D1650909

DEEP INELASTIC SCATTERING

To learn more about the AIP Conference Proceedings, including the Conference Proceedings Series, please visit the webpage
http://proceedings.aip.org/proceedings

DEEP INELASTIC SCATTERING

13th International Workshop on Deep Inelastic Scattering

DIS 2005

Madison, Wisconsin 27 April – 1 May 2005

EDITORS

Wesley H. Smith

Sridhara R. Dasu

University of Wisconsin-Madison

SPONSORING ORGANIZATIONS

Argonne National Laboratory

U.S. Department of Energy

Deutsches Elektronen Synchrotron Laboratory

U.S. National Science Foundation

University of Wisconsin-Madison

Melville, New York, 2005

AIP CONFERENCE PROCEEDINGS ■ VOLUME 792

Editors:

Wesley H. Smith
Sridhara R. Dasu

University of Wisconsin
Physics Department
1150 University Avenue
Madison, WI 53706
USA

E-mail: wsmith@hep.wisc.edu
dasu@hep.wisc.edu

L.C. Catalog Card No. 2005932874
ISBN 0-7354-0283-3
ISSN 0094-243X
Printed in the United States of America

Table of Contents

STRUCTURE FUNCTION WORKING GROUP PRESENTATIONS

DIFFRACTION & VECTOR MESONS WORKING GROUP PRESENTATIONS

ELECTROWEAK AND BEYOND THE STANDARD MODEL WORKING GROUP PRESENTATIONS

HADRONIC FINAL STATES WORKING GROUP PRESENTATIONS

HEAVY FLAVORS WORKING GROUP PRESENTATIONS

SPIN PHYSICS WORKING GROUP PRESENTATIONS

FUTURE OF DIS PRESENTATIONS

Preface

The 13th International Workshop on Deep Inelastic Scattering and QCD (DIS 2005) brought together 280 experimentalists and theorists at the Frank Lloyd Wright Monona Terrace Convention Center in Madison, Wisconsin, U.S.A. The workshop began April 27, 2005 and ran through May 1, 2005. It provided a review of the progress in the field of DIS and QCD and planning for the future. The workshop format involved plenary sessions with review talks and parallel working group sessions with shorter contributions. The Working Groups for DIS 2005 included Structure Functions and Low-x, Diffraction and Vector Mesons Electroweak and Beyond the Standard Model, Hadronic Final States, Heavy Flavors, Spin Physics, and the Future of DIS.

The 2005 workshop was hosted by the High Energy Physics group of the University of Wisconsin–Madison Physics Department. Previous workshops took place in Durham, Eilat, Paris, Rome, Chicago, Brussels, Zeuthen, Liverpool, Bologna, Cracow, St. Petersburg and Strbske Pleso, Solvakia. This is only the second time the conference has been hosted in the United States, with the previous time being 1997 in Chicago by Argonne National Lab.

The workshop conference examined the wealth of new measurements emerging from the HERA, TeVatron, SLAC, CERN, RHIC and Jefferson Lab programs along with new theoretical advances in QCD. There was particular attention brought on the linkages between measurements made in nuclear and particle physics. The timing of the conference at about 2 years before the conclusion of the HERA-II program, also provided an opportunity to consider the planning of the next generation of QCD and DIS experiments.

The organizers wish to thank the International Scientific Advisory Committee for the invitation to host the meeting in Madison and for excellent guidance in preparing for the workshop. We thank the Working Group Convenors for their excellent efforts in assembling an excellent program. We also appreciate the outstanding work by Ms. Aimee Lefkow in making the conference arrangements and her consummate skill in running everything so smoothly. We also wish to acknowledge the superb set up and operation of the web pages and computing for the conference by Steve Rader. We also thank Prof. Brian Foster for arranging a memorable concert with the very talented violinist Jack Liebeck and pianist Inan Barnatan.

The organizers gratefully acknowledge the financial support of the workshop from Argonne National Lab, the U.S. Department of Energy, the Deutsches Elektronen Synchroton Laboratory, the U.S. National Science Foundation, and the University of Wisconsin–Madison, which made this conference possible.

-- Wesley Smith and Sridhara Dasu, *Proceedings Editors*

Dedication

We wish to fondly dedicate the proceedings for DIS05 to our fellow organizer Professor Don Reeder, who continues to guide and mentor us with unerring good judgment and sage perspicacity through our careers in particle physics. Don will retire this year from the University of Wisconsin Physics Department, having served brilliantly as Chairman over two terms for an unmatched total of 11 years. His innovations in so many directions have been extraordinary. He has overseen the renovation of our building, reformed the undergraduate introductory curriculum and steered the hiring of new faculty of exceptional quality.

Don has been a leader in the research celebrated in these proceedings through his direction of the US group on ZEUS during the calorimeter installation and his supervision of many influential physics analyses by his students. Don was also the founding chair of the US program participating in the CMS experiment being built at the Large Hadron Collider at CERN. Don was a co-leader of the experiment at Fermilab in the 1970s that found important experimental evidence for neutral currents in neutrino scattering.

Don has provided exemplary service to the particle physics community through his many years on the Department of Energy High Energy Physics Advisory Panel as a member and as a University representative. He has provided us with a superb role model that we continue to endeavor to follow.

-- Wesley Smith and Sridhara Dasu, *Proceedings Editors*

International Scientific Advisory Committee

Giulio d'Agostini	(Roma)
Guido Altarelli	(CERN)
Violette Brisson	(Orsay)
Dusan Bruncko	(Slovak Acad.)
John Dainton	(Liverpool)
Andrzej Eskreys	(Cracow)
Joel Feltesse	(Saclay)
Robert Klanner	(DESY)
Max Klein	(Zeuthen)
Aharon Levy	(Tel Aviv)
Lev Lipatov	(St. Petersburg)
Pierre Marage	(Brussels)
Rosario Nania	(Bologna)
Jose Repond	(Argonne)
Frank Sciulli	(Columbia)
James Stirling	(Durham)
Albrecht Wagner	(DESY)
Guenter Wolf	(DESY)

Local Organizing Committee

Wesley Smith	(University of Wisconsin)	(Chair)
Sridhara Dasu	(University of Wisconsin)	
Tao Han	(University of Wisconsin)	
Aimee Lefkow	(University of Wisconsin)	
Renee Lefkow	(University of Wisconsin)	
Don Reeder	(University of Wisconsin)	
Jose Repond	(Argonne National Lab)	

OPENING PLENARY PRESENTATIONS

New Results From H1

Olaf Behnke
on behalf of the
H1 Collaboration

Physikalisches Institut, Universität Heidelberg, Philosophenweg 12, 69120 Heidelberg, Germany

Abstract. A summary is given of some of the recent measurements made by the H1 Collaboration in ep collisions at HERA. These include studies of electroweak parameters, of proton charm and beauty structure functions, of α_s from the ratio of tri- to dijet cross sections, of diffractive scattering with dijets in the final state and of searches for strange pentaquarks. Also the first results from the analysis of HERA II e^+p and e^-p data are shown, including studies of events with isolated leptons and large missing transverse momentum and the first measurement of charged current cross sections with left handed polarised e^- beams.

PACS: 10.

INTRODUCTION

Recent measurements are summarised, which were made by the H1 Collaboration in the year leading up to the Deep Inelastic Scattering (DIS) Conference in Madison. First the results are presented which are obtained with the HERA I e^+p and e^-p data taken in the years 1994-2000, comprising an integrated luminosity of up to $\sim 120\,\mathrm{pb}^{-1}$. Second the results are shown which are made with the recent HERA II data with longitudinally polarised lepton beams, comprising up to $\sim 74\,\mathrm{pb}^{-1}$ of integrated luminosity. Many further results, and more detail on the topics discussed here, can be found in the accompanying articles in these Proceedings. The H1 apparatus is described in detail in [1].

FIT OF ELECTROWEAK PARAMETERS

Electron proton scattering at HERA is dominated by photon exchange for the largest part of the covered phase space in momentum transfer Q^2. However one probes at HERA also the region where $\sqrt{Q^2}$ is approaching or exceeding the masses of the W and Z electroweak gauge bosons, and where the exchange of these bosons is of similar strength as the γ exchange. Necessary ingredients for calculating these processes are the quark densities in the proton and the electroweak parameters such as the coupling constants of the quarks to the gauge bosons. Using the e^+p and e^-p charged (W-exchange) and neutral current (γ and Z-exchange) cross sections previously measured by the H1 experiment [2, 3, 4, 5], a combined electroweak and QCD analysis is performed to determine electroweak parameters fully accounting for their correlations with proton parton distributions.

CP792, *Deep Inelastic Scattering, DIS 2005,* edited by W. H. Smith and S. R. Dasu

The *neutral current* measurements are used to determine for the first time at HERA the vector and axial coupling constants of the u and d quarks to the Z boson. The results are shown in figure 1. The sign ambiguities, which are present in the LEP e^+e^- results [6],

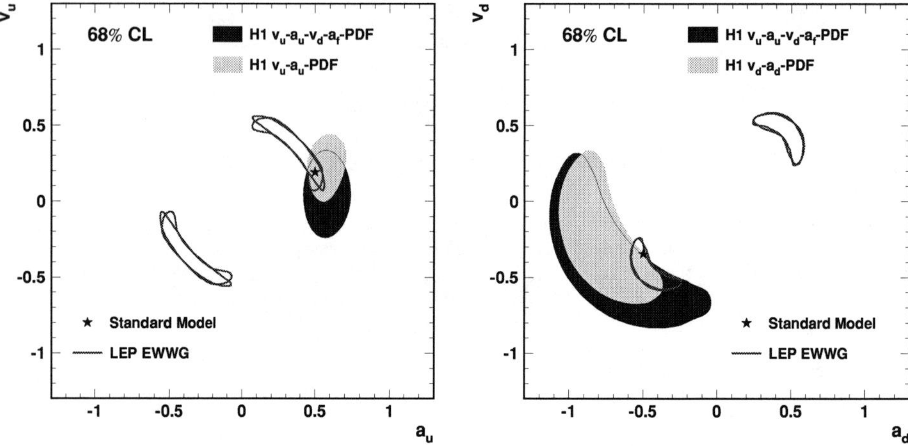

FIGURE 1. Results at 68% confidence level (CL) on the weak neutral current couplings of u (left plot) and d (right plot) quarks to the Z boson determined in the H1 measurement (the shaded contours) in comparison with those determined by the combined LEP data [6] which have both sign and $v-a$ exchange ambiguities. The dark-shaded contours correspond to results of a simultaneous fit of all four couplings whereas the lighted-shaded contours correspond to results of fits where either d or u quark couplings are fixed to their Standard Model (SM) values. The stars show the expected SM values.

as can be seen in figure 1, are unambiguously resolved. This is due to the fact, that at HERA the $\gamma - Z$ interference is probed over a wide range of Q^2. The obtained parameters are in good agreement with the expected Standard Model values.

The observed Q^2 shape of the *charged current* data is exploited to determine a W propagator mass with the result $M_{prop} = 82.87 \pm 1.82_{exp}\,{}^{+0.30}_{-0.16}|_{model}$ GeV. When analysing the data in the context of the Standard model, in the so-called-on-mass-shell (OMS) scheme [7], the Fermi constant G_F can be expressed in terms of M_Z and M_W and hence also the global normalisation of the data becomes sensitive to M_W. The corresponding fit result for M_W is translated into an indirect determination of sin^2_W, yielding $sin^2_W = 0.2151 \pm 0.0040_{exp}\,{}^{+0.0019}_{-0.0011}$.

The W exchange is also sensitive to the top mass through loop corrections. This is exploited to determine for the first time at HERA the top mass with the result $m_t = 108 \pm 44$ GeV, where the error covers only the experimental uncertainty.

PROTON STRUCTURE FUNCTIONS $F_2^{c\bar{c}}$ AND $F_2^{b\bar{b}}$

H1 presents at this conference measurements of inclusive charm and beauty cross sections in ep collisions at HERA over a wide range of momentum transfer $3.5 \leq Q^2 \leq 650\,\mathrm{GeV}^2$. The results at higher values of $Q^2 > 100$ GeV^2 have been recently published [8], the preliminary measurements at lower values of $Q^2 < 100\,\mathrm{GeV}^2$ are presented for the first time at this conference. The fraction of events containing charm and beauty quarks is determined using a method based on the impact parameter, in the transverse plane, of tracks to the primary vertex, as measured by the H1 vertex detector [9]. Values for the structure functions $F_2^{c\bar{c}}$ and $F_2^{b\bar{b}}$ are obtained. This is the first measurement of $F_2^{b\bar{b}}$. The results are shown in figure 2 as function of Q^2 in bins of the Bjorken scaling variable x. The measurements show positive scaling violations which

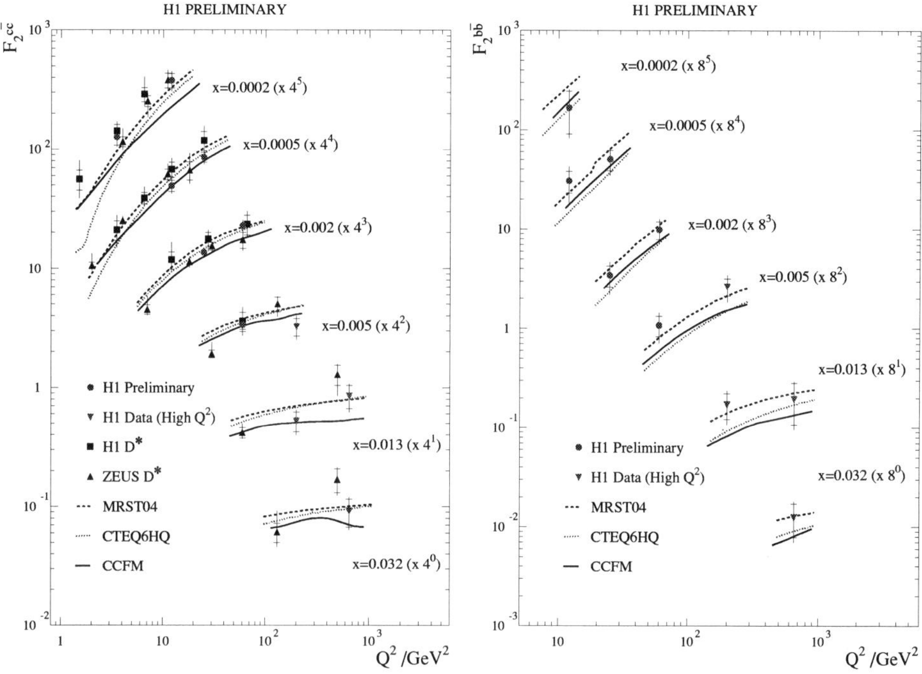

FIGURE 2. The measured $F_2^{c\bar{c}}$ (left) and $F_2^{b\bar{b}}$ (right) shown as a function of Q^2 for various x values. The inner errors bars show the statistical error, the outer error bars represent the statistical and systematic errors added in quadrature. The $F_2^{c\bar{c}}$ measurements obtained of D^* mesons from H1 [10] and ZEUS [11] and the predictions of different QCD model calculations are also shown.

increase with decreasing x. The charm measurements are in good agreement with the previous results extracted from D^* meson measurements by H1 [10] and ZEUS [11], which are also shown in figure 2 (left). The new measurements have the advantage of a high acceptance, which exceeds 70% in each Q^2 and x bin, both for charm and beauty. On the contrary, for the D^* measurements correction factors of typically > 2 have to be applied to extrapolate the data outside the D^* visible kinematic range, in order to obtain an inclusive charm cross section. The charm and beauty data are compared with several QCD model predictions. The next to leading order (NLO) perturbative QCD predictions by MRST [12] and CTEQ [13] use the variable flavour number scheme (VFNS). In this scheme at values of $Q^2 = M^2$ the dominant leading order process is $\gamma g \rightarrow q\bar{q}$, where

$q = c$ or b and where the gluon is emitted from the proton. As Q^2 increases, in the region $Q^2 \gg M^2$ the heavy flavour quarks are treated as massless partons in the proton and the leading order process is $\gamma q \to q$. The calculations by MRST and CTEQ differ in the way the transition between the two regimes is dealt with. This leads to significant differences between the two predictions in the kinematic range of this measurement, especially for beauty as can be seen in figure 2. However, within the experimental uncertainties both calculations describe the new data well, both for charm and beauty.

In the kinematic region of this measurement, the charm contribution to the total ep cross section, i.e. the ratio $F_2^{c\bar{c}}/F_2$ is found to be around 20%-30% and increases slightly with increasing Q^2 and decreasing x. The beauty contribution to the total ep cross section increases rapidly from $\leq 0.4\%$ at $Q^2 = 12\,\mathrm{GeV}^2$ to $\sim 5\%$ at $Q^2 = 650\,\mathrm{GeV}^2$.

QCD TESTS WITH FINAL STATES

Determination of α_s from three to two jet ratio

H1 presents at this conference a new determination of the strong coupling constant α_s via the ratio $R_{3/2}$ of tri- to dijet cross sections in deep inelastic scattering. The measurement is performed over a range of momentum transfer $150 < Q^2 < 150000\,\mathrm{GeV}^2$ and transverse jet energies $5 < E_T < 50$ GeV. Jets are defined by the inclusive $k_\perp$ algorithm in the Breit frame. Figure 3 shows the measured dijet and trijet differential cross sections as function of Q^2. The data are compared to an NLO prediction using the NLOJET++

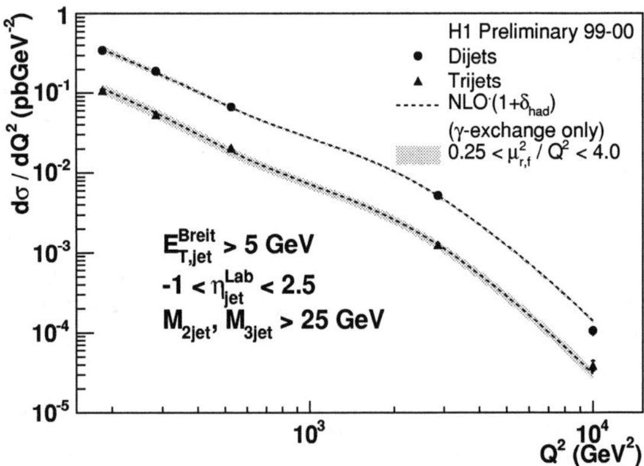

FIGURE 3. Neutral current di- and trijet differential cross sections, with respect to Q^2, shown with NLO pQCD predictions including hadronisation corrections. The shaded bands show the effect of varying the renormalisation/factorisation scale by a factor of two. The highest Q^2 bin is overestimated by the theoretical predictions due to the absence of electroweak effects in the NLO calculations.

program [14], which describes the data very well with the exception of the highest Q^2 bin. The cross section in this bin is overestimated by the theoretical predictions due to

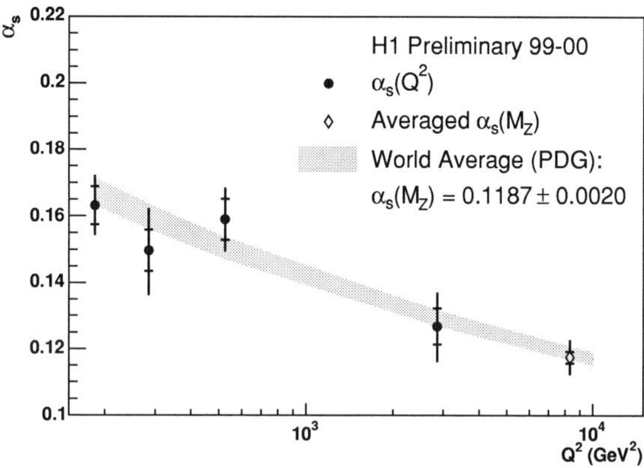

FIGURE 4. $\alpha_s(M_Z)$ values, from each Q^2 bin, as determined from a fit to the ratio of tri- to dijet cross section in that bin, evolved to their values at their respective values of Q^2 (triangles) using the two-loop solution of the renormalisation group equation. Uncertainties shown are statistical up to the horizontal bar, and then the quadratic sum of statistical and systematical uncertainties. The averaged value of $\alpha_s(M_Z)$, found using a χ^2 minimisation fit, is shown at the far right (empty diamond). The evolution of the current world average value of $\alpha_s(M_Z)$ is shown as a shaded band.

the absence of electroweak effects in the NLO calculation. For this reason the highest Q^2 bin is omitted in the current determination of α_s, which is discussed in the following.

For each bin in Q^2 a value of $\alpha_s(M_Z)$ is determined by matching the theoretical prediction from NLOJET++ as function of $\alpha_s(M_Z)$ to the measured ratio $R_{3/2}$ in that bin. Figure 4 shows the $\alpha_S(M_Z)$ values from each Q^2 bin evolved to their values at their respective values of Q^2. The data are in agreement with the predicted evolution by pQCD. An overall value of $\alpha_s(M_Z)$ is calculated from the weighted average of the $\alpha_s(M_Z)$ measurements in the bins and is also shown in figure 4 (right-most point). The result is $\alpha_{s(M_Z)} = 0.1175 \pm 0.00017$ (stat.) ± 0.0050 (syst.) $^{+0.0054}_{-0.0068}$ (th.), which is in good agreement with the world average value $\alpha_s(M_Z) = 0.1187 \pm 0.0020$. The value is competitive with results from other experimental procedures from different processes and is in agreement with the result recently obtained from $R_{3/2}$ by the ZEUS collaboration [15].

Diffraction

QCD predicts that the cross section for diffractive DIS factorises into universal diffractive parton densities (DPFDs) and process-dependent hard scattering coefficients [16] Diffractive parton densities have been determined from QCD fits to inclusive diffractive HERA data [17, 18]. Final state configurations for which a partonic cross section is perturbatively calculable in QCD include dijet production, which is directly sensitive to the gluon component of the diffractive exchange. However, applying this approach to predict diffractive dijet production in $p\bar{p}$ collisions at the Tevatron leads to an overestimation of the observed rate by nearly one order of magnitude [19]. This discrepancy has been attributed to the presence of the additional beam hadron remnant in $p\bar{p}$ collisions, which leads to secondary interactions and a breakdown of factorisation. The transition from DIS to hadron-hadron scattering can be studied at HERA in a comparison of scattering processes in DIS and in photoproduction. While in DIS the virtual photon always directly enters the hard interaction, in photoproduction the real photon can directly interact or be first resolved into partons which then initiate the hard scattering, resembling hadron-hadron scattering. H1 presents at this conference measurements of differential dijet cross sections both in diffractive DIS ($Q^2 > 4$ GeV^2 and in photoproduction ($Q^2 < 0.01$ GeV^2). The DIS results are in good agreement with QCD factorisation. In photoproduction, however, the dijet rate is suppressed by about a factor 0.5 compared to the NLO QCD prediction. This can be seen in figure 5, which shows the differential cross section in photoproduction as function of the quantities $z_{I\!P}^{\rm jets}$ and x_γ^{jets}.

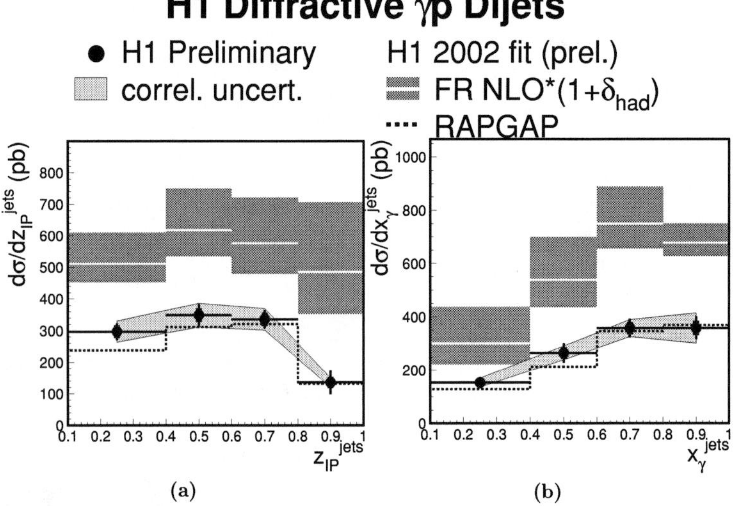

FIGURE 5. Cross section for the diffractive production of two jets in photoproduction as a function of a) $z_{I\!P}^{\rm jets}$ and b) $x_\gamma^{\rm jets}$.The band around the NLO prediction with hadronisation correction indicates the uncertainty resulting from the variation of the renormalisation scale by factors 0.5 and 2.

In the leading order picture $z_{I\!P}^{\rm jets}$ and x_γ^{jets} are the momentum fractions of the diffractive exchange and the photon, respectively, which are transferred into the dijet system. Surprisingly not only the region of $x_\gamma^{jets} < 0.9$, where resolved photon processes are expected to be enhanced but also the complementary region $x_\gamma^{jets} > 0.9$, where the direct photon contributions are expected to dominate, are suppressed at a similar level. Thus these preliminary results are suggestive of a breakdown of factorisation in photoproduction for both direct and resolved photon interactions.

SEARCH FOR STRANGE PENTAQUARK RESONANCE

Recently several experiments have published evidence for the production of a strange pentaquark θ^+ by various reaction processes [20, 21, 22]. It has been observed via its decays into K^+n and into K_s^0p. Negative preliminary results, however, have been reported [23] from pp collisions, e^+e^- annihilation and also from fixed target photoproduction experiments. H1 presents at this conference preliminary results on the search for the θ^+ decaying into K_s^0p in DIS, where the ZEUS experiment has found evidence of a signal at a mass of 1.522 GeV [22].

The identification of the decay protons is based on the measured energy loss in the central jet chambers. The invariant K_s^0p mass distribution obtained by the H1 experiment is shown in figure 6 for three different regions of Q^2 between 5 and 100 GeV^2.

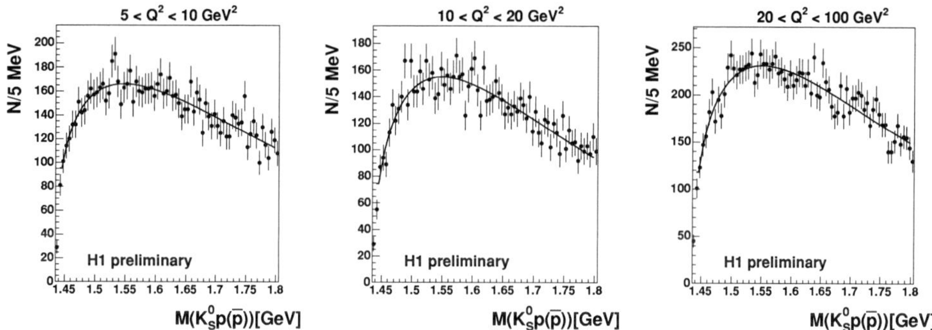

FIGURE 6. Invariant $K_S^0p(\bar{p})$ mass spectra for the standard dE/dx selection, in three bins of Q^2. The full line shows the result from the fit of a background function to the data. The mass spectra show upward and downward fluctuations but no significant peak is observed.

The mass spectra do not show any significant signal in the mass range from threshold up to 1.7 GeV/c^2. With the assumption that pentaquarks are produced by fragmentation, mass dependent upper limits on the visible $\theta^\pm$ production cross sections are obtained for the investigated different kinematic regions. When repeating the analysis with similar cuts as were applied for the ZEUS measurement, no signal is observed either, although the sensitivity is expected to be comparable for both experiments.

RESULTS WITH HERA II DATA

Isolated lepton events with large missing transverse momentum

One of the most exciting results from HERA I is an excess observed in the H1 data [24] for events with an isolated lepton and large missing transverse momentum as this might be a hint for new physics beyond the standard model. The main contribution from the standard model to these events is the production of a real W boson (radiated from a quark) and the subsequent decay into lepton neutrino. An excess over the predicted cross sections was observed both for the electron and muon channel for large hadronic transverse momenta $p_T^X > 25$ GeV. For this conference H1 has updated the measurement with the available HERA II data. Six new electron events are found in these data but no new muon event. The yield of isolated electrons is again in excess over the standard model prediction. One of the new electron events is shown in figure 7.

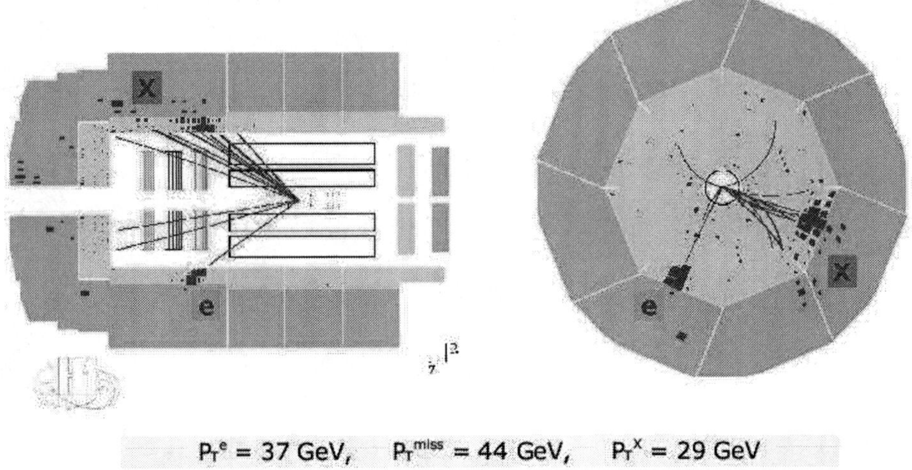

FIGURE 7. Display of an event with an isolated electron, missing transverse momentum and a prominent hadronic jet.

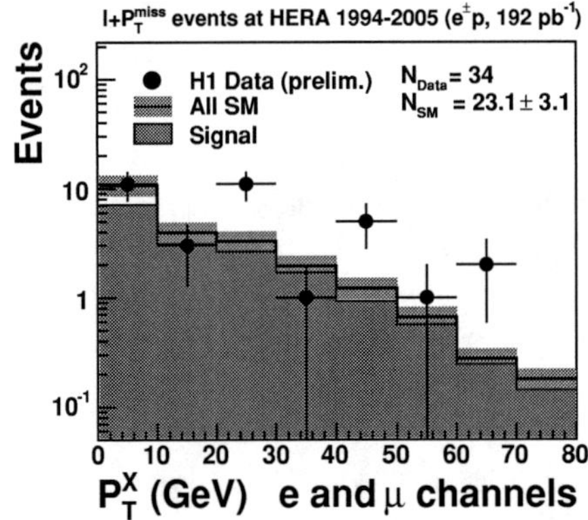

FIGURE 8. The hadronic transverse momentum distribution in the electron and muon channels combined: data (full HERA data set, $\mathscr{L} = 192\ \text{pb}^{-1}$) is compared to the Standard Model expectation (open histogram). The signal component of the SM expectation, dominated by real W production, is given by the hatched histogram. N_{Data} is the total number of data events observed, N_{SM} is the total SM expectation. The total error on the SM expectation is given by the shaded band.

Figure 8 shows the observed event yield, when combining the electron and muon channel and exploiting the full HERA data set of e^+p and e^-p collisions from 1994-2005, as function of p_T^X. The numbers which are given in the figure legend correspond to the total p_T^X range. In the restricted region $p_T^X > 25$ GeV, there are 17 events observed while the SM predicts only 5.8.

Charged current cross sections with polarised e^- and e^+ beams

Since November 2004 the H1 and ZEUS experiments started to take for the first time data with with longitudinally polarised electron beams. This allows for a more detailed test of the electroweak part of the Standard Model at the energy frontier. The Standard model predicts that the cross sections for the charged current (CC) interactions $ep \to \nu X$ should have a linear dependence on polarisation, and furthermore, the cross section for fully right handed electrons should be zero (similarly for fully lefthanded positrons). This follows from the abesence of right handed currents within the framework of the Standard Model. At this conference H1 presents the first measurement of the total CC cross section for longitudinally polarised lefthanded electrons. The data set presented here has a mean lefthanded polarisation of $(-25.40 \pm 0.44)\%$ and comprise an integrated luminosity of ~ 18 pb^{-1}. The measurement requires an excellent control of the energies deposited in the calorimeter, which is obtained by an *in-situ* calibration using a high statistics Neutral Current sample. Figure 9 shows the measured dependence of the e^+p and e^-p CC cross section with the lepton beam polarisation. The new measurement with the lefthanded electron beam enters as the circle point. The figure also shows two data points (triangles) with polarised positron beams and the results for unpolarised beams from the HERA I data [4, 5]. The data point with an average polarisation of -40% is the new and first CC measurement with a lefthanded polarised positron beam. This measurement is based on data from 2004 and was made preliminary [25] in summer 2004. All the data points are in good agreement with the standard model expectation.

CONCLUSION

The H1 Collaboration is completing the analysis of the HERA I data. At this conference many new and even some completely novel measurements with these data are presented such as the first determination at HERA of the axial and vector couplings of u and d quarks to the Z-boson and the measurement of the charm and beauty contribution to the total inclusive ep scattering using a secondary vertex tagging method. First results are obtained with the recent HERA II data. The HERA operation in the last few months leading up to this conference, with left handed polarised electrons, is exploited to make the first measurement of charged current events $ep \to \nu X$ for such polarised electrons. The excess observed in the HERA I data of events with an isolated lepton and missing transverse momentum is continued to be seen in the HERA II data. Six new candidates have been observed in the electron channel but none in the muon channel. To draw further conclusions, a much higher integrated luminosity is needed. This is expected to be provided in the next years of HERA II running and will allow a wealth of new and interesting measurements to be made at HERA.

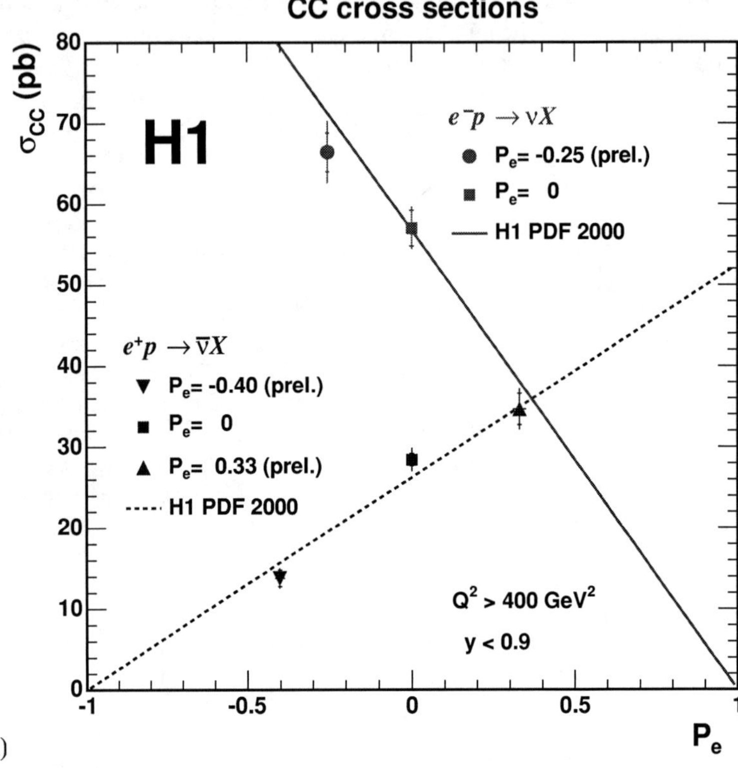

FIGURE 9. The dependence of the e^+p and e^-p CC cross section with the lepton beam polarisation P_e is shown. The data are compared to the prediction from the H1 PDF 2000 fit for polarised positrons (dashed line) and for polarised electrons (full line).

Many more results and details than presented in this summary talk can be found in these proceedings from the parallel sessions contributions.

ACKNOWLEDGEMENTS

I thank all of my H1 colleagues for their hard work which has resulted in the production of the measurements discussed here. Thanks to the H1 colleagues who helped me with the preparation of this talk, especially to Max Klein. I thank the DIS05 organisers for a pleasant workshop.

REFERENCES

1. I. Abt *et al.* [H1 Collaboration], Nucl. Instrum. Meth. A **386** (1997) 310.
2. C. Adloff *et al.* [H1 Collaboration], Eur. Phys. J. C **21**, 33 (2001) [arXiv:hep-ex/0012053].
3. C. Adloff *et al.* [H1 Collaboration], Eur. Phys. J. C **13**, 609 (2000) [arXiv:hep-ex/9908059].
4. C. Adloff *et al.* [H1 Collaboration], Eur. Phys. J. C **30**, 1 (2003) [arXiv:hep-ex/0304003].
5. C. Adloff *et al.* [H1 Collaboration], Eur. Phys. J. C **19**, 269 (2001) [arXiv:hep-ex/0012052].
6. LEP EW working group, http://lepewwg.web.cern.ch/LEPEWWG/plots/summer2004/.
7. A. Sirlin, Phys. Rev. D **22**, 971 (1980) and ibid. D **29**, 89 (1984).
8. A. Aktas *et al.* [H1 Collaboration], Eur. Phys. J. C **40**, 349 (2005) [arXiv:hep-ex/0411046].
9. D. Pitzl *et al.*, Nucl. Instrum. Meth. A **454**, 334 (2000) [arXiv:hep-ex/0002044].
10. C. Adloff *et al.* [H1 Collaboration], Phys. Lett. B **528**, 199 (2002) [arXiv:hep-ex/0108039].
11. S. Chekanov *et al.* [ZEUS Collaboration], Phys. Rev. D **69**, 012004 (2004) [arXiv:hep-ex/0308068].
12. A. D. Martin, R. G. Roberts, W. J. Stirling and R. S. Thorne, Eur. Phys. J. C **39**, 155 (2005) [arXiv:hep-ph/0411040].
13. S. Kretzer, H. L. Lai, F. I. Olness and W. K. Tung, Phys. Rev. D **69**, 114005 (2004) [arXiv:hep-ph/0307022].
14. Z. Nagy and Z. Trocsanyi, Phys. Rev. Lett. **87**, 082001 (2001) [arXiv:hep-ph/0104315].
15. S. Chekanov *et al.* [ZEUS Collaboration], arXiv:hep-ex/0502007.
16. J. C. Collins, Phys. Rev. D **57**, 3051 (1998) [Erratum-ibid. D **61**, 019902 (2000)] [arXiv:hep-ph/9709499].
17. C. Adloff *et al.* [H1 Collaboration], Z. Phys. C **76**, 613 (1997) [arXiv:hep-ex/9708016].
18. H1 Collaboration, paper 980 submitted to Intl.Conf. on High Energy Physics, ICHEP2002, Amsterdam.
19. T. Affolder *et al.* [CDF Collaboration], Phys. Rev. Lett. **84**, 5043 (2000).
20. T. Nakano *et al.*, (LEPS), Phys. Rev. Lett. **91** (2003) 012002;
    S. Stepanyan *et al.*, (CLAS), Phys. Atom. Nuclei **66**, 1715 (2003);
    J. Barth *et al.*, (SAPHIR), Phys. Lett. B **572** (2003) 127;
    V. Kubarovsky *et al.*, (CLAS), Phys. Rev. Lett. **92** (2004) 03201 [erratum-ibid. 92 (2004) 049902].
21. V. V. Barnim *et al.*, (DIANA), Phys. Atom. Nucl. **66** (2003) 1715 [Yad. Fiz. 66 (2003) 1763];
    A. E. Asratyan, A. G. Dolgolenko and M. A. Kubantsev, Phys. Atom. Nucl. **67** (2004) 682;
    A. Aleev *et al.*, (SVD), hep-ex/0401024;
    A. Airapetian *et al.*, (HERMES), Phys. Lett. B **585** (2004) 213;
    M. Abdel-Bary *et al.*, (COSY-TOF),Phys. Lett. B **595** (2004) 127.
22. S. Chekanov *et al.* [ZEUS Collaboration], Phys. Lett. **591** (2004) 7.
23. J. Z. Bai *et al.*, (BES), Phys. Rev. D **70** (2004) 012004;
    BaBar Collaboration, hep-ex/0408064;
    BELLE Collaboration, hep-ex/0409010;
    S. R. Armstrong, hep-ex/0410080; S. Schael *et al.*, (ALPEH), Phys. Lett. B **599**(2004) 1;
    I. Abt *et al.*, (HERA-B), Phys. Rev. Lett. **93** (2003) 212003;
    Yu. M. Antipov *et al.*, (SPHINX), Eur. Phys. J. A **21**, 455 (2004);
    M. J. Longo *et al.*, (HyperCP), Phys. Rev. D **70** (2004) 111101;
    D. O. Litvintsev, (CDF), hep-ex/0410024;
    R. Mizuk *et al.*, (Belle), hep-ex/0411005;
    C. Pinkerton *et al.*, (PHENIX), J. Phys. G **30** (2004) 1201;
    R. De Vita, (CLAS Collaboration), *Search for pentaquarks at CLAS in photoproduction from protons*,American Physical Society Meeting, Tampa, Florida, April 2005.
24. V. Andreev *et al.* [H1 Collaboration], Phys. Lett. B **561**, 241 (2003) [arXiv:hep-ex/0301030].
25. H1 Collaboration , Contributed paper no. 756/758 to ICHEP04, http://www-h1.desy.de/h1/www/publications/htmlsplit/H1prelim-04-141.long.html.

ZEUS Results

Elisabetta Gallo[†]

INFN Firenze, Italy
(On behalf of the ZEUS Collaboration)

Abstract. Several results from the ZEUS Collaboration were presented at this Workshop. The highlights are presented in this summary, and include results from NLO QCD fits and determination of α_S, from forward jets and diffractive final states, from pentaquarks and searches and from heavy flavour production. Also the first results from the analysis of the HERA II e^+p/e^-p data are shown.

Keywords: Document processing, Class file writing, LaTeX 2_ε
PACS: 43.35.Ei, 78.60.Mq

INTRODUCTION

These proceedings report on the highlights of recent results from the ZEUS Collaboration. The published structure functions data have been used to determine the proton parton distribution functions in NLO QCD fits. The fit includes also jet data to constrain the gluon in the middle-x region. The strong coupling constant α_S is determined in various measurements. The diffractive parton distribution functions are also determined from inclusive data and used to obtain predictions for diffractive final states. ZEUS has performed searches for exotic final states like pentaquarks, involving strange or charm quarks. The HERA experiments are still competitive in the search for SUSY with R-parity-violating interactions and the recent search for stop is described. Several results on heavy flavour production were presented in the parallel sections, here the results on beauty cross sections at HERA I and D^* production at HERA II are reported. Finally the first results on the charged current polarized cross sections are shown.

STRUCTURE FUNCTIONS AND JETS

NLO QCD fits

The ZEUS Collaboration has completed the publication of the structure function data from the HERA I (94-00) running. These cross sections have been used in a NLO QCD fit, based on the DGLAP evolution, to determine the proton parton distributions functions (PDFs), using the ZEUS data only [1]. Uncertainties from heavy-target corrections, present in global analyses which include also fixed-target data, are therefore avoided. While the low Q^2 neutral current (NC) data determine the low-x sea and gluon distributions, the high Q^2 NC and charged (CC) cross sections constrain the valence distributions. In addition, published jet cross sections from the 96-97 data were used to constrain the gluon density in the mid-to-high-x region ($x \simeq 0.01 - 0.5$). The predictions for the

CP792, *Deep Inelastic Scattering, DIS 2005,* edited by W. H. Smith and S. R. Dasu

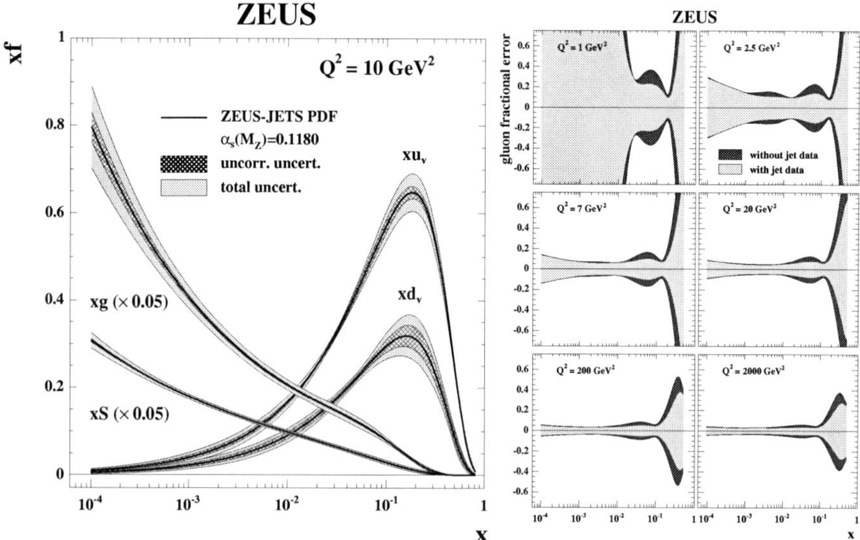

FIGURE 1. On the left: parton distribution functions for the u-valence, d-valence, sea and gluon density obtained from the ZEUS-JETS fit, for a Q^2 value of 10 GeV^2. On the right side: uncertainty on the gluon density for different values of Q^2 and for the two cases in which the jets are included or not in the fit.

two jet cross sections used (DIS jets in the Breit frame and jets in direct photoproduction) were calculated to NLO and used in the fit in a rigourous way. The resulting PDFs in this fit (see figure 1), called ZEUS-JETS fit, give a very good description of both inclusive and jet cross sections, in agreement with QCD factorization. The gain in using the jets is shown in the right part of figure 1, where the yellow (light) band shows the total experiment uncertainty on the gluon density in the fit including the jets, compared to the red (dark) band which is the error for the fit without including the jets. As an example, in the bin at $Q^2 = 7$ GeV^2 and $x \simeq 0.06$, the uncertainty is reduced from 17% to 10% using the jets. A similar decrease of approximately a factor two is visible in the whole Q^2 range in the mid-to-high-x region.

In the inclusive cross sections, there is a strong correlation between α_S and the gluon density. As the jets data depend on α_S and on $xg(x)$ in a different way, the addition of the jets data to the fits permits also a more precise determination of the strong coupling constant, compared to previous fits, when α_S is left as a free parameter. The value resulting from this fit, called ZEUS-JETS-α_S, is

$$\alpha_S(M_Z) = 0.1183 \pm 0.0028(\text{exp.}) \pm 0.0008(\text{model}) \pm 0.005(\text{scale}) \quad (1)$$

which is in very good agreement with the world average of 0.1182 ± 0.0027.

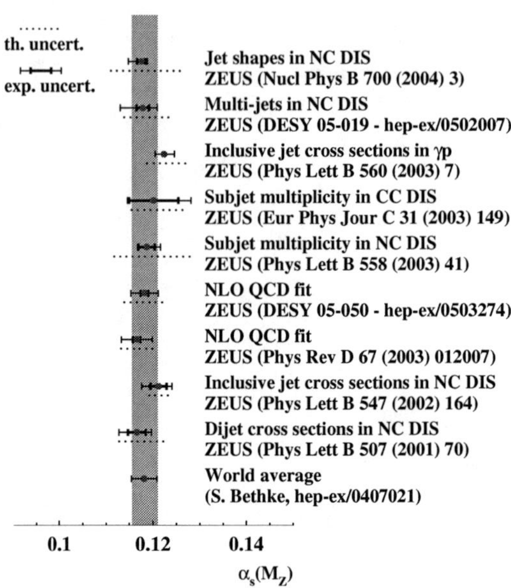

FIGURE 2. Compilation of ZEUS results on the determination of α_S. The band shows the world average as calculated in the reference cited in the figure.

Summary of α_S results

Figure 2 shows a compilation of results on the determination of α_S at ZEUS (see also [2]). The most precise result, from the experimental point of view, comes from the inclusive jets in photoproduction. Although each of the measurements has a precise experimental determination (typically 3%), competitive with the world average, the error is dominated by the theoretical uncertainties (typically 4%), which are shown by the dashed lines. NNLO (next-to-NLO) calculations for jet based variables and for PDFs are therefore needed.

Forward jets at low-x

The DGLAP evolution scheme describes the HERA structure function data down to low-x. In order to disentangle the effect of different evolution schemes, like BFKL or CCFM, it has been suggested long ago by Mueller [3] to look at jets in the forward (proton) region, where the differences between the different parton evolution schemes should be more prominent. For such an analysis, the jets were reconstructed with the k_T cluster algorithm in the longitudinally invariant inclusive mode in the Breit frame and then boosted to the laboratory frame. Thanks to the forward plug calorimeter installed for

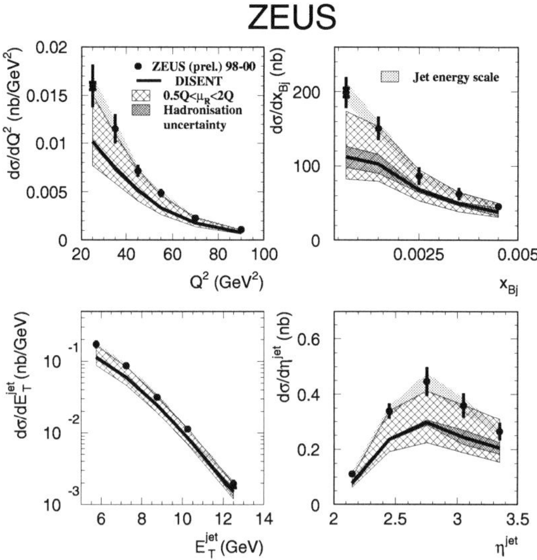

FIGURE 3. Measured differential cross sections compared to the NLO QCD calculations for forward jet production at low x.

the period 98-00, jets could be selected with a pseudorapity coverage in the laboratory frame of $2 < \eta_{\text{jet}} < 3.5$. In order to enhance the contribution of possible BFKL effects, the jets were required to have a $x_{\text{jet}} > x$, where x_{jet} is the ratio of the longitudinal momentum of the jet and the proton momentum, and x is the Bjorken variable. This requirement maximises the phase space for BFKL evolution. In addition the jets were required to have a transverse momentum squared $(E_T^{\text{jet}})^2$ approximately equal to Q^2 in order to suppress DGLAP evolution as this cut leaves no room for evolution in Q^2. The events were selected in the kinematic region $20 < Q^2 < 100$ GeV^2 and $0.0004 < x < 0.005$. The differential cross sections as a function of Q^2, x, E_T^{jet} and η_{jet} are shown in figure 3, compared to NLO QCD calculations based on the DGLAP evolution, with the program DISENT. The data are slightly above the theoretical NLO calculations, especially in the lowest x bin, however the uncertainties on the theory are still very large, as shown by the hatched bands. The variation of the calculations due to the change of renormalization scale is particularly large, which indicates the need for higher-oder calculations.

Very high-x at HERA

While HERA has provided precision results on structure functions at low-x, the high-x region has still to be explored, because of both limited statistics and difficulty in reconstructing the kinematic variables. ZEUS has developed a new method to select

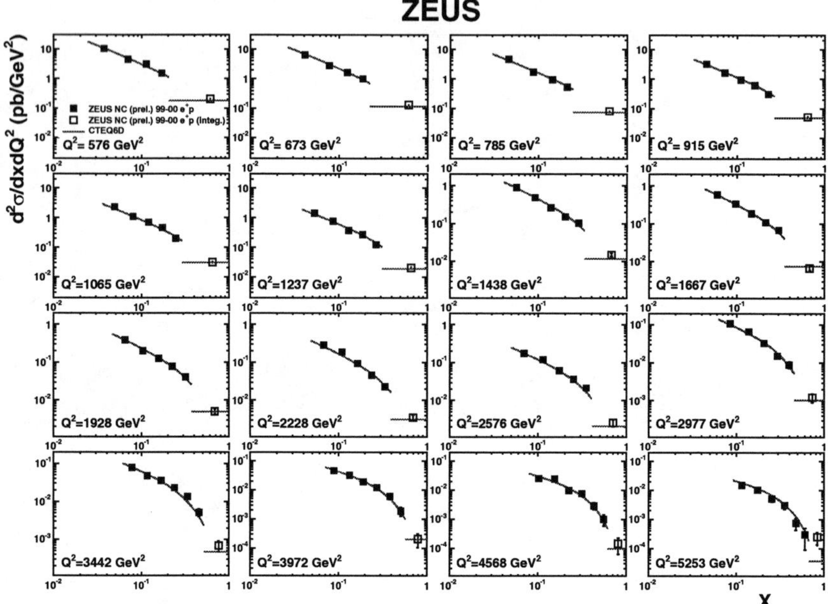

FIGURE 4. Double differential cross sections (solid squares) for the 99-00 e^+p data, as a function of x in Q^2 bins, compared to the CTEQ6D parton distributions. The last bin (open symbol) shows the integrated cross section over x divided by the bin width, i.e. $1/(1-x_{edge}) \cdot \int_{x_{edge}}^{1}(d^2\sigma/dxdQ^2)dx$; the symbol is shown at the centre of the bin. In this bin the prediction is drawn as an horizontal line. The error bars represent the quadratic sum of the statistical and systematic uncertainties.

events at very high-x and in the middle Q^2 region ($Q^2 > 576$ GeV^2). These events are characterized by a well measured scattered electron or positron in the central part of the calorimeter; and by a jet, very forward, close to the proton beampipe. As x increases, the jet is more and more boosted in the forward direction and eventually disappears in the beam pipe; the value of x at which this occurs is Q^2 dependent. The kinematic of the event is reconstructed in this way. First the Q^2 of the event is determined from the electron variables E_e and θ_e, which are measured with good resolution. The events are then separated in those with exactly one good reconstructed jet and those without any jet. For the 1-jet events, the jet information is used to calculate the Bjorken x variable from E_{jet} and θ_{jet} and the double differential cross section $d^2\sigma/dxdQ^2$ is calculated in each x, Q^2 bin. For the 0-jet sample, it is assumed that the events come from very high-x, with a lower value x_{edge} which can be evaluated for each Q^2 bin based on kinematic constraints. In this case the events are collected in a bin $x_{\text{edge}} < x < 1$ and an integrated cross section $\int_{x_{\text{edge}}}^{1}(d^2\sigma/dxdQ^2)dx$ is calculated.

The result is shown in figure 4, where, for each Q^2, the open squares in the last bin are the integrated cross sections up to $x = 1$ and the closed squares show the double differential cross section in x, Q^2. The precision in the last bin is comparable to the other

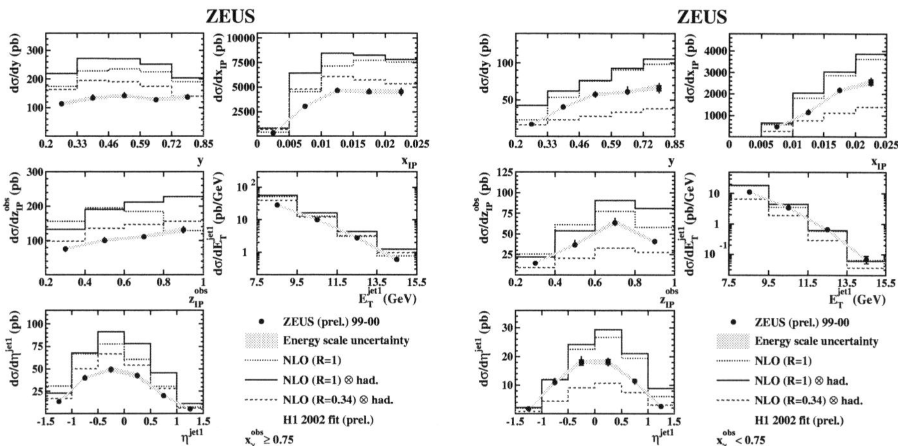

FIGURE 5. Differential cross sections for diffractive dijet photoproduction for a direct-enriched (left) and a resolved-enriched (right) sample. The variable z_{IP} is the longitudinal fractional momentum taken from the diffractive exchange by the dijet system.

bins. For most of these highest x-bins, where there is no previous measurement, the data tend to lie above the expectations from CTEQ6D [4]. Precision measurements at high-x at HERA will allow to constrain the valence parton distributions, where up to now only fixed target experiments at low Q^2 have provided experimental data.

DIFFRACTION

At HERA, NLO QCD fits have also been performed to inclusive diffractive data and diffractive parton distribution functions (dPDFs) have been extracted. Assuming the QCD factorization theorem to be valid, in hard diffractive processes the cross section can be factorized into the partonic cross sections and these dPDFs, which are assumed to be universal. The dPDFs are then used to make predictions for various diffractive final states. Here I will concentrate on diffractive photoproduction of dijets, which were measured in the kinematic range $0.2 < y < 0.85$ and $x_{IP} < 0.025$. Jets were selected with the longitudinally invariant k_T algorithm and with the asymmetric cuts $E_T > 7.5, 6.5$ GeV for the two jets. Double differential cross sections were measured separately in the region $x_\gamma^{obs} < 0.75$, which is the resolved-photon enriched region, and the region $x_\gamma^{obs} > 0.75$, which is the region where direct-photon interactions dominate. The measured distributions in y, x_{IP}, z_{IP} and E_T and η of the highest-E_t jet are shown at the hadron level in figure 5. The data are compared to NLO calculations [5] with diffractive PDFs extracted from the H1 data [6]. In general the NLO calculations (shown by the line $\mathrm{R} = 1$) reproduce the shape of the distributions, but their normalization is too high.

A reduction factor $\mathrm{R} = 0.34$ has been calculated [7] in $p\bar{p}$ diffractive interactions, originating from interactions between spectator partons in the two hadronic beams.

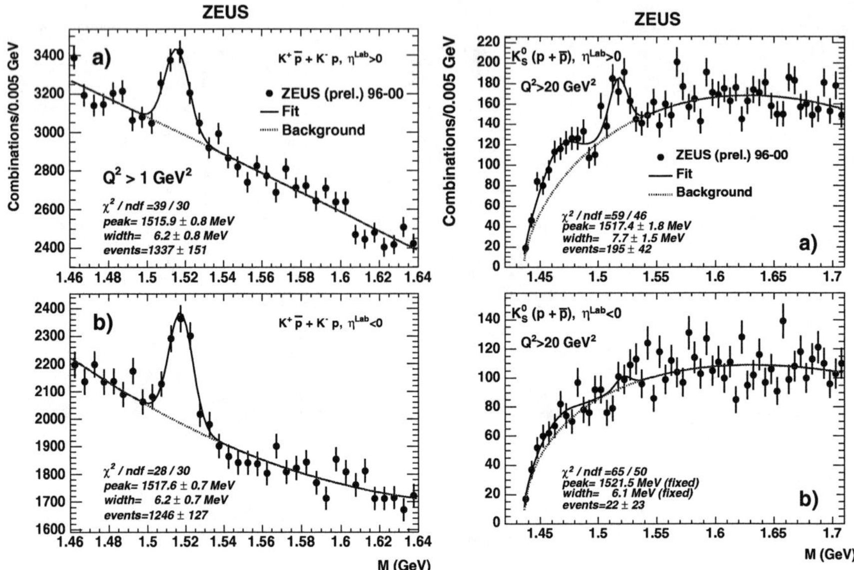

FIGURE 6. Invariant mass spectra for the $\Lambda(1520)$ (left) and for the Θ^+ (right), for $Q^2 > 20\ \text{GeV}^2$ and divided in forward pseudorapidity region (top) and rear region (bottom). The forward region corresponds to the direction of the proton.

Such rescattering processes create additional particles that fill the large rapidity gap characteristic of diffractive processes, causing a suppression of the measured cross sections and a break of factorization. A similar reduction could be expected for resolved processes in which the photon behaves like a hadron. The dashed curve in figure 5, corresponding to NLO predictions with $R = 0.34$, is however too low compared to the measured resolved-photon cross sections and a suppression factor of $R \simeq 0.5$ seems more appropriate. A similar suppression factor seems to be needed also for direct processes, as shown by the left plot of figure 5, although in this case the photon is more point-like and factorization is expected to hold.

EXOTIC FINAL STATES

Recently there has been many experiments reporting evidence of new baryonic states consisting of five quarks. A number of experiments, including ZEUS [8], have observed a narrow resonance decaying either to nK^+ or pK^0_S and with a mass around 1530 MeV, which could correspond to the pentaquark state $\Theta^+ = uudd\bar{s}$. Other experiments have reported negative searches for this state and ZEUS is at the moment the only high energy experiment to have observed the Θ^+ candidate. Recent studies from ZEUS have focused in trying to understand the production mechanism of this state.

The $K^0_S p(\bar{p})$ spectrum was studied for $Q^2 > 20\ \text{GeV}^2$ separately in the forward proton region, selecting events with $\eta > 0$, and in the rear region, $\eta < 0$. The resulting spectra

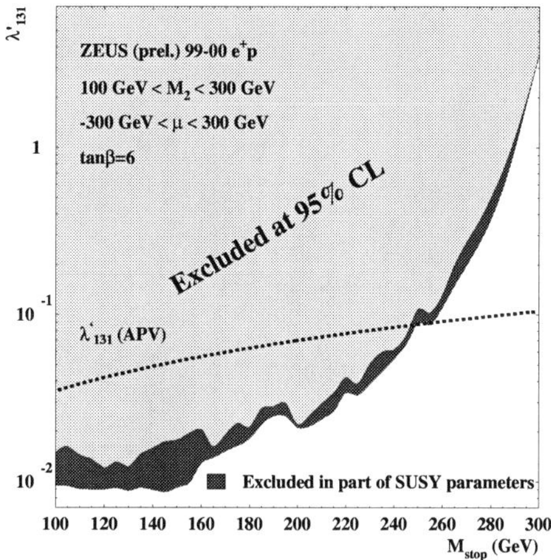

FIGURE 7. Limits on the R-parity-violating coupling λ'_{131} as a function of the stop mass. The light region shows the excluded region for all considered SUSY scenarios, the dark area for a part of them. The dashed line is the limit from atomic-parity violation measurements.

are shown in figure 6: the fitted number of events under the peak is higher in the region closer to the proton remnant, compared to the rear region. This is not what is observed in the production of the $\Lambda(1520) \to K^+p$, as shown in the same figure. This suggests that the production mechanism for the Θ^+ could be different from pure fragmentation in the central rapidity region and could be related to proton-remnant fragmentation.

ZEUS has also performed a search for two other exotic states: the $\theta_C = uudd\bar{c}$ observed by H1 [9] in the HERA I data: and the states $\Xi^{--}_{3/2}$ or $\Xi^{0}_{3/2}$ with a mass around 1860 MeV decaying in $\Xi\pi$, observed by the NA49 Collaboration [10]. The Θ_c was observed by H1 as a narrow resonance at a mass of 3099 GeV decaying to D^*p, both in a $Q^2 > 1$ GeV^2 sample and in a photoproduction sample: approximately 1% of the selected D^* mesons were observed to come from the decay of a Θ_C. A similar analysis was performed by ZEUS [11] in an inclusive sample with approximately 60,000 D^* candidates, but no evidence for this state was found. Also a search [12] for the two states observed by NA49, decaying respectively to $\Xi^-\pi^-$ or $\Xi^-\pi^+$, was performed and no resonance was found in a large mass range nor around the region of 1860 GeV.

R-PARITY-VIOLATING SUSY SEARCHES

HERA is the ideal place to look for possible exotic states originating from the fusion of the initial positron/electron and a valence quark in the proton. Among these, the stop can be produced from $e^+d \to \tilde{t}$ via the R-parity-violating coupling λ'_{131}. The stop can then decay as $\tilde{t} \to e^+d$ or through the decay $\tilde{t} \to b\chi^+$, giving a rich topology of final states. ZEUS has performed a search with the 99-00 e^+p data looking for final state topologies

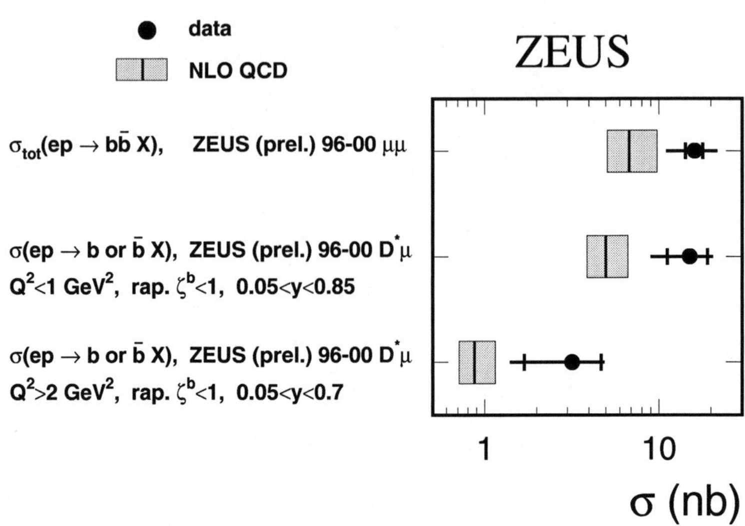

FIGURE 8. Beauty cross section from dimuon and $D^* + \mu$ tagged events. The measurements (dots) are compared to the NLO QCD calculations.

with one positron and one jet (e-J) or multi jets (e-MJ) and one neutrino and multi jets (ν-MJ). The branching ratios to the various channels depend on the SUSY parameters, therefore a scan over a wide range of SUSY parameters was performed. Selection cuts were designed to optimize the signal sensitivity with respect to the main background, which comes from NC and CC interactions at high Q^2. No resonance was observed in the invariant mass of the final state particles in the different topologies, therefore 95% confidence level limits were set on stop production. The results are shown in figure 7 for the stop mass versus the coupling λ'_{131}. The yellow (light) area is the excluded region; the effect of changing the SUSY parameters in a wide range and for $\tan\beta = 6$, as shown in the legend of the figure, is shown by the dark shaded band. The limits for masses up to 250 GeV improve on the low-energy limits from atomic parity violation (APV) measurements and depend weekly on the SUSY parameters.

HEAVY FLAVOURS

Beauty production has been measured by ZEUS both in photoproduction and in DIS. The hard scale provided by the mass of the b-quark allows to compare to NLO QCD calculations. A new measurement based on 121 pb^{-1} of HERA I data in which two muons are observed in the final state was presented at this workshop. Tagging both muons coming from the semileptonic b decays allows a better suppression of the backgrounds from charm and light flavour production. This also means that muons, and therefore b's, can be selected at low p_T and in a large η range, which implies less extrapolation in deriving the total cross section. In addition the normalizations of the various backgrounds can be constrained dividing the sample into high- and low-mass, isolated and non-isolated, like-

and unlike-sign muon pairs.

From the visible cross section, a total cross section for beauty production at HERA is extracted and compared to NLO QCD predictions from the sum of the two calculations FMNR (for photoproduction) and HVQDIS (for DIS), obtaining:

$$\sigma^b_{tot}(ep \to b\bar{b}X)[318\ \text{GeV}] = 16.1 \pm 1.8(stat.)^{+5.3}_{-4.8}(syst.)\ \text{nb}, \tag{2}$$

$$\sigma^b_{tot}(\text{NLO}) = 6.9^{+3.0}_{-1.8}\ \text{nb}, \tag{3}$$

where the uncertainties on the theory were evaluated changing the renormalization and factorisation scales by a factor 2 and varying the b mass between 4.5 and 5 GeV.

The result is shown in figure 8, where also previous results from ZEUS, based on $D^* + \mu$ samples [13], are shown. These samples also allow measurement of the cross section to be made for low transverse momentum of the b's. From the figure it looks like there is the tendency for these cross sections to be above NLO QCD predictions, while for higher Q^2 and higher p_T there is good agreement between theory and measurements (see e.g. [14]).

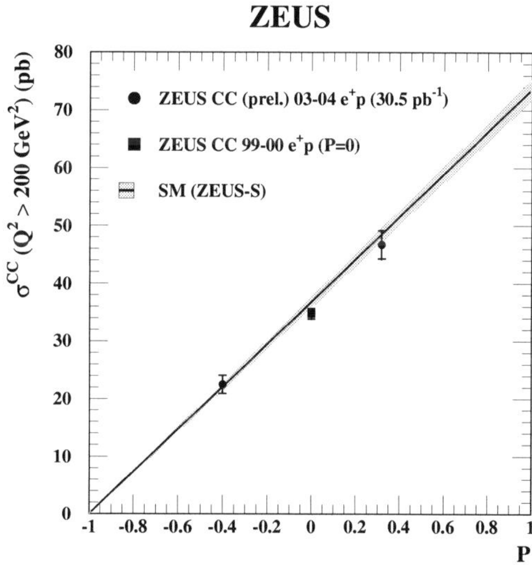

FIGURE 9. Charged current total cross sections as a function of the longitudinal polarization of the positron beam, compared to the SM predictions with the ZEUS-S PDFs. The square symbol shows the HERA I results from the published unpolarized e^+p data.

HERA II

Cross sections at high Q^2

The HERA II running phase has started, e^+p collisions with longitudinally polarized positrons have been collected in 2003-2004 and in 2004-2005 the machine has switched to polarized electron beams. Cross sections at high Q^2 were measured with the first 2003-2004 data with polarized positron beams [15], with a mean luminosity weighted polarization $P = -40.2\%$ for 16.4 pb^{-1} of collected data, and $P = +31.8\%$ for 14.1 pb^{-1} of integrated luminosity. The total charged current cross section is predicted in the Standard Model to have a linear dependence on the polarization, becoming zero for left-handed positrons (or right-handed electrons).

The measured CC cross sections for the two polarization values are shown in figure 9: the linear dependence on P is clearly visible and the points are in good agreement with the SM prediction, which is calculated with the ZEUS-S parton densities for the proton. The effect of the longitudinal polarization in the NC cross sections can only be seen at very high Q^2 with a high integrated luminosity. A marginal effect could anyway already be seen in these data [15].

Charm cross sections

Charm is produced in DIS mainly through the boson-gluon-fusion mechanism and is therefore directly sensitive to the gluon density in the proton. The cross section for charm was measured in 40 pb^{-1} of e^+p and 33 pb^{-1} of e^+p data collected in 2003-2005, as a function of Q^2 in the range $5 < Q^2 < 1000$ GeV^2. Charm was tagged using the $D^* \to D^0\pi_S \to K\pi\pi_S$ decay, where tracks were reconstructed with the central tracking detector and the newly installed silicon microvertex detector. Approximately 1200 D^* candidates were selected in each sample, e^+p and e^-p. The ratio of the cross sections for the two samples was measured, as shown in figure 10, as a function of Q^2. This ratio is particularly interesting as, with the previous published HERA I data [16] (based on 17 pb^{-1} of e^-p and 65 pb^{-1} of e^+p), the D^* production rate was observed to be higher in e^-p than in e^+p, as shown also in the figure. This ratio was found to be 1.67 ± 0.21 (only statistical error) in the range $40 < Q^2 < 1000$ GeV^2 and such phenomenon is not expected from any physics process. With the higher statistics in the e^-p data, collected up to now at HERA II, this effect is not confirmed and the ratio agrees with unity through the whole range of Q^2 measured.

CONCLUSION

The ZEUS Collaboration is completing the analysis of the HERA I data and starting to look at the HERA II data. Recent highlights have been described here; many more results and details can be found in these proceedings from the parallel sessions contributions.

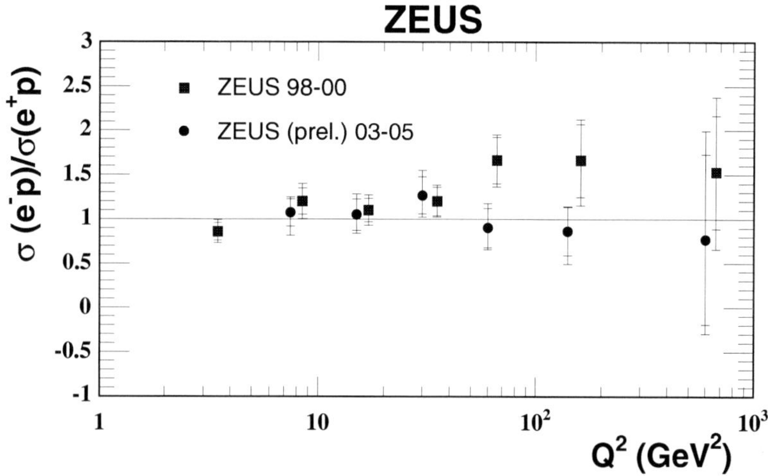

FIGURE 10. The ratio of the cross sections for D^* production in e^-p to the one in e^+p, as measured in the 03-05 data and in the 98-00 data. The inner error bars show the statistical uncertainty, the outer error bar the systematic uncertainty. Many systematic effects cancel in the ratio.

ACKNOWLEDGMENTS

I would like to thank my ZEUS colleagues, and in particular Matthew Wing and Rik Yoshida, for the preparation of this talk; and Prof. Wesley H. Smith for the excellent organization of this Workshop.

REFERENCES

1. ZEUS Collaboration, DESY-05-071 (2005).
2. C. Glasman, these proceedings and references therein (2005).
3. A. H. Mueller, *J. Phys. G*, **19**, 1463–1468 (1993).
4. J. Pumplin et al., *JHEP*, **0207**, 012 (2002).
5. M. Klasen, and G. Kramer, *Eur. Phys. J*, **C38**, 93 (2004).
6. H1 Collaboration, paper 980 submitted to ICHEP02 (2002).
7. A. B. Kaidalov, *Phys. Lett.*, **B567**, 61 (2003).
8. ZEUS Collaboration, *Phys. Lett.*, **B591**, 7–22 (2004).
9. H1 Collaboration, *Phys. Lett.*, **B588**, 17–28 (2004).
10. NA49 Collaboration, *Phys. Rev. Lett.*, **92**, 042003 (2004).
11. ZEUS Collaboration, *Eur. Phys. J.*, **C38**, 29–41 (2004).
12. ZEUS Colaboration, DESY-05-018 (2005).
13. ZEUS Collaboration, paper 5-0342 submitted to ICHEP04 (2004).
14. H1 Collaboration, *Eur. Phys. J.*, **C41**, 453 (2005).
15. ZEUS Collaboration, paper 4-0256 submitted to ICHEP04 (2004).
16. ZEUS Collaboration, *Phys. Rev.*, **D69**, 012004 (2004).

Tevatron Results

Régis Lefèvre [1]
on behalf of the CDF and DØ Collaborations

Institut de Física d'Altes Energies, Universitat Autònoma de Barcelona, Edifici Cn. Facultat Ciènces UAB, E-08193 Bellaterra, Spain

Abstract. Recent results obtained by the CDF and DØ experiments at the Tevatron Run II are presented. A first part is dedicated to QCD physics where inclusive jet production, dijet azimuthal decorrelations and jet shapes measurements are reported. Electroweak physics is then discussed relating measurements of the W and Z bosons productions, of the forward-backward charge asymmetry in W production, of the W width and of the top quarks mass. The extensive Run II exploration program is finally approached reporting about searches for neutral supersymmetric Higgs bosons in multijet events and for sbottom quark from gluino decays.

Keywords: Tevatron, CDF, DØ, QCD, Electroweak, Supersymmetry, MSSM, Jet, W, Z, Top, Higgs, Bottom, Gluino, Sbottom, Production, Shape, Asymmetry, Width, Mass, Search
PACS: 13.85.-t, 13.87.-a, 14.65.Ha, 14.70.Fm, 14.80.Cp, 14.80.Ly

INTRODUCTION

In Run II, the Tevatron $p\bar{p}$ collider is operating at a center of mass energy of 1.96 TeV. The 36 bunches of protons and anti-protons collide at two interaction points every 396 ns. The peak luminosity has greatly improved since the beginning on Run II and is now about 10^{32} cm^{-2}s^{-1}. The record integrated luminosity in a week is 20 pb^{-1}. The long term luminosity plan of the Tevatron for Run II is to deliver between 4.4 fb^{-1} (base goal) and 8.5 fb^{-1} (design goal) by the end of 2009.

The CDF [1] and DØ [2] detectors are located at the interaction points. Those two multi-purpose experiments have been highly upgraded for Run II. Both have new Silicon micro-vertex tracker, new tracking system and upgraded muon chambers. CDF also has integrated a new time-of-flight detector and new plug calorimeters. DØ has integrated a new solenoid and new pre-showers. Both experiments have new trigger and data acquisition systems and are taking data with good efficiency, around 85 %. At the time of the conference, each experiment had already collected about 0.7 fb^{-1} on tape.

CDF and DØ has developed a very broad and exciting physics program [3, 4]. Important precision measurements are carried on in order to test and further constrain the Standard Model. New physics beyond the Standard Model are also extensively explored. Some recent results from Run II are presented in the following. They are based on data sets corresponding to integrated luminosities between 72 pb^{-1} and 385 pb^{-1} depending on the analysis.

[1] Régis Lefèvre is supported by the EU funding under the RTN contract: HPRN-CT-2002-00292, Probe for New Physics.

CP792, *Deep Inelastic Scattering, DIS 2005,* edited by W. H. Smith and S. R. Dasu

QCD PHYSICS

Inclusive jet production

The measurement of the inclusive jet production cross section at the Tevatron provides a stringent test of perturbative QCD (pQCD) predictions over almost nine orders of magnitude. The high p_T tail is sensitive to new physics, it probes distances up to around 10^{-19} m. This measurement can also be used to constrain the Parton Distribution Functions (PDFs) at high x and high Q^2. Thanks to the Tevatron increase of center of mass energy, from 1.8 TeV in Run I to 1.96 TeV in Run II, the jet production rate at high p_T has significantly increased, it has been multiply by a factor five around 600 GeV/c for instance. First Run II measurements have already extended the p_T coverage by about 150 GeV/c compared to Run I. In addition, new jet algorithms are now explored following theoretical prescription suggesting that the cone based algorithm used in Run I is not infrared safe and compromises meaningful comparisons with pQCD calculations.

DØ has measured the inclusive jet production cross section in two regions of jet rapidity, $|y^{jet}| < 0.4$ and $0.4 < |y^{jet}| < 0.8$. Jets were reconstructed using the midpoint algorithm [5]. This iterative seed-based cone algorithm uses midpoints between pair of proto-jets as additional seeds which makes the clusterization procedure infrared safe [6]. The following parameters were used: a cone radius of 0.7 in the $Y - \phi$ space and a merging fraction of 50 %. Results based on 378 pb^{-1} of Run II data are presented in figure 1. The measurements are directly compared to Next-to-Leading-Order (NLO) pQCD calculations computed with NLOJET++ [7] using CTEQ6.1M [8] and MRST2004 [9] PDFs, without any correction for non-perturbative contributions. The renormalization and factorization scales were set to the jet transverse momentum $\mu_R = \mu_F = p_T^{jet}$. The non-perturbative contributions, associated with underlying event and hadronization processes, were investigated using PYTHIA [10] and HERWIG [11]. The two Monte Carlos give consistent results: those contributions were found to be small allowing the direct comparison to NLO calculations presented here to be meaningful within an uncertainty between 10 % for p_T^{jet} lower than 100 GeV/c and 5 % at higher transverse momenta. The measurements agree with the theoretical predictions. Experimental errors are dominated by the uncertainty on the jet energy scale. The largest theoretical error is coming from the PDFs, theoretical predictions especially suffer from the limited knowledge of the gluon distribution at high x.

In CDF, the inclusive jet production cross section has been measured for central jets, $0.1 < |y^{jet}| < 0.7$, using the longitudinally invariant K_T algorithm [12]. Merging pairs of nearby particles in order of increasing relative transverse momentum, as suggested by pQCD gluon emissions, this algorithm is infrared and collinear safe to all orders in pQCD. Unlike cone based algorithms, it does not include any merging/splitting feature that may affect comparisons to pQCD. It contains a parameter D that controls the merging termination and characterizes the approximate size of the resulting jets:

$$d_{ij} = min(p_{T,i}^2, p_{T,j}^2) \cdot \frac{(y_i - y_j)^2 + (\phi_i - \phi_j)^2}{D^2} \; ; \; d_{ii} = p_{T,i}^2$$

To make sure that soft contributions such as the underlying event and multiple $p\bar{p}$

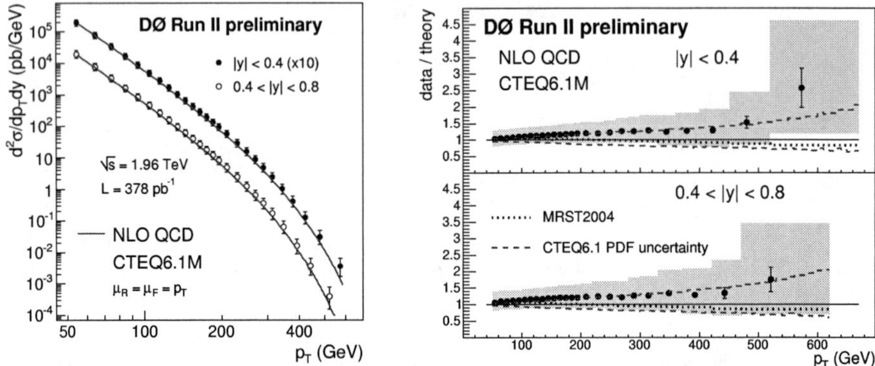

FIGURE 1. *Left:* Inclusive jet cross section measured by DØ in two regions of jet rapidity using the midpoint algorithm. The errors bars indicate the total experimental uncertainty. The data at $|y^{jet}| < 0.4$ are scaled by a factor of ten for presentation purposes. The predictions from NLO pQCD are overlaid on the data. *Right:* Ratio of the measured inclusive jet cross section and the NLO pQCD prediction in the two regions of jet rapidity. The total experimental uncertainty is shown by the shaded bands. The uncertainty due to the proton PDFs is indicated by the dashed lines. The NLO prediction for MRST 2004 PDFs are shown as the dotted line.

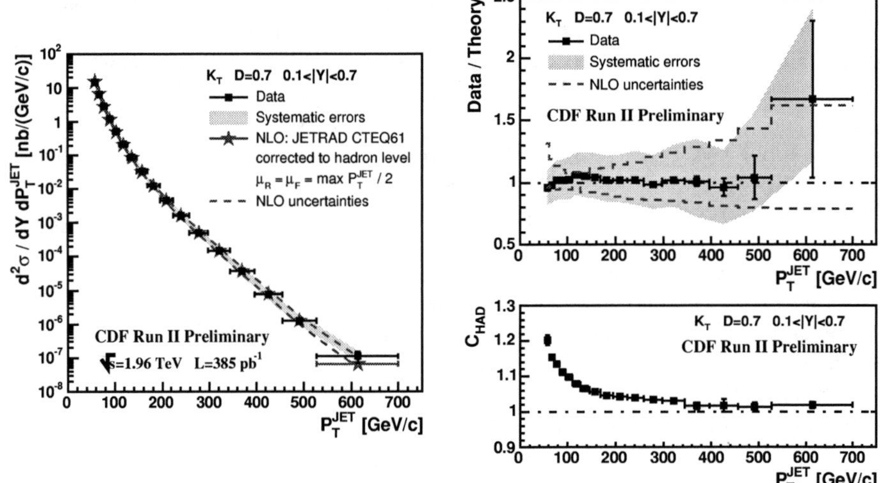

FIGURE 2. Inclusive jet cross section measured by CDF for jets with $0.1 < |y^{jet}| < 0.7$ using the longitudinally invariant K_T algorithm. The measured cross section is compared to NLO pQCD predictions corrected to the particle level: the bottom right plot shows the correction factor that take into account both underlying event and hadronization effects. Data points include the statistical errors, the shaded bands represent the experimental systematic uncertainties. The dashed lines represent the theoretical uncertainties.

interactions are under control, three different values of the D parameter were investigated with 385 pb^{-1} of Run II data: $D = 0.5$, 0.7 and 1.0. The results obtained with $D = 0.7$ are presented in figure 2. They are compared to pQCD NLO predictions computed with JETRAD [13] using CTEQ6.1M PDFs and corrected to the particle level. The renormalization and factorization scales were set to half of the maximum jet transverse momentum in the rapidity region $\mu_R = \mu_F = p_T^{max}/2$. pQCD predictions are corrected to the particle level to take into account the non-perturbative contributions associated with underlying event and hadronization processes, see bottom right plot of figure 2. This correction has been obtained with PYTHIA comparing the predicted jet inclusive cross sections at the particle level with Multiple Parton Interactions (MPI) turned on, and at the parton level with Multiple Parton Interactions turned off:

$$C_{HAD} = \sigma_{Particle\ level}^{PYTHIA-Tune\ A\ with\ MPI\ ON} / \sigma_{Parton\ level}^{PYTHIA-Tune\ A\ with\ MPI\ OFF}$$

for each p_T^{jet} bin. A special set of PYTHIA parameters, tuned on Run I CDF data to reproduce the underlying event activity in the transverse region, denoted as PYTHIA-Tune A [14], was used to evaluate this correction. After correcting pQCD calculations to the particle level, the measurements are in very good agreement with the theoretical predictions over the whole p_T^{jet} range. As for the DØ study previously discussed, systematic errors are dominated by the uncertainty on the jet energy scale while theoretical uncertainties mainly comes from the gluon PDF at high x. Similar very good agreements between data and theory were obtained using a D parameter of 0.5 and of 1.0 showing that soft contributions are well under control as their importance depends a lot on the size of the jets: with respect to $D = 0.7$, the corrections of pQCD to the particle level are for instance about twice smaller using $D = 0.5$ and about twice bigger using $D = 1.0$.

Dijet azimuthal decorrelations

Using an inclusive dijet sample corresponding to 150 pb^{-1} of Run II data, DØ has studied the dijet azimuthal decorrelations measuring the normalized differential dijet cross section in $\Delta\phi_{dijet}$ [15]. This measurement is sensitive to the gluon radiation spectrum.

Jets were reconstructed with the midpoint jet algorithm using a cone radius of 0.7 and a merging fraction of 50 %. Both jets were required to have central rapidities: $|y^{jet}| < 0.5$. Four regions of leading jet transverse momentum were investigated starting at $p_T^{max} > 75$ GeV/c, the second jet was required to have $p_T^{jet} > 40$ GeV/c.

The measurements are reported in figure 3. In the left plot, they are compared to pQCD Leading-Order (LO) and NLO calculations computed with NLOJET++ using CTEQ6.1M PDFs. The renormalization and factorization scales were set to $\mu_R = \mu_F = p_T^{max}/2$. The LO prediction, with at most three partons in the final state, is limited to $\Delta\phi_{dijet} > 2\pi/3$ corresponding to three partons of equal transverse momenta, *Mercedes-star* topology. It presents a prominent peak at $\Delta\phi_{dijet} = \pi$ corresponding to the soft limit for which the third parton is collinear to the direction of the two leading partons. The NLO prediction, up to four partons in the final state, describes the measured distribution except close to $\Delta\phi_{dijet} = \pi$ which is dominated by soft processes

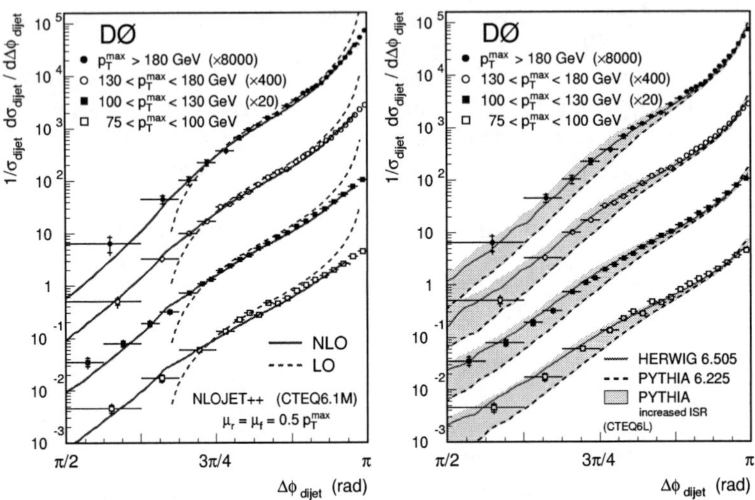

FIGURE 3. $\Delta\phi_{dijet}$ distributions measured in four regions of p_T^{max}. Data and predictions with $p_T^{max} > 100$ GeV/c are scaled by successive factors of 20 for presentation purposes. Predictions are either from NLO (solid lines) and LO pQCD (dashed lines) on the left, either from HERWIG (solid lines) and PYTHIA (dashed lines plus shaded bands, see text) on the right.

and where a resummed calculation is mandatory. A reasonnable approximation to such a calculation is provided by parton shower Monte Carlo programs. In the right plot of figure 3, the measurements are compared to HERWIG and PYTHIA, both with default parameters and CTEQ6L PDFs [8]. HERWIG describes the data. PYTHIA clearly underestimates the gluon radiation at large angles with default parameters but can describes the data if Initial State Radiation (ISR) contributions are increased: the shaded bands in figure 3 (right) indicate the range of variation observed increasing the maximum allowed virtually, directly related to the maximum p_T in the initial-state parton shower, from the default value up to four times higher.

Jet shapes

The internal structure of jets is dominated by multi-gluon emissions from the primary final-state parton. It is sensitive to the relative quark and gluon-jet fraction and receives contributions from soft gluon initial-state radiation and beam-beam remnant interactions. The study of the jet shapes at the Tevatron provides a stringent test of QCD predictions and probes the validity of the models used in the Monte Carlo for parton cascades and soft gluon emissions in $p\bar{p}$ collisions.

Based on an inclusive jet sample corresponding to 170 pb^{-1} of Run II data, CDF has measured the jet shapes for jets with rapidity $0.1 < |y^{jet}| < 0.7$ and transverse momentum $37 < p_T^{jet} < 380$ GeV/c [16]. Jets were reconstructed with the midpoint jet algorithm using a cone radius of 0.7 and a merging fraction of 75 %.

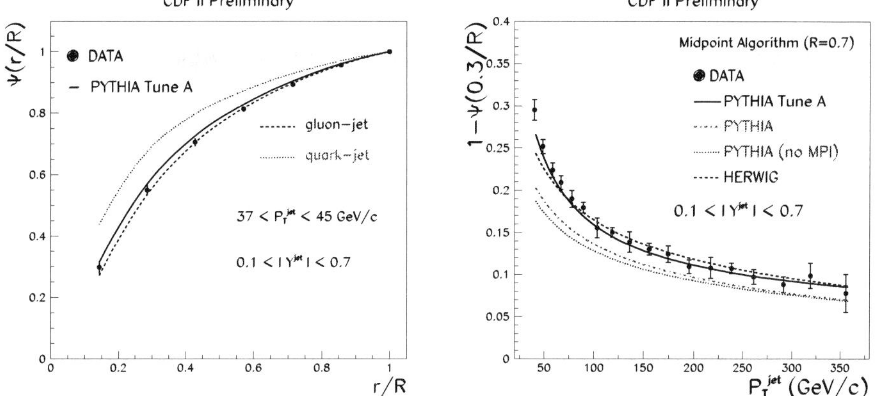

FIGURE 4. *Left:* Measured integrated jet shape for $37 < p_T^{jet} < 45$ GeV/c. The predictions of PYTHIA-Tune A (solid line) and the separate predictions for quark-initiated jets (dotted line) and gluon-initiated jets (dashed line) are shown for comparison. *Right:* $1-\psi(0.3/R)$ versus p_T^{jet}. The predictions of PYTHIA-Tune A (solid line), PYTHIA (dashed-dotted line), PYTHIA-(no MPI) (dotted line) and HERWIG (dashed line) are shown for comparison. In both plots, error bars indicate the statistical and systematic uncertainties added in quadrature.

Figure 4 shows some of the obtained results. The left plot shows the measured integrated jet shape for the first p_T^{jet} bin. It is defined as the average fraction of the jet transverse momentum that lies inside a cone of radius r concentric to the jet cone:

$$\psi(r) = \frac{1}{N_{jet}} \sum_{jets} \frac{p_T(0,r)}{p_T(0,R)}, \quad 0 \leq r \leq R$$

where N_{jet} denotes the number of jets. The right plot of figure 4 shows $1-\psi(0.3/R)$, the average fraction of jet transverse momentum outside an inner cone of fixed radius $r_0 = 0.3$, as a function of p_T^{jet}. The evolution of $1-\psi(0.3/R)$ with p_T^{jet} for gluon-jets only on one hand or for quark-jets only on the other hand is directly related to the running of the strong coupling $\alpha_s(p_t^{jet})$.

The measurements have been compared to the predictions from PYTHIA and HERWIG, using CTEQ5L PDFs [8] in both cases. Tune A parameters were used for PYTHIA, default ones for HERWIG. PYTHIA was also investigated with default parameters and with default parameters but turning off the Multiple Parton Interactions (MPI), the latter solution being denoted as PYTHIA-(no MPI).

With default parameters, PYTHIA produces too narrow jets while PYTHIA-Tune A describes all the data very well. HERWIG gives good predictions for p_T^{jet} above 55 GeV/c but produces too narrow jets bellow. The Monte Carlo predictions indicate that the measured jet shapes are dominated by contributions from gluon-initiated jets at low p_T^{jet} as shown in figure 4 (left), similarly they indicate that contributions from quark-initiated jets dominate at high p_T^{jet}: this is related to the partonic contents of the proton and anti-proton since the quark and gluon mixture in the final state partially reflects the nature on the incoming partons that participate in the hard interaction.

ELECTROWEAK PHYSICS

W and Z productions

In their electron and muon decay modes, the W and Z productions at the Tevatron provide clean, abundant and well know signals that allow stringent tests of the Standard Model. An indirect determination of the W width can for instance be obtained from the ratio of W and Z cross sections. Experimentally more challenging, the tau decay modes are also very interesting as they allow to test the lepton coupling universality.

Figure 5 summarizes the W (left) and Z (right) production measurements that have been performed at the Tevatron. Results are in good agreement with the Standard Model predictions [17]. The main systematic uncertainty arises from the integrated luminosity determination (around 6 %). The second main source of systematic uncertainty comes from the PDFs in the case of the electron and muon decay modes (around 1.5 %). For the tau decay modes, it come from the tau identification (around 5 %).

Making the ratio of the $W \rightarrow \mu\nu$ and $W \rightarrow e\nu$ cross sections on one hand and the ratio of $W \rightarrow \tau\nu$ and $W \rightarrow e\nu$ cross sections on the other hand, CDF has investigated the lepton coupling universality to the W boson and found no violation: $g_\mu/g_e = 0.998 \pm 0.004_{(\mathrm{stat})} \pm 0.011_{(\mathrm{syst})}$, $g_\tau/g_e = 0.99 \pm 0.02_{(\mathrm{stat})} \pm 0.04_{(\mathrm{syst})}$.

As there is no sign of non-universality, CDF has made the ratio of the W and Z production cross sections combining electron and muon decay modes and extracted an indirect determination of the W width. The obtained value is $\Gamma_W = 2.079 \pm 0.041$ GeV, compatible with the previous world average $\Gamma_W = 2.124 \pm 0.041$ GeV [18] and in good agreement with the Standard Model expectation $\Gamma_W = 2.092 \pm 0.003$ GeV [18].

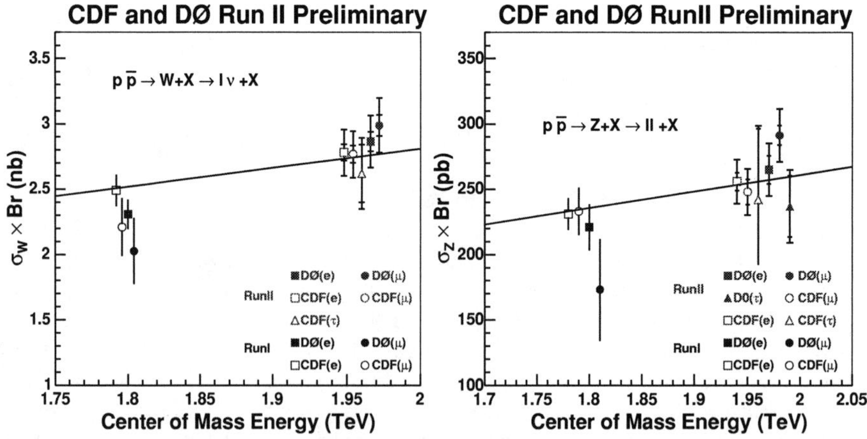

FIGURE 5. Summary of the W (left) and Z (right) production measurements at the Tevatron. Run I (Run II) results are all at 1.8 TeV (1.96 TeV): they are spaced along the x axis for presentation purposes. Inner (outer) error bars on Run II data points exclude (include) the uncertainty from the luminosity measurement. The curves report the Standard Model expectations.

W charge asymmetry

CDF has measured the forward-backward charge asymmetry from $W \longrightarrow e\nu$ production using 170 pb^1 of Run II data [19]. This measurement provides important input on the ratio of the u and d quark components of the PDFs at high momentum transfer, $Q^2 \approx M_W^2$. Since, on average, u quarks carry a higher fraction of the proton momentum than d quarks [20], produced W^+ (W^-) tend to be boosted forward (backward), in the proton (anti-proton) direction. The $W \longrightarrow e\nu$ decays provide a high purity sample however, since the p_Z of the neutrino is not reconstructed, the asymmetry can only be measured with respect to the electron pseudo-rapidity , η_e, as:

$$A(\eta_e) = \frac{d\sigma(e^+)/d\eta_e - d\sigma(e^-)/d\eta_e}{d\sigma(e^+)/d\eta_e + d\sigma(e^-)/d\eta_e}$$

As shown in figure 6, the asymmetry has been measured in two intervals of electron transverse energy that probe different ranges of W rapidities and thus increase sensitivity to the PDFs, especially at $x > 0.3$ where they are currently least constrained. Theoretical predictions from CTEQ6.1M [8] and MRST02 [21] PDFs are shown for comparison. They were obtained using NLO RESBOS [22] Monte Carlo calculation with soft gluon resummation to correctly model the p_T spectrum of the W. Inclusion of those results will further constrain future PDFs fits.

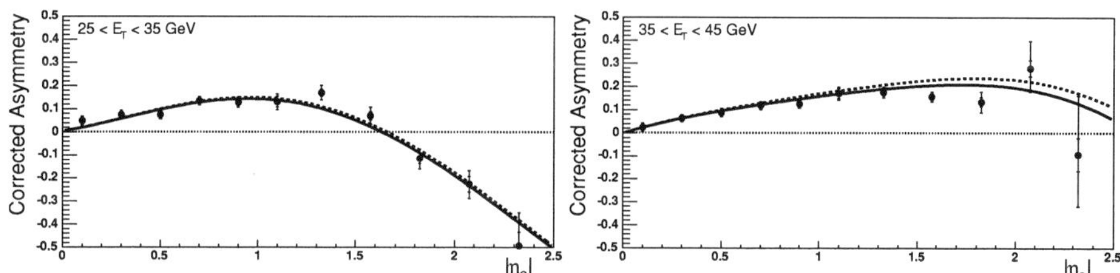

FIGURE 6. Measured forward-backward charge asymmetry from $W \longrightarrow e\nu$ production as a function of the electron pseudo-rapidity, $A(|\eta_e|)$, for 2 intervals of electron transverse energy: $25 < E_T < 35$ GeV (left) and $35 < E_T < 45$ GeV (right). Theoretical predictions from CTEQ6.1M (solid line) and MRST02 (dashed line) PDFs are overlaid on the data.

Direct measurement of the W width

In the Standard Model, the width of the W boson is precisely predicted in terms of the masses and coupling constants of the gauge bosons: $\Gamma_W = 2.092 \pm 0.003$ GeV [18]. DØ has performed a direct measurement of the W width using 177 pb^{-1} of Run II data in the $W \longrightarrow e\nu$ channel. It was determined from a binned maximum likelihood fit to the transverse mass distribution in the region $100 < \mathrm{M_T} < 200$ GeV/c^2. Figure 7 (left) shows the good agreement between data and Monte Carlo for the fitted width. The obtained result is $\Gamma_W = 2.011 \pm 0.093_{(\mathrm{stat})} \pm 0.099_{(\mathrm{syst})}$ GeV, in good agreement with the Standard

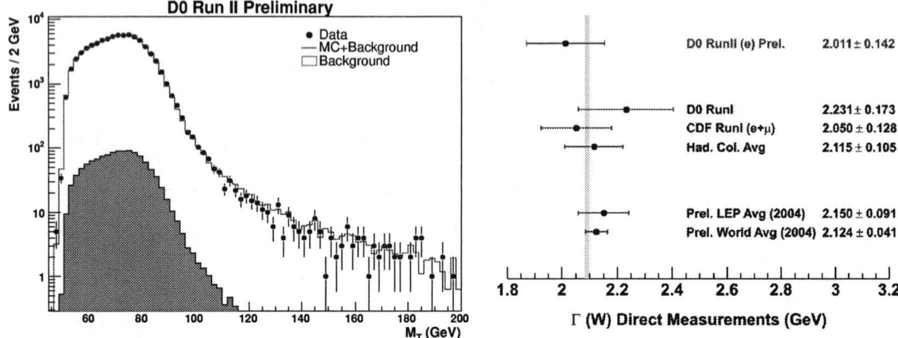

FIGURE 7. *Left:* Comparison of data to Monte Carlo templates for the shape of the transverse mass in $W \to e\nu$ candidates. The dots with error bars are data. The shadowed area is the QCD background. The line corresponds to the sum of the QCD background and of the $W \to e\nu$ and $W \to \tau\nu$ Monte Carlo samples for the fitted value of the W width. The normalization of the Monte Carlo samples is obtained in the transverse mass region $50 \leq M_T \leq 100$ GeV/c^2. The W width is then obtained fitting the tail region $100 \leq M_T \leq 200$ GeV/c^2. *Right:* Comparison of the preliminary result obtained in this analysis to previously published direct measurements of the W boson width. The shaded band indicates the Standard Model prediction.

Model prediction. The main systematic uncertainties come from the modeling of the hadronic response and resolution (64 MeV), the underlying event (47 MeV) and the electromagnetic response (30 MeV). Figure 7 (right) shows that the achieved uncertainty is comparable to Run I ones.

Top quark mass

The top quark mass is a fundamental parameter of the Standard Model. Because radiative corrections are often dominated by the large top quark mass, it plays an important role toward precise prediction of electroweak observables, such as the Higgs boson mass as shown in figure 8 (left). Due to its large mass, the top quark is particularly sensitive to the electroweak symmetry breaking mechanism. A precise measurement of the top quark mass provides a crucial test of the consistency of the Standard Model and help constraining physics beyond the Standard Model.

At the Tevatron, the top quark is mostly pair produced through quark-antiquark annihilation and gluon-gluon fusion. Thanks to the Tevatron increase of center of mass energy, from 1.8 TeV in Run I to 1.96 TeV in Run II, the cross section has increased by about 30 %. Run II experimental results for what concerns the top quark pair production cross section agree with the Standard Model expectation [23].

Because its width ($\Gamma_t > 1$ GeV for $m_t > 120$ GeV/c^2) is quite larger than Λ_{QCD} (≈ 100 MeV), the top quark decays before hadronizing. In the Standard Model, the top quark decays almost exclusively to a W boson and a b quark. $t\bar{t}$ events are then classified with respect to the W decays. The most precise measurements of the top quark mass are

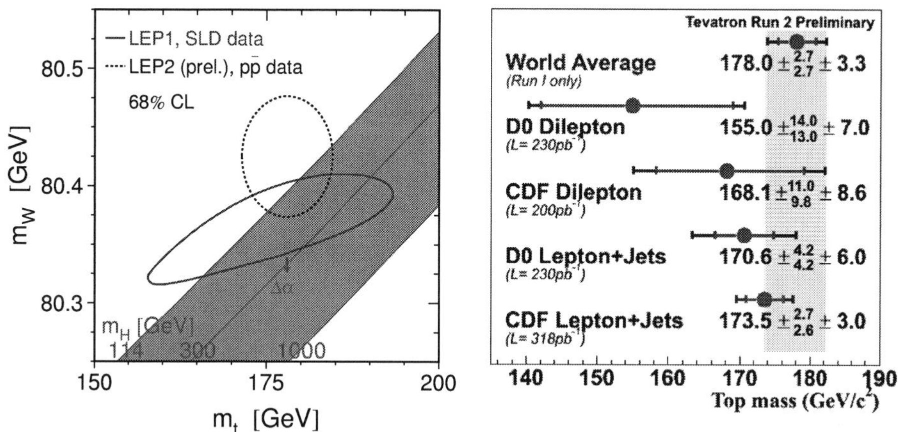

FIGURE 8. *Left:* Mass of the Higgs boson predicted by the electroweak fit as a function of the top quark mass and the W boson mass. Direct and indirect experimental contraints are also reported, they do not include Run II preliminary results. *Right:* Best Run II preliminary measurements of the top quark mass. The shaded band represents the world average from Run I.

obtained in the lepton plus jets sample in which one of the W decays to leptons and the other to quarks. The measurement has also been performed in the dilepton sample in which both W decay to leptons.

Figure 8 (right) summarizes the best Run II preliminary results at the time of the conference. CDF lepton plus jets analysis is the most recent one and achieves the best precision, even better than previous world average from Run I. In this study, the observed invariant mass distribution of the W hadronic decays is used to reduce the largest systematic uncertainty which arises from the jet energy scale. The different measurements agree with Run I combination, the central values tend to be a bit smaller.

SEARCHES

Search for neutral supersymmetric Higgs in multijet events

DØ has searched for neutral supersymmetric Higgs bosons produced in association with bottom quarks using 260 pb^{-1} of data collected in Run II. In the Minimal Supersymmetric extension of the Standard Model (MSSM), the $b\bar{b}\phi(\rightarrow b\bar{b})$ processes are enhanced at large $\tan\beta$. An excess in the distribution of the two leading jets invariant mass in events with at least three b-tagged jets was investigated.

Figure 9 (left) shows that the obtained dijet mass spectrum agrees very well with the estimated background. Limits on signal production cross sections were then set using a modified frequentist method [24]. Expected signal cross sections in the MSSM and their uncertainties were taken from NLO calculations [25, 26]. Figure 9 (right) shows the obtained limits in the $\tan\beta$ versus m_A plane on two MSSM scenarios: "no mixing"

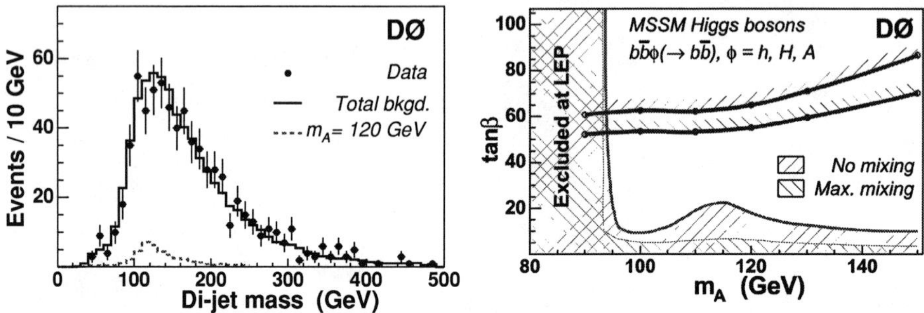

FIGURE 9. *Left:* Invariant mass spectrum of the two leading jets in events with at least three b-tagged jets, estimated background and expected signal for a 120 GeV/c^2 Higgs boson that can be excluded at 95 % CL. *Right:* 95 % CL upper limit on tan β as a function of m_A for the two scenarios of the MSSM, "no mixing" and "maximal mixing". Also shown are the limits obtained by the LEP experiments for the same two scenarios of the MSSM.

in the scalar top quark sector [27, 28], $X_t = A_t - \mu \cot\beta = 0$ where A_t is the tri-linear coupling and $\mu = -0.2$ TeV in the Higgsino mass parameter, and "maximal mixing", $X_t = \sqrt{6} \times M_{SUSY}$ where $M_{SUSY} = 1$ TeV is the mass scale of supersymmetric particles. A significant portion of the parameter space is excluded down to tan $\beta \approx 50$.

Search for sbottom quarks from gluino decays

CDF has searched for sbottom quarks from gluino decays using 156 pb^{-1} of Run II data. For supersymmetric scenarios with large tan β, the lighter sbottom mass eigenstate ($\tilde{b}_1$) can be significantly lighter than other squarks due to the substantial mixing in the sbottom sector [29]. This analysis assumes a scenario where the sbottom quark is lighter than the gluino and relies on the large gluino pair production cross section. R-parity conservation is assumed and the Lighest Supersymmetric Particle (LSP) is supposed to be the lighest neutralino ($\tilde{\chi}_1^0$). Each of the pair produced gluinos decays as follow: $\tilde{g} \to b\tilde{b}_1 \to b\bar{b}\tilde{\chi}_1^0$. The analysis requires four jets, large missing E_T and one or two b-tagged jets. In the more sensitive inclusive double b-tagged sample, see figure 10 (left), 4 events were observed in the signal region, missing $E_T > 80$ GeV, where 2.6 ± 0.7 events were expected from Standard Model processes. Figure 10 (right) shows the obtained limits in the gluino sbottom mass plane. Earlier limits are significantly extended.

ACKNOWLEDGMENTS

I am very grateful to the DIS 2005 organizers for their invitation. I would also like to acknowledge the members the CDF and DØ Collaborations for their works in achieving the results reported here. Finally, I would like to acknowledge the EU for its funding under the RTN contract: HPRN-CT-2002-00292, Probe for New Physics.

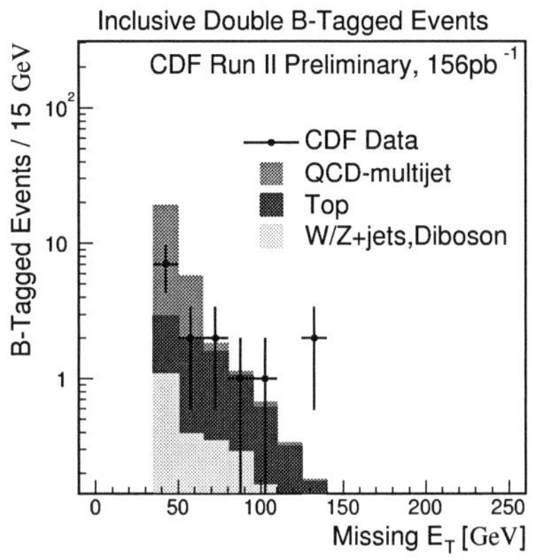

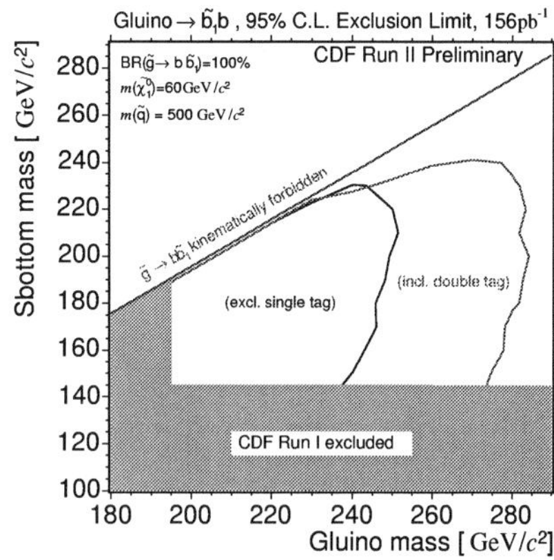

FIGURE 10. *Left:* missing E_T spectrum in the inclusive double b-tagged sample. Estimated backgrounds are overlaid on the data. *Right:* 95 % CL exclusion contours in the gluino sbottom mass plane obtained in the exclusive single b-tagged sample and in the inclusive double b-tagged sample. Previous limits from Run I are also reported.

REFERENCES

1. CDF Collaboration, D. Acosta *et al.*, Phys. Rev. D **71**, 032001 (2005).
2. T. LeCompte and H.T. Diehl, Annu. Rev. Nucl. Part. Sci. **50**, 71 (2000).
3. CDF Collaboration, http://www-cdf.fnal.gov/physics/physics.html
4. DØ Collaboration, http://www-d0.fnal.gov/Run2Physics/WWW/results.htm
5. G.C. Blazey *et al.*, Proceedings of the Workshop: "QCD and Weak Boson Physics in Run II", Batavia, Illinois, 1999, edited by U. Baur, R.K. Ellis and D. Zeppenfeld, 47 (2000).
6. M.H. Seymour, Nucl. Phys. B **513**, 269 (1998).
7. Z. Nagy, Phys. Rev. Lett. **88**, 122003 (2002); Z. Nagy, Phys. Rev. D **68**, 094002 (2003).
8. J. Pumplim *et al.*, JHEP **0207**, 12 (2002); D. Stump *et al.*, JHEP **0310**, 046 (2003).
9. A.D. Martin *et al.*, Phys. Lett. B **604**, 61 (2004).
10. T. Sjöstrand *et al.*, Comp. Phys. Comm. **135**, 238 (2001).
11. G. Marchenisi *et al.*, Comp. Phys. Comm. **65**, 465 (1992); G. Corcella *et al.*, JHEP **0101**, 010 (2001).
12. S.D. Ellis and D.E. Soper, Phys. Rev. D **48**, 3160 (1993).
13. W.T. Giele *et al.*, Phys. Rev. Lett. **73**, 2019 (1994).
14. R.D. Field, "ME/MC Tuning Workshop", Fermilab, October 2002.
15. DØ Collaboration, V.M. Abazov *et al.*, Phys. Rev. Lett. **94**, 221801 (2005).
16. CDF Collaboration, D. Acosta *et al.*, FERMILAB-PUB-05-156-E, accepted by Phys. Rev. D (2005).
17. C.R. Hamberg, W.L. van Neerven and T. Matsuura, Nucl. Phys. B **359**, 343 (1991).
18. S. Eidelman *et al.*, Phys. Lett. B **592**, 1 (2004).
19. CDF Collaboration, D. Acosta *et al.*, Phys.Rev. D **71**, 051104 (2005).
20. H.L. Lai *et al.*, Phys.Rev. D **51**, 4763 (1995).
21. A. Martin, R. Roberts, W. Stirling, and R. Thorne, Eur. Phys. J. C **4**, 463 (1998).
22. F. Landry, R. Brock, P. M. Nadolsky, C. P. Yuan, Phys. Rev. D **67**, 073016 (2003).
23. M. Cacciari *et al.*, JHEP **404**, 068 (2004).
24. T. Junk, Nucl. Instrum. Methods Phys. Res. A **434**, 435 (1999).
25. J. Campbell, R.K. Ellis, F. Maltoni and S. Willenbrock, Phys. Rev. D **67**, 095002 (2003).
26. S. Dawson, C.B. Jackson, L. Reina and D. Wackeroth, Phys. Rev. Lett. **94**, 031802 (2005).
27. M. Carena, S. Mrenna and C.E.M. Wagner, Phys. Rev. D **60**, 075010 (1999).
28. M. Carena, S. Mrenna and C.E.M. Wagner, Phys. Rev. D **62**, 055008 (2000).
29. A. Bartl, W. Majerotto ans W. Porod, Z. Phys. C **64**, 499 (1994); Erratum-ibid. C **68**, 518 (1995).

Lepton-Nucleon Spin Physics

D. Ryckbosch

University of Gent, Belgium
Proeftuinstraat 86, B-9000 Gent

Abstract. An overview of recent experimental results in the field of nucleon spin physics is given. The emphasis is on experiments using polarized deep inelastic scattering, but some important new results from e^+e^--annihilation and pp-scattering are included as well.

Keywords: Nucleon Spin Structure
PACS: 13.60.Hb

INTRODUCTION

Traditionally the topic of spin structure of the nucleon has been studied almost exclusively using polarized deep inelastic scattering (DIS). In the past, most of the experimental work dealt with the determination of the helicity structure of the nucleon, trying to determine the various contributions to its spin. Consequently, most of the data were on the helicity structure function $g_1(x)$ and its associated distribution function $\Delta q(x)$[1] and only very few other subjects received any attention. Two exceptions were measurements of the second polarized structure function $g_2(x)$ and the polarization distribution of the gluon $\Delta G(x)$.

It is only recently that other topics dealing with the nucleon spin structure received experimental attention as it was realized that a complete understanding of the quark-gluon structure of the proton required measurements of other observables as well. Topics like the Collins- and Sivers-effect, orbital motion of the quarks, higher twist, and other parton-correlation effects have been addressed both theoretically and experimentally and the new results have added considerably to our understanding of the structure of the polarized nucleon. At the same time new facilities have become available to probe this structure. Polarized fragmentation functions are now being probed at e^+e^- collider experiments and polarization phenomena in pp scattering are accessible with high accuracy from the RHIC-collider.

EXPERIMENTS

The main fixed-target experiments studying nucleon spin physics are listed in Tables 1 and 2. HERMES has been running the longest and to a large extent compensates its relatively low luminosity by the extended periods of data taking (for almost a decade

[1] Further on we will also use $g_1^q(x)$ to denote this distribution function.

CP792, *Deep Inelastic Scattering, DIS 2005,* edited by W. H. Smith and S. R. Dasu

TABLE 1. Main beam properties and (typical) kinematic coverage of the major lepton-nucleon spin physics experiments.

exp.	E_b (GeV)	x	Q^2 (GeV^2)	P_b
HERMES	27.6 $e^{\pm}$	0.02 - 0.6	0.1 - 15	$\pm$0.55
COMPASS	160 μ	0.003 - 0.6	1 - 100	-0.76
JLAB	<6 e^{-}	0.1 - 0.85	1 - 4.5	$\pm$0.7

TABLE 2. Main target properties of the major lepton-nucleon spin physics experiments.

exp.	P_t	target	$\mathscr{L}$ (cm^{-2}s^{-1})
HERMES	0.85	$\vec{\mathrm{H}},\vec{\mathrm{D}}$	10^{31}
COMPASS	0.50	$\vec{\mathrm{Li}}\vec{\mathrm{D}}$	$5 \cdot 10^{32}$
Hall A	0.35	$^3\vec{\mathrm{He}}$	10^{36}
CLAS	0.8 (0.3)	NH_3 (ND_3)	10^{34}

now) as compared to the other experiments. Moreover, the gaseous targets used by HERMES have no dilution from unpolarized nucleons, which can be a disadvantage in other experiments. At the present time both HERMES and COMPASS have yielded results on transverse degrees of freedom with comparable statistical accuracy. The JLAB experiments run at much higher luminosity but due to the low energy of the electron beam they are usually limited to a rather narrow range in kinematics. With the planned upgrade of the JLAB accelerator to 12 GeV this range will be substantially expanded in the future. Common to most experiments today is their large acceptance enabling the efficient detection of hadrons emitted in the DIS reaction. Only Hall A at JLAB works with (relatively) small acceptance magnetic spectrometers and is limited to inclusive DIS studies, albeit at high luminosity.

HELICITY DISTRIBUTIONS

Inclusive DIS experiments where only the scattered lepton is detected are mainly used to determine the polarized structure function $g_1(x)$ which -in the quark parton model- is related to the helicity distribution of quarks by:

$$g_1(x) = \frac{1}{2}\sum_q e_q^2[q^+(x) - q^-(x)] = \frac{1}{2}\sum_q e_q^2 \Delta q(x) \tag{1}$$

where $q^{+(-)}(x)$ denotes the probability to find a quark of flavour q with momentum fraction x and the same (opposite) helicity to that of the nucleon.

A summary of world data on the helicity structure function for the proton and the deuteron is shown in Figure 1. (See the contribution by D. Reggiani to these proceedings for more details on the HERMES data presented in this figure.) These data from several different experiments show a remarkable degree of agreement. In fact, since the data were taken at quite different values of the scale Q^2 this is important input to pQCD

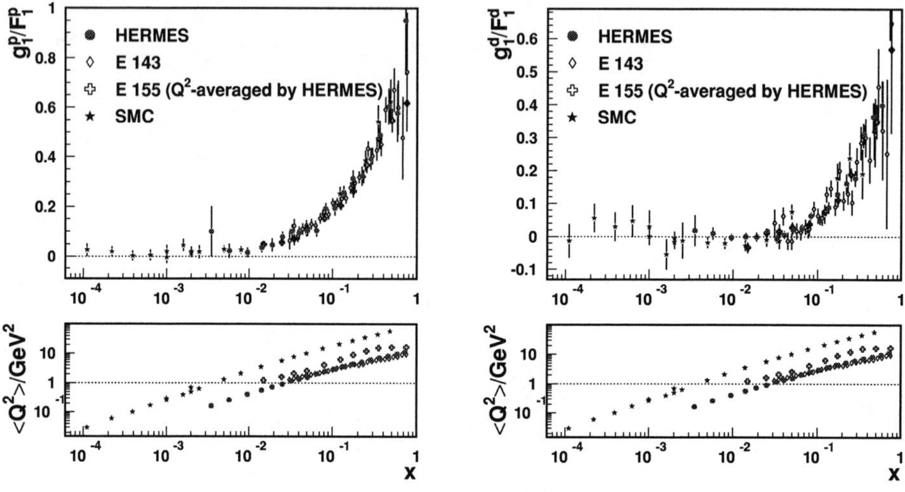

FIGURE 1. Summary of world data on $g_1(x)$

analyses of the helicity structure. Although the statistical accuracy is definitely good enough to allow a detailed analysis of the overall features of the structure function there still remain domains where our present knowledge is insufficient. This is mainly at the highest and lowest ends of the x-range.

At the highest values of x there are clear -and sometimes strikingly different- predictions from various models for g_1 on the proton and the neutron. The data plotted in Figure 1 are not accurate enough to decide which of the models can be ruled out. However, new data from JLAB (see the contributions by e.g. J-P. Chen to these proceedings) have considerably better statistical accuracy and indeed show the capability to distinguish between model predictions.

The lowest values in x are important to determine the sum rule over $g_1(x)$ which is directly proportional to the total quark spin contribution $\Delta\Sigma$ to the nucleon spin. Since the sum rule involves an integration over the unmeasured region at low x its value is highly dependent on the extrapolation to that region. Hence a better knowledge of g_1 at the lowest possible values of x is essential. The COMPASS experiment (see the contribution by J. Hannappel to these proceedings) has recently released results for a deuteron target which show an improvement in statistical accuracy in this x-domain of more than a factor of two over the older SMC data shown in the figure.

GLUON POLARIZATION

From pQCD analyses of the Q^2 dependence of the world data on g_1, and further data on among others the flavour decomposition of the helicity distribution $\Delta q(x)$ obtained by

HERMES [1], it is now clear that the quark spin only contributes a minor part to the total nucleon spin. The issue which most experiments have therefore been addressing lately is which other contributions make up the remainder of the nucleon spin. A prime suspect is the polarization of the gluons ΔG. The pQCD analyses performed so far all indicate a relatively large and positive polarization for the gluons, but the uncertainties on these results are large. This is mainly caused by the limited lever arm in Q^2 that is available in the data. More direct methods are needed. The gluons being electrically neutral ΔG cannot be probed directly in DIS and other ways have been identified to gain access to this quantity. In particular the production of (open) charm and the production of pairs of jets or hadrons at high transverse momentum are seen as promising avenues towards a determination of ΔG. In fact, all existing data up to now come from analyses of high-p_T pairs. Some years ago HERMES has published [2] a significantly positive value for ΔG, albeit with large systematic errors due to the uncertainties in determining the relative contributions of competing production mechanisms. Recently, SMC [3] has released a negative value which is, however, consistent with zero. The COMPASS experiment is ideally placed to remedy this unsatisfactory experimental situation due to its higher center-of-mass energy than e.g. HERMES. In the contribution by C. Bernet to these proceedings the COMPASS experiment shows the most accurate determination to day of ΔG, which is within error bars equal to zero. Further data from COMPASS also from open charm production should help to elucidate this situation in the near future.

THE STRUCTURE FUNCTION $G_2(X)$

In inclusive scattering of longitudinally polarized leptons off longitudinally polarized nucleons a second structure function g_2 arises. This is related to the transverse polarization of the target nucleon with respect to the virtual photon direction, and was long wrongly assumed to represent the transverse spin structure of the nucleon. In fact, it is the best known example of a twist-3 function. As seen in Eq. (2) this structure function contains a twist-2 part, directly related to the standard polarized structure function g_1 and a part stemming from actual twist-3 operators $\tilde{g}_2$.

$$g_2(x) = -g_1(x) + \int_x^1 g_1(x')dx'/x' + \tilde{g}_2(x) = g_2^{WW}(x) + \tilde{g}_2(x) \qquad (2)$$

Several experiments have been performed to determine g_2, and in particular to measure the deviation from the (dominant) twist-2 part. Moments of this deviation can be compared directly to results of lattice QCD calculations. The most precise determination of these moments were given by the E155x experiment [4] at SLAC some years ago. However, in several limited kinematic domains two JLAB experiments have now shown results for g_2 with considerably improved statistical accuracy. In contributions to this workshop the latest results are discussed in detail.

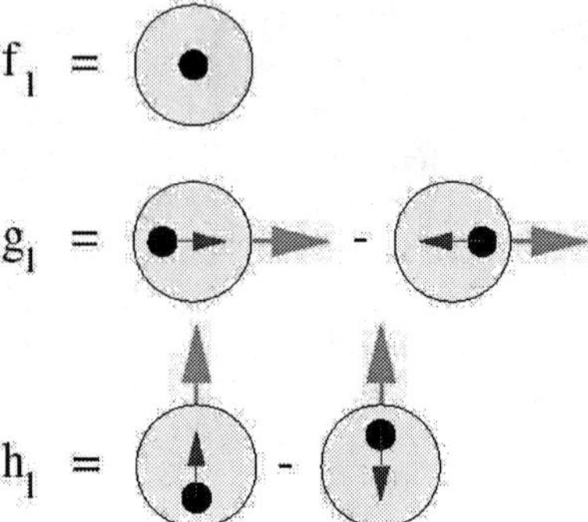

FIGURE 2. Schematic representation of the three twist-2 distribution functions that survive integration over intrinsic transverse momentum: the momentum density f_1, the helicity distribution g_1 and the transversity h_1.

TRANSVERSITY

A full tree-level description of all distribution and fragmentation functions appearing at twist-2 and twist-3 in polarized DIS has been available for some time [5]. Most of these functions are explicitely dependent on intrinsic transverse momentum of the quarks in the nucleon or hadron. Only 3 functions do not disappear when integrated over this intrinsic k_T or p_T. For the distribution functions these are the unpolarized distribution $f_1^q(x)$, the helicity distribution $g_1^q(x)$ and the transversity distribution $h_1^q(x)$. When integrated over x these distributions contain information on the vector charge, the axial charge and the tensor charge, respectively, of the nucleon. While the former two are well known from inclusive DIS measurements, there is no experimental information on the latter one. This is a consequence of the chiral-odd nature of this distribution function, which means that it cannot appear in an inclusive cross section. The transversity distribution is chiral-odd since it involves, in a helicity basis, a simultaneous helicity flip of the target and the quark. For massless quarks this is impossible in a pure inclusive reaction.

The transversity distribution has several features which make it an attractive object of study. As indicated before it is the last forward distribution function to be measured after $f_1(x)$ and $g_1(x)$. It differs from the helicity distribution in a subtle way. Firstly, for relativistic particles the Lorentz boost and rotations do not commute, which means that in this case helicity and transversity are different. Differences between $g_1^q(x)$ and $h_1^q(x)$ thus give information on the relativistic nature of the quark motion. Secondly, and probably more importantly, is the fact that since the transversity distribution involves a spin-flip amplitude which is impossible for (spin 1) gluons in a spin 1/2 target, the transversity distribution decouples from the gluons. This leads to a very different QCD-evolution for $h_1^q(x)$ as compared to $g_1^q(x)$.

As mentioned before it is impossible to observe transversity in inclusive DIS. However, in a semi-inclusive DIS reaction, where at least one of the produced hadrons is

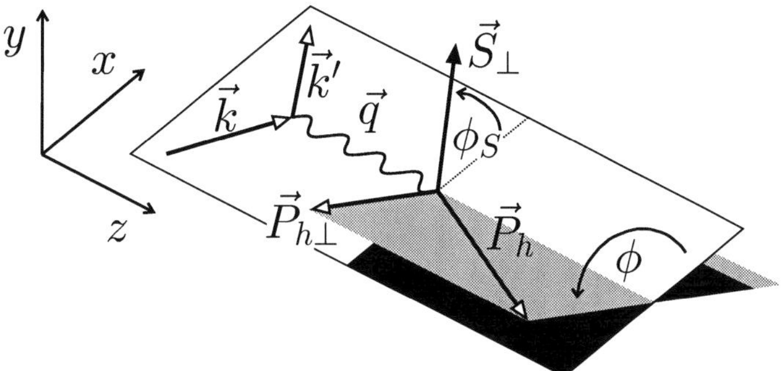

FIGURE 3. Definition of the various azimuthal angles relevant in semi-inclusive DIS on a transversely polarized target. $\vec{S}_\perp$ indicates the direction of the nucleon target spin, while $\vec{k}$, $\vec{k}'$ and $\vec{P}_h$ represent the momentum of the incoming lepton, the scattered lepton and the produced hadron, respectively

detected, the cross section also depends on the fragmentation function:

$$\sigma^{ep\to ehX} = \sum_q f^{p\to q} \otimes \sigma^{eq\to eq} \otimes D^{q\to h}. \tag{3}$$

In this equation $f^{p\to q}$ is the distribution function representing the probability to find a quark q in the proton p (possibly carrying a certain spin polarization), $\sigma^{eq\to eq}$ is the elementary $e-q$ cross section as given by QED, and $D^{q\to h}$ is the fragmentation function describing the probability to find a certain hadron h in a quark q.

This factorized approach to semi-inclusive DIS has been used with great success by the HERMES collaboration [1] to determine the flavour decomposition of the helicity distribution $g_1^q(x)$ (or in a different notation $\Delta q(x)$). In that case the unpolarized fragmentation function $D_1(z)$ was used as a flavour filter to disentangle the contribution of different quark flavours.

Collins effect

In the case of transversity the presence of a second soft object in the form of a fragmentation function allows the actual observation of the distribution function provided that the fragmentation function is also chiral-odd. Such a mechanism is the basis of the Collins effect [6] where the chiral-odd Collins fragmentation function $H_1^\perp(z)$ appears. This fragmentation function describes a correlation between the direction of the outgoing hadron and the transverse spin direction of the initial quark. An important feature of the Collins fragmentation function is that it is also (naive) time-reversal odd (see e.g. the contribution of D. Sivers to these proceedings). As such it causes the appearance of a single spin asymmetry (SSA) in the azimuthal distribution of the outgoing hadrons. Such SSA's have already been observed by HERMES [7] using a longitudinally polarized target, indicating that the Collins fragmentation function is non-zero.

The relevant azimuthal angles are defined in fig. 3. Transversity manifests itself in the Collins effect through a sine modulation in the angle $\Phi = \phi + \phi_S$.

Sivers effect

A completely different mechanism has been suggested that also leads to SSA's in semi-inclusive DIS. A correlation between the intrinsic transverse momentum of an unpolarized quark and the direction of the transverse spin of its parent nucleon can exist and is described by the Sivers distribution function $f_{1T}^{\perp}(x)$ [8]. Part of this correlation may survive the fragmentation process and result again in a correlation between the target spin direction and the direction of the outgoing hadron. This effect was proposed more than a decade ago as a possible explanation for the observed asymmetries in hadron-hadron scattering experiments, but was largely ignored in the context of semi-inclusive DIS. This is because the corresponding distribution function $f_{1T}^{\perp}(x)$ is, like the Collins fragmentation function, a (naive) time-reversal odd object. Hence, the observation of effects related to $f_{1T}^{\perp}(x)$ requires the interference of at least two amplitudes, but this was assumed to be impossible for a distribution function. However, recent insights [9] have shown that restoration of gauge invariance entails the existence of gauge links which appear as a kind of final state interaction involving the exchange of a soft gluon. This extra diagram then enables the existence of the naive time-reversal odd distribution functions.

The importance of such final state interactions could call the universality of distribution and fragmentation functions in doubt. It was, however, shown that up to a non-trivial sign change the distribution functions and fragmentation function are indeed universal between semi-inclusive DIS, Drell-Yan production and e^+e^- annihilation [10].

An interesting feature of the Sivers distribution function is, apart from its T-odd character, the fact that it requires a non-zero orbital angular momentum of the unpolarized quark. It may therefore eventually give access to this elusive component of the nucleon spin structure.

In semi-inclusive DIS the Sivers effect appears as a SSA sine modulation in the angle $\phi - \phi_S$. It can be seen from fig. 3 that this angle does not depend on the orientation of the lepton scattering plane, but only on the angle between the target spin vector and the hadron production plane, as should be the case for the mechanism outlined above.

Single Spin Asymmetries

Both the COMPASS and HERMES collaborations have now released first SSA measurements based on semi-inclusive DIS experiments on transversely polarized targets. While COMPASS used polarized LiD as a deuterium target, HERMES has used a pure hydrogen target (see the contributions by P. Pagano and M. Diefenthaler to these proceedings). The results of HERMES are displayed in figs. 4 and 5. The Collins asymmetries are somewhat surprising. Firstly, their magnitude is smaller than what could be expected on the basis of the published longitudinally polarized target data [11]. Secondly, the absolute magnitude for the π^- asymmetry is actually larger than the one for the π^+ mesons. Keeping in mind that the contribution of u-quarks is usually dominant due to their larger electric charge, one possible explanation is that the disfavoured Collins fragmentation function of a u-quark hadronizing into a π^- has a sizeable magnitude and is of the opposite sign as compared to the favoured fragmentation function. Since the Collins

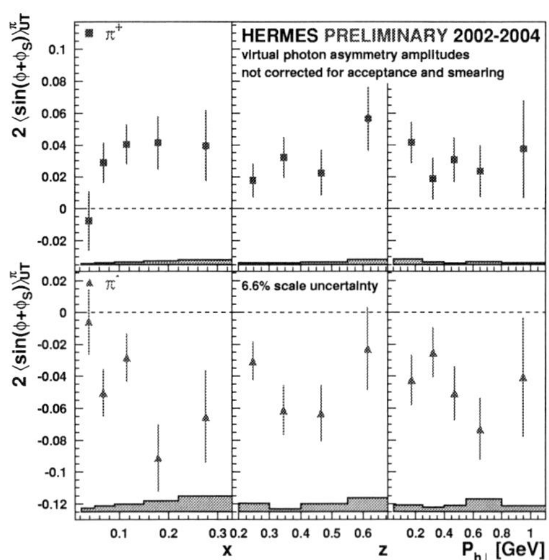

FIGURE 4. HERMES preliminary results for the amplitude of the Collins asymmetries on a transversely polarized hydrogen target, as a function of the kinematic variables x, z and $P_{h\perp}$. The top panels correspond to π^+ production, while the bottom panels represent π^- production.

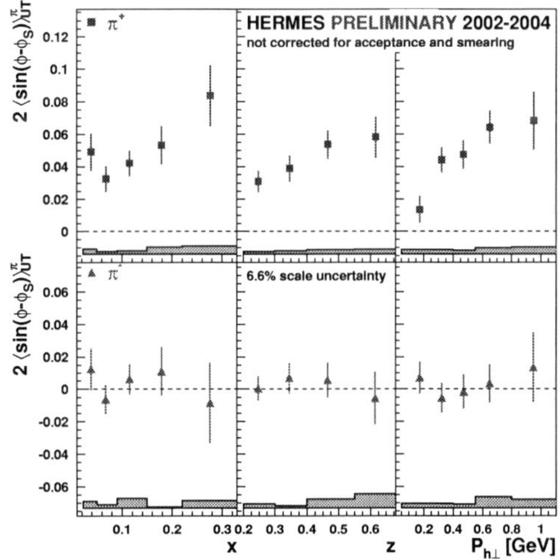

FIGURE 5. HERMES preliminary results for the amplitude of the Sivers asymmetries on a transversely polarized hydrogen target, as a function of the kinematic variables x, z and $P_{h\perp}$. The top panels correspond to π^+ production, while the bottom panels represent π^- production.

fragmentation function actually describes a correlation between two directions there is nothing inherently strange about a negative sign. Such a possibility could also explain the COMPASS result which is consistent with zero on a deuterium target because the proton and neutron fragmentation functions may entail strong cancellations.

In any case the significantly non-zero results from HERMES prove the existence of the Collins fragmentation function and the transversity distribution function. Fortunately the Collins mechanism is not the only way to access transversity. There are other chiral-odd fragmentation functions which can act in conjunction with $h_1(x)$ to produce observable asymmetries. In particular there has been the suggestion to look at asymmetries in 2-hadron production where an interference fragmentation function would be active. This would provide an interesting independent measurement of transversity (see contribution by P. van der Nat to these proceedings).

The Sivers asymmetries from HERMES are also significantly different from zero, whereas the COMPASS results are again consistent with zero. In this case there is only one unknown distribution at work: the Sivers distribution function $f_{1T}^{\perp}(x)$, which in conjunction with the normal unpolarized fragmentation function $D_1(z)$ determines the semi-inclusive DIS cross section. The observation of a non-zero asymmetry is immediate evidence for a non-zero orbital angular momentum for the quarks. This would in particular hold for the u-quark which determines the π^+-asymmetry. Further analysis of both HERMES and COMPASS results should make it possible to have at least a rough idea of the flavour decomposition of the Sivers distribution function.

Collins Fragmentation Function

As mentioned before, the Collins asymmetries observed in semi-inclusive DIS depend on two completely unknown functions: the Collins fragmentation function and the transversity distribution. In order to determine the transversity distribution separately one needs independent information on the fragmentation function. This can come from analysis of high-energy e^+e^- collisions. In the past there have been preliminary analyses of LEP data which gave tentative evidence for a non-vanishing Collins fragmentation mechanism, but they were certainly not conclusive about the magnitude of the function.

Recently there has been an effort going on to analyse the massive amount of fragmentation data available from the high-luminosity asymmetric e^+e^- colliders built for studies of CP violation in B-physics. In particular a group at the BELLE experiment has looked for evidence for polarized fragmentation in their data. In his contribution to these proceedings R. Seidl shows for the first time conclusive evidence for a significant non-zero Collins fragmentation function. Further refined analysis of these data, together with the semi-inclusive data, will eventually enable the full determination of the transversity distribution.

Hadron-hadron scattering

Historically the first non-zero SSA's were observed in pion production in $\vec{p}p$-scattering [12]. There have been many theoretical calculations aiming at a reproduction of these data, basically invoking either the Collins mechanism, the Sivers mechanism or both. A point of uncertainty has, however, always been the rather low scale of the measurements. With the advent of RHIC, and in particular the availability of intense

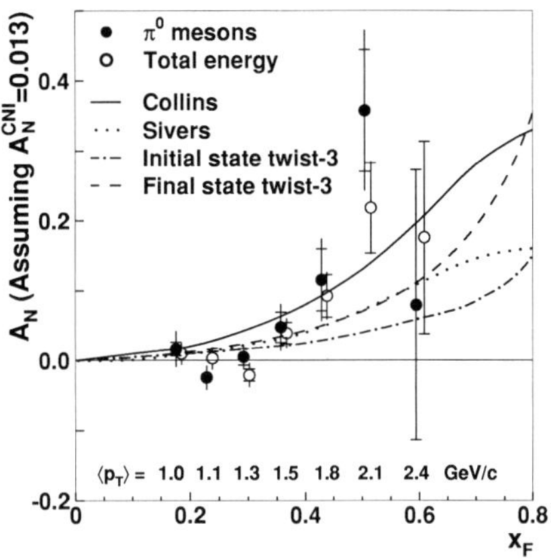

FIGURE 6. Analyzing powers for π^0 production in STAR [13].

proton beams with high polarization, this situation is being remedied. Recent data from both the PHENIX and STAR experiments have shown very good agreement between measured cross sections and NLO pQCD calculations, thus establishing the validity of this framework in the kinematic region covered. The RHIC experiments have now published the first results on single spin asymmetries which confirm and extend the older data. In fig. 6 the analyzing power for π^0 production from STAR [13] is plotted as an example. It is clear from the figure that models based on either Collins or Sivers mechanisms are able to reproduce the data with the present accuracy. However, with the much improved statistics which will become available with continued running of the RHIC-spin programme, a differentiation between the models may be possible. (See the contributions by S. Heppelmann, M. Chiu and F. Videbaek to these proceedings.) This would again give an independent means of determining the transversity distribution, provided that the Collins mechanism proves to be dominant.

In the last runs RHIC actually had both proton beams polarized. This makes the study of double-polarized processes possible. Of particular interest will be the determination of the gluon polarization ΔG. First preliminary results have shown the power of this method (see the contributions by A. Desphande and R. Cadman to these proceedings), which is again completely independent of the measurements in lepton scattering. The statistical accuracy at present is, however, not yet sufficient to edtract a value of ΔG, with a meaningfull significance.

Subleading twist

In a previous section we already discussed the possibility to make quantitative studies of higher twist contributions. Also in the context of SSA measurements this is possible. HERMES has collected a large amount of semi-inclusive DIS data on a longitudinally polarized target. The SSA for such a target contain contributions from both the Collins

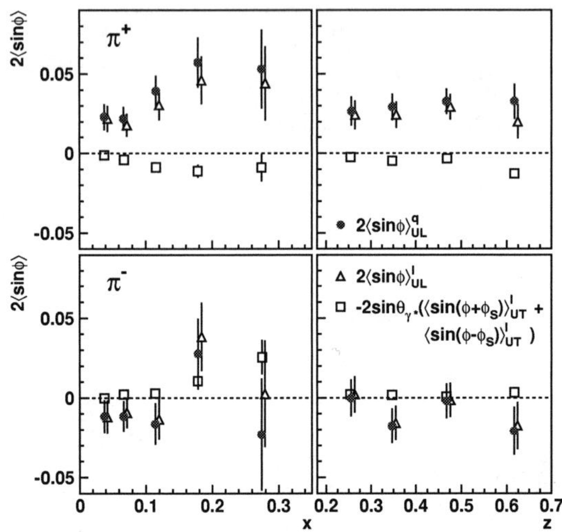

FIGURE 7. The various azimuthal moments appearing in the measurement of the $\sin\phi$ modulations of SSA on a longitudinally polarized proton target. Open triangles: total SSA; open squares: contribution from Collins and Sivers mechanisms; closed circles: subleading twist contributions.

and the Sivers effects, but also contributions from unmeasured twist-3 functions. (In fact, one usually sees a twist-3 distribution function coupled with a twist-2 fragmentation function and vice-versa [5].) Using the asymmetries for Collins and Sivers measured separately on a transverse target (and corrected for different kinematic factors) one can subtract the leading twist contribution to the longitudinal SSA and derive an estimate for the subleading twist part. This was recently done by HERMES [14] and the result is shown in fig. 7. It is observed that at the moderate Q^2 of the experiment the subleading twist terms can be large, certainly in conditions where the leading twist contributions are suppressed (like here).

CONCLUSIONS

The field of nucleon spin physics has seen a proliferation of new topics over the last decade. This reflects our deeper understanding of the complexities involved in the spin degrees of freedom in the nucleon. New sectors of the spin structure are being explored at several, often new, experiments. It is particularly gratifying to see that also facilities outside the traditional field of polarized DIS are now actively contributing to the growing body of experimental data.

ACKNOWLEDGMENTS

It is a pleasure to thank the different experiments for the information they provided me with to prepare this talk. In particular I would like to acknowledge input from J.-P.Chen,

M.Grosse Perdekamp, D.Hasch, S.Kuhn, J.Pretz and W.A.Zajc.

REFERENCES

1. A. Airapetian, et al (HERMES collaboration), *Phys.Rev.*, **D71**, 012003 (2005).
2. A. Airapetian, et al (HERMES collaboration), *Phys.Rev.Lett.*, **84**, 2584-2588 (2000).
3. B. Adeva, et al (SMC collaboration), *Phys.Rev.*, **D70**, 012002 (2004).
4. P.L. Anthony, et al (E155 collaboration), *Phys.Lett.*, **B553**, 18-24 (2003).
5. P.J. Mulders, and R.D. Tangerman, *Nucl.Phys.*, **B461**, 197–237 (1996).
6. J.C. Collins, *Nucl.Phys.*, **B396**, 161 (1993).
7. A. Airapetian, et al (HERMES collaboration), *Phys.Lett.*, **B562**, 182 (2003).
8. D.W. Sivers, *Phys.Rev.*, **D41**, 83 (1990).
9. S.J. Brodsky, D.S. Hwang, and I. Schmidt, *Phys.Lett.*, **B530**, 99 (2002).
10. X. Ji, J-P. Ma, and F. Yuan, *Phys.Rev.*, **D71**, 034005 (2005).
11. A. Airapetian, et al (HERMES collaboration), *Phys.Rev.Lett.*, **94**, 012002 (2005).
12. D. Adams, et al, *Phys.Lett.*, **B264**, 462 (1991).
13. J. Adams, et al (STAR collaboration), *Phys.Rev.Lett.*, **92**, 171801 (2004).
14. A. Airapetian, et al (HERMES collaboration), hep-ex/0505042.

Parton Distributions

Jon Pumplin

Department of Physics and Astronomy, Michigan State University East Lansing MI 48824

Abstract. I present an overview of some current topics in the measurement of Parton Distribution Functions.

Keywords: Parton distributions, QCD, Hadron Colliders
PACS: 12.38.Bx,12.38.Qk,13.60.Hb

INTRODUCTION

Parton distribution functions describe the quark and gluon content of a hadron when the parton-parton correlations and spin structure have been integrated out. By the asymptotic freedom of QCD, the PDFs are the only aspect of initial-state hadron structure that is needed to calculate short-distance hard scattering. Hence PDFs are a *Fundamental Measurement:* a challenge to understand using methods of nonperturbative QCD; and a *Necessary Evil:* essential input to perturbative calculations of signal and background at hadron colliders.

The parton distributions are functions $f_a(x,\mu)$ that tell the probability density for a parton of flavor a in a proton or other hadron, at momentum fraction x and momentum scale μ. An overview is shown in Fig. 1 at $\mu = 2\,\text{GeV}$ and $\mu = 100\,\text{GeV}$. Valence quarks dominate at $x \to 1$, while gluons dominate at small x, especially at large μ—hence their vital importance for the LHC.

The PDFs are measured through a "global analysis" in which a large variety of data from many experiments that probe short distance are fitted simultaneously. The full paradigm consists of the following steps:

1. Parameterize the x-dependence for each flavor at a fixed small μ_0, using functional forms that contain "shape parameters" $A_1,\ldots,A_N$.
2. Compute the PDFs $f_a(x,\mu)$ at $\mu > \mu_0$ by the DGLAP equation.
3. Compute the cross sections for DIS(e,μ,ν), Drell-Yan, Inclusive jets, etc. by perturbation theory.
4. Compute the "χ^2" measure of agreement between the predictions and experiment:
$$\chi^2 = \sum_i \left(\frac{\text{Data}_i - \text{Theory}_i}{\text{Error}_i}\right)^2$$
or generalizations of that formula to include correlated experimental errors.
5. Minimize χ^2 with respect to the parameters $\{A_i\}$ to obtain Best Fit PDFs.
6. Map the PDF uncertainty range as the region in $\{A_i\}$ space where χ^2 is sufficiently close to the minimum.

CP792, *Deep Inelastic Scattering, DIS 2005,* edited by W. H. Smith and S. R. Dasu

7. Make the best fit and uncertainty sets available at `http://durpdg.dur.ac.uk/HEPDATA/`. At that site, you can find the current CTEQ, MRST, Alekhin, H1, and ZEUS sets, along with some older CTEQ, MRST, and GRV sets.

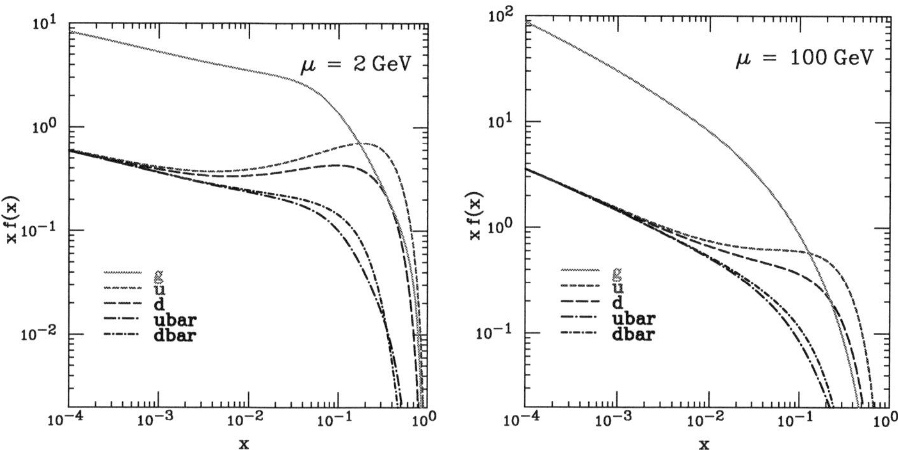

FIGURE 1. Overview of PDF results from CTEQ6.

The global analysis rests on three solid theoretical pillars:

1. *Asymptotic Freedom* $\Rightarrow$ QCD interactions are weak at large scale μ (short distance), so a perturbative expansion in powers of $\alpha_s(\mu)$ at NLO or NNLO can be used to make the calculations;
2. *Factorization Theorem* $\Rightarrow$ PDFs are universal, i.e., the same for all processes;
3. *DGLAP evolution* $\Rightarrow$ the dependence of $f_a(x,\mu)$ on momentum scale μ is perturbatively calculable, so only the dependence on light-cone momentum fraction x for each flavor a at a fixed small μ_0 needs to be measured.

Carrying out the global analysis brings a constant battle against the following challenges:

- Extracting continuous functions from a finite set of measurements is mathematically unclean. In particular, the analysis is carried out by modeling the PDFs at μ_0 by smooth functions containing free parameters. The bias that can result from the choice of these functional forms is known as "parametrization dependence."
- A specific type of parametrization dependence consists of simplifying assumptions that are made in the absence of data, such as strangeness symmetry ($f_s(x,\mu_0) = f_{\bar{s}}(x,\mu_0)$) which was assumed in most older analyses; or the condition of no intrinsic charm ($f_c(x,m_c) = f_{\bar{c}}(x,m_c) = 0$) which is still used.
- Combining data from diverse experiments is frequently made difficult by the presence of unknown errors. This is true even when a single discrete quantity like the mass or lifetime of a particle is measured; but the situation is worse when one is attempting to measure a large number of parameters (the $\{A_i\}$) from many experiments that are sensitive to different combinations of those parameters.

- The quality of the fit to data is measured—by tradition and because there is no obvious better alternative—by a global χ^2 that is based on the reported experimental errors. But unquantified experimental and theoretical systematic errors are found to be almost an order of magnitude larger, based on the level of inconsistencies observed between the "pulls" of various data sets, so the reported experimental errors do not really provide the ideal weighting of the data points.

THE LANDSCAPE IN X AND μ

The kinematic regions of interest in x and μ are shown in Fig. 2 (lifted from a talk by James Stirling). One sees that a large range of scales in μ are connected by DGLAP evolution. The consistency or inconsistency between the different processes that make up the global analysis can be tested only by the global fit, since every experiment depends differently on the PDFs. The LHC will dramatically extend the region of the measurements and their applications—especially at small x.

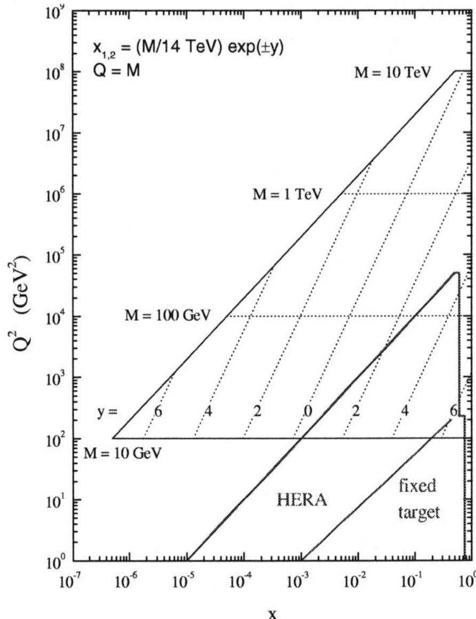

FIGURE 2. Kinematic map in x and $Q \equiv \mu$.

The DGLAP evolution in μ arises from parton branching, so the PDFs at a given x and μ can be thought of as arising from PDFs at smaller μ and larger x. To develop a feel for how this works quantitatively, Fig. 3 shows the regions where one or more of the PDFs changes by $>0.2\%$ (solid) or $>0.05\%$ (dotted) when a 1% change is made in $\bar{u}+\bar{d}$ or $u_v \equiv u-\bar{u}$ or g at $\mu_0 = 1.3\,\text{GeV}$ in a narrow band of x at various values of x. One sees that the valence quarks are unimportant at small x—as expected—and that quark evolution is effectively at constant x, i.e., the quark distributions at a given x and μ are mainly influenced by the quarks at μ_0 at the same x. The gluon at very large x similarly evolves in its own world. But the influence of changes in the input $g(x)$ at moderate x spreads out rapidly.[1] The small-x gluon at $\mu_0 = 1.3\,\text{GeV}$ has little direct influence because gluons at moderate and high μ are mainly generated radiatively.

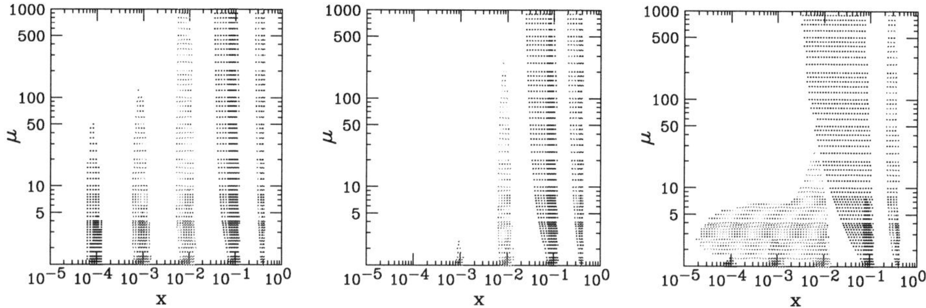

FIGURE 3. Regions of change caused by small changes in $\bar{u}(x) + \bar{d}(x)$ (left) or $u_v(x)$ (center) or $g(x)$ (right) at $\mu = 1.3\,\text{GeV}$ and $x = 10^{-4}, 10^{-3}, 10^{-2}, 10^{-1}, 0.4$.

PDF UNCERTAINTIES

A large effort has been made in recent years to intelligently consider the uncertainties of PDF measurements. This is a more difficult problem than most other error estimates because the objects are functions, rather than discrete values; and because they are extracted from a complicated stew of experiments of different types. Obvious sources of uncertainty are

1. Experimental errors included in χ^2
2. Unknown experimental errors and theory approximations
3. Higher-order QCD corrections + Large Logs
4. Power Law QCD corrections ("higher twist")
5. Parametrization dependence
6. Nuclear corrections for data that are taken on deuterium or other nuclear targets.

Essential difficulties arise from the fact that

- Experiments run until systematic errors dominate, so the remaining systematic error estimates involve guesswork.
- The systematic errors of the theory (e.g. power-law corrections or the approximations of NLO or NNLO) and their correlations are even harder to guess.
- Some combinations of PDFs are unconstrained, like $s - \bar{s}$ was before the NuTeV data.

Empirically, the essence of the uncertainty problem is illustrated by the hypothetical Figure 4. Suppose the quantity θ is measured in two different experiments. What would

[1] This results from the form of the distributions at μ_0—it is not just a property of the DGLAP evolution.

you quote as the central value and the uncertainty? (To play along at home, write down your answer before reading further!) Perhaps you would expand the errors to make the uncertainty range cover both data sets. Or perhaps you would expand it even more, using the difference between experiments as a measure of the uncertainty. Perhaps you would also be suspicious that the spread in the points from each experiment is too small compared to the quoted errors.

In the global analysis, we don't get to see the conflicts so directly as in Figure 4. But if you make a best fit to the points in Figure 4 with different weights assigned to the two experiments, the variation of the best fit value with the choice of weights would map out the information that is needed. That approach can be used in the global fit. In practice to quantify the uncertainties of the PDF global fit, we retain χ^2 as the measure of fit quality, but vary weights of the experiments to estimate a range of acceptable $\Delta\chi^2$ above the minimum value, in place of the classical $\Delta\chi^2 = 1$. Estimates for the current major global analyses are that something like $\Delta\chi^2 = 50 - 100$ corresponds to a $\sim 90\%$ confidence interval.

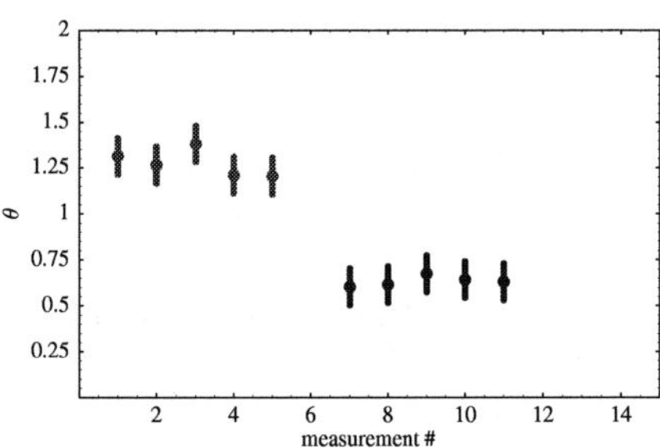

FIGURE 4. Hypothetical measurements of a quantity θ by two experiments.

EIGENVECTOR UNCERTAINTY SETS

A convenient way to characterize the uncertainty of the PDFs is to create a collection of fits by stepping away from the minimum of χ^2 along each eigenvector direction of the quadratic form (the Hessian matrix) that describes the dependence of χ^2 on the fitting parameters $\{A_i\}$ in the neighborhood of the minimum. This stepping is done in both directions along each eigenvector to allow asymmetric errors, so the 20 fitting parameters in CTEQ6 lead to 40 eigenvector uncertainty sets. The PDF uncertainty for any quantity is obtained by evaluating that quantity with each of the eigenvector sets and then applying a simple formula; or more crudely just directly from the spread in eigenvector predictions. This method has proven so useful that generating uncertainty sets should be regarded as an essential part of the job for every general-purpose PDF determination. In order to do this properly, CTEQ has developed an iterative procedure [1] to compute the eigenvector directions in the face of numerical instabilities that arise from the large dynamic range in eigenvalues of the Hessian. (Other PDF groups have adopted the eigenvector method as well, but they avoid the numerical difficulties by keeping substantially fewer free fitting parameters, e.g. 10 – 15, at a cost of greater parametrization bias.)

The uncertainty of the gluon distribution at $\mu = 2\,\text{GeV}$, as calculated by the eigenvector method, is shown in Fig. 5.[2] Also shown are best fits in which the data were reweighted to emphasize DØ (solid) or CDF (dashed) inclusive jet cross section measurements. Note that the uncertainty estimated by the eigenvector method is comparable to the difference between the "pull" of these two experiments. This shows that the eigenvector method is working correctly, and that the jet data are a major source of the information on the gluon distribution.

It is interesting that these two very similar experiments pull so differently, which suggests that the need to allow $\Delta\chi^2 \gg 1$ may be mainly due to unknown systematic errors in the experiments. (In support of that notion, we also find significant differences between the influences of the nominally similar H1 and ZEUS components of DIS data.)

The right-hand side of Fig. 5 demonstrates "convergent evolution:" the fractional uncertainty of the gluon is much smaller at large μ.

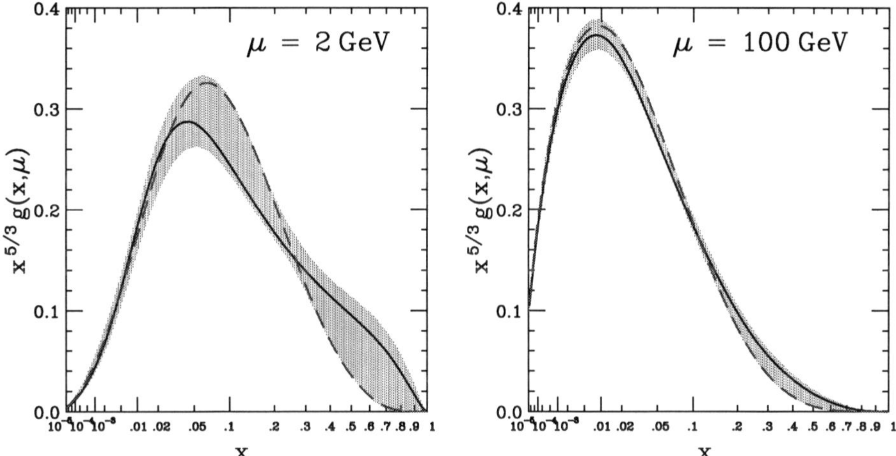

FIGURE 5. Gluon uncertainty at $\mu = 2\,\text{GeV}$ and $\mu = 100\,\text{GeV}$ from CTEQ6, plotted vs. $x^{1/3}$. Solid (dashed) curve has extra weight for DØ (CDF) jet data.

PDF COMPARISONS

The fractional uncertainty of the gluon distribution at an intermediate scale $\mu = 3.16\,\text{GeV}$ relative to CTEQ6 is shown in the left panel of Fig. 6. The dotted curve is CTEQ5, the previous generation of CTEQ PDFs. The dashed curve is CTEQ5HJ, which was an early milestone in the PDF uncertainty business: it accounted for the seemingly

---

[2] The envelope of these uncertainties is not itself an allowed solution, because the area under the curve is equal to the total gluon momentum, which is strongly constrained by DIS data. Hence if $g(x)$ is larger than the central value at $x \approx 0.5$ it must be smaller than the central value at $x \approx 0.05$.

high CDF inclusive jet cross section, relative to the QCD prediction, by an increased gluon distribution at large x that was well within the PDF uncertainty range. The solid curve is CTEQ6.1, which shows very little change from CTEQ6.0.

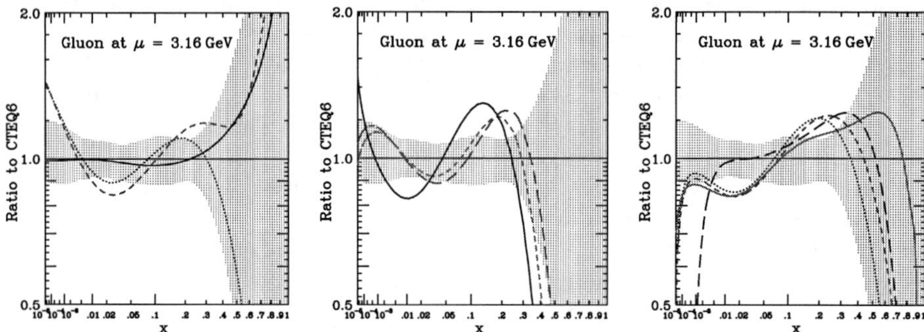

FIGURE 6. Gluon uncertainty at $\mu^2 = 10\,\mathrm{GeV}^2$ relative to cteq6. Left: cteq5 (dotted), cteq5HJ (dashed) and cteq6.1 (solid); Center: zeus2005zj (solid), Alekhin02NLO (long dash) and Alekhin02NNLO (short dash); Right: mrst2001 (dotted), mrst2002 (short dash), mrst2003 (long dash), mrst2004 (solid).

The center panel of Figure 6 shows a comparison with fits by the ZEUS group and by Alekhin. Since those fits are based on only a subset of the available data used in the CTEQ analysis, it is not surprising that they lie outside the CTEQ uncertainty bands. The difference between the Alekhin NLO (long dash) and NNLO (short dash) is seen to be small compared to the PDF uncertainty.

The right panel of Figure 6 shows a comparison with fits by the MRST group. The MRST fits have progressed toward a stronger gluon at large x, which is needed to obtain good fits to the inclusive jet cross sections. The small-x behavior is sensitive to parametrization assumptions that will be discussed later.

It is ironic that the differences between PDF determinations by the various major players are comparable to the estimated uncertainty. For our original motivation to study the uncertainties systematically was the danger that comparing results from different groups might greatly underestimate the uncertainty, since all groups use basically the same method.

NLO AND NNLO

Figure 7 shows a comparison with NLO and NNLO fits by the MRST group. At present, the difference between NLO and NNLO analysis is small compared to the PDF uncertainty. This is also apparent in the Alekhin fits shown in Fig. 6. Hence NNLO fitting, while obviously desirable on theoretical grounds, is not urgent. A reasonable goal would be to have a full set of NNLO global analysis tools—including jet cross sections—in place by the time LHC data taking begins.

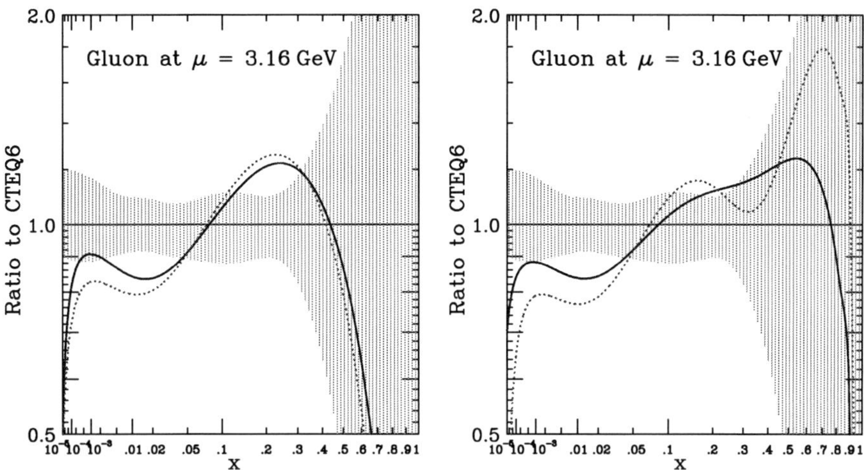

FIGURE 7. Showing that NNLO effects are small for the PDFs: Left is mrst2002 NLO (solid) and NNLO (dotted); Right is mrst2004 NLO (solid) and NNLO (dotted).

NEGATIVE GLUONS AND THE STABILITY OF NLO

It has been hoped that the $W^{\pm}$ production cross section can be used as a "Standard Candle" to measure parton luminosity at the LHC. But a challenge to the belief that σ_W can be reliably predicted arose with the mrst2003c PDFs. The 'c' label refers to 'conservative' cuts ($x > 0.005$, $Q > 3.162\,\mathrm{GeV}$) on the input data set to avoid possible contamination from physics that is missing in the NLO analysis. But—as sometimes also happens in politics—this "conservative" approach led to a "radical" outcome: a much smaller predicted cross section.

The origin of the surprising prediction is a preference of the MRST fit for a much smaller gluon distribution at small x, which is associated with $g(x,\mu)$ actually turning negative, even at a momentum scale as large as $\mu = m_W$. This suppresses the predicted $d\sigma/dy$ for $W^+ + W^-$ at large $|y|$ as shown in Fig. 8.

In agreement with MRST, CTEQ finds that negative gluon PDFs can produce an acceptable fit to the data with the conservative cuts, while predicting a small $d\sigma/dy$ similar to that of mrst2003c. But in disagreement with MRST, CTEQ finds the NLO fits to be stable with respect to variations in the cuts, which leaves no persuasive motivation to make the "conservative" cuts. The disagreement appears to arise from parametrization assumptions, since in the latest MRST NLO fit (mrst2004nlo) a different ansatz for the gluon distribution at μ_0 has led to a different small-x behavior. For details, see [2] and Dan Stump's talk at this conference.

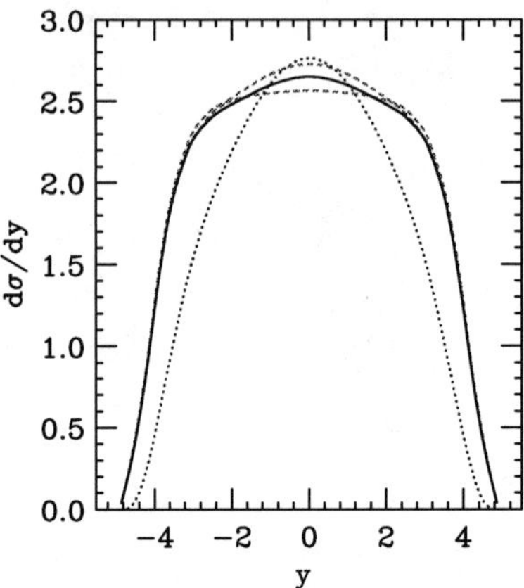

FIGURE 8. Predicted rapidity distribution for $W^{+}+W^{-}$ at the LHC: cteq6.1 (solid); cteq6.1 uncertainty range (dashed); mrst2003cnlo (dotted).

NEW PHYSICS FROM PDF FITS?

The global fit for PDFs relies on lots of Standard Model QCD, so the quality of the fit can be sensitive to Beyond Standard Model physics. As a specific example, the existence of a light gluino would modify PDF evolution and jet production [3]. A contour plot of $\chi^2 - \chi^2_{\text{CTEQ6}}$ vs. $M_{\tilde{g}}$ and $\alpha_s(M_Z)$ is shown in Fig. 9. There is a valley at $5\,\text{GeV} < M_{\tilde{g}} < 20\,\text{GeV}$ with a depth of $\Delta\chi^2 \approx -25$, which could be regarded by a SUSY fanatic as a hint for a light gluino. But a more down-to-earth interpretation is simply as a confirmation of the Standard Model, along with further evidence that a change of at least $50-100$ in χ^2 is necessary to signal a persuasive change in the quality of the current fits.

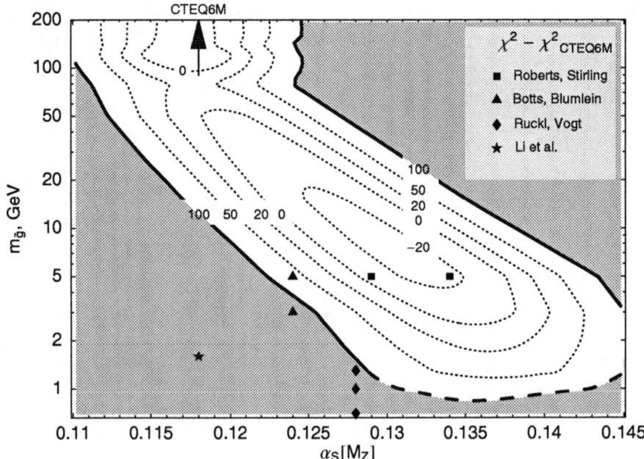

FIGURE 9. Contour plot of $\chi^2 - \chi^2_{\text{CTEQ6}}$ vs. $M_{\tilde{g}}$ and $\alpha_s(M_Z)$.

THE FUTURE

We can expect to see steady progress in the field of parton distribution measurements during the final productive years of HERA and the Tevatron, followed by dramatic developments when LHC data become available.

At the present time, there are a number of improvements underway from the theory side:

- Improved treatment of heavy quarks.
- Complete NNLO calculations.
- Weaker input assumptions.

With regard to the input assumptions, one way to eliminate possible biases caused by a choice of functional forms for the PDfs at μ_0 was discussed at this conference: to replace the parametrizations by neural net methods [4]. Work is underway to open up the assumptions on strangeness: in previous analyses, $s+\bar{s} \propto \bar{d}+\bar{u}$ has been assumed. Work is also underway to allow for possible nonradiatively generated (i.e., "intrinsic") c, $\bar{c}$ and b, $\bar{b}$. At the same time, it is worthwhile to look at possibilities for making *stronger* input assumptions, by incorporating ideas from nonperturbative models or lattice calculations.

There will also be many improvements in the input data set in the near future:

- H1 and ZEUS are taking much more data.
- NuTeV data analysis at NLO is in progress.
- E866 final data are nearly ready.
- CDF and DØ will have improved measurements of inclusive jets and the lepton rapidity asymmetry from W decay.

There are some new types of measurement that could be made at HERA which would be useful for improving the determination of PDFs. Most importantly, they could measure F_L, though it will require them to accept the risk of using part of their remaining running time to switch to a lowered proton energy. Perhaps that risk is actually not so large, since the small increase in statistics to be gained by say the last half year of conventional running is not all that valuable. In a more perfect world, it would also be very nice if HERA would measure DIS on deuterium.

CDF and DØ can also make major new contributions through measurements of

- Inclusive Z^0 and $W^{\pm}$.
- Inclusive jet with c- or b-tag.
- $\gamma/Z^0/W^{\pm}$ + jet with c- or b-tag.

Some data has already been reported for Z^0+bjet (DØ) [5] and γ+bjet (CDF) [6]. For more information on these and other possibilities, see the proceedings of the HeraLHC [7] and TeV4LHC [8] workshops.

Principal efforts in the application of parton distributions in the near term will be to the difficult question of systematic errors of the W mass measurement at the Tevatron, and to All Things LHC—the Standard Model and beyond. It is also worthwhile to mention that important extensions of the PDF approach are going on to study (1) Spin-dependent PDFs, (2) transverse momentum dependent "generalized" PDFs, and (3) parton distributions of nuclei.

ACKNOWLEDGMENTS

I wish to thank the organizers for an excellent conference. I thank Robert Thorne, Stan Brodsky, Wu-Ki Tung, Dan Stump, and Joey Huston for many discussions. This work is supported by the National Science Foundation.

REFERENCES

1. J. Pumplin et al., Phys. Rev. D **65**, 014011 (2002) [arXiv:hep-ph/0008191].
2. J. Huston, et al., "Stability of NLO global analysis and implications for hadron collider physics," [arXiv:hep-ph/0502080] to be published in JHEP.
3. E. L. Berger, et al., Phys. Rev. D **71**, 014007 (2005) [arXiv:hep-ph/0406143].
4. J. Rojo et al., "The neural network approach to parton fitting," [arXiv:hep-ph/0505044] and talk at this conference.
5. V. M. Abazov et al., Phys. Rev. Lett. **94**, 161801 (2005).
6. Reported by Amnon Harel at this conference.
7. `http://www.desy.de/~heralhc/`
8. `http://conferences.fnal.gov/tev4lhc/`

Recent Developments in Perturbative QCD

Lance J. Dixon

Stanford Linear Accelerator Center, Stanford University, Stanford, CA 94309, USA

Abstract. I review recent progress in perturbative QCD on two fronts: extending next-to-next-to-leading order QCD corrections to a broader range of collider processes, and applying twistor-space methods (and related spinoffs) to computations of multi-parton scattering amplitudes.

Keywords: Perturbative QCD calculations
PACS: 12.38.Bx

A method is more important than a discovery,
since the right method will lead to new and even more important discoveries.
– L. D. Landau

INTRODUCTION

Asymptotic freedom [1], for which Gross, Politzer and Wilczek received the 2004 Nobel Prize in Physics, provides the conceptual framework for applying perturbative QCD to short-distance-dominated problems in hadronic physics, such as the deep-inelastic (DIS) scattering process, the focus of this series of workshops. Supplemented with the notion of factorization [2], and the experimental determination of parton distributions, perturbative QCD has become the basis for all quantitative theoretical predictions for large-transverse momentum processes in hadron-hadron and *ep* collisions, as well as jet production in e^+e^- annihilation.

One might have thought that by now, with the aid of computers, perturbative QCD should have been "reduced to quadratures", that is, to a simple exercise in tabulating and numerically evaluating Feynman diagrams. Yet it is often the case that the experimental precision exceeds the theoretical uncertainties, due to unknown higher-order terms in the perturbation series. Here I will cover two topics in computational perturbative QCD, for which there has been a great deal of progress, although much still remains to be done.

The first topic concerns next-to-next-to-leading order (NNLO) corrections to collider processes, also the subject of another talk at this workshop [3]. NNLO computations have been available for a limited number of collider observables for many years, but only now are the prospects becoming good for extending them to a broader range of important precision processes at hadron colliders.

The second topic is a rapidly developing one, in which insights gleaned from the topological string in twistor space proposed by Witten [4], and further developments, promise to provide efficient means for computing tree-level and one-loop QCD amplitudes with a large number of external partons, as well as vector and Higgs bosons. These amplitudes are needed for next-to-leading order corrections to a variety of processes.

CP792, *Deep Inelastic Scattering, DIS 2005,* edited by W. H. Smith and S. R. Dasu

PROGRESS AT NNLO

For most observables, the QCD perturbation series is a slowly converging one. (Technically it is an asymptotic series, but rarely are there enough terms available in the series for the distinction to matter quantitatively.) Typical next-to-leading order (NLO) corrections for collider processes range from 20% to 100%. Clearly any kind of precision measurement, say at the few percent level, will require the NNLO terms in the series as well. Examples where this precision is desirable include the determination of

- α_s via jet production and event shapes in e^+e^- annihilation (as well as in ep collisions)
- parton distributions via DIS, Drell-Yan production, and high-p_T jet production at hadron colliders
- electroweak parameters, such as M_W, via W and Z production at hadron colliders
- the "partonic luminosity" at the LHC [5]
- Higgs couplings.

The progress of NNLO computations for collider processes can be charted in terms of the number of physical scales present in the parton-level cross sections. The more scales, the more difficult the computation, but the more flexible the applications. In perturbative QCD with massless quarks, all relevant scales are associated with the external kinematics. (The dependence of the partonic cross sections on the renormalization and factorization scales can be determined with relatively little effort, so it can be neglected in this counting.) Also, dimensional analysis can be used to remove an overall dimensionful scale from the problem, leaving just the number of dimensionless ratios.

No-scale problems

For example, in the total cross section for e^+e^- annihilation into hadrons, or equivalently the ratio $R_{e^+e^-}(s) = \sigma(e^+e^- \to \text{hadrons})/\sigma(e^+e^- \to \mu^+\mu^-)$, the only physical scale is s, the square of the center-of-mass energy. This scale can be removed trivially, so $R_{e^+e^-}(s)$ is really a no-scale problem, from the computational point of view. That is, each term in the perturbative series for $R_{e^+e^-}$ is (for fixed renormalization scale μ) a pure number. Related to the lack of other scales in the problem is the totally inclusive nature of the observable; that is, it sums over all hadronic final states with no constraints. This sum can be performed using unitarity, or the optical theorem, as illustrated in fig. 1, transforming the problem into the computation of the imaginary part of the virtual photon propagator, or two-point function.

No-scale processes were the first to be computed at NNLO in perturbative QCD, in the early 1990s. Besides $R_{e^+e^-}$ and the closely related problem of the semi-hadronic width of the τ lepton, $\Gamma(\tau \to \nu_\tau + \text{hadrons})$ [6], various DIS sum rules were also evaluated at this order: the Bjorken sum rule for neutrino scattering, $\int_0^1 dx\,[F_1^{\bar{\nu}p}(x,Q^2) - F_1^{\nu p}(x,Q^2)]$ [7], the Bjorken sum rule for polarized electroproduction, and the Gross-Llewellyn Smith sum rule for neutrino scattering [8]. The integrals over x not only remove dependence

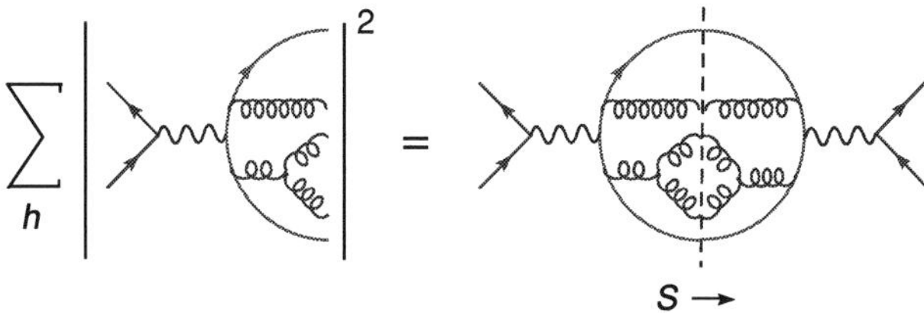

FIGURE 1. Unitarity relates the $e^+e^- \to$ hadrons total cross section to propagator-type loop integrals.

of the observables on parton distribution functions, but they reduce the computation to a no-scale, propagator-type problem very similar to $R_{e^+e^-}$.

The technology underlying all these computations was integration by parts (IBP) [9]. Total derivatives of multi-loop integrals in $D = 4 - 2\varepsilon$ space-time dimensions (*i.e.*, using dimensional regularization), such as

$$\begin{aligned} 0 &= \int d^D p d^D q \ldots \frac{\partial}{\partial q^\mu} \frac{k^\mu}{p^2 q^2 (p+q)^2 \ldots} \\ &= \int d^D p d^D q \ldots \left[-\frac{2k\cdot q}{p^2 [q^2]^2 (p+q)^2 \ldots} - \frac{2k\cdot (p+q)}{p^2 q^2 [(p+q)^2]^2 \ldots} \right], \end{aligned} \tag{1}$$

where p and q are loop momenta, and k is an external momenta, can be re-expressed as linear equations relating loop integrals with propagators raised to different powers. The absorptive parts of four-loop propagators can be related to three-loop propagators via the R^* operation [10]. For the three-loop propagator problem, a recursive solution to the system of linear IBP equations, in terms of a small set of irreducible, "master" integrals, was implemented in the program `MINCER` [11], making possible the previously-mentioned NNLO results.

The pure numbers encountered at NNLO in no-scale problems have a very simple analytic structure. The only algebraic quantities appearing, besides rational numbers, are the Riemann zeta values, $\zeta(n)$, for $n \leq 5$. The results from the early 1990s stood as the computational state-of-the-art for many years. (Very recently, a method for handling the four-loop propagators with all possible topologies has proven successful [12], suggesting that the N^{3}LO results for $R_{e^+e^-}$ may be available before long.) They have led to $\alpha_s(Q^2)$ determinations with the smallest theoretical uncertainty — if Q^2 can be made large enough experimentally, for example, at the Z^0 pole. However, the experimental precision is highly stressed by the leading "1" in $R_{e^+e^-} \propto 1 + \alpha_s/\pi + \cdots$: A 3% measurement of $\alpha_s(M_Z) \approx 0.120$ requires a parts per mil measurement at the Z pole of $\Gamma(Z \to \text{hadrons})/\Gamma(Z \to \mu^+\mu^-)$ [13]. Observables beginning at order α_s, *e.g.*, e^+e^- event-shape variables, are less demanding in this way, motivating their NNLO computation, which is a multi-scale problem, only now approaching completion.

One-scale problems

Also around 1990, the first NNLO computation of a one-scale collider process was carried out, the total cross section for inclusive production in hadronic collisions of a lepton pair via the Drell-Yan process, *i.e.* via a vector boson $V = \gamma^*$, W or Z [14]. At the parton level, the process $pp \to V + X$, where V has mass M_V, introduces the single dimensionless ratio $z \equiv M_V^2/\hat{s}$, where $\hat{s}$ is the squared partonic center-of-mass energy. Whereas the NLO correction to the total cross section was sizable, at NNLO, for W or Z production at the Tevatron or LHC, the perturbative series nicely stabilized.

The NNLO Drell-Yan result was followed quickly by the Wilson coefficient functions $C_i(z)$ for DIS structure functions [15] — except for the longitudinal structure function F_L, which begins at one order higher in α_s, and whose computation was just completed this spring [16].

In the past few years, the NNLO corrections to two additional one-scale collider processes were attacked. First, the total cross section for inclusive production of a Higgs boson in hadronic collisions, $pp \to H + X$ was computed in the large m_t approximation. In this limit, Higgs production is kinematically very similar to the Drell-Yan process, because V and H are both massive color-singlet particles, and no other mass scales remain in the problem — other than the overall Higgs coupling strength, dictated by the operator $C(m_t)H\mathrm{tr}(G_{\mu\nu}G^{\mu\nu})$. Indeed the first Higgs computation, via a high-order expansion in $1-z$ (where now $z = M_H^2/\hat{s}$) [17] was also applied to the Drell-Yan case, revealing a numerically small correction to the original results.

A second Higgs production computation [18] exploited unitarity to express the partonic Higgs cross section as a forward scattering process, as shown in fig. 2. In this case, the state that scatters forward is not a single massive virtual photon, but a pair of massless initial partons. Also, not every cut is considered, but only those that cut through the Higgs particle (or vector boson V). The advantage of this approach is that the large number of phase-space integrals that have to be performed (only one example of which is shown on the left-hand side of fig. 2) can be traded for multi-loop integrals, to which the IBP method can be applied in an automated fashion [19], in order to reduce the integrals to a manageable set of master integrals. In contrast to the no-scale examples, now all the master integrals depend on z. They also depend on the dimensional regularization parameter ε, and have to be expanded in a Laurent expansion around $\varepsilon = 0$, beginning at order $1/\varepsilon^3$, due to infrared divergences in the integrals. Fortunately, the IBP method also provides a way to determine the z-dependence of each coefficient in the Laurent expansion: Taking a derivative with respect to z produces an integral which can also be reduced to master integrals, thus generating a coupled set of differential equations [20] which are readily solved in terms of special functions. In this way the exact dependence of the NNLO partonic Higgs cross section on z was determined [18] (see also ref. [21]).

For the Drell-Yan or Higgs total cross section, the special functions that appear are polylogarithms of the form $\mathrm{Li}_n(z)$, defined by $\mathrm{Li}_1(z) = -\ln(1-z)$,

$$\mathrm{Li}_n(z) = \sum_{j=1}^{\infty} \frac{z^j}{j^n} = \int_0^z \frac{dt}{t} \mathrm{Li}_{n-1}(t), \qquad (2)$$

where $n \leq 3$, and the argument z may be replaced by a few other rational functions of

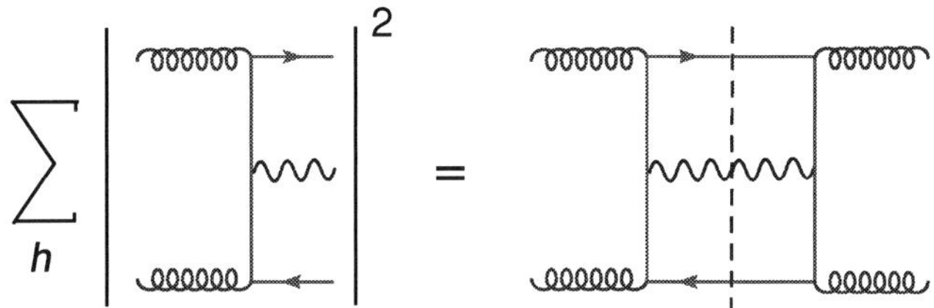

FIGURE 2. Unitarity relates the $gg \to V+X$ cross section to forward-scattering loop integrals.

z. The existence of the single scale z allows the analytical complexity, in terms of the number of different algebraic objects present, to grow significantly with respect to $R_{e^+e^-}$, but it is not yet out of hand.

In principle, no NNLO computation of a collider process is complete without an evolution of the parton distributions from low to high scales at the same NNLO accuracy. Last year saw the long-awaited completion of the NNLO corrections to the DGLAP evolution kernels for parton distribution functions [22], $P_{ij}(x)$. These were also computed by considering a forward scattering problem, with a virtual photon and a quark in the initial and final state. (For the gluon evolution kernel $P_{gg}(x)$, and also for $P_{qg}(x)$, a fictitious Higgs-like scalar ϕ, coupling to gluons via $\phi\mathrm{tr}(G_{\mu\nu}G^{\mu\nu})$, is used instead of a virtual photon.) After renormalization and subtraction of collinear divergences from lower loop orders, the NNLO evolution kernels are identified from the remaining $1/\varepsilon$ poles in the expression, which must be subtracted in the $\overline{\mathrm{MS}}$ scheme for defining parton distribution functions (or equivalently, leading twist operators). As a by-product of the computation, the finite, order ε^0 terms give the N^{3}LO contributions to the DIS structure function F_2 and the NNLO contribution to F_L [16].

For these results, somewhat more complicated special functions are required, such as $\mathrm{Li}_n(x)$ with $n=4$ for $P_{ij}(x)$ and $n=5$ for F_2 and F_L. However, the ordinary polylogarithms are not sufficient; a suitable generalization, harmonic polylogarithms [23], can be used instead. In fact, these computations were *not* performed in terms of the variable x, but rather the variable N appearing in the Mellin transform, $\tilde{f}(N) \equiv \int_0^1 dx\, x^{N-1}\, f(x)$. In N-space, the special functions that appear at NNLO are harmonic sums, such as the one-dimensional sum $S_k(N) = \sum_{i=1}^N i^{-k}$, and the multi-dimensional generalization of it,

$$S_{m,m_1,m_2,\ldots,m_k}(N) = \sum_{i=1}^{N} \frac{S_{m_1,m_2,\ldots,m_k}(i)}{i^m}. \tag{3}$$

Although it might appear that in Mellin-moment space a no-scale problem has been recovered, it is of course an infinite set of no-scale problems. Indeed, at fixed moment N, the program `MINCER` can be used to compute anomalous dimensions at NNLO for $N=2,4,6,8,10$ [24] and even up to $N=16$ [25]. However, for the case of arbitrary N, new integral reduction algorithms had to be developed [22].

Recently, the Mellin moments of the NNLO Drell-Yan and Higgs production cross sections were presented in terms of harmonic sums of the form (3) [26], putting the mathematical structure of the two types of NNLO one-scale problems discussed here on the same footing.

There have been important phenomenological applications of these one-scale results. Also taking into account the resummation of large threshold logarithms from multiple soft-gluon resummation, through next-to-next-to-leading logarithmic accuracy [27], the uncertainty on the total Higgs production cross section at the Tevatron and LHC has been considerably reduced, from perhaps 30–40% at NLO, to perhaps 10–20% at NNLO+NNLL. Some of the immediate phenomenological impact of the NNLO evolution kernels and coefficient functions on "DIS-driven" MRST parton fits [28] was discussed elsewhere at this meeting [3, 29].

Two or more scale problems

Although the total cross sections for inclusive Drell-Yan and Higgs production are now known relatively well theoretically, in practice such quantities are not measurable experimentally. Numerous experimental cuts must be imposed to extract a signal from the background, for example, cuts on the transverse momentum, rapidity, and isolation of leptons or photons visible in the final state. In an ideal world, a flexible hadron-level Monte Carlo program, accurate to NNLO, would allow the effects of such cuts to be assessed. However, even fixed-order computations of generic ("multi-scale") NNLO observables are not yet available.

A special case which can be handled in the style of the previous section is the distribution in rapidity Y_V of a Drell-Yan pair or a Higgs boson. The former process can be used to extract parton distribution information from fixed-target production, or monitor the "partonic luminosity" via W and Z production at the LHC [5], because at leading order it is proportional to $q(x_1)\bar{q}(x_2)$ with $x_{1,2} = (M_V/\sqrt{s})e^{\pm Y_V}$. Compared with the total cross section approach of ref. [18], a δ function needs to be inserted into the phase-space integration, of the form $\delta(Y - Y_V)$. The IBP method still works [30], reducing the phase-space integrals to a set of master integrals depending on z and $u = (x_1/x_2)e^{-2Y_V}$, which can again be determined by integrating differential equations. The NNLO results have much lengthier expressions than the one-scale answers. They involve polylogarithms with arguments which can be irrational functions of z and u, for example $\mathrm{Li}_2[(u - 1 - i\sqrt{(4u^2 - z(1+u)^2)/z}/(2u)]$. The same stability of the perturbative series seen for the total W and Z production cross section at NNLO, holds also bin-by-bin in rapidity [30].

For problems with more than two scales, for example e^+e^- event shapes and generic hadron-collider processes, a flexible, fully numerical approach seems necessary. The major bottleneck at present comes in integrating contributions containing the emission of two extra partons, which have quite complicated singularities as momenta become soft and/or collinear. There has been important recent progress in this direction. For lack of space, and because these developments were described in another talk at this workshop, I refer the reader to that report [3].

TWISTOR SPINOFFS

There has been a great deal of recent interest in novel methods for evaluating QCD tree and loop amplitudes, stimulated by Witten's topological string in twistor space [4]. In general, it is possible to find compact representations for amplitudes for many external particles, and more efficient techniques for computing the amplitudes, by making full use of their analytic structure (which is sometimes hidden). Twistor space [31] is a kind of Fourier transform of the usual momentum-space representation of amplitudes. Very often, a Fourier transform can expose simplicity. Consider the time-dependence of the electric field $E(t)$ associated with light arriving from the Sun. It has a pretty random appearance. However, transforming to energy variables, $E(t) \to E(\omega) = \int dt e^{i\omega t} E(t)$, reveals spectral lines, from which the presence of helium in the Sun could be deduced.

The twistor transform is very well-suited for describing the scattering of massless particles. Traditional scattering variables are the four-momentum vectors k_i^μ — which are null vectors in the massless case, $k_i^2 = 0$ — and their Lorentz-invariant products, $s_{ij} = (k_i + k_j)^2 = 2k_i \cdot k_j$. We can trade the k_i^μ for spinor variables, the right- and left-handed, or $+$ and $-$ chirality, solutions to the Dirac equation, $u_\pm(k_i)$. A shorthand notation for the two-component (Weyl) versions of these spinors is,

$$(\lambda_i)_\alpha \equiv u_+(k_i), \qquad (\tilde{\lambda}_i)_{\dot{\alpha}} \equiv u_-(k_i). \tag{4}$$

The trade is possible thanks to the form of the positive-energy projector for massless spinors, $u(k)\bar{u}(k) = \not{k}$, or in two-component notation,

$$k_i^\mu (\sigma_\mu)_{\alpha\dot{\alpha}} = (\not{k}_i)_{\alpha\dot{\alpha}} = (\lambda_i)_\alpha (\tilde{\lambda}_i)_{\dot{\alpha}}. \tag{5}$$

Instead of Lorentz-invariant products, the natural variables for massless scattering are spinor inner-products [32], defined by

$$\langle jl \rangle = \varepsilon^{\alpha\beta} (\lambda_j)_\alpha (\lambda_l)_\beta = \bar{u}_-(k_j) u_+(k_l), \qquad [jl] = \varepsilon^{\dot{\alpha}\dot{\beta}} (\tilde{\lambda}_j)_{\dot{\alpha}} (\tilde{\lambda}_l)_{\dot{\beta}} = \bar{u}_+(k_j) u_-(k_l). \tag{6}$$

These products are the square roots of the Lorentz products, up to a phase ϕ,

$$\langle jl \rangle = \sqrt{s_{jl}} e^{i\phi_{jl}}, \qquad [jl] = \pm \sqrt{s_{jl}} e^{-i\phi_{jl}}. \tag{7}$$

The utility of these variables was recognized already in the 1980s. For example, the Parke-Taylor tree amplitudes [33, 34] are for the scattering of two negative-helicity gluons, labelled j and l, and $n-2$ positive-helicity gluons. They are termed "maximally helicity-violating" (MHV) amplitudes, because tree amplitudes with fewer (zero or one) negative-helicity gluons vanish. In terms of spinorial variables, they take a remarkably simple form for any n,

$$A_n^{\mathrm{MHV}, jl} \equiv A_n^{\mathrm{tree}}(1^+, 2^+, \ldots, j^-, \ldots, l^-, \ldots, n^+) = i \frac{\langle jl \rangle^4}{\langle 12 \rangle \langle 23 \rangle \cdots \langle n1 \rangle}, \tag{8}$$

depending only on the positive-helicity spinors λ_i, not the $\tilde{\lambda}_i$.

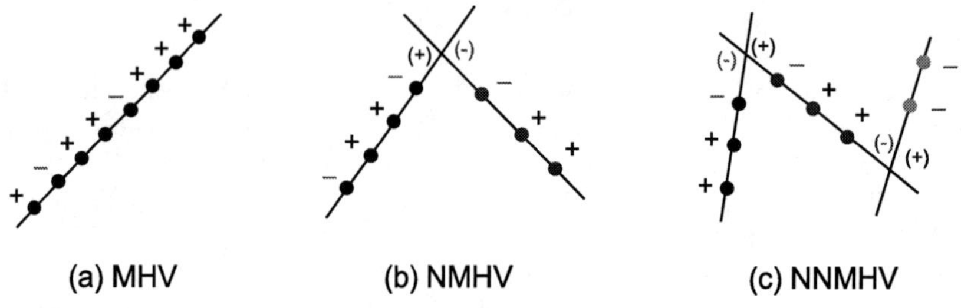

FIGURE 3. Tree amplitudes are supported on networks of intersecting lines in twistor space.

Twistor space and MHV rules

The twistor transform is a Fourier transform of the $\tilde{\lambda}_i$, leaving the λ_i alone. The four coordinates of twistor space, for each of the n particles, are $(\lambda_1, \lambda_2, \mu^{\dot{1}}, \mu^{\dot{2}})$, where $\mu^{\dot{\alpha}}$ is defined by

$$\tilde{\lambda}_{\dot{\alpha}} = i\frac{\partial}{\partial \mu^{\dot{\alpha}}}, \qquad \mu^{\dot{\alpha}} = i\frac{\partial}{\partial \tilde{\lambda}_{\dot{\alpha}}}. \tag{9}$$

In order to transform the MHV amplitudes (8), following ref. [4] we first must multiply them by the momentum-conserving δ-function, which can be written, using eq. (5) as,

$$\delta\Big(\sum_i k_i\Big) = \int d^4x \exp[i x^{\alpha\dot{\alpha}} (\lambda_i)_\alpha (\tilde{\lambda}_i)_{\dot{\alpha}}]. \tag{10}$$

Then the transformed amplitudes are

$$\tilde{A}_n^{\text{MHV},jl}(\lambda_i, \mu_i) = \int \prod_i d\tilde{\lambda}_i \exp(i\mu_i\tilde{\lambda}_i) \int d^4x\, A(\lambda_i) \exp[i x \lambda_i \tilde{\lambda}_i] \propto \prod_i \delta(\mu_i + x\lambda_i). \tag{11}$$

The product of all the linear δ functions means that the amplitude is supported on a line in twistor space, as shown in fig. 3(a).

Investigation of amplitudes with three and four negative helicities (NMHV and NN-MHV amplitudes) revealed the pattern of intersecting lines in fig. 3(b) and fig. 3(c). It also led to a set of “MHV rules” for QCD tree amplitudes, which are simpler than Feynman rules [35]. Each line in fig. 3 corresponds to an “MHV vertex”, which is a clever off-shell continuation of the MHV amplitude (8), labelled with two negative helicities and the rest positive. Many Feynman vertices can be lumped effectively into a single MHV vertex. The MHV vertices are joined with scalar propagators, that is, factors of $1/p^2$, so that no messy contractions of Lorentz indices have to be performed. An example of an MHV-rules diagram, corresponding to the twistor-space structure in fig. 3(c), is given in fig. 4.

The efficiency of the MHV rules for gluonic amplitudes motivated their quick extension to amplitudes with massless external fermions [36], Higgs bosons coupling to gluons via $H\text{tr}(G_{\mu\nu}G^{\mu\nu})$ in the large m_t limit [37], and vector bosons (γ^*, W, Z), including DIS multi-jet processes [38].

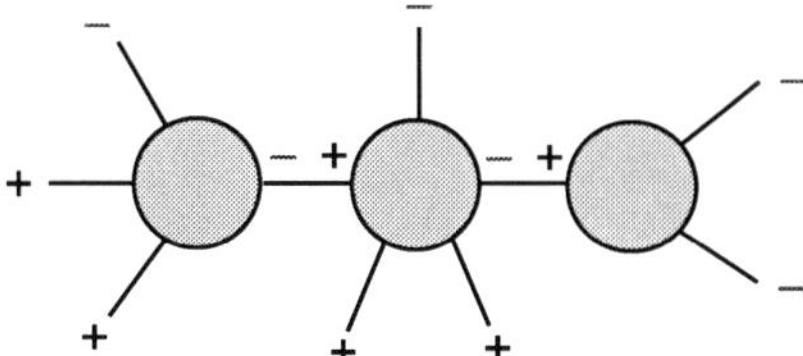

FIGURE 4. Example of an MHV-rules diagram, corresponding to fig. 3(c).

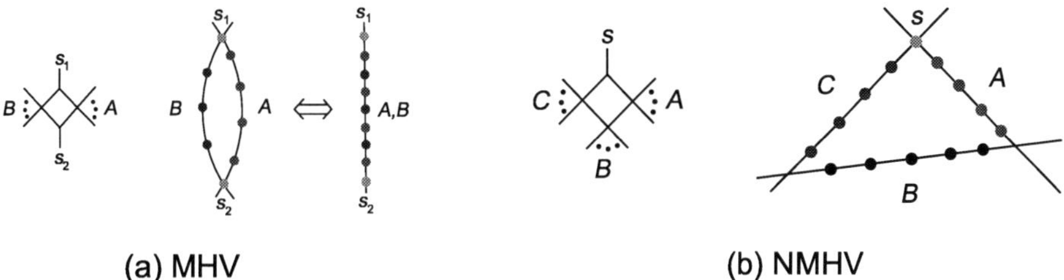

FIGURE 5. Twistor structure of box integral coefficients for one-loop amplitudes in $\mathcal{N} = 4$ supersymmetric Yang-Mills theory.

In a parallel development, the twistor structure of loop amplitudes was explored and exploited [39]. The simplest structure to explain is for gluonic loop amplitudes in a computational "toy model" for QCD, maximally ($\mathcal{N} = 4$) supersymmetric Yang-Mills theory. One-loop amplitudes in this theory can be written as a linear combination of scalar box integrals; no triangles or bubbles are required. The box integrals typically have many legs from the amplitude clustered into a single vertex of the box. In the MHV case, the coefficients of the boxes are either zero, or else equal to the MHV tree amplitudes $A_n^{\mathrm{MHV},jl}$ [40], in which case the vertices all lie on a single line in twistor space, as shown in fig. 5(a). However, this pattern turns out to be a degenerate case, which is resolved in the NMHV amplitudes. Here the simplest non-vanishing box coefficients are those with three clusters A, B, C and a single massless leg s. Their coefficients are supported on three lines intersecting in a ring, as shown in fig. 5(b) [41].

On-shell recursion relations

Quite recently, another approach to tree amplitudes has been developed, on-shell recursion relations [42, 43, 44]. These relations are somewhat more efficient and easier to generalize than the MHV rules [45], and they have very promising implications for loop amplitudes as well [46]. The derivation of the relations [44] is very general, relying just on Cauchy's theorem for functions of a single complex variable, and the factorization properties of amplitudes. The desired amplitude $A_n = A_n(0)$ can be embedded into a family of amplitudes $A_n(z)$, labelled by a complex parameter z characterizing a shift in the momentum flowing through the amplitude. For example, the momenta of the pair of legs 1 and n can be shifted, while respecting overall momentum conservation and masslessness of the external legs, according to $k_1 \to \hat{k}_1(z)$, $k_n \to \hat{k}_n(z)$, where

$$\hat{k}_1(z) + \hat{k}_n(z) = k_1(z) + k_n(z), \qquad \hat{k}_1^2(z) = \hat{k}_n^2(z) = 0. \tag{12}$$

A momentum shift satisfying eq. (12) is best described using spinor variables, as

$$\lambda_1 \to \lambda_1 + z\lambda_n, \quad \tilde{\lambda}_1 \to \tilde{\lambda}_1; \qquad \lambda_n \to \lambda_n, \quad \tilde{\lambda}_n \to \tilde{\lambda}_n - z\tilde{\lambda}_1. \tag{13}$$

This is because a complex massless vector k^μ has $\det(\not{k}) = k^2 = 0$. So the singular 2×2 matrix $\not{k}$ can be factored into a pair of spinors, as $(\not{k})_{\alpha\dot{\alpha}} = \lambda_\alpha \tilde{\lambda}'_{\dot{\alpha}}$, as in eq. (5), where $\tilde{\lambda}'$ is no longer the conjugate spinor to λ.

As z varies over the complex plane, different intermediate states go on shell, generating poles in z. Let $K_{1,l} \equiv k_1 + k_2 + \cdots + k_l$. Then $\hat{K}_{1,l}^2(z) = (K_{1,l} + z\lambda_n\tilde{\lambda}_1)^2$ vanishes at

$$z = z_l = -K_{1,l}^2 / \langle n^- | \not{K}_{1,l} | 1^- \rangle. \tag{14}$$

So long as $A_n(z) \to 0$ as $z \to \infty$, the contour integral over a large circle C, $\frac{1}{2\pi i}\oint_C dz A_n(z)/z$ vanishes. The residue at $z = 0$, which is the desired amplitude $A_n(0)$, is the negative of the sum of the residues at $z = z_l$. Those residues are given by the factorization of the amplitude into two lower-point amplitudes, evaluated in shifted, on-shell kinematics. The resulting recursion relation [43] includes a sum over l, and over the possible intermediate helicities h,

$$A_n(1,2,\ldots,n) = \sum_{h=\pm} \sum_{l=2}^{n-2} A_{l+1}(\hat{1},2,\ldots,l,-\hat{K}_{1,l}^{-h}) \frac{i}{K_{1,l}^2} A_{n-l+1}(\hat{K}_{1,l}^h, l+1,\ldots,n-1,\hat{n}). \tag{15}$$

These relations lead quickly to very compact forms for tree amplitudes. For example, there are 220 Feynman diagrams for the six-gluon amplitude. Using color algebra and symmetries, the information in these diagrams is represented by the MHV amplitudes (8), plus two more helicity amplitudes, $A_6(1^+,2^+,3^+,4^-,5^-,6^-)$ and $A_6(1^+,2^+,3^-,4^+,5^-,6^-)$. Computing the first of them using the shift (13), yields the set of diagrams shown in fig. 6. Diagram (b) vanishes because $A_4(-,+,+,+) = 0$. Diagram (c) is related to diagram (a) by the symmetry $(1 \leftrightarrow 6, 2 \leftrightarrow 5, 3 \leftrightarrow 4)$ (plus spinor conjugation). Diagram (a) can be evaluated in a few steps, using the MHV amplitudes, to give a single-term expression. Adding diagram (c) gives,

$$\begin{aligned} -iA_6(1^+,2^+,3^+,4^-,5^-,6^-) &= \frac{\langle 6^-|(1+2)|3^-\rangle^3}{\langle 61\rangle\langle 12\rangle[34][45]s_{612}\langle 2^-|(6+1)|5^-\rangle} \\ &\quad + \frac{\langle 4^-|(5+6)|1^-\rangle^3}{\langle 23\rangle\langle 34\rangle[56][61]s_{561}\langle 2^-|(6+1)|5^-\rangle}. \end{aligned} \tag{16}$$

The combination $\langle 2^-|(6+1)|5^-\rangle = \langle 26\rangle[65] + \langle 21\rangle[15]$ in the denominator leads to an unphysical singularity in the first term of eq. (16) when $k_6 + k_1$ is a linear combination of k_2 and k_5, which is cancelled by the second term. On the other hand, all of the physical factorization behavior is made manifest, in contrast to Feynman-diagram

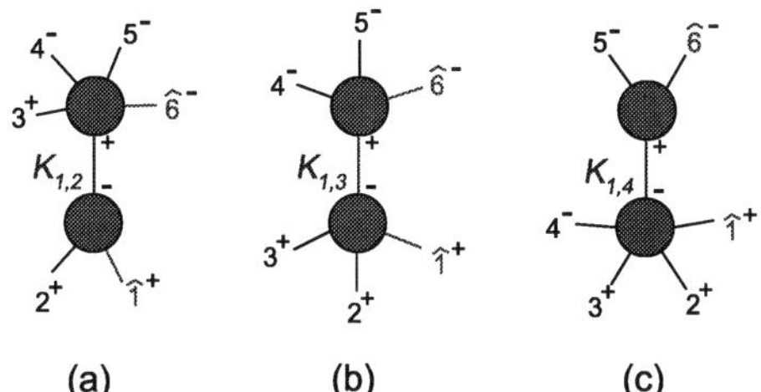

FIGURE 6. On-shell recursive diagrams for $A_6(1^+,2^+,3^+,4^-,5^-,6^-)$.

based representations [32]. The amplitude $A_6(1^+,2^+,3^-,4^+,5^-,6^-)$ has three independent recursive diagrams. Thus the six-gluon calculation is reduced from 220 Feynman diagrams to 4 much simpler ones.

The generality of this approach has led to its rapid application to different tree-level processes, with fermions and massive particles [45], and to one loop [46]. Essentially, amplitudes are being built up directly from their analytic properties. Considering that the heyday of S-matrix analyticity ended with the rise of a gauge theory for the strong interactions, QCD, these recent computational advances may herald the final revenge of the analytic S-matrix.

ACKNOWLEDGMENTS

I am grateful to the organizers of DIS 2005 for the invitation to present this talk, and for arranging such a stimulating meeting. I thank Zvi Bern, Vittorio Del Duca, Michael Klasen and David Kosower for helpful discussions. This work was supported by the US Department of Energy under contract DE–AC02–76SF00515.

REFERENCES

1. D. J. Gross and F. Wilczek, Phys. Rev. Lett. **30**, 1343 (1973); H. D. Politzer, Phys. Rev. Lett. **30**, 1346 (1973).
2. R. K. Ellis, H. Georgi, M. Machacek, H. D. Politzer and G. G. Ross, Nucl. Phys. B **152**, 285 (1979); A. H. Mueller, Phys. Rept. **73**, 237 (1981); J. C. Collins, D. E. Soper and G. Sterman, Nucl. Phys. B **261**, 104 (1985).
3. M. Klasen, these proceedings.
4. E. Witten, Commun. Math. Phys. **252**, 189 (2004) [hep-th/0312171].
5. M. Dittmar, F. Pauss and D. Zürcher, Phys. Rev. D **56**, 7284 (1997) [hep-ex/9705004].
6. S. G. Gorishnii, A. L. Kataev and S. A. Larin, Phys. Lett. B **259**, 144 (1991); L. R. Surguladze and M. A. Samuel, Phys. Rev. Lett. **66**, 560 (1991) [Erratum-ibid. **66**, 2416 (1991)].
7. S. A. Larin, F. V. Tkachov and J. A. M. Vermaseren, Phys. Rev. Lett. **66**, 862 (1991).
8. S. A. Larin and J. A. M. Vermaseren, Phys. Lett. B **259**, 345 (1991).
9. K. G. Chetyrkin and F. V. Tkachov, Nucl. Phys. B **192**, 159 (1981).
10. K. G. Chetyrkin and V. A. Smirnov, Phys. Lett. B **144**, 419 (1984).
11. S. G. Gorishnii, S. A. Larin, L. R. Surguladze and F. V. Tkachov, Comput. Phys. Commun. **55**, 381 (1989).
12. P. A. Baikov, K. G. Chetyrkin and J. H. Kuhn, Nucl. Phys. Proc. Suppl. **144**, 81 (2005).

13. S. Eidelman *et al.* [Particle Data Group], Phys. Lett. B **592**, 1 (2004).
14. R. Hamberg, W.L. van Neerven and T. Matsuura, Nucl. Phys. B **359**, 343 (1991) [Erratum-ibid. B **644**, 403 (2002)].
15. E.B. Zijlstra and W.L. van Neerven, Nucl. Phys. B **383**, 525 (1992).
16. J. A. M. Vermaseren, A. Vogt and S. Moch, hep-ph/0504242.
17. R. V. Harlander and W. B. Kilgore, Phys. Rev. Lett. **88**, 201801 (2002) [hep-ph/0201206].
18. C. Anastasiou and K. Melnikov, Nucl. Phys. B **646**, 220 (2002) [hep-ph/0207004].
19. S. Laporta, Int. J. Mod. Phys. A **15**, 5087 (2000) [hep-ph/0102033]; C. Anastasiou and A. Lazopoulos, JHEP **0407**, 046 (2004) [hep-ph/0404258].
20. T. Gehrmann and E. Remiddi, Nucl. Phys. B **580**, 485 (2000) [hep-ph/9912329].
21. V. Ravindran, J. Smith and W. L. van Neerven, Nucl. Phys. B **665**, 325 (2003) [hep-ph/0302135].
22. S. Moch, J. A. M. Vermaseren and A. Vogt, Nucl. Phys. B **688**, 101 (2004) [hep-ph/0403192]; Nucl. Phys. B **691**, 129 (2004) [hep-ph/0404111].
23. E. Remiddi and J. A. M. Vermaseren, Int. J. Mod. Phys. A **15**, 725 (2000) [hep-ph/9905237].
24. S. A. Larin, T. van Ritbergen and J. A. M. Vermaseren, Nucl. Phys. B **427**, 41 (1994); S. A. Larin, P. Nogueira, T. van Ritbergen and J. A. M. Vermaseren, Nucl. Phys. B **492**, 338 (1997) [hep-ph/9605317].
25. J. Blumlein and J. A. M. Vermaseren, Phys. Lett. B **606**, 130 (2005) [hep-ph/0411111].
26. J. Blumlein and V. Ravindran, Nucl. Phys. B **716**, 128 (2005) [hep-ph/0501178].
27. S. Catani, D. de Florian and M. Grazzini, JHEP **0105**, 025 (2001) [hep-ph/0102227]; S. Catani, D. de Florian, M. Grazzini and P. Nason, JHEP **0307**, 028 (2003) [hep-ph/0306211].
28. A. D. Martin, R. G. Roberts, W. J. Stirling and R. S. Thorne, Phys. Lett. B **604**, 61 (2004) [hep-ph/0410230].
29. R. S. Thorne, these proceedings.
30. C. Anastasiou, L. J. Dixon, K. Melnikov and F. Petriello, Phys. Rev. Lett. **91**, 182002 (2003) [hep-ph/0306192]; Phys. Rev. D **69**, 094008 (2004) [hep-ph/0312266].
31. R. Penrose, J. Math. Phys. 8:345 (1967).
32. M. L. Mangano and S. J. Parke, Phys. Rept. **200**, 301 (1991).
33. S. J. Parke and T. R. Taylor, Phys. Rev. Lett. 56:2459 (1986).
34. F. A. Berends and W. Giele, Nucl. Phys. B294:700 (1987); M. L. Mangano, S. J. Parke and Z. Xu, Nucl. Phys. B298:653 (1988).
35. F. Cachazo, P. Svrček and E. Witten, JHEP **0409**, 006 (2004) [hep-th/0403047].
36. G. Georgiou and V. V. Khoze, JHEP **0405**, 070 (2004) [hep-th/0404072]; J. B. Wu and C. J. Zhu, JHEP **0409**, 063 (2004) [hep-th/0406146]; G. Georgiou, E. W. N. Glover and V. V. Khoze, JHEP **0407**, 048 (2004) [hep-th/0407027].
37. L. J. Dixon, E. W. N. Glover and V. V. Khoze, JHEP **0412**, 015 (2004) [hep-th/0411092]; S. D. Badger, E. W. N. Glover and V. V. Khoze, JHEP **0503**, 023 (2005) [hep-th/0412275].
38. Z. Bern, D. Forde, D. A. Kosower and P. Mastrolia, hep-ph/0412167.
39. A. Brandhuber, B. Spence and G. Travaglini, Nucl. Phys. B **706**, 150 (2005) [hep-th/0407214]; F. Cachazo, P. Svrček and E. Witten, JHEP **0410**, 077 (2004) [hep-th/0409245]; R. Britto, F. Cachazo and B. Feng, Phys. Rev. D **71**, 025012 (2005) [hep-th/0410179]; S. J. Bidder, N. E. J. Bjerrum-Bohr, L. J. Dixon and D. C. Dunbar, Phys. Lett. B **606**, 189 (2005) [hep-th/0410296]; S. J. Bidder, N. E. J. Bjerrum-Bohr, D. C. Dunbar and W. B. Perkins, Phys. Lett. B **608**, 151 (2005) [hep-th/0412023].
40. Z. Bern, L. J. Dixon, D. C. Dunbar and D. A. Kosower, Nucl. Phys. B **425**, 217 (1994) [hep-ph/9403226].
41. Z. Bern, L. J. Dixon and D. A. Kosower, hep-th/0412210.
42. R. Roiban, M. Spradlin and A. Volovich, Phys. Rev. Lett. **94**, 102002 (2005) [hep-th/0412265].
43. R. Britto, F. Cachazo and B. Feng, Nucl. Phys. B **715**, 499 (2005) [hep-th/0412308].
44. R. Britto, F. Cachazo, B. Feng and E. Witten, Phys. Rev. Lett. **94**, 181602 (2005) [hep-th/0501052].
45. M. Luo and C. k. Wen, JHEP **0503**, 004 (2005) [hep-th/0501121]; Phys. Rev. D **71**, 091501 (2005) [hep-th/0502009]; R. Britto, B. Feng, R. Roiban, M. Spradlin and A. Volovich, Phys. Rev. D **71**, 105017 (2005) [hep-th/0503198]; S. D. Badger, E. W. N. Glover, V. V. Khoze and P. Svrcek, hep-th/0504159.
46. Z. Bern, L. J. Dixon and D. A. Kosower, Phys. Rev. D **71**, 105013 (2005) [hep-th/0501240]; hep-ph/0505055.

QCD (&) Event Generators

Peter Skands

Theoretical Physics, Fermilab MS106, Batavia IL-60510-0500, USA

Abstract. Recent developments in QCD phenomenology have spurred on several improved approaches to Monte Carlo event generation, relative to the post–LEP state of the art. In this brief review, the emphasis is placed on approaches for 1) consistently merging fixed–order matrix element calculations with parton showers, 2) improving the parton shower algorithms themselves, and 3) improving the description of the underlying event in hadron collisions.

Keywords: DIS05, QCD, hadron collisions, collider phenomenology, event generators, parton showers, underlying event
PACS: 12.38.-t ; 13.85.Hd ; 13.87.-a

INTRODUCTION

The immediate horizon of accelerator-based high energy experiments is dominated by HERA, its legacy and final years of running at DESY, by the Tevatron, currently in its second run of operations at Fermilab, and by the Large Hadron Collider, under construction at CERN. A common denominator for all three machines is the study of high energy hadronic interactions at unprecedented levels of statistical precision. Thus, for a wide range of measurements, the limiting factors are ultimately systematic and theoretical in nature, rather than purely statistical.

Among the most important challenges is naturally that, while perturbative QCD describes the interactions of quarks and gluons, experiments observe hadrons. In addition, many collider observables involve an interplay between widely separated energy scales, logarithms of which may appear and impact the validity of predictions even at the perturbative level. As a result, the field of QCD phenomenology is experiencing a rapid pace of development, a significant portion of which can be traced to either of two sources that I will focus on here.

Firstly, as the Centre-of-Mass energy increases, the phase space for radiation also becomes larger; high–$\hat{s}$ final states are likely to be accompanied by high jet multiplicities. To accurately predict observables in such processes, lowest order scattering matrix elements are not sufficient. Rather, more sophisticated approaches are called for, which combine the rate of hard wide–angle jets predicted by fixed–order $2 \to n$ matrix elements with a resummation of multiple soft emissions in a consistent way, avoiding both double counting and "dead regions" over all of phase space.

Secondly, hadron collisions involve new intrinsic challenges relative to ee (and to some extent also ep) scatterings, since both of the initial states are here composite and strongly interacting. However, on the theoretical side, the description of beam remnants and underlying events has gone through a long period of relative hibernation, essentially during the LEP era, with few new ideas emerging over the last ~ 20 years. Recently,

CP792, *Deep Inelastic Scattering, DIS 2005,* edited by W. H. Smith and S. R. Dasu

however, interest has been rekindled, largely in response to increased interest from the Tevatron and LHC collaborations.

Staying along the same lines, it should also here be emphasized that, when using LEP, HERA, and Tevatron results to make predictions for the LHC, an extrapolation is performed over orders of magnitude in Q^2 and x, at the same time as approximations that are "safe" at lower energies may be stretched into regions where large corrections are to be expected. As such, a non-trivial and many-faceted issue is how to treat the associated systematic and theoretical uncertainties. Though I will not touch directly on this topic below, some recent progress that partly addresses it is the emergence of parton distributions with intrinsic errors, reported on elsewhere in these proceedings [1–3].

HARD & SOFT – ME/PS MATCHING

The evaluation of tree–level transition amplitudes, involving less than, say, 5–6 partons in the final state, is a procedure which by now has been largely automated. Matrix Element Generators like CompHEP [4], Grace [5, 6], HELAS [7], MadGraph [8, 9], O'Mega [10], and AMEGIC++ [11] provide fast and reliable means of obtaining (more or less) tractable analytic expressions for a broad range of matrix elements, both in the Standard Model and beyond. Combining these with efficient numerical phase space integration algorithms such as BASES/SPRING [12] and others [13], it is possible to further automate the phase space weighting, and hence to generate events corresponding to the chosen amplitude at matrix–element level. A number of more dedicated matrix element evaluation codes also exist, with processes hard-coded one by one, most notably AlpGen [14], but also MCFM [15] and others [16–21].

Common to all these approaches (most at tree level and a few at one loop) is that they represent fixed–order expansions in the electromagnetic, weak, and in particular strong coupling constants. As such, the virtue of these calculations is, briefly stated, that they include the entire helicity and interference structure of the amplitude (as well as virtual corrections to it, to the extent that loops are included), up to the given order calculated. Furthermore, asymptotic freedom implies that the stability of this expansion should improve with energy, due to the gradual vanishing of the strong coupling at large energies. Admittedly, the complexity rapidly grows with the number of particles involved, and so as already mentioned, these approaches are presently limited to fairly inclusive observables, where the number of resolved final state particles does not exceed a handful or so.

On the other hand, from LEP we know that multiple soft gluon emissions are important in building up the full event structure. Mathematically, these corrections correspond to logarithmic enhancements of the amplitude, of the form

$$\alpha_s^N \log^{2N}(Q_{\mathrm{hard}}/Q_{\mathrm{soft}}) + \ldots \qquad (1)$$

where $Q_{\mathrm{soft}}/Q_{\mathrm{hard}}$ is a measure of the softness of the gluon(s) relative to the hard scale(s) in the problem. Moreover, when going to higher energies, the phase space for such emissions increases. Thus, while fixed–order calculations should be able to predict reliably the rate of a few well–separated jets (and other observables at the same level of inclusiveness), it is necessary go beyond the fixed–order approximation to obtain a picture of the full event structure.

To improve the logarithmic accuracy, two dominant approaches exist: resummation calculations, and parton showers. Both are approximations to perturbation theory which work at infinite order in the coupling constant, and which are exact in certain limits.

The former approach, resummation, allows to include not only the terms shown explicitly above (double logs $\sim$ LLA), but also less singular logarithms in a systematic way. However, the formalism can still only be applied to relatively inclusive quantities, and a separate calculation must be performed for each observable, though interesting work has recently been carried out on automating calculations of this type [22].

In this talk, I will concentrate on the parton shower approach. While this description is formally correct only to leading logarithmic accuracy, it has the virtue that a fully exclusive description of the final state is obtained, which can be easily matched onto hadronisation descriptions, and from which in principle any observable can be constructed. Moreover, it is possible (and indeed necessary, as e.g. in the case of momentum conservation) to include at least a subset of higher–order effects, such as angular ordering of emissions, optimizing the scale choice in α_s with respect to higher order kernels [23], and choosing azimuthal angles in the branchings non–isotropically [24]. In practice, such refinements have been introduced in all of the standard shower Monte Carlos, including in particular the ARIADNE dipole shower [25–27], the HERWIG [28–30] and HERWIG++ [31] showers, and both the PYTHIA virtuality–ordered [32–34] and transverse-momentum–ordered [35] showers (the SHERPA generator [36] basically uses a variant of the PYTHIA virtuality–ordered algorithm). It would therefore be grossly misleading to equate leading log analytical calculations, where no such refinements are included, with leading log parton showers.

The parton shower approximation starts from the observation that the collinear limit of QCD (and QED, for that matter) is universal. Thus, a process like $\mathrm{e^+e^-} \to \mathrm{q\bar{q}}$ can be corrected to the process $\mathrm{e^+e^-} \to \mathrm{q\bar{q}g}$ using universal expressions for the $\mathrm{q} \to \mathrm{qg}$ splitting probability. Due to the universality, the same expressions may then be applied again to describe the radiation of further gluons, as well as gluon splittings into quarks and so forth. Since the integrated probability at each step is nominally infinite, an ordering is introduced, whereby the emissions are generated sequentially according to some resolution criterion, like angle, virtuality, or transverse momentum. A lower cutoff on the resolution variable may then be naturally introduced, that regulates the infrared divergences, and at which scale a hadronisation description is supposed to take over.

Thus, the virtues are that final states with an arbitrary number of partons may be built up, with a transition to hadronisation descriptions built in from the start. The down side is that the approximation is only exact in the collinear limit. For hard and/or wide–angle emissions, different parton showers can give widely different answers, reflecting the approximate nature of the approach in those regions. However, as mentioned above these are precisely the regions where the fixed–order calculations are at their best, hence it has been a long–standing wish to join consistently the state of the art of both worlds.

TABLE 1. ME/PS Merging: List of processes, X, for which X+jet merging has been implemented in the HERWIG and PYTHIA Monte Carlos (e^+e^- processes omitted).

	$pp \to h^0$	$pp \to V\ (=\gamma/Z/W)$	**DIS**	**top decay**	**SM decays**	**SUSY decays***
HERWIG	√	√	√	√	-	-
PYTHIA†	√	√	√	√	√	√

* corrections approximate for 2–body RPV modes, absent for RPV 3–body modes [37].
† PYTHIA: applies to both the Q^2 and $p_\perp^2$ ordered shower algorithms in PYTHIA 6.3.

ME/PS Merging

The simplest (and oldest) approaches to join matrix elements and parton showers I will here refer to as matrix–element/parton–shower (ME/PS) "merging", to be contrasted with "matching" below. Essentially, merging improves the parton shower off a hard system, call it X, by re-weighting the position of the hardest jet in phase space to reproduce the matrix element distribution for X+jet. An overview of hadron collider processes for which such corrections have been implemented in the HERWIG and PYTHIA models is given in Tab. 1.

Technically, the way these corrections are implemented can be quite different, depending on the showering algorithm. In HERWIG, the showering algorithm has a "dead zone" in the hard wide–angle region, where no radiation at all is produced. In order to match to the matrix element which does produce jets there, two classes of events are effectively merged, e.g. X and X+jet, with the latter chosen inside the dead region of the former [29]. The detailed procedure is somewhat complicated and has only been worked out for a few cases, most recently for Higgs production [38].

In PYTHIA, the problem is rather the opposite. Too much radiation is generally produced in the hard wide–angle region, as compared to the matrix element answer. It is thereby straightforward to introduce a re-weighting, vetoing some of the extra emissions, to arrive back down at the matrix element rate [33, 39]. Note that these corrections are applied for both the Q^2– and $p_\perp^2$–ordered shower algorithms in PYTHIA 6.3.

So far, so good. However, for both the HERWIG and PYTHIA style merging, the procedure rapidly becomes more involved when attempting to generalize the methods to more complicated final states (see e.g. [40, 41]). Moreover, recent developments along related lines have resulted in a range of more generic approaches which are now being more actively pursued, as will be discussed below.

ME/PS Matching at Leading Order

The problem of consistently adding parton showers to a set of leading–order matrix elements for X, X + jet, X + 2 jets, etc, to obtain an inclusive sample of X production, matched to all available hard radiation matrix elements, has recently been studied in detail by a number of authors, in particular by Mangano (MLM) [42], by Catani, Krauss, Kuhn, and Webber (CKKW) [43, 44], by Lönnblad [45], and most recently by Mrenna and Richardson [46]. (See also talk by Frixione [47].)

All these approaches essentially allow a consistent adding together of events generated with different jet multiplicities at the matrix–element level (e.g. W, W+jet, W+2jets, ...), by re-weighting them and showering them in way so that double counting and empty regions are avoided over all of phase space.

The approach proposed by CKKW [43] is, briefly stated, to first select the jet multiplicity, n, at the matrix element level according to a known probability,

$$P(n) = \frac{\sigma_n}{\sigma_0 + \sigma_1 + \sigma_2 + \sigma_3 + \ldots}, \quad ; \quad \sigma_i \equiv \sigma_i(Q_{\mathrm{cut}}), \tag{2}$$

with Q_{cut} a cutoff scale (in principle arbitrary) that regulates the infrared divergencies of the matrix elements. According to the chosen matrix element, a set of explicit four-momenta p_i are then generated, to which a jet clustering algorithm is applied. A series of 'branchings' is thereby reconstructed, which can be interpreted as a parton shower history. The event as a whole is then re-weighted according to the Sudakov form factors (see below) and α_s values associated with the reconstructed intermediate scales. A parton shower can then be applied as the final step, with emissions above the cut scale vetoed for all except the highest jet multiplicity matrix elements.

Why this works is more technical: recall that the parton shower is formulated in terms of the no–emission probability between two scales, the Sudakov form factor, which in all simplicity is the (singular part of the) probability for an n–jet configuration to remain an n–jet one as a function of the resolution scale. By re-weighting the matrix elements with Sudakov factors for each leg, the leading divergencies which would lead to double–counting between e.g. σ_1 and σ_2 are cancelled. Roughly speaking, the Sudakov re-weighting takes into account that for every 2-jet configuration you gain, you must lose one 1-jet one. It was shown by [43] that this procedure makes the sum stable at least to next-to-leading logarithmic (NLL) accuracy. Note that this stabilisation is to some extent equivalent to the cancellation of real divergencies by virtual ones in a full NLO calculation, with the Sudakovs here playing the role of virtual corrections, that have the same structure (but opposite signs) as the tree–level divergencies.

To further explain what the Sudakovs are doing, note that the parton shower is correct in the limit of strongly ordered emissions, i.e. in the limit that each successive scale is much smaller than the preceding one. In this case, the emission probability is large (i.e. the Sudakov = *no*–emission probability, is small), and the Sudakov re-weighting becomes quite important. At the other extreme are emissions which happen at similar scales; here, the Sudakovs are very close to one (the probability to go from one scale to another without emission becomes unity in the limit that the two scales coincide), and hence the matrix element results remain essentially unaltered here, as desired.

This style of matching has since been implemented in the SHERPA event generator for several processes [36]. It was then noticed by Lönnblad [45] that a better matching could be obtained by replacing the analytical Sudakovs used in the original CKKW prescription by Sudakovs numerically generated by running the actual parton shower (so–called 'pseudo–showers'). This is the approach implemented [48] in the ARIADNE generator [26]. Mrenna and Richardson subsequently made further refinements [46], applying the methodology also to hadronic collisions, using MadGraph and the HERWIG and PYTHIA generators, part of which work is stored in the PATRIOT event database [49].

TABLE 2. LO ME/PS Matching: List of processes, X, for which LO ME/PS matched generators/samples are available.

	$e^+e^- \to q\bar{q}$	$pp \to V\,(=\gamma/Z/W)$	$pp \to VV$	**DIS**
ARIADNE	√	√*	-	√
SHERPA	√	√	√	-
PATRIOT	-	√	-	-

* V = W only.

Mangano's prescription ("MLM matching") [42] is similar but somewhat simpler in spirit than CKKW. In particular, it is based on clustering of events after showering and thus has a much simpler interface between the matrix element and parton shower generators.

Tab. 2 gives an overview of processes for which Leading Order matching is currently implemented/available.

ME/PS Matching at NLO

Above, the aim was essentially to describe real QCD radiation as accurately as possible, over all regions of phase space. This was accomplished by consistently matching on parton shower descriptions of soft radiation to a set of tree–level matrix elements describing as many hard emissions as one cares to calculate, which in practice currently means up to 3–4 extra jets.

However, since virtual corrections are not included, the normalization of the distributions is still only correct to leading order. Quite recently, the problem of matching parton showers to the full NLO theory, i.e. one–leg *and* one–loop corrections to the lowest order, has been addressed by several groups (see also talk by Frixione [47]). Early approaches include the use of phase space slicing to separate resolvable and unresolvable regions [50, 51], which, despite a number of initial successes, suffers from the drawback of not reproducing the perturbative expansion correctly. Important studies have also been carried out by the group [52–54], though so far practical applications have been limited.

Presently, the most mature NLO matching approach is the one put forth by Frixione, Nason, and Webber [55, 56], which has been implemented in the program MC@NLO (essentially a superstructure built onto the HERWIG Monte Carlo). Another promising approach suggested by Krämer and Soper [57, 58] has so far only been applied to e^+e^- observables, though work is in progress to generalise it. A more complete list of processes is given in Tab. 3.

Naturally, the big boon is that one automatically obtains cross sections normalised to NLO precision. A vice, as compared to the leading order matchings, has so far been that the NLO matrix elements only include the first (real) hard emission; subsequent emissions, even when hard, must still be generated by the parton shower, though recent work has been carried out on including also CKKW-style matching for higher jet multiplicities [59].

A related issue is that, since MC@NLO is hard-wired to HERWIG, it is presently

TABLE 3. NLO ME/PS Matching: List of processes, X, for which unweighted NLO ME/PS matched event samples can be generated with the MC@NLO generator (weights = ± 1).

	$e^+e^- \to q\bar{q}$	$pp \to h^0$	$pp \to V\,(=\gamma/Z/W)$	$pp \to VV$	$pp \to QQ$	**single top**
MC@NLO	-*	√	√	√	√†	in progress

* Note that, due to the simple structure of e^+e^-, NLO matching can here effectively be obtained using the existing HERWIG and PYTHIA merging methods, and re-weighting by the loop factor $(1{+}\alpha_s/\pi)$.
† Q: heavy quark.

not possible to vary the parton shower model. Consider for instance the peak position of the Drell–Yan and h^0 $p_\perp$ spectra. This is where the bulk of the cross section sits, and this is, roughly speaking, the region that gets the most enhancement by the loop corrections. However, this region is also highly sensitive to multiple soft emissions, which are resummed by the parton shower. A much awaited future development is thus the creation of tools that allow a more generic interfacing with different shower models.

In a similar vein, note that, while the overall normalisation is improved by going from LO to NLO matching, the same is not necessarily true for the shapes of distributions, as follows. Consider again the Drell–Yan $p_\perp$ spectrum in hadron collisions. As mentioned above, it is dominated by Z + multiple soft emissions in the peak, while Z+jet dominates in the tail. It would now be possible experimentally to define a semi-inclusive Z+jet *fraction* as a function of $p_{\perp Z}$. At NLO, the prediction for Z+jet (and hence for this subsample) is effectively only correct to leading order, since virtual corrections to Z+jet are not included. Since the Z+jet fraction tends towards unity at large $p_{\perp Z}$ values, the overall shape should probably not be regarded as being more precisely determined here than in the leading order matching schemes discussed above. The extreme example would be an observable sensitive to multiple hard emissions, where a leading order matching including several hard jet radiations would clearly be superior to the present NLO matching schemes, where the second jet has to be radiated by the parton shower.

NEW PARTON SHOWER ALGORITHMS

Both HERWIG++ [60] and PYTHIA 6.3 [61] contain new parton shower models. In the HERWIG++ case, refinements have been made [31] on the already existing HERWIG model, while a completely new shower model [35] has been implemented in PYTHIA 6.3. The recently completed program APACIC++ [62] also contains a parton shower model, which essentially is an adaption of the old PYTHIA shower model with the specific implementation of CKKW style matching in mind.

The HERWIG shower is based on a strictly angular–ordered sequence of emissions [28]. This correctly accounts for coherence effects in the emission of soft gluons, but has the disadvantage that it leaves a 'dead zone' in the hard 3–jet region, which has to be filled in separately, as discussed above. The shower evolution is stopped once a fixed low scale is reached, at which time a transition is made to a non–perturbative hadronisation description, the HERWIG one being based on the cluster model [63].

The HERWIG++ algorithm [31] starts from the same basic principles and thus inherits

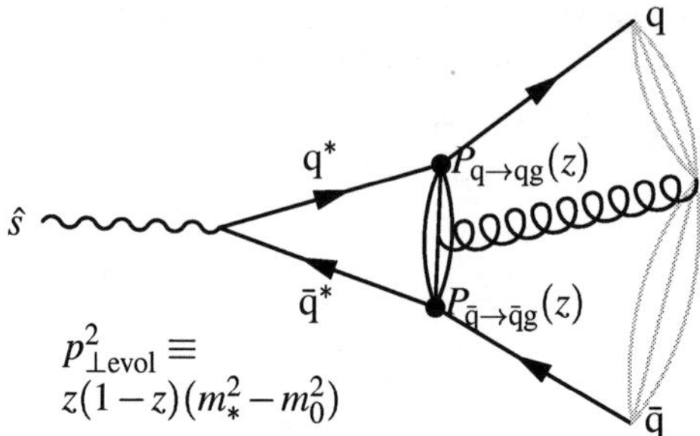

FIGURE 1. Final–state branching in the $p_\perp$–ordered shower. The evolution is performed on each side of the dipole separately, in the variable $p^2_{\perp\mathrm{evol}}$, here expressed in terms of the parton virtuality, m_*, its rest mass, m_0 (for massive partons), and the energy sharing fraction z appearing in the splitting kernels $P(z)$. Kinematics are constructed in the dipole CM frame, conserving energy and momentum inside the system.

most features from the old shower, but the definition of the evolution variable has been slightly changed, to allow a more correct treatment of heavy quark radiation as well as a more consistent behaviour in the soft–gluon limit. Finally, the treatment of cascade decays of resonances interspersed with showers should be improved by a more consistent implementation of multi–scale showers. Recent studies exploring this algorithm can be found in [31, 64, 65]. At present, HERWIG++ is still mainly a tool for e^+e^- collisions, though work is underway to extend it to hadron collisions. Another item on the agenda is the inclusion of a CKKW-style ME/PS matching scheme.

More radical changes have occurred in PYTHIA 6.3, where a new transverse–momentum–ordered shower based on a recently proposed hybrid between the dipole and parton shower formalisms [35] has been implemented (in addition to the old virtuality–ordered shower, which is carried over from previous versions). The choice of $p_\perp$ as evolution variable was made for several reasons. Firstly, it has the dual property of simultaneously being a good measure of hardness while still leading to a natural angular ordering of emissions [66]. Secondly, it is Lorentz invariant under longitudinal boosts, which is not the case e.g. for the HERWIG angular–ordered evolution variable. Thirdly, while both PYTHIA and HERWIG can be tuned to give good descriptions of the LEP data, the ARIADNE $p_\perp$–ordered dipole shower tends to do even better. Note, however, that the $p_\perp$ measure proposed in [35] is different from the one used in the ARIADNE evolution. Fourthly, the underlying–event model in PYTHIA is based on a $p_\perp$–ordered sequence of multiple perturbative interactions, hence a shower also ordered in $p_\perp$ allows for a more unified treatment of the perturbative activity as a whole, as will be discussed below.

The dipole/parton shower hybrid approach implies an evolution of individual parton lines, but with recoils occurring inside dipoles, as illustrated in Fig. 1. For the final–state shower, these dipoles are normally defined by the respective colour partners, with some exceptions involving decays of heavy resonances. For the initial–state shower, where a backwards evolution is performed from the hard scattering down, only a single dipole is relevant, spanned between the two incoming partons.

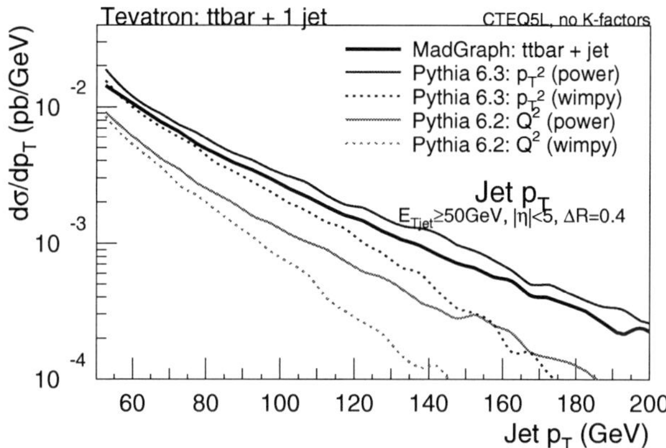

FIGURE 2. The Tevatron $t\bar{t}$+jet rate as a function of jet $p_\perp$. Results are shown for MadGraph, and for hard ('power') and soft ('wimpy') variants of both the Q^2– and $p_\perp^2$–ordered PYTHIA showers, see text.

Studies carried out so far indicate that the new shower leads to an improved description e.g. of event shapes at LEP [67] and of Drell–Yan production at the Tevatron [35, 68]. Fig. 2 gives a preliminary comparison of the $t\bar{t}$+jet rate at the Tevatron, as a function of jet $p_\perp$ [69]. Results are shown for MadGraph (thick black line), for two variants of the old Q^2–ordered shower (green lines, solid and dotted), and for two variants of the new $p_\perp^2$–ordered shower (blue lines, solid and dotted). The 'power' and 'wimpy' versions of each shower represent a reasonable range of variation of the maximum scale for shower emissions, with the former populating all available phase space, and the latter bounded by the scale of the hard interaction. Though it is still too early to draw more general conclusions, it is encouraging that both the shape and normalisation of the distribution appear to be improved, especially in the 'power shower' variant of the $p_\perp$–ordered algorithm.

THE UNDERLYING EVENT

The underlying event may be defined somewhat loosely as the activity in a (single) hadron–hadron collision which does not originate directly from the hard scattering that triggered the event. Currently, this component is not well understood from first principles. What is known, on the other hand, is that it produces omnipresent systematic as well as random fluctuations in activity, which impact isolation criteria, jet energy scales, etc., which can be of significant magnitude for experimental analyses.

During the last few years it has become increasingly clear that the underlying event not only contains soft activity but also has a semi–hard, 'lumpy' component [70]. To explain this, the concept of multiple perturbative interactions [71] appears increasingly attractive. Several recent implementations are built on ideas incorporating multiple interactions in one form or another, including the underlying–event framework in the HERWIG add–on JIMMY [72], the new interleaved model [35, 73] in PYTHIA 6.3, and the underlying–event model being developed for SHERPA [74], this latter being quite similar to the old PYTHIA one [71].

The most recent development has been the interleaved model of [35, 73]. In this picture, the evolution of the initial state radiation and the generation of multiple scatterings are no longer independent. Instead, a successive 'fine-graining' of all the perturbative activity is performed, with radiations interspersed (or 'interleaved') with interactions. This allows correlations to be introduced between all the perturbative activity at successively smaller scales. For instance, the third interaction will know about the presence of a gluon having been radiated off a valence quark in the first, etcetera. This introduces for the first time non–trivial correlations in flavour and x, while all known sum rules are still respected (e.g. momentum conservation and quark counting rules). In addition, the model allows new possibilities for impact parameter dependence, and contains a more refined treatment of beam remnants, based on an extension of the Lund string model to 'baryonic' colour topologies [37].

With the development of more sophisticated physics models, the hope is that it will be possible to ask a range of more meaningful physics questions, among which especially the topic of colour correlations and colour reconnections (in view of a possible difference between the vacua left by e^+e^- and hadronic initial states) is presently the most actively investigated. Finally, the energy dependence of the underlying activity is currently very poorly known; further studies aiming to pin down the scaling behaviour, for instance by including the available RHIC pp data, would be of great interest.

CONCLUSION

The quest for the cause of electroweak symmetry breaking is (hopefully) nearing its end. Should the Tevatron not discover it within a few years, the experimental programme at the LHC promises a comprehensive exploration of the TeV scale, with the emphasis on precisely this question, as well as on more general searches for signals of physics beyond the Standard Model. In addition, the high energies and hadronic environment of both the Tevatron and in particular the LHC will challenge our understanding of QCD, especially our ability to solve it for large numbers of partons, both real and virtual, and our control of phenomena near the borders of its perturbative domain.

It is encouraging, then, to see a truly impressive effort mounted in recent years to address several of the most important issues, among which I have here discussed more precise and realistic simulation of events at higher perturbative orders, improved parton shower models, and some progress towards a better understanding of the underlying event.

On a general note, the emergence of many sophisticated but specialised tools entails an increased need for efficient cross–communication between the programs. For instance, while the traditional Monte Carlo generators have been largely self-contained in the past, the time-consuming process of hard–coding matrix elements for each process by hand can now be left to automatic programs optimised specifically for this task. One should therefore expect an increased reliance on external interfaces, such as provided by the Les Houches Accords [75, 76], in the future.

Finally, though much of the effort reported here is still centred around a relatively few groups, for which simple manpower restrictions often represent a non–trivial problem, the many new and exciting developments do seem to have led to an increased communication, bringing people together from different fields. Hopefully, with continued nourishment, this is a trend that will continue and grow in the future.

REFERENCES

1. J. Pumplin, "Parton Distributions," in *these proceedings*, 2005, transparencies available from www.hep.wisc.edu/dis05/.
2. D. Stump, "Stability and uncertainty of parton distribution functions," in *these proceedings*, 2005, transparencies available from www.hep.wisc.edu/dis05/.
3. R. Thorne, "MRST PDFs and impact on LHC physics," in *these proceedings*, 2005, transparencies available from www.hep.wisc.edu/dis05/.
4. A. Pukhov, et al. (1999), hep-ph/9908288.
5. F. Yuasa, et al., *Prog. Theor. Phys. Suppl.*, **138**, 18–23 (2000), hep-ph/0007053.
6. S. Tsuno, et al., *Comput. Phys. Commun.*, **151**, 216–240 (2003), hep-ph/0204222.
7. H. Murayama, I. Watanabe, and K. Hagiwara (1991), kEK-91-11.
8. T. Stelzer, and W. F. Long, *Comput. Phys. Commun.*, **81**, 357–371 (1994), hep-ph/9401258.
9. F. Maltoni, and T. Stelzer, *JHEP*, **02**, 027 (2003), hep-ph/0208156.
10. M. Moretti, T. Ohl, and J. Reuter (2001), hep-ph/0102195.
11. F. Krauss, R. Kuhn, and G. Soff, *JHEP*, **02**, 044 (2002), hep-ph/0109036.
12. S. Kawabata, *Comp. Phys. Commun.*, **88**, 309–326 (1995).
13. V. A. Ilyin, D. N. Kovalenko, and A. E. Pukhov, *Int. J. Mod. Phys.*, **C7**, 761 (1996), hep-ph/9612479.
14. M. L. Mangano, M. Moretti, F. Piccinini, R. Pittau, and A. D. Polosa, *JHEP*, **07**, 001 (2003), hep-ph/0206293.
15. J. M. Campbell, and R. K. Ellis, *Phys. Rev.*, **D62**, 114012 (2000), hep-ph/0006304.
16. W. Beenakker, R. Hopker, and M. Spira (1996), hep-ph/9611232.
17. F. A. Berends, W. T. Giele, and H. Kuijf, *Phys. Lett.*, **B232**, 266 (1989).
18. F. A. Berends, H. Kuijf, B. Tausk, and W. T. Giele, *Nucl. Phys.*, **B357**, 32–64 (1991).
19. W. T. Giele, E. W. N. Glover, and D. A. Kosower, *Nucl. Phys.*, **B403**, 633–670 (1993), hep-ph/9302225.
20. B. P. Kersevan, and E. Richter-Was, *Comput. Phys. Commun.*, **149**, 142–194 (2003), hep-ph/0201302.
21. Z. Nagy, and Z. Trocsanyi, *Phys. Rev.*, **D59**, 014020 (1999), hep-ph/9806317.
22. A. Banfi, G. P. Salam, and G. Zanderighi, *JHEP*, **03**, 073 (2005), hep-ph/0407286.
23. D. Amati, A. Bassetto, M. Ciafaloni, G. Marchesini, and G. Veneziano, *Nucl. Phys.*, **B173**, 429 (1980).
24. B. R. Webber, *Ann. Rev. Nucl. Part. Sci.*, **36**, 253 (1986).
25. G. Gustafson, and U. Pettersson, *Nucl. Phys.*, **B306**, 746 (1988).
26. L. Lönnblad, *Comput. Phys. Commun.*, **71**, 15–31 (1992).
27. L. Lönnblad, *Nucl. Phys.*, **B458**, 215–230 (1996), hep-ph/9508261.
28. G. Marchesini, and B. R. Webber, *Nucl. Phys.*, **B238**, 1 (1984).
29. M. H. Seymour, *Comp. Phys. Commun.*, **90**, 95–101 (1995), hep-ph/9410414.
30. G. Corcella, et al., *JHEP*, **01**, 010 (2001), hep-ph/0011363.
31. S. Gieseke, P. Stephens, and B. Webber, *JHEP*, **12**, 045 (2003), hep-ph/0310083.
32. M. Bengtsson, and T. Sjöstrand, *Nucl. Phys.*, **B289**, 810 (1987).
33. E. Norrbin, and T. Sjöstrand, *Nucl. Phys.*, **B603**, 297–342 (2001), hep-ph/0010012.
34. T. Sjöstrand, et al., *Comput. Phys. Commun.*, **135**, 238–259 (2001), hep-ph/0010017.
35. T. Sjöstrand, and P. Z. Skands, *Eur. Phys. J.*, **C39**, 129–154 (2005), hep-ph/0408302.
36. T. Gleisberg, et al., *JHEP*, **02**, 056 (2004), hep-ph/0311263.
37. T. Sjöstrand, and P. Z. Skands, *Nucl. Phys.*, **B659**, 243 (2003), hep-ph/0212264.
38. G. Corcella, and S. Moretti (2004), hep-ph/0402149.
39. G. Miu, and T. Sjöstrand, *Phys. Lett.*, **B449**, 313–320 (1999), hep-ph/9812455.

40. J. Andre, and T. Sjöstrand, *Phys. Rev.*, **D57**, 5767–5772 (1998), hep-ph/9708390.
41. S. Mrenna (1999), hep-ph/9902471.
42. M. Mangano, "The so–called MLM prescription for ME/PS matching" (2004), Talk presented at the Fermilab ME/MC Tuning Workshop, October 4, 2004.
43. S. Catani, F. Krauss, R. Kuhn, and B. R. Webber, *JHEP*, **11**, 063 (2001), hep-ph/0109231.
44. F. Krauss, *JHEP*, **08**, 015 (2002), hep-ph/0205283.
45. L. Lönnblad, *JHEP*, **05**, 046 (2002), hep-ph/0112284.
46. S. Mrenna, and P. Richardson, *JHEP*, **05**, 040 (2004), hep-ph/0312274.
47. S. Frixione, "Monte-Carlo generators," in *these proceedings*, 2005, transparencies available from www.hep.wisc.edu/dis05/.
48. N. Lavesson, and L. Lönnblad (2005), hep-ph/0503293.
49. S. Mrenna, Patriot database (2004), see http://cepa.fnal.gov/personal/mrenna/Matched_Dataset_Description.html.
50. M. Dobbs, *Phys. Rev.*, **D65**, 094011 (2002), hep-ph/0111234.
51. B. Potter, and T. Schorner, *Phys. Lett.*, **B517**, 86–92 (2001), hep-ph/0104261.
52. J. C. Collins, *JHEP*, **05**, 004 (2000), hep-ph/0001040.
53. J. C. Collins, and F. Hautmann, *JHEP*, **03**, 016 (2001), hep-ph/0009286.
54. J. C. Collins, and X. Zu, *JHEP*, **03**, 059 (2005), hep-ph/0411332.
55. S. Frixione, and B. R. Webber, *JHEP*, **06**, 029 (2002), hep-ph/0204244.
56. S. Frixione, P. Nason, and B. R. Webber, *JHEP*, **08**, 007 (2003), hep-ph/0305252.
57. M. Kramer, and D. E. Soper, *Phys. Rev.*, **D69**, 054019 (2004), hep-ph/0306222.
58. D. E. Soper, *Phys. Rev.*, **D69**, 054020 (2004), hep-ph/0306268.
59. Z. Nagy, and D. E. Soper (2005), hep-ph/0503053.
60. S. Gieseke, A. Ribon, M. H. Seymour, P. Stephens, and B. Webber, *JHEP*, **02**, 005 (2004), hep-ph/0311208.
61. T. Sjöstrand, L. Lönnblad, S. Mrenna, and P. Skands (2003), hep-ph/0308153.
62. F. Krauss, A. Schalicke, and G. Soff (2005), hep-ph/0503087.
63. B. R. Webber, *Nucl. Phys.*, **B238**, 492 (1984).
64. S. Gieseke (2004), hep-ph/0408034.
65. S. Gieseke, *JHEP*, **01**, 058 (2005), hep-ph/0412342.
66. G. Gustafson, *Phys. Lett.*, **B175**, 453 (1986).
67. G. Rudolph (2003), ALEPH, unpublished.
68. J. Huston, I. Puljak, T. Sjöstrand, and E. Thomé, "Resummation and shower studies," in *The QCD/SM working group: Summary report, 3rd Les Houches Workshop: Physics at TeV Colliders, Les Houches, France, 26 May - 6 Jun 2003, M. Dobbs et al., hep-ph/0403100*, 2004, hep-ph/0401145.
69. T. Plehn, D. Rainwater, and P. Skands (2005), in preparation.
70. R. Field, presentations at the 'Matrix Element and Monte Carlo Tuning Workshop, Fermilab, 4 October 2002 and 29–30 April 2003 (2002–2003), talks available from webpage http://cepa.fnal.gov/CPD/MCTuning/, and further recent talks available from http://www.phys.ufl.edu/~rfield/cdf/.
71. T. Sjöstrand, and M. van Zijl, *Phys. Rev.*, **D36**, 2019 (1987).
72. J. M. Butterworth, J. R. Forshaw, and M. H. Seymour, *Z. Phys.*, **C72**, 637–646 (1996), hep-ph/9601371.
73. T. Sjöstrand, and P. Z. Skands, *JHEP*, **03**, 053 (2004), hep-ph/0402078.
74. S. Schumann, SHERPA: an event generator for the LHC (2004), talk given at HERA/LHC Workshop, CERN, October 2004.
75. E. Boos, et al. (2001), hep-ph/0109068.
76. P. Skands, et al., *JHEP*, **07**, 036 (2004), hep-ph/0311123.

Heavy flavour production

Stefano Frixione

INFN, Sezione di Genova
Via Dodecaneso 33, 16146 Genova, Italy

Abstract. I review theoretical and phenomenological results relevant to heavy flavour production at colliders, and discuss the impact of recent measurements at the Tevatron and at HERA

Keywords: Collider Physics, Heavy Flavours

GENERALITIES

Heavy flavour production is one of the most extensively studied topics in high-energy particle physics. An impressive amount of data is available, for basically all kinds of colliding particles, which renders it possible to test QCD predictions to some accuracy. In open heavy flavour production final-state observables must be defined using either the variables of the heavy quarks or of the hadrons containing at most one heavy quark, and must not contain any reference to quarkonium states. By definition, a quark is heavy when

$$m_Q \gg \Lambda_{QCD}\,. \tag{1}$$

According to this equation, up, down, and strange quarks are definitely not heavy; for the remaining flavours, we have

$$m_t/\Lambda_{QCD} \simeq 800, \quad \Longrightarrow \quad \alpha_S(m_t) \simeq 0.1, \tag{2}$$

$$m_b/\Lambda_{QCD} \simeq 15, \quad \Longrightarrow \quad \alpha_S(m_b) \simeq 0.21, \tag{3}$$

$$m_c/\Lambda_{QCD} \simeq 4, \quad \Longrightarrow \quad \alpha_S(m_c) \simeq 0.33, \tag{4}$$

from which one is entitled to consider the top and the bottom to be heavy, while the case of charm is borderline. I shall treat the charm as heavy in what follows, for the simple technical reason that the condition in eq. (1) allows one to define an open-quark cross section without the need to convolute it with fragmentation functions, and this puts the charm formally on the same footing as the bottom and the top. On the other hand, since the quark mass typically sets the hard scale of the process, the values of α_S reported in eqs. (2)–(4) imply that in the case of charm the perturbative results will be affected by the larger uncertainties, and that non-perturbative effects are liable to play a major role.

In perturbative QCD, the production of a pair of heavy quarks $Q\overline{Q}$ in the collision of two hadrons $H_{1,2}$ is written according to the factorization theorem

$$d\sigma_{H_1H_2\to Q\overline{Q}}(S) = \sum_{ij}\int dx_1 dx_2 f_i^{(H_1)}(x_1) f_j^{(H_2)}(x_2) d\hat{\sigma}_{ij\to Q\overline{Q}}(\hat{s}=x_1x_2S), \tag{5}$$

CP792, *Deep Inelastic Scattering, DIS 2005*, edited by W. H. Smith and S. R. Dasu

and a fairly similar formula holds in the case of γ-hadron collisions (see ref. [1] for a review). As is well known, the parton density functions (PDFs) $f_i^{(H)}$ cannot be computed in perturbation theory, but are universal. On the other hand, the short distance cross sections $d\hat{\sigma}_{ij\to Q\overline{Q}}$ are process-specific, and computable in perturbation theory

$$d\hat{\sigma} = \sum_{i=2}^{\infty} a_i \alpha_S^i = a_2 \alpha_S^2 + a_3 \alpha_S^3 + a_4 \alpha_S^4 + \ldots \tag{6}$$

The coefficients a_2, a_3, and a_4 explicitly indicated in eq. (6) correspond to the LO, NLO, and NNLO contributions respectively. Although the computation of the LO term is almost trivial, that of the NLO term is not, and its achievement represented a breakthrough in heavy flavour physics at the end of the 80's [2, 3]. No exact result beyond NLO is currently available, and it does not seem probable that it will for some time. This is worrisome, since the NLO corrections for c and b production are of the same size as the LO contributions (i.e., the K factor is about 2), and the scale dependence for some observables is very large; NNLO terms may thus be numerically sizable, and quite relevant to the correct predictions of measured quantities.

Even if NNLO (or higher) contributions were available, one must keep in mind that such *fixed-order* results may still be insufficient to obtain phenomenologically sensible predictions. There are two main issues that need be considered.

1) Large logs appear in the perturbative coefficients

$$a_i = \sum_{k=0}^{i-2} a_i^{(i-2-k)} \log^{i-2-k} \mathscr{Q}, \tag{7}$$

where "large" means $\alpha_S \log^2 \mathscr{Q} \gtrsim 1$. $\mathscr{Q}$ *may or may not* depend on the observable. If $\mathscr{Q}$ is large, all terms in the expansion on the r.h.s. of eq. (6) are numerically of the same order, and the convergence of the series is spoiled. The way out is that of keeping only (some of) the leading logs in eq. (7), in such a way that the series of eq. (6) can be summed. This effectively corresponds to rearranging the perturbative expansion; technically, one says that the logs are *resummed*.

2) Bottom and charm quarks, although heavy, cannot be directly observed; therefore, the open heavy flavour cross section in such cases must be supplemented with the description of the quark-to-hadron transition (called fragmentation), which always involves a quantity, the non-perturbative fragmentation function (NPFF), not computable in perturbation theory. For the single-inclusive p_T spectrum, one writes

$$\frac{d\hat{\sigma}(H_Q)}{dp_T} = \int \frac{dz}{z} D^{Q\to H_Q}(z;\varepsilon) \frac{d\hat{\sigma}(Q)}{d\hat{p}_T}, \qquad p_T = z\hat{p}_T, \tag{8}$$

where Q is the heavy quark, H_Q is a given heavy-flavoured hadron, $\hat{p}_T$ (p_T) is the transverse momentum of Q (H_Q), $d\hat{\sigma}(Q)$ is the cross section for open-Q production, and $D^{Q\to H_Q}$ is the NPFF.

I'll start by discussing the case of large perturbative logs. In general, the logs to be resummed can be divided into two broad classes.

Observable-dependent logarithms: these logs depend on the kinematics of the final state (including cuts); a sample of their arguments is given in the equations below

$$\mathscr{Q} = \frac{p_T(Q)}{m_Q}, \qquad p_T(Q) \gg m_Q, \tag{9}$$

$$\mathscr{Q} = \frac{p_T(Q\overline{Q})}{m_Q}, \qquad p_T(Q\overline{Q}) \simeq 0, \tag{10}$$

$$\mathscr{Q} = 1 - \frac{\Delta\phi(Q\overline{Q})}{\pi}, \qquad \Delta\phi(Q\overline{Q}) \simeq \pi. \tag{11}$$

Equation. (9) is relevant to the single-inclusive transverse momentum distributions, whereas eqs. (10) and (11) are relevant to $Q\overline{Q}$ correlations. Analytic resummations for the logs of this class are observable-dependent and technically fairly involved, which renders the resummed cross sections unavailable except for a few simple cases. Even if the resummed observable can be computed, in general it must be matched to the corresponding fixed-order result for the predictions to be physically meaningful. Fortunately, this is the case for the single-inclusive p_T spectrum: the FONLL formalism [4] allows one to consistently combine (i.e., avoiding over-counting) the NLO result with the cross section in which p_T/m logs are resummed to NLL accuracy. Thus, FONLL can describe both the small-p_T ($p_T \sim m$, where resummed results don't make sense) and the large-p_T ($p_T \gg m$, where NLO results are not reliable) regimes.

The technical complications of the analytic resummations can be avoided by letting a parton shower Monte Carlo (PSMC) perform the resummation numerically. This procedure has the advantage that the logs can always be resummed, no matter how complicated the definition of the observable and the final state cuts are. The drawback is that the PSMC resummation is formally less accurate in terms of log accuracy than the analytic resummations, although in practice the difference between the two approaches is almost always negligible. On the other hand, PSMC's are based on a LO computation at the level of matrix elements, which is largely insufficient in the case of heavy flavours. In recent years, however, the problem of the consistent inclusion of NLO matrix elements into a PSMC framework has been successfully solved in QCD (MC@NLO [5, 6]); phenomenological results obtained with this formalism will be presented later.

Observable-independent logs: these logs do not depend on the kinematics of the final state. Those relevant to heavy flavour production are

$$\mathscr{Q} = 1 - 4m_Q^2/\hat{s}, \qquad \hat{s} \simeq 4m_Q^2, \tag{12}$$

$$\mathscr{Q} = m_Q^2/\hat{s}, \qquad \hat{s} \gg m_Q^2, \tag{13}$$

denoted as threshold and small-x logs respectively. Techniques to resum the former logs are rather well established; their effects are rather marginal, however, in c and b physics, except for b production at HERA-B (their role in top production has been mentioned before). On the other hand, small-x logs are theoretically challenging and intriguing. The standard Altarelli-Parisi evolution equations are replaced by those of CCFM; upon doing so, one is forced to introduce the so-called unintegrated PDFs, which have a functional dependence on a transverse momentum in addition to those on x and Q^2 of the standard PDFs.

After all the relevant large logs are properly resummed, one needs to understand whether the description of the fragmentation is physically sensible. I remind the reader that the NPFF is not calculable from first principles, and the free parameters it contains (denoted by ε in eq. (8)) are fitted to data after choosing a functional form in z. This fit is typically performed using eq. (8), identifying the l.h.s. with e^+e^- data. It follows that the value of ε is strictly correlated to the short-distance cross section $d\hat{\sigma}(Q)$ used in the fitting procedure, and thus is *non-physical*. When eq. (8) is used to predict H_Q cross sections, it is therefore mandatory to make consistent choices for ε and $d\hat{\sigma}(Q)$. The failure to do so has been one of the reasons behind the long-standing discrepancy between single-inclusive b measurements at colliders, and QCD predictions.

There is also another peculiar issue relevant to the NPFF: $d\hat{\sigma}(Q)/d\hat{p}_T$ has a power-like form, and one writes

$$\frac{d\hat{\sigma}(Q)}{d\hat{p}_T} \simeq \frac{C}{\hat{p}_T^N} \quad \Longrightarrow \quad \frac{d\hat{\sigma}(H_Q)}{dp_T} = \frac{C}{p_T^N} D_N^{Q\to H_Q}, \tag{14}$$

$$D_N^{Q\to H_Q} = \int dz\, z^{N-1} D^{Q\to H_Q}(z;\varepsilon). \tag{15}$$

It turns out that the approximation for the H_Q cross section given in eq. (14) is in excellent agreement with the exact result [7]. Since typically $N = \mathscr{O}(5)$ (for c and b at the Tevatron) it follows that, in order to have an accurate prediction for the p_T spectrum in hadroproduction, it is mandatory that the first few Mellin moments computed with $D(z)$ agreed with those measured. In ref. [8], where b production is considered, it is pointed out that this is not the case, in spite of the fact that the prediction for the inclusive B cross section in e^+e^- collisions, obtained with *the same* $D(z)$, displays an excellent agreement with the data. There may seem to be a contradiction in this statement: if the shape is reproduced well, why this is not true for the Mellin moments? The reason is that, when fitting $D(z)$, one excludes the region of large z, since it is affected by Sudakov logs and by complex non-perturbative effects which are unlikely to be described by the NPFF. On the other hand, the large-z region is important for the computation of D_N (because of the factor z^{N-1} in the integrand of eq. (15)). Therefore, for the purpose of predicting D- and B-meson spectra at colliders, ref. [8] advocates the procedure of fitting the NPFF directly in the N-space. A fit to the second moment (denoted as $N = 2$ fit henceforth) is found to fit well all the D_N's for N up to 10.

HEAVY FLAVOUR PRODUCTION: PHENOMENOLOGY

As discussed above, a typical state-of-the-art theoretical computation matches an NLO prediction with a cross section in which the relevant logs have been resummed. The ideal testing ground for such computation is top production for which, according to eqs. (2)–(4), we expect the perturbative series to be well-behaved, and that has the extra advantage of being unaffected by problems related to the understanding of the fragmentation mechanism. At present, the only $t\bar{t}$ data available are those from the Tevatron, where the c.m. hadronic energy is never much larger than the $t\bar{t}$ invariant mass, and therefore threshold logs may play an important role. It turns out that the resummation of such logs is seen to increase only marginally the NLO result (+3.5%), while it reduces substantially the scale uncertainty (from about 10% to about 5%). The comparison of preliminary Run II data with NLO predictions matched to the NLL-resummed result of ref. [9] is presented in fig. 1; as can be seen, theory and data are in good agreement, which confirms the findings of Run I. It should be stressed that, although the overall theoretical accuracy (about 12%, a 7% of which is due to PDF uncertainty) is small compared to that one finds in other heavy-flavour production processes, its combination with experimental errors will render it unlikely that any significant discrepancies with QCD will be found in Run II, even if top mass measurements will result in a value smaller than the present world average (CDF presented at this conference the most precise result so far obtained, $m_{top} = 173.5^{+4.1}_{-4.0}$ GeV – see ref. [10]).

Predictions used in fig. 1 have been obtained in ref. [11], and have been called NLO+NLL to remind that they are based on an NLO result matched with a resummed one that includes logs up to the next-to-leading accuracy. I stress that this terminology is not universal, and in particular is different from the one used in ref. [13] and subsequent papers. In those papers, one generally finds the notation NNLO+N^{k}LL, with k=2,3,4. It is important to realize that NNLO in this context does *not* mean that the full NNLO result has been computed (in fact, such result is not available, since the corresponding two-loop amplitudes have not been computed); it means that all of the NNLO terms have been retained, which are obtained upon a re-expansion of the resummed result, matched with the full NLO result. As far as the resummation is concerned, while the logs in ref. [9] are counted upstairs (i.e. in the exponent), in ref. [13] they are counted downstairs (i.e., after exponent expansion). This implies that the two approaches differ

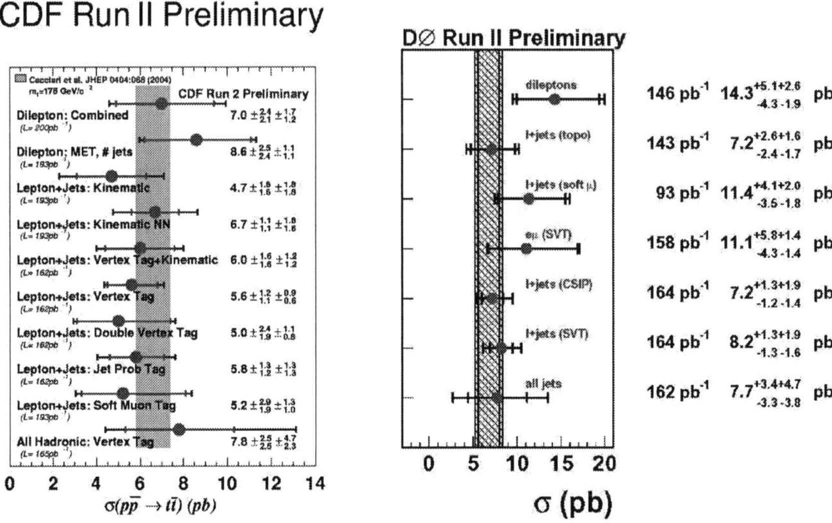

FIGURE 1. Comparison between NLO+NLL QCD predictions [11] and Run II data (see e.g. ref. [12]) for top production at the Tevatron.

by terms of NNLO, and of NNLL (in the exponent), which are in any case beyond control, and therefore that the perturbative accuracies of these results are the same.

Predictions for bottom and charm cross sections are much more involved than those for top, and therefore they represent a tougher test for perturbative QCD. As is well known, bottom data have received a lot of attention in the past. Such data are extremely abundant at hadron colliders, especially so for single-inclusive observables that are relatively well measured also in absence of vertex detectors. It was therefore worrisome that basically all Run I measurements overshot NLO QCD predictions by sizable factors (see e.g. ref. [1]). In general, the shape of single-inclusive p_T spectrum given by QCD is consistent with the data, but if one uses "default" values for the inputs of the computations (such as $m_b = 4.75$ GeV, $\mu = \sqrt{p_T^2 + m_b^2}$, and central PDF sets) the measured rates are larger than NLO results by a factor of 2.84 (for CDF) and 2.12 (for D0) on average. The last data available from Run I confirmed this trend: for the B^+ p_T spectrum, CDF found that the average data/theory ratio is $2.9 \pm 0.2 \pm 0.4$.

The disagreement between b production data at the Tevatron and QCD predictions has been one of the most compelling problems in hadronic physics. It triggered studies on possible beyond-the-SM explanations, which to the best of my knowledge have not survived the tests of LEP data. Although the BSM scenario is still an option, at present it seems premature to adopt it without first reassessing carefully all possible sources of mistakes in the past comparisons between theory and data, and without considering the uncertainties that so-far uncalculated SM contributions can give. To start with, it appears that a better comparison with Tevatron data should be obtained by using FONLL rather than pure NLO predictions, with the fragmentation function obtained with an $N = 2$ fit. The result of such a comparison is shown in the left panel of fig. 2 (taken from ref. [8]), where the data are the same B^+ ones quoted above. The average data/theory ratio is now $1.7 \pm 0.5 \pm 0.5$ i.e. data are within 1σ from the default theoretical prediction. A factor of 20% of the reduction from the formerly claimed discrepancy of 2.9 to the current 1.7 is due to the use of FONLL in place of NLO results; the remaining 45% to the correct treatment of the NPFF.

These findings suggest to recompute the theoretical predictions upon which the vast majority of the past comparisons between theory and data were based. Unfortunately, this would not help much, since most of those data are relevant to b quarks, rather

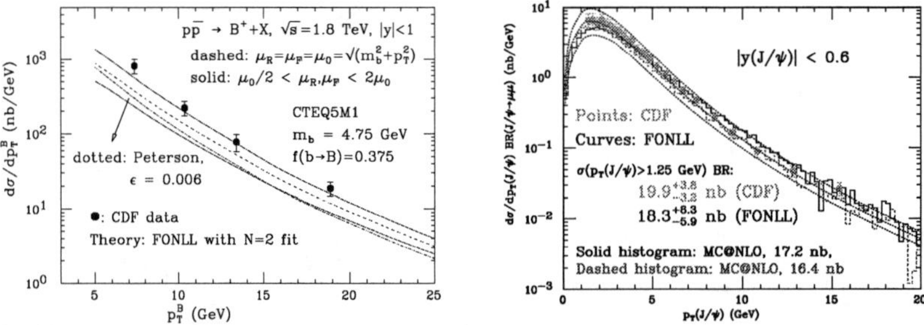

FIGURE 2. Comparisons between b data at the Tevatron, and QCD predictions. See the text for details.

than to B mesons; in other words, experimental collaborations deconvoluted the $b \to B$ fragmentation. This has been typically done using PSMC models, and in general it appears rather difficult to recover the data for B mesons, which is a practical example of a theoretical bias affecting data. On the other hand, there are other ways to understand whether we are on the right track. One interesting possibility consists in considering jets containing b quarks (i.e., any b-hadron species) rather than tagging a specific b-hadron: in this case, the NPFF simply doesn't enter the cross section, and the theoretical predictions are also less prone to develop large p_T logs, since the p_T of the b is not involved in the definition of the observable. The comparison between NLO predictions for b-jets [14] and D0 measurements [15] is indeed satisfactory: data are consistent with theory in the range $25 < E_T^{b-jet} < 100$ GeV. More measurements relevant to b-jets are being performed in Run II.

To fully test the ideas relevant to the NPFF treatment one still needs heavy-flavoured meson data. Fortunately, a lot of these are expected to become available in the near future, thanks to the ongoing Run II. The first results on single-inclusive b-hadron p_T spectrum have been presented by CDF [16], and they are particularly interesting in view of the fact that for the first time they probe the region of $p_T \simeq 0$, which is fairly sensitive to the description of the fragmentation. The comparison of data with FONLL (dotted lines) is presented in the right panel of fig. 2 and displays the best-ever agreement between theoretical predictions and b data at colliders. The same pattern of agreement is obtained by using MC@NLO (histograms), which constitutes a very powerful check on the theoretical results: both the resummation and the $b \to B$ transition are performed in vastly different ways in FONLL and MC@NLO. I should stress that the plot presents the p_T spectrum of the J/ψ's emerging from b-flavoured hadron decays, and thus involves a highly non-trivial combination of short- and long-distance dynamics (see ref. [17] for a more detailed discussion).

Although the tests presented above prove that when comparing modern data sets with up-to-date theoretical predictions heavy flavour production appears to be fairly well predicted by QCD, we must keep in mind that the uncertainties affecting the QCD results are still quite large; thus, the possibility remains that such uncertainties hide physics effects not included in the computations. Within non-BSM physics, one of the most interesting questions is whether b production at the Tevatron is small-x physics. According to ref. [18], small-x resummation would increase the b cross section by a mere 30%. On the other hand, a good description of Tevatron data is obtained by using CASCADE [19], a Monte Carlo code that implements CCFM evolution equations. It should be noted that, since the (LO) matrix elements convoluted with unintegrated PDFs have off-shell partons, they include part of the contributions which are of NLO in the standard collinear approach. This renders the interpretation of the results more complicated, since it is impossible to tell the *pure* small-x effects apart from higher-order corrections to the matrix elements. It seems therefore necessary to compute the NLO corrections rigorously in the context of the small-x approach in order to achieve firmer conclusions. I should also mention the fact that unintegrated PDFs, exactly like standard PDFs, cannot be computed from first principles, and need to be extracted from data. Such an extraction is at present affected by large uncertainties (especially for the gluon density), which must be systematically reduced in order for small-x computations to be as reliable phenomenologically as those based on collinear factorization.

Thanks to the much improved performances in terms of luminosity of the HERA collider, QCD predictions for b production can also be sensibly tested in γp (photoproduction) and ep (DIS) collisions. The first HERA results on b physics gave ambiguous indications. H1 photoproduction measurement [20] was a factor 3.26 ± 0.74 higher than (i.e., 3σ away from) NLO QCD predictions [21]. On the other hand, in the case of ZEUS [22] the ratio data/theory was 2.50 ± 1.18, i.e. a mere 1.3σ away from NLO QCD. Experiments understand much better now these very complicated analyses; furthermore, the increased statistics allow them to quote cross sections in the visible regions, and therefore to be more independent of the (mainly PSMC) biases implicit in the extrapolations from the visible to the invisible regions. In ref. [23], H1 re-analyzed the data of ref. [20]; the theory/data ratio went down to 1.99 ± 0.47. Furthermore, new data presented in that paper are larger than NLO QCD predictions only by a factor 1.61 ± 0.51, i.e. little more than 1σ. As far as ZEUS is concerned, newer photoproduction results [24] (summarized here in the left panel of fig. 3) are also in better agreement with QCD than those of ref. [22]. By a closer inspection of the experimental results, one can see that the bulk of the discrepancy wrt NLO QCD predictions observed by H1 in ref. [23] is concentrated at low transverse momenta, in both photoproduction and DIS regimes. The same effect occurs in a recent ZEUS DIS measurement [25]. A few more glitches can be seen in the comparison between data and theory, which is summarized for total (visible) rates in the right panel of fig. 3. These kind of discrepancies certainly deserve more attention in the future, and improved vertexing techniques and statistics will help to clarify the situation. On the other hand, it appears to me that at present there is a satisfactory agreement between HERA measurements and QCD predictions for b observables. It is worth noting that small-p_T regions have the highest sensitivity to PSMC biases, since standard (i.e. based on a leading-order picture) parton shower Monte Carlo's cannot give sensible predictions for such momenta (see ref. [6] for a discussion on this point). One should not forget the lesson learnt at the Tevatron, that these kind of biases can lead to overestimating the discrepancy between theory and data. An example of (possible) large biases can be found in ref. [26], in which the b sample is obtained entirely through MC methods: it is perhaps not a coincidence that the ratio data/theory turns out to be extremely large in that paper.

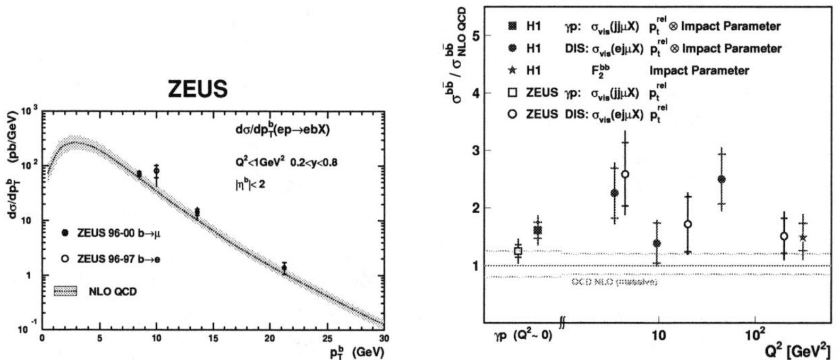

FIGURE 3. Comparisons between b data at HERA, and QCD predictions. See the text for details.

Let me now turn to the case of charm production. Being the lightest of the heavy quarks, QCD predictions for charm cross sections are affected by the largest perturbative uncertainties and non-perturbative effects. That notwithstanding, the overall agreement between charm data and QCD predictions is basically satisfactory, although a few disturbing facts remain. In particular, at HERA data on D^* production cannot be described by QCD in the region $\eta > 0$. Also, the $p_T(D^*)$ photoproduction spectrum measured by ZEUS is harder than FONLL predicts, while that of H1 is softer. In spite of these discrepancies, I believe that open charm production will not pose any serious problem for QCD in the near future. A fairly different situation is that of the $c\bar{c}$ bound states, whose theoretical understanding is less developed than that for open-heavy-quark states, and whose phenomenology poses at present quite a challenge to QCD. I'll give the briefest of the reviews on quarkonium production in what follows.

The hadroproduction cross section of a $c\bar{c}$ or $b\bar{b}$ quarkonium state H can be written in a manner analogous to eq. (5)

$$d\sigma_{H_1H_2\to H}(S) = \sum_{ij} \int dx_1 dx_2 f_i^{(H_1)}(x_1) f_j^{(H_2)}(x_2) d\hat{\sigma}_{ij\to H}(\hat{s} = x_1x_2S). \tag{16}$$

The forms of the short-distance cross sections that appear in eq. (16) depend on the theory or the model adopted to compute them. Among the various approaches, the *only one* that can be mathematically derived from QCD is non-relativistic QCD (NRQCD [27]), an effective theory in which the heavy quarks move at non-relativistic velocities. Within NRQCD, the partonic cross sections are [28]

$$d\hat{\sigma}_{ij\to H} = \sum_n d\hat{\sigma}(ij \to Q\overline{Q}[n]) \langle \mathcal{O}^H[n] \rangle, \tag{17}$$

where $d\hat{\sigma}(Q\overline{Q}[n])$ is the cross section for the production of a $Q\overline{Q}$ pair in a given spin and colour state (symbolically, $n = \{c = (1,8); {}^{2S+1}L_J\}$), and the NRQCD matrix elements $\langle \mathcal{O}^H[n] \rangle$ are analogous to PDFs and NPFFs, since they cannot be computed in perturbation theory, and are universal; loosely speaking, they are proportional to the probability for the $Q\overline{Q}$ pair in the state n to fragment into the quarkonium state H.

Given the fact that NRQCD is derived from QCD, and that pQCD can describe open-Q data, we expect that NRQCD does a good job too. A difficulty, however, is immediately apparent by looking at eq. (17), that features an infinite sum in which each term contains a non-calculable long-distance parameter ($\langle \mathcal{O}^H[n] \rangle$); this implies a complete loss of predictive power. Fortunately, the NRQCD matrix elements obey a (velocity) scaling rule [29]

$$\langle \mathcal{O}^H[n] \rangle \propto v^{f(n,H)}, \qquad v^2 \simeq 0.3, 0.1 \quad \text{for} \quad c\bar{c}, b\bar{b}, \tag{18}$$

where v is the relative velocity of the $Q\overline{Q}$ pair in the quarkonium state, and f is a function that depends in a complicated manner on the quantum numbers of the states $Q\overline{Q}[n]$ and H (see e.g. ref. [30]). Eq. (18), combined with the usual expansion in α_s of the short-distance cross sections, implies that the r.h.s. of eq. (17) can be rewritten as a double series

$$d\hat{\sigma}_{ij\to H} = \sum_{m,k} s_{m,k} \alpha_s^m v^k. \tag{19}$$

This systematic expansion in α_s and v provides a computational framework similar to that relevant to open-Q production. Still, eq. (19) poses some non-trivial computational problems, given the fact that the double series is slowly "convergent", particularly so for charm (owing to eqs. (4) and (18)), and thus one needs to determine a large number of NRQCD matrix elements (some of them can be expressed in terms of others, for example using heavy quark spin symmetry and vacuum saturation approximation). Furthermore, the same problems that affect the short distance coefficients of eq. (6) are also relevant to the present case; most notably, the occurrence of large logs can render the expansion (19) useless.

The first stringent experimental tests of NRQCD predictions at colliders have become available only relatively recently, thanks to the use by CDF of microvertices that allow the precise measurements of the *direct* cross sections (those in which the observed quarkonium state is not obtained through feeddown from more massive bound states). This immediately led to ruling out the so-called colour-singlet model (CSM), which can be obtained from eq. (17) by dropping all but the leading colour-singlet contribution there. In fact, it turns out that no amount of tuning of the NRQCD matrix elements can bring the LO CS contribution in agreement with data, and CO contributions are therefore necessary; this fact holds both for J/ψ and for Υ production.

Although the dominance of CO contributions in quarkonia production at the Tevatron has to be regarded as a highly successful prediction of NRQCD, we must keep in mind that the NRQCD matrix elements are (in part) fitted to the data that the theory is supposed to predict. Such a determination, furthermore, is not only affected by fairly large uncertainties (see ref. [31]), but is also biased by the fact that a definite choice for the PDFs of the colliding hadrons must be made. Within these uncertainties, the NRQCD matrix elements do obey the scaling rules of eq. (18). On the other hand, a more convincing test of NRQCD predictions is the check of the universality of the matrix elements, whose values must be independent of the hard process and therefore, if fitted at the Tevatron, can be used at HERA to predict the quarkonium cross sections there. Unfortunately, the comparison between ep data and NRQCD predictions is largely inconclusive as far as the CO contributions are concerned. The J/ψ energy distribution in photoproduction does not support the growth towards the endpoints of the spectrum which is a consequence of the CO terms; on the other hand, it is known that such regions are strongly affected by large higher-order corrections (in v), and thus a resummation would be necessary in order to draw definite conclusions. On the other hand, the data are in good agreement, for both the energy distribution and the p_T spectrum, with the pure CS prediction if NLO effects [32] are taken into account. The situation appears to be more consistent with the findings at the Tevatron in the case of DIS data, although a few glitches remain there as well (especially in the z distribution). The bottom line is that it is hard to draw any conclusion at present; data of larger statistics must be obtained in the present HERA run phase in order to test NRQCD more thoroughly.

It is interesting to observe that a good agreement with the Tevatron J/ψ and Υ data is also obtained in the context of the colour evaporation model (CEM). When using such a model, the short-distance cross sections in eq. (16) are written as follows

$$d\hat{\sigma}_{ij\to H}^{(\mathrm{CEM})} = F_H \int_{4m_Q^2}^{4m_M^2} dm_{Q\overline{Q}}^2 \frac{d\hat{\sigma}(ij\to Q\overline{Q})}{dm_{Q\overline{Q}}^2}, \tag{20}$$

where m_M is the mass of the lowest-lying Q-flavoured meson state, and F_H is a universal (long-distance) constant. The colour evaporation model can be formally written in the same form as NRQCD, by replacing the original expression for the NRQCD matrix elements

$$\mathscr{O}^H[n] = \chi^* \kappa_n \psi \left(\sum_X |H+X\rangle\langle H+X| \right) \psi^* \kappa_n' \chi \tag{21}$$

with

$$\begin{aligned} \mathscr{O}^H[n] &= F_H \sum_n \chi^* \kappa_n \psi \sum_X |Q\overline{Q}(m_{Q\overline{Q}}^2 < 4m_M^2) + X\rangle \\ &\quad \times \langle Q\overline{Q}(m_{Q\overline{Q}}^2 < 4m_M^2) + X| \psi^* \kappa_n' \chi. \end{aligned} \tag{22}$$

Eq. (22) basically entails a change of the scaling rules, $v^{f(n,H)} \to v^{2L}$; this is interesting, since it implies that a re-organization of the double expansion in eq. (19) can also give satisfactory phenomenological results. It should be noted that in general CEM predictions must be supplemented with a k_T-kick in order to describe the data, which decreases the predictive power of the model. The fact that the only information concerning the quarkonium state is contained in the constant F_H is also troublesome, since for example the ratio $\sigma(\chi_c)/\sigma(J/\psi)$ turns out to be different if measured in hadron-hadron and photon-hadron collisions at fixed-target. Furthermore, a weak p_T dependence is found for the J/ψ decay fractions.

In conclusions, it appears that at present perturbative QCD is doing a satisfactory job in predicting open heavy flavour cross sections measured at colliders. The most substantial improvement in recent years occurred in b physics; the long-standing discrepancy between single-inclusive b data at colliders and theoretical predictions has been settled mainly thanks to a better understanding of the non-perturbative phenomena, since the backbone of the computations is still the NLO result of refs. [2, 3], which is essential to get anywhere close to the measurements. On the other hand, the careful re-analysis of the computations motivated their improvements, through the matching with resummed results (FONLL) or with Monte Carlo techniques (MC@NLO); the flexibility of the latter guarantees that studies with the same accuracy of those performed so far only for single-inclusive observables can be repeated for basically any type of variable. It is reassuring that, by increasing the collected statistics and with a better understanding of the analyses, the measurements performed at HERA are also in a much better agreement with QCD than they used to be. At present, I don't see any serious discrepancies between data and theory there; some problems remain, however, and this will serve as a powerful motivation for obtaining data of yet better quality in the last years of HERA running.

The situation is not as bright in the case of quarkonium production. We do have a theory, NRQCD, whose elegant and compact formulation allows the systematic computation of any observable relevant to quarkonium production. Being a direct consequence of QCD, it appears to be fairly solid. On the other hand, problems remain in the comparisons with collider data, the most serious of which is that of the J/ψ polarization. It should be noted that NRQCD computations are not at the same level of accuracy as those used to predict open heavy flavour cross sections; both the v and the α_s expansions must be considered, and the computation of the observables beyond LO (which appears to be a necessity in heavy flavour physics) is extremely complicated. The success of the CEM may suggest that, at least for the case of charm, velocity scaling rules may not be adequate, and more theoretical work is needed in this direction.

REFERENCES

1. S. Frixione, M. L. Mangano, P. Nason and G. Ridolfi, Adv. Ser. Direct. High Energy Phys. **15** (1998) 609 [arXiv:hep-ph/9702287].
2. P. Nason, S. Dawson and R. K. Ellis, Nucl. Phys. B **303** (1988) 607.
3. W. Beenakker, W. L. van Neerven, R. Meng, G. A. Schuler and J. Smith, Nucl. Phys. B **351** (1991) 507.
4. M. Cacciari, M. Greco and P. Nason, JHEP **9805** (1998) 007 [arXiv:hep-ph/9803400].
5. S. Frixione and B. R. Webber, JHEP **0206** (2002) 029 [arXiv:hep-ph/0204244].
6. S. Frixione, P. Nason and B. R. Webber, JHEP **0308** (2003) 007 [arXiv:hep-ph/0305252].
7. P. Nason *et al.*, arXiv:hep-ph/0003142.
8. M. Cacciari and P. Nason, Phys. Rev. Lett. **89** (2002) 122003 [arXiv:hep-ph/0204025].
9. R. Bonciani, S. Catani, M. L. Mangano and P. Nason, Nucl. Phys. B **529** (1998) 424 [arXiv:hep-ph/9801375].
10. J-F. Arguin, these proceedings.
11. M. Cacciari, S. Frixione, M. L. Mangano, P. Nason and G. Ridolfi, JHEP **0404** (2004) 068 [arXiv:hep-ph/0303085].
12. K. Ranjan, these proceedings.
13. N. Kidonakis, E. Laenen, S. Moch and R. Vogt, Phys. Rev. D **64** (2001) 114001 [arXiv:hep-ph/0105041].
14. S. Frixione and M. L. Mangano, Nucl. Phys. B **483** (1997) 321 [arXiv:hep-ph/9605270].
15. B. Abbott *et al.* [D0 Collaboration], Phys. Rev. Lett. **85** (2000) 5068 [arXiv:hep-ex/0008021].
16. D. Acosta *et al.* [CDF Collaboration], arXiv:hep-ex/0412071.
17. M. Cacciari, S. Frixione, M. L. Mangano, P. Nason and G. Ridolfi, JHEP **0407** (2004) 033 [arXiv:hep-ph/0312132].
18. J. C. Collins and R. K. Ellis, Nucl. Phys. B **360** (1991) 3.
19. H. Jung and G. P. Salam, Eur. Phys. J. C **19** (2001) 351 [arXiv:hep-ph/0012143].
20. C. Adloff *et al.* [H1 Collaboration], Phys. Lett. B **467** (1999) 156 [Erratum-ibid. B **518** (2001) 331] [arXiv:hep-ex/9909029].
21. S. Frixione, M. L. Mangano, P. Nason and G. Ridolfi, Nucl. Phys. B **412** (1994) 225 [arXiv:hep-ph/9306337].
22. J. Breitweg *et al.* [ZEUS Collaboration], Eur. Phys. J. C **18** (2001) 625 [arXiv:hep-ex/0011081].
23. A. Aktas *et al.* [H1 Collaboration], Eur. Phys. J. C **41** (2005) 453 [arXiv:hep-ex/0502010].
24. S. Chekanov *et al.* [ZEUS Collaboration], arXiv:hep-ex/0312057.
25. S. Chekanov *et al.* [ZEUS Collaboration], Phys. Lett. B **599** (2004) 173 [arXiv:hep-ex/0405069].
26. A. Aktas *et al.* [H1 Collaboration], arXiv:hep-ex/0503038.
27. W. E. Caswell and G. P. Lepage, Phys. Lett. B **167** (1986) 437.
28. G. T. Bodwin, E. Braaten and G. P. Lepage, Phys. Rev. D **51** (1995) 1125 [Erratum-ibid. D **55** (1997) 5853] [arXiv:hep-ph/9407339].
29. G. P. Lepage, L. Magnea, C. Nakhleh, U. Magnea and K. Hornbostel, Phys. Rev. D **46** (1992) 4052 [arXiv:hep-lat/9205007].
30. M. Kramer, Prog. Part. Nucl. Phys. **47** (2001) 141 [arXiv:hep-ph/0106120].
31. M. Beneke and M. Kramer, Phys. Rev. D **55** (1997) 5269 [arXiv:hep-ph/9611218].
32. M. Kramer, Nucl. Phys. B **459** (1996) 3 [arXiv:hep-ph/9508409].

Life and Death Among the Hadrons

R. L. Jaffe

Center for Theoretical Physics,
Laboratory for Nuclear Science and Department of Physics
Massachusetts Institute of Technology,
Cambridge, Massachusetts 02139

Abstract.
This talk covers two related topics: a quick review of the present evidence for, and especially against the exotic Θ^+ baryon, and a brief advertisement for the study of diquark correlations in QCD[1].

THE $\Theta^+(1540)$

In January of 2003 evidence was reported of a very narrow baryon resonance with strangeness one and charge one, of mass $\approx$ 1540 MeV, now dubbed the Θ^+, with minimum quark content $uudd\bar{s}$ [2, 3]. The first experiment was followed by evidence for other exotics: a strangeness minus two, charge minus two particle now officially named the Φ^{--} by the PDG, with minimum quark content $ddss\bar{u}$ [5] at 1860 MeV, and an as-yet nameless charm exotic $(uudd\bar{c})$ [6] at 3099 MeV. Many experimental groups published confirmations of the Θ^+. Theorists, myself included, descended upon these reports and tried to extract dynamical insight into QCD[1]. Other experimental groups began searches for the Θ^+ and some reported negative results, especially in higher energy, inclusive production environments. Recently the balance has been tipping toward the negative sightings. Late last year Dzierba, Meyer, and Szczepaniak (DMS)[4] summarized the experimental evidence for (their Table 1) and against (their Table 2) the Θ^+ and its exotic partners.

In April 2005 a new experiment undertaken at Jefferson Lab (G11@JLab) has reported null results[7]. This report is particularly significant because it comes in a channel where a positive signal was seen before. The reaction was $\gamma p \to K^0\ \Theta^+$, followed by $\Theta^+ \to K^+ n$ and $K_S \to \pi^+\pi^-$. The most compelling way to present their data is to compare the G11 signal for the $\Lambda(1520)$, a well known non-exotic resonance, with their null result for the Θ^+, and then contrast the G11 results with the earlier positive sighting from SAPHIR[8] (see Fig. 1). SAPHIR quote a ratio

* Several other negative results have been reported since DIS05, including $\gamma D \to pK^+K^-n$ from G10@JLab [9] and in secondary kaon interactions at Belle[10]. Unofficial reports from BaBar indicate negative results from $\gamma A \to kK_S$ using beam-detector interactions. No *new* positive results were reported at the 2005 Lepton-Photon Symposium at Uppsala[11], although several of the first sightings survived reanalyses of the original data.

CP792, *Deep Inelastic Scattering, DIS 2005*, edited by W. H. Smith and S. R. Dasu

of production rates: $N(\Theta^+)/N(\Lambda(1520)|_{\mathrm{SAPHIR}} \approx 9\%$. In contrast CLAS G11 quote $N(\Theta^+)/N(\Lambda(1520)|_{\mathrm{G11@JLab}} < 0.5\%$, a striking disagreement*

Things do not look good for the Θ^+. Typically in our field, effects seen weakly in discovery experiments are quickly confirmed when they are real, especially in the next round of experiments expressly designed to find them (*e.g.* CLAS G11). While it is probably too early to declare it dead, it is not too early to think about life without the Θ^+. So, for the rest of this talk, I will assume that the Θ^+ is no more.

The absence of a Θ^+ has implications for phenomenological models of QCD. Chiral soliton models (CSM) need to be rethought. Diakonov, Petrov, and Polyakov used a version of the CSM to predict unequivocally a light, narrow, exotic baryon with $Y = 1, I = 0$ and $J^\Pi = 1/2^+$[12]. They defended their prediction against criticism of the formulation of their model[13, 14, 15] and of the accuracy of both the mass[13] and width[16] preditions. What could be wrong with the CSM? Here are some possibilities:

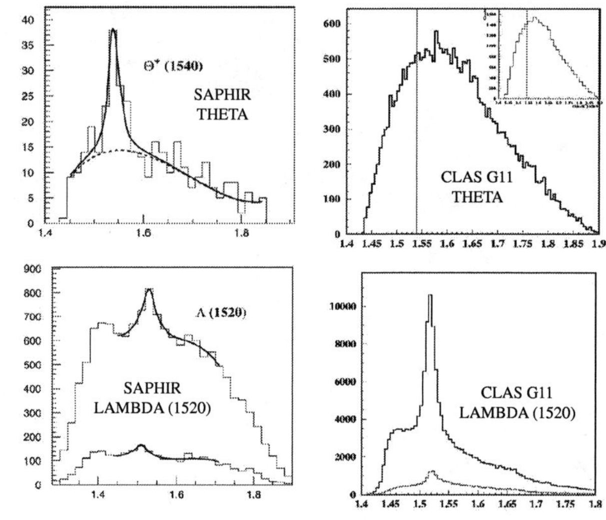

1: SAPHIR (left) and CLAS G11 (right) data on the Θ^+ (upper) and $\Lambda(1520)$ (right). The CLAS G11 upper limit on the $\Theta^+/\Lambda(1520)$ ratio is a factor of 20 smaller than measured by SAPHIR.

- Perhaps baryons are not chiral solitons in the first place! Witten introduced the idea into QCD as a *heuristic*, and it clearly fails in the case of 1-flavor and N_c colors, where there is no chiral sector at all, but there are perfectly good baryons (the analogue of the Δ^{++}).
- Perhaps it was not appropriate to truncate the chiral lagrangian after the "Syrme term". The soliton is stabilized by balancing the dimension-four ($\mathcal{L}_4 = f_\pi^2 \partial_\mu U \partial^\mu U^\dagger$) term against the dimension-six Skyrme term, $\mathcal{L}_6$, ignoring other operators of dimension-six and higher. If two operators of different dimension are equally important in a particular regime, there is no *a priori* reason to ignore operators of arbitrarily high dimension.
- Perhaps collective coordinate quantization fails. Why should the soliton keeps its rigid profile as it is "spun up" in angular momentum and flavor space? Since there is only one scale in QCD, shape excitations should appear at the same scale as rotations.
- Perhaps $SU(3)$ violation cannot he treated perturbatively. This is the subject of Refs. [14, 15] and Refs. [17, 18] before them. Attempts to find and characterize the

Θ^+ on the lattice have been inconclusive: some studies report no K^+n resonance at all, some report a negative parity (presumably s-wave) resonance, and others report positive parity[1].

Most other phenomenological models — large N_c[19], and quark models[1, 20, 21, 22], for example — never predicted the mass of the Θ^+ in the first place. Not as ambitious as the CSM, they generally do not claim to determine the overall mass scale of a new sector of QCD (*eg.* $qqqq\bar{q}$) accurately. Pre-2003 estimates were typically hundreds of MeV heavier than 1540 MeV. After the Θ^+ was reported, its mass was used to tie down the $qqqq\bar{q}$ spectrum and other states were predicted relative to M_Θ. If the Θ^+ is gone, the correlations invoked to stabilize it in these models must be weaker than proposed.Attempts to find and characterize the Θ^+ on the lattice have been inconclusive: some studies report no K^+n resonance at all, some report a negative parity (presumably s-wave) resonance, and others report positive parity[1].

What, then, are the lessons of the "Θ-affair", if it is over? First, there *really are no light, narrow exotics in QCD*. Second, phenomenological models should not be taken too seriously, especially for the overall mass scale in untested regimes (recent surprises in the D_{sJ}-spectrum support this[23]). Third, lattice QCD has not yet got to the stage where it can provide reliable, quantitative insight into novel phenomena. Note, however, that the demise of the Θ^+ would not rule out exotics of a different character in the heavy quark sectors. In particular, the charm exotic baryons proposed in Ref. [24] remain interesting.

DIQUARKS

It is clear that exotics are very rare in QCD. Perhaps they are entirely absent. This remarkable feature of QCD is often forgotten when exotic candidates are discussed. The existence of any exotic has to be understood in a framework that also explains their overall rarity. Along the same line, the *aufbau* principle of QCD differs dramatically from that of atoms and nuclei: to make more atoms add electrons, to make more nuclei, add neutrons and protons. However in QCD the spectrum seems to stop at qqq and $q\bar{q}$.

Thinking about early reports of the Θ^+ in light of early work on multiquark correlations in QCD [25], Frank Wilczek and I [20]† began to re-examine the role of diquark correlations in QCD. Diquarks are not new; they have been championed by a small group of QCD theorists for several decades [26, 27]. We already knew [25] that diquark correlations can naturally explain the general absence of exotics and predict a supernumerary nonet of scalar mesons which seems to exist. They appear in many successful pictures of soft QCD phenomenology. A light Θ^+ can be accommodated, but is not required, by diquark dynamics. Whether or not the Θ^+ survives, diquarks are here to stay.

Spectroscopy was at the cutting edge of high energy physics in the '60's and '70's. A great deal of effort and sophisticated analysis was brought to bear on the study of the hadron spectrum, and the conclusions remain important. In the decade that followed the first conjectures about quarks experimental groups studied meson-baryon and meson-

† Closely related ideas have been explored by Nussinov [21] and by Karliner and Lipkin [22].

meson scattering, and extracted the masses and widths of meson and baryon resonances. Resonances were discovered in nearly all non-exotic meson and baryon channels, but no prominent exotics were found.

The zeroth order summary prior to January 2003 was simple: no exotic mesons or baryons. In fact the only striking anomaly in low energy scattering was the existence of a supernumerary (*i.e.*, not expected in the quark model) nonet of scalar, ($J^{\Pi} = 0^{+}$) mesons with masses below 1 GeV: the $f_0(600)$, $\kappa(800)$, $f_0(980)$, and $a_0(980)$ that is now widely considered to contain important $\bar{q}\bar{q}qq$ components [28].

QCD phenomena are dominated by two well known quark correlations: confinement and chiral symmetry breaking. Confinement hardly need be mentioned: color forces only allow quarks and antiquarks correlated into color singlets. Chiral symmetry breaking can be viewed as the consequence of a very strong quark-antiquark correlation in the color, spin, and flavor singlet channel: $[\bar{q}q]^{\mathbf{1}_c\mathbf{1}_f0}$. The attractive forces in this channel are so strong that $[\bar{q}q]^{\mathbf{1}_c\mathbf{1}_f0}$ condenses in the vacuum, breaking $SU(N_f)_L \times SU(N_f)_R$ chiral symmetry.

The "next most attractive channel" in QCD seems to be the color antitriplet, flavor antisymmetric (which is the $\overline{\mathbf{3}}_f$ for three light flavors), spin singlet with even parity: $[qq]^{\overline{\mathbf{3}}_c\overline{\mathbf{3}}_f0^+}$. This channel is favored by one gluon exchange and by instanton interactions. It will play the central role in the exotic drama to follow.

The classification of diquarks is not entirely trivial. Operators that will create a diquark of any (integer) spin and parity can be constructed from two quark fields and insertions of the covariant derivative. We are interested in potentially low energy configurations, so we omit the derivatives. There are eight distinct diquark multiplets (in color×flavor×spin) that can be created from the vacuum by operators bilinear in the quark field [1]. However, the interesting candidates can be pared down quickly: Color $\mathbf{6}_c$ diquarks would appear to have much larger color electrostatic field energy. Odd parity diquarks require quarks to be excited relative to one another. This leaves only two diquarks consistent with fermi statistics,

$$\begin{aligned}
&\left|\{qq\}\ \overline{\mathbf{3}}_c(A)\ \overline{\mathbf{3}}_f(A)\ 0^+(A)\right\rangle \\
&\left|\{qq\}\ \overline{\mathbf{3}}_c(A)\ \mathbf{6}_f(S)\ 1^+(S)\right\rangle, \qquad (1)
\end{aligned}$$

where A or S denotes the exchange symmetry of the preceding representation. Both of these configurations are important in spectroscopy. In what follows I will refer to them sometimes as the "scalar" and "vector" diquarks, or more suggestively, as the "good" and "bad" diquarks. Remember, though, that there are many "worse" diquarks that we are ignoring entirely.

Models universally suggest that the scalar diquark is lighter than the vector. For example, one gluon exchange evaluated in a quark model gives rise to a color and spin dependent interaction that favors the scalar diquark. The matrix elements of this interaction in the "good" and "bad" diquark states are $-2\mathcal{M}$ and $+2/3\mathcal{M}$ respectively, where $\mathcal{M}$ is model dependent. To set the scale, the Δ–nucleon mass difference is $4\mathcal{M}$, so the energy difference between good and bad diquarks is $\sim \frac{2}{3}(M_\Delta - M_N) \sim 200$ MeV. Not a huge effect, but large enough to make a significant difference in spectroscopy. After all, the nucleon is stable and the Δ is 300 MeV heavier and has a width of 120 MeV!

Characterizing diquarks

The good scalar and bad vector diquarks are our principal subjects. Since the good diquarks are antisymmetric in flavor they lie in the $\overline{\mathbf{3}}$ representation of $SU(3)_f$. We will denote them by $[q_1,q_2]:\{[u,d]\ [d,s]\ [s,u]\}$ when flavor is important and by $\mathbb{Q}$ when it is not. Under flavor $SU(3)$ transformations they behave exactly like antiquarks, $[u,d]\leftrightarrow\bar{s}$, $[d,s]\leftrightarrow\bar{u}$, $[s,u]\leftrightarrow\bar{d}$. The bad diquarks are symmetric in flavor, forming the **6** representation of $SU(3)_f$. The notation $\{q_1,q_2\}:\{\{u,u\}\ \{u,d\}\ \{d,d\}\ \{d,s\}\ \{s,s\}\ \{s,u\}\}$ will do.

Although diquarks are colored states, their properties can be studied in a formally correct, color gauge invariant way on the lattice. To define the non-strange diquarks, introduce an infinitely heavy quark, Q, *i.e.* a Polyakov line. Then study the qqQ correlator with the qq quarks either antisymmetric ($[u,d]Q$) or symmetric ($\{u,d\}Q$) in flavor. The results, $M[u,d]$ and $M\{u,d\}$ — labelled unambiguously — are meaningful in comparison, for example, with the mass of the lightest $\bar{q}Q$ meson, $M(u)=M(d)$. $M\{u,d\}-M[u,d]$ is the good-bad diquark mass difference for massless quarks. It is a measure of the strength of the diquark correlation. The diquark-quark mass difference, $M[u,d]-M(u)$, is another. The same analysis can be applied to diquarks made from one light and one strange quark giving $M[u,s]$ and $M\{u,s\}$. These mass differences are *fundamental* characteristics of QCD, which should be measured carefully on the lattice[29].

In practice we can estimate these masses by replacing the infinitely heavy quark by the physical charm or bottom, or even the strange quark. The analysis is complicated by the fact that the spin interactions between the light quarks and the s, c or b quark are not negligible. Of course the scalar diquark has no spin interaction with the spectator heavy quark (Q), but the vector diquark does. In order to obtain estimates of diquark mass differences, it is necessary to take linear combinations of baryon and meson masses that eliminate these spin interactions [1].

A simple analysis of strange, charm, and bottom hadron masses leads to quite a consistent picture of diquark mass differences. First, for non-strange quarks and diquarks,

$$\begin{aligned} M\{u,d\}|_s-M[u,d]|_s &= 205\text{ MeV} \\ M\{u,d\}|_c-M[u,d]|_c &= 212\text{ MeV}, \end{aligned} \qquad \begin{aligned} M[u,d]|_s-M(u)|_s &= 321\text{ MeV} \\ M[u,d]|_c-M(u)|_c &= 312\text{ MeV} \\ M[u,d]|_b-M(u)|_b &= 310\text{ MeV}, \end{aligned}$$

it appears that the properties of hypothetical non-strange diquarks are the pretty much the same when extracted from the charm and bottom, and even strange, baryon sectors. Second,

$$\begin{aligned} M\{u,s\}|_c-M[u,s]|_c &= 152\text{ MeV} \\ M[u,s]|_c-M(s)|_c &= 498\text{ MeV}, \end{aligned} \tag{2}$$

shows that the diquark correlation decreases when one of the light quarks is strange. This is certainly to be expected, since it originates in spin dependent forces. As the correlation decreases, the mass difference between the scalar and vector diquarks decreases (∼210→∼150 MeV) and the mass difference between the scalar diquark and the antiquark increases (∼310→∼500 MeV).

Diquarks and higher twist

Diquarks need not be pointlike. As we have seen, the energy difference between the good and bad diquarks is only ~ 200 MeV, enough to be quite important in spectroscopy, but corresponding only to a correlation length of 1 fermi, the same as every other mass scale in QCD. It is interesting, nevertheless to ask whether other hadronic phenomena can constrain the correlation. Although many nucleon properties, like form factors, are often discussed in terms of quark correlations, as far as I know, the correspondence can only be made exact for deep inelastic scattering (DIS).

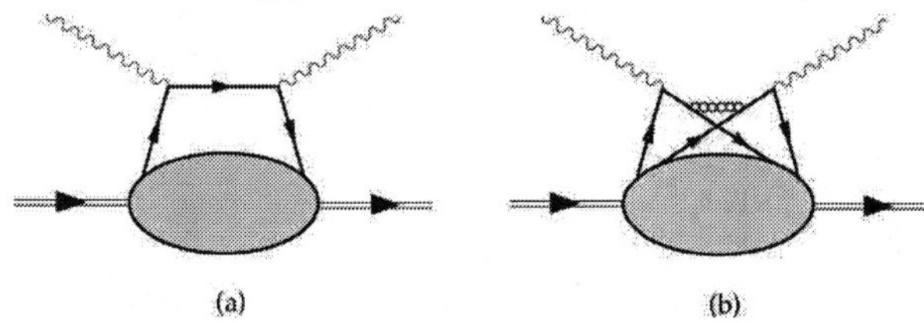

2: (a) Leading twist, single quark contribution to DIS (b) Twist-4, diquark contribution to DIS.

Any kind of quasi-pointlike, (*i.e.*, characterized by a mass scale $\Lambda_{\bar{q}q} \gg \Lambda_{\rm QCD}$) correlation in the nucleon is certainly excluded for $\Lambda_{\bar{q}q}$ ranging from ~ 1 GeV up to the highest scales where deep inelastic data exist ($\sim$100 GeV). Diquarks would be especially obvious because as bosons they would generate an anomalously large longitudinal/transverse inelastic cross section ratio in DIS at scales below $\Lambda_{\bar{q}q}$, which would disappear above $\Lambda_{\bar{q}q}$. Such an effect is certainly ruled out by the early, and apparently permanent, onset of scaling seen in a multitude of experiments.

On the other hand one might think that the absence of large higher twist effects in DIS could be used to place an uncomfortably *low* limit on the mass scale of diquark correlations. This is not the case[30]. In fact measurements of $1/Q^2$ corrections to DIS *place no limits whatsoever* on scalar diquark correlations in the nucleon. To understand this it is necessary to review some of the basics of the twist analysis of deep inelastic scattering. "Twist" refers to the dimension (d) minus the spin (n) of the operators that contribute to DIS, $t = d - n$. The smaller the twist, the more important the contribution to DIS: A given operator contributes like $1/Q^{t-2}$. The leading operators are twist-2 and act on a single quark**. They have the generic structure

$$O_\mu^{(2)} \sim \bar{q}\gamma_\mu \mathcal{D}\mathcal{D}\ldots q \tag{3}$$

The covariant derivatives, their Lorentz indices suppressed, denoted schematically by $\mathcal{D}$, have $d(\mathcal{D}) - n(\mathcal{D}) = 0$, so they are irrelevant for counting twist. The quark fields have $d(q) = 3/2$ and the γ-matrix contributes $n(\gamma) = 1$, so in all, $t = 2(3/2) - 1 = 2$, and these operators' contributions to DIS are independent of Q (modulo logarithmic corrections from perturbative QCD). The $\bar{q}\gamma q$ operators sum up to give the "handbag" diagram shown in Fig. 2(a).

It is easy to write down operators with twist greater than two[31]. The most important are twist-four (twist-three does not contribute to spin average DIS for light quarks),

** I am ignoring gluon operators, which do not figure in the argument.

which contribute corrections of order $1/Q^2$ to deep inelastic structure functions. The factor of $1/Q^2$ is accompanied by some squared mass-scale, M_4^2, in the numerator. Twist-four effects have been studied for years, and the qualitative conclusion is that M_4 is small. How small need not concern us, for we are about to see that it anyway places no limit on the good diquark that interests us.

Twist four operators invariably involve products of more than two quark and gluon fields (again ignoring pure-gluon operators). Examples include quark-gluon operators, $\bar{q}\mathcal{F}q$ and $\bar{q}\mathcal{F}\mathcal{F}q$, and four-quark operators, $\bar{q}q\bar{q}q$. The matrix elements of these operators in the target nucleon determine the magnitude of higher twist effects. The four quark operators are the culprits: they can be Fierz-transformed into diquark-antidiquark operators, $\bar{q}\bar{q}\ldots qq$ and therefore measure the scale of diquark correlations in the nucleon. They can be summed (in a well-defined way) to give diagrams like Fig. 2(b), where two quarks are removed from the nucleon, scattered at high momentum, and then returned. The generic structure of four quark operators is (there are others, but the results are the same),

$$O^{(4)}_{\mu\nu} \sim \bar{q}\gamma_\mu \mathcal{D}\mathcal{D}\ldots q\, \bar{q}\gamma_\nu \mathcal{D}\mathcal{D}\ldots q. \tag{4}$$

The γ-matrices are necessary. With $d(q) = 3/2$ and $d(\mathcal{D}) - n(\mathcal{D}) = 0$ it is easy to see that the twist of $O^{(4)}$ would be six if it were not for the two factors of γ, each of which corresponds to a unit of spin. In other words: when Fierzed, the two diquarks in $O^{(4)}$ must be coupled to spin-2. So *only the vector diquark contributes at twist-four*. Bounds on twist four in DIS tell us that the bad, vector diquark cannot be tightly bound, but they do not constrain the good, scalar diquark at all. It contributes only to twist-six and beyond, where it cannot be separated from the flood of non-perturbative effects that emerge at low Q^2.

We can proceed without concern that correlations of the extent necessary to influence the spectrum are ruled out by deep inelastic phenomena.

Diquarks and the Absence of Exotics

I want to look at exotics assuming little more than that two quarks prefer to form the good, scalar diquark when possible. States dominated by that configuration should be systematically lighter, more stable, and therefore more prominent, than states formed from other types of diquarks. This qualitative rule leads to qualitative predictions — all of which seem to be supported by the present state of experiment. This is clearly an idealization — a starting place for describing exotic spectroscopy. To learn the real extent of $\mathbb{Q}$ dominance will require more models and more information from experiment.

The predictions that follow from $\mathbb{Q}$-dominance are simple, and striking. They capture all the important features of exotic spectroscopy and provide the conceptual basis of a unified description of this sector of QCD.

a) There should be no (light, prominent) exotic mesons: The good diquark, $\mathbb{Q}$, is a flavor $\bar{\mathbf{3}}$, just like the antiquark. Tetraquarks, $\bar{q}\bar{q}qq$, potentially include exotics in **27**, **10**, and $\overline{\mathbf{10}}$ representations of flavor $SU(3)$. However $\overline{\mathbb{Q}} \otimes \mathbb{Q}$ contains *only non-*

exotic representations, **1** and **8**, just like $\bar{q}\otimes q$: $q^3\otimes\bar{q}^{\bar{3}} = (\bar{q}q)^{\mathbf{1}}\oplus(\bar{q}q)^{\mathbf{8}}$ compared with $\overline{\mathcal{Q}}^{\mathbf{3}}\otimes\mathcal{Q}^{\bar{\mathbf{3}}} = (\overline{\mathcal{Q}}\mathcal{Q})^{\mathbf{1}}\oplus(\overline{\mathcal{Q}}\mathcal{Q})^{\mathbf{8}}$. Other diquark-antidiquark mesons are heavier, where they would be buried in the meson-meson continuum. Probably they are not just "broad", but in fact absent[32].

b) The only prominent tetraquark mesons should be an $SU(3)$ nonet with $J^{\Pi} = 0^+$. This prediction — a simple corollary of the one just above — dates back to the late 1970's[25]. Since the good diquarks, $\mathcal{Q}$, are spinless bosons, the spin^{parity} of the lightest nonet is $J^{\Pi} = 0^+$. Over the years evidence has accumulated that the nine 0^+-mesons with masses below 1 GeV (the $f_0(600)$, $\kappa(800)$, $f_0(980)$, and $a_0(980)$) have important tetraquark components[28]. Space does not permit me to present the evidence here. The interested reader can find more in Ref. [1].

c) If there are any exotic pentaquark baryons, they lie in a positive parity $\overline{10}$ of $SU(3)_{\text{f}}$. This is also a simple consequence of combining good diquarks. To make pentaquarks it is necessary to combine two diquarks and an antiquark. The result is $\bar{\mathbf{3}}\otimes\bar{\mathbf{3}}\otimes\bar{\mathbf{3}} = \mathbf{1}\oplus\mathbf{8}\oplus\mathbf{8}\oplus\overline{\mathbf{10}}$. The only exotic is the $\overline{\mathbf{10}}$. Other exotic flavor multiplets, like the **27** and **35**, which occur in the uncorrelated quark picture and/or the chiral soliton models, should be heavier and most likely lost in the meson-baryon continuum.

d) Nuclei will be made of nucleons. To a good approximation, nuclei are made of nucleons — a fact which QCD should explain. If diquark correlations dominate, systems of $3A$ quarks should prefer to form individual nucleons, not a single hadron. The argument is based on statistics: Good diquarks are spinless color anti-triplet bosons. Only one, $[u,d]$, is non-strange. A six-quark system made of three of these, antisymmetrized in color to make a color singlet, would have to have fully antisymmetric space-wavefunction to satisfy Bose statistics. The simplest would be a triple-scalar product, $\vec{p}_1\cdot\vec{p}_2\times\vec{p}_3$, which should be much more energetic than two separate, color-singlet nucleons in an s-wave (*e.g.*, the deuteron). The argument generalizes to heavy nuclei. Of course it does not explain nuclear binding or the rich phenomena of nuclear physics.

CONCLUSIONS

There are two distinct, but related issues at the core of this discussion: first, a question: are there light, prominent exotic baryons, and if so, what is the best dynamical framework in which to study them? and second, a proposal: diquark correlations are important in QCD spectroscopy, especially in multiquark systems, where they account naturally for the principal features.

I believe the case for diquarks is already quite compelling. There are many projects ahead: re-evaluating the qqq spectrum [33]; systematically exploring the role of diquarks in deep inelastic distribution and fragmentation functions, and in scaling violation; seeing if diquarks can help in other areas of hadron phenomenology like form-factors, low p_T particle production, and polarization phenomena; developing a more sophisticated treatment of quark correlations, recognizing that diquarks are far from pointlike inside hadrons; establishing diquark parameters and looking for diquark structure in hadrons using lattice QCD; and — the holy grail of this subject — seeking a more fundamental and quantitative phenomenological paradigm for light quark dynamics at the confinement scale. Diquark advocates have considered many of these issues in the past [27]. No doubt many other important contributions, like the diquark analysis of the $\Delta I = 1/2$-rule [34], have already been accomplished. We can hope eventually to have as sophisticated an understanding of diquark correlations as we have of $\bar{q}q$ correlations, as expressed in chiral dynamics.

The situation with the Θ^+ is less clear. Evidence for the Θ^+ is not growing. Instead two (CLAS G10 & G11) second generation experiments have reported negative results. This is particularly disheartening because these experiments were designed after the initial reports of the Θ^+ and were optimized in light of them. High statistics, high energy experiments also report negative results, although they are too different from the low energy discovery experiments to be conclusive. Also, theorists' attempts to understand the Θ^+ have raised more questions than they have answered. To wit: (a) A negative parity (KN s-wave) Θ^+ is intolerable to theorists, but that is what most lattice studies find, if they find anything at all. (b) No one has come up with a simple, qualitative explanation for the exceptionally narrow width of the Θ^+. (c) The original prediction of a narrow, light Θ^+ in the chiral soliton model does not appear to be robust. (d) Quark models can accomodate the Θ^+, but only by reversing the naive, and heretofore universal, parity of the $q^{n_q}\bar{q}^{n_{\bar{q}}}$ ground state. It is necessary to excite the quarks in order to capture the correlation energy of the good diquarks. This does not sound like a way to make an exceptionally light and stable pentaquark. (e) When models are adjusted to accomodate the Θ^+, they predict the existence of other states that should have been observed by now: The diquark picture wants both a $\Theta^{\frac{1}{2}^+}$ and a $\Theta^{\frac{3}{2}^+}$; the CSM and large N_c want a relatively light **27**, which includes an $I = 1$ triplet: $\Theta^{*0}, \Theta^{*+}, \Theta^{*++}$.

Fortunately, ours is an experimental science, and the situation will eventually become clear — a virtue of working on QCD as opposed to string theory!

ACKNOWLEDGEMENTS

This work is supported in part by the U.S. Department of Energy (D.O.E.) under cooperative research agreement #DF-FC02-94ER40818.

REFERENCES

1. Many details have been omitted due to lack of space. A more complete discussion can be found in R. L. Jaffe, Phys. Rept. **409**, 1 (2005) [Nucl. Phys. Proc. Suppl. **142**, 343 (2005)] [arXiv:hep-ph/0409065].
2. T. Nakano *et al.* [LEPS Collaboration], Phys. Rev. Lett. **91**, 012002 (2003) [arXiv:hep-ex/0301020].
3. For references, see for example, Ref. [4].
4. A. R. Dzierba, C. A. Meyer and A. P. Szczepaniak, arXiv:hep-ex/0412077.
5. C. Alt *et al.* [NA49 Collaboration], Phys. Rev. Lett. **92**, 042003 (2004) [arXiv:hep-ex/0310014]; but see also H. G. Fischer and S. Wenig, arXiv:hep-ex/0401014 and K. T. Knopfle, M. Zavertyaev and T. Zivko [HERA-B Collaboration], J. Phys. G **30**, S1363 (2004) [arXiv:hep-ex/0403020].
6. A. Aktas *et al.* [H1 Collaboration], arXiv:hep-ex/0403017, but see also K. Lipka [H1 Collaboration], arXiv:hep-ex/0405051.
7. R. de Vita, Talk presented at the April Meeting of the APS, Tampa, Fla.

8. J. Barth *et al.* [SAPHIR Collaboration], "Observation of the positive-strangeness pentaquark Theta+ in photoproduction with the SAPHIR detector at ELSA," arXiv:hep-ex/0307083;
9. K. Hicks and S. Stepanyan, for the CLAS Collaboration, Paper-357 submitted to the Lepton-Photon-2005 (Uppsala).
10. Belle Collaboration, Paper-153 submitted to Lepton-Photon-2005 (Uppsala).
11. V. Burket, Plenary talk at Lepton-Photon-2005 (Uppsala).
12. D. Diakonov, V. Petrov and M. V. Polyakov, Z. Phys. A **359**, 305 (1997) [arXiv:hep-ph/9703373].
13. H. Weigel, Eur. Phys. J. A **2**, 391 (1998) [arXiv:hep-ph/9804260].
14. T. D. Cohen, Phys. Lett. B **581**, 175 (2004) [arXiv:hep-ph/0309111], Phys. Rev. D **70**, 014011 (2004), arXiv:math-ph/0407031. For further discussion, see A. Cherman, T. D. Cohen and A. Nellore, arXiv:hep-ph/0408209.
15. N. Itzhaki, I. R. Klebanov, P. Ouyang and L. Rastelli, Nucl. Phys. B **684**, 264 (2004) [arXiv:hep-ph/0309305]; I. R. Klebanov and P. Ouyang, arXiv:hep-ph/0408251.
16. R. L. Jaffe, Eur. Phys. J. C **35**, 221 (2004) [arXiv:hep-ph/0401187]; D. Diakonov, V. Petrov and M. Polyakov, arXiv:hep-ph/0404212; R. L. Jaffe, arXiv:hep-ph/0405268.
17. C. G. Callan and I. R. Klebanov, Nucl. Phys. B **262**, 365 (1985).
18. M. P. Mattis and M. Karliner, Phys. Rev. D **31**, 2833 (1985), Phys. Rev. Lett. **56**, 428 (1986), Phys. Rev. D **34**, 1991 (1986).
19. E. Jenkins and A. V. Manohar, Phys. Rev. Lett. **93**, 022001 (2004) [arXiv:hep-ph/0401190]; JHEP **0406**, 039 (2004) [arXiv:hep-ph/0402024].
20. R. L. Jaffe and F. Wilczek, Phys. Rev. Lett. **91**, 232003 (2003) [arXiv:hep-ph/0307341].
21. S. Nussinov, arXiv:hep-ph/0307357.
22. M. Karliner and H. J. Lipkin, arXiv:hep-ph/0307243.
23. B. Aubert *et al.* [BABAR Collaboration], Phys. Rev. Lett. **90**, 242001 (2003) [arXiv:hep-ex/0304021].
24. I. W. Stewart, M. E. Wessling and M. B. Wise, Phys. Lett. B **590**, 185 (2004) [arXiv:hep-ph/0402076].
25. R. L. Jaffe and K. Johnson, Phys. Lett. B **60**, 201 (1976). R. L. Jaffe, Phys. Rev. **D14** 267, 281 (1977).
26. M. Ida and R. Kobayashi, Prog. Theor. Phys. **36**, 846 (1966); D.B. Lichtenberg and L.J. Tassie, Phys. Rev. **155**, 1601(1967). Lichtenberg has recently discussed exotics using diquark concepts, D. B. Lichtenberg, [arXiv:hep-ph/0406198].
27. For a review an further references, M. Anselmino, E. Predazzi, S. Ekelin, S. Fredriksson and D. B. Lichtenberg, Rev. Mod. Phys. **65**, 1199 (1993), or M. Anselmino, E. Predazzi, eds. *International Workshop on Diquarks and Other Models of Compositeness: Diquarks III, Turin, Italy, 28-30 Oct 1996* (World Scientific, 1998).
28. C. Amsler and N. A. Tornqvist, Phys. Rept. **389**, 61 (2004); F. E. Close and N. A. Tornqvist, J. Phys. G **28**, R249 (2002) [arXiv:hep-ph/0204205]; For a recent reconsideration, see L. Maiani, F. Piccinini, A. D. Polosa and V. Riquer, arXiv:hep-ph/0407017.
29. For an extension of these ideas to more general color-non-singlet states in QCD, see R. L. Jaffe, "Color Non-Singlet Spectroscopy", to be published.
30. R. L. Jaffe and A. Vainshteyn, unpublished.
31. R. L. Jaffe and M. Soldate, Phys. Lett. B **105**, 467 (1981), Coefficient Phys. Rev. D **26**, 49 (1982).
32. R. L. Jaffe and F. E. Low, Phys. Rev. D **19**, 2105 (1979). For a pedagogical introduction, see R. L. Jaffe, "How to analyse low energy scattering", in H. Guth, K. Huang, and R. L. Jaffe, eds. *Asymptotic Realms of Physics, Essays in Honor of Francis E. Low*, (MIT Press, Cambridge, 1983).
33. F. Wilczek, arXiv:hep-ph/0409168; see also, A. Selem, S. B. Thesis (MIT, 2005) unpublished.
34. M. Neubert and B. Stech, Phys. Lett. B **231**, 477 (1989), Phys. Rev. D **44**, 775 (1991).

Summary of the HERA/LHC workshop

Albert De Roeck

CERN, 1211 Geneva 23

Abstract. This report summarizes some of the main results of the one year long workshop on HERA and the LHC.

Keywords: Parton distributions, hadronic final states, LHC, HERA, diffraction, heavy flavors

INTRODUCTION

In roughly two years time from this writing, i.e. in the second half of the year 2007, the Large Hadron Collider (LHC), presently under construction at CERN, Geneva, will come into operation. This collider will produce proton-proton interactions at a centre of mass system (CMS) energy of 14 TeV. Present experimental and theoretical indications are that this energy range, also called the TeV-scale or Terascale, will break new ground in the understanding of particle physics and even the Universe [1].

Specifically, the LHC will unveil the mystery of electro-weak symmetry breaking, either by discovering the Higgs bosons, or otherwise. Furthermore the chances are extremely high that new physics will be discovered, such as supersymmetry, extra dimensions or other. The LHC will also be a precision instrument, allowing for a measurement of the masses of the top quark and W boson to respectively 1 GeV and 15 MeV[2].

However the LHC will also allow for e.g. new measurements in the field of QCD, b and c physics, diffraction etc., in this new energy regime. Many of these measurements will need to be made and understood early on, in order to allow to estimate backgrounds correctly for searches of new phenomena. Precision measurements will also need a good understanding of QCD, both in the perturbative range (parton showering, jets,...) and the nonperturbative range (fragmentation, underlying events, minimum bias event cross sections,...). Since the protons are composite particles, consisting of gluons and quarks, the pp cross sections of hard scattering processes depend on the parton distributions in the proton. The LHC can make some measurements of these quantities, but will rely to a large extend on precision data collected at other colliders, in particular data from HERA.

The HERA ep collider has proven itself in the past years as a precision instrument for QCD measurements. A plethora of precision measurements on jet physics, diffraction, soft scattering and on particular the structure functions of the proton has been released by the two experiments H1 and ZEUS since HERA's start of operation in 1992. The experiments have collected about 100 pb^{-1} of data each during the first run which ended in 2000. Then a luminosity upgrade was initiated which should lead to an additional 500 pb^{-1} per experiment by the middle of 2007, when the HERA data taking program is scheduled to terminate. Given that the end of HERA may be near, and that the physics

CP792, *Deep Inelastic Scattering, DIS 2005*, edited by W. H. Smith and S. R. Dasu

requirements at the LHC are by now sufficiently understood, it seemed like an excellent opportunity to launch a workshop to bring these two communities in direct dialog with each other, to make sure one can extract the maximum information from the HERA data to help the future analyses at the LHC[3]. This workshop will be described here. Another workshop with a similar program, but for the Tevatron-LHC combination, called TeV4LHC[4], was launched as well in 2004.

THE HERA/LHC WORKSHOP

The seeds of the idea to organize this workshop came from two other workshops organized in the year 2003. At CERN a one month workshop took place during the early part of the summer on "Monte Carlo Tools for the LHC" [5], and in Binn (Switzerland) there was a small but very topical workshop on "precision measurements" [6] in the Fall. Both workshops brought –among others– the HERA community in direct contact with the LHC community. These workshops were received enthusiastically, but could only scratch the top of the iceberg and something on longer term was needed to work out the ideas that were generated. Hence the idea for a full fletched one-year long workshop was born.

The goals of the HERA-LHC workshop have been defined as follows

- To identify and prioritize those measurements to be made at HERA which have an impact on the physics reach of the LHC.
- To encourage and stimulate the transfer of knowledge between the HERA and LHC communities and establish an ongoing interaction.
- To encourage and stimulate theory and phenomenological efforts related to the above goals.
- To examine and improve theoretical and experimental tools related to the above goals
- To increase the quantitative understanding of the implications of HERA measurements on LHC physics.

Five working groups have been formed to accomplish these tasks: (WG1) Parton Densities; (WG2) Multi-jet Final States; (WG3) Heavy Quarks; (WG4) Diffraction; (WG5) MC-Tools. The first meeting took place in CERN in March 03 (250 participants), and the final meeting was held at DESY (150 participants).

WG1 PARTON DISTRIBUTIONS

Parton distribution functions are the prime measurements that are made at HERA. The charged weighted quark distributions are measured directly via the structure function F_2. The gluon distributions can be measured indirectly via QCD evolution fits of F_2 or semi-directly in e.g. jet and charm cross section measurements.

The F_2 structure functions at HERA are now measured with a precision of typically 2% or better in large kinematic regions, and are basically limited by systematics. The Run-II high statistics HERA data is expected in particular to improve the region of large

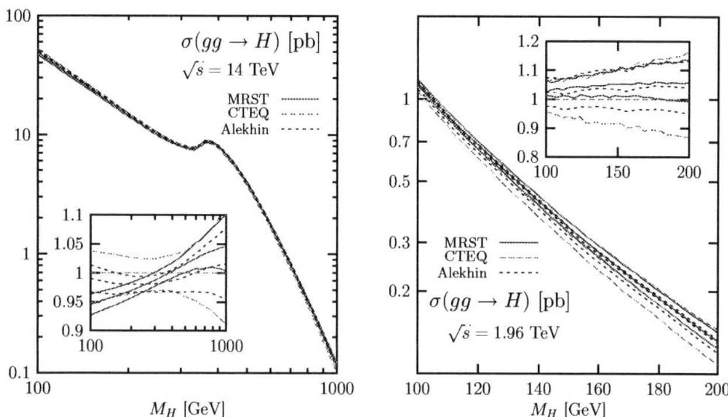

FIGURE 1. The CTEQ, MRST and Alekhin PDF uncertainty bands for the NLO cross sections for the production of the Higgs boson at the LHC (left) and Tevatron (right) for the process $gg \to$ Higgs. The insert shows the spread of the predictions when the NLO cross sections are normalized to the prediction of the reference CTEQ6M set [10].

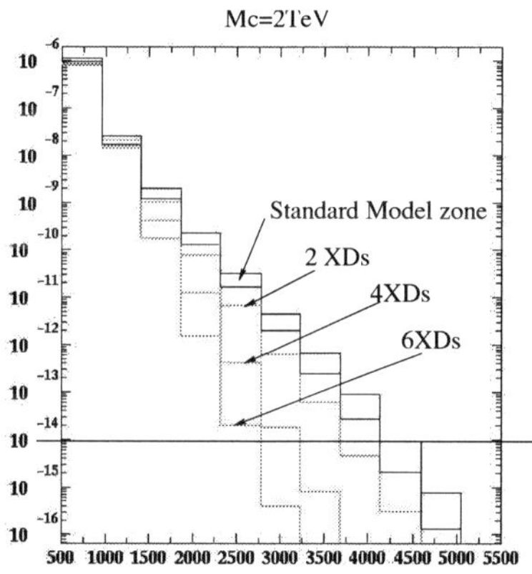

FIGURE 2. The dijet cross section in ADD extra dimensions with compactification scale $M_c = 2$ TeV. The Standard Model zone includes the uncertainties of the PDFs on the cross section prediction.

x and Q^2 which is still statistically limited. The results of HERA are used in PDF fits by either the collaborations themselves or in global fits by groups that try to include as much PDF sensitive data as possible. The current popular PDF sets are the MRST[7], CTEQ[8] and Alekhin[9] ones. The latter differ from the two former sets in that it includes only DIS data. The fits performed by the experimental collaborations themselves include less data but allow for a more easy determination of the error band as the errors and their correlations are often fully under control.

Taking naively the simple spread of the existing PDFs gives up to a 10% uncertainty in the SM Higgs cross section, as demonstrated in Fig. 1 [10]. The message for the workshop is clear: ultimately we have to do better than that. Another example is shown in Fig. 2: the sensitivity of the reach in discovery of ADD extra dimensions can be significantly reduced due to the parton density uncertainties [11]. The example is for virtual graviton exchange in a two jet final state. The two-jet cross section at the LHC gets reduced due to the interference of signal with SM QCD background. However the observation of the effect can be partially blurred by the PDF uncertainties: if one would know the SM di-jet cross section precisely, the sensitivity would be up to 5 TeV, but gets reduced to 2 (3,4) TeV for 2 (4,6) extra dimensions, due to the cross section uncertainty resulting from the PDF uncertainties.

The working group has defined the following program to be studied

- Study and document the potential experimental and theoretical accuracy for various LHC processes (Drell-Yan, W, Z, WW, γ+ jet production...) How can these be used for precision measurements at the LHC and e.g for luminosity determination? Cross sections and distributions will be studied and benchmarked with LHC detector simulation.
- Study of the impact of PDFs on LHC measurements. Here one will try to make the most of the HERA data. Is there a need for F_L and/or eD scattering? Can one judge which PDF is preferred? If so, what are the most precise PDFs and their errors?
- On the more theoretical side: what is the impact of small x and large x resummation and saturation corrections on PDFs? How well is the QCD evolution validated in the different kinematic regimes? How can we verify this at HERA and what is the impact on the LHC?

The systematic study of well measurable LHC final states is ongoing. As an example Table1 shows the summary of the uncertainties for W,Z and di-bosons production with experimental cuts, for the parton distributions and perturbative scale [12].

TABLE 1. Summary of uncertainties for measurments including experimental cuts, for the PDFs and scale of the perturbative calculation.

	W/Z	W/Z+jet	WW/ZZ
$\Delta_{PDF}(\%)$	±5.3	±4.3	±3.7
$\Delta_{Pert}(\%)$	±5.4	±9.1	±3.8

Many of the processes in this study can be used for the extraction of information on the PDFs, but it needs still to be quantified to what precision this can be done.

Fig. 3 shows the plane in x, Q^2 covered presently by HERA and the part that will be covered by the LHC [13]. Extrapolation or rather QCD evolution of the PDFs will be required over about 3 orders of magnitude. Clearly we need to understand as good as we can the evolution in the region where we have precise data at present, to check the uncertainty which is 'tolerated' by these data (e.g. the amount of non-linear effects). In the course of this workshop the NNLO splitting functions for the DGLAP evolution became available [14], so full NNLO fits can be made soon. Low-x resummation is important and was shown that it can lead to differences of about 20% at $x = 10^{-3}$ and

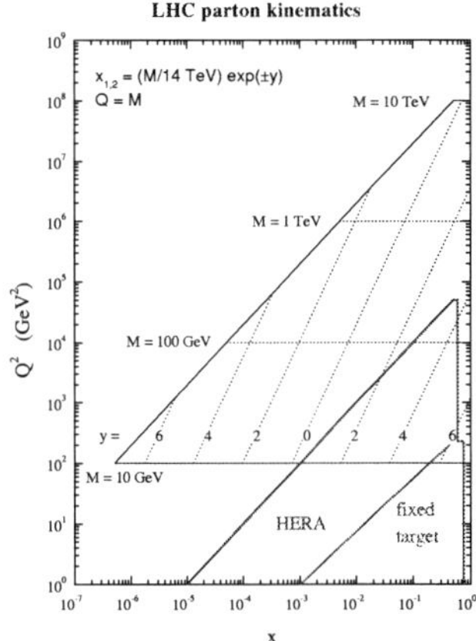

FIGURE 3. The kinematic plane (x, Q^2) and the reach of the LHC, together with that of the existing data (HERA, fixed target). Lines of constant pseudo-rapidity are shown to indicate the kinematics of the produced objects in the LHC centre of mass frame [13].

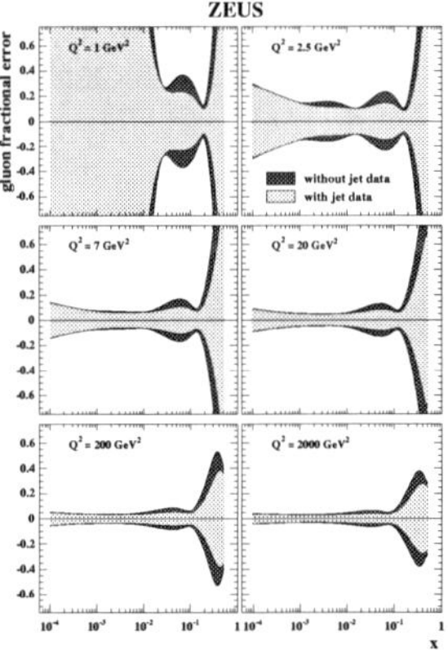

FIGURE 4. The total experimental uncertainty on the gluon PDF for a fit including the jets, compared to a fit not including jet data (outer error bands). The uncertainties are shown as fractional differences from the central values of the fits, for several values of Q^2 [17].

low Q^2 for the gluon distribution extracted by global fits [15]. On the high x side, $x > 0.7$, resummations can lead to 15% changes in the quark distributions [16].

The key issues nowadays for the global fits are the selection of data, a consistent treatement of errors and calculation of error bands. There are some tensions observed between data sets which need to be understood. While several prescriptions are being tried out for the error treatement, one radical way to approach this is to take data of one experiment only, but try to include as much as possible information. ZEUS presented and encouraging study on a combined PDF study using F_2 data and jet cross sections.

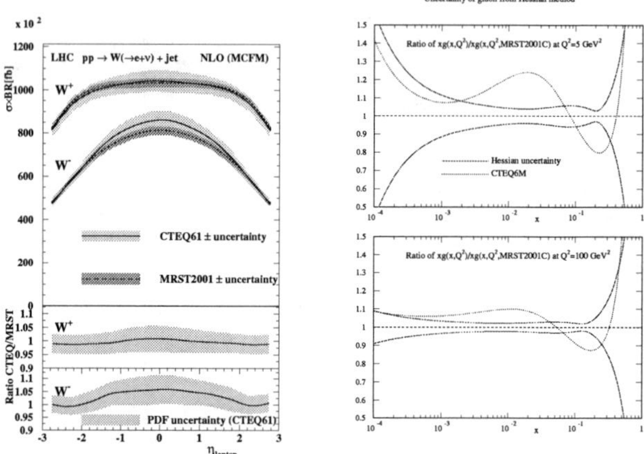

FIGURE 5. (Left) The PDF uncertainties for W^+ and W^- production [12]. (Right) The gluon distribution uncertainty from MRST, compared with the CTEQ central values [15]

Fig. 4 shows the potential gain in the uncertainty of the gluon distribution. Particularly at medium-x one can gain of order of 30% in precision in the gluon determination.

A new initiative that started during this workshop are the first steps towards a creation of combined data sets from HERA, i.e. really combining the experimental data points, rather than using the sets as two independent ones in the fit. The first results are very encouraging: they show that the extracted PDF fit from the combined data set can be much better than the fit to the sum of all the data points. What happens in practice is that one experiment 'calibrates' the other during the combining procedure. Similar improvements have been noted at LEP in combining measurements.

Turning back for a moment to the present PDF uncertainty: Fig. 5 shows the PDF error bands one gets using the present prescriptions of the PDF uncertainties, for W+jet production at the LHC. One notes that the error band of one PDF does not cover the central value of the other. One of the main reasons is the low-x behaviour of the parton distributions which is presently very different for the two sets of PDFs shown in Fig. 5. Both PDFs however are consistent with the HERA low-x data. Clearly nature may have chosen one or the other way, so how can one make progress here? What is needed are measurements that are more directly sensitive to the gluon in that region. The measurement of the longitudinal structure function F_L could do the trick, if it can reach the necessary precision. Better than the F_2^{charm}, F_L is as fundamental as F_2 with little theoretical ambiguity. To make a clean measurement of F_L HERA will have to operate some time at lower energies, and this is not yet on the program. Similarly for a good flavour separation and non-singlet structure function extraction, electron scattering on Deuterons would be needed. This option will however need some modifications in the HERA injection scheme. HERA is a unique machine and if these measurements do NOT happen at HERA, they won't happen for at least a very long time to come. Hence, the physics case should be made and discussed in all detail now, before it becomes irreversibly too late.

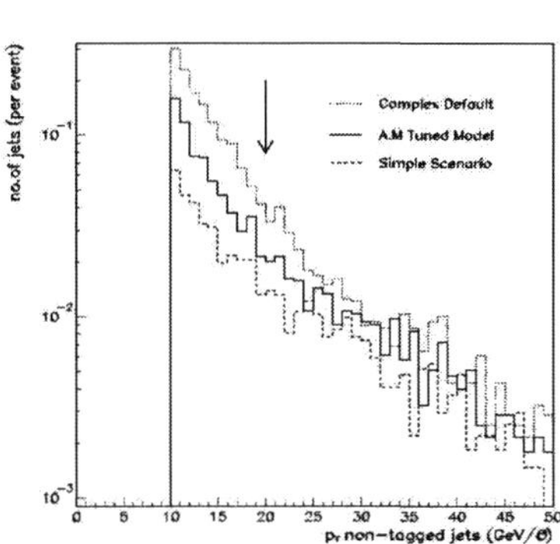

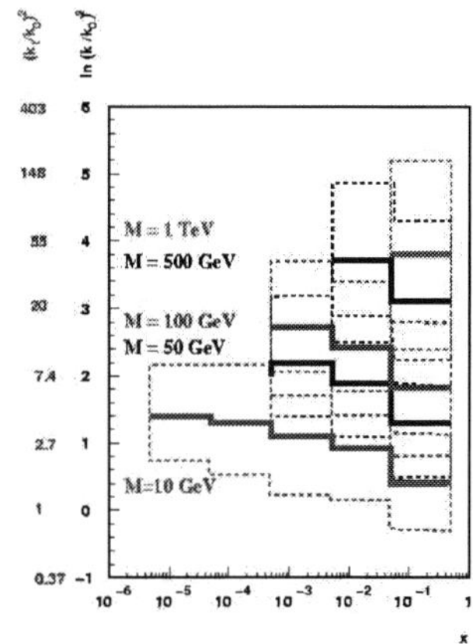

FIGURE 6. (left) Number of central jets per event in an analysis of $H \rightarrow WW^* \rightarrow 2l$ for different models/assumptions of the underlying event. The study was performed with ATLFAST. (Right) the k_T from QCD evolution for different values of the mass of a produced system M in $gg \rightarrow M$.

WG2: MULTI-JET FINAL STATES AND ENERGY FLOW

The following topics were studied by WG2

- The study of the structure of the underlying event, and of minimum bias events. New models were proposed and tested during the workshop. Tunes to existing data were discussed. A task force was installed to study similar observables in *ep* as done in *pp* for the tuning.
- The gap survival probability. The dynamics of gaps void of particles in *pp* and the consequences for the LHC are still poorly understood. New measurements were suggested to make further progress.
- A study of the phenomenology related to the CASCADE Monte Carlo, which shows differences with other QCD generators at the LHC at low-x
- Unintegrated PDFs and their importance e.g. on p_T distributions of the Higgs particle.
- Issues connected with Matrix Element and Parton Shower matching.
- Resummation of event shape variables.
- Future parton shower developments, such as unintegrated parton correlation functions and QEDxQCD exponentation.

Certainly one of the unknowns for studies at the LHC at present is the control of the underlying event and the event shape and number of minimum bias events which will be added to hard scattering event as pile-up: we expect about 4 interactions per bunch crossing on average at the first years luminosity of $2.10^{33}\text{cm}^{-2}\text{s}^{-1}$. Studies of tunes of PYTHIA and to some extend also HERWIG have been made using Tevatron and even lower energy data. These tunes should be validated next with the plethora of available HERA data. New models are now available: a new PYTHIA version,

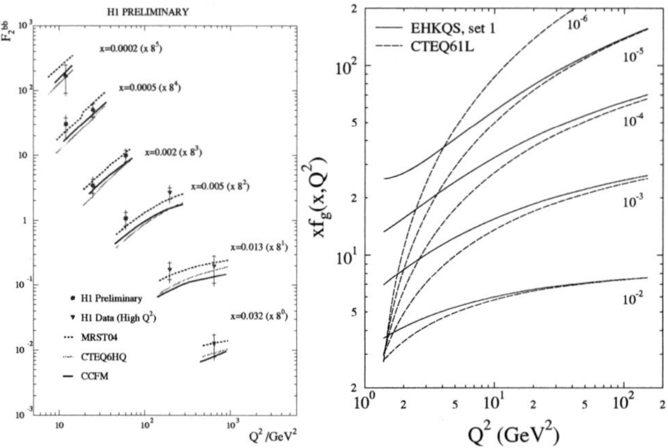

FIGURE 7. (Left) Data on $F_2^b(x,Q^2)$ from the H1 experiment, compared to QCD predictions. (Right) Comparison of EHKQS set 1 (solid line) and CTEQ11 (Dashed line) gluon distributions as a function of Q^2 for various x values.

Jimmy for HERWIG, and the SHERPA underlying event. All these need tuning and validating. The effect of the (importance of) underlying event was demonstrated with the vector boson fusion channel for Higgs production. In this process two forward jets are produced, plus the Higgs, choosen to decay in two W's, which in turn decay leptonically. Hence there is no color flow and hadron activity in the central region, except from the underlying event. To select these events over background a central jet veto is introduced, the efficiency of which will be affected and dependent on the underlying event model. Results in Fig. 6(left) show that there is a 10% variation in the selection efficiency, depending on the model choosen for the underlying event.

A challenge for final state studies will be to predict cross sections and topologies for many-jet events at the LHC, e.g. 8-jets or more. Certain SUSY cascades can lead to such number of jets, and a pure event counting technique will need a solid prediction of the QCD background. This needs good matching between matrix elements and parton showers. Such matching algorithms have been developed over the past year, in particular for *ee* and *pp* scattering, and are now being extended to *ep* such that these can be used to test on HERA multi-jet data.

A very important aspect is the intial k_T in the hard scattering, built up during the parton evolution before, say, the gluon enters in the hard scattering to produce a Higgs in the process $gg \to$ Higgs. The growth in k_T can be large as shown in Fig. 6(right) for a CASCADE calculation, for massive systems, thus affecting the p_T distribution of the produced particle. This means that for such production processes the unintegrated partons will be needed to correctly follow this evolution and provide the expected k_T in the scattering. HERA can test these k_T predictions and their effects with its data, and will allow to measure the unintegrated PDFs via final state measurements.

WG3: HEAVY FLAVOURS

Follows a list of measurements to be done at HERA, proposed by WG3

- The charm and bottom structure functions F_2^c and F_2^b
- Charm exclusive final states in γp and DIS: cross sections, fragmentation universality, contributions from higher charm resonances.
- Charm exclusive final states with jets
- Bottom exclusive final states
- Double quark tags
- Charm and bottom in charged current events
- Quarkonia
- Diffractive production of charm

To have significant impact and improve the already available data, at least 400 pb^{-1} will be needed at HERA-II. The topics listed are of general interst for the study of heavy flavour physics, but several have direct impact on the LHC. A clear case is the measurements of F_2^b, which is important for $bb \to$ Higgs production contribution. This needs a measurement of F_2^b at a scale of $m_H/2$. Fig. 7 shows recent results of a measurement of F_2^b from H1 based on HERA-I data [18]. The HERA-II data could reduce the errors by a factor of 4.

Heavy flavour measurements are also very sensitive to non-linear QCD evolution effects in the parton distributions. Fits to the HERA F_2 data at small x and small Q^2 improve by adding non-linear terms to the gluon evolution, see Fig. 7 [19]. This will lead to more charm production at low p_T [20]. The effects will become visible at the LHC for p_T values below about 2 GeV. ALICE will be best placed to measure these effects in the LHC data, since they can measure p_T values down to almost zero.

WG4: DIFFRACTION

This working group studied the following topics

- Diffractive Higgs production
- Backgrounds to diffractive Higgs
- Diffractive factorization breaking in di-jet, charm and leading neutron production
- Rapidity gap survival
- New measurements eg. F_L^D
- Exclusive diffractive di-jets
- Saturation effects and relation to multiple interactions and the gap survival

A large part of the activities was the transfer of experience and knowledge and design and operation of the detectors for forward physics from HERA to the LHC.

A topic of recent strong interest is the possibility to produce central diffractive Higgs particles in pp collisions, see Fig. 8. The advantages of this channel are [21]: a good missing mass resolution, of order 1-2 GeV via the protons for the Higgs, and low

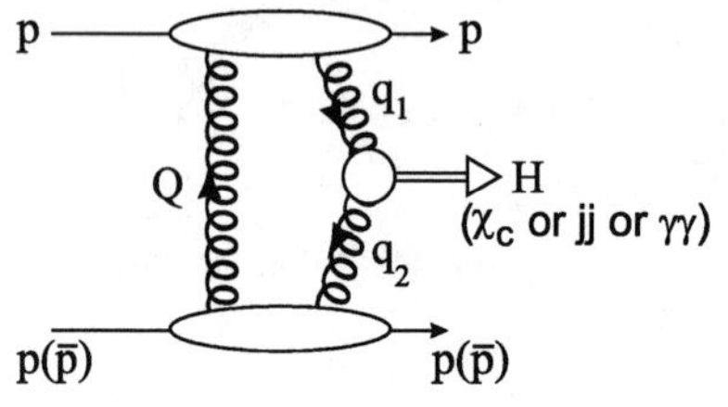

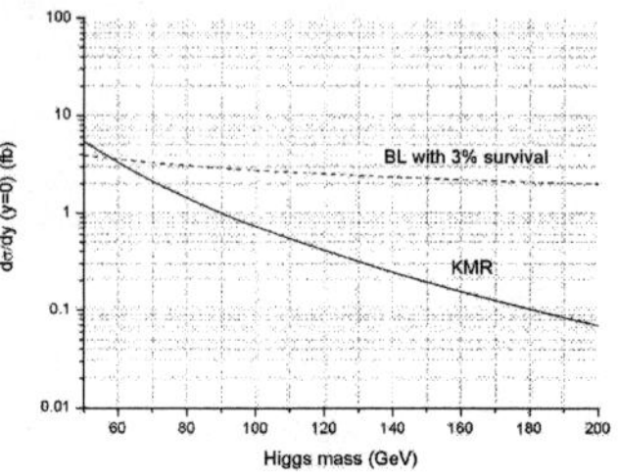

FIGURE 8. (Left) Diagram for exclusive Higgs production. (Right) Evolution of the cross section as function of mass for KMR[23] and the model as proposed in [24].

backgrounds. The cross sections are generally of the order of femtobarns and there has been quite some discussion on the validity of certain calculations. Also Monte Carlo models have been compared with one another in detail. The differences are basically understood as due to Sudakov suppression factors and parton distributions. In particular the Exhume[22] program is considered to give the more natural expected η behaviour. The KMR[23] calculation has been checked by independent groups and found to be ok. In all it means that the perturbative cross section for the Standard Model exclusive Higgs production is likely to stay below 10 fb. There are however alternative model predictions, based on non-perturbative calculations. Fig. 7(right) shows the different energy dependence in the KMR and the model proposed in [24]. It is not excluded that the total exclusive cross section could be larger than the one calculated in [23] if an additional soft component would be present.

It will be important in the coming year to test and measure the ingredients that go in that calculation. An example is the rescattering effects in collisions. It has been suggested to look into events with jets and a leading neutron at HERA [25] and study eg. $x - p_T$ correlations.

An input used in the exclusive Higgs cross section calculations are the generalized unintegrated parton distributions. HERA can measure these distributions via in exclusive J/ψ production. The double pomeron process itself can be measured at HERA in the reaction $\gamma p \to V + X + p$ with V a vector meson and X the centrally produced system. Finally the leading proton spectra as measured at HERA are found not to be described with standard Monte Carlo generators. This has an effect on the background studies to diffractive processes at the LHC, and some tunning based on the HERA leading baryon measurements will be essential.

Diffraction and low-x is part of the LHC physics program and there are plans to equip the central detectors with detectors in the forward region, which also offers new opportunities for groups to join in this activity.

WG5: TOOLS

WG5 had the following program

- Parton distribution library: LHAPDF is now the official carrier of the PDFs. It is used by the LHC experiments in generators. The HERA PDFs have been added recently. LHAPDF allows for uncertainty estimates. The Pion and Photon PDFs have been added to the library. Should the F_2^D parametrizations also be added?
- NLOLIB framework for NLO QCD programs. A uniform user interface is being developed, as well as an interface to HZTOOL. e^+e^-/ep have been included but pp still needs to be added.
- HZTOOL/JetWeb/RunMC/Cedar tools for Monte Carlo tuning. All HERA results have been included, some e^+e^- results. Include more pp data?
- Discussions on RAPGAP and CASCADE monte carlo programs for inclusive and diffractive pp
- Plenty of exchange on other MC tools, leading to new MC tools and comparisons with ep where possible.
- Continuation of the MC@LHC workshop, concerning validation of MC programs.

THE VERDICT AND OUTLOOK

Coming back to the goals that were set at the start of the workshop, one can say items (1) $\rightarrow$ (4) have been achieved. For item (5) many studies are still ongoing, and more quantitative examples/results are expected for the proceedings end of summer '05.

The final meeting is not the end of the workshop, however. The link between the communities is now strong and should not fade away. Therefore it was unanimously decided to continue the workshop but on a "one meeting per year basis". The next meeting will be in March 2006 at CERN. Everybody is invited to continue (or start) participating in the workshop.

ACKNOWLEDGMENTS

It is a pleasure to thank all participants of the workshop for their work, and especially Hannes Jung for the co-organization. My thanks goes also to the organizers of DIS05 for this kind invitation.

REFERENCES

1. J. G. Branson et al., hep-ph/0110021.
2. Fabiola Gianotti, Monica Pepe-Altarelli, Nucl. Phys. Proc. Suppl. 89 (2000) 189,hep-ex/0006016.
3. http://www.desy.de/~heralhc
4. http://conferences.fnal.gov/tev4lhc/
5. http://mlm.home.cern.ch/mlm/mcwshop03/mcwshop.html
6. http://wwweth.cern.ch/WorkShopBinn/home.html
7. A.D. Martin, R.G. Roberts, W.J. Stirling, R.S. Thorne Eur. Phys. J. **C23** (2002) 73 hep-ph/0110215

8. J. Pumplin et al., **JHEP 0207** (2002) 012, hep-ph/0201195
9. Sergey Alekhin, Phys. Rev. **D68** (2003) 014002, hep-ph/0211096
10. Abdelhak Djouadi, Samir Ferrag, Phys. Lett. **B586** (2004) 352, hep-ph/0310209.
11. S. Ferrag, hep-ph/0407303.
12. H. Stenzel, contribution to this workshop.
13. A.D. Martin, R.G. Roberts, W.J. Stirling, R.S. Thorne Eur. Phys. J. **C14** 2000, 133.
14. S. Moch, J.A.M. Vermaseren, A. Vogt Nucl. Phys. **B691** (2004) 129, hep-ph/0404111
15. A.D. Martin, R.G. Roberts, W.J. Stirling, R.S. Thorne, Eur. Phys. J. **C35** (2004) 325, hep-ph/0308087
16. G. Corcella and L. Magnea, hep-ph/0507042
17. ZEUS Collaboration, S. Chekanov et al., hep-ph/0503274.
18. H1 Collaboration , A. Aktas et al., Eur. Phys. J. **C41** (2005) 453, hep-ex/0502010.
19. K.J. Eskola, V.J. Kolhinen, R. Vogt, Phys. Lett. **B582** (2004) 157, hep-ph/031011.
20. A. Dainese, J.Phys. **G30** (2004) 1787-1799, hep-ph/040309.
21. A. de Roeck et al.,Eur. Phys. J. **C25** (2002) 391, 2002, hep-ph/0207042
22. J. Monk, A. Pilkington, hep-ph/0502077.
23. Valery A. Khoze, Alan D. Martin, M.G. Ryskin, Eur.Phys.J. **C14** (2000) 525 hep-ph/0002072.
24. M. Boonekamp, R. Peschanski and C. Royon, Nucl.Phys. **B669** (2003) 277, Erratum-ibid. **B676** (2004) 493 hep-ph/0301244.
25. A. Kaidalov and V. Khoze, private communication.

WORKING GROUP AND WORKSHOP SUMMARIES

Summary of the Structure Functions and Low-x working group

J.E. Cole*, Jianwei Qiu† and Un-ki Yang**

*H.H. Wills Physics Laboratory, University of Bristol, Bristol, BS8 1TL, UK
†Department of Physics and Astronomy Iowa State University, Iowa 50011, USA
**Enrico Fermi Institute, University of Chicago, Chicago, Illinois 60637, USA

Abstract. We report a summary of the structure function working group which covers a wide range of the recent results from HERA, Tevatron, RHIC, and JLab experiments, and many theoretical issues from low x to high x.

Keywords: structure functions, low-x
PACS: 12.38.-t,12.38.Mh,12.38.Qk,13.60.Hb

INTRODUCTION

Much of the predictive power of Quantum Chromodynamics (QCD) is provided by universality of the non-perturbative functions, in particular, the parton distribution functions (PDFs), in factorization theorems for hard processes. With the aid of factorization and perturbative calculation of short-distance dynamics, the universality allows us to extract a set of PDFs from some reactions and then use them to predict observables in other reactions. Knowledge of PDFs is critical for testing QCD dynamics in asymptotic region at existing facilities, as well as, for making predictions for future facilities, like the Large Hadron Collider (LHC). It is also essential for exploring non-perturbative QCD dynamics when the extracted PDFs are compared with what calculated in lattice QCD or in effective field theory approaches.

With only one identified hadron, structure functions of inclusive lepton-hadron deeply inelastic scattering (DIS) are clean observables for extracting PDFs. After more than 30 years of continuous effort, and many generations of machines and detectors, measurements of proton structure functions have become the benchmark tests of QCD dynamics. With the HERA at DESY, we are able to explore the kinematic region with the Bjorken x_B as low as 10^{-5} while staying the DIS regime. The continuous growth of structure functions as x_B decreases raises an urgent question: when such growth will hit the unitarity limit and slow down? The knowledge of structure functions and low x physics is extremely important for testing QCD dynamics and our ability to explore new physics beyond the standard model.

Our working group had a total of 46 talks divided into 9 sessions including one joint session with Electroweak and Beyond the Standard Model working group. In this writeup, we summarize the recent achievements, progresses, and open questions that were presented at our working group meetings. We organize this summary into seven

CP792, *Deep Inelastic Scattering, DIS 2005*, edited by W. H. Smith and S. R. Dasu

parts:

1. Structure function measurements at low x
2. Structure functions and PDFs at high x
3. Progress in the determination of PDFs
4. Toward QCD precision tests
5. Low-x physics: parton evolution and saturation
6. Nuclear structure functions and nuclear PDFs
7. New approaches to PDFs

STRUCTURE FUNCTION MEASUREMENTS AT LOW X

Measurements of the proton structure function, F_2, in neutral current (NC) deep inelastic scattering (DIS) at HERA are vital for testing the predictions of perturbative QCD and in the determination of the parton distribution functions of the proton. Recent results from the two general-purpose detectors, H1 and ZEUS, cover five orders of magnitude in the photon virtually, Q^2, and in the Bjorken scaling variable, x [1].

The possibility afforded by HERA of studying the structure functions down to values of x as low as $\sim 10^{-6}$ is important, as it gives access to partons which have undergone a large number of QCD branching processes. The density of these partons, both gluons and the so-called "sea" quarks has be found to increase dramatically as x decreases, which may indicate the need to taken into account non-linear effects in QCD evolution, such as saturation.

The double-differential cross section for inclusive NC DIS is given by:

$$\frac{xQ^4}{2\pi\alpha^2 Y_+}\frac{d^2\sigma}{dxdQ^2} = \sigma_r = F_2(x,Q^2) - \frac{y^2}{Y_+}F_L(x,Q^2) - \frac{Y_-}{Y_+}xF_3(x,Q^2) \qquad (1)$$

where $Y_\pm = 1 \pm (1-y)^2$, in which $y = Q^2/xs$ is the inelasticity, s is the total squared center-of-mass energy and F_2, F_L and F_3 are the structure functions of the proton. The quantity, σ_r is also defined in this equation, which is known as the reduced cross section.

Although the precision measurements of F_2 exist over such a large kinematic range, the same cannot be said for the longitudinal structure function, F_L, which has not been directly measured at HERA. F_L is directly sensitive to scaling violations and hence to the gluon content of the proton and is therefore crucial to our understanding of proton structure.

The final term in equation (1) contains xF_3, or the parity violating structure function. This structure function is, however, only important at high Q^2 and will not be considered further in this section, where the interest is primarily in low Q^2 structure function measurements.

The rise of F_2 with decreasing x persists down to very low values of Q^2 [2], although it is known that as $Q^2 \to 0$, $F_2 \to$ constant Q^2, as it must in order to satisfy the conservation of the electromagnetic current. It is also known that around $Q^2 \sim 1$ GeV^2, perturbative QCD begins to breakdown and phenomenological models must be invoked to explain the behavior of the F_2 data.

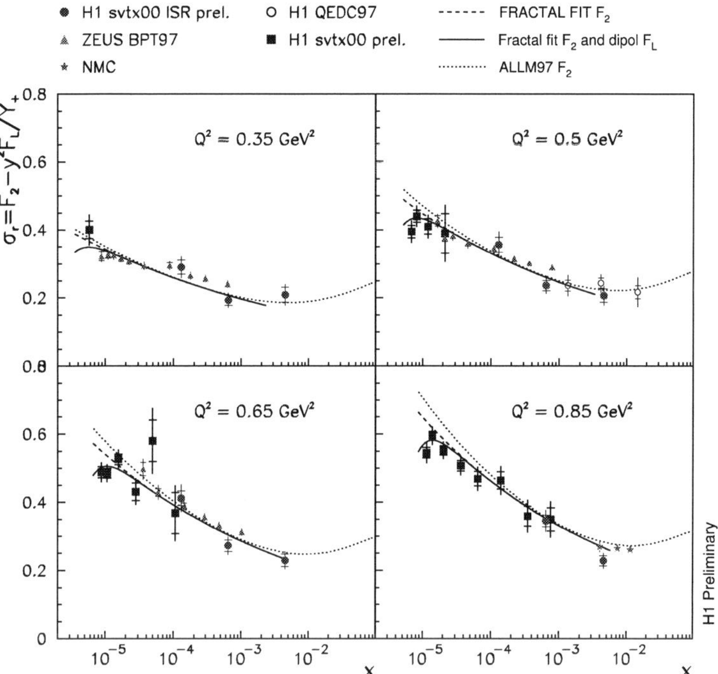

FIGURE 1. Compilation of reduced cross section measurements for $Q^2 < 1$ GeV^2 from the H1, ZEUS and NMC Collaborations.

From an experimental point of view, accessing very low values of Q^2 is technically challenging, but has been achieved by the HERA experiments using a number of different techniques. The H1 Collaboration presented recent measurements of the reduced cross section at low Q^2 using two of these techniques, namely, via the identification of QED Compton events [3] and using a small sample of data in which the interaction vertex was intentionally shifted by $+70$ cm toward the outgoing proton beam direction [4], effectively extending the acceptance of the H1 detector to values of Q^2 down as low as 0.35 GeV^2. Using this so-called "shifted vertex" data sample, they have also specifically identified events in which an energetic photon was emitted by the incoming lepton prior to its interaction with the proton; these initial-state radiative (ISR) events give access not only to even lower values of Q^2, but also to higher values of x, giving a wide coverage in x at low Q^2.

These measurements are shown in figure 1, in which it can be seen that F_2, and hence the reduced cross section, rises with decreasing x, even at low values of Q^2. The only exception to this behavior is at the very lowest values of x, at which the contribution to the reduced cross section from the longitudinal structure function, F_L, becomes significant, causing σ_r to decrease. As can be seen in equation (1), the contribution to the reduced cross section from F_L is suppressed for all but the highest values of y (low x). This be-

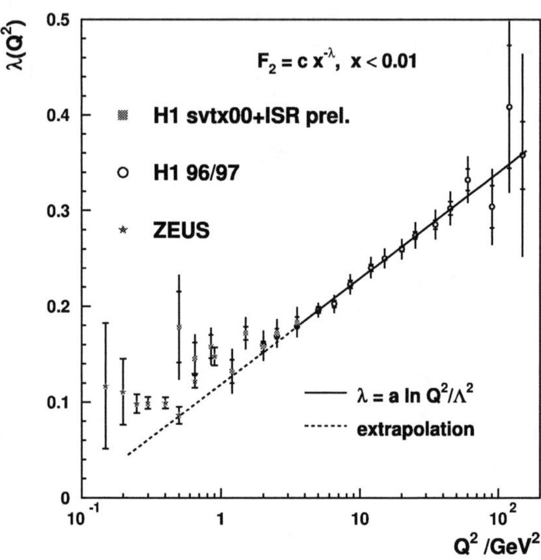

FIGURE 2. Compilation of selected HERA results on the parameter λ, obtained from fits of the form $F_2 = c(Q^2) \cdot x^{-\lambda(Q^2)}$ to low x data

havior can be exploited to perform an extraction of F_L, albeit in a model-dependent way. The resulting F_L points were also shown at this workshop [4] and are already able to discriminate between different PDF parameterizations.

The low Q^2 F_2 data can be fitted in order to quantify the change of the low x slope of F_2 with Q^2. Figure 2 shows the result of just such a fit, performed for $x < 0.01$ by the H1 Collaboration. The expected change in behavior around $Q^2 \sim 1$ GeV^2, is clearly observed.

STRUCTURE FUNCTIONS AT HIGH X

Structure Functions at high x region has brought many attentions at this workshop. Recently, it has been realized that it is very important to understand this region in order to achieve precise elecroweak measurements and to extract new physics signals from HERA, Tevatron and LHC at high Q^2 region. A large uncertainty on the PDFs at very high x and low Q^2 region can make a big impact on the high Q^2 region even at intermediate x due to the effect of DGLAP evolution.

Most precise data on high x come from the traditional fixed target experiments (SLAC/BCDMS/NMC). But their high x data corresponds to low Q^2 region where we face many challenges in understanding all non-perturbative QCD and nuclear effects. One clean way is to probe the structure functions at high x and Q^2 directly. Both H1 and ZEUS [5] showed measurements of the cross sections for neutral and charged-current scattering as a function of Q^2 using polarized beams. The measured cross sections are

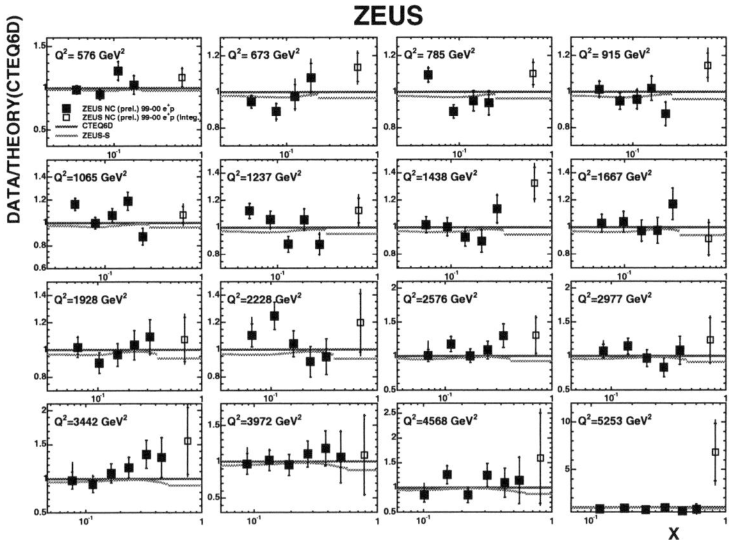

FIGURE 3. [Left] The ratio of the ZEUS differential cross sections data and predictions using CTEQ6D PDFs.

well described by the Standard Model. But more data is required to constrain parton distributions functions at high x. The ZEUS [6] presented a very promising method to probe the PDFs up to $x = 1$ using the jet information (E_{jet} and θ_{jet}) to calculate the value of x. Events with no jets reconstructed within their fiducial volume is assumed to come from very high x_{edge} to 1. The measured cross sections using early dataset show good agreements with the predictions using CTEQ6D PDFs, shown in Figure 3. However, their highest x data tend to be higher than the predictions. Thus, it would be interesting to see their results using a full dataset.

The NuTeV [7] presented their final differential cross sections using neutrino-iron scattering. The extracted F_2 and xF_3 from their differential cross sections are 20% higher than the CCFR measurements, and 10-15% higher than the BCDMS F_2, as shown in Figure 4. They explained that two third of the difference between the NuTeV and CCFR measurements is due to an improved calibration of the magnetic field, and a better modelling in Monte Carlo. This result implies that that nuclear effect in neutrino scattering is different from that in the charged lepton scattering at high x. Thus, we need to resolve this difference before the NuTeV data can be used in a global PDFs analysis to constrain the PDFs at high x. It would be interesting to see their QCD fit results. A possible difference in the nuclear effect can be resolved by the CHORUS data on the lead target, and future Minerva/MINOS results.

The ratio of d and u quarks at high x primary comes from the measurements of F_2(deuterium) and F_2(proton). Because of a large uncertainty of nuclear binding effect on deuterium target, this ratio is poorly known. Figure 5[left] shows the NMC F_2^d/F_2^p with and without nuclear correction [8]. With a nuclear correction, the NMC data favors

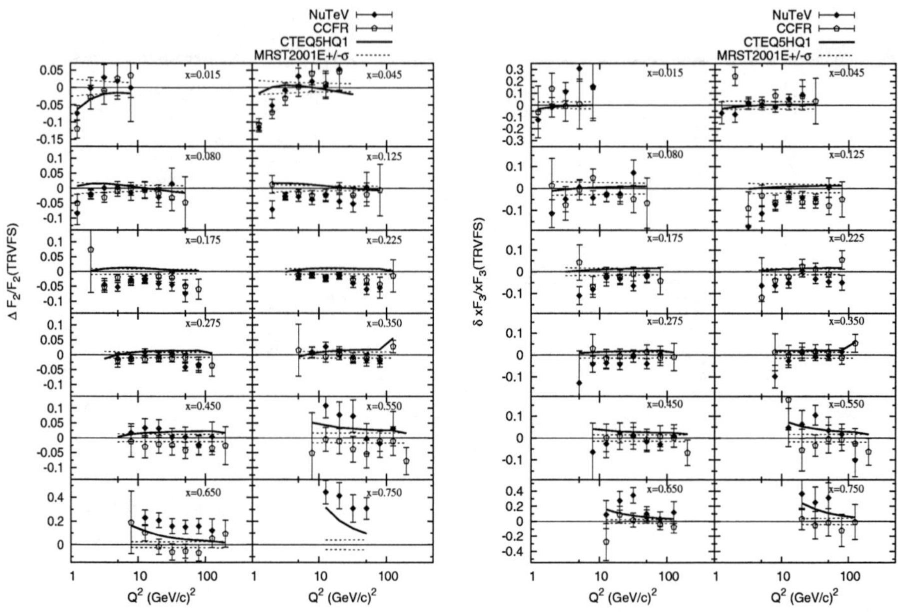

FIGURE 4. NuTeV F_2[left] and xF_3[right] data are compared with CCFR data and the NLO prediction with MRST2001E.

0.2 for d/u as $x \to 1$, which is of theoretical interest for nuclear physics community. However, the size of nuclear binding correction is still controversial. S.Kuhn [9] presented dedicated JLab efforts to study d/u at high x. Their programs are to study the effect of nuclear binding on neutron structure, and to measure the structure functions of a free neutron by detecting a slow spectator proton. Information on d/u can be also extracted from W production data at the Tevatron. The CDF collaboration [10] measured the forward-backward charge asymmetry of electrons from W boson decays. In order to get a better d/u sensitivity on higher x region, they have looked at a higher electron E_T region. Fig 5[right] shows comparisons with the NLO RESBOS predictions using CTEQ6M and MRST 2001 PDFs. At high η, the CDF data tends to favor higher d/u value at high x. Thus, it would be interesting to compare with the PDFs which was extracted, assuming a large nuclear correction on the deuterium target. They expect to have a big improvement on this measurement by reconstructing W rapidity directly.

At the workshop, one of the hot subjects was the phenomenon of a parton-hadron duality which states that the average behavior of the nucleon resonances follows the DIS scaling limit curve. Jlab has very precise data at high x and low Q^2 region (where a resonance production occurs). Besides many theoretical issues discussed by S. Liuti [11], C. Keppel and I. Niculescu [12] demonstrated that the duality holds for F_2 proton, EMC

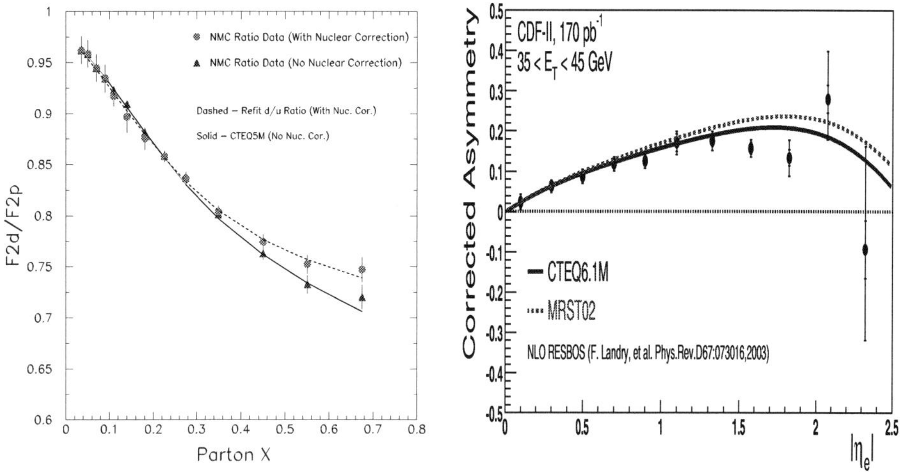

FIGURE 5. [Left] The ratio of F_2 data on the deuterium and hydrogen targets with and without the nuclear corrections. [Right] The CDF lepton charge asymmetry is compared with predictions with CTEQ6M and MRST02 PDFs using a NLO RESBOS calculation.

effect, and even spin structure functions in the region of $Q^2 > 0.5$ GeV^2. Issues are whether non-perturbative power corrections between DIS and resonance region is same, and DGLAP evolution & factorization works in the resonance region too. S. Liuti's studies suggest that the size of the higher twist effect may not be same. However, A. Bodek [13] showed that all DIS F_2 data and JLab's resonance data are well described by his Bodek-Yang leading-order model, as shown in Figure 6. This implies that there is not much difference in the power corrections between two regions. This model uses a new scaling variable ξ_w and Q^2 dependent K factors to all PDFs to describe both pQCD and non-perturbative QCD regions very smoothly.

A community of neutrino oscillation physics have started to pay attention on this non-perturbative QCD region. For precise measurements of mass splitting and mixing angles, neutrino oscillation experiments (MINOS, NOνA, and T2K) need to have a good understanding of neutrino cross section at low energy. This point was well presented by H. Gallagher [14]. Certainly this would be a place where DIS, nuclear physics, and neutrino physics communities need to make a coherent effort.

At the end of the workshop, PDF uncertainties at the Tevatron and impact on various measurements are presented by F. Chlebana, and A. Harel [15]. Chelbana discussed many ideas to constrain the PDFs using the Tevatron data. Figure 7 shows the latest status of the Tevatron jet data with the NLO predictions where there is no observed discrepancy at high E_T region. These measurements are currently dominated by the jet energy scale uncertainties. In future, it would be important to separate PDF effects from any new physics signal. F. Chlebana He also pointed out that it is crucial to measure the size of heavy flavor quarks densities for Higgs and Top physics.

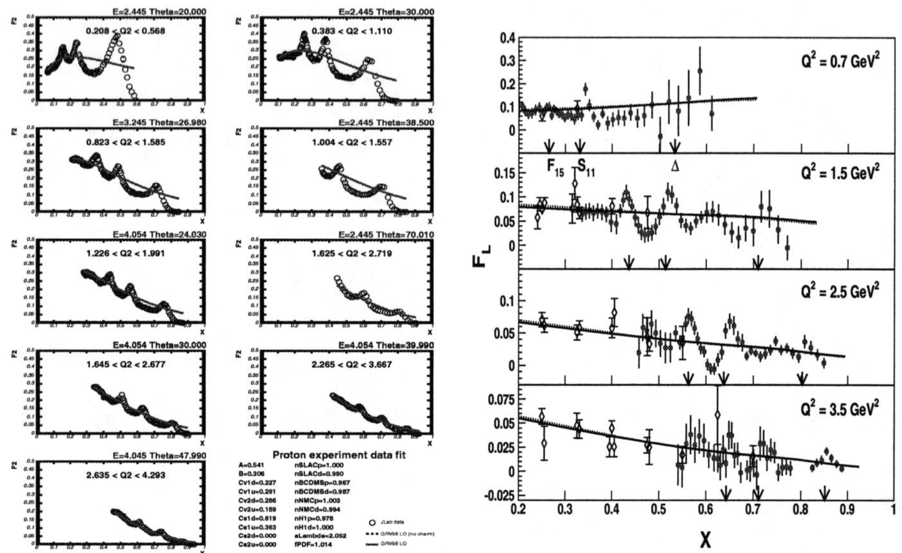

FIGURE 6. Comparisons of the predictions of Bodek-Yang model to resonance electro-production data on proton: JLab F_2 proton [left], and F_L proton[right].

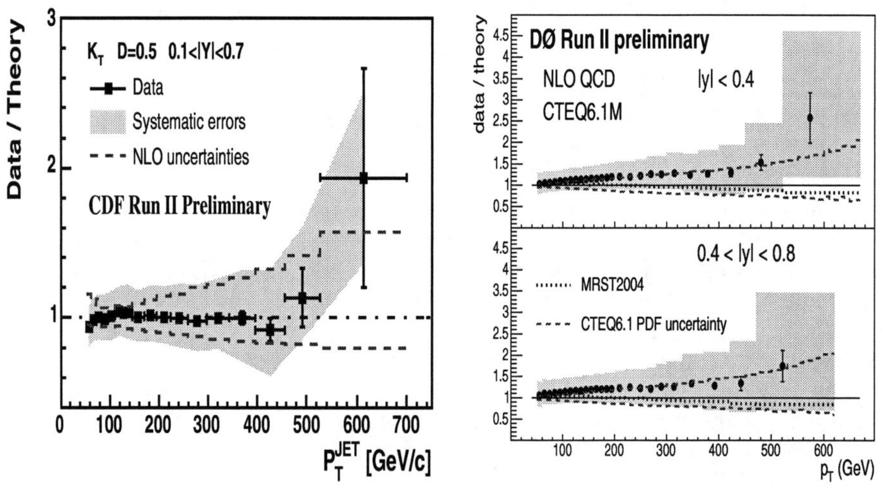

FIGURE 7. The ratios of the inclusive jet cross section data and the NLO predictions: CDF data using K_T algorithm [left] and D0 data using cone algorithm[right].

PROGRESS IN THE DETERMINATION OF PARTON DISTRIBUTION FUNCTIONS

The structure functions discussed in the previous section can be expressed in terms of the parton distribution functions (PDFs) of the proton. The structure function, $F_2(x,Q^2)$ is proportional to the sum of the quark and antiquark PDFs. At low x, F_2 is therefore sensitive to the sea quark distributions and hence is indirectly sensitive to the gluon density. The longitudinal structure function, F_L, is directly sensitive to the gluon density. The parity-violating structure function, $xF_3(x,Q^2)$, is proportional to the difference between the quark and antiquark PDFs, making it sensitive to the valence quark distributions.

The PDFs of the proton may be extracted by fitting, among other things, the HERA structure function data. These fits have been performed by a number of different groups, as well as the experimental collaborations themselves. It is crucial that the best possible understanding of the proton PDFs is achieved, given their central role in predictions for other processes, for example, at the LHC.

The traditional method of determining the PDFs of the proton relies on assuming an x-dependence for each of the different PDFs at some starting scale Q_0^2 and then using the Dokshitzer-Gribov-Lipatov-Altarelli-Parisi (DGLAP) evolution equations [16] to model the Q^2 dependence of the PDFs. This approach is used by both the CTEQ Collaboration [17] and Martin et al. (MRST) [18]; both groups presented progress reports at this workshop. This approach has also been adopted by the ZEUS Collaboration [19], who also presented results of their latest fit at this workshop.

A number of issues have been addressed by both the CTEQ and MRST groups recently. In particular, the compatibility of datasets and the stability of the fit results have been studied.

Both groups have performed studies of fit stability [20, 21], by studying the impact of restricting the fits to data at higher x. The studies were performed by looking at the next-to-leading order (NLO) $W^{\pm}$ production cross section predictions from fits with different lower x limits. The results of both studies are shown in figure 8. The CTEQ group conclude from their studies that the fits are stable. The MRST group conclude that the uncertainties increase significantly as the lower x limit is tightened and that next-to-next-to-leading-order (NNLO) is inherently more stable and these fits should become the standard in the future.

The MRST group also presented the results of other studies, in particular the inclusion of electroweak corrections and QED effects [22]. The latter, in particular, have a negligible effect on the PDFs themselves, as expected. However, the inclusion of QED effects does lead to a small isospin violation, which significantly improves the predictions for prompt photon production at HERA.

The ZEUS Collaboration also presented their latest determination of the proton PDFs using only ZEUS data [23]. In comparison to their previous fits, ZEUS jet cross section data has been included, which is directly sensitive to the gluon density of the proton and benefit from small experimental and theoretical uncertainties. Both NC DIS jet data and direct-enriched dijet photoproduction data, in which the photon behaves as a point-like object, have been included, significantly improving the uncertainty on the gluon PDF in the range $0.01 < x < 0.4$. If a simultaneous fit of both the PDFs and α_s is performed, the

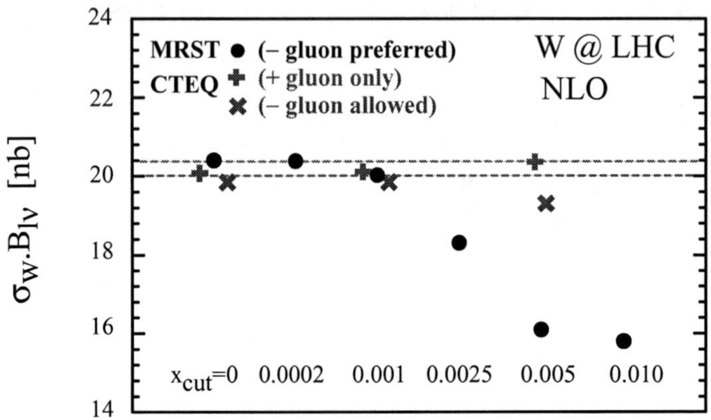

FIGURE 8. Predictions for the $W^{\pm}$ production cross section at the LHC from the CTEQ and MRST Collaborations. The cross sections are plotted as a function of the lower x limit applied to the data used to extract the PDFs. The CTEQ Collaboration have considered two different scenarios: one in which the gluon is forced to be positive-definite and the other in which the gluon is left free. Both are indicated by crosses. The MRST predictions are indicated by the dots and here the gluon is left free.

result for α_s is very precise and in agreement with the world average.

Several presentations were also made at this workshop, in which alternative approaches to PDF determination were explained. One such presentation was made by the NNPDF Collaboration [24], who are developing a neural network approach to PDF fitting. This approach avoids some of the shortcomings of the standard method, such as avoiding any potential bias from the choice of functional form for the PDFs. It should also lead to a better estimation of the PDF uncertainties. So far the structure functions have been successfully determined using this approach, but work is still in progress to successfully determine the PDFs using this method.

The estimation and reduction of PDF uncertainties was a key theme at this workshop. One presentation made at this workshop looked at the possibility of averaging the F_2 data from the H1 and ZEUS experiments, prior to including it in any global PDF fit [25]. This has advantages when it comes to the handling of the systematic uncertainties; it also provides a model-independent method of checking the consistency of the data from the two experiments. It has been found that several contributions to the systematic uncertainties from each experiment are reduced; the experiments are effectively constraining each other. This is an interesting approach which, it is hoped, will be pursued further by the two Collaborations.

Another presentation made at this workshop looked at the impact of future HERA data on the PDF uncertainties [26]. This study was performed using the ZEUS PDF fit as a basis. A number of different scenarios were considered, including simply the expected increase in the amount of luminosity, as well as the inclusion of new cross section measurements which have been optimized to constrain the PDFs as tightly as possible. These improvements would lead to significant improvements in the valence quark distributions, as well as in the high x sea quark and gluon distributions. Other scenarios which were also considered are the inclusion of precision measurements of F_L from HERA data (low-energy proton running) and the possibility of $e-D$ running to constrain the sea quark asymmetries.

TOWARD QCD PRECISION TESTS

New HERA data on unpolarized DIS structure functions, combined with the present world data, allow to reduce the experimental error on the strong coupling constant, $\alpha_s(M_Z^2)$, the fundamental constant of QCD and strong interaction, to the level of 1%. On the theoretial side, the next-to-leading order (NLO) analyses have limitations due to scale variations being present which allow no better than 5% accuracy in the determination of α_s [27]. In order to match the experimental accuracy, it was stressed [27] that analyses of DIS structure functions need to be carried out at the NNLO level. With the recent computation of the 3–loop anomalous dimensions [28], a complete NNLO study of DIS structure functions is now possible. A full NNLO analysis of unpolarized DIS structure functions aiming to obtain a high accuracy determination of α_s was presented at the workshop [27]. It was pointed out that a combination of standard NNLO QCD analysis and fits based on factorization scheme-invariant evolution of DIS structure functions will provide a valuable tool in high-precision analyses aiming at 1% accuracy in the determination of α_s [27]. The factorization scheme-invariant evolution of DIS structure functions can be implemented to different pair of structure functions, such as F_2 and F_L or F_2 and its $t \propto \ln(\alpha_s(Q))$ derivative. In this approach, α_s is determined by performing an one dimensional fit between the evolution of DIS structure functions and the data. Work is still ongoing. A full NNLO accuracy evolution for F_2 and $\partial F_2/\partial t$ have been completely implemented for massless flavors. Inclusion of heavy flavors and the fit to the data, and the one parameter fit to determine Λ_{QCD} are on the way. One interesting result is that comparing the behavior of slopes of $\partial F_2/\partial t$ to the slopes extracted experimentally points toward a positive gluon density in small-x region [27], while NLO global analysis of PDFs points to a negative gluon density in low-x and low Q^2 region [18].

An effort to develop a next generation of event generators was reported at the workshop [29]. The effort was aiming to set up a systematic scheme for developing event generators that are consistent to QCD factorization of differential cross sections up to NLO accuracy. It was argued that in order to achieve this accuracy, one has to use unintegrated PDFs to replace the parton shower in the LO event generators. The basic rules have been established for a DIS event generator, but, there are still works to be done [29].

At hadron colliders, it is the PDFs that determine partonic flux of hard collisions. Full discovery potentials of the LHC and precision tests of QCD are sensitive to PDFs at large x. In the form of QCD factorization, extraction of PDFs depends on short-distance dynamics and perturbatively calculated coefficient functions. The coefficient functions often have high powers of logarithms like $\ln(1-x)$ and $\ln(1/x)$, which become large as x near 1 and 0. Resummation of these large logarithms is necessary for observables dominated by those kinematic regions. A presentation made at this workshop looked at the effect of large-x resummation on the extraction of PDFs [30]. Large-x resummation was performed for coefficient functions of DIS structure functions in massless approximation as well as in an approach that includes heavy quark-mass effects. After performing fits to the fixed target DIS data from NuTeV, BCDMS and NMC collaborations, using NLO and NLL-resummed coefficient functions, it was found that the resummation has a visible impact on the extraction of quark distributions at large x, and was stressed that large-x partonic resummation is needed whenever a high precision is required for cross sections evaluated near partonic threshold [30].

A precise knowledge of PDFs, in particular, gluon distribution at $x \sim 0.001 - 0.01$ are vital for understanding almost all standard production processes at the LHC. When we move away from zero rapidity, much smaller x partons, as small as 10^{-5}, are required for some observables. Although perturbative QCD has been very successful in interpreting data on scaling violation of PDFs in terms of DGLAP evolution in Q^2 [17, 18], the extrapolation of PDFs to smaller x has not been very consistent with the BFKL evolution in x (or in energy) [31]. Although the strong rise of proton structure function F_2 with energy, observed at HERA, can be well described by a simple (3 parameters) LO BFKL fit [32], a much too small effective $\alpha_s < 0.1$ is needed while the world average is $\alpha_s \sim 0.2$ for HERA kinematics. A phenomenological study of confronting NLO BFKL with new HERA data on F_2 structure function was presented at the workshop [33]. A big discrepancy between theory and data, especially at low Q^2, was clearly evident [33]. A further study is needed although more progresses have been made recently [34].

LOW-*X* PHYSICS: PARTON EVOLUTION AND SATURATION

One of the challenging problems in QCD is to understand the behavior of hadronic cross sections in high energy limit. Experimental data on the total cross section show a slow but distinct rise with collision energy $\sqrt{s}$. This rise could be parametrized by a power of s, $\sigma(s) \sim s^{0.08}$, which is consistent with an exchange of soft Pomerons [35]. On the other hand, after resumming leading powers of $\ln(s)$ contributions, perturbative QCD calculation, in the form of BFKL evolution in energy (or in x), predicts a much stronger rise with a much large power of s [31]. As $\sqrt{s} \to \infty$ (or $x \to 0$), the power-like rise is not compatible with the unitarity of the S-matrix in the high energy limit, or in contradiction with the Froissart bound [36], which allows at most a logarithmic increase with collision energy.

BFKL equation is a linear evolution equation and predicts a large number of low-x partons due to the strength of soft gluon radiation in QCD. On the other hand, the large number of soft partons generated by parton radiation are likely to interact and recombine. Parton recombination introduces non-linear terms into the BFKL equation, slows down the small-x evolution, and removes the apparent violation of the unitarity. When parton recombination is strong enough to balance parton radiation, PDFs saturate as $x \to 0$ [37]. The state of saturated partons is sometime referred as the Color Glass Condensate (CGC) [38, 39, 40]. A lot of work, both theoretical and experimental, have been done and many progresses have been made in recent years to understand this saturation phenomenon, especially, in a nuclear environment because of an $A^{1/3}$ length enhancement in parton density at a given impact parameter. Two sessions at this workshop were devoted to the presentations related to this novel phenomenon.

A simple modification to the BFKL equation is Balitsky-Kovchekov (BK) equation [41], which adds a quadratic term to the BFKL equation. The BK equation is a non-linear integro-differential equation for unintegrated PDFs. Its non-linearity leads to many interesting features that could be seen in high energy reactions. It was shown [42] that the BK equation is in the equivalence class of the Fisher–Kolmogorov–Petrovsky–Piscounov (FKPP) non-linear partial differential equation, which has so-called traveling wave solutions. The similarity leads to an interesting point of view that high energy QCD is equivalent to a reaction–diffusion system [43]. A detailed numerical studies of the mean field approximation to the BK equation was presented at the workshop [43]. It was demonstrated that the numerical solutions of the BK equation does show features of traveling wave solutions. It was also confirmed that the influence of the initial condition disappears for large $Y \sim \ln(1/x)$, so that a universal propagation speed is approached, which should help establish statistical interpretations of the phenomena observed in QCD scattering at high energy.

A study of discrete version of the BK equation was presented at this workshop [44]. By noting that the number of gluons in the hadron wave functions is discrete, and their formation in the chain of small x evolution occurs in the discrete intervals of $\ln(1/x)$, a discrete version of BK equation was formulated [44]. It was found that numerical solutions of the discrete BK equation behave chaotically in the phenomenologically interesting kinematic region. It was concluded [44] that the evolution of the scattering amplitude at high energies in the saturation region might be chaotic, while the scattering amplitude in the normal perturbative region is not affected by the discretization. Although the model used in the numerical calculations neglected the diffusion in transverse momentum, stochasticity of gluon emission and the dynamical fluctuations beyond the mean field approximation, it was hoped that at least some of the features of discrete quantum evolution at small x will survive a more realistic treatment. The chaotic features of small x evolution open a new intriguing prospective on the studies of hadron and nuclear interactions at high energies.

Recent developments of CGC theory were reported at the workshop [45, 46]. The CGC theory is an effective theory of the strong interactions at very high energies [40]. The basic equation of CGC theory is the JIMWLK equation [39], which governs the evolution of a weight function of a color medium (or a hadron target) with rapidity. The evolution kernel is often referred as the JIMWLK Hamiltonian. The weight function is needed for calculating physical scattering amplitude when it is averaged over the medium's color charge. In the large N_c limit and in the dipole scattering picture, the JIMWLK equation reduces to the closed and relatively simple BK equation. In a language of Feynman diagrams, the JIMWLK equation includes both BFKL ladder diagrams and the fan diagrams of triple ladder interactions, and it naturally describes the physics of scattering on a dense medium (or a target) with multiple scattering corrections. It naturally interprets the geometric scaling observed in the data [40]. However, the JIMWLK equation does not include Pomeron (or ladder) splittings or all the Pomeron loops. Modifications and improvements to the JIMWLK equation were proposed [45, 46]. In addition, a similar evolution equation was derived for a dilute target [46], while the JIMWLK equation is suitable for scatterings on a dense target. A striking result is that the evolution kernels of these two equations are apparently dual to each other [45, 46]. The selfduality of the kernel is somewhat similar (although different

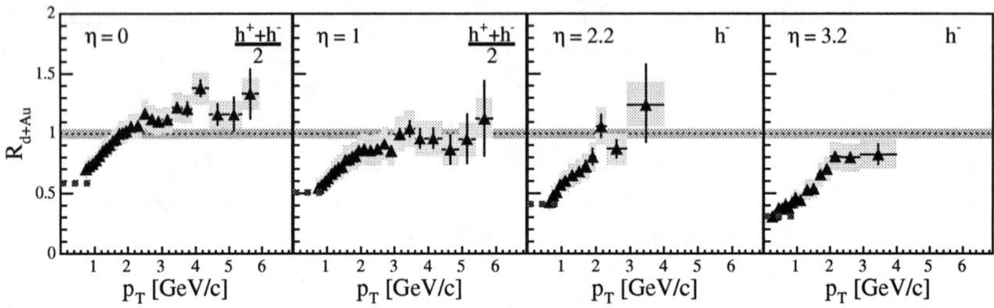

FIGURE 9. Nuclear modification factor for charged hardons at pseudorapidities $\eta = 0, 1.0, 2.2, 3.2$. Statistical errors are shown with error bars. Systematic errors are shown with shaded boxes with widths set by the bin sizes. The shaded band around unity indicates the estimated error on the normalization to $\langle N_{coll} \rangle$. Dashed lines at $p_T < 1$ GeV/c show the normalized charged particle density ratio $\frac{1}{\langle N_{coll} \rangle} \frac{dN/d\eta(dAu)}{dN/d\eta(pp)}$

in detail) to the duality symmetry, $p \to x$, $x \to -p$ in the Hamiltonian of a harmonic oscillator [46].

Relativistic Heavy Ion Collider (RHIC) is a unique place to test the theory of CGC because of the high density of partons involved. In order to probe small x partons and the phenomena of CGC at RHIC, one has to go to extremely forward and backward region in rapidity because of the relatively low colliding energy. Three major experimental collaborations, BRAHMS, PHENIX, and STAR, at RHIC carried out the effort and presented their early results at the workshop [47, 48, 49].

Rapidity dependence of high-p_T particle suppression was measured in d-Au collisions at $\sqrt{s_{NN}} = 200$ GeV by BRAHMS Collaboration and presented at the workshop [47]. The data collected from d-Au collisions at RHIC is compared to p-p in Figure 9 using the nuclear modification factor defined as

$$R_{\mathrm{d-Au}} = \frac{1}{N_{coll}} \frac{\frac{dN^{\mathrm{dAu}}}{dp_T d\eta}}{\frac{dN^{\mathrm{pp}}}{dp_T d\eta}} \tag{2}$$

where N_{coll} is the number of binary collisions estimated to be 7.2 ± 0.6 for minimum biased d+Au collisions. Nuclear modification factors for charged hadrons at pseudorapidities $\eta = 0, 1.0, 2.2$ and 3.2 were shown as a function of hadron transverse momentum p_T. At the central region, or zero rapidity, data confirm the Cronin type enhancement in large p_T region. However, as the pseudorapidity increases, the enhancement vanishes, and the modification factor is less than the unity for entire measured p_T region. The forward region is, the measurement probes target partons with smaller x. The observed rapidity dependence of the suppression, which increases with rapidity, fits naturally into the picture of CGC [50], and can be also interpreted by the recombination model of hadronization [51]. In addition, the suppression is consistent with the perturbative QCD calculation based on resummation of coherent multiple scattering [52].

PHENIX Collaboration reported its measurement of charged hadron production in the same d-Au collisions at RHIC [48]. It covers pseudorapidities from -2.0 to -1.4 and 1.4 to 2.2 with the forward coverage overlaps with some of BRAHMS measurements. PHENIX also observes a suppression in hadron yields in d-Au collision relative to binary

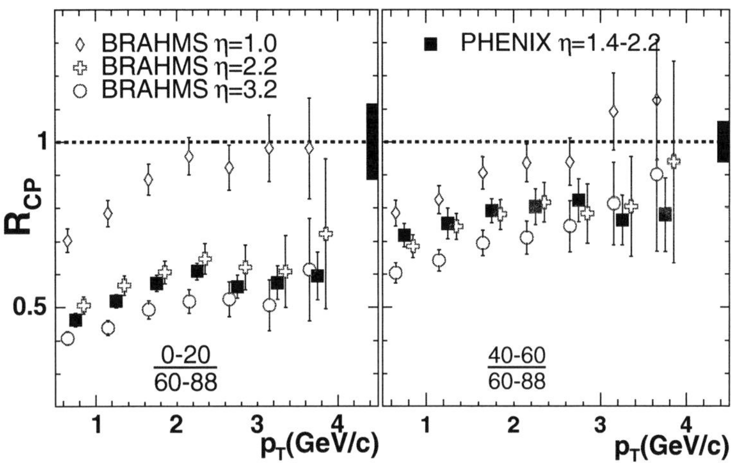

FIGURE 10. PHENIX R_{cp} as a function of p_T at forward rapidities shown as the average of the two methods. Note that the BRAHMS results are for negative hadrons at $\eta = 2.2,3.2$ and their centrality ranges $(0-20\%/60-80\%$ and $30-50\%/60-80\%)$ are somewhat different from ours.

collision. The data was presented in terms of a different nuclear modification factor, R_{cp}, which is defined as the ratio of the particle yield in central collisions to the particle yield in peripheral collisions, each normalized by the averaged number of binary collisions N_{coll}

$$R_{\rm cp} = \frac{\frac{dN^{\rm central}}{dp_T d\eta} \Big/ N_{coll}^{\rm central}}{\frac{dN^{\rm peripheral}}{dp_T d\eta} \Big/ N_{coll}^{\rm peripheral}}. \tag{3}$$

As shown in Figure **??**, PHENIX data are consistent with BRAHMS data, and are in qualitative agreement with theoretical expectation. Quantitatively, the ratio R_{cp} for the most central over the most peripheral collisions is more suppressed than theoretical calculations.

Measurements of the inclusive yields of π^0 mesons in p-p and d-Au collisions at RHIC were presented by STAR Collaboration [49]. With a forward π^0 detector installed at the Solenoidal Tracker at RHIC (STAR), it can detect high energy π^0 mesons with pseudorapidity as large as $3.3 < \eta < 4.1$ [49]. The inclusive yield in p-p collisions at $\sqrt{s} = 200$ GeV are consistent with NLO pQCD calculations. The nuclear modification factor, R_{dAu} in Figure 11, shows a strong suppression at the large pseudorapidity. It was argued that the d-Au yield is consistent with a model calculation treating the Au nucleus as a CGC for forward particle production [49]. Comparisons with other production models will be interesting to perform. Additional measurements with different final-state particles and at different centralities will help elucidate the cause of the observed strong suppression, which covers a broad range of nuclear gluon momentum fraction with a peak value $x \sim 0.02$ [49].

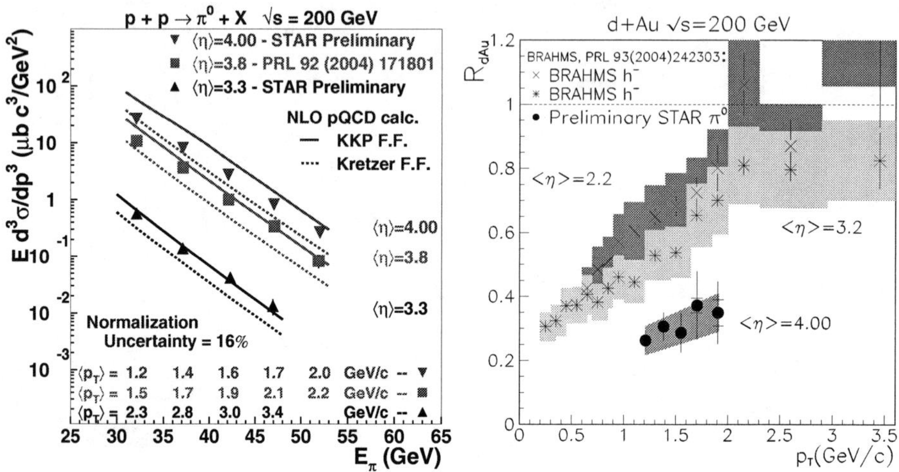

FIGURE 11. Inclusive π^0 yield for p+p [left] and d+Au collisions normalized by p+p [right]. The pion energy (E_π) is correlated with the transverse momentum (p_T), as the FPD was at fixed values of pseudorapidity (η). The inner error bars are statistical, while the outer combine these with the E_π- (p_T-) dependent systematic errors, and are often smaller than the points. The curves (left) are NLO pQCD calculations evaluated at fixed η, using different fragmentation functions. The x's and stars (right) are BRAHMS data for h^- production at smaller η.

Two theory talks were presented at the workshop to specifically address the strong suppression observed in the forward region of d-Au collisions at RHIC [50, 51]. Two completely different pictures were presented on how a leading hadron was produced in the d-Au collisions at RHIC energies. In one approach [50], single hadron production was assumed to be proportional to gluon production. Under this approximation, the nuclear modification factor R_{dAu} is the same for both hadron and gluon production. The gluon production in d-Au collisions was calculated in the framework of CGC physics [50]. In the other approach [51], a single hadron was produced via recombination of partons available during the collisions. It was argued that p-p, p(d)-Au, and Au-Au collisions produce different shapes of parton spectra. Recombination of partons with different spectra naturally leads to different hadron distribution and a nontrivial nuclear modification factor [51]. A striking fact is that both of these approaches provided calculations that are consistent with the observed data.

NUCLEAR STRUCTURE FUNCTIONS AND NUCLEAR PDFS

It was observed about two decades ago that DIS structure functions of nuclei differ from simple sum of those in the free nucleon [53]. As a result, PDFs of a nucleus of atomic weight A also differ from those in the free proton, $f_i^A(x,Q^2) \neq f_i(x,Q^2)$. In order to understand the overwhelming data from the RHIC and make predictions for the heavy ion programs at the future facilities, like the LHC and Electron Ion Collider (EIC), we need precise information of nuclear PDFs (nPDFs), in particular, at small $x(< 0.1)$.

A brief overview of the global DGLAP analyses of nPDFs was presented at the workshop [54]. The nPDFs are defined in terms of the same operators that define the free nucleon PDFs with the free nucleon state replaced by a nuclear state. Therefore, nPDFs and free nucleon PDFs should share the same DGLAP evolution equations, and only difference between nPDFs of different nuclei and free nucleon PDFs is the input distributions to DGLAP equations at a scale Q_0^2. Once a set of the nonperturbative input distributions are chosen, DGLAP evolution equations predict nPDFs at a larger momentum scale Q^2. There are typically two approaches to choose the input distributions: calculated by the nuclear models and determined by fit to the data [54]. The second approach shares the same procedures as that used in the determination of PDFs, and often referred as the global analyses of nPDFs. There are three groups who have been carrying out the global analyses and reanalyses of nPDFs: Eskola *et al.* (usually called as *EKS98*) [55, 56, 57], Hirai *et al.* [58, 59], and de Florian and Sassot [60]. The first two groups use LO DGLAP evolution while the third uses NLO evolution. All analyses, only DIS and Drell-Yan data on nuclear targets were used in the fits. Because of the large error in nuclear data and the lack of direct information on gluon initiated processes, all fits have reasonable constrains and consistencies on quark distributions, but, not on gluon distributions [54].

A hard probe often refers to a scattering process with a large momentum exchange q^μ whose invariant mass $Q \equiv \sqrt{|q^2|} \gg \Lambda_{\rm QCD}$, and it can probe a distance scale much smaller than size of a nucleon at rest, $1/Q \ll$ fm. However, when an active parton's momentum fraction $x < x_c \sim 0.1$, a hard probe might interact with more than one partons of the nucleon coherently [61]. When $x \ll x_c$, the hard probe can cover a whole Lorentz contracted nucleus and interact with partons from different nucleons. Although such coherent multi-parton interactions are power suppressed by hard scales of the scattering, they are enhanced by the nuclear size and could be one of the important sources of nuclear dependence observed in high energy nuclear collisions. A presentation made at this workshop looked at the impact of coherent multiple scattering in DIS on nuclear targets and leading particle production in p(d)-Au collisions [52]. An all power resummation of nuclear enhanced power corrections to DIS structure functions on nuclear targets was achieved. The calculated results for the Bjorken x-, Q^2- and A-dependence of nuclear shadowing in $F_2^A(x,Q^2)$ and the nuclear modifications to $F_L^A(x,Q^2)$ are consistent with the existing data [52]. Predictions were made for the dynamical shadowing from final state interactions in $\nu + A$ reactions for sea and valence quarks in the structure functions $F_2^A(x,Q^2)$ and $xF_3^A(x,Q^2)$, respectively. In addition, calculations for the centrality and rapidity dependent nuclear suppression of single and double inclusive hadron production at moderate transverse momenta in $p+A$ collisions were presented and consistent with the RHIC data [52].

In the Gribov-Glauber picture, nuclear shadowing and antishadowing observed in nuclear structure functions are due to the destructive and constructive interference of amplitudes arising from the multiple-scattering of quarks in the nucleus, respectively. A calculation of shadowing and antishadowing of nuclear structure functions in the Gribov-Glauber picture were presented at the workshop [62]. The coherence of multistep nuclear processes leads to shadowing and antishadowing of the electromagnetic nuclear structure functions in agreement with the data. But, the same picture leads to substantially different antishadowing for charged and neutral current reactions, thus affecting the extraction of the weak-mixing angle θ_W [62]. This is due to the fact that Reggeon couplings depend on the quantum numbers of the struck quark implies non-universality of nuclear antishadowing for charged and neutral currents [62].

NEW APPROACHES TO PDFS

Moments of PDFs are matrix elements of local gauge invariant operators which in principle can be calculated by using lattice QCD. A brief review of recent lattice effort in determining the PDFs was presented at the workshop [63]. Lattice QCD calculations of three representative observables, the transverse quark distribution, momentum fraction, and axial charge, were presented [63]. It was emphasized that lattice calculations of nucleon structure are beginning to realize their promise to elucidate QCD and make contact with the experimental programs. It was concluded that recent calculations are painting a qualitative three dimensional picture of nucleon structure revealing a significant x dependence of the transverse size of the nucleon. Quantitative calculations of moments of PDFs are progressing, in particular, the calculation of g_A may soon reach a few percent accuracy.

An analytical approach to understand the three dimensional picture of nucleon structure was presented at the workshop [64]. A concept of the quantum phase-space (Wigner) distributions for the quarks and gluons in the nucleon was introduced. The quark Wigner functions were related to the transverse-momentum dependent PDFs and generalized PDFs with emphasis on the physical role of the skewness parameter. Any knowledge on the generalized PDFs can be immediately translated into the correlated coordinate and momentum distributions of partons. In particular, the generalized PDFs can be used to visualize the phase-space motion of the quarks, and hence allow studying the contribution of the quark orbital angular momentum to the spin of the nucleon. It was concluded that measurements of generalized PDFs and/or direct lattice QCD calculations of them will provide a fantastic window to the quark and gluon dynamics in the proton [64].

Another presentation made at the workshop looked at quark asymmetries in nucleons [65]. Instead of fitting the data, a physical model for the non-perturbative x-shape of PDFs was developed. The model was based on Gaussian fluctuations in momenta, and quantum fluctuations of the proton into meson-baryon pairs. It was found that the model gives a good description of the proton structure function and a natural explanation of observed quark asymmetries, such as the difference between the anti-up and anti-down sea quark distributions and between the up and down valence distributions [65]. Within this model, there is an asymmetry in the momentum distributions of strange and anti-strange quarks in the nucleon, and the asymmetry is large enough to reduce the NuTeV anomaly to a level which does not give a significant indication of physics beyond the standard model.

Effective field theory was used to investigate the nuclear modification to the PDFs and a recent result was presented at the workshop [66]. It was found that the universality of the shape distortion in nPDFs (the factorization of the Bjorken x and atomic weight A dependence) is model independent and emerges naturally in effective field theory. For a simple parameterization of nonperturbative functions in the approach, fits to the data confirm the factorization [66].

ACKNOWLEDGMENTS

We would like to thank all the members of our working group for the excellent presentations and for lively discussions they provoked. We would also like to thank the conveners of the Electroweak and Beyond the Standard Model working group for their assistance in the joint session on high Q^2 structure function measurements. We would also like to thank all the session chairs for agreeing to be involved. Last, but not least, we would like to thank the organizers of DIS 2005 for interesting and well-organized meeting.

REFERENCES

1. ZEUS Collaboration, S. Chekanov et al. *Eur. Phys. J.* **C21**, 443–471 (2001); H1 Collaboration, C. Adloff et al. *Eur. Phys. J.* **C21**, 33–61 (2001).
2. ZEUS Collaboration, S. Chekanov et al. *Phys. Lett.* **B487**, 1–2, 53–73 (2000).
3. E. Lobodzinska, *these proceedings*; H1 Collaboration, A. Aktas et al. *Phys. Lett.* **B598**, 159–171 (2004).
4. A. Petrukhin, *these proceedings.*
5. H1 Collaboration, A. Nikiforov, *these proceedings*; ZEUS Collaboration, A. Tapper, *these proceedings.*
6. ZEUS Collaboration, Y. Ning, *these proceedings.*
7. NuTeV Collaboration, M. Tzanov, *these proceedings, hep-ex/0507040.*
8. S. Kuhlmann *et al*, *Phys. Lett.* **B476**,291 (2000).
9. S. Kuhn, *these proceedings.*
10. CDF Collaboration, Y.S. Chung, *these proceedings.*
11. S. Liuti, *these proceedings.*
12. C. Keppel, *these proceedings*; I. Niculescu, *these proceedings.*
13. A. Bodek, *these proceedings.*
14. H. Gallagher, *these proceedings.*
15. F. Chlebana, *these proceedings*; A. Harel, *these proceedings.*
16. Yu. Dokshitzer, Soviet Phys. JETP **46**, 641 (1977); V. N. Gribov and L. N. Lipatov, Soviet J. Nucl. Phys. **15**, 438,675, (1972); L. N. Lipatov, Soviet J. Nucl. Phys. *20*, 95, (1975); G. Altarelli and G. Parisi, Nucl. Phys **B126**, 298 (1977).
17. D. Stump, *these proceedings.*
18. R. Thorne, *these proceedings, hep-ph/0507015.*
19. J. Terron, *these proceedings.*
20. J. Huston, J. Pumplin, D. Stump and W. K. Tung, *hep-ph/0502080.*
21. A. Martin, R. Roberts, W. J. Stirling and R. Thorne, *Phys. Lett.* **B604**, 61–68 (2004).
22. A. Martin, R. Roberts, W. J. Stirling and R. Thorne, *Eur. Phys. J.* **C39**, 155–161 (2005).
23. ZEUS Collaboration, S. Chekanov et al. *DESY 05-050.*
24. J. Rojo, *these proceedings, hep-ph/0505044.*
25. A. Glazov, *these proceedings.*
26. C. Gwenlan, *these proceedings.*
27. A. Guffanti, *these proceedings.*
28. S. Moch, J. A. M. Vermaseren and A. Vogt, *Nucl. Phys.* **B688**, 101–134 (2004); **B691**, 129–181 (2004).
29. X. Zu, *these proceedings*; J. C. Collins and X. Zu, *JHEP* **0503**, 059 (2005); **0206**, 018 (2002).
30. G. Corcella, *these proceedings.*

31. L. N. Lipatov, *Sov. J. Nucl. Phys.* **23**, 338 (1976); E. A. Kuraev, L. N. Lipatov and V. S. Fadin, *Sov. Phys. JETP* **45**, 199–204 (1977); I. I. Balitsky and L. N. Lipatov, *Sov. J. Nucl. Phys.* **28**, 822–829 (1978).
32. H. Navelet, R. Peschanski, C. Royon, and S. Wallon, *Phys. Lett.* **B385**, 357–364 (1996); S. Munier and R. Peschanski, *Nucl. Phys.* **B524**, 377–393 (1998).
33. C. Royon, *these proceedings.*
34. G. Altarelli, R. D. Ball and S. Forte, *Nucl. Phys. Proc. Suppl.* **135**, 163-167 (2004).
35. A. Donnachie, P.V. Landshoff, *Phys. Lett.* **B296**, 227 (1992).
36. M. Froissart, *Phys. Rev.* **123**, 1053 (1961).
37. L. V. Gribov, E. M. Levin and M. G. Ryskin, *Phys. Rept.* **100**, 1–150 (1983); A. H. Mueller and J. W. Qiu, *Nucl. Phys.* **B268**, 427 (1986); A. H. Mueller, *Nucl. Phys.* **B558**, 285 (1999).
38. L. D. McLerran and R. Venugopalan, *Phys. Rev.* **D49**, 3352 (1994); *Phys. Rev.* **D49**, 2233 (1994); *Phys. Rev.* **D50**, 2225 (1994).
39. J. Jalilian-Marian, A. Kovner, A. Leonidov and H. Weigert, *Phys. Rev.* **D59**, 014014 (1999); J. Jalilian-Marian, A. Kovner and H. Weigert, *Phys. Rev.* **D59**, 014015 (1999).
40. E. Iancu, A. Leonidov and L. D. McLerran, *Nucl. Phys.* **A692**, 583 (2001); *Phys. Lett.* **B510**, 133 (2001); E. Iancu and L. D. McLerran, *Phys. Lett.* **B510**, 145 (2001).
41. I. Balitsky, *Nucl. Phys.* **B463**, 99 (1996); *Phys. Rev. Lett.* **81**, 2024 (1998); *Phys. Lett.* **B518**, 235 (2001); Y.V. Kovchegov, *Phys. Rev.* **D60**, 034008 (1999); *Phys. Rev.* **D61**, 074018 (2000).
42. S. Munier and R. Peschanski, *Phys. Rev. Lett.* **91**, 232001 (2003); *Phys. Rev.* **D69**, 034008 (2004); *Phys. Rev.* **70**, 077503 (2004).
43. R. Enberg, *these proceedings.*
44. K. Tuchin, *these proceedings.*
45. A. M. Stasto, *these proceedings.*
46. A. Kovner, *these proceedings.*
47. F. Videbaek for BRAHMS Collaboration, *these proceedings.*
48. C. Zhang for PHENIX Collaboration, *these proceedings.*
49. G. Rakness for STAR Collaboration, *these proceedings.*
50. Y. Kovchegov, *these proceedings.*
51. R. Fries, *these proceedings.*
52. I. Vitev, *these proceedings.*
53. J. Aubert *et al.*, *Phys. Lett.* **B123**, 275 (1983).
54. V. Kolhinen, *these proceedings.*
55. K. J. Eskola, V. J. Kolhinen, and P. V. Ruuskanen, *Nucl. Phys.* **B535**, 351–371 (1998).
56. K. J. Eskola, V. J. Kolhinen, and C. A. Salgado, *Eur. Phys. J.* **C9**, 61–68 (1999).
57. K. J. Eskola, V. J. Kolhinen, and C. A. Salgado, *In preparation.*
58. M. Hirai, S. Kumano, and M. Miyama, *Phys. Rev.* **D64**, 034003 (2001).
59. M. Hirai, S. Kumano, and T. H. Nagai, *Phys. Rev.* **C70**, 044905 (2004).
60. D. de Florian, and R. Sassot, *Phys. Rev.* **D69**, 074028 (2004).
61. J. W. Qiu, Nucl. Phys. A **715**, 309 (2003) [arXiv:nucl-th/0211086].
62. S.J. Brodsky, *these proceedings.*
63. D. Renner, *these proceedings.*
64. A. V. Belitsky, *these proceedings*; A. V. Belitsky, X. Ji and F. Yuan, *Phys.Rev.* **D69**, 074014 (2004).
65. J. Alwall, *these proceedings.*
66. W. Detmold, *these proceedings.*

Summary of the "Diffraction & Vector Mesons" working group at DIS05

X. Janssen*, M. Ruspa† and V. A. Khoze**

*DESY, 22607 Hamburg, Germany
†University of Eastern Piedmont, 28100 Novara, Italy
**IPPP, University of Durham, DH1 3LE,UK

Abstract.
We survey the contributions presented in the working group: "Diffraction & Vector Mesons".

Keywords: diffractive deep inelastic scattering, Pomeron, diffractive PDFs, factorization, vector mesons, DVCS, GPDs, diffractive Higgs
PACS: 13.60Hb, 13.60Le, 12.38Bx, 12.39St, 12.40Nn

INTRODUCTION

In diffractive interactions in hadron-hadron or photon-hadron collisions at least one of the beam particles emerges intact from the collision, having lost only a small fraction of its initial energy, and carrying a small transverse momentum. Therefore no color is exchanged in the t-channel. The signature for such processes is the presence of a gap in rapidity between the two hadronic final states. At high energy this is described by the exchange of an object with the quantum numbers of the vacuum, referred to as the Pomeron in the framework of Regge phenomenology [1]. Note that at low energies similar reactions can also proceed when quantum numbers are exchanged through subleading Regge trajectories (Reggeons); however, these contributions are exponentially suppressed as a function of the gap size and are negligible at small values of the longitudinal momentum loss. The understanding and description of diffractive processes is one of the aims of QCD.

Diffractive events are being extensively studied at HERA, TEVATRON, RHIC, JLAB and CERN and there is a growing community planning to continue these studies at the LHC. Updates on the available experimental data and on their theoretical interpretation were given at this workshop; many discussions also took place on the future plans. In the present summary we focus on the path from HERA to the LHC through the TEVATRON.

FROM HERA TO HADRON COLLIDERS

Selection of diffractive processes

Let us first look at the diffractive reaction $ep \rightarrow eXp$ at HERA, depicted in Fig. 1a: a photon of virtuality Q^2 diffractively dissociates interacting with the proton at a center

CP792, *Deep Inelastic Scattering, DIS 2005*, edited by W. H. Smith and S. R. Dasu

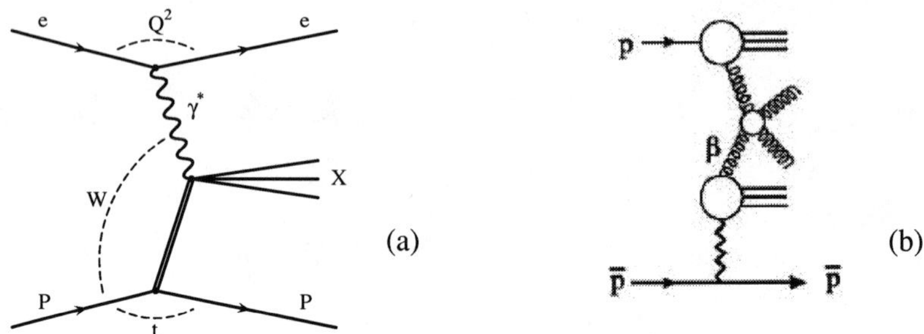

FIGURE 1. (a) Diffraction in ep interactions. (b) Diffraction in $p\bar{p}$ interactions.

of mass energy W and squared four momentum transfer t and produces the hadronic system X with mass M_X in the final state. The fraction of the proton momentum carried by the exchanged object is denoted by $x_{\mathbb{P}}$, while the fraction of the momentum of the exchanged object carried by the struck quark is denoted by β (note that sometimes z is used instead of β). The virtual photon emitted from the lepton beam provides a point-like probe to study the structure of the diffractive exchange, similarly to ordinary DIS probing proton structure. The fact that a large fraction ($\sim 10\%$) of deep inelastic (DIS) events at HERA is diffractive has thus opened the possibility of investigating the partonic nature of the Pomeron and has established a theoretical link between Regge theory and QCD.

At the TEVATRON inclusive diffraction is mainly studied via the reaction $p\bar{p} \rightarrow \bar{p}X$, sketched in Fig. 1b; in the TEVATRON jargon $x_{\mathbb{P}}$ is usually indicated as ξ.

At HERA three methods are used to select diffractive events [2]. The first is based on the measurement of the scattered proton with a spectrometer installed very close to the beam in a region with acceptance for protons which have lost only a small fraction of their initial longitudinal energy. A second method requires the presence of a large rapidity gap (LRG) in the forward region. A third method is based on the different shape of the M_X distribution between diffractive and non diffractive events. At the TEVATRON diffractive interactions are selected by tagging events by either a rapidity gap or a leading antiproton [3].

The proton tagging method has the advantage of excluding the proton dissociation processes $ep \rightarrow eXN$, where the proton also diffractively dissociates into a state N of mass M_N that escapes undetected into the beam pipe. In order to ensure that the scattered proton resulted from a diffractive process one requires $x_{\mathbb{P}} < 0.01$. This cut removes contributions coming from Reggeon exchanges [4].

The large rapidity gap method selects events which include some proton dissociation processes and some Reggeon contributions. The latter can be removed by the same $x_{\mathbb{P}}$ cut as above. If the mass M_N of the dissociative system is large enough to be measured in the forward detector the proton dissociation background can be removed, whereas the contribution of low mass proton dissociation can be estimated with a Monte Carlo simulation (10% of background with $M_N < 1.6$ GeV is quoted from the H1 analysis [5]).

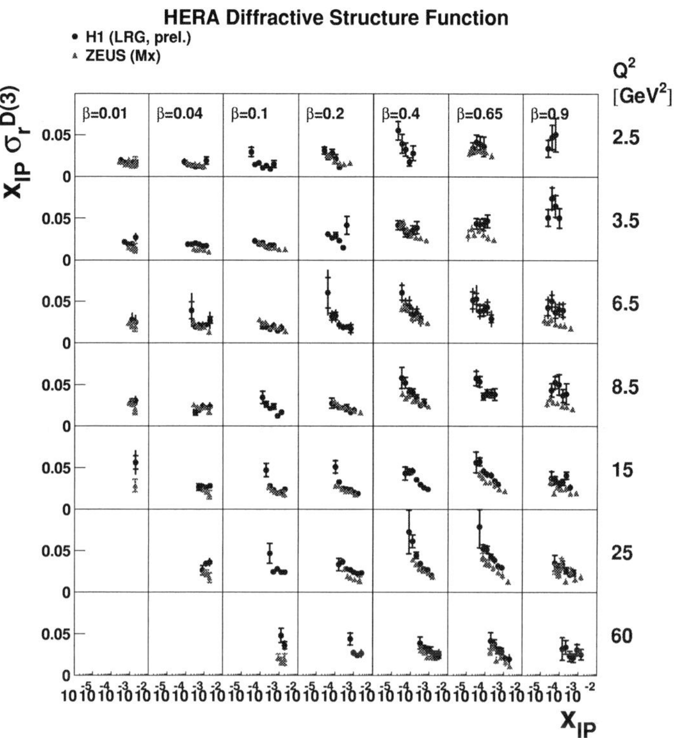

FIGURE 2. ZEUS M_X and H1 LRG measurements of the diffractive structure function.

In the M_X method the statistical subtraction of the non diffractive background eliminates also the the Reggeon contribution, but the selected sample is left with an important contamination from proton dissociative events with masses $M_N < 2.3$ GeV [6]. By comparing the measured cross sections with those coming from the leading proton analysis one can estimate the amount of this background (around 30% [7]) and determine a correction factor.

HERA diffractive structure function and PDFs

H1 and ZEUS have presented recent precise measurements of the diffractive structure function obtained with all three HERA methods and covering a wide kinematic range (proton tagging method: [2, 8], LRG method: [5], M_x method: [6]). In Fig. 2 the diffractive structure function is presented as a function of x_{IP} for fixed values of Q^2 and β. The data points come from two samples analysed by H1 with the LRG method and by ZEUS with the M_X method, respectively. The ZEUS M_X data have been scaled to $M_Y < 1.6$ GeV, the region of dissociative masses included in the H1 data. There is

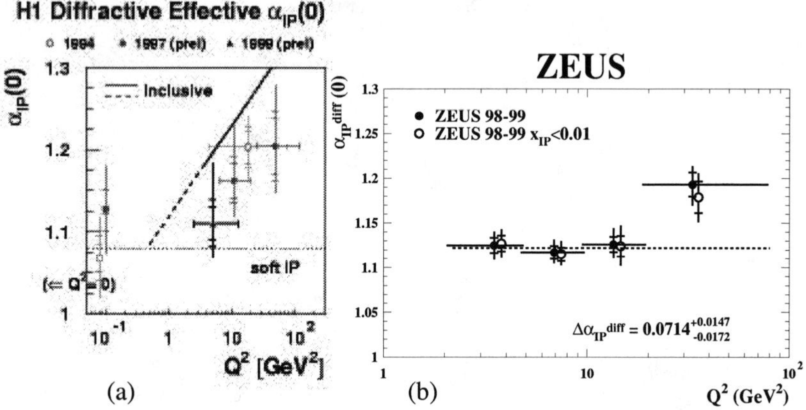

FIGURE 3. Q^2 dependence of the Pomeron intercept $\alpha_{I\!P}(0)$, measured by H1 (a) and by ZEUS (b).

a reasonable agreement between the two data sets, but at a closer inspection it turns out that the Q^2 dependences are different, namely the positive scaling violations in the ZEUS data are smaller than in the H1 data. This discrepancy has been investigated very recently by a combined set of next-to-leading-order (NLO) QCD fits of the diffractive structure function, attempted by two different groups (P. Newman et al. [9] and A. Levy et al. [10] - see also the upcoming proceedings of the HERA-LHC workshop, http://www.desy.de/~heralhc).

Such fits are based on the validity of a collinear factorization theorem in diffractive processes [11], which allows to write F_2^D as a convolution of the usual partonic cross sections as in DIS with diffractive parton distribution functions (DPDFs). The DPDFs, parametrised at a starting scale, are evolved according to the DGLAP equations [12] and fitted to the data. In the ideal case we would evolve in Q^2 for fixed t and $x_{I\!P}$, or at least for fixed $x_{I\!P}$ if t is integrated over, but this is not allowed by the rather limited statistics of the present data. An alternative approach is the assumption, known as "Regge factorization" hypothesis, that F_2^D can be expressed as the product of a flux, depending only on $x_{I\!P}$ and t, and the structure function of a particle-like object. Whether the data support this assumption or not is a controversial problem. It translates into determining whether or not the intercept $\alpha_{I\!P}(0)$ of the Pomeron trajectory $\alpha_{I\!P}(t) = \alpha_{I\!P}(0) + \alpha' t$ depends on Q^2.

Fig. 3a shows $\alpha_{I\!P}(0)$ as a function of Q^2, as measured by H1 : there is a suggestion of a dependence of $\alpha_{I\!P}(0)$ on Q^2, though firm conclusions are not possible with the present uncertainties. In Fig. 3b, where the ZEUS measurement is presented, the Pomeron intercept rises by $\Delta\alpha_{\text{diff}} = 0.0741 \pm 0.0140(\text{stat.})^{+0.0047}_{-0.0100}(\text{syst.})$ between Q^2 of 7.8 GeV^2 and 27 GeV^2, with a significance of 4.2 standard deviations. This scenario suggests a possible violation of Regge factorization and a clear need for more precise data. Nevertheless it has been shown [10] that, when restricting the analysed range to $x_{I\!P} < 0.01$, Regge factorization is a sufficiently good approximation and this is the compromise at the basis of the NLO DGLAP fits discussed in the following.

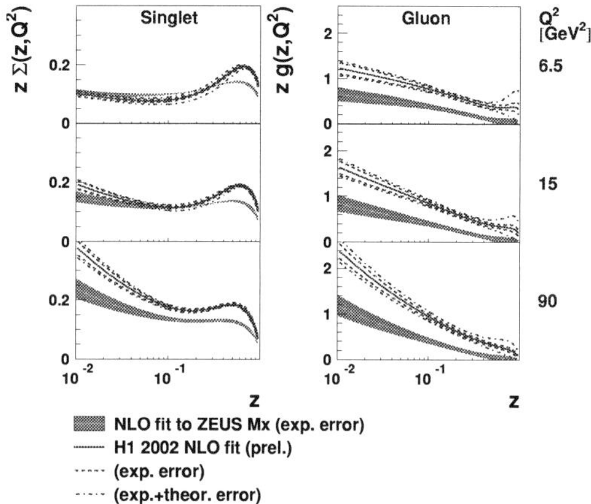

FIGURE 4. The diffractive parton densities resulting from a NLO QCD fit by P. Newman et al. [9] to the ZEUS M_X data (solid line) and to the H1 LRG data (shaded line).

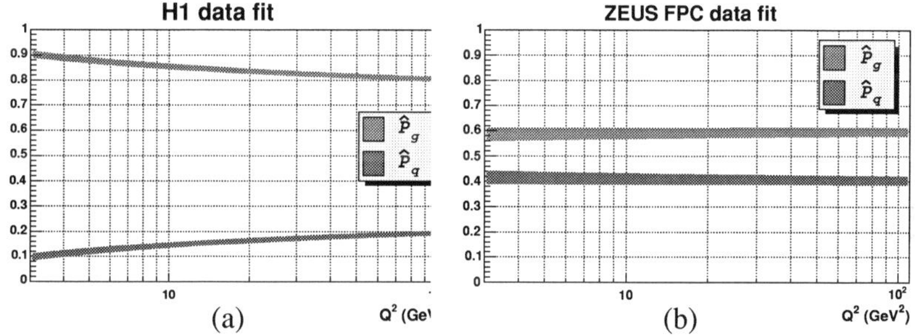

FIGURE 5. The parton momentum fraction as a function of Q^2 from a NLO QCD fit by A. Levy et al. [10] to the H1 LRG data (a) and to the ZEUS M_X data (b).

In Fig. 4 a comparison is shown between the diffractive PDFs extracted from the NLO QCD fit by P. Newman et al. to the ZEUS M_X data (solid line) and from the same fit to the H1 LRG data (shaded line), the latter being essentially the well known H1 fit 2002 [5]. Note that most of the data points from the high β region, where discrepancies arise between the data sets (Fig. 2), have not been included in the fit. As a reflection of the difference in the scaling violations between the two sets of measurements (Fig. 2), the quark density is similar at low Q^2 and evolves differently to higher Q^2; the gluon density is a factor ~ 2 smaller in the ZEUS data than in the H1 sample. This disagreement is confirmed and quantified in Fig. 5, which shows the fraction of the Pomeron momentum carried by quarks (red/dark line) and by gluons (blue/light line), as a function of Q^2, as resulting from the fit by A. Levy et al., similar to the previous one, but completely independent, performed always on the H1 LRG data (Fig. 5a) and on the ZEUS M_X data

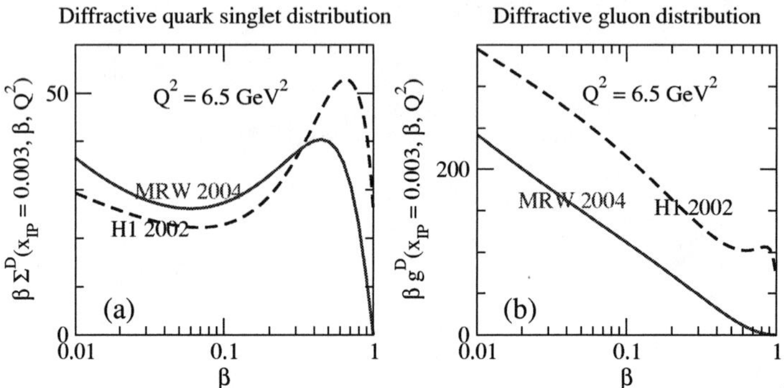

FIGURE 6. The diffractive parton densities resulting from a combined QCD fit by A. Martin et al. [13] to the ZEUS proton tagged, ZEUS M_X and H1 LRG data. The dashed lines are the densities obtained in the H1 fit 2002 [5].

(Fig. 5b). The fraction of the Pomeron momentum carried by gluons turns to be between 70% and 90% in the H1 data and between 55% and 65% in the ZEUS M_X data. The same study has been carried out also on the ZEUS proton tagged data and the resulting integral of the fractional momentum is in agreement with the H1 value.

The same data have also been analysed according to a new approach by A. Martin et al. [13], which does not assume Regge factorization and shows that the collinear factorization theorem, though valid asymptotically in diffractive DIS, has important modifications at the energies relevant at HERA, which can be quantified using perturbative QCD. The DPDFs are shown to satisfy an inhomogeneous evolution equation and the need of including both the gluonic and sea-quark components of the perturbative Pomeron is considered. The DPDFs resulting from a combined fit to the ZEUS proton tagged data and M_X data and to the H1 LRG data are shown in Fig. 6 (solid line), together with H1 fit 2002 (dashed line). While the quark densities are not very different from those of H1, the gluon distribution is significantly lower than H1 one.

The discrepancies between the various DPDFs shown in Figs. 4 and 6 are large and presently not fully understood. They are due to a combination of effects: disagreement in the data, different fit methods and assumptions behind them. Therefore these very differences between the DPDFs are at the moment the only realistic estimate we have of their uncertainties. A precise and consistent determination of the DPDFs is certainly one of the main tasks that the HERA community has to face in the near future. Among other reasons, they are a crucial input for the prediction of any inclusive diffractive cross section at the LHC.

QCD factorization tests

According to the factorization theorem, calculations based on DPDFs extracted from inclusive measurements should allow to predict cross sections for other diffractive pro-

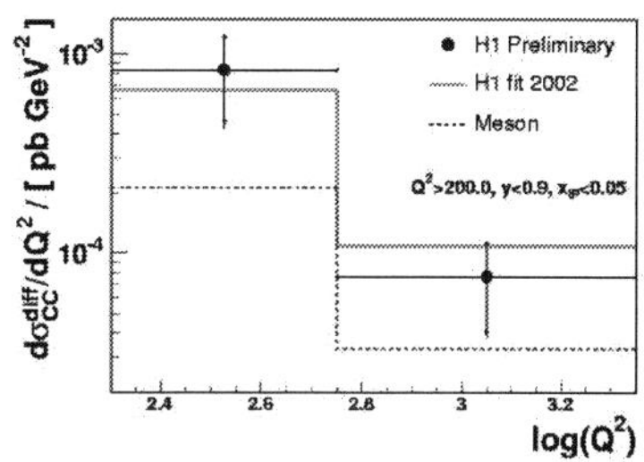

FIGURE 7. H1 charge current differential cross section $d\sigma_{cc}^{diff}/dQ^2$ as a function of $log(Q^2)$.

cesses. Calculations based on H1 fit 2002 agree well with the data on diffractive D^* production in DIS [14] and diffractive dijet production in DIS [15]. A further test of factorization comes from the study of events with a large rapidity gap in charged current interactions at high Q^2: in Fig. 7 the differential cross section $d\sigma_{cc}^{diff}/dQ^2$, as measured by H1 [16], is presented as a function of Q^2 and is well described by a calculation based on H1 fit 2002. A similar result was obtained by ZEUS [17]. However, the important uncertainties on the DPDFs discussed in the previous section make the conclusions on the validity of QCD factorization in DIS rather weak.

The factorization theorem does not hold in the case of diffractive hadron-hadron scattering [11]: indeed it has been known for years that the DPDFs extracted from HERA data overestimate the rate of diffractive dijets at the TEVATRON by one order of magnitude [18]. It was shown in [19] that this breakdown of factorization can be explained by screening (unitarization) effects. In the t-channel Reggeon framework, these effects are described by multi-Pomeron exchange diagrams. Because of the screening, the probability of rapidity gaps in high energy interactions to survive decreases since they may be populated by rescattering processes. The screening corrections are accounted for by the introduction of a suppression factor, which is often called the *survival probability of rapidity gaps*. As shown in [19] and [20], the current CDF diffractive dijet data, with one or two rapidity gaps, are in good quantitative agreement with the multi-Pomeron-exchange model.

In photoproduction at HERA ($Q^2 \sim 0$), the exchanged photon, which is real or quasi real, can either interact directly with the proton or first dissolve into partonic constituents which then scatter off the target (resolved process). In the former case dijet photoproduction is described by a photon gluon fusion process. In the latter case the photon behaves like a hadron. Factorization should then be valid for direct interactions as in the case of DIS with large Q^2, whereas for the resolved contribution it is expected to fail due to rescattering corrections. In the ideal theoretical limit, the suppression factor of 0.34 is evaluated for the resolved process within the multi-Pomeron exchange model [21]. However, in reality there is no clear model independent separation between the direct and resolved processes. In particular, the direct contribution is smeared by the experimental resolution and uncertainties. Moreover, at NLO these contributions are closely related. Recently Klasen and Kramer [22] have performed an analysis of diffractive dijet photoproduction data at NLO where they suppressed the resolved process by a factor 0.34.

Fig. 8 shows the differential cross section, as measured by H1 [15], for the diffractive photoproduction of two jets as a function of x_γ (the fraction of the photon momentum entering the hard scattering), where the NLO prediction has been tested in two different weighting schemes: in Fig. 8a only the resolved part has been scaled by the factor 0.34, while in Fig. 8b a global suppression factor 0.5 is applied to both the direct and resolved components. In Fig. 9 the ratio of the ZEUS data [23] to the NLO predictions of Klasen and Kramer [22] with no suppression factor ($R = 1$) is shown separately for the sample enriched (a) in the direct ($x_\gamma \geq 0.75$) and (b) in the resolved ($x_\gamma < 0.75$) components. Both for resolved and direct photoproduction the ratio is flat, but the data are lower by a factor of ~ 2 compared to the NLO calculations. Deviations are seen in E_T^{jet1} and in η^{jet1}, which are know to be sensitive to the structure function of the photon [24]. The overall message from the data of Figs. 8 and 9 is that, while a suppression of only the resolved contribution at NLO is disfavored by the data, a good agreement is achieved with the global suppression 0.5, which furthermore yields a good description of all measured cross sections.

The fact that the data, apparently against expectations, support suppression of direct photoproduction, has been addressed by M. Klasen [25] and has been related to the critical role of an initial state singularity in the way factorization breaks down and to the need of a modification of the suppression mechanism: separation of direct and resolved photoproduction events is a leading order concept. At NLO they are closely connected. The sum of both cross sections is the only physical relevant observable, which is approximately independent of the factorization scale, M_γ [26]. By multiplying the resolved cross section with the suppression factor $R = 0.34$, the scale dependence of the NLO direct cross section is compensated against that of the LO resolved part [22]. But at NLO collinear singularities arise from the photon initial state, which are absorbed at the factorization scale into the photon PDFs; the latter become in turn M_γ dependent. An equivalent M_γ dependence, just with the opposite sign, is then left in the NLO corrections to the direct contribution. Hence, in order to get a physical cross section at NLO, that is the superimposition of the NLO direct and LO resolved cross section, and to restore the scale invariance, one must multiply the M_γ dependent term of the NLO correction to the direct contribution with the same suppression factor as the resolved cross section.

The situation with the factorization breaking in dijet photoproduction is not completely clear and further experimental and theoretical efforts are needed. As was emphasized in [21], a possible way to study the effects of factorization breaking due to rescattering in diffractive photoproduction is to measure the ratio of diffractive and inclusive dijet photoproduction as a function of x_γ. In such quantity (at least) some of the theoretical and experimental uncertainties will cancel.

The understanding of factorization breaking in hadron-hadron collisions is of fundamental importance for the diffractive physics at the LHC. The rapidity gap survival factor is an essential ingredient of the predictions [27] on exclusive diffractive Higgs production, which will be discussed in the last section.

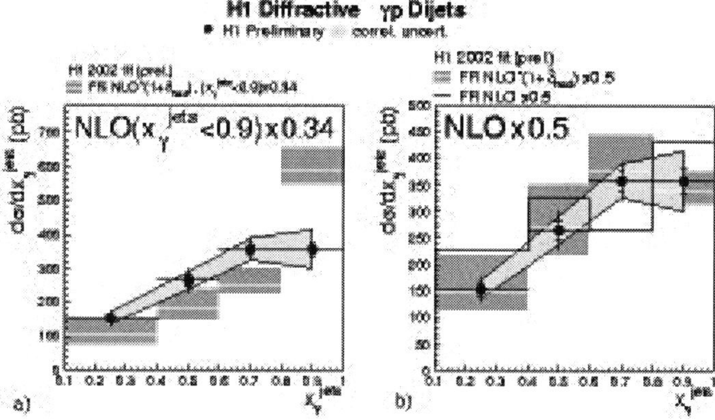

FIGURE 8. H1 cross section for the diffractive production of dijets in photoproduction as a function of x_γ. In (a) only the resolved contribution to the NLO calculation has been scaled by the factor 0.34, while in (b) the complete prediction is multiplied by a factor 0.5.

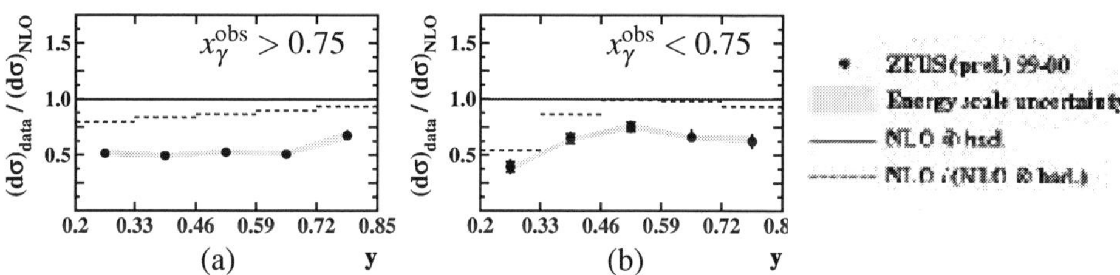

FIGURE 9. Ratio of the ZEUS diffractive dijet data to the NLO QCD predictions [22] of the single differential cross section in y for the sample enriched in direct (a) and resolved (b) photoproduction.

Diffraction at the TEVATRON

As discussed in the previous section, factorization is not expected to hold in hadron-hadron collisions. A strong breakdown of factorization at the TEVATRON has been known for some time from run-I (1992-1995) results [28]: the single-diffractive to non-diffractive ratios for dijets, W, b-quark and J/ψ production, as well as the ratio of double-diffractive to non-diffractive dijet production are all $\sim 1\%$, a factor 10 less than at HERA. However, the ratio of double- to single-diffractive dijets is found to be about a factor 5 larger than the ratio of single- to non-diffractive dijets, suggesting that there is only a small extra suppression when going from one to two rapidity gaps in the event, as confirmed by predictions [20]. In this respect the TEVATRON data are being a very powerful tool to shed light on the factorization breaking mechanism.

One of the major challenges of run-II is the measurement of central exclusive production rates (dijets, χ_c^0, diphotons). By central exclusive, we refer to the process $p\bar{p} \rightarrow p \oplus \phi \oplus \bar{p}$, where $\oplus$ denotes the absence of hadronic activity ('gap') between

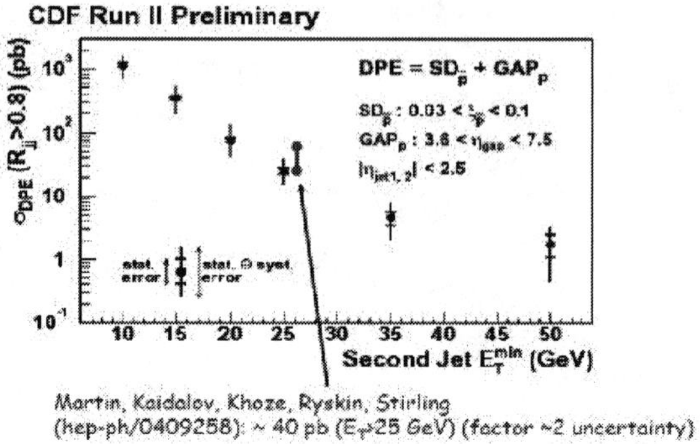

FIGURE 10. CDF dijet production cross section for $R_{jj} > 0.8$ in double Pomeron exchange events as a function of E_T^{min}, the E_T of the lower E_T jet.

the outgoing hadrons and the decay products of the central system ϕ. As we will discuss in the last section, the exclusive Higgs signal is particularly clean and the signal-to-background ratio is especially favorable, in comparison with other proposed selection modes. However, the expected number of events is low. Therefore it is important to check the predictions for exclusive Higgs production by studying processes mediated by the same mechanism, but with rates which are high enough to be observed at the TEVATRON (as well as at the LHC) [29].

The CDF search for exclusive dijet production is based on the reconstruction of the dijet mass fraction R_{jj} in double Pomeron exchange events. R_{jj} is defined as the mass of the two leading jets in an event divided by the total mass measured in all calorimeters. At first sight, we might expect that the exclusive dijets form a narrow peak concentrated at R_{jj} close to 1. In reality, the peak is smeared out due to hadronization and jet searching procedure as well as due to a 'radiative tail' phenomenon [30]. So it is not so surprising that within the CDF selection cuts no peak has been seen. CDF reports production cross sections for events with $R_{jj} > 0.8$, which are interpreted as the upper limits for exclusive production. Fig. 10 [28] shows such cross sections as a function of E_T^{min}, the E_T of the lower E_T jet. These data agree, within errors, with recent predictions for exclusive dijet production [29]. The analysis benefits from using dijet events in which at least one of the jets is b-tagged: presently more data on heavy flavor exclusive dijets are being collected with a special b-tagged dijet trigger.

Diffraction at RHIC

New interesting experimental results from RHIC were presented by Guryn, White and Klein. In particular, Guryn [31] described the results of the measurement of the single spin analyzing power A_N in polarized pp elastic scattering at 200 GeV. The recent results on inelastic diffraction with Au-Au, d-Au and pp beams were reviewed by White [32]. And Klein [33] showed the results of the STAR collaboration for coherent photonuclear ρ and 4 charged pion production.

Updates on theory

Several excellent mini review type theoretical talks were presented. Hard diffraction in DIS and the origin of hard Pomeron from rescattering were discussed by Brodsky [34]. He also reviewed such effects as Color Transparency, Color Opaqueness and Intrinsic Charm. Levin [35] gave a brief review of the current status of high density QCD with its ups and downs. The recent progress in the BFKL studies was covered by Andersen [36]. In particular, he discussed the high-energy limit of diffractive scattering processes in the BFKL resummation framework. He showed that the BFKL equation was solved at full next-to-leading logarithmic accuracy.

EXCLUSIVE MESON PRODUCTION AND DVCS

The dynamics of diffractive interactions can also be studied through exclusive vector meson ($V = \rho^0, \omega, J/\psi, ...$) and photon production, $l^{\pm} + N \longrightarrow l^{\pm} + V + Y$, where Y is either an elastically scattered nucleon or a low-mass state dissociative system. At low transverse momentum transfer at the nucleon vertex, the photoproduction of ρ^0, ω and ϕ mesons is characterized by a "soft" dependence of their cross-sections in the γp center-of-mass energy, W. This can be interpreted in the framework of Regge theory as due to the exchange of a "soft" Pomeron ($I\!P$) resulting in an energy dependence of the form $\mathrm{d}\sigma/\mathrm{dt} \propto \mathrm{W}^{4(\alpha_{\mathrm{P}}(\mathrm{t})-1)}$, where the Pomeron trajectory is parametrised as $\alpha_{I\!P}(t) = \alpha_{I\!P}(0) + \alpha' t \simeq 1.08 + 0.25t$. However, in the presence of a "hard" scale like large values of the photon virtuality Q^2or of the momentum transfer $|t|$ or of the vector meson mass, perturbative QCD (pQCD) is expected to apply. Diffractive vector meson production can then be seen in the nucleon rest frame as a sequence of tree subprocesses well separated in time: the fluctuation of the exchanged photon in a $q\bar{q}$ pair, the hard interaction of the $q\bar{q}$ pair with the nucleon via the exchange of (at least) two gluons in a color singlet state, and the $q\bar{q}$ pair recombination into a real vector meson. This approach results in a stronger rise of the cross section with W, which reflect the strong rise at small x of the gluon density in the nucleon. Such an energy dependence is observed in J/ψ production, where the quark charm mass provides a hard scale. It is of particular interest to study the role of other hard scales like Q^2 and t as well as the transition from a "soft" to "hard" behavior expected for light vector mesons. Furthermore, to take into account the skewing effect, i.e. the difference between the proton momentum fractions carried by the two exchanged gluons, one has to consider generalized parton distributions (GPDs). GPDs are an extension of standard PDFs, which include additional information on the correlations between partons and their transverse motion. There are four different types of GPDs, $H(x,x',t)$ and $E(x,x',t)$, where x and x' are the momentum fraction of the two parton considered, in the unpolarized case to which one should add $\tilde{H}(x,x',t)$ and $\tilde{E}(x,x',t)$ in the polarized case. While E and $\tilde{E}$ have no equivalent in the ordinary PDFs approach, H and $\tilde{H}$ reduce to the usual unpolarized and polarized PDFs respectively in the forward limit ($x = x'$ and $t = 0$).

The COMPASS experiment has presented [37] a study of the diffractive elastic lepto-production of ρ^0 mesons, $\mu + N \longrightarrow \mu + \rho^0 + N$, where N is a quasi-free nucleon from any of the nuclei of their polarized target, at $< W >= 10$ GeV for a wide range of Q^2,

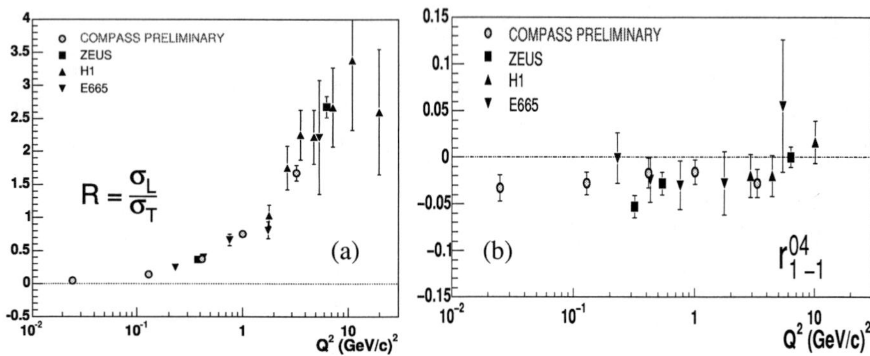

FIGURE 11. Q^2 dependence (a) of the ratio R between the longitudinal (σ_L) and the transverse (σ_T) cross sections and (b) of the r^{04}_{00} matrix element for elastic leptoproduction of ρ^0 as measured by COMPASS.

$0.01 < Q^2 < 10\,\text{GeV}^2$. Several spin density matrix elements (SDME), which carry information on the helicity structure of the production amplitudes, have been extracted from the production and decay ρ^0 angular distributions. The COMPASS data provide a large statistics which allows to extend the previous measurements towards low Q^2. Measurements of the r^{04}_{00} matrix element, which can be interpreted as the fraction of longitudinal ρ^0 in the sample, have been performed as a function of Q^2. If one assumes s-channel helicity conservation (SCHC) between the exchanged photon and the ρ^0 meson, one can obtain the ratio R between the longitudinal (σ_L) and the transverse (σ_T) cross sections (see Fig. 11a). A weak violation of SCHC is observed through the r^{04}_{1-1} matrix element (see Fig. 11b, in agreement with results of previous experiments. It has to be noted that the study of systematic effects is still ongoing and that only the statistical errors are provided.

Elastic electroproduction of ϕ mesons has been studied in $e^{\pm}p$ collisions by the ZEUS experiment [38] in the kinematic range $2 < Q^2 < 70$ GeV^2, $35 < W < 145$ GeV and $|t| < 0.6$ GeV^2. The energy dependence of the $\gamma^* p$ cross section has been measured and can be parametrised as $\sigma \propto W^{\delta}$, with $\delta \simeq 0.4$. This value is between the "soft" diffraction value and the one observed for J/ψ. No Q^2 or t dependence of the slope δ was observed with the present precision. When parametrised as a falling exponential, the t dependence of the cross section leads to b slopes in the range from 6.4 ± 0.4 GeV^{-2} at $Q^2 = 2.4$ GeV^2 to 5.1 ± 1.1 GeV^{-2} at $Q^2 = 19.7$ GeV^2. The values of δ and b were found to scale with respect to other vector mesons results when plotted as a function of $Q^2 + m_V^2$, where m_V is the mass of the vector meson, suggesting that this could be a good approximation of the universal scale in this process. The ratio between the longitudinal (σ_L) and the transverse (σ_T) cross sections, extracted from the ϕ angular distributions, was found to increase with Q^2 and when compared with results obtained for other vector mesons to scale with Q^2/m_V^2.

H1 has presented [39] comprehensive results on elastic J/ψ production in the $\gamma^* p$ center-of-mass energy ranges $40 < W < 305$ GeV in photoproduction and $40 < W < 160$ GeV in electroproduction up to 80 GeV^2 in Q^2 and in both cases for $|t| < 1.2$ GeV^2.

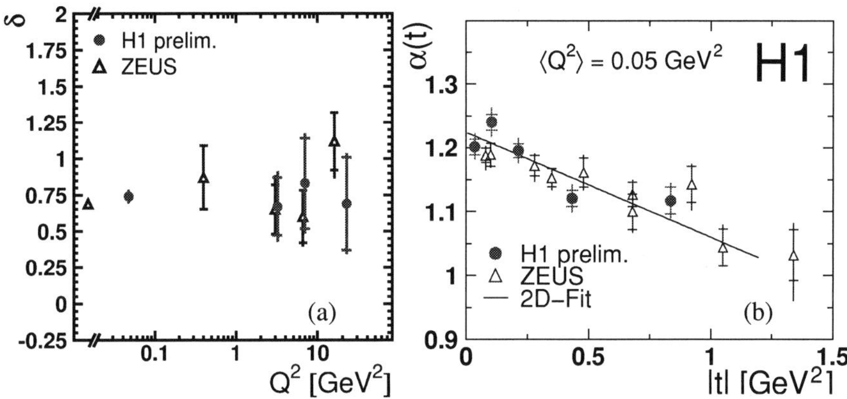

FIGURE 12. (a) The $\sigma \propto W^{\delta}$ fit parameter δ for J/ψ production as a function of Q^2. (b) The effective trajectory $\alpha_{I\!P}(t)$ as a function of t for J/ψ photoproduction.

In such a process, the hard scale provided by the mass of the involved charm quark ensures the validity of a pQCD description. This is even more so in electroproduction where Q^2 can provide a second hard scale. The Q^2 and W dependent $\gamma^* p$ cross-sections have been extracted. A steep rise with energy, $\sigma \propto W^{\delta}$, was observed with values of $\delta \simeq 0.7$ independently of Q^2 (see Fig. 12a). The effective Pomeron trajectories $\alpha_{I\!P}(t) = \alpha_{I\!P}(0) + \alpha' t$ have been extracted from the study of the doubly differential $\mathrm{d}\sigma/\mathrm{dt}$ cross-section as a function of W and t. In photoproduction (see Fig. 12b), a positive value of $\alpha' = 0.164 \pm 0.028 \pm 0.030\ \mathrm{GeV}^{-2}$ was obtained, leading to a shrinkage of the forward scattering peak, even if the effect is smaller than observed in hadron-hadron interactions. In electroproduction, within its large error, the obtained value of α' was found compatible both with the photoproduction result and zero. Finally, the helicity structure has been analyzed as a function of Q^2 and t and no evidence for a violation of SCHC has been observed. Assuming SCHC, the ratio of the longitudinal and the transverse cross sections has been extracted as a function of Q^2.

Teubner [40] has presented a model for vector meson production based on k_T factorization, which uses a parton-hadron duality ansatz to avoid the large uncertainties arising from the poorly known vector meson wave functions. The predictions obtained for J/ψ cross section as a function of W (see Fig. 13) with different sets of gluon distribution show a huge spread. This indicates a possible sensitivity to the gluon at small x and small to intermediate scales, i.e. a kinematic region where fits to the inclusive data do not constrain the gluon with high precision. Getting high precision data on vector meson production at HERA and reducing the remaining theoretical uncertainties might then allow to pin down the gluon at low x.

Kroll [41] presented a LO QCD calculation for light vector meson electroproduction taking into account the transverse momenta of the quark ant the anti-quark as well as Sudakov factors. The GPDs are modeled according to the ansatz of Radyushkin and Gaussian wavefunctions are used for the vector mesons. A fair agreement with the available data on ρ^0 and ϕ production at HERA is obtained between the predictions for

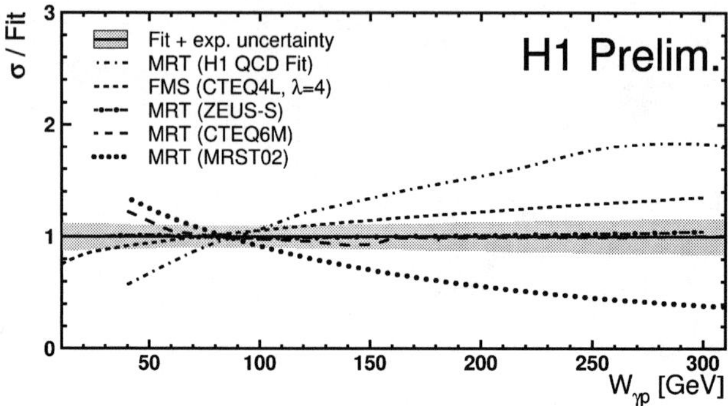

FIGURE 13. Ratio versus W of the MRT [40] theoretical predictions based on several gluon distributions to a parametrization of H1 J/ψ photoproduction preliminary data.

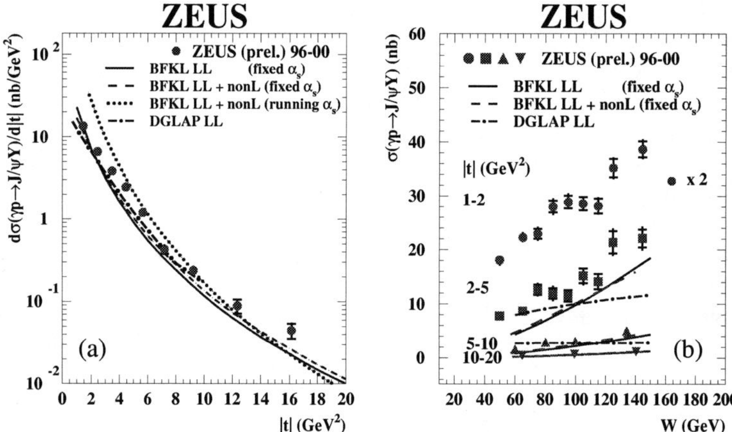

FIGURE 14. The proton-dissociative diffractive photoproduction of J/psi mesons cross sections for $|t| > 1$ GeV^2 (a) as a function of t and (b) as a function of W for four bins in t compared to predictions from based on BFKL and DGLAP.

the transverse and the longitudinal cross section as well as for the spin density matrix elements.

Photoproduction of vector mesons at large $|t|$ is largely studied since a few years as it is expected to be described by perturbative models involving the BFKL dynamics in the exchanged gluon ladder [42]. These models predict a power law behavior of the t dependence of the cross section and a rise with $|t|$ of the steepness of the W dependence.

H1 has presented [43] results on ρ^0 photoproduction in the kinematic range $75 < W < 95$ GeV and $1.5 < |t| < 10$ GeV^2 where the mass of the proton dissociative system Y is limited to $M_Y < 5$ GeV. The measured t dependence of the cross-section is well described by a power law of the form $|t|^{-n}$ with $n = 4.41 \pm 0.07^{+0.07}_{-0.10}$ and can

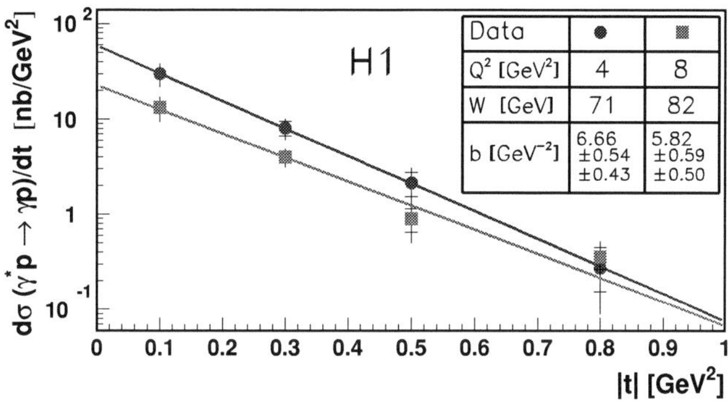

FIGURE 15. The DVCS cross section differential in t, for $Q^2 = 4$ GeV^2 at $W = 71$ GeV and for $Q^2 = 8$ GeV^2 at $W = 82$ GeV as measured by H1. The lines represent the results of fit of the form e^{-bt}.

be reproduced by BFKL model predictions. A study of the helicity structure has been performed and confirms the violation of SCHC in the case of ρ^0 photoproduction at large $|t|$ in contrast to what was observed for high $|t|$ J/ψ production [44, 45]. This is generally attributed to differences in the wave function between ρ and J/ψ.

J/ψ photoproduction at large $|t|$ has been studied by ZEUS [46] in the kinematic range $50 < W < 150$ GeV, $|t| > 1$ GeV^2 and $M_Y < 30$ GeV. Both the t dependence and the W dependence of the cross-section have been extracted, as shown on Fig. 14. Fits of the form W^δ to the W dependence of the cross section lead to values of $\delta \simeq 1$ with an indication for a rise of δ with $|t|$. The model based on BFKL has been found to describe the t dependence.

The opportunity to study Generalized Parton Distributions (GPDs) was discussed in a common session with the Spin Physics working group. Information about GPDs in lepton nucleon scattering can be provided by measurements of exclusive processes in which the nucleon remain intact. The simplest process sensitive to GPDs is Deeply Virtual Compton Scattering (DVCS), i.e. exclusive photon production off the proton $\gamma^* p \longrightarrow \gamma p$ at small $|t|$ but large Q^2, which is calculable in perturbative QCD. Such a final state also receives contributions from the purely electromagnetic Bethe-Heitler process, where the photon is radiated from the lepton. The resulting interference term in the cross section vanishes as long as one integrates over the azimuthal angle between the lepton and the hadron plane. It is then possible to extract the DVCS cross section by subtracting the Bethe-Heitler contribution, as done by H1 and ZEUS. The azimuthal asymmetries resulting from the interference are also sensitive to GPDs and are studied by HERMES, COMPASS and at JLAB. Extracting GPDs from the DVCS process would allow, through the Ji's sum rule, to determine the total angular momentum carried by the quarks which contribute to the proton spin.

A new high statistics analysis of DVCS has been performed by the H1 experiment [47] in the kinematic region $2 < Q^2 < 80$ GeV^2, $30 < W < 140$ GeV and $|t| < 1$ GeV^2. The $\gamma^* p \longrightarrow \gamma p$ cross section has been measured as a function of Q^2 and as a function of W. The W dependence can be parametrised as $\sigma \propto W^\delta$, yielding $\delta = 0.77 \pm 0.23 \pm 0.19$ at

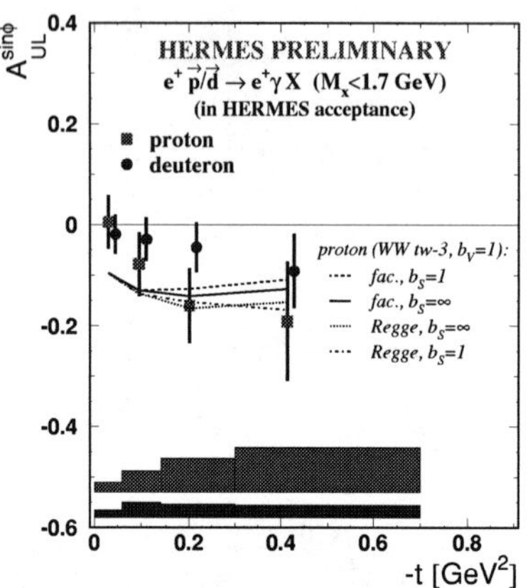

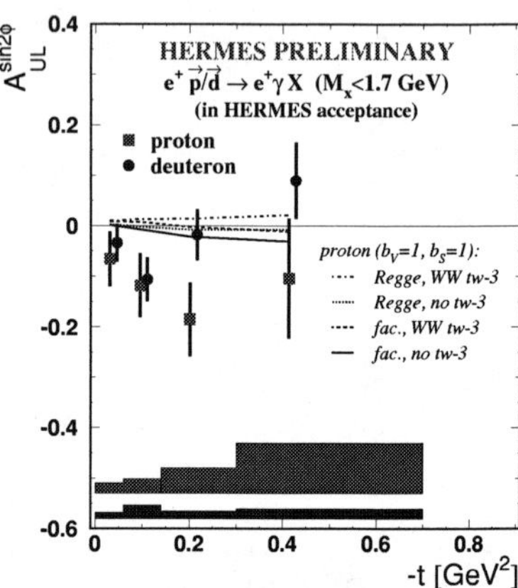

FIGURE 16. The $\sin\phi$ and $\sin 2\phi$ moments of longitudinal target-spin asymmetry on hydrogen and deuterium as function of t as measured by HERMES compared to predictions based on GPDs.

$Q^2 = 8$ GeV^2, i.e. a value similar to J/ψ production indicating the presence of a hard scattering process. For the first time, the DVCS cross section has been measured differentially in t (see Fig. 15) and the observed fast decrease with $|t|$ can be described by the form $e^{-b|t|}$ with $b = 6.02 \pm 0.35 \pm 0.39$ GeV^2 at $Q^2 = 8$ GeV^2. This measurement allows to further constrain the models, as their normalization depends directly on the t slope parameter. NLO QCD calculations using a GPD parametrization based on the ordinary parton distributions in the DGLAP region and where the skewedness is dynamically generated provide a good description of both the Q^2 and the W dependences.

A review of the HERMES results on DVCS [48] has been presented, including new data on polarized targets. On basis of unpolarized target data, one can measure the beam charge asymmetries, which are sensitive to the real part of the DVCS amplitudes, and the beam spin asymmetries, which are sensitive to the imaginary part. These are in fact mainly sensitive to the H GPD. Both asymmetries have been extracted and show the expected $\cos(\phi)$ and $\sin(\phi)$ behavior, respectively. A measurement of the t dependence of the beam charge asymmetry has been performed and comparison with models indicate the possible sensitivity of the data to constrain GPDs. Polarized target have been analyzed and the longitudinal target spin asymmetry, which is sensitive to the $\tilde{H}$ GPD, has been measured for the fist time. The resulting $\sin(\phi)$ and $\sin(2\phi)$ moments are shown as a function of t in Fig. 16, together with prediction based on GPD models. The sizeable $\sin(2\phi)$ moment might indicate a sensitivity to the twist-3 H and $\tilde{H}$ contributions. The installation of a new recoil detector will allow to tag directly the final state proton and to reduce the uncertainties due to the backgrounds arising from the missing mass techniques used up to now to guarantee exclusivity.

Gavalian [49] summarized the previous results on DVCS obtained by the CLAS experiment at JLAB, which measured in particular the beam spin asymmetry. In 2004 a dedicated DVCS experiment has been operated in Hall A and new results are expected soon. He also reviewed the status of the upgrade of the CLAS experiment which would allow to measure DVCS with the expected 12 GeV beam.

Exclusive meson production processes provide as well access to GPDs. HERMES has studied [50] exclusive π^+ production which is sensitive to the $\tilde{H}$ and the $\tilde{E}$ GPDs. The Q^2 dependent cross section has been measured and found to be in good agreement with a GPDs based model. A first measurements of the target spin asymmetry for ρ^0 production, which probes the E GPD, has been performed.

Weiss [51] reviewed the theoretical status of hard electroproduction of pions and kaons and their link to GPDs.

TOWARDS THE LHC

Diffractive physics has provided a rich source of important results from both HERA and the TEVATRON. Within the past few years there has been increasing interest to the study of diffractive processes at the LHC in connection with the proposal to add forward proton detectors to the LHC experiments. Various aspects of physics with forward proton tagging at the LHC have been under discussion in our working group.

Eggert [52] described the status of the TOTEM detector and the prospects of measurements of total and elastic pp - cross section. In particular, the total pp - cross section will be measured with the record (order 1 %) accuracy. This would allow to strongly restrict the range of existing theoretical models. Elastic cross section will be measured in the wide interval of momentum transfer $10^{-3} < -t < 8\,\mathrm{GeV}^2$.

The measurement of the elastic slope $b(t = 0)$ at the LHC is especially important, since it is expected (see for example [53, 54]) that this quantity is much more sensitive to the effects of the multi-Pomeron cuts than the total cross section.

Studies of diffractive physics at TOTEM require integration with CMS. CMS and TOTEM together will provide the largest acceptance detector ever built at a hadron collider. From the point of view of testing different regimes of the asymptotical behavior of the pp -scattering amplitude, it will be very informative to measure accurately the survival probabilities of one, two, three (maybe even four) rapidity gaps [55, 56, 57]. CMS/TOTEM physics menu will include also measurements of the centrally produced low mass systems (χ -bosons, dijets, diphotons). Special attention in his talk Eggert paid to the new ($\beta*$ =172m) optics aimed at optimization of diffractive proton detection at L= $10^{32}\mathrm{cm}^{-2}\mathrm{s}^{-1}$.

Several speakers (Albrow, Cox, Kowalski, Piotrzkowski and Royon) discussed the unique physics potential of forward proton tagging at 420m at the LHC. The use of forward proton detectors as a means to study Standard Model (SM) and New Physics at the LHC has only been fully appreciated within the last few years, (see e.g. [57, 58] and references therein). By detecting protons that have lost less than 2 % of their longitudinal momentum, a rich QCD, electroweak, Higgs and BSM program becomes accessible, with a potential to study phenomena which are unique at the LHC, and difficult even at a future linear collider [59, 60, 61, 62, 63].

It was emphasized by Albrow, Cox and Royon [64] that the so-called central exclusive production (CEP) process might provide a particularly clean environment to search for, and identify the nature of, new particles at the LHC. There is also a potentially rich, more

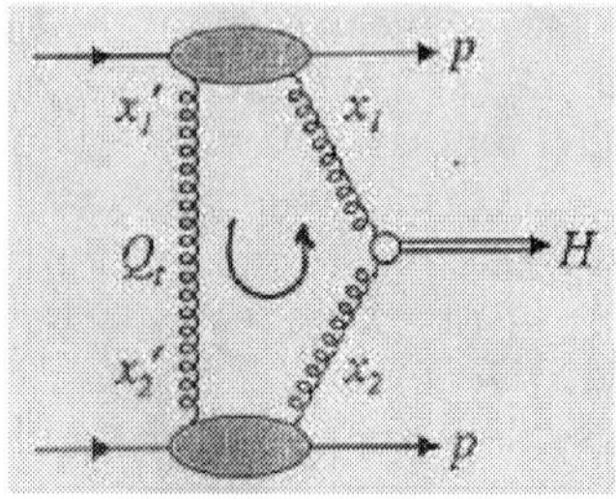

FIGURE 17. Schematic diagram for central exclusive Higgs production at the LHC, $pp \to p + H + p$.

exotic physics menu including (light) gluino and squark production, gluinonia, radions, and indeed any object which has 0^{++} or 2^{++} quantum numbers and couples strongly to gluons [57].

By central exclusive, we refer to the process $pp \to p \oplus \phi \oplus p$, where the symbol meaning has been described in the section "Diffraction at the TEVATRON". The process is attractive for two main reasons. Firstly, if the outgoing protons remain intact and scatter through small angles, then, to a very good approximation, the central system ϕ must be dominantly produced in a spin 0, CP even state, therefore allowing a clean determination of the quantum numbers of any observed resonance. Secondly, as a result of these quantum number selection rules, coupled with the (in principle) excellent mass resolution on the central system achievable if suitable proton detectors are installed, signal to background ratios greater than unity are predicted for SM Higgs production [65], and significantly larger for the lightest Higgs boson in certain regions of the MSSM parameter space [66]. Simply stated, the reason for these large signal to background ratios is that exclusive b quark production, the primary background in light Higgs searches, is heavily suppressed due to the quantum number selection rules. Another attractive feature is the ability to directly probe the CP structure of the Higgs sector by measuring azimuthal asymmetries in the tagged protons [67]. Another strategy to explore the manifestation of the explicit CP violation in the Higgs sector was recently studied by Ellis et al. [68].

The 'benchmark' CEP process for new physics searches is SM Higgs production, sketched in Fig. 17. The cross section prediction for the production of a 120 GeV Higgs at 14 TeV is 3 fb, falling to 1 fb at 200 GeV [27] [1] The simplest channel to observe the SM Higgs in the tagged proton approach from the experimental perspective is the WW decay channel [60, 69]. More challenging from a trigger perspective in the $b\bar{b}$ channel. This mode, however, becomes extremely important in the so-called 'intense coupling regime' of the MSSM, where the CEP is likely to be the discovery channel. In this case it is expected close to 10^3 exclusively produced double-tagged Higgs bosons in 30fb^{-1} of delivered luminosity. About 100 would survive the experimental cuts, with a signal-to-background ratio of order 10.

---

[1] for the discussion of the uncertainties in this calculation, see [66].

Furthermore, as was reported in [32, 33, 63], forward proton tagging will make possible a unique program of high-energy photon interactions physics at the LHC. For example, the two-photon production of W pairs will allow a high precision study of the quartic gauge couplings [63]. Photon interactions are enhanced in heavy ion collisions and studies of such ultra-peripheral collisions were discussed in [32, 33]. In addition, two-photon exclusive production of lepton pairs provides an excellent tool for calibrating both luminosity and the energy scale of the tagged events [63].

Finally, by tagging both outgoing protons, the LHC is effectively turned into a glue-glue collider. This will open up a rich, high rate QCD physics menu (especially in what concerns diffractive phenomena), allowing to study the skewed unintegrated gluon densities and the details of rapidity gap survival [62]. Note that the CEP provides a source of practically pure gluon jets (gluon factory [72]). This can be an ideal laboratory to study the properties of gluon jets, especially in comparison with the quark jets, and even the way to search for the glueballs.

Cox and Kowalski [60, 62] discussed the outline of the FP420 R&D project aimed at assessing whether it is possible to install forward proton detectors with appropriate acceptance at ATLAS and/or CMS, and to fully integrate such detectors within the experimental trigger frameworks [61].

REFERENCES

1. P.D.B. Collins, *An Introduction to Regge Theory and High Energy Physics,* Cambridge University Press, Cambridge, 1977.
2. ZEUS Coll., S. Chekanov et al, Eur. Phys. J C 38 (2004) 43.
3. K. Goulianos, arXiv:hep-ph/0407035.
4. K. Golec Biernat, J. Kwiecinski and A. Szczurek, Phys. Rev. D 56 (1997) 3995.
5. paper 980 submitted to ICHEP 2002; paper 981 submitted to ICHEP 2002; paper 5-090 submitted to EPS 2003.
6. DESY-05-011, to be published in Nucl. Phys. B.
7. ZEUS Coll., S. Chekanov et al, Eur. Phys. J C 25 (2002) 169.
8. paper 984 submitted to ICHEP 2002.
9. P. Laycock, these Proceedings.
10. A. Levy, these Proceedings.
11. J.C. Collins. Phys. Rev. D 557 (1998) 3051; [erratum-ibid. Phys. Rev. D 61 (2000) 019902].
12. V.N. Gribov and L.N. Lipatov., Sov. J. Nucl. Phys. 15 (1972) 438; Yu.L. Dokshitzer, Sov. Phys. JETP 46 (1977) 641; G. Altarelli and G. Parisi, Nucl. Phys. B 126 (1977) 298.
13. A. Martin, these Proceedings and references therein.
14. M. Beckingham, these Proceedings; paper 6-178 submitted to ICHEP 2004.
15. M. Mozer, these Proceedings; paper 6-177 submitted to ICHEP 2004.
16. P. Laycock, these Proceedings; paper 6-821 submitted to ICHEP 2004.
17. L. Adamczyk, these Proceedings; paper 6-0229 submitted to ICHEP 2004.
18. CDF Coll., T. Affolder et al., Phys. Rev. Lett. 84 (2000) 5043.
19. A.B. Kaidalov et al., Eur. Phys. J. C 21 (2001) 521.
20. A.B. Kaidalov et al., Phys. Lett. B 559 (2003) 235.
21. A.B. Kaidalov et al., Phys. Lett. B 567 (2003) 61.
22. M. Klasen and G. Kramer, DESY-04-011, arXiv:hep-ph/0401202; M. Klasen and G. Kramer, Eur. Phys. J. C 38 (2004) 93.
23. R. Renner, these Proceedings; paper 6-0259 submitted to ICHEP 2004.
24. ZEUS Coll., S. Chekanov et al., Eur. Phys. J. C 23 (2002) 615.
25. M. Klasen, these Proceedings; arXiv:hep-ph/0506121.
26. D. Bodeker et al., Z. Phys. C 63 (1994) 471.

27. V. A. Khoze, A. D. Martin and M. G. Ryskin, Eur. Phys. J. C 14 (2000) 525.
28. K. Goulianos, these Proceedings.
29. A.B. Kaidalov et al., arXiv:hep-ph/0409258.
30. V.A.Khoze et al., arXiv:hep-ph/0507040.
31. W. Guryn, these Proceedings.
32. S. White, these Proceedings.
33. S.Klein, these Proceedings.
34. S. Brodsky, these Proceedings.
35. E. Levin, these Proceedings.
36. J. R. Andersen, these Proceedings.
37. N. d'Hose, these Proceedings.
38. ZEUS Coll., S. Chekanov et al., Nucl. Phys. B 718 (2005) 3.
39. C. Kiesling, these Proceedings.
40. T. Teubner, these Proceedings.
41. P. Kroll, these Proceedings.
42. R. Enberg, J. R. Forshaw, L. Motyka and G. Poludniowski, JHEP 0309 (2003) 008
G. G. Poludniowski, R. Enberg, J. R. Forshaw and L. Motyka, JHEP 0312 (2003) 002 [arXiv:hep-ph/0311017].
43. C. Gwilliam, these Proceedings.
44. H1 Coll., A. Aktas et al., Phys. Lett. B 568 (2003) 205.
45. ZEUS Coll., S. Chekanov et al., Eur. Phys. J. C 26 (2003) 389.
46. D. Szuba, these Proceedings.
47. S. Glazov, these Proceedings.
H1 Coll., A. Aktas et al., arXiv:hep-ex/0505061.
48. M. Koputin, these Proceedings.
49. G. Gavalian, these Proceedings.
50. A. Vandenbroucke, these Proceedings.
51. C. Weiss, these Proceedings.
52. K. Eggert, these Proceedings.
53. V. A. Khoze, A. D. Martin and M. G. Ryskin, Nucl. Phys. B (Proc. Suppl.) 99 (2001) 213.
54. V. A. Khoze, A. D. Martin and M. G. Ryskin, Eur. Phys. J. C 18 (2000) 167.
55. K. Goulianos, Phys. Lett. B 358 (1995) 379.
56. K. Goulianos, these Proceedings.
57. V. A. Khoze, A. D. Martin and M. G. Ryskin, Eur. Phys. J. C 23 (2002) 311;
M.G. Ryskin, A. D. Martin and V.A. Khoze, arXiv:hep-ph/0506272;
58. B. Cox, arXiv:hep-ph/0409144.
59. M. Albrow, these Proceedings.
60. B. Cox, these Proceedings.
61. FP-420 :M. Albrow et al., CERN-LHCC-2005-025, LHCC-I-015.
62. H. Kowalski, these Proceedings.
63. K. Piotrzkowski, these Proceedings.
64. C. Royon, these Proceedings.
65. A. De Roeck, V. A. Khoze, A. D. Martin, R. Orava and M. G. Ryskin, Eur. Phys. J. C 25 (2002) 391.
66. A. B. Kaidalov, V. A. Khoze, A. D. Martin and M. G. Ryskin, Eur. Phys. J. C 33 (2004) 261.
67. V. A. Khoze, A. D. Martin and M. G. Ryskin, Eur. Phys. J. C 34 (2004) 327.
68. J. R. Ellis, J.S.Lee and A. Pilaftsis, Phys.Rev. **D71** (2005) 075007.
69. B. E. Cox et al., arXiv:hep-ph/0505220.
70. V. A. Khoze, A. D. Martin and M.G.Ryskin, Phys. Lett. B401 (1997) 30.
71. V. A. Khoze, A. D. Martin and M. G. Ryskin, arXiv:hep-ph/0006005, In Proc. of 8th Int. Workshop on Deep Inelastic Scattering and QCD (DIS2000), Liverpool, eds. J.Gracey and T.Greenshaw (World Scientific, 2001), p.592.
72. V. A. Khoze, A. D. Martin and M. G. Ryskin, Eur. Phys. J. C 19 (2001) 477; Erratum: ibid. C 20 (2001) 599.
73. A. B. Kaidalov, V. A. Khoze, A. D. Martin and M.G.Ryskin, Eur. Phys. J. C 31 (2003) 387.

Summary of the Hadronic Final State Working Group

Claudia Glasman*, Stephen Maxfield† and Pavel Nadolsky**

*Universidad Autónoma de Madrid, Spain
†University of Liverpool, U.K.
**Argonne National Laboratory, U.S.A.

Abstract. Talks given during the Hadronic Final State sessions of the DIS 2005 workshop are summarized. Recent experimental studies on many topics ranging from perturbative and non-perturbative QCD effects in high-energy hadroproduction to hadron spectroscopy and high-energy nuclear matter were presented. Considerable progress achieved in higher-order QCD calculations, development of parton showers, resummations, and other theoretical areas was also reviewed.

Keywords: hadronic final states
PACS: 12.38.-t,13.85.-t,13.87.-a

INTRODUCTION

The study of hadronic final states at particle colliders is one of the most active areas of research, as demonstrated by 38 talks presented during the sessions of this working group. Tests of perturbative QCD (pQCD), studies of non-perturbative QCD models, fragmentation effects, baryon and meson spectroscopy, search for exotic particles (e.g. pentaquarks), study of the photon structure and high-energy nuclear matter are some of the topics discussed during the workshop. The results presented constitute significant progress since the last conference in this series (DIS2004, Slovakia).

JET PHYSICS

Jet physics provides a testing ground for pQCD as well as the possibility of determining the strong coupling constant α_s and QCD colour factors. High-precision measurements of jet cross sections at HERA and Tevatron help constrain nonperturbative QCD inputs (e.g., gluon densities in the proton) and reduce uncertainties in signal and background cross sections in searches for Higgs bosons and other new particles at future accelerators, like LHC. Understanding of jet algorithms, multiparticle interactions and underlying events is a key issue on the experimental side. New experimental results comprise:

- Inclusive jet cross sections in neutral current deep inelastic scattering (ZEUS, J. Standage): measurements in a new kinematic regime show good agreement with next-to-leading-order (NLO) QCD calculations. The calculations present small uncertainties in the region of phase space spanned by the measurements.
- Multijet production in neutral current deep inelastic scattering (H1, T. Kluge): three-jet cross sections in deep inelastic scattering (DIS) are directly sensitive to

CP792, *Deep Inelastic Scattering, DIS 2005,* edited by W. H. Smith and S. R. Dasu

QCD radiation and the gluon density in the proton. Measurements of dijet and three-jet cross sections were presented at high Q^2. In this kinematic regime, the $\mathcal{O}(\alpha_s^3)$ corrections are suppressed by the small QCD coupling strength α_s. The calculations from NLOJET++ give a good description of the data. The ratio of the three-jet to the dijet cross sections as a function of Q^2 was used to make a determination of α_s with accuracy competitive with other more inclusive measurements.

- Color dynamics in photoproduction of jets (ZEUS, J. Terrón): three-jet production in direct photoproduction provides a means of testing the underlying gauge symmetry group of the strong interactions. The observables studied show a sensitivity to the color components. The data are consistent with the mixture of different color configurations predicted by SU(3) color symmetry; the contribution from diagrams with a triple-gluon vertex is $\sim 42\%$ and, thus, these measurements have a potential for extraction of color factors.
- Precision measurements of α_s at HERA (C. Glasman): an average value of $\alpha_s(M_Z)$ from several measurements by the H1 and ZEUS Collaborations was presented. The experimental uncertainty in the combined value is below 1%, whereas the theoretical uncertainty is $\sim 4\%$. Thus, further theoretical work is needed to obtain more precise determinations of α_s from HERA data. The combination of the measurements of α_s at different energies allows one to demonstrate the running of the QCD coupling by using the HERA jet data alone.
- Latest jet results from the DØ Collaboration (B. Davies): new cross sections for high transverse-energy jets in Run II show agreement with NLO QCD calculations within the uncertainties. The experimental uncertainty arising from the absolute energy scale of the jets ($\sim 5\%$) dominates. The azimuthal decorrelation between jets in multijet events also shows good agreement with the theoretical expectations. First measurements of tagged jet cross sections were also presented.
- Jet physics in Run II at CDF (R. Field): jet cross sections in Run II from CDF using the k_T cluster algorithm were presented. Good agreement between data and theory is observed. Measurements of b-tagged jet cross sections also show a good agreement with QCD calculations. Detailed studies of the effects of the "underlying event" were presented. Aside from Tevatron applications and constraints on parton densities, these studies may help understand the "underlying event" at the LHC, which may be quite different from that at the Tevatron as a consequence of the absence of KNO scaling. Hadronization models in the present Monte-Carlo showering programs may be inadequate ("still sort of plug and pray", to use S. Frixione's expression) at the LHC. Consequently substantial re-tuning of Monte-Carlo generators in the first few years at LHC may be needed. Alternatively, data from low energies at RHIC ($\sqrt{s} = 200$ GeV) may be possibly utilized to determine the energy dependence of the underlying event. The relevant data is already becoming available from the RHIC measurements (see J. Jia's talk).

PERTURBATIVE QCD, MONTE-CARLO GENERATORS, AND RESUMMATIONS

Successful theory predictions are obtained in quantum chromodynamics by combining systematic calculations for hard-scattering processes with all-order resummation of semi-hard hadronic activity and reliable modeling of nonperturbative contributions. Each of these three aspects of perturbative QCD theory was amply presented during this workshop.

The status of calculations for hard-scattering processes was reviewed by M. Klasen. Significant advances are being made in computations of two- and three-loop QCD matrix elements, improving accuracy of theory predictions up to next-to-next-to-leading-order (NNLO). The latest NNLO results of relevance to hadroproduction include complete expressions for 3-loop DIS structure functions and anomalous dimensions, qq helicity amplitudes for $pp \to 2$ jets, partial 2-loop expression for the average thrust in e^+e^- hadroproduction, fully differential distributions for Higgs + 2 jets, and 2-loop vertex corrections in e^+e^- heavy-quark production. Two-loop cross sections for semi-inclusive DIS hadroproduction at high p_T have been computed by several groups (see below). The soft and collinear limits of NNLO hard cross sections, where extra care is needed in dealing with kinematical singularities, are also being studied.

The progress in perturbative calculations is based, in part, on the development of efficient methods for evaluation of complex Feynman integrals: color and helicity decomposition, recursive relations suggested by graph topology and unitarity, integration by parts, sector decomposition, etc. An intriguing new connection has been established between the structures of massless multi-leg QCD diagrams and string representations in twistor space, which opens the door for more efficient calculations at tree and one-loop level. Besides our working group session, the technical aspects of QCD calculations and connections to the twistor formalism were covered in the plenary talk by L. Dixon.

With the exception of the simplest one-scale observables [such as the ratio $R = \sigma(e^+e^- \to \text{hadrons})/\sigma(e^+e^- \to \mu^+\mu^-)$], the hard cross sections cannot be immediately accessed in experiments without also including effects of multiple parton radiation and long-distance nonperturbative dynamics. Resummation of large logarithms from emissions of partons in a wide span of energies can be realized by solving evolution equations (such as the DGLAP equations for parton densities and fragmentation functions), or numerically by Monte-Carlo showering programs. A presentation by S. Frixione (complementary to the plenary talk by P. Skands) described the latest achievements in the development of next-to-leading order (NLO) Monte-Carlo generators. The Monte-Carlo programs for leading-logarithm resummation are ubiquitously applied in experimental analyses because of their capability to simulate exclusive final states and flexibility. Until recently, the Monte-Carlo generators included only the leading-order hard-scattering cross sections and did not properly describe high-p_T and multi-jet configurations, or the total rate. On the other hand, fixed-order calculations properly account for hard emissions, estimate reliably the total rates, and can be systematically improved to reduce dependence on arbitrary factorization scales. The powerful features of the higher-order and Monte-Carlo computations are complementary, and significant effort is being dedicated to the combination of the two approaches in one calculation. The talk compared two promising methods for such a merging (matrix element corrections and NLO with parton showers) and presented examples of successful applications in each hybrid approach.

In contrast to the Monte-Carlo methods, analytical resummation often preserves desirable features of the finite-order calculations, including systematical factorization of short- and long-distance scattering, order-by-order expansion in series of α_s, controlled total rates and scale dependence. Analogously to the finite-order calculations, resummation is applicable to limited classes of sufficiently inclusive observables, so that distinct types of resummation must be applied under different circumstances. Resummation issues were discussed in the following contributed talks:

- Soft-gluon expansions through NNNLO (N. Kidonakis): large logarithms enhance QCD cross sections when the scattering energy is barely above the threshold in the considered particle production channel. The leading threshold effects are process-independent, so that a master formula can be devised to describe such effects in a variety of reactions. The master formula has now been derived up to NNNLO, and its application to charged Higgs boson production was described.
- Transverse momentum resummation at small x (F. Olness): logarithms arising at small transverse momenta in Drell-Yan-like processes and semi-inclusive DIS are described by a generalized factorization formalism (Collins-Soper-Sterman resummation). Higher-order corrections will modify this formalism at small momentum fractions x, where Catani-Ciafaloni-Fiorani-Marchesini (CCFM) or Balitsky-Fadin-Kuraev-Lipatov (BFKL) small-x resummations are more appropriate. The talk pointed out that small-x modifications of logarithmic or other nature may have already been observed in q_T resummation for SIDIS at HERA, which would also imply observable modifications in q_T distributions in the comparable x range at the Tevatron and LHC. Possible effects on electroweak boson production at hadron-hadron colliders suggested by the HERA SIDIS data and opportunities to experimentally measure them in the Tevatron Run-2 were discussed.
- The high energy limit of QCD as described by the NNL BFKL equation (J. Andersen): an iterative method for solving the small-x BFKL resummation equation at full NNL accuracy directly in the space of physical rapidity and transverse momentum was described. The method has the advantage of being unaffected by instabilities arising when the NNL BFKL equation is solved in an approximate analytical form, and it gives stable predictions for the gluon Green function and effective intercept at large rapidity gaps. The instabilities in the analytical solution can be traced to the wrong choice of NNL kernel eigenfunctions and remedied by selecting an NNL eigenbasis with correct renormalization scale dependence.
- Gaps between jets in the high energy limit (A. Kyrieleis): dependence of the cross section for production of two jets on the jet separation in rapidity was considered. A combination of leading-log-Q_0 and BFKL resummations, applicable at small and large rapidity gaps respectively,was discussed. Several all-order schemes for interpolation between the leading-log-Q_0 and BFKL resummed cross sections in the whole rapidity range without double counting were worked out, and differences between the interpolation schemes were analyzed. This work contributes to the development of a unified formalism describing the "jet-gap-jet" process at all

rapidity separations, with resummations included in the appropriate kinematical limits.

SMALL X AND FORWARD PARTICLE PRODUCTION

Structure function data from HERA and fixed-target experiments have allowed a good determination of the proton PDFs over a large region in phase space. The evolution of the structure function data with Q^2 is successfully described by the DGLAP equations. Measurements of jet production have provided accurate tests of pQCD. At high Q^2 or $E_T^{\rm jet}$, where $E_T^{\rm jet}$ is the jet transverse energy, NLO calculations using the DGLAP evolution equations have been found to give a good description of the jet data. However, in production of forward, close to the proton direction, jets with $E_T^{\rm jet} \sim Q$ at low x, discrepancies with NLO DGLAP calculations may have been observed. In some cases, the NLO DGLAP cross section agrees with the data only after a large renormalization scale uncertainty (up to 100%) is taken into account. This indicates importance of higher-order corrections at low x and Q^2, possibly arising from small-x effects associated with CCFM or BFKL dynamics. An improved description of the data can alternatively be obtained by assigning a partonic structure to the exchanged virtual photon. New results on forward-jet production from H1 and ZEUS, as well as new theoretical calculations have been presented in the course of this workshop. The contributed talks include:

- Measurement of forward jet production at low x in DIS (H1, A. Knutsson): forward jet cross sections were measured and compared with the predictions of perturbative QCD of $\mathcal{O}(\alpha_s)$ and $\mathcal{O}(\alpha_s^2)$. The predictions using DISENT at NLO give a reasonable description of the data at high $x_{\rm Bj}$, but fail at low values. For dijet production, the predictions of NLOJET++ describe the data within the experimental and theoretical uncertainties. The data were also compared to the predictions of several leading-logarithm parton-shower models. Those models include a color-dipole treatment of the parton shower, which mimics the BFKL regime, or a resolved component for the virtual photon, which adds a second DGLAP evolution ladder on the photon side. They give a good description of the data.
- Forward jet production in DIS (ZEUS, N. Vlasov): inclusive forward jet measurements were presented and compared with theory predictions in different pseudorapidity regions. The leading-logarithm color-dipole parton shower ARIADNE gives the best overall description of the cross sections, while LO DGLAP parton showers fail to describe the forward data. NLO QCD predictions by DISENT give a good description of the measured transverse energy dependence of the jet cross section. The data are well described by DISENT at low pseudorapidity, but some discrepancies are seen in the most forward region, where a large renormalization scale uncertainty is also present. Discrepancies are also observed with the showering program CASCADE in the CCFM k_T-factorization approach, suggesting that adjustments in parameters of unintegrated parton distributions are necessary.
- QCD corrections to the electroproduction of hadrons with high p_T (R. Sassot); DIS production of inclusive hadrons with large p_T at NLO (B. Kniehl): NLO, or $\mathcal{O}(\alpha_s^2)$, cross sections for hadron electroproduction at large p_T were independently

computed by two groups (A. Daleo, D. de Florian, and R. Sassot; B. A. Kniehl, G. Kramer, and M. Maniatis). The NLO calculation describes well most of the HERA phase space, in contrast to the LO predictions, which typically fall below the HERA data. The NLO/LO K-factor is moderate at higher x, p_T and Q^2, but grows up to 10-25 toward the lowest reachable x and Q^2, including the forward production region. A new hadroproduction channel $\gamma^* g \to q\bar{q}g$ opens up at NLO and contributes a rate several times higher in forward π^0 production compared to the other channels (Sassot). Large renormalization scale uncertainty suggests that the NNLO analysis will ultimately be required for reliable interpretation of the forward production data (Kniehl). Reduction of other uncertainties, *e.g.* arising from substantial dependence on the choice of fragmentation functions, may also be needed.

- Small x effects in forward jet production at HERA (C. Marquet): an alternative interpretation of the forward jet data can be made in the BFKL framework, designed to resum the $\ln(1/x)$ logarithms from multiple $g \to g$ splittings as $x \to 0$. Such an analysis was performed in the leading-logarithm BFKL approximation, and the possibility of saturation effects at HERA energies was also considered. Both small and large saturation energy scales were found to be compatible with the forward jet data, suggesting that these data alone cannot distinguish between the weak saturation (essentially a pure BFKL regime) and strong saturation (with the BFKL cross section growth tamed by unitarity). Consequently, more data has to be included in the fit to help discriminate between the different saturation models.

As can be seen from the review of the talks, interpretation of the forward production data remains unsettled, with different (sometimes complementary) explanations presented by many authors. Further experimental and theoretical studies are clearly needed to work out a consistent point of view on the nature of QCD radiative corrections in the HERA forward region, especially given that the relevant region of x will be routinely probed at the LHC.

PARTICLE PRODUCTION

Studies of particle production have become increasingly detailed and flavour specific, providing stringent tests of QCD and our understanding of the hadronisation process. New and improved measurements of azimuthal asymmetries, polarisation in the $\Lambda(\bar{\Lambda})$ system and of Bose-Einstein correlations in kaon production were shown. A puzzling discrepancy between two determinations of electric and magnetic dipole moments of protons may have an explanation, and heavy particle production at the LHC may be modified by onset of the “black body” limit:

- Predictions for azimuthal asymmetries (T. Tymieniecka): the energy flow method for studying azimuthal asymmetries was presented, with the goal of measuring contributions from subleading angular functions in semi-inclusive deep inelastic scattering. This method is more reliable than multiplicity methods, as it is not sensitive to details of fragmentation in the final state.

- Azimuthal asymmetry using energy flow method (ZEUS, A. Ukleja): measurements of azimuthal asymmetries of hadronic energy flow, charged and neutral hadrons in neutral current deep inelastic scattering were presented. Next-to-leading-order QCD calculations show a reasonable description of the asymmetry as a function of the pseudorapidity in the hadronic center-of-mass frame.
- Polarization and asymmetries in neutral strange particle production (ZEUS, A. Cottrell): new measurements of $\Lambda(\bar{\Lambda})$ polarization were presented. The transverse and longitudinal polarizations are consistent with zero. Results on baryon asymmetry and the baryon-meson ratio were also presented. No significant asymmetry was observed. The data give a higher baryon to meson ratio at low p_T than the predictions. This ratio rises steeply as x decreases.
- New possible insight into JLab proton polarization data puzzle by DIS (A. Dubnickova): the talk discussed a disagreement between the ratios of the electric and magnetic form factors determined in polarized ep scattering at JLab and by Rosenbluth technique. The disagreement between the two measurements is explained by different behaviour of the electric form factor, which is known with less certainty than the magnetic form factor as a result of its kinematical suppression in the Rosenbluth method. A new sum rule was proposed to independently constrain the electric form factor from measurement of the difference of proton and neutron differential distributions $d\sigma/dQ^2$ (with x integrated out) in DIS.
- Neutral and charged kaon Bose-Einstein correlations in DIS (ZEUS, A. Galas): results on Bose-Einstein correlations for charged kaons show an r value similar to pions, but a somewhat smaller λ value. The results for r are in good agreement with the results from LEP, but the λ value is smaller. For neutral kaons, the value of r is in good agreement with that for charged kaons and similar to that of pions, and is also in good agreement with the measurements of LEP. The value of λ extracted for neutral kaons is rather large.
- Black body limit in central pp/pA collisions at the LHC (C. Weiss): massive electroweak bosons and other heavy particles will be frequently produced at the LHC in central proton-proton or proton-nucleus collisions. Production of very forward hadrons in such collisions may saturate the "black-body" (unitarity) limit, which will modify properties of the hadronic final state at very large rapidities. Signatures of the "black-body" limit were analyzed in the dipole approximation for DGLAP evolution, and kinematical regions where the "black-body" limit may be reached were identified based on the HERA data.

NON-PERTURBATIVE EFFECTS

Event-shape variables are particularly sensitive to details of the non-perturbative effects of hadronization and can be used to test the models for these effects. Recently, new developments with regard to power-law corrections have prompted a revived interest in the understanding of hadronization from first principles. In this type of analysis, the data are compared to model predictions which combine NLO calculations and theoretical expectations for the power corrections, characterized in this case by an effective coupling

$\bar{\alpha}_0$. Previous results supported the concept of power corrections in the approach of Dokshitzer et al., but a large spread of the results suggested that higher-order corrections were needed. Now, resummed next-to-leading-logarithm calculations matched to NLO are available, and so it is possible to study event-shape distributions instead of only their mean values. New results from ZEUS were presented:

- Event shapes at HERA (ZEUS, A. Everett): measurements of event-shape variables in neutral current deep inelastic scattering were presented both as mean values and differential distributions. The results were used to test the model of power corrections for the hadronization process. This model uses only one universal non-perturbative parameter, $\bar{\alpha}_0$, in addition to α_s to describe fragmentation. For the differential distributions, the NLO calculations compared to the data were complemented by resummed calculations. This led to a good description of the data and extraction of values for α_s and $\bar{\alpha}_0$, which are consistent with measurements in other experiments. New variables based on event shapes with jets were also discussed.

FRAGMENTATION EFFECTS

Studies of the fragmentation process through measurement of particle multiplicities have been expanded at HERA. Tests of the universality of the fragmentation process were extended to higher energy scales, and HERMES presented a detailed and careful analysis of flavour-separated multiplicities at lower energies.

- Recent results on multiplicity from ZEUS (M. Rosin): charged particle multiplicities were measured in the current regions of Breit and hadronic center-of-mass frames at a higher energy scale than in previous *ep* measurements. The results were compared to hadron multiplicities in e^+e^- and *pp* collisions. Similarity between the energy dependences of multiplicities in *ep*, *ee*, and *ep* scattering was found, if the energy scale in *ep* collisions was equated to twice the energy in the current region of the Breit frame.
- The fragmentation process at HERMES (B. Maiheu): the fragmentation function was studied at HERMES, with the focus on flavor separation of multiplicities in semi-inclusive deep inelastic scattering. Cross sections as a function of Q^2 and x_{Bj} were presented. The results obtained are model-independent, and much effort was invested in a proper treatment of uncertainties. No signs of strong factorization breaking were found.

PENTAQUARKS

The experimental status of pentaquarks remains ambiguous, with observation and non-observation of the various states being reported in roughly equal numbers of experiments. Careful comparison of superficially contradictory results can sometimes show that the differences are not really significant (see talks from H1 and ZEUS on the strange pentaquark), and the new results from CLAS show that apparently "solid" observations can sometimes evaporate in the light of higher statistics. The results presented are:

- Search for pentaquarks at HERMES (A. Airapetian): HERMES has searched for Θ^+ and Ξ^{--} states in the decay channels of $\Theta^+ \to K_s^0 p$ and $\Xi^{--} \to \Xi^- \pi^- \to \Lambda^0 \pi^- \pi^-$. A signal with a significance of 3.7σ is observed at $M = 1528 \pm 2.6$ MeV in the $\pi^+\pi^- p$ mass spectrum. No signal for $\Theta^{++} \to pK^+$ is observed, thus Θ^+ cannot be an isotensor state. No signal is observed for a Ξ^{--} state; a limit on the production cross section of 2.1 nb was set at 90% CL.
- Pentaquarks at JLab: new results from CLAS (M. Battaglieri): CLAS has searched the Θ^+ state in high-statistics dedicated runs in exclusive photoproduction on protons. The process studied is $\gamma p \to \Theta^+ \bar{K}^0 \to (nK^+)(K_s \to \pi^+\pi^-)$. Preliminary results with only 1% of the collected data were presented. The observed nK^+ spectrum is smooth; no signal is observed at masses of ~ 1540 MeV. An upper limit of $1-4$ nb at 95% CL was derived for the production cross section of the process $\gamma p \to \Theta^+ \bar{K}^0$, in contrast to the SAPHIR results.
- Analysis of the anti-charmed baryon state at H1 (K. Daum): a report on the observation and analysis of the $D^* p(3100)$ state by the H1 collaboration was made. The corrected fraction of $D^* p(3100)$ to D^* was measured to be $1.59 \pm 0.33^{+0.33}_{-0.45}\%$, whereas the ratio of the production cross section is $\sigma(D^* p(3100))/\sigma(D^*) = 2.48 \pm 0.52^{+0.85}_{-0.64}\%$. The kinematics of the $D^* p(3100)$ state were studied: D^* mesons from $D^* p(3100)$ decay are significantly softer than normal D^*s, and $D^* p(3100)$ production at central pseudorapidities $\eta_{\rm LAB}$ is suppressed. A simple fragmentation approach with isotropic decay for the $D^* p(3100)$ state describes the W and Q^2 spectra, but does not describe the properties of the D^* mesons produced in the $D^* p(3100)$ decay.
- Searches for pentaquarks at BaBar (E. Eckhart): the Θ^+, Ξ^{--}, Ξ^0, and Θ^{*++} states have been searched for by BaBar. No enhancement was observed in electroproduction, hadroproduction, or e^+e^- $K_s p$ mass spectrum at masses of about 1540 MeV corresponding to the Θ^+ state. Also, no enhancement in the $\Xi^-\pi^+$ or $\Xi^-\pi^-$ spectrum was observed at masses corresponding to the $\Xi(1860)$ state. No evidence was found for a Θ^{*++} signal in the pK^+ decay mode.
- The experimental search for charm pentaquarks at ZEUS (Y. Eisenberg): ZEUS has searched for decays of charmed pentaquarks into $D^* p$ using 127 pb^{-1} of integrated luminosity. The D^* mesons were searched for in the $D^{*+} \to D^0 \pi_s^+ \to (K^-\pi^+)\pi_s^+ + (c.c.)$ and $D^{*+} \to D^0 \pi_s^+ \to (K^-\pi^+\pi^+\pi^-)\pi_s^+ + (c.c.)$ decay channels. No evidence for a signal at 3.1 GeV, as reported by H1, was observed in the $D^* p$ spectra studied, even when using a very similar selection to that of H1. An upper limit on the fraction of D^* mesons originating from Θ_c^0 decays was set to 0.37% (full sample) and 0.51% (a sample from deep inelastic scattering at $Q^2 > 1$ GeV^2) at 95% CL.
- Strange pentaquark search with ZEUS (Z. Ren): ZEUS observes the Θ^+ state in the $K_s^0 p(\bar{p})$ decay channel at a mass of 1521.5 ± 1.5 (stat.) $^{+1.28}_{-1.7}$ (syst.) MeV with 4.6σ significance in deep inelastic scattering events with $Q^2 > 20$ GeV^2. The production cross section is $\sigma(ep \to e\Theta + X) = 125 \pm 27(\text{stat.})^{+36}_{-28}(\text{syst.})$ pb. New results on production mechanisms of various states were presented: the production

of $\Lambda(1520)$ is consistent with a pure fragmentation origin; the production of Λ_c can be attributed to boson-gluon fusion processes; the production of Θ^+ may favor a proton-remnant fragmentation origin.

- H1 search for a narrow baryonic resonance decaying to $K_s^0 p(\bar{p})$ (C. Risler): H1 has searched for the Θ^+ state in the $K_s^0 p(\bar{p})$ decay channel. No significant signal was found in the region of $1.48 - 1.7$ GeV of the decay-products mass spectrum in different regions of Q^2. An upper limit of $40 - 120$ pb at 95% CL was set. This limit is not incompatible with the measured production cross section of ZEUS.
- Theoretical aspects of pentaquark searches (A. Szczepaniak): a detailed review on experimental pentaquark results was presented. Most pentaquark sightings seem to come from low statistics, low resolution, low-energy experiments with kinematically constrained final states and complicated cuts. On the other hand, it seems that high-resolution, high-statistics experiments with low- or high-particles multiplicity do not report signals.

PHOTON STRUCTURE AND PROMPT PHOTONS

The photon, being the gauge particle of QED, can couple directly to $q\bar{q}$ pairs. In this light, a hard photon observed in the final state can give direct insight into the mechanism of the hard interaction with the advantage of not undergoing a fragmentation process. Thus, measurements of final-state photons are less affected by the experimental and theoretical uncertainties of hadronization. On the other hand, softer photons in the initial state can also interact via their resolved partonic structure, and photon PDFs are needed to describe these processes. The PDFs of the photon are less constrained at present than the proton PDFs, especially the gluon densities. Inclusion of data directly sensitive to the gluonic content in the photon, such as HERA photoproduction jet data, in principle helps to constrain the gluon density in the PDF fits. Results on these two topics were presented:

- Measurement of prompt photon cross sections in photoproduction at H1 (J. Ferencei): new measurements of prompt photon inclusive cross sections and those accompanied by jets were presented using a method with a likelihood discriminator to isolate the signal. The measured cross sections are reasonably well described by NLO QCD calculations, whereas leading-logarithm parton-shower Monte Carlo models describe the shape, but underestimate the normalization of the data.
- Next-to-leading-order photon PDF (A. Levy): new parametrizations for the photon parton distribution functions at NLO were presented. Analysis of experimental PDF uncertainties with the error matrix method was performed. Several data sets, including measurements of F_2^γ, low-x F_2^p and dijets in photoproduction, were used in the fits. A good description of the data is obtained with these parametrizations. Although the dijet data receives a contribution from the gluon photon density, this contribution is kinematically suppressed compared to the gluon proton density in the region covered by HERA. Consequently, inclusion of the dijet data did not substantially reduce uncertainties in the gluon photon density, contrary to the original anticipations.

HIGH-ENERGY HEAVY-ION COLLISIONS

Analysis of the hadronic final state in RHIC experiments is providing insight into the existence and properties of the hadronic matter produced in the aftermath of heavy-ion collisions. New results provide strong evidence that the observed suppression in hadron yields indeed arises from propagation through a dense hadronic medium and is not caused by intrinsic properties of the colliding nuclei or production mechanism. Experimental tools for analysing jets in this new environment are being developed and tested.

- π, η and direct photon production in pp and $AuAu$ collisions (PHENIX, T. Awes): measurements of high-p_T π^0 and γ cross sections were presented and compared with pQCD calculations. The data on photon production agrees well with the pQCD calculation, in contrast to π^0 production, which is strongly suppressed compared to the perturbative prediction. Direct photons do not interact with the hadronic matter after the hard scattering occurs. The difference between the γ and π^0 yields therefore indicates presence of a dense hadronic medium in the final state, and is not likely to be caused by an initial-state effect.
- Jet properties from di-hadron correlation in pp and dAu (PHENIX, J. Jia): two-particle correlations were described as a useful characteristic of the jets produced in pp and dAu collisions at RHIC. The comparison of proton-proton and deuteron-gold data provides useful information about the impact of the nuclear matter on formation of the jets. Transverse momentum distributions of hadrons in the jets are very similar between dAu and pp collisions, *i.e.* presence of a cold nuclear medium (dAu) does not cause significant broadening of the jets comparatively to pp scattering. The jet-yield distribution is also very similar between these two reactions, thus, no significant increase in jet multiplicity occurs for dAu with respect to pp. The underlying event is larger at small p_T in dAu, but essentially coincides with the pp underlying event at large p_T. Assuming that hard-scattering events in the dAu system happen in independent nucleon-nucleon collisions, the underlying event rates in pp and dAu are related by a simple function depending on the pp minimum bias yield, number of collisions, and dAu nuclear modification factor.

SUMMARY

This short article does not attempt to give more than a flavour of the diverse topics presented during the Hadronic Final State sessions of the conference. The reader can find the full details of each presentation within the individual contributions. Several conclusions can be drawn from this overview, which of course do not cover the whole span of important topics discussed in many interesting talks.

High-Q^2 data now arriving from HERA become increasingly suited to precision studies of QCD: accurate determination of the QCD coupling, color factors, parton densities, etc. Further developments in QCD theory must also be pursued to match the experimental accuracy. For example, percent-level precision in the measurement of the fundamental parameter of QCD, the strong coupling constant, has been achieved at

HERA. Further theoretical work, namely NNLO calculations, is needed to reduce theory uncertainties for this observable to a level comparable with experimental errors.

Despite substantial recent progress, interpretation of the low-x scattering at HERA remains challenging, with none of the existing theoretical frameworks adequately explaining all features of forward hadroproduction. Future precision data may provide additional clues about the mechanisms involved. Improvements in the theoretical frameworks for the large- and small-x limits (DGLAP and BFKL) and development of specialized methods for moderately small x (e.g., CCFM) are required to explain the whole range of x and Q^2 covered by HERA measurements. Both inclusive and exclusive observables must be well understood as a function of x, given that many new physics searches at the LHC will rely on observation of multi-particle final states in the x range accessible at HERA and/or at forward rapidities.

The evidence for pentaquarks is dwindling; high-statistics experiments will be able to provide decisive information on the pentaquark existence or absence in the near future. Remarkable progress is being observed in higher-order QCD computations, development of Monte-Carlo showering programs, and resummations, with combined methods comprising several theoretical approaches (parton showers with NLO; joint resummations) gradually emerging. Details of high-energy hadronic scattering (angular momentum and spin dependence; partonic structure; power-suppressed contributions; fragmentation, hadron multiplicities, and jet formation; underlying event; ...) are systematically explored. Examination of hadronic interactions in heavy-nucleus collisions provides telling clues about properties of nuclear matter at high energies. Taken together, the results presented demonstrate that the physics of hadronic final states remains a leading area of particle research, with many new discoveries awaiting in the future.

ACKNOWLEDGMENTS

We would like to thank the organizers of the conference, and specially Wesley Smith, the chair of the organizing committee, for a well prepared conference and for providing a warm atmosphere that led to many lively physics discussions.

Spin Physics Summary

Pasquale Di Nezza*, Krishna S. Kumar† and Marco Stratmann**

*INFN - Laboratori Nazionali di Frascati, via E.Fermi 40, I-00044 Frascati, Rome, Italy
†Department of Physics, University of Massachusetts, Amherst, MA 01002, USA
**Department of Physics, Institut für Theoretische Physik, Universität Regensburg, D-93040 Regensburg, Germany

Abstract.
We present a brief summary of recent results presented at the Spin Physics Working Group at DIS 2005. The Spin Physics parallel sessions hosted a total of 33 talks equally distributed between experimental and theoretical aspects. An improved theoretical understanding in several recent topics of interest in this field is now evident, especially in the emerging area of transversity distributions.

Keywords: Spin Physics, Polarized Lepton Scattering, Polarized Proton Scattering
PACS: 12.38.Qk, 13.88.+e

Precise results on the spin-dependent structure functions $g_1^p(x,Q^2)$ and $g_1^d(x,Q^2)$ were presented by the HERMES Collaboration while a new measurement of A_1^d and $g_1^d(x,Q^2)$ from the COMPASS experiment extends this measurement to lower values of x_{Bj} and higher Q^2. HERMES showed the comprehensive dataset on g_1 including results from proton, deuterium and neutron targets. The latter comes both from the ^{3}He measurement and from the difference of the proton data from the deuterium data. These results adopt a multidimensional unfolding to take into account both radiative background and intra-bin migration. The technique implies the elimination of the systematic correlation between different kinematic bins [1]. COMPASS combined the 2002 and 2003 data taking for an integrated luminosity of $\sim$1500 pb^{-1} covering a Q^2 range up to 100 GeV^2. Data significantly improved statistical accuracy in the low x region (down to 10^{-3}, with good agreement with previous experiments [2].

HERMES all showed the final results of the first measurement of the tensor asymmetry A_{zz} and the tensor structure function $b_1^d(x,Q^2)$ [3]. These results show that. as a nuclear-polarized deuteron target always carries a large tensor polarization, it is a priori not justified to neglect the effect of the tensor asymmetry while analyzing g_1^d measurements. In practice, b_1 represents the difference in the quark distribution between the helicity-0 and the averaged non-zero helicity states of the deuteron.

Our understanding of the high x region has significantly improved with new data of unprecedented precision from experiemnts at Jefferson Laboratory. This facilitates access not only to the valence quark region, but also the more elusive spin structure function g_2 in order to study the higher twist effects. Results on A_1^n [4] show, for the first time, a significant positive value for for high x, as would be expected with $SU(6)$ breaking. In particular, the results are in significant disagreement with the predictions from leading-order perturbative QCD models that assume hadron helicity conservation. This suggests that effects beyond leading-order perturbative QCD, such as the quark

CP792, *Deep Inelastic Scattering, DIS 2005*, edited by W. H. Smith and S. R. Dasu

orbital angular momentum, may play an important role in this particular kinematic region. In summary, although the statistical accuracy is definitely good enough to allow a detailed analysis of the overall behavior of the helicity structure functions, there still remain regions at high and low x where our present knowledge is insufficient to distinguish between various models.

The resonance region for both proton and neutron have been investigated by precision data on spin structure provided by several experiments in all three experimental Halls A, B and C in JLab[5]. These results indicate that for Q^2 less than $\sim$1.5 GeV^2, quark-hadron duality is violated for g_1, mostly due to the strong negative contribution of $\Delta(1232)$.

The transversity distribution $h_1(x)$ completes the mapping of the spin structure of the nucleon at leading twist. Several features (i.e. the chiral-odd nature) makes transversity an attractive subject for detailed experimental investigation. As is well known, two main mechanisms affect azimuthal single-spin asymmetries on transversely polarized targets. The so-called Collins effect is probably the most representative of the different observables involving trasversity. Here a chiral-odd fragmentation function allows the observation of chiral-odd $h_1(x)$, which is otherwise impossible to access directly in a DIS process. On the other hand, a correlation between the intrinsic transverse momentum of an unpolarized quark and the direction of the transverse spin of its parent nucleon can exist and is described by the (naive) t-odd Sivers distribution function $f_{1T}^{\perp}(x)$. Probably the most interesting qualitative feature is that transversity distributions require a non-zero orbital angular momentum of the unpolarized quark.

HERMES, after the publication [6] of the first evidence for azimuthal single spin asymmetries in the semi-inclusive production of charged pions on transversely polarized target, has shown even more significant signals for both Collins and Sivers mechanisms analyzing the data collected during the period 2002-2004 with a pure hydrogen target. Asymmetries have been shown as a function of x, the hadron momentum fraction z and the hadron transverse momentum $P_{h\perp}$. The average Collins moment is positive for π^+ and negative for π^-. Significantly, the magnitude of the π^- moment appears to be similar in magnitude to that for π^+. The average Sivers moment is significantly positive for π^+, showing for the first time that a non-vanishing orbital angular momentum of quarks inside the nucleon is required. For the π^- the averaged Sivers moment is consistent with zero. The COMPASS has devoted 20% of the running time to transversely polarized ^6LiD (deuteron) target and has shown the data collected during the year 2002 corresponding to an integrated luminosity of about 200 pb^{-1}. Results have been shown for positive and negative charged hadrons selecting all hadrons or the leading one only [7]. Within the accuracy of the measurement, both the Collins and Sivers asymmetries turned out to be small and compatible with zero, with a marginal indication of a Collins effect at large z for both positive and negative charges. Concerning the Sivers asymmetry, the work presented by A.Prokudin [8] has demonstrated that HERMES results for protons and COMPASS results do not contradict each other. In 2006 the COMPASS collaboration plans to take data on a transversely polarized proton target (NH_3).

As previously explained, due to the chiral-odd nature, both the transversity and the Collins fragmentation function cannot be measured directly in conventional DIS processes. The results from HERMES and COMPASS are a convolution of these two objects. However, in order to measure transversity one needs that particular fragmentation function (FF) to be precisely known. BELLE, using hadron production from e^+e^- interactions, has shown the first signature of the Collins fragmentation function using a combined measurement of a quark and an anti-quark distribution. In fact, the number density for producing two unpolarized hadrons from transversely polarized quarks contains two times the chiral-odd FF, resulting in a chiral-even object accessible in electromagnetic interaction directly. A clear signature is visible, with the data also showing an increase with rising fractional energy z. This result can be consider a milestone in the study of the chiral-odd objects connected to the transversity measurement.

In case of $\vec{p}p$ scattering the situation shows some complication because, together with the Collins and Sivers mechanism, there is the contribution of the higher twist initial-state or final state interactions. However, in the hadron-hadron scattering, the first non zero single spin asymmetries were observed in pion production by E704 [9]. They have generated much interest since leading-twist collinear factorized pQCD predicts these asymmetries should be small. Much of the rising interest now is at RHIC with the idea that a full QCD description of the proton requires the inclusion of transverse momentum. In general, the performance of the polarized RHIC collider has been steadily improving. Transverse polarization data sets from 2003 and 2004 were both obtained with less than half Picobarn^{-1} of beam. The average proton beam polarization in 2003 was less than 20%. In the current year (2005), online polarization values in excess of 50% have been seen. Moreover a luminosity about one order of magnitude greater than the one used to extract the presented results have been collected. A much higher statistics measurement of these transverse asymmetries can be expected in the near future. STAR has shown the first measurements at $\sqrt{s}$=200 GeV indicating that the π^0 production exhibits a large single spin asymmetry which increases rapidly for x_F above about 0.3 [10]. These results are similar to those observed at lower energy by the aforementioned E704 experiment. Furthermore, for backward production, relative to the polarized beam, no significant asymmetry is observed. This result for this k_T measurement is exactly what the Sivers model predicts. If events are selected with a pair of π^0's separated by approximately 180° in azimuthal angle and produced within the very forward kinematic region, the momentum of an observed π^0 represents a very large fraction of the jet energy. The typical fragmentation fraction for these events is about 90%. The STAR collaboration has established, for the first time, that these dramatic transverse spin asymmetries persist at the higher energies of the RHIC collider. It is worth mentioning that, while the asymmetries are similar, the cross sections at lower energy did not agree with perturbative calculations. With improving forward calorimetry working in the rapidity range $-1< \eta <4$, used in conjunction with the existing more central STAR detector system, this experiment will be able to distinguish the signatures of the Collins and Sivers effects. PHENIX, with a limited dataset of 0.15 pb^{-1} had calculated the transverse single spin asymmetry for π^0 in the central arm of the spectrometer ($|\eta| <0.35$ and non-identified charged hadrons at x_F=0 and transverse hadron momentum up to $p_\perp =$ 5 GeV. In the results there is an overall scale uncertainty of $\pm$35% from uncertainty in the absolute polarization, and an absolute uncertainty of 0.2% on all points. The asymmetry at mid-rapidity is consistent with zero, to the few percent level, for all $p_\perp$ for both the π^0 and charged hadrons. In a near future PHENIX will access the measurement of back-to-back di-hadron azimuthal correlation to decouple contributions to SSA from the Sivers function. The BRAHMS experiment is well suited to study single spin asymmetries for identified pions at moderate x_F because of the PID coverage up to momenta of 40 GeV and the option to measure at $\eta \simeq 4$. The π^+ measured asymmetries are positive while the π^- are negative i.e. the same sign dependence as seen in the E704 data at lower energy. In addition the protons are found to have an asymmetry consistent with zero. BRAHMS has also compared the data vs. extrapolations of twist 3 (initial state) calculations [11]. The pQCD is apriori not valid at the lower values of $p_\perp$ covered in the presented measurement. Nevertheless it gives a good estimate how kinematic cuts may effect predictions as to give rise to a near constant asymmetry in a limited range of x_F. Both the magnitude and the x_F dependence is in reasonable agreement with the data. All the RHIC data show the clear potential to provide an independent means of determining the transversity distribution, provided that the Collins mechanism proves to be dominant.

A complete different way to approach transversity has been show by HERMES and COMPASS concerning the single spin asymmetries in two-hadron production in semi inclusive DIS. Although this method comes at the expense of a large statistical uncertainty, it has the advantage to relate, at leading twist, directly to the product of h_1 and the fragmentation function, whereas the single-hadron method show a a convolution of that product with the transverse momentum of the hadron. Moreover the two-hadron measurement involves a completely different fragmentation function as compared to the single-hadron one providing an independent method of measuring transversity. Specifically the fragmentation functions involved describe the interference between different production channels of the pion pair. HERMES has measured this asymmetry on transversely polarized hydrogen target in the invariant mass region $0.51 < M_{\pi+\pi-} < 0.97$ GeV, showing significant non zero results. The asymmetry is clearly positive over the mentioned entire invariant mass range and largest in the region of the ρ^0 mass. At the end of 2005 HERMES expects to collect a full data sample to lead to a decrease of the uncertainty on the asymmetry with approximately a factor $\sqrt{2}$ allowing a multi-dimensional analysis studying the x and z dependence. COMPASS has investigated the same asymmetry choosing hadron pairs selecting all combinations of positive and negative hadrons. The measurements has been performed as a function of M_{hh}, x and z showing a very small or no significant signal. Especially the asymmetry vs M_{hh} does not at all show any strong dependence. COMPASS expects new results including hadron identification using RICH information resulting in a cleaner hadron sample and a complementary measurement on a proton target from the planned run in 2006.

Besides the transversity distribution h_1 itself, there are other objects describing the transverse polarization of quarks but this time in longitudinally polarized targets. Hall B at Jefferson Lab has shown single-spin and double-spin asymmetries in semi-inclusive electroproduction of pions with a polarized NH_3 target using the CEBAF 6 GeV polarized electron beam [12]. In particular for such kind of process the only azimuthal asymmetry arising in leading order is the $sin2\phi$ moment involving the transverse momentum dependent Collins fragmentation function and the Mulders distribution function $h_{1L}^{\perp}$. One of the key results shown was a factorization test coming from examining the z-dependence of the double spin asymmetry for π^+, π^- and π^0. If factorization holds, the asymmetries should be approximately independent of z, broken by the different weights given to the polarized u and d quarks by the favored and unfavored fragmentation functions. These studies suggested that factorization works at CEBAF 5.7 GeV for $W>2$ GeV, $Q^2>1.1$ GeV^2, $0.15<x<0.5$ and $0.3<z<0.7$. The data for π^+ show a clear $sin\phi$ and $sin2\phi$ modulations and the x-dependence is consistent with the expectation. No sign of a large unfavoured Collins function is seen. The π^+ asymmetry is dominated by u-quarks, therefore with some assumption about the ratio of unfavoured and favoured Collins functions, it can provide a first glimpse of the twist-2 function $h_{1L}^{\perp}$. This extraction, however, still suffers from low statistics and significant systematics coming from the theoretical assumptions.

Another important topic addressed in the Spin Physics sessions was the measurement of the gluon polarization, with new results shown by PHENIX and COMPASS. The PHENIX experiment has started making measurement of double spin asymmetries that would eventually lead to the polarised gluon distribution. The measurements were done for π^0 production in longitudinally polarised proton-proton collisions. Moreover, PHENIX has reported the unpolarised cross section for π^0 production at mid-rapidity which is described extremely well by the next-to-leading-order pQCD calculations over eight orders of magnitude. This becomes the basis on which the measured asymmetries will be interpretable in terms of polarised gluon distribution in the pQCD formalism directly sensitive to the polarised gluon distribution function in the proton through gluon-gluon and gluon-quark scattering subprocesses. The data to date are consistent with $\Delta G/G=0$, albeit with large errors. Due to the uncertainties of theoretical nature, it is still not possible to rule out large values of gluon polarization. Significant results will be be available soon, with improved statistics and the beam polarization. In particular after the installation of a superconducting "Siberian Snake" system, the beam polarization is expected to be above 60%.

A more direct way to measure the gluon polarization has been presented by COMPASS based on the helicity asymmetry of the photon-gluon fusion cross-section. The analysed data are still unsufficient to access the gluon contribution directly (i.e. by D^0 or D^* production) so what has been measured was the spin asymmetry of quasi-real photopruduction events for which a pair of large trasverse momentum hadrons is produced. However the asymmetry has many competitive processed that must be subtracted by using a Monte Carlo. In this case the PYTHIA generator has been used. This introduces large systematic uncertainties which substancially decrease the significance of the results. Analysing both the kinematical regions for $Q^2<1$ and $Q^2>1$ GeV^2 data show no significant gluon polarization for an avarage x_g=0.095. An agreement in that specific x_g, at the level of 1.5 σ, has been shown with the models [13, 14]. It is worth mentioning that a large positive gluon polarization has been shown previously by HERMES [15] using the same method. However this result is not in contradiction with the COMPASS one because of the different kinematical region where this result was extracted and because it suffers from the same large uncertainties from theoretical modelling.

Finally, a special session was devoted to new proposals for future projects, with topics ranging from improvements in electron and proton polarimetry to ideas for new experiments. The PAX collaboration has presented a rich and innovative physics program to be realised in the upcoming FAIR hadron facility at the GSI laboratory. The storage of polarised antiprotons at HESR will open unique possibilities to test QCD in an unexplored kinematic domain. The idea is to have polarised antiprotons produced by spin-filtering with an internal polarised gas target providing access to a wealth of single- and double-spin observables. This includes a first measurement of the transversity distribution of the valence quarks in the proton as well as a first measurement of the moduli and the relative phase of the time-like electric and magnetic form factors of the proton.

In conclusion, the physics goals in the field of spin physics related to deep inelastic scattering were clearly outlined in the parallel sessions. The evolution of a new topic and the steady improvements to topics that have been studied previously show the ability of this subfield to explore missing elements and address new theoretical concepts. Our understanding of the spin structure of hadronic matter continues to improve and we are looking forward to an exciting future.

REFERENCES

1. HERMES Coll., A. Airapetian et al., Phys. Rev. D71 (2005) 012003.
2. COMPASS Coll., Phys. Lett. B612 (2005) 154.
3. HERMES Coll., A. Airapetian et al, submitted to Phys. Rev. Lett., hep-ex/0506018.
4. X.Zeng et al., Phys. Rev. Lett. 92 (2004) 012004; Phys. Rev. C70 (2004) 065207.
5. K.Kramer et al., submitted to Phys. Rev. Lett., nucl-ex/0506005.
6. HERMES Coll., A. Airapetian et al., Phys. Rev. Lett. 94 (2005) 012002; Physics Letters B (in press), hep-ex/0505042.
7. COMPASS Coll, V.Yu.Alexakhin et al., Phys. Rev. Lett. 94 (2005) 202002.
8. M.Anselmino et al., Phys. Rev. D71 (2205) 074006.
9. D.L.Adams et al., Phys. Rev. D53 (1996) 4747.
10. STAR Coll., J.Adams et al., Phys. Rev. Lett. 92 (2004) 171801.
11. J.Qiu and G.Sterman, Phys. Rev. D59 (1998) 014064.
12. CLAS Coll., H.Avakian et al., Phys. Rev. D69 (2004) 112004.
13. M.Hirai et al., Phys. Rev D69 (2004) 054021.
14. E.Leader et al., Eur. Phys. J. C23 (2002) 479.
15. HERMES Coll.,A. Airapetian et al., Phys. Rev. Lett. 84 (2000) 2584.

Heavy flavours: theory summary

G. Corcella

CERN, Department of Physics, Theory Division, CH-1211 Geneva 23, Switzerland

Abstract. I summarize the theory talks given in the Heavy Flavours Working Group. In particular, I discuss heavy-flavour parton distribution functions, threshold resummation for heavy-quark production, progress in fragmentation functions, quarkonium production, heavy-meson hadroproduction.

Keywords: Heavy quarks, heavy hadrons, parton distributions, fragmentation functions
PACS: 12.38.Bx, 14.40.Lb, 14.40.Nd

INTRODUCTION

Heavy-flavour physics is currently one of the main fields of investigation, in both theoretical and experimental particle physics. In this talk, I summarize the main theoretical issues that were presented in the Heavy Flavour session of DIS 2005. Among the topics discussed, we had updates on heavy-flavour parton distributions, large-x resummation for heavy quark production in DIS, heavy-quark fragmentation functions to next-to-next-to-leading order, progress in heavy-meson and quarkonium production.

HEAVY-QUARK PARTON DISTRIBUTIONS

Relevant work was carried out on the subject of heavy-quark parton distribution functions. Progress was reported by R. Thorne, from the MRST collaboration, concerning the formulation of the Variable Flavour Number Scheme (VFNS) for heavy quarks at NNLO.

In fact, there are problems in defining VFNS for heavy quarks. When switching naively from a given order in the coupling constant α_S with n_f flavours, to the same order, but with $n_f + 1$ flavours, one would get a discontinuity when Q^2 is equal to the heavy quark mass. At NNLO, for example, the contribution to a heavy-quark structure function $F_2^H(x,Q^2)$ with n_f flavours is $\sim \alpha_S^3 C_2^{\mathrm{FF},3} \otimes f^{n_f}$, where $C_2^{\mathrm{FF},3}$ is the fixed-flavour (FF) coefficient function and f the parton distribution function. The corresponding contribution, in the region of $n_f + 1$ flavours, is instead $\sim \alpha_S^2 C_2^{\mathrm{VF},2} \otimes f^{n_f+1}$, where $C_2^{\mathrm{VF},2}$ is the variable-flavour (VF) coefficient function. As a result, $F_2^H(x,Q^2)$ would be discontinuous through the heavy-quark mass threshold $Q^2 = m_H^2$.

The Thorne–Roberts (TR) prescription handles this discontinuity by freezing higher-order terms when crossing the value $Q^2 = m_H^2$. At LO, for instance, the TR prescription reads, in terms of gluon (g) and heavy quark (h, $\bar{h}$) densities:

CP792, *Deep Inelastic Scattering, DIS 2005*, edited by W. H. Smith and S. R. Dasu

$$\frac{\alpha_S(Q^2)}{4\pi} C^{\mathrm{FF},1}_{2,Hg}\left(\frac{Q^2}{m_H^2}\right) \otimes g^{n_f}(Q^2) \rightarrow \frac{\alpha_S(M^2)}{4\pi} C^{\mathrm{FF},1}_{2,Hg}(1) \otimes g^{n_f}(m_H^2)$$
$$+ \ C^{\mathrm{VF},0}_{2,HH}\left(\frac{Q^2}{m_H^2}\right) \otimes (h+\bar{h})(Q^2). \tag{1}$$

In Eq. (1), we note the 'frozen' term $\sim \alpha_S(M^2) C^{\mathrm{FF},1}_{2,Hg}(1)$. In order to apply this prescription to NNLO, we would need the FF $\mathcal{O}(\alpha_S^3)$ heavy-quark coefficient functions for $Q^2 \leq m_H^2$, which have not been computed yet. However, a reliable approximation of such functions can be obtained, gaining information from the available calculations which resum threshold logarithms and leading $\ln(1/x)$ terms, and from the known NNLO coefficient functions and space-like splitting functions.

Using this method, parton densities are still discontinuous at $Q^2 = m_H^2$, but structure functions are now continuous. Results were presented for the charm-quark structure function $F_2^c(x,Q^2)$: the NNLO VFNS prediction fits rather well the ZEUS and H1 data, while the NLO result is always below the data.

F. Olness, from the CTEQ collaboration, also presented recent work on heavy-quark parton distributions, implemented in the CTEQ6HQ set. Mainly, the new set resums logarithms of the heavy-quark mass $\ln(m_H^2/\mu_F^2)$ in the full kinematic range, by reabsorbing them in the heavy-quark distribution function. This yields a small, but visible difference with respect to the previous set CTEQ6M. The predictions obtained using the CTEQ6HQ parton distributions fit the HERA data on $F_2^c(x,Q^2)$ quite well.

Progress was reported on the determination of the strange-quark density, which, in the previous CTEQ sets, was tied to the $\bar{u}$ and $\bar{d}$ distributions via the relation $s = \bar{s} = \kappa(\bar{u}+\bar{d})/2$. This assumption might lead to an underestimate of the uncertainty on the s-quark density. Rather, the s quark can be treated as an additional set, described by a more general parametrization, which can be fitted to data. Preliminary results with an independent s-quark density give reasonable χ^2 when comparing with DIS and Tevatron data.

Moreover, the talk by F. Olness also discussed soft-resummation for b production in DIS, which might also be included in future fits of CTEQ parton distribution functions.

THRESHOLD RESUMMATION FOR HEAVY-QUARK PRODUCTION IN DIS

Soft-gluon resummation for heavy-quark production in DIS was discussed in the talk by A. Mitov. In fact, the DIS $\overline{\mathrm{MS}}$ coefficient functions present, at $\mathcal{O}(\alpha_S)$, terms $\sim 1/(1-x)_+$ and $\sim [\ln(1-x)/(1-x)]_+$, which become large for $x \to 1$ and need to be resummed (threshold resummation). Such contributions correspond to collinear- or soft-gluon radiation.

It was pointed out that the large-x behaviour of the coefficient functions crucially depend, via the ratio m/Q, on the mass of the final-state quark. In fact, a light quark emits both soft- and collinear-divergent radiation, while gluon radiation off a heavy quark can only be soft-enhanced. As a result, the two regimes $m/Q \simeq 1$ and $m/Q \ll 1$ must be treated separately when performing large-x resummation.

The threshold resummation reported by A. Mitov was performed in the next-to-leading logarithmic approximation (NLL), which corresponds to keeping in the Sudakov exponent terms $\sim \alpha_S^n \ln^{n+1} N$ (LL) and $\sim \alpha_S^n \ln^n N$ (NLL), where N is the Mellin moment variable. In fact, soft resummation is analytically performed in N-space, and the results are then inverted to x-space. Predictions were presented for the charm-quark structure function $F_2^c(x,Q^2)$, for charged-current interactions, in the environment of the HERA and NuTeV experiments. The results showed that threshold resummation is relevant, especially at small Q^2. Moreover, resumming large-x terms in the $\overline{\rm MS}$ coefficient function yields a milder dependence on both factorization and renormalization scales, which corresponds to a reduction of the theoretical uncertainty.

NNLO PERTURBATIVE FRAGMENTATION FUNCTION

A. Mitov also reported on progress in heavy-quark fragmentation functions. The energy distribution of a heavy quark presents, at fixed order, terms $\sim \alpha_S^n \ln^k(Q^2/m^2)$ $(k \leq n)$, where Q is the hard scale of the process, that are large for $m \ll Q$, which is often the case. Such logarithms can be resummed using the approach of perturbative fragmentation functions, which expresses the energy spectrum of a heavy quark as the convolution of a coefficient function, describing the emission of a massless parton, and a perturbative fragmentation function $D(m,\mu_F)$, associated with the fragmentation of a massless parton (a light quark or a gluon) into a massive quark. The dependence of $D(m,\mu_F)$ on the factorization scale μ_F is determined by solving the Dokshitzer–Gribov–Altarelli–Parisi (DGLAP) evolution equations, once an initial condition at a scale μ_{0F} is given. Neglecting powers $(m/Q)^p$, the initial condition of the perturbative fragmentation function was proved to be process independent and calculated several years ago to NLO. This talk discussed the recent calculation of NNLO contributions, i.e. up to $\mathcal{O}(\alpha_S^2)$.

Denoting by Q and q heavy and light quarks respectively, NNLO corrections to the initial condition of the perturbative fragmentation function come from the elementary processes: $Q \to Qgg$, $Q \to Qq\bar{q}$, $Q \to QQ\bar{Q}$, $\bar{Q} \to Q\bar{Q}\bar{Q}$, $q(\bar{q}) \to Q\bar{Q}q(\bar{q})$ and $g \to Q\bar{Q}g$, which have now been fully computed.

Solving the DGLAP equations, for an evolution from μ_{0F} to μ_F, allows us to resum terms $\sim \ln(\mu_F^2/\mu_{0F}^2)$, i.e. $\sim \ln(Q^2/m^2)$ if we choose $\mu_{0F} \simeq m$ and $\mu_F \simeq Q$ (collinear resummation). In particular, the leading logarithms are $\alpha_S^n \ln^n(Q^2/m^2)$, the NLLs $\alpha_S^n \ln^{n-1}(Q^2/m^2)$, the NNLLs $\alpha_S^n \ln^{n-2}(Q^2/m^2)$. In principle, the calculation of the initial condition of the perturbative fragmentation function to NNLO would allow one to study the spectrum of heavy quarks in the NNLO approximation, with NNLL collinear resummation. However, for this level of accuracy to be achieved, one would also need NNLO Altarelli–Parisi time-like splitting functions, which are currently known to NLO. The computation of NNLO corrections to such functions is in progress.

HADROPRODUCTION OF HEAVY MESONS IN A MASSIVE VFNS

We had a presentation from I. Schienbein on hadroproduction of heavy mesons in a Massive Variable Flavour Number Scheme (MVFNS). Considering, for example, D-meson production at the Tevatron ($p\bar{p} \to DX$), the MVFNS subtracts and resums logarithms of the heavy-quark mass $\ln(\mu_F^2/m^2)$. Unlike the perturbative fragmentation function approach discussed above, it keeps powers of m/Q in the hard-scattering cross section. This prescription is equivalent to $\overline{\rm MS}$ mass factorization in a scheme where the heavy-quark mass regularizes the collinear divergences. It was numerically checked that the short-distance coefficient function corresponds to the $\overline{\rm MS}$ one in the limit where the heavy-quark mass tends to zero. As a result, predictions can be obtained still using $\overline{\rm MS}$ parton distributions and fragmentation functions, but convoluted with a massive hard-scattering cross section.

The considered partonic subprocesses that were calculated are $gg \to c\bar{c}$ and $q\bar{q} \to c\bar{c}$, at LO; $qq \to c\bar{c}g$, $q\bar{q} \to c\bar{c}g$ and $gq \to c\bar{c}q$, at NLO. If the heavy (charm) quark is in the initial state, it is treated in the massless approximation.

Using the CTEQ6M parton distribution function set and the Binnewies–Kniehl–Krämer (BKK) NLO fragmentation functions, fitted to the OPAL data, predictions were given on the transverse momentum of D^0, D^{*+}, D^+ and D_s^+ mesons at the Tevatron. It was considered just prompt charm production, while D production from B-meson decays was not taken into account. Within the error range, agreement was found with the CDF data, although the ratio of the central values of theoretical predictions and data is about 1.5–1.8, which may warrant further investigation. A prediction was finally given for the transverse momentum distribution of c-flavoured baryons Λ_c^+.

This approach will be in the near future extended to B-meson production at the Tevatron, and both D and B mesons in Deep Inelastic Scattering.

HEAVY-QUARKONIUM PRODUCTION

The talk by J.–P. Lansberg discussed the production of heavy quarkonium ($Q\bar{Q}$) in a new model, which consists of an extension of the Colour Singlet Model (CSM) and the Colour Octet Model (COM). The naive CSM factorizes heavy-quarkonium production in a hard and a soft part. In the hard process, Q and $\bar{Q}$ are assumed to be on-shell, in a colour-singlet state, with zero relative momentum, in a $^3S_1$ angular-momentum state (for J/ψ, ψ' and Υ). As far as the soft part is concerned, the amplitude for the quark-binding probability is a wave function, solution of the Schrödinger equation. This model is, however, unable to reproduce the Tevatron Run I data from CDF on J/ψ and ψ' direct production.

The COM proposes instead that quarkonium states are produced by the fragmentation of a gluon, which is transversely polarized, so that, according to Non-Relativistic QCD (NRQCD), the quarkonium is to have itself transverse polarization. Nevertheless, this prediction disagrees with CDF measurements of unpolarized or slightly longitudinally polarized quarkonium states.

The new model goes beyond the static and on-shell approximations of the CSM. $^3S_1$

quarkonium ($\mathscr{Q} = Q\bar{Q}$) is produced via gluon fusion $gg \to \mathscr{Q}g$, but it includes new contributions with respect to the usual CSM (see J.–P. Lansberg's presentation in these proceedings for details).

To describe the soft, non-perturbative part, two phenomenological vertex functions are chosen: $\psi(\vec{p}_{\rm rel}) \sim \exp[-\vec{p}_{\rm rel}^{\,2}/\Lambda^2]$ and $\psi(\vec{p}_{\rm rel}) \sim \left(1+\vec{p}_{\rm rel}^{\,2}/\Lambda^2\right)^{-2}$, where $\vec{p}_{\rm rel}$ is the relative $Q\bar{Q}$ momentum and Λ is a free size parameter. The results yielded by this model are in good agreement with RICH data on J/ψ, and Tevatron data on J/ψ, ψ' and $\Upsilon(1S)$ production. Fragmentation contributions are then taken from the COM, which gives transverse polarization, and included in the model. The agreement with polarization measurements at the Tevatron is shown to be now pretty good.

NLO CHARMONIUM PRODUCTION IN $\gamma\gamma$ COLLISIONS

The talk given by B. Kniehl discussed NLO charmonium production in photon–photon collisions. The computation of NLO corrections to such processes is a relevant improvement, since it reduces renormalization and factorization scale dependence and allows a test of NRQCD.

New results were presented for processes $\gamma\gamma \to J/\psi X$, with direct photons and prompt J/ψ's, within the framework of NRQCD. In the considered processes, X can be a purely hadronic state, or a hadronic state with a prompt photon. Phenomenological results were presented for e^+e^- colliders, with characteristics similar to the TESLA Linear Collider project, and a centre-of-mass energy $\sqrt{s} = 500$ GeV. Cuts on transverse momentum and rapidity of final-state photons were set to $p_T^\gamma > 3$ GeV and $|y^\gamma| < 2.79$.

It was shown that, unlike the LO, the NLO prediction yields very little dependence on the phase-space slicing parameters adopted to cancel soft and collinear singularities. Results were presented for the J/ψ rapidity and transverse momentum spectrum at LO and NLO, along with the NLO K-factor. NLO corrections exhibit a remarkable impact on both shape and normalization of the distributions which were shown.

In fact, for $\gamma\gamma \to J/\psi X$, the K factor is large because of the subprocesses $\gamma\gamma \to c\bar{c}[^3S_1^{(8)}]g$ and $\gamma\gamma \to c\bar{c}[^3S_1^{(8)}]q\bar{q}$, which are mediated by the gluon splitting $g \to c\bar{c}$, where the $c\bar{c}$ pair is in a $^3S_1^{(8)}$ state.

The presented analysis will be extended to electron–proton photoproduction and hadroproduction, which will yield predictions that could be compared with data from HERA II, the Tevatron Run II and, ultimately, the LHC.

CONCLUSIONS

The heavy-flavour session of DIS 2005 had a number of interesting theory talks.

The reported progress in heavy-quark parton distribution functions will be a key ingredient for precision studies of heavy-flavour physics at present and future high-energy colliders, such as the LHC.

Large-x resummation in the coefficient function for heavy-quark production in DIS will allow us to extract resummed parton densities as well. The NNLO calculation of the initial condition of heavy-quark perturbative fragmentation functions could allow the promotion of the perturbative fragmentation function approach to NNLO/NNLL accuracy if time-like Altarelli–Parisi splitting functions were to be known at NNLO. Given the process independence of the perturbative fragmentation function, we shall be able to apply such results to any process whose coefficient functions are known to NNLO.

An alternative approach to address heavy-quark (hadron) production is the MVFNS, which was used to predict charm-flavoured hadron production at the Tevatron. Some discrepancies between theory and CDF data in the central values of the D transverse-momentum distribution may require further investigation. It may also be worthwhile comparing MVFNS and perturbative fragmentation function approaches.

Quarkonium production studies were also discussed. A new model was proposed, which goes beyond the static approximation of the CSM, uses the COM for the fragmentation, and is able to reproduce Tevatron and RICH data on J/ψ, ψ' and Υ production.

NLO charmonium production in two-photon collisions was also discussed. The results, shown for J/ψ production at an e^+e^- collider with $\sqrt{s} = 500$ GeV, exhibit a relevant effect of the inclusion of NLO corrections.

In summary, all the given talks show active work and progress on heavy-flavour phenomenology, which will allow the performance of increasingly more accurate measurements in Deep Inelastic Scattering experiments, as well as at any present and future hadron collider facility.

Heavy flavours: experimental summary

Andrew Mehta* and Massimo Corradi†

*Liverpool University, Oliver Lodge Lab., Oxford Street, Liverpool, L69 7ZR, UK.
†INFN Bologna, via Irnerio 46, Bologna, Italy

Abstract. New measurements of charm and beauty production presented in the heavy flavour working group are summarized and discussed in comparison with QCD predictions.

Keywords: DIS, Heavy Flavour, Beauty, Charm
PACS: 12.38.Qk,13.60.-r,13.60.Hb,13.90.+i

INTRODUCTION

The measurement of heavy flavours in lepton-hadron interactions provides a test of many aspects of QCD and a probe of the gluon content of the proton or, in charged currents, of its content of strange quarks. In the previous years, the charm measurements at HERA focused on the inclusive production of charmed mesons. At this workshop many new results have been presented that look beyond single heavy-meson production, by measuring fragmentation parameters, looking at di-jet correlations or at heavy-quark-jet characteristics and extending into unexplored kinematic regions. New measurements of $F_2^{c\bar{c}}$and $F_2^{b\bar{b}}$, based on an inclusive lifetime tagging technique, have been presented by H1. NuTeV has shown results on charm production in νA interactions from its full data sample. New data on heavy-quarkonium production in DIS and in hadronic interactions have also been presented. Various results on b production at HERA are now available, obtained from different experimental techniques and extended to a wide range of the b-production phase space.

CHARM FRAGMENTATION

The extremely large charm cross-section at HERA (up to 25% of the total cross-section for DIS) has allowed precise determinations of the of the probability that a charm quark will produce a specific charm hadron (fragmentation fraction) and also the relative momentum carried by that hadron (fragmentation function). These measurements may be compared with those from e^+e^- collisions in order to test the hypothesis of a universal charm fragmentation. Z. Rurikova and R. Walsh showed new measurements of the the fragmentation fraction from DIS at HERA for D^0, $D^\pm$, $D_s^\pm$, $D^{*\pm}$ mesons and the $\Lambda_c^\pm$ baryon. The measurements show in many cases comparable precision to e^+e^- measurements. A good agreement is observed between the two HERA experiments, with HERA photoproduction measurements and with e^+e^- measurements, supporting a universal charm fragmentation.

Measurements were also made of related quantities. The ratio of twice the cross section for mesons with a s quark to that for mesons with a u or d quark (strangeness suppression factor) is found to be about 0.25 in agreement with e^+e^-. The ratio of neutral to charged meson production is found to be consistent with 1 as expected. The ratio of vector c meson production to vector plus pseudoscalar meson production is found to be significantly less than the value of 0.75, which would be naively predicted from spin counting. Measurements of the fragmentation function, although not as precise as e^+e^- determinations, show a range of preferred values of fragmentation parameters for DIS.

These results on charm fragmentation are complemented by the huge amount of data on charm spectroscopy being produced at B factories, as was shown by C. Chen and B. Yabsley.

CHARMONIUM AND UPSILON PRODUCTION

Inelastic J/ψ production has long been described in pp collisions by non-relativistic quantum chromo-dynamics (NRQCD), which incorporates colour singlet (CS) and colour octet (CO) contributions. New measurements of J/ψ production from RHIC ($\sqrt{s} = 200$ GeV), presented by K. Read, have confirmed the good description of NRQCD in pp collisions. New measurements, presented by H. Wahl, of Tevatron ($\sqrt{s} = 2$ TeV) Υ data showed good agreement with NRQCD calculations.

In contrast to the good description in pp collisions NRQCD has not been able to describe ep collisions very well. New results, presented by A. Antonov, of inelastic J/ψ production at HERA show that, although there is good agreement between experiments, the data are substantially below predictions of NRQCD. Much better agreement is observed if only CS contributions are taken.

CHARM AND STRANGE MEASUREMENTS

New measurements of charm production at HERA, tagged using D^* were presented by G. Aghuzumtsyan, for the very low Q^2 region of $0.05 < Q^2 < 0.7$ GeV^2. Even at these low Q^2 values next-to-leading-order (NLO) QCD is found to describe the data well. New D^* measurements have also been made at higher Q^2. R. Hall-Wilton showed results for $Q^2 > 5$ GeV^2 using HERA II e^+p and e^-p data. The new e^+p and e^-p data are found to agree well as expected. There is no confirmation of the slight excess seen with the HERA I e^-p data.

R. Hall-Wilton also showed the huge improvement in signal to background possible when using the new ZEUS micro-vertex detector. As an example the background under $D^\pm$ peak can be reduced by a factor of 45, whilst the signal is reduced by only 2.7 if a cut on track significance is made. It is expected that much more precise charm and beauty results will be obtained from the HERA II data than existing measurements, due to the increased integrated luminosity and the new vertex detectors installed in H1 and ZEUS.

D. Mason showed results on the forward di-muon cross section in νN charged current interactions from NuTeV. Since the sign of the beam can be determined at NuTeV it is possible to extract the strange and anti-strange sea quark distributions from the di-muon cross section. It is found that at leading order the data favour a slight positive strange asymmetry. It should be pointed out, however, this effect is not sufficient to wholely explain the observed difference of NuTeV's measurement of $\sin^2(\theta_W)$ with the world average.

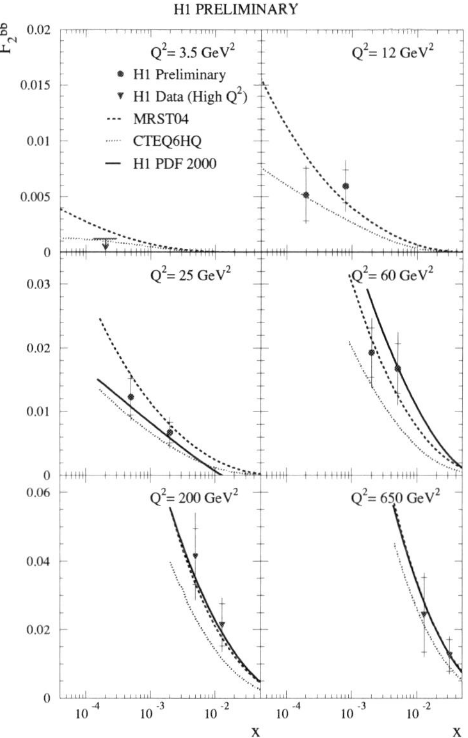

FIGURE 1. The measured $F_2^{b\bar{b}}$ shown as a function of Q^2 for various x values. The inner error bars show the statistical error, the outer error bars represent the statistical and systematic errors added in quadrature. The predictions of QCD are also shown.

INCLUSIVE CHARM AND BEAUTY MEASUREMENTS

New measurements, shown by T. Klimkovich, have been made at HERA using the vertex detector of the differential c and b cross sections ($\mathrm{d}^2\sigma^{c\bar{c}}/\mathrm{d}x\mathrm{d}Q^2$ and $\mathrm{d}^2\sigma^{b\bar{b}}/\mathrm{d}x\mathrm{d}Q^2$) for values of $Q^2 \geq 3.5$ GeV^2. The cross sections are converted into the structure functions $F_2^{c\bar{c}}$ and $F_2^{b\bar{b}}$.

The measurements of $F_2^{c\bar{c}}$ are compared to determinations of $F_2^{c\bar{c}}$ obtained from D^* cross sections where QCD predictions are used to extrapolate over unmeasured phase space. Both determinations agree well. The measurements also agree well with the next

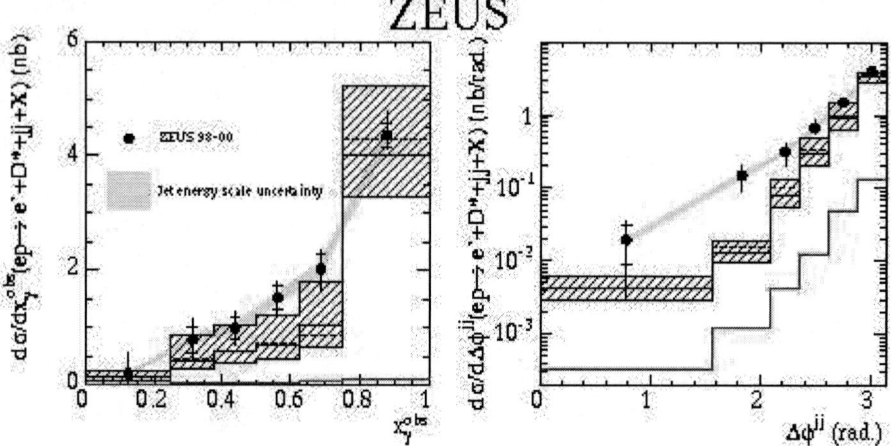

FIGURE 2. ZEUS results on jet-jet correlations in D^* photoproduction. The plots show the differential cross sections in x_γ^{obs}, the fraction of the photon energy carried by the two-jet system, and $\Delta\phi^{jj}$, the azimuthal angle between the two jets. An excess over the NLO QCD predictions is observed at low $\Delta\phi^{jj}$.

to leading order QCD calculations of MRST and CTEQ, which use a variable flavour number scheme (VFNS). These measurements of $F_2^{b\bar{b}}$ (shown in figure 1)are the first to have been made. The measurements agree well with the VFNS calculations of MRST and CTEQ, although the difference between MRST and CTEQ is large at low Q^2 and x, where it can reach a factor of 2.

HEAVY-QUARK JETS

Jet cross sections are a powerful tool to study the parton dynamics in heavy quark production since jets are less influenced by fragmentation details and can be measured in a wider angular range than heavy hadrons.

T. Kohno and G.Fluke presented new results on jet photoproduction at HERA in events with a D^*. The differential cross sections in the jet transverse momentum and rapidity were found to agree well with NLO-QCD predictions based on the FMNR program, similarly to what was found in previous analyses of inclusive D^* photoproduction.

Further constraints on the theory can be obtained from dijet variables. The fraction of the photon's longitudinal momentum taken by the two-jet system, x_γ^{obs}, can be used to distinguish the contributions of the leading-order direct-photon process ($\gamma g \to c\bar{}c$) that peaks at $x_\gamma^{obs} \sim 1$ from resolved-photon processes in which the photons behaves as a source of partons and from higher-order radiative processes that dominate the low-x_γ^{obs} region. The azimuthal angle between the two jets, $\Delta\phi^{jj}$, and the transverse momentum of the dijet system, p_T^{jj}, are direct probes of high-order QCD effects since the leading order diagrams contribute only at $\Delta\phi^{jj} = \pi$ and $p_T^{jj} = 0$.

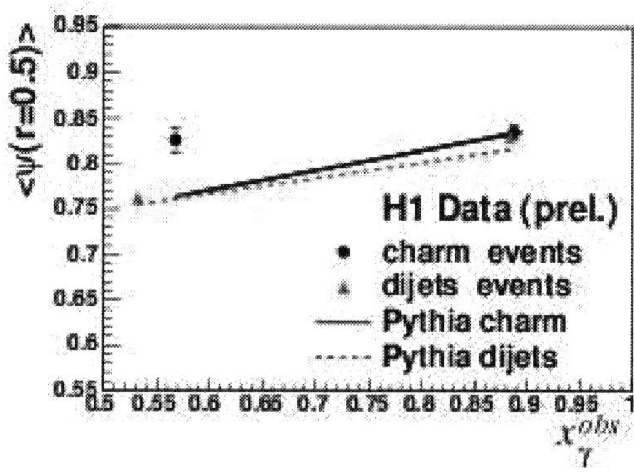

FIGURE 3. H1 results on the mean integrated jet shape as a function of x_γ^{obs} for an inclusive dijet sample (triangles) and for a charm-enriched dijet sample (points), compared to predictions from the PYTHIA MC.

Differential cross sections as a function of x_γ, $\Delta\phi^{\text{jj}}$ and p_T^{jj} were measured by ZEUS in events with two jets and a D^*. The NLO QCD is able to describe the cross section in x_γ^{obs} but fails to reproduce the tails at small $\Delta\phi^{\text{jj}}$ (Fig. 2) and large p_T^{jj}. Parton-shower MC models can instead reproduce reasonably well the shape of these distributions, though with an incorrect overall normalisation. The H1 collaboration studied D^*-jet pairs in which the jet was explicitly required not to contain the reconstructed D^*. Also in this case the NLO-QCD agrees in general with the data but fails to reproduce the region of small azimuthal separation between the D^* and the jet.

To describe jet-jet correlations in heavy-quark photoproduction the theory should probably merge ingredients from NLO-QCD and parton shower MC as in the case of the MC@NLO program that is currently available only for hadron-hadron collisions. Results on correlations between b-tagged jets in $p\bar{p}$ collisions at the Tevatron were presented by R. Lefevre. All the distributions, including $\Delta\phi^{\text{jj}}$ are well described by MC@NLO.

N. Parua presented the first measurement of Z^0 production associated to a b-tagged jet at Tevatron. The rate for this process, which is directly sensitive to the b PDF, was found in agreement with the NLO prediction.

The study of jet shapes is another tool to understand the production of heavy flavours in ep collisions at HERA. M. Martisikova reported about the measurement of jet shapes in the photoproduction of two jets, one of which tagged as charm. The integrated jet shape $\psi(r)$, defined as the fraction of the jet transverse energy contained in a cone of radius r, was measured for the untagged jet and compared to the PYTHIA MC as a function of different kinematic variables. The PYTHIA MC comprises flavour-creation (FC) processes ($\gamma g \to c\bar{c}$, $gg \to c\bar{c}$, ...), in which the two hard jets come from charm quarks and flavour-excitation (FE) processes ($cg \to cg$, $cq \to cq$), in which a charm quark from the photon interacts with a parton from the proton to produce a charm-jet plus a gluon or light-quark jet. Since gluon jets are broader than quark jets (including charm jets), the shape of the untagged jet is sensitive to the relative amount of FC and FE events. For $x_\gamma^{\text{obs}} > 0.75$, were FC processes are expected be dominate, the untagged-

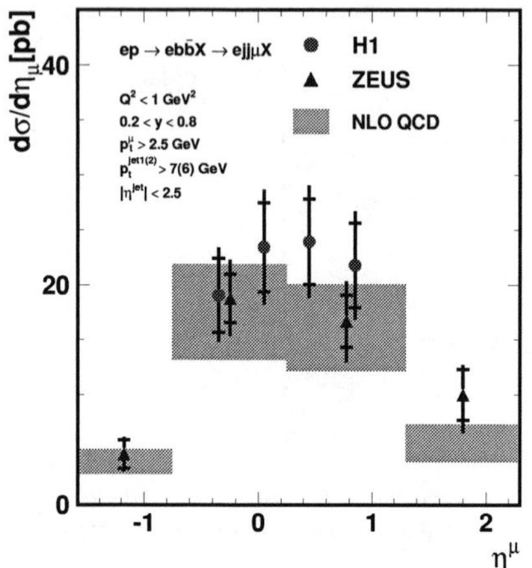

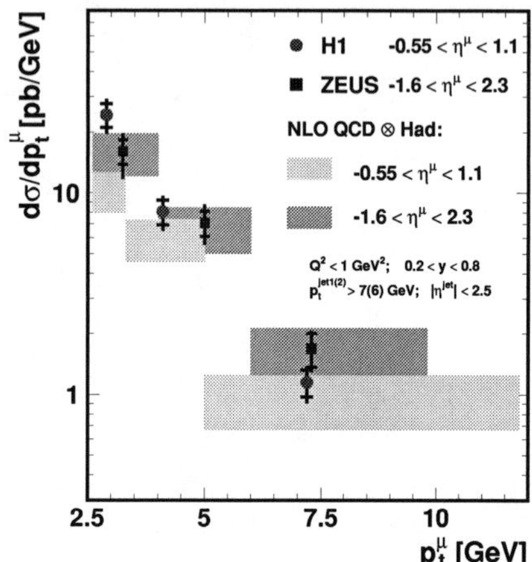

FIGURE 4. Beauty production cross section at HERA in events with two muons and a jet as a function of the muon pseudorapidity (η^μ) and of transverse momentum (p_T^μ). H1 and ZEUS results are compared to NLO-QCD theory (FMNR). In the left plot the ZEUS and H1 data are given for two different η^μ ranges and should be therefore compared to the the theoretical prediction for the corresponding range.

jet shape was found to be well described by PYTHIA. Conversely, for $x_\gamma^{\rm obs} < 0.75$, were most of the contribution from FE is expected, PYTHIA predicts significantly broader jets than those found in the data as show in figure 3. The charm sample is indeed compatible with being composed of two charm jets even at low $x_\gamma^{\rm obs}$.

An interesting QCD prediction is that, in heavy-quark jets, the gluon radiation is suppressed in a dead cone of size $\alpha_0 = M_Q/E_Q$ around the quark direction. A. Peireanu presented a study of jet shapes and subjet multiplicities in H1 DIS data containing a D^* meson, with a first attempt to measure α_0 in DIS. At the present stage the value of α_0 found for the charm jets is statistically compatible with that for non-charm jets. Further studies and more statistics are needed to obtain conclusive results on the dead cone effect.

BEAUTY PRODUCTION AT HERA

QCD predictions are expected to be more reliable for beauty than for charm since the b-quark mass is large enough to assure a reliable perturbative approach. In past years the comparison between NLO-QCD and HERA data gave some problem, while more recent results have found substantial agreement. O. Behnke presented the final H1 measurement of beauty production in events with jets and muons from HERA-I data. This analysis exploited a combination of the muon impact parameter and of the muon transverse momentum with respect to the associated jet, to distinguish the beauty signal from the background from charm and fake muons. The the cross sections for the photoproduction

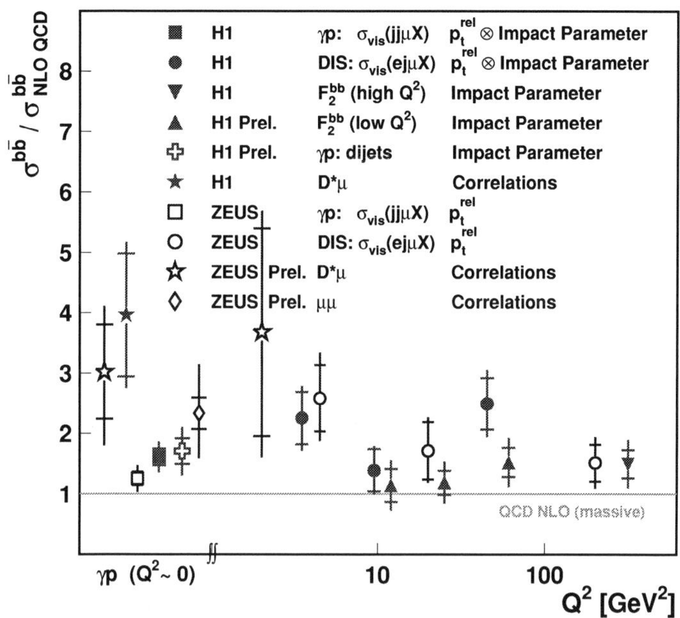

FIGURE 5. Ratio of different beauty cross-section measurements at HERA to the corresponding NLO QCD predictions. The theoretical uncertainty on the NLO Theory ranges from 40% for the total cross-section to 15% at large Q^2.

of two jets and a muon $ep \to e'b\bar{b}X \to e'jj\mu X'$ are found in agreement with previous ZEUS data and are close to the upper limit of the uncertainty band of the NLO-QCD prediction, as shown in Figure 4. Some discrepancy with the theory was found at low muon transverse momentum, were the data are ~ 3 standard deviations above the theory. In the same p_T^μ range the ZEUS data are in agreement with the theory. A similar excess at low muon p_T, though less statistically compelling, was found in DIS ($Q^2 > 2$ GeV^2).

Beauty photoproduction in dijet events $ep \to e'b\bar{b}X \to e'jjX'$ was also measured by H1 with a multi-track impact parameter technique, similar to that used for charm and beauty in DIS, which provided a measurement of the b and c components of an inclusive dijet sample without any muon selection. As explained by L. Finke, the b cross section was found somewhat larger but still compatible with the NLO-QCD theory.

Measurements of beauty productions in events with jets at HERA have reached good precision but are sensitive only to a relatively small region of the $b-$quark production phase space, namely large p_T and central rapidities. Measurements based on a double b tags give the opportunity to cover a larger phase space, including sensitivity for b-quarks produced with zero transverse momentum and a larger rapidity range. As shown by A.

Longhin, the recent ZEUS double-tag analysis based on dimuon events has a sizeable acceptance for about 90% of produced $b\bar{b}$ pairs, allowing for a measurement of the total $b\bar{b}$ cross section at HERA with small extrapolation uncertainties. The result,

$$\sigma_{\text{tot}}(ep \to e'b\bar{b}X) = 16.1 \pm 1.8(\text{stat.})^{+5.3}_{-4.8}(\text{syst.})\,\text{nb}$$

is about two standard deviations larger than the NLO QCD predicted value $\sigma_{\text{tot}}^{NLO}(ep \to e'b\bar{b}X) = 6.8^{+3.8}_{-1.7}$ nb. A similar result was found in the recently published H1 analysis of $D^*\mu$ correlations presented by N. Malden. The visible beauty cross-section was found again about 2 standard deviations larger than expected by NLO-QCD. A new measurement of the total $b\bar{b}$ cross section in pA collision was presented by U. Husemann. In this case the NLO theory has been found in good agreement with the data.

Figure 5 summarizes in one plot the status of the agreement between data and theory for beauty production at HERA by showing the data/NLO ratio for all recent measurements. The theory predictions are obtained with FMNR for $Q^2 < 1$ GeV^2 and with HVQDIS for $Q^2 > 1$ GeV^2. The uncertainties on the theory, that ranges from $\sim$40% for the total cross section to $\sim$15% at large Q^2, are not shown. At large Q^2 the inclusive $F_2^{b\bar{b}}$ results from H1 are in agreement with the NLO and with ZEUS and H1 data from jet-plus-muon ($ej\mu X$) analyses. For Q^2 around few GeV^2, the points from the $ej\mu X$ analyses are somewhat above the theory but still compatible. In photoproduction ($Q^2 < 1$ GeV^2) the most precise measurements are those based on jets and muons (H1 and ZEUS $jj\mu X$) and the H1 dijet analysis with inclusive impact parameter tag. They are substantially in agreement with the theory. The double-tag analyses (ZEUS and H1 $D^*\mu$ correlations and ZEUS $\mu\mu$ correlations) that are sensitive to lower b-quark momenta, but are in general less precise than the dijet ones, are about a factor two above the predictions, with a 2-3 sigma significance. It will be therefore very interesting to see if these hints for an inaccuracy of the theory at low transverse momentum will be confirmed from high statistics analyses at HERA-II. It is also interesting to note the large difference between VFNS calculations of CTEQ and MRST for the inclusive DIS b measurements (see figure 1) in certain regions of phase space. This may indicate that the theory errors that have been assumed so far are under estimated. It should also be noted that the final state observables are compared with theory based on the fixed flavour number scheme, which is expected to be valid only for p_T and Q^2 of not much larger than the quark mass. Unfortunately at the present time there are no VFNS calculations of final state heavy quark observables. The participants in the heavy flavour session expressed a strong wish to theorists to provide a VFNS that could be used to compare with final state observables.

CONCLUSIONS

The huge heavy flavour datasets, provided by HERA, have been used to make measurements of charm fragmentation that are comparible in precision with the world's best. Good agreements is observed between ep and e^+e^- supporting a universal charm fragmentation.

Despite several new precision measurements of inelastic charmonium and upsilon producion and good agreement between independent measurements there is still no good description of *ep* data with non-relativistic QCD calculations. This contrats with *pp* data where similar calculations give a good description of the data.

One of the few methods to constrain the strange quark density of the nucleon is from the di-muon cross section of neutrino-nucleon scattering. New measurements of the strange-antistrange asymmetry were demonstated to be insufficient to wholely explain the deviation of NuTeV's measurement of $\sin^2(\theta_W)$ compared to the world average.

There have been many new measurments of charm production, inclusive or in conjunction with a muon and or jets. All such measurements are broadly in agreement with next to leading order QCD predictions in both *pp* and *ep* scattering. The situation is less clear in beauty cross sections. Measurements made in *pp* collsions show good agreement with calculations. In *ep* collisions many measurements lie above the QCD predictions. It is interesting that predictions of MRST and CTEQ for the inclusive DIS cross section differ by as much as a factor of 2 in some regions of phase space. Perhaps if the reasons for these differences could be understood a better description of beauty production in *ep* scattering might be had.

AKNOWLEDGMENTS

The authors would like to that all speakers and particpants in the heavy flavour session for the interesting talks and lively discussion. We would also express our thanks to the local organising commitee for providing an expertly run and highly enjoyable conference.

Electroweak and beyond the Standard Model physics

B. Heinemann*, A. Tapper† and C.-P. Yuan**

*The University of Liverpool, Oliver Lodge Laboratory, Oxford Street, Liverpool L67 7ZE, U.K.
†Imperial College London, The Blackett Laboratory, Prince Consort Road, London SW7 2BW, U.K.
**Department of Physics and Astronomy, Michigan State University, East Lansing, MI 48823, USA.

Abstract. The status of the electroweak Standard Model and searches for physics beyond the Standard Model are reviewed. We focus on recent results from the HERA and Tevatron accelerators and also discuss measurements at the B factories and the future at the LHC.

Keywords: LEP, HERA, Tevatron, PEP II, KEK-b, LHC, CDF, DØ, H1, ZEUS, BaBar, Belle, Atlas, CMS, Standard Model, Electroweak, Higgs, SUSY, MSSM, GMSB, mSUGRA, Top, Leptoquark
PACS: 12.15.Ff, 12.15.Ji, 12.15.Lk, 12.15.Mm, 12.60.Cn, 12.60.Fr, 12.60.Jv, 13.60.Hb, 13.66.Fg, 13.66.Hk, 13.66.Jn, 13.88.+e, 14.65.Ha, 14.70.Fm, 14.80.Bn, 14.80.Cp, 14.80.Ly

INTRODUCTION

The Standard Model has been extraordinarily successful in describing all electroweak measurements to date. However, the pivotal particle in the explanation of electroweak symmetry breaking, the Higgs boson, has not yet been observed. Additionally there are theoretical problems with the Standard Model, the hierarchy problem, the lack of unification of the electroweak and strong forces and the lack of a candidate for the dark matter observed in the Universe.

In this report we summarise the latest results on electroweak precision measurements and outline the future prospects. We report the latest results on searches for the Higgs boson within and beyond the Standard Model. Finally we present a number of searches for physics beyond the Standard Model such as supersymmetry, leptoquarks and lepton-flavour violation, large extra dimensions and searches for new physics in the flavour sector.

The experimental measurements have been made at a variety of accelerators: the HERA $e^{\pm}p$ collider ($\sqrt{s} = 319$ GeV), the Tevatron $p\bar{p}$ collider ($\sqrt{s} = 1.96$ TeV) and the KEK-b and PEP-II e^+e^- colliders ($\sqrt{s} = 10.6$ GeV). Future prospects at the LHC ($\sqrt{s} = 14$ TeV) are also discussed.

CP792, *Deep Inelastic Scattering, DIS 2005*, edited by W. H. Smith and S. R. Dasu

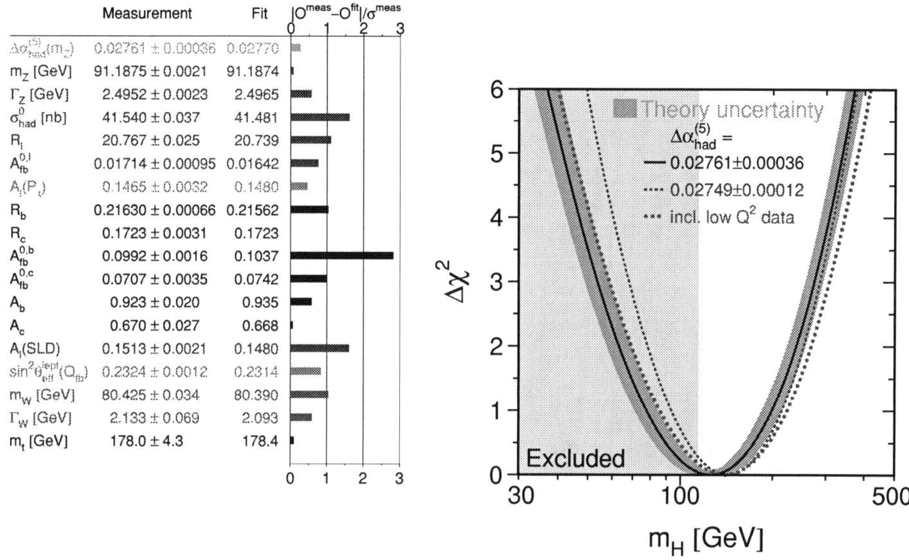

FIGURE 1. Left: The measurements used for the electroweak precision fit. The measurements and the pulls resulting from the fit are shown. Right: the χ^2 of the fit versus the Higgs boson mass. The shaded area is excluded by the direct Higgs boson searches at LEP.

ELECTROWEAK PHYSICS

Electroweak precision measurements

The electroweak sector of the Standard Model is examined to high precision by many direct measurements made at the LEP and SLC colliders in e^+e^- collisions and at the Tevatron in $p\bar{p}$ collisions [1]. A summary of these measurements is shown in Figure 1.

It is seen that all data agree to within 3σ with the fit result. The largest discrepancy arises from the measurement of the forward-backward asymmetry for b-quarks, $A_{FB}^{0,b}$.

Within the electroweak theory these measurements can be used to constrain the mass of the Higgs boson. The χ^2 of the fit is shown in Figure 1 versus the Higgs boson mass. The minimum is at 129 GeV with an uncertainty of $+74$ GeV and -49 GeV at 68% confidence level. The 95% upper limit on the Higgs boson mass is 285 GeV.

Most of these measurements are final and are not expected to be improved in the near future. The only measurements that are expected to change in the near future are the W-boson mass and the top-quark mass. The W-boson mass measurement at LEP is still being finalised and further improvements are expected from the Tevatron experiments. The top-quark mass measurement is also expected to improve substantially due to the new data at the Tevatron.

A new analysis of the W-boson mass using Run II data at the Tevatron of luminosity 200 pb^{-1} was performed and yielded a systematic uncertainty of 76 MeV [2]. However, the measurement is still ongoing and the central value is not yet known. The width of the W boson was measured by the DØ experiment to be $\Gamma_W = 2.011 \pm 0.09(\text{stat.}) \pm 0.107(\text{stat.})$ GeV. This is consistent with the theoretical prediction of $\Gamma_W = 2.099 \pm 0.003$ GeV.

The most recent precise measurement of the top-quark mass is not yet included in the electroweak fit described above. The best precision is currently achieved in the $t\bar{t} \to WWbb \to l\nu jjbb$ decay channel, where one W boson decays leptonically and one decays hadronically. It has recently been made by the CDF collaboration with a luminosity of 319 pb^{-1} of Run II data and yields $m_{top} = 173.5\,^{+3.7}_{-3.6}\,(\text{stat.+JES}) \pm 1.7(\text{syst.})$ GeV where JES denotes the uncertainty coming from the jet energy scale [3]. This measurement was made using a new technique to minimise the dominant systematic uncertainty coming from the jet energy scale uncertainty. Instead of making a one-dimensional fit for the top mass, the top mass and the jet energy scale are fitted simultaneously. It exploits the fact that within $t\bar{t}$ events the jets from the W decay can be used to gain additional knowledge of the jet energy scale. The major advantage of this method is that the jet energy scale uncertainty will improve directly with increasing statistics. The total uncertainty of $^{+4.1}_{-4.0}$ GeV is smaller than that of the Run I world average of ± 4.3 GeV.

The best result from the DØ collaboration in Run II with a luminosity of 230 pb^{-1} yields $170.0 \pm 4.2(\text{stat.}) \pm 6.0(\text{syst.})$ GeV and the uncertainty is dominated by the uncertainty of the jet energy scale. Additional measurements were made in events where both W-bosons decay leptonically and yield $m_{top} = 165.3 \pm 6.3(\text{stat.}) \pm 3.6(\text{syst.})$ GeV (CDF with $\mathscr{L}dt$ = 340 pb^{-1}) and $m_{top} = 155\,^{+14}_{-13}\,(\text{stat.}) \pm 7(\text{syst.})$ GeV (DØ with $\mathscr{L}dt$ = 230 pb^{-1}).

By the end of 2006 the Tevatron experiments will have accumulated an integrated luminosity of about 2 fb^{-1} per experiment. With this luminosity a precision of $\approx$ 35 MeV is expected for the W-boson mass and $\approx$ 2 GeV for the top-quark mass. These measurements will further test the electroweak sector of the Standard Model and indirectly constrain the mass of the Higgs boson within the Standard Model.

At the LHC the precision is expected to be further improved due to the large statistics available: the goal is to measure the W-boson mass to 15 MeV and the top mass with a precision of $1-2$ GeV.

The Standard Model predicts that the cross sections for charged and neutral current ep deep inelastic scattering should exhibit dependence on the longitudinal polarisation of the incoming lepton beam. In the charged current case the dependence is predicted to be linear with the cross section becoming zero for right-handed (left-handed) electron (positron) beams, due to the chiral nature of the Standard Model. Figure 2 shows measurements of the cross section for charged current scattering as a function of the longitudinal polarisation of the lepton beam from the H1 collaboration [4]. The measurements are in agreement with the Standard Model. Similar measurements have also been made by the ZEUS collaboration [5].

The neutral current scattering process at HERA is sensitive to Z-boson exchange at large virtualities and can be used to constrain the couplings of the light quarks to the Z

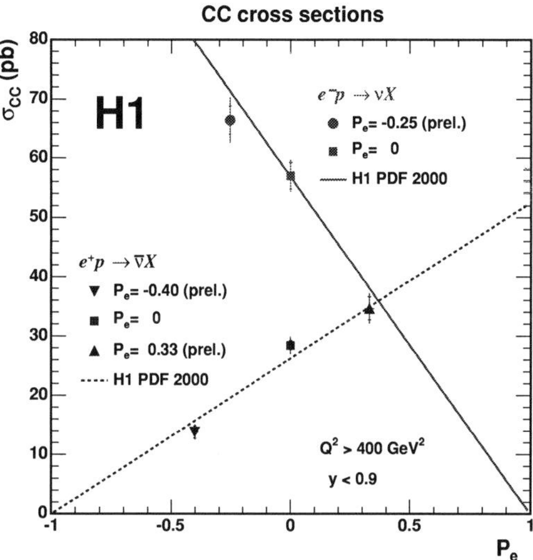

FIGURE 2. Measurements of the cross section for $e^{\pm}p$ charged current deep inelastic scattering as a function of the longitudinal polarisation of the lepton beam from the H1 collaboration.

boson. The results are shown in Figure 3 for the axial and vector couplings of u- and d-quarks using data from H1 corresponding to a luminosity of $\mathscr{L}dt = 100$ pb^{-1} [6].

It is seen that the precision is better for up-quarks as expected due to the larger u-quark density in the proton. It is also seen that the measurements resolve the ambiguity in the LEP data but are of inferior precision to the LEP data. It is expected that the precision will be improved substantially with the HERA II data and with the running with polarised lepton beams.

Top-quark and W-boson production

Due to the large mass of the top quark it can only be produced at the Tevatron $p\bar{p}$ collider (singly and in pairs) and at the HERA ep collider (singly). However, within the Standard Model the rate of top production at HERA is too small to be observed while at the Tevatron the cross section of 6.7 pb [7] for pair production and 3.9 pb for single top production is large enough, and the top quark was observed in pair production in 1995.

The Tevatron collaborations have measured the top quark pair production cross section in many different modes and using many different analysis techniques [8]. A summary of all these measurements is shown in Figure 4.

All cross section measurements agree with the theoretical expectation. The highest precision measurements are the kinematic measurement using a neural network and the measurement using a secondary vertex tag in the lepton + jets channels. Similar measurements from the DØ collaboration are also in good agreement with the theoretical

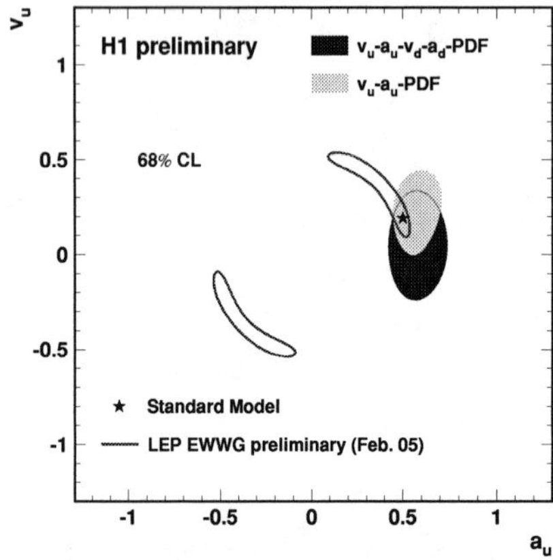

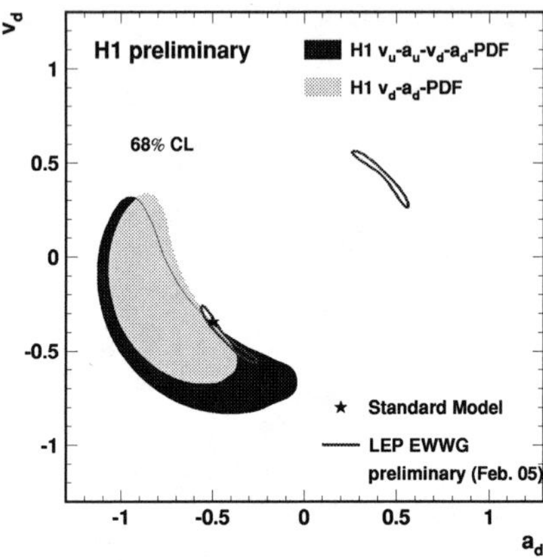

FIGURE 3. Vector versus axial coupling to the Z boson for up-quarks (left) and down-quarks (right). The open contours show the 68% C.L. contours for LEP data and the dark (blue) and light (yellow) contours show the regions allowed by H1. The dark contour shows the result of a fit for all couplings simultaneously and the light contour show the result fixing either the u- or the d-quark couplings.

prediction.

The production of single top quarks can go via two different processes. The s-channel process $q\bar{q} \to W^* \to tb$ where W is produced in the s-channel and decays into a t and b quark. In the t-channel process a W boson is exchanged in the t-channel between a b-quark and another quark, resulting in a final state of a t-, a b- and a 3rd quark. The cross section for the s-channel process is 0.88 ± 0.14 pb, and for the t-channel process it is 1.98 ± 0.30 pb [9].

CDF and DØ have searched for single top production but neither the s- or t-channel process has yet been observed. The best sensitivity is currently achieved by a neural network based analysis from the DØ collaboration with $\mathscr{L}dt = 230$ pb^{-1} [10]. Two separate neural networks are trained to discriminate against the main backgrounds: $t\bar{t}$ and $W + b\bar{b}$ production. An example of the output of neural network which discriminates between $t\bar{t}$ and single top is shown in Figure 4. The data are well described by the background predictions and no evidence for single top production is seen. Thus upper limits at 95% C.L. are set on the production cross sections, separately for the s-channel, $\sigma_s^{95} < 6.4$ pb, and t-channel, $\sigma_t^{95} < 5.0$ pb. CDF obtains upper limits of 13.6 pb in the s-channel and 10.1 pb in the t-channel with $\mathscr{L}dt = 162$ pb^{-1} [11].

With the expected integrated luminosity of Run II at the Tevatron, single-top production as predicted by the Standard Model will be discovered and the W-t-b coupling can be measured. A global analysis using measurements of W-boson polarisation from top-quark decays (in $t\bar{t}$ events) and the s- and t-channel single-top production cross sections can provide the most general determination of W-t-b couplings [12]. Furthermore, different models for the electroweak symmetry breaking mechanism could be distinguished by applying such a global analysis to the LHC data when it becomes available.

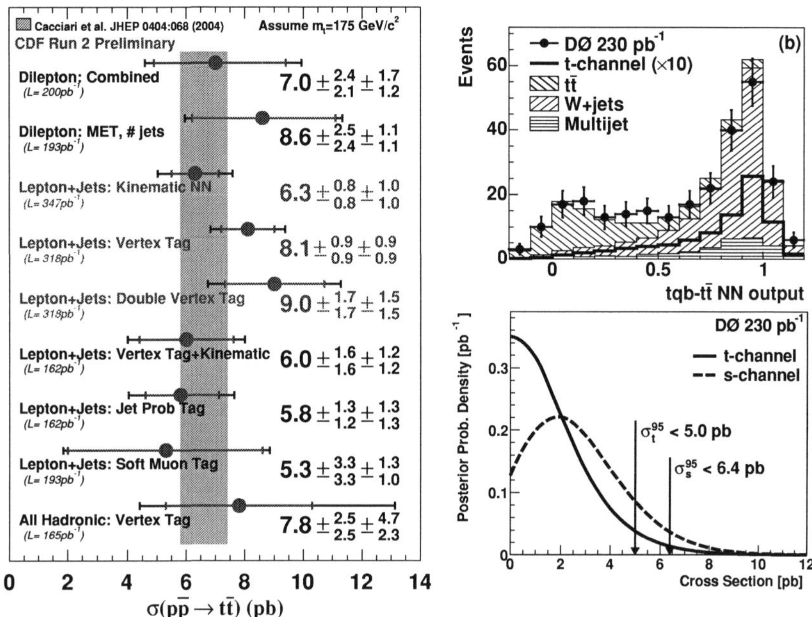

FIGURE 4. Left: top quark production cross section for several CDF measurements using different decay channels: the "Dilepton" channel ($t\bar{t} \to WWbb \to l\nu l\nu bb$), the "Lepton+jets" channel ($t\bar{t} \to WWbb \to l\nu jjbb$) and the "All Hadronic" channel ($t\bar{t} \to WWbb \to jjjjbb$). Right: Neural network output for single top production to discriminate against $t\bar{t}$ production (top) and probability density function for single top production cross section for the t- and s-channel production mechanisms (bottom). The 95% C.L. upper limits are indicated.

At HERA the H1 and ZEUS experiments have searched for direct W production. H1 searches for events with an isolated high momentum lepton and large missing transverse momentum. In the HERA I e^+p data ($\mathscr{L}dt = 104.7\ \text{pb}^{-1}$) an excess of these events was observed [13]: 18 events were observed compared to a Standard Model prediction of 12.4 ± 1.7. The excess was more pronounced at large p_T of the hadronic recoil, $p_T^X > 25$ GeV: 10 events were observed and 2.9 ± 0.5 expected. No such excess is observed by the ZEUS collaboration [14]: 36 events were observed and $32.5\ ^{+1.8}_{-4.7}$ were expected. At high p_T^X ZEUS observe 7 events and expect $5.7\ ^{+0.7}_{-0.4}$. The purity of the samples for W production are about 80% for H1 and 45% for ZEUS.

H1 have analysed $\mathscr{L}dt = 53\ \text{pb}^{-1}$ of HERA II data and repeated the analysis using identical cuts [15]: 10 events are observed and 6.1 ± 0.9 events are expected. At high $p_T^X > 25$ GeV there are 5 events observed and 1.7 ± 0.3 events expected. ZEUS have not yet analysed the data from HERA II. It remains to be seen whether this excess continues in H1 and whether the two experiments agree with the increased luminosity expected during the next few years.

At Run II of the Tevatron, a more precise theory calculation that includes the effects from not only the initial state multiple QCD gluon emission but also final state NLO QED corrections to the production and decay of the W boson is needed in order to determine the mass (M_W) of the W boson to an accuracy of about 40 MeV. Such a calculation has recently been done [16]. Since the modeling of the transverse momentum (q_T) distribution of the W boson is also important for the determination of M_W, a possible broadening in the q_T distribution, as suggested by semi-inclusive DIS energy flow data, would yield a different M_W value. Hence, it requires the measurement of the q_T distribution of the Z boson produced in the large rapidity region at the Tevatron to calibrate the q_T distribution of the W boson [17].

Searches for the Higgs boson

The Higgs boson is the only particle predicted by the Standard Model which has not yet been observed. In addition, many extensions of the Standard Model predict additional Higgs bosons that are searched for at colliders.

The most stringent limit on the Standard Model Higgs boson comes from the LEP collaborations which excluded a Higgs boson with $m_H < 114.4$ GeV at 95% C.L.[18]. After the end of LEP the Tevatron is now the only collider where the Standard Model Higgs boson could be observed.

At the Tevatron the main discovery modes are $WH \to l\nu b\bar{b}$, $ZH \to \nu\bar{n}ub\bar{b}$ and $ZH \to l^+l^-b\bar{b}$ if the mass of the Higgs boson is below about 135 GeV since at low mass the Higgs boson decays primarily to b-quarks. At higher masses the higgs boson decays mostly to W^+W^- and this decay is used to search for this Higgs boson.

The search results for $WH \to l\nu b\bar{b}$ (CDF) and $ZH \to \nu\bar{\nu}bb$ (DØ) are shown in Figure 5 where the invariant mass distribution of the b-jets is shown. For Higgs production a resonance is expected while for the backgrounds a continuous spectrum is expected. For the WH analysis only one of the two jets is required to be b-tagged with a secondary vertex algorithm while for the ZH analysis two b-tagged jets are required. In both channels the data agree well with the Standard Model prediction and an upper limit on the cross section times branching ratio is set. The cross section limits of all Standard Model Higgs searches are shown in Figure 5 and compared to the theoretical prediction.

Currently the cross section limits are about a factor of 20 higher than the Standard Model predictions. With increasing luminosity and improved analysis techniques the Tevatron expects to be able to set 95% C.L. limits up to $m_H = 135$ GeV and make a 3σ observation if $m_H < 120$ GeV with a luminosity of 8 fb^{-1}.

The LHC will give the definitive answer to the Standard Model Higgs boson after three years of low luminosity running: the 5σ discovery reach is shown in Figure 6 for $\mathscr{L}dt = 30\ \text{fb}^{-1}$ for ATLAS and CMS. A discovery will be possible for Higgs masses between 100 and 800 GeV. In a wide mass range a discovery can already be made after one year of LHC data taking with $\mathscr{L}dt = 10\ \text{fb}^{-1}$.

The success of the Higgs boson search in the di-photon channel depends on how well we know about its background rates, i.e. di-photon rates produced via the $q\bar{q}$, qg and gg fusion processes. A theory calculation that includes the effects of multiple QCD gluon emission in the initial state has been performed to compare with Tevatron Run II data

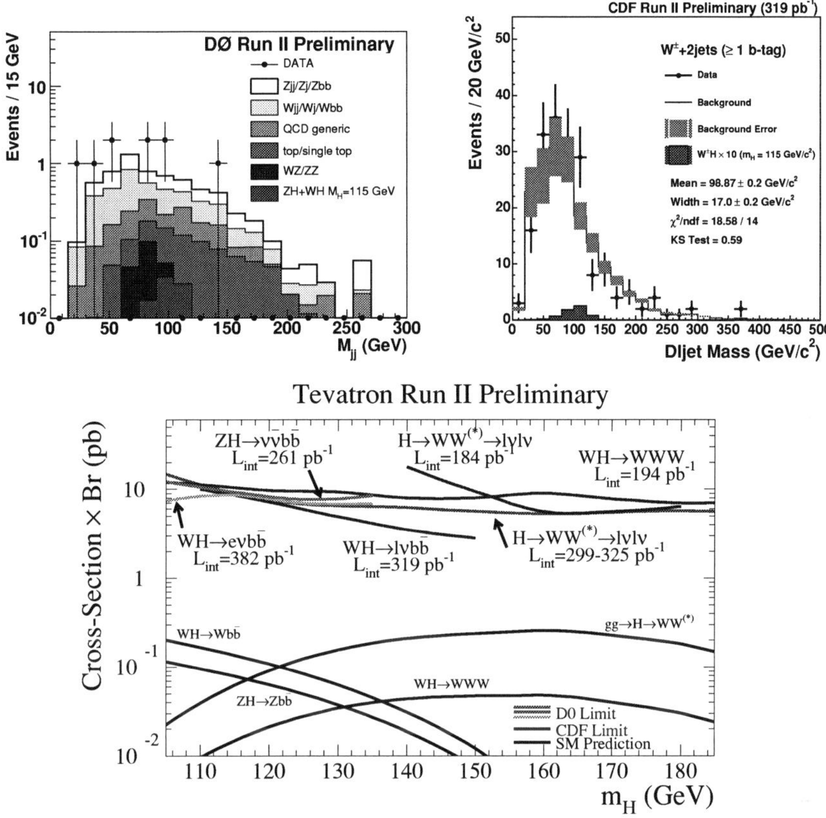

FIGURE 5. Top: $b\bar{b}$ invariant mass distribution for $ZH \rightarrow \nu\bar{\nu}b\bar{b}$ (left) and $WH \rightarrow l\nu b\bar{b}$ production (right). Bottom: Higgs Production cross section times the branching ratio: the experimental upper limits at 95% C.L. and the theoretical expectations for various production and decay modes are shown.

and the complete NLO calculation. It was found that the resummed calculation agrees well with the NLO calculation in the invariant mass distribution of the di-photon pair. The q_T distribution of the $\gamma\gamma$ pair agrees better with data than the NLO calculation in the low q_T region [19].

SEARCHES FOR PHYSICS BEYOND THE STANDARD MODEL

Higgs bosons beyond the Standard Model

In many extensions of the Standard Model additional Higgs bosons are predicted.

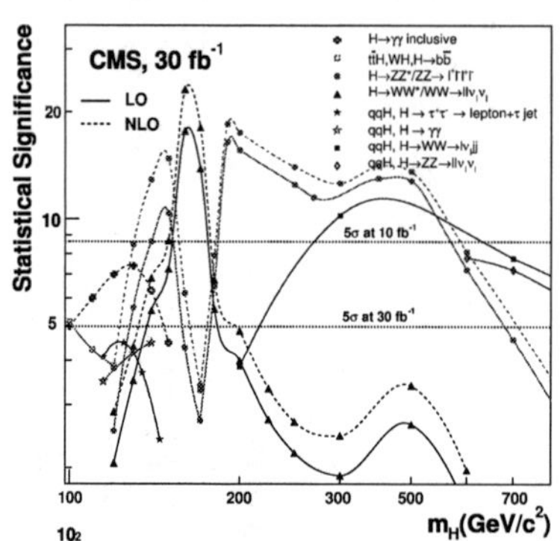

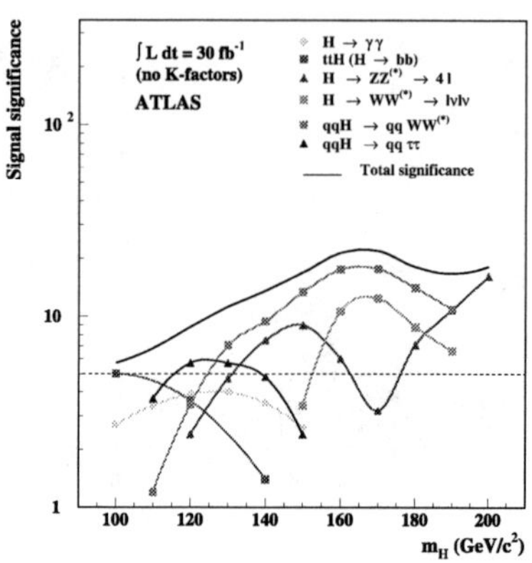

FIGURE 6. LHC Discovery reach for the Standard Model Higgs boson at CMS (left) and ATLAS (right) for $\mathscr{L}dt = 30\ \text{fb}^{-1}$.

In supersymmetry (see section below) there are two Higgs doublets, leading to the existence of 5 Higgs bosons: the neutral scalars h and H, the neutral pseudo-scalar A and two charged Higgs bosons $H^{\pm}$. The main parameter that governs the Higgs sector is $\tan\beta$ which is the ratio of the vacuum expectation values of the two Higgs fields. At low $\tan\beta$ the h behaves very similarly to the Standard Model Higgs. However, at large $\tan\beta$ the couplings to down-type quarks are enhanced ($\propto \tan^2\beta$) and the A is degenerate in mass with either the h or the H.

DØ have made a dedicated search for Higgs bosons at high $\tan\beta$ in the associated production of a Higgs boson with a b-quark: $p\bar{p} \to \Phi b + X \to b\bar{b}b + X$ where $\Phi = h, A, H$. No evidence was found for Higgs production and the data are used to constrain the allowed parameter space [20]. The result is shown in Figure 7 where $\tan\beta$ is shown versus the mass of the Higgs boson. It is seen that a large region of the parameter space is excluded by these data.

In left-right symmetric models doubly-charged Higgs bosons appear naturally. These decay into two leptons and several searches have been made by the Tevatron and HERA collaborations in the channels $H^{\pm\pm} \to e^{\pm}e^{\pm}$, $H^{\pm\pm} \to e^{\pm}\mu^{\pm}$, $H^{\pm\pm} \to \mu^{\pm}\mu^{\pm}$ and $H^{\pm\pm} \to \tau^{\pm}\tau^{\pm}$.

H1 observe a slight excess in high mass, $m_{ee} > 100$ GeV, dielectron production: 6 events are observed compared to a Standard Model background of 0.75 ± 0.13 events. However, these events are not consistent with doubly-charged Higgs production. Furthermore, that interpretation is also ruled out by the CDF and OPAL experiments as can be seen from Figure 7 where the coupling is shown versus the mass of the $H^{\pm\pm}$. Limits are also available for the other decay modes [21].

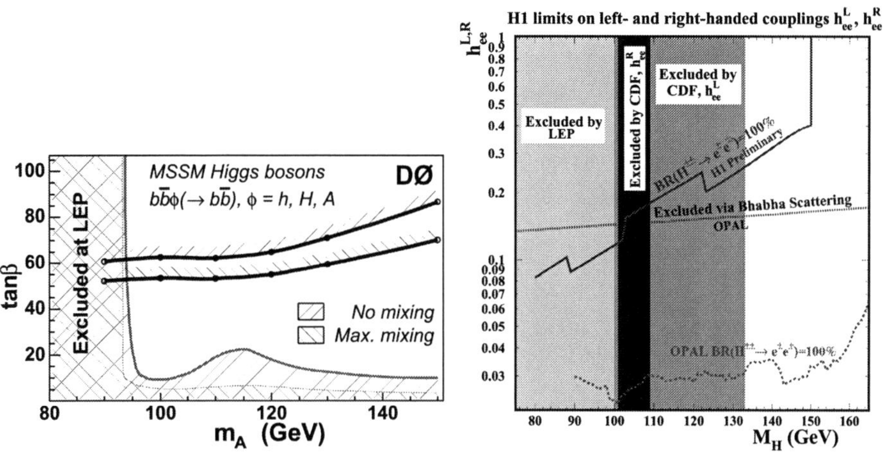

FIGURE 7. Left: Excluded regions of $\tan\beta$ versus m_A for the DØ and the LEP experiments. Shown are two scenarios ("no mixing" and "max. mixing") which result in different sensitivities due to different radiative corrections. Right: excluded region in coupling versus $m(H^{\pm\pm})$ plane. The H1, CDF, OPAL and LEP regions are shown.

Supersymmetry

Supersymmetry (SUSY) is one of the most promising candidates for a theory beyond the Standard Model. (A comprehensive introduction to SUSY models can be found in the talk given by H. Baer in these proceedings [22].) Through the introduction of superpartners, differing in spin by half a unit, for each of the Standard Model particles it provides solutions to many of the problems associated with the Standard Model and may provide a dark matter candidate particle. None of the superpartners of the Standard Model particles have been observed indicating that supersymmetry is a broken symmetry which predicts additional particles in the Higgs sector.

R-parity is a multiplicative quantum number defined as $R_p = (-1)^{3B+L+2S}$ where B is baryon number, L is lepton number and S is spin. It follows that Standard Model particles have $R_p = 1$ and SUSY particles $R_p = -1$. If R_p is a conserved quantity then SUSY particles may only be pair-produced and the lightest SUSY particle is stable and may be a dark matter candidate. Conversely in R-parity violating scenarios SUSY particles may decay to Standard Model particles. Searches for SUSY at HERA generally consider scenarios in which R-parity is violated, which yield the most competitive limits, while at the Tevatron both R-parity violating and conserving scenarios are explored.

The R_p violating Yukawa coupling λ'_{131} (where the indices refer to generation) allows the resonant production of the stop squark in eq fusion at HERA. The stop squark is of particular interest since due to the large top-quark mass, large mixing can occur in the third generation of squarks, which can lead to low stop masses for high $tan\beta$ SUSY scenarios. Stop quark decays have been searched for by the ZEUS collaboration [23]. No evidence for Stop production was observed and Fig. 8 shows the upper limits on

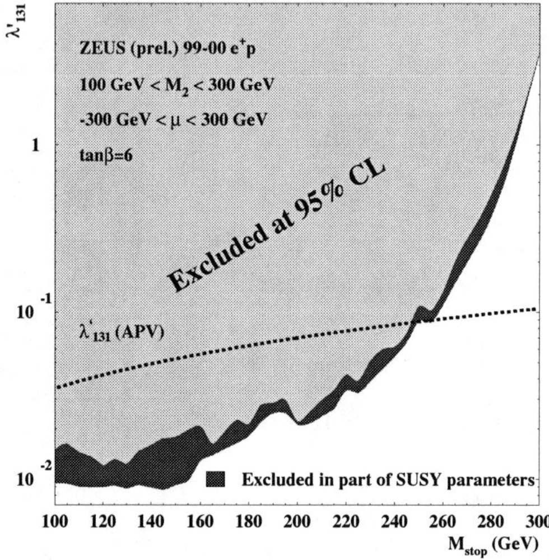

FIGURE 8. Limits on λ'_{131} and the stop mass from the ZEUS collaboration.

the coupling and mass of the Stop that were derived in the framework of the Minimal Supersymmetric Standard Model (MSSM). For a coupling of electromagnetic strength stop masses up to 265 GeV are excluded by this search. Comparable limits are obtained by the H1 collaboration in a similar scenario [24].

A search for light gravitinos in the Gauge Mediated Supersymmetry Breaking (GMSB) model has been performed by the H1 collaboration [24]. In this model the lightest SUSY particle is the gravitino, and decays of a neutralino to a photon and a gravitino were considered. In this way the search is independent of the squark sector. Events with an isolated photon, jet and missing E_T were selected and since no signal was observed limits were set on the neutralino masses and Yukawa couplings. For Yukawa couplings equal to one, neutralino masses up to 112 GeV can be excluded

R-parity conserving decays are characterised by large missing E_T from the lightest SUSY particle which remains undetected. The CDF and DØ collaborations have both searched for events in which a chargino and neutralino are produced and decay through channels with three leptons and missing E_T [25]. No signal is observed but the search yields lower limits on the mass of the chargino of 113-128 GeV depending on the parameters of the minimal supergravity (mSUGRA) model chosen. The DØ collaboration set a limit on the cross section times branching ratio to three leptons, for production of neutralino and chargino of around 0.2 pb. This extends the limit set by LEP, but unlike the LEP limit is model dependent. Similarly the Tevatron experiments have searched for chargino and neutralino production in the GMSB framework through the decay to two photons and missing E_T. The results from the two experiments have been combined to yield a lower limit on the chargino mass of 209 GeV. A large variety of other SUSY searches are also being explored by the Tevatron experiments and their discovery potential will benefit from increased accumulated luminosity in the future.

With a 14 TeV centre-of-mass energy and an expected integrated luminosity of 10 fb^{-1} per year in the first three years the LHC has vast potential to discover supersymmetry at the TeV scale. Detailed studies to develop reconstruction methods and algorithms to best measure various SUSY parameters are well advanced [26]. Full simulation studies show it will be possible to accurately determine the masses of the SUSY particles using a variety of complementary techniques and also to determine the spins of the new particles in order to demonstrate that they are indeed the predicted superpartners of the Standard Model particles.

To fully explore the discovery reach of the Tevatron and the LHC for supersymmetric Higgs bosons produced via bottom quark fusion, the parton distribution function (PDF) uncertainty in its production rate needs to be studied [27]. It was found that at the Tevatron the PDF uncertainty dominates the higher order uncertainty.

Recently, the production cross section for $gb \to H^- t$ at the LHC was calculated to higher orders with less scale dependence than the exact NLO calculation [28].

Leptoquarks and lepton-flavour violation

The symmetry between quarks and leptons in the Standard Model suggests some more fundamental theory may exist allowing direct interactions between quarks and leptons. Leptoquarks are hypothetical bosons which couple to a lepton and a quark and are suggested in many theories that extend the standard model to mediate such interactions. It is usually assumed that leptoquarks couple to fermions of the same generation. Models in which this is not the case introduce flavour changing neutral currents and lepton flavour violation.

Searches by the HERA experiments for leptoquarks produced in eq fusion yield limits on the masses of first generation leptoquarks that depend on the Yukawa coupling of the leptoquark to the lepton and quark. In contrast in searches by the Tevatron experiments for pair-production of leptoquarks the production mechanism is via the strong interaction through $q\bar{q}$ annihilation and gg fusion and is therefore independent of the leptoquark coupling.

Searches at HERA have been made for s-channel resonant leptoquark production through the electron-quark and neutrino-quark decay channels [29, 30]. For a coupling of electromagnetic strength, first generation leptoquark masses below 275-325 GeV can be excluded depending on the leptoquark type. In addition searches for lepton-flavour violation through the interactions $ep \to \mu(\tau)X$ have been performed. Such interactions could be mediated by leptoquarks and also by squarks in R_p violating supersymmetry models. No evidence for lepton-flavour violation was found and the results were interpreted in terms of limits on leptoquark and squark masses. For couplings of electromagnetic strength masses up to around 300 GeV can be excluded. Despite stringent limits on lepton-flavour violation from low energy experiments HERA improves on some of these limits, particularly in the $e - \tau$ transition when second or third generation quarks are involved.

Searches for pair-produced leptoquarks at the Tevatron have been made in both the charged lepton and neutrino decay channels [31]. Results from Run II include only the first and second generations so far and depend on the branching ratio to the charged lepton and quark final state, β. First generation leptoquarks with masses below 256 GeV and 234 GeV can be excluded for values of β of 1 and 0.5, respectively. The corresponding limits for second generation leptoquarks are around 225 GeV and 210 GeV.

Future leptoquark searches should benefit from increased luminosity at HERA and the Tevatron and in the case of HERA also longitudinal lepton-beam polarisation, should lead to more stringent limits.

Searches for new physics in the flavour sector

A complementary method to direct searches for new particles is the high precision study of complex low energy decays in which new physics can manifest itself through virtual loops and corrections to Standard Model processes. The SLAC and KEK B factories are ideal for this purpose since very high luminosities, clean signal topologies and low backgrounds make the BaBar and Belle experiments sensitive to rare processes at a level never before achieved. The large number of B mesons produced at the Tevatron and high precision tracking detectors also make the CDF and DØ experiments sensitive to such decays.

Searches for new physics through gluonic penguin decays of B mesons have been performed by the BaBar and Belle collaborations [32]. In these processes the b-quark decays to an s-quark and a gluon via an internal loop. By comparing these decays with those to charmonium which proceed without an internal loop, there is sensitivity to new particles contributing virtually. Figure 9 compares the values of $\sin 2\beta$ obtained from gluonic penguin processes to that measured in charmonium decays. A discrepancy of 3.7σ significance is observed between the penguin and charmonium decays, however improved statistics and a better theoretical understanding are necessary before a signature for new physics can be established. Similar measurements have been made by the CDF collaboration [33], however higher statistics are necessary to make the measurements competitive with the B factories.

Similarly, radiative penguin decays in which a b-quark decays to an s-quark and a photon are sensitive to new physics. The branching fractions for the fully inclusive process $B \to X_s\gamma$ and a semi-inclusive sum over many final states have been measured at the B factories and found to be in agreement with the Standard Model predictions. In addition the direct CP asymmetry has also been measured in these channels and found to be compatible with the Standard Model prediction of zero.

Decays of B mesons to di-leptons have very small branching fractions in the Standard Model since they are suppressed by CKM, GIM and helicity effects. However, many models for physics beyond the Standard Model predict large enhancements in these branching fractions, in some cases up to three orders of magnitude, for example in SUSY scenarios with high $\tan\beta$. Such decays have been searched for by the B-factory and Tevatron experiments and stringent limits set on the branching ratios. For example

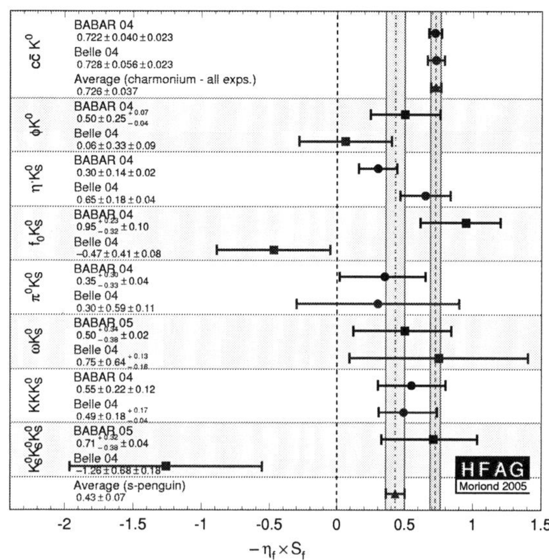

FIGURE 9. Measurements of $sin2\beta$ from gluonic penguin and charmonium decays by the Babar and Belle collaborations.

the CDF collaboration sets a limit of $Br(B_d \to \mu^+\mu^-) < 3.8 \cdot 10^{-8}$ at 90% CL.

Continued high luminosity running at the B factories and Tevatron offers the prospect of higher precision tests of the Standard Model in the flavour sector in the future.

Many new physics models could be compatible with current data and yield interesting new phenomena in future data. One such example is the MSSM with U(2) flavour symmetry in the quark sector [34]. For example, the apparent deviation in the decay branching ratio of $B_d \to \phi K_s$ could be explained together with all known flavour physics data.

Other searches

Many other searches for physics beyond the Standard Model have been performed, most notably at the Tevatron [31]. Direct searches for new particles in the di-lepton invariant mass spectra allow limits to be set on Randall-Sundrum gravitons, additional vector bosons such as the Z', technicolour mesons, little Higgs scenarios etc. Indirect searches yield limits on large extra dimension scales in excess of 1 TeV and quark-lepton compositeness scales from 4.2-9.8 TeV depending on the model. A direct search for magnetic monopoles was performed by the CDF collaboration using a novel time-of-flight trigger and dedicated track reconstruction algorithm, which resulted in a lower mass limit of 350 GeV for Drell-Yan pair production.

For a more extensive discussion on new physics models with extra bosons and fermions, see the talks given by L. Wang [35] and T. Tait [36], respectively, in these proceedings. It was argued that extra bosons with electroweak gauge quantum numbers are well motivated by the "little hierarchy problem" [35]. Similarly, vector-like quarks are motivated by the precision electroweak measurement of A^b_{FB} [36].

SUMMARY

The electroweak Standard Model continues to give an extraordinarily good description of the wide range of measurements made at LEP, HERA, the Tevatron and the *B* factories. The expected increases in luminosity at HERA, the Tevatron and the B factories in the next few years will test the accuracy of the Standard Model to further precision. The LHC will extend the potential for precision measurements and searches for physics beyond the Standard Model into new territory.

ACKNOWLEDGMENTS

It's a pleasure to be able to thank all the participants for their interesting talks and lively discussion, and the organisers for an exciting and perfectly organised workshop.

REFERENCES

1. P. Renton, in these proceedings.
2. C. Hays, in these proceedings.
3. J.-F. Arguin, in these proceedings.
4. A. Nikiforov, in these proceedings.
5. A. Tapper, in these proceedings.
6. B. Portheault, in these proceedings.
7. R. Bonciani *et al.*, Nucl. Phys. **B529**, 424 (1998); N. Kidonakis and R. Vogt, Phys. Rev. D **68**, 114014 (2003); M. Cacciari *et al.*, JHEP **404**, 68 (2004).
8. K. Ranjan, in these proceedings.
9. B. W. Harris *et al.* Phys. Rev. D **66**, 054024 (2002), Z. Sullivan, Phys. Rev. D **70**, 114012 (2004).
10. V. Abazov *et al.*, DØ Collaboration, preprint hep-ex/0505063, submitted to Phys. Lett. B.
11. D. Acosta *et al.*, CDF Collaboration, Phys. Rev. D**71** 012005 (2005).
12. F. Larios, in these proceedings.
13. V. Andreev *et al.*, H1 Collaboration, Phys. Lett. B**561** (2003) 241.
14. K. Piotrzkowski, in these proceedings.
15. C. Veelken, in these proceedings.
16. Q.-H. Cao, in these proceedings.
17. P. Nadolsky, in these proceedings.
18. R. Barate *et al.*, ALEPH, DELPHI, L3 and OPAL Collaborations and LEP Working Group for Higgs boson searches, Phys. Lett. B**565**:61-75, 2003.
19. C. Balazs, in these proceedings.
20. C. Hensel, in these proceedings.
21. A. Schoening, in these proceedings.
22. H. Baer, in these proceedings.
23. C. Horn, in these proceedings.
24. D. South, in these proceedings.
25. J. Zhou, in these proceedings.
26. N. Ozturk, in these proceedings.
27. A. Belyaev, in these proceedings.

28. N. Kidonakis, in these proceedings.
29. L. Lindfeld, in these proceedings.
30. G. Barbagli, in these proceedings.
31. R. Illingworth, in these proceedings.
32. T. Meyer, in these proceedings.
33. M. Herndon, in these proceedings.
34. S. Gopalakrishna, in these proceedings.
35. L.T. Wang, in these proceedings.
36. T. Tait, in these proceedings.

Summary of DIS05

Allen Caldwell

Max-Planck-Institut für Physik, Munich, Germany

Keywords: Deep-inelastic Scattering
PACS: 13.60

APOLOGIES

I wish to issue the usual apologies - this talk is not really a summary of the conference, but rather a discussion of some selected topics. No attempt is made to summarize in a balanced way all the material presented in the different sessions. These sessions were expertly summarized already by the conveners of the sessions. In particular, I will not have anything to say about the status of Pentaquark searches. The situation is quite confusing and requires a review of its own. For this topic, I refer the reader to the summaries of R. Jaffe and S. Maxfield and to the individual presentations. I will also not cover any topics in electroweak physics and searches for physics beyond the Standard Model. These are clearly interesting and will be a major focus of HERA II. However, they are outside the scope of my selected topics. I will not do justice to the theoretical developments, nor to the work at JLAB, nor the spin community, ... Rather, this 'summary' is really a commentary on where we stand in our understanding of the structure of hadrons and a discussion of possibilities for future research. It should be read in the right spirit - as a qualitative look at the state of DIS physics.

THE KEY TOPICS

In my view, the key topics addressed in experimental and theoretical studies of deep-inelastic scattering are:

- the study of the structure of hadronic matter;
- small-x physics, or the universal QCD fuzz;
- the value and running of the strong coupling constant, α_S;
- the precision to which we can predict cross sections involving strong interactions;
- the decomposition of the spin of the proton into its different components.

Many other topics are addressed in deep-inelastic scattering, such as measuring electroweak parameters, testing the Standard Model, searches for new physics such as Leptoquark production, searches for exotic physics such as Pentaquark production, etc. These are all important aspects of the research program, but DIS makes a more unique contribution to the topics listed above.

CP792, *Deep Inelastic Scattering, DIS 2005,* edited by W. H. Smith and S. R. Dasu

In the following, I will give my view on selected topics in the list, giving more weight to the topics where I consider myself more expert. I will conclude with a discussion of possible future projects.

HADRONIC STRUCTURE

How would we describe a proton to non-experts ? Would we resort to the infinite momentum frame, use light-cone variables, ... ? We would clearly like to describe the proton, or any particle, in its rest frame. The goal should be something like this: the 6-D (position and momentum) distribution of the quarks and gluons. We immediately run into the problem that the usual variables we use to describe proton structure, Q^2 and x need to be translated into a space-time picture. For Q^2, this is straightforward - the transverse distance scale probed is conjugate to Q, and we have the simple relation $r \sim 0.2\text{fm}/Q$, with Q in GeV. What about x ? The Bjorken-frame identification of x as a longitudinal momentum fraction is certainly not applicable in the proton rest frame, where the longitudinal momentum of a parton can be positive or negative and averages to zero. We would therefore like a different, more intuitive, variable to describe the proton scattering data at small-x. One possibility is to view the scattering as an (evolved) dipole-proton interaction.

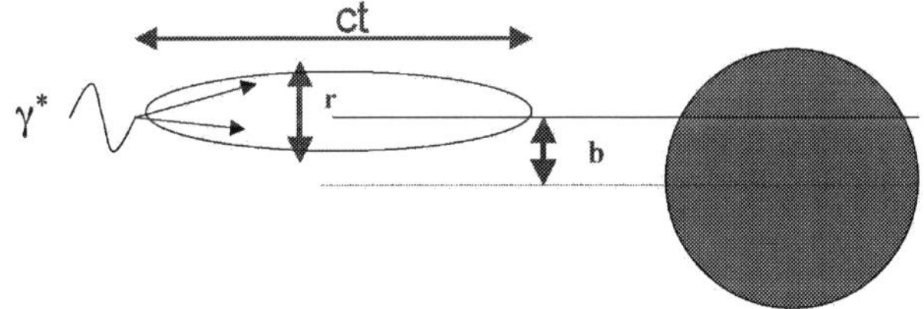

The distance scale over which the photon fluctuations survive scale as $1/x$, and this coherence length can be considered as a possible variable to describe the scattering. To ensure that we are studying proton structure, we should require $x > 0.1$, which corresponds to a coherence length of $ct < 1$ fm. For $x < 0.01$, we are measuring something else - the general properties of radiation in QCD or the time structure of the fluctuations. More on this later. We therefore make the classification:

- hadron structure $x \geq 0.1$;
- small-x physics $x \leq 0.01$.

The region in between is some mix of the two.

Recent Results

We have seen many interesting results on hadronic structure in this workshop. For example, S. Kuhn presented results on the structure functions of bound neutrons measured

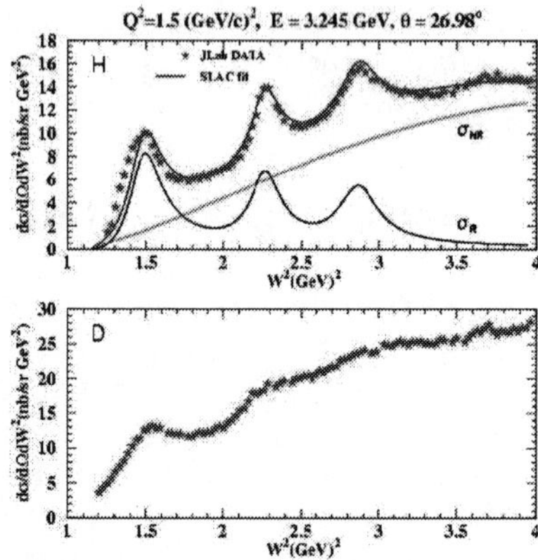

FIGURE 1. Proton and deuteron structure at small Q^2 and small W measured at JLAB.

at JLAB. An example of the extensive data available is given in Fig. 1. Here the data are plotted as a function of W, the center-of-mass energy of the final hadronic system, and the familiar resonances are very clearly seen when the scattering takes place on a proton, but are strongly suppressed when the scattering takes place on a deuteron.

New results were also presented for much high Q^2 and W. The final NuTeV structure functions were shown by M. Tzanov, and are now seen to be in good agreement with CDHS measurements. The discrepancy with previous CCFR results at the higher values of x is understood. These data represent the most precise data on neutrino-nucleon scattering we are likely to have for quite some time. Y. Ning presented a new technique for measuring cross sections up to $x = 1$ at HERA. The technique relies on the observation that the scattered electron is measured with full acceptance at high-x for $Q \geq 500$ GeV^2. At the highest x, the value of x cannot be determined, but an integrated cross section up to $x = 1$ can be extracted. The ZEUS results using this technique are shown in Fig. 2. Precise nucleon structure data is available over a wide range of Q^2 and provides a genuine challenge to models purporting to describe this structure.

Our final understanding of hadronic structure will likely come from the lattice. Interesting results were shown in the workshop by D. Renner on quark densities as a function of x and distance from the center of the proton, indicating that steady progress is being made on this front. A. Belitsky presented beautiful pictures on quark imaging via quantum phase space distributions, albeit in a fast-moving frame. One of his pictures is reproduced in Fig. 3. Maybe we will eventually have a 6-D picture of hadrons !

SMALL-X PHYSICS

Small-x physics means long radiation chains. Given that every subsequent fluctuation has a shorter duration than the parent, this implies that small-x physics can be interpreted as the study of QCD radiation at very short time scales. The higher the energy available in the scattering, the shorter the time fluctuations which can be observed. The source

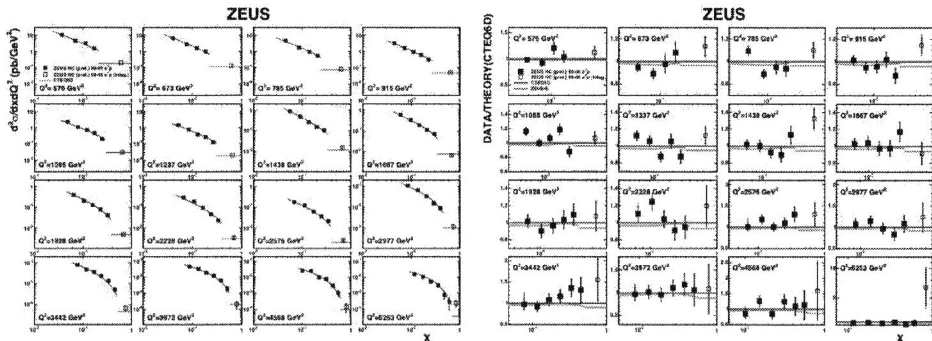

FIGURE 2. Differential cross sections measured by ZEUS as a function of x in Q^2 bins. The right-hand plot shows the ratio of the measurements to expectations based on the CTEQ6D pdf. The cross section in the highest x bin is $\int \frac{d^2\sigma}{dxdQ^2} dx/\Delta x$, where Δx is the size of the bin. This value is plotted at the center of the bin.

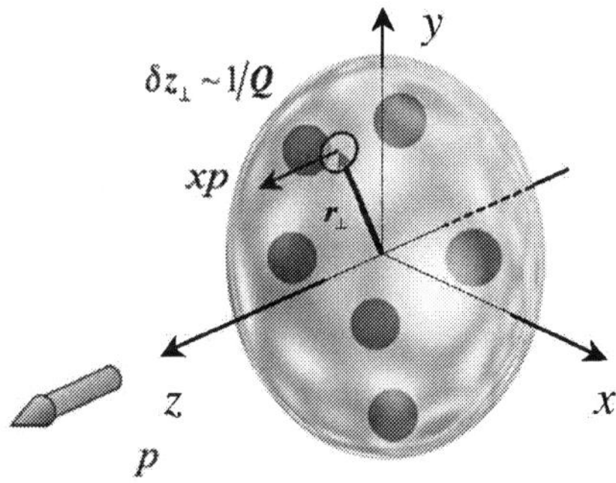

FIGURE 3. Quark imaging in the proton, from A. Belitsky.

of the radiation in a scattering process cannot be identified uniquely, and the intuition depends on the reference frame chosen. However, in a long chain one can suppose that the knowledge of the initial conditions has been forgotten and the behavior of the scattering cross sections with energy should become universal. This appears to be supported by the data. As Donnachie and Landshoff pointed out, hadronic total cross sections can be described by a universal energy dependence at high energies. This same energy dependence fits the HERA photoproduction data. It therefore appears that for soft scattering, this universality is present. What about at larger Q^2 values ? HERA discovered the rise of the structure functions at small-x, and this rise is clearly Q^2 dependent (see Fig. 4). A transition from the soft scattering behavior to a steeper dependence of the cross section on energy is seen around $Q^2 \approx 0.5$ GeV^2. However,

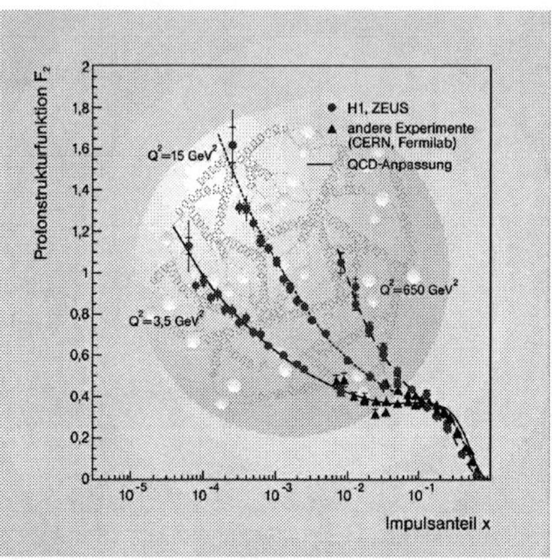

FIGURE 4. Sample of proton F_2 data, showing the change in the x-dependence as Q^2 changes.

it appears that, for fixed Q^2, the energy dependence is again universal. For example, the ratio of the diffractive cross section to the total DIS cross section at HERA is flat vs energy (see Fig. 5). The structure function data for events with a leading neutron, dominated by electron-pion scattering, also shows the same behavior as the total cross section. The implication is that whether the electron is scattering on a quark from the proton, or from a constituent of the proton such as a Pomeron or pion, the energy dependence of the cross section is the same. I.e., the quark and gluon densities evolve in a universal way.

The behavior of the total cross section at small-x is strikingly simple. The cross section rises as a power of W which is Q^2 dependent, and is otherwise featureless. This is in contrast to the large-x behavior, where the real structure of the proton is seen. However, the simplicity of the data has not led to a simple theoretical modelling. Despite the efforts of a large number of theorists over a period covering twenty years, we still do not have a satisfactory theoretical understanding of the high energy behavior of the scattering cross sections. In the language used earlier, we do not have a description of the QCD fluctuations in the time domain, in particular for very short times. The Golec-Biernat, Wüsthoff model has helped tremendously, in that it provided a framework for explaining simultaneously the total as well as diffractive cross sections. Further developments of this model have allowed for better fits and for the mapping of the hadronic matter profile in the proton. The discovery that this model contained a new type of scaling - geometric scaling - has prompted many theorists to look for new ways to attack the small-x physics, and we heard interesting talks, e.g., from A. Stasto and R. Enberg about the relationship of the small-x physics to travelling wave solutions to known nonlinear differential equations. The Color Glass Condensate is a related approach to the small-x physics which also provides an appealing physical picture of small-x physics. The point made by L. McLerran, R. Venugopalan et al., is that there should exist a kinematical region where perturbative calculations are possible although the Q^2 scale is small. The large gluon density at small-x introduces a new scale, the saturation scale, Q_S, which

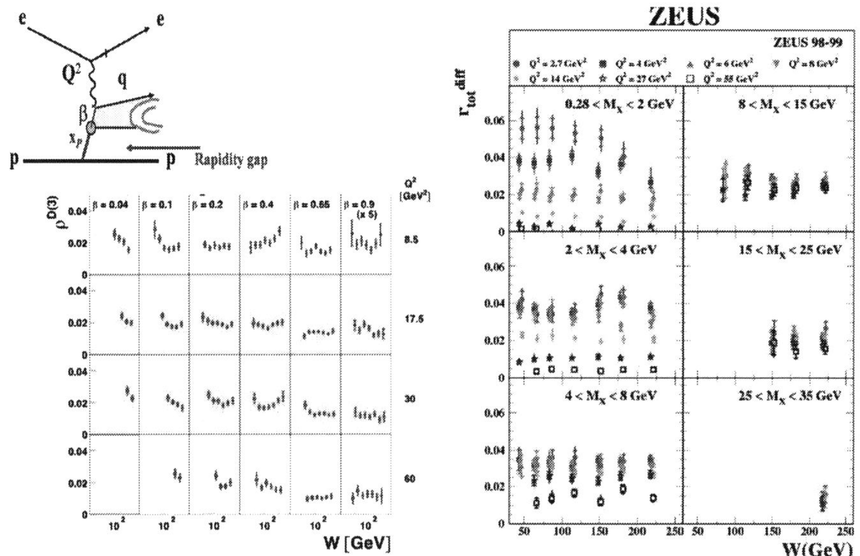

FIGURE 5. Ratio of diffractive to total scattering cross sections versus W and binned in different variables. The left plot shows H1 data, whereas the right plot shows ZEUS data.

can be large enough so that the tools of classical field theory can be used. J. Jalilian-Marian made the point that the CGC made predictions for RHIC which have indeed been validated. Nevertheless, I think it is still fair to say that we are still some ways from a satisfactory understanding of the small-x physics.

The Gluon Density

Understanding small-x processes means understanding **the** gluon density. The **the** is in bold face because I refer to the universal gluon density present around all particles (in all interactions). In DIS, we customarily call this the gluon density of the proton, but as discussed we could also view the scattering as an evolved dipole scattering on the proton. We would then be talking about the gluon density of the electron or photon. Figure 6 gives an indication of our present knowledge of the gluon density. There are clearly large uncertainties at both small- and large-x. The ZEUS collaboration has shown in this workshop that adding jet production data can help reduce the strong correlation between the gluon density and α_S found in fits to structure function data. This is clearly an important step and will result in more reliable parton densities from HERA.

I will focus here on the small-x gluon density. Eleven years ago, we held a workshop at Nevis Labs, Columbia University, dedicated to investigating the different possibilities for measuring the gluon density at HERA. Many processes were considered, from a measurement of F_L, to vector meson production, jet production, etc. At the conclusion of the workshop, the mood was rather gloomy - although many techniques were shown

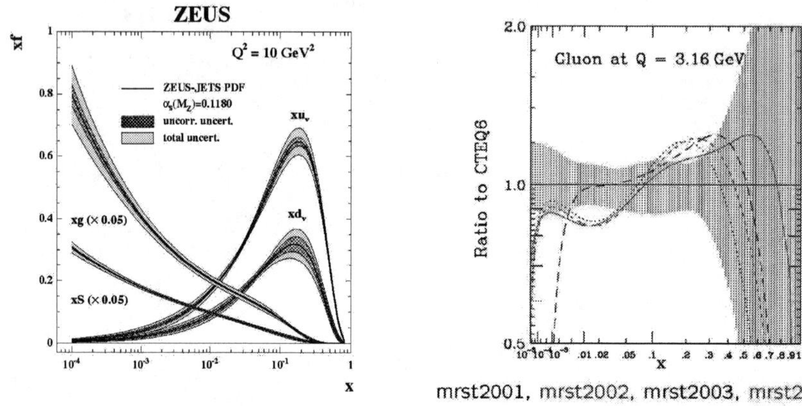

FIGURE 6. Recent result on parton densities from ZEUS (left). The right-hand plot shows the ratio of different gluon densities to that from CTEQ6.

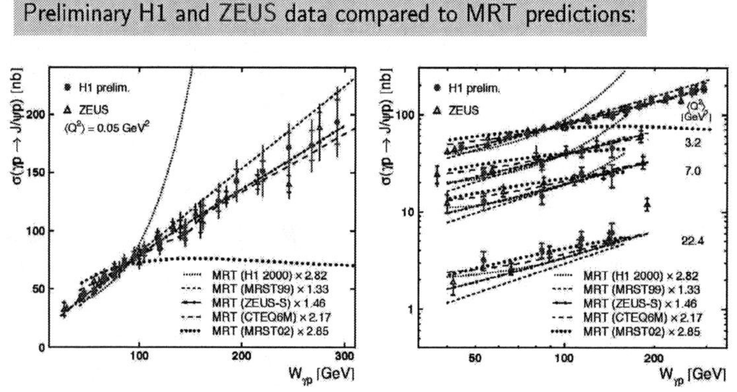

FIGURE 7. W dependence of the J/ψ cross section compared to calculations by Martin, Ryskin and Teubner using different gluon densities.

to in principle give access to the gluon density, the theory was clearly not advanced enough to make quantitative statements. It appears we are now on the brink of a new era, where exclusive processes can be used to help pin down the gluon. As mentioned above, HERA jet data are now used in pdf extraction fits. In addition, M. Teubner showed at this conference that predictions for exclusive J/ψ production depend very strongly on the gluon density. The calculations are compared to H1 data in Fig. 7. While the normalization of the calculations is still very uncertain, the shapes are less so. If the curves in the figure are to be taken seriously, then it is clear that many gluon density parametrizations could immediately be ruled out. The claim, and hope, is that we are now close to this goal.

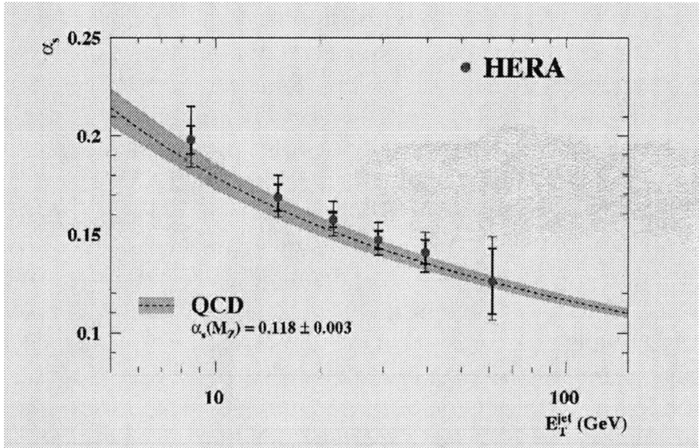

FIGURE 8. The strong coupling constant, α_S, measured from jet rates.

The gold-plated measurement for the gluon density has always been the measurement of F_L. This is a difficult measurement, requiring running HERA with different beam energies. A study by M. Klein and C. Gwenlan-Barr indicates that a limited low energy run at HERA would indeed be able to discriminate between different gluon densities. This running option should certainly be considered very seriously. The small-x behavior of the parton densities, and in particular the gluon, are likely to be the legacy of HERA.

α_S

The strong coupling constant, α_S, is the least well known of the fundamental coupling constants, and therefore merits all the attention it receives. Not only is the absolute value important, but the running of the coupling is also important for understanding the unification of the different forces. Many different techniques are available in DIS to extract this parameter, from the evolution of the structure functions with Q^2, to the measurement of jet rates, events shapes, etc. Getting consistent answers from these different approaches is a valuable test of how well we can calculate using QCD. A summary of α_S measurements at HERA using jet rate data, presented by C. Glasman, is shown in Fig. 8. The HERA data provides not only precise information on the value of α_S, but also provides clear evidence for the 'running' of α_S as a function of scale. The dominant uncertainty on α_S has been, and continues to be, the theoretical uncertainty - in particular the definition of the scale at which α_S is measured. Further progress on our knowledge of α_S is contingent on getting better theoretical control of the scales involved.

PRECISION QCD TESTS

X. Zu quoted a translation (by J. Legge) of Confucius: "... The mechanic, who wishes to do his work well, must first sharpen his tools ...". Now that there is general agreement that QCD is the correct theory to describe the strong interactions, the task is to understand what QCD tells us. We still do not have a first-principles understanding of some of our most basic observations, such as how color confinement comes about, why the constituent quark model works so well, the chiral phase transition, how spin is built up in a hadron, or why the high-energy behavior of the cross sections are as they are.

In addition to trying to understand what is implied by QCD, we need precise calculations so we can distinguish new physics from 'known' physics. Our standard calculational tool is perturbative QCD, and the standard approach has been to make more and more precise perturbative calculations. L. Dixon presented a summary of where theorists stand in this task. Steady progress is visible, despite the fact that the complexity of the calculations is growing enormously. On the non-pQCD front, lattice QCD is becoming ever more reliable, but also here the calculational barrier is tremendous. Meanwhile, the search goes on for different calculational approaches. A recent favorite is the application of the twistor-space approach to multi-parton scattering which has brought string theorists and QCD experts closer together. On the phenomenological side, there is continued progress on producing more reliable parton densities, in particular with respect to the uncertainties on the pdfs. J. Pumplin showed us the pitfalls awaiting those who attempt to produce global fits to world data to extract the pdfs. The standard rules of statistics are thrown out the window and pragmatic approaches are required - i.e., this is the business of experts. A very important element for practically all predictions involving QCD at colliders is the event simulator. Models are employed to simulate the QCD radiation processes and the later recombination of the partons into hadrons. New developments here are always welcome. In an interesting contribution, X. Zu described her efforts at making event generators more true to QCD by including correct parton kinematics.

Beauty Production

As an example of the difficulty in trying to discern new physics from 'known' QCD, consider the status of open beauty production. The HERA results are summarized in Fig. 9. According to the standard rules of statistics, one would say that the theory (QCD at NLO) does not reproduce the data. I.e., we either need higher order calculations or we have found new physics. However, we were told on at least two occasions (S. Frixione in his opening plenary talk, M. Corradi in his summary talk on heavy flavors) that this level of disagreement does not indicate a problem with the calculations. In a Bayesian language, we could phrase the situation as follows:

$$P(theory+model|data)P(data) = P(data|theory+model)P(theory+model)$$

where *theory* is the prediction of NLO QCD and *model* contains the fragmentation and hadronization. The data is Fig. 9 is clearly at odds with the prediction, i.e., *P(data | theory)* is very small. Saying that we are happy with the result implies that either *P(data)* is quite small - i.e., we have a problem with the measurements (in this case, we would be using QCD as a guide for making proper measurements), or *P(theory+model | data)* is quite small. But we believe that QCD is correct. So, the problem must be in the modelling or the lack of higher order calculations. It would certainly be useful to have

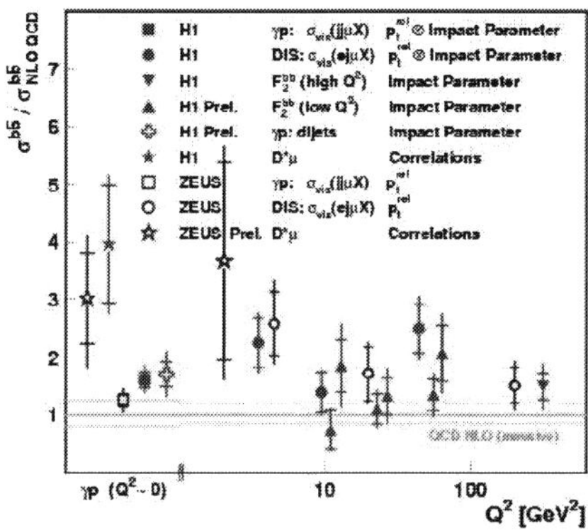

FIGURE 9. Summary of open beauty production at HERA.

some idea of our degree of belief in each of 1) the data; 2) the NLO predictions; 3) the modelling of the fragmentation and hadronization.

Jet Rates at the Tevatron

The example given in the last section is far from unique. There is the famous story of the high-E_T jet rates observed at the Tevatron and trying to puzzle out whether or not new physics was being observed. In the end, the discrepancy between data & theory was ascribed to uncertainties in the gluon density of the proton. The current status of the comparison of measured to predicted cross sections is shown in Fig. 10. The data and predictions have become much more precise, and currently no significant deviation is observed.

Parton Density Functions

The lesson from the jet rates was that we needed to understand the uncertainties in the parton density functions, and a tremendous effort was made by CTEQ and MRST to incorporate uncertainties with the pdfs. This effort has been taken up by other groups (H1 &ZEUS, Alekhin). There are also completely new approaches under study, such as the use of neural nets to extract pdfs and their uncertainties. J. Chacon presented the latest results from the NNPDF collaboration, and made the point that neural networks are the least biased approach, requiring only the assumption of smoothness of the pdfs. As an exercize for the method, the knowledge of the pdfs using only data prior to HERA was shown (see Fig. 11), and it was clear that all solutions, from rising structure functions at small-x to falling structure functions were compatible with existing data and QCD. This example shows that: 1) we cannot predict the x-dependence of structure functions, 2) the

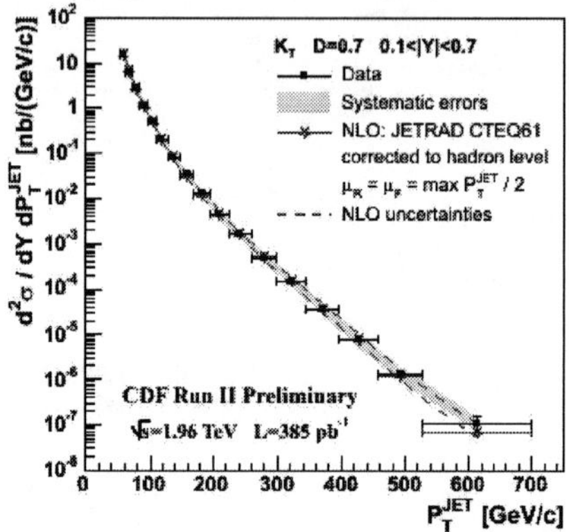

FIGURE 10. Jet differential cross section measured at the Tevatron by the CDF collaboration.

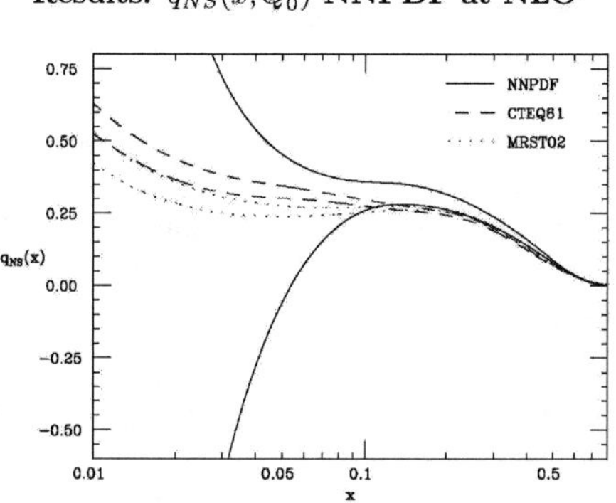

FIGURE 11. Neural-net results on the non-singlet structure function at small-x resulting from using pre-HERA data only.

rise of the structure functions with decreasing x at small x is therefore an experimental discovery, and 3) the NN approach appears to indeed produce sensible results.

J. Pumplin reiterated that the most serious problem in determining uncertainty bands on the pdfs is the problem of incompatible data sets. The point was made using with the plot shown in Fig. 12. Two different data sets are shown which are clearly incompatible given the quoted errors. All analysis methods assume Gaussian distributions of the observed values around the true values, and the toy data shown in Fig. 12 clearly violate this assumption. Someone made a mistake - to paraphrase Donald Rumsfled, there are ‘unknown unknowns’. All pdf fitting groups have to deal with this issue, and different choices are made on how to modify standard statistical methods to extract a best fit and uncertainty for the pdfs. I.e., the results are necessarily subjective and require expert

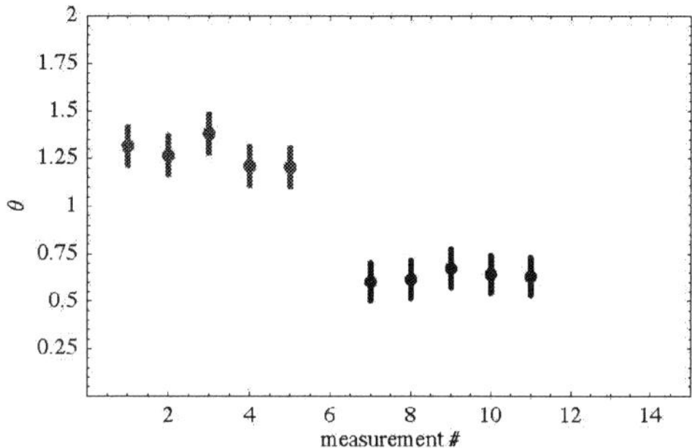

FIGURE 12. Representation of incompatible data sets used in pdf extraction, from J. Pumplin.

knowledge. Efforts to extract pdfs from a single or a limited number of data sets reduce this problem, but at the price of having less discerning data.

SPIN

We have no idea how the spin of a hadron is built up out of the constituent parton spins and orbital angular momenta. This is probably too strong a statement - the constituent quark model does very well at explaining the magnetic moments, and must contain some truth. So perhaps it is better to say that we have no idea how to build up the spin of a constituent quark from the partons. While the question seems rather straightforward, the techniques used to experimentally address this issue get ever more complicated. As D. Ryckbosch pointed out, traditionally lepton-nucleon spin physics meant measuring the structure function $g_1(x)$, and using this to extract the net spin carried by quarks and antiquarks. This resulted in the now well-known fact that we cannot build up the spin of a nucleon simply by adding the spins of the quarks and antiquarks, and the focus turned toward the spin carried by gluons and the spin contained in orbital angular momentum. A summary of the data on g_1/F_1 on protons and deuterons is shown in Fig. 13. The data are measured at many different Q^2, and yet match beautifully when plotted as a function of x. This implies that g_1 and F_1 have a very similar Q^2 dependence.

The (sparse) data on the spin carried by gluons is shown in Fig. 14. There are different approaches discussed to accessing the gluon content of the spin structure. One is to study the Q^2 evolution of g_1. This is limited by the relatively small lever arm available in Q^2. Other approaches mentioned include heavy flavor production or jet production. I note that similar processes have also been discussed for some time in the unpolarized community as techniques for extracting the gluon density, and only recently have been found to be useful. It is very difficult to improve on the scaling violations as a technique for extracting the gluon density.

Newer topics have arisen in spin physics in the last several years. It has become clear

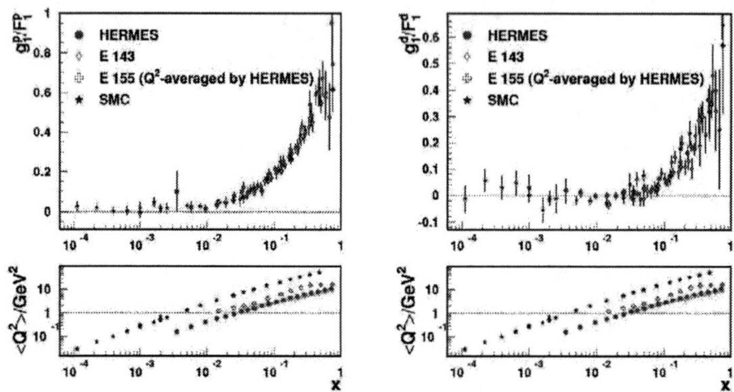

FIGURE 13. Summary of the world's data on the structure function g_1, scaled to F_1.

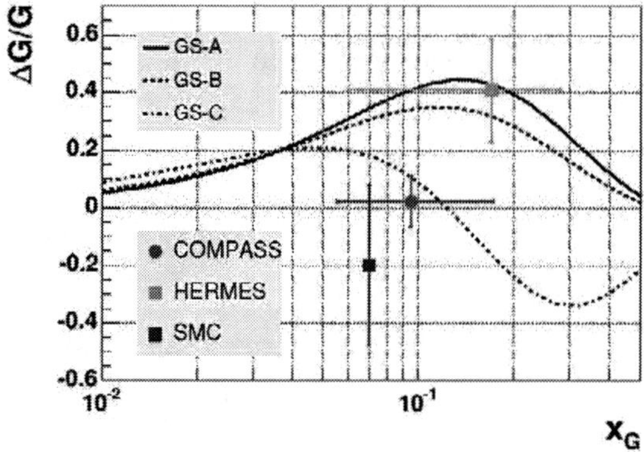

FIGURE 14. Available data on the net spin carried by gluons, compared with various calculations.

that a complete description will require also an understanding of the transverse spin distributions. New effects, such as the Collins or Sivers effects involving the transverse polarization of quarks have been introduced and are under study. One thing is obvious - new experiments over a broad kinematic range will be necessary to measure the required quantities over a large enough phase space to really understand how spin is built up from the constituents.

FUTURE PROSPECTS

There are several ongoing and proposed experiments to further our understanding of hadron structure (which I roughly defined as $x > 0.1$). These are

- The JLAB upgrade to 12 GeV electron beam energy and the subsequent experiments (described by C. Keppel);
- The MINERVA experiment at FNAL to study nuclear structure with a high-intensity neutrino beam (J. Morfin);
- Continued operation of COMPASS and future fixed target experiments at CERN (E. Rondio).

These experiments will primarily study the spin and flavor structure of the nucleon in the valence region. Very large data sets will be available so that multiply differential cross sections can be measured, and a genuine 3-D picture of hadronic matter extracted. E. Rondio also stressed the possibility for Compass to make precision measurements in the Q^2 range where HERA has observed a transition from a soft energy dependence to a Q^2 dependent variation of the W-dependence of the cross section. Decisions on going ahead with these projects, in particular the JLAB upgrade, are expected in the near future.

The eRHIC program, described by A. Deshpande, would cover the major topics of DIS: the mapping out of the flavor and spin structure of hadronic matter, and the detailed study of small-x physics, particularly in the realm of heavier nuclei. Given that RHIC exists, the expensive part of eRHIC has already been built. What is needed is an electron beam - the concept is for a ring to intersect RHIC at one interaction region. The strong points of eRHIC in relation to HERA would be the much larger expected luminosity, the ability to run with a variety of nuclei, the availability of polarized proton beam, and the variability of the beam energies over a wide range. The disadvantage relative to HERA is the considerably smaller center-of-mass energy. As B. Surrow has discussed, one detector will not be enough to get all the physics out of eRHIC: for the small-x physics and QCD studies, the ideal detector has acceptance over the largest possible rapidity range. This detector would require a lot of free space along the beam line. A design for such a detector is shown in Fig. 15, and follows in the lines of ideas from J. Bjorken regarding a full acceptance spectrometer. For spin structure function measurements, however, the key is luminosity, which is maximized by putting accelerator elements as near the interaction point as possible. If only one IP is possible, then a staged approach could be considered: first the low luminosity, large acceptance detector is built, and then the high luminosity version of the accelerator and detector are introduced.

Continuing on the road to higher energies and smaller-x, three other experimental possibilities were discussed. H. Kowalski presented calculations showing the acceptance and physics which could be reached with a very forward (420m) detector at the LHC. The basic reaction studied is shown in Fig. 16. This type of reaction has many advantages, as repeatedly stressed by V. Khoze: the quantum numbers of the produced system can be disentangled and the signal-to-background ratio for some processes, such as Higgs production, can be considerably improved. These reactions also allow the continued study of the gluon in the proton in an interesting kinematic region, and could allow the measurement of the energy dependence of pp diffractive reactions with a hard scale.

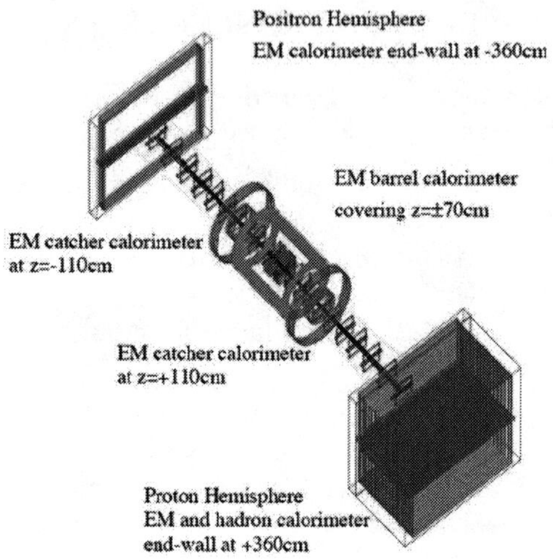

FIGURE 15. Conceptual design for a large acceptance detector for eRHIC.

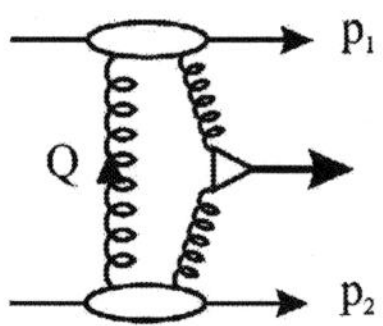

FIGURE 16. Diagram for a $pp \to ppX$ reaction at the LHC mediated by gluon exchange.

In principle, the ILC will also give access to small-x. However, the e^+e^- option is not ideal, since the rates will be low, and forward detectors would be difficult to implement. A better option for DIS type physics is an $e\gamma$ collider.

At the very highest energies, one could also consider building a linear electron accelerator to collide with the LHC. One could extend the HERA reach by one order of magnitude in x and thereby pursue the study of small-x physics in a completely new domain. The time scale for such a project is clearly very long.

There are clearly many ideas being discussed for future projects. The choice(s) of which of these projects to follow should obviously be driven by the physics. However, the time scale on which such projects could be realized should be considered in conjunction with the physics arguments, and a reasonable balance should be struck between the ultimate physics program and a realizeable program.

ACKNOWLEDGMENTS

I would like to thank the Wesley Smith and his team for organizing such an informative and pleasant workshop. I would also like to thank all those I had the pleasure of learning from and discussing with during DIS05.

STRUCTURE FUNCTION
WORKING GROUP PRESENTATIONS

Measurement of the proton structure function F_2 at low Q^2 in QED Compton scattering at HERA

Ewelina Maria Łobodzińska

DESY, Zeuthen Germany and INP, Cracow Poland

Abstract. The proton structure function F_2 is measured using inelastic QED Compton events. The data were collected by H1 experiment at the HERA in 1997 and correspond to a luminosity of 9.25 pb^{-1}. QED Compton events allow to access very low Q^2 region, down to 0.5 GeV^2 and Bjorken x up to ~ 0.06, a region that has not been covered by previous inclusive measurements at HERA. The results are in agreement with measurements from fixed targed lepton-nucleon scattering experiments.

Keywords: proton structure function F_2, inelastic QED Compton, low Q^2, medium x, low y, H1, HERA, *ep* scattering
PACS: 13.85.Lg

INTRODUCTION

Radiative processes in *ep* scattering - depicted in Fig.1 - are of special interest, since the photon emission from the lepton line gives rise to event kinematics which open new ways of investigating proton structure . In the present analysis [1] the QED Compton (QEDC) process is considered, which is characterised by low viruality of the exchanged photon and high virtuality of the exchanged electron. To account for the additional photon in the final state the standard kinematic variables x and Q^2 have to be redefined:

$$Q^2 = -q^2 = -(l - l' - k)^2, \qquad x = \frac{Q^2}{2P \cdot (l - l' - k)},$$

where k,l,l' represent the four-momenta of the radiated photon, incoming and outgoing electron, respectively.

In this analysis inelastic QEDC events were used. For these events the proton breaks up, so the $\gamma^* p$ cross section is defined through the proton structure functions F_2 and F_L. In the kinematical range studied in this analysis (low y region) the contribution from F_L can be neglected, such that the cross section is proportional to F_2.

QEDC EVENT SIMULATION

In order to investigate inelastic QEDC events an improved version of the COMPTON generator was developed [2]. For the simulation of the hadronic final state, the SOPHIA program [3] was used in the range of low Q^2 ($Q^2 < 2$ GeV^2) or low masses W ($W <$ 5 GeV). The SOPHIA model provides an accurate description of photon-hadron interactions reproducing large sets of available data. The simulation includes the production

CP792, *Deep Inelastic Scattering, DIS 2005,* edited by W. H. Smith and S. R. Dasu

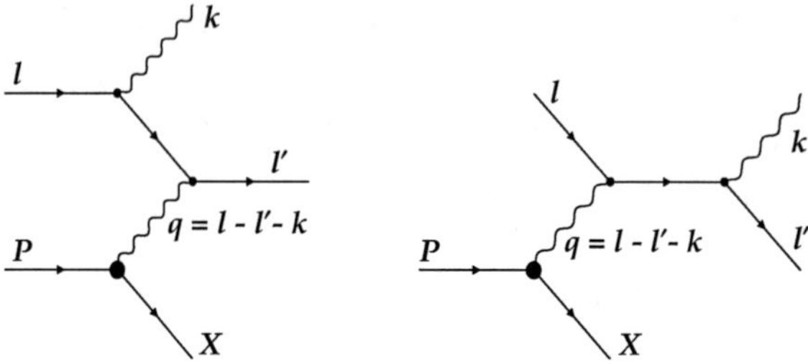

FIGURE 1. Lowest order Feynman diagrams for the radiative process $ep \to e\gamma X$ with photon emission from the electron line. The four-momenta of incoming electron and proton, outgoing electron, photon and hadronic final state are represented by l, P, l', k and X respectively.

of the major baryon resonances, direct pion production, multiparticle production based on the Dual Parton Model with subsequent Lund string fragmentation, as well as the diffractive production of the light vector mesons ρ and ω.

BACKGROUND REJECTION

The dominant background to the QEDC process arises from inclusive DIS events in which one particle from the hadronic final state (typically a π^0) fakes the outgoing photon. At high y, where the hadronic final state lies mostly in the backward region, this process is hampering a clean QEDC event selection. Therefore, the analysis had to be constrained to the low y values. Remaining background coming from this source is modelled using an inclusive DIS Monte Carlo simulation. Another source of significant background comes from the inelastic Deeply Virtual Compton Scattering (DVCS) events, in which the final state photon is diffractively produced in the virtual photon proton collision. This source of background is estimated to contribute 5.5% to the measured cross section. Elastic QEDC and DVCS events may also contribute to the measured cross section if the noise in the calorimeter is missidentified as hadronic activity. This source of background contributes 0-2% to the measured signal. Other background sources are even less significant [1].

EVENT SELECTION

To select QEDC events, two energetic clusters in the backward electromagnetic calorimeter are required. The sum of both energies must be close to the electron beam energy and the azimuthal angle between both clusters is supposed to be close to π. In addition, a well reconstructed interaction vertex is required. In order to remove elastic events at least one particle from the hadronic final state has to be detected in the calorimeter.

As mentioned above, to suppress the DIS background, the analysis is constrained to the low y region. The additional suppression of the DIS, photoproduction and dielectron

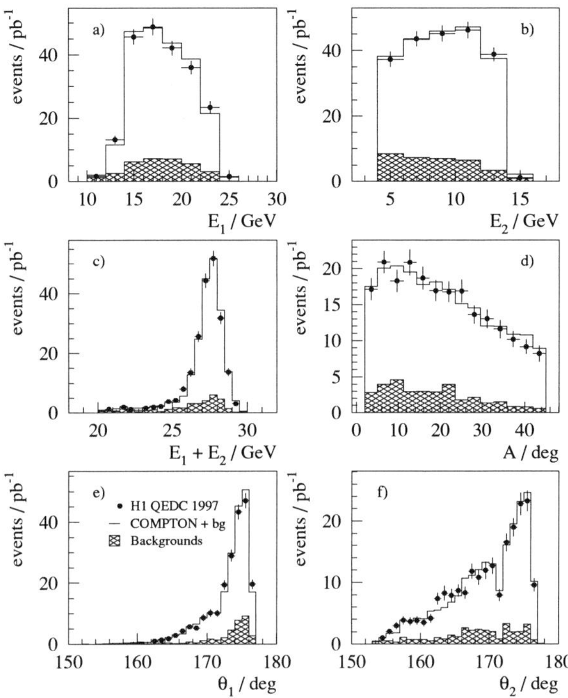

FIGURE 2. Control distributions for the measured electron and photon final state: a.) energy of the particle with the higher energy; b.) energy of the particle with the lower energy; c.) sum of both energies; d.) $e\gamma$ acoplanarity; e.) polar angle of the particle with the higher energy and f.) polar angle of the particle with the lower energy.

background is performed by requiring the residual energy in the electromagnetic Spacal to be below 1 GeV. Furthermore, to separate electrons and photons from hadrons, cuts on the shower shape estimators are performed. The control distributions shown in Fig.2 illustrate the good description of the electron-photon final state provided by the simulation.

RESULTS

In order to extract the structure function F_2 the data sample is divided into subsamples corresponding to a grid in y and Q^2. The bin sizes are adapted to the resolution in the maesured kinematic quantities such that the stability and purity in all bins shown are greater than 30%. [1] The statistical errors lie in the range 6 - 10%, while the systematic

[1] Here, the stability (purity) is defined as the number of simulated QEDC events originating from and reconstructed in a specific bin to the number of generated (reconstructed) events in the same bin.

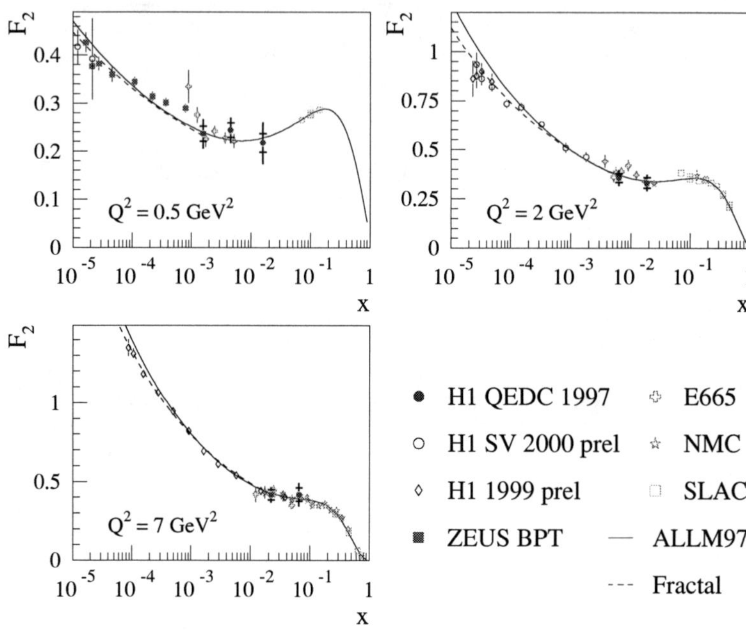

FIGURE 3. F_2 measurements from QED Compton scattering by H1 compared to other measurements at HERA and fixed targed experiments.

uncertainties are typically 9 - 12%, rising to 18% in the lowest y region. The total errors are obtained by adding the statistical and systematic uncertainties.

The F_2 values as measured with the QED Compton process are depicted in Fig.3 as a function of x for fixed Q^2 and compared to other HERA [4] and fixed targed data [5].

The present analysis extends the kinematic range of HERA measurements at low Q^2 towards higher x values, thus complementing standard inclusive and shifted vertex measurements. QEDC F_2 data are consistent with the results from fixed targed experiments.

REFERENCES

1. H1 Collab., A. Aktas et al., *Phys Lett B* **598** (2004) 159-171 hep-ex/0406029
2. V. Lendermann, Doctoral Thesis University of Dortmund, 2002 DESY-THESIS-02-004.
3. A. Mücke et al., *Comput. Phys. Commun.*, **124**, 290 (2000) astro-ph/9903478.
4. H1 Collab. C. Adloff et al., *Eur. Phys. J. C*, **21** (2001) 33 hep-ex/0012053,
   H1 Collab. C. Adloff et al. *Nucl. Phys. B*, **497** (1997) 3 hep-ex/9703012,
   ZEUS Collab. S. Chekanov et al., *Phys Lett B*, **487** (2000) 53 hep-ex/0005018
5. L. W. Whitlow et al., *Phys Lett B*, **282** (1992) 475 SLAC-PUB-5445,
   M. Arneodo et al. New Muon Collaboration, *Nucl. Phys. B*, **483** (1997) 3 hep-ph/9610231,
   M. R. Adams et al. E665 Collaboration, *Phys Rev. D* **54** (1996) 3006.

Measurements of the Structure Functions $F_2(x,Q^2)$ and $F_L(x,Q^2)$ at low Q^2

A. Petrukhin

DESY, Notkestrasse 85, 22607 Hamburg, Germany
ITEP, Bolshaja Cheremushkinskaja 25, 117218 Moscow, Russia

Abstract. Recent measurements of the proton structure function F_2 and determination of the structure function F_L in e^+p deep inelastic scattering (DIS) by the H1 Collaboration are summarized. The F_2 results presented here extend the previous measurements at HERA into the Q^2 region below 1 GeV^2 by also using Initial State Radiation (ISR) events. The F_L structure function determined in the current analysis is more precise and accesses lower Q^2 values as compared to previous determinations published by H1.

Keywords: F_2 proton structure function, longitudinal proton structure function, DIS, HERA, H1, initial state radiation, low Q^2, low x
PACS: 13.85.Lg

INTRODUCTION

At low Q^2 the DIS double-differential cross section can be expressed as:

$$\frac{d^2\sigma}{dx\,dQ^2}\cdot\frac{Q^4x}{2\pi\alpha^2Y_+}=\sigma_r=F_2(x,Q^2)-\frac{y^2}{Y_+}\cdot F_L(x,Q^2), \qquad (1)$$

where Q^2 is the squared four-momentum transfer, x is the Bjorken scaling variable, $Y_+ = 1+(1-y)^2$. The inelasticity $y = Q^2/sx$ is a fraction of the lepton's energy loss and s denotes the center of mass energy squared of the lepton-nucleon system. For $y < 0.6$ the contribution of F_L is supressed by the kinematic factor y^2/Y_+. Therefore the measurement of σ_r at low y determines F_2 with a small correction for F_L. At high y the contribution of the longitudinal structure function becomes significant which allows to determine F_L.

MEASUREMENT OF THE CROSS SECTION

The data used in the analysis are taken by the H1 detector during special running periods with open triggers for inclusive DIS events. In order to access low values of Q^2 around 1 GeV^2 the interaction vertex for the 2000 data was shifted by approximately 70 cm along the proton beam direction. This data sample is also used for accessing larger values of x with ISR events. Data collected in 1999 with a nominal position of the interaction vertex are used for measurements in the high y region and 1997 data for accessing of low values of y using QED Compton events (QEDC). A new ISR analysis method is

CP792, *Deep Inelastic Scattering, DIS 2005*, edited by W. H. Smith and S. R. Dasu

applied where the radiative photon is not required to be measured in the photon detector. Details concerning the ISR analysis can be found in [1].

The kinematic reconstruction was done with two methods. The sigma method [2] has been used for medium and low y and the electron method was prefered at highest y due to better y resolution in this region. The H1 and ZEUS inclusuve cross sections measurements at low Q^2 are shown in Fig. 1 together with the fixed target data.

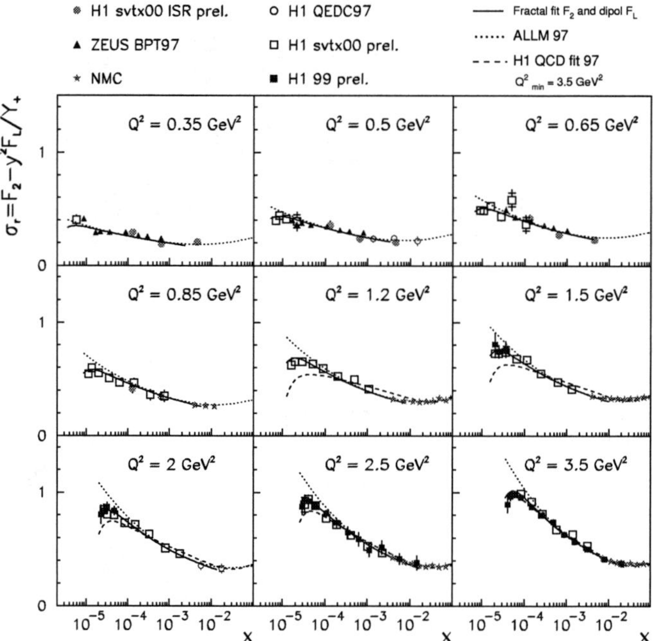

FIGURE 1. Measurements of the cross section - H1 shifted vertex ISR data 2000 (closed points) [1], data from ZEUS BPT97 (closed triangles) [3], NMC (stars) [4], H1 QEDC97 (open points) [5], H1 shifted vertex preliminary data 2000 (open boxes) [6] and H1 preliminary data 1999 (closed boxes) [7]. The data are compared to the Fractal fit F_2 [8] and dipol model F_L [9] (solid curve), to the ALLM97 [10] parameterization (dotted curve) and also to the H1 QCD fit [11] (dashed curve)

The preliminary results of the ISR analysis and the QEDC results from H1 extend the measured kinematic range at low Q^2 towards higher x values.

STRUCTURE FUNCTION F_2 AT LOW Q^2

The proton structure function F_2 obtained from the inclusive cross section measurements rises towards low x for all measured Q^2 bins. The detailed analysis of the x dependence of F_2 is described in [12]. It was found that the x dependence of F_2 at fixed Q^2 exhibits a power law behaviour $F_2 = c \cdot x^{-\lambda(Q^2)}$. The result of fitting this form to the shifted vertex data [1] and [6], compared to published H1 [13] and ZEUS [3] data, is presented in Fig. 2.

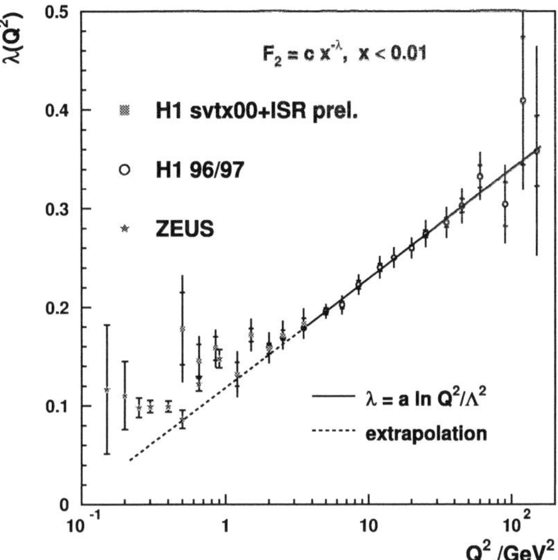

FIGURE 2. The slope $\lambda(Q^2)$ from the fit $F_2 = c \cdot x^{-\lambda(Q^2)}$ to the H1 and ZEUS data.

One can see that the shifted vertex results are in agreement with previously published data in the regions of overlap. Deviation of $\lambda(Q^2)$ from logarithmic dependence $a \cdot \ln[Q^2/\Lambda^2]$ was observed around Q^2=1 GeV^2.

THE EXTRACTION OF F_L

Fig. 1 shows the rise of the cross section with decreasing x for fixed Q^2. However, for the lowest x, a turnover of the cross section can be seen. Lowest x corresponds to highest y and the contribution of $F_L(x,Q^2)$ to the cross section is not negligible at such values of y. An extraction of F_L was done with a "shape method" [14]. It assumes the power behaviour of F_2 and the difference in the shape between the cross section and F_2 to be mostly driven by kinematic factor y^2/Y_+. Using the cross section, parametrised as $\sigma_r = F_2 - y^2/Y_+ \cdot F_L$, F_L for a given Q^2 was extracted by fitting. The results of such an F_L extraction, together with previously published H1 results and different QCD fits, are presented in Fig. 3.

Good agreement of the data and F_L determined by H1 [14], Alekhin's [17] and H1's QCD [11] fits is observed. However, MRST [16] and ZEUS [15] fits tend to be lower.

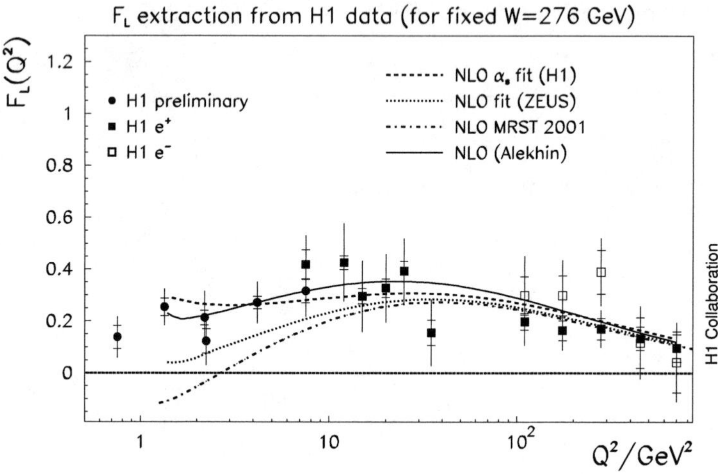

FIGURE 3. F_L obtained from H1 data compared to H1 QCD fit (dashed curve) [11], ZEUS NLO fit (dotted curve) [15], MRST NLO fit (dash-dotted curve) [16], Alekhin's fit (solid curve) [17]

SUMMARY

The inclusive cross section measurements presented here extend the kinematic region covered by HERA at low Q^2. The ISR and QEDC data allow access medium and high x values. The F_L values determined in this analysis are significantly above zero everywhere and already now can discriminate between different predictions.

REFERENCES

1. H1 Collaboration: contributed paper to the International Conference on High Energy Physics, ICHEP2004, Beijing, Abstract **170**, Parallel session **5**.
2. U. Bassler, and G. Bernardi, *Nucl. Instrum. Meth.* **A361** (1995) 197.
3. J. Breitweg *et al.* [ZEUS Collaboration], *Phys. Lett.* **B487** (2000) 53.
4. M. Arneodo *et al.* [NMC Collaboration], *Nucl. Phys.* **B483** (1997) 3.
5. H1 Collaboration: contributed paper to the International Erophysics Conference on High Energy Physics, EPS 2003, Aachen, Abstract **084**, Parallel session **4**.
6. T. Laštovička, *Acta Phys. Pol.* **B33** (2002), 2835
7. H1 Collaboration: contributed paper to the International Erophysics Conference on High Energy Physics, EPS 2001, Budapest, Abstract **799**
8. T. Laštovička, *Eur. Phys. J.* **C24** (2002) 529-533.
9. K. Golec-Biernat and M. Wüsthoff, *Phys. Rev.* **D59** (1999) 014017.
10. H. Abramowicz and A.Levy, hep-ph/9712415.
11. C. Adloff *et al.* [H1 Collaboration], *Eur. Phys. J.* **C21** (2001) 33-61.
12. C. Adloff *et al.* [H1 Collaboration], *Phys. Lett.* **B487** (2001) 183.
13. C. Adloff *et al.* [H1 Collaboration], *Eur. Phys. J.* **C21** (2001) 33.
14. H1 Collaboration: contributed paper to the International Conference on High Energy Physics, ICHEP2004, Beijing, Abstract **161**, Parallel session **5**.
15. S. Chekanov *et. al.* [ZEUS Collaboration], *Phys. Rev.* **D67** (2003) 012007.
16. A. Martin, R. G. Roberts, W. J. Stirling and R. S. Thorne, *Eur.Phys. J.* **C23** (2002) 73
17. S. Alekhin, *Phys. Rev.* **D68** (2003) 114002

Averaging of DIS Cross Section Data

A. Glazov

DESY, Notkestrasse 85, Hamburg, D 22603 Germany

Abstract. A method to combine measurements of the structure functions performed by several experiments in a common kinematic domain is presented. This method generalises the standard averaging procedure by taking into account point-to-point correlations which are introduced by the systematic uncertainties of the measurements. The method is applied to the neutral and charged current deep inelastic scattering cross section data published by the H1 and ZEUS collaborations. The averaging improves accuracy owing to the cross calibration of the H1 and ZEUS measurements.

Keywords: Proton Structure Functions
PACS: 13.60.-r

Modern QCD fit procedures (Alekhin [1], CTEQ [2], MRST [3], H1 [4], ZEUS [5]) use data from a number of individual experiments directly to extract the parton distribution functions (PDF). All modern programs use both the central values of measured cross section data as well as information about the correlations among the experimental data points.

This direct extraction procedure has some drawbacks. Firstly the number of input datasets is large consisting of many individual publications. The data points are correlated through common systematic uncertainties, within and also across the publications. Handling of the experimental data without additional expert knowledge often becomes very difficult. In addition, the treatment of the correlations produced by the systematic errors is not unique [6]. In the Lagrange Multiplier method [7] each systematic error is treated as a parameter and thus fitted to QCD. Error propogation is then used to estimate resulting uncertainties on PDFs. In the so-called "offset" method (see e.g. [5]) the datasets are shifted in turn by each systematic error before fitting. The resulting fits are used to form an envelope function to estimate the PDF uncertainty. Each method has its own advantages and shortcomings, and it is difficult to select the standard one. Finally, some global QCD analyses use non-statistical criteria to estimate the PDF uncertainties ($\Delta\chi^2 >> 1$). This is driven by the apparent discrepancy between different experiments which is often difficult to quantify. Without a model independent consistency check of the data it might be the only safe procedure.

These drawbacks can be significantly reduced by averaging of the input structure function data in a model independent way before performing a QCD analysis of that data. One combined dataset of deep inelastic scattering (DIS) cross section measurements is much easier to handle compared to a scattered set of individual experimental measurements, while retaining the full correlations between data points. The averaging method proposed here is unique and removes the drawback of the offset method, which fixes the size of the systematic uncertainties. In the averaging procedure the correlated systematic uncertainties are floated coherently allowing in some cases reduction of the uncertainty. In addition, study of a global χ^2/dof of the average and distribution of the

CP792, *Deep Inelastic Scattering, DIS 2005,* edited by W. H. Smith and S. R. Dasu

pulls allows a model independent consistency check between the experiments. In case of discrepancy between the input datasets, localised enlargement of the uncertainties for the average can be performed.

A standard way to represent a cross section measurement of a single experiment is given in the case of the F_2 structure function by:

$$\chi^2_{exp}\left(\left\{F_2^{i,true}\right\},\{\alpha_j\}\right) = \Sigma_i \frac{\left[F_2^{i,true} - \left(F_2^i + \Sigma_j \frac{\partial F_2^i}{\partial \alpha_j}\alpha_j\right)\right]^2}{\sigma_i^2} + \Sigma_j \frac{\alpha_j^2}{\sigma_{\alpha_j}^2}. \quad (1)$$

Here F_2^i (σ_i^2) are the measured central values (statistical and uncorrelated systematic uncertainties) of the F_2 structure function[1], α_j are the correlated systematic uncertainty sources and $\partial F_2^i/\partial\alpha_j$ are the sensitivities of the measurements to these systematic sources. Eq. 1 corresponds to the correlated probability distribution functions for the structure function $F_2^{i,true}$ and for the systematic uncertainties α_j.

The χ^2 function Eq. 1 by construction has a minimum $\chi^2 = 0$ for $F_2^{i,true} = F_2^i$ and $\alpha_j = 0$. One can show that the total uncertainty for $F_2^{i,true}$ determined from the formal minimisation of Eq. 1 is equal to the sum in quadrature of the statistical and systematic uncertainties. The reduced covariance matrix $cov(F_2^{i,true}, F_2^{j,true})$ quantifies the correlation between experimental points.

In the analysis of data from more than one experiment, the χ^2_{tot} function is taken as a sum of the χ^2 functions Eq. 1 for each experiment. The QCD fit is then performed in terms of parton density functions which are used to calculate predictions for $F_2^{i,true}$.

Before performing the QCD fit, the χ^2_{tot} function can be minimised with respect to $F_2^{i,true}$ and α_j. If none of correlated sources is present, this minimisation is equivalent to taking an average of the structure function measurements. If the systematic sources are included, the minimisation corresponds to a generalisation of the averaging procedure which contains correlations among the measurements.

Being a sum of positive definite quadratic functions, χ^2_{tot} is also a positive definite quadratic and thus has a unique minimum which can be found as a solution of a system of linear equations. Although this system of the equations has a large dimension it has a simple structure allowing fast and precise solution.

A dedicated program has been developed to perform this averaging of the DIS cross section data (`http://www.desy.de/~glazov/f2av.tar.gz`). This program can calculate the simultaneous averages for neutral current (NC) and charged current (CC) electron- and positron-proton scattering cross section data including correlated systematic sources. The output of the program includes the central values and uncorrelated uncertainties of the average cross section data. The correlated systematic uncertainties can be represented in terms of (i) covariance matrix, (ii) dependence of the average cross section on the original systematic sources together with the correlation matrix for the systematic sources, (iii) and finally the correlation matrix of the systematic sources

[1] The structure function is measured for different Q^2 (four momentum transfer squared) and Bjorken-x values which are omitted here for simplicity.

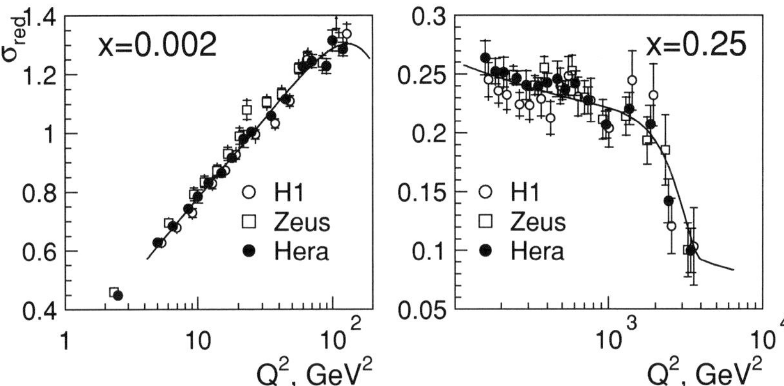

FIGURE 1. Q^2 dependence of the NC reduced cross section for $x = 0.002$ and $x = 0.25$ bins. H1 data is shown as open circles, ZEUS data is shown as open squares and the average of H1 and ZEUS data is shown as filled circles. The line represents the expectation from the H1 PDF 2000 QCD fit.

can be diagonalised, in this case the form of χ^2 for the average data is identical to Eq. 1 but the original systematic sources are not preserved.

The first application of the averaging program has been a determination of the average of the published H1 and ZEUS data [4,10-18]. Nine individual NC and CC cross section measurements are included from H1 and seven are included from ZEUS. Several sources of systematic uncertainties are correlated between datasets, the correlations among H1 and ZEUS datasets are taken from [4] and [8], respectively. No correlations are assumed between H1 and ZEUS systematic uncertainties apart from a common 0.5% luminosity measurement uncertainty. The total number of data points is 1153 (552 unique points) and the number of correlated systematic sources, including normalisation uncertainties, is 43.

The averaging can take place only if most of the data from the experiments are quoted at the same Q^2 and x values. Therefore, before the averaging the data points are interpolated to a common Q^2,x grid. This interpolation is based on the H1 PDF 2000 QCD fit [4]. The interpolation of data points in principle introduces a model dependency. For H1 and ZEUS structure function data both experiments employ rather similar Q^2,x grids. About 20% of the input points are interpolated, for most of the cases the correction factors are small (few percent) and stable if different QCD fit parametrizations [2, 3] are used.

The cross section data have also been corrected to a fixed center of mass energy squared $S = 101570$ GeV^2. This has introduced a small correction for the data taken at $S = 90530$ GeV^2. The correction is based on H1-2000 PDFs, it is only significant for high inelasticity $y > 0.6$ and does not exceed 6%.

The HERA data sets agree very well: χ^2/dof for the average is 521/601. The distribution of pulls does not show any significant tensions across the kinematic plane. Some systematic trends can be observed at low $Q^2 < 50$ GeV^2, where ZEUS NC data lie systematically higher than the H1 data, although this difference is within the normalisation uncertainty. An example of the resulting average DIS cross section is shown in Fig. 1, where the data points are displaced in Q^2 for clarity.

A remarkable side feature of the averaging is a significant reduction of the correlated systematic uncertainties. For example the uncertainty on the scattered electron energy measurement in the H1 backward calorimeter is reduced by a factor of three. The reduction of the correlated systematic uncertainties thus leads to a significant reduction of the total errors, especially for low $Q^2 < 100$ GeV^2, where systematic uncertainties limit the measurement accuracy. For this domain the total errors are often reduced by a factor two compared to the total errors of the individual H1 and ZEUS measurements.

The reduction of the correlated systematic uncertainties is achieved since the dependence of the measured cross section on the systematic sources is significantly different between H1 and ZEUS experiments. This difference is due mostly to the difference in the kinematic reconstruction methods used by the two collaborations, and to a lesser extent to the individual features of the H1 and ZEUS detectors. For example, the cross section dependence on the scattered electron energy scale has a very particular behaviour for H1 data which relies on kinematic reconstruction using only the scattered electron in one region of phase space. ZEUS uses the double angle reconstruction method where the pattern of this dependence is completely different leading to a measurement constraint.

In summary, a generalised averaging procedure to include point-to-point correlations caused by the systematic uncertainties has been developed. This averaging procedure has been applied to H1 and ZEUS DIS cross section data. The data show good consistency. The averaging of H1 and ZEUS data leads to a significant reduction of the correlated systematic uncertainties and thus a large improvement in precision for low Q^2 measurements. The goal of the averaging procedure is to obtain HERA DIS cross section set which takes into account all correlations among the experiments.

I would like to thank E. Rizvi, M. Klein, M. Cooper-Sarkar and C. Pascaud for help and many fruitful discussions.

REFERENCES

1. S. I. Alekhin, Phys. Rev. **D68**, 014002 (2003).
2. J. Pumplin *et al.* [CTEQ Collaboration], JHEP **0207**, 012 (2002).
3. A.D. Martin *et al.* [MRST Collaboration], Eur. Phys. J. **C28**, 455 (2002).
4. C. Adloff *et al.* [H1 Collaboration], Eur. Phys. J. **C30** (2003) 1
5. S. Chekanov *et al.* [ZEUS Collaboration], Phys. Rev. **D67**, 012007 (2003).
6. A.M. Cooper-Sarkar, J. Phys. **G28**, 2717 (2002).
7. C. Pascaud and F. Zomer, LAL preprint, LAL/95-05 (1995)
8. S. Chekanov *et al.* [ZEUS Collaboration], DESY-05-050, Submitted to Eur. Phys. J. **C**
9. C. Adloff *et al.* [H1 Collaboration], Eur. Phys. J. **C21** (2001) 33
10. C. Adloff *et al.* [H1 Collaboration], Eur. Phys. J. **C13** (2000) 609
11. C. Adloff *et al.* [H1 Collaboration], Eur. Phys. J. **C19** (2001) 269
12. S. Chekanov *et al.* [ZEUS Collaboration] Eur. Phys. J. **C21**, 443 (2001).
13. J. Breitweg *et al.* [ZEUS Collaboration] Eur. Phys. J. **C12**, 411 (2000).
14. S. Chekanov *et al.* [ZEUS Collaboration] Eur. Phys. J. **C28**, 175 (2003).
15. S. Chekanov *et al.* [ZEUS Collaboration] Phys. Lett. **B539**, 197 (2002).
16. S. Chekanov *et al.* [ZEUS Collaboration] Phys. Rev. D. **D70**, 052001 (2004).
17. S. Chekanov *et al.* [ZEUS Collaboration] Eur. Phys. J. **C32**, 16 (2003).

NuTeV Structure Function Measurement

M. Tzanov for the NuTeV Collaboration

University of Pittsburgh, Pittsburgh, PA 15260

Abstract. The NuTeV experiment obtained high statistics samples of neutrino and antineutrino charged current events during the 1996-1997 Fermilab fixed target run. The experiment combines sign-selected neutrino and antineutrino beams and the upgraded CCFR iron-scintillator neutrino detector. A precision continuous calibration beam was used to determine the muon and hadron energy scales to a precision of 0.7% and 0.43% respectively. The structure functions $F_2(x,Q^2)$ and $xF_3(x,Q^2)$ obtained by fitting the y-dependence of the sum and the difference of the ν and $\overline{\nu}$ differential cross sections are presented.

Keywords: NuTeV, Structure function, Experimental results
PACS: 12.38.Qk, 13.15.+g, 13.60.Hb

Neutrino deep inelastic scattering (DIS) provides a unique information for the structure of the proton and QCD, allowing the measurement of two structure functions (SF): $F_2(x,Q^2)$, and the parity-violating $xF_3(x,Q^2)$,which is accessible only by neutrino DIS [1]. The NuTeV experiment is a high-energy fixed target $\nu - Fe$ scattering experiment, which combines two new features: Separate high-purity neutrino and antineutrino beams, used to tag the primary lepton in charged-current interactions, and a continuous precision calibration beam, which improves the experiment's knowledge of the absolute energy scale for hadrons and muon, produced in neutrino interactions, to a precision of 0.43% and 0.7% respectively [2]. NuTeV took data during 1996-97 and collected 8.6×10^5 ν and 2.4×10^5 $\overline{\nu}$ charged-current (CC) interactions that passed analysis cuts.

ν-FE CHARGE CURRENT DIFFERENTIAL CROSS SECTION

The differential cross section is determined from

$$\frac{d^2\sigma^{\nu(\overline{\nu})}}{dxdy} = \frac{1}{\Phi(E)}\frac{d^2N^{\nu(\overline{\nu})}(E)}{dxdy}, \tag{1}$$

where $\Phi(E)$ is the $\nu(\overline{\nu})$ flux in energy bins. The cross section event sample is required to pass fiducial volume cuts, μ track reconstruction quality cuts, a minimum muon energy threshold $E_\mu > 15$ GeV, a minimum hadronic energy threshold $E_{HAD} > 10$ GeV, and a minimum neutrino energy threshold $E_\nu > 30$ GeV. Selected events are binned in x, y, and E_ν bins, and corrected for acceptance and smearing using a fast detector simulation. $Q^2 > 1$ GeV^2 is required to minimize the non-perturbative contribution to the cross section. NuTeV data ranges from 10^{-3} to 0.95 in x, 0.05 to 0.95 in y, and from 30 GeV to 360 GeV in E_ν.

The flux is determined from data with $E_{HAD} < 20$ GeV using the "fixed ν_0" relative flux extraction method [1]. The integrated number of events in this sample is propor-

CP792, *Deep Inelastic Scattering, DIS 2005,* edited by W. H. Smith and S. R. Dasu

tional to the flux as $y = \frac{E_{HAD}}{E_\nu} \to 0$. Corrections up to order y^2, determined from the data sample, are applied to determine the relative flux to about the 1% level. Flux is normalized using the world average ν-Fe cross section $\frac{\sigma^\nu}{E_\nu} = 0.677 \times 10^{-38} cm^2/GeV$ [3].

The fast detector simulation, which takes into account acceptance and resolution effects, uses an empirically determined set of PDFs extracted by fitting the differential cross section [4]. The procedure is then iterated until convergence is achieved (within 3 iterations). Detector response functions are parameterized from the NuTeV calibration beam data samples [2].

STRUCTURE FUNCTIONS

The structure function $F_2(x,Q^2)$ is determined from a fit to the y-dependence of the sum of the $\nu, \overline{\nu}$ differential cross sections:

$$\left(\frac{d^2\sigma}{dxdy}^{\nu} + \frac{d^2\sigma}{dxdy}^{\overline{\nu}}\right) = \frac{G_F^2 ME}{\pi}\left[2\left(1-y-\frac{Mxy}{2E}+\frac{y^2}{2}\frac{1+4M^2x^2/Q^2}{1+R_L}\right)F_2 + y\left(1-\frac{y}{2}\right)\Delta xF_3\right], \tag{2}$$

where $F_2 = \frac{F_2^\nu + F_2^{\overline{\nu}}}{2}$, $R_L(x,Q^2)$ is the ratio of the cross section for scattering from longitudinally to transversely polarized W-bosons,and $\Delta xF_3 = xF_3^\nu - xF_3^{\overline{\nu}}$. Cross sections are corrected for QED radiative effects and for 5.67% excess of neutrons over protons in our iron target before the sum is formed [5]. To extract $F_2(x,Q^2)$ we use ΔxF_3 from a NLO QCD model as input (TRVFS) [6]. The input value of $R_L(x,Q^2)$ comes from a fit to the world's measurements [7]. NuTeV $F_2(x,Q^2)$ for neutrino scattering on iron is shown on Fig. 1 (left) compared with previous ν-Fe scattering measurements (CDHSW [8], CCFR [9]). NuTeV F_2 is in reasonable agreement with CDHSW and CCFR for $x < 0.4$. At high-x NuTeV F_2 is systematically above CCFR: 4% at $x = 0.45$, 9% at $x = 0.55$, 18% at $x = 0.65$.

Similarly, the structure function $xF_3(x,Q^2)$ is determined from a fit to the y-dependence of the difference of the $\nu, \overline{\nu}$ differential cross sections:

$$\left[\frac{d^2\sigma^\nu}{dxdy} - \frac{d^2\sigma^{\overline{\nu}}}{dxdy}\right] = \frac{2G_F^2 ME}{\pi}\left(y - \frac{y^2}{2}\right) xF_3^{AVG}(x,Q^2), \tag{3}$$

where $xF_3^{AVG} = \frac{1}{2}(xF_3^\nu + xF_3^{\overline{\nu}})$. $F_2^\nu(x,Q^2) \approx F_2^{\overline{\nu}}(x,Q^2)$ are nearly identical so no additional model input is required. Cross sections are corrected for QED radiative effects and for 5.67% excess of neutrons over protons in our iron target before the difference is formed [5]. Fig. 1 (right) shows the NuTeV measurement of $xF_3(x,Q^2)$ compared to previous ν-Fe results (CDHSW [8], CCFR(97) [3]). NuTeV xF_3 agrees with CCFR(97) and CDHSW for $x < 0.4$. For $x > 0.4$ NuTeV result is systematically higher than CCFR(97) [3].

We have determined that the largest contribution to the discrepancy with CCFR at high-x is due to a mis-calibration of the magnetic field map of the muon spectrometer in CCFR. NuTeV and CCFR used the same muon spectrometer. Hence, the radial dependence of the magnetic field should be the same. NuTeV mapped the entire surface of the

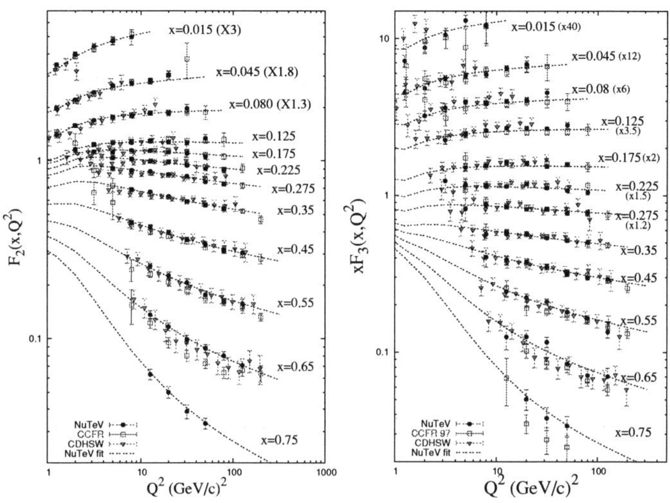

FIGURE 1. NuTeV F_2 (left) and xF_3 (right) in comparison with previous ν-Fe scattering experiments.

muon spectrometer with calibration beam of muons, which provided precise calibration of the magnetic field [2], while CCFR used a model for the magnetic field map and one high statistics calibration muon run, aimed at a single point of the spectrometer, to set the overall normalization [10]. The difference of the two magnetic field maps is an effective 0.8% shift of the muon energy scale, which accounts for a third of the discrepancy. Additional contributions to the discrepancy are the different cross section models used by NuTeV and CCFR (3% of the 18%), and the NuTeV's improved muon and hadron energy smearing models (2% of the 18%). All of the above differences account for two thirds of the discrepancy.

A comparison with TRVFS(MRST2001E) [6, 11] and ACOT(CTEQ5) [12, 13] for F_2 and xF_3 is shown on Fig.2. Both theoretical curves are corrected for nuclear target [1, 3] and target mass effects [14]. NuTeV agrees with both theoretical calculations for $0.06 < x < 0.5$. For $x < 0.06$ both NuTeV and CCFR measure different Q^2-dependence than the theoretical predictions. At high-x both theoretical predictions are systematically higher than the NuTeV F_2 and xF_3.

The nuclear correction used to correct the theory curves is independent of Q^2 and based on a fit to charged-lepton data on nuclear targets. NuTeV perhaps indicates that neutrino scattering favors smaller nuclear effects at high-x than are found in charged-lepton scattering. At small x, new theoretical calculations show that in the shadowing region the nuclear correction has Q^2 dependence [15, 16]. The standard nuclear correction obtained from a fit to charged lepton data implies a suppression of 10% independent of Q^2 at $x = 0.015$, while for $x = 0.015$ reference [16] finds a suppression of 15% at $Q^2 = 1.25\text{GeV}^2$ and a suppression of 3.4% at $Q^2 = 7.94\text{GeV}^2$. This effect improves agreement with data at low-x.

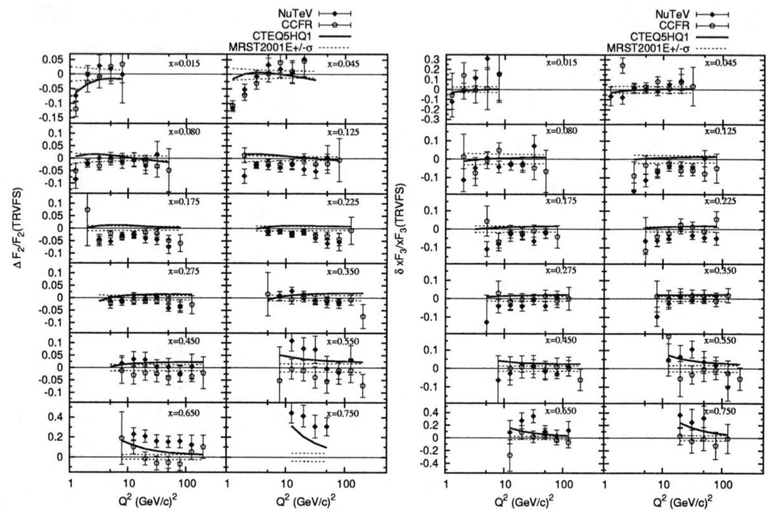

FIGURE 2. NuTeV and CCFR F_2(left) and xF_3(right) compared with TRVFS(MRST2001E) and ACOT(CTEQ5HQ1).

CONCLUSIONS

In conclusion, NuTeV has measured F_2 and xF_3 structure functions. This is the most precise measurement from neutrino scattering experiment to date. NuTeV result is in good agreement with previous ν-Fe results over the intermediate x region. At high-x NuTeV result is higher than the theoretical predictions. Perhaps, the nuclear correction is different for neutrino scattering.

REFERENCES

1. J. M. Conrad, M. H. Shaevitz, and T. Bolton, *Rev. Mod. Phys.* **70**, 4 (1998).
2. D. A. Harris *et al.*, *Nucl. Instrum. Methods* A **447**, 377-415 (2000).
3. W. Seligman, Ph. D. Thesis (Columbia University), Nevis Reports 292, (1997).
4. A. J. Buras and K. L. F. Gaemers, *Nucl. Phys.* B **132**, 2109 (1978).
5. Bardin, D. Y. and Dokuchaeva, JINR-E2-86-260 (1986).
6. R. S. Thorne and R. G. Roberts,*Phys. Lett.* B **421**, 303 (1998).
7. L. W. Whitlow *et al.*,*Phys. Lett.* B **250**, 193 (1990).
8. P. Berge *et al.*, *Z. Phys.* C **49**, 187 (1991).
9. U. K. Yang , Ph. D. Thesis, University of Rochester, (2001), UR-1583
10. B. King *et al.* Nucl. Instrum. Methods **A302** (1991) 254.
11. A. D. Martin, R. G. Roberts, W. J. Stirling, R. S. Thorne, Eur.Phys.J.**C 28** (2003) 455-473.
12. M. A. G. Aivazis, J. C. Collins, F. I. Olness, and W. K. Tung Phys Rev **D50**, 3102 (1994).
13. H. L. Lai *et. al.*, Eur.Phys.J.**C 12** (2000) 375-392.
14. H. Georgi and H. D. Politzer, Phys Rev **D**14 (1976) 1829.
15. S.A. Kulagin, R. Petti, [arXiv:hep-ph/0412425]
16. J. W. Qiu and I. Vitev, Phys. Lett. B **587**, (2004) 52 [arXiv:hep-ph/0401062].

W Asymmetry and Z Rapidity Measurements

Y. S. Chung

Department of Physics and Astronomy, University of Rochester, Rochester, New York 14627-0171

Abstract. The measurements of W charge asymmetry and Z rapidity distributions are inputs to constrain the parton distribution functions at high Q^2. The CDF and DØ experiments at the Tevatron have analyzed up to 340 pb^{-1} of Run 2 data to measure the W charge asymmetry and Z rapidity. The measurements are in generally good agreement with predictions.

Keywords: asymmetry rapidity PDF
PACS: 12.38.Qk

INTRODUCTION

The W and Z bosons are primarily produced by a quark and anti-quark annihilation in $p\bar{p}$ collisions at the Tevatron. Many precision measurements at the Tevatron are limited by the uncertainties on the parton distribution functions (PDFs). These include the W mass, cross section, transverse momentum distributions of W and Z bosons, and so on. For the W mass measurement in the CDF experiment, the current systematic uncertainty from PDF uncertainties is about 15 MeV/c^2 [1] and will become more significant as data accumulated.

The measurement of the W charge asymmetry provides input to the ratio of u and d quark components of PDFs, especially at the medium/high x. In the $p\bar{p} \to Z$ process, the momentum fraction of a parton is related to the Z rapidity (y)[1]. Analysis of the high y region probes the PDFs at both high and low momentum fraction.

W ASYMMETRY MEASUREMENT

The $W^+(W^-)$ bosons at the Tevatron are produced primarily by the annihilation of $u(d)$ quarks in the proton and $\bar{d}(\bar{u})$ quarks in the anti-proton. Because u quarks carry, on average, more momentum than d quarks, the W^+ tend to follow the direction of the incoming proton and the W^- that of the anti-proton.

The $W^{\pm}$ charge asymmetry

$$A(y_W) = \frac{d\sigma_+/dy_W - d\sigma_-/dy_W}{d\sigma_+/dy_W + d\sigma_-/dy_W}, \tag{1}$$

where the subscript (+,-) denotes the charge of the W and y_W is the rapidity of the W bosons is a sensitive probe of the momentum fraction difference between u and d quarks

---

[1] $y = \frac{1}{2} \ln \frac{(E+P_z)}{(E-P_z)}$

CP792, *Deep Inelastic Scattering, DIS 2005*, edited by W. H. Smith and S. R. Dasu

in the $Q^2 \approx M_W^2$ region. A precise measurement of the W asymmetry serves as a valuable constraint on the u and d quark momentum distributions. For $p\bar{p}$ collisions in leading-order parton model, $A(y_W)$ is given approximately by

$$A(y_W) \approx \frac{u(x_1)d(x_2) - d(x_1)u(x_2)}{u(x_1)d(x_2) + d(x_1)u(x_2)}, \tag{2}$$

where $x_{1,2} = x_0 e^{\pm y_W}$, $x_0 = M_W/\sqrt{s}$, and s is the center of mass energy. The asymmetry, $A(y_W)$, is related to the slope of d/u at $Q^2 = M_W^2$ and in the moderate x region of W production.

The rapidity of the W boson is not measured because the longitudinal momentum of the neutrino can not be experimentally determined. Therefore, we measure the lepton asymmetry, which is a convolution of the W production charge asymmetry and the V-A couplings in the W decay. The lepton asymmetry is still sensitive to PDFs and is defined as:

$$A(y_l) = \frac{d\sigma_+/dy_l - d\sigma_-/dy_l}{d\sigma_+/dy_l + d\sigma_-/dy_l}. \tag{3}$$

The CDF experiment measured the forward-backward charge asymmetry of electrons from W boson decays in $p\bar{p}$ collisions at $\sqrt{s}$ = 1.96 TeV using a data sample of 170 pb^{-1} of Run 2 data [2]. Candidate W $\to e\nu$ events were required to have exactly one electron candidate of E_T> 25 GeV, with $\not{E}_T$>25 GeV and transverse mass in the range 50 GeV/c^2 < M_T < 100 GeV/c^2.

The CDF experiment improved its sensitivity to the PDFs by reducing the decay asymmetry. The measurement separates the asymmetry measurement into two bins: 25 GeV < E_T < 35 GeV and 35 GeV < E_T < 45 GeV (Figure 1 and 2). This is done for the first time and the separation into different bins is a novel addition to the previous measurements. Predictions from CTEQ [3] and MRST [4] PDFs using a NLO RESBOS calculation [5] are shown for comparison.

The CDF Collaboration is additionally developing a new method to measure the W asymmetry directly by partially reconstructing W rapidity using a W mass.

Z RAPIDITY DISTRIBUTION

Measurement of the rapidity distribution of Drell-Yan pairs in the Z boson mass region provides a stringent test of PDFs. At leading order (LO), the momentum fraction, $x_+(x_-)$, carried by the parton from the proton (anti-proton) is related to the rapidity of the Z boson via the equation:

$$x_\pm = \frac{M_Z}{\sqrt{s}} e^{\pm y} \tag{4}$$

where M_Z is the mass of Z boson. Therefore the analysis of the high y region probes the high x as well as low x region.

The DØ experiment measures the rapidity distribution, $d\sigma/dy$, of the Drell-Yan process in the dielectron's mass range between 71 and 111 GeV/c^2 [6]. The data were collected with the DØ detector in Tevatron $p\bar{p}$ collisions at $\sqrt{s}$ = 1.96 TeV using a data

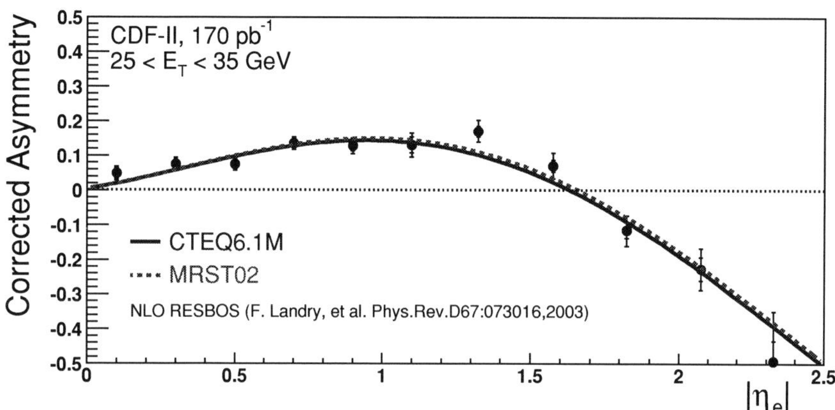

FIGURE 1. The measured lepton charge asymmetry for 25 GeV < E_T < 35 GeV with predictions from CTEQ6.1M (solid) and MRST02 (dashed) PDFs using a NLO RESBOS calculation. Both statistical and total (statistical+systematic) uncertainties are shown.

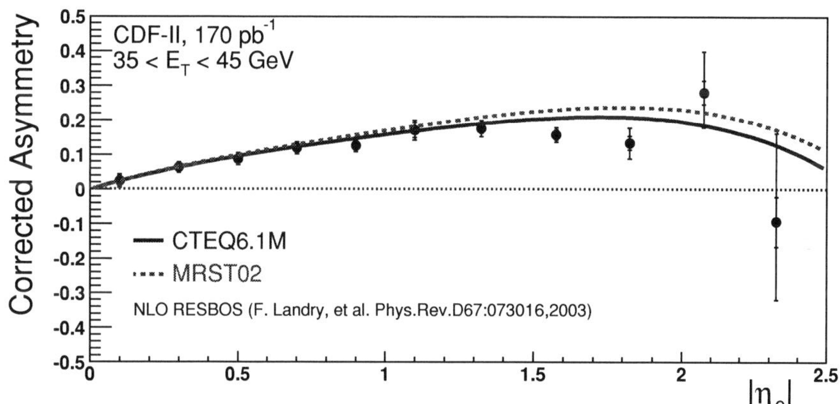

FIGURE 2. The measured lepton charge asymmetry for 35 GeV < E_T < 45 GeV with predictions from CTEQ6.1M (solid) and MRST02 (dashed) PDFs using a NLO RESBOS calculation. Both statistical and total (statistical+systematic) uncertainties are shown.

sample of 337 pb^{-1} of Run 2 data. The final result, with statistical and systematic errors, is shown in Figure 3. The measurement of $d\sigma/dy$ is compared with a prediction using an NNLO calculation based on the MRST 2001 PDFs [7]. The DØ forward calorimeters provide data over almost the entire rapidity range accessible at Tevatron. Measurements are in generally good agreement with the prediction.

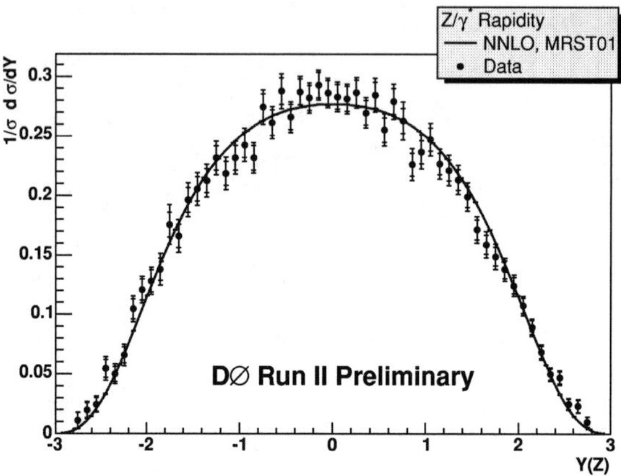

FIGURE 3. The DØ measurement of the Z rapidity distribution. The outer error bars show the total error (statistical+systematic), while the inner error bars indicate the statistical error alone. The solid line shows the NNLO prediction based on the MRST 2001 PDFs.

CONCLUSIONS

The CDF experiment have measured the W charge asymmetry. Since the previous measurements upon which the current predictions are based are least constraining for $|\eta|>1$ and do not separate the E_T dependence, inclusion of the new W asymmetry results will further constrain future fits and improve the predictions. The DØ measurement of the Z rapidity distribution is in generally good agreement with theory prediction. We expect that Tevatron Z and W measurements will provide important input to constrain the PDFs, especially in the high x region.

ACKNOWLEDGMENTS

I would like to thank the workshop organizers and working group leaders for their invitation and a wonderful workshop. Also many thanks to the electroweak conveners of both the CDF and DØ Collaborations for providing material and valuable comments.

REFERENCES

1. CDF 2 Collaboration, *CDF Run 2 W mass*, hep-ex/0505064 (2005), hep-ex/0506016 (2005).
2. CDF 2 Collaboration, D. Acosta *et al*, *Phys. Rev. D* **71**, 051104(R) (2005).
3. J. Pumplin, D. R. Stump, J. Huston, H. L. Lai, P. Nadolsky, and W. K. Tung, *JHEP* **0207** 012 (2002).
4. A. Martin, R. Roberts, W. Stirling, and R. Thorne, *Eur. Phys. J. C* **4**, 463 (1998).
5. C. Balazs and C.P. Yuan, *Phys. Rev. D*, **56**, 5558 (1997).
6. The DØ Collaboration, http://www-d0.fnal.gov/Run2Physics/WWW/results/ew.htm.
7. C. Anastasiou, L. Dixon, K. Melnikov and F. Petriello, *Phys. Rev. D* **69**, 094008 (2004).

Structure Functions of Bound Neutrons

Sebastian Kuhn

Old Dominion University, Norfolk, VA 23529

Abstract. We describe an experiment measuring electron scattering on a neutron bound in deuterium with coincident detection of a fast, backward-going spectator proton. Our data map out the relative importance of the pure PWIA spectator mechanism and final state interactions in various kinematic regions, and give a first glimpse of the modification of the structure function of a bound neutron as a function of its off-shell mass. We also discuss a new experimental program to study the structure of a free neutron by extending the same technique to much lower spectator momenta.

Keywords: deuterium, off-shell, neutron, structure functions
PACS: 24.85.+p, 25.30.-c, 21.45.+v

INTRODUCTION

The structure functions of the proton have been measured over a truly impressive kinematic range, with high precision. These data have given insight into the internal quark structure of the proton and into the properties of QCD. Similar data for the neutron are highly desirable, but are lacking due to the absence of a free neutron target (or beam) of sufficient density. Instead, data taken on the deuteron are used to infer information on the neutron structure; however, because of Fermi motion, binding effects [1], possible modifications of bound nucleons in the medium [2, 3] and non-nucleonic degrees of freedom in deuterium like 6-quark configurations [4, 5], this extraction is non-trivial and model-dependent, especially at high x and in the resonance region.

One possible approach to this problem is a detailed study of the effect of binding on nucleon structure. If we can model the structure modifications of a bound neutron confidently, we can correct the deuteron data and extract neutron information. For comparison with various models of binding effects, it is useful to emphasize scattering off a nucleon which is far off its mass shell. This can be done in the case of deuterium by "tagging" the struck neutron with a fairly high-momentum backward-moving proton. While on average the nucleons in deuterium are only loosely bound, if we require a "spectator" momentum of at least 300 MeV/c, we can emphasize the part of the deuteron wave function corresponding to short distances, where binding effects are largest. This is the goal of the "Deeps" experiment, which studied the reaction $D(e,e'p_s)X$. The experiment and some of its results are discussed in the next section.

A more direct approach would use the same technique – spectator proton tagging – to select neutrons which are only weakly bound, close to their mass shell, and little affected by final state interactions. The additional advantage of measuring the spectator proton kinematics is that one can correct for the Fermi motion of the struck neutron. This method, "Barely Off-shell NUcleon Scattering", is the goal of the "BONUS" program which will be discussed in the last section.

CP792, *Deep Inelastic Scattering, DIS 2005,* edited by W. H. Smith and S. R. Dasu

THE "DEEPS" EXPERIMENT

The data for the Deeps experiment were collected in spring 2002 with a 5.7 GeV electron beam ($6-9$ nA current) in experimental hall B of the Thomas Jefferson National Accelerator Facility (Jefferson Lab, formerly CEBAF). The scattered electrons and the backward-moving protons were detected with the CEBAF Large Acceptance Spectrometer (CLAS) [6]. Electrons were identified using Cherenkov counters and by matching their momentum (measured by three layers of drift chambers) to the energy deposited in the electromagnetic calorimeter. The large acceptance of CLAS allowed us to collect data simultaneously from the quasi-elastic to deep inelastic kinematics ($W < 2.7$ GeV) and over a range in momentum transfer of $1.2\ \text{GeV}^2 \leq Q^2 \leq 5.5\ \text{GeV}^2$. Coincident protons were detected at angles up to 140° and with momenta from 0.3 GeV/c to 0.7 GeV/c. The missing mass of the unobserved debris in $\text{D}(e,e'p_s)\text{X}$ can be reconstructed from the electron and proton kinematics, without Fermi smearing or binding shifts. In the spectator picture, the usual Bjorken variable x can also be corrected for the initial motion of the struck neutron.

Our main goal in this experiment was to test the spectator approximation in this reaction, where the backward-moving proton is an uninvolved spectator to the reaction that occurs on the off-shell neutron. We developed a simple PWIA model of the reaction in this framework, using the Paris wave function for the momentum distribution of the nucleons in the initial state. By correcting for acceptance, efficiency and backgrounds and by dividing out trivial kinematic factors, we can express our results as the product of the probability to find the proton with initial momentum p_s, $P(p_s)$, and a (possibly) modified structure function F_{2n} of the neutron, which depends on the relativistically rescaled variables x or W and Q^2, as well as (possibly) the off-shell mass of the struck neutron.

In the top half of Fig. 1, we show the product $P(p_s) \times F_{2n}$ for a final state mass around $W = 1.5$ GeV and $Q^2 \approx 1.8\ \text{GeV}^2$, in two bins of p_s. While the agreement with our simple PWIA spectator picture is rather good at backward angles relative to the momentum transfer vector $\vec{q}$ ($\cos(\theta_{pq}) \leq -0.3$), the data are clearly rising above the model curve at angles around 90°, especially for higher proton momenta. This is in good agreement with models of final state interaction (FSI) effects based on the reinteraction of the debris from the struck neutron with the spectator proton [7]. The FSI effect is larger for large W, since one can expect more hadronic particles in the final state in this case. It is maximal for light cone fractions around $\alpha_s \approx 1$, which corresponds to angles θ_{pq} slightly forward of 90°. FSI are much reduced at very forward and very backward angles, as well as lower momenta. All of these general trends are confirmed by our data.

If we concentrate on the backward region $\cos(\theta_{pq}) \leq -0.3$, the spectator model works relatively well and we can extract the "off-shell" neutron structure function F_{2n}^{eff}. Our results (lower part of Fig. 1) indicate that at the highest momenta (above 400 MeV/c) there may be some depletion of F_{2n}^{eff} relative to the on-shell neutron structure function, especially at moderate to high x. However, a full calculation with proper inclusion of FSI effects is needed to quantify this statement.

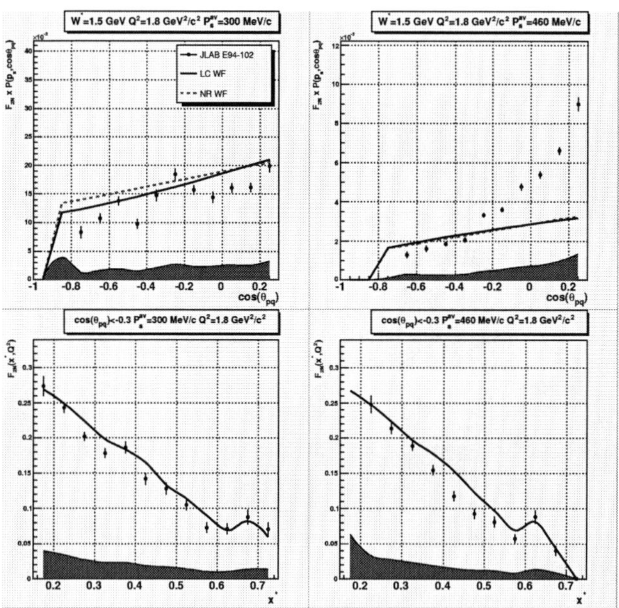

FIGURE 1. Angular distribution (reduced cross section vs. $\cos(\theta_{pq})$) of "spectator" protons relative to the momentum transfer vector $\vec{q}$ (top two panels). The invariant mass of the unobserved final state of the struck neutron is in the region of the S_{11} and D_{13} resonances (around 1.5 GeV). The solid line is from a PWIA calculation within a strict spectator model, using a light cone wave function for the deuteron. The dashed line corresponds to the same calculation with a non-relativistic deuteron wave function. Bottom two panels: The effective structure function F_{2n}^{eff} of the neutron extracted from our data for two different momentum bins and backward spectator proton kinematics. The solid line shows a simple parametrization of the free neutron structure function.

THE "BONUS" EXPERIMENT

According to nearly all theoretical models of the D$(e,e'p_s)$X reaction (and corroborated by our "Deeps" data), both FSI and binding effects should be minimal if one selects spectator protons going backwards at momenta below 0.1 GeV/c. The "BONUS" collaboration has prepared an experiment (to run in Fall 2005) that will use this method to extract the structure of the (nearly) free neutron, both in the resonance region and in the DIS region. In particular, we will measure the DIS structure function of the neutron out to $x = 0.6$ without the nuclear uncertainties that become sizable around that point.

The scattered electron will again be detected with CLAS. The main new ingredient of this experiment is a novel recoil detector in the form of a radial time projection chamber (RTPC) surrounding a gaseous deuteron target. This detector is shown in Fig. 2. After exiting from the thin (4-6 mm diameter) target tube (50 μm Kapton), recoil protons with momenta down to 70 MeV/c traverse a buffer volume filled with helium where Möller electrons are curled up by a 3 Tesla solenoid field (not shown). The protons then enter a drift volume which extends from a radius of 3 cm to 6 cm, surrounding the beam line. The ionization electrons released in this volume are drifted

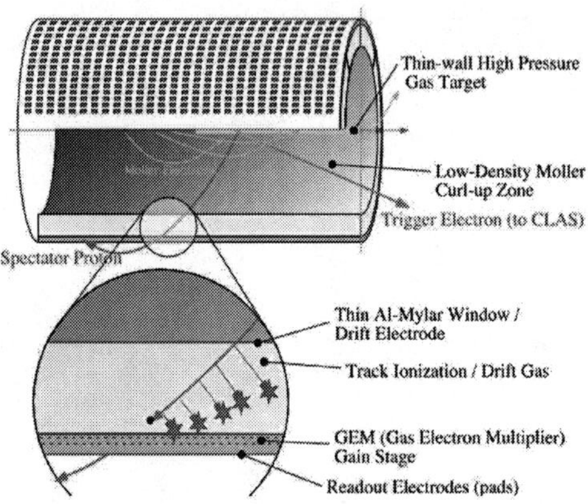

FIGURE 2. The general layout of the BoNuS RTPC. See text for explanation

to a series of three gaseous electron multiplier (GEM) foils that will amplify the signal more than 1000 fold. The signal is read out on a two-dimensional pattern of readout pads connected to continuously sampling ADCs (using the ALTRO chip developed at CERN). From the two spatial dimensions and the radial distance inferred from drift time, a three-dimensional track can be reconstructed. Proton momenta and identity can be reconstructed from the curvature of this track in the solenoid field and the energy deposited in the drift volume.

The detector system has been assembled and just passed a first engineering run with CLAS. The data to be collected in Fall will improve our knowledge of F_{2n}, especially in the resonance region and at high x. At the same time, an extension of both experiments described here to 11 GeV (after the anticipated Jefferson Lab upgrade) will generate a wealth of truly fundamental data not attainable anywhere else.

ACKNOWLEDGMENTS

We express our thanks to the CLAS collaboration and the technical staff at Jefferson Lab for their help and support.

REFERENCES

1. W. Melnitchouk, M. Sargsian, and M. I. Strikman, *Z. Phys.*, **A359**, 99–109 (1997), nucl-th/9609048.
2. L. L. Frankfurt, and M. I. Strikman, *Phys. Rept.*, **76**, 215–347 (1981).
3. F. E. Close, R. L. Jaffe, R. G. Roberts, and G. G. Ross, *Phys. Rev.*, **D31**, 1004 (1985).
4. C. E. Carlson, and K. E. Lassila, *Phys. Rev.*, **C51**, 364–370 (1995), hep-ph/9401307.
5. C. E. Carlson, J. Hanlon, and K. E. Lassila, *Phys. Rev.*, **D63**, 117301 (2001), hep-ph/9902281.
6. B. A. Mecking, et al., *Nucl. Instrum. Meth.*, **A503**, 513–553 (2003).
7. C. Ciofi degli Atti, L. P. Kaptari, and B. Z. Kopeliovich, *Eur. Phys. J.*, **A19**, 145–151 (2004), nucl-th/0307052.

Structure functions at large Bjorken x and the Transition between perturbative and non-perturbative QCD

S. Liuti*, N. Bianchi† and A. Fantoni†

*University of Virginia, Charlottesville, Virginia 22901, USA
†Laboratori Nazionali di Frascati dell'INFN, Via E. Fermi 40, 00044 Frascati (RM), Italy

Abstract. We study both polarized and unpolarized proton structure functions in the kinematical region of large Bjorken x and four-momentum tranfer of few GeV^2, characterized by the phenomenon of parton-hadron duality between the smooth continuation of the deep inelastic scattering curve and the average of the nucleon resonances which dominate this region. We present results on a perturbative-QCD analysis using recent accurate data aimed at extracting the infrared behavior of the nucleon structure functions.

PACS: 13.60.Hb, 13.88.+e

INTRODUCTION

Parton-hadron duality, or the similarity between hadronic cross sections in the Deep Inelastic Region (DIS) and in the resonance region, encompasses a range of phenomena where one expects to observe a trasmogrification from partonic to hadronic degrees of freedom. It lies, therefore, at the very heart of Quantum ChromoDynamics (QCD), as the theory of strong interactions. A number of experiments were conducted, in fact, in the early days of QCD where it was shown that the *continuation* of the smooth curve describing different observables from a wide variety of reactions – structure functions, sum rules, $R(s)$, heavy meson decays... – at large four-momentum transfers/energies into the low momentum/energy region characterized by resonances, could be considered as an *average* of the resonances trend. A fully satisfactory theoretical description of this phenomenon, that became to be accepted as a "natural"feature of hadronic interactions, is still nowadays very difficult to obtain. Recent progress both on the theoretical and experimental side [1, 2], has however renovated and reinforced the hadronic physics comunity's interest in this subject [3].

In this contribution, by conducting an analysis of the most recent polarized and unpolarized inclusive electron scattering data, we present evidence that standard Perturbative QCD (PQCD) approaches might not be adequate in order to describe parton-hadron duality. In particular, we unravel a discrepancy in the behavior of the extracted power corrections in the DIS and resonance regions, respectively.

CP792, *Deep Inelastic Scattering, DIS 2005,* edited by W. H. Smith and S. R. Dasu

COMPARISON BETWEEN DIS AND RESONANCE REGIONS

In this Section we define the key concepts and quantities in our analysis, namely what is meant by: *i) Continuation* of DIS curve into the resonance region; *ii) Average* over the resonances. Although these concepts are equivalently found in a number of different reactions, and in different channels (see *e.g.* [4] for a review), we concentrate on the proton structure functions for polarized, g_1, and unpolarized, F_2, electron scattering.

Continuation of DIS Curve

It is important to define exactly what one means by "continuation" of the DIS curve, in order to be able to define whether parton-duality can be considered to be fulfilled. The accuracy of current data allows us, in fact, to address the question of *what extrapolation from the large Q^2, or asymptotic regime the cross sections in the resonance region should be compared to*. In principle any extrapolation from high to low Q^2 is expected to be fraught with theoretical uncertainties ranging from the propagation of the uncertainty on $\alpha_S(M_Z^2)$ into the resonance region to the appearance of different types of both perturbative and power corrections in the low Q^2 regime. All of these aspects need therefore to be evaluated carefully. In our approach we considered as essential ingredients for the analysis which is centered on the *large Bjorken x* behavior:

- Non-Singlet (NS) Parton Distribtion Functions (PDFs) evolved at Next to Leading Order (NLO).
- The correct scale for the transverse momentum integration yielding the leading log approximation result [5, 6]
- Target Mass Corrections (TMCs) [7]

In Ref.[8] we performed an extensive study using all available parametrizations of PDFs which are pure DIS, extended to the measured x and Q^2 ranges by pQCD evolution. Definitions for both F_2 and g_1 in terms of PDFs are given *e.g.* in Ref.[8]. As shown in [8], the uncertainty due to the use of different parametrizations can be taken into account by a band that is currently smaller than the experimental one in the region of interest. A potential theoretical error in the extrapolation of the ratios to low Q^2 could be, however, generated by the error in $\alpha_S(M_Z^2)$. However, also in this case, because at large x DGLAP evolution involves only Non-Singlet (NS) distributions, there is very little uncertainty in the extrapolation of the initial pQCD distribution evn to the low values of W^2 considered.

As noticed in a pioneering paper [5], the problem of resumming the large logarithm terms arising at large x can be accounted for by considering the correct definition of the upper limit of integration for the transverse momentum in the ladder diagrams defining the leading log approximation. This implies replacing Q^2 with $\approx \widetilde{W}^2$, an invariant mass, in the evolution equations. Such a procedure was taken into account to obtain our results both in Refs.[8, 9], and in the current contribution.

Finally, TMCs need to be implemented. As a word of caution, we notice that since they apply as a series in the parameter: $4M^2x^2/Q^2$, one has to ensure that the kinematical region considered is consistent with such an expansion (this question was addressed explicitly in [9].

Averaging Procedure

Resonant data can be averaged over, according to different procedures We considered the following complementary methods:

$$I(Q^2) = \int_{x_{\rm min}}^{x_{\rm max}} F_2^{\rm res}(x,Q^2)\, dx \tag{1}$$

$$M_n(Q^2) = \int_0^1 dx\, \xi^{n-1} \frac{F_2^{\rm res}(x,Q^2)}{x}\, p_n \tag{2}$$

$$F_2^{\rm ave}(x,Q^2(x,W^2)) = F_2^{\rm Jlab}(\xi,W^2) \tag{3}$$

where $F_2^{\rm res}$ is evaluated using the experimental data in the resonance region [1]. In Eq.(1), for each Q^2 value: $x_{\rm min} = Q^2/(Q^2+W^2_{\rm max}-M^2)$, and $x_{\rm max} = Q^2/(Q^2+W^2_{\rm min}-M^2)$. $W_{\rm min}$ and $W_{\rm max}$ delimit either the whole resonance region, *i.e.* $W_{\rm min} \approx 1.1$ GeV^2, and $W^2_{\rm max} \approx 4$ GeV^2, or smaller intervals within it. In Eq.(2), ξ is the Nachtmann variable [10], and $M_n(Q^2)$ are Nachtmann moments [10]; p_n is a kinematical factor [10]. The r.h.s. of Eq.(3), $F_2^{\rm Jlab}(\xi,W^2)$, is a smooth fit to the resonant data [1], valid for $1 < W^2 < 4$ GeV^2; $F_2^{\rm ave}$ symbolizes the average taken at the $Q^2 \equiv (x,W^2)$ of the data.

After describing our program to address quantitatively all sources of theoretical errors started in [9, 8], we finally, in Fig. 1 we present our main result, namely extraction of the dynamical Higher Twist (HT) terms from the resonance region, and we compare them to results obtained in the DIS region [11, 12]. A clear discrepancy marking perhaps a *breakdown of the twist expansion* at low values of W^2 is seen for the unpolarized structure function, F_2 (upper panel). More data at large x are needed in order to draw conclusion for the polarized structure function, g_1.

ACKNOWLEDGMENTS

This work is supported by the U.S. Department of Energy grant no. DE-FG02-01ER41200.

REFERENCES

1. I. Niculescu *et al.*, Phys. Rev. Lett. **85**, 1186 (2000).
2. A. Airapetian *et al.* [HERMES Collaboration], Phys. Rev. Lett. **90**, 092002 (2003)
3. `http://www.lnf.infn.it/conference/duality05/`

[1] Similar formulae hold for the polarized structure function, g_1.

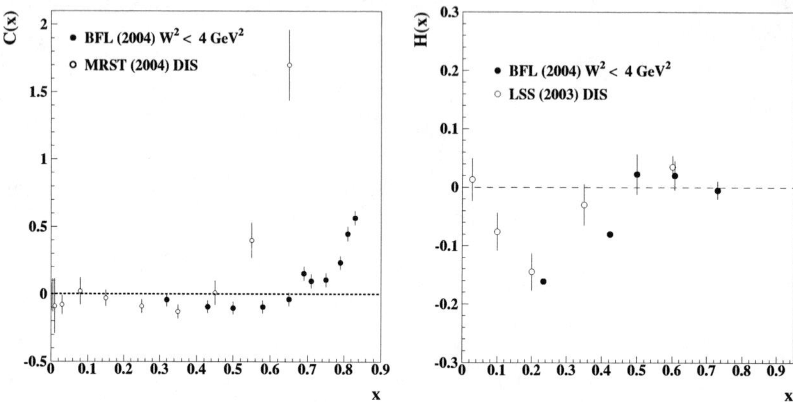

FIGURE 1. Comparison of HT contributions for both the structure function F_2 (left panel) and the polarized structure function g_1 (right panel) in the DIS and resonance regions, respectively. The full circles are the values obtained in the resonance region [8]. For F_2 these are compared with extractions using DIS data, from [11]. For g_1 they are compared to the extraction from [12]. Notice that we show our results in a factorized model for F_2, and in a non-factorized one for g_1 for a consistent comparison with [11, 12].

4. W. Melnitchouk, R. Ent and C. Keppel, Phys. Rept. **406**, 127 (2005)
5. S. J. Brodsky and G. P. Lepage, SLAC-PUB-2447, *Presented at Summer Inst. on Particle Physics, SLAC, Stanford, Calif., Jul 9-20, 1979*
6. R. G. Roberts, Eur. Phys. J. C **10**, 697 (1999)
7. H. Georgi and H. D. Politzer, Phys. Rev. D **14**, 1829 (1976);
8. N. Bianchi, A. Fantoni and S. Liuti, Phys. Rev. D **69**, 014505 (2004)
9. S. Liuti, R. Ent, C. E. Keppel and I. Niculescu, Phys. Rev. Lett. **89**, 162001 (2002)
10. O. Nachtmann, Nucl. Phys. B **63**, 237 (1973).
11. A. D. Martin, R. G. Roberts, W. J. Stirling and R. S. Thorne, Eur. Phys. J. C **35**, 325 (2004)
12. E. Leader, A. V. Sidorov and D. B. Stamenov, Phys. Part. Nucl. Lett. **1**, 229 (2004), *and references therein*

A Unified Model for inelastic $e-N$ and $\nu-N$ cross sections at all Q^2

Arie Bodek* and Un-ki Yang†

*Department of Physics and Astronomy, University of Rochester, Rochester, New York 14618, USA
†Enrico Fermi Institute, University of Chicago,Chicago, Illinois 60637, USA

Abstract. We present results using a new scaling variable, ξ_w in modeling electron- and neutrino-nucleon scattering cross sections with effective leading order PDFs. Our model uses all inelastic charged lepton F_2 data (SLAC/BCDMS/NMC/HERA), and photoproduction data on hydrogen and deuterium. We find that our model describes all inelastic scattering charged lepton data, the average of JLAB resonance data, and neutrino data at all Q^2. This model is currently used by current neutrino oscillation experiments in the few GeV region.

Keywords: structure functions, low-x
PACS: 12.38.-t,12.38.Mh,12.38.Qk,13.60.Hb

The field of neutrino oscillation physics has progressed from the discovery of neutrino oscillation [1] to the era of precision measurements of mass splitting and mixing angles. Currently, there are only poor measurements of differential cross sections for neutrino interactions in the few GeV region. This results in large systematic uncertainties in the extraction of mass splitting and mixing parameters (e.g. by the MINOS, NOνA , K2K and T2K experiments). Therefore, reliable modeling of neutrino cross sections at low energies is essential for precise (next generation) neutrino oscillations experiments. In the few GeV region, there are three types of neutrino interactions: quasi-elastic, resonance, and inelastic scattering. It is very challenging to disentangle each contribution separately, especially, resonance production versus deep inelastic scattering (DIS) contributions. There are large non-perturbative QCD corrections to the DIS contributions in this region.

Our approach is to relate neutrino interaction processes using a quark-parton model to precise charged-lepton scattering data. In a previous communication [2], we showed that our effective leading order model using an improved scaling variable ξ_w describes all deep inelastic scattering charged lepton-nucleon scattering data including resonance data (SLAC/BCDMS/NMC/HERA/Jlab) [3, 4] from very high Q^2 to very low Q^2 (down to photo-production region), as well as high energy CCFR neutrino data [5].

The proposed scaling variable, ξ_w is derived using energy momentum conservation, assuming massless initial state quarks bound in a proton of mass M.

$$\xi_w = \frac{2x(Q^2+M_f{}^2+B)}{Q^2[1+\sqrt{1+(2Mx)^2/Q^2}]+2Ax}, \tag{1}$$

here, M_f is the final quark mass (zero except for charm-production in neutrino processes). The parameter A accounts for the higher order (dynamic higher twist) QCD terms in the form of an enhanced target mass term (the effects of the proton target mass

CP792, *Deep Inelastic Scattering, DIS 2005*, edited by W. H. Smith and S. R. Dasu

are already taken into account using the exact form in the denominator of ξ_w). The parameter B accounts for the initial state quark transverse momentum and final state quark effective $\Delta M_f{}^2$ (originating from multi-gluon emission by quarks). This parameter also allows us to describe the data also in the photoproduction limit (all the way down to Q^2=0).

A brief summary of our effective leading order (LO) model is given as follows;

- The GRV98 LO PDFs [6] are used to describe the F_2 data at high Q^2 region.
- The scaling variable x is replaced with the improved scaling variable ξ_w (Eq. 1).
- All PDFs are modified by K factors to describe low Q^2 data in the photoproduction limit.

$$K_{sea}(Q^2) = \frac{Q^2}{Q^2+C_s}, \qquad K_{valence}(Q^2) = [1-G_D^2(Q^2)]\left(\frac{Q^2+C_{v2}}{Q^2+C_{v1}}\right), \tag{2}$$

  where $G_D = 1/(1+Q^2/0.71)^2$ is the proton elastic form factor. At low Q^2, $[1-G_D^2(Q^2)]$ is approximately $Q^2/(Q^2+0.178)$. Different values of the K factor are obtained for u and d quarks
- The evolution of the GRV98 PDFs is frozen at a value of $Q^2 = 0.80$ which is the minimum Q^2 value of this PDFs. Below this Q^2, F_2 is given by;

$$F_2(x,Q^2<0.8) = K(Q^2)\times F_2(\xi,Q^2=0.8) \tag{3}$$

- Finally, using these effective GRV98 LO PDFs (ξ_w) we fit to all inelastic charged lepton scattering data (SLAC/BCDMS/NMC/H1) and photoproduction data on hydrogen and deuterium. Note that no resonance data is included in the fit. We obtain excellent fits with; A=0.538, B=0.305, C^d_{v1}=0.202, C^u_{v1}=0.291, C^d_{v2}=0.255, C^u_{v2}=0.189, C^d_{s1}=0.621, C^u_{s1}=0.363, and χ^2/DOF =1874/1574. Because of the K factors to the PDFs, we find that the GRV98 PDFs need to be multiplied by a factor of 1.015.

The measured structure functions data are corrected for the relative normalizations between the SLAC, BCDMS, NMC and H1 data. The deuterium data are corrected for nuclear binding effects [7]. We also add a separate charm pair production contribution to lepton-nucleon scattering using the photon-gluon fusion model. This component is not necessary at low energies, and is only needed to describe the highest ν HERA F_2 and photoproduction data (since the GRV98 LO PDFs do not include a charm sea).

Our effective LO model describes various DIS (SLAC, BCDMS, NMC, and HERA) and photo-production data down to the $Q^2 = 0$ limit. Fig. 1 shows some of comparisons. Furthermore, based on duality arguments [8], it appears that this model also provides a reasonable description of the average value of F_2 for SLAC and Jlab data in the resonance region, as shown in Fig. 2[left]. Although no resonance data has been included in our fit, our model gives a good description of the most recent $2xF_1$ electron-proton Rosenbluth separated data in the resonance region from Jefferson Lab Hall C E94-110 Collaboration [9], and also data from the first phase of the JUPITER program at Jlab. Our predictions for $2xF_1$ are obtained using our F_2 model and R_{1998} [10]. For heavy targets,

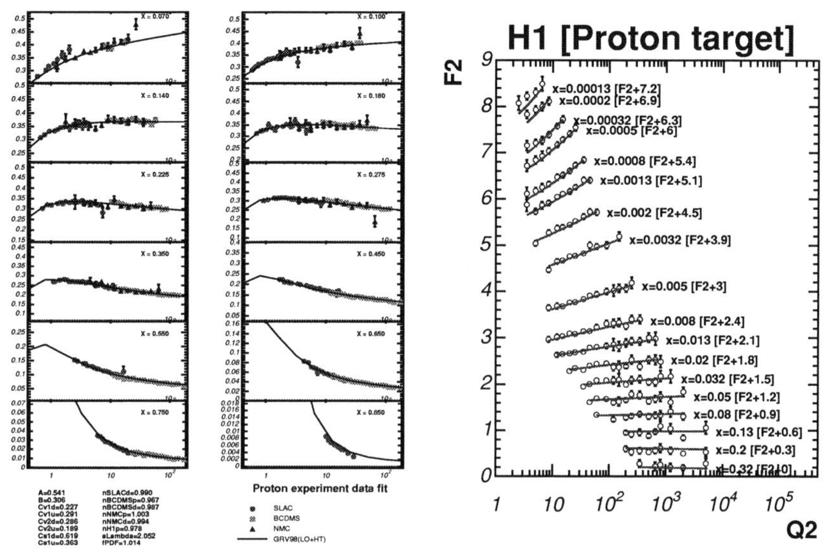

FIGURE 1. Comparisons of the predictions of our effective LO model for F_2 to charged lepton inelastic scattering data. [left] DIS F_2 proton data (SLAC, BCDMS, NMC), [right] H1 F_2 proton.

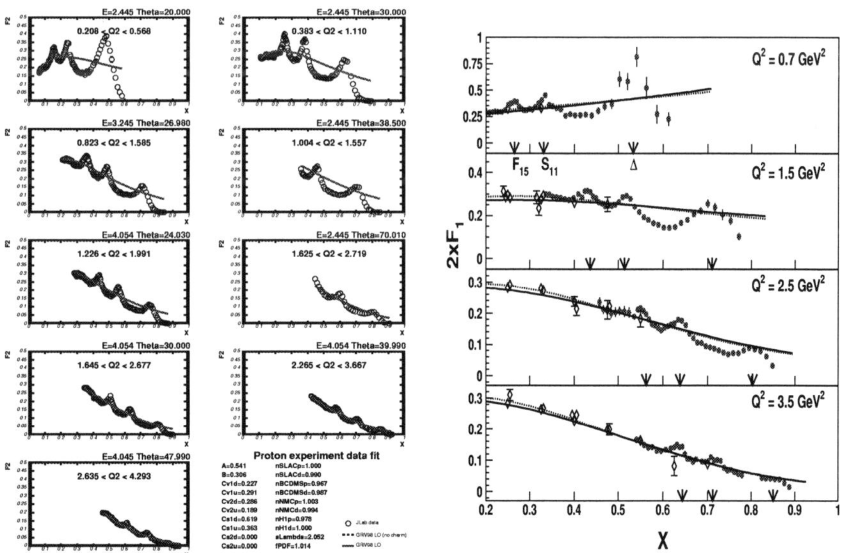

FIGURE 2. Comparisons of the predictions of our effective LO model to resonance electro-production data on protons (which was not included in our fit). Shown are F_2 proton [left], and $2xF_1$ proton data [right] (from the Jefferson Lab Hall C E94-110 Collaboration). The predictions for $2xF_1$ are obtained from our model for F_2 with R_{1998}. In the right plot, the solid line uses GRV98 PDFs, and the dashed line is our previous model using GRV94 LO PDFs.

nuclear effects are important, especially at low Q^2. Recent results from Jlab indicate that the Fe/D ratio in the resonance region is the same as the Fe/D ratio from DIS data for the same value of ξ (or ξ_w). Future Jlab experiments with deuterium and heavy nuclear targets (e.g. JUPITER) will provide a high statistics data in the resonance region which will be very important to improve our model at Q^2 region.

In neutrino scattering, in addition to the vector structure function, there is an axial vector structure function contribution. At the $Q^2 = 0$ limit, the vector structure function goes to zero, while the axial-vector part has a finite contribution. At high Q^2, these two structure functions are expected to be same. Thus, it is important to understand the axial-vector contribution at low Q^2 by comparing to future low energy neutrino data (e.g. MINERνA [14]). As a preliminary step, we compare the CCFR and CDHSW [11] high energy neutrino data with our model, assuming that the vector contribution is the same as the axial vector contribution. We find that the CCFR/CDHSW neutrino data are well described by our model.

We are currently working on constraining the low Q^2 axial vector contribution using low energy CDHSW and CHORUS [12] data. The form of the fits we plan to use is motivated by the Adler sum rule [13] for the axial vector contribution as follows:

$$K_{sea-ax}(Q^2) = \frac{Q^2 + C_{2s-ax}}{Q^2 + C_{1s-ax}}, \qquad K_{valence}(Q^2) = [1 - F_A^2(Q^2)] \left(\frac{Q^2 + C_{2v-ax}}{Q^2 + C_{1v-ax}} \right), \tag{4}$$

where $F_A(Q^2) = -1.267/(1 + Q^2/1.00)^2$. Nuclear effects for heavy target are also important and may be different for the vector and axial vector structure functions. Future measurements on the axial vector contribution from the MINERνA experiment [14] will be important in constraining this model.

REFERENCES

1. S. Fukuda *et al.*, Phys. Rev. Lett. **85**, 3999 (2000); T. Toshito, hep-ex/0105023.
2. A. Bodek and U. K. Yang, hep-ex/0308007.
3. L. W. Whitlow *et al.* (SLAC-MIT), Phys. Lett. B **282**, 433 (1995); A. C. Benvenuti *et al.* (BCDMS), Phys. Lett. B**237**, 592 (1990); M. Arneodo *et al.* (NMC), Nucl. Phys. B**483**, 3 (1997).
4. C. Keppel, Proc. of the Workshop on Exclusive Processes at High P_T, Newport News, VA, May (2002).]
5. U. K. Yang, Ph.D. thesis, Univ. of Rochester, UR-1583 (2001).
6. M. Gluck, E. Reya, A. Vogt, Eur. Phys. J **C5**, 461 (1998).
7. U. K. Yang and A. Bodek, Phys. Rev. Lett. **82**, 2467 (1999), Eur. Phys. J. C**13**, 241 (2000).
8. E. D. Bloom and F. J. Gilman, Phys. Rev. Lett. **25**, 1140 (1970).
9. Y. Liang *et al.*, nucl-ex/0410027.
10. K. Abe *et al.*, *Phys. Lett.* **B452**, 194 (1999).
11. P. Berge *et al.* (CDHSW), Zeit. Phys. **C49**, 607 (1991).
12. R. Oldeman, Proc. of 30th International Conference on High-Energy Physics (ICHEP 2000), Osaka, Japan, 2000.
13. S. Adler, Phys. Rev. **143**, 1144 (1966); F. Gillman, Phys. Rev. **167**, 1365 (1968).
14. MINERνA Proposal, D. Drakoulakos *et al.*, hep-ex/0405002

Scheme-Invariant Evolution to NNLO

J. Blümlein* and A. Guffanti*

*DESY, Platanenallee 6, D-15738 Zeuthen, Germany

Abstract. The (Factorization)-Scheme-Invariant analysis of unpolarized DIS structure functions is presented as a method to reduce theoretical errors on the determination of the strong coupling constant α_s, in order to match the accuracy foreseen for determinations coming from future high statistics measurements.

Keywords: Unpolarized DIS, structure functions, scheme-invariant evolution
PACS: 12.38.Bx

1. INTRODUCTION

The final HERA-II data on unpolarized DIS structure functions, combined with the present world data, will allow to reduce the experimental error on the strong coupling constant, α_s, to the level of 1% [1]. On the theoretical side the NLO analyzes have limitations due to the scale variations being present which allow no better than 5% accuracy in the determination of α_s. In order to match the expected experimental accuracy analyzes of DIS structure functions need then to be carried out to NNLO level.

To perform a full NNLO analysis the knowledge of the 3-loop β-function coefficient, β_2, the 2-(resp. 3-)loop Wilson coefficients and the 3-loop anomalous dimensions is required. With the calculation of the latter [2], the whole scheme-independent set of quantities is known, thus allowing a complete NNLO study of DIS structure functions.

While pushing the analysis one order further in perturbation theory will help reducing the theoretical errors, we think that other ways of reducing theoretical and conceptual uncertainties should be seeked. In this perspective we are persuaded that combining the standard QCD analysis and fits based on scheme-invariant evolution will provide a valuable tool in high-precision analyzes aiming to 1% accuracy in the determination of α_s. The advantage of scheme-invariant evolution consists in the fact that the input distributions are *physical observables* extracted from data at a reference scale Q_0^2. They are then evolved through evolution equations with physical anomalous dimensions and in the end a one-parameter fit to the data is performed to determine α_s.

The aim of our work is to perform a full NNLO analysis of unpolarized DIS structure functions in order to obtain a high-accuracy determination of α_s and to extract a set of parton distribution functions (PDFs) with fully correlated errors.

A complete study of structure functions data includes taking into account singlet and non-singlet evolution. We refer the reader interested in the non-singlet analysis to [3] and in the present letter we will concentrate on the singlet sector with particular reference to the scheme-invariant approach.

CP792, *Deep Inelastic Scattering, DIS 2005,* edited by W. H. Smith and S. R. Dasu

2. PHYSICAL ANOMALOUS DIMENSIONS

The coupled evolution of the singlet quark and gluon parton distribution functions can be mapped into the evolution of a pair of structure functions, related to the PDFs via convolution with the Wilson coefficients:

$$\begin{pmatrix} F_A^N \\ F_B^N \end{pmatrix} = \begin{pmatrix} C_{A,\Sigma}^N & C_{A,g}^N \\ C_{B,\Sigma}^N & C_{B,g}^N \end{pmatrix} \begin{pmatrix} \Sigma^N \\ G^N \end{pmatrix}. \tag{1}$$

The observables satisfy the matrix evolution equation:

$$\frac{d}{dt}\begin{pmatrix} F_A^N \\ F_B^N \end{pmatrix} = -\frac{1}{4}\mathbf{K}^N \begin{pmatrix} F_A^N \\ F_B^N \end{pmatrix}, \qquad t = -\frac{2}{\beta_0}\ln\frac{a_s(Q^2)}{a_s(Q_0^2)}, \qquad a_s = \frac{\alpha_s}{4\pi}. \tag{2}$$

The physical anomalous dimensions K_{IJ} are given in terms of the Wilson coefficients $C_{I,k}^N$ and the unpolarized anomalous dimensions γ_{ij}^N by

$$K_{IJ}^N = \left[-4\frac{\partial C_{I,m}^N(t)}{\partial t}\left(C^N\right)_{m,J}^{-1}(t)\frac{\beta_0 a_s(Q^2)}{2\beta(a_s(Q^2))}C_{I,m}^N(t)\gamma_{mn}^N(t)\left(C^N\right)_{n,J}^{-1}(t)\right]. \tag{3}$$

While the anomalous dimensions and the Wilson coefficients are, separately, factorization scheme dependent quantities, the physical anomalous dimensions are factorization scheme invariants. Equation (2) relates physical quantities, namely the structure functions and their slopes. This means that the physical kernels are also physical quantities and thus free of soft or collinear divergences. They can be expanded perturbatively in powers of the coupling constant $a_s(\mu^2)$ and, when computed in fixed-order perturbation theory they retain a dependence on the renormalization scale which can be used to estimate the theoretical error due to higher order corrections.

Different pairs of structure functions can be taken into consideration:

- F_2 and $\partial F_2/\partial t$ [4];
- F_2 and F_L [5, 6].

In [7] we presented the physical anomalous dimensions for the coupled evolution of the structure functions F_2 and $\partial F_2/\partial t$ up to NNLO. We also computed the ones for the coupled evolution of F_2 and F_L but, due to space limitations, we refer the interested reader to [8], where the complete expressions will be given.

3. HEAVY FLAVOURS CONTRIBUTION

It is known that the heavy flavour contribution to DIS structure functions in the kinematical regime of HERA is sizable. The electromagnetic structure function F_2, for example, receives contributions from heavy flavours of 20 – 40%, depending on the actual event kinematics, which requires a careful account for the heavy flavours contributions.

Since we solve the evolution equation for scheme-invariant evolution in Mellin space we implement heavy flavour contributions using the parametrization derived in [9]. The

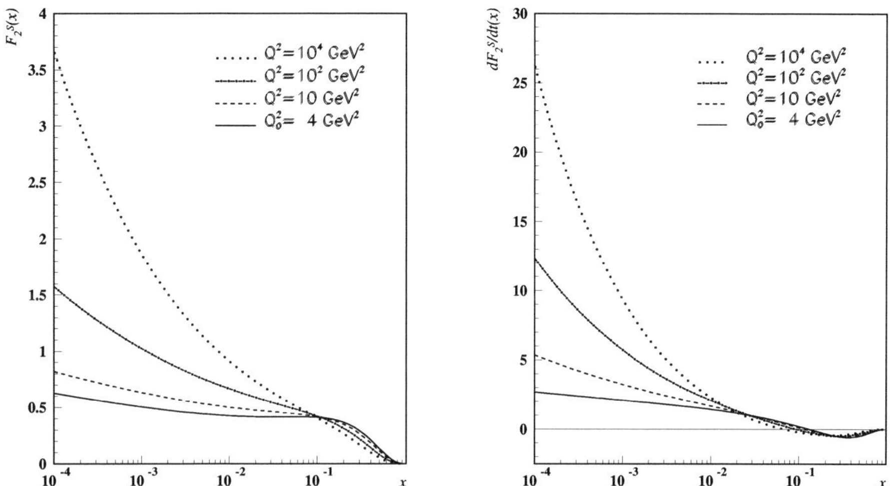

FIGURE 1. NNLO scheme invariant evolution for the singlet part of the structure function F_2 and its slope $\partial F_2/\partial t$ for four massless flavours.

inclusion into a scheme-invariant analysis requires, beyond the one of the heavy flavours Wilson coefficients, the knowledge of their derivatives with respect to the evolution variable t, defined in (2).

4. NUMERICAL RESULTS

As far as the numerical implementation of factorization scheme invariant evolution is concerned, we concentrated on the F_2 and $\partial F_2/\partial t$ system. We implemented the evolution to NNLO for massless flavours and we are on the way to include heavy flavours contributions.

In Fig. 1 we present the scheme invariant evolution for the structure functions F_2 and $\partial F_2/\partial t$ to NNLO. The input distribution at the reference scale are not extracted from data, but rather built up as a convolution of Wilson coefficients and PDFs, the latter being parametrised according to [10]. Comparing the behaviour of the slopes in Fig. 1b to the slopes extracted experimentally in Fig. 12 of [11] points towards a positive gluon density in the small x region.

5. CONCLUSIONS

The future high precision HERA-II data will allow a reduction of the experimental error on the determination of the strong coupling constant α_s to the level of 1%. On the theoretical side, the inclusion of NNLO corrections is therefore mandatory to cope with such a level of accuracy.

In view of a high accuracy determination of α_s we think that combining standard $\overline{MS}$ analysis with fits based on factorization-scheme invariant evolution could provide a method to have a better control on theoretical and conceptual errors.

ACKNOWLEDGMENTS

This work is supported in part by DFG Sonderforschungbereich Transregio 9, Computergestüzte Theoretische Physik.

REFERENCES

1. M. Botje, M. Klein and C. Pascaud, arXiv:hep-ph/9609489.
2. S. Moch, J. A. S. Vermaseren and A. Vogt, Nucl. Phys. B **688** (2004) 101, [arXiv:hep-ph/0403192] and Nucl. Phys. B **691** (2004) 129, [arXiv:hep-ph/0404111].
3. J. Blümlein, H. Böttcher and A. Guffanti, arXiv:hep-ph/0407089.
4. W. Furmanski and R. Petronzio, Z. Phys. C **11** (1982) 293.
5. S. Catani, Z. Phys. C **75** (1997) 665, [arXiv:hep-ph/9609263].
6. J. Blümlein, V. Ravindran and W. L. van Neerven, Nucl. Phys. B **586** (2000) 349, [arXiv:hep-ph/0004172].
7. J. Blümlein and A. Guffanti, arXiv:hep-ph/0411110.
8. J. Blümlein and A. Guffanti, in preparation.
9. E. Laenen, S. Riemersma, J. Smith, and W.L. van Neerven, Nucl. Phys. **B392** (1993) 162, 229; S. Riemersma, J. Smith, and W.L. van Neerven, Phys. Lett. **B347** (1995) 143; S. I. Alekhin and J. Blümlein, Phys. Lett. B **594**, 299 (2004) [arXiv:hep-ph/0404034].
10. W. Giele *et al.*, arXiv:hep-ph/0204316.
11. C. Adloff *et al.* [H1 Collaboration], Eur. Phys. J. C **21**, 33 (2001) [arXiv:hep-ex/0012053].

NLO BFKL tests using the proton structure function F_2 measured at HERA

C. Royon

Service de physique des particules, CEA/Saclay, 91191 Gif-sur-Yvette cedex, France
Fermilab, Batavia, USA

Abstract. We propose a phenomenological study of the next-to-leading Balitsky-Fadin-Kuraev-Lipatov (BFKL) approach applied to the data on the proton structure function F_2 measured at HERA in the small-x_{Bj} region.

INTRODUCTION

Precise phenomenological tests of QCD evolution equations are one of the main goals of deep inelastic scattering phenomenology. For the Dokshitzer-Gribov-Lipatov-Altarelli-Parisi (DGLAP) evolution in Q^2 [1], it has been possible to test it in various ways with NLO (next-to-leading $\log Q^2$) and now NNLO accuracy and it works quite well in a large range of Q^2 and x_{Bj}. Testing precisely the Balitsky-Fadin-Kuraev-Lipatov (BFKL) evolution in energy [2] (or x_{Bj}) beyond leading order appears more difficult.

The first experimental results from HERA confirmed the existence of a strong rise of the proton structure function F_2 with energy which, in the BFKL framework, can be well described by a simple (3 parameters) LO-BFKL fit [3]. The main issue of Ref.[3] was that not only the rise with energy but also the scaling violations observed at small x_{Bj} are encoded in the BFKL framework through the Q^2 variation of the effective anomalous dimension. However one was led [3] to introduce an effective but unphysical value of the strong coupling constant $\alpha \sim .07 - .09$ instead of $\alpha \sim .2$ in the Q^2-range considered for HERA small-x_{Bj} physics, revealing the need for NLO corrections. Indeed, the running of the strong coupling constant is not taken into account.

In fact, the theoretical task of computing these corrections appears to be quite hard. It is now in good progress but still under completion. For the BFKL kernel, they have been calculated after much efforts [5]. It was realized [4] that the main problem comes from the existence of spurious singularities brought together with the NLO corrections, which ought to be cancelled by an appropriate resummation at all orders of the perturbative expansion [5, 6, 7], resummation required by consistency with the QCD renormalization group.

CP792, *Deep Inelastic Scattering, DIS 2005*, edited by W. H. Smith and S. R. Dasu

NLO BFKL PHENOMENOLOGY

Saddle point approximation

Following the successful BFKL-LO parametrisation of the proton structure F_2 at HERA, we perform the same saddle point approximation as at LO using χ_{NLO} given by resummed NLO BFKL kernels [8].

$$F_2 = C e^{\alpha_{RGE}\chi_{eff}(\gamma_c,\alpha_{RGE})Y} \left(Q^2/Q_0^2\right)^{\gamma_c} e^{-\frac{log^2(Q/Q_0)}{2\alpha_{RGE}\chi''_{eff}(\gamma_c,\alpha_{RGE})Y}} \tag{1}$$

where γ_C and χ_{eff} come directly from the properties of the NLO BFKL equation if the small-x structure function is dominated by the perturbative Green function:

$$\frac{d\chi_{eff}}{d\gamma}(\gamma_C,\alpha_{RGE}(Q^2)) = 0 \quad ; \quad \chi_{eff}(\gamma,\alpha_{RGE}) = \omega(\gamma,\alpha_{RGE})/\alpha_{RGE}. \tag{2}$$

Instead of getting a 3-parameter formula like at LO (normalisation, α_S, and Q_0), we get only two free parameters at NLO since the value of α_S and its Q^2 evolution are imposed by the renormalisation group equations (RGE). The delicate aspect of the problem comes for the fact that χ is now scheme dependent.

Strategy for NLO fits

The strategy for BFKL-NLO is the following [8]:

- The first step is the knowledge of $\chi_{NLO}(\gamma,\omega,\alpha)$ from the BFKL equation and different resummation schemes
- The second step is to use the implicit equation $\chi(\gamma,\omega) = \omega/\alpha$ to compute numerically ω as a function of γ for different schemes and values of α
- The third step is to determinate numerically the saddle point values γ_C as a function of α as well as the values of χ and χ''
- The fourth step is to perform the BFKL-NLO fit to HERA F_2 data with two free parameters C and Q_0^2

Details about the numerical results can be found in Ref. [8].

The results of the NLO BFKL fit to the H1 and ZEUS data [9] using the saddle point approximation for two different schemes (CSS and S3, see Ref [4, 6]) are given in Fig 1 where the data over theory ratio is displayed.

We see a big dicrepancy between data and theory especially at lower Q^2. To understand further the reason of that discrepancy, we performed an analysis in the Mellin space where the formulation of the BFKL NLO resummed kernels is easier.

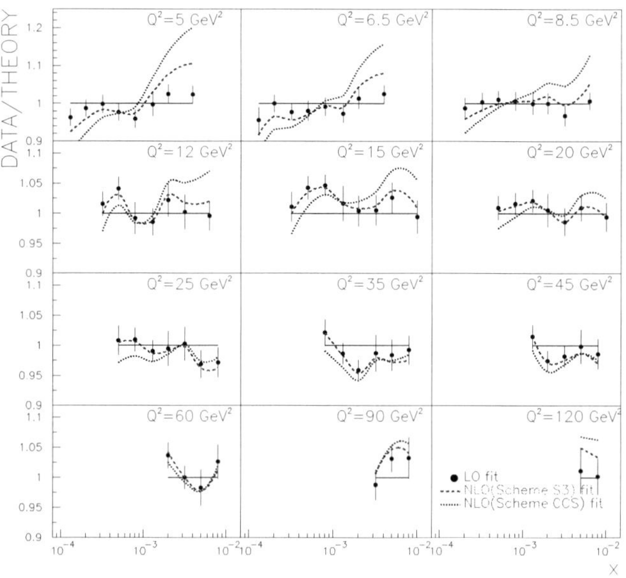

FIGURE 1. Data/Theory ratios for the proton structure function F_2. The points show the results of the LO fit as a reference. The dashed (resp. dotted) lines show the results of the BFKL-NLO fits including the S3 (resp. CCS) resummation scheme.

ANALYSIS IN MELLIN SPACE

In this section we want to analyze in more detail the features of the BFKL parametrizations and in particular the reasons of the still quantitatively unsatisfactory results of the NLO fits. For this sake, it is important to come back to the key ingredient of our analysis, i.e. the dominance of the hard Pomeron singularity expressed by the relation (2). Equality (2) can be checked at NLO using the GRV98 [10], MRS2001 [11], CTEQ6.1 [12] and ALLM [13] parametrisations. These four parametrisations give a fair description of the proton structure functions measured by the H1 and ZEUS collaborations over a wide range of x_{Bj} and Q^2, as well as fixed target experiment data. The three first parametrisations correspond to a DGLAP NLO evolution whereas the ALLM one corresponds to a Regge analysis of proton structure function data.

We notice in Fig.2 that the linear property of relation (2), namely for $\chi_{eff}(\gamma^*, \alpha_{RG})$ as a function of ω is well verified. We indeed can describe the GRV and MRS parametrisations using a linear fit with a good precision. However the predicted zero at the origin $\omega = 0$ is not obtained, even if the value at the origin remains small. The fit does not go through the origin and we would need to add a constant term to the linear fit formula, and the slope is not equal to α. Small but sizeable effects give phenomenological deviations from the expected theoretical properties of the NLO kernels.

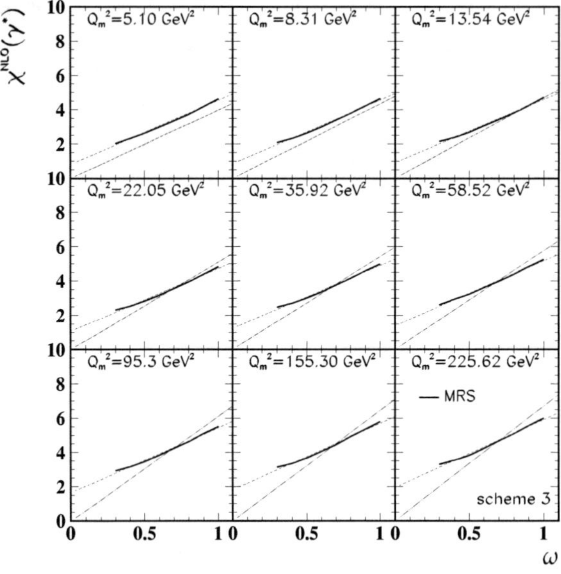

FIGURE 2. Test of $\chi(\omega, Q^2)$ for scheme $S3$. The result for the MRS parametrization is shown in black in the different bins in Q^2 together with a linear fi t, and the expectation if formula (2) is fulfi lled. We notice the discrepancy between the MRS result and formula (2).

ACKNOWLEDGMENTS

There results come from a fruitful collaboration with R. Peschanski and L. Schoeffel.

REFERENCES

1. G.Altarelli and G.Parisi, *Nucl. Phys.* **B126** 18C (1977) 298. V.N.Gribov and L.N.Lipatov, *Sov. Journ. Nucl. Phys.* (1972) 438 and 675. Yu.L.Dokshitzer, *Sov. Phys. JETP.* **46** (1977) 641.

2. L.N.Lipatov, *Sov. J. Nucl. Phys.* **23** (1976) 642; V.S.Fadin, E.A.Kuraev and L.N.Lipatov, *Phys. lett.* **B60** (1975) 50; E.A.Kuraev, L.N.Lipatov and V.S.Fadin, *Sov.Phys.JETP* **44** (1976) 45, **45** (1977) 199; I.I.Balitsky and L.N.Lipatov, *Sov.J.Nucl.Phys.* **28** (1978) 822.
3. H Navelet, R.Peschanski, Ch. Royon, S.Wallon, *Phys. Lett.* **B385** (1996) 357. S.Munier, R.Peschanski, *Nucl.Phys.* **B524** (1998) 377.
4. G.P. Salam, *JHEP 9807* (1998) 019
5. V.S. Fadin and L.N. Lipatov, Phys. Lett. B429 (1998) 127; M.Ciafaloni, Phys. Lett. B429 (1998) 363; M. Ciafaloni and G. Camici, Phys. Lett. B430 (1998) 349.
6. M. Ciafaloni, D. Colferai, G.P. Salam, *Phys.Rev.* **D60** 114036, , *JHEP 9910* (1999) 017; M. Ciafaloni, D. Colferai, G.P. Salam,A.M. Stasto, *Phys.Lett.* **B541** (2002) 314.
7. Stanley J. Brodsky, Victor S. Fadin, Victor T. Kim, Lev N. Lipatov, Grigorii B. Pivovarov, *JETP Lett.* **70** (1999) 155.
8. R.Peschanski, Ch. Royon, L.Schoffel, *Nucl. Phys.* **B716** (2005) 401.
9. H1 Collab., C. Adloff et al, *Eur.Phys.J.* **C21** (2001) 33; ZEUS Collab., S. Chekanov et al., *Eur.Phys.J.* **C21** (2001) 443.
10. M. Gluck, E. Reya, A. Vogt, *Eur.Phys.J.* **C5** (1998) 461, for updated parametrizations.
11. A.D. Martin, R.G. Roberts, W.J. Stirling, R.S. Thorne, *Eur.Phys.J.* **C23** (2002) 73.
12. D. Stump, J. Huston, J. Pumplin, W.-K. Tung, H.L. Lai, S. Kuhlmann, J. F. Owens, *JHEP* **0310** (2003) 046.
13. H. Abramowicz, E. Levin, A. Levy, U. Maor, *Phys. Lett. B* **269** (1991) 46.

Global Analyses of Nuclear PDFs

V.J. Kolhinen

Helsinki Institute of Physics, P.O. Box 64, FIN-00014 University of Helsinki, Finland
Department of Physics, P.O.Box 35 (YFL), FIN-40014 University of Jyväskylä, Finland

Abstract. A brief overview of the global DGLAP analyses of the nuclear parton distribution functions is given. Although all the current global nPDF sets describe $R^A_{F_2}(x,Q^2)$ well in the large-x region where the data exist, variations between their parton distributions can be substantial.

Keywords: Global analysis, nuclear PDF, shadowing
PACS: 24.85.+p

Ever since it was observed about two decades ago that parton distribution functions (PDFs) of nuclei differ from those in the free proton, $f_i^A(x,Q^2) \neq f_i(x,Q^2)$, several analyses of the nuclear effects have been presented. Similarly to the case of the free proton PDFs, the nuclear PDFs (nPDFs) at an initial scale provide the nonperturbative input for the perturbative QCD analysis. Once they are known the Dokshitzer-Gribov-Lipatov-Altarelli-Parisi (DGLAP) evolution equations predict their behaviour at larger scales.

Although several DGLAP analyses [1, 2, 3, 4, 5, 6] exist, I will concentrate here only on the recent global ones where the initial distributions are based on the fit to the data, not on a model. Only three such analyses and their reanalyses currently exist, namely the ones by us, Eskola *et al.* (usually called as *EKS98* [7, 8]), Hirai *et al.* (*HKM* [9] and *HKN* [10]) and by de Florian and Sassot (*nDS* [11]). Along with these I will also present preliminary results of the reanalysis of our nPDFs (*EKS05* [12]).

The analyses of the nPDFs are performed much in the same way as those of the free PDFs. However, lack of data especially in the small-x region has kept the nPDF analyses less constrained than the free PDF ones. For example, whereas the recent PDF analyses have been performed in next-to-leading order (NLO) and some are currently being calculated in NNLO, only one global nPDF analysis, by de Florian and Sassot (nDS), is currently calculated in NLO.

Since the initial states of the global nPDF sets are based on the data they naturally describe well the structure function $F_2^A(x,Q^2)$ at large x where the most of the data lie. However, their nuclear modifications for the different parton flavours can vary greatly in some regions.

Nuclear effects are commonly defined through a ratio

$$R_i^A(x,Q^2) = \frac{f_i^A(x,Q^2)}{f_i(x,Q^2)}, \tag{1}$$

where f_i stands for parton distribution function for a parton type, $i = u, \bar{u}, d, \bar{d}, \ldots, g$. However, due to lack of data one usually defines only some 3-5 ratios for the initial distributions: for valence (u and d together (EKS98), or separately (HKM,HKN,nDS)),

CP792, *Deep Inelastic Scattering, DIS 2005*, edited by W. H. Smith and S. R. Dasu

sea ($\bar{u}$ and $\bar{d}$ together (EKS98,HKM,HKN), or separately (nDS)) and gluon. These ratios are usually given for a bound proton.

The $R_{F_2}^A$ ratios are fairly well constrained in large-x region by the data from lepton-nucleus deep inelastic scattering (DIS). As valence quarks dominate the F_2 in this region, they also become well determined there. At mid-x the Drell-Yan (DY) dilepton data constrain the sea quark distributions together with the DIS data. For gluon distribution the only data constraints arise from the $\log Q^2$ slopes of the NMC data for $F_2^{\mathrm{Sn}}/F_2^{\mathrm{C}}$ [13] as shown in Refs. [7, 14]:

$$\frac{\partial R_{F_2}^A(x,Q^2)}{\partial \log Q^2} \approx \frac{10\alpha_s}{27\pi} \frac{xg(2x,Q^2)}{F_2^D(x,Q^2)} \left\{ R_G^A(2x,Q^2) - R_{F_2}^A(x,Q^2) \right\}. \tag{2}$$

In addition to the data further constraints for the fits arise from the momentum, charge and baryon number conservation. Let us next take a look at these three analyses individually.

EKS98, EKS05: In the nPDF analysis by us, Eskola, Kolhinen, Ruuskanen and Salgado [7, 8] the $R_{F_2}^A(x,Q_0^2)$ distribution was first parametrized piecewise for each A and fitted to the data. The $R_{F_2}^A$ was then split to valence and sea part which were constrained by the DY data and baryon number conservation. Finally, the gluon distribution was constructed from the $R_{F_2}^A$ fit. The actual fits to the DIS and DY data were done by eye. However, later calculations with EKS98 prove $\chi^2 \approx 390$ for 503 data points, though the data set is slightly different than in the original analysis.

As a continuation for this work, we are currently performing a reanalysis of the nPDFs, with some more recent data included [12]. We have now also included a proper χ^2 analysis and use a Hessian method for the error estimates. Instead of parametrizing the $R_{F_2}^A$ as in EKS98, we now parametrize directly the nuclear effects in the initial valence, sea and gluon distributions. Although the results are still preliminary, they seem to resemble much the EKS98 ones and giving $\chi^2 \approx 390-400$ for 503 data points.

HKM, HKN: In their first analysis Hirai, Kumano and Miyama (HKM) [9] use the DIS, but not DY, data from several experiments. They composed two different fits, "quadratic" and "cubic" referring to the polynomial in the fit. These fits are performed for each A separately. The resulting valence ratios R_u^A and R_d^A are not given for bound proton but for an average nucleon in a nucleus. The calculated χ^2 of the fit is 583.7 (quadratic) and 546.6 (cubic) for 309 data points, or $\chi^2/d.o.f. = 1.93$ and 1.82, respectively. The nuclear effects show small antishadowing for the valence at small x. Sea and gluons are shadowed in small-x region, but antishadowed at larger values of x. Only valence shows an EMC effect (shadowing) at $x \sim 0.7$.

In the subsequent analysis by Hirai, Kumano and Nagai (HKN) [10], the DY data have been included along with some more DIS data. The statistical error analysis is also performed using a Hessian method. The general form of the fit is similar to the "cubic" form in HKM. The resulting distributions fit the small x region better, obviously due to the DY data and improved description of the sea quarks. Whereas valence and gluon behave much in the same way as in the HKM, the sea quarks now have a valence-like EMC effect at $x \sim 0.7$. The resulting $\chi^2 = 1489.8$ for 951 data points.

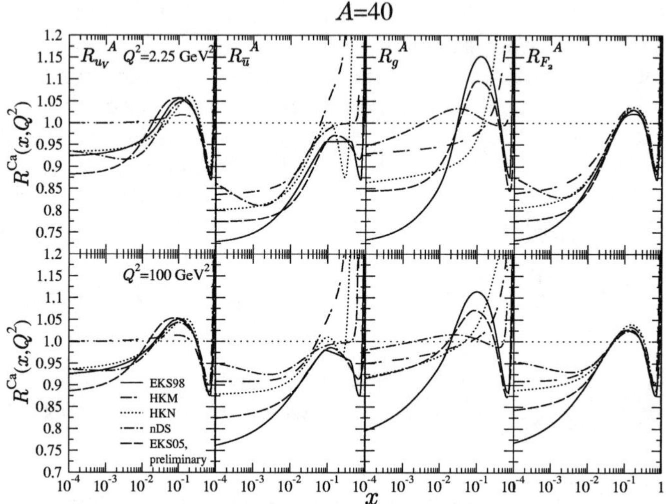

FIGURE 1. Ratios $R^A_{u_v}(x,Q^2)$, $R^A_{\bar{u}}(x,Q^2)$, $R^A_g(x,Q^2)$ and $R^A_{F_2}(x,Q^2)$ for $Q^2 = 2.25$ GeV^2 (upper panels) and $Q^2 = 100$ GeV^2 (lower panels) for $A = 40$ given by EKS98 (solid), HKM (double dashed), HKN (dotted) and nDS [NLO] (dotted-dashed). Preliminary EKS05 results are also shown (dashed).

nDS: The global nPDF analysis by de Florian and Sassot (nDS) [11] is so far the only one performed in NLO. In this analysis PDFs are defined using the convolution method, $f_i^A(x,Q_0^2) = \int_x^A \frac{dy}{y} w_i(y,A) f_i(\frac{x}{y},Q_0^2)$, which enables evolution in the Mellin space. The advantages of this approach are that the calculations are faster and more straightforward, as well as that the x dependence of nPDFs is strongly correlated to that of free PDFs. The total χ^2 obtained is 316.35 for LO and 300.15 for NLO for 420 data points. The main difference between LO and NLO results are in sea and gluon distributions at small x, small Q^2 and large A. As the χ^2 values suggest, LO fit describes the data almost equally well as the NLO one. This fact is reported to arise from the rather restricted Q^2 range of the data as well as the absence of the data strongly dependent to gluon distribution.

Compared to the other analyses, the largest difference is in the gluon distribution, which is much less shadowed than e.g. in EKS98. The authors have tried another parametrization with stronger gluon shadowing, but they report the results to be worse.

Comparison between the nuclear effects of different sets are shown in Fig. 1 for $A = 40$ and for $Q^2 = 2.25$ and 100 GeV^2. As seen in the figure, $R^A_{F_2}$'s calculated using different nPDF sets coincide in the large x region. Also valence quarks become well determined in large x. However, in other regions the differences between the sets can be large. In order to constrain the fits more properly, especially at small x, more data would be needed. As pointed out earlier and shown in Fig. 2, currently only the $\log Q^2$ slopes of NMC data for $F_2^{\rm Sn}/F_2^{\rm C}$ give constraints to gluon distributions. Probes sensitive to nuclear gluon PDFs, such as the charm production, would thus be crucial for more accurate analysis.

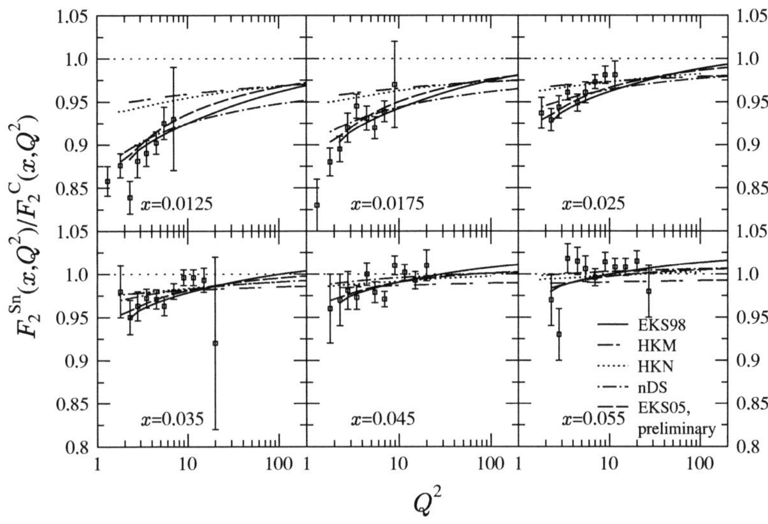

FIGURE 2. Calculated F_2^{Sn}/F_2^{C} ratios compared to the NMC data [13] for a few small x values.

In future analyses data on structure function F_3 could also provide more information on the u_v^A vs d_v^A ratio and the valence shadowing.

Acknowledgements. This project was funded by the Academy of Finland, projects nos. 80383 and 206024.

REFERENCES

1. J.-w. Qiu, *Nucl. Phys.*, **B291**, 746 (1987).
2. L. L. Frankfurt, M. I. Strikman, and S. Liuti, *Phys. Rev. Lett.*, **65**, 1725–1728 (1990).
3. K. J. Eskola, *Nucl. Phys.*, **B400**, 240–266 (1993).
4. D. Indumathi, and W. Zhu, *Z. Phys.*, **C74**, 119–129 (1997), `hep-ph/9605417`.
5. D. Indumathi, *Z. Phys.*, **C76**, 91–97 (1997), `hep-ph/9609361`.
6. L. Frankfurt, V. Guzey, and M. Strikman, *Phys. Rev.*, **D71**, 054001 (2005), `hep-ph/0303022`.
7. K. J. Eskola, V. J. Kolhinen, and P. V. Ruuskanen, *Nucl. Phys.*, **B535**, 351–371 (1998), `hep-ph/9802350`.
8. K. J. Eskola, V. J. Kolhinen, and C. A. Salgado, *Eur. Phys. J.*, **C9**, 61–68 (1999), `hep-ph/9807297`.
9. M. Hirai, S. Kumano, and M. Miyama, *Phys. Rev.*, **D64**, 034003 (2001), `hep-ph/0103208`.
10. M. Hirai, S. Kumano, and T. H. Nagai, *Phys. Rev.*, **C70**, 044905 (2004), `hep-ph/0404093`.
11. D. de Florian, and R. Sassot, *Phys. Rev.*, **D69**, 074028 (2004), `hep-ph/0311227`.
12. K. J. Eskola, V. J. Kolhinen, and C. A. Salgado, *In preparation.*
13. M. Arneodo, et al., *Nucl. Phys.*, **B481**, 23–39 (1996).
14. K. Prytz, *Phys. Lett.*, **B311**, 286–290 (1993).

Coherent Power Corrections to Structure Functions

Ivan Vitev

Los Alamos National Laboratory, Theory Division and Physics Division, Mail Stop H846, Los Alamos, NM 87544

Abstract.
We calculate and resum a perturbative expansion of nuclear enhanced power corrections to the structure functions measured in deeply inelastic scattering of leptons on a nuclear target. Our results for the Bjorken x-, Q^2- and A-dependence of nuclear shadowing in $F_2^A(x,Q^2)$ and the nuclear modifications to $F_L^A(x,Q^2)$, obtained in terms of the QCD factorization approach, are consistent with the existing data. We predict the dynamical final state shadowing in $\nu+A$ reactions for sea and valence quarks in the structure functions $F_2^A(x,Q^2)$ and $F_3^A(x,Q^2)$, respectively. In $p+A$ collisions we calculate the centrality and rapidity dependent nuclear suppression of single and double inclusive hadron production at moderate transverse momenta.

Keywords: Power corrections, High twist shadowing
PACS: 12.38.Cy; 12.39.St; 24.85.+p

Dynamical high twist shadowing

Under the approximation of one-photon exchange, the lepton-hadron DIS cross section $d\sigma_{\ell h}/dx dQ^2 \propto L_{\mu\nu} W^{\mu\nu}(x,Q^2)$, with Bjorken variable $x = Q^2/(2p\cdot q)$ and virtual photon's invariant mass $q^2 = -Q^2$. The hadronic tensor can be expressed in terms of structure functions based on the polarization states of the exchange virtual photon: $W^{\mu\nu}(x,Q^2) = \varepsilon_T^{\mu\nu} F_T(x,Q^2) + \varepsilon_L^{\mu\nu} F_L(x,Q^2)$. In DIS the exchange photon γ^* of virtuality Q^2 and energy $\nu = Q^2/(2xm_N)$ probes an effective volume of transverse area $1/Q^2$ and longitudinal extent $\Delta z_N \times x_N/x$, where Δz_N is the nucleon size, $x_N = 1/(2r_0 m_N) \sim 0.1$ and $r_0 \sim 1.2$ fm. When Bjorken $x \ll x_N$ the lepton-nucleus DIS covers several nucleons in longitudinal direction while it is localized in the transverse plane.

In the lightcone $A^+ = 0$ gauge and the Breit frame we identify the natural short and long distance separation of the multiple final state interactions from the propagator structure of the struck quark, $i(\gamma^+/2p^+)/(x_i - x \pm i\varepsilon)$ (pole term) and $i(xp^+/Q^2)\gamma^-$ (contact term) [1]. The two gluon contact exchange is therefore evaluated in a single nucleon state. Resumming the $A^{1/3}$-enhanced power corrections we find [1]:

$$F_T^A(x,Q^2) \approx A F_T^{(\mathrm{LT})}\left(x + \frac{x\xi^2(A^{1/3}-1)}{Q^2}, Q^2\right), \tag{1}$$

$$F_L^A(x,Q^2) \approx A F_L^{(\mathrm{LT})}(x,Q^2) + \frac{4\xi^2}{Q^2} F_T^A(x,Q^2). \tag{2}$$

CP792, *Deep Inelastic Scattering, DIS 2005,* edited by W. H. Smith and S. R. Dasu

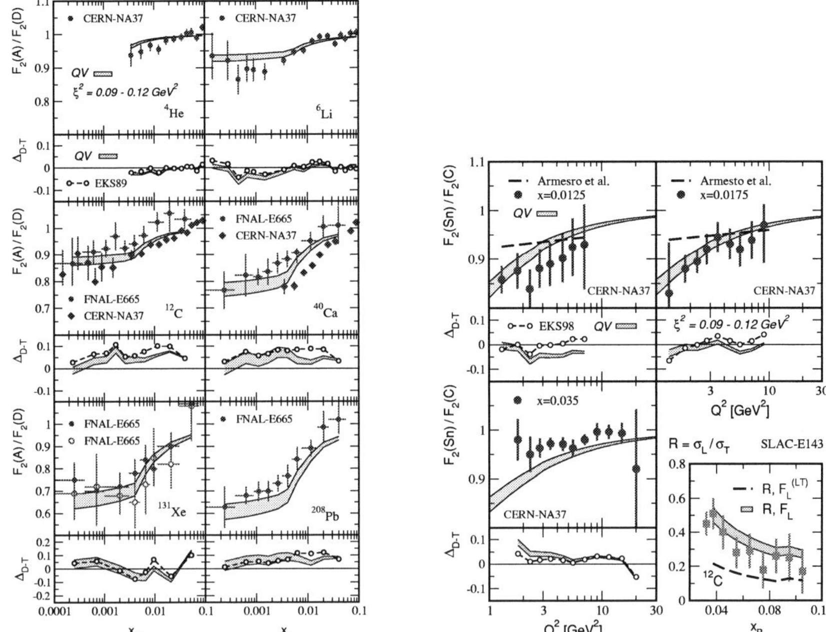

FIGURE 1. Left panel: $F_2(A)/F_2(D)$ calculation of resummed power corrections versus nuclear A and Bjorken-x [1]. Right panel: $F_2(Sn)/F_2(C)$ show evidence for a power-law in $1/Q^2$ behavior consistent with the all-twist resummed calculation [1]. The bottom right panel illustrates the role of higher twist contribution to F_L on the example of $R = \sigma_L/\sigma_T$.

Here ξ^2 represents the characteristic scale of higer twist per nucleon to $\mathcal{O}(\alpha_s)$:

$$\xi^2 = \frac{3\pi\alpha_s(Q^2)}{8\,r_0^2}\langle p|\hat{F}^2(\lambda_i)\,|p\rangle\,, \qquad \langle p|\hat{F}^2(\lambda_i)\,|p\rangle = \lim_{x\to 0}\frac{1}{2}xG(x,Q^2)\,.$$

The x- and A-dependence of $F_2(A)/F_2(D)$, calculated for $\xi^2 = 0.09 - 0.12$ GeV^2, is given in the left panel of Fig. 1. Comparison to a leading twist shadowing parameterization [2] is also shown. The right panel of Fig. 1 indicates the power law nature of the nuclear modification to the structure functions. The physical gluon exchange leads to a high twist contribution to the longitudinal structure function F_L and enhances the ratio $R = \sigma_L/\sigma_T$. We emphasize that both leading twist [3] and high twist shadowing [1] have their origin in the *final state* coherent scattering. This provides a natural explanation of the apparent *lack* of gluon shadowing in the NLO global analysis [4] which is the only one directly sensitive to gluons.

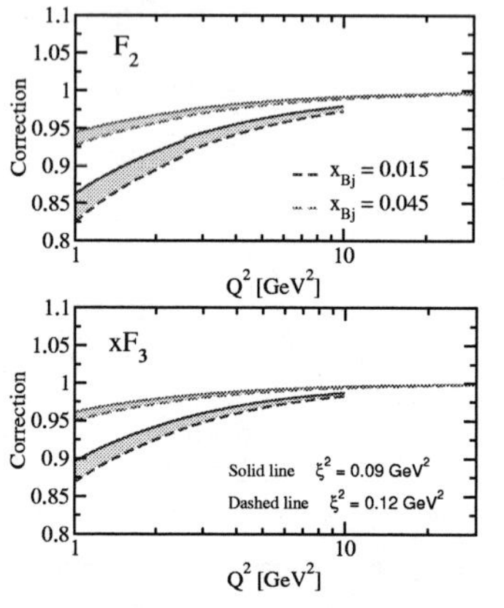

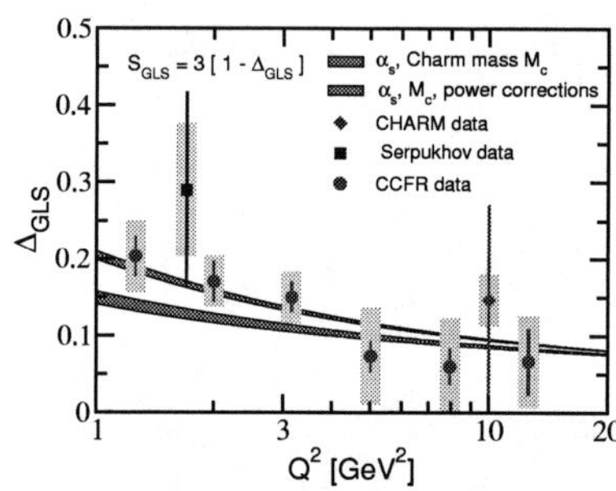

FIGURE 2. Left panel: power corrections to the structure functions $F_2(x,Q^2)$ and $xF_3(x,Q^2)$ [5] for two values of x_B corresponding to NuTeV measurements [7]. Right panel: high twist modification to the Gross-Llewellyn-Smith sum rule Δ_{GLS} [5].

Neutrino-nucleus scattering

Neutrino-nucleus scattering provides the unique opportunity to separately study the effect of coherent power corrections for sea and valance quarks [5] through the structure functions:

$$\begin{aligned} F^{\nu A}_{1(3)}(x_B,Q^2) \approx\ & A(2)\left[\sum_{D,U}|V_{DU}|^2\phi^A_D\left(x_B+x_B\frac{\xi^2(A^{1/3}-1)}{Q^2}+x_B\frac{M^2_U}{Q^2},Q^2\right)\right. \\ & \left. +(-)\sum_{\bar{U},\bar{D}}|V_{\bar{U}\bar{D}}|^2\phi^A_{\bar{U}}\left(x_B+x_B\frac{\xi^2(A^{1/3}-1)}{Q^2}+x_B\frac{M^2_{\bar{D}}}{Q^2},Q^2\right)\right] . \end{aligned} \tag{3}$$

Here V_{DU} are the CKM matrix elements. Eq. (3) identifies the nuclear enhanced high twist corrections with dynamical mass $m^2_{dyn} = \xi^2(A^{1/3}-1)$ generated by the final state parton scattering through direct comparison to $M^2_{U,D}$.

The modification to the structure functions $F_2(x,Q^2)$ and $xF_3(x,Q^2)$ for two select values of x_B are shown in the left panel of Fig. 2. These give a good description of the observed power law deviation of the reduced cross sections measured by NuTeV [6, 7] from the leading twist pQCD at small values of Q^2. Note the difference in the "shadowing" of F_2 and xF_3 due to the different steepness of sea and valence quark PDFs (in x). The right panel of Fig. 2 demonstrates the improved agreement between data and theory for the Gross-Llewellyn-Smith sum rule [5]:

$$\Delta_{\rm GLS} \equiv \frac{1}{3}(3-S_{\rm GLS}) = \frac{\alpha_s(Q^2)}{\pi} + \frac{\mathscr{G}}{Q^2} + \mathscr{O}(Q^{-4}) . \tag{4}$$

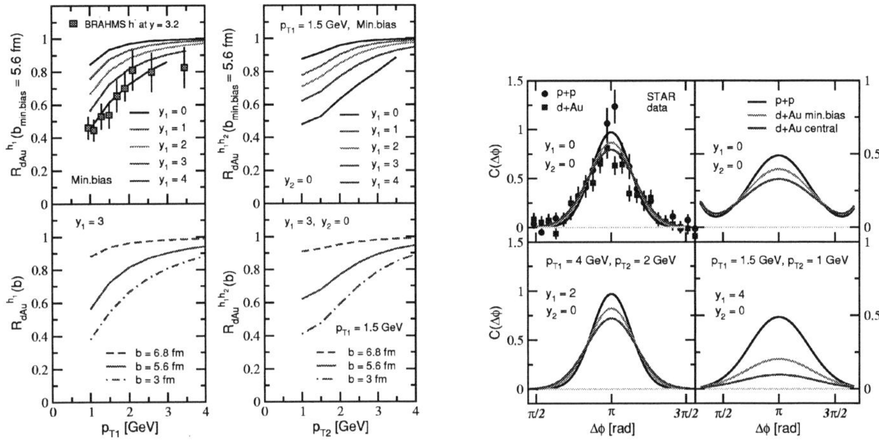

FIGURE 3. Left panel: upper limit on the suppression of the single inclusive particle production $R^{(1)}_{pA}(p_{T_1})$ from coherent power corrections versus rapidity and centrality [8]. Data is from BRAHMS [9]. Right panel: suppression of the double inclusive cross section $R^{(2)}_{pA}(p_{T_1}, p_{T_2})$ for different rapidity gaps, p_{T_1}, p_{T_2} ranges and centrality.

Proton-nucleus collisions

The $p+A$ analogue of the DIS coherent power corrections is the final state interactions of the small x_b parton in the $|\hat{t}| \ll |\hat{s}|, |\hat{u}|$ regime. Here $\hat{t} = q^2 = (x_a P_a - P_c/z_1)^2$ and the x_b rescaling in the lowest order pQCD formalism reads [8]:

$$F_{ab\to cd}(x_b) \;\Rightarrow\; F_{ab\to cd}\left(x_b \left[1 + C_d \frac{\xi^2}{-t}(A^{1/3} - 1) \right] \right) . \tag{5}$$

In Eq.(5) $F_{ab\to cd}(x_b) = |M_{ab\to cd}|^2 \phi(x_b)/x_b$ and C_d is a color factor, $C_{q(\bar{q})} = 1$ and $C_g = C_A/C_F = 9/4$ for quark (antiquark) and gluon, respectively.

The left panel of Fig. 3 shows the *upper limit* on the centrality and rapidity dependent suppression $R^{(1)}_{pA}$ of single inclusive hadron production at RHIC. Data is from BRAHMS [9]. Additional nuclear suppression arises form the energy loss in cold nuclei [10]. The right panel shows the suppression of away side dihadron correlations $R^{(2)}_{pA}$ versus transverse momentum, rapidity and centrality on the example of the area of the correlation function $C(\Delta\phi) = dN^{h_1,h_2}/d\Delta\phi$ The pronounced p_{T_2} dependence is consistent with STAR preliminary data [11].

Acknowledgments: Useful discussion with S. J. Brodsky, R. Jaffe, J. W. Qiu and M. Tzanov is acknowledged. This work is supported by the J. Robert Oppenheimer Fellowship of the Los Alamos National Laboratory and by the US Department of Energy.

REFERENCES

1. J. W. Qiu and I. Vitev, Phys. Rev. Lett. **93**, 262301 (2004).
2. K. J. Eskola, V. J. Kolhinen and C. A. Salgado, Eur. Phys. J. C **9**, 61 (1999); V. Kolhinen, these proceedings.
3. S. J. Brodsky, P. Hoyer, N. Marchal, S. Peigne and F. Sannino, Phys. Rev. D **65**, 114025 (2002); these proceedings.
4. D. de Florian and R. Sassot, Phys. Rev. D **69**, 074028 (2004).
5. J. W. Qiu and I. Vitev, Phys. Lett. B **587**, 52 (2004).
6. V. A. Radescu [NuTeV Collaboration], arXiv:hep-ex/0408006.
7. M. Tzanov *et al.* [NuTeV Collaboration], arXiv:hep-ex/0306035; these proceedings.
8. J. W. Qiu and I. Vitev, hep-ph/0405068.
9. I. Arsene *et al.* [BRAHMS Collaboration], Phys. Rev. Lett. **93**, 242303 (2004).
10. B. Z. Kopeliovich, J. Nemchik, I. K. Potashnikova, M. B. Johnson and I. Schmidt, hep-ph/0501260.
11. A. Ogawa [STAR collaboration], nucl-ex/0408004.

Novel Nuclear Effects in QCD: The Non-Universality of Nuclear Antishadowing and the Implications of Hidden Color

Stanley J. Brodsky

Stanford Linear Accelerator Center, Stanford University, Stanford, CA, 94309

Abstract. The shadowing and antishadowing of nuclear structure functions in the Gribov-Glauber picture is due to the destructive and constructive interference of amplitudes arising from the multiple-scattering of quarks in the nucleus, respectively. The diffractive contributions to deep inelastic scattering includes Pomeron and Odderon contributions from multi-gluon exchange as well as Reggeon quark-exchange contributions. The coherence of multi-step nuclear processes leads to shadowing and antishadowing of the electromagnetic nuclear structure functions in agreement with measurements. This picture also leads to substantially different antishadowing for charged and neutral current reactions, thus affecting the extraction of the weak-mixing angle θ_W. The fact that Reggeon couplings depend on the quantum numbers of the struck quark implies non-universality of nuclear antishadowing for charged and neutral currents as well as a dependence of antishadowing on the polarization of the beam and target. The implications of hidden color degrees of freedom in the nuclear wavefunction is also briefly discussed.

Keywords: Quantum Chromodynamics, Deep Inelastic Lepton-Hadron Interactions, Nuclear Processes
PACS: 24.85.+p,12.38.-t,12.38.Aw,13.60.-r,13.15.+g

ANTISHADOWING OF NUCLEAR STRUCTURE FUNCTIONS

One of the novel features of QCD involving nuclei is the *antishadowing* of the nuclear structure functions which is observed in deep inelastic lepton scattering and other hard processes. Empirically, one finds $R_A(x,Q^2) \equiv \left(F_{2A}(x,Q^2)/(A/2)F_d(x,Q^2)\right) > 1$ in the domain $0.1 < x < 0.2$; *i.e.*, the measured nuclear structure function (referenced to the deuteron) is larger than than the scattering on a set of A independent nucleons. For many years the only theoretical guidance to this phenomenon was that given by Nikolaev and Zakharov [1], who argued that antishadowing was needed to restore the momentum sum rule in nuclei, compensating the shadowing and EMC regimes. However, this argument does not explain the dynamical mechanism which creates antishadowing nor the location in x_{bj} where it occurs.

The shadowing of the nuclear structure functions: $R_A(x,Q^2) < 1$ at small $x < 0.1$ can be readily understood in terms of the Gribov-Glauber theory. Consider the two-step process illustrated in Fig. 1 in the nuclear target rest frame. The incoming $q\overline{q}$ dipole first interacts diffractively $\gamma^* N_1 \to (q\overline{q})N_1$ on nucleon N_1 leaving it intact. This is the leading-twist diffractive deep inelastic scattering (DDIS) process which has been measured at HERA to constitute approximately 10% of the DIS cross section at high energies. The $q\overline{q}$ state then interacts inelastically on a downstream nucleon $N_2 : (q\overline{q})N_2 \to X$. The phase of the pomeron-dominated DDIS amplitude is close to imaginary, and the

CP792, *Deep Inelastic Scattering, DIS 2005,* edited by W. H. Smith and S. R. Dasu

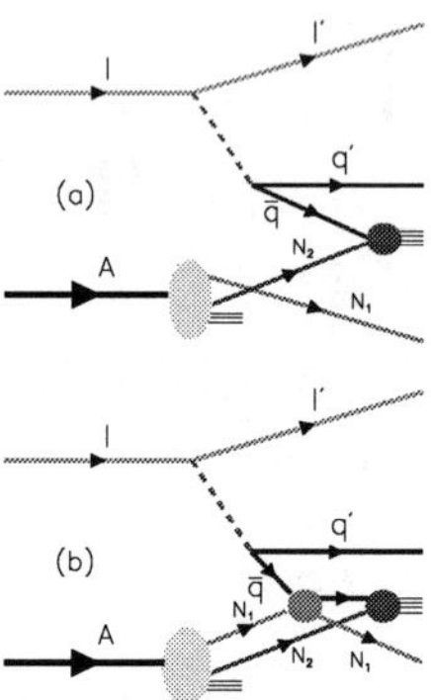

FIGURE 1. Illustration of one-step and two-step processes.

Glauber cut provides another phase i, so that the two-step process has opposite phase and destructively interferes with the one-step DIS process $\gamma * N_2 \to X$ where N_1 acts as an unscattered spectator. The one-step and-two step amplitudes can coherently interfere as long as the momentum transfer to the nucleon N_1 is sufficiently small that it remains in the nuclear target; *i.e.*, the Ioffe length [2] $L_I = 2M\nu/Q^2$ is large compared to the inter-nucleon separation. In effect, the flux reaching the interior nucleons is diminished, thus reducing the number of effective nucleons and $R_A(x,Q^2) < 1$.

As noted by Hung Jung Lu and myself [3], there are also leading-twist diffractive contributions $\gamma^* N_1 \to (q\overline{q})N_1$ arising from Reggeon exchanges in the t-channel. For example, isospin–non-singlet $C = +$ Reggeons contribute to the difference of proton and neutron structure functions, giving the characteristic Kuti-Weisskopf $F_{2p} - F_{2n} \sim x^{1-\alpha_R(0)} \sim x^{0.5}$ behavior at small x. The x dependence of the structure functions reflects the Regge behavior $\nu^{\alpha_R(0)}$ of the virtual Compton amplitude at fixed Q^2 and $t = 0$. The phase of the diffractive amplitude is determined by analyticity and crossing to be proportional to $-1+i$ for $\alpha_R = 0.5$, which together with the phase from the Glauber cut, leads to *constructive* interference of the diffractive and nondiffractive multi-step nuclear amplitudes. Furthermore, because of its x dependence, the nuclear structure function is enhanced precisely in the domain $0.1 < x < 0.2$ where antishadowing is empirically observed. The strength of the Reggeon amplitudes is fixed by the fits to the nucleon structure functions, so there is little model dependence.

The origin of the diffractive contributions to DIS was shown in Ref. [4] to be due to the rescattering of the struck quark after it is struck in the usual parton model frame $q^+ \leq 0$, an effect induced by the Wilson line connecting the currents. Thus one cannot attribute DDIS to the physics of the target nucleon computed in isolation. It is an effect resulting from the $\gamma^* p$ collision. Similarily, since shadowing and antishadowing arise from the physics of diffraction, we cannot attribute these phenomena to the structure of the nucleus itself: shadowing and antishadowing arise because of the $\gamma^* A$ collision and the history of the $q\overline{q}$ dipole as it propagates through the nucleus.

In a recent paper, Ivan Schmidt, Jian-Jun Yang, and I [5] have extended this analysis to the shadowing and antishadowing of all of the electroweak structure functions. Quarks

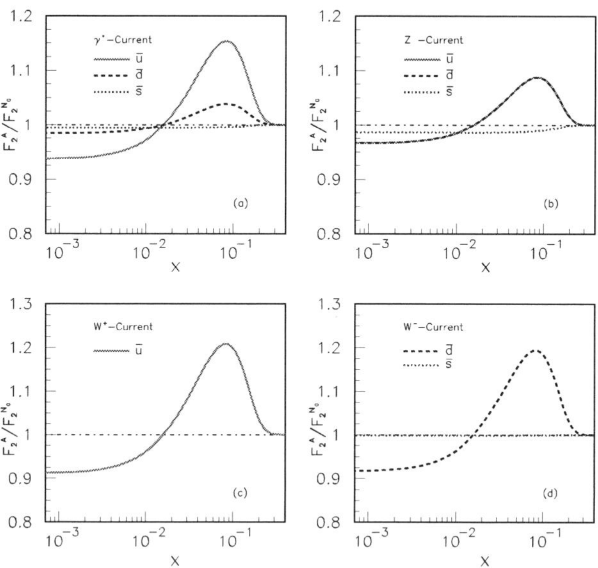

FIGURE 2. Model predictions [5] for interactions of electroweak interactions on antiquarks in nuclear targets. The antishadowing effect is not as large for quark currents.

of different flavors will couple to different Reggeons; this leads to the remarkable prediction that nuclear antishadowing is not universal; it depends on the quantum numbers of the struck quark. This picture leads to substantially different antishadowing for charged and neutral current reactions, thus affecting the extraction of the weak-mixing angle θ_W. See Fig. 2. We find that part of the anomalous NuTeV result [6] for θ_W could be due to the nonuniversality of nuclear antishadowing for charged and neutral currents. Detailed measurements of the nuclear dependence of individual quark structure functions are thus needed to establish the distinctive phenomenology of shadowing and antishadowing and to make the NuTeV results definitive. Schmidt, Yang, and I have also identified contributions to the nuclear multi-step reactions which arise from odderon exchange and also hidden color degrees of freedom in the nuclear wavefunction. There are other ways in which this new view of antishadowing can be tested; antishadowing can also depend on the target and beam polarization.

HIDDEN COLOR

One of the most important distinctions between traditional nuclear physics and QCD descriptions of nuclei are the hidden color degrees of freedom of the nuclear wavefunction [7]. For example, there are five color-singlet combinations of six color-triplet quarks in the deuteron valence Fock state: $3_C \times 3_C \times 3_C \times 3_C \times 3_C \times 3_C = 1_C + 1_C + 1_C + 1_C + 1_C + \cdots$, only one of which is identified with the $n-p$ degrees of freedom at large distances. At short distances the 5-component deuteron distribution amplitude

$\phi(x_i, Q)_I$ evolves at $Q^2 \gg \Lambda^2_{QCD}$ until at asymptotic momenta all five components have equal weight. The asymptotic large-momentum-transfer behavior of the deuteron form factor and the form of the deuteron distribution amplitude at short distances can be rigorously derived from perturbative QCD. The fact that the six-quark state is 80% hidden color at small transverse separation implies that the deuteron form factors cannot be described at large Q^2 by meson-nucleon degrees of freedom, and that the nucleon-nucleon potential is repulsive at short distances.

The observed $Q^{10}F_d(Q^2)$ scaling of the helicity-conserving deuteron form factor [8] and the fixed CM angle scaling of the deuteron photodisintegration cross section [9] $s^{11}\frac{d\sigma}{dt}(\gamma d \to np) = F(t/s)$ provide remarkable tests of the conformal properties of QCD at short distances in a nuclear system as predicted by QCD and the AdS/CFT correspondence. The measured reduced form factor of the deuteron [10] $f_d(Q^2) \equiv F_d(Q^2)/F_N^2(Q^2/4)$ falls quickly at small $Q^2R_d^2 < 1$, and then scales as the pion monopole form factor: $f_d(Q^2) \sim 0.15 \times F_\pi(Q^2)$ at $Q^2 >> 1$ GeV^2, suggesting that the hidden color degrees of freedom dominate the deuteron wavefunction at high momentum transfer $k_\perp^2 > 1$ GeV^2 with an approximate normalization of 15%.

The hidden color $(uud)_{8C}(duu)_{8C}$ Fock states of the deuteron have a sizable overlap with the $\Delta^{++}(uuu)\Delta^-(ddd)$ di-isobar state. Thus one expects from QCD hidden color that the $\frac{d\sigma}{dt}(\gamma d \to \Delta^{++}\Delta^-)$ and $\frac{d\sigma}{dt}(\gamma d \to np)$ cross sections become comparable in the fixed angle scaling regime.

ACKNOWLEDGMENTS

The work on antishadowing reported here is based on collaborations with H. J. Lu, I. A. Schmidt, and Jian-Jun Yang, and the work on hidden color is based on collaborations with C. R Ji, and G. P. Lepage. This work was supported by the Department of Energy, contract No. DE-AC02-76SF00515.

REFERENCES

1. N. N. Nikolaev and V. I. Zakharov, Phys. Lett. B **55**, 397 (1975).
2. B. L. Ioffe, Phys. Lett. B **30**, 123 (1969).
3. S. J. Brodsky and H. J. Lu, Phys. Rev. Lett. **64**, 1342 (1990).
4. S. J. Brodsky, P. Hoyer, N. Marchal, S. Peigne and F. Sannino, Phys. Rev. D **65**, 114025 (2002) [arXiv:hep-ph/0104291].
5. S. J. Brodsky, I. Schmidt and J. J. Yang, Phys. Rev. D **70**, 116003 (2004) [arXiv:hep-ph/0409279].
6. G. P. Zeller *et al.* [NuTeV Collaboration], Phys. Rev. Lett. **88**, 091802 (2002) [Erratum-ibid. **90**, 239902 (2003)] [arXiv:hep-ex/0110059].
7. S. J. Brodsky, C. R. Ji and G. P. Lepage, Phys. Rev. Lett. **51**, 83 (1983).
8. R. J. Holt, Phys. Rev. C **41**, 2400 (1990).
9. P. Rossi *et al.* [CLAS Collaboration], arXiv:hep-ph/0405207.
10. S. J. Brodsky and B. T. Chertok, Phys. Rev. D **14**, 3003 (1976).

Rapidity Dependence of High-p_T suppression

F.Videbæk for the BRAHMS Collaboration

Physics Department, Brookhaven National Laboratory

Abstract. The rapidity dependence of nuclear modifications factor in d-Au collisions at $\sqrt{s_{NN}}$ = 200 GeV at RHIC is discussed.

Keywords: Nuclear Modification Factor, gluon saturation
PACS: 25.75.Dw,13.85Hd.

INTRODUCTION

Particle production at forward rapidities in p(d)-A reactions probes partons in the target at low-x values. At sufficient high energy, large rapidities or large nucleii the initial gluon distribution will saturate, and is expected to modify the particle pseudo rapidity densities, as well as the p_T spectra of hadrons. A theory based on QCD has been developed for dense low-x systems, termed the Color Glass Condensate (CGC) [1]. This description has inspired much theoretical and experimental work, and was also a motivation for the BRAHMS measurements of charged hadrons at forward rapidities in 200 GeV d-Au collisions at RHIC.

RESULTS

The data reported here were all obtained with the BRAHMS spectrometers. The BRAHMS forward spectrometer consists of 4 dipole magnets, 5 tracking chambers, two Time-Of-Flight systems and a Ring Imaging Chrenkov Detector (RICH) for particle identification. The angular coverage of the spectrometer extends from 2.3° to 15° with solid angle of 0.8 msr. The mid-rapidity spectrometer covers angles from 40° and 90° . Details of experimental setup can be found in [2]. The collision vertex is determined from timing measurements done with a set of symmetricaly placed scintillator counters around the beam pipe at 1.5, 4.15 and 6.7 meters [3]. The resolution of the vertex determination is $\approx$ 10 cm. This set of counters also provides the minimum bias normalization. It is estimated that for pp collisions they record 70% of the inelastic cross section. Additional details of the setup as well as the analysis can be found in [3, 4] where most of the data discussed here were first published. Spectra of charged hadrons in d-Au and pp collisions at 200 GeV are presented in Fig. 1 for several pseudorapidities. The data at $\eta = 0$ and 1 are for the average of the positive and negatives charges, while the high rapidity data are for negative only. Both the pp and dAu cross sections becomes steeper with increasing η.

CP792, *Deep Inelastic Scattering, DIS 2005,* edited by W. H. Smith and S. R. Dasu

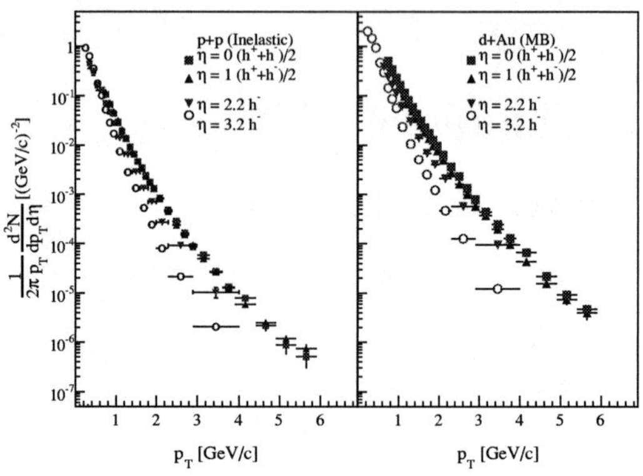

FIGURE 1. Spectra for dAu and pp

The data collected from d-Au collisions is compared to p-p using the nuclear modification factor defined as: $R_{dAu} = \frac{1}{N_{coll}} \frac{\frac{dN^{dAu}}{dp_T d\eta}}{\frac{dN^{pp}}{dp_T d\eta}}$. where N_{coll} is the number of binary collisions estimated to be equal to 7.2 ± 0.6 for minimum biased d+Au collisions. The pt dependence of the factor is shown in Fig.2. Each panel shows the ratio calculated at a different η value. At mid-rapidity ($\eta = 0$), the nuclear modification factor exceeds 1 for transverse momenta greater than 2 GeV/c in a similar, although less pronounced way as Cronin's p+A measurements performed at lower energies [5].

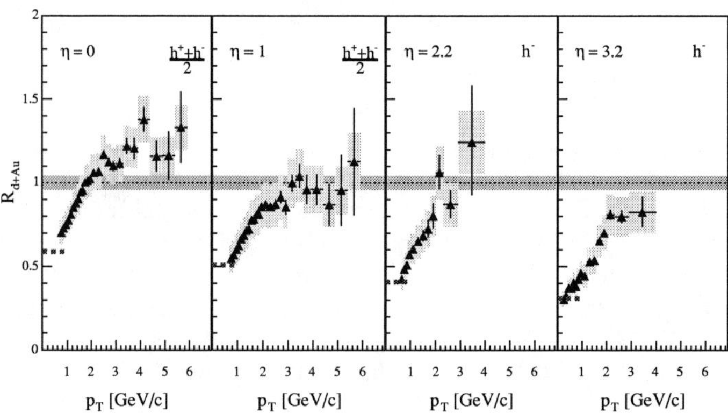

FIGURE 2. Nuclear modification factor for charged hadrons at pseudorapidities $\eta = 0, 1.0, 2.2, 3.2$. Statistical errors are shown with error bars. Systematic errors are shown with shaded boxes with widths set by the bin sizes. The shaded band around unity indicates the estimated error on the normalization to $\langle N_{coll} \rangle$. Dashed lines at $p_T < 1$ GeV/c show the normalized charged particle density ratio $\frac{1}{\langle N_{coll} \rangle} \frac{dN/d\eta(d+Au)}{dN/d\eta(pp)}$.

A shift of one unit of rapidity is enough to make the Cronin type enhancement disappear, and as the measurements are done at higher rapidities, the ratio becomes consistently smaller than 1 indicating a suppression in dAu collisions compared to scaled pp systems at the same energy. In all four panels, the statistical errors, shown as error bars (vertical lines), are dominant specially in the most forward measurements.

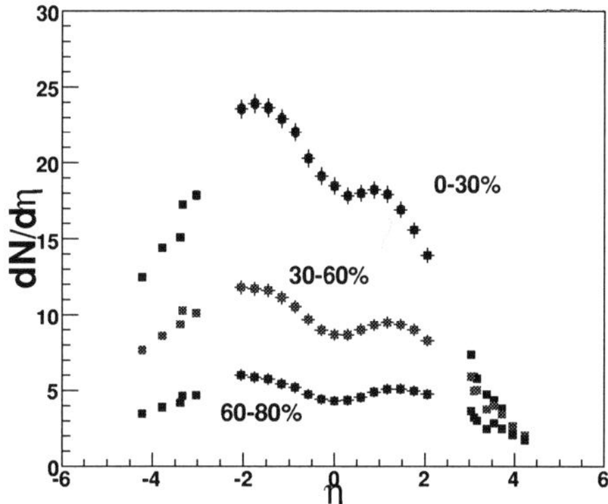

FIGURE 3. Pseudo rapidity dependence of dN/deta in dAu for 3 centrality bins.

These results have been described within the context of the Color Glass Condensate [1]; the evolution of the nuclear modification factor with rapidity and centrality is consistent with a description of the Au target where the rate of gluon fusion becomes comparable with that of gluon emmission as the rapidity increases and it slows down the overall growth of the gluon density. The measured nuclear modification factor compares the slowed down growth of the numerator to a sum of incoherent p+p collisions, considered as dilute systems, whose gluon densities grow faster with rapidity because of the abscence of gluon fusion in dilute systems [6]. Other explanations for the measured suppression have been proposed and they also reproduce the data [9, 10, 11].

Some of these other explanations particluar focus on the observation that the charged particle pseudo-rapidity density distributions exhibits a change in shape vs. centrality. This is illustrated in Fig.2 by the dashed lines at low p_T where the ratios where obtained from the BRAHMS data[4] (shown in Fig.3) and the UA5 pp results[7]. Thus already the overall soft spectrum shows suppression when going to forward angles.

The nuclear modification factor of baryons is different from the one calculated with mesons, whenever the factor shows the so called Cronin enhancement, baryons show a stronger enhancement. Such difference, seen at lower energies, has also been found at RHIC energies at all rapidities, in particular, Fig. 4 presents the minimum bias nuclear modification R_{dAu} for anti-protons and negative pions at $\eta = 3.2$. These ratios were obtained making use of ratios of raw counts of identified particles compared to those of charged particles in each p_T bin. This nuclear modification factor was calculated from the measured ratios between anti-protons and pions to changed particles for dAu and pp and applying these factors to the minimum bias nuclear modification factor for negatively charged hadrons[8]. No attempt was made to estimate the contributions from anti-lambda feed down to the anti-proton result. The remarkable difference between

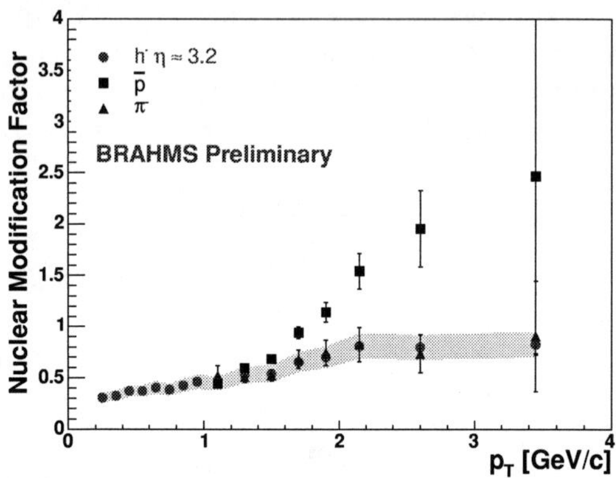

FIGURE 4. The nuclear modification factor R_{dAu} calculated for anti-protons (filled squares) and negative pions (filled triangles) at $\eta = 3.2$. The same ratio calcutated for negative particles is shown with filled circles, and the systematic error for that measurement is shown as grey band.

baryons and mesons has been related to parton recombination [9] for heavy ion reactions. It is though surprising that this effect in the dA system at forward rapidities where only a small soft parton component should be able to account for this increase.

In summary, particle production from dAu and pp collisions at $\sqrt{s_{NN}} = 200$ GeV and at different rapidities with the BRAHMS setup offers a window to the small-x components of the Au wave function. The suppression found in the particle production at high rapidities from d+Au collisions may be the first indication of the onset of saturation in the gluon distribution function of the Au target.

This work is supported by the Division of Nuclear Physics of the Office of Science of the U.S. Department of energy under contract DE-AC02-98-CH10886, the Danish Natural Science Research Council, the Research Council of Norway, the Jagiellonian University Grants and the Romanian Ministry of Education and Research. I thank R.Debbe for providing valuable input to this talk and conbtribution.

REFERENCES

1. L.McLerran and R.Venugopalan,*Phys. Rev.* **D49** 1994, 2233.
2. J.Adamczyk et.al. *Nucl. Instr. Meth.* **A499** 2003, 437
3. I. Arsene *et al.* [BRAHMS Collaboration], *Phys. Rev. Lett.* **93**, 2004, 242303
4. I. Arsene at.al. *Phys. Rev. Letter*, **94** 2005, 032301
5. D. Antreasyan et. al. *Phys. Rev.* **D19** 1979, 764
6. D. Kharzeev, Y. V. Kovchegov and K. Tuchin *Phys. Rev.* **D68**,2003, 094013 ; hep-ph/0307037;
7. G.J.Almer et.al *Z.Phys.***C33** 1986, 1
8. R. Debbe. proceedings CINPP05; nucl-ex/0506022
9. R. Hwa,C. .B .Yang and R. .J. Fries *Phys. Rev.* **C71** 2004, 024902
10. Jian-Wei Qiu and Ivan Vitev hep-ph/0410218; Ivan Vitev hep-ph/0506039.
11. B.Z. Kopeliovich *et al.* hep-ph/0501260.

Nuclear Modification Factors for Hadrons At Forward and Backward Rapidities in Deuteron-Gold Collisions at $\sqrt{s_{NN}} = 200$ GeV

Chun Zhang, for the PHENIX Collaboration

Oak Ridge National Laboratory, Oak Ridge, TN 37831, USA

Abstract. We report on charged hadron production in deuteron-gold reactions at $\sqrt{s_{NN}} = 200$ GeV. Our measurements in the deuteron-direction cover $1.4 < \eta < 2.2$, referred to as forward rapidity, and in the gold-direction $-2.0 < \eta < -1.4$, referred to as backward rapidity, and a transverse momentum range $p_T = 0.5-4.0$ GeV/c. We compare the relative yields for different deuteron-gold collision centrality classes. We observe a suppression relative to binary collision scaling at forward rapidity, sensitive to low momentum fraction (x) partons in the gold nucleus, and an enhancement at backward rapidity, sensitive to high momentum fraction partons in the gold nucleus.

Keywords: Hadron, Rapidity, Nuclear Modification
PACS: 25.75.DW

INTRODUCTION

In 2003 the Relativistic Heavy Ion Collider (RHIC) collided deuteron and gold nuclei at $\sqrt{s_{NN}} = 200$ GeV. At this energy, most hadrons with $p_T > 2.0$ GeV/c arise from parton-parton interactions and can be used as a probe of nuclear partonic structure. Hadrons with $p_T > 2.0$ GeV/c at forward rapidity $1.4 < \eta < 2.2$ are sensitive to low x partons in the gold nucleus $0.001 < x < 0.03$. Hadrons at backward rapidity $-2.0 < \eta < -1.4$ are sensitive to high x partons in the gold nucleus $0.04 < x < 0.5$. It has been predicted that gluon saturation at small x will suppress hadronic yields at forward rapidity [1] with the transverse momentum scale for the onset of the gluon saturation set by $Q_s^2[\text{GeV}^2] = 0.13\, N_{coll} e^{\lambda y}$ [2] for $d+Au$ collisions at RHIC. Here $\lambda \sim 0.3$ is determined from HERA data [3] and N_{coll} is the number of nucleon-nucleon inelastic collisions. Thus, for central collisions and within our forward rapidity coverage Q_s^2 is expected to be of order $2-4$ GeV^2 and may have observable consequences. Other hadron production mechanisms, such as quark recombination [4], can also result in an effective suppression in the forward rapidity region.

Results on charged hadron yields at forward rapidity from the BRAHMS experiment have shown a suppression of the yield of hadrons in central, compared to peripheral, $d+Au$ collisions [7]. At mid-rapidity, PHENIX has reported a modest enhancement of the yield of hadrons with $p_T > 1.5$ GeV/c [8]. This enhancement, generally referred to as the "Cronin effect" is often ascribed to initial state scattering of the parton traversing the nucleus prior to the high Q^2 scattering [5]. At backward rapidities (large x), anti-shadowing and other effects of the surrounding nuclear medium (e.g. the EMC effect) [9] may compete, making predictions challenging.

CP792, *Deep Inelastic Scattering, DIS 2005*, edited by W. H. Smith and S. R. Dasu

RESULTS

We measured hardon via two independed methods [6]. First method directly identify hadron based on matching between hadron momentum and where it stops in the detector. The second method measures muons from light hadron decays.

We divide deuteron-gold collisions into four centrality classes based on the number of particle hits in the backward BBC counter covering $-3.9 < \eta < -3.0$. Using a Glauber model [8] and simulation of the BBC, we determine the average number of binary collisions in each centrality class. The classes are categorized as follows: $60-88\%$ ($\langle N_{coll}\rangle = 3.1 \pm 0.3$), $40-60\%$ ($\langle N_{coll}\rangle = 7.0 \pm 0.6$), $20-40\%$ ($\langle N_{coll}\rangle = 10.6 \pm 0.7$),and $0-20\%$ ($\langle N_{coll}\rangle = 15.4 \pm 1.0$).

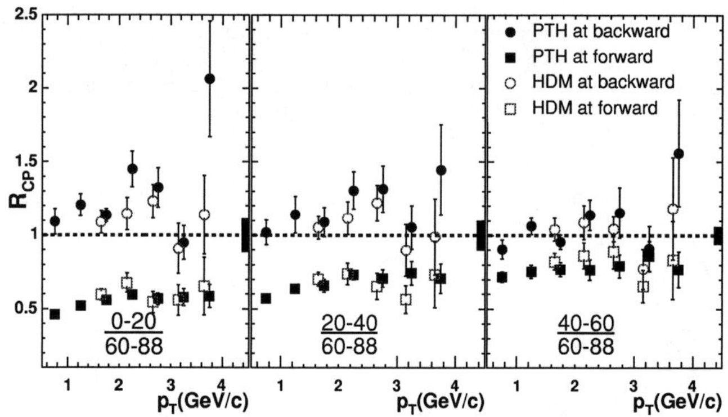

FIGURE 1. (color online). R_{cp} as a function of p_T at forward rapidity (squares) and backward rapidity (circles) for different centrality classes.

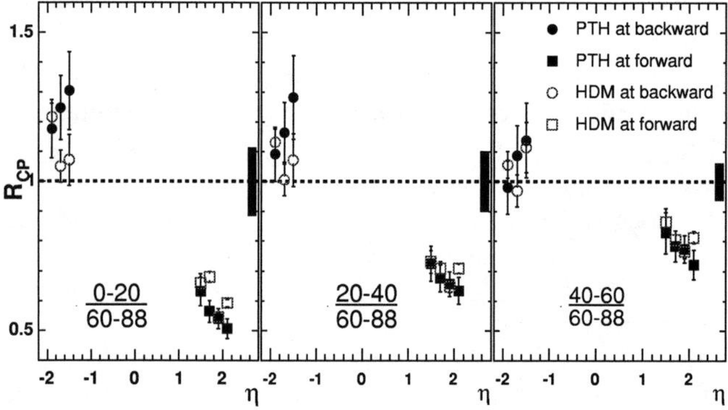

FIGURE 2. (color online). R_{cp} as a function of η for $1.5 < p_T < 4.0\,\mathrm{GeV}/c$ for different centrality classes.

The *nuclear modification factor* R_{cp} is defined as the ratio of the particle yield in central collisions to the particle yield in peripheral collisions, each normalized by the average number of nucleon-nucleon binary collisions ($\langle N_{coll}\rangle$):

$$R_{cp} = \frac{\langle \left(\frac{dN}{d\eta dp_T}\right)^{Central} \rangle / \langle N_{coll}^{Central} \rangle}{\langle \left(\frac{dN}{d\eta dp_T}\right)^{Peripheral} \rangle / \langle N_{coll}^{Peripheral} \rangle} \tag{1}$$

The hadron R_{cp}, using the most peripheral centrality class ($60-88\%$) for normalization, is shown in Figure 1 as a function of p_T at forward and backward rapidities. The results from the punch-through hadron (PTH) and hadron decay muon (HDM) techniques are both shown and are in quite good agreement. We also show the results integrated over $1.5 < p_T < 4.0\,\mathrm{GeV}/c$ as a function of pseudorapidity in Figure 2.

We observe that R_{cp} shows a suppression at forward rapidity that is largest for the most central events. The opposite trend is observed at backward rapidity where R_{cp} shows an enhancement that is also largest for the most central events. We observe a weak p_T dependence with slightly smaller R_{cp} values at lower p_T. We observe a clear pseudorapidity dependence at forward rapidity with R_{cp} dropping further at larger η values. Within our current uncertainties we are unable to discern any pseudorapidity dependence at backward rapidity.

In Figure 3 we compare results from the BRAHMS experiment [7] with our results at forward rapidity. The PHENIX data and the BRAHMS data are in agreement within systematic uncertainties.

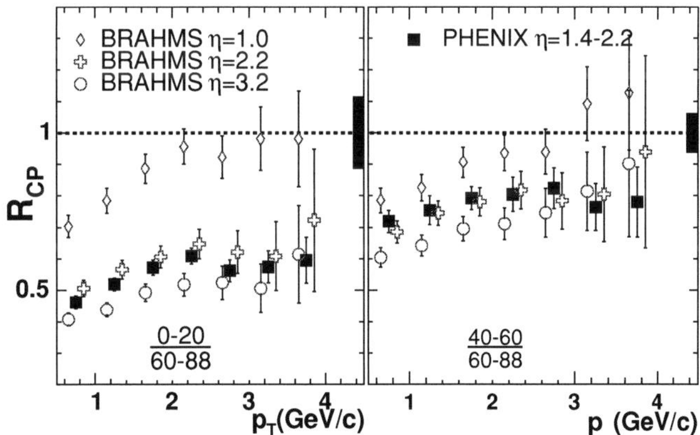

FIGURE 3. (color online). PHENIX R_{cp} as a function of p_T at forward rapidities shown as the average of the two methods. Note that the BRAHMS results are for negative hadrons at $\eta = 2.2, 3.2$ and their centrality ranges ($0-20\%/60-80\%$ and $30-50\%/60-80\%$) are somewhat different from ours.

The suppression of hadron yields relative to binary collision scaling at forward rapidity is expected from initial state nuclear effects. However, detailed comparisons with various theoretical approaches is necessary in order to discriminate between different models. In particular the lack of a strong p_T depedence at both forward and backward rapidities must be understood as the physics processes transition from "soft" to "hard" physics scales.

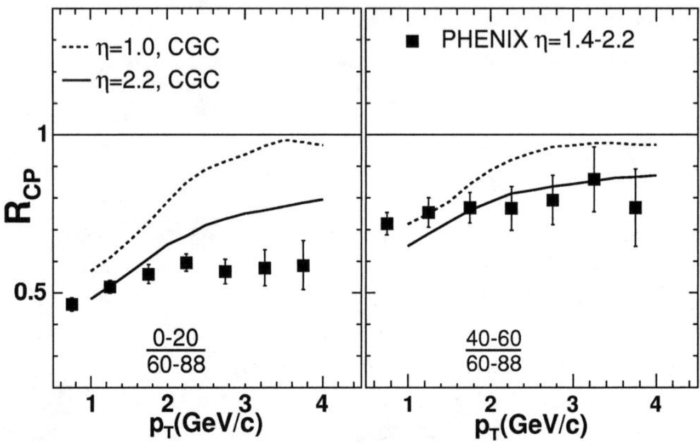

FIGURE 4. (color online). PHENIX R_{cp} as a function of p_T at forward rapidities ploted with CGC ca. Note that the CGC calculations have two free parameters fitted by BRAHMS results and their centrality ranges ($0-20\%/60-80\%$ and $30-50\%/60-80\%$) are somewhat different from ours.

SUMMARY AND DISCUSSIONS

To summarize, we observe a suppression in hadron yields relative to binary collision scaling at forward rapidities and an enhancement at backward rapidity for central relative to peripheral $d+Au$ reactions at $\sqrt{s_{NN}} = 200\,\text{GeV}$. The forward rapidity suppression is in qualitative agreement with the expectation of shadowing and saturation effects in the small x region in the gold nucleus. In figure5, we give the comparison between our measured nuclear modification from forward rapidities and the CGC calculations [2]. However, other physics effects must also be considered in understanding the full p_T and η dependence. The source of the backward rapidity enhancement, and the possible contribution of anti-shadowing of large x partons, has yet to be understood.

REFERENCES

1. A. Dumitru and J. Jalilian-Marian, *Phys. Lett.* **B547**, *15 (2002)*; F. Gelis and J. Jalilian-Marian, *Phys. Rev. D* **66**, *014021 (2002)*.
2. D. Kharzeev, Y.V. Kovchegov and K. Tuchin, arXiv:hep-ph/0405045.
3. K. Golec-Biernat and M. Wusthoff, *Phys. Rev. D* **59**, *014017 (1999)*; K. Golec-Biernat and M. Wusthoff, *Phys. Rev. D* **60**, *114023 (1999)*.
4. R. Hwa, C.B. Yang, R.J. Fries, arXiv:nucl-th/0410111.
5. D. Antreasyan, et al., *Phys. Rev. D* **19**, *764 (1979)*.
6. S. S. Adler, et al., *Phys. Rev. Lett.* **94**, *082302 (2005)*.
7. I. Arsene, et al., *Phys. Lett.* **B595**, *209 (2004)*.
8. S. S. Adler, et al., *Phys. Rev. Lett.* **91**, *072303 (2003)*.
9. M. Arneodo, *Phys. Repts.* **240**, *301 (1994)*.

Neutral Pion Suppression at Forward Rapidities in d+Au Collisions at STAR

G. Rakness for the STAR Collaboration

Penn State University/Brookhaven National Lab

Abstract. Measurements of the inclusive yields of π^0 mesons in p+p and d+Au collisions at center of mass energy $\sqrt{s_{NN}} = 200\,\text{GeV}$ and pseudorapidity $\langle\eta\rangle = 4.00$ (d beam direction) are reported. The yield for p+p collisions is in general agreement with perturbative QCD calculations. The d+Au yield is in agreement with a calculation which models the Au nucleus as a Color Glass Condensate for forward particle production. The nuclear modification factor derived from the inclusive yields is qualitatively consistent with models which suppress the gluon density in nuclei.

Keywords: low-x, saturation, Color Glass Condensate, particle production, shadowing
PACS: 24.85.+p,25.75.-q,13.85.Ni

In contrast to the nucleon, very little is known about the density of gluons in nuclei [1]. For protons, the gluon parton distribution function (PDF) is constrained primarily by scaling violations in deeply-inelastic lepton scattering (DIS) measured at the HERA collider [2]. The data are accurately described by QCD evolution equations that allow the determination of the gluon PDF. As the momentum fraction of the parton (x_{Bj}) decreases, the gluon PDF is found to increase. At sufficiently low x_{Bj}, the increase in gluon splitting is expected to become balanced by gluon-gluon recombination, resulting in gluon saturation. In nuclei, the density of gluons per unit transverse area is expected to be larger than in nucleons, and so is a natural environment in which to establish if, and under which conditions, gluon saturation occurs. Quantifying if gluon saturation occurs at RHIC energies is important because most of the matter created in heavy ion collisions is expected to evolve from an initial state produced by the collisions of low-x_{Bj} gluon fields in the nuclei [3]. Fixed target nuclear DIS experiments are restricted in the kinematics available and have determined the nuclear gluon PDF for $x_{Bj} \gtrsim 0.02$ [1]. In a conventional pQCD description of d(p)+A collisions at $\sqrt{s_{NN}} = 200\,\text{GeV}$, inclusive forward hadroproduction probes the nuclear gluon density over a broad distribution of x_{Bj} peaked around 0.02 and extending well below $x_{Bj} \approx 0.005$ [4].

Using factorization in a perturbative QCD (pQCD) framework, the PDF's and fragmentation functions (FF's) measured in electromagnetic processes can be used in the description of hadronic scattering processes. In p+p collisions at $\sqrt{s} = 200\,\text{GeV}$, factorized leading twist pQCD calculations have been shown to quantitatively describe inclusive π^0 production over a broad rapidity window [5, 6]. In pQCD, forward π production in p+p collisions probes low-x_{Bj} gluons (g) in one proton using the valence quarks (q) of the other. Recently, the yield of forward negatively charged hadrons (h^-) in the d-beam direction of d+Au collisions was found to be reduced when normalized to p+p collisions [7]. The reduction is especially significant since isospin effects are expected to suppress h^- production in p+p collisions, but not in d+Au collisions [4].

CP792, *Deep Inelastic Scattering, DIS 2005,* edited by W. H. Smith and S. R. Dasu

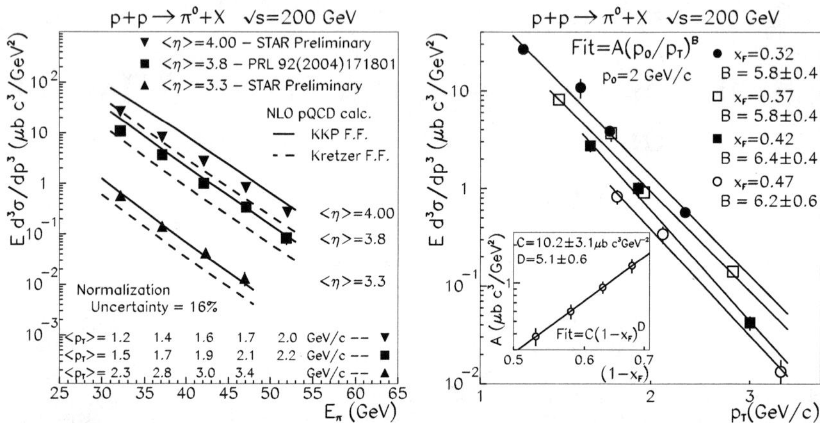

FIGURE 1. Inclusive π^0 production cross section for p+p collisions. The error bars, often smaller than the points, combine the statistical and point-to-point systematic errors. (Left) Versus pion energy (E_π) at fixed pseudorapidity (η). The curves are NLO pQCD calculations using two sets of fragmentation functions (FF). (Right) Versus transverse momentum (p_T) at fixed Feynman-x (x_F). Curves here and in the inset are fits to the data of the form shown on the plots. (Inset) Versus $(1-x_F)$ at $p_T = 2\,\mathrm{GeV/c}$.

Many models attempt to describe forward hadroproduction from nuclear targets. Saturation models [8] include a QCD based theory called the Color Glass Condensate (CGC) [9, 10]. Another approach models quarks scattering coherently from multiple nucleons, leading to an effective shift in the x_{Bj} probed [11]. Shadowing models suppress the nuclear gluon PDF in a standard factorization framework [12]. Parton recombination models modify the fragmentation of a quark passing through many gluons [13]. Other descriptions include factorization breaking in heavy nuclei [14]. More data are needed to constrain the mechanisms for forward hadroproduction in nuclear collisions.

In this paper we present the inclusive yields of high energy π^0 mesons at $\langle\eta\rangle = 4.00$ ($\eta = -\ln[\tan(\theta/2)]$) in p+p and d+Au collisions at $\sqrt{s_{NN}} = 200\,\mathrm{GeV}$. The results are compared with theoretical predictions. The η dependence of the normalized cross section ratio is also presented. Analysis of the azimuthal correlations of the forward π^0 with coincident hadrons at midrapidity is presented elsewhere [15].

A forward π^0 detector (FPD) was installed at the Solenoidal Tracker At RHIC (STAR) to detect high energy π^0 mesons with $3.3 < \eta < 4.1$. In the 2002 run, p+p collisions at $\sqrt{s} = 200\,\mathrm{GeV}$ were studied with a prototype FPD [5]. In the 2003 run, p+p collisions were studied with the complete FPD and exploratory measurements were performed for d+Au collisions at $\sqrt{s_{NN}} = 200\,\mathrm{GeV}$. Details of the FPD performance can be found in Ref. [16]. The luminosity was determined by measuring the transverse size of the colliding beams and the number of colliding ions, giving a normalization error of $\approx 11\%$.

The cross sections for $p+p \to \pi^0 + X$ at $\langle\eta\rangle = 3.3$, 3.8, and 4.00 are presented in Fig. 1 (left) [5, 16]. The E_π-dependent systematic error at $\langle\eta\rangle = 4.00$ is $10-20\%$, dominated by the energy calibration of the FPD. The absolute η uncertainty contributes 11% to the normalization error [16]. The curves are NLO pQCD calculations using the KKP and Kretzer sets of FF's differing primarily in the gluon-to-pion FF, D_g^π. At

$\eta = 3.3$ and 3.8, the data are consistent with KKP. As p_T decreases further, the data begin to undershoot KKP and approach consistency with Kretzer. This is consistent with the trend seen at midrapidity [6]. At low p_T, the dominant contribution to π^0 production becomes gg scattering, making D_g^π the dominant FF [17]. In Fig. 1 (right), the cross section is shown versus p_T at fixed Feynman-x ($x_F \approx 2E_\pi/\sqrt{s}$), and versus $(1-x_F)$ at $p_T = 2\,\mathrm{GeV/c}$ (inset). As was reported at the CERN ISR [18], the forward rapidity yields are rapidly changing functions of both x_F and p_T.

The cross section for $d+Au \to \pi^0 + X$ at $\sqrt{s_{NN}} = 200\,\mathrm{GeV}$ and $\langle\eta\rangle = 4.00$ is presented in Fig. 2 (left). The E_π-dependent systematic error is $\approx 20\%$, dominated by the background correction because, on average, 0.5 more photons are observed in d+Au than in p+p collisions per event with $> 30\,\mathrm{GeV}$ detected in the FPD. The curves are LO calculations from Ref. [10], using CTEQ5 PDF's and KKP FF's convoluted with a dipole-nucleus cross section which models parton scattering from a CGC in the Au nucleus. The "MV" and "No DGLAP" curves neglect QCD evolution of the Au wave function and the PDF/FF, respectively. The difference in the slopes with and without evolution is greater than the slope change from LO to NLO for $p+p \to \pi^0 + X$ at $\eta = 3.8$. The calculations in Fig. 2 are normalized by a p_T-independent K-factor of 0.8. This is smaller than the K-factor used to normalize the theory to the $d+Au \to h^-$ yield at the nominal value of $\eta = 3.2$, in the same direction as the renormalization needed to scale the NLO calculations with KKP FF to $p+p \to \pi^0$ data at $\eta = 4.00$. Note from Fig. 1 that a change of $\Delta\eta \approx 0.05$ at these kinematics results in $\Delta\sigma/\sigma \approx 35\%$ at fixed x_F. The slope of the π^0 yield is consistent with the calculation where the PDF and FF evolve à la DGLAP, and includes small-x_{Bj} evolution of the Au wave function.

Nuclear effects on particle production are quantified by the nuclear modification factor, R^Y_{dAu}, which can be defined for minimum-bias events by the ratio,

$$R^Y_{\mathrm{dAu}} = \frac{\sigma^{pp}_{\mathrm{inel}}}{\langle N_{\mathrm{bin}}\rangle \sigma^{dAu}_{\mathrm{hadr}}} \frac{Ed^3\sigma/dp^3(d+Au \to Y+X)}{Ed^3\sigma/dp^3(p+p \to Y+X)}. \tag{1}$$

We adopt $\sigma^{pp}_{\mathrm{inel}} = 42\,\mathrm{mb}$ for the inelastic p+p cross section, while the non-elastic d+Au cross section, $\sigma^{dAu}_{\mathrm{hadr}} = 2.21 \pm 0.09\,\mathrm{b}$, and the mean number of binary collisions, $\langle N_{\mathrm{bin}}\rangle = 7.5 \pm 0.4$, are determined by a Glauber model calculation [19]. Normalization systematic errors mostly cancel in the ratio. At $\eta \approx 0$, $R^{h^\pm}_{\mathrm{dAu}} \gtrsim 1$, with a Cronin peak at $p_T > 2\,\mathrm{GeV/c}$ [19]. In contrast, at $\langle\eta\rangle = 4.00$, $R^{\pi^0}_{\mathrm{dAu}} \ll 1$, as seen in Fig. 2 (right). The decrease of R_{dAu} with η is qualitatively consistent with models that suppress the nuclear gluon density [9, 11, 12, 13]. $R^{\pi^0}_{\mathrm{dAu}}$ is significantly smaller than a linear extrapolation of $R^{h^-}_{\mathrm{dAu}}$ to $\eta = 4$, consistent with expectations of isospin suppression of $p+p \to h^- + X$ [4].

In summary, inclusive yields of forward π^0 mesons from p+p collisions at $\sqrt{s} = 200\,\mathrm{GeV}$ are consistent with NLO pQCD calculations. The cross sections show rapid variation with both x_F and p_T. The d+Au yield is consistent with a model which treats the Au nucleus as a CGC for forward particle production. Comparisons with other models will be interesting to perform. Normalizing to equal numbers of binary collisions, the d+Au yield is significantly smaller than the p+p yield. The η dependence of the reduction is consistent with models which suppress the gluon density in nuclei, in addition to exhibiting isospin effects at these kinematics. Additional measurements

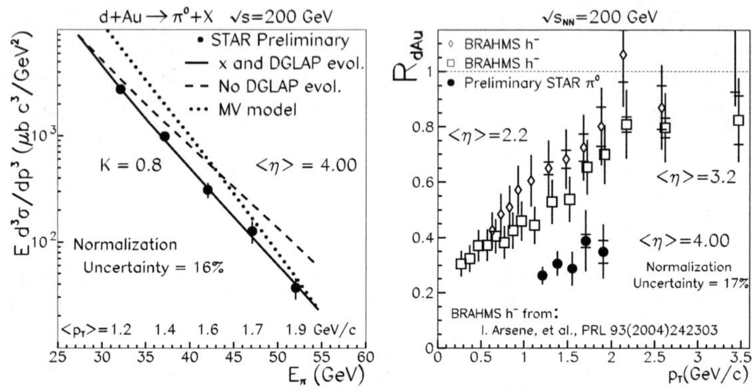

FIGURE 2. (Left) Inclusive π^0 production cross section for d+Au collisions, displayed as in Fig. 1. The curves are from models which treat the Au nucleus as a CGC, normalized with a p_T-independent K-factor. (Right) Nuclear modification factor for minimum-bias d+Au collisions versus p_T. The solid circles are data for π^0 mesons. The open diamonds and squares are data for h^- at smaller η. The inner error bars are statistical, while the outer combine these with the point-to-point systematic errors.

at different centralities and with other final states will help elucidate the cause of the suppression. In addition, both di-hadron correlation data and quantitative theoretical understanding thereof are needed to facilitate tests of a possible CGC.

REFERENCES

1. M. Hirai, S. Kumano, and T.-H. Nagai, Phys. Rev. C **70**, 044905 (2004).
2. C. Adloff, *et al.*, Eur. Phys. J. **C21**, 33 (2001); S. Dhekanov, *et al.*, Eur. Phys. J. **C21**, 443 (2001).
3. M. Gyulassy and L. McLerran, Nucl. Phys. A **750**, 30 (2005).
4. V. Guzey, M. Strikman, and W. Vogelsang, Phys. Lett. B **603**, 173 (2004).
5. J. Adams, *et al.*, Phys. Rev. Lett. **92**, 171801 (2004).
6. S. S. Adler, *et al.*, Phys. Rev. Lett. **91**, 241803 (2003).
7. I. Arsene, *et al.*, Phys. Rev. Lett. **93**, 242303 (2004); **91**, 072305 (2003).
8. L. Gribov, E. Levin, and M. Ryskin, Phys. Rep. **100**, 1 (1983); A. H. Mueller and J. Qiu, Nucl. Phys. B **268**, 427 (1986); L. McLerran and R. Venugopalan, Phys. Rev. D **49**, 3352 (1994); A. Dumitru and J. Jalilian-Marian, Phys. Rev. Lett. **89**, 022301 (2002); E. Iancu, K. Itakura, and D. N. Triantafyllopoulos, Nucl. Phys. A **742**, 182 (2004).
9. J. Jalilian-Marian, Nucl. Phys. A **748**, 664 (2005); D. Kharzeev, Yu. V. Kovchegov, and K. Tuchin, Phys. Lett. B **599**, 23 (2004); Phys. Rev. D **68**, 094013 (2003).
10. A. Dumitru, A. Hayashigaki, and J. Jalilian-Marian, hep-ph/0506308.
11. J. Qiu and I. Vitev, Phys. Rev. Lett. **93**, 262301 (2004); hep-ph/0410218.
12. R. Vogt, Phys. Rev. C **70**, 064902 (2004); N. Armesto, C. A. Salgado, and U. A. Wiedemann, Phys. Rev. Lett. **94**, 022002 (2005).
13. R. C. Hwa, C. B. Yang, and R. J. Fries, Phys. Rev. C **71**, 024902 (2005).
14. B. Kopeliovich, *et al.*, hep-ph/0501260; N. Nikolaev and W. Schäfer, Phys. Rev. D **71**, 014023 (2005).
15. A. Ogawa, for the STAR Collab., nucl-ex/0408004.
16. G. Rakness, for the STAR Collab., hep-ex/0505062; Nucl. Phys. B (Proc. Suppl.) **146**, 73 (2005) (nucl-ex/0501026); D. Morozov, for the STAR Collab., hep-ex/0505024.
17. S. Kretzer, hep-ph/0410219.
18. J. Singh, *et al.*, Nucl. Phys. B **140**, 189 (1978).
19. J. Adams, *et al.*, Phys. Rev. Lett. **91**, 072304 (2003).

Particle Production in p(d)A Collisions

Yuri V. Kovchegov

Department of Physics, The Ohio State University
Columbus, OH 43210

Abstract. We discuss particle production in $p(d)A$ collisions in the saturation/Color Glass Condensate physics framework. We describe how predictions of saturation physics can be tested by studying energy/rapidity dependence of particle spectra in p(d)A collisions. We concentrate on the nuclear modification factor R^{pA} for gluon production. We show that at moderately high energy/rapidity the nuclear modification factor R^{pA} exhibits Cronin enhancement. As the energy/rapidity increases, R^{pA} decreases. At sufficiently high energy/rapidity R^{pA} becomes less than 1 for all values of p_T indicating the onset of suppression of gluon production due to quantum small-x evolution effects. These predictions were confirmed by RHIC data.

Keywords: saturation, Color Glass Condensate, particle production, BFKL evolution, shadowing
PACS: 12.38.Bx, 12.38.Cy, 24.85.+p

INTRODUCTION

This talk is based on work done in collaboration with Dima Kharzeev and Kirill Tuchin [1, 2].

We will discuss single inclusive particle production in $p(d)A$ collisions calculated in the framework of saturation/Color Glass Condensate physics. In $p(d)A$ collisions, due to absence of fi nal state interactions, predictions of saturation physics can be tested directly. Single gluon production cross section in pA collisions was found in [3] in the quasi-classical approximation and the effects of quantum small-x evolution were included in it in [4]. Concentrating on the resulting nuclear modifi cation factor R^{pA} we will argue that the quasi-classical gluon production cross section calculated in [3] leads to Cronin enhancement [1, 5]. This regime is probably relevant for mid-rapidity particle production in dAu collisions at RHIC. At higher (forward) rapidities the effects of small-x evolution would lead to suppression of R^{pA} and disappearance of Cronin effect [6, 1, 7]. The suppression predicted in [6, 1, 7] was observed experimentally in [8, 9], confi rming the expectation of saturation/Color Glass physics: it may also be regarded as experimental evidence of BFKL evolution. We will demonstrate that a quantitative saturation-inspired model from [2] describes the data of [8] rather well.

SINGLE GLUON PRODUCTION IN PA COLLISIONS

We start by discussing single inclusive particle production cross section and transverse momentum spectra in dAu collisions. The gluon production cross section in pA in the

CP792, *Deep Inelastic Scattering, DIS 2005,* edited by W. H. Smith and S. R. Dasu

quasi-classical approximation was constructed in [3] yielding

$$\frac{d\sigma^{pA}}{d^2k\,dy} = \int d^2b\,d^2x\,d^2y\,\frac{1}{(2\pi)^2}\,\frac{\alpha_s C_F}{\pi^2}\,\frac{\underline{x}\cdot\underline{y}}{\underline{x}^2\underline{y}^2}\,e^{-i\underline{k}\cdot(\underline{x}-\underline{y})}$$

$$\times\left[1-e^{-\underline{x}^2 Q_{s0}^2\ln(1/x_T\Lambda)/4}-e^{-\underline{y}^2 Q_{s0}^2\ln(1/y_T\Lambda)/4}+e^{-(\underline{x}-\underline{y})^2 Q_{s0}^2\ln(1/|\underline{x}-\underline{y}|\Lambda)/4}\right], \tag{1}$$

where $\underline{k}$ and y are the produced gluon's transverse momentum and rapidity, $\underline{b}$ is the proton's impact factor, Q_{s0} is the saturation scale in McLerran-Venugopalan model and $\underline{x}$, $\underline{y}$ are two-dimensional gluon transverse position vectors which are integrated over.

Eq. (1) can be used to construct the nuclear modifi cation factor [1]

$$R^{pA}(\underline{k},y) = \frac{\frac{d\sigma^{pA}}{d^2k\,dy}}{A\,\frac{d\sigma^{pp}}{d^2k\,dy}}. \tag{2}$$

The ratio $R^{pA}(k_T)$ is plotted in Fig. 1 for $\Lambda = 0.2\,Q_{s0}$. It clearly exhibits an enhancement at high-k_T characteristic of Cronin effect. Similar conclusions have been reached by other authors [5]. Note that the position and height of the Cronin peak are increasing functions of the centrality of pA collisions [1].

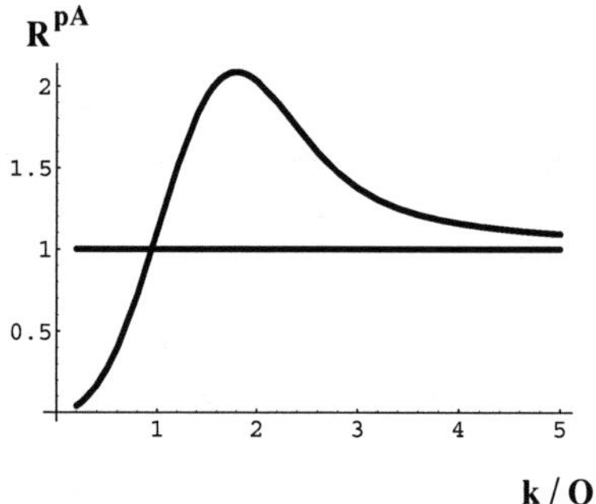

FIGURE 1. Nuclear modification factor R^{pA} plotted as a function of k_T/Q_{s0} for gluon production in the quasi-classical McLerran-Venugopalan model as found in [3]. The cutoff is $\Lambda = 0.2\,Q_s$.

Before including quantum evolution into Eq. (1), let us fi rst note that Eq. (1), even though it includes multiple rescatterings, can still be written in k_T-factorized form (!) [4, 1]

$$\frac{d\sigma^{pA}}{d^2k\,dy} = \frac{2\,\alpha_s}{C_F}\,\frac{1}{\underline{k}^2}\int d^2q\,\phi_p(\underline{q},0)\,\phi_A(\underline{k}-\underline{q},0), \tag{3}$$

with the unintegrated "gluon distributions" given by

$$\phi_A(x,\underline{k}^2) = \frac{C_F}{\alpha_s\,(2\pi)^3}\int d^2b\,d^2r\,e^{-i\underline{k}\cdot\underline{r}}\,\nabla_r^2 N_G(\underline{r},\underline{b},y=\ln 1/x), \tag{4}$$

and

$$\phi_p(x,\underline{k}^2) = \frac{C_F}{\alpha_s\,(2\pi)^3}\int d^2b\,d^2r\,e^{-i\underline{k}\cdot\underline{r}}\,\nabla_r^2 n_G(\underline{r},\underline{b},y=\ln 1/x). \tag{5}$$

Here $N_G(\underline{r},\underline{b},y=\ln 1/x)$ and $n_G(\underline{r},\underline{b},y=\ln 1/x)$ are forward scattering amplitudes of a gluon dipole of size r located at impact parameter $\underline{b}$ on a nucleus and a proton correspondingly. Eq. (1) is reproduced by using $N_G(\underline{r},\underline{b},0)=1-e^{-\underline{r}^2 Q_{s0}^2 \ln(1/r_T\Lambda)/4}$ and $n_G(\underline{r},\underline{b},0)=\underline{r}^2\Lambda^2\ln(1/r_T\Lambda)$ [1].

As was shown in [4], Eq. (3) makes inclusion of quantum small-x evolution straightforward. A tedious analysis shows that the inclusion of evolution preserves k_T-factorization of Eq. (3) yielding the full answer for inclusive cross section [4]

$$\frac{d\sigma^{pA}}{d^2k\,dy}=\frac{2\alpha_s}{C_F}\frac{1}{\underline{k}^2}\int d^2q\,\phi_p(\underline{q},Y-y)\,\phi_A(\underline{k}-\underline{q},y), \tag{6}$$

where Y is the full rapidity interval between the proton and the nucleus. The gluon distribution functions in Eq. (6) are still given by Eqs. (4) and (5), but now with n_G given by the solution of the BFKL equation and with N_G given by the solution of the non-linear evolution equation.

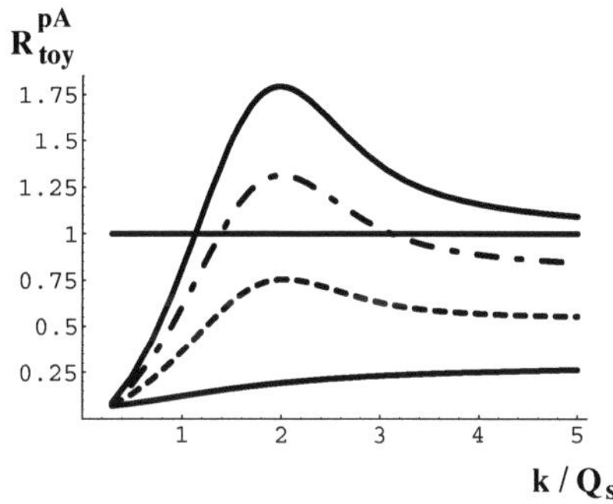

FIGURE 2. R^{pA} plotted as a function of k_T/Q_s for (i) McLerran-Venugopalan model, which is valid for moderate energies/rapidities (upper solid line); (ii) our toy model for very high energies/rapidities from [1] (lower solid line); (iii) an interpolation to intermediate energies/rapidities (dash-dotted and dashed lines).

The analysis of nuclear modifi cation factor resulting from Eq. (6) was carried out in [6, 1, 7]. The result is that $R^{pA}\sim A^{-1/6}$ at very large y for all p_T. Note that R^{pA} will become a decreasing function of centrality at high energies/rapidities.

Our conclusions regarding the variation of R^{pA} with energy/ centrality are summarized in Fig. 2. The top solid curve in Fig. 2 is the same as in the quasi-classical approximation shown in Fig. 1. It corresponds to moderately high energy/rapidity. As energy/rapidity increases R^{pA} decreases (dash-dotted and dashed lines) and the Cronin peak flattens, eventually approaching a flat curve (lower solid line) which has suppression at all p_T. Similar conclusions have been reached in [7].

THE DATA

The data on the nuclear modifi cation function R^{dAu} for $d+Au$ collisions reported by BRAHMS collaboration [8] is shown in Fig. 3, along with predictions of a saturation-inspired model constructed in [2], which includes valence quark contribution as well. As one can see from Fig. 3, the onset of suppression at higher rapidities, predicted by CGC approach, is confirmed by RHIC data.

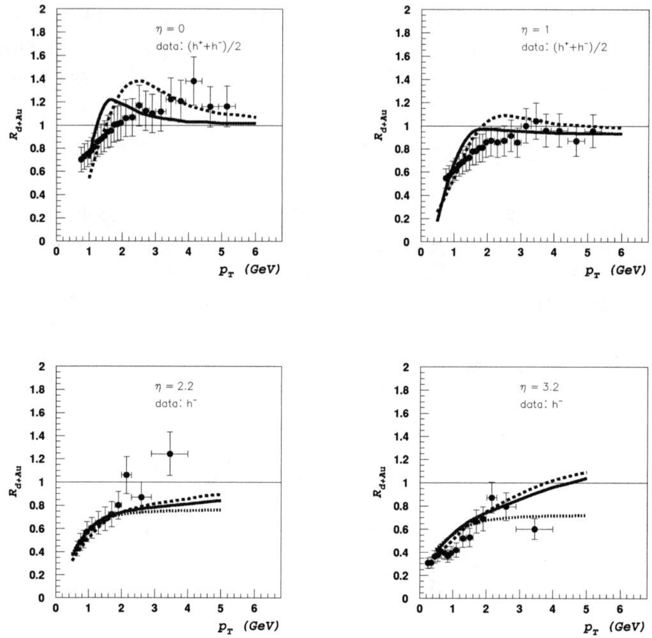

FIGURE 3. Nuclear modification factor R_{dAu} of charged particles for different rapidities. The fit uses the model described in [2]. Data is from [8].

REFERENCES

1. D. Kharzeev, Y. V. Kovchegov and K. Tuchin, Phys. Rev. D **68**, 094013 (2003) [arXiv:hep-ph/0307037].
2. D. Kharzeev, Y. V. Kovchegov and K. Tuchin, Phys. Lett. B **599**, 23 (2004) [arXiv:hep-ph/0405045].
3. Yu. V. Kovchegov and A. H. Mueller, Nucl. Phys. B **529**, 451 (1998) [arXiv:hep-ph/9802440].
4. Yu. V. Kovchegov and K. Tuchin, Phys. Rev. D **65**, 074026 (2002) [arXiv:hep-ph/0111362].
5. R. Baier, A. Kovner and U. A. Wiedemann, Phys. Rev. D **68**, 054009 (2003) [arXiv:hep-ph/0305265]; J. Jalilian-Marian, Y. Nara and R. Venugopalan, Phys. Lett. B **577**, 54 (2003) [arXiv:nucl-th/0307022]; B. Z. Kopeliovich, J. Nemchik, A. Schafer and A. V. Tarasov, Phys. Rev. Lett. **88**, 232303 (2002) [arXiv:hep-ph/0201010]; A. Accardi and M. Gyulassy, arXiv:nucl-th/0308029.
6. D. Kharzeev, E. Levin and L. McLerran, Phys. Lett. B **561**, 93 (2003) [arXiv:hep-ph/0210332].
7. J. L. Albacete, N. Armesto, A. Kovner, C. A. Salgado and U. A. Wiedemann, Phys. Rev. Lett. **92**, 082001 (2004) [arXiv:hep-ph/0307179].
8. I. Arsene *et al.* [BRAHMS Collaboration], Phys. Rev. Lett. **93**, 242303 (2004) [arXiv:nucl-ex/0403005]; R. Debbe [BRAHMS Collaboration], arXiv:nucl-ex/0403052.
9. B. B. Back *et al.* [PHOBOS Collaboration], Phys. Rev. C **70**, 061901 (2004) [arXiv:nucl-ex/0406017]; S. S. Adler *et al.* [PHENIX Collaboration], Phys. Rev. Lett. **94**, 082302 (2005) [arXiv:nucl-ex/0411054]; G. Rakness [STAR Collaboration], these proceedings.

Forward Production in d+Au Collisions by Parton Recombination

R. J. Fries

School of Physics and Astronomy, University of Minnesota, Minneapolis, MN 55455

Abstract. We discuss parton recombination as a hadronization mechanism for d+Au collisions. We show that several features of hadron production measured at the Relativistic Heavy Ion Collider (RHIC) can be explained by recombination, including the Cronin effect at midrapidity and the suppression of hadrons in forward direction.

Keywords: p+A collisions;parton recombination
PACS: 13.85.Ni,13.87.Fh,25.75.Nq

Recombination of partons has been suggested as a possible hadronization mechanism in high energy scattering reactions with final state hadrons almost 30 years ago [1]. Quarks pick other partons in their neighborhood and cluster to objects that have the quantum numbers of QCD bound states. This happens after the virtuality of the scattered partons has been reduced by radiation to values well in the non-perturbative regime $\sim \Lambda_{QCD}$ and before the system is so dilute that colored partons have an average distance of more than the confinement radius.

Of course, a full description of this process would be equivalent to solving the QCD equations of motion in the non-perturbative regime, so it is not surprising that the recombination idea sketched above was only implemented in the form of simplified models. Nevertheless, these models have turned out to be very useful in understanding certain aspects of hadronization, e.g. for the leading particle effect [2] or, most recently, in high energy nuclear collisions [3, 4, 5, 6].

It has been realized that if thermalization of partons is achieved in nuclear collisions, this acts like a reset button that wipes out a lot of the complicated dynamics present in other collisions. Recombination of thermal partons has become an extremely successful concept to describe the relative behavior of hadron species, e.g. pions compared with protons, kaons compared with Lambdas etc. This is true for most single inclusive hadron observables in high energy nuclear collisions. E.g. in order to describe the transverse momentum spectrum or elliptic flow of the dozen or so *hadron* species identified in Au+Au collisions at the Relativistic Heavy Ion Collider (RHIC), it is sufficient to start from one universal parameterization of the corresponding *parton* quantity just before hadronization [7].

It was the unexpected baryon excess observed in Au+Au at RHIC that triggered the success of recombination models. Because the ejection probability of clusters of particles with a given relativistic momentum p from a thermal ensemble is independent of the number of particles in the cluster, the production probability for baryons is intrinsically equal to that of mesons, leading to a much higher baryon yield than expected from vacuum fragmentation. Surprisingly, when first data on hadron production in $d+Au$

CP792, *Deep Inelastic Scattering, DIS 2005*, edited by W. H. Smith and S. R. Dasu

from RHIC was presented, a similar, though smaller effect could be observed [8]. The modification factor R^h_{dAu} is defined as the ratio of the yield of a certain hadron species h in $d+Au$ over the yield of h in $p+p$ scaled by the number of binary nucleon-nucleon collisions in d+Au. It is unity if a d+Au collision is only a superposition of individual nucleon-nucleon collisions. One expects to see the typical Cronin enhancement at intermediate p_T, i.e. between roughly 2 and 6 GeV/c [9]. Indeed a slight Cronin effect was seen for pions, with $R^\pi_{dAu} \approx 1.1$. However, for protons the nuclear modification seems to be much larger in the same momentum range. This is completely at odds with the explanation of the Cronin effect as a broadening by multiple initial state scatterings in the nucleus [10]. There must be a final state effect that can have a much larger influence on the momentum spectrum than the initial state multiple scattering.

Hwa and Yang suggested that although there is no thermalization of partons in a rather "cold" d+Au collision, the exponential shape of the low and intermediate p_T spectra could still make the production of baryons through the recombination channel so important that it significantly increases the overall baryon yield at intermediate p_T. Indeed in their calculation they can explain the Cronin enhancement for pions and the even larger effect for protons entirely by additional recombination of partons which is not present in $p+p$ collisions, without having to invoke any conventional initial state multiple scattering at all [11].

Let us pause here and add a brief review of the recombination formalism in the version of Hwa and Yang [6]. The yield of, e.g., a π^+ is given by

$$\frac{dN_\pi}{p_T dp_T} = \frac{1}{p_T^2} \int \frac{dq_1}{q_1} \frac{dq_2}{q_2} F_{u\bar{d}}(q_1, q_2) R_\pi(q_1, q_2; p_T) \tag{1}$$

with the recombination function R_π. The two particle distribution $F_{u\bar{d}}$ for the u and $\bar{d}$ quarks is given by four different contributions, schematically written as

$$F_{u\bar{d}} = TT + TS + (SS) + SS'. \tag{2}$$

The nomenclature is as follows: TT are pairs where both partons are from the "soft" exponential part of the parton spectrum; TS are pairs where one parton is soft and the other is from the shower induced by a jet; (SS) is the situation where both partons are from the shower (of the same jet) and SS' is the contribution where both partons are from different jets.

The (SS) contribution, the recombination of 2 partons from the shower of a jet is nothing else than the fragmentation of a pion from a jet. Therefore the shower parton distributions have been parameterized to fit the known fragmentation functions. While the SS' contribution is negligible, TS and TT are small for $p+p$ but have increasing importance when going to d+Au and Au+Au collisions.

The BRAHMS experiment has reported that at forward rapidities in d+Au the enhancement of the nuclear modification factor changes to suppression for (unidentified) hadrons [12]. This has been seen as a possible earmark of gluon saturation in the Au nucleus at small Bjorken x which would be an exciting first piece of evidence for the existence of a color glass condensate [13, 14, 15] . However, any alternative explanations for this phenomenon have to be discussed thoroughly [16].

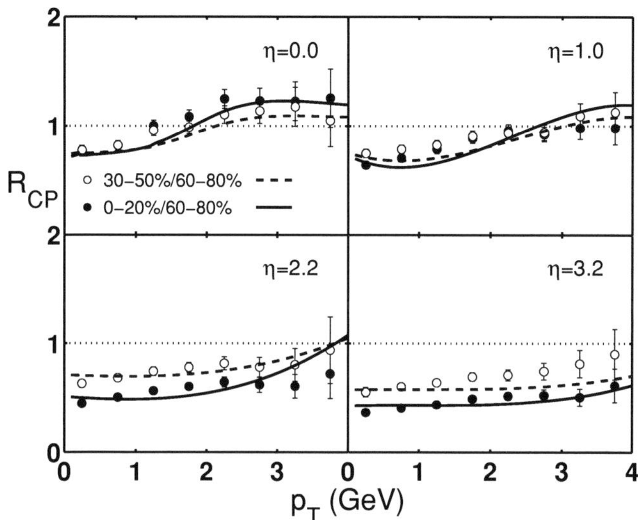

FIGURE 1. R_{CP} for 0-20%/60-80% (filled circles and solid lines) and 30-50%/60-80% (open circles and dashed lines) for four pseudorapidities. Data are from [12].

It was noted that in very forward direction the phase space for hard jet production is very small. The kinematic limit for the jet p_T is 8.1 GeV/c for rapidity $\eta = 3.2$. This leads to a dramatic decrease in the (SS) term in Eq. (2). Now the scaling property depends on the competing TT and TS channels. A first principle calculation of the soft parton spectrum T in the event is not available and the shape of the spectrum at forward rapidity has not been published yet. We assume here that the slope of the exponential $T = Cp_T \exp(-\beta p_T)$ does not vary with η and take the value obtained from a fit at midrapidity ($\beta^{-1} = 0.21$ GeV). The normalization C can be fixed by the measured charged hadron multiplicity $dN_{ch}/d\eta$. See [16] for details.

Our results for the ratio R_{CP}, a relative of R_{dAu} comparing different centrality classes for d+Au, are shown in Fig. 1 for pions. We compare with BRAHMS data for four different values of rapidity. Our results show the characteristic suppression at forward rapidity. This is indeed due to the fact that hadron production, even for pions, is dominated by parton recombination at very forward rapidities. The developing dominance of the soft component with increasing η can also be seen in Fig. 2 where the predicted pion spectrum is shown for 4 different rapidities and 2 centralities. The spectrum above $p_T = 2$ GeV/c cleary evolves from a power-law (hard) spectrum for $\eta = 0$ to an exponential (soft) spectrum at $\eta = 3.2$.

To conclude, we have reported about an alternative way to understand the suppression of hadrons at forward rapidities in d+Au collisions. It is based on the assumption that recombination of soft partons is dominating in this kinematic region. This is motivated by the observation that recombination effects have likely been observed at midrapidity through the particular dependence of the Cronin effect on the hadron species. It should be emphasized that in this framework nothing can be said about the origin of the soft

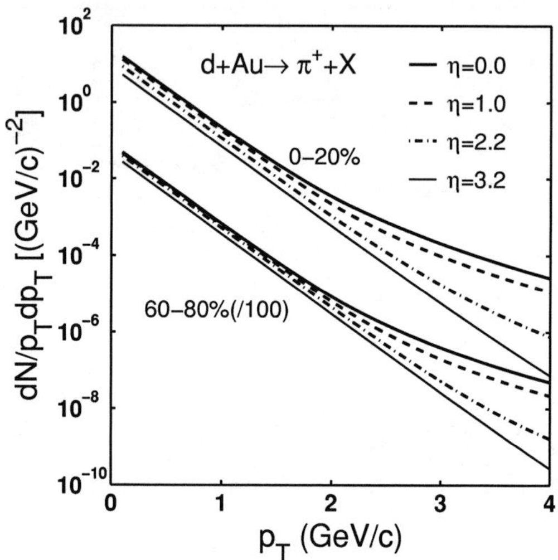

FIGURE 2. The spectrum $dN/(p_T dp_T)$ for pions for different centralities and rapidities.

parton spectrum.

ACKNOWLEDGMENTS

I want to thank my collaborators R. C. Hwa and C. B. Yang. This work was supported by DOE grant DE-FG02-87ER40328.

REFERENCES

1. K. P. Das and R. C. Hwa, *Phys. Lett. B* **68**, 459 (1977); *Erratum-ibid.* **73B**, 504 (1978).
2. E. Braaten, Y. Jia and T. Mehen, *Phys. Rev. Lett.* **89**, 122002 (2002) and references therein.
3. R. C. Hwa and C. B. Yang, *Phys. Rev. C* **67**, 034902 (2003).
4. R. J. Fries, B. Müller, C. Nonaka and S. A. Bass, *Phys. Rev. Lett.* **90**, 202303 (2003); *Phys. Rev. C* **68**, 044902 (2003).
5. V. Greco, C. M. Ko, and P. Lévai, *Phys. Rev. Lett.* **90**, 202302 (2003); *Phys. Rev. C* **68**, 034904 (2003).
6. R. C. Hwa and C. B. Yang, *Phys. Rev. C* **70**, 024905 (2004).
7. R. J. Fries, *J. Phys. G* **30**, S853 (2004).
8. F. Matathias [PHENIX Collaboration], *preprint* nucl-ex/0504019.
9. J. W. Cronin *et al.*, *Phys. Rev. D* **11**, 3105 (1975).
10. A. Accardi, and M. Gyulassy, *J. Phys. G* **30**, S969 (2004).
11. R. C. Hwa and C. B. Yang, *Phys. Rev. Lett.* **93**, 082302 (2004).
12. I. Arsene *et al.* (BRAHMS Collaboration), *Phys. Rev. Lett.* **93**, 242303 (2004).
13. L. McLerran, *preprint* hep-ph/0402137.
14. J.-P. Blaizot and F. Gelis, *Nucl. Phys. A* **750**, 148–165,(2005).
15. D. Kharzeev, Y. Kovchegov, and K. Tuchin, *Phys. Rev. D* **68**, 094013 (2003); *Phys. Lett. B* **599**, 23–31 (2004).
16. R. C. Hwa, C. B. Yang and R. J. Fries, *Phys. Rev. C* **71**, 024902 (2005).

Impact of large-x resummation on parton distribution functions

G. Corcella* and L. Magnea†

*CERN, Department of Physics, Theory Division, CH-1211 Geneva 23, Switzerland
† Università di Torino and INFN, Sezione di Torino, Via P. Giuria 1, I-10125, Torino, Italy

Abstract. We investigate the effect of large-x resummation on parton distributions by performing a fit of Deep Inelastic Scattering data from the NuTeV, BCDMS and NMC collaborations, using NLO and NLL soft-resummed coefficient functions. Our results show that soft resummation has a visible impact on quark densities at large x. Resummed parton fits would therefore be needed whenever high precision is required for cross sections evaluated near partonic threshold.

Keywords: Resummation, parton distribution functions
PACS: 12.38.Bx, 12.38.Cy

A precise knowledge of parton distribution functions (PDF's) at large x is important to achieve the accuracy goals of the LHC and other high energy accelerators. We present a simple fit of Deep Inelastic Scattering (DIS) structure function data, and extract NLO and NLL-resummed quark densities, in order to establish qualitatively the effects of soft-gluon resummation.

Structure functions $F_i(x,Q^2)$ are given by the convolution of coefficient functions and PDF's. Finite-order coefficient functions present logarithmic terms that are singular at $x=1$, and originate from soft or collinear gluon radiation. These contributions need to be resummed to extend the validity of the perturbative prediction. Large-x resummation for the DIS coefficient function was performed in [1, 2] in the massless approximation, and in [3, 4] with the inclusion of quark-mass effects, relevant at small Q^2.

Soft resummation is naturally performed in moment space, where large-x terms correspond, at $\mathcal{O}(\alpha_s)$, to single ($\alpha_s \ln N$) and double ($\alpha_s \ln^2 N$) logarithms of the Mellin variable N. In the following, we shall consider values of Q^2 sufficiently large to neglect quark-mass effects. Furthermore, we shall implement soft resummation in the next-to-leading logarithmic (NLL) approximation, which corresponds to keeping terms $\mathcal{O}(\alpha_s^n \ln^{n+1} N)$ (LL) and $\mathcal{O}(\alpha_s^n \ln^n N)$ (NLL) in the Sudakov exponent.

To gauge the impact of the resummation on the DIS cross section, we can evaluate the charged-current (CC) structure function F_2 convoluting NLO and NLL-resummed $\overline{\mathrm{MS}}$ coefficient functions with the NLO PDF set CTEQ6M [5]. We consider $Q^2 = 31.62\ \mathrm{GeV}^2$, since it is one of the values of Q^2 at which the NuTeV collaboration collected data [6]. In Fig. 1 we plot $F_2(x)$ with and without resummation (Fig. 1a), as well as the normalized difference $\Delta = (F_2^{\mathrm{res}} - F_2^{\mathrm{NLO}})/F_2^{\mathrm{NLO}}$ (Fig. 1b). We note that the effect of the resummation is an enhancement of F_2 for $x > 0.6$. Such an enhancement is compensated by a decrease at smaller x: the resummation, in fact, does not change the first moment of F_2, since we include in the Sudakov exponent only terms $\sim \ln^k N$, which vanish for $N = 1$. Our predictions for F_2 at different values of Q^2 can be compared with

CP792, *Deep Inelastic Scattering, DIS 2005*, edited by W. H. Smith and S. R. Dasu

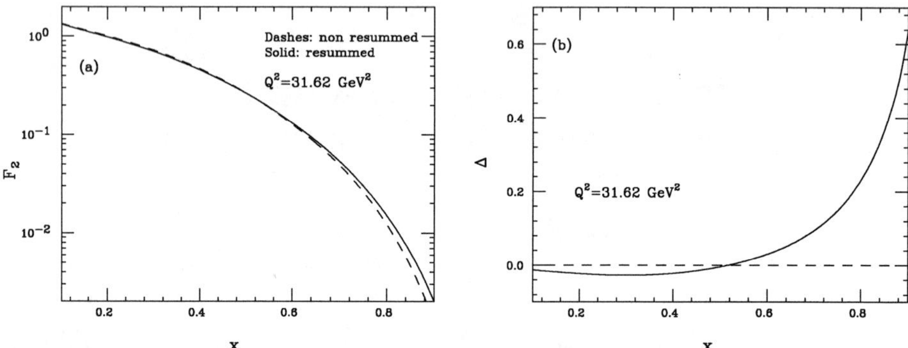

FIGURE 1. (a): CC structure function $F_2(x)$ using NLO (dashes) and NLL-resummed (solid) coefficient functions, at $Q^2 = 31.62$ GeV^2; (b): relative difference $\Delta = (F_2^{\rm res} - F_2^{\rm NLO})/F_2^{\rm NLO}$

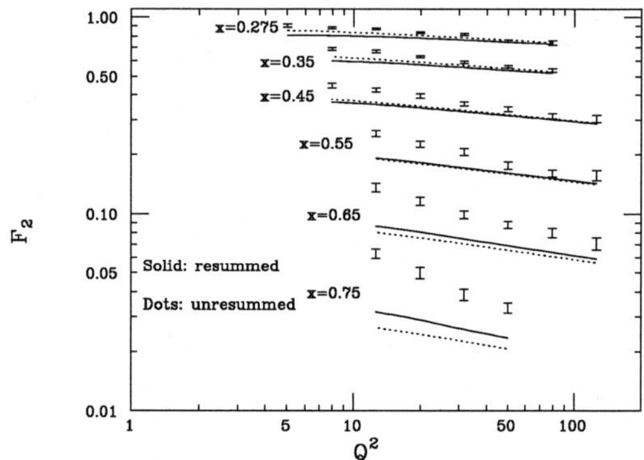

FIGURE 2. Comparison of NuTeV data on the CC structure function $F_2(x,Q^2)$ with a theoretical prediction using CTEQ6M PDF's and NLO (dots) or NLL-resummed (solid) coefficient functions.

NuTeV data at large x. The results of the comparison are shown in Fig. 2: although the resummation moves the prediction towards the data, we are still unable to reproduce the large-x data. Several effects are involved in the mismatch: at very large values of x, power corrections will certainly play a role. Moreover, we have used so far a parton set (CTEQ6M), extracted by a global fit which did not account for the NuTeV data. Rather, data from the CCFR experiment [7], which disagree at large x with NuTeV [6], were used. The discrepancy has recently been described as understood [8]; however, it is not possible to draw any firm conclusion from our comparison.

We wish to reconsider the CC data in the context of an indipendent fit. We shall use NuTeV data on $F_2(x)$ and $xF_3(x)$ at $Q^2 = 31.62$ GeV^2 and 12.59 GeV^2, and extract NLO and NLL-resummed quark distributions from the fit. F_2 contains a gluon-initiated contribution F_2^g, which is not soft-enhanced and is very small at large x: we can therefore safely take F_2^g from a global fit, e.g. CTEQ6M, and limit our fit to the quark-initiated term F_2^q. We choose a parametrization of the form $F_2^q(x) = F_2(x) - F_2^g(x) = Ax^{-\alpha}(1-$

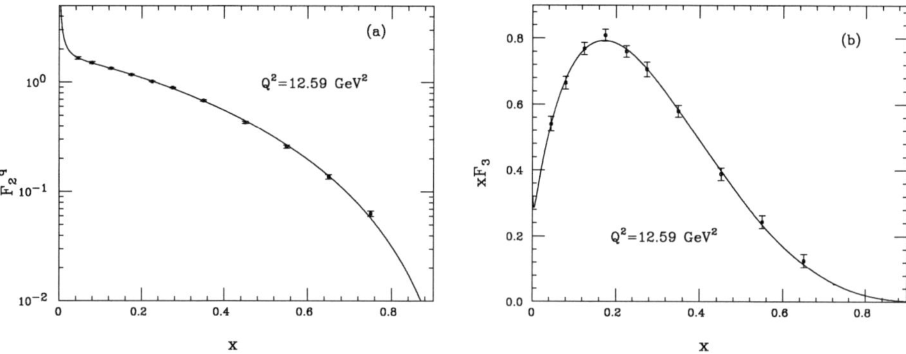

FIGURE 3. NuTeV data and best-fit curves at $Q^2 = 12.59$ GeV^2 for F_2^q (a) and xF_3 (b).

$x)^{\beta}(1+bx)$; $xF_3(x) = Cx^{-\rho}(1-x)^{\sigma}(1+kx)$. The best-fit parameters and the χ^2 per degree of freedom are quoted in [9]. In Fig. 3, we present the NuTeV data on $F_2(x)$ and $xF_3(x)$ at $Q^2 = 12.59$ GeV^2, along with the best-fit curves. Similar plots at $Q^2 = 31.62$ GeV^2 are shown in Ref. [9].

In order to extract individual quark distributions, we need to consider also neutral current data. We use BCDMS [10] and NMC [11] results, and employ the parametrization of the nonsinglet structure function $F_2^{\rm ns} = F_2^p - F_2^D$ provided by Ref. [12]. The parametrization [12] is based on neural networks trained on Monte-Carlo copies of the data set, which include all information on errors and correlations: this gives an unbiased representation of the probability distribution in the space of structure functions.

Writing F_2, xF_3 and $F_2^{\rm ns}$ in terms of their parton content, we can extract NLO and NLL-resummed quark distributions, according to whether we use NLO or NLL coefficient functions. We assume isospin symmetry of the sea, i.e. $s = \bar{s}$ and $\bar{u} = \bar{d}$, we neglect the charm density, and impose a relation $\bar{s} = \kappa\bar{u}$. We obtain a system of three equations, explicitly presented in [9], that can be solved in terms of u, d and s. We begin by working in N-space, where the resummation has a simpler form and quark distributions are just the ratio of the appropriate structure function and coefficient function. We then revert to x-space using a simple parametrization $q(x) = Dx^{-\gamma}(1-x)^{\delta}$.

Figs. 4–5 show the effect of the resummation on the up-quark distribution at $Q^2 = 12.59$ and 31.62 GeV^2, in N- and x-space respectively. The best-fit values of D, γ and δ, along with the χ^2/dof, can be found in [9]. The impact of the resummation is noticeable at large N and x: there, soft resummation enhances the coefficient function and its moments, hence it suppresses the quark densities extracted from structure function data. In principle, also d and s densities are affected by the resummation; the errors on their moments, however, are too large for the effect to be statistically significant. In [9] it was also shown that the results for the up quark at 12.59 and 31.62 GeV^2 are consistent with NLO perturbative evolution.

In summary, we have presented a comparison of NLO and NLL-resummed quark densities extracted from large-x DIS data. We found a suppression of valence quarks in the $10-20\%$ range at $x > 0.5$, for moderate Q^2. We believe that it would be interesting

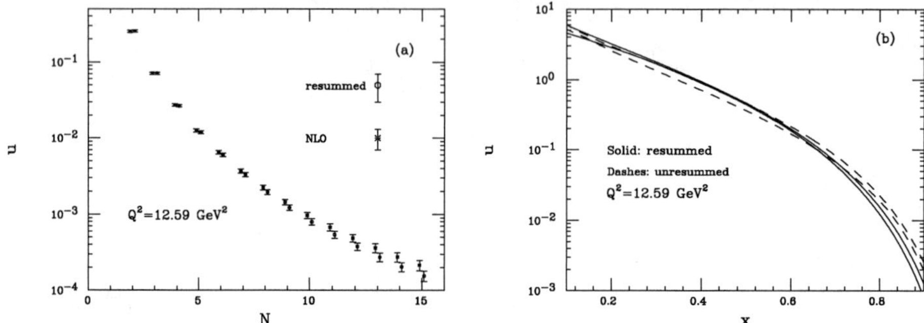

FIGURE 4. NLO and resummed up quark distribution at $Q^2 = 12.59$ GeV^2 in moment (a) and x (b) spaces. Following [9], in x space, we have plotted the edges of a band corresponding to a prediction at one-standard-deviation confidence level (statistical errors only).

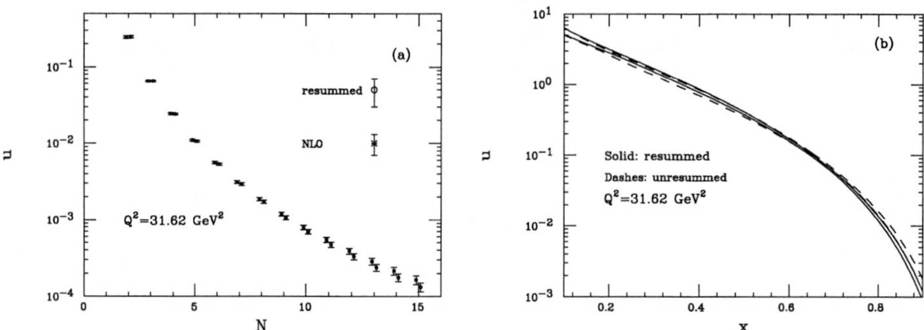

FIGURE 5. The same as in Fig. 4, but at $Q^2 = 31.62$ GeV^2.

and fruitful to extend this analysis and include large-x resummation in the toolbox of global fits. Our results show in fact that this would be necessary to achieve precisions better than 10% in processes involving large-x partons.

REFERENCES

1. G. Sterman, *Nucl. Phys. B* **281** (1987) 310.
2. S. Catani and L. Trentadue, *Nucl. Phys. B* **327** (1989) 323.
3. E. Laenen and S. O. Moch, *Phys. Rev. D* **59** (1999) 034027.
4. G. Corcella and A. D. Mitov, *Nucl. Phys. B* **676** (2004) 346.
5. J. Pumplin, D. R. Stump, J. Huston, H. L. Lai, P. Nadolsky and W. K. Tung, *JHEP* **0207** (2002) 012.
6. D. Naples *et al.* [NuTeV Collaboration], hep-ex/0307005.
7. U. K. Yang *et al.* [CCFR/NuTeV Collaboration], *Phys. Rev. Lett.* **86** (2001) 2742.
8. M. Tzanov *et al.* [NuTeV Collaboration], these proceedings.
9. G. Corcella and L. Magnea, hep-ph/0506278.
10. A. C. Benvenuti *et al.* [BCDMS Collaboration], *Phys. Lett. B* **237** (1990) 592.
11. M. Arneodo *et al.* [New Muon Collaboration], *Nucl. Phys. B* **483** (1997) 3.
12. L. Del Debbio, S. Forte, J. I. Latorre, A. Piccione and J. Rojo, *JHEP* **0503** (2005) 080.

Saturation, traveling waves and fluctuations

Rikard Enberg

Theoretical Physics Group, Lawrence Berkeley National Laboratory, CA 94720, USA
Centre de Physique Théorique, École Polytechnique, 91128 Palaiseau, France

Abstract. In this talk I discuss the high energy asymptotics of QCD scattering, and its similarity to a reaction–diffusion process. I also discuss detailed numerical studies of the mean field approximation to this picture, i.e., the Balitsky–Kovchegov equation.

Keywords: Saturation, high energy scattering, stochastic processes
PACS: 12.38.-t, 12.38.Aw, 12.40.Ee

Perturbative QCD, in the form of the BFKL equation, predicts that in the high energy (small-x) limit, cross sections rise as a power of the center of mass energy. This rise is not compatible with unitarity of the S-matrix when extrapolated to very high energies. The BFKL equation is a linear evolution equation; the limits set by unitarity requires the introduction of non-linear terms in the evolution equation. The simplest such evolution equation is the Balitsky–Kovchegov (BK) equation [1].

More involved QCD evolution equations exist; they are known as the Balitsky–JIMWLK equations. Very recently, it has been discovered that there are important effects that are not accounted for by this set of equations (see [2] and references therein).

A different point of view has recently been emerging, in which high energy QCD is equivalent to a reaction–diffusion system. First the BK equation was shown [3] to be in the equivalence class of the Fisher–Kolmogorov–Petrovsky–Piscounov (FKPP) non-linear partial differential equation, which has so-called traveling wave solutions. This allowed deriving results for the amplitude and saturation scale as a function of energy. Then it was realized [4] that the effects neglected in the BK equation can be described within the equivalence class of the *stochastic* version of the FKPP equation.

This talk is based on our paper [5] where we wanted first to understand in more detail how the statistical interpretation of high-energy QCD comes about and then to make detailed numerical studies of the properties of the mean field approximation (the BK equation) and of a toy model that captures the important features of high energy QCD.

Let us first discuss the mean field approximation of the full evolution, which yields the BK equation. The first equation in Balitsky's hierarchy of coupled equations can be written as

$$\partial_{\bar{\alpha}_s Y} \langle T(k,Y) \rangle = K \otimes \langle T(k,Y) \rangle - \langle T^2(k,Y) \rangle, \tag{1}$$

where $L = \ln(k^2/\Lambda^2)$, and $K\otimes$ means the action of the BFKL integral kernel. The next equation in the hierarchy is an equation for the evolution of the correlator $\langle T^2(k,Y) \rangle$. If one approximates this correlator as the factorized form $\langle T^2(k,Y) \rangle \approx \langle T(k,Y) \rangle^2 \equiv A^2(k,Y)$ one gets the BK equation for the mean-field amplitude A:

$$\partial_{\bar{\alpha}_s Y} A(k,Y) = K \otimes A(k,Y) - A^2(k,Y) \tag{2}$$

CP792, *Deep Inelastic Scattering, DIS 2005*, edited by W. H. Smith and S. R. Dasu

Munier and Peschanski showed that this equation is in the equivalence class of the FKPP equation, $\partial_t u(x,t) = \partial_x^2 u(x,t) + u(x,t) - u^2(x,t)$, where x corresponds to L and t corresponds to Y. This equation has traveling wave solutions for certain conditions on the initial conditions, see [3]. This means that there is a solution which has a more or less fixed shape, and under the evolution in time (rapidity), the position of the wave front moves in space (momentum). Using results from the study of the FKPP equation, they obtained an expression for the saturation scale as a function of rapidity, and analytical expressions for the shape of the wave front for momenta above the saturation scale.

In [5], we solved the BK equation numerically and studied the properties of the solution. In particular we compared the numerical results to the analytical results. In Fig. 1, we see that there are indeed traveling wave solutions, and that the analytical results agree very well with the full solution for large rapidities. We also confirm the prediction that the influence of the initial condition disappears for large Y, so that a universal propagation speed is approached.

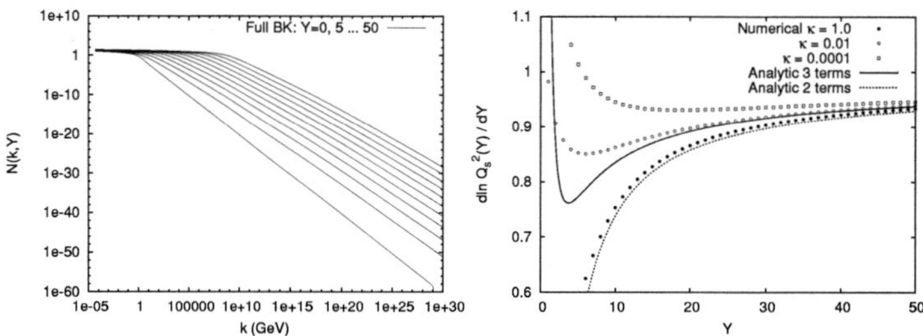

FIGURE 1. Left: Evolution of the initial condition using the full BK equation. Right: Logarithmic derivative of the saturation scale $d\ln Q_s^2(Y)/dY$ obtained from numerical simulations, compared to the analytical results to two different levels of accuracy.

The stochastic FKPP equation describes physical situations in which objects evolve by multiplying and diffusing, up to a limiting threshold. The crucial point is that the QCD parton model has such dynamics [4, 6, 5]. Once this is realized, all mathematical results obtained in statistical physics can be transposed to high energy QCD [6].

In the dipole picture of high energy scattering, a QCD dipole scatters off a hadronic target by interacting with one of its quantum fluctuations. The probe effectively "counts" the partons in this Fock state of the target with transverse momenta k: the amplitude $T(k)$ for the scattering is proportional to the number of partons $n(k)$.

The wave function of a hadronic object is built up from successive splittings of partons starting from the valence structure. As one increases the rapidity Y, the phase space for parton splittings grows and makes the probability for high occupation numbers larger. In the initial stages of the evolution, the parton density grows diffusively from these splittings. On the other hand, the number of partons in each cell of transverse phase space is limited to a maximum number N that depends on the strength of the interaction of the probe, i.e., on the relationship between T and n. This property is necessary from unitarity, which imposes an upper bound on the amplitude (e.g. $T(k) \leq 1$) which, in turn, results in an upper bound on $n(k)$. This is parton saturation.

Viewed in this way, scattering in the parton model is a reaction–diffusion process. The rapidity evolution of the Fock states of the target hadron is like the time evolution of a set

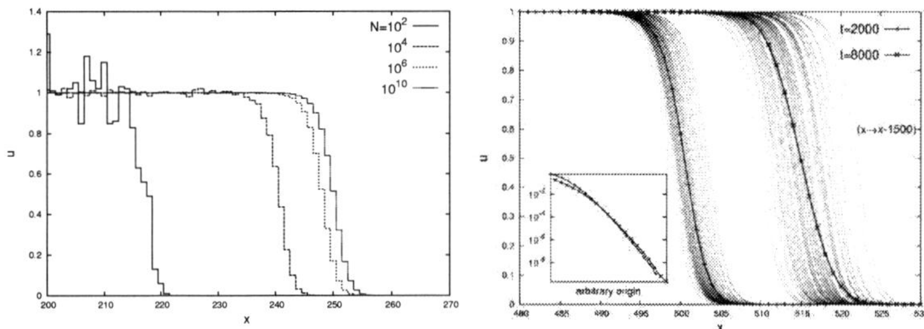

FIGURE 2. Left: Numerical integration of the toy model over 1000 units of time for different values of N. Right: 1000 realizations of the evolution of the toy model between time 0 and $t_1 = 2000$ (left bunch of curves), and $t_2 = 8000$ (right bunch of curve). *Insert:* the average front for these two times.

of particles that diffuse in space, multiply and recombine so that on average there are no more than N particles on each site. The QCD amplitude T corresponds to the fractional occupation number $u = n/N$ in the reaction–diffusion model. In the continuum limit, u obeys the so-called Reggeon field theory equation (which is in the same equivalence class as the stochastic FKPP equation)

$$\partial_t u(x,t) = D\partial_x^2 u(x,t) + \lambda u(x,t) - \mu u^2(x,t) + \sqrt{(2/N)u(x,t)}\,\eta(x,t)\,, \qquad (3)$$

where η is a Gaussian white noise function. We take this as a model of QCD scattering at high energy. The model does not give the exact behavior of the scattering, but should describe its gross features in the saturation limit. The second order differential operator should be replaced with the BFKL kernel in real QCD. See [5, 7] for a detailed discussion of the correspondence to QCD.

Eq. (3) describes the scattering off one partonic realization of the target. The amplitude u is a random variable, which fluctuates between different realizations of a scattering. To get the physical amplitude one must take the average of all realizations.

Taking into account the properties of the noise term leads to a hierarchy of equations for correlators similar to the Balitsky hierarchy, which, however, has some extra boundary terms that are important (see [5]). The mean field approximation leads, in this case, to the FKPP equation, in analogy with the BK case.

To study this numerically, we construct a toy model with a number N of particles on a one-dimensional lattice, that are allowed to jump to the neighboring sites, to multiply, and to disappear with certain probabilities. This type of model has been extensively studied in statistical physics [8] and allows straightforward simulation. For the exact description of the model, see [5]. Here, we just note that the number N corresponds to $1/\alpha_s^2$ in QCD, so the many-particle limit corresponds to small α_s.

The left plot in Fig. 2 shows the evolution of one realization of the numerical model for different N. The fronts have fluctuations around the steady front shape without fluctuations, with smaller fluctuations for larger N.

The large time velocity of the wave front for a system described by the model (3) has been shown to be [9] $v = v_0 - c/\ln^2 N$, where v_0 is the velocity in the absence of fluctuations, and c is a constant. Fig. 3 shows the velocity of one realization as a function

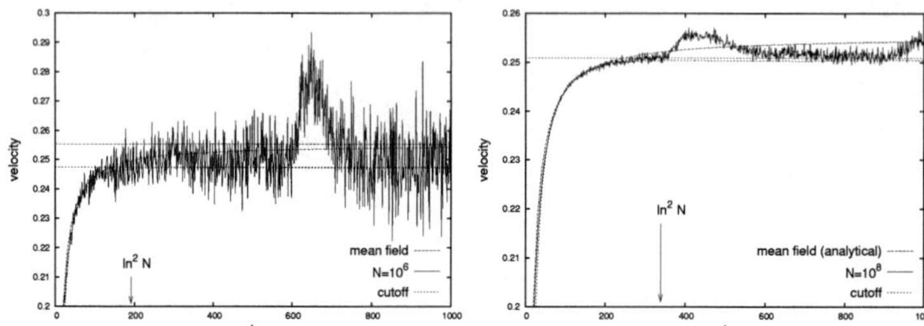

FIGURE 3. Instantaneous velocity of the front as a function of time (full fluctuating line) from a numerical solution of the stochastic evolution equation for $N = 10^6$ and $N = 10^8$ particles per site.

of time for two different N, showing appreciable fluctuations. The dashed curves show the analytical predictions; for large times the simulation gets closer to the asymptotic value; for smaller times the mean field prediction is still good. This also means that geometric scaling is expected to hold for intermediate times (or rapidities in QCD).

The right plot in Fig. 2, finally, shows the breaking of geometric scaling that arises from the averaging. Since the position of the front at a fixed time fluctuates between different realizations, the average will have a different shape than each individual curve. This is illustrated in the plot at two different times. At the earlier time the breaking is not so large, but at the larger time it is much more pronounced, as is the spread in position.

Although the mean field approximation seems to remain useful for intermediate, phenomenologically accessible energies (see also [10]), it is important to understand the corrections to this picture. The main interest of the statistical approach described above is to get a simple physical understanding of the fluctuations in the (extended) Balitsky-JIMWLK equations, and to make it possible to make a simple derivation of the universal asymptotics of the scattering amplitude at high energy. The simplicity comes from the reaction–diffusion nature of QCD scattering at high energy.

Acknowledgment This talk is based on work done in collaboration with Krzysztof Golec-Biernat and Stéphane Munier [5].

REFERENCES

1. I. Balitsky, Nucl. Phys. **B463** (1996) 99; Phys. Rev. Lett. **81** (1998) 2024; Phys. Lett. **B518** (2001) 235; Y.V. Kovchegov, Phys. Rev. **D60** (1999) 034008; Phys. Rev. **D61** (2000) 074018.
2. A. M. Stasto, these proceedings, hep-ph/0506103; A. Kovner, these proceedings, hep-ph/0506279.
3. S. Munier and R. Peschanski, Phys. Rev. Lett. **91** (2003) 232001; Phys. Rev. **D69** (2004) 034008; Phys. Rev. D **70**, 077503 (2004).
4. S. Munier, BNL-73263-2004; hep-ph/0501149.
5. R. Enberg, K. Golec-Biernat and S. Munier, hep-ph/0505101.
6. E. Iancu, A. H. Mueller and S. Munier, Phys. Lett. B **606**, 342 (2005).
7. E. Iancu and D. N. Triantafyllopoulos, Nucl. Phys. A **756**, 419 (2005); Phys. Lett. B **610**, 253 (2005).
8. For a recent review, see D. Panja, Phys. Rept. **393** (2004) 87.
9. E. Brunet and B. Derrida, Phys. Rev. **E56** (1997) 2597; Comp. Phys. Comm. **121-122** (1999) 376; J. Stat. Phys. **103** (2001) 269.
10. G. Soyez, "Fluctuations effects in high-energy evolution of QCD," hep-ph/0504129.

Nonlinear evolution equations in QCD and effective Hamiltonian at high energy

Anna M. Staśto

Physics Department, Brookhaven National Laboratory, Upton, NY 11973, USA
and
H. Niewodniczański Institute of Nuclear Physics, 31-342 Kraków, Poland

Abstract. In this talk I briefly present recent developments in the theory of the Color Glass Condensate. The duality between the dense and dilute regimes of the gluon field is discussed as well as the effective selfdual Hamiltonian which includes both Pomeron merging and Pomeron splitting.

Keywords: small x, Pomeron, high gluon density, JIMWLK equation
PACS: 12.38-t,12.38.Bx

Introduction: high energy limit and the Pomeron

One of the central problems in the theory of strong interactions is the understanding of the behaviour of the hadronic cross sections in the limit of high energies. The experimental data on the total cross section show slow but distinct increase with energy. It has been suggested that this rise, which can be parametrized by a power-like form,

$$\sigma(s) \sim s^{0.08}, \tag{1}$$

is mediated by the 'soft Pomeron' [1]. Perturbative calculation summing the leading logarithmic contributions of large logs $\ln s/t$ results in the BFKL Pomeron, which however leads to a stronger rise of the cross section with energy [2]

$$\sigma(s) \sim s^{4\ln 2\,\alpha_s N_c/\pi}, \tag{2}$$

with the leading exponent being much larger than in (1) when evaluated at physically interesting values of the strong coupling constant ($\alpha_s \simeq 0.2$). In any of these cases, the power-like increase of the cross section is in contradiction with the Froissart bound [3] which allows at most logarithmic increase with energy

$$\sigma(s) \sim \sigma_0 \ln^2 s, \tag{3}$$

with a normalization coefficient σ_0 related to the inverse of the pion mass squared. Froissart bound stems from the very general principles: unitarity and the finite range of the strong interactions. Thus, the important question arises whether it is possible to identify and calculate other type of Feynman diagrams which will lead to the restoration of the unitarity.

CP792, *Deep Inelastic Scattering, DIS 2005,* edited by W. H. Smith and S. R. Dasu

Color Glass Condensate and the JIMWLK equation

The Color Glass Condensate is an effective theory of the strong interactions at very high energies. Apart from the standard BFKL evolution of the gluon density it also contains the recombination diagrams which are important when the gluon density becomes very high. It therefore describes the phenomenon known as perturbative parton saturation. When the density of gluons becomes very high due to their enhanced splitting described by the BFKL Pomeron, the recombination effects start to become important and reduce the growth. It is believed that this phenomenon is important for the restoration of the unitarity. The Color Glass Condensate theory describes the scattering of the projectile off a target which constitutes a dense system of soft gluons. The S-matrix for the scattering of the quark-antiquark dipole in the target field α is described by

$$S(\boldsymbol{x},\boldsymbol{y}) \;=\; \frac{1}{N_c}\mathrm{Tr}(V_{\boldsymbol{x}}V_{\boldsymbol{y}}^{\dagger})\;, \tag{4}$$

where Wilson line

$$V_{\boldsymbol{x}} \;=\; P\exp\Big(ig\int dx^-\,\alpha_a(x^-,\boldsymbol{x})\,t^a\Big)\;, \tag{5}$$

is the path ordered exponential along the trajectory of the projectile. The physical scattering matrix between the $q\bar{q}$ dipole and the hadron target is obtained by taking the average of (4) over the color field of the target

$$\langle S(\boldsymbol{x},\boldsymbol{y})\rangle_\tau \;=\; \int D[\rho]\,\mathscr{Z}[\rho]_\tau\,S(\boldsymbol{x},\boldsymbol{y})\;, \tag{6}$$

where ρ is a color charge which generates field α. One cannot compute the weight function $\mathscr{Z}[\rho]$ since it contains nonperturbative information about the hadron but one can compute the evolution of this weight function with increasing rapidity $\tau \sim \ln s$. The basic equation of the Color Glass Condensate theory is the JIMWLK equation [4] which governs the evolution of the weight function $\mathscr{Z}[\rho]_\tau$ with rapidity

$$\frac{\partial\mathscr{Z}[\rho]_\tau}{\partial\tau} \;=\; H_{JIMWLK}\,\mathscr{Z}[\rho]_\tau\;, \tag{7}$$

where H is the JIMWLK Hamiltonian. The JIWMLK equation is a very complicated functional evolution equation, which however reduces to one closed and relatively simple equation, the Balitsky-Kovchegov equation, in the large N_c limit and in the dipole picture. Diagramatically JIMWLK equation contains BFKL Pomeron ladder diagrams as well as the triple Pomeron interaction diagrams which reduce the growth of the gluon density. More precisely it contains Pomeron merging diagrams, however it misses the Pomeron splittings. One can understand it by looking at the structure of the operator H_{JIMWLK}

$$H_{JIMWLK} \;=\; \frac{1}{2\pi}\int_{\boldsymbol{x},\boldsymbol{y},\boldsymbol{z}} K_{\boldsymbol{xyz}}\,\frac{\delta}{\delta\alpha^a(\boldsymbol{x})}\Big[1+\tilde{V}_{\boldsymbol{x}}^{\dagger}\tilde{V}_{\boldsymbol{y}}-\tilde{V}_{\boldsymbol{z}}^{\dagger}\tilde{V}_{\boldsymbol{y}}-\tilde{V}_{\boldsymbol{x}}^{\dagger}\tilde{V}_{\boldsymbol{z}}\Big]^{ab}\,\frac{\delta}{\delta\alpha^b(\boldsymbol{y})}\,, \tag{8}$$

which is second order in derivatives $\frac{\delta}{\delta\alpha}$ and all orders (through $\tilde{V}$'s)[1] in the field α. That means, that the evolution of the correlation functions in the field α can be (by using (6,7,8)) schematically represented as

$$\frac{\partial \overbrace{\langle \alpha \ldots \alpha \rangle}^{n}}{\partial \tau} = \sum_{m \geq n} \mathscr{K}_m \otimes \overbrace{\langle \alpha \ldots \alpha \rangle}^{m} . \tag{9}$$

The evolution of n point functions is coupled to the higher order correlation functions but not to the lower correlation functions.

Dual version of JIMWLK and effective Hamiltonian

In order to include the additional diagrams corresponding to the Pomeron splittings in the Color Glass Condensate formalism, terms which are higher order in derivatives have to be incorporated [5]. By including these terms in the evolution, the Pomeron loops are generated, which are known to be important contributions at very high energy. It has been suggested [6], that in general there is a duality relation between the evolution of the dense system, governed by the JIWMLK equation, and the dilute regime. In particular, the Pomeron mergings and splittings are believed to be the dominant diagrams in the dense and the dilute regimes, respectively, of the gluon field. In particular the evolution in the dilute regime is governed by the dual equation to the JIMWLK evolution. Formally, the duality can be expressed as the transformation [6, 7]

$$x^- \leftrightarrow x^+, \quad \frac{\delta}{\delta\alpha^a(x^-,\boldsymbol{x})} \leftrightarrow i\rho^a(x^+,\boldsymbol{x}), \quad \alpha^a(x^-,\boldsymbol{x}) \leftrightarrow -i\frac{\delta}{\delta\rho^a(x^+,\boldsymbol{x})} , \tag{10}$$

where ρ is the charge of the target and the α is the field generated by this charge measured by the projectile. The dual version of the JIMWLK equation is then

$$\frac{\partial \mathscr{Z}[\rho]_\tau}{\partial \tau} = \tilde{H}\, \mathscr{Z}[\rho]_\tau , \tag{11}$$

where the dual Hamiltonian

$$\tilde{H} = -\frac{1}{2\pi}\int_{\boldsymbol{x},\boldsymbol{y},\boldsymbol{z}} K_{\boldsymbol{xyz}}\, \rho^a(\boldsymbol{x}) \left[1 + \tilde{W}_{\boldsymbol{x}}^\dagger \tilde{W}_{\boldsymbol{y}} - \tilde{W}_{\boldsymbol{z}}^\dagger \tilde{W}_{\boldsymbol{y}} - \tilde{W}_{\boldsymbol{x}}^\dagger \tilde{W}_{\boldsymbol{z}}\right]^{ab} \rho^b(\boldsymbol{y}) . \tag{12}$$

The Wilson line $\tilde{W}$ is given by

$$\tilde{W}_{\boldsymbol{x}} = P\exp\left(g\int dx^+ \frac{\delta}{\delta\rho_a(x^+,\boldsymbol{x})} T^a\right) , \tag{13}$$

where now the integration and ordering is in the other light-cone variable x^+. The above evolution equation thus includes the process of Pomeron splitting by evolving

[1] $\tilde{V}$ denotes the Wilson line in the adjoint representation.

the weight function of the target. The duality can be also rexpressed in term of the symmetry between the projectile and target, what is the splitting from the target side, can be interpreted as a Pomeron merging from the projectile side. The general Hamiltonian at high energy should therefore include both of these processes at the same time. It can be shown [7] that it is entirely expressed in terms of Wilson lines along x^- and x^+ directions

$$H_{\rm eff} = \frac{1}{2\pi g^2 N_c} \int_{\boldsymbol{x}} {\rm Tr}\big[\tilde{V}_\infty(\partial^i\tilde{W}_\infty)(\partial^i\tilde{V}^\dagger_{-\infty})\tilde{W}^\dagger_{-\infty} + \tilde{V}_\infty\tilde{W}_\infty(\partial^i\tilde{V}^\dagger_{-\infty})(\partial^i\tilde{W}^\dagger_{-\infty}) + (\partial^i\tilde{W}^\dagger_{-\infty})(\partial^i\tilde{V}_\infty)\tilde{W}_\infty\tilde{V}^\dagger_{-\infty} + \tilde{W}^\dagger_{-\infty}(\partial^i\tilde{V}_\infty)(\partial^i\tilde{W}_\infty)\tilde{V}^\dagger_{-\infty}\big] \,, \qquad (14)$$

where the infinity limits in the Wilson lines $\tilde{V}$ and $\tilde{W}$ correspond to the limits in the x^+ and x^- lightcone variables correspondingly. This Hamiltonian is reminiscent of the nonlinear sigma model for the effective theory at high energy, proposed in [8]. The important point to note here however, is that $\tilde{V}$ and $\tilde{W}$ are interpreted as operators which are the functions of fields and derivatives of fields, respectively, with very nontrivial commutation relations. The effective Hamiltonian (14) has an interesting symmetry property, namely it is invariant under the following transformation

$$\tilde{W}_{-\infty} \to \tilde{V}_{-\infty}, \qquad \tilde{V}_{-\infty} \to \tilde{W}^\dagger_\infty, \qquad \tilde{W}^\dagger_\infty \to \tilde{V}^\dagger_\infty, \qquad \tilde{V}^\dagger_\infty \to \tilde{W}_{-\infty}, \qquad (15)$$

which is refferred to as the selfduality property. The effective Hamiltonian reduces to JIMWLK (8) and its dual (12) when the Wilson lines $\tilde{W}$ and $\tilde{V}$ are expanded to the first nontrivial order in derivatives and fields, respectively.

ACKNOWLEDGMENTS

The results presented in this talk have been obtained in the collaboration with Y.Hatta, E.Iancu, L.McLerran and D.Triantafyllopoulos [7]. This research has been supported by the U. S. Department of Energy, Contract No. DE-AC02-98CH10886 and by the Polish Committee for Scientific Research, KBN Grant No. 1 P03B 028 28.

REFERENCES

1. A. Donnachie, P.V. Landshoff, *Phys. Lett.* **B296** (1992) 227.
2. L. N. Lipatov, *Sov. J. Nucl. Phys.* **23** (1976) 338;
   E. A. Kuraev, L. N. Lipatov and V. S. Fadin, *Sov. Phys. JETP* **45** (1977) 199;
   I. I. Balitsky and L. N. Lipatov, *Sov. J. Nucl. Phys.* **28** (1978) 338.
3. M. Froissart, *Phys. Rev.* **123** (1961) 1053.
4. J. Jalilian-Marian, A. Kovner, A. Leonidov and H. Weigert, *Nucl. Phys.* **B 504** (1997) 415; *Phys. Rev.* **D 59** (1999) 014014. J. Jalilian-Marian, A. Kovner and H. Weigert, *Phys. Rev.* **D 59** (1999) 014015; E. Iancu, A. Leonidov and L. McLerran, *Nucl.Phys.* **A692** (2001) 583.
5. A.H. Mueller, A.I. Shoshi and S.M.H. Wong, *Nucl. Phys.* **B715** (2005) 440.
6. A. Kovner and M. Lublinsky, *Phys. Rev. Lett.* **94** (2005) 181603.
7. Y. Hatta, E. Iancu, L. McLerran, A.M. Stasto, D.N. Triantafyllopoulos, hep-ph/0504182.
8. H. Verlinde and E. Verlinde, hep-th/9302104.

From dense-dilute duality to self duality in high energy evolution

Alex Kovner

Physics Department, University of Connecticut, Storrs, CT 06269, USA

Abstract. I describe recent work on inclusion of Pomeron loops in the high energy evolution. In particular I show that the complete eikonal high energy evolution kernel must be selfdual

Keywords: Pomeron loops, high energy evolution
PACS: 12.38Lg, 12.40Nn

Last year has seen renewed attempts to understand Pomeron loop contributions to the high energy evolution of hadronic cross sections in QCD. In recent years the study of the high energy scattering has centered around the so called JIMWLK evolution equation [1, 2, 3]. It describes the approach of the scattering amplitude to saturation due to multiple scattering corrections on dense hadronic targets, or in the diagrammatic language, the fan diagrams. The JIMWLK equation however only partially takes into account the processes whereby the gluons emitted in the projectile wave function at an early stage of the evolution, are "bleached" by subsequently emitted gluons, or the so-called Pomeron loops[4],[5],[6],[7],[8].

Recently we have calculated corrections to the JIMWLK equation, which take into account some finite density effects in the projectile wave function (or equivalently, resum certain corrections away form the dense limit of the the target)[9]. We have also derived the evolution equation valid for dilute target, which is the opposite limit to that considered in JIMWLK[7]. The most striking feature of the two results, is that they appear to be dual to each other. The improved JIMWLK equation is given by[9]

$$\begin{aligned}\chi^{\rm JIMWLK+} &= \frac{1}{2\pi}\int_z\Big\{b_i^a(z,0,[\frac{\delta}{\delta\alpha}])b_i^a(z,0,[\frac{\delta}{\delta\alpha}])+b_i^a(z,1,[\frac{\delta}{\delta\alpha}])b_i^a(z,1,[\frac{\delta}{\delta\alpha}])\\ &- 2\,b_i^a(z,0,[\frac{\delta}{\delta\alpha}])\left[\mathscr{P}e^{i\int_0^1 dx^- T^c\alpha^c(x^-,z)}\right]^{ab} b_i^b(z,1,[\frac{\delta}{\delta\alpha}])\Big\}\end{aligned} \quad (1)$$

where $\mathscr{P}$ denotes path ordering with respect to x^- and the field b_i^a satisfies the "classical" equation of motion[9]. The low density limit evolution kernel (KLWMIJ) including the same type of corrections but in the target wave function derived in [7] is[1]:

$$\chi^{\rm KLWMIJ+} = -\frac{1}{2\pi}\int_z\{b_i^a(z,0,[\rho])b_i^a(z,0,[\rho])+b^a(z,1,[\rho])b^a(z,1,[\rho])$$

[1] Following the work [7], the same expression has also been obtained in [10] using the effective action techniques.

CP792, *Deep Inelastic Scattering, DIS 2005,* edited by W. H. Smith and S. R. Dasu

$$- \quad 2b_i^a(z,0,[\rho]) \left[\mathscr{P} e^{\int_0^1 dx^- T^c \frac{\delta}{\delta \rho^c(x^-,z)}} \right]^{ab} b_i^b(z,1,[\rho])\} . \qquad (2)$$

The two kernels are strikingly similar which suggests an intriguing duality between the high and the low density limits of the evolution kernel.

In this contribution I follow [8] and show that indeed the full eikonal kernel for the high energy evolution must satisfy the property of self duality. The requirement that the evolution of the projectile and the target wave functions has the same functional form coupled with the requirement of Lorentz invariance of the scattering matrix, leads to the condition that the kernel of the evolution $\chi[\rho, \frac{\delta}{\delta\rho}]$ must satisfy

$$\chi[\alpha, \frac{\delta}{\delta\alpha}] = \chi[-i\frac{\delta}{\delta\rho}, i\rho] . \qquad (3)$$

where ρ is the charge density in the target wave function and α is defined by[2, 3]

$$\begin{aligned} \alpha^a(x,x^-)T^a &= \frac{1}{\partial^2}(x-y) \left\{ U^\dagger(y,x^-)\, \rho^a(y,x^-) T^a\, U(y,x^-) \right\}, \\ U(x,x^-) &= \mathscr{P} \exp\{ i \int_{-\infty}^{x^-} dy^- T^a \alpha^a(x,y^-) \} . \end{aligned} \qquad (4)$$

I note that from the functional integral point of view this duality has been discussed earlier in [11].

Consider the general expression for the S-matrix of a projectile with the wave function $|P\rangle$ scattering on a target with the wave function $|T\rangle$[8], where the total rapidity of the process is Y. The projectile is assumed to be moving to the left with rapidity $Y - Y_0$ (and thus has sizeable color charge density ρ^-), while the target is moving to the right with rapidity Y_0 (and has large ρ^+). We assume that the projectile and the target contain only partons with large k^- and k^+ momenta respectively: $k^- > \Lambda^-$ and $k^+ > \Lambda^+$. The eikonal expression for the S-matrix reads

$$\mathscr{S}_Y = \int D\rho^{+a}(x,x^-)\, W^T_{Y_0}[\rho^+(x^-,x)]\, \Sigma^P_{Y-Y_0}[\alpha] , \qquad (5)$$

where Σ^P is the S-matrix averaged over the projectile wave function

$$\Sigma^P[\alpha] = \langle P| \mathscr{P} e^{i\int dx^- \int d^2x \hat{\rho}^{-a}(x)\alpha^a(x,x^-)} |P\rangle . \qquad (6)$$

where $W^T[\alpha]$ is the weight function representing the target, which is related to the target wave function in the following way: for an arbitrary operator $\hat{O}[\hat{\rho}^+]$

$$\langle T| \hat{O}[\hat{\rho}^+(x)] |T\rangle = \int D\rho^{+a}\, W^T[\rho^+(x^-,x)]\, O[\rho^+(x,x^-)] . \qquad (7)$$

The field $\alpha(x)$ is the A^+ component of the vector potential in the light cone gauge $A^- = 0$. This is the natural gauge from the point of view of partonic interpretation of the projectile wave function. In the formulae above we use hats to denote quantum operators.

Note that the quantum operators $\hat{\rho}^{-a}(x)$ and $\hat{\rho}^{+a}(x)$ do not depend on longitudinal coordinates, but only on transverse coordinates x. The "classical" variables α and ρ^+ on the other hand do depend on the longitudinal coordinate x^-. This dependence, as discussed in detail in [7] arises due to the need to take correctly into account the proper ordering of noncommuting quantum operators. Thus the ordering of the quantum operators $\hat{\rho}^+$ in the expansion of $\hat{O}$ in the lhs of eq.(7) translates into the same ordering with respect to the longitudinal coordinate x^- of $\rho^+(x^-)$ in the expansion of $O[\rho^+(x^-)]$ in the rhs of eq.(7).

As shown in [7] the functional $W^T[\alpha]$ cannot in general be interpreted as probability density, as it contains a complex factor. This factor - the Wess-Zumino term, ensures correct commutators between the quantum operators $\hat{\rho}^a$. In the present derivation we do not require an explicit form of this term, but the following property which is implicit in eq. (7) is crucial to our discussion. The "correlators" of the charge density $\langle \rho^{a_1}(x_1,x_1^-)...\rho^{a_n}(x_n,x_n^-)\rangle$ do not depend on the values of the longitudinal coordinates x_i^-, but only on their ordering[7].

Note that one can define an analog of W^T for the wave function of the projectile via

$$\langle P|\hat{O}[\hat{\rho}^-(x)]\,|P\rangle = \int D\rho^{-a}\,W^P[\rho^-]\,O[\rho^-(x,x^-)]\,. \tag{8}$$

With this definition it is straightforward to see that Σ^P and W^P are related through a functional Fourier transform. To represent Σ as a functional integral with weight W^P we have to order the factors of the charge density $\hat{\rho}^-$ in the expansion of eq.(6), and then endow the charge density $\hat{\rho}^-(x)$ with an additional coordinate t to turn it into a classical variable. This task is made easy by the fact that the ordering of $\hat{\rho}$ in eq.(6) follows automatically the ordering of the coordinate x^- in the path ordered exponential. Since the correlators of $\rho(x,t_i)$ with the weight W^P depend only on the ordering of the coordinates t_i and not their values, we can simply set $t = x^-$. Once we have turned the quantum operators $\hat{\rho}$ into the classical variables $\rho(x^-)$, the path ordering plays no role anymore, and we thus have

$$\Sigma^P(\alpha) = \int D\rho^a\,W^P[\rho]\,e^{i\int dx^-\int d^2x\rho^a(x,x^-)\alpha^a(x,x^-)}\,. \tag{9}$$

We now turn to the discussion of the evolution. The evolution to higher energy can be achieved by boosting either the projectile or the target. The resulting S-matrix should be the same. This is required by the Lorentz invariance of the S-matrix. Consider first boosting the projectile by a small rapidity δY. This transformation leads to the change of the projectile S-matrix Σ of the form

$$\frac{\partial}{\partial Y}\Sigma^P = \chi^\dagger[\alpha,\frac{\delta}{\delta\alpha}]\,\Sigma^P[\alpha] \tag{10}$$

Substituting eq.(10) into eq.(5) we have

$$\begin{aligned}\frac{\partial}{\partial Y}\mathscr{S}_Y &= \int D\rho^{+a}(x,x^-)\,W^T_{Y_0}[\rho^+(x^-,x)]\,\left\{\chi^\dagger[\alpha,\frac{\delta}{\delta\alpha}]\,\Sigma^P_{Y-Y_0}[\alpha]\right\}\\ &= \int D\rho^{+a}(x,x^-)\,\left\{\chi[\alpha,\frac{\delta}{\delta\alpha}]\,W^T_{Y_0}[\rho^+(x^-,x)]\right\}\,\Sigma^P_{Y-Y_0}[\alpha]\,.\end{aligned} \tag{11}$$

Where the second equality follows by integration by parts. We now impose the requirement that the S-matrix does not depend on Y_0 [12]. Since Σ in eq.(5) depends on the difference of rapidities, requiring that $\partial\mathscr{S}/\partial Y_0 = 0$ we find that W should satisfy

$$\frac{\partial}{\partial Y}W^T = \chi[\alpha, \frac{\delta}{\delta\alpha}]\, W^T[\rho^+] \tag{12}$$

Thus we have determined the evolution of the target eq.(12) by boosting the projectile and requiring Lorentz invariance of the S-matrix. On the other hand the extra energy due to boost can be deposited in the target rather than in the projectile. How does W^T change under boost of the target wave function? To answer this question we consider the relation between Σ and W together with the evolution of Σ. Referring to eqs.(9) and (10) it is obvious that multiplication of Σ^P by α is equivalent to acting on W^P by the operator $-i\delta/\delta\rho$, and acting on Σ^P by $\delta/\delta\alpha$ is equivalent to multiplying W^P by $i\rho$. Additionally, the action of $i\rho$ and $-i\delta/\delta\rho$ on W^P must be in the reverse order to the action of $\delta/\delta\alpha$ and α on Σ^P. This means that the evolution of the functional W^P is given by

$$\frac{\partial}{\partial Y}W^P = \chi[-i\frac{\delta}{\delta\rho}, i\rho]\, W^P[\rho]\,. \tag{13}$$

Although eq.(13) refers to the weight functional representing the projectile wave function, clearly the functional form of the evolution must be the same for W^T. Comparing eq.(12) and eq.(13) we find that the high energy evolution kernel must, as advertised, satisfy the selfduality relation eq.(3). This is the main result.

The selfduality of the kernel is somewhat similar (although different in detail) to the duality symmetry of a harmonic oscillator Hamiltonian $p \to x,\ x \to -p$. One thus hopes that it may eventually be of help in solving the complete evolution equation, once it is derived.

REFERENCES

1. I. Balitsky, *Nucl. Phys.* **B463**:99,1996; *Phys. Rev. Lett.* **81**:2024-2027,1998; Phys. Rev. D60:014020,1999.
2. J. Jalilian Marian et.al. *Nucl. Phys.* **B504** (1997) 415; *Phys. Rev.* **D59** (1999) 014014; *Phys. Rev.* **D59** (1999) 014015; A. Kovner and J.G. Milhano, *Phys. Rev.* **D61** (2000) 014012. A. Kovner, J.G. Milhano and H. Weigert, *Phys. Rev.* **D62**:114005,2000; H. Weigert, Nucl. Phys. A703 (2002) 823.
3. E. Iancu et.al., *Phys. Lett.* **B510** (2001) 133; *Nucl. Phys.* **A692** (2001) 583; *Nucl. Phys.* **A703** (2002) 489.
4. E. Iancu and A. H. Mueller, *Nucl. Phys.* **A730** (2004) 460, 494.
5. A. Mueller and A. Shoshi, Nucl. Phys. B692 (2004) 175.
6. E. Iancu and D. N. Triantafyllopoulos, *Phys.Lett.* **B610**:253-261,2005; A. Mueller, A. Shoshi and S. Wong, *Nucl.Phys.* **B715**:440-460,2005 ; E. Levin and M. Lublinsky, hep-ph/0501173.
7. A. Kovner and M. Lublinsky, *Phys.Rev.***D.71**:085004,2005
8. A. Kovner and M. Lublinsky, *Phys.Rev.Lett.* **94**:181603,2005
9. A. Kovner and M. Lublinsky, *JHEP* **0503**:001,2005
10. Y. Hata et.al. hep-ph/0504182.
11. I. Balitsky , In *Shifman, M. (ed.): At the frontier of particle physics, vol. 2* 1237-1342, hep-ph/0101042
12. E. Levin and M. Lublinsky, *Phys. Lett.***B607**:131-138,2005l;

Development of Chaos in the Color Glass Condensate

Kirill Tuchin

Nuclear Theory Group,
Physics Department,
Brookhaven National Laboratory,
Upton, NY 11973-5000, USA

Abstract. Noting that the number of gluons in the hadron wave function is discrete, and their formation in the chain of small x evolution occurs over discrete rapidity intervals of $\Delta y \simeq 1/\alpha_s$, we formulate the discrete version of the Balitsky–Kovchegov evolution equation and show that its solution behaves chaotically in the phenomenologically interesting kinematic region.

Keywords: small-x, evolution equation, chaos
PACS: 24.60.Lz,12.38.-t

EVOLUTION AS A DISCRETE PROCESS

The color field of an ultra–relativistic hadron is a quasi-classical non-Abelian Weizsacker-Williams field [1, 2]. It emerges when the occupation number of the bremsstrahlung gluons emitted at a given impact parameter exceeds unity and eventually saturates at $\sim 1/\alpha_s$. It has been argued that in a big nucleus, such that $\alpha_s A^{1/3} \gg 1$, and not very high energies the mean-field treatment is a reasonable approximation to the evolution equations.

As a result of broken scale symmetry of QCD there exist the dimensional scale Λ which is the infrared cutoff on the gluon's momenta. An introduction of an infrared cutoff Λ on the momentum of the emitted gluons amounts to imposing the boundary condition. This is equivalent to the quantization of the gluon modes in a box of size $L \sim \Lambda^{-1}$, in which case the spectrum of the emitted gluons and their number become discrete. The formation of a gluon occurs over a rapidity interval of $\Delta y \simeq 1/\alpha_s$. Therefore, the evolution in rapidity can be considered as a discrete quantum process, where each subsequent step occurs when $\Delta y\, \alpha_s \simeq 1$ [3].

Assuming that $\Delta y\, \alpha_s$ is a certain number for all steps in evolution process neglects a stochastic nature of quantum evolution. Full treatment of the discrete BK equation requires taking these effects into account. However, unfortunately BK equation is known to resist all attempts of analytical solution, and our hope at present is to develop a meaningful approximation. Thus, we suggest an approximation in which the gluons are emitted over a fixed "time" defined by $\bar{\alpha}_s \Delta y = C$ with $C = 1$. To justify this assumption, let us note that BFKL takes into account only fast gluons, i.e. those with $C \sim 1$. It is beyond the leading logarithmic (LL) approximation to take into account slow gluons. Moreover, it is known that an account of NLL corrections effectively leads to imposing a rapidity veto [4] on the emission of gluons with close rapidities, which restricts

CP792, *Deep Inelastic Scattering, DIS 2005,* edited by W. H. Smith and S. R. Dasu

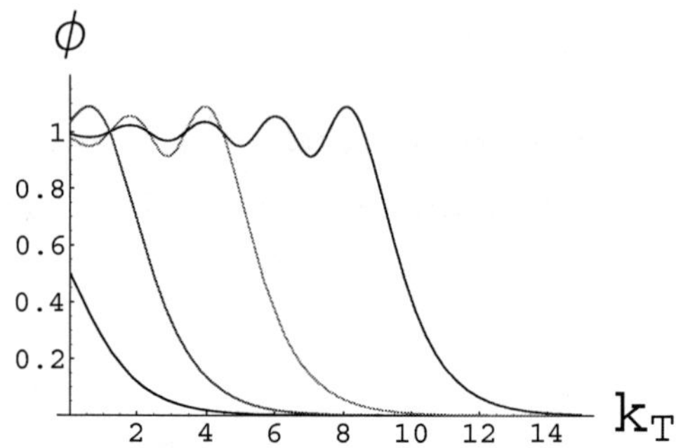

FIGURE 1. Discrete BK equation at $\omega = 2.8$. Different lines from left to right correspond to $n = 1,4,7,10$ (black, red, green, blue).

production of gluons with small C (this is due to an effective repulsion between the emitted gluons induced at the NLL level). Therefore C is bounded from below by a number close to one. On the other hand the probability that no gluon is emitted when C becomes larger than one is very small if we choose the high density initial condition, such as the one given by the McLerran-Venugopalan model[1]. Therefore, C takes random values around 1, but the effective dispersion can be expected quite small.

DISCRETE EVOLUTION EQUATION

Equation which describes the gluon evolution in the high parton density regime of QCD is the Balitsky-Kovchegov equation [5, 6]. This equation is formulated for the forward scattering amplitude of a color dipole of transverse momentum $\underline{k}$ and rapidity y to scatter of a big nucleus at impact parameter $\underline{b}$. Assumption that the dispersion of the dipole transverse momenta with y is a slow process one can develop a diffusion approximation to the BK equation. This approximation is meaningful in the saturation region. The discrete version of BK equation takes the form (see also [7])

$$\tilde{N}_{n+1}(\underline{k},y) = (1 + \chi(\gamma_0))\tilde{N}_n(\underline{k},y) - \tilde{N}_n^2(\underline{k},y)\,, \tag{1}$$

where we introduced the discrete variable n to enumerate emitted gluons and $\chi(\gamma_0) = 4\ln 2 \equiv \omega - 1$. It is convenient to re-scale the scattering amplitude $\phi_n = \chi(\gamma_0)\tilde{N}_n$ so that the corresponding continuous amplitude is normalized to $\tilde{N}_n(\underline{k},y) \leq 1$:

$$\phi_{n+1} = \omega\,\phi_n - (\omega - 1)\phi_n^2\,. \tag{2}$$

Numerical solution to the discrete BK equation is shown in Figures 1–4.

Let us now consider how does the evolution proceeds for various values of ω. In Fig. 1 the case of $1 < \omega < 3$ is shown. In that case, for any k there is one stable fixed point at $\phi_n = 1$ and one unstable fixed point at $\phi_n = 0$. The fixed points are determined from the condition $\phi_{n+1} = \phi_n$.

In Fig. 2 we consider the case $3 < \omega < 3.442\ldots$. The point $\phi_n = 1$ ceases to be unstable. Instead two new stable points appear. These can be determined from the

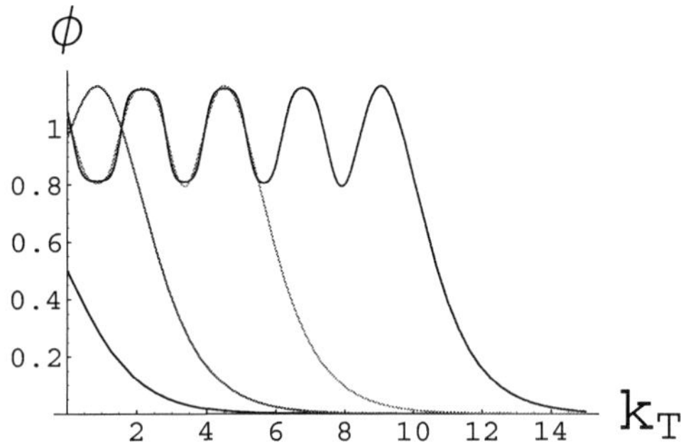

FIGURE 2. Discrete BK equation at $\omega = 3.1$. Different lines from left to right correspond to $n = 1,4,7,10$ (black, red, green, blue).

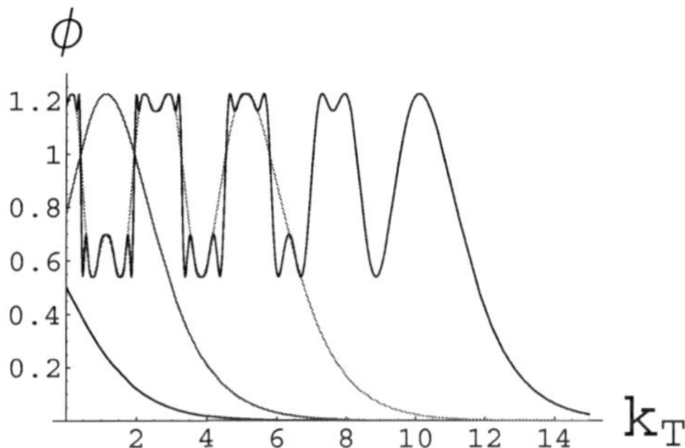

FIGURE 3. Discrete BK equation at $\omega = 3.495$. Different lines from left to right correspond to $n = 1,4,7,10$ (black, red, green, blue).

condition $\phi_{n+2} = \phi_n$. The described multiplication of stable points is called in general *bifurcation*. In our particular case it is referred to as the *period doubling scenario*.

When $3.442\ldots < \omega < 3.56\ldots$ there are four new fixed points, see Fig. 3. It is important to emphasize that at $n \to \infty$ the value of ϕ_n settles to a given set of fixed points (specified by the value of ω) independently of the initial condition. In other words, at very high energies the scattering amplitude in the saturation regime $k_T < Q_s(n)$ is independent of the initial condition.

ONSET OF CHAOS

The period doubling proceeds at increasingly smaller increments of ω until the *accumulation point* $\omega_F = 3.569\ldots$ known also as the *Feingenbaum's number*. At this point there is no more universal limiting behavior at large n. Instead small change in the initial condition leads to large change in the final state. The onset of this chaotic behavior can be observed in Fig. 4. Note now that at first, $\omega_{BFKL} = 1 + 4\ln 2 = 3.77 > \omega_F$ and at second, ω_{BFKL} is the absolute minimum of the function $1 + \chi(\gamma)$. Thus, we conclude

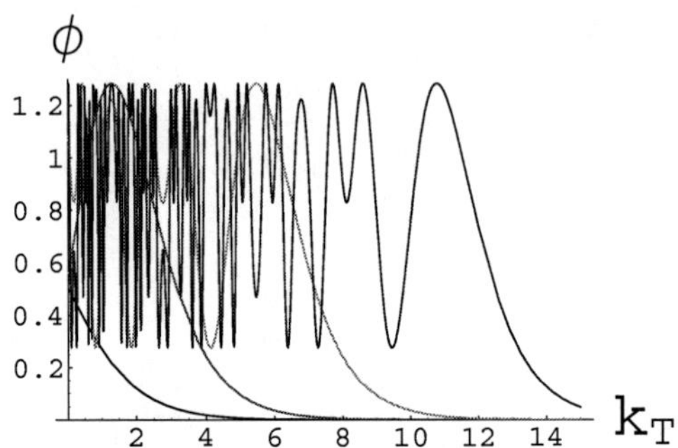

FIGURE 4. Discrete BK equation at $\omega = 3.77$. Different lines from left to right correspond to $n = 1, 4, 7, 10$ (black, red, green, blue).

that the evolution of the scattering amplitude at high energies in the saturation region may be chaotic. At the same time, it is evident from Figs. 1–4 that the high k_T tail of the scattering amplitude which describes the perturbative regime is not affected by the peculiar behavior of the discrete equation.

By averaging over all events one can define the mean value of the scattering amplitude. However, this procedure hides a lot of interesting physics. The most obvious example of this is diffraction, which measures the strength of fluctuations in the inelastic cross section. Figs. 1–4 imply that diffraction is a significant part of the total inelastic cross section at very high energies, and is universal (independent of the properties of the target).

The model used in this letter is admittedly oversimplified: we neglected the diffusion in transverse momentum, stochasticity of gluon emission and the dynamical fluctuations beyond the mean field approximation. Nevertheless, we hope that at least some of the features of discrete quantum evolution at small x will survive a more realistic treatment. The chaotic features of small x evolution open a new intriguing prospective on the studies of hadron and nuclear interactions at high energies.

ACKNOWLEDGMENTS

This research was done in the collaboration with Dima Kharzeev. I would like to thank Yuri Kovchegov, Alex Kovner, Misha Kozlov, Genya Levin, Larry McLerran, Al Mueller and Anna Stasto for informative and helpful discussions and comments. This research was supported by the U.S. Department of Energy under Grant No. DE-AC02-98CH10886. BNL preprint number is BNL-NT-05/24.

REFERENCES

1. L. D. McLerran and R. Venugopalan, Phys. Rev. D **49**, 3352 (1994) [arXiv:hep-ph/9311205]; Phys. Rev. D **49**, 2233 (1994) [arXiv:hep-ph/9309289]; Phys. Rev. D **50**, 2225 (1994) [arXiv:hep-ph/9402335].

2. Y. V. Kovchegov, Phys. Rev. D **54**, 5463 (1996) [arXiv:hep-ph/9605446]; Phys. Rev. D **55**, 5445 (1997) [arXiv:hep-ph/9701229].
3. D. Kharzeev and K. Tuchin, "Chaos in the color glass condensate," arXiv:hep-ph/0501271.
4. C. R. Schmidt, Phys. Rev. D **60**, 074003 (1999) [arXiv:hep-ph/9901397];
   J. R. Forshaw, D. A. Ross and A. Sabio Vera, Phys. Lett. B **455**, 273 (1999) [arXiv:hep-ph/9903390];
   G. Chachamis, M. Lublinsky and A. Sabio Vera, Nucl. Phys. A **748**, 649 (2005) [arXiv:hep-ph/0408333].
5. I. Balitsky, Nucl. Phys. B **463**, 99 (1996) [arXiv:hep-ph/9509348].
6. Y. V. Kovchegov, Phys. Rev. D **60**, 034008 (1999) [arXiv:hep-ph/9901281].
7. A. Bialas and R. Peschanski, arXiv:hep-ph/0502187.

Gluon Distributions and Fits Using Dipole Cross-Sections

Robert S. Thorne

Cavendish Laboratory, University of Cambridge, Madingley Road, Cambridge, CB3 0HE, UK

Abstract. I carry out a comparison between the gluon distribution obtained from a dipole picture analysis with structure function data and the standard DGLAP gluon. The former is smaller and steeper, and I explain the approximations that have resulted in this difference.

Keywords: <QCD,Structure Functions>
PACS: 12.38.Bx,13.60.Hb

There has recently been a lot of work on calculating/modelling dipole cross-sections and using the dipole model [1, 2, 3, 4] to fit to structure function data, and a variety of approaches match data very well. However, the picture of steeply growing quantities at small x tamed by saturation conflicts with the DGLAP picture of a small/negative gluon distribution at small x and Q^2 [5, 6]. We need to understand why this happens. My approach is based on the assumption that QCD factorization theory is correct and quantitative at high Q^2 (and not too small x), and then I work back to the dipole cross-sections. In this way I examine whether the dipole motivated fits are truly quantitative and whether a large/steep dipole cross-section means a large/steep gluon distribution. Any evidence for saturation is only a side issue. More details may be found in [7].

Within LO k_T-factorization theory the $\gamma^\star p$ cross-section can be written as [8]

$$\sigma(x,Q^2) \propto \int dz \int \frac{d^2k}{k^4} \int d^2p\tilde{\sigma}(z,p^2,Q^2)f(x,k^2)$$

where $f(x,k^2)$ is the unintegrated gluon distribution. In the limit $x \to 0$, i.e. LO in the k_T-factorization theory [9], this formula can be simplified. Integrating over z and p

$$F(x,Q^2) = \int \frac{d^2k}{k^2} \frac{\alpha_S 2N_f}{6\pi} h(k^2/Q^2)f(x,k^2).$$

Taking the double Mellin transformation $\int dQ^2 Q^{2-2\gamma}$ and $\int dx x^N$ we obtain,

$$\tilde{F}(N,\gamma) = \frac{\alpha_S 2N_f}{6\pi}\tilde{h}(\gamma)\tilde{f}(N,\gamma)/\gamma \equiv \alpha_S\tilde{h}(\gamma)\tilde{g}(N,\gamma).$$

$g(x,Q^2)=\int_0^{Q^2}\frac{d^2k}{k^2}f(x,k^2)$ is the integrated gluon distribution. If $g(N,Q^2)\sim(Q^2)^{\gamma(\alpha_S,N)}$

$$F(N,Q^2)=\frac{\alpha_S 2N_f}{6\pi}h(\gamma(\alpha_S,N))g(N,Q^2) \to F(x,Q^2)=\frac{\alpha_S 2N_f}{6\pi}h(\gamma(\alpha_S,\ln(1/x)))\otimes g(x,Q^2),$$

where $\gamma(\alpha_S/N)$ is a positive quantity, e.g. in LO BFKL [10] $\gamma(x) = x + 2.4x^4 + 2x^6 + 17x^7 + \cdots$.

CP792, *Deep Inelastic Scattering, DIS 2005,* edited by W. H. Smith and S. R. Dasu

Alternatively, taking a Fourier transformation with respect to p, with r the conjugate variable, integrating over p^2 and z, and letting $x \to 0$ one can equivalently write

$$\sigma = \frac{2\pi}{3}\int_0^1 dz \int d^2r |\Psi(r,z,Q)|^2 \int \frac{d^2k}{k^4}\alpha_S f(x,k^2)(1-J_0(kr)) \equiv \int_0^1 dz \int d^2r |\Psi(r,z,Q)|^2 \hat{\sigma}(x,r^2).$$

Taking the Mellin transformations leads to

$$F_i(N,Q^2) = \frac{\alpha_S 2N_f}{6\pi} h_{id}(\gamma(\alpha_S,N)) h_{dg}(\gamma(\alpha_S,N) g(N,Q^2) \equiv \frac{\alpha_S 2N_f}{6\pi} h_i(\gamma(\alpha_S,N)) \otimes g(N,Q^2).$$

So the effective coefficient function for the hard cross-section $h_i(\gamma(\alpha_S,N))$ is the product of a photon-dipole part $h_{id}(\gamma(\alpha_S,N))$ and a dipole-gluon part $h_{dg}(\gamma(\alpha_S,N))$, both of which are calculable. For $dF_2/d\ln Q^2$ [11], expanding in powers of γ

$$h_2(\gamma(\bar{\alpha}_S/N) = 1 + 2.17\gamma + 2.30\gamma^2 + 5.07\gamma^3 + 3.58\gamma^4 + 8.00\gamma^5 + \cdots.$$

$$h_{dg}(\gamma(\bar{\alpha}_S/N) = 1 + 2.23\gamma + 3.49\gamma^2 + 3.95\gamma^3 + 4.22\gamma^4 + 4.06\gamma^5 + \cdots$$

$$h_{2d}(\gamma(\bar{\alpha}_S/N) = 1 - 0.07\gamma - 1.05\gamma^2 + 3.77\gamma^3 - 4.94\gamma^4 + 6.53\gamma^5 + \cdots.$$

This leads to a steep growth of $dF_2/d\ln Q^2$ relative to the gluon, but this is all generated by the dipole-gluon cross-section, rather than the photon-dipole coefficient. Indeed, taking the simple Golec-Biernat Wüsthoff model [12] within this picture we find that a flat $\hat{\sigma}(x,r^2)$ comes from a valence-like $xg(x,Q^2)$ and $f_g(x,k^2)$.

In order to see how this works out in practice I perform a fit to data from the starting point of a well-defined gluon distribution that has the same shape in x and Q^2 as a standard LO or NLO gluon distribution and for $Q^2 \gg Q_0^2$ evolves in a quantitatively correct way. Clearly for $Q^2 \sim Q_0^2$ the evolution needs to be modified. In essence I just replace Q^2 by $Q^2 + Q_0^2$. $\alpha_S(\mu^2)$ is also modified in the same way. This expression for the gluon is converted into a dipole cross-section using

$$\hat{\sigma}(x,r^2) = \frac{2\pi}{3}\int \frac{d^2k}{k^4}\alpha_S(k^2) f(x,k^2)(1-J_0(kr)),$$

and put into a fit to data. The comparison of the gluon and the resulting dipole cross-section is shown in fig. 1. Clearly the dipole cross-section is generally steeper, particularly comparing $Q^2 = 0.2\text{GeV}^2$ with $r = 10\text{GeV}^{-1}$.

There are various details of the fit to consider. In reality 3 types of diagram contribute. A gluon can radiate gluons before entering into the scattering, or a quark can fluctuate into a gluon which enters, or a quark can scatter directly off the photon. In the LO k_T-factorization theorem the first two processes combine as $f_g(x,k^2) + 4/9 f_S(x,k^2)$, i.e. the quark also contributes to the dipole cross-section. For the last process I include a term $f \times Q^2/(Q^2+Q_0^2)$, where f is free (and in practice very small). Another very important issue is heavy quarks. These are often ignored in dipole fits, but charm constitutes about 40% of $dF_2/d\ln Q^2$ for $Q^2 > m_c^2$. Its omission leads to $\hat{\sigma}(x,r^2)$ and $g(x,Q^2)$ being up to 5/3 times too big. Since saturation corrections are $\propto g^2(x,Q^2)$ they can be enormously

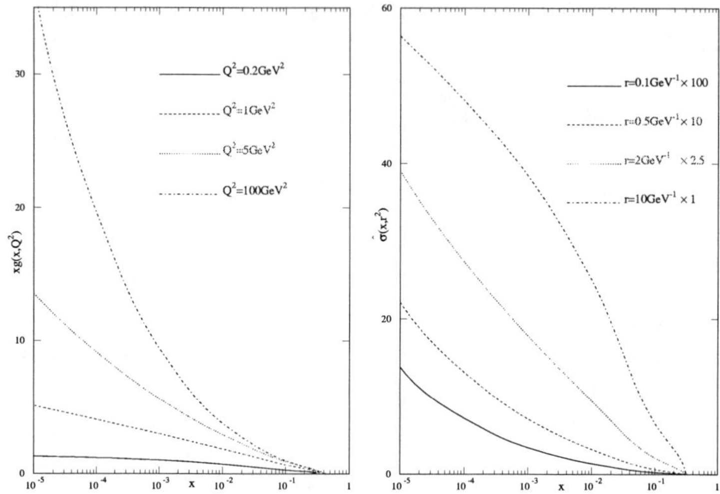

FIGURE 1. Comparison of $xg(x,Q^2)$ at various Q^2 with $\hat{\sigma}(x,r^2)$ at various r

exaggerated. To check the consequences of ignoring charm I performed a DGLAP fit without the charm contribution to $F(x,Q^2)$. The gluon and α_S are bigger, but also the global fit is terrible: $\chi^2 = 2$ per point for 2000 points. One cannot get $dF_2/d\ln Q^2$ consistently correct at all. Hence, NLO and NNLO DGLAP are good enough and constraining enough to determine that charm has to be there. This suggests that one should be suspicious of good qualitative results present without heavy quarks. They should really be wrong until corrected. In my fit I use $m_c = 1.3\text{GeV}$.

I fit H1 [13], ZEUS [14] and E665 [15] data from $0.5\text{GeV}^2 \leq Q^2 \leq 50\text{GeV}^2$. The fit does not work for $Q^2 < 0.5\text{GeV}^2$, perhaps suggesting saturation. The quality of fit, $\chi^2 = 1.1$ per point, is very good for 3 different data sets. In fig. 2 the resulting dipole fit gluon is compared to the MRSTNLO gluon. It is approximately $0.65-0.75$ of the size. It should really be compared to $g(x,Q^2)+4/9f_S(x,Q^2)$. The biggest change is at high x, and the factor is now $0.5-0.65$. The dipole gluon does not match onto the standard DGLAP gluon at high Q^2. At low Q^2 it is much smaller than the DGLAP gluon at moderate x but eventually becomes bigger at very small x.

This relative behaviour comes from the effective coefficient functions/splitting functions. Consider $dF_2/d\ln Q^2$ controlled by $\gamma^{DIS}(\alpha_S(Q^2),N)$. For my model $\gamma_{gg}(\alpha_s(Q^2),N) = \bar{\alpha}_s(Q^2)(1/N-1)$ (i.e. correct for DGLAP). This means that

$$\gamma^{DIS}_{dip}(\alpha_S(Q^2),N) \approx \frac{\alpha_S(Q^2)2N_f}{6\pi}\left(1+2.17\bar{\alpha}_S(Q^2)\left(\frac{1}{N}-1\right)+2.30\bar{\alpha}_S^2(Q^2)\left(\frac{1}{N}-1\right)^2+\cdots\right).$$

At highish Q^2 and moderate x the first two terms are dominant. The exact result is

$$\gamma^{DIS}_{exact}(\alpha_S(Q^2),N) = \frac{\alpha_S(Q^2)2N_f}{6\pi}\left(1-1.08N+\cdots+2.17\bar{\alpha}_S(Q^2)\left(\frac{1}{N}-3.05+\cdots\right)+\cdots\right).$$

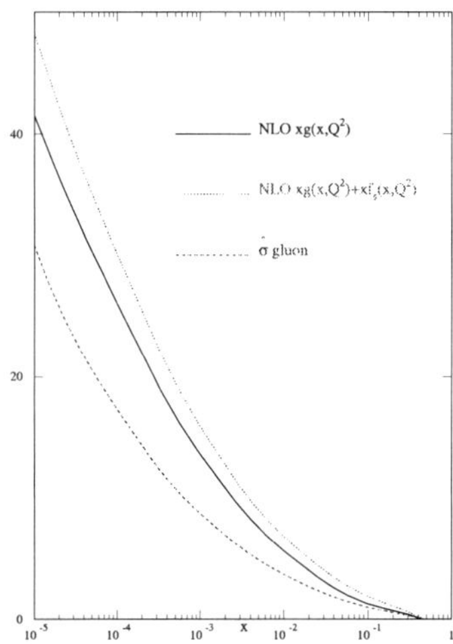

FIGURE 2. Comparison of dipole inspired gluon and standard gluon at $Q^2 = 50\text{GeV}^2$

The first terms are much bigger in the dipole approach, leading to a smaller gluon at all x, and the difference in dipole and DGLAP gluons is mainly due to these effective splitting functions. This is verified by modifying the DGLAP splitting functions in a normal global fit, resulting in a dipole-like gluon [7]. Hence, the good fit at $Q^2 \sim 20-50\text{GeV}^2$ using the dipole model is achieved with a wrong gluon. But the terms in the splitting function are by no means negligible at $Q^2 \sim 1\text{GeV}^2$ – the error in the $\mathcal{O}(\alpha_S)$ and $\mathcal{O}(\alpha_S^2)$ evolution is more important as gluons get flatter. The DGLAP gluons are not accurate here, but the dipole fits are also incorrect. More sophisticated (beyond LO k_t-factorization theory) calculations are needed for semi-quantitative results. But the simple dipole picture also needs extension beyond LO k_t-factorization theory.

REFERENCES

1. L. L. Frankfurt and M. I. Strikman, *Phys. Rept.* **1**60, 235 (1988).
2. A. H. Mueller, *Nucl. Phys.* **B**335, 115 (1990).
3. N. N. Nikolaev and B. G. Zakharov, *Z. Phys.* **C**49, 607 (1991).
4. N. N. Nikolaev and B. G. Zakharov, *Phys. Lett.* **B**332, 184 (1994); Z. Phys. **C**64, 631 (1994).
5. CTEQ Collaboration: J. Pumplin *et al.*, *JHEP* **0**207:012 (2002).
6. A. D. Martin, R. G. Roberts, W. J. Stirling and R. S. Thorne, *Phys. Lett.* **B**604 61 (2004).
7. R. S. Thorne, *Phys. Rev.* **D**71 054024 (2005).
8. A. Bialas, H. Navalet and R. Peschanski, *Nucl. Phys.* **B**593, 438 (2001).
9. S. Catani, M. Ciafaloni and F. Hautmann, *Nucl. Phys.* **B**366, 135 (1991); J. C. Collins and R. K. Ellis, *Nucl. Phys.* **B**360, 3 (1991).
10. L. N. Lipatov, *Sov. J. Nucl. Phys.* **2**3 338 (1976); E. A. Kuraev, L. N. Lipatov, V. S. Fadin, *Sov. Phys. JETP* **4**5 199 (1977); I. I. Balitsky, L. N. Lipatov, *Sov. J. Nucl. Phys.* **2**8 338 (1978).
11. S. Catani and F. Hautmann, *Nucl. Phys.* **B**427, 475 (1994).
12. K. Golec-Biernat and M. Wusthoff, *Phys. Rev.* **D**59, 014017 (1999).
13. H1 Collaboration: C. Adloff *et al.*, *Eur. Phys. J.* **C**21 33 (2001).
14. ZEUS Collaboration: S. Chekanov *et al.*, *Eur. Phys. J.* **C**21 443 (2001): ZEUS Collaboration: J. Breitweg *et al.*, *Phys. Lett.* **B**487 53 (2001).
15. M. R. Adams *et al.*, *Phys. Rev.* **D**54 3006 (1996).

The EMC effect in effective field theory

William Detmold

Department of Physics, Box 351560, University of Washington, Seattle, WA 98195, USA

Abstract. Using effective field theory, we investigate nuclear modification of nucleon parton distributions (for example, the EMC effect). We show that the universality of the shape distortion in nuclear parton distributions (the factorisation of the Bjorken x and atomic number (A) dependence) is model independent and emerges naturally in effective field theory. We present simple fits to experimental data that incorporate this factorisation.

Keywords: Nuclear modification; parton distributions
PACS: 25.30.Mr

The 1983 European Muon Collaboration's (EMC) observation [1] of the deviation of the ratio $R_{Fe}(x)$ of $F_2(x)$ structure functions in iron and deuterium in deep inelastic scattering (DIS) has provoked much analysis and discussion over the years [2] since it shows that the parton distributions are modified in the nuclear environment. Here [3] we employ effective field theory (EFT) to investigate the EMC effect by studying nuclear matrix elements of the twist-two operators which are related to parton distributions and structure functions via the operator product expansion. We find that the universality of the shape distortion of the EMC effect (the factorisation of the x and A dependencies) is a model independent result, arising from the symmetries of QCD and the separation of the relevant scales. The x dependence of $R_A(x)$ is governed by short distance physics, while the overall magnitude (the A dependence) of the EMC effect is governed by long distance matrix elements calculable using traditional nuclear physics.

Recently EFT has been applied to the computation of hadronic matrix elements of twist-two operators in the meson and single nucleon sectors [4]. The approach has also been extended to analyse the deuteron system [5] and is readily generalised to the multi-nucleon case. To described the EMC effect observed in F_2 data on isoscalar nuclei, we consider the normalised, spin singlet, isoscalar twist-two operators,

$$\mathcal{O}_q^{\mu_0\cdots\mu_n} = \overline{q}\gamma^{(\mu_0} iD^{\mu_1}\cdots iD^{\mu_n)}q/\left(2M^{n+1}\right), \tag{1}$$

where $(\ldots)$ indicates that enclosed indices have been symmetrised and made traceless, $D^\mu = (\overrightarrow{D}^\mu - \overleftarrow{D}^\mu)/2$ is the covariant derivative and M is the nucleon mass. The matrix elements of $\mathcal{O}_q^{\mu_0\ldots\mu_n}$ in an unpolarised nucleon of momentum P can be parametrised as

$$\langle P|\mathcal{O}_q^{\mu_0\ldots\mu_n}|P\rangle = \langle x^n\rangle_q \widetilde{v}^{\mu_0}\cdots\widetilde{v}^{\mu_n}, \tag{2}$$

where the nucleon velocity $\widetilde{v}^\mu = P^\mu/M$. It is well known that the coefficients $\langle x^n\rangle_q$ correspond to moments of the isoscalar combination of parton distribution functions, $\langle x^n\rangle_q = \int_{-1}^{1} dx x^n q(x)$ where $q(x)$ is the isoscalar quark distribution and $q(-x) = -\overline{q}(x)$.

We first consider only nucleonic degrees of freedom (*i.e.*, assume that pions are integrated out of the EFT – they will be reintroduced below) and perform the standard

CP792, *Deep Inelastic Scattering, DIS 2005,* edited by W. H. Smith and S. R. Dasu

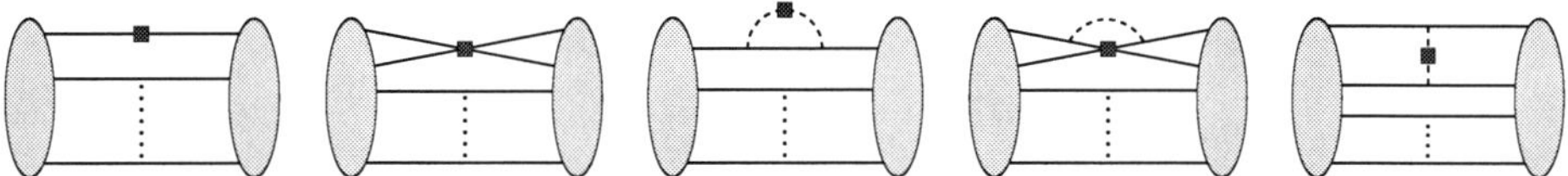

FIGURE 1. Contributions to nuclear matrix elements, labeled (a) to (e) left to right. The dark square represents the various operators in Eq. (3) and the light shaded ellipse corresponds to the nucleus, A. The solid and dashed lines represent nucleons and pions respectively. The dots in the lower part of the diagrams indicate the spectator nucleons.

matching procedure in EFT, equating the quark level twist-two operators to the most general combinations of hadronic operators of the same symmetries [4]. The leading one- and two-body hadronic operators in the matching are

$$\mathcal{O}_q^{\mu_0\cdots\mu_n} = \langle x^n\rangle_q v^{\mu_0}\cdots v^{\mu_n} N^\dagger N[1+\alpha_n N^\dagger N]+\cdots, \tag{3}$$

where $v^\mu = \tilde{v}^\mu + \mathcal{O}(1/M)$ is the velocity of the nucleus. Operators involving additional derivatives are suppressed by powers of M in the EFT power-counting and we have only kept the SU(4) (spin and isospin) singlet two-body operator in the above equation as the other independent two-body operator, $v^{\mu_0}\cdots v^{\mu_n}\left(N^\dagger \tau N\right)^2$, is suppressed [3]. Three- and higher- body operators also appear in Eq. (3), however numerical evidence from other EFT calculations indicates that these contributions are generally small [6].

Nuclear matrix elements of $\mathcal{O}_q^{\mu_0\cdots\mu_n}$ give the moments of the isoscalar nuclear parton distributions, $q_A(x)$. The leading order (LO) and the next-to-leading order (NLO) contributions to these matrix elements are shown in Fig. 1(a) and 1(b), respectively. For an unpolarized, isoscalar nucleus,

$$\begin{aligned}\langle x^n\rangle_{q|A} &\equiv v^{\mu_0}\cdots v^{\mu_n}\langle A|\mathcal{O}_q^{\mu_0\cdots\mu_n}|A\rangle \\ &= \langle x^n\rangle_q[A+\langle A|\alpha_n(N^\dagger N)^2|A\rangle],\end{aligned} \tag{4}$$

where we have used $\langle A|N^\dagger N|A\rangle = A$. Notice that if there were no EMC effect, the α_n would vanish for all n. Also $\alpha_0 = 0$ by charge conservation. From Eq. (4) we see that

$$\left(\frac{\langle x^n\rangle_{q|A}}{A\langle x^n\rangle_q}-1\right)\Big/\left(\frac{\langle x^m\rangle_{q|A}}{A\langle x^m\rangle_q}-1\right) = \frac{\alpha_n}{\alpha_m} \tag{5}$$

is independent of A which has powerful consequences. In all generality, the isoscalar nuclear quark distribution can be written as

$$q_A(x) = A\left[q(x)+\tilde{g}(x,A)\right]. \tag{6}$$

Taking moments of Eq. (6), Eq. (5) then demands that the x dependence and A dependence of $\tilde{g}$ factorise,

$$\tilde{g}(x,A) = g(x)\mathcal{G}(A), \tag{7}$$

with

$$\mathcal{G}(A) = \langle A|(N^\dagger N)^2|A\rangle/A\Lambda_0^3, \tag{8}$$

and $g(x)$ satisfying

$$\alpha_n = \frac{1}{\Lambda_0^3 \langle x^n \rangle_q} \int_{-A}^{A} dx x^n g(x) . \tag{9}$$

Λ_0 is an arbitrary dimensionful parameter . Crossing symmetry dictates that the even and odd α_n separately determine the nuclear modifications of valence and total quark distributions. These results apply to any isoscalar combination of parton distributions including $F_2(x)$ for isoscalar nuclei. Thus our result implies that

$$R_A(x) = \frac{F_2^A(x)}{AF_2^N(x)} = 1 + g_{F_2}(x)\mathscr{G}(A), \tag{10}$$

which says that the EMC effect (the deviation of $R_A(x)$ from unity) has an universal shape described by $g_{F_2}(x)$ while the magnitude of deviation, $\mathscr{G}(A)$, only depends on A.

The above analysis gives a simple explanation of the observed universal shape of the EMC effect, or equivalently, the factorisation of $\widetilde{g}(x,A)$. The key to establishing this factorisation is that other sources of nuclear modification contributing to the right-hand side of Eq. (3) must be suppressed (higher order in the EFT) such that the A independence of Eq. (5) can be established. We stress that the factorisation persists when pions are included in our analysis. In Fig. 1, examples of the leading pionic contributions are shown. The various single-nucleon diagrams, such as Fig. 1(c), simply renormalise the nucleon moments, $\langle x^n \rangle_q$, without contributing to the EMC effect. Two- and more-nucleon diagrams such as those in Fig. 1(d) and 1(e) contribute to the EMC effect, but only at N^3LO and higher (see Ref. [3] for explicit calculations). Other contributions that could upset the factorization include a two-body operator which is similar to that in Eq. (3) but with two more derivatives. However this operator also contributes at N^3LO. Consequently, the universality of the EMC effect is preserved to good accuracy. For large x it is clear that the factorisation must break down (simply consider the region $x > 2$ in which only three- and higher- body operators contribute) though the structure function is very small in this region anyway.

It is clear from Eq. (8) that $\mathscr{G}(A)$ is governed by long distance physics which can be computed using a traditional, non-relativistic nuclear physics approach. Information on the shape distortion function $g(x)$ is encoded in the short distance parameters α_n associated with the strength of the two-body currents. One can either fix the α_n from experimental data (to determine all α_n, data on $F_2^A(x)$ and $F_3^A(x)$ are required) or calculate $\langle NN|\mathscr{O}_q^{\mu_0 \ldots \mu_n}|NN\rangle$ in two nucleon systems to extract them. The latter approach, however, is intrinsically non-perturbative and thus requires lattice QCD [7, 3].

In Figure 2 we present simple fits to the world data on the ratio $R_A(x)$. We choose the simple parameterisation

$$g_{F_2}(x) = (a + b\sqrt{x} + cx + dx^2)(1-x)^f , \qquad \mathscr{G}(A) = 1 - A^{-1/3} \tag{11}$$

though other similar forms also work. This five parameter fit describes the data well in both the small and large x regions, giving a χ^2 per degree of freedom of ~ 1.4. Whilst these fits do not include the (weak) scale dependence of the data, they show consistency with factorisation.

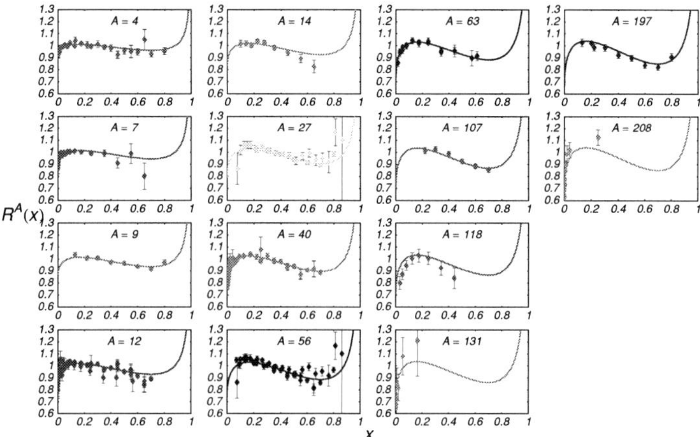

FIGURE 2. Fits to nuclear $R_A(x)$ data.

Similar techniques can also be used to study nuclear modifications of the isovector and spin-dependent parton distributions. Comparable factorisations are expected in EFT. In the isovector case this factorisation can be tested; one can either consider the difference between F_2's in $(Z,N) = (n+m,n)$ and $(n,n+m)$ mirror nuclei [8] and compare it with $F_2^p - F_2^n$, or disentangle $u_A(x)$ and $d_A(x)$ with the upcoming neutrino-nucleus experiment, MINERνA [9]. For spin dependent PDFs, experimental tests are unlikely. Finally, EFT analysis of off-forward matrix elements of the same twist-two operators leads to information about nuclear effects in generalised parton distributions [3].

ACKNOWLEDGMENTS

I thank J.-W. Chen with whom this work was performed. This work was supported by the DOE contract DE-FG03-97ER41014.

REFERENCES

1. J. Aubert *et al.*, Phys. Lett. B **123**, 275 (1983).
2. M. Arneodo, Phys. Rept. **240**, 301 (1994); D. F. Geesaman, K. Saito and A. W. Thomas, Ann. Rev. Nucl. Part. Sci. **45**, 337 (1995); G. Piller and W. Weise, Phys. Rept. **330**, 1 (2000); P. R. Norton, Rept. Prog. Phys. **66**, 1253 (2003).
3. W. Detmold and J. W. Chen, arXiv:hep-ph/0412119; *In preparation.*
4. W. Detmold *et al.*, Phys. Rev. Lett. **87**, 172001 (2001); D. Arndt and M. J. Savage, Nucl. Phys. **A697**, 429 (2002); J. W. Chen and X. Ji, Phys. Lett. **B523**, 107 (2001); *ibid*, 73 (2001); Phys. Rev. Lett. **87**, 152002 (2001). W. Detmold and C.-J. D. Lin, Phys. Rev. D **71**, 054510 (2005).
5. S. R. Beane and M. J. Savage, arXiv:nucl-th/0412025.
6. K. Kubodera and T.-S. Park, Ann. Rev. Nuc. Part. Sci. **54**, 19 (2004);
7. W. Detmold, Phys. Rev. D **71**, 054506 (2005).
8. K. Saito *et al.*, Phys. Lett. B **493**, 288 (2000).
9. D. Drakoulakos *et al.*, hep-ex/0405002.

Understanding Parton Distributions from Lattice QCD

Dru B. Renner

Department of Physics, University of Arizona, 1118 E 4th Street, Tucson, AZ 85721, USA

Abstract. I examine the past lattice QCD calculations of three representative observables, the transverse quark distribution, momentum fraction, and axial charge, and emphasize the prospects for not only quantitative comparison with experiment but also qualitative understanding of QCD.

Keywords: Hadron structure, Generalized parton distributions, Lattice matrix elements
PACS: 12.38.Gc,14.20.Dh,13.60.Fz

INTRODUCTION

Lattice QCD calculations provide the opportunity for both quantitative comparison with experimental measurements and for advancing our qualitative understanding of QCD. I examine several observables which exemplify this range of opportunities. The lattice calculation of the moments of parton distributions is an essential step in achieving quantitative agreement with experimental results. Additionally our understanding of QCD can be further expanded by calculating the remaining moments of the generalized parton distributions which determine the three dimensional distribution of the transverse position and longitudinal momentum of quarks and gluons within the nucleon.

Generalized Form Factors. The generalized form factors provide an alternative but equivalent language to the generalized parton distributions. They encode the ordinary form factors and parton distributions as well as the nucleon spin decomposition [1] and transverse quark distributions [2] and hence provide a unifying language to describe calculations of nucleon structure. Each tower of twist two operators has a corresponding set of generalized form factors. As an example the unpolarized operators $O_q^{\mu_1\cdots\mu_n} = \overline{q}iD^{(\mu_1}\cdots iD^{\mu_{n-1}}\gamma^{\mu_n)}q$ define the generalized form factors A^q_{ni}, B^q_{ni}, and C^q_n via

$$\langle P'|O_q^{\mu_1\cdots\mu_n}|P\rangle = \overline{U}(P')[\sum_{i=0,\,\text{even}}^{n-1}\left(A^q_{ni}(t)K^A_{ni}+B^q_{ni}(t)K^B_{ni}\right)+\delta^n_{\text{even}}C^q_n(t)K^C_n\,]U(P) \qquad (1)$$

where K are known functions of P and P'. A complete set of results can be found in [3].

Lattice Calculations. There have been several full QCD calculations of nucleon structure to date. Results from these and other calculations [4-14] were shown at the conference. Each calculation uses different actions as well as differing lattice spacings and volumes. However the dominate systematic error in lattice calculations of nucleon observables is still due to the chiral extrapolation.

CP792, *Deep Inelastic Scattering, DIS 2005,* edited by W. H. Smith and S. R. Dasu

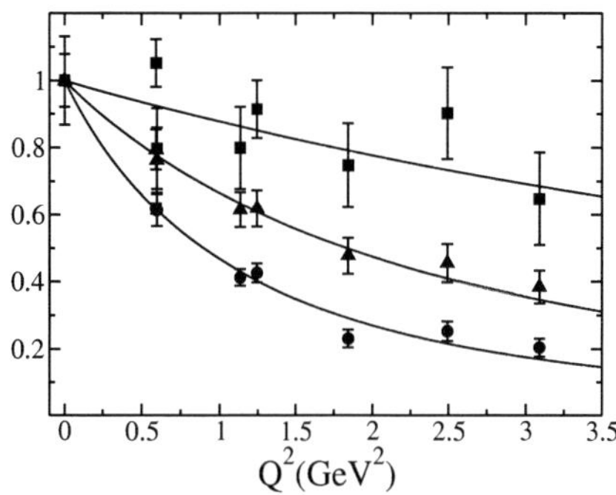

FIGURE 1. squares are A_{30}^{u-d}, triangles are A_{20}^{u-d}, circles are A_{10}^{u-d} [4]

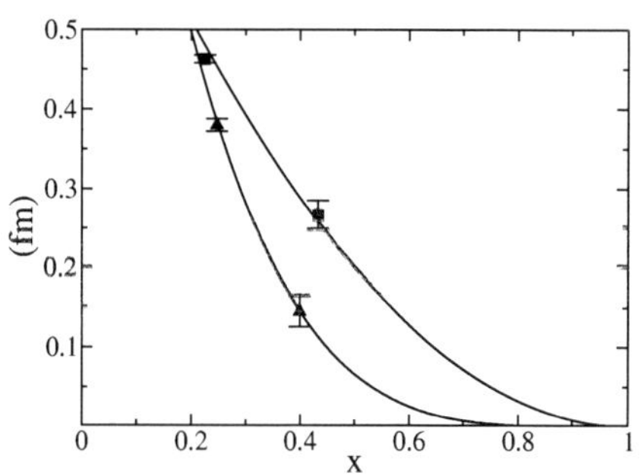

FIGURE 2. squares are $\langle b_\perp^2 \rangle_{(n)}^{u+d}$, triangles are $\langle b_\perp^2 \rangle_{(n)}^{u-d}$ for $n = 1,3$ [4, 6]

TRANSVERSE QUARK DISTRIBUTIONS

The transverse quark distribution $q(x,\vec{b}_\perp)$ gives the probability to find a quark of flavor q carrying a fraction x of the nucleon's longitudinal momentum at a displacement $\vec{b}_\perp$ from the center of the nucleon. The moments of the transverse quark distribution are given by

$$q_n(\vec{b}_\perp) = \int_{-1}^{1} dx\, x^{n-1}\, q(x,\vec{b}_\perp) = \int \frac{d^2\Delta_\perp}{(2\pi)^2} e^{-i\vec{b}_\perp \cdot \vec{\Delta}_\perp} A_{n0}^q(-\vec{\Delta}_\perp^2). \tag{2}$$

Current calculations are restricted to the lowest three moments, however the following shows several ways to examine the transverse structure using just the low moments.

Q^2 Dependence. The slope of the generalized form factors in the forward limit is of particular interest because it determines the rms radius of the nth moment of $q(x,\vec{b}_\perp)$,

$$\langle b_\perp^2 \rangle_n = \frac{\int d^2 b_\perp\, b_\perp^2 \int_{-1}^{1} dx x^{n-1} q(x,\vec{b}_\perp)}{\int d^2 b_\perp \int_{-1}^{1} dx x^{n-1} q(x,\vec{b}_\perp)} = \frac{-4}{A_{n0}^q(0)} \frac{dA_{n0}^q(0)}{dQ^2}. \tag{3}$$

The moments in Eq. 3 are dominated by x near 1 for large n. In such a limit the quark carries all the longitudinal momentum and is kinematically constrained to reside at the center of the nucleon [4]. Thus higher moments determine the transverse size of larger x quarks which are distributed more narrowly in $b_\perp$. Consequently the qualitative expectation, illustrated in Fig. 1, is that the slopes should decrease as n increases for large enough n. That an expectation for large n is so clearly seen for the lowest three moments demonstrates that the transverse distribution of quarks within the nucleon depends strongly on the momentum fraction at which the quarks are probed.

x Dependence. The transverse rms radius of the nucleon at a fixed x is

$$\langle b_\perp^2 \rangle_x = \frac{\int d^2 b_\perp\, b_\perp^2\, q(x,\vec{b}_\perp)}{\int d^2 b_\perp\, q(x,\vec{b}_\perp)},$$

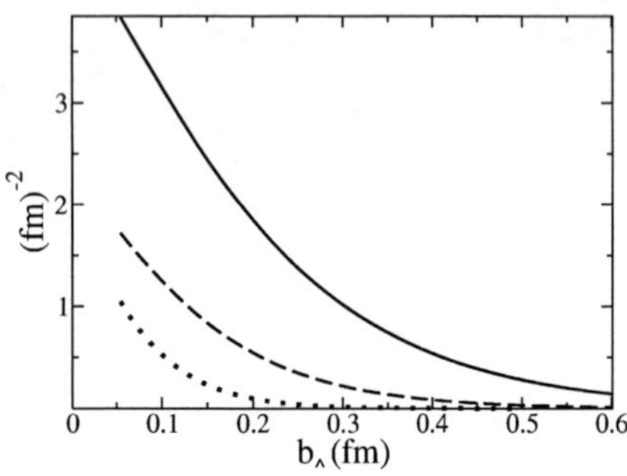

FIGURE 3. solid, dashed, dotted are $q_n(\vec{b}_\perp)$ for $n = 1,2,3$, $q = u - d$ [6]

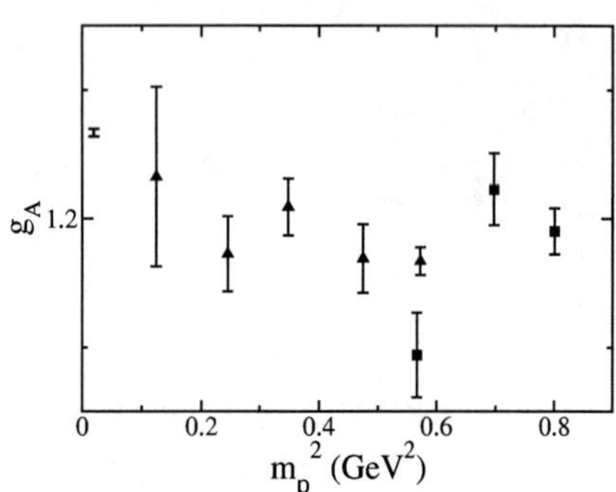

FIGURE 4. g_A, triangles [12], squares [15]

whereas lattice calculations determine the transverse radius at a fixed moment n as shown in Eq. 3. To understand the meaning of the transverse radius at a fixed moment we can think of $\langle b_\perp^2 \rangle_n$ as a coarse grained transverse size of the nucleon corresponding to a region centered on the average x of the nth moment,

$$\langle x \rangle_n = \frac{\int d^2b_\perp \int_{-1}^{1} dx\, |x| x^{n-1} q(x,\vec{b}_\perp)}{\int d^2b_\perp \int_{-1}^{1} dx x^{n-1} q(x,\vec{b}_\perp)} = \frac{\langle x^n \rangle + 2(-1)^n \int d^2b_\perp \int_0^1 dx x^n \overline{q}(x)}{\langle x^{n-1} \rangle}. \tag{4}$$

Lattice QCD is not currently capable of calculating the anti-quark contribution in $\langle x \rangle_n$, however the phenomenologically determined parton distributions indicate it is small enough that it does not affect the following qualitative conclusions. Fig. 2 shows the transverse radius versus corresponding momentum fraction for the lowest moments illustrating, as above, that the transverse size of the nucleon depends significantly on the longitudinal momentum of its constituents.

$b_\perp$ *Dependence.* By assuming a dipole ansatz for the generalized form factors the $b_\perp$ dependence of each moment can be determined from Eq. 2. The details are given in [6], and the results are shown in Fig. 3. Of particular importance are the lowest two moments which determine the transverse distribution of quarks ($n = 1$) and of longitudinal momentum ($n = 2$) within the nucleon.

MOMENTS OF PARTON DISTRIBUTIONS

Moments of parton distributions provide an opportunity for quantitative comparison between experimental measurements and lattice calculations of nucleon structure. The axial charge and momentum fraction of the nucleon represent the state of affairs with the former observable providing an example of a potential success of current calculations and the latter an example of a challenge to future calculations.

$\langle x \rangle_{u-d}$. Extensive quenched calculations of $\langle x \rangle_{u-d}$ [7] have shown very little dependence on m_π while overestimating the experimental result by nearly a factor of two. The first unquenched calculations [15] confirmed the earlier quenched results and lead to the suggestion that sizable chiral corrections could accommodate both lattice calculations and experimental measurements [8]. Recent calculations with lighter quark masses [5, 9] have not resolved this discrepancy, however one recent quenched calculation [16] shows a significant but unconfirmed shift toward the experimental result.

g_A. Lattice calculations of the nucleon axial coupling are beginning to mature. In particular recent calculations with chiral actions allow for a non-perturbative renormalization of g_A. This observable has been shown to have sizable finite size corrections for light quark masses [10, 11], however current calculations have reached large enough volumes that such effects appear under control. Furthermore simple linear extrapolations, in m_π^2, of the lightest calculations [12] give estimates of 1.23(2) to 1.26(2) (using 3 to 6 of the lightest points) the latter of which agrees with the experimental measurement within the statistical errors. However detailed study of the chiral behavior is needed to reliably estimate the systematic errors in such extrapolations.

CONCLUSIONS

Lattice calculations of nucleon structure are beginning to realize their promise to elucidate QCD and make contact with the experimental programs. Recent calculations are painting a qualitative three dimensional picture of nucleon structure revealing a significant x dependence of the transverse size of the nucleon. Quantitative calculations of moments of parton distributions are progressing, in particular the calculation of g_A may soon reach a few percent accuracy.

REFERENCES

1. X. D. Ji, *Phys. Rev. Lett.*, **78**, 610–613 (1997), hep-ph/9603249.
2. M. Burkardt, *Phys. Rev.*, **D62**, 071503 (2000), hep-ph/0005108.
3. M. Diehl, *Phys. Rept.*, **388**, 41–277 (2003), hep-ph/0307382.
4. P. Hagler, et al., *Phys. Rev. Lett.*, **93**, 112001 (2004), hep-lat/0312014.
5. M. Gockeler, et al., *Nucl. Phys. Proc. Suppl.*, **140**, 399–404 (2005), hep-lat/0409162.
6. D. B. Renner (2005), hep-lat/0501005.
7. M. Gockeler, et al., *Nucl. Phys. Proc. Suppl.*, **119**, 32–40 (2003), hep-lat/0209160.
8. W. Detmold, et al., *Phys. Rev. Lett.*, **87**, 172001 (2001), hep-lat/0103006.
9. S. Ohta, and K. Orginos, *Nucl. Phys. Proc. Suppl.*, **140**, 396–398 (2005), hep-lat/0411008.
10. S. Sasaki, K. Orginos, S. Ohta, and T. Blum, *Phys. Rev.*, **D68**, 054509 (2003), hep-lat/0306007.
11. A. A. Khan, et al., *Nucl. Phys. Proc. Suppl.*, **140**, 408–410 (2005), hep-lat/0409161.
12. D. B. Renner, et al., *Nucl. Phys. Proc. Suppl.*, **140**, 255–260 (2005), hep-lat/0409130.
13. P. Hagler, et al., *Phys. Rev.*, **D68**, 034505 (2003), hep-lat/0304018.
14. M. Gockeler, et al., *Phys. Rev. Lett.*, **92**, 042002 (2004), hep-ph/0304249.
15. D. Dolgov, et al., *Phys. Rev.*, **D66**, 034506 (2002), hep-lat/0201021.
16. M. Gurtler, et al. (2004), hep-lat/0409164.

Quark Asymmetries in Nucleons

Johan Alwall

High Energy Physics, Uppsala University, Box 535, S-75121 Uppsala, Sweden

Abstract. We have developed a physical model for the non-perturbative x-shape of parton density functions in the proton, based on Gaussian fluctuations in momenta, and quantum fluctuations of the proton into meson-baryon pairs. The model describes the proton structure function and gives a natural explanation of observed quark asymmetries, such as the difference between the anti-up and anti-down sea quark distributions and between the up and down valence distributions. We also find an asymmetry in the momentum distribution of strange and anti-strange quarks in the nucleon, large enough to reduce the NuTeV anomaly to a level which does not give a significant indication of physics beyond the standard model.

Keywords: quark asymmetries, parton density distributions, s-sbar asymmetry, NuTeV anomaly
PACS: 12.39.Ki,11.30.Hv,12.40.Vv,13.15.+g,13.60.Hb

The low-scale parton density functions give a description of the hadron at a non-perturbative level. The conventional approach to these functions is to make parameterizations using some more or less arbitrary functional forms, based on data from deep inelastic scattering and hadron collision experiments. Another approach, however, is to start from some ideas of the behavior of partons in the non-perturbative hadron, and build a model based on that behavior. The advantage with this approach is that the successes and failures of such a model allows us to get insight into the non-perturbative QCD dynamics. The model presented here, and described in detail in [1, 2], describes the F_2 structure function of the proton, as well as sea quark asymmetries of the nucleon. Most noteworthy, our model predicts an asymmetry between the momentum distributions of strange and anti-strange quarks in the nucleon of the same order as the newly reported results from NuTeV [3].

This work extends the model previously presented in [4]. The model gives the four-momentum k of a single probed valence parton (see Fig. 1a for definitions of momenta) by assuming that, in the nucleon rest frame, the shape of the momentum distribution for a parton of type i and mass m_i can be taken as a Gaussian $f_i(k) = N(\sigma_i, m_i)\exp\{-\left[(k_0-m_i)^2+k_x^2+k_y^2+k_z^2\right]/2\sigma_i^2\}$ which may be motivated as a result of the many interactions binding the parton in the nucleon. The width of the distribution should be of order hundred MeV from the Heisenberg uncertainty relation applied to the nucleon size, *i.e.* $\sigma_i = 1/d_N$. The momentum fraction x of the parton is then defined as the light-cone fraction $x = k_+/p_+$. We impose constraints on the final-state momenta in order to obtain a kinematically allowed final state, which also ensures that $0 < x < 1$. Using a Monte Carlo method these parton distributions are integrated numerically without approximations.

To describe the dynamics of the sea partons, we note that the appropriate basis for the non-perturbative dynamics of the bound state nucleon should be hadronic. Therefore we

CP792, *Deep Inelastic Scattering, DIS 2005*, edited by W. H. Smith and S. R. Dasu

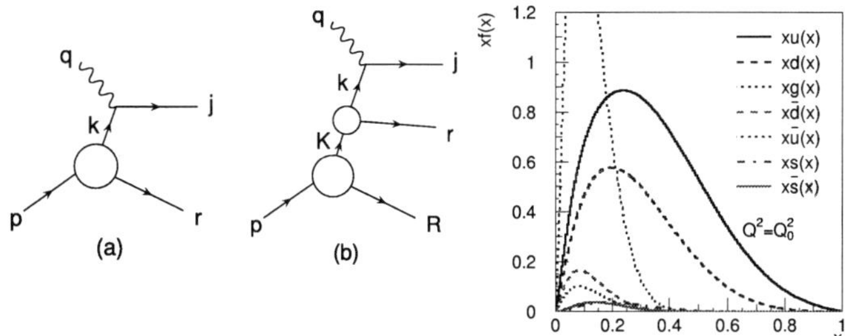

FIGURE 1. Illustration of the processes (a) probing a valence parton in the proton and (b) a sea parton in a hadronic fluctuation (letters are four-momenta). (c) shows the resulting parton distributions at the starting scale Q_0^2.

consider hadronic fluctuations, for the proton

$$|p\rangle = \alpha_0|p_0\rangle + \alpha_{p\pi^0}|p\pi^0\rangle + \alpha_{n\pi^+}|n\pi^+\rangle + \ldots + \alpha_{\Lambda K}|\Lambda K^+\rangle + \ldots \tag{1}$$

Probing a parton i in a hadron H of a baryon-meson fluctuation $|BM\rangle$ (see Fig. 1b) gives a sea parton with light-cone fraction $x = x_H x_i$ of the target proton. The momentum of the probed hadron is given by a similar Gaussian, but with a separate width parameter σ_H. Also here, kinematic constraints ensure that we get a physically allowed final state. The procedure gives $x_H \sim M_H/(M_B + M_M)$, *i.e.* the heavier baryon gets a harder spectrum than the lighter meson. The normalization of the sea distributions is given by the amplitude coefficients α_{BM}^2 of Eq. (1). These cannot be calculated from first principles in QCD and are therefore taken as free parameters to be fitted using experimental data.

The resulting valence and sea parton x-distributions apply at a low scale Q_0^2, and the distributions at higher Q^2 are obtained using perturbative QCD evolution at next-to-leading order.

The model has in total four shape parameters and three normalization parameters, plus the starting scale, to determine the parton densities u, d, g, $\bar{u}$, $\bar{d}$, s, $\bar{s}$. These are (with values resulting from fits to experimental data as described below):

$$\begin{array}{cccc} \sigma_u = 230\,\text{MeV} & \sigma_d = 170\,\text{MeV} & \sigma_g = 77\,\text{MeV} & \sigma_H = 100\,\text{MeV} \\ \alpha^2_{p\pi^0} = 0.45 & \alpha^2_{n\pi^+} = 0.14 & \alpha^2_{\Lambda K} = 0.05 & Q_0 = 0.75\,\text{GeV} \end{array} \tag{2}$$

The resulting parton densities are shown in Fig. 1(c).

In order to fix the values of the model parameters, we make a global fit using several experimental data sets: Fixed-target F_2 data to fix large-x (valence) distributions (Fig. 2a); HERA F_2 data for the gluon distribution width and the starting scale Q_0^2; $\bar{d}/\bar{u}$-asymmetry data for the normalizations of the $|p\pi^0\rangle$ and $|n\pi^+\rangle$ fluctuations (Fig. 3); and strange sea data to fix the normalization of fluctuations including strange quarks (Fig. 4a). We have also compared with $W^\pm$ charge asymmetry data as a cross-check on the ratio of Gaussian widths for the u and d valence quark distributions (Fig. 2b). It is interesting to note that this simple model can describe such a wealth of different data with just one or two parameters per data set.

In our model, the shape difference between the valence u and d distributions in the proton, apparent from the $W^\pm$ charge asymmetry data, is described as different Gaussian

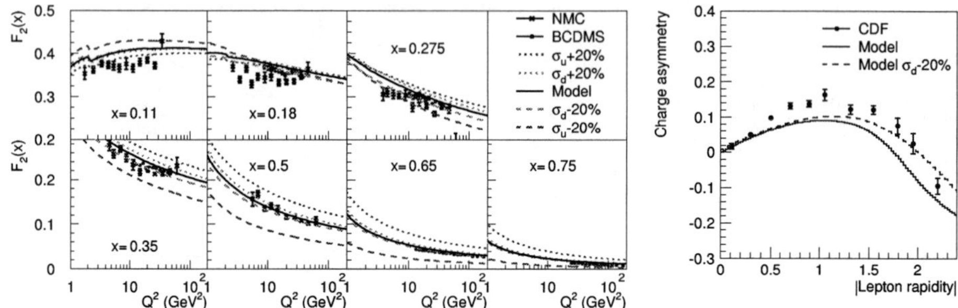

FIGURE 2. Left: The proton structure function $F_2(x,Q^2)$ for large x values; NMC and BCDMS data [5, 6] compared to our model, also showing the results of $\pm 20\%$ variations of the width parameters σ_u and σ_d for the u and d valence distributions. Right: The charge asymmetry for leptons from $W^\pm$-decays in $p\bar{p}$ collisions at the Tevatron [7] compared to our model, with best-fit parameters and a 20% reduced width of the valence d quark distribution.

widths. This would correspond to a larger effective volume in the proton for d quarks than for u quarks, an effect which could conceivably be explained by Pauli blocking of the u quarks.

Since the proton can fluctuate to π^0 and π^+ by $|p\pi^0\rangle$ and $|n\pi^+\rangle$, but to π^- only by the heavier $|\Delta^{++}\pi^-\rangle$, we get an excess of $\bar{d}$ over $\bar{u}$ in the proton sea. Interestingly, the fit to data improves when we use a larger effective pion mass of 400 MeV (see Fig. 3). This might indicate that we have a surprisingly large coupling to heavier ρ mesons, or that one should use a more generic meson mass rather than the very light pion.

The lightest strange fluctuation is $|\Lambda K^+\rangle$. If we let this implicitly include also heavier strange meson-baryon fluctuations, we can fit the normalization $\alpha^2_{\Lambda K}$ to strange sea data (see Fig. 4a). The result corresponds to $\int_0^1 (xs + x\bar{s})dx / \int_0^1 (x\bar{u} + x\bar{d})dx \approx 0.5$, in agreement with standard parton density parameterizations. We note that this indicates a normalization $\propto 1/\Delta M_{BM} = 1/(M_B + M_M - M_p)$ rather than the expected $\propto 1/\Delta M^2_{BM}$.

Since the s quark is in the heavier baryon Λ and the $\bar{s}$ quark is in the lighter meson K^+, we get a non-zero asymmetry $S^- = \int_0^1 dx[xs(x) - x\bar{s}(x)]$ in the momentum distribution of the strange sea, as seen in Fig. 4b and 5. Depending on details of the model, we get the range $0.0010 \leq S^- \leq 0.0023$ for this asymmetry.

This is especially interesting in connection to the NuTeV anomaly [9]. NuTeV found, based on the observable $R^- = \frac{\sigma(\nu_\mu N \to \nu_\mu X) - \sigma(\bar{\nu}_\mu N \to \bar{\nu}_\mu X)}{\sigma(\nu_\mu N \to \mu^- X) - \sigma(\bar{\nu}_\mu N \to \mu^+ X)} = g_L^2 - g_R^2 = \frac{1}{2} - \sin^2\theta_W$ a 3σ deviation of $\sin^2\theta_W$ compared to the Standard Model fit: $\sin^2\theta_W^{\mathrm{NuTeV}} = 0.2277 \pm 0.0016$ compared to $\sin^2\theta_W^{\mathrm{SM}} = 0.2227 \pm 0.0004$. However, an asymmetric strange sea would

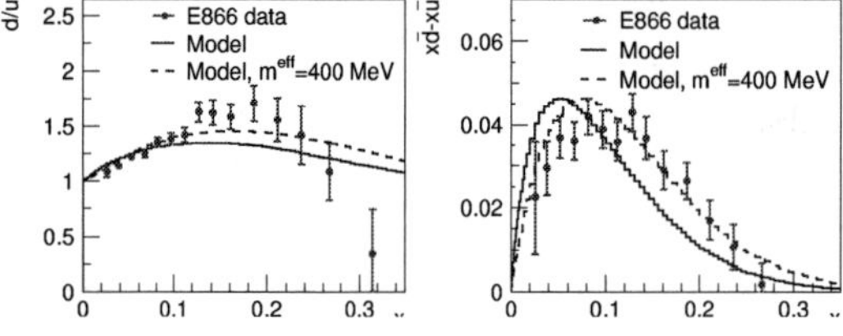

FIGURE 3. Comparison between our model and data from the E866/NuSea collaboration [8]: (a) $\bar{u}(x)/\bar{d}(x)$ (b) $xd(x) - xu(x)$. The full line uses the physical pion mass, while the dashed line uses an effective pions mass $m^{\mathrm{eff}} = 400$ MeV as discussed in the text.

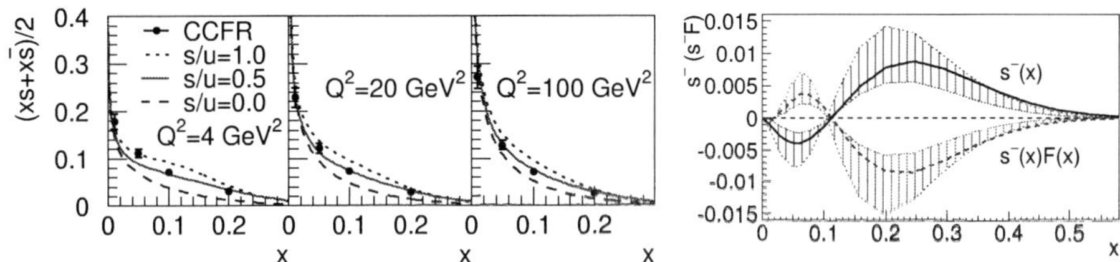

FIGURE 4. Left: CCFR data [10] on the strange sea distribution $(xs(x)+x\bar{s}(x))/2$ compared to our model based on $|\Lambda K\rangle$ fluctuations with different normalizations. Right: The strange sea asymmetry $s^-(x)=xs(x)-x\bar{s}(x)$ (at $Q^2=20\,\mathrm{GeV}^2$) from the model and combined with the function $F(x)$ accounting for NuTeV's analysis giving $\Delta\sin^2\theta_W=\int_0^1 dx s^-(x)F(x)=-0.0017$. The uncertainty bands correspond to the uncertainties for S^- and $\Delta\sin^2\theta_W$ quoted in the text.

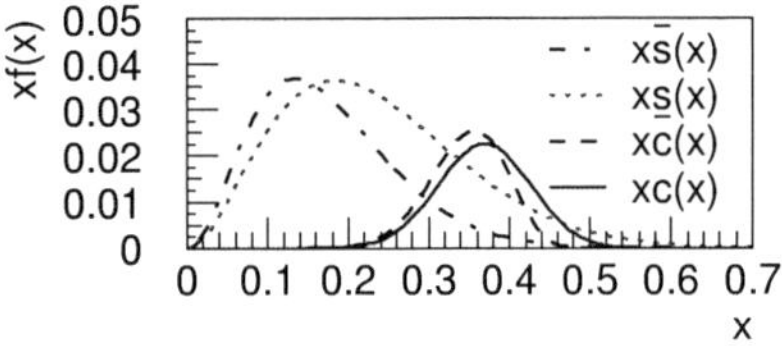

FIGURE 5. Comparison between the strange and charm sea obtained from our model with the inclusion of the $\Lambda_C\overline{D}$ fluctuation. The normalization is here taken to be $\propto 1/(M_B+M_M-M_p)$ as suggested by strange sea data.

change their result, since ν only have charged current interactions with s and $\bar{\nu}$ with $\bar{s}$. Using the folding function provided by NuTeV to account for their analysis, the s-$\bar{s}$ asymmetry from our model gives a shift $-0.0024 \le \Delta\sin^2\theta_W = \int_0^1 dx s^-(x)F(x) \le -0.00097$, *i.e.* the discrepancy with the Standard Model result is reduced to between 1.6σ and 2.4σ, leaving no strong hint of physics beyond the Standard Model.

We have also considered charmed fluctuations. The lightest charmed baryon-meson fluctuation $|\Lambda_C\overline{D}\rangle$ gives c and $\bar{c}$ distributions as in Fig. 5, where the normalization $\alpha^2_{\Lambda_C D}$ is taken to be $\propto 1/\Delta M_{BM}$, as suggested by the strange sea normalization. However, in order to conform to the EMC F_2^c data at large x, a normalization close to $1/\Delta M^2_{\Lambda_C\overline{D}}$ seems to be enough [11].

Acknowledgments: We would like to thank the organizers for the opportunity to talk at DIS'05, and Stan Brodsky for very interesting discussions.

REFERENCES

1. J. Alwall, and G. Ingelman, *Phys. Rev.*, **D71**, 094015 (2005), hep-ph/0503099.
2. J. Alwall, and G. Ingelman, *Phys. Rev.*, **D70**, 111505 (2004), hep-ph/0407364.
3. D. Mason, "NuTeV strange/antistrange sea measurements from neutrino charm production" (2005), presented at this conference.
4. A. Edin, and G. Ingelman, *Phys. Lett.*, **B432**, 402–410 (1998), hep-ph/9803496.
5. M. Arneodo, et al., *Phys. Lett.*, **B364**, 107–115 (1995), hep-ph/9509406.
6. A. C. Benvenuti, et al., *Phys. Lett.*, **B223**, 485 (1989).
7. F. Abe, et al., *Phys. Rev. Lett.*, **81**, 5754–5759 (1998), hep-ex/9809001.
8. R. S. Towell, et al., *Phys. Rev.*, **D64**, 052002 (2001), hep-ex/0103030.
9. G. P. Zeller, et al., *Phys. Rev. Lett.*, **88**, 091802 (2002), hep-ex/0110059.
10. A. O. Bazarko, et al., *Z. Phys.*, **C65**, 189–198 (1995), hep-ex/9406007.
11. J. Alwall, and G. Ingelman (work in progress).

Polarisation Dependence of the Total CC $e^{\pm}p$ Cross Section

Andrei Nikiforov
on behalf of the H1 collaboration

Max Planck Institute for Physics, Munich, Germany
E-mail: nikifor@mail.desy.de

Abstract. Data taken with the H1 detector, with longitudinally polarised electrons and positrons in collision with unpolarised protons at HERA, are used to measure the charged current cross section σ_{CC}^{tot} for $Q^2 > 400\,\text{GeV}^2$ and inelasticity $y < 0.9$. The polarisation dependence of the measured cross section is in agreement with the Standard Model predictions.

Keywords: Charged current, electroweak, cross section, polarisation, deep inelastic scattering
PACS: 13.60.Hb, 13.88.+e

INTRODUCTION

Data taking of the second, high luminosity phase of the HERA program (HERA II) started in October 2003. An increase of specific luminosity after the HERA upgrade has been achieved by placing strong super-conducting focusing magnets inside the H1 and ZEUS detectors, close to the interaction point. A major success at HERA II is a longitudinal polarisation of the electron beam which collides with protons inside the H1 and ZEUS detectors.

Both, neutral current (NC), $ep \rightarrow eX$, and charged current (CC) interactions, $ep \rightarrow \nu X$, can be measured at HERA and provide complementary information. In the Standard Model, there is clear prediction on polarisation dependence of these cross sections. Specifically, the e^+p CC cross section for a left handed positron and e^-p CC cross section for right handed electron should be zero and should have full strength for opposite helicities. This follows from absence of right handed weak charged currents in the Standard Model. Thus, for unpolarised electrons (positrons), the cross section is a half of the one for left handed electron (right handed positron).

The incident electron (positron) beam energy is 27.6 GeV, whilst the upolarised proton beam energy is 920 GeV. This yields a center-of-mass energy of $\sqrt{s} = 319$ GeV. The e^+p data collected in 2003 and 2004 consist of two periods with positive and negative polarisations. The first period has a mean polarisation of (33.0±0.7)% and integrated luminosity of (15.3±0.4) pb^{-1}. The second period, with negative polarisation, has (-40.2±0.6)% and (21.7±0.6) pb^{-1}. The 2005 e^-p data set has a mean polarisation of (-25.4±0.4)% and an integrated luminosity of (17.8±0.2) pb^{-1}.

The double differential CC cross section for collisions of polarised leptons with upolarised protons corrected for QED radiative effects can be written as

CP792, *Deep Inelastic Scattering, DIS 2005,* edited by W. H. Smith and S. R. Dasu

$$\frac{d^2\sigma_{CC}^{e^\pm p}}{dxdQ^2} = [1 \pm P_e]\frac{G_F^2}{2\pi x}[\frac{M_W^2}{Q^2+M_W^2}]^2\Phi_{CC}^\pm, \tag{1}$$

where G_F is the Fermi coupling constant, M_W is the mass of the W boson. $\Phi_{CC}^\pm$ is the term depending on the partonic content of the proton and can be expressed as follows:

$$\Phi_{CC}^+ = \bar{u} + \bar{c} + (1-y)^2(d+s+b), \tag{2}$$

$$\Phi_{CC}^- = u + c + (1-y)^2(\bar{d}+\bar{s}+\bar{b}). \tag{3}$$

Here, P_e is the degree of longitudinal polarisation defined as $P_e = (N_R - N_L)/(N_R + N_L)$ with $N_R(N_L)$ being the number of right (left) handed polarised leptons in the beam. The cross section has a linear dependence on the polarisation of the lepton beam. For positrons (electrons) with positive (negative) polarisation it is enlarged, whilst the cross section for negative (positive) polarised positrons (electrons) is diminished. For $P_e = -1$ ($P_e = +1$) the e^+p (e^-p) CC cross section is identically equal to zero in the Standard Model.

EXPERIMENTAL TECHNIQUE

The H1 detector components most relevant for this analysis are the LAr calorimeter, which measures energies of charged and neutral particles over the polar[1] angular range $4^\circ < \theta < 154^\circ$, and the inner tracking detectors which measure the angles and momenta of charged particles over the range $7^\circ < \theta < 165^\circ$. A full description of the detector can be found in [3].

Transverse polarisation of the electron beam at HERA arises naturally through synchrotron radiation via the Sokolov-Ternov effect [4]. In 2000 pairs of spin rotators were installed in the beamline around the H1 detector transforming transversely polarised positrons into longitudinally polarised. The polarisation is continuously measured using two independent polarimeters TPOL [5] and LPOL [6].

In order to determine acceptance corrections and background contributions for the CC cross section measurements, the detector response for events generated by various Monte Carlo programs is simulated in detail [2].

The selection and analysis of deep inelastic scattering processes with charged current interactions follows closely that of published unpolarised data [2]. The CC events are identified as having missing transverse momentum, $P_{T,h}$, where $P_{T,h} = \sqrt{(\sum_i p_{x,i})^2 + (\sum_i p_{y,i})^2}$ and is summed over all particles of the hadronic final state.

The CC kinematics quantities are determined using hadronic final state [7]:

$$y_h = \frac{\Sigma}{2\,E_e}, \qquad Q_h^2 = \frac{P_{T,h}^2}{1-y_h}, \qquad x_h = \frac{Q_h^2}{s\,y_h}, \tag{4}$$

[1] The polar angle θ is defi ned with respect to the direction of the outgoing proton beam.

where $\Sigma \equiv \sum_i E_i - p_{z,i}$ and E_e is the incident electron (positron) beam energy.

MEASUREMENT PROCEDURE

Candidate for CC interactions are selected by requiring $P_{T,h} > 12\,\text{GeV}$. In order to ensure high efficiency of the trigger and kinematic resolution the analysis is further restricted to the domain of $Q_h^2 > 200\ \text{GeV}^2$ and $0.03 < y_h < 0.85$. The ep background is dominantly due to photoproduction events in which the scattered electron (positron) escapes undetected in the beam pipe and large missing transverse momentum is faked by fluctuations in the detector response or undetected particles. This background is suppressed using the correlation between the ratio V_{ap}/V_p and $P_{T,h}$.

The quantities V_p and V_{ap} are the transverse energy flow parallel and anti-parallel to the direction in which points the transverse momentum vector of the hadronic final state respectively. The residual ep background is negligible for most of the measured kinematic domain. The simulation is used to estimate this contribution, which is subtracted statistically from the CC data sample with a systematic uncertainty of 30% on the number of subtracted events.

Finally, remaining non-ep background is rejected by identifying event topologies characteristic for cosmic ray and beam-induced background.

RESULTS

The total polarised e^+p and e^-p CC cross sections are measured in the range $Q^2 > 400$ GeV^2 and $y < 0.9$. The results are:

$$\sigma_{CC}^{e^+p}(P_e = +0.33) = 34.67 \pm 1.94_{\text{stat}} \pm 1.66_{\text{sys}}\ \text{pb}, \tag{5}$$

$$\sigma_{CC}^{e^+p}(P_e = -0.40) = 13.80 \pm 1.04_{\text{stat}} \pm 0.94_{\text{sys}}\ \text{pb}, \tag{6}$$

$$\sigma_{CC}^{e^-p}(P_e = -0.25) = 66.42 \pm 2.39_{\text{stat}} \pm 2.99_{\text{sys}}\ \text{pb}. \tag{7}$$

The polarisation dependence of the CC cross section is shown in Fig. 1. The polarisation uncertainty, which does not directly enter in the cross section measurement, is shown in the figure as horizontal error bars. The measurements of the unpolarised CC cross section (in the same phase space regions) are also shown in Fig. 1. They are based on HERA I data with a luminosity of 65.2 pb^{-1} for e^+p and 16.4 pb^{-1} for e^-p:

$$\sigma_{CC}^{e^+p}(P_e = 0) = 28.44 \pm 0.77_{\text{stat}} \pm 1.22_{\text{sys}}\ \text{pb}, \tag{8}$$

$$\sigma_{CC}^{e^-p}(P_e = 0) = 57.03 \pm 2.21_{\text{stat}} \pm 1.38_{\text{sys}}\ \text{pb}. \tag{9}$$

The measured cross sections agree well the Standard Model predictions based on the H1 PDF 2000 fit. A linear fit to the polarisation dependence of the three measured $\sigma_{CC}^{e^+p}$ cross sections is performed. The fit takes into account a correlation of systematic errors of the

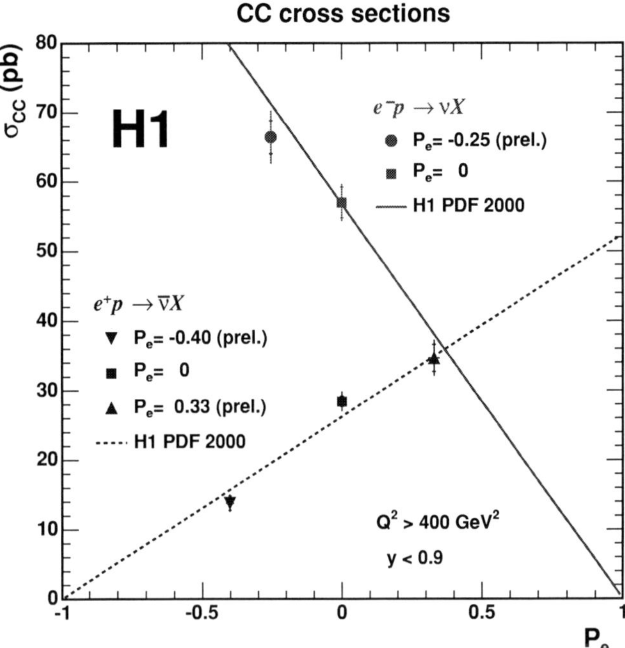

FIGURE 1. The dependence of the e^+p and e^-p cross section with the lepton beam polarisation P_e is shown. The data are compared to the Standard Model prediction based on the H1 PDF 2000 fi t.

measurements. The fit provides a reasonable description of the data with a $\chi^2 = 2.45$. The cross section extrapolation to $P_e = -1.0$ yields a value:

$$\sigma_{CC}(P_e = -1.00) = -3.7 \pm 2.4_{\mathrm{stat}} \pm 2.7_{\mathrm{sys}}\ \mathrm{pb}. \tag{10}$$

This value is consistent with zero and indicates the absence of right handed weak charged currents, in agreement with the Standard Model.

REFERENCES

1. M. Klein and T. Riemann, Z. Phys. C **24** (1984) 151.
2. C. Adloff *et al.* [H1 Collaboration], Eur. Phys. J. C **30** (2003) 1-32 [hep-ex/0304003].
3. I. Abt *et al.* [H1 Collaboration], Nucl. Instrum. Meth. A **386** (1997) 310 and 348; R. D. Appuhn *et al.* [H1 SPACAL Group], Nucl. Instrum. Meth. A **386** (1997) 397.
4. A.A. Sokolov and I.M. Ternov, Sov. Phys. Dokl. **8** No. 12 (1964) 1203.
5. D.B. Barber *et al.*, Nucl. Instrum. Meth. A **329** (1993) 79.
6. M. Beckmann *et al.*, Nucl. Instrum. Meth. A **479** (2002) 334.
7. A. Blondel and F. Jacquet, Proceedings of the Study of an *ep* Facility for Europe, ed. U. Amaldi, DESY 79/48 (1979) 391.

High Q^2 DIS cross sections at HERA with longitudinally polarised positron beams

A. Tapper

Imperial College London, The Blackett Laboratory, Prince Consort Road, London SW7 2BW, U.K.

Abstract. Measurements of the cross sections for neutral and charged current deep inelastic scattering in e^+p collisions with longitudinally polarised positron beams are presented. The total cross section for e^+p charged current deep inelastic scattering is presented at positive and negative values of positron beam longitudinal polarisation. In addition, single differential cross sections are presented for charged and neutral current deep inelastic scattering in the kinematic region $Q^2 > 200$ GeV^2. The measurements are based on data of integrated luminosity 30.5 pb^{-1} collected with the ZEUS detector in 2003 and 2004 at a centre-of-mass energy of 318 GeV. The measured cross sections are compared with the predictions of the Standard Model.

Keywords: HERA, DIS
PACS: 13.88.+e, 13.60.Hb

INTRODUCTION

Deep inelastic scattering (DIS) of leptons off nucleons probes the structure of matter at small distance scales. Two types of DIS interactions are possible at HERA: neutral current (NC) reactions $e^-p \to e^-X$ and $e^+p \to e^+X$, where a photon or Z^0 boson is exchanged and charged current (CC) interactions $e^-p \to \nu X$ and $e^+p \to \bar{\nu}X$, where a $W^{\pm}$ boson is exchanged.

The Standard Model predicts that the cross sections for charged and neutral current DIS should exhibit dependence on the longitudinal polarisation of the incoming lepton beam. In the charged current case the dependence is predicted to be linear with the cross section becoming zero for right-handed (left-handed) electron (positron) beams, due to the chiral nature of the Standard Model.

The kinematics of charged current and neutral current deep inelastic scattering processes are defined by the four-momenta of the incoming lepton (k), the incoming proton (P), the outgoing lepton (k') and the hadronic final state (P'). The four-momentum transfer between the electron and the proton is given by $q = k - k' = P' - P$. The square of the centre-of-mass energy is given by $s = (k+P)^2$. The description of DIS is usually given in terms of three Lorentz invariant quantities, which may be defined in terms of the four-momenta k, P and q:

- $Q^2 = -q^2$, the negative square of the four-momentum transfer,
- $x = \frac{Q^2}{2P \cdot q}$, the Bjorken scaling variable,
- $y = \frac{q \cdot P}{k \cdot P}$, the fraction of the energy transferred to the proton in its rest frame.

CP792, *Deep Inelastic Scattering, DIS 2005,* edited by W. H. Smith and S. R. Dasu

These variables are related by $Q^2 = xys$, when the masses of the incoming particles can be neglected.

This paper presents measurements of the cross sections for e^+p CC and NC DIS with longitudinally polarised positron beams. The measurements are based on 16.4 pb^{-1} of data collected at a mean luminosity weighted polarisation of -40.2%, and 14.1 pb^{-1} collected at a polarisation of 31.8% with the ZEUS detector in 2003 and 2004. During this time HERA collided protons of energy 920 GeV with positrons of energy 27.5 GeV, yielding collisions at a centre-of-mass energy of 318 GeV. The measured cross sections are compared to the Standard Model predictions.

CROSS SECTIONS

The electroweak Born-level cross-section for the CC reaction, $e^+p \rightarrow \bar{\nu}X$, with a longitudinally polarised positron beam (defined in Eqn. (2)), can be expressed as

$$\frac{d^2\sigma^{\rm CC}}{dxdQ^2} = (1+\mathscr{P})\frac{G_F^2}{4\pi x}\left(\frac{M_W^2}{M_W^2+Q^2}\right)^2\left[Y_+F_2^{\rm CC}(x,Q^2) - Y_-xF_3^{\rm CC}(x,Q^2) - y^2F_L^{\rm CC}(x,Q^2)\right], \quad (1)$$

where G_F is the Fermi constant, M_W is the mass of the W boson and $Y_\pm = 1 \pm (1-y)^2$. The structure functions $F_2^{\rm CC}$ and $xF_3^{\rm CC}$ contain sums and differences of the quark and anti-quark parton density functions (PDFs) and $F_L^{\rm CC}$ is the longitudinal structure function. The longitudinal polarisation of the positron beam is defined as

$$\mathscr{P} = \frac{N_R - N_L}{N_R + N_L}, \quad (2)$$

where N_R and N_L are the numbers of right and left-handed positrons in the beam. Similarly the cross section for the NC reaction, $e^+p \rightarrow e^+X$, can be expressed as

$$\frac{d^2\sigma^{\rm NC}(e^+p)}{dxdQ^2} = \frac{2\pi\alpha^2}{xQ^4}[H_0^+ + \mathscr{P}H_{\mathscr{P}}^+], \quad (3)$$

where α is the QED coupling constant and H_0^+ and $H_{\mathscr{P}}^+$ contain the unpolarised and polarised structure functions, respectively.

RESULTS

The total cross section for e^+p CC DIS in the kinematic region $Q^2 > 200$ GeV^2 was measured to be

$$\sigma^{\rm CC}(\mathscr{P} = +0.318 \pm 0.009) = 46.7 \pm 2.4(\text{stat.}) \pm 1.0(\text{syst.})\ \text{pb}, \quad (4)$$

and

$$\sigma^{\rm CC}(\mathscr{P} = -0.402 \pm 0.011) = 22.5 \pm 1.6(\text{stat.}) \pm 0.5(\text{syst.})\ \text{pb}. \quad (5)$$

The contribution to the systematic uncertainty of 5% from the luminosity measurement is not included in the quoted systematic uncertainty. The total cross section is shown as a function of

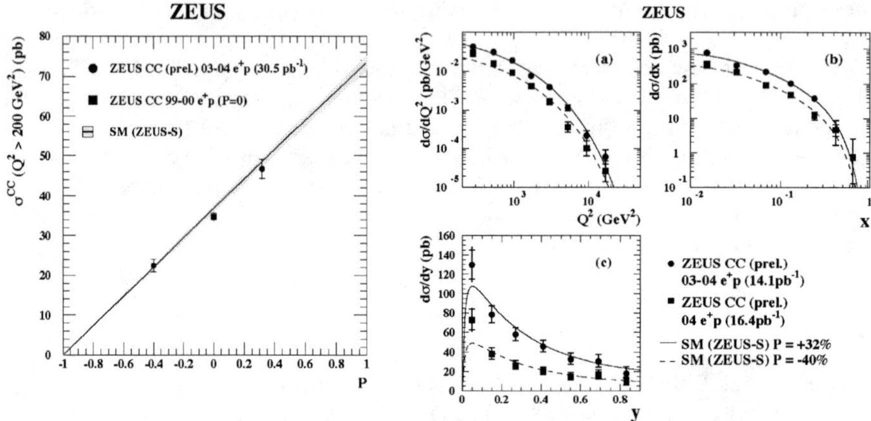

FIGURE 1. The total cross section for e^+p CC DIS as a function of the longitudinal polarisation of the positron beam is shown on the left. On the right the differential cross-sections (a) $d\sigma/dQ^2$, (b) $d\sigma/dx$ and (c) $d\sigma/dy$ are shown. The curves show the SM predictions evaluated using the ZEUS-S PDFs.

the longitudinal polarisation of the positron beam in Fig. 1 including the unpolarised ZEUS measurement from the 1999-2000 data [1]. The data are compared to the Standard Model prediction evaluated using the ZEUS-S PDFs [2]. The SM prediction describes the data well.

The single-differential cross-sections, $d\sigma/dQ^2$, $d\sigma/dx$ and $d\sigma/dy$ for charged current DIS are also shown in Fig. 1. A clear difference is observed between the measurements for positive and negative longitudinal polarisation, which is well described by the Standard Model evaluated using the ZEUS-S PDFs.

Figure 2 shows the cross-section $d\sigma/dQ^2$ for NC DIS with positively and negatively polarised positron beams. Ratios of the cross sections for positive and negative longitudinal polarisations to the ZEUS unpolarised measurements from the 1999-2000 data [3] are also shown in Fig. 2. In addition the ratio of the cross sections for positive and negative longitudinal polarisations is shown. Only statistical uncertainties were considered when taking ratios of the positively and negatively polarised cross sections. In taking ratios to the unpolarised cross sections the systematic uncertainties were considered uncorrelated with those of the polarised cross sections. The measurements are well described by the SM evaluated using the ZEUS-S PDFs and consistent with the expectations of the electroweak Standard Model for polarised NC DIS, although the statistical precision of the current data set does not allow the polarised effect to be conclusively observed.

SUMMARY

The cross sections for neutral and charged current deep inelastic scattering in e^+p collisions with longitudinally polarised positron beams have been measured. The measurements are based on data of integrated luminosity 30.5 pb^{-1} collected with the ZEUS detector in 2003 and 2004 at a centre-of-mass energy of 318 GeV. The total cross section for e^+p charged current deep inelastic scattering is presented at positive and negative values of positron beam longitudinal

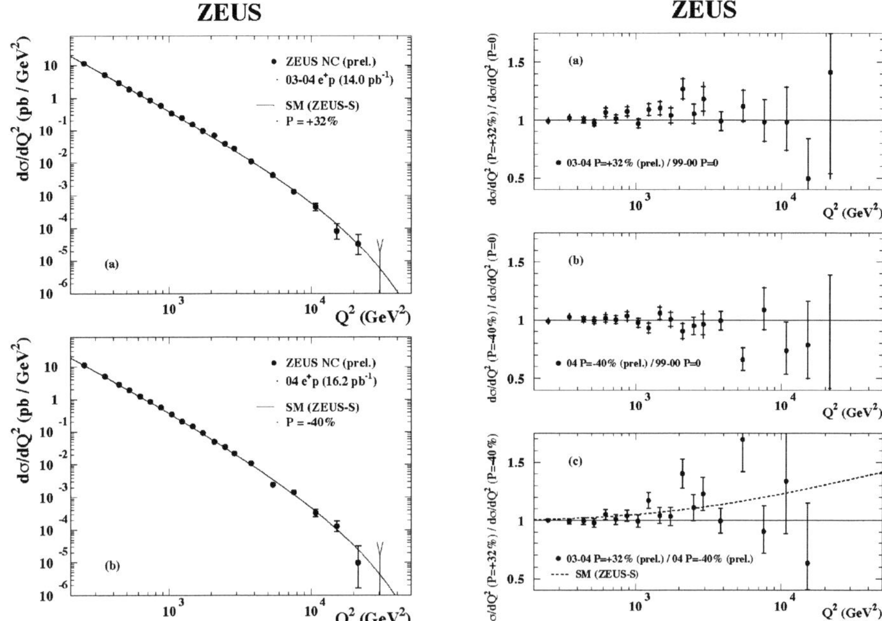

FIGURE 2. The NC DIS cross-sections $d\sigma/dQ^2$ on the left for (a) positively and (b) negatively polarised data. The curves show the SM prediction evaluated using the ZEUS-S PDFs. On the right the ratios of the cross sections for (a) positively polarised data to unpolarised data, (b) negatively polarised to unpolarised data and (c) positively polarised to negatively polarised data are shown. The dashed curve shows the SM prediction evaluated using the ZEUS-S PDFs.

polarisation. In addition, single differential cross sections are presented for charged and neutral current deep inelastic scattering in the kinematic region $Q^2 > 200$ GeV^2. The measured cross sections are well described by the predictions of the Standard Model.

REFERENCES

1. ZEUS Collab., S. Chekanov *et al.*, Eur. Phys. J. **C32**, 1 (2003).
2. ZEUS Collab., S. Chekanov *et al.*, Phys. Rev. **D67**, 012007 (2003).
3. ZEUS Collab., S. Chekanov *et al.*, Phys. Rev. **D70**, 052001 (2004).

Neutral current cross sections at high x

Y. Ning on behalf of the ZEUS collaboration

DESY ZEUS Columbia, Notke Str. 85, 22607 Hamburg, Germany
Pupin 5283, 538 W. 120th Str., New York, NY 10027, USA

Abstract. A new method is employed to measure the neutral current cross section up to Bjorken-x values equal to one with the ZEUS detector at HERA using an integrated luminosity of 65 pb^{-1}. Cross sections have been extracted for $Q^2 > 576\ \text{GeV}^2$ and are compared to predictions using different parton density functions (PDFs). The data produce new constraints on the PDFs at the high values of x.

Keywords: neutral current cross section, high x, ZEUS, HERA
PACS: 13.60.Hb

INTRODUCTION

The electron-proton deep inelastic scattering (DIS) cross section is written in terms of the proton structure functions, which can be writen in terms of PDFs and electroweak parameters. The PDFs are found to decrease very quickly for $x \geq 0.3$. The form of the PDF is typically parameterized as $(1-x)^{\eta}$ near $x \to 1$, as expected from counting rule arguments [1], and this form follows the data quite well [2, 3]. However, a direct confrontation with data has not been possible to date for $x \to 1$ due to limitations in beam energies and measurement techniques. The highest measured points in the DIS regime are for $x = 0.75$ [4]. Data at higher x exist [5, 6] but these are in the resonance production region and cannot be easily interpreted in terms of parton distributions. The highest x value for HERA structure function data is $x = 0.65$.

EXPERIMENTAL SETUP AND PROPOSED METHOD

At HERA, proton beams of 920 GeV (820 GeV prior to 1998), collide with either electron or positron beams of 27.5 GeV. ZEUS is a multipurpose detector described in detail elsewhere [7]. A schematic depiction of the ZEUS detector is given in Fig. 1. The components most relevant in this analysis are the uranium–scintillator calorimeter (CAL) [8–11], and the central tracking detector (CTD) [12–14]

Figure 1 includes a schematic depiction of a NC event: a scattered electron and a jet are outlined in the CAL, while the proton remnant largely disappears down the forward beam pipe. E'_e and E_{jet} are the energies of the scattered electron and jet; θ_e and θ_{jet} are the polar angles with respect to the proton beam direction. As x increases, the jet is boosted in the forward direction and θ_{jet} decreases. When x is too high, a part of the jet is lost in the beam pipe. The value of x at which this occurs is Q^2 dependent: the x values for which jets are well contained increases as Q^2 increases. At the Q^2 values considered in

CP792, *Deep Inelastic Scattering, DIS 2005*, edited by W. H. Smith and S. R. Dasu

this analysis, the scattered electron is at large angles and well contained in the detector. The new method employed in this analysis combines electron and jet information to

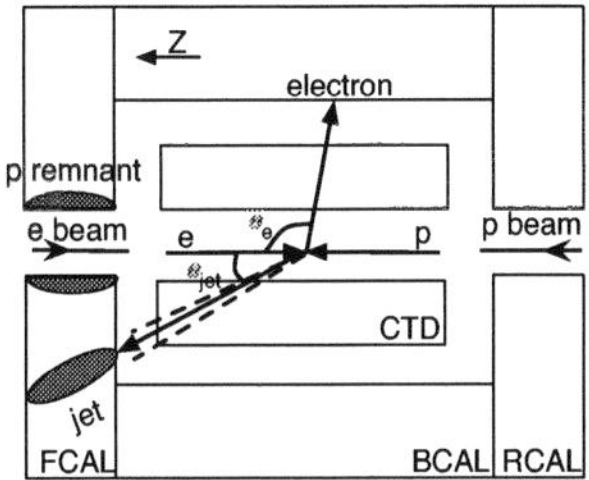

Figure 1. A schematic depiction of the ZEUS detector with the main components used in this analysis labeled. Also shown is a typical topology for events studied in this analysis. The electron is scattered at large angles and is reconstructed using the central tracking chamber (CTD) and the barrel calorimeter (BCAL), while the scattered jet is typically reconstructed in the forward calorimeter (FCAL). The jet of particles from the proton remnant largely disappears down the beam pipe.

allow a measurement of the differential cross section up to $x = 1$. Events are first sorted into Q^2 bins using information from the electron only: $Q^2 = 2E_eE'_e(1 + \cos\theta_e)$, where E_e is the electron beam energy. The jet information is then used to calculate x from E_{jet} and θ_{jet} for events with a well reconstructed jet. These events are sorted into x bins to allow a measurement of the double differential cross section $d^2\sigma/dxdQ^2$. Events with no jet reconstructed within the fiducial volume are assumed to come from high x and are collected in a bin with $x_{Edge} < x < 1$. Since these bins are generally large and the form of the PDF is not well known in this region, an integrated cross section is calculated; $\int_{x_{Edge}}^{1}(d^2\sigma/dxdQ^2)dx$. Events with more than one high energy jet are discarded.

The features of this method are:

- good resolution in Q^2 for all x;
- good resolution in x in events where a jet can be reconstructed;
- cross section measurements possible up to $x = 1$.

DATA SET

The measurement is based on the data collected by ZEUS from 1999 to 2000. The data corresponds to an integrated luminosity of 65.1 pb^{-1} for e^+p collisions. Events with high energy electron with strict fiducial cuts and zero or one jet with high transverse energy and $\theta_{jet} > 0.12$ rad are selected. The simulated MC events were used to evaluate the efficiency for the event selection and to determine the accuracy of the kinematic reconstruction. A sufficient number of events was used to ensure the statistical uncertainties from the MC samples were negligible compared to those in the data.

MC distributions are compared with those from the data. The MC distributions have been normalized to the measured luminosity. Good agreement between data and MC simulation is observed for both zero and one jet events, and there is no indication of residual backgrounds. For zero jet events, about 10 % more data events are observed than expected in the simulation.

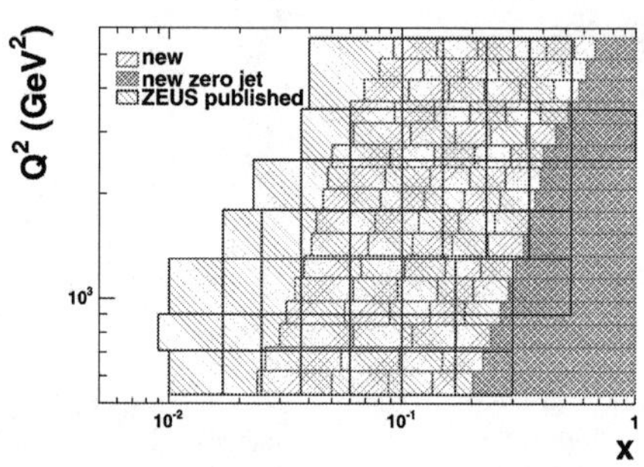

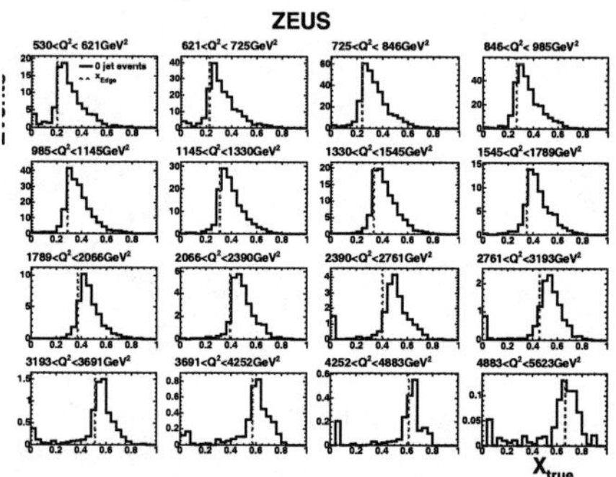

Figure 2. Definition of the bins as used in this analysis. The magenta bins extending to $x = 1$ are for the zero jet events. The blue bins show the bin structure ZEUS published [15].

Figure 3. The true x distribution from NC MC simulations for zero jet events in different Q^2 bins as indicated in the plots. The dashed lines represent the lower edge of the bins, x_{Edge}. The MC distributions are normalized to the luminosity of the data.

The bin definitions used in this analysis are given in Fig. 2. The bin widths for the double differential cross section measurements were chosen to correspond to three times the resolution of the reconstructed kinematic variables and the definition of the x bin boundaries vary with Q^2 since x_{Edge} is strongly Q^2 dependent. The x resolution of the new method is better than that of the double angle method (DA) which is usually used by ZEUS, which allows a more accurate measurement and smaller bins as shown in the figure.

The MC simulation was used to study the x distribution of the zero jet events which are assigned to the highest x bin. Figure 3 shows the true x distribution for these MC events in different Q^2 bins. As can be seen in this figure, the zero jet events originate predominantly from the interval $x_{Edge} < x < 1$. The purity in these bins is high and comparable to the purity in mid-x bins.

RESULTS AND CONCLUSION

The measured cross sections are shown in Fig. 4 and compared to SM expectations at NLO using the CTEQ6D parton distributions [16]. The double differential cross sections are represented by solid points, and generally agree well with the expectations. The cross section in the highest x bin is given as

$$\frac{1}{1-x_{Edge}} \int_{x_{Edge}}^{1} \frac{d^2\sigma}{dxdQ^2} dx \ .$$

In this bin, the expected cross section is drawn as a horizontal line, while the measured cross section is displayed as the open symbol. The measured data is plotted at the center of the bin, but it should be understood to be an integrated cross section for the bin.

The error bars represent the quadratic sum of the systematic and statistical uncertainties, where the statistical uncertainties are calculated from the square root of the number of observed events.

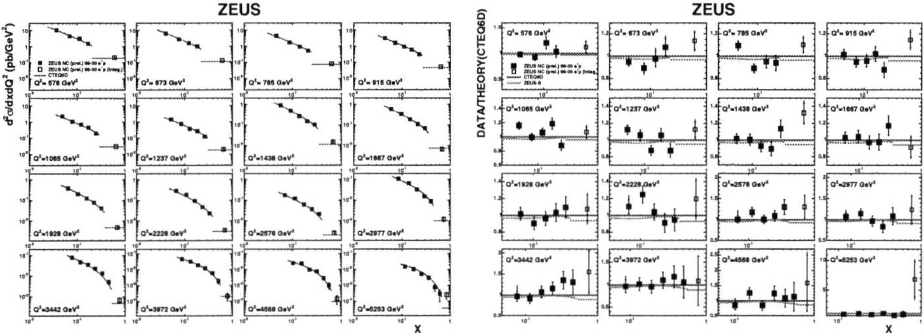

Figure 4. The double differential cross section (solid squares) and the integral of the double differential cross section divided by the bin width (open squares) compared to the Standard Model (SM) expectations evaluated using CTEQ6D PDFs (lines). The inner error bars show the statistical uncertainty, while the outer ones show the statistical and systematic uncertainties added in quadrature.

Figure 5. Ratio of the double differential cross section (solid squares) and the integral of the double differential cross section divided by x bin width (open squares) to the SM expectation evaluated using the CTEQ6D PDFs. The inner error bars show the statistical uncertainty, while the outer ones show the statistical and systematic uncertainties added in quadrature. The ratio of the expectations using the CTEQ6D PDFs to those using the ZEUS-S predictions is shown as the green (grey) lines.

The ratio of the measured cross sections to SM expectation using the CTEQ6D PDFs is shown in Fig. 5. The ratio of the expectation using the CTEQ6D PDFs to that using ZEUS-S PDFs [17] is also shown. The measured double differential cross sections generally agree well with both sets of expectations. For the highest x bins, which are in previously unmeasured kinematic ranges, the data has a tendency to lie above the expectations. These data are expected to have an impact on the extraction of the PDFs at the highest values of x, and via sum rules, also the PDFs at smaller x.

REFERENCES

1. R. Blankenbecler, and S. Brodsky, *Phys. Rev.*, **D 10**, 2973 (1974).
2. A. Martin, et al., NNLO global parton analysis (2002), hep-ph/0201127.
3. J. Pumplin, et al., New generation of parton distributions with uncertainties from global QCD analysis (2002), hep-ph/0201195.
4. A. Benvenuti, et al., *Phys. Lett.*, **B 223**, 485 (1989).
5. L. Whitlow, et al., *Phys. Lett.*, **B 282**, 475 (1992).
6. M. Osipenko, et al., *Phys. Rev.*, **D 67**, 092001 (2003).
7. The ZEUS detector, Status Report (unpublished), DESY (1993).
8. M. Derrick, et al., *NIM*, **A 309**, 77 (1991).
9. A. Andresen, et al., *NIM*, **A 309**, 101 (1991).
10. A. Caldwell, et al., *NIM*, **A 321**, 356 (1992).
11. A. Bernstein, et al., *NIM*, **A 336**, 23 (1993).
12. N. Harnew, et al., *NIM*, **A 279**, 290 (1989).
13. B. Foster, et al., *NPPS*, **B 32**, 181 (1993).

14. B. Foster, et al., *NIM*, **A 338**, 254 (1994).
15. S. C. ZEUS Collab., et al., *Phys. Rev.*, **D 70**, 052001 (2004), hep-ex/0401003.
16. J. Pumplin, et al., *JHEP*, **0207**, 012 (2002).
17. S. Chekanov, et al., *Phys. Rev.*, **D 67**, 12007 (2003).

Parity Violation in Møller Scattering

Krishna S. Kumar[†]

Department of Physics, University of Massachusetts, Amherst, MA 01002, USA.
Representing the E158 Collaboration

Abstract. We have carried out a precision of the parity-violating asymmetry A_{PV} in the scattering of longitudinally polarized electrons off electrons in a liquid hydrogen target. The measurement was performed with the 50 GeV beam line at SLAC. The final result with the full data set collected in three production runs is $A_{PV} = -131 \pm 14$ (stat) ± 10 (syst) parts per billion. The result leads to new limits on possible contact interactions at the TeV scale.

Keywords: Parity Violation, Møller Scattering, Weak Neutral Current Interactions
PACS: 11.30.Er, 12.15.Lk, 12.15.Mm, 13.66.Lm, 13.88.+e, 14.60.Cd

INTRODUCTION

Precision measurements of weak neutral current (WNC) interactions, mediated by the Z boson, play a central role in tests of the electroweak theory and in the search for new dynamics at very high energy scales. One class of WNC experiments involves measurements of the fractional difference in the cross-section for longitudinally polarized electrons scattering off unpolarized nuclear targets. A non-zero asymmetry (A_{PV}) is a signature of parity violation and arises from the interference between the WNC and electromagnetic amplitudes[1]. A_{PV} thus rises linearly with the 4-momentum transfer Q^2. For the range of Q^2 typical for fixed target scattering, A_{PV} ranges from 10^{-4} to as small as 10^{-7}, depending on the WNC couplings of the particles involved.

Measurements of WNC amplitudes can test the standard model by comparing the extracted value of the weak mixing angle $\sin^2\theta_W$ in each case (evolved to the same energy scale) to the precise value obtained in high energy collider expectation. A deviation would be a signature of new contact interactions at the TeV scale, such as a new Z′ boson or compositeness. Since fixed target WNC measurements are carried out at $Q^2 \ll M_Z^2$, such new physics amplitudes can interfere with the electromagnetic amplitude, unlike the case of WNC measurements at the Z resonance[2]. In order for such low Q^2 WNC measurements to provide new information, it is necessary to measure $\sin^2\theta_W$ to fractional accuracy better than 1%[3].

Two measurements have reached such sensitivity: the weak charge measurement in atomic Cesium[4] and the NuTeV neutrino deep-inelastic scattering measurement[5]. In this paper we discuss a measurement of A_{PV} in electron-electron (Møller) scattering[6], a purely leptonic reaction with little theoretical uncertainty[7]. The E158 experiment at the Stanford Linear Accelerator Center (SLAC) was designed to use the longitudinally polarized 50 GeV electron beam to measure A_{PV} to a relative accuracy of 10%, which enabled a measurement of $\sin^2\theta_W$ to an accuracy of 0.001.

CP792, *Deep Inelastic Scattering, DIS 2005*, edited by W. H. Smith and S. R. Dasu

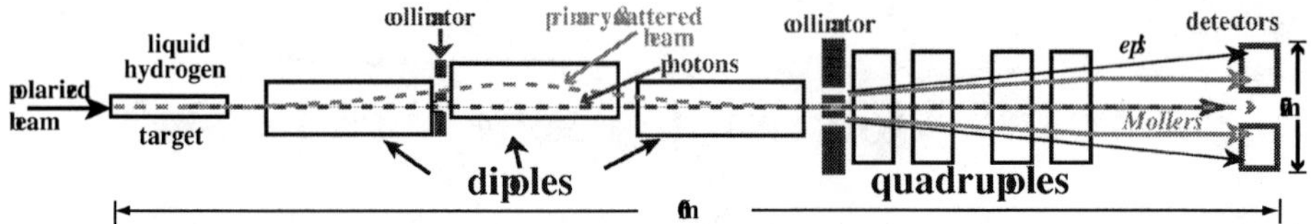

FIGURE 1. Schematic Overview of the E158 Apparatus

EXPERIMENTAL OVERVIEW

Target electrons, in a 1.54 m long cell of liquid hydrogen (10.5 gm/cm^2), are bombarded by a 48 GeV electron beam, the longitudinal polarization of which is changed pseudo-randomly, keeping all other beam parameters unchanged. Møller electrons, i.e., beam electrons scattering from target electrons, are isolated by a forward magnetic spectrometer consisting of 3 dipole "chicane" and 4 quadrupole magnets, oriented with their magnetic axes along the primary beam direction. Møller electrons of interest in the full range of the azimuth (spanning the polar angular range 4.5 mrad $< \theta_{lab} <$ 8 mrad) traverse through the bores of the quadrupoles and are brought to focus in a ring on a calorimeter located 60 m downstream of the target. A schematic overview of the apparatus is shown in Fig. 1.

The experimental asymmetry is measured by averaging the fractional difference in the cross-section over many complementary pairs of beam pulses of opposite helicity. In order to achieve the desired statistical precision of 10 parts per billion (ppb) in a reasonable length of time, the intergrated signal of more than 20 million electrons must be detected each beam pulse.

The calorimeter provides both radial and azimuthal segmentation. Four radial rings are uniformly covered in the azimuth by a total of 60 photomultiplier tubes (PMTs). The three inner rings are predominently sensitive to Møller electrons while the outermost ring intercepts the bulk of the flux from electron-proton (*ep*) scattering. The radiative tail of the *ep* flux is the main background in the Møller rings, totalling $\approx 8\%$. Other charged and neutral particles contribute less than 1%.

Data were collected over three run periods in 2002 and 2003. The longitudinally polarized electron beam is generated via photoemission of a GaAs photocathode by circularly polarized laser light. This facililated rapid reversal of the electron beam polarization. In addition, spurious asymmetries were suppressed by passively reversing the sign of the helicity asymmetry by two independent methods. First, the state of a half-wave plate in the laser line was toggled each day. Second, spin precession in the 24.5° bend after beam acceleration created opposite helicity orientation at 45 GeV and 48 GeV beam energies. Roughly equal statistics were accumulated with opposite signs of the measured asymmetry.

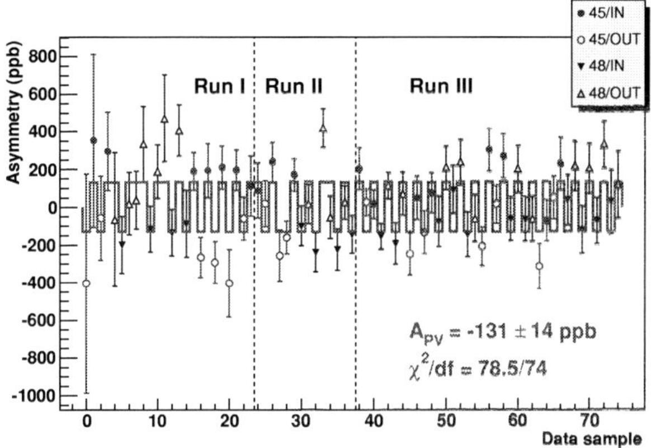

FIGURE 2. A_{PV} for 75 data samples

RESULTS AND IMPLICATIONS

Figure 2 shows the parity-violating asymmetry as a function of data-set number. Each data-set constitutes about two days of data, after which either the beam energy or the state of the half-wave plate were changed to flip the sign of the measured asymmetry. The grand average result for the parity-violating asymmetry in Møller scattering at $Q^2 = 0.03$ GeV^2 was found to be:

$$A_{\text{PV}} = -131 \pm 14(\text{stat}) \pm 10(\text{syst}) \quad (\text{ppb}). \tag{1}$$

From this measurement, the value of $\sin^2\theta_W$ can be extracted within the context of the Standard Model. Using a definition which reproduces the effective leptonic couplings at the Z pole, we determine:

$$\sin^2\theta_{\text{W}}^{\text{eff}} = 0.2397 \pm 0.0010(\text{stat}) \pm 0.0008(\text{syst}). \tag{2}$$

Figure 3 shows the E158 result, which establishes the "running" of $\sin^2\theta_W$[8] by more than 6 standard deviations. Also shown are the two other precise low energy $\sin^2\theta_W$ determinations mentioned earlier. It can be seen that the atomic Cesium result and the E158 result are consistent with the standard model expectation. The deviation of the NuTeV result thus implies that either there are new contact interactions specific to neutrino interactions or that there are additional strong interaction effects that are unaccounted for in neutrino deep inelastic scattering. One leading candidate is charge symmetry violation in the parton distribution functions[9].

The E158 measurement can be used to set limits on the size of possible new contributions beyond the standard model. Assuming a new contact interaction scale[10] characterized by Λ_{LL}, the 95% C.L. limit is 7 TeV or 16 TeV depending on the sign of the contact interaction term.

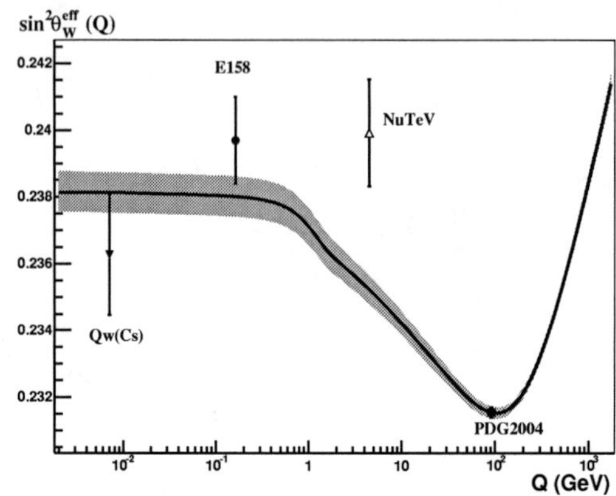

FIGURE 3. $\sin^2\theta_W$ as a function of Q^2

FUTURE PROSPECTS

The figure of merit on the achievable precision on the WNC amplitude in Møller scattering rises with incident beam energy. It is therefore possible to contemplate improved new measurements of $A_{\rm PV}$ at future facilities[11]. One possibility is to carry out a new measurement at Jefferson Laboratory after it is upgraded to 12 GeV, where it is potentially possible to achieve a factor of 5 improvement over the reported measurement. The ultimate measurement could be carried out at the proposed International Linear Collider, using the electron beam downstream of the primary collider interaction region. More than an order of magnitude improvement is possible [12], which competes very favorably with future collider determinations of the weak mixing angle and measurements of the W boson mass.

REFERENCES

1. Ya.B. Zel'dovich, *Sov.Phys.JETP* **94**, 1959, pp. 262-263.
2. K.S. Kumar et. al, *Mod.Phys.Lett.* **A10**, 1995, 2979.
3. M.J. Ramsey-Musolf, *Phys.Rev.* **C60**, 1991, 015501.
4. S.C. Bennett and C.E. Wieman, *Phys.Rev.Lett.* **82**, 1999, 2484-2487.
5. G.P. Zeller et. al, *Phys.Rev.Lett.* **88**, 2002, 0918802.
6. A. Czarnecki and W.J.Marciano, *Phys.Rev.* **D53**, 1996, 1066.
7. P.L. Anthony et.al, *Phys.Rev.Lett.* **92**, 2004, 181602.
8. A. Czarnecki and W.J. Marciano, *Int.J.Mod.Phys.* **A15**, 2000, 2365; J. Erler, A. Kurylov and M.J. Ramsey-Musolf, *Phys.Rev.* **D68**, 2003, 016006; A. Ferroglia, G. Ossola, and A. Sirlin, *Eur.Phys.J.* **C34**, 2004, 165; J. Erler and M.J. Ramsey-Musolf, hep-ph/0409169, 2004; F.J. Petriello, *Phys.Rev.* **D68**, 2003, 033006.
9. J.T. Londergan and A.W. Thomas, hep-ph/0407247, 2004, and references therein.
10. E.J. Eichten, K.D. Lane, and M.E. Peskin, *Phys.Rev.Lett* **50**, 1983, 811.
11. K.S. Kumar, *Eur.Phys.J.* **A24**, s2, 2005, 191-195.
12. K.S. Kumar, in *DPF/DPB Summer Study on New Directions in High Energy Physics*, econf **C960625**, 1996, NEW168.

Parity Violation in ep scattering at JLab

P. A. Souder

Department of Physics, Syracuse University, Syracuse, NY13244

Abstract. We review the program of parity violation in the scattering of polarized electrons at JLab. Results are presented from recent experiments measuring the weak form factors, which in turn measure the contribution of strange quarks to the elastic form factors. In addition, we discuss the physics of parity violation in deep inelastic scattering, which will become possible with the upgrade of the JLab energy to 12 GeV.

Keywords: Parity violation, Form Factors, Structure Functions
PACS: 11.30.Er, 134.0.Gp, 13.60.Hb

INTRODUCTION

By measuring the parity-violating asymmetry in the scattering of polarized electrons $A_{PV} = (\sigma_\uparrow - \sigma_\downarrow)/(\sigma_\uparrow + \sigma_\downarrow)$, we determine the electroweak interference between the photon and the Z boson. From this information, we can extract weak amplitudes from electron scattering. The first experiment of this type was done by Prescott et al. [1], who in 1978 measured a 10^{-4} asymmetry in deep inelastic scattering, thereby establishing that the interactions of the Z are parity violating.

Recently, there has been a large number of experimental results published on parity violation in electron scattering from SLAC, Mainz, Bates, and JLab. The asymmetries measured have been as small as 10^{-7}. The goals of these experiments have been either to probe hadronic structure or to test the Standard Model [2].

ELASTIC SCATTERING

Elastic scattering from the proton, deuteron, and He targets is a classic example of using the weak interaction to probe hadronic structure [3, 4]. Elastic scattering is determined by two functions of Q^2, G_E and G_M. Both form factors have contributions from the various quarks, including u, d, and possibly s. The strangeness contribution to the form factors, possibly caused by a "kaon cloud" around the nucleon, can be differentiated from the larger u and d contributions by measuring the weak form factor.

Before this year, data on parity violation were consistent with the absence of significant strange contribution to the form factors. Then the A4 experiment at Mainz published a 2 σ effect at $Q^2 = 0.1(\mathrm{GeV/c})^2$ [5]. With the publication of data from the HAPPEX [6, 7] and G_0 [8] collaborations at JLab, the significance of the world's data at $Q^2 = 0.1(\mathrm{GeV/c})^2$ remains at about the 2 σ level. However, data from G_0, which covers a wide range of Q^2, indicates that the observed effect of strangeness in the range $Q^2 = 0.2 - 0.3(\mathrm{GeV/c})^2$ is quite small. Thus strange form factors, if they are large, have

CP792, *Deep Inelastic Scattering, DIS 2005*, edited by W. H. Smith and S. R. Dasu

a rapid Q^2 dependence. Further experiments are planned that will clarify this potentially exciting situation.

DEEP INELASTIC SCATTERING

In the limit of deep inelastic scattering (DIS), the parity-violating asymmetry is given by

$$A^{PV} = Q^2 \frac{G_F}{2\sqrt{2}\pi\alpha}\left[a(x) + \frac{1-(1-y)^2}{1+(1-y)^2} b(x)\right],$$

with $a(x) = \sum_i f_i(x) C_{1i} Q_i / \sum_i f_i(x) Q_i^2$, and $b(x) = \sum_i f_i(x) C_{2i} Q_i / \sum_i f_i(x) Q_i^2$. Here $f_i(x)$ is the probability of a parton with flavor i having a fraction x of the momentum of the nucleon. Also Q_i is the electromagnetic charge and $C_{1i}(C_2 i)$ are the weak vector (axial vector) charges for the ith flavor. For an isoscalar target such as deuterium, we have

$$a(x) = \frac{6}{5}\left[\left(C_{1u} - \frac{1}{2}C_{1d}\right) + \frac{2}{15}\frac{s(x)+\bar{s}(x)-c(x)-\bar{c}(x)}{D}\right] \tag{1}$$

The key feature is that large $a(x)$ term is constant and independent of hadron structure at moderately large x where the heavy quarks make small contributions. Thus the deuteron makes an ideal target for tests of the Standard Model[9].

For hydrogen targets, the $a(x)$ term is given by

$$a(x) \approx \frac{3}{4}\left[\frac{6C_{1u}u(x) - 3C_{1d}d(x)}{u(x)+\frac{1}{4}d(x)}\right] \sim \left[\frac{u(x)+0.912d(x)}{u(x)+0.25d(x)}\right]$$

In this case the asymmetry depends on the ratio $d(x)/u(x)$, a quantity of considerable interest for very large x[10, 11, 12, 13].

Possible Hadronic Corrections

Since any interesting corrections to Eq. 1 are at most 5-10%, the experiments must attain precision at the 1-2% level. However, the surprising results of the NuTeV experiment [14] have led to a critical examination of whether all hadronic effects are understood to this high precision. We discuss below two possible effects which might be interesting in themselves, higher twist and charge symmetry.

Higher Twist

One interesting candidate for hadronic corrections are the higher twist (HT) effects, which might add Q^2 dependence to the asymmetry [15]:

$$A^{PV}(x, Q^2) = A^{PV}(x)(1 + C(x)/Q^2).$$

The size of the coefficients $C(x)$ is not known. However, the effect might be about the same size as the corresponding coefficients in the formula

$$F_2(x,Q^2) = F_2(x)(1+D(x)/Q^2).$$

extracted from GLAP fits to structure function data $F_2(x,Q^2)$ [16]. The sizes of the $D(x)$ for $x < 0.4$ are negligible for GLAP fits at the NNNLO level. However, the effects may be substantial and easy to measure for $x > 0.5-0.6$. A_{PV} makes an ideal laboratory for studying HT because the dominant $a(x)$ term is constant and requires no GLAP evolution.

Charge Symmetry Violation

One of the assumptions in deriving Eq. 1 is charge symmetry: $u^p = d^n$ and $d^p = u^n$. If we allow charge symmetry violation (CSV), new structure functions are required:

$$\delta u(x) = u^p(x) - d^n(x); \;\; \delta d(x) = d^p(x) - u^n(x).$$

The effect of the CSV parameters on R^{PW}, the quantity measured by NuTeV, is[17]:

$$\frac{\delta R^{PW}}{R^{PW}} \sim 0.85 \frac{\delta u - \delta d}{u+d}.$$

For parity, we have

$$\frac{\delta A^{PV}}{A^{PV}} = 0.28 \frac{\delta u - \delta d}{u+d}. \tag{2}$$

Knowledge about $(\delta u - \delta d)/(u+d)$ is limited. Empirical limits from the MRST group allow CSV effects to be large enough to explain the NuTeV result[18].

The asymmetry A^{PV} is less sensitive to CSV than is R^{PW}. However, for the NuTeV experiment with its large kinematic acceptance, the data is averaged over a large range in x. With the spectrometers at JLab, data may be obtained over a narrow kinematic range and in particular focus on the region of large x. Then if the CSV ratio in Eq. 2 is larger at high x, perhaps because of the falling values of the structure functions, measurements of APV at JLab might observe a significant effect. Any experimental demonstration of an x-dependent CSV would be extremely important.

Required Apparatus

A DIS point can be done with existing facilities at JLab at low x and low Q^2. With the advent of the 12 GeV upgrade and the SHMS spectrometer, additional points can be reached. Also, useful data with $W < 2$ GeV can be obtained. However, to reach points with the highest Q^2 and x values with sufficient statistics, a spectrometer with at least 50% acceptance in azimuth operating at angles of 20°-40° is required. We do not know how to achieve this goal economically with a conventional magnetic spectrometer.

We suggest that a calorimeter such as the one used for A4 at Mainz might be more suitable. At the higher beam energies at JLab, the calorimeter must be shielded from photons and low energy pions originating in the target. Thus a sweeping magnet would be required. Such an apparatus will also be useful for many other physics topics, such as measurements of $A_1(x)$.

Summary of the DIS Program

With the proposed spectrometer, we can plan to measure A_{PV} for hydrogen and deuterium with <2% relative precision over the range $0.3 < x < 0.7$ with Q^2 varying by a factor of 2 for each point except at $x = 0.7$. Unexpected variation of A_{PV} for the deuteron with Q^2 or x would be a signature for HT effects or CSV respectively. If these effects are demonstrated to be under control, we then can use deuterium as a test of the Standard Model and measure d/u at high x for hydrogen.

ACKNOWLEDGMENTS

I would like to thank A. Afanesev, D. Beck, P. Bosted, J. T. Londergan, K. Kumar, K. McFarland, F. Maas, Z. Meziani, W. Melnitchouk, M. J. Ramsey-Musolf, P. Reimer, A. Thomas, and M. Vanderhaeghen for useful discussions. This work is supported by the U. S. Department of Energy under Grant No. DE-FG020-84ER40146.

REFERENCES

1. C. Y. Prescott, *et al., Phys. Lett.* B 84, 524 (1979) .
2. M. J. Musolf, T. W. Donnelly, J. Dubach, S. J. .. Pollock, S. Kowalski and E. J. Beise, Phys. Rept. **239**, 1 (1994).
3. K. S. Kumar and P. A. Souder, Prog. Part. Nucl. Phys. **45**, S333 (2000).
4. D. H. Beck and B. R. Holstein, Int. J. Mod. Phys. E **10**, 1 (2001) [arXiv:hep-ph/0102053].
5. F. E. Maas *et al.*, Phys. Rev. Lett. **94**, 152001 (2005).
6. K. A. Aniol *et al.* [HAPPEX Collaboration], arXiv:nucl-ex/0506011.
7. K. A. Aniol *et al.* [HAPPEX Collaboration], arXiv:nucl-ex/0506010.
8. D. S. Armstrong *et al.* [G0 Collaboration], arXiv:nucl-ex/0506021.
9. A. Kurylov, M. J. Ramsey-Musolf and S. Su, Phys. Lett. B **582**, 222 (2004).
10. W. Melnitchouk, I. R. Afnan, F. Bissey and A. W. Thomas, Phys. Rev. Lett. **84**, 5455 (2000).
11. W. Melnitchouk and A. W. Thomas, Phys. Lett. B **377**, 11 (1996).
12. S. I. Alekhin, Phys. Rev. D **63**, 094022 (2001).
13. S. Kuhlmann *et al., Phys. Lett.* B 476, 297 (2000) .
14. G. P. Zeller *et al.* [NuTeV Collaboration], Phys. Rev. Lett. **88**, 091802 (2002) [Erratum-ibid. **90**, 239902 (2003)].
15. S. Brodsky, in *"Proceedings from the JLab/Temple University HiX2000 Workshop,"*, 2000.
16. A. D. Martin, R. G. Roberts, W. J. Stirling and R. S. Thorne, Eur. Phys. J. C **35**, 325 (2004).
17. J. T. Londergan and A. W. Thomas, Phys. Rev. D **67**, 111901 (2003).
18. J. T. Londergan, Eur. Phys. J. A **24S2**, 85 (2005).

Parton Uncertainties and the Stability of NLO Global Analysis

Daniel R. Stump

Department of Physics and Astronomy
Michigan State University
East Lansing, MI 48824

Abstract. In global analysis of QCD, both experimental and theoretical errors contribute to the uncertainty of the results. A recent study of the stability of NLO global analysis is described.

Keywords: Parton Distribution Functions, QCD
PACS: 12.38.-t, 12.39.St, 13.60.Hb

In the global analysis of QCD and parton distribution functions (PDFs), data from many different experiments are combined to extract the PDFs and to test the predictions of perturbative QCD. Both experimental and theoretical errors contribute to the uncertainties of the results. Assessment of the uncertainties of PDFs is a nontrivial issue [1-4]: a standard statistical analysis is not adequate because of systematic errors.

The parton structure of the nucleon is a nonperturbative aspect of QCD. The PDFs $f_i(x,Q)$ are not calculated from theory, but must be extracted from data. The functions are parameterized at a low momentum scale Q_0 of order 1 GeV, using functional forms that have reasonable behavior as $x \to 0$ and $x \to 1$, and with a set of free parameters $\{a_n\}$ that can be adjusted to fit the full set of data chosen for the global analysis. The dependence of the PDFs on momentum scale Q is assumed to be well-described by the DGLAP evolution equations in next-to-leading order (NLO) perturbation theory. (The NNLO approximation is also used in some cases [5-7].)

GLOBAL ANALYSIS AND UNCERTAINTIES OF PDFs

Data from disparate processes are used in the global analysis of QCD. Deep-inelastic scattering (DIS) gives crucial information on the PDFs. The Drell-Yan process yields complementary information. Measurements of inclusive jet production at the Tevatron collider provide important additional constraints on the gluon distribution. By combining experimental data from these processes, and theoretical calculations of the cross sections, we construct a consistent set of PDFs.

In global analysis, we inevitably face two questions: of compatibility and of stability. Are data sets from different experiments compatible? Are the final results of the global analysis stable and robust? A simple parable illustrates the question of compatibility. Suppose two experimental groups have measured a quantity θ. Each

CP792, *Deep Inelastic Scattering, DIS 2005*, edited by W. H. Smith and S. R. Dasu

experiment has both statistical and systematic errors. Because of the errors, there is a systematic difference between the experiments. The data sets are consistent, provided that the systematic errors are taken into account. But the combined result must be a compromise, with a large uncertainty associated with the systematic errors. This simple example illustrates what happens in global analysis of QCD. Data from different experiments are only consistent when systematic errors are taken into account. Therefore the PDFs that are extracted from the analysis must be a compromise between experiments with systematic differences. The best fit to one data set will not be the best fit to another data set. The final PDFs must fit all data acceptably. They then have large uncertainties comparable to the systematic differences between experiments.

To assess the uncertainties of the PDFs constructed from a global analysis is a painstaking process. Complete methods have been devised to study the PDF uncertainties [1,2]. These methods are applied to the comparison between perturbative QCD theory and current data, and to predictions for future experiments.

THE CTEQ STABILITY STUDY

In order for the results of a global analysis to be trustworthy, they must be robust. The fitting parameters are adjusted such that the theory matches all chosen data sets acceptably, i.e., within the experimental errors. "Stability" means that small changes in the inputs, e.g., the selected data or the theoretical assumptions, will not produce large changes in the outputs, e.g., the PDF parameters or PDF-dependent predictions.

The stability of next-to-leading-order (NLO) global analysis has been challenged in an interesting study by the MRST group [4]. They imposed cuts on the data selected for the global analysis, requiring $Q>Q_{cut}$ and $x>x_{cut}$, and asked whether the resulting PDFs are stable with respect to small variations of the cutoffs, Q_{cut} or x_{cut}. They found surprisingly large changes in the PDFs, for their parameterization, as the cutoff x_{cut} on x was raised from 0 to 0.005. For example, the central prediction for the cross section σ_W(LHC) (for inclusive production of $W^{\pm}$ at the LHC) decreased by 20 percent from x_{cut}=0 (the default MRST PDFs) to x_{cut}=0.005 (the "conservative" PDF analysis). Since the default PDF uncertainty on σ_W(LHC) is estimated to be approximately ±5%, the large dependence on x_{cut} raises a question of the stability of the NLO analysis. Is the apparent instability a breakdown of the NLO approximation, or a consequence of increased PDF uncertainty, or an artifact of the parameterization, or due to some other reason?

The CTEQ global analysis group has carried out a study of the stability of the NLO global analysis for the CTEQ parameterization of PDFs, in order to clarify the question of stability [8]. Table 1 shows the results. Three choices of exclusionary cuts are compared: standard and strong (similar to the MRST default and conservative cuts, respectively) and an intermediate case. N_{pts} is the number of data points used in each global analysis, equal to 1926 for the standard cuts and 1588 for the strong cuts. The value of χ^2 for the data that is included, changes very little from standard to strong cuts; χ^2 decreases only from 1583 to 1573 for the 1588 data points that pass the strong cuts. Hence the NLO global analysis is in fact stable with respect to the exclusion of low-x (x < 0.005) data. Table 1 also shows the central predictions for the cross section σ_W(LHC) (times the branching ratio B for $W^{\pm}$ to decay to leptons) which changes by only 1.5 percent from standard to strong cuts.

Table 1: Results of the CTEQ stability study							
Cuts	Q_{min}	x_{min}	N_{pts}	$\chi^2(1926)$	$\chi^2(1770)$	$\chi^2(1588)$	σ_W B
Standard	2 GeV	0	1926	2023	1850	1583	20.02 nb
Intermed	2.5 GeV	0.001	1770	-	1849	1579	20.10 nb
Strong	3.16 "	0.005	1588	-	-	1573	20.34 nb

Figure 1 shows σ_W.B graphically, from the CTEQ (+) and MRST (•) studies. (Also shown (×) are the CTEQ results for a parameterization in which the gluon distribution is allowed to be negative for small x at $Q=Q_0$.) The NLO prediction of σ_W(LHC) is stable for the CTEQ parameterization.

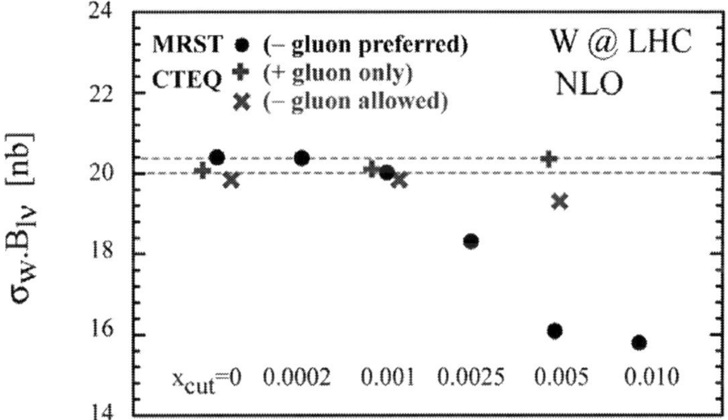

FIGURE 1. The cross section for $W^{\pm}$ production at the LHC, based on global analyses of the PDFs, as a function of the cutoff x_{cut} on x.

LAGRANGE MULTIPLIER METHOD AND THE GLUON PDF

To gain more insight into the stability of the NLO global analysis, we also used the Lagrange Multiplier method (LM) to study the uncertainty of σ_W(LHC) as a function of exclusionary cuts on input data [8]. In general, the LM method calculates the minimum χ^2 as a function of any chosen constrained variable X that depends on the PDFs [1]. We applied the method to the cross section σ_W(LHC) for $W^{\pm}$ production at the LHC, separately for the three choices of exclusionary cuts in Table 1. Thus for each case we obtained the parabola of the best χ^2 versus σ_W(LHC).

The results of the LM analysis are (i) that the position of the absolute minimum of the χ^2 parabola changes very little from standard to strong cuts; but (ii) that the width of the χ^2 parabola increases significantly from standard to strong cuts. In other words, the central prediction, which is the value of σ_W at the absolute minimum of the χ^2

parabola, is stable; that result is already seen in Table 1 and Figure 1. However, the uncertainty of the prediction increases significantly from standard to strong cuts. The latter result makes sense. By excluding 338 data points (those with $x < 0.005$) we have lost a lot of information about the PDFs. Any prediction that is sensitive to the parton structure at small x will have a much larger uncertainty for the strong cuts. From this viewpoint, the stability of the central prediction is not such a significant issue; more important is the growth of uncertainty.

It is easy to see that σ_W(LHC) is sensitive to the gluon distribution function. In the LO parton process $pp \rightarrow u\bar{d} \rightarrow W^+$, the $\bar{d}$ is a sea quark of the second proton; sea quarks are closely related to the gluon distribution. Or, the NLO process $ug \rightarrow dW^+$ depends directly on the gluon distribution. The gluon distribution still has a large uncertainty at present, so the cross section σ_W(LHC) must be uncertain and sensitive to unconstrained assumptions about the form of the gluon PDF—a theoretical uncertainty. The CTEQ and MRST parameterizations for the gluon PDF are quite different [8]. The difference between CTEQ and MRST on the stability of the NLO global analysis, illustrated in Figure 1, must originate in the difference between their parameterizations, i.e., part of the theoretical uncertainty of PDFs.

ACKNOWLEDGEMENTS

The work described here was done in collaboration with J. Huston, J. Pumplin and W.K. Tung. We thank Robert Thorne for useful discussions concerning the MRST results and suggestions to clarify the comparison between CTEQ and MRST .

REFERENCES

1. D. Stump et al., Phys. Rev. D **65**, 014012 (2002) [arXiv:hep-ph/0101051].
2. J. Pumplin et al., Phys. Rev. D **65**, 014013 (2002) [arXiv:hep-ph/0101032].
3. A. D. Martin, R. G. Roberts, W. J. Stirling and R. S. Thorne, Eur. Phys. J. C **28**, 455 (2003).
4. A. D. Martin, R. G. Roberts, W. J. Stirling and R. S. Thorne, Eur. Phys. J. C **35**, 325 (2004).
5. A. D. Martin, R. G. Roberts, W. J. Stirling and R. S. Thorne, Phys. Lett. B **531**, 216 (2002).
6. S. Alekhin, Phys. Rev. D **68**, 014002 (2003); arXiv:hep-ph/0311184.
7. A. Vogt, S. Moch and J. A. M. Vermaseren, Nucl. Phys. B **691**, 129 (2004).
8. J. Huston, J. Pumplin, D. Stump and W.K. Tung, Stability of NLO Global Analysis and Implications for Hadron Collider Physics, to appear in JHEP [arXiv:hep-ph/0502080].

Recent Progress in Parton Distributions and Implications for LHC Physics

Robert S. Thorne*, A. D. Martin†, R. G. Roberts** and W. J. Stirling†

*Cavendish Laboratory, University of Cambridge, Madingley Road, Cambridge, CB3 0HE, UK
†Institute for Particle Physics Phenomenology, University of Durham, DH1 3LE, UK
**Rutherford Appleton Laboratory, Chilton, Didcot, Oxon, OX11 0QX, UK

Abstract. I outline some of the most recent developments in the global fit to parton distributions performed by the MRST collaboration.

Keywords: <QCD,Structure Functions>
PACS: 12.38.Bx,13.60.Hb

At present there is a great deal of interest in the importance of parton distributions for studies at the LHC. This necessarily involves a certain kinematic range for the partons. The kinematic range for particle production at the LHC is shown in fig. 1. Parton distributions at $x \sim 0.001 - 0.01$ are vital for understanding the standard production processes at the LHC. However, even smaller (and higher) x partons are required when one moves away from zero rapidity, e.g. when calculating the total production cross-section of the heavy boson. As well as the central values one needs the uncertainties on the partons, and there has been a lot of work on this [1]–[9]. This uncertainty is shown for the $\bar{u}$ and $\bar{d}$ quarks in fig. 2. Central rapidity production of W,Z Higgs at

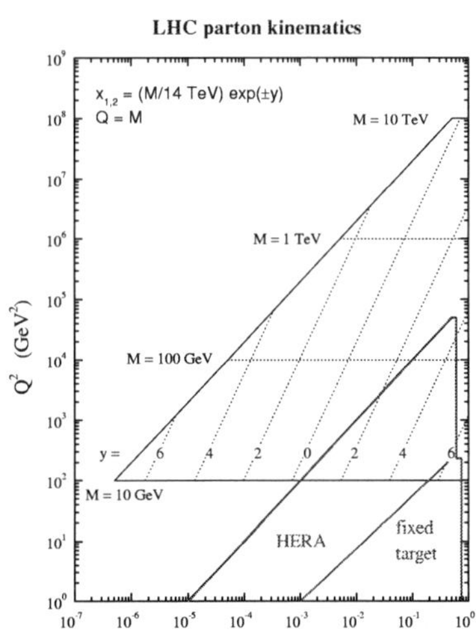

FIGURE 1. The kinematic range for particle production at the LHC

CP792, *Deep Inelastic Scattering, DIS 2005*, edited by W. H. Smith and S. R. Dasu

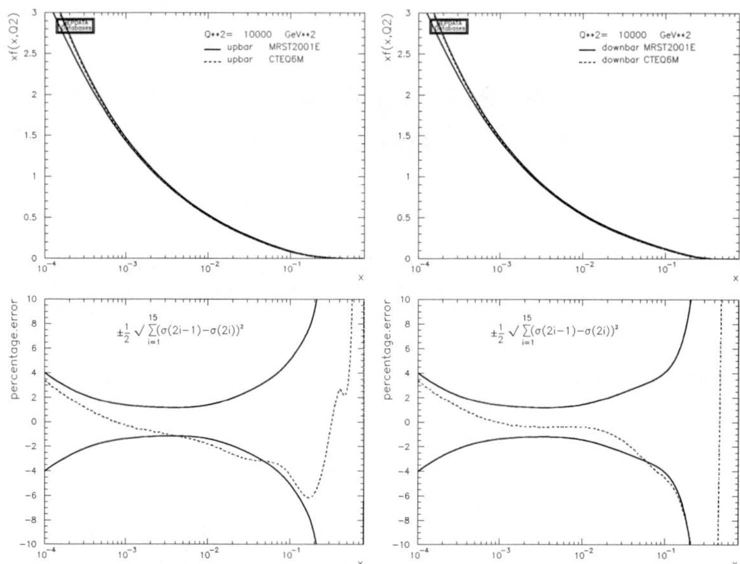

FIGURE 2. Uncertainty on MRST $\bar{u}$ and $\bar{d}$ distributions, along with CTEQ6

the LHC probes $x = 0.006$, which is ideal for the MRST partons. The current best estimate for the uncertainty due to experimental errors is $\delta\sigma_{W,Z}^{\rm NLO}(\text{expt. pdf}) = \pm 2\%$, but we note that there is a theoretical uncertainty, which is potentially large due to possible problems at small x. This is because the large rapidity W and Z cross-sections sample very small x. However, the ratio $\sigma(W^+)/\sigma(W^-)$ is a *gold-plated* prediction, where $R_\pm = \frac{\sigma(W^+)}{\sigma(W^-)} \simeq \frac{u(x_1)\bar{d}(x_2)}{d(x_1)\bar{u}(x_2)} \simeq \frac{u(x_1)}{d(x_1)}$ and using the MRST2001E partons $\delta R_\pm(\text{expt. pdf}) = \pm 1.4\%$. Assuming all other uncertainties cancel, this is probably the most accurate SM cross-section test at LHC.

This suggests that $\sigma(W)$ or $\sigma(Z)$ could be used to calibrate other cross-sections, e.g. $\sigma(WH)$, $\sigma(Z')$. As an example we consider W plus Higgs production. $\sigma(WH)$ is more precisely predicted than $\sigma(W)$ because it samples quark pdfs at higher x and scale. However, the ratio shows no improvement in uncertainty, and can be worse, see fig. 3. This is because partons in different regions of x are often anti-correlated rather than correlated, partially due to sum rules. Similarly, there is no obvious advantage in using $\sigma(t\bar{t})$ as a calibration SM cross-section, except maybe for very particular, and rather large, M_H, where the gluon is probed in the same region for both. However, a light (SM or MSSM) Higgs is dominantly produced via $gg \to H$, and the cross-section has a small pdf uncertainty because $g(x)$ at small x is well constrained by HERA DIS data. The current best MRST estimate, for $M_H = 120\,\text{GeV}$, is $\delta\sigma_H^{\rm NLO}(\text{expt pdf}) = \pm 2-3\%$ with less sensitivity to small x than $\sigma(W)$. This is a much smaller uncertainty than that from higher-order corrections, for example [10], $\delta\sigma_H^{\rm NNLL}(\text{scale variation}) = \pm 8\%$. In constrast, the error on predictions for very high-E_T jets at the LHC is dominated by the parton uncertainties, because it is sensitive to the relatively poorly known high-x gluon.

Different approaches to fits generally lead to similar uncertainties for measured quantities, but can lead to different central values [9]. For the true uncertainty one must consider the effect of assumptions made during the fit and the correctness of fixed order QCD. The failings of NLO QCD are indicated by some areas where the fit quality could

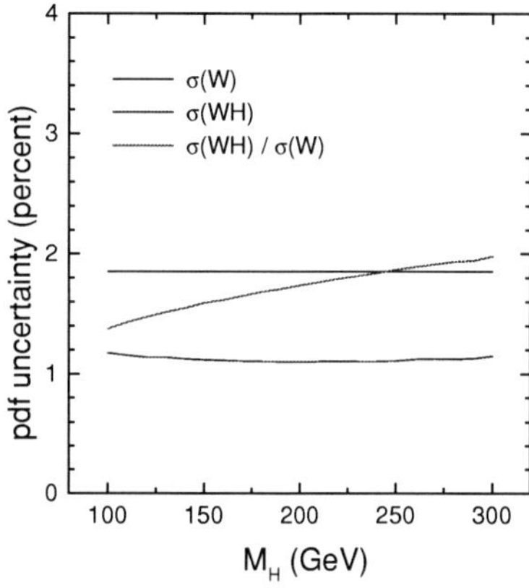

FIGURE 3. The uncertainty on W and Higgs production

be improved. There is a good fit to HERA data, but there are some problems at the highest Q^2 at moderate x, i.e. in $dF_2/d\ln Q^2$. Also the data require the gluon to be small or negative at low Q^2 and x, and this is needed by all data (e.g. Tevatron jets), not just low-Q^2, low-x data. Other groups find similar problems with the gluon at low x. CTEQ have a valence-like input gluon at $Q_0^2 = 1.69\text{GeV}^2$ which would marginally prefer to be negative [11]. There is also instability in the physical, gluon-dominated quantity $F_L(x,Q^2)$ going from LO to NLO to NNLO, seen in fig. 4. The exact NNLO coefficient function [12] has a very large effect, a possible sign of $\ln(1/x)$ corrections being required.

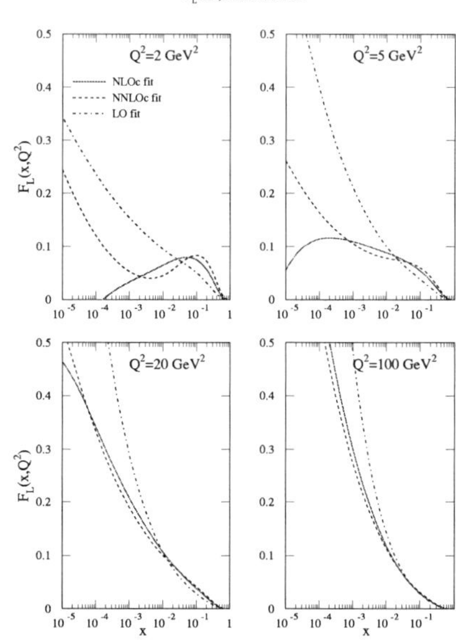

FIGURE 4. The MRST prediction for $F_L(x,Q^2)$ at LO, NLO and NNLO

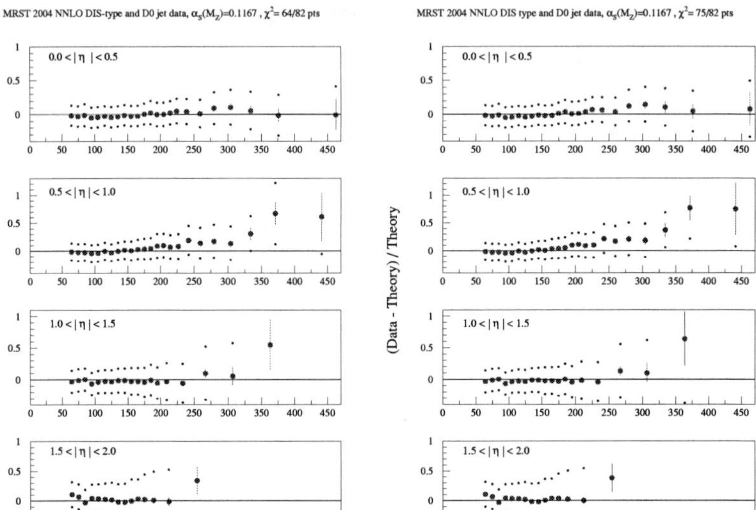

FIGURE 5. Change in fit to D0 data with weak corrections

As an example of the effect assumptions can make to the fit, MRST found only a reasonable fit to jet data [13, 14], but needed to use the large systematic errors, while the result is better for CTEQ6 [6] due to different cuts on other data, and a different type of high-x parameterization. However, for the CTEQ6.1M partons, which give a good fit to the jet data, the gluon is very hard as $x \to 1$. MRST have recently utilised the fact that, under a change of scheme from $\overline{MS}$ to DIS schemes, the scheme transformation will dominate the high-x gluon if valence quarks are naturally biggest at high x [15]. This allows a high-x gluon in the $\overline{MS}$ scheme which is determined from the quarks. At NLO the χ^2 for jets reduces from 154 to 116. This prescription works even better at NNLO – χ^2 for the jets goes from 164 to 117, and the total $\Delta\chi^2 = -79$.

Regarding high E_T jets, there has recently been a calculation of weak corrections [16], which implies that $\sigma_{QCD} \to \sigma_{QCD}(1 - \frac{2}{3}C_F \frac{\alpha_W}{\pi} \log^2(E_T^2/M_W^2))$. This is a 12% correction at $E_T = 450$GeV, and the authors question the validity of recent partons due to this. We have studied the phenomenological impact, and the movement of both CDF and D0 data is relatively small, as shown in fig. 5. The total χ^2 changes by ~ 15 without refitting, which is significant but not a disaster. The correction is more important at higher E_T, but there are positive real corrections to be added which depend on the jet definitions.

There has also been a study of the inclusion of QED effects by MRST [17]. The overall effect is small, but does lead to small isospin violation because $u_V^p(x)$ quarks radiate more photons than $d_V^n(x)$ quarks. This is in the correct direction to reduce the NuTeV $\sin^2\theta_W$ anomaly [18], and with current quark masses it is halved. Our approach is supported by data on wide-angle photon scattering, i.e. $ep \to e\gamma X$ [19] where the final state electron and photon have equal and opposite large transverse momentum.

A much more important correction is NNLO QCD. The NNLO splitting functions are now complete [20], but very similar to the average of previous best estimates, so lead to no large change in our previous NNLO partons [21]. The NNLO corrections improve the quality of the fit slightly, and reduce α_S. However, to perform an absolutely correct NNLO fit we need not only exact NNLO splitting functions, but also require a rigorous treatment of heavy quark thresholds [22], NNLO Drell-Yan cross-sections [23], and a complete treatment of uncertainties. All this is in hand, and an essentially full NNLO determination of partons will appear very soon. Only the NNLO jet cross-section is missing. This is probably not too important – the NLO corrections themselves are not large, except at high rapidities, being $\sim 10\%$ at central rapidities. There are also good NNLO estimates, i.e. the threshold correction logarithms, which are expected to be a major component of the total NNLO correction [24]. These give a flat $3-4\%$ correction, i.e. smaller than the systematic errors on the data. Hence, the mistakes from ignoring jets in the fits are bigger than the mistakes made at NNLO by not knowing the exact hard cross-section. There is reasonable stability order by order for (quark-dominated) W and Z cross-sections, but this fairly good convergence is largely guaranteed because the quarks are fit directly to the data. This is much worse for gluon-dominated quantities, e.g. $F_L(x,Q^2)$, which is unstable at small x and Q^2, as seen in fig. 4.

Given this theoretical uncertainty, we devised an approach to look for the safe theoretical regions, i.e. change Q^2_{cut} and x_{cut}, re-fit and see if the quality of the fit to the remaining data improves and/or the input parameters change dramatically [25]. Raising Q^2_{cut} from 2GeV^2 in steps, there is a slow, continuous and significant improvement for higher Q^2 up to $> 10\text{GeV}^2$. Raising x_{cut} from 0 to 0.005, there is a continuous improvement, and at each step the moderate x gluon becomes more positive. This led to the MRST2003 conservative partons, which should be the most reliable method of parton determination, but are *only* applicable for a restricted range of x and Q^2. We also have NNLO conservative partons, with similar cuts and improvement in fit quality, but the change in the partons is considerably less in this case because NNLO includes important theoretical corrections lacking at NLO. The variation in predictions with the cuts indicates the range of possible theoretical errors. There is a large change in σ_W at the LHC since this is sensitive to the low x region. The prediction is much more stable at NNLO, and LHC uncertainties are $\sim 3-4\%$ including the theoretical uncertainty. Hence, σ_W is a good candidate for luminosity determination. CTEQ have repeated this type of analysis and see a similar type of behaviour with cuts [11], although much less dramatic. With conservative cuts on data their input gluon again marginally prefers to have a negative component, confirming that a negative/small gluon at low x and Q^2 is not due to the data at low x and Q^2. They also find that the prediction for σ_W at the LHC moves down, but only a little, as more cuts are imposed. However, the loss of data with more cuts leads to larger errors, and the χ^2 profile is very flat indeed in the downwards direction, as seen in fig. 6. There is not really any inconsistency with MRST. If one is cautious about the accuracy of theory at low x and Q^2, the conclusion that the uncertainty is large on small x-sensitive quantities holds. CTEQ claim no reason to be cautious. This theoretical uncertainty is not so much of an issue at NNLO though, as discussed above.

In conclusion, we determine the parton distributions and predict cross-sections by performing global fits, and the fit quality using NLO or NNLO QCD is fairly good. There are various ways of looking at uncertainties due to errors on data, and they are $1-5\%$ for most LHC quantities. Ratios often do not reduce uncertainties. QED corrections are small, but introduce important isospin asymmetry. The uncertainty from

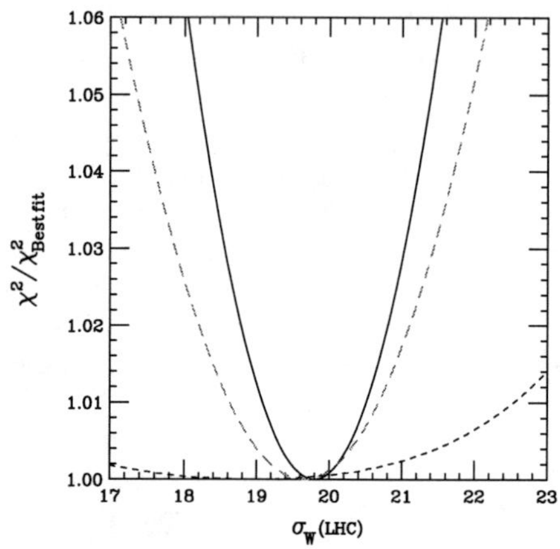

FIGURE 6. CTEQ χ^2 profile for σ_W [11], where the wide profile is for conservative cuts

using different approaches is often comparable to or even larger than deriving from the errors on the data. For example, a model for the input form of the gluon can solve the apparent high-E_T jet problem. Errors from higher orders/resummation are potentially large. Conservative cuts on x and Q^2 allow an improved fit to the remaining data, and altered partons. CTEQ see some effects from this type of study, but these are much smaller. NNLO is much more stable than NLO, and more theoretically reliable. For MRST full NNLO fits are imminent, and should become the new standard.

REFERENCES

1. M. Botje, *E*ur. Phys. J. **C**14 285 (2000).
2. W. T. Giele, S. Keller and D. A. Kosower, `hep-ph/0104052`.
3. S. I. Alekhin, *P*hys. Rev. **D**68 014002 (2003).
4. CTEQ Collaboration: D. Stump *et al.*, *P*hys. Rev. **D**65 014012 (2002).
5. CTEQ Collaboration: J. Pumplin *et al.*, *P*hys. Rev. **D**65 014013 (2002).
6. CTEQ Collaboration: J. Pumplin *et al.*, JHEP 0207:012 (2002).
7. H1 Collaboration: C. Adloff *et al.*, *E*ur. Phys. J. **C**30 1 (2003).
8. ZEUS Collaboration: S. Chekanov *et al.*, `hep-ph/0503274`.
9. A. D. Martin, R. G. Roberts, W. J. Stirling and R. S. Thorne, *E*ur. Phys. J. **C**28 455 (2003).
10. S. Catani,D. de Florian and M. Grazzini, JHEP 0307:028 (2003).
11. J. Huston, J. Pumplin, D. Stump and W. K. Tung, `hep-ph/0502080`.
12. S. Moch, J. A. M. Vermaseren and A. Vogt, *P*hys. Lett. **B**606 123 (2005); `hep-ph/0504242`.
13. D0 Collaboration: B. Abbott *et al.*, *P*hys. Rev. Lett. **8**6 1707 (2001).
14. CDF Collaboration: T. Affolder *et al.*, *P*hys. Rev. **D**64 032001 (2001).
15. A. D. Martin, R. G. Roberts, W. J. Stirling and R. S. Thorne, *P*hys. Lett. **B**604 61 (2004).
16. S. Moretti, M. R. Nolten and D. A. Ross, `hep-ph/0503152`.
17. A. D. Martin, R. G. Roberts, W. J. Stirling and R. S. Thorne, *E*ur. Phys. J. **C**39 155 (2005).
18. G. P. Zeller *et al.*, *P*hys. Rev. Lett. **8**8 091802 (2002).
19. ZEUS collaboration: S. Chekanov *et al.*, *P*hys. Lett. **B**595 86 (2004).
20. S. Moch, *et al.*, *N*ucl. Phys. **B**688 101 (2004); *N*ucl. Phys. **B**691 129 (2004).
21. A. D. Martin, R. G. Roberts, W. J. Stirling and R. S. Thorne, *P*hys. Lett. **B**531 216 (2002).
22. R. S. Thorne, `hep-ph/0506251`, these proceedings.
23. C. Anastasiou, *et al*, *P*hys. Rev. Lett. **9**1, 182002 (2003); *P*hys. Rev. **D**69, 094008 (2004).
24. N. Kidonakis and J. F. Owens, *P*hys. Rev. **D**63 054019 (2001).
25. A. D. Martin, R. G. Roberts, W. J. Stirling and R. S. Thorne, *E*ur. Phys. J. **C**35 325 (2004).

Improving the determination of the gluon density in the proton using jet data from ZEUS

Juan Terrón (on behalf of the ZEUS Collaboration)

Universidad Autónoma de Madrid, Spain

Abstract. A combined analysis of the structure function measurements in neutral- and charged-current deep inelastic $e^{\pm}p$ scattering and jet cross-section measurements in ep collisions at HERA to determine the parton distribution functions in the proton is presented. The inclusion of the jet measurements in such analysis provides a sizeable reduction of the uncertainty on the gluon density in the mid- to high-x region as well as a precise determination of the strong coupling constant.

Keywords: QCD, parton distribution functions in the proton
PACS: 12.38.Qk

INTRODUCTION

The determination of the parton distribution functions (PDFs) in the proton has primarily used inclusive measurements of deep inelastic lepton-nucleon scattering. This procedure has the advantage that only the final-state lepton is tagged and, thus, neither QCD nor hadronisation corrections are associated to the final-state signature. However, the gluon density contributes indirectly to these observables and, as a result, is affected by larger uncertainties, specially at high x. On the other hand, jet cross sections in hadron-hadron or lepton-hadron collisions can directly probe the gluon content of the proton. Nevertheless, observables based on jets have hardly been used in a determination of the proton PDFs since large QCD and hadronisation corrections affected this type of measurements. An improved determination of the proton PDFs would have, in addition to its intrinsic importance, an impact on future research with proton-proton collisions at very high energies, such as those foreseen in the Large Hadron Collider.

In the last few years, measurements of jet cross sections have been made in ep collisions at HERA which directly probe the gluon content of the proton and have small experimental and theoretical uncertainties. The precision achieved in these measurements has been accomplished through a reduction of the uncertainty on the jet energy scale (the main experimental uncertainty) and a careful choice of the jet algorithm, phase-space region and jet selection criteria, which allowed a reduction of the theoretical uncertainties.

The ZEUS Collaboration has pioneered a new determination [1] of the proton PDFs in which the inclusive measurements of neutral- (NC) and charged-current (CC) deep inelastic $e^{\pm}p$ scattering (DIS) [2] have been used in combination with the measurements of jet cross section in DIS [3] and γp collisions [4]. As a result of this new procedure, an improved determination of the gluon density in the proton in the mid- to high-x region has been obtained. In this contribution, the analysis method and the results are presented.

CP792, *Deep Inelastic Scattering, DIS 2005,* edited by W. H. Smith and S. R. Dasu

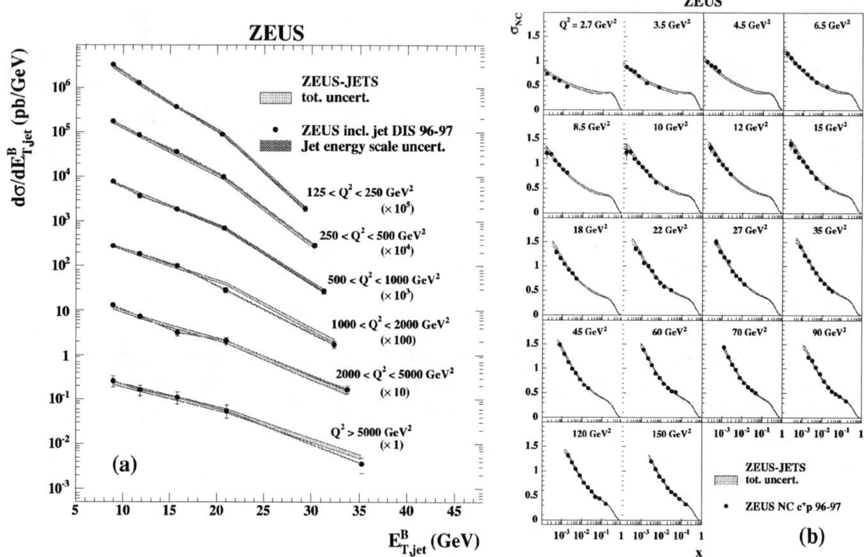

FIGURE 1. (a) Measurements of $d\sigma/dE^B_{T,\text{jet}}$ for inclusive jet production in NC DIS in different regions of Q^2; (b) measurements of the reduced double differential cross section in NC DIS; for comparison, NLO QCD calculations using the best determination of the proton PDFs are also shown.

JET MEASUREMENTS AND CALCULATIONS

Two sets of jet cross-section measurements have been used in the analysis. The first set is that of the measurements of the differential cross section $d\sigma/dE^B_{T,\text{jet}}$ for the inclusive production of jets with transverse energy $E^B_{T,\text{jet}} > 8$ GeV in the Breit frame of NC DIS in different regions of Q^2 [3]. The results are shown in Fig. 1a. The dominant experimental uncertainty is that on the jet energy scale, which is known to $\pm 1\%$, and amounts to $\pm 5\%$ on the differential cross sections. The leading-order (LO) QCD contribution to the production of high-E_T jets in the Breit frame comes from the photon-gluon fusion (PGF, $\gamma^* g \to q\bar{q}$) and QCD Compton (QCDC, $\gamma^* q \to qg$) processes. Therefore, this observable is directly sensitive to the gluon density in the proton. Next-to-leading-order (NLO) QCD calculations are available in the form of flexible programmes [5]. The dominant theoretical uncertainty arises from terms beyond NLO and was estimated by varying the renormalisation scale μ_R by a factor of two up and down; it amounts to $\sim \pm 5\%$.

The second set is that of the measurements of the differential cross section $d\sigma/dE_T^{\text{jet},1}$ for dijet production with $E_T^{\text{jet},1} > 14$ GeV and $E_T^{\text{jet},2} > 11$ GeV in photoproduction [4]. Photoproduction at HERA is studied by means of ep scattering at low four-momentum transfers ($Q^2 \approx 0$). In photon-proton reactions, two types of QCD processes contribute to jet production at LO: either the photon interacts directly with a parton in the proton (the direct process) or the photon acts as a source of partons which scatter off those in the proton (the resolved process). Direct-photon dijet events directly probe the gluon density in the proton since the LO QCD contribution comes from the PGF and QCDC processes. In order to suppress the resolved-photon contribution, so as to reduce the dependence on the knowledge of the photon PDFs, differential cross sections for dijet production with $x_\gamma^{obs} > 0.75$ have been used in the analysis, where x_γ^{obs} is the fraction of the photon momentum participating in the production of the two jets with highest E_T^{jet}. NLO QCD calculations are available [6]. The dominant theoretical uncertainty arises from terms beyond NLO and was estimated by varying μ_R by a factor of two up and down; it amounts to $\sim \pm 10\%$.

For both data sets, the corresponding NLO QCD calculations are so much time consuming that their straightforwad inclusion in a global analysis to determine the proton PDFs is prevented. A method has been developed to include in a rigorous way the NLO QCD calculations for jet cross sections; it basically consists of deconvoluting the proton PDFs and α_s dependencies from the matrix elements in the calculations and building grids. Using this method, jet-cross-section calculations can be performed for any set of PDFs and any value of α_s in a fast way and with an accuracy better than 0.5%.

INCLUSIVE MEASUREMENTS AND COMBINED ANALYSIS

The inclusive measurements used in the analysis were those of the reduced double differential cross sections in x and Q^2 from NC and CC $e^\pm p$ scattering [2]. The kinematic region covered by these data is $6.3 \cdot 10^{-5} < x < 0.65$ and $2.7 < Q^2 < 3 \cdot 10^4$ GeV^2. As an example, the measurements of the reduced double differential cross sections in NC DIS in the region $2.7 < Q^2 < 150$ GeV^2 are shown in Fig. 1b.

The proton PDFs have been determined by simultaneously fitting the inclusive and jet measurements with NLO QCD calculations. In these calculations, the proton PDFs have been evolved using the DGLAP equations at NLO in the $\overline{MS}$ scheme as implemented in the programme QCDNUM [7]. The contribution from heavy quarks has been evaluated using the general-mass variable flavour-number scheme of Roberts and Thorne [8]. The PDFs for u valence, d valence, total sea, gluon and the difference between the d and u contributions to the sea, are each parametrised at the input scale $Q_0^2 = 7$ GeV^2 by the form: $xf(x) = p_1 x^{p_2}(1-x)^{p_3}(1+p_4 x)$. Some constraints were imposed on the parameters p_i, such as the number and momentum sum rules (for details, see [1]). In total, there are 11 free parameters to describe the PDFs at the input scale Q_0^2. In a first instance, the strong coupling constant $\alpha_s(M_Z)$ was fixed to the world average $\alpha_s(M_Z) = 0.118$; alternatively, it was considered as a free parameter (see next Section). Full account has been taken of the correlated experimental systematic uncertainties using the offset method [9].

RESULTS

A good description of all the data sets is obtained by the NLO QCD calculations using the proton PDFs as determined by the best fit. This is illustrated in Figs. 1a and 1b. The resulting gluon and sea distributions in the proton as well as their estimated uncertainties are shown in Fig. 2a. At low x these distributions are as well determined as in any global analysis of the proton PDFs which makes use of inclusive measurements at HERA. The improvement which is brought by the inclusion of the jet measurements in the combined analysis is illustrated in Fig. 2b, where the total experimental uncertainty on the gluon distribution is compared for the case in which the jet data was included with that in which they were excluded. A reduction by about a factor of two in the uncertainty on the gluon distribution over the range $0.01 \lesssim x \lesssim 0.4$ has been accomplished by the use of the jet-cross-section measurements.

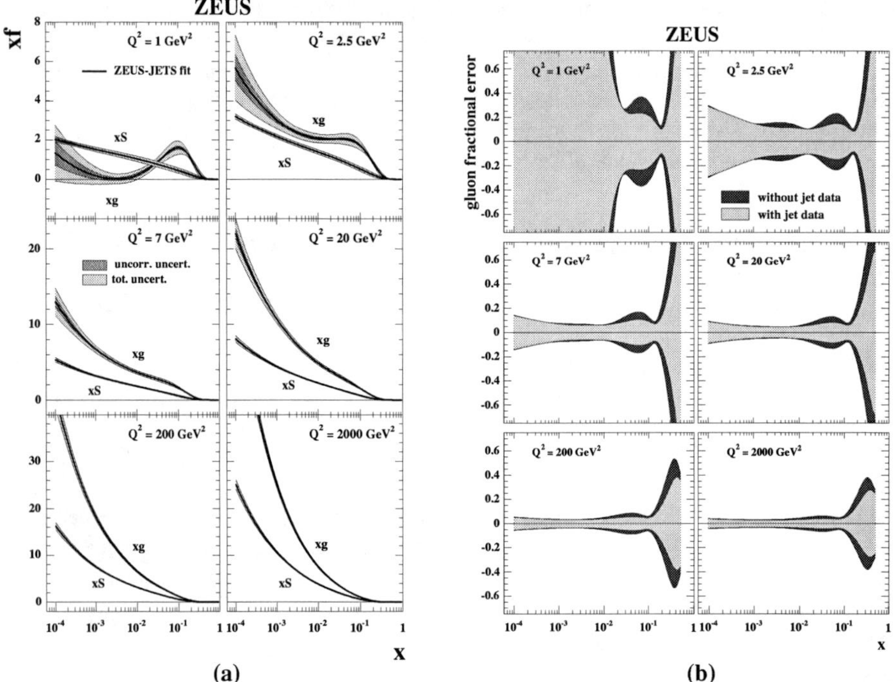

FIGURE 2. (a) Gluon and sea distributions in the proton as determined by the best fi t; (b) the total experimental uncertainty on the gluon distribution as determined either including (inner band) or excluding (outer band) the jet measurements in the combined analysis.

Furthermore, the use of the inclusive and jet measurements allowed a simultaneous determination of the proton PDFs and α_s. The jet cross sections are directly sensitive to α_s and provide, via the QCDC process, a tight contrain on α_s which is not strongly correlated with the gluon density in the proton. In this analysis, the parameters to describe the PDFs as well as α_s were set free. The fitted value is

$$\alpha_s(M_Z) = 0.1183 \pm 0.0007(\text{uncorr.}) \pm 0.0022(\text{corr.}) \pm 0.0016(\text{norm.}) \pm 0.0008(\text{model}),$$

where the uncertainties arise from the following sources: statistical and other uncorrelated sources; experimental correlated systematic sources excluding normalisation uncertainties; normalisation uncertainties; and model uncertainty. The theoretical uncertainty that arises from terms beyond NLO has been estimated by varying μ_R by a factor of $\sqrt{2}$ up and down and amounts to $\Delta\alpha_s(M_Z) \sim \pm 0.0050$. This determination of α_s has the advantage with respect to previous extractions using jet-cross-section measurements at HERA that the correlation between α_s and the proton PDFs is fully taken into account in the present analysis. For this reason, this determination can be considered the first one that is based on HERA data alone.

SUMMARY

Due to the precision and kinematic coverage of the ZEUS measurements on structure functions and jet cross sections, it is now possible to determine the proton PDFs within a single experiment with minimal external input. The feasibility of including, in a rigorous way, jet-cross-section measurements in a global analysis has been demonstrated. The advantages of including jet cross sections which are directly sensitive to the gluon density in the proton and have small experimental and theoretical uncertainties are: a sizeable reduction of the uncertainty on the gluon distribution in the mid- to high-x region as well as a precise determination of the strong coupling constant from HERA data alone.

ACKNOWLEDGMENTS

I would like to thank the organisers of the conference for the superb location, the warm hospitality and the fruitful atmosphere.

REFERENCES

1. ZEUS Coll., S. Chekanov et al., DESY preprint DESY-05-050, 2005.
2. ZEUS Coll., J. Breitweg et al., *Eur. Phys. Jour.* **C** 12 (2000) 411 (erratum in *Eur. Phys. Jour.* **C** 27 (2003) 305); ZEUS Coll., S. Chekanov et al., *Eur. Phys. Jour.* **C** 21 (2001) 443, *Phys. Lett.* **B** 539 (2002) 197 (erratum in *Phys. Lett.* **B** 552 (2003) 308), *Eur. Phys. Jour.* **C** 28 (2003) 175, *Eur. Phys. Jour.* **C** 32 (2003) 16 and *Phys. Rev.* **D** 70 (2004) 052001.
3. ZEUS Coll., S. Chekanov et al., *Phys. Lett.* **B** 547 (2002) 164.
4. ZEUS Coll., S. Chekanov et al., *Eur. Phys. Jour.* **C** 23 (2002) 615.
5. S. Catani and M.H. Seymour, *Nucl. Phys.* **B** 485 (1997) 291.
6. S. Frixione and G. Ridolfi , *Nucl. Phys.* **B** 507 (1997) 315.
7. M. Botje, QCDNUM version 16.12 (unpublished), available on `http://www.nikhef.nl/ h24/qcdcode/index.html`.
8. R.G. Roberts and R.S. Thorne, *Phys. Rev.* **D** 57 (1998) 6871.
9. ZEUS Coll., S. Chekanov et al., *Phys. Rev.* **D** 67 (2003) 012007,

The neural network approach to parton fitting[1]

The NNPDF Collaboration: Joan Rojo*, Luigi Del Debbio†, Stefano Forte**, José I. Latorre* and Andrea Piccione‡

* *Departament d'Estructura i Constituents de la Matèria, Universitat de Barcelona*
†*Theory Division, CERN*
** *Dipartimento di Fisica, Università di Milano and INFN, Sezione di Milano*
‡*Dipartimento di Fisica Teorica, Università di Torino, and INFN Sezione di Torino*

Abstract. We introduce the neural network approach to global fits of parton distribution functions. First we review previous work on unbiased parametrizations of deep-inelastic structure functions with faithful estimation of their uncertainties, and then we summarize the current status of neural network parton distribution fits.

Keywords: QCD
PACS: 12.38.-t

1. INTRODUCTION

The requirements of precision physics at hadron colliders have recently led to a rapid improvement in the techniques for the determination of parton distributions of the nucleon [1]. Specifically it is now mandatory to determine accurately the uncertainty on these quantities. The main difficulty is that one is trying to determine the uncertainty on a function, that is, a probability measure in a space of functions, and to extract it from a finite set of experimental data, a problem which is mathematically ill-posed

The shortcomings of the standard approach to global parton fits are well known: the bias introduced by choosing fixed functional forms to parametrize the parton distributions (also known as *model dependence*), the problems to assess faithfully their uncertainties, the combination of inconsistent experiments, and the lack of general, process-independent error propagation techniques. Although the problem of quantifying the uncertainties in pdfs has seen a huge progress since its paramount importance was raised some years ago, until now no unambiguous conclusions have been obtained.

Here we present a novel strategy to address the problem of constructing unbiased parametrizations of parton distributions with a faithful estimation of their uncertainties, based on a combination of two techniques: Monte Carlo methods and neural networks. First we review recent work on the related problem of the construction of bias-free parametrizations of structure functions from experimental data, and then we turn to the application of our strategy to parton distributions.

[1] Talk given at Deep Inelastic Scattering 2005 Workshop (Madison) by J. R., on behalf of the NNPDF Collaboration.

CP792, *Deep Inelastic Scattering, DIS 2005*, edited by W. H. Smith and S. R. Dasu

2. STRUCTURE FUNCTIONS

The strategy presented in [2, 3] to address to problem of parametrizing deep-inelastic structure functions $F(x,Q^2)$ is a combination of two techniques: first we construct a Monte Carlo sampling of the experimental data (generating artificial data replicas), and then we train neural networks in each data replica, to construct a probability measure in the space of structure functions $\mathscr{P}\left[F(x,Q^2)\right]$. The probability measure constructed in this way contains all information from experimental data, including correlations, with the only assumption of smoothness. Expectation values and moments over this probability measure are then evaluated as averages over the trained network sample,

$$\left\langle \mathscr{F}\left[F(x,Q^2)\right]\right\rangle = \int \mathscr{D}F\,\mathscr{P}\left[F(x,Q^2)\right]\mathscr{F}\left[F(x,Q^2)\right] = \frac{1}{N_{\rm rep}}\sum_{k=1}^{N_{\rm rep}} \mathscr{F}\left(F^{({\rm net})(k)}(x,Q^2)\right) . \tag{1}$$

where $\mathscr{F}[F]$ is an arbitrary function of $F(x,Q^2)$.

The first step is the Monte Carlo sampling of experimental data, generating $N_{\rm rep}$ replicas of the original $N_{\rm dat}$ experimental data,

$$F_i^{({\rm art})(k)} = \left(1+r_N^{(k)}\sigma_N\right)\left[F_i^{({\rm exp})} + r_i^{s,(k)}\sigma_i^{stat} + \sum_{l=1}^{N_{sys}} r^{l,(k)}\sigma_i^{sys,l}\right], \qquad i=1,\ldots,N_{\rm dat}\,, \tag{2}$$

where r are gaussian random numbers with the same correlation as the respective uncertainties, and $\sigma^{stat}, \sigma^{sys}, \sigma_N$ are the statistical, systematic and normalization errors. The number of replicas $N_{\rm rep}$ has to be large enough so that the replica sample reproduces central values, errors and correlations of the experimental data.

The second step consists on training a neural network on each of the data replicas. A neural network [4] (see fig. 1) is a highly nonlinear mapping between input and output patterns as a function of its parameters (the so-called *weights* $\omega_{ij}^{(l)}$ and *thresholds* $\theta_i^{(l)}$). Neural networks are specially suitable to parametrize parton distributions since they are unbiased, robust approximants and interpolate between data points with the only assumption of smoothness. The neural network training consist on the minimization for each replica of the χ^2 defined with the inverse of the experimental covariance matrix,

$$\chi^{2(k)} = \frac{1}{N_{\rm dat}}\sum_{i,j=1}^{N_{\rm dat}}\left(F_i^{({\rm art})(k)} - F_i^{({\rm net})(k)}\right){\rm cov}_{ij}^{-1}\left(F_j^{({\rm art})(k)} - F_j^{({\rm net})(k)}\right) . \tag{3}$$

Our minimization strategy is based on Genetic Algorithms [5], which are specially suited for finding global minima in highly nonlinear minimization problems.

The set of trained nets, once is validated through suitable statistical estimators, becomes the sought-for probability measure $\mathscr{P}\left[F(x,Q^2)\right]$ in the space of structure functions. Now observables with errors and correlations can be computed from averages over this probability measure, using eq. (1). For example, the average and error of a structure function $F(x,Q^2)$ at arbitrary (x,Q^2) can be computed as

$$\left\langle F(x,Q^2)\right\rangle = \frac{1}{N_{\rm rep}}\sum_{k=1}^{N_{\rm rep}} F^{({\rm net})(k)}(x,Q^2), \quad \sigma(x,Q^2) = \sqrt{\left\langle F(x,Q^2)^2\right\rangle - \left\langle F(x,Q^2)\right\rangle^2} . \tag{4}$$

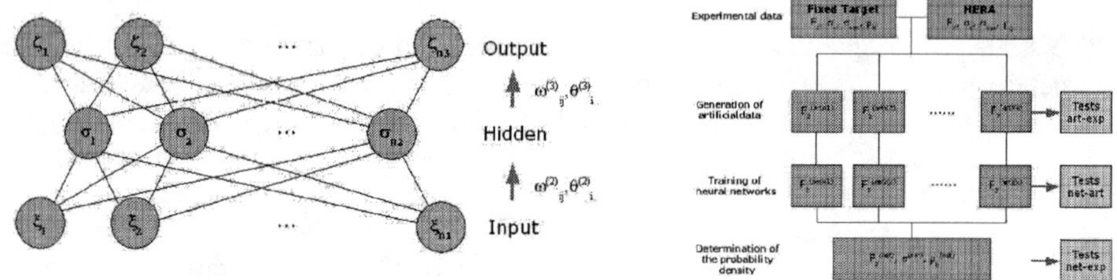

FIGURE 1. Left: a neural network. Right: Summary of the NNPDF approach.

Our strategy is summarized in fig. 1. In fig. 2 we show our results[2] for the proton structure function $F_2^p(x,Q^2)$, both in our original fit without HERA data [2] and in the latest fit [3] including HERA data.

3. PARTON DISTRIBUTIONS

The strategy presented in the above section can be used to parametrize parton distributions, provided one now takes into account Altarelli-Parisi QCD evolution. Therefore we need to define a suitable evolution formalism, and we will consider for the sake of simplicity nonsinglet parton evolution. Since complex neural networks are not allowed, we must use the convolution theorem to evolve parton distributions in $x-$space using the inverse $\Gamma(x)$ of the Mellin space evolution factor $\Gamma(N)$, defined as

$$q(N,Q^2) = q(N,Q_0^2)\Gamma\left(N,\alpha_s\left(Q^2\right),\alpha_s\left(Q_0^2\right)\right) \ , \tag{5}$$

$$\Gamma\left(x,\alpha_s\left(Q^2\right),\alpha_s\left(Q_0^2\right)\right) \equiv \frac{1}{2\pi i}\int_{c-i\infty}^{c+i\infty} dN\, x^{-N}\Gamma\left(N,\alpha_s\left(Q^2\right),\alpha_s\left(Q_0^2\right)\right) \ . \tag{6}$$

The only subtlety is that eq. (6) defines a distribution, which must therefore be regulated at $x=1$, yielding the final evolution equation,

$$q(x,Q^2) = q(x,Q_0^2)\int_x^1 dy\,\Gamma(y) + \int_x^1 \frac{dy}{y}\Gamma(y)\left(q\left(\frac{x}{y},Q_0^2\right) - yq(x,Q_0^2)\right) \ , \tag{7}$$

where in the above equation $q(x,Q_0^2)$ is parametrized using a neural network. At higher orders in perturbation theory coefficient functions $C(N)$ are introduced through a modified evolution factor, $\tilde{\Gamma}(N) \equiv \Gamma(N)C(N)$. We have benchmarked our evolution code with the Les Houches benchmark tables [6]. The evolution factor $\Gamma(x)$ and its integral are computed and interpolated before the neural network training in order to have a faster fitting procedure.

---

[2] The source code, driver program and graphical web interface for our structure function fits is available at **http://sophia.ecm.ub.es/f2neural**.

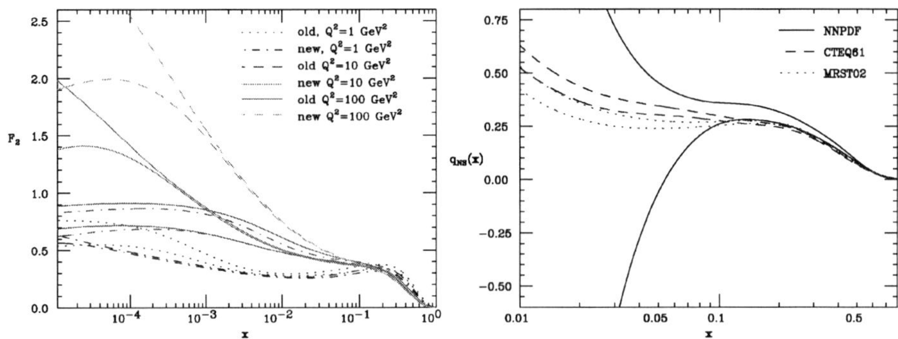

FIGURE 2. Left: The proton structure function, both the old version [2] without HERA data and the new version [3] which included HERA data. Right: Preliminary results on the nonsinglet parton distribution $q_{NS}(x, Q_0^2)$ compared with other global fits, for $Q_0^2 = 2$ GeV^2

As a first application of our method, we extract the nonsinglet parton distribution $q_{NS}(x, Q_0^2) = \left(u + \bar{u} - d - \bar{d}\right)(x, Q_0^2)$ from the nonsinglet structure function $F_2^{NS}(x, Q^2)$ measured by the NMC [7] and BCDMS [8, 9] collaborations. The preliminary results of a NLO fit with fully correlated uncertainties can be seen in fig. 2. Our result is consistent within the error bands with the results from other global fits [10, 11, 12] in almost all the range of Bjorken-x. It is clear that the large uncertainties at small x do not allow, within the current experimental data, to determine if $q_{NS}(x, Q^2)$ grows at small x, as supported by different theoretical arguments as well as by other global parton fits. Only additional nonsinglet structure function data at small x can settle this issue.

Summarizing, we have described a general technique to parametrize experimental data in an bias-free way with a faithful estimation of their uncertainties, which has been successfully applied to structure functions and that now is being implemented in the context of global parton distribution fits.

REFERENCES

1. W.-K. Tung, *AIP Conf. Proc.*, **753**, 15–29 (2005), hep-ph/0410139.
2. S. Forte, L. Garrido, J. I. Latorre, and A. Piccione, *JHEP*, **05**, 062 (2002), hep-ph/0204232.
3. L. Del Debbio, S. Forte, J. I. Latorre, A. Piccione, and J. Rojo (2004), hep-ph/0501067.
4. G. Stimpfl-Abele, and L. Garrido, *Comput. Phys. Commun.*, **64**, 46–56 (1991).
5. J. Rojo, and J. I. Latorre, *JHEP*, **01**, 055 (2004), hep-ph/0401047.
6. W. Giele, et al. (2002), hep-ph/0204316.
7. M. Arneodo, et al., *Nucl. Phys.*, **B483**, 3–43 (1997), hep-ph/9610231.
8. A. C. Benvenuti, et al., *Phys. Lett.*, **B223**, 485 (1989).
9. A. C. Benvenuti, et al., *Phys. Lett.*, **B237**, 592 (1990).
10. A. D. Martin, R. G. Roberts, W. J. Stirling, and R. S. Thorne, *Eur. Phys. J.*, **C28**, 455–473 (2003), hep-ph/0211080.
11. D. Stump, et al., *JHEP*, **10**, 046 (2003), hep-ph/0303013.
12. S. Alekhin, *Phys. Rev.*, **D68**, 014002 (2003), hep-ph/0211096.

Tevatron Measurements and PDF Uncertainties

Frank Chlebana

Fermi National Accelerator Lab, P.O. Box 500, Batavia IL, 60150 USA

Abstract. The impact of PDF uncertainties on recent Tevatron measurements is explored. One of the most poorly constrained PDFs is the gluon distribution which is seen to be the dominant source of uncertainty for many interesting calculations. Tevatron measurements that can be used to better constrain PDFs are highlighted. Recent techniques to quantify the error on measured distributions resulting from PDF uncertainties are discussed.

Keywords: QCD, Parton Distributions, Hadronic Colliders
PACS: 12.38.-t, 12.38.Bx, 12.38.Qk

Parton density functions (PDFs) are essential input in the calculation of production cross sections for many lepton-hadron and hadron-hadron processes. Once the PDFs are specified as a function of the kinematic variable, x, and for a given Q^2, the DGLAP [1] equations can be used to calculate cross sections in any region of phase space. The validity of the extrapolation depends on theoretical assumptions as well as the uncertainties of the PDFs. Refining theory predictions is an iterative process and as new data are incorporated into the global fits, more precise predictions can be made. Calculations have an uncertainty arising from both experiment and theory. Theoretical errors include the choice of parameterization, input parameters such as flavor threshold and α_s and uncertainties on the modeling (scale errors, nonperterbative effects). Experimental errors originate from the statistical and systematic errors of the data that are used in the global fits. PDF errors propagate to the measurement when calculating the acceptance, luminosity, event selection and background estimate.

Recently new methods to estimate PDF uncertainties based on the Hessian and Lagrange multiplier techniques [2] have been developed taking into account the statistical and correlated systematic errors. N_{PDF} eigenvectors are calculated from the error matrix and their upper and lower deviations result in $2 \times N_{PDF}$ new sets of error PDFs. The $\Delta\chi^2$ which best describes the error tolerance is somewhat intuitive and different conventions have been adopted by different groups, for example the CTEQ group uses $\Delta\chi^2 = 100$ while MRST uses $\Delta\chi^2 = 50$. Figure 1 shows the CDF Run I inclusive jet data with the error PDFS determined for the CTEQ6.1M PDF [2]. The dominant error arises from the uncertainty of the gluon distribution. The higher precision Run II data will lead to improved PDF sets with reduced uncertainties. In order to quantify the error arising from PDF uncertainties, the error PDFs have to be available in a form that can be generally used. The Les Houches Accord Parton Density Function Interface (LHAPDF) provides a standard interface enabling PDF error sets to be used with MC generators.

The increased center-of-mass energy available in Run II ($1.8 \rightarrow 1.96$ TeV) yields a cross section larger by $\sim 2\times$ at 400 GeV and $\sim 5\times$ at 600 GeV compared to Run I. Preliminary measurements are now available based on about $350 pb^{-1}$, already extending

CP792, *Deep Inelastic Scattering, DIS 2005,* edited by W. H. Smith and S. R. Dasu

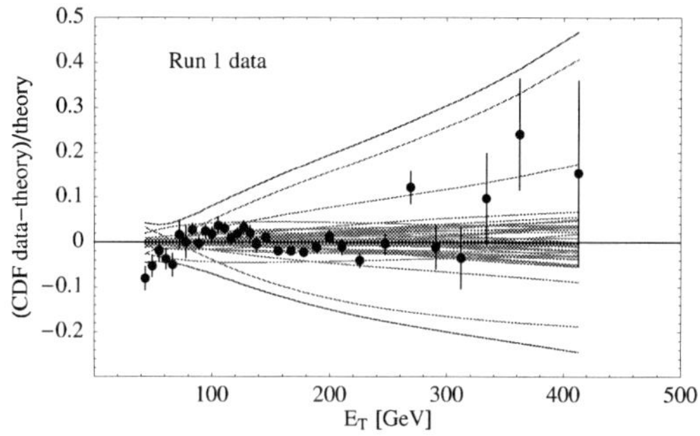

FIGURE 1. The CDF Run I inclusive jet cross section compared to the CTEQ6.1M error PDFs.

the Run I inclusive jet results by ~ 150 GeV as is shown in Figure 2.

In Run I, the measured inclusive jet cross section from the Tevatron was larger than predicted using then current PDFs. New Physics could show up as a deviation from the Standard Model predictions at high E_T. Global PDF fits including the Run I Tevatron jet data were able to accommodate the excess by an enhanced gluon PDF at high x. In order to separate PDF effects from new physics one can study the angular distribution between the leading jets or more generally include measurements of inclusive jet production over a wide rapidity region in global fits [2]. Preliminary Run II results from DØ in two rapidity bins are shown in Figure 2. The dominant systematic error originates from the jet energy scale and can be expected to be improved.

The W cross section has been proposed to determine the luminosity at the LHC since is it theoretically well understood and well measured. Calculation of the cross section depends on the data included in the global PDF fit while the choice of $\Delta\chi^2$ influences the error on calculation. The $\sigma_{\text{NLO}}(W)$ at LHC energies was recently calculated using the following PDF sets $\sigma_{\text{NLO}}(W) = 204 \pm 4$ (nb) (MRST2002), 205 ± 8 (nb) (CTEQ6),

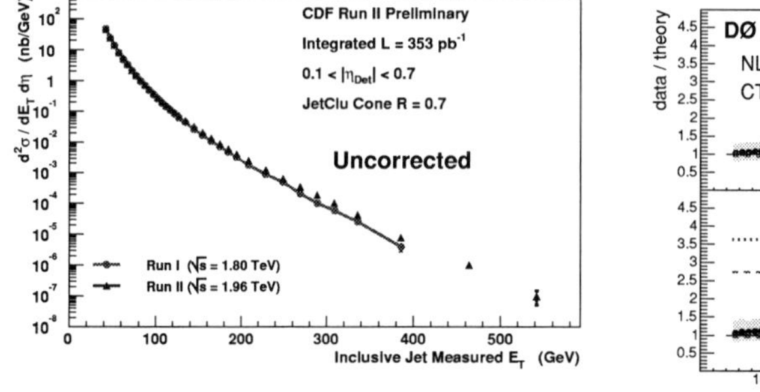

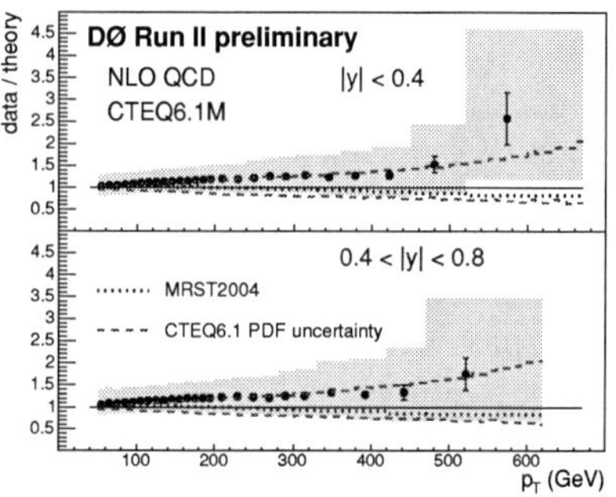

FIGURE 2. Left: Uncorrected inclusive jet distribution showing the extended reach in Run II using $\sim 350 pb^{-1}$ of data. Right: Preliminary DØ results for the inclusive jet cross section in two rapidity bins.

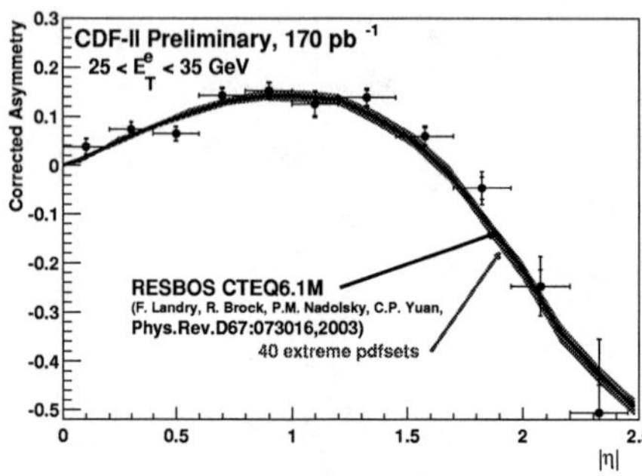

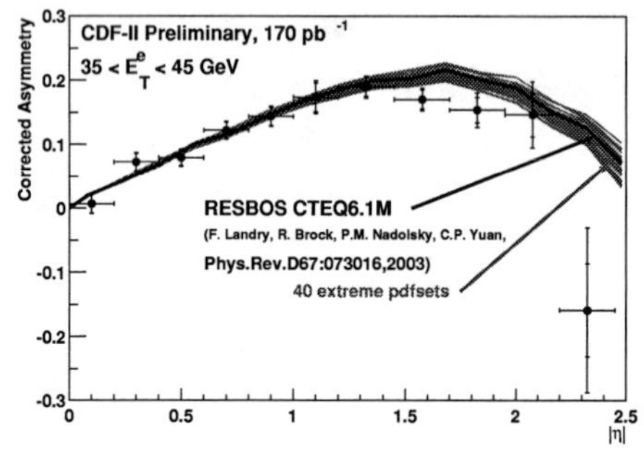

FIGURE 3. *W* charge asymmetry for two bins in E_T. The 40 PDF error sets from CTEQ6.1M are shown as the band around the central value.

215 ± 6 (nb) (Alekhin02). The error on the luminosity is reflected in any cross section measurement and it is important to understand the source of the error as well as to reduce it. The impact of PDF related uncertainties on the determination of the luminosity can be further reduced to $\sim 1\%$ by using cross section ratios [4].

More data and enhanced detector capabilities allow for the possibility of including new as well as more precise measurements in the PDF fits. Measurement of the W charge asymmetry constrains the ratio $d(x,M_W)/u(x,M_W)$ as $x \to 1$. Global QCD fits include the W charge asymmetry results from Run I which was averaged over E_T. In Run II we now have two E_T bins allowing us to explore the E_T dependence. Preliminary CDF results are shown in Figure 3 where predictions from the error PDFs are overlayed.

There is very little direct experimental input on the intrinsic heavy flavor of the proton. All c and b distributions in existing PDF sets are radiatively generated. Tevatron measurements with tagged final states $(W/Z/\gamma + c/b)$ probe the sea quark distributions. As more data is collected these measurements may provide insight into the heavy flavor content of the proton which would have a impact on several important production channels including the Higgs and single top.

Inclusion of full PDF systematics leads to a more realistic estimate of the top cross section uncertainty [5]. For $m_t = 175$ GeV, $\sigma = 6.70 \pm 0.45 pb$ (CTEQ6M) and $\sigma = 6.76 \pm 0.21 pb$ (MRST2001). The error on the top cross section is dominated by PDF and α_s uncertainties where the $\pm 3 - 6\%$ error mainly arises from the uncertainty of large-x gluons. Comparison of the data to the theory is shown in Figure 4 where it can be seen that the measurement error is approaching the magnitude of the error on the prediction. The inclusion of Run II data in PDF fits will help reduce the theoretical uncertainties.

The uncertainty on the Standard Model Higgs cross section was calculated for the main production processes [6]; associate production with W/Z $(q\bar{q} \to VH)$, massive vector boson fusion $(qq \to Hqq)$, gluon fusion $(gg \to H)$ and associate production with top quarks $(gg, q\bar{q} \to \bar{t}H)$. The different results obtained when using different PDF sets is attributed to the choice of data used as input to fits and the treatment of errors. For a given PDF set there is $\sim 5\%$ uncertainty while there is $\sim 15\%$ spread between PDF sets at Tevatron and LHC energies.

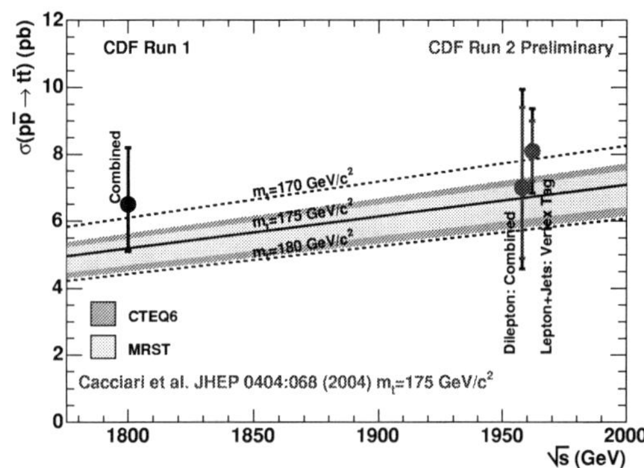

FIGURE 4. The top cross section measured by CDF is shown at different center of mass energies. The predictions determined using two PDF sets are shown as the bands.

CONCLUSION

PDFs are essential input in the calculation of hard scattering lepton-hadron and hadron-hadron cross sections and directly influence the precision of measurements. New techniques to estimate PDF related errors together with a standard interface between PDF sets and MC generators allow possibilities to better understand the impact on measured observables. The increased luminosity of the Tevatron and the upgraded detectors of CDF and DØ greatly extend the kinematic reach and precision of measurements that can be used in global QCD fits. The new data will lead to refined PDF sets with reduced errors. Uncertainty of the gluon distribution at high x is the dominant error on many interesting measurements such as top and Higgs production. Inclusive jet measurements will provide an important constraint on the gluon distribution at high x. Significantly different predictions can be obtained when using different PDF sets which is attributed to the choice of the input data in the fits as well as the treatment of errors. PDFs are universal and efforts should be made to use as much of the available data in the fits as possible.

REFERENCES

1. V.N. Gribov and L.N. Lipatov, *Sov. J. Nucl. Phys.*, **15**, 438, 675, (1972); L.N. Lipatov, *Sov. J. Nucl. Phys.*, **20**, 95, (1975); Yu.L. Dokshitzer, *Sov. J. Phys. JETP*, **46**, 641, (1977); G. Altarelli and G. Parisi, *Nucl. Phys.*, **B126**, 298, (1977).
2. Stump et al., *JHEP* 0310: 046, (2003).
3. Pumplin et al., *JHEP* 0207: 012, (2002).
4. M. Dittmar et al., *Phys.Rev.*, **D56**, pp. 7284–7290 (1997).
5. Cacciari et al., *JHEP* 0404: 068, (2004).
6. A. Djouadi and S. Ferrag, *Phys.Lett.*, **B586**, pp. 345–352, (2004).
7. A.D. Martin et al., *Eur. Phys. J.*, **C28**, 455, (2003).
8. S.I. Alekhin, *Phys.Rev.*,**D68**, :014002, (2003).

Precision Measurements of W and Z Boson Production at the Tevatron

Jonathan Hays for the CDF and DØ Collaborations

Northwestern University, Evanston, IL, USA

Abstract. Measurements of the inclusive W and Z boson production cross section times leptonic branching ratio in proton anti-proton collisions at $\sqrt{s} = 1.96$ TeV are presented. The ratio of the W and Z cross sections is used to derive an indirect measurement of the W boson width. CDF results[5] derive from 72pb^{-1} of integrated luminosity. Preliminary DØ results presented here use 177pb^{-1} integrated luminosity for the electron channels and 148 pb^{-1} and 96 pb^{-1} for the $Z \rightarrow \mu\mu$ and $W \rightarrow \mu\nu$ channels respectively.

Keywords: W boson, Z boson, production cross-section, cross-section ratio, W Boson Width
PACS: 01.30.Cc, 12.38.Qk, 13.38.Be, 13.38.Dg, 13.85.Qk, 14.70.Fm

INTRODUCTION

The measurements of the inclusive W and Z boson production cross section times leptonic branching ratios at the upgraded Run II Tevatron are presented. The Run II Tevatron collides proton and anti-proton beams with a centre of mass energy of $\sqrt{s} = 1.96$ TeV. Previous measurements carried out during Run I at $\sqrt{s} = 1.8$ TeV have been reported in [1, 2, 3, 4]. The CDF results presented here are derived from a data sample representing a total of 72pb^{-1} of integrated luminosity. For the DØ results the data comprises an integrated luminosity of 177 pb^{-1} for the electron channels and 148 pb^{-1} and 96 pb^{-1} for the $Z \rightarrow \mu\mu$ and $W \rightarrow \mu\nu$ channels respectively.

W AND Z PRODUCTION

Precise measurements of the inclusive W and Z boson production cross sections allow tests of the Standard Model predictions and represent benchmark analyses for the CDF and DØ collaborations. Assuming the predicted values for these cross sections the measurement can also be turned into a standard candle for measuring or making a cross check of the luminosity - a procedure which may prove valuable at the LHC. Taking the Standard Model predictions for the ratio of the total cross sections; the partial width of the W decaying to leptons and the LEP measurements of the Z boson width, an indirect measurement of the total W width can be derived from the measured ratio of the W and Z production cross section times leptonic branching ratios.

Events are selected by requiring an isolated high transverse momentum $pT > 25$ (20) GeV electron (muon) which must have fired the trigger. Electrons are identified by their deposits in the calorimeter and by the presence of a matching charged track reconstructed in the tracking detectors. Outside the coverage of the central tracking

CP792, *Deep Inelastic Scattering, DIS 2005*, edited by W. H. Smith and S. R. Dasu

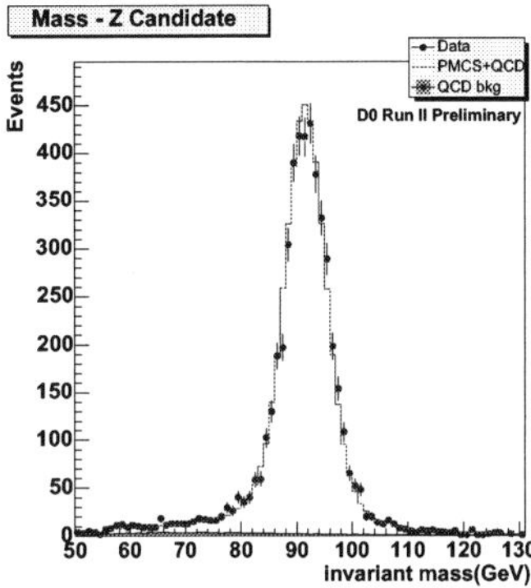

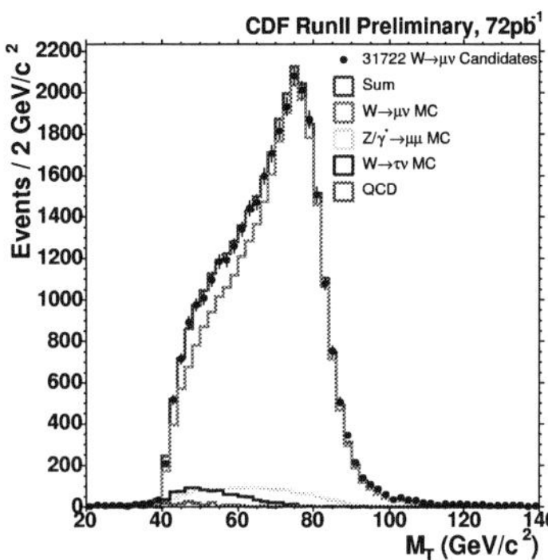

FIGURE 1. Distributions of di-electron invariant mass in the $Z \rightarrow e^+e^-$ channel at DØ (left) and transverse mass in the $W \rightarrow \mu\nu$ channel at CDF (right)

detectors at CDF, pseudo-rapidity $|\eta| > 1$ calorimetry alone is used to identify electrons. Candidate electrons are required to have a large fraction of their energy deposited in the electromagnetic calorimeter and the shower shape is required to be consistent with that expected for an electron. Muons are identified by matching reconstructed stubs in the muon detector with central charged tracks and by having energy deposits in the calorimeter consistent with the passage of a minimum ionising particle. Further selection criteria are applied to the track quality and timing to remove muons from decays in flight and from cosmic rays. For muon reconstruction at DØ the allowed fiducial region extends out to $|\eta| < 1.8$.

To select the W sample, in addition to a high pT lepton reconstructed in the central region $|\eta| < 1$, a large imbalance in the transverse momentum of the event - measured using the calorimetry - is required $\not{E}_T > 25$ (20) GeV for the electron (muon) channel. This arises from the momentum carried away by the undetected neutrino.

Z candidates are selected by requiring an additional oppositely charged muon candidate or loose electron candidate for the muon and electron channels respectively. The selection criteria on the second electron are relaxed - for example, the track matching requirement is dropped - in order to maintain a high efficiency. CDF requires at least one electron to be reconstructed in the central region, allowing the second to be reconstructed in the plug calorimeter. In the results presented here DØ requires both electrons to be reconstructed in the central cryostat of the calorimeter.

Single lepton trigger, reconstruction and identification efficiencies are measured directly using $Z \rightarrow l^+l^-$ events in the data. Events are selected with standard cuts applied to one leg of the decay leaving the other unbiased leg to be tested to see if it passes the selection criteria. Biases have been investigated in full Monte-Carlo simulations and found to be small.

Primary backgrounds come from multi-jet events or from those events containing W and Z decays such as: $Z \rightarrow l^+l^-$ where one lepton is not reconstructed, and $W \rightarrow \tau\nu$ in W decays and $W \rightarrow l\nu$ in Z decays. Additionally, muons from cosmic rays contribute to backgrounds in the muon channels. Electroweak boson backgrounds are estimated using full Monte-Carlo simulation and estimates of the multi-jet background are extracted from the data.

Figure 1 shows the reconstructed di-electron invariant mass distribution from the $Z \to e^+e^-$ analysis at DØ (left) and the W transverse mass distribution in the $W \to \mu\nu$ channel at CDF (right). In the left plot electroweak backgrounds have been subtracted from the data (points) and the QCD multi-jet backgrounds are shown as the shaded area beneath the peak. The right hand plot shows the various estimated signal and background contributions as histograms and the measured data as points. In both cases good agreement between measurement and expectation is demonstrated.

RESULTS

TABLE 1. Summary of published CDF cross-section, ratio and indirect W width results [5]

Channel		stat	sys	lum	
$\sigma_W(e+\mu)$	2775	$\pm$ 10	$\pm$ 53	$\pm$ 167	pb
$\sigma_Z(e+\mu)$	254.9	$\pm$ 3.3	$\pm$ 4.6	$\pm$ 15.2	pb
$R(e+\mu)$	10.92	$\pm$ 0.15	$\pm$ 0.14		
Γ_W	2.079	$\pm$ 0.042			GeV

TABLE 2. Summary of preliminary DØ cross section and ratio results

Channel		stat	sys	pdf	lum	
$\sigma_W \times Br(W \to e\nu)$	2865.2	$\pm$8.3	$\pm$62.8	$\pm$ 40.4	$\pm$ 186.2	pb
$\sigma_W \times Br(W \to \mu\nu)$	2989	$\pm$15	$\pm$81		$\pm$ 194	pb
$\sigma_Z \times Br(Z \to e^+e^-)$	264.9	$\pm$3.9	$\pm$ 8.5	$\pm$ 5.1	$\pm$ 17.2	pb
$\sigma_Z \times Br(Z \to \mu^+\mu^-)$	291	$\pm$3.0	$\pm$ 6.9		$\pm$ 18.9	pb
R(electron)	10.82	$\pm$0.15	$\pm$ 0.25	$\pm$ 0.13		

Tables 1 and 2 summarise the published results from CDF and the preliminary results from DØ respectively. For the CDF Z cross section results the quoted value represents the cross section including virtual photon effects within a di-electron mass window $66 < M_{ee} < 116$ GeV. The corresponding DØ results represent the pure Z cross section corrected for the full acceptance over the whole mass range. Seeing no evidence for breaking of lepton universality CDF have produced combined muon and electron cross section results. Work to combine the equivalent analyses at DØ is still underway. Separate columns are used for each of the contributing errors to the measurements. Where estimates of the uncertainties arising from the choice of parton density function (PDF) used in determining the acceptance corrections were available separately these have been quoted - in all other cases such effects are included in the total systematic error. These results are systematics limited at around the 2-3 % level, ignoring the uncertainty in the luminosity which contributes at around 6%. The uncertainties in the determination of the efficiency and acceptance corrections dominate the non-luminosity systematic errors. The contribution to systematic error deriving from the uncertainty in the PDFs used in the acceptance correction is between 1.7 and 2.0 % depending on the channel. The uncertainty on the luminosity is correlated between the two experiments with approximately 4% coming from understanding the performance of the luminosity detectors and around 4.5% from the error on the total $p\bar{p}$ cross section.

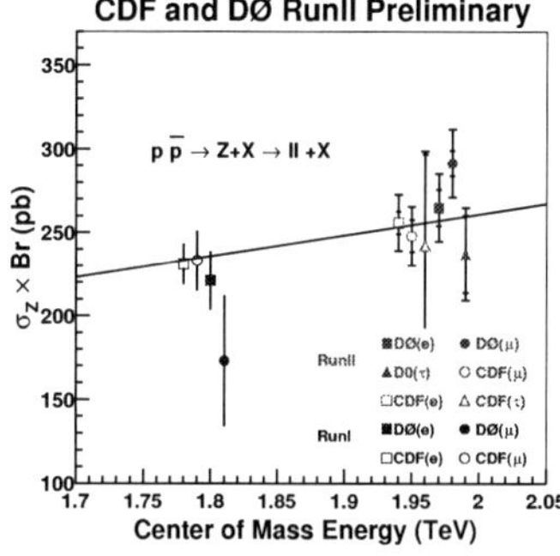

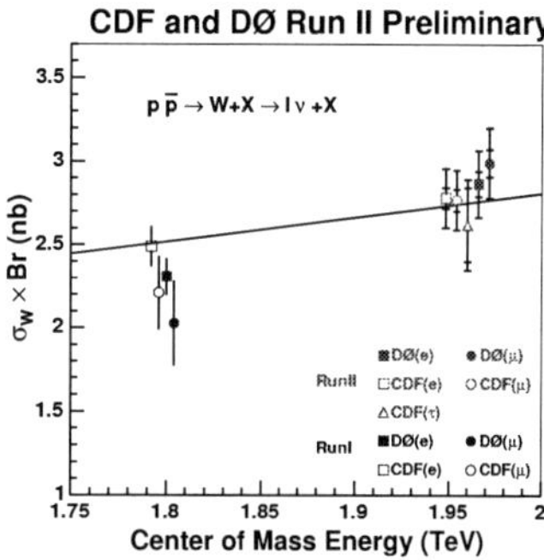

FIGURE 2. Summary of Z (left) and W (right) boson production cross-section times leptonic branching ratio results from Run I and Run II

Figure 2 summarises graphically the results from Run I and the results reported here. The points have been displaced horizontally for clarity but represent results from $\sqrt{s} =$ 1.8 TeV and $\sqrt{s} = 1.96$ TeV. The solid curve shows the Standard Model predictions.

The ratio of the cross sections times branching fractions can be defined as in Equation 1. From the measured value of $Br(Z \to l^+l^-) = 0.033658 \pm 0.000023$ [7] and a theoretical calculation at NLO of the production cross section ratio [6] CDF extract a measurement of the W leptonic branching ratio: $Br(W \to l\nu) = 0.1089 \pm 0.0022$. With these numbers and the theoretical value of the W partial width, $\Gamma(W \to l\nu) = 226.4 \pm 0.3$MeV [7] the total width of the W can be extracted: $\Gamma_W = 2079 \pm 42$ MeV.

$$R = \frac{\sigma_W \times Br(W \to l\nu)}{\sigma_Z \times Br(Z \to ll)} = \frac{\sigma_W}{\sigma_Z}\frac{\Gamma_Z}{\Gamma_{Z\to ll}}\frac{\Gamma_{W\to l\nu}}{\Gamma_W} \tag{1}$$

CONCLUSIONS

CDF and DØ have measured the W and Z boson production times leptonic branching fractions and their ratios. These are high precision, systematics limited measurements at the 2-3% level with an uncertainty from the luminosity of around 6%. Good agreement is found between data and Standard Model predictions.

REFERENCES

1. T. Affolder *et al.* (CDF Collab.), Phys. Rev. Lett., **84**, 845, (2000)
2. F. Abe *et al.*, (CDF Collab.), Phys. Rev., **D59**, 052002, (1999)
3. B. Abbott *et al.* (DØ Collab.), Phys. Rev., **D60**, 05203, (1999)
4. F. Abe *et al.*, (CDF Collab.), Phys. Rev. Lett., **76**, 3070, (1996)
5. D. Acosta *et al.*, (CDF Collab.), Phys. Rev. Lett., **94**, 091803, (2005)
6. See [5] and references therein.
7. K. Hagiwara *et al.*, Phys. Rev., **D66**, 010001, (2002)

Jet Physics and PDFs at the Tevatron

Amnon Harel

On behalf of the CDF and DØ Collaborations
Wuppertal University, 42097 Wuppertal, Germany

Abstract. During Run II of Fermilab's Tevatron collider, the CDF and DØ experiments have collected over 1 fb^{-1} of data. The collaborations made several jet measurements relevant to determining the proton's particle density functions, which are presented here.

Keywords: QCD, Jet, Tevatron, PDF
PACS: 13.87.Ce; 13.85.-t

Run II of Fermilab's Tevatron collider and its upgraded detectors started on April 2001, with a proton-antiproton center of mass energy of 1.96 TeV. The detectors are CDF and DØ[1], both general-purpose detectors containing silicon micro-vertex detectors, a central tracking system, and extensive calorimetry and muon systems. The Tevatron allows us to probe the proton's parton density functions (PDFs) in regions of phase space inaccessible elsewhere. Hard collisions often result in collimated particle sprays, which are clustered into jets using various jet algorithms. The jet algorithms are applied to partons, particles, tracks and calorimeter towers.

The primary impact of PDFs on jet physics is through the gluon PDF, which at the Tevatron is strongly related to the inclusive jet and dijet cross sections. These results are presented in the next section. It is also possible to probe the b-quark PDF at the Tevatron, and such measurements from both collaborations are presented. It should be noted that these are the first experimental probes of the b-quark PDF in this region of phase space. Also briefly mentioned are recent Tevatron results that can serve to verify the hard-scattering matrix elements and aspects of the simulation used to extract the PDFs from the measured cross sections.

Due to space limitations, only a few of the plots shown at the conference will be included here. Further plots, details and updates are available on CDF's [2] and DØ's [3] public results web pages, and on the conference's web page [4].

JET CROSS SECTION MEASUREMENTS

Both the CDF and the DØ collaborations have recently updated their inclusive cross sections measurements as a function of the transverse jet momentum, p_T. The most recent measurement is from CDF and uses the $k_\perp$ jet algorithm as described in [6]. The measurement includes jets with $p_T > 54$ GeV in the rapidity region $0.1 < |y_{jet}| < 0.7$. The jets were reconstructed with the $k_\perp$ algorithm's D parameter set to 0.5, 0.7 and 1.0. The results for D=0.5 are shown in Figure 1. The measured cross sections were corrected to the hadron level and compared to perturbative QCD (pQCD) calculations in next-to-leading order (NLO) as implemented in JETRAD [5], and additional parton-to-hadron

CP792, *Deep Inelastic Scattering, DIS 2005,* edited by W. H. Smith and S. R. Dasu

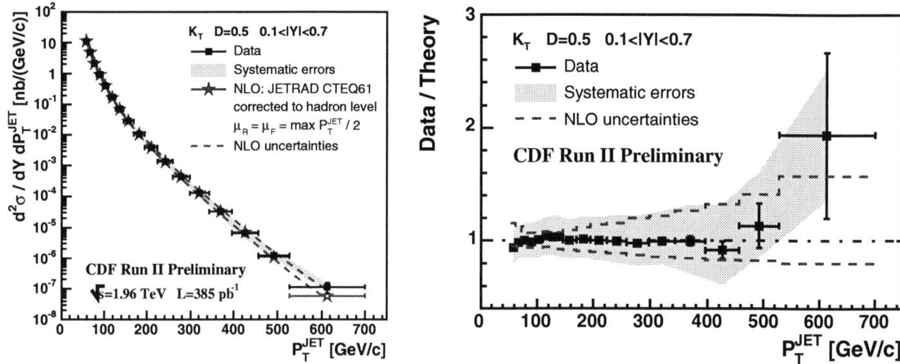

FIGURE 1. CDF's $k_{\perp}$-jet inclusive cross section measurement and its ratio over the NLO prediction

corrections were applied. Larger values of D include particles at larger distances into the jets and hence make the jet cross section more sensitive to soft contributions from the underlying event. This measurement achieved a precision similar to that of the NLO calculations. The NLO calculations and the measured cross section are consistent.

The DØ collaboration measured the inclusive jet cross section using the Run II Cone Algorithm [7] with $R_{cone} = 0.7$. The cross section was measured in two rapidity regions: $|y_{jet}| < 0.4$ and $0.4 < |y_{jet}| < 0.8$, and the results are shown in Figure 2. The measured cross sections were corrected to the hadron level and compared to pQCD NLO calculations made using NLOJET++ [8] and the PDFs from CTEQ6.1M [9] and MRST2004 [10]. Parton-to-hadron corrections were evaluated using PYTHIA [11] and HERWIG [12] and result in uncertainties on the ratio between the observed and the calculated cross sections between 10% (below $p_T = 100$ GeV) and 5% (for $p_T > 100$ GeV). The NLO calculations and the results are consistent.

CDF also measured the inclusive cross section using cone jets in the rapidity region $0.1 < |y_{jet}| < 0.7$, using data with an integrated luminosity of $177\,\text{pb}^{-1}$; DØ also measured the dijet cross section using cone jets in the rapidity region $|y_{jet}| < 0.5$, using data with an integrated luminosity of $143\,\text{pb}^{-1}$. In both measurements the pQCD NLO calculations and the data were consistent. Further details can be found in [2, 3, 4].

B-JET MEASUREMENTS

In current PDF fits, the b-PDF is derived from the gluon PDF using theoretical predictions which assume no intrinsic b content and often use the massless gluon splitting approximation. The Tevatron collaborations made the direct experimental probes of the b-PDF necessary to verify these approximations at Tevatron kinematics, using vector-boson plus b-jet final states.

DØ has recently published [13] a measurement of the ratio of inclusive cross sections $\frac{\sigma(\text{Z+bjet})}{\sigma(\text{Z+jet})} = 0.021 \pm 0.004(stat)^{+0.002}_{-0.003}(syst)$, using data with an integrated luminosity of $\approx 180\,\text{pb}^{-1}$. The cross sections were measured in the $Z \to e^+e^-$ and the $Z \to$

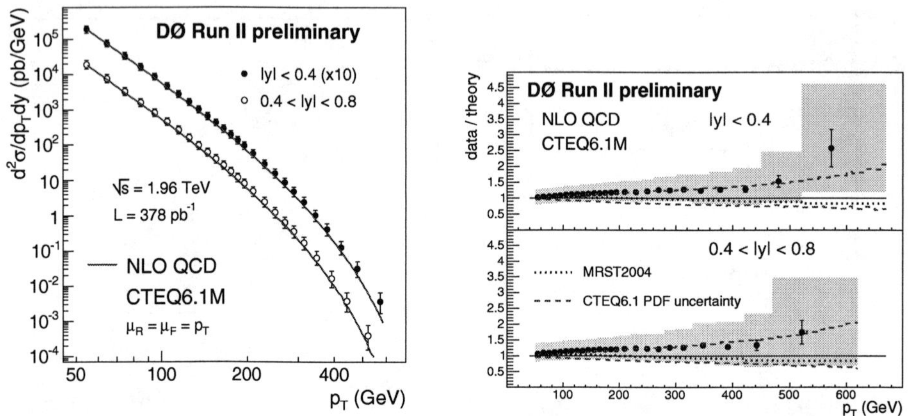

FIGURE 2. DØ's cone jet inclusive cross section measurement and its ratio over the NLO prediction

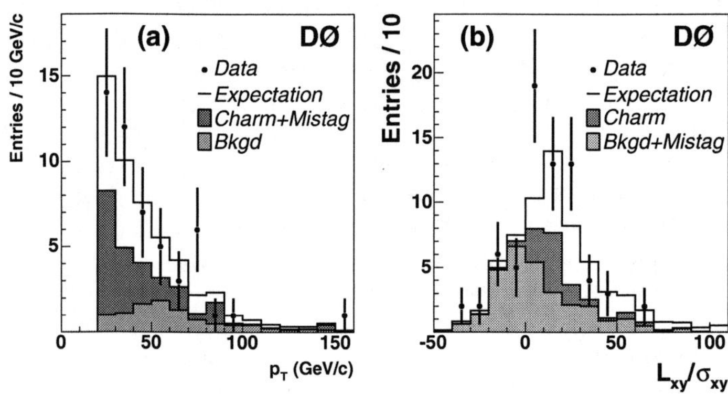

FIGURE 3. Decay length significance and p_T distributions from DØ's cross section ratio measurement

$\mu^+\mu^-$ modes, for jets with a pseudo-rapidity $|\eta| < 2.5$ and $p_T > 20$ GeV. The b-jets were tagged with a secondary vertex algorithm, Figure 3 shows the distribution of the transverse decay length significance before requiring $L_{xy}/\sigma_{xy} > 7$. The measurement is in good agreement with the NLO prediction of 0.018 ± 0.004, made using the CTEQ6 [9] PDFs.

CDF measured the γ b-jet production cross section $\sigma(\gamma + b\text{jet}) = 40.6 \pm 19.5(stat)^{+7.4}_{-7.8}(syst)$pb, using data with an integrated luminosity of $\approx 67\,\text{pb}^{-1}$. The cross sections were measured for electrons with $E_T > 25$ GeV and for jets with $E > 30$ GeV that were tagged with a secondary vertex algorithm. The measurement is in agreement with the LO prediction made using the CTEQ5L PDFs.

SUPPORTING MEASUREMENTS

To extract information on the PDFs from a jet measurement, the particle-level cross sections must be derived from the measured (detector level) observables. These "hadronic corrections" include the effects of the proton and anti-proton remnants and their subsequent interactions, as well as fragmentation effects. CDF measured the jet shapes for cone jets, in p_T bins from 37 to 380 GeV. This measurement is sensitive to the hadron remnants and to fragmentation effects, and shows that both are well simulated.

Hard scattering matrix elements, computed using pQCD, are used to constrain the PDFs from the particle-level cross sections. DØ published [14] a measurement of dijet azimuthal decorrelations at central rapidities ($|y_{jet}| < 0.5$), which directly probes the matrix elements at orders $O(\alpha_s^3)$ and $O(\alpha_s^4)$. The distribution of the azimuthal angle measured, ϕ is given by the differential cross section in ϕ normalized by the total cross section. This reduces the systematic effects and uncertainties from the PDFs and jet energy response and allows a check of the matrix elements. The measurements was presented in detail elsewhere in this conference, but is mentioned here since it serves to verify the extraction of PDFs from jet cross section measurements.

ACKNOWLEDGMENTS

It is a pleasure to thank the organizing committee of DIS 2005 for a pleasant and productive meeting. I would also like to thank D. Alton, M. Begel, F. Chlebana, C. Royon and M. Wobisch for their help in preparing the talk.

REFERENCES

1. CDF II Collab., FERMILAB-PUB-96/390-E (1996);
   D. Acosta et al., Phys. Rev. D **71**,032001 (2005); DØ Collab., V. Abazov et al. (to be published);
   T. LeCompte and H. T. Diehl, Annu. Rev. Nucl. Part. Sci. **50**, 71 (2000).
2. http://www-cdf.fnal.gov/physics/pub_run2/
3. http://www-d0.fnal.gov/www_buffer/pub/Run2_publications.html
4. http://www.hep.wisc.edu/dis05/
5. W. T. Giele, E. Q. N. Glover, D. A. Kosower, Phys. Rev. Lett. **73**, 2019 (1994).
6. S. D. Ellis, D. E. Soper, Phys. Rev. D **48**, 3160 (1993) [hep-ph/9305266].
7. G. C. Blazey et al., FERMILAB-Conf-00/092-E . See Section 3.5 for details.
8. Z. Nagy, Phys. Rev. Lett. **88**, 122003 (2002); Z. Nagy, Phys. Rev. D **68**, 094002 (2003).
9. J. Pumplin et al., JHEP **0207**, 12 (2002); D. Stump et al., JHEP **0310**, 46 (2003).
10. A. D. Martin et al., Phys. Lett. B **604**, 61 (2004).
11. T. Sjöstrand et al., Comp. Phys. Comm. **135**, 238 (2001).
12. G. Marchesini et al., Comp. Phys. Comm. **67**, 465 (1992); G. Corcella et al., JHEP **0101**, 10 (2001).
13. V. M. Abazov et al., Phys. Rev. Lett. **94**, 161801 (2005).
14. V. M. Abazov et al., Phys. Rev. Lett. **94**, 221801 (2005).

Neutrino Oscillation Experiments and Cross Section Modeling

Hugh Gallagher

Tufts University, Medford, MA 02155

Abstract. In this paper I will discuss the modeling of neutrino interaction physics for neutrino oscillation experiments, focusing in particular on the cross section modeling for the MINOS experiment.

NEUTRINO OSCILLATION EXPERIMENTS

The discovery of neutrino oscillations in the solar and atmospheric fluxes has opened a new window of exploration into the standard model. Precision measurements of the lepton mixing matrix will require intense neutrino beams and large detectors at remote locations. The MINOS experiment, which began taking oscillation data in March 2005, uses a conventional neutrino beam from the Fermilab Main Injector and a 5.4 ton iron calorimeter located 730 km away in northern Minnesota [1]. A number of additional experiments are either in the construction, advanced planning, or proposal stages. Conventional "super-beams" produced by megawatt-scale proton drivers, neutrino factories, or even neutrino beams produced from radioactive ion beams have all been considered. Generally speaking the energy range of interest to future experiments is 0.5 - 10 GeV with detectors consisting primarily of carbon, oxygen, or iron.

Through the study of neutrino interactions over several decades we have learned a great deal about electroweak unification and the QCD structure of the nucleon. With the advent of high-statistics oscillation experiments the study of neutrino interactions has entered a new phase, where the ability to accurately model the cross sections and nuclear physics is important in an "engineering" sense, i.e. as the backdrop against which any oscillation signatures will play out.

NEUTRINO INTERACTION PHYSICS UNCERTAINTIES

One of the clearest challenges in modeling neutrino interactions for an experiment like MINOS is incorporating physics models over a broad range of kinematics and nuclear targets. Figure 1 shows the kinematic coverage in x and Q^2 of the NuMI (Neutrinos at the Main Injector) low energy beam, the standard configuration for the MINOS experiment. The contours indicate regions of kinematic space including 50, 75, 90, and 99% of the events. The dashed lines indicate rough boundaries for common theoretical assumptions. The line at $Q^2 = 1$ GeV^2 is the approximate minimum value for which the Plane Wave Impulse Approximation (PWIA), a key assumption in the treatment of scattering from

CP792, *Deep Inelastic Scattering, DIS 2005*, edited by W. H. Smith and S. R. Dasu

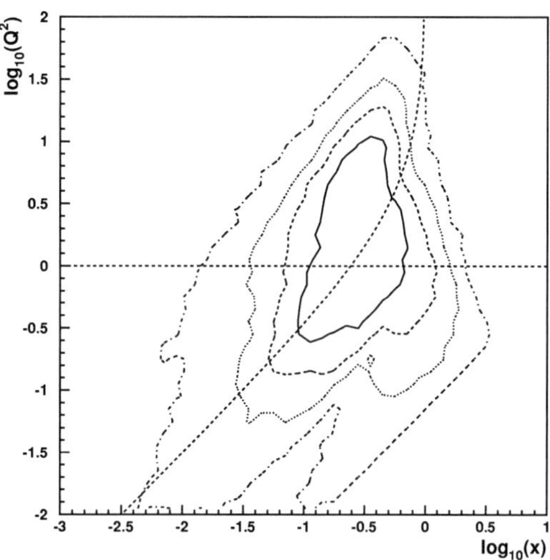

FIGURE 1. Kinematic coverage of the NuMI low energy beam.

nuclear targets, is considered valid. The other corresponds to W=2 GeV, the canonical transition into the DIS regime. Neutrino analyses historically avoided inelastic data below W=2 GeV and above the $\Delta(1232)$ because of the large higher twist corrections and difficulty in identifying a clean theoretical approach for this region which stands at the transition between perturbative and non-perturbative regimes.

In addition to uncertainties in modeling the cross section itself, uncertainties in the hadronization process and in the intranuclear rescattering of produced hadrons can have a large impact on an oscillation search. Part of the goal of a new generation of experiments, like the MINERνA [2] experiment at Fermilab, is to measure these effects with high precision so that they do not become limiting systematics for future oscillation measurements [3].

TUNING TO ELECTRON SCATTERING DATA

There is much to be gained by tuning neutrino interaction models to electron scattering data [4]. The kinematic region of primary importance to MINOS largely overlaps the kinematic range explored by electron scattering experiments at the Jefferson lab and other facilities going back many years. This data is both far more precise than neutrino data and exists over the entire resonance region.

The standard treatment for neutrino cross section modeling is now centered around the modified leading order DIS model of Bodek and Yang [5] which uses the GRV98LO parton distributions. This model describes higher twist effects through the use of new scaling variables and is able to describe electron scattering data down to the photopro-

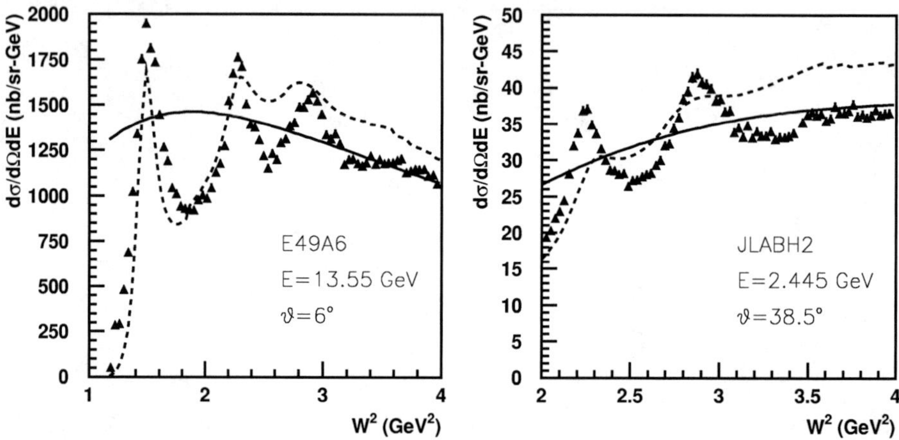

FIGURE 2. Comparison between electron scattering data and predictions from two versions of the NEUGEN3 program. The dashed curve is from the default 2004 version of the program which is dominated in this regime by the Rein-Seghal model. The solid line is the prediction of the Bodek-Yang model. Data are from [8].

duction limit. This model has been comprehensively compared against F_2 and xF_3 data from both charged lepton and neutrino scattering experiments.

Tuning cross section models in the resonance region (1.2 GeV $<$ W $<$ 2 GeV) has always posed a challenge because of the lack of precise neutrino data. Many neutrino programs have, in the past, relied on the Rein-Seghal model [6] which implements the Feynman-Kislinger-Ravndal model [7] of baryon resonances and attempts to describe data up to W=2 GeV in terms of 18 hadronic resonances and an incoherent background. This model has been a favorite of neutrino simulators for decades as it attempts to describe scattering over a broad range of kinematics historically avoided by perturbative QCD models. While the Bodek-Yang model gets this region right on average, some experiments may require more accurate modeling of the resonance structure, in particular the $\Delta(1232)$. Tying the Bodek-Yang model to an explicit resonance model is an area of current work.

Figure 2 shows a comparison between the 2004 NEUGEN3 [9] prediction (which was based on the Rein-Seghal model), the Bodek-Yang model and a representative sample of electron scattering data. While the Bodek-Yang model describes the data in an average sense, the Rein-Seghal model attempts to fully describe the resonance structure. For other kinematics the Rein-Seghal based predictions are not nearly as good, and can differ from the data by as much as 100%.

With the high statistics recorded in the MINOS near detector a number of additional, non-oscillation measurements will be possible. Figure 3 shows the expected statistical precision from the MINOS experiment extraction of F_2 on iron with 5 years data [10].

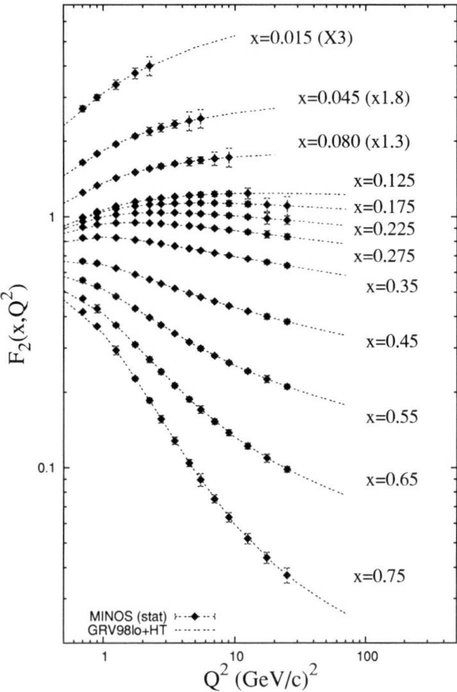

FIGURE 3. Structure function measurement capability of the MINOS experiment.

Systematic errors are not included.

ACKNOWLEDGMENTS

This work was supported by the Department of Energy grant DE-FG02-92ER40702. I would like to thank C. Andreopoulos, A. Bodek, D. Harris, C. Keppel, J. Morfin, D. Naples and S. Wood for many helpful conversations.

REFERENCES

1. D. Michael, *Nucl. Phys. Proc. Suppl.*, **118**, 189–196 (2003).
2. D. Drakoulakos, et al. (2004), `hep-ex/0405002`.
3. D. A. Harris, et al. (2004), `hep-ex/0410005`.
4. H. Gallagher, *AIP Conf. Proc.*, **698**, 153–157 (2004).
5. A. Bodek, and U. K. Yang, *AIP Conf. Proc.*, **721**, 358–362 (2004).
6. D. Rein, and L. M. Sehgal, *Ann. Phys.*, **133**, 79 (1981).
7. R. P. Feynman, M. Kislinger, and F. Ravndal, *Phys. Rev.*, **D3**, 2706–2732 (1971).
8. S. Wood, http://hallcweb.jlab.org/resdata.
9. H. Gallagher, *Nucl. Phys. Proc. Suppl.*, **112**, 188–194 (2002).
10. D. Naples, private communication.

Impact of future HERA data on the determination of proton parton distribution functions using the ZEUS QCD fit

CLAIRE GWENLAN

Nuclear & Astrophysics Laboratory, University of Oxford, Keble Road, Oxford, OX1 3RH. UK.
e-mail: c.gwenlan1@physics.ox.ac.uk

Abstract. The high precision and large kinematic coverage of the data from the HERA-I running period (1994-2000) have already allowed precise extractions of proton parton distribution functions (PDFs). The HERA-II running program is now underway and is expected to provide a substantial increase in the luminosity collected at HERA. In this contribution, a study is presented which investigates the potential impact of future data from HERA on the proton PDF uncertainties, within the currently planned running scenario. In addition, the effect of a possible future measurement of the longitudinal structure function, F_L, on the gluon distribution is investigated.

Keywords: HERA, proton structure, PDF, QCD
PACS: 12.38.-t

INTRODUCTION

Since the advent of HERA, much progress has been made in determining the Parton Distribution Functions (PDFs) of the proton. The PDFs must be known as precisely as possible in order to make reliable predictions for any process involving protons, and to maximise the discovery potential for new physics at both current and future colliders.

HERA is now in its second stage of operation. With the measurements that can now be expected from HERA-II, our knowledge of PDFs should be further improved. In this paper, first studies of the potential impact of future measurements from HERA, on the PDF uncertainties, are presented.

HERA PHYSICS AND KINEMATICS

Lepton-proton Deep Inelastic Scattering (DIS) can proceed either via the Neutral Current (NC) interaction (through the exchange of a γ^* or Z^0), or via the Charged Current (CC) interaction (through the exchange of a $W^{\pm}$). The kinematics of lepton-proton DIS are described in terms of the Bjorken scaling variable, x, the negative invariant mass squared of the exchanged vector boson, Q^2, and the fraction of energy transferred from the lepton to the hadron system, y. The three quantities are related by $Q^2 = sxy$, where s is the centre-of-mass energy squared.

At leading order (LO) in the electroweak interaction, the double differential cross section for the NC DIS process is given in terms of proton structure functions,

$$\frac{d^2\sigma^{NC}(e^{\pm}p)}{dxdQ^2} \sim Y_+ F_2(x,Q^2) - y^2 F_L(x,Q^2) \mp Y_- xF_3(x,Q^2) \quad (1)$$

CP792, *Deep Inelastic Scattering, DIS 2005*, edited by W. H. Smith and S. R. Dasu

TABLE 1. The data-sets included in the ZEUS PDF fits. The last two columns give the integrated luminosities of the HERA-I measurements and those assumed in the HERA-II projection. Note that the 96-97 NC and the 94-97 CC measurements have not had their luminosity scaled.

data sample	kinematic coverage	HERA-I $\mathscr{L}$ (pb^{-1})	HERA-II $\mathscr{L}$ (pb^{-1}) (assumed)
96-97 NC e^+p [3]	$2.7 < Q^2 < 30000\ \text{GeV}^2;\ 6.3\cdot 10^{-5} < x < 0.65$	30	30
94-97 CC e^+p [6]	$280 < Q^2 < 17000\ \text{GeV}^2;\ 6.3\cdot 10^{-5} < x < 0.65$	48	48
98-99 NC e^-p [4]	$200 < Q^2 < 30000\ \text{GeV}^2;\ 0.005 < x < 0.65$	16	350
98-99 CC e^-p [7]	$280 < Q^2 < 17000\ \text{GeV}^2;\ 0.015 < x < 0.42$	16	350
99-00 NC e^+p [5]	$200 < Q^2 < 30000\ \text{GeV}^2;\ 0.005 < x < 0.65$	63	350
99-00 CC e^+p [8]	$280 < Q^2 < 17000\ \text{GeV}^2;\ 0.008 < x < 0.42$	61	350
96-97 inc. DIS jets [9]	$125 < Q^2 < 30000\ \text{GeV}^2;\ E_T^{Breit} > 8\ \text{GeV}$	37	500
96-97 dijets in γp [10]	$Q^2 \lesssim 1\ \text{GeV}^2;\ E_T^{jet1,2} > 14, 11\ \text{GeV}$	37	500
optimised jets [12]	$Q^2 \lesssim 1\ \text{GeV}^2;\ E_T^{jet1,2} > 20, 15\ \text{GeV}$	-	500

where $Y_\pm = 1 \pm (1-y)^2$. The structure functions are directly related to the PDFs and their Q^2 dependence is predicted in perturbative QCD. In particular, F_2 and xF_3 depend directly on the quark distributions. For $x < 10^{-2}$, F_2 is dominated by sea quarks and the Q^2 dependence is driven by gluon radiation. At high $Q^2 \gtrsim M_Z^2$, the contribution from xF_3 is significant. The longitudinal structure function, F_L, is directly sensitive to the gluon, but is only important at high-y. At LO, the CC cross sections are given by,

$$\frac{d^2\sigma^{CC}(e^+p)}{dx dQ^2} \sim \bar{u}+\bar{c}+(1-y)^2(d+s); \qquad \frac{d^2\sigma^{CC}(e^-p)}{dx dQ^2} \sim u+c+(1-y)^2(\bar{d}+\bar{s})$$

so that a measurement of the e^+p and e^-p cross sections at high-x, provide information on the d- and u-valence quarks, respectively, thereby allowing the separation of flavour.

The QCD scaling violations in the inclusive cross section data, namely the QCD Compton ($\gamma^* q \to gq$) and boson-gluon-fusion ($\gamma^* g \to q\bar{q}$) processes, may also give rise to distinct jets in the final state. Jet cross sections therefore provide a direct constraint on the gluon through the boson-gluon-fusion process.

RESULTS AND DISCUSSION

In this section, the results of two separate studies are presented. The first study, provides an estimate of how well the PDF uncetrainties may be known by the end of HERA-II, within the currently planned running scenario, while the second study investigates the impact of a future HERA measurement of F_L on the gluon uncertainty.

All results presented, are based on the recent ZEUS-JETS PDF analysis [1], which is performed in the conventional next-to-leading-order (NLO) DGLAP [2] framework of QCD. The fit includes the full set of ZEUS inclusive NC [3, 4, 5] and CC [6, 7, 8] data from HERA-I, as well as ZEUS inclusive jet data in DIS [9] and dijets in γp collisions [10]. The ZEUS-JETS fit uses the offset method [11] for the evaluation of PDF uncertainties.

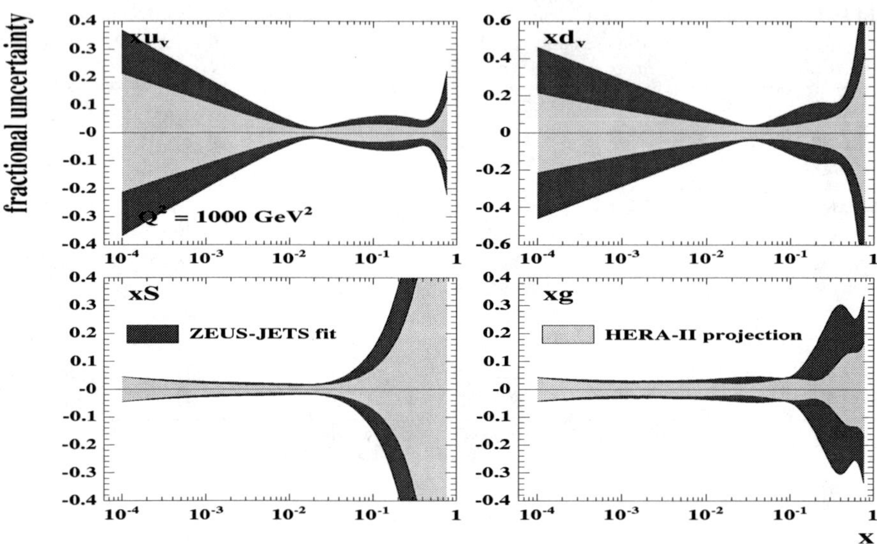

FIGURE 1. The fractional PDF uncertainties, as a function of x, for the u-valence, d-valence, sea-quark and gluon distributions at $Q^2 = 1000$ GeV^2. The red shaded bands show the results of the ZEUS-JETS fit and the yellow shaded bands show the results of the HERA-II projected fit.

1. PDF uncertainty estimates for the end of HERA running

The data from HERA-I (1994-2000) are already very precise and cover a wide kinematic region. However, HERA-II is now running efficiently and is expected to provide a substantial increase in luminosity. Current estimates suggest that, by the end of HERA running, an integrated luminosity of 700 pb^{-1} should be achievable. This will allow more precise measurements of cross sections that are curently statistically limited: in particular, the high-Q^2 NC and CC data, as well as high-Q^2 and/or high-E_T jet data. In addition to the simple increase in luminosity, recent studies [12] have shown that future jet cross section measurements, in kinematic regions optimised for sensitivity to PDFs, should have a significant impact on the gluon uncertainties.

In this contribution, the effect on the PDF uncertainties, of both the higher precision expected from HERA-II and the possibility of optimised jet cross section measurements, has been estimated in a new QCD fit. This fit will be referred to as the 'HERA-II projection', In the HERA-II projected fit, the statistical uncertainties on the currently available HERA-I data have been reduced. For the high-Q^2 inclusive data, a total integrated luminosity of 700 pb^{-1} was assumed, equally divided between e^+ and e^-. For the jet data, an integrated luminosity of 500 pb^{-1} was assumed. The central values and systematic uncertainties were taken from the published data in each case. In addition to the assumed increase in precision of the measurements, a set of optimised jet cross sections were also included, for forward dijets in γp collisions, as defined in a recent study [12]. Since no real data are yet available, simulated points were calculated using the NLO QCD program of Frixione-Ridolfi [13], with statistical uncertainties corresponding to 500 pb^{-1}. For this study, systematic uncertainties on the optimised jet cross sections were ignored. Table 1 summarises the data-sets included in the fit, and gives the luminosities of the (real) HERA-I measurements compared to those assumed in the HERA-II projection.

The results are summarised in Fig. 1, which shows the fractional PDF uncertainties, for the u- and d-valences, sea-quark and gluon distributions, at $Q^2 = 1000$ GeV^2, for the ZEUS-JETS fit compared to the HERA-II projection. Note that the same general features are observed for all values of Q^2. In fits to only HERA data, the information on the valence quarks comes from the high-Q^2 NC and CC cross sections. The increased statistical precision of the high-Q^2 data, as assumed in the HERA-II projected fit, gives a significant improvement in the valence uncertainties over the whole range of x. For the sea quarks, a significant improvement in the uncertainties at high-x is also observed. In contrast, the low-x uncertainties are not visibly reduced. This is due to the fact that the data constraining the low-x region tends to be at lower-Q^2, which are already systematically limited. This is also the reason why the low-x gluon uncertainties are not significantly reduced. However, the mid-to-high-x gluon, which is constrained by the jet data, is much improved in the HERA-II projected fit. Note that about half of the observed reduction in the gluon uncertainties is due to the inclusion of the simulated optimised jet cross sections. The improvement to the high-x partons observed in the HERA-II projected fit, is particularly relevant for high-scale physics at the LHC e.g. high-E_T jets, new particle searches.

2. Impact of a future HERA measurement of F_L on the gluon PDF

The longitudinal structure function, F_L, is directly related to the gluon density in the proton. In principle, F_L can be extracted by measuring the NC DIS cross section at fixed x and Q^2, for different values of y (see Eqn. 1). A precision measurement could be achieved by varying the centre-of-mass energy, since $s = Q^2/xy \approx 4E_eE_p$, where E_e and E_p are the electron and proton beam energies, respectively. Studies [14] have shown that this would be most efficiently achieved by changing the proton beam energy. However, such a measurement has not yet been performed at HERA.

For the present study, the impact of a possible future HERA measurement of F_L on the gluon PDF uncertainties has been investigated, using a set of simulated F_L data-points [14]. The simulation was performed using the GRV94 [15] proton PDF for the central values, and assuming $E_e = 27.6$ GeV and $E_p = 920, 575, 465$ and 400 GeV, with luminosities of 10, 5, 3 and 2 pb^{-1}, respectively. Under such conditions, and assuming the luminosity scales as E_p^2, this scenario would nominally cost 35 pb^{-1} of luminosity under standard HERA conditions. However, this takes no account of time taken for optimisation of the machine with each change in E_p, which could be considerable. The systematic uncertainties on the simulated data-points were calculated assuming a $\sim 2\%$ precision on the inclusive NC cross section measurement. A more comprehensive description of the simulated data is given elsewhere [14].

The simulated data were included in the ZEUS-JETS fit. Figure 2 shows the gluon distribution and fractional uncertainties for fits with and without inclusion of the simulated F_L data. The results indicate that the gluon uncertainties are reduced at low-x, but the improvement is only significant at relatively low Q^2.

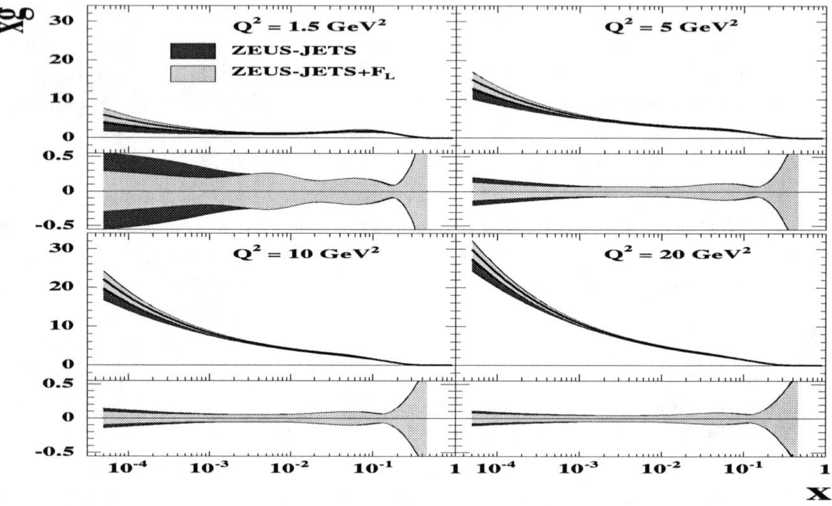

FIGURE 2. The gluon PDFs, showing also the fractional uncertainty, for fits with and without inclusion of the simulated F_L data, for $Q^2 = 1.5$, 5, 10 and 20 GeV^2. The red shaded bands show the results of the ZEUS-JETS fit and the yellow shadeds band show the results of the ZEUS-JETS+F_L fit.

ACKNOWLEDGEMENTS

I thank C. Targett-Adams for providing the NLO QCD jet cross section predictions and the machinery to allow their inclusion in the ZEUS fit. I also thank M. Klein and R. Thorne for providing F_L predictions. This work was supported by PPARC.

REFERENCES

1. ZEUS Coll., S. Chekanov *et al.*, Eur. Phys. J. **C**, 050364 (2005).
2. V. N. Gribov and L. N. Lipatov, Sov. J. Nucl. Phys. **15**, 438 (1972); L. N. Lipatov, Sov. J. Nucl. Phys. **20**, 94 (1975); G. Altarelli and G. Parisi, Nucl. Phys. **B 126**, 298 (1977).
3. ZEUS Coll., S. Chekanov *et al.*, Eur. Phys. J. **C 21**, 443 (2001).
4. ZEUS Coll., S. Chekanov *et al.*, Eur. Phys. J. **C 28**, 175 (2003).
5. ZEUS Coll., S. Chekanov *et al.*, Preprint hep-ex/0401003, (2004).
6. ZEUS Coll., J. Breitwig *et al.*, Eur. Phys. J. **C 12**, 411 (2000).
7. ZEUS Coll., S. Chekanov *et al.*, Phys. Lett. **B 539**, 197 (2002).
8. ZEUS Coll., S. Chekanov *et al.*, Eur. Phys. J. **C 32**, 16 (2003).
9. ZEUS Coll., S. Chekanov *et al.*, Phys. Lett. **B 547**, 164 (2002).
10. ZEUS Coll., S. Chekanov *et al.*, Eur. Phys. J. **C 23**, 615 (2002).
11. A. Cooper-Sarkar, hep-ph/0205153, (2002).
12. C. Targett-Adams, private communication.
13. S. Frixione and G. Ridolfi, Nucl. Phys **B 507**, 315 (1997); S. Frixione, Nucl. Phys. **B 507**, 295 (1997).
14. M. Klein, "Future F_L at HERA", in proceedings, DIS04 (2004).
15. M.Gluck, E. Reya and A. Vogt, Z. Phys. **C 67**, 433 (1995).

DIFFRACTION & VECTOR MESONS
WORKING GROUP PRESENTATIONS

New Data on elastic J/ψ Production from H1 at HERA

Christian Kiesling, for the H1 Collaboration

Max-Planck-Insitute for Physics
Föhringer Ring 6, D-80805 München, Germany

Abstract. New data on cross sections for elastic production of J/ψ mesons in photoproduction and electroproduction are presented for photon virtualities Q^2 up to $80\,\mathrm{GeV}^2$. The photon-proton center-of-mass energy $W_{\gamma p}$ covers the ranges $40 \leq W_{\gamma p} \leq 305\,\mathrm{GeV}$ in photoproduction and $40 \leq W_{\gamma p} \leq 160\,\mathrm{GeV}$ in electroproduction. The steep rise of the cross sections with $W_{\gamma p}$ is confirmed. Differential cross sections $d\sigma/dt$, where t is the squared four-momentum transfer at the proton vertex, are measured in the range $|t| < 1.2\,\mathrm{GeV}^2$ as functions of $W_{\gamma p}$ and Q^2 . An analysis of the J/ψ production and decay angular distributions yields no evidence for a violation of s-channel helicity conservation. The ratio of the cross sections for longitudinal and transverse photons is measured as a function of Q^2 and is found to be described by perturbative QCD models.

Keywords: Total cross sections, differential cross section, meson production
PACS: 13.60Hb, 13.60Le

INTRODUCTION

Quantum Chromodynamics (QCD) is expected to describe the strong force between hadrons, or, at a more fundamental level, between their constituents, the quarks and gluons. QCD is a very successful theory in the limit of short distances, corresponding to small values of the strong coupling constant α_s, where perturbative methods can be applied (perturbative QCD, pQCD). The bulk of the scattering cross section of hadrons, however, is dominated by long-range forces ("soft interactions"), where a satisfactory understanding of QCD still remains a challenge. A large fraction of these soft interactions are mediated by vacuum quantum number exchange and are termed "diffractive". In hadronic interactions, diffraction is well described by Regge theory, which is formulated as a t-channel exchange of a leading trajectory with vacuum quantum numbers, called the "Pomeron" trajectory. In the high energy limit, Pomeron exchange dominates over all other contributions to the scattering amplitude and predicts an almost energy-independent (diffractive) cross section. Elastic scattering is a particular example for such a diffractive process. Measurements of elastic photoproduction $\gamma p \to \mathrm{VM}\ p$ of light vector mesons VM (ρ, ω, and ϕ), which carry the quantum numbers of real photons, $J^{PC} = 1^{--}$, in low Q^2 electron-proton collisions at HERA as function of the photon-proton center-of-mass energy $W_{\gamma p}$ [1, 2] have beautifully verified the expected universal Regge behaviour.

The cross section for elastic photoproduction of J/ψ mesons, $\gamma p \to J/\psi p$, on the other hand, was observed [3, 4, 5, 6] to rise steeply with $W_{\gamma p}$, incompatible with a universal Pomeron. Due to its large mass, providing a "hard" scale (equivalent to a short

CP792, *Deep Inelastic Scattering, DIS 2005,* edited by W. H. Smith and S. R. Dasu

range of the forces involved), the elastic photoproduction of J/ψ mesons is expected to be described by pQCD. This is even more so in electroproduction, where the photon virtuality Q^2 can provide a second hard scale. The presence of two potentially hard scales makes J/ψ production particularly interesting for comparisons with pQCD.

In the past, J/ψ cross sections have been measured (see, e.g., [4, 6, 7, 8, 9, 10, 11]) over a wide range of the photon-proton center-of-mass energy $W_{\gamma p}$ in photoproduction, and in electroproduction spanning the range of photon virtualities up to values of $Q^2 < 100\,\mathrm{GeV}^2$. In this talk more precise data are presented on the Q^2, $W_{\gamma p}$ and t dependence of the cross sections. The data contain up to a factor 3 more integrated luminosity compared to our previous publication [4] for photoproduction and a factor of 2 for electroproduction [7]. The kinematic range has been extended to values of $W_{\gamma p}$ up to 305 GeV in photoproduction, while in electroproduction the range covered now is $40 < W_{\gamma p} < 160\,\mathrm{GeV}$. Furthermore, the angular distributions for production and decay of the J/ψ mesons are determined to extract the cross sections of longitudinal and transverse photons and to test the hypothesis that the helicity of the J/ψ in the final state is the same as that of the initial photon (s-channel helicity conservation, SCHC).

DATA ANALYSIS

The data were recorded by the H1 detector in the years 1999 and 2000 when HERA was operated mostly with positrons. The data sets are selected from an integrated luminosity of 55 pb^{-1}. The J/ψ mesons are detected via their decays into $\mu^+\mu^-$ or e^+e^- pairs in the central and forward tracking detectors (CTD and FTD), as well as in the backward silicon tracker (BST). The scattered positron is detected in the SpaCal calorimeter, while the decay electrons from the J/ψ meson are identified in the LAr and SpaCal calorimeters. Muons are identified as minimum ionizing particles in the LAr calorimeter or in the instrumented iron return yoke (CMD) of the solenoidal magnet which surrounds the central detector. Details on H1 can be found in [12].

In electroproduction the event kinematics is reconstructed using the double angle method. In photoproduction, where the positron is not observed in the central detector, the kinematics is reconstructed from the observed decay particles using the Jaquet-Blondel method [13]. The momentum transfer t is approximated by the transverse momentum of the J/ψ as $t \simeq -\vec{p}_{t,\psi}{}^2$.

Elastic J/ψ production is selected by requiring two leptons (muons or electrons), and a scattered positron in the case of electroproduction. Four data sets are defined covering different regions of Q^2 and $W_{\gamma p}$, corresponding to different signatures of the J/ψ decay leptons and/or the Q^2 range. For the electroproduction and photoproduction at low $W_{\gamma p}$ ($40 < W_{\gamma p} < 160\,\mathrm{GeV}$) the decay channel $J/\psi \rightarrow \mu^+\mu^-$ is selected. Exactly two oppositely charged particles must be present in the CTD, with transverse momenta (with respect to the beamline) $p_t > 0.8\,\mathrm{GeV}$. At least one particle must be identified as a muon in the central calorimeter or in the CMD. The details of this analysis can be found in [14]. For the higher $W_{\gamma p}$ energies the electronic decay of the J/ψ is selected. The use of the BST tracks leads to an improved mass reconstruction at the highest values of $W_{\gamma p}$. The details of this analysis can be found in [15].

Monte Carlo simulations are used to calculate the acceptances and efficiencies for triggering, track reconstruction, event selection and lepton identification, both for the signal and for the various QED backgrounds. Systematic uncertainties of the cross sections arise from detector effects which are not perfectly modeled in the simulation of the physics processes. Most uncertainties are obtained by comparisons of data and simulation after tuning the latter with independent data sets.

RESULTS

Cross sections are calculated from the number of selected events after correction for non resonant, proton dissociative and $\psi(2S)$ backgrounds. The MC simulation is used to determine the efficiencies ε for the event selection. Using the integrated luminosity $\mathscr{L}$ and the branching ratio BR for the decay of the J/ψ mesons, the γp cross section is determined in the Weizsäcker-Williams approximation. Cross sections are given at 'bin centers' $\langle W_{\gamma p}\rangle$ and $\langle Q^2\rangle$, which take into account the measured Q^2 and $W_{\gamma p}$ dependences within the bins. The differential cross sections $d\sigma/dt$ are quoted at the weighted mean $\langle t\rangle$ for each t bin, assuming an exponential dependence of the differential cross section with t.

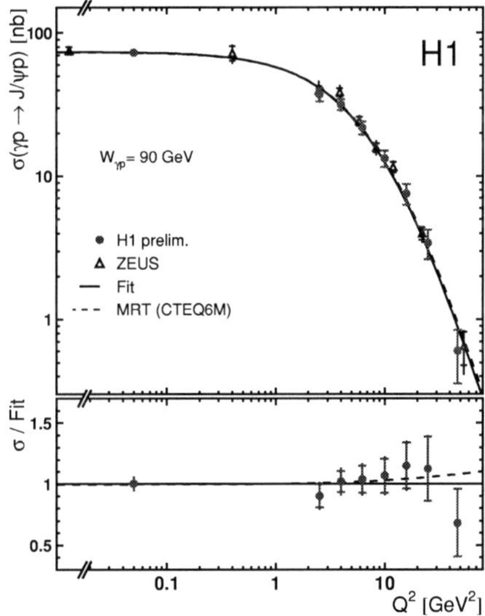

FIGURE 1. Total cross sections for elastic J/ψ production as a function of Q^2 in the range $|t| < 1.2\,\mathrm{GeV}^2$ at $W_{\gamma p} = 90\,\mathrm{GeV}$. The inner error bars show the statistical error, while the outer error bars show the statistical and systematic uncertainties added in quadrature. The solid line is a fit to the H1 data of the form $\sigma_{\gamma p} \propto (M_\psi^2 + Q^2)^{-n}$. Data from the ZEUS experiment [6, 11] are also shown.

A phenomenological fit of the form $\sigma_{\gamma p} \propto (M_\psi^2 + Q^2)^{-n}$ to the H1 data yields a value of $n = 2.486 \pm 0.080(\mathrm{stat.}) \pm 0.068(\mathrm{syst.})$. This result confirms, with smaller errors, the Q^2 dependence observed previously by H1 [7]. The quality of the fit is good (χ^2/ndf$=0.5$) and within errors the value of n is independent of the Q^2 range used in the fit. Recent results from the ZEUS collaboration [6, 11] are also shown in figure 1a, which agree well with the present data in the entire range of Q^2.

In figure 1b the pQCD calculations by Martin et al. (MRT [16]) are compared to the fit result quoted above. The predictions are obtained separately for the contributions from transversely and longitudinally polarized photons which are expected to have a different dependence on Q^2. Results using a gluon density distribution (CTEQ6M [17]) derived from global fits to current inclusive F_2 measurements and other data are shown.

$W_{\gamma p}$ **Dependence**: The γp cross section for elastic J/ψ production is presented as a function of $W_{\gamma p}$ in figure 2 for photoproduction. The data are shown with the result of a model independent fit of the form $\sigma_{\gamma p} \propto W_{\gamma p}^{\delta}$ with an empirical parameter δ. The fit yields a value of $\delta = 0.754 \pm 0.033 \pm 0.032$. The first error is obtained using only the statistical uncertainties in the fit. The second quoted error corresponds to the quadratic difference of the two. The total error is obtained by including also the uncorrelated systematic uncertainties added in quadrature. The fit result is in agreement with our previous measurements [4]. Similar data from the ZEUS collaboration [6] (also shown in figure 2) agree well with the present measurements. Two theoretical predictions are compared with the data: The MRT predictions are normalized with the same coefficients as obtained from the comparison of the Q^2 distributions (figure 1). The $W_{\gamma p}$ dependences have also be studied in electroproduction, but are not shown here. Similar conclusions as for photoproduction can be drawn. More comparisons of the MRT model are given in the talk by T. Teubner in these proceedings [18].

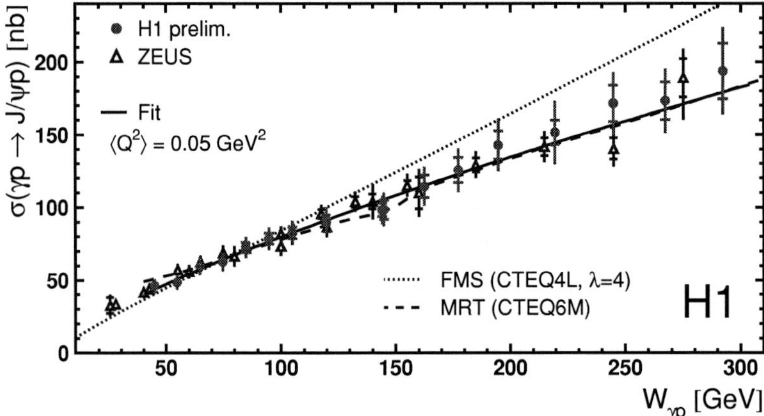

FIGURE 2. Cross sections for elastic J/ψ production as a function of $W_{\gamma p}$ in the range $|t| < 1.2\,\mathrm{GeV}^2$ in photoproduction The solid line shows a fit to the H1 data of the form $\sigma \propto W_{\gamma p}^{\delta}$. Results from the ZEUS experiment [6] in a similar kinematic range are also shown. Predictions from Martin et al. [16] and Frankfurt et al. [19] based on different gluon distributions [17, 20] are shown using the normalization factors determined from the Q^2 distributions.

Differential Cross Sections: The t dependence of the elastic cross section for J/ψ meson production has been studied for photoproduction and for electroproduction in the range $40 < W_{\gamma p} < 160\,\mathrm{GeV}$ for different Q^2 bins. For photoproduction, the measurement of the t dependence has been extended to significantly higher $W_{\gamma p}$ (up to $W_{\gamma p} = 305$

GeV). The differential cross sections $d\sigma/dt$ are well described by single exponentials ($\chi^2/\text{ndf} = 0.25$, e.g. in the case of photoproduction). In the context of developing the calculations using generalized parton densities, Frankfurt and Strikman [21] have proposed an alternative shape. It is based on a dipole function with a t dependent two-gluon form factor $d\sigma/dt \propto (1 - t/m_{2g}^2)^{-4}$. In a fit to the photoproduction data the two-gluon invariant mass m_{2g} is left as a free parameter. This form is strongly disfavored by the data ($\chi^2/\text{ndf} = 5.48$).

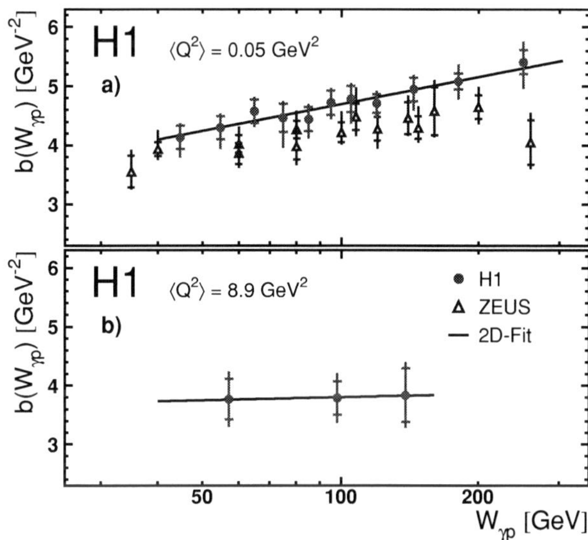

FIGURE 3. The values of the t slope $b(W_{\gamma p})$ as a function of $W_{\gamma p}$in the range $|t| < 1.2\,\text{GeV}^2$ for a) photoproduction and b) electroproduction. $\langle Q^2 \rangle$ indicates the bin center value in the Q^2 range considered. The data points are the results of one-dimensional fits of the form $d\sigma/dt \propto e^{bt}$ in $W_{\gamma p}$ bins. The solid lines show the results of the two-dimensional fits. In figure a) the data are compared to results by the ZEUS collaboration [6].

Exponential fits of the form e^{bt} to the measured differential cross sections $d\sigma/dt$ in bins of $W_{\gamma p}$ are performed and the resulting values for b are displayed in figure 3a and b for photo- and electroproduction, respectively. For photoproduction the b values are seen to increase with $W_{\gamma p}$. These b slope values are independent of normalization uncertainties between data sets. The curves in figure 3a and b show the corresponding result $b(W_{\gamma p})$ from a two-dimensional fit in $W_{\gamma p}$ and t. They agree well with the data.

In figure 3a photoproduction results for the slope parameter from the ZEUS experiment [6] in a similar kinematic region are also shown. They show a similar dependence on $W_{\gamma p}$ but are on average 0.5 GeV^{-2} lower. This difference in the absolute size of b may be due to differences in the handling of the background from proton dissociative events, which has a much shallower b slope (1.6 GeV^{-2}[9]).

Studies of the production and decay angles of the J/ψ yield consistency with s-channel helicity conservation (SCHC) within experimental errors (the corresponding new data are not shown). In the case of SCHC a direct measurement of R, the ratio of

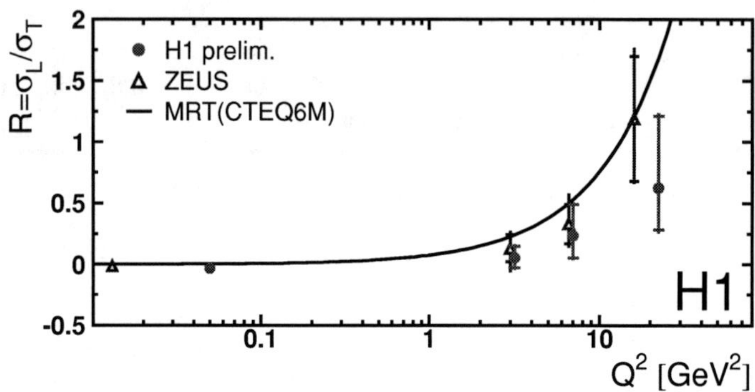

FIGURE 4. Ratio $R = \sigma^L/\sigma^T$ as a function of Q^2 (deduced from the measurement of r_{00}^{04} from the $\cos\theta^*$ distribution) for the range $40 < W_{\gamma p} < 160\,\mathrm{GeV}$ and $|t| < 5\,\mathrm{GeV}^2$. The data are compared to the result of a calculation by Martin et al. [16] based on the CTEQ6M [17] gluon distribution. Also shown are results from the ZEUS collaboration [6, 11].

the cross sections for longitudinal and transverse photons, σ^L and σ^T respectively, can be performed (see figure 4). For comparison the prediction from MRT [16] is shown, which depends only weakly on the particular gluon density. Here, the gluon density from CTEQ6M is chosen, which gives the best description of the Q^2 and $W_{\gamma p}$ dependences of the cross sections. The data are described reasonably well. Similar measurements from [6, 11] agree well with the present data.

SUMMARY

New measurements of elastic J/ψ photoproduction and electroproduction in the range $40 < W_{\gamma p} < 305\,\mathrm{GeV}$ have been presented. The cross section $\sigma(\gamma p \to J/\psi p)$ is measured as a function of Q^2 in the range $0 < Q^2 < 80\,\mathrm{GeV}^2$, and a fit of the form $\sigma_{\gamma p} \propto (M_\psi^2 + Q^2)^{-n}$ yields $n = 2.486 \pm 0.080(\mathrm{stat.}) \pm 0.068(\mathrm{syst.})$. The shape of the Q^2 distribution is well described by a perturbative QCD calculation by Martin et al. [16], almost independent of the gluon density distribution used.

The photoproduction cross section is measured as a function of the photon-proton center-of-mass energy $W_{\gamma p}$ in the range $40 < W_{\gamma p} < 305\,\mathrm{GeV}$, and is parameterized as $\sigma_{\gamma p} \propto W_{\gamma p}^{\delta}$ with $\delta = 0.754 \pm 0.033 \pm 0.032$. The results for δ in electroproduction, measured in the range $40 < W_{\gamma p} < 160\,\mathrm{GeV}$, are consistent with those in photoproduction and no change with Q^2 is observed within experimental errors.

Predictions of the $W_{\gamma p}$ dependence of the cross section in pQCD-based models seem to depend strongly on the form of the gluon distribution, as was observed explicitely in the model of Martin, Ryskin and Teubner: A good description of the shape of the data can currently be achieved only with some gluon distributions. This may demonstrate the high potential of constraining the gluon distribution with the elastic J/ψ data in a kinematic region (low x, low Q^2) where fits from inclusive data yield gluon distributions with very large uncertainties.

The differential cross section $d\sigma/dt$ for elastic J/ψ photoproduction for $|t| \leq 1.2\,\text{GeV}^2$ has been extended to the full $W_{\gamma p}$ range of HERA, $40 - 305\,\text{GeV}$, in photoproduction. The differential cross sections $d\sigma/dt$ are well parameterized as a single exponential, dipole forms are strongly disfavored. The slope parameter shows a dependence on $W_{\gamma p}$, which is weaker than expected from soft Pomeron phenomenology, but is clearly positive, leading to a shrinkage of the forward scattering peak. The slope parameters in electroproduction agree with photoproduction within errors, but have a tendency to decrease with Q^2 .

Finally, the helicity structure of diffractive J/ψ production is analyzed as a function of Q^2 and $|t|$. No evidence is found for a violation of s-channel helicity conservation (SCHC). Assuming SCHC, the ratio of the longitudinal to the transverse cross section is determined as a function of Q^2 and is found to be reasonably well described by QCD calculations.

ACKNOWLEDGMENTS

We very much appreciated the pleasant and stimulating atmosphere of this workshop, as well as the splendid organisation by the local committee from Madison University. We also thank T. Teubner and M. Strikman for valuable discussions and for making their theoretical predictions available.

REFERENCES

1. S. Aid *et al.*, H1 Collaboration, *Nucl. Phys. B*, **468**, 3 (1996).
2. M. Derrick *et al.*, ZEUS Collaboration, *Z. Phys. C*, **69**, 39 (1995).
3. S. Aid *et al.*, H1 Collaboration, *Nucl. Phys. B*, **472**, 3 (1996).
4. C. Adloff *et al.*, H1 Collaboration, *Phys. Lett. B*, **483**, 23 (2000).
5. J. Breitweg *et al.*, ZEUS Collaboration, *Z. Phys. C*, **75**, 215 (1997).
6. S. Chekanov *et al.*, ZEUS Collaboration, *Eur. Phys. J. C*, **24**, 345 (2002).
7. C. Adloff *et al.*, H1 Collaboration, *Eur. Phys. J. C*, **10**, 373 (1999).
8. C. Adloff *et al.*, H1 Collaboration, *Phys. Lett. B*, **541**, 251 (2002).
9. A. Aktas *et al.*, H1 Collaboration, *Phys. Lett. B*, **568**, 205 (2003).
10. J. Breitweg *et al.*, ZEUS Collaboration, *Eur. Phys. J. C*, **6**, 603 (1999).
11. S. Chekanov *et al.*, ZEUS Collaboration, *Nucl. Phys. B*, **695**, 3 (2004).
12. I. Abt *et al.*, H1 Collaboration, *Nucl. Instrum. Meth. A*, **386**, 310 (1996).
13. F. Jacquet, and A. Blondel, "The Kinematics of *ep* Interactions," in *Study for an ep Facility for Europe*, edited by U. Amaldi, DESY-79-048, 1979.
14. P. Fleischmann, *Elastic J/ψ Production at HERA*, Ph.D. thesis, Universität Hamburg, Hamburg (2004).
15. L. Janauschek, *Elastic Photoproduction of J/ψ Vector Mesons at High Photon-Proton Center of Mass Energies at the H1 Experiment at HERA*, Ph.D. thesis, Ludwig-Maximilians-Universität, München (2004).
16. A. D. Martin, M. G. Ryskin, and T. Teubner, *Phys. Rev. D*, **62**, 014022 (1999).
17. J. Pumplin *et al.*, *JHEP*, **0207**, 012 (2002).
18. T. Teubner, *these proceedings* (2005).
19. L. Frankfurt, M. McDermott, and M. Strikman, *JHEP*, **0103**, 045 (2001).
20. A. D. Martin, R. G. Roberts, W. J. Stirling, and R. S. Thorne, *Eur. Phys. J. C*, **23**, 73 (2002).
21. L. Frankfurt, and M. Strikman, *Phys. Rev. D*, **66**, 031502 (2002).

Vector meson production at HERA

Dorota Szuba
on behalf of the ZEUS Collaboration

H. Niewodniczański Inst. of Nuclear Physics
Polish Academy of Sciences
ul. Radzikowskiego 152, 31-342 Cracow, Poland

Abstract. Exclusive ϕ electroproduction and proton-dissociative diffractive photoproduction of J/ψ mesons has been studied in *ep* collisions with the ZEUS detector at HERA, using an integrated luminosity of 65 pb^{-1} and 112 pb^{-1}, respectively. The ϕ cross section was measured as a function of Q^2, W, t and helicity angle θ_h in the kinematic range of $2 < Q^2 < 70$ GeV^2, $35 < W < 145$ GeV and $|t| < 0.6$ GeV^2. The J/ψ photoproduction cross section was measured for $|t|$ up to 20 GeV^2 in the range of $50 < W < 150$ GeV as a function of $|t|$ and W. Both measurements are compared to other vector meson cross sections and predictions from theoretical models.

Keywords: diffraction, vector meson, electroproduction, photoproduction, high t
PACS: 14.40Cs, 14.40.Lb

INTRODUCTION

The ZEUS experiment at the HERA electron-proton collider has made extensive studies of vector meson (VM) production [1–3]. The ZEUS data cover a wide kinematic region, in which perturbative QCD (pQCD) is expected to be applicable, as well as regions in which no hard scale is present and non-perturbative processes dominate. The diffractive production of vector mesons gives the opportunity to explore the partonic nature of the diffractive exchange and to investigate the transition region between soft physics, already explored by fixed targed experiments, and the hard, pQCD physics.

The production of ϕ and J/ψ vector mesons has been studied in the diffractive process, $ep \to e\ V\ Y$, with Y being either a scattered proton (exclusive production of ϕ), or a proton remnant system (proton dissociative production of J/ψ).

The characteristic variables for these processes are: the mass of the vector meson, M_{VM}, the centre-of-mass energy of the $\gamma^* p$ system, W, the photon virtuality, Q^2, and the square of the four momentum, $|t|$, transferred at the proton vertex.

In QCD, the diffractive production of vector mesons can be viewed, in the proton rest frame, as a sequence of processes which are well separated in time. The incoming electron emits a virtual photon, which then fluctuates into a $q\bar{q}$ pair. This pair scatters elastically off the proton, and long after the interaction forms a vector meson. Due to this time separation, the cross section can be factorised [4] into terms representing the $q\bar{q}$ coupling to the photon, the dipole scattering cross section on the proton, and the final state formation of the vector meson.

The non-perturbative approach, based on Regge phenomenology [5] and the Vector Dominance Model (VDM) gives a successful description of light vector meson (ρ, ω, ϕ) production at low values of Q^2 and $|t|$. In this approach diffractive interactions are

CP792, *Deep Inelastic Scattering, DIS 2005*, edited by W. H. Smith and S. R. Dasu

mediated by the exchange of the vacuum quantum numbers which corresponds to soft Pomeron exchange. This approach predicts a weak energy dependence of the cross section ($\sigma \propto W^{0.22}$), shrinkage of the diffractive peak with increasing W, $d\sigma/dt \propto e^{-bt}(W/W_0)^{4(\alpha_{IP}(t)-1)}$, with $\alpha_{IP}(t) = \alpha_{IP}(0) + \alpha'_{IP}t$ and $b = b_0 + 4\alpha' \ln(W/W_0)$ and S-Channel Helicity Conservation, SCHC, (the VM retains the helicity of the photon).

For large values of Q^2, M_{VM} or $|t|$ the process is expected to be 'hard' and hence pQCD should be applicable. In this approach the colour dipole interacts with the proton (or a single parton in it) by the exchange of a colour singlet state, which is assumed to be a two-gluon system or gluon ladder (the latter is often called the hard or pQCD pomeron). The $q\bar{q}$ scattering off the proton is described by pQCD, while the transitions $\gamma \to q\bar{q}$ and $q\bar{q} \to VM$ are modelled with the respective wave functions. The cross section is proportional to the square of the gluon density $g(x,Q^2)$ in the proton: $\sigma \propto \alpha_s^2(Q^2)/Q^6[xg(x,Q^2)]^2$, with $x \approx Q^2/W^2$ (valid for small x) [4]. Thus, in the presence of a hard scale, a steep rise of the cross section with W is predicted, along with a weak t dependence, little shrinkage with increasing W, dominant longitudinal γ^* polarisation and possible SCHC violation.

Exclusive ϕ production allows the study of the transition region between soft and hard dynamics and data to be confronted with theoretical predictions. These results are an update of previous measurements and have been performed using a significantly larger data sample.

In proton-dissociative J/ψ photoproduction $|t|$ serves as a hard scale. This process, with an extended $|t|$ range and larger statistics, compared to previous results, is a very promising test of the BFKL based models.

In this paper these new results from ZEUS are reported.

EXCLUSIVE ϕ IN DIS

Analysis. The exclusive production of ϕ mesons, $ep \to e\phi p$, was measured in the kinematic range of $2 < Q^2 < 70$ GeV^2, $35 < W < 145$ GeV and $|t| < 0.6$ GeV^2. The data correspond to an integrated luminosity of 65.1 pb^{-1} and were collected during the 1998-2000 running period. The signature of exclusive ϕ electroproduction consists of the scattered electron and two oppositely charged kaons from the ϕ decay. The scattered proton is deflected through a small angle and escapes undetected down the beam pipe. The identification and energy measurement of the scattered electron is based on the information from the main calorimeter. The ϕ signal was reconstructed from the properties of the decay kaons. Non-exclusive events and the events with the proton dissociation were rejected if there were additional energy deposits in the calorimeter or other detectors above the noise level.

Results. The cross sections were measured as a function of W, Q^2, $|t|$ and the helicity angle, θ_h. The ratio $R = \sigma_L/\sigma_T$ was studied as a function of Q^2 and W. The data are compared to the cross sections for other vector mesons as well to predictions from theoretical models.

The $\gamma^* p$ cross section rises with W as $\sigma \propto W^{\delta}$, where δ does not depend on Q^2 or t. The measured value of $\delta \approx 0.4$ lies between the 'soft' diffraction value and that observed

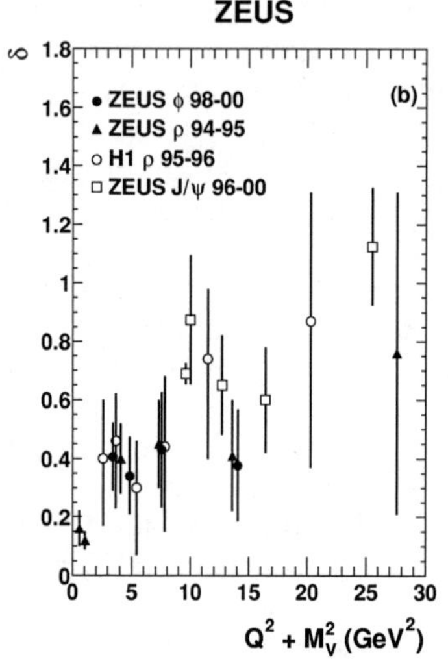

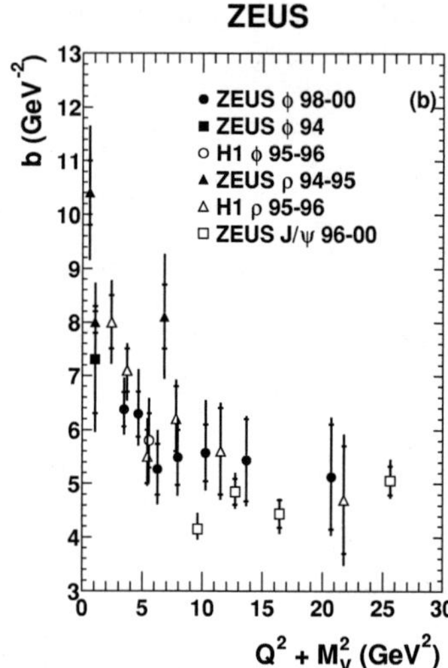

FIGURE 1. The extracted values of δ and b for ϕ as a function of $Q^2 + M_{VM}^2$ compared with results from other vector mesons. The error bars represent the quadratic sum of the statistic and systematic uncertainties.

in exclusive J/ψ production.

The Q^2 dependence of the cross section, fitted with a function of the form $(Q^2 + M_\phi^2)^{-n}$, shows that there is no a single value of n which is valid over the whole range of Q^2. The fit parameter n varies from $n = 2.087 \pm 0.055(\text{stat.}) \pm 0.050(\text{syst.})$ for $2.4 \leq Q^2 \leq 9.2$ GeV^2 to $n = 2.75 \pm 0.13(\text{stat.}) \pm 0.07(\text{syst.})$ for $9.2 \leq Q^2 \leq 70$ GeV^2. The longitudinal, σ_L, and transverse, σ_T, components of the cross section have been extracted from the measured ratio R. Both cross sections show different Q^2 dependence.

The t dependence is well described over the whole range of Q^2 and at $W = 75$ GeV by an exponential form $d\sigma/dt = d\sigma/dt|_{t=0} \cdot e^{bt}$, where b shows no the Q^2 dependence. The Pomeron trajectory was extracted from the W dependence of the differential cross section, $d\sigma/dt$, for $Q^2 = 5$ GeV^2 yielding an intercept of $\alpha_{IP}(0) = 1.10 \pm 0.02(\text{stat.}) \pm 0.02(\text{syst.})$ and a slope of $\alpha'_{IP} = 0.08 \pm 0.09(\text{stat.}) \pm 0.08(\text{syst.})$. The measured value of α'_{IP} is closer to the values measured in hard processes than in soft ones.

The data were compared to the available measurements of $ep \rightarrow e\,Vp$ [6–12] and the parameters δ and b were found to scale as a function of $Q^2 + M_V^2$ (Fig. 1). The Q^2 dependence of R is well described by the expression $a(Q^2/M_\phi^2)^b$, yielding $a = 0.51 \pm 0.07(\text{stat.}) \pm 0.05(\text{syst.})$ and $b = 0.86 \pm 0.11(\text{stat.}) \pm 0.05(\text{syst.})$. The value of R was found also to scale with Q^2/M_V^2.

The measured quantities were compared with two different leading order theoretical

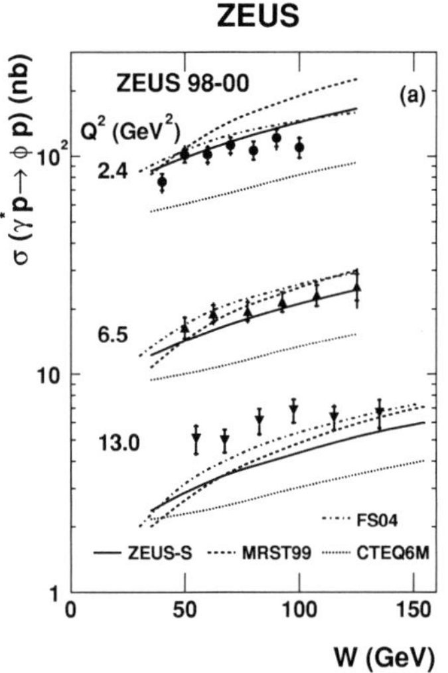

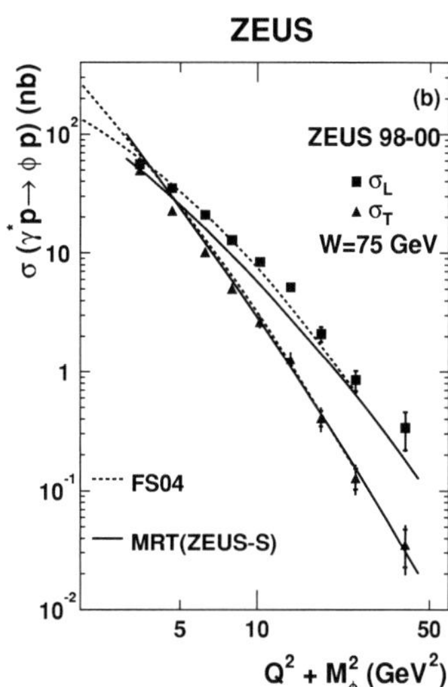

FIGURE 2. Exclusive ϕ cross sections are shown as a function of W and $Q^2 + M_\phi^2$. The data are compared to the MRT and FS04 predictions. The inner bars correspond to the statistical uncertainty and the outer to the statistical and systematic uncertainty added in quadrature.

predictions (Fig. 2): the Martin, Ryskin and Teubner (MRT) [13] model using three different gluon distributions: ZEUS-S, CTEQ6M and MRST99 and predictions by Forshaw and Shaw (FS04) [14]. The W dependence of the $\gamma^* p$ cross section is reproduced well by the FS04 predictions, especially for higher values of Q^2, while the MRT model using the ZEUS-S and CTEQ6M parametrisations is compatible with the data. The predictions based on MRST99 are too steep. Both models generally describe the Q^2 dependence, although for smaller values of Q^2 the normalisation is not well reproduced. The power-law dependence of R with Q^2 is reproduced by the theoretical predictions but is systematically low. The value of R is weakly dependent on W, this observation being confirmed by both the MRT and FS04 models.

PROTON-DISSOCIATIVE J/ψ PHOTOPRODUCTION

Analysis. The data used in this analysis were collected in 1996 - 2000 and correspond to an integrated luminosity of 112 pb^{-1}. The signature of proton dissociative J/ψ photoproduction, $ep \to eJ/\psi Y$, consists of two oppositely charged muons from the J/ψ decay, the dissociated proton and the lack of the scattered electron in the main detector. The electron is scattered at small angle and escapes undetected down the beam pipe. The low mass hadronic states originating from the scattered proton are registered in

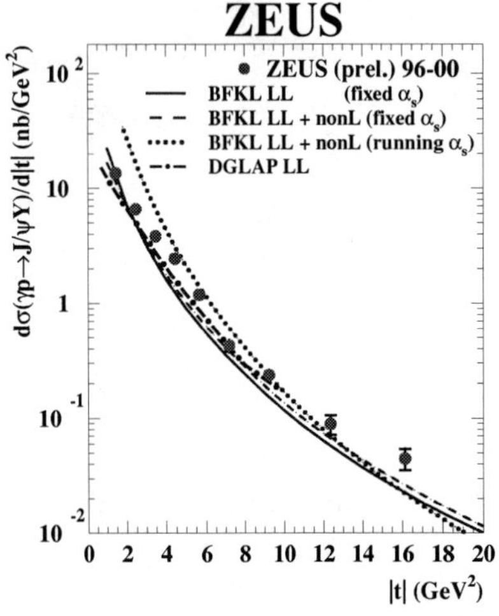

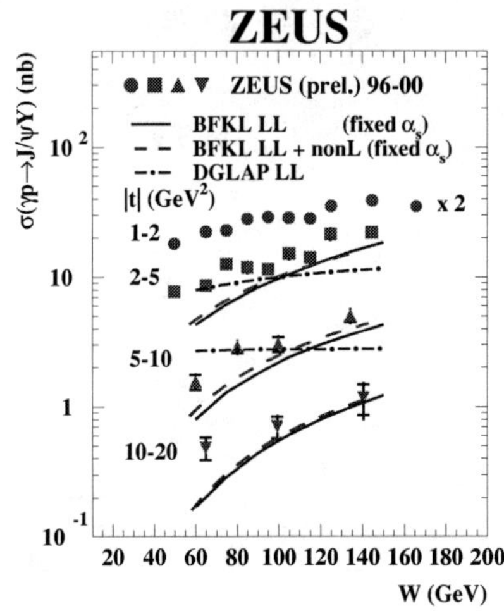

FIGURE 3. The differential cross section $d\sigma/d|t|$ and the cross section as a function of W for the process $\gamma p \rightarrow J/\psi Y$. The inner bars correspond to the statistical uncertainty and the outer to the statistical and systematic uncertainty added in quadrature. The solid line represents the results of BFKL LL with fixed α_s and the dashed (dotted) line – including non-leading corrections (nonL) with fixed (running) α_s. The dashed-dotted curve is a prediction based on DGLAP LL calculations.

the forward part of the calorimeter or in the forward detectors. The J/ψ mesons were reconstructed from the decay muons measured in the central tracking detector. Events with additional energy deposits above the noise level and not associated with lepton candidates or with products of dissociated proton were rejected.

In order to compare the cross sections with theoretical predictions and previous measurements, an additional cut of $z > 0.95$ was applied. This ensures the diffractive nature of the interaction and also restricts the invariant mass of the system Y to be $M_Y < 30$ GeV through the relation $z = 1 - (M_Y^2 - t)/W^2$, where $z = (P \cdot v)/(P \cdot q)$ is the event elasticity, i.e. the fraction of virtual photon energy transferred to the J/ψ in the proton rest frame.

Events were required to be in a kinematic range where the properties of the final state particles are properly measured and the acceptance is well defined, which is satisfied for $1 < |t| < 20$ GeV^2, $40 < W < 160$ GeV and $M_Y < 30$ GeV.

Results. The cross section was measured as a function of $|t|$ and W. The energy dependence was fitted to the form $\sigma \propto W^{\delta}$ and the measured values of δ are found to rise with $|t|$. The Pomeron trajectory determined from δ, with an intercept of $\alpha_{IP}(0) = 1.153 \pm 0.048(\text{stat.}) \pm 0.039(\text{syst.})$ and a slope of $\alpha'_{IP} = -0.020 \pm 0.014(\text{stat.}) \pm 0.010(\text{syst.})$, is quite different from that of the 'soft' Pomeron [15] but is consistent with the prediction for that of the BFKL Pomeron[16, 17]. The data are in good agreement with previous measurements.

The data were compared with pQCD models (Fig. 3) based on BFKL [18–20] and DGLAP [21] evolution performed in the leading logarithmic (LL) approximation. The

BFKL model also provides calculations which include non-leading corrections (nonL) and for fixed and running α_s. The parameters of both pQCD models are those which best describe previous HERA measurements[22]. The BFKL LL and BFKL nonL predictions with fixed α_s as well as the DGLAP LL calculations generally describe the $|t|$ dependence of the cross section but rather underestimate its magnitude, while the BFKL nonL calculations with running α_s give a steeper $|t|$ dependence and are unable to describe the data across the whole range of $|t|$.

For the W dependence of the cross section the predictions are only available for $|t| > 2$ GeV^2 and, in the case of the BFKL-based model, only for fixed α_s. The DGLAP LL predictions are not able to describe the W dependence while the BFKL LL predictions quantitatively reproduce the rise of the cross section.

REFERENCES

1. J. A. .Crittenden, Springer Tracts in Modern Physics, Volume 140
2. H. Abramowicz and A. Caldwell, Rev. Mod. Phys., 71, 1999, 1275
3. I. P. Ivanov, N. N. Nikolaev, A. A. Savin, preprint hep-ph/0501034, 2005
4. S. J. Brodsky et al., Phys. Rev., D50, 1994, 3134
5. P.D.B.Collins, An Introduction to Regge Theory and High Energy Physics, Cambridge University Press, 1977
6. ZEUS Coll., J. Breitweg et al., *Eur. Phys. J.*, **C 2**, 1998, 247
7. ZEUS Coll., J. Breitweg et al., *Eur. Phys. J.*, **C 6**, 1999, 603
8. H1 Coll., C. Adloff et al., *Eur. Phys. J.*, **C 13**, 2000, 371
9. ZEUS Coll., J. Derrick et al., *Phys. Lett.*, **B 377**, 1996, 259
10. H1 Coll., C. Adloff et al., *Phys. Lett.*, **B 483**, 2000, 360
11. ZEUS Coll., S. Checkanov et al., *Eur. Phys. J.*, **C 24**, 2002, 345
12. ZEUS Coll., S. Checkanov et al., *Nucl. Phys.*, **B 695**, 2004, 3
13. A. D. Martin, M. G. Ryskin, T. Teubner, *Phys. Rev.*, **D 62**, 2000, 14022
14. J. R. Forshaw and G. Shaw, *JHEP*, **0412**, 2004, 052
15. A. Donnachie and P.V. Landshoff, *Phys. Lett.*, **B 348**, 1995, 213
16. S.J. Brodsky et al., *JETP Lett.*, **70**, 1999, 155
17. N.N. Nikolaev, B.G. Zakharov and V.R. Zoller, *Phys. Lett.*, **B 366**, 1996, 337
18. J. Bartels et al., *Phys. Lett.*, **B 375**, 1996, 301
19. J. R. Forshaw and G. Poludniowski, *Eur. Phys. J.*, **C 26**, 2003, 411
20. R. Enberg et al., *Eur. Phys. J.*, **C 26**, 2002, 219
21. E. Gotsman et al., *Phys. Lett.*, **B 532**, 2002, 37
22. H1 Coll., A. Aktas et al., *Phys. Lett*, **B 568**, 2003, 205

Diffractive Production of Vector Mesons and the Gluon at small x

Thomas Teubner

The University of Liverpool, Liverpool L69 3BX, England, U.K.

Abstract. The theory of diffractive production of vector mesons at HERA is briefly reviewed. Perturbative QCD calculations within the MRT model are discussed. The uncertainties of the predictions are scrutinized, and the strong sensitivity of the diffractive cross section to the gluon distribution at small x is emphasized.

Keywords: Diffraction, meson production, parton distribution functions
PACS: 12.38Bx, 12.38Aw, 13.60Le

Introduction

Diffractive high energy scattering is an unique laboratory to study QCD at various scales. Depending on the final state and kinematic variables diffraction can be a purely soft, semi-hard or even hard process, thus enabling us to study the transition from soft to hard QCD. While more and more precise data for different final states are analysed at various colliders, the theoretical description of diffraction has to be further refined and predictions must become quantitative. The motivation for this ranges from understanding the dynamics of QCD and measuring structure functions and parton distributions, to better predicting the structure of underlying events in hadron collisions or even the exciting possibility to first finding the Higgs in exclusive events at the LHC. In the following I will briefly review the prediction of diffractive vector meson (VM) production at HERA and discuss the main uncertainties. It will be demonstrated that the QCD predictions strongly depend on the gluon distribution required as input, which are, however, not well constrained in the region of small x.

Vector meson production in QCD beyond leading order

In leading order QCD, diffractive production of VMs through colourless two gluon exchange (in the forward limit) is given by [1]

$$\frac{\mathrm{d}\sigma}{\mathrm{d}t}(\gamma^* p \to V p)\Big|_{t=0} = \frac{\Gamma_{ee}^V M_V^3 \pi^3}{48\alpha} \frac{\alpha_s(\overline{Q}^2)^2}{\overline{Q}^8} \left[xg(x,\overline{Q}^2)\right]^2 \left(1 + \frac{Q^2}{M_V^2}\right), \qquad (1)$$

where M_V and Γ_{ee}^V are the mass and electronic width of the VM, $\overline{Q}^2 = (Q^2 + M_V^2)/4$ is the effective scale, and $x = (Q^2 + M_V^2)/(Q^2 + W^2)$ (W the c.m.s. energy of $\gamma^* p$). Eq. (1) is valid in the high energy (small x) regime, where quark contributions are negligible.

CP792, *Deep Inelastic Scattering, DIS 2005,* edited by W. H. Smith and S. R. Dasu

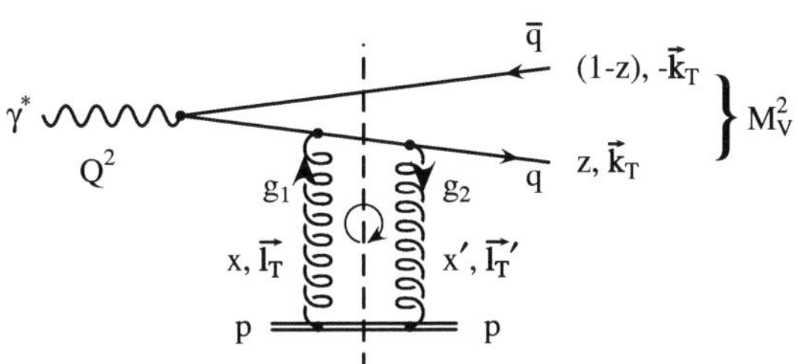

FIGURE 1. One of four leading order amplitudes for the diffractive process $\gamma^* p \to q\bar{q}p$.

The improvement of several other approximations as done e.g. in MRT predictions [2] will be discussed in the following.

Non-relativistic limit. Eq. (1) assumes that q and $\bar{q}$ forming the VM are produced with zero transverse momentum. This can be overcome by introducing the quark's transverse momentum k_T (see Fig. 1) and convoluting the k_T dependent amplitude with a suitable VM wave function, thus accounting for Fermi motion inside the VM. This introduces considerable uncertainties due to the limited knowledge of the form of the wave function. An alternative approach is to assume Parton Hadron Duality and to integrate *open* $q\bar{q}$ production over a given mass interval, $M_{q\bar{q}} \sim M_V$.[1] The size of the interval is a free parameter (though chosen universally for different VM in the MRT predictions) which results in an uncertainty of the overall normalization.

l_T of the gluons and skewing. The identification of the two gluons with the integrated, forward gluon is only valid in the limit $l_T^2 \ll \overline{Q}^2 + k_T^2$ and $x \simeq x'$, respectively. (Here l_T is the transverse momentum of the gluons and x, x' are the longitudinal momentum fractions of the gluons g_1, g_2, see Fig. 1.) However, beyond the leading $\log Q^2$ and $\log(1/x)$ approximations the amplitudes depend on generalized (unintegrated, non-forward) parton distributions (GPDs). In the predictions from MRT normal (integrated, forward) gluons $g(x,\mu^2)$ are used as input to derive unintegrated distributions, $f(x, l_T^2)$, thus replacing collinear with l_T factorization. For $t \sim 0$[2] and small x the effect of skewing comes mainly from evolution and can be determined from the unskewed (forward) gluon [3] (see also [4] for a different approach). Note that both effects, performing the l_T integral explicitely and applying skewing corrections, enhance the cross section considerably (typically by about a factor of two for electroproduction of J/ψ at HERA).

---

[1] In addition, a projection of the $\gamma^*_{L,T} p \to q\bar{q}p$ amplitudes on the correct quantum numbers of the VM, $J^P = 1^-$, is performed which is crucial for the suppression of endpoint-singularities for transverse (T) photons, see [2] for details.

[2] As usual, the cross section is calculated in the forward limit $t \sim 0$, assuming an exponential t dependence $\sigma \sim \exp(-bt)$ with a slope b determined experimentally or by phenomenological models.

Real part and K factor corrections. The actual diagrammatic calculations of the amplitudes corresponding to Fig. 1 take into account only the imaginary parts (leading $\log(1/x)$ approximation). Assuming a power-like behaviour $\mathscr{A} \sim x^{-\lambda}$ of the amplitudes $\mathscr{A}_{L,T}$, the real part can be corrected for via the relation $\mathrm{Re}\mathscr{A} = \tan(\pi\lambda/2)\,\mathrm{Im}\mathscr{A}$ numerically and enhance the cross section for J/ψ by about 20%. Still missing at next-to-leading order (NLO) are full one-loop corrections to the coupling of the two gluons to the $q\bar{q}$ pair (impact factor at NLO). Such NLO corrections are not expected to significantly alter the Q^2 or W dependence of the cross section, or the ratio of longitudinal to transverse cross section, but could well lead to a sizeable change of the normalization (K factor). In [5] the K factor was estimated from π^2 enhanced terms, and MRT predictions are incorporating these contributions. First results for a full calculation of the NLO impact factors became available recently [6], so further improvement towards a complete NLO prediction seems in reach.

Results and discussion

Fig. 2 displays data from H1 [7] and ZEUS [8] for the cross section of diffractive photo- and electroproduction of J/ψ as a function of the $\gamma^* p$ c.m.s. energy. Also shown are predictions from MRT for different input gluon distributions, multiplied by (Q^2 independent) factors to fit the normalization which is not well predicted within the MRT model. Obviously the predictions strongly depend on the gluon distribution, as expected already from Eq. (1) with the quadratic dependence on g, further enhanced through skewing and real part corrections. The W behaviour of the cross section reflects the x dependence of the gluon in the range $x \sim 10^{-4} \ldots 10^{-2}$ at relatively small scales. Note that the scale assigment in $g(x, \mu^2 = \overline{Q}^2)$ only holds at leading order; applying l_T factorization and using the unintegrated gluon distribution $f(x, l_T^2)$ one samples all scales $l_T^2 = 0 \ldots \infty$ in the loop integral, see Fig. 1.[3] In diffractive J/ψ production at HERA typically 70% of the total cross section come from $l_T^2 < 10\ \mathrm{GeV}^2$, with a considerable contribution from low scales, where the gluon is not well constrained.[4] The HERA J/ψ data are therefore sensitive to scales relevant for diffractive (exclusive) Higgs production at the LHC, a topic discussed in detail at this workshop.

As pointed out, current fits of the gluon PDF using DIS and jet data are not well constrained at small x and scales, and allow for a wide range of shapes at small x, including non-monotonic behaviour or a falling gluon which even turns negative. As demonstrated, such a behaviour leads to QCD predictions incompatible with the diffractive data. Also note that the differences when using different gluon parametrizations as input well exceed the uncertainties as obtained through the error estimates of single parametrizations. Diffractive data therefore have a strong potential to constrain the gluon at small x. Due

---

[3] In the infrared regime with scales l_T^2 smaller than the minimal value allowed in the gluon PDF fits (typically $\sim 1.5 - 2\ \mathrm{GeV}^2$) one has to extrapolate the gluon.[2] This introduces an additional uncertainty, which is, however, small compared to the uncertainties from the PDFs.

[4] For larger W and Q^2 (or M_V^2) the l_T^2 spectrum gets harder and the relative contribution from the regime of soft scales is suppressed.

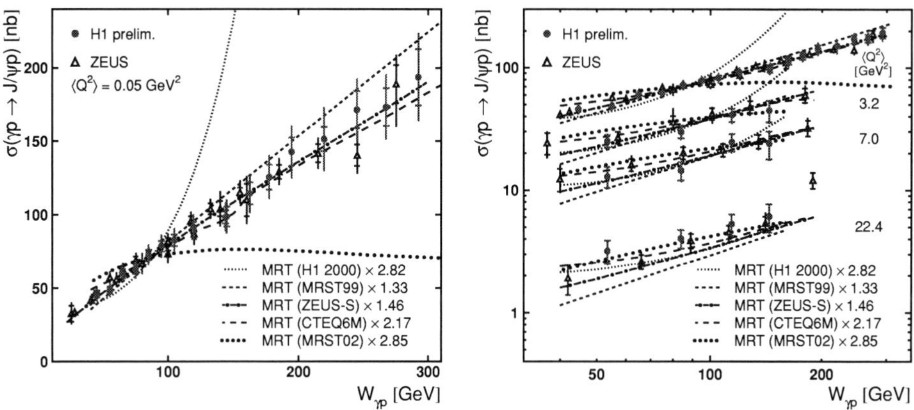

FIGURE 2. H1 [7] and ZEUS [8] data for diffractive J/ψ production compared to MRT predictions. (Figure provided by P. Fleischmann, H1 Collaboration.)

to the complicated relation between diffractive cross sections and the gluon distribution the determination of the gluon from diffractive data is non-trivial. However, with data for different VMs $(\rho, \omega, \phi, J/\psi, \Upsilon)$ and different distributions (W, Q^2, t dependence, ratio σ_L/σ_T) one has a handle to disentangle the correlations between different scales and parameters, and we are looking forward to quantitative studies to improve PDFs at small x.

Acknowledgments

It is my pleasure to thank Alan Martin and Misha Ryskin for enjoyable collaboration on the subject of this talk and numerous members of the ZEUS and H1 Collaborations, especially Emmanuelle Perez, for very fruitful discussions.

REFERENCES

1. M. G. Ryskin, *Z. Phys.* **C57**, 89 (1993).
2. A. D. Martin, M. G. Ryskin and T. Teubner, *Phys. Rev.* **D55**, 4329 (1997); *Phys. Lett.* **B454**, 339 (1999); *Phys. Rev.* **D62**, 014022 (2000).
3. A. G. Shuvaev, K. J. Golec-Biernat, A. D. Martin and M. G. Ryskin, *Phys. Rev.* **D60**, 014015 (1999)
4. P. Kroll, in these proceedings (2005).
5. E. M. Levin, A. D. Martin, M. G. Ryskin and T. Teubner, *Z. Phys.* **C74**, 671 (1997).
6. D. Yu. Ivanov, M. I. Kotsky, A. Papa, *Eur. Phys. J.* **C38**, 195 (2004), and references therein.
7. C. Kiesling, for the H1 Collaboration, in these proceedings (2005), and references therein.
8. ZEUS Collaboration, S. Chekanov et al., *Nucl. Phys.* **B695**, 3 (2004); *Eur. J. Phys.* **C24**, 345 (2002). ZEUS Collaboration, J. Breitweg et al., *Eur. J. Phys.* **C6**, 603 (1999); *Z. Phys.* **C75**, 215 (1997).

New Measurement of DVCS Cross Section at HERA

A. Glazov

DESY, Notkestrasse 85, Hamburg, D 22603 Germany

Abstract. A new measurement is presented of elastic deeply virtual Compton scattering based on data taken by the H1 detector at HERA [1]. For the first time the cross section dependence is reported on the momentum transfer squared at the proton vertex, t. The data is well described by the QCD based calculations.

Keywords: Deeply virtual Compton scattering
PACS: 13.60.Fz

INTRODUCTION

Compton scattering processes, $ep \to e\gamma p$, is one of the most used reactions at HERA. The QED elastic Compton scattering, Bethe-Heitler (BH) process, is utilized to determine the luminosity. The inelastic Compton scattering process is employed to measure the F_2 structure function in extended kinematic range [2, 3].

Recently the focus has been on measuring of the deeply virtual Compton scattering process (DVCS). One of the remarkable properties of this process is factorisation: in the presence of a hard scale, here photon virtuality Q^2, the DVCS scattering amplitude factorises [4, 5, 6] into a hard part, which can be calculated in perturbative QCD, and so-called generalized parton distribution functions (GPDs) which contain the non-perturbative proton structure effects. GPDs correspond to a natural extension of the usual parton density functions (PDFs), they add information about correlations between partons and partons transverse motion [7, 8, 4]. The new degrees of freedom, additional to the Bjorken-x, are so-called skewedness ξ, which measures the fractional momenta difference between emitted and absorbed parton and t.

This paper presents a measurement of DVCS cross section based on 46.5 pb^{-1} of data collected with the H1 detector at HERA in years 1996 to 2000 [1]. The cross section is presented as a function of Q^2, of the invariant mass of $\gamma^* p$ system, W, and of t.

DATA ANALYSIS

The elastic Compton scattering events have very distinct features making them very precious experimental commodity, see Figure 1. The event kinematics can be redundantly determined using the two reconstructed electromagnetic clusters. The fraction of DVCS events with respect to ordinary QED Compton scattering is enhanced requiring the scattered photon to be reconstructed in the central region. A typical signal to background ratio for the DVCS analysis is 10 to 1.

CP792, *Deep Inelastic Scattering, DIS 2005,* edited by W. H. Smith and S. R. Dasu

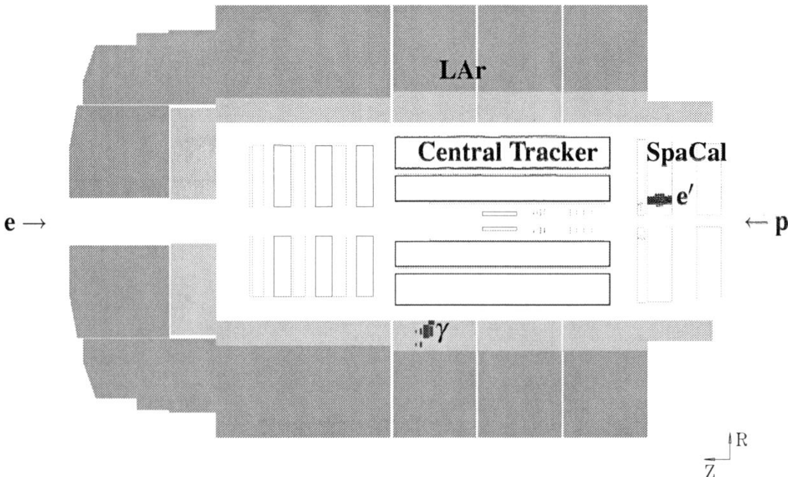

FIGURE 1. A typical DVCS event candidate recorded by the H1 detector. The Z axis is defined by the proton beam directions. The beam pipe which is located in the center of the detector (not shown in the figure) is surrounded by the central silicon tracker, central drift chamber tracker and the liquid argon calorimeter (LAr). The particles scattered in the backward direction are measured in the backward silicon tracker, backward drift chamber and the lead scintillating fiber calorimeter (SpaCal). Both calorimeters contain electromagnetic and hadronic compartments. The dark (red) areas in the calorimeters represent two electromagnetic clusters. The cluster in the LAr calorimeter is associated with the scattered photon (γ) since there is no matching track in the central tracker. The cluster in SpaCal us associated with the scattered electron (e'). The scattered proton escapes in the forward beam pipe.

The selected DVCS sample contains 1243 events. To extract the cross section, the data are corrected for detector acceptance and initial state radiation using the Monte Carlo simulation (MC). The generation of MC is based on MILOU program [9] and then the events are passed through a detailed simulation of the H1 detector response.

The measured $e^+p \to e^+\gamma p$ cross section is converted to the $\gamma^* p \to \gamma p$ cross section using equivalent photon approximation:

$$\frac{d^3\sigma(ep \to e\gamma p)}{dydQ^2dt}(Q^2,y,t) = \Gamma(Q^2,y)\frac{d\sigma(\gamma^* p \to \gamma p)}{dt}(Q^2,y,t). \tag{1}$$

Here inelasticity y is calculated as $y = (W^2+Q^2)/s$ (s is the square of the ep centre-of-mass energy) and the transverse photon flux Γ is [10]

$$\Gamma = \frac{\alpha(1-y+y^2/2)}{\pi y Q^2}. \tag{2}$$

RESULTS

For the triggering reasons, the cross sections are measured separately in 1996-1997 and 1999-2000, covering different Q^2 ranges, and are then combined. The differential

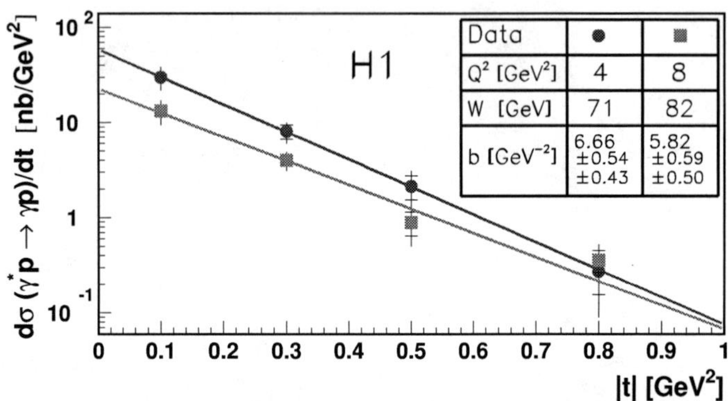

FIGURE 2. The differential cross section $d\sigma(\gamma^* p \to \gamma p)/d|t|$ for $Q^2 = 4$ GeV^2 at $W = 71$ GeV and $Q^2 = 8$ GeV^2 at $W = 82$ GeV. The inner error bars represent the statistical and the full error bars the total uncertainties. The lines show the results of fits $e^{-b|t|}$, the corresponding values of b are given in the insert.

$d\sigma(\gamma^* p)/d|t|$ cross section is shown in Fig. 2. The data points are fitted with the exponential form $e^{-b|t|}$, which gives $b = 6.66 \pm 0.54_{\text{stat}} \pm 0.43_{\text{sys}}$ GeV^{-2} at $Q^2 = 4$ GeV^2 and $b = 5.82 \pm 0.59_{\text{stat}} \pm 0.50_{\text{sys}}$ GeV^{-2} at $Q^2 = 8$ GeV^2. Since the measurements agree within the experimental errors, they are averaged to obtain $b = 6.02 \pm 0.35_{\text{stat}} \pm 0.39_{\text{sys}}$ GeV^{-2} for $Q^2 = 8$ GeV^2 and $W = 82$ GeV.

Fig. 3 shows Q^2 and W dependence of the $\gamma^* p$ cross section. Fitting the Q^2 dependence with a form $(1/Q^2)^n$ returns $n = 1.54 \pm 0.09_{\text{stat}} \pm 0.04_{\text{sys}}$. Fitting the W dependence with a form W^δ gives $\delta = 0.77 \pm 0.23_{\text{stat}} \pm 0.19_{\text{sys}}$. The steep rise of the cross section with W indicates the presence of a hard scattering process, the value of δ is similar to the value measured in exclusive J/ψ production [11, 12].

Fig. 3 also compares the data with the NLO QCD based calculation [14]. In this model, the DVCS cross section is calculated using two different GPD parameterisations. The t dependence is assumed to follow $e^{-b|t|}$ with the coefficient b taken from the H1 measurement. The two parametrisations use either MRST [16] or CTEQ [15] standard PDFs for the region $|x| > \xi$. The generalized structure functions $H^{q,g}$ are given at a starting scale μ by $H^q(x,\xi,t) = q(x)e^{-b|t|}$ for the quarks and $H^g(x,\xi,t) = xg(x)e^{-b|t|}$ for the gluons. For the larger scales, both Q^2 dependence and skewing are generated dynamically. For $|x| < \xi$, the parameterisations are modified preserving smooth transition into $|x| > \xi$ regime [14].

The theoretical estimations agree well with the data for both shape and absolute normalization. The uncertainty in the normalization for the theory is significantly reduced owing to the H1 measurement of the $\gamma^* p$ cross section t slope; this uncertainty becomes smaller than the input PDF uncertainty which is quantified comparing MRST and CTEQ PDF set based predictions.

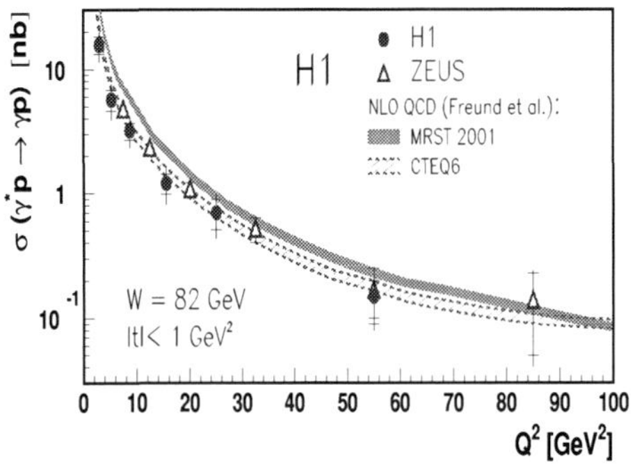

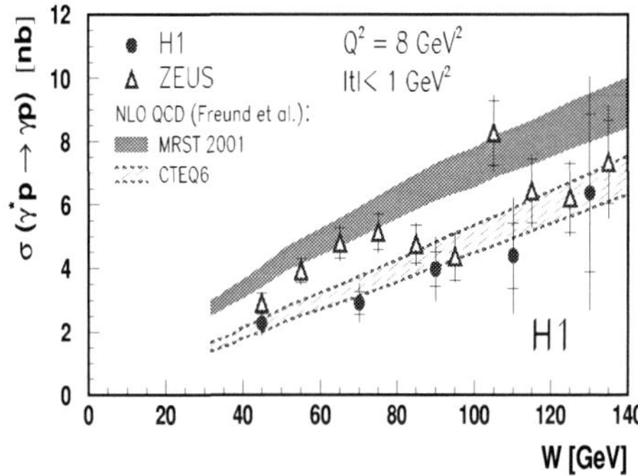

FIGURE 3. Q^2 (left) and W (right) dependence of the $\gamma^* p \to \gamma p$ cross section measured by H1 (filled circles) and ZEUS [13] (open triangles) collaborations compared to NLO QCD based calculations [14] which are performed for two different PDF sets, CTEQ6 [15] (dashed band) and MRST [16] (filled band). The inner error bars represent the statistical and the full error bars the total uncertainties. The band associated with each prediction corresponds to the uncertainty in the measured t-slope.

SUMMARY

The DVCS process are measured in the kinematic domain $30 < W < 140$ GeV, $2 < Q^2 < 80$ GeV^2 and $|t| < 1$ GeV^2 by the H1 collaboration using data collected in years 1996-2000. The $\gamma^* p \to \gamma p$ cross section is reported as a function of Q^2, W and for the first time as a function of t. The measurement of the the t dependence allows to reduce normalization uncertainties of the theoretical predictions. NLO QCD calculations provide a good description of the data.

REFERENCES

1. A. Altas *et al.* [H1 collaboration] DESY 05-065, Submitted to Eur. Phys. J. C.
2. A. Aktas *et al.* [H1 Collaboration], Phys Lett B **598** (2004) 159.
3. E. Lobodzinska, DIS-05 proceedings, Madison 2005.
4. A,V, Radyushkin, Phys. Rev. **D56** (1997) 5524.
5. J.C. Collins and A. Freund, Phys. Rev. **D59** (1999) 074009.
6. X. Ji and J. Osborne, Phys. Rev. **D58** (1998) 094018.
7. D. Muller *et al.*, Fortsch. Phys. **42** (1994) 101.
8. X. Ji, Phys. Rev. Lett. **78** (1997) 610.
9. E. Perez, L. Schoeffel and L. Favart, DESY-04-228.
10. L.N. Hand, Phys. Rev. **129** (1963) 1834.
11. C. Adolf *et al.* [H1 Collaboration], Eur.Phys. J. C **10** (1999) 373.
12. S. Chekanov *et al.* [ZEUS Collaboration], Eur.Phys. J, C **24** (2002) 345.
13. S. Chekanov *et al.* [ZEUS Collaboration], Phys. Lett. B **573** (2003) 46.
14. A. Freund, M.F. McDermott and M. Strikman, Phys. Rev. **D67** (2003) 036001.
15. D. Stump *et al.* [CTEQ Collaboration], JHEP **0310** (2003) 046
16. A.D. Martin *et al.* [MRST Collaboration], Eur. Phys. J. C **28** (2003) 455.

Measurement of Deeply Virtual Compton Scattering at HERMES

M. Kopytin on behalf of the HERMES collaboration

DESY, 15738 Zeuthen, Germany. email: Mikhail.Kopytin@desy.de

Abstract. The measurement of azimuthal cross section asymmetries from deeply virtual Compton scattering on the proton and deuteron at HERMES is discussed. In particular results on the longitudinal target spin asymmetry as a function of the azimuthal angle and the Mandelstam t are given. The t-dependence of the asymmetry is compared with calculations based on generalized parton distribution models.

Keywords: GPD, DVCS, azimuthal asymmetry, target-spin asymmetry, HERMES
PACS: 13.60.Fz

1. INTRODUCTION

Generalized Parton Distributions (GPD) were introduced a decade ago as a unified description of hard exclusive processes in the Bjorken regime [1, 2]. As a generalization of the usual Parton Distribution Functions (PDF) they give additional information about quark and gluon properties in the nucleon. Because of their off-forward nature GPDs contain information about both PDF and nucleon form factors.

Strong interest in the physics properties of GPDs was triggered by the work of Ji [3] who demonstrated that in the forward limit GPDs can give information about the total angular momentum carried by quarks (and gluons) in the nucleon.

2. DEEPLY VIRTUAL COMPTON SCATTERING

The presently cleanest way to access GPDs is to study Deeply Virtual Compton Scattering (DVCS), the hard exclusive electroproduction of a real photon. This process occurs always together with the Bethe-Heitler process where the photon is radiated from one of the involved leptons. Both processes have identical final states, making them experimentally indistinguishable, and leading to interference between amplitudes. This yields the following amplitude for real photon production

$$\frac{d\sigma}{dx_B dQ^2 d|t| d\phi} \propto |\tau_{BH}|^2 + |\tau_{DVCS}|^2 + \overbrace{\tau_{DVCS}\tau_{BH}^* + \tau_{DVCS}^*\tau_{BH}}^{I}, \tag{1}$$

where x_B represents the Bjorken scaling variable, $-Q^2$ the virtual-photon four-momentum squared and t the square of the four-momentum transfer to the target. The azimuthal angle ϕ is defined by the lepton scattering plane and by the photon production plane.

CP792, *Deep Inelastic Scattering, DIS 2005*, edited by W. H. Smith and S. R. Dasu

In leading twist the dependence of the interference term I on the azimuthal angle ϕ can be written as [4]

$$I \propto \frac{e_l}{P(cos\phi)}\Big(cos\phi\,\mathrm{Re}\widehat{M}_{++} - P_l\sqrt{1-\varepsilon^2}sin\phi\,\mathrm{Im}\widehat{M}_{++} - \\ - S_L\big[sin\phi\,\mathrm{Im}\widehat{M}^L_{++} - P_l\sqrt{1-\varepsilon^2}cos\phi\,\mathrm{Re}\widehat{M}^L_{++}\big]\Big), \qquad (2)$$

where $e_l = \pm 1$ is the charge of the lepton beam with polarization P_l scattered on a target with longitudinal polarization S_L. Here ε is the ratio of fluxes of longitudinal to transverse initial photons in the DVCS process and the factor $P(cos\phi)$ coming from the lepton propagators in the Bethe-Heitler process gives an additional ϕ dependence.

In case of an *unpolarized* target ($S_L = 0$) the ϕ dependence of the cross section asymmetry with respect to the charge (spin) of the lepton beam gives access to the real (imaginary) part of the DVCS amplitude $\widehat{M}_{++}$ which is a linear combination of the so-called Compton form factors (CFFs) $\mathcal{H}$, $\tilde{\mathcal{H}}$ and $\mathcal{E}$:

$$\widehat{M}_{++} = \sqrt{1-\xi^2}\frac{\sqrt{t_0-t}}{2M_p}\Big[F_1\mathcal{H} + \xi(F_1+F_2)\tilde{\mathcal{H}} - \frac{t}{4M_p^2}F_2\mathcal{E}\Big], \qquad (3)$$

where $\xi \simeq \frac{x_B}{2-x_B}$ in Bjorken limit is the skewedness parameter, t_0 is the minimum possible value of $-t$ at a given ξ, and F_1 and F_2 are the Dirac and Pauli form factors, respectively.

In case of a *polarized* target ($S_L \neq 0$) the dependence of the asymmetry with respect to the target polarization state gives access to the imaginary part of $\widehat{M}^L_{++}$ which is defined by another linear combination of the CFFs $\mathcal{H}$, $\tilde{\mathcal{H}}$, $\mathcal{E}$ and $\tilde{\mathcal{E}}$:

$$\widehat{M}^L_{++} = \sqrt{1-\xi^2}\frac{\sqrt{t_0-t}}{2M_p}\Big[F_1\tilde{\mathcal{H}} + \xi(F_1+F_2)\big(\mathcal{H} + \frac{\xi}{1+\xi}\mathcal{E}\big) - \big(\frac{\xi}{1+\xi}F_1 + \frac{t}{4M_p^2}F_2\big)\xi\tilde{\mathcal{E}}\Big]. \qquad (4)$$

These complex CFFs are flavor sums of convolutions of corresponding leading-twist GPDs H, $\tilde{H}$, E and $\tilde{E}$ with the hard scattering amplitudes that are available up to NLO in pQCD [5].

3. DVCS AT HERMES

HERMES is a fixed-target experiment at the 27.6 GeV electron or positron beam of HERA [6]. Its gas target provides polarized H, D, ^{3}He as well as unpolarized H, D, Ne, Kr and Xe targets. In this paper only result for H and D targets are discussed.

The selected DVCS events are required to have a detected photon in addition to one charged track identified as the scattered lepton. The kinematic requirements imposed on the scattered lepton were $Q^2 > 1$ GeV^2, $W^2 > 8$ GeV^2 and $\nu < 23$ GeV. The polar angle $\theta_{\gamma^*\gamma}$ between the virtual and the real photon is required to be between 5 and 45 mrad. Since the recoiling proton was not detected the exclusive events were selected with a cut on the missing mass M_x of the reaction $ep \to e\gamma X$ that requires M_x to correspond to the proton mass. Due to the limited detector resolution the missing mass range $-1.5 < M_x < 1.7$ GeV was selected using an MC simulation for the optimum separation of exclusive events from the semi-inclusive background.

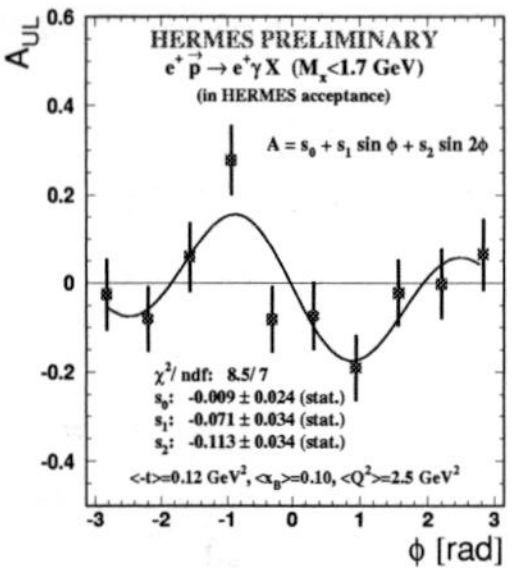

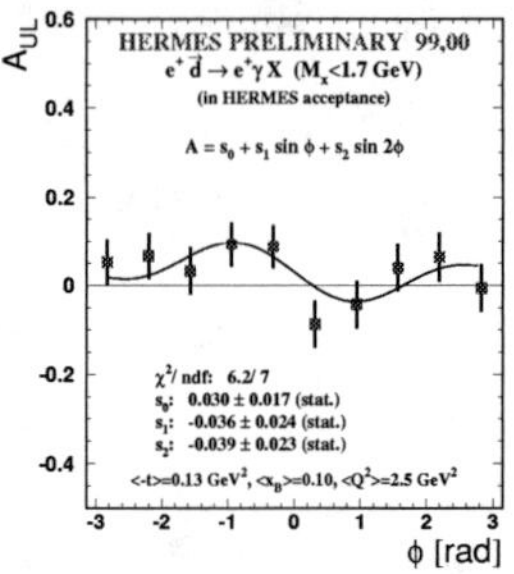

FIGURE 1. Longitudinal target spin asymmetry A_{UL} for hard electroproduction of photons off the proton (left) and the deuteron (right) as a function of the azimuthal angle ϕ for the exclusive sample. The solid curves show the results of the indicated fi ts with the values given in the plots.

4. AZIMUTHAL CROSS SECTION ASYMMETRIES

The discussion on the beam-spin asymmetry (BSA) and the beam-charge asymmetry (BCA) is beyond the scope of this paper, since the HERMES results on these two asymmetries for the proton and the deuteron have been reported before (the BSA dependence on ϕ is discussed in [7, 8], the BCA dependence on ϕ and t in [9]).

4.1. Longitudinal Target Spin Asymmetry

The single-spin asymmetry w.r.t. the polarization of a *longitudinally* (L) polarized target (LTSA) as a function of ϕ is calculated as

$$A_{UL}(\phi) = \frac{1}{<|S_L|>} \frac{(N^{\leftarrow\Leftarrow} + N^{\rightarrow\Leftarrow}) - (N^{\leftarrow\Rightarrow} + N^{\rightarrow\Rightarrow})}{(N^{\leftarrow\Leftarrow} + N^{\rightarrow\Leftarrow}) + (N^{\leftarrow\Rightarrow} + N^{\rightarrow\Rightarrow})}, \tag{5}$$

where $\Leftarrow$ ($\Rightarrow$) or $\leftarrow$ ($\rightarrow$) denote target spin or beam helicity antiparallel (parallel) to the beam direction, respectively, and $N^{\leftarrow(\rightarrow)\Leftarrow(\Rightarrow)}$ represents the single photon yield normalized to the acquired luminosity for the corresponding target spin and beam helicity states. The imposed requirement on the beam polarization $<|P_l^{\leftarrow\Leftarrow(\Rightarrow)}|>=<|P_l^{\rightarrow\Leftarrow(\Rightarrow)}|>$ excludes a possible influence of the double-spin asymmetry on A_{UL}.

Since the main contribution to the DVCS amplitude $\widehat{M}^L_{++}$ at HERMES kinematics originates from the CFFs $\mathscr{H}$ and $\tilde{\mathscr{H}}$, measurements of the LTSA can constrain models for the imaginary parts of GPDs H and $\tilde{H}$. The asymmetries on proton and deuteron as function of ϕ are shown in Fig. 1. These results demonstrate the expected $sin\phi$ dependence [9]. A sizeable $sin2\phi$ amplitude $A_{UL}^{sin2\phi}$ is found which is expected to be kinematically suppressed w.r.t. $A_{UL}^{sin\phi}$; it can be sensitive to the next-to-leading contributions to the asymmetry (e.g. twist-3 GPDs H^3, $\tilde{H}^3$). In Fig. 2 the dependences of the $sin\phi$ and $sin2\phi$ amplitudes of the LTSA on proton and deuteron are shown as function of $-t$ derived from fits for every $-t$ bin. No difference is observed between the asymmetries on

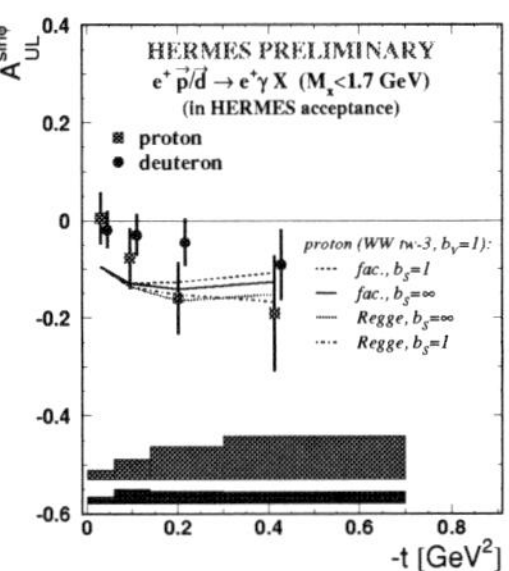

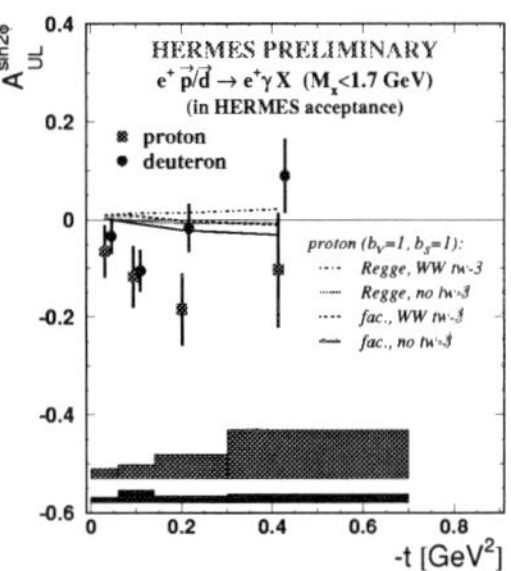

FIGURE 2. The $sin\phi$ (left) and $sin2\phi$ (right) amplitudes of the longitudinal target-spin asymmetry on the proton and the deuteron as a function of $-t$. The GPD model calculations use a factorized or a Regge-inspired t-dependence with or without a Wandzura-Wilczek (WW) term.

H and D targets in the first t-bin, where effects from coherent scattering on the deuteron may be expected. The difference between the two targets in higher t-bins might be due to incoherent scattering on the neutron.

Calculations based on a GPD model developed in [10] were carried out at the average kinematics of every t bin for the proton. Although the model describes the $sin\phi$ amplitude well, it does not agree with the data for $sin2\phi$. This can be due to the fact that the twist-3 GPDs are modeled only by the Wandzura-Wilczek term, hence no information on quark-gluon correlations is included.

5. OUTLOOK

The measurement of azimuthal asymmetries in DVCS performed at HERMES can give access to a variety of GPDs. The BSA and BCA constrains both *imaginary* and *real* part of the GPD H, respectively. The longitudinal TSA is sensitive to the imaginary part of the GPD $\tilde{H}$. The present HERMES data taking with a transversely polarized target allows to measure the transverse TSA that is sensitive to the GPD E. These results may allow to constrain the total angular momentum of u-quark J^u for the first time [11].

REFERENCES

1. D. Mueller et al., Fortsch. Phys. **42** (1994) 101.
2. A.V. Radyushkin, Phys. Lett. **B385** (1996) 333.
3. X. Ji, Phys. Rev. Lett. **78** (1997) 610, Phys. Rev. **D55** (1997) 7114.
4. M. Diehl, Physics Reports **388** (2003) 41.
5. A.V. Belitsky, D. Müller, A. Kirchner, Nucl. Phys. **B629** (2002) 323-392.
6. HERMES Coll., K. Ackerstaff et al., Nucl. Instr. and Meth. **A417** (1998) 230.
7. A. Airapetian et al., Phys. Rev. Lett. **87** (2001) 182001.
8. HERMES Coll., F. Ellinghaus et al., hep-ex/0212019.
9. W.-D. Nowak, hep-ex/0503010
10. K. Goeke, M.V. Polyakov and M. Vanderhaegen, Prog. Part. Nucl. Phys. **47** (2001) 401.
11. F. Ellinghaus et al., hep-ph/0506264.

Exclusive meson production at HERMES

A. Vandenbroucke*
On behalf of the HERMES Collaboration

*Department of Subatomic and Radiation Physics, Proeftuinstraat 86, B-9000 Gent

Abstract. Generalized Parton Distributions (GPDs) provide a new level of insight into the quark structure of the nucleon. Experimentally they can be probed by hard exclusive electroproduction of both scalar and vector mesons. Results for the cross section for the reaction $ep \to en\pi^+$, and a first result for the asymmetry A_{UT} for exclusive ρ^0 production are presented.

Keywords: Proton Structure, GPD, Exclusive processes
PACS: 13.60.Le

INTRODUCTION

It has been shown [1, 2] that in the case of large Q^2 and for longitudinally polarised virtual photons a factorization theorem can be applied to exclusive meson production. The theorem states that the meson production amplitude can be written as a product of a hard part, describing the interaction of the virtual photon, a meson wave function, and a soft part. The latter can be expressed in terms of Generalised Parton Distribution functions, describing the nucleon. These are related to the unknown quark orbital angular momentum L_q.

For every quark flavor q there are four GPD's, two of them are polarized ($\tilde{E}^q$ and $\tilde{H}^q$), whereas the other two are unpolarized (E^q and H^q). Pseudoscalar (vector) meson production is sensitive to the polarized (unpolarized) GPD's. Different observables like cross-sections, or single-spin asymmetries select different combinations of GPD's. The experimental separation between γ_L^* and γ_T^* is in principle possible for vector mesons, for pseudoscalar mesons this separation will be tested by the overall Q^2 dependence.

In this paper, the cross section for exclusive π^+ production, and the target related single spin asymmetry A_{UT} for exclusive ρ^0 production will be covered. The data is coming from the HERMES experiment at the HERA storage ring at DESY. Only the scattered lepton and the produced meson are detected in the HERMES spectrometer [3], however, by putting constraints to missing mass or missing energy, exclusive reactions can be selected.

CROSS SECTION FOR EXCLUSIVE π^+ PRODUCTION

The cross section for exclusive pion leptoproduction on a proton target can be related to $(\tilde{H}+\tilde{E})^2$. The selection of exclusive π^+ candidates in the reaction $e+p \to e'+\pi^+ +X$ was done by calculating the missing mass $M_X^2 = (P_e+P_p-(P_{e'}+P_{\pi^+}))^2$, and applying a missing mass cut M_X^2 around the squared proton mass. Unfortunately the M_X^2 resolution

CP792, *Deep Inelastic Scattering, DIS 2005*, edited by W. H. Smith and S. R. Dasu

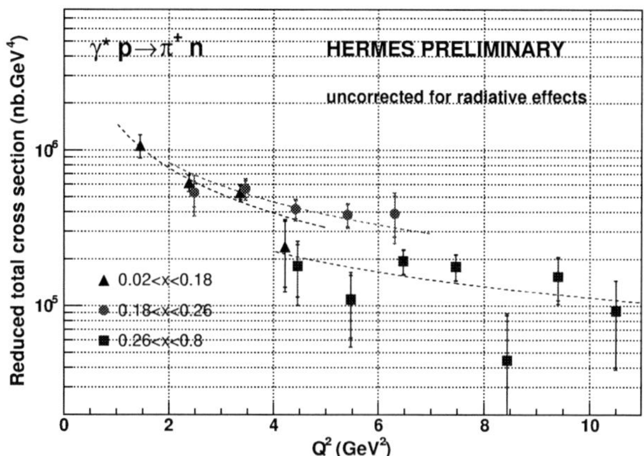

FIGURE 1. Reduced cross section for exclusive π^+ production as a function of Q^2 in different x bins. A polynomial of the order $\frac{1}{Q^2}$ has been fit to every bin.

doesn't allow the separation of the exclusive events from the non-exclusive background. One can however use the normalised π^- yield as an estimate for the background, as there's no exclusive channel in π^- production on the proton with the creation of a nucleon. Applying this background subtraction yields a peak in the M_X^2 spectrum around the squared nucleon mass, whose width and mean could be reproduced by a Monte Carlo based on a GPD model.

The exclusive π^+ cross section was determined by analysing HERMES data sets from 1996 up to 2000. Out of these, approximately 3500 exclusive π^+ events were selected. The cross section was evaluated by applying the following formula:

$$\sigma^{\gamma^* p \to n+\pi^+}(x, Q^2) = \frac{N_{\text{excl}}^{\pi^+}}{L \cdot \Delta x \Delta Q^2 \cdot \kappa(x, Q^2) \cdot \Gamma(<x>, <Q^2>)}, \tag{1}$$

where L is the integrated luminosity, κ the probability to detect the scattered lepton and the produced π^+, and Γ the virtual photon flux factor, calculated following the description described in [4]. The detection probability has been determined using two different Monte Carlo samples based on GPD calculations [5, 6]. The agreement between data and Monte Carlo was good. More details can be found in [7]

The factorization theorem predicts a $\frac{1}{Q^6}$ dependence for σ_L at fixed x and t. We can write σ_L as:

$$\frac{d\sigma_L}{dt} = \frac{1}{16\pi} \frac{x^2}{1-x} \frac{1}{Q^4} \frac{1}{\sqrt{1+\frac{4m^2x^2}{Q^2}}} \sum_{\text{spin}} |\mathbf{A}(\gamma^* p \to pM)|^2, \tag{2}$$

where the sum term is the so-called *reduced* cross section, which is independent of x and should have a $\frac{1}{Q^2}$ behavior. The reduced cross section for exclusive π^+ production is shown in figure 1. The remaining x-dependence as seen in that figure can be explained by the finite size of the x-bins. For every x-bin a function $\frac{1}{Q^p}$ has been fitted to the spectrum. The fit parameters $p = 1.9 \pm 0.5$ (triangles), $p = 1.7 \pm 0.6$ (circles), and $p = 1.5 \pm 1.0$ (squares) agree with the predicted Q^2 dependence.

A_{UT} FOR EXCLUSIVE ρ^0 PRODUCTION

Since 2002 HERMES is running with a transversely polarized proton target. The transverse target spin asymmetry for exclusive ρ^0 production is directly related to the GPD E, and is the most direct probe of that GPD. The asymmetry is sensitive to the total angular momentum of quarks as shown in [8].

For the study of A_{UT}, ρ^0's are reconstructed from h^+h^- pairs. Exclusivity is achieved by requiring the missing energy ΔE to be zero. Subtracting the non-exclusive background as described by Monte Carlo from the ΔE spectrum leaves a clear peak around $\Delta E = 0$, which is evidence for exclusive ρ^0 production on a proton target.

Generally, a single spin asymmetry $\mathrm{A_{UT}}$ is measured by subtracting the number of observed events when the target is in one state from the number of events when the target is in the other state. The subtraction occurs in every bin $(\phi - \phi_S)$:

$$\mathrm{A_{UT}}(\phi - \phi_S) = \frac{1}{|P|} \frac{N^\uparrow - N^\downarrow}{N^\uparrow + N^\downarrow}.$$

ϕ (ϕ_S) is the angle betweeen the lepton scattering plane and the hadron production plane (target spin), as shown in figure 2. The average polarization $< |P| >$ over the combined 2002-2004 data sample was 0.755 ± 0.049.

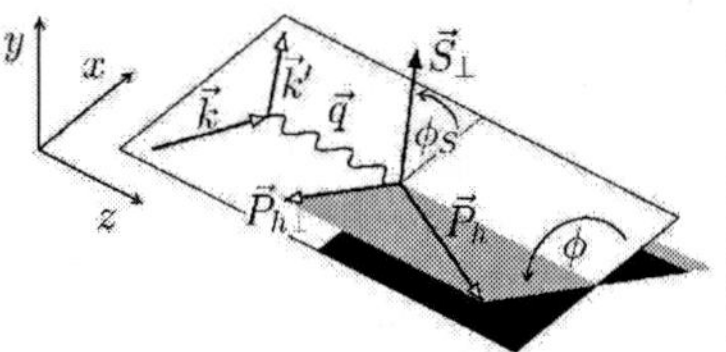

FIGURE 2. Definition of the angles ϕ and ϕ_S.

The relevant observable is the $\sin(\phi - \phi_s)$ amplitude of the asymmetry and is depicted in figure 3 as a function of x and t', the latter variable being the difference of t and t_0, where t_0 is the maximum kinematically allowed value of t. Despite the large error bars a large asymmetry is visible for low x and high t'.

HERMES will run with a transversely polarized proton target up to November 2005. Statistics thus are expected to increase, allowing for a full $\sigma_L - \sigma_T$ separation in the future.

OUTLOOK

Analysis of exclusive processes with HERMES data is still continuing. The transverse target spin asymmetry for exclusive π^+ production for example is an interesting ob-

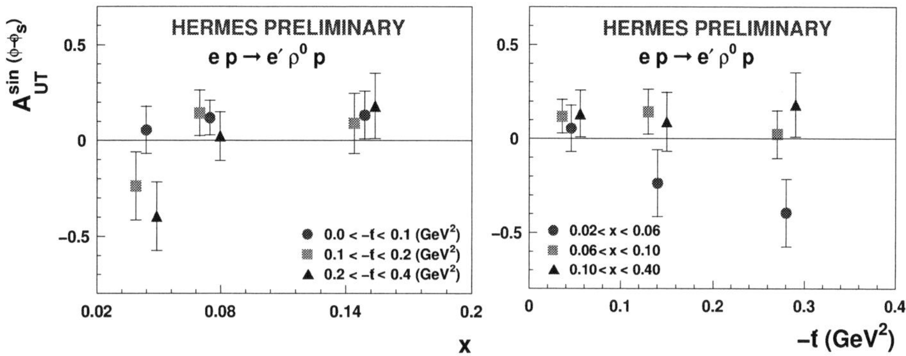

FIGURE 3. A_{UT} for exclusive ρ^0 production on a proton target. Left (right) as a function of x (-t').

servable. The asymmetry is related to $\tilde{E} \cdot \tilde{H}$, and can be used together with the cross section information to disentangle $\tilde{E}$ and $\tilde{H}$. Moreover the scaling region is expected to be reached at lower Q^2 in the case of asymmetries. In [9] the sensitivity of A_{UT} to the pion form factor is shown.

Looking at π^+-electroproduction we see that the pseudoscalar contribution $\tilde{E}$ is dominated by the pion-pole exchange, and therefore is related to the pion form factor. In the case of exclusive π^0 production there's no pion-pole contribution. The production ratio of π^+ to π^0 mesons thus promises to be an interesting observable in the quest for the understanding of the nucleon's structure.

CONCLUSIONS

Exclusive meson production is related to GPDs and can be used to explore the structure of the nucleon. It was shown that the Q^2 dependence of the reduced cross section for exclusive π^+ production on a proton target is in agreement with theoretical predictions.

The exclusive ρ^0 production shows target spin asymmetries which are related th the GPD E. First results for this asymmetry were presented. HERMES is continuing to take data and more results from a refined analysis are expected soon.

REFERENCES

1. J. C. Collins, L. Frankfurt, and M. Strikman (1997), hep-ph/9709336.
2. J. C. Collins (1999), hep-ph/9907513.
3. K. Ackerstaff, et al., *Nucl. Instrum. Meth.*, **A417**, 230–265 (1998).
4. L. N. Hand, *Phys. Rev.*, **129**, 1834–1846 (1963).
5. M. Vanderhaeghen, P. A. M. Guichon, and M. Guidal, *Phys. Rev.*, **D60**, 094017 (1999).
6. L. Mankiewicz, G. Piller, and A. Radyushkin, *Eur. Phys. J.*, **C10**, 307–312 (1999).
7. C. Hadjidakis, D. Hasch, and E. Thomas, *Int. J. Mod. Phys.*, **A20**, 593–595 (2005).
8. K. Goeke, M. V. Polyakov, and M. Vanderhaeghen, *Prog. Part. Nucl. Phys.*, **47**, 401–515 (2001).
9. L. L. Frankfurt, P. V. Pobylitsa, M. V. Polyakov, and M. Strikman, *Phys. Rev.*, **D60**, 014010 (1999).

Diffractive photoproduction of ρ mesons with large momentum transfer at HERA

Carl Gwilliam,
on behalf of the H1 collaboration

School of Physics and Astronomy, The University of Manchester, Manchester M13 9PL, ENGLAND

Abstract. The diffractive photoproduction of ρ mesons with large momentum transfer, $ep \to e\rho Y$, is studied at HERA in the kinematic range $Q^2 < 0.01$ GeV^2, $75 < W < 95$ GeV, $1.5 < |t| < 10.0$ GeV^2 and $M_Y < 5$ GeV. The t dependence of the cross section is measured, as well as the spin density matrix elements. All results are compared to BFKL model predictions.

Keywords: H1, DIFFRACTION, PHOTOPRODUCTION, VECTOR MESON, RHO, HELICITY ANGLES, HIGH T
PACS: 14.40.C5

INTRODUCTION

Results are presented on the diffractive photoproduction of ρ mesons at large momentum transfer in high energy ep collisions: $ep \to e\rho Y$; $\rho \to \pi^+\pi^-$. The scattered proton is mainly excited into a system Y of mass M_Y, which is much lower than the γp centre of mass (cms) energy W ("proton dissociative" scattering). There is also a small contribution from "proton elastic" scattering where the proton remains intact.

For $|t|$ larger than a few GeV^2, t being the square of the four momentum transferred at the proton vertex, perturbative QCD (pQCD) is expected to apply and diffractive ρ production is viewed, in the proton rest frame, as a sequence of three processes well separated in time: the photon fluctuates into a $q\bar{q}$ pair; the $q\bar{q}$ pair is involved in a hard interaction with the proton via the exchange, at lowest order, of two gluons in a colour singlet state and the $q\bar{q}$ pair recombine to form a bound ρ meson. In the leading logarithm approximation the process is represented by the effective exchange of a gluonic ladder, which, in the low x region of interest, is described by BFKL evolution.

The data are taken with the H1 detector [1] in the year 2000 and correspond to an integrated luminosity of 20.1 pb^{-1}. A selection is performed based on the requirement of two tracks (pion candidates) in the central H1 detector along with an energy deposit in the electron tagger situated 44 m along the beam pipe in the electron direction (electron candidate). The kinematic range is restricted to is $1.5 < |t| < 10.0$ GeV^2, $75 < W < 95$ GeV and $Q^2 < 0.01$ GeV^2, where Q^2 is the modulus squared of the four momentum carried by the intermediate photon, in order to ensure high acceptance over all the distributions studied. The further requirement of no additional energy deposits, not associated to the two decay pions, detected within the liquid argon calorimeter (LAr) limits the dissociative proton system to $M_Y \lesssim 5$ GeV.

CP792, *Deep Inelastic Scattering, DIS 2005,* edited by W. H. Smith and S. R. Dasu

THEORETICAL MODEL

The process of interest is described by the non-forward BFKL equation for which a complete analytical solution, in the leading logarithm approximation, is presented by Poludniowski *et al.* in [2, 3]. Their prescription is to factorise the meson production from the hard subprocess using a set of meson wavefunctions expanded on the light-cone. All expansions are performed up to twist-3, i.e. next-to-leading twist, which is the lowest order able to accommodate a non-zero r^{04}_{10} matrix element (see below). The constituent, as opposed to current, quark mass is used to enhance the coupling of the photon to chiral odd $\bar{q}q$ configurations; thereby producing a strong transverse polarisation as supported by previous data [4, 5].

At leading log accuracy there are three free parameters α_s^{IF} (0.21) , α_s^{BFKL} (0.2) and γ (1.0): where α_s^{IF} is the coupling of the two gluons to each impact factor, α_s^{BFKL} is the coupling within the BFKL ladder and γ determines the scale $\Lambda^2 = m_\rho^2 - \gamma t$. The numbers in parenthesis refer to the parameter values chosen to provide the best description of the previous ZEUS measurements [4] and are used to describe the results that follow.

The spin density matrix elements characterise the helicity states of the ρ meson and the photon. They are defined as bilinear combinations of the helicity amplitudes $M_{\lambda_\rho \lambda_\gamma}$, where $\lambda_\rho, \lambda_\gamma = -, 0, +$ represent the respective helicities of the ρ meson and the photon [6]. In the assumption of *s*-channel helicity conservation (SCHC), whereby the ρ meson retains the helicity of the photon, both the single-flip and double-flip dependent matrix elements are expected to be zero. In contrast, the model of Poludniowski *et al.* predicts violation of SCHC with a hierarchy of helicity amplitudes given by: non-flip (M_{++}) > double-flip (M_{+-}) > single-flip (M_{+0})[1].

The theoretical challenge is to provide a simultaneous description of both the t spectra of the vector mesons and the spin density matrix elements; in particular the largeness of the double-flip dependent matrix element r^{04}_{1-1} and the smallness of the single-flip dependent matrix element r^{04}_{00}.

DEPENDENCE ON t

The t dependence of the $\gamma p \to \rho Y$ cross section is presented in Fig. 1. Both the experimental result and the theoretical prediction are normalised to unity by dividing by their respective integrated cross-section over the range of interest. The dependence is well described by the BFKL model of Poludniowski *et al.* Further, it can be fitted by a power law distribution of the form $d\sigma/dt \propto |t|^{-n}$, where $n = 4.41 \pm 0.07$ (stat.) $^{+0.07}_{-0.05}$ (syst.), over the measured range of t. It should be noted that the dependence is steeper than that observed in [4] as is expected theoretically due to the different M_Y range.

---

[1] There are also the corresponding amplitudes M_{--}, M_{-+} and M_{-0}, which satisfy $M_{--} = M_{++}$, $M_{+-} = M_{-+}$ and $M_{-0} = M_{+0}$.

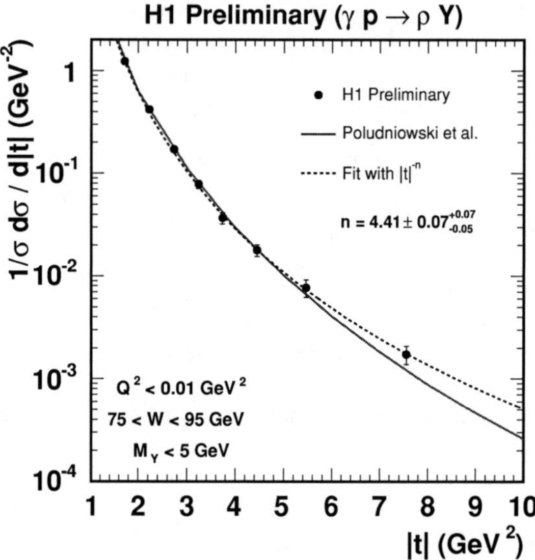

FIGURE 1. The t dependence of the $\gamma p \to \rho Y$ cross section. The inner error bars show the statistical errors, while the outer ones represent the sum of the statistical and non-correlated systematic errors added in quadrature. The dashed line represents a power law fit $|t|^{-n}$, which results in a power $n = 4.41 \pm 0.07\,(\text{stat.})\,^{+0.07}_{-0.05}\,(\text{syst.})$. The full line shows the result from the BFKL model of Poludniowski *et al.*

SPIN DENSITY MATRIX ELEMENTS

The measurement of the production and decay angular distributions provides access to the spin density matrix elements. Here the angles $\cos\theta^*$ and ϕ^* (where θ^* and ϕ^* represent, respectively, the polar and azimuthal angles of the π^+ in the ρ rest frame, the quantisation axis being taken as the direction opposite to that of the scattered photon) are fitted to extract the matrix elements r^{04}_{00} and r^{04}_{1-1} respectively. The fitting procedure is performed separately in three bins of t and the corresponding t dependencies are presented in Fig. 2. Also shown are the results of the ZEUS collaboration [4], with which there is excellent agreement. The single-flip dependent element r^{04}_{00} is consistent with zero indicating that the production is dominated by transversely polarised ρ mesons. The non-zero double-flip dependent matrix element r^{04}_{1-1} confirms the expected SCHC violation. The matrix element r^{04}_{00} agrees well with the BFKL model of Poludniowski *et al.*, while the prediction for r^{04}_{1-1}, although qualitatively able to reproduce the data, is too large at low values of $|t|$.

CONCLUSIONS

The photoproduction of ρ mesons with large momentum transfer has been studied using the H1 detector at HERA. The t dependence of the γp cross sec-

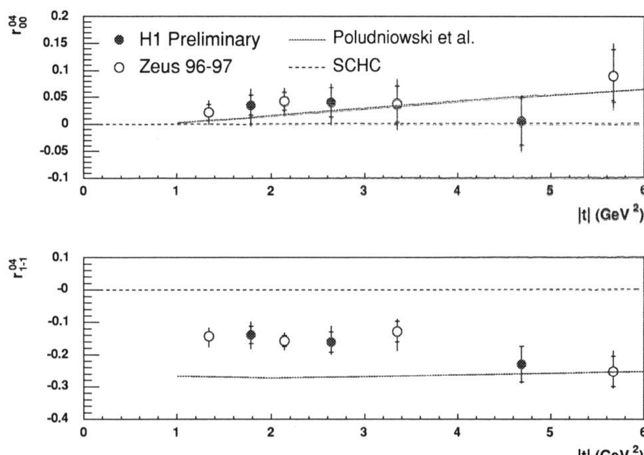

FIGURE 2. Measurements of r^{04}_{00} and r^{04}_{1-1} as a function of $|t|$ (full points) together with the previous measurement [4] (open points). The inner error bars show the statistical errors, while the outer ones represent the sum of the statistical and systematic errors added in quadrature. The full line shows the result from the BFKL model of Poludniowski *et al.*, while the dashed line shows the prediction from SCHC

tion is measured and fitted with a power law of the form $|t|^{-n}$, which results in $n = 4.41 \pm 0.07$ (stat.) $^{+0.07}_{-0.05}$ (syst.). Moreover, it is well described by BFKL model predictions. The spin density matrix elements r^{04}_{00} and r^{04}_{1-1} are measured as a function of t. The r^{04}_{00} matrix element is consistent with zero while the r^{04}_{1-1} matrix element differs significantly from zero and thus confirms violation of SCHC. BFKL model predictions well describe r^{04}_{00} but the prediction for r^{04}_{00}, although qualitatively able to reproduce the data, is too large at low $|t|$.

ACKNOWLEDGMENTS

We would like to thank J. Forshaw and R. Enberg for providing the theoretical model calculations used throughout.

REFERENCES

1. I. Abt *et al.* [H1 Collaboration], *Nucl. Instr. Meth. A*, **386** 310 and 348 (1997);
2. R. Enberg *et al.*, *JHEP*, **309**, 008-024 (2003).
3. G. G. Poludniowski *et al.*, *JHEP*, **312**, 002-041 (2003).
4. S. Chekanov *et al.* [ZEUS Collaboration], *Eur. Phys. J. C*, **26**, 389-409 (2003).
5. A. Aktas *et al.* [H1 Collaboration], *Phys. Lett. B.*, **586**, 205-218 (2003).
6. K. Schilling and G. Wolf, *Nucl. Phys. B*, **61**, 381-413 (1973).

Diffractive ρ^0 production at COMPASS

N. d'Hose,
on behalf of the COMPASS collaboration

CEA-Saclay, DSM/DAPNIA/SphN, F-91191 Gif-sur-Yvette Cedex

Abstract. Diffractive leptoproduction of ρ^0 mesons, $\mu + N \to \mu + N + \rho$ is measured at COMPASS at $< W >$ = 10 GeV over a wide range of Q^2, $0.01 < Q^2 < 10$ GeV^2. Angular distributions allow to determine spin density matrix elements. Preliminary results from COMPASS 2002 data are presented. They are consistent with a substantial increase of $R = \sigma_L/\sigma_T$ with Q^2 and a weak violation of SCHC, in agreement with other high energy experiments.

Keywords: COMPASS, DIFFRACTION, HELICITY, SCHC, SPIN DENSITY MATRIX
PACS: 13.60.Le; 14.40.Cs

PHYSICS MOTIVATIONS

Exclusive production of vector mesons is part of the COMPASS physics program. Here the reaction, $\mu + N \to \mu + N + \rho$, where N is a quasi-free nucleon from any of the nuclei of the COMPASS polarised target is studied in the diffractive regime at small $|t|$, $< W >$ = 10 GeV over a wide range of Q^2, $0.01 < Q^2 < 10$ GeV^2.

Diffractive ρ^0 production in lepton-nucleon scattering is often described as the fluctuation of the virtual photon emitted by the lepton into an intermediate $q\bar{q}$ state or off-shell ρ^0 meson. The intermediate state is scattered onto the mass shell by a diffractive interaction leaving the target nucleon intact. In Regge phenomenology the process is described by the exchange in the t channel of an intermediate object (Reggeon at low energy ($W < 5$ GeV^2) and Pomeron at higher energy). Experimental data obtained at E665 [1], ZEUS [2] and H1 [3] have indicated that the exchange in the t channel of an object of natural parity (with J^P such as $P = (-1)^J$) dominates such diffractive processes, and that the helicity of the photon in the $\gamma^* N$ centre-of-mass system is approximatively retained by the vector meson, a phenomenon known as s-channel helicity conservation (SCHC).

The angular distributions of the reaction give access to spin density matrix elements, which are bilinear combinations of the helicity amplitudes $T_{\lambda_{VM}\lambda_\gamma}$, where $\lambda_{VM}(\lambda_\gamma)$ is the vector meson (virtual photon) helicity. Assuming natural parity exchange to hold $T_{-\lambda_{VM}-\lambda_\gamma} = (-1)^{\lambda_{VM}-\lambda_\gamma} T_{\lambda_{VM}\lambda_\gamma}$, the 9 helicity amplitudes reduce to 5 independent complex amplitudes. At $Q^2 > 2$ GeV^2, the following hierarchy has emerged from previous measurements: the longitudinal T_{00} helicity non-flip amplitude dominates over the transverse T_{11} helicity non-flip amplitude, which dominates over the helicity single flip amplitude from transverse photon to longitudinal vector meson T_{01} which finally dominates over the single T_{10} and double T_{1-1} helicity flip amplitudes.

The goal of the COMPASS experiment is to quantify any violation of SCHC and to

CP792, *Deep Inelastic Scattering, DIS 2005,* edited by W. H. Smith and S. R. Dasu

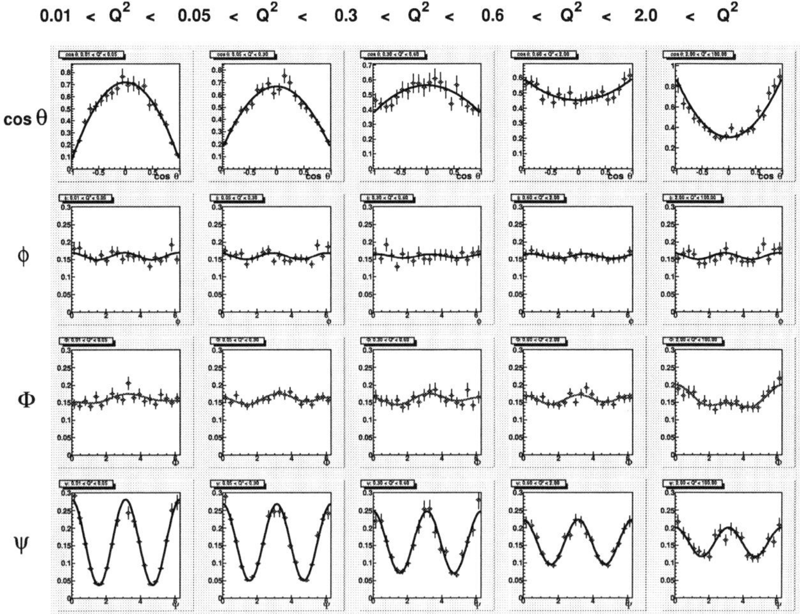

FIGURE 1. Angular distributions as a function of $cos\theta$, ϕ, Φ and $\psi = \phi - \Phi$ for 5 intervals in Q^2.

allow for a extraction of $R = \sigma_L/\sigma_T$. This will provide an easy determination of the longitudinal cross section which can be described at high Q^2 in terms of Generalized Partons Distributions.

SELECTION OF DIFFRACTIVE EVENTS

The COMPASS experiment uses the 160 GeV/c polarised muon beam of the CERN SPS. Muons are scattered off polarised nucleons in a 1.2m long frozen ^{6}LiD target.The scattered particles and the decay products of the ρ are detected in two high resolution magnetic spectrometers equipped with tracking detectors, calorimetry, RICH and muon identification (see ref. [4] for more detailed description of the COMPASS apparatus). The present analysis concerns the full data sample taken in 2002 with a longitudinally polarised target. For an event to be selected we required it to originate inside the target, to have reconstructed beam and scattered muon tracks and to have only two additional tracks, which correspond to charged pions from the decay of the ρ^0. A cut on the invariant mass of two pions, $0.5 < M_{\pi\pi} < 1$ GeV, is applied to identify the ρ^0. In order to select exclusive events as the slow recoiling target particles are not detected, we use cuts on the missing energy, $-2.5 < E_{miss} < 2.5$ GeV and on the transverse momentum of ρ^0 with respect to the virtual photon direction, $p_t^2 < 0.5$ GeV^2. Here $E_{miss} = (M_X^2 - M_p^2)/2M_p$ where M_X is the mass of the undetected system and M_p the proton mass. Coherent interactions on the target nuclei are removed by a cut $p_t^2 > 0.15$ GeV^2. Finally to avoid large corrections for acceptance, smearing and misidendification of events additional cuts are applied: $Q^2 > 0.01$ GeV^2, $\nu > 30$ GeV and $E_{\mu'} > 20$ GeV. After all selections the sample consists of about 700 000 events, of which about 6000 events at $Q^2 > 2$ GeV^2. The remaining non-exclusive background in the whole sample is about 12%.

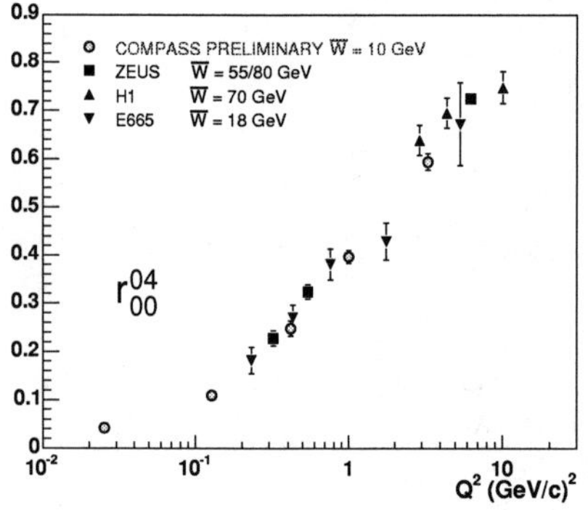

Figure 2. Q^2 dependence of r^{04}_{00}.

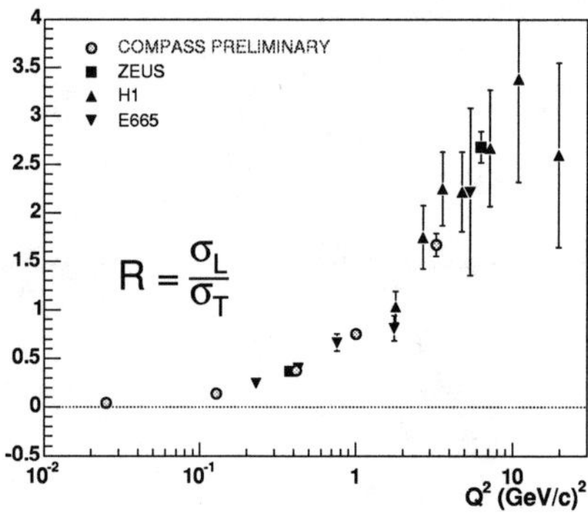

Figure 3. Q^2 dependence of $R = \sigma_L/\sigma_T$.

EXTRACTION OF SPIN DENSITY MATRIX

The angular distribution $W(\cos\theta, \phi, \Phi)$ is studied in the s-channel helicity frame [5]. The ρ^0 direction in the $\gamma^* N$ centre-of-mass system is taken as the quantization axis. The angle θ is the polar angle and ϕ the azimuthal angle of the positive decay meson in the ρ^0 centre-of-mass system. ϕ is then the angle between the production plane and the decay plane and Φ is the angle between the production plane and the lepton scattering plane. In principle 23 spin density matrix elements (SDME) can be extracted with a longitudinally polarised beam. However in this preliminary analysis, we consider only single dimensional projections of the angular distribution leading to the determination of a few SDME. For cos θ et ϕ the angular distributions are:

$$W(\cos\,\theta) = \frac{3}{4}[(1 - r^{04}_{00}) + (3\,r^{04}_{00} - 1)\cos^2\theta]$$

$$W(\phi) = \frac{1}{2\pi}[1 - 2r^{04}_{1-1}\cos 2\phi + P_\mu\sqrt{1-\varepsilon^2}\,2\Im m\, r^{3}_{1-1}\sin 2\phi]$$

where the r's are the SDME, P_μ is the polarisation of the muon beam and ε is the virtual photon polarisation parameter. If the SCHC holds the complete angular distribution reduces to $W(\cos\theta, \psi)$ where $\psi = \phi - \Phi$ is the angle between the ρ^0 decay plane and the lepton scattering plane:

$$W(\psi) = \frac{1}{2\pi}[1 + 2\varepsilon\, r^{1}_{1-1}\cos 2\psi]$$

RESULTS

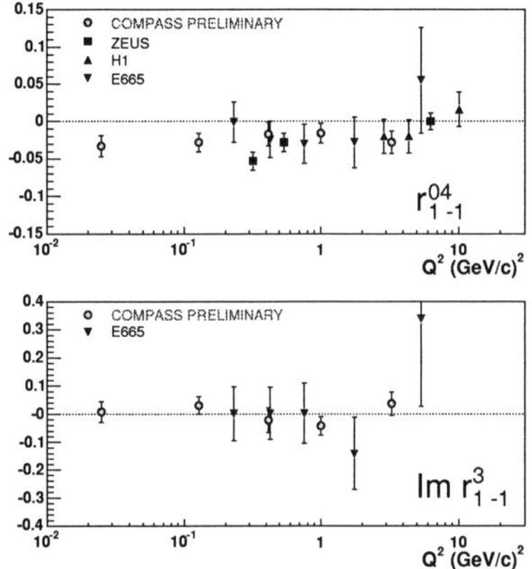

Figure 4. Q^2 dependence of r^{04}_{1-1} and $\Im m\, r^3_{1-1}$.

The angular distributions presented in Fig. 1 are corrected for acceptance, smearing and efficiency using a full MC simulation of the apparatus and the DIPSI event generator. Fits of these distributions allow to extract some SDME. r^{04}_{00} determined from $cos\theta$ distribution, is displayed as a function of Q^2 in Fig. 2 where only statistical errors have been drawn. The COMPASS data, with its good statistical precision, cover a wide range of Q^2 from quasi-real photoproduction to the hard scattering regime ($0.01 < Q^2 < 10$ GeV^2). The results are in fair agreement with the other experiments [1, 2, 3].

$$r^{04}_{00} \propto (|T_{00}|^2 + |T_{01}|^2)/\sigma_{tot} \overset{if\ SCHC}{\rightarrow} \sigma_L/\sigma_{tot}$$

Then if SCHC holds, $R = \sigma_L/\sigma_T$ can be determined (see Fig. 3). At small Q^2 the production by transverse photons dominates while when $Q^2 > 2$ GeV^2 the production by longitudinal photons takes overs. From ϕ distribution the SDME r^{04}_{1-1} and $\Im m\, r^3_{1-1}$ are extracted and compared to other experiments in Fig. 4. $r^{04}_{1-1} \propto (\Re e\ (T_{11}T^*_{1-1}) + |T_{10}|^2)/\sigma_{tot}$ should be 0 if SCHC holds. Its non-zero value indicates a small contribution of amplitudes with helicity flip. Note, that $\Im m\, r^3_{1-1}$ could be accessed only with polarised lepton beams. It is consistent with 0.

SUMMARY AND OUTLOOK

Preliminary results from the analysis of angular distributions for exclusive incoherent ρ^0 production from COMPASS 2002 data are presented. The data measured at $< W >= 10$ GeV over a wide range of Q^2, $0.01 < Q^2 < 10$ GeV^2 are consistent with a substantial increase of R with Q^2 and a weak violation of SCHC in agreement with other high energy experiments. These new COMPASS data will be completed by the 2003 and 2004 sets. Although matching previous measurements, they are of much higher statistical significance and cover a larger range in Q^2. At $Q^2 > 1$ GeV^2 they consist of a very first step towards the study of Generalized Parton Distributions.

REFERENCES

1. E665 Collab., M.R. Adams *et al., Z. Phys.* **C 74** (1997) 237.
2. ZEUS Collab., J. Breitweg *et al., Eur. Phys. J* **C 12** (2000) 393.
3. H1 Collab., C. Adloff *et al., Eur. Phys. J* **C 13** (2000) 371 and *Phys. Lett.* **B 539** (2002) 25.
4. G. Mallot, *Nucl. Instr. and Meth.* **A 518** (2004) 121.
5. K. Schilling and G. Wolf, *Nucl. Phys.* **B 61** (1973) 381.

Deeply virtual vector meson electroproduction at small Bjorken-x

P. Kroll

Fachbereich Physik, Universität Wuppertal, D-42097 Wuppertal, Germany

Abstract. It is reported on an analysis of vector meson electroproduction at small Bjorken-x ($x_{\rm Bj}$) within the handbag approach. Using a model for the generalized parton distributions (GPDs) and calculating the partonic subprocess, electroproduction off gluons, within the modified perturbative approach, cross sections and spin density matrix elements (SDME) are evaluated. The numerical results of this analysis agree fairly well with recent HERA data.

Keywords: meson electroproduction, generalized parton distributions
PACS: 12.38Bx,12.39St,13.60Le

It has been shown [1] that, at large photon virtuality Q^2, meson electroproduction factorizes in a partonic subprocess, electroproduction off gluons or quarks, $\gamma^* g(q) \to Vg(q)$, and GPDs, representing soft proton matrix elements (see Fig. 1). At small $x_{\rm Bj}$ ($\lesssim 10^{-2}$) the quark subprocesses can be ignored. In the following I am going to report on an analysis of vector meson electroproduction within this handbag approach [2] in the kinematical regime of large Q^2 and large energy W in the photon-proton c.m.s. but small $x_{\rm Bj}$ and Mandelstam t. An exploratory study of the longitudinal cross section σ_L for $\gamma^* p \to Vp$ has been performed by Mankiewicz et al. [3] within this approach. Effects of the GPDs have been estimated by Martin et al [4].

The structure of the proton is rather complex. In correspondence to its four form factors there are four gluon GPDs H^g, E^g, $\widetilde{H}^g$ and $\widetilde{E}^g$ and four for each quark flavour. All GPDs are functions of three variables, t, skewness ξ and the average momentum fraction $\bar{x}$, the latter two are defined by

$$\xi = \frac{(p-p')^+}{(p+p')^+}, \qquad \bar{x} = \bar{k}^+ / \bar{p}^+ . \tag{1}$$

These parameters are related to the usual momentum fractions the gluons carry with respect to their parent proton, by $x^{(\prime)} = (\bar{x} \pm \xi)/(1 \pm \xi)$. The skewness is kinematically fixed to $\xi \simeq x_{\rm Bj}/2$ in a small $x_{\rm Bj}$ approximations. Hence, $x \neq x'$. This is to be contrasted with the leading $\log(1/x_{\rm Bj})$ approximation [5] where $x \simeq x' \simeq x_{\rm Bj}$ is assumed and the GPD replaced by the usual gluon distribution $g(x)$.

The handbag approach leads to the following proton helicity non-flip amplitude

$$M^V_{\mu'+,\mu+} = \frac{e}{2}\mathscr{C}_V \int_0^1 \frac{d\bar{x}}{(\bar{x}+\xi)(\bar{x}-\xi+i\varepsilon)} \left[H^V_{\mu'+,\mu+} + H^V_{\mu'-,\mu-}\right] H^g(\bar{x},\xi,t). \tag{2}$$

Contributions from other GPDs can be neglected at small $x_{\rm Bj}$ and for unpolarized protons. The photon and meson helicities are denoted by μ and μ', respectively. The

CP792, *Deep Inelastic Scattering, DIS 2005*, edited by W. H. Smith and S. R. Dasu

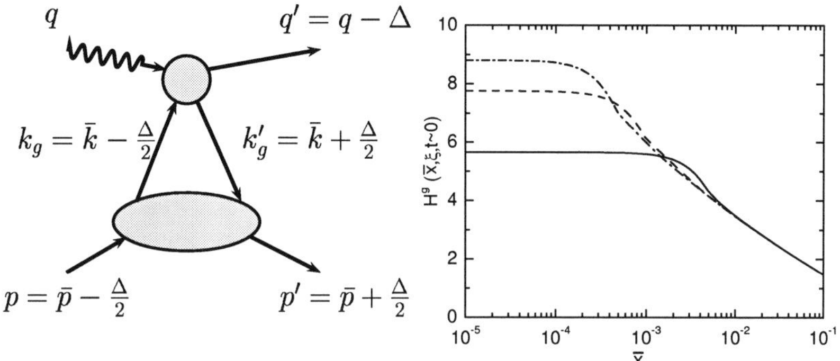

FIGURE 1. Left: The handbag diagram for meson electroproduction off protons. The large blob represents a GPD while the small one stands for the subprocess. The momenta of the involved particles are specified. Right: Model results for the GPD H^g at $t \simeq 0$ and for the case $n = 1$. The results for $n = 2$ are similar. The solid (dashed, dash-dotted) line represents the GPD at $\xi = 5\ (1\,,\ 0.5) \cdot 10^{-3}$ and at a scale of 2 GeV.

explicit labels in the full (subprocess) amplitude, M^V (H^V) refer to the helicities of the protons (gluons).

The GPDs are controlled by non-perturbative QCD. In the absence of an GPD analysis in analogy to those of the usual PDFs (see however [6]) one has to rely on a model. Its contruction is however not an easy matter since the GPDs are functions of three variables. Factorising the t dependence from the $\bar{x}, \xi$ one is probably incorrect. We therefore restrict ourselves to the forward direction and exploit the ansatz for a double distribution proposed in Ref. [7] ($n = 1, 2$)

$$f(\beta, \alpha, t \simeq 0) = g(\beta) \frac{\Gamma(2n+2)}{2^{2n+1}\Gamma^2(n+1)} \frac{[(1-|\beta|)^2 - \alpha^2]^n}{(1-|\beta|)^{2n+1}} . \tag{3}$$

The GPDs is then obtained by an integral over f

$$H^g(\bar{x}, \xi) = \Big[\Theta(0 \leq \bar{x} \leq \xi) \int_{x_3}^{x_1} d\beta + \Theta(\xi \leq \bar{x} \leq 1) \int_{x_2}^{x_1} d\beta\Big] \frac{\beta}{\xi} f(\beta, \alpha = \frac{\bar{x} - \beta}{\xi}) . \tag{4}$$

Using the NLO CTEQ5M [8] result as input we obtain the GPD H^g shown in Fig. 1.

The last item of the amplitude (2) to be discussed is the subprocess amplitude. Its treatment is rather standard, it only differs in detail from versions to be found in the literature [4, 9]. In the modified perturbative approach invented by Sterman and collaborators [10], in which quark transverse momenta are retained and gluonic radiative corrections in the form of a Sudakov factor are taken into account, it reads

$$\mathscr{H}^V = \int \frac{d\tau dk_\perp^2}{\sqrt{2}16\pi^2} \Psi_V \,\mathrm{Tr}\left\{ (\not{q}' + m_V)\not{\varepsilon}_V^* T_0 - \frac{k_\perp^2 g_\perp^{\alpha\beta}}{2M_V} \{(\not{q}' + m_V)\not{\varepsilon}_V^*, \gamma_\alpha\} \Delta T_\beta \right\} . \tag{5}$$

Higher order terms in this expansion are not shown. Gaussians for the meson's wavefunctions, $\Psi_V = \Psi(\tau, k_\perp^2)$, are used which may depend on the polarization of the vector

meson. There are two parameters specifying the wavefunction, the meson's decay constant and a transverse size parameter. For longitudinally polarized vector mesons these parameters are fairly well-known. The first term in (5) dominates for V_L while it is approximately zero for transversally polarized vector mesons (V_T). In the latter case the second term is the dominant one. Note that the soft physics parameter M_V in this term is of order of the vector meson mass m_V. As can be seen from Eq. (5) the $L \to L$ transition is dominant while the $T \to T$ one is of relative order $\langle k_\perp^2 \rangle^{1/2}/Q$ and the $T \to L$ one of order $\sqrt{-t}/Q$. The latter amplitude is tiny and only noticeable in some of the SDMEs. All other transitions are negligible. Eventual infrared singularities that may occur for transitions to V_T, are regularized in the modified perturbative approach.
Before comparing the results to experiment I have to comment on the t dependence of the amplitudes. Exponentials in t are assumed with slopes $B^V_{LL(TT)}$ taken from experiment. Combined with the calculated forward amplitudes one can evaluate the integrated cross sections and the SDME at small t. From (5) one sees that the size of the $T \to T$ amplitude is controlled by the following product of parameters

$$\left| M^V_{TT} \right| \propto \left(\frac{f^V_T}{M_V} \right)^2 \frac{1}{B^V_{TT}} . \tag{6}$$

Without precise t-dependent data at disposal only this product is probed. One can therefore, for instance, assume $B^V_{TT} \simeq B^V_{LL}/2$. Combined with $M_V = m_V$ and $f^\rho_T = 250$ MeV this assumption provides reasonable results for vector meson electroproduction, see Fig. 2. An alternative choice is $B^V_{TT} \simeq B^V_{LL}$, $M_V = m_V$, $f^\rho_T = 170$ MeV which leads to practically the same results for the cross sections. Only the t dependence of the SDME differs in both the cases. Given the accuracy of the present data [11, 12] both the scenarios are in agreement with experiment.
In Fig. 2 the cross section σ_L and the ratio $R = \sigma_L/\sigma_T$ are displayed. The data on R are extracted from the SDME measurements. This extraction is problematic if the slopes are different. For comparison the ratio of the corresponding differential cross sections is also shown in Fig. 2 (at $t \simeq -0.15$ GeV^2). Results for the SDME of ρ and ϕ mesons are also presented in [2] in fair agreement with experiment.
Finally I want to comment on the W dependence of the dominant longitudinal cross section. It is given by the imaginary part of the $L \to L$ amplitude with a correction of about 10% from the real part. The cross section is therefore approximately proportional to $|H^g(\xi,\xi)|^2$. Through the model (4) the low-x behaviour of the PDF $\bar{x}g(\bar{x}) \sim \bar{x}^{-\delta(Q^2)}$ is tranferred to the GPD and one finds

$$\sigma_L \propto W^{-4\delta(Q^2)} . \tag{7}$$

The Q^2 dependence of δ is a consequence of evolution. Comparison with experiment reveals that this behaviour is in remarkable agreement with the data within admittedly large errors. A last remark: The expression for σ_L the GPD approach provides, is also obtained in the leading $\log 1/x_{\rm Bj}$ approximation given that the subprocess is treated equally and that $H^g(\xi,\xi)$ is replaced by $2\xi g(2\xi)$. The quality of this approximation is rather good for $\xi \lesssim 10^{-2}$, there is only an enhancement of the GPD by about 18%, the skewing effect [4]. For increasing ξ the approximation becomes gradually worse.

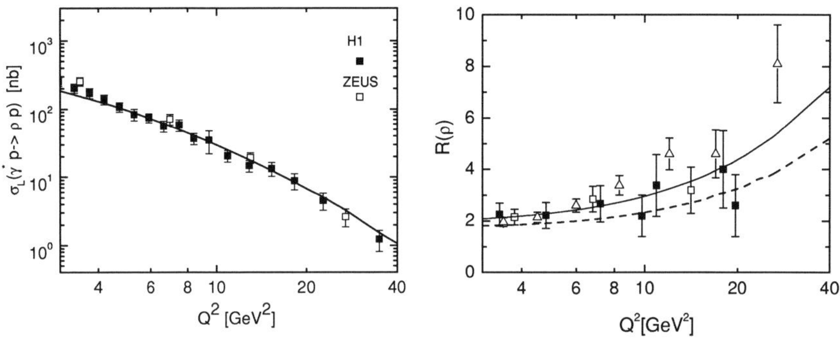

FIGURE 2. Left: The integrated cross section for $\gamma_L p \to \rho p$ versus Q^2 at $W \simeq 75$ GeV. Right: The ratio of longitudinal and transverse cross sections for ρ production versus Q^2 at $W \simeq 75$ GeV. Data taken from [11] (filled squares) and [12] (open symbols). The solid (dashed) lines are the results of Ref. [2] for the ratio of differential (integrated) cross sections.

I summarize: Vector meson electroproduction off unpolarized protons at small $x_{\rm Bj}$ and small t probes the GPD H^g. Calculating the partonic subprocess within the modified perturbative approach (using gaussian wavefunctions) fair agreement with HERA data on the integrated cross sections for longitudinally and transversally polarized virtual photons and the spin density matrix elements are obtained for electroproduction of ρ and ϕ mesons. It is to be stressed that only the forward amplitudes are caluclated within the GPD approach. Their t dependencies are assumed to be exponentials with slopes taken from experiment. The present data do, however, not fix the slope of the $T \to T$ amplitude precisely. This treatment of the t dependence is unsatisfactory and improvements are required. In principle the GPD approach has the potential to do better but the GPDs as a function of t are needed for that. It is also possible to go to larger values of $x_{\rm Bj}$ with it. Some results on ϕ production at COMPASS kinematics are presented in [2].

REFERENCES

1. A.V. Radyushkin, *Phys. Lett. B 385, 333 (1996)*; J.C. Collins *et al.*, *Phys. Rev. D* **5**6, 2982 (1997).
2. S. V. Goloskokov and P. Kroll, hep-ph/0501242, to be published in Eur. Phys. J. C.
3. L. Mankiewicz, G. Piller and T. Weigl, *Eur. Phys. J. C* **5**, 119 (1998).
4. A. D. Martin, M. G. Ryskin and T. Teubner, *Phys. Rev. D* **6**2 (2000) 014022
5. S. J. Brodsky, *at al.*, *Phys. Rev. D* **5**0, 3134 (1994).
6. M. Diehl, T. Feldmann, R. Jakob and P. Kroll, *Eur. Phys. J. C* **3**9, 1 (2005).
7. I. V. Musatov and A. V. Radyushkin, *Phys. Rev. D* **6**1, 074027 (2000).
8. J. Pumplin, D. R. Stump, J. Huston, H. L. Lai, P. Nadolsky and W. K. Tung, *JHEP* **0**207, 012 (2002).
9. L. Frankfurt, W. Koepf and M. Strikman, *Phys. Rev. D* **5**4, 3194 (1996)
10. J. Botts and G. Sterman, Nucl. Phys. B **3**25, 62 (1989).
11. C. Adloff *et al.* [H1 Collaboration], *Eur. Phys. J. C13, 371 (2000)*; S. Aid *et al.* [H1 Collaboration], *Nucl. Phys. B* **4**68, 3 (1996).
12. J. Breitweg *et al.* [ZEUS Collaboration], *Eur. Phys. J. C* **6**, 603 (1999) S. Chekanov [ZEUS Collaboration], hep-ex/0504010.

Diffractive Dijet Photoproduction

Michael Klasen* and Gustav Kramer†

*Institute for Nuclear Theory, University of Washington, Box 351550, Seattle, WA 98195, USA and Laboratoire de Physique Subatomique et de Cosmologie, Université Joseph Fourier/CNRS-IN2P3, 53 Avenue des Martyrs, 38026 Grenoble, France
†II. Institut für Theoretische Physik, Universität Hamburg, Luruper Chaussee 149, 22761 Hamburg, Germany

Abstract. We have calculated diffractive dijet production in deep-inelastic scattering (DIS) at low-Q^2 and next-to-leading order (NLO) of perturbative QCD, including contributions from direct and resolved photons. We study how the cross section depends on the factorization scheme and scale M_γ at the virtual photon vertex for the occurance of factorization breaking. The strong M_γ-dependence, which is present when only the resolved cross section is suppressed, is tamed by intodrucing the suppression also in the initial-state NLO correction of the direct part.

Keywords: Perturbative QCD calculations, Factorization, Regge Theory
PACS: 12.38.Bx, 12.39.St, 12.40.Nn

INTRODUCTION

From a perturbative QCD (pQCD) point of view, the central question for hard diffractive scattering events, characterized by a large rapidity gap devoid of particles in high-energy collisions, is whether they can be factorized into non-perturbative diffractive parton density functions (PDFs) of a colorless object (*e.g.* a pomeron) and perturbative partonic cross sections. This concept, based on a long-standing proposal by Ingelman and Schlein [1], is believed to hold for the scattering of point-like electromagnetic probes off a hadronic target, such as deep-inelastic scattering (DIS) or *direct* photoproduction [2], whereas it has been shown to fail for purely hadronic collisions [2, 3].

Factorization is thus expected to fail also in *resolved* photoproduction, where the photon first dissolves into partonic constituents, before these scatter off the hadronic target. The separation of these two types of photoproduction events is, however, a leading order (LO) concept. At next-to-leading order (NLO) of pQCD, they are closely connected by an initial state singularity originating from the splitting $\gamma \to q\bar{q}$ [4], which may play a crucial role in the way factorization breaks down in diffractive photoproduction [5]. Factorization breaking effects are therefore expected to show up first in observables that distinguish between direct and resolved photoproduction, such as distributions in the longitudinal momentum fraction x_γ of partons in the photon [6], the photon virtuality Q^2 [7], or the dependence of the predicted cross sections on the factorization scale M_γ [5]. It is clear that this M_γ-dependence is unphysical and must be remedied also for the case of factorization breaking of the resolved part of the cross section. A proposal how to achieve this has been worked out in our previous work [5] and will be described in the next Section. For demonstrative purposes we restrict ourselves to low-Q^2 DIS using the kinematic framework of our earlier publication [7].

CP792, *Deep Inelastic Scattering, DIS 2005*, edited by W. H. Smith and S. R. Dasu

Three groups have recently tried to extract diffractive parton densities from inclusive diffractive DIS data at DESY HERA, treating the Pomeron either as a hadronic object within Regge factorization [8, 9]

$$f_a^D(x,Q^2;x_{I\!P},t) = f_{I\!P/p}(x_{I\!P},t) f_{a/I\!P}(\beta = x/x_{I\!P},Q^2) \tag{1}$$

or perturbative QCD [10], but so far only the H1 set has been tested for factorization in a different scattering environment, *i.e.* in photoproduction of dijets at HERA [11].

FACTORIZATION SCHEME AND SCALE DEPENDENCE

A factorization scheme for virtual photoproduction has been defined and the full NLO corrections for inclusive dijet production have been calculated in [12]. They have been implemented in the NLO Monte Carlo program JETVIP [13]. We have adapted this NLO framework to diffractive dijet production. According to [12], the subtraction term, which is absorbed into the PDFs of the virtual photon $f_{a/\gamma}(x_\gamma, M_\gamma)$, is of the form as given in [5]. The main term is proportional to $\ln(M_\gamma^2/Q^2)$ times the splitting function

$$P_{q_i \leftarrow \gamma}(z) = 2N_c Q_i^2 \frac{z^2 + (1-z)^2}{2}, \tag{2}$$

where $z = p_1 p_2 / p_0 q \in [x;1]$ and Q_i is the fractional charge of the quark q_i. p_1 and p_2 are the momenta of the two outgoing jets, and p_0 and q are the momenta of the ingoing parton and virtual photon, respectively. Since $Q^2 = -q^2 \ll M_\gamma^2$, the subtraction term is large and is therefore resummed by the DGLAP evolution equations for the virtual photon PDFs. After this subtraction, the finite term $M(Q^2)_{\overline{\rm MS}}$, which remains in the matrix element for the NLO correction to the direct process [12], has the same M_γ-dependence as the subtraction term, *i.e.* $\ln M_\gamma$ is multiplied with the same factor. As already mentioned, this yields the M_γ-dependence before the evolution is turned on. In the usual non-diffractive dijet photoproduction these two M_γ-dependences cancel, when the NLO correction to the direct part is added to the LO resolved cross section [14]. Then it is obvious that the approximate M_γ-independence is destroyed, if the resolved cross section is multiplied by a suppression factor R to account for the factorization breaking in the experimental data. To remedy this deficiency, we propose to multiply the $\ln M_\gamma$-dependent term in $M(Q^2)_{\overline{\rm MS}}$ with the same suppression factor as the resolved cross section. This is done in the following way: We split $M(Q^2)_{\overline{\rm MS}}$ into two terms using the scale p_T^* in such a way that the term containing the slicing parameter y_s, which was used to separate the initial-state singular contribution, remains unsuppressed. In particular, we replace the finite term after the subtraction by

$$\begin{aligned} M(Q^2,R)_{\overline{\rm MS}} &= \left[-\frac{1}{2N_c} P_{q_i \leftarrow \gamma}(z) \ln\left(\frac{M_\gamma^2 z}{p_T^{*2}(1-z)} \right) + \frac{Q_i^2}{2} \right] R \\ &\quad - \frac{1}{2N_c} P_{q_i \leftarrow \gamma}(z) \ln\left(\frac{p_T^{*2}}{zQ^2 + y_s s} \right), \end{aligned} \tag{3}$$

where R is the suppression factor. This expression coincides with the finite term after subtraction (see Ref. [5]) for $R = 1$, as it should, and leaves the second term in Eq. (3) unsuppressed. In Eq. (3) we have suppressed in addition to $\ln(M_\gamma^2/p_T^{*2})$ also the z-dependent term $\ln(z/(1-z))$, which is specific to the $\overline{\rm MS}$ subtraction scheme as defined in [12]. The second term in Eq. (3) must be left in its original form, *i.e.* being unsuppressed, in order to achieve the cancellation of the slicing parameter (y_s) dependence of the complete NLO correction in the limit of very small Q^2 or equivalently very large s. It is clear that the suppression of this part of the NLO correction to the direct cross section will change the full cross section only very little as long as we choose $M_\gamma \simeq p_T^*$. The first term in Eq. (3), which has the suppression factor R, will be denoted by $\rm DIR_{IS}$ in the following.

To study the left-over M_γ-dependence of the physical cross section, we have calculated the diffractive dijet cross section with the same kinematic constraints as in the H1 experiment [15]. Jets are defined by the CDF cone algorithm with jet radius equal to one and asymmetric cuts for the transverse momenta of the two jets required for infrared stable comparisons with the NLO calculations [16]. The original H1 analysis actually used a symmetric cut of 4 GeV on the transverse momenta of both jets [17]. The data have, however, been reanalyzed for asymmetric cuts [15].

For the NLO resolved virtual photon predictions, we have used the PDFs SaS1D [18] and transformed them from the $\rm DIS_\gamma$ to the $\overline{\rm MS}$ scheme as in Ref. [12]. If not stated otherwise, the renormalization and factorization scales at the pomeron and the photon vertex are equal and fixed to $p_T^* = p_{T,jet1}^*$. We include four flavors, *i.e.* $n_f = 4$ in the formula for α_s and in the PDFs of the pomeron and the photon. With these assumptions we have calculated the same cross section as in our previous work [7]. First we investigated how the cross section $\mathrm{d}\sigma/\mathrm{d}Q^2$ depends on the factorization scheme of the PDFs for the virtual photon, *i.e.* $\mathrm{d}\sigma/\mathrm{d}Q^2$ is calculated for the choice SaS1D and SaS1M. Here $\mathrm{d}\sigma/\mathrm{d}Q^2$ is the full cross section (sum of direct and resolved) integrated over the momentum and rapidity ranges as in the H1 analysis. The results, shown in Fig. 2 of Ref. [5] demonstrate that the choice of the factorization scheme of the virtual photon PDFs has negligible influence on $\mathrm{d}\sigma/\mathrm{d}Q^2$ for all considered Q^2. The predictions agree reasonably well with the preliminary H1 data [15].

We now turn to the M_γ-dependence of the cross section with a suppression factor for $\rm DIR_{IS}$, which is the main part of this Report. To show this dependence for the two suppression mechanisms, (i) suppression of the resolved cross section only and (ii) additional suppression of the $\rm DIR_{IS}$ term as defined in Eq. (3) in the NLO correction of the direct cross section, we consider $\mathrm{d}\sigma/\mathrm{d}Q^2$ for the lowest Q^2-bin, $Q^2 \in [4,6]$ GeV^2. In the left part of Fig. 1, this cross section is plotted as a function of $\xi = M_\gamma/p_T^*$ in the range $\xi \in [0.25;4]$ for the cases (i) (light full curve) and (ii) (full curve). We see that the cross section for case (i) has an appreciable ξ-dependence in the considered ξ range of the order of 40%, which is caused by the suppression of the resolved contribution only. With the additional suppression of the $\rm DIR_{IS}$ term in the direct NLO correction, the ξ-dependence of $\mathrm{d}\sigma/\mathrm{d}Q^2$ is reduced to approximately less than 20%, if we compare the maximal and the minimal value of $\mathrm{d}\sigma/\mathrm{d}Q^2$ in the considered ξ range. The remaining ξ-dependence is caused by the NLO corrections to the suppressed resolved cross section and the evolution of the virtual photon PDFs. How the compensation of the

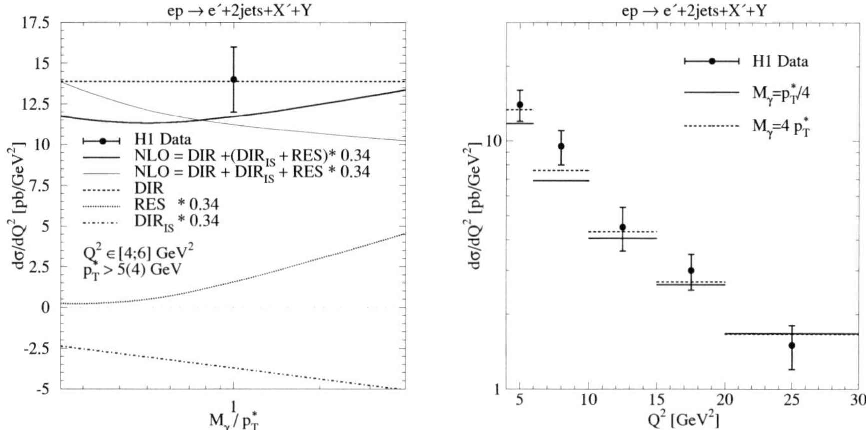

FIGURE 1. Left: Photon factorization scale dependence of resolved and direct contributions to $d\sigma/dQ^2$ together with their weighted sums for (i) suppression of the resolved cross section and for (ii) additional suppression of DIR_{IS}, using SaS1D virtual photon PDFs [18]. Right: Q^2-dependence of the dijet cross section for $M_\gamma = p_T^*/4$ (full) and $M_\gamma = 4p_T^*$ (dashed) and comparison with preliminary H1 data using SaS1D virtual photon PDFs [18].

M_γ-dependence between the suppressed resolved contribution and the suppressed direct NLO term works in detail is exhibited by the dotted and dashed-dotted curves in Fig. 1 (left). The suppressed resolved term increases and the suppressed direct NLO term decreases by approximately the same amount with increasing ξ. In addition we show also $d\sigma/dQ^2$ in the DIS theory, *i.e.* without subtraction of any $\ln Q^2$ terms (dashed line). Of course, this cross section must be independent of ξ. This prediction agrees very well with the experimental point, whereas the result for the subtracted and suppressed theory (full curve) lies slightly below. We notice, that for $M_\gamma = p_T^*$ the additional suppression of DIR_{IS} has only a small effect. It increases $d\sigma/dQ^2$ by 5% only.

In order to get an idea about the M_γ scale dependence of $d\sigma/dQ^2$ for the other Q^2 bins we have computed this cross section for two choices of M_γ, namely $M_\gamma = p_T^*/4$ and $M_\gamma = 4p_T^*$ corresponding to the lowest and highest ξ in Fig. 1 (left). The result for the $d\sigma/dQ^2$ is shown on the right side of Fig. 1. We see that the M_γ-dependence in the considered range decreases with increasing Q^2. This is to be expected since the resolved contribution diminishes with increasing Q^2, so that the NLO corrections to the resolved cross section and the effect of the evolution of the photon PDF diminish as well.

CONCLUSION

In Summary, we described in this Report a new factorization scheme for diffractive production of jets in low-Q^2 deep inelastic scattering. By suppressing not only the resolved photon contribution, but also the unresummed logarithm as well as scheme-dependent finite terms in the NLO direct initial state correction, factorization scheme and scale invariance is restored up to higher order effects, while at the same time the cut-off invariance required in phase space slicing methods is preserved.

For pedagogical reasons, we have chosen in this Report the kinematic region of finite, but low photon virtuality Q^2, which exposes and regularizes a logarithmic virtual photon initial state singularity. We do, however, not rely on the finiteness of Q^2, but rather separate suppressed and unsuppressed terms using the hard transverse momentum scale p_T^*, so that our scheme is equally valid for real photoproduction.

The scheme- and scale invariance has been demonstrated numerically using the kinematics of a recent H1 analysis, differential in Q^2 and parton momentum fraction in the pomeron $z_{I\!P}$ (not shown). Very good stability with respect to scheme- and scale variations and good agreement with the experimental data has been found.

ACKNOWLEDGMENTS

M.K. thanks the working group convenors for the kind invitation and the DoE's INT at the University of Washington for kind hospitality and partial financial support during preparation of this work.

REFERENCES

1. G. Ingelman and P. E. Schlein, Phys. Lett. B **152**, 256 (1985).
2. J. C. Collins, Phys. Rev. D **57**, 3051 (1998) [Erratum-ibid. D **61**, 019902 (2000)].
3. T. Affolder *et al.* [CDF Collaboration], Phys. Rev. Lett. **84**, 5043 (2000).
4. M. Klasen, Rev. Mod. Phys. **74**, 1221 (2002).
5. M. Klasen and G. Kramer, DESY 05-095, LPSC 05-053, hep-ph/0506121, submitted to J. Phys. G.
6. M. Klasen and G. Kramer, contribution to DIS 2004, hep-ph/0401202; Eur. Phys. J. C **38**, 93 (2004).
7. M. Klasen and G. Kramer, Phys. Rev. Lett. **93**, 232002 (2004).
8. H1 Collaboration, Abstract 980, contributed to the 31st International Conference on High Energy Physics (ICHEP 2002), Amsterdam, July 2002.
9. S. Chekanov *et al.* [ZEUS Collaboration], Eur. Phys. J. C **38**, 43 (2004).
10. A. D. Martin, M. G. Ryskin and G. Watt, Eur. Phys. J. C **37**, 285 (2004).
11. M. Mozer and R. Renner for the H1 and ZEUS Collaborations, these proceedings.
12. M. Klasen, G. Kramer and B. Pötter, Eur. Phys. J. C **1**, 261 (1998).
13. B. Pötter, Comput. Phys. Commun. **133**, 105 (2000).
14. D. Bödeker, G. Kramer and S. G. Salesch, Z. Phys. C **63**, 471 (1994).
15. S. Schätzel, hep-ex/0408049, to appear in the proceedings of the 12th International Workshop on Deep Inelastic Scattering (DIS 2004), Strbske Pleso, April 2004; H1 Collaboration, Abstract 6-0176, contributed to the 32nd International Conference on High Energy Physics (ICHEP 2004), Beijing, August 2004.
16. M. Klasen and G. Kramer, Phys. Lett. B **366**, 385 (1996).
17. C. Adloff *et al.* [H1 Collaboration], Eur. Phys. J. C **20**, 29 (2001).
18. G. A. Schuler and T. Sjöstrand, Phys. Lett. B **376**, 193 (1996).

ZEUS results on inclusive diffraction

Heuijin Lim (on behalf of the ZEUS Collaboration)

DESY Notkestr 85. D-22607 Hamburg, Germany

Abstract. Deep inelastic diffractive scattering, $ep \to e'\gamma^* p \to e'XN$, has been studied at HERA with the ZEUS detector in a wide kinematic range in the $\gamma^* p$ centre-of-mass energy W, the photon virtuality Q^2 and the mass of the system X, M_X. ZEUS results on diffraction have been obtained using the M_X method and by identifying leading protons which carry a large fraction of the incoming proton beam energy. They are presented in terms of the diffractive cross section, $d\sigma(M_X, W, Q^2)/dM_X$ and the diffractive structure function, $x_{IP}F_2^{D(3)}(\beta, x_{IP}, Q^2)$.

Keywords: diffractive deep inelastic scattering, QCD
PACS: 13.60.Hb, 12.38.-t

INTRODUCTION

In the Quantum Chromodynamics (QCD) picture, diffraction, where the proton or a low-mass nucleonic system emerges from the interaction with almost the full energy of the incident proton, is mediated by the exchange of a colour singlet carrying the quantum numbers of the vacuum, called the Pomeron, which in lowest order could be a two-quark or two-gluon system.

ZEUS diffractive results measured in two different analyses have recently been published. One uses data taken in 1998-1999 with the forward plug calorimeter (FPC) installed in the $20 \times 20\,\text{cm}^2$ beam hole of the forward uranium calorimeter [1]. The FPC increased the calorimetric coverage in the outgoing proton beam direction by about one unit in pseudorapidity and expanded the accessible range in M_X by a factor of about 1.7. It also substantially limited the contribution from the nucleon dissociation to $M_N < 2.3$ GeV. Diffractive events were selected using the M_X method which is based on the different characteristics of the M_X distributions in diffractive and non-diffractive processes. Results are presented for $\gamma^* p \to XN$, $M_N < 2.3$ GeV in the range $2.2 < Q^2 < 80$ GeV^2, $37 < W < 245$ GeV and $M_X < 35$ GeV. The other analysis uses the leading proton spectrometer (LPS) to detect the scattered protons, carrying at least 90 % of the incoming proton momentum. Based on data taken in 1997, results are presented for $\gamma^* p \to Xp$ in the range $0.03 < Q^2 < 100$ GeV^2, $25 < W < 240$ GeV and $M_X > 1.5$ GeV [2].

DIFFRACTIVE CROSS SECTION

The LPS analysis shows that the diffractive cross section falls steeply with the squared four-momentum transfer at the proton vertex, t. Fitting the distribution with a function of the form, $d\sigma_{\gamma^* p \to Xp}/dt \propto e^{b|t|}$, yields $b = 7.9 \pm 0.5(\text{stat.})^{+0.8}_{-0.5}(\text{syst.})$ GeV^{-2}. This value is compatible with b-value measured for elastic pion-proton scattering [3].

CP792, *Deep Inelastic Scattering, DIS 2005,* edited by W. H. Smith and S. R. Dasu

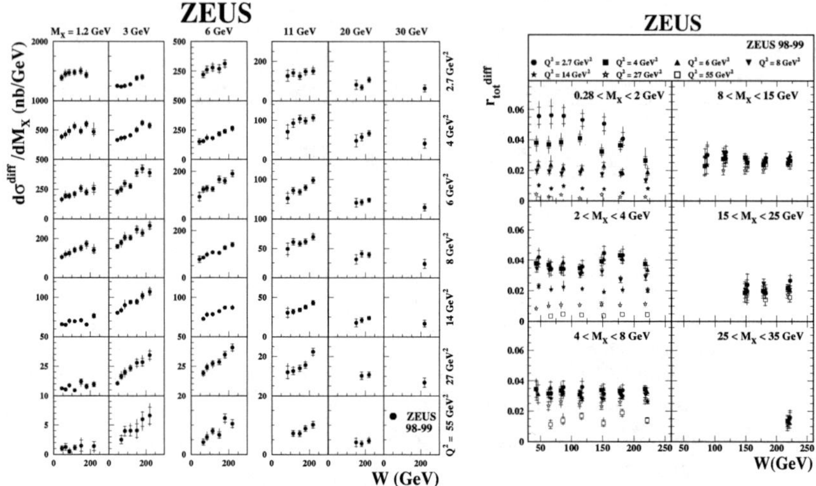

FIGURE 1. (left) Diffractive cross section from FPC. (right) Ratio $r_{\text{tot}}^{\text{diff}} = \sigma^{\text{diff}}/\sigma_{\gamma^* p}^{\text{tot}}$ from FPC.

Figure 1-(left) shows the diffractive cross section $d\sigma^{\text{diff}}_{\gamma^* p \to XN, M_N < 2.3\text{GeV}}/dM_X$ measured in the FPC analysis as a function of W for different Q^2 and M_X. For $M_X < 2$ GeV, the diffractive cross section is rather constant with W while at higher M_X, a strong rise with W is observed for all values of Q^2. This rise was quantified by fitting the data in the range $2 < M_X < 15$ GeV and in different Q^2 intervals with the form, $d\sigma^{\text{diff}}/dM_X = c(M_X, Q^2) \cdot (W/W_0)^{4(\overline{\alpha_{IP}}(Q^2)-1)}$ where $c(M_X, Q^2)$ and $\overline{\alpha_{IP}}(Q^2)$ were treated as free parameters. Assuming $d\sigma/dt \propto e^{b|t|}$ and $\alpha_{IP}(t) = \alpha_{IP}(0) + \alpha'_{IP} \cdot t$, it leads to $\alpha_{IP}(0) = \overline{\alpha_{IP}} + \alpha'_{IP}/b$. The Pomeron intercept rises by $\Delta\alpha_{IP}^{\text{diff}} = 0.0741 \pm 0.0140(\text{stat.})^{+0.0047}_{-0.0100}(\text{syst.})$ between Q^2 of 7.8 and 27 GeV^2. This establishes a Q^2 dependence of the Pomeron intercept and shows that diffractive DIS like inclusive DIS processes, cannot be interpreted as resulting from single Pomeron exchange combined with the assumption of Regge factorisation.

The ratio $r_{\text{tot}}^{\text{diff}} \equiv (\int_{M_a}^{M_b} dM_X d\sigma^{\text{diff}}_{\gamma^* p \to XN, M_N < 2.3\text{GeV}}/dM_X)/\sigma_{\gamma^* p}^{\text{tot}}$ was determined in all M_X bins and is shown in fig. 1-(right). For $M_X < 2$ GeV, $r_{\text{tot}}^{\text{diff}}$ falls with W, while at the higher M_X, it is approximately constant. It agrees with the conclusion that for $M_X < 2$ GeV, $\alpha_{IP}^{\text{diff}}(0)$ is compatible with the soft Pomeron and for larger M_X, $\alpha_{IP}^{\text{diff}}(0)$ increases with Q^2, invalidating the idea of single Pomeron exchange. The low M_X bins exhibit a strong decrease of $r_{\text{tot}}^{\text{diff}}$ with increasing Q^2, while for $M_X > 4$ GeV, this decrease becomes less dramatic and for $M_X > 8$ almost disappears. The diffractive cross section for $0.28 < M_X < 2$ GeV shows a much stronger decrease than $1/Q^2$, which is characteristic of a higher twist behavior. For $M_X > 8$ GeV, the cross section decreases as $1/Q^2$, consistent with a leading twist behavior. For the highest W bin ($200 < W < 245$ GeV), the ratio σ^{diff} ($0.28 < M_X < 35$ GeV, $M_N < 2.3$ GeV)$/\sigma^{\text{tot}}$ reaches $15.8^{+1.2}_{-1.0}$ % at $Q^2 = 4$ GeV^2 and decreases slowly with Q^2, reaching $9.6^{+0.7}_{-0.7}$ % at $Q^2 = 27$ GeV^2.

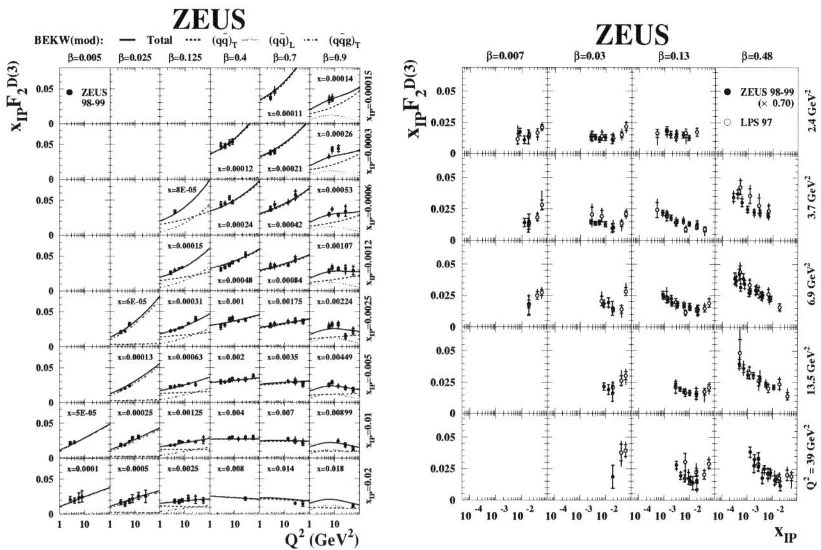

FIGURE 2. (left) $x_{IP}F_2^{D(3)}$ from the FPC analysis. The lines are the results of BEKW(mod) fit. (right) $x_{IP}F_2^{D(3)}$ from the FPC analysis compared with the results of LPS measurement.

Diffractive processes account for a substantial part of the total deep inelastic cross section.

DIFFRACTIVE STRUCTURE FUNCTION

The diffractive analogue to the proton structure function F_2, the diffractive structure function $F_2^{D(3)}$ is parametrized in terms of Q^2, the momentum fraction $x_{IP} = (M_X^2 + Q^2)/(W^2 + Q^2)$ of the proton carried by the Pomeron, and the momentum fraction $\beta = Q^2/(M_X^2 + Q^2)$ of the Pomeron carried by the struck quark. If $F_2^{D(3)}$ is interpreted in terms of quark densities, it specifies the probability to find, in a proton undergoing a diffractive reaction, a quark carrying a fraction $x = \beta x_{IP}$ of the proton momentum.

Figure 2-(left) shows $x_{IP}F_2^{D(3)}$ as a function of Q^2 for different values of β and x_{IP}. For $\beta = 0.9$, the region dominated by diffractive production of states with $M_X < 2$ GeV, $x_{IP}F_2^{D(3)}$ is constant or slowly decreasing with Q^2 due to higher twist effects in longitudinal diffractive $\gamma^* p$ scattering. For $\beta \leq 0.7$ and $x = \beta x_{IP} < 0.002$, positive scaling violations are observed presumably due to perturbative effects such as gluon emission. For fixed β, the Q^2 dependence of $x_{IP}F_2^{D(3)}$ changes with x_{IP}. This is inconsistent with the hypothesis that single Pomeron exchange is responsible for these data.

Comparison with the LPS data shows that about 30 % of the FPC cross section comes from nucleon dissociation with $M_N < 2.3$ GeV. Figure 2-(right) compares the LPS data with the FPC data multiplied by a factor of 0.7. Good agreement between two

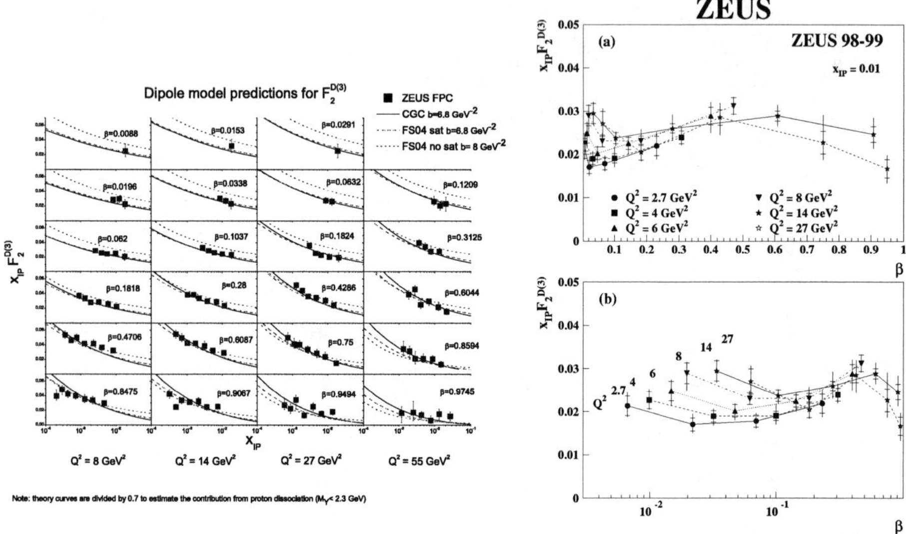

FIGURE 3. (left) $x_{IP}F_2^{D(3)}$ compared with FS04 model without saturation (dotted line), FS04 model with saturation (dashed line) and CGC model (solid line). (right) $x_{IP}F_2^{D(3)}$ for $x_{IP} = 0.01$ from FPC.

measurements is observed, with the possible exception of the region of $x_{IP} > 0.01$, where LPS data include contributions from Reggeon exchange.

The data were compared with a colour dipole model assuming that the virtual photon fluctuates into a colour dipole ($q\bar{q}$ or $q\bar{q}g$) which interacts with the proton by the exchange of a colourless object. The BEKW(mod) fit describes the data well (fig. 2-(left)). As shown in fig. 3-(left), the Forshaw and Shaw (FS04) model without saturation overestimates the diffractive contribution somewhat for the low β region where the contribution of gluon is dominant. The FS04 model with saturation and and the Colour Glass Condensate (CGC) model give good descriptions of the data [4].

Figure 3-(right) shows $x_{IP}F_2^{D(3)}$ for fixed $x_{IP} \equiv x_0 = 0.01$ as a function of β for different values of Q^2. The $x_{IP}F_2^{D(3)}$ have a maximum near $\beta = 0.5$, consistent with a $\beta(1-\beta)$ variation which is explained in the dipole models of diffraction by $\gamma^* \to q\bar{q}$ splitting and two-gluon exchange. The rise of $x_{IP}F_2^{D(3)}$ as $\beta \to 0$ and its increase with Q^2 is reminiscent of the logarithmic scaling violations observed in F_2 at low x, which arises from QCD evolution.

REFERENCES

1. ZEUS Collaboration; S. Chekanov *et al.*, Nucl. Phys. **B 713**, 3 (2005).
2. ZEUS Collaboration; S. Chekanov *et al.*, Eur. Phys. J. **C 38**, 43 (2004).
3. K. Goulianos, Phys. Rep. **101**, 169 (1983).
4. J. R. Forshaw and G. Shaw, JHEP 0412 (2004) 052 and hep-ph/0411337.

H1 F_2^D and Diffractive Charged Current Results

Paul Laycock

FH1 Liverpool 10b, DESY, Notkestrasse 85, 22603 Hamburg, Germany
email: laycock@mail.desy.de

Abstract. The high precision measurements of inclusive diffractive deep inelastic scattering are presented with the NLO DGLAP QCD fits to this data. The first differential measurements of the diffractive charged current cross-section are also presented and compared with the predictions of the fit to the neutral current data.

Keywords: Diffraction
PACS: 13

INTRODUCTION

At H1 inclusive diffractive DIS events are selected on the basis of a large rapidity gap being present in the final state due to the diffractive exchange [1]. Figure 1 shows a diagram of the generic diffractive DIS process at HERA. The kinematics of DIS are given in equation 1.

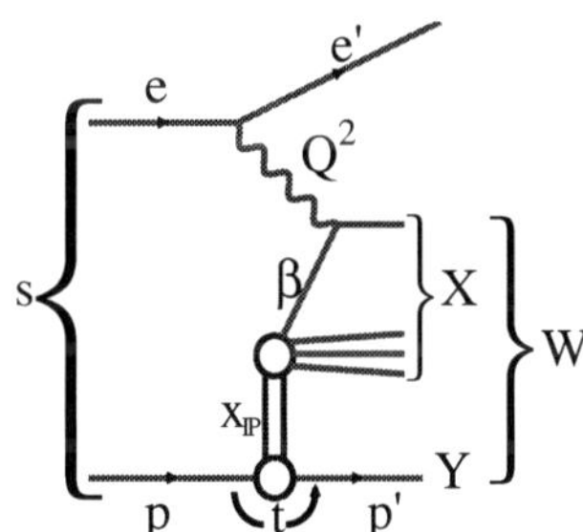

FIGURE 1. A schematic of the generic diffractive DIS process.

$$Q^2 = -q^2 = -(k-k')^2,\ x = \frac{Q^2}{2p\cdot q},\ y = \frac{p\cdot q}{p\cdot k} \tag{1}$$

Here Q^2 is the virtuality of the exchange boson, x is the Bjorken scaling variable and y is the inelasticity. They are related to the ep centre of mass energy $\sqrt{s}$ by the equation $Q^2 = sxy$. In addition to these standard DIS variables and the Mandelstam variables (t,s) the kinematic variables x_{IP} and β are useful in describing the diffractive DIS interaction. They are defined as:

$$\beta = \frac{Q^2}{Q^2 + M_X^2 - t}, \qquad x_{IP} = \frac{Q^2 + M_X^2 - t}{Q^2 + W^2} \tag{2}$$

CP792, *Deep Inelastic Scattering, DIS 2005,* edited by W. H. Smith and S. R. Dasu

where M_X is the invariant mass of the hadronic system X and $W^2 = (q+p)^2$ is the square of the centre of mass of the photon-proton system. Assuming that the diffractive exchange can be attributed to a QCD object, i.e. a pomeron, then x_{IP} is the fractional momentum of the pomeron with respect to the proton and β is the fractional momentum of the struck parton with respect to the pomeron.

The diffractive reduced cross-section $\sigma_r^{D(3)}(\beta, Q^2, x_{IP})$ can be defined using:

$$\frac{d^3\sigma_{ep\to eXp}}{d\beta dQ^2 dx_{IP}} = \frac{4\pi\alpha_{em}^2}{\beta Q^4}\left(1-y+\frac{y^2}{2}\right)\sigma_r^{D(3)}(\beta, Q^2, x_{IP}) \tag{3}$$

and can be related to the structure functions by:

$$\sigma_r^{D(3)}(\beta, Q^2, x_{IP}) = F_2^{D(3)}(\beta, Q^2, x_{IP}) - \frac{y^2}{1+(1-y)^2}F_L^{D(3)}(\beta, Q^2, x_{IP}). \tag{4}$$

and for most of the measured phase space $\sigma_r^{D(3)} = F_2^{D(3)}$ is a good approximation.

Measurements of the inclusive diffractive DIS process now cover most of the accessible kinematic range of HERA, as shown in figure 2. Also shown in figure 2 is the good agreement between the cross-sections obtained using the large rapidity gap method and measurements made by H1 [1] and ZEUS [2] where the complete final state, including the elastically scattered proton, is detected. The agreement is seen to be very good across the overlapping kinematic range[1].

Factorisation and NLO QCD

It has been proven by Collins [3] that the diffractive $\gamma^* p$ cross-section can be written in terms of diffractive PDFs, $q(x, Q^2, x_{IP}, t)$, which are dependent on four kinematic variables, convoluted with a hard-scattering cross-section:

$$\sigma(\gamma^* p \to Xp) \;\sim\; q(x_{IP}, t, x, Q^2) \;\otimes\; \hat{\sigma}_{\gamma^* q}(x, Q^2) \tag{5}$$

At fixed x_{IP} and t these PDFs will evolve with Q^2 and x in exactly the same way as the inclusive proton PDFs. Such a proof allows a full QCD fit to the data with no additional assumptions. At the present time the necessary technical knowledge is as yet unavailable to make such a fit. Instead the H1 Collaboration proceed via an assumption supported by the data that the x_{IP} dependence of the data can be modelled using a Regge-motivated parameterisation [1] (flux factor). This allows the extraction of the diffractive PDFs shown in figure 3. Also shown in figure 3 is the diffractive reduced cross-section divided by the flux factor as a function of β for different Q^2 and x_{IP} bins. The agreement between overlaid x_{IP} bins supports the ansatz that the flux factor contains all of the x_{IP} dependence of the data.

---

[1] A factor of 1.1 is applied to the elastic proton data to scale it to the same M_Y range as the large rapidity gap data

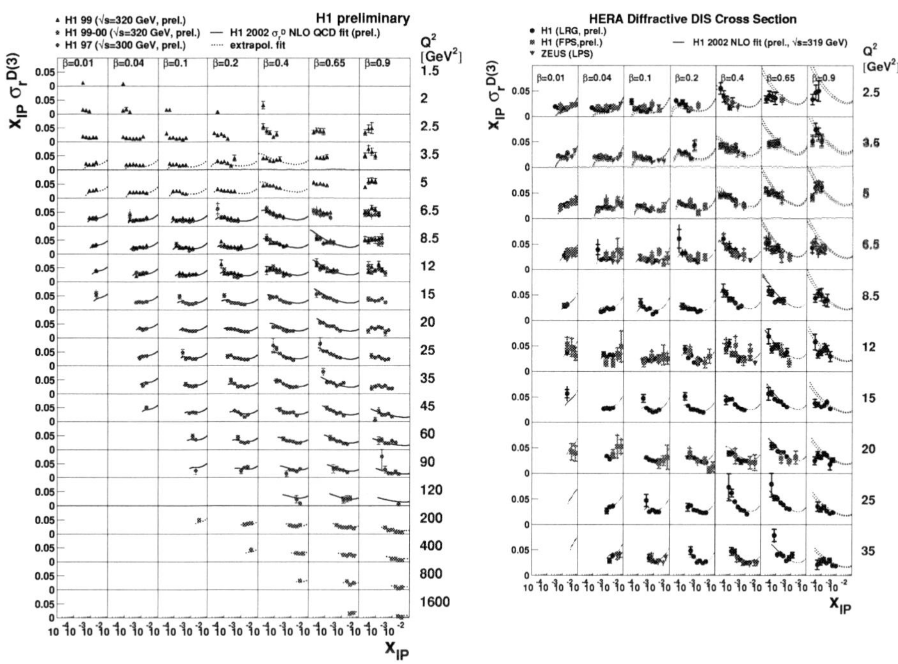

FIGURE 2. Measurements from H1 of the inclusive diffractive cross-section across the full kinematic range at HERA (left) and (right) a comparison of the large rapidity gap data with elastic proton data from both H1 and ZEUS; good ageement is seen between the two experimental techniques.

DIFFERENTIAL DIFFRACTIVE CHARGED CURRENT CROSS-SECTIONS

The large rapidity gap method can also be applied to charged current data to select a sample of events with a large rapidity gap in the final state [4]. The total cross-section for this process, for the kinematic range $Q^2 > 200\,\mathrm{GeV}^2$, $y < 0.9$ and $x_{IP} < 0.05$ is $0.42 \pm 0.13(\mathrm{stat}) \pm 0.09(\mathrm{sys})$pb, in good agreement with the prediction of 0.43pb obtained from the NLO DGLAP QCD fit to neutral current data described above. The ratio of this diffractive charged current process to the inclusive charged current process (for $x_{Bjorken} < 0.05$) for the kinematic range defined above is $2.5 \pm 0.8 \pm 0.6\%$.

The differential cross-section as a function of x_{IP} is shown for this event class in figure 4 compared to the prediction of the fit. Figure 4 also shows the differential cross-sections as a function of Q^2 and β. Reasonable agreement is seen between the data and the fit prediction for all three kinematic variables.

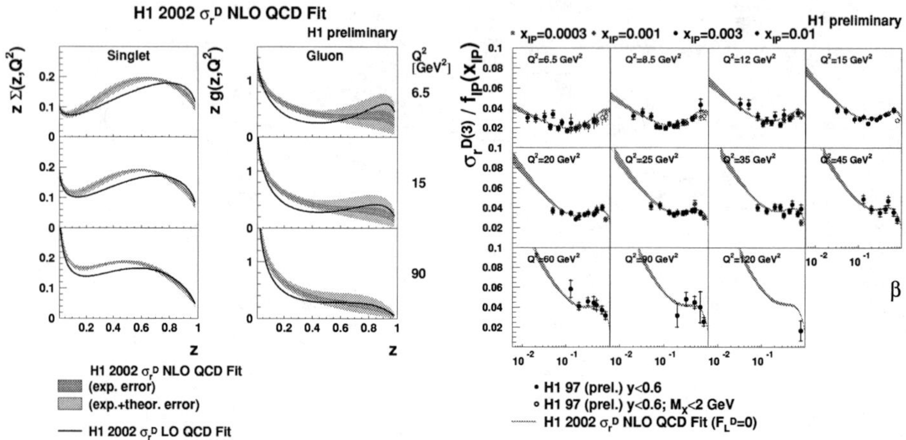

FIGURE 3. Diffractive PDFs extracted by H1 from their inclusive cross-section measurements (left) and (right) the diffractive reduced cross-section divided by the flux factor as a function of β for different Q^2 and x_{IP} bins.

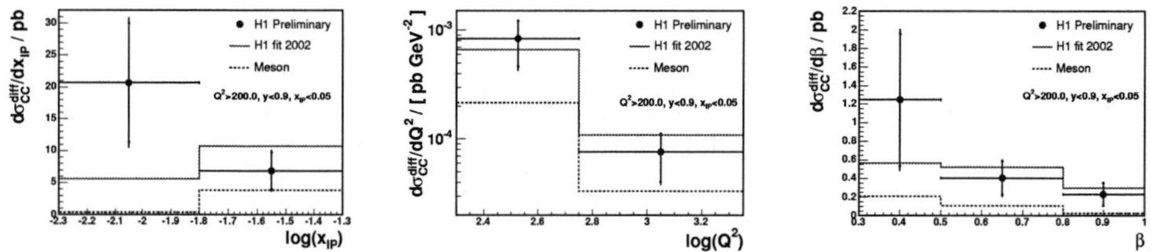

FIGURE 4. The Differential Diffractive Charged Current Cross-section as a function of x_{IP}, Q^2 and β

REFERENCES

1. H1 Collab., Adloff, C. et al, Measurement and NLO DGLAP QCD Interpretation of Diffractive Deep-Inelastic Scattering at HERA, Submitted to ICHEP 2002, Abstract 980 (2002).
2. Study of Deep Inelastic Inclusive and Diffractive Scattering with the ZEUS Forward Plug Calorimeter. ZEUS Collab. (S. Chekanov et al.). DESY-05-011, Jan 2005. 87pp. Published in Nucl.Phys.B713:3-80,2005
3. John C. Collins, Proof of Factorization for Diffractive Hard Scattering, Phys. Rev., D57, 3051-3056 (1998).
4. H1 Collab., Adloff, C. et al, Measurement of the Diffractive Cross Section in Charged Current Interactions at HERA, Submitted to ICHEP2004.

ZEUS results on rapidity gap events in charged and neutral current processes at large Q^2

Leszek Adamczyk (on behalf of the ZEUS Collaboration)

AGH - University of Science and Technology, Al. Mickiewicza 30, 30-059 Cracow

Abstract. The observation of large rapidity gap(LRG) events in charged and neutral current deep inelastic scattering using e^+p data collected with the ZEUS detector is reported. These events are compared with phenomenological models. The ratio of LRG to inclusive cross sections have been determined at $Q^2 > 200$ GeV^2, where Q^2 is the negative square of the four-momentum of the exchanged boson, both for neutral and charged current processes.

Keywords: charrged current DIS, diffraction, large rapidity gap, neutral curent DIS
PACS: 10

INTRODUCTION

A significant fraction of the neutral current (NC) deep inelasitic scattering (DIS) measured at HERA was found to result from diffractive processes [1, 2]. The exchange of colour singlet leads to formation of a large rapidity gap (LRG) in the hadronic final state, located between the exchanged boson and the proton fragmentation regions.

ZEUS has observed LRG events also in charged current (CC) DIS, $e^+p \to \bar{\nu}_e XY$, at $Q^2 > 200$ GeV^2 [3]. This analysis presents the corresponding measurement for NC DIS, $e^+p \to e^+XY$, in the same kinematic region. Both CC and NC results are based on the same data sample collected during the running periods of 1999 and 2000 when HERA collided 27.5 GeV positrons with 920 GeV protons, yielding a centre-of-mass energy of $\sqrt{s} = 318$ GeV.

MONTE CARLO MODELS AND DATA SELECTION

Diffractive NC and CC events were modelled with the RAPGAP 2.08/06 [4] generator, in which colour singlet exchange occurs between virtual boson and proton.

Non-diffractive NC and CC events were produced with the DJANGOH 1.1 [5] generator interfaced to the colour-dipole model of ARIADNE 4.10 [6] for the fragmentation.

Soft Colour Interaction (SCI) model interfaced with the MEPS version of LEPTO 6.5 [7] gives rise to LRG events without introducing the concept of diffractive exchange. This model was used instead of the combination of the ARIADNE and RAPGAP samples.

The reconstruction and selection criteria for CC and NC event candidates are identical to those of inclusive CC [8] and NC [9] measurements. The kinematic requirements $Q^2 > 200$ GeV^2, $x_{\mathrm{Bj}} < 0.05$ and $y < 0.9$ were imposed, where x_{Bj} is the Bjorken scaling

CP792, *Deep Inelastic Scattering, DIS 2005*, edited by W. H. Smith and S. R. Dasu

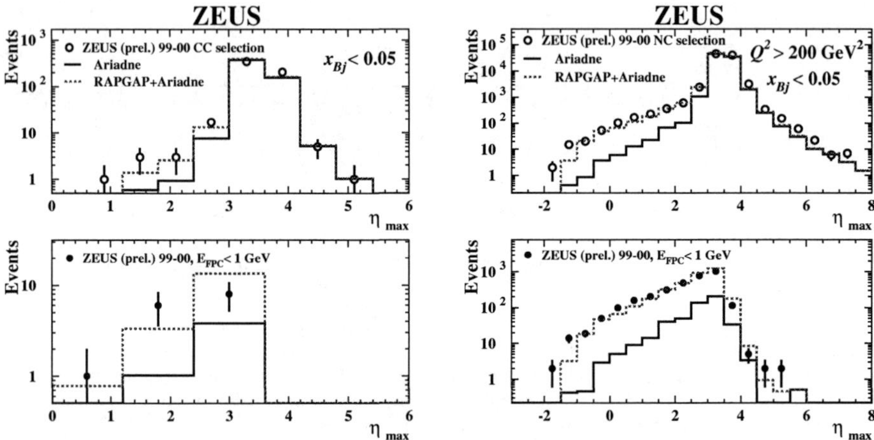

FIGURE 1. The distribution of $\eta_{\rm max}$ is shown in the upper left(right) plot for CC(NC) DIS events in the kinematic range $Q^2 > 200$ GeV^2, $y < 0.9$ and $x_{\rm Bj} < 0.05$. The circles are the data points, the solid histogram is the result of the non-diffractive (ARIADNE) MC simulation and the dashed histogram is that of the sum of non-diffractive and diffractive (RAPGAP) MC sample. The bottom plots show the same distributions with an additional requirement of $E_{\rm FPC} < 1$ GeV.

variable, and y is the inelasticity parameter. To observe events with a LRG, the variable $\eta_{\rm max}$, defined as the pseudorapidity of the energy deposit in the calorimeter closest to the proton direction, was considered. The distributions of $\eta_{\rm max}$ for the CC and NC candidates are shown in Fig. 1. The following criteria were applied to both the NC and CC samples to select LRG events:

- $E_{\rm FPC} < 1.0$ GeV, where $E_{\rm FPC}$ is the energy deposited in the forward plug calorimeter (FPC). The FPC increased the forward calorimeter coverage by about 1 unit of pseudo-rapidity to $\eta < 5$;
- $\eta_{\rm max} < 2.9$. This cut together with $E_{\rm FPC} < 1.0$ GeV cut ensures a gap of at least two units of rapidity;
- $x_{\rm IP} < 0.05$, where $x_{\rm IP}$ is the fraction of proton four-momentum carried by the exchanged colour singlet (Pomeron).

RESULTS

Figure 2 shows the distributions of relevant variables for CC and NC DIS events for data and the sum of non-diffractive (ARIADNE) and diffractive (RAPGAP) MC simulations. The MC simulations describe well both the inclusive sample and the LRG subsample for both the CC and NC.

Figure 3 shows the distributions of relevant variables for CC and NC DIS events together with the results of the MEPS with SCI included MC simulation. The rate of

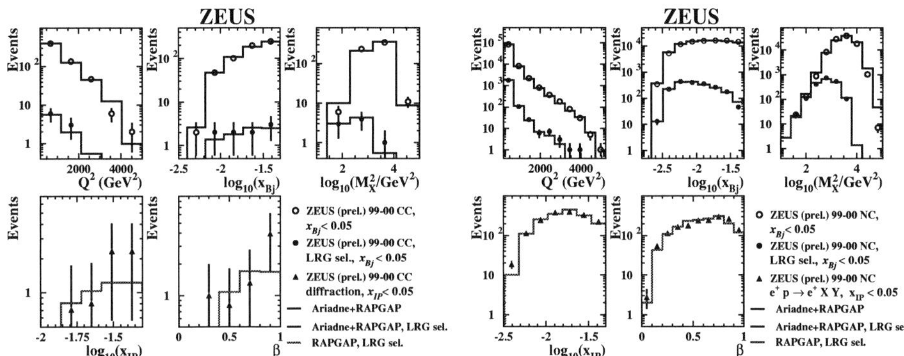

FIGURE 2. The upper left(right) plots show the distributions of Q^2, $\log x_{\mathrm{Bj}}$ and $\log M_X^2$, where M_X is the mass of the hadronic system X measured in the main detector, for CC(NC) DIS events in the kinematic range of $Q^2 > 200$ GeV^2, $y < 0.9$ and $x_{\mathrm{Bj}} < 0.05$. The circles are the data points and the solid histograms are the sum of results of non-diffractive (ARIADNE) and diffractive (RAPGAP) MC simulations. The lower data points and MC histograms on each plot correspond to the subsamples satisfying the LRG selection criteria. The bottom plots show the $\log x_{\mathrm{IP}}$ and β, distributions of the DIS events satisfying the LRG selection criteria. The triangles are the data points after statistical subtraction of non-diffractive (ARIADNE) contribution. The histogram is the result of diffractive (RAPGAP) MC simulation.

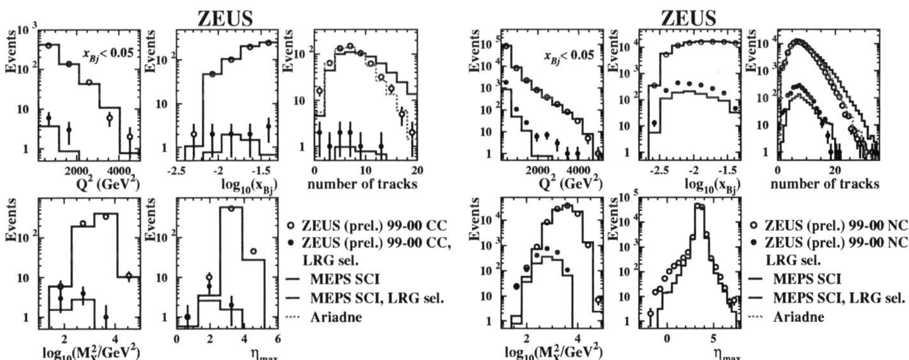

FIGURE 3. The left(right) plots show the distributions of Q^2, $\log x_{\mathrm{Bj}}$, number of tracks, $\log M_X^2$ and η_{max} for CC(NC) DIS events in the kinematic range $Q^2 > 200$ GeV^2, $y < 0.9$ and $x_{\mathrm{Bj}} < 0.05$. The circles are the data points and the solid histograms are the results of MEPS with SCI included MC simulations. The dashed histogram superimposed on the distribution of the number of tracks is the result of ARIADNE simulation. The lower data points and MC histograms on each plot correspond to the subsamples satisfying the LRG selection criteria.

events with a LRG expected by this MC model is smaller than observed in the data.

The ratio of LRG cross section to the total cross section was measured in the kinematic region of $Q^2 > 200$ GeV^2 and $x_{\mathrm{Bj}} < 0.05$ for both NC and CC processes. These ratios were found to be compatible as shown in Fig. 4.

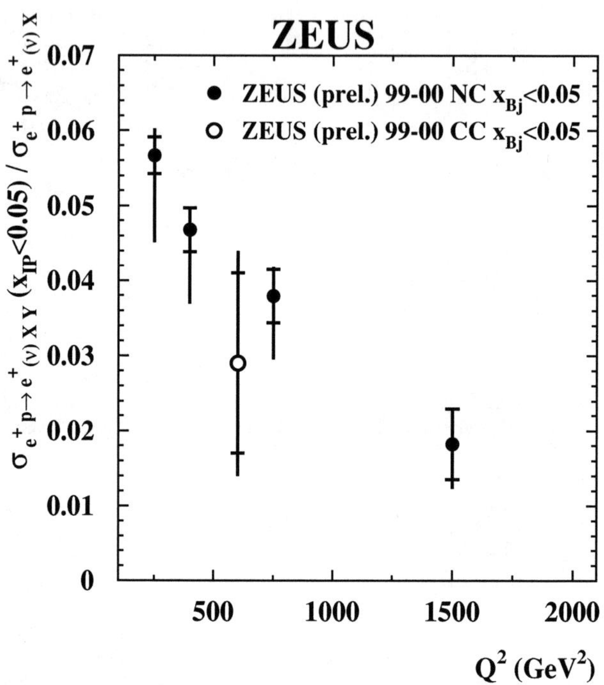

FIGURE 4. The ratio of the LRG cross section to the total cross section measured in the kinematic region $x_{\mathrm{Bj}} < 0.05$ as a function of Q^2 is presented both for NC and CC processes.

REFERENCES

1. ZEUS Coll., M. Derrick et al., Phys. Lett. **B 315**, 481 (1993)
2. H1 Coll., T. Ahmed et al., Nucl. Phys. **B 429**, 477 (1994)
3. ZEUS Coll., *Observation of Large Rapidity Gap Events in Charged Current High Q^2 DIS at HERA*. Abstract 229, International Conference on High Energy Physics, Beijing, China, 2004
4. H. Jung, Comp. Phys. Comm. **86**, 147 (1995)
5. H. Spiesberger, HERACLES *and* DJANGOH*: Event Generation for ep Interactions at HERA Including Radiative Processes*, 1998, available on http://www.desy.de/~ hspiesb/djangoh.html
6. L. Lönnblad, Comp. Phys. Comm. **71**, 15 (1992)
7. G. Ingelman, A. Edin and J. Rathsman, Comp. Phys. Comm. **101**, 108 (1997)
8. ZEUS Coll., S. Chekanov et al., Eur. Phys. J. **C 32**, 1 (2003)
9. ZEUS Coll.; S. Chekanov et al., Phys. Rev. **D 70** (2004)

The Pomeron Structure and Diffractive Parton Distributions

Halina Abramowicz[1], Michael Groys and Aharon Levy

School of Physics and Astronomy, Raymond and Beverly Sackler Faculty of Exact Sciences, Tel Aviv University, Tel Aviv, Israel

Abstract. Measurements of the diffractive structure function, F_2^D, of the proton at HERA are used to extract the partonic structure of the Pomeron. Regge Factorization is tested and is found to describe well the existing data within the selected kinematic range. The analysis is based on the next to leading order QCD evolution equations. The results obtained from various data sets are compared. An analysis of the uncertainties in determining the parton distributions is provided. The probability of diffraction is calculated using the obtained results.

Keywords: Diffractive DIS, diffractive structure functions, Pomeron structure, diffractive parton distribution functions, Regge factorization
PACS: 13.60.Hb, 12.38.Bx, 12.38.Qk, 12.40.Nn

INTRODUCTION

In the last 10 years a large amount of diffractive data was accumulated at the HERA collider [1, 2, 3]. There are three methods used at HERA to select diffractive events. One uses the Leading Proton Spectrometer (LPS) [3] to detect the scattered proton and by choosing the kinematic region where the scattered proton looses very little of its initial longitudinal energy, it ensures that the event was diffractive. A second method [2] simply requests a large rapidity gap (LRG) in the event and fits the data to contributions coming from Pomeron and Reggeon exchange. The third method [1] relies on the distribution of the mass of the hadronic system seen in the detector, M_X, to isolate diffractive events and makes use of the Forward Plug Calorimeter (FPC) to maximize the phase space coverage. We will refer to these three as ZEUS LPS, H1 and ZUES FPC methods.

The experiments [4, 5, 6] provide sets of results for inclusive diffractive structure function, $x_{I\!P}F_2^{D(3)}$, in different regions of phase space. In extracting the initial Pomeron parton distribution functions (pdfs), the data are fitted assuming the validity of Regge factorization.

In the present study, Regge factorization is tested. New fits, based on a NLO QCD analysis, are provided and include the contribution of the longitudinal structure function. The obtained PDFs are systematically analyzed. A comparison of the different experimental data sets is provided. Additional quantities derived from the fit results are also presented.

In order to make sure that diffractive processes are selected, a cut of $x_{I\!P} < 0.01$ was

---

[1] also at Max Planck Institute, Munich, Germany, Alexander von Humboldt Research Award.

CP792, *Deep Inelastic Scattering, DIS 2005,* edited by W. H. Smith and S. R. Dasu

performed, where $x_{I\!P}$ is the fraction of the proton momentum carried by the Pomeron. It was shown [7] that this cuts ensures the dominance of Pomeron exchange. In addition, a cut of $Q^2 > 3$ GeV^2 was performed on the exchanged photon virtuality for applying the NLO analysis. Finally, a cut on $M_X > 2$ GeV was used so as to exclude the light vector meson production.

REGGE FACTORIZATION

The Regge Factorization assumption can be reduced to the following,

$$F_2^{D(4)}(x_{I\!P}, t, \beta, Q^2) = f(x_{I\!P}, t) \cdot F(\beta, Q^2), \tag{1}$$

where $f(x_{I\!P}, t)$ represents the Pomeron flux which is assumed to be independent of β and Q^2 and $F(\beta, Q^2)$ represents the Pomeron structure and is β and Q^2 dependent. In order to test this assumption, we check whether the flux $f(x_{I\!P}, t)$ is indeed independent of β and Q^2 on the basis of the available experimental data.

The flux is assumed to have a form $\sim x_{I\!P}^{-A}$ (after integrating over t which is not measured in the data) . A fit of this form to the data was performed in different Q^2 intervals, for the whole β range, and for different β intervals for the whole Q^2 range.

Figure 1 shows the Q^2 dependence of the exponent A for all three data sets, with the $x_{I\!P}$ and M_X cuts as described in the introduction. The H1 and the LPS data show no Q^2 dependence. The ZEUS FPC data show a small increase in A at the higher Q^2 region. It

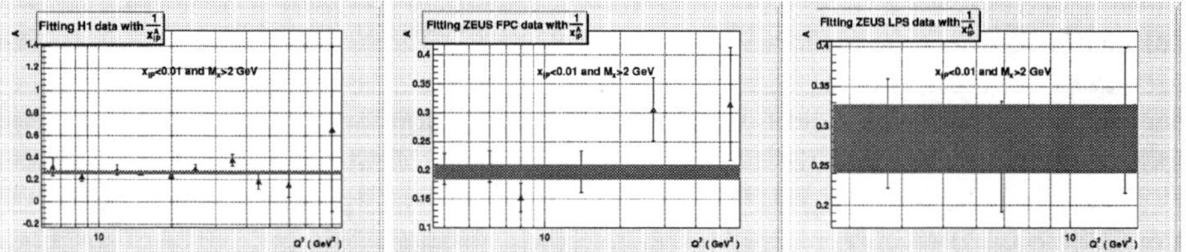

FIGURE 1. A as a function of Q^2 for $x_{I\!P} < 0.01$ and $M_X > 2$ GeV, for the three data sets, as indicated in the figure. The line corresponds to a fit over the whole Q^2 region

should be noted that while for the H1 and LPS data, releasing the $x_{I\!P}$ cut to 0.03 seems to have no effect, the deviation of the ZEUS FPC data from a flat dependence increases from a 2.4 standard deviation (s.d.) to a 4.2 s.d. effect (not shown).

The β dependence of A is shown in figure 2. All three data sets seem to show no β dependence, within the errors of the data. Note however, that by releasing the $x_{I\!P}$ cut to higher values, a strong dependence of the flux on β is observed (not shown).

We thus conclude that for $x_{I\!P} < 0.01$, the Pomeron flux seems to be independent of Q^2 and of β and thus the Regge factorization hypothesis holds.

NLO QCD FITS

We parameterized the parton distribution functions of the Pomeron at $Q_0^2 = 3$ GeV^2 in a simple form of $Ax^b(1-x)^c$ for u and d quarks (and anti-quarks) and set all other quarks

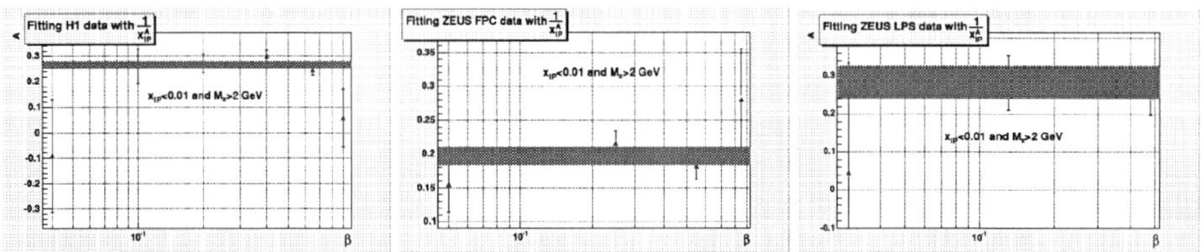

FIGURE 2. A as a function of β for $x_{I\!P} < 0.01$ and $M_X > 2$ GeV, for the three data sets, as indicated in the figure. The line corresponds to a fit over the whole β region

to zero at the initial scale. The gluon distribution was also assumed to have the same mathematical form. We thus had 3 parameters for quarks, 3 for gluons and an additional parameter for the flux, expressed in terms of the Pomeron intercept $\alpha_{I\!P}(0)$. Each data set was fitted to 7 parameters and a good fit was achieved for each. The H1 and ZEUS FPC had χ^2/df ≈ 1, while for the LPS data, the obtained value was 0.5. The data together with the results of the fits are shown in figure 3. The following values were obtained for $\alpha_{I\!P}(0)$,

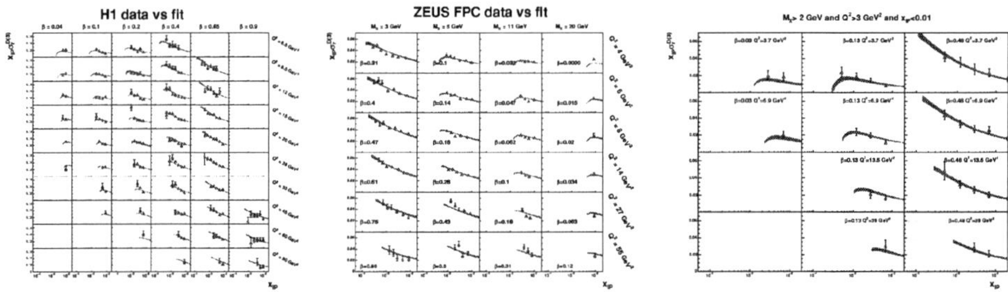

FIGURE 3. The diffractive reduced cross section of the proton multiplies by $x_{I\!P}$, as a function of $x_{I\!P}$ for the different data sets (the most right plot is for the LPS data) in different bins of Q^2 and β, as indicated in the figure. The bands are the results of the fits including uncertainties.

for each of the three data sets: $\alpha_{I\!P}(0) = 1.138 \pm 0.011$, for the ZEUS FPC data, $\alpha_{I\!P}(0) = 1.189 \pm 0.020$, for the ZEUS LPS data, $\alpha_{I\!P}(0) = 1.178 \pm 0.007$, for the H1 data.

The parton distribution functions are shown in figure 4 for the H1 and the ZEUS FPC data points. Because of the limited β range covered by the LPS data, the resulting pdfs uncertainties are large and are not shown here. For the H1 fit one sees the dominance of the gluons in all the β range. For the ZEUS FPC data, the quark constituent of the Pomeron dominates at high β while gluons dominate at low β. We can quantify this by calculating the Pomeron momentum carried by the gluons. Using the fit results one gets for the H1 data 80-90%. while for the ZEUS FPC data, 55-65%.

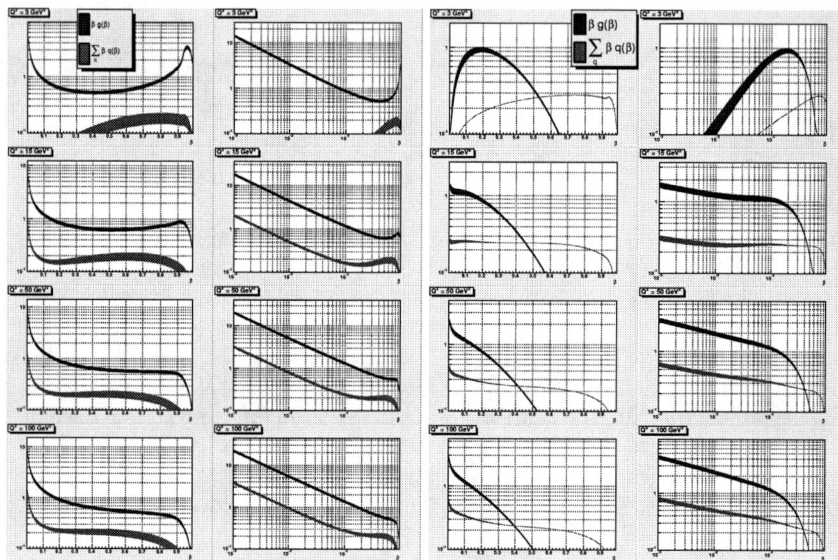

FIGURE 4. Quark and gluon pdfs of the Pomeron as obtained from the H1 data fit (left two figures) and from the ZEUS FPC data fit (two rightmost figures) as a function of β, at different values of Q^2.

COMPARISON OF THE DATA SETS

One way of checking the compatibility of all three data sets is to make an overall fit for the whole data sample. Since the coverage of the β range in the LPS data is limited, we compare only the H1 and the ZEUS FPC data. A fit with a relative overall scaling factor of the two data sets failed. Using the fit results of one data sets superimposed on the other shows that the fit can describe some kinematic regions, while failing in other bins. This leads to the conclusion that there seems to be some incompatibility between the two data sets.

PROBABILITY OF DIFFRACTION

It is of interest to calculate the probability that a certain parton is produced in a diffractive process [8]. The probability of diffraction on quarks and gluons, as a function of Bjorken x at different values of Q^2 are shown in figure 5, using the results of the H1 and the ZEUS FPC data fits. The ZEUS FPC data shows that throughout the whole kinematic range shown in the figures, the probability for diffraction is not bigger than 0.15, far from the Pumplin [9] limit of 0.5. This is not the case for the H1 data for which, at small x and low Q^2, the probability of diffraction induced by gluons becomes greater than 0.5 and thus unphysical. Note however that the results for $x < 2 \cdot 10^{-4}$ are in a region where H1 has no data and thus the calculated probability in this region is an extrapolation based on the fit parameters. In order to get physical results, some process, like saturation, must

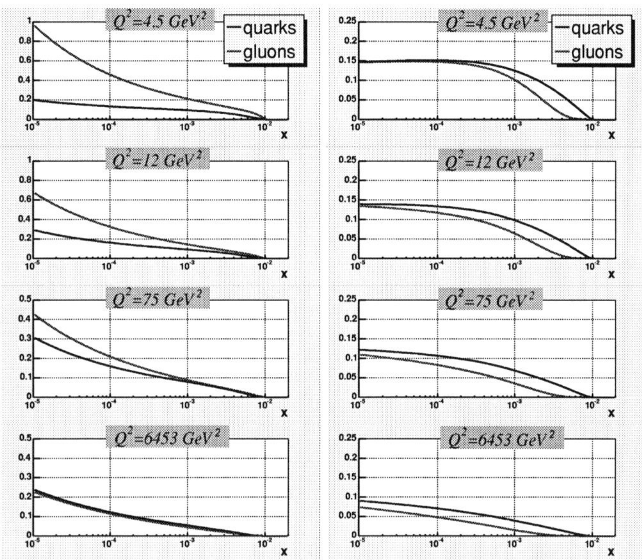

FIGURE 5. Probability of diffraction as a function of x, at different values of Q^2, calculated from the results of the H1 data fit (left figure) and from those of the ZEUS FPC data fit (right figure).

lower the expected value.

ACKNOWLEDGMENTS

We would like to thank Prof. John Collins for providing the program to calculate the NLO QCD equations for the diffractive data. This work was supported in part by the Israel Science Foundation (ISF).

REFERENCES

1. ZEUS Collab., M. Derrick et al., *Phys. Lett.* **315** (1993) 481; J. Breitweg et al., *Eur. Phys. J.* **C6** (1999) 43.
2. H1 Collab., C. Adloff et al., *Zeit. Phys.* **C76** (1997) 613.
3. ZEUS Collab., S. Chekanov et al., *Eur. Phys. J.* **C25** (2002) 169.
4. ZEUS Collaboration, S. Chekanov et al., *Nucl. Phys.* **B713** (2005) 3.
5. H1 Collaboration, "Measurement and NLO DGLAP QCD Interpretation of Diffractive Deep-Inelastic Scattering at HERA," paper 089 submitted to EPS 2003, Aachen.
6. ZEUS Collaboration, S. Chekanov et al., *Eur. Phys. J.* **C38** (2004) 43.
7. K. Golec-Biernat, J. Kwiecinski and A. Szczurek, *Phys. Rev.* **D56** (1997) 3955.
8. L. Frankfurt and M. Strikman, "Future small x physics with *ep* and *eA* colliders," hep-ph/9907221.
9. J. Pumplin, *Phys. Rev.* **D8** (1973) 2899.

HERA Diffractive Structure Function Data and Parton Distributions

Paul Laycock*, Paul Newman† and Frank-Peter Schilling**

*FH1 Liverpool 10b, DESY, Notkestrasse 85, 22603 Hamburg, Germany
email: laycock@mail.desy.de
†School of Physics and Astronomy, University of Birmingham, B15 2TT, UK
email: newmanpr@mail.desy.de
**CERN/PH-CMT (CMS Collaboration) Office 40-1-B06, mail E27710, CH-1211 Geneva 23
email:fpschill@mail.cern.ch

Abstract. The high precision diffractive DIS data from the H1 and ZEUS collaborations discussed elsewhere in these proceedings are compared. NLO DGLAP QCD fits are performed separately to the H1 and ZEUS data samples and the resulting diffractive PDFs are compared.

Keywords: Diffraction
PACS: 13

INTRODUCTION

The H1 collaboration has extracted diffractive PDFs from neutral current data as discussed in [1, 2]. Further to the proof of Collins [3] that the process $\gamma^* p \to Xp$ is factorisable an additional assumption is required. This assumption, which is supported by the data, is that the x_{IP} dependence of the data can be modelled using a Regge-motivated parameterisation [2]. Following this assumption the diffractive PDFs shown in figure 1 can be extracted. Also shown in figure 1 is the diffractive dijet cross-section compared with the predictions of a leading order Monte Carlo using the H1 diffractive PDFs. The prediction does rather well.

COMPARISON OF DIFFRACTIVE DIS DATA

The different experimental techniques for selecting diffractive DIS events imply different kinematic ranges for the various datasets considered here, in particular the range of M_Y varies. In order to compare the various datasets they have all been corrected to the same phase space as the H1 large rapidity gap measurements [2, 4, 5], i.e. $M_Y < 1.6$ GeV. The H1 [6] and ZEUS [7] leading proton data are scaled up by a global factor of 1.1 [8]. A factor of 0.7 [9] is used to correct the ZEUS M_X data [9] from the measured $M_Y < 2.3$ GeV to an elastic proton. The same factor of 1.1 is then used to correct from an elastic proton to $M_Y < 1.6$ GeV, resulting in an overall global scale factor of 0.77 being applied to the ZEUS M_X data.

Shown in figure 2 is a comparison of the H1 large rapidity gap data and the ZEUS M_X data. There is in general good agreement but differences are observed at low M_X (high

CP792, *Deep Inelastic Scattering, DIS 2005*, edited by W. H. Smith and S. R. Dasu

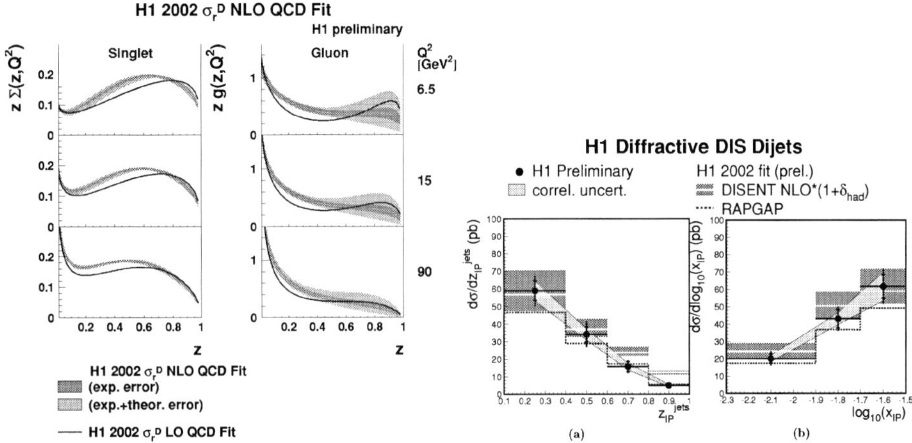

FIGURE 1. Diffractive PDFs extracted by H1 from their inclusive cross-section measurements (left) and (right) the diffractive dijet cross-section compared with the predictions using the H1 PDFs.

β) and in the Q^2 dependence of the data. Figure 3 is a comparison of the H1 rapidity gap measurements and the two leading proton measurements. The two leading proton measurements agree well with each other. There is also good agreement between the leading proton analyses and the H1 large rapidity gap measurement.

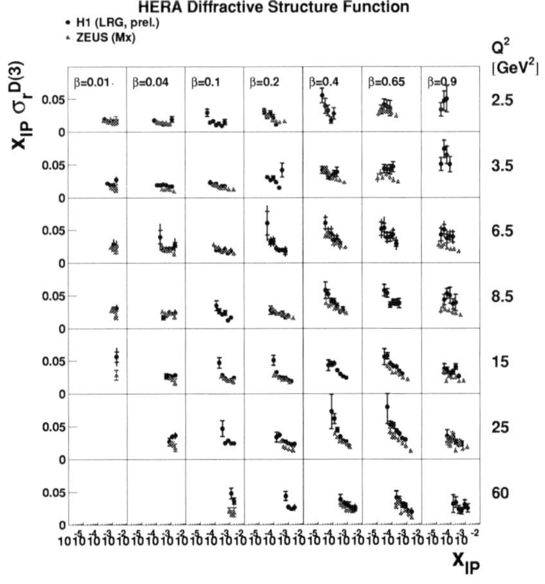

FIGURE 2. A comparison of the H1 large rapidity gap measurement and the ZEUS M_X data.

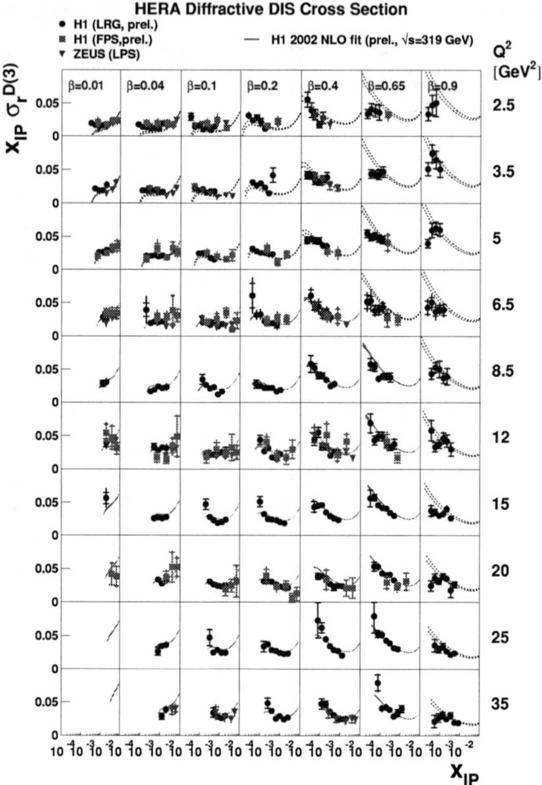

FIGURE 3. A comparison of the H1 large rapidity gap measurement and the leading proton analyses of H1 and ZEUS.

NLO DGLAP QCD FIT COMPARISONS

The same NLO DGLAP QCD fit procedure as described in [2] was used to fit the ZEUS M_X data, with only a few minor modifications. The modifications were:

- All M_x data for $Q^2 > 4\,\mathrm{GeV}^2$ were included in the fit (H1: $Q^2 > 6.5\,\mathrm{GeV}^2$)
- Only the total error of the data was considered
- No meson component was included in the fit
- The Pomeron intercept was fitted at the same time as the PDFs

and otherwise the fit was the same as for the H1 large rapidity gap data. The resulting fit gives a good desciption of the data.

Figure 4 shows a comparison of the two sets of diffractive PDFs extracted from the H1 and ZEUS fits. The two singlet distributions are similar at low Q^2, evolving differently to higher Q^2. As a result the ZEUS gluon is a factor of ≈ 2 smaller than the H1 gluon.

The different Q^2 evolution of the H1 and ZEUS datasets results in the differences in the extracted PDFs. Note that the fits remain largely unaffected by the differences seen at low M_X between the two datasets because of a cut of $M_X > 2$ GeV.

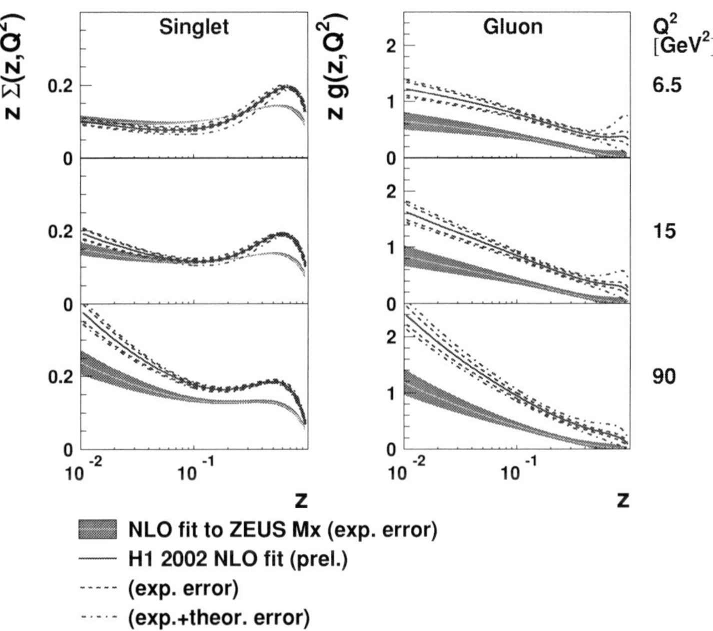

FIGURE 4. A comparison of the diffractive PDFs extracted from the NLO DGLAP QCD fits to H1 and ZEUS data.

REFERENCES

1. H1 F_2^D and Diffractive Charged Current Results, Paul Laycock, these proceedings.
2. H1 Collab., Adloff, C. et al, Measurement and NLO DGLAP QCD Interpretation of Diffractive Deep-Inelastic Scattering at HERA, Submitted to ICHEP 2002, Abstract 980 (2002).
3. John C. Collins, Proof of Factorization for Diffractive Hard Scattering, Phys. Rev., D57, 3051-3056 (1998).
4. Measurement of the Diffractive DIS Cross Section at low Q^2, Paper 981 subm. to ICHEP 2002.
5. Measurement of the Inclusive Diffractive Cross Section $\sigma_r^D(3)$ at high Q^2, Paper 5-090 subm. to EPS 2003.
6. Measurement of semi-inclusive diffractive deep-inelastic scattering with a leading proton at HERA, Paper 6-984 subm. to ICHEP 2002.
7. Dissociation of virtual photons in events with a leading proton at HERA, Eur. Phys. J C38 (2004) 43.
8. Deep Inelastic Scattering Events with a Large Rapidity Gap at HERA, H1 Collab., T. Ahmed et al., Nucl. Phys. B429 (1994) 477
9. Study of Deep Inelastic Inclusive and Diffractive Scattering with the ZEUS Forward Plug Calorimeter. ZEUS Collab. (S. Chekanov et al.). DESY-05-011, Jan 2005. 87pp. Published in Nucl.Phys.B713:3-80,2005

Diffractive deep inelastic scattering

A.D. Martin*, M.G. Ryskin* and G. Watt†

*IPPP, Physics Department, University of Durham, DH1 3LE, UK
†DESY, 22607 Hamburg, Germany

Abstract. A new approach to the analysis of diffractive deep inelastic data is presented. We show that the collinear factorisation theorem, which holds for diffractive DIS, has important modifications in the sub-asymptotic HERA regime, which can be quantified by using perturbative QCD. In fact the diffractive parton densities are shown to satisfy an inhomogeneous evolution equation. Moreover it is necessary to include both the gluonic and sea-quark t-channel components of the perturbative Pomeron.

Keywords: diffractive deep inelastic scattering, QCD Pomeron
PACS: 12.38.Bx, 13.60.Hb

INTRODUCTION

A notable feature of deep-inelastic scattering is that some $10-20\%$ of the events are diffractive, $\gamma^* p \to X + p$, in which the slightly deflected proton and the cluster X of outgoing hadrons are well-separated in rapidity. At high energies, the large rapidity gap is believed to be associated with Pomeron, or vacuum-quantum-number, exchange. Some secondary Reggeons also have vacuum quantum numbers, but these contributions are exponentially suppressed as a function of the gap size, and are negligible at small $x_{\mathbb{P}}$.

It is common to perform analyses of diffractive DIS (DDIS) data based on two levels of factorisation. First the diffractive structure function $F_2^{\rm D}$ may be written as the convolution of the usual coefficient functions as in DIS with diffractive parton distribution functions $a^{\rm D}$ [1]:

$$F_2^{\rm D} = \sum_{a=q,g} C_{2,a} \otimes a^{\rm D}, \tag{1}$$

with factorisation scale μ_F, where $a^{\rm D} = \beta q^{\rm D}$ or $\beta g^{\rm D}$ satisfy DGLAP evolution in μ_F. The *collinear factorisation* theorem (1) applies when μ_F is made large, therefore it is correct up to power-suppressed corrections. In a second stage, *Regge factorisation* is usually assumed [2], such that the diffractive parton densities $a^{\rm D}$ are written as a product of the Pomeron flux factor $f_{\mathbb{P}}(x_{\mathbb{P}}, t)$ and the Pomeron parton densities $a^{\mathbb{P}} = \beta q^{\mathbb{P}}$ or $\beta g^{\mathbb{P}}$. Taking $\mu_F = Q$, the t-integrated form is

$$a^{\rm D}(x_{\mathbb{P}}, \beta, Q^2) = f_{\mathbb{P}}(x_{\mathbb{P}})\, a^{\mathbb{P}}(\beta, Q^2), \tag{2}$$

where the Pomeron flux factor is taken from Regge phenomenology,

$$f_{\mathbb{P}}(x_{\mathbb{P}}) = \int_{t_{\rm cut}}^{t_{\rm min}} {\rm d}t \, \frac{{\rm e}^{B_{\mathbb{P}} t}}{x_{\mathbb{P}}^{2\alpha_{\mathbb{P}}(t)-1}}, \tag{3}$$

CP792, *Deep Inelastic Scattering, DIS 2005,* edited by W. H. Smith and S. R. Dasu

with $\alpha_{\mathbb{P}}(t) = \alpha_{\mathbb{P}}(0) + \alpha'_{\mathbb{P}} t$. For simplicity of presentation we omit the contribution of secondary Reggeons to the right-hand side of (2). Strictly speaking, the parameters $\alpha_{\mathbb{P}}(0)$, $\alpha'_{\mathbb{P}}$, and $B_{\mathbb{P}}$ must be taken from fits to soft hadron data. However, then one fails to describe the $x_{\mathbb{P}}$ dependence of the $F_2^{\mathrm{D}(3)}$ diffractive data. Therefore, as a rule (see [3]), $\alpha'_{\mathbb{P}}$ and $B_{\mathbb{P}}$ are fixed by the analyses of soft hadron data, while $\alpha_{\mathbb{P}}(0)$, and the parameters describing the input Pomeron parton distributions at some scale μ_0, are determined from the fit to the DDIS data. In these analyses, the Pomeron is treated as an effective pole in the complex angular momentum plane, and regarded as a hadron-like object of more or less fixed size. This Regge factorisation takes place in the non-perturbative region at some low scale μ, with $\mu < \mu_0$, of the order of the (inverse) size of the hadron. However, in such fits, the value of $\alpha_{\mathbb{P}}(0)$ extracted from DDIS data (for example, 1.17 in Ref. [3]) lies significantly above the value of 1.08 obtained from soft hadron data [4]. This indicates that there is a contribution coming from the small-size component of the Pomeron, that is, coming from the perturbative QCD region where the vacuum singularity has a larger intercept.

The Regge factorisation approach is a simplified phenomenological model. We shall not assume Regge factorisation for the whole a^{D}, but instead study the impact of applying perturbative QCD to the analysis of DDIS data. Moreover, although collinear factorisation holds in DDIS in the asymptotic limit, we will show that at the relevant HERA energies there exist important modifications.

EVOLUTION OF THE DIFFRACTIVE PARTON DENSITIES

The LO Feynman graph for the DDIS amplitude is shown by (the left-hand side of) Fig. 1, where Pomeron exchange is described by a gluon ladder. The kinematics of the process are fixed by the momentum $\tilde{k}$ of the first emitted parton at the end of the rapidity gap. Indeed, the virtuality of the first t-channel parton in the upper ladder fixes the scale $\mu^2 = k_t^2/\tilde{\beta}$, where $\tilde{\beta}$ is the (light-cone) momentum fraction of the Pomeron carried by parton $\tilde{k}$ and k_t is its transverse momentum. This scale μ plays the rôle of the lowest factorisation scale for the usual DGLAP evolution in the upper part of the diagram. *Simultaneously*, it is the upper scale for the lower part of the diagram. Indeed, the integral over the transverse momentum l_t of the t-channel gluon has the logarithmic form for $l_t^2 \ll \mu^2$, whereas for $l_t^2 \gg \mu^2$ it converges as $\mathrm{d}l_t^2/l_t^4$ and its contribution may be neglected as usual. After accounting for the evolution in the lower part of the diagram, and integrating over l_t, the probability amplitude to find the appropriate t-channel gluons is given, at LO, by the conventional integrated gluon distribution of the proton, $x_{\mathbb{P}}g(x_{\mathbb{P}}, \mu^2)$, which is known from global analyses of DIS and related hard-scattering data. To be precise, the skewing effect is incorporated using Ref. [5].

The evolution of a^{D} is a little subtle. Before we integrate over the momentum k_t — that is, selecting events with a fixed transverse momentum of the lowest parton — we can indeed see that the scale μ is the lowest possible factorisation scale for the diffractive parton densities a^{D}. However, in inclusive DDIS, we must integrate over k_t. The integral

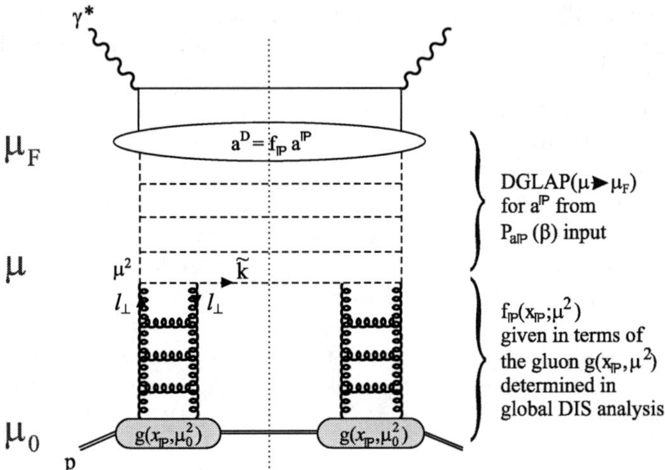

FIGURE 1. A ladder-type diagram showing a contribution to the diffractive parton densities $a^{\rm D}(x_{\mathbb{P}},\beta,\mu_F^2)$ in the perturbative region, $\mu > \mu_0 \sim 1$ GeV. DGLAP evolution for the Pomeron densities ($a^{\mathbb{P}} = \beta q^{\mathbb{P}}, \beta g^{\mathbb{P}}$) is performed from μ to μ_F for each component μ of the Pomeron in the perturbative interval $\mu_0 < \mu < \mu_F$ and then the sum is taken. The factorisation scale μ_F^2 is usually taken to be Q^2.

over k_t translates into an integral over μ, which is of the form

$$a^{\rm D} = \int_{\mu_0^2}^{Q^2} \frac{{\rm d}\mu^2}{\mu^2}\frac{1}{x_{\mathbb{P}}}\left[\frac{\alpha_S}{\mu}x_{\mathbb{P}}g(x_{\mathbb{P}},\mu^2)\right]^2 a^{\mathbb{P}} \equiv \int_{\mu_0^2}^{Q^2} \frac{{\rm d}\mu^2}{\mu^2}\, f_{\mathbb{P}}\, a^{\mathbb{P}}. \tag{4}$$

where the Pomeron flux, $f_{\mathbb{P}}$, is now given in terms of the integrated gluon, unlike (3).

Differentiating (4) with respect to $\ln Q^2$, we see that the evolution equations for the diffractive parton densities ($a^{\rm D} = \beta q^{\rm D}, \beta g^{\rm D}$) are

$$\begin{aligned}\frac{\partial a^{\rm D}}{\partial \ln Q^2} &= \int_{\mu_0^2}^{Q^2} \frac{{\rm d}\mu^2}{\mu^2}\, f_{\mathbb{P}}(x_{\mathbb{P}};\mu^2)\frac{\partial a^{\mathbb{P}}}{\partial \ln Q^2} + f_{\mathbb{P}}(x_{\mathbb{P}};\mu^2)\, a^{\mathbb{P}}(\beta,Q^2;\mu^2)\big|_{\mu^2=Q^2} \\ &= \frac{\alpha_S}{2\pi}\sum_{a'=q,g} P_{aa'}\otimes a'^{\rm D} + f_{\mathbb{P}}(x_{\mathbb{P}};Q^2)\, P_{a\mathbb{P}}(\beta).\end{aligned} \tag{5}$$

Here, $a^{\mathbb{P}}(\beta,Q^2;\mu^2)$ are the Pomeron parton densities DGLAP-evolved from a starting scale μ^2 up to Q^2, from input distributions $a^{\mathbb{P}}(\beta,\mu^2;\mu^2) = P_{a\mathbb{P}}(\beta)$. The quantities $P_{a\mathbb{P}}(\beta)$, with $a = q,g$, are the splitting functions of the perturbative Pomeron into quarks and gluons, whose LO forms are known from perturbative QCD [6, 7]. To allow for higher-order effects, we multiply the LO splitting functions by constant K factors, which are taken as free parameters in a fi t to the DDIS data [8].

If we were to assume that $f_{I\!P}$ were independent of μ^2, as is the case for BFKL asymptotics where $g(x_{I\!P},\mu^2) \sim (\mu^2)^{0.5}$, then (5) would be an inhomogeneous DGLAP equation exactly analogous to that for the evolution of the parton densities of the photon [9]. At first sight this appears strange. The Pomeron, unlike the photon, is an extended object, and we might have anticipated a form factor dependence $f_{I\!P} \sim 1/(\mu^2 R_{I\!P}^2)$. For such an extreme behaviour, the inhomogeneous term would be a power correction, which conventionally is neglected in the evolution. However, as $x_{I\!P} \to 0$, the proton gluon density $g(x_{I\!P},\mu^2)$ grows as $(\mu^2)^{0.5}$, and compensates the form factor power-like suppression. The HERA domain is an intermediate region: the anomalous dimension γ is not small, but is less than 0.5, with $g \sim (\mu^2)^\gamma$. Thus, although the integral in (4) is convergent at large μ^2, we still cannot neglect the inhomogeneous term in (5).

In practice, rather than to solve the inhomogeneous equation (5) directly, it is more convenient to use a separate standard homogeneous DGLAP equation for each component μ of the Pomeron. Then we can sum up (that is, integrate over μ) the different μ^2 contributions, in which the standard DGLAP evolution of each component starts from its own scale μ, provided that μ is in the pQCD domain ($\mu > \mu_0$), and continues up to the (collinear) factorisation scale μ_F. The contribution coming from $\mu < \mu_0$ must be treated non-perturbatively. This component of the diffractive densities $a^{\rm D}$ is included in the starting distributions at μ_0 whose parameters are obtained by fitting to data.

SUB-ASYMPTOTIC MODIFICATION OF DDIS FACTORISATION

From the formal viewpoint, for any fixed $x_{I\!P}$, the inhomogeneous term in the evolution equation (5) for $a^{\rm D}$ dies out as $Q^2 \to \infty$. Thus if we evolve from a large enough starting scale, μ_S, we are entirely in the perturbative regime and the diffractive parton densities, $a^{\rm D}$, are described by the usual (homogeneous) DGLAP equations and satisfy the collinear factorisation theorem [1].

However, this is not true in the HERA domain. For a low starting scale μ_S, but still satisfying $\mu_S \gg \Lambda_{\rm QCD}$, we can no longer omit the inhomogeneous term in the evolution (5) of the diffractive parton densities, $a^{\rm D}$. One consequence is that we will obtain a *smaller* diffractive gluon density $g^{\rm D}$. Indeed, it is known from the global analyses of DIS data that the gluon is mainly driven by $\partial F_2/\partial \ln Q^2 \sim \alpha_S\, g$. However, for DDIS, the perturbative Pomeron contribution to $F_2^{\rm D(3)}$ is [7, 8]

$$F_{2,\rm pert.}^{\rm D(3)}(x_{I\!P},\beta,Q^2) = \int_{\mu_0^2}^{Q^2} \frac{{\rm d}\mu^2}{\mu^2}\, f_{I\!P}(x_{I\!P};\mu^2)\, F_2^{I\!P}(\beta,Q^2;\mu^2) \tag{6}$$

$$\Longrightarrow \quad \frac{\partial F_{2,\rm pert.}^{\rm D(3)}}{\partial \ln Q^2} = \int_{\mu_0^2}^{Q^2} \frac{{\rm d}\mu^2}{\mu^2}\, f_{I\!P}(x_{I\!P};\mu^2) \frac{\partial F_2^{I\!P}(\beta,Q^2;\mu^2)}{\partial \ln Q^2} + f_{I\!P}(x_{I\!P};Q^2) F_2^{I\!P}(\beta,Q^2;Q^2). \tag{7}$$

Here, $F_2^{I\!P}(\beta,Q^2;\mu^2)$ is the Pomeron structure function, evaluated from the DGLAP-evolved Pomeron parton densities $a^{I\!P}(\beta,Q^2;\mu^2)$. The first term of (7) is roughly $\alpha_S\, g^{\rm D}$. Therefore, since part of the derivative $\partial F_2^{\rm D(3)}/\partial \ln Q^2$ comes from the upper limit Q^2 of the integral over the Pomeron scale μ, this results in a smaller diffractive gluon density than if the second term of (7) was neglected.

In addition, contributions associated with the charm quark and F_L^D violate factorisation. For moderate Q^2 it is convenient to use the fixed flavour number scheme, where the charm contribution arises from photon-gluon fusion $\gamma^* g^{I\!P} \to c\bar{c}$. However, in this case we will miss the diagram where the Pomeron directly produces the c quark, that is, when c is the lowest parton $\tilde{k}$ in the upper ladder of Fig. 1. In this case there is no evolution and no factorisation. This contribution should be added separately. Also, for large β and the Q^2 values typical at HERA, we cannot neglect the twist-four $F_L^{\mathrm{D}(3)}$ contribution [6, 7].

THE SEA-QUARK COMPONENT OF THE POMERON

So far we have taken the Pomeron to be a parton ladder where the two uppermost t-channel partons are gluons. However, at small scales μ^2, the gluon densities have a valence-like structure. They decrease with decreasing x already from $x \sim 0.01$ for $\mu^2 \sim 2$ GeV^2. On the other hand, the sea quark density $S = 2(\bar{u} + \bar{d} + \bar{s})$ increases as $xS \sim x^{-0.2}$ with decreasing x. As a consequence we have to include another contribution to the Pomeron in which the two uppermost t-channel partons in the lower ladders in Fig. 1 are a sea quark–antiquark pair. Thus we must introduce a sea-quark Pomeron flux $f_{I\!P=S}$ given by (4) with $x_{I\!P} g$ replaced by $x_{I\!P} S$. To avoid confusion we denote the flux in (4) by $f_{I\!P=G}$. In addition, we must include the $I\!P = GS$ interference term. In general, the two lower ladders in Fig. 1, shown as gluon ladders, contain both quarks and gluons.

PQCD ANALYSIS OF DDIS DATA

The pQCD approach described above has been used to analyse the new HERA DDIS data [8]. This was accomplished by summing up the parton densities given by a series of homogeneous DGLAP equations with different values of the starting scale μ. Suppressing the $x_{I\!P}$ and β dependence, the general structure is

$$a^{\mathrm{D}}(\mu_F;\mu_0) = a^{\mathrm{D}}_{\text{non-pert.}}(\mu_F;\mu_0) + \int_{\mu_0^2}^{\mu_F^2} \frac{\mathrm{d}\mu^2}{\mu^2} f_{I\!P}(\mu^2)\, a^{I\!P}(\mu_F;\mu), \tag{8}$$

where $a^{I\!P}(\mu_F;\mu)$ results from DGLAP evolution up to μ_F starting from the input $a^{I\!P}(\mu;\mu) = P_{aI\!P}(\beta)$. The non-perturbative term, $a^{\mathrm{D}}_{\text{non-pert.}}$, is exactly analogous to the procedure used in the Regge factorisation analysis described in the Introduction. The important new ingredient is the pQCD contribution given by the second term on the right-hand side of (8), which has contributions from the gluonic and sea-quark Pomeron ($I\!P = G,S$), and their interference ($I\!P = GS$). Setting $\mu_F = Q$, it is clear that (8) satisfies the inhomogeneous evolution equation (5).

A good description of both the ZEUS [10] and H1 [3] data was obtained. Diffractive parton densities resulting from the fit [8] to these combined data are shown in Fig. 2.

In summary, we have shown how to obtain *universal* diffractive parton densities a^{D} which can be used in the description of different diffractive processes. Of course, for

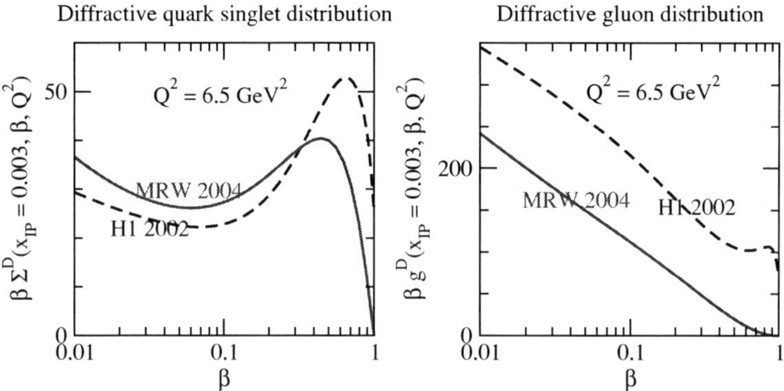

FIGURE 2. The diffractive parton densities at $Q^2 = 6.5$ GeV^2 obtained [8] from a fit to ZEUS [10] and H1 [3] DDIS data. The dashed curves are the densities obtained in the H1 fit [3].

diffractive production in high energy hadron-hadron collisions, we have to take care that the rapidity gaps are not populated by secondaries produced during the soft interaction of spectators. This well-known rapidity gap survival factor, usually denoted by S^2, may be calculated from phenomenological models tuned to describe elastic and related 'soft' hadron-hadron processes [11]. Also, in addition to the hard subprocesses originating from the collision of quarks and gluons from the *resolved* Pomeron, we must include the matrix element where the perturbative Pomeron *directly* participates in the hard interaction.

ACKNOWLEDGMENTS

ADM thanks the Leverhulme Trust for an Emeritus Fellowship and the Royal Society for a Joint Project Grant with the former Soviet Union.

REFERENCES

1. J. C. Collins, Phys. Rev. D **57** (1998) 3051 [Erratum-ibid. D **61** (2000) 019902].
2. G. Ingelman and P. E. Schlein, Phys. Lett. B **152** (1985) 256.
3. H1 Collaboration, paper 089 submitted to EPS 2003, Aachen.
4. A. Donnachie and P. V. Landshoff, Phys. Lett. B **296** (1992) 227.
5. A. G. Shuvaev, K. J. Golec-Biernat, A. D. Martin and M. G. Ryskin, Phys. Rev. D **60** (1999) 014015.
6. M. Wüsthoff, Phys. Rev. D **56** (1997) 4311.
7. A. D. Martin, M. G. Ryskin and G. Watt, arXiv:hep-ph/0504132.
8. A. D. Martin, M. G. Ryskin and G. Watt, Eur. Phys. J. C **37** (2004) 285.
9. E. Witten, Nucl. Phys. B **120** (1977) 189.
10. S. Chekanov *et al.* [ZEUS Collaboration], Eur. Phys. J. C **38** (2004) 43; Nucl. Phys. B **713** (2005) 3.
11. V. A. Khoze, A. D. Martin and M. G. Ryskin, Eur. Phys. J. C **18** (2000) 167.

Multiplicity structure in inclusive and diffractive deep inelastic e^+p collisions at HERA

Tinne Anthonis (on behalf of the H1 collaboration)

Physics Department, University of Antwerpen
Universiteitsplein 1, B-2610 Antwerpen, Belgium

Abstract. The multiplicity structure of the hadronic final state system X produced in diffractive deep-inelastic e^+p collisions of the type $e^+p \to e^+XY$, in which the photon dissociation system X is separated from a leading low mass baryonic state Y by a large rapidity gap, has been measured. Results are presented on charged particle multiplicity distributions and rapidity spectra. We investigate the kinematical dependences of the mean multiplicity and the particle density in rapidity space on the diffractive variables. The comparison with non-diffractive deep-inelastic e^+p collisions has been performed.

Keywords: Charged particle multiplicity, diffractive deep-inelastic e^+p collision.
PACS: 24.85.+p, 25.75.Gz

INTRODUCTION

The observation of "Large Rapidity Gap" (LRG) events in deep-inelastic (DIS) *ep* scattering at HERA renewed the interest in diffraction. To develop an understanding of the diffractive process, it is tempting and traditional to start from the popular pomeron picture. The pomeron is regarded as a color-singlet hadronic component in the target proton and the virtual photon in diffractive DIS (DDIS) probes the quark content of the pomeron. Just as in the case of the proton, the pomeron could then be characterized by diffractive structure functions. The pomeron structure functions should be universal if the pomeron were indeed an intrinsic part of the target proton wave function. However, striking similarities between inclusive and diffractive data in DIS have led to the insight that "the pomeron in the proton" is a dynamical effect of the interaction and thus not universal. The data provide strong hints that the underlying short-time, hard scattering sub-processes are identical in inclusive and diffractive DIS. The formation of a rapidity gap would then be a soft process happening on a longer time scale.

MULTIPLICITY DISTRIBUTION AND MOMENTS

Q^2 dependence of $\langle n \rangle$ in DIS and DDIS at fixed W

In the simple quark-parton model the properties of the hadronic system produced in DIS depend on the kinematical variables x_{Bj} and Q^2 only through the invariant mass W of the hadronic system. In QCD, weak scaling violations of the quark fragmentation

CP792, *Deep Inelastic Scattering, DIS 2005,* edited by W. H. Smith and S. R. Dasu

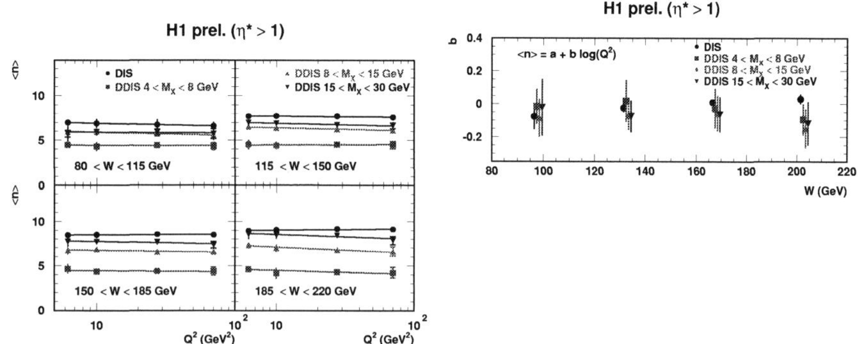

FIGURE 1. The Q^2 dependence of $\langle n\rangle$ for bins in W for either DIS and DDIS data (at fixed M_X).

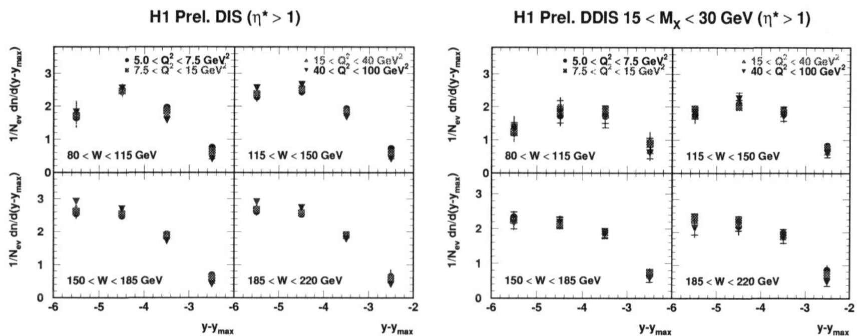

FIGURE 2. Rapidity densities in DIS and DDIS as a function of $y - y_{max}$.

functions and of the parton distributions introduce a Q^2 dependence even at fixed W. Figure 1 (left) shows the Q^2 dependence of $\langle n\rangle$ for bins in W for either DIS and DDIS data (at fixed M_X). The data were fitted to a form $\langle n\rangle = a + b\log(Q^2)$. The slopes b are given in Figure 1 (right). The data show no dependence on Q^2.

The Q^2 dependence is examined in more detail in Figure 2, where the rapidity densities in DIS and DDIS, respectively, are plotted as a function of $y - y_{max}$; $y_{max} = \ln(W/m_\pi)$ is the maximum rapidity at a given W and is calculated event-by-event. With W fixed, the spectra for DIS show little, if any Q^2 dependence. The same holds within errors for DDIS spectra at fixed M_X. We further note that the particle density is essentially independent of W. The mean multiplicity in DIS is predominantly a function of W and not of Q^2 and x separately.

β dependence of $\langle n\rangle$ at fixed M_X in DDIS

Various models view DDIS as elastic scattering of the virtual photon Fock states ($q\bar{q}$, $q\bar{q}g$, ...) off the target, with subsequent hadronisation, leading to the diffractive system X.

In such models, the relative fraction of quark and gluon fragmentation depends strongly on β.

The results for $\langle n\rangle$ are shown in Figure 3. Within errors, no dependence on β is observed when M_X is kept fixed. Combining the observation of the previous section with the results of this section, we conclude that the mean multiplicity of the diffractive system X is essentially a function of M_X only.

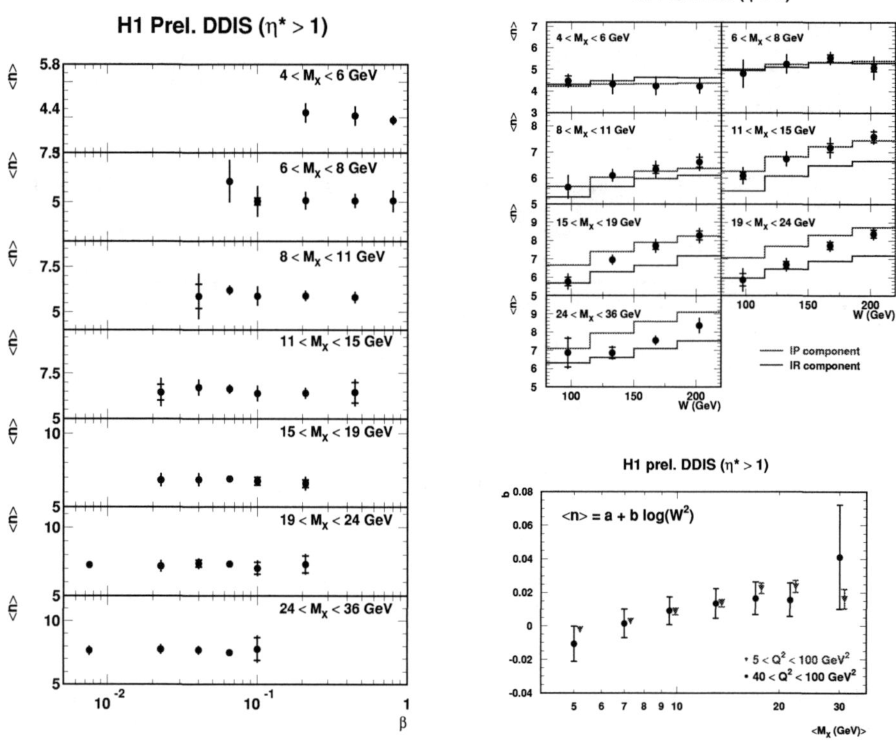

FIGURE 3. Dependence of $\langle n\rangle$ on β.

FIGURE 4. Mean multiplicity as a function of W ($x_{I\!P}$) for fixed bins in M_X.

W dependence at fixed M_X in DDIS

Based on experimental evidence, one often assumes Regge factorisation, whereby F_2^D is decomposed into a pomeron flux factor and the structure function of the pomeron. Regge factorisation states that the diffractive parton densities are independent of $x_{I\!P}$ and/or t.

Figure 4 (top) shows the mean multiplicity as a function of W ($x_{I\!P}$) for fixed bins in M_X and $5 < Q^2 < 100$ GeV^2. The curves are the predictions from the pomeron and reggeon component in the resolved pomeron picture.

Figure 4 (bottom, dots) shows the parameter b from a fit $\langle n\rangle = a + b\log(W^2)$. We observe the strongest W dependence for diffractive masses $M_X > 10$ GeV.

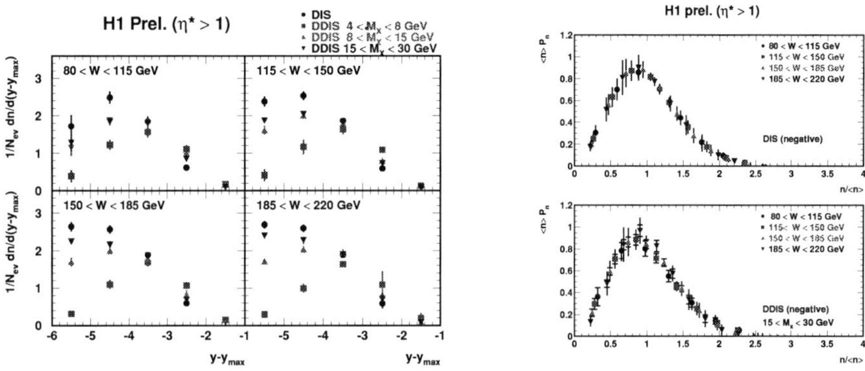

FIGURE 5. Particle density in rapidity space (left) and KNO scaling (right).

In models which assume Regge factorisation no W dependence at fixed values of M_X is expected. On the contrary, in models in which the rapidity gap formation is due to multiple soft rearrangements of colour configurations and where diffraction is a dynamical effect of the DIS interaction itself, such a W dependence occurs naturally as the energy evolution of the particle density in perturbative QCD is controlled by the anomalous multiplicity dimension $\gamma = \mathrm{d}\ln\langle n\rangle/\mathrm{d}\ln W^2$.

Regge factorisation breaking is also expected in multiple scattering (absorption) models. For DIS these effects are predicted to diminish with increasing Q^2. Figure 4 (bottom, triangles) shows the mean multiplicity only in the highest Q^2 bin ($40 < Q^2 < 100\,\mathrm{GeV}^2$). Factorisation breaking is, within the errors, not dependent on Q^2.

COMPARISON OF DIS AND DDIS

Figure 5 (left) shows the particle density in rapidity space as a function of $y - y_{max}$ for DIS and DDIS data (at fixed M_X).

In rapidity regions sufficiently far away from the gap (right side), we observe that the particle density is remarkably similar for DIS and DDIS at the highest M_X.

To demonstrate the energy scaling of the multiplicity distribution in DIS and DDIS, Figure 5 (right) shows the KNO distributions $\langle n\rangle P(n)$ versus $n/\langle n\rangle$ for different values of W, for negative particles in the domain $\eta^* > 1$. We observe, within the errors, KNO scaling for DIS and DDIS. The shape of the KNO distribution is similar for DIS and DDIS.

CONCLUSIONS

The charged particle multiplicity structure has been studied for DIS and DDIS events. The main kinematical variable for DIS is W and for DDIS it is M_X. The similarities for DIS and DDIS are in favor of models which incorporate the scattering of the virtual photon on the proton itself. The formation of a rapidity gap would then be a soft process happening on a longer time scale.

Measurement of Dijets with a Leading Neutron in ep Interactions at HERA

Armen Bunyatyan

Max-Planck-Institut für Kernphysik,
Postfach 103980, 69029 Heidelberg, Germany
and Yerevan Physics Institute, Armenia
E-mail: bunar@mail.desy.de

on behalf of the H1 Collaboration

Abstract. Measurements are reported of the production of dijet events with a leading neutron in ep interactions at HERA. Differential cross sections for photoproduction and DIS are presented. LO QCD simulation programs and NLO perturbative QCD calculations are compared with the measurements. Models in which the real or virtual photon interacts with a parton of an exchanged pion are found to be in good agreement with the measured cross sections. The fraction of leading neutron dijet events with respect to all dijet events is also determined.

Keywords: leading neutron, pion exchange, jets
PACS: 13.60.-r, 13.60.Rj, 13.87.-a

INTRODUCTION

Previous HERA measurements [1, 2] show that the cross section for the semi-inclusive ep scattering process, $e+p \to e+n+X$, where the leading neutron carries a large fraction of the proton energy, is described by the pion exchange mechanism [3] in which the virtual photon interacts with a parton from the pion. In the present analysis [4], the leading neutron production mechanism is investigated further in both photoproduction ($0.3 < y < 0.65$, $Q^2 < 10^{-2}$ GeV^2) and DIS ($2 < Q^2 < 80$ GeV^2, $0.1 < y < 0.7$) regimes by requiring that the hadronic final state contain two jets with transverse energies above 7 GeV and 6 GeV, respectively, measured in the $\gamma^* p$ centre-of-mass frame. Events with a leading neutron are selected from the inclusive dijet samples by requiring an energetic cluster in the forward neutron calorimeter (FNC).

RESULTS

Figure 1 shows the energy spectra as measured in the FNC, together with the predictions from Monte Carlo models. In addition to the models for inclusive jet production, PYTHIA [5] for photoproduction and RAPGAP [6] and LEPTO [7] for DIS, RAPGAP is also used to model the hard interaction assuming that it proceeds via π-exchange.

For neutron energies $E_n > 400$ GeV, the photoproduction data are described by the π–exchange model RAPGAP-π, as well as by PYTHIA without multiple interactions. If multiple interactions are included, PYTHIA predicts a rate which is too high for lower neutron energies. In DIS, the RAPGAP-π model describes the distribution, while the

CP792, *Deep Inelastic Scattering, DIS 2005*, edited by W. H. Smith and S. R. Dasu

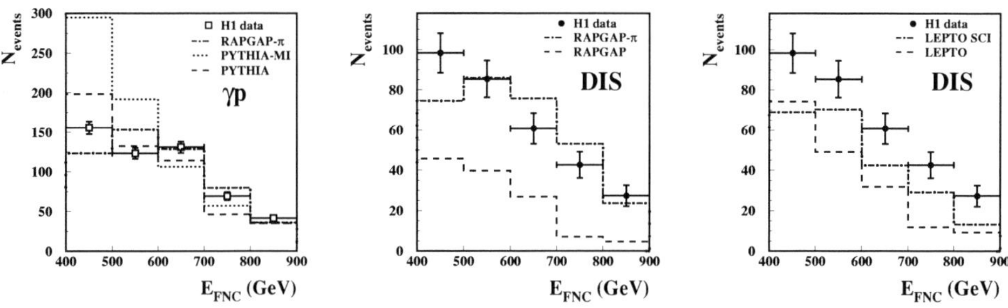

FIGURE 1. Energy distributions observed in the FNC for dijet events compared with the Monte Carlo model predictions normalized to the integrated luminosity of the corresponding data samples.

rates predicted by the standard DIS programs- LEPTO and RAPGAP- are too low. The LEPTO predictions are somewhat increased if soft colour interactions [8] are included.

The differential cross sections are measured at the hadron level for neutron energies $E_n > 500$ GeV. In Fig. 2 the cross sections are shown in the DIS regime. The RAPGAP-π model describes the measured distributions well, while the standard DIS processes, as simulated by LEPTO, lie below the data. In Fig. 3 the jet cross sections are shown in the photoproduction regime. Here, x_γ^{jet} and x_π^{jet} are the fractions of the 4-momentum of the photon and of the pion which participate in the hard interaction. Within the 20% normalization uncertainties the data are well described by the RAPGAP-π model, as well as by the NLO QCD calculation of dijet production in photon–pion collisions [9]. PYTHIA without multiple interactions, which does not include pion exchange, also provides a good description of the photoproduction data. Predictions of PYTHIA with multiple interactions are too high for low values of E_T^{jet} and for $x_\gamma^{jet} < 0.6$.

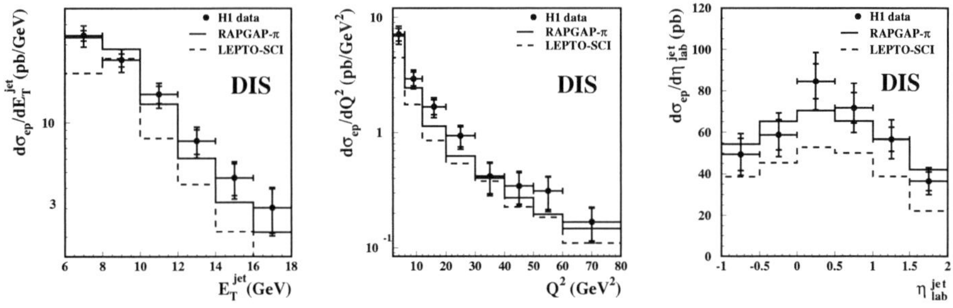

FIGURE 2. The measured differential ep cross sections as a function of E_T^{jet}, Q^2 and η_{lab}^{jet} for dijet events with a leading neutron in DIS.

The measured ratios of the dijet cross sections with and without the requirement of a leading neutron, f_{ln}, are shown in Figs.4 and 5. If the hard interaction is independent of the neutron production, f_{ln} should be essentially independent of the jet kinematics which reflect the hard process. For the photoproduction data, f_{ln} is, within errors, independent

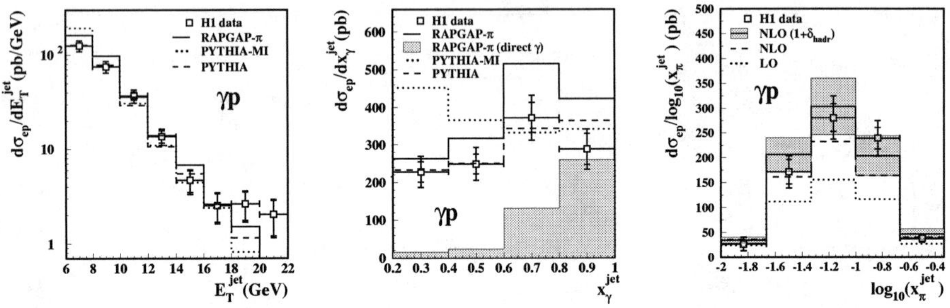

FIGURE 3. The measured differential *ep* cross sections as a function of E_T^{jet}, x_γ^{jet} and x_π^{jet} for dijet events with a leading neutron in photoproduction.

of E_T^{jet} and has an average value of about 2.3%. However, f_{ln} shows a dependence on other variables. These dependences can only partly be reproduced by the PYTHIA model, which provides some estimate of the size of possible phase space effects. A better description of the ratio in Fig. 4 is possible if the leading neutron data are described by the π-exchange model, RAPGAP-π, and the inclusive dijet data by PYTHIA-MI.

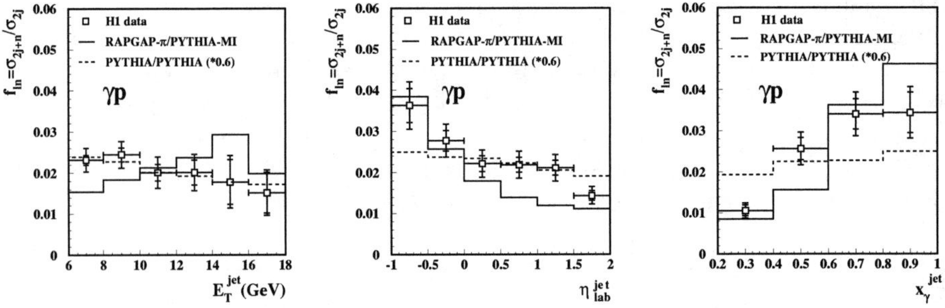

FIGURE 4. The ratio of the cross section for dijet photoproduction with a leading neutron to that for inclusive dijet photoproduction, as a function of E_T^{jet}, η^{jet} and x_γ^{jet}. Monte Carlo predictions for the ratios are obtained by using either RAPGAP-π for the leading neutron cross sections and PYTHIA-MI for the inclusive cross sections, or by using PYTHIA in both cases.

The Q^2 dependence of the ratio f_{ln} is shown in Fig. 5. Within errors, the RAPGAP model describes the measured ratio, when the leading neutron and the inclusive dijet data are represented by RAPGAP-π and standard RAPGAP, respectively. However, there is some tendency for the measured ratio to increase with Q^2, for Q^2 below 20 GeV^2.

CONCLUSIONS

It was observed in [1, 2] that pion exchange provides a good description of the semi-inclusive DIS process in which a leading neutron is produced. The present results demonstrate that this is also the case for the small subsample of leading neutron events

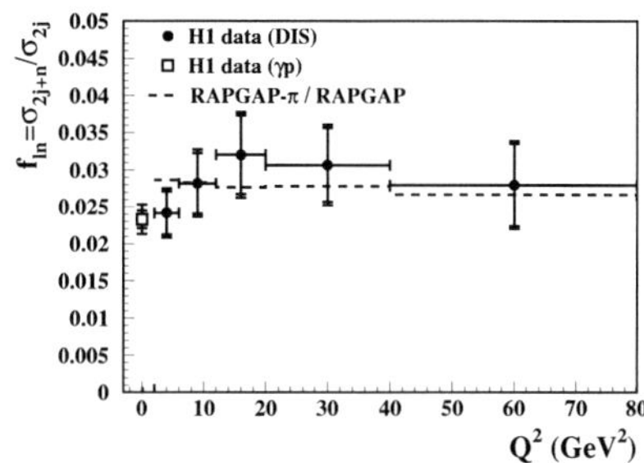

FIGURE 5. The ratio of the cross section for dijet production with a leading neutron to that for inclusive dijet production, as a function of Q^2. The Monte Carlo prediction, shown only for DIS, is obtained by using RAPGAP-π for the leading neutron cross section and RAPGAP for the inclusive cross section.

in which a dijet system is produced, in both DIS and photoproduction. Both the cross section measurements and the neutron energy spectrum are reasonably well described by pion exchange models.

For DIS, the standard Monte Carlo models for the simulation of the hadronic final state, such as LEPTO and RAPGAP, predict cross sections for the production of dijets with a leading neutron which are too small. The increase of the leading neutron rate in dijet production caused by the introduction of non-perturbative soft colour interactions in LEPTO-SCI is not large enough to provide a good description of the measurements.

For the photoproduction sample, the predictions of PYTHIA-MI clearly fail to describe the leading neutron data. This discrepancy between the data and PYTHIA-MI predictions, which on the other hand describe the inclusive jet production [10], suggests that the leading neutron dijet data have a lower fraction of resolved photon processes than do the inclusive dijet data. The decrease of the ratios of the dijet cross sections with and without the leading neutron requirement with decreasing x_γ^{jet} can be explained by the contribution of multiple interactions in inclusive dijet production. There is no evidence for a strong dependence of the ratio on Q^2.

REFERENCES

1. C. Adloff *et al.* [H1 Collaboration], *Eur. Phys. J.* **C 6** (1999) 587 [hep-ex/9811013].
2. S. Chekanov *et al.* [ZEUS Collaboration], *Nucl. Phys.* **B 637** (2002) 3 [hep-ex/0205076].
3. J.D. Sullivan, *Phys. Rev.* **D 5** (1972) 1732; H. Holtmann *et al.*, *Phys. Lett.* **B 338** (1994) 363; B. Kopeliovich, B. Povh and I. Potashnikova, *Z. Phys.* **C 73** (1996) 125 [hep-ph/9601291]; A. Szczurek, N.N. Nikolaev and J. Speth, *Phys. Lett.* **B 428** (1998) 383 [hep-ph/9712261].
4. A. Aktas *et al.* [H1 Collaboration], *Eur. Phys. J.* **C 41** (2005) 273, [hep-ex/0501074].
5. T. Sjöstrand, *Comp. Phys. Commun.* **82** (1994) 74.
6. H. Jung, *Comp. Phys. Commun.* **86** (1995) 147.
7. G. Ingelman, A. Edin and J. Rathsman, *Comp. Phys. Commun.* **101** (1997) 108 [hep-ph/9605286].
8. A. Edin, G. Ingelman and J. Rathsman, *Phys. Lett.* **B 366** (1996) 371 [hep-ph/9508386].
9. M. Klasen and G. Kramer, *Phys. Lett.* **B 508** (2001) 259 [hep-ph/0103056].
10. S. Aid *et al.* [H1 Collaboration], *Z. Phys.* **C 70** (1996) 17 [hep-ex/9511012].

Diffractive $D^{*\pm}$ Meson Production in Deep-Inelastic Scattering at HERA

Matthew Beckingham for the H1 Collaboration

DESY, Notkestrasse 85, 22607 Hamburg, Germany.

Abstract. A new measurement is presented of $D^{*\pm}$ meson production in deep-inelastic scattering at HERA. Cross sections are measured for the process $ep \to eXY$, where the system X contains at least one $D^{*\pm}$ meson and is separated by a large rapidity gap from a low mass proton remnant system Y. The cross sections are measured in the kinematic region $2 < Q^2 < 100$ GeV^2, $0.05 < y < 0.7$, $x_{I\!P} < 0.04$, $M_Y < 1.6$ GeV and $|t| < 1$ GeV^2. The $D^{*\pm}$ mesons are restricted to the range $p_T(D^*) > 2$ GeV and $|\eta(D^*)| < 1.5$. The data are compared with NLO QCD calculations using recent H1 diffractive PDFs as well as with a model of two gluon exchange.

Keywords: H1, diffraction, DIS, charm
PACS: 12.38.Qk,12.39.St

INTRODUCTION

This paper presents a new measurement of diffractive $D^{*\pm}$ production in deep inelastic scattering at HERA. This process is characterised by two distinct hadronic systems X and Y separated by a large rapidity gap. The system X contains at least one $D^{*\pm}$ meson. The system Y consists of the elastically scattered proton or a low mass diffractive state. In addition to the hard scale from the large photon virtuality, the hard scale from the charm mass means diffractive charm production is an interesting process to test different perturbative QCD approaches to diffraction. More details of this analysis can be found in [1].

Hard diffractive processes may be described theoretically using the collinear factorisation or perturbative two gluon approaches. The collinear approach [2] uses diffractive parton densities obtained from next-to-leading order (NLO) fits to the reduced diffractive cross section. Charm production then proceedes dominantly via the Boson Gluon Fusion process, where the factorisation theorem states that the hard scattering matrix elements are the same as for inclusive heavy flavour production. Hence diffractive charm production is directly sensitive to the diffractive gluon density, which is only indirectly constrained in inclusive diffraction through scaling violations. The pertubative two gluon approach [3] uses non-diffractive un-integrated gluon densities of the proton to combine two gluons into a colourless exchange. This exchange can then couple directly to a $c\bar{c}$ pair ($\gamma^* p \to c\bar{c}\ p$) or to a $c\bar{c}g$ system ($\gamma^* p \to c\bar{c}g\ p$).

CP792, *Deep Inelastic Scattering, DIS 2005,* edited by W. H. Smith and S. R. Dasu

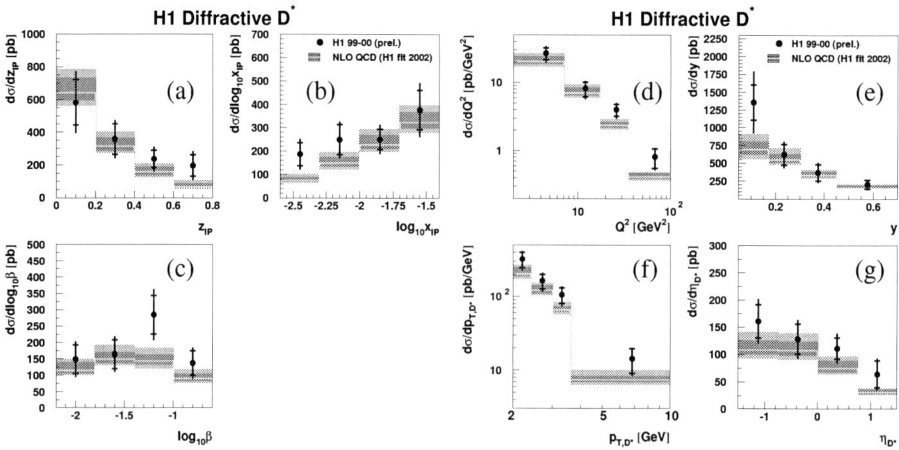

FIGURE 1. Differential cross sections for $D^{*\pm}$ meson production in diffractive DIS as a function of (a) $z_{I\!P}$, (b) $x_{I\!P}$, (c) β, (d) Q^2, (e) y, (f) $p_T(D^*)$ and (g) $\eta(D^*)$. The inner error bars show the statistical error and the outer error bars the total error. The data are compared to a NLO QCD prediction from the diffractive version of HVQDIS using NLO diffractive PDFs. The inner error band represents the renormalisation scale uncertainty and the outer error band the total uncertainty.

EVENT SELECTION AND KINEMATIC RECONSTRUCTION

This analysis is based on an integrated luminosity of 42.6 pb^{-1} collected in the years 1999 and 2000 by the H1 detector [4], from collisions of 27.5 GeV positrons with 920 GeV protons at HERA. DIS events are selected in the kinematic range $2 < Q^2 < 100$ GeV^2 and $0.05 < y < 0.7$. Diffractive events are selected by requiring no activity above noise thresholds in the forward detectors, thus limiting the X system to $\eta_{max} < 3.2$. Monte Carlo simulations are then used to correct to the range of the proton remnant system mass $M_Y < 1.6$ GeV and the squared four-momentum transfer at the proton vertex to $|t| < 1$ GeV^2. The kinematic range is restricted to $x_{I\!P} < 0.04$ to supress contributions from non-diffractive scattering and secondary reggeon exchanges. Further diffractive kinematic quantites are β, the fractional longitudinal momentum of the colourless exchange carried by the struck quark, and $z_{I\!P}$, the fraction of of the colour singlet exchange that enters the hard scattering.

The $D^{*\pm}$ mesons are reconstructed through the decay channel $D^{*+} \to D^0\pi^+_{slow} \to (K^-\pi^+)\pi^+_{slow}(+c.c.)$, with the decay products within the CTD acceptence ($20^\circ < \theta < 160^\circ$) and with transverse momentum p_T of at least 120 MeV for the π_{slow}, 300 MeV for the other pion and 500 MeV for the K meson. The invariant mass of the $K\pi$ combination must be within ± 80 MeV of the D^0 mass. Additionally the $D^{*\pm}$ must be in the pseudorapidity range $|\eta(D^{*\pm})| < 1.5$ and have a transverse momentum $p_T(D^{*\pm}) > 2$ GeV. The number of reconstructed $D^{*\pm}$ mesons is determined by fitting the mass difference $\Delta M = M(K^\mp\pi^\pm\pi^\pm_{slow}) - M(K^\mp\pi^\pm)$ for all selected events and track combinations. A total of 140 ± 16 diffractive $D^{*\pm}$ mesons are found.

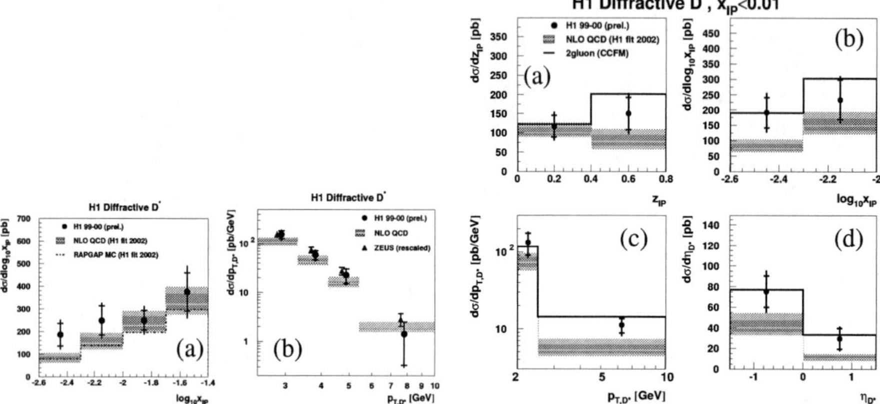

FIGURE 2. Differential cross sections for $D^{*\pm}$ meson production in diffractive DIS as a function of (a) $x_{I\!P}$ and $p_T(D^*)$. The data are compared to a NLO QCD prediction (see Fig.1). In (a) the data are also compared to the LO hadron level from RAPGAP (dot-dashed curve). In (b) the data are compared to the measurement from ZEUS (see text).

FIGURE 3. Differential cross sections for $D^{*\pm}$ meson production in diffractive DIS, in the kinematic region $x_{I\!P} < 0.01$, as a function of (a) $z_{I\!P}$, (b) $x_{I\!P}$ (c) $p_T(D^*)$ and (d) $\eta(D^*)$. The data are compared to predictions from NLO QCD (see Fig.1) and from the perturbative two gluon approach of BJKLW.

RESULTS AND COMPARISONS WITH CALCULATIONS

The differential cross sections are obtained from fits to the ΔM distribution in each kinematic bin. The data are then corrected for detector effects using the RAPGAP Monte Carlo generator [5], using diffractive parton densities. The visible cross sections are measured in the kinematic region $2 < Q^2 < 100$ GeV^2, $0.05 < y < 0.7$, $x_{I\!P} < 0.04$, $M_Y < 1.6$ GeV^2, $|t| < 1$ GeV^2, $p_T(D^*) > 2$ GeV and $|\eta(D^*)| < 1.5$.

The diffractive $D^{*\pm}$ cross sections are compared to two models based on NLO (LO) diffractive parton distributions from H1 [6]. For all models the charm mass was set to $m_c = 1.5$ GeV, $\Lambda_{QCD} = 0.2$ and the number of quarks set to $N_f = 4$. For the NLO calculation the diffractive version of the HVQDIS [7, 8] program with NLO diffractive parton densities (PDFs) was used. The renormalisation, μ_r, and fragmentation, μ_f, scales were set to $\mu_f^2 = \mu_r^2 = Q^2 + 4m_c^2$ and the Peterson fragmentation function was set to $\varepsilon = 0.078$. Uncertainties were evaluated by varying μ_r by factors of $1/4$ and 4, m_c within 1.35-1.65 GeV and ε between 0.035 and 0.1. For the LO calculation $O(\alpha_s)$ matrix elements and LO diffractive PDFs [6] were used in the RAPGAP Monte Carlo event generator. The D^* meson was fragmented using the Lund string model [9] and μ_f and μ_r were set to $\mu_f^2 = \mu_r^2 = Q^2 + p_T^2 + 4m_c^2$. The data are also compared to a prediction from the perturbative two gluon approach 'BJKLW' [3, 10, 11] using unintegrated inclusive gluon PDFs [12] evolved using the CCFM evolution equations. The comparison is made only for small $x_{I\!P} < 0.01$, so quark exchange may be neglected, and with a cut on the gluon p_T in the $c\bar{c}g$ system of $p_T > 1.5$ GeV, to ensure perturbative QCD is applicable.

Figure 1 shows the diffractive $D^{*\pm}$ cross sections differentially as a function of $z_{I\!P}$, $x_{I\!P}$, β, Q^2, y, $p_T(D^*)$ and $\eta(D^*)$. The differential cross sections are reproduced within the errors by the NLO HVQDIS calculation. Figure 2 (a) shows the differential $x_{I\!P}$ cross section again, compared to the predictions from the NLO HVQDIS calculation and to the LO calculation as implemented in RAPGAP. The prediction from RAPGAP is in agreement with the full NLO calculation. Figure 2 (b) shows the differential $p_T(D^*)$ cross section, compared to the measurement from the ZEUS [13] collaboration, rescaled to the phase space of this measurement using RAPGAP and with an additional 10% correction for different M_Y ranges. Both measurements are in good agreement. The description of the shapes of the differential cross sections by the LO and NLO calculations supports the validity of hard scattering factorisation.

Figure 3 shows the diffractive $D^{*\pm}$ cross sections for $x_{I\!P} < 0.01$ differentially as a function of $z_{I\!P}$, $x_{I\!P}$, $p_T(D^*)$ and $\eta(D^*)$ compared to both the NLO HVQDIS calculation and the prediction from the perturbative two gluon calculation of BJKLW. The NLO HVQDIS calculation falls below the data in all differential bins, however is still able to describe the data within errors. The data are also described in all differential bins by the two gluon prediction.

SUMMARY

A new measurement of diffractive $D^{*\pm}$ production in DIS at HERA has been presented, based on an integrated luminosity of 42.6pb^{-1}. Cross sections have been measured differentially as a function of various kinematic variables. LO and NLO QCD calculations using the colinear factorisation approach with diffractive PDFs are able to describe the shapes of the differential cross sections within errors. This agreement supports the validity of hard scattering factorisation. In the range $x_{I\!P} < 0.01$, the data are described by both a prediction from the perturbative two gluon model and the NLO colinear factorisation prediction for all differential cross sections.

REFERENCES

1. H1 Collaboration, paper 178 ICHEP04, H1prelim-04-111
2. J. C. Collins, Phys. Rev. D **57** (1998) 3051, Erratum-ibid. D **61** (2000) 019902, hep-ph/9709499.
3. J. Bartels, H. Jung and A. Kyrieleis, Eur. Phys. J. C **24** (2002) 555, hep-ph/0204269.
4. I. Abt *et al.* [H1 Collaboration], Nucl. Instrum. Meth. A **386** (1997) 310.
5. H. Jung, Comput. Phys. Commun. **86** (1995) 147.
6. H1 Collaboration, paper 980 ICHEP02, H1prelim-02-012
7. B. W. Harris and J. Smith, Nucl. Phys. B **452** (1995) 109, hep-ph/9503484.
8. L. Alvero, J. C. Collins and J. J. Whitmore, hep-ph/9806340.
9. T. Sjostrand, Comput. Phys. Commun. **82** (1994) 74.
10. J. Bartels, H. Lotter and M. Wusthoff, Phys. Lett. B **379** (1996) 239, Erratum-ibid. B **382** (1996) 449, hep-ph/9602363.
11. J. Bartels, H. Jung and M. Wusthoff, Eur. Phys. J. C **11** (1999) 111, hep-ph/9903265.
12. M. Hansson and H. Jung, hep-ph/0309009.
13. S. Chekanov *et al.* [ZEUS Collaboration], Nucl. Phys. B **672** (2003) 3, hep-ex/0307068.

Diffractive Dijets in DIS and Photoproduction

Matthias U. Mozer
on behalf of the
H1 Collaboration

Physikalisches Institut, Universität Heidelberg, Philosophenweg 12, 69120 Heidelberg, Germany

Abstract. The factorization of diffractive ep cross sections into a (perturbatively accessible) matrix element and a diffractive structure function was proven for the case of DIS by Collins in 1998. This factorization is not expected to hold for Photoproduction. Diffractive dijet production provides a valuable probe to testing factorization and characterize possible mechanisms of factorization breaking. Measurements of dijet cross-sections in DIS and photoproduction made with the H1 detector at the HERA accelerator are presented. Diffractive events were identified by a rapidity gap selection. The resulting differential cross sections are compared to QCD calculations in NLO, based on structure functions extracted from inclusive diffraction. The diffractive dijet cross sections in DIS agree well with the QCD predictions, supporting the QCD factorization for DIS. This is not the case for the diffractive dijet cross sections in photoproduction, which are significantly suppressed with respect to the expectations from QCD factorization.

Keywords: diffraction, factorization
PACS: 12.38.Bx, 12.38.Qk, 12.39.St, 12.40.Nn, 13.87.-a

INTRODUCTION

It can be shown [1] in Quantum Chromodynamics (QCD) that the cross section for diffractive deep-inelastic *ep* scattering (DIS) factorises into universal diffractive parton densities of the proton and process-dependent hard scattering cross sections (QCD factorisation). Diffractive parton densities have been determined from QCD fi ts to inclusive diffractive corss sections [2, 3].

Previous measurements of diffractive dijet and D* meson production in DIS have been found to be described by leading order (LO) Monte Carlo QCD calculations including parton showers (PS) based on diffractive parton densities [5, 6]. However, applying this approach in LO QCD calculations to predict diffractive cross sections for dijet production in $p\bar{p}$ collisions at the Tevatron leads to an overestimation of the observed rate by approximately one order of magnitude [7]. This discrepancy has been attributed to the presence of the additional beam hadron remnant in $p\bar{p}$ collisions, which leads to secondary interactions and a breakdown of factorisation. The suppression, often characterised by a 'rapidity gap survival probability,' cannot be calculated perturbatively and has been parameterised in various ways (see for example [8]).

The transition from deep-inelastic scattering to hadron-hadron scattering can be studied at HERA in a comparison of scattering processes in DIS and in photoproduction. Processes in which a real photon participates directly in the hard scattering are expected to be similar to the deep-inelastic scattering of highly virtual photons. By contrast, processes in which the photon is fi rst resolved into partons which then initiate the hard scat-

CP792, *Deep Inelastic Scattering, DIS 2005,* edited by W. H. Smith and S. R. Dasu

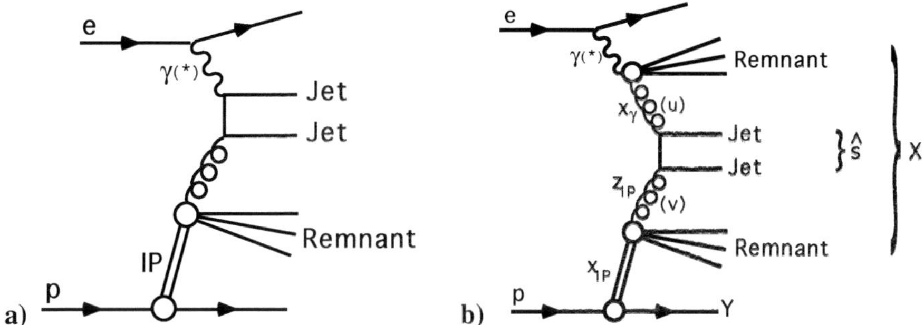

FIGURE 1. Leading order diagrams for diffractive dijet production at HERA. a) Direct (point-like) photon process (photon-gluon fusion), b) resolved (hadron-like) photon process with gluon-gluon scattering. This process can also lead to quark-quark and quark-gluon scattering.

tering resemble hadron-hadron scattering. Via resolved photon processes in hard photoproduction, gluon-gluon and gluon-quark final states are accessible, which are present in the equivalent $p\bar{p}$ collisions but negligible in DIS. Different models for diffractive scattering can therefore be tested in the regime of hard diffractive DIS and photoproduction.

In this article, measurements of diffractive dijet cross sections in DIS and photoproduction are presented, based on data collected with the H1 detector at HERA. Apart from the different range for the photon virtuality, the two measurements are performed in the same kinematic ranges to allow the closest possible comparison of the cross sections. The cross sections are compared with next to leading order (NLO) QCD predictions based on diffractive parton densities.

Examples of hard diffractive processes leading to jets in the final state at HERA are given by the diagrams of Figure 1 for the photon-gluon fusion production mechanism ("direct" photon process) and a process in which the photon develops a hadronic structure from which a single parton undergoes the hard scatter ("resolved" photon process). In photoproduction, resolved processes give a large contribution, whereas in DIS they are suppressed due to the large photon virtuality. The diffractive exchange in these diagrams is modeled by the exchange of the Pomeron with momentum fraction $x_{I\!P}$.

EXPERIMENTAL PROCEDURE

A detailed description of the H1 detector can be found in [9]. The data used in the present analysis were taken in the 1996 and 1997 running periods, in which HERA collided 820 GeV protons with 27.5 GeV positrons, corresponding to an integrated luminosity of 18 pb^{-1}. The photoproduction data are collected using a trigger which requires the scattered positron to be measured in the small angle positron detector, while DIS events are collected using a trigger which requires the scattered positron to be detected in the backward electromagnetic calorimeter. Rapidity gap events are selected by requiring an absence of activity in the direction of the outgoing proton. Jets are formed from the hadronic final state using the inclusive k_T cluster algorithm with a distance parameter of

TABLE 1. The kinematic ranges defining the measured cross sections.

Photoproduction	DIS
$Q^2 < 0.01$ GeV^2	$4 < Q^2 < 80$ GeV^2
$165 < W < 242$ GeV	
inclusive k_T jet algorithm, distance parameter=1	
$N_{\text{jet}} \geq 2$	
$E_T^{*,\text{jet1}} > 5$ GeV	
$E_T^{*,\text{jet2}} > 4$ GeV	
$-1 < \eta_{\text{jet}(1,2)} < 2$	$-3 < \eta^*_{\text{jet}(1,2)} < 0$
$x_{I\!P} < 0.03$	

unity in the γp rest frame. The final jet cross sections are given at the hadron level. The measured distributions are corrected for detector inefficiencies and migrations between measurement intervals during the reconstruction. The kinematic regions in which the cross sections are measured are given in Table 1. The kinematic variables $x_{I\!P}, M_Y$ and t are defined on the basis of the largest rapidity gap in the final state hadron distribution which divides the hadrons into the systems X and Y (the variables are defined as in [5]).

NEXT-TO-LEADING ORDER QCD CALCULATIONS

To calculate diffractive dijet cross sections to NLO in QCD for deep-inelastic positron-proton scattering, the DISENT [10] program is used. To obtain NLO cross sections for diffractive dijet photoproduction, the program by Frixione et al. [11] is used. Both programs are interfaced to the 'H1 2002 fit' (prel.) NLO diffractive parton distributions obtained in [3]. For the comparison with the measured hadron level jet cross sections, the calculated NLO parton jet cross sections were corrected for the effects of hadronisation using the RAPGAP Monte Carlo generator with Lund string fragmentation.

RESULTS

Figure 2 shows the dijet production cross section in diffractive DIS for $z_{I\!P}^{\text{jets}}$ (an estimator for the fraction of the momentum of the diffractive exchange entering the hard scatter) and $\log_{10}(x_{I\!P})$ in comparison to the QCD predictions calculated with the DISENT program. Good agreement is found within the given uncertainties for $\log_{10}(x_{I\!P})$. Some discrepancy between data and prediction can be observed for $z_{I\!P}^{\text{jets}} > 0.6$. The uncertainty shown in Figure 2 for the NLO prediction does not include the uncertainty of the gluon density as determined from F_2^D which is very large at high $z_{I\!P}$.

The situation is quite different for the dijet production cross sections in diffractive photoproduction (Figure 3). The QCD calculations are higher by a factor ≈ 2 compared with the experimental results. Therefore QCD factorisation seems to be broken in photoproduction.

Figure 4 shows the differential cross section for the diffractive production of two jets in photoproduction as a function of x_γ^{jets}(the fraction of the photon momentum entering

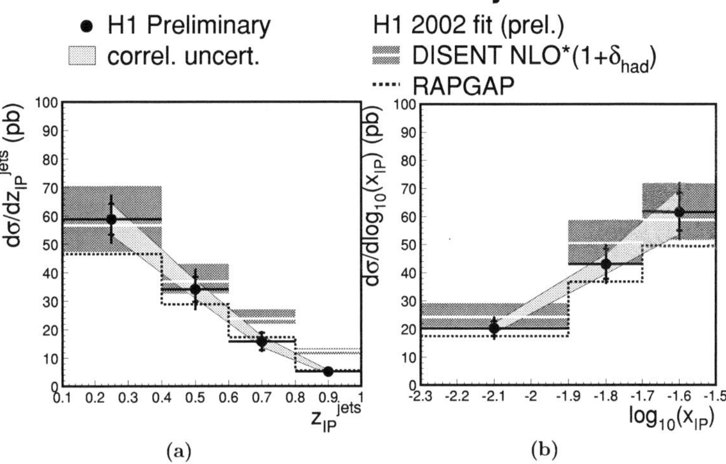

FIGURE 2. Cross section for the diffractive production of two jets in the DIS as a function of a) $z_{I\!P}^{\rm jets}$and b) $\log_{10}(x_{I\!P})$.The band around the NLO prediction with hadronisation correction indicates the uncertainty resulting from the variation of the renormalisation scale by factors 0.5 and 2. Not shown is the uncertainty resulting from the uncertainty on the diffractive parton densities and the uncertainty due to the imperfect knowledge of the hadronisation corrections.

the hard scatter, $x_\gamma^{\rm jets} = \frac{\sum_{\rm jets}(E-p_z)}{2yE_e}$), where the QCD prediction has been modifi ed by two different weighting schemes. In Figure 4 a) the "resolved" part for which $x_\gamma^{\rm jets} < 0.9$ at the parton level is scaled by a factor 0.34 as proposed by Kaidalov et al. [8]. The calculation for $x_\gamma^{\rm jets} > 0.9$ is left unscaled.

In Figure 4 b) the whole QCD calculation was globally scaled down by a factor 0.5. The global scaling of the QCD prediction describes the differential cross section much better than the QCD prediction weighted according to Kaidalov et al. [8].

The global suppression of the QCD calculation by a factor 2 yields a good description of all measured cross sections (see also Figure 5) and is favored over the suppression of only the resolved contribution.

SUMMARY

QCD factorisation in diffraction is tested with diffractive dijet production at HERA. While in the DIS case the measurement supports factorisation, the calculation in photoproduction has to be suppressed by a global factor 2.

REFERENCES

1. J. Collins, *Phys. Rev.* **D 57** (1998) 3051 and erratum ibid. **D61** (2000) 019902.
2. H1 Collaboration, C. Adloff *et al.*, *Z. Phys.* **C 76** (1997) 613.

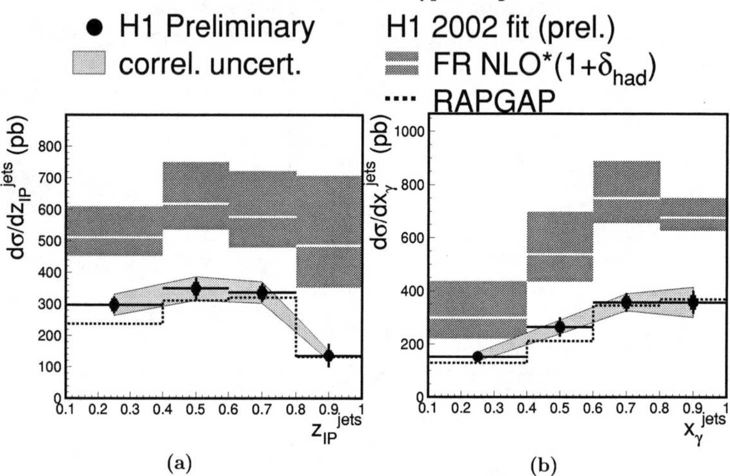

FIGURE 3. Cross section for the diffractive production of two jets in photoproduction as a function of a) z_{IP}^{jets} and b) x_{γ}^{jets}.The band around the NLO prediction with hadronisation correction indicates the uncertainty resulting from the variation of the renormalisation scale by factors 0.5 and 2.

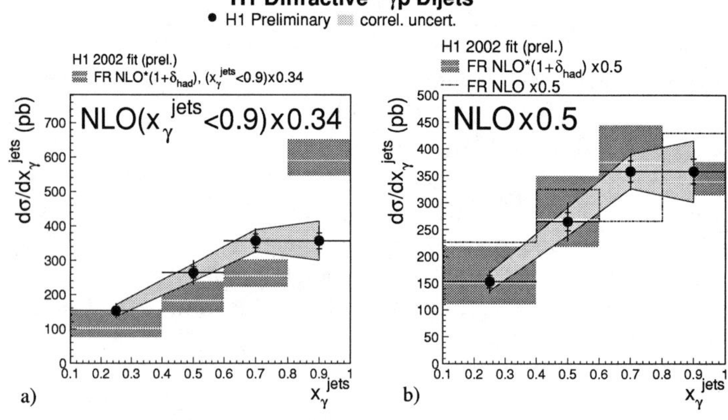

FIGURE 4. Cross section for the diffractive production of two jets in photoproduction as a function of x_{γ}^{jets}. In a) the resolved contributions to the QCD calculation ($x_{\gamma}^{\text{jets}} < 0.9$) is multiplied by 0.34 as proposed by Kaidalov et al. [8], while in b) the complete prediction is multiplied by a factor 0.5.

3. H1 Collaboration, paper 980 submitted to 31st Intl. Conf. on High Energy Physics, ICHEP 2002, Amsterdam.
4. H1 Collaboration, C. Adloff *et al.*, *Eur. Phys. J.* **C 6** (1999) 421.
5. H1 Collaboration, C. Adloff *et al.*, *Eur. Phys. J.* **C 20** (2001) 29.
6. H1 Collaboration, C. Adloff *et al.*, *Phys. Lett.* **B 520** (2001) 191.
7. CDF Collaboration, T. Affolder *et al.*, *Phys. Rev. Lett.* **84** (2000) 5043.

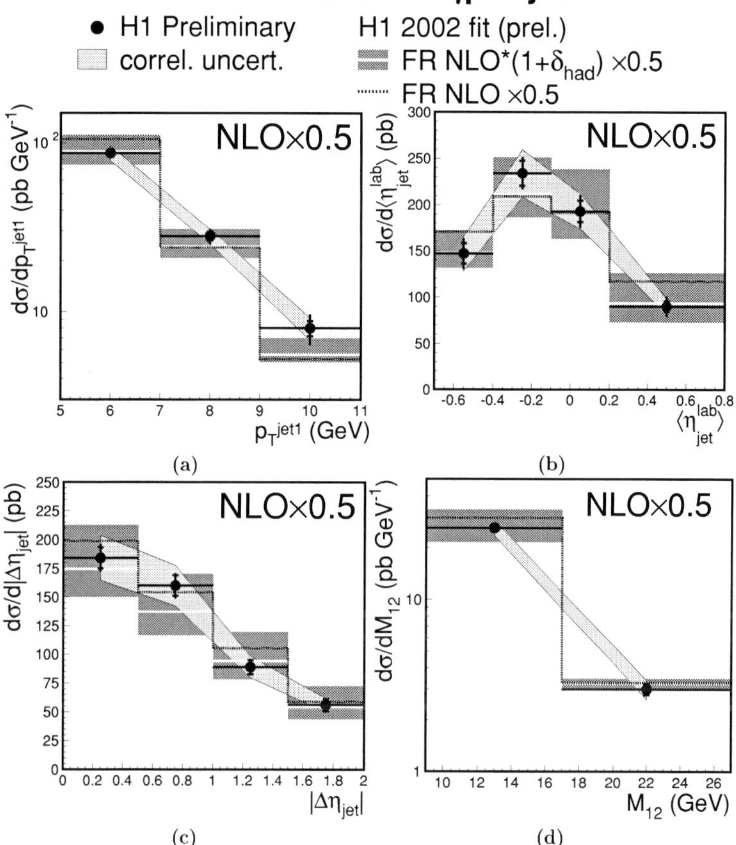

FIGURE 5. Cross section for the diffractive production of two jets in photoproduction as a function of a) p_T^{jet1}, b) $\langle\eta_{\text{jet}}^{\text{lab}}\rangle$, c) $\left|\Delta\eta_{\text{jet}}\right|$ and d) M_{12}.

8. A. Kaidalov, V. Khoze, A. Martin, M. Ryskin, *Phys. Lett.* **B 567** (2003) 61
9. H1 Collaboration, I. Abt *et al.*, *Nucl. Instrum. Methods* **A 386** (1997) 310 and 348.
10. S. Catani, M. H. Seymour, *Nucl. Phys.* **B 485** (1997) 29 [erratum ibid. **B510** (1997) 503].
11. S. Frixione, Z. Kunszt and A. Signer, *Nucl. Phys.* **B 467** (1996) 399;
S. Frixione, *Nucl. Phys.* **B 507** (1997) 295.

Diffractive photoproduction of dijets at ZEUS

Roger Renner (on behalf of the ZEUS Collaboration)* and S. Kagawa†

*Physikalisches Institut, University of Bonn
†Department of Physics, University of Tokyo

Abstract. The diffractive photoproduction of dijets has been studied using 77.6 pb^{-1} of data taken by the ZEUS detector at HERA. The measurements have been made in the kinematic range $0.2 < y < 0.85$ and $x_{\mathbb{P}} < 0.035$, where y is the inelasticity and $x_{\mathbb{P}}$ is the fraction of the proton momentum taken by the diffractive exchange. The jets are reconstructed using the k_T algorithm. The two highest transverse energy jets are required to satisfy $E_T > 7.5$ and 6.5 GeV, respectively, and to lie in the pseudorapidity range $-1.5 < \eta < 1.5$. Differential cross sections have been measured and are confronted with the predictions from leading order Monte Carlo models and next-to-leading order QCD calculations.

Keywords: diffractive photoproduction dijets ZEUS
PACS: 13.60.-r

INTRODUCTION

Diffractive dijets provide an important means to test whether factorisation holds in photoproduction (PHP) as it has been shown to hold in deep inelastic scattering (DIS) [1]. In PHP, the exchanged photon emitted from the incoming electron or positron[1] can either interact directly or as a source of quarks and gluons (resolved PHP). In the direct process, the photon is point-like and mimics the DIS process. In the resolved process, rescattering between the hadronic content of the photon and the proton may lead to a breakdown of factorisation, as was observed and discussed in [2, 3]. In one model, that takes rescattering into account, contributions from resolved PHP were predicted to be suppressed by a factor 0.34 [4].
In this presentation, data enriched in direct and resolved PHP are compared separately with leading order (LO) Monte Carlo as well as next-to-leading (NLO) QCD calculations to test for such factorisation breaking.

EXPERIMENTAL SET-UP

The analysis is based on data with an integrated luminosity of 77.6 pb^{-1} which was collected with the ZEUS detector [5] at the HERA collider in the years 1999/2000. The energies of the incoming protons and electrons were 920 GeV and 27.6 GeV, respectively. The jets are reconstructed using information from the central tracking detector and the compensating uranium-scintillator calorimeter. Diffractive dijet events are selected

[1] Hereafter, "electron" is used to refer to both electron and positron.

CP792, *Deep Inelastic Scattering, DIS 2005,* edited by W. H. Smith and S. R. Dasu

by requiring a rapidity gap between the direction of the incoming proton and the most-forward energy deposits[2] in the forward calorimeter (FCAL) or the forward plug calorimeter (FPC), which is located around the beampipe inside the FCAL, extending the coverage in pseudorapidity up to $\eta \simeq 5$.

KINEMATICS AND DATA SELECTION

Diffractive dijet events are described by the process $ep \to ep + X(jet + jet + X')$. In PHP both the scattered electron and the scattered proton escape down the beampipe and remain undetected. The hadronic system X with invariant mass M_X contains the dijet system and any other hadronic activity X'. The kinematic variables used in this analysis are depicted in Fig. 1 and are defined as follows:

- Q^2, the photon virtuality;
- y, the energy fraction of the electron carried by the exchanged photon;
- x_γ, the longitudinal momentum fraction of the photon carried by the parton;
- $x_{I\!P}$, the longitudinal momentum fraction of the proton entering the diffractive exchange, also referred to as Pomeron($I\!P$)-exchange;
- $z_{I\!P}$, the longitudinal momentum fraction of the diffractive exchange carried by the parton;
- $E_T^{jet\,1,2}$ and $\eta^{jet\,1,2}$, the transverse energy and pseudorapidity of the two jets with the highest E_T as reconstructed by the k_T algorithm [6] run in the longitudinally invariant inclusive mode in the laboratory frame [7].

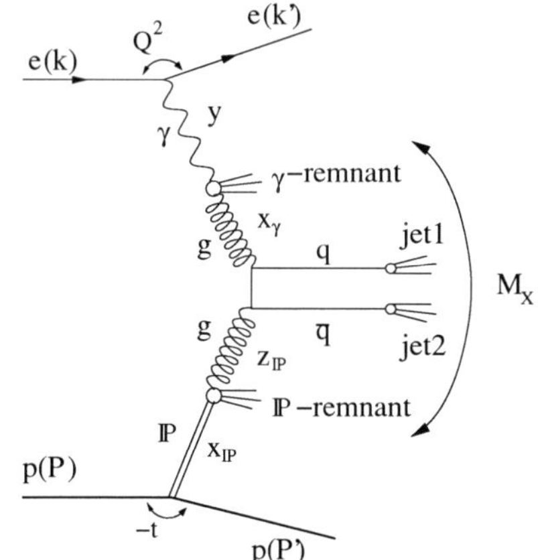

FIGURE 1. Depiction of the production of diffractive dijets in resolved PHP

[2] above noise threshold

For x_γ and $z_{I\!P}$, the observable estimators x_γ^{obs} and $z_{I\!P}^{obs}$ are calculated as defined below:

$$x_\gamma^{obs} = \frac{\sum_{jet1,2} E_T^{jet} e^{-\eta^{jet}}}{2yE_e} \qquad\qquad z_{I\!P}^{obs} = \frac{\sum_{jet1,2} E_T^{jet} e^{\eta^{jet}}}{2x_{I\!P}E_p}$$

The observable x_γ^{obs} is an estimator of the photon momentum fraction entering the dijet system. The range $x_\gamma^{obs} > 0.75$ ($x_\gamma^{obs} \leq 0.75$) is chosen to select data enriched with direct (resolved) PHP events [8]. The observable $z_{I\!P}^{obs}$ is sensitive to the parton densities in the diffractive exchange.

The cross sections are measured in the following kinematic region:

- $Q^2 < 1.0\,\mathrm{GeV}^2, \quad 0.2 < y < 0.85;$
- $x_{I\!P} < 0.035;$
- $E_T^{jet\,1} > 7.5\,\mathrm{GeV}, \quad E_T^{jet\,2} > 6.5\,\mathrm{GeV}, \quad -1.5 < \eta^{jet\,1,2} < 1.5$

PHP events are selected by requiring that no scattered electron candidate is found in the detector. For diffractive events a rapidity gap covering the pseudorapidity range $3 < \eta < 5$ was required. Background from cosmic events was rejected by a cut based on the timing of the two jets with highest E_T. After applying all cuts, 10673 events remain. The background from events with an undetected low-mass dissociative proton system is estimated to be $(16 \pm 4)\%$ [9] and independent from the measured variables. This fraction is statistically subtracted from all cross sections. Background from non-diffractive events is expected to be less than 10% and is not subtracted.

THEORETICAL PREDICTIONS

The data are compared to LO(+shower) Monte Carlo and NLO QCD calculations at hadron level.

The LO Monte Carlo is generated with RAPGAP [10] using the photon structure function GRV-G-HO and H1-fit2 [11] for the diffractive PDFs. Subsequently, the parton shower model MEPS [12] and the hadronisation model JETSET [13, 14] are applied. The LO Monte Carlo sample is also used to correct the data to the hadron level to account for acceptance losses.

NLO calculations on hadron level were provided by M. Klasen and G. Kramer [15] and are compared to data in two different ways. In one model resolved PHP is suppressed by a factor of $R = 0.34$ as motivated by [4], whereas in the other model no suppression of resolved PHP is applied ($R = 1$). In both cases no suppression was applied on direct PHP. The NLO predictions use the preliminary H1-2002 fit [16] for the diffractive PDFs. The renormalisation and factorisation scales were varied from $\frac{1}{2}E_T^{jet1}$ to $2E_T^{jet1}$ to account for theoretical uncertainties [15]. LO Monte Carlo is used to correct the predictions from the parton level[3] to the hadron level.

---

[3] after parton showers

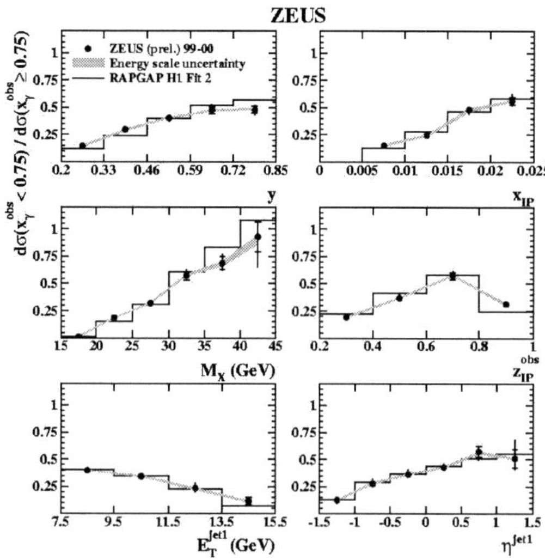

FIGURE 2. The ratio of single differential cross sections in y, x_{IP}, M_X, z_{IP}, $E_T^{jet\,1}$ and $\eta^{jet\,1}$ of data and LO Monte Carlo. The data are shown as dots, with the corresponding energy scale uncertainty shown as a band; the inner error bars indicate the statistical uncertainty and the outer error bars indicate the statistical and systematic uncertainties added in quadrature. The solid line show the prediction of the LO RAPGAP Monte Carlo. The H1 Fit2 diffractive PDFs are used.

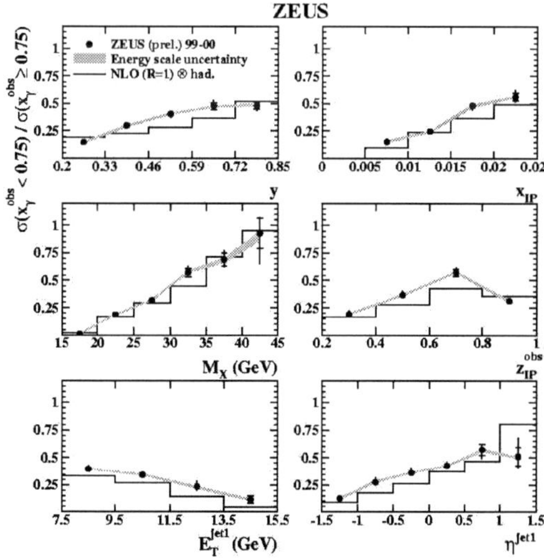

FIGURE 3. The ratio of single differential cross sections in y, x_{IP}, M_X, z_{IP}, $E_T^{jet\,1}$ and $\eta^{jet\,1}$ of data and NLO QCD calculations. For details of the data, see caption of Fig. 2. The solid line show the prediction of the NLO QCD calculation. The diffractive PDFs are from the H1 2002 fit.

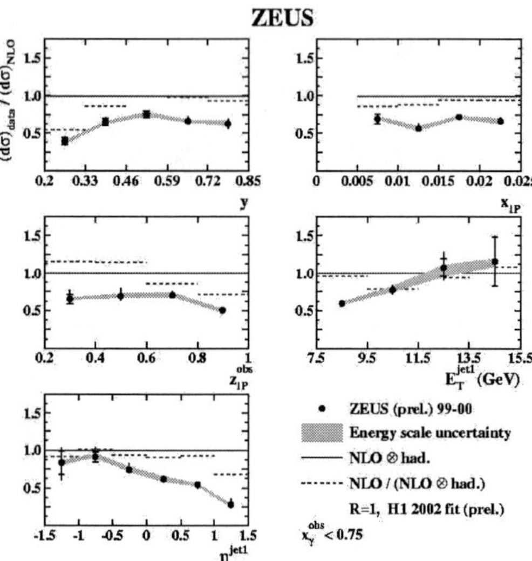

FIGURE 4. The ratio of data to the NLO QCD predictions of the single differential cross sections in y, $x_{I\!P}$, M_X, $z_{I\!P}^{obs}$, $E_T^{jet\,1}$ and $\eta^{jet\,1}$ for the sample enriched in resolved PHP ($x_\gamma^{obs} < 0.75$). For details of the data, see caption of Fig. 2. The solid lines indicate the expectations for $R = 1$. The dashed lines show the ratio of parton level to hadron level for NLO ($R = 1$). The diffractive PDFs are from the H1 2002 fit.

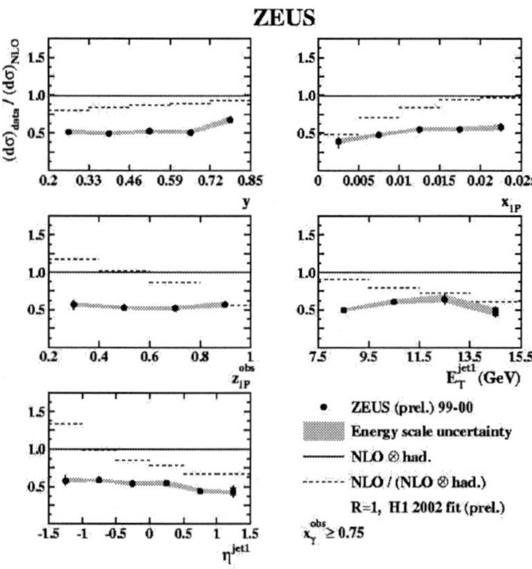

FIGURE 5. The ratio of data to the NLO QCD predictions of the single differential cross sections in y, $x_{I\!P}$, M_X, $z_{I\!P}^{obs}$, $E_T^{jet\,1}$ and $\eta^{jet\,1}$ for the sample enriched in direct PHP ($x_\gamma^{obs} \leq 0.75$). For details of the data, see caption of Fig. 2. For details of NLO predictions, see caption of Fig. 4.

RESULTS

The cross sections are measured as a function of y, x_{IP}, M_X, x_γ^{obs}, z_{IP}^{obs}, $E_T^{jet\,1}$ and $\eta^{jet\,1}$ and presented separately for the samples enriched with direct and resolved PHP.
The LO Monte Carlo describes the data well as can be seen in Fig. 2, which shows the ratio of the direct vs. resolved enriched sample for both data and MC events and has the benefit that certain uncertainties, like the normalisation factor for MC, cancel out.
The ratio is also fairly well described by the NLO predictions (Fig. 3) of Klasen and Kramer. The ratio of data to NLO ($R = 1$) is shown separately for the resolved and direct enriched samples in Figs. 4 and 5. For resolved PHP ($x_\gamma^{obs} < 0.75$) the ratio is flat but data are lower by a factor of 2 compared to NLO. Deviations are seen at small $E_T^{jet\,1}$ and high $\eta^{jet\,1}$, which are known to be sensitive to the structure function of the photon [17]. For direct PHP ($x_\gamma^{obs} \geq 0.75$) the ratio is also uniformly flat and data are also lower by a factor of 2. This is in contradiction to theoretical considerations which predict a suppression factor only for resolved PHP [4] and could indicate that a global suppression of both direct and resolved PHP is more likely. However, uncertainties in the diffractive PDFs need to be evaluated before stronger conclusions can be made.

REFERENCES

1. J.C. Collins, *Proof of factorisation for diffractive hard scattering.* Phys. Rev. **D 57**, 3051 (1998), hep-ph/9709499.
2. CDF Coll., T. Affolder et al., *Diffracitve dijets with a leading antiproton in $p\bar{p}$ collisions at $\sqrt{s} = 1800$ GeV.* Physi. Rev. Lett. **84**, 5043 (200).
3. A.B. Kaidalov, V.A. Khoze, A.D. Martin and M.G. Ryskin, *Probabilities of rapidity gaps in high energy interactions.* Eur. Phys. J. **C 21**, 521 (2001), hep-ph/0105145
4. A.B. Kaidalov, V.A. Khoze, A.D. Martin and M.G. Ryskin, Eur. Phys. J. **C21**, 521 (2003).
5. ZEUS Coll., U. Holm (ed.), *The ZEUS Detector.* Status Report (unpublished), DESY (1993), available on `www-zeus.desy.de/bluebook/bluebook.html`.
6. S. Catani et al., Nucl. Phys. **B 406**, 187 (1993).
7. S.D. Ellis and D.E. Soper, Phys. Rev. **D48**, 3160 (1993).
8. ZEUS Coll., S. Chekanov et al., Eur. Phys. J. **C23**, 615 (2001).
9. ZEUS Coll., S. Chekanov et al., Nucl. Phys. **B672**, 3 (2003).
10. H. Jung, *The RAPGAP Monte Carlo for Deep Inelastic Scattering version 2.08/00* (unpublished). 2001.
11. H1 Coll., C. Adloff et al., Z. Phys. **C 76**, 613 (1997).
12. R.K. Ellis, W.J. Stirling and B.R. Webber, *QCD and Collider Physics*, Cambridge Monographs on Particle Physics, Nuclear Physics and Cosmolgy, Vol.8.
13. M. Bengtsson and T. Sjöstrand, Comp. Phys. Comm. **46**, 43 (1987).
14. T. Sjöstrand, Comp. Phys. Comm. **82**, 74 (1994).
15. M. Klasen and G. Kramer, Preprint DESY-04-011 (hep-ph/0401202), 2004.
16. H1 Coll., *paper 980 submitted to 31. Intl. Conf. on High Energy Physics ICHEP 2002, Amsterdam, and paper 089 submitted to the EPS 2003 Conf., Aachen* (unpublished).
17. ZEUS Coll., S. Chekanov et al., Eur. Phys. J. C 23, 615 (2002).

DØ RESULTS IN DIFFRACTION

LUIZ MUNDIM
(for the DØ Collaboration)

High Energy & Nuclear Physics Department
State University of Rio de Janeiro,
R. São Francisco Xavier, 524
Rio de Janeiro, RJ, 20550-900, Brazil

Abstract. The first search for diffractively produced Z bosons in the muon decay channel is presented, using a data set collected by the DØ detector at the Fermilab Tevatron at $\sqrt{s} = 1.96$ TeV between April and September 2003, corresponding to an integrated luminosity of approximately 110 pb^{-1}. The current status of the FPD commissioning is also presented.

INTRODUCTION

QCD models single diffractive scattering of hadrons as proceeding via the exchange of a colour singlet object. In single diffraction, where one proton[1] remains intact with a small momentum loss and the other dissociates, there may be an area devoid of activity (rapidity gap) in the region of the outgoing intact proton. We present here the first ever search for diffractively produced Z bosons in the muon decay channel at the Fermilab Tevatron.

DIFFRACTIVE Z BOSON PRODUCTION

Event selection and data analysis

Z bosons produced via single diffraction are identified by demanding a rapidity gap near the beampipe in either the outgoing proton or antiproton direction. The data set was collected between April and September 2003 by the DØ detector at the Fermilab Tevatron, corresponding to an integrated luminosity of approximately 110 pb^{-1}. The DØ detector is described in detail elsewhere [1]. The Z boson is selected via its decay into two oppositely charged muons each with $p_T > 15$ GeV. At least one muon must be isolated in the central tracking detector and the calorimeter: Σp_T of tracks within a cone of radius 0.5 around the muon is required to be less than 3.5 GeV, and in the calorimeter (ΣE_T in a cone of radius of 0.5 around the muon) - (ΣE_T in a cone of radius of 0.1 around the muon) is required to be less than 2.5 GeV, where the cone radius is defined in pseudorapidity η and azimuthal angle ϕ as $\Delta R = \sqrt{\Delta\eta^2 + \Delta\phi^2}$. Cosmic ray muon events

[1] Here the term 'proton' is used to refer to both protons and antiprotons.

CP792, *Deep Inelastic Scattering, DIS 2005*, edited by W. H. Smith and S. R. Dasu

are vetoed by requiring that the distance of closest approach of muon tracks to the beam position is less than 0.02 cm for tracks with hits in both the Silicon Micro-vertex Tracker (SMT) and Central Fiber Tracker (CFT), or less than 0.2 cm for tracks with hits only in the CFT. In addition, the muon tracks are required to fulfil $|\Delta\phi_{\mu\mu} + \Delta\theta_{\mu\mu} - 2\pi| > 0.05$ radians, where θ is polar angle.

The rapidity gap search makes use of two detectors, the Luminosity Monitor (LM) and the end calorimeter. The LM comprises two scintillating detectors, one on each side of the interaction region, which cover the pseudorapidity range $2.7 < |\eta| < 4.4$. The total output charge is discriminated to give an on/off signal for each detector. The end calorimeter is divided into three regions: (1) four electromagnetic layers closest to the beam, (2) four fine hadronic layers and (3) one coarse hadronic layer furthest from the beam. Each layer is divided into cells in the $\eta - \phi$ plane. For this analysis, the energy is summed separately on each side (outgoing proton and antiproton) in the range $2.6 < |\eta| < 5.3$, using electromagnetic cells with $E_{\text{cell}} > 100$ MeV and fine hadronic cells with $E_{\text{cell}} > 200$ MeV.

The log of the energy sum on the outgoing antiproton side is plotted in Figure 1 for bunch crossings in which there are no visible interactions. These are selected from a randomly triggered sample with the requirements that both LM detectors are off and there is no vertex with greater than two associated tracks. These events are used to approximate rapidity gap events, in which there is no activity in the outgoing antiproton direction. The log of the energy sum on the outgoing antiproton side is also shown for a sample of minimum bias events in the figure. These are selected requiring hits in both detectors of the LM within a small time window. A third (25 GeV jet) sample is selected by requiring a vertex with at least three tracks, and at least one jet with $p_T > 25$ GeV that passes jet quality cuts. Jet events in which the highest p_T jet lies in the region $|\eta| > 2.4$ are excluded. The minimum bias and jet samples are dominated by events in which both protons dissociate.

Events with no interaction and events with antiproton dissociation are separated by applying a cut at an energy sum of 10 GeV. This is also the case in the outgoing proton direction. To select single diffractive candidates in the Z boson sample the LM detector is required to be off and the energy sum less than 10 GeV on one side, and the LM detector is required to be on and the energy sum greater than 10 GeV on the other side.

Results

Figure 2 shows the di-muon invariant mass distribution for two samples. Fig. 2(left) shows those events that fail the two rapidity gap cuts on both the outgoing proton and antiproton sides. These are strong candidates for non-diffractive production of Z bosons. A resonant peak is observed together with a small background contribution, arising mainly from the $(Z/\gamma)^*$ continuum. Fig. 2(right) shows those events that pass both rapidity gap cuts on one side and fail both on the other. These are candidates for single diffractively produced Z bosons, where one proton is intact and the other dissociates.

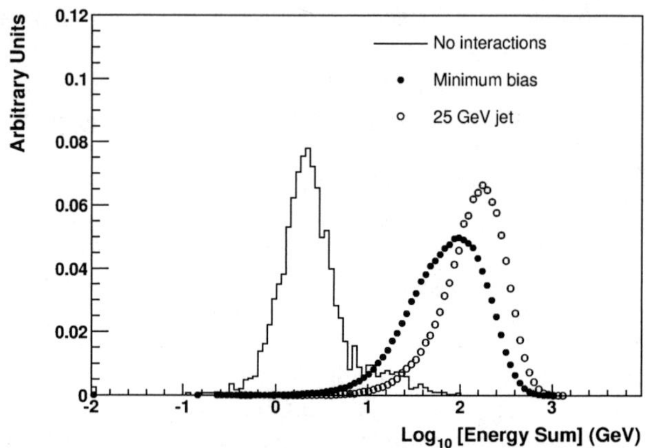

FIGURE 1. Log of energy sum in the outgoing antiproton direction ($-5.3 < \eta < -2.6$), comparing events with no visible interactions with events in which both protons dissociate. Areas are normalised to unity. An energy sum cut is applied at 10 GeV for rapidity gap candidates.

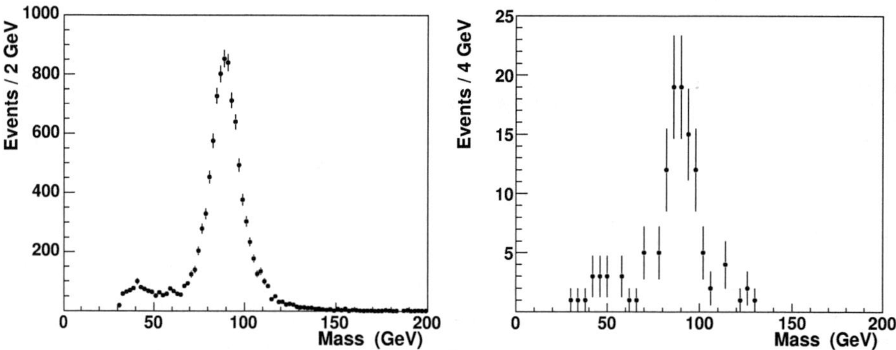

FIGURE 2. The di-muon invariant mass distribution for Z boson candidates with no rapidity gap (left) and a single rapidity gap (right). A rapidity gap is defined as one LM detector off and energy sum less than 10 GeV in the same region (see text for details).

FORWARD PROTON DETECTOR

The Forward Proton Detector was designed to study diffractive events produced at the DØ interaction point at Tevatron. It is an ensemble of 9 momentum spectrometers comprised of 18 Roman Pots, inside which a detector based on scintillating fibers can be brought as close to the beam as 6 mm, allowing the reconstruction of the trajectories of the scattered protons. This approach provides a way to measure the scattering angle and momentum fraction lost with a much better resolution than using the rapidity gap method alone. For a full description of the FPD, see [2].

The first 10 Roman pots were installed between 2001 and 2002 and since January

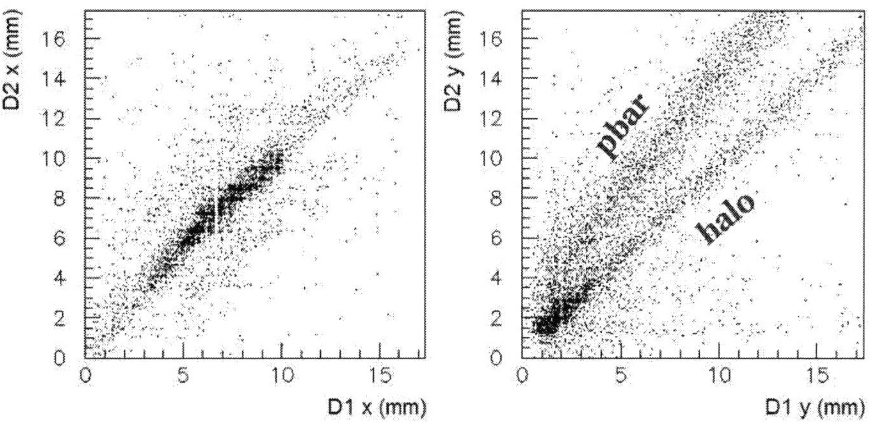

FIGURE 3. Hit correlation for the dipole spectrometer. On the left, the x coordinate of a hit on the first detector is plotted against the one on the second detector; on the right, the same correlation is plotted for the Y coordinate

of 2004, all of them are taking data integrated with the DØ detector and functioning as designed.

Each one of the 18 detectors has a sensitive area of about 17.5x17.5 mm^2, 6 layers of fibers in 3 planes (u,x and v) and a trigger scintillator. The 3 planes can measure the hit of a proton passing through it without ambiguity. Combining the hit measured in two detectors of a spectrometer (see [2]), it is possible to track the particle from the interaction point.

The hit correlation on both detectors of a spectrometer can be used to separate the scattered proton from the halo particles. Since the halo particles have basically the same energy as the beam ones, they are affected the same way by the electromagnetic fields along the path between both detectors, however, the scattered proton, after loosing energy has its path more affected by the fields along its way from the first detector to the second one. Thus, the scattered proton has its path deviated towards the center of the ring (increasing Y coordinate), making them to be off diagonal in the correlation plot.

This effect can be seen in figure 3, where it is shown the X (left) and the Y (right) coordinate correlations for the dipole spectrometer. On these plots, it has been used a simple selection criteria where only events with only one hit per detector were selected.

SUMMARY

A search for diffractively produced Z bosons in the muon channel has been presented. The sample is large enough to allow a study of the kinematic properties of the Z bosons for the first time. It has also been shown that the Forward Proton Detector is under commissioning and working as expected. Future diffractive studies at DØ will be done using both the FPD signals and the rapidity gap technique.

ACKNOWLEDGMENTS

We thank the staffs at Fermilab and collaborating institutions, and acknowledge support from the Department of Energy and National Science Foundation (USA), Commissariat à l'Energie Atomique and CNRS/Institut National de Physique Nucléaire et de Physique des Particules (France), Ministry of Education and Science, Agency for Atomic Energy and RF President Grants Program (Russia), CAPES, CNPq, FAPERJ, FAPESP and FUNDUNESP (Brazil), Departments of Atomic Energy and Science and Technology (India), Colciencias (Colombia), CONACyT (Mexico), KRF (Korea), CONICET and UBACyT (Argentina), The Foundation for Fundamental Research on Matter (The Netherlands), PPARC (United Kingdom), Ministry of Education (Czech Republic), Natural Sciences and Engineering Research Council and WestGrid Project (Canada), BMBF (Germany), A.P. Sloan Foundation, Civilian Research and Development Foundation, Research Corporation, Texas Advanced Research Program, and the Alexander von Humboldt Foundation.

REFERENCES

1. V. Abazov *et al.* (DØ collaboration), to be submitted to Nucl. Instrum. Methods A; T. LeCompte and H. T. Diehl, Ann. Rev. Nucl. Part. Sci. **50**, 71 (2000).
2. A. Brandt *et al.* (DØ collaboration), FERMILAB-PUB-97-377.

Update of CDF Results on Diffraction

Konstantin Goulianos

The Rockefeller University, 1230 York Avenue, New York, NY 10021, U.S.A.
(Presented on behalf of the CDF Collaboration)

Abstract. The diffractive program of the CDF Collaboration at the Fermilab Tevatron $\bar{p}p$ Collider is reviewed with emphasis on recent results from Run-II and future prospects.

Diffractive $\bar{p}p$ interactions are characterized by the presence of at least one large rapidity gap, defined as a region of pseudorapidity [1] devoid of particles. A diffractive rapidity gap, which may be forward (adjacent to a leading nucleon) or central, is presumed to be formed by the exchange of a *Pomeron* [2], which in QCD is a color singlet quark/gluon object with vacuum quantum numbers. Diffraction in which there is a high momentum-transfer partonic scattering in the event in addition to the rapidity gap is referred to as *hard diffraction*. In this paper, we briefly review the results on diffraction obtained by the Collider Detector at Fermilab (CDF) in Run-I (1992-1995), present an update of results from Run-II, which is in progress, and discuss future prospects.

RUN-I RESULTS

In addition to measuring $\bar{p}p$ elastic, single diffraction (SD), and total cross sections at $\sqrt{s} = 540$ and 1800 GeV, CDF studied several soft and hard diffraction processes at 1800 GeV, and in some cases at $\sqrt{s} = 630$ GeV [3]. Soft processes studied include:

DD	Double Diffraction	$\bar{p}p \rightarrow X + \text{gap} + Y$
DPE	Double Pomeron Exchange	$\bar{p}p \rightarrow \bar{p} + \text{gap} + X + \text{gap} + p$
SDD	Single $\oplus$ Double Diffraction	$\bar{p}p \rightarrow \bar{p} + \text{gap} + X + \text{gap} + Y$

In hard diffraction CDF measured SD dijet, W, b-quark and J/ψ, DD dijet, and DPE dijet production. Schematic diagrams and event topologies for representative processes are shown in Fig. 1.

SOFT DIFFRACTION HARD DIFFRACTION

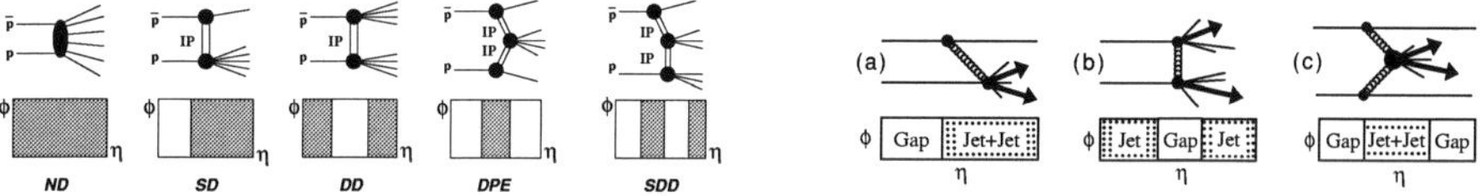

FIGURE 1. Schematic diagrams and η-ϕ topologies of representative diffractive processes studied by CDF. The shaded areas represent regions of pseudorapidity in which there is particle production.

CP792, *Deep Inelastic Scattering, DIS 2005,* edited by W. H. Smith and S. R. Dasu

Two types of hard diffraction results were obtained in Run-I: diffractive to non-diffractive cross section ratios using the rapidity gap signature to select diffractive events and diffractive to non-diffractive structure function ratios using a Roman Pot Spectrometer (RPS) to trigger on leading antiprotons. The results exhibit regularities in normalization and factorization properties that point to the QCD character of diffraction (see [3]).

At $\sqrt{s}$ =1800 GeV, the SD/ND ratios (gap fractions) for dijet, W, b-quark, and J/ψ production, as well the ratio of DD/ND dijet production, are all $\approx 1\%$. These ratios are suppressed relative to standard QCD inspired theoretical expectations (*e.g.* 2-gluon exchange) by a factor of $\sim$10, which is comparable to that observed in soft diffraction relative to Regge theory expectations. This suppression represents a severe breakdown of QCD factorization. It is, however, interesting to note that except for the overall suppression in normalization factorization approximately holds at fixed $\sqrt{s}$.

Another interesting aspect of the Run-I results is that ratios of two-gap to one-gap cross sections for both soft and hard processes appear to obey factorization. This feature of the data provides both a clue to understanding diffraction and a tool for diffractive studies using processes with multiple rapidity gaps [4].

RUN-II PROGRAM

The goal of the Run-II diffractive program of CDF is twofold: (a) to obtain results that could help decipher the QCD nature of the Pomeron, such as dependence of the diffractive structure function (DSF) on Q^2, x_{Bj}, t, and ξ (fractional momentum loss of the diffracted nucleon), and (b) to measure exclusive production rates (dijet, χ_c^0, $\gamma\gamma$), which could to be used to establish benchmark calibrations for exclusive Higgs production at LHC [5]. Preliminary results from data collected at $\sqrt{s} = 1.96$ GeV confirm the Run-I DSF results [3, 7]. New in Run-II are the measurement of the Q^2 dependence of the DSF obtained from dijet production and limits on exclusive production rates.

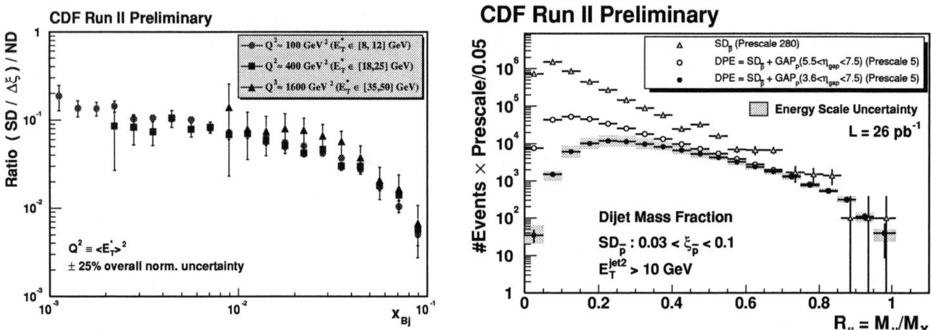

FIGURE 2. (*left*) Ratio of SD/$\Delta\xi_{\bar{p}}$ over ND rates obtained from dijet data at various Q^2 ranges; (*right*) ratio of dijet mass to total mass "visible" in the calorimeters for dijet production in events with a leading antiproton within $0.3 < \xi_{\bar{p}} < 0.1$ and various gap requirements on the proton side: (triangles) no gap requirement, (open circles) gap in $5.5 < \eta < 7.5$, and (filled circles) gap in region $3.5 < \eta < 7.5$.

The diffractive structure function

In Fig. 2 (*left*), the ratio of SD/ND rates, which in LO QCD and at fixed x_{Bj} is equal to the ratio of the corresponding structure functions, shows no appreciable Q^2 dependence. This result was foreseen in the renormalization model [8]. in which the diffractive structure function is basically the low-x ($x < \xi$) structure function of the diffracted nucleon. More data are currently being analyzed to improve the statistics of this measurement.

Data are at hand and analyses are in progress for the measurement of the t, ξ, and flavor dependence of the DSF using dijet, W, and J/ψ production. In addition, factorization will be tested more accurately than in Run-I by comparing the DSFs obtained from dijet production in SD and DPE.

Exclusive production

Exclusive dijet production

The search for exclusive dijet production is based on measuring the dijet mass fraction M_{jj}, defined as the mass of the two leading jets in an event divided by the total mass reconstructed from all the energy observed in all calorimeters. Fig. 2 (*right*) shows M_{jj} distributions for events with different selection criteria. The signal from exclusive dijets is expected to be concentrated in the region of $R_{jj} > 0.8$, with values of $R_{jj} < 1$ being caused by measurement resolution effects and final state radiation. Of course, background events from inclusive DPE production, $\bar{p}p \to (\bar{p}+gap)+JJ+X+gap$, are expected to contribute to the entire M_{jj} region.

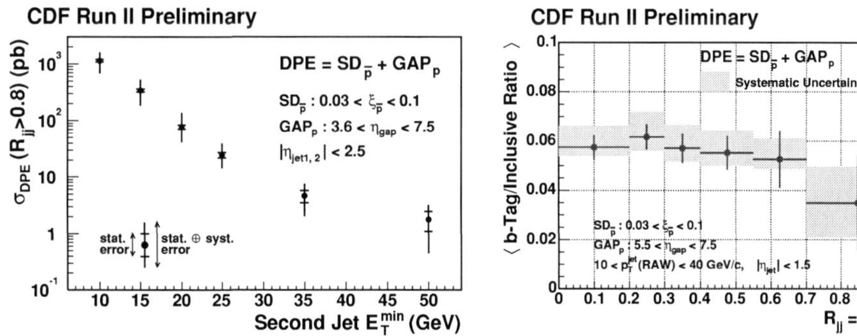

FIGURE 3. (*left*) Dijet production cross sections for $R_{jj} > 0.8$ in DPE events as a function of E_T^{min}, the E_T of the next to the highest E_T jet; (*right*) the ratio of b-tagged to all jets in the DPE dijet event sample versus the dijet mass fraction.

Since no peak is observed at $R_{jj} > 0.8$ in Fig. 2 (*right*), CDF reports production cross sections for events with $R_{jj} > 0.8$, which could be used as upper limits for exclusive production. Figure 3 (*left*) shows such cross sections for various kinematic cuts plotted versus E_T^{min}, the next to leading jet E_T. These cross sections agree, within errors, with recent predictions for exclusive dijet production [5]. Thus, for these predictions to be correct, the background would have to vanish as $R_{jj} \to 1$. While this is guaranteed by the $J_z = 0$ selection rule for leading order $gg \to q\bar{q}$ jets of $m_q << M_{jet}$, Monte Carlo (MC) simulations are used to deal with the dominant $gg \to gg$ process. To avoid using simulations $q\bar{q}$ events could be used to estimate the background and this could be done using dijet events in which at least one of the jets is b-tagged. Figure 3 (*right*) shows the ratio of b-tagged to inclusive dijet events versus dijet mass fraction. A suppression is observed as $M_{jj} \to 1$, as would be expected if there were exclusive dijets in the sample. However, background still may exist from the gluon splitting process $gg \to g + g(\to b\bar{b})$. This background could be practically eliminated if both jets were required to be b-tagged. Presently, more data are being collected with an unprescaled b-tagged dijet trigger to yield a large sample of double-b-tagged dijet events to measure the rate for exclusive production in a low background environment.

Exclusive χ_c^0 production

CDF has reported an upper limit of 49 ± 18 (stat) ± 39 (syst) pb for exclusive χ_c^0 production from a search for $J/\psi + \gamma$ events from $\bar{p}p \to \bar{p} + \chi_c^0(\to J/\psi + \gamma \to \mu\mu + \gamma) + \bar{p}$. Theoretical predictions of ~ 70 pb have recently been revised to ~ 50 pb [5]. More data, collected with a dedicated trigger, are currently being analyzed.

CONCLUSIONS

A comprehensive program of measurements of the diffractive structure function and of exclusive diffractive production is currently under way at CDF aiming at deciphering the QCD nature of diffraction and at providing benchmark calibrations for estimating rates for diffractive Higgs production at the LHC.

REFERENCES

1. We use rapidity, $y = \frac{1}{2}\frac{E+p_L}{E-p_L}$, and pseudorapidity, $\eta = -\ln\tan\frac{\theta}{2}$, interchangeably, since in the kinematic region of interest they are approximately equal.
2. V. Barone and E. Predazzi, "High-Energy Particle Diffraction," Springer Press, 2001.
3. K. Goulianos, "Hadronic Diffraction: Where do we Stand?," in *La Thuile 2004, Results and Perspectives in Particle Physics*, edited by M. Greco, Proc. of Les Rencontres de Physique de la Vallé d'Aoste, La Thuile, Aosta Valley, Italy, February 29 - March 6, 2004, pp. 251-274; e-Print Archive: hep-ph/0407035.
4. K. Goulianos, "Multigap Diffraction at the LHC," in these proceedings.
5. V. A. Khoze, A. D. Martin, and M. G. Ryskin, *Eur. Phys. J. C*, **34**, 327 (2004); A. B. Kaidalov, V. A. Khoze, A. D. Martin, and M. G. Ryskin, *Eur. Phys. J. C*, **33**, 261 (2004).
6. V. A. Khoze, A. B. Kaidalov, A. D. Martin, M. G. Ryskin, and W. J. Stirling, IPPP/05/36, DCPT/05/72, to appear in Proc. of *Gribov-75 Memorial Wprkshop*, Budapest, Hungary, May 2005.
7. M. Gallinaro, "New Diffractive Results from the Tevatron;"e-Print Archive: hep-ph/0505159.
8. K. Goulianos, *J.Phys. G*, **26**, 716 (2000).

Double Pomeron Physics at the LHC

Michael G. Albrow

Fermi National Accelerator Laboratory, Batavia, IL 60510, USA

Abstract. I discuss central exclusive production, also known as Double Pomeron Exchange, $D\mathbb{P}E$, from the ISR through the Tevatron to the LHC. There I emphasize the interest of exclusive Higgs and W^+W^-/ZZ production.

Keywords: Double Pomeron, Central Exclusive Production, Diffractive Higgs
PACS: 13.85.-t, 14.70.Fm, 14.80.Bn, 14.80.Cp

INTRODUCTION

In 1973, shortly after the CERN Intersecting Storage Rings (ISR) provided the first colliding hadron beams, "high mass" diffraction was discovered by the CERN- Holland-Lancaster- Manchester collaboration [1]. In this context "high mass" meant $\approx$ 10 GeV, much larger than the $\approx$ 2 GeV diffractive states seen hitherto. Then Shankar [2] and D.Chew and G.Chew [3] predicted in the framework of Triple-Regge theory double pomeron exchange, $D\mathbb{P}E$,where both beam hadrons are coherently scattered and a central hadronic system is produced. Later experiments, in particular at the Split Field Magnet [4] and the Axial Field Spectrometer (AFS) [5] discovered the processes: $\mathbb{P}\,\mathbb{P} \rightarrow \pi^+\pi^-, K^+K^-, p\bar{p}, 4\pi$ at $\sqrt{s}$ up to 63 GeV. In the case of the AFS we added very forward proton detectors to the large central high-p_T detector, motivated largely by a search for glueballs. Structures were indeed found in the $\pi^+\pi^-$ mass spectrum, not all understood and not, unfortunately, studied at higher $\sqrt{s}$. The absence of a ρ signal verified that $D\mathbb{P}E$ is indeed dominant at this energy, but not at lower (SPS) energies. Measuring the (coherently scattered) forward protons allowed a partial wave analysis to select $J = 0,2$ central states.

Now we want to do a similar experiment on a much grander scale, adding very small forward proton detectors to the large central high-p_T detectors: CMS and ATLAS. At $\sqrt{s}$ = 14,000 GeV rather than 63 GeV we will be measuring W^+W^- and ZZ rather than $\pi^+\pi^-$ and looking for Higgs bosons or other phenomena (perhaps even more interesting, such as anomalous EWK-QCD couplings). What will the $M(W^+W^-), M(ZZ)$ spectra look like? As at the ISR, measurements of the (coherently scattered) forward protons will enable one to determine the quantum numbers of the central states, picking out the S-wave (scalars), D-wave (spin 2), etc. This is very powerful; even if, for example, a Higgs boson is discovered another way it may take central exclusive production to prove that it is a scalar. There will be forward roman pots around CMS at 220m for the TOTEM experiment to measure (in special runs) $\sigma_{TOT}, \frac{d\sigma}{dt}$ and other diffractive processes. To study central masses below 200 GeV (the favored Higgs region) in normal

CP792, *Deep Inelastic Scattering, DIS 2005,* edited by W. H. Smith and S. R. Dasu

high luminosity low-β running we need to measure protons even farther from the collision point, at 420m. Physicists from ATLAS, CMS and TOTEM have joined forces on an R&D project called FP420 to develop common technical solutions; we hope both large detectors will have this proton tagging capability.

In symmetric colliding beams the beam rapidity $y_{BEAM} = ln\frac{\sqrt{s}}{M_p}$, and a central produced state of mass M_{CEN} spans approximately $\Delta y_{CEN} = 2ln\frac{M_{CEN}}{M_0}; M_0 \approx 1$ GeV. Pomeron $I\!P$ exchanges begin to dominate (exceeding Reggeon exchanges) when a rapidity gap exceeds about 3 units, which is a good "rule-of-thumb", although 4 units is safer. Requiring two gaps of > 3 units, the maximum central mass follows from the above as simply $M_{CEN}(max) \approx \frac{\sqrt{s}}{20}$, which gives nominal limits of 3 GeV at the ISR (less at the SPS fixed target, which is therefore very marginal), 100 GeV at the Tevatron and 700 GeV at the LHC. The central exclusive mass spectra did indeed extend to $\approx$ 3 GeV at the ISR [5], and the Tevatron experiment CDF finds [6] *DIPE* di-jets with masses up to $\approx$ 100 GeV. The Tevatron would be a perfect place for low mass *DIPE* spectroscopy (glueballs, hybrids, odderon search) but this has not yet been done. At the $Sp\bar{p}S$ collider, with $\sqrt{s}$ = 630 GeV, a few *DIPE* studies were done. UA1 had no forward proton detection but studied [7] charged multiplicity $n^{\pm}$ and p_T distributions up to $M_{CEN} \approx$ 60 GeV using rapidity gaps. UA8 had roman pots, but studied mostly single diffraction, with some low mass *DIPE* [8]. At the Tevatron ($\sqrt{s}$ = 630, 1800, 1960 GeV) CDF has forward proton (FP) detection (roman pots) on the $\bar{p}$ side only, and uses the gap criterion on the p side. As well as jet physics, searches are underway for exclusive χ_c and exclusive central $\gamma\gamma$ without, unfortunately, detecting the protons. D0 in Run 1 had no FP detection but studied jets with gaps. In Run 2 they now have FP detection on both sides but have not presented *DIPE* data yet.

The extension of the *DIPE* mass range from $\approx$100 GeV at the Tevatron to $\approx$700 GeV at the LHC is exciting, as it takes us into the $W, Z, H, t\bar{t}$ domain.

CENTRAL EXCLUSIVE PRODUCTION AT THE LHC

The main channel for Higgs boson production at the LHC is gg-fusion. Another gluon exchange can cancel the color and can even leave the protons intact: $pp \rightarrow p+H+p$ where the $+$ denote large rapidity gaps and there are *no* other particles produced (i.e. it is *exclusive*). If the outgoing protons are well measured, the mass $M_{CEN} = M_H$ can be determined by the missing mass method [9] with $\sigma_M \approx 2$ GeV, and its quantum numbers can be determined. Theoretical uncertainties in the cross section involve skewed gluon distributions, gluon k_T, gluon radiation, Sudakov form factors, etc. Probably [10, 11] for a Standard Model (SM) Higgs, $\sigma_{SMH} \approx 0.2$ fb at the Tevatron, which is not detectable, but at the LHC $\sigma_{SMH} \approx 3$ fb (within a factor 2-3) and with the higher luminosity (30-100 fb^{-1}) there should be enough events to be valuable. Some of the uncertainties in the cross section can be addressed by measuring related processes at the Tevatron. The process $gg \rightarrow H$ proceeds through a top loop. The same diagram with instead a $b(c,u)$ loop can give exclusive $\chi_b(\chi_c, \gamma\gamma)$, which can therefore be used to "calibrate" the theory now at the Tevatron and then in the early days of the LHC. There are predictions [12, 10] for exclusive $pp \rightarrow p+\chi_c+p \approx 600$ nb at the Tevatron, $\approx$ 20/sec! In reality requiring decay to a useful channel ($\chi_c \to J/\psi\gamma \to \mu^+\mu^-\gamma$), no other interaction (for cleanliness), trigger efficiency and acceptance reduces this to effectively a few pb (still, thousands of events in 1 fb^{-1}). Candidates have been seen (also candidates for exclusive J/ψ which may be from photoproduction ($\gamma I\!P$)). Exclusive χ_b may also be possible but is marginal; the cross section is 5000 times smaller. It will be valuable to measure this early at the LHC (with TOTEM+CMS at high β^*?). Unfortunately in CDF we cannot detect the associated protons, which would provide a quantum number filter, selecting mainly $I^G J^{PC} = 0^+0^{++}$; $J^P = 2^+$ is forbidden at $t = 0$ for a $q\bar{q}$ state. D0 *may*, but they have $|t_{min}| \approx 0.7$ GeV^2 which limits the statistics.

The process $pp \to p + H + p$ with $H \to b\bar{b}$ with no other activity (e.g. no gluon emission) would have two and only two central jets. We can also have $pp \to p + gg + p$ or $pp \to p + b\bar{b} + p$ which we call "exclusive dijets", although it is clear that both experimentally and theoretically that is not a well defined state (unlike exclusive χ_c or exclusive W^+W^- production). Nonetheless we look in CDF for signs of "exclusive dijets" which we can define, with some arbitrariness, as events where two central jets as defined by a jet algorithm (again, not unique) have $R_{jj} = \frac{M_{jj}}{M_{CEN}} > 0.8$. (The events selected have a forward $\bar{p}$ detected and a rapidity gap on the p-side.) There is no R_{jj} = 1 "exclusive" peak, and probably none is expected; there may be a broad high R_{jj} enhancement but with respect to what? CDF look to see if at $R_{jj} > 0.8$ there is a depletion of quark (specifically b) jets as expected [12]; we can also look at the g/q-jet ratio using internal jet features vs R_{jj}. At the LHC, one could get very large samples (early, with low luminosity, tagging the protons) of exclusive dijets with $M_{CEN} = M_{jj} \approx$ 100-200 GeV. These should be very pure gluon jets, which could be used to study QCD (think of the large samples of quark jets studied at LEP on the Z).

A difficult issue with exclusive SMH(120-130 GeV) is that the 420m p-detectors are too far away to be included in the 1st level trigger, L1, and the central jets from the H-decay are completely overwhelmed by QCD jet production. Putting forward rapidity gaps in the L1 trigger can be done but only works with single interactions/low luminosity. The total integrated luminosity if only single interactions can be used is expected to be $\approx$ 2-3 fb^{-1} which is not enough for a SM Higgs, although it might be for some MSSM scenarios which can have a much bigger (factor $\approx$ 50) cross section. [J.Ellis, J.Lee and A.Pilaftsis discussed [13] diffractive production of MSSM Higgs at the LHC.] A solution might be to have a L1 trigger based on a 220m pot track and 2 jets with specific kinematics, such as 100 GeV $< M_{jj} <$150 GeV, small $\sum \vec{E_T}$ (the forward protons will have $\sum \vec{p_T} <\approx 2$ GeV and the jets balance that), and with the jets in the same rapidity hemisphere as the 220m proton. Better, for the desired process $pp \to p + J_1 J_2 + p$ there is a relation between the rapidities of the jets y_1, y_2 and the momentum loss fractions ξ_1, ξ_2 of the forward protons: $\xi_{1(2)} = \frac{E_T}{\sqrt{s}}[e^{-(+)y_1} + e^{-(+)y_2}]$. If a (even a few-bit) measurement of ξ from the 220m pot track can be combined with the jets' (E_T, η, ϕ) at L1 it should help, but the technical feasibility (and value) remains to be studied. The 420m detectors can be included at L2. If the Higgs boson mass is 140 - 200 GeV W^+W^- and eventually ZZ decays come in, and can provide L1 triggers, so the forward detectors can be part of L2. They can again be of great value for quantum number determinations and for a good mass measurement ($\sigma(M_H)$ = 2 GeV *per event*) even in the mode $W^+W^- \to l^+\nu l^-\bar{\nu}$. These events are clean even with pile-up, as the l^+l^- vertex has no other particles (the distribution of $n^\pm$ on the l^+l^- vertex will be broad but with a peak at 0). Two photon processes $\gamma\gamma \to W^+W^-$ give a continuum background for WW (not for ZZ) and the $|t|$ of the protons is smaller which helps the rejection.

PROMPT VECTOR BOSON PAIR PRODUCTION

By "prompt" I mean not from $t\bar{t}$ (a most prolific source) and not from Higgs, and by "vector boson" V I mean γ, W, Z. There are several production mechanisms. Approximately 90% of prompt W^+W^- are from $q\bar{q}$ annihilation with t-channel q exchange. This can produce any $Q = 0, 1$ pair. $q\bar{q}$ annihilation with an s-channel V can produce only $\gamma W, WW$ and WZ; it is important as a probe of the VVV vertex. Virtual V emission from quarks, rescattering to a real pair (any pair, even W^+W^+) is negligible at the Tevatron, but is $\approx 10\%$ at the LHC (WW-scattering with a possible Higgs pole, something else, or unitarity violation!). Two photon production $\gamma\gamma \to W^+W^-$ is about 100 fb at the LHC, is well known and has the characteristic feature of very small t_1, t_2 for the protons. *DIPE* production of W-pairs has not been calculated, but whether inclusive or exclusive it should be very small in the SM. I address later the possibility of this being dramatically wrong. Note that in these various processes for VV production there are different color flows (color triplet annihilation, color singlet exchange, high p_T forward jets, low-p_T protons) which can give different hadronic activity. For single interactions that might be of interest.

At the Tevatron the (non-diffractive) cross sections agree with CTEQ NLO which predict: $\sigma(WW) = 12.4$ pb, $\sigma(WZ) = 3.65$ pb and $\sigma(ZZ) = 1.39$ pb (the latter has not yet been measured). At the LHC the cross sections should be $\approx 10\times$ higher. At the Tevatron we found the following "rules-of-thumb" for diffractive production of hard final states (jets, W): about 1% (within a factor 2) are produced by single diffraction, and about 10^{-3} are produced by *DIPE*. This would imply 120 fb for SDE $\to W^+W^-$ and 12(1) fb for *DIPE* $\to WW(ZZ)$ (+ anything).

The WW decay mode of the SM Higgs rises through 10% at 120 GeV, through 50% at 140 GeV and is about 98% above 160 GeV. Let us consider three WW event classes at the LHC. In all cases consider only the $e\nu$ and $\mu\nu$ decay modes, which unfortunately gives a factor $(4 \times 0.106^2) = 0.045$ (later we will relax this). The *DIPE* $WW \to l^+l'^-\nu\bar{\nu} + X$ cross section is ≈ 0.5 fb, small but perhaps not impossible to see; in any case this might be considered a background to the following more interesting signals. Exclusive W^+W^- with the two forward protons and nothing else can come from exclusive Higgs production or from $\gamma\gamma$ collisions. The former is predicted to be, for a 170 GeV Higgs, $\approx$ 3 fb $\times$ 0.045 (BR) $\approx$ 0.13 fb. The latter is larger, $\approx$ 100 fb $\times$ 0.045 = 4.5 fb. *However* (a) the $\gamma\gamma$ data is a mass continuum while the Higgs events are localised with the missing mass method in a $\approx$ 4 GeV bin (b) the t_1 and t_2 of the protons is more peaked at low values in the $\gamma\gamma$ case. For both classes of exclusive events, with the $pWWp$ missing mass method one can probably use also the $\tau\nu$ decay mode and even the dominant $W \to q\bar{q}$ decay mode for one of the W's. Note that there are potentially useful missing mass games one can play, e.g. in $p_1 p_2 \to p_3 + WW + p_4 \to p_3 + l^\pm \nu j_1 j_2 + p_4$ the missing mass squared: $MM^2 = (p_1 + p_2 - p_3 - p_4 - p_e - p_{j_1} - p_{j_2})^2 = M_\nu^2 = 0$. Ability to use $W \to q\bar{q}$ modes for one W would increase the statistics by a factor 7.4 over $e\nu, \mu\nu$ only. In $H \to ZZ \to \mu^+\mu^-\nu\bar{\nu}$ the invisible missing mass from $MM^2 = (p_1 + p_2 - p_3 - p_4 - p_{\mu_1} - p_{\mu_2})^2 = M_Z^2$ should help distinguish this from the $WW \to \mu^+\mu^-\nu\bar{\nu}$ state (of course we also have $(p_{\mu_1} + p_{\mu_2})^2 = M_Z^2$), as well as measuring $M(ZZ)$.

We cannot expect to see $DI\!PE \to W^+W^-$ at the Tevatron, but it may still be very interesting to study the associated hadronic activity in VV and also single V events. CDF and D0 each have around 20 $WW/WZ/ZZ$ events in Run 2 based on the first 0.2 fb^{-1}, with a factor $\approx$25 more to come. Counting associated hadrons in the CDF events we find a very large spread, with $n_{ass}^{\pm}$ in $p_T > 400$ MeV/c, $|\eta| < 1.0$ ranging from 0 to 34! More statistics and more studies are needed to say if there is anything anomalous, and the "super-clean" event cannot be called diffractive, but it is likely that the high $n_{ass}^{\pm}$ event was a small impact parameter collision and the super-clean event had large impact parameter *and yet produced a W-pair*.

DIFFERENT POMERONS

To 0th order soft (low $|t|$, low Q^2) diffractive interactions are due to a pair of gluons in a color singlet ... a classical Low-Nussinov soft pomeron. There can be a small (ggg) component which becomes relatively more important at larger $|t|$. These exchanges are equivalent to a sum over towers of virtual glueballs. As Q^2 increases, $q\bar{q}$ evolve in. Reggeons are predominantly towers of virtual $q\bar{q}$ mesons, summed over spins. There has been an ambitious attempt to calculate the pomeron in QCD as a "reggeized gluon ladder" ... the BFKL pomeron. It is known that the exchange of a single gluon between quark lines, the leading order qq-scattering QCD diagram, is 'sick"; it is not gauge invariant. A summing procedure over diagrams can result in a gauge invariant exchange, the "reggeized gluon". In the BFKL pomeron two reggeized gluons cancel each other's color. This "pomeron exchange between quarks" diagram enhances jet production in the forward direction (low $|t|$, high s). In the "White pomeron" [14] the color of the reggeized gluon is cancelled instead by an infinite number of wee gluons (they have no momentum even in the infinite momentum frame). The wee gluons have the properties of the vacuum; in a sense they *are* the vacuum. In White's theory asymptotic freedom requires a pair of very heavy color sextet quarks, which couple strongly to the pomeron *and* to the W and Z once the energy is high enough. Consequently at the LHC diffractive W,Z production should be prolific, including $pp \to p + WW/ZZ + p$ exclusive states.There should also be an *effective* $\gamma Z I\!P$ coupling through color sextet quark loops, and hence photoproduction of single Z seen as $pp \to p + Z + p$, which would be another surprise (effectively an anomalous EWK-QCD coupling).

FP420

The potentially rich physics program at the LHC with $DI\!PE$, especially with central states $WW, ZZ, H, jj, t\bar{t}, X$, needs the big central detectors CMS and ATLAS together with very forward proton detection and precision measurement. This can be partially provided by the TOTEM detectors with CMS, but it is necessary to supplement them with detectors at 420m. At this place the relevant protons have been deflected out of the beam by $\approx$ 3-25 mm where they can be detected in small precision pixel tracking detectors. An international consortium of CMS, TOTEM and ATLAS physicists has been formed to develop this proposal, and a LOI for support for R&D will be sent to the LHCC in June.

The proposed precise very forward proton detectors may have a side benefit of calibrating the energy scale of the hadronic calorimetry. (At the Tevatron this gives the largest uncertainty in e.g. the top quark mass.) During a special run at low luminosity with less than one interaction per crossing, trigger on events with two forward protons and nothing else beyond (say) $\eta = 4.0$ ($\theta = 2°$). The total central mass (e.g. $\approx$ 200 GeV) is contained in the main CMS/ATLAS detctors and is known to $\approx 1\%$. The electromagnetic calorimetry should already be well calibrated with $Z \to e^+e^-$, so this calibrates the hadronic energy scale, for jet or non-jet events. This should be competitive with other approaches (γ-jet balancing and $W \to$ jet+jet in top events).

CLOSING REMARKS

There are many other related talks at this meeting (e.g. Cox, Eggert, Klein, Kowalski, Piotrzowski, Royon, ...) demonstrating the interest in the field. This is sure to be a very exciting field at the LHC, *whether or not the Higgs boson is in reach.* It it exists and we see it, central exclusive production will be important for measurements of the mass, quantum numbers, couplings and other properties. If it does not exist, exotic new physics may manifest itself through this process. We will have come a long way from $pp \to p + \pi^+\pi^- + p$ at $\sqrt{s} = 63$ GeV to $pp \to p + W^+W^- + p$ at $\sqrt{s} = 14{,}000$ GeV!

REFERENCES

1. M.G.Albrow et al., *Nucl.Phys.* **B51** 388 (1973).
2. R.Shankar, *Nucl.Phys.* **B63** 168 (1974).
3. D.Chew and G.Chew, *Phys.Lett.* **53B** 191 (1974).
4. L.Baksay et al.,*Phys.Lett.* **61B** 89 (1976); A.Breakstone et al., *Z.Phys.C* **31** 185 (1986) and references therein.
5. T.Akesson et al., *Phys.Lett.* **133B** 268 (1983); T.Akesson et al., *Nucl.Phys.* **B264** 154 (1986).
6. T.Affolder et al., *Phys.Rev.Lett.* **85** 4215 (2000); D.Acosta et al., *Phys.Rev.Lett.* **93** 141601 (2004) and hep-ex/0311023.
7. D.Joyce et al., *Phys.Rev.* **D48** 48 (1993).
8. A.Brandt et al., *Eur.Phys.J.* **C25** 361 (2002)
9. M.G.Albrow and A.Rostovtsev, hep-ph/0009336 (2000)
10. V.Khoze, A.Martin and M.Ryskin, *Eur.Phys.J.* **C23** 311 (2002)
11. V.Khoze, A.Martin and M.Ryskin, *Eur.Phys.J.* **C26** 229 (2002)
12. V.Khoze, A.Martin and M.Ryskin, *Eur.Phys.J.* **C19** 477 (2001)
13. J.R.Ellis, J.S.Lee and A.Pilaftsis, *Phys.Rev* **D71**:075007 (2005), hep-ph/0502251.
14. A.R.White, The Physics of a Sextet Quark Sector, hep-ph/0412062. See M.G.Albrow, hep-ph/0409308 for an introduction.

Multigap Diffraction at LHC

Konstantin Goulianos

The Rockefeller University, New York, NY 10021, USA

Abstract. The large rapidity interval available at the Large Hadron Collider (LHC) offers an arena in which the QCD aspects of diffraction may be explored in an environment free of gap survival complications using events with multiple rapidity gaps.

SOFT DIFFRACTION

Diffractive processes are characterized by large rapidity gaps, defined as regions of (pseudo)rapidity [1] in which there is no particle production. Diffractive gaps are presumed to be produced by the exchange of a color singlet quark/gluon object with vacuum quantum numbers referred to as the *Pomeron* [2, 3] (the present paper contains excerpts from these two references).

Traditionally diffraction had been treated in Regge theory using an amplitude based on a simple Pomeron pole and factorization. This approach was successful at $\sqrt{s}$ energies below $\sim$ 50 GeV [4], but as collision energies increased to reach $\sqrt{s}$ =1800 GeV at the Fermilab Tevatron the SD cross section was found to be suppressed by a factor of $\sim \mathcal{O}(10)$ relative to the Regge-based prediction [5]. This blatant breakdown of factorization was traced back to the energy dependence of the Regge theory $\sigma_{sd}^{tot}(s)$,

$$d\sigma_{sd}(s,M^2)/dM^2 \sim s^{2\varepsilon}/(M^2)^{1+\varepsilon}, \tag{1}$$

which is faster than that of $\sigma^{tot}(s) \sim s^{\varepsilon}$, so that at high $\sqrt{s}$ unitarity would have to be violated if factorization held.

In contrast to the Regge theory prediction of Eq. (1), the measured SD M^2-distribution shows no explicit s-dependence (M^2-scaling) over a region of s spanning six orders of magnitude [6]. Thus, factorization appears to *yield* to M^2-scaling. This is a property built into the *Renormalization Model* of hadronic diffraction, in which the Regge theory Pomeron flux is renormalized to unity [7].

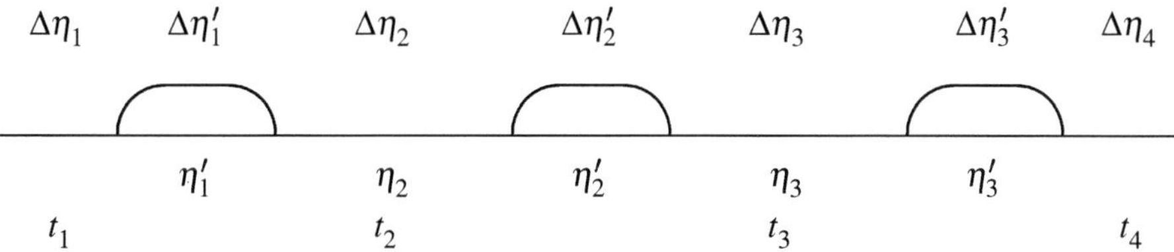

FIGURE 1. Average multiplicity $dN/d\eta$ versus η for a process with four rapidity gaps $\Delta\eta_{i=1-4}$.

CP792, *Deep Inelastic Scattering, DIS 2005,* edited by W. H. Smith and S. R. Dasu

In a QCD inspired approach, the renormalization model was extended to central and multigap diffractive processes [8], an example of which is the four-gap process shown schematically in Fig 1. In this approach cross sections depend on the number of wee partons [9] and therefore the pp total cross section is given by

$$\sigma_{pp}^{tot} = \sigma_0 \cdot e^{\varepsilon \Delta\eta'}, \tag{2}$$

where $\Delta\eta'$ is the rapidity region in which there is particle production. Since, from the optical theorem, $\sigma^{tot} \sim \mathrm{Im\,f^{el}}(t=0)$, the full parton model amplitude may be written as

$$\mathrm{Im\,f^{el}}(t,\Delta\eta) \sim e^{(\varepsilon+\alpha' t)\Delta\eta}, \tag{3}$$

where $\alpha' t$ is a simple parameterization of the t-dependence of the amplitude. On the basis of this amplitude, the cross section of the four-gap process of Fig. 1 takes the form

$$\frac{d^{10}\sigma^D}{\Pi_{i=1}^{10} dV_i} = N_{gap}^{-1} \underbrace{F_p^2(t_1) F_p^2(t_4) \Pi_{i=1}^4 \left\{ e^{[\varepsilon+\alpha' t_i]\Delta\eta_i} \right\}^2}_{\text{gap probability}} \times \kappa^4 \left[\sigma_0 e^{\varepsilon \sum_{i=1}^3 \Delta\eta_i'} \right], \tag{4}$$

where the term in square brackets is the pp total cross section at the reduced s-value, defined through $\ln(s'/s_0) = \sum_i \Delta\eta_i'$, κ (one for each gap) is the QCD color factor for gap formation, the gap probability is the amplitude squared for elastic scattering between two diffractive clusters or between a diffractive cluster and a surviving proton with form factor $F_p^2(t)$, and N_{gap} is the (re)normalization factor defined as the gap probability integrated over all 10 independent variables t_i, η_i, η_i', and $\Delta\eta \equiv \sum_{i=1}^4 \Delta\eta_i$.

The renormalization factor N_{gap} is a function of s only. The color factors are $c_g = (N_c^2-1)^{-1}$ and $c_q = 1/N_c$ for gluon and quark color-singlet exchange, respectively. Since the reduced energy cross section is properly normalized, the gap probability is (re) normalized to unity. The quark to gluon fraction, and thereby the Pomeron intercept parameter ε may be obtained from the inclusive parton distribution functions (PDFs) [2]. Thus, normalized differential multigap cross sections at $t=0$ may be fully derived from inclusive PDFs and QCD color factors without any free parameters.

The exponential dependence of the cross section on $\Delta\eta_i$ leads to a renormalization factor $\sim s^{2\varepsilon}$ independent of the number of gaps in the process. This remarkable property of the renormalization model, which was confirmed in two-gap to one-gap cross section ratios measured by the CDF Collaboration (see [2]), suggests that multigap diffraction can be used as a tool for exploring the QCD aspects of diffraction in an environment free of rapidity gap suppression effects. The LHC with its large rapidity coverage provides the ideal arena for such studies.

HARD DIFFRACTION

Hard diffraction processes are those in which there is a hard partonic scattering in addition to the diffractive rapidity gap. SD/ND ratios for W, dijet, b-quark, and J/ψ production at $\sqrt{s}$ =1800 GeV measured by the CDF Collaboration are approximately equal ($\sim 1\%$), indicating that the rapidity gap formation probability is largely *flavor independent*. However, the SD structure function measured from dijet production is suppressed by $\sim \mathcal{O}(10)$ relative to expectations based on diffractive PDFs measured from diffractive DIS at HERA.

A modified version of our QCD approach to soft diffraction can be used to describe hard diffractive processes and has been applied to diffractive DIS at HERA, $\gamma^* + p \to p + Jet + X$, and diffractive dijet production at the Tevatron, $\bar{p} + p \to \bar{p} + \text{dijet} + X$ in [3]. The hard process generally involves several color "emissions" from the surviving proton, the sum of which comprises a color singlet exchange with vacuum quantum numbers. Two of these emissions are of special interest, one at $x = x_{Bj}$ from the proton's hard PDF at scale Q^2, which causes the hard scattering, and another at $x = \xi$ (fractional momentum loss of the diffracted nucleon) from the soft PDF at $Q^2 \approx 1$ GeV^2, which neutralizes the exchanged color and forms the rapidity gap. Neglecting the t-dependence, the diffractive structure function could then be expressed as the product of the inclusive hard structure function and the soft parton density at $x = \xi$,

$$F^D(\xi, x, Q^2) = \frac{A_{\text{norm}}}{\xi^{1+\varepsilon}} \cdot c_{g,q} \cdot F(x, Q^2) \Rightarrow \frac{A_{\text{norm}}}{\xi^{1+\varepsilon+\lambda(Q^2)}} \cdot c_{g,q} \cdot \frac{C(Q^2)}{\beta^{\lambda(Q^2)}}, \tag{5}$$

where $c_{g,q}$ are QCD color factors, λ is the parameter of a power law fit to the hard structure function in the region $x < 0.1$, A_{norm} is a normalization factor, and $\beta \equiv x/\xi$.

At high Q^2 at HERA, where factorization is expected to hold [7, 12], A_{norm} is the nominal normalization factor of the soft PDF. This factor is constant, leading to two important predictions, which are confirmed by the data:

i) The Pomeron intercept in diffractive DIS (DDIS) is Q^2-dependent and equals the average value of the soft and hard intercepts:

$$\alpha_{I\!P}^{DIS} = 1 + \lambda(Q^2), \quad \alpha_{I\!P}^{DDIS} = 1 + \frac{1}{2}\left[\varepsilon + \lambda(Q^2)\right] \tag{6}$$

ii) The ratio of DDIS to DIS structure functions at fixed ξ is independent of x and Q^2:

$$R\left[\frac{F^D(\xi, x, Q^2)}{F^{ND}(x, Q^2)}\right]_{\text{HERA}} = \frac{A_{\text{norm}} \cdot c_q}{\xi^{1+\varepsilon}} = \frac{\text{const}}{\xi^{1+\varepsilon}} \tag{7}$$

At the Tevatron, where high soft parton densities lead to saturation, A_{norm} must be renormalized to

$$A_{\text{renorm}}^{\text{Tevatron}} = 1 \Big/ \int_{\xi_{min}}^{\xi=0.1} \frac{d\xi}{\xi^{1+\varepsilon+\lambda}} \propto \left(\frac{1}{\beta \cdot s}\right)^{\varepsilon+\lambda}, \tag{8}$$

where $\xi_{min} = x_{min}/\beta$ and $x_{min} \propto 1/s$. Thus, the diffractive structure function acquires a term $\sim (1/\beta)^{\varepsilon+\lambda}$, and the ratio of the diffractive to inclusive structure functions a term $\sim (1/x)^{\varepsilon+\lambda}$. This prediction is confirmed by CDF data, where the x-dependence of the diffractive to inclusive ratio was measured to be $\sim 1/x^{0.45}$ (see [2]).

A comparison [1] between the diffractive structure function measured on the proton side in events with a leading antiproton to expectations from diffractive DIS at HERA showed approximate agreement, indicating that factorization is largely restored for events that already have a rapidity gap. Thus, as already mentioned for soft diffraction, events triggered on a leading proton at LHC provide an environment in which the QCD aspects of diffraction may be explored without complications arising from rapidity gap survival.

PROPOSED PROGRAM OF MULTIGAP DIFFRACTION AT LHC

The rapidity span at LHC running at $\sqrt{s} = 14$ TeV is $\Delta\eta = 19$ as compared to $\Delta\eta = 15$ at the Tevatron. This suggests the following program for studies of non-suppressed diffraction:

- Trigger on two forward rapidity gaps of $\Delta\eta_F \geq 2$ (one on each side of the interaction point), or equivalently on forward protons of fractional longitudinal momentum loss $\xi = \Delta p_L / p_L \leq 0.1$, and explore the central rapidity region of $|\Delta\eta| \leq 7.5$, which has the same width as the entire rapidity region of the Tevatron. In such an environment, the ratio of the rate of dijet events with a gap between jets to that without a gap, *gap*+[jet-gap-jet]+*gap* to *gap*+[jet-jet]+*gap*, should rise from its value of $\sim 1\%$ at the Tevatron to $\sim 5\%$.
- Trigger on one forward gap of $\Delta\eta_F \geq 2$ or on a proton of $\xi < 0.1$, in which case the rapidity gap available for non-suppressed diffractive studies rises to 17 units.

REFERENCES

1. We use *pseudorapidity*, $\eta = -\ln\tan\frac{\theta}{2}$, and *rapidity*, $y = \frac{1}{2}\frac{E+p_L}{E-p_L}$, interchangeably.
2. K. Goulianos, "Hadronic Diffraction: Where do we Stand?," in *La Thuile 2004, Results and Perspectives in Particle Physics*, edited by M. Greco, Proc. of Les Rencontres de Physique de la Vallé d'Aoste, La Thuile, Aosta Valley, Italy, February 29 - March 6, 2004, pp. 251-274; e-Print Archive: hep-ph/0407035.
3. K. Goulianos, *Nucl. Phys. (Proc. Suppt.) B*, **146**, 166 (2005).
4. K. Goulianos, *Phys. Reports*, **101**, 171 (1983).
5. F. Abe et al. (CDF Collaboration), *Phys. Rev. D*, **50**, 5535 (1994).
6. K. Goulianos and J. Montanha, *Phys. Rev. D*, **59**, 114017 (1999).
7. K. Goulianos, *Phys. Lett. B*, **358**, 379 (1995); Erratum-*ib.* **363**, 268 (1995).
8. K. Goulianos, in *Diffraction in QCD*, Corfu Summer Institute on Elementary Particle Physics, Corfu, Greece, 31 Aug - 20 Sep 2001; e-print Archive: hep-ph/0203141.
9. E. Levin, "An Introduction to Pomerons," Preprint DESY 98-120.
10. A. Petrukhin (H1 Collaboration), "Measurement of the Inclusive DIS Cross Section at Low Q^2 and High x Using Events with Initial State Radiation," presented at DIS2004, 14-18 April 2004, Slovakia.
11. F. P. Schilling (H1 Collaboration), "Measurement and NLO DGLAP QCD Interpretation of Diffractive Deep-Inelastic Scattering at HERA," submitted to 31st International Conference on High Energy Physics, ICHEP02, Amsterdam, The Netherlands, Jul. 24—31, 2001 (abstract 980).
12. J. Collins, *J. Phys. G*, **28**, 1069 (2002); e-Print Archive: hep-ph/0107252.

---

[1] Performed by the author and K. Hatakeyama (see [2]) using CDF published data and preliminary H1 diffractive parton densities [11].

Hard Diffraction in QCD

Stanley J. Brodsky

Stanford Linear Accelerator Center, Stanford University, Stanford, CA, 94309

Abstract. Gluon exchange between the outgoing quarks and the target spectators affects the structure functions measured in deep inelastic scattering in a profound way, leading to diffractive leptoproduction processes, the shadowing and antishadowing of nuclear structure functions, and target spin asymmetries – leading-twist physics not incorporated in the light-front wavefunctions of the target computed in isolation. I also discuss the diffraction dissociation of hadrons into jets as a tool for resolving fundamental hadron substructure.

Keywords: Quantum Chromodynamics, Deep Inelastic Lepton-Hadron Interactions, Diffractive Processes
PACS: 12.38.-t,13.60.Hb,13.60.-r,13.88.+e

DIFFRACTIVE DEEP INELASTIC SCATTERING

A remarkable feature of deep inelastic lepton-proton scattering at HERA is that approximately 10% events are diffractive [1, 2]: the target proton remains intact, and there is a large rapidity gap between the proton and the other hadrons in the final state. These diffractive deep inelastic scattering (DDIS) events can be understood most simply from the perspective of the color-dipole model: the $q\overline{q}$ Fock state of the high-energy virtual photon diffractively dissociates into a diffractive dijet system. The exchange of multiple gluons between the color dipole of the $q\overline{q}$ and the quarks of the target proton neutralizes the color separation and leads to the diffractive final state. The same multiple gluon exchange also controls diffractive vector meson electroproduction at large photon virtuality [3]. This observation presents a paradox: if one chooses the conventional parton model frame where the photon light-front momentum is negative $q+ = q^0 + q^z < 0$, the virtual photon interacts with a quark constituent with light-cone momentum fraction $x = k^+/p^+ = x_{bj}$. If one chooses light-cone gauge $A^+ = 0$, then the gauge link associated with the struck quark (the Wilson line) becomes unity. Thus the struck "current" quark experiences no final-state interactions. Since the light-front wavefunctions $\psi_n(x_i, k_{\perp i}$ of a stable hadron are real, it appears impossible to generate the required imaginary phase associated with pomeron exchange, let alone large rapidity gaps.

This paradox was resolved by Paul Hoyer, Nils Marchal, Stephane Peigne, Francesco Sannino and myself [4]. Consider the case where the virtual photon interacts with a strange quark – the $s\overline{s}$ pair is assumed to be produced in the target by gluon splitting. In the case of Feynman gauge, the struck s quark continues to interact in the final state via gluon exchange as described by the Wilson line. The final-state interactions occur at a light-cone time $\Delta\tau \simeq 1/\nu$ shortly after the virtual photon interacts with the struck quark. When one integrates over the nearly-on-shell intermediate state, the amplitude acquires an imaginary part. Thus the rescattering of the quark produces a separated color-singlet $s\overline{s}$ and an imaginary phase. In the case of the light-cone gauge $A^+ = \eta \cdot A = 0$, one

CP792, *Deep Inelastic Scattering, DIS 2005,* edited by W. H. Smith and S. R. Dasu

must also consider the final-state interactions of the (unstruck) $\bar{s}$ quark. The gluon propagator in light-cone gauge $d^{\mu\nu}_{LC}(k) = (i/k^2 + i\varepsilon)[-g^{\mu\nu} + (\eta^\mu k^\nu + k^\mu \eta^\nu / \eta \cdot k)]$ is singular at $k^+ = \eta \cdot k = 0$. The momentum of the exchanged gluon k^+ is of $\mathcal{O}(1/\nu)$; thus rescattering contributes at leading twist even in light-cone gauge. The net result is gauge invariant and is identical to the color dipole model calculation. The calculation of the rescattering effects on DIS in Feynman and light-cone gauge through three loops is given in detail for an Abelian model in Ref. [4]. The result shows that the rescattering corrections reduce the magnitude of the DIS cross section in analogy to nuclear shadowing.

A new understanding of the role of final-state interactions in deep inelastic scattering has thus emerged. The multiple scattering of the struck parton via instantaneous interactions in the target generates dominantly imaginary diffractive amplitudes, giving rise to an effective "hard pomeron" exchange. The presence of a rapidity gap between the target and diffractive system requires that the target remnant emerges in a color-singlet state; this is made possible in any gauge by the soft rescattering. The resulting diffractive contributions leave the target intact and do not resolve its quark structure; thus there are contributions to the DIS structure functions which cannot be interpreted as parton probabilities [4]; the leading-twist contribution to DIS from rescattering of a quark in the target is a coherent effect which is not included in the light-front wave functions computed in isolation. One can augment the light-front wave functions with a gauge link corresponding to an external field created by the virtual photon $q\bar{q}$ pair current [5, 6]. Such a gauge link is process dependent [7], so the resulting augmented LFWFs are not universal [4, 5, 8]. We also note that the shadowing of nuclear structure functions is due to the destructive interference between multi-nucleon amplitudes involving diffractive DIS and on-shell intermediate states with a complex phase. In contrast, the wave function of a stable target is strictly real since it does not have on-energy-shell intermediate state configurations. The physics of rescattering and shadowing is thus not included in the nuclear light-front wave functions, and a probabilistic interpretation of the nuclear DIS cross section is precluded.

Rikard Enberg, Paul Hoyer, Gunnar Ingelman and I [9]. have shown that the quark structure function of the effective hard pomeron has the same form as the quark contribution of the gluon structure function. The hard pomeron is not an intrinsic part of the proton; rather it must be considered as a dynamical effect of the lepton-proton interaction. Our QCD-based picture also applies to diffraction in hadron-initiated processes. The rescattering is different in virtual photon- and hadron-induced processes due to the different color environment, which accounts for the observed non-universality of diffractive parton distributions. This framework also provides a theoretical basis for the phenomenologically successful Soft Color Interaction (SCI) model [10] which includes rescattering effects and thus generates a variety of final states with rapidity gaps.

SINGLE-SPIN ASYMMETRIES FROM FINAL-STATE INTERACTIONS

Among the most interesting polarization effects are single-spin azimuthal asymmetries in semi-inclusive deep inelastic scattering, representing the correlation of the spin of the proton target and the virtual photon to hadron production plane: $\vec{S}_p \cdot \vec{q} \times \vec{p}_H$. Such asymmetries are time-reversal odd, but they can arise in QCD through phase differences in different spin amplitudes. In fact, final-state interactions from gluon exchange between the outgoing quarks and the target spectator system lead to single-spin asymmetries in semi-inclusive deep inelastic lepton-proton scattering which are not power-law suppressed at large photon virtuality Q^2 at fixed x_{bj} [11] In contrast to the SSAs arising from transversity and the Collins fragmentation function, the fragmentation of the quark into hadrons is not necessary; one predicts a correlation with the production plane of the quark jet itself. Physically, the final-state interaction phase arises as the infrared-finite difference of QCD Coulomb phases for hadron wave functions with differing orbital angular momentum. The same proton matrix element which determines the spin-orbit correlation $\vec{S} \cdot \vec{L}$ also produces the anomalous magnetic moment of the proton, the Pauli form factor, and the generalized parton distribution E which is measured in deeply virtual Compton scattering. Thus the contribution of each quark current to the SSA is proportional to the contribution $\kappa_{q/p}$ of that quark to the proton target's anomalous magnetic moment $\kappa_p = \sum_q e_q \kappa_{q/p}$ [11, 12]. The HERMES collaboration has recently measured the SSA in pion electroproduction using transverse target polarization [13]. The Sivers and Collins effects can be separated using planar correlations; both contributions are observed to contribute, with values not in disagreement with theory expectations [13, 14] .

A related analysis also predicts that the initial-state interactions from gluon exchange between the incoming quark and the target spectator system lead to leading-twist single-spin asymmetries in the Drell-Yan process $H_1 H_2^{\updownarrow} \to \ell^+ \ell^- X$ [7, 15]. The SSA in the Drell-Yan process is the same as that obtained in SIDIS, with the appropriate identification of variables, but with the opposite sign. Initial-state interactions also lead to a $\cos 2\phi$ planar correlation in unpolarized Drell-Yan reactions [16]. There is no Sivers effect in charged-current reactions since the W only couples to left-handed quarks [17].

DIFFRACTION DISSOCIATION AS A TOOL TO RESOLVE HADRON SUBSTRUCTURE

Diffractive multi-jet production in heavy nuclei provides a novel way to resolve the shape of light-front Fock state wave functions and test color transparency [18]. For example, consider the reaction [19, 20] $\pi A \to \mathrm{Jet}_1 + \mathrm{Jet}_2 + A'$ at high energy where the nucleus A' is left intact in its ground state. The transverse momenta of the jets balance so that $\vec{k}_{\perp i} + \vec{k}_{\perp 2} = \vec{q}_\perp < R^{-1}{}_A$. Because of color transparency, the valence wave function of the pion with small impact separation will penetrate the nucleus with minimal interactions, diffracting into jet pairs [19]. The $x_1 = x$, $x_2 = 1 - x$ dependence of the di-jet distributions will thus reflect the shape of the pion valence light-cone wave function in x; similarly, the $\vec{k}_{\perp 1} - \vec{k}_{\perp 2}$ relative transverse momenta of the jets gives key information on the second transverse momentum derivative of the underlying shape of the valence pion wavefunction [20, 21]. The diffractive nuclear amplitude extrapolated to $t = 0$ should be linear in nuclear number A if color transparency is correct. The integrated diffractive rate will then scale as $A^2/R_A^2 \sim A^{4/3}$. This is in fact what has been observed by the E791 collaboration at FermiLab for 500 GeV incident pions on nuclear targets [22].

The measured momentum fraction distribution of the jets is found to be approximately consistent with the shape of the pion asymptotic distribution amplitude [23, 24, 25]. $\phi_\pi^{\rm asympt}(x) = \sqrt{3} f_\pi x(1-x)$ [26]. The concept of high energy diffractive dissociation can be generalized to provide a tool to materialize the individual Fock states of a hadron, photon, or nuclear projectile; *e.g.*, the diffractive or Coulomb dissociation of a high energy proton $pA \to qqqA'$ or $pe \to qqqe$ can be used to measure the valence light-front wavefunction of the proton.

ACKNOWLEDGMENTS

This work is based on collaborations with Rikard Enberg, Paul Hoyer, Dae Sung Hwang, Gunnar Ingelman, and Ivan Schmidt. This work was supported by the Department of Energy, contract No. DE-AC02-76SF00515.

REFERENCES

1. C. Adloff *et al.* [H1 Collaboration], Z. Phys. C **76**, 613 (1997) [arXiv:hep-ex/9708016].
2. J. Breitweg *et al.* [ZEUS Collaboration], Eur. Phys. J. C **6**, 43 (1999) [arXiv:hep-ex/9807010].
3. S. J. Brodsky, L. Frankfurt, J. F. Gunion, A. H. Mueller and M. Strikman, Phys. Rev. D **50**, 3134 (1994) [arXiv:hep-ph/9402283].
4. S. J. Brodsky, P. Hoyer, N. Marchal, S. Peigne and F. Sannino, Phys. Rev. D **65**, 114025 (2002) [arXiv:hep-ph/0104291].
5. A. V. Belitsky, X. Ji and F. Yuan, Nucl. Phys. B **656**, 165 (2003) [arXiv:hep-ph/0208038].
6. J. C. Collins and A. Metz, Phys. Rev. Lett. **93**, 252001 (2004) [arXiv:hep-ph/0408249].
7. J. C. Collins, Phys. Lett. B **536**, 43 (2002) [arXiv:hep-ph/0204004].
8. J. C. Collins, Acta Phys. Polon. B **34**, 3103 (2003) [arXiv:hep-ph/0304122].
9. S. J. Brodsky, R. Enberg, P. Hoyer and G. Ingelman, Phys. Rev. D **71**, 074020 (2005) [arXiv:hep-ph/0409119].
10. A. Edin, G. Ingelman and J. Rathsman, Phys. Lett. B **366**, 371 (1996) [arXiv:hep-ph/9508386].
11. S. J. Brodsky, D. S. Hwang and I. Schmidt, Phys. Lett. B **530**, 99 (2002) [arXiv:hep-ph/0201296].
12. M. Burkardt, Nucl. Phys. Proc. Suppl. **141**, 86 (2005) [arXiv:hep-ph/0408009].
13. A. Airapetian *et al.* [HERMES Collaboration], Phys. Rev. Lett. **94**, 012002 (2005) [arXiv:hep-ex/0408013].
14. H. Avakian and L. Elouadrhiri [CLAS Collaboration], AIP Conf. Proc. **698**, 612 (2004).
15. S. J. Brodsky, D. S. Hwang and I. Schmidt, Nucl. Phys. B **642**, 344 (2002) [arXiv:hep-ph/0206259].
16. D. Boer, S. J. Brodsky and D. S. Hwang, Phys. Rev. D **67**, 054003 (2003) [arXiv:hep-ph/0211110].
17. S. J. Brodsky, D. S. Hwang and I. Schmidt, Phys. Lett. B **553**, 223 (2003) [arXiv:hep-ph/0211212].
18. S. J. Brodsky and A. H. Mueller, Phys. Lett. B **206**, 685 (1988).
19. G. Bertsch, S. J. Brodsky, A. S. Goldhaber and J. F. Gunion, Phys. Rev. Lett. **47**, 297 (1981).
20. L. Frankfurt, G. A. Miller and M. Strikman, Found. Phys. **30**, 533 (2000) [arXiv:hep-ph/9907214].
21. N. N. Nikolaev, W. Schafer and G. Schwiete, Phys. Rev. D **63**, 014020 (2001) [arXiv:hep-ph/0009038].
22. E. M. Aitala *et al.* [E791 Collaboration], Phys. Rev. Lett. **86**, 4773 (2001) [arXiv:hep-ex/0010044].
23. G. P. Lepage and S. J. Brodsky, Phys. Lett. B **87**, 359 (1979).
24. A. V. Efremov and A. V. Radyushkin, Theor. Math. Phys. **42**, 97 (1980) [Teor. Mat. Fiz. **42**, 147 (1980)].
25. G. P. Lepage and S. J. Brodsky, Phys. Rev. D **22**, 2157 (1980).
26. E. M. Aitala *et al.* [E791 Collaboration], Phys. Rev. Lett. **86**, 4768 (2001) [arXiv:hep-ex/0010043].

A_N measurement in the CNI region, at $\sqrt{s} = 200GeV$ in polarized pp elastic scattering

Wlodek Guryn for pp2pp collaboration [1]

Brookhaven National Laboratory Upton, NY 11973, USA [1]

Abstract. We describe the first measurement of the single spin analysing power (A_N) at $\sqrt{s} = 200$ GeV in the four momentum transfer t range $0.01 \leq |t| \leq 0.03$ $(\text{GeV}/c)^2$, obtained by the pp2pp experiment using polarized proton beams at the Relativistic Heavy Ion Collider (RHIC). The results presented are preliminary.

Keywords: Polarization, Elastic Scattering
PACS: 13.85.Dz and 13.88.+e

THE EXPERIMENT

The layout of the experiment is shown in Fig. 1. More details can be found in [2, 3]. The identification of elastic events is based on the collinearity criterion, which requires the simultaneous detection of the scattered protons in the pair of Roman Pot (RP) detectors [4] on either side of the IP.

The silicon strip detectors (SSD) in the RPs were used to record the x, y coordinates of the scattered protons. They are made of 0.40 mm thick n-type silicon with p^+-type implanted strips of 0.07 mm width and a strip pitch of 0.10 mm. Each strip is capacitively coupled to an input channel of a SVXIIe [5].

The elastic trigger scintillators are 8 mm thick, 80×50 mm^2 in area, and each is viewed by two photomultiplier tubes. The elastic event trigger is a coincidence between signals in the RP's scintillators, belonging either to arm A or arm B, see Fig. 1. For each arm two closest to the collision point RP's were used: RP1 and RP3. The overall trigger was a logical OR of a coincidence between up and down pots: (RP3U and RP1D) OR (RP3D and RP1U) in coincidence with the beam crossing signal derived from the RHIC master clock. For each event, TDC and ADC information for the trigger scintillation counters was recorded.

SELECTION OF ELASTIC EVENTS

The detectors in RP1 and RP3 were used for elastic event reconstruction, as this provided the highest acceptance for the experiment. Particle hits in the silicon detector were identified for each strip requiring that the energy deposited (ΔE) was $\Delta E \geq 5\sigma$ of

---

[1] Supported under Prime Contract between Brookhaven Science Associates and the Department of Energy No. DE-AC02-98CH10886.

CP792, *Deep Inelastic Scattering, DIS 2005,* edited by W. H. Smith and S. R. Dasu

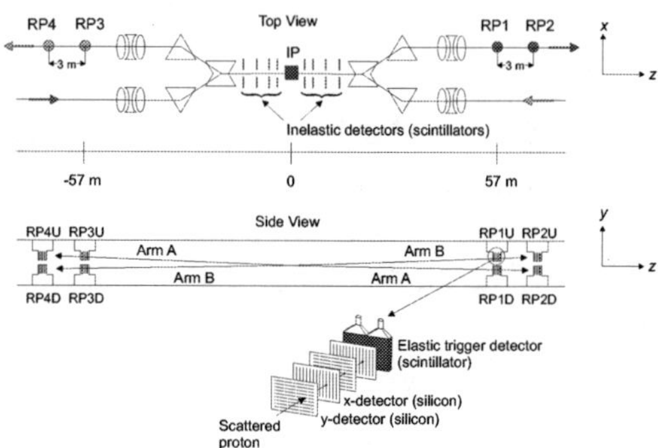

FIGURE 1. Layout of the pp2pp experiment. Note the detector pairs RP1, RP2 and RP3, RP4 lie in different RHIC rings. Scattering is detected in either one of two arms: Arm A is formed from RP3U and RP1D. Conversely, Arm B is formed from RP3D and RP1U.

its pedestal value. From those hits a cluster of consecutive strips was formed and the coordinate for that cluster was calculated as an energy-weighted average of the positions of the strips. The cluster size was limited to no more than five consecutive strips and its ΔE required to be larger than 20 ADC counts, where one ADC count is about 700 electrons of charge. The average silicon detector plane efficiency was better than 0.98, and the signal-to-noise ratio was better than 22.

The collinearity of the scattered protons results in correlation between coordinates measured on each side of the IP. Hence the main criterion to select the elastic scattering events was the hit coordinate correlation in the corresponding silicon detectors on the opposite sides of the IP.

DETERMINATION OF ANALYZING POWER A_N

After the cuts, the sample of 1.14 million events in the t-interval $0.010 \leq -t < 0.030$, subdivided into three intervals $0.010 \leq -t < 0.015$, $0.015 \leq -t < 0.020$, $0.020 \leq -t < 0.030$, was used to determine A_N. In each t-interval the asymmetry was calculated as a function of azimuthal angle ϕ using 5°-bins. Azimuthal angle dependence of the cross section for the elastic collision of the vertically polarized protons is given by

$$2\pi \frac{d^2\sigma}{dt d\phi} = \frac{d\sigma}{dt} \cdot (1 + (P_B + P_Y) A_N \cos\phi + P_B P_Y (A_{NN} \cos^2\phi + A_{SS} \sin^2\phi)), \quad (1)$$

where P_B and P_Y are the beam polarizations and A_{NN}, A_{SS} are double spin asymmetries (see Ref.[6] for definitions). Given beam poalrizations, see below, an upper constraint is 0.028 for the term $P_B P_Y (A_{NN} \cos^2\phi + A_{SS} \sin^2\phi)$, even if both double-spin asymmetries

A_{NN} and A_{SS} were as large as 0.15. This term is small in comparison to the systematic errors on A_N and was therefore neglected in Eq. (1) but included in the systematic error, as described below.

Then the square root formula [7] for the single spin raw asymmetry $\varepsilon(\phi)$ can be written as Eq. (2). A cosine fit to the raw asymmetry $\varepsilon(\phi)$ was used to determine values of A_N.

$$\varepsilon(\phi) \approx (P_B + P_Y)A_N \cos\phi = \frac{\sqrt{N^{\uparrow\uparrow}(\phi)N^{\downarrow\downarrow}(\pi-\phi)} - \sqrt{N^{\downarrow\downarrow}(\phi)N^{\uparrow\uparrow}(\pi-\phi)}}{\sqrt{N^{\uparrow\uparrow}(\phi)N^{\downarrow\downarrow}(\pi-\phi)} + \sqrt{N^{\downarrow\downarrow}(\phi)N^{\uparrow\uparrow}(\pi-\phi)}} \tag{2}$$

Equation (2), from which the asymmetry is calculated has important features; namely, luminosities of the differently polarized proton beam bunches cancel as do the relative detection efficiencies, including geometrical acceptance, for each t and ϕ. Two other contributions to the systematic error were considered: backgrounds, which affect the asymmetry value, and sensitivity to the transport matrix parameters and to the beam position with respect to the detectors that affect the determination of t and ϕ.

The error in A_N due to uncertainty in the transport is 1.4%. The systematic error due to an uncertainty of beam positions at the detectors is 1.8% and due to the variation in L_{eff} was also studied and estimated to be 6.4%. the upper limit of the systematic error due to the background is 4.5%. Since all the above errors are not correlated adding them in quadrature results in the systematic error of $\Delta A_N/A_N = 8.4\%$. This error is smaller than the statistical errors of the measurement.

The polarization values of the proton beams was obtained from the Collider–Accelerator Department (C–AD). For our running period the beam polarizations were $P_Y = 0.346 \pm 0.0731$ and $P_B = 0.532 \pm 0.0988$. The errors include the contribution of the systematic part of the error due to the calibration of pC polarimeter of 13%, which is correlated for both beams and the statistical errors of the measurement. This results in the sum of the polarizations and its error $P_Y + P_B = 0.877 \pm 0.146$.

The total systematic error is comprised of A_N scale error of 16.6% mostly due to the systematic error of the polarization measurement, and 8.4% error due to the experimental systematic effects as described above.

RESULTS AND CONCLUSIONS

The values of A_N obtained in this experiment and their statistical errors are shown in Fig. 2 for the three t-intervals.

The solid curve in Fig. 2 corresponds to the calculation without hadronic spin flip. Recent measurements of A_N at substantially lower cms energies than the one reported here indicate small but significantly different from zero contribution of spin-flip amplitude in case of proton-carbon scattering and are consistent with no spin-flip contribution for proton-proton scattering at $\sqrt{s}$ = 13.7 GeV [8].

Our results, as well as A_N measurements at lower energies, provide the much needed input for the theoretical calculations of the exchange process. They also underline a need for further measurements to be able to reconcile the differences for a more complete picture to emerge and also to extend the measurements to higher energies. In addition,

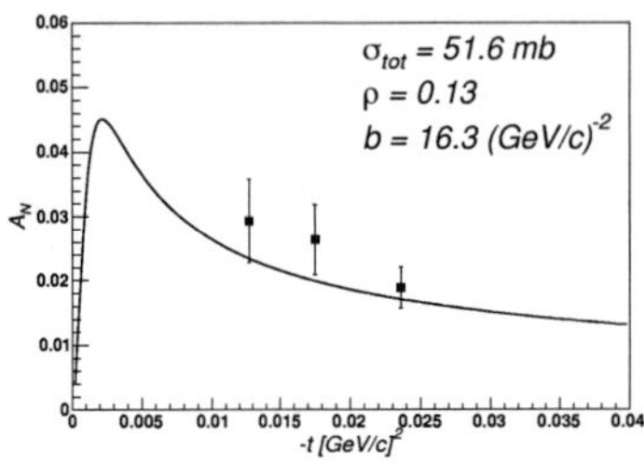

FIGURE 2. The single spin analyzing power A_N for three t intervals. Vertical error bars show statistical errors. The solid curve corresponds to theoretical calculations without hadronic spin flip.

an extension of the t-range will allow us to constrain both the magnitude and the shape of the analyzing power as a function of t, and higher statistics will permit measurements of A_{NN} and A_{SS}. This will help establish the role of multigluon exchanges in near–forward polarized proton-proton scattering.

ACKNOWLEDGMENTS

The research reported here has been performed in part under the US DOE contract DE-AC02-98CH10886, and was supported by the US National Science Foundation and the Polish Academy of Sciences. The authors are grateful for the help of N. Akchurin, D. Alburger, P. Draper, R. Fleysher, D. Morse, Y. Onel, A. Penzo, and P. Schiavon at various stages of the experiment and the support of the BNL Physics Department, Instrumentation Division, and C-A Department at the RHIC-AGS facility.

REFERENCES

1. S. Bültmann, I. H. Chiang, R. E. Chrien, A. Drees, R. L. Gill, W. Guryn, J. Landgraf, T. A. Ljubičič, D. Lynn, C. Pearson, P. Pile, A. Rusek, M. Sakitt, S. Tepikian, K. Yip (*BNL, Upton, NY, USA*); J. Chwastowski, B. Pawlik (*INP, Cracow, Poland*); M. Haguenauer (*Ecole Polytechnique, Palaiseau, France*); A. A. Bogdanov, S. B. Nurushev, M. F. Runzo, M. N. Strikhanov (*MEPhI, Moscow, Russia*); I. G. Alekseev, V. P. Kanavets, L. I. Koroleva, B. V. Morozov, D. N. Svirida (*ITEP, Moscow, Russia*); A. Khodinov, M. Rijssenbeek, L. Whitehead (*Stony Brook University, NY, USA*); K. De, J. Li, N. Öztürk (*University of Texas at Arlington, TX, USA*); A. Sandacz (*INS, Warsaw, Poland*).
2. W. Guryn *et al.*, RHIC Proposal R7 (1994) (unpublished).
3. S. Bültmann *et al.*, Nucl. Instr Meth. **A535**, 415 (2004).
4. R. Battiston *et al.*, Nucl. Instr. Meth. **A238**, 35 (1985)
5. R. Lipton, Nucl. Instr. Meth. **A418**, 85 (1998)
6. N. H. Buttimore *et al.*, Phys. Rev. **D59**, 114010 (1999)
7. G.G. Ohlsen and P.W. Keaton, Jr., *Nucl. Instr. Meth.* **109**, 41 (1973).
8. A. Bravar *et al.*, these proceedings and references therein.

Diffraction Dissociation - 50 Years later

Sebastian N. White

Brookhaven National Laboratory,
Upton, N.Y. 11973, USA

Abstract. The field of Diffraction Dissociation, which is the subject of this workshop, began 50 years ago with the analysis of deuteron stripping in low energy collisions with nuclei. We return to the subject in a modern context- deuteron dissociation in $\sqrt{s_{NN}} = 200$ GeV d-Au collisions recorded during the 2003 RHIC run in the PHENIX experiment. At RHIC energy, d→n+p proceeds predominantly (90%) through Electromagnetic Dissociation and the remaining fraction via the hadronic shadowing described by Glauber. Since the dissociation cross section has a small theoretical error we adopt this process to normalize other cross sections measured in RHIC.

Keywords: RHIC,ATLAS, heavy-ion
PACS: 25.75.Nq –25.75.Dw

INTRODUCTION

When deuteron beams were first accelerated to 190 MeV and collided with internal targets in the Berkeley cyclotron, experiments found a very collimated forward beam of neutrons which were identified with the process of absorptive stripping originally proposed by Serber[1]. Glauber[2] then showed that deuteron breakup could also proceed via a process he called "free dissociation", which has no classical analog. Absorption of part of the deuteron occurs even when neither nucleon strikes the target nucleus (treated as a black disc) and this absorption mixes unbound states of the proton and neutron which can then appear in the outgoing beam. The calculated cross section for this process is large (60% of the absorptive stripping cross section).

At RHIC collision energy ($\sqrt{s_{NN}} = 200$ GeV) a second mechanism for free dissociation of the deuteron becomes dominant. The intense Coulomb field of the target nucleus appears to the incident deuteron as a beam of photons whose flux can be calculated by the Equivalent Photon Approximation originally due to Fermi[9, 8]. Since the spectrum extends well above the deuteron photodissociation energy ($E_\gamma = 2.23$ MeV) this becomes the dominant process for free dissociation. A recent calculation[5] for RHIC yields 1.38 ($\pm 5\%$) barn of which 0.14 barn is due to the original nuclear dissociation process calculated by Glauber.

In 2 companion papers in the early 60's Good and Walker[7] observed that both Coulomb dissociation and nuclear dissociation should have analogs in diffractive excitation of elementary particles in high energy collisions with nuclei. Diffractive processes play a significant role in both Coulomb interactions (in Ultraperipheral Collisions) and hadronic collisions of Heavy Ions. This should be contrasted with e-p (HERA) and $\bar{p}p$ (Tevatron) where either one or the other process is studied. A unique aspect of the Heavy Ion program in ATLAS is that HERA and Tevatron measurements can be extended to higher energies and nuclear targets[10, 11].

CP792, *Deep Inelastic Scattering, DIS 2005,* edited by W. H. Smith and S. R. Dasu

D-AU CROSS SECTIONS

In addition to the dissociation cross section the total d-Au inelastic cross section is of interest for the RHIC program. The inelastic cross section is sampled in the experiments ("min-bias trigger") for use as a luminosity monitor. Once the min-bias cross section is known those of other processes recorded during the same luminosity interval can be calculated in the usual way. There are 2 approaches to this cross section normalization. In the first it is derived from known, elementary, NN inelastic cross sections using a Glauber model with a Woods-Saxon distribution parametrization of the p,n distributions in the nuclei. Calculations done for d-Au at RHIC energy range from 2.26$\pm$0.1 barn [12] to 1.93 barn[6]. It is difficult to assign an overall error to this calculation since one may find discussions in the literature of whether or not to include the diffractive part of NN cross sections and whether n and p should have the same matter distributions in the Au nucleus, etc.

The second approach, which we adopt here, is to directly determine the min-bias trigger cross section by comparison to the reliably calculated [5] deuteron dissociation process which was also measured in the PHENIX 2003 data.

INSTRUMENTATION

The four RHIC experiments have mid-rapidity spectrometers with different characteristics but all share identical Zero Degree Calorimeters (ZDC's) located at $\pm$18 m. The ZDC's cover $\pm$5 cm (in x and y) about the forward beam direction and have an energy resolution of $\sigma_E/E < 21\%$ for 100 GeV neutrons within x,y$\leq$ 4.5 cm[3, 4]. Essentially, all non-interaction ("spectator") neutrons are detected in the ZDC's, while charged particles are mostly swept out of the ZDC region by strong ($16T \cdot m$) accelerator dipoles at z= $\pm$11m.

The same dipoles sweep spectator protons from d-dissociation beyond the outgoing beam trajectory (since they have twice the deuteron charge-to-mass ratio) and in PHENIX they are detected in a proton calorimeter("fCal")[14].

The PHENIX experiment used two additional hodoscopes (BBC's)[4], located at z= $\pm$1.5m and covering $3.0 \leq |\eta_{BBC}| \leq 3.9$, as its main min-bias trigger. Events with one or more charged particles hitting both the +z and -z BBC fired this trigger. Determining σ_{BBC} is equivalent to determining the luminosity of the PHENIX data. d-Au events occuring well within the z= $\pm$1.5 m interval between the BBC's fire this trigger with an efficiency of $88 \pm 4\%$ [13] but the efficiency falls off for $|z_{event}| \geq 40$ cm. For this reason we will determine σ_{BBC} using only events within this interval. A correction is then applied for the fraction of all RHIC events within this interval. The actual distribution of events along z within the data can be measured using time-of-flight measurements between the ZDC's for events with a ZDC coincidence (with single event resolution of $\sigma_Z \sim 2$ cm).

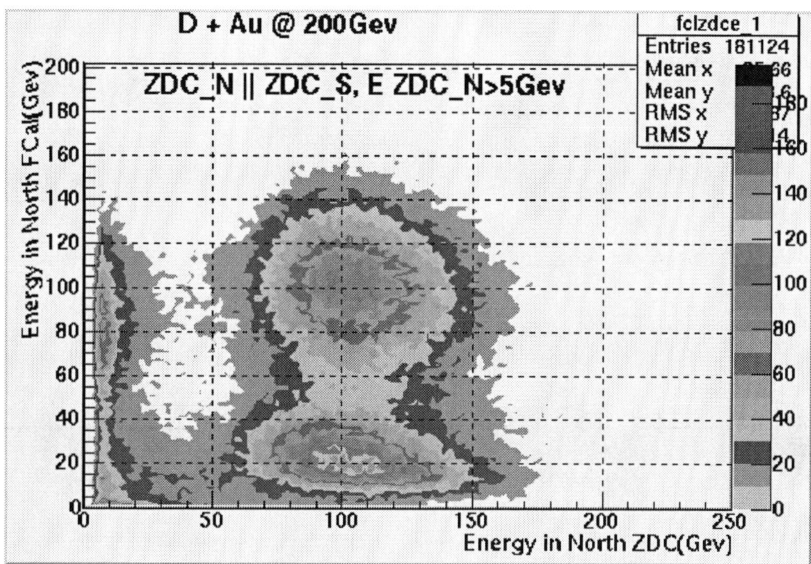

FIGURE 1. Energy deposition in proton calorimeter vs. ZDC (neutron calorimeter) for events with some activity in the ZDC in the deuteron beam direction. This sample includes absorptive stripping as well as d→ n+p.

DATA ANALYSIS

Typical event rates were several kHz ($\mathscr{L} \sim 1-4\times 10^{28} cm^{-2}s^{-1}$) for all processes considered here. Therefore this analysis is based on a representative data sample with

$$N_{BBC}^{trig} = 230k \tag{1}$$

$$N^{trig}(ZDC_{Au}\text{"}or\text{"}ZDC_d) = 460k \tag{2}$$

events, where subscripts Au and d represent $E_{ZDC} > 10$ GeV in the Au or d direction. The second trigger is sensitive to d-dissociation, characterized by a 100 GeV neutron in ZDC_d and a 100 GeV proton in fCal with no activity at mid-rapidity.

Additional data samples were recorded with one of the RHIC beams intentionally displaced by up to 1 mm to measure the fraction of triggers due to d-Au collisions (as opposed to beam gas background). The largest background was <3% and quoted rates have been corrected for the measured background.

The BBC rate, corrected for accelerator interaction distribution is

$$N_{BBC}^{corrected} = 228634\,\text{events}\,(\pm 0.5\%) \tag{3}$$

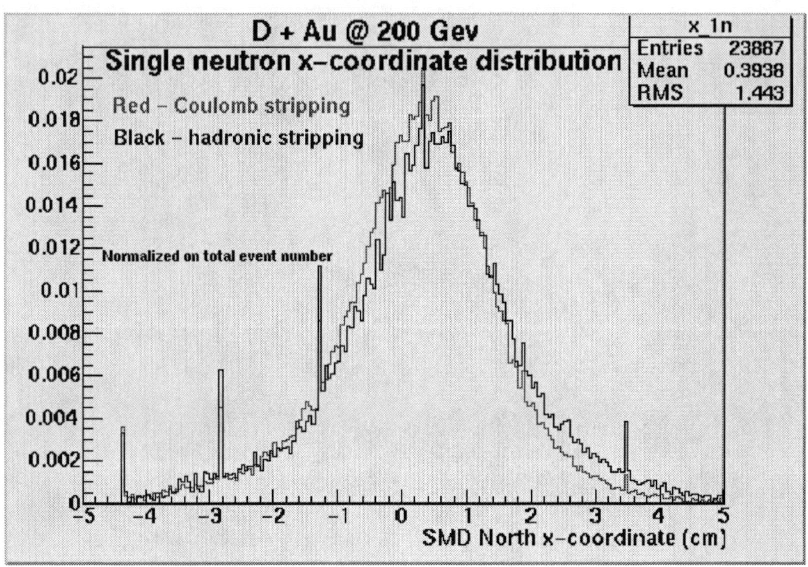

FIGURE 2. Neutron impact distribution for both absorptive stripping and dissociation events.

D-DISSOCIATION ANALYSIS

As stated above, d-dissociation events have a clear signature in the PHENIX experiment. This is illustrated in Figs 1 where we display the forward (in the deuteron direction) neutron vs. proton calorimeter energy for ZDC-trigger events.

The neutron impact parameter distribution (measured by the PHENIX Shower maximum detector) is displayed in Fig. 2 both for events with the d-dissociation signature ("Coulomb") and BBC trigger ("hadronic"). The latter correspond exactly to Serber's absorptive stripping proccess and, as noted in ref.[6], the neutron angular distribution should have an interesting correlation with event centrality. This will be discussed in a future note.

In any case one can see that neutrons have a small angular divergence and consequently there is only a small correction for ZDC acceptance. Instead, the dominant correction is for absorptive stripping events which feed into the dissociation sample. In order to extract the dissociation event yield we used an iterative procedure, fitting the sum of ZDC_d+fCal total energy to the sum of 100 +200 GeV lineshapes and correcting for calculated efficiency as successive cuts on activity in other detectors were applied. The first 2 iterations yield

$$N(d \rightarrow n+p) = 157149 \,\text{and}\, 156951 \tag{4}$$

events, so the procedure is clearly stable.

Our final result is :

$$\sigma_{BBC} = N(BBC)/N(d \rightarrow n+p) \times \sigma(d \rightarrow n+p) = 228634/158761 \times 1.38(\pm 0.5\%) \tag{5}$$

$$= 1.99\,(\pm 1.6\% \pm 5.0\%)\,\text{barn}. \tag{6}$$

This is the quantity needed for luminosity normalization.

In order to compare with Glauber calculations in the literature we then correct for the BBC detector efficiency given above:

$$\sigma(inelastic_{d-Au}) = \sigma_{corrected}(BBC) = 1.99/0.88 \tag{7}$$

$$= 2.26(\pm 1.6\% \pm 5.0\% \pm 4.5\%)\,\text{barn} \tag{8}$$

where the last 2 errors reflect the theoretical error on $\sigma(d \to n + p)$ and the BBC inefficiency uncertainty.

A similar analysis yields the cross section for the selection $ZDC_{Au}(E > 10GeV)$, also used as a min-bias trigger:

$$\sigma(ZDC_{Au}) = 2.06(\pm 1.7\% \pm 5.0\%)\,\text{barn}. \tag{9}$$

ACKNOWLEDGMENTS

I would like to thank the organizers of this workshop and particularly Dr. Khoze for a very stimulating and enjoyable meeting. I would like to thank my collaborator in this work, A. Denisov and also R. Glauber for helpful discussions. This work was supported in part under DOE Contract number DE-AC02-98CH10886.

REFERENCES

1. R. Serber, Phys. Rev. *72*(1947) 1008.
2. R. J. Glauber, *Deuteron Stripping Processes at High Energy*,Phys. Rev. *99*(1955) 1515.
3. C. Adler et al., Nucl. Instr. And Meth. *A470* (2001) 488.
4. M. Chiu et al. Phys. Rev. Lett. *89* (02) 012302 and nucl-ex/0109018.
5. S. Klein and R. Vogt, Phys. Rev. C 68 (2003) 017902. and nucl-ex/0303013.
6. B. Kopeliovich, Phys. Rev. C 68 (2003) 044906 and nucl-th/0306044.
7. M. L. Good and W. D. Walker, Phys.Rev.*120* (1960) pp.1855-1856 and 1857-1858.
8. S. White "Applications of the Equivalent Photon Approximation to Heavy Ion Collisions" in "Electromagnetic Probes of Fundamental Physics" W. Marciano and S. White, eds. World Scientific (2003).
9. E. Fermi (translation by M. Gallinaro and S. White) "On the Theory of Interactions between Atoms and Electrically Charged Particles", ibid and hep-th/0205086.
10. G. Bauer et al. "Hot Topics in Ultraperipheral Heavy Ion Collisions", ibid.
11. M. Strikman, R.V ogt and S. White, manuscript in preparation. To be published in CERN yellow report on UltraPeripheral Collision physics and R. Vogt ,hep-ph/0407298.
12. D. Kharzeev, E. Levin and M. Nardi, hep-ph/0212316.
13. S. S. Adler et al., (PHENIX Collaboration), Phys. Rev. Lett. 91, 072303 (2003).
14. The fCal calorimeters were re-used from BNL experiment E-864. See T.A. Armstrong et al., Nucl. Instr. and Meth. A 406 (1998) 227.

Photoproduction at Hadron Colliders

Spencer R. Klein, for the STAR Collaboration

Lawrence Berkeley National Laboratory, Berkeley, CA, 94720, USA

Abstract.
Photoproduction can be studied at hadron colliders by using the virtual photons associated with the hadron beams. The LHC will reach photonucleon energies 10 times higher than that available elsewhere. These reactions are already being studied at RHIC. After introducing photoproduction at hadron colliders, I will discuss recent results from STAR on ρ^0, $\pi^+\pi^-\pi^+\pi^-$ and e^+e^- production.

Keywords: photoproduction, ρ^0, ultra-peripheral collisions
PACS: 13.60.Le, 25.20.Lj, 12.20.-m

INTRODUCTION

The upcoming Large Hadron Collider (LHC) will reach proton-proton energies an order of magnitude higher than any existing accelerator. Because relativistic protons and heavier nuclei accompanied by fields of virtual photons, the LHC can be used to study photonuclear and two-photon interactions at energies far beyond those accessible at HERA or other accelerators.

Photoproduction is of interest in both pp and heavy-ion collisions. Proton-proton collisions produce photons with the highest energies, and, because of the very high pp luminosities, good rates. However, for many channels, the signal to noise ratio may be lower than in ion collisions. Heavy ions are accompanied by very high photon fluxes, and, because of the very strong fields, a single ion-ion collision can induce multiple electromagnetic interactions. The correlations between the multiple photons lead to the ability to "tune" the photon beam, by selecting different photon energy spectra and polarizations. Since these reactions take place at large impact parameters, where no hadronic interactions occur, they are often known as "ultra-peripheral collisions" (UPCs).

UPCs can be used to study a variety of topics [1][2][3]. Low$-x$ gluon distributions can be probed via heavy quark (including quarkonium) and jet production. UPCs can be used for many other studies of QCD. At the LHC, photonuclear interactions can be used to search for new physics. The strong fields allow for many tests of quantum electrodynamics in the very strong field regime, where perturbation theory may be expected to fail. Many of these topics are already being studied at RHIC.

PHOTOPRODUCTION AT HADRON COLLIDERS

For most reactions, the photon flux from protons or nuclei is well described by the Weizsäcker-Williams method of virtual photons. The photon flux per unit area for an

CP792, *Deep Inelastic Scattering, DIS 2005*, edited by W. H. Smith and S. R. Dasu

energy ω at a distance b from a relativistic nucleus with charge Z is [4]

$$N(\omega,b) = \frac{Z^2\alpha\omega^2}{\pi^2\gamma^2\hbar^2}K_1^2(x) \tag{1}$$

where $x = \omega b/\gamma$, γ is the Lorentz boost of the nucleus, $\alpha \approx 1/137$ is the fine structure constant, and K_1 is a modified Bessel function. The total photon flux from an ion with radius R_A is

$$n(\omega) = \int d^2bN(\omega,b). \tag{2}$$

The constrant $b > R_A$ is usually imposed to eliminate the photon flux inside the nucleus (where Eq. 1 fails, and, in any case, most of the flux is not usable). For photonuclear or two-photon interactions to be visible, the two nuclei must not interact hadronically, requiring $b > 2R_A$. This flux is calculated numerically, but can be approximated within about 15% by requiring $b > 2R_A$ in Eq. 2 [1].

The cross section for photonuclear interactions can be written [6]

$$\sigma(A+A \rightarrow A+A+X) = \int d^2bP(b) \tag{3}$$

where $P(b)$ is the probability for a photonuclear interaction, $P(b) = \int d\omega N(\omega,b)\sigma_{\gamma A}(\omega)$ where $\sigma_{\gamma A}(\omega)$ is the cross section for the photonuclear interaction in question. This formulation is easy to generalize to include multiple interactions between a single ion pair:

$$\sigma(A+A \rightarrow X_1+X_2+...) = \int d^2P_1(b)P_2(b). \tag{4}$$

In general, $P(b) \approx 1/b^2$, so the integrand for a $n-$photon reaction goes as $1/b^{2n}$ and the more photons involved in a reaction, the smaller the average impact parameters [5]. For example, with gold at RHIC, the median impact parameter drops from 46 fm for unselected ρ^0 production to 18 fm for ρ^0 accompanied by mutual Coulomb excitation [6]. The smaller impact parameters harden the photon spectrum, from $1/\omega$ to independent of ω. For some reactions, $P(2R_A) > 1$; in this case $P(b)$ is the mean number of reaction at that b.

Factorization can be used to simplify triggering on UPCs. One reaction can serve as a 'trigger' for another. STAR has studied the reactions $Au+Au \rightarrow Au^*+Au^*+\rho^0$ and $Au+Au \rightarrow Au^*+Au^*+e^+e^-$, using signals from the neutrons emitted in the Au^* decays to trigger the detector, providing ρ^0 and e^+e^- samples without trigger bias.

The photon polarizations are also correlated. The electric field of the photon-emitting nucleus parallels the impact parameter vector, so photons are linearly polarized along the impact parameter vector. For multiple interactions between a single ion-pair, the parallel polarizations can lead to observable angular correlations between decay products [5].

RESULTS FROM STAR AT RHIC

The STAR collaboration has produced final results on ρ^0 production [7] and on two-photon production of e^+e^- pairs [8]. Events were selected with two types of triggers:

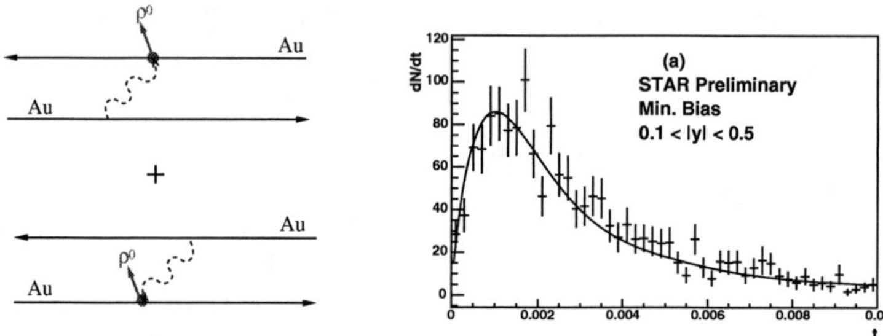

FIGURE 1. (a) Diagrams for the two interferening production mechanisms, at two spatially separated locations. (b) $t_\perp = p_T^2$ spectrum for ρ^0 accompanied by mutual Coulomb excitation. The drop at low $t_\perp$ is due to interference. The solid histogram is a fit to the data, using a model that includes the interference.

minimum bias triggers that select events with mutual Coulomb dissociation, taking advantage of factorization, and topological triggers, that select low multiplicity events with appropriate topologies in the central detector [7].

The ρ data is well described by the soft Pomeron model, and the previously discussed factorization holds. In the soft Pomeron model, the incident photon fluctuates to a quark-antiquark pair, which then elastically scatters (via Pomeron exchange) from the target nucleus [9]. Because the scattering is coherent, the momentum transfer is limited to order $\hbar/R_A$. This low p_T is a distinctive experimental signature; for gold, most of the signal occurs for $p_T < 150$ MeV/c.

Figure 1 shows the $t_\perp = p_T^2$ spectrum of ρ^0 with rapidity $0.1 < |\eta| < 0.6$, selected with stringent cuts to minimize the background [12]. At moderate and high t, the spectrum is well fit by an exponential, $dN/dt = a\exp(-bt)$. However, for $t < 0.0015\text{GeV}^2$, the data drops off. This drop can be explained by interference between two indisinguishable possibilities: nucleus 1 emits a photon which interacts with nucleus 2, or vice-versa [10]. In pp or AA collisions, these two possibilities are related by a parity transformation. Since the ρ^0 is negative parity, the interference is destructive. At mid-rapidity,

$$\sigma = \sigma_0[1 - \cos(p_T b)] \tag{5}$$

Of course, b is unknown, and the overall interference depends on the integral over all b. Away from $y = 0$, the interference is reduced because the photon energies, fluxes, amplitudes *etc.* for the two directions are different. The solid curve in Fig. 1 shows a fit to a functional form based on these factors; for this sample, the interference is $101 \pm 8(\text{stat.}) \pm 15(\text{syst.})\%$ of that expected [12]. Because the two sources are spatially separated, the final state $\pi^+\pi^-$ wave function does not factorize into single-particle wave functions, and the system exhibits the Einstein-Podolsky-Rosen paradox [11]. For $\overline{p}p$ collisions, the transformation between the two possibilies is a charge-parity transformation; vector mesons are CP positive, so the interference in Eq. 5 is positive [13]; this may be studied at the Fermilab Tevatron.

STAR has also studied ρ^0 production in dAu collisions. The photon is usually emitted by the gold nucleus, and the deuteron is the target. Both coherent (deuteron stays intact)

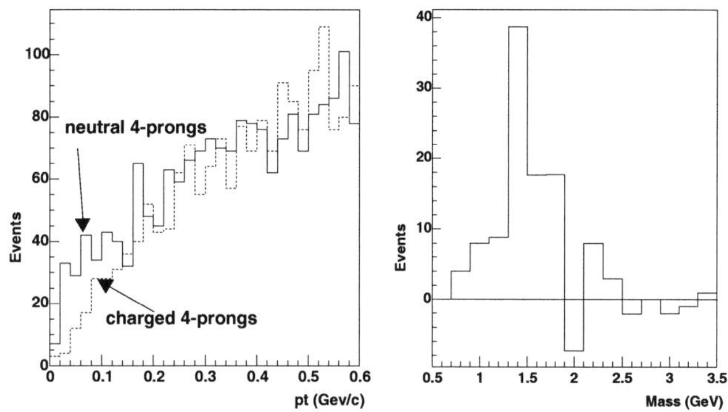

FIGURE 2. (a)p_T spectrum (dN/dp_T) for 4-prong events for neutral and net-charged combinations. (b) The mass spectrum for the neutral excess is peaked around 1.5 GeV.

and incohent (deuteron dissociates) interactions have been observed. The $t_\perp$ spectrum for the incoherent interactions is similar to that observed in *eA* collisions at HERA [14].

The STAR e^+e^- data is well described by lowest order quantum electrodynamics and factorization [8]. The p_T spectrum of the e^+e^- pairs is not well described by the virtual photon paradigm - the photon virtuality is required to fit the data.

STAR has also studied 4-prong final states, like $\pi^+\pi^-\pi^+\pi^-$. Fig. 2 compares the p_T spectrum of 4-prong events with net charge 0 with those of net charge 2. This data was taken in 2002 with the minimum-bias trigger. A neutral excess is present for $p_T < 150$ MeV/c, with the mass spectrum of the excess centered around 1.5 GeV/c^2.

This work was supported by the U.S. D.O.E. under Contract No. DE-AC-03076SF00098.

REFERENCES

1. C. A. Bertulani, S. R. Klein and J. Nystrand, nucl-ex/0502005.
2. G. Baur *et al.*, Phys. Rept. **364**, 359 (2002).
3. F. Krauss, M. Greiner and G. Soff, Prog. Nucl. Part. Phys. **39**, 503 (1997).
4. J.D. Jackson, *Classical Electrodynamics*, 2nd ed. (Wiley, 1975).
5. G. Baur *et al.*, Nucl. Phys. **A729**, 787 (2003).
6. A.J. Baltz, S. Klein and J. Nystrand, Phys. Rev. Lett. **89**, 012301 (2002).
7. C. Adler *et al.* (STAR Collaboration), Phys. Rev. Lett. **89**, 272302 (2002).
8. J. Adams *et al.* (STAR Collaboration), Phys. Rev. **C70**, 031902 (2004); V. B. Morozov, nucl-ex/0403002.
9. S. Klein and J. Nystrand, Phys. Rev. **C60**, 014903 (1999); L. Frankfurt, M. Strikman and M. Zhalov, Phys. Rev. **C67**, 034901 (2003); V. P. Goncalves and M. V. T. Machado, hep-ph/0501099.
10. S. Klein and J. Nystrand, Phys. Rev. Lett. **84**, 2330 (2000); K. Hencken, G. Baur and D. Trautmann, hep-ph/0506014.
11. S. Klein and J. Nystrand, Phys. Lett. **A308**, 323 (2003).
12. S. Klein, nucl-ex/0310020; S. Klein, nucl-ex/0402007.
13. S. Klein and J. Nystrand, Phys. Rev. Lett. **92**, 142003 (2004).
14. S. Timoshenko, nucl-ex/0501010.

Saturation 2005 (mini-review)

Eugene Levin

Particle Physics Department, School of Physics,
Tel Aviv University, Tel Aviv 69978, Israel

Abstract. This talk is a brief review of ups and downs of high density QCD during the past year.

Keywords: Saturation, Colour Glass Condensate, non-linear evolution, stochastic processes
PACS: 11.15.-q; 12.20.-m; 12.28.Lg

SATURATION AT HERA AND RHIC

Before discussing the high density QCD news we would like to summarize what we have learned about saturation at HERA and RHIC.

HERA:

- The power - like growth of $xG(x,Q^2)$ at low x ($xG(x,Q^2) \propto x^{-\lambda}$ with $\lambda \approx 0.3$;
- The geometrical scaling behaviour for $x \leq 10^{-2}$;
- Fit of all HERA data for $Q^2 = 0 \div 500 GeV^2$ with $\chi^2/d.o.f. \leq 1$ based on non-linear equation [1, 2];

RHIC:

- Saturation approach for dN/dy versus y, energy and number of participants predicted and led to a reasonable description of the experimental data [3];
- Prediction for suppression of the hadron production in dA collision and confirmation in the experimental data [4, 5].

The only consistent explanation all these observations is to assume that at HERA we have started to approach a new phase of QCD, with large gluon density but still with small coupling constant. The regime of high parton density at HERA is reached due to the QCD emission of gluons that was incorporated in the QCD evolution equations. The independent check of the effects of high gluon density at HERA was performed by RHIC experiment in heavy ion-ion collisions. In this reaction the energies are much lower than at HERA, but the large values of the parton densities were achieved due to the large number of nucleons in a nucleus. Based on these experimental observations we can anticipate that the LHC will be a machine for discovery a new phase of QCD: colour glass condensate with saturated gluon density.

CP792, *Deep Inelastic Scattering, DIS 2005*, edited by W. H. Smith and S. R. Dasu

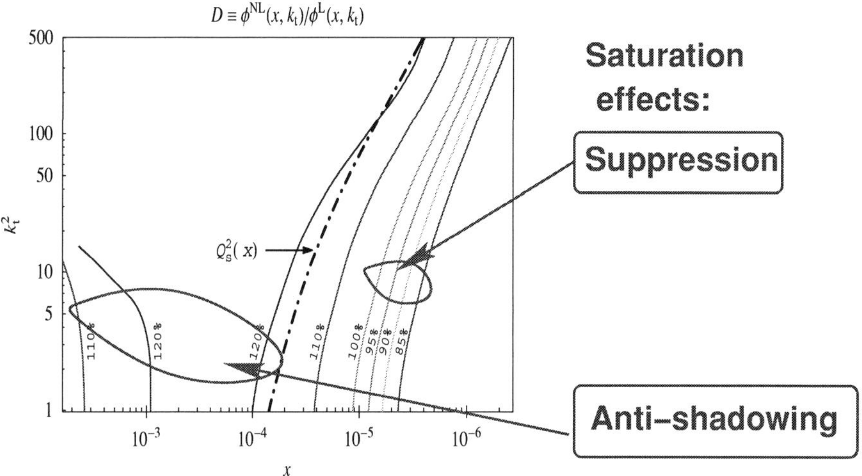

FIGURE 1. The ratio ϕ^{NL}/ϕ^{L} which shows the influence of non-linear correction on the predictions for inclusive gluon jet production at LHC energies

PREDICTIONS FOR THE LHC RANGE OF ENERGIES

Our main challenge is to provide reliable estimates for the influence of high density QCD (saturation) effects in the LHC range of energies. The first such estimates have been discussed [6, 7], and the results for the ratio of the unintegrated structure functions $D = \phi^{NL}/\phi^{L}$ are plotted in Fig.1 where

$$\frac{d\sigma}{dyd^2p_t} \propto \frac{\alpha_S}{p_t^2} \int d^2k_t\, \phi(k_t^2)\, \phi((\vec{p}-\vec{k})_t^2) \qquad (1)$$

and $\phi^{NL}(\phi^{L})$ is solution of the non-linear (linear) equation.

It should be stressed that non-linear evolution predicts not only suppression in the saturation region, but also the anti-shadowing effect which results in an increase of the value of ϕ for $Q^2 > Q_s^2(x)$, where Q_s is the saturation scale. One can see that the suppression and increase could be rather large leading to an inclusive cross section twice as large or twice as small, as the predictions based on routine linear evolution.

THEORETICAL DEVELOPMENT

B- JIMWLK approach $\longleftrightarrow$ BFKL Pomeron Calculus

The good news is that it turns out that Balitsky-JIMWLK approach [8] can be reduced to BFKL Pomeron calculus [9], and JIMWLK effective Lagrangian give us possibility to calculate all multi-Pomeron vertices. For the first time, we can do such calculations using operator formalism without spending years to obtain result just summing Feyman diagrams. Since the colour dipoles are the 'wee' partons of the BFKL equation the Balitsky-JIMWLK formalism can be discussed in terms of the dipole approach.

The bad news is that we have not achieved any progress in Pomeron calculus.

Probabilistic interpretation

Our last hope is the probabilistic approach to Pomeron interaction. The best way to express our optimism is to cite Grassberger and Sundermeyer [10] who proposed this interpretation: " *Reggeon field theory is equivalent to a chemical process where a radical can undergo diffusion, absorption, recombination, and autocatalytic production. Physically, these "radicals" are wee partons (*colour dipoles*)*".

It turns out that B-WLKJIM approach can be written as a typical death-birth process (Markov's chain)[11, 12]

$$\frac{\partial P_n}{\partial Y} = -\sum_i \Gamma(1 \to 2) \bigotimes \left(P_n(...x_i, y_i...;Y) - P_{n-1}(...x_i, y_i...;Y)\right) \tag{2}$$

where P_n - probability to find n-dipoles at rapidity Y, $\Gamma(1 \to 2)$ describe the decay of one dipole into two dipoles and $\bigotimes$ denotes all needed integration. This equation can be a basis for the Monte Carlo code which will be able to solve high density QCD equations, and which will lead to theoretical treatment of the multiparticle production.

Hunt for Pomeron loops

The process of two Pomeron merging into one Pomeron is naturally included in Pomeron calculus with the same vertex as the process of Pomeron splitting. However, we need correctly normalize this process if we wish to use the probabilistic interpretation. Such normalization was suggested in Ref. [13] and this vertex $\Gamma(2 \to 1)$ has been calculated [13, 14, 12]. Using this vertex, we can generalize Eq.(2) which takes the form

$$\frac{\partial P_n}{\partial Y} = \text{Eq.(2)} - \sum_i \Gamma(2 \to 1) \bigotimes \left(P_n(...x_i, y_i...;Y) - \sum_k P_{n+1}(...x_i, y_i...x_k, y_k;Y)\right) \tag{3}$$

Solution

Attempts to solve Eqs.(3) have been made in Refs.[15, 16, 17]. The result is surprisingly unexpected, namely,

- Asymptotic solution leads to a *gray* disc (*not black!!!*);
- Using the large parameters of our theory ($\Gamma(1 \to 2)/\Gamma(2 \to 1) \approx N_c^2/\alpha_S^2$ and $\Gamma(1 \to 2)/\Gamma(2 \to 3) \approx N_c^2$) the semiclassical approach can be developed for searching for both the asymptotic solution and the corrections to it, at high energy;

- The corrections to the asymptotic solution decrease at large values of Y, and can be found from the Liouville-type linear equation;
- The important role in searching for high energy asymptotic behaviour of the amplitude plays the role of t-channel unitarity constraint, which specifies the value of the typical amplitude for dipole-dipole interaction.

Topics which I have no room to discuss

This brief review is my personal view on news in low x (high density) QCD. Unfortunately, I had no room even to express my point of view. It is pity since I think that a more microscopic approach, related to the new effective Lagrangian, and to a search for a Bogolubov transformation between dipole and quarks (antiquark) and gluon degrees of freedom [9, 18, 19], looks very interesting. It is very attractive approach and I hope that my references provide the reader with names of active players in this field. However, I must admit that the theory becomes dangerously complicated and reminds me more and more my nightmare that Lipatov [20] is correct with his effective action, which is not easier to solve than the full QCD Lagrangian.

Acknowledgments: I am very grateful to E. Gotsman for everyday discussions on the subject of this talk.

REFERENCES

1. E. Gotsman, E. Levin, M. Lublinsky and U. Maor, *Eur. Phys. J.*, **C27**, 411-425 (2003).
2. E. Iancu, K. Itakura and S. Munier, *Phys. Lett.*, **B590**, 199 -208 (2004).
3. D. Kharzeev, E. Levin and M. Nardi, *Nucl. Phys.*, **A730**, 448-459 (2004); D. Kharzeev and E. Levin, *Phys. Lett.*, **B523**, 79-87 (2001.
4. D. Kharzeev, E. Levin and L. McLerran, *Phys. Lett.*, **B561**,93-101 (2003).
5. D. Kharzeev, Y. V. Kovchegov and K. Tuchin, *Phys. Lett.*, **B599**, 23-31 (2004).
6. E. Gotsman, E. Levin, U. Maor and E. Naftali, arXiv:hep-ph/0504040.
7. K. J. Eskola, V. J. Kolhinen and R. Vogt, *Phys. Lett.*, **B582**, 157-166 (2004); K. J. Eskola, H. Honkanen, V. J. Kolhinen, J. w. Qiu and C. A. Salgado, *Nucl. Phys.*, **B660**,211-224 (2003).
8. I. Balitsky, *Nucl. Phys.*, **B463**,99-160 (1996); J. Jalilian-Marian, A. Kovner, A. Leonidov and H. Weigert, *Phys. Rev.*, **D59**,014014 - 014019 (1999); *Nucl. Phys.*,**B504**, 415-431 (1997); E. Iancu, A. Leonidov and L. D. McLerran, *Phys. Lett.*, **B510**,133-144 (2001); *Nucl. Phys.*,**A692**, 583-645 (2001); H. Weigert,*Nucl. Phys.*, **A703**, 823-860 (2002).
9. A. Kovner and M. Lublinsky, *Phys. Rev.*, **D71**, 085004- 085011 (2005); *JHEP*, **0503**, 001 - 008 (2005).
10. P. Grassberger and K. Sundermeyer, *Phys. Lett.*, **B77**, 220-222 (1978).
11. J. P. Blaizot, E. Iancu and H. Weigert, *Nucl. Phys.*, **A713**, 441-469 (2003).
12. E. Levin and M. Lublinsky, *Phys. Lett.*, **B607**, 131-138 (2005); arXiv:hep-ph/0501173.
13. E. Iancu and D. N. Triantafyllopoulos, *Phys. Lett.*, **B610**, 253-261 (2005); *Nucl. Phys.*, **A756**, 419-467 (2005).
14. A. H. Mueller, A. I. Shoshi and S. M. H. Wong, *Nucl. Phys.*, **B715**, 440-460 (2005).
15. K. G. Boreskov,arXiv:hep-ph/0112325.
16. E. Levin, arXiv:hep-ph/0502243.
17. P. Rembiesa and A. M. Stasto, arXiv:hep-ph/0503223.
18. Y. Hatta, E. Iancu, L. McLerran, A. Stasto and D. N. Triantafyllopoulos arXiv:hep-ph/0504182; arXiv:hep-ph/0505235.
19. C. Marquet, A. H. Mueller, A. I. Shoshi and S. M. H. Wong, arXiv:hep-ph/0505229.
20. L. N. Lipatov, *Nucl. Phys.*, **B452**, 69-400 (1995).

The FP420 R&D Project: Forward Proton Tagging at the LHC as a Means to Discover New Physics

B. E. Cox

School of Physics and Astronomy, The University of Manchester, Oxford Road, Manchester, M139PL, UK

Abstract. We review the theoretical and experimental motivations behind recent proposals to add forward proton tagging detectors to the LHC experiments as a means to search for new physics.

Keywords: Higgs, MSSM Higgs, Proton Tagging, LHC
PACS: 14.80.Bn,14.80.Cp,14.70.Fm,14.70.Hp,12.38.Aw

INTRODUCTION

There has been increasing interest in the past few years in the possibility of using diffractive interactions as a search tool for new physics (see for example [1, 2, 3, 4, 5] and references therein). In particular, it has been suggested that the so-called central exclusive production process might provide a particularly clean environment to search for, and identify the nature of, new particles at the LHC. By central exclusive, we refer to the process $PP \to P \oplus \phi \oplus P$, where $\oplus$ denotes the absence of hadronic activity ('gap') between the outgoing protons and the decay products of the central system ϕ. An example would be standard model Higgs Boson production, where the central system could consist of 2 b-quark jets, or the decay products of 2 W bosons, and no other activity.

There are three primary reasons why this process is attractive. Firstly, if the outgoing protons remain intact and scatter through small angles, then, to a very good approximation, the central system ϕ must be produced in a spin 0, CP even state, therefore allowing a clean determination of the quantum numbers of any observed particle. Secondly, the mass of the central system can be determined very accurately from a measurement of the transverse and longitudinal momentum components of the outgoing protons alone. The mass of any new particles produced in this way can therefore be precisely determined irrespective of their decay mode. Thirdly, because of the accurate mass measurement, the spin selection rules and the cleanness of the events in the central detectors, excellent signal to background rations are achievable for a wide range of Standard Model and MSSM Higgs production scenarios. We discuss these two possibilities in more detail in section 2. Another attractive feature is the ability to directly probe the CP structure of the Higgs sector by measuring azimuthal asymmetries in the tagged protons (a measurement previously proposed only at a future linear collider) [6].

In order to make use of the central exclusive production process at the LHC, proton tagging detectors must be installed on both sides of either the ATLAS and / or CMS

CP792, *Deep Inelastic Scattering, DIS 2005,* edited by W. H. Smith and S. R. Dasu

detectors, with an acceptance that covers the appropriate mass range for Standard Model and MSSM Higgs bosons, and other possible new particles. For central systems in the 120 GeV mass range, the outgoing protons emerge in the region 420m away from the interaction points. The FP420 R&D project [7] aims to assess the feasibility of installing such detectors. We review the key technical issues to be addressed by this project in the following section .

There is also a wealth of QCD and 2-photon physics that becomes accessible if 420m proton detectors are installed. Of particular interest is the study of the quartic gauge couplings $\gamma\gamma WW$. It is estimated that approximately 1000 events could be collected in the semi and fully leptonic decay channels with 30 fb^{-1} of delivered luminosity, delivering a sensitivity to anomalous quartic couplings a factor of 10,000 better than the current LEP2 limits. There is similar sensitivity to the anomalous production of Z pairs in the process $\gamma\gamma \to ZZ$. The rich QCD program includes precision measurements of the diffractive structure functions of the proton in the HERA kinematic range, allowing detailed studies of rapidity gap survival probability (and hence the contribution of multi-parton interactions to the underlying event), and the off-diagonal unintegrated gluon distributions of the proton. For more details, see [7] and references therein.

THE FP420 R&D PROJECT

The 420m region at the LHC consists of a 15m drift space (i.e. no magnets), and is at present enclosed in a 'connection cryostat' which maintains a series of superconducting bus-bars, and the beam pipes themselves, at a temperature of 1.7K. The first goal of the FP420 project is to assess the feasibility of replacing the 420m interconnection cryostat to facilitate access to the beam pipes and therefore allow proton tagging detectors to be installed.The first opportunity to install such detectors would be the planned LHC shutdown in autumn 2008.

The resolution and acceptance of the proposed detectors is essentially fixed by the LHC high-luminosity beam optics. For the case of a 140 GeV central system, the acceptance for tagging both outgoing protons in the 420m detectors is 22%, and the mass resolution is expected to be ~ 1 %. The acceptance rises to 60 % if one of the protons is tagged using the proposed 220m detectors of CMS / TOTEM or the ATLAS luminosity system. The mass resolution for these asymmetric events is ~ 6 % [7].

A key challenge for the FP420 project is level 1 triggering. The 420m detectors are too far away from the interaction point to be included in the level 1 trigger systems of ATLAS or CMS without increasing the level 1 trigger latency. It is therefore necessary to save the interesting physics at level 1 using the central detectors, the forward detectors (T1, T2 and CASTOR in CMS / TOTEM and LUCID in Atlas), or the CMS / TOTEM or ATLAS Roman Pots at 220 m, if available. Central systems that contain high-pT leptons, such as $H \to WW$, are not a problem. The challenge arises in the case of low-mass states, such as the Standard Model Higgs boson, decaying into relatively low E_T jets.

A preliminary study of Level 1 triggering in this case was carried out in [7]. The conclusions were that L1 triggering is not a problem as long as there is no pile-up, since the forward detectors can be used to veto events in which there are no forward rapidity gaps. For a 25 ns bunch structure, it is possible to collect ~ 6 fb^{-1} of clean (no pile-up) events within 3 years of LHC running. This is enough to detect signals with central exclusive cross sections in the 50 fb range. An example would be the high $\tan\beta$ MSSM scenario, to be discussed in the following section. At higher luminosities, up to $\sim 2\times 10^{33}$ cm^{-2}s^{-1}, events in which one proton is detected at 220m can be saved. If the Roman Pot detectors at 420 m could be used, requiring that a proton be seen on both sides would yield a signal efficiency of $\sim$ 15 % and an L1 trigger rate at a luminosity of 10^{34}cm^{-2}s^{-1} of $\sim$ 7 kHz [7]. No trigger problems are foreseen for final states containing high-p_T leptons, such as the WW decay modes of the Standard Model Higgs boson.

THE CENTRAL EXCLUSIVE PRODUCTION OF SM AND MSSM HIGGS BOSONS

The benchmark central exclusive production process for new physics searches is Standard Model (SM) Higgs production. The cross section for $pp \to p \oplus H \oplus p$ was calculated in [8, 9] to be 3 fb for $M_H = 120$ GeV, falling to $\sim$ 1 fb at $M_H = 200$ GeV. The simplest channel to observe the SM Higgs from an experimental perspective is the WW decay channel. For $M_H = 140$ GeV, 19 exclusive $H \to$ WW events are expected to have double proton tags using both 220m and 420m detectors (none using 220m detectors alone), for an LHC luminosity of 30 fb^{-1}. This rises to 25 at 160 GeV. Of these, approximately 25 % will be taken by the standard ATLAS and CMS level 1 leptonic triggers, although it is expected that with further optimisation of the trigger thresholds this efficiency should rise to close to 50 % [9]. In the gold plated semi-leptonic channels, the signal to background ratio will be in excess of unity, and observation of SM Higgs in this channel will cleanly establish its quantum numbers with 30 fb^{-1} of delivered luminosity.

More challenging from a trigger perspective is the b-jet decay channel. That this channel is possible to observe at all is a consequence of the spin-0 selection rules for central exclusive production [10], which heavily suppresses exclusive b-jet production; in conventional channels this signal is swamped by the copious QCD background. For $M_H = 120$ GeV, we expect 60 exclusive $H \to b\bar{b}$ events to have double proton tags using both 220m and 420m detectors. A recent study [1] found that, after taking into account losses due to b-tagging efficiencies and kinematic cuts to reduce backgrounds, and the likely achievable mass resolution of the proton tagging detectors, 11 signal events remain with a signal to background ratio of order unity for a luminosity of 30 fb^{-1}.

The b-jet channel becomes extremely important in the so-called intense coupling regime of the MSSM. This is a region of MSSM parameter space in which the couplings of the Higgs to the electroweak gauge bosons are strongly suppressed, making discovery challenging at the LHC by conventional means [11]. The rates for central exclusive production of the two scalar MSSM Higgs bosons are enhanced by an order of magnitude in these models, however. We expect close to 1000 exclusively produced double-tagged h and H bosons with 220m and 420m detectors in 30 fb^{-1} of delivered luminosity, for $M_{h,H} \sim 125$ GeV and $\tan\beta = 50$ [12]. Under the same assumptions as for the SM Higgs, approximately 100 would survive the experimental cuts, with a signal to background ratio of order 10. It is also worth noting that the pseudo-scalar Higgs (A) is practically not produced in the central exclusive channel, allowing for a clean separation of the scalar Higgs bosons which is impossible in conventional channels. For such regions of the MSSM, central exclusive production is likely to be the discovery channel.

SUMMARY

Installing proton tagging detectors in the 420m region at the LHC will open up a rich QCD, electroweak, Higgs and BSM program, with the potential to make measurements which are unique at LHC, and difficult even at a future linear collider. The FP420 R&D project aims to assess the feasibility of installing such detectors, as an upgrade to either ATLAS and / or CMS. Full details of the R&D proposal can be found in [7].

ACKNOWLEDGMENTS

We would like to thank all the authors of the FP420 LOI, whose contributions have been summarised in this article. This work was supported in the UK by PPARC.

REFERENCES

1. A. De Roeck, V. A. Khoze, A. D. Martin, R. Orava and M. G. Ryskin, Eur. Phys. J. C **25** (2002) 391 [arXiv:hep-ph/0207042].
2. K. A. Assamagan *et al.* [Higgs Working Group Collaboration], arXiv:hep-ph/0406152.
3. M. Boonekamp, R. Peschanski and C. Royon, Phys. Lett. B **598** (2004) 243 [arXiv:hep-ph/0406061].
4. V. A. Petrov, R. A. Ryutin, A. E. Sobol and J. P. Guillaud, arXiv:hep-ph/0409118.
5. B. E. Cox, AIP Conf. Proc. **753** (2005) 103 [arXiv:hep-ph/0409144].
6. V. A. Khoze, A. D. Martin and M. G. Ryskin, Eur. Phys. J. C **34** (2004) 327 [arXiv:hep-ph/0401078].
7. FP420 LOI, CERN-LHCC-2005-025
8. V. A. Khoze, A. D. Martin and M. G. Ryskin, Eur. Phys. J. C **23** (2002) 311 [arXiv:hep-ph/0111078].
9. B. E. Cox *et al.*, arXiv:hep-ph/0505240.
10. V. A. Khoze, A. D. Martin and M. G. Ryskin, Eur. Phys. J. C **19** (2001) 477 [Erratum-ibid. C **20** (2001) 599] [arXiv:hep-ph/0011393].
11. E. Boos, A. Djouadi and A. Nikitenko, Phys. Lett. B **578**, 384 (2004) [arXiv:hep-ph/0307079].
12. A. B. Kaidalov, V. A. Khoze, A. D. Martin and M. G. Ryskin, Eur. Phys. J. C **33** (2004) 261 [arXiv:hep-ph/0311023].

High Energy Photon Interactions at the LHC

K. Piotrzkowski

Université catholique de Louvain, B-1348 Louvain-la-Neuve, Belgium

Abstract. Experimental prospects for studying high-energy photon interactions at the LHC are discussed. Rôle of the forward proton detectors in selection and reconstruction of such events is briefly described. Physics cases for the two-photon fusion and photon-proton interactions are introduced. Several examples, as the associated WH photo-production, or two-photon production of supersymmetric pairs are given. Finally, a possibility of studying the photon interactions in ion collisions is shortly discussed.

Keywords: photon-photon and photon-proton interactions, electroweak processes, Higgs boson, searches for supersymmetry.
PACS: 13.40.-f, 13.60.-r, 14.70.-e,14.80.Bu, 14.80.Ly

INTRODUCTION

The proton beam energy of the LHC will be so high that not a negligible fraction of the *pp* collisions will involve high energy two-photon (and photon-proton) processes, where the protons emitting a photon will survive the interaction. Such protons are scattered at very small angles, comparable to the beam angular divergence at the interaction point (IP). The scattered protons can however be measured when a significant fraction of the initial proton energy is carried away by a photon. In such a case these protons are more strongly deflected by the beam-line magnets and can be detected in the so-called Roman pots installed far away from the IP and close to the proton beam. Tagging photon interactions using forward proton detectors opens up a novel domain of research at the LHC [1]. The *double tagging* corresponds to the case when the two scattered protons are detected, whereas the *single tagging* occurs when only one proton is detected. In this case, however also those two-photon events are tagged where one proton does not survive the interaction. In fact, these *inelastic* two-photon events have even higher effective luminosity than the nominal, *elastic* events. The effective luminosity, available for the tagged two-photon collisions, reaches 1% of the *pp* luminosity for the $\gamma\gamma$ center of mass energy W > 100 GeV. One should note also that the luminosity spectrum extends to large values of W, even beyond 1 TeV.

The ultimate resolution of the scattered proton momentum is determined by the beam properties at the IP, its lateral size and angular divergence [1]. Good resolution of p_T is essential in case when a good separation of the two-photon and pomeron-pomeron events is needed.

Similarly, one can identify photo-production processes. Here, the effective luminosity of photon-quark (or photon-gluon) collisions is much higher and extending

CP792, *Deep Inelastic Scattering, DIS 2005,* edited by W. H. Smith and S. R. Dasu

to even higher energies. In Figure 1 the γq luminosity S_γ (and its integral) is shown as a function of the photon-quark *cms* energy. This quantity can be used, assuming factorization, to convert γp to *pp* cross sections as follows: $\sigma_{pp} = \int \sigma_{\gamma p}(W) S_\gamma(W) dW$. Similar values of S_γ have been obtained for the effective photon-gluon luminosity.

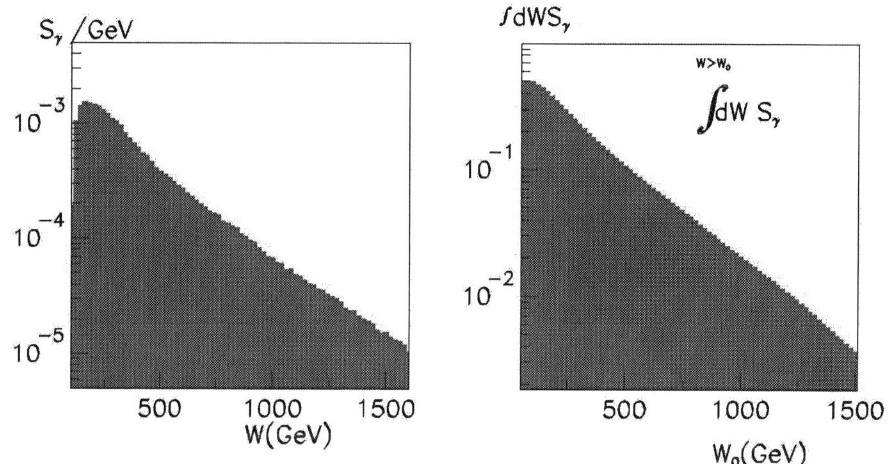

FIGURE 1. Effective luminosity spectrum S_γ of the photon-quark collisions (left plot), using MRST2001 pdfs; and its integral between an initial cms energy W_0 and the kinematical limit (right plot).

PHYSICS CASE

$\gamma\gamma$ collisions

Possibility of using $\gamma\gamma$ collisions at the LHC to search for new particles was already considered theoretically some time ago – see, for example [2]. The exclusive production of the Standard Model (SM) Higgs boson has been an especially attractive case, but the corresponding cross section is small [3]. This is quite opposite for the W boson pair production via the $\gamma\gamma$ fusion – the cross section at the *pp* level is large, above 100 fb, and the average *cms* energy is high, above 300 GeV. Detecting these events would allow for very significant improvement (with respect to the present limits), in sensitivity to anomalous quartic gauge couplings [4]. Similar results are obtained for the $\gamma\gamma \to ZZ$ process (in the SM strongly suppressed).

If the super-symmetric particles exist and are not too heavy they could be also studied in a very complementary way using the exclusive $\gamma\gamma$ production. In Figure 2 the tagged production rates of pairs of charginos, charged Higgses and sleptons are shown. Clearly, if the sparticle masses are in the 150–250 GeV range, the number of events becomes interesting already for the low luminosity running of the LHC. One should note that contrary to the inclusive case, one is dealing here with very simple production mechanism and final states, as well as much better constrained initial state.

γp collisions

The Higgs boson associated photo-production at the LHC, $\gamma p \rightarrow WHX$, yields a cross section above 50 fb for $M_H < 170$ GeV. This is naturally much smaller than for the inclusive case but has much more favorable background conditions. For example, the cross section of the main reducible $\bar{t}t$ background is almost thousand times bigger than the WH signal in the inclusive case, but only about 50 times in case of the photo-production [5]. Similarly, the irreducible background $\bar{b}bW$, for $H \rightarrow \bar{b}b$, is more than 100 times bigger for the inclusive case and is only a couple of times bigger for photo-production. The higher signal-to-background ratio and better event reconstruction (thanks to photon energy measurement and less busy event 'environment') in the WH photo-production, should allow to provide a useful and complementary information for not too heavy Higgs boson.

In addition, the $\bar{t}t$ photo-production is interesting by its own. It has a respectable cross-section of about 3 pb and proceeds mainly via the photon-gluon fusion. The high statistics samples should be therefore collected and used for precision measurement of the top mass and its charge, for example.

Finally, the single W photo-production, $pp \rightarrow pWX$, at high transverse momentum will be studied in a very similar way as the related process, $ep \rightarrow eWX$, at HERA. High statistics samples will be again available and should allow for stringent tests of the SM.

Ion collisions

In ion collisions at the LHC photon interactions are strongly enhanced due to coherent coupling to the large ion charge Z. For example, two-photon interactions in the coherent domain scale as Z^4. This usually leads to improved signal to background ratio, as the latter often scales like A^2. This opens up new, very interesting field of research including both photon-photon and photon-nucleon interactions, and studies of low-x phenomena, heavy quark production, or of the electroweak processes as the W pair production via $\gamma\gamma$ fussion, for example [6].

Recently proposed forward detectors will be crucial for these studies allowing for detection of protons in pA, and light ions in AA interactions [7].

CONCLUSIONS

Preliminary results of novel studies of the photon-induced processes at the LHC strongly indicate that it could provide valuable and complementary information to the nominal measurements, in particular in the electro-weak sector. Forward proton detectors are instrumental in this context [7].

ACKNOWLEDGEMENTS

I would like to thank wholeheartedly my colleagues in Louvain-la-Neuve, V. Lemaître, M. vander Donckt, J. de Favereau, Y. Liu, S. Ovyn, T. Pierzchała and X. Rouby for fruitful collaboration on the photon physics at the LHC.

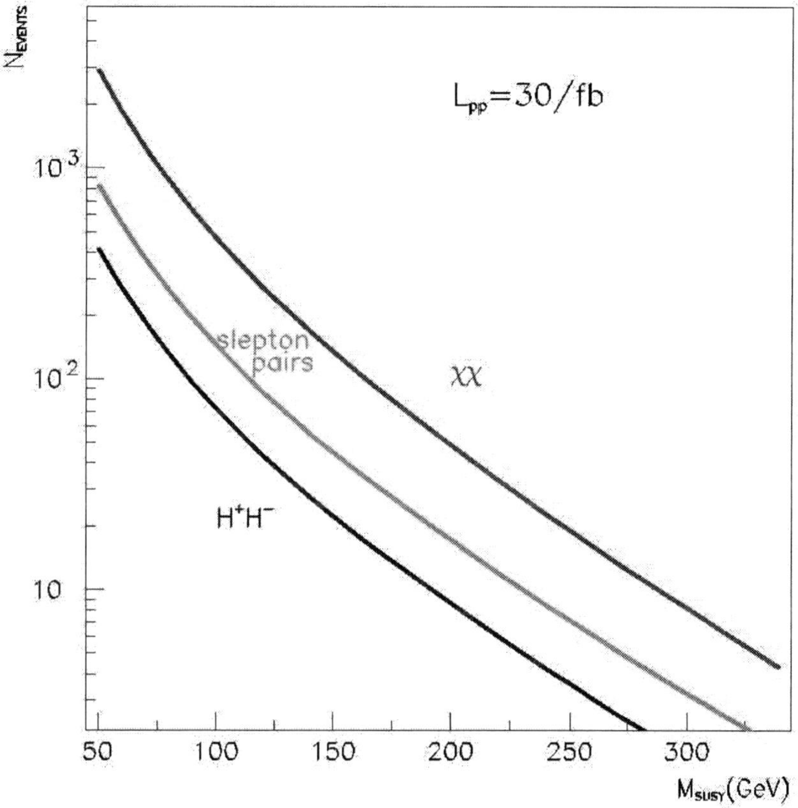

FIGURE 2. Number of super-symmetric pairs produced, assuming single tagging and the 30 fb^{-1} integrated luminosity, as a function of the chargino, slepton and charged Higgs mass, respectively. The 70–700 GeV photon tagging range is assumed.

REFERENCES

1. K. Piotrzkowski, *Phys. Rev. D* **63** (2001) 071502(R).
2. J. Ohnemus *et al.*, *Phys. Lett. B* **328** (1994) 369.
3. E. Papageorgiu, *Phys. Lett.* B **352** (1995) 394.
4. T. Pierzchala *et al.,* publication in preparation.
5. M. vander Donckt *et al.*, publication in preparation.
6. K. Piotrzkowski *et al.*, *Hot topics in ultra-peripheral ion collision*s, hep-ph/0201034.
7. A.G. Albrow *et al.*, FP420: An R&D Proposal, CERN-LHCC-2005-025;
   LoI of CMS and Totem, in preparation.

High mass diffraction at the LHC

C. Royon

Service de physique des particules, CEA/Saclay, 91191 Gif-sur-Yvette cedex, France
Fermilab, Batavia, USA

Abstract. We use a Monte Carlo implementation of recently developped models of exclusive diffractive W, top, Higgs and stop productions to assess the sensitivity of the LHC experiments.

THEORETICAL FRAMEWORK

The proposed model for $pp \to p + H + p$, the Bialas-Landshoff (BL) [1] model, is based on a summation of two-gluon exchange Feynman graphs coupled to Higgs production by the top quark loop. The non-perturbative character of diffraction at the proton vertices relies on the introduction of "non-perturbative" gluon propagators which are modeled on the description of soft total cross-sections within the additive constituent quark model.

More details about the theoretical model and its phenomenological applications can be found in Refs. [2] and [3]. In the following, we use the BL model for exclusive Higgs production recently implemented in a Monte-Carlo generator [2].

EXPERIMENTAL CONTEXT

The analysis is based on a fast simulation of the CMS detector at the LHC (similar results would be obtained using the ATLAS simulation). The calorimetric coverage of the CMS experiment ranges up to a pseudorapidity of $|\eta| \sim 5$. The region devoted to precision measurements lies within $|\eta| \leq 3$, with a typical resolution on jet energy measurement of $\sim 50\%/\sqrt{E}$, where E is in GeV, and a granularity in pseudorapidity and azimuth of $\Delta\eta \times \Delta\Phi \sim 0.1 \times 0.1$.

In addition to the central CMS detector, the existence of roman pot detectors allowing to tag diffractively produced protons, located on both p sides, is assumed [4]. The ξ acceptance and resolution have been derived for each device using a complete simulation of the LHC beam parameters. The combined ξ acceptance is $\sim 100\%$ for ξ ranging from 0.002 to 0.1, where ξ is the proton fractional momentum loss. The acceptance limit of the device closest to the interaction point is $\xi > \xi_{min}$ =0.02.

In exclusive double Pomeron exchange, the mass of the central heavy object is given by $M^2 = \xi_1 \xi_2 s$, where ξ_1 and ξ_2 are the proton fractional momentum losses measured in the roman pot detectors.

CP792, *Deep Inelastic Scattering, DIS 2005,* edited by W. H. Smith and S. R. Dasu

TABLE 1. Exclusive Higgs production cross section for different Higgs masses, number of signal and background events for 100 fb^{-1}, ratio, and number of standard deviations (σ).

M_{Higgs}	cross section	signal	backg.	S/B	σ
120	3.9	27.1	28.5	0.95	5.1
130	3.1	20.6	18.8	1.10	4.8
140	2.0	12.6	11.7	1.08	3.7

RESULTS ON DIFFRACTIVE HIGGS PRODUCTION

Results are given in Fig. 1 for a Higgs mass of 120 GeV, in terms of the signal to background ratio S/B, as a function of the Higgs boson mass resolution.

In order to obtain an S/B of 3 (resp. 1, 0.5), a mass resolution of about 0.3 GeV (resp. 1.2, 2.3 GeV) is needed. The forward detector design of [4] claims a resolution of about 2.-2.5 GeV, which leads to a S/B of about 0.4-0.6. Improvements in this design would increase the S/B ratio as indicated on the figure. As usual, this number is enhanced by a large factor if one considers supersymmetric Higgs boson production with favorable Higgs or squark field mixing parameters.

The cross sections obtained after applying the survival probability of 0.03 at the LHC as well as the S/B ratios are given in Table 1 if one assumes a resolution on the missing mass of about 1 GeV (which is the most optimistic scenario). The acceptances of the roman pot detectors as well as the simulation of the CMS detectors have been taken into account in these results.

Let us also notice that the missing mass method will allow to perform a W mass measurement using exclusive (or quasi-exclusive) WW events in double Pomeron exchanges, and QED processes. The advantage of the QED processes is that their cross section is perfectly known and that this measurement only depends on the mass resolution and the roman pot acceptance. In the same way, it is possible to measure the mass of the top quark in $t\bar{t}$ events in double Pomeron exchanges.

The diffractive SUSY Higgs boson production cross section is noticeably enhanced at high values of $\tan\beta$ and since we look for Higgs decaying into $b\bar{b}$, it is possible to benefit directly from the enhancement of the cross section contrary to the non diffractive case. A signal-over-background up to a factor 50 can be reached for 100 fb^{-1} for $\tan\beta \sim 50$ [6].

THRESHOLD SCAN METHOD: W, TOP AND STOP MASS MEASUREMENTS

We propose a new method to measure heavy particle properties via double photon and double pomeron exchange (DPE), at the LHC [5]. In this category of events, the heavy objects are produced in pairs, whereas the beam particles often leave the interaction region intact, and can be measured using very forward detectors.

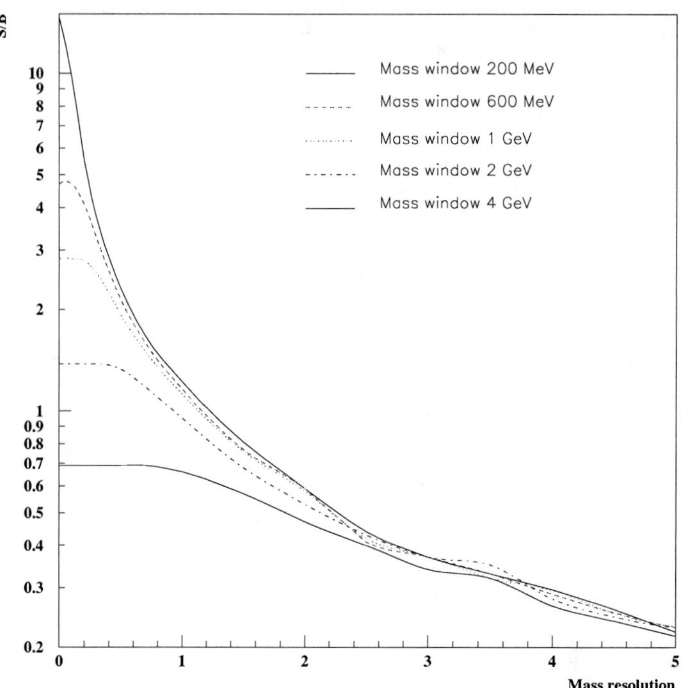

FIGURE 1. Standard Model Higgs boson signal to background ratio as a function of the resolution on the missing mass, in GeV. This figure assumes a Higgs boson mass of 120 GeV.

Pair production of WW bosons and top quarks in QED and double pomeron exchange are described in detail in this section. WW pairs are produced in photon-mediated processes, which are exactly calculable in QED. There is basically no uncertainty concerning the possibility of measuring these processes at the LHC. On the contrary, $t\bar{t}$ events, produced in exclusive double pomeron exchange, suffer from theoretical uncertainties since exclusive diffractive production is still to be observed at the Tevatron, and other models lead to different cross sections, and thus to a different potential for the top quark mass measurement. However, since the exclusive kinematics are simple, the model dependence will be essentially reflected by a factor in the effective luminosity for such events.

Explanation of the methods

We study two different methods to reconstruct the mass of heavy objects double diffractively produced at the LHC. The method is based on a fit to the turn-on point of the missing mass distribution at threshold.

One proposed method (the "histogram" method) corresponds to the comparison of the mass distribution in data with some reference distributions following a Monte Carlo simulation of the detector with different input masses corresponding to the data luminosity. As an example, we can produce a data sample for 100 fb^{-1} with a top mass of 174 GeV, and a few MC samples corresponding to top masses between 150 and 200 GeV by steps of. For each Monte Carlo sample, a χ^2 value corresponding to the population difference in each bin between data and MC is computed. The mass point where the χ^2 is minimum corresponds to the mass of the produced object in data. This method has the advantage of being easy but requires a good simulation of the detector.

The other proposed method (the "turn-on fit" method) is less sensitive to the MC simulation of the detectors. As mentioned earlier, the threshold scan is directly sensitive to the mass of the diffractively produced object (in the WWW case for instance, it is sensitive to twice the WW mass). The idea is thus to fit the turn-on point of the missing mass distribution which leads directly to the mass of the produced object, the WW boson. Due to its robustness, this method is considered as the "default" one in the following.

Results

To illustrate the principle of these methods and their achievements, we apply them to the WW boson and the top quark mass measurements in the following, and obtain the reaches at the LHC. They can be applied to other threshold scans as well. The precision of the WW mass measurement (0.3 GeV for 300 fb^{-1}) is not competitive with other methods, but provides a very precise calibration of the roman pot detectors. The precision of the top mass measurement is however competitive, with an expected precision better than 1 GeV at high luminosity. The resolution on the top mass is given in Fig. 2 as a function of luminosity for different resolutions of the roman pot detectors.

The other application is to use the so-called "threshold-scan method" to measure the stop mass in *exclusive* events. The idea is straightforward: one measures the turn-on point in the missing mass distribution at about twice the stop mass. After taking into account the stop width, we obtain a resolution on the stop mass of 0.4, 0.7 and 4.3 GeV for a stop mass of 174.3, 210 and 393 GeV for a luminosity (divided by the signal efficiency) of 100 fb^{-1}. We notice that one can expect to reach typical mass resolutions which can be obtained at a linear collider. The process is thus similar to those at linear colliders (all final states are detected) without the initial state radiation problem.

The caveat is of course that production via diffractive *exclusive* processes is model dependent, and definitely needs the Tevatron data to test the models. It will allow to determine more precisely the production cross section by testing and measuring at the Tevatron the jet and photon production for high masses and high dijet or diphoton mass fraction.

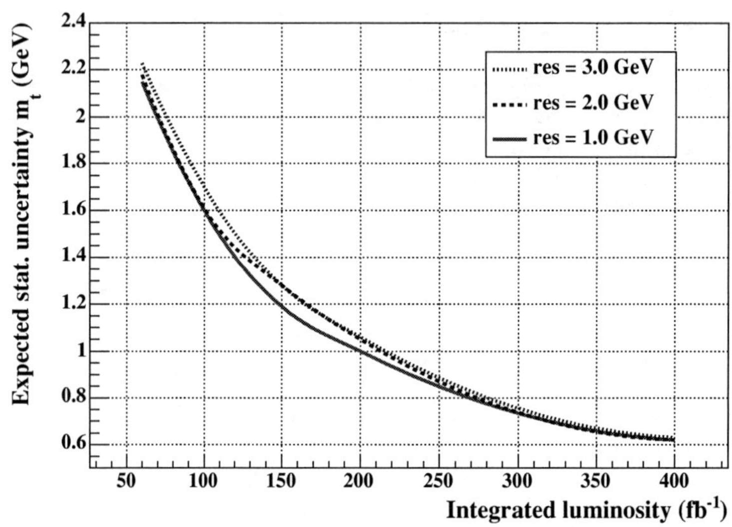

FIGURE 2. Expected statistical precision of the top mass as a function of the integrated luminosity for various resolutions of the roman pot detectors (full line: resolution of 1 GeV, dashed line: 2 GeV, dotted line: 3 GeV).

ACKNOWLEDGMENTS

There results come from a fruitful collaboration with M. Boonekamp, J. Cammin, S. Lavignac and R. Peschanski.

REFERENCES

1. A. Bialas, P.V. Landshoff, Phys. Lett. **B256** (1990) 540.
2. M. Boonekamp, R. Peschanski, and C. Royon, Phys. Lett. **B598** (2004) 243.
3. M. Boonekamp, R. Peschanski, C. Royon, Phys. Rev. Lett. **87** (2001) 251806; M. Boonekamp, A. De Roeck, R. Peschanski, C. Royon, Phys. Lett. **B550** (2002) 93; M. Boonekamp, R. Peschanski, C. Royon, Nucl. Phys. **B669** (2003) 277, Err-ibid **B676** (2004) 493; for a general review see C. Royon, Mod. Phys. Lett. **A18** (2003) 2169.
4. J. Kalliopuska, T. Mäki, N. Marola, R. Orava, K. Österberg, M. Ottela, HIP-2003-11/EXP.
5. M. Boonekamp, J. Cammin, R. Peschanski, C. Royon, hep-ph/0504199.
6. M. Boonekamp, J. Cammin, S. Lavignac, R. Peschanski, C. Royon, hep-ph/0506275.

ELECTROWEAK AND BEYOND THE STANDARD MODEL WORKING GROUP PRESENTATIONS

Electroweak Fits and the Higgs Mass

Peter B. Renton

University of Oxford, Oxford OX1 3RH, UK; p.renton@physics.ox.ac.uk

Abstract. The current electroweak data and the constraints on the Higgs mass are discussed.

Keywords: electroweak, Higgs
PACS: 12.15.Lk

This report contains an update on the values of the precision electroweak properties and fits within the context of the SM, with respect to [1], where more details can be found. The e^+e^- data are from the ALEPH, DELPHI, L3 and OPAL experiments at LEP, and from the SLD experiment at SLAC. The $p\bar{p}$ data come from the CDF and D0 experiments from both Run 1 ($\sqrt{s}$=1.8 TeV) and Run 2 ($\sqrt{s}$=1.96 TeV).

The Z-lepton couplings (see [1]) are extracted from the τ polarisation (A_e, A_μ), the SLAC polarised electron asymmetry $A_{\rm LR}$ (A_e) and the forward-backward asymmetries for leptons (A_ℓ, ℓ=e,μ, τ). The results are reasonably compatible with lepton universality and, assuming this, give $A_e = 0.1501 \pm 0.0016$. Within the context of the SM this favours a light Higgs mass. The invisible width of the Z boson allows the number of light neutrinos to be extracted (assuming Γ_ν/Γ_l from the SM), and gives $N_\nu = 2.9841 \pm 0.0083$, which is 1.9 σ below 3.

The current combined LEP and SLD heavy flavour results are given in Table 1. There are downward shifts of 0.0001 and 0.0006, since Summer 2004, in $A_{\rm FB}^{0,c}$ and $A_{\rm FB}^{0,b}$ respectively due to changes in the theory corrections applied to offpeak data. There is good internal consistency in the determinations of R_b, R_c, $A_{\rm FB}^{0,b}$ and $A_{\rm FB}^{0,c}$. The largest correlation is -0.18, between R_b and R_c. The value of $A_{\rm FB}^{0,b}$ (and also $A_{\rm FB}^{0,c}$) favours a rather heavy Higgs mass.

TABLE 1. Combination of Z heavy flavour results

quantity	value	error
R_b	0.21630	0.00066
R_c	0.1723	0.0031
$A_{\rm FB}^{0,b}$	0.0992	0.0016
$A_{\rm FB}^{0,c}$	0.0707	0.0035
A_b	0.923	0.020
A_c	0.670	0.027

The overall χ^2 probability for the compatibility of the six methods used for $\sin^2\theta_{\rm eff}^{\rm lept}$ is reasonable (3.8%), but the value obtained from purely leptonic processes

CP792, *Deep Inelastic Scattering, DIS 2005,* edited by W. H. Smith and S. R. Dasu

($\sin^2\theta_{\rm eff}^{\rm lept}$=0.23113 ± 0.00021) is some 3.2σ different to that obtained using heavy quarks ($\sin^2\theta_{\rm eff}^{\rm lept}$=0.23222 ± 0.00027). This comes mostly from the 3.2σ difference in the SLD $A_{\rm LR}$ and $A_{\rm FB}^{0,\rm b}$ values.

The W boson is produced singly at the Tevatron (eg $u+\bar{d} \rightarrow W^+$). The leptonic decays W$\rightarrow \ell\nu$ (with $\ell = e, \mu$) are used to determine the W mass and width, using the transverse mass or p_T^ℓ. From Run 1 the values $M_{\rm W}$ = 80.433 ± 0.079 GeV (CDF) and 80.483 ± 0.084 GeV (D0) were obtained. Taking into account common systematics, the combined Run 1 values are $M_{\rm W}$ = 80.452 ± 0.059 GeV and $\Gamma_{\rm W}$ = 2.102 ± 0.106 GeV [2]. Run 2 analyses are currently underway.

At LEP2 the W bosons are pair-produced in $e^+e^- \rightarrow W^+W^-$. The final analyses are still in progress. The statistical uncertainties from the $\ell\nu q\bar{q}'$ and $q\bar{q}'q\bar{q}'$ channels are similar. However, there is at present a large systematic uncertainty (97 MeV) in the $q\bar{q}'q\bar{q}'$ channel, due to final-state interaction effects (colour reconnection and Bose Einstein correlations). This means that the $q\bar{q}'q\bar{q}'$ channel carries only 10% of the weight in the LEP2 average. The preliminary LEP2 values are $M_{\rm W}$ = 80.412 ± 0.042 GeV and $\Gamma_{\rm W}$ = 2.152 ± 0.091 GeV.

The combined Tevatron and LEP2 values are $M_{\rm W}$ = 80.425 ± 0.034 GeV and $\Gamma_{\rm W}$ = 2.133 ± 0.069 GeV. $\Gamma_{\rm W}$ is compatible with the SM value of 2.097 ± 0.003 GeV. The world average $M_{\rm W}$ value favours a low Higgs mass in the context of the SM.

The SM parameters

The SM parameters are taken to be $M_{\rm Z}$, G_F, $\alpha(M_{\rm Z})$ and $\alpha_{\rm s}(M_{\rm Z})$, and the top-quark mass $m_{\rm t}$. Through loop diagrams measurements of the precision electroweak quantities are sensitive to $m_{\rm t}$ and, the 'unknown' in the SM, $m_{\rm H}$. The SM computations use the programs TOPAZ0 and ZFITTER. The latter program (version 6.41) incorporates the recent fermion 2-loop corrections to $\sin^2\theta_{\rm eff}^{\rm lept}$ and full 2-loop, and leading 3-loop, corrections to $M_{\rm W}$ [3].

The D0 Collaboration have improved their original Run 1 measurement using a weighting method based on the matrix element, giving $m_{\rm t}$ = 179.0 ± 3.5 (stat) ± 3.8 (syst) GeV. The CDF Run 1 value is $m_{\rm t}$ = 176.1 ± 4.2 (stat) ± 5.1 (syst) GeV. Taking into account common systematic uncertainties the combined value is [4] $m_{\rm t}$ = 178.0 ± 4.3 GeV, with statistical and systematic error components of 2.7 and 3.3 GeV respectively. The previous value was $m_{\rm t}$ = 174.3 ± 5.1 GeV.

Run 2 values have been obtained by both the CDF and D0 Collaborations, but these have not yet been included in the average.

The value of α at the scale $M_{\rm Z}$ requires the use of data on $e^+e^- \rightarrow$hadrons at low energies and the use of perturbative QCD at higher energies. The various estimations of $\alpha(M_{\rm Z})$ differ in the extent to which QCD is used, as well as in the data used in the evaluation. The quantity needed is the hadronic contribution $\Delta\alpha_{\rm had}^{(5)}$ and the value used by the LEP EWWG [1] is $\Delta\alpha_{\rm had}^{(5)}(M_{\rm Z})$ = 0.02761 ± 0.00036.

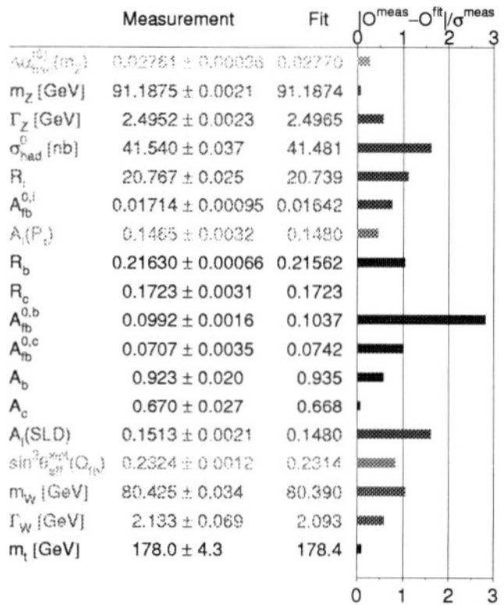

FIGURE 1. Measured and SM fitted values of electroweak quantities.

Electroweak fits

The measurements used in the global SM electroweak fits, and the fitted values, are shown in figure 1. The SM fit to these high Q^2 data gives

$$m_t = 178.4 \pm 3.9 \text{ GeV}$$
$$m_H = 126\ ^{+73}_{-48} \text{ GeV}$$
$$\alpha_s(M_Z) = 0.1188 \pm 0.0027.$$

The χ^2/df is 18.3/13, giving a probability of 15%. The variation of the fit χ^2, compared to the minimum value, is shown in the 'blue-band' plot of figure 2, as a function of m_{H}. Also shown is the direct search limit of 114 GeV. The one-sided 95% upper limit is $m_{\mathrm{H}} \leq 280$ GeV. This includes the theoretical uncertainty (blue-band) which is evaluated by considering the uncertainties in the new 2-loop calculations [3]. If the more theory driven value $\Delta\alpha^{(5)}_{\mathrm{had}}(M_Z) = 0.02749 \pm 0.00012$ is used, then m_{H} increases to 143 GeV.

Since 2004[1] the main change is from the modified result for $A^{0,\mathrm{b}}_{\mathrm{FB}}$. The direct versus indirect values of m_{t} and M_{W} is a powerful test of the SM; see figure 3. The contours shown are for the 68% cl. It can be seen that there is a reasonable degree of overlap and that the data prefer a light Higgs mass.

The above fits use only high Q^2 data. There are also low Q^2 data from Atomic Parity Violation in ^{133}Cs (Q_W = -72.74 $\pm$ 0.46), the SLAC polarised electron Moller scattering experiment E158 ($\sin^2\theta^{\mathrm{lept}}_{\mathrm{eff}} = 0.2333 \pm 0.0016$) and the deep-inelastic $\nu(\bar{\nu})$ experiment NuTeV ($\sin^2\theta_{\mathrm{W}} = 0.2277 \pm 0.0016$). The NuTeV value can be used to extract M_{W}, and gives a value 3.1σ below that from direct measurement. Including all these low Q^2 data in the SM fit increases m_{H} by 13 GeV to 139 GeV, and the χ^2 probability drops to 3.7%, essentially due to the NuTeV result.

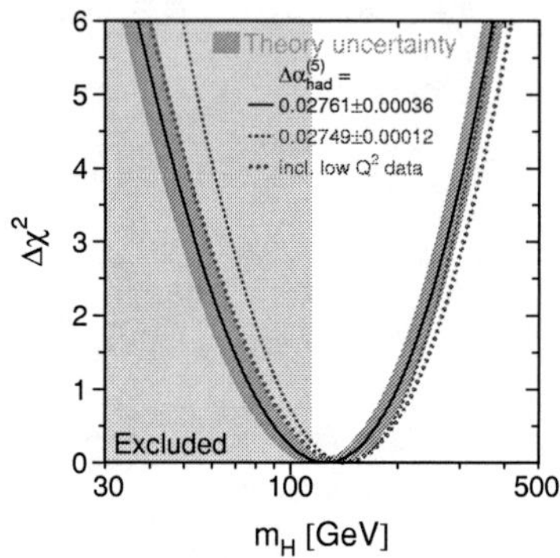

FIGURE 2. Variation of χ^2 versus m_{H}.

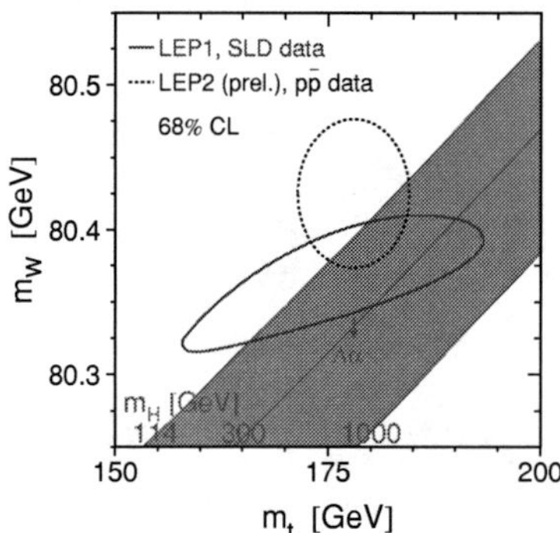

FIGURE 3. Direct versus indirect m_{t} and M_{W} measurements.

In conclusion, the SM fits favour a relatively light Higgs mass, $m_{\mathrm{H}} = 126\,^{+73}_{-48}$ GeV, and a 95% cl upper limit of 280 GeV. Improved measurements of both m_{t} and M_{W} at the Tevatron, and then the LHC, will significantly improve the precision of the indirect estimation of m_{H}.

REFERENCES

1. The LEP Collaborations and EWWG; CERN-PH-EP/2004-069 and hep-ex/0412015 (2004).
2. Combination of CDF and D0 results on W boson mass and width, hep-ex/0311039 (2003).
3. M. Awramik et al., hep-ph/0311148, 0407317; M. Faisst et al. Nucl. Phys B665 (2003) 649.
4. Combination of CDF and D0 results on the top-quark mass, hep-ex/0404010 (2004).

W Boson Mass and Properties

Christopher P. Hays

Department of Physics, Duke University, Durham, North Carolina 27708

Abstract. Measurements of the W boson determine fundamental parameters of the standard model and test its internal consistency at the level of radiative corrections. In particular, the W boson mass constrains the mass of the unobserved Higgs boson through loop corrections, and the W boson width indirectly determines elements of the CKM matrix defining quark generational mixing in the electroweak interactions. The CDF Collaboration has analyzed $\approx$200 pb^{-1} of $\sqrt{s} = 1.96$ TeV $p\bar{p}$ collision data and determined the uncertainty on its W boson mass measurement to be 76 MeV. The DØ Collaboration has directly measured the W boson width to be $\Gamma_W = 2.011 \pm 0.09$(stat.)$\pm 0.107$(syst.) GeV using 177 pb^{-1} of $W \to e\nu$ data.

Keywords: W, mass; W, width; CDF; DZERO
PACS: 12.15.Lk; 12.20.Fv; 13.38.Be; 14.17.Fm

INTRODUCTION

Many properties of the W boson are precisely predicted by the standard model (SM), which describes electroweak interactions as an $SU(2)_L \times U(1)_Y$ gauge symmetry. Ongoing $\sqrt{s} = 1.96$ TeV $p\bar{p}$ collisions at the Fermilab Tevatron in Batavia, Illinois, have produced the largest W boson data samples in the world to test these predictions. Important W boson measurements include those of its mass (m_W) [1], width (Γ_W) [1], cross section [2], couplings to other bosons [3], and angular distributions (including charge asymmetries [4]). The m_W and Γ_W measurements are particularly important for understanding the mechanism of $SU(2)_L \times U(1)_Y \to U(1)_{EM}$ symmetry breaking and the generational fermionic structure, respectively. The CDF and DØ Collaborations have analyzed the first $\approx$200 pb^{-1} of Run 2 collisions and CDF has determined its m_W measurement uncertainties, while DØ has performed the first Run 2 direct Γ_W measurement.

W BOSON MASS MEASUREMENT

Electroweak Symmetry Breaking

The SM predicts m_W in terms of the Z boson mass (m_Z) and the electromagnetic (α_{EM}) and weak (G_F) couplings [5]:

$$m_W^2 = \frac{\pi \alpha_{EM}}{\sqrt{2} G_F (1 - m_W^2/m_Z^2)(1 - \Delta r)}, \tag{1}$$

where Δr is the radiative correction to the mass. This correction includes a term proportional to the square of the top quark mass (m_t) and the logarithm of the Higgs boson mass (m_H). The unobserved Higgs boson provides the mechanism for electroweak symmetry

CP792, *Deep Inelastic Scattering, DIS 2005*, edited by W. H. Smith and S. R. Dasu

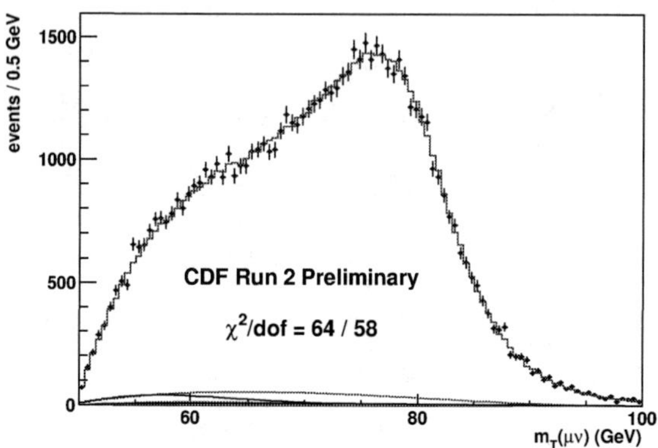

FIGURE 1. The measured (points) and simulated (histogram) m_T distribution in $W \to \mu\nu$ events. The m_T distribution has the best statistical power for the W mass fit.

breaking in the SM. The m_W measurement constrains m_H in the context of the SM, and could provide evidence for new particles coupling to the W boson when the Higgs is discovered. One promising scenario is the existence of supersymmetric particles, which would be favored if m_W is measured to be larger than predicted by the SM [6].

CDF Run 2 Measurement

At hadron colliders m_W is measured in the $e\nu$ and $\mu\nu$ decay channels, since these channels can be cleanly separated from background and the lepton momenta can be calibrated to a precision better than a part per thousand. The mass is fit using templates of the transverse mass (m_T) distribution (Fig. 1), which has the greatest statistical power. The transverse mass is defined as the two-dimensional mass ignoring the unmeasured z component of the neutrino momentum:

$$m_T = \sqrt{2p_T^l p_T^\nu(1-\cos(\Delta\phi))}. \tag{2}$$

The mass measurement consists of the following main components: simulation of the W boson production and decay; lepton momentum calibration; and hadronic recoil energy measurement and modelling.

The distributions of W boson p_T and rapidity affect the momentum distributions of the decay leptons, which in turn affect the m_T distribution. The p_T distribution is simulated using a next-to-leading-log QCD event generator (RESBOS [7]), which uses Tevatron Run 1 $Z \to ll$ data to parametrize the boson p_T distribution at low momentum. The preliminary Run 2 data show good agreement with this parametrization (Fig. 2).

The W boson rapidity distribution is affected by the momentum distributions of the valence and sea quarks and gluons in the proton and antiproton. These distributions are parametrized by the parton distribution functions (PDFs) using a global data fit.

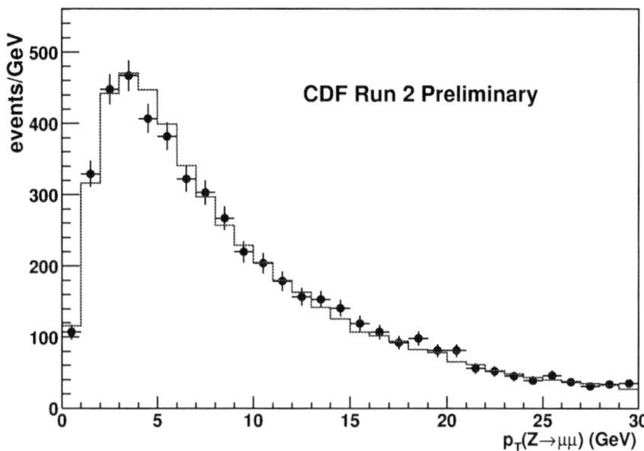

FIGURE 2. The measured (points) and simulated (histogram) p_T^Z distribution in $Z \to \mu\mu$ events. The simulated distribution is based on the RESBOS generator, which parametrizes the non-perturbative QCD region of low p_T^Z using Tevatron Run 1 data.

The uncertainty on these functions results in a 15 MeV uncertainty on m_W, and will be reduced by constraints from the new W production charge asymmetry measurement [4].

Photon radiation off the decay lepton reduces the measured m_T of a given W event. An event generator based on a NLO QED calculation simulates this radiation. The NNLO two-photon radiation process, which is not simulated, results in a measured W mass uncertainty of 15 (20) MeV in the e (μ) channel.

CDF bases its lepton momentum calibration on the charged-particle tracker. The high-statistics $J/\Psi \to \mu\mu$ and $\Upsilon \to \mu\mu$ samples provide a precise tracker momentum calibration, which can be extrapolated to high momenta. The electron energy measurement in the calorimeter is scaled to match the momentum of the electron track (after simulating effects such as photon radiation in the passive material).

The hadronic recoil energy is not simulated from first principles; rather, a model based on $Z \to ll$ data is developed. The data provide a response function for the ratio of measured to produced hadronic energy. The energy resolution model has two components: resolution from the residual energy from the underlying interaction; and the resolution of the higher momentum hadrons recoiling off the Z boson.

Table 1 summarizes the uncertainties associated with the CDF m_W measurement.

W BOSON WIDTH MEASUREMENT

Electroweak Generational Mixing

The difference between the quark electroweak and mass eigenstates results in generational mixing in the electroweak interactions. This mixing arises from the couplings of W bosons to quarks and can be probed by the W decay width. Since the SM predicts the W leptonic width to within $\approx$0.1%, a measurement of the total W width probes the W

TABLE 1. The uncertainties on the W boson mass measurement in MeV/c^2 using 0.2 fb^{-1} of Run 2 CDF data [8]. The combined uncertainty is 76 MeV.c^2. The CDF Run 1B uncertainties are shown for comparison [9].

Sytematic Uncertainty	Electrons (Run 1B)	Muons (Run 1B)
Production and Decay Model	30 (30)	30 (30)
Lepton E Scale and Resolution	70 (80)	30 (87)
Recoil Scale and Resolution	50 (37)	50 (35)
Backgrounds	20 (5)	20 (25)
Statistics	45 (65)	50 (100)
Total	105 (110)	85 (140)

decays to quarks:

$$\Gamma_W = \Gamma(W \to l\nu_l)(3 + f_{QCD}\sum_{ij}|V_{ij}|^2), \tag{3}$$

where f_{QCD} is a QCD correction factor and V_{ij} are the elements of the CKM quark mixing matrix. The unitarity of the CKM matrix in the SM requires $\sum_{ij}|V_{ij}|^2 = 2$.

DØ Run 2 Measurement

As with the m_W measurement, the direct Γ_W measurement uses templates of the m_T distribution in lepton decays. The DØ Collaboration has measured the W width using 625 $W \to e\nu$ candidates in the region 100 GeV $< m_T <$ 200 GeV. Since the intrinsic width is convoluted by detector resolution, the dominant uncertainties arise from uncertainties in the resolution model. The model is constrained using $Z \to ee$ data and results in Γ_W uncertainties of 51 (69) MeV for the electron (hadron) resolution. The fit result is $\Gamma_W = 2.011 \pm 0.093 \pm 0.107$ GeV, from which $\sum_{ij}|V_{ij}|^2 = 1.89 \pm 0.20$ can be extracted. This result is consistent with the SM prediction of $\Gamma_W = 2.099 \pm 0.003$ GeV [10] and the more precise CKM measurements using the world average of the leptonic branching ratio of the W, $|V_{ij}|^2 = 2.044 \pm 0.024$.

REFERENCES

1. CDF and DØ Collaborations, V. M. Abazov *et al.*, *Phys. Rev. D*, **70**, 092008 (2004).
2. CDF Collaboration, D. Acosta *et al.*, *Phys. Rev. Lett.*, **94**, 091803 (2005).
3. CDF Collaboration, D. Acosta *et al.*, *Phys. Rev. Lett.*, **94**, 041803 (2005); *ibid*, 211801 (2005); *ibid*, *Phys. Rev. D*, **71**, 091105 (2005); and DØ Collaboration, V. M. Abazov *et al.*, *Phys. Rev. D*, **71**, 091108 (2005).
4. CDF Collaboration, D. Acosta *et al.*, *Phys. Rev. D*, **71**, 051104 (2005).
5. S. Eidelman *et al.*, *Phys. Lett. B*, **592**, 1 (2004), and references therein.
6. S. Heinemeyer, W. Hollik, and G. Weiglein, *hep-ph/0412214* (2004).
7. C. Balazs and C.P. Yuan, *Phys. Rev. D*, **56**, 5558 (1997).
8. C. Hays, "W Boson Mass and Width at the Tevatron," in *Proceedings of the 32nd International Conference on High Energy Physics*, edited by H. Chen *et al.*, World Scientific, 2005.
9. CDF Collaboration, T. Affolder *et al.*, *Phys. Rev. D*, **63**, 032003 (2001).
10. P. B. Renton, *Rep. Prog. Phys.*, **65**, 1271-1330 (2002), and references therein.

Determination of electroweak parameters at HERA with the H1 experiment

Benjamin Portheault

Laboratoire de l'Accélérateur Linéaire
IN2P3-CNRS et Université de Paris Sud
F-91898 Orsay Cedex
E-mail: portheau@lal.in2p3.fr

Abstract. Using deep inelastic $e^{\pm}p$ charged and neutral current scattering cross sections previously published, a combined electroweak and QCD analysis is performed to determine electroweak parameters accounting for their correlation with parton distributions. The data used have been collected by the H1 experiment in 1994-2000 and correspond to an integrated luminosity of 117.2 pb^{-1}. The W boson mass is determined via the propagator and in the on-mass-shell scheme. A first measurement at HERA is made of the light quark weak couplings to the Z^0 boson.

Keywords: deep inelastic scattering, QCD analysis, electroweak analysis, W boson mass, quarks weak neutral current couplings
PACS: 12.38.-t, 12.15.Ji

THE HERA INCLUSIVE DIS DATA

The H1 experiment has collected 117.2 pb^{-1} of data during the first phase of HERA operation (HERA I) with unpolarised $e^{\pm}$ beams. The inclusive deep inelastic scattering (DIS) cross section [1, 2, 3, 4] have been measured in neutral current (NC) and charged current (CC). The precise low x, Q^2 data became the cornerstone of any QCD analysis (the so called QCD fits) aiming at extracting the parton distribution functions (PDFs). The addition of CC data allowed the flavor separation of PDFs and thus a QCD analysis based solely on H1 data has been performed [4], with the advantage of a consistent treatment of correlated systematic errors.

A COMBINED ELECTROWEAK-QCD ANALYSIS

The combined EW-QCD analysis aims at determining EW and PDFs parameters in the same fit to NC and CC data. This removes a bias due to the fact that PDFs parameters are obtained with assumptions on EW parameters, so the use of fixed PDFs parameters to fit the EW ones is not a consistent procedure. The combined fitting allows to consistently take into account the uncertainty coming from the proton's structure.

CP792, *Deep Inelastic Scattering, DIS 2005,* edited by W. H. Smith and S. R. Dasu

THE H1PDF2000 QCD ANALYSIS

The basis for the combined EW-QCD analysis is the H1 QCD analysis of HERA I data, the so-called H1PDF2000 [4, 5]. In this analysis five combinations are parameterized and fitted, the gluon g, the up-type quarks $U = u + c$, the down-type quarks $D = d + s$ and the anti-quarks distribution $\overline{U}, \overline{D}$. The parameterization space is explored with a systematic procedure until a χ^2 saturation is reached. The resulting paramterization has a total of 10 free parameters.

DESCRIPTION OF THE ELECTROWEAK SCHEMES

The choice of a renormalization scheme defines the set of theory parameters to be used as inputs for the cross sections calculations. For the CC and NC cross sections, two main schemes are used : the On Mass Shell scheme (OMS) which uses the weak boson masses as fundamental parameters, and the Modified On Mass Shell scheme (MOMS) in which the W boson mass is replaced by the Fermi constant G_F. Since the Fermi constant includes by definition the radiative corrections due to the W self energy, the CC cross section reads

$$\frac{d^2\sigma_{CC}^{e^\pm p}}{dxdQ^2} = \frac{G_F^2}{4\pi x}\left[\frac{M_W^2}{Q^2+M_W^2}\right]^2 \Phi_{\mathrm{PDFs}}^{\pm}(x,Q^2) \qquad (1)$$

in which $\Phi_{\mathrm{PDFs}}^{\pm}$ is the structure function part and M_W has to be computed from G and the other parameters. To use of the OMS scheme, one has to replace G by its expressions as a function of masses and one needs to compute the relatively large ($\sim 3\%$) radiative correction which is a function of all the Standard Model parameters $\Delta r(\alpha, M_Z, M_W, m_t, M_H)$. In particular Δr has a quadratic dependency upon the top quark mass m_t and a logarithmic dependency upon the Higgs mass M_H. The CC cross section reads

$$\frac{d^2\sigma_{CC}^{e^\pm p}}{dxdQ^2} = \frac{\pi\alpha^2}{4M_W^2\left(1-\frac{M_W^2}{M_Z^2}\right)^2}\frac{1}{(1-\Delta r)^2}\left[\frac{M_W^2}{Q^2+M_W^2}\right]^2 \Phi_{\mathrm{PDFs}}^{\pm}(x,Q^2). \qquad (2)$$

The W mass entering in Eq. (1) is called the *propagator mass* as it enters only in the Q^2 dependency of the cross section. In the OMS scheme, the W mass enters also in the normalization of the cross section, yielding an increased sensitivity to this parameter.

RESULTS FOR THE PROPAGATOR MASS

In a first analysis, we consider the propagator mass M_{prop} and the normalization constant G in Eq. (1) as independent. The result of a 12 parameter fit G-M_{prop}-PDFs is shown on Fig. 1. Then the parameter G is fixed to the well measured Fermi constant G_F and the propagator mass is fitted together with the PDFs. The result is

$$M_W = 82.87 \pm 1.83(\mathrm{exp})^{+0.30}_{-0.16}(\mathrm{mod})\ \mathrm{GeV} \qquad (3)$$

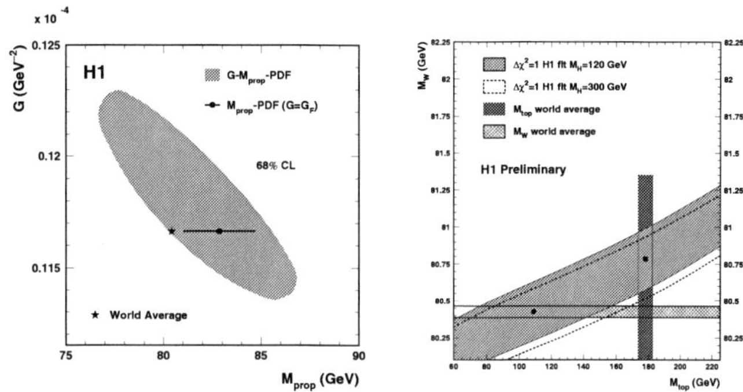

FIGURE 1. Left figure: result of the fit to G and M_{prop} at 68% CL. Fixing G to G_F, the fit results in a measurement of the propagator mass M_{prop} shown as the dot with the horizontal error bar. Right figure: result of the m_t, M_W and PDFs fit in the m_t, M_W plane for $M_H = 120$ GeV (shaded area) and for $M_H = 300$ GeV (dashed area). The vertical shaded band and the horizontal hatched band represent respectively the world average values of the top and W masses.

where the experimental uncertainty accounts for all statistical and systematic errors. The second error is due to model uncertainties, accounting for variations on the parameters entering in the QCD fit modelisation[4]. Using fixed PDFs from H1PDF2000 would introduce a systematic bias of 0.5 GeV on M_{prop}. This is the first coherent determination of the propagator mass in ep collisions.

OMS ANALYSIS

In the OMS scheme one needs to compute the radiative correction to the W self energy. This is done using EPRC program [6], and terms of order $\mathcal{O}(\alpha)$, $\mathcal{O}(\alpha\alpha_s)$ and leading $\mathcal{O}(\alpha^2)$ are included. First, a 12 parameter fit of M_W, m_t and PDFs is performed, and the result in the M_W, m_t plane is shown in Fig. 1. The choice of the Higgs mass shifts the allowed region. Since the top mass has been directly measured at the TeVatron, one can use its value to constrain the W mass. The result for $m_t = 178$ GeV, $M_H = 120$ GeV is

$$M_W = 80.786 \pm 0.207(\text{exp})^{+0.048}_{-0.029}(\text{mod}) \pm 0.025(\text{top}) \pm 0.033(\text{th}) \text{ GeV}. \qquad (4)$$

In addition to the experimental and model uncertainties, uncertainties from the top mass and theoretical uncertainty on Δr calculation have been taken into account[7]. Using a Higgs mass of 300 GeV instead of 120 GeV changes M_W by -0.084 GeV. One needs to emphasize that this is not a measurement of the W mass but a model dependent determination of a Standard Model parameter, because the validity of Standard Model is assumed is to compute the radiative correction. Given that the result is 1.7σ from the world average value of the W mass, the H1 data are consistent with the Standard Model of strong and electroweak interactions. This result can be translated into a determination of $\sin^2\theta_W = 1 - M_W^2/M_Z^2$ in the OMS scheme, using the Z^0 mass world average. The result is $\sin^2\theta_W = 0.2151 \pm 0.0040(\text{exp})^{+0.0019}_{-0.0011}(\text{th})$. It is also possible to fit other parameters via the radiative correction, such as the top or the Higgs mass. The result

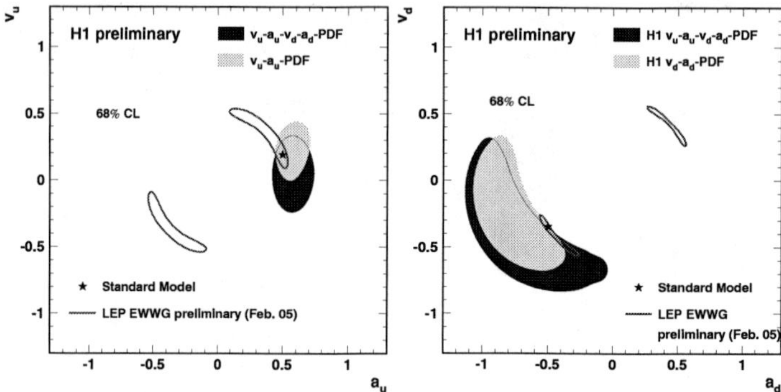

FIGURE 2. Results at 68% CL on the weak neutral current couplings of U (left plot) and D (right plot) quarks to the Z^0 boson determined in this analysis. The dark-shaded contours correspond to results of a simultaneous fit of all four couplings. The light-shaded contours correspond to results of fits where either d or u quark couplings are fixed to their SM values. The open contours are from the combined LEP data[8].

is $m_t = 108 \pm 44$ GeV, and $\log_{10} M_H = 3.9 \pm 2.2(\text{exp})$. One thus obtains a top mass consistent with direct measurement and a Higgs mass larger than 50 GeV at 68% CL via the radiative corrections. This is the first time such a determination is obtained at HERA.

QUARKS COUPLINGS TO THE Z^0

The axial a_q and vector v_q couplings of quarks to the Z have been measured precisely for heavy quarks in e^+e^- collisions, but are less well measured for light quarks. In DIS, combinations weighted by partons densities of the four couplings a_U, a_D, v_U, v_D appear in the $\gamma - Z^0$ and pure Z^0 terms of F_2 and xF_3. It is possible to extract unambiguously these couplings together with the parton densities in a combined fit. The results are shown in Fig. 2. This is the first HERA result on this topic, and a significant future improvement is expected with the HERA II polarised data.

REFERENCES

1. C. Adloff *et al.* [H1 Collaboration], Eur. Phys. J. C **13**, 609 (2000) [arXiv:hep-ex/9908059].
2. C. Adloff *et al.* [H1 Collaboration], Eur. Phys. J. C **21** (2001) 33 [arXiv:hep-ex/0012053].
3. C. Adloff *et al.* [H1 Collaboration], Eur. Phys. J. C **19**, 269 (2001) [arXiv:hep-ex/0012052].
4. C. Adloff *et al.* [H1 Collaboration], Eur. Phys. J. C **30** (2003) 1 [arXiv:hep-ex/0304003].
5. C. Pascaud and F. Zomer, LAL preprint, LAL 95-05 (1995); C. Pascaud and F. Zomer, [hep-ph/0104013].
6. H. Spiesberger, *Prepared for Workshop on Future Physics at HERA (Preceded by meetings 25-26 Sep 1995 and 7-9 Feb 1996 at DESY), Hamburg, Germany, 30-31 May 1996* `http://www.desy.de/~hspiesb/eprc.html`
7. B. Portheault, Ph.D. thesis (March 2005),LAL 05-05 (IN2P3/CNRS), Université de Paris Sud Orsay.
8. LEP and SLD Electroweak Working Groups, `http://lepewwg.web.cern.ch/LEPEWWG/plots/winter2005/`

Top Quark Mass and Properties at the Tevatron

Jean-François Arguin
on behalf of the CDF and DØ Collaborations

Department of Physics, University of Toronto, 60 St. George St., Toronto, M5S 1A7, ON, Canada

Abstract. We present recent analyses of top quark properties performed at Run II of the Tevatron. Measurements of the top quark mass, branching ratios and W boson helicity inside top quark decays are covered.

Keywords: Particle physics; hadron colliders; top quark; electroweak physics; heavy quark
PACS: 12.15.Ff; 12.15.Lk; 14.65.Ha

TOP PHENOMENOLOGY AT THE TEVATRON

The top quark has been discovered only recently [1] due to its very large mass: $M_{top} \approx 175$ GeV$/c^2$. Indeed, the top quark is easily the heaviest particle in the Standard Model (SM). This peculiar property brings several reasons to study it. First, the Yukawa coupling of the top quark is nearly one which could be a sign that it plays a special role in the origin of mass beyond the Standard Model. Second, radiative corrections due to top quark loops are often dominant in the prediction of precision observables like the W or Higgs boson mass. Finally, its large mass implies it has a very short lifetime ($\sim 10^{-25}$s), about an order of magnitude smaller than the hadronization time, which provides a unique opportunity to study a bare quark.

The top quarks are produced by the Tevatron, a $p\bar{p}$ collider operating at a center-of-mass energy of 1.96 TeV for the current period of data-taking (Run II). They are produced predominantly in pairs ($t\bar{t}$) via the strong interaction with a predicted cross-section of $6.7^{+0.7}_{-0.9}$ pb [2]. Because the V_{tb} CKM matrix element is nearly one under the 3 families SM assumption, the top quarks are predicted to decay $> 99.8\%$ of the time to a real W boson and a b-quark. The two resulting W bosons in turn decay either hadronically or leptonically, defining the three channels of $t\bar{t}$ events (with branching ratios in parenthesis): "all-hadronic" for two hadronic decays (46%), "lepton+jets" for one leptonic and hadronic decays (29%) and "dilepton" for two leptonic decays (5%). By "leptons" we refer only to electrons and muons, taus being generally too difficult to identify to be usually considered in top quark properties measurements.

The $t\bar{t}$ events reconstruction are performed by the CDF and DØ detectors that are described in detail elsewhere [3]. Typical event selections include identification of isolated leptons with high transverse momentum (p_T), large transverse missing energy ($\not{E}_T$) due to the undetected neutrino(s) from W boson decays, several high-p_T jets and identification of b-jets (b-tagging) generally using secondary vertex tagging. The integrated luminosity ($\int \mathscr{L} dt$) used for the measurements presented here vary from $\approx$ 200–350 pb^{-1}.

CP792, *Deep Inelastic Scattering, DIS 2005,* edited by W. H. Smith and S. R. Dasu

MASS MEASUREMENTS

The current information on M_{top} comes from measurements performed during Run I of the Tevatron (with a luminosity of $\approx$ 100 pb^{-1}). The world average is $M_{top} = 178.0 \pm 4.3$ GeV$/c^2$ [4]. The final goal in Run II is to collect between 4-8 fb^{-1} and reduce the uncertainty on M_{top} to ≈ 2 GeV$/c^2$.

A precise measurement of M_{top} is strongly motivated in the SM because the radiative corrections to many precision observables are dominated by top quark loops. A famous example is the indirect constraint on the Higgs boson mass (M_H). The current constraint is $M_H = 126^{+73}_{-48}$ GeV$/c^2$ [5]. More detailed discussions on the precision electroweak observables can be found in these proceedings (see contribution from P. Renton).

Before describing specific analyses, we first discuss general considerations. One challenge of the M_{top} measurement is to solve the combinatorics problem in the event reconstruction. For example there are four jets in the final state in the lepton+jets channel ($t\bar{t} \to l\nu q\bar{q}'b\bar{b}$), resulting in 12 possible jet-parton assignments (since the two W daughter jets are interchangeable in the reconstruction). The problem is simplified by tagging b-jets as described above. Another limitation of the measurement is the large uncertainty in the modeling of the jet energy response, referred to as the jet energy scale uncertainty (JES). This is generally the dominant systematic uncertainty, and is currently about 3 GeV$/c^2$ for CDF and 5–6 GeV$/c^2$ for DØ[1].

CDF recently reported the most precise measurement of M_{top} to date. It has been performed in the lepton+jets channel using $\int \mathscr{L} dt = 318$ pb^{-1} of data. One novelty of this analysis is the usage of the hadronic W boson decays ($W \to jj$) to improve the jet energy scale uncertainty. Templates of reconstructed top quark and W boson mass are created in MC events as a function of the true top quark mass and the jet energy scale and compared with the data. The fit to the data yields $M_{top} = 173.5\ ^{+3.7}_{-3.6}$ (stat. + JES) GeV$/c^2$ where both the statistical and JES uncertainties are included. The systematic uncertainties not including the JES are small (1.7 GeV$/c^2$). The final result is $M_{top} = 173.5\ ^{+4.1}_{-4.0}$ GeV$/c^2$. Figure 1 shows the reconstructed top quark mass in the data with the best fits from the MC templates overlaid. Analyses using the template method in the lepton+jets channel have also been performed by the DØ collaboration. The best measurement so far corresponds to a luminosity of 230 pb^{-1} and yields $M_{top} = 170.0 \pm 4.2$ (stat.) ± 6.0 (syst.) GeV/c^2.

A different class of analyses uses a matrix-element method that consists of computing a probability for each event to be signal with a given top quark mass. The probability is computed using the full SM production and decay matrix-elements. This method has been shown to be very powerful statistically, as demonstrated by the best measurement in Run I [6]. The CDF collaboration has recently reported a measurement using this method: $M_{top} = 173.8\ ^{+2.7}_{-2.6}$ (stat.) $\pm$ 3.3 (syst.) GeV/c^2. The result is in very good agreement with the template analysis above. A similar measurement is expected to be released by the DØ collaboration very soon.

The top quark mass can also be measured in the dilepton channel but with larger

[1] The DØ JES uncertainty is expected to go down near the CDF level in the next few months.

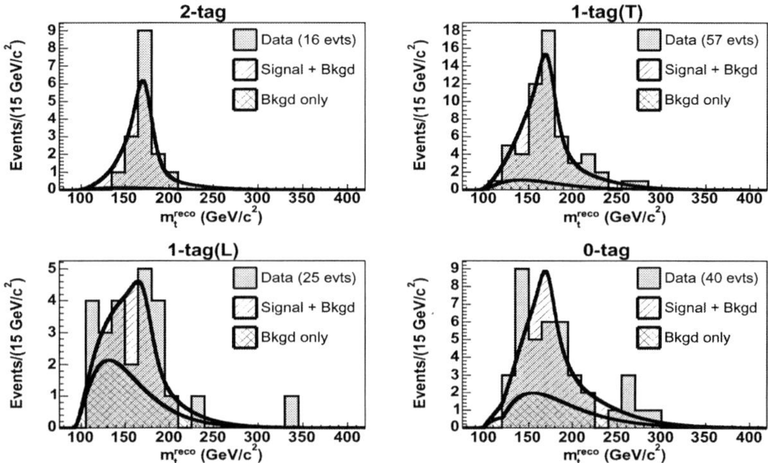

FIGURE 1. The reconstructed top quark mass for various subsamples of the lepton+jets sample collected by CDF ($\sim$ 318 pb^{-1}). Overlaid are the best fit from signal and background templates.

statistical uncertainties due to the smaller branching ratio and the two neutrinos in the final state that complicate the event reconstruction. The CDF collaboration has recently reported the best measurement in this channel to date using the matrix-element method: $M_{top} = 165.3 \pm 6.3$ (stat.) ± 3.6 (syst.) GeV/c^2 (340 pb^{-1}). The best measurement at DØ so far is: $M_{top} = 155\ ^{+14}_{-13}$ (stat.) ± 7 (syst.) GeV/c^2 (230 pb^{-1}).

OTHER PROPERTIES

It is very important to study in detail the properties of the top quark since it might be more closely related to new physics than other SM particles due to its large mass. For instance, one can study the polarization of the W boson inside the top decays. In the SM, 70%, 30% and 0% of W bosons are predicted to have a longitudinal, left-handed and right-handed helicity, respectively. The W helicity is sensitive to the angle between the charged lepton in W rest frame and the b-jet angle, denoted as $\cos\theta^*$. The DØ collaboration measured the fraction of right-handed W bosons in top decays: $f^+ < 0.25$ at 95% confidence level (C.L.) [7]. Figure 2 shows the distribution $\cos\theta^*$ in data. The W helicity is also sensitive to the charged lepton p_T spectrum from W boson decays that has been used at CDF to measure the fraction of longitudinal W bosons: $f^0 = 0.27^{+0.35}_{-0.21}$.

The assumption that the top quark decays nearly 100% of the time to $t \to Wb$ is checked by measuring $R = B(t \to Wb)/B(t \to Wq)$. This is done by counting the number of b-tags in the sample. Such a measurement was performed at DØ and yielded $R > 0.64$ at 95% C.L.. A consistent result has been obtained by CDF: $R > 0.61$ at 95% C.L. [8].

Other properties of the top quark will be measured in Run II as more data is accumulated like the electric charge, $t\bar{t}$ spin correlation and rare decays.

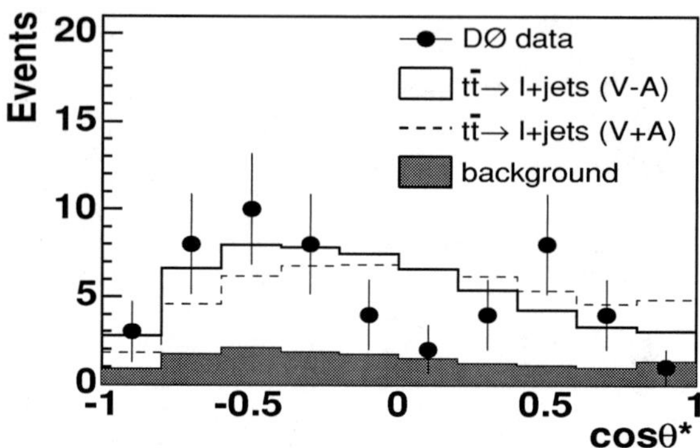

FIGURE 2. Distribution of $\cos\theta^*$ for data (points), signal assuming pure V-A interaction (full line), pure V+A interaction (dashed line) and background for events with b-tag (DØ experiment [7]).

CONCLUSION

The precise determination of top quark properties is already well underway in Run II at the Tevatron. The best measurement of the top quark mass to date has been performed at CDF: $M_{top} = 173.5\,^{+4.1}_{-4.0}$ GeV$/c^2$. A very competitive measurement is expected to be released by DØ very soon. Measurements of other properties of top quarks show a good agreement with the SM using only the small $t\bar{t}$ samples collected so far ($\mathcal{O}(100)$ events per experiment). We can expect these datasets to grow by a factor of ten or so by the end of Run II, which will greatly improve our knowledge of the peculiar particle that is the top quark.

We note that there was no time to present all analyses of top quark properties performed at the Tevatron. More details can be found on the public web pages of the CDF [9] and DØ [10] experiments.

REFERENCES

1. F. Abe *et al.*, CDF Collaboration, Phys. Rev. Lett. **74**, 2626 (1995);
   S. Abachi *et al.*, DØ Collaboration, Phys. Rev. Lett. **74**, 2632 (1995).
2. M. Cacciari *et al.* JHEP **0404**, 068 (2004);
   N. Kidonakis and R. Vogt, Phys. Rev. **D68**, 114014 (2003).
3. R. Blair *et al.* (CDF-II), PUB 96-390-E, FERMILAB (1996);
   V. Abazov *et al.*, DØ Collaboration (to be published), T. LeCompte and H.T. Diehl, Annu. Rev. Nucl. Part. Sci. **50**, 71 (2000).
4. CDF and DØ Collaborations, hep-ex/0404010.
5. LEP Electroweak Working Group, `http://lepewwg.web.cern.ch/LEPEWWG/`, To be published (2005).
6. V.M. Abazov *et al.*, DØ Collaboration, Nature **429**, 638 (2004).
7. DØ Collaboration, Submitted to Phys. Rev. **D**, hep-ex/0505031.
8. CDF Collaboration, Submitted to Phys. Rev. Lett., hep-ex/0505091.
9. CDF Collaboration, `http://www-cdf.fnal.gov/physics/new/top/top.html`.
10. DØ Collaboration, `http://www-d0.fnal.gov/Run2Physics/WWW/results/top.htm`.

Universality of q_T resummation for electroweak boson production

Anton V. Konychev* and Pavel M. Nadolsky†

*Department of Physics, Indiana University, Bloomington, IN 47405-7105, U.S.A.
†High Energy Physics Division, Argonne National Laboratory, Argonne, IL 60439-4815, U.S.A.

Abstract. We perform a global analysis of transverse momentum distributions in Drell-Yan pair and Z boson production in order to investigate universality of nonperturbative contributions to the Collins-Soper-Sterman resummed form factor. Our fit made in an improved nonperturbative model suggests that the nonperturbative contributions follow universal nearly-linear dependence on the logarithm of the heavy boson invariant mass Q, which closely agrees with an estimate from the infrared renormalon analysis.

Transverse momentum distributions of heavy Drell-Yan lepton pairs, W, or Z bosons produced in hadron-hadron collisions present an interesting example of factorization for multi-scale observables. If the transverse momentum q_T of the electroweak boson is much smaller than its invariant mass Q, $d\sigma/dq_T$ at an n-th order of perturbation theory includes large contributions of the type $\alpha_s^n \ln^m(q_T^2/Q^2)/q_T^2$ $(m = 0, 1 \ldots 2n-1)$, which must be summed through all orders of α_s to reliably predict the cross section [1]. Such resummation is realized in the Collins-Soper-Sterman (CSS) formalism [2], which describes soft and collinear QCD radiation in a wide range of energies by introducing a resummed form factor $\widetilde{W}(b)$ in impact parameter (b) space.

While the short-distance contributions ($b \lesssim 1$ GeV^{-1}) to the CSS form factor $\widetilde{W}(b)$ can be calculated in perturbative QCD, long-distance nonperturbative contributions from $b > 1$ GeV^{-1} are not yet fully computable, even though their basic form can be deduced from the infrared renormalon analysis [3]. The factorization theorem behind the CSS formalism predicts that the nonperturbative contributions are universal in unpolarized Drell-Yan-like and semi-inclusive DIS processes. Consequently the function $\mathcal{F}_{NP}(b,Q)$ that describes the nonperturbative terms can be constrained in a global fit to the hadronic q_T data, just as the k_T-integrated parton densities are constrained with the help of inclusive scattering data. $\mathcal{F}_{NP}(b,Q)$ must be known precisely in order to successfully measure the W boson mass, because uncertainties in $\mathcal{F}_{NP}(b,Q)$ may affect the measured value of M_W at the level comparable to the targeted accuracy of the measurement, $\delta M_W \approx 30$ MeV at the Tevatron and 15 MeV at the LHC. It is therefore interesting to investigate if $\mathcal{F}_{NP}(b,Q)$ found in the q_T fit is consistent with the universality hypothesis, and whether its preferred form is compatible with the renormalon analysis.

These issues were explored recently in Ref. [4], where a global analysis of q_T data from fixed-target Drell-Yan pair production and Tevatron Z boson production was performed in the context of an improved model for the nonperturbative contributions. Although $\mathcal{F}_{NP}(b,Q)$ primarily parametrizes the "power-suppressed" terms, *i.e.*, terms pro-

CP792, *Deep Inelastic Scattering, DIS 2005*, edited by W. H. Smith and S. R. Dasu

portional to positive powers of b, its form found in the fit is correlated with the assumed behavior of the leading-power terms (logarithmic in b terms) at $b < 2$ GeV^{-1}. The exact behavior of $\widetilde{W}(b)$ at $b > 2$ GeV^{-1} is of reduced importance, as $\widetilde{W}(b)$ is strongly suppressed at such b. For these reasons, we closely followed the procedure of the previous global q_T analysis [5], while paying close attention to the model of the leading-power terms at perturbative and moderately nonperturbative transverse distances, $b < 2$ GeV^{-1}.

The large-b contributions were introduced by using the b_* model [2], as

$$\widetilde{W}(b) = \widetilde{W}_{pert}(b_*)\, e^{-\mathcal{F}_{NP}(b,Q)}. \tag{1}$$

Here $\widetilde{W}_{pert}(b_*)$ is the perturbative part of $\widetilde{W}(b)$, *i.e.*, its leading-power part evaluated at a finite order of α_s. $\widetilde{W}_{pert}(b_*)$ depends on the variable $b_* \equiv b/(1+b^2/b_{max}^2)^{1/2}$ and serves as an approximation for all leading-power terms. Its shape is varied at all b by adjusting a single parameter b_{max}. The b_* model with a relatively low $b_{max} = 0.5$ GeV^{-1} was a choice of the previous q_T fits [5, 6]. However, it is natural to consider b_{max} above 1 GeV^{-1} in order to avoid *ad hoc* modifications of $\widetilde{W}_{pert}(b)$ in the b region where perturbation theory is still applicable. In Ref. [4], we proposed a modification in the b_* model that allowed us to increase b_{max} at least up to ≈ 3 GeV^{-1}, while preserving correct resummation of the large logarithms at small b and numerical stability of the Fourier-Bessel transform. If a very large b_{max} comparable to $1/\Lambda_{QCD}$ is taken, $\widetilde{W}_{LP}(b)$ essentially coincides with $\widetilde{W}_{pert}(b)$, extrapolated to large b by using the known, although not always reliable, dependence of $\widetilde{W}_{pert}(b)$ on $\ln b$. Hence, the new prescription can be also used to test viability of extrapolation of $\widetilde{W}_{pert}(b)$ to large b, reminiscent of similar extrapolations introduced in the alternative models [7, 8].

Following the renormalon analysis and Ref. [5], we assumed a Gaussian form of the nonperturbative function, $\mathcal{F}_{NP}(b,Q) \equiv a(Q)b^2$, with

$$a(Q) \equiv a_1 + a_2 \ln[Q/(3.2\ \text{GeV})] + a_3 \ln[100 x_1 x_2]. \tag{2}$$

The dependence of $\mathcal{F}_{NP}$ on $\ln Q$ is a consequence of renormalization-group invariance of the soft-gluon radiation. The coefficient a_2 of the $\ln Q$ term has been related to the vacuum average of the Wilson loop operator and evaluated within lattice QCD as $0.19^{+0.12}_{-0.09}$ GeV^2 [9]. To see if the universal Gaussian behavior is consistent with the data, we first examined the values of $a(Q)$ that are independently preferred by each bin of Q in 5 examined experimental data sets. Fig. 1(a) shows the best-fit values of $a(Q)$ obtained in independent fits to the data in each bin of Q for $b_{max} = 1.5$ GeV^{-1}. The best-fit $a(Q)$ follow a nearly linear dependence on $\ln Q$, and the slope $a_2 \equiv da(Q)/d(\ln Q)$ is close to the renormalon analysis expectation of 0.19 GeV^2 [9]. Such nearly linear behavior of $a(Q)$ is observed in the entire range $b_{max} = 1-2$ GeV^{-1}, and it less pronounced at b_{max} outside of the interval 1-2 GeV^{-1}. Since the best-fit $a(Q)$ in each Q bin are essentially independent, we conclude that the data support the universality of $\mathcal{F}_{NP}$, when b_{max} lies in the range $1-2$ GeV^{-1}. In addition, each experimental data set individually prefers a nearly quadratic dependence on b, $\mathcal{F}_{NP} = a(Q)b^{2-\beta}$, with $|\beta| < 0.5$ in all experiments.

Next, we performed a simultaneous fit of our model to all the data. Fig. 1(b) shows the dependence of the best-fit χ^2, a_1, a_2, and a_3 on b_{max}. As b_{max} is increased above

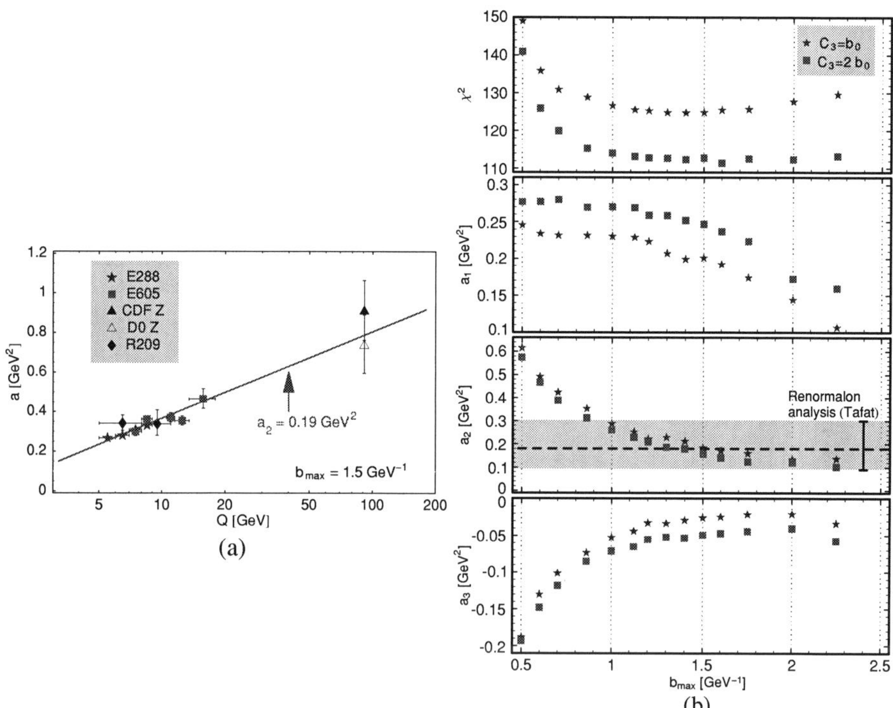

Figure 1. (a) The best-fit values of $a(Q)$ obtained in independent scans of χ^2 for the contributing experiments. The vertical error bars correspond to the increase of χ^2 by unity above its minimum in each Q bin. The slope of the line is equal to the central-value prediction from the renormalon analysis [9]. (b) The best-fit χ^2 and coefficients a_1, a_2, and a_3 in $\mathscr{F}_{NP}(b,Q)$ for different values of b_{max}. The size of the symbols approximately corresponds to 1σ errors for the shown parameters.

$0.5\ \mathrm{GeV}^{-1}$ assumed in the studies [5, 6], χ^2 rapidly decreases, becomes relatively flat at $b_{max} = 1-2\ \mathrm{GeV}^{-1}$, and grows again at $b_{max} > 2\ \mathrm{GeV}^{-1}$. The global minimum of χ^2 is reached at $b_{max} \approx 1.5\ \mathrm{GeV}^{-1}$, where all data sets are described equally well, without major tensions among the five experiments. The magnitudes of a_1, a_2, and a_3 are reduced when b_{max} increases from 0.5 to $1.5\ \mathrm{GeV}^{-1}$. In the whole range $1 \leq b_{max} \leq 2\ \mathrm{GeV}^{-1}$, a_2 agrees with the renormalon analysis estimate. The coefficient a_3, which parametrizes deviations from the linear $\ln Q$ dependence, is considerably smaller (< 0.05) than both a_1 and a_2 (~ 0.2). This behavior supports the conjecture in [7] that a_3 is small if the exact form of $\widetilde{W}_{pert}(b)$ is maximally preserved.

The preference for the values of b_{max} between 1 and 2 GeV^{-1} indicates, first, that the data do favor the extension of the b range where all leading-power terms are approximated by their finite-order expression $\widetilde{W}_{pert}(b)$. In Z boson production, this region extends up to $3-4\ \mathrm{GeV}^{-1}$ as a consequence of the strong suppression of the large-b tail by the Sudakov exponent. The fit to the Z data is actually independent of b_{max} within the experimental uncertainties for $b_{max} > 1\ \mathrm{GeV}^{-1}$. In the low-$Q$ Drell-Yan process, the continuation of $b\widetilde{W}_{pert}(b)$ far beyond $b \approx 1\ \mathrm{GeV}^{-1}$ is disfavored because of large higher-order corrections to $b\widetilde{W}_{pert}(b)$ at b around 1.5 GeV^{-1}. To summarize, the extrapolation of $\widetilde{W}_{pert}(b)$ to $b > 1.5$ GeV^{-1} is disfavored by the low-Q data sets, if a purely Gaussian form of $\mathscr{F}_{NP}$ is assumed. The Gaussian approximation is adequate, on the other hand, in the b_* model with b_{max} in the range $1-2$ GeV^{-1}.

In Z boson production, our best-fit $a(M_Z) = 0.85 \pm 0.10$ GeV^2 agrees with 0.8 GeV^2 found in the extrapolation-based models [7, 8], and it is about a third of 2.7 GeV^2 predicted by the BLNY parametrization. In the low-Q Drell-Yan case, our $a(Q) = 0.2-0.4$ GeV^2 is close to the average $\langle a \rangle = 0.19-0.28$ GeV^2 in four Q bins of the E288 and E605 data found in the model [7]. To describe the low-Q data, Ref. [7] allowed a large discontinuity in the first derivative of $\widetilde{W}(b)$ at b equal to the separation parameter $b_{max}^{QZ} = 0.3-0.5$ GeV^{-1}, where switching from the exact $\widetilde{W}_{pert}(b)$ to its extrapolated form occurs. In the revised b_* model, such discontinuity does not happen, and $\widetilde{W}_{LP}(b)$ is closer to the exact $\widetilde{W}_{pert}(b)$ in a wider b range than in Ref. [7].

The best-fit parameters in $\mathscr{F}_{NP}$ found in the new model are quoted in Ref. [4]. The global fit places stricter constraints on $\mathscr{F}_{NP}$ at $Q = M_Z$ than the Tevatron Run-1 Z data alone. Theoretical uncertainties from a variety of sources may be substantial in the low-Q Drell-Yan process, which is indicated, in particular, by the dependence of the agreement with the low-Q data on an arbitrary factorization scale C_3 in $\widetilde{W}_{pert}(b)$. The low-Q uncertainties do not substantially affect predictions at the electroweak scale. The $\mathscr{O}(\alpha_s^2)$ corrections and scale dependence are smaller in W and Z production, and, in addition, the term $a_2 \ln Q$, which arises from the soft factor $\mathscr{S}(b,Q)$ and dominates $\mathscr{F}_{NP}$ at $Q = M_Z$, shows little variation with C_3. Consequently, the revised b_* model with $b_{max} \approx 1.5$ GeV^{-1} increases our confidence in the transverse momentum resummation at electroweak scales by exposing the soft-gluon origin and universality of the dominant nonperturbative contributions at collider energies.

We express our gratitude to C.-P. Yuan for his crucial help with the setup of the fitting program. This work was supported in part by the US Department of Energy, High Energy Physics Division, under Contract W-31-109-ENG-38, and by the U.S. National Science Foundation under grants PHY-0100348 and PHY-0457219.

REFERENCES

1. Yu. L. Dokshitzer, D. I. D'yakonov, S. I. Troyan, Phys. Lett. **B79**, 269 (1978); G. Parisi, R. Petronzio, Nucl. Phys. **B154**, 427 (1979).
2. J. C. Collins, D. E. Soper, Nucl. Phys. **B193**, 381 (1981); **B213**, 545(E) (1983); **B197**, 446 (1982); J. C. Collins, D. E. Soper, and G. Sterman, Nucl. Phys. **B250**, 199 (1985); J. C. Collins, A. Metz, Phys. Rev. Lett. **93**, 252001 (2004).
3. G. P. Korchemsky, G. Sterman, Nucl. Phys. **B437**, 415 (1995).
4. A. V. Konychev, P. M. Nadolsky, arXiv:hep-ph/0506225.
5. F. Landry, R. Brock, P. M. Nadolsky, and C.-P. Yuan, Phys. Rev. D**67**, 073016 (2003).
6. C. Davies, B. Webber, W. Stirling, Nucl. Phys.**B256**, 413 (1985); G. Ladinsky, C.-P. Yuan, Phys. Rev. D**50**, 4239 (1994); F. Landry, R. Brock, G. Ladinsky, C.-P. Yuan, *ibid.* **63**, 013004 (2001); R. K. Ellis, D. A. Ross, S. Veseli, Nucl. Phys. **B503**, 309 (1997).
7. J.-W. Qiu, X.-F. Zhang, Phys. Rev. Lett. **86**, 2724 (2001); Phys. Rev. D**63**, 114011 (2001).
8. A. Kulesza, G. Sterman, and W. Vogelsang, Phys. Rev. D**66**, 114011 (2002).
9. S. Tafat, JHEP **05**, 004 (2001).

Search for Events with Isolated Leptons and Large Missing Transverse Momentum

Christian Veelken

Deutsches Elektronen Synchrotron
Notkestrasse 85, 22607 Hamburg, Germany
E-mail: veelken@mail.desy.de

Abstract.
Events with isolated leptons and large missing transverse momentum are searched for in the e^+p and e^-p data recorded by the H1 experiment at HERA. In previous analyses of the HERA I e^+p data ($\mathcal{L} = 104.7$ pb^{-1}), an excess of events with isolated electrons or muons and large missing transverse momentum was found. No significant excess was found in the HERA I e^-p data ($\mathcal{L} = 13.6$ pb^{-1}). In the following, new results of the search for events with isolated leptons and large missing transverse momentum in the HERA II e^+p and e^-p datasets are presented. The analysed HERA II data correspond to integrated luminosities of 53 pb^{-1} and 21 pb^{-1} in e^+p and e^-p collisions, respectively.

Keywords: ep, isolated lepton
PACS: 13.60.-r

HERA I RESULTS

In previous H1 analyses of the HERA I e^+p dataset [1, 2], an excess of events with isolated electrons or muons and large missing transverse momentum has been observed.

Within the Standard Model, the dominant process yielding such kind of events is the production of real W bosons via neutral current interactions $ep \to eW^{\pm}X$ with subsequent leptonic decay $W \to \ell\nu$ of the W (henceforth denoted as "signal"). The cross-section for W boson production in ep collisions is expected to be about 1.2 pb at HERA centre-of-mass energies of $\sqrt{s} = 318$ GeV [3]. Small additional contributions to the signal arise from the equivalent charged current process $ep \to \nu W^{\pm}X$ and from the production of real Z bosons with subsequent decay of the Z into two neutrinos, in events where the beam electron is scattered into the main detector. The event selection of the analysis is tuned to optimize the acceptance for the rare signal and to reduce the contribution from all other SM processes (henceforth denoted as "background").

In the HERA I e^+p dataset ($\mathcal{L} = 104.7$ pb^{-1}), 18 events with isolated electrons or muons and large missing transverse momentum are selected, exceeding a SM expectation of 12.4 ± 1.7, which is dominated by the contribution from signal processes (9.4 ± 1.6 events). The observed excess is concentrated in the region of large hadronic transverse momenta (abbreviated as P_T^X in the following). In the region $P_T^X > 25$ GeV, 10 events are observed, while only 2.9 ± 0.5 are expected from all SM processes.

In the HERA I e^-p dataset ($\mathcal{L} = 13.6$ pb^{-1}), 1 event passes the event selection, in agreement with a SM expectation of 2.1 ± 0.3. From the HERA I data alone, no conclusion can be made whether the excess observed in e^+p collisions is also present in

CP792, *Deep Inelastic Scattering, DIS 2005,* edited by W. H. Smith and S. R. Dasu

e^-p collisions or not, due to the small integrated luminosity of the HERA I e^-p dataset, however.

Preliminary results of the search for events with isolated tau leptons have been presented [4]. The analysis of the tau channel is based on HERA I e^+p and e^-p datasets recorded between 1996 and 2000 ($\mathscr{L} = 108$ pb^{-1}). In this dataset, events with collimated jets containing a single isolated track as expected for hadronic "one-prong" tau decays are selected. In total, 5 candidate events pass the event selection, in agreement with a SM expectation of 5.8 ± 1.4. The SM expectation is dominated by the contributions from background processes at small P_T^X and by signal processes in the region of large hadronic transverse momenta. The events observed in the data are all concentrated at small P_T^X, where the contribution from background processes is expected to be dominant. In the region $P_T^X > 25$ GeV, where the excess observed in the electron and muon channels is concentrated, no event is observed in the tau channel. Taking the significantly lower selection efficiency in the tau channel into account, the non-observation of an excess in this channel is compatible with the excess observed in the electron and muon channels.

NEW HERA II RESULTS

The new HERA II results presented at the DIS 2005 conference are based on an analysis of the e^+p and e^-p datasets of H1 recorded between October 2003 and March 2005. In this dataset, events with isolated electrons or muons are searched for. In the e^+p dataset, corresponding to an integrated luminosity of 53 pb^{-1}, 10 events (9 events with isolated electrons and 1 event with an isolated muon) are selected, exceeding a SM expectation of 6.1 ± 0.9. The excess observed in the HERA II e^+p data is concentrated in the region of large hadronic transverse momenta, as was the case in the HERA I e^+p data. In the region $P_T^X > 25$ GeV, 5 events are observed, while only 1.7 ± 0.3 are expected from all SM processes. In the e^-p dataset, corresponding to an integrated luminosity of 21 pb^{-1}, 5 events (all containing isolated electrons) pass the event selection, while 2.8 ± 0.4 are expected from all Standard Model processes. In the region $P_T^X > 25$ GeV, in which the excess of events in the e^+p data is observed, only 1 event is found, in agreement with a SM expectation of 0.7 ± 0.1.

The present status of the search for events with isolated leptons and large missing transverse momentum is summarized in Table 1, in which the number of events with isolated electrons, muons or tau leptons selected in the HERA I and II e^+p and e^-p datasets are compared with the SM expectation. The signal component is given as percentage of the expectation for all SM processes. Note the small contribution from background processes, especially at large $P_T^X > 25$ GeV. The current significance of the excess observed at large hadronic transverse momenta may be inferred from Figure 1, which shows the P_T^X spectrum of the observed events in comparison to the SM expectation.

TABLE 1. Summary of the H1 results of searches for events with isolated leptons and large missing transverse momentum. The results are shown for the full selected sample and for the subsample at large $P_T^X > 25$ GeV. The number of observed events is compared to the SM expectation, the signal fraction of which is given in percent in parentheses. The quoted errors contain statistical and systematic uncertainties added in quadrature.

H1 Preliminary 1994-2005 $e^{\pm}p$ 192 **pb**$^{-1}$	**Electron obs./exp. (Signal contribution)**	**Muon obs./exp. (Signal contribution)**	$e+\mu$ **Combined obs./exp. (Signal contribution)**	**Tau*** **obs./exp. (Signal contribution)**
Full Sample	$25/18.4 \pm 2.5$ (70%)	$9/4.9 \pm 0.8$ (85%)	$34/23.3 \pm 3.2$ (73%)	$5/5.8 \pm 1.4$ (15%)
$P_T^X > 25$ GeV	$11/2.9 \pm 0.6$ (81%)	$6/2.9 \pm 0.6$ (86%)	$17/5.8 \pm 1.1$ (84%)	$0/0.5 \pm 0.1$ (49%)

* Results for 1996-2000 $e^{\pm}p$ Data only ($\mathscr{L} = 108$ pb^{-1})

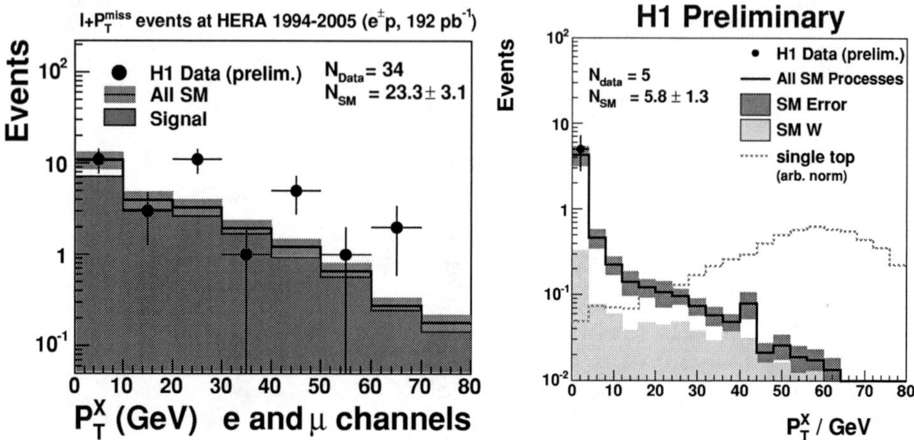

FIGURE 1. The hadronic transverse momentum spectrum of the events with isolated electrons or muons (left) and tau leptons (right) observed in the HERA I and II e^+p and e^-p datasets compared to the SM expectation. In the tau channel, P_T^X is defined as the transverse momentum of all hadrons not contained within the collimated jet associated with the hadronic tau decay. The dashed grey line shown in the plot on the right represents the P_T^X distribution expected for a particular process beyond the Standard Model that may explain the excess observed in the electron and muon channels [5], the production of single top quarks by flavour-changing neutral current interactions.

SUMMARY

Results of the search for events with isolated leptons and large missing transverse momentum in the full HERA I and II e^+p and e^-p datasets of H1 have been presented. The excess of events with isolated electrons or muons observed at large hadronic transverse momenta in the HERA I e^+p dataset continues to be seen in the HERA II e^+p data. In the HERA I and II e^-p datasets, there is no evidence for an excess of events in the region $P_T^X > 25$ GeV, where the excess observed in the e^+p datasets is concentrated.

The integrated luminosity of the e^-p datasets is not yet sufficiently large to conclude whether the excess observed in the e^+p data is also present in e^-p collisions.

The continuing analysis of the HERA II data, in particular that of the increasing e^-p dataset, will hopefully provide an insight into the nature of the excess observed at high P_T^X in the HERA e^+p data.

REFERENCES

1. H1 Collab., C. Adloff *et al.*, *Eur. Phys. J.* **C5**, 575–584 (1998).
2. H1 Collab., V. Andreev *et al.*, *Phys. Lett.* **B561**, 241–257 (2003).
3. K. P. Diener, C. Schwanenberger and M. Spira, *Eur. Phys. J.* **C25** 405–411 (2002).
4. G. Brandt, *Events with τ Leptons at HERA*, talk presented at the DIS 2004 conference.
5. H1 Collab., A. Aktas *et al.*, *Eur. Phys. J.* **C33**, 9–22 (2004).

Single W Boson Production at HERA

K. Piotrzkowski, for of the ZEUS collaboration

Université catholique de Louvain, B-1348 Louvain-la-Neuve, Belgium

Abstract. A search for the process $e^+p \rightarrow WX$, with the subsequent decay $W \rightarrow e\nu_e$, was performed using 66 pb^{-1} of e^+p collision data taken with the ZEUS detector during the 1999-2000 running period. The process leads to final states characterized by isolated high transverse energy electrons and large missing transverse momentum P_T.

Keywords: W boson production, isolated leptons, electron-proton collisions.
PACS: 13.38.Be, 13.60.-r, 14.70.Fm

INTRODUCTION

We report the results of an investigation of the production of W bosons in positron-proton collisions at center-of-mass energy of 318 GeV at HERA. Single W production is a rare Standard Model (SM) process and an important source of background to searches for physics beyond the Standard Model [1, 2]. Investigations of the process $ep \rightarrow eWX$, $W \rightarrow l\nu$, where $l = \mu$, e, have been performed at HERA by both the H1 [2,3] and ZEUS [1, 4] collaborations. H1 observes an excess of events with isolated muons or electrons[1] and high missing transverse momentum over the SM prediction, dominated by single W production. The ZEUS results based on searches for isolated electrons and muons at center-of-mass energy of 300 GeV do not confirm this excess.

This paper reports the results of a similar search using the larger sample of data taken by ZEUS in the 99-00 running period. In the electron decay-channel of the W, the event topology contains an isolated electron with high transverse momentum (P_T) and missing total P_T arising from the escaping neutrino. If the negative of the four-momentum transfer squared (Q^2) is greater than a few GeV^2, the scattered electron can be observed in the detector. The study was performed by selecting events containing well isolated high p_T electrons, and large missing total P_T. The data set corresponds to the 1999-2000 running period, with a total integrated luminosity of 66 pb^{-1}.

SIMULATION OF THE SIGNAL AND OF THE BACKGROUND

The leading order (LO) cross section for $e^+p \rightarrow e^+WX$ has been calculated using the EPVEC generator [5]. EPVEC calculates the cross section in two regions, corresponding to photoproduction and deep inelastic scattering. The photon (proton) structure functions used in the calculation are GRV-G(LO) (CTEQ4D). The final state

[1] In this paper "electron" refers both to electrons and positrons unless specified.

CP792, *Deep Inelastic Scattering, DIS 2005,* edited by W. H. Smith and S. R. Dasu

simulation does not include hard gluon radiation. Such calculations yield a total cross section of 0.945 pb for $\sqrt{s}$ =300 GeV and 1.09 pb for $\sqrt{s}$ = 320 GeV. The uncertainties on this value are approximately 5% for the choice of boundary between the two regions, 5% for the choice of proton structure function, 10% for photon structure function and 10% from the choice of Q^2 scale used in EPVEC. Next-to-leading order corrections were calculated in [6] and were found to be of the order of 10%; they were however neglected in this analysis.

The most important background to W production in the electron decay-channel arises from high Q^2 charged and neutral current deep inelastic scattering (DIS) events. These DIS events have been simulated using the generator DJANGO6, an interface to the Monte Carlo (MC) programs HERACLES 4.5 and LEPTO 6.5. Leading order QCD and electroweak radiative corrections were included and higher order QCD effects were simulated via parton cascades using the color-dipole model ARIADNE or parton shower based on a leading-logarithm approximation (MEPS). The hadronization of the partonic final state was performed by JETSET. Minor contributions to the background come from two-photon processes, which were simulated using the GRAPE dilepton generators, and direct and resolved photoproduction processes which were simulated using the HERWIG 6.1 event generator.

The generated events were passed through the GEANT-based ZEUS detector and trigger simulation programs [7]. They were reconstructed and analyzed by the same program chain as the data.

EVENT RECONSTRUCTION AND DATA SELECTION

The missing transverse momentum is defined as: $P_T = \sqrt{\{(\Sigma\, p_x^i)^2 + (\Sigma\, p_y^i)^2\}}$, where $p_x^i = E^i \sin\theta^i \cos\phi^i$ and $p_y^i = E^i \sin\theta^i \sin\phi^i$ are calculated using the energies (E^i) of individual calorimeter cells that are above noise thresholds. The angles θ^i and ϕ^i are estimated from the geometric cell centers and the event vertex. In $W \to e\nu$ events, P_T defined above is an estimate of the missing transverse momentum carried by the neutrino. Electron (hadron) transverse momenta are defined as sums over those calorimeter cells that are (are not) assigned to the electron candidate cluster.

Longitudinal momentum conservation ensures that $E-p_z$ (δ), defined as: $E-p_z = \Sigma\, E^i (1-\cos\theta^i)$, peaks at twice the electron beam energy E^e for fully contained events. Small values of δ are expected for proton-gas interactions. Only events with $5 < \delta < 60$ GeV were chosen in the pre-selection.

The acoplanarity angle Φ_{ACOP} is the azimuthal separation of the outgoing lepton and the vector in the {X, Y} plane that balances the hadronic-P_T vector. For well-measured neutral current events, the acoplanarity angle is close to zero, while large acoplanarity angles indicate large missing energies.

The transverse mass is defined as: $M_T = \sqrt{\{2P_T^l\, P_T^{\nu}(1-\cos\Phi^{l\nu})\}}$, where P_T^l is the lepton transverse momentum, P_T^{ν} is the magnitude of the P_T, and $\Phi^{l\nu}$ is the azimuthal separation of the lepton and the P_T vectors.

Events that pass the trigger requirements are further required to have a P_T greater than 9 GeV and a $M_T > 10$ GeV. Other pre-selection cuts are the requirement that the

Z-coordinate of the tracking vertex be reconstructed within 50 cm of the nominal interaction. Cuts on the calorimeter timing and algorithms based on the pattern of tracks reject beam-gas, cosmic ray and halo-muon events.

SEARCH FOR W PRODUCTION AND THE DECAY W → eν

A neural-network-based algorithm to identify electrons, trained on Monte Carlo events and optimized for maximum electron-finding efficiency and electron-hadron separation, selects candidate electromagnetic clusters in the calorimeter [8]. A cut on the electromagnetic cluster energy is made at 8 GeV, above which the neural network is fully efficient, except at the boundaries between different calorimeter parts. The background from fake electrons is reduced by requiring that the energy not associated with the electron in an $\{\eta,\phi\}$ cone of radius 0.8 around the electron direction, be less than 4 GeV. In addition, since most fake electrons are misidentified hadrons close to jets, the background is further reduced by requiring that the electron track be separated by at least 0.5 units in $\{\eta,\phi\}$ space, from the other tracks associated with the event vertex. The data are compared to the expectation from the Monte Carlo simulation in Figure 1, after requiring that the transverse momentum of the electron P_T^e be greater than 5 GeV, and the polar angle of the electron measured in the calorimeter, θ^e, be less that 2.0 rad.

Neutral current background events dominate the sample at this stage of the selection, as is evident from the steeply falling P_T spectrum and the concentration of events at mall acoplanarity angles. A Jacobian peak structure is visible in the transverse mass distribution for the Monte Carlo simulation of the signal events. The neutral current background is strongly suppressed by requiring the P_T to be greater than 20 GeV and the acoplanarity angle to be greater than 0.3 rad. The latter cut is applied only to events with a hadronic transverse momentum (P_T^X) in excess of 4 GeV, for which the acoplanarity is well defined. Electrons in the final event sample must have $P_T^e > 10$ GeV and $\theta^e < 1.5$ rad. Finally, requiring that the matching electron track have a transverse momentum greater than 5 GeV, removes most of the remaining fake electrons.

Table 1. Five W candidates.

Candidate	1	2	3	4	5
P_T^e (GeV)	20.3	33.5	52.8	28.0	30.6
Hadronic P_T^X (GeV)	9.3	17.0	35.2	2.82	29.4
Missing P_T (GeV)	25.2	20.0	25.1	29.2	48.4

Five data events survive these final cuts, their properties are outlined in Table 1. The expectation for W production from EPVEC Monte Carlo is 3.2±0.1(Stat.)+1.1−1.0(Syst.) events corresponding to an efficiency of selecting decays of the W in the electron channel of 39%. From other SM processes 3.5±0.6(Stat.)+1.7−1.6(Syst.) events are expected. A program using Feldman-Cousins ordering, was used to extract a 95% confidence limit on the total cross section for single W production of: $\sigma(e^+p \rightarrow e^+WX) < 2.8$ pb; a stronger limit than that obtained in [4].

The systematic effects on the background and signal expectations considered were: Using the electron finder described in [9], rather than the neural network based finder; using the MEPS model instead of the Ariadne model for hadronization in the background simulation; varying the absolute energy scale of the calorimeter by ±1.5%. The dominant systematic effect on signal efficiency arose from using the alternative electron finder. In the case of the background, the dominant systematic uncertainty arose from the choice of hadronization model.

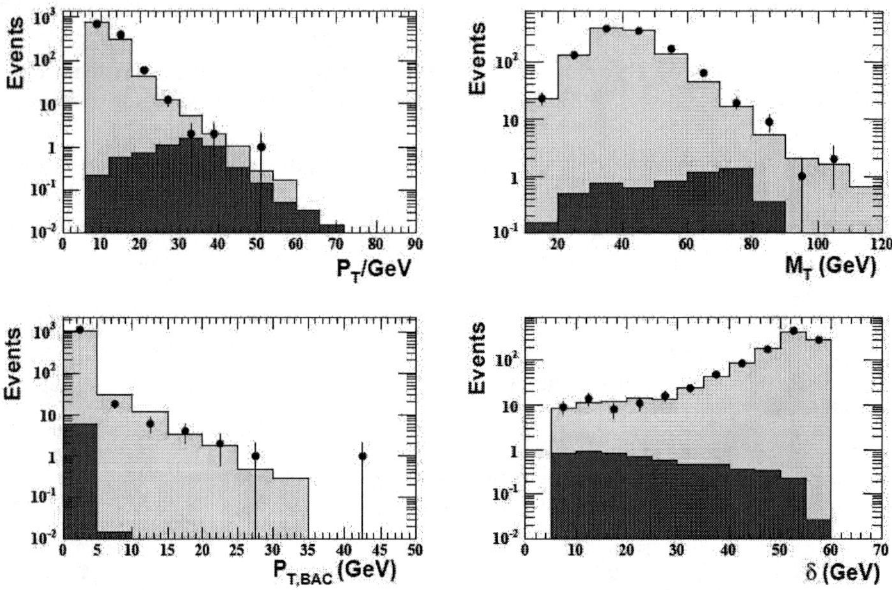

FIGURE 1. Data (points) compared to luminosity-normalized Monte Carlo. The light shaded histogram represents the Standard Model (SM) MC prediction, and the dark shaded area the signal ($e^+p \rightarrow e^+ WX$) prediction.

REFERENCES

1. ZEUS Coll., S. Chekanov *et al.*, Phys. Lett. **559**, 153 (2003).
2. H1 Coll., Preprint DESY-02-022 (hep-ex/0301030), DESY, 2002.
3. H1 Coll., C. Adloff *et al.,* Eur. Phys. J. C **5**, 575 (1998).
4. ZEUS Coll., J. Breitweg *et al.,* Phys. Lett. B **471**, 411 (2000).
5. U. Baur, J.A.M. Vermaseren and D. Zeppenfeld, Nucl. Phys. B **375**, 3 (1992).
6. Diener,Schwanenberger and Spira, Preprint DESY-02-035 (hep-ph/0203269), 2002.
7. R. Brun *et al.*, Geant3, Technical Report CERN-DD/EE/84-1, CERN, 1987; ZEUS Coll., U. Holm (ed.), The ZEUS Detector. Status Report (unpublished), DESY (1993), available on http://www-us.desy.de/bluebook/bluebook.html.
8. H. Abramowicz, A. Caldwell and R. Sinkus, Nucl. Inst. Meth. A **365**, 508 (1995).
9. ZEUS Coll., J. Breitweg *et al.*, Eur. Phys. J. C **11**, 427 (1999).

Top Quark Production Cross-Section at the Tevatron Collider

Kirti Ranjan

For the DØ and the CDF collaborations
University of Delhi, Delhi - 110007, India and Fermilab, Batavia, Illinois - 60510, USA

Abstract. We present the preliminary results of the $t\bar{t}$ pair production cross-section measurements and the single top quark exclusion limits carried out by the DØ and the CDF collaborations in Run II of the Tevatron. The dataset for the various measurements ranges from 140 pb^{-1} to 350 pb^{-1}.

Keywords: top quark, production cross-section
PACS: 14.65.Ha, 13.85.Qk, 13.85.Lg

1. INTRODUCTION

The top quark was discovered jointly by the DØ and the CDF collaborations in 1995 [1]. At the Tevatron $p\bar{p}$ collider (center of mass energy, $\sqrt{s}$ = 1.96 TeV), top quarks are dominantly produced in pairs via strong interaction: $q\bar{q}$ annihilation ($\sim$ 85 %) and gluon fusion ($\sim$ 15 %). Existing theoretical predictions for the $t\bar{t}$ pair production cross-section at NLO in QCD ranges from 6.7 - 7.5 pb [2] (for m_t = 175 GeV), representing a 30 % increase in cross-section over Tevatron's RunI center of mass energy of 1.8 TeV.

Within the Standard Model (SM), the top quark decays almost exclusively to a W-boson and a b-quark. Decay of the two top quarks, $t \to Wb$, are characterized by the three distinct final state signatures depending on the decay modes of the W-boson: the dilepton final states $(ee, e\mu, \mu\mu)$ where both the Ws decay leptonically is the cleanest channel but accounts only for $\sim$ 5 % of the $t\bar{t}$ pairs; the all-hadronic channel where both the Ws decay hadronically constitutes $\sim$ 44 % of the signal but suffers from a huge QCD multijet background; and the lepton + jets ($e + jets$ or $\mu + jets$) channel where one W decays leptonically and the other one decays hadronically accounts for $\sim$ 30 % of the $t\bar{t}$ events and can be considered the best compromise between statistics and purity.

The SM also predicts the production of single top quark via electroweak interaction with a production cross-section of $\sim$ 0.88 pb in the s-channel and $\sim$ 1.98 pb in the t-channel at $\sqrt{s}$ = 1.96 TeV [3] . This production mechanism has not yet been observed and is the subject of current active searches at the Tevatron.

2. TOP QUARK PAIR PRODUCTION

2.1. Di-lepton channel

A dileptonic final state is characterized by the presence of two isolated, high p_t leptons, two high p_t b-jets and large $\not{E}_T$ from the two neutrinos. The background

CP792, *Deep Inelastic Scattering, DIS 2005,* edited by W. H. Smith and S. R. Dasu

contribution comes mainly from instrumental effects (estimated from data), sources of which include QCD multijet, W+jets, and $Z \to l^+l^-$ events with mismeasured $\not{E}_T$ or misidentified leptons. The additional contribution comes from irreducible physics background (estimated from Monte Carlo simulations), mostly from $Z \to \tau^+\tau^-$ and $WW/WZ \to l^+l^-$ processes.

CDF has performed two complementary analyses using 200 pb^{-1} of the data. In one of the analysis, two leptons are explicitly identified as e or μ. In the other analysis, CDF requires the 2nd leading lepton to be an isolated, high p_t track and does not require its explicit identification as e or μ. This increases the signal acceptance and also extends the sensitivity to $W \to \tau\nu$ (with single prong decay of τ), however, at the cost of higher background contamination. The DØ experiment has done separate analyses in the ee, $e\mu$, and $\mu\mu$ final states with about 140 pb^{-1} of the data and then combined the final results. The measured cross-sections for the CDF and DØ analyses are $\sigma_{t\bar{t}} = 7.0^{+2.4}_{-2.1}(stat)^{+1.7}_{-1.2}(syst+lumi)pb$ and $\sigma_{t\bar{t}} = 14.3^{+5.1}_{-4.3}(stat)^{+2.6}_{-1.9}(syst) \pm 0.9(lumi)pb$ respectively. It can be observed that the dominant uncertainties in this channel are statistical in nature. CDF also pursues an alternative approach where only two leptons are selected and a likelihood fit has been performed in the $\not{E}_T$ - N_{jets} plane to simultaneously determine the contributions from $t\bar{t}$, WW, and $Z \to \tau\bar{\tau}$. The measured $t\bar{t}$ cross-section for an integrated luminosity of 200 pb^{-1} is $\sigma_{t\bar{t}} = 8.6^{+2.5}_{-2.4}(stat) \pm 1.1(syst)pb$

2.2. Lepton + Jets channel

The signature of the lepton+jets channel consists of one isolated, high p_t lepton, at least 3 or 4 high p_t jets and $\not{E}_T$ from the neutrino. The dominant background comes from the W+jets processes with an additional component from the QCD multijet events. To discriminate signal from the backgrounds, which are significantly higher than dilepton analyses, two approaches are used: topological selection and b-tagging selection.

Topological analysis: The general approach is to choose various topological and kinematical variables, which can provide maximum separation between signal and background events. CDF combines these variables into a 7-input Neural Network (NN) discriminant requiring at least 3 jets and then performs template fit to the NN output (figure 1) and measures the $t\bar{t}$ cross-section with 347 pb^{-1} of the data to be $\sigma_{t\bar{t}} = 6.0 \pm 0.8(stat) \pm 1.0(syst)$ pb for m_t = 178 GeV. In a similar manner DØ has combined these variables into an event likelihood discriminant with the requirement of at least 4 jets in the events and a template fit to the discriminant is performed to extract the cross-section (figure 1). Measured DØ cross-section for 230 pb^{-1} of the data is $\sigma_{t\bar{t}} = 6.7^{+1.4}_{-1.3}(stat)^{+1.6}_{-1.1}(syst) \pm 0.4(lumi)$ pb. In this channel the systematic uceratinties are comparable to the statistical ones.

b-tagging analysis: The $t\bar{t}$ decays contain two high p_t b-jets. B-hadrons are long lived and travel a few mm distance away from the primary vertex before decay. Using the silicon microvertex sub-detectors in both the DØ and the CDF experiments, secondary vertex can be explicitly identified (referred to as SVX in CDF and SVT in DØ). Applying b-tagging is a very powerful tool in suppressing the backgrounds. DØ has pursued the b-tagging analysis separately for exactly 1 b-tagged jet events and at least 2 b-tagged jet

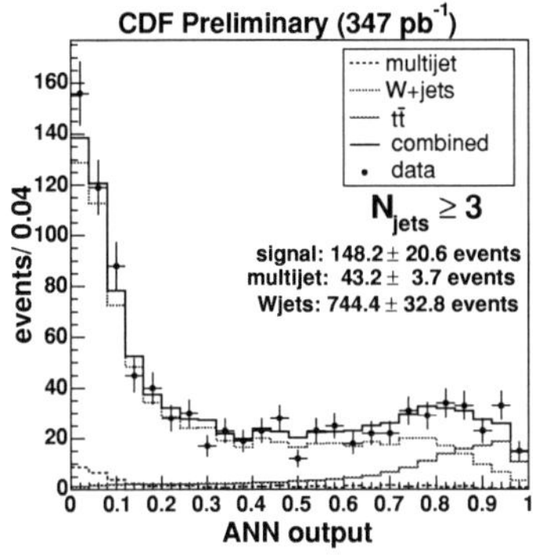

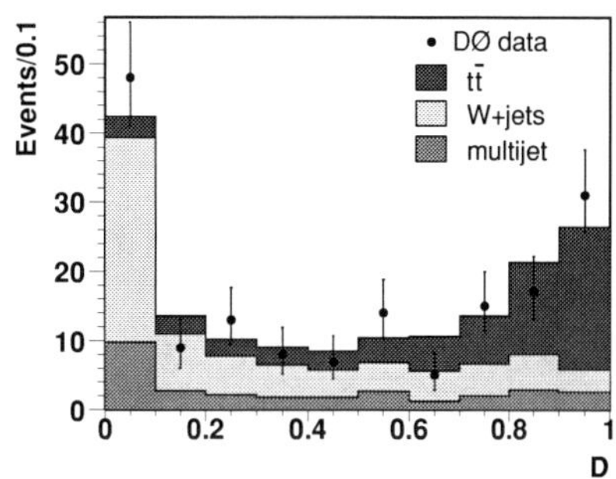

FIGURE 1. Result of the fi t to the NN-output shape for the CDF (left plot) and the fi t to the likelihood discriminant for the DØ (right plot) lepton+jet analysis.

events for lepton+$\not{E}_T$+ $\geq$ 4jets selection and the combined measured cross-section for 230 pb^{-1} of the data is $\sigma_{t\bar{t}} = 8.6^{+1.6}_{-1.1}(stat)^{+1.1}_{-1.0}(syst) \pm 0.6(lumi)pb$. CDF has applied the b-tagging requirement in lepton+$\not{E}_T$+$\geq$3jets events and performed the analysis for $\geq$ 1 b-tags and $\geq$ 2 b-tags. The measured cross-section for the two analyses for 318 pb^{-1} of the data are $\sigma_{t\bar{t}} = 7.9 \pm 0.9(stat) \pm 0.9(syst)pb$ and $\sigma_{t\bar{t}} = 8.7 \pm 1.7(stat) \pm 1.5(syst)pb$ respectively. CDF has also performed separate analysis using the semileptonic decay of the B-hadrons and hence identifying the b-jet by the presence of "soft" muon inside a jet. CDF has extracted the $t\bar{t}$ cross-section using soft muon tagging for 194 pb^{-1} of the data to be $\sigma_{t\bar{t}} = 5.2^{+2.9}_{-1.9}(stat)^{+1.3}_{-1.0}(syst)pb$.

2.3. All hadronic channel

Owing to an overwhelming QCD multijet background, the cross-section measurement in all hadronic decay mode is relatively difficult. Both topological cuts and b-tagging are used to suppress the background. The CDF experiment selects events with at least six jets and requires at least one jet to be b-tagged. It then employs several cuts on the kinematical variables to further reduce the multijet background. The measured $t\bar{t}$ cross-section for an integrated luminosity of 165 pb^{-1} is $\sigma_{t\bar{t}} = 7.8 \pm 2.5(stat)^{+4.7}_{-2.3}(syst)pb$. In the DØ analysis, events with at least 6 jets are selected requiring at least one SVT b-tag and it combines topological variables in 3 successive NNs at various stages. The distribution of 3rd NN discriminant is fit to extract the $t\bar{t}$ cross-section. The DØ cross-section measurement yields $\sigma_{t\bar{t}} = 7.7^{+3.4}_{-3.3}(stat)^{+4.7}_{-3.7}(syst) \pm 0.5(lumi)pb$.

3. SINGLE TOP QUARK SEARCHES

Both the CDF and the DØ experiments have performed single top quark searches in the s- and t-channels. No evidence for the single top quark signal has yet been found

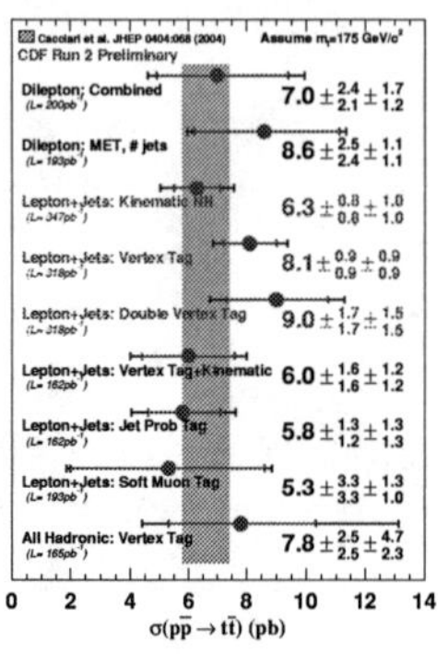

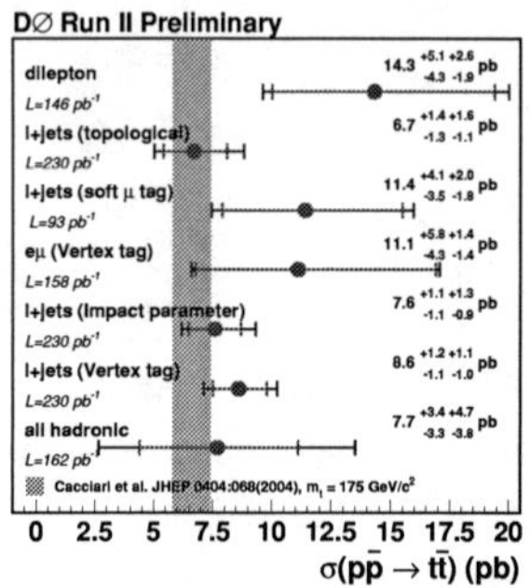

FIGURE 2. Summary of the $t\bar{t}$ pair production cross-section measurements from the CDF and the DØ collaborations. For comaparison, the NLO predictions for $m_t = 175GeV$ including uncertainties on top quark mass is also shown.

in these analyses. Single top quark events are selected by requiring one isolated, high p_t lepton, large $\not{E}_T$ and two high p_t jets. CDF has performed channel specific searches by requiring exactly 1 b-tagged jet events for the t-channel and at least 2 b-tagged jet events for the s-channel. CDF has also pursued the combined channel searches using a discriminating variable which can separate single top signal from the $t\bar{t}$ and the non-top background events. Cross-section limits at the 95% confidence level for the s-channel, t-channel and the combined channel searches are placed at $\sigma_s < 13.6$ pb, $\sigma_t < 10.1$ pb and $\sigma_{s+t} < 17.8$ pb respectively for 162 pb^{-1} of the data. DØ has performed three separate analyses using NNs, decision trees and cut-based techniques. 95% confidence level Bayesian upper limits on the production cross-section using the binned likelihood fits to the NN output distributions are σ_s< 6.4 pb in the s-channel, and σ_t< 5.0 pb in the t-channel using 230 pb^{-1} of the data.

4. SUMMARY

The measurement of the top quark pair production cross-section has been performed by both the DØ and the CDF collaborations in various channels using differnt approaches for an integrated luminosity ranging from 140 pb^{-1} to 350 pb^{-1} (figure 2). The measured cross-sections are consistent with the SM predictions. Upper limits on the production cross-section of the single top quark have also been placed by both the collaborations.

REFERENCES

1. F. Abe et al., The CDF Collaboration, *Phys. Rev. Lett.*, **74**, 2626 (1995).
   S. Abachi et al., The DØ Collaboration, *Phys. Rev. Lett.*, **74**, 2632 (1995).
2. N. Kidonakis, and R. Vogt, *Phys. Rev. D*, **68**, 114014 (2003).
   M. Cacciari, S. Frixione, G. Ridolfi , M. Mangano, P. Nason, *JHEP*, **404**, 68 (2004).
3. B. W. Harris et al., *Phys. Rev. D*, **66**, 054024(2002), and Z. Sullivan, *Phys. Rev. D*, **70**, 114012(2004).

Precision Electroweak Measurements at LHC

K. Mazumdar

EHEP Group, Tata Institute of Fundamental Research, Mumbai, India.

Abstract. The LHC experiments will perform a large number of precision measurements. Only W physics and top physics are discussed here. Lepton pair production through Drell Yan process, measurements in Higgs and SUSY sectors, determination of triple gauge boson couplings are also important for precision physics at LHC. Limited knowledge of physics may dictate the final precision.

Keywords: LHC, Precision, Electroweak model, Top quark
PACS: 14.70.-e, 12.38.-b, 14.65.Ha,13.87.-a

LHC AND PRECISION MEASUREMENTS

The precision measurements of the Standard Model (SM) lies at the heart of the physics programme of the LHC, probing the physics at TeV energy scale. The main physics motivation of the LHC is to understand the concept of Electroweak Symmetry Breaking (EWSB) and to look for *New Physics* (NP) beyond SM. Due to the wide range of cross-sections and the lack of knowledge for the unknown, the discovery potential of LHC will be realised or strengthened after re-establishing SM through various precision measurements which also provide indirect constraints on NP.

The centre-of-mass energy of the LHC is 14 TeV with initial low luminosity of 2×10^{33} cm^{-2} s^{-1} to final luminosity of 10^{34} cm^{-2} s^{-1}. It is expected to have the first proton-on-proton collision in 2007 and by 2010 the accumulated luminosity will be 30 fb^{-1}/year/experiment. Thus, LHC will be a factory for many particles: W, Z, $b\bar{b}$, $t\bar{t}$ etc. Table 1 gives an idea of the event rates and the sample sizes of physics processes during low luminosity phase of LHC.

TABLE 1. Cross-section and approximate event rate per experiment for some of the LHC physics processes with low luminosity.

Process	σ (pb)	Events/sec.	Events/year
$W \to e\nu$	1.5×10^4	15	10^8
$Z \to e^+e^-$	1.5×10^3	1.5	10^7
$t\bar{t}$	830	0.8	10^7
$b\bar{b}$	5×10^8	5×10^5	10^{12}
H ($m_H = 200$ GeV)	10	0.01	10^5
Inclusive jets ($p_T > 200$ GeV)	10^5	100	10^9

The general-purpose experiments, ATLAS (A Toroidal Lhc ApparatuS) and CMS (Compact Muon Solenoid), are both capable of precision measurements of physics objects: e, γ, μ, τ, b-jets within rapidity range ($|\eta|$) of about 2.5 and jets, missing transverse energy upto $|\eta| \leq 5$. Both the experiments have similar detector components

CP792, *Deep Inelastic Scattering, DIS 2005*, edited by W. H. Smith and S. R. Dasu

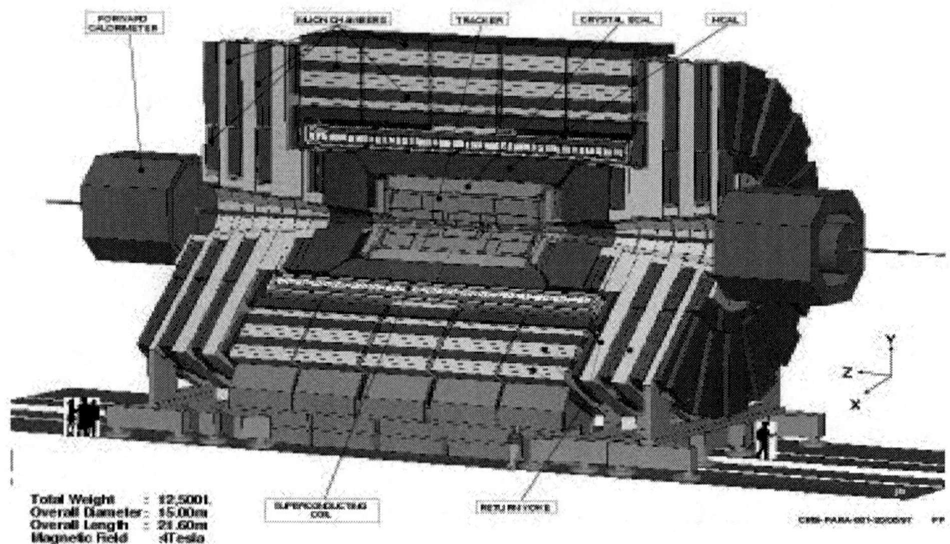

FIGURE 1. CMS detector.

with comparable resolutions and are well under way of construction. Fig. 1 shows the CMS detector, which is much smaller in size but heavier than the ATLAS detector.

The knowledge of absolute luminosity affects all cross-section measurements. As evident from Table 1, most of the measurements will have negligible statistical error and the systematics component which scales as $1/\sqrt{N}$. On the other hand, the uncertainty will be dominated by the experimental environment of the LHC imparting heavy challenges from the physics, the machine, as well as the detector. Considering W, Z production rates as the luminosity monitor will help in getting rid of several systematics since the rates are well known as well as the branching ratios of these particles. The accuracy of the cross-sections are limited by the uncertainties in the parton distributions and will be known eventually to 2% level.

ELECTROWEAK PHYSICS

The key to understand EWSB is in constraining the Higgs boson mass through precision measurements of the W-mass (M_W) and the top-quark mass (m_t) since they are related through radiative corrections as illustarted in Fig. 2. The status of electroweak precision from experimental measurements of LEP and Tevatron and the expectations from LHC are summarised in Table 2.

For LHC, the lepton energy and momentum scale dominates the error on M_W and the jet energy scale affects the estimate of m_t. The transverse W-mass can be reconstructed from $W \to \ell\nu$ decay by deducing the missing energy of the neutrino in the transverse direction. The trailing edge of this mass distribution is sensitive to the value of M_W.

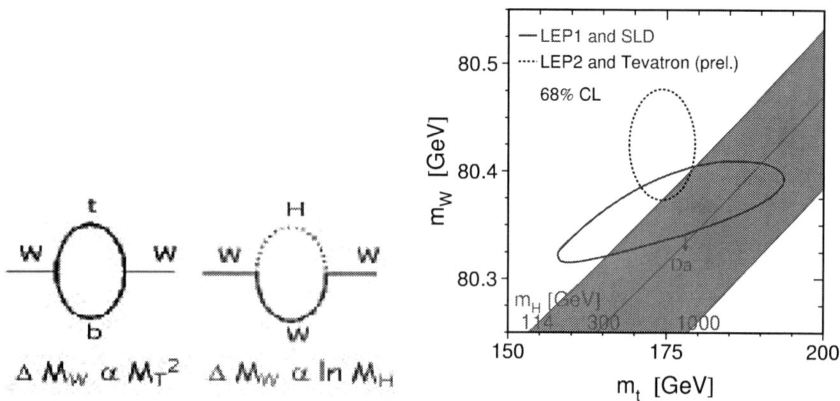

FIGURE 2. Radiative corrections to W mass and the constraints on Higgs mass from experiments.

TABLE 2. current situation about precision measuremnets and LHC expectations

exp. error	$\delta sin^2\theta_{eff}(\times 10^5)$	δM_W [MeV/c^2]	δm_t [GeV/c^2]	δM_H [MeV/c^2]
today	17	34	5.1	-
LHC	14-20	15	1-2	200

Using a control sample of $Z \to \ell\ell$, the lepton energy scale and the detector response the W-recoil, the p_T spectrum of the W can be modelled. The expected errors with 10 fb^{-1} are presented in Table 3. For an ultimate precision of 20 MeV/c^2, the lepton scale needs to be estimated to a level of 0.02% which is very difficult but not impossible.

The large values of top quark mass ($m_t = 173.5^{+4.1}_{-4.0}$ GeV/c^2), width ($\Gamma_t \sim 1.4$ GeV/c^2) and its role in electroweak radiative corrections make the precision measurements in top-sector very important to probe the origin of the fermion mass-hierarchy. Top physics is a very sensitive window for NP scenarios and top events are also major backgrounds for many searches; moreover they will be used extensively to understand the detector. e.g., trigger, b-tagging, calibration of calorimeters etc.. The pair-production, as depicted in Fig. 3, has cross-section value at NLO: $\sigma(t\bar{t}) \sim 830 \pm 10$pb of which $\approx 87\%$ is from gluon fusion and $\approx 13\%$ is due to $q\bar{q}$ annihilation. The estimated cross-section is sensitive to scales of renormalisation, factorisation and PDF, which can be given as

TABLE 3. components of errors in W-mass measurement

Source	ΔM_W (MeV/c^2)	Source	ΔM_W (MeV/c^2)
lepton E/p scale	15	lepton E/p resolution	5
structure functions	< 10	p_T^W	5
recoil model	5	W-width	7
radiative deacys	< 10		

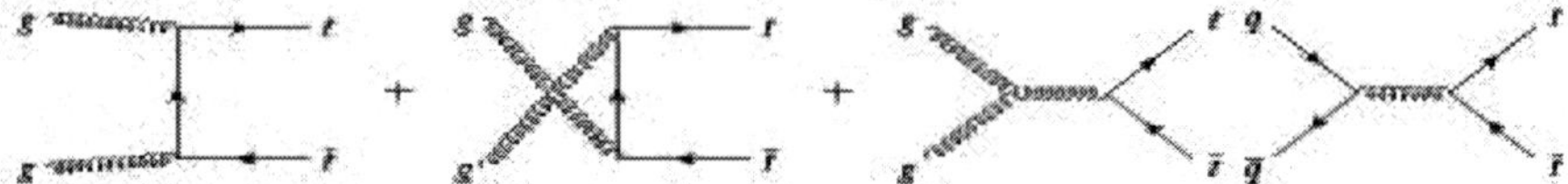

FIGURE 3. Feynman diagrams for top-pair production at LHC

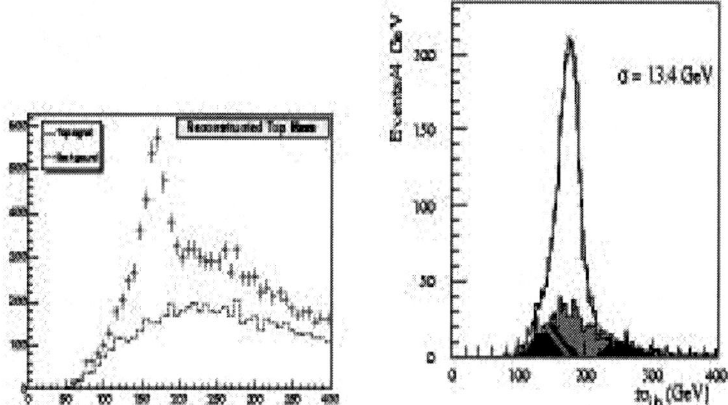

FIGURE 4. Top-peak in W+4 jets channel at LHC start-up and the ultimate reconstructed mass.

$\Delta\sigma \sim 5\%$ for $\Delta m_t = 2$ GeV/c^2. This can be compared with the statistical uncertainty on rate in a week with low lumi, which is $\sim 2.5\%$. Top peak will be visible above background from early periods ($\sim$ 150 pb $^{-1}$ after LHC startup with simple analysis (*ie.,* no b-tagging) and ultimately with a precision of 1.5 GeV/c^2 as shown in Fig. 4.

Single top quark production due only to electroweak physics is not yet observed experimentally. Tree level processes at LHC with their rates can be classified as: Wg fusion (245 pb), Wt (62 pb) and W* (10 pb). Each process is sensitive to different kinds of NP effects. With 30 fb^{-1} data, V_{tb} is expected to be determined to an accuracy of 5%. Single top studies provide independent measurements of mass, top spin polarisation, FCNC processes etc.. Anomalous couplings via FCNC decays like $t \to Vq$, $V = Z, \gamma, g$ are highly suppressed in SM, but get enhanced by NP. Very good improvements on branching ratio limits are expected with 100 fb^{-1}.

ACKNOWLEDGEMENT

I would like to thank the organizers of the conference for the exciting and beautiful time we had in Madison. My sincere gratitude to Prof. Wesley Smith for providing me the local hospitality without which I could not attend the conference. I also thank my colleagues in CMS and ATLAS collaborations for the material I have used here.

General Analysis of Single Top Production and W Helicity in Top Decay

Chuan-Ren Chen*, F. Larios*† and C.-P. Yuan*

*Department of Physics and Astronomy, Michigan State University, East Lansing, Michigan 48824, USA
†Departamento de Física Aplicada, CINVESTAV-Mérida, A.P. 73, 97310 Mérida, Yucatán, México

Abstract. We provide a framework for the analysis of the W boson helicity in the decay of the top quark that is based on a general effective tbW coupling. Four independent coupling coefficients can be uniquely determined by the fractions of longitudinal and transverse W boson polarizations as well as the single top production rates for the t-channel and the s-channel processes.

INTRODUCTION

The measurement of single top production cross section has turned out to be a challenging task and no single top events have been observed so far [1]. Concerning the W boson helicity in the $t \to bW$ decay, there are three modes depending on the polarization state of the W boson. Each mode is associated with a fraction: f_0, f_+ or f_- that corresponds to the longitudinal, right-handed or left-handed polarization, respectively. By definition, we have the restriction $f_0 + f_+ + f_- = 1$. Recent reports by the DØ and CDF collaborations at Fermilab give the following (95% C.L.)results for the longitudinal and right-handed fraction of $t \to bW$ in the $t\bar{t}$ pair events [2]:

$$\begin{aligned} f_0 &= 0.91 \pm 0.38\,(\mathrm{CDF}), \quad f_0 = 0.56 \pm 0.32\,(\mathrm{D}), \\ f_+ &\leq 0.18\,(\mathrm{CDF}), \qquad\quad f_+ \leq 0.24\,(\mathrm{D}). \end{aligned}$$

In this work we want to propose a new strategy to use the measurements on the single top production cross section and on the polarization of the W boson in the $t \to bW$ decay in order to determine the general effective tbW vertex. Our strategy consists of using four measurements: a) σ_s and σ_t, the cross sections of the two most important modes of single top quark production at the Tevatron, referred to as s-channel and t-channel [3] as well as, b) two of the three decay ratios f_0, f_- and f_+, to determine the four independent couplings that define the general effective tbW vertex.

CP792, *Deep Inelastic Scattering, DIS 2005,* edited by W. H. Smith and S. R. Dasu

Therefore, the effects of our general effective Lagrangian to the processes considered here can be completely described by the following tbW vertex:

$$\begin{aligned}\mathscr{L}_{tbW} &= \frac{g}{\sqrt{2}} W_\mu^- \bar{b}\gamma^\mu \left(f_1^L P_L + f_1^R P_R\right) t \\ &- \frac{g}{\sqrt{2}M_W} \partial_\nu W_\mu^- \bar{b}\sigma^{\mu\nu} \left(f_2^L P_L + f_2^R P_R\right) t \;\; + h.c.,\end{aligned} \tag{1}$$

where we have changed the mass scale Λ to m_W to keep the same notation used in the literature [4, 5].

In the SM the values of the form factors are $f_1^L = V_{tb} \simeq 1$, $f_1^R = f_2^L = f_2^R = 0$. To focus on deviations from SM values, let us define $f_1^L \equiv 1 + \varepsilon_L$.

It is well known that $b \to s\gamma$ can impose a strong constraint on f_1^R and f_2^L to be less than 0.004 [6, 7]. Also, $b \to s l^+ l^-$ can be sensitive to f_2^R and impose constraint of order 0.03 [7]. For ε_L, the LEP precision data imposes some constraint: assuming no deviations from the SM ttZ vertex we would have that $\varepsilon_L \leq 0.02$ [6]. These constraints do not use the combined effects of other couplings. For the dimension 5 couplings we will consider values at most of order 0.5 in order to satisfy the unitarity condition [8].

Previous studies of $f_2^{L,R}$ in connection with the single top quark production at hadron colliders have shown that a sensitivity of order 0.2 (0.05) might be achieved at the Tevatron (LHC) [9]. Information on the helicity of the W boson in $t \to bW$ can be obtained by measuring a forward-backward asymmetry (A_{FB}) based on the angle between the charged lepton and the b-jet of the observed decay process [10]. Preliminary studies show that if A_{FB} is measured with 20% accuracy at the Tevatron, it may be sensitive to values of order $f_2^{L,R} \sim 0.3$; similarly, if A_{FB} is measured with 1% accuracy at the LHC this may be translated to a sensitivity of order $f_2^L \sim 0.03$ and $f_2^R \sim 0.003$ [5]. We would like to point out that, since the observable A_{FB} is only proportional to the difference between f_+ and f_- [10], it is clear that it does not provide any more information than the separate measurements of (two of) the ratios f_0, f_- and f_+.

In Table 1 we show the leading order (LO) and the next-to-leading order (NLO) SM predictions for σ_t and σ_s at the Tevatron and at the LHC [3]. For the LO predictions the CTEQ6L1 parton distribution function (PDF) has been used [11]. For the NLO predictions the CTEQ6M PDF has been used [3]. We are taking the mass of the top quark as $m_t = 178\,\text{GeV}$ and the mass of the W boson as $m_W = 80.4\,\text{GeV}$.

TABLE 1. SM single top production cross section predictions in units of pb [3]. The mass of the top quark is taken as $m_t = 178\,\text{GeV}$.

Channel	Tevatron (t LO)	(t NLO)	LHC (t LO)	(t NLO)	LHC ($\bar{t}$ LO)	($\bar{t}$ NLO)
t-channel	0.827	0.924	146.0	150.0	84.9	88.5
s-channel	0.27	0.405	4.26	6.06	2.59	3.76

Neglecting terms proportional to the bottom mass, the Born level values of the top quark width and its W-polarization ratios are $\Gamma_t = 1.65\,\text{GeV}$, $f_0 = 0.71$, $f_- = 0.29$ and

$f_+ = 0$. In the SM, including terms proportional to m_b, order α_s^2 QCD, electroweak, and finite W width corrections produce a 10% decrease in the top's width ($\Gamma_t = 1.49$) and a small $\sim 1\%$ variation for decay ratios ($f_0 = 0.701$, $f_- = 0.297$ and $f_+ = 0.002$) [10].

In this work we will be interested in deviations from the SM values (up to the NLO) that come from the effects of the anomalous ε_L, f_1^R, f_2^L and f_2^R couplings cf. Eq. (1), induced by heavy new physics effects. In the following, we will write down the Born level contributions of these couplings on the observables f_0, f_+, f_-, σ_t and σ_s.

SINGLE TOP PRODUCTION AND W HELICITY IN $T \to BW$ DECAY

The tree level $t \to bW$ decay width of the top quark with the general tbW vertex can be easily obtained with the helicity amplitude method, and it is given by [4]:

$$
\begin{aligned}
\Gamma_t &= \Gamma_0 + \Gamma_- + \Gamma_+ \\
&= \frac{g^2 m_t}{64\pi} \frac{(x_t^2 - 1)^2}{x_t^4} \left(x_t^2 (1 + x_0) + 2(1 + x_m) + 2x_p \right) , \\
x_0 &\equiv (f_1^L + f_2^R / x_t)^2 + (f_1^R + f_2^L / x_t)^2 - 1 , \\
x_m &\equiv (f_1^L + x_t f_2^R)^2 - 1 , \\
x_p &\equiv (f_1^R + x_t f_2^L)^2 , \\
x_t &\equiv \frac{m_t}{m_W} .
\end{aligned}
\tag{2}
$$

As the notation suggests, x_0, x_m and x_p are the effective terms that originate the contribution to f_0, f_- and f_+, respectively. Below, we will write down the explicit expressions for these decay ratios.

The contributions of the effective tbW couplings to the observables of interest are:

$$
f_0 = \frac{x_t^2 (1 + x_0)}{x_t^2 (1 + x_0) + 2(1 + x_m + x_p)} , \tag{3}
$$

$$
f_+ = \frac{2x_p}{x_t^2 (1 + x_0) + 2(1 + x_m + x_p)} , \tag{4}
$$

$$
f_- = \frac{2(1 + x_m)}{x_t^2 (1 + x_0) + 2(1 + x_m + x_p)} ,
$$

$$
\Delta\sigma_t = a_0 x_0 + a_m x_m + a_p x_p + a_5 x_5 , \tag{5}
$$

$$
\Delta\sigma_s = b_0 x_0 + b_m x_m + b_p x_p + b_5 x_5 , \tag{6}
$$

$$
x_5 \equiv x_t^2 ({f_2^L}^2 + {f_2^R}^2) ,
$$

where $\Delta\sigma$ stands for the variation from the SM NLO prediction. The numerical values of the a_i and b_i coefficients are given in Table 2 for the Tevatron and the LHC. They have been obtained by integrating over the parton luminosities which are evaluated using the PDF CTEQ6L1 [11].

TABLE 2. The single top production cross section coefficients of Eqs. (5-6). In units of pb.

t-channel:	a_0	a_m	a_p	a_5
Tevatron	0.896	-0.069	-0.153	0.247
LHC (t)	165.2	-19.1	-34.2	62.5
LHC ($\bar{t}$)	105.8	-20.9	-12.5	38.6
s-channel:	b_0	b_m	b_p	b_5
Tevatron	-0.081	0.352	0.352	0.230
LHC (t)	-1.41	5.67	5.67	6.34
LHC ($\bar{t}$)	-0.836	3.43	3.43	3.38

Eqs. (3)-(6) can be used to make a general analysis of the effective tbW vertex. We note that in case a new light resonance is found, like a scalar or vector boson, the s-channel process could be significantly enhanced and its production rate may not be dominated by a virtual W-boson s-channel diagram [12]. Here, we do not include possible new production channels for the s-channel single top events. We want to emphasize that in general all four observables of Eqs. (3)-(6) are needed to determine the four couplings of the tbW vertex and to make a complete analysis that could test the different models of EWSB.

ACKNOWLEDGMENTS

We thank A. Belyaev and H.-J. He for discussions. The work of F.L. has been supported in part by the Fulbright-Garcia Robles grant and by Conacyt. This work was supported in part by the NSF grant PHY-0244919.

REFERENCES

1. R. Schweinhorst, for the DØ and CDF collaborations, hep-ex/0411039.
2. By the DØ collaboration, hep-ex/0404040 and references there in; by the CDF collaboration, hep-ex/0411070 and references therein.
3. Z. Sullivan, Phys. Rev. D**70** (2004), 114012. and references therein; J. Campbell, R.K. Ellis and F. Tramontino, Phys. Rev. D**70**(2004), 094012.
4. G.L. Kane, G.A. Ladinsky and C.-P. Yuan, Phys. Rev. D**45** (1992), 124.
5. F. del Aguila and J.A. Aguilar-Saavedra, Phys. Rev. D**67** (2003), 014009.
6. F. Larios, M.A. Pérez and C.-P. Yuan, Phys. Lett. B**457** (1999), 334. and references therein.
7. G. Burdman, M.C. Gonzalez-Garcia and S.F. Novaes, Phys. Rev. D**61** (2000), 114016.
8. G.J. Gounaris, F.M. Renard and C. Verzegnassi, Phys. Rev. D**52** (1995), 451.
9. E. Boos, L. Dudko and T. Ohl, Eur. Phys. J. C**11** (1999) 473.
10. H.S. Do, S. Groote, J.G. Korner and M.C. Mauser, Phys. Rev. D**67** (2003), 091501. and references therein.
11. J. Pumplin, D.R. Stump, J. Huston, H.L. Lai, P. Nadolsky and W.K. Tung, JHEP **0207** (2002) 012.
12. H.-J. He, L. Diaz-Cruz and C.-P. Yuan, Phys. Lett. B**530** (2002), 179.

Supersymmetry: Theory Overview

Howard Baer

Dep't of Physics, Florida State University, Tallahassee, FL 32306

Abstract. I present here a brief overview of supersymmetry theory as it pertains to collider physics. After a discussion of supersymmetric models and their constraints, I discuss prospects for supersymmetry at the DESY HERA, Fermilab Tevatron and CERN LHC colliders. I also compare with expectations for direct and indirect searches for neutralino dark matter.

Keywords: Supersymmetry,Hadron Colliders
PACS: 12.60.Jv,11.30.Pb,14.80.Ly

SUPERSYMMETRIC MODELS

There has been a recent flurry of new theoretical ideas for physics beyond the Standard Model (SM). In spite of these novel approaches, it still seems likely that the most promising avenue for physics beyond the SM to be revealed is weak scale supersymmetry[**?**]. If light fundamental scalars exist, then SUSY is needed to stabilize quantum corrections to scalar masses. SUSY is an integral ingredient in superstring theory, and has a deep connection to gravity via supergravity. Still, it seems diabolical that all the SM particles (apart from the Higgs boson) have been discovered, while no superpartners have. From the theory side, this is easy to understand: the SM particles are those that remain massless upon supersymmetry breaking, and only gain mass upon electroweak symmetry breaking (EWSB). Still, will broken supersymmetry be revealed at or around the TeV (weak) scale? Here, we appeal to data, and cite indirect evidence in the form of 1. gauge coupling unification, 2. precision measurement of electroweak parameters imply a Higgs mass $\lesssim 200$ GeV, in accord with MSSM predictions, 3. the fact that EWSB can occur *radiatively* in SUSY models, but only if $m_t \sim 150-200$ GeV (as is measured), 4. many SUSY models offer an explanation of cold dark matter (CDM) in the universe, 5. the see-saw neutrino mass mechanism coupled to GUT theories suggests a high mass scale $M_N \sim 10^{15}$ GeV. The presence of such a high mass scale will require SUSY to stabilize the weak scale vs. M_N or M_{GUT}. 6. The SM contains no viable mechanism for baryogenesis in the universe, while SUSY theories conatin at least three: electroweak baryogenesis (nearly excluded), leptogenesis (well-motivated in light of theories with massive neutrinos), and Affleck-Dine baryogenesis. In spite of all this indirect evidence, as Dirac once said, "That could all be wrong!". And so we must calculate as well as possible all observable consequences of weak scale supersymmetry, to test teh theory against experiment.

The Minimal Supersymmetric Standard Model, or MSSM, provides a simple supersymmetric extension of the SM[**?**]. It is straightforward to write down the Lagrangian for globally supersymmetric, renormalizable gauge theories. To construct the MSSM, we then adopt the SM gauge symmetry, and elevate all SM fermions to left-chiral super-

CP792, *Deep Inelastic Scattering, DIS 2005,* edited by W. H. Smith and S. R. Dasu

fields, and all SM gauge fields to gauge superfields. Two Higgs doublets are necessary in the MSSM for triangle anomaly cancellation, and to give mass to all SM fermions. An R-parity conserving (or violating) superpotential may be chosen, and explicit soft SUSY breaking terms can be added by hand. The result is a 124 parameter model which provides the masses, mixings and Feynman rules needed for calculating physical processes[**?**].

The MSSM is expected to be the low energy effective field theory of some more unified theory based on (a combination of) strings, supergravity and GUT ideas. Since spontaneous SUSY breaking is not phenomenologically viable in the MSSM, a hidden sector needs to be introduced to serve as an arena for SUSY breaking. In gauge-mediated SUSY breaking (GMSB), SUSY is broken in the hidden sector, and communicated to the visible sector via messenger fields. The lightest SUSY particle (LSP) is a gravitino of very low mass, so that these models have a hard time explaining CDM in the universe. In anomaly-mediated SUSY breaking, the LSP is wino-like, and again it is difficult to accommodate CDM. In supergravity models (SUGRA), gravity acts as the messenger for SUSY breaking, and the LSP is usually the lightest neutralino, or a TeV scale gravitino. The SUGRA theories most easily accomodate CDM in the universe.

In SUGRA theories, the gravitino mass $m_{3/2} \sim M_s^2/M_{Pl} \sim 10^3$ GeV for $M_s \sim 10^{11}$ GeV. In minimal renditions of SUSY GUT models, it is assumed that the MSSM is valid between scales $Q = M_{GUT}$ and $Q = M_{weak}$. The soft SUSY breaking masses are stipulated at $Q = M_{GUT}$, and their value at M_{weak} is calculated via renormalization group evolution (RGE). A minimal choice, with some motivation from string theory is that all scalars have a common mass m_0, all gauginos a common mass $m_{1/2}$ and all trilinears have a common mass A_0. Finally, a bilinear SB term B is also stipulated at $Q = M_{GUT}$. The gauge and Yukawa couplings and soft terms are evolved from M_{GUT} to M_{weak} and EW symmetry is broken radiatively, allowing B to be traded for the ratio of Higgs vevs $\tan\beta$, and the magnitude (but not the sign) of the superpotential μ term can be determined. The parameter space of the minimal supergravity model (mSUGRA) is thus given by

$$m_0,\ m_{1/2},\ A_0,\ \tan\beta,\ sign(\mu). \tag{1}$$

This simple model offers a common baseline for examining observable consequences of SUSY. But it is important to keep in mind that nature may be considerably more complicated. Some possibilities include non-universal scalar and/or gaugino masses, off diagonal SSB masses, large CP violating phases, and additional fields beyond those of the MSSM at scales below (and maybe far below) M_{GUT}.

Once a parameter space point has been selected, several publicly available computer codes are available which calculate sparticle masses and mixings, so that observable processes can be calculated. There are several constraints on the mSUGRA parameter space: there must be successful REWSB, the lightest SUSY particle (LSP) must be electrically and color neutral, and one must fulfill limits from LEP2 on the chargino mass ($m_{\widetilde{W}_1} > 103.5$ GeV) and the Higgs mass ($m_h > 114.4$ GeV for a SM-like Higgs boson). In addition, there are constraints from the measured decay $BF(b \to s\gamma)$ and muon anomalous magnetic moment a_μ. Lately, the most stringent constraint comes from the neutralino relic density, where $\Omega_{\widetilde{Z}_1} h^2 = 0.113 \pm 0.009$ from WMAP and other measurements. The WMAP constraint leads to only a few allowed regions of mSUGRA parameter space: 1. the bulk region at low m_0 and low $m_{1/2}$ where neutralino annihilation in the early universe is enhanced by light t-channel slepton exchange, 2. the stau co-annihilation region at low m_0 where $m_{\tilde{\tau}_1} \simeq m_{\widetilde{Z}_1}$, and neutralino (co)-annihilation is enhanced, 3. the hyperbolic branch/focus point region (HB/FP) at large m_0 where μ becomes small and we have mixed higgsino dark matter, 4. the A-annihilation funnel at large $\tan\beta$ where neutralino annihilation is enhanced by the heavy Higgs resonances since $2m_{\widetilde{Z}_1} \sim m_A$. Also, there is a less prominent light Higgs annihilation region (where $2m_{\widetilde{Z}_1} \simeq m_h$) and a top squark co-annihilation region at particular A_0 values.

SUSY AT COLLIDERS

SUSY is being searched for currently at the HERA ep collider and the Tevatron $p\bar{p}$ collider. At HERA, one may look for $ep \to \tilde{e}\tilde{q}$. Usually, this search has been neglected because of combined bounds on 1. the selectron mass from LEP2 and 2. the squark mass from the Tevatron. However, the latter limit assumes 5 degenerate generations of left- and right- squarks. Recently, SUGRA models with split Higgs masses at the GUT scale have been shown to give rise to just two light squarks: the $\tilde{u}_R$ and $\tilde{c}_R$, while $\Omega_{\widetilde{Z}_1}h^2$ is consistent with WMAP. Thus, HERA experiments should take this possibiliy into account. In addition, a strength ofthe HERA collider is to search for R-parity violating decays via the λ'_{ijk} couplings which lead to s-channel squark production via the $ue\tilde{d}$ or $ed\tilde{u}$ couplings. In this case, squark production is identical to s-channel spin-0 leptoquark production.

At the Fermilab Tevatron collider, the best signature for mSUGRA is clean trilepton production from $p\bar{p} \to \widetilde{W}_1\widetilde{Z}_2$ production followed by $\widetilde{W}_1 \to \ell\nu\widetilde{Z}_1$ and $\widetilde{Z}_2 \to \ell\bar{\ell}\widetilde{Z}_1$ decays. Also, as for HERA, Tevatron experiments should search for single squark production, in case one of $\tilde{u}_R$ or $\tilde{c}_R$ is quite light. Searches are also ongoing for top and bottom squark production, and gluino pair production.

At the CERN LHC, SUSY can be searched for in a variety of jets plus leptons plus missing $\not{E}_T$ events at levels above SM backgrounds, where the events will originate from squark and gluino production followed by cascade decays. The LHC with 100 fb^{-1} can probe to $m_{\tilde{g}} \sim 1.8$ TeV for large m_0, and to $m_{\tilde{g}} \sim 3$ TeV for low $m_{1/2}$: see the Figure. A variety of scenarios have been examined by Frank Paige et al. as to how to extract fundamental SUSY parameters from SUSY events at the LHC.

There are plans to build a TeV scale International Linear e^+e^- Collider (ILC), starting at $\sqrt{s} = 0.5$ TeV, and ultimately increasing CM energy to 1 TeV. In most of parameter space, the LHC has a better reach for SUSY than the ILC. However, in the HB/FP region, squarks and sleptons, and possibly gluinos, may all be heavy, and beyond LHC reach. In this case, μ is small if the WMAP constraint is respected, and then charginos remain quite light and possibly accessible to ILC searches. In fact, in this region the ILC reach can exceed that of LHC. In addition, if SUSY is discovered, then the well defined beam energy, beam polarization, variable CM beam energy and clean environment all allow a bevy of precision measurements to be made on the SUSY particles.

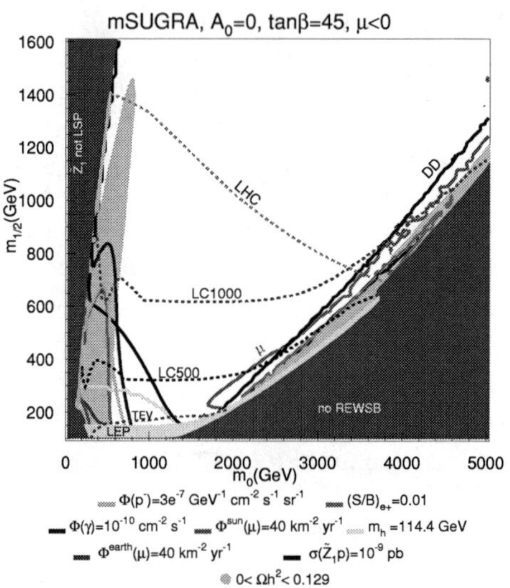

FIGURE 1. Reach of various collider and direct and indirect DM detection experiments for SUSY in the mSUGRA model for $\tan\beta = 45, A_0 = 0$ and $\mu < 0$.

DIRECT AND INDIRECT SEARCH FOR NEUTRALINO DM

Neutralinos can also be seached for directly as relics from the Big Bang in underground cryogenic detectors such as CDMS2, Edelweiss2 and Zeplin2, which should all probe spin-independent neutralino-nucleon scattering cross sections of $\sim 10^{-8}$ pb. Third generation detectors being planned such as Genius, Xenon and Zeplin4 should be able to probe to $\sim 10^{-9}$ pb, which allows coverage of the interesting HB/FP region with mixed higgsino DM.

Neutralinos can also be collected gravitationally in the solar core. At high density, they will annihilate at a high rate to SM particles, which will decay to muon neutrinos, which can be searched for at the IceCube and Antares neutrino telecopes. Prospects are especially good for the HB/FP region. Relic neutralinos in the galactic halo can annihilate to gamma rays or anti-matter. Searches for gamma rays from the galactic center by GLAST and for e^+, $\bar{p}$ or $\bar{d}$ by Pamela or AMS can probe the HB/FP region *and* the A-annihilation funnel. Prospects for direct or indirect DM detection are generally poor for SUSY in the stau co-annihilation region. Thus, the detection of supersymmetric signals at colliders and at direct and indirect detection experiments may help to pinpoint the identity of cold dark matter in the universe.

REFERENCES

1. For an overview, see *Weak Scale Supersymmetry: From Superfields to Scattering Events*, by H. Baer and X. Tata (Cambridge University Press, 2006).

SUSY Searches at H1

David South

Deutsches Elektronen Synchrotron
Notkestrasse 85, 22607, Hamburg, Germany.
email: David.South@desy.de

Abstract.
The deep inelastic collisions produced at HERA provide and ideal environment to search for new particles and physics Beyond the Standard Model (BSM) and three such analyses performed by the H1 experiment are presented here. A search for bosonic stop decays in R-parity violating SUSY finds no significant deviation from the Standard Model (SM) and sets competitive limits, excluding the existence of stop quarks at 95% CL with masses up to 275 GeV for a Yukawa coupling λ' of electromagnetic strength. Secondly, a search for light gravitinos in events with photons and missing transverse momentum sets the first constraints from HERA on SUSY models independent of the squark sector, ruling out neutralino masses up to 112 GeV at 95% CL for $\lambda' = 1$. Finally, a model independent general search for new phenomena finds good agreement between the H1 data and the SM, the most significant deviation found being a previously reported observation.

Keywords: H1, Supersymmetry, Searches, ep
PACS: 11.30.Pb

A SEARCH FOR BOSONIC STOP DECAYS IN R-PARITY VIOLATING SUSY IN E$^+$P COLLISIONS AT HERA

H1 has previously performed a search for squarks ($\tilde{q}$), the scalar supersymmeteric (SUSY) partners of quarks, in models with R–parity violation ($\not{R}_p$) [1]. This analysis, which uses H1 e^+p data corresponding to an integrated luminosity of 106 pb^{-1}, complements the previous analysis and focuses on resonant stop quark production in e^+q fusion which proceeds via an $\not{R}_p$ coupling λ', with the subsequent bosonic stop decay $\tilde{t} \to \tilde{b}W$ [2]. The R–parity violating decay of the sbottom quark $\tilde{b} \to d\bar{\nu}_e$ and leptonic and hadronic W decays are considered. The $\not{R}_p$ decay $\tilde{t} \to eq$ is also examined in order to cover all possible stop decay modes. The bosonic stop decay leads to three different final state topologies depending on the decay of the W boson; a jet, a lepton (electron or muon)[1] and missing transverse momentum ($je\not{P}_\perp$ channel and $j\mu\not{P}_\perp$ channel) or, if the W decays to jets, three jets and missing transverse momentum ($jjj\not{P}_\perp$ channel).

A slight excess of events compared to the SM prediction is observed in the $j\mu\not{P}_\perp$ channel, confirming the results of a previous H1 analysis [3]. All of the other channels are found to be in good agreement with the SM. Assuming the presence of a stop mass $M_{\tilde{t}}$ decaying bosonically, the observed event yields are used to determine the allowed range for a stop production cross section $\sigma_{\tilde{t}}$, as illustrated in figure 1 (left). It can

---

[1] The W decay into $\nu_\tau\tau$, where $\tau \to$ hadrons $+\nu$ is not investgated in this analysis.

CP792, *Deep Inelastic Scattering, DIS 2005,* edited by W. H. Smith and S. R. Dasu

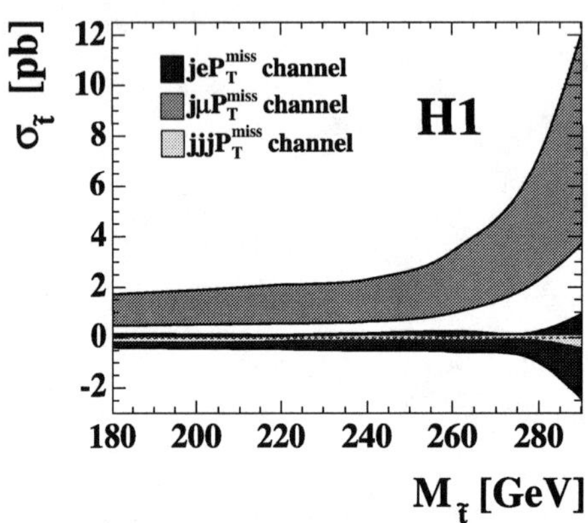

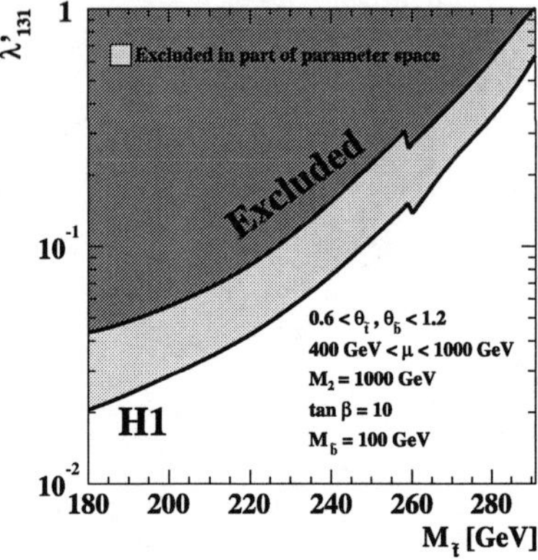

FIGURE 1. Left: Bands representing the allowed stop cross section regions $\sigma_{\tilde{t}} \pm \Delta\sigma_{\tilde{t}}$ as a function of the stop mass as obtained from the analysis of each bosonic stop decay channel. Right: Exclusion limits at the 95% CL on the $\not{R}_p$ coupling λ'_{131} as a function of the stop mass for $M_{\tilde{b}} = 100$ GeV.

be seen that the excess in the $j\mu\not{P}_\perp$ channel is not supported by the other channels, and hence no evidence for stop production is observed. The results from the different channels are combined to derive constraints on stop quarks decaying bosonically in the Minimal Supersymmetric Standard Model (MSSM). The resulting limits projected on the $(M_{\tilde{t}}, \lambda'_{131})$ plane for $M_{\tilde{b}} = 100$ GeV are shown in figure 1 (right). In a large part of the model parameter space the existence of stop quarks coupling to an e^+d pair with masses up to 275 GeV is excluded at 95% CL for a Yukawa coupling λ'_{131} of electromagnetic strength.

A SEARCH FOR LIGHT GRAVITINOS IN EVENTS WITH PHOTONS AND MISSING TRANSVERSE MOMENTUM AT HERA

The fermion–boson symmetry in SUSY models leads to an extention of the particle spectrum by associating each SM particle to a super symmetric partner, with different spin by half a unit. The masses of the new particles are related to the symmetry breaking mechanism and in Gauge Mediated Supersymmetry Breaking (GMSB) models, new "messenger" fields are introduced which couple to the source of the symmetry breaking. The gravitino, $\tilde{G}$ is the lightest supersymmetric particle (LSP) and the next–to–lightest supersymmetric particle (NLSP) is generally either the lightest neutralino $\tilde{\chi}_1^0$ or a slepton $\tilde{l}$, which decays accordingly to the stable gravitino. This analysis, which uses H1 $e^\pm p$ data corresponding to an integrated luminosity of 78 pb^{-1}, investigates $\not{R}_p$ SUSY in a GMSB scenario and searches for resonant single neutralino production $\tilde{\chi}_1^0$ via t–channel selectron exchange, $e^\pm p \to \tilde{\chi}_1^0 q'$, where it is assumed the $\tilde{\chi}_1^0$ is the NLSP and the subsequent decay $\tilde{\chi}_1^0 \to \gamma\tilde{G}$ occurs with an unobservably small lifetime [4]. The experimental signature is thus a photon, a jet from the struck quark and missing

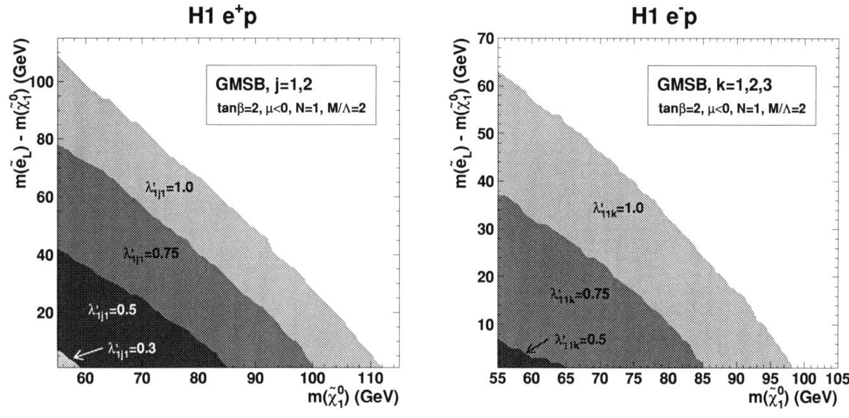

FIGURE 2. Excluded regions at the 95% CL in the $\Delta m = m(\tilde{e}_L) - m(\tilde{\chi}_1^0)$ and $m(\tilde{\chi}_1^0)$ plane for various values of λ'_{1j1} $(j = 1,2)$ and λ'_{11k} $(k = 1,2,3)$.

transverse momentum due to the gravitino. The main SM background is due to radiative charged current deep inelastic scattering.

In the search for light gravitinos in events with photons and missing transverse momentum no significant deviation from the SM is observed. The results are used to derive constraints on GMSB models for different values of the $R\!\!\!/_p$ coupling λ' for fixed values of the SUSY parameters $tan\beta$ (the ratio of the Higgs vacuum expectation values), N (the number of sets of messenger particles) and μ (the sign of the Higgs sector mixing parameter). Figure 2 displays excluded regions in the ($m(\chi_1^0)$, $m(\tilde{e}_L)$–$m(\chi_1^0)$) plane derived from the H1 e^+p (left) and e^-p (right) data for various values of λ'. It can be seen that for small mass differences between the neutralino and selectron, neutralino masses up to 112 GeV are ruled out at 95% CL for $R\!\!\!/_p$ coupling $\lambda' = 1$. Furthermore, for neutralino masses close to 55 GeV, λ'_{1j1} Yukawa couplings of electromagnetic strength are excluded. These are the first constraints from HERA on SUSY models independent of the squark sector.

A GENERAL SEARCH FOR NEW PHENOMENA IN EP SCATTERING AT HERA

A model independent general search for deviations from the SM is performed using the complete H1 HERA I data sample (1994-2000), corresponding to an integrated luminosity of 117 pb^{-1} [5]. All high transverse momentum (P_T) final state configurations involving electrons (e), muons (μ), jets (j), photons (γ) or neutrinos (ν) are considered. All final state configurations containing at least two such objects with $P_T > 20$ GeV in the central region of the detector are investigated and classified into exclusive event classes, for example e–j, μ–j–ν, j–j–j and so on.

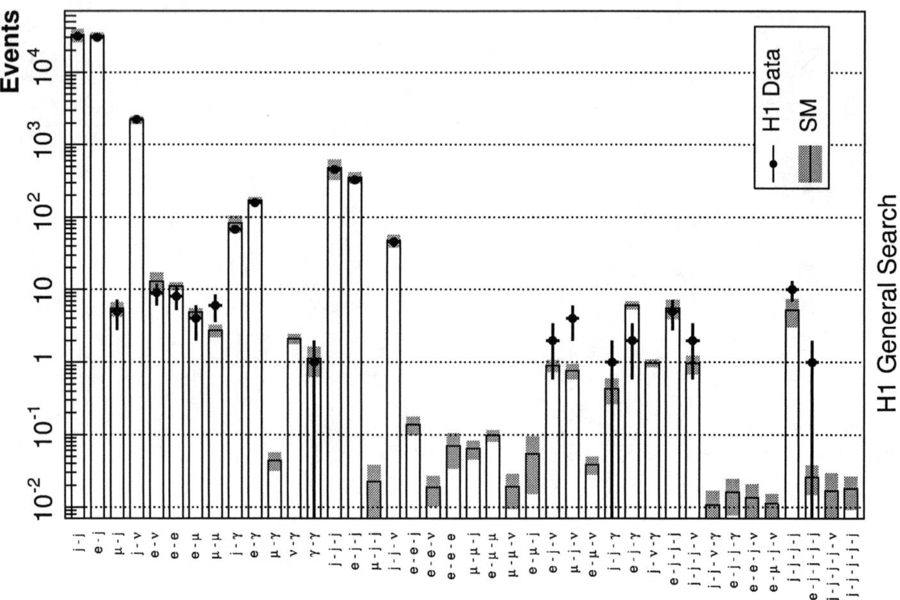

FIGURE 3. The data and the SM expectation for all event classes with a SM expectation greater than 0.01 events. The error bands on the predictions include model uncertainties and experimental systematic errors added in quadrature.

Data events are found in 22 such event classes and good agreement is observed between data and the SM expectation in most event classes, as can be seen in figure 3. The dominant high P_T processes at HERA, namely photoproduction of jets (j–j class) and neutral and charged current deep inelastic scattering (e–j and j–ν classes respectively) are seen to dominate the observed rates. The observed data excess in the μ–j–ν event class, where 4 data events are observed compared to a SM expectation of 0.8 ± 0.2, was previously reported in [3]. Additionally, in the e–j–j–j–j event class 1 event is observed in the data compared to a SM prediction of 0.026 ± 0.011.

REFERENCES

1. H1 Collab., A. Aktas et al., *Eur. Phys. J.* **C36** 425 (2004).
2. H1 Collab., A. Aktas et al., *Phys. Lett.* **B599** 159 (2004).
3. H1 Collab., V. Andreev et al., *Phys. Lett.* **B561** 241 (2003).
4. H1 Collab., A. Aktas et al., *Phys. Lett.* **B616** 31 (2005).
5. H1 Collab., A. Aktas et al., *Phys. Lett.* **B602** 14 (2004).

Searches for R-parity violating Supersymmetry at ZEUS

Claus Horn
(On behalf of the ZEUS Collaboration)

DESY, Notkestrasse 85, D-22607 Hamburg, Germany

Abstract. Two searches for R_P-violating supersymmetry were performed by the ZEUS experiment at HERA. No deviation from Standard Model expectations was found in either case. Limits were calculated for the Yukawa couppling λ'_{131} and the MSSM parameters M_2 and μ.

Keywords: SUSY;MSSM;RPV
PACS: 11.30.Pb;14.80.Ly

1. INTRODUCTION

Supersymmetry (SUSY) is one of the most promising candidates for a theory beyond the Standard Model (SM). The two analyses presented here are placed in the framework of the Minimal Supersymmetric Standard Model (MSSM) with R-parity violation. R-parity, defined as $R_P = (-1)^{3B+L+2S}$ (here B, L and S are the Baryon number, the Lepton number and the spin of the particle, respectively), is a multiplicative, discrete symmetry, often considered to be conserved as in such models the lightest supersymmetric partice (LSP) is stable and can be a candidate for dark matter. However, in the most general supersymmetric theory that is renormalisable and gauge invariant under the SM gauge transformations the superpotential contains three trelinear terms of Yukawa couplings which allow for R_P-violating interactions:

$$W_{RPV}^{tril} = \lambda_{ijk} L_i L_j \bar{E}_k + \lambda'_{ijk} L_i Q_j \bar{D}_k + \lambda''_{ijk} \bar{U}_i \bar{D}_j \bar{D}_k + h.c. \tag{1}$$

Because of its unique initial state which provides both baryonic and leptonic quantum numbers the *ep* collider HERA is an ideal place to look for processes which proceed through the second term characterised by the Yukawa coupling λ'_{ijk} (here ijk denote family indices).

Both s-channel and t-channel processes have been considered in searches for stop and gaugino production as described in Sect. 2 and 3, respectively. The Feynman diagrams for these processes are illustrated in Fig. 1.

2. SEARCH FOR STOP PRODUCTION

In R_P-violating scenarios the Yukawa coupling λ'_{131} allows for the resonant production of the stop ($e^+d \to \tilde{t}$). The stop squark is especially interesting since large mixings can be significant in the third family, which can lead to low stop masses for high tanβ. The

CP792, *Deep Inelastic Scattering, DIS 2005,* edited by W. H. Smith and S. R. Dasu

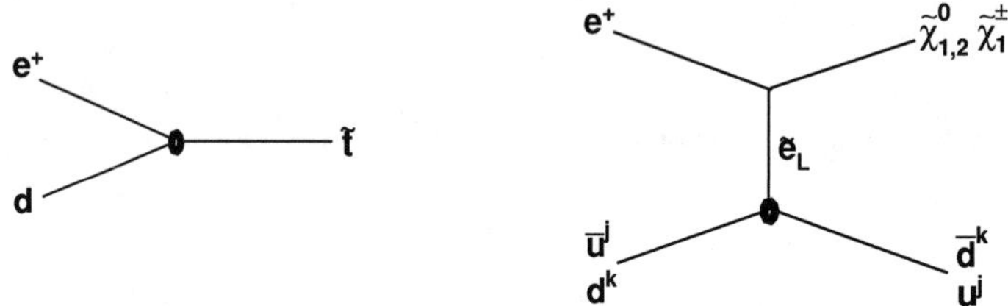

FIGURE 1. Feynman diagrams of production processes proceeding through the Yukawa coupling λ' (vertices marked by a circle), resonant stop production (left) and t-channel gaugino production (right).

subsequent decay can proceed again via $\tilde{t} \to e^+d$ or the R_P-conserving decay $\tilde{t} \to b\tilde{\chi}_1^+$ followed by the decay of the $\tilde{\chi}_1^+$. Here, the decay channels of a positron and one jet, a positron and multiple jets and a neutralino and multiple jets have been analysed which dominate in the considered MSSM parameter region (100 GeV $< M_2 <$ 300 GeV, $|\mu| <$ 300 GeV and tan$\beta = 6$). Scenarios where the lightest neutralino, $\tilde{\chi}_1^0$, is not the lightest supersymmetric particle (LSP) or it has a mass less than 35 GeV were discarded. Furthermore, the gluino was assumed to be heavier than the lightest stop mass eigenstate so that the decay $\tilde{t} \to t\tilde{g}$ is kinematically forbidden.

In the preselection, a reconstructed stop mass of $m_{\tilde{t}} >$ 100 GeV for events with a positron in the final state and $m_{\tilde{t}} >$ 80 GeV for events with a neutralino in the final state was required. To pass the final selection, events had to fulfill the conditions of $Q^2 > 3000$ GeV^2 and $y > 0.2-0.6$, depending on the stop mass. Additionally, different cuts on P_T/E_T, which measures how equally an event is distributed over the azimuthal angle ϕ, were made in order to select either single or multiple jet events.

Since no evidence for stop production was observed in any of the three channels, upper limits on the Yukawa coupling λ'_{131} were derived as a function of the stop mass (see Fig. 2). It was found that a variation of the MSSM parameters M_2 and μ has only a small influence on the resulting limit (dark region in Fig 2). For values of λ'_{131} of electromagnetic strength stop masses up to 265 GeV can be excluded. Comparable constraints were obtained by the H1 collaboration using similar SUSY scenarios [1, 2].

3. SEARCH FOR GAUGINO PRODUCTION

If squarks are much heavier than sleptons resonant squark production is suppressed and t-channel processes become dominant. At HERA the initial electron or positron can exchange a slepton with a quark from the proton leading to the production of neutralinos and charginos. They decay predominantly in two decay channels resulting in two quarks and an electron or positron, or two quarks and a neutrino or anti-neutrino.

It was assumed that the single Yukawa coupling λ'_{111} associated with pure first generation particle interaction dominates and the contributions from the other Yukawa couplings were neglected.

The presented results are based only on the analysis of events with an electron or

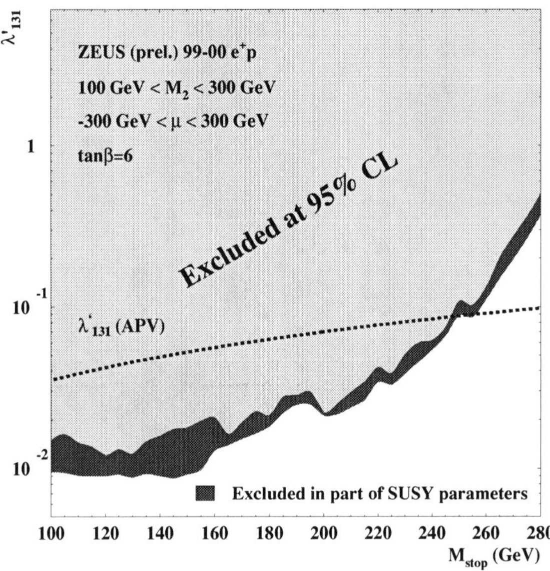

FIGURE 2. Limit on λ'_{131} as a function of the stop mass for different points in the SUSY parameter space. The Low energy limit from APV measurements is also shown (dashed line).

positron in the final state. Branching ratios for this channel vary between 30% and 70% depending on the chosen SUSY parameters. Signal events are characerised by high transverse energy, at least two jets and an electron candidate. As main selection criteria $E_T > 75$ GeV was required.

For a better signal to background separation a multivariate discriminant method was used [3] with the variables E_T, circularity, $E - P_z$, y_{JB}, and E_T^{jet} of the highest and second highest E_T jets. As no deviation from the SM was found in the signal region, the results were used to calculate limits in the M_2-μ plane (see Fig. 3) for $tan\beta = 30$, $M_{\tilde{q}}$= 1 TeV, $M_{\tilde{l}}$ = 100 GeV and $\lambda'_{111} = 1$. Existing limits from the ALEPH and DELPHI experiments [4, 5, 6] were improved by around 40GeV at high $|\mu|$.

4. SUMMARY AND OUTLOOK

Searches for R_P-violating SUSY have been performed by the ZEUS experiment at HERA. Exclusion limits have been derived which extend the existing limits for the MSSM.

Further studies will include the investigation of the decay channel with a neutralino in the final state for the gaugino analysis and the investigation of the neutralino decay in other SUSY models.

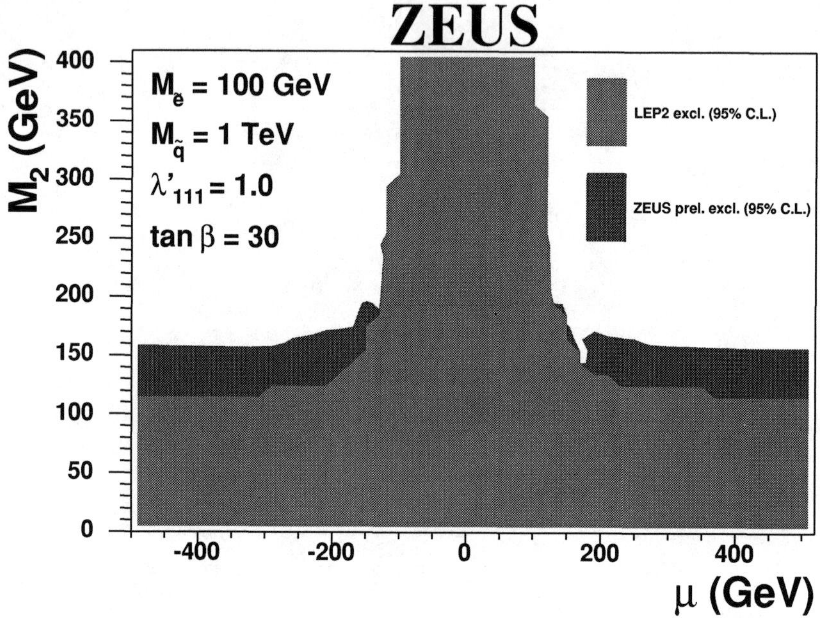

FIGURE 3. Regions in the μ-M_2 plane which can be excluded at 95% CL by the ALEPH and DELPHI experiments at LEP (lightly-shaded area) and by the gaugino analysis (darkly-shaped area). The ALEPH/DELPHI exclusion region is derived from the lower limit for the $\tilde{\chi}_1^{\pm}$ mass of 103 GeV [4, 5, 6].

REFERENCES

1. H1 Coll. A. Aktas et al. DESY-04-025, Accepted by Eur. Phys. J
2. H1 Coll., A. Aktas et al. DESY-04-084, Submitted to Phys. Lett. B
3. T. Carli and B. Koblitz, Nucl. Inst. Meth. **A 501**, 576 (2003)
4. DELPHI Coll., J. Abdallah et al., Eur. Phys. J. **C 36**, 1 (2004)
5. DELPHI Coll., J. Abdallah et al., Eur. Phys. J. **C 37**, 129 (2004)
6. ALEPH Coll., A. Heister et al., Eur. Phys. J. **C 31**, 1 (2003)

Search for Supersymmetry at the Tevatron DIS'05, Wisconsin, USA

John Zhou
on behalf of the CDF and DØ collaborations

P.O.Box 500
MS 318 (Rutgers University)
Fermi National Accelerator Laboratory
Batavia, IL 60510

Abstract. We report at the DIS'05 conference the latest results of search for supersymmetry (SUSY) at the Fermilab Tevatron. No evidence of Supersymmetry is found and limits are set accordingly.

Keywords: Supersymmetry, Tevatron, Run2, CDF, DØ
PACS: 12.60.Jv

Introduction

Tevatron Run2 has been taking data smoothly since 2003. The analyses reported here are based on 200–390 pb^{-1} of CDF and DØ data. They include results of Supersymmetry searches for R-parity conserving and violating scenarios under the framework of minimal supergravity (mSUGRA), gauge mediated SUSY breaking (GMSB), or anomaly mediated SUSY breaking (AMSB).

Search for $\widetilde{\chi}_1^{\pm}\widetilde{\chi}_2^0$ Pair Production via the Tri-lepton Final States

The chargino-neutralino ($\widetilde{\chi}_1^{\pm}\widetilde{\chi}_2^0$) pair decaying to 3 isolated leptons is consider the "golden" channel for SUSY search because there is little standard model (SM) background which is dominated by fakes and $W\gamma$. For the results in this section, the underlying framework is mSUGRA.

DØ carried out 6 analyses based on the final states listed in Table 1. After all the optimized cuts and a $p_T^{l3} > 4$ GeV cut on the third lepton are made, 4 data events are observed in 320 pb^{-1} of data with 3.85 ± 0.75 expected. Since no SUSY signal is observed, the results of the six channels are combined to set limits on the mass of the chargino: $m_{\widetilde{\chi}_1^{\pm}} > 117$ (132) GeV/c^2 for the *3l-max* (*heavy-squarks*) scenario at the 95% confidence level. The leptonic branching fraction is maximally enhanced in the *3l-max* scenario and the scalar mass unification is relaxed in the *heavy-squarks* scenario. These limits are beyond those by LEP although the LEP limits are model independent.

CP792, *Deep Inelastic Scattering, DIS 2005*, edited by W. H. Smith and S. R. Dasu

TABLE 1. Left: Number of expected background and observed events in 320 pb^{-1} of data in the six individual DØ tri-lepton analyses and in total. Right: The 95% confidence limits on $\sigma \times BR$ as a function of chargino mass.

Channel	Expected	Observed
$ee+l$	0.21 ± 0.12	0
$e\mu+l$	0.31 ± 0.13	0
$\mu\mu+l$	1.75 ± 0.57	2
$\mu^{\pm}\mu^{\pm}$	0.64 ± 0.38	1
$e\tau+l$	0.58 ± 0.14	0
$\mu\tau+l$	0.36 ± 0.11	1
All	3.85 ± 0.75	4

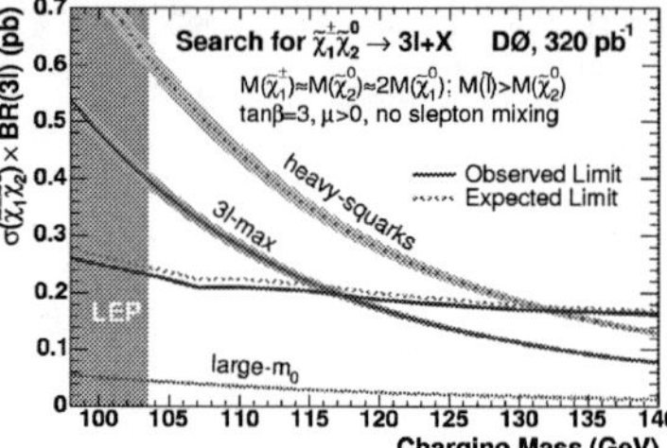

CDF also completed two analyses in the high and low p_T $ee+l$ channel using data from different electron triggers. The low p_T analysis includes hadronic τ contribution by allowing just an isolated track in the detector in addition to the two electrons. The luminosity used, the number of expected and observed data events are listed in Table. 2. Results in the $\mu\mu+l$ and $e\mu+l$ channels, and all channels combined are expected soon.

TABLE 2. Number of expected signal, background and observed events in 346 and 224 pb^{-1} of data in the two individual CDF tri-lepton analyses. The signal is mSUGRA with $m_0 = 100$ GeV, $m_{1/2} = 180$ GeV, $\tan\beta = 5$, $\mu > 0$, and $A_0 = 0$.

Channel	Luminosity	Expected Signal	Expected Background	Observed
$ee+l$ high p_T	346 pb^{-1}	0.5	0.16 ± 0.07	0
$ee+l$ low p_T	224 pb^{-1}	0.5	0.36 ± 0.27	0

Search for $\tilde{\chi}_1^{\pm}\tilde{\chi}_2^0$ Pair Production via the Diphoton Final States

In the GMSB model, the gravitino can be the lightest SUSY particle and the neutralino, $\tilde{\chi}_1^0$, can be the next lightest SUSY particle. A $\tilde{\chi}_1^{\pm}\tilde{\chi}_2^0$ pair production may result in diphoton + $\not\!\!E_T$ final state with $\tilde{\chi}_1^0$ decaying to a photon and a gravitino. CDF and DØ searched in this final state in 200 and 263 pb^{-1} of data, respectively. The analysis cuts, background expectation, and observation are listed in Table. 3. Background is dominated by $e\gamma$ and non-collision events. Combining the CDF and DØ results, we set a limit: $m_{\tilde{\chi}_1^{\pm}} > 209$ GeV/c^2 at the 95% confidence level as shown in the left plot in Fig. 1.

TABLE 3. Number of expected SM background and observed events in 263 and 202 pb^{-1} of data in the DØ and CDF diphoton + $\not\!\!E_T$ analyses.

Experiment	Luminosity	E_T^{γ} cut	$\not\!\!E_T$ cut	Expected Background	Observed
DØ	263 pb^{-1}	20 GeV	45 GeV	3.7 ± 0.6	2
CDF	202 pb^{-1}	13 GeV	40 GeV	0.3 ± 0.1	0

Search of Gluino-Squark in Hadronic Final States

If they exist, squarks ($\tilde{q}$) and gluinos ($\tilde{g}$) are copiously produced at the Tevatron because they couple strong to gluons and quarks. The typical signatures are hard jets and large $\not{E}_T$ resulted from chain decays of squarks and/or gluinos. The largest background come from QCD multijet events in which the jet energies are mis-measured giving rise to large $\not{E}_T$.

DØ divided the final states into 2, 3, and 4 jets + $\not{E}_T$, each optimized for $\tilde{q}\tilde{q}$, $\tilde{q}\tilde{g}$, and $\tilde{g}\tilde{g}$ pair production, respectively. The cuts on jet p_T, event $H_T = \sum p_T^{\text{jet}}$, and $\not{E}_T$, and the resulting expected background and observed number of events based on 310 pb^{-1} of data are listed in Table. 4. Since no evidence of squark or gluino production is found, a limit is set as a function of squark and gluino mass as shown in the middle plot in Fig. 1.

TABLE 4. Left: Cuts on the number of jets, p_T, event H_T, and $\not{E}_T$; the resulting expected number of background events and observed number of events in 310 pb^{-1} of data for the DØ analyses to search for squarks and gluinos.

# jets (p_T (GeV))	H_T (GeV)	$\not{E}_T$ (GeV)	Expected	Obs.
2 jets (60, 50)	250	175	12.8 ± 5.4	12
3 jets (60,40,25)	325	100	6.1 ± 3.1	5
4 jets (60,40,30,25)	175	75	9.3 ± 0.5	10

CDF also performed a similar analysis in a 3 jet + $\not{E}_T$ channel. The 3 jets are required to have $E_T > 125$, 75, and 25 GeV, respectively. In addition, the event must pass the cuts: $H_T = E_T^{j1} + E_T^{j2} + E_T^{j3} > 350$ GeV and $\not{E}_T > 165$ GeV. 4.2 ± 1.1 background events are expected with 3 events observed in 254 pb^{-1} of data. A limit on the masses of gluino and squark is being set.

Search for SUSY in the R-parity violating processes

DØ searched for slepton in the mSUGRA framework in four different channels. The coupling vertices, the corresponding final states considered, and the results are shown in Table. 5. In each analysis, only the coupling under investigation is assumed to dominate.

TABLE 5. Left: R-parity violating slepton search results. Right: the limit contour resulted from the $\mu\mu$+2jet analysis on the λ'_{211} coupling.

Coupling	λ'_{211}	λ_{121}	λ_{122}	λ_{133}
μ	< 0	> 0	> 0	> 0
$\tan\beta$	2	5	5	10
Channel	$\mu\mu$+2jets	$eel+\not{E}_T$	$\mu\mu l+\not{E}_T$	$ee\tau+\not{E}_T$
Lum (pb^{-1})	154	238	160	199
# Expected	1.1 ± 0.4	0.5 ± 0.4	0.6 ± 1.9	1.0 ± 1.4
# Observed	2	0	2	0
$m_{\tilde{\chi}_1^0}$ (GeV/c^2)	Fig.	>95	>90	>66
$m_{\tilde{\chi}_1^\pm}$ (GeV/c^2)	Fig.	>181	>165	>118

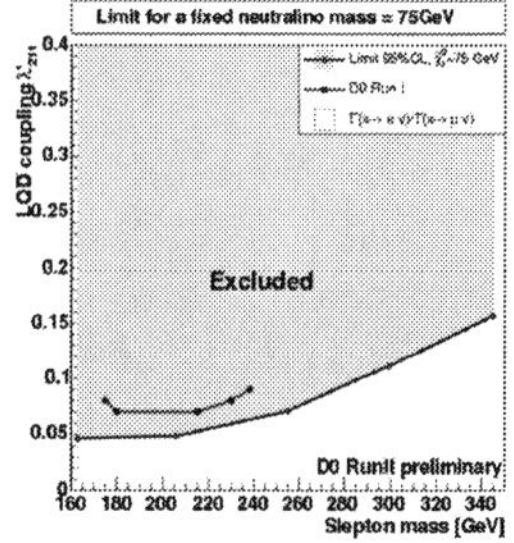

Using 200 pb^{-1} of data CDF investigated the coupling λ'_{333} in the process of stop pair production and decaying into 2 τ leptons and 2 b-jets. The mSUGRA framework and $BR(\tilde{t}_1 \to b\tau) = 100\%$ are assumed. It is required that one τ decays hadronically and the other decays leptonically. There is no b-tagging required in order to obtain maximal acceptances. After all optimal cuts we expect 4.8 ± 0.7 events and observe 5. From that we are able to extract a limit on $m_{\tilde{t}_1} > 129$ GeV/c^2, as shown in the right plot in Fig. 1.

Search for Charged Mass Particles (CHAMP) at DØ

Charginos in the AMSB models or the $\tilde{\tau}$ in the GMSB models may be long lived and decay outside of the detector ($c\tau > 10$m). They leave their trace as minimum-ionizing particles moving at a speed much less than the speed of light in the detector. Using 390 pb^{-1} of data, DØ searched for the pair production of CHAMPs in the dimuon channel. The muons are required to have $p_T > 15$ GeV and that they must have a large speed significance defined as: $S_v = (1 - \text{speed})/\sigma_{\text{speed}}$. After optimization for the signal in a 2D plane in di-muon mass and $S_v^1 \times S_v^2$, 0.66 ± 0.06 background events are expected and 0 is observed. A limit is set on the chargino masses in the framework of AMSB: 140 (174) GeV/c^2 if the chargino is higgsino (gaugino) like.

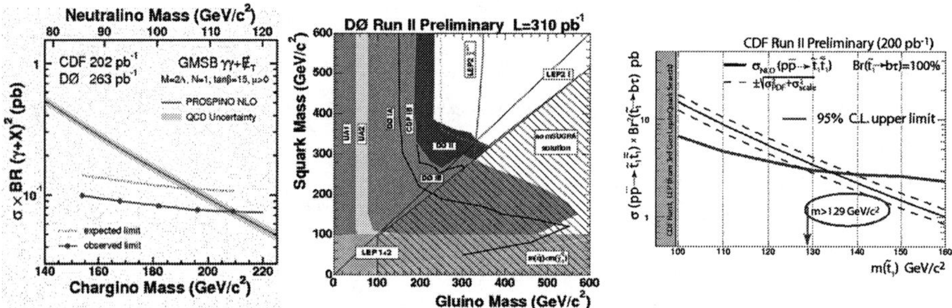

FIGURE 1. Left: combined CDF/DØ limit on $m_{\tilde{\chi}_1^\pm}$ in GMSB model in the diphoton+$\not\!\!E_T$ analyses. Middle: limit as a function of masses of squark and gluino from DØś jets+$\not\!\!E_T$ analysis. Right: limit on $m_{\tilde{t}}$ from CDF di-τ+di-jet analysis.

Conclusion

With no SUSY signal observed, limits on the SUSY model parameters have been placed. Some limits, for example, the masses of the gluino and squarks are already beyond the world's current limit. With delivered luminosity just surpassing 1 fb^{-1}, new limits and even discoveries will become a reality soon. References to the results reported here as well as the latest results can be obtained at `http://www-cdf.fnal.gov/physics/exotic/exotic.html` and `http://www-d0.fnal.gov/Run2Physics/WWW/results/np.html`.

SUSY Searches at the LHC

Nurcan Öztürk

Department of Physics, University of Texas at Arlington, Arlington, TX 76109, USA

Abstract. Supersymmetry is a strong candidate for theories beyond the Standard Model. Discovering supersymmetry is one of the motivations for building the LHC. I report on the searches for supersymmetry from inclusive and leptonic signatures within the mSUGRA framework at the LHC.

Keywords: Supersymmetry, LHC, mSUGRA, Data Challenges
PACS: 11.30.Pb, 12.60.Jv

INTRODUCTION

The Large Hadron Collider (LHC) is a 14-TeV proton-proton collider which is scheduled to start taking data in 2007 at CERN. Luminosity goals of the collider are 10 fb^{-1}/year for the first three years and 100 fb^{-1}/year subsequently. Supersymmetry (SUSY) will be explored primarily in the ATLAS (A Torodial LHC ApparatuS) and CMS (Compact Muon Solenoid) experiments as documented in Ref. [1] and [2] respectively.

Heavy strongly interacting SUSY particles (gluinos and squarks) are expected to be produced in the proton-proton collision. Due to long decay chains and large mass differences between SUSY particles, many high transverse momentum (p_T) objects; leptons, jets and b-jets will be observed. If R-parity is conserved, cascade decays of SUSY particles into the stable undetected LSP (lightest SUSY particle), which escapes the detector, will give a clear signature containing jets plus large missing energy (E_T^{miss}).

In the minimal Supergravity (mSUGRA) framework which is used as a standard benchmark model, SUSY particle masses and couplings are calculated by five parameters; m_0 and $m_{1/2}$ (universal scalar and gaugino mass parameters), A_0 (universal trilinear coupling), $\tan(\beta)$ (the ratio of the vacuum expectation values of the scalar fields) and $\text{sgn}(\mu)$ (the sign of the higgsino mass term). The sensitivity of the experiments to inclusive signatures is typically mapped in the m_0-$m_{1/2}$ parameter space for set values of the other parameters as shown in Figure 1. Multiple signatures for most of the parameter space are available; E_T^{miss} (dominant signature), E_T^{miss} with lepton veto, with one lepton, with two same sign (SS) leptons, and with two opposite sign (OS) leptons.

DATA CHALLENGE ACTIVITIES

Data challenge (DC) activities aim to provide simulated data to optimize the detectors and to validate the Computing model. Analyzing SUSY events in the simulated data is important to test and evaluate the reconstruction software since typical SUSY events contain the complete set of physics objects that can be reconstructed in the detector. Several mSUGRA points that are favored by the WMAP data (as discussed in [3]) have

CP792, *Deep Inelastic Scattering, DIS 2005,* edited by W. H. Smith and S. R. Dasu

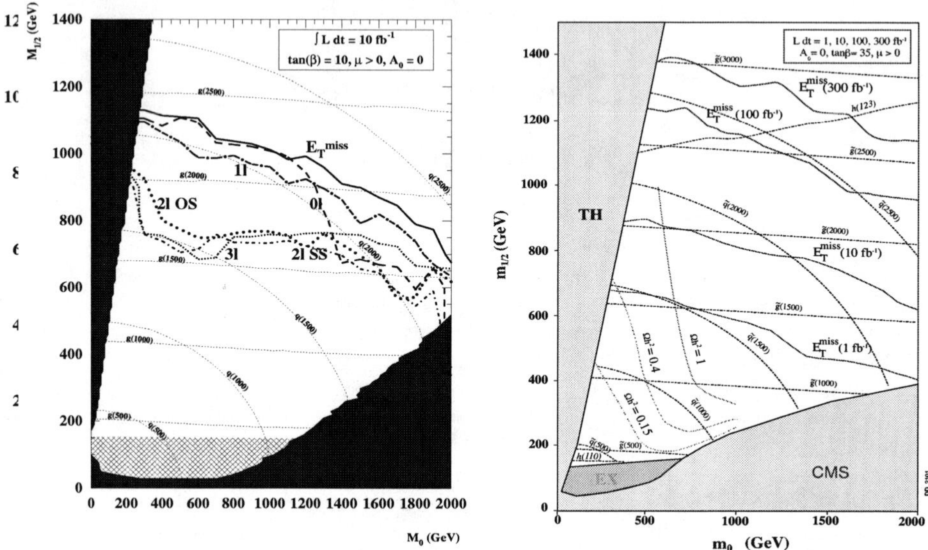

FIGURE 1. Maps of discovery potential corresponding to a 5σ excess above background in the mSUGRA parameter space for the ATLAS (left plot) and CMS (right plot) experiments.

been chosen for data production as summarized in Table 1. Similar studies have been carried out in CMS Data Challenges [4].

TABLE 1. mSUGRA points chosen for data production in ATLAS Data Challenges

ATLAS Data Challenges	mSUGRA Point m_0 (GeV), $m_{1/2}$ (GeV), A_0 (GeV), tan(β), sgn(μ)	Cross Section	Events Produced
DC1 July 2002-March 2003	100, 300, -300, 6, + (bulk region point)	19.3 pb	100k
DC2 June-December 2004	100, 300, -300, 6, + (bulk region point)	19.3 pb	50k
	70, 350, 0, 10, + (coannihilation region point)	6.8 pb	200k
Rome Production January-June 2005	100, 300, -300, 6, + (bulk region point)	19.3 pb	100k
	70, 350, 0, 10, + (coannihilation region point)	6.8 pb	200k
	3550, 300, 0, 10, + (focus region point)	4.9 pb	100k
	320, 375, 0, 50, + (funnel region point)	4.5 pb	100k
	200, 160, -400, 10, + (low mass point)	280 pb	100k
	130, 600, 0, 10, + (scan point in parameter space)	0.4 pb	10k
	250, 600, 0, 10, + (scan point in parameter space)	0.4 pb	10k
	500, 600, 0, 10, + (scan point in parameter space)	0.3 pb	10k

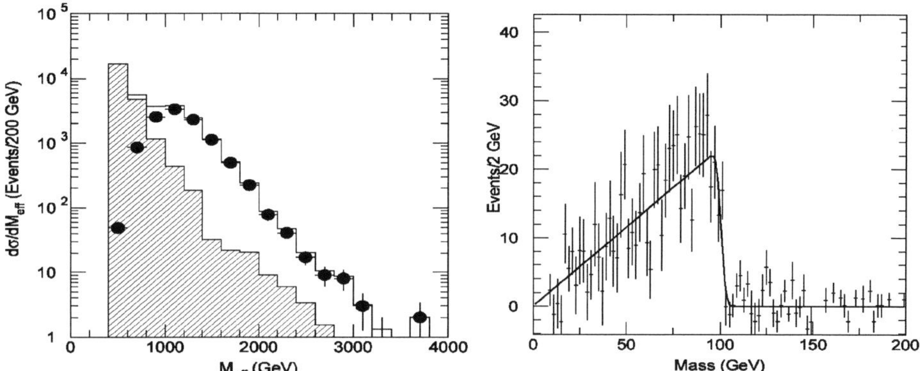

FIGURE 2. Results from the DC1 mSUGRA point for an integrated luminosity of 5.13 fb^{-1}. Left plot: The effective mass distribution; SUSY signal events (circles), Standart Model background events (hatched histogram), and the sum of signal and background (histogram). The background events were produced by fast simulation. Right plot: The dilepton invariant mass distribution ($e^+e^- + \mu^+\mu^- - e^\pm\mu^\mp$), solid line shows a triangular function convoluted with a Gaussian to estimate the edge position.

FULL SIMULATION RESULTS FROM DATA CHALLENGES

Inclusive searches for SUSY will be one of the most important task for ATLAS and CMS experiments during the first few years of operation. The SUSY mass scale, defined as the cross section weighted mean of the masses of the two SUSY particles initially produced, is the first parameter likely to be measured using jets plus E_T^{miss} signature. The effective mass variable defined as $M^{\text{eff}} = E_T^{\text{miss}} + \sum_{\text{jet}} E_{T,\text{jet}}$ peaks at twice the SUSY mass scale (see [5]). Events are selected by requiring at least four jets with $E_T > 100, 100, 50, 50$ GeV, $E_T^{\text{miss}} > \max(100\text{ GeV}, 0.20\,M^{\text{eff}})$ and no muons or isolated electrons in $|\eta| < 2.5$. The left plot in Figure 2 shows the M^{eff} distribution for the DC1 mSUGRA point defined in Table 1: it peaks around 1120 GeV which is consistent with twice the expected SUSY mass scale of 590 GeV.

If R-parity is conserved, all SUSY events contain two invisible neutralinos ($\tilde{\chi}_1^0$) which escape the detector, therefore no mass peaks can be reconstructed directly and the kinematic endpoint methods are used. For instance from the decay chain of $\tilde{\chi}_2^0 \to \tilde{l}_R^\mp l^\pm \to \tilde{\chi}_1^0 l^\mp l^\pm$ the mass difference between $\tilde{\chi}_2^0$ and $\tilde{\chi}_1^0$ is measured from the endpoint of the dilepton invariant mass distribution. The background from $t\bar{t}$ and $\tilde{\chi}^\pm$ decays cancel in the combination $e^+e^- + \mu^+\mu^- - e^\pm\mu^\mp$. Events are selected by requiring electrons and muons with $p_T \geq 10$ GeV and separating leptons from jets by $\Delta\text{R} > 0.4$. Reconstruction efficiencies for electrons and muons are taken into account, resulting in the distribution shown in the right plot of Figure 2. The endpoint fit gives a value of 100.25±1.14 GeV which is consistent with the expected value of 100.31 GeV within the error.

The mass of the right-handed squark can also be determined by using the endpoint technique. From the decay of $\tilde{q}_R \to \tilde{\chi}_1^0 q$ the signal events for $\tilde{q}_R$ contain two hard jets plus large E_T^{miss}. The stransverse mass (defined in [6]) distribution (using the mass of

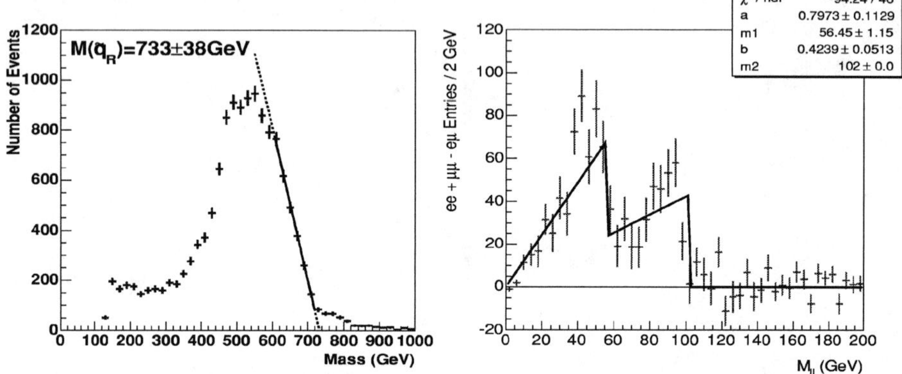

FIGURE 3. Preliminary results from the mSUGRA point in the coannihilation region from Rome production for an integrated luminosity of 20.6 fb^{-1}. Left plot: The stransverse mass distribution. Right plot: The dilepton invariant mass distribution ($e^+e^- + \mu^+\mu^- - e^\pm\mu^\mp$), solid line shows a triangular function fit to estimate the two edge positions.

$\widetilde{\chi}_1^0$ as input) gives an endpoint at the mass of $\widetilde{q}_R$. The left plot of Figure 3 shows this distribution for the events produced in Rome production from the mSUGRA point in the coannihilation region in Table 1. Events are selected to have two jets with $p_T > 200$ GeV and $E_T^{\text{miss}} > 400$ GeV. The endpoint fit gives a value of 733±38 GeV which is consistent with the expected value of 735 GeV within the error. The right plot of Figure 3 shows the dilepton invariant distribution from the same mSUGRA point, two mass edges are clearly seen as expected at 58.2 GeV and 100.9 GeV from the decays of $\widetilde{\chi}_2^0 \to \widetilde{l}_L^\pm l^\mp \to \widetilde{\chi}_1^0 l^\pm l^\mp$ and $\widetilde{\chi}_2^0 \to \widetilde{l}_R^\pm l^\mp \to \widetilde{\chi}_1^0 l^\pm l^\mp$ respectively.

Detailed analysis of the mSUGRA points from DC1 and Rome Production including Standard Model backgrounds, inclusive, leptonic and tau signatures as well as b-jet signatures can be found in [7]. These studies show a rich discovery potential for SUSY at the LHC.

REFERENCES

1. ATLAS Collaboration, *ATLAS TDR*, CERN/LHCC/99-14, `http://atlasinfo.cern.ch/Atlas/GROUPS/PHYSICS/TDR/access.html`.
2. S. Abdullin et al., [CMS Collaboration], "Discovery Potential for Supersymmetry in CMS," *J. Phys.*, **G28**, 469 (2002), arXiv:hep-ph/9806366.
3. Talk by H. Baer, in this proceedings.
4. URL `http://cmsdoc.cern.ch/cms/PRS/susybsm/`.
5. D. R. Tovey, *Eur. Phys. J. directC* , **4**, N4 (2002).
6. A. Barr et al., *J. Phys.*, **G29**, 2343 (2003).
7. M. Biglietti et al., *ATLAS internal note*, ATL-PHYS-2004-011. SUSY talks at the Rome ATLAS Physics Workshop, URL `http://agenda.cern.ch/fullAgenda.php?ida=a044738`.

Multi-Lepton Events at H1 and Search for Doubly-Charged Higgs Bosons

André Schöning

ETH Hönggerberg Zürich, Switzerland

Abstract.
Events with two or more leptons (electrons or muons) with high transverse momentum are measured in electron-proton collisions at HERA using the data sample collected in the period 1994-2005. Multi-lepton events at high transverse momenta are of special interest as these signature might reveal new physics beyond the Standard Model. An example is the single production of doubly-charged Higgs bosons $H_{L,R}^{\pm\pm}$, which couple to leptons of the i'th and j'th generation via Yukawa couplings $h_{ij}^{L,R}$. Results from a search for doubly-charged Higgs bosons in the decays into electrons, muons and taus are presented using data taken in the period 1994-2000. No evidence for doubly-charged Higgs production is found and we derive limits on the $h_{ee}^{L,R}$ and $h_{e\mu}^{L,R}$ Yukawa couplings as a function of the $H_{L,R}^{\pm\pm}$ mass.

Keywords: Higgs, Doubly-Charged Higgs, Leptons, H1, HERA
PACS: 12.60.Fr; 13.60.r; 13.85.Rm; 14.80.Cp

INTRODUCTION

The H1 experiment has reported an excess of di-electron and tri-electron events with invariant masses above 100 GeV observed in ep collisions at HERA I in the period 1994-2000 [1]. In this talk results from a recent multi-lepton search at high transverse momentum are presented using the HERA I+II data samples collected in 1994-2005.

Multi-lepton events at high transverse momenta are of special interest as this signature might reveal new physics beyond the Standard Model. An example is the single production of doubly-charged Higgs bosons $H_{L,R}^{\pm\pm}$, which couples to leptons of the i'th and j'th generation via Yukawa couplings $h_{ij}^{L,R}$.

Doubly-charged Higgs bosons ($H^{\pm\pm}$) appear naturally in various extensions of the Standard Model (SM) in which the usual Higgs sector is extended by one or more triplet(s) [2, 3]. Examples are provided by some Left-Right Symmetric (LRS) models [4, 5] with a spontaneously broken $SU(2)_L \times SU(2)_R \times U(1)_{B-L}$ symmetry. These models are of particular interest as Higgs triplets can give Majorana masses to neutrinos, which are known to be massive from recent experimental data. Results from a search for doubly-charged Higgs bosons in the decays into electrons, muons and taus are presented using data taken in the period 1994-2000.

CP792, *Deep Inelastic Scattering, DIS 2005*, edited by W. H. Smith and S. R. Dasu

TABLE 1. H1 preliminary event yields for 1994-2004 $e^{\pm}p$ data (L=163 pb^{-1}) of the different multi-lepton channels for $M_{ll'} > 100\ GeV$. The numbers are compared with the expectations from the Standard Model and from pair production only.

channel	data events	Standard Model	Pair Production
ee	3	0.4 ± 0.1	0.32
$\mu\mu$	0	0.04 ± 0.02	0.04
$e\mu$	0	0.31 ± 0.03	0.31
eee	3	0.31 ± 0.08	0.31
$e\mu\mu\ (M_{e\mu})$	1	0.04 ± 0.01	0.04
$e\mu\mu\ (M_{\mu\mu})$	1	0.02 ± 0.01	0.02

EVENTS WITH MULTI-LEPTONS

Events with clearly identified isolated electrons and muons at high transverse momentum (p_T) are selected and classified into the following final state topologies: ee, eee, $e\mu$, $\mu\mu$, $e\mu\mu$, etc.). Within the Standard Model such events are mainly produced via photon-photon collisions. The invariant mass distribution of the two highest p_T electrons are shown for the ee and eee classes in Fig. 1 using 163 pb^{-1} of HERA I+II data. The event yields for all classes after the cut $M_{\ell\ell'} > 100$ GeV are shown in Tab. 1. An excess of data events over the SM expectation is seen in the ee and eee classes.

For some event classes the same analysis was repeated using the most recent e^-p dataset from 2004/05 (21 pb^{-1}), see Fig. 1. No new event was found for $M_{\ell\ell'} > 100$ GeV in the analysed $ee, e\mu, eee, e\mu\mu$ event classes in agreement with the SM expectation.

DOUBLY-CHARGED HIGGS PRODUCTION

The search for single doubly-charged Higgs production is based on studies of dilepton production ee, $\mu\mu$ [1, 6], and new multi-lepton searches for the $e\mu$ and $\tau\tau$ final states at high p_T. The analyses use data collected at HERA I from 1994-2000 (ee, $\mu\mu$ and $e\mu$) and 1999/2000 ($\tau\tau$). The selections involving electrons and muons are identical to those described in the previous section. Tau pairs are searched for taking into account leptonic and hadronic τ-decays in the event classes $\tau\tau \to e\mu, ej, \mu j, jj$ plus missing energy from the unobserved tau neutrinos. The invariant mass of the $H^{\pm\pm}$ is fitted by exploiting energy and momentum conservation resulting in a resolution of $2.5 - 4.0$ GeV.

Further $H^{\pm\pm}$ selection criteria are applied to all multi-lepton final states. Events where the charges of the highest p_T leptons are measured to be inconsistent with the $H^{\pm\pm}$ hypothesis are rejected. For the $H^{\pm\pm} \to e^{\pm}e^{\pm}$ decay both electrons are required to have large transverse energies in the calorimeter. After the final selections only one candidate event ($H^{\pm\pm} \to e^{\pm}e^{\pm}$) is found for $M_H > 100$ GeV. Various limits are derived as function of M_H at 95% CL on the product $\sigma \times$ BR for each decay topology, see Fig 2 left, and on the coupling $h_{ee}^{L,R}$ assuming a model with democratic couplings BR$(H^{\pm\pm} \to l^{\pm}l^{\pm}) = 1/3$ (right). In Fig. 3 exclusion limits on $h_{ee}^{L,R}$ (left) and $h_{ee}^{L,R}$ (right) are shown as function of M_H assuming a decay branching ratio of 100% for each case. For comparison direct and indirect limits from other experiments are shown [7, 8, 9].

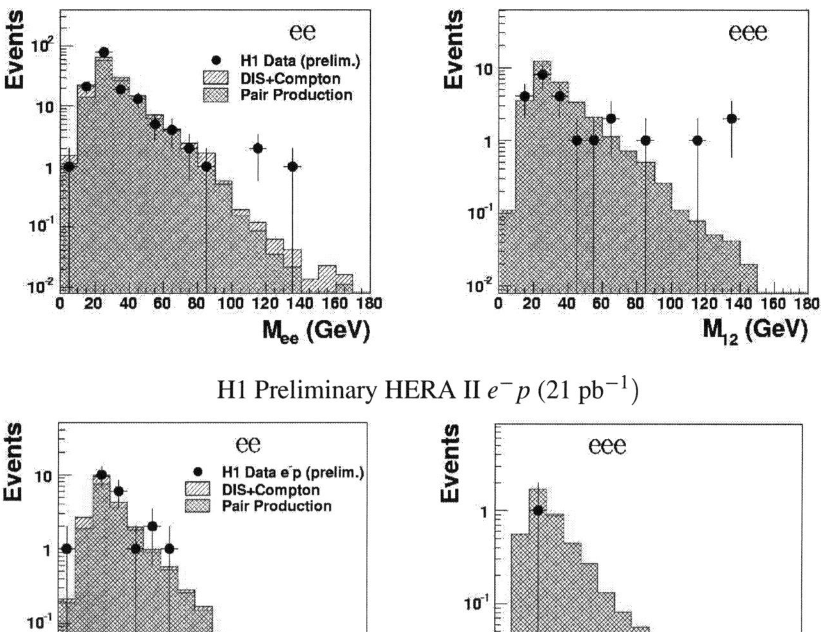

FIGURE 1. Invariant masses M of the two highest p_T leptons compared to expectations for ee and eee classifi ed event. The data samples correspond to an integrated luminosity of 163 pb^{-1} (1994+2004, top fi gures) and to 21 pb^{-1} of the recent e^-p data taken in 2004/05 (bottom).

CONCLUSION

Multi-lepton events with electrons and muons are studied using HERA I+II data. No new di-electron and tri-electron events are observed with invariant masses above 100 GeV in the most recent e^-p dataset. More integrated luminosity is required to enhance the sensitivity at high masses. No evidence for doubly-charged Higgs production is found and various limits on $H^{\pm\pm}_{L,R}$ Yukawa couplings $h^{L,R}_{\ell\ell'}$ were derived.

REFERENCES

1. A. Aktas *et al.* [H1 Collaboration], *Eur. Phys. J. C* **31** (2003) 17 [hep-ex/0307015].
2. G. B. Gelmini and M. Roncadelli, *Phys. Lett. B* **99** (1981) 411.
3. J. C. Pati and A. Salam, *Phys. Rev. D* **10** (1974) 275; R. E. Marshak and R. N. Mohapatra, *Phys. Lett. B* **91** (1980) 222; R. N. Mohapatra and G. Senjanovic, *Phys. Rev. Lett.* **44** (1980) 912.

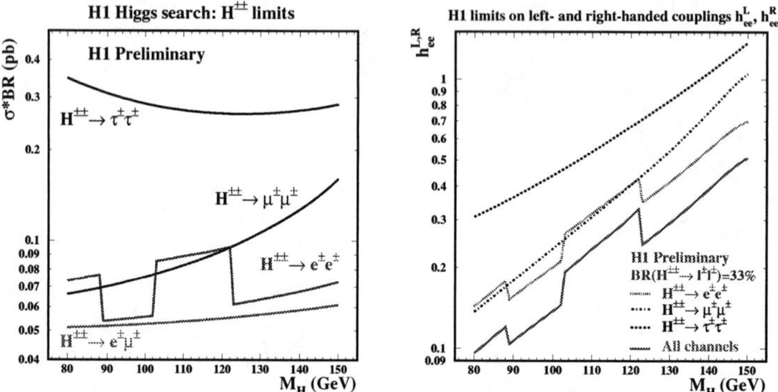

FIGURE 2. Left: upper limits at 95% CL on $\sigma(e^{\pm}p \to e^{\mp}H^{\pm\pm}X) \times \text{BR}(H^{\pm\pm} \to \ell^{\pm}\ell'^{\pm})$ as a function of the doubly-charged Higgs mass. Right: exclusion limits on the coupling $h_{ee}^{L,R}$ at 95% CL as function of M_H for the decay channels $H^{\pm\pm} \to e^{\pm}e^{\pm}, \mu^{\pm}\mu^{\pm}, \tau^{\pm}\tau^{\pm}$ and their combination (full curve) assuming democratic couplings, ie. $\text{BR}(H^{\pm\pm} \to \ell^{\pm}\ell^{\pm}) = 1/3$.

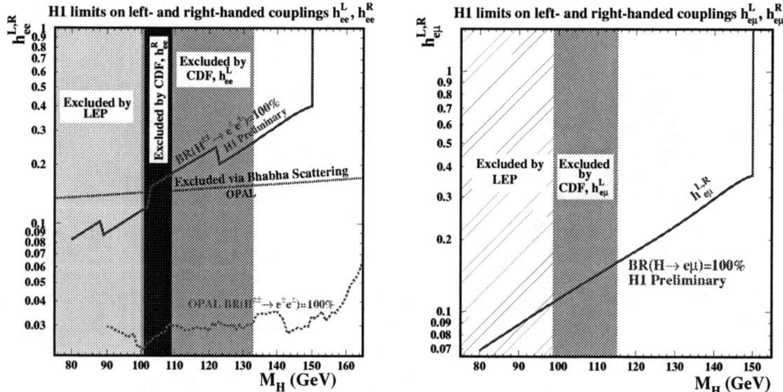

FIGURE 3. Left: exclusion limits on the coupling $h_{ee}^{L,R}$ at 95% CL as a function of the doubly-charged Higgs mass for the decay $H^{\pm\pm} \to e^{\pm}e^{\pm}$ assuming $\text{BR}(H^{\pm\pm} \to e^{\pm}e^{\pm}) = 1$. Right: exclusion limits on the coupling $h_{e\mu}^{L,R}$ at 95% CL as a function of M_H for the decay $H^{\pm\pm} \to e^{\pm}\mu^{\pm}$ assuming $\text{BR}(H^{\pm\pm} \to e^{\pm}\mu^{\pm}) = 1$. The results are compared to direct and indirect limits (Bhabba scattering) obtained by OPAL (single production), and LEP and CDF (pair production).

4. G. Senjanovic and R. N. Mohapatra, *Phys. Rev. D* **12** (1975) 1502.
5. R. N. Mohapatra and R. E. Marshak, *Phys. Rev. Lett.* **44** (1980) 1316 [Erratum-ibid. **44** (1980) 1643].
6. A. Aktas *et al.* [H1 Collaboration], *Phys. Lett. B* **583** (2004) 28 [hep-ex/0311015].
7. G. Abbiendi *et al.* [OPAL Collaboration], *Phys. Lett. B* **577** (2003) 93 [hep-ex/0308052].
8. J. Abdallah *et al.* [DELPHI Collaboration], *Phys. Lett. B* **552** (2003) 127 [hep-ex/0303026];
P. Achard *et al.* [L3 Collaboration], *Phys. Lett. B* **576** (2003) 18 [hep-ex/0309076];
G. Abbiendi *et al.* [OPAL Collaboration], *Phys. Lett. B* **526** (2002) 221 [hep-ex/0111059].
9. D. Acosta *et al.* [CDF Collaboration], *Phys. Rev. Lett.* **93** (2004) 2218802 [hep-ex/0406073].

Higgs Searches at the Tevatron

Carsten Hensel

University of Kansas, Lawrence, Kansas 66045

Abstract. We report on Higgs searches performed by the CDF and DØ collaboration in Tevatron RunII p$\bar{\text{p}}$ collisions at $\sqrt{s} = 1.96\,\text{TeV}$. No Higgs signal has been observed and limits on the Standard Model Higgs and on Higgs bosons in models beyond the Standard Model have been derived.

Keywords: Tevatron, CDF, DØ, Electroweak, Higgs, MSSM
PACS: 13.85.-t, 14.80.Bn, 14.80.Cp

INTRODUCTION

The Standard Model (SM) of elecroweak interactions is extremely successful in describing interactions of elementary particles. One of the key elements of the SM is the Higgs boson. In interactions with other particles, particle masses are generated and the theory is kept renormalizable at the electroweak scale. The Higgs boson is the remaining particle of the SM particle zoo yet to be experimentally observed. The discovery of the Higgs boson will provide further insight into the electroweak symmetry breaking mechanism and thus is one of the most important goals in high energy physics.

In this note we review some of the most recent Higgs search results obtained by the CDF and DØ collaborations. The note is divided in two parts. The first one focuses on results of searches for the SM Higgs boson. In most theories beyond the SM the potential for a Higgs boson discovery is much larger. The second part covers examples for the search for the Higgs boson in the minimal supersymmetric extension of the Standard Model (MSSM).

SM HIGGS SEARCHES

While the LEP experiments currently set the most stringent lower limit on the SM Higgs mass of $\text{M}_\text{H} > 114.4\,\text{GeV}$ at 95% CL[1], electroweak fits favor a light SM Higgs boson at $\text{M}_\text{H} = 126^{+73}_{-48}\,\text{GeV}$. As shown in Figure 1 the expected production cross-section in this mass window is rather small and pose a real challenge to the experimental search of the Higgs boson. The dominant production mechanism in p$\bar{\text{p}}$ collisions is the gluon fusion ($\text{gg} \rightarrow \text{H}$). However, the irreducible QCD background ($\text{gg}, \text{q}\bar{\text{q}} \rightarrow \text{b}\bar{\text{b}}$) is too large to allow a direct observation in this channel. Therefor processes in which the Higgs boson is produced in association with an additional vector boson are more promising.

For Higgs boson masses in the range of 100 - 140 GeV the cross-section $\sigma(\text{p}\bar{\text{p}} \rightarrow \text{WH})$ varies from 0.3 to 0.1 pb, and high-p_T leptons from W decays allow for an efficient trigger. The CDF collaboration searched for the associate WH production requiring at least one lepton with $p_{\text{T},1} > 20\text{GeV}$ and a second one with $p_{\text{T},2} > 16\text{GeV}$. A dataset

CP792, *Deep Inelastic Scattering, DIS 2005*, edited by W. H. Smith and S. R. Dasu

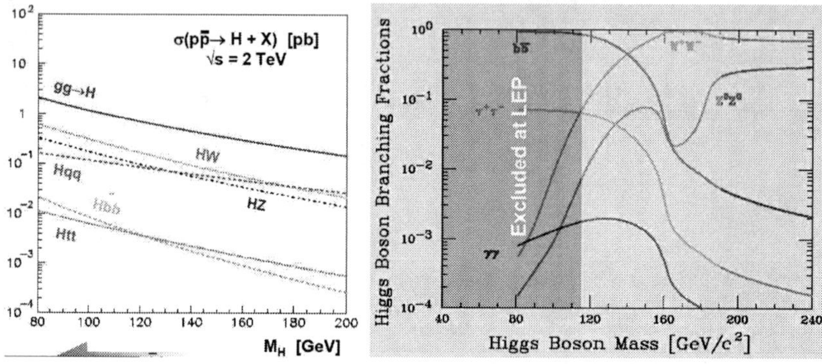

FIGURE 1. Left: Production cross-section in $p\bar{p}$ collisions of the SM Higgs boson as a function of the Higgs bsoson mass. Right: SM Higgs branching ratios as a function of the Higgs boson mass.

corresponding to $\mathscr{L}_{\text{int}} = 193.5\text{pb}^{-1}$ was analyzed. While a total of 0.95 background events were expected, 0 candidates were observed. The derived upper limits on the Higgs branching ratio times cross-section are summarized in Figure 2.

DØ shows in Tevatron's first RunII search for ZH production that also the Z boson provides a clean and simple signature. Assuming the Z boson decays into thwo neutrinos, signal events can be tagged by missing transverse energy. The actual Higgs search is then performed in the dijet mass spectrum of the two b-jets. In a dataset of $\mathscr{L}_{\text{int}} = 261\text{pb}^{-1}$ no excess over the SM background was observed and the DØ collaboration sets the upper limits on the production cross-section of ZH as shown in Figure 3.

As these examples show the searches for the SM Higgs boson at Tevatron are not yet competitive with the LEP results. But with only a fraction of the collected data analyzed and more data yet to come the CDF and DØ collaborations have a good starting position for the search for the Higgs boson. As described in the the following section theories

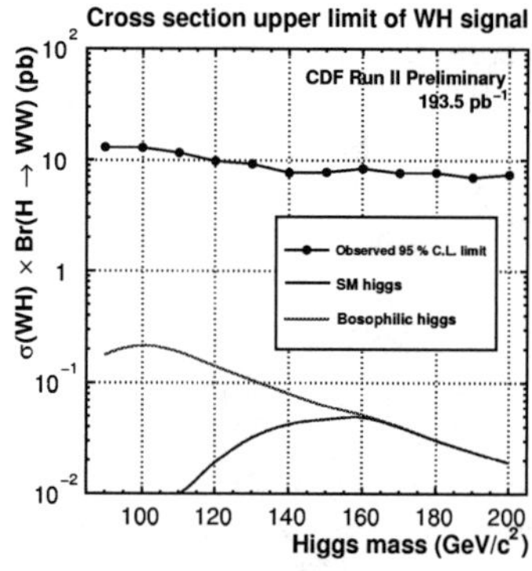

FIGURE 2. Cross-section of WH production times branching ratio of Higgs into W W* upper limit at 95% CL.

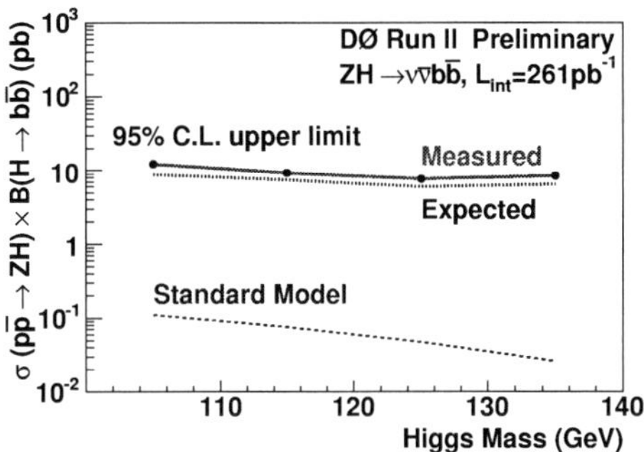

FIGURE 3. 95% CL upper limits on the cross-section of ZH production times the branching ratio for $H \rightarrow b\bar{b}$

beyond the SM like the MSSM currently offer a much higher discovery potential at the Tevatron.

MSSM HIGGS SEARCHES

One possible extension to the SM is the introduction of an additional doublet of scalar fields in the Higgs sector. In the framework of the MSSM this results in five Higgs boson mass eigenstates - three neutral (h/H/A) and two charged ($H^{\pm}$). The MSSM allows for larger couplings of fermions to the MSSM Higgs bosons and thus puts the model within the current reach of the Tevatron RunII experiments.

CDF reports results on a search for neutral MSSM Higgs bosons assuming the Higgs subsequently deacys into a pair of τ leptons. In a dataset corresponding to an integrated Luminosity of $310\,\text{pb}^{-1}$. No significant excess of events above the SM backgrounds was observed. The measurements are used to set exclusion limits on production cross-section times branching fraction to τ pairs for Higgs masses in the range from 115 to 250 GeV. The observed and expected upper limits are shown in Figure 4

That the Tevatron collaborations are already able to significantly contribute to the reduction of the MSSM phase space is illustrated in Figure 5. Based on the DØ results for the search for the neutral MSSM Higgs boson in the channel $p\bar{p} \rightarrow Hb\bar{b} \rightarrow b\bar{b}b\bar{b}$ the plot shows the 95% CL exclusion region in the $\tan\beta$- M_A plane.

SUMMARY

The search for the Higgs boson is one of the main scientifical goals of the Tevatron experiments. The start of the RunII phase increased the discovery potential significantly. With more and more data collected the CDF and DØ collaborations are approaching quickly the sensitivity region for a SM Higgs discovery. The efficiency of the their analyzes could already be proven in non-SM Higgs searches.

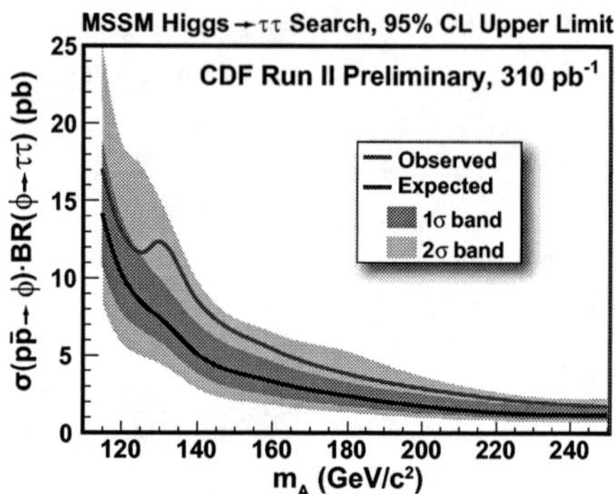

FIGURE 4. Observed and expected upper limits at 95% CL for Higgs production cross-section times branching fraction into τ pairs.

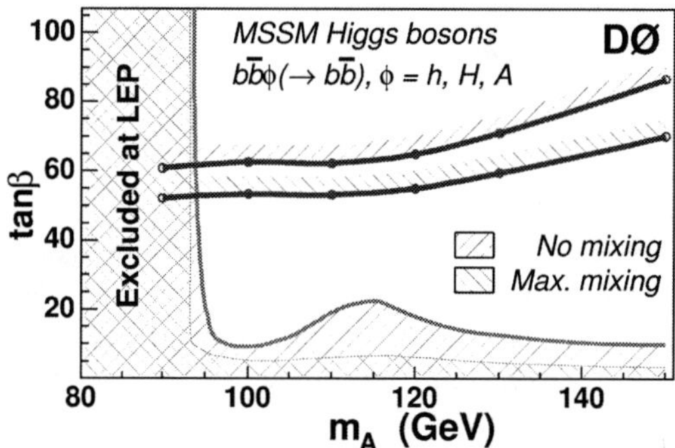

FIGURE 5. The 95% CL upper limit on $\tan\beta$ as a function of M_A for two scenarios of the MSSM, "no mixing" and "maximal mixing". Also shown are the limits obtained by the LEP experiments for the same two scenarios of the MSSM[2]

ACKNOWLEDGMENTS

I would like to thank the CDF and DØ collaborations for providing me with their results. I thank the organizers of DIS'05 for their hospitality and extremely well organized conference.

REFERENCES

1. R. Barate, et al., *Phys. Lett.*, **B565**, 61–75 (2003), `hep-ex/0306033`.
2. The Lep Working Group for Higgs Boson Searches (2004), `LHWG-Note2004-01`.

Higgs Searches at the LHC

M. Escalier

LPNHE, Paris, France

Abstract. The search for the Higgs boson, introduced to give a dynamical mass to Standard Model particles, is one of the crucial tasks of LHC experiments. This proceeding reviews and evaluates the potential discovery of the main channels for the Standard Model Higgs boson at this accelerator.

Keywords: Higgs, LHC, Standard Model
PACS: 14.80.Bn

INTRODUCTION

Over the last 30 years, the Standard Model has been very successful to describe particle physics. However, one of its key ingredient, the Higgs boson [1] has not yet been discovered. The Large Hadron Collider (LHC), expected to start operations in summer 2007, will play an important role in his discovery by colliding two 7 TeV proton beams every 25 ns. Two optimized detectors, ATLAS (A Toroidal LHC ApparatuS) [2], and CMS (Compact Muon Solenoid) [3], are expected to cover all the Higgs mass range from 80 GeV to 1 TeV, larger than the allowed range expected by the theoretical and experimental constraints of previous experiments [4, 5]. After a 3-year luminosity period equivalent to 10 fb^{-1} of data taking per year, the design luminosity is expected to reach 100 fb^{-1} per year. While the mode of production (Fig. 1 left) of the Higgs boson keeps a hierarchy over the whole range of mass, dominated by the gluons fusion and vector boson fusion, seconded by associated production with a W, Z or heavy quarks, the branching ratio changes as a function of the Higgs mass (Fig. 1 right).

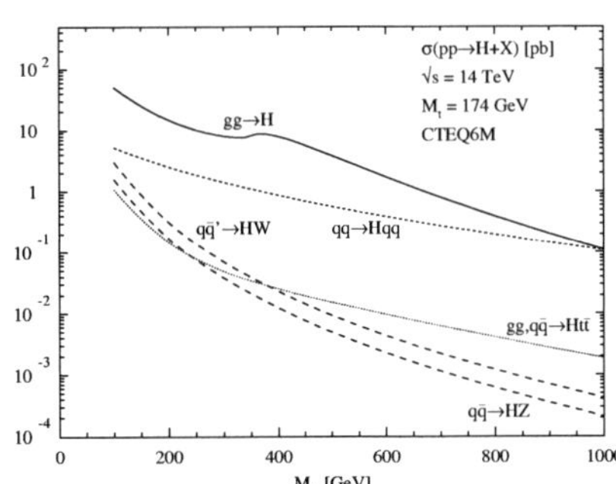

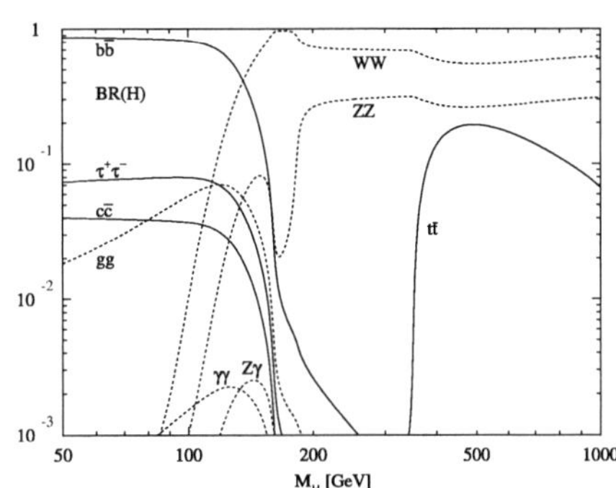

FIGURE 1. Production cross sections at NLO (left panel [6]) and branching ratios (right panel [6]) for the main channels of the Standard Model Higgs boson at LHC as a function of its mass.

CP792, *Deep Inelastic Scattering, DIS 2005,* edited by W. H. Smith and S. R. Dasu

LOW MASSES RANGE

In the low masses region up to 140 GeV, although the $H \rightarrow b\bar{b}$ favoured branching ratio may not be considered in an inclusive mode because of the huge QCD background, its decay in $t\bar{t}H$ production allows to increase the sensitivity. This associated production is the most promising channel when requesting $t \rightarrow W(\rightarrow l\nu)b$ and $t \rightarrow W(\rightarrow q\bar{q}')b$ decays respectively for the two top quarks. The lepton decays is the key for the trigger. This channel with at least 6 jets needs a good (60 % efficiency) b-tagging capability of the inner detector and a pairing likelihood to reduce the combinatorics from the b quarks. The other top quark is identified by reconstructing the neutrino impulsion from m_W. With these strategies to reduce the $t\bar{t}jj$ background, the significance would be of about 3 for a 120 GeV Higgs boson with an integrated luminosity of 30 fb^{-1}, and up to 2 σ in one year [7].

Despite its low branching ratio, $H \rightarrow \gamma\gamma$ is an important channel for low Higgs masses between 80 and 150 GeV with a signal on background ratio of 1/20. This requires an excellent energy and angular resolutions in order to extract the signal from the irreducible $\gamma\gamma$ continuum and the large γ-jet and jet-jet reducible background. This motivates the choice for LAr (ATLAS) and $PbWO_4$ (CMS) electromagnetic calorimeters. The Standard Model Higgs boson may be discovered (left panel of Fig. 2 is at Leading Order only) in this channel with an integrated luminosity of 30 fb^{-1} up to 140 GeV [8], [9].

An another interesting channel in the region $110 < m_H < 140$ GeV is the $H \rightarrow \tau\tau$ decay in vector boson fusion production, in association with two energetic forward jets. Two decays of the τ may be considered : the double leptonic decay mode $qqH \rightarrow qq\ l\nu\nu\ l\nu\nu$ and the lepton-hadron decay mode $qqH \rightarrow qq\ l\nu\nu\ had\nu$. Because the H boson is produced with an important transverse momentum, the τ decay products are emitted nearly collinear in the laboratory frame so that the collinear approximation is considered in the reconstruction of the $\tau\tau$ invariant mass. The dominant background is the $Z+$jet production with $Z \rightarrow \tau\tau$. The absence of colorflow between the partons for the signal[1] gives rise to suppressed jet production in the central region. Forward jet tagging and central jet veto are requested to reject the background. The resolution on the missing transverse momentum will be crucial for the Higgs resolution. A significance of about 2.5 σ can be reached during the first year of the LHC. A integrated luminosity of about 30 fb^{-1} is needed for the discovery [10], and allows a direct measurement of the $H\tau\tau$ Yukawa coupling.

INTERMEDIATE AND HIGH MASS RANGES

In the range $120 < m_H < 700$ GeV, the so-called gold-plated channel[2] $H \rightarrow ZZ(^*) \rightarrow 4l$, made of $4e$, 4μ and $2e2\mu$ decays, provides a clean signature. The ZZ irreducible back-

[1] On the contrary the t-channel contribution of the background has a color flow.

[2] The two real Z bosons channel opens up above 2 m_Z.

ground is made of four leptons decays and of a smaller (10 %) contribution from one $Z \to l\bar{l}$ decay and from the second decaying into a pair of τ, each of them decaying into a lepton and a neutrino. The reducible background comes from $t\bar{t}$ and $Zb\bar{b}$ production. Despite that $t\bar{t}$ production cross section is 5 times smaller than the second one, the lepton pair from the b decays gives raise to larger invariant mass which is unfortunately better accepted in the Higgs mass range. Nevertheless, the use of the peaking invariant mass, the harder transverse momentum for the signal and the CP and spin quantum numbers of the leptons allow to extract Higgs signal in that channel. A multivariate analysis using all these informations is used to improve rejection of the background. With an integrated luminosity of 30 fb^{-1}, this channel may allow discovery above 5 σ in the range $130 < m_H < 180$ GeV [11], with an exception near 170 GeV where this branching ratio is reduced due to the opening of the $H \to WW$ decay. Above 180 GeV until 700 GeV, this decay mode is the most promising channel and allow discovery in less than one year of low luminosity for $200 < m_H < 600$ GeV. For larger masses, the large width of the Higgs boson makes the rate to drop quickly. The $H \to ZZ \to ll\nu\nu$ and $H \to ZZ \to lljj$ decay modes allow to extend the observability of the Higgs up to 1 TeV.

In the region around 170 GeV, the discovery potential is enhanced by the opening of the $H \to WW^* \to l\nu l\nu$ channel. Due to the missing energy, no mass peak can be reconstructed. The irreducible background is made of the WW continuum, 5 to 7 times larger than the signal, and of the $W(\to l\nu)Z(\to ll)$ and $Z(\to ll)Z(\to ll)$ contributions. The reducible background comes from $t\bar{t}$, Wt production with isolated leptons in the final state, and from $Wb\bar{b}$ and $b\bar{b}$ production with one or two leptons from semileptonic b-decay and $W+$jet production. Requesting central jet veto, strong angular correlations between the leptons and high missing transverse mass allows to discriminate between signal and background. With an integrated luminosity of 10 fb^{-1}, a significance larger than 5 σ may be obtained in the region $150 < m_H < 190$ GeV [9]. The $H \to WW^* \to ll$ decay can also be searched in the vector boson fusion production [10] for $130 < m_H < 180$ GeV. Despite the lower signal rate, this channel has a better signal to background ratio, thanks to the use of the forward jet tagging. It is thus less sensitive to systematic uncertainties on the background and has a significant discovery potential (Fig. 2). In the region above 800 GeV, the decay mode $H \to WW \to l\nu jj$ in the same production mode allows to isolate the signal from the background and to extend the Higgs search to 1 TeV.

CONCLUSION

An overview of the ATLAS and CMS capabilities to discover the Standard Model Higgs boson has been presented. The figure 2 presents the statistical significances for the Standard Model Higgs boson in good agreements by the ATLAS (using leading order cross sections for the signal) and CMS experiments (at leading and next-to-leading order). Combining these results for all the channels should allow a 7 σ discovery over the whole mass spectrum with 30 fb^{-1}, provided systematics on the background are under control. In the MSSM sector, most of the $m_A - \tan\beta$ is covered with 10 fb^{-1}.

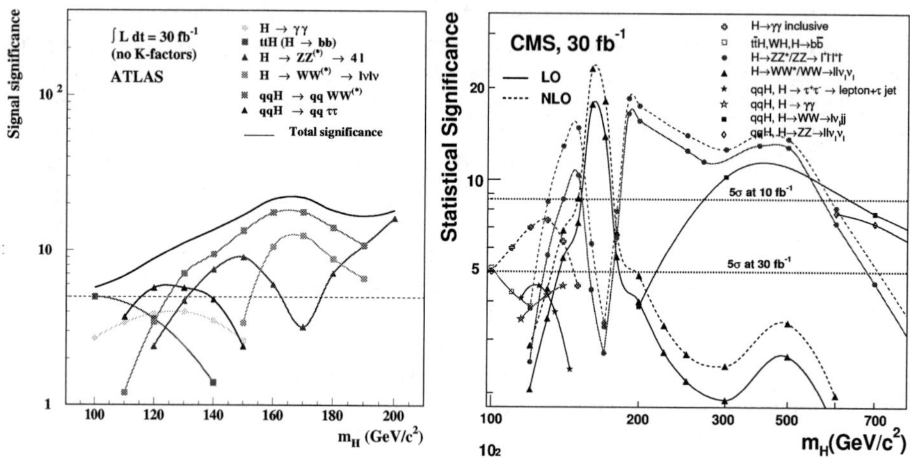

FIGURE 2. Potential of discovery (significances) at the LHC for the ATLAS (left panel) [10] and the CMS (right panel) [12] experiments.

ACKNOWLEDGMENTS

This work is the result of the combined effort from the ATLAS and the CMS collaborations. The author is grateful to the ATLAS speakers Commitee for the invitation to this conference, especially M. Cobal, G. Polesello and G. Unal. Thanks to the organizers of DIS05 for the fruitful conference.

REFERENCES

1. P. W. Higgs, *Phys. Lett.*, **12**, 132–133 (1964).
2. ATLAS (1994), `CERN-LHCC-94-43`.
3. CMS (1994), `CERN-LHCC-94-38`.
4. P. B. Renton (2004), `hep-ph/0410177`.
5. R. Barate, et al., *Phys. Lett.*, **B565**, 61–75 (2003), `hep-ex/0306033`.
6. V. Büscher, and V. Jakobs, *hep-ph/0504099* (2005).
7. J. Camin, and M. Schumacher (2003), `ATL-PHYS-20003-024`.
8. ATLAS (1999), `CERN-LHCC-99-15`.
9. S. Abdullin, and al. (2003), `CMS-Note-2003/033`.
10. S. Asai, et al., *Eur. Phys. J.*, **C32S2**, 19–54 (2004), `hep-ph/0402254`.
11. M. Sani (2004), `CMS-CR-2004/035`.
12. S. Abdullin et al., *CMS-NOTE-2003/033* (2003).

The role of PDF uncertainty for inclusive Higgs boson production at the Tevatron and LHC

Alexander Belyaev[1]

Department of Physics and Astronomy, Michigan State University, East Lansing, MI 48824, USA

Abstract. We perform a detailed study of the inclusive Higgs boson production uncertainty due to the parton distribution functions. We find that the Lagrange Multiplier and Hessian methods for determination of PDF uncertainty are in good agreement. We demonstrate that the PDF uncertainty may be comparable to or even much larger than the theoretical scale uncertainty. This reveals the crucial role of PDF uncertainty for correct signal interpretation and understanding the underlying electroweak symmetry breaking mechanism.

Keywords: Parton density functions, Higgs boson, hadron colliders, cross section uncertainty
PACS: 13.85.-t,14.80.Cp

INTRODUCTION

The search for the Higgs boson – the *only* Standard Model (SM) particle that remains undiscovered – has become one of the central problem in high energy physics. It is important not only to find the Higgs boson(s), it is just as important to understand its properties, including its couplings. These will reveal the underlying electroweak symmetry breaking mechanism, be it the Standard Model (SM) or beyond the SM. In order to distinguish between the different physical mechanisms, it is essential to assess the inherent uncertainties of theoretical predictions based on the various underlying physics scenarios.

A lot of work has been done on higher order QCD corrections to the Higgs cross section in SM and in Supersymmetry models. These have reduced considerably the theoretical uncertainties on these cross sections, usually estimated by the scale variation of the calculated cross sections. (We shall refer to this type of uncertainties, in short, as *scale uncertainties*.) However, the total uncertainty of the predicted Higgs production cross sections must also include uncertainty due to input parton distribution functions (PDFs). (These will be refered to, in short, as *PDF uncertainties*.)

We perform a detailed study of the PDF uncertainty for inclusive Higgs boson production at the Tevatron and LHC and compare them with existing estimates of the scale uncertainty. Particular attention is given to the b-quark initiated Higgs boson production mechanism. Whereas the $b\bar{b} \to H$ process is relatively insignificant compared to $gg \to H$ (via a top quark loop) in the SM, it can be significantly enhanced in the models beyond the SM which have several vacuum expectation values v_i, like Supersymmetry. We show that PDF uncertainty are comparable or even significantly dominate the scale uncertainty which has a very important consequenses for the reasons mentioned above.

---

[1] The work reported in this talk is done in collaboration with Jon Pumplin, Wu-Ki Tung and C.-P. Yuan

CP792, *Deep Inelastic Scattering, DIS 2005*, edited by W. H. Smith and S. R. Dasu

HIGGS PRODUCTION IN SM AND SUPERSYMMETRY

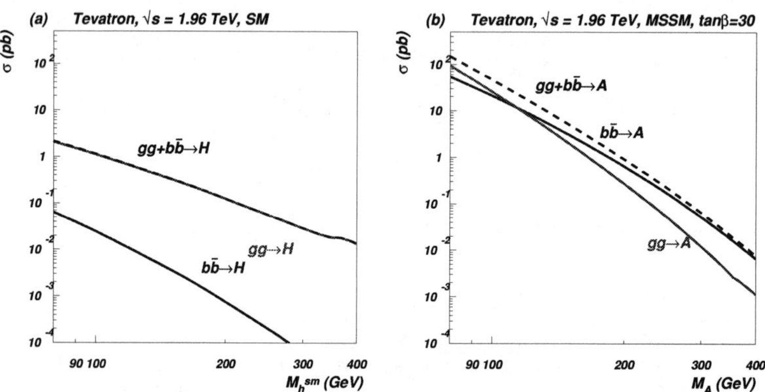

FIGURE 1. NLO cross sections for $b\bar{b} \to H$ and $gg \to H$ processes and their sum for the case of SM (a), and Supersymmetry (b) ($\tan\beta = 30$, CP-odd Higgs production).

While in the Standard Model, $b\bar{b} \to H$ production is highly suppressed compared to $gg \to H$, the ratio of $b\bar{b} \to H$ to $gg \to H$ could be drastically different in models beyond the SM because of the large enhancement of Yukawa couplings which could take place in models with several vacuum expectation values. In particular, in Supersymmetry, interactions of the Higgs bosons to bottom- and top- quarks relative to SM interactions, take the following form:

$$\begin{array}{lll} Y_{ht\bar{t}} = \cos\alpha/\sin\beta Y^{SM}_{ht\bar{t}}, & Y_{Ht\bar{t}} = \sin\alpha/\sin\beta Y^{SM}_{ht\bar{t}}, & Y_{At\bar{t}} = \cot\beta\gamma_5 Y^{SM}_{ht\bar{t}} \\ Y_{hb\bar{b}} = -\sin\alpha/\cos\beta Y^{SM}_{hb\bar{b}}, & Y_{Hb\bar{b}} = \cos\alpha/\cos\beta Y^{SM}_{hb\bar{b}}, & Y_{Ab\bar{b}} = \tan\beta\gamma_5 Y^{SM}_{ht\bar{t}} \end{array} \quad (1)$$

where α is the mixing angle of two CP-even Higgs bosons and $\tan\beta$ is the ratio of v_u/v_d vacuum expectation values. To avoid dependence on many SUSY parameters we study CP-odd Higgs production in $b\bar{b}$ or gg fusion which is independent of the Higgs mixing angle α. Figure 1a(b) presents NLO cross sections for $b\bar{b} \to H(A)$ and $gg \to H(A)$ (and their sum) for the case of SM (Supersymmetry, $\tan\beta = 30$) respectively. The cross section was obtained using the HIGLU program [1] for $gg \to A$ process and the code developed in [2] for evaluation of $b\bar{b} \to A$ process at NLO. Contrary to the SM case, the contribution from $b\bar{b} \to A$ becomes important even for moderate values of $\tan\beta \sim 10$. For $M_H < 110-115$ GeV the contribution from the $gg \to A$ process is bit bigger than the one from $b\bar{b} \to A$ while for $M_H > 115$ GeV values $b\bar{b} \to A$ production starts dominate over the $gg \to A$ one. The ratio of $b\bar{b} \to A$ and $gg \to A$ processes at the Tevatron and LHC is very similar, while the absolute values of the production rate at the LHC is about two orders of magnitude higher then at the Tevatron.

RESULTS FOR PDF UNCERTAINTIES

The best theoretical control of the QCD scale uncertainties for the inclusive $b\bar{b} \to A+X$ processes was reported in [3], where this process was calculated at next-to-next-to-leading order. It was found that for this process the scale uncertainty decreases from 15%

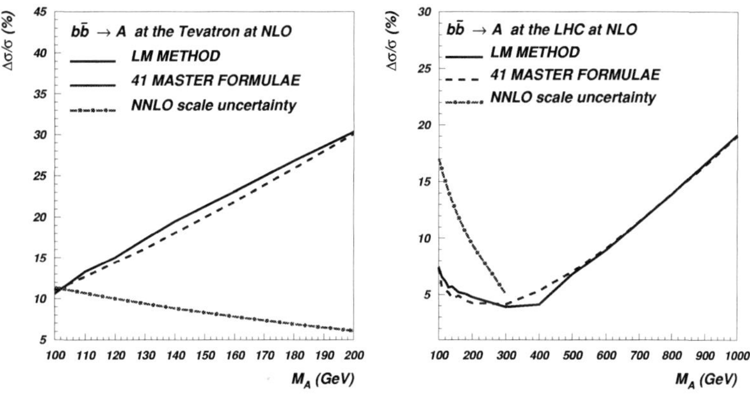

FIGURE 2. PDF uncertainty of $b\bar{b} \to A$ process for Hessian method(41 CTEQ PDF set) and for LM method for Tevatron(left) and LHC (right). Scale uncertainties from [3] are presented by red line.

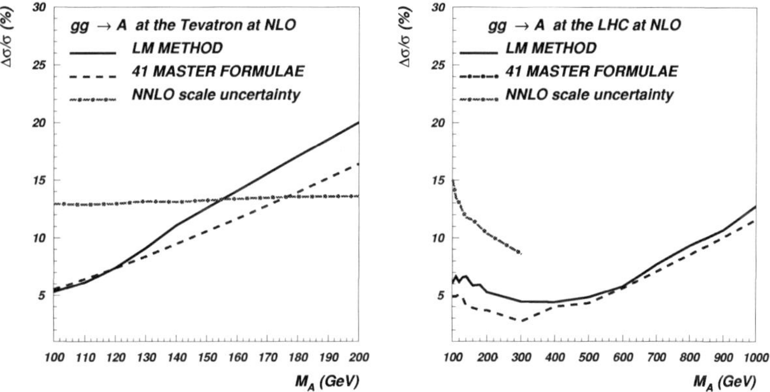

FIGURE 3. PDF uncertainty of $gg \to A$ process. Scale uncertainties are taken from Ref. [5]

to 5% for the LHC and from 10% to 3% for the Tevatron with increasing Higgs mass in the 120 $\to$ 300 GeV range. Here we perform a detailed study of PDF uncertainties for inclusive Higgs production processes to understand the total uncertainty for the Higgs boson production rate. We cross check results by using two methods for estimating the PDF uncertainty – Lagrange multiplier(LM) and Hessian(HS) (publicly available as or 41 set CTEQ PDF set) methods [4]. We find the PDF variations of cross section ($\sigma_{\pm} = \sigma_0 \pm \Delta\sigma$) at 90% confidence level. Our main results on PDF uncertainties as a function of Higgs boson mass are presented in Figure 2(left) and 2(right) for Tevatron and LHC respectively. Results for LM and HS methods – solid and dash lines respectively – are in a good agreement. QCD scale uncertainties available for the $M_A \leq 300$ GeV from Ref. [3] are presented by dot-dash line. Comparison $\Delta\sigma_{bb}^{PDF}$ and scale uncertainties ($\Delta\sigma_{bb}^{SC}$) reveals the opposite trend in their dependence on the Higgs boson mass. At the Tevatron: $\Delta\sigma_{bb}^{PDF}$ goes *up* as 11% $\rightarrow$ 30% for M_A in 100 $\rightarrow$ 200 GeV range, while $\Delta\sigma_{bb}^{SC}$ decreases from 11% to 3% in this mass range. For high M_A values, $\Delta\sigma_{bb}^{PDF}$ could be one order of magnitude bigger then $\Delta\sigma_{bb}^{SC}$! At the LHC, both $\Delta\sigma_{bb}^{PDF}$ and $\Delta\sigma_{bb}^{SC}$, decrease with M_A increase in 100 $\rightarrow$ 200 GeV mass range and $\Delta\sigma_{bb}^{SC}$ dominates $\Delta\sigma_{bb}^{PDF}$. For higher values of M_A $\Delta\sigma_{bb}$ is again dominated by PDF uncertainties similarly to the situation at the Tevatron.

There is a qualitative difference between Tevatron and LHC in the behavior of $\Delta\sigma^{PDF}$ as a function of Higgs mass: $\Delta\sigma^{PDF}$ always increases with the Higgs mass increase at the Tevatron, while at the LHC it has a minimum for Higgs mass around 300 GeV. The reason for this fact is that at the Tevatron, the value of x entering the PDF is big: $\simeq 0.05$ ($x = M_A/\sqrt{S} \geq 100/1986$) while at the LHC, for low $M_A \simeq 100$ GeV, x is fairly small: $\simeq 0.007$. The PDF uncertainty has a minimum at $x_{min} \simeq 0.02$ which never takes place at the Tevatron (for $M_A > 100$ GeV) but happens at the LHC for $M_A \simeq 300$ GeV. In Figure3 we present analogous results for the $gg \rightarrow A$ process which are qualitatively similar to those for $b\bar{b} \rightarrow A$ production. PDF uncertainties for $gg \rightarrow A$ process are about factor two lower compared to $b\bar{b} \rightarrow A$ one. For details, we refer reader to Ref. [6], where $\Delta g/g$ and $\Delta b/b$ uncertainties and their correlation were studied in detail. Our results on $\Delta\sigma_{gg}^{PDF}$ PDF uncertainties are in agreement with results presented in Refs. [7].

CONCLUSIONS

The role of $b\bar{b} \rightarrow A$ and $gg \rightarrow A$ processes could be central for the Higgs boson search. Therefore the correct understanding of uncertainties of their production rate is crucial. We show that $b\bar{b} \rightarrow A$ is equally important or even dominant over the $gg \rightarrow A$. The PDF uncertainty of $b\bar{b} \rightarrow A$ is about a factor two bigger then the PDF uncertainty for $gg \rightarrow A$ process; therefore the PDF uncertainty of $b\bar{b} \rightarrow A$ basically determines the total PDF uncertainty of the $b\bar{b} + gg \rightarrow A$ rate. It was found that at the Tevatron, PDF uncertainty dominates the scale uncertainty for M_A >130 GeV and could be as big as 30% for M_A=200 GeV which is one order of magnitude bigger then the NNLO scale uncertainty. At the LHC the scale uncertainty is dominant and could be as big as 15% for M_A <300 GeV region. In this region one could expect high Higgs production rates from the new physics, which would require higher order corrections for better theoretical control of the cross section. We have also found that Lagrange Multiplier and Hessian methods are in good agreement.

To conclude, we have shown that PDF uncertainties plays a crucial role in understanding the sensitivity of inclusive Higgs production process to new physics.

REFERENCES

1. M. Spira, *Nucl. Instrum. Meth.*, **A389**, 357–360 (1997), hep-ph/9610350.
2. C. Balazs, H.-J. He, and C. P. Yuan, *Phys. Rev.*, **D60**, 114001 (1999), hep-ph/9812263.
3. R. V. Harlander, and W. B. Kilgore, *Phys. Rev.*, **D68**, 013001 (2003), hep-ph/0304035.
4. J. Pumplin, D. R. Stump, and W. K. Tung, *Phys. Rev.*, **D65**, 014011 (2002), hep-ph/0008191.
5. R. V. Harlander, and W. B. Kilgore, *Phys. Rev. Lett.*, **88**, 201801 (2002), hep-ph/0201206.
6. Z. Sullivan, and P. M. Nadolsky, *eConf*, **C010630**, P511 (2001), hep-ph/0111358.
7. A. Djouadi, and S. Ferrag, *Phys. Lett.*, **B586**, 345–352 (2004), hep-ph/0310209.

Search for Leptoquarks and Lepton Flavor Violation at the H1 experiment

Linus Lindfeld

Universität Zürich, Physik-Institut, Winterthurerstr. 190, 8057 Zürich, Switzerland

Abstract.
We review recent preliminary results of searches for leptoquarks (LQs) with the H1 experiment at HERA. A search for first generation leptoquarks was performed using the e^-p and e^+p scattering data collected between 1994-2000 corresponding to an integrated luminosity $\mathscr{L}_{\text{int}} = 117\,\text{pb}^{-1}$. In addition, the e^+p H1 data set of the period 1999-2000 ($\mathscr{L}_{\text{int}} = 66\,\text{pb}^{-1}$) was used to search for lepton flavor violating (LFV) processes mediated by second and third generation LQs. No evidence for direct or indirect production of LQs was found in either analysis. The analyses established constraints on the Yukawa coupling of LQs and LFV processes.

Keywords: H1, Leptoquark, LQ, LFV, BRW
PACS: 13.85.Rm, 12.60.-i, 13.85.Qk, 14.80.-j

INTRODUCTION

Leptoquark (LQ) color triplet bosons appear naturally in many extensions of the standard model (SM). The ep-collider HERA provides a unique possibility to search for such new particles which couple to lepton-quark pairs. By the fusion of the initial state lepton of energy 27.6 GeV with a quark from the incoming proton of energy 820 GeV (1994-1997) or 920 GeV (1998-2000), single LQs of masses up to the centre-of-mass energy $\sqrt{s} = 319\,\text{GeV}$ can be resonantly produced. The phenomenology of LQs is described in detail in [1].

Singly produced LQs could also decay to a muon-quark pair or a tau-quark pair. Therefore, LQs are a convenient concept to explain such hypothetical exotic signatures inducing LFV at HERA.

SEARCH FOR LEPTOQUARKS (LQ)

Processes in ep-scattering involving a LQ decaying to an electron-quark or neutrino-quark pair lead to similar final states to those of NC and CC DIS at very high Q^2, the negative four-momentum tranfer squared. The differential cross section involves LQ, interference and SM contributions. The search is based on inclusive NC and CC DIS data in the kinematic domain $Q^2 > 2500\,\text{GeV}$ and $0.1 > y > 0.9$, where the ineleasticity variable y is defined as $y = Q^2/M_{\text{LQ}}^2$. The NC selection including an identified electron with transverse momentum above 15 GeV is decribed in [1]. The selection of CC-like events with a missing transverse momentum exceeding 25 GeV follows closely that presented in [1, 2].

CP792, *Deep Inelastic Scattering, DIS 2005,* edited by W. H. Smith and S. R. Dasu

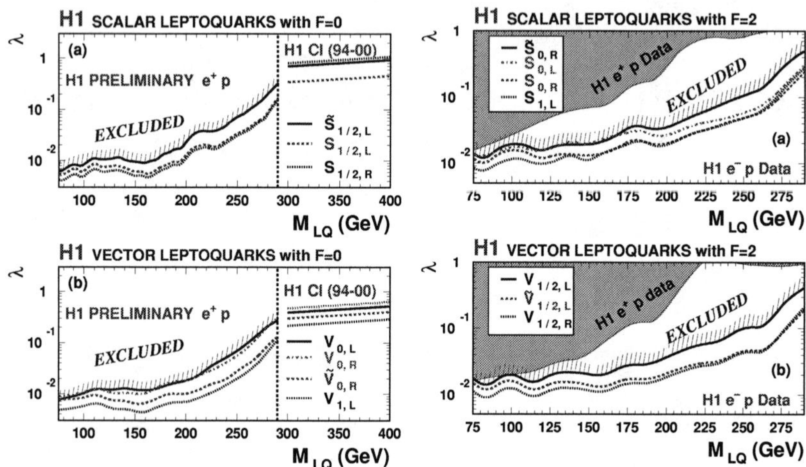

FIGURE 1. 95% CL exclusion limits on the Yukawa coupling λ as a function of the LQ mass for each of the 14 LQs described in the BRW model. Figures on the left-hand side show preliminary limits on $F = 0$ LQs up to a LQ mass of 290 GeV derived from e^+p data. Above the centre-of-mass energy of $\sqrt{s} = 319$ GeV, published limits for contact interaction (CI) processes are shown in addition. On the right-hand side published limits refer to $F = 2$ LQs derived from e^-p data.

The SM predictions, obtained using a DJANGO Monte Carlo [3] with CTEQ5D parametrisation [4] of the parton densities, describe the mass spectra of both NC- and CC-like data events very well. Since no evidence for LQ production is observed, constraints on LQs are set. To exploit fully the sensitivity to a LQ signal, all selected events are analysed in bins in the $M-y$ plane with varying size adapted to the experimental resolution and optimised with respect to the signal selection efficiency. Assuming poisson distributions for the SM expectation and the LQ signal, upper limits at 95% confidence level (CL) on the LQ production are deduced following a modified frequentist approach [5] to constrain LQ models.

Figure 1 shows limits on the Yukawa coupling λ of an electron[1]-quark pair to a LQ derived within the phenomenological model proposed by Buchmüller, Rückl and Wyler (BRW) [6]. In addition to the BRW model, where the branching ratio β_e (β_ν) is fixed to 0.5, constraints are derived on generic models where β_e (β_ν) is not fixed (see figure 2).

SEARCH FOR LEPTON FLAVOR VIOLATION (LFV)

In these analyses, searches for LFV in e^+p collisions mediated by LQs are performed. Only LQs with fermion number $F = |L+3B| = 0$ are considered here[2]. For the de-

[1] electron or positron depending on the data taking period

[2] where L is the lepton number and B is the baryon number

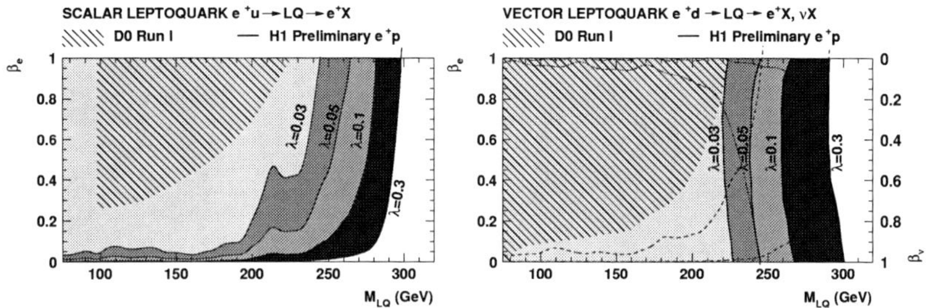

FIGURE 2. Mass dependent exclusion limits (95% CL) on generic models where the branching ratio β_e (β_ν) is not fixed. Left: Constraints on β_e of a scalar LQ which couples to e^+u (same quantum numbers as $S_{1/2,L}$). Right: Domains ruled out by combination of the NC (alone: lower dashed lines) and CC analyses (alone: upper dashed lines), for a vector LQ which couples to e^+d (same quantum numbers as $V_{0,L}$) and decays to eq and νq final states. Both figures show the resulting limits for four different Yukawa couplings λ. Hatched: Superceded coupling-independent limit of the DØ experiment (see latest results in [7]).

termination of the signal detection efficiencies the LEGO [8] event generator is used and the complete H1 detector response is simulated. The contributions from several SM background processes which may mimic the signal through measurement fluctuations are evaluated in this analysis. These processes include NC DIS, lepton pair production, W-production, photoprodution and CC DIS.

LQs with couplings to second generation leptons, leading to $\mu + q$ final states, exhibit a clear signature in the detector. Namely, an isolated high P_T muon "back-to-back" in the azimuthal angle, ϕ, with respect to the hadronic final state (HFS) is balanced in P_T with the HFS. In general, a muon deposits only a very small fraction of its energy in the LAr calorimeter, so the signal is expected to exhibit large P_T^{Calo}, where P_T^{Calo} is the measured P_T reconstructed from all clusters recorded in the calorimeter. A detailed description of the selection criteria can be found in [9].

The search for LQs possessing couplings to a third generation lepton leading to $\tau + q$ final states is restricted to hadronic decays of the tau. Hadronic decays of a high P_T tau lead to a typical signature of a "pencil-like" τ-jet. This τ-jet is characterised by a narrow shape in the calorimeter and low track multiplicity, i.e. one to three tracks in the identification cone of the jet. The signal topology is a dijet event with no leptons. The missing transverse momentum in the event carried by the neutrino(s) from the tau decay is aligned with the second highest P_T jet (jet2), which has to fulfil the τ-jet criteria. A detailed description of the selection criteria can be found in [10].

The results of the selections in the muon and tau channel are summarised in table 1. In order to set limits on the signal cross section, the mass spectra are scanned for signals using a sliding mass window with optimised borders. Within this window the number of data events, background events and the selection efficiency are used to calculate an upper limit on the signal at a 95% CL [5]. Applying the BRW model to second and third generation leptons, these limits are converted into limits on the couplings $\lambda_{\mu q}$ and $\lambda_{\tau q}$. The obtained limits are shown in figure 3 for one of the scalar and one of the vector LQs with an assumed LFV branching ratio of $BR_{\mathrm{LQ}\to\mu q,\tau q} = BR_{\mathrm{LQ}\to eq}$. The constraints on LQs coupling to first generation leptons only are shown as comparison.

TABLE 1. Results of the search for LFV. Statistical and systematic errors are added in quadrature.

$\mathcal{L} = 66\,\text{pb}^{-1}$	H1 Data*	SM Total	NC DIS	Lepton Pairs	Photo-Production	W-Production
$\text{LQ} \to \mu + q$	0	0.74 ± 0.25	0.09 ± 0.05	0.50 ± 0.24	0.06 ± 0.07	0.08 ± 0.02
$\text{LQ} \to \tau + q$	1	0.56 ± 0.16	0.37 ± 0.13	< 0.01	0.13 ± 0.04	0.06 ± 0.02

* preliminary

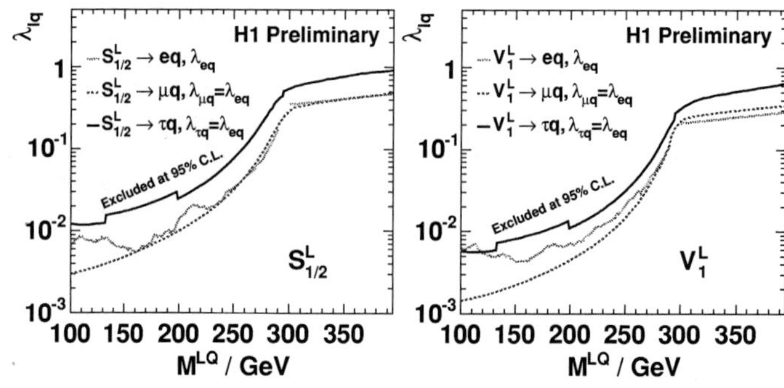

FIGURE 3. Limits on the coupling strength λ_{lq} at 95% C.L. for $S^L_{1/2}$ and V^L_1.

CONCLUSIONS

In recent searches for LQs with the H1 experiment at HERA no signal has been observed and constraints have been set. LQs up to a mass of 290 GeV and a Yukawa coupling of electromagnetic strength could be ruled out.

Also, a search for LFV mediated by LQs coupling to second and third generation leptons in e^+p collisions has been performed. No evidence has been found for lepton flavor violation in neither the muon nor the tau decay channel. Assuming a Yukawa coupling of electromagnetic strength, couplings of scalar (vector) LQs with masses up to 275-300 (288-330) GeV to second generation leptons, and couplings of scalar (vector) LQs with masses up to 260-284 (278-300) GeV to third generation leptons are excluded.

REFERENCES

1. C. Adloff *et al.* [H1 Coll.], Eur. Phys. J. C**11** (1999) 447 [Er.-ibid. C**14** (1999) 553][hep-ex/9907002].
2. C. Adloff *et al.* [H1 Coll.], Eur. Phys. J. C**19** (2001) 269 [hep-ex/0012052].
3. G.A. Schuler and H. Spiesberger, Workshop Proc.'Physics at HERA" (1991), Vol. 3, p. 1419.
4. H.L. Lai *et al.* [CTEQ Coll.], Eur. Phys. J. C **12** (2000) 375 [hep-ph/9903282].
5. T. Junk, Nucl. Instrum. Meth. A **434** (1999) 435.
6. W. Buchmüller, R. Rückl and D. Wyler, Phys. Lett. B **191** (1987) 442.
7. V.M. Abazov *et al.* [DØ Coll.], Phys. Rev. D**71** (2005) 071104 [hep-ex/0412029].
8. K. Rosenbauer, Dissertation, RWTH Aachen (in German), PITHA 95/16, Germany (1995).
9. I. Mudhahir, Dissertation, Univ. of Manchester, UK (2005).
10. H1 Coll., Conference paper subm. to 32nd ICHEP '04, Beijing, Aug 16-22 2004.

Search for lepton flavor violation at ZEUS

G. Barbagli
(On behalf of the ZEUS Collaboration)

INFN Firenze Via G.Sansone 1, 50019 Sesto Fiorentino (FI),Italy

Abstract. A search for lepton-flavor violating interactions has been performed at ZEUS using the whole luminosity collected in the HERA I phase. No event was found. Limits have been set on leptoquarks which could mediate such interactions.

Keywords: lepton,flavor,violation, leptoquark, ZEUS, BRW
PACS: 12.60.-i; 13.85.Qk; 13.85.Rm; 14.80.-j

INTRODUCTION

Lepton flavor conservation is not implied by any universal symmetry. After the evidence of neutrino oscillations one can think that lepton flavor could be violated also for charged leptons. Lepton flavor violation (LFV) is foreseen in many extensions of the Standard Model (SM), for example GUT, Leptoquark, SUSY models with R-Parity violation. It could be mediated by leptoquarks (LQs) (or s-quarks in R-Parity violating SUSY).
As framework for interpreting the results of the search described in this paper the phenomenological Buchmüller-Rück-Wyler (BRW) leptoquark model [1] was chosen. It has very general assumptions about the interaction Lagrangian. It is invariant under $SU(3)_c \times SU(2)_L \times U(1)_Y$. It couples to left handed (LH) or right handed (RH) leptons but not to both. It has fixed branching ratios to eq and νq. It includes 14 leptoquarks , with both lepton(L) and baryon(B) number different from zero, classified according to fermion number $F = 3B+L = 0, \pm 2$. There 7 are scalar and 7 vector LQs, with isospin 0, 1/2, 1, helicity L or R. The model is non-diagonal in lepton families and so it can accomodate LFV.
Electron-proton collisions at HERA are ideal for this kind of searches. The cross section for LFV processes has in general both s- channel and u- channel contributions (see Fig. 1). Two different approximations for the cross section can be used depending on the

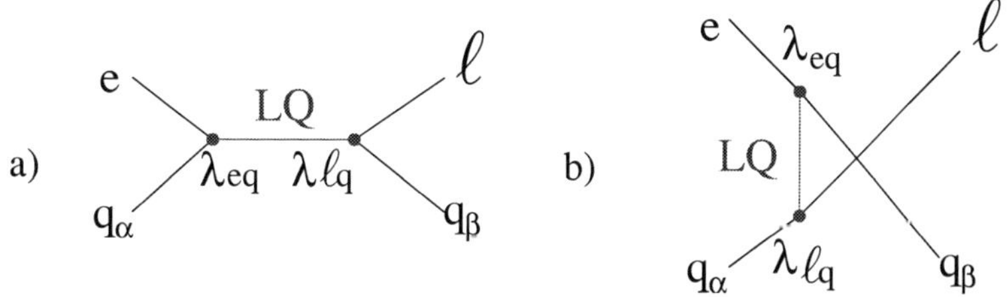

FIGURE 1. Contributions from (a) s-channel and (b) u-channel to LFV processes at HERA. α and β denote quark generations, ℓ is either a μ or a τ.

CP792, *Deep Inelastic Scattering, DIS 2005,* edited by W. H. Smith and S. R. Dasu

leptoquark mass [2]. Low mass LQs can be resonantly produced. The s-channel process dominates and the production occurs at $x = M_{LQ}^2/s$. When $M_{LQ}^2 < s$ we can use the so called Narrow Width Approximation (NWA):

$$\sigma_{NWA} \propto \lambda_{eq}^2 B_{\ell q_\beta} \tag{1}$$

where λ_{eq} is the coupling and $B_{\ell q_\beta}$ is the branching ratio into the quark generation β. The experimental signature for such resonances is a high p_t isolated lepton and a peak in the invariant mass of the lepton-jet system. Due to the dominance of valence quark contributions only $F = 0$ LQs are considered for e^+p data and only $|F| = 2$ LQs are considered for e^-p data.

For high mass LQs both s- and u- channel contribute. When $M_{LQ}^2 >> s$ we can use the so called Contact Interaction (CI) Approximation:

$$\sigma_{CI} \propto [\frac{\lambda_{eq_\alpha}\lambda_{\ell q_\beta}}{M_{LQ}^2}]^2 \tag{2}$$

The spectrum of the isolated leptons is softer compared to the previous case. Both $F = 0$ and $|F| = 2$ LQs are considered for both e^+p and e^-p samples.

FEATURES OF LFV EVENTS, BACKGROUND AND SIMULATION

When $e - \mu$ or $e - \tau$ transitions take place the electron or positron in the initial state is replaced by an isolated muon, or an isolated tau, which can be identified, according to the decay channel, via a high p_t-e or a high-p_t μ or narrow jet. For the $e - \tau$ transition tau leptonic and hadronic channels were considered, and a multivariate technique (a tau finder based on a discriminant), already used in the isolated tau search, was employed [3].

Background from SM includes deep inelastic scattering (neutral and charged currents), direct and resolved photoproduction and lepton pair production. Selection criteria and efficiencies have been studied using Monte Carlo simulation of the signal and of the different background sources.

RESULTS

The whole HERA I luminosity has been analyzed by ZEUS looking for LFV: 112.8 pb^{-1} taken with e^+p in the years 1994-2000 and 16.7 pb^{-1} taken with e^-p in the years 1998-1999, at the two center-of-mass energies of 300 and 318 GeV.

In the search for $e - \mu$ transitions no event was found while 0.87 ± 0.15 were expected from simulation of SM processes. In the search for $e - \tau$ transitions no event was found looking for the e, μ and hadronic decay channel of the τ, while 2.3 ± 0.5 were expected. Such results were interpreted within the BRW model and limits were set on couplings

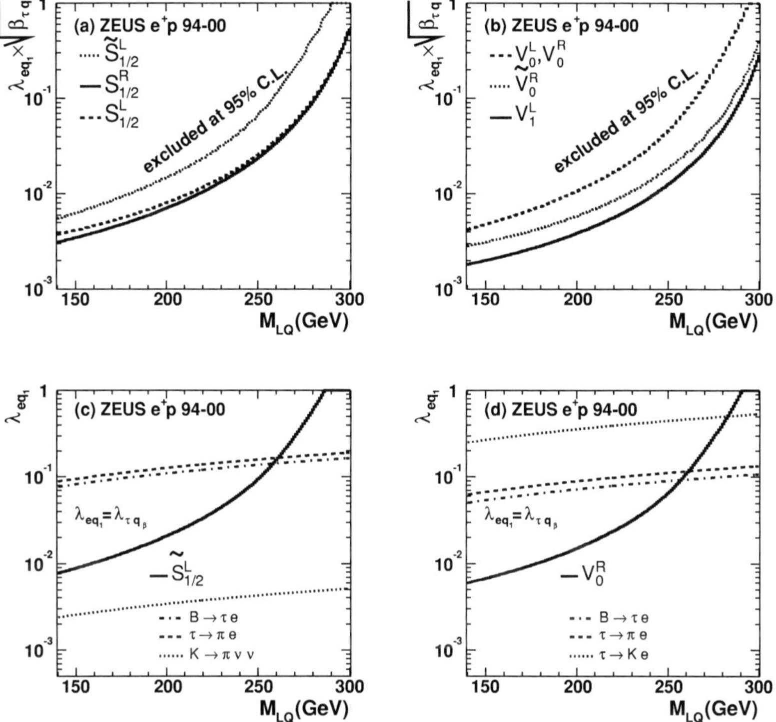

FIGURE 2. Limits for $F = 0$ low-mass LQs for $e - \tau$ transitions obtained from e^+p collisions. The upper plots correspond to 95 % C.L. limits on $\lambda_{eq_1} \cdot \sqrt{\beta\tau q}$ vs. M_{LQ} (β is the branching ratio) for (a) scalar and (b) vector LQs. In the lower plots ZEUS limits on λ_{eq_1} for a representative (c) scalar and (d) vector LQ are compared to indirect constraints from low energy experiments, assuming $\lambda_{eq_1} = \lambda_{\tau q_\beta}$.

and masses both for low-mass and high mass LQs [4, 5]. Many of the limits also apply to R-Parity violating SUSY squarks. Figures 2 and 3 show, as an example, contraints for $e - \tau$ transitions, for $F = 0$ and $F = 2$ low-mass scalar and vector LQs. The limits set by ZEUS are competitive with low energy experiments and often improve them, especially for the $e - \tau$ transition (since existing limits from rare decays are less stringent than for muons), and when heavy quarks are involved.

CONCLUSIONS

ZEUS has searched for the lepton flavor violating processes $ep \to \mu X$ and $ep \to \tau X$ using the full HERA I data set. No evidence was found for $e - \mu$ or of $e - \tau$ transitions. Limits have been set for leptoquarks mediating LFV. Some of these limits are the most

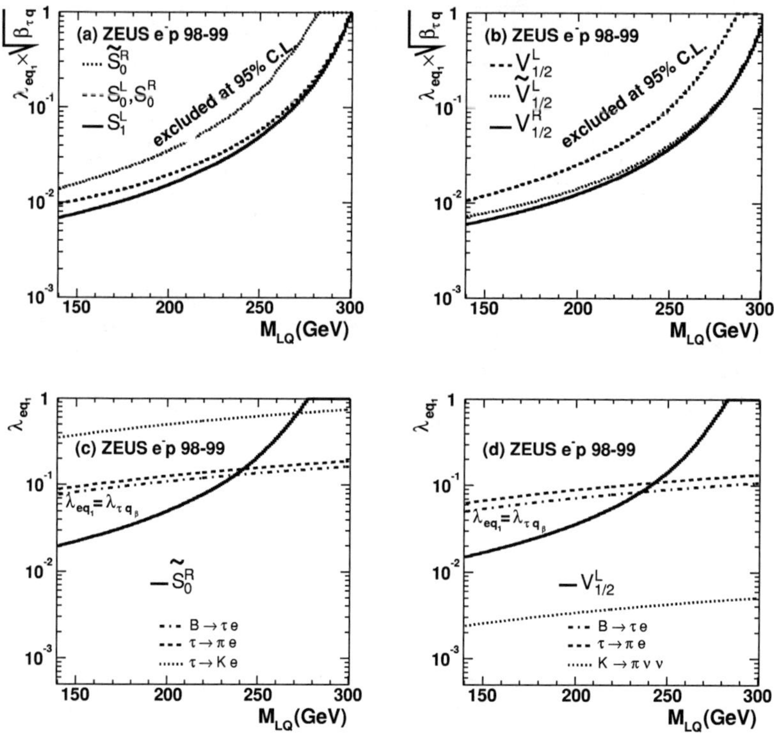

FIGURE 3. Limits for $F = 2$ low-mass LQs for $e - \tau$ transitions obtained from $e^- p$ collisions. The upper plots correspond to 95 % C.L. limits on $\lambda_{eq_1} \cdot \sqrt{\beta \tau q}$ vs. M_{LQ} (β is the branching ratio) for (a) scalar and (b) vector LQs. In the lower plots ZEUS limits on λ_{eq_1} for a representative (c) scalar and (d) vector LQ are compared to indirect constraints from low energy experiments, assuming $\lambda_{eq_1} = \lambda_{\tau q_\beta}$.

stringent up to date. The new higher luminosity phase HERA II should increase the sensitivity to LFV. Polarized electron and positron beams should permit the study of LH or RH LQs.

REFERENCES

1. W. Buchmüller, R. Rück, D. Wyler, *Phys. Letters*, **B191**, 442(1987) Erratum *Phys. Letters*, **B448**, 320(1999).
2. ZEUS Collaboration, M.Derrick et al., *Zeitschrift.f.Physik*, **C73**, 4,613(1997).
3. ZEUS Collaboration, S,Chekanov et al., *Phys. Letters*, **B583**, 41(2004).
4. ZEUS Collaboration, S,Chekanov et al., *Desy Report*, **DESY-05-016**, (2005) and references therein.
5. C. Genta, *Search for New Physics in Events with high p_t Leptons at HERA*, Ph.D. thesis, Florence University, January 2005, also available as , DESY-THESIS-2005-017 June 2005.

Searches at the Tevatron

Robert Illingworth
(on behalf of the CDF and DØ collaborations)

Fermi National Accelerator Laboratory, P.O. Box 500, Batavia, Illinois 60510, USA

Abstract. The CDF and DØ collider experiments at Fermilab have searched for evidence of physics beyond the Standard Model of particle physics. We report on the results of searches for extra heavy gauge bosons, quark-lepton compositeness, leptoquarks, and magnetic monopoles in $p\bar{p}$ collisions at $\sqrt{s} = 1.96$ TeV. No evidence of new signals were found, therefore we derive limits on the model parameters.

Keywords: New phenomena; Beyond the Standard Model
PACS: 13.85.Rm;14.70.Pw;14.80.-j

INTRODUCTION

The Standard Model of particles and interactions has been an extremely successful theory for describing high-energy physics data. However, although its predictions have been verified to great accuracy, it leaves many fundamental questions unanswered. Among these are why there are three generations of quarks and leptons and why their masses are what they are, why gravity seems to be so much weaker than the other forces, and whether all the forces can be unified into one single fundamental interaction. Many models predicting new physics beyond the Standard Model have been proposed. These models predict new signatures that can potentially be detected at high-energy colliders like the Tevatron. These searches can be challenging: the production cross-sections tend to be small and the SM backgrounds very large. Processes involving jets are usually overwhelmed by the background from QCD, so lepton final states are used even though the rates are often suppressed.

These results are based on up to 400 pb^{-1} of data delivered to the CDF and DØ experiments at Fermilab between 2002 and 2004. All limits quoted are at the 95% confidence level.

NEW CHARGED HEAVY VECTOR BOSON

The W' is an additional charged heavy vector boson that appears in theories based on extending the gauge group of the Standard Model. For example left-right symmetric models[1, 2] feature new gauge bosons including a heavy, right-handed W'. CDF performed a search using 200 pb^{-1} of data for a W' decaying to an electron-neutrino pair where the neutrino is assumed to be light and does not decay inside the detector volume. Events were selected which contained an isolated electron with $p_T > 25$ GeV and missing transverse energy $\not\!\!E_T > 25$ GeV. The backgrounds to W' production are from

CP792, *Deep Inelastic Scattering, DIS 2005,* edited by W. H. Smith and S. R. Dasu

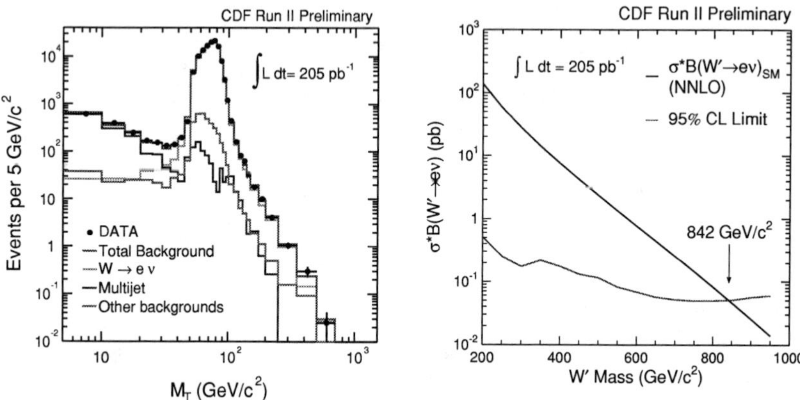

FIGURE 1. *(left)* Transverse mass distribution from selected W' sample, *(right)* 95% confidence limit on the W' cross-section, assuming the Standard Model coupling

real W decays, sources producing real electrons such as $Z/\gamma \to ee$, $t\bar{t} \to eX$, and QCD multijet events where a jet is misidentified as an electron and the jet energy is mismeasured to create significant $\not{E}_T$. The data show no evidence of an excess of events over the predicted background, which leads to a limit on the W' mass of $M_{W'} > 842$ GeV (Figure 1).

QUARK-LEPTON COMPOSITENESS

The Standard Model predicts neither the masses nor the number of families of quarks and leptons, which suggests there may be a more fundamental basis. In one scenario quarks, leptons and heavy bosons are formed from constituents called preons interacting through a new gauge interaction called metacolour[3]. Below the characteristic energy scale Λ the interaction binds the preons into composite states such as quarks and leptons. The experimental signature is highly model dependent, but exhibits itself as a deviation from the SM cross-section for dilepton production through the Drell-Yan process at high invariant masses. These deviations are highly model dependent. DØ studied 400 pb^{-1} of data in the dimuon channel. Selected events were required to have two isolated muons with $p_T > 50$ GeV. No deviations from the Standard Model prediction were found (Figure 2), so lower limits on the compositeness scale were set, ranging from $\Lambda = 4.2$ TeV to $\Lambda = 9.8$ TeV depending on the model.

LEPTOQUARKS

Several extensions of the Standard Model contain particles which possess colour, electric charge and both lepton and quark quantum numbers, and which decay into a lepton and a quark[4]. At the Tevatron leptoquarks would be produced predominantly through

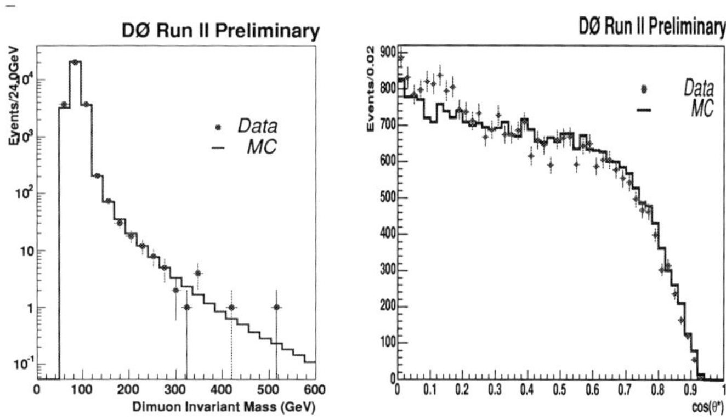

FIGURE 2. Dimuon invariant mass and cosine of scattering angle distributions from the DØ compositeness search

$q\bar{q}$ annihilation and gluon fusion, and decay to either a charged lepton and a quark (with branching fraction β), or a neutrino and a quark. DØ and CDF have searched for first generation scalar leptoquarks in the $\ell q \ell q$ and $\ell q \nu q$ channels. These have an experimental signature of two jets and either two charged leptons, or one charged lepton and missing transverse energy respectively. Additionally, CDF has included results from the $\nu j \nu j$ channel, which cannot distinguish between the different generations and has an experimental signature of two jets and missing transverse energy. No excesses over the Standard Model predictions were found. The mass limits set depend on the branching ratio β. For $\beta = 1$ first generation limits are $M > 230$ GeV (CDF) and $M > 256$ GeV (DØ), and the second generation limit is $M > 224$ GeV (CDF). First generation limits as a function of β from DØ are shown in Figure 3.

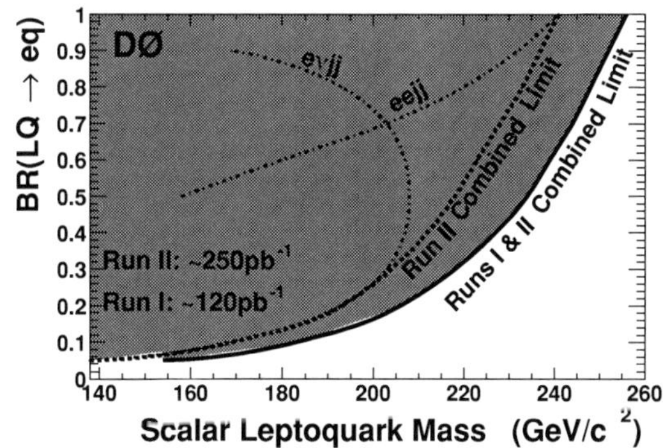

FIGURE 3. DØ first generation leptoquark exclusion region

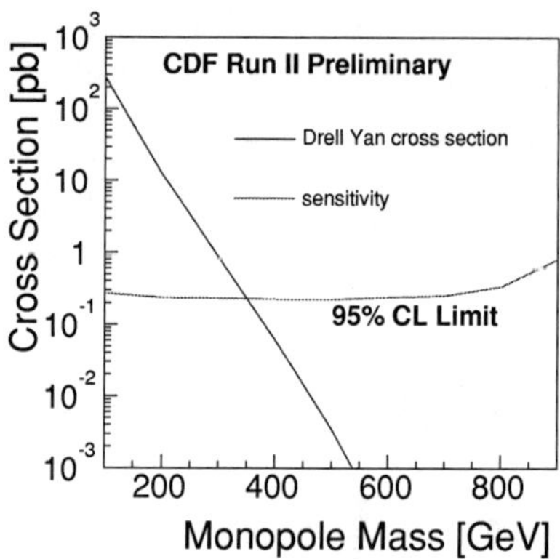

FIGURE 4. 95% confidence limit on the Dirac monopole cross-section, assuming Drell-Yan production

MAGNETIC MONOPOLES

A Dirac magnetic monopole[5] is a particle with no electric charge, no hadronic interactions and whose magnetic charge satisfies the quantization condition $g \approx 68.5 \cdot n \cdot e$. CDF conducted a search for Dirac monopoles. These have a unique signature in a detector like CDF: they are accelerated down the solenoidal magnetic in the tracking region field rather than bending in the transverse plane. Additionally, the very large magnetic charge of the monopole causes huge amounts of ionization, so CDF was able to use a very efficient dedicated trigger requiring large light pulses in the Time-Of-Flight detector scintillator bar. This trigger collected 25 pb^{-1} of data which were reconstructed with a special tracking algorithm. The unique path of the monopole means that the background is negligible. No events were found, which gives a mass limit of $M > 350$ GeV, assuming the monopoles are produced by a Drell-Yan like process (Figure 4).

SUMMARY

The CDF and DØ collaborations have analyzed data from up to 400 pb^{-1} of $p\bar{p}$ collisions at $\sqrt{s} = 1.96$ TeV. No deviations from the predictions of the Standard Model have been found, and limits have been set on the parameters of charged heavy gauge bosons, quark-lepton compositeness, leptoquarks, and magnetic monopoles.

REFERENCES

1. J. C. Pati, and A. Salam, *Phys. Rev.*, **D10**, 275–289 (1974).
2. R. N. Mohapatra, and J. C. Pati, *Phys. Rev.*, **D11**, 566–571 (1975).
3. E. Eichten, K. D. Lane, and M. E. Peskin, *Phys. Rev. Lett.*, **50**, 811–814 (1983).
4. D. E. Acosta, and S. K. Blessing, *Ann. Rev. Nucl. Part. Sci.*, **49**, 389–434 (1999).
5. P. A. M. Dirac, *Proc. R. Soc. (London)*, **A133** (1931).

Top quark and charged Higgs production at hadron colliders

Nikolaos Kidonakis

Kennesaw State University, Physics #1202, 1000 Chastain Rd, Kennesaw, GA 30144-5591, USA

Abstract. I present a brief theoretical update on top quark pair production at the Tevatron and give values of the NNLO-NNNLL cross section for both $m_t = 175$ and 178 GeV. I then present a calculation of the cross section for charged Higgs production in association with a top quark at the LHC, including NNLO soft-gluon corrections.

Keywords: top quark;Higgs
PACS: 12.38.Bx;13.85.-t;14.65.Ha;14.80.Cp

TOP QUARK PRODUCTION AT THE TEVATRON

The properties of the top quark, in particular its mass and production cross section, are subjects of intense study at the Tevatron [1, 2]. The most accurate theoretical prediction [3] for top quark pair production at the Tevatron includes soft-gluon corrections [4, 5, 6] through next-to-next-to-next-to-leading logarithmic (NNNLL) accuracy at next-to-next-to-leading order (NNLO), denoted as NNLO-NNNLL [3]. These corrections are sizable and provide a dramatic decrease in the scale dependence of the cross section. Results have been derived in both single-particle-inclusive (1PI) kinematics and pair-invariant-mass (PIM) kinematics. There are differences in the results in the two kinematics due to subleading terms, and the best estimate is given by the average of the two kinematics.

For a top quark mass $m_t = 175$ GeV the theoretical value of the cross section is [3]

$$\sigma_{t\bar{t}}^{NNLO-NNNLL}(\sqrt{S} = 1.8 \text{ TeV}, m_t{=}175 \text{ GeV}) = 5.24 \pm 0.31 \text{ pb} \quad \text{and}$$

$$\sigma_{t\bar{t}}^{NNLO-NNNLL}(\sqrt{S} = 1.96 \text{ TeV}, m_t{=}175 \text{ GeV}) = 6.77 \pm 0.42 \text{ pb}$$

at Run I and Run II, respectively. The uncertainty indicated is due to the kinematics ambiguity; the scale uncertainty is much smaller.

Some recent data from the Tevatron suggest a value for the top quark mass around $m_t = 178$ GeV. For that value of top mass the theoretical cross sections become

$$\sigma_{t\bar{t}}^{NNLO-NNNLL}(\sqrt{S} = 1.8 \text{ TeV}, m_t{=}178 \text{ GeV}) = 4.76 \pm 0.28 \text{ pb} \quad \text{and}$$

$$\sigma_{t\bar{t}}^{NNLO-NNNLL}(\sqrt{S} = 1.96 \text{ TeV}, m_t{=}178 \text{ GeV}) = 6.15 \pm 0.38 \text{ pb.}$$

Results for the top quark transverse momentum distributions at NNLO-NNNLL are also available [3].

CP792, *Deep Inelastic Scattering, DIS 2005,* edited by W. H. Smith and S. R. Dasu

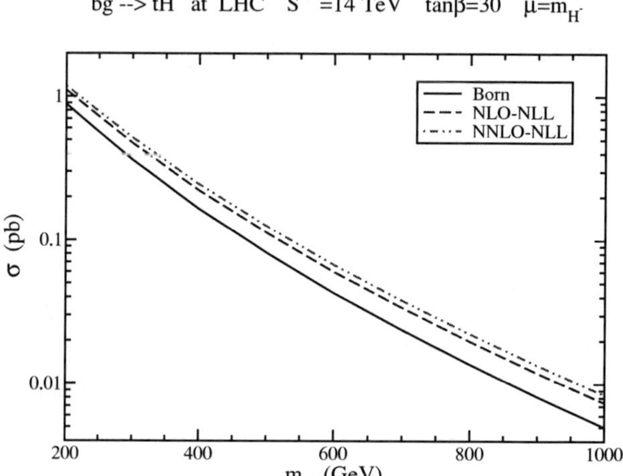

FIGURE 1. The total cross section for charged Higgs production at the LHC.

CHARGED HIGGS PRODUCTION VIA $BG \to TH^-$

A future discovery of a charged Higgs boson would be an umistakable sign of new physics beyond the Standard Model [7]. The LHC has good potential for such a discovery through the partonic process $bg \to tH^-$. The Born cross section is proportional to $\alpha\alpha_s(m_b^2 \tan^2\beta + m_t^2 \cot^2\beta)$, where $\tan\beta = v_2/v_1$ is the ratio of the vacuum expectation values (vev's) of two Higgs doublets in the MSSM.

Full NLO calculations have recently become available [8, 9], and they show that the NLO corrections are big. Since charged Higgs production will be a near-threshold process at the LHC, given the expected large mass of this particle (hundreds of GeV), threshold soft-gluon corrections can provide significant enhancements of the cross section. A next-to-leading logarithm (NLL) calculation of these corrections at NNLO, denoted as NNLO-NLL [10], showed that indeed the soft-gluon corrections are substantial and they decrease the scale dependence of the cross section, thus providing a better theoretical prediction.

For the process $b(p_b) + g(p_g) \longrightarrow t(p_t) + H^-(p_{H^-})$ we define $s = (p_b + p_g)^2$, $t = (p_b - p_t)^2$, $u = (p_g - p_t)^2$, and $s_4 = s + t + u - m_t^2 - m_{H^-}{}^2$. At threshold $s_4 \to 0$. The soft-gluon corrections take the form $[(\ln^l(s_4/m_{H^-}^2))/s_4]_+$. For the order α_s^n corrections, $l \leq 2n-1$. The leading logarithms (LL) are those with $l = 2n-1$, while for the NLL $l = 2n-2$. We calculate NLO and NNLO corrections at NLL accuracy.

In Figure 1 we plot the cross section versus charged Higgs mass for pp collisions at the LHC with $\sqrt{S} = 14$ TeV. We use the MRST2002 approximate NNLO parton distributions functions (PDF) [11] with the respective three-loop evaluation of α_s. We set the factorization scale equal to the renormalization scale and denote this common scale by μ. We show results for the Born, NLO-NLL, and NNLO-NLL cross sections, all with a choice of scale $\mu = m_{H^-}$. In our calculations we use $\tan\beta = 30$. The NLO and

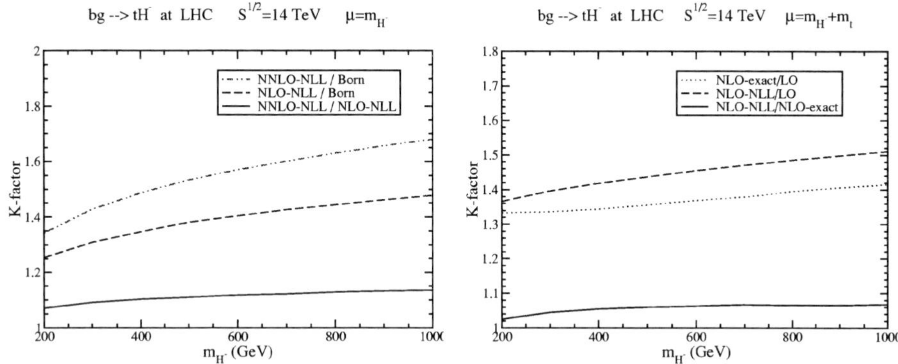

FIGURE 2. *K*-factors for charged Higgs production at the LHC.

NNLO threshold corrections are positive and provide a significant enhancement to the lowest-order result. We note that the cross sections for the related process $\bar{b}g \to \bar{t}H^+$ are exactly the same.

In Figure 2 we plot K-factors, i.e. ratios of cross sections at various orders. On the left-hand side, the NLO-NLL / Born curve shows that the NLO threshold corrections enhance the Born cross section by approximately 25% to 50% depending on the mass of the charged Higgs. The NNLO-NLL / Born curve shows that if we include the NNLO threshold corrections we get an enhancement over the Born result of approximately 35% to 70% in the range of masses shown. Finally, the NNLO-NLL / NLO-NLL curve shows clearly the further enhancement over NLO that the NNLO threshold corrections provide, between 7% and 14%. On the right-hand side we compare our NLO-NLL results with the exact results that have been derived in [8]. To make the comparison with [8], the NLO-NLL result is calculated here for $\mu = m_{H^-} + m_t$, the choice of scale used in that reference, and also using a two-loop α_s. Also the use of K-factors removes any discrepancies arising from different choices of parton distribution functions. The NLO-NLL / NLO-exact curve is very close to 1 (only a few percent difference), and this shows that the NLO-NLL cross section is a remarkably good approximation to the exact NLO result. As noted before, we might have expected this on theoretical grounds since this is near-threshold production, and also from prior experience with many other near-threshold hard-scattering cross sections [3, 5, 12].

In Figure 3, we plot the scale dependence of the cross section for a fixed charged Higgs mass $m_{H^-} = 500$ GeV. We plot a large range in scale, $0.1 \leq \mu/m_{H^-} \leq 10$, and see indeed that the threshold corrections greatly decrease the scale dependence of the cross section. The NNLO-NLL curve is relatively flat. For comparison we also plot the results using only a leading logarithm (LL) approximation. We see that the LL results display a large scale dependence at both NLO and NNLO, and are not an improvement over the Born result. The NLL terms are essential in diminishing the scale dependence. The difference between the LL and NLL results at both NLO and NNLO can be very substantial. Thus having a complete NLL calculation, as provided here, is crucial in providing stable theoretical predictions.

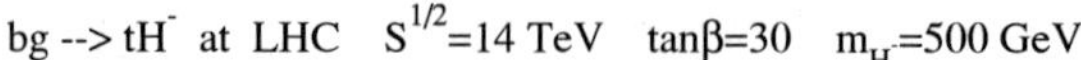

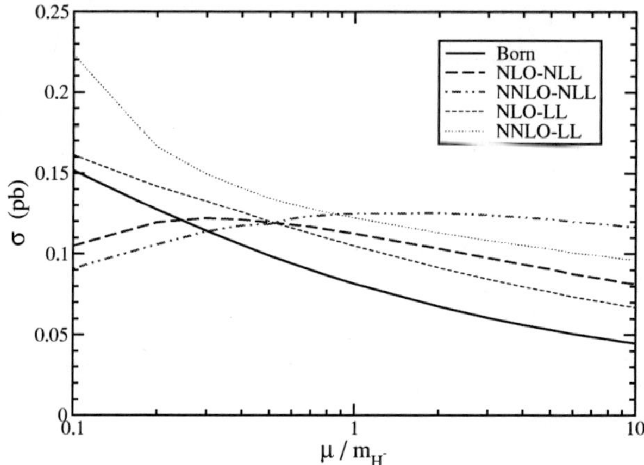

FIGURE 3. The scale dependence of the charged Higgs cross section.

Finally, we note that even higher-order corrections may provide sizable contributions to hard-scattering cross sections. In particular current calculations of next-to-next-to-next-to-leading order (NNNLO) soft-gluon corrections indicate a non-negligible enhancement of the cross section for charged Higgs production.

REFERENCES

1. CDF Collaboration, *Phys. Rev. Lett.*, **93**, 142001 (2004); hep-ex/0410041.
2. D0 Collaboration, *Phys. Lett. B*, **606**, 25 (2005); hep-ex/0504043; hep-ex/0504058; hep-ex/0505082.
3. N. Kidonakis and R. Vogt, *Phys. Rev. D*, **68**, 114014 (2003); *Eur. Phys. J. C*, **33**, s466 (2004); in *DPF 2004*, hep-ph/0410367.
4. N. Kidonakis and G. Sterman, *Phys. Lett. B*, **387**, 867 (1996); *Nucl. Phys. B*, **505**, 321 (1997); N. Kidonakis, *Int. J. Mod. Phys. A*, **15**, 1245 (2000).
5. N. Kidonakis, *Phys. Rev. D*, **64**, 014009 (2001); *Int. J. Mod. Phys. A*, **16S1A**, 363 (2001); in *Snowmass 2001*, hep-ph/0110145; N. Kidonakis, E. Laenen, S. Moch, and R. Vogt, *Phys. Rev. D*, **64**, 114001 (2001); *Nucl. Phys. A*, **715**, 549c (2003); *Phys. Rev. D*, **67**, 074037 (2003); N. Kidonakis and R. Vogt, *Eur. Phys. J. C*, **36**, 201 (2004); in *DIS 2004*, hep-ph/0405212.
6. N. Kidonakis, *Int. J. Mod. Phys. A*, **19**, 1793 (2004); in *DIS 2003*, hep-ph/0306125, hep-ph/0307207; *Mod. Phys. Lett. A*, **19**, 405 (2004); in *DPF 2004*, hep-ph/0410116;
7. The Higgs Working Group: Summary Report, in *Les Houches 2003*, hep-ph/0406152.
8. S.H. Zhu, *Phys. Rev. D*, **67**, 075006 (2003).
9. T. Plehn, *Phys. Rev. D*, **67**, 014018 (2003).
10. N. Kidonakis, *JHEP*, **05**, 011 (2005); in *DIS 2004*, hep-ph/0406179.
11. A.D. Martin, R.G. Roberts, W.J. Stirling, and R.S. Thorne, *Eur. Phys. J. C*, **28**, 455 (2003).
12. N. Kidonakis and J.F. Owens, *Phys. Rev. D*, **61**, 094004 (2000); *Int. J. Mod. Phys. A*, **19**, 149 (2004); *Phys. Rev. D*, **63**, 054019 (2001); A. Belyaev and N. Kidonakis, *Phys. Rev. D*, **65**, 037501 (2002); N. Kidonakis and A. Belyaev, *JHEP*, **12**, 004 (2003); in *DIS 2004*, hep-ph/0407032; N. Kidonakis and A. Sabio Vera, *JHEP*, **02**, 027 (2004); in *DPF 2004*, hep-ph/0409206; hep-ph/0409337.

Searches for New Physics in the Flavor Sector at the *B* Factories

W. T. Meyer

Iowa State University, Ames, Iowa

Abstract. Very high luminosities and uniquely favorable kinematics make the SLAC and KEK *B* factories promising places to look for physics beyond the standard model. This presentation summarizes results from three of the many searches that have been performed by the Belle and BaBar collaborations. Two of the searches give results consistent with the standard model, the third one gives results that may indicate new physics, but which require more data and a better understanding of possible competing processes before a conclusion can be drawn.

Keywords: *B* meson, penguin diagram, leptonic decay, new physics search.
PACS: 13.20.He, 13.25.Hw, 13.66.Hk

INTRODUCTION

The outstanding performance of the KEK-B and PEP-II colliders has opened up many areas in which to search for new physics. Between them, the BaBar and Belle detectors have recorded data from more than 650 fb^{-1} of luminosity, permitting the study of rare processes at a level never before achieved. Added to this is the uniquely clean situation of the Y(4S) resonance, where about one-fourth of the hadronic events are *B*-pairs, where even topologies are clean, and where backgrounds are low.

Because of the large number of searches for new physics that the two experiments have performed, I have used personal discretion to choose three topics of interest[1].

A Review of the Weak Interaction of Quarks

When a W-boson interacts with a quark-antiquark pair, the vertex contains a factor V_{pq}, which is an element of the Cabbibo-Kobayashi-Maskawa (CKM) matrix. The CKM matrix is a 3x3 unitary matrix, where the elements are complex numbers. Since the matrix element for $q \to W p$ contains a factor of V_{pq} while that *for* $\bar{q} \to W^{+} \bar{p}$ contains a factor of V_{pq}^{*}, there is a source of CP-violation possible. In the standard model, the complex behavior of the CKM matrix can be represented by a single phase. Using the property of unitary triangles that one column multiplied by the complex conjugate of another column gives zero, we can construct a so-called Unitarity Triangle using the three complex terms in the sum formed by taking the product to form the sides of the triangle. The angles, called [α,β,γ] by BaBar and [ϕ_1,ϕ_2,ϕ_3] by Belle, are the parameters of most interest. In particular, the angle β/ϕ_2 is the one most readily studied at *B* factories. For conciseness, I will us the BaBar notation in the rest of this report.

CP792, *Deep Inelastic Scattering, DIS 2005*, edited by W. H. Smith and S. R. Dasu

The measurement of β is done by measuring the amplitude of the oscillation in neutral B-meson decays to states which are CP eigenstates. This oscillation is measured by identifying ("tagging") one of a pair of B^0 mesons as either a B or $\overline{B}$ from its decay properties. By comparing the magnitude of the asymmetry between the decay rates when the tagging particle is a B and when it is a $\overline{B}$, we determine sin(2β).

GLUONIC PENGUIN PROCESSES

The first results are from processes where a b-quark coverts to an s-quark and a gluon via an internal loop, the so-called "gluonic penguin" or "s-penguin" diagram. The tree level diagram is shown in figure 1a.

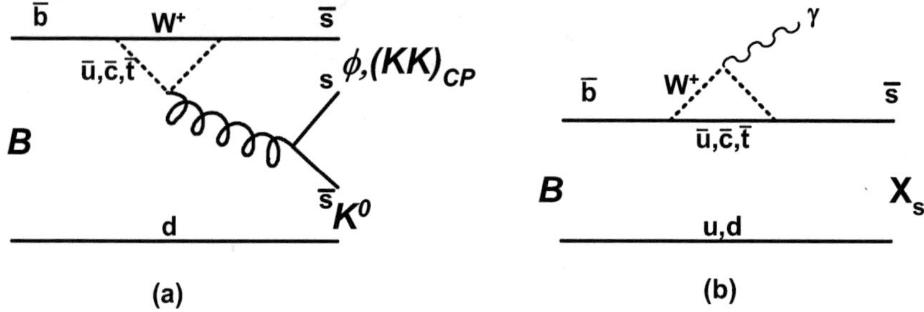

FIGURE 1. The tree-level diagrams for (a) $b\rightarrow sg$ and (b) $b\rightarrow s\gamma$ processes

In the standard model the value of sin(2β) from this process should be the same as for processes where the b-quark converts to a c-quark without an internal loop. An example of such a "golden mode" process is $B\rightarrow J/\psi K^0$, where experiments give a world average result of sin(2β) = 0.726±0.037. If there is new physics, e.g. new heavy particles in the internal loop, sin($2\beta_{eff}$) could differ from this value.

BaBar and Belle have measured sin(2 β_{eff}) for seven gluonic penguin processes. The results are shown in figure 2. The narrow vertical band on the right represents the golden mode result cited above. The wider band to the left is the naïve average of the seven modes measured by both experiments. Taken at face value there appears to be a 3.7σ disagreement between the two bands, but this result is subject to some uncertainty due to the contributions from sub-leading order diagrams, such as multiple gluons being emitted from the internal loop. Moreover, it may not be valid to average over all seven processes if the sub-leading diagrams for them differ. It is worth noting that when analyzed separately the results from Belle and BaBar each give a disagreement of 2.9σ. While this is very suggestive, improved statistics and a better theoretical understanding are needed before any new physics can be claimed from these processes.

RADIATIVE PENGUIN PROCESSES

The second search for new physics is in the radiative penguin processes, as shown at tree level in figure 1b. As before, new physics could show up in the form of additional particles in the internal loop.

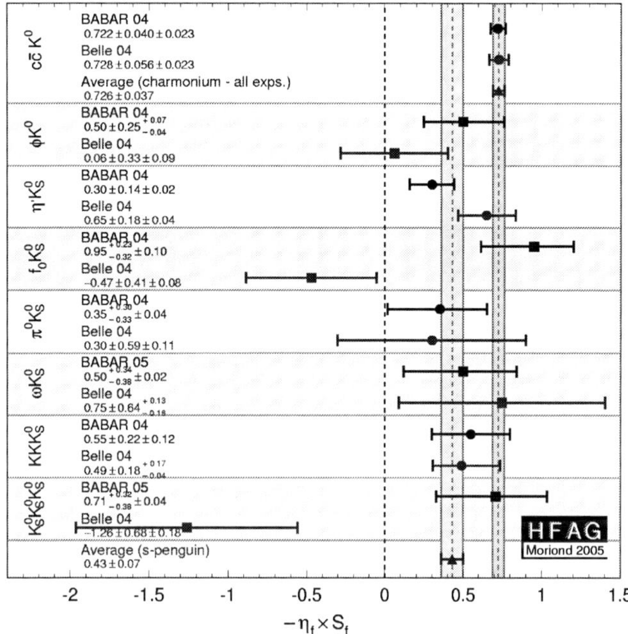

FIGURE 2. The combined $\sin(2\beta_{eff})$ results from gluonic penguin processes.

Two methods have been used to study this process: a fully inclusive $B \rightarrow X_s\gamma$ measurement and a semi-inclusive sum over a large number of final states. Two parameters have been measured: the inclusive branching fraction and the direct CP asymmetry. The results are presented in tables 1 and 2. The results agree with the standard model prediction to within one sigma or less.

TABLE 1. $B \rightarrow X_s\gamma$ Inclusive Branching Fraction

Experiment	Integrated Luminosity	Result
CLEO-01	9.1 fb^{-1}	$(3.21\pm0.43(stat)\pm0.27(sys)^{+0.18}_{-0.10}(th))\times10^{-4}$
BABAR-02	55 fb^{-1}	$(3.90\pm0.36(stat)\pm0.37(sys)^{+0.44}_{-0.25}(th))\times10^{-4}$
BELLE-04	140 fb^{-1}	$(3.55\pm0.32(stat)\pm0.31(sys)^{+0.11}_{-0.07}(th))\times10^{-4}$

The world average for this branching ratio is $(3.34\pm0.38)\times10^{-4}$, which compares well to the standard model prediction of $(3.57\pm0.30)\times10^{-4}$. Note that in this comparison there is an implicit cutoff on the photon energy. The value of the cutoff varies slightly between experiments and for the standard model calculation and is approximately 1.6 GeV. The uncertainty introduced by this is much less than the quoted uncertainty.

TABLE 2. $B \rightarrow X_S \gamma$ Direct CP Asymmetry

Experiment	Integrated Luminosity	Result
CLEO-01	9.1 fb^{-1}	(-0.079±0.108(stat) ±0.022(sys)(1.0±0.030)
BABAR-04	80 fb^{-1}	-0.002±0.050(stat) ±0.030(sys)
BELLE-04	140 fb^{-1}	-0.025±0.050(stat) ±0.015(sys)

All three experiments are consistent with the standard model expectation of zero.

DI-LEPTON DECAYS OF B MESONS

The final result is a measurement of the branching fraction for $B^0 \rightarrow l^+ l$, where l is either an electron or a muon. The standard model predictions for these modes are very close to zero. The results at the 90% confidence level, including those from CLEO and CDF (presented at this workshop) are given in table 3.

TABLE 3. $B^0 \rightarrow l^+ l$ branching fractions, 90% C>L> limits

Mode	Std Model	Belle (78 fb^{-1})	BaBar(111 fb^{-1})	CLEO (9.1 fb^{-1})	CDF (364 pb^{-1})
e^+e^-	2.4 x 10^{-15}	1.9 x 10^{-7}	6.1 x 10^{-8}	8.3 x 10^{-7}	-
$\mu^+\mu^-$	1.0 x 10^{-10}	1.6 x 10^{-7}	8.3 x 10^{-8}	6.1 x 10^{-7}	3.8 x 10^{-8}
$e^{\pm}\mu^{\mp}$	~0	1.7 x 10^{-7}	18 x 10^{-8}	15 x 10^{-7}	-

CONCLUSION

Results from three searches for new physics show one with interesting, but inconclusive, results and two which agree well with standard model expectations. Planned luminosity upgrades offer the prospect of results from a combined 2000 fb^{-1} of data taking by 2008.

ACKNOWLEDGMENTS

Both the BaBar and Belle experiments are deeply grateful to the KEK-B and PEP-II staffs which have worked hard to obtain the excellent performance of both machines. The results presented here are those of large collaborations, the author is merely a presenter of this work by many people. The author thanks his collaborators at BaBar and his friends at Belle for their help in collecting these results.

REFERENCES

1. Space restrictions prevent a detailed list of references. Readers wishing to obtain the results from the Belle and BaBar experiments can do so at the "publications" or "documentation" links on their web sites: http://belle.kek.jp/ and http://www.slac.stanford.edu/BF/ .

Searches for New Physics in the Flavor Sector

M. Herndon for the CDF and D0 Collaborations

Johns Hopkins University

Abstract. Looking for deviations from the Standard Model in measurements from the flavor sector can be a powerful probe for the indications of new physics. In this proceeding we discuss the potential of lifetime measurements, CP asymmetry measurements and searches for rare decays of B hadrons as probes for new physics and present results from the Tevatron experiments.

Keywords: HEP, Flavor Physics, B Physics

INTRODUCTION

Traditionally searches for particles predicted by extensions of or alternatives to the Standard Model(SM) have been performed by looking for direct production of the particles. The simplest example of a direct search is particle anti-particle annihilation leading to the production of single or pairs of new particles. However, another way to search for the evidence of new particles is to look at the decay properties of hadrons. In this scenario the new physics particles occur virtually in the decay diagrams and can lead to branching ratios or decay distributions not predicted by the SM. The best place to look for non SM effects is in decays that are low probability. For instance, in weak decays B hadron decays that can only occur via loop diagrams the predicted contribution from the SM can be on order the contributions from new physics models.

Searching for these rare decays or small deviations from the SM distributions required large statistics. The Tevatron experiments, CDF and D0, are acquiring very large samples of B decays using dedicated triggers. Promising areas for looking for new physics effects include examining B lifetimes, measuring direct charge parity(CP) asymmetries and searching for very rare decays.

In the following sections we will briefly comment on the Tevatron detectors and the properties that make them well suited for the indirect searches discussed above. Then we will review current and in progress measurements of interest.

THE CDF AND D0 DETECTORS

The CDF and D0 detectors are typical high energy physics multipurpose devices. They consist inner and outer trackers immersed in a magnetic field and designed for precession interaction or decay vertex finding and high efficiency track finding; calorimeter systems for measuring the energy of electromagnetic and hadronic particles; and muon chambers. The CDF tracking detector has a large radius which allows for high precision measurements of the mass of B hadrons while the D0 detector is lower radius but has efficient track finding to higher pseudorapidity which is well matched to the large

CP792, *Deep Inelastic Scattering, DIS 2005,* edited by W. H. Smith and S. R. Dasu

coverage of its muon chambers. The CDF detector collects interesting physics events by selecting possible B events based on finding muons and displaced tracks. D0 primarily relies on it's large muon coverage though they are upgrading their trigger to also include lifetime information.

NEW PHYSICS IN $\Delta\Gamma_{B_s^0}$

Particle-antiparticle oscillation occurs in the B_s^0 meson system resulting in two eigenstates with definite masses and widths. Also in the SM the CP eigenstates of the B_s^0 meson are expected to be nearly identical to the mass eigenstates. This makes it possible to directly measure the decay width difference, $\Delta\Gamma_{B_s^0}$, by measuring the lifetime of states with known CP content. In the SM the mass difference, Δm_s, which can be measured in a B_s^0 oscillation analysis, is related to $\Delta\Gamma_{B_s^0}$ by a simple ratio [1]. Where observing B_s oscillations may be challenging at the Tevatron for higher oscillation frequencies, a $\Delta\Gamma_{B_s}$ measurement would be feasible. In new physics models $\Delta\Gamma_{B_s}$ is related to the SM value by the expression $\Delta\Gamma_{B_s} = \Delta\Gamma_{B_s}^{(CP\ conserving)} cos(\phi^{(SM)} + \phi^{(New\ physics)})$. In this expression the SM phase is expected to be zero and new physics contributions would reduce $\Delta\Gamma_{B_s}$ from the SM expectation.

There are several interesting ways to probe $\Delta\Gamma_{B_s^0}$. Examples of such analysis are: measuring the CP eigenstate lifetimes by disentangling the eigenstates using angular information in $B_s^0 \to J/\psi\phi$ decays; measuring the lifetime in modes that are expected to be primarily one eigenstate such as the decay $B_s^0 \to KK$ which is 97% CP even; or considering a decays such as $B_s^0 \to D_sD_s$ which is expected to account for most of the decay width and lifetime difference.

The $\Delta\Gamma_{B_s^0}$ analysis using $B_s^0 \to J/\psi\phi$ has been performed by both CDF and D0. The CDF analysis is performed in the transversity angle basis [2], which defines three decay amplitudes corresponding to linear combinations of the two eigenstates [3]. The D0 analysis is similar except in that they integrates over two of the three transversity angles [4]. The CDF and D0 experiments extract values of $\Delta\Gamma_{B_s^0} = 00.65^{+0.25}_{-0.33} \pm 0.01$ and $\Delta\Gamma_{B_s^0} = 00.21^{+0.33}_{-0.45}$ respectively. The measured values of $\Delta\Gamma_{B_s^0}$ are plotted relative to the SM value and world average constraints in Figure 1. Both values are high, which would not be expected in new physics scenarios, but are compatible with each other and the SM expectations.

Measurements of the lifetime in $B_s^0 \to KK$ decays or the lifetime and decay branching fraction in $B_s^0 \to D_sD_s$ have not been performed yet. However, the CDF experiment has observed the decay $B_s^0 \to KK$ [5] as well as the first B decay to two charmed hadrons at a proton anti-proton collider $B^0 \to D_sD^+$ [6] and is working on extending these analysis.

NEW PHYSICS IN CP ASYMMETRIES

Many models of new physics predict enhancements of the CP asymmetries (A_{CP}) of B decays [7]. At hadron colliders the most promising measurements are of direct A_{CP} or

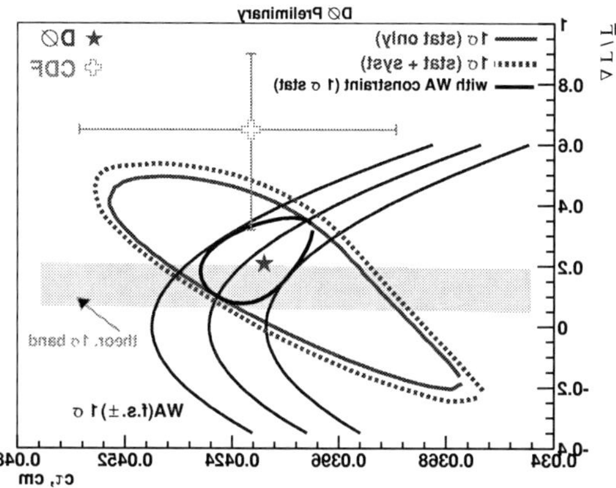

FIGURE 1. $\Delta\Gamma_{B_s^0}$ vs. B_s^0. Also plotted is the standard models predicted range for $\Delta\Gamma_{B_s^0}$ and the range allowed by the world average lifetime results.

differences in the time integrated decay rate of the CP eigenstates. These measurements are simplest when the decays of the eigenstates are flavor specific(when the eigenstate can be identified by the decay products). Examples of flavor specific decays are $B^+ \to J/\psi K^\pm$ or $B^0 \to \pi^+K^-$ vs. $\bar{B}^0 \to \pi^-K^+$. Another possibility is to look at angular information in decays such as $B_s^0 \to \phi\phi$.

The CDF experiment has recently performed CP asymmetry measurements in the $B^+ \to J/\psi K^+$ [8] and $B^0 \to \pi^+K^-$ [5] modes. The measured asymmetries, $A_{CP}(B^+ \to J/\psi K^+) = -0.07 \pm 0.17^{+0.03}_{-0.02}$ and $A_{CP}(B^0 \to \pi^+K^-) = -0.04 \pm 0.08 \pm 0.006$, are of comparable accuracy to world average measurements.

In the mode $B_s^0 \to \phi\phi$ CDF has published an observation[8] based on 12 events with 1.98 ± 0.62 background and preliminarily found that 44 events are present in the data set up to August 2004. This larger amount of events approaches the number necessary to perform an A_{CP}.

NEW PHYSICS IN RARE DECAYS

Searching for rare decays can give one of the least ambiguous signals for new physics. For instance, the decays $B_s^0 \to \mu^+\mu^-$and $B_d^0 \to \mu^+\mu^-$are highly suppressed in the SM with expected branching ratios of $\mathrm{BR}(B_s^0 \to \mu^+\mu^-) = 3.5 \times 10^{-9}$ and $\mathrm{BR}(B_d^0 \to \mu^+\mu^-) = 1.0 \times 10^{-10}$. However, it has been noted [9] that the decay $B_s^0 \to \mu^+\mu^-$can be enhanced by up to 3 orders of magnitude in supersymetric extensions to the SM(SUSY) making it observable at the Tevatron. An observation of this decay would be a clear indication of new physics. In addition, the enhancement of this decay is proportional to $tan^6\beta/m_A^4$ and an observation would give interesting information on the $tan\beta$, the ratio of the vacuum expectation values of the SUSY Higgs' and the mass of the pseudoscaler Higgs [10].

The D0 and CDF collaboration have recently put limits on these processes. The D0 experiment expected 4.3 events and observed 4 and sets a limit of of $\mathrm{BR}(B_s^0 \to \mu^+\mu^-) = 3.7 \times 10^{-9}$ at 95% confidence level(CL) [11]. The CDF experiment expects 1.47 evens and observes none and sets limits of $\mathrm{BR}(B_s^0 \to \mu^+\mu^-) = 2.0 \times 10^{-9}$ and $\mathrm{BR}(B_s^0 \to \mu^+\mu^-) = 4.9 \times 10^{-10}$ at 95% CL [12]. These measurements are twice as sensitive as previous published measurements [13]. In addition, CDF and D0 they have recently produced a combined limit of $\mathrm{BR}(B_s^0 \to \mu^+\mu^-) = 1.6 \times 10^{-9}$ at 95% CL [14]. This combined limit starts to severely constrain the phase space of SUSY variants such as SO(10) gauge unification models [15].

CONCLUSION

Indirect searches for the evidence of new physics in the flavor sector is a promising avenue of investigation. The Tevatron experiments, CDF and D0, have performed a number of new measurements using the large samples of B hadron decays they have collected. No evidence of new physics is yet seen and limits are set on various new physics scenarios.

ACKNOWLEDGMENTS

I would like to acknowledge the work of the CDF and D0 collaborations, the Fermi National Accelerator Laboratory staff and the contributions of the various funding agencies to the experiments and the laboratory

REFERENCES

1. M.Beneke *et al.* Phys. Lett. B **459**, 631 (1999).
2. A.S.Dighe, I.Dunietz, H.J.Lipkin and J.L.Rosner, Phys. Lett. B **369**, 144 (1996).
3. CDF Collaboration, D. Acosta *et al.*, Phys. Rev. Lett. **94**, 101803 (2005).
4. D0 Collaboration, D0 Conference Note 4557 (2005).
5. CDF Collaboration, CDF Public Note 7142 (2004).
6. CDF Collaboration, CDF Public Note 7495 (2005).
7. X.Q.Li, G.R. Lu and Y.D. Yang, Phys. Rev. D **68**, 114015 (2004).
8. CDF Collaboration, D. Acosta *et al.*, Submitted to Phys. Rev. Lett., hep-ex/0502044 (2005).
9. S.Choudhury and N.Gaur, Phys. Lett. B **451**, 86 (1999); K.S.Babu and C.Kolda, Phys. Rev. Lett. **84**, 228 (2000).
10. A.Dedes and B.T.Huffman, Phys. Lett. B **600**, 261 (2004); G.L. Kane, C.Kolda and J.E.Lennon, hep-ph/0310042 (2003);
11. D0 Collaboration, D0 Conference Note 4733 (2005).
12. CDF Collaboration, CDF Public Note 7670 (2005).
13. CDF Collaboration, D.Acosta *et al.*, Phys. Rev. Lett. **93** (2004) 032001; D0 Collaboration, V. M.Abzov *et al.*, Phys. Rev. Lett. **94** (2005) 071802; BABAR Collaboration, Submitted to Phys. Rev. Lett., hep-ex/0408096 (2004).
14. D0 Collaboration, D0 Conference Note 4856 (2005) and CDF Collaboration, CDF Public Note 7705 (2005).
15. R.Dermisek, S.Raby, L.Roszkowski, and R. Ruis de Austri, JHEP **0304**, 037 (2003).

New Physics in the Flavor Sector

Shrihari Gopalakrishna* and C.-P. Yuan†

*Dept. of Physics and Astronomy, Northwestern University, Evanston, IL - 60202.
†Dept. of Physics and Astronomy, Michigan State University, East Lansing, MI - 48824.

Abstract.
The standard model suffers from the hierarchy problem and the flavor problem. Various extensions have been proposed to alleviate these problems, supersymmetry and U(2) flavor symmetry being among them. Flavor changing neutral current rare decay modes of the B-meson is an excellent probe of such extensions. Recent preliminary data from the B-factories indicate that the CP asymmetry in $b \to s$ penguin modes has a discrepancy with standard model predictions, and if established to a higher confidence level would indicate new physics beyond the standard model. We show that a supersymmetric U(2) flavor model can explain such a discrepancy, while being consistent with all other K and B-meson data.

Keywords: Supersymmetry, Flavor symmetry, B physics
PACS: 12.60.Jv, 11.30.Hv, 13.20.-v, 13.25.-k

INTRODUCTION

The standard model (SM) describes impressively the data accumulated to date.[1] The SM, however, suffers from a few aesthetic problems, for example, the gauge hierarchy problem and the flavor problem. The first is the fine tuning required to maintain a low electroweak mass scale (M_{EW}) in the presence of a very high scale, the Planck Scale (M_{Pl}). The second problem is a lack of explanation of the mass hierarchy and mixings of the quarks and leptons. Various beyond the SM (BSM) extensions have been proposed to cure these ills of the SM. A very well motivated extension of the SM that eliminates the hierarchy problem is supersymmetry (SUSY). To address the flavor problem, various flavor symmetries have been proposed, one of which is a nonabelian U(2) "horizontal" symmetry acting on the first 2 generations.

A sensitive probe of BSM physics is flavor changing neutral current processes (FCNC). This is because, they are absent at tree level in the SM and GIM suppressed at loop level making their amplitudes small. A vigorous experimental program is underway to continue testing the SM to glean hints of BSM physics. The recent discrepancy between the SM predictions for $b \to s$ penguin modes and data from BaBar and Belle, if solidified to a higher confidence level, might be signs of BSM physics. In this paper we show that a supersymmetric U(2) flavor theory can accommodate such a deviation while being consistent with all other K and B-meson data. Details are presented in Ref. [1].

[1] There is now a very strong evidence that neutrinos have nonzero masses. The correct ultraviolet finite theory that is responsible for neutrino masses is presently the subject of intense theoretical work. Nonzero neutrino masses also imply mixing in the lepton sector like in quarks.

CP792, *Deep Inelastic Scattering, DIS 2005,* edited by W. H. Smith and S. R. Dasu

THE MODEL

We consider a situation in which in general there is no alignment of the quark/lepton flavor structure with that of the scalar-quark/lepton sector, leading to a non-minimal flavor violation (NMFV) scenario. We consider a spontaneously broken U(2) flavor symmetry [2, 3] in the framework of "effective supersymmetry" [4], in which the first two generation scalars are relatively heavy (a few TeV mass), thereby satisfying electric dipole moment constraints, while still allowing large CP violating phases in the scalar sector.

Consider that the first and second generation superfields (ψ_a, a=1,2) transform as a U(2) doublet while the third generation superfield (ψ) is a singlet [3]. The most general U(2) symmetric superpotential can be written as

$$\mathscr{W} = \psi\alpha_1 H\psi + \frac{\phi^a}{M}\psi\alpha_2 H\psi_a + \frac{\phi^{ab}}{M}\psi_a\alpha_3 H\psi_b + \frac{\phi^a\phi^b}{M^2}\psi_a\alpha_4 H\psi_b + \frac{S^{ab}}{M}\psi_a\alpha_5 H\psi_b + \mu H_u H_d\,, \tag{1}$$

where M is the cutoff scale below which such an effective description is valid, the α_i are O(1) constants, ϕ^a is a U(2) doublet, ϕ^{ab} and S^{ab} are second rank antisymmetric and symmetric U(2) tensors respectively. If U(2) is broken spontaneously by the vacuum expectation values (VEV)

$$\langle\phi^a\rangle = \begin{pmatrix} 0 \\ V \end{pmatrix};\quad \left\langle\phi^{ab}\right\rangle = v\varepsilon^{ab};\quad \left\langle S^{11,12,21}\right\rangle = 0, \left\langle S^{22}\right\rangle = V,$$

with $\frac{V}{M} \equiv \varepsilon \sim 0.02$ and $\frac{v}{M} \equiv \varepsilon' \sim 0.004$, and if U(2) is broken below the SUSY breaking scale, the SUSY breaking masses would also have a structure dictated by U(2). The resulting quark and scalar down-type masses are

$$\mathscr{M}_d = v_d \begin{pmatrix} O & -\lambda_1\varepsilon' & O \\ \lambda_1\varepsilon' & \lambda_2\varepsilon & \lambda_4\varepsilon \\ O & \lambda_4'\varepsilon & \lambda_3 \end{pmatrix}, \quad \mathscr{M}^2_{RL} = v_d \begin{pmatrix} O & -A_1\varepsilon' & O \\ A_1\varepsilon' & A_2\varepsilon & A_4\varepsilon \\ O & A_4'\varepsilon & A_3 \end{pmatrix}, \tag{2}$$

$$\mathscr{M}^2_{LL} = \begin{pmatrix} m_1^2 & i\varepsilon' m_5^2 & 0 \\ -i\varepsilon' m_5^2 & m_1^2 + \varepsilon^2 m_2^2 & \varepsilon m_4^{2*} \\ 0 & \varepsilon m_4^2 & m_3^2 \end{pmatrix}_{LL}, \quad \mathscr{M}^2_{RR} = \begin{pmatrix} m_1^2 & i\varepsilon' m_5^2 & 0 \\ -i\varepsilon' m_5^2 & m_1^2 + \varepsilon^2 m_2^2 & \varepsilon m_4^{2*} \\ 0 & \varepsilon m_4^2 & m_3^2 \end{pmatrix}_{RR},$$

where $v_d = \langle h_d\rangle$ is the VEV of the Higgs field, the λ_i's are O(1) coefficients, and, m_i and A_i (complex in general) are determined by the SUSY breaking mechanism. It has been shown [3] that such a pattern of the quark mass matrix explains the quark masses and CKM elements.

For our study, we consider the following values for the various SUSY parameters: $m_{\tilde{b}_R,\tilde{t}_R} = 100\,\text{GeV}$, the other squark masses given by $m_0 = 1000\,\text{GeV}$, $A = 1000\,\text{GeV}$, $\tan\beta = 5$, $|\mu| = 200\,\text{GeV}$, $M_2 = 250\,\text{GeV}$, $M_{\tilde{g}} = 300\,\text{GeV}$ and $m_{H^\pm} = 250\,\text{GeV}$. ($m_0$ and A denote generic SUSY breaking mass scales.)

Here, we consider processes that go through the $b \to s$ quark level transition, and in our framework the dominant SUSY contributions are due to $\delta^{RL,RR,LL}_{32,23} \equiv \frac{(\mathscr{M}^2_{RL,RR,LL})_{32,23}}{m_0^2}$.

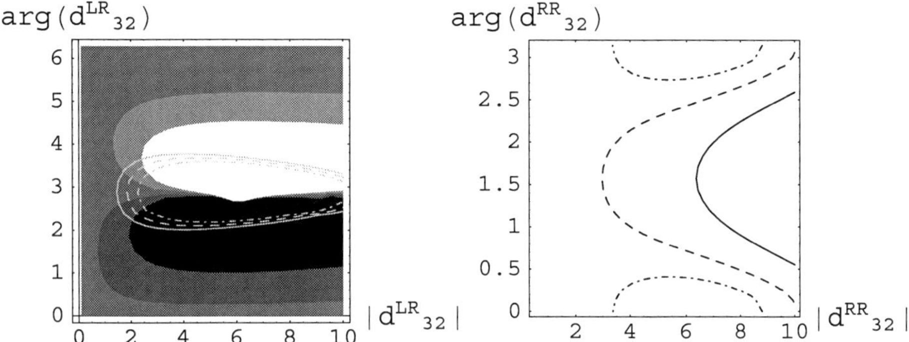

FIGURE 1. (Left) The shaded regions, darkest to lightest are (-7,-3,3,7)% contours of $A_{CP}^{B_d \to X_s\gamma}$ with 2σ contours of B.R.($B_d \to X_s\gamma$) superimposed. (Right) (15, 25, 40 ps^{-1}) contours of Δm_{B_s}.

For the chosen values of the parameters, we find $|\delta_{32,23}^{RL}| \sim \frac{v_d A\varepsilon}{\tilde{m}_0^2} = 6.8 \times 10^{-4} d_{32,23}^{RL}$, and, $|\delta_{32}^{LL,RR}| \sim \varepsilon\frac{m_4^2}{m_0^2} = 0.02 d_{32}^{LL,RR}$, where the $d_{32,23}^{RL,LR}$ are $O(1)$ coefficients determined by SUSY and U(2) breaking dynamics, unspecified in our effective theory.

RESULTS

The SUSY U(2) theory described has new FCNC contributions which can be searched for experimentally. In the theory described, we compute ε_K ($K^0\bar{K^0}$ mixing), Δm_d ($B_d\bar{B}_d$ mixing), $\sin 2\beta$ (CP violation in $B_d \to \psi K_s$), and Branching ratios (B.R.) of $B_d \to X_s\gamma$, $B_d \to X_s\ell^+\ell^-$and $B_d \to \phi K_s$, and select regions of SUSY U(2) parameter space that are consistent with experiment. For this region of parameter space allowed by all the preceding processes, we then obtain expectations for the CP violation in $B_d \to X_s\gamma$ and $B_d \to \phi K_s$, and, Δm_s ($B_s\bar{B}_s$ mixing). We show, as an example, in Fig. 1 (left), the region in d_{23}^{LR} parameter space allowed by B.R.($B_d \to X_s\gamma$) and also show the CP asymmetry in $B_d \to X_s\gamma$ expected. The boundaries between the shaded regions show (-7,-3,3,7)% (darkest to lightest) contours of $A_{CP}^{B_d \to X_s\gamma}$ with the experimentally allowed 2σ contours of B.R.($B_d \to X_s\gamma$) superimposed. We note here that at 95% C.L. the present limit from BaBar and Belle is $A_{CP}^{B_d \to X_s\gamma} < 7\%$. We see that a significant enhancement above the SM prediction (which is $< 1\%$) is possible in our SUSY U(2) model. Fig. 1 (right) shows the expectations for Δm_{B_s} which in some regions of parameter space is much larger than the SM prediction (which is $\lesssim 25\ \text{ps}^{-1}$). The current experimental lower bound is $14.4\ \text{ps}^{-1}$, and in the future will be improved upon by the Tevatron.

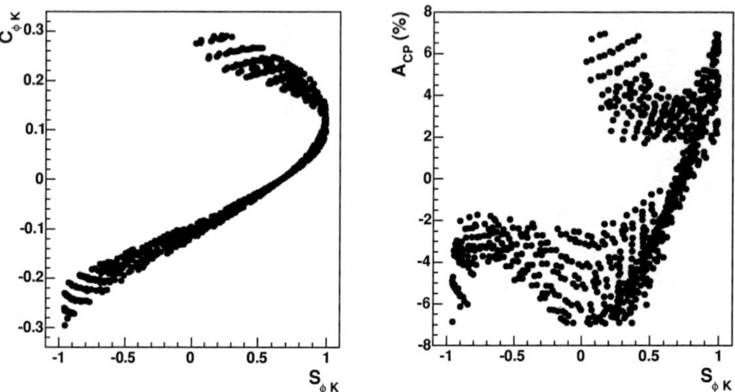

FIGURE 2. Correlations between $A_{CP}^{B_d \to X_s \gamma}$, $S_{\phi K}$ and $C_{\phi K}$ for points that satisfy all experimental constraints (within 2σ), resulting from a scan over d_{32}^{LR} and $\arg(\mu)$.

The CP asymmetry in $B_d \to \phi K_s$ can be written as

$$A_{CP}^{B_d \to \phi K_s} = -C_{\phi K} \cos(\Delta m_{B_d} t) + S_{\phi K} \sin(\Delta m_{B_d} t) \; . \tag{3}$$

In the SM, $S_{\phi K}$ is equal to $S_{\psi K} \equiv \sin 2\beta$, and the present value is $\sin 2\beta = 0.725 \pm 0.037$. The recent heavy flavor averaging group (HFAG) [5] average of BaBar and Belle $B_d \to \phi K_s$ data ($S_{\phi K} = 0.34 \pm 0.2, C_{\phi K} = -0.04 \pm 0.17$) shows about a 2σ deviation [2] from the SM expectation, and if this discrepancy is established to a higher confidence limit with more data, it could be due to BSM physics. We show in Fig. 2 that a SUSY U(2) theory of the kind we are considering can accommodate this lower $S_{\phi K}$. We show correlations of $C_{\phi K}$ versus $S_{\phi K}$ (left), and $A_{CP}^{B_d \to X_s \gamma}$ versus $S_{\phi K}$ (right) that result from a scan over d_{32}^{LR} and $\arg(\mu)$, and satisfy all other K and B-meson constraints. As we collect more data, using such plots, we can test the validity of our framework and the parameter choices, or shrink down the allowed parameter space. We thus await with interest more data from the B-factories and the Tevatron. Before the LHC can probe BSM physics directly, flavor physics might provide the first evidence indirectly.

REFERENCES

1. S. Gopalakrishna and C. P. Yuan, Phys. Rev. D **71**, 035012 (2005).
2. A. Pomarol and D. Tommasini, Nucl. Phys. B **466**, 3 (1996).
3. R. Barbieri, G. R. Dvali and L. J. Hall, Phys. Lett. B **377**, 76 (1996); R. Barbieri, L. J. Hall and A. Romanino, Phys. Lett. B **401**, 47 (1997).
4. A. G. Cohen, D. B. Kaplan and A. E. Nelson, Phys. Lett. B **388**, 588 (1996).
5. The heavy flavor averaging group, URL: `http://www.slac.stanford.edu/xorg/hfag/`

[2] When all the $b \to s$ penguin modes are averaged [5], the discrepancy with the SM is about $3.5\,\sigma$.

HADRONIC FINAL STATES
WORKING GROUP PRESENTATIONS

H1 search for a narrow baryonic resonance decaying to $K^0_S p(\bar{p})$

Christiane Risler

DESY, Notkestrasse 85, D-22607 Hamburg, Germany

Abstract. Preliminary results from the H1 experiment on the search for the production of a candidate for the strange pentaquark in the decay channel $\Theta^+ \to K^0_s p$ and its antiparticle in the invariant mass combinations of K^0_s mesons with protons and antiprotons in deep-inelastic *ep*-scattering at HERA are presented.

Keywords: deep-inelastic scattering, strange pentaquark
PACS: 12.39.Mk

INTRODUCTION

Recently several experiments have observed narrow baryonic resonances in various reaction processes [1] which can be interpreted as strange pentaquark Θ^+. However, negative results have been reported [1] from *pp* and *ep* collisions, e^+e^- annihilation and also from fixed target photo-production experiments. Preliminary results on the search for the strange pentaquark Θ^+ [1] decaying into $K^0_s p$ in deep-inelastic *ep* scattering (DIS) with the H1 detector are presented here. Despite the lack of understanding of the pentaquark production mechanism in high energy processes it is assumed that pentaquark formation is part of the fragmentation process. Since no significant signal was found upper limits are derived on the visible Θ^+ cross section in DIS at HERA.

ANALYSIS OF K^0_S P COMBINATIONS

The analysed data was collected with the H1 detector in the years 1996 to 2000 and corresponds to an integrated luminosity of $71\,\text{pb}^{-1}$. A detailed description of the H1 detector can be found elsewhere [4]. DIS events were selected by requiring a reconstructed scattered electron in the backward calorimeter of H1 and an exchanged photon virtuality of $Q^2 > 5\,\text{GeV}^2$. The condition $0.1 < y < 0.6$ ensures substantial hadronic final state energies in the central detector region. K^0_s meson decays are reconstructed via the decay mode $K^0_s \to \pi^+\pi^-$ by requiring a radial displacement of the decay vertex of at least 2 cm from the primary interaction point. Only those K^0_s candidates having transverse momenta $p_T(K^0_s) \geq 0.3$ GeV and pseudorapidities $|\eta(K^0_s)| \leq 1.5$ in the laboratory frame are accepted. About $142000 K^0_S$ mesons are reconstruced. Candidate K^0_s mesons within

[1] The charge conjugate state is always implied if not otherwise stated expliciely.

CP792, *Deep Inelastic Scattering, DIS 2005*, edited by W. H. Smith and S. R. Dasu

$\pm 2\sigma$ of the measured K_s^0 mass are further combined with tracks originating from the primary vertex assigned the proton mass. These proton tracks are selected using the measurement of the ionisation loss $\mathrm{d}E/\mathrm{d}x$ in the central drift chambers of H1 yielding a resolution of about 8% for minimum ionizing particles. From the difference of the measurement and a Bethe-Bloch-like parameterisation the likelihoods for different particle hypotheses are calculated which are used for particle identification.

The selected K_s^0 meson and proton candidates are combined and the invariant mass $M(K_s^0 p)$ of these combinations is formed by fixing the K_s^0 mass to its nominal value [5]. For the $K_s^0 p$ system $p_t(K_s^0 p) > 0.5\ \mathrm{GeV}^2$ and $|\eta(K_s^0 p)| < 1.5$ is required. The $M(K_s^0 p)$ distributions for three different bins in Q^2, $5 < Q^2 < 10\,\mathrm{GeV}^2$, $10 < Q^2 < 20\,\mathrm{GeV}^2$ and $20 < Q^2 < 100\,\mathrm{GeV}^2$ are shown in Fig. 1 together with a fit of a background function. No significant structure is observed in neither of the Q^2 bins. The $M(K_s^0 p)$

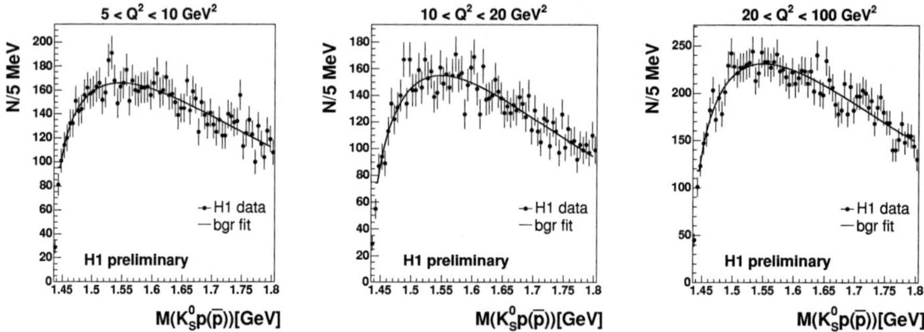

FIGURE 1. Invariant $K_S^0 p(\bar{p})$ mass spectra for the standard dE/dx selection in bins of Q^2, which are used for the limit extraction. The full line shows the result from the fi t of a background function to the data. The mass spectra show upward fluctuations at different masses but no signifi cant peak is observed.

distributions are used to derive mass dependent upper limits at 95 % confidence level (C.L.) on the visible Θ^+ production cross section $\sigma(ep \to e\Theta^+ X \to K^0 pX)$. In order to set limits, it is assumed that strange pentaquarks are produced by fragmentation. The acceptances were calculated using the RAPGAP 3.1 [6] event generator incorporating fragmentation according to the Lund string model [7] implemented in PYTHIA 6.2 [8]. The experimental resolution of a possible state with zero width decaying to $K_s^0\ p$ is expected to be $\sigma(M(K_s^0 p)) = 5$ MeV.

The experimental systematic uncertainties comprise contributions from e.g. the DIS event selection, uncertainties of the efficiencies of track reconstruction and $\mathrm{d}E/\mathrm{d}x$ selection and variations of the fitting method. The total systematic uncertainty is 18%.

The resulting upper limits on the Θ^+ cross section $\sigma_{u.l.}(ep \to e\Theta^+ X \to K^0 pX)$ in the visible range defined by $5 < Q^2 < 100\,\mathrm{GeV}^2$, $0.1 < y < 0.6$, $pt(K_s^0 p) > 0.5\,\mathrm{GeV}$ and $|\eta(K_s^0 p)| < 1.5$ at 95% confidence level are shown in Fig. 2 in different Q^2 bins. The upper limit on the number of Θ^+ is derived assuming a width of a possible signal of 5 and 8 MeV, using signal mass windows of ± 10 and ± 16 MeV, respectively, shown as full and dashed lines. In the lowest Q^2 bin an upward fluctuation of the upper limit in the interesting mass region 1.52 to 1.54 GeV is observed. However, a different shape of the fluctuations of the limits for the different Q^2 bins is found. The 95% C.L. upper

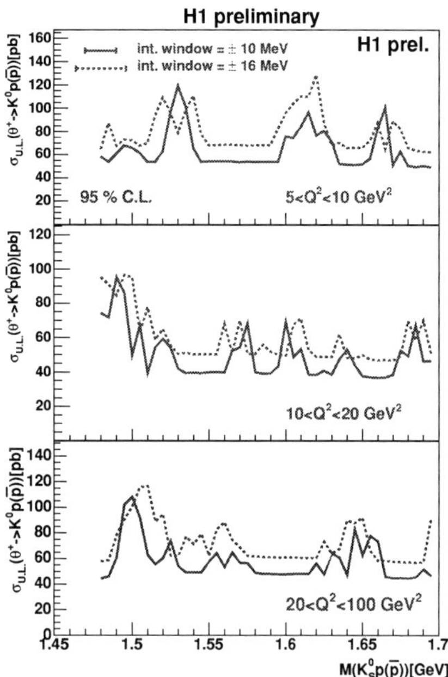

FIGURE 2. Upper limits on the cross section $\sigma_{U.L.}(ep \to e\theta^+X \to eK^0p(\bar{p})X)$ at 95% confidence level in bins of Q^2 in the visible range $p_T(K_s^0p) > 0.5$ GeV and $|\eta(K_S^0p)| < 1.5$. The full and dashed line represents the limit using ± 10 and ± 16 MeV mass windows, respectively.

limits vary between 40 and 120 *pb* for the different Q^2 regions and over the mass range from 1.48 to 1.7 GeV. The upper limits for the decay $\Theta^+ \to K_s^0 p$ and its charge conjugate $\bar{\Theta}^- \to K_s^0 \bar{p}$ were found to be of comparable size and the fluctuations of the corresponding limits were found to be at different masses.

A positive Θ^+ observation in DIS at HERA has been reported at a mass of 1.522 GeV by the ZEUS experiment [2] with an observed visible cross section of $\sigma(e^\pm p \to e^\pm\Theta^+X \to e^\pm K^0 pX) = 125 \pm 27(stat.)^{+36}_{-28}(syst.)$ pb in the kinematic range $Q^2 > 20\,\text{GeV}^2, 0.04 < y < 0.95$ using a data sample with an integrated luminosity of 121 pb^{-1}. In order to be able to compare the upper limits on the Θ^+ production more directly to these results, the analysis was repeated using a proton selection more similar to that used in [2], in the following called "low momentum dE/dx selection". The dE/dx-likelihood proton selection was replaced by a visual selection, requiring $\mathrm{d}E/\mathrm{d}x > 1.15$, $p(p) < 1.5$ GeV and $f_1 < \mathrm{d}E/\mathrm{d}x < f_2$, where f_i are functions enclosing 98 % of the proton dE/dx band. The invariant $K_s^0 p(\bar{p})$ mass spectrum for $20 < Q^2 < 100\,\text{GeV}^2$ and the resulting upper limits on the Θ^+ cross section at 95 % confidence level are shown in Fig. 3. Around the mass of 1.52 GeV an upper limit on the cross section of roughly 100 pb is found, which does not contradict the cross section of the ZEUS experiment quoted above. Also for this low momentum dE/dx selection the upper limits for the positive and nevative combinations were investigated separetely and the up- and down-

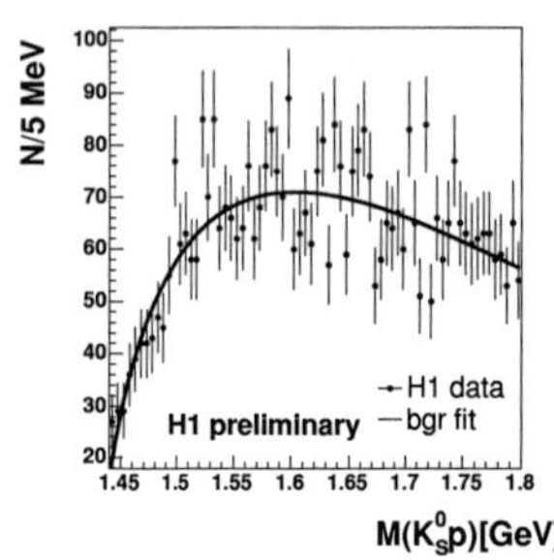

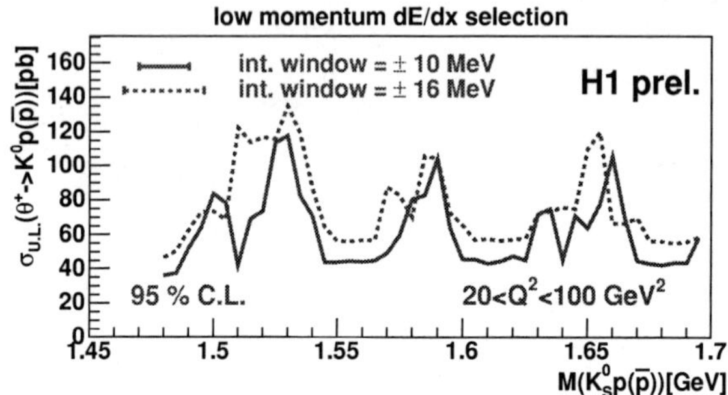

FIGURE 3. Invariant $K^0_S p$ mass spectra in the highest Q^2 bin, $20 < Q^2 < 100$ GeV^2, for the low momentum dE/dx selection (left) and the corresponding upper limit at 95 % confidence level on the Θ^+ cross section $\sigma_{U.L.}(ep \to e\theta^+X \to eK^0p(\bar{p})X)$ (right).

ward fluctuations were found to be at different masses in the respective upper limits on the Θ^+ ($\bar{\Theta}^-$) cross section.

CONCLUSIONS

The preliminary results of the search for the strange pentaquark Θ^+ in deep-inelastic ep scattering has been presented. No significant signal for Θ^+ production in the decay mode $\Theta^+ \to K^0_s p(\bar{p})$ has been observed in the $K^0_s\, p(\bar{p})$ mass distribution for different regions in Q^2 between 5 and 100 GeV^2. With the assumption that pentaquarks are produced by fragmentation mass dependent upper limits on the cross section $\sigma(ep \to e\Theta^+X \to K^0p(\bar{p})X)$ are derived as a function of Q^2 and found to vary between 40 and 120 pb over the mass range of 1.48 to 1.7 GeV. In order to compare to the previous measurement of the visible Θ^+ cross section by the ZEUS collaboration, the analysis was repeated with a low momentum dE/dx selection. The resulting upper limit does not exclude the previosly observed cross section [9]. The present statistical precision in the HERA I data sample is not sufficient to draw a stronger conclusion.

REFERENCES

1. K. Hicks,*Experimental Search for Pentaquarks*, [hep-ex/0504027], and references therein.
2. S. Chekanov *et al.* [ZEUS Collaboration], Phys. Lett. **591** (2004) 7.
3. A.Aktas *et al.* [H1 Collaboration] Phys. Lett **B 588** (2004) 17.
4. I. Abt *et al.* [H1 Collaboration], Nucl. Inst. Meth. **A386** (1997) 310;
   I. Abt *et al.* [H1 Collaboration], Nucl. Inst. Meth. **A386** (1997) 348.
5. Particle Data Group, H. Hagiwara *et al.*, Phys. Rev. **D66** (2002) 010001.
6. H. Jung, Comp. Phys. Comm. **71**, 15 (1992).
7. B. Andersson, G. Gustafson, G. Ingelman and T. Sjöstrand, Phys. Rept. **97** (1983) 31.
8. T. Sjöstrand *et al.*, Comp. Phys. Commun. **135** (2001) 238 [hep-ph/0010017].
9. S. Chekanov *et al.* [ZEUS Collaboration], Contributed paper to 32nd International Conference on High-Energy Physics, Beijing(2004), abstract no. 10-0273.

Analysis of the Anti-charmed Baryon State at H1

Karin Daum

Bergische Universität Wuppertal, Gaußstrasse 20,D-42097 Wuppertal, Germany [1]

Abstract. The measurement of acceptance corrected ratios $\sigma(D^*p(3100))/\sigma(D^*)$ for electro-production of the anti-charmed baryon state $D^*p(3100)$ decaying into D^* and p is presented. The analysis based on the 1996-2000 data is performed in the deep inelastic scattering region $1 < Q^2 < 100\ \mathrm{GeV}^2$, $0.05 < y_e < 0.7$.

Keywords: deep-inelastic scattering, charmed pentaquark
PACS: 12.39.Mk

INTRODUCTION

Recently the H1 experiment has reported the observation of a narrow resonance decaying to $D^{*-}p$ [2] with a mass of 3099 MeV in deep inelastic ep scattering at HERA [1]. This resonance is a candidate for the charmed pentaquark Θ_c^0. Subsequent searches by other high energy physics experiments did not confirm this observation [2]. To facilitate further comparisons of experimental results and to investigate the production mechanism of the $D^*p(3100)$ resonance[3] its production phase space is explored in this paper. The data presented here include acceptance corrections assuming pentaquark production as part of the fragmentation process.

ANALYSIS OF $D^{*-}P$ COMBINATIONS

The data were collected with the H1 detector in the years 1996 to 2000 and corresponds to an integrated luminosity of $76\,\mathrm{pb}^{-1}$. A detailed description of the H1 detector is given in [3]. DIS events are selected by requiring a reconstructed scattered electron in the backward calorimeter of H1 in the kinematic range $Q^2 > 1\,\mathrm{GeV}^2$ and $0.05 < y < 0.7$. The selection of D^* mesons and proton candidates is the same as in [1]. The decay channel $D^* \to D^0\pi_s \to K\pi\pi_s$ is used to reconstruct D^* mesons. D^* candidates in the visible region $p_T(D^*) > 1.5$ GeV, $-1.5 < \eta(D^*) < 1$ and $z(D^*) > 0.2$ having a mass difference $\Delta M_{D^*} = m(K\pi\pi_s) - m(K\pi)$ within ± 2.5 MeV of the nominal value $\Delta M_{D^*} = 145.4$ MeV are combined with oppositely charged proton candidates selected according to the proton likelihood based on the particles energy loss $\mathrm{d}E/\mathrm{d}x$ in the central trackers.

---

[1] Permanent address: DESY, Notkestrasse 85, D-22607 Hamburg, Germany; e-mail:daum@mail.desy.de
[2] The charge conjugate state is always implied if not otherwise stated explicitly.
[3] Since the spin of the resonance is unknown the term $D^*p(3100)$ is used through out this paper.

CP792, *Deep Inelastic Scattering, DIS 2005,* edited by W. H. Smith and S. R. Dasu

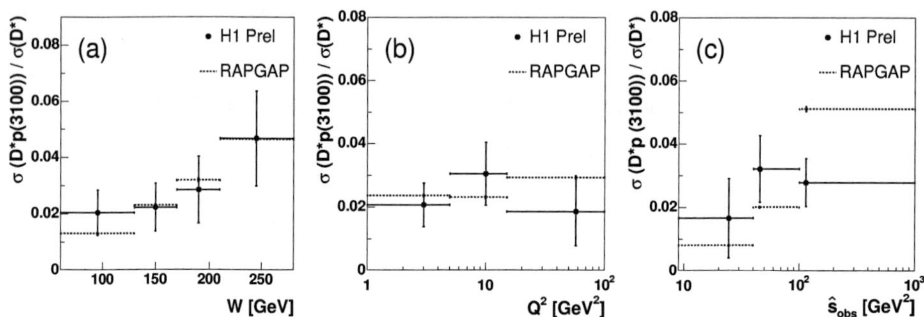

FIGURE 1. $\sigma_{vis}(D^*p(3100))/\sigma_{vis}(D^*)$ as a function of the kinematic variables (a) W, (b) Q^2 and (c) $\hat{s}_{obs}$. Data (closed symbols) are compared with the expectation (dashed line) of RAPGAP 3.1 which assumes the same mechanism for D^* and $D^*p(3100)$ production. Only statistical errors are shown.

The acceptances for the D^* meson and for the $D^*p(3100)$ baryon are calculated by Monte Carlo methods using the RAPGAP 3.1 [4] event generator incorporating fragmentation according to the Lund string model [5] implemented in PYTHIA 6.1 [6]. Pentaquarks are assumed to be produced by fragmentation. The generated events are passed through the full detector simulation using GEANT 3.15 [7] and are subsequently subjected to the same reconstruction and analysis chain as the data.

For the visible range of the $D^*p(3100)$: $p_t(D^*p(3100)) > 1.5$ GeV, $-1.5 < \eta(D^*p(3100)) < 1.0$ and of the D^* meson: $p_t(D^*) > 1.5$ GeV, $-1.5 < \eta(D^*) < 1.0$, $z(D^*) > 0.2$ a total acceptance corrected yields ratio of

$$R_{cor}(D^*p(3100)/D^*) = \left(1.59 \pm 0.33(stat.)^{+\,0.33}_{-\,0.45}(syst.)\right)\ \% \tag{1}$$

has been observed. The same D^* visibility cuts are required for the D^* meson originating from $D^*p(3100)$ baryon decay and for those from the inclusive D^* mesons sample. If acceptance corrections to the $D^*p(3100)$ signal are applied by extrapolating to the full D^* phase space from $D^*p(3100)$ decay the visible cross section ratio is

$$\sigma_{vis}(D^*p(3100))/\sigma_{vis}(D^*) = \left(2.48 \pm 0.52(stat.)^{+0.85}_{-0.64}(syst.)\right)\ \%. \tag{2}$$

In figure 1 the acceptance corrected ratio $\sigma_{vis}(D^*p(3100))/\sigma_{vis}(D^*)$ is shown as a function of the hadronic mass W, the four momentum transfer squared of the virtual photon Q^2 and the invariant mass of the $c\bar{c}$ system, $\hat{s}_{obs}$ in comparison with the expectations of the fragmentation production model. Since the absolute normalization of the $D^*p(3100)$ rate is not fixed in the model, the $D^*p(3100)$ yield is normalized such to reproduce the ratio R_{cor} in (1). The observed dependence on W and on Q^2 is well described by this model, while it is significantly above the data at large $\hat{s}_{obs}$.

In order to investigate the properties of the D^* mesons contributing to the $D^*p(3100)$ resonance the ratio $\sigma_{vis}(D^*p(3100))/\sigma_{vis}(D^*)$ is shown in figure 2 as a function of the pseudorapidity $\eta(D^*)$ and the transverse momentum $p_t(D^*)$, both in the laboratory frame, the inelasticity $z(D^*)$ and the pseudorapidity $\eta^*(D^*)$ in the hadronic centre-of-mass system. Also shown are the expectations from the model. The most striking feature

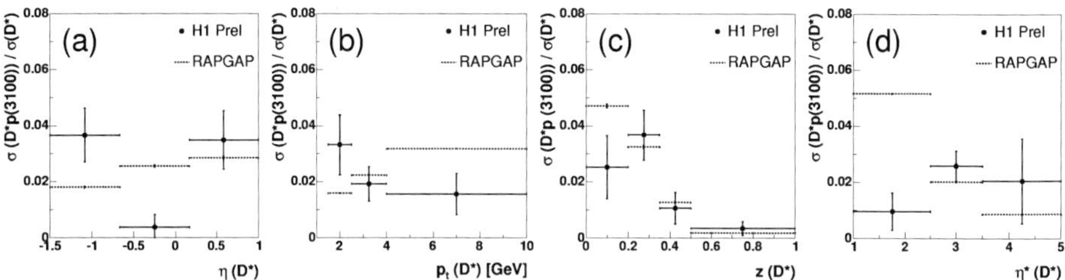

FIGURE 2. $\sigma_{vis}(D^*p(3100))/\sigma_{vis}(D^*)$ as a function of D^* variables (a) $\eta(D^*)$ and (b) $p_t(D^*)$ both in the laboratory frame, (c) $z(D^*)$ and (d) $\eta^*(D^*)$ in the hadronic centre-of-mass system. See fig.1 for details.

in the data is the suppression of the $D^*p(3100)$ baryon relative to D^* meson production in the near to central region in both frames. Such a dependence is not predicted by the fragmentation production model. The data indicate that $D^*p(3100)$ baryon production is closer to the photon direction than normal D^* meson production.

In figure 3 $D^*p(3100)$ differential cross sections are presented as a function of $\eta(D^*p)$, $p_t(D^*p)$, $z(D^*p)$ and $\eta^*(D^*p)$. The $D^*p(3100)$ production cross section shows the same features as a function of $\eta(D^*p)$ and $\eta^*(D^*p)$ than observed for the ratio $\sigma_{vis}(D^*p(3100))/\sigma_{vis}(D^*)$ as a function of the D^* variables. Within the quite large statistical errors the shapes of the $z(D^*p)$ and of the $p_t(D^*p)$ distributions are consistent with the fragmentation production model. These two distributions are suggesting that boson gluon fusion is the source for the production of $D^*p(3100)$ baryons while the pseudorapidity distributions are not evidently supporting this picture.

Finally, information on the $D^*p(3100)$ fragmentation process and the D^* hadronization of D^* mesons from $D^*p(3100)$ decay has been extracted from the data. In figure 4 the cross section ratio $\sigma_{vis}(D^*p(3100))/\sigma_{vis}(D^*)$ as a function of the D^* hadronization variable $x_{obs}(D^*)$ and the differential $D^*p(3100)$ cross section as a function of fragmentation variable $x_{obs}(D^*p)$ are shown together with the predictions from the model. The ratio $\sigma_{vis}(D^*p(3100))/\sigma_{vis}(D^*)$ increases with decreasing $x_{obs}(D^*)$ value which means that D^* mesons originating from $D^*p(3100)$ decay are significantly softer than inclusive D^* mesons. This is expected in decay of a real $D^*p(3100)$ particle. In figure 4b the differential cross section $d\sigma_{vis}(D^*p(3100))/dx_{obs}(D^*p)$ is shown as a function of $x_{obs}(D^*p)$.

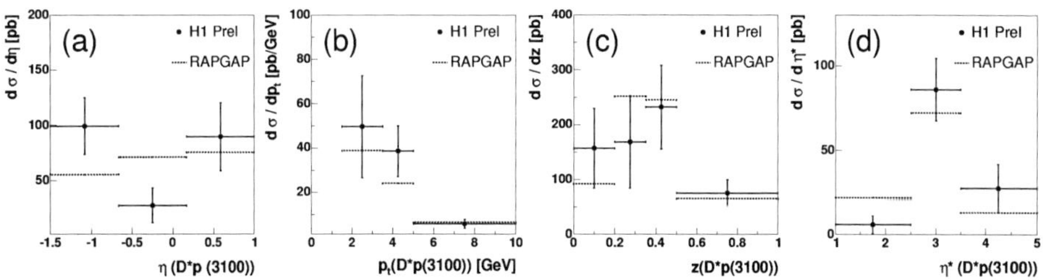

FIGURE 3. Differential $(D^*p(3100))$ cross sections as a function of D^*p variables (a) $\eta(D^*p)$ and (b) $p_t(D^*p)$, (c) $z(D^*p)$ and (d) $\eta^*(D^*p)$. See fig.1 for details.

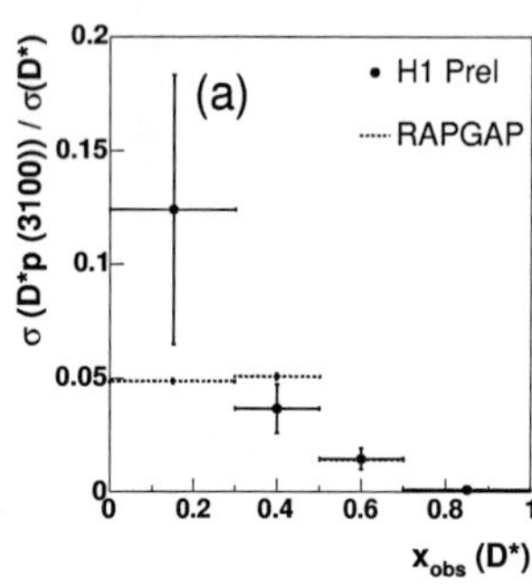

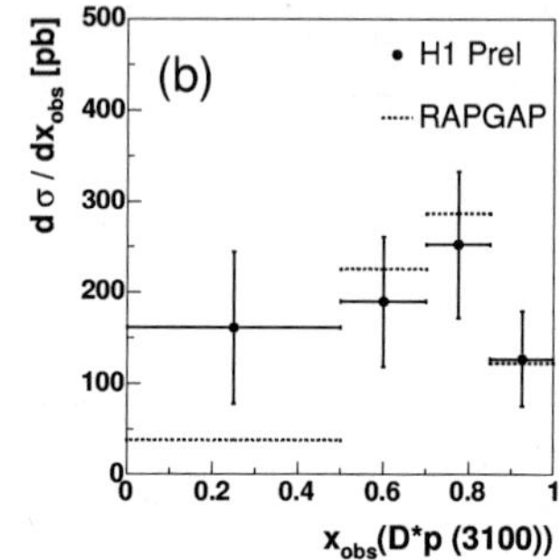

FIGURE 4. $\sigma_{vis}(D^*p(3100))/\sigma_{vis}(D^*)$ as a function of the D^* hadronization fraction $x_{obs}(D^*)$ in (a) and $d\sigma_{vis}(D^*p(3100))/dx_{obs}(D^*p)$ in (b). See fig.1 for details.

The $D^*p(3100)$ fragmentation function is very hard compared to the D^* hadronization function of figure 4a. Such hard fragmentation is expected for charmed hadrons.

CONCLUSION

A detailed analysis of the exotic $D^*p(3100)$ baryon has been presented. An acceptance corrected yields ratio $R_{cor}(D^*p(3100)/D^*) = \left(1.59 \pm 0.33(stat.)^{+\,0.33}_{-\,0.45}(syst.)\right)$ % for the visible $D^*p(3100)$ and D^* range has been observed.

Differential distributions of $\sigma_{vis}(D^*p(3100))/\sigma_{vis}(D^*)$ as a function of event kinematics and D^* quantities as well as differential $D^*p(3100)$ cross sections as a function of $D^*p(3100)$ variables have been presented. In general the fragmentation production model leads to a reasonable description of the data with some exceptions. Compared to inclusive D^* production the $D^*p(3100)$ production seems to be suppressed in the close to central rapidity region. The $D^*p(3100)$ fragmentation function is hard, as expected for charmed hadrons. The hadronization function of D^* mesons from the $D^*p(3100)$ resonance is much softer than observed in inclusive D^* mesons production.

REFERENCES

1. A.Aktas *et al.* [H1 Collaboration] Phys. Lett **B 588** (2004) 17.
2. S. R. Armstrong, hep-ex/0410080; S. Schael *et al.*, (ALPEH), Phys. Lett. B **599**(2004) 1; D. O. Litvintsev, (CDF), hep-ex/0410024; K. Stenson (FOCUS), hep-ex/0412021; R. Mizuk *et al.*, (Belle), hep-ex/0411005. S. Chekanov *et al.* [ZEUS Collaboration], Eur. Phys. J. **C 38** (2004) 29.
3. I. Abt *et al.* [H1 Collaboration], Nucl. Inst. Meth. **A386** (1997) 310;
I. Abt *et al.* [H1 Collaboration], Nucl. Inst. Meth. **A386** (1997) 348.
4. H. Jung, Comp. Phys. Comm. **71** (1992) 15.
5. B. Andersson, G. Gustafson, G. Ingelman and T. Sjöstrand, Phys. Rept. **97** (1983) 31.
6. T. Sjöstrand *et al.*, Comp. Phys. Commun. **135** (2001) 238 [hep-ph/0010017].
7. R. Brun *et al.*, GEANT3, Technical Report CERN-DD/EE/84-1, CERN, 1987.

The search for strange pentaquarks at ZEUS

Z.Ren on behalf of ZEUS collaboration

DESY ZEUS Columbia, Notke Str. 85, 22607 Hamburg, Germany
Pupin 5283, 538 W. 120th Str., New York, NY 10027, USA

Abstract. A study of states and enhancements reconstructed using the invariant-mass spectra associated with strange baryons has been performed in ep collisions with the ZEUS detector at HERA using an integrated luminosity of 121 pb^{-1}. The invariant-mass spectra were reconstructed in several kinematic regions with the main emphasis on the spectra which are sensitive to the production of pentaquarks. The candidate Θ^+ signal was found to be produced predominantly in the forward hemisphere in the laboratory frame. This is unlike the case for the $\Lambda(1520)$ or the Λ_c, and indicates that the Θ^+ may have an unusual production mechanism related to proton-remnant fragmentation.

Keywords: pentaquark, ZEUS, HERA
PACS: 14.80.-j

INTRODUCTION

This paper describes an inclusive search for new states whose decay products include strange hadrons. The searches involving Ξ baryons have already been published [1]. The invariant-mass spectra were reconstructed in several kinematic regions with the main emphasis on spectra which are sensitive to the production of pentaquarks. The invariant masses of known states with similar decay channels as for pentaquarks were also studied in order to show the sensitivity to new resonances and to compare their respective production mechanisms.

DATA SAMPLE

The data sample for this analysis was taken during the 1996–2000 running period of HERA, and corresponds to an integrated luminosity of 121 pb^{-1}. In this analysis, the data sample was divided into two categories: photoproduction (PHP) and DIS. The latter sample was studied at low ($Q^2 > 1$ GeV^2) and medium ($Q^2 > 20$ GeV^2) values of the exchanged photon virtuality, Q^2.

The present analysis was based on tracks measured in the CTD. The energy-loss measurement in the CTD, dE/dx, was used for particle identification [2].

The K_S^0 mesons and Λ baryons were identified by their charged decay mode, $K_S^0 \to \pi^+\pi^-$ and $\Lambda \to p\pi^-$ using the pairs of tracks originating from secondary vertices. The resulting $\pi^+\pi^-$ and $p\pi^-$ invariant-mass spectra are shown in previous publications. About 4.4M K_S^0 and 840k $\Lambda(\bar{\Lambda})$ candidates were reconstructed in both PHP and DIS samples [1, 2].

CP792, *Deep Inelastic Scattering, DIS 2005*, edited by W. H. Smith and S. R. Dasu

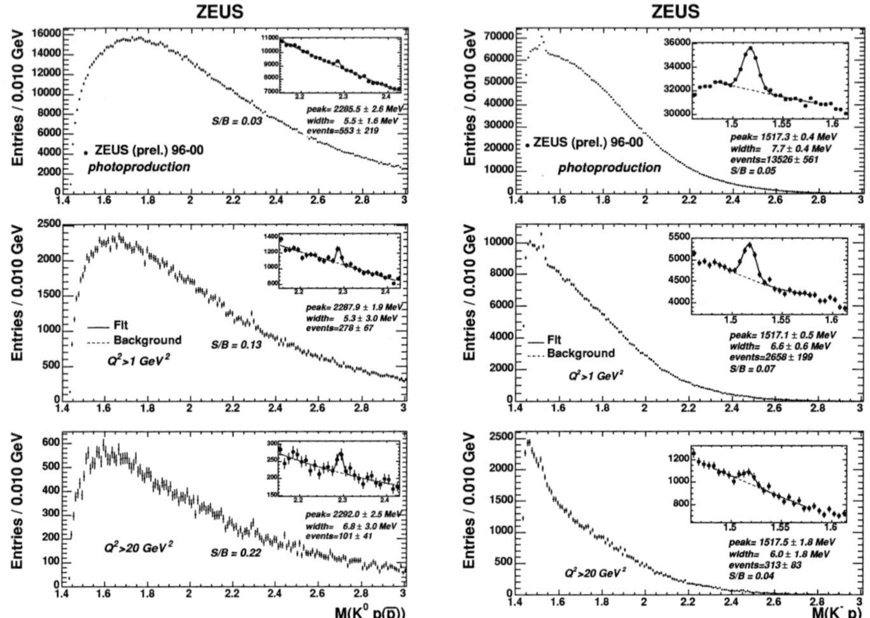

FIGURE 1. The $K^0_S p$ invariant-mass spectra in PHP and DIS ($Q^2 > 1$ GeV^2 and $Q^2 > 20$ GeV^2). The insets show the invariant-mass distribution near the Λ_c mass region. The solid line shows a fit using a Gaussian plus a second-order polynomial background function using 5 MeV bins. The signal-over-background ratio (S/B) for the Λ_c peak is indicated on each figure.

FIGURE 2. The $K^- p$ $(K^+ \bar{p})$ invariant-mass spectra in PHP and DIS ($Q^2 > 1$ GeV^2 and $Q^2 > 20$ GeV^2). The solid line in the insets shows the result of a fit using a Gaussian plus a second-order polynomial background function. The signal-over-background ratio (S/B) is indicated on each figure.

RESULTS

Figure 1 shows the $K^0_S p(\bar{p})$ invariant-mass spectra for three data samples: 1) all data which passed the ZEUS trigger chain after removing DIS events. This data sample is dominated by PHP events as explained in Section ; 2) DIS events with $Q^2 > 1$ GeV^2. This sample represents the largest DIS data sample taken; 3) DIS events at medium $Q^2 > 20$ GeV^2. The latter sample was used in the previous ZEUS publication [2]. The $K^0_S p(\bar{p})$ distribution has two peaks, at around 1522 MeV ($Q^2 > 20$ GeV^2) and at around 2286 MeV (PHP and DIS). The first peak is attributed to the Θ^+ state, which has been discussed in detail in the previous ZEUS publication [2]. The second peak, which corresponds to the established Λ_c, was fitted using a Gaussian with a second-order polynomial function for the background. Both peaks are best seen at $Q^2 > 20$ GeV^2. This is a region where the Λ_c peak has largest signal-over-background ratio (S/B=0.22). The signal-over-background ratio in PHP is rather small, since the average charged-track multiplicity in PHP is larger by 50% than in DIS at $Q^2 > 1$ GeV^2. Since the combinatorial background is larger for low Q^2 DIS and PHP, this may explain the non-observation of the Θ^+ for these two samples [2]. The Λ_c peak was further studied in the forward ($\eta > 0$) and in the rear ($\eta < 0$) pseudorapidity regions, shown in Fig. 3, and for protons and antiprotons. The fit was performed using a Gaussian with a second-order polynomial function. Due to low statistics, the peak position and the width were fixed from the fit to the overall mass spectra for $Q^2 > 1$ GeV^2 (shown in Fig. 1) in order to obtain stable fits. The numbers of the extracted Λ_c candidates are about the same for all four mass distributions. This indicates that the production of Λ_c is consistent with the fragmentation of $c(\bar{c})$ quarks produced by the boson-gluon-fusion mechanism, $\gamma^* g \to c\bar{c}$.

Figure 2 shows the invariant mass of $K^- p$ (and $K^+ \bar{p}$) combinations for PHP and DIS ($Q^2 > 1$ GeV^2 and $Q^2 > 20$ GeV^2). The fits of the invariant-mass distributions shown in this figure lead to significant numbers of reconstructed $\Lambda(1520)$ baryons (13500 for PHP, 2600 for DIS). Unlike Λ_c, the signal-over-background ratio for $\Lambda(1520)$ is similar for PHP and DIS. This is possible if the production rate of the $\Lambda(1520)$ baryons is proportional to the energy available for the fragmentation process. This indicates that $\Lambda(1520)$ can be produced by pure fragmentation mechanism, without partons from the hard interaction. The $\Lambda(1520)$ was also studied in the forward ($\eta > 0$) and in the rear$\eta < 0$ pseudorapidity regions, shown in Fig. 4. The fit was performed using a Gaussian with a second-order polynomial function. There are similar numbers of $\Lambda(1520)$ candidates for these two pseudorapidity regions. Also the fits give similar numbers of $\Lambda(1520)$ baryons and antibaryons. Both observations strengthen the conclusion that the dominant production mechanism of $\Lambda(1520)$ is pure fragmentation.

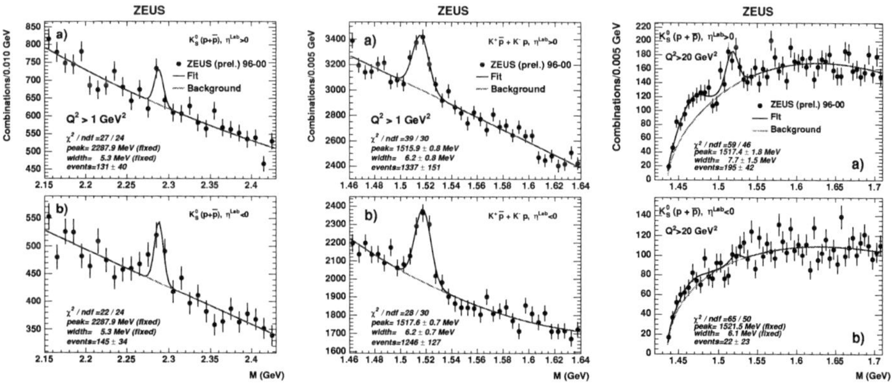

FIGURE 3. The $K^0_S p(\bar{p})$ invariant-mass distribution near the Λ_c mass for $\eta > 0$ and $\eta < 0$ regions in DIS for $Q^2 > 1$ GeV^2.

FIGURE 4. The $K^- p(K^+ \bar{p})$ invariant-mass distribution near the $\Lambda(1520)$ mass in the forward ($\eta > 0$) and in the rear ($\eta < 0$) regions for $Q^2 > 1$ GeV^2.

FIGURE 5. The $K^0_S p(\bar{p})$ invariant-mass distribution near the 1520 MeV mass for the forward ($\eta > 0$) and rear ($\eta < 0$) region in DIS for $Q^2 > 20$ GeV^2.

The Θ^{++} state was searched in the $K^+ p$ ($K^- \bar{p}$) invariant-mass spectrum in PHP and DIS. However, no signal was found in either distribution. This indicates that Θ^+ is not an isotensor. If the peak in the $K^0_S p$ invariant mass spectrum near 1520 MeV corresponds to a new Σ state rather than to the strange pentaquark, the decay $\Theta^+ \to \Lambda\pi^+$ should also be allowed. Peaks due to the known PDG states, $\Xi(1320)$ and $\Sigma(1385)$ baryons, are clearly observed in the reconstructed $\Lambda\pi$ invariant mass. However, no statistically significant peak is seen near 1520 MeV.

If the observed $K^0_S p(\bar{p})$ peak near 1520 MeV corresponds to a new state, then the studies of this peak in different pseudorapidity regions, as well as for proton and antiproton samples, can help to qualify the production mechanism of this baryonic state. The results of the fits of the $K^0_S p$ invariant mass near 1520 MeV are shown in Fig. 5 for a region where the contribution from the proton remnant is expected to be more pronounced (the upper figure), and a region dominated by pure fragmentation (the bottom figure). The fit was performed using a double Gaussian with the threshold function: $P_1(M - m_p - m_{K^0_S})^{P_2} \times (1 + P_3(M - m_p - m_{K^0_S}))$, where M is the $K^0_S p(\bar{p})$ candidate mass, m_p and $m_{K^0_S}$ are the masses of the proton and the K^0_S, respectively, and P_1, P_2 and P_3 are parameters. For the fit in the $\eta < 0$ region, the peak position and the Gaussian width were fixed to the sum of these two distributions. The signal is found to occur predominantly in the first pseudorapidity region. Also the fits give relatively larger number of Θ^+ candidates than Θ^- candidates. However, no strong conclusion can be made since the background spectra for $K^0_S \bar{p}$ is too complicated near 1480 MeV region.

CONCLUSIONS

The invariant-mass spectra sensitive to possible baryonic states decaying to strange hadrons have been investigated. The candidate Θ^+ signal reported earlier [2] was found to be produced predominantly in the forward pseudorapidity region, unlike the well-established baryons, $\Lambda(1520)$ and Λ_c. This indicates that the Θ^+ may have an unusual production mechanism related to proton-remnant fragmentation. The production of Θ^- has lower rate relative to Θ^+, however, the statistical significance of this observation is low, and the background for antiproton spectra is too complicated to draw a strong conclusion.

As for the previous ZEUS studies [2], the candidate Θ^+ signal exists predominantly at medium Q^2 events. The studies of Λ_c indicate that combinatorial background for this Q^2 region is rather favorable for the reconstruction of a state whose production is not driven by pure fragmentation mechanism.

A significant number of $\Lambda(1520)$ baryons were reconstructed in the K^-p decay channel. There is a strong indication that the main production mechanism of this baryon is pure fragmentation. A search was performed for $\Theta^{++} \to K^+p$ signal, however, the results were negative.

The $K^0_S p$ peak near 1520 MeV could also be a new Σ state, which can decay to $\Lambda\pi$. To check this, $\Lambda\pi$ mass combinations were investigated for PHP and DIS. No significant peak near the Θ^+ mass region was found.

REFERENCES

1. S. Chekanov, et al., *Phys. Lett.*, **B 610**, 212 (2005), `hep-ex/0501069`.
2. S. Chekanov, et al., *Phys. Lett.*, **B 591**, 7–22 (2004), `hep-ex/0403051`.

The experimental search for CHARM pentaquarks in the ZEUS detector at HERA[1]

Yehuda Eisenberg[2]

Weizmann Institute of Science,
Rehovot, Israel

Abstract. Using the full 1996 - 2000 ZEUS data at HERA (121 pb^{-1}) we have searched for the $\Theta_c^0(3100)$ pentaquark in the photoproduction and DIS regime. The search has yielded negative results. The 95% C.L. upper limits on the visible rate $R(\Theta_c^0 \to D^* p/D^*)$ is 0.23% (0.35% for DIS).

Keywords: charmed pentaquarks
PACS: 12.39

INTRODUCTION: PENTAQUARKS

Observation of a narrow exotic baryon with strangeness +1 around 1530 MeV decaying into K^+n was reported during the last 2 years by a few low energy experiments [1]. It was suggested that these are due to the $\Theta^+ = uudd\bar{s}$ pentaquark candidate predicted by Diakonov et al. [2], at the top of a $SU(3)$ spin 1/2 anti-decuplet of baryons. Narrow peaks were also seen at a similar mass in the final state $K_S^0 p$, which is exotic if the K_S^0 strangeness is +1. In high energy experiments ZEUS searched for the $\Theta^+(1530) \to K_S^0 p$ [3] in the DIS regime ($Q^2 > 20$ GeV^2) and reported observation of the $\Theta^+(1530)$.

It should be noted however that other high energy experiments at LEP, BaBar, CDF and FOCUS searched and have not seen the $\Theta^+(1530)$ pentaquark [5]. Recently some theoretical arguments have been presented [6] suggesting possible reasons why the $\Theta^+(1530)$ can be seen only in very specific experiments.

SEARCH FOR CHARM PENTAQUARK DECAYING TO $D^{*\pm}P^{\mp}$

The existence of the strange pentaquark Θ^+ suggests that charmed pentaquarks, $\Theta_c^0 = uudd\bar{c}$, may also exist.

The H1 Collaboration reported [9] observation of a narrow signal in the $D^{*\pm}p^{\mp}$ at 3.1 GeV with a width consistent with the detector resolution. The signal was seen in a DIS sample of $\approx 3400\ D^{*\pm} \to D^0\pi^\pm \to (K^\mp\pi^\pm)\pi^\pm$ with a rate of $\approx 1.5\%$ of the visible

[1] This work is supported by the Israel Science Foundation and the U.S.-Israel Bi-national Science Foundation
[2] On behalf of the ZEUS Collaboration

CP792, *Deep Inelastic Scattering, DIS 2005,* edited by W. H. Smith and S. R. Dasu

D^* production. A less clean signal of a comparable rate was seen also in the H1 PHP sample (see previous talk by K.Daum in this session)

The Θ_c^0 search of ZEUS in the $D^{*\pm}p^{\mp}$ mode was performed with the full HERA-I data [10]. Clean $D^{*\pm}$ signals were seen in the $\Delta M = M(D^{*\pm}) - M(D^0)$ plots (Fig. 1 left). Two $D^{*\pm} \rightarrow D^0\pi^\pm$ decay channels were used with $D^0 \rightarrow K^\mp\pi^\pm$ and $D^0 \rightarrow K^\mp\pi^\pm\pi^+\pi^-$. The Θ_c^0 search was performed in the kinematic range $|\eta(D^*)| < 1.6$ and $p_T\ (D^*) > 1.35\ (2.8)$ GeV and with ΔM values between $0.144 - 0.147(0.1445 - 0.1465)$ GeV for the $K\pi\pi$ $(K\pi\pi\pi\pi)$ channel. The shaded bands in Fig.1a,b contains a total of ≈ 62000 D^*'s after subtracting wrong-charge combinations with charge ± 2 for the D^0 candidate. Selecting DIS events with $Q^2 > 1$ GeV^2 yielded smaller, but cleaner D^* signals with a total of ≈ 13500 D^*'s (Fig. 1c-d left).

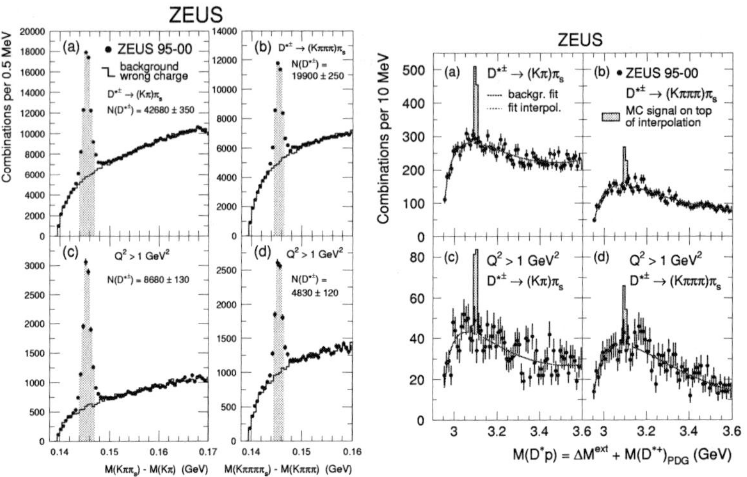

FIGURE 1. Left: ΔM distributions (dots) for (a) $D^* \rightarrow K\pi\pi$ and (b) $D^* \rightarrow K\pi\pi\pi\pi$ candidates. Events with $Q^2 > 1$ GeV^2 for the two channels, respectively, are shown in (c) and (d). The histograms are for wrong charge combinations. Right: $M(D^{*\pm}p^{\mp})$ distributions (dots) for the same samples. Solid curves are fi ts to a background function (see text). Shaded historgams are MC Θ_c^0 signals, normalised to $\Theta_c^0/D^* = 1\%$, on top of the background fi t.

Protons were selected with $p_T(p) > 0.15$ GeV. To reduce the pion and kaon background, a parameterisation of the expected dE/dx as a function of P/m was obtained using tagged protons from Λ decays and tagged pions from K_S^0 decays. The χ^2 probability of the proton hypothesis was required to be above 0.15. Fig. 2 shows the $M(D^*p) = M(K\pi\pi p) - M(K\pi\pi) + M(D^*)_{PDG}$ distributions for the $K\pi\pi$ channel for the full (left) and the DIS (right) samples, where $M(D^*)_{PDG}$ is the $D^{*\pm}$ mass [4]. In the low-P selection (Fig. 2b), a clean proton sample separated from the π and K dE/dx bands was obtained by taking only tracks with $P < 1.35$ GeV and $dE/dx > 1.3$ mips. In the high-P selection (Fig. 2c) only tracks with $P(p) > 2$ GeV were used. The latter selection was prompted by the H1 observation [9] of a better Θ_c^0 signal-to-background ratio for high proton momenta. No narrow signal is seen in the $K\pi\pi$ (Fig. 2) as well as in the $K\pi\pi\pi\pi$ (Fig. 1b,d right,Fig.3a,b left) channel. The $K\pi\pi$ analysis was repeated using very similar selection criteria as in the H1 analysis [9]. No indication of a narrow resonance was found in either the DIS or the PHP event sample [10].

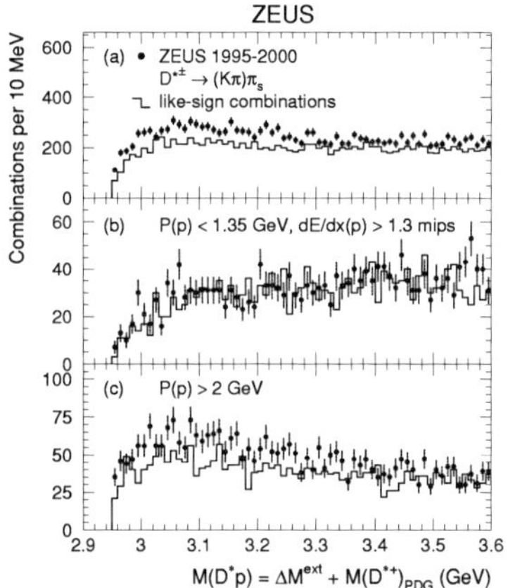

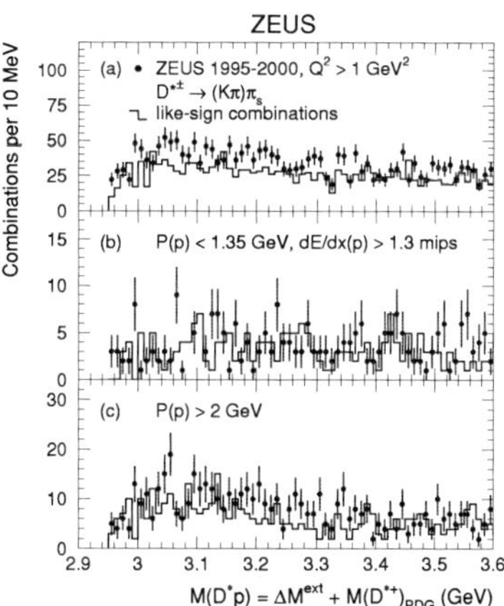

FIGURE 2. Left: $M(D^{*\pm}p^{\mp})$ distributions for the $K\pi\pi$ channel (dots) with (a) all proton candidates, (b) candidates with $P(p) < 1.35$ GeV and $dE/dx > 1.3$, and (c) candidates with $P(p) > 2$ GeV. Histograms show the $M(D^{*\pm}p^{\pm})$ like-sign combinations. Right: Same for DIS events with $Q^2 > 1$ GeV^2.

95% C.L. upper limits on the fraction of D^* mesons originating from Θ_c^0 decays, $R(\Theta_c^0 \to D^*p/D^*)$, were calculated in a signal window $3.07 < M(D^*p) < 3.13$ GeV for the $K\pi\pi$ and $K\pi\pi\pi\pi$ channels. A visible rate of 1% for this fraction (Fig. 1 right), as claimed by H1 [9], is excluded by 9σ (5σ) for the full (DIS) combined sample. The $M(D^*p)$ distributions were fitted to the form $x^a e^{-bx+cx^2}$, where $x = M(D^*p) - M(D^*) - m_p$ (Fig. 1 right). The number of reconstructed Θ_c^0 baryons was estimated by subtracting in the signal window the background function from the observed number of events, yielding $R(\Theta_c^0 \to D^*p/D^*) < 0.23\%$ and $< 0.35\%$ for the full and DIS combined two channels. The acceptance-corrected rates are, respectively, 0.37% and 0.51%. The 95% C.L. upper limit on the fraction of charm quarks fragmenting to Θ_c^0 times the branching ratio $\Theta_c^0 \to D^*p$ for the combined two channels is $f(c \to \Theta_c^0) \cdot B_{\Theta_c^0 \to D^*p} < 0.16\%$ ($< 0.19\%$) for the full (DIS) sample.

We conclude that the ZEUS data are not compatible with the H1 result of $\approx 1.5\%$ rate of $\Theta_c^0 \to D^*p/D^*$. Such a rate is excluded by more then 9σ for the full data and 5σ for the ZEUS DIS data. Further experiments perhaps with HERA II may help resolve this discrepancy. No other experiment reported the observation of Θ_c^0 [5].

REFERENCES

1. CLAS Coll., V. Kubarovsky et al, Phys. Rev. Lett. 92, 032001 (2004); SAPHIR Coll., J.Narth et al, Phys. Lett. B572,127 (2003).
2. D. Diakonov, V. Petrov and M.V. Polyakov, *Z. Phys.* **A359**, 305 (1997).
3. ZEUS Coll., S. Chekanov et al., *Phys. Lett.* **B591**, 7 (2004). See update in the talk of Zhenhai Ren ,on behalf of ZEUS at this conference.
4. Particle Data Group, K. Hagiwara et al., *Phys. Rev.* **D66**, 10001 (2002).
5. See reports in this conference: Christiane Risler DESY H1 results; Eric Eckhart, BaBar results; Marco Battaglieri, Pentaquark at JLab: new results from CLAS. Aleph Coll.,S. Schael et al, Phys. Lett. B

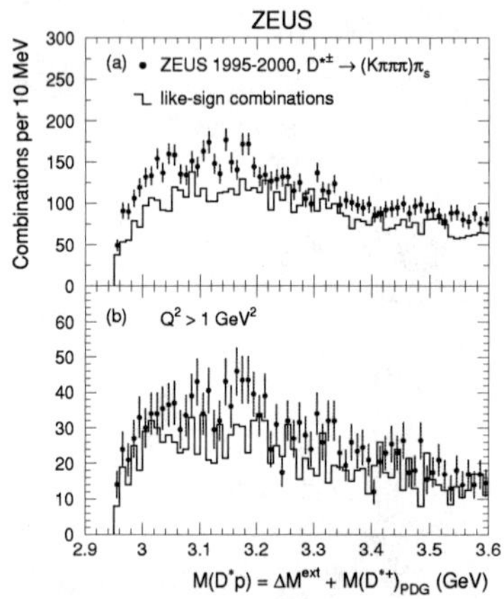

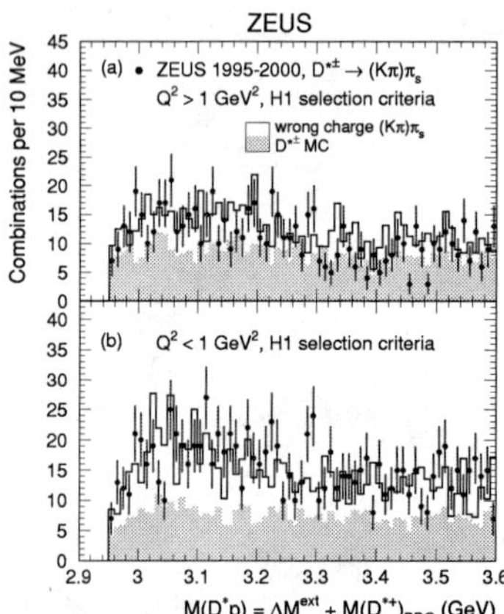

FIGURE 3. Left: $M(D^{*\pm}p^{\mp})$ distributions for the $K\pi\pi\pi\pi$ channel (dots) with (a) full sample, (b) DIS sample, $Q^2 > 1GeV^2$. Right: same distributions for the $K\pi\pi$ channel for DIS ($Q^2 > 1$ GeV^2) and photoproduction using the H1 event selection criteria.

599,1 (2004); CDF Coll., I. V. Gorelov Preprint (hep-ex/0408025). See also references in K.Hicks,hep-ex//0504027 and C. Risler talk,this session

6. M. Karliner and H. J .Lipkin, preprint (hep=ph/0506084)
7. A. Jaffe and F. Wilczek, , *Phys. Rev. Lett.* **91**, 232003 (2003).
8. M. Karliner and H.J. Lipkin, Preprint hep-ph/0307343 (2003).
9. H1 Coll., C. Atkas et al., , *Phys. Lett.* **B588**, 17 (2004).
10. ZEUS Coll., S. Chekanov et al., *Eur.Phys. J.* **C38** (2004)29 .

Pentaquark searches at HERMES

Avetik Airapetian
on behalf of HERMES Collaboration

*Randall Laboratory of Physics, University of Michigan,
Ann Arbor Michigan 48109-1040,USA
E-Mail: Avetik.Airapetian@desy.de*

Abstract. An experimental search for exotic baryons was performed with the HERMES experiment at DESY in quasi-real photoproduction. Positive evidence is presented for a Θ^+ at a mass of $1528 \pm 2.6(stat) \pm 2.1(sys)$ MeV. No evidence for possible Θ^{++}, $\Xi^{--}(1860)$ and $\Xi^0(1860)$ resonances was found, instead, upper limits for their production cross section are given. In addition, photoproduction crosssections for the $\Lambda(1520)$ and $\Xi^0(1530)$ resonances are presented, because they have similar decay modes as the Θ^{++} and $\Xi^{--}(1860)$, respectively.

Keywords: Exotic baryon production,pentaquarks,nonstandard multi-quark states
PACS: 12.39.Mk,13.60.-r,13.60.Rj,14.20.-c

INTRODUCTION

A recent prediction of the existence of narrow exotic baryon resonances [1], based on the Chiral Soliton Model, has triggered an intensive search for the exotic members of an anti-decuplet. In this anti-decuplet all three vertices are manifestly exotic. The lightest exotic member, lying at its apex, was predicted to have a mass of 1530 MeV and a narrow width. It corresponds to a $uudd\bar{s}$ configuration, and decays through the channels $\Theta^+ \to pK^0$ or $\Theta^+ \to nK^+$. Other approaches, based on the constituent quark model [2], or on the Chiral Soliton model [3], predict that rather than an isosinglet Θ^+, an isotriplet or an isovector Θ particle should exist.

The first experimental evidence for the Θ^+ came from the LEPS experiment [4] in Japan, which reported the observation of a narrow resonance at $1540 \pm 10(\text{sys})$ MeV by analyzing the K^- missing mass spectrum in the reaction $\gamma n \to K^- K^+ n$ on ^{12}C. The decay mode corresponds to an S=+1 resonance, containing an $\bar{s}$ quark with baryon number +1, signaling a manifestly exotic pentaquark state with minimum quark content ($uudd\bar{s}$). Confirmation came quickly from a series of experiments [5, 6], with the observation of narrow peaks in pK^0 or nK^+ mass spectra near 1530 MeV, in each case with a width consistent with the experimental resolution. Doubts concerning the validity of these observations have been raised recently, because of the failure to observe a signal in many other experiments [7].

Experimental evidence for a second exotic member of the anti-decuplet came from the reported observation of a $S=-2$, $Q=-2$ baryon resonance in proton-proton collisions

[1] This work is supported by the U. S. National Science Foundation under grant 0244842

CP792, *Deep Inelastic Scattering, DIS 2005*, edited by W. H. Smith and S. R. Dasu

at $\sqrt{s} = 17.2$ GeV at the CERN SPS [8]. A narrow peak at a mass of about 1862 MeV in the $\Xi^-\pi^-$ invariant mass spectrum was proposed as a candidate for the predicted exotic $\Xi^{--}_{3/2}$ baryon with S=−2, I=$\frac{3}{2}$ and a quark content of ($ddss\bar{u}$). At the same mass, a peak was observed that is a candidate for the $\Xi^0_{3/2}$ member of this isospin quartet. This state has not been confirmed by other experimental searches [9, 10, 11].

EXPERIMENT

The HERMES collaboration has performed an experimental search for the Θ^+, Θ^{++}, Ξ^{--} and Ξ^0 particles in quasi-real photoproduction on a deuterium target. The data were obtained using the 27.6 GeV positron beam of the HERA storage ring at DESY. The analysis searched for inclusive photo-production of the Θ^+, either off a proton or a neuteron, followed by the decay $\Theta^+ \to pK^0_S \to p\pi^+\pi^-$. The search for the Θ^{++} was pursued via the possible decay mode $\Theta^{++} \to pK^+$. The search for inclusive photo-production of $\Xi(1860)$ pentaquarks was performed assuming the decay modes: $\Xi^{--} \to \Xi^-\pi^- \to \Lambda\pi^-\pi^- \to p\pi^-\pi^-\pi^-$ or $\Xi^0 \to \Xi^-\pi^+ \to \Lambda\pi^-\pi^+ \to p\pi^-\pi^-\pi^+$.

Identification of charged pions and protons was accomplished with a Ring-Imaging Čerenkov (RICH) detector [12]. The data from the simulation indicated that cross contaminations in the search of the Θ^+ and Θ^{++} resonances is negligible if protons are restricted to a momentum range of 4–9 GeV/c, kaons to 2–15 GeV/c and pions to 1–15 GeV/c. In the search of the $\Xi(1860)$ baryons, the requirements on the proton and pion momenta were relaxed to a momentum range of 2–15 GeV/c and 0.25–15 GeV/c respectively, because the intermediate Λ and $\Xi^-(1321)$ particles were clearly identified in the particle reconstruction.

The event selection included constraints on the event topology to maximize the yield of the K^0_S, Λ or $\Xi^-(1321)$ peaks in the $M_{\pi^+\pi^-}$, $M_{p\pi^-}$ and $M_{\Lambda\pi^-}$ spectra, respectively, while minimizing their background.

RESULTS

To search for $\Xi^{--}_{3/2}$ ($\Xi^0_{3/2}$) candidates, first events were selected with an invariant mass $M_{p\pi^-}$ within $\pm 3\,\sigma$ of the centroid of the Λ peak. These events were combined with a π^- to form the Ξ^-. In the next step, events were selected with a $M_{\Lambda\pi^-}$ invariant mass within $\pm 3\,\sigma$ of the centroid of the Ξ^- peak. The resulting spectrum of the invariant mass of the $p\pi^-\pi^-\pi^-$ and $p\pi^-\pi^-\pi^+$ system is displayed in Fig. 1. While no peak structure is observed near 1862 MeV, one appears at the mass of the known $\Xi^0(1530)$ resonance.

To search for the Θ^+ candidates, events were selected with a $M_{\pi^+\pi^-}$ invariant mass within $\pm 2\sigma$ of the centroid of the K^0_S peak. The resulting $p\pi^+\pi^-$ invariant mass spectrum is shown in Fig. 2 (left panel). A narrow peak is observed at $1528.0 \pm 2.6 \pm 2.1$ MeV with a Gaussian width of $\sigma = 8 \pm 2$ MeV and a statistical significance of $N_s/\delta N_s = 3.7\,\sigma$. The state observed here may be interpreted as the predicted exotic Θ^+ pentaquark S=+1 baryon.

In view of the speculation that the observed resonance is isotensor [2, 3], the possibility that the Θ^{++} partner is present in the M_{pK^+} spectrum was explored. Although Fig. 2 (right panel) shows a clear peak for the $\Lambda(1520)$ in the M_{pK^-} spectrum, there is no peak

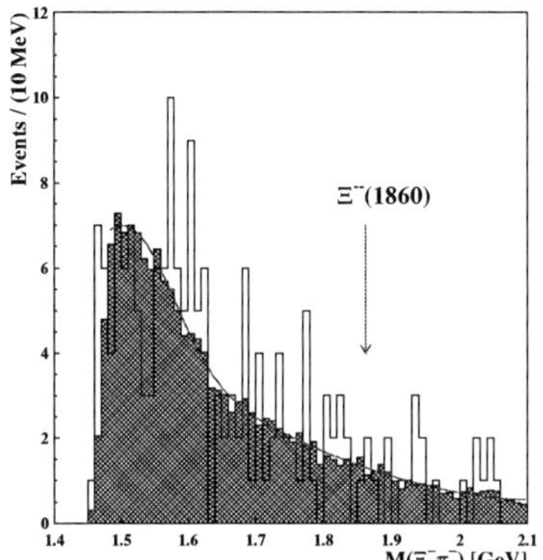

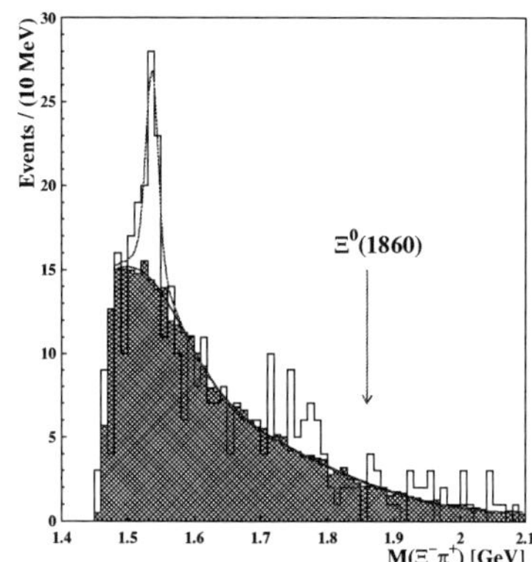

FIGURE 1. Invariant mass distribution of the $p\pi^-\pi^-\pi^-$(plus c.c.) system (left panel) and $p\pi^-\pi^-\pi^+$(plus c.c.) (right panel). The mixed-event background is represented by the gray shaded histogram, which is normalized to the background component of the fitted curve. The arrow shows the hypothetical $\Xi^{--}_{3/2}$ mass. The excess near $1770\,MeV$(right panel) has a statistical significance of only $1.8\,\sigma$. The peak near $1530MeV$ represents the $\Xi^0(1530)$ resonance.

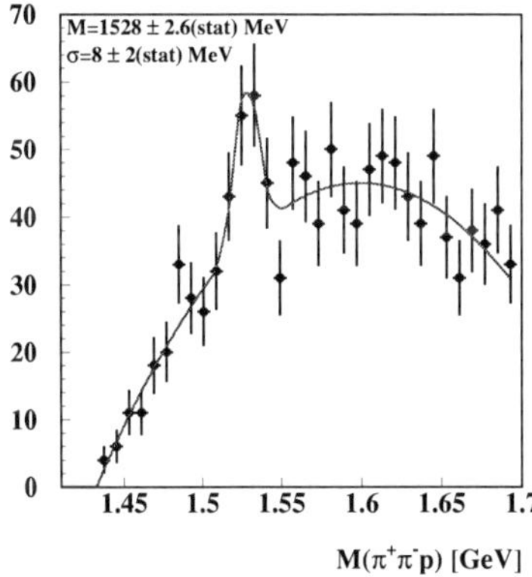

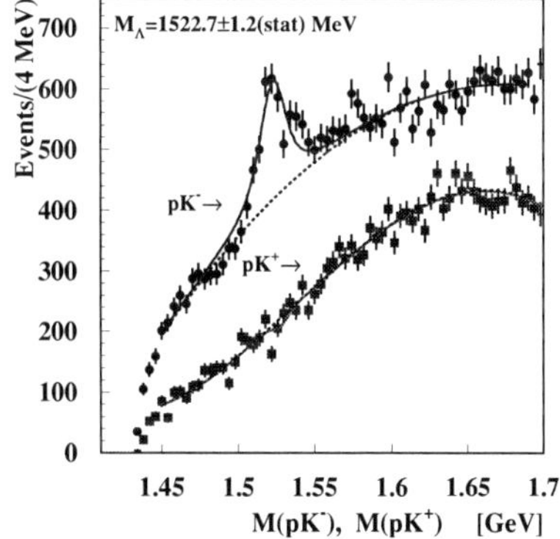

FIGURE 2. Distribution in invariant mass of the $p\pi^+\pi^-$ system (left panel). The smooth curve results from a fit to the data of a Gaussian plus a third-order polynomial. Spectra of invariant mass (right panel) M_{pK^-}(top) and M_{pK^+}(bottom).

structure observed in the M_{pK^+} mass distribution at that mass. This suggests that if the observed peak is the Θ^+, it is likely to be isoscalar.

A Breit-Wigner form convoluted with a Gaussian representing the simulated instrumental resolution was used to estimate the intrinsic width of the observed resonance ($\Gamma = 17 \pm 9(\text{stat}) \pm 3(\text{sys})\,\text{MeV}$[6]). Estimates of the total spectrometer acceptance from detector simulations have been used to extract the inclusive cross section in the reaction $\gamma^* D \to \Theta^+ X$. The result varies between 100 and 220 nb $\pm 25\%$(stat) depending on the model for the background and the functional form fitted to the peak. An additional factor of two uncertainty is due to the unknown initial kinematic distributions. The cross section for photo-production of the $\Lambda(1520)$ is found to be 62 ± 11 (stat) nb. If the branching ratio(BR) for the $\Xi^0(1530) \rightarrow \Xi^- \pi^+$ decay is taken to be $2/3$ [13], its photoproduction cross section is found to be between 8.8 and 24 nb[14]. The results for the upper limits for the $\Xi^{--}(\Xi^0)(1860)$ photo production cross section times BR is found to be 1.0 to 2.1 nb (1.2 to 2.5 nb) at the 90% C.L.

In summary it is important to note that the existence of pentaquark baryons is still unsettled. The experiments reporting evidence for the Θ^+ are confronted with a growing number of experiments that fail to see a signal. Given the present ambiguous experimental situation, opposing views can be found in the recent literature on this subject [15, 16], in efforts to resolve the apparent experimental inconsistencies. All this brings us to the conclusion that only dedicated, high-statistics and high-resolution measurements can settle the question whether or not exotic baryons such as the hypothetical Θ^+ pentaquark state do exist.

ACKNOWLEDGMENT

I wish to thank my colleagues from the HERMES Collaboration. I highly acknowledge Wolfgang Lorenzon for constant support and fruitful discussions.

REFERENCES

1. D. Diakonov, V. Petrov, and M. Polyakov, Z. Phys. A **359**, 305 (1997).
2. S. Capstick *et al.*, Phys. Lett. B **570**, 185 (2003).
3. H. Gao and B.-Q. Ma, Mod. Phys. Lett. A **14**, 2313 (1999).
4. LEPS Collaboration, T. Nakano *et al.*, Phys. Rev. Lett. **91**, 012002 (2003).
5. DIANA Collaboration, V.V. Barmin *et al.*, Yad. Fiz. **66**, 1763 (2003), CLAS Collaboration, S. Stepanyan *et al.*, Phys. Rev. Lett. **91**, 252001 (2003), SAPHIR Collaboration, J. Barth *et al.*, Phys. Lett. B **572**, 127 (2003), A. E. Asratyan *et al.*, Yad. Fiz. **67**, 704 (2004), CLAS Collaboration, V. Kubarovsky *et al.*, Phys. Rev. Lett. **92**, 032001 (2004); *erratum ibid.* Phys. Rev. Lett. **92**, 049902 (2004), ZEUS Collaboration, S. Chekanov *et al.*, Phys. Lett. B **591**, 7 (2004), SVD Collaboration, A. Aleev *et al.*, archive:hep-ex/0401024, COSY-TOF Collaboration, M. Abdel-Bary *et al.*, Phys. Lett. B **595**, 127 (2004), P.Zh. Aslanyan *et al.*, archive:hep-ex/0403044, Yu. A. Troyan *et al.*, archive:hep-ex/0404003.
6. HERMES Collaboration, A. Airapetian *et al.*, Phys. Lett. B **585**, 213 (2004).
7. B. Aubert *et al.*, (Babar Collaboration), hep-ex/0408064; K. Stenson (FOCUS Collaboration), hep-ex/0412021, BES Collaboration, J.Z. Bai *et al.*, Phys. Rev. D **70**, 012004 (2004), PHENIX collaboration, C. Pinkenburg *et al.*, J. Phys. G **30**, S1201 (2004), HERA-B Collaboration, K.T. Knöpfle *et al.*, J. Phys. G **30**, S1363 (2004), SPHINX Collaboration, Yu.M. Antipov *et al.*, Eur. Phys. J. A **21**, 455 (2004), ALEPH Collaboration, S. Schael *et al.*, Phys. Lett. B **599**, 1 (2004), HyperCP Collaboration, M.J. Longo *et al.*, Phys. Rev. D **70**, 111101 (2004).
8. NA49 Collaboration, C. Alt *et al.*, Phys. Rev. Lett. **92**, 042003 (2004).
9. WA89 Collaboration, M.I. Adamovich *et al.*, Phys. Rev. C **70**, 022201(R) (2004).
10. HERA-B Collaboration, I. Abt *et al.*, Phys. Rev. Lett. **93**, 212003 (2004).
11. S.V. Chekanov (ZEUS Collaboration), hep-ex/0405013; I. Gorelov (CDF Collaboration), hep-ex/0408025; D. Christian (E690 Collaboration), Quarks and Nuclear Physics 2004, Bloomington, Indiana, 2004, http://www.qnp2004.org/; K. Stenson (FOCUS Collaboration), hep-ex/0412021.
12. N. Akopov *et al.*, Nucl. Instr. Meth. A **479**, 511 (2002).
13. WA89 Collaboration, M.I. Adamovich *et al.*, Eur. Phys. J. C **11**, 271 (1999).
14. HERMES Collaboration, A. Airapetian *et al.* Phys. Rev. D **71**, 032004 (2004).
15. A.R. Dzierba *et al.*, Phys. Rev. D **92**, 042003 (2004).
16. A.I. Titov, A. Hosaka, S. Date and Y. Ohashi, Phys. Rev. C **70**, 035210 (2004).

Status Report of NNLO QCD Calculations

Michael Klasen[1]

Institute for Nuclear Theory, University of Washington, Box 351550, Seattle, WA 98195-1550, USA

Abstract. We review recent progress in next-to-next-to-leading order (NNLO) perturbative QCD calculations with special emphasis on results ready for phenomenological applications. Important examples are new results on structure functions and jet or Higgs boson production. In addition, we describe new calculational techniques based on twistors and their potential for efficient calculations of multiparticle amplitudes.

Keywords: Perturbative QCD calculations
PACS: 12.38.Bx

INTRODUCTION

Despite the fact that the theory of strong interactions, quantum chromodynamics (QCD), is today a well-established part of the Standard Model of particle physics, it continues to be an extremely active field of experimental and theoretical research. This is as true for non-perturbative (and often lattice) determinations of meson and baryon spectra, decay constants, or form factors for electric dipole moments or B-meson decays, as it is for perturbative calculations. Today, the experimental precision of high-energy collider data always requires calculations beyond the leading order (LO) in the strong coupling constant, $\alpha_s(\mu)$, and often even beyond the next-to-leading order (NLO) for reliable comparisons and determinations of unknown Standard Model parameters.

After a relatively slow start due to large technical difficulties in the mid-1990s, QCD calculations at next-to-next-to-leading order (NNLO) have recently attained their goals for several phenomenological applications. The aim of this Report is therefore to review recent progress in this field with special emphasis on phenomenologically relevant results, including structure functions, jet, and Higgs boson production. Very recently, twistor methods have received particular attention, as they may offer a route to efficient multiparticle calculations, even beyond tree-level. They will therefore be briefly discussed, before we present our conclusions.

STRUCTURE FUNCTIONS

Since its start-up in 1992, the DESY ep collider HERA has provided a wealth of data on the structure functions of the proton, and thus its parton densities (PDFs), in a large region of Bjorken-x and Q^2 [1]. The PDFs represent at the same time an important

[1] Permanent address: Laboratoire de Physique Subatomique et de Cosmologie, Université Joseph Fourier/CNRS-IN2P3, 53 Avenue des Martyrs, 38026 Grenoble, France

CP792, *Deep Inelastic Scattering, DIS 2005*, edited by W. H. Smith and S. R. Dasu

ingredient in the search for new physics, that is currently underway at the Fermilab Tevatron and is an important research goal at the CERN LHC [2]. In particular, QCD uncertainties in PDF determinations, *e.g.* from renormalization and factorization scale and scheme [3, 4] variations, have to be under control before disagreement between theory and experiment can be interpreted as evidence for new physics.

It is thus fortunate that, after completion of the three-loop singlet splitting functions (obtained from the $1/\varepsilon$-poles in dimensional regularization) and longitudinal coefficient function (obtained from the finite terms) [5], a full NNLO calculation of the structure functions $F_{1,2}(x,Q^2)$ (and thus also of $F_L = F_2 - 2xF_1$), is now available. These results have been obtained using a large variety of techniques, such as application of the optical theorem, transformation to Mellin space, mapping of diagrams with composite topologies to basic building blocks, and integration by parts. They also relied on recent advances in mathematics (harmonic sums) and computer algebra (QGRAF, FORM). Fortunately, the results can also be applied to other physical observables such as photon structure functions and total cross sections in e^+e^--annihilation.

While the numerical effects of the NNLO QCD corrections are well visible in splitting functions and coefficient functions *separately* and show clear differences in normalization, although not in shape, from earlier leading-ln(x) estimates (see Fig. 1), the total

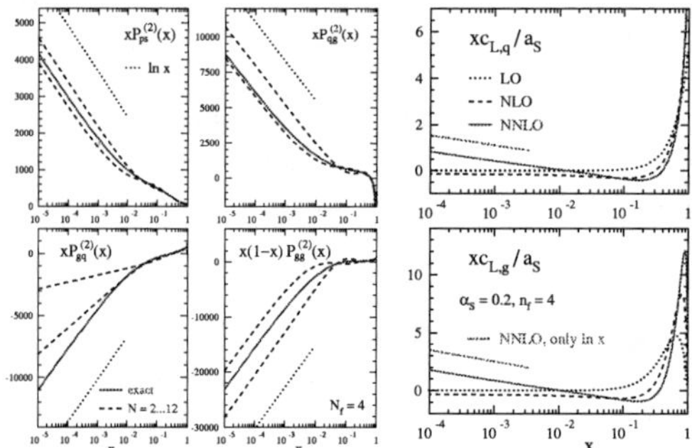

FIGURE 1. Splitting functions (left) and longitudinal coefficient functions (right) at NNLO [5].

effect in the *convoluted* structure functions is only important for very small or large x and small Q^2. For this reason, the NNLO corrections have not yet been included in the global CTEQ analysis, as the authors find sufficient stability of their fit at NLO [6], whereas the MRST collaboration have now implemented the exact NNLO splitting functions, but find minor effects w.r.t. a DIS-scheme motivated variation of the input gluon density [4] or an earlier analysis using approximate NNLO splitting functions [7].

JET PRODUCTION IN ELECTRON-POSITRON ANNIHILATION

Calculations of observables less inclusive than deep inelastic scattering (DIS) structure functions and total e^+e^- cross sections do not allow for a straight-forward application of the optical theorem, as they require multiparticle cuts in the three-loop photon propagator. For this reason, NNLO jet calculations in e^+e^--annihilation have been lagging behind, although recently considerable progress has also been achieved in this field. The interference terms of $1 \to 3$ tree-level and two-loop amplitudes are now known, as are the single soft and/or collinear regions of $1 \to 4$ squared one-loop amplitudes. What remains to be calculated are the maximally double-soft and triple-collinear regions of $1 \to 5$ squared tree-level amplitudes. Here, antenna functions have successfully been used for quark-gluon and gluon-gluon final states [8].

A first preliminary numerical result has been obtained for the NNLO contribution to the C_F^2 color class of the average thrust [9]

$$\langle 1-T \rangle = \int (1-T) \frac{1}{\sigma_0} \frac{d\sigma}{dT} = C_F \left[\left(\frac{\alpha_s}{2\pi} \right) A + \left(\frac{\alpha_s}{2\pi} \right)^2 B + \left(\frac{\alpha_s}{2\pi} \right)^3 C + \ldots \right],$$

where $A = 1.57$ and $B = 32.3$ were known and now $C(C_F^2) = -20.4 \pm 4$. The full result will be important for reliable estimates of higher-twist effects in e^+e^- event shapes. Furthermore, analytic continuation allows to apply the results obtained for three-jet production in e^+e^--annihilation also to dijet production in DIS.

HIGGS BOSON PRODUCTION AT HADRON COLLIDERS

The inclusive NNLO cross section for light/heavy scalar and pseudo-scalar Higgs production in gg scattering at hadron colliders has been known for several years, but the relevance of these calculations has been limited by the need for experimental cuts on associated photons, jets or b-quarks. Fully differential cross sections, such as a recent NNLO calculation for Higgs plus dijet production [10], allow for transverse energy vetos on the additional jets and thus for a suppression of the QCD background in the diphoton (or other) Higgs decay channels (see Fig. 2).

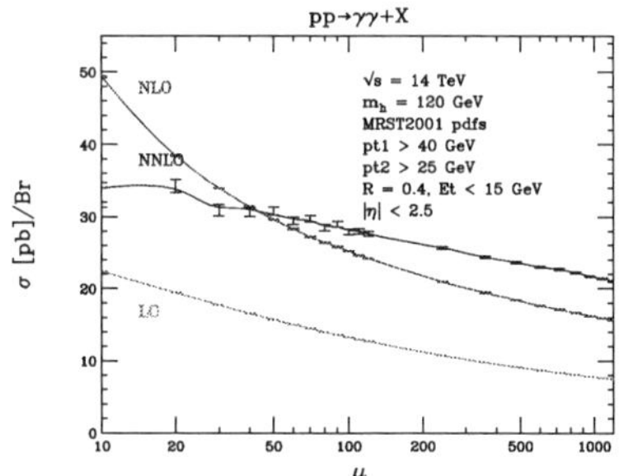

FIGURE 2. Higgs production at the LHC with diphoton decay and no additional hard jets [10].

JET HADROPRODUCTION AND TWISTOR METHODS

For hadron-hadron scattering, the two-loop $2 \to 2$ parton amplitudes have been known for some time, and the result for polarized $qq \to qq$ scattering has recently been confirmed using supersymmetric methods [11]. Since the soft/collinear singular regions of one-loop $2 \to 3$ amplitudes are also known, recent work has focused on multiple soft/collinear emissions in tree-level multiparton amplitudes and on extending the dipole subtraction formalism to NNLO, where two alternatives to previously known methods have been proposed [12, 13]. Although the subtraction terms are in principle known, the singular regions often overlap and need to be disentangled to avoid double-counting. Furthermore, phase space factorization and integration of the singular matrix elements still needs to be resolved. For this reason, further progress towards a full NNLO calculation for jet hadroproduction has been slow, and numerical integration methods using sector decomposition [14] may have to be developped further.

Multiparticle amplitudes may be calculated efficiently by relating perturbative gauge theory to the D-instanton expansion for a topological string in twistor space. While originally motivated by $N = 4$ super Yang-Mills theory, this approach has been shown to work at tree-level for generic QCD amplitudes. $N = 4$ supersymmetric and cut-constructible one-loop diagrams have also been calculated, and extensions to non-supersymmetric theories and two-loop amplitudes are underway [15].

CONCLUSION

Although this brief Report is necessarily far from complete, it should be clear that NNLO QCD calculations represent a theoretically challenging, phenomenologically important, and fast-moving field of research allowing for interesting advances in the fields of mathematics, computer science and last, but not least, elementary particle physics.

ACKNOWLEDGMENTS

The author thanks the working group convenors for the kind invitation, the DoE's INT at the University of Washington for kind hospitality and partial financial support during preparation of this work, and L. Dixon and T. Teubner for useful discussions.

REFERENCES

1. O. Behnke and E. Gallo for the H1 and ZEUS Collaborations, these proceedings.
2. A. de Roeck for the HERA-LHC Workshop, these proceedings.
3. M. Klasen and G. Kramer, Phys. Lett. B **386**, 384 (1996).
4. A. D. Martin, R. G. Roberts, W. J. Stirling and R. S. Thorne, Phys. Lett. B **604**, 61 (2004).
5. S. Moch, J. A. M. Vermaseren and A. Vogt, Nucl. Phys. B **691**, 129 (2004); Phys. Lett. B **606**, 123 (2005); hep-ph/0504242.
6. J. Huston, J. Pumplin, D. Stump and W. K. Tung, hep-ph/0502080.
7. A. D. Martin, R. G. Roberts, W. J. Stirling and R. S. Thorne, Phys. Lett. B **531**, 216 (2002).
8. A. Gehrmann-De Ridder, T. Gehrmann and E. W. N. Glover, Phys. Lett. B **612**, 36 (2005); **612**, 49 (2005); hep-ph/0505111.
9. A. Gehrmann-De Ridder, T. Gehrmann and E. W. N. Glover, Nucl. Phys. Proc. Suppl. **135**, 97 (2004).
10. C. Anastasiou, K. Melnikov and F. Petriello, hep-ph/0409088 and hep-ph/0501130.
11. A. De Freitas and Z. Bern, JHEP **0409**, 039 (2004).
12. S. Frixione and M. Grazzini, hep-ph/0411399.
13. G. Somogyi, Z. Trocsanyi and V. Del Duca, hep-ph/0502226.
14. T. Binoth and G. Heinrich, Nucl. Phys. B **693**, 134 (2004).
15. L. Dixon, these proceedings.

Monte Carlo event generators

Stefano Frixione

INFN, Sezione di Genova
Via Dodecaneso 33, 16146 Genova, Italy

Abstract. I review recent progress in the physics of parton shower Monte Carlos, emphasizing the ideas which allow the inclusion of higher-order matrix elements into the framework of event generators

Keywords: Collider Physics, Event Generators

Reliable predictions of cross sections and final-state distributions for QCD processes are a crucial ingredient in high-energy collider experiments, not only as a test of QCD but also for new particle searches. All systematic approaches to this problem are based on fixed-order (FO) results in perturbation theory, and yield (usually at the next-to-leading order, NLO) the best available results for sufficiently inclusive observables. However, in many cases a more exclusive description of final states is needed. In such cases, in which one also combines perturbative calculations with a model for the conversion of partonic final states into hadrons, Monte Carlo (MC) simulations are generally adopted. MC's operate on partonic states with high multiplicity and low relative transverse momenta, which are obtained from a parton shower or dipole cascade approximation to QCD dynamics. This has to be confronted with FO results, which can describe the complementary region of small multiplicities, and large relative transverse momenta.

The lack of large transverse momentum emissions, and the fact that total rates are computed to leading order accuracy only, are serious problems in MC simulations, especially when the c.m. energies are in the TeV range. These problems can be solved by a suitable combination of MC and FO methods. Given the flexibility of MC's, it is actually desirable to embed as much as possible of FO information into the framework of MC simulations, since the other way around would just prove to be too complicated. In order to explain how this could be done, it is useful to briefly remind how an MC works: for a given process, which at the LO receives contribution from $2 \to n_0$ reactions, $(2+n_0)$-particle configurations are generated, according to exact tree-level matrix element (ME) computations. The quarks and gluons (partons henceforth) among these primary particles are then allowed to emit more quarks and gluons, which are obtained from a parton shower or dipole cascade approximation to QCD dynamics. To lessen the impact of this approximation on physical observables, one can devise two strategies. The first aims at having n_E extra hard partons in the final state; thus, in the example given above, the number of final-state hard particles would increase from n_0 to n_0+n_E. This approach is usually referred to as *matrix element corrections* (MEC), since the MC must use the $(2+n_0+n_E)$-particle ME's to generate the correct hard kinematics. The second strategy also aims at simulating the production of n_0+n_E hard particles, but

CP792, *Deep Inelastic Scattering, DIS 2005,* edited by W. H. Smith and S. R. Dasu

improves the computation of rates as well, to N^{n_E}LO accuracy. I'll generally denote the resulting MC as N^{n_E}LOwPS.

There are basically two major problems in the implementation of MEC. The first problem is that of achieving a fast computation of the ME's themselves for the largest possible $n_0 + n_E$, and an efficient phase-space generation. A variety of solutions exist nowadays for this problem, implemented in packages which I'll denote as ME generators; popular ones include AlpGEN [1], CompHEP [2], Grace [3], and MadEvent [4]. The second problem stems from the fact that multi-parton ME's are IR divergent. Clearly, in hard-particle configurations IR divergences don't appear; however, the definition of what hard means is, to a large extent, arbitrary. In practice, hardness is achieved by imposing some cuts on suitable partonic variables, such as p_T's and (η, φ)-distances. I collectively denote these cuts by δ_{sep}. One assumes that n hard partons will result (after the shower) into n jets; but, with a probability depending on δ_{sep}, a given n-jet event could also result from $n+m$ hard partons. This means that, when generating events at a fixed $n_0 + n_E$ number of primary particles, physical observables in general depend upon δ_{sep}; I refer to this as the δ_{sep}-bias problem. Any solution to the δ_{sep}-bias problem implies a procedure to combine consistently ME's with different $n_0 + n_E$'s. It should be stressed that, in presence of a δ_{sep} bias, the interface of an ME generator (which is responsible for producing the hard configurations, i.e. the initial conditions for the shower), and a parton shower code is *not*, strictly speaking, an event generator (EvG), since the events depend somehow on the value of δ_{sep}. In practice, the dependence is of the order of 20%, which is acceptable if one considers that, without MEC, multi-jet configurations predicted by standard MC's are completely unreliable. A solution to the δ_{sep}-bias problem has been presented, for e^+e^- collisions, in ref. [5] (CKKW henceforth), and subsequently extended (without formal proof) to hadronic collisions in ref. [6]; an alternative method for colour-dipole cascades has been presented in ref. [7]. Loosely speaking, CKKW achieve the following: if an n-jet observable is affected by the δ_{sep} bias in the following way

$$\sigma_n \sim \alpha_S^{n-2} \sum_k a_k \alpha_S^k \log^{2k} \delta_{sep}, \tag{1}$$

by applying the CKKW prescription one gets

$$\alpha_S^{n-2} (\delta_{sep}^a + \sum_k b_k \alpha_S^k \log^{2k-2} \delta_{sep}). \tag{2}$$

There is a considerable freedom in the implementation of the CKKW prescription in the case of hadronic collisions. This freedom is used to tune (some of) the EvG's parameters in order to reduce as much as possible the δ_{sep} dependence, which typically manifests itself in the form of discontinuities in the derivative of the physical spectra. A discussion on these issues, with practical examples of the implementation of CKKW in HERWIG and PYTHIA, can be found in ref. [8]. CKKW has also been implemented in SHERPA [9]; an alternative procedure, proposed by Mangano, is being implemented in AlpGEN.

I stress that the complete independence of δ_{sep} cannot be achieved; this would be possible only by including all diagrams (i.e., also the virtual ones) contributing to a given order in α_S. This fact appears to be pretty obvious: it is well known, and formally established by the BN and KLN theorems, that the infrared and collinear singularities of the real matrix elements are cancelled by the virtual contributions. One may in fact be surprised by the mild δ_{sep} dependence left in the practical implementation of CKKW (see for example ref. [8]); however, we should keep in mind that parton showers do contain part of the virtual corrections, thanks to the unitarity constraint which is embedded in the Sudakov form factors. However, to cancel exactly the δ_{sep} dependence there is no alternative way to that of inserting the exact virtual contributions to the hard process considered. In doing so, one is also able to include consistently in the computation the K factor. It is important to realize that this is *the only manner* to obtain this result in a theoretically consistent way. The procedure of reweighting the EvG's results to match those obtained with fixed-order codes for certain observables must be considered a crude approximation (since no fixed-order computation can keep into account all the complicated final-state correlations that are present when defining the cuts used in experimental analyses).

The desirable thing to do would be that of adding the virtual corrections of the same order as all of the real contributions to CKKW implementations. Unfortunately, this is unfeasible, for practical and principle reasons. The practical reason is that, at variance with real corrections, we don't know how to automatize efficiently the computations of loop diagrams in the Minkowskian kinematic region. The principle reason is that there's no known way of achieving the cancellation of infrared and collinear divergences in an universal and observable-independent manner beyond NLO. We have thus to restrict ourselves to the task of including NLO corrections in EvG's, i.e. $n_E = 1$ in the notation used above.

The fact that only one extra hard emission can be included in NLOwPS's is the reason why such codes must be presently seen as complementary to MEC. When one is interested in a small number of extra emissions, then NLOwPS's must be considered superior to MEC; on the other hand, for studying processes with many hard legs involved, such as SUSY signals or backgrounds, MEC implementations should be used. A realistic goal for the near future is that of incorporating the complete NLO corrections to all the processes with different n_E's in CKKW.

The striking feature of an NLOwPS is the computation of loop diagrams (which are necessary in order to compute total rates to NLO accuracy); this in general implies the presence of negative weights. This is a new feature in MC's, which however doesn't spoil their probabilistic nature. In fact, in NLOwPS the distributions of positive and negative weights are *separately* finite, at variance with what happens in NLO computations; thus, each of them can be unweighted and evolved separately, since no cancellation between large numbers is involved in this procedure. On the other hand, the contribution of loop diagrams implies that the δ_{sep}-bias problem which affects ME corrections is simply not present. This advantage comes at a price: KLN cancellation cannot be achieved any longer solely at the level of Sudakov form factors through unitarity, since virtual and real matrix elements are now explicitly present in the computation. Technically, this complicates enormously the problem wrt the case of MEC: KLN cancellation is inclusive by nature, and one wants to achieve it here in the context of a parton shower MC, whose final state is fully exclusive thanks to the use of a hadronization model. A solution to this problem has been presented for the first time in ref. [10]. It is based on the observation that, upon formally expanding an MC result in α_S, the first non-trivial order obtained in this way must match the behaviour of the fixed-order computations at the same order, and in the collinear limit. Thus, the $\mathscr{O}(\alpha_S)$ MC result can be used effectively to cancel locally the matrix element singularities. It can be shown [10] that, after the subtracted matrix elements are matched to the shower, the singularity cancellation achieved in this way is equivalent to the KLN one for inclusive observables, up to power-suppressed terms (which are not correctly included anyhow in results based on collinear factorization theorems).

The strategy outlined above has been implemented in MC@NLO [10, 11], which features a steadily-growing number of production processes in hadronic collisions, such as single vector and Higgs bosons, vector boson pairs, heavy quark pairs, lepton pairs, and Higgs boson in association with a W or Z [12]. Apart from MC@NLO there are at present only a couple of NLOwPS hadronic codes, Φ-veto [13] and GRACE_LLsub [14], which feature only single-Z production. On the other hand, there has been a substantial theoretical activity in the field in the past few years, which will certainly lead to more practical implementation of NLOwPS in the future. Nason [15] has proposed a method for constructing NLOwPS's that should result in a smaller number of negative weights wrt those obtained with MC@NLO. Collins and Zu [16] have defined a framework in which the shower can be improved beyond the LL accuracy; at the moment, the method cannot work in QCD, since it does not include a proper treatment of soft emissions. Soper and Nagy [17] attempt to introduce a formalism in which NLOwPS techniques are embedded in a CKKW framework, thus potentially improving the latter by adding an extra $\mathscr{O}(\alpha_S)$ accuracy for each emission. Finally, one should not forget that a lot of work is being done on different aspects of standard MC's: see ref. [18]).

REFERENCES

1. M. L. Mangano, M. Moretti, F. Piccinini, R. Pittau and A. D. Polosa, JHEP **0307** (2003) 001 [arXiv:hep-ph/0206293].
2. A. Pukhov *et al.*, arXiv:hep-ph/9908288.
3. J. Fujimoto *et al.*, Comput. Phys. Commun. **153** (2003) 106 [arXiv:hep-ph/0208036].
4. F. Maltoni and T. Stelzer, JHEP **0302** (2003) 027 [arXiv:hep-ph/0208156].
5. S. Catani, F. Krauss, R. Kuhn and B. R. Webber, JHEP **0111** (2001) 063 [arXiv:hep-ph/0109231].
6. F. Krauss, JHEP **0208** (2002) 015 [arXiv:hep-ph/0205283].
7. L. Lonnblad, JHEP **0205** (2002) 046 [arXiv:hep-ph/0112284].
8. S. Mrenna and P. Richardson, JHEP **0405** (2004) 040 [arXiv:hep-ph/0312274].
9. T. Gleisberg, S. Hoeche, F. Krauss, A. Schaelicke, S. Schumann, J. Winter and G. Soff, arXiv:hep-ph/0407365.
10. S. Frixione and B. R. Webber, JHEP **0206** (2002) 029 [arXiv:hep-ph/0204244].
11. S. Frixione, P. Nason and B. R. Webber, JHEP **0308** (2003) 007 [arXiv:hep-ph/0305252].
12. S. Frixione and B. R. Webber, arXiv:hep-ph/0506182.
13. M. Dobbs, Phys. Rev. D **65** (2002) 094011 [arXiv:hep-ph/0111234].
14. Y. Kurihara, J. Fujimoto, T. Ishikawa, K. Kato, S. Kawabata, T. Munehisa and H. Tanaka, Nucl. Phys. B **654** (2003) 301 [arXiv:hep-ph/0212216].
15. P. Nason, JHEP **0411** (2004) 040 [arXiv:hep-ph/0409146].
16. J. C. Collins and X. Zu, JHEP **0503** (2005) 059 [arXiv:hep-ph/0411332].
17. Z. Nagy and D. E. Soper, arXiv:hep-ph/0503053.
18. P. Skands, these proceedings.

Precision Measurements of α_s at HERA

Claudia Glasman[1]

Universidad Autónoma de Madrid

Abstract. The precision measurements of the strong coupling constant, α_s, and its energy-scale dependence carried out at HERA by the H1 and ZEUS Collaborations are reviewed. An average value of

$$\overline{\alpha}_s(M_Z) = 0.1186 \pm 0.0011\,(\text{exp.}) \pm 0.0050\,(\text{th.})$$

is obtained from these measurements. The combined HERA determinations of the energy-scale dependence of α_s clearly show the running of α_s from jet data alone and are in agreement with the running of the coupling as predicted by QCD.

Keywords: strong coupling constant
PACS: 12.38.Qk

INTRODUCTION

The strong coupling constant, α_s, is one of the fundamental parameters of QCD. However, its value is not predicted by the theory and must be determined by experiment. Many precise and consistent determinations of α_s from diverse phenomena underlie the success of perturbative QCD (pQCD). At HERA, α_s has been determined from many observables, which include jet cross sections and structure functions, by the H1 and ZEUS Collaborations. All the available determinations [1–11] are shown in Fig. 1a. They are in good agreement with each other and are consistent with the current world average ($(\overline{\alpha}_s(M_Z))^{WA} = 0.1182 \pm 0.0027$ [12]). These determinations, most of which come from observables which involve jet algorithms, lead to determinations of α_s some of which are as precise as those from more inclusive measurements. The uncertainty in these determinations is dominated by the theoretical contributions, which amount to 4% for jet cross sections and fits of structure functions and 8% for the internal structure of jets, whereas the experimental uncertainties amount to $\sim 3\%$.

AVERAGING THE α_S DETERMINATIONS FROM HERA

To make a proper average of all these diverse measurements, the correlation among the different determinations has to be taken into account. The experimental contribution to the uncertainty due to that of the energy scale of the jets, which is the dominant source in the jet measurements, is correlated among the determinations from each experiment. On the theoretical side, the uncertainty coming from the proton parton distribution functions (PDFs) is certainly correlated whereas that coming from the hadronisation

[1] Ramón y Cajal Fellow.

CP792, *Deep Inelastic Scattering, DIS 2005,* edited by W. H. Smith and S. R. Dasu

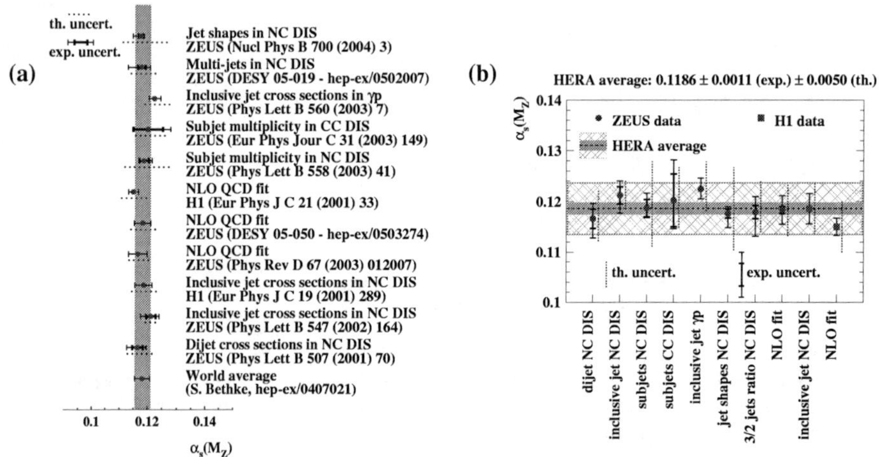

FIGURE 1. (a) Summary of $\alpha_s(M_Z)$ determinations at HERA compared with the world average; (b) $\alpha_s(M_Z)$ determinations at HERA compared with the HERA average.

corrections is only partially correlated. The uncertainty coming from the terms beyond NLO is correlated up to a certain, a priori unknown, degree; since these uncertainties are dominant, special care must be taken in the treatment of these uncertainties when making an average of the HERA determinations.

Several methods have been used to obtain an average value of α_s from the HERA measurements and its uncertainty. Using a naive method in which all uncertainties are assumed to be uncorrelated, the average value and its uncertainty are:

$$\overline{\alpha}_s(M_Z) = 0.1188 \pm 0.0020 \text{ (ZEUS + H1)}.$$

The second method used is that developed by M. Schmelling [13] to average correlated data when correlations are present but hard to quantify. In this method, an error-weighted average and an optimised correlation error were obtained from the error covariance matrix by assuming an overall correlation factor between the total errors of all measurements; the overall factor was determined by the condition that the overall χ^2/dof is equal to unity. First, an error-weighted average was done separately for the ZEUS and H1 measurements, and then the two averages were combined:

$$\overline{\alpha}_s(M_Z) = 0.1196 \pm 0.0060 \text{ (ZEUS)} \quad \text{and} \quad \overline{\alpha}_s(M_Z) = 0.1166 \pm 0.0053 \text{ (H1)},$$
$$\overline{\alpha}_s(M_Z) = 0.1188 \pm 0.0057 \text{ (ZEUS + H1)}.$$

The averages from the ZEUS and H1 determinations are compatible within the uncertainties. The uncertainty of the combined average is $\sim 5\%$. This procedure gives rise to relatively large uncertainties when there are large correlations among some of the measurements, as it is the case here. To overcome this effect, the method has been repeated by restricting to the most accurate measurements [14]. The result of applying the procedure to those measurements with a total error $\Delta\alpha_s(M_Z) < 0.006$ [2, 3, 5, 6, 9] is:

$$\overline{\alpha}_s(M_Z) = 0.1192 \pm 0.0047 \ (\Delta\alpha_s^i < 0.006) \ (\text{ZEUS} + \text{H1}).$$

Finally, a more reliable, but conservative, approach has been used in which the known correlations from the determinations of α_s coming from the same experiment were taken into account ("correlation method"). The theoretical uncertainties arising from the terms beyond NLO were assumed to be (conservatively) fully correlated. Error-weighted averages were obtained separately for the ZEUS and H1 measurements:

$$\overline{\alpha}_s(M_Z) = 0.1200 \pm 0.0023 \ (\text{exp.})^{+0.0058}_{-0.0049} \ (\text{th.}) \ (\text{ZEUS}),$$

$$\overline{\alpha}_s(M_Z) = 0.1160 \pm 0.0016 \ (\text{exp.})^{+0.0048}_{-0.0049} \ (\text{th.}) \ (\text{H1}).$$

A HERA average was obtained by using the error-weighted average method on the ZEUS and H1 averages, assuming the experimental uncertainties to be uncorrelated and taking the overall theoretical uncertainty as the linear average of its contribution in each experiment. As a result, the average of the HERA measurements and its uncertainty are:

$$\overline{\alpha}_s(M_Z) = 0.1186 \pm 0.0011 \ (\text{exp.}) \pm 0.0050 \ (\text{th.}) \ (\text{ZEUS} + \text{H1}).$$

This average, together with the individual values considered, is shown in Fig. 1b. It is found to be in good agreement with the current world average, which does not include any of these determinations. The results of applying Schmelling's and the correlation methods are very similar, giving confidence on the average obtained and its estimated uncertainty.

ENERGY-SCALE DEPENDENCE OF α_S

The H1 and ZEUS Collaborations have tested the pQCD prediction for the energy-scale dependence of the strong coupling constant by determining α_s from the measured differential jet cross sections at different E_T^{jet} [1–3, 9]. Figure 2a shows the determinations of the energy-scale dependence of α_s as a function of E_T^{jet} from H1 and ZEUS. The determinations are consistent with the running of α_s as predicted by pQCD over a large range in E_T^{jet}.

The determinations of $\alpha_s(E_T^{\text{jet}})$ from H1 and ZEUS at similar E_T^{jet} have been combined using the correlation method explained above. The combined HERA determinations of the energy-scale dependence of α_s are shown in Fig. 2b, in which the running of α_s from HERA jet data alone is clearly observed.

SUMMARY

A comprehensive average of α_s and its energy-scale dependence from HERA data has been performed taking into account the known correlations in each experiment and assuming conservatively that the theoretical uncertainties arising from the terms beyond NLO are fully correlated. The HERA average is

$$\overline{\alpha}_s(M_Z) = 0.1186 \pm 0.0011 \ (\text{exp.}) \pm 0.0050 \ (\text{th.}).$$

The experimental uncertainty of this average is $\sim 0.9\%$ and the theoretical uncertainty amounts to $\sim 4\%$. There is still room for improvement when the next-to-NLO (NNLO)

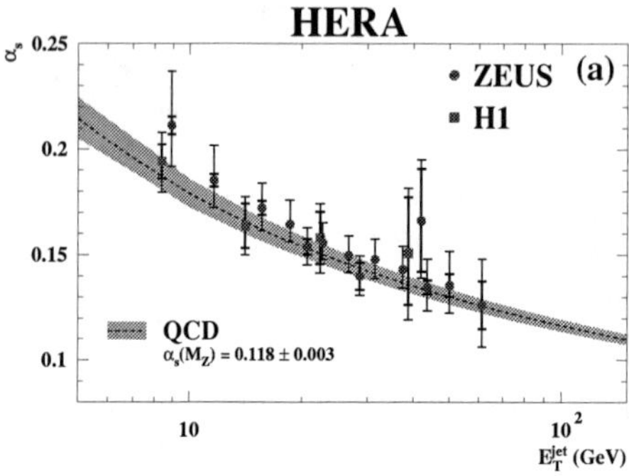

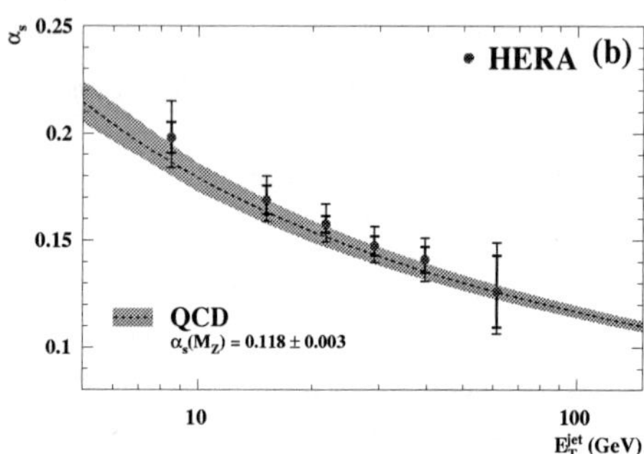

FIGURE 2. (a) α_s as a function of $E_T^{\rm jet}$ from H1 and ZEUS; (b) combined α_s as a function of $E_T^{\rm jet}$ from HERA jet data. In both fi gures, the QCD prediction for the running of α_s is also shown. In (a), the inner error bars display the statistical uncertainties and the outer error bars display the systematic and theoretical uncertainties added in quadrature. In (b), the inner (outer) error bars show the experimental (theoretical) uncertainties.

calculations needed for the determination of the PDFs are included and when the NNLO calculations needed for jet-based observables are finished.

ACKNOWLEDGMENTS

This work has been carried out in collaboration with Juan Terrón. I would like to thank my colleagues from H1 and ZEUS for their help in the preparation of this report.

REFERENCES

1. ZEUS Coll., J. Breitweg et al., *Phys. Lett.* **B** 507 (2001) 70.
2. ZEUS Coll., S. Chekanov et al., *Phys. Lett.* **B** 547 (2002) 164.
3. H1 Coll., C. Adloff et al., *Eur. Phys. Jour.* **C** 19 (2001) 289.
4. ZEUS Coll., S. Chekanov et al., *Phys. Rev.* **D** 67 (2003) 012007.
5. ZEUS Coll., S. Chekanov et al., DESY 05-050 (March 2005), hep-ph/0503274.
6. H1 Coll., C. Adloff et al., *Eur. Phys. Jour.* **C** 21 (2001) 33.
7. ZEUS Coll., S. Chekanov et al., *Phys. Lett.* **B** 558 (2003) 41.
8. ZEUS Coll., S. Chekanov et al., *Eur. Phys. Jour.* **C** 31 (2003) 149.
9. ZEUS Coll., S. Chekanov et al., *Phys. Lett.* **B** 560 (2003) 7.
10. ZEUS Coll., S. Chekanov et al., DESY 05-019 (January 2005), hep-ex/0502007.
11. ZEUS Coll., S. Chekanov et al., *Nucl. Phys.* **B** 700 (2004) 3.
12. S. Bethke, hep-ex/0407021.
13. M. Schmelling, *Phys. Scr.* 51 (1995) 676.
14. S. Bethke, *J. Phys.* **G** 26 (2000) R27.

JET PHYSICS IN RUN 2 AT CDF

Rick Field
(for the CDF Collaboration)

Department of Physics, University of Florida, Gainesville, Florida, 32611, USA

Abstract: New CDF Run 2 results on the inclusive jet cross section (K_T algorithm) and the b-jet cross section (MidPoint algorithm) are presented and compared with theory. We also study the "underlying event" by using the direction of the leading jet to isolate regions of η-ϕ space that are very sensitive to the "beam-beam" remnants and to multiple parton interactions.

Keywords: QCD, Jets, Hadron Collider.
PACS: 12.38.-t, 12.38.Bx, 12.38.Qk.

The study of proton-antiproton collisions in Run 2 at CDF is teaching us a lot about how QCD works. Comparing data with theory will lead to improved QCD Monte-Carlo models and to more precise parton distribution functions. In the CDF-QCD group we are studying the inclusive jet cross section using both the MidPoint cone algorithm and the K_T algorithm [1]. We are studying heavy flavor jets (*i.e.* b-jets) and jets produced in association with photons, W bosons, and Z bosons. We are studying jet fragmentation (jet shapes, momentum distributions, two-particle correlations) and we are making good progress in understanding and modeling the "underlying event" in hard scattering processes. Here I will only be able to show a little bit of what we have learned.

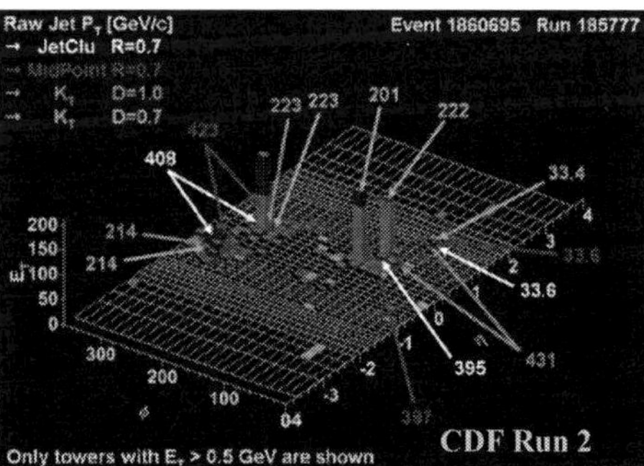

FIGURE 1. Shows the transverse energy of calorimeter towers with $E_T > 0.5$ GeV for an event in the CDF detector. The MidPoint algorithm combines the two clusters into one "jet" with $p_T = 423$ GeV/c while the K_T algorithm (D = 0.7) finds two "jets" with $p_T = 223$ GeV/c and 214 GeV/c.

Experimentally we measure "jets" at the detector (*i.c.* calorimeter) level by observing the energy in each calorimeter cell as illustrated in Fig. 1. Of course the "jet" cross section depends on ones choice of jet algorithm. Each jet algorithm is a different observable and comparing the results of different jet algorithm teaches us

CP792, *Deep Inelastic Scattering, DIS 2005,* edited by W. H. Smith and S. R. Dasu

about QCD. Most theorists prefer the K_T algorithm over cone algorithms, however, one must demonstrate that the K_T algorithm will work in the collider environment where there is an "underlying event".

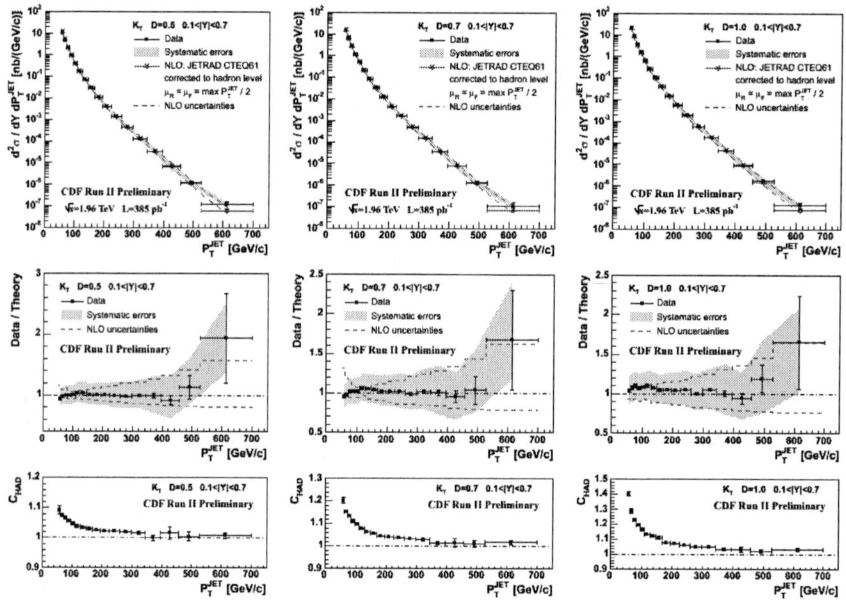

FIGURE 2. The CDF Run 2 jet cross section using the K_T algorithm with D = 0.5, 0.7, and 1.0. The data have been corrected to the particle level (with an "underlying event") and the NLO parton level (CTEQ61M) has been corrected for fragmentation and the "underlying event" (with correction factors C_{HAD}).

What we measure in the calorimeter must be corrected for detector efficiency which is done by comparing the QCD Monte-Carlo models at the particle (*i.e.* generator level) with the result after detector simulation (*i.e.* CDFSIM). At CDF we correct the data using PYTHIA Tune A [2,3] and use differences between PYTHIA and HERWIG [4] as a measure of the systematic uncertainty in correcting to the particle level. To compare with a NLO parton level theory calculation requires removing the "underlying event" and correcting the data to the parton level. I believe that experimenters should publish what they measure (*i.e.* observables at the particle level with an "underlying event") and one should correct the NLO parton level theory by adding in the effects of fragmentation and the "underlying event". Fig. 2 shows the CDF Run 2 jet cross section using the K_T algorithm. The correction factors for the "underlying event" are large for p_T(jet) < 300 GeV/c. Nonetheless, the agreement between the theory and data is very good. The highest p_T bin contains only a few events and the resulting large error makes it consistent with the theory. Within the errors the data and theory agree, however, we will continue to pay close attention to the highest p_T bins.

We identify b-jets by studying the invariant mass of the charged particles emanating from the secondary vertex which is displaced slightly from the primary interaction vertex due to the long lifetime of the heavy b-quark. As shown in Fig. 3, the fraction of b-tagged jets is determined by fitting (on a bin-by-bin bases) the secondary vertex invariant mass to templates determined from PYTHIA Tune A. Fig. 4 shows the resulting CDF b-jet inclusive cross section at 1.96 TeV compared with PYTHIA Tune A.

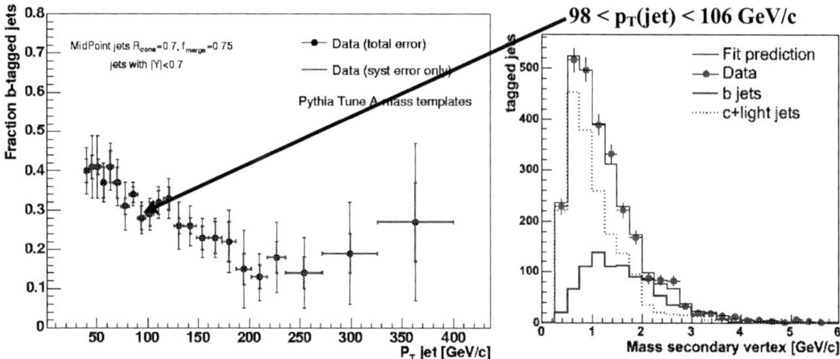

FIGURE 3. (*left*) Shows the fraction of b-tagged jets as a function of the jet p_T. (*right*) Shows the fit to the secondary vertex mass for the bin $98 < p_T(\text{jet}) < 106$ GeV/c.

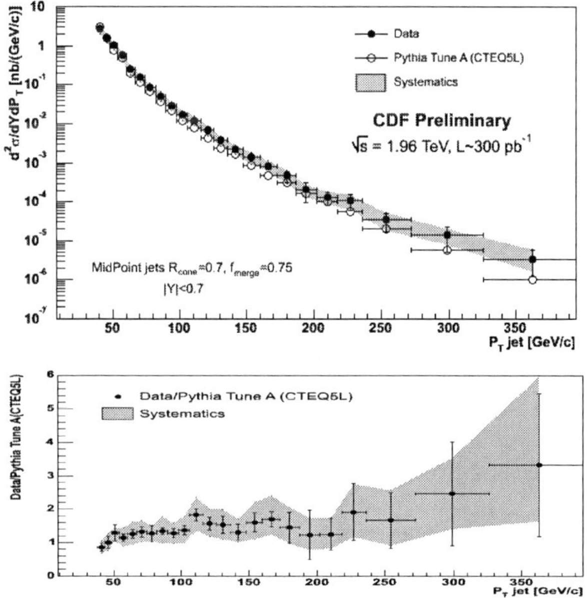

FIGURE 4. (*top*) Shows the CDF Run 2 b-jet inclusive cross section at 1.96 TeV compared with PYTHIA Tune A. (*bottom*) Shows the ratio data/theory for PYTHIA Tune A.

Both the inclusive jet cross section and the b-jet cross section depend sensitively on the "underlying event" and we are working to understand and model the "underlying event" in Run 2 at CDF. We study the "underlying event" using the direction of the leading calorimeter jet (JetClu, R = 0.7) to isolate regions of η-ϕ space that are sensitive to the "underlying event". The angle $\Delta\phi = \phi - \phi_{jet\#1}$ is the relative azimuthal angle between a charged particle and the direction of jet#1. As illustrated in Fig. 5, the "transverse" region is defined by $60^o < |\Delta\phi| < 120^o$ and $|\eta| < 1$. The "transverse" region is perpendicular to the plane of the hard 2-to-2 scattering and is therefore very sensitive to the "underlying event". We restrict ourselves to charged particles in the range $p_T > 0.5$ GeV/c and $|\eta| < 1$, but allow the leading jet that is used to define the "transverse" region to have $|\eta(\text{jet\#1})| < 2$. In our Run 2 analysis we consider two classes of events. We refer to events in which there are no restrictions placed on the second and third highest E_T jets (jet#2 and jet#3) as "leading jet" events. Events with at least two jets with $E_T > 15$ GeV where the leading two jets are nearly "back-to-back" ($|\Delta\phi_{12}| > 150^o$) with $E_T(\text{jet\#2})/E_T(\text{jet\#1}) > 0.8$ and $E_T(\text{jet\#3}) < 15$ GeV are referred to as "back-to-back" events. "Back-to-back" events are a subset of the "leading jet" events. The idea here is to suppress hard initial and final-state radiation thus increasing the sensitivity of the "transverse" region to the "beam-beam remnant" and the multiple parton scattering component of the "underlying event".

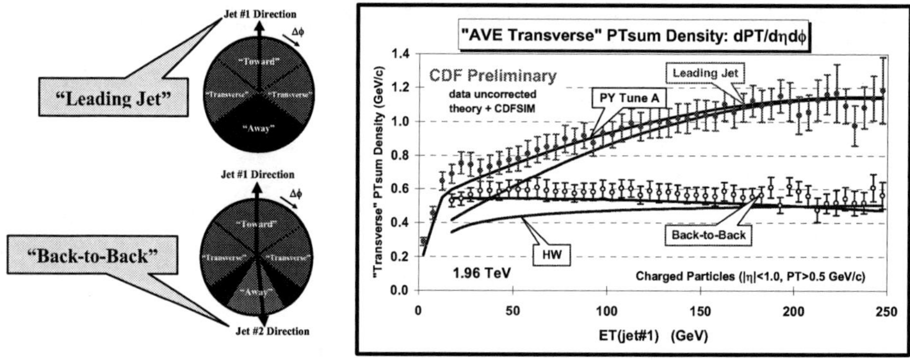

FIGURE 5. Data on the average PTsum density, $dPT_{sum}/d\eta d\phi$, for charged particles with $p_T > 0.5$ GeV/c and $|\eta| < 1$ in the "transverse" region for "leading jet" events and for "back-to-back" events as a function of the leading jet E_T compared with PYTHIA Tune A and HERWIG at 1.96 TeV after CDFSIM..

Fig. 5 compares PYTHIA Tune A and HERWIG with the "leading jet" and "back-to-back" data on the average PT_{sum} density, $dPT_{sum}/d\eta d\phi$, in the "transverse" region as a function of the leading jet E_T. The multiple parton interaction parameters of PYTHIA were adjusted to agree with Run 1 data [2,3]. The "back-to-back" data show a slight decrease in the "transverse" density with increasing $E_T(\text{jet\#1})$ which is described well by PYTHIA Tune A (with multiple parton interactions) but not by HERWIG (without multiple parton interactions).

CONCLUSIONS

We have measured the inclusive jet cross section using the K_T algorithm. The data agree well with the NLO theory after the theory is corrected for fragmentation effects and for the "underlying event". Our results show that the K_T algorithm works fine at the Tevatron collider, which has implications for the LHC.

We have also measured the b-jet production cross section at 1.96 TeV. Measurements of the b-jet production at hadron colliders provide an important test of QCD. Past Tevatron measurements of b-quark production indicated a possible "excess" with respect to QCD predictions. However, the b-jet cross is in agreement expectations. The data are about a factor of 1.4 larger than the prediction of PYTHIA Tune A, however, this is to be expected since PYTHIA is a "leading log order" model. One cannot expect it correctly predict the precise amount of "flavor excitation" and "gluon splitting". We are working on the comparisons with MC@NLO [5].

Both the inclusive jet cross section and the b-jet cross section depend sensitively on the "underlying event". In Run 2 at CDF we use the direction of the leading calorimeter jet in each event to define the "transverse" regions of η-ϕ space that is very sensitive to the "underlying event". In addition, by selecting events with at least two jets that are nearly back-to-back ($\Delta\phi_{12} > 150^o$) and with E_T(jet#3) < 15 GeV we are able to look closer at the "beam-beam remnant" and multiple parton interaction components of the "underlying event". PYTHIA Tune A (with multiple parton interactions) does a good job in describing the "underlying event" (*i.e.* "transverse" regions) for both "leading jet" and "back-to-back" events. HERWIG (without multiple parton interactions) does not have enough activity in the "underlying event" for E_T(jet#1) less than about 150 GeV, which was also observed in our published Run 1 analyses [6].

JIMMY [7] is a model of multiple parton interaction which can be combined with HERWIG or MC@NLO to enhance the "underlying event" thereby improving the agreement with data. We are working on tunes of JIMMY that fit the CDF Run 2 "underlying event" data [8]. We are also studying the energy in the "underlying event".

REFERENCES

1. S. D. Ellis and D. E. Soper, Phys. Rev. **D48**, 3160 (1993).
2. T. Sjostrand, Phys. Lett. **157B**, 321 (1985); M. Bengtsson, T. Sjostrand, and M. van Zijl, Z. Phys. **C32**, 67 (1986); T. Sjostrand and M. van Zijl, Phys. Rev. **D36**, 2019 (1987).
3. *Min-Bias and the Underlying Event at the Tevatron and the LHC*, talk by R. Field at the Fermilab ME/MC Tuning Workshop, Fermilab, October 4, 2002. *Toward an Understanding of Hadron Collisions: From Feynman-Field until Now*, talk by R. Field at the Fermilab Joint Theoretical Experimental "Wine & Cheese" Seminar, Fermilab, October 4, 2002.
4. G. Marchesini and B. R. Webber, Nucl. Phys **B310**, 461 (1988); I. G. Knowles, Nucl. Phys. **B310**, 571 (1988); S. Catani, G. Marchesini, and B. R. Webber, Nucl. Phys. **B349**, 635 (1991).
5. *The MC and NLO 3.1 Event Generator*, Stefano Frixione and Bryan R. Webber, CAVENDISH-HEP-05-09, hep-ph/0506182 (2005). *Matching NLO QCD and Parton Showers in Heavy Flavor Production*, Stefano Frixione, Paolo Nason, and Bryan R. Webber, JHEP 0308:007 (2003).
6. *Charged Jet Evolution and the Underlying Event in Proton-Antiproton Collisions at 1.8 TeV*, The CDF Collaboration (T. Affolder et al.), Phys. Rev. **D65**, 092002 (2002). *The Underlying Event in Hard Interactions at the Tevatron Proton-Antiproton Collider*, CDF Collaboration (D. Acosta et al.), Phys. Rev. **D70**, 072002 (2004).
7. *Multiparton Interactions in Photoproduction at HERA*, J.M. Butterworth, J.R. Forshaw, and M.H. Seymour, Z. Phys. **C7**, 637-646 (1996).
8. *HERWIG, JIMMY, and PYTHIA Tune A*, talk by R. Field at the TeV4LHC Workshop, Fermilab, December 1, 2004.

Soft-gluon expansions through NNNLO

Nikolaos Kidonakis

Kennesaw State University, Physics #1202, 1000 Chastain Rd, Kennesaw, GA 30144-5591, USA

Abstract. I present universal master formulas for soft-gluon corrections to hard-scattering cross sections through next-to-next-to-next-to-leading order (NNNLO). I also briefly discuss applications to some processes where these corrections enhance the cross section and decrease the scale dependence.

Keywords: soft gluons;high-order corrections;resummation
PACS: 12.38.Bx;12.38.Cy;13.85.-t

SOFT-GLUON RESUMMATION

Cross sections in perturbative QCD can be calculated by employing factorization theorems as $\sigma = \sum_f \int [\prod_i dx_i \, \phi_{f/h_i}(x_i,\mu_F)] \, \hat{\sigma}(s,t,u,\mu_F,\mu_R)$, where $\hat{\sigma}$ is the perturbatively calculable hard-scattering cross section, and the parton distributions ϕ are determined from experiment. The renormalization and factorization scales are denoted by μ_R and μ_F respectively, and s,t,u are standard kinematical invariants formed from the momenta of the partons in the hard scattering.

Near threshold for the production of a specified system, such as a top quark pair or a Higgs boson, there is restricted phase space for real gluon emission. The incomplete cancellation of infrared divergences between real and virtual graphs results in the appearance of large logarithms. If we define $s_4 = s+t+u-\sum m^2$, with m the masses of the particles in the scattering, then $s_4 \to 0$ at threshold and these soft-gluon logarithmic corrections take the form of plus distributions, $\mathcal{D}_l(s_4) \equiv [\ln^l(s_4/M^2)/s_4]_+$, where M is a relevant hard scale, such as the mass of a heavy quark or the transverse momentum of a jet, and $l \leq 2n-1$ for the n-th order corrections.

If we define moments of the cross section $\hat{\sigma}(N) = \int_0^\infty ds_4 \, e^{-Ns_4/M^2} \, \hat{\sigma}(s_4)$ then the soft corrections are transformed as $[\ln^l(s_4/M^2)/s_4]_+ \to [(-1)^{l+1}/(l+1)] \ln^{l+1} N + \cdots$. We can formally resum these logarithms $\ln N$ to all orders in α_s by factorizing the soft gluons from the hard scattering [1, 2]. Although the formal resummation in moment space is well defined, when inverting back to momentum space we encounter ambiguities due to the infrared singularity which require a prescription. Unfortunately different prescriptions can give different numerical results as well as have dubious theoretical underpinnings (see discussion in Ref. [3]). However, fixed-order expansions can provide us with solid, prescription-independent, theoretical and numerical results [3, 4].

At next-to-leading order (NLO) in α_s, the cross section includes $\mathcal{D}_1(s_4)$ terms which are the leading logarithms (LL), and $\mathcal{D}_0(s_4)$ terms which are the next-to-leading logarithms (NLL). At next-to-next-to-leading order (NNLO), we have $\mathcal{D}_3(s_4)$ (LL), $\mathcal{D}_2(s_4)$ (NLL), $\mathcal{D}_1(s_4)$ (NNLL), and $\mathcal{D}_0(s_4)$ (NNNLL) terms. At next-to-next-to-next-to-leading order (NNNLO), we have $\mathcal{D}_5(s_4)$ (LL), $\mathcal{D}_4(s_4)$ (NLL), $\mathcal{D}_3(s_4)$ (NNLL),

CP792, *Deep Inelastic Scattering, DIS 2005,* edited by W. H. Smith and S. R. Dasu

$\mathscr{D}_2(s_4)$ (NNNLL), $\mathscr{D}_1(s_4)$ (NNNNLL), and $\mathscr{D}_0(s_4)$ (NNNNNLL) terms.

The threshold resummation formalism has been applied by now to many processes in hadron-hadron and lepton-hadron colliders, for both total and differential cross sections, in both single-particle-inclusive (1PI) and pair-invariant-mass (PIM) kinematics, for both simple and complex color flows, and in both $\overline{\mathrm{MS}}$ and DIS factorization schemes [4].

Specific processes for which soft-gluon corrections have been calculated at NNLO include top quark pair hadroproduction [3, 5], beauty and charm production [6], jet production [7], direct photon production [8], large-p_T W production [9], FCNC top production [10], and charged Higgs production [11]. Numerical results show that usually the soft corrections are a good approximation of the full NLO result. In all cases the higher-order corrections are sizable and produce a dramatic decrease of the scale dependence of the cross section.

The resummed cross section can be written for an arbitrary process as [4]

$$\begin{aligned}\hat{\sigma}^{res}(N) \;=\; & \exp\left[\sum_i E_i(N_i)\right] \exp\left[\sum_j E'_j(N_j)\right] \exp\left[2 d_{\alpha_s} \int_{\mu_R}^{\sqrt{s}} \frac{d\mu'}{\mu'} \beta\left(\mu'\right)\right] \\ & \times \exp\left[\sum_i 2 \int_{\mu_F}^{\sqrt{s}} \frac{d\mu'}{\mu'} \left(\frac{\alpha_s(\mu')}{\pi}\gamma_i^{(1)} + \gamma'_{i/i}\left(\mu'\right)\right)\right] \\ & \times \mathrm{Tr}\left\{H\left(\mu_R\right) \exp\left[\int_{\sqrt{s}}^{\sqrt{s}/\tilde{N}_j} \frac{d\mu'}{\mu'} \Gamma_S^{\dagger}\left(\mu'\right)\right] S\left(s/\tilde{N}_j^2\right) \exp\left[\int_{\sqrt{s}}^{\sqrt{s}/\tilde{N}_j} \frac{d\mu'}{\mu'} \Gamma_S\left(\mu'\right)\right]\right\} \quad (1)\end{aligned}$$

where $E_i(N_i)$ denotes contributions from the incoming partons and is given in the $\overline{\mathrm{MS}}$ scheme by

$$E_i(N_i) = -\int_0^1 dz \frac{z^{N_i-1}-1}{1-z} \left\{\int_{(1-z)^2 s}^{\mu_F^2} \frac{d\mu'^2}{\mu'^2} A_i(\mu') + \nu_i\left[(1-z)^2 s\right]\right\} \quad (2)$$

with $A_i = C_i[\alpha_s/\pi + (\alpha_s/\pi)^2 K/2] + \cdots$, $\nu_i = (\alpha_s/\pi)C_i + (\alpha_s/\pi)^2 \nu_i^{(2)} + \cdots$; $E_j(N_j)$ denotes contributions from massless final-state partons (if any) and is given by

$$E'_j(N_j) = \int_0^1 dz \frac{z^{N_j-1}-1}{1-z} \left\{\int_{(1-z)^2}^{1-z} \frac{d\lambda}{\lambda} A_j(\lambda s) - B'_j[(1-z)s] - \nu_j\left[(1-z)^2 s\right]\right\} \quad (3)$$

where $B'_j = (\alpha_s/\pi)B'^{(1)}_j + (\alpha_s/\pi)^2 B'^{(2)}_j + \cdots$; γ_i are parton anomalous dimensions; H are hard scattering matrices, independent of soft-gluon radiation; S are soft matrices which describe noncollinear soft-gluon emission; and Γ_S are soft anomalous dimension matrices which appear in the evolution of the S matrices. H, S, and Γ_S are matrices in the space of color exchanges; they become simple functions for processes with simple color structure, such as Drell-Yan production.

Expansions of the resummed cross section through NNLO were given in Ref. [4], where master formulas were derived and then used in calculations for specific processes [5-11]. Here we extend this expansion to NNNLO.

Expanding Eq. (1) to NLO, gives us the master formula for the NLO corrections

$$\begin{aligned}\hat{\sigma}^{(1)} &= \sigma^B \frac{\alpha_s(\mu_R^2)}{\pi}\{c_3\,\mathscr{D}_1(s_4)+c_2\,\mathscr{D}_0(s_4)+c_1\,\delta(s_4)\} \\ &+ \frac{\alpha_s^{d_{\alpha_s}+1}(\mu_R^2)}{\pi}[A^c\,\mathscr{D}_0(s_4)+T_1^c\,\delta(s_4)] \end{aligned} \tag{4}$$

with $c_3=\sum_i 2C_i-\sum_j C_j$, where for quarks $C_F=(N_c^2-1)/(2N_c)$ and for gluons $C_A=N_c$; $c_2=c_2^\mu+T_2$, with $c_2^\mu=-\sum_i C_i\ln(\mu_F^2/M^2)$ and

$$T_2=-\sum_i\left[C_i+2C_i\ln\left(\frac{-t_i}{M^2}\right)+C_i\ln\left(\frac{M^2}{s}\right)\right]-\sum_j\left[B_j'^{(1)}+C_j+C_j\ln\left(\frac{M^2}{s}\right)\right]; \tag{5}$$

and $c_1=c_1^\mu+T_1$, with

$$c_1^\mu=\sum_i\left[C_i\ln\left(\frac{-t_i}{M^2}\right)-\gamma_i^{(1)}\right]\ln\left(\frac{\mu_F^2}{M^2}\right)+d_{\alpha_s}\frac{\beta_0}{4}\ln\left(\frac{\mu_R^2}{M^2}\right), \tag{6}$$

where for quarks $B_q'^{(1)}=\gamma_q^{(1)}=3C_F/4$ and for gluons $B_g'^{(1)}=\gamma_g^{(1)}=\beta_0/4$. The term A^c involves matrices and is given by $A^c=\mathrm{tr}\left(H^{(0)}\Gamma_S'^{(1)\dagger}S^{(0)}+H^{(0)}S^{(0)}\Gamma_S'^{(1)}\right)$. Finally T_1 and T_1^c can be read off a complete NLO calculation for a specific process.

Expanding the resummed cross section through NNLO and matching with the NLO result gives us the master formula for the NNLO corrections [4]

$$\begin{aligned}\hat{\sigma}^{(2)} &= \sigma^B\frac{\alpha_s^2(\mu_R^2)}{\pi^2}\,\frac{1}{2}c_3^2\,\mathscr{D}_3(s_4) \\ &+\sigma^B\frac{\alpha_s^2(\mu_R^2)}{\pi^2}\left\{\frac{3}{2}c_3c_2-\frac{\beta_0}{4}c_3+\sum_j C_j\frac{\beta_0}{8}\right\}\mathscr{D}_2(s_4)+\frac{\alpha_s^{d_{\alpha_s}+2}(\mu_R^2)}{\pi^2}\,\frac{3}{2}c_3A^c\,\mathscr{D}_2(s_4) \\ &+\sigma^B\frac{\alpha_s^2(\mu_R^2)}{\pi^2}\left\{c_3c_1+c_2^2-\zeta_2c_3^2-\frac{\beta_0}{2}T_2+\frac{\beta_0}{4}c_3\ln\left(\frac{\mu_R^2}{M^2}\right)+c_3\frac{K}{2}-\sum_j\frac{\beta_0}{4}B_j'^{(1)}\right\}\mathscr{D}_1(s_4) \\ &+\frac{\alpha_s^{d_{\alpha_s}+2}(\mu_R^2)}{\pi^2}\left\{\left(2c_2-\frac{\beta_0}{2}\right)A^c+c_3T_1^c+F^c\right\}\mathscr{D}_1(s_4)+\mathscr{O}(\mathscr{D}_0(s_4)),\end{aligned} \tag{7}$$

where we show terms explicitly through NNLL. Here $F^c=\mathrm{tr}[H^{(0)}\left(\Gamma_S'^{(1)\dagger}\right)^2S^{(0)}+H^{(0)}S^{(0)}(\Gamma_S'^{(1)})^2+2H^{(0)}\Gamma_S'^{(1)\dagger}S^{(0)}\Gamma_S'^{(1)}]$.

Finally, expanding the resummed formula through NNNLO and matching with the NLO and NNLO results gives us the master formula for the NNNLO corrections

$$\hat{\sigma}^{(3)} = \sigma^B\frac{\alpha_s^3(\mu_R^2)}{\pi^3}\,\frac{1}{8}c_3^3\,\mathscr{D}_5(s_4)$$

$$+\sigma^B \frac{\alpha_s^3(\mu_R^2)}{\pi^3}\left\{\frac{5}{8}c_3^2 c_2-\frac{5}{2}c_3 X_3\right\}\mathscr{D}_4(s_4)+\frac{\alpha_s^{d_{\alpha_s}+3}(\mu_R^2)}{\pi^3}\frac{5}{8}c_3^2 A^c\,\mathscr{D}_4(s_4)$$

$$+\sigma^B \frac{\alpha_s^3(\mu_R^2)}{\pi^3}\left\{c_3 c_2^2+\frac{1}{2}c_3^2 c_1-\zeta_2 c_3^3+(\beta_0-4c_2)X_3+2c_3X_2-\sum_j C_j\frac{\beta_0^2}{48}\right\}\mathscr{D}_3(s_4)$$

$$+\frac{\alpha_s^{d_{\alpha_s}+3}(\mu_R^2)}{\pi^3}\left\{\frac{1}{2}c_3^2T_1^c+\left[2c_3c_2-\frac{\beta_0}{2}c_3-4X_3\right]A^c+c_3F^c\right\}\mathscr{D}_3(s_4)+\mathscr{O}(\mathscr{D}_2(s_4)) \quad (8)$$

where again we show terms explicitly through NNLL. Here $X_3=(\beta_0/12)c_3-\sum_j C_j\beta_0/24$ and $X_2=-(\beta_0/4)T_2+(\beta_0/8)c_3\ln(\mu_R^2/M^2)+c_3K/4-\sum_j(\beta_0/8)B'^{(1)}_j$.

This calculation has recently been applied to charged Higgs production with a top quark via bottom gluon fusion at the LHC through NNLO [11], where for a charged Higgs mass of 500 GeV the NLO-NLL soft-gluon corrections provide a enhancement of 38% over the leading-order cross section and the NNLO-NLL corrections provide a enhancement of 11% over the NLO-NLL cross section. A new calculation of the NNNLO-NLL corrections shows that they provide an additional 7% enhancement over the NNLO-NLL cross section and further stabilize the scale dependence of the cross section. Similarly, for top quark pair production at the Tevatron the scale dependence is considerably decreased. More details will be given in a forthcoming paper.

REFERENCES

1. N. Kidonakis and G. Sterman, *Phys. Lett. B*, **387**, 867 (1996); *Nucl. Phys. B*, **505**, 321 (1997); N. Kidonakis, G. Oderda, and G. Sterman, *Nucl. Phys. B*, **525**, 299 (1998); *Nucl. Phys. B*, **531**, 365 (1998); N. Kidonakis, *Int. J. Mod. Phys. A*, **15**, 1245 (2000).
2. E. Laenen, G. Oderda, and G. Sterman, *Phys. Lett. B*, **438**, 173 (1998).
3. N. Kidonakis, *Phys. Rev. D*, **64**, 014009 (2001); *Int. J. Mod. Phys. A*, **16S1A**, 363 (2001).
4. N. Kidonakis, *Int. J. Mod. Phys. A*, **19**, 1793 (2004); in *DIS 2003*, hep-ph/0306125, hep-ph/0307207; *Mod. Phys. Lett. A*, **19**, 405 (2004); in *DPF 2004*, hep-ph/0410116.
5. N. Kidonakis and J. Smith, *Phys. Rev. D*, **51**, 6092 (1995); N. Kidonakis, E. Laenen, S. Moch, and R. Vogt, *Phys. Rev. D*, **64**, 114001 (2001); N. Kidonakis and R. Vogt, *Phys. Rev. D*, **68**, 114014 (2003); *Eur. Phys. J. C*, **33**, s466 (2004); in *DPF 2004*, hep-ph/0410367.
6. N. Kidonakis, E. Laenen, S. Moch, and R. Vogt, *Phys. Rev. D*, **67**, 074037 (2003); *Nucl. Phys. A*, **715**, 549c (2003); N. Kidonakis and R. Vogt, *Eur. Phys. J. C*, **36**, 201 (2004); in *DIS 2004*, hep-ph/0405212.
7. N. Kidonakis and J.F. Owens, *Phys. Rev. D*, **63**, 054019 (2001).
8. N. Kidonakis and J.F. Owens, *Phys. Rev. D*, **61**, 094004 (2000); *Int. J. Mod. Phys. A*, **19**, 149 (2004).
9. N. Kidonakis and A. Sabio Vera, *JHEP*, **02**, 027 (2004); hep-ph/0409337.
10. A. Belyaev and N. Kidonakis, *Phys. Rev. D*, **65**, 037501 (2002); N. Kidonakis and A. Belyaev, *JHEP*, **12**, 004 (2003); in *DIS 2004*, hep-ph/0407032.
11. N. Kidonakis, *JHEP*, **05**, 011 (2005); in *DIS 2004*, hep-ph/0406179; in *DIS 2005*, hep-ph/0505271.

Multi Jet Production at High Q^2

Thomas Kluge [1]

DESY, Notkestr. 85, 22607 Hamburg, Germany
E-mail: thomas.kluge@desy.de

Abstract. Deep-inelastic e^+p scattering data, taken with the H1 detector at HERA, are used to investigate jet production over a range of four-momentum transfers $150 < Q^2 < 15000\,\text{GeV}^2$ and transverse jet energies $5 < E_T < 50\,\text{GeV}$. The analysis is based on data corresponding to an integrated luminosity of $\mathscr{L}_{\text{int}} = 65.4\,\text{pb}^{-1}$ taken in the years 1999-2000 at a centre-of-mass energy $\sqrt{s} \approx 319\,\text{GeV}$. Jets are defined by the inclusive k_t algorithm in the Breit frame of reference. Dijet and trijet jet cross sections are measured with respect to the exchanged boson virtuality and in addition the ratio of the trijet to the dijet cross section $R_{3/2}$ is investigated. The results are compared to the predictions of perturbative QCD calculations in next-to-leading order in the strong coupling constant α_s. The value of $\alpha_s(m_Z)$ determined from the study of $R_{3/2}$ is $\alpha_s(m_Z) = 0.1175 \pm 0.0017(\text{stat.}) \pm 0.0050(\text{syst.})^{+0.0054}_{-0.0068}(\text{theo.})$.

INTRODUCTION AND OBSERVABLES

Deep-inelastic scattering (DIS) at HERA is a precision tool for studies of Quantum Chromo Dynamics (QCD). Jet cross sections at high transverse momenta are in particular attractive, because perturbative calculations (pQCD) are precise and non-perturbative hadronisation effects are weak. In the past inclusive jet cross sections [1, 2], as well as multijet cross sections [3, 4] have been studied at HERA.

The aim of the present analysis is to check quantitatively pQCD predictions for di- and trijet cross sections and to determine the value of the strong coupling at different values of the relevant hard scale.

In the following, jets are defined by the inclusive k_t cluster algorithm [5, 6] in the Breit frame of reference. Accepted jets are required to have a transverse energy of more than 5 GeV. These jets, boosted back to the laboratory frame, have to fulfill the pseudorapidity requirement $-1 < \eta_{\text{lab}} < 2.5$ to ensure good detector acceptance. Events with at least two (three) accepted jets are assigned to the dijet (trijet) event sample. In order to ensure stability of the perturbative calculations, cuts on the invariant jet masses are applied: $M_{\text{dijet}} > 25\,\text{GeV}$ and $M_{\text{trijet}} > 25\,\text{GeV}$ for the di- and trijet sample, respectively.

DATA SAMPLE AND ANALYSIS METHODS

The data this analysis is based on were taken with the H1 detector in the years 1999-2000 and correspond to an integrated luminosity of $\mathscr{L}_{\text{int}} = 65.4\,\text{pb}^{-1}$. A DIS selection

[1] on behalf of the H1 Collaboration

CP792, *Deep Inelastic Scattering, DIS 2005,* edited by W. H. Smith and S. R. Dasu

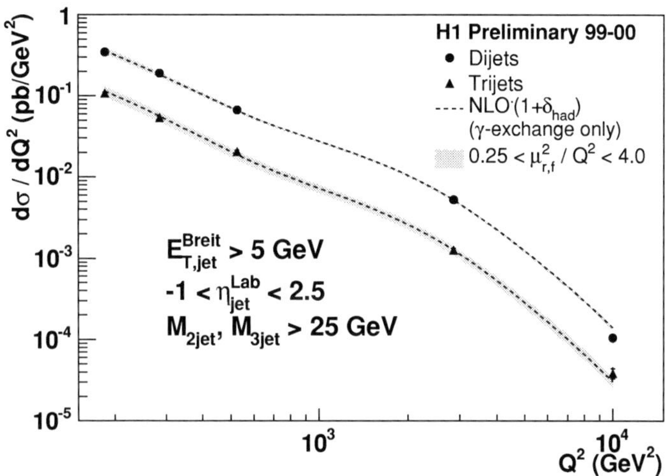

FIGURE 1. NC dijet and trijet differential cross-sections, with respect to Q^2, shown with NLO pQCD predictions including hadronisation corrections. The shaded bands show the effect of varying the renormalisation/factorisation scale by a factor of two.

in the phase space of four-momentum transfers $150 < Q^2 < 15000\,\text{GeV}^2$ and inelasticity $0.2 < y < 0.6$ leaves 5460 dijet- and 1757 trijet events.

To account for limited detector acceptance and resolution, correction factors were applied, which were determined with the Monte Carlo event generators DJANGOH [7] (including ARIADNE [8]) and RAPGAP [9]. The cross sections were corrected for QED radiative effects with HERACLES [10].

Relevant experimental uncertainties include: the electromagnetic energy scale, the hadronic energy scale, the scattering angle of the positron, a model uncertainty in the acceptance correction and the luminosity measurement, with the hadronic energy scale uncertainty being the dominant contribution.

RESULTS

Fig. 1 shows the differential di- and trijet cross sections as a function of Q^2. The data are compared to a calculation carried out with NLOJET++ [11], using parton density functions of the proton from the CTEQ5M1 [12] set and assuming a value of the strong coupling $\alpha_s(m_Z) = 0.118$. The calculation includes matrix elements at next-to-leading order (NLO) for two and three partons in the final state, corresponding to orders of the strong coupling up to $\mathscr{O}(\alpha_s^3)$. Hadronisation corrections are applied to the parton level prediction, determined with the help of DJANGOH and RAPGAP event samples. An uncertainty estimate for the prediction is provided by variations of the renormalisation and factorisation scale by a factor of two. The cross sections span several orders of magnitude, where a good description of the data by the prediction is observed. At the

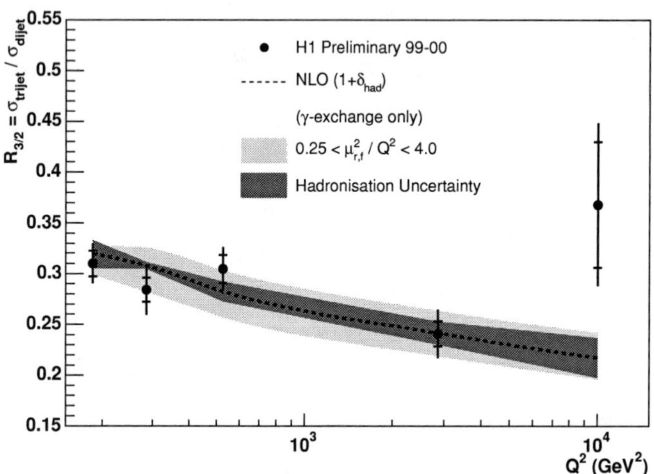

FIGURE 2. Measured values of $R_{3/2}$ against Q^2 compared with a NLO pQCD prediction with hadronisation corrections. The dark shaded band shows the uncertainty associated with the hadronisation corrections, while the light shaded band shows the effect of varying the renormalisation/factorisation scale by a factor of two.

highest Q^2 bin electroweak effects, which are not present in the calculation, cannot be neglected. Hence this point will not be used in the following α_s fit.

The ratio of the tri- and dijet cross section, $R_{3/2} = \sigma_{\text{trijet}}/\sigma_{\text{dijet}}$ is shown in Fig. 2, where again a good description by the perturbative calculation is observed. Based on this measurement, a fit of the strong coupling is performed: for each Q^2 value at which $R_{3/2}$ is measured (but the highest one) the pQCD calculation is repeated with the five variations of the proton p.d.f.s, available within the CTEQ4A [13] set. This yields five predictions of $R_{3/2}$ as a function of $\alpha_s(m_Z)$, to which a function of the form $C_1\alpha_s(m_Z) + C_2\alpha_s^2(m_Z)$ is fitted. Consequently, the function is used to obtain from the measured value of $R_{3/2}$ and its uncertainty the corresponding value and uncertainty of $\alpha_s(m_Z)$.

The fit results are shown as points in Fig. 3. From this diagram the running of the renormalised strong coupling is clearly evident. In addition an average value is obtained from this values by χ^2 minimisation, shown as a diamond at the reference of $Q^2 = m_Z^2$. The presented result is well compatible with the world average from the PDG [14], which is shown as a band for comparison.

CONCLUSION

Differential di- and trijet cross sections measured at high Q^2 have been shown. The distributions are well described by NLO pQCD with hadronisation corrections, except for the highest Q^2 data point, where electroweak corrections cannot be neglected. A fit

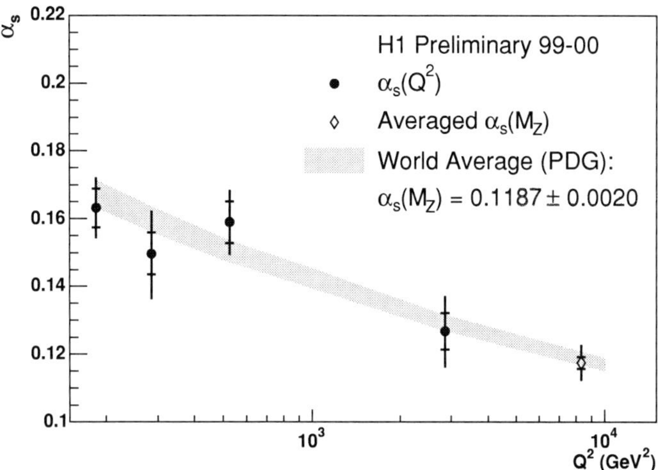

FIGURE 3. $\alpha_s(m_Z)$ values from each Q^2 bin evolved to values at their respective values of Q^2 (points) using the two-loop solution of the renormalisation group equation. The averaged value of $\alpha_s(m_Z)$, found using a χ^2 minimisation fit, is shown at the far right (empty diamond). The inner error bars show the statistical errors alone and the outer error bars denote the quadratic sum of statistical and systematical errors. The evolution of the world average value of $\alpha_s(m_Z)$ is shown as a shaded band.

of the strong coupling yields

$$\alpha_s(m_Z) = 0.1175 \pm 0.0017(\text{stat.}) \pm 0.0050(\text{syst.})^{+0.0054}_{-0.0068}(\text{theo.}),$$

well in agreement with the world average. Future improvements in the understanding of the hadronic energy scale will significantly reduce the experimental uncertainty.

REFERENCES

1. C. Adloff, et al., *Eur. Phys. J.*, **C19**, 289–311 (2001), hep-ex/0010054.
2. S. Chekanov, et al., *Phys. Lett.*, **B547**, 164–180 (2002), hep-ex/0208037.
3. C. Adloff, et al., *Phys. Lett.*, **B515**, 17–29 (2001), hep-ex/0106078.
4. S. Chekanov, et al. (2005), hep-ex/0502007.
5. S. D. Ellis, and D. E. Soper, *Phys. Rev.*, **D48**, 3160–3166 (1993), hep-ph/9305266.
6. S. Catani, Y. L. Dokshitzer, M. H. Seymour, and B. R. Webber, *Nucl. Phys.*, **B406**, 187–224 (1993).
7. K. Charchula, G. A. Schuler, and H. Spiesberger, *Comput. Phys. Commun.*, **81**, 381–402 (1994).
8. L. Lonnblad, *Comput. Phys. Commun.*, **71**, 15–31 (1992).
9. H. Jung, *Comp. Phys. Commun.*, **86**, 147–161 (1995).
10. A. Kwiatkowski, H. Spiesberger, and H. J. Mohring, *Comp. Phys. Commun.*, **69**, 155–172 (1992).
11. Z. Nagy, and Z. Trocsanyi, *Phys. Rev. Lett.*, **87**, 082001 (2001), hep-ph/0104315.
12. H. L. Lai, et al., *Eur. Phys. J.*, **C12**, 375–392 (2000), hep-ph/9903282.
13. H. L. Lai, et al., *Phys. Rev.*, **D55**, 1280–1296 (1997), hep-ph/9606399.
14. S. Eidelman, et al., *Phys. Lett.*, **B592**, 1 (2004).

INCLUSIVE JET CROSS SECTIONS IN NEUTRAL CURRENT DIS EVENTS IN THE BREIT FRAME

Jeff Standage

York University
(On behalf of the ZEUS Collaboration)
Email: standage@mail.desy.de

Abstract. Inclusive jet differential cross sections have been measured in neutral current deep inelastic scattering e^+p collisions for photon virtualities $Q^2 > 125$ GeV^2 with the ZEUS detector at HERA using an integrated luminosity of 65 pb^{-1}. Jets were identified in the Breit frame using the longitudinally invariant k_T-cluster algorithm. Measurements of differential inclusive jet cross sections are presented as a function of jet transverse energy, E_T^B(jet), jet pseudorapidity, η^B, and Q^2, for jets with E_T^B(jet) > 8 GeV. Next-to-leading-order (NLO) QCD calculations describe well the measurements. An NLO QCD analysis of the differential cross sections allows a precise determination of $\alpha_s(M_Z)$.

Keywords: DIS Inclusive Jets Breit
PACS: 12.38.Qk

1. Introduction

Measurements of jet cross sections are a well-established tool for QCD studies. In particular, inclusive jet cross sections at high photon virtuality, Q^2, are sensitive to the strong coupling constant, α_s, and can be used to determine this quantity. Several measurements of this quantity have already been performed at HERA [1] with results competitive with the world average [2]. ZEUS has also started to include jet cross sections in their global QCD fits, reducing substantially the uncertainty on the gluon density in the high-x region [3].

In this paper, a report is made of a new ZEUS measurement of the inclusive jet cross section at high $Q^2 > 125$ GeV in 1999-2000 e^+p data. Jets were reconstructed using the longitudinally invariant inclusive k_T-cluster algorithm [4] in the Breit frame. The measurement was made differentially in the transverse energy of the jet in the Breit frame, E_T^B, in Q^2 and in the jet pseudorapidity in the Breit frame, η^B. The measurements are compared to next-to-leading order QCD calculations.

2. Theoretical Predictions

The data are compared to NLO QCD calculations using the program DISENT [5]. These calculations interface NLO matrix elements with proton parton density functions (PDFs), in this case the MRST99 [6] parameterization was used, to make parton level predictions for the cross sections. The factorization scale was set to Q and

CP792, *Deep Inelastic Scattering, DIS 2005,* edited by W. H. Smith and S. R. Dasu

the renormalization scale was set to E_T^B(jet). The number of flavours was set to five and $\alpha_s(M_Z)$ was set to 0.1175. The calculations do not include any modeling of higher order corrections. Parton level cross sections were then corrected to the hadron level using Monte Carlo (MC) simulations.

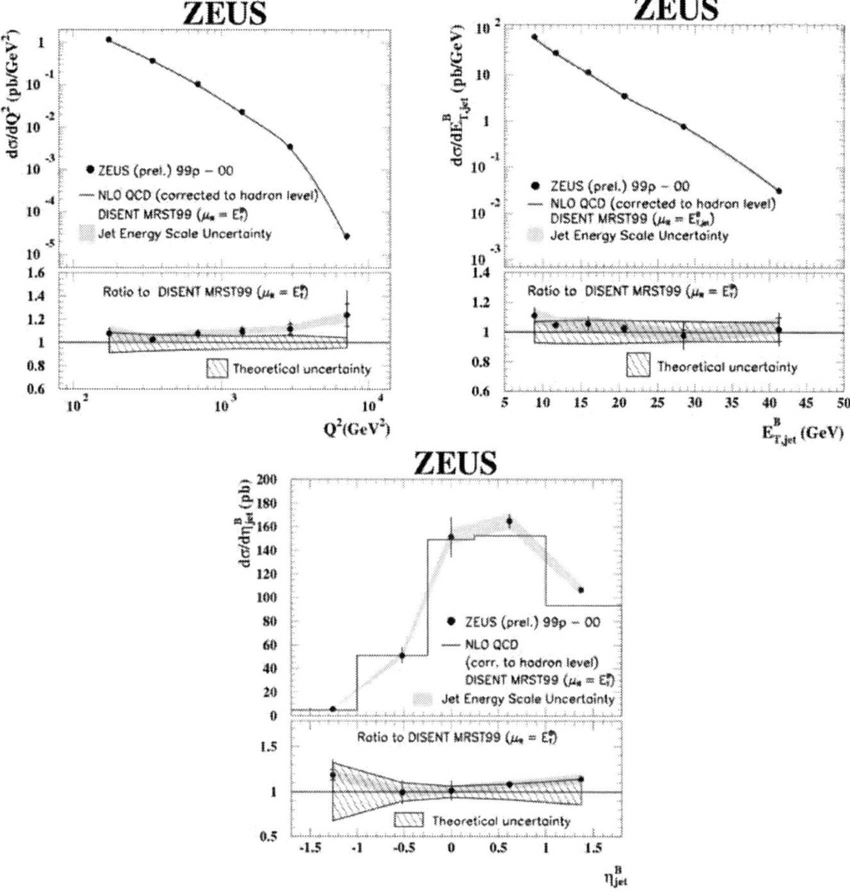

FIGURE 1. Measured inclusive differential jet cross sections in Q^2, E_T^B(jet), η^B (upper plots). Lower plots show ratio of measurement to theoretical calculations.

3. Event Selection and Monte Carlo Simulations

Neutral current (NC) deep inelastic scattering (DIS) events were selected from data collected with the ZEUS detector during 1999-2000 when HERA operated with protons of 920 GeV and positrons of 27.5 GeV. An isolated high energy positron was required (E' > 10 GeV) in the final state and cuts made using the uranium-scintillator calorimeter and central tracking detectors to reject charged current events and beam-gas and cosmic background. Kinematic variables were reconstructed using the angles of the scattered positron and the hadronic system [7]. The hadronic angle, γ_h, which corresponds to the direction of the outgoing struck quark in the quark-parton model, was reconstructed from the calorimeter measurement of the hadronic final state.

The jet search was made from the calorimeter cell deposits (excluding those assigned to the scattered positron) in the Breit frame using the longitudinally invariant inclusive k_T-cluster algorithm.

Events selected for the measurement satisfied $Q^2 > 125$ GeV^2 and $-0.7 < \cos\gamma_h < 0.5$ and jets were selected to satisfy $E_T^B > 8$ GeV and $-2.0 < \eta^B < 1.8$.

Monte Carlo simulations were used to correct the data for detector acceptance, resolution and efficiency. Two MC models were used to generate DIS events; ARIADNE v4.0 [8] and LEPTO v6.5 [9]. ARIADNE uses the colour-dipole model (CDM) [10] and LO QCD diagrams to simulate the QCD cascade. LEPTO, which was used as a systematic check, uses the exact matrix elements to generate the hard subprocess and parton shower algorithms to simulate higher order processes. Both models use the LUND string model [11] of JETSET v7.4 [12] for the hadronization of the final state. To take into account first-order electroweak corrections, ARIADNE and LEPTO were both interfaced to HERACLES v4.6 [13] using the DJANGO v1.10 [14] program. The response of the ZEUS detector was simulated using a program based on the GEANT 3.13 package [15]. Simulated events were passed through the same selection process as the data.

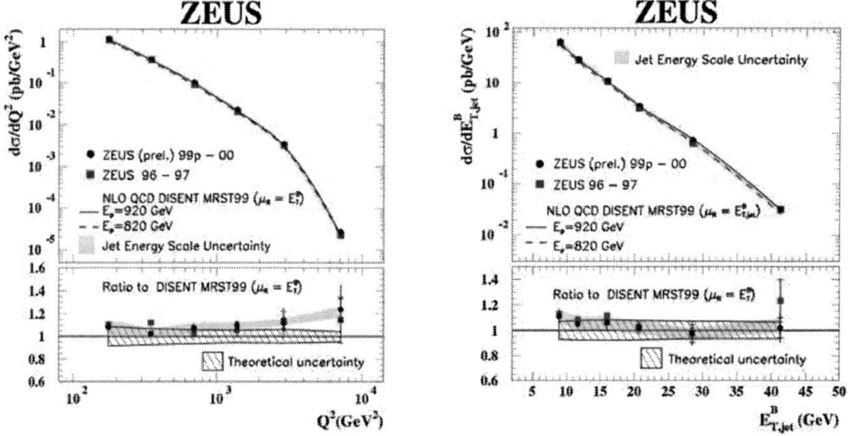

FIGURE 2. Comparison with previously published results. Measured inclusive differential jet cross sections in Q^2, E_T^B(jet) (upper plots). Lower plots show ratio of measurement to theoretical calculations.

4. Results

The inclusive jet differential cross sections as a function of Q^2, E_T^B(jet) and η^B(jet) are shown in Figure 1. In the top part of the plots the data (corrected to the hadron level and for first-order electroweak effects) are compared to NLO QCD calculations (corrected for hadronization effects). The lower part of the plots shows the ratio of data to the NLO QCD calculations. The hashed band indicates the theoretical error. For all three distributions the data is described well by the theory. The solid band indicates the uncertainty in jet energy scale.

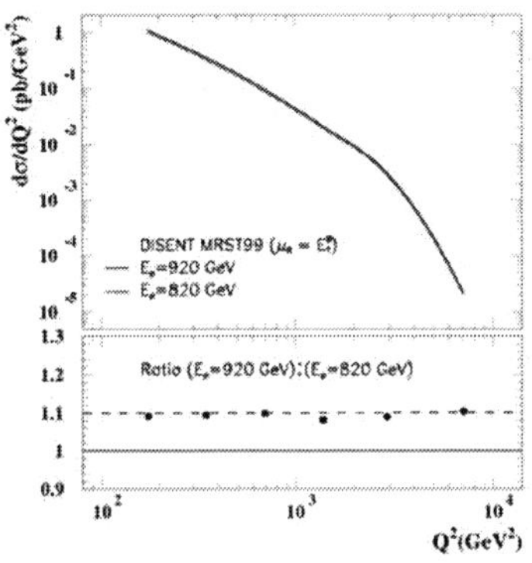

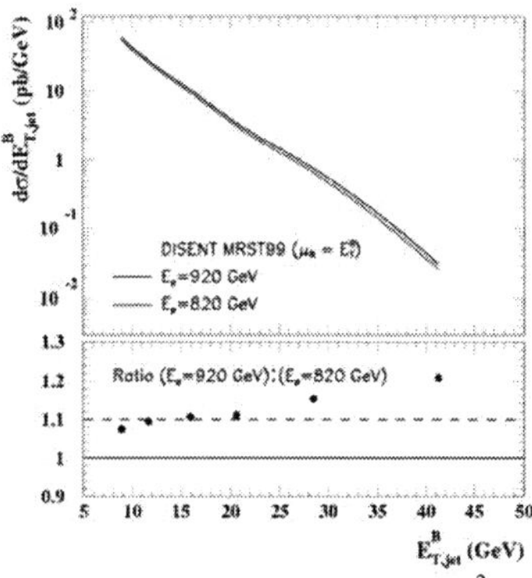

FIGURE 3. Comparison of theoretical calculations. Inclusive differential cross sections in Q^2, $E_T{}^B$(jet) (upper) for E_P = 820 (920) GeV. Lower plots show ratio of calculations with E_P = 820 to E_P = 920 GeV.

Figure 2 shows the same distributions for Q^2 and $E_T{}^B$(jet) directly compared with equivalent distributions from a former ZEUS analysis done for lower proton energy (E_P = 820 GeV). It can be seen that there is comparable agreement between data and theory for both analyses. The reduced statistical uncertainties in the new measurement are clearly visible.

Figure 3 shows the effect of the increased beam energy on the NLO QCD predictions for Q^2 and $E_T{}^B$(jet) inclusive differential jet cross sections. For the Q^2 distribution the effect of the increased beam energy is to increase the differential cross section by ~10% consistently across the range calculated. For the $E_T{}^B$(jet) distribution, this enhancement itself increases relatively with rising $E_T{}^B$(jet).

I would like to thank the members of the ZEUS collaboration who helped with this research.

REFERENCES

1. ZEUS Collaboration, S. Chekanov et al., *Phys. Lett* **B** 560 (2003) 7; 558 (2003) 41; 531 (2002) 9.
   ZEUS Collaboration, S. Chekanov et al., *Eur. Phys. J.* **C** 31 (2003) 149; 23 (2002) 615; 23 (2002) 13.
   H1 Collaboration, A. Aktas et al., *Eur. Phys. J.* **C** 33 (2004) 477; 29 (2003) 497; 25 (2002) 13; 19 (2001) 289.
   H1 Collaboration, C. Adloff et al., *Phys. Lett.* **B** 542 (2002) 193.
2. S. Bethke, *Nucl. Phys. Proc. Suppl.* 121 (2003) 74.
3. ZEUS Collaboration, S.Chekanov et al., DESY-05-050 (March 2005), submitted to *Eur. Phys. J.*
4. S. Catani et al., *Nucl. Phys.* **B** 406 (1993) 187.
5. S. Catani and M.H. Seymour, . *Phys.* **B** 485 (1997) 291; *Erratum in Nucl. Phys.* B510 (1998) 503.
6. A.D. Martin et al., *Eur. Phys. J.* **C** 4 (1998) 463; 14 (2000) 133.
7. S. Bentvelsen, J. Engelen and P. Kooijman, *Proc. Workshop on Physics at HERA;* W. Buchmüller and G. Ingelman (eds.), Vol. 1, p. 23, Hamburg, Germany, DESY (1992).
8. L. Lönnblad, *Comp. Phys. Comm.* 71 (1992) 15; *Z. Phys.* **C** 65 (1995) 285.
9. G. Ingelman, A. Edin and J. Rathsman, *Comp. Phys. Comm.* 101 (1997) 108.
10. Y. Azimov et al, *Phys. Lett.* **B** 165 (1985) 147;
    G. Gustafson, *Phys. Lett.* **B** 175 (1986) 453; 306 (1988) 746;
    B. Andersson et al., *Z. Phys.* **C** 43 (1989) 625.
11. B. Andersson et al., *Phys. Rep.* 97 (1983) 31.
12. T. Sjöstrand, *Comp. Phys. Comm.* 39 (1986) 347;
    T. Sjöstrand and M. Bengtsson, *Comp. Phys. Comm.* 43 (1987) 367.
13. A. Kwistkowski, H. Spiesberger and H.-J. Möhring, *Comp. Phys. Comm.* 69 (1986) 155.
14. K. Charchula, G.A. Schuler and H. Spiesberger, *Comp. Phys. Comm.* 81 (1994) 381.
15. R. Brun et al., GEANT3, Technical Report CERN-DD/EE/84-1, CERN, 1987.

Study of color dynamics in photoproduction at HERA

Juan Terrón (on behalf of the ZEUS Collaboration)

Universidad Autónoma de Madrid, Spain

Abstract. Measurements of normalised differential cross sections for the photoproduction of three-jet events are presented as functions of jet angular variables to highlight the contributions from the different color configurations. Fixed-order calculations were compared to the results and used to study the underlying gauge-group symmetry.

Keywords: QCD
PACS: 12.38.Qk

INTRODUCTION

The gauge theory of the strong interactions, Quantum chromodynamics (QCD), is based on the non-abelian group SU(3). In general, the physical manifestation of the underlying group structure is encapsulated in the color factors C_F, C_A and T_F. In strong interactions, they represent the relative strengths of the processes $q \to qg$, $g \to gg$ and $g \to q\bar{q}$, respectively. Investigations of the color factors have been carried out at LEP using angular correlations in four-jet events from Z° hadronic decays.

At HERA, the effects of the different color configurations arising from the underlying gauge-group structure should manifest themselves in three-jet events in photoproduction. Photoproduction at HERA is studied by means of ep scattering at low four-momentum transfers ($Q^2 \approx 0$, where Q^2 is the virtuality of the exchanged photon). In photon-proton reactions, two types of QCD processes contribute to jet production at leading order (LO): either the photon interacts directly with a parton in the proton (the direct process) or the photon acts as a source of partons which scatter off those in the proton (the resolved process). Direct-photon events provide a clean way to study the effects of the different color configurations. An illustrative diagram for each color configuration is shown in Fig. 1: (A) double-gluon bremsstrahlung from a quark line (C_F^2), (B) the splitting of a virtual gluon into a pair of final-state gluons ($C_F C_A$), (C) the production of a $q\bar{q}$ pair through the exchange of a virtual gluon emitted by an incoming quark ($C_F T_F$) and (D) the production of a $q\bar{q}$ pair through the exchange of a virtual gluon arising from the splitting of an incoming gluon ($T_F C_A$). The variables that have been devised to highlight the contributions from the different color configurations are based on the angular correlations between the final-state jets and the beam and make use of the three jets with highest transverse energy (E_T^{jet}) in an event:

- θ_H, the angle between the plane determined by the highest-E_T^{jet} jet and the beam and the plane determined by the two lowest-E_T^{jet} jets [1];

CP792, *Deep Inelastic Scattering, DIS 2005,* edited by W. H. Smith and S. R. Dasu

- α_{23}, which is inspired by the variable $\alpha_{34}^{e^+e^-}$ [2] for $e^+e^- \to 4$ jets, is defined as the angle between the two lowest-$E_T^{\rm jet}$ jets;
- $\beta_{\rm KSW}$, which is inspired by the Körner-Schierholz-Willrodt angle $\Phi_{\rm KSW}^{e^+e^-}$ [3] for $e^+e^- \to 4$ jets, is defined as $\cos(\beta_{\rm KSW}) = \cos\left[\frac{1}{2}\left(\angle[(\vec{p}_1 \times \vec{p}_3),(\vec{p}_2 \times \vec{p}_B)] + \angle[(\vec{p}_1 \times \vec{p}_B),(\vec{p}_2 \times \vec{p}_3)]\right)\right]$, where $\vec{p}_i$, $i = 1,...,3$ is the momentum of jet i and $\vec{p}_B$ is a unit vector in the direction of the beam; the jets are ordered according to decreasing transverse energy.

In this contribution, measurements of angular distributions [4] in photoproduced three-jet events are presented and compared to fixed-order $O(\alpha\alpha_s^2)$ perturbative calculations [5].

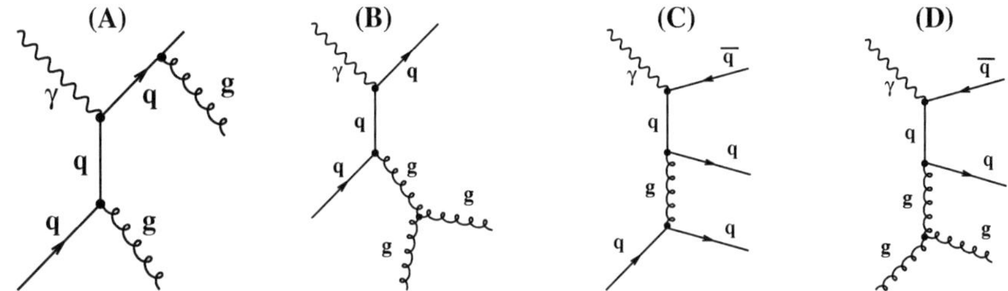

FIGURE 1. Examples of diagrams for the photoproduction of three-jet events through direct-photon processes in each color configuration: (A) C_F^2; (B) $C_F C_A$; (C) $C_F T_F$; (D) $T_F C_A$.

DATA SELECTION AND JET SEARCH

The data sample was collected with the ZEUS detector at HERA and corresponds to an integrated luminosity of 45.0 ± 0.7 (65.5 ± 1.5) pb^{-1} for e^+p collisions taken during 1995-97 (1999-2000) and 16.7 ± 0.3 pb^{-1} for e^-p collisions taken during 1998-99. During 1995-97 (1998-2000), HERA operated with protons of energy $E_p = 820$ GeV (920 GeV) and positrons or electrons of energy $E_e = 27.5$ GeV, yielding a centre-of-mass energy of $\sqrt{s} = 300$ GeV (318 GeV).

Events from collisions between protons and quasi-real photons were selected offline using similar criteria as reported in a previous publication [6].

The k_T cluster algorithm [7] was used in the longitudinally invariant inclusive mode [8] to reconstruct jets in the hadronic final state. The axis of the jet was defined according to the Snowmass convention, where $\eta^{\rm jet}$ ($\varphi^{\rm jet}$) is the transverse energy-weighted mean pseudorapidity (azimuth) of all the particles belonging to that jet. Events with at least three jets of $E_T^{\rm jet} > 14$ GeV and $-1 < \eta^{\rm jet} < 2.5$ were retained. Direct-photon events were selected by requiring $x_\gamma^{\rm obs} > 0.7$, where $x_\gamma^{\rm obs}$, the fraction of the photon momentum participating in the production of the three jets with highest $E_T^{\rm jet}$, is defined as $x_\gamma^{\rm obs} = \sum_{i=1}^{3}(E_T^{\rm jet,i} e^{-\eta^{\rm jet,i}})/(2yE_e)$, where y is the inelasticity variable. The remaining contribution from resolved-photon events was estimated by Monte Carlo techniques to be $\approx 34\%$. It was checked that the angular distributions of the events from resolved processes with $x_\gamma^{\rm obs} > 0.7$ were similar to those from direct processes and, therefore, no subtraction was performed when comparing to the fixed-order calculations described in the next Section. The selected sample consisted of events from ep interactions with $Q^2 < 1$ GeV^2 and a median of $Q^2 \approx 10^{-3}$ GeV^2. The event sample was restricted to the kinematic range $0.2 < y < 0.85$. There remained 2233 events after all selection criteria were applied.

FIXED-ORDER CALCULATIONS

The calculations of direct-photon processes used in this analysis are based on the program by Klasen, Kleinwort and Kramer [5]. The MRST99 [9] parameterisations of the parton distribution functions (PDFs) of the proton were used as defaults for the comparisons with the measured cross sections. These calculations are $O(\alpha\alpha_s^2)$ and represent the lowest-order contribution to three-jet photoproduction. The LO calculation of three-jet cross sections for direct-photon processes can be expressed in terms of the color factors C_A, C_F and T_F as follows [10]: $\sigma_{ep\rightarrow 3\text{jets}} = C_F^2 \cdot \sigma_A + C_F C_A \cdot \sigma_B + C_F T_F \cdot \sigma_C + T_F C_A \cdot \sigma_D$, where $\sigma_A, ..., \sigma_D$ are the cross sections for the different contributions (see Fig. 1). The component which contains the contribution from the triple-gluon vertex in quark-induced processes, σ_B, has a very different shape than the other components for the three angular variables considered here (not shown here, see [4]). The other components are best separated by the distribution of $\cos\alpha_{23}$. Thus, these variables are sensitive to the different color configurations. The predicted relative contributions in QCD (i.e. SU(3)) are A: 13%, B: 10%, C: 45% and D: 32%. The full prediction for a given gauge group was obtained by adding up σ_i ($i = A, ..., D$) weighted with the corresponding color factors.

RESULTS

Using the selected data sample, normalised three-jet differential cross sections were measured for $Q^2 < 1$ GeV^2, $0.2 < y < 0.85$ and $x_\gamma^{\text{obs}} > 0.7$. The cross sections were determined for jets of $E_T^{\text{jet}} > 14$ GeV and $-1 < \eta^{\text{jet}} < 2.5$. The normalised cross-sections $(1/\sigma)(d\sigma/d\theta_H)$, $(1/\sigma)(d\sigma/d\cos\alpha_{23})$ and $(1/\sigma)(d\sigma/d\cos\beta_{\text{KSW}})$ are presented in Fig. 2. The experimental systematic uncertainties were added in quadrature to the statistical uncertainty (thin error bars) of the data and are shown as thick error bars in the figure. The calculations based on SU(3), shown in Fig. 2, give a good description of the data for $\theta_H > 35°$, $\cos\alpha_{23} > -0.6$ and $-0.8 < \cos\beta_{\text{KSW}} < 0.6$.

To illustrate the sensitivity of the measurements to the color factors, calculations based on different symmetry groups are also compared to the data in Fig. 2. In this figure, the color components have been combined in such a way as to reproduce the color structure of a theory based on the non-abelian group SU(N) in the limit of large N ($C_F = (N^2-1)/2N, T_F = 1/2, C_A = N$), the abelian group U(1)^3 ($C_F = 1, T_F = 3, C_A = 0$) and, as an extreme choice, a calculation with $C_F = 0$. The shapes of the distributions predicted by U(1)^3 are very similar to those by SU(3) due to the smallness of the component σ_B and the difficulty to distinguish the component σ_D. The data clearly disfavour a theory in which $T_F/C_F \approx 0$ such as predicted by SU(N) in the limit of large N or $C_F = 0$.

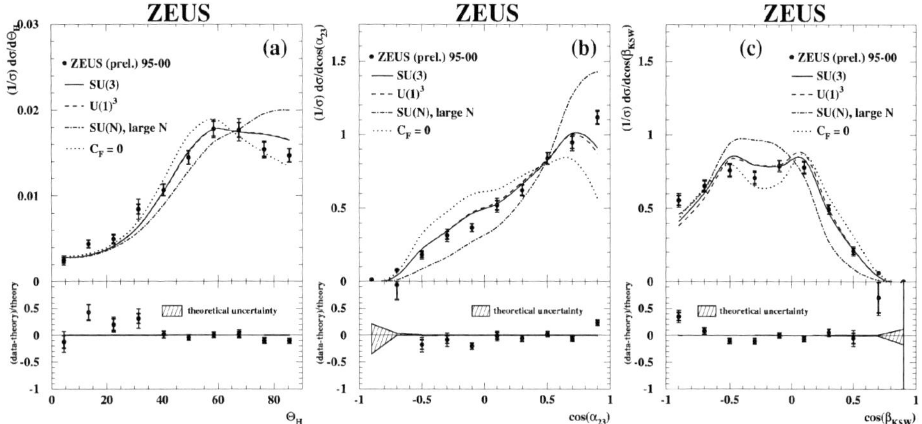

FIGURE 2. (a) Normalised differential *ep* cross sections for three-jet photoproduction as functions of (a) θ_H, (b) $\cos\alpha_{23}$ and (c) $\cos\beta_{\mathrm{KSW}}$. For comparison, the $\mathrm{O}(\alpha\alpha_s^2)$ calculations for direct-photon processes based on SU(3) (solid lines), $\mathrm{U}(1)^3$ (dashed lines), SU(N) in the limit of large N (dot-dashed lines) and $C_F = 0$ (dotted lines) are included. The lower part of the figures displays the fractional difference between the measured normalised cross section and the calculation based on SU(3).

SUMMARY

Measurements of angular correlations in three-jet photoproduction have been made in *ep* collisions using 127 pb^{-1} of data collected with the ZEUS detector at HERA. Normalised differential cross sections were measured as functions of θ_H, $\cos\alpha_{23}$ and $\cos\beta_{\mathrm{KSW}}$. Fixed-order ($\mathrm{O}(\alpha\alpha_s^2)$) calculations for three-jet photoproduction through direct-photon processes separated according to the color configurations were used to study the sensitivity of the angular distributions to the underlying gauge-group structure. The measurements are found to be consistent with the admixture of color configurations as predicted by SU(3). The data clearly disfavour a theory in which $T_F/C_F \approx 0$, as predicted by SU(N) in the limit of large N, or $C_F = 0$.

REFERENCES

1. R. Muñoz-Tapia and W.J. Stirling, *Phys. Rev.* **D** 52 (1995) 3894.
2. DELPHI Coll., P. Abreu et al., *Phys. Lett.* **B** 255 (1991) 466.
3. J.G. Körner, G. Schierholz and J. Willrodt, *Nucl. Phys.* **B** 185 (1981) 365.
4. ZEUS Coll., contributed paper 5-0271, 32nd International Conference on HEP, Beijing, China, 2004.
5. M. Klasen, T. Kleinwort and G. Kramer, *Eur. Phys. Jour. direct* **C** 1 (1998) 1.
6. ZEUS Coll., S. Chekanov et al., *Phys. Lett.* **B** 560 (2003) 7.
7. S. Catani et al., *Nucl. Phys.* **B** 406 (1993) 187.
8. S.D. Ellis and D.E. Soper, *Phys. Rev.* **D** 48 (1993) 3160.
9. A.D. Martin et al., *Eur. Phys. Jour.* **C** 4 (1998) 463 and *Eur. Phys. Jour.* **C** 14 (2000) 133.
10. P. Aurenche et al., *Nucl. Phys.* **B** 286 (1987) 553.

Event shapes in deep inelastic $ep \rightarrow eX$ scattering at HERA

A. Everett

On behalf of the ZEUS Collaboration
University of Wisconsin - Madison, 1150 University Ave., Madison WI 53706, USA
E-mail: adam.everett@desy.de

Abstract. The study of energy flow in high energy particle collisions is a powerful tool for testing QCD predictions. Event shape variables extend this study to the non-perturbative region of QCD.

Keywords: Event Shapes, Power Correction
PACS: 13.85.Hd

INTRODUCTION

Hadronic final states in *ep* collisions at HERA provides a unique opportunity to test perturbative QCD predictions over a large range of the four-momentum transfer squared. Q^2 is the virtuality of the photon in the region where the experimental uncertainties and non-perturbative effects are expected to be reduced. However, understanding of the nature of the non-perturbative contribution, hadronization, is one of the most important subjects of high energy physics. Recent revival of interest in event shapes has been prompted by theoretical developments in the understanding of hadronization. This gives a chance to go beyond the phenomenological models of hadronization that are currently used in Monte Carlo event generators.

EVENT SHAPES

The event shapes studied at HERA by ZEUS[1] are thrust T, broadening B, jet mass M, and the C-parameter C. The measurements were performed in the kinematic range $80 < Q^2 < 2 \cdot 10^4\,\mathrm{GeV}^2$ and $0.0024 < x < 0.6$, in the current region of the Breit frame, in order to facilitate comparison with e^+e^- experiments.

According to the theoretical model of power corrections (PC) introduced by Y. Dokshitzer and B. Webber[2], event-shape values, $\langle F \rangle$, can be described as the sum of perturbative, $\langle F \rangle_{pert}$, and non-perturbative , $\langle F \rangle_{PC}$, terms. The perturbative contribution is then calculated using QCD next-to-leading-order (NLO) programs such as DISENT and DISASTER[3], while the non-perturbative term is an analytical function of two parameters $(\alpha_s, \overline{\alpha_0})$, where $\overline{\alpha_0}$ is an effective coupling. By fitting the measured Q^2 dependence of the mean event-shape values to the NLO + PC prediction, values for α_s and $\overline{\alpha_0}$ were extracted.

CP792, *Deep Inelastic Scattering, DIS 2005,* edited by W. H. Smith and S. R. Dasu

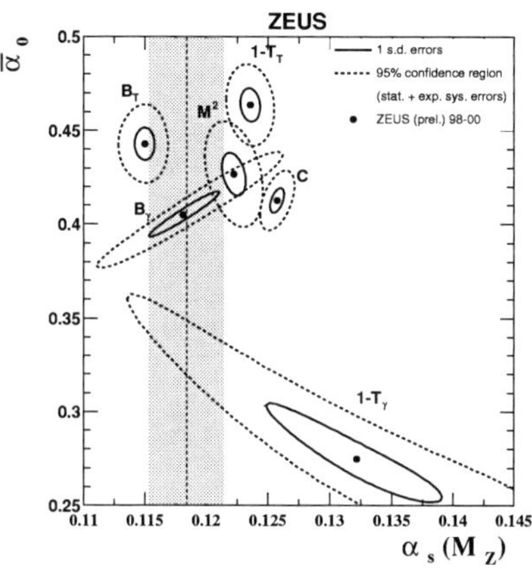

FIGURE 1. ZEUS extractions of α_s and $\overline{\alpha_0}$ shown in the (α_s,$\overline{\alpha_0}$) plane from fits to the event-shape means. The experimental systematic errors were determined using the Hessian method which takes into account bin-by-bin correlations. The solid curve is the statistical plus experimental systematic 1 σ error contour; the dashed curve represents the 95% confidence limit for α_s and $\overline{\alpha_0}$. The shaded band is the world average of α_s.

The extraction of α_s and $\overline{\alpha_0}$ from each mean event-shape value provided values of α_s and $\overline{\alpha_0}$ that are consistent overall with each other and with the world average. Figure 1 shows the extracted values $(\alpha_s, \overline{\alpha_0})$ for each event-shape mean studied. There is a slight spread between most of the values and with the values extracted from thrust with respect to the photon axis in the Breit frame and broadening with respect to the thrust axis in the Breit frame.

The dispersion of the α_s and $\overline{\alpha_0}$ values could be due to higher-order terms that are not present in the NLO + PC calculations. Studying the differential event-shape distributions provides a way to analyze the effects of those higher-order terms. Due to the lack of convergence of the perturbative series for the differential distributions, the NLO + PC model used for the mean is not expected to describe the distributions. To study the event-shape distributions, a resummation[4] of the next-to-leading logarithm (NLL) is required. The effect of adding the PC terms to the NLO + NLL prediction cause a shift in the distributions.

Power-correction theory is not expected to be accurate in all regions of the event shape distributions because the lack of terms beyond NLO in the perturbative series causes a deficit of very broad events and an excess of strongly collimated events. Therefore, the predictions do not match the data at each end of the distributions. By studying the differential distributions, one can identify regions in phase space where the model is performing well, and restrict the fits to these regions. The missing higher-order, resummation terms still make small, but significant, contributions, even in this restricted phase space.

ZEUS has recently made measurements of the differential distributions. As described above, the NLO + NLL + PC predictions are fit to the measurements in a restricted range. The NLO + NLL + PC predictions are shown together with the measured distributions

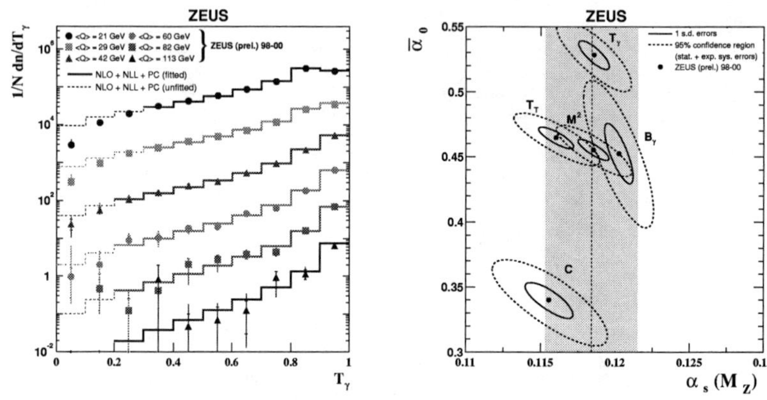

FIGURE 2. ZEUS measurement of T with respect to the photon axis in the Breit frame in different bins of Q^2 compared to the prediction of NLO + NLL + PC, as described in the text. Errors are shown. Solid lines for the prediction represent the fi t range, and the dashed prediction line represents the region not fi t to the data.

FIGURE 3. ZEUS extractions of α_s and $\overline{\alpha_0}$ shown in the $(\alpha_s,\overline{\alpha_0})$ plane from fi ts to the event shape differential distributions. The experimental systematic errors were determined using the Hessian method, which takes into account bin-by-bin correlations. The solid curve is statistical plus experimental systematic 1 σ contour; the dashed curve represents the 95% confi dence limit for α_s and $\overline{\alpha_0}$.

in Fig. 2, for the thrust distribution.

Based on the fits to the event-shape distributions, values for α_s and $\overline{\alpha_0}$ are extracted from each event shape. As with the values extracted from the measurement of the means, the values extracted from the measured distributions provide consistent values for α_s (see Fig. 3). The values extracted for $\overline{\alpha_0}$ are a little lower than for other experiments, and notably the ZEUS value from the C-parameter is lower.

To study further the non-perturbative region of QCD, it is possible to investigate event shapes which are more sensitive to higher-order effects. In particular, the study of dijet events provides a sample which is sensitive to $\mathcal{O}(\alpha_s^2)$ contributions. The out-of-plane momentum and jet rates are event shapes that are sensitive to these higher order effects. The out-of-plane momentum, K_{OUT}, is the sum of momentum out of the event plane formed by the two highest energy jets, and the jet rate, y_2, is the (2+1)-jet resolution defined by the k_T cluster algorithm[5] in the longitudinally invariant inclusive mode in the Breit frame. Figure 4 shows the distributions of the multijet event shape K_{OUT}. For these shapes, no fits were performed up to now since only lowest order K_{OUT} calculations are available. Figure 4 shows K_{OUT} in two Q^2 bins; the available LO + NLL + PC prediction[6] is in acceptable agreement with K_{OUT} data. For y_2, NLOJET[7] gives a good description of y_2 for higher values of Q.

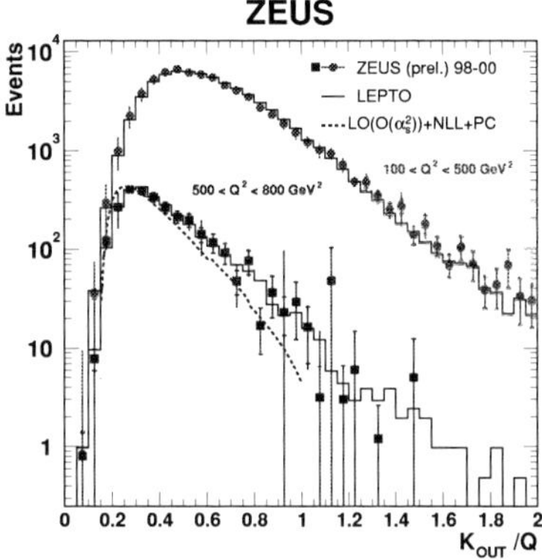

FIGURE 4. K_{OUT} distribution compared to LEPTO shown in bins of $100 < Q^2 < 500\,\mathrm{GeV}^2$ and $500 < Q^2 < 800\,\mathrm{GeV}^2$. The LO + NLL + PC calculation is shown in the higher Q^2 bin.

SUMMARY

Existing ZEUS event-shape measurements [1] are complemented by a new measurement of the means, using a larger data sample, and by measurements of differential distributions. For the kinematic ranges and cuts employed by this analysis, the NLO + PC calculation is unable to extract consistent results for all of the mean values and differential distributions of the event-shape variables. When matched NLL resummations are added to the model, good fits to the differential distributions are obtained that yield $\alpha_s(M_Z)$ values that are consistent with the world average. For the first time, ZEUS has measured a two jet event-shape variable, K_{OUT}.

ACKNOWLEDGMENTS

This work has been conducted in collaboration with S. Hanlon, A. Savin, and I. Skillicorn. I would like to thank my colleagues from ZEUS for their help in the preperation of this report.

REFERENCES

1. S. C. ZEUS Coll., *Eur. Phys. Jour.*, **C 27**, 531 (2003).
2. Y. Dokshitzer, and B. Webber, *Phys. Lett.*, **B 352**, 451 (1997), hep-ph/9504219.
3. D. Graudenz (1997), hep-ph/9710244.
4. M. Dasgupta, and G. Salam, *JHEP*, **0208**, 032 (2002), hep-ph/0208073.
5. S. Catani, Y. L. Dokshitzer, M. H. Seymour, and B. R. Webber, *Nucl. Phys.*, **B 406**, 187–224 (1993).
6. G. Zanderighi, *Personal Communication* (2004).
7. Z. Nagy, and Z. Trocsanyi, *Phys. Rev. Lett.*, **87**, 082001 (2001).

Neutral- and charged-kaon Bose-Einstein correlations in DIS

Anna Galas

on behalf of the ZEUS Collaboration
The Henryk Niewodniczanski Institute of Nuclear Physics, Polish Academy of Sciences
ul.Radzikowskiego 152, 31-342 Cracow, Poland

Abstract. The results of the measurements of Bose-Einstein correlations (BEC) between identified kaons are presented for deep inelastic scattering (DIS) at HERA. An integrated luminosity of 121 pb^{-1} of $e^{\pm}p$ collisions data have been analysed. The two-particle correlation function was studied as a function of the $Q_{12} = \sqrt{-(p_1 - p_2)^2}$ - four momenta difference of the kaon-pair, assuming a Gaussian shape of BEC source. The values of radius, r, of the production volume and as a coherence strength factor, λ, were obtained for both, neutral and charged kaon-pairs. The results of r and λ are disccused and compared with the LEP measurements.

Keywords: Bose-Einstein correlations, neutral kaons, charged kaons
PACS: PACS numbers: 13.90.+i

Two identical bosons emitted from the same source show the tendency to have similar energy-momentum characteristic. This enhancement in the production of identical bosons with almost equal momenta is called Bose-Einstein effect and was first observed by Goldhaber *et al.*[1] in $p\bar{p}$ collisions.An important information on non-perturbative QCD and hadronisation processes can be obtained from analysys of Bose-Einstein effect of hadronic final states produced in particle reactions. Moreover, space-time characteristic of the particle emission region provides some knowledge about emission source size in different reactions.

The study of the size of BEC of identical bosons pairs in DIS has beed extended to pairs of $K_s^0 K_s^0$ and $K^{\pm}K^{\pm}$. From LEP results for BEC and FDC (Fermi-Dirac correlations) a hierarchy of the size of the emission volumes is observed. This hierarchy seems to decrease as the hadron mass increases [2]. New DIS results may bring new information about this behaviour.

The DIS data sample was taken during the 1996-2000 period from $e^{\pm}p$ collisions for the Q^2 range from 2 to 15000 GeV^2 for 121 pb^{-1} of integrated luminosity. The ZEUS detector is described in detail in [3]. The analysed data comes from the central tracking detector (CTD) and CALorimeter (CAL).

The two-particle correlation function $R(Q_{12})$ was calculated using double ratio method and divided by correction coefficient to remove correlations other than BEC:

$$R(Q_{12}) = \frac{P(Q_{12})^{\text{data}}}{P_{\text{mix}}(Q_{12})^{\text{data}}} / \frac{P(Q_{12})^{\text{MC,noBEC}}}{P_{\text{mix}}(Q_{12})^{\text{MC,noBEC}}} \,, \tag{1}$$

CP792, *Deep Inelastic Scattering, DIS 2005,* edited by W. H. Smith and S. R. Dasu

where the so-called mixed-event sample $P_{\rm mix}$ contains pairs of bosons coming from different events. Q_{12} is given by $Q_{12} = \sqrt{-(p_1 - p_2)^2} = \sqrt{M^2 - 4m_{boson}}$, where M is the invariant mass of the two particles with four-momenta p_1 and p_2 and mass m_{boson}.

Assuming a Gausian shape of emission source, $R(Q_{12})$ can be described by the standard Goldhaber-like function:

$$R(Q_{12}) = \alpha(1 + \lambda e^{-Q_{12}^2 r^2})(1 + \delta Q_{12}) \,, \tag{2}$$

where the most important parameters are r - a geometrical radius of the boson emitting source and λ - a coherence strength factor. The remaining parameters, α and δ, describe the overall normalization and non-linearity in the behavior of background. This parametrisation is introduced for a spherically symmetric boson emitting source and was used to fit data and extract the values of r and λ.

Figure 1a shows the measured two-particle correlation function of $K^{\pm}K^{\pm}$ pairs with fit. The extracted values of the BEC parameters for charged kaon-pairs are in Table 1.

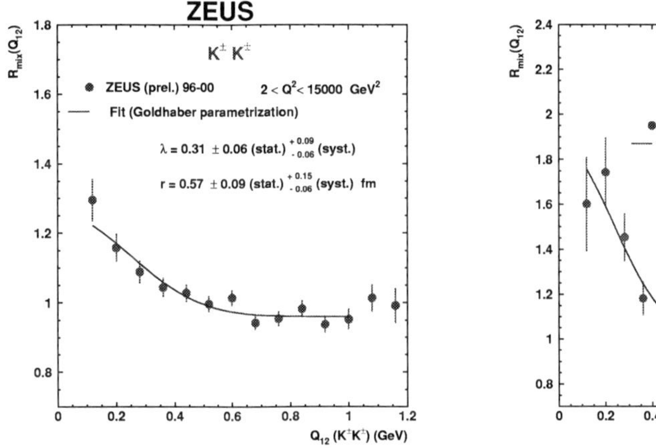

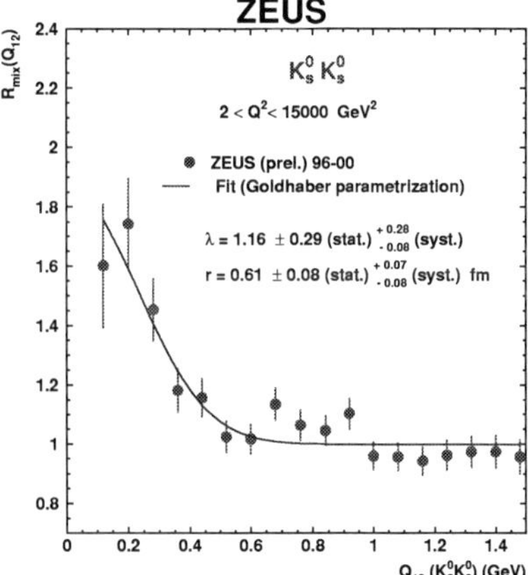

FIGURE 1. Bose-Einstein correlation function of (a) $K^{\pm}K^{\pm}$ and (b) $K^0_s K^0_s$ pairs.

The radius value for charged kaons is consistent with that for charged pions [4]. However, the λ value for kaons is smaller than for pions. Figure 2a shows the comparison of DIS and LEP results [5,6] for r and λ. The good agreement with LEP for radius is observed. The λ value for DIS is smaller than for LEP. This is due to the fact that the DIS data populate mostly proton fragmentation region, the number of non prompt kaons in the final state may significantly increase.

Like for charged kaons, the extracted values of BEC parameters for neutral kaons are presented in Table 1. Figure 1b shows the correlation function of $K^0_s K^0_s$ pairs with the Gaussian fit. The radius for $K^0_s K^0_s$ is in good agreement with that for $K^{\pm}K^{\pm}$ but the λ value is larged than for charged kaons. This may be explained by $f_0(980)$ resonance impact on the low Q_{12} region which is not well described in the ZEUS Monte Carlo. It was veryfied that small contribution of such resonance can significantly decrease λ value. The corresponding change in the value of r is small.

TABLE 1. The two-particle emitter size and the strength factor obtained from BEC. The first and the second errors are the statistical and systematic respectivly.

	r **[fm]**	λ
$K^{\pm}K^{\pm}$	$0.57 \pm 0.09 ^{+0.15}_{-0.06}$	$0.31 \pm 0.06 ^{+0.09}_{-0.06}$
$K_s^0 K_s^0$	$0.61 \pm 0.08 ^{+0.07}_{-0.08}$	$0.57 \pm 0.09 ^{+0.28}_{-0.08}$
pions*	$0.666 \pm 0.009 ^{+0.022}_{-0.036}$	$0.475 \pm 0.007 ^{+0.011}_{-0.003}$

* charged pions published by ZEUS [4]

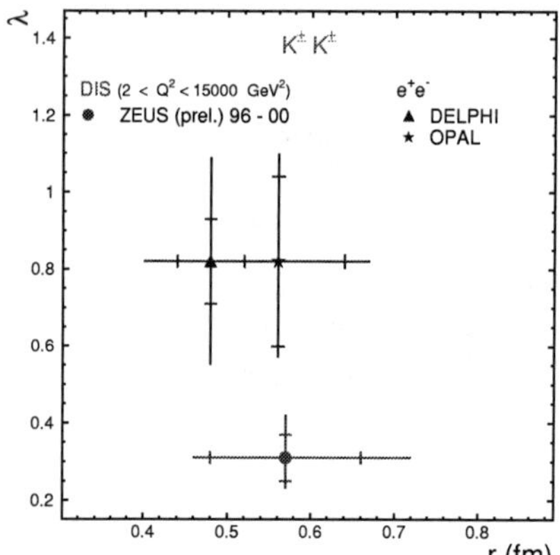

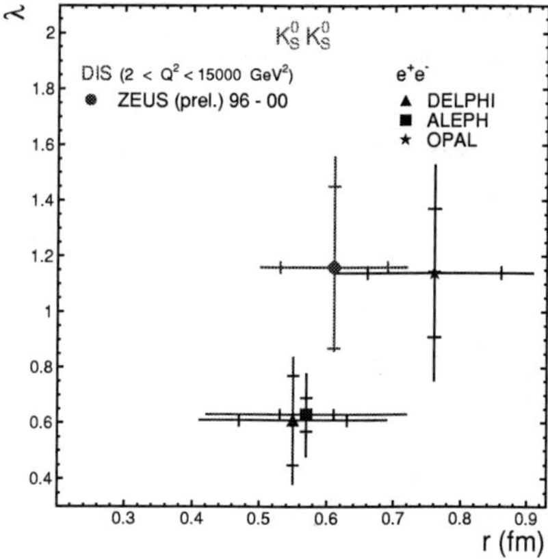

FIGURE 2. Comparison of DIS and e^+e^- LEP results for r and λ obtained from BEC of (a) $K^{\pm}K^{\pm}$ and (b) $K_s^0 K_s^0$ pairs.

Figure 2b shows the comparison between DIS and LEP results [5,7,8] for neutral kaons. The radius value obtained in DIS agrees with the measurements from LEP. Although the λ value is larger for DIS than for ALEPH and DELPHI is similar to the OPAL measurements within the uncertainties. An influence of $f_0(980)$ resonance was not removed from our analysys and from OPAL measurements but was excluded from ALEPH and DELPHI studies.

The LEP results on BEC and FDC suggest that radius depends on the hadron mass. This results as well as the recent DIS results for pions and kaons are plotted in Figure 3. The $r(m)$ behaviour is compared with Heisenberg uncertainty relation and QCD approach based on the the virial theorem [2]. Both theoretical expectations well describe the data. However, string (LUND) model does not predict such dependece. This requires more studies on heavier particles in DIS.

Bose-Einstein correlations of charged and neutral kaons have been studied in DIS

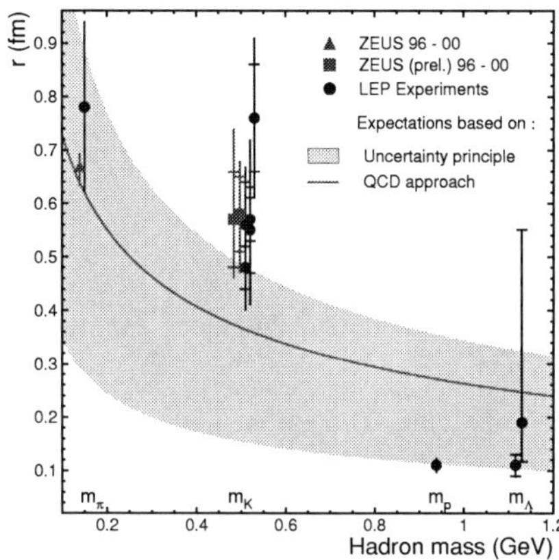

FIGURE 3. The dependence of the radius on the hadron mass. The results obtained in DIS and LEP are compared with the thoretical predictions.

at HERA. The values of radius for charged and neutral kaons and also for charged pions agree within errors. They are also compatible with LEP measurements. The difference between λ values for charged and neutral kaons can be due to possible influence of the proton fragmentations mechanism and production of $f_0(980)$ resonance.

In order to confirm the $r(m)$ dependence observed at LEP, more studies on Bose-Einstein and Fermi-Dirac correlations for identified baryons are needed.

ACKNOWLEDGMENTS

I would like to thank the following people for their help, many discussions and suggestions on BEC analysys: K. Olkiewicz, L. Zawiejski and Coordinators from QCD HFS group in ZEUS.

REFERENCES

1. Goldhaber *at al.*, *Phys. Rev. Lett* **3**, 181 (1959),
   Goldhaber *at al.*, *Phys. Rev.* **D 120**, 300 (1960).
2. G. Alexander *Rep. Prog. Phys.* **66**, 481-522 (2003);
   G. Alexander, I. Cohen, E. Levin, *Phys. Lett.* **B 452**, 159 (1999).
3. ZEUS Collab., U. Holm (ed.), *The ZEUS Detector. Status Report* (unpublished), DESY (1993), available on `http://www-zeus.desy.de/bluebook/bluebook.html`
4. ZEUS Collab., S. Chekanov *at al.*, *Phys. Lett.* **B 583**, 231 (2004).
5. DELPHI Collab., P. Abreu *at al.*, *Phys. Lett.* **B 379**, 330 (1996).
6. OPAL Collab., G. Abbiendi *at al.*, *Eur. Phys. J.* **C 21**, 23 (2001).
7. OPAL Collab., R. Akers *at al.*, *Z. Phys.* **C 67**, 389 (1995).
8. ALEPH Collab., S. Schael *at al.*, *Phys. Lett.* **B 611**, 66 (2005).

q_T Uncertainties for W and Z Production[1]

Stefan Berge*, Pavel M. Nadolsky†, Fredrick I. Olness* and C.-P. Yuan**

*Department of Physics, Southern Methodist University, Dallas, Texas 75275-0175, U.S.A.
†High Energy Physics Division, Argonne National Laboratory, Argonne, IL 60439-4815, U.S.A.
**Michigan State University, Department of Physics and Astronomy, East Lansing, MI 48824-1116, U.S.A.

Abstract. Analysis of semi-inclusive DIS hadroproduction suggests broadening of transverse momentum distributions at small x below $10^{-3} \sim 10^{-2}$, which can be modeled in the Collins-Soper-Sterman formalism by a modification of impact parameter dependent parton densities. We investigate these consequences for the production of electroweak bosons at the Tevatron and the LHC. If substantial small-x broadening is observed in forward Z^0 boson production in the Tevatron Run-2, it will strongly affect the predicted q_T distributions for $W^{\pm}$ and Z^0 boson production at the LHC.

PACS: 12.15.Ji, 12.38 Cy, 13.85.Qk

INTRODUCTION: As we move from the Tevatron collider at 1.96 TeV to the Large Hadron Collider (LHC) at 14 TeV, we encounter an entirely unexplored kinematic regime. There is good reason to believe that in this regime we will discover phenomena which may have significant consequences for precision measurements and searches for new physics.

In this paper,[2] we analyze the consequences of anomalous transverse momentum (q_T) broadening driven by possible small-x effects in W and Z boson production at the Tevatron and LHC. Such effects have been observed at the HERA ep collider [2, 3, 4, 5], and we use these results to predict the effects in hadron-hadron processes. The resulting modifications in the transverse momentum distributions will affect the measurements of the W boson mass and width, as well as the W and Z boson background in the search for new gauge bosons. The q_T broadening may also affect the detection of the Higgs boson at the LHC by altering its q_T distribution and the relevant QCD background.

At present, the magnitude of the q_T broadening corrections to W, Z, and Higgs boson production is largely unknown, in part because limited experimental data on q_T distributions is available in Drell-Yan-like processes at small x. If we turn to the crossed deep inelastic scattering (DIS) process, q_T broadening was observed at the HERA ep collider in the small x region: $x = 10^{-4} \sim 10^{-2}$ [2, 3, 4, 5].

PARAMETERIZING THE BROADENING: We now characterize the q_T broadening which was observed in semi-inclusive DIS processes, and consider implications for the (crossed) Drell-Yan process. We examine the resummed transverse momentum distribu-

---

[1] Presented by Fredrick Olness.

[2] Note, the results presented here are based on Ref. [1]; refer to this reference for a detailed description of the process and more extensive references.

CP792, *Deep Inelastic Scattering, DIS 2005*, edited by W. H. Smith and S. R. Dasu

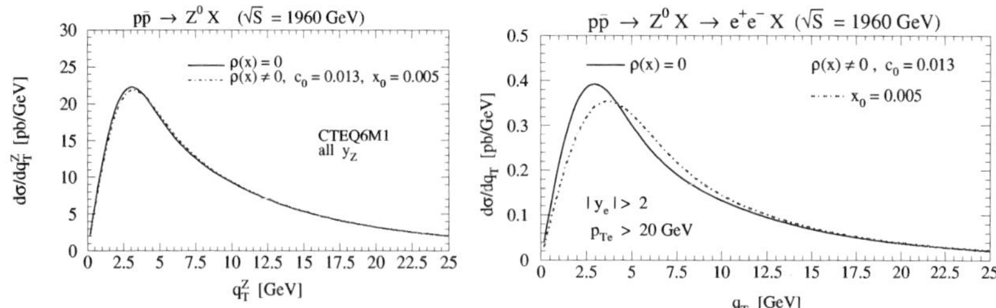

Figure 1. Transverse momentum distributions of Z bosons in the Tevatron Run-2: (A) integrated over the full range of Z boson rapidity y; (B) for events with both decay electrons registered in the forward ($y_{e^+} > 2$, $y_{e^+} > 2$) or backward ($y_{e^+} < -2$, $y_{e^+} < -2$) detector regions. The solid curve is the standard CSS cross section, calculated using the BLNY parametrization[6] of the non-perturbative Sudakov factor. The dashed curve includes the additional term responsible for the q_T broadening in the small-x region.

tion for the Drell-Yan process, following the notations of Refs. [1, 7]

$$\frac{d\sigma}{dydq_T^2} = \frac{\sigma_0}{S} \int \frac{d^2b}{(2\pi)^2} \, e^{-i\vec{q}_T \cdot \vec{b}} \, \widetilde{W}(b,Q,x_A,x_B) + Y(q_T,Q,x_A,x_B). \tag{1}$$

Here $x_{A,B} \equiv Qe^{\pm y}/\sqrt{S}$, y is the rapidity, the integral is the Fourier-Bessel transform of a resummed form factor $\widetilde{W}$ given in impact parameter (b) space, and Y is the regular part of the fixed-order cross section (Y is small at $q_T \to 0$). The form factor $\widetilde{W}$ is given by a product of a Sudakov exponent $e^{-S(b,Q)}$ and generalized parton distributions $\overline{\mathscr{P}}(x,b)$:

$$\widetilde{W}(b,Q,x_A,x_B) = \frac{\pi}{S} \sum_{a,b} \sigma_{ab}^{(0)} \, e^{-S(b,Q)} \, \overline{\mathscr{P}}(x_A,b) \, \overline{\mathscr{P}}(x_B,b) \quad .$$

In the limit of small b, we can write the generalized parton distributions, $\overline{\mathscr{P}}(x,b)$, in the form: $\overline{\mathscr{P}}(x,b) \simeq (\mathscr{C} \otimes f)(x, b_0/b) \, e^{-\rho(x)b^2}$, where $\mathscr{C}(x, b_0/b)$ are coefficient functions, $f(x,\mu)$ are integrated parton distributions, and $b_0 = 2e^{-\gamma_E}$.

The expressions for $\overline{\mathscr{P}}(x,b)$ differer from the conventional form by the introduction of the term $e^{-\rho(x)b^2}$ which will provide an additional q_T broadening with an x dependence specified by $\rho(x)$. This phenomenological characterization of the q_T broadening follows the corresponding analysis of the effect which was observed at HERA [2, 3, 4, 5]. This q_T broadening may approximate x-dependent higher-order contributions that are not included in the finite-order expression for $(\mathscr{C} \otimes f)$. We parametrize $\rho(x)$ in the following functional form:

$$\rho(x) = c_0 \left(\sqrt{\frac{1}{x^2} + \frac{1}{x_0^2}} - \frac{1}{x_0} \right)$$

such that $\rho(x) \sim c_0/x$ for $x \ll x_0$, and $\rho(x) \sim 0$ for $x \gg x_0$. This parameterization ensures that the formalism reduces to the usual CSS form for large x ($x \gg x_0$) and introduces an additional source of q_T broadening (growing as $1/x$) at small x ($x \ll x_0$). The parameter c_0 determines the magnitude of the broadening for a given x, while x_0

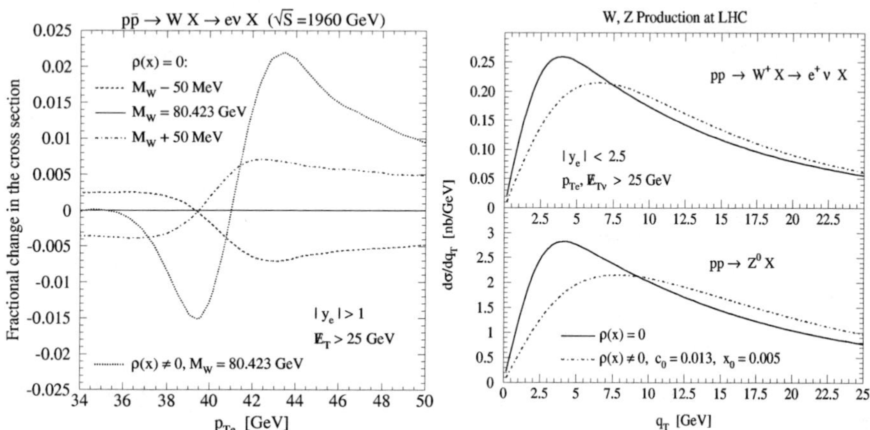

Figure 2. A) The fractional difference in the distribution $d\sigma/dp_{Te}$ for the forward-rapidity sample of electrons ($|y_e| > 1$) at the Tevatron. B) Transverse momentum distributions of (i) W^+ bosons and (ii) Z^0 bosons at the Large Hadron Collider.

specifies the value of x below which the broadening effects become important. Based on the observed dependence $\rho(x) \sim 0.013/x$ at $x \lesssim 10^{-2}$ in SIDIS energy flow data, we choose $c_0 = 0.013$ and $x_0 = 0.005$ as a representative choice for our calculations.

KINEMATICS: The small-x broadening occurs when one or both longitudinal momentum fractions $x_{A,B} \approx M_V e^{\pm y}/\sqrt{S}$ are $\lesssim x_0 = 0.005$. Lower values of x_A can be reached at the price of pushing x_B closer to unity, and vice versa; hence, we anticipate this effect will be enhanced as we move to either large boson rapidity $|y|$, or to small $M_V/\sqrt{S}$. The rate in the forward $|y|$ region is suppressed by the decreasing parton densities at $x \to 1$.

Z BOSONS AT THE TEVATRON: The broadening may be observed in the di-lepton channel in Z boson production in the Tevatron Run-2. For the Z boson cross section, the dominant contributions come from $x \sim M_Z/\sqrt{S} \sim 0.046 \gg x_0$, where the broadening function $\rho(x)$ is negligible. Consequently, the strategy here is to exclude contributions from the central-rapidity Z bosons, which are almost insensitive to the broadening. If no distinction between the central and forward Z bosons is made (e.g., as in the Run-1 analysis), the small-x broadening contributes at or below the level of the other uncertainties in the resummed form factor. Fig. 1(A) shows the Z boson distribution $d\sigma/dq_T$, integrated over the Z boson rapidity y without selection cuts on the decay leptons. The cross section with the broadening term (dashed line) essentially coincides with the cross section without such a term (solid line).

In contrast, the small-x broadening visibly modifies $d\sigma/dq_T$ at forward rapidities ($|y| < 2$), where one of the initial-state partons carries a small momentum fraction ($x \lesssim 0.005$). Fig. 2(B) displays the cross sections with the acceptance cuts $y_{e^\pm} > 2$ or $y_{e^\pm} < -2$ simultaneously imposed on both decay leptons. The cuts exclude central Z contributions and retain a fairly large cross section (≈ 3.4 pb), most of which falls within the experimental acceptance region. Run-2 can discriminate between the curves in Fig. 1(B) given the improved acceptance and increased luminosity of the upgraded

Tevatron collider; this result will have important implications for the W boson measurements, as we illustrate in the following section.

M_W MEASUREMENT: The q_T distribution of the W boson is important as this influences the extraction of M_W from the distribution $d\sigma/dp_{Te}$. The distribution $d\sigma/dp_{Te}$ exhibits the typical Jacobian peak located at $p_{Te} \sim M_W/2 \approx 40$ GeV. To better visualize percent-level changes in $d\sigma/dp_{Te}$ associated with the broadening, we plot in Fig. 2(A) the fractional difference $\left(d\sigma^{mod}/dp_{Te}\right) / \left(d\sigma^{std}/dp_{Te}\right) - 1$ of the "modified" (*mod*) and "standard" (*std*) theory cross sections. The broadening of $d\sigma/dq_T$ shifts the Jacobian peak in the positive direction. At $|y_e| > 1$, the small-x broadening is large and exceeds the other theoretical uncertainties, and is comparable with a variation of M_W by more than 50 MeV. For $|y_e| < 1$ (not shown), the effect is comparable with a variation of M_W by ~20 MeV. In either case, if these effects are present then they must be taken into account for precision measurements.

W & Z BOSONS AT THE LHC: At the LHC, the small-x broadening can be observed in W and Z boson production at all rapidities because $x < 0.005$ for all y. Fig. 2(B-i) displays the q_T shift for the production of W bosons with experimental cuts for ATLAS. The shift is slightly larger in W^+ boson production as compared in W^- boson production (not shown) because of the flatter y distribution for W^+ bosons. The observed q_T broadening propagates into the leptonic transverse mass and transverse momentum distributions. The q_T shifts for the production of Z bosons is comparable, and is displayed in Fig. 2(B-ii). A measurement of the rapidity dependence of q_T distributions at the LHC will test for the presence of such effects.

CONCLUSIONS: For the choice of parameters extracted from the fit to the HERA data, the q_T broadening may be discovered via the analysis of forward produced Z bosons at the Tevatron Run-2. Then the q_T broadening will shift the measured W boson mass in the p_{Te} method by ~ 20 MeV in the central region ($|y_e| < 1$), and more than 50 MeV in the forward region ($|y_e| > 1$). At the LHC, these effects produce a much harder q_T distribution for W and Z bosons, and has important implications for the measurement of the W boson mass.

REFERENCES

1. S. Berge, P. Nadolsky, F. Olness, and C. P. Yuan (2004), hep-ph/0410375.
2. P. Nadolsky, D. R. Stump, and C.-P. Yuan, *Phys. Rev.*, **D61**, 014003 (2000), erratum: *ibid.*, **D64**, 059903 (2001).
3. P. M. Nadolsky, D. R. Stump, and C.-P. Yuan, *Phys. Rev.*, **D64**, 114011 (2001).
4. C. Adloff, et al., *Eur. Phys. J.*, **C12**, 595–607 (2000).
5. S. Aid, et al., *Phys. Lett.*, **B356**, 118–128 (1995).
6. F. Landry, R. Brock, P. M. Nadolsky, and C. P. Yuan, *Phys. Rev.*, **D67**, 073016 (2003), hep-ph/0212159.
7. S. Berge, P. Nadolsky, F. Olness, and C.-P. Yuan (2004), hep-ph/0401128.

A Closer Look at the Analysis of NLL BFKL

Jeppe R. Andersen

Cavendish Laboratory, University of Cambridge, Madingley Road, CB3 0HE, Cambridge, UK

Abstract. The initial analyses of the next-to-leading logarithmic corrections to the BFKL kernel were very discouraging. Encouraged by the success of new methods in the analysis of the BFKL equation at full NLL accuracy we demonstrate in this talk how some of the initial conclusions were based on a breakdown of the tools used in the analysis rather than the framework itself.

INTRODUCTION

The Balitsky–Fadin–Kuraev–Lipatov[1] (BFKL) framework systematically resums the class of logarithms originating from the kinematics that dominate the total cross section in the Regge limit of scattering amplitudes, where the centre of mass energy $\sqrt{\hat{s}}$ is large and the momentum transfer $\sqrt{-\hat{t}}$ is fixed. In this limit the scattering of two gluons $p_A p_B \to p_{A'} p_{B'}$ will be dominated by multi–particle production leading to final states described by momenta $k_0 = p_{A'}, k_1, \ldots, k_n, k_{n+1} = p_{B'}$ satisfying

$$s \gg 2k_{i-1}k_i \gg t_i = q_i^2, q_i = p_A - \sum_{r_0}^{i-1} k_r, \prod_{i=1}^{n+1} s_i = s \prod_{i=1}^{n} \mathbf{k}_i^2, k_\perp^2 = -\mathbf{k}^2, |k_{i\perp}| \simeq |p_{A'\perp}| \quad (1)$$

The Regge limit is therefore suitable for describing the production of multiple hard partons from e.g. gluon scattering (and in fact the large-rapidity limit of any process that includes a t-channel gluon exchange). We will in this talk focus entirely on processes within the multi-Regge kinematics of Eq. (1). In this limit the partonic cross section can be approximated by

$$\hat{\sigma}(\Delta) = \int \frac{d^2\mathbf{k}_a}{2\pi\mathbf{k}_a^2} \int \frac{d^2\mathbf{k}_b}{2\pi\mathbf{k}_b^2} \Phi_A(\mathbf{k}_a)\, f(\mathbf{k}_a, \mathbf{k}_b, \Delta)\, \Phi_B(\mathbf{k}_b), \quad (2)$$

where $\Phi_{A,B}$ are the impact factors characteristic of the particular scattering process, and $f(\mathbf{k}_a, \mathbf{k}_b, \Delta)$ is the gluon Green's function describing the interaction between two Reggeised gluons exchanged in the t–channel with transverse momenta $\mathbf{k}_{a,b}$ spanning a rapidity interval of length Δ. The leading and next-to-leading logarithmic contributions to this gluon Green's function can be resummed by solving the BFKL equation to the required accuracy

$$\omega\, f_\omega(\mathbf{k}_a, \mathbf{k}_b) = \delta^{(2+2\varepsilon)}(\mathbf{k}_a - \mathbf{k}_b) + \int d^{2+2\varepsilon}\mathbf{k}\, \mathscr{K}(\mathbf{k}_a, \mathbf{k}+\mathbf{k}_a)\, f_\omega(\mathbf{k}+\mathbf{k}_a, \mathbf{k}_b), \quad (3)$$

where w is the Mellin-conjugated variable to Δ, and the BFKL kernel $\mathscr{K}(\mathbf{k}_a, \mathbf{k}+\mathbf{k}_a)$ is presently known to next-to-leading logarithmic accuracy.

CP792, *Deep Inelastic Scattering, DIS 2005*, edited by W. H. Smith and S. R. Dasu

SOLUTIONS OF THE BFKL EQUATION

The solution to integral equations of the form in Eq. (3) can be written in terms of the eigenfunctions $\phi_i(k_a)$ and eigenvalues λ_i as

$$f_\omega(k_a,k_b) = \sum\!\!\!\!\!\!\int_i \frac{\phi_i(k_a)\ \phi_i^*(k_b)}{\omega-\lambda_i} \tag{4}$$

leading to

$$f(k_a,k_b,\Delta) = \sum\!\!\!\!\!\!\int_i \frac{1}{2\pi i}\, e^{\Delta\lambda_i}\ \phi_i(k_a)\ \phi_i^*(k_b) \tag{5}$$

Leading Logarithmic Accuracy

At leading logarithmic accuracy the BFKL kernel is conformal invariant, since the running of the coupling only enters at higher logarithmic orders. The eigenfunctions of the angular averaged kernel are of the form $k^{2(\gamma-1)}, \gamma = 1/2+i\nu$, which means that to this accuracy, the BFKL evolution can be solved analytically, with the transverse momentum of emitted gluons integrated to infinity, by analysing the Mellin transform of the kernel. One finds

$$\int \mathrm{d}^{D-2}\mathbf{k}\ \mathscr{K}^{\mathrm{LL}}(\mathbf{k}_a,\mathbf{k})\ \left(\mathbf{k}^2\right)^{\gamma-1} = \frac{\alpha_s N}{\pi}\chi^{\mathrm{LL}}(\gamma)\left(\mathbf{k}_a^2\right)^{\gamma-1}, \tag{6}$$

with N being the number of colours and

$$\chi^{\mathrm{LL}}(\gamma) = 2\psi(1)-\psi(\gamma)-\psi(1-\gamma), \qquad \psi(\gamma) = \Gamma'(\gamma)/\Gamma(\gamma). \tag{7}$$

Since both the eigenfunctions and eigenvalues are known, the angular averaged gluon Green's function can now be obtained as

$$\bar{f}(k_a,k_b,\Delta) = \frac{1}{k_b^2}\int_{\frac{1}{2}-i\infty}^{\frac{1}{2}+i\infty}\frac{\mathrm{d}\gamma}{2\pi i}\ e^{\Delta\omega^{\mathrm{LL}}(\gamma)}\left(\frac{k_b^2}{k_a^2}\right)^{\gamma}, \tag{8}$$

where

$$\omega^{\mathrm{LL}}(\gamma) \equiv \int \mathrm{d}^{D-2}\mathbf{k}\ \mathscr{K}^{\mathrm{LL}}(\mathbf{k}_a,\mathbf{k})\ \left(\frac{\mathbf{k}^2}{\mathbf{k}_a^2}\right)^{\gamma-1} = \frac{\alpha_s(k_a^2)N}{\pi}\chi^{\mathrm{LL}}(\gamma). \tag{9}$$

We stress that at LL the coupling is formally fixed, and so the regularisation scale is completely arbitrary. In Fig. 1 we have plotted $\omega^{\mathrm{LL}}(\frac{1}{2}+i\nu)$ for $\alpha_s = 0.2$ and it is seen that there is a maximum at $\nu = 0$. Therefore, the behaviour of the gluon Green's function in the limit of $\Delta\to\infty$ is determined by the value of $\omega^{\mathrm{LL}}(\frac{1}{2}) = 4\ln 2\alpha_s N/\pi$. A saddle point approximation based on the second order Taylor polynomial around $\nu = 0$ will correctly describe the asymptotic exponential growth in Δ, since the polynomial attains the correct value at the maximum.

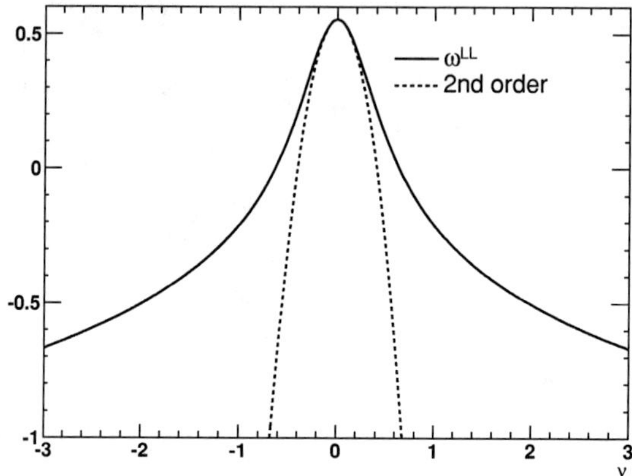

FIGURE 1. $\omega^{\mathrm{LL}}(\frac{1}{2}+i\nu)$ and the second order Taylor polynomial around $\nu = 0$.

Next-to-Leading Logarithmic Accuracy

When trying to extend this analysis to NLL accuracy one is immediately faced with the complications introduced by the breaking of the conformal invariance by the running coupling terms. This effect will necessarily change the eigenfunctions, and thus far the eigenfunctions for the full NLL kernel in QCD have not been constructed. Traditionally, the kernel at NLL has been studied using the projection on the Born level eigenfunctions as in Eq. (9). One finds [2]

$$\begin{aligned}\omega^{\mathrm{NLL}}(\gamma) &\equiv \int \mathrm{d}^{D-2}\mathbf{k}\, \mathscr{K}^{\mathrm{NLL}}(\mathbf{k}_a,\mathbf{k}) \left(\frac{\mathbf{k}^2}{\mathbf{k}_a^2}\right)^{\gamma-1} \\ &= \frac{\alpha_s(\mathbf{k}_a^2)N}{\pi}\left(\chi^{\mathrm{LL}}(\gamma)+\chi^{\mathrm{NLL}}(\gamma)\frac{\alpha_s(\mathbf{k}_a^2)N}{\pi}\right)\end{aligned} \tag{10}$$

with

$$\begin{aligned}\chi^{\mathrm{NLL}}(\gamma) = -\frac{1}{4}\Bigg[&\left(\frac{11}{3}-\frac{2}{3}\frac{n_f}{N}\right)\frac{1}{2}\left(\chi^{\mathrm{LL}}(\gamma)-\psi'(\gamma)+\psi'(1-\gamma)\right) \\ &-6\zeta(3)+\frac{\pi^2\cos(\pi\gamma)}{\sin^2(\pi\gamma)(1-2\gamma)}\left(3+\left(1+\frac{n_f}{N^3}\right)\frac{2+3\gamma(1-\gamma)}{(3-2\gamma)(1+2\gamma)}\right) \\ &-\left(\frac{67}{9}-\frac{\pi^2}{3}-\frac{10}{9}\frac{n_f}{N}\right)\chi^{\mathrm{LL}}-\psi''(\gamma)-\psi''(1-\gamma)-\frac{\pi^3}{\sin(\pi\gamma)}+4\phi(\gamma)\Bigg],\end{aligned} \tag{11}$$

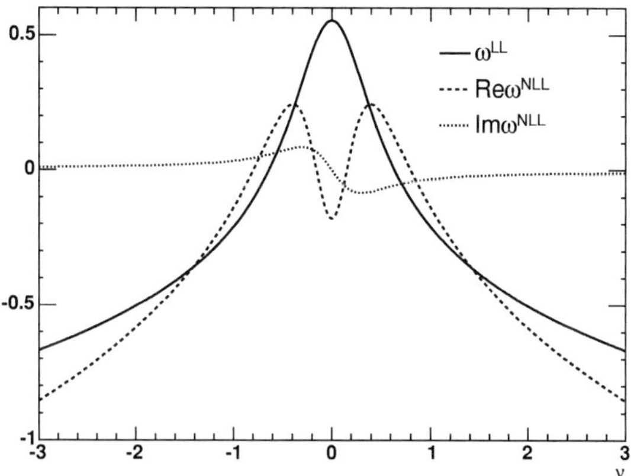

FIGURE 2. $\omega^{\mathrm{LL}}(\frac{1}{2}+i\nu)$ and $\omega^{\mathrm{NLL}}(\frac{1}{2}+i\nu)$ for $\alpha_s = 0.2$, and $N = n_f = 3$.

where

$$\begin{aligned}\phi(\gamma) &= -\int_0^1 \frac{\mathrm{d}x}{1+x}(x^{\gamma-1}+x^{-\gamma})\int_x^1 \frac{\mathrm{d}t}{t}\ln(1-t)\\ &= \sum_{n=0}^{\infty}(-1)^n\left[\frac{\psi(n+1+\gamma)-\psi(1)}{(n+\gamma)^2}+\frac{\psi(n+2-\gamma)-\psi(1)}{(n+1-\gamma)^2}\right].\end{aligned} \tag{12}$$

An approximation to the gluon Green's function at NLL can then be constructed by use of ω^{NLL} in place of ω^{LL} in Eq. (8). We have in Fig. 2 plotted the real and imaginary part of ω^{NLL} compared with ω^{LL} for $\alpha_s = 0.2$. The double hump structure of the real part of ω^{NLL} is potentially a disaster for the gluon Green's function. At asymptotically large Δ the behaviour of the gluon Green's function is determined by the position and value of the maxima of ω^{NLL}. Since there are two such distinct maxima located at $\gamma_1 = 1/2 - i\nu_0, \gamma_2 = 1/2 + i\nu_0$ the asymptotic estimate of the NLL gluon Green's function based on the LL eigenfunctions will have an oscillatory behaviour in $\ln k_a/k_b$. This is a problem of matching to the DGLAP evolution, and is a problem strictly outside the Regge kinematics.

The initial observation of the large NLL corrections was based on the difference $\omega^{\mathrm{LL}} - \omega^{\mathrm{NLL}}$ evaluated at $\nu = 0$. Indeed, for reasonable values of the coupling ω^{NLL} is even negative at this point. However, this is not what determines the intercept. If indeed the solution of the BFKL equation at NLL was obtained using ω^{NLL} of Eq. (10), then the asymptotic intercept at NLL would again be determined by the maximum value attained by the real part of $\omega^{\mathrm{NLL}}(\gamma)$ along the contour $\gamma = 1/2 + i\nu$. We see from Fig. 2 that this maximum value is roughly halved compared to the LL asymptotic intercept.

However, even within the Regge kinematics, this analysis leads to severe problems in the very limit, where the resummed logarithmic terms are meant to dominate the scattering matrix. The non-zero imaginary part of $\omega^{\mathrm{NLL}}(\frac{1}{2}+i\nu_0)$ leads to oscillations with increasing rapidity! This clearly signals a breakdown of the approach in the very limit it is meant to describe.

The solution to this apparent problem is the realisation that what one has studied with this method is indeed not the real solution to the BFKL equation at NLL. The LL eigenfunctions are *not* the eigenfunctions at NLL (for non-zero β_0). Indeed, the only contribution to the imaginary part of $\omega^{\rm NLL}(\gamma)$ for $\gamma = 1/2 + i\nu$ comes from the term $-\psi'(\gamma) + \psi(1-\gamma)$ in Eq. (11), which contributes to $\omega^{\rm NLL}$ with a factor proportional to β_0 (that is, it vanishes in the limit where the LL eigenfunctions diagonalises the NLL kernel). It was observed already in Ref.[2] that this part of the NLL corrections is the only one that is not symmetric under $\gamma \leftrightarrow 1-\gamma$, and that if one expands the kernel on the LL eigenfunctions rescaled by a square root of the coupling, i.e. $(k^2)^{\gamma-1}\left(\frac{\alpha_s(k^s)}{\alpha_s(\mu^2)}\right)^{-1/2}$ then these and only these terms would disappear from Eq. (11). Since this is the only contribution to the imaginary part of $\omega^{\rm NLL}$ this would simultaneously cure the problem of oscillations within the Regge-limit. It should be emphasised though that these rescaled functions still are not the true eigenfunctions at NLL, but it is straightforward to check numerically that the “off-diagonal elements” of $\omega^{\rm NLL}$ (i.e. those obtained with a different γ for k_a and k) are far smaller in this case than when using the pure LL eigenfunctions. With the advance of new approaches to the solution of the BFKL equation at full NLL accuracy[3, 4, 5, 6] it has also been possible to check explicitly how well the two approximations fare with the full solution. We find that the approximation using the rescaled LL eigenfunctions is much closer to the true solution than the one using the pure LL ones. It should be noted that a saddle point approximation based around $\nu = 0$ would have to use an extremely large order approximation in order to describe correctly the asymptotic intercept. Even the 16th order Taylor polynomial would fail to reach the maximum value for $\omega^{\rm NLL}$ in Fig. 2. It would be far better to base a saddle point approximation around ν_0.

Using the guess obtained from these rescaled eigenfunctions it is possible to calculate the intercept as a function of the rapidity. This is depicted on Fig. 3. We see that although the NLL correction amounts to roughly a factor of two, it is stable. Also, it should be remembered that the study of both the LL and NLL intercept has been performed without constraining the phase space of the BFKL resummation to such which is attainable at a given collider. The effects of such a constrain are known to be large[7, 8, 9, 10, 11, 12] and reduce the LL evolution significantly (and presumably the NLL intercept to a slightly lesser extent).

CONCLUSIONS

We have shown that the NLL corrections to the BFKL intercept are large but stable. The instability observed in initial analyses is a direct result of using the conformal set of leading log eigenfunctions as if they were eigenfunctions at NLL. Indeed, for $\beta_0 = 0$ even this instability disappear, and this the case also for non-zero β_0 if one applies rescaled eigenfunctions. Although these still do not diagonalise the kernel fully at NLL, the results obtained using the rescaled eigenfunctions do compare favourably when compared directly with a full solution obtained numerically. In the conformal limit

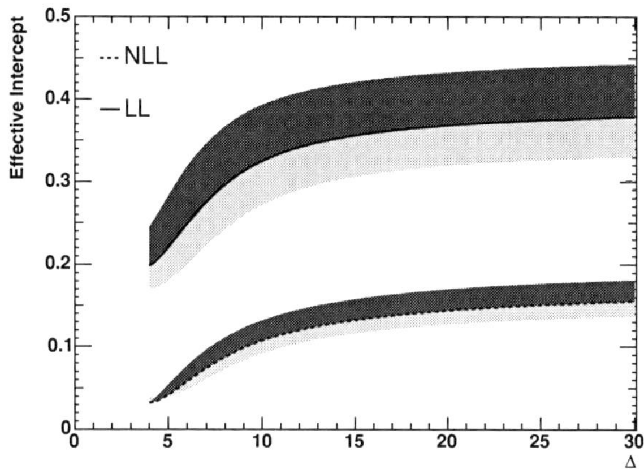

FIGURE 3. The effective intercept (of $f(k_a = 20\text{GeV}, k_b = 30\text{GeV}, \Delta)$ with $\alpha_s(30\text{GeV}) = 0.15$), as a function of the rapidity at LL (upper line) and NLL (lower line). The uncertainty due to a renormalisation scale variation of a factor of two is indicated by the colour band.

of $\beta_0 = 0$, where the NLL corrections can be studied exactly using the projections on the LL eigenfunctions, any modification of the NLL kernel to match better to the DGLAP region (like the ones of Ref.[13, 14]) must move the position of the maximum of ω^{NLL} to $\nu = 0$ while not changing the maximum value itself significantly (since this would lead to a change in the asymptotic intercept obtained within the Regge kinematic).

ACKNOWLEDGMENTS

I would like to thank Agustín Sabio Vera for lively discussions. This research was supported by PPARC (postdoctoral fellowship PPA/P/S/2003/00281).

REFERENCES

1. L. N. Lipatov, Sov. J. Nucl. Phys. **23** (1976) 338 [Yad. Fiz. **23** (1976) 642],
   E. A. Kuraev, L. N. Lipatov and V. S. Fadin, Sov. Phys. JETP **45** (1977) 199 [Zh. Eksp. Teor. Fiz. **72** (1977) 377],
   I. I. Balitsky and L. N. Lipatov, Sov. J. Nucl. Phys. **28** (1978) 822 [Yad. Fiz. **28** (1978) 1597]
2. V. S. Fadin and L. N. Lipatov, Phys. Lett. B **429** (1998) 127
3. J. R. Andersen and A. Sabio Vera, Phys. Lett. B **567** (2003) 116
4. J. R. Andersen and A. Sabio Vera, Nucl. Phys. B **679** (2004) 345
5. J. R. Andersen and A. Sabio Vera, Nucl. Phys. B **699** (2004) 90
6. M. Ciafaloni, D. Colferai, G. P. Salam and A. M. Stasto, Phys. Rev. D **68** (2003) 114003
7. J. R. Andersen and W. J. Stirling, JHEP **0302** (2003) 018
8. L. H. Orr and W. J. Stirling, Phys. Rev. D **56** (1997) 5875
9. L. H. Orr and W. J. Stirling, Phys. Lett. B **429** (1998) 135
10. L. H. Orr and W. J. Stirling, Phys. Lett. B **436**, 372 (1998)
11. J. R. Andersen, V. Del Duca, F. Maltoni and W. J. Stirling, JHEP **0105** (2001) 048

12. J. R. Andersen, V. Del Duca, S. Frixione, C. R. Schmidt and W. J. Stirling, JHEP **0102** (2001) 007
13. G. P. Salam, JHEP **9807** (1998) 019
14. A. S. Vera, "An all-poles approximation to collinear resummations in the Regge limit of perturbative QCD," arXiv:hep-ph/0505128.

Azimuthal asymmetries in deep inelastic scattering at HERA

Artur Ukleja

Hoża 69, 00681 Warszawa, Poland

Abstract. The distribution of the azimuthal angle of charged and neutral hadrons has been studied in the hadronic centre-of-mass system for neutral current deep inelastic *ep* scattering with the ZEUS detector at HERA using an integrated luminosity of 45.21 pb^{-1}. Measurements of the dependence of the moments of the azimuthal distribution on the pseudorapidity and minimum transverse energy of the final state hadrons are presented using the energy flow method.

Keywords: Deep inelastic scattering, azimuthal asymmetry
PACS: 13.60-r, 12.38Qk

INTRODUCTION

The investigation of the semi-inclusive process $ep \rightarrow ehX$ in deep inelastic scattering (DIS), where h is an observed hadron, addresses an important prediction of perturbative Quantum Chromodynamics (pQCD) in the description of hadron production. It is of interest to investigate the distribution of the azimuthal angle of the detected hadrons around the virtual photon direction in the hadronic centre-of-mass frame (HCM). The azimuthal angle ϕ, is defined as the angle between the hadron production plane and the lepton scattering plane (Figure 1a).

The azimuthal dependence of hadron production has the form [1, 2, 3] as

$$\frac{d\sigma}{d\phi} = \mathscr{A} + \mathscr{B}\cos\phi + \mathscr{C}\cos 2\phi \tag{1}$$

where the azimuthal asymmetries, defined as parameters $\mathscr{B}$ and $\mathscr{C}$, can be evaluated experimentally. They are extracted from experimental data by calculating statistical moments for experimental distributions of the respective trigonometrical functions of ϕ:

$$\langle\cos\phi\rangle = \frac{\mathscr{B}}{2\mathscr{A}} \qquad \langle\cos 2\phi)\rangle = \frac{\mathscr{C}}{2\mathscr{A}} \tag{2}$$

For neutral current deep inelastic scattering interactions NC DIS with an unpolarised lepton beam the $\langle\cos\phi\rangle$ and $\langle\cos 2\phi\rangle$ values are of the order of a few percent [4, 5].

Azimuthal asymmetries, (2) exist only if the the final state hadron has transverse momentum. The higher order QCD processes such as QCD-Compton QCDC and boson-gluon fusion BGF are the main source of these hadrons. These two processes have a different ϕ behaviour as described in [6] as well as a different rapidity dependence. The rapidity or pseudorapidity η^{HCM} is defined here w.r.t. the incoming proton direction. Hadrons from BGF and QCDC dominate over hadrons from the zeroth order DIS process

CP792, *Deep Inelastic Scattering, DIS 2005,* edited by W. H. Smith and S. R. Dasu

in the region $-4 < \eta^{\mathrm{HCM}} < 0$. In addition the gluons and quarks from the QCD Compton process populate different region of rapidity. The coefficient $\mathscr{B}$ has an opposite sign for gluons and quarks, thus motivating a study of the azimuthal asymmetry as a function of rapidity η^{HCM}.

Chay, Ellis and Stirling [2] proposed analysing the asymmetry as a function of the detected hadron's transverse momentum cutoff $p_{T\,\mathrm{cut}}$. This is equivalent to removal of the zeroth order QCD processes and a selection of leading hadrons produced directly from the scattered partons. Consequently at higher $p_{T\,\mathrm{cut}}$ values a better agreement should be obtained with the perturbative QCD predictions that suggest the coefficient $\mathscr{C}$ to be always positive and larger for higher $p_{T\,\mathrm{cut}}$ values.

Asymmetries in the ϕ distribution arise whenever a non-zero transverse momentum is presented in the scattering process in the HCM frame. Thus the perturbative azimuthal asymmetry originates in the first and higher order QCD processes and is observable in single particle production because the high-energy hadrons are produced close to the direction of the scattered hard partons. The transverse momenta arising from parton hadronisation does not contribute to the asymmetry but smears the observed distribution. The energy flow method enhances the contribution of the hard hadrons and is used here to calculate the mean values (2). In this method the direction of each particle h in the final state is weighted with its transverse energy. The range of the investigated phase space is increased with respect to the previous studies [4, 5].

DATA SAMPLE

The experimental results are based on the data collected in 1995-97 with the ZEUS detector at the HERA collider with protons of energy 820 GeV. Electrons of energy 27.5 GeV are longitudinally unpolarised. Neutral current deep inelastic scattering events have been selected from the data corresponding to an integrated luminosity of 45.21 pb^{-1}. ZEUS is a multipurpose detector described in detail elsewhere [7].

Particles in the final state were reconstructed by combining information from tracking and calorimeter in the ZEUS detector, as energy flow objects [8]. The selection criteria were based on the earlier ZEUS investigation [4]. The main cuts were:

- the event had an identified scattered positron with energy $E_{e'} > 10$ GeV;
- in order to define the phase space of the measurement, the event was required to have $100 < Q^2 < 8000\ \mathrm{GeV}^2$, $0.2 < y < 0.8$ and $0.01 < x < 0.1$. The double angle method was used to reconstruct these variables [9];
- the reconstructed hadrons (charged and neutral particles) were required to have their transverse momenta $p_T^{\mathrm{LAB}} > 150$ MeV. These cuts excluded hadrons contained within the beam pipe or failing to traverse sufficient layers of the tracker to ensure good reconstruction.

CORRECTION PROCEDURE

Monte Carlo (MC) events were used to correct the data for detector inefficiences. The detector simulation is based on the GEANT 3.13 program [10]. Neutral current

(NC) events with electroweak radiative corrections came from the LEPTO 6.5.1 code interfaced to HERACLES 4.6.1 [11] via the DJANGOH 1.1 code [12]. High order QCD processes were simulated using the MEPS option of LEPTO.
A second sample of NC DIS Monte Carlo events was generated with ARIADNE 4.10 [13] where the QCD cascade came from the colour-dipole model. In all cases, the events have been generated using the CTEQ4D next-to-leading order parton density parametrization of the proton. The final state parton system was hadronised using the LUND string model as implemented in JETSET 7.4.10 [14].

The correction factor for ϕ was defined as the ratio of energy flow of hadrons, $E(\phi^{MC}_{had})$, to energy flow detected, $E(\phi^{MC}_{det})$, i.e. $F^{MC}(\phi) = \frac{E(\phi^{MC}_{had})}{E(\phi^{MC}_{det})}$. The corrected integrated energy flow $E(\phi)$ was determined separately bin-by-bin for each region in the η-ϕ plane as $E(\phi^{DATA}) = F^{MC}(\phi) \cdot E(\phi^{DATA}_{det})$.

RESULTS

The measured azimuthal asymmetries in terms of the mean values of the trigonometrical functions which appear in the functional form (1) for the differential cross section for $ep \rightarrow ehX$ are presented in Figure 1b as a function of pseudorapidity η^{HCM} and in Figure 2 as the minimum transverse energy $\mathrm{E}_{\mathrm{T}}^{\mathrm{HCM}}$ (min) into three regions of η^{HCM}: $-5 < \eta^{\mathrm{HCM}} < -2.5$, $-2.5 < \eta^{\mathrm{HCM}} < -1$ and $-1 < \eta^{\mathrm{HCM}} < 0$.

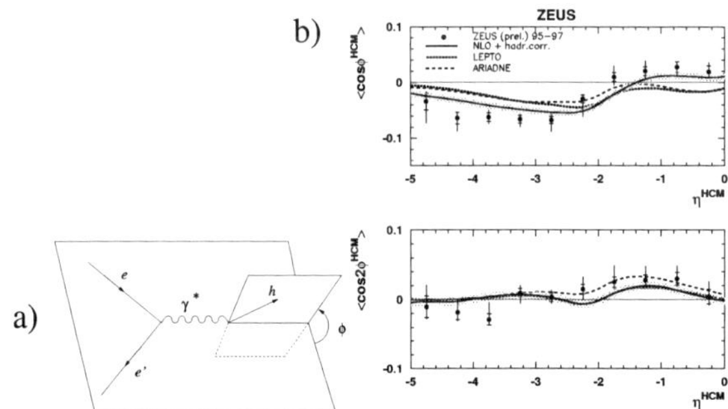

FIGURE 1. a) The definition of the azimuthal angle ϕ; b) The values of $\langle\cos\phi^{\mathrm{HCM}}\rangle$ and $\langle\cos 2\phi^{\mathrm{HCM}}\rangle$ are shown as a function of hadron η^{HCM} obtained using the energy flow method.

Figure 1b shows that the mean value of $\langle\cos\phi^{\mathrm{HCM}}\rangle$ is negative for $\eta^{\mathrm{HCM}} < -2$ and becomes positive for larger η^{HCM}. This is in disagreement with the LO predictions that are negative throughout the measured η^{HCM} range. The measured $\langle\cos 2\phi^{\mathrm{HCM}}\rangle$ values are consistent with zero for for $\eta^{\mathrm{HCM}} < -2$ and are positive for higher values of η^{HCM}. This is consistent with the LO expectations from both LEPTO and ARIADNE.

In region $-5 < \eta^{\mathrm{HCM}} < -2.5$ (Figure 2a) the main contribution to azimuthal asymmetry comes from QCD Compton $\gamma q \rightarrow gq$ and arises from hadrons coming from quark fragmentation. This analysis confirms that the value of $\langle\cos\phi^{\mathrm{HCM}}\rangle$ is more negative than expected from the LO predictions. The $\langle\cos 2\phi^{\mathrm{HCM}}\rangle$ values are small and in agreement with both LEPTO and ARIADNE.

The region $-2.5 < \eta^{\mathrm{HCM}} < -1$ (Figure 2b) is that with an increasing contribution from boson-gluon fusion. The results presented here confirm a small value of $\langle \cos\phi^{\mathrm{HCM}} \rangle$ and positive values for $\langle \cos 2\phi^{\mathrm{HCM}} \rangle$ for all $\mathrm{E}_{\mathrm{T}}^{\mathrm{HCM}}$(min). The LO predictions of LEPTO and ARIADNE are in good agreement with data.

The third region $-1 < \eta^{\mathrm{HCM}} < 0$ (Figure 2c) is populated equally by hadrons from QCD Compton and from boson-gluon fusion processes. The $\langle \cos\phi^{\mathrm{HCM}} \rangle$ values are positive, contrary to LO predictions, whereas the $\langle \cos 2\phi^{\mathrm{HCM}} \rangle$ values are positive and in agreement with LO predictions.

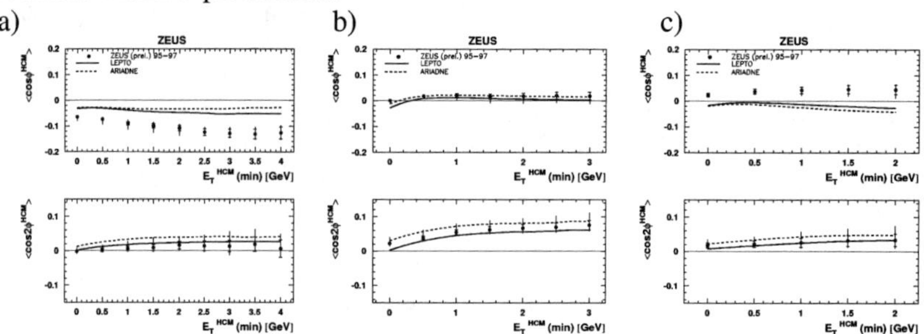

FIGURE 2. The values of $\langle \cos\phi^{\mathrm{HCM}} \rangle$ and $\langle \cos 2\phi^{\mathrm{HCM}} \rangle$ are shown as a function of hadron minimum transverse energy $\mathrm{E}_{\mathrm{T}}^{\mathrm{HCM}}$ (min) in the HCM for a) $-5 < \eta^{\mathrm{HCM}} \leq -2.5$; b) $-2.5 < \eta^{\mathrm{HCM}} \leq -1$; c) $-1 < \eta^{\mathrm{HCM}} \leq 0$

CONCLUSIONS

Azimuthal asymmetries are investigated as a function of hadron pseudorapidity in the hadronic centre-of-mass frame. The $\langle \cos\phi^{\mathrm{HCM}} \rangle$ values are not described by LO predictions. The $\langle \cos 2\phi^{\mathrm{HCM}} \rangle$ values are only significant in the region $\eta^{\mathrm{HCM}} > -2.5$ when high minimum transverse momentum is selected for hadrons.

REFERENCES

1. A. Méndez, *Nucl. Phys.*, **B 145**, 199 (1978).
2. J. Chay, S. D. Ellis, and W. J. Stirling, *Phys. Lett.*, **B 269**, 175 (1991).
3. M. Ahmed, and T. Gehrmann, *Phys. Lett.*, **B 465**, 297 (1999).
4. J. Breitweg, et al., *Phys. Lett.*, **B 481**, 199 (2000).
5. S. Chekanov, et al., *Phys. Lett.*, **B 551**, 3 (2003).
6. V. Hedberg, et al., *Physics at HERA, Workshop*, **1**, 331 (1991).
7. The ZEUS detector, Status Report (unpublished), DESY (1993).
8. J. Breitweg, et al., *Eur. Phys. J.*, **C 6**, 43 (1999).
9. S. Bentvelsen, J. Engelen, and P. Kooijman, "Reconstruction of (x, Q^2) and Extraction of Structure Functions in Neutral Current Scattering at HERA," in *Proc. Workshop on Physics at HERA*, edited by W. Buchmüller, and G. Ingelman, DESY, Hamburg, Germany, 1992, vol. 1, p. 23.
10. R. Brun, et al., GEANT3, Tech. Rep. CERN-DD/EE/84-1, CERN (1987).
11. A. Kwiatkowski, H. Spiesberger, and H. J. Möhring, *Comp. Phys. Comm.*, **69**, 155 (1992), also in *Proc. Workshop Physics at HERA*, 1991, DESY, Hamburg.
12. K. Charchula, G. Schuler, and H. Spiesberger, *Comp. Phys. Comm.*, **81**, 381 (1994).
13. L. Lönnblad, *Comp. Phys. Comm.*, **71**, 15 (1992).
14. T. Sjöstrand, and M. Bengtsson, *Comp. Phys. Comm.*, **43**, 367 (1987).

Azimuthal asymmetries in deep inelastic $e^+p \to e^+X$ scattering at HERA

Teresa Tymieniecka

Warsaw University, Warsaw, Poland

Abstract. A Monte Carlo study of the azimuthal angular distribution around the virtual boson-proton beam axis at HERA is presented. In the presence of typical acceptance and selection criteria in the laboratory frame the azimuthal distribution is investigated for different kinematical variables. The best measureable dependence of $\langle\cos\phi\rangle$ and $\langle\cos 2\phi\rangle$ is found to be for rapidity $\mathscr{Y}$ or pseudorapidity η using the energy flow method.

Keywords: Deep inelastic scattering, azimuthal asymmetry, boson polarisation, HERA
PACS: 13.60-r, 12.38Aw, 12.38Bx

INTRODUCTION

The high-luminosity data on deep inelastic scattering collected at HERA allow the investigation of hadronic final state features going beyond hadron multiplicities or jet structures. One of the possible observables is related to the angular correlation between the lepton scattering plane and the hadron production plane in the hadronic centre-of-mass (HCM) created by the exchanged boson and proton.

This paper presents an investigation of azimuthal asymmetries by counting the emitted hadrons and with energy flow to optimize observables and the kinematical region for precise measurements. The experimental data are presented elsewhere [1]. The Z-axis directed along the incoming proton direction defines the transverse and longitudinal quantities of the emitted hadrons. Around this axis the azimuthal angle ϕ is measured from the lepton scattering plane to the hadron/parton production plane (Fig. 1a).

The semi-inclusive cross section for neutral current can be decomposed according to the dependence on ϕ e.g. [2]:

$$\frac{d^5\sigma^{ep\to ehX}}{dxdQ^2dp_\perp^2\dots d\phi} = \mathscr{A} + \mathscr{B}\cos\phi + \mathscr{C}\cos 2\phi \tag{1}$$

The terms with $\mathscr{B}$ and $\mathscr{C}$ quantities exist only if the final hadron or parton carries some transverse momentum. In the zeroth order QCD process the incoming quark is backscattered along the Z-axis. The parton shower development and hadronisation cause the appearance of hadrons with some transverse energy E_T^{HCM}, which on average, is rather small ($E_T^{\mathrm{HCM}} < 0.5$ GeV). The first order QCD processes such as QCD Comptons ($\gamma^* q \to qg$) and boson-gluon fusion ($\gamma^* g \to q\bar{q}$) are the main source of hadrons with large transverse momenta. Hadronisation which is essentially symmetric around the parton direction smears the primary events shape.

CP792, *Deep Inelastic Scattering, DIS 2005*, edited by W. H. Smith and S. R. Dasu

The $\mathscr{B}$ and $\mathscr{C}$ values can be extracted within any kinematical cuts by calculating statistical moments for the obtained distributions of the respective trigonometrical functions of ϕ. Some simple relations exist for the averaged asymmetries:

$$\langle\cos\phi\rangle = \frac{\mathscr{B}}{2\mathscr{A}} \qquad\qquad \langle\cos 2\phi\rangle = \frac{\mathscr{C}}{2\mathscr{A}} \tag{2}$$

EXPERIMENTAL ASPECTS

The investigations are done for 820 GeV protons and 27.5 GeV positrons. The kinematical region is the same as used in the high-Q^2 ZEUS analysis [3, 4] namely the exchanged boson fourvector is $100 < Q^2 < 8000$ GeV^2, the Bjorken variable $0.01 < x < 0.1$ and inelasitcity $0.2 < y < 0.8$. The main feature of the experimental data included in the MC generation are losses of the hadrons in the beam pipe ($\mathrm{P_T^{LAB}} < 150$ MeV).

In the HERA experiments calorimeters are the dominant suppliers of information on emitted hadrons covering 99% of 4π. They do not measure individual hadrons but are designed to measure clusters of energy for jet algorithms. The detection efficiencies of charged and neutral particles are similar. Tracking detectors cover a narrow region of phase space, i.e. $|\eta^{\mathrm{LAB}}| < 1.75$ for the ZEUS detectors which limits investigations based on charged hadrons [3].

The idea of energy flow method was introduced by theorists [5, 6] to avoid soft and collinear singularities in perturbative QCD. By weighting the directions of the emitted hadrons or partons with their energies any hadrons/partons emitted together are treated as being equivalent to one object carrying the sum of the energies. Experimentally two hadrons emitted nearby in the HCM frames are also detected as one object on the detector level.

The energy flow method permits the use of charged and neutral hadrons as well as the enhancement of hard hadrons/partons with weights thus avoiding dependence on jet algorithms discussed elsewhere [7]. The method diminishes the contributions from QCD Compton and boson-gluon fusions emitted along the Z-axis where some theoretical uncertainties exist [8]. The method is sensitive to energy scale uncertainties but the investigation of mean values (2) leads to cancellation.

Here an individual hadron/parton transverse energy E_T^{HCM} is used as a weight.

MONTE CARLO MODELS

The azimuthal distributions are generated using the LEPTO 6.5.1 program [9] which includes photon and Z exchanges and the QCD processes described by the first order QCD matrix elements (ME). If needed, parton shower (MEPS) in the leading logarithm approximation is switched on. As an alternative, the QCD cascade is modelled with the ARIADNE 4.12 program [10] modified in 1997 to include BGF. In both cases, fragmentation into hadrons is performed according to the JETSET 7.410 code [11] based on the Lund string model. The next-to-leading order (NLO) prediction is obtained with DISENT using the dipole factorization formulae [12].

PREDICTIONS

Using the LO predictions of LEPTO and ARIADNE azimuthal asymmetries are investigated as a function of pseudorapidity η^{HCM} of the emitted hadrons/partons for the $-5 < \eta^{\mathrm{HCM}} < 0$ region because, (see Fig. 1b) the zeroth order QCD process (QPM) contribution is eliminated ($\eta^{\mathrm{HCM}} > -5$) as well as target fragments ($\eta^{\mathrm{HCM}} < 0$). The relative contribution from boson-gluon fusion (BGF) with respect to QCD Compton (QCDC) also changes with rapidity (Fig. 1b) and in addition quarks from QCDC populate mainly $\eta^{\mathrm{HCM}} < -2.8$ whereas gluons populate $\eta^{\mathrm{HCM}} > -2.8$ (Fig. 1c).

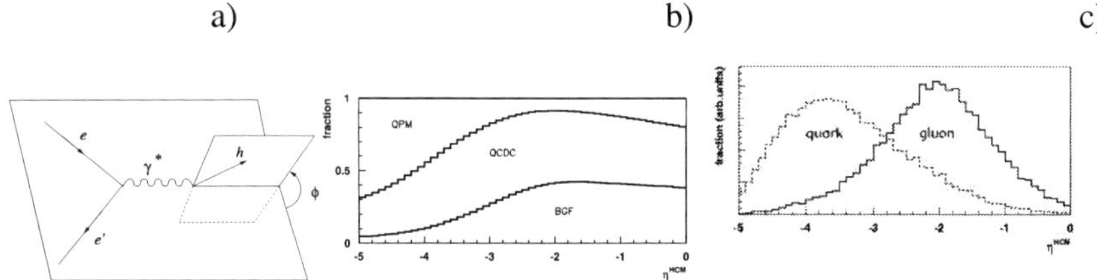

FIGURE 1. a) The definition of the azimuthal angle ϕ; b) the fraction of boson-gluon fusion (BGF), QCD Compton (QCDC) and the zeroth order QCD process (QPM) as a function of pseudorapidity η^{HCM} in the hadronic centre-of-mass frame for the energy flow method; c) for the QCD Compton process the guark and gluon contributions as a function of pseudorapidity η^{HCM}. These predictions are taken from LEPTO 6.5.1.

Two methods are used: the first method is based on energy flow; the second method — on multiplicity. Figure 2 (left side) presents asymmetries in form of $\langle\cos\phi\rangle$ and $\langle\cos 2\phi\rangle$ obtained from LO predictions of LEPTO and ARIADNE on the hadron level but without detector losses. No significant difference is found between predictions from the different MC models and methods. However asymmetries from energy flow are slightly larger than the ones from multiplicity.

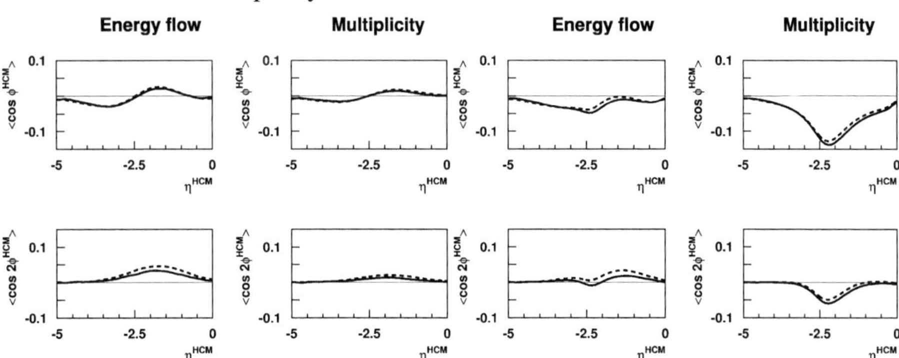

FIGURE 2. Predictions for azimuthal asymmetries in form of $\langle\cos\phi\rangle$ and $\langle\cos 2\phi\rangle$ for the energy flow method and for the multiplicity method generated with the LEPTO 6.5.1 code (solid line) and the ARIADNE 4.12 (dashed line) code. All emitted hadrons are detected (left side) and with the detection effects included (right side) i.e. losses of hadrons in the HERA beampipe ($P_T^{LAB} < 150$ GeV).

Detection of hadrons modify the conclusion. With the present colliding beams the reconstruction excludes hadrons with the low transverse momenta in the laboratory frame

$P_T^{LAB} < 150$ MeV. Loss of hadrons modifies slightly asymmetries from the energy flow in the region of pseudorapidity, $-2.5 < \eta^{HCM} < -1$ but the ones from the multiplicity method are completely different (Fig. 2, right side).

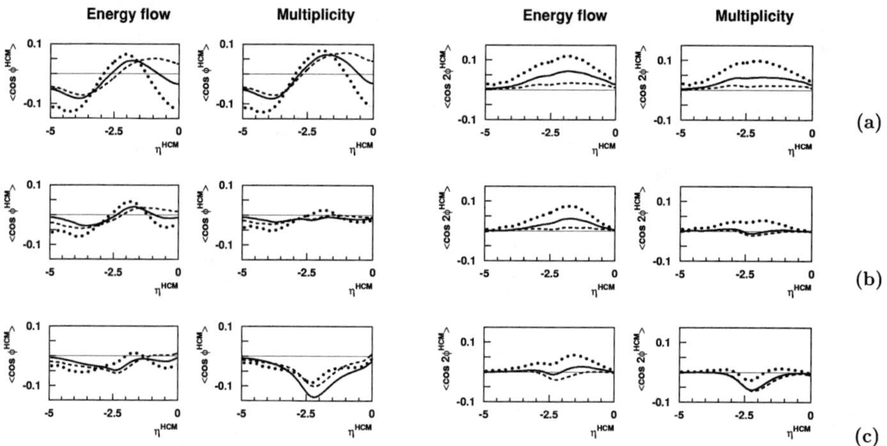

FIGURE 3. Predictions for azimuthal asymmetries for $\langle\cos\phi\rangle$ and $\langle\cos 2\phi\rangle$ generated by the LEPTO 6.5.1 code as a function of pseudorapidity η^{HCM} in the hadronic centre-of-mass system for the energy flow method (left side) and for the multiplicity method (right side). The results are presented for all emitted partons together (solid line) and separately for main processes: for QCD Compton (dashed line) and for boson-gluon fusion (BGF) (dotted line): (a) for hard processes only, i.e. for the matrix element option (ME) without hadronisation; (b) with parton shower included, i.e. for the MEPS option without hadronisation; (c) with hadronisation included, i.e. for the MEPS option with hadronisation. The detection effects in form of losses in the HERA beam pipe ($P_T^{LAB} < 150$ GeV) are included but are insignificant without hadronisation. Line at $\langle\cos\phi\rangle = 0$ is to guide an eye.

Using LEPTO the step-by-step contributions to azimuthal asymmetries are investigated including detector effects: for hard scattering of partons described by the QCD matrix elements (ME) followed by parton showers (PS) and hadronisation (JETSET). This is shown for all partons from QPM + QCDC + BGF together, and separately for QCDC and BGF. Figure 3a is equivalent to the jet study; the results are expected to be nearly the same for both methods. One can see that the asymmetry is large for hard processes and larger for the multiplicity method; therefore the multiplicity method is better for jets but experimentally can be used in the narrower region of phase space [4]. Although losses of particles affect predictions on the parton level the main effect is visible for the multiplicity method after hadronisation where many low P_T^{LAB} hadrons are lost.

Figure 4 shows the LO and NLO predictions of DISENT and the LO predictions with the LEPTO and ARIADNE codes. Some differences come from a lack of correlation in generation of the hadronic part of the LEPTO/ARIADNE events with amount of the longitudinal structure function F_L; helicity conservation for the longitudinally polarized exchanged boson excludes QPM but allows for QCDC. In the region $-5 < \eta^{HCM} < -2.5$ the main contribution to the azimuthal asymmetry comes from the QCD Compton process and arises from hadrons coming from quark fragmentation (Fig. 1c). The region $-2.5(-3) < \eta^{HCM} < -1$ is that with an increasing contribution from boson-gluon fusion (Fig. 1b); this region is strongly affected by detection losses. The third region $-1 < \eta^{\mathrm{HCM}} < 0$ is populated equally by hadrons from QCD Compton process and from boson-gluon fusions (Fig. 1b); the $\langle \cos \phi^{\mathrm{HCM}} \rangle$ values are sensitive to NLO corrections; they are positive whereas the LO predictions give the negative values (Fig. 4).

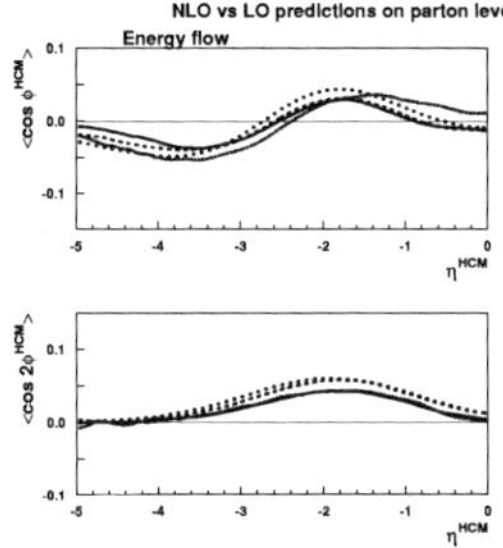

FIGURE 4. The LO predictions for azimuthal asymmetries generated with the LEPTO 6.5.1, ARIADNE 4.12 and DISENT codes (solid, dashed and dotted line respectively) on the parton level. The NLO predictions of DISENT is also imposed (dashed-dotted line). No detection effects included.

SUMMARY AND CONCLUSIONS

Azimuthal asymmetries are proposed to be analyzed as a function of hadron rapidity in the HCM frame using the energy flow method. This novel approach provides precise measurements and small systematical uncertainties in the wider interval of phase space [1]. It permits the investigation of the contribution of boson-gluon fussion with respect to the QCD Compton process as well as contribution from quarks produced in the QCD Compton process with respect to the gluon. Calculation of mean values minimizes the experimental uncertainties by cancellation of some systematical effects like the calorimeter energy scale, uncertainties of the parton density functions and partially of the fragmentation functions to some extent. The latter means that hadron spectra and consequently their detection depend on the feature of fragmentation function. Detection inefficiency modifies the values of asymmetries.

REFERENCES

1. A.Ukleja, *these preceedings* (2005).
2. K. Hagiwara, K. Hikasa, and N. Kai, *Phys. Rev.*, **D27**, 84 (1983).
3. J. Breitweg, et al., *Phys. Lett.*, **B 481**, 199 (2000).
4. S. Chekanov, et al., *Phys. Lett.*, **B 551**, 3 (2003).
5. R. D. Peccei, and R. Rückl, *Phys. Lett.*, **B 84**, 95 (1979).
6. G. Ingelman, B. Andersson, G. Gustafsona, and T. Sjöstrand, *Nucl. Phys.*, **B 206**, 239 (1982).
7. E. Mirkes, and S. Willfahrt, *Phys. Lett.*, **B 414**, 205 (1997).
8. V. Hedberg, et al., *Physics at HERA, Workshop*, **1**, 331 (1991).
9. G. Ingelman, A. Edin, and J. Rathsman, *Comp. Phys. Comm.*, **101**, 108 (1997).
10. L. Lönnblad, *Comp. Phys. Comm.*, **71**, 15 (1992).
11. T. Sjöstrand, *Comp. Phys. Comm.*, **39**, 347 (1986).
12. S. Catani, and M. H. Seymour, *Nucl. Phys.*, **B 485**, 401 (1997).

Pentaquark at JLab: the g11 experiment in CLAS

M.Battaglieri*, R. De Vita*, V.Kubarovsky†,** and the CLAS Collaboration

*Istituto Nazionale di Fisica Nucleare - Via Dodecaneso 33 16139 Genova ITALY
†Rensselaer Polytechnic Institute - Troy New York 12180-359
**Jefferson Laboratory - 12000 Jefferson Avenue Newport News 23606 Virginia

Abstract. After the claiming of the possible discovery of a pentaquark state, many experiments reported positive and negative results opening a discussion about the pentaquark existence. New experiments with high resolution and high statistics are needed to solve the controversy. Jefferson Lab started a comprehensive program to search for pentaquark in photoproduction at threshold on proton and deuteron targets, collecting more than 10 times the existing statistics. The first experiment on the proton (g11) just finished to analyze the data and in a short time will be able to report about the pentaquark search.

Keywords: Pentaquark, photoproduction, proton target.
PACS: 13.60 Rj; 25.20 Lj

INTRODUCTION

The possible existence of a narrow resonance with strangeness quantum number S=+1 and valence quark structure $udud\bar{s}$ was reported by the LEPS Collaboration [1] and then confirmed by many other experiments using a wide variety of probes (photon, electron, protons) and targets (proton, deuteron, nuclei). This state, known as Θ^+ was predicted to be few MeV wide with a mass of about 1540 MeV, in very good agreement with the experimental findings. If it exists, this would be the first example of a baryon state that is not made up of a simple 3-quark (qqq) valence configuration. However, the low statistics of the observed structures and many null results obtained from the reanalysis of high energy experiments questioned the pentaquark existence. The experimental evidence, both positive or negative, was obtained from data previously collected for other purposes in many reaction channels and very different kinematic conditions, which may involve different production mechanisms. Thus a direct comparisons of the results of the different experiments are very difficult, preventing a definitive conclusion. A second generation of dedicated experiments, optimized for the pentaquark search, was undertaken at Jefferson Laboratory (JLab). These experiments cover the few GeV region where most of the positive evidence were reported, and collected at least an order of magnitude more statistics than the previous measurements. The mass resolution is of the order of few MeV and the accuracy in the mass determination is of 1-2 MeV, allowing a precise determination of any possible narrow peaks in the spectra. In this talk we report about the so called g11 experiment run with real photon on a proton target.

CP792, *Deep Inelastic Scattering, DIS 2005,* edited by W. H. Smith and S. R. Dasu

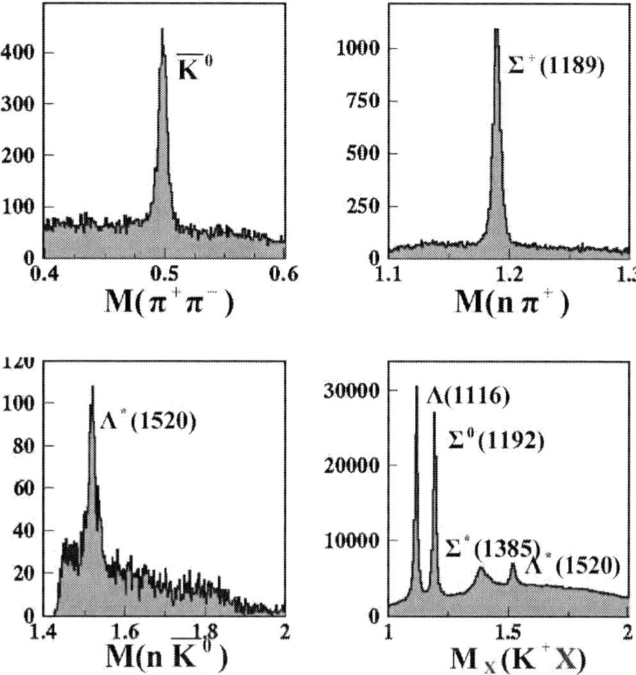

FIGURE 1. Missing and invariant masses for different final states measured in CLAS. Error on mass determination of known baryons and mesons is within 1-2 MeV of the nominal PDG value.

THE G11 EXPERIMENT AT JLAB

The g11 experiment aimed to measure two production channels on the proton: $\bar{K}^0\Theta^+$ and $K^*\Theta^+$, each using two decay modes of the Θ^+: K^+n and K^0p, for a total of four final states. In particular, the reaction $\gamma p \to \bar{K}^0K^+n$ was previously investigated at ELSA by the SAPHIR collaboration [2] in a similar photon energy range, finding a positive evidence for a narrow Θ^+ state with M=1540 MeV, FWHM $\Gamma < 25$ MeV, and a production cross section of the order of 300 nb. The observed peak had 55 events in it over a background of 56 events. This result called for a confirmation from an high resolution and high statistics experiment.

The measurement was performed using the CLAS detector in Hall-B with a bremsstrahlung photon beam produced by a continuous 60 nA electron beam of E_0 = 4.0 GeV impinging on a gold foil 8×10^{-5} radiation lengths thick. A bremsstrahlung tagging system with a photon energy resolution of 0.1% E_0 was used to tag photons in the energy range from $1.6-3.8$ GeV. A liquid hydrogen target was contained in a mylar cylinder cell 4 cm in diameter and 40 cm long. Outgoing hadrons were detected in the CLAS [3] spectrometer. Momentum information for charged particles was obtained via tracking through three regions of multi-wire drift chambers immersed in a toroidal

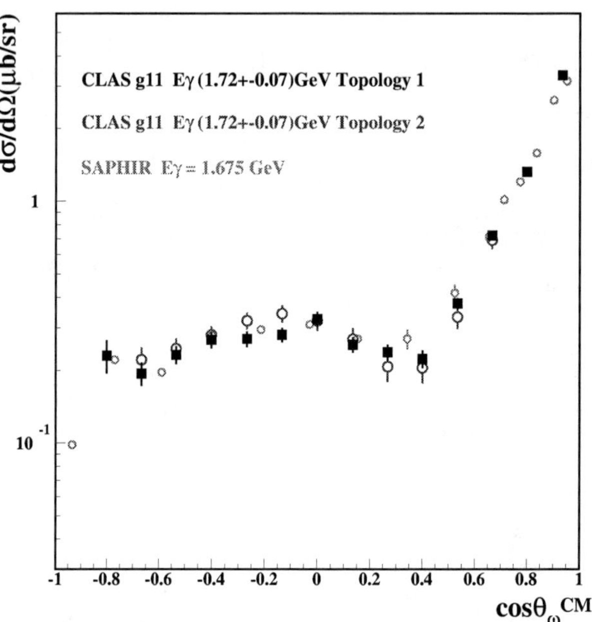

FIGURE 2. The differential cross section obtained by the two topologies and compared to existing data

magnetic field (~ 0.5 T), which was generated by six superconducting coils. The field was set to bend the positive particles away from the beam into the acceptance region of the detector. Time-of-flight scintillators (TOF) were used for hadron identification. The interaction time between the incoming photon and the target was measured by the Start Counter (ST), consisting of a set of 24, 2.2 mm thick plastic scintillators surrounding the hydrogen cell. The CLAS momentum resolution is of the order of 0.5-1% depending on the kinematics. The detector geometrical acceptance for each positive particle in the relevant kinematic region is about 40%. It is somewhat less for low energy negative hadrons, which can be lost at forward angles because they are bent out of the acceptance by the toroidal field. Coincidences between the photon tagger and the CLAS detector triggered the recording of the events. The trigger in CLAS was defined requiring the coincidence between the TOF system and the ST in at least two sectors. We took data for 50 days during June and July 2004 collecting more than 7G triggers corresponding to an integrated luminosity of 70 pb^{-1}. This is probably the highest statistics ever collected in experiments with tagged photons.

FIRST RESULTS

Data analysis searching for pentaquark is still underway. Here we report some preliminary results showing the quality of our data. Exploiting the capability of CLAS to measure simultaneously different final states we were able to check detector calibrations. Fig.1 shows some missing and invariant mass spectra for different final states detected in CLAS. Starting from the left-upper panel: $(\pi^+\pi^-)$ invariant mass with a clear K_S peak; on the right-up, $(n\pi^+)$ invariant mass centered and the Σ^+ peak; in the bottom-left panel $(n\bar{K}^0)$ invariant mass and the $\Lambda^*(1520)$ peak, and finally, the missing mass M_X when a K^- is measured in CLAS with peaks corresponding to the hyperons $\Lambda(1116)$, $\Sigma(1192)$, and their excited states. These spectra were produced using only 1% of the g11 statistics!

As a check of the entire analysis procedure we derived differential (and total) cross sections for some known reactions. The luminosity was monitored and recorded during data taking while the CLAS efficiency was evaluated by mean of realistic Monte Carlo simulations. In particular, the reaction $\gamma p \to p\omega \to p\pi^+\pi^-\pi^0$ was studied measuring two different final states: $p\pi^+$ and $p\pi^+\pi^-$ detected. In both cases missing mass technique was used to close the kinematics. Since the two topologies mainly involve different parts of the CLAS detector a comparison of the differential cross sections extracted in the two cases is a good cross check of the entire analysis procedure. Fig.2 shows the differential cross section measured for the two topologies and compared to the existing published data. The good agreement shows that we control the systematics errors at few percent level.

The analysis of the the reaction $\gamma p \to \bar{K}^0 K^+ n$ in search of evidence of the Θ^+ pentaquark in the nK^+ decay channel is currently in progress. The final channel is isolated detecting the K^+ and $\bar{K}^0 \to \pi^+\pi^-$ and identifying the neutron with the missing mass technique. The direct measurement of the K^+ allows one to define the strangeness of any resonance observed in this final state. Thanks to CLAS capability of detecting many particles in a wide kinematic range, the background hyperons decaying into this same final can be clearly identified and cut in the analysis. Using the full g11 data set a total number of 100k $\Lambda^*(1520)$ were collected. When compared to the 630 $\Lambda^*(1520)$ seen by SAPHIR, this gives a feeling of the statistical significance of g11 experiment. Results about pentaquark search will be published shortly.

REFERENCES

1. T. Nakano *et al.* (LEPS Collaboration), Phys. Rev. Lett. **91**, 012002 (2003).
2. J. Barth *et al.* (SAPHIR Collaboration), Phys. Lett. B **572**, 127 (2003).
3. B. Mecking et al.,*Nucl. Instrum. and Meth.* **A503**, 513 (2003).

Theoretical aspects of pentaquark searches

A.P. Szczepaniak

Physics Department and Nuclear Theory Center,
Indiana University, Bloomington IN, 47405, USA E-mail: aszczepa@indiana.edu

Abstract. The experimental evidence for the Θ^+ pentaquark was examined. We reviewed old duality arguments against pentaquarks based on the observed exchange degeneracy of Regge trajectories, and reminded about past null searches and kinematical effects. We shoed how various kinematical effects could also be responsible for the present sightings of peaks in the K^+n and K_Sp spectra. The work presented was based on papers written in collaboration with A. Dzierba and C. Meyer [1, 2]

Keywords: pentaquarks, resonances
PACS: 11.80.Cr, 13.60.Le, 13.60Rj

REFERENCES

1. A. R. Dzierba *et al.* Phys. Rev. **D** 69, 051901(R) (2004).
2. A. R. Dzierba, C. A. Meyer and A. P. Szczepaniak, "Reviewing the evidence for pentaquarks," arXiv:hep-ex/0412077.

CP792, *Deep Inelastic Scattering, DIS 2005,* edited by W. H. Smith and S. R. Dasu

Forward jets in DIS at ZEUS

Nikolai Vlasov

Physikalisches Institut, University of Freiburg,
on behalf of the ZEUS Collaboration

Abstract. Jet cross sections in neutral current deep inelastic scattering at low x_{Bj} have been measured with the ZEUS detector in the forward region towards the proton direction. Hadronic final-state measurements in this region are expected to be particularly sensitive to QCD evolution effects. The measurements have been compared with leading-logarithm parton-shower Monte Carlo models and perturbative QCD calculations.

Keywords: forward jets DIS BFKL DGLAP ZEUS
PACS: 13.60.-r

INTRODUCTION

In deep inelastic scattering (DIS) a parton in the proton can induce a QCD cascade consisting of several subsequent parton emissions before a quark absorbs the virtual photon. The scattered quark then radiates partons until hadronization sets in. Several different models of parton evolution dynamics have been proposed.

One set of parton evolution equations, derived on the basis of the collinear factorization theorem is that of the DGLAP evolution equations. They are characterized by resummation of the terms of $\alpha_s \ln(Q^2)$, where Q^2 is the virtuality of exchanged boson. This approach assumes that the dominant contribution to the evolution comes from subsequent gluon emissions which are strongly ordered in transverse momenta k_T. DGLAP is applicable for the region $\ln(Q^2) \gg \ln(1/x)$, where x is the fraction of the proton's momentum carried by the struck parton.

The BFKL evolution equations allow resummation of terms independently of $\ln(Q^2)$. Therefore, the gluon ladder need not be ordered in k_T. The BFKL equations can be applied in the region $\ln(1/x) \gg \ln(Q^2)$ in the leading-logarithmic approximation, so that they are expected to primarily contribute to the evolution at low x.

The CCFM evolution equations are based on the idea of coherent gluon radiation, which leads to angular ordering of gluon emissions in the gluon ladder such that $\theta_i > \theta_{i-1}$, where θ_i is the i-th gluon with respect to the incoming parton. The CCFM approach incorporates both types of evolution, DGLAP and BFKL, so that it should be applicable across the whole kinematic plane.

Differences between these different approaches are expected to be most prominent in the phase-space region towards the proton-remnant (forward) direction. The results of two ZEUS analyses [1] which study forward jet production in DIS, are presented here.

CP792, *Deep Inelastic Scattering, DIS 2005,* edited by W. H. Smith and S. R. Dasu

FORWARD JET PRODUCTION

Jets with transverse energy $E_T^{\rm jet} > 6$ GeV and pseudorapidity $0 < \eta^{\rm jet} < 3$ were selected in the laboratory frame with the longitudinally-invariant k_T cluster algorithm in the inclusive mode. A phase-space region, called "BFKL", was defined by the additional condition $0.5 < \frac{(E_T^{\rm jet})^2}{Q^2} < 2$. The requirement on $(E_T^{\rm jet})^2/Q^2$ restricts the jet kinematics to the region where BFKL effects are expected to be large. The jet cross sections were measured as functions of $E_T^{\rm jet}$, $\eta^{\rm jet}$, Q^2 and x.

Only events with $Q^2 > 25$ GeV^2 and $y > 0.04$ were considered in the measurement. The analysis was performed with $\cos\gamma_h < 0$, where γ_h is hadronic angle which corresponds to the direction of the outgoing quark. The restriction was made to reject quark-parton model events.

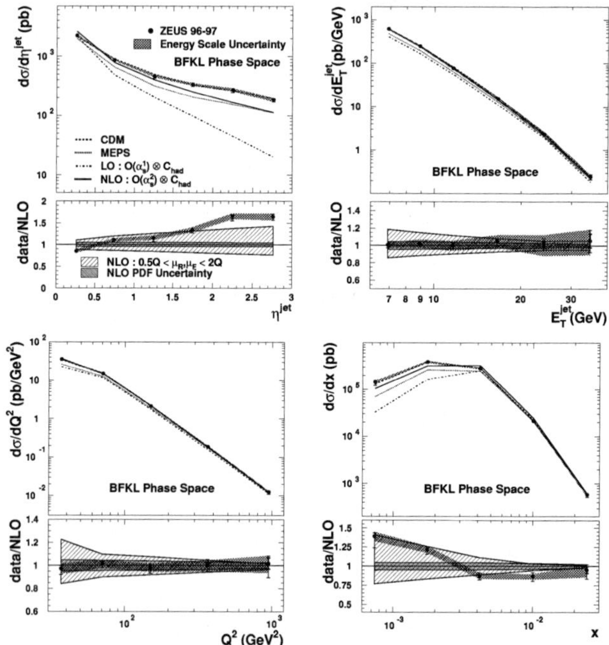

FIGURE 1. Differential cross sections (dots) in the BFKL phase space for inclusive jet production in DIS as functions of $\eta^{\rm jet}$, $E_T^{\rm jet}$, Q^2 and x. The calculations of CDM (dashed lines), MEPS (dotted lines), $\mathcal{O}(\alpha_s^1)$ (dot-dashed lines) and $\mathcal{O}(\alpha_s^2)$ (solid lines) QCD calculations are shown.

The data are compared to predictions from the Monte Carlo (MC) programs ARIADNE and LEPTO, and to perturbative QCD calculations using the program DISENT. The ARIADNE MC is based on the BFKL-like Color Dipole Model (CDM) which produces a cascade of gluons not strongly ordered in transverse momentum. LEPTO MC is a pure DGLAP type MC based on the first-order QCD matrix elements plus parton showers (MEPS).

The program DISENT, using DGLAP-evolved proton PDFs, allows calculations that sum up to two orders of the perturbation series (LO = $\mathcal{O}(\alpha_s^1)$ and NLO = $\mathcal{O}(\alpha_s^2)$). The uncertainty on the calculations due to higher-order terms was estimated by changing the renormalisation scale.

The measurements are presented in Fig. 1. The predictions of ARIADNE describe well all data distributions. The predictions of LEPTO fail to describe the data, especially in the $\eta^{\rm jet}$ distribution and low-x region. Fixed-order QCD calculations describe the data well for Q^2, $E_T^{\rm jet}$ and x, but underestimate the data at high values of $\eta^{\rm jet}$.

FORWARD JET PRODUCTION IN EXTENDED $\eta^{\rm JET}$ REGION

During the 1998-2000 running period, the forward plug calorimeter (FPC) [2] was installed with a small hole of radius 3.15 cm in the center to accommodate the beam pipe. The FPC increased the forward calorimetric coverage by about 1 unit of pseudorapidity to $\eta \leq 5$. This new component allowed to extend the reconstruction of jets by 0.5 unit of pseudorapidity.

In the following, cross sections are presented as functions of the variables Q^2, x, $E_T^{\rm jet}$ and $\eta^{\rm jet}$. The differential jet cross sections were determined in the kinematic region $20 < Q^2 < 100$ GeV^2, $0.04 < y < 0.7$ and $0.0004 < x < 0.005$. The jet search was performed using the k_T cluster algorithm in the longitudinally invariant inclusive mode in the Breit frame. The reconstructed jets were then boosted to the laboratory frame. The following jet selection cuts were applied in the laboratory frame: $E_T^{\rm jet} > 5$ GeV, $2 < \eta^{\rm jet} < 3.5, 0.5 < (E_T^{\rm jet})^2/Q^2 < 2$; the scaled longitudinal momentum was required to satisfy $x_{\rm jet} = p_z^{\rm jet}/p > 0.036$, where p is the proton momentum and $p_z^{\rm jet}$ is the longitudinal jet momentum, which selected forward jets with a large energy.

The left side of Figure 2 compares the measured cross sections as function of Q^2, x, $E_T^{\rm jet}$ and $\eta^{\rm jet}$ with the predictions of different MC models. The predictions of ARIADNE reproduce the shapes and the normalizations of the differential cross sections. LEPTO describes well the shapes of the data but is lower than the data by a factor of two.

The CASCADE MC based on CCFM parton evolution uses k_T-factorization of the cross section into an off-shell matrix element and an unintegrated gluon density function. The following sets of CCFM unintegrated gluon density function have been tried for comparison with the data: J2003 set 1 and J2003 set 2. The latter version includes non-singular terms in the splitting function and lowers the cross sections at low x. CASCADE with J2003 set 2 describes the absolute values of the cross sections better than J2003 set 1. However it does not reproduce the shapes of the distributions in x and $\eta^{\rm jet}$.

The right side of the Figure 2 compares the measurements with the predictions of $\mathcal{O}(\alpha_s^2)$ QCD calculations as it is implemented in DISENT. The calculations describe the measurement within the theoretical uncertainties. The variation of the calculations with the renormalization scale is large, emphasizing the need for higher-order calculations.

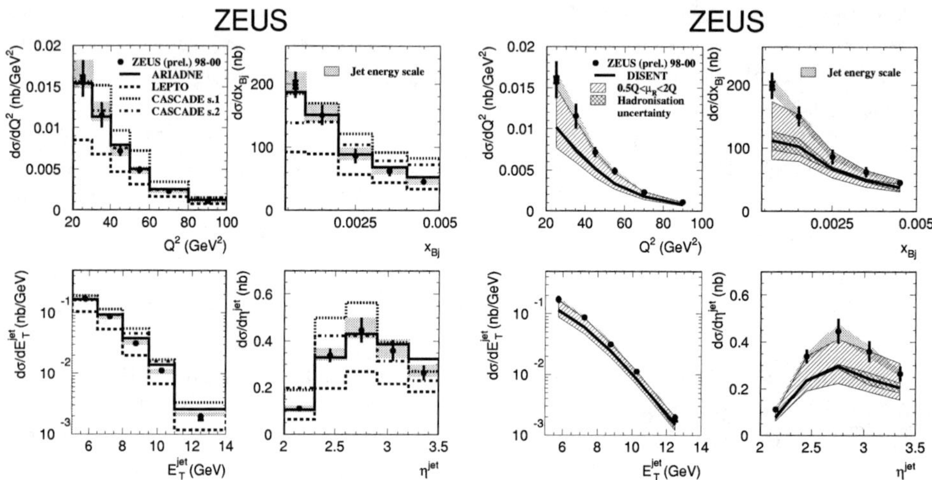

FIGURE 2. Differential cross sections for inclusive jet production for the data (dots) compared with ARIADNE (solid histogram), LEPTO (dashed histogram) and CASCADE (dotted and point-dashed histograms) predictions (left side), and compared with the NLO QCD calculations (solid line, right side). The cross sections are shown as a function of Q^2, x, $E_{T,jet}$, and η^{jet}.

CONCLUSIONS

Two ZEUS measurements of differential inclusive jet cross sections were performed in the forward region, in Q^2, x, $E_T^{\rm jet}$ and $\eta^{\rm jet}$. The ARIADNE MC model gives the best overall description of the cross sections. CASCADE with J2003 set 2 reproduces the normalisation of the measured cross sections, but fails to describe the shapes of the differential cross sections in x and $\eta^{\rm jet}$. These experimental results may be used to adjust the parameters of the intrinsic k_T distribution. The perturbative QCD calculations are consistent with the data within the large theoretical uncertainties, which prevent us to draw any strong conclusion.

REFERENCES

1. ZEUS Coll., S. Chekanov et al., Report DESY-05-117, DESY, 2005.
2. A. Bamberger et al., Nucl. Inst. Meth. **A 450**, 235 (2000).

QCD corrections to the electroproduction of hadrons with high p_T

A. Daleo*, D. de Florian† and R. Sassot†

*Institut für Theoretische Physik, Universität Zürich, Winterthurerstrasse 190, CH-8057 Zürich, Switzerland
†Departamento de Física, Universidad de Buenos Aires, Ciudad Universitaria, Pab.1 (1428) Buenos Aires, Argentina

Abstract. We compute the order α_s^2 corrections to the one particle inclusive electroproduction cross section of hadrons with non vanishing transverse momentum. We compare our results with H1 data on forward production of π^0, and conclude that the data is well described by the DGLAP approach. within the theoretical uncertainties.

Keywords: Semi-Inclusive DIS; perturbative QCD.
PACS: 12.38.Bx,13.85.Ni

INTRODUCTION

The precise measurement of final state hadrons in lepton nucleon deep inelastic scattering constitutes an excellent benchmark for different features of perturbative quantum chromodynamics. Among them, the calculation of higher order corrections, which have been explored and validated for most processes up to next to leading order (NLO) accuracy. For the one particle inclusive processes only very recently there has been progress beyond the leading order (LO) [1, 2, 3, 4, 5]. and up to now there were no analytic computation of the $\mathcal{O}(\alpha_s^2)$ corrections for the electroproduction of hadrons with non vanishing transverse momentum. The analytic computation of the $\mathcal{O}(\alpha_s^2)$ corrections allows us to check factorization in a direct way, which means that collinear singularities showing up in the partonic cross section factorize into parton densities (PDFs) and fragmentation functions (FFs). As a consequence of this explicit cancellation, the resulting cross section is finite and can be straightforwardly convoluted with PDFs and FFs in a fast and stable numerical codes. The analytical result is still sufficiently exclusive and keeps the dependence on the rapidity and the transverse momentum of the produced hadron, allowing a detailed comparison with the experimental data. In the following we summarize the results obtained in Ref. [6]

$\mathcal{O}(\alpha_S^2)$ QCD CORRECTIONS

We consider the process

$$l(l) + P(P) \longrightarrow l'(l') + h(P_h) + X, \tag{1}$$

CP792, *Deep Inelastic Scattering, DIS 2005,* edited by W. H. Smith and S. R. Dasu

where a lepton of momentum l scatters off a nucleon of momentum P with a lepton of momentum l' and a hadron h of momentum P_h tagged in the final state. Omitting target fragmentation at zero transverse momentum, which has been discussed at length in [1, 2], the cross section for this process can be written as

$$\frac{d\sigma^h}{dx_B\, dQ^2} = \sum_{i,j,n} \int_0^1 d\xi \int_0^1 d\zeta \int \mathrm{dPS}^{(n)} \left[f_i(\xi) D_{h/j}(\zeta) \frac{d\sigma_{ij}^{(n)}}{dx_B\, dQ^2\, \mathrm{dPS}^{(n)}} \right] \quad (2)$$

where $\sigma_{ij}^{(n)}$ is the partonic level cross section corresponding to the process and is calculated order by order in perturbation theory through the related parton-photon squared matrix elements $\overline{H}_{\mu\nu}^{(n)}(i,j)$ for the $i+\gamma \to j+X$ processes

$$\frac{d\sigma_{ij}^{(n)}}{dx_B\, dQ^2\, dy\, dz} = \frac{\alpha_{em}^2}{e^2} \frac{1}{\xi\, x_B^2\, S_H^2} \left(Y_M(-g^{\mu\nu}) + Y_L \frac{4x_B^2}{Q^2} P^\mu P^\nu \right) \sum_n \overline{H}_{\mu\nu}^{(n)}(i,j)\,. \quad (3)$$

in terms of the standard kinematical variables [6].

At order-α_s^2, the partonic cross sections receive contributions from the following reactions:

$$\begin{array}{ll} \text{Real contributions} & \left\{ \begin{array}{lcl} \gamma+q(\bar{q}) & \to & g+g+q(\bar{q}) \\ \gamma+q_i(\bar{q}_i) & \to & q_i(\bar{q}_i)+q_j+\bar{q}_j \;\; (i \neq j) \\ \gamma+q_i(\bar{q}_i) & \to & q_i(\bar{q}_i)+q_i+\bar{q}_i \\ \gamma+g & \to & g+q+\bar{q} \end{array} \right. \\ \text{Virtual contributions} & \left\{ \begin{array}{lcl} \gamma+q(\bar{q}) & \to & g+q(\bar{q}) \\ \gamma+g & \to & q+\bar{q} \end{array} \right. \end{array} \quad (4)$$

where any of the outgoing partons can fragment into the final state hadron h.

At variance with the $|p_T| = 0$ case, where the integration over final states leads to overlapping singularities along various curves in the residual phase space, here the only remaining singularities are found at $z = 0$ and thus they can be dealt with the standard method. After combining real and virtual contributions to a given partonic process, the cross section can be written as

$$\frac{d\sigma_{ij}^{(2)}}{dx_B\, dQ^2\, dy\, dz} = \frac{c_q C_\varepsilon^2}{\xi\, x_B^2\, S_H^2} \left\{ \frac{1}{\varepsilon} \mathscr{P}_{1ij}^{(2)}(\rho,y,z) + C_{ij}^{(2)}(\rho,y,z) + \mathscr{O}(\varepsilon) \right\}, \quad (5)$$

where the coefficient of the single poles, $\mathscr{P}_{1ij}^{(2)}(\rho,y,z)$, as well as the finite contributions $C_{ij}^{(2)}(\rho,y,z)$, include 'delta' and 'plus' distributions in z. The IR double poles present in the individual real and virtual contributions cancel out in the sum, providing the first straightforward check on the angular integration of real amplitudes and the loop integrals in the virtual case. In the real terms, the above mentioned double poles come from the product of a pole arising in the integration over the spectators phase space and a single pole coming from the expansion of $z^{-1+\varepsilon}$ factors. Double poles in the virtual contributions always arise from loop integrals.

The remaining singularities, contributing to the single pole, are of UV and collinear origin. The former are removed by means of coupling constant renormalization, whereas the latter have to be factorized in the redefinition of parton densities and fragmentation functions.

PHENOMENOLOGY

In Figure 1 we show the LO and NLO predictions for the electroproduction of neutral pions as a function of x_B and p_T, respectively, in the kinematical range of the H1 experiment [7], together with the most recent data for the range $p_T \geq 3.5$ GeV. The cross sections are computed as described in the previous sections, applying H1 cuts and using MRST02 parton densities [8]. Similar results are found using other sets of modern PDFs. For the input fragmentation functions, we use two different sets, the ones from reference [9] denoted as KKP and those from [10] referenced as K. We set the renormalization and factorization scales as the average between Q^2 and p_T^2, and we compute α_s at NLO(LO) fixing Λ_{QCD} as in the MRST analysis. The plots clearly show

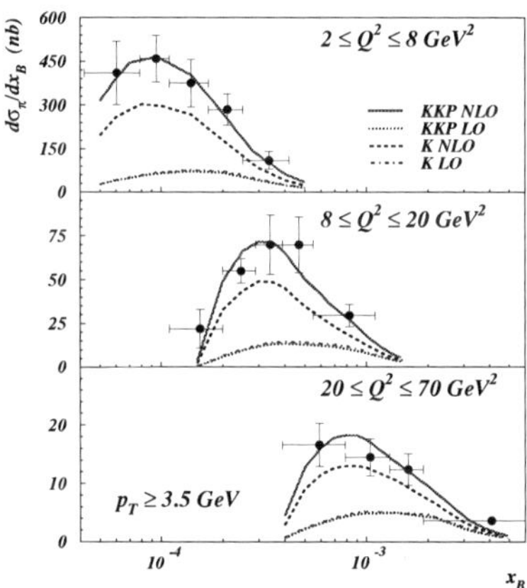

FIGURE 1. LO and NLO cross sections, including experimental cuts as explained in the text, as a function of x_B. H1 data [7] for the range $p_T \geq 3.5$ GeV are also shown.

that the NLO cross sections are much larger than the LO ones, even by the required order of magnitude in certain kinematical regions. Another interesting feature is that the uncertainty due to the choice of a fragmentation functions set is also quite noticeable, this fact driven by the different gluon content of the two sets considered here. Low Q^2 bins seem to prefer KKP set, which have a larger gluon-fragmentation content, whereas for larger Q^2 both sets agree with the data within errors. LO estimates show a much smaller sensitivity on the choice of fragmentation functions, since gluon fragmentation does not contribute significantly to the cross section at this order.

The rather large size of the K-factor can, then, be understood as a consequence of the opening of a new dominant ('leading-order') channel, and not to the 'genuine' increase in the partonic cross section that might otherwise threaten perturbative stability. The dominance of the new channel is due to the size of the gluon distribution at small x_B and to the fact that the H1 selection cuts highlight the kinematical region dominated by the $\gamma + g \rightarrow g + q + \bar{q}$ partonic process.

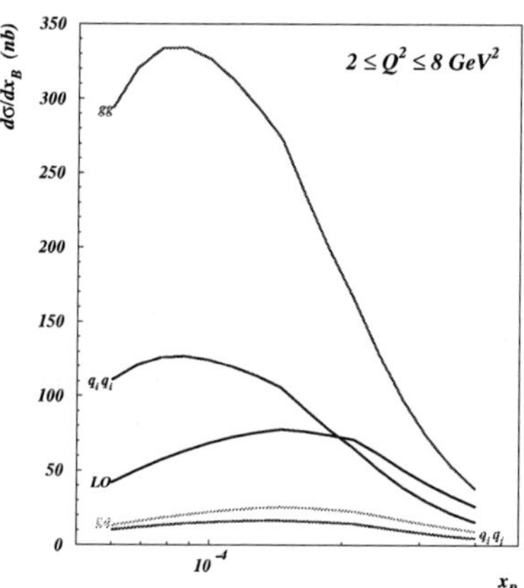

FIGURE 2. Contributions to the cross section for the lowest Q^2 bin of Figure 1

In Figure 2 we show the different contributions to the cross section discriminated by the underlying partonic process. Notice that at very small x_B the gg term can be by itself several times larger than the LO contribution, remaining larger or comparable even for higher x_B values. The forward selection is also responsible of the scale sensitivity of the cross section, as it supresses large components with small scale dependence whereas it stresses components as gg whose scale dependence would be partly canceled only at NNLO

ACKNOWLEDGMENTS

Partially supported by CONICET, Antorchas, UBACYT and ANPCyT, Argentina, and the Swiss National Science Foundation (SNF) through grant No. 200021-101874.

REFERENCES

1. A. Daleo, C. García Canal, R. Sassot, *Nucl. Phys.B* **662** 334 (2003).
2. A. Daleo, R. Sassot, *Nucl. Phys.B* **673** 357 (2003).
3. P. Aurenche, R. Basu, M. Fontannaz and R. M. Godbole, *Eur. Phys. J. C* **34** 277 (2004).
4. M. Fontannaz,arXiv:hep-ph/0410021.
5. M. Maniatis,arXiv:hep-ph/0403002.
6. A. Daleo, D. de Florian and R. Sassot, *Phys. Rev. D* **71**, 034013 (2005).
7. A. Aktas *et al.* [H1 Collaboration], *Eur. Phys. J. C* **36** 441 (2004). [arXiv:hep-ex/0404009].
8. A. D. Martin, R. G. Roberts, W. J. Stirling and R. S. Thorne, *Eur. Phys. J. C* **28** 455 (2003).
9. B. A. Kniehl, G. Kramer and B. Potter, *Nucl. Phys. B* **582** 514 (2000).
10. S. Kretzer, *Phys. Rev.D* **62** 054001 (2000). [arXiv:hep-ph/0003177].

Inclusive electroproduction of light hadrons with large p_T at next-to-leading order

Bernd A. Kniehl

II. Institut für Theoretische Physik, Universität Hamburg, Luruper Chaussee 149, 22761 Hamburg, Germany

Abstract. We review recent results on the inclusive electroproduction of light hadrons at next-to-leading order in the parton model of quantum chromodynamics implemented with fragmentation functions and present updated predictions for HERA experiments based on the new AKK set.

Keywords: Quantum chromodynamics, parton model, radiative corrections, inclusive hadron production, deep-inelastic scattering
PACS: 12.38.Bx, 12.39.St, 13.87.Fh, 14.40.Aq

1. INTRODUCTION

In the framework of the parton model of quantum chromodynamics (QCD), the inclusive production of single hadrons is described by means of fragmentation functions (FFs) $D_a^h(x,\mu)$. At lowest order (LO), the value of $D_a^h(x,\mu)$ corresponds to the probability for the parton a produced at short distance $1/\mu$ to form a jet that includes the hadron h carrying the fraction x of the longitudinal momentum of a. Analogously, incoming hadrons and resolved photons are represented by (non-perturbative) parton density functions (PDFs) $F_{a/h}(x,\mu)$. Unfortunately, it is not yet possible to calculate the FFs from first principles, in particular for hadrons with masses smaller than or comparable to the asymptotic scale parameter Λ. However, given their x dependence at some energy scale μ, the evolution with μ may be computed perturbatively in QCD using the timelike Dokshitzer-Gribov-Lipatov-Altarelli-Parisi (DGLAP) equations. Moreover, the factorization theorem guarantees that the $D_a^h(x,\mu)$ functions are independent of the process in which they have been determined and represent a universal property of h. This entitles us to transfer information on how a hadronizes to h in a well-defined quantitative way from e^+e^- annihilation, where the measurements are usually most precise, to other kinds of experiments, such as photo-, lepto-, and hadroproduction. Recently, FFs for light charged hadrons with complete quark flavour separation were determined through a global fit to e^+e^- data from LEP, PEP, and SLC [1] thereby improving a previous analysis [2].

The QCD-improved parton model should be particularly well applicable to the inclusive production of light hadrons carrying large transverse momenta (p_T) in deep-inelastic lepton-hadron scattering (DIS) with large photon virtuality (Q^2) due to the presence of two hard mass scales, with $Q^2, p_T^2 \gg \Lambda^2$. In Fig. 1, this process is represented in the parton-model picture. The hard-scattering (HS) cross sections, which include colored quarks and/or gluons in the initial and final states, are computed in perturbative QCD. They were evaluated at LO more than 25 years ago [3]. Recently, the next-to-

CP792, *Deep Inelastic Scattering, DIS 2005,* edited by W. H. Smith and S. R. Dasu

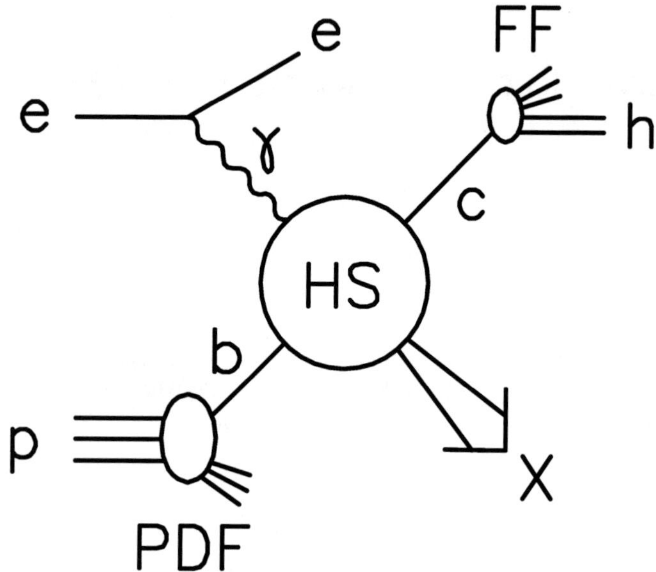

FIGURE 1. Parton-model representation of $ep \to eh + X$.

leading-order (NLO) analysis was performed independently by three groups [4, 5, 6]. A comparison between Refs. [5, 6] using identical input yielded agreement within the numerical accuracy.

The cross section of $e^+p \to e^+\pi^0 + X$ in DIS was measured in various distributions with high precision by the H1 Collaboration at HERA in the forward region, close to the proton remnant [7, 8]. This measurement reaches down to rather low values of Bjorken's variable $x_B = Q^2/(2P\cdot q)$, where P and q are the proton and virtual-photon four-momenta, respectively, and $Q^2 = -q^2$, so that the validity of the DGLAP evolution might be challenged by Balitsky-Fadin-Kuraev-Lipatov (BFKL) dynamics.

In Ref. [5], the H1 data [7, 8] were compared with NLO predictions evaluated with the KKP FFs [2]. In Section 2, we present an update of this comparison based on the new AKK FFs [1]. Our conclusions are summarized in Section 3.

2. COMPARISON WITH H1 DATA

We work in the modified minimal-subtraction ($\overline{\text{MS}}$) renormalization and factorization scheme with $n_f = 5$ massless quark flavors and identify the renormalization and factorization scales by choosing $\mu^2 = \xi[Q^2 + (p_T^*)^2]/2$, where the asterisk labels quantities in the $\gamma^* p$ center-of-mass (c.m.) frame and ξ is varied between 1/2 and 2 about the default value 1 to estimate the theoretical uncertainty. At NLO (LO), we employ set CTEQ6M (CTEQ6L1) of proton PDFs [9], the NLO (LO) set of AKK FFs [1], and the two-loop (one-loop) formula for the strong-coupling constant $\alpha_s^{(n_f)}(\mu)$ with $\Lambda^{(5)} = 226$ MeV

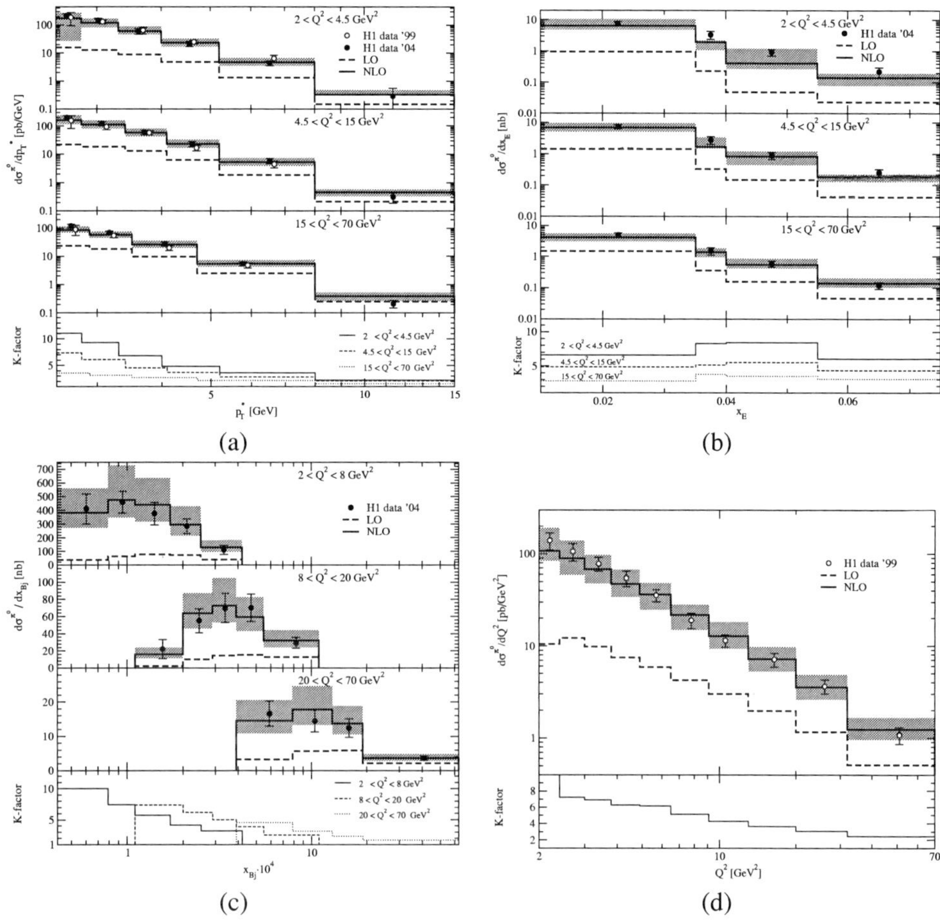

FIGURE 2. H1 data on (a) $d\sigma/dp_T^*$, (b) $d\sigma/dx_E$, and (c) $d\sigma/dx_B$ for $2 < Q^2 < 4.5$ GeV^2, $4.5 < Q^2 < 15$ GeV^2, or $15 < Q^2 < 70$ GeV^2, and on (d) $d\sigma/dQ^2$ from Refs. [7] (open circles) and [8] (solid circles) are compared with our default LO (dashed histograms) and NLO (solid histograms) predictions including theoretical uncertainties (shaded bands). The QCD-correction (K) factors are also shown.

(165 MeV) [9].

The H1 data [7, 8] were taken in DIS of positrons with energy $E_e = 27.6$ GeV on protons with energy $E_p = 820$ GeV in the laboratory frame, yielding a c.m. energy of $\sqrt{S} = 2\sqrt{E_e E_p} = 301$ GeV. The DIS phase space was restricted to $0.1 < y < 0.6$ and $2 < Q^2 < 70$ GeV^2, where $y = Q^2/(x_B S)$. The π^0 mesons were detected within the acceptance cuts $p_T^* > 2.5$ GeV, $5^\circ < \theta < 25^\circ$, and $x_E > 0.01$, where θ is their angle with respect to the proton flight direction and $E = x_E E_p$ is their energy in the laboratory frame. The comparisons with the our updated LO and NLO predictions are displayed in Figs. 2(a)–(d).

3. CONCLUSIONS

We calculated the cross section of $ep \to e\pi^0 + X$ in DIS for finite values of p_T^* at LO and NLO in the parton model of QCD [5] using the new AKK FFs [1] and compared it with a precise measurement by the H1 Collaboration at HERA [7, 8].

We found that our LO predictions always significantly fell short of the H1 data and often exhibited deviating shapes. However, the situation dramatically improved as we proceeded to NLO, where our default predictions, endowed with theoretical uncertainties estimated by moderate unphysical-scale variations, led to a satisfactory description of the H1 data in the preponderant part of the accessed phase space. In other words, we encountered K factors much in excess of unity, except towards the regime of asymptotic freedom characterized by large values of p_T^* and/or Q^2. This was unavoidably accompanied by considerable theoretical uncertainties. Both features suggest that a reliable interpretation of the H1 data within the QCD-improved parton model ultimately necessitates a full next-to-next-to-leading-order analysis, which is presently out of reach, however. For the time being, we conclude that the successful comparison of the H1 data with our NLO predictions provides a useful test of the universality and the scaling violations of the FFs, which are guaranteed by the factorization theorem and are ruled by the DGLAP evolution equations, respectively.

Significant deviations between the H1 data and our NLO predictions only occurred in certain corners of phase space, namely in the photoproduction limit $Q^2 \to 0$, where resolved virtual photons are expected to contribute, and in the limit $\eta \to \infty$ of the pseudorapidity $\eta = -\ln[\tan(\theta/2)]$, where fracture functions are supposed to enter the stage. Both refinements were not included in our analysis. Interestingly, distinctive deviations could not be observed towards the lowest x_B values probed, which indicates that the realm of BFKL dynamics has not actually been accessed yet.

ACKNOWLEDGMENTS

The author thanks G. Kramer and M. Maniatis for their collaboration. This work was supported in part by BMBF Grant No. 05 HT1GUA/4.

REFERENCES

1. S. Albino, B. A. Kniehl, and G. Kramer, Report No. DESY 05-022 and hep-ph/0502188, *Nucl. Phys. B* (in press).
2. B. A. Kniehl, G. Kramer, and B. Pötter, *Nucl. Phys. B* **582**, 514 (2000); *Phys. Rev. Lett.* **85**, 5288 (2000); *Nucl. Phys. B* **597**, 337 (2001).
3. A. Mendez, *Nucl. Phys. B* **145**, 199 (1978).
4. P. Aurenche, R. Basu, M. Fontannaz, and R. M. Godbole, *Eur. Phys. J. C* **34**, 277 (2004).
5. B. A. Kniehl, G. Kramer, and M. Maniatis, *Nucl. Phys. B* **711**, 345 (2005); **720**, 231(E) (2005).
6. A. Daleo, D. de Florian, and R. Sassot, *Phys. Rev. D* **71**, 034013 (2005); R. Sassot, in these proceedings.
7. H1 Collaboration, C. Adloff et al., *Phys. Lett. B* **462**, 440 (1999).
8. H1 Collaboration, A. Aktas et al., *Eur. Phys. J. C* **36**, 441 (2004).
9. J. Pumplin, D. R. Stump, J. Huston, H.-L. Lai, P. Nadolsky, and W.-K. Tung, *JHEP* **0207**, 012 (2002).

Small-x effects in forward-jet production at HERA

Cyrille Marquet

Service de physique théorique, CEA/Saclay, 91191 Gif-sur-Yvette cedex, France
URA 2306, unité de recherche associée au CNRS

Abstract. We investigate small$-x$ effects in forward-jet production at HERA in the two-hard-scale region $k_T \sim Q \gg \Lambda_{QCD}$. We show that, despite describing different energy regimes, both a BFKL parametrization and saturation parametrizations describe well the H1 and ZEUS data for $d\sigma/dx$ published a few years ago. This is confirmed when comparing the predictions to the latest data.

INTRODUCTION

Forward-jet production is a process in which a virtual photon strongly interacts with a proton and a jet is detected in the forward direction of the proton. The virtuality of the photon Q^2 and the squared transverse momentum of the jet k_T^2 are hard scales of about the same magnitude. In the Regge limit of perturbative QCD, *i.e.* when the centre-of-mass energy in a collision is much bigger than the fixed hard scales of the problem, the scattering amplitudes grow with increasing energy as described by the BFKL equation [1]. The forward-jet measurement was originally proposed [2] to test the BFKL equation because, if the energy in the photon-proton collision W is large enough, it lies in the kinematic regime corresponding to the Regge limit ($W^2 \gg Q^2$).

The question is whether the BFKL equation is relevant at the present energies, or if usual perturbative QCD in the Bjorken limit is still sufficient to describe the data. We adress that problem by computing the forward-jet cross-section in the high-energy regime and by comparing the BFKL predictions with the available data. We also adress the problem of saturation [3]: it is well-known that the BFKL growth is damped by saturation effects when energies become too high and the scattering amplitudes approach the unitatity limit. We implement saturation effects in a very simple way, inspired by the Golec-Biernat and Wüsthoff approach [4] and check the consistency with the data.

FORMULATION

The QCD cross-section for forward-jet production in a lepton-proton collision reads

$$\frac{d^{(4)}\sigma}{dx dQ^2 dx_J dk_T^2} = \frac{\alpha_{em}}{\pi x Q^2}\left\{\left(\frac{d\sigma_T^{\gamma * p\rightarrow JX}}{dx_J dk_T^2} + \frac{d\sigma_L^{\gamma * p\rightarrow JX}}{dx_J dk_T^2}\right)(1-y) + \frac{d\sigma_T^{\gamma * p\rightarrow JX}}{dx_J dk_T^2}\frac{y^2}{2}\right\}, \quad (1)$$

where x and y are the usual kinematic variables of deep inelastic scattering and Q^2 is the virtuality of the intermediate photon that undergoes the hadronic interaction.

CP792, *Deep Inelastic Scattering, DIS 2005*, edited by W. H. Smith and S. R. Dasu

$d\sigma_{T,L}^{\gamma * p \to JX}/dx_J dk_T^2$ is the cross-section for forward-jet production in the collision of this transversally (T) or longitudinally (L) polarized virtual photon with the target proton. k_T is the jet transverse momentum and x_J its longitudinal momentum fraction with respect to the proton.

Let us now consider the high-energy regime: $x = \log(Q^2/(Q^2+W^2)) \ll 1$. In an appropriate frame called the dipole frame, the virtual photon undergoes the hadronic interaction via a fluctuation into a colorless $q\bar{q}$ pair, a dipole. The squared wavefunctions ϕ_T^γ and ϕ_L^γ describing the splitting of the virtual photon onto a dipole are well-known. The dipole then interacts with the target proton and one has the following factorization

$$\frac{d\sigma_{T,L}^{\gamma * p \to JX}}{dx_J dk_T^2} = \int_0^\infty 2\pi r dr\, \phi_{T,L}^\gamma(r,Q) \frac{d\sigma_{q\bar{q}}}{dx_J dk_T^2}(r)\,. \tag{2}$$

$d\sigma_{q\bar{q}}(r)/dx_J dk_T^2$ is the cross-section for forward-jet production in the dipole-proton collision. The integration variable r represents the size of the intermediate dipole.

It was shown in [5] that the emission of the forward jet can be described through the interaction of an effective gluonic (gg) dipole:

$$\frac{d\sigma_{q\bar{q}}}{dx_J dk_T^2}(r) = \frac{\pi N_c}{16 k_T^2} f_{eff}(x_J,k_T^2) \int_0^\infty d\bar{r}\, J_0(k_T \bar{r}) \frac{\partial}{\partial \bar{r}}\left(\bar{r}\frac{\partial}{\partial \bar{r}}\sigma_{(q\bar{q})(gg)}(r,\bar{r},Y)\right) \tag{3}$$

with $Y = \log(x_J/x)$ the rapidity assumed very large. $\sigma_{(q\bar{q})(gg)}(r,\bar{r},Y)$ is the $q\bar{q}$ dipole (size r)-gg dipole (size $\bar{r}$) total cross-section with rapidity Y. As usual, the dipoles emerge as the effective degrees of freedom at high energies: $\sigma_{(q\bar{q})(gg)}$contains any number of gluon exchanges and therefore this formulation goes beyond k_T-factorization which assumes only a two-gluon exchange. The effective parton distribution function f_{eff} is given by: $f_{eff}(x_J,k_T^2) = g(x_J,k_T^2) + C_F\left(q(x_J,k_T^2) + \bar{q}(x_J,k_T^2)\right)/N_c$ where g (resp. q, $\bar{q}$) is the gluon (resp. quark, antiquark) distribution function in the incident proton.

BFKL parametrization

The BFKL $q\bar{q}-$dipole $gg-$dipole cross-section reads

$$\sigma_{(q\bar{q})(gg)}^{BFKL}(r,\bar{r},Y) = 2\pi\alpha_s^2 r^2 \int \frac{d\gamma}{2i\pi}\left(\frac{\bar{r}}{r}\right)^{2\gamma} \frac{\exp\left(\frac{\alpha_s N_c}{\pi}\chi(\gamma)Y\right)}{\gamma^2(1-\gamma)^2} \tag{4}$$

with the complex integral running along the imaginary axis from $1/2-i\infty$ to $1/2+i\infty$ and with the BFKL kernel given by $\chi(\gamma) = 2\psi(1)-\psi(1-\gamma)-\psi(\gamma)$ where $\psi(\gamma)$ is the logarithmic derivative of the Gamma function. It comes about when the interaction between the $q\bar{q}-$dipole and the $gg-$dipole is restricted to a two-gluon exchange. One can easily show, putting (4) in (2) and (3), that this formulation is equivalent to using k_T-factorization. We are going to perform a fit of the parametrization (4) to the data. The parameters are $\lambda = 4\alpha_s N_c \log(2)/\pi$ and a normalization.

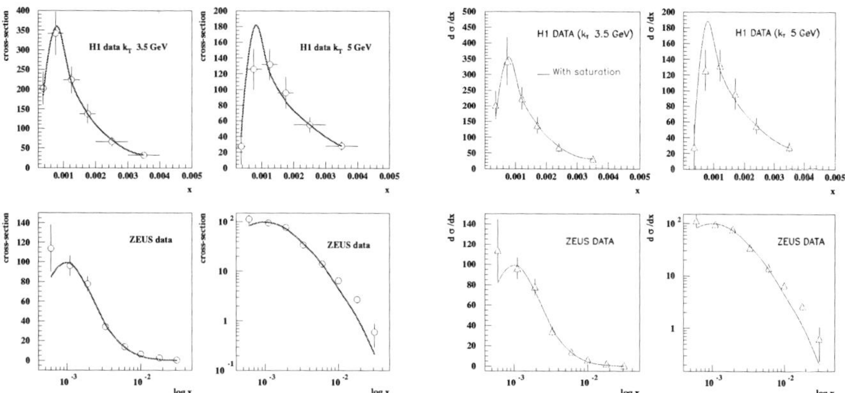

FIGURE 1. Fits to the H1 and ZEUS forward-jet old data for $d\sigma/dx$. The left plot shows the BFKL fit and the right plot shows one of the saturation fits (called sat. in the text).

Saturation parametrization

To take into account saturation effects, let us consider the following parametrization:

$$\sigma^{sat}_{(q\bar{q})(gg)}(r,\bar{r},Y) = 4\pi\alpha_s^2\sigma_0\left(1-\exp\left(-\frac{r_{\rm eff}^2(r,\bar{r})}{4R_0^2(Y)}\right)\right). \tag{5}$$

The dipole-dipole *effective* radius $r_{\rm eff}^2(r,\bar{r})$ is defined through the two-gluon exchange:

$$4\pi\alpha_s^2 r_{\rm eff}^2(r,\bar{r}) \equiv \sigma^{BFKL}_{(q\bar{q})(gg)}(r,\bar{r},0) = 4\pi\alpha_s^2\min(r^2,\bar{r}^2)\left\{1+\log\frac{\max(r,\bar{r})}{\min(r,\bar{r})}\right\} \tag{6}$$

while the saturation radius is parametrized by $R_0(Y)=e^{-\frac{\lambda}{2}(Y-Y_0)}/Q_0$ with $Q_0\equiv 1\ GeV$. The parameters for the fit are λ, Y_0 and the normalization σ_0.

PHENOMENOLOGY

To compare the cross-section (1) with the data for $d\sigma/dx$, the three remaining integration are carried out taking into account the different sets of cuts provided by the different experiments. Fits have been performed to the old sets of data [6, 7] for the BFKL parametrization [8] and the saturation parametrization [9]. With all χ^2 values of about 1, the BFKL fit gives $\lambda=0.430$ and the saturation fit shows two χ^2 minima for $(\lambda=0.402, Y_0=-0.82)$ (sat.) and $(\lambda=0.370, Y_0=8.23)$ (weak sat.). The plots are shown on Fig1. Despite describing different energy regimes, both a BFKL parametrization and saturation parametrizations describe well the data. The first saturation minima corresponds to a strong saturation effect as, for typical values of Y, the saturation scale $1/R_0$ is 5 Gev which is the value of a typical k_T. The second saturation minima corresponds to small saturation effets and rather describes BFKL physics.

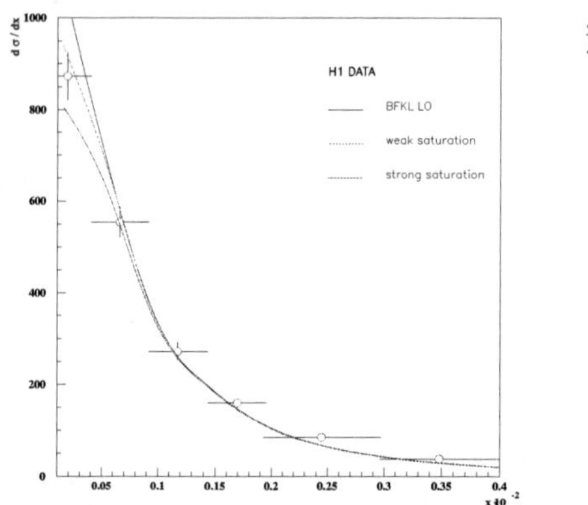

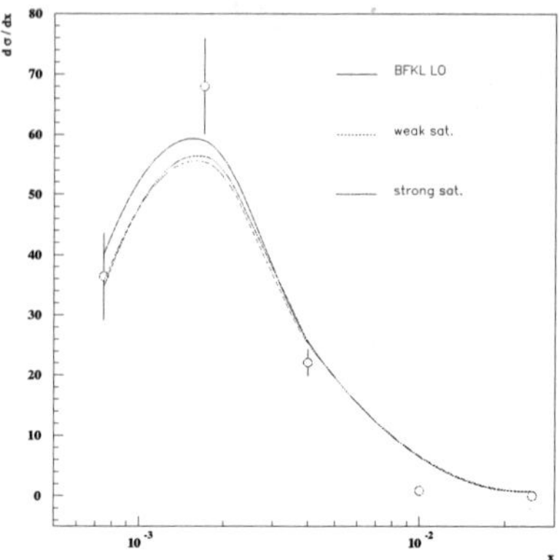

FIGURE 2. Comparisons between the H1 (left plot) and ZEUS (right plot) forward-jet new data for $d\sigma/dx$ and the BFKL and saturation parametrizations.

Let us now look at the new data [10, 11] for $d\sigma/dx$ which go to lower x. Without performing any new fit of the parameters, but rather by taking the values already obtained, the three parametrizations describe very well the new data, as shown on Fig2. One cannot really distinguish between the three curves even if at small values of x, one starts to see the difference between them. At the lowest values of x, NLOQCD predictions are about a factor 1.5 to 2.5 below the data depending on the experiment and the error bars. However, it could be that adding a resolved-photon component to the NLO predictions will pull them within the uncertainties which moderates the conclusion that the BFKL resummation is needed to describe those data. The fact that two saturation parametrizations are consistent with the data also asks for further study.

We intend to complete our analysis [12] by considering the other measurements $d\sigma/dQ^2$ and $d\sigma/dk_T$ by ZEUS and $d\sigma/dxdQ^2dk_T^2$, by H1 and Mueller-Navelet jets [13] at Tevatron or LHC. These could help clarifing the situation [14].

REFERENCES

1. L. N. Lipatov, *Sov. J. Nucl. Phys.* **23**, (1976) 338; E. A. Kuraev, L. N. Lipatov and V. S. Fadin, *Sov. Phys. JETP* **45**, (1977) 199; I. I. Balitsky and L. N. Lipatov, *Sov. J. Nucl. Phys.* **28**, (1978) 822.
2. A. H. Mueller, *Nucl. Phys. Proc. Suppl.* **B18C** (1990) 125; *J. Phys.* **G17** (1991) 1443.
3. L. V. Gribov, E. M. Levin and M. G. Ryskin, *Phys. Rep.* **100** (1983) 1.
4. K. Golec-Biernat and M. Wüsthoff, *Phys. Rev.* **D59** (1999) 014017; *Phys. Rev.* **D60** (1999) 114023.
5. C. Marquet *Nucl. Phys.* **B705** (2005) 319.
6. H1 Collaboration, C. Adloff *et al*, *Nucl. Phys.* **B538** (1999) 3.
7. ZEUS Collaboration, J. Breitweg *et al*, *Eur. Phys. J.* **C6** (1999) 239.
8. J. G. Contreras, R. Peschanski and C. Royon, *Phys. Rev.* **D62** (2000) 034006.
9. C. Marquet, R. Peschanski and C. Royon, *Phys. Lett.* **B599** (2004) 236.
10. A. Knutsson for the H1 Collaboration, these proceedings.
11. ZEUS Collaboration, hep-ex/0502029; N. Vlasov for the ZEUS collaboration, these proceedings.
12. C. Marquet and C. Royon, in preparation.
13. A. H. Mueller and H. Navelet, *Nucl. Phys.* **B282** (1987) 727.
14. C. Marquet and R. Peschanski, *Phys. Lett.* **B587** (2004) 201; C. Marquet, hep-ph/0406111.

Measurement of prompt photon cross sections in photoproduction at H1

Jozef Ferencei
for the H1 Collaboration

Institute of Experimental Physics SAS, Watsonova 47, 04353 Košice, Slovakia

Abstract. Cross section measurements of isolated prompt photons, inclusively and associated with jets, have been made at the HERA *ep* collider with the H1 detector, using the data taken in the years 1996-2000 corresponding to an integrated luminosity of 105 pb^{-1}. The results are compared to a perturbative QCD calculations in next to leading order and to predictions of the event generators PYTHIA and HERWIG.

Keywords: prompt photons, photoproduction
PACS: 13.60.Hb

INTRODUCTION

An inclusive prompt photon measurement, tagging only the final-state hard photon, may be directly compared with perturbative QCD calculations with no need for a jet definition matching theory and experiment and without hadronisation uncertainties. The photon's momentum reflects the collision kinematics since such photons are produced at the elementary interaction vertex. This contrasts with jet production, where the hadronisation process obscures the measurement of energy and direction of the outgoing parton. Therefore prompt photon events provide an interesting tool to test QCD, in a way which is complementary to jets. Further understanding of the production process may be obtained by detecting the prompt photon together with a jet.

The main experimental difficulty is the separation of the prompt photons from hadronic background, in particular from signals due to π^0 mesons and, to less extent, due to η's when at high energies the two energetic decay photons cannot be resolved in the detector. In order to suppress this background one looks for energetic and isolated photons.

RESULTS

In the H1 analysis [1] an isolation cone was imposed around the photon candidate: within a cone of unit radius in the (η, ϕ) plane, the total transverse energy E_T from other particles was required not to exceed 10% of the transverse energy E_T^γ of the photon. A typical signal is a compact energy deposition in the calorimeter, with no track pointing to it. The prompt photon fractions are determined in a likelihood analysis using shower shape observables for prompt photons, and for photons from π^0 and η decays. In the following the results are presented as bin averaged *ep* cross sections in the kinematic

CP792, *Deep Inelastic Scattering, DIS 2005*, edited by W. H. Smith and S. R. Dasu

region:
$\sqrt{s} = 319\,\text{GeV},\ 0.2 < y < 0.7,\ Q^2 < 1\,\text{GeV}^2,\ 5 < E_T^\gamma < 10\,\text{GeV}$ and $-1 < \eta^\gamma < 0.9$.

Inclusive prompt photons. Differential cross sections for inclusive prompt photons as a function of E_T^γ and pseudorapidity η^γ are shown in Fig. 1 and compared with the predictions of PYTHIA [2] and HERWIG [3].

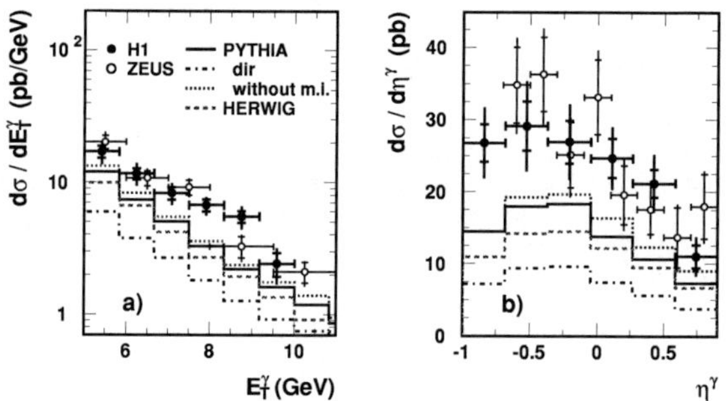

FIGURE 1. The prompt photon cross section measured by H1 [1] and ZEUS [4] are compared with the predictions of HERWIG and PYTHIA including multiple interactions. The contribution of direct interactions (dir) and the full PYTHIA predictions without multiple interactions (without m.i.) are also shown.

The distributions are also compared to the NLO pQCD calculations (Fig.2a,b) by Fontannaz, Guillet and Heinrich (FGH) [5] and Krawczyk and Zembruski (K&Z) [6, 7].

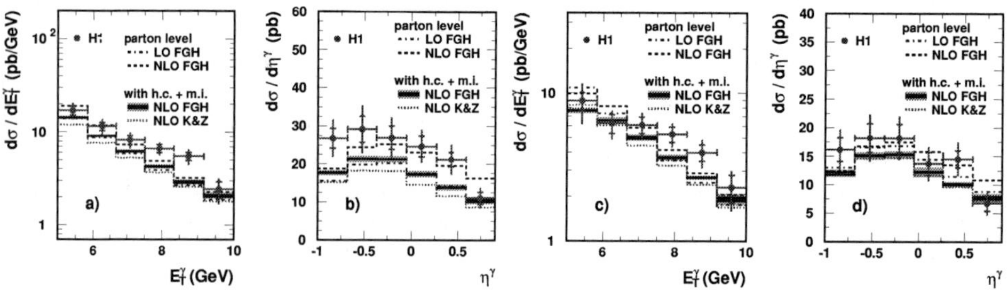

FIGURE 2. Inclusive prompt photon cross sections (a,b) and with an additional jet requirement (c,d) compared with NLO pQCD calculations: FGH [5] and K&Z [7]. The NLO results are corrected for hadronisation and multiple interactions (h.c. + m.i.). The FGH results are also shown without corrections for hadronisation and multiple interactions (parton level).

Prompt photons with jets. Cross sections for the production of a prompt photon associated with a jet ($E_T^{jet} > 4.5$ GeV, $-1 < \eta^{jet} < 2.3$) are presented in Figs. 2c,d as a function of the variables E_T^γ and η^γ and in Fig. 3 as function of $E_T^{jet}, \eta^{jet}, x_\gamma^{LO}$ and x_p^{LO}. The estimators $x_\gamma^{LO} = E_T^\gamma(e^{-\eta^{jet}} + e^{-\eta^\gamma})/2yE_e$ and $x_p^{LO} = E_T^\gamma(e^{\eta^{jet}} + e^{\eta^\gamma})/2E_p$ for the momentum fractions of constituents of the incident photon and proton, respectively, participating in the hard process make explicit use only of the photon energy and are most easily interpreted in leading (LO) order approximation [8].

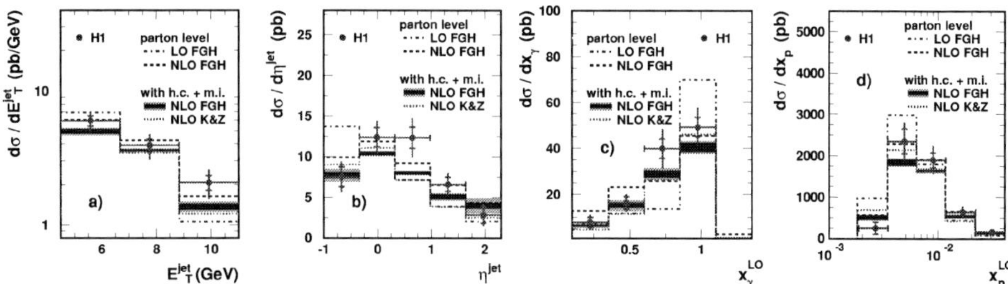

FIGURE 3. Prompt photon cross sections with an additional jet requirement differential in E_T^{jet}, η^{jet}, x_γ^{LO} and x_p^{LO}. The data are compared with NLO pQCD calculations by K&Z [7] and FGH [5, 8]. The NLO results are corrected for hadronisation and multiple interaction (h.c.+ m.i.) effects. The FGH results are also shown without corrections for h.c. and m.i. at NLO and LO.

In the absence of higher order processes and of intrinsic transverse momentum (k_T) of the incoming parton inside the proton, the photon and jet would be balanced in p_T. The observation of p_T imbalance between the photon and the jet thus allows a measurement of the mixed influence of the higher order processes and the intrinsic k_T of partons in the photon [9]. The distribution of the component of the prompt photon's momentum perpendicular to the jet direction in the transverse plane, $p_\perp \equiv | \vec{p}_T^{\,\gamma} \times \vec{p}_T^{\,jet} | / | \vec{p}_T^{\,jet} | = E_T^\gamma \cdot \sin(\Delta\phi)$, where $\Delta\phi$ is the difference in azimuth between the photon and the jet, is therefore sensitive to effects beyond LO. The normalised $p_\perp$ distribution is shown in Fig. 4 separately for the regions $x_\gamma^{LO} > 0.85$ and $x_\gamma^{LO} < 0.85$, where direct and resolved photon induced processes dominate, respectively.

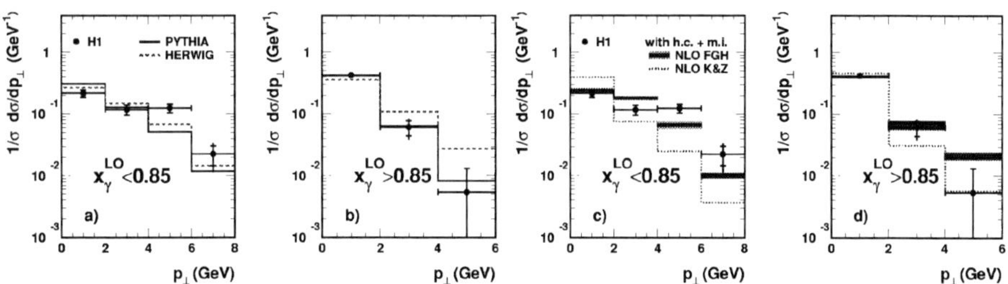

FIGURE 4. Distribution of the prompt photon momentum component, perpendicular to the jet direction in the transverse plane, for $x_\gamma^{LO} < 0.85$, a) and c), and $x_\gamma^{LO} > 0.85$, b) and d). In a) and b) the data are compared with PYTHIA and HERWIG. In c) and d) the data are compared with NLO pQCD calculations by K&Z [7] and FGH [5, 8]. The NLO results are corrected for hadronisation and multiple interactions.

CONCLUSIONS

The production of inclusive prompt photons and associated with jets has been studied in γp interactions. The cross sections predicted by the PYTHIA and HERWIG event generators describe the distributions well in shape but the normalisations are by about $40\% - 50\%$ too low. Multiple interactions tend to reduce the cross section due to the isolation requirement. H1 and ZEUS measurements are consistent within the errors. The η^γ and E_T^γ distributions of the inclusive prompt photons are reasonably well described in shape by NLO pQCD calculations, but after corrections for hadronisation and multiple interactions the predictions are $30\%(40\%)$ below the data for FGH (K&Z) calculations.

The distributions for prompt photons associated with a jet are better described by the NLO calculations including corrections for hadronisation and multiple interactions than the inclusive prompt photon distributions. The NLO corrections are smaller on average than in the inclusive case, which suggests that contributions beyond NLO are less important if an energetic jet is required together with the prompt photon.

PYTHIA describes the normalised $p_\perp$ distributions well. In contrast HERWIG predicts too hard $p_\perp$ distribution at large x_γ^{LO}, where direct photon interactions dominate. At low x_γ^{LO}, the $p_\perp$ distribution is better described by the NLO calculations if NLO QCD corrections are also applied for the resolved photon interactions, as is the case in the FGH calculations.

The large differences between the predictions of the various NLO calculations and Monte Carlo models in the present comparisons preclude a reliable conclusion on the intrinsic k_T of initial state partons in the proton.

ACKNOWLEDGMENTS

Attendance to the conference was supported by DESY Hamburg. I would like to thank Joerg Gayler for a careful reading of the manuscript.

REFERENCES

1. A. Aktas *et al.* [H1 Collaboration], Eur. Phys. J. C **38**, 437 (2005) [hep-ex/0407018].
2. T. Sjöstrand, P. Edén, C. Friberg, L. Lönnblad, G. Miu, S. Mrenna and E. Norrbin, Comput. Phys. Commun. **135** (2001) 238 [hep-ph/0010017].
3. G. Corcella *et al.*, "HERWIG 6.1 release note", hep-ph/9912396;
G. Marchesini *et al.*, Comput. Phys. Commun. **67** (1992) 465.
4. J. Breitweg *et al.* [ZEUS Collaboration], Phys. Lett. B **472** (2000) 175 [hep-ex/9910045].
5. M. Fontannaz, J. P. Guillet and G. Heinrich, Eur. Phys. J. C **21** (2001) 303 [hep-ph/0105121].
6. M. Krawczyk and A. Zembrzuski, Phys. Rev. D **64** (2001) 114017 [hep-ph/0105166].
7. A. Zembrzuski and M. Krawczyk, "Photoproduction of isolated photon and jet at the DESY HERA", IFT-2003-27 [hep-ph/0309308].
8. M. Fontannaz, J. P. Guillet and G. Heinrich, Eur. Phys. J. C **22** (2001) 303 [hep-ph/0107262].
9. S. Chekanov *et al.* [ZEUS Collaboration], Phys. Lett. B **511** (2001) 19 [hep-ex/0104001].

NLO PHOTON PARTON PARAMETRIZATION USING *ee* and *ep* DATA

Halina Abramowicz[1], Aharon Levy* and Wojtek Slominski†

*School of Physics and Astronomy, Raymond and Beverly Sackler Faculty of Exact Sciences, Tel Aviv University, Tel Aviv, Israel
†M. Smoluchowski Institute of Physics, Jagellonian University, Reymonta 4, 30-059, Cracow, Poland

Abstract. An NLO photon parton parametrization is presented based on the existing F_2^γ measurements from e^+e^- data and the low-x proton structure function from ep interactions. Also included in the extraction of the NLO parton distribution functions are the dijets data coming from $\gamma p \to j_1 + j_2 + X$. The new parametrization is compared to other available NLO parametrizations.

Keywords: Photon structure function, QCD, parton distribution functions, jets
PACS: 14.70.Bh, 13.60.Hb, 13.66.Bc, 12.38.Bx

INTRODUCTION

A new parametrization of the parton distributions in the photon is extracted in next-to-leading order (NLO) of perturbative QCD. It differs from other NLO parametrizations [1, 2, 3, 4, 5] in that the data used in the fitting procedure include the expected behaviour of F_2^γ at low-x, as derived from F_2^p measurements [6] under Gribov factorization assumption [7], as suggested in [8] and, in addition, the measurements of the dijet photoproduction cross sections [9].

GRIBOV FACTORIZATION

It was suggested [8] that for low Bjorken x ($x < 0.01$) one can use the relation based on Gribov factorization [7], to find a simple relation between F_2^γ and F_2^p. Gribov factorization relates the total $\gamma\gamma$ cross section to those of γp and pp. For low x one can thus obtain

$$F_2^\gamma(x,Q^2) = F_2^p(x,Q^2)\frac{\sigma_{\gamma p}(W)}{\sigma_{pp}(W)}. \quad (1)$$

Here Q^2 is the virtuality of the probing photon and W is the center of mass energy. Using the parameterization of Donnachie and Landshoff [10], which gives a good representation of the data, one obtains at large W

$$F_2^\gamma/\alpha = 0.43 F_2^p, \quad (2)$$

[1] also at Max Planck Institute, Munich, Germany, Alexander von Humboldt Research Award.

CP792, *Deep Inelastic Scattering, DIS 2005*, edited by W. H. Smith and S. R. Dasu

where α is the electromagnetic coupling constant. In extracting parton distributions in the photon, this last relation allows the use of the precise F_2^p data to constrain the low-x region, where F_2^γ data are very scarce.

THE PARAMETRIZATION

Our parametrization of the initial parton distributions, defined at $Q_0^2 = 2\,\text{GeV}^2$, aims at describing the experimental data below the charm threshold. Thus we explicitly parametrize only the u,d,s quarks and the gluon. The c,b and t quarks are generated radiatively once their respective thresholds are crossed.

All quark distributions in the photon are parametrized as a sum of point-like and hadron-like contributions,

$$f_q(x) = f_{\bar{q}}(x) = e_q^2 A^{\text{PL}} \frac{x^2 + (1-x)^2}{1 - B^{\text{PL}} \ln(1-x)} + f_q^{\text{HAD}}(x) . \tag{3}$$

Apart from the e_q^2 factor, the point-like contribution is the same for all quarks. The hadron-like contribution is assumed to depend on the quark mass only. For u and d quarks we parametrize it as

$$f_u^{\text{HAD}}(x) = f_d^{\text{HAD}}(x) = A^{\text{HAD}} x^{B^{\text{HAD}}} (1-x)^{C^{\text{HAD}}} , \tag{4}$$

and for the s quark we fix it to be

$$f_s^{\text{HAD}}(x) = 0.3 \, f_d^{\text{HAD}}(x) . \tag{5}$$

The gluons in the photon are assumed to have hadron-like behaviour

$$f_G(x) = A_G^{\text{HAD}} x^{B_G^{\text{HAD}}} (1-x)^{C_G^{\text{HAD}}} . \tag{6}$$

As there are no data at x close to 1 we fix $C^{\text{HAD}} = 1$ and $C_G^{\text{HAD}} = 3$ as suggested by counting rules [11, 12] based on dimensional arguments. Thus we are left with 6 free parameters.

THE FIT PROCEDURE AND THE DATA

We use the DIS_γ scheme to relate F_2^γ to the parton densities. We use the zero mass variable-flavor-number-scheme (VFNS) for the DGLAP evolution of heavy flavor parton distribution functions (pdfs). For the heavy quark contribution to F_2^γ we adopt a phenomenological parametrization as a weighted sum of the Bethe-Heitler and pdf contributions [13]. The weights are defined so as to avoid double counting. The following masses of heavy quarks were used: m_c = 1.5 GeV, m_b = 4.5 GeV and m_t = 174 GeV.

For fitting the parameters we used all published data on the photon structure function F_2^γ, from LEP, PETRA and TRISTAN [14]. We also used the Gribov factorization relation in order to produce F_2^γ 'data' at low x from the proton structure function data measured by ZEUS [6]. In addition the dijet photoproduction measurements were taken from the ZEUS experiment [9]. All in all we used 164 points of F_2^γ measurements coming from e^+e^- reactions, 122 proton structure function data points from ep interactions and 24 points of dijet photoproduction reactions.

RESULTS

The fit to the 286 structure function data points gave a value of 1.06 for the χ^2 per degree of freedom. This increased to 1.63 when the additional 24 dijets points were added. Nevertheless, it had only a minor effect on the overall fit results and their errors. The best fit expectations (denoted as the SAL parametrization), using all the 310 data points, are shown in figure 1, where F_2^γ is plotted as a function of x in bins of Q^2. The

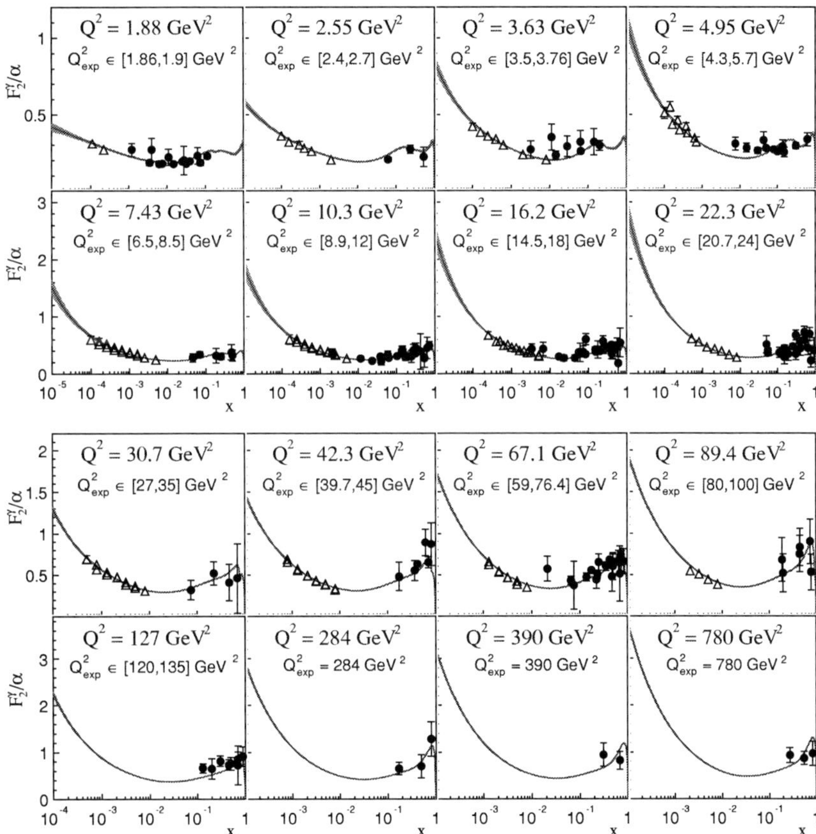

FIGURE 1. The SAL expectations for $F_2^\gamma(x,Q^2)$ as a function of x at selected Q^2 values, as denoted in th fi gure. The plotted data (dots for F_2^γ measured directly and triangles for F_2^γ deduced from F_2^p) are from the range Q^2_{exp} presented in the fi gure

real F_2^γ data and the ones deduced from F_2^p are shown with different symbols. Note that wherever available, the two data sets overlap within errors. To limit the number of plots without loss of information, the data are shown within a range of Q^2, while the corresponding curve is calculated for the average Q^2 of that bin. The shaded error band is calculated according to the final error matrix of the fitted parameters as returned by MINUIT. The uncertainty becomes smaller with increasing Q^2, due to the expected loss of sensitivity to the initial pdf parametrization.

The dijet data gave a poor fit and did not help to constrain the photon pdfs. The main reason is that the data are in a kinematical region where the gluons in the proton dominate and thus may need to be adjusted in order to get a better fit.

PARTON DISTRIBUTIONS

The SAL parton distributions in the photon are shown in figure 2. The features to be

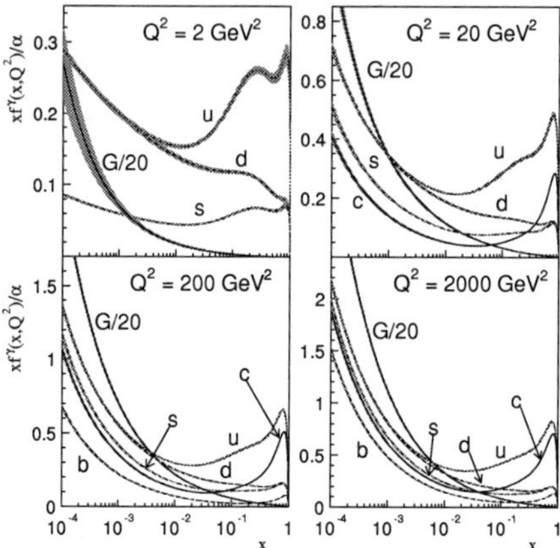

FIGURE 2. Parton distributions in the photon for different values of Q^2, as denoted in the figure.

noted are the behaviour of quarks at large x, typical of the point-like contribution of the photon, and the dominance of the gluon distribution at low x.

The comparison of the SAL pdfs and the other available NLO DIS$_\gamma$ photon parametrizations, GRV [1], GRS[2] [4], and CJK [5], is shown in figure 3 for Q^2 = 2.5 GeV^2. There are big differences between the various pdfs[3]. They are especially pronounced for $x < 10^{-3}$, where no F_2^γ data are available and the result is subject to additional theoretical assumptions. The SAL parametrization has the lowest gluon distribution down to $x \sim 10^{-4}$, below which value we observe a steep rise, steeper than other pdfs. At higher Q^2, where the sensitivity to initial conditions is diminished, there are still noticeable differences [13].

---

[2] This parametrization uses Fixed Flavor Number Scheme (FFNS), where only u, d and s pdfs exist.

[3] A non-vanishing b-quark density at Q^2 = 2.5 GeV^2 is a feature of the CJK parametrization.

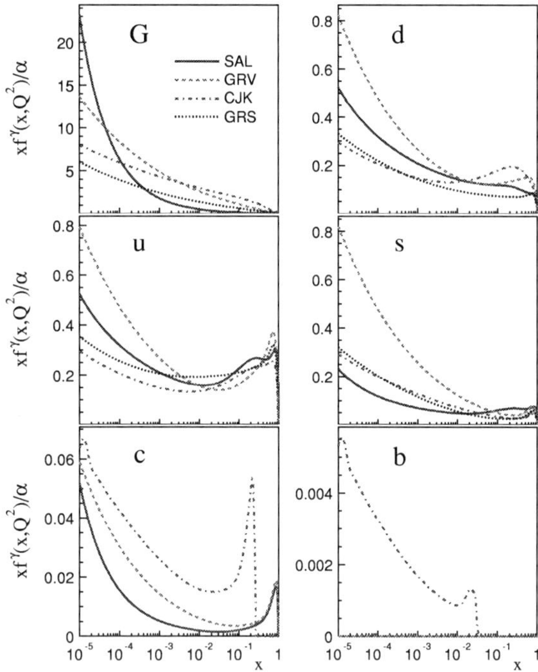

FIGURE 3. Comparison of SAL to other NLO parametrization at $Q^2 = 2.5$ GeV^2.

ACKNOWLEDGMENTS

This work was supported in part by the Israel Science Foundation (ISF).

REFERENCES

1. M. Gluck, E. Reya, A. Vogt, *Phys. Rev.* **D45** (1992) 3986, *Phys. Rev.* **D46** (1992) 1973.
2. P. Aurenche, J.-Ph. Guillet, M. Fontannaz, *Zeit. Phys.* **C64** (1994) 621.
3. L.E. Gordon and J.K. Storrow, *Nucl. Phys.* **B489** (1997) 405.
4. M. Gluck, E. Reya and I. Schienbein, *Phys. Rev.* **D60** (1999) 054019; Erratum *Phys. Rev.* **D62** (2000) 019902.
5. F. Cornet, P. Jankowski and M. Krawczyk, *Phys. Rev.* **D70** (2004) 093004.
6. ZEUS Collaboration, *Eur. Phys. J.* **C7** (1999) 609; *Eur. Phys. J.* **C21** (2001) 443.
7. V. N. Gribov, L. Ya. Pomeranchuk, *Phys. Rev. Lett.* **8** (1962) 343.
8. A. Levy, *Phys. Lett.* **B404** (1997) 369.
9. ZEUS Collaboration, *Eur. Phys. J.* **C23** (2002) 615.
10. A. Donnachie and P. Landshoff, *Phys. Lett.* **B296** (1992) 227.
11. R. Blankenbecler and S. J. Brodsky, *Phys. Rev.* **D10** (1974) 2973.
12. G. R. Farrar and D. R. Jackson, *Phys. Rev. Lett.* **35** (1975) 1416.
13. W. Slominski, H. Abramowicz and A. Levy, *NLO photon parton parametrization using ee and ep data*, arXiv:hep-ph/0504003.
14. See e.g. M. Krawczyk, A. Zembrzuski and M. Staszel, *Phys. Rep.* **345** (2001) 265;

Charged multiplicity distributions in deep inelastic scattering at HERA

Michele Rosin
on behalf of the ZEUS collaboration

DESY-ZEUS Wisconsin, Notestrasse 85, 22607 Hamburg, Germany

Abstract.
The hadronic final state has been investigated in inclusive neutral current deep inelastic *ep* scattering with the ZEUS detector at HERA, using an integrated luminosity of 38.6 pb^{-1}. The mean charged multiplicity has been measured for the hadrons belonging to the current region of the Breit frame, as well as for those belonging to the photon (current) fragmentation region of the hadronic centre-of-mass frame (HCM). The results are compared to leading-logarithm parton-shower Monte Carlo predictions as well as to results of e^+e^- and pp measurements.

Keywords: multiplicity distributions, hadronic final states
PACS: 13.85Hd

INTRODUCTION

The measurements of multiplicities of charged particles at colliders have yielded insights into hadronisation mechanisms. It has been found that charged particle multiplicities measured as a function of the centre-of-mass (cms) energy, $\sqrt{s}$, at e^+e^- colliders are the same as those measured at pp colliders as a function of $\sqrt{q_{\mathrm{tot}}^{\mathrm{had}}} = \sqrt{[(q_1^{\mathrm{inc}} - q_1^{\mathrm{leading}}) + (q_2^{\mathrm{inc}} - q_2^{\mathrm{leading}})]^2}$, where $q_{1,2}^{\mathrm{inc}}$ and $q_{1,2}^{\mathrm{leading}}$ are the four-momenta of the incoming protons and leading particles that escape down the beampipe, respectively [1]. It is therefore interesting to study charged particle multiplicities in *ep* collisions. At the HERA *ep* collider, because of the large asymmetry between the *e*-beam and *p*-beam energies, a large part of the fi nal hadronic system produced near the proton falls outside the region of acceptance of the detectors. Therefore, to perform studies similar to e^+e^- and pp also in *ep* collisions at HERA, only hadrons belonging to the photon (current) fragmentation region of the hadronic centre-of-mass (HCM) frame and to the current region of the Breit frame were used.

DATA SELECTION

The data were collected with the ZEUS detector during the 1996 and 1997 running periods, using protons of energy $E_p = 820$ GeV and positrons of energy $E_e = 27.5$ GeV, and correspond to an integrated luminosity of $38.6 \pm 0.6\ \mathrm{pb}^{-1}$.

The ZEUS detector is described in detail elsewhere [2]. The most important components used in the current analysis are the central tracking detector (CTD) which was

CP792, *Deep Inelastic Scattering, DIS 2005,* edited by W. H. Smith and S. R. Dasu

used to measure the charged-particle multiplicity and the uranium-scintillator calorimeter (CAL) which measured the energy of the final-state hadrons, charged and neutral.

Neutral current deep inelastic scattering (DIS) events were selected in the range $Q^2 > 25$ GeV^2 and $70 < W < 225$ GeV , where Q^2 is the photon virtuality and W is the $\gamma^* P$ centre-of-mass energy. Diffractive contributions were reduced by rejecting events where the pseudorapidity of the most forward energy deposit in the CAL is large. The reconstructed tracks were required to be associated with the primary vertex and have a sufficient p_T to ensure good reconstruction. A cut on the pseudorapidity of the measured tracks restricts the analysis to a region of high CTD acceptance where detector response and systematics are well understood.

In the Breit frame, the exchanged virtual boson is purely spacelike, with three-momentum $\mathbf{q}$ = (0,0,-Q). The particles produced in the interaction can be assigned to one of two regions: the current region if their longitudinal momentum in the Breit frame is negative, and the target region if their logitudinal momentum is positive. The hadronic system of the current region used in this analysis is almost fully (about 95%) contained within the acceptance of the CTD.

For the HCM frame the final particles are separated into the photon (current) and proton (target) fragmenation regions. The target region of the HCM can not be observed in the CTD and only 60-80% of the hadrons belonging to the photon (current) region are detected.

The boost to the corresponding frames was performed using the positron four-momentum as reconstructed using the scattered-positron angle, θ_e, and the angle γ_H, where in the naive quark parton model,γ_H is the angle of the scattered massless quark in the laboratory frame.

MONTE CARLO MODELS, ACCEPTANCE CORRECTIONS AND SYSTEMATIC ERRORS

Samples of neutral current DIS events were generated using the HERACLES 4.6.1 [3, 4] MC program with the DJANGOH 1.1 [5, 6] interface to the hadronisation programs. The QCD cascade is simulated using the colour-dipole model as implemented in ARIADNE 4.08 [7] or with the MEPS model of LEPTO 6.5 [8]. Both ARIADNE and LEPTO use the Lund string model [9] for the hadronisation. All event samples were generated using the CTEQ4D parameterisation of the parton distribution functions in the proton. The generated hadron distributions do not include charged particles produced from weak decays with lifetimes below $3 \cdot 10^{-10}$ seconds. The charged-particle decay products of K^0_S and Λ were excluded.

The MC event samples were passed through reconstruction and selection procedures identical to those for the data. Monte Carlo studies were used to determine the event and track acceptances in the selected kinematic region of Q^2 and W^2.

The corrections applied to the data accounted for the effects of acceptance and resolution of the detector, event selection cuts, QED-radiative effects, track reconstruction, track selection cuts, the decay products of K^0_S and Λ which were assigned to the primary vertex, and energy losses in the inactive material in front of the calorimeter in the case of the energy measurement.

Two different correction procedures for the multiplicity distributions were used. One of them is based on the matrix unfolding method as used in earlier studies [10]. The second is a bin-by-bin method. Both methods give similar results. The difference between them is included as a systematic uncertainty.

The dominant sources of systematic uncertainties were investigated. The main uncertainty for the W measurement arises from different MC models used in the correction procedure and it amount to up to 5%. To reduce the uncertainty the average correction factor between ARIADNE and LEPTO was taken, thus reducing the uncertainty by a factor of 2. For the Breit frame measurement this uncertainly is typically below 0.5%. Other sources of uncertainties are (typical values of the uncertainties are shown in brackets): uncertainty in the CAL energy scale ($0.5-1.1\%$), event reconstruction and selection ($< 0.5\%$), track reconstruction and selection ($< 0.5\%$) and method of correction (matrix or modifi ed bin-by-bin, $0.6-1.3\%$). The contaminations due to migrations from $Q^2 < 25$ GeV^2 and from diffractive events are negligible.

RESULTS

The current region of the Breit frame is analogous to a single hemisphere of e^+e^- annihilation. In $e^+e^- \rightarrow q\bar{q}$ the two quarks are produced with equal and opposite momenta, $\pm\sqrt{s_{ee}}/2$. The fragmentation of these quarks can be compared to that of the quark struck from the proton in DIS. This quark has an outgoing momentum $-Q/2$ in the Breit frame. The multiplicity of the current region of the Breit frame is expected to have a dependence on Q similar to that in e^+e^- annihilation versus the energy $\sqrt{s_{ee}} = Q$. To take into account contributions from soft and hard QCD processes that lead to a decrease of the energy and the number of particles in the current region of the Breit frame, in this analysis $2E_{\text{current}}$ was used instead of Q, where E_{current} is the energy of all the particles in the current region of the Breit frame.

For the photon (current) fragmentation region of the HCM frame the total number of charged particles can be studied as a function of W. Additional studies of the energy in this region demonstrated that there are no signifi cant migrations from the photon (current) to the target region and the variable W can be safely used.

Figure 1 shows the measured mean charged multiplicity, $\langle n_{\text{ch}} \rangle$, in the photon (current) region of the HCM frame as a function of W. The measurement as a function of W can be only performed at high W, because of the acceptance of the ZEUS detector. The predictions of ARIADNE and LEPTO are also shown.

To compare the results of measurements in the Breit and HCM frame with results of the e^+e^- and pp experiments the mean charge multiplicity was multiplied by 2. Figure 2 shows twice the measured mean charged multiplicity, $2 \cdot \langle n_{\text{ch}} \rangle$, in the current region of the Breit frame plotted versus $2 \cdot E_{\text{current}}$ and twice the measured mean charged multiplicity in the photon (current) region of the HCM frame plotted versus W. Also shown are the predictions of ARIADNE, LEPTO and the measurements from e^+e^- and pp

experiments together with a previous ZEUS measurement in the current region of the Breit frame [11]. For the previous ZEUS measurement, $2 \cdot \langle n_{ch} \rangle$ is measured as a function of Q.

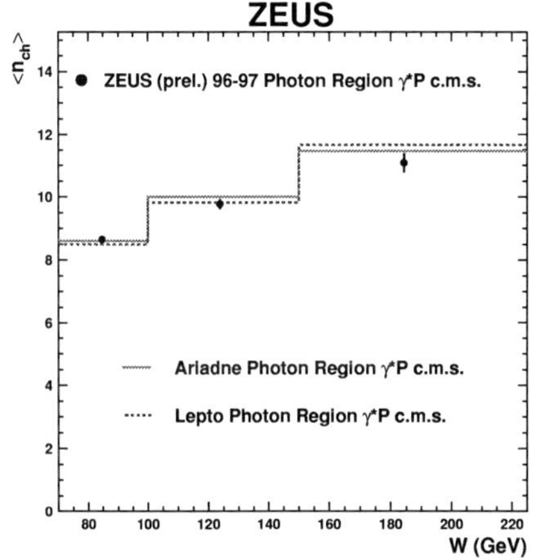

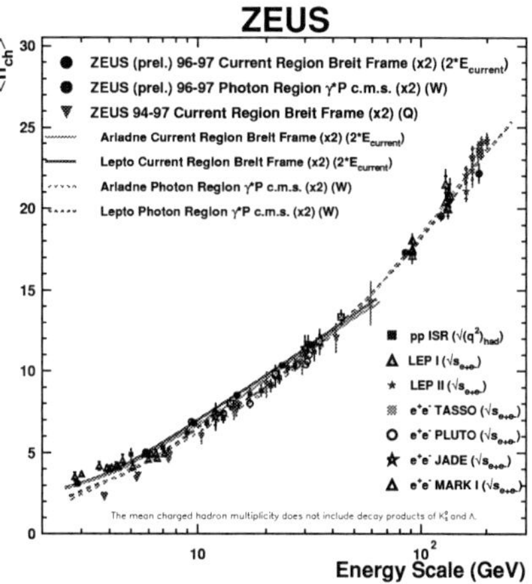

FIGURE 1. Mean charged multiplicity, $\langle n_{\text{ch}} \rangle$, in the photon (current) fragmentation region of the HCM as a function of W. Also shown are the predictions from ARIADNE and LEPTO.

FIGURE 2. Mean charged multiplicity, $\langle n_{\text{ch}} \rangle$, in the current region of the Breit frame multiplied by 2 plotted versus $2E_{\text{current}}$, where E_{current} is the sum of the energies of the particles (charged and neutral) in the current region. The blue dots represent the $\langle n_{\text{ch}} \rangle$ in the photon (current) region of the HCM frame, multiplied by 2, plotted versus W. The downward triangles show the ZEUS measurement of ref. [11]. The prediction of the ARIADNE and the LEPTO, as well as the measurements of pp and e^+e^- interactions are also shown.

The ZEUS measurements presented in this paper agree with the e^+e^- and pp measurements. At low values of energy, the measurement as a function of $2 \cdot E_{\text{current}}$ agrees better with e^+e^- than the measurement as a function of Q, since the migrations of final state particles out of the current region are larger at the low values of energy and they are properly taken into account in E_{current}.

REFERENCES

1. M. Basile, et al., *Nuovo Cim.*, **A65**, 400 (1981).
2. The ZEUS detector, Status Report (unpublished), DESY (1993).
3. H. S. A. Kwiatkowski, and H.-J. Möhring, *Comp. Phys. Commun.*, **69**, 155 (1992), also in *Proc. Workshop Physics at HERA*, 1991, DESY, Hamburg.
4. H. Spiesberger, *An Event Generator for ep Interactions at HERA Including Radiative Processes (Version 4.6)* (1996).
5. H. Spiesberger, HERACLES *and* DJANGOH*: Event Generation for ep Interactions at HERA Including Radiative Processes* (1998).
6. G. S. K. Charchula, and H. Spiesberger, *Comp. Phys. Commun.*, **81**, 381 (1994).
7. L. Lönnblad, *Comp. Phys. Commun.*, **71**, 15 (1992).
8. A. E. G. Ingelman, and J. Rathsman, *Comp. Phys. Commun.*, **101**, 108 (1997).
9. B. A. et al., *Preprint*, **97**, 31 (1983).
10. M. D. e. a. ZEUS Coll., *Z. Phys.*, **C 67**, 93 (1995), `hep-ex/9501012`.
11. J. B. e. a. ZEUS Coll., *EUR. PHYS. J.*, **C 11**, 251 (1999), `hep-ex/9903056`.

The fragmentation process at HERMES

B. Maiheu[1]

On behalf of the HERMES collaboration

Abstract. In semi-inclusive deep-inelastic scattering one can study the fragmentation process of partons by extracting multiplicity distributions for the resulting hadrons. At the HERMES experiment a unique possibility exists to study this hadronization at a $\sqrt{s}$ of 7.2 GeV. Making full use of the hadron identification capabilities of a Ring Imaging Čerenkov (RICH) detector, we were able to extract charge-separated pion and kaon multiplicities. Significant effort was put reducing the model dependence of the result by tuning the used MC generator and using an unfolding method to correct for experimental inefficiencies and migration of events due to both radiative and detector smearing.

Keywords: HERMES, fragmentation, hadron multiplicity
PACS: 13.87.Fh

INTRODUCTION

The confi nement principle states that partons cannot exist as free particles. Therefore when a highly energetic probe hits one of the partons inside a nucleon, it will break free from the nucleon, forming new hadrons from its color fi eld link with the nucleon remnant. It is interesting to study this process at moderate energies at which perturbative QCD and factorization are not yet fully established and tested. With a positron beam energy of 27.6 GeV and a fi xed, gaseous hydrogen target, the HERMES experiment [1] provides an ideal setup to undertake this study at $\sqrt{s}$ of about 7 GeV. It is located on the HERA storage ring at DESY, Hamburg. The analysis briefly presented here displays the extraction of flavor separated multiplicity distributions, defi ned in this way :

$$\frac{1}{\sigma^{DIS}} \cdot \frac{d\sigma^{ep\to e'hX}(z,Q^2)}{dz} = \frac{\sum_f e_f^2 \int_0^1 dx\, q_f(x,Q^2) D_f^h(z,Q^2)}{\sum_f e_f^2 \int_0^1 dx\, q_f(x,Q^2)}. \tag{1}$$

In this equation $q_f(x,Q^2)$ represent the parton distribution functions and $D_f^h(z,Q^2)$ the fragmentation function of the parton of flavor f into a hadron of type h. The observable $z = E_h/\nu$ represents the energy fraction carried away by the hadron considered. Of course Q^2 is the exchanged photon virtuality and x the momentum fraction of the struck quark.

---

[1] University of Gent, Belgium, email: bino@inwfsun1.ugent.be

CP792, *Deep Inelastic Scattering, DIS 2005,* edited by W. H. Smith and S. R. Dasu

EXPERIMENTAL HADRON MULTIPLICITIES

At the first stage in the analysis a hadron PID correction was applied that accounts for misidentification of hadrons by the RICH [2] detector. From a GEANT description of the RICH, a matrix **P** was extracted in which each element $\mathbf{P}_t^i$ represents the probability that a particle of type t was identified as type i, thus $\vec{I} = \mathbf{P} \cdot \vec{N}$ in which $\vec{I}$ and $\vec{N}$ are respectively the vectors with identified particles and real particles. Inversion of this matrix $\mathbf{P}^{-1}$ results in a weight to obtain the true particle flux for each identified particle type i . The **P** matrix depends on the momentum of the hadrons and the event topology. The contribution to the systematic uncertainty from this correction results from the non-perfect knowledge of the response of the RICH detector. Two different settings were used in the RICH Monte Carlo description, one for $\beta = 1$ particles tuned the Čerenkov angle and photon yield distributions, while the second uses information from decaying particles (ϕ, K_S, Λ^0) from both data and Monte Carlo. The difference between the two resulting **P** matrices is taken as a measure for the systematic uncertainty.

Next, two sources of background were identified and corrected for. Positrons from e^+e^- pair production that ended up in the DIS sample were corrected for by subtracting the oppositely[2] charged leptons within the DIS sample. This correction is small and increases total experimental hadron multiplicity by 1 to 1.5 %. The second identified source of background comes from the decay of diffractively produced vector mesons like ρ^0 or ϕ. For this the PYTHIA6 generator [3] was taken and adapted with QED radiative corrections and tuned to fit HERMES kinematic distributions. Especially at low Q^2 and high z contributions can reach 40 to 50 % of pions mainly coming from diffractively produced ρ^0, whereas for kaons the contribution is limited to about 10 %. This is shown in figure 1. The gray area is not used in the rest of the analysis. This correction was applied following the philosophy that the production process for these hadrons is totally different from quark fragmentation. However the final results will be presented with and without this background subtracted.

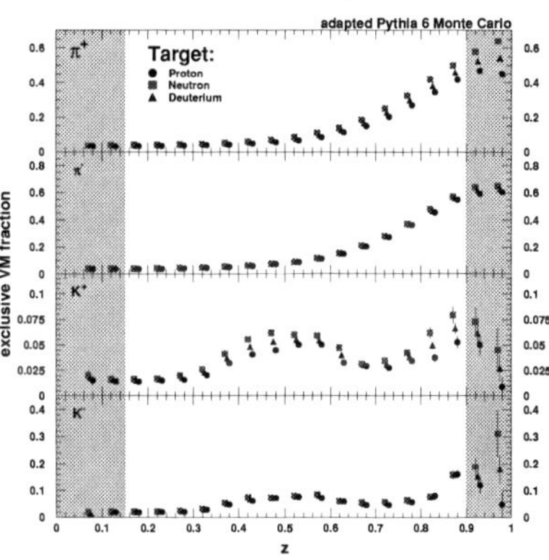

1: Exclusive VM fraction vs. z

SMEARING AND ACCEPTANCE CORRECTION

The following step is to correct these experimental multiplicity distributions for smearing and acceptance. In order to have a clear treatment of the uncertainties involved, we keep track of the smearing between kinematical bins by means of a smearing matrix

---

[2] With respect to the beam charge.

$S(i,j) = n(i,j)/n^B(j)$, of which each element represents the flow of events from a Born level bin j to an experimental bin i. It is defined as the migration matrix $n(i,j)$ normalized by the Born level distribution $n^B(j)$. Since both of these quantities are extracted from Monte Carlo using the same model, the dependence on this input model is greatly reduced. Figure 2 displays the migration matrix $n(i,j)$ for the 15 z bins from 0.15 to 0.9 as used in the analysis. The last j bins represents events smeared from outside the Born level sample into the experimental sample. One can clearly see the asymmetric shape of the migration matrix. This is due to QED radiative corrections which only work in one direction of ν (and thus z). The change in kinematics due to detector smearing is much smaller in size and works in both directions. The model dependence of this approach was checked by using different tunes for the JETSET fragmentation model in the Monte Carlo. The variation was found to be well inside the assigned systematic uncertainty. Normalizing to a 4π Born distribution $n^B(j)$ also takes into account the HERMES acceptance. Being a forward detector with an gap of 80 mrad around the beam pipe, the HERMES experiment is limited in its acceptance, it nevertheless samples the kinematical phase space well enough to fix the dynamics.

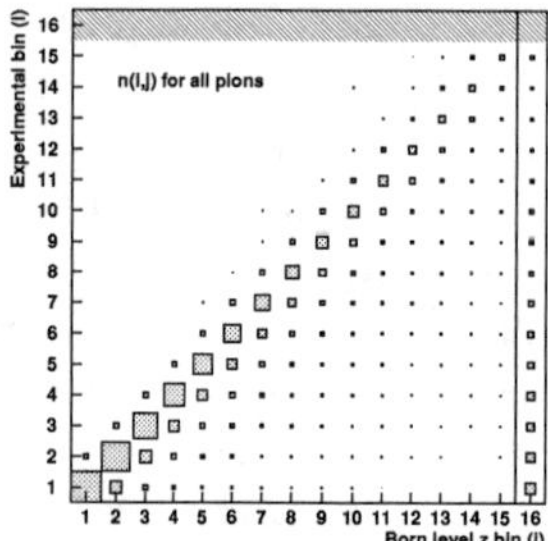

2: The migration matrix

RESULTS & CONCLUSION

The resulting hadron multiplicities were evolved to a Q_0^2 of 2.5 $\mathrm{GeV}^2/\mathrm{c}^2$. The multiplicity distributions versus z are displayed in figure 3, both with and without the correction for exclusive vector meson production. When evolving to a scale of 25 $\mathrm{GeV}^2/\mathrm{c}^2$ this data is found well in agreement with previously published fragmentation functions by the EMC experiment [4]. Finally the Q^2 dependence of the multiplicities was investigated. Figure 4 shows this compared to a parameterization by S. Kretzer [5]. One can see that the agreement is good. The other plot the same figure shows the residual x dependence compared to data. This analysis observed a much weaker residual x dependence than the EMC data or previous HERMES data [6]. This can be attributed to a much better understanding of the Monte Carlo and the corrections involved. From this analysis no signs of strong factorization breaking can be observed for the HERMES kinematical regime.

REFERENCES

1. K. Ackerstaff *et al.* Nucl. Inst. Meth., **A 417** (1998) 230-265
2. N. Akopov *et al.* Nucl. Inst. Meth., **A 479** (2002) 511-530
3. T. Sjöstrand *et al.* Comp. Phys. Commun., **135** (2001) 238
4. EMC Collaboration, Nucl. Phys. **B 321** (1989) 541-560
5. S. Kretzer, Phys. Rev. **D 62** (2000) 054001
   S. Kretzer, *private communication*
6. A. Airapetian, *et al.* Eur. Phys. J., **C 21** (2001) 599-606

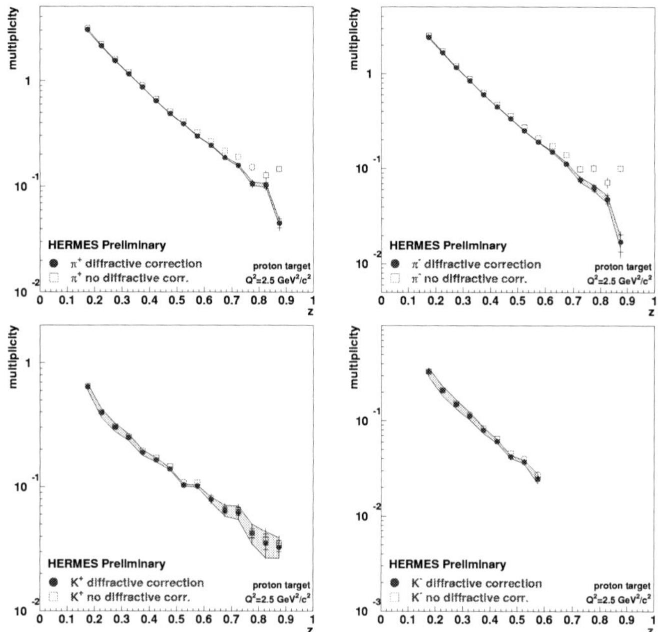

FIGURE 3. Multiplicity distributions on a proton target for π^+, π^-, K^+ and K^- versus z. All the data has been evolved to $Q^2 = 2.5$ GeV^2/c^2, the closed symbols represent the data with the exclusive VM correction in.

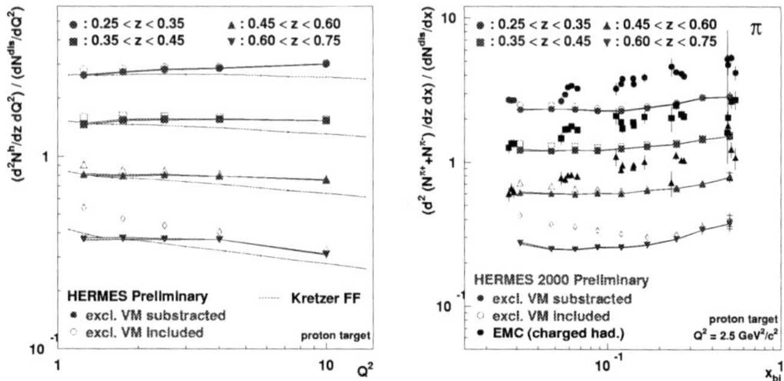

FIGURE 4. The Q^2 dependence for the total hadron multiplicity compared to the Kretzer parameterization (left). Right is shown the residual x dependence of the total pion multiplicity, compared to previously released EMC data [4] evolved to Q^2 of 2.5 GeV^2/c^2. The HERMES points are connected to the systematic uncertainty bands.

Gaps between Jets: Matching two Approaches

A. Kyrieleis

University of Manchester, Oxford Road, Manchester M13 9PL, U.K.
email: kyrieleis@hep.man.ac.uk

Abstract. We calculate the parton level cross section for the production of two jets that are far apart in rapidity, subject to a limitation on the total transverse momentum Q_0 in the interjet region. We specifically address the question of how to combine the approach which sums all leading logarithms in Q/Q_0 (where Q is the jet transverse momentum) with the BFKL approach, in which leading logarithms of the scattering energy are summed. Using an "all orders" matching, we obtain results for the cross section which correctly reproduce the two approaches in the appropriate limits.

Keywords: qcd, jet
PACS: 12.38.Bx, 12.38.Cy

INTRODUCTION

Final states with high-p_T jets separated by large rapidity gaps at hadron colliders offer the possibility to better understand QCD in the high energy limit and also to understand QCD radiation in "gap" events. There are two major approaches to the production of two gap-separated jets. In the BFKL [1] approach, parton-parton elastic scattering with a QCD colour singlet exchange is regarded as providing the leading contribution to the cross-section. The leading-Y terms (Y is the rapidity interval between the jets) are summed i.e. terms $\sim \alpha_s^n Y^n$ [2, 3]. The observable calculated in this approach does not consider any radiation into the interjet region. Experiments though, impose an upper bound on this radiation by necessity. In the second approach soft radiation with transverse energy below Q_0 is allowed in the interjet region. This gives rise to logarithms of Q/Q_0 where Q is the transverse momentum of the jets. The global leading logarithms of Q/Q_0 (LLQ_0) have been summed for various jet definitions [4, 5] i.e terms $\sim \alpha_s^n Y^m L^n$ ($m \leq n$) where $L = \ln Q^2/Q_0^2$. Non-global effects have been considered in [5, 6]. In order to get a better understanding of the gaps-between-jets processes at colliders it is desirable to combine the two approaches. This is the main issue in this contribution, for details see [7]

SUMMING LOGARITHMS IN Q_0

As the first step we recalculate the cross section for two-jet production in the high-energy (i.e. high rapidity separation) limit, with limited total scalar transverse momentum in the interjet region. We require this transverse momentum to be below Q_0 and consider the region $Q_0^2 \ll Q^2 \ll \hat{s} = e^Y Q^2$. Since we are not sensitive to collinear emission, we work at the parton level and calculate the all-orders gap cross section $\sigma \equiv \frac{d\sigma(\hat{s},Q_0,Y)}{dQ^2}$ for the

CP792, *Deep Inelastic Scattering, DIS 2005,* edited by W. H. Smith and S. R. Dasu

process $qq' \to qq'$. $\sigma^{(n)}$ denotes the cross section at $\mathcal{O}(\alpha_s^n)$. Our approximation implies the eikonal (soft gluon) approximation. To generate the leading logs in Q_0, we make the approximation of strongly-ordered transverse momenta for real and virtual gluons.

As the basis for the calculation to all orders we employ the following theorem. Let us denote by $\mathcal{A}_1^{(n)}(Q_0)\mathbf{C}_1$ ($\mathcal{A}_8^{(n)}(Q_0)\mathbf{C}_8$) the singlet (octet) component of the $\mathcal{O}(\alpha_s^n)$ $qq' \to qq'$ amplitude in the approximation defined above, with the phase space for the gluons constrained to the gap region in rapidity and with transverse momentum above Q_0 ($\mathbf{C}_{1,8}$ are the colour factors). With $\mathcal{B}(Q_0)$ denoting the production amplitude (including colour factor) for more than 2 particles, the theorem reads

$$\sigma^{(k)} = |\mathcal{A}_1(0)|^2\mathbf{C}_1^2 + |\mathcal{A}_8(0)|^2\mathbf{C}_8^2 + |\mathcal{B}(Q_0)|^2 = |\mathcal{A}_1(Q_0)|^2\mathbf{C}_1^2 + |\mathcal{A}_8(Q_0)|^2\mathbf{C}_8^2 \tag{1}$$

where the squares are to be read symbolically representing the sums over $\mathcal{A}^{(n)*}\mathcal{A}^{(m)}$ (and $\mathcal{B}(Q_0)$, respectively). This is clearly a major simplification, since it means that we never have to calculate any real emission or triple-gluon-vertex diagrams. This theorem provides the basis for the matching with BFKL. We calculate $\mathcal{A}(Q_0)_{1,8}$ and hence σ to all orders. Besides the double-leading-logarithmic (*DLL*) terms we include those terms sub-leading in Y that arise from the imaginary parts of the loop integrals.

MATCHING WITH BFKL

To combine the gap cross section with the BFKL approach order-by-order we need to prevent double counting and make sure the divergences arising from the BFKL approach (at each order in α_s) cancel in the jet cross section. To this end we calculate the leading-Y approximation of the singlet component $\mathcal{A}_1^{(n)}(Q_0)$:

$$\mathcal{A}_{1,S}^{(n)}(Q_0) \equiv \mathcal{A}_1^{(n)}(Q_0)\Big|_{LY}. \tag{2}$$

$\mathcal{A}_{1,S}^{(n)}(0)$ is divergent at each order and it is this contribution to σ that is also included in the BFKL result.

Fixed order matching. We denote by $\mathcal{A}_{BFKL}^{(n)}\mathbf{C}_1$ the $\mathcal{O}(\alpha_s^n)$ elastic quark scattering amplitude with colour singlet exchange in the leading-Y approximation. We want σ to include $\mathcal{A}_{BFKL}^{(n)}$. However, $\mathcal{A}_1^{(n)}(0)$ also includes terms sub-leading in Y which we have to keep; they are given by $(\mathcal{A}_1(0) - \mathcal{A}_{1,S}(0))^{(n)}$. We therefore define the following fixed order gap cross section (again omitting the sum over indices in the first line).

$$\sigma_{gap}^{(k)} \equiv |\mathcal{A}_{BFKL} + \mathcal{A}_1(0) - \mathcal{A}_{1,S}(0)|^2\mathbf{C}_1^2 + |\mathcal{A}_8(0)|^2\mathbf{C}_8^2 + |\mathcal{B}(Q_0)|^2 \tag{3}$$

$$= \sigma^{(k)} + \sum_{m+n=k}\left[2\mathrm{Im}\mathcal{A}_1^{(m)}(0)\cdot(-i\delta^{(n)}) + \delta^{(m)}\delta^{(n)*}\right]\mathbf{C}_1^2 \tag{4}$$

$$\text{with} \quad \delta^{(n)} = \mathcal{A}_{BFKL}^{(n)} - \mathcal{A}_{1,S}^{(n)}(0) \tag{5}$$

where, in the last line we have invoked the theorem (1). This cross section combines the two approaches without double counting. However, not surprisingly, the strong ordering approximation cannot cancel the divergence in the BFKL amplitude at any order. The second term in (4) and hence $\sigma_{gap}^{(k)}$ is divergent for $k \geq 6$. Via (4) we can therefore combine the all-orders cross section σ with the BFKL result up to $\mathcal{O}(\alpha_s^5)$.

The theorem (1) holds beyond the high energy approximation, the matching with BFKL can therefore be extended to full (global) LLQ_0 accuracy in a straightforward way [7].

All orders matching. Although the order-by-order combination of the LLQ_0 and the BFKL result can only work for the first few orders it is possible to construct an all-orders cross section that does smoothly interpolate the LLQ_0 and BFKL results, agreeing with each in its region of validity and avoiding any double-counting. Central to this are the following two observations. First, the amplitude $\mathcal{A}_{1,S}^{(n)}(Q_0)$ summed to all orders reads:

$$\mathcal{A}_{1,S}(Q_0) = -i\frac{N_c^2-1}{2N_c^3}\frac{\pi}{Y}\mathcal{A}_8^{(1)} \cdot \left[1 - \exp\left(-\frac{N_c\alpha_s}{2\pi}YL\right)\right]. \tag{6}$$

The exponential vanishes as $Q_0 \to 0$. In contrast to the fixed order result, $\mathcal{A}_{1,S}(0)$ is therefore finite. Secondly, we find the following relation between the (finite) all-orders results for the BFKL $2 \to 2$ cross section σ_{BFKL} [3] and the gap cross section σ:

$$\sigma_{BFKL}|_{Y\to 0} = \sigma|_{Y\to\infty} = \sigma_S \equiv |\mathcal{A}_{1,S}(0)|^2\mathbf{C}_1^2 = \sigma^{(2)}\frac{N_c^2-1}{N_c^4}\frac{\pi^2}{Y^2} \tag{7}$$

which implies $\mathcal{A}_{BFKL}|_{Y\to 0} = \mathcal{A}_{1,S}(0)$. Using these two remarkable results we construct three different matched cross sections (δ is given by (5) summed to all orders).

Simple matching: $\sigma_{gap} = \sigma + N_c^2|\delta|^2$
Cross section matching: $\sigma_{gap} = \sigma + \sigma_{BFKL} - \sigma_S$
Amplitude matching: $\sigma_{gap} = \frac{1}{4}(N_c^2-1)|\mathcal{A}_8(Q_0)|^2 + N_c^2|\mathcal{A}_1(Q_0)+\delta|^2$

In the first scheme we have replaced all expressions in (4) with the (finite) all-orders results and exploited the fact that $\mathcal{A}_1(0)$ is zero. In all three cases we subtract from the sum of the LLQ_0 and $BFKL$ amplitudes (cross sections) the double-counted term $\mathcal{A}_{1,S}(0)$ (σ_s). In all schemes $\sigma_{gap} \to \sigma$ for $Y \to 0$ since $\delta \to 0$ ($\sigma_{BFKL} - \sigma_S \to 0$), see (7). As $Y \to \infty$ we have $\sigma, \sigma_S \to 0$ and $\mathcal{A}_{1,8}(Q_0), \mathcal{A}_{1,S}(0) \to 0$ (i.e. $\delta \to \mathcal{A}_{BFKL}$) and hence $\sigma_{gap} \to \sigma_{BFKL}$. Each scheme therefore achieves our goal of having a smooth matching of the two all-orders cross sections, in that for small and large Y it agrees with the LLQ_0 and BFKL cross sections respectively avoiding any double-counting.

As a measure of the uncertainty inherent in the matching procedure fig. 1 shows numerical results of all three schemes. Indeed, they all match the two cross sections in the small and large Y limits and the differences are not large in between.

CONCLUSION

Working in the high energy limit we have calculated the (partonic) cross section for the production of two jets distant in rapidity and with limited transverse energy flow into the

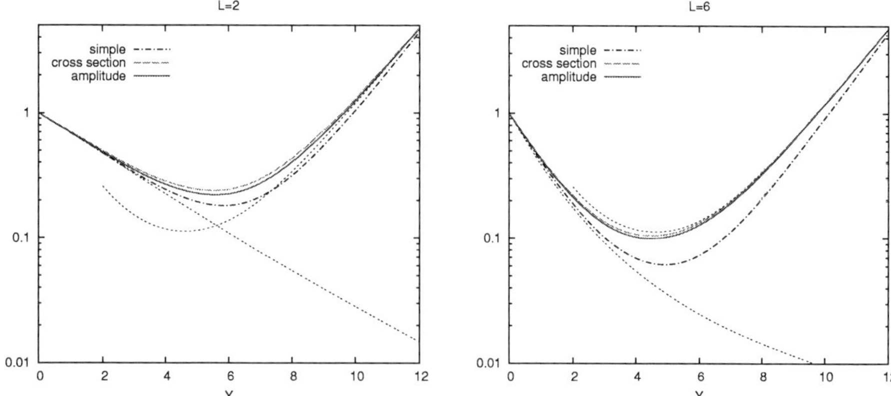

FIGURE 1. The gap cross section in the three matching schemes for $L = 2$ and 6 ($\alpha_s = 0.2$) compared to σ_{BFKL} (dots) and σ (double-dots)

region between the jets. Besides the *DLL* terms, we have summed terms sub-leading in Y stemming from the imaginary parts of the loop integrals. This allowed us to consistently combine the terms of the LLQ_0 series and the BFKL series to $\mathcal{O}(\alpha_s^5)$ accuracy without double counting. In the LLQ_0A, the inclusion of higher orders of the BFKL cross section in this way is not possible since it implies a divergent cross section.

We have also studied several "all order" matching schemes that effectively interpolate between the LLQ_0 and BFKL results. Although they all yield similar results, the differences between them cannot be resolved without further work, specifically understanding the role of real-emission contributions in the high energy limit. We have made a first step towards the unification of the two main approaches to the "jet–gap–jet" process.

ACKNOWLEDGMENTS

The presented work was done in collaboration with J. R. Forshaw and M. H. Seymour.

REFERENCES

1. E. A. Kuraev, L. N. Lipatov and V. S. Fadin, *Sov. Phys. JETP* **45** (1977) 199 [*Zh. Eksp. Teor. Fiz.* **72** (1977) 377]; I. I. Balitsky and L. N. Lipatov, *Sov. J. Nucl. Phys.* **28** (1978) 822 [*Yad. Fiz.* **28** (1978) 1597].
2. A. H. Mueller and W. K. Tang, *Phys. Lett. B* **284** (1992) 123; B. Cox, J. Forshaw and L. Lonnblad, *JHEP* **9910** (1999) 023; R. Enberg, L. Motyka and G. Ingelman, Presented at the 9th International Workshop on Deep Inelastic Scattering (DIS 2001), Bologna, Italy. arXiv:hep-ph/0106323; R. Enberg, G. Ingelman and L. Motyka, *Phys. Lett. B* **524** (2002) 273
3. L. Motyka, A. D. Martin and M. G. Ryskin, *Phys. Lett. B* **524** (2002) 107
4. G. Oderda and G. Sterman, *Phys. Rev. Lett.* **81**, 3591 (1998); G. Oderda, *Phys. Rev. D* **61** (2000) 014004; C. F. Berger, T. Kucs and G. Sterman, *Phys. Rev. D* **65** (2002) 094031
5. R. B. Appleby and M. H. Seymour, *JHEP* **0309** (2003) 056
6. R. B. Appleby and M. H. Seymour, *JHEP* **0212** (2002) 063
7. J. R. Forshaw, A. Kyrieleis and M.H. Seymour, arXiv:hep-ph/0502086, to be published in *JHEP*.

Polarization and asymmetries in neutral strange particle production

Andrew Cottrell
for the ZEUS Collaboration

Denys Wilkinson Building, Department of Physics, Oxford, OX1 3RH

Abstract. Inclusive Λ, $\bar{\Lambda}$ and K_s^0 production in deep inelastic *ep* scattering has been studied with the ZEUS detector at HERA using an integrated luminosity of 120pb^{-1}. Differential cross sections, baryon to antibaryon production asymmetry and baryon to meson production ratios have been measured in the laboratory system for $Q^2 > 25\text{GeV}^2$.

Keywords: Lambda Polarization, Baryon Number, Hadronization
PACS: 14.20.Jn, 13.85.Ni, 13.87.Fh, 13.60.Rj

INTRODUCTION

This paper discusses the production of the neutral strange particles Λ, $\bar{\Lambda}$[1] and K_s^0 in deep inelastic scattering at HERA. As well as measuring differential cross sections for their production, the baryon-antibaryon asymmetry and the baryon to meson ratio are investigated, and a first ZEUS measurement of the transverse and longitudinal Λ polarization is made.

All of these measurements endeavour to clarify in different ways the transition from quark to hadron. For example the Λ to K_s^0 ratio has been measured in e^+e^- colliders [1] and in heavy ion collisions [2]. The ZEUS measurement in *ep* collisions adds more information in trying to understand when a baryon is produced, and when a meson.

Λ transverse polarization also gives information on hadron production. The DeGrand-Miettinen model [3] explains the Λ spin as being carried predominantly by the s quark which picks up polarization via Thomas Precession when it is accelerated. Hence a measurement of the transverse polarization will give information on the initial direction of the s quark that ends up in the Λ.

Observing these particles also allows an investigation into baryon number and how it is transported. Significant baryon number transport over several units of rapdidity has been observed in heavy ion collisions [4]. Various models have been developed to explain this, including associating baryon number with valence quarks and moving it through rapidity by multiple scattering [5] or associating baryon number with a gluonic junction [6]. In HERA *ep* collisions initially a baryon number of +1 exists as the proton moving down the beampipe. The possibility of this baryon number being observed in the Λ system in the central rapidity region is investigated.

[1] Hereafter,both Λ and $\bar{\Lambda}$ are referred to as Λ, unless explicit comparisons are made.

CP792, *Deep Inelastic Scattering, DIS 2005,* edited by W. H. Smith and S. R. Dasu

EVENT SELECTION AND ANALYSIS

This analysis uses an inclusive sample of neutral current deep inelastic scattering (DIS) events collected by ZEUS in the 1996-2000 HERA running period, corresponding to an integrated luminosity of 120 pb^{-1}. The kinematic region was $Q^2 > 25\text{GeV}^2$ and $0.02 < y < 0.95$.

Λ and K_s^0 are detected in the $p\pi$ and $\pi^+\pi^-$ decay channels respectively. A secondary vertex is observed and the mass is reconstructed from the momenta of two oppositely charged tracks coming from the vertex. Both tracks are assumed to have the mass of a π^+ (for K_s^0) or the track with the most momentum has the proton mass and the other the mass of a π^+ (for Λ). Combinatorial background is removed with a bin-by-bin sideband subtraction method.

The Λ polarization is measured via the angular distribution of the decay products:

$$\frac{dN}{d\Omega} \propto \frac{1}{4\pi}(1 \pm \alpha P \cos\theta) \tag{1}$$

in the Λ rest frame, where $\alpha = 0.642 \pm 0.013$[1] is the decay asymmetry parameter and P is the polarization. θ is the angle between the decay proton momentum, $\vec{p}$ and the Λ momentum, $\vec{P}_\Lambda$ (longitudinal polarization) or between $\vec{p}$ and $\mathbf{n} = \vec{P}_{beam} \times \vec{P}_\Lambda$, where $\vec{P}_{beam}$ is the momentum of the electron beam (transverse polarization).

MONTE CARLO SIMULATION

Data were corrected to hadron level by using the ARIADNE 4.08[7] Monte Carlo (MC) interfaced to HERACLES via DJANGOH 1.1[8]. The parton density functions were taken from the CTEQ4D set. The strange suppression factor, λ_s was set to 0.3. ARIADNE is based on the Color Dipole Model and the LUND string model[9] is used to simulate the fragmentation of the partons. The Ariadne prediction of the cross sections is also shown on the results plots.

RESULTS

The results of the $\Lambda/\bar{\Lambda}$ polarization are presented in Fig. 1. Averaged over the full kinematic range, the transverse polarization of Λ and $\bar{\Lambda}$ were observed to be $+1.4 \pm 4.5(\text{stat})^{+4.1}_{-1.9}(\text{syst})\%$ and $-1.8 \pm 4.4(\text{stat})^{+3.1}_{-1.3}(\text{syst})\%$ respectively. The longitudinal polarization of Λ and $\bar{\Lambda}$ was also measured to be consistent with zero within the total uncertainties. The longitudinal polarization was observed to be $+0.3 \pm 10.3(\text{stat})^{+0.1}_{-7.8}(\text{syst})\%$ for Λ and $+19.8 \pm 10.8(\text{stat})^{+4.2}_{-12.6}(\text{syst})\%$ for $\bar{\Lambda}$.

The differential Λ and K_S^0 cross sections, $\frac{\Lambda-\bar{\Lambda}}{\Lambda+\bar{\Lambda}}$ asymmetry and Λ/K_S^0 ratio as a function of p_T, η, x and Q^2 are shown in Fig. 2 and Fig. 3. The MC generally describes the cross sections in the data. However, the MC tends to overestimate the K_s^0 production, particularly at low P_T.

The ratio $\frac{\Lambda-\bar{\Lambda}}{\Lambda+\bar{\Lambda}}$ as a function of all kinematic variables is consistent with zero.

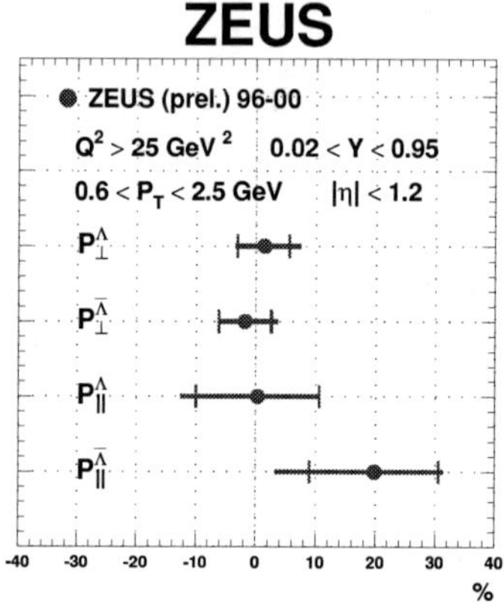

FIGURE 1. Transverse and longitudinal Λ and $\bar{\Lambda}$ polarization.

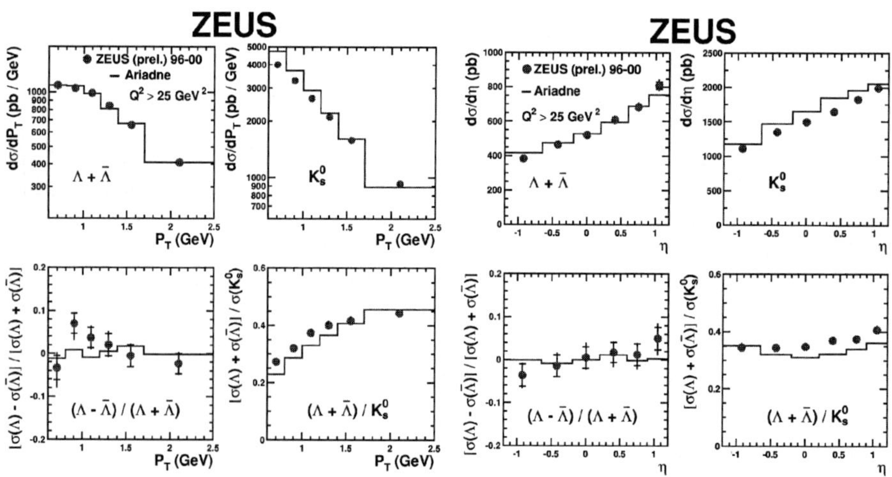

FIGURE 2. Λ and K_s^0 cross sections, $\frac{\Lambda-\bar{\Lambda}}{\Lambda+\bar{\Lambda}}$ ratio and $\frac{\Lambda+\bar{\Lambda}}{K_s^0}$ ratio as a function of P_T and η

The Λ/K_S^0 ratio is generally described by the MC. However, although the steep rise as x decreases is modelled by the MC to some extent, it is not to the same degree. Other differences between the MC and the data can be seen as a function of P_T, where the Λ/K_s^0 ratio is in excess of the MC at low P_T, and as a function of η where the MC is symmetric about $\eta = 0$, whereas the data shows an increase in this ratio as η increases.

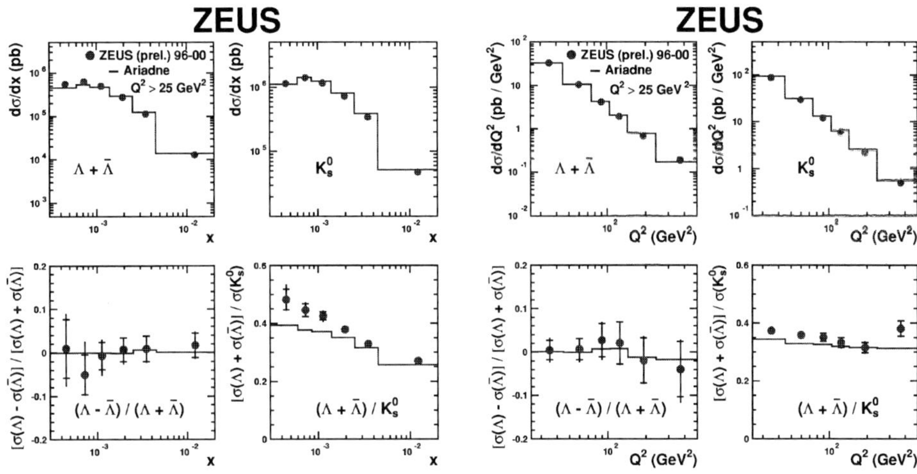

FIGURE 3. Λ and K_s^0 cross sections, $\frac{\Lambda-\bar{\Lambda}}{\Lambda+\bar{\Lambda}}$ ratio and $\frac{\Lambda+\bar{\Lambda}}{K_s^0}$ ratio as a function of x and Q^2

CONCLUSIONS

Using the model of DeGrand and Miettinen [3], the lack of any transverse polarization suggests that the strange quarks in the Λ baryons observed by ZEUS do not come from any particular direction. The longitudinal polarization being consistent with zero is as expected with HERA-I data, but gives a measure of the potential to measure any polarization transfer from the electron beam to the Λ in HERA-II. The Λ/K_S^0 ratio has been measured, and the data is generally described by ARIADNE. Areas of the phase space do exist where the MC is not sufficient to describe the data, particularly at low P_T and low x. No significant $\Lambda - \bar{\Lambda}$ asymmetry is seen.

REFERENCES

1. K. Hagiwara et al., *Phys. Rev. D*, **66**, 010001+ (2002), URL `http://pdg.lbl.gov`.
2. H. Huang, and J. Rafelski, *hep-ph/0501187*, 2005.
3. T. DeGrand, and H. Miettinen, *Phys. Rev. D*, **24**, 2419 (1981).
4. S. Bass et al., *Phys. Rev. Lett.* , **91**, 052302 (2003).
5. S. Bass et al., *J. Phys. G: Nucl. Part. Phys.*, **30**, 1283 (2004).
6. S. Vance et al., *Phys. Lett. B*, **443**, 45 (1998).
7. L. Lönnblad, *Comp. Phys. Comm.*, **71**, 15 (1992).
8. H. Spiesberger, HERACLES and DJANGOH: Event Generation for *ep* Interactions at HERA Including Radiative Processes, 1998, URL `http://www.desy.de/~hspiesb/djangoh.html`.
9. B. Andersson et al., *Phys. Rep.*, **97**, 31 (1983).

New possible insight into JLab proton polarization data puzzle by DIS

E. Bartoš*, S Dubnička†, A. Z. Dubničková* and E. A. Kuraev**

*Dept. of Theor. Physics, Comenius Univ., Bratislava, Slovak Republic
†Inst. of Physics, Slovak Academy of Sciences, Bratislava, Slovak Republic
**Bogolyubov Lab. of Theor. Physics, JINR, Dubna, Russian Federation

Abstract. It is demonstrated that the JLab proton polarization data puzzle could be solved by the new sum rule giving into a relation proton and neutron Dirac and Pauli form factors in the space-like region with a difference of the differential proton and neutron cross sections describing just Q^2 distribution in deep inelastic electron scattering process.

Keywords: proton electric form factor; polarization data; structure of nucleon
PACS: 12.40.Vv, 13.40Bp

The proton electromagnetic (EM) structure is described by two form factors (FF's) dependent on the squared four-momentum transfer $t = -Q^2$. The most natural are Dirac $F_{1p}(t)$ and Pauli $F_{2p}(t)$ FF's obtained in a parametrization of the matrix element of the EM current

$$\langle p'|J_\mu^{EM}|p\rangle = \bar{u}(p')\{\gamma_\mu F_{1p}(t) + i\frac{\sigma_{\mu\nu}(p'-p)_\nu}{2m_p^2}F_{2p}(t)\}u(p) \tag{1}$$

according to a maximum number of linearly independent covariants to be constructed from proton momenta and spin parameters.

The most suitable in extracting of experimental information are Sachs electric $G_{Ep}(t)$ and magnetic $G_{Mp}(t)$ FF's, giving in the Breit frame charge and magnetization distributions within the proton, respectively. Both sets of FF's are related

$$G_{Ep}(t) = F_{1p}(t) + \frac{t}{4m_p^2}F_{2p}; \quad G_{Mp} = F_{1p} + F_{2p}, \tag{2}$$

similarly for the neutron

$$G_{En}(t) = F_{1n}(t) + \frac{t}{4m_n^2}F_{2n}; \quad G_{Mn} = F_{1n} + F_{2n}. \tag{3}$$

The proton EM FF data in the space-like region ($t < 0$) have been obtained (see Fig. 1) from the measured cross-section (in DESY, SLAC and Bonn)

$$\frac{d\sigma^{lab}(e^-p \to e^-p)}{d\Omega} = \frac{\alpha^2}{4E^2}\frac{\cos^2(\theta/2)}{\sin^4(\theta/2)\frac{1}{1+(\frac{2E}{m_p})}\sin^2(\theta/2)}\left[A(t) + B(t)\tan^2(\theta/2)\right], \tag{4}$$

where $\alpha = 1/137$ and E - incidental energy

CP792, *Deep Inelastic Scattering, DIS 2005*, edited by W. H. Smith and S. R. Dasu

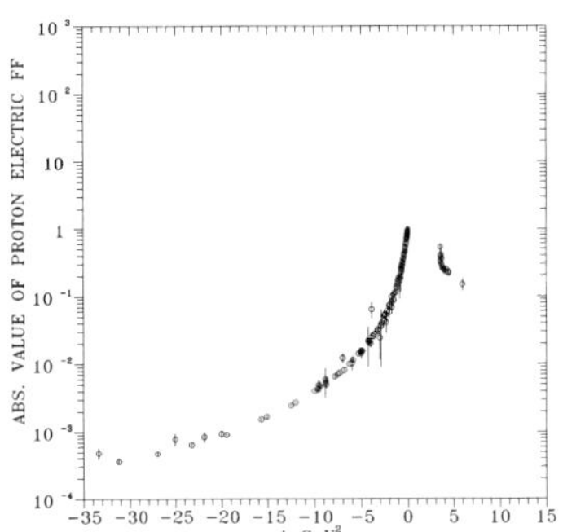

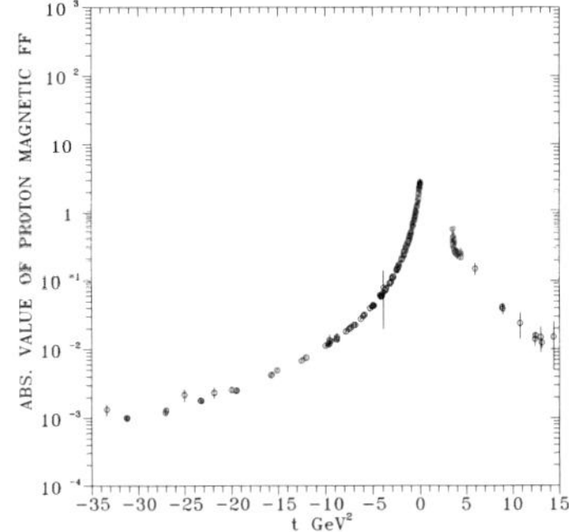

FIGURE 1. Compiled proton electric and magnetic FF data.

$$A(t) = \frac{G_{Ep}^2(t) - t/4m_p^2 G_{Mp}^2(t)}{1 - t/4m_p^2} \qquad B(t) = -2 \cdot t/4m_p^2 G_{Mp}^2(t) \tag{5}$$

by so-called Rosenbluth technique, employing a linear $\tan^2(\theta/2)$ dependence of (4).

Large progress has been recently done in the obtaining of the ratio (see Fig. 2)

$$\frac{G_{Ep}}{G_{Mp}} = -\frac{P_t}{P_l}\frac{(E+E')}{2m_p}\tan(\theta/2). \tag{6}$$

in the space-like ($t = -Q^2 < 0$) region by measuring [1, 2] simultaneously transverse

$$P_t = \frac{h}{I_0}(-2)\sqrt{\tau(1+\tau)}G_{Ep}G_{Mp}\tan(\theta/2) \tag{7}$$

and longitudinal

$$P_l = \frac{h(E+E')}{I_0 m_p}\sqrt{\tau(1+\tau)}G_{Mp}^2\tan^2(\theta/2), \tag{8}$$

components of the recoil proton's polarization in the electron scattering plane of the polarization transfer process $\vec{e}^- p \to e^- \vec{p}$, where h is the electron beam helicity, I_0 is the unpolarized cross-section excluding σ_{Mott} and $\tau = Q^2/4m_p^2$. As one can see from Fig. 2, these ratio data are in strong disagreement with the data obtained by Rosenbluth technique.

Due to the fact that $G_{Mp}^2(t)$ in (5) is multiplied by $-t/4m_p^2$ factor, i.e. as $-t$ increases, the measured cross-section becomes dominant by $G_{Mp}^2(t)$ part contribution making the extraction of $G_{Ep}^2(t)$ more and more difficult, the independent determination of $G_{Mp}(t)$ and $G_{Ep}(t)$ by Rosenbluth technique has been done [3] only up to $8.7 GeV^2$ and the extraction of G_{Mp} at higher values of Q^2 up to $\sim 31\ GeV^2$ assumes G_{Ep}=G_{Mp}/μ_p. As a result one could believe more to experimental data on G_{Mp} in space-like region than to experimental data on G_{Ep} and the disagreement of ratios in Fig. 2 is caused by contradicting behaviours of $G_{Ep}(Q^2)$ and on no account by $G_{Mp}(Q^2)$.

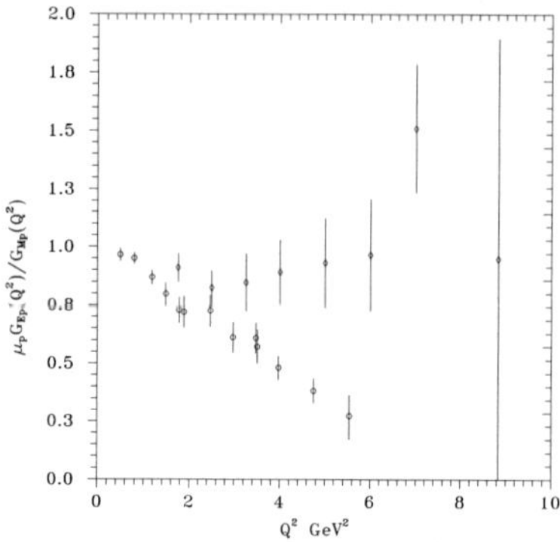

FIGURE 2. The JLab proton polarization data on ratio $\mu_p G_{Ep}/G_{Mp}$ (circles) and the same ratio calculated from data on G_{Ep} and G_{Mp} obtained by Rosenbluth technique (diamonds).

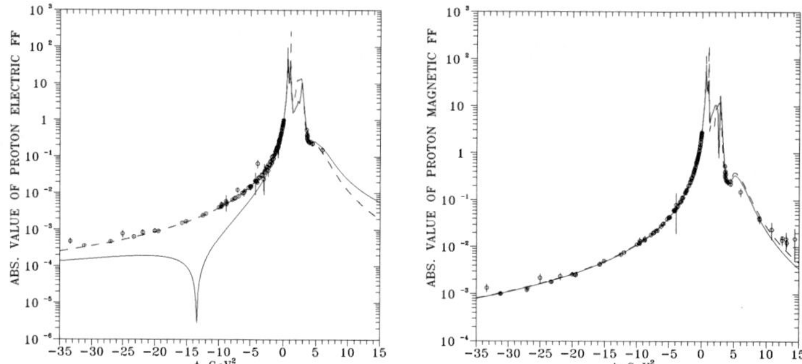

FIGURE 3. The predicted different behaviours of G_{Ep} in $t < 0$ region dependent on the fact if Rosenbluth technique data (dashed line) or JLab proton polarization data (full line) are used in the analysis

We have carried out a test of this hypothesis in framework of the ten-resonance Unitary and analytic model of nucleon EM structure [4], which is formulated in the language of isoscalar $F^s_{1,2}(t)$ and isovector $F^v_{1,2}(t)$ parts of the Dirac and Pauli FF's and comprises all known nucleon FF properties.

First, we have carried out the analysis of all proton and neutron data obtained by Rosenbluth technique together with all proton and neutron data in time-like region.

Then all $|G_{Ep}(t)|$ space-like data obtained by Rosenbluth technique were excluded and the new JLab proton polarization data on $\mu_p G_{Ep}(Q^2)/G_{Mp}(Q^2)$ for 0.49 GeV^2 $\leq Q^2 \leq 5.54$ GeV^2 were analyzed together with all electric proton time-like data and all space-like and time-like magnetic proton, as well as electric and magnetic neutron data.

The results of the analysis are presented in Fig. 3 from where it is seen that almost nothing is changed in a description of $G_{Mp}(t)$, $G_{En}(t)$ and $G_{Mn}(t)$ in both space-like and time-like regions, and also $|G_{Ep}(t)|$ in the time-like region. There is only a difference in behaviours of $G_{Ep}(t)$ in $t < 0$ region dependent on the fact if old data obtained by

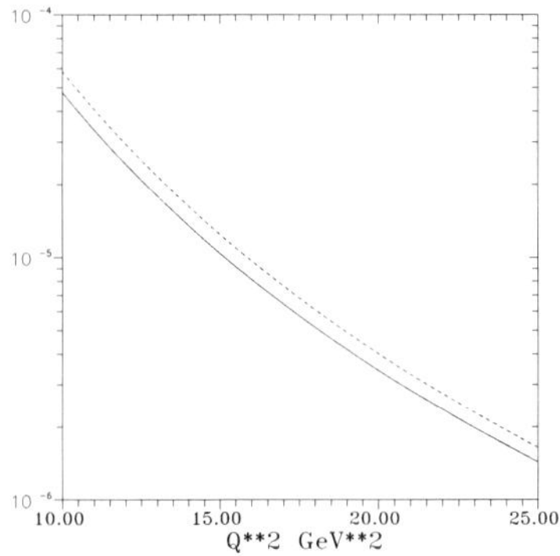

FIGURE 4. A prediction of two different behaviours of the right-hand side in (9) following from two different behaviours of G_{Ep} in Fig. 3.

Rosenbluth technique are used (dashed line) or the new JLab proton polarization data are analysed (full line).

In order to distinguish which of these behaviours of $G_{Ep}(t)$ in the space-like ($t < 0$) region is correct we suggest to employ new sum rule [5]

$$\begin{aligned} & F_{1p}^2(-Q^2) + \frac{Q^2}{4m_p^2} F_{2p}^2(-Q^2) - \\ & -F_{1n}^2(-Q^2) - \frac{Q^2}{4m_n^2} F_{2n}^2(-Q^2) = \\ = 1 - 2\frac{(Q^2)^2}{\pi\alpha^2} & \left(\frac{d\sigma^{e^- p \to e^- X}}{dQ^2} - \frac{d\sigma^{e^- n \to e^- X}}{dQ^2} \right), \end{aligned} \tag{9}$$

giving into a relation proton and neutron Dirac and Pauli FF's in the space-like region with a difference of the differential proton and neutron cross-sections describing Q^2 distribution in DIS. Evaluating Dirac and Pauli FF's on the left-hand side corresponding to the old (dashed line) and new (full line) space-like behaviour of $G_{Ep}(t)$ in Fig. 3 one predicts the corresponding behaviours of the difference of deep inelastic cross sections in Fig. 4. By a measurement of the latter the true $t < 0$ behaviour of $G_{Ep}(t)$ can be chosen.

The work was in part supported by the Slovak Grant Agency for Sciences VEGA, Gr. No. 2/4099/24.

REFERENCES

1. M. K. Jones et al, *Phys. Rev. Lett.* **84** (2000) 1398.
2. O. Gayon et al, *Phys. Rev. Lett.* **88** (2002) 092301.
3. L. Andivakis et al, *Phys. Rev.* **D50** (1994) 5491.
4. S. Dubnička, A. Z. Dubničková, P. Weisenpacher, *J. Phys.* **G29** (2003) 405.
5. E. Bartoš, S. Dubnička, E. A. Kuraev, *Phys. Rev.* **D70** (2004) 117901.

Production of Direct Photons, π^0's, and η's in p+p and Au+Au Collisions at RHIC

Terry C. Awes, for the PHENIX Collaboration

Oak Ridge National Laboratory, Oak Ridge, TN 37831, USA

Abstract. The PHENIX experiment at RHIC has measured neutral pion, eta, and direct photon production in $\sqrt{s_{NN}} = 200$ GeV p+p and Au+Au collisions. The neutral pion and direct photon yields in p+p collisions are well described by NLO pQCD predictions. In Au+Au collisions the neutral pion yield is suppressed as compared to expectations from p+p collisions, with a suppression which increases with increasing nuclear overlap. However, the direct photon yield shows no centrality dependent suppression, indicating that the pion suppression is due to a final state effect, such as parton energy loss in the excited final state.

Keywords: Pion Production, Direct Photons, Heavy Ions, Nuclear Effects, Quark Gluon Plasma
PACS: 25.75.Dw

INTRODUCTION

The most dramatic observation from experiments at the Relativistic Heavy Ion Collider (RHIC) is the strong suppression of the yield of hadrons at large transverse momenta ($p_T > 2$ GeV/c) in central Au+Au collisions, as compared to measured yields in $p+p$ collisions scaled by the number of binary nucleon-nucleon collisions [1, 2, 3, 4]. This effect was predicted to result from the energy loss of hard-scattered partons propagating through the high density matter created in heavy ion collisions [5]. It was later proposed that the observed hadron suppression could be an initial-state effect due to saturation of the initial parton distributions in large nuclei [6]. Measurement of direct photon production allows a rather definitive discrimination between initial- and final-state suppression due to the fact that photons, once produced, are essentially unaffected by the surrounding matter. Hence photons produced directly in initial parton scatterings will not be quenched unless the initial parton distributions are suppressed in the nucleus. In fact, the direct photon yield may be enhanced in *AA* collisions [7] due to various processes such as momentum broadening of the incoming partons, additional fragmentation contributions [8, 9], or additional scatterings in the thermalizing dense matter of the final state.

The results described here were obtained with the PHENIX experiment at RHIC. Photons were measured with the PHENIX electromagnetic calorimeter subsytem located in the central arms of PHENIX. Each arm covers 90° in ϕ with $|y| < 0.35$. The neutral pion and eta yields were extracted from their two-photon decay branch via a photon-pair invariant mass analysis. Further details may be found in [3, 10, 11, 12].

CP792, *Deep Inelastic Scattering, DIS 2005,* edited by W. H. Smith and S. R. Dasu

RESULTS

The completely corrected inclusive photon yields are compared to the expected yields of background photons from hadronic decays in the left panel of Fig. 1 for minimum bias Au+Au collisions (0-92% of the geometric cross section) and for five centrality selections. The decay photon calculations are based on the measured π^0 and η spectra [3, 12] assuming m_T-scaling for all other radiative decays (η',K_s^0,ω). The comparison is made as the ratio of measured (inclusive) γ/π^0 and calculated background γ/π^0 since many uncertainties, such as the energy scale, cancel to varying extent in the ratio. Since the π^0 spectra of the background calculations are taken to be the same as the measured spectra any significant deviation of the double ratio above unity indicates a direct photon excess. In Fig. 1 an excess is observed at high p_T with a magnitude that increases with increasing centrality of the collision. The extracted direct photon spectra as a function of centrality are shown in the right panel of Fig. 1. The curves shown in Fig. 1 are NLO pQCD $p+p$ predictions [12] scaled by the number of binary nucleon-nucleon collisions corresponding to the chosen centrality selection. The NLO pQCD predictions are found to be consistent with the observed direct photon yields.

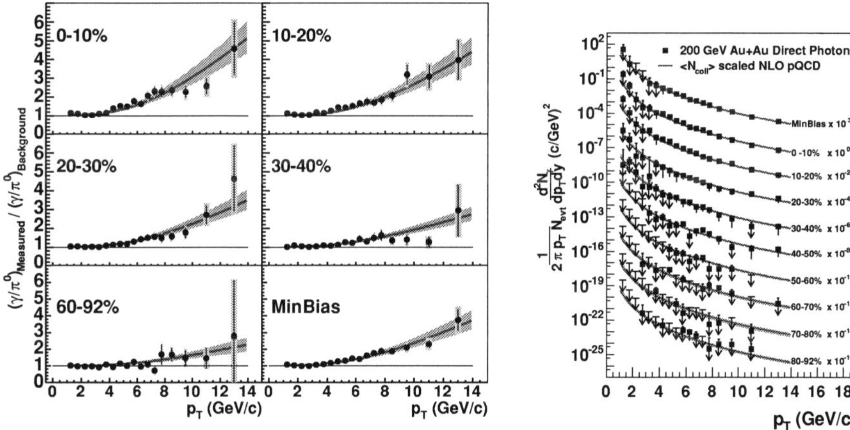

FIGURE 1. Left: Double ratio of measured $(\gamma/\pi^0)_{\text{Measured}}$ invariant yield ratio to the background decay $(\gamma/\pi^0)_{\text{Background}}$ ratio as a function of p_T for minimum bias and for five centralities of Au+Au collisions at $\sqrt{s_{NN}}$ = 200 GeV (0-10% is the most central). Statistical and total errors are indicated separately on each data point by the vertical bar and shaded region, respectively. The solid curves are the ratio of NLO pQCD predictions to the background photon invariant yield based on the measured π^0 yield for each centrality class. The shaded region around the curves indicates the variation of the pQCD calculation for scale changes from $p_T/2$ to $2p_T$, plus the $\langle N_{coll} \rangle$ uncertainty. Right: Direct γ invariant yields as a function of transverse momentum for 9 centrality selections and minimum bias Au+Au collisions at $\sqrt{s_{NN}}$ = 200 GeV. The vertical error bar on each point indicates the total error. Arrows indicate measurements consistent with zero yield with the tail of the arrow indicating the 90% confidence level upper limit. The solid curves are NLO pQCD predictions.

Medium effects in *AA* collisions are often presented using the *nuclear modification factor* given as the ratio of the measured *AA* invariant yields to the yield per $p+p$ collision times the average number of nucleon-nucleon collisions $\langle N_{coll} \rangle$ in the centrality

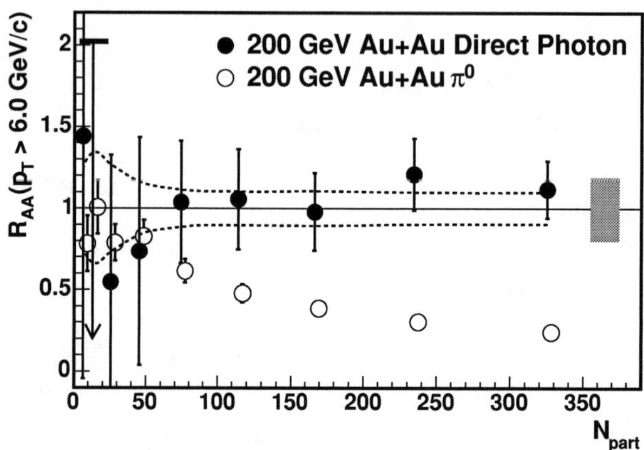

FIGURE 2. Ratio of Au+Au yield to $p+p$ yield normalized by the number of binary nucleon collisions as a function of the number of participating nucleons N_{part} for direct γ (closed circles) and π^0 (open circles) yields integrated above 6 GeV$/c$. The $p+p$ direct photon yield is taken as the NLO pQCD prediction. The error bars indicate the total error excluding the error on $\langle N_{coll}\rangle$ shown by the dashed lines and the scale uncertainty of the NLO calculation shown by the shaded region at the right.

bin under consideration:

$$R_{AA}(p_T) = \frac{(1/N_{AA}^{evt})\, d^2N_{AA}/dp_T dy}{\langle N_{coll}\rangle/\sigma_{pp}^{inel} \times d^2\sigma_{pp}/dp_T dy}. \tag{1}$$

The deviation of $R_{AA}(p_T)$ from unity provides a measure of the deviation of AA data from an incoherent superposition of NN collisions.

The centrality dependence of $R_{AA}(p_T > 6$ GeV$/c)$ for high transverse momentum direct photons are shown in Fig. 3 and compared to the result for π^0 production. Unlike the π^0 production which shows a suppression which increases with Au+Au collision centrality, the high p_T direct photon yield is consistent with the NLO pQCD predictions scaled by the number of nucleon-nucleon collisions.

The p_T dependence of the direct photon $R_{AA}(p_T)$ for central Au+Au collisions is shown in Fig. 3 and compared to the π^0 $R_{AA}(p_T)$. Unlike the suppression observed for π^0 production, the direct photon yield is observed to scale with the number of collisions, within errors, over the full p_T range.

SUMMARY AND CONCLUSION

The PHENIX experiment at RHIC has measured π^0, η, and direct photon production in $p+p$ and Au+Au collisions at $\sqrt{s_{NN}} = 200$ GeV. The π^0 and direct photon yields in $p+p$ collisions are well-described by NLO pQCD predictions. As a function of collision centrality in Au+Au collisions, the π^0 yield is suppressed in comparison to the yield expected from $p+p$ collisions scaled by the number of binary nucleon-nucleon collisions, with a suppression which increases with increasing centrality and

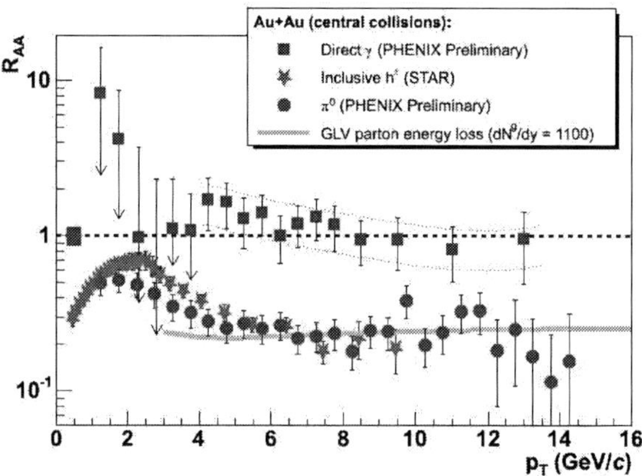

FIGURE 3. Ratio of Au+Au yield to $p+p$ yield normalized by the number of binary nucleon collisions as a function of p_T for direct γ (squares) and π^0 (circles) yields for the 10% most central Au+Au collisions. The $p+p$ direct photon yield is taken as the NLO pQCD prediction with scale variation indicated by the dashed lines. The error bars indicate the total error excluding the error on $\langle N_{coll} \rangle$ shown by the shaded region $R = 1$. The scale uncertainty of the NLO calculation is shown by the dashed lines about the γ result. Results for non-identified hadrons from STAR [13] are also shown (stars). Hadron energy loss calculations [14, 15, 16] are shown by the solid curve.

that is essentially independent of p_T at high p_T. In contrast, the direct photon yield is observed to scale with the number of binary nucleon-nucleon collisions with a yield that is consistent with the NLO pQCD predictions. The results clearly indicate that the hadron suppression observed in central Au+Au collisions is a consequence of the dense matter produced in the final state.

REFERENCES

1. K. Adcox, et al., *Phys. Rev. Lett.*, **88**, 022301 (2002).
2. C. Adler, et al., *Phys. Rev. Lett.*, **89**, 202301 (2002).
3. S. S. Adler, et al., *Phys. Rev. Lett.*, **91**, 072301 (2003).
4. S. S. Adler, et al., *Phys. Rev.*, **C69**, 034909 (2004).
5. M. Gyulassy, and M. Plumer, *Phys. Lett.*, **B243**, 432–438 (1990).
6. D. Kharzeev, E. Levin, and L. McLerran, *Phys. Lett.*, **B561**, 93–101 (2003).
7. T. Peitzmann, and M. H. Thoma, *Phys. Rept.*, **364**, 175–246 (2002).
8. R. J. Fries, B. Muller, and D. K. Srivastava, *Phys. Rev. Lett.*, **90**, 132301 (2003).
9. B. G. Zakharov, *JETP Lett.*, **80**, 1–6 (2004).
10. S. S. Adler, et al., *Phys. Rev. Lett.*, **91**, 241803 (2003).
11. S. S. Adler, et al., *Phys. Rev.*, **D71**, 071102 (2005).
12. S. S. Adler, et al. (2005), `nucl-ex/0503003`.
13. J. Adams, et al., *Phys. Rev. Lett.*, **91**, 172302 (2003), `nucl-ex/0305015`.
14. M. Gyulassy, P. Levai, and I. Vitev, *Phys. Rev. Lett.*, **85**, 5535–5538 (2000), `nucl-th/0005032`.
15. I. Vitev, *J. Phys.*, **G30**, S791–S800 (2004), `hep-ph/0403089`.
16. D. d'Enterria (2005), `nucl-ex/0504001`.

Jet properties from $\pi^{\pm}$ - $h^{\pm}$ correlation in p+p and d+Au collisions

Jiangyong Jia, for the PHENIX Collaboration

Columbia University, New York, NY 10027 and Nevis Laboratories, Irvington, NY 10533, USA

Abstract. We discuss results on the charged pion - charged hadron correlation in $p+p$ and d + Au collisions as measured by the PHENIX Collaboration. Properties of di-jet system, such as the jet shape, associated hadron yield per trigger pion, and the underlying event are extracted statistically from the $\pi^{\pm}-h^{\pm}$ correlation function in $\Delta\phi$ and $\Delta\eta$. For jet triggered with high p_T pions ($p_T > 5$ GeV/c), no apparent differences in the jet properties are seen between $p+p$ and d + Au.

Keywords: Two particle azimuth correlation, conditional yield, nuclear effect, underlying event
PACS: 25.75.Dw

INTRODUCTION

The technique of two particle correlation in relative azimuth ($\Delta\phi$) and pseudorapidity ($\Delta\eta$) is an useful tool to access the (di-)jet properties in heavy-ion collisions. Comparing with the traditional full jet reconstruction method, the two particle correlation method is relatively insensitive to the level of the underling event, thus can probe soft jets ($\lesssim$5 GeV/c); combining with event mixing technique, it can also be used for detectors with limited acceptance.

To leading order in QCD, high p_T jets are produced back-to-back in azimuth. This back-to-back correlation, however, is smeared by the fragmentation process and the initial and final state radiation, which lead to a typical di-hadron correlation function in $\Delta\phi$ as shown schematically in Figure.1. The associated hadron yield per trigger $\pi^{\pm}$ (conditional yield or CY) can be parameterized by a constant plus a double gauss function,

$$\frac{1}{N^0_{\rm trig}}\frac{dN_0}{d\Delta\phi} = B + \frac{\rm N_S}{\sqrt{2\pi}\sigma_N}e^{\frac{-\Delta\phi^2}{2\sigma_N^2}} + \frac{\rm N_A}{\sqrt{2\pi}\sigma_F}e^{\frac{-(\Delta\phi-\pi)^2}{2\sigma_F^2}}, \quad (1)$$

In this analysis, everything about the (di-)jet is extracted from this parameterization. The peaks in the same side ($\Delta\phi = 0$) and the away side ($\Delta\phi = \pi$) represent the intra-jet and di-jet correlation, respectively. The widths of the peaks are controlled by the jet fragmentation momentum j_T and the parton transverse momentum k_T [1, 2]: $\sigma_{same} \propto j_{Ty}$, $\sqrt{\sigma^2_{away} - \sigma^2_{same}} \propto k_{Ty}$, where the subscript "$y$" represent the 1D projection in transverse plane; The integrals of the peaks, N_S and N_A give the total number of hadrons associated with the trigger hadrons in the same side and the away side; The pedestal level beneath the jet structure, B, represents contributions from the underlying event.

We focus the physics discussions on three aspects of the $\pi^{\pm} - h^{\pm}$ correlations in Figure.1: jet shape, jet yield and the underlying event. Further details on the method,

CP792, *Deep Inelastic Scattering, DIS 2005,* edited by W. H. Smith and S. R. Dasu

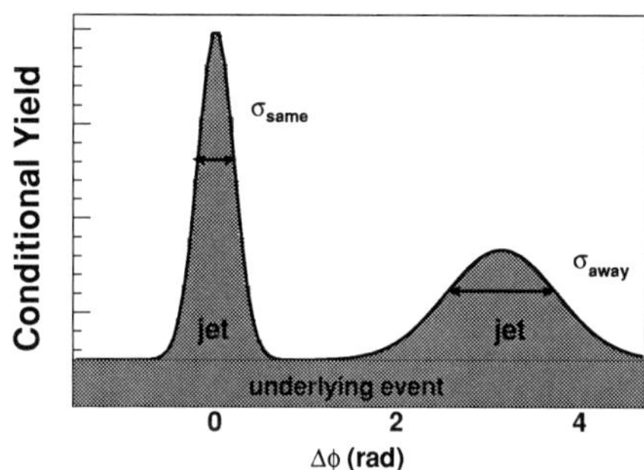

FIGURE 1. Cartoon of the two particle $\Delta\phi$ correlation. The yield of hadron per trigger (Conditional Yield) has a di-jet part and a part corresponding to the underlying event.

analysis, and physics results can be found in[2, 3].

RESULTS

Left panel of Figure.2 shows the $\pi^{\pm} - h^{\pm}$ $\Delta\phi$ distribution from $p+p$ and d + Au collisions for several ranges of associated hadron transverse momentum, $p_{T,\mathrm{assoc}}$, with trigger $\pi^{\pm}$ p_T $5 < p_{T,\mathrm{trig}} < 10$ GeV/c. The widths decrease with increasing $p_{T,\mathrm{assoc}}$, which is consistent with narrowing of the jet cone for larger $p_{T,\mathrm{assoc}}$. It is interesting to notice that a large fraction of all hadrons in the event are associated with the trigger, thus are originated from the hard-scattered partons. Even for $p_{T,\mathrm{assoc}}$ as low as $0.4-1$ GeV/c, about 51% hadron yield in $p+p$ (27% in d + Au) comes from the jet fragmentation. Right panel of Figure.2 shows the extracted jet widths in $\Delta\phi$, comparing with those in $\Delta\eta$ for various $p_{T,\mathrm{assoc}}$. The overall agreement between the jet widths in $\Delta\eta$ and $\Delta\phi$ is pretty good, except at small $p_{T,\mathrm{assoc}}$, where the width in $\Delta\eta$ is systematically lower than that in $\Delta\phi$. The fact that this discrepancy exist in both $p+p$ and d + Au collisions indicates that this deviation is likely due to the systematics of the fitting in a limited $\Delta\eta$ range [1] rather than any real physics effect in d + Au.

The same side and away side p_T distributions of the charged hadrons associated with trigger pions are plotted in Fig3, comparing between $p+p$ and d + Au collisions. The same side yield is related to the di-hadron fragmentation, since both particles comes from the same jet, while the away side yield depends on two independent fragmentation functions: one parton fragments to produce the trigger, while the second parton produces the associated hadron. No apparent differences are seen between $p+p$ and d + Au; this observation is in contradiction to some recombination model prediction [4], in which a significant difference is expected due to shower-thermal contribution.

Events triggered by high p_T hadrons not only contain particles originated from the two hard-scattered partons, but also those come from soft multiple interaction and the beam

[1] PHENIX coverage in η is $|\eta| < 0.35$, which gives a pair coverage of $|\Delta\eta| < 0.7$.

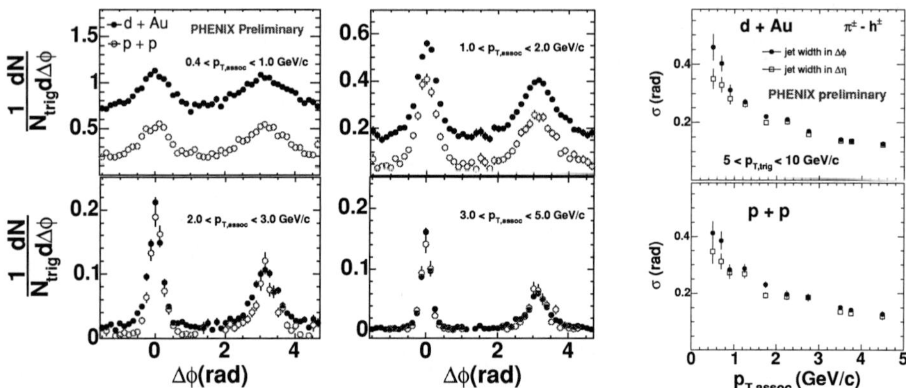

FIGURE 2. Left: Per-trigger pair distributions in $p+p$ and minimum bias d+Au collisions. The trigger $\pi^{\pm}$ are correlated with hadrons with $p_{T,\text{assoc}}$ $0.4-1.0$ GeV/c, $1.0-2.0$ GeV/c, $2.0-3.0$ GeV/c and $3.0-5.0$ GeV/c (from top to bottom and left to right). Right: The comparison of jet width as function of $p_{T,\text{assoc}}$ in $\Delta\phi$ (solid circles) and $\Delta\eta$ (open boxes) from $\pi^{\pm}-h^{\pm}$ correlation. Top and bottom panels show the results for d + Au and $p+p$, respectively.

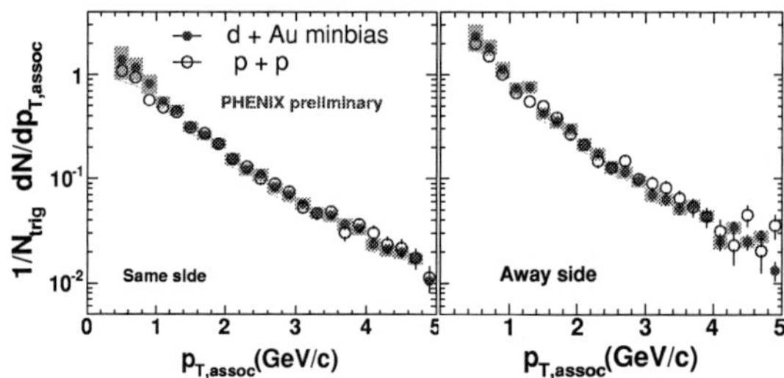

FIGURE 3. Jet pair distribution as function of $p_{T,assoc}$ for same side (right panel) and away side (left panel) in $p+p$ and d + Au.

remnants. Underlying event in $p+p$ and d + Au collisions refers to all hadrons except those from the two outgoing hard-scattered partons, which includes contributions from the beam remnants and initial and final state radiation [5]. It has been studied extensively at the Tevatron energy [6, 5]. Similar studies at the RHIC are useful in understanding it's dependence on $\sqrt{s}$, and can provide valuable constrains on the underlying event physics at the LHC.

Figure.4 shows the jet pair distribution in $p+p$ collisions, reproduced from Figure.2 but plotted under semi-log scale. The pedestal in the $\Delta\phi$ correlation, which represents the underlying event contribution, decreases quickly and becomes negligible at $p_{T,\text{assoc}} > 2$ GeV/c. However, the level corresponding to minimum bias $p+p$ events, denoted by the thick horizontal line, seems to decrease even faster. Since minimum bias event has small hard-scattering contribution, the relative abundance of the pedestal in triggered events over the minimum bias events indicates that most of the underlying event comes from the initial or final state radiation of the hard-scattered partons.

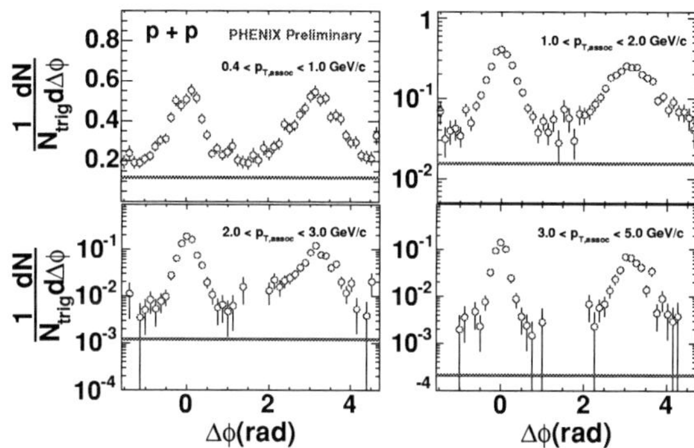

FIGURE 4. Condition yield in $\Delta\phi$ for $p+p$ collisions (from Figure.2). The thick solid line represents the average level for minimum bias events, i.e. it is equal to $\text{Yield}_{\text{pp}}/(2\pi)$.

In d + Au case, Figure.2 indicates that the underlying event levels are larger than those in $p+p$, although the properties of the jets are quite similar. Under the assumption that the hard-scattering happens in one nucleon-nucleon collision and that the ambient particle production scale as the nuclear modification factor, R_{dAu} measured in d + Au [7]. The underlying event yields in $p+p$ and d + Au, U_{dAu} and U_{pp} are connected to each other through the following simple relation,

$$U_{\text{dAu}} = U_{\text{pp}} + R_{\text{dAu}}\,(N_{\text{coll}} - 1)\,\text{Yield}_{\text{pp}} \tag{2}$$

where Yield_{pp} represents the hadron yield per event in minimum bias $p+p$ collisions. Divide both side by Yield_{pp}, we get,

$$\lambda_{\text{dAu}} = \lambda_{\text{pp}} + R_{\text{dAu}}\,(N_{\text{coll}} - 1) \tag{3}$$

$$\lambda_{\text{dAu}} = U_{\text{dAu}}/\text{Yield}_{\text{pp}}, \lambda_{\text{pp}} = \text{U}_{\text{pp}}/\text{Yield}_{\text{pp}} \tag{4}$$

note λ_{pp} denotes the ratio of underlying event yield to minimum bias event in $p+p$, which should be larger than 1 according to Figure.4.

In summary, we have demonstrated that the two particle correlation method is a powerful tool in accessing basic jet properties. The extracted jet shape and associated yield in $p+p$ and d+Au are very similar, indicating little modification in cold nuclear medium. These measurements serve as important baselines for jet correlation studies in heavy-ion collisions.

REFERENCES

1. J.Rak, J. Phys. **G30**, S1309 (2004), hep-ex/0403038.
2. J. Jia, J. Phys. **G31**, S521 (2005), nucl-ex/0409024.
3. J. Jia, nucl-ex/0506009.
4. R. C. Hwa and C. B. Yang, Phys. Rev. **C70**, 054902 (2004), nucl-th/0407081.
5. T. Affolder et al (CDF), Phys. Rev. **D65**, 092002 (2002).
6. D. Acosta et al (CDF), Phys. Rev. **D70**, 072002 (2004), hep-ex/0404004.
7. F. Matathias, (PHENIX) (2005), nucl-ex/0504019.

HEAVY FLAVORS

WORKING GROUP PRESENTATIONS

Charm Production at Low Q^2 With the ZEUS Detector

Gayane Aghuzumtsyan (on behalf of the ZEUS Collaboration)

Physikalisches Institut, University of Bonn

Abstract. The production of $D^*(2010)$ mesons in deep inelastic scattering at low Q^2 has been measured with the ZEUS detector at HERA using an integrated luminosity of 81.9 pb^{-1}. The D^* mesons have been reconstructed from their decay into D^0 and π_s with the decay $D^0 \to K^-\pi^+$ and corresponding antiparticle decay. Differential D^* cross sections as functions of exchanged photon virtuality, Q^2, inelasticity, y, transverse momentum of the D^* meson, $p_T(D^*)$, and pseudorapidity of the D^* meson, $\eta(D^*)$, have been measured, using the beam-pipe calorimeter of ZEUS. The kinematic region of the measurement is $0.05 < Q^2 < 0.7$ GeV^2, $0.02 < y < 0.85$, $1.5 < p_T(D^*) < 9.0$ GeV and $|\eta(D^*)| < 1.5$. The measured differential cross sections are compared with the predictions of next-to-leading-order QCD.

Keywords: Charm, production, D*, low Q^2, cross sections, QCD
PACS: 12.38.Bx, 12.38.Qk, 13.25.Ft, 13.60.Le, 13.85.Hd, 14.40.Lb

INTRODUCTION

Charm quarks have been measured both in photoproduction and in deep inelastic scattering (DIS) at HERA [1, 2, 3, 4, 5]. The measurements are reasonably described by perturbative QCD calculations where the charm is produced mainly by boson-gluon fusion (BGF).

Charm production in the transition region from DIS to photoproduction is probed. Event containing D^* mesons were selected in which the virtuality of the exchanged photon , Q^2, lies in the range $0.05 < Q^2 < 0.7$ GeV^2. The measurement was performed using the beam pipe calorimeter (BPC) [6, 7]. Differential D^* cross sections have been measured as a function of Q^2, y, $p_T(D^*)$ ans $\eta(D^*)$ and compared to NLO predictions using the HVQDIS program.

EVENT SELECTION AND DATA ANALYSIS

The data from the years 1998-2000 with an integrated luminosity of 82 pb^{-1} have been analysed with the ZEUS [8] detector at HERA, where protons of energy $E_p = 920$ GeV collided with the electrons or positrons of energy $E_e = 27.5$ GeV.

Events which fulfil the following conditions were selected: characteristic energy deposit of an electron within the fiducial area of the BPC with $E_{\rm BPC} > 4$ GeV; BPC timing measurement consistent with an ep interaction $| < \tau_{\rm BPC} > | < 3$ ms; a primary vertex with $|Z_{\rm vertex}| < 50$ cm was reconstructed; the ratio of the transverse momentum of the D^* to the total tranverse CAL energy deposit was $p_T(D^*)/E_T > 0.1$ and $35 < \delta_{\rm BPC} < 65$ GeV, where $\delta_{\rm BPC} = \delta + E_{\rm BPC}(1 - \cos(\Theta_{\rm BPC}))$, $\delta = \sum_i (E - p_z)_i$, the index i runs over

CP792, *Deep Inelastic Scattering, DIS 2005,* edited by W. H. Smith and S. R. Dasu

the CAL clusters, and Θ_{BPC} is the angle of the scattered electron w.r.t. the proton beam axis. Events with an additional reconstructed electron in the calorimeter are suppressed.

The selected kinematic region was $0.05 < Q^2 < 0.7$ GeV^2 and $0.02 < y < 0.85$.

D^* mesons were reconstructed from tracks in the decay channel $D^{*+} \to D^0 \pi_s^+$ (+c.c.) with $D^0 \to K^- \pi^+$ (+c.c.). Pairs of well-reconstructed tracks with $p_T > 0.45$ GeV were combined to form a D^0 candidate. A third track with $p_T > 0.12$ GeV and charge opposite to that of the kaon track was combined with the D^0 candidate to form a D^* candidate, and kept if its charge was opposite to the kaon track. A different mass window for the D^0 was used for each bin of $p_T(D^*)$ from $1.82 < M(K\pi) < 1.91$ GeV, to $1.79 < M(K\pi) < 1.94$ GeV. To allow the background to be determined, D^0 candidates with wrong-sign combinations, in which both tracks forming the D^0 candidates have the same charge and third track has the opposite charge, were also retained.

D^* mesons were selected in the kinematic region $1.5 < p_T(D^*) < 9$ GeV and $|\eta(D^*)| < 1.5$.

Figure 1 shows the distribution of $\Delta M = M(D^*) - M(D^0)$ for the reconstructed events. A clear signal is seen around the nominal value of ΔM. The number of D^* mesons, extracted by an unbinned fit, was $N(D^*) = 253 \pm 25$.

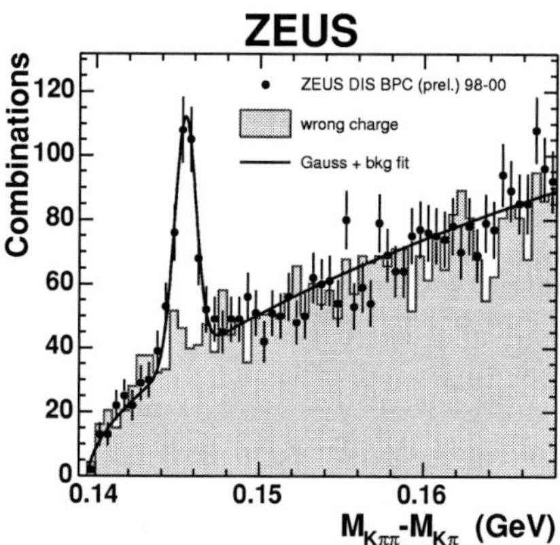

FIGURE 1. The distribution of the mass difference, $\Delta M = M(K\pi\pi_s) - M(K\pi)$, for $D^{*\pm}$ candidates from BPC measurements. The histogram shows the ΔM distribution for wrong charge combinations. The solid curve represents the fi t.

CROSS SECTIONS

The inclusive D^* cross sections at low Q^2 were measured in the kinematic region $0.05 < Q^2 < 0.7$ GeV^2, $0.02 < y < 0.85$, $1.5 < p_T(D^*) < 9$ GeV and $|\eta(D^*)| < 1.5$. The HERWIG [9] Monte Carlo program was used to correct the data for detector effects and calculate acceptances. The measured cross section is

$$\sigma(e^{\pm}p \to e^{\pm}D^*X) = 10.1 \pm 1.0\ (\text{stat})\ ^{+1.1}_{-0.8}(\text{syst})\ \text{nb}. \tag{1}$$

The replacement of HERWIG by RAPGAP [10] for acceptance corrections was the main source of systematic error (8%).

The NLO prediction of the $c\bar{c}$ cross section was obtained using the program HVQDIS. The fragmentation of the charm quarks was performed according to the Peterson model with the parameter $\varepsilon = 0.035$. The nominal mass of the charm quark was set to $m_c = 1.35$ GeV. The normalisation and factorisation scales were set to $\mu = \sqrt{Q^2 + 4m_c^2}$. The ZEUS NLO QCD fit and CTEQ5F3 were used as the parametrisation of the proton PDFs. The NLO predicted total cross section is

$$\sigma_{\mathrm{HVQDIS}}(e^{\pm}p \to e^{\pm}D^{*}X) = 8.6^{+1.9}_{-1.8}(\mathrm{syst.})\,\mathrm{nb} \tag{2}$$

To estimate the theoretical uncertainty, the ZEUS PDF fit was used and the scale μ, the mass of the charm quark and the parameter ε in the Peterson fragmentation function were varied in the range: $(Q^2 + m_c^2) < \mu^2 < 4(Q^2 + 4m_c^2)$, $1.2 < m_c < 1.5$ GeV, $0.02 < \varepsilon < 0.05$, respectively.

Figure 2 shows the single differential cross sections as a function of Q^2, y, $p_T(D^*)$ and $\eta(D^*)$ compared to the NLO QCD predictions. In general, shape and normalization of the distributions are described by the NLO predictions. A comparison of $d\sigma/dQ^2$ with previous ZEUS results is shown in figure 3. For this figure, in order to have data comparable to older measurements, the y range was restricted to $0.02 < y < 0.7$. The unbinned fit in this restricted kinematic region yielded $N(D^*) = 239 \pm 23$. The transition from high to low Q^2 is well described by the predictions, and therefore well understood.

SUMMARY

The production of D^* mesons in DIS at HERA was measured with the ZEUS detector in the kinematic region $0.05 < Q^2 < 0.7$ GeV^2, $0.02 < y < 0.85$, $1.5 < p_T(D^*) < 9$ GeV, $|\eta(D^*)| < 1.5$, probing the transition region to photoproduction regime. The theoretical NLO QCD calculation of BGF charm production is consistent with the measured cross sections at low Q^2. A comparison to data at higher Q^2 shows that this transition region is well understood.

REFERENCES

1. H1 Coll., C. Adolf et el. *Phys. Lett.*, **B 528**, 199 (2002).
2. ZEUS Coll., J. Breitweg et al., *Phys. Lett.*, **B 481**, 213 (2000).
3. ZEUS Coll., J. Breitweg et al., *Eur. Phys. J.*, **C 12**, 35 (2000).
4. H1 Coll., C. Adolf et el. *Nucl. Phys.*, **B 545**, 21 (1999).
5. ZEUS Coll., J. Breitweg et al., *Eur. Phys. J.*, **C 6**, 67 (1999).
6. C. Amelung, *Measurement of Proton Structure Function F_2 at Very Low Q^2 at HERA*, Ph.D. Thesis, Universität Bonn, Bonn, (Germany), Report BONN-IR-99-15, DESY-THESIS-2000-002,1999.
7. B. Surrow, *Measurement of the Proton Structure Function F2at Low Q^2 and Very Low x with the ZEUS Beam Pipe Calorimeter at HERA*, Ph.D. Thesis, University Hamburg, Hamburg, (Germany), DESY Report DESY-THESIS-1998-004.
8. ZEUS Coll., U. Holm et al.,*The ZEUS Detector*, DESY Status Report (unpublished), 1993, *www-zeus.desy.de/bluebook/bluebook.html*.
9. ZEUS Coll., G. Marchesini et al.,*Comp.Phys. Comm.*, **67**, 465 (1992).
10. H. Jung,*Comp. Phys. Commun.*, **86**, 147 (1995).

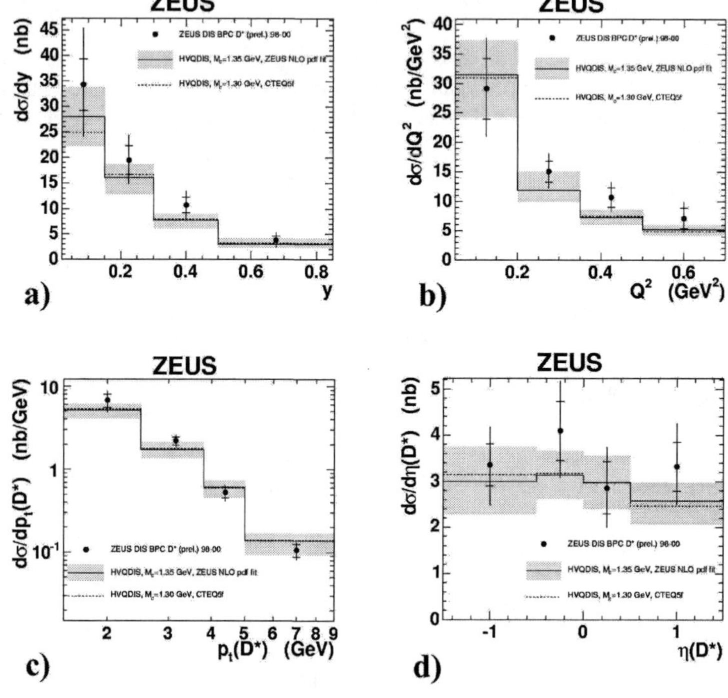

FIGURE 2. Differential D^* cross sections as a function of y (a), Q^2 (b), $p_T(D^*)$ (c) and $\eta(D^*)$ for low Q^2 compared to the NLO predictions from HVQDIS.

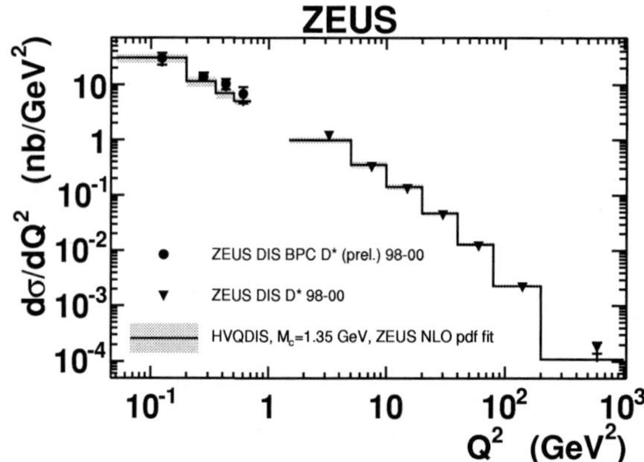

FIGURE 3. Differential D^* cross sections as a function of Q^2 for low Q^2 and from previous results on D^* production in DIS compared to the NLO predictions from HVQDIS.

The Structure of Charm Jets in Deep-Inelastic Scattering

Adrian Perieanu

on behalf of the H1 collaboration
DESY, Hamburg, Germany, email: perieanu@mail.desy.de

Abstract. The structure of charm jets in deep-inelastic scattering (DIS) is studied with the H1 detector at HERA using an integrated luminosity of 50 pb^{-1}. The charm events are tagged by a D^*-meson. The events are required to have at least one jet containing the D^*-meson (D^*Jet). A second jet (*Other*Jet) is also studied, if present. The structure of these jets is investigated by measuring jet shapes and subjet multiplicities as well as the angular distribution of a subjet defined to give a correlation with a corresponding gluon at the parton level. The study of the latter distribution and its dependence on the jet energy is motivated by the suppression of soft gluon radiation from heavy quarks, as expected from QCD. All data are reasonably well described by a QCD model for charm production as implemented in RAPGAP and JETSET.

Keywords: Jet Shape, Charm, Soft Gluons, Dead Cone
PACS: 13.85.Hd, 14.40.Lb, 14.65.Dw

INTRODUCTION

Jets initiated by heavy quarks are expected to be different from those initiated by light quarks. In the case of heavy quarks, the phase space available for gluon emissions is reduced due to kinematic effects of the heavy quark mass [1].

In this analysis the structure of charm jets is investigated by measuring jet shape variables and subjet multiplicities, as well as a specific angular distribution and its dependence on the jet energy, which might be expected to be sensitive to the suppression of soft gluons from heavy quarks close to their direction of flight. For such gluons the differential cross section can be approximated by [1]:

$$\frac{d\sigma_{Q\to Q+g}}{d\alpha} \approx K \frac{\alpha^3}{(\alpha_0^2+\alpha^2)^2}\,, \tag{1}$$

where

$$\alpha_0 = M_Q/E_Q\,, \tag{2}$$

with M_Q being the mass of the heavy quark and E_Q its energy. The angle between the radiated gluon and the primary heavy quark momentum is denoted by α. It follows from Eq. 1 that in the region of small α the emission of soft gluons is suppressed, and therefore this region is called the "dead cone". It is characterized by an opening angle α_0 which depends on the mass and energy of the heavy quark. Here an attempt is made to define an experimentally accessible angle which should approximate the parton level angle α.

Indirect evidence for the presence of the dead cone, based on comparing multiplicities in light-quark and heavy-quark jets has been reported by the DELPHI collaboration [2].

CP792, *Deep Inelastic Scattering, DIS 2005,* edited by W. H. Smith and S. R. Dasu

EXPERIMENTAL METHOD AND DATA SELECTION

At HERA, in ep collisions at $\sqrt{s} = 318$ GeV, the charm quarks are produced in DIS mainly via the boson-gluon fusion process [3]. They are produced in pairs and, in contrast to most e^+e^- experiments, close to the kinematic threshold. This allows to study the effect of the charm mass on jet structure at low jet momenta, in a region of phase space where these effects are expected to be measurable with the detector resolution of the H1 experiment. The data presented here were collected during the running periods of 1999 and 2000. A detailed description of the H1 detector is given in [4], and the components most relevant for this analysis are discussed in [3].

The scattered positron is detected in the backward calorimeter (SpaCal). In order to ensure high acceptance for the entire kinematic region, the square of the four momentum transfer[1] is restricted to $2 \leq Q^2 \leq 100$ GeV^2 and the inelasticity to $0.05 \leq y \leq 0.7$. The energy of the scattered positron is required to be larger than 8 GeV. In order to suppress background, cuts on the position of the event vertex ($|z_{\rm vtx}| < 35$ cm) and on the quantity $E - p_z$, summed over all hadronic particles and the scattered positron, ($40 < E - p_z < 75$ GeV) are applied.

The events with charm are tagged by reconstructing a D^*-meson[2] using the channel $D^{*+} \to D^0\pi_s^+ \to K^-\pi^+\pi_s^+$. As soon as a D^*-meson is found in an event, it is treated as one particle and its decay products are discarded. The range of the transverse momentum and the pseudorapidity[3] of the D^*-meson is restricted to $p_{{\rm t},D^*} > 1.5$ GeV and $|\eta_{D^*}| < 1.5$. Kinematic cuts, applied to individual charged particle tracks are: $p_{{\rm t},\pi_s} > 0.12$ GeV, $p_{{\rm t},K,\pi} > 0.25$ GeV, $|\eta_{K,\pi,\pi_s}| < 1.5$ and $p_{{\rm t},K} + p_{{\rm t},\pi} > 2$ GeV.

The D^*-signal is identified using the mass difference $\Delta m = m_{K\pi\pi_s} - m_{K\pi}$ for a D^0-meson mass window $|m_{K\pi} - m_{D^0}| < 0.07$ GeV. The background is subtracted using the wrong-charge (WrCh)[4] events in the Δm range $0.143 < \Delta m < 0.148$ GeV. The normalization of the WrCh background is determined from a simultaneous fit to both the right-charge (RiCh) and WrCh Δm distributions.

Jets are found using the inclusive $k_{\rm t}$ cluster algorithm [5], applied in the laboratory frame, and events with at least one jet having $p_{\rm t,jet} > 1.5$ GeV and containing a D^*-meson are accepted. The properties of this jet, the D^*Jet, and of a second jet defined as the largest $p_{\rm t}$ jet not containing the reconstructed D^*-meson, the *Other*Jet[5], if observed within fiducial limits, are studied.

The observables used to study the energy flow inside a jet are the jet shape variables, averaged over all jets in data, defined as:

$$< \psi(r/R) >= \frac{1}{N_{\rm jets}} \sum_{\rm jets} \frac{E_{\rm t,jet}(r)}{E_{\rm t,jet}(r=R)}, \quad < \rho(r/R) >= \frac{d\psi}{dr} = \frac{1}{N_{\rm jets}} \sum_{\rm jets} \frac{E_{\rm t,jet}(r, r+\Delta r)}{E_{\rm t,jet}(r=R)}, \quad (3)$$

---

1 $Q^2 = 4E_eE'_e\cos^2\left(\frac{\theta_e}{2}\right)$; $y = 1 - \frac{E'_e}{E_e}\sin^2\left(\frac{\theta_e}{2}\right)$.

2 D^* refers to both D^{*+} and D^{*-} states.

3 Here and later in this paper η denotes the pseudorapidity of a particle: $\eta = -\ln\left[\tan\left(\frac{\theta}{2}\right)\right]$.

4 $D^{*+} \to K^+\pi^+\pi_s^-$ and charge conjugate.

5 The azimuthal opening angle between the D^*Jet and the *Other*Jet shows the expected back-to-back topology in the $\gamma^* p$-frame, with a small tail extending to lower angles.

where $E_{t,jet}(r)$ is the total transverse energy of particles inside a cone with the size $r = \sqrt{\Delta\phi^2 + \Delta\eta^2}$ around the jet axis, $E_{t,jet}(r, r+\Delta r)$ is the total transverse energy of particles between a cone with size r and one with $r+\Delta r$. The radius R is set to one.

The mean subjet multiplicity is defined as the number of subjets inside the jet, found by the inclusive k_t jet finder as a function of the jet resolution parameter y_{cut}.

$$< n_{sbj}(y_{cut}) >= \frac{1}{N_{jets}} \sum_{i=1}^{N_{jets}} n^i_{sbj}(y_{cut}) \qquad (4)$$

Fig. 1 shows that the observed distributions are well described by the RAPGAP 2.8/JETSET 7.4 MC model [6] predictions. The observables are shown at the detector level[6].

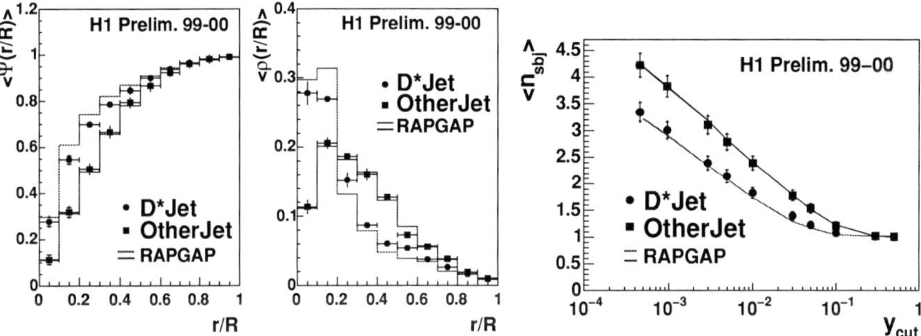

FIGURE 1. The jet shape variables compared to the QCD model for the D^*Jet and the *Other*Jet.

The method to approximate the angular distribution of the soft gluons with respect to the quark direction measures the "gluon" subjet angle, α, in different bins of $E_{D^*\mathrm{Jet}}$ and $E_{Other\mathrm{Jet}}$. The subjets are found by re-running the k_t algorithm on particles inside the D^*Jet and the *Other*Jet until two subjets are found. The "gluon" subjet angle α is defined as the angle between the direction of the D^*Jet (*Other*Jet) and the direction of the subjet not containing the D^*-meson (the less energetic subjet of the *Other*Jet).

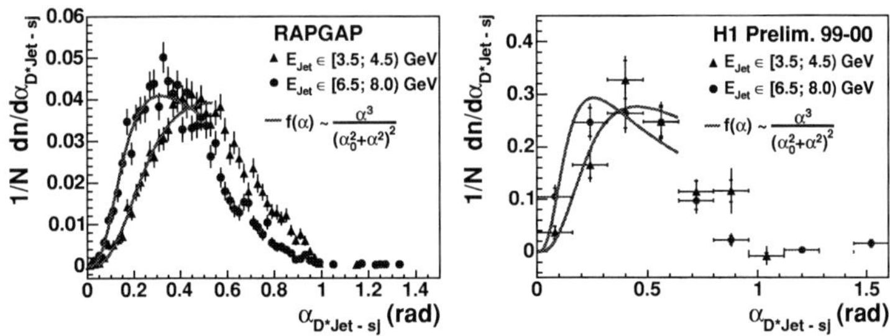

FIGURE 2. The distribution of the angle α between the "gluon" subjet and the D^*Jet for two bins in jet energy for data (right) and the QCD model RAPGAP/JETSET (left). The solid lines indicate the results of a fit using the function $f(\alpha) \sim \alpha^3/(\alpha_0^2+\alpha^2)^2$ predicted by the QCD theory.

---

[6] The α distributions have been calculated also at the parton and hadron level in the MC.

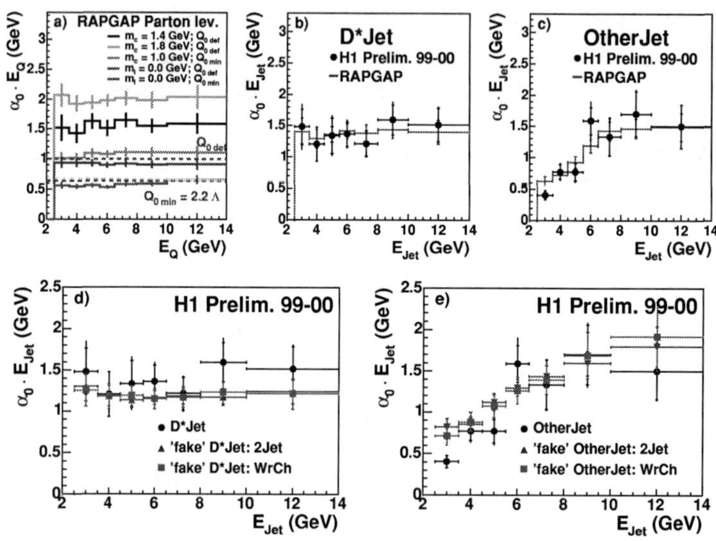

FIGURE 3. The $\alpha_0 E_{\rm jet(Q)}$ distributions. Q_0 is the invariant mass cut-off ($Q_{0\rm def} = 1$ GeV, $\Lambda = 0.29$ GeV).

In Fig. 2 one observes a shift of the α distribution to smaller angles with increasing $E_{D^*\rm Jet}$ as expected from Eq.1, which provides a reasonable fit at small angles to the data and MC. It allows to determine values for α_0 for different $E_{\rm jet}$. From Eq. 2 one expects that the distribution $\alpha_0 \cdot E_{\rm jet}$ is independent of the $E_{\rm jet}$, i. e. a constant of size M_Q. In Fig. 3a) the QCD model predictions, including the soft gluon suppression, are shown at parton level for different values of m_c, Q_0 and for $m_{l=u,d,s} = 0$. In Fig. 3b) and c) the distributions of $\alpha_0 E_{\rm jet}$ for the D^*Jet and the $Other$Jet are shown as a function of $E_{\rm jet}$. They are well described by the QCD model predictions; a flat behavior is not expected for $E_{Other\rm Jet}$ below about 5 GeV due to a deteriorating reconstruction of the charm quark or even misidentifying it. In Fig. 3d) and e) the D^*Jet and $Other$Jet data are compared with "fake" D^*Jet: WrCh and 2Jet[7], respectively "fake" $Other$Jet.

The distributions look somewhat different from the ones from genuine charm jets but 'agree' within the errors[8]. Further studies and more statistics are needed to clarify these differences.

REFERENCES

1. Y. L. Dokshitzer, V. A. Khoze and S. I. Troian, J. Phys. G **17** (1991) 1602.
2. P. Abreu et al. [DELPHI collaboration], *Phys. Lett.* B**479** (2000) 118.
3. C. Adloff et al. [H1 Collaboration], Z. Phys. C **72** (1996) 593 [hep-ex/9607012].
4. I. Abt et al. [H1 collaboration], *Nucl. Instrum. Methods* A**386** (1997) 310/348.
5. S. D. Ellis and D. E. Soper, Phys. Rev. D **48**, 3160 (1993) [hep-ph/9305266].
6. H. Jung, *Comp. Phys. Commun.* **86** (1995)147.

[7] Both samples are dominated by light quarks WrCh($\sim 70\%$) and 2Jet($\sim 80\%$).

[8] The individual systematic errors have been added in quadrature.

Jet Cross Sections in D^* Photoproduction with ZEUS

Takanori Kohno for the ZEUS Collaboration

Denys Wilkinson Building Keble Road, Oxford, OX1 3RH, United Kingdom
Email: t.kohno1@physics.ox.ac.uk

Abstract. Charm photoproduction in D^* photoproduction have been studied using 78.6 pb^{-1} of data collected by the ZEUS detector. The measurement of inclusive jet cross section with a D^* in the final state was performed in the kinematic region, $Q^2 < 1$ GeV, $130 < W < 280$ GeV, $p_T^{D^*} > 3$ GeV, $|\eta^{D^*}| < 1.5$, $E_T^{\mathrm{jet}} > 6$ GeV and $-1.5 < \eta^{\mathrm{jet}} < 2.4$. Differential cross sections as a function of E_T^{jet} and η^{jet} were compared to the NLO QCD predictions. There is a general agreement with the NLO QCD preditions. In addition to above requirements, a dijet sub-sample having at least two jets with $E_T^{\mathrm{jet}} > 6$ GeV and the leading jet with $E_T^{\mathrm{jet}} > 7$ GeV was used to measure dijet correlations. Dijet correlations are compared to the NLO QCD predictions and leading-order (LO) Monte Carlo (MC) models with parton showers (PS). Dijet correlations are described well by the LO+PS models, while the NLO QCD prediction underestimates the measurement in the region where higher-order effects are expected to become significant.

Keywords: charm photoproduction, inclusive jet cross section, dijet correlation
PACS: 12.38.Qk,13.60.Le,13.85.Hd,14.40.Lb

INTRODUCTION

Charm jet production in ep collision should be calculable in perturbative Quantum Chromodynamics (pQCD) since the charm-quark mass, m_c, and the transverse energy of the jet, E_T^{jet}, provide hard scales. In photoproduction, where the virtuality of the photon emitted from the incoming electron is small, there are two main processes contributing to high E_T scattering. In direct processes, the photon interacts directly with the parton in the proton, while in resolved processes, the photon acts as a source of partons such that partons from both photon and the proton participates in the hard scattering. Predictions of charm production can be made using leading order (LO) matrix element followed by the parton shower (PS) algorithm as implemented in Monte Carlo (MC) generators. Next-to-leading order (NLO) QCD calculations are available in two schemes. In the "massive" scheme [1], the charm mass is considered to be large such that only light quarks are the active flavors in the initial state. Therefore, charm is only produced dynamically. In the "massless" approach [2], the charm quark is treated similarly to the light quarks.

The inclusive jet cross section with respect to the transverse energy, E_T^{jet}, and the pseudo-rapidity, η^{jet}, of the jet provides a fundamental test of the parton-level dynamics in a wide range in pseudo-rapidity. In dijet events, more detailed information of the event can be reconstructed. The azimuthal diffence of the two highest transverse energy jets, $\Delta\phi^{\mathrm{jj}}$, and the transverse momentum of the dijet system, p_T^{jj} are particularly interesting.

CP792, *Deep Inelastic Scattering, DIS 2005,* edited by W. H. Smith and S. R. Dasu

In LO $2 \to 2$ process, the two jets are produced back-to-back such that $\Delta\phi^{\rm jj} = \pi$ and $p_T^{\rm jj} = 0$. Large deviations from these values come from the higher-order QCD radiations. In addition, the fractional momentum of the photon contributing to the dijet production, $x_\gamma^{\rm obs}$, can be measured which allows to study the direct-enriched ($0.75 < x_\gamma^{\rm obs} < 1$) and the resolved-enriched ($0 < x_\gamma^{\rm obs} < 0.75$) samples separately.

EVENT SELECTION

The measurement is based on 78.6 pb^{-1} of $e^\pm p$ collisions collected by the ZEUS detector [3] from 1998 to 2000.

The measurement was performed in the photoproduction regime characterised by $Q^2 < 1$ GeV^2, where Q^2 is the virtuality of the photon emitted from the electron. The centre-of-mass energy of the photon-proton system was restricted to $130 < W < 280$ GeV.

Charm events were tagged by reconstructing a $D^{*\pm}$ meson in its decay mode, $D^{*+} \to D^0\pi_s^+ \to (K^-\pi^+)\pi_s^+$ and the charge conjugate. The D^0 meson was formed using tracks in the pseudo-rapidity region $|\eta^{track}| < 1.75$ with a transverse momentum cut of $p_T > 0.4$ GeV. The mass of a pion and a kaon was assumed for each track in turn, to calculate the invariant mass, $m(K\pi)$, which was required to be within $1.80 < m(K\pi) < 1.92$ GeV. After forming a D^* candidate, the slow pion track was selected in the same pseudo-rapidity range as above and $p_T > 0.12$ GeV to form the D^* candidate. The mass difference between the D^* and the D^0, ΔM was required to be within $0.145 < \Delta M < 0.148$ GeV. In order to keep high efficiency and purity, $p_T^{D^*} > 3$ GeV and $|\eta^{D^*}| < 1.5$ were applied to the reconstructed D^*.

Jets were reconstructed by the k_T clustering algorithm over energy flow objects (EFO), which were reconstructed using both tracking and calorimeter information to represent final state particles. Events having at least one jet with $E_T^{\rm jet} > 6$ GeV and $-1.5 < \eta^{\rm jet} < 2.4$ were selected for the inclusive cross section measurement and for the dijet sub-sample, events with two such jets and a further satisfying the requirement on the leading jet, $E_T^{\rm jet1} > 7$ GeV, were used. Jets were divided into two categories; the D^*-tagged jet which is defined as a jet associated with the D^* and the untagged jet referring to all other jets. Experimentally, the D^*-tagged jet was obtained as the jet closest to the D^* in the $\eta - \phi$ space satisfying $\Delta R = \sqrt{(\eta^{\rm jet} - \eta^{D^*})^2 + (\phi^{jet} - \phi^{D^*})^2} < 0.6$.

Results

Jet cross sections for the process $ep \to D^* + {\rm jet} + X$ are measured. With this definition, D^* from beauty production is also included. The beauty contribution is found to be around 3% at low $E_T^{\rm jet}$ and increases up to 15% at high $E_T^{\rm jet}$.

The measurements are compared to the NLO QCD predictions and also to the HERWIG [4] and PYTHIA [5] MC models for dijet correlations. The parton-level cross section provided by the NLO QCD calculations were corrected for the hadronisation effects to be compared with data. A bin-by-bin correction was adopted for this, where the correc-

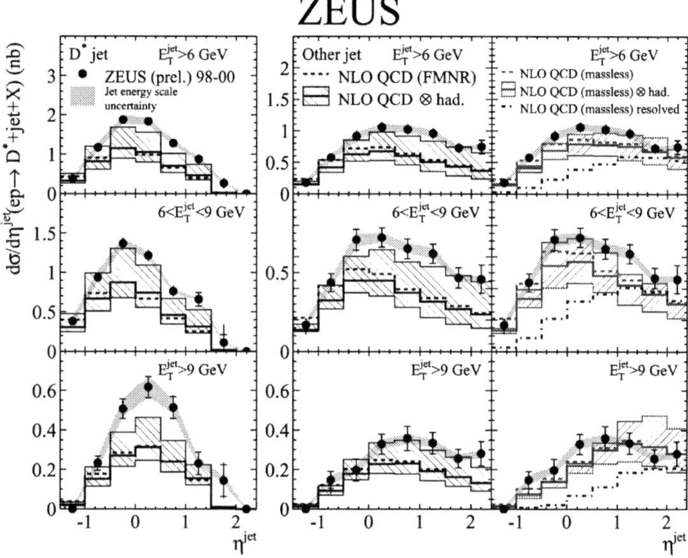

FIGURE 1. The cross-section $d\sigma/d\eta^{\text{jet}}$ for D^*-tagged jet (left) and untagged jets (center, right) in different E_T^{jet} ranges compared to the "massive" and "massless" NLO predictions.

tion factors were obtained as the ratio of the hadron-level cross section to the parton-level cross section using MC simulations. The average value obtained by HERWIG and PYTHIA was used for the correction with their difference used as the uncertainty. The hadronisation uncertainty was added in quadrature to other uncertainties in the NLO QCD calculations estimated by changing the charm mass and the renormalisation scale.

In figure 1, the cross-section $d\sigma/d\eta^{\text{jet}}$ in different ranges of E_T^{jet} are shown for D^*-tagged jets and untagged jets seperately. Measurements are compared to the "massive" NLO QCD prediction. For untagged jet distributions, comparisons to the "massless" NLO QCD predictions are also available. The shapes of the distributions, in general, agree well with the upper bound of both NLO QCD predictions. A better agreement is obtained after applying the hadronisation correction.

Dijet cross sections, $d\sigma/dx_\gamma^{\text{obs}}$, $d\sigma/d\Delta\phi^{\text{jj}}$, $d\sigma/d(p_T^{\text{jj}})^2$ and $d\sigma/dM^{\text{jj}}$ are compared to the "massive" NLO QCD predictions in figure 2. The contribution from beauty production was estimated by reweighting the PYTHIA prediction to the "massive" NLO prediction according to the p_T of the two stable B-hadrons, which is also shown in the figure. A good agreement is observed for $d\sigma/dx_\gamma^{\text{obs}}$ and $d\sigma/dM^{\text{jj}}$. However, the NLO prediction underestimates the cross section at low $\Delta\phi^{\text{jj}}$ and large $(p_T^{\text{jj}})^2$. In figure 2, the cross-section $d\sigma/d(p_T^{\text{jj}})^2$ is shown separately for direct-enriched ($0.75 < x_\gamma^{\text{obs}} < 1$) and resolved-enriched ($0 < x_\gamma^{\text{obs}} < 0.75$) samples. The direct-enriched sample shows a softer spectrum in both data and the NLO prediction. A significant discrepancy between data and the NLO prediction is observed for the resolved-enriched sample, while HERWIG

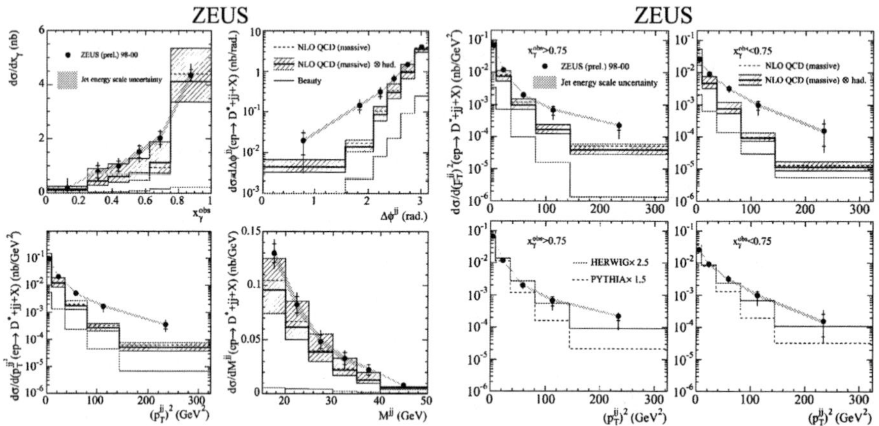

FIGURE 2. The dijet cross-sections $d\sigma/dx_\gamma^{\rm obs}$, $d\sigma/d\Delta\phi^{\rm jj}$, $d\sigma/d(p_T^{\rm jj})^2$ and $d\sigma/dM^{\rm jj}$ compared to the "massive" NLO predictions (left). The dijet cross-section $d\sigma/d(p_T^{\rm jj})^2$ for direct- and resolved-enriched samples compared to the "massive" NLO predictions and MC models (right).

MC gives an excellent agreement in shape.

The NLO QCD prediction alone is not sufficient to reproduce the distribution at low $\Delta\phi^{\rm jj}$ or large $(p_T^{\rm jj})^2$ where higher-order corrections are important. The fact that LO+PS MC model gives a better description of the data in these regions indicates that the PS algorithm is effectively simulating the higher-order effects in charm photoproduction. Matching the NLO calculation with a PS algorithm such as in the MC@NLO program [6] is expected to give a better description of the data.

ACKNOWLEDGMENTS

The author is financially supported by the Japan Society for the Promotion of Science (2004).

REFERENCES

1. S. Frixione, M. Mangano, P. Nason and G. Ridolfi, Nucl. Phys. **B 412** (1994) 225;
   M. Mangano, P. Nason and G. Ridolfi, Nucl. Phys. **B 373** (1992) 295.
2. G. Heinrich and B. A. Kniehl, Phys. Rev. **D 70**, 094035 (2004).
3. ZEUS Collaboration, Phys. Lett. **B 293** (1992) 465;
   ZEUSCollaboration, The ZEUS Detector. Status Report, (unpublished), DESY (1993), Avaiable on http://www-zeus.desy.de/bluebook/bluebook.html.
4. G. Marchesini et al., Comp Phys. Comm. **67** (1992) 465;
   G. Corcella et al., JHEP **0101** (2001) 010.
5. T. Sjœstrand, Comp. Phys. Comm. **82** (1994) 74.
6. S. Frixione and B. R. Webber, JHEP **0206**, 029 (2002);
   S. Frixione, P. Nason and B. R. Webber, JHEP **0308**, 007 (2003).

Photoproduction of D^* Mesons and Jets with H1

Gero Flucke (on behalf of the H1 collaboration)

DESY, Hamburg, Germany

Abstract. Cross sections are measured for photoproduction events containing a $D^{*\pm}$ meson and a jet. They are determined for photon-proton centre-of-mass energies $171 < W < 256$ GeV and photon virtualities $Q^2 < 0.01$ GeV^2. A jet that does not contain the D^* meson is required. Differential cross sections are compared with perturbative QCD predictions in collinear and k_t-factorisation.

Keywords: Charm, Jets, Photoproduction, QCD, Cross sections
PACS: 12.38.Qk,13.60.Le,13.85.Hd,14.40.Lb

INTRODUCTION

Inclusive D^* photoproduction at HERA is interesting as a general test of the charm production mechanism in high energy ep collisions. The measurements [1, 2, 3, 4] are consistent with the assumption that heavy quarks are predominantly produced via a photon-gluon fusion mechanism. The measurement of a D^*+jet pair gives the possibility to investigate further details of the charm production mechanism.

DATA SELECTION

Data of the years 1999 and 2000 with an integrated luminosity of $\mathscr{L} = 51.1$ pb^{-1} are analysed where protons of $E_p = 920$ GeV have been collided with positrons of $E_e = 27.6$ GeV. $D^{*\pm}$ mesons are reconstructed from tracks in the decay channel[1] $D^{*+} \rightarrow D^0\pi_s^+ \rightarrow K^-\pi^+\pi_s^+$ as in [3]. Photoproduction is selected through the reconstruction of the scattered positron under a small angle $(\pi - \theta) < 5$ mrad[2], leading to photon virtualities of $Q^2 < 0.01$ GeV^2 and photon-proton centre-of-mass energies of $171 < W < 256$ GeV.

Jets are defined by the inclusive k_t-algorithm [5]. The input of the jet algorithm are hadronic-final-state objects (HFS) from combining tracks and calorimeter clusters. The HFS objects of the D^* decay tracks are replaced by the D^* candidate itself. The minimal required transverse momentum is $p_t(\text{jet}) > 3$ GeV. A satisfying jet reconstruction even at these low values of p_t is achieved for jets in the central detector region $|\eta(\text{jet})| < 1.5$ where well measured tracks dominate the HFS objects. The aim is to tag a second object besides the D^* from the hard process. Therefore the jet with the highest transverse momentum *not* containing the D^* meson is selected. In total a signal of 588 ± 46 D^*+jet combinations is observed. Details of the D^* and jet selection are described in [6].

---

[1] Charge conjugate states are always implicitly included.
[2] The polar angle θ is measured with respect to the direction of the colliding protons.

CP792, *Deep Inelastic Scattering, DIS 2005,* edited by W. H. Smith and S. R. Dasu

TABLE 1. *Parameters of the QCD calculations: renormalisation (μ_r) and factorisation (μ_f) scales, parton density parametrisations and charm fragmentation treatment.*

	PYTHIA 6.1	CASCADE 1.2	FMNR	ZMVFNS
μ_r^2	$m_c^2+p_{t,c}^2$	$4m_c^2+p_{t,c}^2$	$m_c^2+(p_{t,c}^2+p_{t,\bar{c}}^2)/2$	$m_c^2+p_{t,D^*}^2$
μ_f^2	$m_c^2+p_{t,c}^2$	$Q_t^2+\hat{s}$	$4(m_c^2+(p_{t,c}^2+p_{t,\bar{c}}^2)/2)$	$4(m_c^2+p_{t,D^*}^2)$
p-PDF	CTEQ5L[12]	A0 [13]	CTEQ5M [12]	
γ-PDF	GRV-G LO [14]	–	GRV-G HO [14]	
Fragm.	Lund with $\varepsilon_{pet}=0.078$		$\varepsilon_{pet}=0.035$ [15]	BKK O [16]

QCD CALCULATIONS

The results will be compared with two leading order calculations supplemented with parton showers, the collinear factorising PYTHIA 6.15 [7] and the k_t-factorising CASCADE 1.2 [8], as well as with collinear next-to-leading order (NLO) calculations in the massive (FMNR [9]) and massless (ZMVFNS [10]) scheme. The charm pole mass is set to $m_c = 1.5$ GeV and the fragmentation fraction $f(c \to D^*) = 0.235$ [11] is applied. The main parameters of all calculations are listed in table 1. For both NLO calculations corrections are applied to account for the transition from parton level to hadron level jets. They are obtained as bin-by-bin corrections calculated between parton and hadron level jets in PYTHIA. Uncertainties of the CASCADE, FMNR and ZMVFNS calculations are estimated by varying the charm mass and the factorisation and renormalisation scales.

CROSS SECTIONS

To determine bin averaged differential cross sections, the numbers of reconstructed D^*+jet combinations, determined by fits to the $\Delta m = m(K\pi\pi_s) - m(K\pi)$ distributions, is corrected for the branching ratio $BR(D^* \to K\pi\pi_s) = 0.0257$ [17], limited acceptances, reconstruction and trigger inefficiencies. Systematic uncertainties amount to $15-17\%$. Their main contributions are the uncertainties in the track reconstruction efficiency.

The D^*+jet cross section are shown in figure 1 and compared with the predictions. The $p_t(D^*)$ (fig. 1a+c) and $p_t(\text{jet})$ (fig. 1e+h) distributions are rapidly falling towards higher transverse momenta and are reasonably well described by all QCD calculations, but the CASCADE prediction shows a slightly harder spectrum than observed in the data.

The cross sections as a function of the pseudorapidity of the jet and the D^* differ: The $\eta(D^*)$ (fig. 1b+d) distribution falls off with increasing values of η whereas $\eta(\text{jet})$ (fig. 1f+i) is almost flat. The difference is not caused by the different kinematic cuts for the D^* and the jet, but suggests that the jet cross section contains contributions from non-charm jets. They could arise from hard gluon radiation from the initial state, which mainly populates the forward (large η) region. This hypothesis is also supported from comparison with the PYTHIA calculations: for direct photon processes ($\gamma g \to c\bar{c}$) the η spectra of the jet and the D^* are found to be similar (figures 1b+f). Only after inclusion of the charm excitation processes, which effectively simulate processes like $\gamma g \to c\bar{c}g$, the η spectrum of the jet can be described. A similar feature is obtained with k_t factorisation as implemented in CASCADE, but also with the NLO calculations.

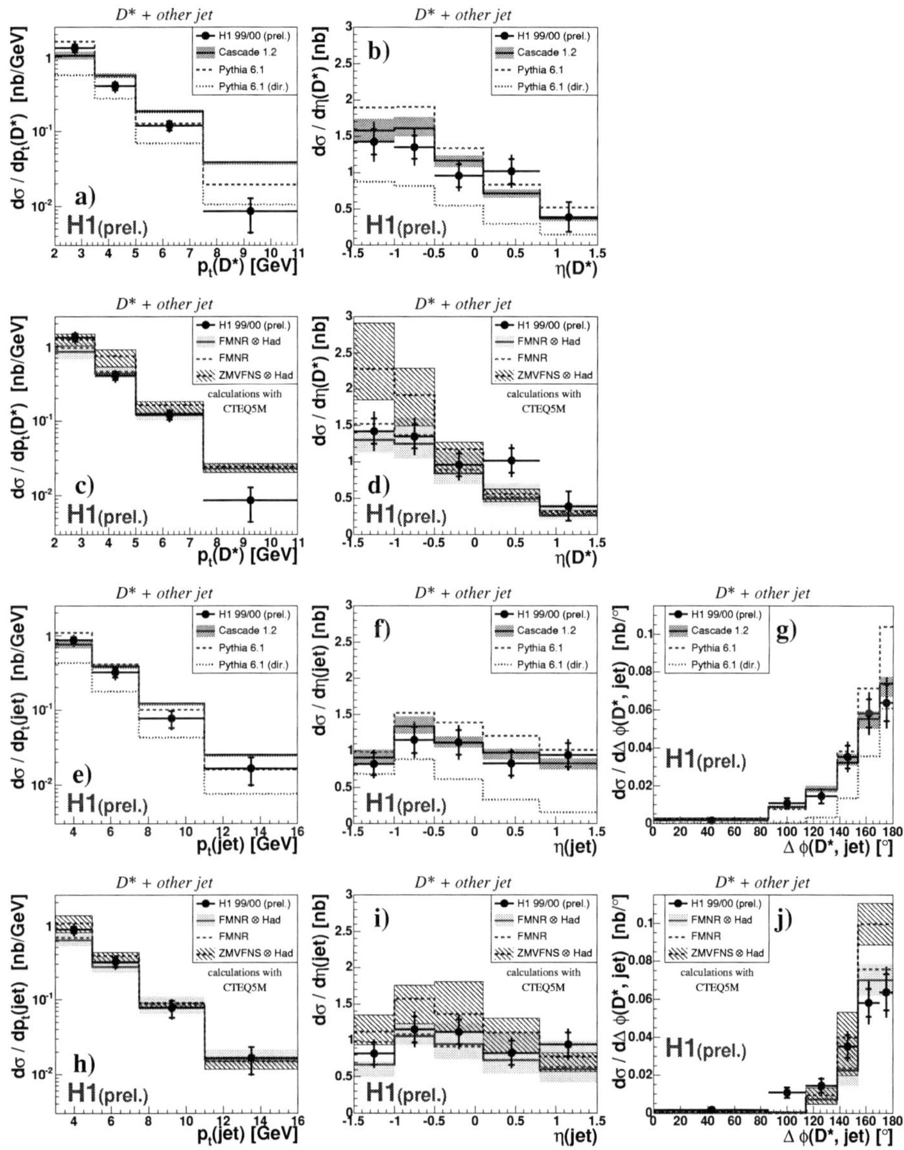

FIGURE 1. *D^*+jet cross sections in bins of the transverse momentum (left column) and pseudorapidity (middle column) of the jet and the D^* and their azimuthal distance $\Delta\phi(D^*,jet)$ (right column) compared with the predictions of PYTHIA and CASCADE (a-b,e-g) and of the next-to-leading order calculations FMNR and ZMVFNS (c-d,h-j). For PYTHIA the direct contribution of the prediction is shown separately and labelled as* "dir.". *The central FMNR prediction is shown before and after applying the hadronisation corrections. The inner error bars of the data indicate the statistical uncertainties, the outer error bars indicate the statistical and systematic uncertainties added in quadrature.*

In the figures 1g+j the cross section is presented as a function of $\Delta\phi(D^*,\text{jet})$. For processes like $\gamma g \rightarrow c\bar{c}$ a back-to-back configuration is expected. The large fraction of events where the D^* and the jet are not back-to-back suggests a significant contribution of higher order QCD radiation.

CONCLUSIONS

Photoproduction of D^* mesons has been analysed at relatively low transverse momenta. Cross sections have been determined for events with a D^* meson and a jet not containing the D^*. The results have been compared with four approaches of QCD calculations.

The measured transverse momentum distributions of the D^* and jet are in general in agreement with the prediction within the theoretical uncertainties.

Comparing the pseudorapidity distributions of the D^* and the jet it is striking that the $\eta(D^*)$ distribution falls in forward direction whereas $\eta(\text{jet})$ is almost constant. This observation suggests that the jet cross section contains not only a charm jet but also a significant contribution form a further parton, most likely a gluon jet.

The azimuthal difference $\Delta\phi(D^*,\text{jet})$ is sensitive to the amount of gluon radiation. The measured cross section shows that higher orders beyond the LO process $\gamma g \rightarrow c\bar{c}$ with a collinear gluon contribute significantly. The $\Delta\phi(D^*,\text{jet})$ is reasonably well described by calculations applying leading log parton showers in the collinear factorisation ansatz, or by using un-integrated gluon densities in the k_t-factorisation ansatz, but not by fixed order NLO calculations.

ACKNOWLEDGMENTS

We would like to thank S. Frixione for the FMNR code and G. Heinrich for the ZMVFNS calculations.

REFERENCES

1. C. Adloff, et al., *Nucl. Phys.*, **B545**, 21–44 (1999), hep-ex/9812023.
2. J. Abbiendi, et al., *Eur. Phys. J.*, **C6**, 67–83 (1999), hep-ex/9807008.
3. H1, Photoproduction of D^* Mesons at HERA, EPS03, H1prelim-03-071.
4. ZEUS, Measurement of D^* Photoproduction at HERA, ICHEP02, ZEUS-prel-02-004.
5. S. Ellis, and D. Soper, *Phys. Rev.*, **D48**, 3160–3166 (1993), hep-ph/9305266.
6. G. Flucke, Ph.D. thesis, Universität Hamburg (2005), DESY-THESIS-2005-006.
7. T. Sjöstrand, et al., *Comput. Phys. Commun.*, **135**, 238–259 (2001), hep-ph/0010017.
8. H. Jung, *Comput. Phys. Commun.*, **143**, 100–111 (2002), hep-ph/0109102.
9. S. Frixione, P. Nason, and G. Ridolfi, *Nucl. Phys.*, **B454**, 3–24 (1995), hep-ph/9506226.
10. G. Heinrich, and B. A. Kniehl, *Phys. Rev.*, **D70**, 094035 (2004), hep-ph/0409303.
11. L. Gladilin, Charm hadron production fractions (1999), hep-ex/9912064.
12. H. L. Lai, et al., *Eur. Phys. J.*, **C12**, 375–392 (2000), hep-ph/9903282.
13. H. Jung, "Un-integrated Parton Density Functions in CCFM," (DIS 2004), hep-ph/0411287.
14. M. Glück, E. Reya, and A. Vogt, *Phys. Rev.*, **D46**, 1973–1979 (1992).
15. P. Nason, and C. Oleari, *Nucl. Phys.*, **B565**, 245–266 (2000), hep-ph/9903541.
16. J. Binnewies, B. A. Kniehl, and G. Kramer, *Phys. Rev.*, **D58**, 014014 (1998), hep-ph/9712482.
17. S. Eidelman, et al., *Phys. Lett.*, **B592**, 1 (2004).

Study of Jet Shapes in Charm Photoproduction at HERA

Mária Martišíková
for the H1 Collaboration

DESY, Notkestrasse 85, 22603, Hamburg, Germany
E-mail: maria.martisikova@desy.de

Abstract. Jet shapes in charm photoproduction events in *ep* collisions at HERA are studied. The main goal is to distinguish jets initiated by quarks and gluons in order to shed light on the production mechanism of charm events.

Keywords: charm photoproduction, direct and resolved processes, excitation processes, jet shape
PACS: 14.65.Dw

INTRODUCTION

Charm production in *ep* collisions at HERA is dominated by photon gluon fusion processes (fig. 1). In photoproduction in addition to direct photon processes (fig. 1a), resolved photon processes (fig. 1b-d) can contribute significantly. A key question is to which extent the charm production can be attributed to the so called c-excitation processes (fig. 1c-d). In perturbative QCD at LO only those processes produce a hard gluon in addition to a charm quark, while the direct photon and other resolved processes (fig. 1b) lead to the production of a $c\bar{c}$ pair. The aim of the present study is the distinction of the $c\bar{c}$ component from the cg events. Due to differences in the jet formation processes, gluon jets are expected to be broader than quark jets [1]. Such differences have been observed previously between jets initiated by light quarks and by gluons [2].

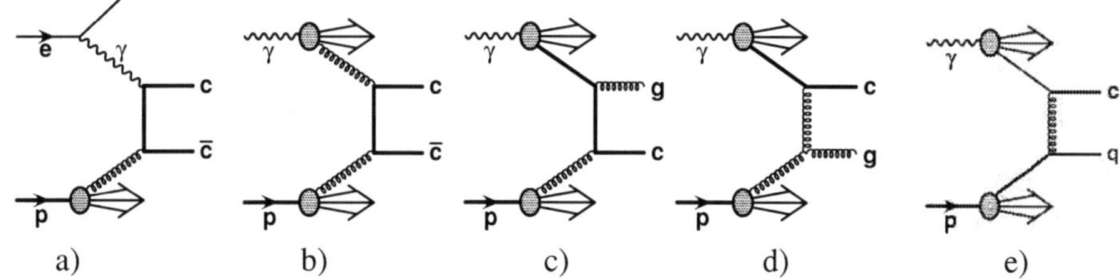

FIGURE 1. Charm quark production processes in leading order pQCD.

EXPERIMENTAL METHOD

The jet shapes are studied using the variable 'integrated jet shape' $\psi(r)$. It is defined as the fraction of the jet transverse momentum (with respect to the beam axis) deposited

CP792, *Deep Inelastic Scattering, DIS 2005,* edited by W. H. Smith and S. R. Dasu

within a cone of the radius r around the jet axis relative to the transverse momentum of the jet, contained in a cone of radius R=1 (fig. 2)

$$\psi(r) = \frac{p_t^{cone}(r)}{p_t^{cone}(R)}$$

The radius is defined as $r = \sqrt{\Delta\eta^2 + \Delta\phi^2}$, where $\Delta\eta$ and $\Delta\phi$ are the distances of the particles to the jet axis in pseudorapidity η and azimuthal angle ϕ respectively. The mean $\langle\psi(r)\rangle$ over events is studied as a function of r/R.

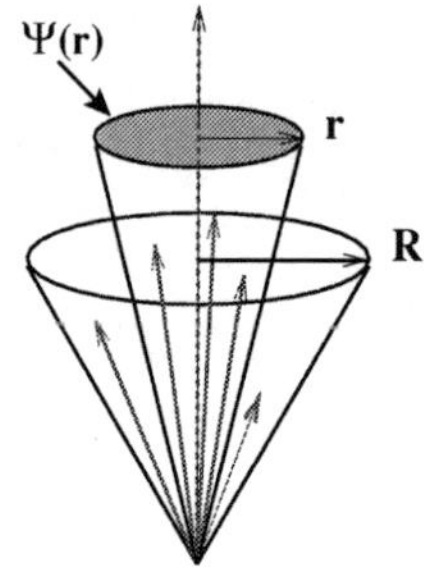

FIGURE2. Definition of $\psi(r)$

EVENT SELECTION

The analysis is based on the event sample selected for the recent determination of the beauty cross section [3]. The data were recorded in 1999 and 2000 at a center-of-mass energy of 318 GeV and correspond to an integrated luminosity of 48 pb^{-1}. The kinematic range is restricted to $Q^2 < 1$ GeV^2 and $0.2 < y < 0.8$. Massless jets are reconstructed by the inclusive k_t algorithm in the p_t recombination scheme using a combination of tracks and calorimeter energy deposits. The selection requires at least two jets with transverse momentum $p_t > 7(6)$ GeV, of which at least one contains a muon candidate. Muons are selected in the angular range $35° < \theta(\mu) < 130°$, with $p_t^{\mu} > 2.5$ GeV. In order to enrich the sample in charm events a cut on the transverse momentum of the muon track relative to the associated jet, $p_t^{rel} < 1$ GeV, is imposed. The final sample contains 800 events with a charm purity of about 75% . The remaining background due to b and light quark production is subtracted statistically, using the fractions as predicted by the inclusive PYTHIA Monte Carlo simulation, which were checked to describe the data.

For comparison a dijet event sample is selected which fulfills the same jet selection criteria as the charm sample except for the requirement of a high p_t muon. In this case the kinematic range is restricted to $Q^2 < 0.01$ GeV^2 and $0.3 < y < 0.65$. This event sample is dominated by the production of light quarks ($\sim$ 75%).

RESULTS OF THE JET SHAPE MEASUREMENTS

The fraction x_γ^{obs} of the photon energy entering the hard interaction is estimated using the observable

$$x_\gamma^{obs} = \frac{\sum_{Jet_1}(E - p_z) + \sum_{Jet_2}(E - p_z)}{\sum_{all\ hadrons}(E - p_z)}$$

For the direct process x_γ^{obs} approaches unity, in resolved processes x_γ^{obs} can be small.

The distributions of the averaged integrated jet shapes $\langle\psi(r)\rangle$ as functions of r/R are presented for two separate regions of x_γ^{obs}. Distributions for the charm sample are shown in fig. 3. Only the jet without muon is studied. The PYTHIA 6.1 predictions, using CTEQ5L and GRVG-LO the parton densities for proton and photon, respectively, are shown with the data. Direct and resolved photon processes are simulated, including excitation processes. The curves for the direct and resolved photon events are shown separately. The data are described well by the PYTHIA prediction at high x_γ^{obs}, but a disagreement is observed in the region of $x_\gamma^{obs} \leq 0.75$. In the PYTHIA simulation the rise of $\langle \psi(r) \rangle$ is softened due to the presence of gluon jets in the resolved photon sample. Such a slow rise is not observed in the data.

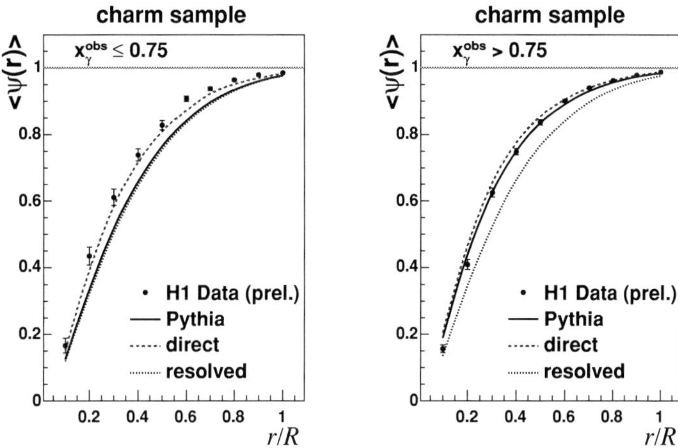

FIGURE 3. $\langle \psi(r) \rangle$ for the charm event sample in two different regions of x_γ^{obs}. The data are compared with the prediction from PYTHIA. The expected curves for direct and resolved photon processes are shown separately.

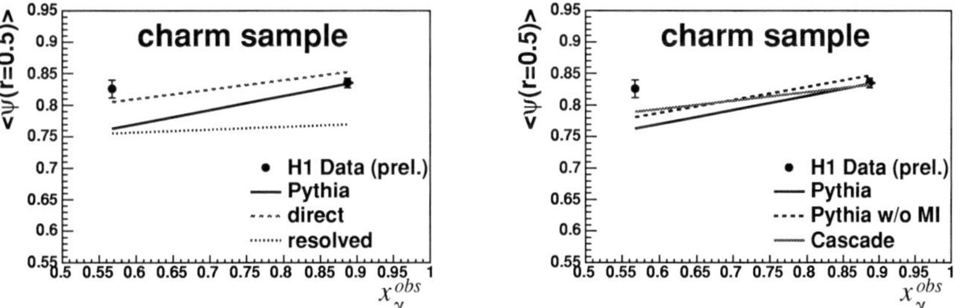

FIGURE 4. $\langle \psi(r/R = 0.5) \rangle$ as a function of x_γ^{obs} for the charm sample. The same points are compared with PYTHIA (left), CASCADE and PYTHIA without multiple interactions (right).

In fig. 4 the averaged integrated jet shape at a fixed value $r/R = 0.5$ is shown for the charm sample as a function of x_γ^{obs} and compared with different model calculations. PYTHIA describes the data at high x_γ^{obs}, while deviations are seen at low x_γ^{obs}. PYTHIA without multiple interactions (MI) is closer to the data at low x_γ^{obs}, while the description at high x_γ^{obs} is only marginally worse. The CASCADE 1.0 model using an unintegrated parton density (JS2001) describes the data well in the high x_γ^{obs} region and is closer to the data at low x_γ^{obs}.

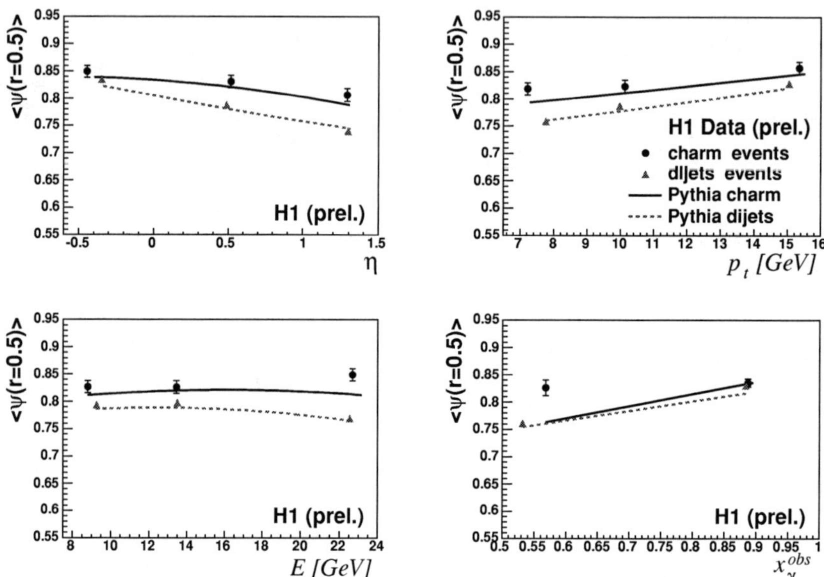

FIGURE 5. Differential distributions of $\langle\psi(r/R=0.5)\rangle$ as functions of the jet quantities η, p_t, E and x_γ^{obs} for the charm and dijet samples. The predictions from PYTHIA are shown for both data samples.

It was checked that variation of the background due to b quarks and light quarks or of the Peterson fragmentation parameter ε_c, did not change the distributions significantly.

In fig. 5 the jet shapes in charm events are directly compared to dijet events which are dominated by light quarks. For dijets both selected jets enter the distributions. PYTHIA describes the dijet data well everywhere.

CONCLUSIONS

Jet shapes are studied in a charm dijet photoproduction sample with the H1 detector at HERA. The resolved component in PYTHIA is dominated by charm excitation, where in addition to the charm quark jet a gluon initiated jet is expected, which is broader than the charm jet. In the charm data sample the jet shapes are however found to be similar in direct and resolved photon enriched samples. For comparison a light quark dominated dijet sample was studied. The jet shapes are in agreement with PYTHIA. The observed discrepancy between the charm data and PYTHIA indicates a lack of understanding of the charm production process in the resolved photon region.

REFERENCES

1. R.K. Ellis, W.J. Stirling and B.R. Webber, QCD and Collider Physics, Cambridge University Press (1996)
2. G. Alexander *et al.* [OPAL Collaboration], Z. Phys. C **69** (1996) 543.
3. A. Aktas *et al.* [H1 Collaboration], arXiv:hep-ex/0502010, Accepted by Eur Phys J, 02/05.

Heavy Quarkonium Production: Extending CSM and COM

J.P. Lansberg

Physique théorique fondamentale, Département de Physique, Université de Liège, allée du 6 Août 17, bât. B5, B-4000 Liège 1, Belgium
E-mail: JPH.Lansberg@ulg.ac.be

Abstract. By questioning the applicability of the static approximation of the Colour-Singlet Model, we have seen that the production amplitude receives contributions from two different cuts. The first one in its static limit gives the colour-singlet mechanism. The second one has not been considered so far. We treat it in a gauge-invariant manner by introducing necessary new 4-point vertices, suggestive of the colour-octet mechanism. This new contribution can be as large as the colour-singlet mechanism at high p_T, however these vertices are not totally constrained and when the freedom in their determination is fully exploited, we are able to reproduce the production cross-sections at the Tevatron for the J/ψ, ψ' and $\Upsilon(1S)$ and at RHIC for the J/ψ.

Keywords: heavy quarkonium production, vector-meson production, gauge invariance, relativistic effects, non-static extension
PACS: 14.40.Gx, 13.85.Ni, 11.10.St, 13.20.Gd

1. INTRODUCTION

Ten years after the discovery of the "ψ' anomaly" by the CDF collaboration [1, 2], no totally conclusive solution has been proposed so far (for a comprehensive and up-to-date review on the subject, see [3]). Even though the Colour-Octet Mechanism (COM), coming from the application of NRQCD to heavy quarkonium, is a good candidate, it appears clearly that as long as fragmentation is the dominant production contribution and the velocity scaling rules of NRQCD hold, it cannot accommodate the polarisation measurements of CDF [4], which show a non-polarised, if not slightly longitudinal, production.

In that context, we have felt the necessity to reconsider the appropriateness of the static and on-shell approximation of the Colour-Singlet Model (CSM) [5], which is still the most natural model from QCD. These approximations are also implicit in the COM, therefore any feature arising from this study should have some implication for the COM.

In order to study properly non-static and off-shell effects, we have used a vertex function as an input for the bound state characteristics, whereas the Schrödinger wave function at the origin is used in the CSM and Long Distance Matrix Elements (LDME) of NRQCD enter the COM. We emphasise again that we probe all the internal phase space of the quarkonium, and thus need a function, where the two models simply need a constant factor.

CP792, *Deep Inelastic Scattering, DIS 2005,* edited by W. H. Smith and S. R. Dasu

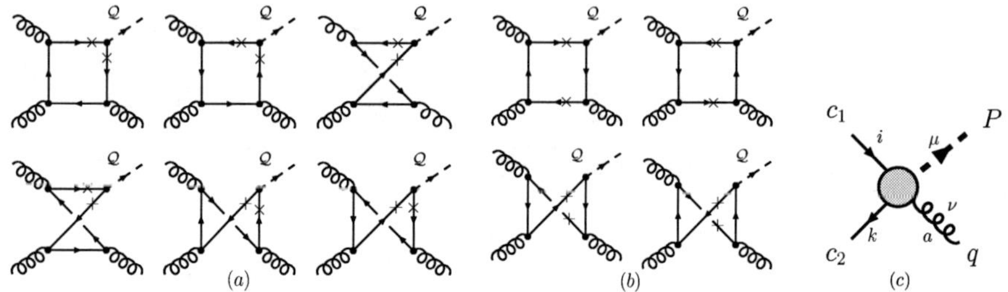

FIGURE 1. The first family (a) has 6 diagrams and the second family (b) 4 diagrams contributing the discontinuity of $gg \to {}^3S_1 g$ at LO in QCD.(c): the gauge-invariance restoring vertex, $\Gamma^{(4)}$.

2. OUR MODEL

In the case of ${}^3S_1$ quarkonium (noted $\mathcal{Q}$) production in high-energy hadronic collisions, we are to consider gluon fusion $gg \to \mathcal{Q}g$. Using the Landau equations [6], we have shown in [7] that there are two families of contributions (see Fig. 1 (a) and (b)): the first is the usual colour-singlet mechanism, where in the context of our model, we use a 3-point function $\Gamma^{(3)}_\mu(p,P) = \Gamma(p,P)\gamma_\mu$ at the $Q\bar{Q}\mathcal{Q}$ vertex; the second family was never considered before. To simplify the study, we set $m > M/2$ so that the first cut does not contribute.

For the functional form of $\Gamma(p,P)$, we neglect possible cuts, and choose two opposite scenarios: a dipolar form which decreases gently with its argument, and a Gaussian form: $\Gamma(p,P) = N(1+\frac{\vec{p}^2}{\Lambda^2})^{-2}$ and $\Gamma(p,P) = Ne^{-\frac{\vec{p}^2}{\Lambda^2}}$, both with a free size parameter Λ, and a normalisation N. In [8], we have shown how to fix the normalisation N of $\Gamma(p,P)$ as a function of Λ.

In addition to the second family, one is driven – to preserve gauge invariance (GI) – to introduce new contributions arising from the presence of 4-point vertices. Besides restoring GI, these vertices have to satisfy specific constraints [7, 9, 10]. For the following simple choice for $\Gamma^{(4)}_{\mu\nu}(c_1,c_2,P,q)$

$$-ig_s T^a_{ki}\left(\Gamma(2c_1-P,P)-\Gamma(2c_2-P,P)\right)\left[\frac{c_{1\mu}}{(c_2+P)^2-m^2}+\frac{c_{2\mu}}{(c_1-P)^2-m^2}\right]\gamma_\nu, \quad (1)$$

where the momenta and indices are as in Fig. 1 (c), we got for the J/ψ and ψ' production at the Tevatron the results shown in Fig. 2. In the ψ' case, we employed the ambiguity upon the vertex function normalisation due to the node position a_{node} to describe the data at low P_T. Note that the slope is not that different from that of the data. This is at variance with what is widely believed since fragmentation (with a typical $1/P_T^4$ behaviour) processes describe the data.

However there exist different choices for the GI restoring vertex (GIRV). In the following, we present some interesting results obtained by studying the effects of autonomous vertices. The latter link different suitable choices of GIRV: they are GI alone and a priori unconstrained in normalisation. For a first study, let us

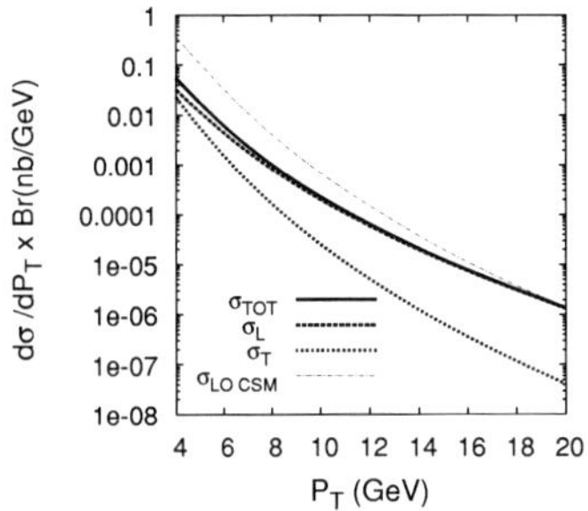

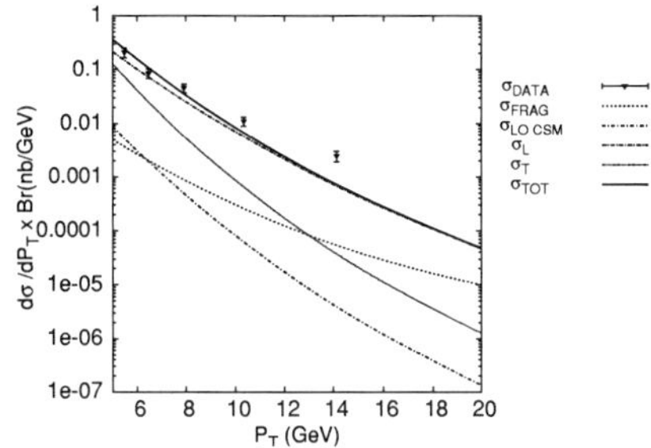

FIGURE 2. Polarised (σ_T and σ_L) and total (σ_{TOT}) cross sections obtained with a Gaussian vertex functions On the left, for J/ψ with $m = 1.87$ GeV and $\Lambda = 1.8$ GeV; on the right for ψ' with $a_{node} = 1.333$ GeV, $m = 1.87$ GeV and $\Lambda = 1.8$ GeV to be compared with the data from CDF [1, 2].

restrict the choice to the three simplest possible ones [9, 11] (omitting the factor $-ig_s T^a_{ki}(\Gamma(2c_1 - P, P) - \Gamma(2c_2 - P, P)))$

$$(a)\ \alpha/(\sqrt{\hat{s}}m_Q)\gamma^\mu q^\nu \qquad (b)\ \beta'(c_1+c_2)^\mu(c_1+c_2)^\nu \qquad (c)\ \xi/m_Q g^{\mu\nu} \tag{2}$$

The factors α, β' and ξ are free constant. If we introduce these contributions in the amplitude calculation, we see in Fig. 3 that we can fit the data for some set of values for (α, ξ).

3. CONCLUSION

We have shown that it is possible to go beyond the static approximation of the CSM. It may also be possible to extend the COM in the same manner. This necessitates the introduction of 4-point vertices due to the non-local 3-point vertex relevant for the non-static and off-shell contributions.

By going deeper in the analysis, we see that the form of these 4-point vertices is not absolutely constrained even after imposing necessary conditions to conserve crossing symmetry and the analytic structure of the amplitude. When this lack of constraint is used, we are able to reproduce the cross-section for the J/ψ, ψ' and $\Upsilon(1S)$ as measured at the Tevatron by CDF (and also at RHIC by PHENIX for J/ψ).

In our framework, cross-sections are dominated by longitudinal $\mathcal{Q}$, therefore by combining our approach with COM fragmentation they could [11] agree with the polarisation measurements of CDF.

ACKNOWLEDGMENTS

J.P.L. is an IISN Postdoctoral Researcher, this work has been done in collaboration with J.R. Cudell and Yu.L. Kalinovsky [7].

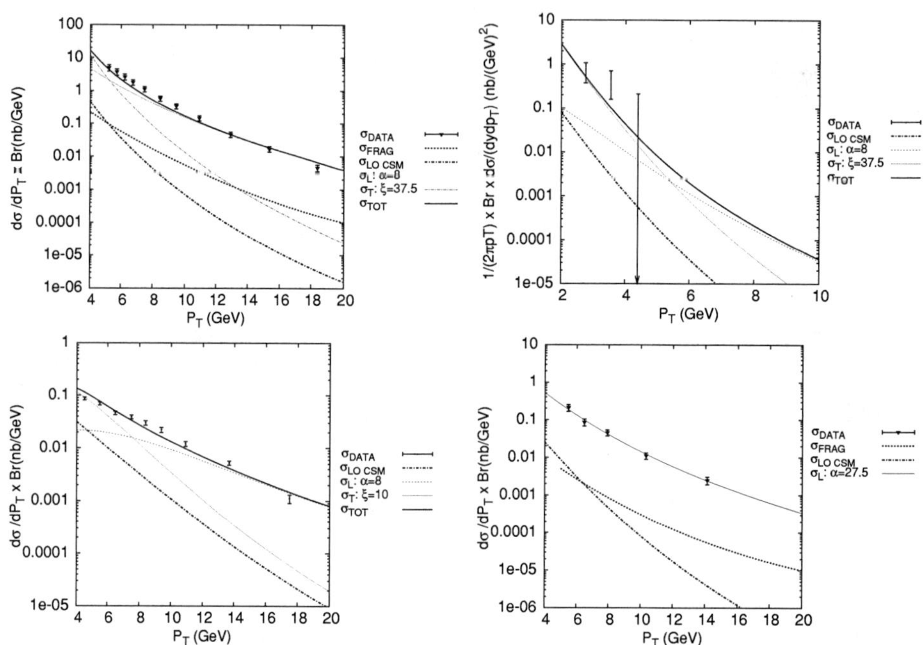

FIGURE 3. Polarised (σ_T and σ_L) and total (σ_{TOT}) cross sections obtained: upleft – for J/ψ at $\sqrt{s} = 1800$ GeV with $\alpha = 8$ and $\xi = 37.5$ to be compared with LO CSM, the fragmentation in the CSM and with the data of CDF [2]; upright – for J/ψ at $\sqrt{s} = 200$ GeV with $\alpha = 8$ and $\xi = 37.5$ to be compared with LO CSM and with the data of PHENIX [13]; downleft – for $\Upsilon(1S)$ at $\sqrt{s} = 1800$ GeV with $\alpha = 8$ and $\xi = 10$ to be compared with LO CSM and with the data of CDF [12]; downright – for ψ' at $\sqrt{s} = 1800$ GeV with $\alpha = 27.5$ and to be compared with LO CSM, the fragmentation in the CSM and with the data of CDF [1].

REFERENCES

1. F. Abe *et al.* [CDF Collaboration], Phys. Rev. Lett. **79** (1997) 572.
2. F. Abe *et al.* [CDF Collaboration], Phys. Rev. Lett. **79** (1997) 578.
3. N. Brambilla *et al.*, arXiv:hep-ph/0412158.
4. T. Affolder *et al.* [CDF Collaboration], Phys. Rev. Lett. **85** (2000) 2886 [arXiv:hep-ex/0004027].
5. C-H. Chang, Nucl. Phys. **B 172** (1980) 425;
   R. Baier and R. Rückl, Phys. Lett. B **102** (1981) 364;
   R. Baier and R. Rückl, Z. Phys. **C 19** (1983) 251.
6. L. D. Landau, Nucl. Phys. **13** (1959) 181.
7. J. P. Lansberg, J. R. Cudell and Yu. L. Kalinovsky, arXiv:hep-ph/0507060.
8. J. P. Lansberg, AIP Conf. Proc. **775** (2005) 11.
9. J. P. Lansberg, *Quarkonium Production at High-Energy Hadron Colliders*, Ph.D. Thesis, ULg, Liège, Belgium, 2005.
10. S. D. Drell and T. D. Lee, Phys. Rev. D **5** (1972) 1738.
11. J. R. Cudell, Yu. L. Kalinovsky and J. P. Lansberg, (in preparation).
12. D. Acosta *et al.* [CDF Collaboration], Phys. Rev. Lett. **88** (2002) 161802.
13. S. S. Adler *et al.* [PHENIX Collaboration], Phys. Rev. Lett. **92** (2004) 051802 [arXiv:hep-ex/0307019].

ZEUS Measurement of Inelastic $J/\psi \rightarrow \mu^+\mu^-$ Production in DIS

Alexei Antonov for the ZEUS Collaboration

Moscow State Engineering Physics Institute. Kashirskoe shosse 31 115409 Moscow now at Deutsches Elektronen-Synchrotron. Notkestrasse 85, 22607 Hamburg Germany.

Abstract. The inelastic production of J/ψ in ep collisions has been studied with the ZEUS detector at HERA using an integrated luminosity of $109\,pb^{-1}$. The J/ψ mesons were identified using the decay channel $J/\psi \rightarrow \mu^+\mu^-$. The measured cross sections are compared to theoretical predictions within the non-relativistic QCD framework including colour–singlet and colour–octet contributions and the k_T–factorisation approach. Calculations of the colour–singlet process only generally agree with the data whereas inclusion of colour–octet terms spoils this agreement.

Keywords: Deep inelastic scattering,HERA,Inelastic J/ψ Electroproduction
PACS: 13.20.Gd, 13.60.Le, 14.40.Gx

INTRODUCTION

Inelastic production of charmonium is described in two steps. The first step is the creation of a $c\bar{c}$ quark pair which can be calculated in perturbative Quantum Chromodynamics (QCD). The second step is the formation of the J/ψ bound state, which occurs at long distances and is described by phenomenological models. The J/ψ bound state can be considered to be formed by a $c\bar{c}$ pair in either a colour singlet (CS) or colour octet (CO) state. In the colour singlet model (CSM)[1] only CS contribution is assumed. In the framework of non-relativistic QCD (NRQCD)[2] both CS and CO contributions exist and the latter contribution is parametrised using a set of long distance matrix elements tuned to describe hadroproduction data.

In the semi-hard or k_T-factorisation approach[3], based on non-collinear parton dynamics governed by the BFKL or CCFM evolution equations, effects of non-zero initial gluon virtuality (transverse momentum) are taken into account.

The inelasticity, z, which is the fraction of the virtual photon energy transferred to the J/ψ in the proton rest frame, is sensitive to the various production mechanisms. CS processes are expected to contribute to the region of medium z values, whereas CO (and diffractive) processes populates the high-z region. Resolved-photon processes, in which the photon acts as a source of incoming partons, populate low values of z.

This paper presents a measurement of inelastic J/ψ production in DIS with the ZEUS detector at HERA. The reaction $ep \rightarrow eJ/\psi X$ is studied for $Q^2 > 2$ GeV^2.

CP792, *Deep Inelastic Scattering, DIS 2005,* edited by W. H. Smith and S. R. Dasu

INELASTIC ELECTROPRODUCTION OF J/ψ

The measurements were performed in the kinematic range $2 < Q^2 < 80\,\mathrm{GeV}^2$, $50 < W < 250\,\mathrm{GeV}$, $0.2 < z < 0.9$ and $-1.6 < Y_{lab} < 1.3$, where Q^2 is the virtuality of the exchanged photon, W is the photon–proton centre–of–mass energy, z is the fraction of the photon energy carried by the J/ψ meson in the proton rest frame and Y_{lab} is the rapidity of the J/ψ in the laboratory frame. The cross section for the process $ep \to eJ/\psi X$ in the kinematic region is

$$302 \pm 23\ (\mathrm{stat.})\ ^{+28}_{-20}\ (\mathrm{syst.})\ \mathrm{pb},$$

where the first uncertainty is statistical and the second systematic. In Fig. 1 the differential cross sections as a function of z, Q^2, W, p_T^{*2}, Y^* and the invariant mass of the hadronic system X, $\log M_X^2$, are shown. They are compared to the predictions of a NRQCD model[4], a CSM with k_T factorisation (LZ)[5] and to the CASCADE MC program[6].

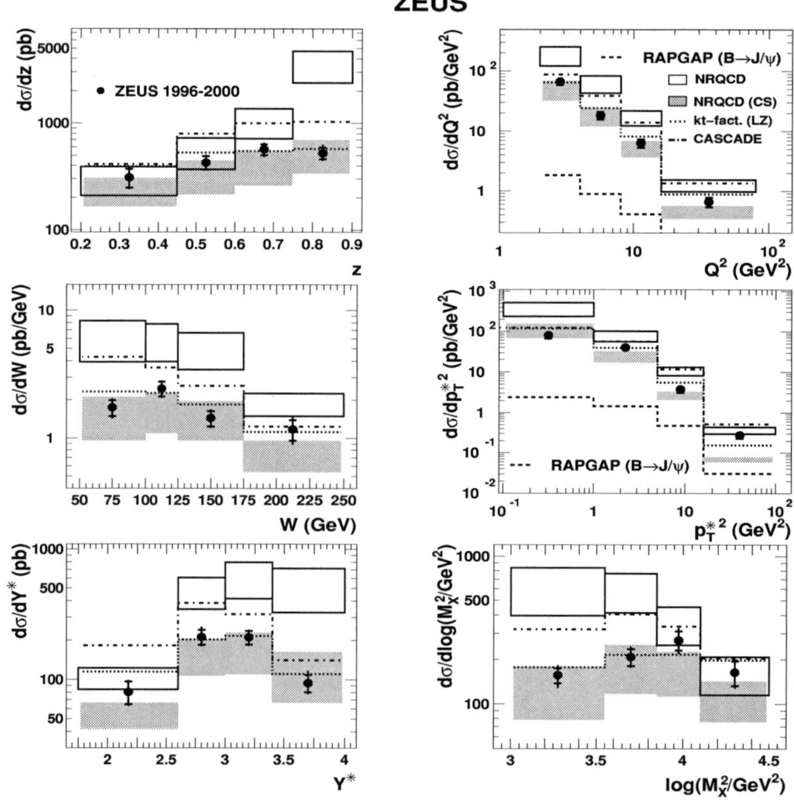

FIGURE 1. Differential cross sections as a function of z, Q^2,W,p_T^{*2},Y^* and $\log M_X^2$. The inner error bars of the data points show the statistical uncertainty; the outer bars show statistical and systematic uncertainties added in quadrature.

The uncertainties for the CS and CO NRQCD predictions correspond to variations of the charm–quark mass ($m_c = 1.5 \pm 0.1$ GeV) and of the renormalisation and factorisation scales from $\frac{1}{2}\sqrt{Q^2 + M_\psi^2}$ to $2\sqrt{Q^2 + M_\psi^2}$. The uncertainty on the long–distance matrix elements and the effect of different choices of parton distribution functions (default set is MRST98LO) are also taken into account. The bands in the figures shows all these uncertainties added in quadrature.

In general, the CSM is consistent with the data. The predictions including both CS and CO contributions are higher than the data, especially at high z and low p_T^{*2}. At high values of p_T^{*2} the agreement with the data is reasonable. The prediction does not describe the shapes of the z, Y^* and $\log M_X^2$ distributions. Previous photoproduction results[7] showed that the agreement between data and theory at high z can be improved using resummed LO NRQCD predictions [8]. It should be noted that, in photoproduction, inclusion of the NLO corrections to the CSM, not available for DIS, significantly improved the description of the data.

For the LZ k_T–factorisation predictions, the parametrisation, KMS, of the unintegrated gluon density was used. The charm–quark mass was set to $m_c = 1.4$ GeV, which is the mass used in the KMS parametrisation. The renormalisation and factorisation scales were both set to $\mu = k_T$ for $k_T > 1$ GeV. For $k_T \leq 1$ GeV the scales were fixed at 1 GeV. Calculations based on the k_T-factorisation approach give a reasonable description of the data both in shape and normalisation.

The data are also compared with the predictions of the CASCADE MC using the k_T-factorisation approach, where gluons are treated according to the CCFM evolution equations. These predictions were obtained by setting the charm–quark mass to 1.5 GeV, the evolution scale of the strong coupling constant to the J/ψ transverse mass, $\sqrt{M_\psi^2 + p_T^2}$, and using the unintegrated gluon–density parametrisation "J2003 set 2". The CASCADE MC is above the data for $z > 0.45$ and for $W < 175$ GeV.

In order to compare the present measurements directly to the H1 results[9], differential cross sections were determined in the kinematic range $2 < Q^2 < 100$ GeV^2, $50 < W < 225$ GeV, $0.3 < z < 0.9$ and $p_T^{*2} > 1$ GeV^2. The results of this comparison are shown in Fig. 2. The present results are in agreement with those from H1. In Fig. 1, the ZEUS data are in better agreement with the CSM prediction than in Fig. 2. This is a consequence of the $p_T^{*2} > 1$ GeV^2 cut used in Fig. 2 combined with the fact that the CS prediction underestimate the data at high p_T^{*2}, as seen in Fig. 1.

CONCLUSIONS

Inelastic J/ψ production in DIS has been measured in the kinematic region $2 < Q^2 < 80$ GeV^2, $50 < W < 250$ GeV, $0.2 < z < 0.9$ and $-1.6 < Y_{lab} < 1.3$. The data are in agreement with the H1 results in the kinematic region $2 < Q^2 < 100$ GeV^2, $50 < W < 225$ GeV, $0.3 < z < 0.9$ and $p_T^{*2} > 1$ GeV^2. The data are compared with the predictions of a LO NRQCD model including both CS and CO contributions, as well as with those of a CS model with k_T–factorisation. Calculations of the CS process generally agree with the data, whereas inclusion of CO terms spoils this agreement.

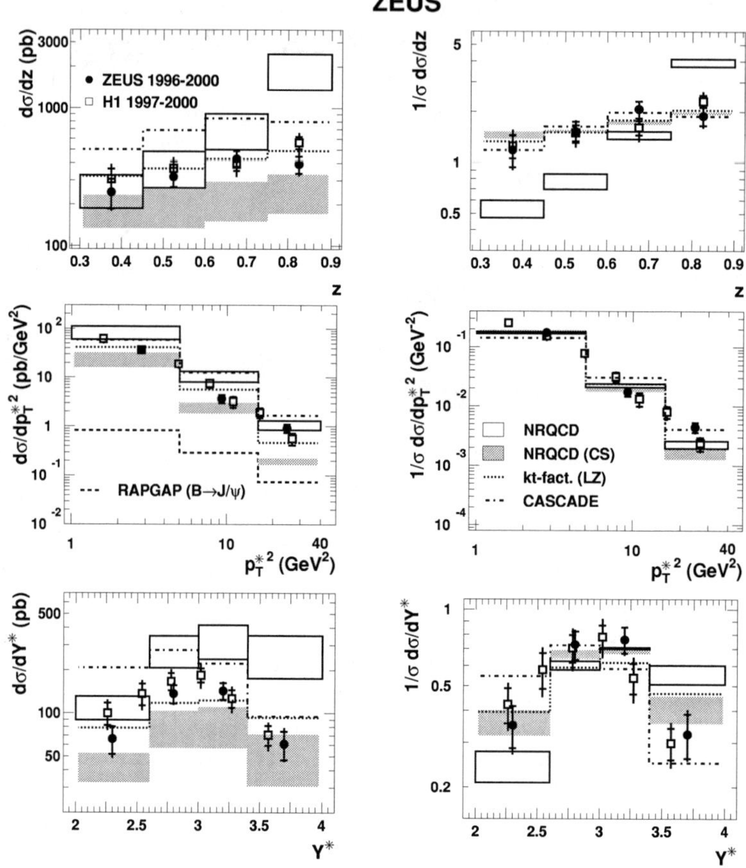

FIGURE 2. Differential cross sections as a function of z, p_T^{*2} and Y^*. Right plots show the data and the theoretical predictions normalised to unit area.

ACKNOWLEDGMENTS

The author would like to thank the DIS05 crew for the well-organized and interesting conference.

REFERENCES

1. E. L. Berger and D. Jones, Phys. Rev. **D 23**, 1521 (1981).
2. G. T. Bodwin, E. Braaten and G. P. Lepage, Phys. Rev. **D 51**, 1125 (1995). Erratum-ibid **D 55**, 5853 (1997).
3. L.V. Gribov, E.M. Levin and M.G. Ryskin, Phys. Rep. **100**, 1 (1983).
4. B. A. Kniehl and L. Zwirner, Nucl. Phys. **B 621**, 337 (2002).
5. A. V. Lipatov and N. P. Zotov, Eur. Phys. J. **C 27**, 87 (2003).
6. H. Jung and G. P. Salam, Eur. Phys. J. **C 19** (2001).
7. ZEUS Coll., S. Chekanov et al., Eur. Phys. J. **C 27**, 173 (2003).
8. M. Beneke, G. A. Schuler and S. Wolf, Phys. Rev. **D 62**, 34004 (2000).
9. H1 Coll., C. Adloff et al., Eur. Phys. J. **C 25**, 41 (2002).

Soft Resummation for Heavy-Quark Production

Alexander Mitov

University of Hawaii, Honolulu, HI-96822, USA

Abstract. We review the need for soft-gluon resummation for heavy quark production in charged- and neutral-current deep-inelastic scattering and explain its effect on observables like the charm structure function.

Keywords: Heavy Quarks, Resummation , Higher Order Perturbative Calculations
PACS: 12.38.Cy , 12.38.Bx , 13.85.Ni , 14.65.Dw

INTRODUCTION: LARGE LOGS OF KINEMATICAL ORIGIN

In perturbative calculations one often encounters numerically large logarithms. One can imagine dividing these logs into "parametrical" and "kinematical". The former are simply large numbers that appear at any order in the coupling. A typical example is $\sim \ln(m/Q)$ where m is a mass and Q stands for some fixed large kinematical scale. More details on the origin of these logarithms and ways to deal with them can be found elsewhere [1]. The subject of the present discussion are what I call "kinematical" logs. These logs can be numerically large, but unlike the "parametrical" ones, they depend on kinematical scales and variables. Therefore, in certain kinematical regimes or corners of phase space these logs can be large, while in others they can be small. A typical example for a kinematical log is $\sim \ln^k(N)$ where N is the Mellin variable conjugated to particular observable, for example, the energy fraction of an observed particle.

It is well known that presence of large logarithms represents a problem for the perturbative expansion since the appearance of numerically large terms spoils the convergence of the perturbation series. A truncated perturbative expansion becomes completely unreliable in such case. The way to deal with this problem is to restore the applicability of the expansion by resuming infinite series of terms of the type $\sim \ln^k(N)\alpha_s^n$ where $k-n$ is a fixed integer.

It is by now well understood what is the origin of the kinematical logs and how to resum them [2]. The reason why such all-order resummation is even possible lies in the universal origin of the soft logs. Their appearance is related to uncontrolled (by fixed order perturbative expansion) predominant soft gluon radiation in particular parts of the phase space. A classical example is the large $x \to 1$ region of the light quark structure functions (here x is the usual Bjorken variable).

It becomes clear from the definition of x that before the hard scattering process between the on-shell quark and the virtual photon has taken place, only soft gluon radiation from the quark is allowed. This follows from the limited phase space $1-x \approx 0$ available for real radiation. However, in the fully inclusive measurement, after that hard scattering has taken place the final state quark has no such restriction and can radiate energy only subjected to a virtuality restriction. Clearly, if we are to probe the structure

CP792, *Deep Inelastic Scattering, DIS 2005*, edited by W. H. Smith and S. R. Dasu

function at not-so-high x, hard radiation will be also allowed and it can be seen to significantly decrease the value of the soft terms $\sim \ln^k(1-x)/(1-x)$.

SOFT RESUMMATION IN HEAVY-QUARK PRODUCTION

In this talk I would like to describe the extension of the methods for soft-gluon resummation for the case of massive quark production. As I will demonstrate, the inclusion of the quark mass is an important effect. Let us begin by a close examination of the known NLO QCD correction for heavy quark production in charged-current DIS. For the case of neutrino-nucleon scattering one has:

$$\frac{d^2\sigma^{\nu(\bar{\nu})}}{dxdy} = \frac{G_F^2 ME}{\pi(1+Q^2/m_W^2)^2}\left\{y^2 xF_1 + \left[1-\left(1+\frac{Mx}{2E}\right)y\right]F_2 \pm y\left(1-\frac{y}{2}\right)xF_3\right\}, \quad (1)$$

where the scalar form-factors F_i are the corresponding structure functions, G_F is the Fermi constant, E the neutrino energy in the target rest frame and y is the inelasticity variable.

After an appropriate normalization, the structure functions are given as a convolution of the usual parton distributions and perturbatively calculable coefficient functions:

$$F_i(x,Q^2) \sim \int_\chi^1 \frac{d\xi}{\xi}\left[C_i^q(\xi,\mu^2,\mu_F^2,\lambda)q_1\left(\frac{\chi}{\xi},\mu_F^2\right) + C_i^g(\xi,\mu^2,\mu_F^2,\lambda)g\left(\frac{\chi}{\xi},\mu_F^2\right)\right]. \quad (2)$$

The variable χ takes values on the interval $(0,1)$ and incorporates also the effect of the target mass in addition to the non-zero mass m of the final quark and $\lambda = Q^2/(Q^2+m^2)$.

The coefficient functions are known through NLO QCD:

$$\begin{aligned} C_i^q(z,\mu^2,\mu_F^2,\lambda) &= \delta(1-z) + \frac{\alpha_S(\mu^2)}{2\pi}H_i^q(z,\mu_F^2,\lambda), \\ C_i^g(z,\mu^2,\mu_F^2,\lambda) &= \frac{\alpha_S(\mu^2)}{2\pi}H_i^g(z,\mu_F^2,\lambda), \end{aligned}$$

and they describe the corrections to the tree-level partonic processes $q+W^* \to \mathcal{Q}$ and $g+W^* \to \mathcal{Q}+\bar{q}$ respectively ($\mathcal{Q}$ and q stand for heavy and light quark resp.).

From the explicit form of these functions one can verify that the gluon initiated contribution is not enhanced in the soft limit $z \to 1$ while the quark initiated component is:

$$\begin{aligned} H_i^q(z\to 1,\mu_F^2,\lambda) &= H^{\text{soft}}(z,\mu_F^2,\lambda)\ ;\ i=1,2,3, \\ H^{\text{soft}}(z,\mu_F^2,\lambda) &= 2C_F\left\{2\left[\frac{\ln(1-z)}{1-z}\right]_+ - \left[\frac{\ln(1-\lambda z)}{1-z}\right]_+\right. \\ &+ \left.\frac{1}{4}\left[\frac{1-z}{(1-\lambda z)^2}\right]_+ + \frac{1}{(1-z)_+}\left(\ln\frac{Q^2+m^2}{\mu_F^2}-1\right)\right\}. \end{aligned} \quad (3)$$

The most important feature of Eq.(3) is the interplay between the limits $z \to 1$ and $m \to 0$. It is easy to see that if the mass m is non-zero, terms depending on z and λ through

$(1-\lambda z)$ are finite in the limit $z \to 1$, i.e. are not soft-enhanced. Therefore we are led to the conclusion that the behavior of the coefficient function at large z in the massive case is very different from the one in the massless case; in fact, that difference is apparent even at the leading log level.

To perform the soft-gluon resummation one can apply the usual techniques for evaluation of the corresponding partonic cross-section in the eikonal approximation. However, a more instructive method is to use the available results for the process $t \to W + b$ [3]. It is easy to see that the two are related by the simple replacement:

$$m_t \to m \ ; \ m_W^2 \to -Q^2. \tag{4}$$

The effect of the mass in the Sudakov exponent Δ_N (evaluated in Mellin space) with next-to-leading logarithmic (NLL) accuracy is to set the corresponding characteristic scale:

$$\ln \Delta_N = \int_0^1 dz \frac{z^{N-1}-1}{1-z} \left\{ \int_{\mu_F^2}^{\mathcal{M}^2(1-z)^2} \frac{dk^2}{k^2} A\left[\alpha_S(k^2)\right] + S\left[\alpha_S\left(\mathcal{M}^2(1-z)^2\right)\right] \right\}, \tag{5}$$

and

$$\mathcal{M}^2 = m^2 \left(1 + \frac{Q^2}{m^2}\right)^2. \tag{6}$$

Note that this equation is valid only in the limit when the mass m is not very small compared to the scale Q. If that is not the case, then one can neglect the mass m and obtain the well known massless result with the identification $\mathcal{M}^2 = Q^2$ and replacing the function S characteristic for processes with heavy quarks with the “massless" function $1/2B$ (see [4] for further details). The latter change accounts for the enhancement due to collinear radiation when the final state quark is massless.

APPLICATIONS

The implementation of the soft gluon resummation is based on matching of the perturbative spectrum which is valid away from the point $z \approx 1$ and the Sudakov exponent (with the coefficients of the $\sim \delta(1-z)$ terms added multiplicatively) which is only valid at large $z \approx 1$. In order to study the effect of the soft-gluon resummation on observables like the structure functions F_i, one has to convolute the matched coefficient function with the next-to-leading order parton distributions. For more details about this procedure and its numerical impact on observables and on the extraction of parton distributions see the talk by G. Corcella [5] as well as [4]. Here we will only briefly mention that the effect of the resummation on the charm structure function is rather large at $Q^2 \sim 2-5$ GeV^2 and can reach a factor of 2 at $x \sim 0.5$ or factor of 5 for $x \sim 0.6$. Another important feature of the resummed result is that the predictions with and without soft-resummation are distinguishable above $x \sim 0.4$ as can be established by studying that stability of both results with respect to variation of the factorization and renormalization scales [4].

SOFT-RESUMMATION IN THE NEUTRAL CURRENT CASE

The soft-limit behavior for massive quark production is different in the neutral current case. Clearly, at low $Q^2 \sim m^2$ only the gluon initiated component $g+\gamma^* \to \mathscr{Q}+\overline{\mathscr{Q}}$ will have significant contribution. However, as we already mentioned, this component is not soft divergent; in the limit $N \to \infty$ (which is the Mellin space analog of $z \to 1$) it behaves only as $\sim \ln(N)/N$. In order to uncover a sensitivity to soft gluon radiation in that process one need to consider observables that are more differential with respect to the final state. Two cases that have been studied in the literature involve one-particle differential distributions of the type:

$$g+\gamma^* \to \mathscr{Q}+X(\overline{\mathscr{Q}}).$$

Since the heavy flavor is now being directly measured, one can select the so-called kinematical thresholds, i.e. parts of the phase space that leads to soft gluon radiation. The corresponding conditions are $q_T/Q \to 0$ [6] and $m_X^2 - m^2 \to 0$ and $m^2_{\mathscr{Q}\overline{\mathscr{Q}}} - s \to 0$ [7]. The authors of [6] have studied the effect of the resummation on the angular distribution of bottom mesons at HERA. They work in terms of the polar angle with respect to the direction of the photon momentum in the $\gamma^* p$ rest frame. The conclusion is that the effect of the resummation at small angles is important. The authors of [7] have considered the effect of the soft-gluon resummation on the p_T dependence of the charm structure function. They also find important effect due to the resummation at low p_T. This effect might be important for the determination of the gluon distribution function.

SUMMARY

In this talk I have reviewed the need for performing soft-gluon resummation in heavy quark production in both CC and NC DIS. In the cases that have been previously considered, the soft-gluon resummation has an important numerical effect. In particular, it is interesting to study its impact on the parton distributions.

ACKNOWLEDGMENTS

It is my pleasure to thank Gennaro Corcella for fruitful collaboration on this subject. Research supported by the US DOE under contract DE-FG03-94ER-40833 and by the start up funds of the University of Hawaii.

REFERENCES

1. Alexander Mitov, these proceedings.
2. G. Sterman, Nucl. Phys. B 281 (1987) 310; S. Catani and L. Trentadue, Nucl. Phys. B 327 (1989) 323.
3. M. Cacciari, G. Corcella and A. Mitov, JHEP 0212 (2002) 015 (hep-ph/0209204).
4. G. Corcella and A. Mitov, Nucl.Phys. B 676 (2004) 346 (hep-ph/0308105).
5. Gennaro Corcella, these proceedings.
6. P. M. Nadolsky, N. Kidonakis, F. I. Olness and C.-P. Yuan, Phys.Rev. D 67 (2003) 074015.
7. E. Laenen and S-O. Moch, Phys. Rev. D 59 (1999) 034027 (hep-ph/9809550).

Charmonium production in two-photon collisions at next-to-leading order

Bernd A. Kniehl

II. Institut für Theoretische Physik, Universität Hamburg, Luruper Chaussee 149, 22761 Hamburg, Germany

Abstract. We review recent results on the production of prompt charmonium in association with a hadron jet or a prompt photon in two-photon collisions at next-to-leading order in the factorization framework of nonrelativistic quantum chromodynamics.

Keywords: Nonrelativistic quantum chromodynamics, radiative corrections, charmonium, two-photon scattering
PACS: 12.38.Bx, 12.39.St, 13.66.Bc, 14.40.Gx

1. INTRODUCTION

The factorization formalism of nonrelativistic quantum chromodynamics (NRQCD) [1] provides a rigorous theoretical framework for the description of heavy-quarkonium production and decay that is renormalizable and predictive. Theoretical predictions are decomposed into sums over products of short-distance coefficients, which can be calculated perturbatively as expansions in the strong-coupling constant α_s, and long-distance matrix elements (MEs), which are subject to relative-velocity (v) scaling rules and must be extracted from experiment, and are so organized as double expansions in α_s and v. This formalism takes into account the complete structure of the $Q\overline{Q}$ Fock space, which is spanned by the states $n = {}^{2S+1}L_J^{(a)}$ with definite spin S, orbital angular momentum L, total angular momentum J, and color multiplicity $a = 1, 8$, and so predicts the existence of color-octet (CO) processes in nature.

The greatest triumph of the NRQCD factorization formalism was its ability to correctly describe the cross section of inclusive charmonium hadroproduction at the Tevatron, which exceeds the color-singlet-model prediction by more than one order of magnitude. In order to convincingly establish the phenomenological significance of the CO processes, it is indispensable to identify them in other kinds of high-energy experiments as well. The verification of the NRQCD factorization hypothesis is presently hampered both from the theoretical and experimental sides. On the one hand, the theoretical predictions to be compared with existing experimental data are, apart from very few exceptions, of lowest order (LO) and thus suffer from considerable uncertainties. The measurement of charmonium polarization at the Tevatron currently presents a challenge for NRQCD factorization, but any conclusions are premature in the absence of a full-fledged next-to-leading-order (NLO) analysis. It is, therefore, mandatory to calculate the NLO corrections to the hard-scattering cross sections and to include the composite operators that are suppressed by higher powers in v. Apart from the usual reduction of the renor-

CP792, *Deep Inelastic Scattering, DIS 2005,* edited by W. H. Smith and S. R. Dasu

malization and factorization scale dependences, sizeable effects, e.g. due to the opening of new partonic production channels, are expected at NLO. On the other hand, the experimental errors are still rather sizeable. The latter are being significantly reduced by HERA II and run II at the Tevatron, and will be dramatically more so by the LHC and hopefully a future e^+e^- linear collider such as the TeV-Energy Superconducting Linear Accelerator (TESLA), which is presently being designed and planned at DESY.

Recently, $2 \to 2$ processes of heavy-quarkonium production were for the first time studied at NLO in the NRQCD factorization formalism [2, 3]. Specifically, the production of prompt charmonium, which is produced either directly or through the decay of heavier charmonia, with finite transverse momentum (p_T) in association with a hadron jet [2] or a prompt photon [3] via direct photoproduction in two-photon collisions was considered. In this presentation, we review the most important conceptional issues and phenomenological results of Refs. [2, 3], in Sections 2 and 3, respectively.

2. CONCEPTIONAL ISSUES

We focus attention on the process $\gamma\gamma \to J/\psi + X$, where J/ψ is promptly produced at finite value of p_T and X is a purely hadronic remainder. Since the incoming photons can interact either directly with the quarks participating in the hard-scattering process (direct photoproduction) or via their quark and gluon content (resolved photoproduction), this process receives contributions from the direct, single-resolved, and double-resolved channels, which are formally of the same order in the perturbative expansion. At LO, the bulk of the cross section is due to single-resolved photoproduction, and the NRQCD prediction [4] based on the MEs determined from fits [4] to Tevatron data nicely agrees with a recent measurement by the DELPHI Collaboration at LEP2 [6].

Here, we consider direct photoproduction at NLO [2]. At LO, there is only one partonic subprocess, namely

$$\gamma + \gamma \to c\bar{c}[{}^3S_1^{(8)}] + g. \tag{1}$$

At NLO, virtual corrections to process (1) and $\langle \mathcal{O}^H[{}^3S_1^{(8)}] \rangle$, where $H = J/\psi, \chi_{cJ}, \psi'$, and real corrections to

$$\gamma + \gamma \to c\bar{c}[n] + g + g, \qquad n = {}^3P_J^{(1)}, {}^1S_0^{(8)}, {}^3S_1^{(8)}, {}^3P_J^{(8)}, \tag{2}$$

$$\gamma + \gamma \to c\bar{c}[n] + u + \bar{u}, \qquad n = {}^3S_1^{(8)}, \tag{3}$$

$$\gamma + \gamma \to c\bar{c}[n] + q + \bar{q}, \qquad n = {}^1S_0^{(8)}, {}^3S_1^{(8)}, {}^3P_J^{(8)}, \tag{4}$$

where u and $\bar{u}$ denote the Faddeev-Popov ghosts of the gluon, contribute.

The virtual corrections to process (1) receive contributions from self-energy, triangle, box, and pentagon diagrams. The self-energy and triangle diagrams are in general ultraviolet (UV) divergent; the triangle, box, and pentagon diagrams are in general infrared (IR) divergent; and the pentagon diagrams without three-gluon vertex also contain Coulomb divergences. As for the light-quark loops, the triangle diagrams vanish by Furry's theorem, while the box diagrams form a finite subset. The virtual corrections to

$\langle \mathcal{O}^H[^3S_1^{(8)}]\rangle$ also produce UV, IR, and Coulomb divergences. The UV and IR divergences are extracted using dimensional regularization in $d = 4 - 2\varepsilon$ space-time dimensions, leading to poles in $\varepsilon_{\rm UV}$ and $\varepsilon_{\rm IR}$, respectively, while the Coulomb singularities are regularized by a small value of v. The UV divergences are removed by the renormalization of $\langle \mathcal{O}^H[^3S_1^{(8)}]\rangle$, α_s, the charm-quark mass and field, and the gluon field, which is performed in the modified minimal-subtraction ($\overline{\rm MS}$) scheme for the former two quantities, rendering them dependent on the renormalization scales λ and μ, respectively, and in the on-mass-shell scheme for the residual three quantities. The IR divergences cancel among the virtual and real corrections, the wave-function renormalizations, and $\langle \mathcal{O}^H[^3S_1^{(8)}]\rangle$. The Coulomb divergences cancel between the virtual corrections and $\langle \mathcal{O}^H[^3S_1^{(8)}]\rangle$.

The real corrections are plagued by IR divergences, which come as collinear divergences from the initial state and collinear and/or soft ones from the final state. They are identified by appropriately slicing the three-particle phase space using small parameters δ_i and δ_f, respectively. The collinear and/or soft regions of phase space are integrated over analytically in d dimensions, while the hard region is integrated over numerically in four dimensions. The sum of these contributions is, to very good approximation, independent of δ_i and δ_f. The initial-state collinear divergences are factorized at some factorization scale M and absorbed into the parton density functions (PDFs) of the q and $\overline{q}$ quarks inside the resolved photon. The M dependence thus introduced is approximately compensated by the LO single-resolved contribution.

Combining the contributions arising from the virtual corrections (vi), the parameter and wave-function renormalization (ct), the operator redefinition (op), the initial-state (is) and final-state (fs) collinear configurations, the soft-gluon radiation (so), and the hard-parton emission (ha) as

$$d\sigma(\mu,\lambda,M) = d\sigma_0(\mu,\lambda)[1+\delta_{\rm vi}(\mu;\varepsilon_{\rm UV},\varepsilon_{\rm IR},v)+\delta_{\rm ct}(\mu;\varepsilon_{\rm UV},\varepsilon_{\rm IR})+\delta_{\rm op}(\mu,\lambda;\varepsilon_{\rm IR},v) + \delta_{\rm fs}(\mu;\varepsilon_{\rm IR},\delta_f)] + d\sigma_{\rm is}(\mu,\lambda,M;\delta_i) + d\sigma_{\rm so}(\mu,\lambda;\varepsilon_{\rm IR},\delta_f) + d\sigma_{\rm ha}(\mu,\lambda;\delta_i,\delta_f), \quad (5)$$

the regulators $\varepsilon_{\rm UV}$, $\varepsilon_{\rm IR}$, v, δ_i, and δ_f drop out and the μ and λ dependences formally cancel up to terms beyond NLO, while the M dependence is unscreened at NLO.

3. PHENOMENOLOGICAL RESULTS

We consider two-photon collisions at TESLA operating at a center-of-mass energy of 500 GeV, where the photons arise from electromagnetic initial-state bremsstrahlung, with antitagging angle $\theta_{\rm max} = 25$ mrad, and beamstrahlung, with effective beamstrahlung parameter $\Upsilon = 0.053$. The J/ψ, χ_{cJ}, and ψ' MEs are adopted from Ref. [5] and the photon PDFs from Ref. [7].

In Fig. 1, we study $d^2\sigma/dp_T dy$ (a) for rapidity $y = 0$ as a function of p_T and (b) for $p_T = 5$ GeV as a function of y, comparing the LO (dashed lines) and NLO (solid lines) results of direct photoproduction with the LO result of single-resolved photoproduction (dotted lines). From Fig. 1(a), we observe that, with increasing value of p_T, the NLO result of direct photoproduction falls of considerably more slowly than the LO one. In fact, the QCD correction (K) factor, defined as the NLO to LO ratio, rapidly increases with p_T, exceeding 10 for $p_T \gtrsim 10$ GeV. This feature may be understood by observing that so-called *fragmentation-prone* [8] partonic subprocesses start to contribute to direct photoproduction at NLO, while they are absent at LO. Such subprocesses contain a gluon with small virtuality, $q^2 = 4m_c^2$, that splits into a $c\bar{c}$ pair in the Fock state $n = {}^3S_1^{(8)}$ and thus generally generate dominant contributions at $p_T \gg 2m_c$ due to the presence of a large gluon propagator. In single-resolved photoproduction, a fragmentation-prone partonic subprocess already contributes at LO. This explains why the solid and dotted curves in Fig. 1(a) run parallel in the upper p_T range. At low values of p_T, the fragmentation-prone partonic subprocesses do not matter, and the relative suppression of direct photoproduction is due to the fact that, at LO, this is a pure CO process.

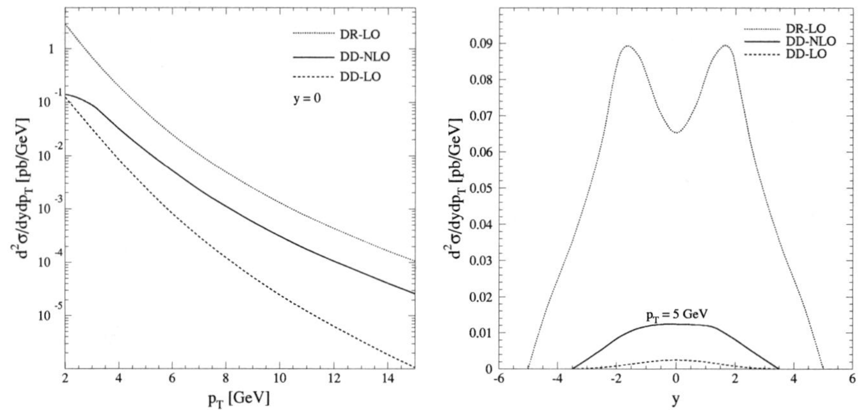

FIGURE 1. LO single-resolved (dotted lines), LO direct (dashed lines), and NLO direct (solid lines) contributions to $d^2\sigma/dp_T dy$ (a) for $y = 0$ as a function of p_T and (b) for $p_T = 5$ GeV as a function of y.

In the case of $\gamma\gamma \to J/\psi + X_\gamma$, where X_γ contains a prompt photon, the K factor was found to decrease fast with increasing value of p_T, falling below 01 for $p_T \gtrsim 14$ GeV [3].

ACKNOWLEDGMENTS

The author thanks M. Klasen, L.N. Mihaila, and M. Steinhauser for their collaboration. This work was supported in part by BMBF Grant No. 05 HT1GUA/4.

REFERENCES

1. G. T. Bodwin, E. Braaten, and G. P. Lepage, *Phys. Rev. D* **51**, 1125 (1995); **55**, 5853(E) (1997).
2. M. Klasen, B. A. Kniehl, L. N. Mihaila, and M. Steinhauser, *Nucl. Phys. B* **713**, 487 (2005).
3. M. Klasen, B. A. Kniehl, L. N. Mihaila, and M. Steinhauser, *Phys. Rev. D* **71**, 014016 (2005).
4. M. Klasen, B. A. Kniehl, L. N. Mihaila, and M. Steinhauser, *Phys. Rev. Lett.* **89**, 032001 (2002).
5. E. Braaten, B. A. Kniehl, and J. Lee, *Phys. Rev. D* **62**, 094005 (2000).
6. DELPHI Collaboration, J. Abdallah et al., *Phys. Lett. B* **565**, 76 (2003).
7. M. Glück, E. Reya, and I. Schienbein, *Phys. Rev. D* **60**, 054019 (1999); **62**, 019902(E) (1999).
8. B. A. Kniehl, C. P. Palisoc, and L. Zwirner, *Phys. Rev. D* **66**, 114002 (2002); **69**, 115005 (2004).

Measurements of Charm and Charmonium Production by PHENIX

K.F. Read, for the PHENIX Collaboration[1]

Physics Division, Oak Ridge National Laboratory, Oak Ridge, TN 37831, USA
Department of Physics and Astronomy, University of Tennessee, Knoxville, TN 37996, USA

Abstract. The PHENIX Experiment at RHIC has measured charmonium production using dileptons and open heavy flavor production via semileptonic decays in $p+p$, $d+Au$, and $Au+Au$ collisions at $\sqrt{s_{NN}} = 200$ GeV. A nuclear dependence affecting J/ψ production in $d+Au$ collisions is observed. For electrons from heavy flavor decay in $Au+Au$ collisions, the transverse momentum spectrum is observed to be strongly modified compared to scaled results from $p+p$ collisions. An initial measurement of the azimuthal anisotropy parameter, v_2, for such electrons is reported.

Keywords: Charm, charmonium, heavy ions
PACS: 25.75.-q; 25.75.Dw; 25.75.Ld; 13.20.Fc

INTRODUCTION

The PHENIX Experiment [1, 2] at RHIC has collected $p+p$, $d+Au$, $Cu+Cu$, and $Au+Au$ collision data at $\sqrt{s_{NN}} = 200$ GeV. Measurements of open charm production in $p+p$ collisions can be used to make important tests of pQCD predictions at $\sqrt{s} = 200$ GeV and establish a baseline for total charm production for the heavy ion program [3]. Analogous measurements for $Au+Au$ collisions can be used to study medium modification effects such as charm quark energy loss and collective flow as well as establish a baseline for J/ψ production. Such measurements can also be used to study potential thermal production of charm from a possible Quark Gluon Plasma. The J/ψ production cross section is predicted to be modified by the medium produced in heavy ion collisions. Measurements of open heavy flavor and J/ψ production in $d+Au$ collisions can be used to study modification of the gluon structure function in nuclei.

PHENIX EXPERIMENT

The PHENIX experiment consists of two separate central arms with 90° coverage in azimuth and pseudorapidity coverage of $|\eta| < 0.35$. Two separate muon spectrometers at forward and backward rapidity cover the range $1.2 < |\eta| < 2.4$, with full azimuthal coverage. Electrons are measured in the central arm detectors by matching charged particle tracks to clusters in an electromagnetic calorimeter and rings in a ring imaging Čerenkov detector. Muons are measured in the forward/backward arms using Iarocci

[1] For the full PHENIX Collaboration author list, see Ref. [1].

CP792, *Deep Inelastic Scattering, DIS 2005,* edited by W. H. Smith and S. R. Dasu

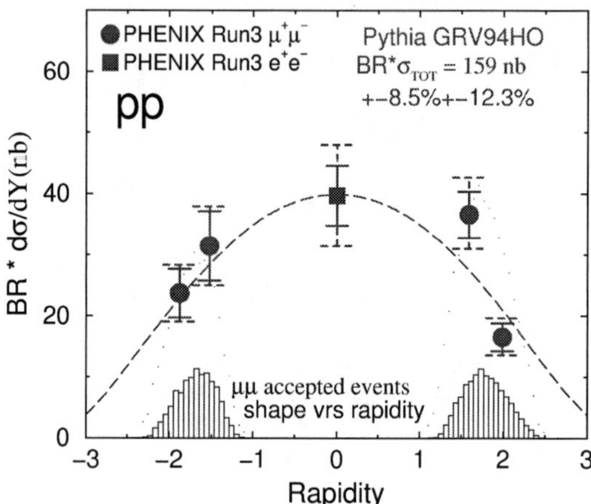

FIGURE 1. J/ψ differential cross section, multiplied by the dilepton branching ratio, versus rapidity for dimuons (circles) and dielectrons (square) [4], compared to a theoretical prediction (curve) from PYTHIA using the GRV94HO parton distribution.

tubes sandwiched between steel absorber planes for identification and cathode strip chambers for momentum measurement. Experimental details are provided in Ref. [2].

J/ψ PRODUCTION

Using $p+p$ collisions at $\sqrt{s} = 200$ GeV, PHENIX has measured the J/ψ differential production cross section at both mid-central and forward/backward rapidities (Fig. 1) using dielectrons and dimuons, respectively. The preliminary value of the measured [4] branching ratio times cross section is 159 nb $\pm$ 8.5% $\pm$ 12.3%. Comparisons have been made to different theoretical predictions [4]. Future runs with increased statistics will be used to discriminate between predictions, extend the measurements to higher transverse momentum, and allow measurement of the J/ψ polarization.

The ratio between the J/ψ differential production cross section in $d+Au$ collisions and the same quantity for $p+p$ collisions scaled by 2×197 has been measured [4]. The ratio (nuclear modification factor) is near unity at backward rapidity and significantly lower at forward rapidity. Such measurements can be used to test theoretical predictions for the modification of the gluon structure function in nuclei (shadowing).

PHENIX has measured the cross section for J/ψ production in $Au+Au$ collisions as a function of centrality [5]. Such measurements provide important tests of theoretical predictions for modification of J/ψ production due to the creation of a possible Quark Gluon Plasma. The present statistics are insufficient to make strong conclusions, although models with strong enhancement appear to be disfavored. Results based on data from subsequent runs with greater statistics as well as different colliding species ($Cu+Cu$) are in preparation.

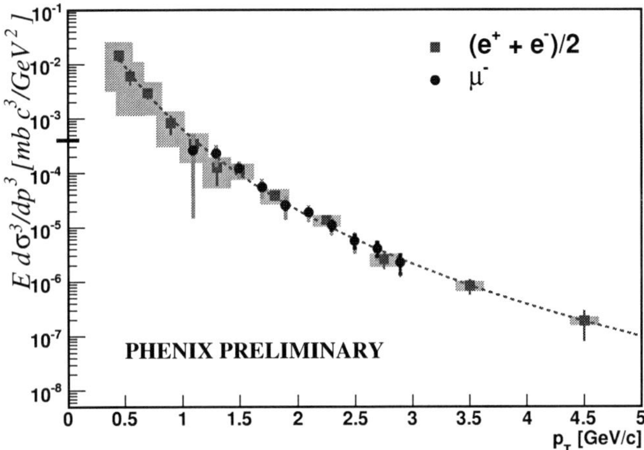

FIGURE 2. Transverse momentum distribution of invariant differential production cross section for negative muons (circles) and the average of electrons and positrons (squares) from heavy flavor decay for $p+p$ collisions at $\sqrt{s} = 200$ GeV. The pseudorapidity interval is centered around $|\eta| = 1.65$ for muons and $\eta = 0$ for electrons and positrons. Systematic errors are indicated as bands. The systematic error associated with overall normalization of the muon data is separately indicated by the horizontal band on the vertical axis. The dashed curve is a power law fit to the electron data points.

OPEN HEAVY FLAVOR PRODUCTION

After appropriate subtraction of physics backgrounds, PHENIX measurements of inclusive single electron and muon production can be used to extract the yield of leptons resulting from the semileptonic decay of heavy quarks, and thereby obtain the cross section for open heavy flavor production.

Hadrons that punch through the absorber of the muon spectrometer and mimic muons and real muons resulting from the decay of light hadrons must be subtracted on a statistical basis from the measured inclusive single muon yield, in order to obtain the signal for muons resulting from heavy flavor decay. Data collected from successive layers in the Muon Identifier are used to measure an effective nuclear absorption length which is used in a Monte Carlo simulation to estimate the punch-through contribution. The measured vertex dependence of the inclusive muon yield is used to obtain the contribution due to the decay in flight of light hadrons. The production cross section for negative single muons from heavy flavor semileptonic decay for $p+p$ collisions at $\sqrt{s} = 200$ GeV is shown in Fig. 2 as a function of transverse momentum.

The contributions of electrons resulting from photonic sources such as photon conversions and Dalitz decays of π^0 and η mesons must be subtracted from the inclusive single electron yield, in order to obtain the signal for non-photonic electrons which primarily results from semileptonic heavy flavor decay [6, 7]. PHENIX has measured the transverse momentum distribution of the cross section for non-photonic single electrons resulting from heavy flavor decay in $p+p$ (Fig. 2), $d+Au$, and $Au+Au$ collisions

[6, 7] at $\sqrt{s_{NN}} = 200$ GeV. Preliminary measurements indicate that in $p+p$ collisions the shape of the transverse momentum distribution of the differential production cross section for these electrons is harder than that of leading-order PYTHIA predictions.

Recent preliminary measurements with increased statistics indicate that the production of electrons from heavy flavor decay in $Au+Au$ collisions is suppressed at high p_T relative to $p+p$ results scaled by the number of binary collisions. The spectral shape is modified by the medium in a pattern which is consistent with theoretical models incorporating quark energy loss.

PHENIX has measured the second harmonic, v_2, of the azimuthal distribution of electrons from heavy flavor decay in $Au+Au$ collisions at $\sqrt{s_{NN}} = 200$ GeV as a function of transverse momentum [8]. The measured v_2 is nonzero at 90% confidence level; however, the present uncertainties are too large to draw conclusions concerning flow of the heavy quark. Analysis of a subsequent data set with significantly greater statistics is underway which should provide more detailed information concerning charm flow.

SUMMARY AND CONCLUSIONS

Heavy flavor production has been measured via leptonic decays in $p+p$, $d+Au$, and $Au+Au$ collisions at $\sqrt{s_{NN}} = 200$ GeV by the PHENIX Experiment at RHIC. As discussed above, such results have been used to test theoretical predictions concerning topics such as the momentum spectrum of open heavy flavor production, shadowing in nuclei, and modifications of the J/ψ production cross section in heavy ion collisions. Strong modification of the spectra for electrons from heavy flavor decay is observed in $Au+Au$ collisions. An initial measurement of the azimuthal anisotropy, v_2, for electrons from heavy flavor decay has been made. Results with higher statistics and other colliding species are in preparation.

ACKNOWLEDGMENTS

This research was sponsored by the Division of Nuclear Physics, U.S. Department of Energy, under contract DE-AC05-00OR22725 with UT-Battelle, LLC (Oak Ridge National Laboratory) and contract DE-FG02-96ER40982 with the University of Tennessee. The complete list of PHENIX acknowledgements is provided in Ref. [1].

REFERENCES

1. S. Adler, et al., *J. Phys. G: Nucl. Part. Phys.*, **30**, S1415–1418 (2004).
2. K. Adcox, et al., *Nucl. Instrum. Meth. A*, **499**, 469–602 (2003).
3. S. Adler, et al., *Phys. Rev. Lett.*, **92**, 051802 (2004).
4. R. Granier de Cassagnac, et al., *J. Phys. G: Nucl. Part. Phys.*, **30**, S1341–1345 (2004).
5. S. Adler, et al., *Phys. Rev. C*, **69**, 014901 (2004).
6. S. Kelly, et al., *J. Phys. G: Nucl. Part. Phys.*, **30**, S1189–1192 (2004).
7. K. Adcox, et al., *Phys. Rev. Lett.*, **94**, 082301 (2005).
8. S. Adler, et al., (2005), nucl-ex/0502009.

Heavy Quark Parton Distribution Functions[1], [2]

Stefan Kretzer*,† and Fredrick I. Olness**

*Physics Department, Brookhaven National Laboratory, Upton, New York 11973
† RIKEN-BNL Research Center, Brookhaven National Laboratory, Upton, New York 11973 – 5000
**Department of Physics, Southern Methodist University, Dallas, Texas 75275-0175

Abstract. We present the CTEQ6HQ parton distribution set which is determined in the general variable flavor number scheme which incorporates heavy flavor mass effects; hence, this set provides advantages for precision observables which are sensitive to charm and bottom quark masses. We describe the analysis procedure, examine the predominant features of the new distributions, and compare with previous distributions. We also examine the uncertainties of the strange quark distribution and how the the recent NuTeV dimuon data constrains this quantity.

Parton distributions functions (PDFs) provide the essential link between the theoretically calculated partonic cross-sections, and the experimentally measured physical cross-sections involving hadrons and mesons. This link is crucial if we are to make incisive tests of the standard model, and search for subtle deviations which might signal new physics. The choice of the renormalization scheme is an important issue to address if we are to make the most efficient use of our fixed-order perturbation expansion. Separately, it is important to know the uncertainty range of the PDFs, and properly fold these into the overall uncertainty estimates. We will address both of these issues in turn.

GLOBAL FITTING AND THE CTEQ6HQ PDFS: The CTEQ6HQ (or C6HQ for short) PDFs[1] are obtained by performing a global analysis using the generalized (non-zero quark-mass) $\overline{\text{MS}}$ perturbative QCD framework of Refs. [2, 3], which we label the general-mass variable-flavor-number scheme (GM-VFNS). When matched to the corresponding hard-scattering cross-sections calculated in the same scheme, the combination should provide a more accurate description of the precision DIS structure function data, as well as other processes which are sensitive to charm and bottom mass effects.

The C6HQ global fitting follows the same procedure as that of the earlier CTEQ6 analysis.[4] The data sets used before are supplemented by the H1 and ZEUS data sets for the structure function F_2^c with tagged charm particles in the final state. The F_2^c data sets are quite relevant for this analysis since F_2^c is sensitive to the charm and gluon distributions, which are tightly coupled in the generalized $\overline{\text{MS}}$ formalism.

The C6HQ set is the best fit obtained with these inputs. A broad measure of the quality of this fit is provided by the overall χ^2 of 2008 for a total number of 1925 data points (χ^2/DOF = 1.04). This is to be compared to a χ^2 of 2037 for 1925 points

---

[1] The research presented here was performed in collaboration with W.K. Tung, P. Nadolsky, J.F. Owens, J. Pumplin, D. Stump, J. Huston, & H.L. Lai.
[2] Presented by Fredrick Olness.

CP792, *Deep Inelastic Scattering, DIS 2005*, edited by W. H. Smith and S. R. Dasu

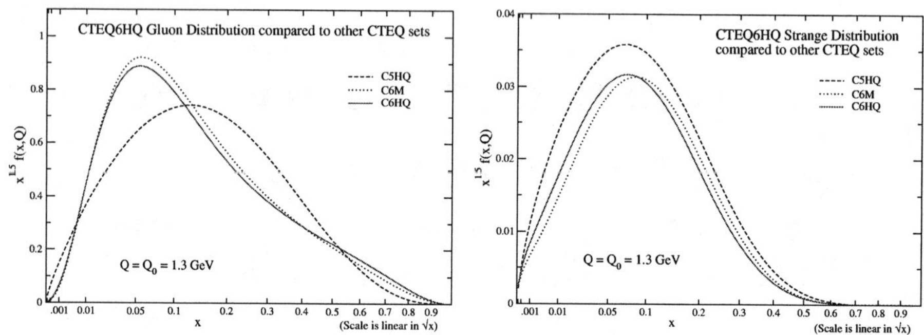

Figure 1. Comparison of the a) gluon and b) strange-quark distributions at $Q_0 = m_c = 1.3\,\mathrm{GeV}$ for the CTEQ5HQ, CTEQ6M, and CTEQ6HQ sets. The axes are scaled to highlight the valence components of these distributions.

(χ^2/DOF = 1.07) in the case of CTEQ6M (or C6M for short). The new C6HQ fit reduces the overall χ^2 by 29 out of ~2000 as compared to the C6M fit. The improvement of this generalized $\overline{\text{MS}}$ result over the zero-mass $\overline{\text{MS}}$ result is encouraging, since the generalized $\overline{\text{MS}}$ formalism represents a more accurate formulation of PQCD. However, a difference of χ^2 of 29 is within the current estimated range of uncertainty of PDF analysis. [4] Therefore, the significance of this difference is arguable. We also note that the improvement in χ^2 is spread over most of the data sets: there is no "smoking gun" for the overall difference.

Since perturbative calculations are renormalization scheme dependent, it is important to use properly matched hard-scattering cross-sections and PDFs when evaluating factorized cross-sections for physical applications. This issue is particularly relevant for applications involving heavy quarks, since the heavy quark introduces a new mass scale which leads to complications of the PQCD formalism. To illustrate this point, we compare the above results with two possible uses of the PDFs that represent a *mis-use* of PQCD in principle, but occur frequently in the literature in practice, perhaps out of necessity. These involve using PDFs obtained in the general-mass scheme convoluted with hard-scattering cross-sections (Wilson coefficients) defined in the zero-mass scheme, and vice versa. For example, if we use the C6M PDFs which are derived in the ZM-VFN scheme with the GM-VFN hard-scattering cross-sections, we obtain a total χ^2 of 2431 for 1925 data points (χ^2/DOF = 1.26). Conversely, if we use the C6HQ PDFs which are derived in the GM-VFN scheme with the ZM-VFN hard-scattering cross-sections, we obtain a total χ^2 of 2496 for 1925 data points (χ^2/DOF = 1.30). For the same data sets, these mis-matched schemes result in a overall χ^2 difference of 420~490 compared to the C6HQ set (χ^2 of 2008). These are quite large differences relative to the tolerances discussed in Refs. [5, 6, 7, 4], and result in obvious discrepancies with some of the precision DIS data sets. Clearly, for quantitative applications it is imperative to maintain consistency between the PDFs and the hard-scattering cross-sections.

COMPARISON WITH RELATED PDFS: The C6HQ and C6M fits provide comparable descriptions of the global QCD data in two *different* schemes. Some of the differences in the PDFs arise purely from the choice of scheme. We are particularly interested in

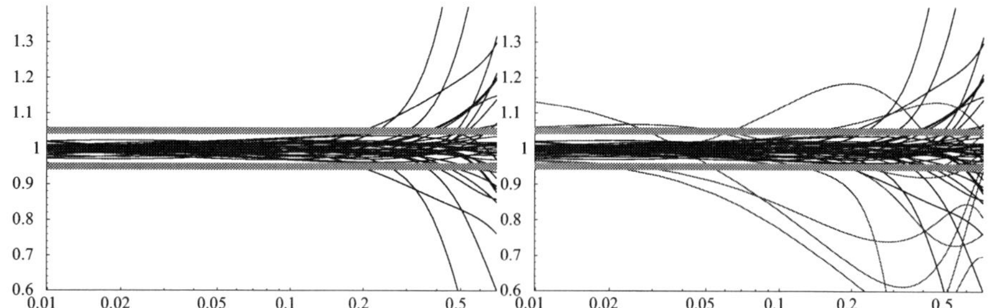

Figure 2. a) Ratio of the 40 CTEQ6M PDF sets compared to the central set for the strange-quark vs x. b) Same as previous, with additional sets included (with a more general $s(x)$ parameterization). In both figures, guide-lines are drawn at $\pm5\%$.

the differences for the gluon distribution, which will strongly influence the closely correlated charm distribution (since it is generated via the $g \to c\bar{c}$ process). It is also interesting to compare the differences between the earlier CTEQ5HQ (C5HQ) distributions with the new C6HQ distributions; differences between these PDFs are attributable both to new data, and to minor differences in the way the theoretical inputs are implemented.

Fig. 1a) shows the comparison of the gluon distribution at Q_0. Here the difference between C5HQ and the CTEQ6 generation of gluon distributions is pronounced. The change in this least-well-determined parton distribution is due to the recent precision DIS data (most influential in the small x region) in conjunction with the greatly improved inclusive jet data from the Tevatron (critical for the medium to large x regions). The differences between C6HQ and C6M gluons at large x are due to a combination of scheme-dependence, and the inherent uncertainty range of the current analysis.

Fig. 1b) shows the comparison of the strange distributions at the same Q_0. The noticeable difference between the C5HQ curve and the others, in this case, is largely the result of different theoretical inputs: namely, the κ parameter which determines the ratio of strange to non-strange sea quarks at the initial scale Q_0. This κ factor, known only approximately, was chosen to be $1/2$ both in the CTEQ5 and CTEQ6 analyses, but for slightly different values of Q_0: 1.0 GeV for C5HQ, and 1.3 GeV for the CTEQ6 sets. We now look at the uncertainty of the strange quark in further detail; this has received increased attention recently since this is an important ingredient in the NuTeV determination of $\sin\theta_W$.

STRANGE-QUARK PDF AND UNCERTAINTIES: Using the set of 40 C6M PDF sets, we can produce a band of of distributions which, in principle, should characterize the uncertainty of the s-quark, cf., Fig. 2a). Based on this plot, one would be tempted to conclude that the uncertainty on the strange quark is better than 5% over much of the x range. Because the parameterization of $s(x)$ in the global fit is constrained to be $\kappa[u(x)+\bar{d}(x)]$ at Q_0, the figure actually reflects the uncertainty not on $s(x)$ but instead on $[\bar{u}(x)+\bar{d}(x)]$. This constraint is imposed because none of the data in the global analysis directly measures the $s(x)$ distribution; hence, Fig. 2a) is by no means a representation of the true uncertainty.

Recent measurements of the charged-current charm production ($\nu s \to c\mu \to \mu^{\pm}\mu^{\mp}X$)

from CCFR and NuTeV [8, 9] have the potential to determine $\{s(x), \bar{s}(x)\}$ with much more precision. In Fig. 2b) we again show the 40 C6M PDF sets combined with a number of additional fits that relax the parameterization of $s(x)$. This is not an exhaustive collection designed to span the full parameter space, but rather an illustration that the implied uncertainty of Fig. 2a) is much too conservative; the dimuon data will go a long way toward improving our knowledge of the strange PDF. An accurate determination of the strangeness of the proton, as well as the strangeness asymmetry $[s(x) - \bar{s}(x)]$ will have important implications for a number of measurements, including the NuTeV $\sin\theta_W$ measurement.

CONCLUDING REMARKS: The C6HQ PDFs presented here complement the CTEQ6 sets by providing distributions which can be used in the generalized $\overline{\text{MS}}$ scheme with massive partons. This analysis includes the complete set of NLO processes including the real and virtual quark-initiated terms.

While the zero-mass parton scheme is sufficient for many purposes, the fully massive scheme can be important when physical quantities are sufficiently sensitive to heavy quark contributions. This is evident when comparing the C6HQ and C6M fits to the mis-matched sets where the precise DIS data from HERA highlights the discrepancies.

The C6HQ fits also provide the basis for a series of further studies involving more quantitative analysis of strange, charm, and bottom quark distributions inside the nucleon. For example, the C6HQ PDFs are necessary for a consistent analysis of resumed differential distributions for heavy quark production.[10] Using the full range of data from both the charged and neutral current processes, these distributions can reduce the uncertainties in the calculations; hence, they have significant implications for charm and bottom production, and can help resolve questions about intrinsic heavy quark constituents inside the proton, the $\Delta x F_3$ structure function, and the extraction of $\sin\theta_W$.

ACKNOWLEDGMENTS: F.I.O. acknowledges the hospitality of Fermilab and BNL, where a portion of this work was performed. This work was supported by RIKEN, BNL, the U.S. DoE under grant DE-AC02-98CH10886 & DE-FG03-95ER40908, and the Lightner-Sams Foundation.

REFERENCES

1. S. Kretzer, H. L. Lai, F. I. Olness, and W. K. Tung, *Phys. Rev.*, **D69**, 114005 (2004), hep-ph/0307022.
2. M. A. G. Aivazis, J. C. Collins, F. I. Olness, and W.-K. Tung, *Phys. Rev.*, **D50**, 3102–3118 (1994), hep-ph/9312319.
3. W.-K. Tung, S. Kretzer, and C. Schmidt, *J. Phys.*, **G28**, 983–996 (2002), hep-ph/0110247.
4. J. Pumplin, et al., *JHEP*, **07**, 012 (2002), hep-ph/0201195.
5. A. D. Martin, R. G. Roberts, W. J. Stirling, and R. S. Thorne, *J. Phys.*, **G26**, 663–665 (2000).
6. A. D. Martin, R. G. Roberts, W. J. Stirling, and R. S. Thorne, *Eur. Phys. J.*, **C28**, 455–473 (2003), hep-ph/0211080.
7. A. D. Martin, R. G. Roberts, W. J. Stirling, and R. S. Thorne, *Eur. Phys. J.*, **C35**, 325–348 (2004), hep-ph/0308087.
8. A. O. Bazarko, et al., *Z. Phys.*, **C65**, 189–198 (1995), hep-ex/9406007.
9. M. Goncharov, et al., *Phys. Rev.*, **D64**, 112006 (2001), hep-ex/0102049.
10. P. M. Nadolsky, N. Kidonakis, F. I. Olness, and C. P. Yuan, *Phys. Rev.*, **D67**, 074015 (2003), hep-ph/0210082.

A Variable-Flavour-Number Scheme at NNLO

Robert S. Thorne

Cavendish Laboratory, University of Cambridge, Madingley Road, Cambridge, CB3 0HE, UK

Abstract. I present a formulation of a Variable Flavour Number Scheme for heavy quarks that is implemented up to NNLO in the strong coupling constant and may be used in NNLO global fits for parton distributions.

Keywords: QCD, Structure Functions, Heavy Quarks
PACS: 12.38.Bx,13.60.Hb

While up, down and strange quarks are treated as effectively massless partons, charm, bottom and top have to be regarded as heavy partons. There are two distinct regimes for these types of quarks. At low scales, $Q^2 \sim m_H^2$, they are only created in the final state and described using the Fixed Flavour Number Scheme (FFNS)

$$F_i(x,Q^2) = C_{i,k}^{FF}(Q^2/m_H^2) \otimes f_k^{n_f}(Q^2).$$

However, for $Q^2 \gg m_H^2$, the coefficient functions contain large $\ln(Q^2/m_H^2)$ terms, spoiling the perturbative expansion. In this regime it is more appropriate to treat the quarks like massless partons, and the large $\ln(Q^2/m_H^2)$ terms are summed via the DGLAP evolution equations. The simplest recipe involving this regime is the Zero Mass Variable Flavour Number Scheme (ZMVFNS). This ignores all $\mathcal{O}(m_H^2/Q^2)$ corrections, i.e.

$$F_i(x,Q^2) = C_{i,j}^{ZMVF} \otimes f_j^{n_f+1}(Q^2).$$

The partons in different flavour-number regions are related perturbatively,

$$f_k^{n_f+1}(Q^2) = A_{jk}(Q^2/m_H^2) \otimes f_k^{n_f}(Q^2),$$

where the perturbative matrix elements $A_{jk}(Q^2/m_H^2)$ containing $\ln(Q^2/m_H^2)$ terms guarantee the correct evolution for both descriptions. At LO, i.e. zeroth order in α_S, the relationship between the two descriptions is trivial – $q(g)_k^{n_f+1}(Q^2) \equiv q(g)_k^{n_f}(Q^2)$. At NLO, i.e. first order in α_S ($h^+(Q^2) = (h+\bar{h})(Q^2)$),

$$h^+(Q^2) = \frac{\alpha_S}{4\pi} P_{qg}^0 \otimes g^{n_f}(Q^2) \ln\left(\frac{Q^2}{m_H^2}\right), \quad g^{n_f+1}(Q^2) = \left(1 - \frac{\alpha_S}{6\pi} \ln\left(\frac{Q^2}{m_H^2}\right)\right) g^{n_f}(Q^2),$$

i.e. the heavy flavour evolves from zero at $Q^2 = m_H^2$ and the gluon loses corresponding momentum. It is natural to choose $Q^2 = m_H^2$ as the transition point. At NNLO, i.e. second order in α_S, there is much more complication

$$f_i^{n_f+1}(Q^2) = \left(\frac{\alpha_S}{4\pi}\right)^2 \sum_{ij} (A_{ij}^{2,0} + A_{ij}^{2,1} \ln(Q^2/m_H^2) + A_{ij}^{2,2} \ln^2(Q^2/m_H^2)) \otimes f_j^{n_f}(Q^2),$$

CP792, *Deep Inelastic Scattering, DIS 2005,* edited by W. H. Smith and S. R. Dasu

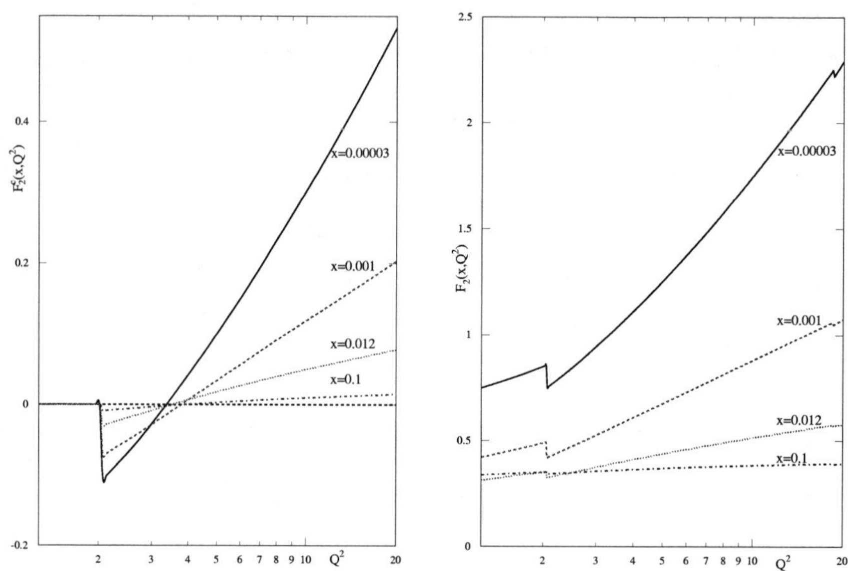

FIGURE 1. NNLO $F_2^c(x,Q^2)$ and $F_2(x,Q^2)$ in zero-mass VFNS

where $A_{ij}^{2,0}$ is generally non-zero [2]. There is no longer a smooth transition at this order, and in fact the heavy parton begins with a negative value at small x.

This leads to discontinuities in the partons and, without the correct treatment, also in the structure functions. ZMVFNS coefficient functions also lead to discontinuities at the transition point due to a sudden change in the flavour number in the coefficient functions. (This is already true at NLO, i.e. $F_2^H(x,Q^2) = 0 \quad Q^2 < m_H^2, = \frac{\alpha_S}{4\pi} C_{2,g} \otimes g^{n_f+1}(Q^2) Q^2 > m_H^2$, but the effect is very small.) This is a large effect at NNLO and is also negative at smallish x ($x \sim 0.001$). Hence, ZMVFNS is not really feasible at NNLO, leading to a huge discontinuity in $F_2^c(x,Q^2)$, which is significant in $F_2^{Tot}(x,Q^2)$, as shown in Fig. 1.

Hence we need a general Variable Flavour Number Scheme (VFNS) interpolating between the two well-defined limits of $Q^2 \leq m_H^2$ and $Q^2 \gg m_H^2$. The VFNS can be defined by demanding equivalence of the n_f and $n_f + 1$-flavour descriptions at all orders,

$$
\begin{aligned}
F_i(x,Q^2) &= C_{i,k}^{FF}(Q^2/m_H^2) \otimes f_k^{n_f}(Q^2) = C_{i,j}^{VF}(Q^2/m_H^2) \otimes f_j^{n_f+1}(Q^2) \\
&\equiv C_{i,j}^{VF}(Q^2/m_H^2) \otimes A_{jk}(Q^2/m_H^2) \otimes f_k^{n_f}(Q^2) \\
\rightarrow \quad C_{i,k}^{FF}(Q^2/m_H^2) &= C_{i,j}^{VF}(Q^2/m_H^2) \otimes A_{jk}(Q^2/m_H^2).
\end{aligned}
$$

At $\mathcal{O}(\alpha_S)$ this gives

$$
C_{2,g}^{FF,1}(Q^2/m_H^2) = C_{2,HH}^{VF,0}(Q^2/m_H^2) \otimes P_{qg}^0 \ln(Q^2/m_H^2) + C_{2,g}^{VF,1}(Q^2/m_H^2).
$$

The VFNS coefficient functions tend to the massless limits as $Q^2/m_H^2 \to \infty$, as demonstrated to all orders in [3], and if we use the zeroth order cross-section for photon-heavy quark scattering we obtain the original ACOT scheme [1].

However, $C^{VF,0}_{2,HH}(Q^2/m_H^2)$ is only uniquely defined as $Q^2/m_H^2 \to \infty$, i.e. one can swap $\mathcal{O}(m_H^2/Q^2)$ terms between $C^{VF,0}_{2,HH}(Q^2/m_H^2)$ and $C^{VF,1}_{2,g}(Q^2/m_H^2)$. Similar reasoning holds for $C^{VF,n}_{2,HH}(Q^2/m_H^2)$. The ACOT prescription violated the threshold $W^2 = Q^2(1-x)/x > 4M^2$ since only one quark was needed in final state. The Thorne-Roberts variable flavour number scheme (TR-VFNS) [4] recognized this ambiguity and removed it by imposing continuity of $(dF_2/d\ln Q^2)$ at the transition point. This guaranteed smoothness at $Q^2 = m_H^2$, but was complicated and cumbersome when extended to higher orders.

There have been other alternatives, and most recently the ACOT(χ) prescription [5] defines $F_2^{H,0}(x,Q^2) = h^+(x/x_{max},Q^2)$, where $x_{max} = Q^2/(Q^2+4m_H^2)$. The coefficient functions tend to the massless limit for $Q^2/m_H^2 \to \infty$ but also respect the threshold requirement $W^2 \geq 4m_H^2$ for quark-antiquark production. Moreover it is very simple. For the VFNS to remain simple (and physical) at all orders I choose $C^{VF,n}_{2,HH}(Q^2/m_H^2,z) = C^{ZM,n}_{2,HH}(z/x_{max})$.[1] Adopting this convention then at NNLO we have, for example,

$$C^{VF,2}_{2,Hg}\left(\frac{Q^2}{m_H^2}\right) = C^{FF,2}_{2,Hg}\left(\frac{Q^2}{m_H^2}\right) - C^{ZM,1}_{2,HH}\left(\frac{z}{x_{max}}\right) \otimes A^1_{Hg}\left(\frac{Q^2}{m_H^2}\right) - C^{ZM,0}_{2,HH}\left(\frac{z}{x_{max}}\right) \otimes A^2_{Hg}\left(\frac{Q^2}{m_H^2}\right).$$

Since $A^2_{Hg}(1,z) \neq 0$, $C^2_{2,Hg}(Q^2/m_H^2,z)$ is discontinuous at $Q^2 = m_H^2$, and this compensates exactly for the discontinuity in the heavy flavour parton distribution.[2]

There is one more issue in defining the VFNS: the ordering for $F_2^H(x,Q^2)$, i.e.

	n_f-flavour	n_f+1-flavour
LO	$\frac{\alpha_S}{4\pi} C^{FF,1}_{2,Hg} \otimes g^{n_f}$	$C^{VF,0}_{2,HH} \otimes h^+$
NLO	$\left(\frac{\alpha_S}{4\pi}\right)^2 (C^{FF,2}_{2,Hg} \otimes g^{n_f} + C^{FF,2}_{2,Hq} \otimes \Sigma^{n_f})$	$\frac{\alpha_S}{4\pi}(C^{VF,1}_{2,HH} \otimes h^+ + C^{FF,1}_{2,Hg} \otimes g^{n_f+1})$.

Switching directly when going from n_f to n_f+1 flavours leads to a discontinuity. We must decide how to deal with this. Up to now ACOT have used e.g. at NLO

$$\frac{\alpha_S}{4\pi} C^{FF,1}_{2,Hg} \otimes g^{n_f} \to \frac{\alpha_S}{4\pi}(C^{VF,1}_{2,HH} \otimes h^+ + C^{FF,1}_{2,Hg} \otimes g^{n_f+1}) + C^{VF,0}_{2,HH} \otimes h^+,$$

i.e. the same order of α_S above and below, but LO below and NLO above. The Thorne-Roberts scheme proposed e.g. at LO

$$\frac{\alpha_S(Q^2)}{4\pi} C^{FF,1}_{2,Hg}\left(\frac{Q^2}{m_H^2}\right) \otimes g^{n_f}(Q^2) \to \frac{\alpha_S(m_H^2)}{4\pi} C^{FF,1}_{2,Hg}(1) \otimes g^{n_f}(m_H^2) + C^{VF,0}_{2,HH}\left(\frac{Q^2}{m_H^2}\right) \otimes h^+(Q^2)$$

i.e. the higher order α_S term is frozen when going upwards through $Q^2 = m_H^2$. This difference in choice is extremely important at low Q^2.

---

[1] It is also important to choose $C^{VF,n}_{L,HH}(Q^2/m_H^2,z) \propto C^{ZM,n}_{L,HH}(z/x_{max})$.

[2] At NNLO there are also contributions due to heavy flavours in loops away from the photon vertex. These are included within the VFNS and lead to a discontinuity in the coefficient functions for light flavours cancelling that in the light quark distributions. Strictly, part of this contribution should be interpreted as light flavour structure functions, while part of it contributes to $F_2^H(x,Q^2)$ [8].

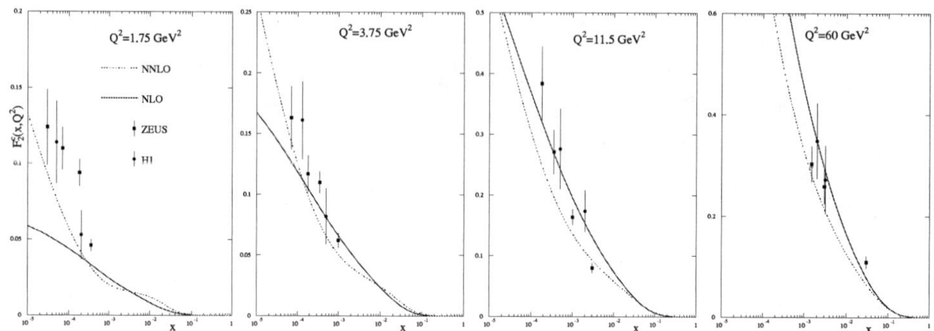

FIGURE 2. Comparison of NLO and NNLO predictions for $F_2^c(x,Q^2)$

Making this choice, in order to define the VFNS at NNLO we need the $\mathcal{O}(\alpha_S^3)$ heavy flavour coefficient functions for $Q^2 \leq m_H^2$. However, these are not yet calculated (making a NNLO FFNS problematic). We know the leading threshold logarithms [6], and can derive the leading $ln(1/x)$ term from k_T-dependent impact factors [7],

$$C_{2,Hg}^{FF,3,lowx}(Q^2/m_H^2,z) = 96\frac{\ln(1/z)}{z}f(Q^2/m_H^2), \qquad f(1) \approx 4,$$

and $C_{2,Hq}^{FF,3,lowx}(Q^2/m_H^2,z) = 4/9\, C_{2,Hg}^{FF,3,lowx}(Q^2/m_H^2,z)$. By analogy with the known NNLO coefficient functions and splitting functions I hypothesize that

$$C_{2,Hg}^{FF,3,lowx}(Q^2/m_H^2,z) = \frac{96}{z}(\ln(1/z)-4)(1-z/x_{max})^{20}f(Q^2/m_H^2),$$

i.e. the leading $\ln(1/z)$ term is always accompanied by ~ -4, and the effect of the small z term is damped as $z \to 1$. Using the full (if slightly approximate) VFNS one can produce NNLO predictions for charm with discontinuous partons, but a continuous $F^c(x,Q^2)$. NNLO clearly improves the match to lowest Q^2 data [9, 10], where NLO is generally too low, as seen in Fig. 2.

Hence, I have devised a full NNLO VFNS, with a small amount of necessary modelling. This seems to improve the fit to the lowest x and Q^2 data greatly. It also guarantees continuity of the physical observables, such as structure functions, despite the discontinuity in NNLO parton distributions. It can now be used in a full NNLO global analysis.

REFERENCES

1. M. Aivazis, F. Olness and W. K. Tung, *P*hys. Rev. **D**50 3102 (1994).
2. M. Buza, *et al.*, *E*ur. Phys. J. **C**1 301 (1998).
3. J. C. Collins, *P*hys. Rev. **D**58 2000 (1998).
4. R. S. Thorne and R. G. Roberts, *P*hys. Lett. **B**421 303 (1998); *P*hys. Rev. **D**57 6871 (1998).
5. W. K. Tung *et al.*, *J*. Phys. **G**28 983 (2002); S. Kretzer *et al.*, *P*hys. Rev. **D**69 114005 (2004).
6. E. Laenen and S. Moch, *P*hys. Rev. **D**59 034027 (1999).
7. S. Catani, M. Ciafaloni and F. Hautmann, *N*ucl. Phys. **B**366 135 (1991).
8. A. Chuvakin, J. Smith and W. L. van Neerven, *P*hys. Rev. **D**61 096004 (2000).
9. ZEUS collaboration: S. Chekanov *et al.*, *P*hys. Rev. **D**69 012004 (2004).
10. H1 collaboration: C. Adloff *et al.*, *P*hys. Lett. **B**528 199 (2002).

NuTeV Strange/Antistrange Sea Measurements from Neutrino Charm Production

D. Mason for the NuTeV Collaboration

University of Oregon, Eugene OR 97403

Abstract. An updated forward dimuon cross section from $\nu - N$ DIS charm production at the NuTeV experiment at FNAL is presented. Charged current interactions in neutrino-nucleon scattering provide a unique means of studying nucleon structure. Additionally, charged current events with two oppositely charged muons in the final state allow direct study of charm production and measurement of the strange sea. NuTeV's sign selected beam gives it the ability to extract the strange and antistrange seas independently, for which an asymmetry has been predicted in some theoretical models, and which is currently of intense interest in interpreting neutrino electroweak results. The results presented here represent a re-analysis of the NuTeV data utilizing the full inclusive CC data sample for normalization. New preliminary leading order strange asymmery results are also presented.

Keywords: Strange asymmetry,Neutrino charm production,NuTeV,Experimental results
PACS: 01.30.Cc,12.38.Qk,13.15.+g,14.65.Dw,13.87.Fh,13.60.Hb,11.30.Hv

INTRODUCTION

Events containing two oppositely signed muons (dimuons) from muon neutrino deep inelastic scattering experiments provide a unique window into the strange quark content of the nucleon[1]. These events occur predominantly when a charmed particle is produced in a charged current (CC) interaction with a strange or down quark. Approximately 10% of the time the charmed hadrons decay semi-muonically, yielding a final state with two oppositely charged muons, one from the weak vertex, one from the charm decay. Charm production from down quarks is Cabibbo suppressed, making dimuons most sensitive to the strange sea. Dimuons are clearly distinguishable in a large neutrino detector such as that employed by NuTeV[2].

The NuTeV experiment, which ran during FNAL's 96-97 fixed target run, recorded 5163 dimuon events from CC ν_μ, and 1380 from CC $\overline{\nu}_\mu$ scattering in its iron target. The detector was calibrated throughout data running by muon, electron, and hadron beams so its response is well understood. NuTeV's beamline was constructed to be able to select ν_μ or $\overline{\nu}_\mu$ beams with very high purity. This a priori knowledge of whether the event was the result of a neutrino or antineutrino interaction allows separate analysis of dimuon events for each polarity, which in turn makes it possible to measure the strange sea independently from the antistrange sea.

Whether the strange sea is different from the anti-strange sea is of particular interest. Over the years several models which predict an asymmetry between the strange and antistrange seas have been proposed[3, 4, 5, 6, 7]. More recently this possibility has been entertained[8, 9, 10, 11, 12] as an explanation for the almost 3σ difference between NuTeV's $\sin^2\theta_W$ result[13] and global electroweak fits. An asymmetry, as defined by the

CP792, *Deep Inelastic Scattering, DIS 2005*, edited by W. H. Smith and S. R. Dasu

momentum weighted integral: $S^- \equiv \int x[s(x) - \bar{s}(x)]dx$, would need to be positive and as large as +0.0068[14] to completely bring NuTeV into agreement with standard model expectations.

FORWARD DIMUON CROSS SECTION

NuTeV's dimuon data sample provides the ability to measure the nature and size of any asymmetry directly. This data has been made available in the form of a model independent forward dimuon cross section table[15], which is defined to be the cross section of charm produced dimuon events such that the muon from charm decay is greater than 5 GeV. The table was extracted in E_ν, x, and y bins for both ν and $\overline{\nu}$ modes. Presenting the data in this way, rather than as a differential charm production cross section, eliminates subjecting the users of the data to NuTeV's assumptions of charm production, fragmentation and semileptonic decay models.

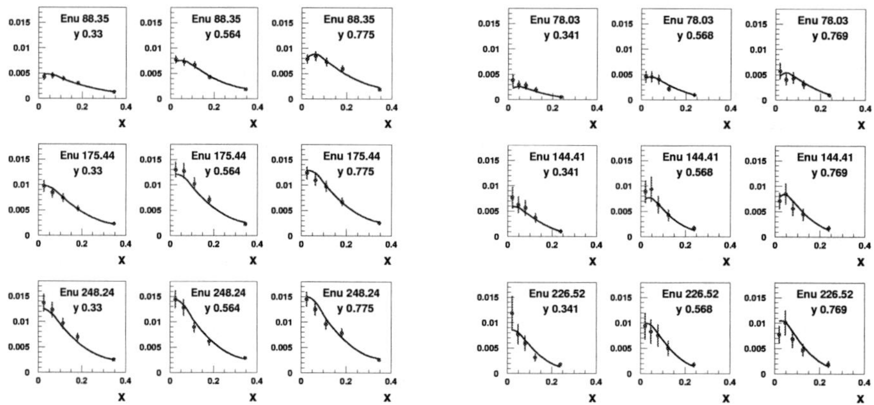

FIGURE 1. Forward dimuon cross section, ν mode on left, $\overline{\nu}$ mode on right. Table points are plotted as a function of x, along with LO Buras-Gaemers parameterization fit (curve)

In preparing for finalization of NuTeV's NLO dimuon analysis[16] this table was re-extracted utilizing the complete inclusive CC event sample for normalization. Due to storage space constraints at the time, the original NuTeV measurement used a 5% prescaled normalization subsample. It was found in the re-analysis however, that a small population of CC events which had also fired the toroid spectrometer shower pedestal trigger "leaked" into this sample unprescaled. Normalizing relative to the full inclusive sample corrects this error, which shifted both the overall normalization and, due to the tendency of the shower events to have lower values of y, the relative normalizations of ν and $\overline{\nu}$ dimuon events. This error only affected the previous NuTeV dimuon analysis referenced above.

Figure 1 shows the corrected dimuon cross section points plotted as a function of x for each E_ν and y bin. The plotted curve is the result of a LO QCD fit to the dimuon data utilizing a Buras-Gaemers[17] parameterization of the final NuTeV inclusive CC cross section[18]. The errors displayed are statistical only. The systematic errors are small, and are in the process of finalization.

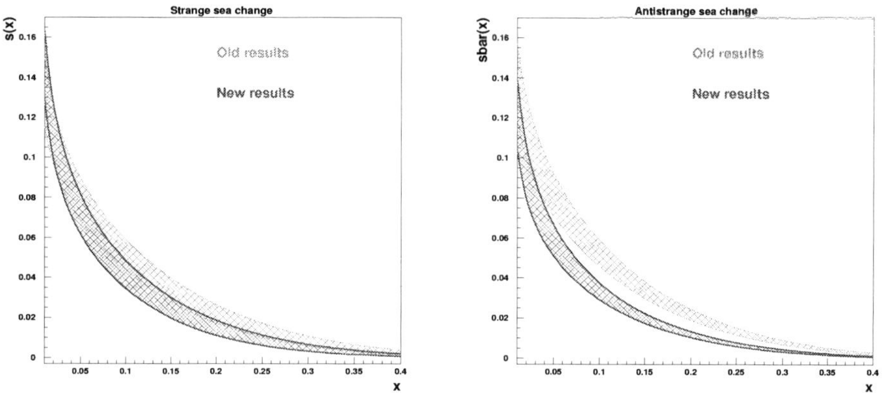

FIGURE 2. Effect of normalization sample fix on strange(left) and antistrange(right) seas

PRELIMINARY LO FITS

Figure 2 shows the shift from correcting the normalization error in extracted strange and antistrange seas with the CTEQ5L pdf set[19]. The light blue (upper) bands in each plot represent fits to the old table, while the pink (lower) bands come from the corrected table. Both are plotted at a Q^2 scale of 10 GeV^2. As can be seen in the figure, the overall effect on the seas is to reduce them slightly, antistrange moreso than strange.

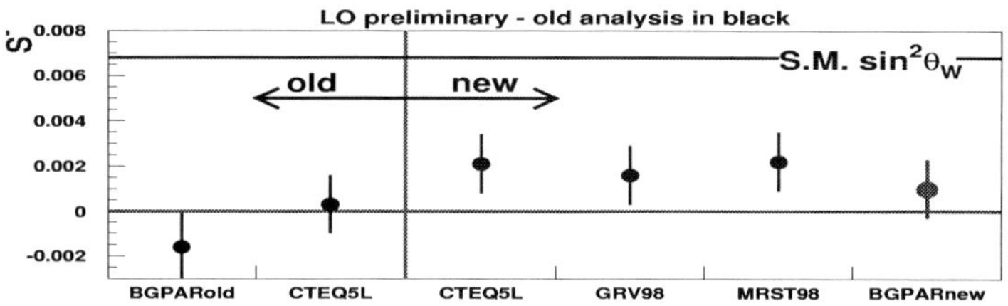

FIGURE 3. Preliminary LO extractions of S^- for old and new tables

Preliminary LO momentum weighted strange asymmetries (S^-), extracted from fits with several internal and external pdf sets are compared in figure 3. The two leftmost points represent fits to the old table, using the CCFR based Buras Gaemers parameterization from the original dimuon analysis, and then with the CTEQ5L pdf set. The four on the right are fits to the corrected table, where the rightmost point comes from a fit using the final NuTeV cross section parameterization, which best fits the NuTeV inclusive charged current data. The remaining three points are from fits using external pdf sets, CTEQ5L, GRV98[20], and MRST98[21]. Unlike the internal NuTeV iron pdfs, the external pdfs require the application of nuclear corrections for NuTeV's iron target[22].

The errors on the points in figure 3 are not yet final, however are near their expected final values. The line on the top of the plot is the level of asymmetry one would need to completely eliminate the discrepancy between NuTeV's $\sin^2\theta_W$ measurement and the world average. It is more than 4 sigma away from the LO best fit value.

CONCLUSIONS

An updated NuTeV dimuon cross section has been presented, utilizing the full inclusive CC sample for normalization. Moving from the 5% normalization subsample used in the old analysis to the full sample led to the discovery and resolution of a normalization error in the original dimuon table. Based on preliminary LO fits to the corrected dimuon cross section table, there may be evidence for a slight positive strange asymmetry. However, eliminating the discrepency between NuTeV's $\sin^2\theta_W$ measurement and standard model expectations solely by means of a strange asymmetry appears excluded at the 3-4 sigma level. The NLO dimuon analysis is progressing well using the updated table, with results expected soon.

REFERENCES

1. J. M. Conrad, M. H. Shaevitz and T. Bolton, Rev. Mod. Phys. **70**, 1341 (1998)
2. D. A. Harris, J. Yu *et al.* [NuTeV Collaboration], Nucl. Instrum. Meth. A **447**, 377 (2000)
3. A. I. Signal and A. W. Thomas, ADP-87-54-T47 *Contributed to School on Quarks and Mesons in Nuclei, Erice, Italy, Jul 16-25, 1987*
4. M. Burkardt and B. Warr, Phys. Rev. D **45**, 958 (1992).
5. S. J. Brodsky and B. Q. Ma, Phys. Lett. B **381**, 317 (1996)
6. G. Rodrigo, S. Catani, D. de Florian and W. Vogelsang, Nucl. Phys. Proc. Suppl. **135** (2004) 188
7. J. Alwall and G. Ingelman, Phys. Rev. D **71**, 094015 (2005)
8. S. Davidson, S. Forte, P. Gambino, N. Rius and A. Strumia, JHEP **0202**, 037 (2002)
9. G. P. Zeller *et al.* [NuTeV Collaboration], Phys. Rev. D **65**, 111103 (2002) [Erratum-ibid. D **67**, 119902 (2003)]
10. F. G. Cao and A. I. Signal, Phys. Lett. B **559**, 229 (2003)
11. S. Kretzer, F. Olness, J. Pumplin, D. Stump, W. K. Tung and M. H. Reno, arXiv:hep-ph/0312322.
12. J. Alwall, these proceedings
13. G. P. Zeller *et al.* [NuTeV Collaboration], Phys. Rev. Lett. **88**, 091802 (2002) [Erratum-ibid. **90**, 239902 (2003)]
14. K. S. McFarland and S. O. Moch, arXiv:hep-ph/0306052.
15. M. Goncharov *et al.* [NuTeV Collaboration], Phys. Rev. D **64**, 112006 (2001)
16. D. Mason [NuTeV Collaboration], arXiv:hep-ex/0405037.
17. A.J.Buras and K.L.F.Gaemers, Nucl. Phys. B **132** (1978) 2109.
18. M. Tzanov, these proceedings
19. H. L. Lai *et al.* [CTEQ Collaboration], Eur. Phys. J. C **12**, 375 (2000)
20. M. Gluck, E. Reya and A. Vogt, Eur. Phys. J. C **5**, 461 (1998)
21. A. D. Martin, R. G. Roberts, W. J. Stirling and R. S. Thorne,
22. B. Seligman PHD thesis, Columbia University, page 250

Heavy Quark Fragmentation Function at NNLO

Alexander Mitov

University of Hawaii, Honolulu, HI-96822, USA

Abstract. We examine not-completely inclusive production of heavy flavors in DIS and other collider processes. We review the Perturbative Fragmentation Functions formalism as the appropriate tool for such studies and detail the extension of this formalism at Next-to-Next-to-Leading order. Various collider applications are considered.

Keywords: Heavy Quarks, Fragmentation, Higher Order Perturbative Calculations
PACS: 12.38.Bx , 12.38.Cy , 13.85.Ni , 13.87.Fh

INTRODUCTION: PRODUCTION OF HEAVY FLAVORS

Production of heavy flavors is an important part of the Deep-Inelastic scattering experiments. Specifically, the subject of our discussion in this talk will be not-completely inclusive differential observables like the transverse momentum distribution of heavy flavored mesons. Such observables have intricacies that are not present in the more familiar, fully inclusive case. For example, the p_T distribution of heavy flavored hadrons cannot be adequately described by fixed order perturbative calculations at very large p_T.

A closer inspection of the structure of the spectrum after the higher order perturbative corrections have been added shows that the perturbative spectrum contains terms of the type $\sim \ln(Q/m)$ at all orders in the coupling constant. Here we consider the general case where Q stands for the typical hard scale of the process (like $\sqrt{p_T^2+m^2}$ for the case of p_T-distributions) and m is the mass of the "observed' final quark. For example, in $e^+e^- \to b\bar{b}$, the scale Q is identified with the center of mass energy $\sqrt{s}$ and m with the pole mass of the b-quark.

It is also clear, however, that heavy flavors can behave as massless in reactions where the typical hard scale Q is much larger than the mass $Q >> m$. To illustrate the point, one can imagine taking formally the limit $m/Q \to 0$; then the corresponding differential distribution diverge due to logarithmic terms $\sim \ln(m/Q)$, i.e. we get situation similar to the one in the pure massless case but with the collinear divergences regulated with small quark mass rather than dimensionally. In such situation one may wish to find a way to compute hard cross-sections with massless quarks (because it is simpler) and at the same time to keep the relevant information on the mass in the differential distribution.

THE PERTURBATIVE FRAGMENTATION FUNCTION

The method that combines the convenience of massless, dimensionally regularized calculations while keeping the relevant information about thc mass of the heavy quark was proposed by Mele and Nason [1] and is known as the perturbative Fragmentation Func-

CP792, *Deep Inelastic Scattering, DIS 2005,* edited by W. H. Smith and S. R. Dasu

tion (PFF) formalism. The method relies on the factorization of long and short distance effects in QCD by making extensive use of the DGLAP evolution equation to resum the above mentioned large logs $\sim \ln^k(m/Q)$ to all orders in the strong coupling constant. The resummation of classes of terms $\sim \ln^k(m/Q)\alpha_s^n$ is possible because the logs $\sim \ln^k(m/Q)$ are of universal collinear origin and can all be predicted with the help of the QCD evolution equations.

Let us consider the production of a heavy quark of flavor $\mathcal{Q}$ with mass m and a definite value of energy $E_{\mathcal{Q}}$ in a hard scattering process. According to the QCD factorization theorems the heavy quark energy spectrum can be computed as a convolution of the energy distribution of massless partons produced in the hard process, and the fragmentation function that describes the probability that the massless parton fragments into a massive quark with a definite energy. If the energy fraction $E_{\mathcal{Q}}/E_{\mathcal{Q},max}$ of the heavy quark is denoted by z, then the energy distribution of that quark can be written as:

$$\frac{d\sigma_{\mathcal{Q}}}{dz}(z,Q,m) = \sum_a \int_z^1 \frac{dx}{x}\frac{d\hat{\sigma}_a}{dx}(x,Q,\mu)D_{a/\mathcal{Q}}\left(\frac{z}{x},\frac{\mu}{m}\right) + \mathcal{O}\left(\frac{m}{Q}\right)^p. \tag{1}$$

Here, the sum runs over all partons (i.e. quarks, antiquarks and gluons) that can be produced in the hard process and μ denotes the factorization scale. The coefficient function $d\hat{\sigma}_a/dx$ is the differential cross-section for producing a massless parton a and with collinear singularities subtracted in the $\overline{\mathrm{MS}}$ scheme. Important feature of Eq.(1) is that power corrections $(m/Q)^p$, i.e. all terms vanishing in the strict limit $m/Q \to 0$, are neglected. We next explain what is the effect of these terms and why we may (not) want to neglect them.

There are basically two reasons why such power corrections are omitted. The first one is that the non-logarithmic terms do not have universal origin and are therefore not controlled by a Renormalization Group (RG) equation. They are instead process dependent and must be calculated perturbatively in any particular process. The second reason has to do with the size of these terms. Clearly, when $Q >> m$, such terms will have negligible numerical effect. Of importance will be only the logarithmic terms that are resumed with the help of Eq.(1) and terms that are finite in the limit $m/Q \to 0$ (often referred to as constant term). A very important implication of the PFF formalism is that while the constant term cannot be predicted and must be obtained from a process-dependent, fixed order calculation, it is sufficient to do that calculation in a massless fashion, i.e. by setting $m = 0$ from the outset. Of course, it is also possible to encounter situation where the condition $Q >> m$ does not really hold; a typical case is again the p_T distribution of hadrons where the power corrections are negligible at large p_T and dominant at low p_T. In such situations one may want to consider a "mix" of resummation and fixed order calculation of the power suppressed terms as is done for example in the p_T spectrum of hadrons implemented in the so-called FONLL formalism [3].

The functions $D_{a/\mathcal{Q}}(x,\mu/m)$ in Eq.(1) are the perturbative fragmentation functions [1]. They satisfy the DGLAP evolution equation and can be fully reconstructed from it, if the initial condition at a scale $\mu = \mu_0$ is known. When we take $\mu_0 \sim m$ the initial condition D_a^{ini} cannot contain large logarithms and can be derived from fixed order perturbative calculations. The initial condition for the PFF is now known through order $\mathcal{O}(\alpha_s^2)$ and was calculated in [2] in a process-independent approach. This approach has the advantage that it fits best the process independent nature of the PFF. The method relies on the universal factorization properties of the QCD amplitudes in the collinear limit and utilizes a light-cone gauge. The computational method relies heavily on recent developments and techniques developed in higher order perturbative calculations. We refrain from presenting here the results in full since they are rather large.

APPLICATIONS

The formalism described in the previous Sections is indispensable in studies of not-completely inclusive observables where heavy flavored hadrons (typically mesons) are measured. Such observables are more complicated and at the same time very interesting because they are very sensitive to the non-perturbative structure of the observed hadrons. The usual way of treating such processes is the following: one assumes that a heavy flavored meson is produced at a scale set by the mass of the heavy quark from the non-perturbative hadronization of a heavy quark with the corresponding flavor. In hard reactions where the hard scale Q is much larger than the quark's mass one can split the production of the heavy flavor in a convolution of perturbative and non-perturbative parts. The perturbative part describes the production of heavy quark and correctly accounts for all the radiation at scales down to $\sim m$. Naturally, that function is correct up to power corrections $\sim (m/Q)^p$. If necessary, these corrections can be incorporated from a fixed order calculation .

The non-perturbative part of the production of the observed meson is described by the so-called non-perturbative fragmentation function $D^{n.p.}(z)$. At that stage the fragmenting quark can radiate only small, of the order of the hadronic scale, transverse energy. Therefore, it is natural to identify z with the fraction of the large component of the momentum of the observed hadron. For example, that can be its longitudinal momentum fraction $p_H^{||} = zp_{\mathscr{Q}}^{||}$.

Similarly to the parton distributions, the non-perturbative heavy quark fragmentation function must be extracted from the experiment. The cleanest process studied so far is the b-energy spectrum in $e^+e^- \to b\bar{b}$. The most important property of the function $D^{n.p.}(z)$ is its process independence i.e. it can be applied to any other hard process subject only to the restriction $\mathscr{O}(m/Q)^p$.

There are many processes where this formalism can be successfully applied, like the high p_T spectrum of hadrons in hadron collisions, photoproduction, b-fragmentation in t-decay etc. The later process, for example, presents an interesting method for the precise determination of the top-quark mass at the LHC. The knowledge of the initial condition for the PFF through order $\mathscr{O}(\alpha_s^2)$ permits resummation of quasi-collinear logs with NNLL accuracy. That will, however, be possible only after the time-like splitting kernels of the DGLAP equation have been evaluated.

It seems that after the perturbative part has been promoted to the NNLO/NNLL level, the dominant uncertainty will be associated with the extracted non-perturbative function. Unfortunately, the only source presently available is the data on b-fragmentation from LEP which, in this respect, has been already analyzed, i.e. no significant further improvements can be expected. Therefore, the only viable source of improvement in the quality of the extracted non-perturbative component lies with the future International

Linear Collider. A Giga-Z option can supply new data with much higher quality that will permit new level of precision in the extraction of the non-perturbative fragmentation function. This in turn will be very important for the precise understanding of perturbative and non-perturbative aspects of the strong interactions involving heavy flavors.

SUMMARY

Production of heavy flavors is becoming an integral part of the precision physics program of the present and future colliders. Although the presence of masses leads to theoretical complications, it is by now well understood how to deal with them in a systematic fashion. Putting larger effort into the study of processes of heavy flavor production at colliders will be rewarding. In particular, differential distributions that are very sensitive to the non-perturbative hadronization effects are a very powerful way to probe the dynamics of heavy mesons.

The successful realization of such program depends crucially on two factors: improvement of the quality of the perturbative corrections and precise and systematic extraction and application of the non-perturbative fragmentation component. On the perturbative side the future is bright. Many important ingredients are already known at the very impressive NNLO level and the advances in the higher order perturbative calculations makes plausible the calculation of all relevant pieces.

However, the present state of the non-perturbative components is not as exciting. The knowledge of that function is based on analyzes of LEP data which results in relatively large uncertainty. The situation is somewhat better for the lowest order moments of the non-perturbative fragmentation function (for b-mesons); the first moment is measured with $\sim 1\%$ accuracy. However, future prospects for making significant progress in this direction exist. They are based on the Giga-Z option of the International Linear Collider. In view of the impressive state of the perturbative side, improvement in the quality of the extracted non-perturbative component is mandatory. Achieving such precision can bring to a qualitatively new level our understanding of the processes involving heavy flavors. Our community can make this happen.

ACKNOWLEDGMENTS

It is my pleasure to thank Kirill Melnikov for fruitful collaboration on this project. Research supported by the US DOE under contract DE-FG03-94ER-40833 and by the start up funds of the University of Hawaii.

REFERENCES

1. B. Mele and P. Nason, Nucl. Phys. **B 361** (1991) 626.
2. K. Melnikov and A. Mitov, Phys. Rev. D **70**, 034027 (2004) (hep-ph/0404143); A. Mitov, Phys. Rev. D **71**, 054021 (2005) (hep-ph/0410205).
3. M. Cacciari, M. Greco and P. Nason, JHEP 9805 (1998) 007 (hep-ph/9803400); M. Cacciari, S. Frixione and P. Nason, JHEP 0103 (2001) 006 (hep-ph/0102134).

Charm Fragmentation Fractions and the Charm Fragmentation Function

Zuzana Rúriková
for the H1 collaboration

Max-Planck-Institut für Physik, München, Germany
E-mail: rurikova@mail.desy.de

Abstract.
The fragmentation of a charm quark to charmed hadrons in deep-inelastic scattering has been studied using the H1 detector at HERA.

The visible cross sections for inclusive D^+, D^0, D_s^+ and D^{*+} meson production were determined making use of the secondary vertex displacement measured by the central silicon detector of H1 and fragmentation fractions and fragmentation ratios were extracted.

To investigate the fraction of the charm quark's momentum transferred to a charmed hadron, the differential cross section of D^*-meson has been measured as a function of two scaling observables.

The parameters of the Peterson and Kartvelishvili fragmentation functions, as implemented in a model with leading-order matrix elements, parton showers, string fragmentation and particle decays, were extracted from a fit to the measured distributions. The fits, using both observables, lead to the following ranges for the fragmentation parameters: $0.014 < \varepsilon < 0.036$ (Peterson) and $4 < \alpha < 6.8$ (Kartvelishvili).

Keywords: Charm, Fragmentation Fractions, Fragmentation Function
PACS: 12.39.St,13.87.Fh,14.40.Lb,14.65.Dw

INTRODUCTION

The production of a charm quarks is expected to be well described by perturbative QCD (pQCD) calculations due to the hard scale provided by the charm mass. However, the hadronization of a charm quark into a cluster of hadrons involves non-perturbative processes. Therefore, a theoretical description of the production of charmed hadrons contains a phenomenological, non-perturbative part, which, if used consistently, is expected to be process independent. In this article, we investigate the probability of a charm quark to hadronize to various charmed mesons and to transfer a fraction of its momentum to a D^{*+}-meson (charge conjugated states are always included). The analyses cover kinematic regions of photon virtuality $2 < Q^2 < 100$ GeV^2 and inelasticity $0.05 < y < 0.7$.

MEASUREMENT OF FRAGMENTATION FRACTIONS

The visible cross sections of D^+, D^0, D_s^+ and D^{*+} mesons were measured by the H1 collaboration [1] making use of the following decay channels: $D^+ \to K^-\pi^+\pi^+$, $D^0 \to K^-\pi^+$, $D_s^+ \to \Phi\pi^+ \to (K^+K^-)\pi^+$ and $D^{*+} \to D^0\pi^+ \to K^-\pi^+\pi^+$.

The relative large life time of weakly decaying D-mesons makes it possible to reconstruct the displacement of their decay point with respect to the primary vertex, using the

CP792, *Deep Inelastic Scattering, DIS 2005,* edited by W. H. Smith and S. R. Dasu

H1 central silicon tracker [3].

After cuts on secondary vertex quantities, the number of signal events is determined for each D-meson individually, by fitting the invariant mass distributions with a Gaussian to describe the signal and an appropriate background shape.

The fragmentation fractions $f(c \to D)$ are defined as the ratio of the total cross section of a given charmed meson to the one for a charm quark. Monte Carlo (MC) model was used to perform the extrapolation from visible to the total cross section for the charmed meson and to predict the cross section for charm production. The resulting values for $f(c \to D)$ are listed in Table 1. They agree well within errors with the world average values.

TABLE 1. Fragmentation fractions

Frag. fractions	D^+	D^0	D_s^+	D^{*+}
H1: $f(c \to D)$	0.203 ± 0.026	0.560 ± 0.046	0.151 ± 0.055	0.263 ± 0.032
World Average: $f(c \to D)$	0.232 ± 0.018	0.549 ± 0.026	0.101 ± 0.027	0.235 ± 0.010

The fragmentation fractions include in addition to the directly produced D-mesons all possible decay chains resulting in a particular charmed meson. The fragmentation ratios are used to characterize distinct aspects of fragmentation, such as the proportions of light quark flavours u, d and s created in the fragmentation process and the formation of different angular momentum states. The definition of these include only directly produced mesons. The measured ratios are summarized in Table 2. The detailed information about the definition of these parameters and their experimental determination can be found in [1].

TABLE 2. Fragmentation ratios

Ratio	**H1 measurement**				**e^+e^- experiments**	
	value	**stat.err.**	**syst.err.**	**theo.err.**	**value**	**err.**
P_V^d	0.693	± 0.045	± 0.004	± 0.009	0.595	± 0.045
P_V^{u+d}	0.613	± 0.061	± 0.033	± 0.008	0.620	± 0.014
$R_{u/d}$	1.26	± 0.20	± 0.11	± 0.04	1.02	± 0.12
γ_s	0.36	± 0.10	± 0.01	± 0.08	0.31	± 0.07

The ratio P_V of cross sections for a given combination of quarks to be produced in various spin states has been evaluated separately for the $c\bar{d}$ quark combination and the sum of $c\bar{u}$ and $c\bar{d}$ quark combinations. The ratio $R_{u/d}$ of probabilities for a charm quark to hadronize together with a u or d quark agrees well with the value of unity (isospin invariance). The suppression of s-quarks with respect to u and d quarks is measured by the factor γ_s and is found to be almost three.

These observations agree well with the values measured in e^+e^- and thus support the assumption of a universal fragmentation.

MEASUREMENT OF FRAGMENTATION FUNCTION

The momentum transfer from a charmed quark to a charmed hadron is usually described by a phenomenological fragmentation function. Widely used parametrizations are the

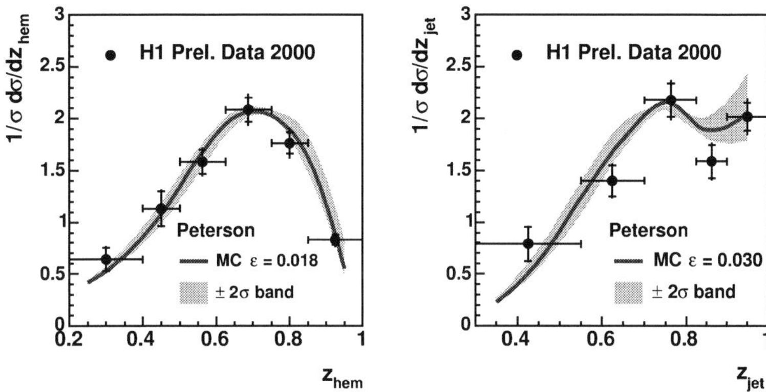

FIGURE 1. Measured D^{*+} z-distributions for the hemisphere method (left) and jet method (right) compared with a QCD model prediction (RAPGAP/PYTHIA), using the Peterson parametrization for the non-perturbative fragmentation function.

ones by Peterson et al. [4]. and by Kartvelishvili et al. [6], Their free parameters ε and α determine the "hardness" of the respective fragmentation function.

The H1 collaboration has measured the fragmentation function of charm to the D^{*+}-meson, reconstructing a D^{*+}-meson close to the kinematic threshold ($p_t(D^{*+}) > 1.5$ GeV) from the decay into $K^-\pi^+\pi_s^+$, with the $K^-\pi^+$ originating from the decay of the D^0-meson [2]. The definition of the variable z describing the fragmentation of a charm quark into a D^{*+}-meson in DIS is more ambiguous than for example in e^+e^- annihilation, where typically it is defined as $z = E_{D^*}/(\sqrt{s}/2)$. In this analysis we investigate two definitions based on different approaches.

In the first approach, the *jet method*, the charm quark momentum is approximated by the momentum of the jet which contains the reconstructed D^{*+}-meson. For this purpose the inclusive k_t-cluster algorithm [7] was applied in the γp-frame. The observable z is defined as $z = (E+p_{\rm L})_{D^*}/(E+p)_{\rm D^*jet}$, where p_L is the D^{*+}-momentum parallel to the jet direction. The $p_t(D^{*+}jet)$ is required to be greater than 3 GeV.

The second approach, the *hemisphere method*, is analogous to the method used in e^+e^- measurements. The event is divided into two hemispheres using the thrust axis calculated in a plane perpendicular to the photon direction in the γp-frame. The energy and the momentum of particles which are found in the hemisphere which contains the D^{*+}-meson is in first order a good approximation of that of the charm quark. Thus one defines $z = (E+p_{\rm L})_{D^*}/(E+p)_{\rm D^*hem}$.

The data were corrected to hadron level using the RAPGAP 3.1 MC interfaced with PYTHIA 6.2 and the full detector simulation and reconstruction package of H1. Radiative corrections were calculated using HERACLES 4.63.

The normalized D^{*+} fragmentation distributions $\frac{1}{\sigma}\frac{d\sigma}{dz}$ are used to extract the parameters of the Peterson and Kartvelishvili non-perturbative fragmentation functions, using the RAPGAP/PYTHIA MC. For steering the production of excited heavy states and other hadronization parameters, a set tuned [5] by ALEPH to their data has been used.

The fraction of resolved photon interactions contributing to D^{*+} production in our range of Q^2 was taken to be 33 % as predicted by RAPGAP.

The normalized cross section as a function of z_{hem} and z_{jet} on hadron level was generated for various values of the respective fragmentation parameter, and the resulting χ^2, comparing data and model predictions as a function of the fragmentation parameter, was calculated. The full correlation matrix was used, taking statistical and uncorrelated systematic errors as well as correlations into account.

TABLE 3. The extracted fragmentation parameters for Peterson and Kartvelishvili parametrizations.

Parametrization		Hemisphere Method	Jet Method	Range
Peterson	ε	$0.018^{+0.004}_{-0.004}$	$0.030^{+0.006}_{-0.005}$	$0.014 < \varepsilon < 0.036$
Kartvelishvili	α	$5.9^{+0.9}_{-0.6}$	$4.5^{+0.5}_{-0.5}$	$4 < \alpha < 6.8$

If the QCD based Monte Carlo model provides a correct description of all aspects of D^{*+}-production, then the jet and hemisphere methods should lead to the same extracted fragmentation parameter. The discrepancy ($< 3\sigma$) between the parameters may signal inadequacies in one or more aspects of the model. Therefore, at this stage of the analysis H1 prefers to suggest the following ranges for the fragmentation parameters: $0.014 < \varepsilon < 0.036$ (Peterson) and $4 < \alpha < 6.8$ (Kartvelishvili).

CONCLUSION

The production of a charm quarks is expected to be well described by perturbative QCD calculations. However, the hadronization of a charm quark involves a non-perturbative process which is not described by pQCD. The H1 collaboration has measured cross sections of D^+, D^0, D_s^+ and D^{*+}-mesons and used them to extract fragmentation fractions and ratios. The results of the measurement are in agreement with the world averages. The differential cross section for D^{*+}-mesons as a function of two scaling variables has been measured and used to extract parameters for the Peterson and Kartvelishvili parametrizations of fragmentation function.

REFERENCES

1. A. Aktas et al. [H1 collaboration], Eur.Phys.J. C38 (2005) 447-459.
2. A. Aktas et al. [H1 collaboration], H1prelim-05-074.
3. D. Pitzl et al., Nucl. Instrum. Meth. A **454** (2000)334. [arXiv:hep-ex/0002044]
4. C. Peterson, D. Schlatter, I. Schmitt and P. M. Zerwas, Phys. Rev. D **27** (1983) 105.
5. G. Rudolph [ALEPH collaboration], private communication.
6. V. G. Kartvelishvili, A. K. Likhoded and V. A. Petrov, Phys. Lett. B **78** (1978) 615.
7. S. D. Ellis and D. E. Soper, Phys. Rev. D **48**, 3160 (1993) [arXiv:hep-ph/9305266]; S. Catani, Y. L. Dokshitzer, M. H. Seymour and B. R. Webber, Nucl. Phys. B **406**, 187 (1993).

Measurements of charmed hadrons production in deep inelastic scattering with ZEUS

Roberval Walsh

On behalf of the ZEUS Collaboration
McGill University, Physics Department, 3600 University Street, Montreal, Quebec H3A 218

Abstract.
Charm production in deep inelastic scattering has been measured with the ZEUS detector at HERA using an integrated luminosity of approximately 82 pb^{-1}. Charm has been tagged by reconstructing $D^{*\pm}$, D^0, $D^\pm$, $D_s^\pm$ and $\Lambda_c^\pm$ charm hadrons in the kinematic region $1.5 < Q^2 < 1000$ GeV^2, $0.02 < y < 0.7$, $p_T(D,\Lambda_c) > 3$ GeV and $|\eta(D,\Lambda_c)| < 1.6$. The charm fragmentation ratios and fractions are measured in the kinematic range considered.

Keywords: Heavy flavour, charm fragmentation, ZEUS, HERA
PACS: 13.60.-r, 13.60.Le, 13.60.Rj

INTRODUCTION

This paper presents the measurements of charm production in DIS using D^0, $D^\pm$, $D_s^\pm$ and $D^{*\pm}$ mesons and $\Lambda_c^\pm$ baryons. Measurements of the charm fragmentation ratios and fractions were performed in ep scattering at HERA in the DIS regime with $1.5 < Q^2 < 1000$ GeV^2.

The data presented in this analysis were collected with the ZEUS detector at HERA. The data sample corresponds to an integrated luminosity of 82 pb^{-1}. A detailed description of the ZEUS detector can be found elsewhere [1].

EVENT SELECTION AND RECONSTRUCTION OF HADRONS

An event was selected if it satisfied the following criteria: a scattered electron was identified with energy $E_e' > 10$ GeV; $y_e \leq 0.95$; $y_{\mathrm{JB}} \geq 0.02$; $1.5 < Q_\Sigma^2 < 1000$ GeV^2; $40 < E - p_z < 65$ GeV; a reconstructed primary vertex with $|Z_{\mathrm{vertex}}| < 50$ cm is required; the impact point (X,Y) for the scattered lepton in the RCAL must lie outside the region 26×14 cm^2 centred on $X = Y = 0$. The selected kinematic region was $1.5 < Q^2 < 1000$ GeV^2 and $0.02 < y < 0.7$.

The charm hadrons were reconstructed in the kinematic region $p_T(D,\Lambda_c) > 3.0$ GeV and $|\eta(D,\Lambda_c)| < 1.6$. Details of the reconstruction of the charm hadrons[1] can be found in [2]. Figure 1 shows the mass distributions of the charm hadrons.

---

[1] Additionaly to the cuts in [2], a cut $p_T > 0.25$ GeV on the soft pion in the reconstruction of D^0 tag and additional D^* candidates is applied for a data subsample, for which the track reconstruction efficiency at low-momentum was smaller due to the operating conditions of the CTD [3].

CP792, *Deep Inelastic Scattering, DIS 2005,* edited by W. H. Smith and S. R. Dasu

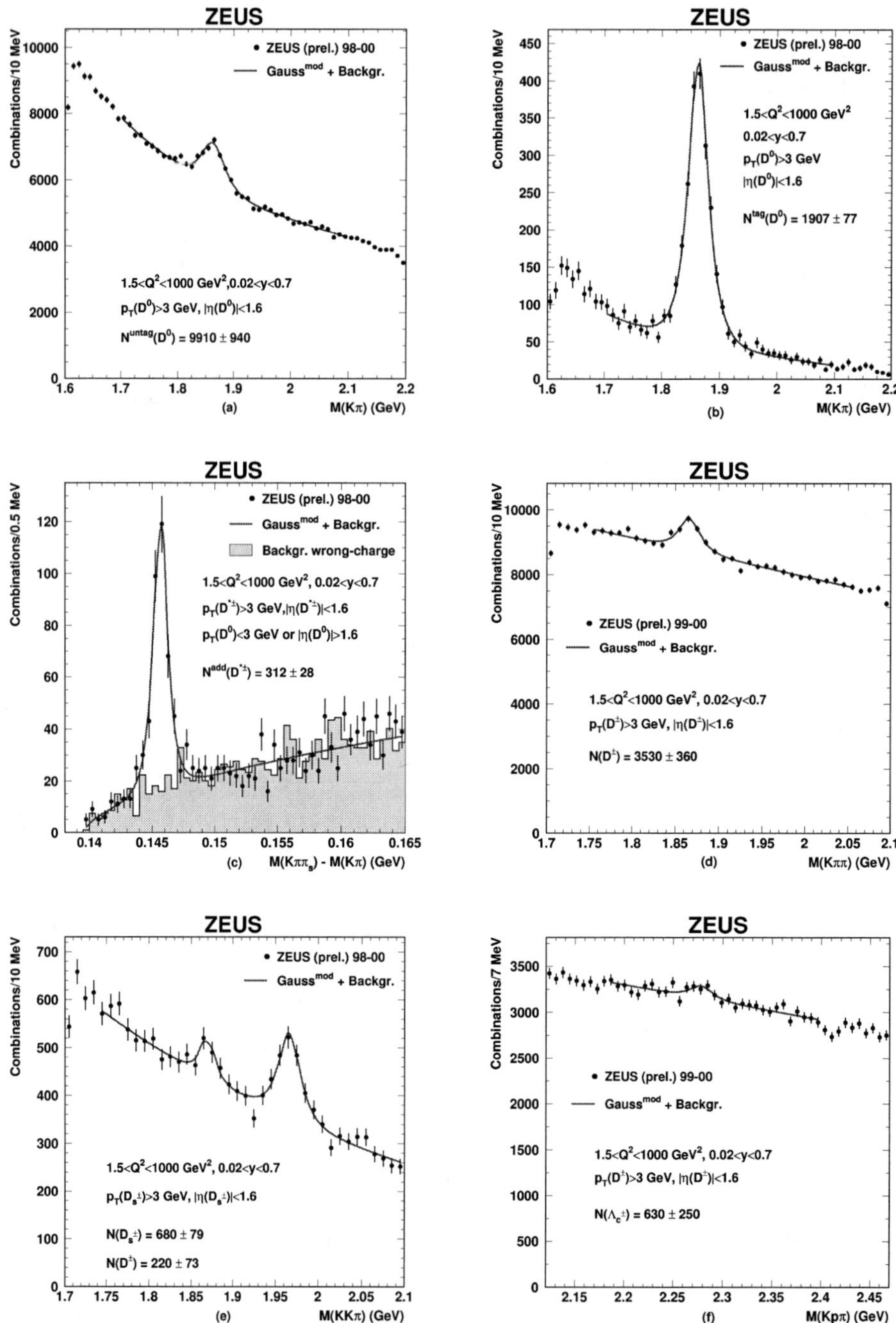

FIGURE 1. The mass distributions for the (a) untagged D^0, (b) tagged D^0, (c) additional $D^{*\pm}$, (d) $D^\pm$, (e) D_s and (f) Λ_c candidates. The solid curves represent fits to the sum modified Gaussian functions and a background functions.

CHARM FRAGMENTATION RATIOS AND FRACTIONS

The ratio of neutral and charged D meson production rates $R_{u/d}$ is the ratio of the sum of direct neutral mesons (D^{*0}, D^0) production cross sections to the sum of the charged mesons ($D^{*\pm}$, $D^{\pm}$) production cross sections. The measured ratio is

$$R_{u/d} = 1.46 \pm 0.17^{+0.10}_{-0.34}.$$

The measured ratio is consistent with one. The large systematic effect comes mainly from the signal extraction procedure which also affects the other measurements of ratios and fractions.

The strangeness suppression factor γ_s is the ratio of twice the cross sections for charmed meson containing a strange quark divided by the cross sections for those containing an up or down quark. The strangeness suppression factor obtained is

$$\gamma_s = 0.265 \pm 0.035^{+0.039}_{-0.048}.$$

The fraction of D mesons produced in a vector state is given by the ratio of vector charm-meson production cross sections to the sum of vector and direct pseudoscalar charm-meson production cross sections. The measured fractions for charged P_V^d and for charged and neutral P_V charm-mesons are, respectively,

$$P_V^d = 0.590 \pm 0.037^{+0.022}_{-0.018}, \qquad P_V = 0.490 \pm 0.032^{+0.071}_{-0.019}.$$

It can be seen from the measured fractions P_V^d and P_V that naive spin counting, which predicts a value of 0.75, does not hold for charm production.

The fraction of c quarks hadronising into a particular charm hadron, $f(c \to D, \Lambda_c)$, is the ratio of the charmed-hadron production rate to the sum of the production rate of all charm ground states. The measured fragmentation fractions are shown in Table 1.

The measurements are compared with other values obtained at HERA [2, 4, 5] and with those from e^+e^- annihilations [6, 7, 8, 9] in Figure 2. All measurements agree within experimental uncertainties.

SUMMARY

The production of D^0, $D^{\pm}$, $D_s^{\pm}$ and $D^{*\pm}$ charm mesons and $\Lambda_c^{\pm}$ charm baryons has been measured in DIS at HERA in the kinematic region $1.5 < Q^2 < 1000$ GeV^2, $0.02 < y < 0.7$, $p_T(D, \Lambda_c^{\pm}) > 3.0$ GeV and $|\eta(D, \Lambda_c^{\pm})| < 1.6$ with the ZEUS detector.

TABLE 1. Fractions of c quarks hadronising as a particular charm hadron.

	ZEUS prel. (DIS)		**ZEUS prel. (DIS)**
$f(c \to D^0)$	$0.584 \pm 0.039^{+0.024}_{-0.050}$	$f(c \to D^{*\pm})$	$0.190 \pm 0.014^{+0.023}_{-0.009}$
$f(c \to D^{\pm})$	$0.194 \pm 0.020^{+0.023}_{-0.011}$	$f(c \to D_s^{\pm})$	$0.103 \pm 0.013^{+0.012}_{-0.017}$
$f(c \to \Lambda_c^{\pm})$	$0.104 \pm 0.048^{+0.018}_{-0.010}$		

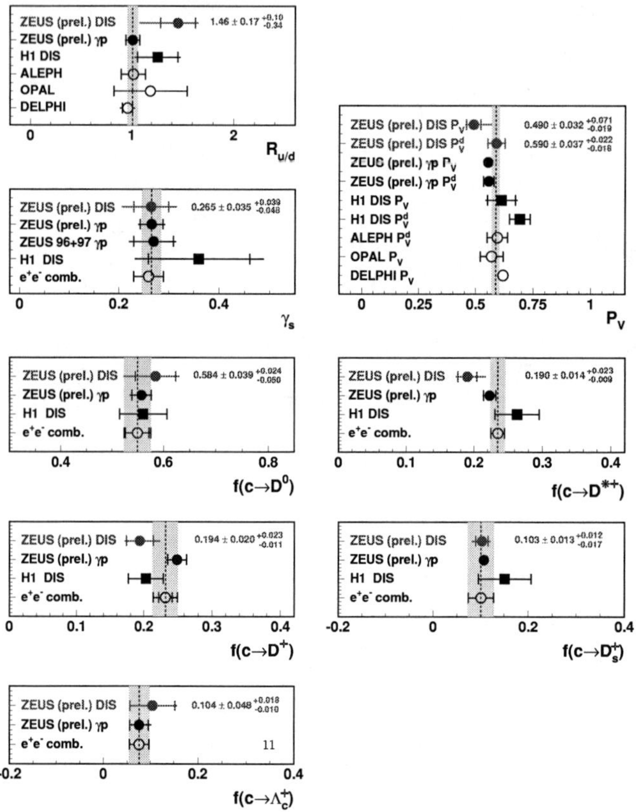

FIGURE 2. Comparison of the charm fragmentation ratios and fractions measurements with those obtained in other experiments.

Charm fragmentation ratios and fractions were determined in the above kinematic range. All fragmentation ratios and fractions agree with those obtained in charm production at HERA and in e^+e^- annihilations, confirming the universality of charm fragmentation.

REFERENCES

1. ZEUS Coll., U. Holm (ed.), *The ZEUS Detector.* Status Report (unpublished), DESY (1993).
2. ZEUS Coll., *Measurement of charm fragmentation ratios and fractions in γp collisions at HERA*, Abstract 564, International Europhysics Conf. on High Energy Physics, Aachen, Germany (2003).
3. D. Bailey and R. Hall-Wilton, *Nucl. Inst. Meth.* **A 515**, 37-42 (2003).
4. ZEUS Coll., J. Breitweg *et al.*, *Phys. Lett.* **B 481**, 213-227 (2000).
5. H1 Coll., A. Aktas *et al.*, *Eur. Phys. J.* **C 38**, 447-459 (2005).
6. L. Gladilin, *Preprint hep-ex/9912064* (1999).
7. DELPHI Coll., P. Abreu *et al.*, *Eur. Phys. J.* **C 12**, 225-241 (2000).
8. ALEPH Coll., R. Barate *et al.*, *Eur. Phys. J.* **C 16**, 597-611 (2000).
9. OPAL Coll., K. Ackerstaff *et al.*, *Eur. Phys. J.* **C 5**, 1-17 (1998).

Hadroproduction of D and B mesons in a massive VFNS

B. A. Kniehl, G. Kramer, I. Schienbein* and H. Spiesberger†

*II. Institut für Theoretische Physik, Universität Hamburg, Luruper Chaussee 149, 22761 Hamburg, Germany
†Institut für Physik, Johannes-Gutenberg-Universität, Staudinger Weg 7, 55099 Mainz, Germany

Abstract. We present a calculation of the next-to-leading order cross section for the inclusive hadroproduction of D and B mesons as a function of the transverse momentum and the rapidity in a massive variable flavor number scheme. We compare our numerical results with recent data from the CDF Collaboration at the Fermilab Tevatron for the production of D^0, $D^{\star+}$, D^+, and D_s^+ mesons at center-of-mass energy $\sqrt{S} = 1.96$ TeV and find reasonably good agreement with the measured cross sections.

Keywords: QCD, Heavy Quarks, Heavy Mesons, Heavy Flavor Schemes
PACS: 12.38.Bx,12.39.St,13.85.Ni,14.40.Lb

Various approaches for next-to-leading order (NLO) calculations in perturbative QCD have been applied to one-particle inclusive hadroproduction of D or B mesons. For definiteness, we shall consider here D mesons. However, all results can easily be carried over to any other heavy-flavored hadron.

A basic approach is the fixed flavor number scheme (FFNS) [1], in which the number of active flavors in the initial state is fixed to $n_f = 3$ and the charm quark appears only in the final state. The charm mass m is explicitly taken into account together with the transverse momentum p_T of the observed meson. In this scheme the charm mass acts as a cutoff for the initial- and final-state collinear singularities and collinear logarithms $\ln(p_T^2/m^2)$ are kept in the hard scattering cross sections. However, for $p_T \gg m$, these logarithms become large and spoil the convergence of the perturbation series.

Therefore, in the regime $p_T \gg m$, it is more appropriate to treat charm quarks like massless partons and to absorb the collinear logarithms into scale dependent parton distribution functions (PDFs) and fragmentation functions (FFs). As is well-known, by this procedure the large logarithms $\ln(p_T^2/m^2)$ are summed via the DGLAP evolution equations and the hard scattering cross sections are finite (infrared safe) in the limit $m \to 0$. If the power-like charm mass terms $\mathcal{O}(m^2/p_T^2)$ are neglected this is just the conventional parton model or zero-mass variable flavor number scheme (ZM-VFNS). Usually, in the ZM-VFNS the charm mass is neglected from the beginning and the collinear singularities appear in dimensional regularization as poles in ε where $d = 4 - 2\varepsilon$ is the number of space-time dimensions. Conventionally, these poles are removed in the modified-minimal-subtraction ($\overline{MS}$) scheme. If, on the other hand, the collinear singularities have been regularized with help of a mass m it is necessary also to subtract finite terms along with the collinear logarithms $\ln m^2$ in order to recover the hard scattering cross sections in the $\overline{MS}$ scheme.

CP792, *Deep Inelastic Scattering, DIS 2005,* edited by W. H. Smith and S. R. Dasu

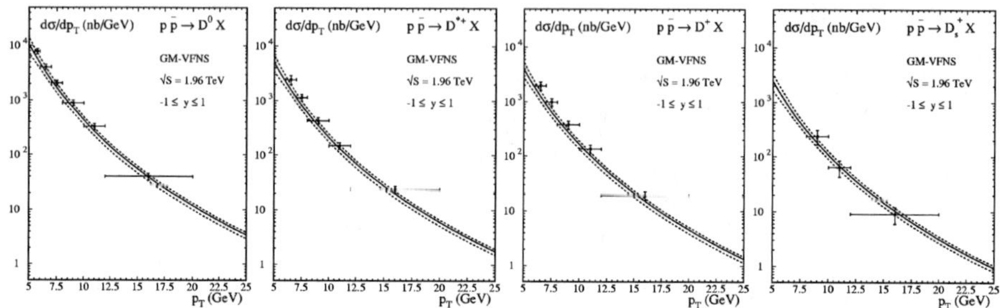

FIGURE 1. QCD predictions for one-particle inclusive production of charmed mesons, $X_c = D^0, D^{\star+}, D^+, D_s^+$, at the Tevatron Run II. In each case, the results are shown for the average of the observed meson with its antiparticle $(X_c + \overline{X}_c)/2$. The solid lines have been obtained with $\mu_{\rm R} = \mu_{\rm F} = \mu_{\rm F}' = m_T$. The upper and lower dashed curves represent the maximum and minimum cross sections found by varying $\mu_{\rm R}$, $\mu_{\rm F}$, and $\mu_{\rm F}'$ independently within a factor of 2 up and down relative to the central values while keeping their ratios $0.5 \leq \mu_{\rm F}/\mu_{\rm R}, \mu_{\rm F}'/\mu_{\rm R}, \mu_{\rm F}/\mu_{\rm F}' \leq 2$. CDF data [9] are shown for comparison.

On top of these two basic approaches, schemes have been devised which combine the two features, non-zero charm mass and resummation of $\ln(p_T^2/m^2)$-terms. One such scheme, which has been applied already to inclusive charmed meson production for the Tevatron experiment is the so-called fixed-order next-to-leading-logarithm (FONLL) scheme. This scheme smoothly interpolates between the traditional cross section in the FFNS and a suitably modified cross section in the ZM-VFNS approach with perturbative FFs with the help of a p_T dependent weight function [2, 3]. In both non-zero-charm-mass approaches, FFNS and FONLL, the theoretically calculated cross sections are convoluted with a scale-independent non-perturbative FF extracted from e^+e^- data describing the transition from the produced charm quark to the observed D meson.

Recently, a general mass variable flavor number scheme (GM-VFNS) has been worked out by us [4, 5, 6, 7] which is closely related to the ZM-VFNS, but keeps all m^2/p_T^2 terms in the hard-scattering cross sections in order to achieve better accuracy in the intermediate region $p_T \geq m$. The massive hard scattering cross sections have been constructed in a way that the conventional hard scattering cross sections in the $\overline{MS}$ scheme are recovered in the limit $p_T \to \infty$ (or $m \to 0$). The requirement to adjust the massive theory to the ZM-VFNS with $\overline{MS}$ subtraction is necessary since all commonly used PDFs and FFs for heavy flavors are defined in this particular scheme. In this sense this subtraction scheme is a consistent extension of the conventional ZM-VFNS for including charm-quark mass effects. It should be noted that our implementation of a GM-VFNS is similar to the ACOT scheme which has been extended to 1-particle inclusive production of B mesons a few years ago [8]. There are small differences concerning the collinear subtraction terms [5]. Further, in [8], the resummation of the final state collinear logarithms has been performed only to leading logarithmic accuracy.

To calculate the cross section $d^2\sigma/dp_T dy$ for the reactions $p + \bar{p} \to D + X$, FFs are needed which describe the fragmentation of the charm quarks, the light quarks, and the gluon into the observed D mesons. Fragmentation functions for the $D^\star$ meson

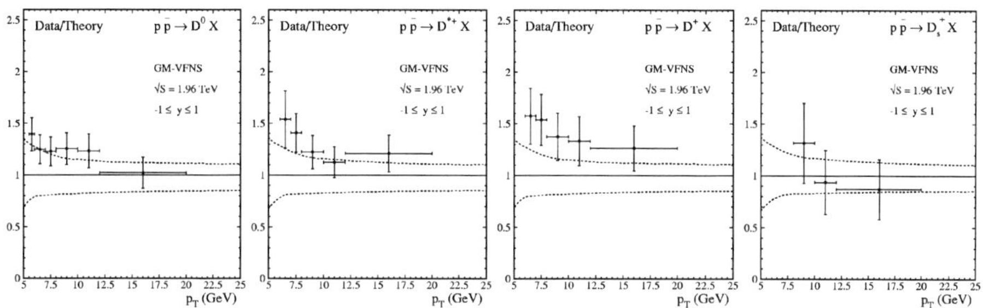

FIGURE 2. Ratios of experimental results and our central theoretical predictions (solid lines in Fig. 1). In addition, the theoretical uncertainty bands are shown, obtained as the ratio of the upper (lower) QCD prediction and the central curve.

have been extracted at leading and next-to-leading order already some time ago [10], using experimental data from the OPAL [11] and ALEPH [12] collaborations at LEP1. Recently, using the same procedure as in [10], also the FFs for $X_c = D^0, D^+, D_s^+, \Lambda_c$ have been determined [13] using OPAL data for $e^+e^- \to X_c + X$ [14]. In Refs. [10, 13] the fits for the X_c FFs have been performed using a starting scale $\mu_0 = 2m$ for the gluon and the u, d, s and c quarks and their antiquarks, while $\mu_0 = 2m_b$ ($m_b = 5$ GeV) was chosen for the FFs of the bottom quark and antiquark. The FFs of the gluon and the first three flavors were assumed to be zero at this μ_0. At larger scale μ, these FFs are generated through the usual DGLAP evolution. Since the effect of the gluon FF is important at Tevatron energies as was found for $D^\star$ production in [4] we decided to repeat the fits for the X_c FFs with the lower starting scales $\mu_0 = m$ and $\mu_0 = m_b$, respectively. This changes the FFs of the c quark only marginally but has a sizable effect on the gluon FF. The details of these new FFs will be presented elsewhere [15].

Next we show our predictions for the cross sections $d\sigma/dp_T$ for $D^0, D^{\star+}, D^+$ and D_s^+ production obtained in the GM-VFNS. For a comparison with the ZM-VFNS we refer to Ref. [16]. The partonic cross sections are convoluted with the (anti-)proton PDFs and the FFs for $c \to X_c$, $u, d, s \to X_c$ and $g \to X_c$. We use CTEQ6M PDFs [17] and the FF sets for D^0, $D^{\star+}$, D^+ and D_s^+ from [15]. [1] Results are shown for the average of the observed X_c mesons with their antiparticles. We consider $d\sigma/dp_T$ at $\sqrt{S} = 1.96$ TeV as a function of p_T with y integrated over the range $-1.0 < y < 1.0$. For the charm mass we take $m = 1.5$ GeV and evaluate $\alpha_s^{(n_f)}(\mu_R)$ with $n_f = 4$ and scale parameter $\Lambda_{\overline{MS}}^{(4)} = 328$ MeV, corresponding to $\alpha_s^{(5)}(m_Z) = 0.1181$. The results are presented in Figs. 1 and 2. The solid lines correspond to the central scale choice $\mu_R = \mu_F = \mu_F' = m_T = (p_T^2 + m^2)^{1/2}$, where μ_R is the renormalization, μ_F the initial-state and μ_F' the final-state factorization scale, respectively. To investigate the scale

[1] It should be noted that the results presented at the DIS05 have been obtained with the FFs from [10, 13].

variation of our predictions, we independently vary the renormalization and factorization scales by a factor of two: $0.5 \leq \mu_R/m_T, \mu_F/m_T, \mu_F'/m_T \leq 2$ while keeping their ratios $0.5 \leq \mu_F/\mu_R, \mu_F'/\mu_R, \mu_F/\mu_F' \leq 2$ [16]. Our theoretical results are compared with the experimental data from CDF [9]. As can be seen, the data are in good agreement with the upper curve of the uncertainty band whereas they are a factor of about 1.5(1.2) above our central prediction at low(high) p_T.

Residual sources of theoretical uncertainty include the variations of the charm mass and the assumed PDF and FF sets. A variation of the value of the charm-quark mass does not contribute much to the theoretical uncertainty. Also the use of other up-to-date NLO proton PDF sets produces only minor differences. Concerning the choice of the NLO FF sets we obtain results reduced by a factor of 1.2–1.3 when we use the NLO sets obtained by fitting with the initial scale choice $\mu_0 = 2m, 2m_b$.

In conclusion, we have presented a NLO perturbative QCD calculation of D meson production at the Tevatron in a GM-VFNS [4, 5] which provides the best description of these experimental results obtained so far. It completes earlier work in this scheme on D meson production in $\gamma\gamma$ and γp collisions [18]. This approach will be applied next to B meson production at the Tevatron. Furthermore, it is planned to extend this scheme to heavy meson production in deep inelastic scattering.

ACKNOWLEDGMENTS

This work was supported in part by the Bundesministerium für Bildung und Forschung through Grant No. 05 HT4GUA/4. The work of I. S. was supported by DESY.

REFERENCES

1. P. Nason, S. Dawson, and R. K. Ellis, *Nucl. Phys.*, **B303**, 607 (1988); **B327**, 49 (1989); **B335**, 260(E) (1990); W. Beenakker, H. Kuijf, W. L. van Neerven, and J. Smith, *Phys. Rev.*, **D40**, 54 (1989); W. Beenakker, W. L. van Neerven, R. Meng, G. A. Schuler, and J. Smith, *Nucl. Phys.*, **B351**, 507 (1991); I. Bojak, and M. Stratmann, *Phys. Rev.*, **D67**, 034010 (2003).
2. M. Cacciari, M. Greco, and P. Nason, *JHEP*, **05**, 007 (1998).
3. M. Cacciari, and P. Nason, *JHEP*, **09**, 006 (2003).
4. B. A. Kniehl, G. Kramer, I. Schienbein, and H. Spiesberger, *Phys. Rev.*, **D71**, 014018 (2005).
5. B. A. Kniehl, G. Kramer, I. Schienbein, and H. Spiesberger, *Eur. Phys. J.*, **C41**, 199 (2005).
6. I. Schienbein, hep-ph/0408036.
7. I. Schienbein, Open heavy-flavour photoproduction at NLO, Proceedings of the Ringberg Workshop, *New Trends in HERA Physics 2003*, edited by G. Grindhammer, B. A. Kniehl, G. Kramer and W. Ochs, World Scientific, 2004, p. 197.
8. F. I. Olness, R. J. Scalise, and W.-K. Tung, *Phys. Rev.*, **D59**, 014506 (1999).
9. D. Acosta, et al., *Phys. Rev. Lett.*, **91**, 241804 (2003).
10. J. Binnewies, B. A. Kniehl, and G. Kramer, *Phys. Rev.*, **D58**, 014014 (1998).
11. K. Ackerstaff, et al., *Eur. Phys. J.*, **C1**, 439 (1998).
12. R. Barate, et al., *Eur. Phys. J.*, **C16**, 597 (2000).
13. B. A. Kniehl, and G. Kramer, *Phys. Rev.*, **D71**, 094013 (2005).
14. G. Alexander, et al., *Z. Phys.*, **C72**, 1 (1996).
15. B. A. Kniehl, G. Kramer, I. Schienbein, and H. Spiesberger, in preparation.
16. See the Proceedings of the HERA-LHC Workshop.
17. J. Pumplin, et al., *JHEP*, **07**, 012 (2002).
18. G. Kramer, and H. Spiesberger, *Eur. Phys. J.*, **C22**, 289 (2001); **C28**, 495 (2003); **C38**, 309 (2004).

Charm Physics at BABAR

Chunhui Chen

Department of Physics, University of Maryland
College Park, Maryland 20742-4111, U.S.A
(for the BABAR Collaboration)

Abstract. Large production of the $c\bar{c}$ pairs and high integrated luminosity make the PEP-II B Factory an excellent place for studying the charm hadrons. In this paper, we present a few of the most recent results from the *BABAR* collaboration in the charm sector.

Keywords: *BABAR*, charm, Standard Model
PACS: 13.20.Fc,13.25.Ft,13.30.Eg,14.20.Lq

INTRODUCTION

The *BABAR* detector [1] is a general purpose detector designed to collect data at PEP-II asymmetric e^+e^- collider, operating at the center-of-mass energy corresponding to the $\Upsilon(4S)$ resonance or ~ 40 MeV below it. With copious production of $c\bar{c}$ pairs from the continuum and high integrated luminosity, *BABAR* is not only a B Factory, it is also an excellent laboratory to study the charm production and decays. In this paper, we present a few of the most recent charm analysis results from *BABAR*.

$\mathbf{D^0 - \bar{D}^0}$ MIXING

Charm mixing is characterized by two dimensionless parameters, $x \equiv \Delta m/\Gamma$ and $y \equiv \Delta\Gamma/2\Gamma$, where Δm ($\Delta\Gamma$) is the mass (width) difference between the two neutral D mass eigenstates, and Γ is the average width. If either x or y is nonzero, then the $D^0 - \bar{D}^0$ mixing will occur. In the Standard Model (SM), $D^0 - \bar{D}^0$ mixing rate is heavily suppressed by the Glashow-Iliopoulos-Maiani (GIM) mechanism[2]. However, the SM mixing rate can be enhanced by the non-perturbative effects and possible new physics beyond SM.

Based on a sample of $87\,\text{fb}^{-1}$ data, *BABAR* performed a search of $D^0 - \bar{D}^0$ mixing [3] to measure the overall time-integrated mixing rate $R_{mix} = (x^2+y^2)/2$ using the decay chain $D^{*+} \to D^0\pi^+$, $D^0 \to K^{\pm}e^{\mp}\nu$ [4]. The charge of the pion daughter of the charged D^* identifies the production flavor of the neutral D, while the charge of the electron identifies the decay flavor. These charges are equal for unmixed decays and opposite for mixed decay. Using event selection and reconstruction based on neural networks and charged kaon and electron particle identification, we obtained 49620 ± 265 unmixed events and 114 ± 61 mixed events. This results in

$$\begin{aligned} R_{mix} &= 0.0024 \pm 0.0012(\text{stat}) \pm 0.0004(\text{syst}) \\ R_{mix} &< 0.0042 \text{ at } 90\,\% \text{ CL.} \end{aligned} \qquad (1)$$

CP792, *Deep Inelastic Scattering, DIS 2005,* edited by W. H. Smith and S. R. Dasu

SEARCH FOR $D^0 \to \ell^+\ell^-$

In the SM, the flavor-changing neutral current (FCNC) decays $D^0 \to e^+e^-$ and $D^0 \to \mu^+\mu^-$ are highly suppressed by the GIM mechanism. The lepton-flavor violating (LFV) decay $D^0 \to e^\pm\mu^\mp$ is strictly forbidden in the SM. Some extensions to the Standard Model [5], such as the R-parity violating supersymmetry, can enhance the FCNC processes by many orders of magnitude and can also permit the LFV decays.

BaBar performed a search for the decays of $D^0 \to e^+e^-$, $D^0 \to \mu^+\mu^-$, and $D^0 \to e^\pm\mu^\mp$ based on a sample of 122 fb^{-1} data [6]. To ensure as clean a sample as possible, the reconstructed $D^0 \to \ell^+\ell^-$ candidate is required to originate from a $D^{*+} \to D^0\pi^+$ decay. A minimum value of 2.4 GeV/c is imposed on the center-of-mass momentum of each D^0 candidate to further reduce the combinatorial background involving the decay products of B mesons. Tight particle selection criteria are also applied to the daughters of $D^0 \to \ell^+\ell^-$ decays.

We observed no significant signals in all three decay modes. As a result, the branching fraction upper limits (UL) have been calculated using the $D^0 \to \pi^+\pi^-$ decay as the normalization mode. We obtain

$$\begin{array}{lcl} \mathrm{Br}(D^0 \to e^+e^-) & < & 1.2\times10^{-6} \text{ at } 90\,\% \text{ CL.,} \\ \mathrm{Br}(D^0 \to \mu^+\mu^-) & < & 1.3\times10^{-6} \text{ at } 90\,\% \text{ CL.,} \\ \mathrm{Br}(D^0 \to \mu^\pm e^\mp) & < & 8.1\times10^{-7} \text{ at } 90\,\% \text{ CL..} \end{array} \quad (2)$$

These results represent significant improvements on the previous limits [7, 8].

SEARCH FOR $D_{sJ}(2632)^+$

The SELEX Collaboration at FNAL has recently reported the observation of a narrow state [9] at a mass of 2632 MeV/c^2 that decays to $D_s^+\eta$. Evidence for the same state in the D^0K^+ mass spectrum was also presented. BaBar has searched for this resonance in the final states $D_s^+\eta$, D^0K^+ and $D^{*+}K_S^0$ [10] produced in $e^+e^- \to c\bar{c}$ events, using 125 fb^{-1} data. As shown in Fig. 1, no signal is observed in $D_s^+\eta$ decay channel; similarly, no evidence of $D_{sJ}(2632)^+$ was found in the D^0K^+ and $D^{*+}K_S^0$ final states, although large and clean signals for the decay $D_{s2}(2573)^+ \to D^0K^+$ and $D_{s1}(2536)^+ \to D^{*+}K_S^0$ are seen.

Ξ_C^0 PRODUCTION AND DECAYS

Using a sample of 116 fb^{-1} data, *BABAR* has performed a branching fraction ratio measurement of the Ξ_c^0 decaying to Ω^-K^+ and $\Xi^-\pi^+$ [11]. The result

$$\frac{\mathrm{Br}(\Xi_c^0 \to \Omega^-K^+)}{\mathrm{Br}(\Xi_c^0 \to \Xi^-\pi^+)} = 0.294 \pm 0.018\,(\mathrm{stat}) \pm 0.016\,(\mathrm{syst}) \quad (3)$$

is a significant improvement over the previous measurement by CLEO [12] and is consistent with a spectator quark model prediction [13].

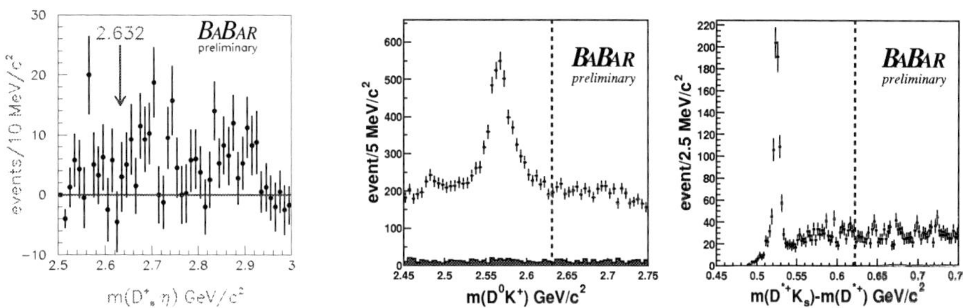

FIGURE 1. (Left:) The $D_s^+\eta$ invariant mass distribution after background subtraction. The arrow indicates the mass location of the expected $D_{sJ}(2632)^+$ state. (Middle:) The D^0K^+ invariant mass distribution. The red histogram is the invariant mass distribution of D^0K^- combinations, and the blue line indicates the mass location of the expected $D_{sJ}(2632)^+$ state. (Right:) The distribution of the difference in invariant mass of the $D^{*+}K_S^0$ combination and D^{*+} candidate. The blue line indicates the mass location of the expected $D_{sJ}(2632)^+$ state.

Although copious production of Ξ_c^0 in B decays has been predicted, such process has been only observed by CLEO [14] with a significance of $\sim 3\sigma$ in the $\Xi_c^0 \to \Xi^-\pi^+$ decay mode and $\sim 4\sigma$ in the $\Xi_c^+ \to \Xi^-\pi^+\pi^+$ decay mode. We studied the Ξ_c^0 production by measuring the spectrum of the Ξ_c^0 momentum p^* in the e^+e^- center-of-mass frame. The Ξ_c^0 produced by the B decays tends to have a smaller momentum; its p^* distribution peaks below 1.5 GeV/c and has a kinematic limit of $p^* = 2.135$ GeV/c at *BABAR*. As for the Ξ_c^0 from continuum production, its momentum distribution peaks at a much higher p^* value. By examining the p^* distribution of the Ξ_c^0 from on-resonance and off-resonance sample, we found

$$\mathrm{Br}(B \to \Xi_c^0 X) \times \mathrm{Br}(\Xi_c^0 \to \Xi^-\pi^+) = 2.11 \pm 0.19\,(\mathrm{stat}) \pm 0.25\,(\mathrm{syst}) \times 10^{-4}, \tag{4}$$

and

$$\sigma(e^+e^- \to \Xi_c^0 X) \times \mathrm{Br}(\Xi_c^0 \to \Sigma^-\pi^+) = 388 \pm 39\,(\mathrm{stat}) \pm 41\,(\mathrm{syst})\,\mathrm{fb}, \tag{5}$$

where both Ξ_c^0 and $\bar{\Xi}_c^0$ are included in the cross-section.

MEASUREMENT OF Λ_C^+ MASS

The invariant masses of the charm hadron ground states are currently reported by the Particle Data Group (PDG) with a precision of about 0.5–0.6 MeV/c^2 [15]. The best individual measurements have a statistical and systematic precision of about 0.5 MeV/c^2 and use data samples of a few hundred events. The *BABAR* data sample contains a large amount of different charm hadron decays and, due to the excellent momentum and vertex resolution in *BABAR*, many of the decay modes can be reconstructed with an event-by-event mass uncertainty of a few MeV/c^2. We can therefore provide significantly improved estimate of the charm hadron masses.

With a sample of $232\,\text{fb}^{-1}$ data, *BABAR* performed a precision measurement of the Λ_c^+ mass. The measurement is based on the reconstruction of the decay modes $\Lambda_c^+ \to \Lambda \bar{K}^0 K^+$ and $\Lambda_c^+ \to \Sigma^0 \bar{K}^0 K^+$. Because almost all of the Λ_c^+ invariant mass in these decays tied to the well-known rest masses of the Λ_c^+ decay products, the systematic uncertainty in the reconstructed mass is significantly reduced compared to the measurements that try to use the more common decay modes. Combining the results from those two modes, the measured Λ_c^+ mass is

$$m(\Lambda_c^+) = 2286.46 \pm 0.14\,\text{MeV}/c^2. \tag{6}$$

This result is in agreement with the mass values measured in other much large sample of Λ_c^+ decays, including $\Lambda_c^+ \to pK^-\pi^+$ and $\Lambda_c^+ \to pK_S^0$ decays, although these are subject to large systematic uncertainties.

This Λ_c^+ mass measurement is the most precise measurement of an open charm hadron mass to date and is an improvement in precision by more than a factor of four over the current PDG value of $2284.9 \pm 0.6\,\text{MeV}/c^2$. Our result is about 2.5σ higher than the PDG value, which is based on several high Q-value decay modes, mainly $\Lambda_c^+ \to pK^-\pi^+$ decays.

CONCLUSION

BABAR has a very rich and active charm physics program. In this paper we discussed only a few most recent results from *BABAR*. Given the excellent luminosity achieved by PEP-II, much more high precision charm physics results are expected in the near future.

REFERENCES

1. *BABAR* Collaboration, B. Aubert *et al.*, Nucl. Instr. Methods Phys. Res., Sect. A **479**, 1 (2002).
2. S. L. Glashow, J. Iliopoulos and L. Maiani, Phys. Rev. D **2**, 1285 (1970).
3. *BABAR* Collaboration, B. Aubert *et al.*, Phys. Rev. D **70**, 091102 (2004).
4. We imply charged conjugate modes throughout the paper.
5. G. Burdman, E. Golowich, J. Hewett and S. Pakvasa, Phys. Rev. D **66**, 014009 (2002).
6. *BABAR* Collaboration, B. Aubert *et al.*, Phys. Rev. Lett. **93**, 191801 (2004).
7. E791 Collaboration, E. M. Aitala *et al.*, Phys. Lett. B **462**, 401 (1999).
8. HERA-B Collaboration, I. Abt *et al.*, Phys. Lett. B **596**, 173 (2004).
9. SELEX Collaboration, A. V. Evdokimov *et al.*, Phys. Rev. Lett. **93**, 242001 (2004).
10. *BABAR* Collaboration, B. Aubert *et al.*, arXiv:hep-ex/0408087.
11. *BABAR* Collaboration, B. Aubert *et al.*, arXiv:hep-ex/0504014.
12. CLEO Collaboration, S. Henderson *et al.*, Phys. Lett. B **283**, 161 (1992).
13. J. G. Korner and M. Kramer, Z. Phys. C **55**, 659 (1992).
14. CLEO Collaboration, B. Barish *et al.*, Phys. Rev. Lett. **79**, 3599 (1997).
15. Particle Data Group, S. Eidelman *et al.*, Phys. Lett. B **592**, 1 (2004).

Charm physics at Belle

Bruce Yabsley,[†]

Virginia Polytechnic Institute and State University, Blacksburg VA 24061
(for the Belle Collaboration)

Abstract. This talk reviews an unrepresentative selection of Belle's open-charm and charmonium analyses, focussing on new developments and topics of interest to the DIS community. Highlights include an $X(3872)$ analysis favoring $J^{PC} = 1^{++}$, and the $D^0\overline{D}^{*0}$ bound-state interpretation.

Keywords: charm,charmonium,spectroscopy,pentaquarks,fragmentation
PACS: 13.66.Bc,14.20.Lq,14.40.Lb,14.40.Gx

INTRODUCTION

A talk of this length does not allow even a representative survey of open-charm and charmonium analyses at Belle, so I've made a selection favoring the most interesting recent developments—concerning the exotic $X(3872)$ state—and topics of interest to the deep inelastic scattering community. Due to length limitations, the writeup is even more cursory than the talk. Interested readers should consult the references.

The aim of the Belle collaboration is to study violation of the CP symmetry, using the time-dependence of decays of $B\overline{B}$ pairs. Open-charm and charmonium studies are an active sideline. The KEKB collider [1] produces $e^+e^- \to \Upsilon(4S) \to B\overline{B}$ and $e^+e^- \to q\bar{q}$ continuum events with unprecedented luminosity: both B-decays and the continuum are copious sources of charmed and charmonium states. The Belle detector [2], at the KEKB interaction point, is a general-purpose detector with good particle ID capabilities.

THE $X(3872)$: QUANTUM NUMBERS AND INTERPRETATION

The $X(3872)$, a narrow state decaying to $\pi^+\pi^- J/\psi$, was discovered in $B \to K\pi^+\pi^- J/\psi$ decays by Belle [3], and confirmed by three other groups [4]. In subsequent analysis, it has not been possible to match the properties of the X with those of an expected $c\bar{c}$ state [5]. Belle has recently reported the observation of $X(3872) \to \gamma J/\psi$ and $\omega J/\psi$ decays [6], confirming that the C-parity of the X must be even. An angular analysis of $X(3872)$ decays has also been performed [7], exploting zeroes in predicted distributions [8] to test various J^{PC} hypotheses. An example is shown in Fig. 1.

The $X \to \pi^+\pi^- J/\psi$ dipion mass distribution is shown in Fig. 2. The rate near the kinematic boundary is sensitive to the parity of the X: for $C = +1$ and even parity, $q^*_{J/\psi}$ dependence is expected (ρ and J/ψ in S-wave, ignoring D-wave admixture); for odd parity, $(q^*_{J/\psi})^3$ (P-wave; ignoring F-wave). Fits to the two cases find $\chi^2 = 43.1$ and 71.0, for 39 degrees of freedom, favoring J^{++} hypotheses. $J^{PC} = 0^{++}$ is disfavored by angular distributions, and $J^{PC} = 2^{++}$ by preliminary evidence for decays to $D^0\overline{D}^0\pi^0$ [9].

CP792, *Deep Inelastic Scattering, DIS 2005,* edited by W. H. Smith and S. R. Dasu

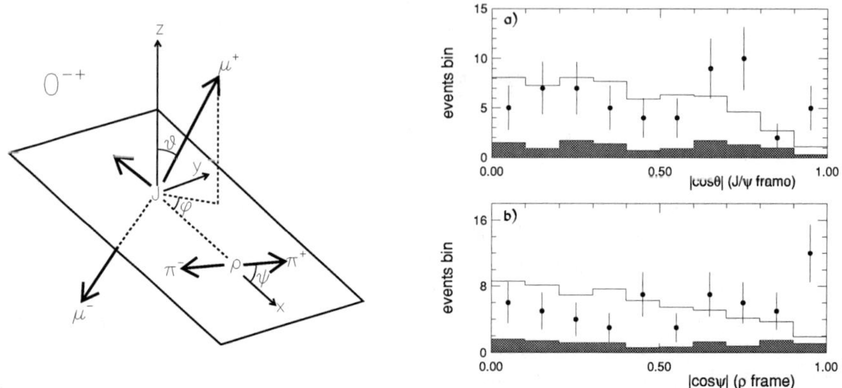

FIGURE 1. $X(3872) \to \pi^+\pi^- J/\psi$ angular distributions for data (points), and for the $J^{PC} = 0^{-+}$ hypothesis (histogram), including background estimated from X-mass sidebands (shaded). The definition of the angles is shown in the sketch on the left. The χ^2 of the fits are (a) 17.7 and (b) 34.2 for 9 degrees of freedom, disfavoring 0^{-+}. Note the concentration of events in the final bins, contrary to expectation.

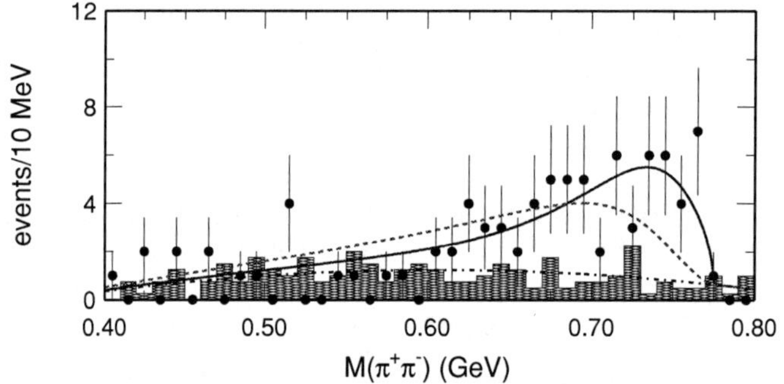

FIGURE 2. $M(\pi^+\pi^-)$ distribution for events in the $X(3872)$ signal region (points) and sideband (shaded). Fits to J^{++} (solid) and J^{+-} (dashed) hypotheses are also shown: see the text.

The $J^{PC} = 1^{++}$ hypothesis is consistent with available data; all other assignments are disfavored by at least one test. However, the identification of the $X(3872)$ with the 1^{++} charmonium state χ'_{c1} is unlikely: potential model predictions for the χ'_{c1} mass are 3953–3990 MeV, and shift upward when coupling to open charm is taken into account [10]. The isospin-violating $\chi'_{c1} \to \pi^+\pi^- J/\psi$ decay would presumably have a small partial width, similar to $\Gamma(\psi(2S) \to \pi^0 J/\psi) = (0.27 \pm 0.06)\,\text{keV}$ [11], to be compared with a total width $\Gamma > 1\,\text{MeV}$ [10]. This contradicts BaBar's 90% confidence limit $\mathscr{B}(X(3872) \to \pi^+\pi^- J/\psi) > 4.3\%$ [12]. The low ratio of radiative and hadronic partial widths $\Gamma(X \to \gamma\psi)/\Gamma(X \to \pi^+\pi^-\psi) = 0.14 \pm 0.05$ [6] also disfavors χ'_{c1}.

By contrast, the observed properties of the $X(3872)$ are consistent with those of a $D^0\overline{D}^{*0}$ bound state [13]: the mass is within errors of $D^0\overline{D}^{*0}$ threshold, $(+0.6 \pm 1.1)\,\text{MeV}$;

as the mass difference $M(D^+D^{*-}) - M(D^0\overline{D}^{*0}) = 8.1\,\text{MeV}$ is large by comparison, isospin violation is natural for such a state, explaining the observation of $X(3872) \to \omega J/\psi$ *and* $\rho J/\psi$ decays. These decays are natural within the model of Swanson [13], where $|\omega J/\psi\rangle$ and $|\rho J/\psi\rangle$ appear as admixtures to the $|D^0\overline{D}^{*0}\rangle$ wavefunction. A small branching ratio $\Gamma(X \to \gamma J/\psi)/\Gamma(X \to \pi^+\pi^- J/\psi)$ is also expected for such a state.

DOUBLE CHARMONIUM PRODUCTION AND THE $X(3940)$

The process $e^+e^- \to c\bar{c}c\bar{c}$ was discovered by Belle in both double charmonium ($J/\psi\eta_c$) and associated charm ($J/\psi D^{(*)}X$) production; both processes have unexpectedly large rates [15]. Various proposed alternative explanations of the data have been contradicted by further tests, including angular analysis and full reconstruction of $e^+e^- \to J/\psi\eta_c$ events [16]. The principal results have recently been confirmed by BaBar [17].

Evidence for a new state, $X(3940)$, seen in the recoil mass (M_X) spectrum in $e^+e^- \to J/\psi X$ events, was presented in 2004 at the Beijing conference [18]. Decays of this state favor $D\overline{D}^*$, based on a study of events with reconstructed J/ψ and D mesons. An updated analysis confirming these results is being prepared for publication this summer [19].

OTHER RESULTS (MOSTLY SPECTROSCOPY)

An enhancement at $\omega J/\psi$ threshold has been seen in $B \to K\omega J/\psi$ [15]. Interpreted as a particle ($M = (3943 \pm 11 \pm 13)\,\text{MeV}$, $\Gamma = (87 \pm 22 \pm 26)\,\text{MeV}$), this "$Y(3940)$" would be exotic: a $c\bar{c}$ state at this mass would be expected to decay to $D\overline{D}^{(*)}$, with very small branching fractions for $\omega J/\psi$ and other hadronic charmonium transitions.

Belle observed the $D^*_{sJ}(2317)$ and $D_{sJ}(2460)$ in both continuum production [20] and B decays [21], confirming the observations by BaBar and CLEO [22, 23], establishing the $D_{sJ}(2460)^+ \to \gamma D_s^+$ decay, and favoring $J^P(D_{sJ}(2460)) = 1^+$, based on the γD_s^+ helicity angle distribution. Study of $D^*_{sJ}(2317)^+ \to \pi^0 D_s^+$ [24] likewise favors $J^P = 0^+$.

Searches for the $D_{sJ}(2632)^+ \to D_s^+\eta$ and D^0K^+ state of SELEX [25] find no evidence of production in B decays or the continuum at Belle [26].

Amongst other charmed baryon results, a new isospin triplet $\Sigma_c(2800)$ decaying to $\Lambda_c^+\pi^{-,0,+}$ has been observed in the continuum [27]. It is tentatively identified as the Σ_{c2} ($J^P = 3/2$), with some admixture of the Σ_{c1} (with the same quantum numbers).

Of the Belle pentaquark searches reported in 2004 [28], the most important uses interactions of kaons (from e^+e^- annihilation) with the material of the detector. This study placed a limit on production of the $\Theta(1540)$ relative to the $\Lambda(1520)$: an updated analysis, to be published in the summer of 2005 [29], also bounds the rate of exclusive production $K^+n \to \Theta(1540)^+ \to pK_S^0$, with similar sensitivity to that of DIANA [30].

A major study of charm fragmentation in $e^+e^- \to c\bar{c}$ at $\sqrt{s} \simeq 10.6\,\text{GeV}$ will also be submitted for publication this summer [31]. Fractional momentum distributions for D, D_s, and D^* mesons, and the Λ_c^+, are measured with much greater precision than in previous studies; a comparison of fragmentation functions is also presented.

SUMMARY

Recent Belle analyses include a study of $X(3872)$ decays and properties, favouring $J^{PC} = 1^{++}$ and the $D^0\overline{D}^{*0}$ molecular model. Other contributions to spectroscopy include the $Y(3940)$, double charmonium production (including $e^+e^- \to J/\psi X(3940)$), and pentaquark searches. A study of charm fragmentation has also been performed. Many other results in charm and charmonium studies lie outside the scope of this talk.

ACKNOWLEDGMENTS

I'd like to thank my colleagues at Belle for their help in preparing this talk, and the organisers of the Deep Inelastic Scattering workshop. The work of Madison's fine local breweries was also greatly appreciated.

REFERENCES

1. S. Kurokawa and E. Kikutani, *Nucl. Instrum. Meth.* **A 499**, 1–7 (2003), and other papers in this Volume.
2. A. Abashian *et al.* (Belle Collaboration), *Nucl. Instrum. Meth.* **A 479**, 117–232 (2002).
3. S. K. Choi, S. L. Olsen *et al.* (Belle Collaboration), *Phys. Rev. Lett.* **91**, 262001 (2003).
4. D. Acosta *et al.* (CDF-II), *Phys. Rev. Lett.* **93**, 072001 (2004); V. M. Abazov *et al.* (D0), *Phys. Rev. Lett.* **93**, 162002 (2004); B. Aubert *et al.* (BaBar), *Phys. Rev.* **D 71**, 071103 (2005).
5. S. L. Olsen (for the Belle Collaboration), *Int. J. Mod. Phys.* **A 20**, 240–249 (2005).
6. K. Abe *et al.* (Belle Collaboration), BELLE–CONF–0540, arXiv:hep-ex/0505037.
7. K. Abe *et al.* (Belle Collaboration), BELLE–CONF–0541, arXiv:hep-ex/0505038.
8. J. Rosner, *Phys. Rev.* **D 70**, 094023 (2004).
9. Paper in preparation. Such decays would proceed in D-wave for $J^{PC} = 2^{++}$.
10. T. Barnes and S. Godfrey, *Phys. Rev.* **D 69**, 054008 (2004); E. J. Eichten, K. Lane, and C. Quigg, *Phys. Rev.* **D 69**, 094019 (2004).
11. S. Eidelman *et al.* (Particle Data Group), *Phys. Lett.* **B 592**, 1 (2004).
12. J. Coleman (for the BaBar Collaboration), talk at the 2005 Rencontre de Moriond on QCD and Hadronic Interactions, March 2005.
13. E. S. Swanson, *Phys. Lett.* **B 588**, 189–195 (2004); **598**, 197–202 (2004).
14. S. K. Choi, S. L. Olsen *et al.* (Belle Collaboration), *Phys. Rev. Lett.* **94**, 182002 (2005).
15. K. Abe *et al.* (Belle Collaboration), *Phys. Rev. Lett.* **94**, 142001 (2002).
16. K. Abe *et al.* (Belle Collaboration), *Phys. Rev.* **D 70**, 071102(R) (2004).
17. B. Aubert *et al.* (BaBar Collaboration), arXiv:hep-ex/0506062, submitted as a Rapid Communication to *Phys. Rev.* **D.**
18. P. Pakhlov (for the Belle Collaboration), arXiv:hep-ex/0412041.
19. K. Abe *et al.* (Belle Collaboration), BELLE–CONF–0517, to be submitted to *Phys. Rev. Lett.*
20. Y. Mikami *et al.* (Belle Collaboration), *Phys. Rev. Lett.* **92**, 012002 (2004).
21. P. Krokovny *et al.* (Belle Collaboration), *Phys. Rev. Lett.* **91**, 262002 (2003).
22. B. Aubert *et al.* (BaBar Collaboration), *Phys. Rev. Lett.* **90**, 242001 (2003).
23. D. Besson *et al.* (CLEO Collaboration), *Phys. Rev.* **D 68**, 032002 (2003).
24. K. Abe *et al.* (Belle Collaboration), BELLE–CONF–0461.
25. A. V. Evdokimov *et al.* (SELEX Collaboration), *Phys. Rev. Lett.* **93**, 242001 (2004).
26. Paper in preparation.
27. R. Mizuk *et al.* (Belle Collaboration), *Phys. Rev. Lett.* **94**, 122002 (2005).
28. R.Mizuk (for the Belle Collaboration), arXiv:hep-ex/0411005.
29. K. Abe *et al.* (Belle Collaboration), BELLE–CONF–0518, to be submitted to *Phys. Lett.* **B.**
30. V. Barmin *et al.* (DIANA Collaboration), *Phys. Atom. Nucl.* **66**, 1715–1718 (2003).
31. R. Seuster *et al.* (Belle Collaboration), arXiv:hep-ex/0506068, submitted to *Phys. Rev.* **D.**

Results from ZEUS on Charm and Beauty Production at HERA II

Richard John Hall-Wilton (on behalf of the ZEUS collaboration)

University Collage London, Gower Street, London, UK.

Abstract. Recent results from the ZEUS collaboration on charm and beauty production using the HERA II data are presented.

Keywords: HERA, ZEUS, charm, beauty, heavy quarks, DIS
PACS: 12.38.Qk,13.85.Hd,14.40.Lb,14.65.Dw,14.65.Fy

INTRODUCTION

The production of heavy quarks in ep collisions at HERA is a stringent test of pQCD since the heavy quark provides a hard scale that should enable reliable predictions.

HERA has been upgraded to provide higher luminosity (HERA II). The ZEUS detector has been upgraded as well, including the installation of the silicon Micro Vertex Detector (MVD) inside the Central Tracking Detector. These detector upgrades, in conjunction with higher luminosities improve the potential for heavy quark physics.

Results are presented here make use of the MVD for tagging charm and beauty. In a previous ZEUS publication [1], there were indications that the production rate of D* mesons differed in e^+p and e^-p collisions, with the ratio $\sigma(e^-p)/\sigma(e^+p)$ increasing towards higher Q^2. Data from HERA II is used to check whether this surprising observation was a statistical fluctuation [2].

D* MESON PRODUCTION IN DEEP INELASTIC SCATTERING

An inclusive selection of neutral current (NC) DIS events was made; this was chosen to be directly comparable to the previous measurement [1]. Good neutral current DIS events were selected in the kinematic region with $5 < Q^2 < 1000$ GeV^2, and $0.02 < y < 0.7$. Studies of the rate of inclusive NC DIS events in the same kinematic region as the D^* candidate events were made to ensure there was no bias introduced between the 2003-4 e^+p and 2004-5 e^-p running periods.

The selection of D^* mesons also followed the strategy used in the previous publication [1]. The D^* mesons were identified using the decay channel $D^{*+} \to D^0\pi_s$ with the subsequent decay $D^0 \to K^-\pi^+$ and the corresponding antiparticle decay.

The signal regions for the reconstructed masses, $M(D^0)$ and $\Delta M = (M_{K\pi\pi_s} - M_{K\pi})$, were $1.80 < M(D^0) < 1.92$ GeV and $0.143 < \Delta M < 0.148$ GeV, respectively. Figure 1 shows the ΔM distributions for the D^* candidates, in e^+p and e^-p scattering. The number of D^* candidates was determined in the signal regions by subtracting the wrong-charge

CP792, *Deep Inelastic Scattering, DIS 2005*, edited by W. H. Smith and S. R. Dasu

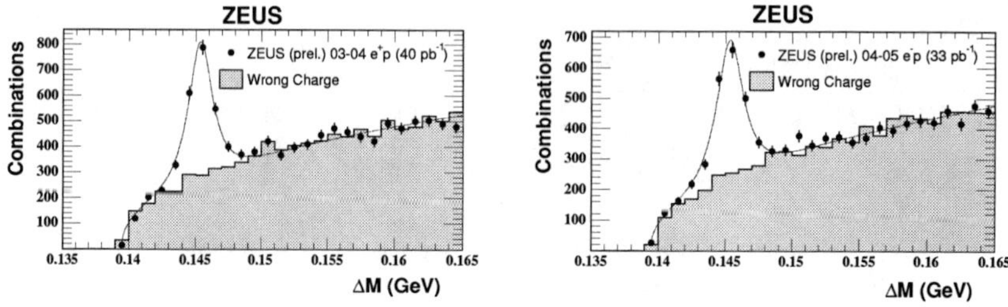

FIGURE 1. The distribution of the mass difference, ΔM, for D^* candidates in e^+p scattering (left) and e^-p scattering (right). The ΔM distribution from wrong-charge combinations is shown as the histogram. The solid line shows the result of a fit.

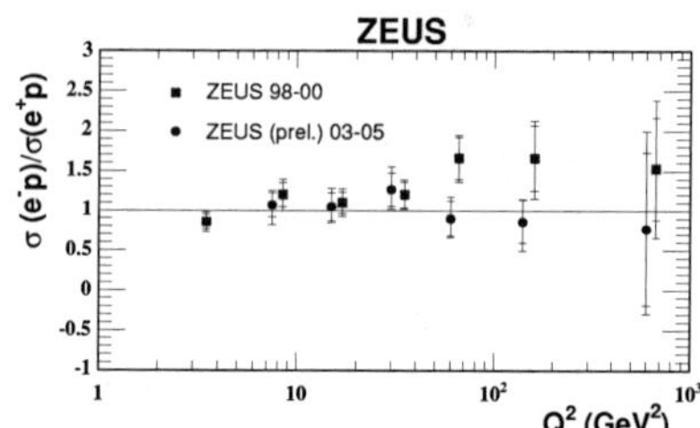

FIGURE 2. The ratio r^{e^-p}/r^{e^+p} for D* Mesons for the 2004-5 data compared to the 1998-00 data.

background from the correct-charge candidates in the signal region.

In total 1240 ± 64 (1118 ± 60) D* candidates in the kinematic region $5 < Q^2 < 1000$ GeV^2, $0.02 < y < 0.7$, $1.5 < p_T(D^*) < 15$ GeV and $|\eta(D^*)| < 1.5$ were found in e$^+$p (e$^-$p) data corresponding to a rate of $30.7 \pm 1.8/\text{pb}^{-1}$ ($33.5 \pm 1.8/\text{pb}^{-1}$). The rate for events with $Q^2 > 40$ GeV^2 was 6.4/pb^{-1} (6.3/pb^{-1}) for e$^+$p (e$^-$p) data. The ratio r^{e^-p}/r^{e^+p} in bins of Q^2 is shown in Fig.2, compared to the ZEUS HERA I values and the theoretical expectation of unity. At all values of Q^2 the value of the ratio is consistent with unity. The observation of the increase of this ratio with increasing Q^2, found previously [1], is not confirmed.

CHARM PRODUCTION — D$^{\pm}$ MESONS

The MVD has also been used to tag D$^{\pm}$ mesons in both photoproduction and DIS, using 15 pb^{-1} of data collected during the 2003-4 running period [3]. The decay channel considered was the three-body decay, $D^{\pm} \rightarrow K^{\mp}\pi^{\pm}\pi^{\pm}$. Only well-measured tracks with a transverse momentum greater than 0.8 GeV were considered.

The three decay tracks were refitted to a secondary vertex and the D$^{\pm}$ invariant mass was determined from the refitted tracks. Badly reconstructed secondary vertices were rejected. The remaining tracks were refitted to a new primary vertex. The distance between the D$^{\pm}$ vertex and the new primary vertex was then used to tag D$^{\pm}$ decays.

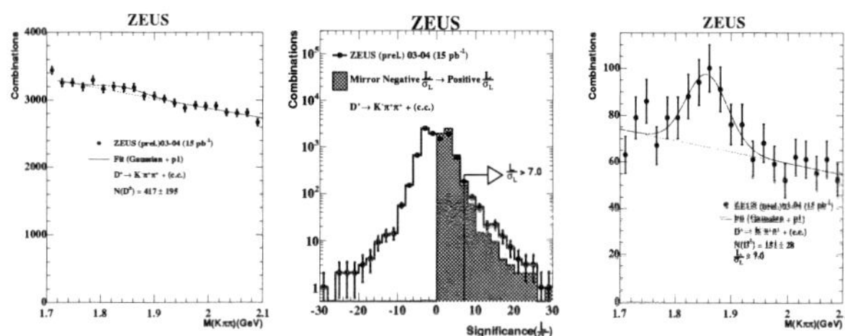

FIGURE 3. $M(K\pi\pi)$ invariant-mass distribution prior to the decay-length significance cut (left). Significance $s_L = \frac{L}{\sigma_L}$ distribution for events satisfying $1.845 < M(D^{\pm}) < 1.905$GeV (middle). The selection cut subsequently used ($s_L > 7$) is indicated. The shaded area represents the negative part of the distribution reflected around s_L=0. The M$(K\pi\pi)$ invariant-mass distribution with decay length significance $L/\sigma_L > 7$ for the inclusive event sample (right).

The decay length, L, can then be assigned a sign with respect to the direction of the D$^{\pm}$ meson: in the case that L is positive, the decay was downstream of the primary interaction in the direction of the $D^{\pm}$; if L is negative, the decay occurred upstream of the primary vertex. The decay length significance is then defined as $s_L = L/\sigma_L$, where σ_L, is the error on the decay length. A heavy-meson decay with a large decay length would lead to an excess at higher positive values of the significance. The negative values of significance can therefore be used as an estimate of the resolution.

Figure 3 shows the invariant-mass distribution $M(K\pi\pi)$ for the inclusive data selection, before the cuts on the decay-length significance were applied. The kinematic region of the measurement was $p_T(D^{\pm}) > 3.7$GeV and $|\eta(D^{\pm})| < 1.5$. A small excess is observed in the mass region expected for the $D^{\pm}$ meson, over a huge combinatorial background. The fit gave a signal of 417$\pm$195 $D^{\pm}$'s.

Figure 3 also shows the significance distribution in the mass range around the nominal mass of the $D^{\pm}$ meson. The excess on the positive side of the distribution indicates the presence of long-lived decays. Figure 3 also shows the mass distribution after the significance cut of $s_L > 7.0$ is applied. A clear peak corresponding to 151$\pm$28 D$^{\pm}$ mesons is observed at the D$^{\pm}$ mass with a width compatible with the detector resolution.

BEAUTY PRODUCTION — MUON ASSOCIATED DIJETS

A study has been performed on muon associated dijet production in photoproduction using 31 pb^{-1} of 2003-4 data [4]. Two jets were required in the massive scheme with p$_T^{jet1,2} > 7, 6$ GeV and $|\eta^{jet}| < 2.5$. The muon is required to have p$_T^{\mu} > 0.75$ GeV. This selection gives 8010 muon candidates associated to a jet.

These muon candidates are used to investigate the beauty fraction using two methods: the P$_T^{rel}$ method as used by a previous ZEUS publication [5]; and by using the impact parameter of the muon with the primary vertex.

The P$_T^{rel}$ distribution of the muon candidates is shown in Figure 4. The fit yields a beauty fraction of 16.1$\pm$2.1%. Also shown in Figure 4 is the impact parameter distribution of the muon with the primary vertex. This has been assigned a sign based upon

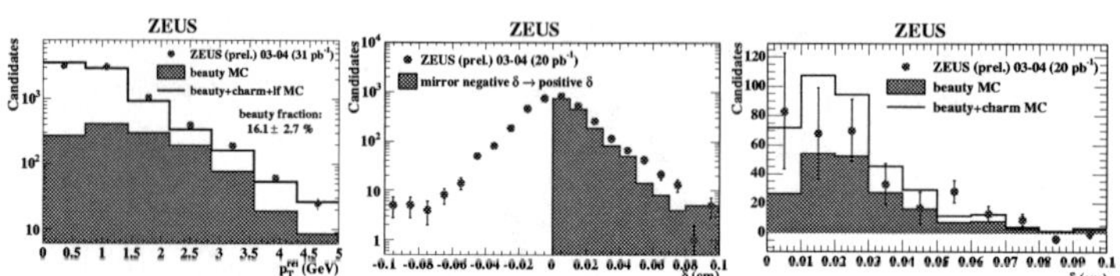

FIGURE 4. P_T^{rel} distribution for the muon associated dijet events (left). The data are compared to a mixture of beauty and charm+light flavour MC given by the fit. The impact parameter distribution (centre). The negative contribution is mirrored to the positive side to show the observed assymetry. The subtracted impact parameter distribution (right). The data are compared to similar subtracted MC distributions from beauty and charm.

the muon direction in an analagous way to that used for D^+ mesons. The distribution is enhanced towards positive values of impact parameter, an indication of the presence of a beauty component. The signal distribution for the impact parameter is also shown in Figure 4 where the negative values of impact parameter have been subtracted from positive values. Superimposed on this is the charm and beauty Monte Carlo (MC), where the beauty fraction in the MC was fixed to be that obtained in the P_T^{rel} method fit. A good description of the data by the charm and beauty MC is observed, indicating that the beauty fractions obtained from the two independent methods are compatible with each other.

SUMMARY

A study of the ratio, r^{e^-p}/r^{e^+p}, of the production rate of D^* mesons in NC DIS in $e^{\pm}$ collisions was made using 73 pb^{-1} data taken by ZEUS during the 2003-5 running period. The sample used was directly comparable to a previous result. Twice as large a sample of e^-p data was available from HERA II. The ratio was observed to be consistent with unity at all values of Q^2, as would be expected from the standard model. The results support the assertion that the higher value of the ratio observed previously was due to a statistical fluctuation.

Initial studies on charm and beauty production using the MVD have been presented, establishing that these new techniques can be used. These methods will be extended in future to tag many other charm and beauty hadrons, with the aim of gaining further understanding of the production dynamics of heavy quarks in electron-proton interactions. As there is more than 120 pb^{-1} data available already, further exciting results can be expected soon from the HERA II programme.

REFERENCES

1. ZEUS, Phys. Rev. D 69 012004 (2004).
2. ZEUS, Lepton-Photon Conf. contributed paper, Uppsala, Abstract 271 (2005).
3. ZEUS, ICHEP Conf. contributed paper, Beijing, Abstract 347 (2004).
4. ZEUS, ICHEP Conf. contributed paper, Beijing, Abstract 354 (2004).
5. ZEUS, Phys. Rev. D 70 012008 (2004).

Inclusive and dijet b productions at CDF

Régis Lefèvre [1]
on behalf of the CDF and Collaboration

Institut de Física d'Altes Energies, Universitat Autònoma de Barcelona, Edifici Cn. Facultat Ciènces UAB, E-08193 Bellaterra, Spain

Abstract. This contribution reports recent CDF measurements of the inclusive b-jet and $b\bar{b}$ dijet production cross sections obtained at the Tevatron Run II in $p\bar{p}$ collisions at $\sqrt{s} = 1.96$ TeV. Preliminary results are in reasonable agreement with QCD predictions.

Keywords: Tevatron, CDF, QCD, b-jet, inclusive, dijet, production
PACS: 13.85.Ni, 13.87.Ce

INTRODUCTION

At the Tevatron, the total $b\bar{b}$ cross section is about 50 μb which results to an event rate of few kHz and very high statistics. Beauty production measurements thus provide stringent test of QCD predictions.

In Run I, CDF and D0 have reported large discrepancies between observed and predicted beauty cross sections [1, 2]. This led to many developments both in the theoretical calculations and the experimental approach [3]. From the theoretical side, a major improvement was the implementation of the Fixed-Order with Next-to-Leading-Log (FONLL) calculations [4] in which the resummation of logarithmic terms, with Next-to-Leading-Logarithmic accuracy (NLL), is matched with the Fixed-Order (FO), exact Next-to-Leading-Order (NLO) calculation for massive quarks. There have also been substantial changes in the Parton Distribution Functions (PDFs) and the bottom fragmentation function as extracted from HERA and LEP data. From the experimental side, there have been many improvements in the data treatment for instance avoiding deconvolution and extrapolation to the quark level with Monte Carlo simulations and using only real observables such as b-hadrons and b-jets. CDF has recently reported a measurement of the b-hadron production cross section in Run II in very good agreement with latest theoretical predictions [5].

The b-jet production is also very interesting because it allows to investigate perturbative QCD (pQCD) with rather small theoretical uncertainties from fragmentation processes as b-jets include most of the b quark fragmentation remnants. CDF preliminary measurements of the inclusive b-jet and $b\bar{b}$ dijet production cross sections in $p\bar{p}$ collisions at $\sqrt{s} = 1.96$ TeV are presented in the following.

[1] Régis Lefèvre is supported by the EU funding under the RTN contract: HPRN-CT-2002-00292, Probe for New Physics.

CP792, *Deep Inelastic Scattering, DIS 2005*, edited by W. H. Smith and S. R. Dasu

INCLUSIVE B-JET PRODUCTION

CDF has measured the inclusive b-jet production cross section for jets in the central rapidity region, $|y^{jet}| < 0.7$. The measurement is based on about 300 pb^{-1} of Run II data. Jets were reconstructed with the midpoint algorithm [6]. This iterative seed-based cone algorithm uses midpoints between pair of jets as additional seeds which makes the clusterization procedure infrared safe [7]. The cone radius in the $Y - \phi$ space was set to $R_{cone} = 0.7$, the merging fraction to $f_{merge} = 75$ %.

The analysis exploits the good tracking capabilities of the CDF detector [8] and relies on the reconstruction of secondary vertexes to identify b-jets using displaced tracks within the jet cone. Taking advantage of the long lifetime of the b hadrons, the tagging is based on the significance of the impact parameter and of the decay length L_{xy}. Furthermore, the sign of L_{xy} is used to reject mis-tagged jets.

To extract the flavor content of tagged jets the shape of the secondary vertex mass distribution is used. Although a full reconstruction of the hadron invariant mass is in general not possible because of the presence of neutral particles, the invariant mass of the tracks used to find the secondary vertex provides a good discrimination between b-jets and c or light flavor jets. The fraction of b-jets is obtained fitting to the data Monte Carlo templates of b-jets on one hand and of c-jets plus light flavor jets on the other hand. The fit is performed independently for each jet transverse momentum bin considered in the cross section measurement.

Figure 1 (left) shows the distribution of the mass of the secondary vertex for jets with $82 < p_T^{jet} < 90$ GeV/c as an example. The b-jet contribution has a quite different shape than the one corresponding to c-jets plus light flavor jets. For the fitted fraction of b-jets, 34 % in this p_T^{jet} bin, the Monte Carlo reproduces very well the data summing both contributions.

Figure 1 (right) shows the measured inclusive b-jet production cross section. The measurement is fully corrected to the particle level to compensate for inefficiencies, energy losses at the calorimeter level and detector resolution. It extends over p_T^{jet} between 38 and 400 GeV/c. This is a considerable improvement with respect to DØ Run I measurement which was limited to $p_T^{jet} < 100$ GeV/c [9].

The main systematic error comes from the jet energy scale for which a conservative uncertainty of 5 % have been considered, although on going studies are investigating the possibility to reduce it to 3 %. On the last bins, an important contribution to the systematic uncertainties comes from the statistics of the Monte Carlo templates used to fit the data in the extraction of the b-jet fraction: bigger Monte Carlo samples are being generated to reduce this error.

The measured b-jet cross section is here compared to the prediction from PYTHIA Monte Carlo [10]. This prediction was obtained using CTEQ5L PDFs [11] and a special set of PYTHIA parameters, tuned on Run I data to reproduce the underlying event activity in the transverse region, denoted as PYTHIA-Tune A [12]. A reasonable agreement is observed considering the fact that PYTHIA integrates matrix element calculations at Leading-Order only. A comparison to NLO pQCD [13] is in progress.

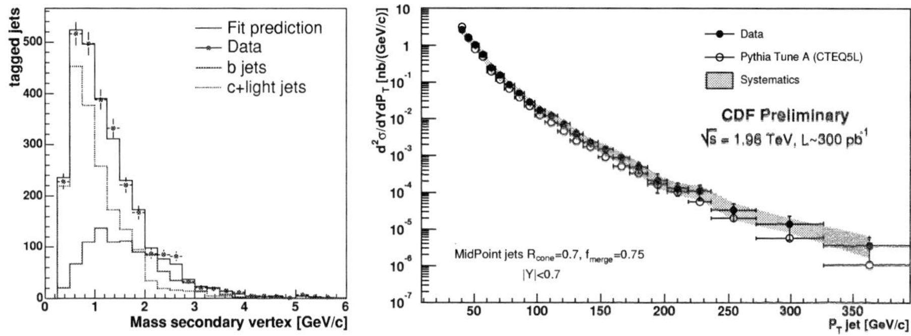

FIGURE 1. *Left:* Mass of the secondary vertex distribution for jets with $82 < p_T^{jet} < 90$ GeV/c. Fitted prediction is overlaid on the data. b-jet contribution as well as the one from c-jets plus light flavor jets are also reported. *Right:* Differential inclusive b-jet production cross section as a function of p_T^{jet}. The measurement is fully corrected to the particle level. Data points include the statistical errors, the shade band represents the systematic uncertainties. Particle level predictions from PYTHIA-Tune A are overlaid on the data.

INCLUSIVE DIJET B PRODUCTION

A preliminary measurement of the $b\bar{b}$ dijet production cross section has been carried out using a small sample of CDF Run II data. Jets were reconstructed using the JetClu Run I cone algorithm [14] with a cone radius of $R_{cone} = 0.7$ in the $\eta - \phi$ space and a merging fraction of $f_{merge} = 75$ %. As in the previous analysis, the b-jet identification is based on the reconstruction of secondary vertexes using displaced tracks within the jet cone and the b-jet content is extracted fitting the secondary vertex mass distribution to the data. Two tagged jets with pseudo-rapidities $|\eta| < 1.2$ and transverse energies $E_T > 30$ GeV for the leading jet and $E_T > 20$ GeV for the second one were required.

Figure 2 shows the differential inclusive $b\bar{b}$ dijet production cross sections measured as a function of the angle between the two jets in the transverse plane $\Delta\phi_{jj}$ and of the invariant mass of the two jets M_{jj}. The $\Delta\phi_{jj}$ distribution shows that the event selection preferentially picks out the leading order flavor creation process when asking for two central b-jets. NLO contributions are however not negligible at small opening angles.

The measurement is compared to predictions from a small MC@NLO [15, 16] sample. In this program, NLO pQCD calculations are matched with parton showers, HERWIG [17] event generator is used. Default MRST2001 PDFs [18] were used. The multiple parton interactions generator JIMMY [19] with default parameters was linked to HERWIG to better take into account the underlying event. Data and theory agree reasonably well.

In the future, the jet clusterization will be done with the midpoint algorithm instead of the JetClu Run I cone algorithm which may compromises meaningful comparisons with pQCD calculations as it is not infrared safe. Work is in progress to include the full data sample as well as bigger MC@NLO samples in order to test QCD predictions more accurately.

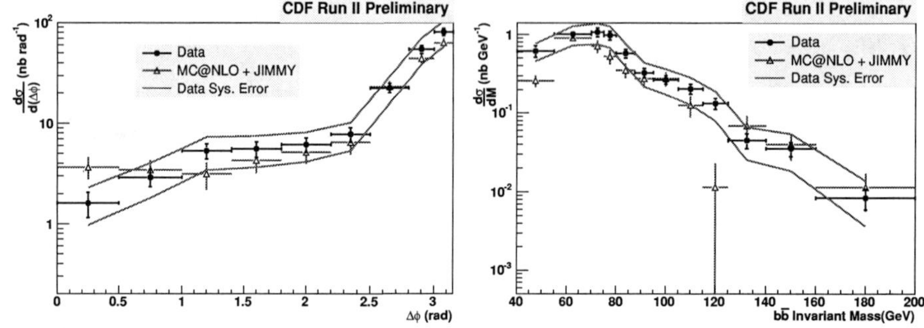

FIGURE 2. Differential inclusive $b\bar{b}$ dijet production cross sections as a function of $\Delta\phi_{jj}$ (left) and of M_{jj} (right). The measurement is fully corrected to the particle level. Data points include the statistical errors, the lines represent the systematic uncertainties. Particle level predictions from MC@NLO using JIMMY in conjunction with HERWIG are overlaid on the data.

ACKNOWLEDGMENTS

I am very grateful to the DIS 2005 organizers for their invitation. I would also like to acknowledge the members of the QCD and b-tagging Groups of the CDF Collaboration, especially Monica D'Onofrio and Anant Gajjar, for their works in achieving the results reported here. Finally, I would like to acknowledge the EU for its funding under the RTN contract: HPRN-CT-2002-00292, Probe for New Physics.

REFERENCES

1. CDF Collaboration, D. Acosta *et al.*, Phys. Rev. D **65**, 052005 (2002).
2. DØ Collaboration, B. Abbott *et al.*, Phys. Lett. B **487**, 264 (2000).
3. M.L. Mangano, Proceedings of the 15th Topical Conference on Hadron Collider Physics (HCP2004), East Lansing, Michigan, 14-18 Jun 2004, AIP Conf. Proc. **753**, 247 (2005).
4. M. Cacciari, M. Greco and P. Nason, JHEP **9805**, 007 (1998).
5. CDF Collaboration, D. Acosta *et al.*, Phys. Rev. D **71**, 032001 (2005).
6. G.C. Blazey *et al.*, Proceedings of the Workshop: "QCD and Weak Boson Physics in Run II", Batavia, Illinois, 1999, edited by U. Baur, R.K. Ellis and D. Zeppenfeld, 47 (2000).
7. M.H. Seymour, Nucl. Phys. B **513**, 269 (1998).
8. CDF Collaboration, D. Acosta *et al.*, Phys. Rev. D **71**, 032001 (2005).
9. DØ Collaboration, B. Abbott *et al.*, Phys. Rev. Lett. **85**, 5068 (2000).
10. T. Sjöstrand *et al.*, Comp. Phys. Comm. **135**, 238 (2001).
11. CTEQ Collaboration, H.L. Lai *et al.*, Eur. Phys. J. C **12**, 375 (2000).
12. R.D. Field, "ME/MC Tuning Workshop", Fermilab, October 2002.
13. S. Frixione and M.L. Mangano, Nucl. Phys. B **483**, 321 (1997).
14. CDF Collaboration, F. Abe *et al.*, Phys. Rev. D **45**, 1448 (1992).
15. S. Frixione and B.R. Webber, JHEP **0206**, 029 (2002).
16. S. Frixione, P. Nason and B.R. Webber, JHEP **0308**, 007 (2003).
17. G. Marchenisi *et al.*, Comp. Phys. Comm. **65**, 465 (1992); G. Corcella *et al.*, JHEP **0101**, 010 (2001).
18. A.D. Martin, R.G. Roberts, W.J. Stirling and R.S. Thorne, Eur. Phys. J. C **23**, 73 (2002).
19. J.M. Butterworth, J.R. Forshaw and M.H. Seymour, Z. Phys. C **72**, 637 (1996); http://jetweb.hep.ucl.ac.uk/JIMMY/index.html

Measurement of beauty production with $\mu\mu$ correlations

A. Longhin on behalf of the ZEUS Collaboration

Dipartimento di Fisica "G.Galilei", via Marzolo 8, 35131 Padova, Italy

Abstract. Beauty production with events in which two muons are observed in the final state has been measured with the ZEUS detector at HERA using an integrated luminosity of 121 pb^{-1}. A low p_T threshold for muon identification, in combination with the large rapidity coverage of the ZEUS muon system, gives access to essentially the full phase space for beauty production. The dimuon selection suppresses backgrounds from charm and light flavor production. Separation of the sample into high and low–mass, isolated and non–isolated, like and unlike–sign muon pairs offers redundancy which is used to further constrain the backgrounds. A total cross section for beauty production at HERA is obtained and compared to QCD predictions.

Keywords: beauty, production, double tag, dimuon, ZEUS, HERA
PACS: 13.85.-t; 13.85.Qk; 14.65.Fy; 13.60.Hb

In recent years *b*–analyses using double tagging techniques ($\mu\mu$, $D^*\mu$) have started to be employed profiting of the increasing amount of data provided by HERA. The redundancy of two heavy flavor tags allows a significant reduction of the intrinsic background with respect to semi–inclusive analyses (i.e. those using the lepton+jets signature). The possibility to employ softer cuts on the muon p_T results in an enhanced sensitivity to the kinematic region of *B* mesons produced at very low transverse momenta where the bulk of the *b* cross section is concentrated. Further advantages, in the dimuon channel especially, come from the wide rapidity coverage of muon detectors in ZEUS. Under these conditions a direct measurement of the total $b\bar{b}$ cross section becomes possible without the necessity to introduce large model-dependent extrapolation factors. Tagging muons arising from different *b*–quarks will also allow to explicitly measure $b\bar{b}$ correlations. Beauty production results in the $D^*\mu$ channel were already presented recently [1, 2]. Selecting unlike–sign $D^*\mu$ pairs produced in the same hemisphere (arising from the same *b*–quark i.e. mainly via $B^0 \to D^*\mu\nu_\mu$) yields a quite pure *b* sample. The signal is cleanly separated from the charm background which generates unlike–sign $D^*\mu$ pairs in a characteristic back–to–back configuration (D^* and μ from different *c*–quarks).

Let us now consider the available $\mu\mu$ final state configurations arising from *b*–events. Muons coming from the same *b*–quark ($b \to c\mu^- \to \mu^+$, $b \to$ charmonium $\to \mu^+\mu^-$) are unlike–sign while, if they arise from different *b*–quarks ($b \to c\mu^- \to \mu^+$, $\bar{b} \to \bar{c}\mu^+ \to \mu^-$), they can be either unlike–sign, if both come from *b* or *c*, or like–sign if only one comes directly from a *b*–quark. Dimuons generated from the decay of the same quark tend to have low invariant mass ($m_{\mu\mu}$). In order to disentangle the various signal topologies from background it is convenient to separate the data into four sub–samples depending on $m_{\mu\mu}$ and the $\mu\mu$ charge relation. Unlike and like–sign samples were further divided into a high and low–mass region ($m_{\mu\mu} \gtrless 4$ GeV).

CP792, *Deep Inelastic Scattering, DIS 2005,* edited by W. H. Smith and S. R. Dasu

The relevant sources of background in this channel are: open–c production, di–μ QED production from "Bethe–Heitler" processes (*BH*), charmonium production and fake–muons from light flavor (*LF*) production. The first three sources introduce unlike–sign dimuons only, while the last contributes in both configurations. Monte Carlo (*MC*) samples for beauty and charm were generated using the RAPGAP and PYTHIA programs. Muon efficiency corrections were estimated directly from data.

Events were selected at online level using a combination of muonic, hadronic charm, di–jet and DIS triggers. The difference between the scalar transverse energy in the calorimeter (out of a cone of 10° around the p–remnant direction) and the scattered electron energy was required to be $E_T^{-10^\circ} - E_T^e > 8$ GeV ($\sim 2m_b$ - unmeasured energy). The μ–finding procedure integrates information from various detectors: the inner and outer barrel, rear and forward muon chambers and the instrumented iron yoke. These detectors were supplemented by the main calorimeter providing the muon minimum ionizing particle signature. Muon candidates were classified into two quality classes depending on the number and type of firing detectors and p_T^μ cuts were chosen accordingly: $p_T^\mu >$ 0.75 or 1.5 GeV for high– and low–quality respectively. For both muons $-2.5 < \eta^\mu <$ 2.2 [1] was required. The $m_{\mu\mu}$ distributions for the above mentioned four data samples are shown in the left hand side plots of Fig. 1.

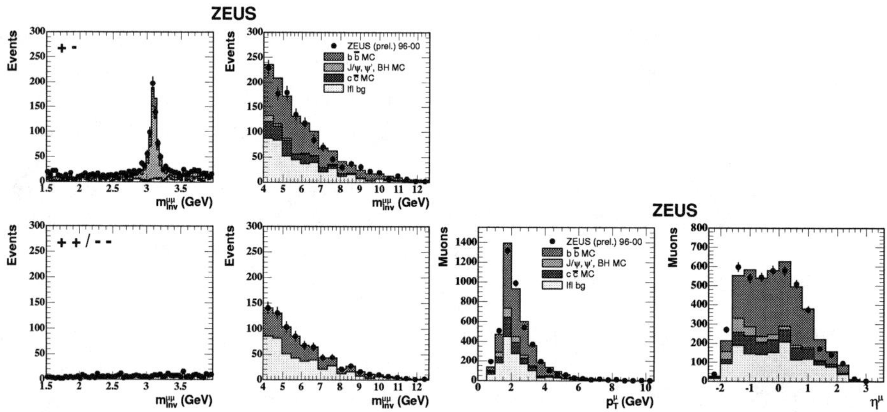

FIGURE 1. Left: $m_{\mu\mu}$ distributions for unlike–sign (top) and like–sign (bottom), low–mass (left), high–mass (right). Right: Final p_T^μ (left) and η^μ (right) distributions in the unlike–sign non–isolated sample.

In order to determine the contribution from BH and charmonium, isolation criteria were applied. For each μ the energy $I_{1,2}$ deposited in a cone with radius $\Delta R \equiv \sqrt{\Delta\phi^2 + \Delta\eta^2} < 1$ around the μ flight direction was calculated, excluding the other

[1] The pseudorapidity η is defined as $\eta = -\ln\tan\frac{\theta}{2}$, θ being the polar angle taking the z–axis along the proton direction.

muon. The quadratic sum of the two energy deposits I was required to be more than 250 MeV. For events in the J/ψ-ψ' mass region where this background is larger, the cut was tightened to 2 GeV. The unlike–sign isolated sample is dominated by BH and charmonium events. Tuning the MC to reproduce the normalization of this sample is used then to constrain these backgrounds in the non–isolated signal sample.

As previously mentioned, muonic decays of different c–quarks constitute a background affecting particularly the high–mass unlike–sign sample. This contribution is irreducible within the di–muon framework. On the other hand the sample of non–isolated high–mass unlike–sign $D^*\mu$ is dominated by charm production and well separated from beauty which mostly populates the low–mass region (top plots of Fig. 4 in [1]). It must be noted that at production level the process producing high–mass $\mu\mu$ and $D^*\mu$ unlike–sign pairs is the same. The charm background was therefore normalized to reproduce the charm signal in the $D^*\mu$ sample.

Dedicated studies performed on light flavor background–enriched samples showed that the normalisation and shapes of the like and unlike sign background distributions agree to a rather good approximation. The total like and unlike–sign beauty contribution was then extracted from the difference between the unlike and like-sign distributions using the formula:

$$N_{b\bar{b}\to\mu\mu} = (N_{data}^{unl} - N_{data}^{like} - (N_{charm} + N_{J/\psi} + N_{BH})) \times \left(\frac{N_{b\bar{b}}^{unl} + N_{b\bar{b}}^{like}}{N_{b\bar{b}}^{unl} - N_{b\bar{b}}^{like}}\right)_{MC}$$

Corrections were applied in order to account for small (4% on average) mass-dependent asymmetries between the like and unlike sign light flavour background distributions.

The final p_T^μ and η^μ distributions in the non–isolated unlike–sign sample, after combination of high and low–mass components, are shown in the right hand plots of Fig. 1. A total beauty contribution of about 1800 events was obtained with a purity of about 50%. The low muon p_T threshold translates into sensitivity to b–quark production down to $p_T^b = 0$ (left plot of Fig. 2). In combination with the large pseudorapidity coverage, this allows the extraction of the total cross section for beauty production. In order to

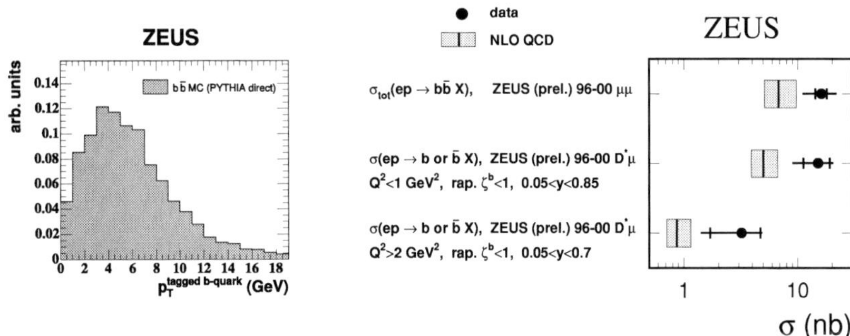

FIGURE 2. Left: p_T^b for tagged b quarks. Right: total b cross sections compared to NLO predictions.

agree with the data, the normalisation of the PYTHIA + RAPGAP MC prediction for the beauty contribution had to be scaled up by a factor 2.06. The total cross section for $b\bar{b}$ pair production for was obtained to be:

$$\sigma_{tot}(ep \to b\bar{b}X, \sqrt{s} = 318 \text{ GeV}) = 16.1 \pm 1.8(\text{stat.})^{+5.3}_{-4.8}(\text{syst.}) \text{ nb}$$

To investigate how much the extracted total cross section depends on details of the $b\bar{b}$ production kinematics , an intermediate visible cross section was extracted for the maximum possible muon phase space region allowed by the preselection and the detector acceptance. The definition of the visible cross section was guided by detector efficiency, to yield minimum extrapolation factor. Both muons were asked to lie in the -2.2 $< \eta^{\mu} <$ 2.5 region. One muon was then required to have $p_T^{\mu} > 1.5$ GeV while for the other muon $p^{\mu} > 1.8$ GeV (for $\eta^{\mu} < 0.6$), $p^{\mu} > 2.5$ or $p_T > 1.5$ GeV (for $\eta^{\mu} > 0.6$) and $p_T^{\mu} > 0.75$ GeV was required. Correcting for muon acceptance, a visible cross section for dimuon production from beauty decays in this phase space $\sigma_{vis}(ep \to b\bar{b}X) = 63 \pm 7(\text{stat.})^{+20}_{-18}(\text{sys.})$ pb was measured. Dividing this cross section by the probability of a $b\bar{b}$ pair to yield a muon pair in this kinematic range, the total cross section already quoted above can again be obtained. This probability (0.38% on average) is quite small. However, it is almost entirely dominated by quantities measured with good precision at e^+e^- colliders (fragmentation functions, branching ratios, decay spectra). Only about 10% of the total phase space remains unmeasured.

One of the main contributions to the systematic error arises from the μ-efficiency correction. Other uncertainties include an adequate variation of the estimated backgrounds, branching ratios, decay spectra and the $B^0\bar{B}^0$ mixing parameters. The NLO predicted p_T^b shape was checked to be well reproduced by PYTHIA and RAPGAP. Nevertheless a related uncertainty was estimated by drastically varying the direct–resolved relative contributions in both efficiency and extrapolation evaluation ($\sim$ 10 % variation in σ).

NLO QCD predictions were obtained in the massive approach by adding the FMNR[3] and HVQDIS[4] results for $Q^2 <,> 1$ GeV^2, respectively. The CTEQ5M and CTEQ5F4 parton density functions were used. Theoretical uncertainties were estimated by varying the beauty mass ($4.5 < m_b < 5.0$ GeV) and the renormalization and factorization default scales used in these calculations by a factor two. The resulting cross section $\sigma_{tot}^{NLO}(ep \to b\bar{b}X) = 6.8^{+3.0}_{-1.7}$ nb is a factor 2.4 lower than the measured value, although still compatible within the large uncertainties (about two standard deviations). The right hand plot of Fig. 2 shows a comparison of this cross section to the slightly less inclusive cross sections from the $D^*\mu$ final state obtained by ZEUS in earlier measurements [1]. These measurements, as well as similar measurements by H1 [2] show the same trend to be larger than the corresponding QCD predictions.

I would like to thank my ZEUS colleagues and especially I. Bloch and A. Geiser for the profitable discussion we had during the preparation of this talk.

REFERENCES

1. ZEUS Coll. Abst. 5-0342,11-0343, ICHEP, Beijing, China, August 16–22,2004.
2. H1 Coll., A. Aktas et al.,hep-ex/0503038 (2005).
3. Frixione *et al.*, Nucl.Phys. B454, 3 (1995); Phys.Lett. B348, 633 (1995).
4. Harris,Smith. Nucl.Phys. B452, 109 (1995); Phys.Lett. B353, 535 (1995); Phys.Rev. D57,2806 (1998);

Measurement of Beauty and Charm Photoproduction at H1 using inclusive lifetime tagging

L. Finke

University of Hamburg, Germany

Abstract. A measurement of the charm and beauty photoproduction cross sections at the *ep* collider HERA is presented. The lifetime signature of c and b-flavoured hadrons is exploited to determine the fractions of events in the sample containing charm or beauty. Differential cross sections as a function of the jet transverse momentum, the rapidity and x_γ^{obs} are measured in the photoproduction region $Q^2 < 1$ GeV^2, with inelasticity $0.15 < y < 0.8$. The results are compared with calculations in next-to-leading order perturbative QCD and Monte Carlo models as implemented in PYTHIA and CASCADE.

Keywords: Heavy Quarks, Charm, Beauty Production, Jets
PACS: 12.38Qk, 13.85.Hd, 14.65.Fy, 14.65.Dw

INTRODUCTION

A measurement of differential charm and beauty dijet photoproduction cross sections in *ep* collisions at HERA is presented here. The analysis covers the photoproduction region, where the virtuality of the photon emitted from the incoming positron is small, $Q^2 \sim 0$. In this process, the production of heavy quarks is expected to be dominated by photon-gluon fusion, $\gamma g \rightarrow c\bar{c}$ or $\gamma g \rightarrow b\bar{b}$, where the photon interacts with a gluon from the proton to produce heavy quarks in the final state. The measurements are compared to calculations in perturbative QCD (pQCD) at next-to-leading order (NLO) in which the mass of the heavy quarks provides a hard scale.

The charm and beauty cross sections are determined using a fit to the lifetime signature of charged particles in jets. This inclusive method yields measurements of differential cross sections that extend to larger values of transverse momenta than in previous HERA analyses in which leptons from beauty quark decays were used to measure beauty cross sections [1, 2, 3, 4, 5, 6].

EXPERIMENTAL METHOD

The analysis is based on a photoproduction dijet sample, corresponding to an integrated luminosity of 57.7pb^{-1}, taken in the years 1999-2000, when HERA was operated in unpolarised e^+p mode, with an *ep* centre of mass energy of $\sqrt{s}$=319 GeV. Events with two jets and large transverse momenta, $p_t^{jet_{1(2)}} > 11(8)$ GeV, in the central rapidity range, $-0.88 < \eta^{jet} < 1.3$ are selected.

CP792, *Deep Inelastic Scattering, DIS 2005,* edited by W. H. Smith and S. R. Dasu

For the final sample only those events which have at least 1 reconstructed track with hits from the central silicon tracker CST [7] with polar angle $30° < \Theta_{track} < 150°$ and a minimum transverse momentum of 0.5 GeV, and which are associated to one of the two highest p_t jets are used. In this analysis, the signed track impact parameter (DCA) with respect to the event vertex is used to separate the different quark flavours. The signed impact parameter is defined as positive if the angle between the jet axis and the line between the vertex and distance of closest approach of the track to the vertex is less than 90°, and is defined as negative otherwise. Tracks from the decays of long lived particles will mainly have positive true DCA, whilst those produced at the event vertex will have zero true DCA. Tracks reconstructed with negative DCA values mainly result from detector resolution.

The separation of charm and beauty events is further enhanced by using different significance distributions for events with different track multiplicities. The first significance S_1 is defined for events with exactly one CST track associated to a jet and is simply the significance of this track. The second significance S_2 is defined for events with two or more CST tracks associated to one of the two jets and is the significance of the track with the second highest absolute significance. Events in which the tracks with the first and second highest absolute significance in a jet have different signs are removed from the S_2 distribution. This latter condition removes around 50% of events from the S_2 distribution, predominantly from the light quark event sample. In order to considerably reduce the uncertainty due to the DCA resolution and the light quark normalisation, the negative bins in the S_1 and S_2 distributions are subtracted from the positive.

The b, c and light quark fractions in the data are extracted by simultaneously fitting the subtracted S_1 and S_2 distributions and the total number of events with the Monte Carlo beauty, charm and light quark distributions used as templates in each interval of the measurement. This procedure was originally proposed by a recent H1 measurement in [8].

Consistent results in all bins are obtained using an alternative method to separate the quark flavours also based on the use of the significance distributions of the selected tracks. The method was employed by the ALEPH collaboration [9].

RESULTS

From the fit results scale factors for charm and beauty are determined for the samples in each bin.

In each bin of the measurement the differential cross section is obtained by multiplying the bin-averaged cross section predicted by the Monte Carlo simulation by the scale factor and dividing by the bin size. The total dijet charm photoproduction cross section in the range $Q^2 < 1\ \mathrm{GeV}^2$, $0.15 < y < 0.8$, $p_t^{jet_{1,2}} > 11(8)$ GeV and $-0.88 < \eta^{jet_{1,2}} < 1.3$ is measured to be

$$\sigma(ep \to ec\bar{c}X \to ejjX) = 694 \pm 69(stat.) \pm 96(sys.)\mathrm{pb}.$$

For beauty, the measurement yields a cross section for the same kinematic range

$$\sigma(ep \to eb\bar{b}X \to ejjX) = 145 \pm 18(stat.) \pm 30(sys.)\mathrm{pb}.$$

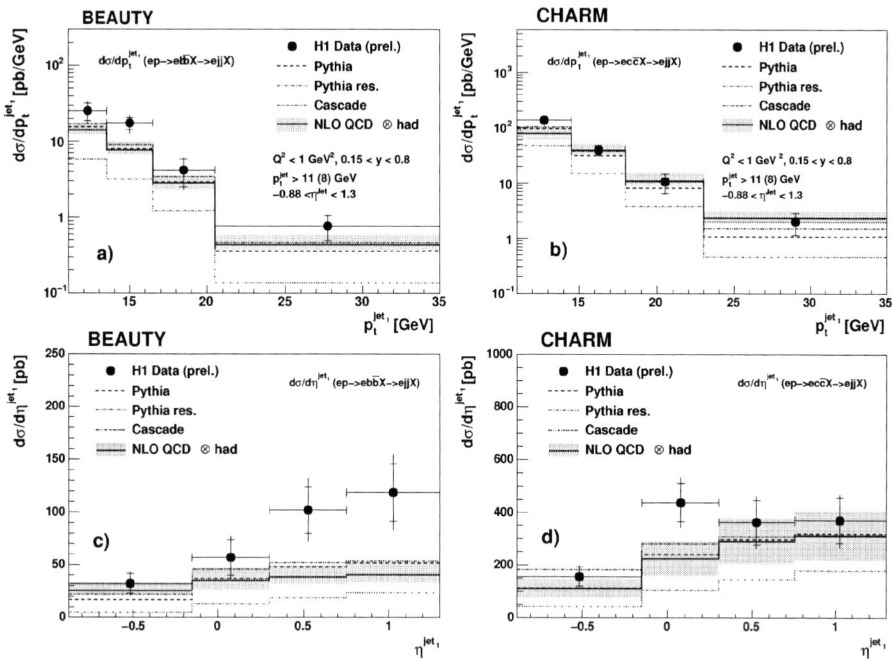

FIGURE 1. Differential charm and beauty cross sections as a function of the transverse momentum $p_t^{jet_1}$ (a and c) and of the rapidity η^{jet_1} (b and d) of the leading jet. The data is compared to the prediction from CASCADE (dotted line) and PYTHIA (dashed line). The contribution in PYTHIA from processes in which the photon is resolved is shown separately (dashed-dotted line). The solid line indicates the prediction from a NLO QCD calculation and the shaded band describes the scale uncertainty of the calculation.

Figures 1 and 2 show the measured differential cross sections for both, charm and beauty, as functions of $p_t^{jet_1}$, η^{jet_1} and x_γ^{obs}, respectively. Here, x_γ^{obs} is defined as the fraction of the $(E-p_z)$ of the hadronic system that is carried by the two highest p_T jets:

$$x_\gamma^{obs} = \frac{(E-p_z)_{jet_1} + (E-p_z)_{jet_2}}{(E-p_z)_h}.$$

The data are compared with predictions from the NLO QCD calculation FMNR as well as from the Monte Carlo programs PYTHIA and CASCADE. The latter implements the CCFM evolution equation using off-shell matrix elements convoluted with k_t-unintegrated parton distributions in the proton.

While the charm data are reasonably well described in both normalization and shape, the beauty data are found to be somewhat higher than the prediction from the NLO QCD calculation. The main difference between the beauty data and the NLO calculation appears to originate in the region of positive values of rapidity, as can be seen in figure 1c, and small values of x_γ^{obs}, where the prediction lies significantly below the

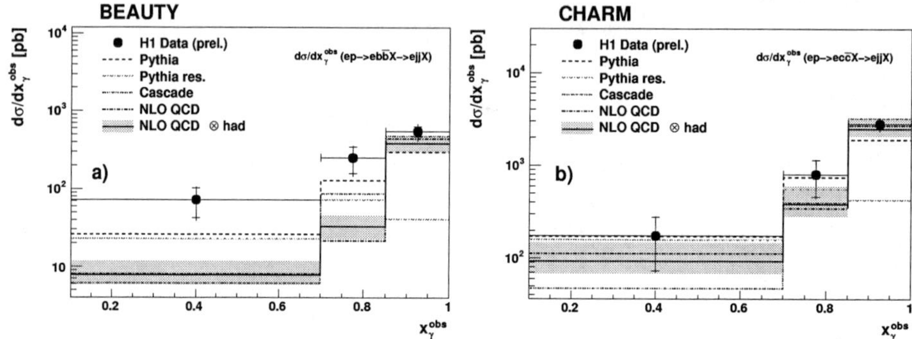

FIGURE 2. Differential charm and beauty cross section as a function of x_γ^{obs}. The data is compared to the prediction from CASCADE (dotted line) and PYTHIA (dashed line). The prediction from a NLO QCD calculation is shown before (dashed-dotted line) and after (solid line) hadronisation corrections, and the shaded band describes the scale uncertainty of the calculation.

data (figure 2a). In these regions the contribution to the cross section from events with resolved photons is particularly large. The prediction from PYTHIA for this contribution is indicated by the dashed-dotted line.

PYTHIA and CASCADE give a good description of the shapes of the data distributions. However, the beauty data are generally higher in normalisation than the PYTHIA (CASCADE) prediction by a factor ~ 1.8 (~ 1.6), respectively.

CONCLUSION

A measurement of differential charm and beauty dijet photoproduction cross sections at HERA has been presented. The measurement makes use of the precise tracking information available from the H1 vertex detector. The heavy quark cross sections are determined by making use of the lifetime distributions of charm and beauty hadrons.

While the charm cross sections are reasonably well described in both normalization and shape, the beauty cross sections are found to be somewhat higher than a calculation in perturbative QCD to next-to-leading order.

REFERENCES

1. C. Adloff *et al.* [H1 Collaboration], Phys. Lett. B **467** (1999) 156 [Erratum-ibid. B **518** (2001) 331] [hep-ex/9909029].
2. J. Breitweg *et al.* [ZEUS Collaboration], Eur. Phys. J. C **18** (2001) 625 [hep-ex/0011081].
3. S. Chekanov *et al.* [ZEUS Collaboration], DESY-03-212, submitted to Physical Review D
4. S. Chekanov *et al.* [ZEUS Collaboration], DESY-04-070 (May 2004), submitted to Phys. Letters B
5. A. Aktas *et al.* [H1 Collaboration], hep-ex/0502010.
6. A. Aktas *et al.* [H1 Collaboration], hep-ex/0503038.
7. D. Pitzl *et al.*, Nucl. Instrum. Meth. A **454**, 334 (2000) [hep-ex/0002044].
8. A. Aktas *et al.* [H1 Collaboration], to appear in Eur. Phys. J. C., [hep-ex/0411046].
9. D. Buskulic *et al.* [ALEPH Collaboration], Phys. Lett. B **313** (1993) 535.

Measurement of $F_2^{c\bar{c}}$ and $F_2^{b\bar{b}}$ at Low and High Q^2 using the H1 Vertex Detector

Tatsiana Klimkovich

DESY, Notkestr. 85, 22607 Hamburg, Germany
Institut für Experimentalphysik, Universität Hamburg, Luruper Chaussee 149, 22761 Hamburg

Abstract. Measurements are presented of inclusive charm and beauty cross sections in e^+p collisions at HERA for values of photon virtuality $3.5 \leq Q^2 \leq 60$ GeV^2 and of the Bjorken scaling variable $0.0002 \leq x \leq 0.005$ using a method based on the impact parameter, in the transverse plane, of tracks to the primary vertex, as measured by the H1 vertex detector. Values for the structure functions $F_2^{c\bar{c}}$ and $F_2^{b\bar{b}}$ are obtained and presented together with recently published high Q^2 measurements [1]. This is the first measurement of $F_2^{b\bar{b}}$ in this kinematic range. The results are found to be compatible with the predictions of perturbative quantum chromodynamics and with previous measurements of $F_2^{c\bar{c}}$.

Keywords: H1, structure function, charm, beauty, heavy quarks
PACS: 13.85.Hd, 14.65.Dw, 14.65.Fy

1. INTRODUCTION

Measurements of the charm (c) and beauty (b) contributions to the inclusive proton structure function F_2 have been made recently in Deep Inelastic Scattering (DIS) at HERA, using information from the H1 vertex detector, for values of the negative square of the four momentum of the exchanged boson $Q^2 > 150$ GeV^2 [1]. In this high Q^2 region a fraction of $\sim 18\%$ ($\sim 3\%$) of DIS events contain c (b) quarks. It was found that perturbative QCD (pQCD) calculations at next-to-leading order (NLO) gave a good description of the data. In this paper a similar method is employed, using data from the same running period, to extend the measurements to the range $3.5 \leq Q^2 \leq 60$ GeV^2 and $0.000197 \leq x \leq 0.005$. Events containing heavy quarks can be distinguished from light quarks events by the long lifetimes of charm and beauty hadrons, which lead to displacements of tracks from the primary vertex.

In the framework of NLO QCD analyses of global inclusive and jet cross section measurements the production of heavy flavours is described using the variable flavour number scheme (VFNS) which aims to provide reliable pQCD predictions over the whole kinematic range in Q^2 [2, 3]. At values of $Q^2 \simeq M^2$ the effects of the quark mass must be taken into account and the heavy flavour partons are treated as massive quarks. The dominant LO process in this region is photon gluon fusion (PGF) and the NLO diagrams are of order α_s^2 [4, 5]. As Q^2 increases, in the region $Q^2 \gg M^2$, the heavy flavour quark may be treated as a massless parton in the proton. Different approaches have been developed which deal with the transition from the heavy quark mass effects at low Q^2 to the asymptotic massless parton behaviour at high Q^2.

CP792, *Deep Inelastic Scattering, DIS 2005,* edited by W. H. Smith and S. R. Dasu

2. EXPERIMENTAL METHOD

The analysis is based on a low Q^2 sample of e^+p neutral current scattering events corresponding to an integrated luminosity of 57.4 pb^{-1}, taken in the years 1999-2000, at an *ep* centre of mass energy $\sqrt{s} = 319$ GeV, with a proton beam energy of 920 GeV. The low Q^2 events are selected in a similar manner to that described in [6].

In order to determine a signed impact parameter (δ) for each track, the azimuthal angle of the struck quark ϕ_{quark} must be determined for each event. To do this, jets with a minimum p_T of 4 GeV, in the angular range $25° < \theta < 155°$, are reconstructed exploiting the invariant k_T algorithm [7, 8] in the laboratory frame using all reconstructed hadronic final state (HFS) particles. The angle ϕ_{quark} is defined as the ϕ of the jet with the highest transverse momentum or, if there is no jet reconstructed in the event, as $180° - \phi_{\text{elec}}$, where ϕ_{elec} is the azimuthal angle of the scattered electron. Approximately 33% (55%) of c (b) events have ϕ_{quark} reconstructed from a jet, as determined from a Monte Carlo simulation.

If the angle between the quark axis and the line joining the primary vertex to the point of DCA is less than 90°, δ is defined as positive, and is defined as negative otherwise. Tracks with azimuthal angle outside $\pm 90°$ of ϕ_{quark} are rejected. The δ distribution, shown in FIGURE 1, is seen to be asymmetric with positive values in excess of negative values indicating the presence of long lived particles. Tracks with $|\delta| > 0.1$ cm are rejected from the analysis to suppress light quark events containing long lived strange particles.

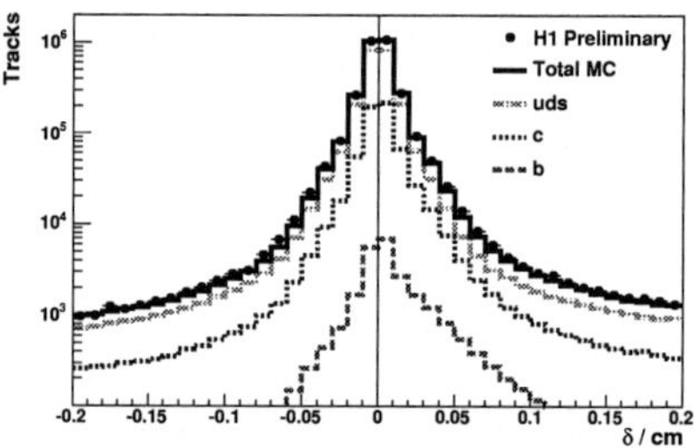

FIGURE 1. The distribution of the signed DCA δ of a track to the primary vertex in the x–y plane.

The method used in [1] to distinguish between the c, b and light quark flavours has been modified in the present analysis because here the fraction of b quarks is smaller. The quantities S_1, S_2 and S_3 are defined as the significance ($\delta/\sigma(\delta)$) of the track with the highest, second highest and third highest absolute significance, respectively, where $\sigma(\delta)$ is the error on δ. Distributions of each of these quantities are made. The events contributing to the S_2 distribution also contribute to the S_1 distribution. Similarly, those contributing to the S_3 distribution also contribute to the S_2 and S_1 distributions. Events in which S_1 and S_2 have opposite signs are excluded from the S_2 distribution. Events in which S_1, S_2 and S_3 do not all have the same sign are excluded from the S_3 distribution.

The fractions of c, b and light quarks of the data are extracted in several x–Q^2 intervals using a least squares simultaneous fit to the subtracted S_1, S_2 and S_3 distributions and the total number of inclusive events before any CST track selection. The c, b and *uds*

Monte Carlo simulation samples are used as templates. The Monte Carlo c, b and uds contributions in each x–Q^2 interval are scaled by factors P_c, P_b and P_l, respectively to obtain the best fit to the observed subtracted S_1, S_2, S_3 and total distributions. Only the statistical errors of the data and Monte Carlo simulation are considered in the fit. The fit to the subtracted significance distributions mainly constrains P_c and P_b, whereas the overall normalization constrains P_l. The fit gives a good description of all the significance distributions, with values of $P_c = 1.34 \pm 0.06$, $P_b = 1.43 \pm 0.17$ and $P_l = 1.16 \pm 0.01$.

The results of the fit in each x–Q^2 interval are converted to a measurement of the 'reduced c cross section' using:

$$\tilde{\sigma}^{c\bar{c}}(x,Q^2) = \tilde{\sigma}(x,Q^2)\frac{P_c N_c^{\mathrm{MCgen}}}{P_c N_c^{\mathrm{MCgen}} + P_b N_b^{\mathrm{MCgen}} + P_l N_l^{\mathrm{MCgen}}}\delta_{\mathrm{BCC}}, \tag{1}$$

where $\tilde{\sigma}(x,Q^2)$ is the measured inclusive reduced cross section from H1 [6], N_c^{MCgen}, N_b^{MCgen} and N_l^{MCgen} are the generated number of c, b and light quark events from the Monte Carlo in each bin, respectively, δ_{BCC} is a bin centre correction. The differential b cross section is evaluated in the same manner.

The structure function $F_2^{c\bar{c}}$ may be evaluated from the reduced cross section expression:

$$\tilde{\sigma}^{c\bar{c}} = F_2^{c\bar{c}} - \frac{y^2}{1+(1-y)^2}F_L^{c\bar{c}}, \tag{2}$$

where the longitudinal structure function $F_L^{c\bar{c}}$ is estimated from the NLO QCD expectation. The structure function $F_2^{b\bar{b}}$ is evaluated in the same manner.

3. RESULTS

The measurements of $F_2^{c\bar{c}}$ and $F_2^{b\bar{b}}$ are shown as a function of x for various Q^2 values in FIGURE 2. In the lowest Q^2 bin $F_2^{b\bar{b}}$ is shown as a limit because it is consistent with zero. The data are compared with the results extracted from D^* meson measurements by H1 [9] and ZEUS [10] obtained using a NLO program based on DGLAP evolution to extrapolate the measurements outside the visible D^* range. The data are also compared with two VFNS predictions of NLO QCD from MRST [2] and CTEQ [3]. The predictions provide a reasonable description of the present data.

4. CONCLUSION

The differential charm and beauty cross sections in deep inelastic scattering are measured at low Q^2 using a technique based on the impact parameters of tracks from decays of long lived c and b hadrons. The measurements are done using all events containing tracks with vertex detector information. The cross sections and derived structure functions $F_2^{c\bar{c}}$ and $F_2^{b\bar{b}}$ are found to be well described by predictions of perturbative QCD.

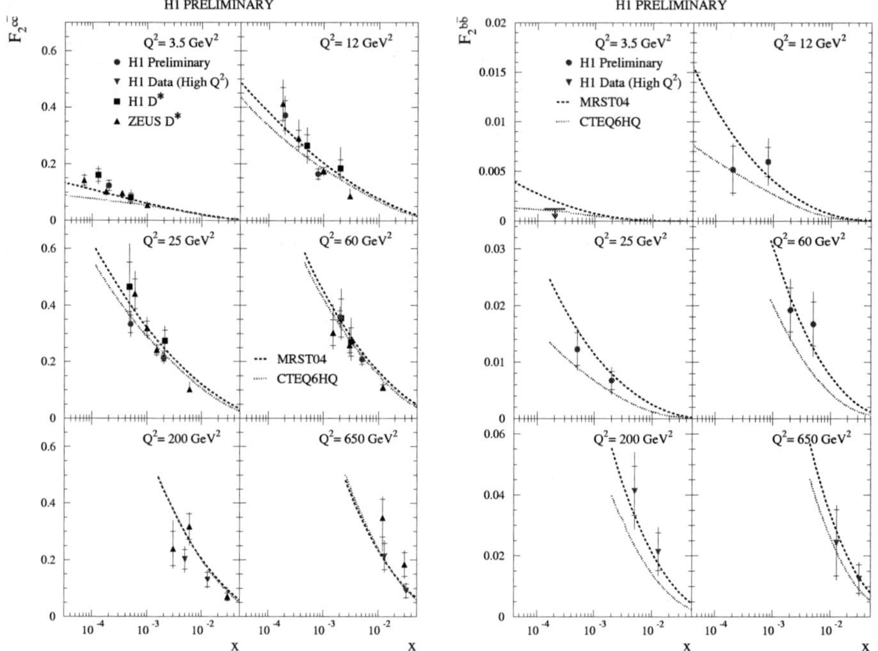

FIGURE 2. The measured $F_2^{c\bar{c}}$(left) and $F_2^{b\bar{b}}$ (right) shown as a function of x for various Q^2 values. The inner error bars show the statistical error, the outer error bars represent the statistical and systematic errors added in quadrature. The $F_2^{c\bar{c}}$and $F_2^{b\bar{b}}$ from H1 at higher values of Q^2 [1], the measurements of obtained from D^* mesons from H1 and ZEUS [9, 10] and the predictions of QCD are also shown.

This is the first measurement of $F_2^{b\bar{b}}$ in the low Q^2 kinematic region. The charm and beauty cross sections contribute on average 22% and 0.8% of the total ep cross section in this kinematic range.

REFERENCES

1. A. Aktas, et al., *Eur. Phys. J.*, **C40**, 349–359 (2005), [hep-ex/0411046].
2. A. D. Martin, R. G. Roberts, W. J. Stirling, and R. S. Thorne, *Eur. Phys. J.*, **C39**, 155–161 (2005), [hep-ph/0411040].
3. S. Kretzer, H. L. Lai, F. I. Olness, and W. K. Tung, *Phys. Rev.*, **D69**, 114005 (2004), [hep-ph/0307022].
4. E. Laenen, S. Riemersma, J. Smith, and W. L. van Neerven, *Nucl. Phys.*, **B392**, 229–250 (1993).
5. E. Laenen, S. Riemersma, J. Smith, and W. L. van Neerven, *Nucl. Phys.*, **B392**, 162–228 (1993).
6. C. Adloff, et al., *Eur. Phys. J.*, **C21**, 33–61 (2001), [hep-ex/0012053].
7. S. D. Ellis, and D. E. Soper, *Phys. Rev.*, **D48**, 3160–3166 (1993), [hep-ph/9305266].
8. S. Catani, Y. L. Dokshitzer, M. H. Seymour, and B. R. Webber, *Nucl. Phys.*, **B406**, 187–224 (1993).
9. C. Adloff, et al., *Phys. Lett.*, **B528**, 199–214 (2002), [hep-ex/0108039].
10. S. Chekanov, et al., *Phys. Rev.*, **D69**, 012004 (2004), [hep-ex/0308068].

Charm and Beauty Production at HERA–B

Ulrich Husemann for the HERA–B Collaboration

Fachbereich Physik, Universität Siegen, D–57068 Siegen, Germany
now at University of Rochester, Rochester, New York 14627

Abstract. The HERA–B experiment at DESY has acquired a data-set of approximately 300,000 decays $J/\psi \to \ell^+\ell^-$ during its 2002/2003 data-taking period. These data are used to analyze the production of heavy quarks in proton-nucleus interactions at a center-of-mass energy of 41.6 GeV. In this article, preliminary results of two measurements are discussed, a measurement of nuclear effects in the production of J/ψ mesons and a measurement of the $b\bar{b}$ production cross section.

Keywords: Heavy quarks, J/ψ production, nuclear effects, beauty production
PACS: 13.20.Gd, 13.20.He, 13.85.Qk,14.40.Gx, 24.85.+p

HERA–B: DETECTOR, TRIGGER AND DATA-SET

The HERA–B detector, depicted in Fig. 1, is a fixed-target spectrometer with large angular acceptance. Protons from the halo of the HERA proton beam are collided with an internal wire target. The target consists of up to eight thin wires of different materials, distributed in two target stations. Each wire can be moved independently to adjust the interaction rate.

The HERA–B detector consists of the following sub-detectors: A silicon micro-strip vertex detector is used to reconstruct and separate primary and secondary vertices. The tracking detectors comprise micro-strip gaseous chambers with gas electron multiplier foils in the inner acceptance region of high particle flux and honeycomb drift chambers in the outer region. Particle identification is performed by a ring-imaging Čerenkov counter, an electromagnetic calorimeter, and a four-layer muon detector.

Lepton pairs from J/ψ decays are enriched by a multi-level trigger system. The first trigger level is implemented as a hardware track trigger in the outer tracking detector. Starting points for the track search are provided by pretriggers in the muon detector and the calorimeter. At the second trigger level, a software trigger running on a PC farm, track candidates are extrapolated to the vertex detector, and a two-prong vertex fit is performed. Accepted events are reconstructed online.

The HERA–B experiment has taken data between October 2002 and February 2003. The data sample recorded with the lepton pair trigger contains approximately 150,000 decays into each of the channels $J/\psi \to e^+e^-$ and $J/\psi \to \mu^+\mu^-$, 90,000 of which were taken with two target wires of different materials simultaneously. Based on this data set, the HERA–B collaboration has analyzed the production of charmonium states such as J/ψ, $\psi(2S)$ and χ_c, as well as the production of open and hidden beauty, e.g. $b\bar{b}$ and Υ.

CP792, *Deep Inelastic Scattering, DIS 2005,* edited by W. H. Smith and S. R. Dasu

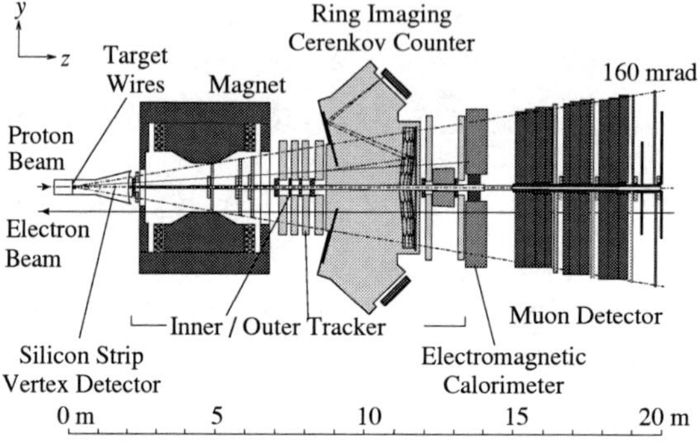

FIGURE 1. Elevation view of the HERA–B detector.

NUCLEAR EFFECTS IN J/ψ PRODUCTION

Many theoretical models of J/ψ production in proton-nucleus interactions predict modifications of the production cross section by nuclear effects, e.g. due to the absorption of final state particles in the nuclear environment [1, 2], or due to coherent proton-nucleus scattering [3, 4]. The nuclear effects are commonly parametrized by the power law

$$\sigma_{pA} = \sigma_{pN} \cdot A^{\alpha(x_F, p_T)}. \tag{1}$$

Here, σ_{pA} is the proton-nucleus cross section, σ_{pN} is the proton-nucleon cross section, and A is the atomic mass number of the target. A value of the nuclear suppression parameter $\alpha(x_F, p_T)$ smaller than unity corresponds to the suppression of J/ψ production. The cross section σ is related to the number of detected particles N, the integrated luminosity $\mathscr{L}$, and the detection efficiency ε by $\sigma = N/(\mathscr{L}\varepsilon)$. Using this relation for a tungsten target and a carbon target operated at the same time, Eq. (1) can be solved for α:

$$\alpha = \frac{1}{\log(A_W/A_C)} \cdot \log\left(\frac{N_W}{N_C} \cdot \frac{\mathscr{L}_C}{\mathscr{L}_W} \cdot \frac{\varepsilon_C}{\varepsilon_W}\right). \tag{2}$$

The measurement hence includes measuring three ratios, the ratio of J/ψ yields, the ratio of luminosities per wire, and the ratio of efficiencies. The analysis thus benefits from the advantages of a relative measurement, in which many systematic uncertainties cancel out. The J/ψ yields are determined by fits to the J/ψ invariant mass spectrum in bins of the kinematic variables. To measure the luminosities, the number of primary vertices per target wire is counted in data recorded with a zero-bias trigger in parallel to the lepton pair trigger. A detailed Monte Carlo (MC) simulation of the HERA–B detector and trigger is performed to calculate the ratio of efficiencies. Details of the analysis in the decay channel $J/\psi \to \mu^+\mu^-$ can be found in [5].

The measured nuclear suppression parameter α in the decay channel $J/\psi \to \mu^+\mu^-$ is shown in Fig. 2 as a function of x_F and p_T. The data show good agreement with previous measurements in the overlap region and extend the existing data to negative x_F. In the covered range of $-0.375 \leq x_F < 0.125$, a small constant suppression of J/ψ production is observed, with an average suppression parameter of $\overline{\alpha} = 0.969 \pm 0.003(\text{stat.}) \pm 0.021(\text{syst.})$.

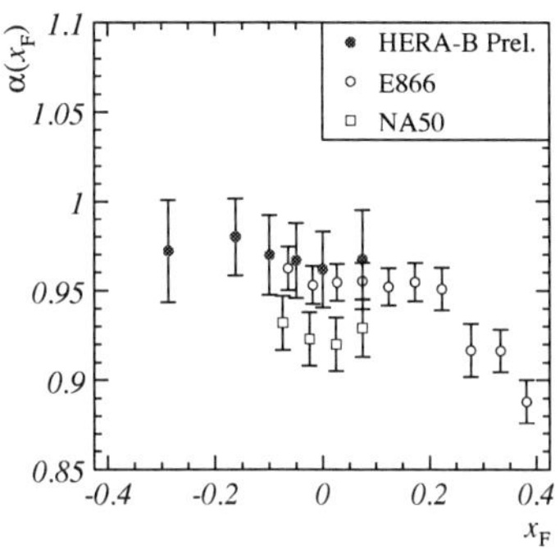

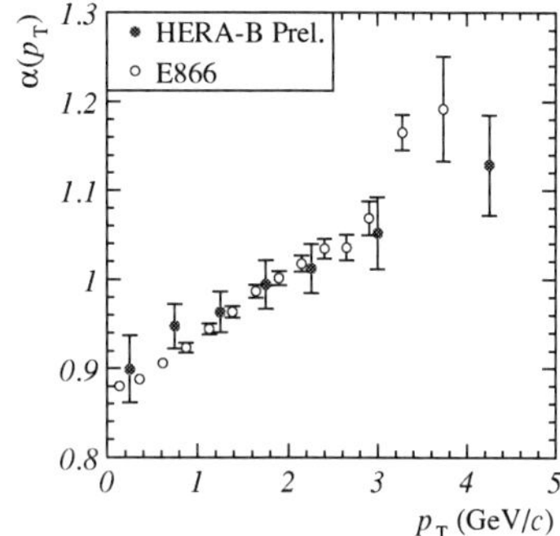

FIGURE 2. Nuclear suppression parameter α in the decay channel $J/\psi \to \mu^+\mu^-$ as a function of x_F (left) and p_T (right), compared to previous measurements by the E866 [6] and the NA50 [7] collaborations. All data points include statistical and systematic uncertainties.

THE $B\overline{B}$ PRODUCTION CROSS SECTION

At the HERA–B center-of-mass energy of 41.6 GeV, beauty production is close to its kinematic threshold. This results in large theoretical uncertainties due to the resummation of soft gluons and increased sensitivity to the mass of the b quark. Previous measurements of the $b\overline{b}$ production cross section in a similar energy range suffered from small statistics and large systematic uncertainties.

In HERA–B, the $b\overline{b}$ production cross section is measured in the inclusive decay channel $b\overline{b} \to J/\psi X \to \ell^+\ell^- X$, where $\ell = e, \mu$. The average flight distance of B mesons in the HERA–B detector amounts to approximately 8 mm. Therefore, B candidate events can be selected based on the separation of the J/ψ decay vertex from the primary vertex on the target wire. The best separation from background decays such as prompt J/ψ decays and semileptonic b and c decays is achieved by a combined cut on the significance of the J/ψ detached vertex and the impact parameter to the target wire. Furthermore, combinatorial background is controlled by comparing the background upstream and downstream of the target with respect to the proton flight direction. A fit to the B lifetime and the requirement of a third track at the J/ψ decay vertex serve as independent confirmations of the B flavor. For the final $b\overline{b}$ cross section result, the consistent results in the dielectron and the dimuon channels using the 2002/2003 data-sample are combined with the previously published result [8] from the 2000 data-taking.

To minimize the influence of external input, the result of the measurement is first presented relative to the J/ψ production cross section and restricted to the kinematic acceptance of the HERA–B detector:

$$R_{b\overline{b}} = \frac{\Delta\sigma_{b\overline{b}}}{\Delta\sigma_{J/\psi}} = \varepsilon \frac{N_{b\overline{b}\to J/\psi X}}{N_{\mathrm{prompt}\, J/\psi}} = 0.033 \pm 0.005(\mathrm{stat.}) \pm 0.004(\mathrm{syst.}). \quad (3)$$

Here, $\Delta\sigma_{b\overline{b}}$ and $\Delta\sigma_{J/\psi}$ are the production cross sections for $b\overline{b}$ and J/ψ within the HERA–B acceptance, $N_{b\overline{b}\to J/\psi X}$ and $N_{\mathrm{prompt}\, J/\psi}$ are the number of $b\overline{b}$ candidates and the number of prompt J/ψ. The efficiency parameter ε includes the branching fractions,

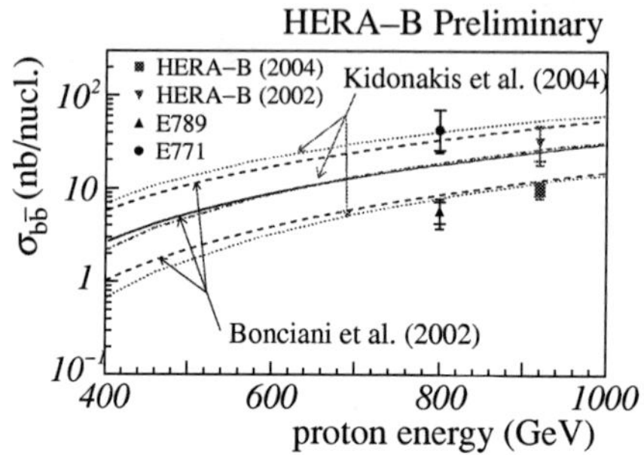

FIGURE 3. Preliminary cross section for $b\bar{b}$ production as a function of the proton energy, compared to theoretical predictions [12, 13] and a previous HERA–B measurement based on the 2000 data-set [8].

the efficiency of the detached vertex selection and the nuclear dependences for $b\bar{b}$ and J/ψ events. The efficiencies are determined from a detailed MC simulation. The result is extrapolated to the full phase-space with the help of PYTHIA [9]. Since to date, there is no published HERA–B measurement of the J/ψ production cross section, a value of $\sigma_{J/\psi} = 352 \pm 2(\text{stat.}) \pm 26(\text{syst.})\,\text{nb/nucl.}$ is assumed [8]. This value is obtained by averaging $\sigma_{J/\psi}$ as measured by the E789 and E771 experiments [10, 11] and extrapolating the result to HERA–B energies. Note that a significant upward revision of this value is anticipated from the HERA–B measurement. The resulting $b\bar{b}$ production cross section is depicted in Fig. 3, along with previous measurements and theoretical predictions. The preliminary numerical result is $\sigma_{b\bar{b}} = 9.9 \pm 1.5(\text{stat.}) \pm 1.4(\text{syst.})\,\text{nb/nucl.}$

ACKNOWLEDGMENTS

The author would like to thank the DIS05 crew for the well-organized and interesting conference. This work was supported by the German Bundesministerium für Bildung und Forschung under the contract number 5HB1PEA/7. Special thanks to the HERA–B charmonium and beauty working groups for their help in preparing this conference contribution and to DESY for the kind financial support.

REFERENCES

1. R. Vogt, *Phys. Rev.*, **C61**, 035203 (2000).
2. R. Vogt, *Nucl. Phys.*, **A700**, 539–554 (2002).
3. B. Z. Kopeliovich, A. Tarasov, and J. Hüfner, *Nucl. Phys.*, **A696**, 669–714 (2001).
4. K. G. Boreskov, and A. B. Kaidalov, *JETP Lett.*, **77**, 599–602 (2003).
5. U. Husemann, *Measurement of Nuclear Effects in the Production of* J/ψ *Mesons with the HERA–B Detector*, Ph.D. thesis, Universität Siegen, Germany (2005), DESY-THESIS-2005-05.
6. M. J. Leitch, et al., *Phys. Rev. Lett.*, **84**, 3256–3260 (2000).
7. B. Alessandro, et al., *Eur. Phys. J.*, **C33**, 31–40 (2004).
8. I. Abt, et al., *Eur. Phys. J.*, **C26**, 345–355 (2003).
9. T. Sjöstrand, *Comput. Phys. Commun.*, **82**, 74–90 (1994).
10. M. H. Schub, et al., *Phys. Rev.*, **D52**, 1307–1315 (1995).
11. T. Alexopoulos, et al., *Phys. Lett.*, **B374**, 271–276 (1996).
12. R. Bonciani, S. Catani, M. L. Mangano, and P. Nason, *Nucl. Phys.*, **B529**, 424–450 (1998).
13. N. Kidonakis, and R. Vogt, *Eur. Phys. J.*, **C36**, 201–213 (2004).

Measurement of beauty production at HERA using events with muons and jets

Olaf Behnke
on behalf of the
H1 Collaboration

Physikalisches Institut, Universität Heidelberg, Philosophenweg 12, 69120 Heidelberg, Germany

Abstract. Several new measurements of beauty production at HERA have been presented at this conference. In this talk we report about the H1 measurement using events with a muon associated to a jet. This is the first beauty analysis at HERA, where both the long lifetime and the large mass of b-flavoured hadrons are exploited to identify the beauty events, leading to an improved signal separation. Differential cross sections are measured both in photoproduction and in deep inelastic scattering. The measured data are found to be somewhat higher then perturbative QCD calculations to next-to-leading order. A significant excess is observed in certain corners of the kinematic phase space. At the end of this report new and recent beauty measurements are summarised.

PACS: 10.

INTRODUCTION

The dominant beauty production mechanism at HERA is the process $\gamma g \rightarrow b\bar{b}$, where the photon is emitted from the electron and the gluon from the proton. Since the large b mass provides a hard scale, rendering a small α_s, it would be expected that beauty production can be accurately calculated using perturbative QCD (pQCD). In contrast, the first measurements at HERA [1, 2] revealed significant excesses of data over pQCD calculations performed to next-to-leading order (NLO). Similar observations were made in hadron-hadron collisions [3] and also in two-photon interactions [4]. Newer beauty production measurements from the H1 [5] and ZEUS Collaborations [6] are in better agreement with QCD predictions or again somewhat higher [7]. In the recent article [8] by Matteo Cacciari with the title 'Rise and Fall of the Bottom Quark Production Excess' it is argued, that improvements both in the theoretical models and in the way the data to theory comparisons are handled, have lead to a much better data description, especially for new measurements from hadron-hadron collisions, as discussed in [9].

At this conference several new measurements are reported on beauty production at HERA. In this paper the H1 measurement [10] is described, where events are used with a muon in the central pseudo rapidity range, which is associated to a jet. Differential cross sections are measured both in photoproduction with photon virtualities $Q^2 < 1$ GeV^2 and in deep inelastic scattering with $2 < Q^2 < 100$ GeV^2. The data were collected in the years 1999-2000. To separate beauty events from charm and light quark background, two observables are used which exploit the large mass and the long lifetime of the b-quark, respectively:

1. The transverse momentum p_T^{rel} of the muon with respect to the axis of the associated jet: For muons from b-decays the p_T^{rel} spectrum extends to much larger values than for the other sources.
2. The signed impact parameter δ of the muon track with respect to the primary event vertex: For muons from b-decays this takes larger values as compared to the other sources.

Finally, the relative contribution of beauty and background in the data is determined from a likelihood fit to the two-dimensional distribution of p_T^{rel} and δ . The ZEUS measurements [6, 7], which are also discussed in the following, rely on the p_T^{rel} observable alone.

CP792, *Deep Inelastic Scattering, DIS 2005*, edited by W. H. Smith and S. R. Dasu

RESULTS

Photoproduction. For the photoproduction analysis at least two jets are required in the final state with transverse momenta $p_t^{jet_{1(2)}} > 7(6)$ GeV. Figure 1 shows the differential cross-sections as function of the muon pseudorapidity (left) and transverse momentum (right). The

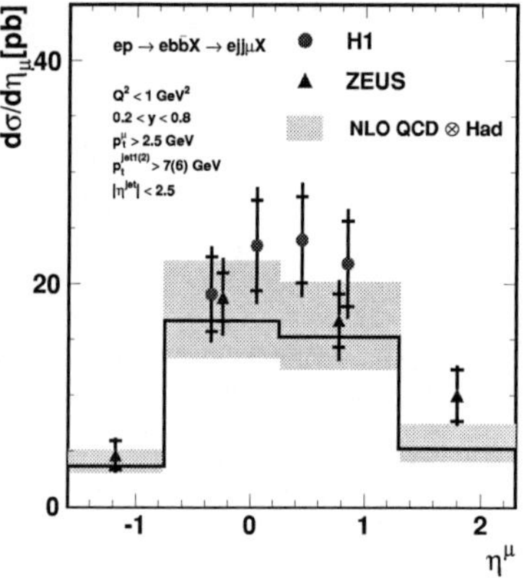

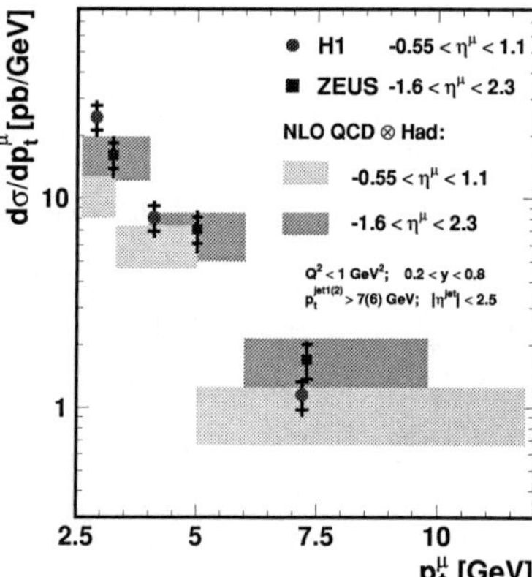

FIGURE 1. Differential beauty cross sections in photoproduction as a function of (left) muon pseudo-rapidity and (right) muon transverse momentum. Shown are the new H1 measurement [10] and the ZEUS measurement [6].

figure also shows the ZEUS measurement [6], which covers a wider muon pseudorapidity range. The H1 and ZEUS measurements agree well in the overlapping region. The data are also compared to an NLO calculation, using the program [11]. For the calculation, fragmentation is performed using the Peterson function [13]. The errors of the theory prediction are dominated by the uncertainties of the renormalisation scale and the b-quark mass. The data tend to lie slightly above this calculation, however, within the errors the calculation describes all the H1 and ZEUS data points. The measured cross sections as function of the muon transverse momentum are compared in figure 1 (right) to the NLO predictions in the respective kinematic ranges of the H1 and ZEUS measurements. In the lowest bin from 2.5 GeV to 3.3 GeV the H1 measurement exceeds the prediction by a factor of ~ 2.5, while at higher transverse momenta a better agreement is observed. Such an excess is not seen in the ZEUS data. This discrepancy needs to be clarified in the future.

Deep inelastic scattering. For the DIS analysis the jet algorithm is applied in the Breit frame and at least one jet with transverse momentum $p_{t,jet}^{Breit} > 6$ GeV is required. Figure 2 shows the differential cross sections of the new H1 measurement (top) and the ZEUS measurement [7] (bottom) as function of muon transverse momentum (left) and pseudo-rapidity (right). The data are compared to an NLO calculation using the program [12]. The H1 and ZEUS measurements are made in similar kinematic regions and also the observations and conclusions are very similar:

1. An excess of data over NLO prediction is observed towards smaller muon transverse momenta below 4 GeV.
2. A rise of the differential cross sections is observed towards more positive muon pseudorapidities, (i.e. more close to the proton direction) which is not reproduced by the NLO calculation.

SUMMARY

Figure 3 presents a summary of the very new and recent HERA beauty cross section measurements as a function of the photon virtuality Q^2. The figure shows the ratios of the measured cross

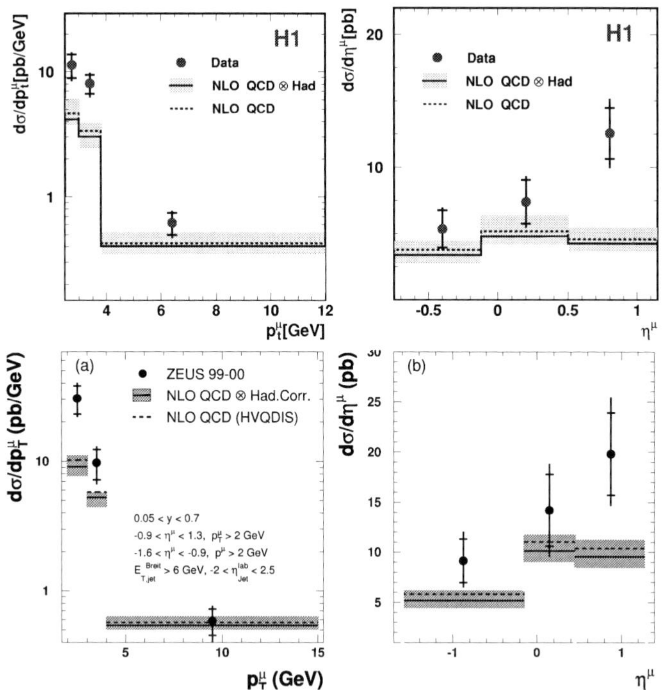

FIGURE 2. Differential beauty cross sections in DIS as a function of (left) muon transverse momentum and (right) muon pseudorapidity for the H1 measurement [10] (top) and the ZEUS measurement [7] (bottom).

sections and the corresponding NLO predictions based on the programs [11, 12]. Uncertainties of the NLO calculations are not taken into account. The new H1 measurement presented in this paper enters the plot as the full square point (photoproduction) and the full circle points (DIS). The following observations are made from all the different measurements:

- Most of the data points are above the predictions.
- There is no clear trend visible for the ratio as function of Q^2.
- There are indications that the data exceed the predictions more significantly towards small b-quark transverse momenta, i.e. closer to the threshold region. This is seen in the new H1 measurement presented here (figures 1 and 2) and also in the H1 and ZEUS $D^*\mu$ measurements [16, 17]. Further evidence comes from another new beauty measurement [18] by ZEUS which is not shown in figure 3. This measurement uses pairs of muons with applying very low momentum cuts and determines the total beauty production cross-section, integrating over photoproduction and DIS, to be higher than the NLO prediction by a factor ~ 2.4. On the contrary, the new inclusive beauty measurements [14, 5] by H1 in the DIS regime, using a vertex tagging method, are in reasonable or good agreement with the NLO predictions as can be seen in figure 3.
- There are indications for an excess of beauty production in the more forward (i.e. proton) direction as can be seen for the DIS measurements in figure 2 and also in the ZEUS photoproduction measurement [6] for muons identified in the forward muon detector (not included in figure 3).

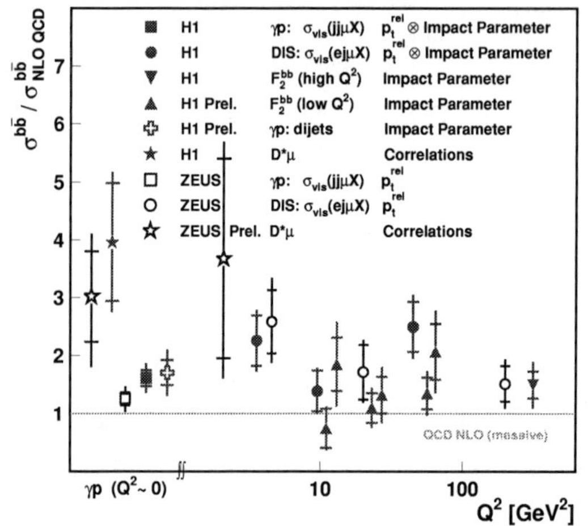

Key	Ref.	Signature	p_T cuts
Photoproduction			
☆	[17]	$D^*\mu$	low
★	[16]	$D^*\mu$	low
□	[6]	2 jets+μ	medium
■	[10]	2 jets+μ	medium
✙	[15]	tracks	high
DIS			
☆	[17]	$D^*\mu$	low
▲	[14]	tracks	low
▼	[5]	tracks	low
●	[10]	1 jet+μ	medium
○	[7]	1 jet+μ	medium

FIGURE 3. Ratio of beauty production cross section measurements at HERA to NLO QCD predictions using the programs [11, 12].

It will be quite interesting to see how this picture develops in the future with the new HERA II data, which will allow to make measurements with much higher precision. It would be very desirable if the theory model improvements reported in [8], which are available for hadron-hadron collisions, would be also made available for the HERA processes.

REFERENCES

1. C. Adloff *et al.* [H1 Collaboration], Phys. Lett. B **467** (1999) 156 [hep-ex/9909029]; [Erratum-ibid. B **518** (2001) 331].
2. J. Breitweg *et al.* [ZEUS Collaboration], Eur. Phys. J. C **18** (2001) 625 [hep-ex/0011081].
3. F. Abe *et al.* [CDF Collaboration], Phys. Rev. Lett. **71** (1993) 2396; Phys. Rev. D **53** (1996) 1051 [hep-ex/9508017];
   S. Abachi *et al.* [D0 Collaboration], Phys. Rev. Lett. **74** (1995) 3548; Phys. Lett. B **370** (1996) 239.
4. M. Acciarri *et al.* [L3 Collaboration], Phys. Lett. B **503** (2001) 10 [hep-ex/0011070].
5. A. Aktas *et al.* [H1 Collaboration], to appear in Eur. Phys. J. C., [hep-ex/0411046].
6. S. Chekanov *et al.* [ZEUS Collaboration], Phys. Rev. D **70** (2004) 012008 [hep-ex/0312057].
7. S. Chekanov *et al.* [ZEUS Collaboration], Phys. Lett. B **599** (2004) 173 [hep-ex/0405069].
8. M. Cacciari, arXiv:hep-ph/0407187.
9. M. Cacciari, S. Frixione, M. L. Mangano, P. Nason and G. Ridolfi, JHEP **0407** (2004) 033 [arXiv:hep-ph/0312132].
10. A. Aktas *et al.* [H1 Collaboration], Eur. Phys. J. C **41**, 453 (2005) [arXiv:hep-ex/0502010].
11. S. Frixione, P. Nason and G. Ridolfi, Nucl. Phys. B **454** (1995) 3 [hep-ph/9506226].
12. B.W. Harris and J. Smith, Nucl. Phys. B **452** (1995) 109.
13. C. Peterson, D. Schlatter, I. Schmitt, and P.M. Zerwas, Phys. Rev. **D27** (1983) 105.
14. T. Klimkovich [H1 Collaboration], these Proceedings.
15. L. Finke [H1 Collaboration], these Proceedings.
16. A. Aktas *et al.* [H1 Collaboration], arXiv:hep-ex/0503038.
17. ZEUS Collaboration, contributed papers no. 783 and 784 to ICHEP2002, Amsterdam 2002.
18. Andrea Longhin [ZEUS Collaboration], these Proceedings.

Heavy quark production in conjunction with Z boson at DØ

Nirmalya Parua
On behalf of the DØ collaboration

State University of New York, Stony Brook, NY 11794-4433; USA

Abstract. We report on the measurement of the ratio of the inclusive cross sections for $Z+b$ jet to $Z+$ jets in $p\bar{p}$ collisions at the center of mass energy of 1.96 TeV. Dataset of 180 pb^{-1} luminosity collected by the DØdetector is used for the measurement. Decays of Z to both $\mu^+\mu^-$ and e^+e^- are studied. The ratio is measured to be 0.021 ± 0.005 for jets with $p_T > 20$ GeV/c within pseudorapidities of 2.5. This is consistent with the next-to-leading order predictions of the standard model. We also briefly report on the measurement of the ratio of the $Z/\gamma^*(\rightarrow e^+e^-)+ \geq n$ jet (where n = 1, 2, 3, 4, 5) to inclusive $Z/\gamma^*(\rightarrow e^+e^-)+$ jet cross sections using 343 pb^{-1} luminosity.

INTRODUCTION

Measurement of the inclusive $Z+b$ jet cross section is important at DØ, as it is expected to be a major background for $p\bar{p} \rightarrow ZH$, where the Higgs boson decays into bb. Also the study of the $Z+b$ jets can be used to probe b quark density in protons as the process $bg \rightarrow Zb$ contributes to two third of the cross section, and the initial b is from the proton parton distribution. Also as the b-quark density of proton influences the rate for single top quark production, and the final state hb, where h is supersymmetric Higgs boson, a clear understanding of $Z+b$ jet production is important.We present the result as the ratio of production cross sections of inclusive $Z+b$ jets to $Z+$ jets. This is beneficial for precise comparison with theory as many systematic uncertainties, like the 6.5 % uncertainty on the measurement of luminosity cancels out.
The DØdetector at the Fermilab Tevatron is a multipurpose detector comprising of central tracking system immersed in a 2T magnetic field, preshower detector, calorimeter and Muon detector [1]. The central tracking system comprises of Silicon Microstrip tracker (SMT) and central fiber tracker (CFT).The system has vertexing capabilities up to pseudorapidities $|\eta| < 3$, where $\eta = -ln(tan(\theta/2))$. θ is the polar angle with direction of proton as the positive Z axis. Covering the tracking system is the uranium-Liquid argon calorimeter. The calorimeter is housed in 3 separate cryostats, Central Cryostat (CC) covering $|\eta| < 1.1$ and two end cryostats (ECs) providing coverage up to $|\eta| < 4.2$. Preshower detector is interspersed between Calorimeter cryostat wall and the solenoid.The muon system comprises of 3 layers, one before the 1.8T Toroid and two after the toroid. Each layer comprises of drift tubes and scintillator detectors.

CP792, *Deep Inelastic Scattering, DIS 2005,* edited by W. H. Smith and S. R. Dasu

RESULTS

We first describe the study for $Z/\gamma^* + \geq n$ jets production. Events for this analysis are required to have good reconstructed vertex within $|Z| < 60$ cm. Two electrons that are reconstructed with simple cone algorithm, satisfy shower shape, isolation and electromagnetic fraction requirements and have $p_T > 25$ GeV/c within $|\eta| < 1.1$ are used to identify Z boson candidate events. Additionally at least one electron is required to have a matching track. Finally invariant mass of the electron pair is required to be within 75-105 GeV/c^2. In 343 pb^{-1} of data a total of 13,893 events passed these criteria. Jets are reconstructed using "Run II cone algorithm" [2] with cone size of 0.5. Jets are required to have $p_T > 20$ GeV/c and $|\eta| < 2.5$, and not to overlap with electrons from Z bosons within $\Delta R = \sqrt{(\Delta\eta)^2 + (\Delta\phi)^2} \leq 0.4$. The main source of background for this analysis comes from QCD events, and are estimated for each jet multiplicity separately. The cross sections for each jet multiplicity is corrected for the jet reconstruction, and identification efficiencies and event migration due to finite jet energy resolution, and are normalized with respect to the inclusive Z/γ^* cross section in the dielectron invariant mass region of 75-105 GeV/c^2. Figure 1 shows the measured cross section ratios for the $Z/\gamma^* + \geq n$ jets as a function of jet multiplicity. The measurement is compared with two QCD predictions 1) MCFM [3], a NLO calculation up to $Z+2$ parton process using CTEQ6M as the parton distribution function and the normalization scale set to $\mu^2_{F/R} = M^2_Z + p^2_{Tz}$, and 2) ME-PS theory based on MADGRAPH [4] $Z+n$ LO matrix element predictions using PYTHIA [5] for parton showering and hadronization. It is to be noted that both QCD predictions agree well with our data. $Z+b$ jet analysis is done

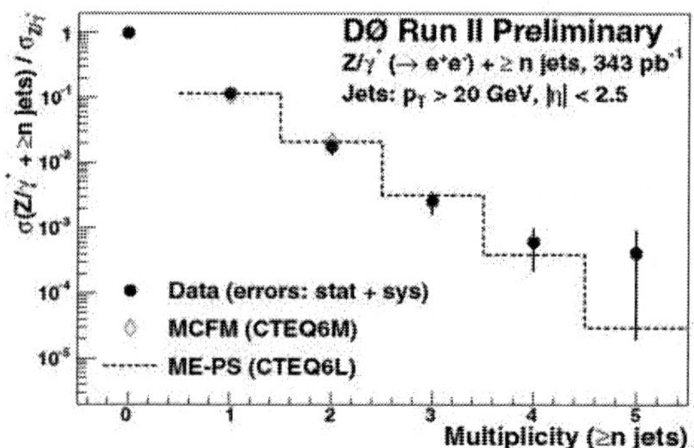

FIGURE 1. Ratio of $Z/\gamma^*(\rightarrow e^+e^-)+ \geq n$ jet cross sections to the inclusive $Z/\gamma^*(\rightarrow e^+e^-)$ cross section versus the jet multiplicity. Theoretical prediction of LO Matrix Element (ME) calculations using Pythia for parton shower is represented by the dotted line.The open diamond represents the MCFM predictions

for Z decaying into both dielectron and dimuon pair. The dielectron sample is selected by two electromagnetic energy clusters at the trigger level. In the offline the EM clusters are required to be isolated from hadronic activity, have shower shape consistent with that expected of an electron and have $p_T > 15$ GeV/c. In addition at least one of the EM cluster is required to have a matching track and the ratio of the energy estimated from the calorimeter to the ratio of the momentum estimated from the tracking detector

consistent with that expected of an electron. The Z candidates are selected by requiring invariant mass of the dielectron within 80-100 GeV/c^2. To select $Z+$ jet sample at least one hadronic jet, reconstructed with Run II cone algorithm with radius 0.5 and with $p_T > 20$ GeV/c is required.In order to reduce background from noise in the calorimeter, jets are required to be "taggable", meaning jets are required to be associated with track cluster, found by applying a $\Delta R < 0.5$ cone algorithm. Total number of $Z(\rightarrow ee)+$ jet candidate events are 1658. Principal background to this channel is from QCD multijet production where two jets mimic electrons. Background for this channel is estimated by side band technique to be 121 ± 4 events.
The dimuon sample is selected by requiring at least 1 muon candidate at the trigger level. In the offline, two oppositely charged muons with $p_T > 15$ GeV/c within $|\eta| < 2$, and isolated from hadronic activities are required. The Z candidates are selected by requiring the invariant mass of the dimuon to be within 65-115 GeV/c^2. Total number of $Z(\rightarrow \mu\mu)+$ taggable jet candidate events is 1406. Main background for this channel is from $b\bar{b}$ production where both b-jet contains muons that satisfy isolation requirement. Total number of background events for $Z(\rightarrow \mu\mu)+$ taggable jet channel is estimated to be 17.5 ± 4.1 events.
A b-jet tagging algorithm for secondary vertices is used to identify heavy quark jets. A jet is considered b-tagged when it is taggable and has at least one secondary vertex associated with it with decay lenghth significance, i.e, the ratio of the decay length in the plane transverse to the beam line (L_{xy}) to its uncertainty σ_{xy}, $L_{xy}/\sigma_{xy} > 7$. After applying b tagging 27 $Z(\rightarrow ee) + b-$jet and 22 $Z(\rightarrow \mu\mu) + b-$jet events remain in the data. Number of estimated background events are 4.2 ± 1.4 and 5.0 ± 1.1 respectively for dielectron and dimuon channels. Finally the ratio of $Z + b$ jet cross section to the Z +jet inclusive cross section is related to the number of events with b (N_b), c (N_c), and light quark jets (N_l) by the relation $\sigma(p\overline{p} \rightarrow Z + bjet)/\sigma(p\overline{p} \rightarrow Z + jet) = N_b/(N_b + N_c + N_l)$ After subtracting the background contributions we can write two equations, one before b-tagging, and the other after b-tagging.

$$N_{before\ b-tag} = t_b N_b + t_c N_c + t_l N_l \tag{1}$$

$$N_{b-tagged} = \varepsilon_b t_b N_b + \varepsilon_c t_c N_c + \varepsilon_l t_l N_l \tag{2}$$

where N_b, N_c, N_l are the number of events with b, c and light jets, respectively. t s and ε s are taggabilities and event-tagging efficiencies for respective jet types as indicated by subscripts.To solve equations (1) and (2) that have 3 unknown, we use theoretical prediction of $N_c = 1.69 N_b$. The ratio $\sigma(p\overline{p} \rightarrow Z + bjet)/\sigma(p\overline{p} \rightarrow Z + jet)$ is found to be 0.023 ± 0.007 for the dielectron channel and 0.019 ± 0.005 for the dimuon channel. Combining these two channels with appropriate statistical weight one obtains the ratio to be 0.021 ± 0.004, where the uncertainty is purely statistical. Figure 2 shows p_T the distribution of b-tagged jet, and the significance of decay length of secondary vertices. For comparison sum of the background and $Z + b$ expectation from MC is also shown. The distribution of decay length significance for secondary vertices shows clear evidence for heavy flavor component in the b-tagged candidate events.Several sources of systematic uncertainties on the measurement are considered. Total systematic uncertainty is estimated to be 10.4% upward and 11.8% downward, with dominant contributions coming from jet energy scale and background estimation. Folding all uncertainties together the ratio of the cross section is found to be $0.021 \pm 0.004(stat)^{+0.002}_{-0.003}(syst)$ This result is in good agreement with the next-to-leading order prediction of 0.018 ± 0.004 [7] using CTEQ6 parton distributions. In the future, with reduced uncertainties for both theoretical prediction and experimental measurement, the result can be used to put constraint on b-quark density of protons.

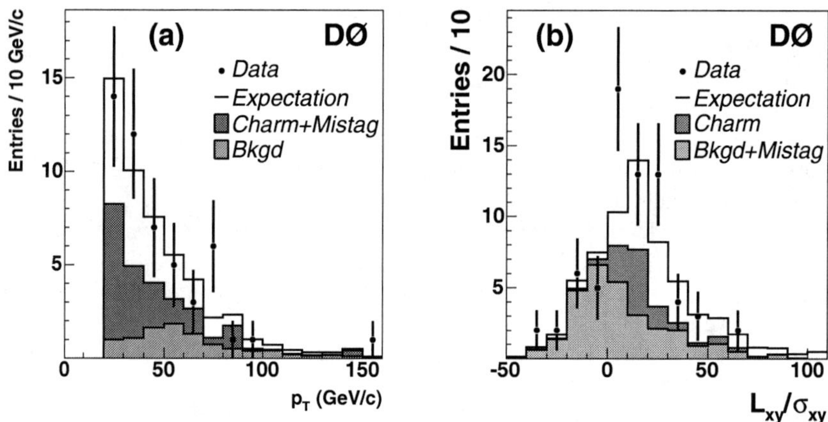

FIGURE 2. (a) p_T spectrum for b-tagged jets. (b) Distribution of decay length significance of secondary vertices in the plane transverse to the beam.

ACKNOWLEDGMENTS

The author would like to acknowledge the conference organizers for an enjoyable conference.

REFERENCES

1. V. Abazov *et al.*, in preparation for submission to *Nucl. Instru. Methods Phys. Res. A* **B18C** (1990) 125. S. Abachi *et al.*, Nucl. Instru. Meth. **A338**, 185 (1994).
2. G.C Blazey *et al.*, Proccedings of the workshop: QCD and weak Boson Physics in Run II, edited by U. Bauer, R.K. Ellis, and D.Zeppenfeld, Batavia, Illinois (2000)
3. J. Campbell and R. K. Ellis, Phys. Rev. D **65**, 113007 (2002)
4. F. Maltoni and T. Stelzer, JHEP **0302**, 027 (2003)
5. T. Sjostrand *et al.*, Comput. Phys. Commun. **135,** 238 (2001)
6. M. L. Mangano, M. Moretti, F. Piccinini, R. Pittau and a. Polosa, J. High Energy Phys. **0307**, 001 (2003)
7. J.M. Campbell, R.K. Ellis, F. Maltoni and S. Willenbrock, Phys. Rev. D **69**, 074021 (2004)

SPIN PHYSICS
WORKING GROUP PRESENTATIONS

Precise Results on $g_1^{\rm p}$ and $g_1^{\rm d}$ and First Measurement of the Tensor Structure Function $b_1^{\rm d}$ with the HERMES-Experiment

Davide Reggiani (on behalf of the HERMES Collaboration)

Physikalisches Institut, Universität Erlangen-Nürnberg, Erwin-Rommel-Str. 1, D-91058 Erlangen, Germany.
reggiani@mail.desy.de

Abstract. Precise results on the spin-dependent structure functions $g_1^{\rm p}(x,Q^2)$ and $g_1^{\rm d}(x,Q^2)$ over the kinematic range $0.0021 < x < 0.9$ and $0.1 < Q^2 < 20$ GeV^2 are presented. The analyses base on data collected at the HERMES experiment at DESY in polarized deep-inelastic scattering and make use of an unfolding algorithm in order to correct for radiation and detector smearing. Furthermore, final results of the first measurement of the tensor asymmetry A_{zz} and the tensor structure function $b_1^{\rm d}(x,Q^2)$ are provided.

Keywords: HERMES at DESY, spin-dependent structure functions, tensor structure function.
PACS: 13.60.-r, 13.88.+e, 14.20.Dh, 14.65.-q

At leading twist, polarized inclusive deep-inelastic scattering (DIS) of charged leptons off a proton (spin 1/2) is described by the spin-dependent structure function $g_1(x,Q^2)$. In case of a deuteron target (spin 1), the parametrization of the hadronic part of the interaction contains the additional structure function $b_1(x,Q^2)$ [1].

In the Quark-Parton Model the structure function g_1 describes the imbalance in the distribution of quarks with the same or opposite helicity with respect to that of the parent hadron and can be measured when both beam and target are polarized. The function g_1 is extracted through the ratio g_1/F_1, which is approximately equal to the virtual photon asymmetry A_1 that is measured via the longitudinal cross section asymmetry $A_{\|}$ [2].

The tensor structure function b_1 represents the difference in the quark distribution between the helicity-0 and the averaged non-zero helicity states of the deuteron. Its measurement does not require a polarized beam and it is obtained from the ratio b_1/F_1 determined via the tensor asymmetry A_{zz} [3]. As a nuclear-polarized deuteron target always carries a large tensor polarization, it is a priori not justifi ed to neglect the effect of the tensor asymmetry during the measurement of $g_1^{\rm d}$.

HERMES is a fi xed-target experiment exploiting the 27.6 GeV lepton beam of the HERA storage ring. The beam is transversely self-polarized, for the reported data reaching an average polarization of 0.53; longitudinal beam polarization in the interaction region is obtained by employing a pair of spin rotators. The HERMES polarized internal hydrogen/deuterium gas target [4] consists of an Atomic Beam Source (ABS) to generate and a Breit Rabi Polarimeter to measure nuclear polarization; a tubular open-ended, cooled storage cell is mounted inside the beam pipe confi ning the gas fed from the ABS. Gas targets offer signifi cant advantages over solid state targets for spin experiment as they provide a high degree of polarization ($|P_z| \simeq 0.85$ for most of the data considered in

CP792, *Deep Inelastic Scattering, DIS 2005,* edited by W. H. Smith and S. R. Dasu

the g_1 analyses), they are free of dilution and they can invert the spin of the atoms within milliseconds, thereby substantially reducing the systematic uncertainties of asymmetry measurements. Moreover, the HERMES deuterium ABS allowed the injection of a pure tensor polarized gas ($|P_{zz}| \simeq 0.83$ with $|P_z| \simeq 0.01$), making it possible to decouple the b_1 from the g_1 measurement.

HERMES employs a conventional forward magnetic spectrometer [5] including tracking chambers and particle identifi cation detectors. The achieved momentum resolution is of the order of 2%, while a very effi cient discrimination between electrons and hadrons, carried out by means of a likelihood method based on the combination of four detectors [6], brings the hadron contamination of the inclusive sample down to less than 1%.

The fi nal HERMES g_1 results base on data taken with longitudinally polarized hydrogen (1996-97) and deuterium (2000) targets (4 resp. $11.5 \cdot 10^6$ DIS events), the b_1 results on a separate data set with a tensor-polarized deuterium target ($2.4 \cdot 10^6$ DIS events).

During these analyses an unfolding method has been applied which corrects within one step both for the (quasi)-elastic radiative background and for the intra-bin migration caused by radiative DIS events and detector smearing. This procedure makes use of a Monte Carlo model in order to keep track of how events migrate from a Born bin j to an experimental bin i due to changes in their kinematics and describes this effect by means of a (jxi) matrix. The Born Asymmetry, retrieved by inverting this matrix and multiplying it by the experimental data, depends only on the experimentally measured asymmetry, on the known unpolarized cross sections and on the Monte Carlo models for background and detector behavior. Unfolding implies the elimination of the systematic correlations between different kinematic bins introducing at the same time statistical correlations between them. Therefore, the HERMES results presented here must always be considered along with their correlation matrices when performing QCD fi ts or moments calculations. The true statistical power of a data set can be seen from the uncertainties of the moments. More analysis details can be found in [7] and [8].

Fig. 1 shows the smearing-unfolded g_1/F_1 and xg_1 results and their statistical correlations. The kinematic range covers $0.0021 < x < 0.9$ and $0.1 < Q^2 < 20$ GeV^2. In the range $0.03 < x < 0.8$ the measured moment for $g_1^{\rm p}$ ($g_1^{\rm d}$) is $0.1174 \pm 0.0027_{\rm stat}$ ($0.0433 \pm 0.0013_{\rm stat}$). A compilation of world data on g_1/F_1 and xg_1 is shown in Fig. 2 [9, 10, 11].

The tensor asymmetry A_{zz}, as measured for the fi rst time by HERMES (in the range $0.01 < x < 0.45$ and $0.5 < Q^2 < 5$ GeV^2), is shown in the left panel of Fig. 3. A_{zz} is different from zero and of the order of 1%, implying the quadrupole contribution to the measurement of g_1 to be negligible. The right plot displays the structure functions $b_1^{\rm d}$ and $xb_1^{\rm d}$ along with the associated average Q^2. The shape of $b_1^{\rm d}$ is in qualitative agreement with coherent double scattering models predicting contributions at low x via the nuclear shadowing mechanism [12, 13, 14]. In the measured range the $b_1^{\rm d}$ moment is $1.05 \pm 0.34_{\rm stat} \pm 0.35_{\rm sys}$.

ACKNOWLEDGMENTS

I thank my HERMES colleagues, in particular Marco Contalbrigo, Lara De Nardo, Markus Ehrenfried and Caroline Riedl. I gratefully acknowledge the German Bundesministerium für Bildung und Forschung for the financial support.

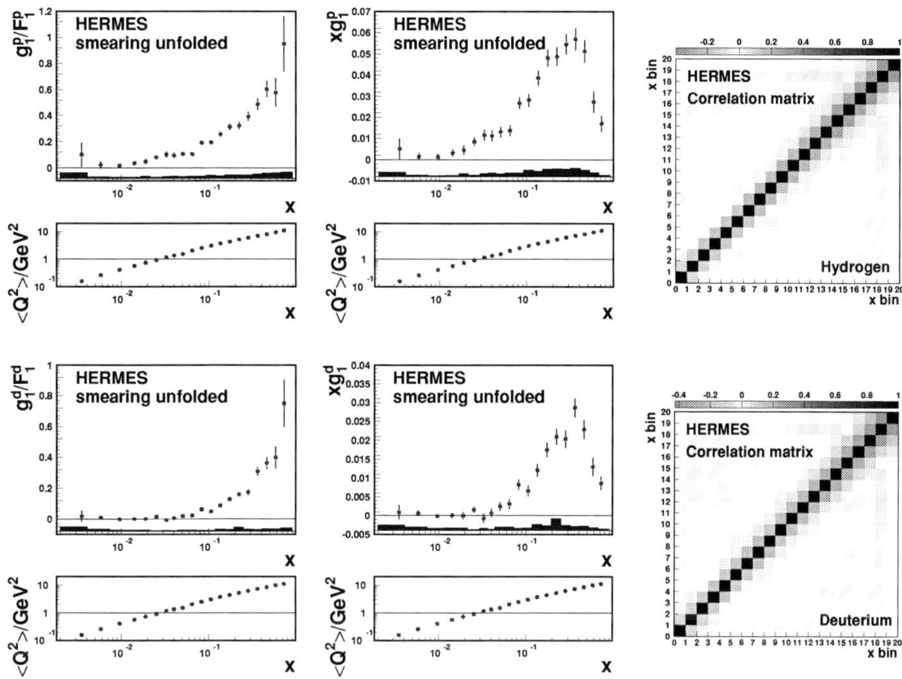

FIGURE 1. Smearing-unfolded g_1/F_1 (left) and xg_1 (middle) as measured by HERMES and the correlation matrix between statistical errors (right), for the proton (top row) and for the deuteron (bottom row). Error bars are statistical, displaying only the diagonal element of the correlation matrix, the shaded bands show the estimated systematic uncertainties

REFERENCES

1. P. Hoodbhoy *et al.*, Nucl. Phys. B**312** (1989) 571.
2. HERMES Collaboration, A. Airapetian *et al.*, Phys. Lett. B**442** (1998) 484.
3. M. Contalbrigo, for the HERMES Collaboration, *Proceedings from SPIN 2002, BNL, USA, Sep 9 - 14, 2002*, hep-ex/0211014
4. HERMES Collaboration, A. Airapetian *et al.*, Nucl. Instr. and Meth. A **540** (2005), 68, hep-ex/0408013.
5. HERMES Collaboration, K. Ackerstaff *et al.*, Nucl. Instr. A **417** (1998) 230.
6. HERMES Collaboration, A. Airapetian *et al.*, Phys. Rev. D **71** (2005) 012003, hep-ex/0407032.
7. HERMES Collaboration, publication on g_1 in preparation.
8. HERMES Collaboration, A. Airapetian *et al.*, First Measurement of the Tensor Structure Function of the Nucleon, submitted to *Phys. Rev. Lett.*, hep-ex/0506018.
9. SMC Collaboration, B. Adeva *et al.*, Phys. Rev. D**60** (1999) 072004.
10. E143 Collaboration, K. Abe *et al.*, Phys. Rev. D**58** (1998) 112003.
11. E155 Collaboration, P.L. Anthony *et al.*, Phys. Lett. B**463** (1999) 339.
12. N.N. Nikolaev and W. Schäfer, *Phys. Lett.* B **398** (1997) 245.
13. J. Edelmann, G. Piller and W. Weise, *Phys. Rev.* C **57** (1998) 3392.
14. K. Bora and R.L. Jaffe, *Phys. Rev.* D **57** (1998) 6906.

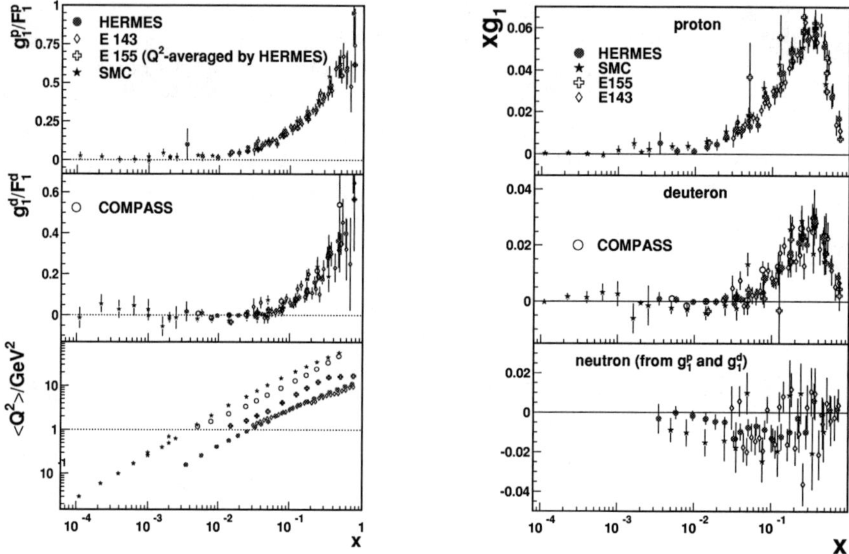

FIGURE 2. HERMES g_1/F_1 (left) and xg_1 data (right) in comparison to SLAC and CERN data for the proton (top panels) and the deuteron (middle panels) and xg_1 for the neutron, as extracted from p and d data (right bottom panel) at the measured Q^2 (left bottom). Error bars show the quadratic sum of statistical and systematic uncertainties.

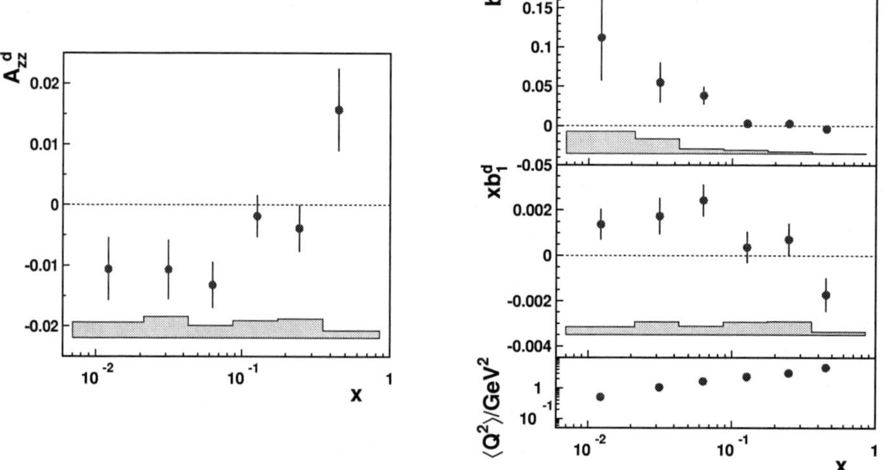

FIGURE 3. The HERMES tensor asymmetry A_{zz} (left) and tensor structure functions $b_1^{\rm d}$ and $xb_1^{\rm d}$ (right upper and middle panel respectively; lower panel: average Q^2 values). The error bars are statistical only, the shaded bands show the estimated systematic uncertainties.

Measurement of the spin structure of the deuteron at COMPASS

Jürgen Hannappel[*,†] and COMPASS collaboration[**]

[*]*Institut für Kernphysik, Johannes Gutenberg-Universität Mainz, Johann-Joachim-Becher-Weg 45, D-55099 Mainz*
[†]*CERN, CH-1211 Genève 23, Switzerland*
[**]`http://wwwcompass.cern.ch/compass/organisation/members.html`

Abstract. A new measurement of the longitudinal spin asymmetry A_1^d and the spin dependent structure function g_1^d of the deuteron is presented in the Q^2 range from $1\mathrm{GeV}^2$ to $100\mathrm{GeV}^2$ and the x range from 0.004 to 0.7. The data were taken in 2002 and 2003 with the COMPASS experiment at CERN, scattering 160 GeV^2 polarised muons off a large polarised ^{6}LiD target. While significantly improving statistical accuracy in the low x region the data agree nicely with previous experiments.

Keywords: Deep inelastic scattering; Structure functions
PACS: 13.60.Hb; 13.88.+e

INTRODUCTION

Most of the work presented here is described in more detail in [1]. First results on the deuteron spin asymmetry A_1^d and the spin dependent structure finction g_1^d are presented.

The data were taken with the COMPASS spectrometer, which is located on the M2 beam line of the SPS at CERN. A polarized muon beam (naturally polarized to $\approx -75\%$) of $160\mathrm{GeV}/c$ (the momentum of each individual muon is measured in the beamline) impinges on a ^{6}LiD polarized target, the reaction products are analyzed in a two stage spectrometer, as seen in Fig. 1 and described in more detail in [2].

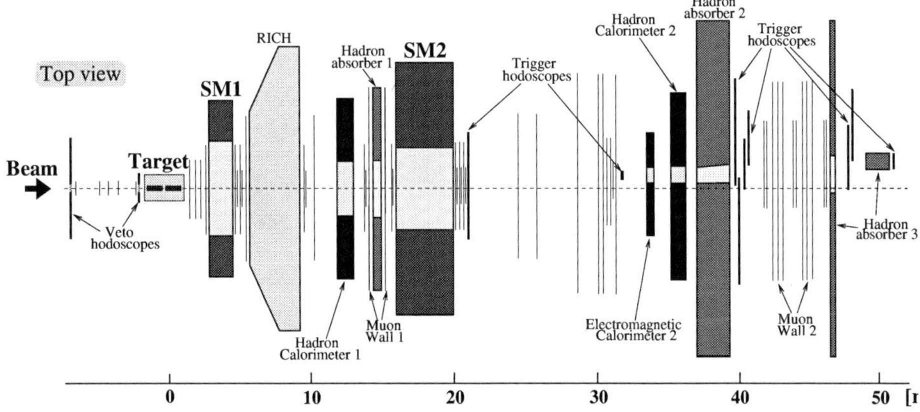

FIGURE 1. Layout of the COMPASS spectrometer in 2002 and 2003

CP792, *Deep Inelastic Scattering, DIS 2005,* edited by W. H. Smith and S. R. Dasu

The target consists of two cells of 60cm length each and 3cm diameter, separated by 10 cm. Each of the cells has its own microwave system for dynamic nuclear polarisation (DNP)[3], so that the two cells are polarized in opposite direction, reaching +54% and -50% of polarisation. Every 8 hours the spin directions in both cells are reversed by rotating the magnetic field direction. This is done to ensure that the variations of flux and acceptance cancel out in the asymmetry calculations. This method is carried further by reversing the sign of the polarisation in each cell several times a year by changing the DNP microwave frequencies.

Data recording is triggered by the detection of the scattered muons. This is done by several sets of hodoscopes, which cover different kinematical ranges. For the large Q^2 range this inclusive triggering is sufficient, but for smaller Q^2 it was necessary to have also triggers that also take into account the muon energy loss in the target and additionally require some hadronic energy to be detected in the hadron calorimeters, while the very large Q^2 range is covered by a trigger on a hadron signal in the calorimeters alone. For more details see [4].

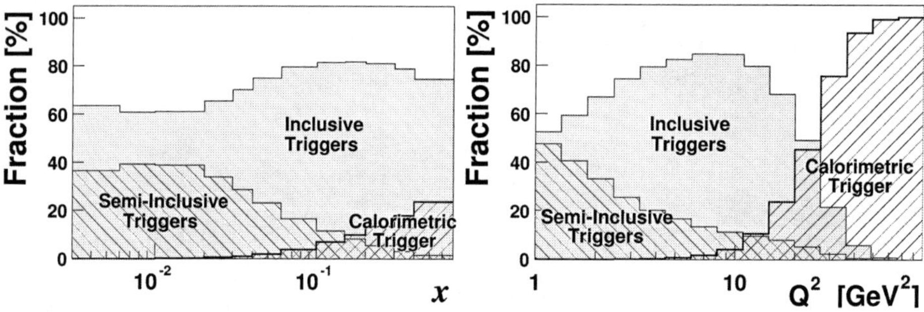

FIGURE 2. Fraction of inclusive, semi-inclusive and calorimetric triggers in the data sample as a function of x (left) and Q^2 (right)

Fig. 2 shows the different contributions of the different kinds of trigger to the data sample. Monte Carlo studies [5] show that the selection of hadronic events due to the semi-inclusive triggering method produces no large effects on the extracted A_1, so the asymmetries were calculated for both hadronic and purely inclusive data sets separately and afterward merged.

Since the measured number of events N_i for the different beam and target spin combinations is related to the asymmetry A_1^d by

$$N_i = a_i \phi_i n_i \bar{\sigma}(1 + P_T P_B f D A_1^d)$$

with a_i being the acceptance for configuration i, ϕ_i the muon flux, n_i the number of target nucleons, P_T the target polarisation, P_B the beam polarisation, f the dilution factor and D the depolarisation factor, we weight each event with $P_B f D$, each of which calculated from the event kinematics. This weighting allows us to decrease the statistical error by 10%.

To go from the asymmetry A_1^d to the structure function g_1^d we use their relation through the structure function F_2^d, which is well known from fixed target experiments [5], and the cross section ratio $R = \sigma_L/\sigma_T$, also from fixed target data [6],[7].

RESULTS

Here the combined results from the beam times in the years 2002 and 2003 are presented, which correspond to integrated luminosities of about 600 pb$^-1$ and 900 pb$^-1$.

The kinematic range covers photon virtualities in the range $1\text{GeV}^2 < Q^2 < 100\text{GeV}^2$, while the fractional energy of the virtual photon was limited to $0.1 < y < 0.9$.

Fig. 3 shows both the asymmetry A_1^d and the structure function g_1^d, the inset on the left showing more clearly the increase of precision reached in the low x region.

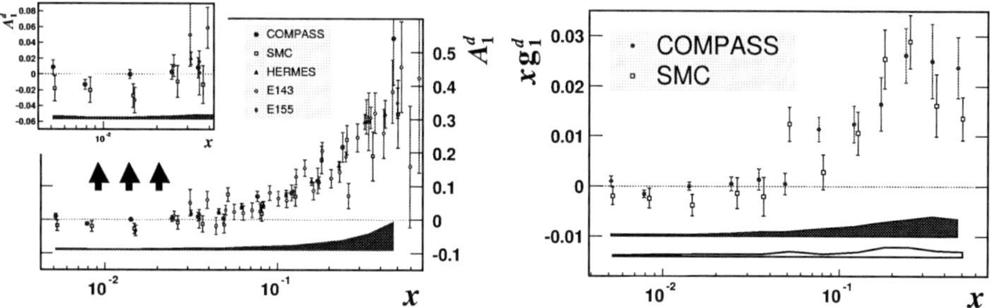

FIGURE 3. Inclusive asymmetry A_1^d (left) and structure function g_1^d (right). For g_1^d the data points are shown at the measured Q^2.

A QCD fit to the world data ([8], [5], [1], [9], [10], [11], [12], [13], [14], [15], [16]) was made to study the impact of the new COMPASS data on our understanding of the nucleon structure. In this fit (which uses the program "2", as it was called in the SMC notation) while DGLAP equations are solved via numerical integration a NLO calculation in the $\overline{\text{MS}}$ scheme is performed. $\Delta\Sigma$, Δq_3,Δq_8 and ΔG are parametrized as

$$\Delta f = \frac{\eta}{\int_0^1 x^\alpha (1-x)^\beta (1+\gamma x) \mathrm{d}x} x^\alpha (1-x)^\beta (1+\gamma x)$$

and then a MINUIT minimization of the deviation between measured and calculated g_1 is performed.

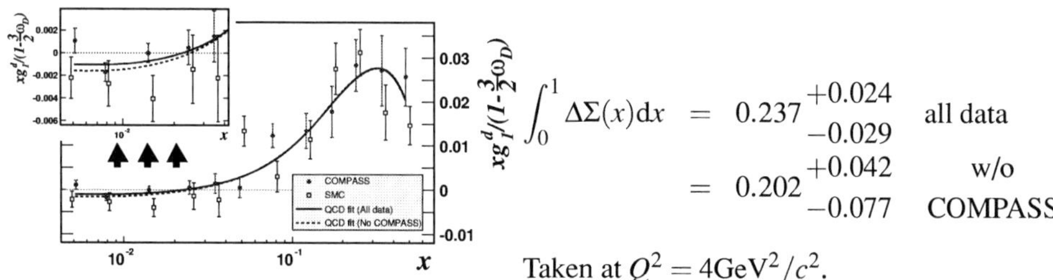

$$\int_0^1 \Delta\Sigma(x)\mathrm{d}x = 0.237 {}^{+0.024}_{-0.029} \quad \text{all data}$$
$$= 0.202 {}^{+0.042}_{-0.077} \quad \text{w/o COMPASS}$$

Taken at $Q^2 = 4\text{GeV}^2/c^2$.

FIGURE 4. QCD fit to word data, shown at measured Q^2

As the improved precision of data due to the COMPASS points at low x changes the curve in that region, there is also a significant change in the $\int_0^1 \Delta\Sigma(x)\mathrm{d}x$ integral, the error of which is also decreased by a factor of 2.

COMPASS being a full spectrometer with a large acceptance it is not only able to measure the scattered muon, but also to identify and measure the other reaction products. Thus asymmetries were also calculated for events in which the leading hadron has positive charge and for negative charge, represented in Fig. 5

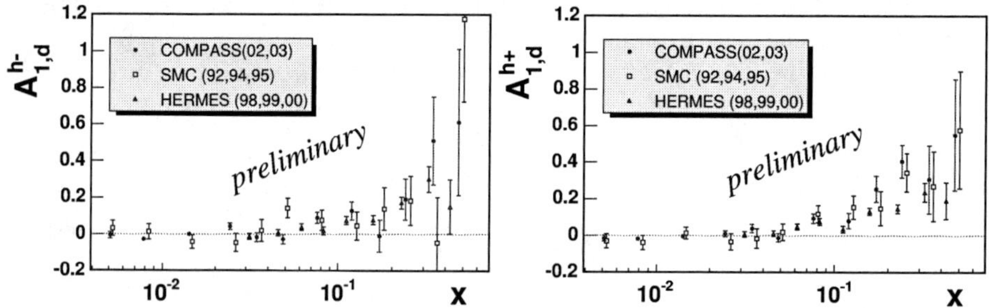

FIGURE 5. semi-inclusive asymmetries, left for negative leading hadrons, right for positive

ACKNOWLEDGMENTS

We gratefully acknowledge the support of the CERN management and staff and the skill and effort of the technicians of our collaborating institutes. Special thanks are due to V. Anosov, J.M. Demolis and V. Pesaro for their technical support during the installation and the running of this experiment. This work was made possible by the financial support of our funding agencies.

REFERENCES

1. COMPASS Collaboration, *Phys. Lett. B*, **612** (2005), 154–164.
2. G.K. Mallot, *Nucl. Instrum. Meth. A*, **518** (2004), 121
3. J. Ball *et al.,Nucl. Instrum. Meth. A* **498** (2003), 101.
4. J. Hannappel *et al., Nucl. Instrum. Meth. A*, submitted for publication
5. B. Adeva *et al., Phys. Rev. D* **58** (1998), 112001
6. NMC, M. Arneodo *et al., Nucl. Phys. B* **483** (1997) 3
7. L. Whitlow *et al., Phys. Lett. B* **250** (1990) 193
   K. Abe *et al., Phys. Lett. b* **452** (1999) 194
8. EMC, *Nucl. Phys. B* **328** (1989) 1
9. E143, *Phys. Rev. D* **58** (1998) 112003
10. E155, *Phys. Lett. B* **463** (1999) 339
11. E155, *Phys. Lett. B* **493** (2000) 19
12. JLAB, *Phys. Rev. Lett.* **92** (2004) 012004
13. E142, *Phys. Rev. D* **54** (1996) 6620
14. E154, *Phys. Rev. Lett.* **79** (1997) 26
15. HERMES, *Phys. Lett. B* **404** (1997) 383
16. HERMES, *Phys. Rev. D* **75** (2005) 012003

Extraction of polarized parton densities from polarized DIS and SIDIS.

D. de Florian*, G. A Navarro* and R. Sassot*

*Departamento de Física, Universidad de Buenos Aires, Ciudad Universitaria, Pab.1 (1428) Buenos Aires, Argentina

Abstract. We present results on the quark and gluon polarization in the nucleon obtained in a combined next to leading order analysis to the available inclusive and semi-inclusive polarized deep inelastic scattering data.

Keywords: polarized PDF, DIS, SIDIS
PACS: 12.38.Bx,13.85.Ni

INTRODUCTION

For more than fifteen years, polarized inclusive deep inelastic scattering (pDIS) has been the main source of information on how the individual partons in the nucleon are polarized at very short distances. Many alternative experiments have been conceived to improve this situation. The most mature among them are those based on polarized semi-inclusive deep inelastic scattering (pSIDIS).

In the following we present results obtained in a combined next to leading order analysis to the recently updated set of pDIS and pSIDIS data [1]. Specifically, we focused on the extraction of sea quark and gluon densities, analyzing the constraining power of the data on the individual densities. As result, we found not only a complete agreement between pDIS and pSIDIS data, but a very useful complementarity, leading to rather well constrained densities.

Using the Lagrange multiplier approach [2], we explored the profile of the χ^2 function against different degrees of polarization in each parton flavor. In this way we obtained estimates for the uncertainty in the net polarization of each flavor, and in the parameters of the pPDFs. We compared results obtained with the two most recent sets of fragmentation functions. The differences are found to be within conservative estimates for the uncertainties. Nevertheless, there is a clear preference for a given set of FF over the other, shown in a difference of several units in the χ^2 of the respective global fits. In NLO global fits the overall agreement between theory and the full set of data is sensibly higher than in LO case.

GLOBAL FIT

In our analysis [1], we followed the same conventions and definitions for the polarized inclusive asymmetries and parton densities adopted in references [3, 4], however we used more recent inputs, such as unpolarized parton densities [5] and the respective values for

CP792, *Deep Inelastic Scattering, DIS 2005,* edited by W. H. Smith and S. R. Dasu

TABLE 1. Inclusive and semi-inclusive data used in the fit.

Collaboration	Target	Final state	# points	Refs.
EMC	proton	inclusive	10	[8]
SMC	proton, deuteron	inclusive	12, 12	[8]
E-143	proton, deuteron	inclusive	82, 82	[8]
E-155	proton, deuteron	inclusive	24, 24	[8]
Hermes	proton,deuteron,helium	inclusive	9, 9, 9	[8]
E-142	helium	inclusive	8	[8]
E-154	helium	inclusive	17	[8]
Hall A	helium	inclusive	3	[8]
COMPASS	deuteron	inclusive	12	[8]
SMC	proton,deuteron	h^+, h^-	24, 24	[8]
Hermes	proton, deuteron, helium	$h^+, h^-, \pi^+, \pi^-, K^+, K^-, K^T$	36,63,18	[8]
	Total		478	

α_s. Fragmentation functions were taken from either [6] or [7], respectively. We also used the flavor symmetry and flavor separation criteria proposed in [6], at the respective initial scales Q_i^2. The data sets analyzed include only points with $Q^2 > 1$ GeV^2, listed in Table 1, and totaling 478 data points.

In Table 2, we summarize the results of the best NLO and LO global fits to all the data listed in Table 1. We present fits obtained using alternatively fragmentation functions from reference [6], labeled as KRE, and from reference [7], labeled as KKP. Since the fit involves 20 parameters, the number of degrees of freedom for these fits is 478-20=458. Consequently, the χ^2 values obtained are excellent for NLO fits and very good for LO. The better agreement between theory and experiment found at NLO, highlights the importance of the corresponding QCD corrections, for the present level of accuracy achieved by the data.

In NLO fits there seems to be better agreement when using KRE fragmentation functions. The difference between the total χ^2 values between KRE and KKP NLO fits comes mainly from the contributions related to pSIDIS data, while those associated to inclusive data are almost the same, as one should expect in a fully consistent scenario.

Table 2 includes also the first moment of each flavor distribution at $Q^2 = 10$ GeV^2, and that for the singlet distribution $\delta\Sigma$, as reference. Most noticeably, while the KRE NLO fit favors the idea of a SU(3) symmetric sea, KKP NLO finds $\overline{u}$ polarized opposite to $\overline{d}$ and to $\overline{s}$. Gluon and strange sea quark polarization are similar in both fits and the total polarization carried by quarks is found to be around 30%.

TABLE 2. χ^2 values and first moments for distributions at $Q^2 = 10$ GeV^2

set		χ^2	χ^2_{DIS}	χ^2_{SIDIS}	$\delta\overline{u}$	$\delta\overline{d}$	$\delta\overline{s}$	δg	$\delta\Sigma$
NLO	KRE	430.91	206.01	224.90	-0.0487	-0.0545	-0.0508	0.680	0.284
	KKP	436.17	205.66	230.51	0.0866	-0.107	-0.0454	0.574	0.311
LO	KRE	457.54	213.48	244.06	-0.0136	-0.0432	-0.0415	0.121	0.252
	KKP	448.71	219.72	228.99	0.0497	-0.0608	-0.0365	0.187	0.271

UNCERTAINTIES

Many strategies have been implemented in order to assess the uncertainties in PDFs and their propagation to observables, specially those associated with experimental errors in the data. The Lagrange multiplier method [2] probes the uncertainty in any observable or quantity of interest relating the range of variation of one or more physical observables dependent upon PDFs to the variation in the χ^2 used to judge the goodness of the fit to data. In Figure 1 we show the outcome of varying the χ^2 of the NLO fits to data against

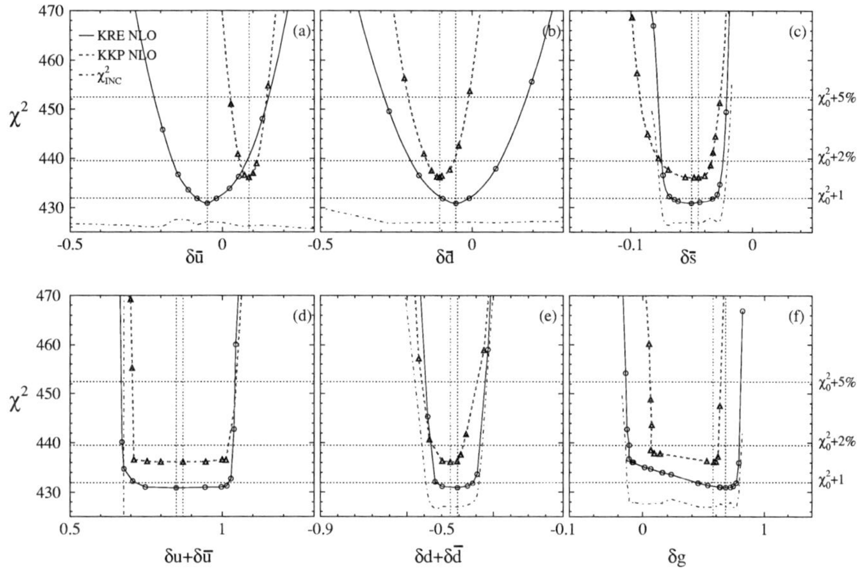

FIGURE 1. χ^2 profiles for NLO fits obtained using Lagrange multipliers at $Q^2 = 10$ GeV^2.

the first moment of the respective polarized parton densities δq at $Q^2 = 10$ GeV^2, one at a time. This is, to minimize

$$\Phi(\lambda_q, a_j) = \chi^2(a_j) + \lambda_q\,\delta q(a_j) \qquad q = u, \bar{u}, d, \bar{d}, s, g. \tag{1}$$

In order to see the effect of the variation in χ^2 on the parton distributions themselves, in Figure 2, we show KRE best fit densities together with the uncertainty bands corresponding to $\Delta\chi^2 = 1$ (darker band) and $\Delta\chi^2 = 2\%$ (light shaded band). As expected, the relative uncertainties in the total quark densities and those strange quarks are rather small. For gluon densities the $\Delta\chi^2 = 1$ band is also small, but the most conservative $\Delta\chi^2 = 2\%$ estimate is much more significative. For light sea quarks the $\Delta\chi^2 = 1$ bands are moderate but the $\Delta\chi^2 = 2\%$ are much more larger.

Two programmed experiments, the one based on the PHENIX detector already running at RHIC [9], and the E04-113 experiment at JLab [10] will be able to reduce dra-

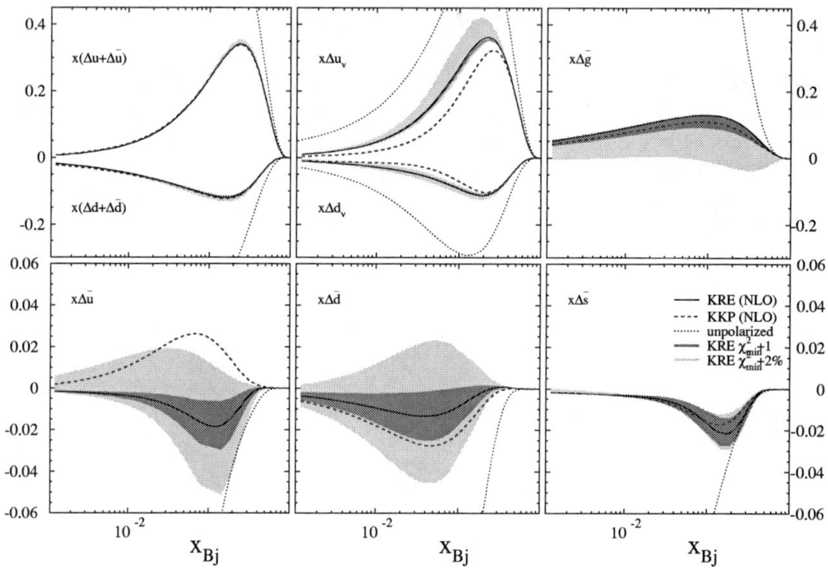

FIGURE 2. Parton densities at $Q^2 = 10$ GeV^2, and the uncertainty bands corresponding to $\Delta\chi^2 = 1$ and $\Delta\chi^2 = 2\%$

matically the uncertainty in both the gluon and the light sea quark densities respectively, the latter providing also an even more stringent test on fragmentation functions.

ACKNOWLEDGMENTS

Partially supported by CONICET, Fundación Antorchas, UBACYT and ANPCyT, Argentina.

REFERENCES

1. D. de Florian, G. A. Navarro and R. Sassot, *Phys. Rev. D* **71**, 094018 (2005).
2. J. Pumplin, D. R. Stump and W. K. Tung, *Phys. Rev. D* **65**, 014011 (2002), D. Stump *et al.*, *Phys. Rev. D* **65**, 014012 (2002).
3. D. de Florian, R. Sassot, *Phys. Rev. D* **62** 094025 (2000).
4. G. A. Navarro and R. Sassot, *Eur. Phys. J. C* **28**, 321 (2003).
5. A. D. Martin, R. G. Roberts, W. J. Stirling and R. S. Thorne, *Eur. Phys. J. C* **28** 455 (2003).
6. S. Kretzer, *Phys. Rev. D* **62** (2000) 054001.
7. B. A. Kniehl, G. Kramer and B. Potter, *Nucl. Phys. B* **582** (2000) 514
8. See [1] and references therein.
9. Y. Fukao [PHENIX Collaboration], hep-ex/0501049.
10. X. Jiang *et al.* hep-ex/0412010.

Synthesis of DGLAP and total resummation of leading logarithms for the non-singlet spin structure function g_1

B.I. Ermolaev, M. Greco[*] and S.I. Troyan[†]

[*]Dipartimento di Fisica and INFN, University Rome III, Rome, Italy
[†]St.Petersburg Institute of Nuclear Physics, 188300 Gatchina, Russia

Abstract. The explicit expressions for the non-singlet DIS structure function g_1 at small x are obtained by resumming the leading logarithmic contributions. The role played by the fits for the initial parton densities currently in the DGLAP on the small-x behavior of the non-singlet g_1 is discussed. Explicit expressions combining DGLAP with our results are presented.

Keywords: deep-inelastic scattering, spin structure functions
PACS: 12.38.Cy

INTRODUCTION

The non-singlet component of the spin structure function g_1 have been investigated in great detail in deep inelastic scattering (DIS) experiments. The standard theoretical framework for studying the DIS structure functions is provided by DGLAP[1]. In this approach, $g_1^{NS}(x,Q^2)$ can be represented as a convolution of the coefficient functions and the evolved quark distributions. Combining these results with appropriate fits for the initial quark distributions, provides a good agreement with the available experimental data.

However, the DGLAP evolution eqs. were originally applied in a range of large x values, where higher-loop contributions to the coefficient functions and the anomalous dimensions are small. Such corrections are becoming essential when x is decreasing, so DGLAP should not work so well at $x \ll 1$. Nevertheless, DGLAP predictions are in a good agreement with available experimental data. It leads to the conclusion that the impact of the higher-order corrections is negligibly small for the available values of x. Below we use our results [2] to show that the impact of the high-order corrections on the Q^2 and x -evolutions of the non-singlet structure functions is quite sizable and bounds the region of strict applicability of DGLAP to $x > 10^{-2}$. We also show that the reason for the success of DGLAP at $x < 10^{-2}$ is related to the sharp x -dependence assumed for the initial parton densities, which is able to mimic the role of high-order corrections.

The paper is organized as follows: In Sect. 2 we discuss the difference of our approach with DGLAP. Then we compare our and the DGLAP formulae for asymptotics of g_1. In Sect. 3 we suggest a method to combine DGLAP with our approach in order to obtain equally correct expressions for both large and small values of x. Sect. 4 contains our conclusions.

CP792, *Deep Inelastic Scattering, DIS 2005*, edited by W. H. Smith and S. R. Dasu

COMPARISON OF DGLAP AND OUR APPROACH

As the DGLAP -expressions for the non-singlet structure functions are well-known, we discuss them briefly only. In this approach, $g_{1\,DGLAP}^{NS}(x,Q^2)$ can be represented as a convolution

$$g_{1DGLAP}^{NS}(x,Q^2) = \int_x^1 \frac{dy}{y} C(x/y) \Delta q(y,Q^2) \tag{1}$$

of the coefficient functions $C(x)$ and the evolved quark distributions $\Delta q(x,Q^2)$. Similarly, $d\Delta q(x,Q^2)/d\ln(Q^2)$ can be expressed through the convolution of the splitting functions and the initial quark densities $\delta q(x \approx 1, Q^2 \approx \mu^2)$ where μ^2 is the starting point of the Q^2 -evolution. It is convenient to represent $f(x,Q^2)$ in the integral form, using the Mellin transform:

$$g_{1\,DGLAP}^{NS}(x,Q^2) = (e_q^2/2) \int_{-\imath\infty}^{\imath\infty} \frac{d\omega}{2\imath\pi} (1/x)^{\omega} C(\omega) \delta q(\omega) \exp\left[\gamma(\omega) \int_{\mu^2}^{Q^2} \frac{dk_\perp^2}{k_\perp^2} \alpha_s(k_\perp^2)\right] \tag{2}$$

where $C(\omega)$ are the non-singlet coefficient functions, $\gamma(\omega)$ the non-singlet anomalous dimensions and $\delta q(\omega)$ the Mellin transforms of the initial non-singlet quark densities. The standard DGLAP fits $\delta q(x)$ for the non-singlet parton densities (see e.g. Refs. [3, 4]) consist of the terms singular when $x \to 0$ and the regular in x part. For example, the fit A of Ref.[3] is chosen as follows:

$$\begin{aligned} \delta q(x) &= N\eta x^{-\alpha} \phi(x) , \\ \phi(x) &\equiv (1-x)^{\beta}(1+\gamma x^{\delta}) , \end{aligned} \tag{3}$$

with N, η being the normalization, $\alpha = 0.576$, $\beta = 2.67$, $\gamma = 34.36$ and $\delta = 0.75$. As the first term $N\eta x^{-\alpha}$ in the rhs of Eq. (3) is singular when $x \to 0$ whereas the second one, $\phi(x)$ is regular, we will address them as the singular and regular parts of the fit respectively. Obviously, in the ω -space Eq. (3) is a sum of pole contributions:

$$\delta q(\omega) = N\eta \left[(\omega-\alpha)^{-1} + \sum_{k=1}^{\infty} m_k \Big((\omega+k-\alpha)^{-1} + \gamma(\omega+k+1-\alpha)^{-1} \Big) \right] , \tag{4}$$

with $m_k = \beta(\beta-1)..(\beta-k+1)/k!$, so that the first term in Eq. (4) (the leading pole) corresponds to the singular term $x^{-\alpha}$ of Eq. (3) and the second term, i.e. the sum of the poles, corresponds to the interference between the singular and regular terms. In contrast to the leading pole position $\omega = \alpha$, all other poles in Eq. (4) have negative values because $k - \alpha > 0$. An alternative approach was used in Refs. [2], by introducing and solving infrared evolution equations with fixed α_s. This approach was improved in Refs. [2], where single-logarithmic contributions were also accounted for and the QCD coupling was running in the Feynman graphs contributing to the non-singlet structure functions. In contrast to the DGLAP parametrization $\alpha_s = \alpha_s(k_\perp^2)$, we used in Refs. [2] another parametrization where the argument of α_s in the quark ladders is given by the time-like virtualities of the intermediate gluons. Refs. [2] suggest the following formulae for the non-singlet structure functions:

$$g_1^{NS}(x,Q^2) = (e_q^2/2)\int_{-\imath\infty}^{\imath\infty}\frac{d\omega}{2\pi\imath}(1/x)^{\omega}C_{NS}(\omega)\delta q(\omega)\exp\left(H_{NS}(\omega)y\right), \quad (5)$$

with $y=\ln(Q^2/\mu^2)$ so that μ^2 is the starting point of the Q^2 -evolution. The new coefficient function C_{NS} is expressed in terms of new anomalous dimensions H_{NS} and the new anomalous dimensions H_{NS} account for the resummation of the double- and single- logarithmic contributions (see Ref. [2] for details).

COMPARISON OF DGLAP AND OUR SMALL-X ASYMPTOTICS

When $x \to 0$, one can use the saddle point method in order to estimate the integrals in Eq. (5) and derive much simpler expressions for the non-singlet structure functions:

$$g_1^{NS} \sim e_q^2\delta q(\omega_0)\xi^{\omega_0}, \quad (6)$$

with $\xi = \sqrt{Q^2/(x^2\mu^2)}$ and with the intercept $\omega_0 = 0.42$. Eq. (6) predicts the asymptotic scaling for the non-singlet structure functions: Asymptotically, g_1^{NS} depends on one argument ξ.

When the standard DGLAP fits, e.g. the fit of Eq. (3), are used, the asymptotics of $g^{NS}_{1\,DGLAP}(x,Q^2)$ is also the Regge-like:

$$g^{NS}_{1\,DGLAP} \sim (e_q^2/2)C(\alpha)(1/x)^{\alpha}\left((\ln(Q^2/\Lambda^2))/(\ln(\mu^2/\Lambda^2))\right)^{\gamma(\alpha)/b}, \quad (7)$$

with $b = (33-2n_f)/12\pi$.

Comparison of Eq. (6)and Eq. (7) demonstrates that both DGLAP and our approach lead to the Regge asymptotic behavior in x. However, it is important that our intercept ω_0 is obtained by the total resummation of the leading logarithmic contributions and without assuming singular fits for δq whereas the DGLAP intercept α in Eq. (7) is generated by the phenomenological factor $x^{-0.57}$ of Eq. (3) which mimics the total resummation. In other words, the impact of the higher-loop radiative corrections on the small-x behavior of the non-singlets is, actually, incorporated into DGLAP phenomenologically, through the fits. It means that the singular factors can be dropped from such fits when the coefficient function includes the total resummation of the leading logarithms and therefore fits for δq can be chosen as regular functions of x in this case.

COMBINING DGLAP WITH OUR HIGHER-LOOP CONTRIBUTIONS

Eq. (5) accounts for the resummation of the double- and single logarithmic contributions to the non-singlet anomalous dimensions and the coefficient functions that are leading when x is small. However, the method we have used does not allow us to account for other contributions which can be neglected for x small but become quite important when x is not far from 1. On the other hand, such contributions are naturally included in

DGLAP, where the non-singlet coefficient function C_{DGLAP} and anomalous dimension γ_{DGLAP} are known with the two-loop accuracy:

$$C_{DGLAP} = 1 + \frac{\alpha_s(Q^2)}{2\pi} C^{(1)}, \tag{8}$$
$$\gamma_{DGLAP} = \frac{\alpha_s(Q^2)}{4\pi}\gamma^{(0)} + \Big(\frac{\alpha_s(Q^2)}{4\pi}\Big)^2 \gamma^{(1)}$$

Therefore, we can borrow from the DGLAP formulae the contributions which are missing in Eq. (5) by adding C_{DGLAP} and γ_{DGLAP} to the coefficient function and anomalous dimension of Eq. (5). It is important to avoid a double counting DL and SL terms common for these expressions.

In order to do so, let us consider the region of $x \sim 1$ where the effective values of ω in Eqs. (2,5) are large. In this region we can expand H_{NS} and C_{NS} into a series in $1/\omega$. Retaining the first two terms in each series, we arrive at $C_{NS} = \widetilde{C}_{NS} + O(\alpha_s^2)$, $H_{NS} = \widetilde{H}_{NS} + O(\alpha_s^3)$, with (see Ref. [2] for details)

$$\widetilde{C}_{NS} = 1 + \frac{A(\omega)C_F}{2\pi}\left[1/\omega^2 + 1/2\omega\right], \tag{9}$$
$$\widetilde{H}_{NS} = \frac{A(\omega)C_F}{4\pi}\left[2/\omega + 1\right] + \Big(\frac{A(\omega)C_F}{4\pi}\Big)^2 (1/\omega)\left[2/\omega + 1\right]^2 + D[1/\omega + 1/2].$$

Now let us define the new coefficient functions $\hat{C}_{NS}$ and new anomalous dimensions $\hat{C}_{NS}$ as follows:

$$\hat{H}_{NS} = \left[H_{NS} - \widetilde{H}_{NS}\right] + \frac{A(\omega)}{4\pi}\gamma^{(0)} + \Big(\frac{A(\omega)}{4\pi}\Big)^2 \gamma^{(1)}, \tag{10}$$
$$\hat{C}_{NS} = \left[C_{NS}^{(\pm)} - \widetilde{C}_{NS}\right] + 1 + \frac{A(\omega)}{2\pi} C^{(1)}.$$

These new, "synthetic" coefficient functions and anomalous dimensions of Eq. (10) include both the total resummation of the leading contributions and the DGLAP expressions in which $\alpha_s(Q^2)$ is replaced by $A(\omega)$ defined in Refs. [2]. The main point is that the factorization of the phase space into transverse and longitudinal spaces used in DGLAP is a good approximation for large x only.

REFERENCES

1. G. Altarelli and G. Parisi, *Nucl. Phys.*, **B126** (1977) 297; V.N. Gribov and L.N. Lipatov, *Sov. J. Nucl. Phys.* textbf15 (1972) 438; L.N.Lipatov, *Sov. J. Nucl. Phys.* **20** (1972) 95; Yu.L. Dokshitzer, *Sov. Phys. JETP* **46** (1977) 641.
2. B.I. Ermolaev, M. Greco and S.I. Troyan, *Nucl.Phys.* **B594**B (2001)71; ibid **B571**(2000)137; *Phys.Lett.* **B579**, 321,(2004); hep-ph/0503019.
3. G. Altarelli, R.D. Ball, S. Forte and G. Ridolfi, *Nucl. Phys.* **B496** (1997) 337; *Acta Phys. Polon.* **B29**(1998)1145;
4. A. Vogt. hep-ph/0408244.

Polarized Structure Functions from Lattice QCD

G. Schierholz

John von Neumann-Institut für Computing NIC,
Deutsches Elektronen-Synchrotron DESY, D-15738 Zeuthen
and
Deutsches Elektronen-Synchrotron DESY, D-22607 Hamburg

Abstract. In this talk I present recent lattice results on polarized structure functions obtained by the QCDSF Collaboration.

Keywords: Lattice QCD, structure functions, higher twist, angular momentum
PACS: 11.15.Ha, 12.38.Gc, 13.60.Hb

INTRODUCTION

Structure functions probe how hadrons are made up from quarks and gluons. The basis for theoretical investigation is the operator product expansion (OPE), which connects moments of structure functions with hadronic matrix elements of local operators. A complete theoretical understanding of the underlying dynamics of quarks and gluons thus requires the calculation of an appropriate set of matrix elements in QCD. This is a nonperturbative problem, and lattice QCD holds the tools to solve it [1].

Polarized structure functions are of particular interest, ever since it was discovered that only a small fraction of the spin of the nucleon is carried by the spin of the quarks. They contain a wealth of information on the distribution of spin and transversity in the fast moving nucleon, and their derivation provides a challenge, both experimentally and theoretically.

In this talk I will concentrate on the axial and tensor charge of the nucleon, on the nucleon's second spin dependent structure function g_2, as well as on the orbital angular momentum of the quarks.

The lattice simulations are done with $N_f = 2$ flavors of light dynamical quarks. To reduce cut-off effects, we use non-perturbatively $O(a)$ improved Wilson fermions. We work on lattices as large as $24^3\,48$ and lattice spacings as small as 0.07 fm. The operators are renormalized non-perturbatively as well throughout this talk.

AXIAL AND TENSOR CHARGE

The nucleon's tensor charge g_T measures the net number of transversely polarized valence quarks in the transversely polarized nucleon, while the axial charge g_A measures the number of longitudinally polarized valence quarks in the longitudinally polarized nucleon. One could argue that the two charges should be the same by rotational invariance. This would be the case if the nucleon was made of free quarks. However, in the

CP792, *Deep Inelastic Scattering, DIS 2005,* edited by W. H. Smith and S. R. Dasu

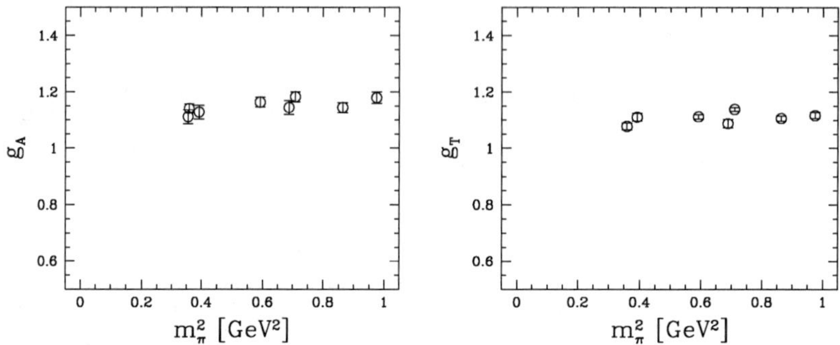

FIGURE 1. The axial and tensor charge of the nucleon.

infinite momentum frame rotational invariance is highly nontrivial and the rotation operators involve interactions. Thus, the difference of axial and tensor charges tells us about the interactions of quarks in the fast moving nucleon. In Fig. 1 I plot the axial and tensor charge of the nucleon as a function of the pion mass squared. The tensor charge refers to the $\overline{MS}$ scheme at 4 GeV^2. We find little difference between g_A and g_T.

$G_2(X, Q^2)$ AND HIGHER TWIST

The nucleon's second spin dependent structure function $g_2(x, Q^2)$ is of considerable phenomenological interest because at leading order in Q^2 it receives contributions from both twist-2 and twist-3 operators. Here we shall be interested in the second moment of g_2 only, and in particular in its twist-3 contribution d_2:

$$d_2 = \int_0^1 dx x^2 g_2(x, Q^2) + \frac{2}{3} \int_0^1 dx x^2 g_1(x, Q^2) . \tag{1}$$

In Fig. 2 I plot d_2 as a function of the pion mass squared for proton and neutron target. The lattice data involve different lattice spacings. For an analysis lattice spacing by spacing and an attempt of a continuum extrapolation see [2]. In the chiral limit d_2 turns out to be consistent with zero, both for proton and nucleon. For the twist-3 contribution to the first moment we find $d_1^q = (2m_q/m_N)\delta q$ (m_q being the quark mass), which vanishes in the chiral limit as well. This suggest that [3]

$$g_2(x, Q^2) = \int_x^1 \frac{dy}{y} g_1(y, Q^2) - g_1(x, Q^2) . \tag{2}$$

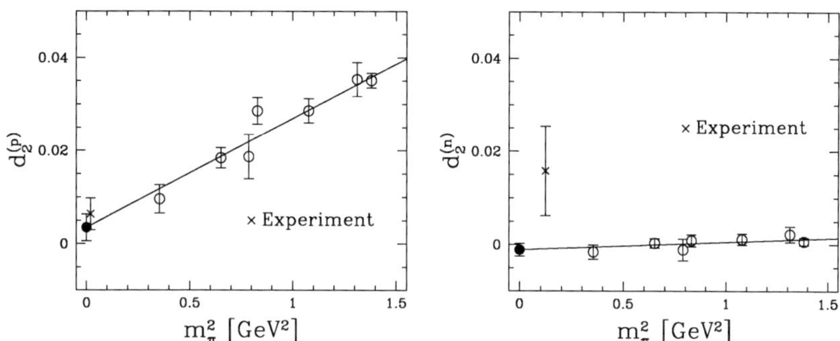

FIGURE 2. The twist-3 contribution to the second moment of g_2 in the $\overline{MS}$ scheme at 5 GeV^2.

ORBITAL ANGULAR MOMENTUM

The spin of the nucleon decomposes into the following contributions:

$$\frac{1}{2} = \frac{1}{2}\Delta\Sigma + \Delta G + L^q + L^g , \tag{3}$$

where $\Delta\Sigma$ (ΔG) is the quark (gluon) spin contribution and L_q (L_g) the contribution of the orbital angular momentum of the quarks (gluon). The angular momentum $J^q = L^q + \Delta q/2$, $\sum_q \Delta q = \Delta\Sigma$, can be computed from the nucleon matrix of the energy-momentum tensor:

$$\begin{aligned} \frac{\mathrm{i}}{2}\langle p'|\bar{q}\gamma_{\{\mu}\overleftrightarrow{D}_{\nu\}}q|p\rangle &= A_2^q(\Delta^2)\,\bar{u}(p')\gamma_{\{\mu}\bar{p}_{\nu\}}u(p) - B_2^q(\Delta^2)\,\frac{\mathrm{i}}{2m_N}\bar{u}(p')\Delta^\alpha\sigma_{\alpha\{\mu}\bar{p}_{\nu\}}u(p) \\ &+ C_2^q(\Delta^2)\,\frac{1}{m_N}\bar{u}(p')u(p)\Delta_{\{\mu}\Delta_{\nu\}} , \qquad (4) \\ J^q &= \frac{1}{2}\left(A_2^q(0) + B_2^q(0)\right) , \qquad (5) \end{aligned}$$

where $\bar{p} = (p+p')/2$ and $\Delta = p' - p$. In Fig. 3 I show the generalized form factors (GFFs) A_2, B_2 and C_2 together with a dipole fit, and in Fig. 4 I show the GFFs extrapolated to $\Delta^2 = 0$, from which we can read off the total angular momentum J. All numbers given refer to valence quarks. If we subtract the contribution of Δq, which is known from an independent calculation [4], we obtain

$$L^{u+d} = 0.03(7) \quad L^{u-d} = -0.45(6) . \tag{6}$$

While the total contribution of u and d quarks appears to be consistent with zero, this is not the case for the individual contributions.

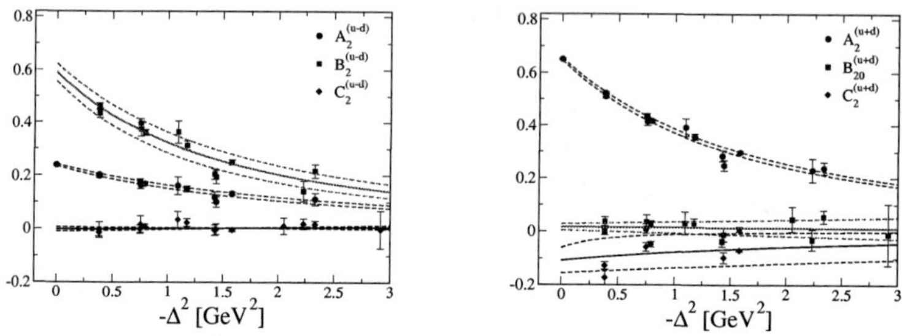

FIGURE 3. The generalized form factors A_2, B_2 and C_2 on the $24^3 48$ lattice at $\beta = 5.4$, $\kappa = 0.135$.

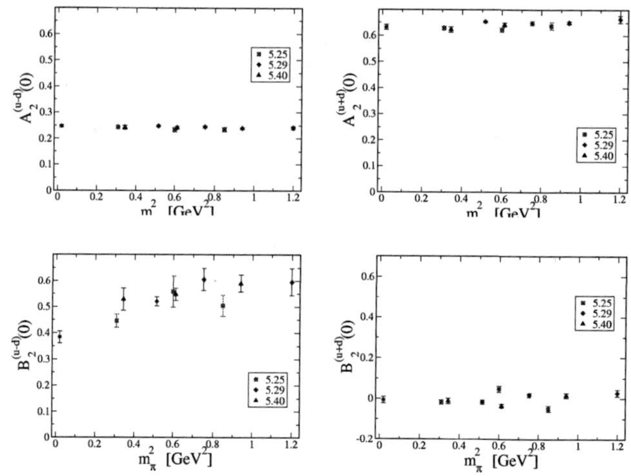

FIGURE 4. The generized form factors A_2 and B_2 extrapolated to $\Delta^2 = 0$.

SUMMARY

Due to space limitations I could show only a selection of results. For more details I refer the interested reader to recent talks and publications of the QCDSF collaboration.

REFERENCES

1. M. Göckeler *et al.*, Phys. Rev. D **53**, 2317 (1996).
2. M. Göckeler *et al.*, arXiv:hep-lat/0506017.
3. S. Wandzura and F. Wilczek, Phys. Lett. B **72**, 195 (1977).
4. G. Schierholz, arXiv:hep-ph/0411191.

Transversity measurements at HERMES

Markus Diefenthaler (on behalf of the HERMES collaboration)

Physikalisches Institut II, Friedrich-Alexander-Universität Erlangen-Nürnberg, Erwin-Rommel-Straße 1, 91058 Erlangen, Germany

Abstract. Azimuthal single-spin asymmetries (SSA) in semi-inclusive electroproduction of charged pions in deep-inelastic scattering (DIS) of positrons on a transversely polarised hydrogen target are presented. Azimuthal moments for both the Collins and the Sivers mechanism are extracted. In addition the subleading-twist contribution due to the transverse spin component from SSA on a longitudinally polarised hydrogen target is evaluated.

Keywords: transversity distribution, azimuthal single-spin asymmetries (SSA), Collins mechanism, Sivers mechanism, subleading-twist effects in SSA on a longitudinally polarised target
PACS: 13.60.-r,13.88.+e,14.20.Dh,14.65.-q

Recently the HERMES collaboration published first evidence for azimuthal single-spin asymmetries (SSA) in the semi-inclusive production of charged pions on a transversely polarised target [1]. Significant signals for both the Collins and Sivers mechanisms were observed in data recorded during the 2002–2003 running period of the HERMES experiment. Below we present a preliminary analysis of these data combined with additional data taken in the years 2003 and 2004. All data was recorded at a beam energy of 27.6GeV using a transversely nuclear-polarised hydrogen-target internal to the HERA positron storage ring at DESY.

At leading twist, the momentum and spin of the quarks inside the nucleon are described by three parton distribution functions: the well-known momentum distribution $q\left(x,Q^2\right)$, the known helicity distribution $\Delta q\left(x,Q^2\right)$ [2] and the unknown *transversity distribution* $\delta q\left(x,Q^2\right)$ [3, 4, 5, 6]. In the helicity basis, transversity is related to a quark-nucleon forward scattering amplitude involving helicity flip of both nucleon and quark $(N^{\Rightarrow}q^{\leftarrow} \rightarrow N^{\Leftarrow}q^{\rightarrow})$. As it is chiral-odd, transversity cannot be probed in inclusive measurements. At HERMES transversity in conjunction with the chiral-odd Collins fragmentation function [7] is accessible in SSA in semi-inclusive DIS on a transversely polarised target (*Collins mechanism*). The Collins fragmentation function describes the correlation between the transverse polarisation of the struck quark and the transverse momentum $P_{\mathrm{h}\perp}$ of the produced hadron. As it is also odd under naive time reversal (T-odd) it can produce a SSA, i.e. a left-right asymmetry in the momentum distribution of the produced hadrons in the directions transverse to the nucleon spin [8].

The *Sivers mechanism* can also cause a SSA: The T-odd Sivers distribution function [9] describes the correlation between the transverse polarisation of the nucleon and the transverse momentum k_T of the quarks within. A non-zero Sivers mechanism provides a non-zero Compton amplitude involving nucleon helicity flip without quark helicity flip $(N^{\Rightarrow}q^{\leftarrow} \rightarrow N^{\Leftarrow}q^{\leftarrow})$, which must therefore involve orbital angular momentum of the quark inside the nucleon [8, 10].

With a transversely polarised target, the azimuthal angle ϕ_S of the target spin direction

CP792, *Deep Inelastic Scattering, DIS 2005*, edited by W. H. Smith and S. R. Dasu

in the "⇑" state is observable in addition to the azimuthal angle ϕ of the detected hadron. Both azimuthal angles are defined with respect to the lepton scattering plane. The additional degree of freedom ϕ_S, not available with a longitudinally polarised target, results in distinctive signatures: $\sin(\phi+\phi_S)$ for the Collins mechanisms and $\sin(\phi-\phi_S)$ for the Sivers mechanism [11]. Therefore, for all detected charged pions and for each bin in x, z or $P_{h\perp}$ the cross section asymmetry for unpolarised beam (U) and transversely polarised target (T) was determined in the two dimensions ϕ and ϕ_S:

$$A_{UT}^{\pi^\pm}(\phi,\phi_S) = \frac{1}{|P_z|} \frac{N_{\pi^\pm}^{\Uparrow}(\phi,\phi_S) - N_{\pi^\pm}^{\Downarrow}(\phi,\phi_S)}{N_{\pi^\pm}^{\Uparrow}(\phi,\phi_S) - N_{\pi^\pm}^{\Downarrow}(\phi,\phi_S)}.$$

Here $N_{\pi^\pm}^{\Uparrow(\Downarrow)}(\phi,\phi_S)$ represents the semi-inclusive normalised yield in the target spin state "$\Uparrow$ ($\Downarrow$)", and $|P_z| = 0.754 \pm 0.050$ denotes the average degree of the target polarisation.

To avoid cross-contamination, the azimuthal moments for the Collins mechanism $\langle\sin(\phi+\phi_S)\rangle_{UT}^{\pi^\pm}$ and the Sivers mechanism $\langle\sin(\phi-\phi_S)\rangle_{UT}^{\pi^\pm}$ were extracted simultaneously. Recent studies showed that the terms for $\sin\phi_S$ and $\sin(2\phi-\phi_S)$ have to be added in the two-dimensional fit for the asymmetry (the kinematic factors $A(\langle x\rangle,\langle y\rangle)$ and $B(\langle y\rangle)$ are defined in [1]):

$$\begin{aligned} A_{UT}^{\pi^\pm}(\phi,\phi_S) \;=\; & 2\langle\sin(\phi+\phi_S)\rangle_{UT}^{\pi^\pm} \frac{B(\langle y\rangle)}{A(\langle x\rangle,\langle y\rangle)} \sin(\phi+\phi_S) + \\ & 2\langle\sin(\phi-\phi_S)\rangle_{UT}^{\pi^\pm} \sin(\phi-\phi_S) + \\ & 2\langle\sin(2\phi-\phi_S)\rangle_{UT}^{\pi^\pm} \sin(2\phi-\phi_S) + 2\langle\sin\phi_S\rangle_{UT}^{\pi^\pm} \sin\phi_S. \end{aligned}$$

The virtual-photon Collins and Sivers moments as a function of x, z and $P_{h\perp}$ are plotted in figure 1 (see caption for the systematic uncertainties). In addition the simulated fraction of charged pions originating from diffractive vector meson production and decay is shown, to estimate the possible contribution from the poorly known asymmetry of this process. The average values of the kinematic variables in the experimental acceptance are $\langle x\rangle = 0.10$, $\langle y\rangle = 0.53$, $\langle Q^2\rangle = 2.43\,\text{GeV}^2$, $\langle z\rangle = 0.36$, $\langle P_{h\perp}\rangle = 0.40\,\text{GeV}$.

This preliminary result is based on nearly five times more statistics than that in the publication [1] and is consistent with the published result: The average Collins moment is positive for π^+ and negative for π^-. Also, the magnitude of the π^- moment appears to be not smaller than the one for π^+. The averaged Sivers moment is significantly positive for π^+ and implies a non-vanishing orbital angular momentum of the quarks inside the nucleon. For π^- the averaged Sivers moment is consistent with zero.

The extracted Collins and Sivers moments allow the evaluation of the subleading-twist contribution to the previously measured SSA on a longitudinally polarised hydrogen target [13, 14, 15]. These subleading moments $\langle\sin\phi\rangle_{UL}^{q}$ are due to the longitudinal component of the target spin along the virtual photon direction ("q"). As shown in figure 2 these moments are almost the same as the previously published moments $\langle\sin\phi\rangle_{UL}^{l}$, where the longitudinal axis was defined along the lepton beam momentum ("l"). However, the maximum contribution of these subleading longitudinal asymmetries to the leading-twist Collins and Sivers moments in figure 1 is 0.004, which is negligible compared to the statistical uncertainty.

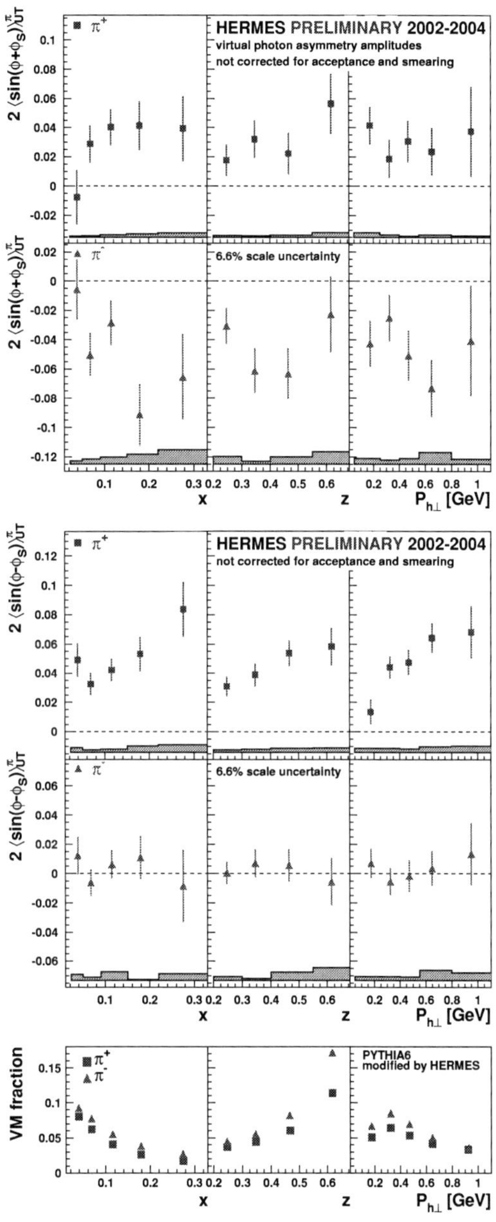

FIGURE 1. Collins moments (upper panel) and Sivers moments (middle panel) for charged pions (as labelled) as a function of x, z and $P_{h\perp}$, multiplied by two to have the possible range ± 1. The error bands represent the maximal systematic uncertainty due to acceptance and detector smearing effects and due to a possible contribution from the $\langle \cos\phi \rangle_{UU}$ moment in the spin-independent cross section. The common overall 6.6% scaling uncertainty is due to the target polarisation uncertainty. The lower panel shows the fraction of charged pions produced in vector meson decay simulated by PHYTHIA6 [12] (tuned for HERMES kinematics).

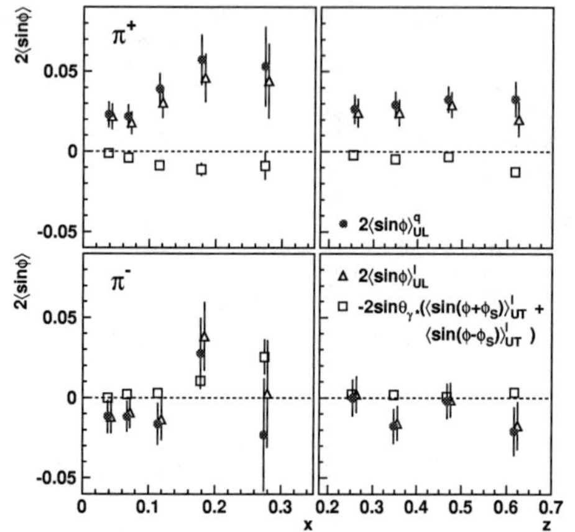

FIGURE 2. The azimuthal moment $\langle \sin \phi \rangle^q_{\mathrm{UL}}$ (•, multiplied by two) shows the subleading-twist contribution to the measured asymmetries on a longitudinally polarised hydrogen target for charged pions as a function of x and z. In addition the measured lepton-axis azimuthal moments are plotted (△ and □). There is an overall systematic error of 0.003. The superscript "q" and "l" distinguishes between moments with respect to the photon-axis and lepton-axis taking into account that the measured asymmetries contain contributions from both transverse and longitudinal polarisation components with respect to the virtual photon direction [15].

ACKNOWLEDGMENTS

This work has been supported by the German Bundesministerium für Bildung und Forschung (BMBF) (contract nr. 06 ER 125I) and the European Community-Research Infrastructure Activity under the FP6 "Structuring the European Research Area" program (HadronPhysics I3, contract nr. RII3-CT-2004-506078).

REFERENCES

1. A. Airapetian, et al., *Physical Review Letters*, **94**, 012002 (2005).
2. A. Airapetian, et al., *Physical Review*, **D71**, 012003 (2005).
3. J. P. Ralston, and D. E. Soper, *Nuclear Physics*, **B152**, 109 (1979).
4. X. Artru, and M. Mekhfi , *Z. Phys.*, **C45**, 669 (1990).
5. R. L. Jaffe, and X.-D. Ji, *Physical Review Letters*, **67**, 552–555 (1991).
6. J. L. Cortes, B. Pire, and J. P. Ralston, *Z. Phys.*, **C55**, 409–416 (1992).
7. J. C. Collins, *Nuclear Physics*, **B396**, 161–182 (1993).
8. M. Burkardt, *Physical Review*, **D69**, 057501 (2004).
9. D. W. Sivers, *Physical Review*, **D41**, 83 (1990).
10. S. J. Brodsky, D. S. Hwang, and I. Schmidt, *Physics Letters*, **B530**, 99–107 (2002).
11. D. Boer, and P. J. Mulders, *Physical Review*, **D57**, 5780–5786 (1998).
12. T. Sjöstrand, et al., *Comput. Phys. Commun.*, **135**, 238–259 (2001).
13. A. Airapetian, et al., *Physical Review Letters*, **84**, 4047–4051 (2000).
14. M. Diehl, and S. Sapeta (2005), `hep-ph/0503023`.
15. A. Airapetian, et al. (2005), accepted by Physics Letters **B**, `hep-ex/0505042`.

Collins and Sivers asymmetries on the deuteron from the COMPASS data

Paolo Pagano
(on behalf of the COMPASS collaboration)

INFN - Sezione di Trieste
via Valerio, 2 - 34127 Trieste (I)

Abstract. COMPASS is a fixed target experiment presently running at CERN. In 2002, 2003, and 2004 it used a 160 GeV polarized muon beam coming from SPS and scattered off a ^{6}LiD (deuteron) target. The nucleons in the target can be polarized either longitudinally or transversely with respect to the muon beam and 20% of the running time has been devoted to transverse polarization. Hereby the final results for the Collins and the Sivers asymmetries calculated from the data taken in transverse polarization in 2002 are presented. In the forthcoming 2006 run, COMPASS plans to run with a NH_3 (proton) target. Projections for the statistical accuracy which will be ultimately achieved on both the proton and the deuteron asymmetries are also given.

Keywords: polarized DIS, transversity, SSA
PACS: 13.60.-r, 13.88.+e, 14.20.Dh, 14.65.-q

THE THEORETICAL FRAMEWORK

The cross-section for polarized deep inelastic scattering[1] of leptons off spin 1/2 hadrons can be expressed, at the leading twist, as a function of three independent quark distribution function: $q(x)$, $\Delta q(x)$ and $\Delta_T q(x)$. The latter is chiral-odd and can be measured in nucleon - (anti) nucleon hard scattering or in semi-inclusive deep inelastic scattering (SIDIS). In SIDIS $\Delta_T q(x)$ can be probed in combination with the Collins fragmentation function, $\Delta D_a^h(z, p_T^h)$, chiral-odd as well, via azimuthal single spin asymmetries (SSA) in the hadronic end-product[2]. A similar effect can arise from the dependence of the nucleon structure on the intrinsic quark transverse momentum k_T[3]; such an effect is described by the so called Sivers distribution function, $\Delta_0^T q(x, k_T)$.

Leptoproduction on transversely polarized nucleons is a favourable setting to disentangle the Collins and Sivers effects since they are function of linearly independent kinematic variables.

According to Collins, the fragmentation function of a quark of flavour a in a hadron h can be written as[4]:

$$D_a^h(z, \mathbf{p_T^h}) = D_a^h(z, p_T^h) + \Delta D_a^h(z, p_T^h) \cdot sin\Phi_C$$

where $\mathbf{p_T^h}$ is the final hadron transverse momentum with respect to the virtual photon direction and $z = E_h/(E_l - E_{l'})$ is the fraction of available energy carried by the hadron (E_h is the hadron energy, E_l is the incoming lepton energy and $E_{l'}$ is the scattered lepton energy). The angle appearing in the fragmentation function, known as "Collins angle" and noted as Φ_C, is conveniently defined in the system where the z-axis is the

CP792, *Deep Inelastic Scattering, DIS 2005*, edited by W. H. Smith and S. R. Dasu

virtual photon direction and the x-z plane is the muon scattering plane. In this frame $\Phi_C = \Phi_h - \Phi'_s$, where Φ_h is the hadron azimuthal angle, and Φ'_s is the azimuthal angle of the transverse spin of the struck quark. Since $\Phi'_s = \pi - \Phi_s$, with Φ_s the azimuthal angle of the transverse spin of the initial quark (nucleon), the relation $\Phi_C = \Phi_h + \Phi_s - \pi$ is also valid. The fragmentation function $\Delta D^h_a(z, p^h_T)$ couples to transverse spin distribution function $\Delta_T q(x)$ and gives rise to SSA (denoted as A_{Coll}) dependent on x, z and p^h_T kinematic variables.

Following the Sivers hypothesis, the difference in the probability of finding an unpolarised quark of transverse momentum $\mathbf{k_T}$ and $-\mathbf{k_T}$ inside a polarised nucleon can be written as [5]:

$$\mathrm{P}_{q/p^\uparrow}(x, \mathbf{k_T}) - \mathrm{P}_{q/p^\uparrow}(x, -\mathbf{k_T}) = \sin\Phi_S \, \Delta_0^T q(x, k_T^2)$$

where $\Phi_S = \Phi_k - \Phi_s$ is the azimuthal angle of the quark with respect to the nucleon transverse spin orientation. It has been demonstrated by theoretical arguments [6, 7], that SSA (denoted as A_{Siv}) coming from the coupling of the Sivers function with the un-polarised fragmentation function $D^h_a(z, p^h_T)$ can be observed at the leading twist from polarised Semi-Inclusive DIS.

PHYSICS RESULTS AND PERSPECTIVES

The COMPASS [8, 9] experiment makes use of a high energy (160 GeV), intense (10^8/spill), muon beam naturally polarised by the $\pi-$ decay mechanism. The polarized target consists of two ^{6}LiD cells, each 60 cm long, located along the beam one after the other in two separate RF cavities. Data are taken simultaneously on the two target cells which are oppositely polarized.

Hereby we discuss the analysis of the data collected in year 2002 with target polarization oriented transversely to the beam direction. This sample, about 200 pb^{-1} in integrated luminosity and divided in two separate periods, is about 20% of the total beam time. Within each period, after 4–5 days of data taking, a polarization reversal was performed by changing the RF frequencies in the two cells. The full description of the analysis and results can be found in [10].

Events were selected in which a primary vertex (with identified beam and scattered muon) was found in one of the two target cells with a least one outgoing hadron. A clean separation of muon and hadron samples was achieved by cuts on the amount of material traversed in the spectrometer. In addition, the kinematic cuts $Q^2 > 1(GeV/c)^2$, $W > 5$ GeV/c^2 and $0.1 < y < 0.9$ were applied to the data to ensure a deep-inelastic sample above the region of the resonances and within the COMPASS trigger acceptance. The upper bound on y also serves to keep radiative corrections small. SSA have been looked for both the leading hadron in the event, and for all the hadrons.

In transverse polarisation, one can write the number of events as follows:

$$N(\Phi_{C/S}) = \alpha(\Phi_{C/S}) \cdot N_0 (1 + \varepsilon_{C/S} \sin\Phi_{C/S}),$$

where ε is the amplitude of the experimental asymmetry and α is a function containing the apparatus acceptance. The former amplitude can be expressed as a function of the

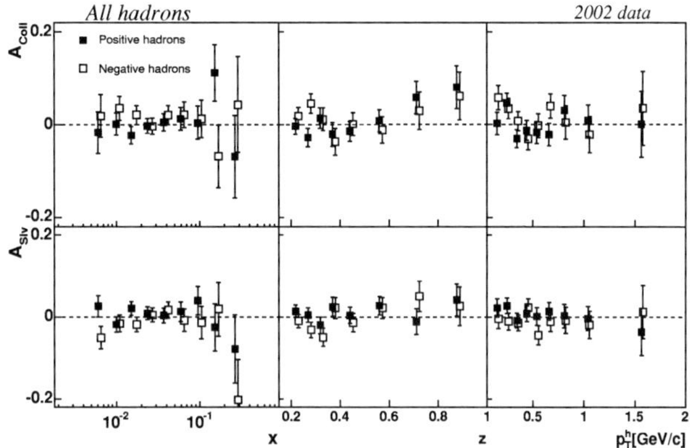

FIGURE 1. Collins and Sivers asymmetry for positive (full points) and negative (open points) hadrons as a function of x, z and p_T^h.

Collins and Sivers asymmetries through the expressions:

$$\varepsilon_C = A_{Coll} \cdot P_T \cdot f \cdot D_{NN} \qquad\qquad \varepsilon_S = A_{Siv} \cdot P_T \cdot f,$$

where P_T ($\simeq 0.45$) is the polarisation of the target, D_{NN} is the spin transfer coefficient, and f ($\simeq 0.40$) is the target dilution factor. To eliminate systematic effects due to acceptance, in each period the asymmetry ε_C (ε_S) is fitted separately for the two target cells from the event flux with the two target orientations using the expression:

$$\varepsilon_{C/S} \sin\Phi_{C/S} = \frac{N_h^{\uparrow}(\Phi_{C/S}) - R \cdot N_h^{\downarrow}(\Phi_{C/S} + \pi)}{N_h^{\uparrow}(\Phi_{C/S}) + R \cdot N_h^{\downarrow}(\Phi_{C/S} + \pi)}$$

where $R = N_{h,tot}^{\uparrow}/N_{h,tot}^{\downarrow}$ is the ratio of the total number of events in the two target polarisation orientations. The results of the asymmetries plotted against the kinematic variables x, z and p_T^h are shown in Fig. 1 for positive (full points) and negative (open points) hadrons. Similar results have been obtained selecting only the leading hadrons from the reactions.

Possible sources for systematic errors have been deeply investigated as described in [10]: the conclusion from those studies is that systematic errors affecting the measurements are smaller than the quoted statistical errors.

Within the accuracy of the measurements, both the Collins and Sivers asymmetries turned out to be small and compatible with zero, with a marginal indication of a Collins effect at large z for both positive and negative charges.

Concerning the Sivers asymmetries, a recent phenomenological work[11] has demonstrated that HERMES results for protons[12] and COMPASS results do not contradict each other.

In 2003 and 2004 COMPASS has been collecting data on a transversely polarized deuteron target with the same share with longitudinal configuration as for 2002. Due

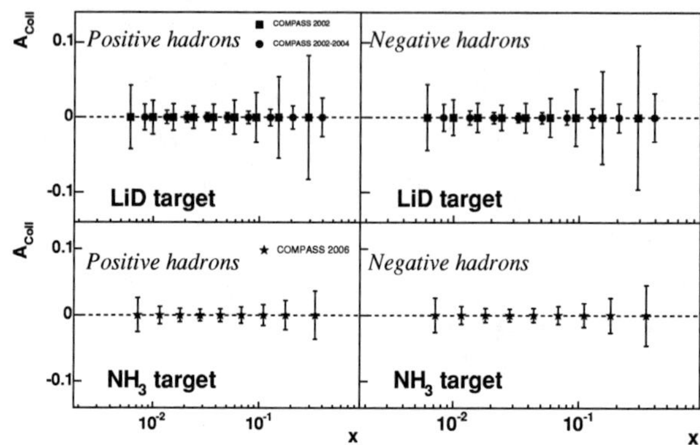

FIGURE 2. Estimation of statistical errors for A_{Coll} as a function of x for deuterium (top) and proton (bottom) targets and for positive (left) and negative hadrons (right).

to the major improvements done on the trigger system, and on the DAQ hardware and software, the total integrated lumonisity taken with a transversely polarized target now amounts to more than 2 nb^{-1}. After the 2005 stop of all the accelerators in CERN, the COMPASS collaboration plans to take data on a transversely polarized proton target (NH_3) in 2006. The statistical accuracy for the measurement of A_{Coll} on both the deuteron and proton targets is shown in fi gure 2 as a function of x. For the proton, it was assumed to have 30 days of data taking.

COMPASS has shown that A_{Coll} and A_{Siv}, if non vanishing, are tiny. The COMPASS proton data, available after the 2006 run, together with the new measurements of the Collins function recently shown by the BELLE [13] collaboration, will permit a flavor decomposition of $\Delta_T q(x)$ achieving an accurate look at the "transverse" spin structure of the nucleon.

REFERENCES

1. V. Barone, A. Drago and P. G. Ratcliffe, Phys. Rept. **359** (2002) 1.
2. J. C. Collins, Nucl. Phys. B **396** (1993) 161.
3. D. W. Sivers, Phys. Rev. D **41** (1990) 83.
4. G. Baum *et al.* [COMPASS Collaboration], CERN-SPSLC-96-14.
5. M. Anselmino, V. Barone, A. Drago, F. Murgia [arXiv:hep-ph/0209073].
6. S. J. Brodsky, D.S. Hwang, I. Schmidt, Phys. Lett. B **530** (2002) 99.
7. J. C. Collins, Phys. Lett. B **536** (2002) 43.
8. G.K. Mallot Nucl. Instrum. Methods Phys. Res., A: 518 (2004) 121
9. F. Bradamante [arXiv:hep-ex/0411076] and references therein.
10. V. Yu. Alexakhin *et al.* [COMPASS Collaboration], Phys. Rev. Lett. **94** (2005) 202002.
11. M. Anselmino *et al.*, Phys. Rev. D **71**, 074006 (2005) [arXiv:hep-ph/0501196].
12. A. Airapetian *et al.* [HERMES Collaboration], Phys. Rev. Lett. **94** (2005) 012002.
13. R. Seidl *et al.*, 'Measurements of chiral-odd fragmentation functions at BELLE", these proceedings.

Transversity Properties of Quarks and Hadrons in SIDIS and Drell-Yan

Leonard P. Gamberg* and Gary R. Goldstein†

*Division of Science, Penn State Berks, Reading, PA 19610, USA
†Department of Physics and Astronomy, Tufts University, Medford, MA 02155, USA

Abstract. We consider the leading twist T-odd contributions as the dominant source of the azimuthal and transverse single spin asymmetries in SIDIS and dilepton production in Drell-Yan Scattering. These asymmetries contain information on the distribution of quark transverse spin in (un)polarized protons. In the spectator framework we estimate these asymmetries at HERMES kinematics and at 50 GeV for the proposed experiments at GSI, where an anti-proton beam is ideal for studying the transversity properties of quarks due to the dominance of *valence* quark effects.

Keywords: Transverse Single Spin Asymmetries, T-Odd Effects
PACS: 13.88.+e, 13.60.-r, 13.15.+g, 13.85.Ni

One of the persistent challenges confronting the QCD parton model is to provide a theoretical basis for the experimentally significant azimuthal and transverse spin asymmetries that emerge in inclusive and semi-inclusive processes. Generally speaking, the spin dependent amplitudes for the scattering will contribute to non-zero transverse single spin asymmetries (SSA) if there are imaginary parts of bilinear products of those amplitudes that have overall helicity change. In perturbative QCD (PQCD), applicable to the hard scattering region, to obtain an imaginary contribution to quark and/or gluon scattering processes demands introducing higher order corrections to tree level processes. One approach incorporates the requisite phases through interference of tree level and one-loop contributions in PQCD in an attempt to explain up-down polarization asymmetry in Λ production [1]. On general grounds the helicity conservation property of massless QCD predicts that such contributions are small, going like $\alpha_s m/Q$, where α_s is the strong coupling, m represents a non-zero quark mass and Q represents the hard QCD scale [1, 2]. Such contributions have failed to account for the large SSA observed in Λ production [3].

However, considering the soft contributions to hadronic processes opens up the possibility that there are non-trivial transversity parton distributions that can contribute to transverse spin asymmetries [4]. For transverse SSA in SIDIS, transverse momentum must be acquired to lead to appropriate helicity changes at leading twist. In describing transverse asymmetries this is particularly relevant when the transverse momentum can arise from intrinsic quark momenta. Here the effects are associated with non-perturbative transverse momentum distribution functions [5] (TMD), where transverse SSAs indicate so called T-odd correlations between transverse spin and longitudinal and intrinsic quark transverse momentum. The T-odd distributions [6, 7] are of importance as they possess both transversity properties and the necessary phases to account for SSA and azimuthal asymmetries [8, 9]. Formally, these phases can be generated from the gauge

CP792, *Deep Inelastic Scattering, DIS 2005*, edited by W. H. Smith and S. R. Dasu

invariant definitions of the T-odd quark distribution functions [10, 11, 12]. In contrast to the transverse SSAs generated from the interference of tree-level and one loop correction in PQCD, such effects go like $\alpha_s < k_\perp > /M$, where now M plays the role of the chiral symmetry breaking scale and $k_\perp$ is characteristic of quark intrinsic motion.

Here we consider the leading twist T-odd contributions as the dominant source of the $\cos 2\phi$ azimuthal asymmetry and $\sin(\phi \pm \phi_s)$ transverse SSAs in SIDIS [13] and azimuthal asymmetry ν in dilepton production in Drell-Yan Scattering [14]. Among other interesting properties, these asymmetries contain information on the distribution of quark transverse spin in an unpolarized proton, $h_1^\perp(x,k_\perp)$ [7]. In a parton-spectator framework we estimate these asymmetries at HERMES kinematics [15] and for Drell-Yan scattering at 50 GeV center of mass energy. The latter azimuthal asymmetry is interesting in light of proposed experiments at GSI, where an anti-proton beam will ideal for studying the transversity properties of quarks due to the dominance of *valence* quark effects [16].

The leading twist contributions to the factorized cross-section for a transversely polarized nucleon target in lepton-proton scattering are

$$\frac{d^6\sigma_{UT}^{\ell N^\uparrow \to \ell\pi X}}{dx_H dy dz_h d\phi_S d^2P_{h\perp}} = \frac{2\alpha^2}{Q^2 y}\Bigg\{|S_T|(1-y)\sin(\phi_h+\phi_S)\sum_q e_q^2 \mathscr{F}\left[\frac{p_\perp\cdot\hat{h}}{M_h}h_1^q H_1^{\perp q}\right]$$
$$+|S_T|\frac{(1+(1-y)^2)}{2}\sin(\phi_h-\phi_S)\sum_q e_q^2\mathscr{F}\left[\frac{k_\perp\cdot\hat{h}}{M}f_{1T}^{\perp q}D_1^q\right]\Bigg\}, \quad (1)$$

where $\mathscr{F}$ is the convolution integral [7]. The twist two T-even and odd distribution and fragmentation functions appearing in Eq. (1) are projected from the correlation functions for the transverse momentum dependent distribution and fragmentation correlators, $\Phi(x,P)$ and $\Delta(p,P_h)$ respectively,

$$\Phi(x,p_\perp) = \frac{1}{2}\Big\{f_1(x,p_\perp)\not{n}_+ + h_1^\perp(x,p_\perp)\frac{\sigma_{\mu\nu}p_\perp^\mu n_+^\nu}{M} + f_{1T}^\perp(x,p_\perp)\frac{\varepsilon_{\mu\nu\rho\sigma}\gamma^\mu n_+^\nu p_\perp^\rho S_T^\sigma}{M}\cdots\Big\}$$
$$\Delta(z,k_\perp) = \frac{1}{4}\Big\{D_1(z,zk_\perp)\not{n}_- + H_1^\perp(z,zk_\perp)\frac{\sigma_{\mu\nu}k_\perp^\mu n_-^\nu}{M_h}+\cdots\Big\},$$

where for example $\int dp^- \mathrm{Tr}\left(\sigma^{\perp +}\gamma_5\Phi\right) = \frac{2\varepsilon_{+-\perp j}p_{\perp j}}{M}h_1^\perp(x,p_\perp)\ldots$. We use the parton inspired quark-diquark spectator framework to model the quark-hadron interactions that enter the T-odd and even TMDs and fragmentation functions contributing to $\Phi(x,P)$ and $\Delta(z,P_h)$ [13]. Noting that parton intrinsic transverse momentum yields a natural regularization for the moments of these distributions, we incorporated a Gaussian from factor into our model. The resulting scalar diquark contribution is $h_1^\perp(x,p_\perp) = \mathscr{N}\alpha_s M\frac{(1-x)(m+xM)}{p_\perp^2\Lambda(p_\perp^2)}\mathscr{R}(p_\perp^2;x)$ where $\mathscr{R}(p_\perp^2;x) = \exp^{-2b(p_\perp^2-\Lambda(0))}\left(\Gamma(0,2b\Lambda(0))-\Gamma(0,2b\Lambda(p_\perp^2))\right)$ is the regularization function. $\Lambda(k_\perp^2)$ is the spectral function and $\mathscr{N}$ is a normalization factor determined with respect to the unpolarized u-quark distribution, obtained from the zeroth moment of $f_1^{(u)}(x,p_\perp)$ normalized with respect to valence distributions. Our regulated expression of the Collins function is given by $H_1^\perp(z,k_\perp) = \mathscr{N}'\alpha_s\frac{1}{4z}\frac{(1-z)}{z}\frac{\mu}{\Lambda'(k_\perp^2)}\frac{M_\pi}{k_\perp^2}\mathscr{R}(z,k_\perp^2)$ where μ is

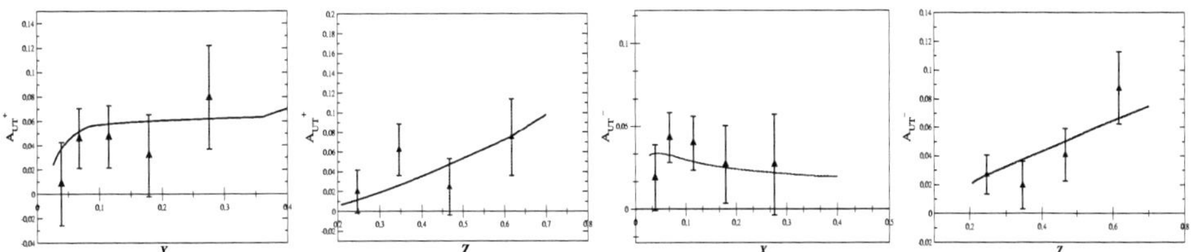

FIGURE 1. Left two Panel: The $\langle\sin(\phi+\phi_s)\rangle_{UT}$ asymmetry for π^+ production as a function of x and z compared to the HERMES data [15] Right two Panels: The $\langle\sin(\phi-\phi_S)\rangle_{UT}$ as a function of x and z.

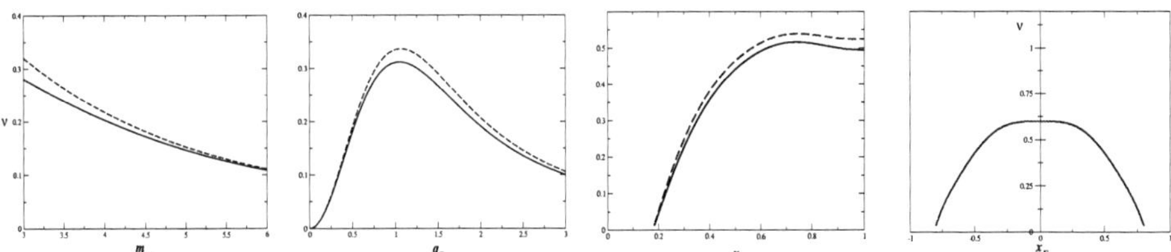

FIGURE 2. Left two Panels: ν plotted as a function of q_T and $q=m_{\mu\mu}$ for $s=50$ GeV^2, x in the range $0.2-1.0$. Right two panels: ν plotted as a function of x and x_F for $s=50$ GeV^2 q_T ranging from 3 to 6 GeV/c and q from 0 to 3 GeV/c.

the quark spectator mass and $\mathcal{N}'$ is determined from the normalization on the unpolarized fragmentation function $D_1(z)$. The Collins and Sivers weighted asymmetries are projected from the differential cross sections, Eq. (1)

$$\langle\frac{P_{h\perp}}{M_\pi}\sin(\phi+\phi_s)\rangle_{UT}=\frac{\int d\phi_s d^2P_{h\perp}\frac{P_{h\perp}}{M_\pi}\sin(\phi+\phi_s)\left(d\sigma^{\uparrow}-d\sigma^{\downarrow}\right)}{\int d\phi_s\int d^2P_{h\perp}\left(d\sigma^{\uparrow}+d\sigma^{\downarrow}\right)}=\frac{|S_T|2(1-y)\sum_q e_q^2 h_1(x)zH_1^{\perp(1)}(z)}{(1+(1-y)^2)\sum_q e_q^2 f_1(x)D_1(z)}$$

and $\langle\frac{|P_{h\perp}|}{M}\sin(\phi-\phi_S)\rangle_{UT}=|S_T|\frac{(1+(1-y)^2)\sum_q e_q^2 f_{1T}^{\perp(1)}(x)zD_1^q(z)}{(1+(1-y)^2)\sum_q e_q^2 f_1(x)D_1(z)}$. We have re-analyzed these asymmetries including both the scalar and vector diquark contributions to the TMDs for the central values of our parameter set, and compared the transverse SSAs to the the HERMES data [15] for π^+ production in Fig. 1. The unweighted asymmetries are approximated as $A_{UT}^{\sin(\phi+\phi_s)}\approx\frac{M_\pi}{\langle P_{h\perp}\rangle}\langle\frac{P_{h\perp}}{M_\pi}\sin(\phi+\phi_s)\rangle$ and $A_{UT}^{\sin(\phi-\phi_s)}\approx\frac{M}{\langle P_{h\perp}\rangle}\langle\frac{P_{h\perp}}{M}\sin(\phi\pm\phi_s)\rangle$. These results agree to within the errors displayed.

An unpolarized double T-odd azimuthal asymmetry enters the Drell-Yan process [17]. For the Drell-Yan process the angular dependence [18] can be expressed as

$$\frac{dN}{d\Omega}\equiv=\frac{3}{4\pi}\frac{1}{\lambda+3}\left(1+\lambda\cos^2\theta+\mu\sin^2\theta\cos\phi+\frac{\nu}{2}\sin^2\theta\cos2\phi\right),\tag{2}$$

where $\frac{dN}{d\Omega}\equiv\left(\frac{d\sigma}{dQ^2dydq_T^2}\right)^{-1}\frac{d\sigma}{dQ^2dydq_T^2d\Omega}$. The solid angle Ω refers to the lepton pair orientation in the pair rest frame relative to the boost direction, and λ,μ,ν are functions that depend on $x,m_{\mu\mu}^2,q_T$, the fraction of quark momentum in the hadron, the invariant mass of the produced lepton pair, and the transverse momentum of the dimuon pair. All of the asymmetry functions, μ, λ and ν, have parton model contributions which at next to leading order predict $1-\lambda-2\nu=0$, the so called Lam-Tung relation [19]. Experimental measurements of $\pi p \to \mu^+\mu^- X$ discovered unexpectedly large values of these asymmetries [20] compared to parton-model expectations resulting in a serious violation of this relation. It has been suggested [17] that there is a dominant leading twist contribution to ν coming from the T-odd transversity distributions $h_1^\perp(x,k_\perp)$ for both hadrons which dominates in the kinematic range, $q_T \ll Q$. The $\cos 2\phi$ azimuthal asymmetry in unpolarized $p\bar{p} \to \mu^+\mu^- X$ involves the convolution of the leading twist T-odd function, $h_1^\perp\ \nu_2 = \frac{\sum_a e_a^2 \mathscr{F}\left[w_2 h_1^\perp(x,k_\perp)\bar{h}_1^\perp(\bar{x},p_\perp)/(M_1 M_2)\right]}{\sum_a e_a^2 \mathscr{F}\left[f_1(x,k_\perp)\bar{f}_1(\bar{x},p_\perp)\right]}$ where $w_2 = (2\hat{h}\cdot k_\perp \cdot \hat{h}\cdot p_\perp - p_\perp \cdot k_\perp)$ is the weight in the convolution integral, $\mathscr{F}$. In addition it is known that there is a non-leading T-even contribution to the $\cos 2\phi$ asymmetry[18] $\nu_4 = \frac{\frac{1}{Q^2}\sum_a e_a^2 \mathscr{F}\left[w_4 f_1(x,k_\perp)\bar{f}_1(\bar{x},p_\perp)\right]}{\sum_a e_a^2 \mathscr{F}\left(f_1(x,k_\perp)\bar{f}_1(\bar{x},p_\perp)\right)}$, where $w_4 = 2(\hat{h}\cdot(k_\perp - p_\perp))^2 - (k_\perp - p_\perp)^2$ and $\hat{h} = q_T/Q_T$. Fig. 2 shows that the $\cos 2\phi$ azimuthal asymmetry ν is not small at center of mass energies of 50 GeV^2. However the T-odd portion dominates with an additional $3-5\%$ from the sub-leading T-even piece. Thus, aside from the competing T-even effect, the experimental observation of a strong x-dependence would indicate the presence of T-odd structures in *unpolarized* Drell-Yan scattering, implying that novel transversity properties of the nucleon can be accessed *without invoking beam or target polarization.*

G.R.G. is supported by U.S. DOE (DE-FG02-92ER40702).

REFERENCES

1. W. G. D. Dharmaratna and G. R. Goldstein, Phys. Rev. D **41**. 1731 (1990).
2. G. L. Kane, J. Pumplin, and K. Repko, Phys. Rev. Lett. **41**, 1689 (1978).
3. K. Heller *et al.*, Phys. Rev. Lett. **41**, 607 (1978); Phys. Rev. Lett. **51**, 2025 (1983).
4. J. Ralston and D. E. Soper, Nucl. Phys. **B152**, 109 (1979).
5. D. E. Soper, Phys. Rev. Lett. **43**, 1847 (1979); R. D. Tangerman and P. J. Mulders, Nucl. Phys. **B461**, 197 (1996).
6. D. Sivers, Phys. Rev. D **41**, 83 (1990); M. Anselmino, *et al.*, Phys. Lett. B **362**, 164 (1995);
7. D. Boer and P. J. Mulders, Phys. Rev. D **57**, 5780 (1998).
8. J. C. Collins, Nucl. Phys. B **396**, 161 (1993).
9. S. J. Brodsky, D. S. Hwang, and I. Schmidt, Phys. Lett. B **530**, 99 (2002).
10. J. C. Collins, Phys. Lett. B **536**, 43 (2002).
11. X. Ji and F. Yuan, Phys. Lett. B **543**, 66 (2002); A.V. Belitsky *et al.*, Nucl. Phys. B **656**, 165 (2003); Daniel Boer *et al.* Nucl.Phys. **B667**, 201 (2003).
12. G. R. Goldstein and L. P. Gamberg, arXiv:hep-ph/0209085, Proceedings ICHEP 2002, Ed. by S. Bentvelsen *et al.*, Amsterdam, The Netherlands (North-Holland 2003), pg. 452.
13. L. P. Gamberg, G. R. Goldstein and K.A. Oganessyan, Phys. Rev. D **67** , 071504 (2003); Phys. Rev. D **68**, 051501 (2003).
14. L. P. Gamberg and G. R. Goldstein arXiv:hep-ph/0506127.
15. A. Airapetian *et al.*, Phys. Rev. Lett. **84**, 4047 (2000); Phys. Rev. Lett. **94** (2005) 012002.
16. *"Antiproton Proton Scattering Experiments with Polarization*, PAX Letter of Intent (2003); M. Maggiora *et al.* [ASSIA Collaboration], arXiv:hep-ex/0504011.
17. D. Boer, Phys. Rev. D **60**, 014012 (1999); D. Boer, S. J. Brodsky, and D. S. Hwang, Phys. Rev D **67**, 054003 (2003).
18. J. C. Collins and D. E. Soper, Phys. Rev. D**16**, 2219 (1977).
19. C.S. Lam and W.K. Tung, Phys. Rev. D **21** (1980) 2712.
20. E615 Collaboration: J. S. Conway *et al.*, Phys. Rev. D **39**, 92 (1989); S. Falciano *et al.* [NA10 Collaboration], Z. Phys. C **31** (1986) 513.

New results on SIDIS SSA from Jefferson Lab

H. Avakian, P. Bosted, V. Burkert, L. Elouadrhiri for the CLAS Collaboration

Jefferson Lab, Newport News, VA 23606, USA

Abstract. We present studies of single-spin and double-spin asymmetries in semi-inclusive electroproduction of pions using the CEBAF 6 GeV polarized electron beam. Kinematic dependences of single and double spin asymmetries have been measured in a wide kinematic range at CLAS with a polarized NH_3 target. Significant target-spin $\sin 2\phi$ and $\sin\phi$ asymmetries have been observed. The hypothesis of factorization has been tested with z-dependence of the double spin asymmetry.

Keywords: quarks,single spin asymmetries,TMD parton distributions
PACS: 13.60.-r, 13.88.+e, 14.20.Dh, 14.65.-q

Single-Spin Asymmetries (SSAs) in azimuthal distributions of final state particles in semi-inclusive deep inelastic scattering play a crucial role in the study of transverse momentum distributions of quarks in the nucleon and provide access to the orbital angular momentum of quarks. Large SSAs, observed for decades in hadronic reactions have been among the most difficult phenomena to understand from first principles in QCD. Recently, significant SSAs were reported in semi-inclusive DIS (SIDIS) by the HERMES collaboration at HERA [1, 2] for longitudinally and transversely polarized targets, and by the CLAS collaboration with a polarized beam [3].

Two fundamental mechanisms have been identified leading to SSAs in hard processes, the Sivers mechanism [4, 5, 6, 7, 8], which generates an asymmetry in the distribution of quarks due to orbital motion of partons, and the Collins mechanism [9, 10], which generates an asymmetry during the hadronization of quarks.

The HERMES Collaboration has recently measured a transverse spin asymmetry in SIDIS providing the cleanest evidence to date for the existence of a non-zero Collins function [2], which describes the fragmentation of a transversely polarized quark into pions. This finding is supported by the preliminary data from BELLE [11] indicating a non-zero Collins effect. The large target SSA in semi-inclusive pion production measured at CLAS and analyzed in terms of the Collins fragmentation [12], also indicate a significant Collins function.

For a longitudinally polarized target the only azimuthal asymmetry arising in leading order is the $\sin 2\phi$ moment [15, 10, 13], involving the transverse momentum dependent (TMD) Collins fragmentation function $H_1^\perp$ [9] and the Mulders distribution function $h_{1L}^\perp$ [14, 10], describing the transverse polarization of quarks in a longitudinally polarized proton [15, 10, 13]. The same distribution function is accessible in double polarized Drell-Yan, where it gives rise to a $\cos 2\phi$ azimuthal moment in the cross section [16].

Single and double spin asymmetries in SIDIS have been measured using the CLAS [17] in Hall B at Jefferson Lab, a 5.7 GeV longitudinally polarized electron beam, a longitudinally polarized proton (NH_3) target. The average beam polarization was 0.73 ± 0.03 and the average target polarization was 0.72 ± 0.05. The open acceptance of

CP792, *Deep Inelastic Scattering, DIS 2005*, edited by W. H. Smith and S. R. Dasu

CLAS and a single electron trigger ensured event recording for a large sample of SIDIS π^+, π^0, and π^- events.

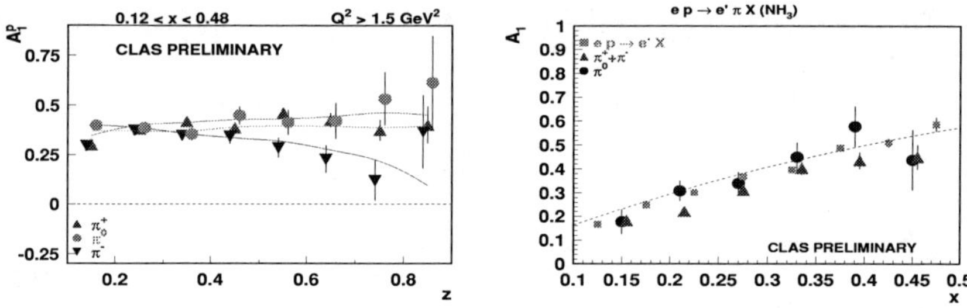

FIGURE 1. The double spin asymmetry for SIDIS π^+, π^-, and π^0 data as a function of z (left) and the comparison of inclusive A_1^p with the A_1^p for the SIDIS sum of π^+ and π^- and A_1^p for SIDIS π^0 (right).

The validity of factorization is crucial to the interpretation of target SSAs in terms of TMDs, which is the main goal of this contribution. A factorization test comes from examining the z-dependence of the double spin asymmetries (A_1^p) for all three pion flavors, as shown in the left panel of Fig. 1. The data cuts included $W > 2$ GeV and $Q^2 > 1.1$ GeV^2 to ensure DIS kinematics, and the average value of x is approximately 0.3. If factorization holds, the asymmetries should be approximately independent of z, broken by the different weights given to the polarized u and d quarks by the favored and unfavored fragmentation functions. We therefore expect the largest z-dependence for the π^- asymmetries. This is indeed born out by the data, which are in excellent agreement in both magnitude and z-dependence up to $z = 0.7$ with predictions of the polarized Lund Monte Carlo using GRSV polarized PDFs as input.

The dependence of the double spin asymmetry on Bjorken x for different pion flavors obtained from the CLAS data for the same kinematic range is presented in Fig. 1 (right panel). The π^0 double spin asymmetry as well as A_1^p for the sum of charged pions are consistent with the inclusive A_1^p as expected in a simple partonic picture.

These studies suggest that factorization works for $W > 2$ GeV, $Q^2 > 1.1$ GeV^2, $0.15 < x < 0.5$, and $0.3 < z < 0.7$ for a 5.7 GeV electron energy.

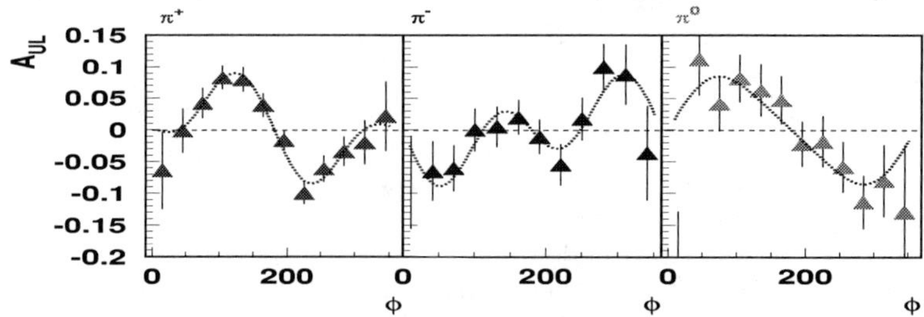

FIGURE 2. The target SSA as a function of azimuthal angle ϕ from data at 5.7 GeV.

The spin-dependent moments ($\sin\phi, \sin 2\phi$) of the semi-inclusive cross section have been extracted in a fit of the normalized-yield asymmetry

$$A_{UL}(\phi) = \frac{1}{P_T}\frac{N^+ - N^-}{N^+ + N^-}. \tag{1}$$

Here $N^{\pm}$ is the number of events for target polarizations antiparallel/parallel to the incoming beam direction and P_T is the target polarization.

Measurements of the $\sin 2\phi$ SSA allow the study of the Collins effect with no contamination from other mechanisms. A recent measurement of $\sin 2\phi$ moment of σ_{UL} by HERMES [1] is consistent with zero. A measurably large asymmetry has been predicted only at large x ($x > 0.2$), a region well-covered by JLab [18].

The data for π^+ (Fig. 2) show a clear $\sin\phi$ and $\sin 2\phi$ modulations from which a $\sin\phi$ moment of $0.058 \pm 0.011(stat)$ and $\sin 2\phi$ moment of $-0.041 \pm 0.011(stat)$ have been determined.

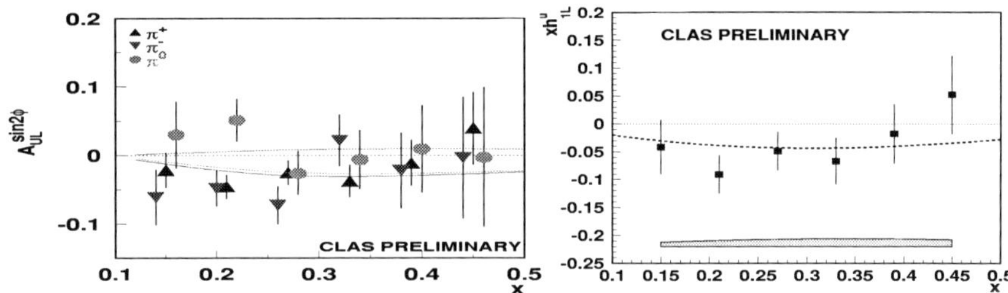

FIGURE 3. The leading twist SSA for the $\sin 2\phi$ moment for π^+, π^0, and π^- as a function of x (left plot). The $h_{1L}^{\perp}$ from the π^+ SSA (right). The contribution from the unfavored production is included in the systematic error band. The variation for the ratio of unfavored to favored Collins functions is from -2.5 to 0. The curves are from [18] using the χQSM calculations of $h_{1L}^{\perp}$.

The x dependence of the SSA for π^+ (Fig. 3) is consistent with predictions [18]. No sign of a large unfavored Collins fragmentation (large π^- SSA with a corresponding π^+ SSA of opposite sign) is seen. The π^+ SSA is dominated by the u-quarks; therefore with some assumption about the ratio of unfavored to favored Collins fragmentation functions, it can provide a first glimpse of the twist-2 TMD function $h_{1L}^{\perp}$ (Fig. 3.) The curve is the calculation by Efremov et al. [18], using $h_{1L}^{\perp}$ from the chiral quark soliton model evolved to Q^2=1.5 GeV^2. The extraction, however, suffers from low statistics and has a significant systematic error from the unknown ratio of the Collins favored and unfavored fragmentation functions, the unknown ratio of $h_{1L}^{\perp d}/h_{1L}^{\perp u}$, as well as from background from exclusive vector mesons.

The $\sin\phi$ moment of the cross section measured with CLAS at 5.7 GeV is in agreement with the HERMES measurement at 27.5 GeV for a longitudinal target [1], indicating a bigger asymmetry for the higher twist contribution compared to the leading twist $\sin 2\phi$ moment. The $P_{\perp}$ dependence of the $\sin\phi$ moment for π^+ (see Fig. 4) is consistent with an increase with increasing $P_{\perp}$, as expected for the TMD function [19]. The $\sin\phi$ moment, for π^- is also plotted in Fig. 4, showing first evidence for a non-zero SSA asymmetry for π^- on a longitudinally polarized target.

The $\sin\phi$ moment of the SIDIS cross section itself can be an important source of independent information on the Collins fragmentation mechanism. Several other contributions to the $\sin\phi$ moment were identified recently [20, 21, 22], involving different unknown distribution and fragmentation functions. A global analysis of beam and target SSA may be required to separate different contributions. Drell-Yan process with longitudinally and transversely polarized protons scattering on longitudinally polarized protons will also provide additional information on these distribution functions [23].

In conclusion CLAS target SSAs were measured using the 5.7 GeV data from the CLAS experiment using a polarized NH_3 target. The twist-2 distribution function $h_{1L}^{\perp}$ was extracted for the first time using the π^+ target SSA. No evidence has been found with CLAS for the large unfavored Collins fragmentation indicated by HERMES [2].

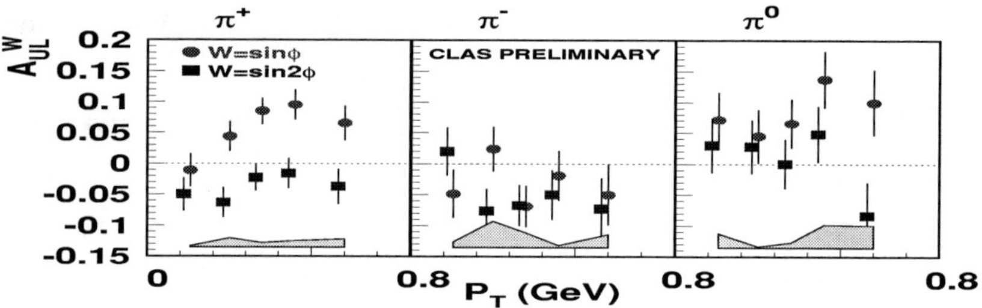

FIGURE 4. The A_{UL} SSA dependence on $P_{\perp}$. The band represents the systematic uncertainty from fi t in the measurement of the $sin2\phi$ moment.

REFERENCES

1. HERMES collaboration (A. Airapetyan *et al.*), Phys.Rev.Lett. **84**, 4047 (2000); hep-ex/0505042.
2. HERMES collaboration (A. Airapetyan *et al.*), Phys.Rev.Lett. **94**, 012002 (2005).
3. CLAS Collaboration (H. Avakian *et al.*) Phys.Rev. **D69**, 112004 (2004).
4. D. Sivers, Phys.Rev. **D43**, 261 (1991).
5. M. Anselmino and F. Murgia, Phys. Lett. B **442**, 470 (1998).
6. S. Brodsky *et al.*, Nucl. Phys. **B 642**, 344 (2002).
7. J. Collins, Phys. Lett. B **536**, 43 (2002).
8. X. Ji, F. Yuan, Phys. Lett. B **543**, 66 (2002); Nucl. Phys. **B 656**, 165 (2003).
9. J. Collins, Nucl. Phys. **B396**, 161 (1993).
10. P.J. Mulders and R.D. Tangerman, Nucl. Phys. **B 461**, 197 (1996).
11. R. Seidl, this conference.
12. A. Efremov Annalen Phys. 13, 651 (2004)
13. A.M. Kotzinian and P.J. Mulders, Phys. Rev. D**54** 1229 (1996).
14. J. Ralston and D. Soper, Nucl. Phys. **B152**, 109 (1979)
15. A. Kotzinian, Nucl. Phys. **B 441**, 234 (1995). **B461**, 197 (1996).
16. R.D. Tangerman and P.J. Mulders, Phys.Rev. D**51** 3357 (1995).
17. B. Mecking *et al.*, Nucl. Inst. & Meth. **503**, 513 (2003).
18. A. V. Efremov *et al.*, Phys. Rev. D **67** (2003) 114014;hep-ph/0412420.
19. X. Ji, J-P. Ma and F. Yuan, Phys. Lett. B **597**, 299 (2004); Nucl. Phys. B **652**, 383 (2003).
20. D. Boer, P. J. Mulders and F. Pijlman, Nucl. Phys. B **667**, 201 (2003).
21. A. Bacchetta, P. J. Mulders, and F. Pijlman, Phys. Lett. **B595**, 309 (2004).
22. K. Goeke, A. Metz, M. Schlegel, hep-ph/0504130.
23. R.L. Jaffe and X. Ji, Nucl.Phys. **B375** (1992) 527.

Spin dependent fragmentation functions at BELLE

A. Ogawa*, D.Gabbert†,*, M. Grosse-Perdekamp†,*, R. Seidl†,* and K. Hasuko**

*RIKEN Brookhaven Research Center
Upton, NY 11973-5000, USA
†University of Illinois at Urbana-Champaign
1100 W Green Street,
Urbana, IL 61801, USA
**RIKEN
Wako, Saitama,351-0198, Japan
E-mail: dgabbert@uiuc.edu, mgp@uiuc.edu, akio@bnl.gov, rseidl@uiuc.edu

Abstract. The measurement of the so far unknown chiral-odd quark transverse spin distribution in either semi-inclusive DIS(SIDIS) or inclusive measurements in pp collisions at RHIC has an additional chiral-odd fragmentation function appearing in the cross section. This chiral-odd fragmentation functions (FF) can for example be the so-called Collins FF or the Interference FF. HERMES has given a first hint that these FFs are nonzero, however in order to measure transversity one needs these FFs to be precisely known. At the Belle e^+e^- collider at the KEK-B factory a data set of 29.0 fb^{-1} has been used to obtain the Collins function.

Keywords: spin transversity e+e- collins
PACS: 13.88.+4,13.66.-a,14.65.-q,14.20.-c

INTRODUCTION

At leading twist 3 quark distribution functions (DF) in the nucleon exist; the well known unpolarized quark DF, the somewhat known quark helicity DF and the so far unknown transversity DF. The latter cannot be measured in inclusive DIS due to its chiral-odd nature, since all possible interactions are chiral-even for nearly massless quarks. Therefore one needs an additional chiral-odd function in the cross section to access transversity. This can be either achieved by an anti quark transversity DF in double transversely polarized Drell-Yan processes or one can have a chiral-odd fragmentation function in SIDIS or hadroproduction.

THE BELLE EXPERIMENT

The Belle [2] experiment at the asymmetric e^+e^- collider KEK-B at Tsukuba, Japan, is mainly dedicated to study CP violation in B meson decays. Its center of mass energy is tuned to the Υ(4S) resonance at $\sqrt{s} = 10.58$ GeV and part of the data was recorded 60 MeV below the resonance. These off-resonance events are studied in order to measure spin dependent and also to perform precise measurements of spin independent fragmentation functions. At present an integrated luminosity of 29.0 fb^{-1} has been accumulated

CP792, *Deep Inelastic Scattering, DIS 2005*, edited by W. H. Smith and S. R. Dasu

in the off-resonance data sample. The aerogel Čerenkov counter (ACC) , time-of-flight (TOF) detector and the central drift chamber (CDC) enable a good particle identification and tracking, which is crucial for these measurements. Using the information from the silicon vertex detector (SVD), one selects tracks originating from the interaction region and thus reducing the contribution of hadrons from heavy meson decays.

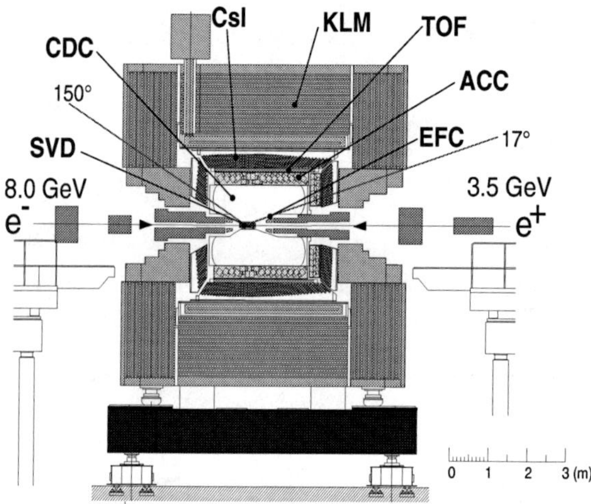

FIGURE 1. A schematic side view of the Belle detector.

To reduce the amount of hard gluon radiative events a cut on the kinematic variable thrust of $T > 0.8$ is applies. This enhances the typical 2-jet topology and the thrust axis is used as approximation of the original quark direction. To ensure that the pions did not originate in the decay of a vector meson and might be mistakenly put in the wrong hemisphere a lower cut on the fractional energy of 0.2 is performed.

COLLINS FF

The Collins effect occurs in the fragmentation of a transversely polarized quark with polarization $\mathbf{S_q}$ and 3-momentum $\mathbf{k}$ into an unpolarized hadron of transverse momentum $\mathbf{P}_{h\perp}$ with respect to the original quark direction. According to the Trento convention [3] the number density for finding an unpolarized hadron h produced from a transversely polarized quark q is defined as:

$$D_{hq^\uparrow}(z,P_{h\perp}) = D_1^q(z,P_{h\perp}^2) + H_1^{\perp q}(z,P_{h\perp}^2)\frac{(\hat{\mathbf{k}} \times \mathbf{P}_{h\perp})\cdot \mathbf{S}_q}{zM_h}, \tag{1}$$

where the first term describes the unpolarized FF $D_1^q(z,P_{h\perp}^2)$, with $z \stackrel{CMS}{=} \frac{2E_h}{Q}$ being the fractional energy the hadron carries relative to half of the CMS energy Q. The second term, containing the Collins function $H_1^{\perp q}(z,P_{h\perp}^2)$, depends on the spin of the quark and thus leads to an asymmetry as it changes sign under flipping the quark spin. The vector product can accordingly be described by a $\sin(\phi)$ modulation, where ϕ is the

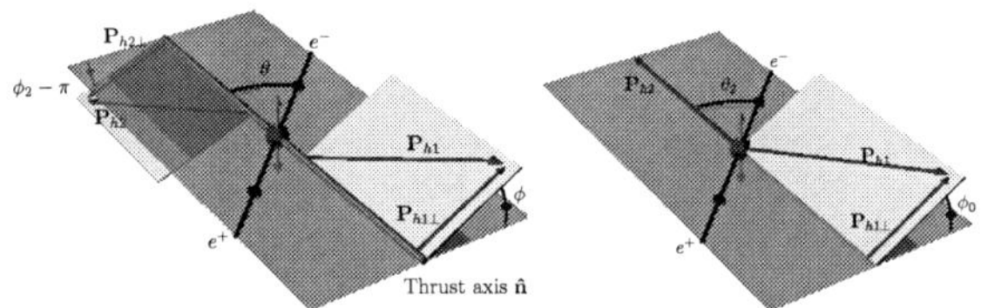

FIGURE 2. Description of the azimuthal angles ϕ_0, ϕ_1 and ϕ_2 relative to the scattering plane defined by the lepton axis and either the thrust axis $\hat{n}$ or the momentum of the 2nd hadron P_{h2}.

azimuthal angle spanned by the transverse momentum and the plane defined by the quark spin and its momentum. In e^+e^- hadron production the Collins effect can be observed by a combined measurement of a quark and an anti quark fragmentation. Combining two hadrons from different hemispheres in jetlike events, with azimuthal angles ϕ_1 and ϕ_2 as defined in Fig. 2, would result in a $\cos(\phi_1 + \phi_2)$ modulation. In the CMS these azimuthal angles are defined between the transverse component of the hadron momenta with regard to the thrust axis $\hat{n}$ and the plane spanned by the lepton momenta and $\hat{n}$. The comparison of the thrust axis calculations using reconstructed and generated tracks in the MC sample shows an average angular separation between the two of 75 mrad with a a root mean square of 74 mrad. Due to that small biases in one of the reconstruction methods used could arise and were studied as discussed later. Following reference[4] one either computes the azimuthal angles of each pion relative to the thrust axis which results in a $\cos(\phi_1 + \phi_2)$ modulation or one calculates the azimuthal angle relative to the axis defined by the 2nd pion which results in a $\cos(2\phi_0)$ modulation 2. While the first method directly accesses moments of the Collins functions the second method also contains a convolution integral of the Collins FF over possible transverse momenta of the hadrons.

Measured asymmetries

We measure the azimuthal asymmetries $N(2\phi)/N_0$, where $N(2\phi)$ denotes the number of hadron pairs in bins of either $2\phi_0$ or $\phi_1 + \phi_2$ and N_0 is the average number of hadron pairs in the whole angle interval. The main background, producing similar azimuthal asymmetries as the Collins effect, is the radiation of soft gluons. This gluonic contribution is proportional to the unpolarized FF and is independent of the charge of the hadrons. Consequently taking the ratio of the normalized distributions for unlike-sign over like-sign pairs the gluonic distributions drop out in leading order:

$$R := \frac{\frac{N(2\phi_0)}{N_0}|_{unlikesign}}{\frac{N(2\phi_0)}{N_0}|_{likesign}} \approx 1 + \frac{\sin^2\theta}{1+\cos^2\theta}\left(F(\frac{H_1^{\perp,fav}}{D_1^{fav}}, \frac{H_1^{\perp,disfav}}{D_1^{disfav}}) + \mathcal{O}f(Q_T, \alpha_S)^2\right)\cos(2\phi_0) \quad , \quad (2)$$

where θ is the angle between the colliding leptons and the produced hadron. Favored and disfavored FF describe the fragmentation of a light quark into a pion of same or opposite charge sign. A similar relation also holds for the $\cos(\phi_1 + \phi_2)$ method. Those double ratios are then fit by the sum of a constant term and a $\cos(2\phi_0)$ or $\cos(\phi_1 + \phi_2)$ modulation. Preliminary results for the cosine fits to the double ratios can be seen in Fig.3 for charged pion pairs, where the combined z-bins are obtained by adding the symmetric bins of the 4×4 $z \in [0.2, 0.3, 0.5, 0.7, 1.0]$ bins. A clear nonzero asymmetry is visible. Additionally the data shows a rising behavior with rising fractional energy z. The systematic errors are obtained by taking the differences of the double ratio results compared with results obtained by subtracting the unlike from the like sign asymmetries. Also the constant fit to the double ratios obtained in MC (without a Collins contribution) together with its statistical error and a similar fit to double ratios of positively charged over negatively charged pion pair data were assigned as systematic error. Also the differences to the results when fitting the double ratios also with higher order azimuthal modulations were added to the systematic errors.

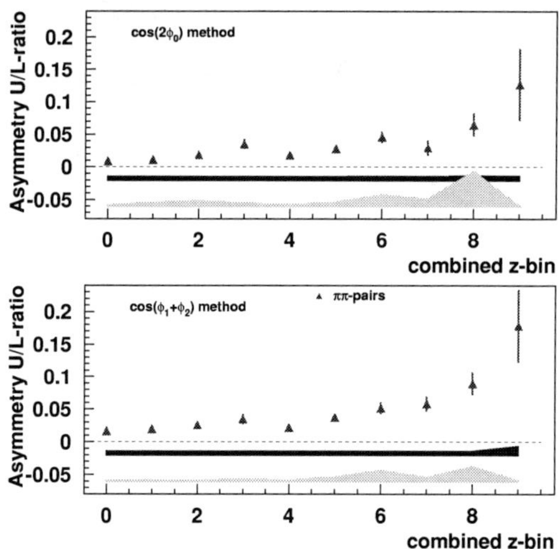

FIGURE 3. Double ratio results for the $\cos(2\phi_0)$ and the $\cos(\phi_1 + \phi_2)$ method. The upper error bars correspond to systematic errors, the lower error bars to possible contributions by charm quarks.

REFERENCES

1. J. C. Collins: Nucl. Phys **B396**(1993):161.
2. A. Abashian et al.(Belle) Nucl. Instrum. Meth.**A479**(2002)117.
3. A. Bacchetta, U. D'Alesio, M. Diehl, A. Miller Phys. Rev. **D70**(2004):117504.
4. D. Boer, R. Jakob, P. J. Mulders: Phys. Let. **B 424**(1998):143
5. R. L. Jaffe, X. m. Jin and J. a. Tang: Phys. Rev. **D57**(1998):5920
6. X. Artru,J. Collins: Z. Phys. **C69**(1996):1166
7. M. Radici, R. Jakob, and A. Bianconi: Phys. Rev.**D65**(2002):074031.

First measurement of interference fragmentation on a transversely polarized hydrogen target

P.B. van der Nat
(on behalf of the HERMES collaboration)

Nationaal Instituut voor Kernfysica en Hoge-Energiefysica (NIKHEF), P.O. Box 41882, 1009 DB Amsterdam, The Netherlands

Abstract. The HERMES experiment has measured for the first time single target-spin asymmetries in semi-inclusive two-pion production using a transversely polarized hydrogen target. These asymmetries are related to the product of two unknowns, the transversity distribution function and the interference fragmentation function. In the invariant mass range 0.51 GeV $< M_{\pi^+\pi^-} <$ 0.97 GeV the measured asymmetry deviates significantly from zero, indicating that two-pion semi-inclusive deep-inelastic scattering can be used to probe transversity.

Keywords: transversity, interference fragmentation
PACS: 13.60.-r, 13.88.+e, 13.87.Fh

INTRODUCTION

An important missing piece in our understanding of the spin structure of the nucleon is the transversity distribution $h_1(x)$. It is the only one of the three leading-twist quark distribution functions, $f_1(x)$, $g_1(x)$ and $h_1(x)$, that so-far remains unmeasured. The function $h_1(x)$ describes the distribution of transversely polarized quarks in a transversely polarized nucleon. It is quite diffi cult to measure $h_1(x)$, since it is a chiral-odd function, which can only be probed in combination with another chiral-odd function. This can be done in semi-inclusive DIS, where the second chiral-odd object is a fragmentation function, describing the fragmentation of the struck quark into one or more fi nal-state hadrons.

HERMES is one of the pioneering experiments on this subject. The structure function $h_1(x)$ is probed by measuring various single-spin asymmetries. First, a longitudinally polarized target [1] was used and more recently a transversely polarized target was used [2]. In these experiments, single spin asymmetries (SSA's) were only measured for *single-hadron* semi-inclusive DIS (SIDIS). However, already in 1993 Collins et al. [3] and in 1998 Jaffe et al. [4] suggested to study transversity in two-hadron SIDIS. Although this comes at the expense of a larger statistical uncertainty, there is a good reason for looking at SSA's in two-hadron SIDIS: the measured SSA's relate directly to the product of $h_1(x)$ and the fragmentation function, whereas in single-hadron SIDIS this product is convoluted with the transverse momentum of the hadron. Also measuring SSA's in two-hadron SIDIS provides an independent method of measuring $h_1(x)$, since it involves a different fragmentation function as compared to single-hadron SIDIS.

In order to fi nally extract the structure function $h_1(x)$, one needs to know the value of the involved fragmentation function. Although this function is also still unknown, it can be cleanly measured in e^+e^- experiments, such as Belle and Babar.

CP792, *Deep Inelastic Scattering, DIS 2005*, edited by W. H. Smith and S. R. Dasu

SINGLE SPIN ASYMMETRY

The transversity distribution can be accessed experimentally by measuring the single target-spin asymmetry, defined as:

$$A_{UT}(\phi_{R\perp},\phi_S,\theta) = \frac{1}{|S_T|}\frac{N^{\uparrow}(\phi_{R\perp},\phi_S,\theta)/N^{\uparrow}_{\text{DIS}} - N^{\downarrow}(\phi_{R\perp},\phi_S,\theta)/N^{\downarrow}_{\text{DIS}}}{N^{\uparrow}(\phi_{R\perp},\phi_S,\theta)/N^{\uparrow}_{\text{DIS}} + N^{\downarrow}(\phi_{R\perp},\phi_S,\theta)/N^{\downarrow}_{\text{DIS}}} = \frac{\sigma_{UT}}{\sigma_{UU}}, \quad (1)$$

where UT refers to Unpolarized beam and Transversely polarized target. The asymmetry is evaluated as a function of the angles $\phi_{R\perp}$, ϕ_S and θ which are defined in Fig. 1[1]. The azimuthal angle ϕ_S represents the spin direction of the target "↑" state and $N^{\uparrow(\downarrow)}(\phi_{R\perp},\phi_S,\theta)$ is the number of semi-inclusive $\pi^+\pi^-$-pairs in the target ↑(↓) spin state. These numbers are normalized to the corresponding number of DIS events, $N^{\uparrow}_{\text{DIS}}$ and $N^{\downarrow}_{DIS}$, respectively. The quantity $|S_T|$ indicates the average target polarization. The asymmetry is equal to the ratio of σ_{UT} and σ_{UU}, which are the polarized and unpolarized cross sections, respectively. According to Bacchetta et al. [6] σ_{UT} can be written at leading-twist[2] as:

$$\begin{aligned}\sigma_{UT} &= -\sum_q \frac{\alpha^2 e_q^2}{2\pi Q^2 y}(1-y)|\vec{S}_\perp|\frac{|\vec{R}|}{M_{\pi\pi}}\sin(\phi_{R\perp}+\phi_S)\sin\theta\, h_{1,q}(x) \\ &\quad \times \left[H_{1,q}^{\sphericalangle,sp}(z,M_{\pi\pi}^2) + \cos\theta H_{1,q}^{\sphericalangle,pp}(z,M_{\pi\pi}^2)\right] \quad (2)\end{aligned}$$

where $|\vec{R}| = \frac{1}{2}\sqrt{M_{\pi\pi}^2 - 4M_\pi^2}$ with $M_{\pi\pi}$ the invariant mass of the pion pair, M_π the pion mass and x, y and z the standard scaling variables used in semi-inclusive DIS. The transversity distribution $h_1(x)$ couples to a combination of two-hadron interference fragmentation functions, $H_1^{\sphericalangle,sp}$ and $H_1^{\sphericalangle,pp}$. These functions describe the interference between different production channels of the pion pair; $H_1^{\sphericalangle,sp}$ relates to the interference between s- and p-wave states and $H_1^{\sphericalangle,pp}$ to the interference between two p-wave states.

A two-dimensional fit function of the form

$$f(\phi_{R\perp}+\phi_S,\theta) = p_0 + p_1\sin(\phi_{R\perp}+\phi_S)\sin\theta \quad (3)$$

was used to extract from the measured asymmetry the part related to the product $h_1 H_1^{\sphericalangle,sp}$, where $p_1 \equiv A_{UT}^{\sin(\phi_{R\perp}+\phi_S)\sin\theta}$.

RESULTS

The present results are based on data taken in the period from 2002 until 2004 using a transversely polarized hydrogen target in the HERMES experiment at DESY. The average target polarization, $|S_T|$, was 75.4 ± 5.0 %.

---

[1] The angle definitions are consistent with the "Trento Conventions" [5].

[2] See [6] for the sub-leading twist expression.

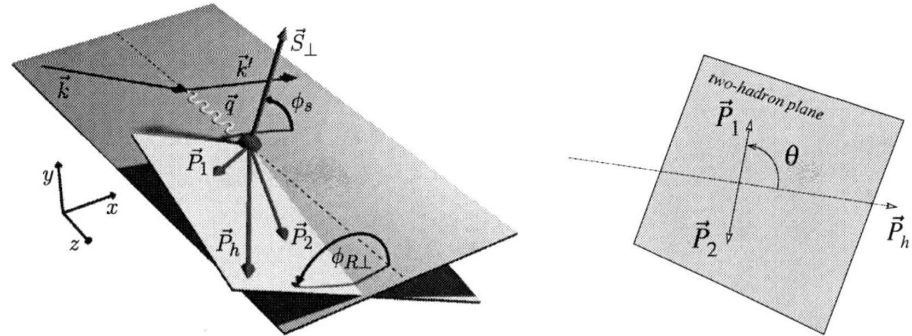

FIGURE 1. Left: kinematic planes, where $\phi_{R\perp}$ is the angle between the plane spanned by the incident ($\vec{k}$) and scattered lepton ($\vec{k}'$) and the plane spanned by the two detected pions $\vec{P}_1$ (π^+) and $\vec{P}_2$ (π^-) with $\vec{P}_h \equiv \vec{P}_1 + \vec{P}_2$. Right: description of the polar angle θ, in the center-of-mass frame of the two pions. The vector $\vec{P}_h$ is evaluated in the hadronic center-of-mass system.

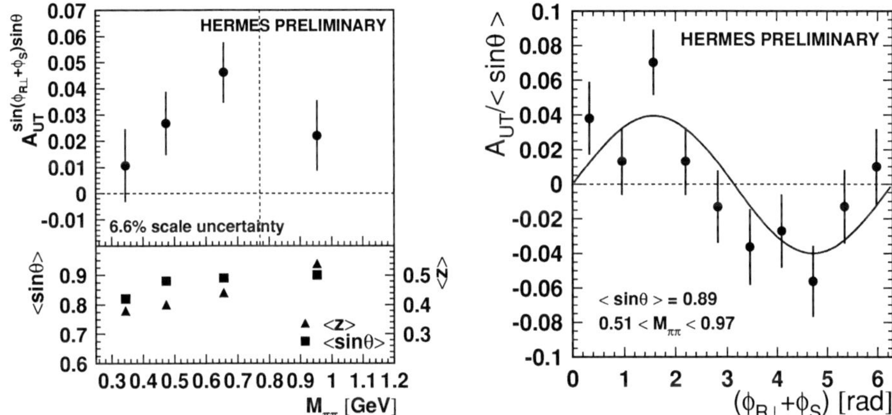

FIGURE 2. On the left the asymmetry $A_{UT}^{\sin(\phi_{R\perp}+\phi_S)\sin\theta}$ is shown versus the invariant mass of the $\pi^+\pi^-$-pair (using mass binning, with the bin boundaries at 0.25, 0.40, 0.55, 0.77, 2.0 GeV) and on the right the asymmetry A_{UT} divided by the average <sin θ> is shown versus the angle combination $(\phi_{R\perp} + \phi_S)$.

In the left plot of Fig. 2 the data for $A_{UT}^{\sin(\phi_{R\perp}+\phi_S)\sin\theta}$ are shown versus the invariant mass of the $\pi^+\pi^-$-pair. The asymmetry is clearly positive over the entire invariant mass range and largest in the region of the ρ^0 mass. The corresponding invariant mass distribution is shown in the left plot of Fig. 3. Whereas the results on SSA's in two-hadron fragmentation using a *longitudinally* polarized deuterium target [8] gave a hint of a sign change of the asymmetry at the ρ^0 mass (770 GeV) as predicted in [4], the new results presented here are clearly inconsistent with such behavior.

In the right plot of Fig. 2 the raw asymmetry is shown in bins of $\phi_{R\perp} + \phi_S$, integrated over the invariant mass range 0.51 GeV $< M_{\pi\pi} <$ 0.97 GeV. This plot shows that a clear $\sin(\phi_{R\perp} + \phi_S)$ behavior is present in the data. The plot includes a curve resulting from

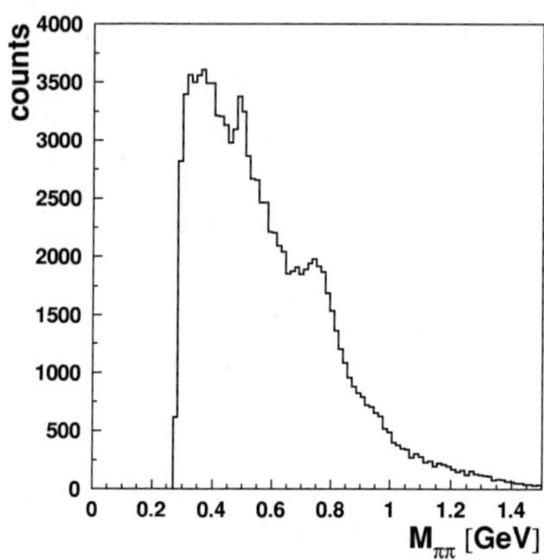

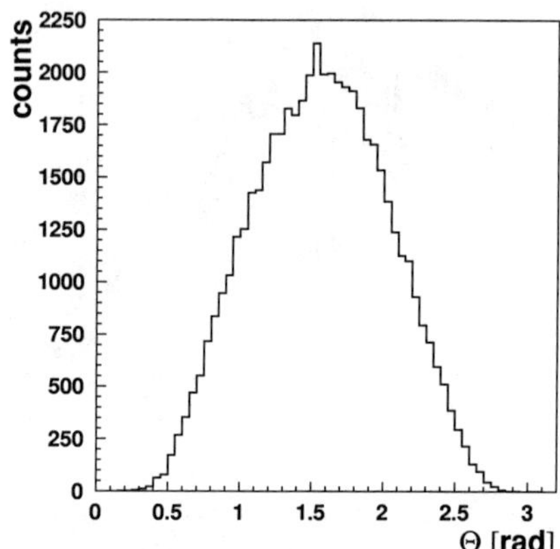

FIGURE 3. The left plot shows the distribution of the invariant mass of the $\pi^+\pi^-$-pairs and the right plot shows the distribution of the angle θ (for the invariant mass range 0.51 GeV $< M_{\pi\pi} <$ 0.97 GeV.

fi tting the data with $f(\phi_{R\perp}+\phi_S) = p_0 + p_1 \sin(\phi_{R\perp}+\phi_S)$, where $p_1 \equiv A_{UT}^{\sin(\phi_{R\perp}+\phi_S)\sin\theta} = 0.040 \pm 0.009$ (stat) $\pm$ 0.003 (syst). Due to the peaked shape of the θ-distribution (right plot in Fig. 3) the asymmetry is mostly evaluated around $\theta = \frac{\pi}{2}$. Therefore the value of $A_{UT}^{\sin(\phi_{R\perp}+\phi_S)\sin\theta}$ is insensitive to whether one uses this one-dimensional fi t function, integrating over θ, or a two-dimensional one, like eq. 3.

Data taking with a transversely polarized hydrogen target will continue until November 2005 after which the analysis of the full data sample is expected to lead to a decrease of the uncertainty on the asymmetry with approximately a factor of $\sqrt{2}$. Further steps in the analysis include looking at the part of the asymmetry coupling to $H_1^{\sphericalangle,pp}$ and studying the x and z dependence of the asymmetries.

ACKNOWLEDGEMENTS

We acknowledge the support of the Dutch Foundation for Fundamenteel Onderzoek der Materie (FOM) and the European Community-Research Infrastructure Activity under the FP6 "Structuring the European Research Area" program (HadronPhysics, contract number RII3-CT-2004-506078).

REFERENCES

1. A. Airapetian *et al.* (HERMES), *Phys. Lett.* **B562**, 182 (2003).
2. A. Airapetian *et al.* (HERMES), *Phys. Rev. Lett.* **94**, 012002 (2005).
3. J. C. Collins, S. F. Heppelmann and G. A. Ladinsky, *Nucl. Phys.* **B420** , 565-582 (1994).
4. R.L. Jaffe, X. Jin, and J. Tang. *Phys. Rev. Lett.* **80**, 1166 (1998).
5. A. Bacchetta *et al.*, hep-ph/0410050 (2004)
6. A. Bacchetta and M. Radici, *Phys. Rev.* D **67**, 094002 (2003), hep-ph/0212300.
7. A. Bacchetta and M. Radici, *Phys. Rev.* D **69**, 074026 (2004), hep-ph/0311173.
8. P. B. van der Nat and K. Griffioen, *Proceedings of SPIN'2004* (2004), hep-ex/0501009.

Transversity signals in two hadron correlation at COMPASS

Rainer Joosten
on behalf of the COMPASS collaboration

Helmholtz Institut für Strahlen- und Kernphysik
Nußallee 14-16, 53115 Bonn, Germany
E-mail: Rainer.Joosten@cern.ch

Abstract.
Measurement of two hadron production introducing the chiral odd interference fragmentation function $H_1^{\sphericalangle}$ is considered a new probe of the transverse spin distribution $\Delta_T q(x)$. COMPASS is a fixed target experiment on the SPS M2 beamline at CERN. Its target can be polarised both longitudinally and transversally with respect to the polarised 160 GeV/c μ^+ beam. In 2002, 2003, and 2004, 20% of the beam-time was spent in the transverse configuration on a ^{6}LiD target, allowing the measurement of transversity effects. First results of the analysis of two hadron production will be reported.

Keywords: Transverse Spin Physics, Interference Fragmentation Functions, Transversity
PACS: 14.20.Dh, 13.60.Hb

THEORETICAL BACKGROUND

The cross-section for deep inelastic scattering off spin 1/2 hadrons can be parametrised, in leading order, in terms of three quark distribution functions: the helicity averaged distribution $q(x)$, the longitudinal helicity distribution $\Delta q(x)$, and the transverse spin distribution $\Delta_T q(x)$. This last distribution function, referred to as transversity, is chiral-odd and can only be measured in combination with another chiral-odd function. So far, attempts were made to measure $\Delta_T q(x)$ in combination with the Collins fragmentation-function $\Delta D_a^h(z, p_T^h)$, requiring the partial detection of the hadronic products (semi-inclusive measurement) [1][2]. Another suggested and very promising probe is the measurement of two hadron production introducing the chiral odd interference fragmentation function $H_1^{\sphericalangle}(z, M_h^2)$. The properties of interference fragmentation functions are described in Refs. [3][4][5][6][7][8].

At leading twist, the fragmentation function of a quark q into a pair h of two hadrons h_1 and h_2 can be written as:

$$D_q^h(z, M_h^2) + H_1^{\sphericalangle}(z, M_h^2) sin(\phi_{RS}) \tag{1}$$

with $\phi_{RS} = \phi_R - \phi_{S'} = \phi_R + \phi_S - \pi$, where $\phi_{S'}$ is the azimuthal angle of the struck quark spin, ϕ_S is the azimuthal angle of the initial quark spin and $\phi_{S'} = \pi - \phi_S$.

ϕ_R is the angle between the lepton scattering plane and the plane spanned by the virtual photon momentum $\mathbf{q}$ and the component $\mathbf{R}_T$ of the relative hadron momentum $\mathbf{R} = \frac{1}{2}(\mathbf{P}_1 - \mathbf{P}_2)$ which is perpendicular to the summed hadron momentum $\mathbf{P}_h = \mathbf{P}_1 + \mathbf{P}_2$.

CP792, *Deep Inelastic Scattering, DIS 2005*, edited by W. H. Smith and S. R. Dasu

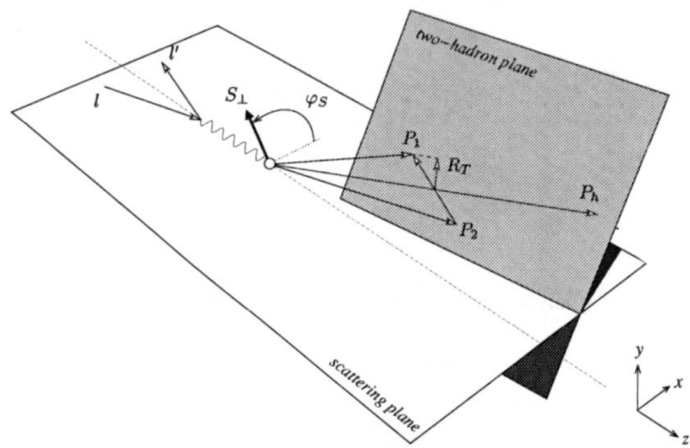

FIGURE 1. Description of the angles involved in the measurement of single spin asymmetries in deep-inelastic production of two hadrons (from Ref. [7])

The angles are defined according to Ref. [9] (see Fig. 1), which follows the so-called *Trento conventions* by:

$$\cos\phi_R = \frac{(\hat{\mathbf{q}}\times\mathbf{l})}{|\hat{\mathbf{q}}\times\mathbf{l}|}\cdot\frac{(\hat{\mathbf{q}}\times\mathbf{R_T})}{|\hat{\mathbf{q}}\times\mathbf{R_T}|}, \qquad \sin\phi_R = \frac{(\mathbf{l}\times\mathbf{R_T})\cdot\hat{\mathbf{q}}}{|\hat{\mathbf{q}}\times\mathbf{l}|\,|\hat{\mathbf{q}}\times\mathbf{R_T}|}, \tag{2}$$

Additionally, $z = z_1 + z_2 = (E(h_1)+E(h_2))/(E_l - E_{l'})$, E_l being the incoming and $E_{l'}$ the scattered lepton energy, is the fraction of the transfered energy carried by the two hadrons, and M_h^2 is their invariant mass squared.

As a result, an asymmetry is expected in the azimuthal angle of the hadron plane which depends on ϕ_{RS}. This asymmetry, which gives information on the transversity distribution, has not been measured so far on a transversely polarised target. One model even predicts a strong dependance of the fragmentation function $H_1^{\sphericalangle}(z,M_h^2)$ on the invariant mass of the two-hadron system in the region of the ρ-mass (Ref. [5]) due to an interference term in two pion production. Another model [6] does not see this strong dependance.

The measured raw asymmetry $A_{UT}^{sin\phi_{RS}}$ is connected to the physically relevant asymmetry $A_{\phi_{RS}}$ by

$$\frac{A_{UT}^{sin\phi_{RS}}}{D_{NN}fP} = A_{\phi_{RS}} = \frac{\Sigma_i e_i \Delta_T q_i(x) H_1^{\sphericalangle h}(z,M_h^2)}{\Sigma_i e_i q_i(x) D_i^h(z,M_h^2)} \tag{3}$$

where $f(\approx 0.4)$ is the dilution factor, $P(\approx 0.45)$ the target polarisation and $D_{NN} = (1-y)/(1-y+y^2/2)$ the depolarisation factor. Here, $y = (E_l - E_{l'})/E_l$ is the fraction of the incoming lepton energy transfered to the hadronic system.

RESULTS FROM THE COMPASS 2002 AND 2003 RUNS

The COMPASS experiment [10][11] uses the high intensity 160 GeV secondary μ^+-beam from π-decay in the CERN SPS M2 beamline. This beam is naturally longitudinally polarised with a polarisation of ≈-76%.

The polarised target consists of two subsequent target cells filled with ^{6}LiD, each 60 cm long, which can be individually polarised using separate RF-cavities. This allows to take data simultaneously on two target cells of opposite polarisation. The target can be polarised longitudinally or transversely with respect to the beam axis.

The data discussed here was taken in 2002 and 2003 with a transversely polarised target. The sample consists of three independent data taking periods, where each period was split in two subperiods with opposite spin orientation in the individual target cells.

The event selection is analogous to the analysis of the Collins and Sivers asymmetries [1][2]. The primary vertex, with identified incoming and scattered muon, is required to be in either of the two target cells. At least two hadrons are required to origin from the same vertex. The separation of muons and hadrons is primarily done by cutting on the amount of traversed material in the spectrometer and on the energy loss in the two hadronic calorimeters. Moreover, the kinematic cuts $Q^2 > 1\,(GeV/c)^2$, $W > 5GeV/c^2$ and $0.1 < y < 0.9$ were applied to ensure a deep-inelastic scattering sample above nuclear resonances and within the COMPASS trigger acceptance. The final data sample had average values for $x = 0.035$, $y = 0.33$, and $Q^2 = 2.4\,(GeV/c)^2$. The mean hadron multiplicity of the events selected by these kinematic cuts is 1.9 hadrons/event.

Based on this sample, hadron pairs are selected by choosing all combinations of positive (h_1) and negative (h_2) hadrons fulfilling the requirements $z_1 = E(h_1)/(E_l - E_{l'}) > 0.1$ and $z_2 = E(h_2)/(E_l - E_{l'}) > 0.1$ as well as $z = z_1 + z_2 < 0.9$. The first two conditions reject the target fragmentation region while the later suppresses the contamination with exclusively produced ρ-mesons. The resulting sample contains 2.810^6 hadron combinations.

From the data, for each target cell and polarisation, the property

$$N(\phi_{RS}) = N_0 \cdot (1 + A_{UT}^{sin\phi_{RS}} \cdot sin\phi_{RS}) \cdot F_{acc}(\phi_{RS}) \tag{4}$$

can be derived, where $F_{acc}(\phi_{RS})$ is the (unknown) angle dependant acceptance function of the detector. However, by comparing the subperiods with opposite target spin, this acceptance function cancels, resulting in

$$A_{UT}^{sin\phi_{RS}} \cdot sin\phi_{RS} = \frac{N^{\Uparrow}(\phi_{RS}) - RN^{\Downarrow}(\phi_{RS} + \pi)}{N^{\Uparrow}(\phi_{RS}) + RN^{\Downarrow}(\phi_{RS} + \pi)} \tag{5}$$

where $R = N_{tot}^{\Uparrow}/N_{tot}^{\Downarrow}$ is the ratio of the events with opposite target polarisation. (For a more detailed description please refer to [2]). This procedure is repeated for both target cells and all periods. Finally, the weighted mean of the results is calculated.

Figure 2 shows the preliminary results from the COMPASS 2002 and 2003 data. The upper plot shows the asymmetries vs. the invariant mass M_h of the hadron pair, while the two lower plots show the asymmetries vs. x and z respectively.

The observed asymmetries are very small and no significant signal can be observed. Especially the asymmetry vs. M_h does not at all show a strong dependance on the hadron invariant mass. The fluctuations for the signal vs. x and z are very small and still compatible with zero. Including the data of the COMPASS 2004 run on a ^{6}LiD target will double the total statistics and improve the sensitivity by a factor 1.4.

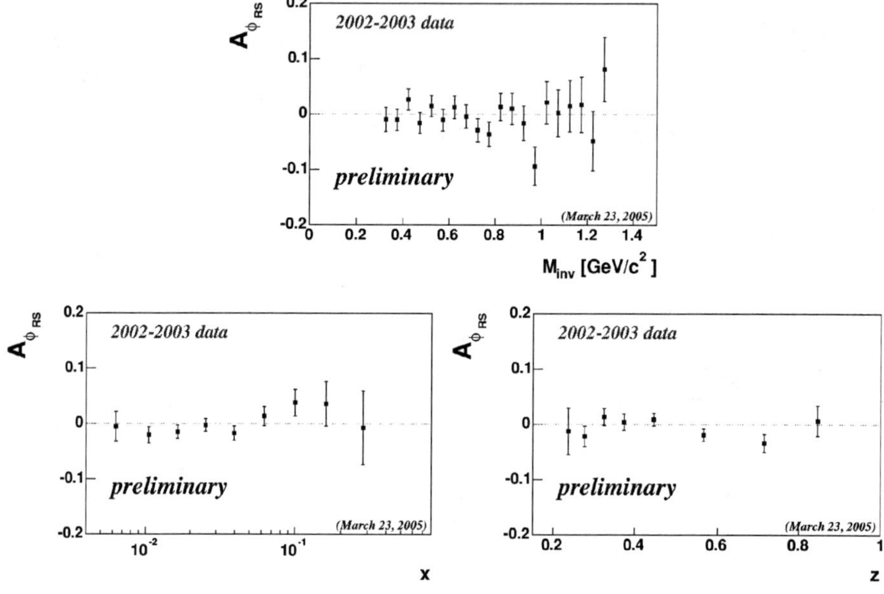

FIGURE 2. Asymmetries $A_{\phi_{RS}}$ for the 2002 and 2003 data vs. invariant mass of the hadron pair (top) and x (bottom left) and $z = z_1 + z_2$ (bottom right).

Furthermore, including hadron identification using the RICH information will clean the hadron sample. Additionally, a complementary measurement on a proton target is planned for the COMPASS 2006 run.

REFERENCES

1. P. Pagano (COMPASS), these proceedings.
2. V.Yu. Alexakhin *et al.* [COMPASS Collaboration], *Phys. Rev. Lett.* **94**, 202002 (2005).
3. J.R. Collins, S.F. Heppelmann and G.A. Ladinsky,*Nucl. Phys.* **B420**, 565 (1994).
4. X. Artru and J. C. Collins, *Z. Phys.* **C69**, 277 (1996).
5. R. L. Jaffe, X. Jin and J. Tang, *Phys. Rev. Lett.* **80**, 1166 (1998).
6. M. Radici, R. Jakob and A. Bianconi, *Phys. Rev.* **D65** 074031 (2002).
7. A. Bianconi, S. Boffi, R. Jakob and M. Radici, *Phys. Rev.* **D62**, 034008 (2000).
8. A. Bacchetta and M. Radici, *Phys. Rev.* **D69**, 074026 (2004).
9. A. Bacchetta and M. Radici, Proceeding of DIS 2004, hep-ph/0407345 (2004).
10. G. Baum *et al.* [COMPASS Collaboration], CERN-SPSLC-96-14.
11. G.K. Mallot, Nucl. Instrum. Meth. A **518**, 121 (2004).

Latest Results on g_1 and g_2 at high x

Jian-ping Chen

Jefferson Lab, Newport News, Virginia 23606, USA

Abstract. Recent progress from Jefferson Lab has significantly improved our understanding of the nucleon spin structure in the high-x region. Results of a precision measurement of the neutron spin asymmetry, A_1^n, in the high-x (valence quark) region are discussed. The up and down quark spin distributions in the nucleon were extracted. A_2^n was also measured. The results were used, in combination with existing data, to extract the second moment, d_2^n. Preliminary results on A_1^p and A_1^d in the high-x region have also become available. Finally, the results of a precision measurement of the g_2 structure function to study higher twist effects will be presented.

Keywords: Spin Structure, high x, higher twist, JLab
PACS: 13.60.Hb

Introduction and Motivation

Recently, the high polarized luminosity available at Jefferson Lab (JLab) has allowed the study of the nucleon spin structure with an unprecedented precision, enabling us to access the valence quark (high-x) region and also to expand the study to the second spin structure function, g_2.

The high-x region is of special interest, because this is where the valence quark contributions are expected to dominate. With sea quark and explicit gluon contributions expected not to be important, it is a clean region to test our understanding of nucleon structure. Relativistic constituent quark models [1] should be applicable in this region and perturbative QCD [2] can be used to make predictions in the large x limit.

To first approximation, the constituent quarks in the nucleon are described by the SU(6) wavefunctions. SU(6) symmetry leads to the following predictions:

$$A_1^p = 5/9; \;\; A_1^n = 0; \;\; \Delta u/u = 2/3; \;\; \Delta d/d = -1/3. \tag{1}$$

Relativistic Constituent Quark Models (RCQM) with broken SU(6) symmetry, e.g., the hyperfine interaction model [1], lead to a dominance of a 'diquark' configuration with the diquark spin $S = 0$ at high x. This implies that as $x \to 1$:

$$A_1^p \to 1; \;\; A_1^n \to 1; \;\; \Delta u/u \to 1; \;\; \text{and} \;\; \Delta d/d \to -1/3. \tag{2}$$

In the RCQM, relativistic effects give rise to the quark orbital angular momentum and reduce the valence quark contributions to the nucleon spin from 1 to $0.6 - 0.75$.

Another approach is leading-order pQCD [2], which assumes the quark orbital angular momentum to be negligible and leads to hadron helicity conservation. It yields:

$$A_1^p \to 1; \;\; A_1^n \to 1; \;\; \Delta u/u \to 1; \;\; \text{and} \;\; \Delta d/d \to 1. \tag{3}$$

CP792, *Deep Inelastic Scattering, DIS 2005,* edited by W. H. Smith and S. R. Dasu

Not only are the limiting values as $x \to 1$ important, but also the behavior in the high-x region. How A_1^n and A_1^p approach their limiting values when x approaches 1, is sensitive to the dynamics in the valence quark region.

g_2, unlike g_1 and F_1, cannot be interpreted in the simple quark-parton model. To understand g_2 properly, it is best to start with the operator product expansion method (OPE) [3]. In the OPE, neglecting quark masses, g_2 can be cleanly separated into a twist-2 and a higher twist term:

$$g_2(x,Q^2) = g_2^{WW}(x,Q^2) + g_2^{H.T.}(x,Q^2)\,. \tag{4}$$

The leading-twist term can be determined from g_1 as [4]

$$g_2^{WW}(x,Q^2) = -g_1(x,Q^2) + \int_x^1 \frac{g_1(y,Q^2)}{y} dy\,, \tag{5}$$

and the higher-twist term arises from the quark-gluon correlations. Therefore g_2 provides a clean way to study higher-twist effects. In addition, at high Q^2, the x^2-weighted moment, d_2, is a twist-3 matrix element and is related to the color polarizabilities [5]:

$$d_2 = \int_0^1 x^2[g_2(x) - g_2^{WW}(x)]dx. \tag{6}$$

Predictions for d_2 exist from various models and lattice QCD.

Recent results from Jefferson Lab

In 2001, JLab experiment E99-117 [6] was carried out in Hall A to measure A_1^n with high precision in the x region from 0.33 to 0.61 (Q^2 from 2.7 to 4.8 GeV^2). Asymmetries from inclusive scattering of a highly polarized 5.7 GeV electron beam on a high pressure (> 10 atm) (both longitudinally and transversely) polarized ^{3}He target were measured. Parallel and perpendicular asymmetries were extracted for ^{3}He. After taking into account the beam and target polarization and the dilution factor, they were combined to form $A_1^{^3He}$. Using the most recent model [7], nuclear corrections were applied to extract A_1^n. The results on A_1^n are shown in the left panel of Fig. 1.

The experiment greatly improved the precision of data in the high-x region, providing the first evidence that A_1^n becomes positive at large x, showing clear SU(6) symmetry breaking. The results are in good agreement with the LSS 2001 pQCD fit to previous world data [8] (solid curve) and the statistical model [9] (long-dashed curve). The trend of the data is consistent with the RCQM predictions (the shaded band). The data disagree with the predictions from the leading-order pQCD models (short-dashed and dash-dotted curves).

In the leading-order approximation, the polarized quark distribution functions $\Delta u/u$ and $\Delta d/d$ were extracted from our neutron data combined with the world proton data. The results are shown in the right panel of Fig. 1, along with predictions from the RCQM (dot-dashed curves), leading-order pQCD (short-dashed curves), the LSS 2001

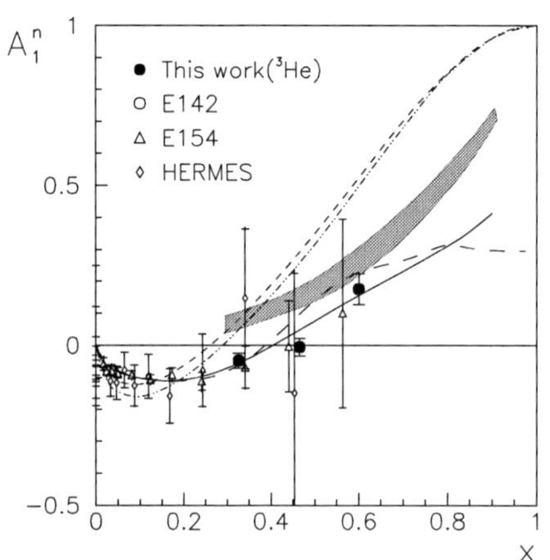

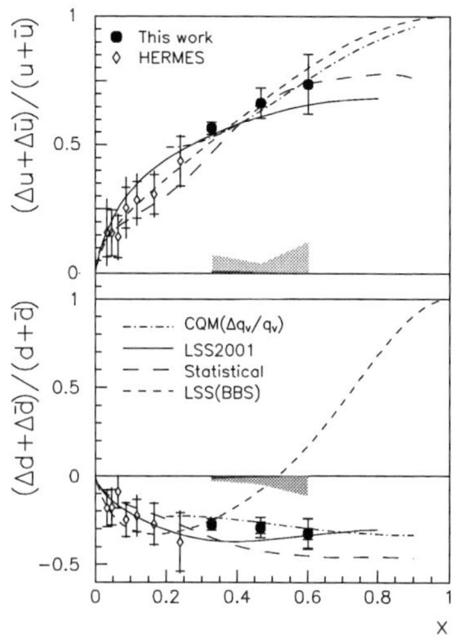

FIGURE 1. A_1^n, $\Delta u/u$ and $\Delta d/d$ results compared with the world data and theoretical predictions.

fits (solid curves) and the statistical model (long-dashed curves). The results agree well with RCQM predictions as well as the LSS 2001 fits and statistical model but are in significant disagreement with the predictions from the leading-order pQCD models assuming hadron helicity conservation. This suggests that effects beyond leading-order pQCD, such as the quark orbital angular momentum, may play an important role in this kinematic region.

A_2^n was also obtained from the same experiment. The precision of the A_2^n data is comparable to that of the best existing world data [10] at high x. Combining these results with the world data, the second moment d_2^n was extracted at an average Q^2 of 5 GeV^2:

$$d_2^n = 0.0062 \pm 0.0028. \tag{7}$$

Compared to the previously published result [10], the uncertainty on d_2^n has been improved by about a factor of 2. The d_2 moment at high Q^2 has been calculated by Lattice QCD and a number of theoretical models. While a negative or near-zero value was predicted by Lattice QCD and most models, the new result for d_2^n is positive.

Preliminary results of A_1^p and A_1^d from the Hall B eg1 experiment [11] have recently become available. The data cover the Q^2 range of 1.4 to 4.5 GeV^2 for x from 0.2 to 0.6 with an invariant mass larger than 2 GeV. The precision of the data improved significantly over that of the existing world data.

A precision measurement of g_2^n from JLab E97-103 [12] covered five different Q^2 values from 0.58 to 1.36 GeV^2 at x $\approx$ 0.2. Results for g_2^n as well as g_1^n are given in Fig. 2. The light-shaded area in the two plots gives the leading-twist contribution to these two quantities, respectively, obtained by fitting world data and evolving to the Q^2 values of this experiment. The systematic errors are shown as the dark-shaded area near the horizontal axes.

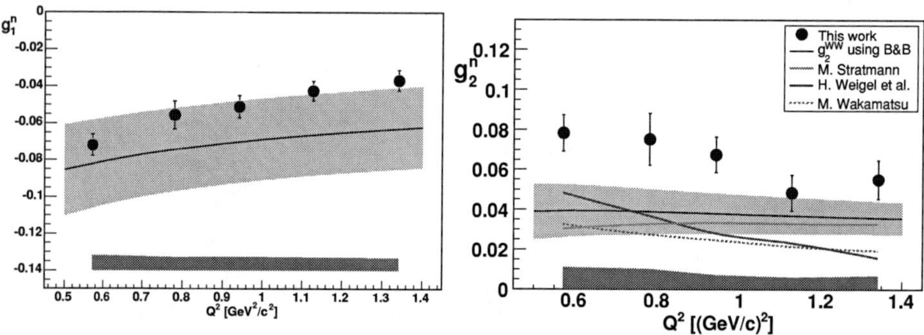

FIGURE 2. Fig. 1: results for g_1^n (left) and g_2^n (right) from E97103.

The precision reached is more than an order of magnitude improvement over that of the best world data. The difference of g_2 from the leading twist part (g_2^{WW})[4] is due to higher twist effects and is sensitive to quark-gluon correlations. The g_2^{WW} values were obtained from a fit [13] to the world high Q^2 data, then evolved to the Q^2 values of this experiment. The measured g_2^n values are consistently higher than g_2^{WW}. For the first time, there is a clear indication that higher twist effects become important at the level of precision of these data. The new g_1^n data agree with the leading-twist calculations within the uncertainties.

In summary, the high polarized luminosity available at JLab, has provided high-precision data to study the nucleon spin structure in the high-x region and higher twist effects, which shed light on the valence quark structure and help to understand quark-gluon correlations.

The work presented was supported in part by the U. S. Department of Energy (DOE) contract DE-AC05-84ER40150 Modification NO. M175, under which the Southeastern Universities Research Association operates the Thomas Jefferson National Accelerator Facility.

REFERENCES

1. N. Isgur, Phys. Rev. D**59**, 034013 (1999).
2. S. Brodsky, M Burkhardt and I. Schmidt, Nucl. Phys. B**441**, 197 (1995).
3. K. Wilson, Phys. Rev. **179**, 1499 (1969).
4. S. Wandzura and F. Wilczek, Phys. Lett. B 72 (1977).
5. X. Ji and P. Unrau, Phys. Lett. B **333**, 228 (1994).
6. X. Zheng, *et al.*, Phys. Rev. Lett. **92**, 012004 (2004); Phys. Rev. C **70**, 065207 (2004).
7. F. Bissey, *et al.*, Phys. Rev. C **65**, 064317 (2002).
8. E. Leader, A. V. Sidorov and D. B. Stamenov, Eur. Phys. J. **C 23**, 479 (2002).
9. C. Bourrely, J. Soffer and F. Buccella, Eur. Phys. J. **C 23**, 487 (2002).
10. K. Abe, *et al.*, E155 collaboration, Phys. Lett. B **493**, 19 (2000).
11. S. Kuhn, private communication.
12. K. Kramer, *et al.*, nucl-ex/0506005, submitted to Phys. Rev. Lett.
13. J. Blümlein and H. Bottcher, Nucl. Phys. B **636**, 225 (2002).

$g_1(x)$ and $g_2(x)$ in the Meson Cloud Model

A. I. Signal

Institute of Fundamental Sciences PN461
Massey University
Private Bag 11 222, Palmerston North
New Zealand

Abstract. We calculate the spin dependent structure functions $g_1(x)$ and $g_2(x)$ of the proton and neutron. Our calculation uses the meson cloud model of nucleon structure and includes the effects of kinematic terms which mix transverse and longitudinal spin components. We find small corrections to the nucleon structure functions, however these are significant for the neutron.

Keywords: Nucleon structure, Polarization, Meson cloud
PACS: 14.20.Dh, 13.88+e, 11.30Hv, 12.39.Ba, 13.60.Hb

The spin dependent structure functions of the nucleon are the subject of much theoretical and experimental interest. As deep inelastic (and other) experiments become more precise it is hoped that it may be possible to make an unequivocal measurement of a higher twist component in the structure function g_2 of the nucleon. This would give new information on the gluon field inside the nucleon, and its relationship with the quark fields.

In order to make such an unequivocal identification it is necessary to understand the relationship between the structure functions g_1 and g_2. In particular there are leading twist contributions to g_2 which arise from scattering from the components of the meson cloud of the physical nucleon. These contributions are of the order of 10% of the structure function, and need to be taken into account when calculating the twist-2 part of g_2.

The Meson Cloud Model (MCM) arises from the crucial observation [1] that the contribution of scattering from the pion cloud of the nucleon scales in the Bjorken limit. This implies that the parton distributions of the nucleon are modified via a convolution between the parton distribution of the meson and the momentum distribution of the meson in the proton, viz.

$$\delta q^p(x) = \int_x^1 \frac{dy}{y} f_{p\pi}(y) q^\pi\left(\frac{x}{y}\right). \tag{1}$$

As well as pions, the MCM takes into account scattering from the other baryon plus meson components in the Fock expansion of the wavefunction i.e.

$$|N\rangle_{\text{physical}} = \sqrt{Z}|N\rangle_{\text{bare}} + \sum_{MB} \int dy\, d^2\mathbf{k}_\perp\, \phi(y, k_\perp^2)\, |M(y, \mathbf{k}_\perp); B(1-y, -\mathbf{k}_\perp)\rangle. \tag{2}$$

The other ingredients of the model are the interaction Lagrangians $\mathscr{L}_{int}$ describing the $N \to BM$ vertices and the form factors for these vertices. The small probability of finding

CP792, *Deep Inelastic Scattering, DIS 2005*, edited by W. H. Smith and S. R. Dasu

high mass states in this model leads to quick convergence of the sum over baryon-meson states for structure function calculations.

The MCM has been applied successfully in spin independent DIS, giving a good description of the HERA data on semi-inclusive DIS with a leading neutron [2, 3], and also dijet events with a leading neutron [4, 5]. In addition the MCM gives a good description of the observed violation of the Gottfried sum rule [6, 7].

To extend the model to spin dependent DIS requires the contributions of both pseudoscalar and pseudovector mesons, particularly the ρ meson. The pseudoscalar contributions mainly 'dilute' the bare spin dependent pdfs, however the importance of $L \neq 0$ amplitudes in the cloud cannot be ignored [8, 9]. The pseudovector mesons can contribute directly to the spin dependent pdfs. In earlier work we calculated the spin dependent sea distributions $\Delta\bar{u}(x)$, $\Delta\bar{d}(x)$, $s(x)$ and $\Delta\bar{s}(x)$ [10]. Our results are in good agreement with the HERMES data [11].

The structure functions $g_1(x)$ and $g_2(x)$ are dominated by valence rather than sea distributions, so the most important contributions in the MCM are those affecting the valence quarks, which are $N \to N\pi$ and $N \to \Delta\pi$, with $\mathscr{L}_{int} = ig_{NN\pi}\bar{\psi}\gamma_5\pi\psi$, $f_{N\Delta\pi}\bar{\psi}\partial_\mu\pi\chi^\mu + \text{h.c.}$ respectively.

At finite Q^2 the spin of the struck hadron from the cloud, in this case either a nucleon or Δ, is not parallel with the initial spin of the target nucleon. This implies,as shown by Kumano and Miyama [12], that both longitudinal and transverse spin structure functions of the cloud hadrons contribute to the observed structure functions. For a spin $1/2$ baryon component of the cloud we have

$$\delta g_1(x,Q^2) = \frac{1}{1+\gamma^2}\int_x^1 \frac{dy}{y} \sum_{i=1,2} (-1)^{i+1}[\Delta f_{iL}(y)+\Delta f_{iT}(y)] g_i^B(\frac{x}{y},Q^2) \tag{3}$$

$$\delta g_2(x,Q^2) = \frac{1}{1+\gamma^2}\int_x^1 \frac{dy}{y} \sum_{i=1,2} (-1)^{i}\left[\Delta f_{iL}(y)+\frac{\Delta f_{iT}(y)}{\gamma^2}\right] g_i^B(\frac{x}{y},Q^2). \tag{4}$$

where $\gamma^2 = 4x^2m_N^2/Q^2$ and $\Delta f_{iL,T}(y)$ are the diferences between spin up and spin down fluctuation functions projected longitudinally along or transverse to the baryon 3-momentum. Similar expressions exist for higher spin components of the cloud [12, 13]. The fluctuations are calculated using standard techniques in time-ordered perturbation theory in the infinite momentum frame [8, 12, 13]. We find that for longitudinal fluctuation functions $\Delta f_{iL}(y)$ the nucleon and Δ contributions are of similar size, with the $s = 3/2$ state of the Δ being important. For transverse fluctuations $\Delta f_{iT}(y)$ the nucleon contributions are much larger than those from the Δ.

In order to estimate the size of the MCM contributions to g_1 and g_2, we need also to calculate the structure functions of the 'bare' hadrons. We use the MIT bag model and the methods developed by the Adelaide group [9, 14] to calculate the spin dependent pdfs. We also add 'by hand' a phenomenological $\Delta g(x)$ such that the integral of $g_1^p(x)$ agrees with experiment. The resulting $g_1^p(x)$ and $g_1^n(x)$ give a reasonable description of the experimental data. To calculate the bare $g_2(x)$ we simply use the leading twist Wandzura-Wilczek term

$$g_2(x) = -g_1(x) + \int_x^1 \frac{dy}{y} g_1(y), \tag{5}$$

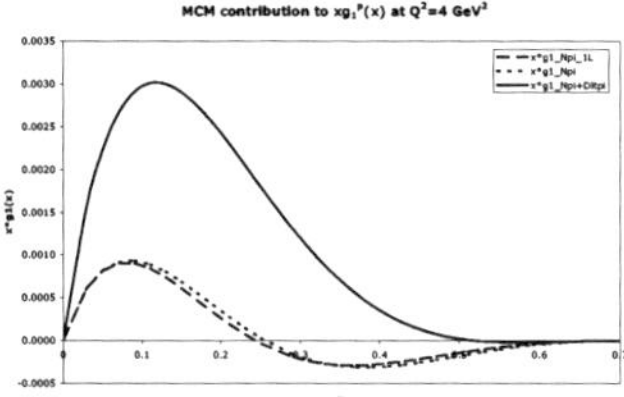

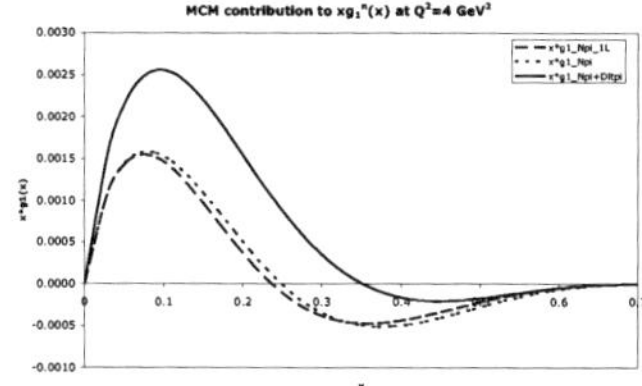

FIGURE 1. Meson cloud Model contributions to g_1 of the proton and neutron. The dashed line is the contribution from longitudinally projected nucleon fluctuations, the dotted line is the total contribution from nucleon fluctuations and the solid line is the total from nucleon and Δ fluctuations.

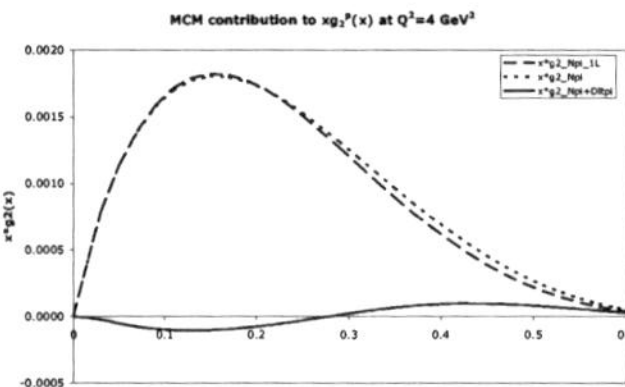

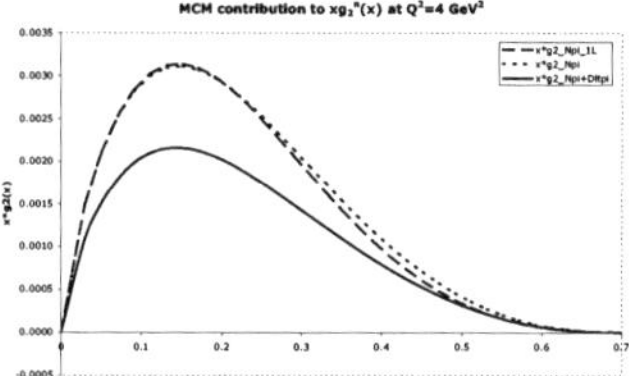

FIGURE 2. Meson cloud Model contributions to g_2 of the proton and neutron. The dashed line is the contribution from longitudinally projected nucleon fluctuations, the dotted line is the total contribution from nucleon fluctuations and the solid line is the total from nucleon and Δ fluctuations.

which also gives a good description of the available experimental data on $g_2^p(x)$ and $g_2^n(x)$.

Our calculations of the contributions to $g_1(x)$ and $g_2(x)$ for both the proton and neutron are shown in figures 1 and 2. For the proton we note that the magnitudes of these contributions are much smaller than the size of the experimental data. We see that the contributions from transversely projected cloud baryons are small, and that the contributions of nucleons and Δ baryons are of similar magnitude, though not necessarily the same sign. For the neutron structure functions the MCM contributions are around 10% of the size of the experimental data. These corrections will be important to consider in any extraction of higher twist components to the neutron structure functions, as they have similar magnitude to these components. Also these corrections have a weak scale dependence which can mimic that of twist-3 contributions at low Q^2. The correction to $g_2^n(x)$ is positive, and may be able to account for the deviation of the JLab E97-103 data around $Q^2 = 1$ GeV^2 from the Wandzura-Wilczek term [15].

ACKNOWLEDGMENTS

A.S. acknowledges the hospitality and support of the Institute for Particle Physics Phenomenology, Durham University, where portions of this work were done.

REFERENCES

1. J. D. Sullivan, *Phys. Rev. D*, **5**, 1732 (1972).
2. C. Adloff et al., *Eur. Phys. J. C*, **6**, 587 (1999).
3. M. Derrick et al., *Phys. Lett. B*, **384**, 388 (1996); S. Chekanov et al., *Nucl. Phys. B*, **637**, 3 (2002).
4. J. Breitweg et al., *Nucl. Phys. B*, **596**, 3 (2001).
5. A. Aktas et al., hep-ex/0501074, *submitted to Eur. Phys. J. C.*
6. P. Amaudraz et al., *Phys. Rev. Lett.*, **66**, 2712 (1991); M. Arneodo et al., *Phys. Rev. D*, **50**, 1 (1994); M. Arneodo et al., *Phys. Lett. B*, **364**, 107 (1995).
7. E. A. Hawker et al., *Phys. Rev. Lett.*, **80**, 3715 (1998); *Phys. Rev. D*, **64**, 052002 (2001).
8. H. Holtmann, A. Szczurek, and J. Speth, *Nucl. Phys. A*, **569**, 631 (1996).
9. C. Boros and A. W. Thomas, *Phys. Rev. D*, **60**, 074017 (1999).
10. F. G. Cao and A. I. Signal, *Phys. Rev. D*, **68**, 074002 (2003).
11. A. Airapetian et al., *Phys. Rev. Lett.*, **92**, 012005 (2004).
12. S. Kumano and M. Miyama, *Phys. Rev. D*, **65**, 034012 (2002).
13. F. Bissey, F. G. Cao and A. I. Signal, *in preparation.*
14. A. I. Signal and A. W. Thomas, *Phys. Lett. B*, **211**, 481 (1988); *Phys. Rev. D*, **40**, 2832 (1989); A. W. Schreiber, A. W. Thomas, and J. T. Londergan, *Phys. Rev. D*, **42**, 2226 (1990).
15. J. P. Cheng, *presentation at DIS05.*

Jefferson Lab's results on the Q^2-evolution of moments of spin structure functions

A. Deur

Thomas Jefferson National Accelerator Facility, Newport-News, VA 23606, USA

Abstract. We present the recent JLab measurements on moments of spin structure functions at intermediate and low Q^2. The Bjorken sum and Burkhardt-Cottingham sum on the neutron are presented. The later appears to hold. Higher moments (generalized spin polarizabilities and d_2^n) are shown and compared to chiral perturbation theory and lattice QCD respectively.

Keywords: nucleon spin structure, moment, sum rule, higher twists, polarizability
PACS: 14.20.Dh, 13.40.-f 13.60.Hb

MOMENTS OF SPIN STRUCTURE FUNCTIONS

Polarized DIS has provided a testing ground for the study of the strong force. Moments of spin structure functions (SSF), among them the Bjorken sum, has played an important rôle in this study. The n-th (Cornwall-Norton) moment of SSF is the integral of the $x^n g_{1,2}(x,Q^2)$ SSF over x. Moments are specially useful because sum rules relate them to other quantities. Such sum rules for Γ_1, the first moment of g_1, are the Ellis-Jaffe [1] and the Bjorken sum rules [2], derived at large Q^2, and the related Gerasimov-Drell-Hearn (GDH) sum rule [3] at $Q^2 = 0$. The first moment of g_2, Γ_2, is given by the Burkhardt-Cottingham (BC) sum rule [4]. Rules can be also derived for higher moments, e.g., spin polarizability or d_2 sum rules.

These relations are useful in many ways: checks the theory on which the rule is based (e.g. QCD and the Bjorken sum rule); checks hypotheses used in the sum rule derivation (e.g. the Ellis-Jaffe sum rules); or checks calculations such as chiral perturbation theory (χpt), lattice QCD or Operator Product Expansion (OPE). If a sum rule rests on solid grounds or is well tested, it can be used to extract quantities otherwise hard to measure (e.g. generalized spin polarizabilities). Because $\Gamma_{1,2}$ are calculable at any Q^2 using either χpt, lattice QCD or OPE, they are particularly suited to study the transition between the hadronic to partonic descriptions of the strong force. Measurements in the transition region (intermediate Q^2) have recently been made at Jefferson Lab (JLab).

MEASUREMENTS AT JEFFERSON LAB

At moderate Q^2, resonances saturate moments. JLab's accelerator delivers CW electron beam with a maximum energy up to 6 GeV. This makes JLab the suited place to measure moments up to Q^2 of a few GeV^2. The beam current can reach 200 μA with a polarization now reaching 85% although at the time of the experiments reported here, it was typically 70%. The beam is sent simultaneously to three halls (A, B and C), all of them equipped with polarized targets. In this talk, we report on results from halls A and B.

Hall A [5] contains a polarized ^{3}He gaseous target and two high resolution spectrometers

CP792, *Deep Inelastic Scattering, DIS 2005,* edited by W. H. Smith and S. R. Dasu

(HRS) with 6 mSr acceptance. The target can be polarized longitudinally or transversally at typically 40% polarization with 10-15 μA of beam. The target's $\sim$ 10 atm. of ^{3}He gives a luminosity greater than 10^{36}cm^{-2}s^{-1}. Hall B [6] luminosity is typically 5×10^{33}cm^{-2}s^{-1} but is compensated by the large acceptance (about 2.5π) of the CLAS spectrometer. Cryogenic polarized targets (NH_3 and ND_3) are well suited for the low beam currents ($\sim$nA) utilized in Hall B. The target is longitudinally polarized with average 75% (NH_3) and 40% (ND_3) polarizations. Both halls can cover the large region of Q^2 and x needed to extract moments at various Q^2, either because of the large CLAS acceptance (Hall B) or because of large luminosity allowing to quickly gather data at various beam energies and HRS settings (Hall A).

I report here on the Hall A E94010 [7] and Hall B EG1 experiments. EG1 was split in two runs: EG1a (1998) which results are published [8], and EG1b (2000) that is still being analyzed. SSF are extracted differently in halls A and B. In Hall A, *absolute* cross sections asymmetries $\Delta\sigma^{\parallel(\perp)}$ were measured for longitudinal (transverse) target spin orientations. g_1 and g_2 are linear combinations of these $\Delta\sigma$ and are extracted without external input. Furthermore, unpolarized contributions, e.g. target cell windows or the (mostly) unpolarized protons in the ^{3}He nucleus, cancel out. The *relative* longitudinal asymmetry $A_{\parallel}$ is measured in Hall B. Models for F_1, g_2 and $R=\sigma_L/\sigma_T$ are then used to extract g_1. F_1 and R are constrained at low Q^2 by recent Hall C data [9]. g_2 is estimated using models (resonance region) or its leading twist part g_2^{ww} (DIS domain). The unmeasured low-x part of the moment is estimated using a parametrization developed by the EG1 collaboration, while the E94010 group used a Regge-type fit of DIS data [10]. Results on Γ_1^p, Γ_1^n and Γ_1^d are shown in Fig. 1, together with χpt calculations [16, 17] models [12, 15] and leading twist OPE prediction. HERMES [13] and SLAC [14] results are also shown. The halls A and B data, reanalyzed at matched Q^2 points and with a consistent low-x estimate [10] were used to form the Bjorken sum Γ_1^{p-n} [11]. Preliminary Γ_1^{p-n} from EG1b is also shown. Γ_1^{p-n} is a unique quantity to study parton-hadron transition because its non-singlet structure makes it an easier quantity to handle for χpt, lattice QCD and OPE. These data form, for both nucleons, an accurate mapping at intermediate Q^2 that connects to SLAC, HERMES and CERN DIS data. At low Q^2, χpt disagrees with the data above $Q^2=0.2$ GeV^2, while models based on different physics reproduce equally well the data. Twist 2 description also works well down to low Q^2, indicating an overall suppressed higher twist rôle. Indeed, in OPE analysis results [11, 18, 19], twist 4 and 6 coefficients are either small or canceling each others at Q^2=1 GeV^2.

The availability of transverse data in Hall A allows us to form Γ_2^n and thereby check the BC sum rule ($\Gamma_2=0$) on the neutron (fig. 2). The sum rule is based on dispersion relations and is Q^2-invariant. A striking feature is the almost perfect cancellation between elastic and resonance contributions leading to the verification of the sum rule. Other sum rules link SSF moments to the generalized spin polarizabilities γ_0 and δ_{LT}:

$$\gamma_0=\frac{4e^2M^2}{\pi Q^6}\int_0^{1^-}x^2(g_1-\frac{4M^2}{Q^2}x^2g_2)dx \ ; \ \delta_{LT}=\frac{4e^2M^2}{\pi Q^6}\int_0^{1^-}x^2(g_1+g_2)dx$$

Results from Hall A can be seen in fig. 2 [20]. δ_{LT} is interesting because the Δ_{1232} rôle is suppressed. Hence δ_{LT} is easier to access by χpt. However, calculations and data disagree for both γ_0 and δ_{LT}. The MAID model [21], however, well reproduces the data.

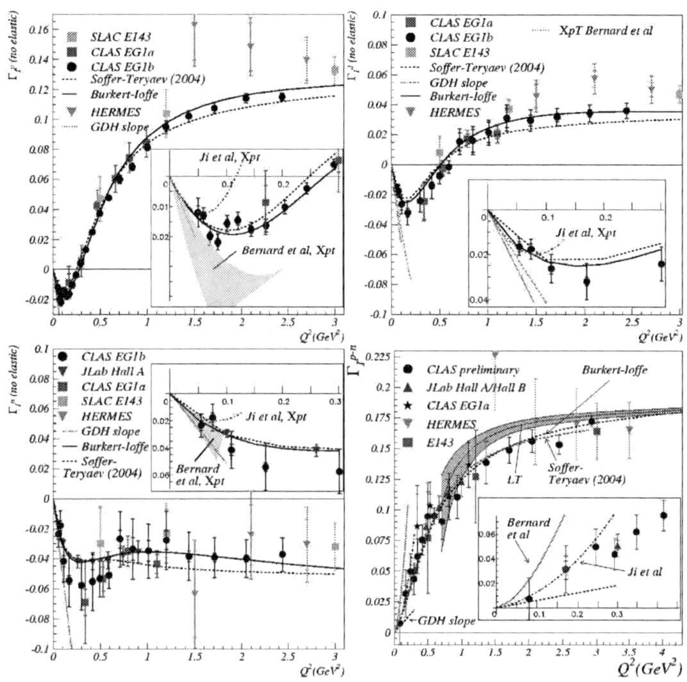

FIGURE 1. First moments Γ_1^p, Γ_1^n, Γ_1^d and the Bjorken sum Γ_1^{p-n}. The elastic contribution is excluded.

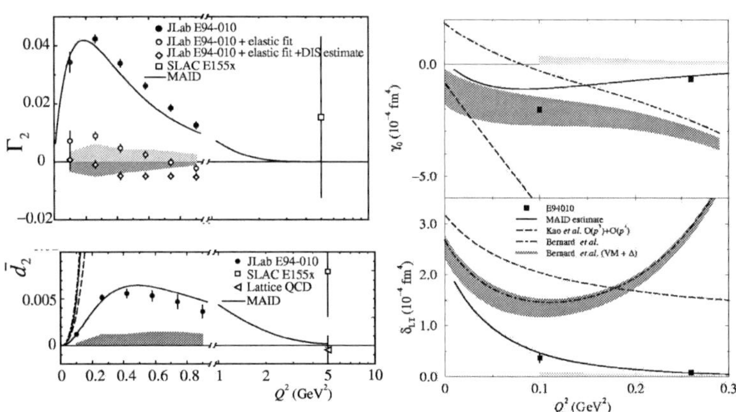

FIGURE 2. Moments Γ_2^n and $\overline{d}_2^n$ (left), and generalized spin polarizabilities γ_0 and δ_{LT} (right)

Another higher moment that can be formed is d_2^n, the integral of $x^2(g_2 - g_2^{ww})$ where g_2^{ww} is the leading twist part of g_2. Thus d_2 is sensitive to twist 3 and higher. The measured $\overline{d}_2^n$ (the bar indicates the exclusion of $x = 1$) trends toward the lattice QCD results, although larger Q^2 data are necessary to establish a possible agreement.

SUMMARY AND PERSPECTIVES

The hadron-parton transition region is covered by data of the SSF moments from JLab. These can be calculated at any Q^2, thus providing a ground for studying the link between hadronic and partonic descriptions of the strong force. An OPE analysis reveals that in this domain, high twist effects are small. The BC sum rule was shown on the neutron and found to be valid. Data and sum rules were used to extract neutron generalized spin polarizabilities. Those disagree with the present χpt calculations. Further data from Hall A E01-012 [22], Hall B EG1b, and Hall C RSS [23] will be available shortly in the resonance region. New data at very low Q^2 have been taken on the neutron in Hall A [24] and will be gathered early 2006 for the proton in Hall B [25]. The 12 GeV upgrade of JLab will allow us to access both larger-x and lower-x. This will allow for more precise measurements of the moments, in particular by addressing the low$x$ issue.

ACKNOWLEDGMENTS

This work is supported by the U.S. Department of Energy (DOE) and the U.S. National Science Foundation. The Southeastern Universities Research Association operates the Thomas Jefferson National Accelerator Facility for the DOE under contract DE-AC05-84ER40150.

REFERENCES

1. J. Ellis and R. L. Jaffe, Phys. Rev. **D9** 1444, (1974)
2. J. D. Bjorken, Phys. Rev. **148**, 1467 (1966)
3. S. Drell and A. Hearn, Phys. Rev. Lett. **16**, 908 (1966); S. Gerasimov, Sov. J. Nucl. Phys. **2**, 430 (1966)
4. H. Burkhardt and W. N. Cottingham, Ann. Phys. 16, 543 (1970)
5. Hall A collaboration: J. Alcorn *et al.*, Nucl. Inst. Meth. **A522**, 294 (2004)
6. CLAS collaboration: B. A. Mecking *et al.*, Nucl. Inst. Meth. **A503**, 513 (2003)
7. E94-010 collaboration: M. Amarian *et al.*, Phys. Rev. Lett. **89**, 242301 (2002); Phys. Rev. Lett. **92**, 022301 (2004); Phys. Rev. Lett. **93**, 152301 (2004)
8. R. Fatemi *et al.*, Phys. Rev. Lett. **91**, 222002 (2003); J. Yun *et al.*, Phys. Rev. C **67**, 055204 (2003)
9. Y. Liang *et al.*, nucl-ex/0410027
10. N. Bianchi and E. Thomas, Nucl. Phys. Proc. Suppl. **82**, 256 (2000)
11. A. Deur *et al.*, Phys. Rev. Lett. **93**, 212001 (2004)
12. V. D. Burkert and B. L. Ioffe, Phys. Lett. **B296**, 223 (1992); J. Exp. Theor. Phys. **78**, 619 (1994)
13. HERMES collaboration: A. Airapetian *et al.*, Eur. Phys. J. **C 26**, 527 (2003)
14. E143 collaboration: K. Abe *et al.*, Phys. Rev. Lett. **78**, 815 (1997)
15. J. Soffer and O. V. Teryaev, Phys. Lett. **B545**, 323 (2002); J. Soffer, hep-ph/0409333
16. V. Bernard, T. R. Hemmert and Ulf-G. Meiner, Phys. Rev. D **67**, 076008 (2003)
17. X. Ji, C. W. Kao and J. Osborne, Phys. Lett. **B472**, 1 (2000)
18. M. Osipenko *et al.*, Phys. Lett. **B609**:258 (2005)
19. Z-E Meziani *et al.*, Phys. Lett. **B613**:148 (2005)
20. M. Amarian et al., Phys. Rev. Lett. **93** 152301 (2004)
21. D. Drechsel, O. Hanstein, S.S. Kamalov and L. Tiator. Nucl. Phys. A **645** 145. (1999)
22. See N. Liyanage contribution to these proceedings
23. JLab E01-006, O. Rondon and M. Jones spokespersons
24. JLab E97-110, J-P Chen, A. Deur and F. Garibaldi spokespersons
25. JLab E03-006, M. Ripani, M. Battaglieri, A. Deur and R. De Vita spokespersons

QCD Factorization for Semi-inclusive Deep Inelastic Scattering

Feng Yuan

RIKEN/BNL Research Center, Building 510A, Brookhaven National Laboratory, Upton, NY 11973

Abstract. In this talk, we will present a QCD factorization theorem for the semi-inclusive deep-inelastic scattering with hadrons in the current fragmentation region detected at low transverse momentum.

Introduction

In recent years, there has been considerable experimental and theoretical interest in semi-inclusive hard processes. Rigorous theoretical studies in this direction started from the classical work on semi-inclusive processes in e^+e^- annihilation by Collins and Soper [1], where a QCD factorization was proved, and non-perturbative transverse-momentum-dependent (TMD) parton distributions and fragmentation functions were introduced [1, 2]. In the past few years, gauge properties of the TMD parton distributions have been investigated [3, 4, 5]. More recently, the factorization theorems for the semi-inclusive deep inelastic scattering (SIDIS) and Drell-Yan processes have been re-examined in the context of the gauge-invariant definitions [6, 7]. In this talk, I will present the theoretical results of [6].

The main result of [6] is a QCD factorization theorem for the SIDIS cross section at low transverse momentum, accurate up to the power corrections $(P_{h\perp}^2/Q^2)^n$ and to all orders in perturbation theory. For example, the leading spin-independent structure function for SIDIS can be factorized as follows,

$$\begin{aligned} F(x_B, z_h, P_{h\perp}, Q^2) &= \sum_{q=u,d,s,\ldots} e_q^2 \int d^2\vec{k}_\perp d^2\vec{p}_\perp d^2\vec{\ell}_\perp \\ &\times q\left(x_B, k_\perp, \mu^2, x_B\zeta, \rho\right) \hat{q}_T\left(z_h, p_\perp, \mu^2, \hat{\zeta}/z_h, \rho\right) S(\vec{\ell}_\perp, \mu^2, \rho) \\ &\times H\left(Q^2, \mu^2, \rho\right) \delta^2(z_h\vec{k}_\perp + \vec{p}_\perp + \vec{\ell}_\perp - \vec{P}_{h\perp}) , \end{aligned} \tag{1}$$

where μ is a renormalization (and collinear factorization) scale; ρ is a gluon rapidity cut-off parameter; *the μ and ρ dependence cancels among various factors*. In a special system of coordinates in which $x_B\zeta = \hat{\zeta}/z_h$, one has $\zeta^2 x_B^2 = \hat{\zeta}^2/z_h^2 = Q^2\rho$. The physical interpretation of the factors are as follows: q is TMD quark distribution function; $\hat{q}$ is the TMD quark fragmentation function depending on; H represents the contribution of parton hard scattering and is a perturbation series in α_s; and, finally, the soft factor S

CP792, *Deep Inelastic Scattering, DIS 2005,* edited by W. H. Smith and S. R. Dasu

comes from soft gluon radiations and is defined by a matrix element of Wilson lines in QCD vacuum.

The Transverse Momentum Dependent Parton Distributions

Consider a hadron, a nucleon for example, with four-momentum P. Let $(xP^+,\vec{k}_\perp)$ represent the momentum of a parton (quark or gluon) in the hadron. In a non-singular gauge (e.g. Feynman gauge), the TMD parton distributions can be defined through the following density matrix [1, 4],

$$\mathcal{M}^{\pm}(x,k_\perp,\mu,x\zeta,\rho) = p^+\int\frac{d\xi^-}{2\pi}e^{-ix\xi^-P^+}\int\frac{d^2\vec{b}_\perp}{(2\pi)^2}e^{i\vec{b}_\perp\cdot\vec{k}_\perp} \times\frac{\left\langle PS\left|\overline{\psi}_q(\xi^-,\vec{b}_\perp)\mathcal{L}_v^\dagger(\pm\infty;\xi^-,\vec{b}_\perp)\mathcal{L}_v(\pm\infty;0)\psi_q(0)\right|PS\right\rangle}{S^{\pm}(\vec{b}_\perp,\mu^2,\rho)} . \quad (2)$$

The $+(-)$ superscript is appropriate for DIS (Drell-Yan) process [4, 5]. v^μ is a time-like dimensionless ($v^2>0$) four-vector with zero transverse components $(v^-,v^+,\vec{0})$ and $v^-\gg v^+$. Thus the v^μ is a quasi light-cone vector, approaching n^μ. The variable ζ^2 denotes the combination $(2P\cdot v)^2/v^2=\zeta^2$. $\mathcal{L}_v$ is a gauge link along v^μ,

$$\mathcal{L}_v(\pm\infty;\xi)=\exp\left(-ig\int_0^{\pm\infty}d\lambda v\cdot A(\lambda v+\xi)\right) . \quad (3)$$

Here the non-light-like gauge link is introduced to regulate the light-cone singularities. In the above definition, we have derived a soft factor defined as [6]:

$$S^{\pm}(\vec{b}_\perp,\mu^2,\rho)=\frac{1}{N_c}\langle 0|\mathcal{L}^\dagger_{\tilde{v}il}(\vec{b}_\perp,-\infty)\mathcal{L}^\dagger_{vlj}(\pm\infty;\vec{b}_\perp)\mathcal{L}_{vjk}(\pm\infty;0)\mathcal{L}_{\tilde{v}ki}(0;-\infty)|0\rangle , \quad (4)$$

where i,j,k,l are color indices and new quasi light-cone vector $\tilde{v}^\mu=(\tilde{v}^-,\tilde{v}^+,\vec{0})$ has been introduced with $\tilde{v}^-\ll\tilde{v}^+$. The ρ parameter is defined as $\rho=\sqrt{v^-\tilde{v}^+/v^+\tilde{v}^-}\gg 1$.

The TMD parton distribution is defined as such to absorb the collinear divergence in the partonic processes. This has been checked by an explicit calculation at one-loop order [6], where the soft divergence associated with soft gluons in the TMDs has been cancelled out in the total result, and we are left with only the collinear singularity.

Factorization at One-loop Order

To demonstrate the factorization at one-loop order, one needs to calculate the TMDs at one-loop order. Then, we have to show that the SIDIS cross section can be written in terms of these TMDs plus the soft and hard factors.

The semi-inclusive DIS cross section under one-photon exchange is

$$\frac{d\sigma}{dx_Bdydz_hd^2\vec{P}_{h\perp}} = \frac{4\pi\alpha_{em}^2 s}{Q^4}\left[(1-y+y^2/2)x_BF(x_B,z_h,P_{h\perp},Q^2)+\cdots\right] , \quad (5)$$

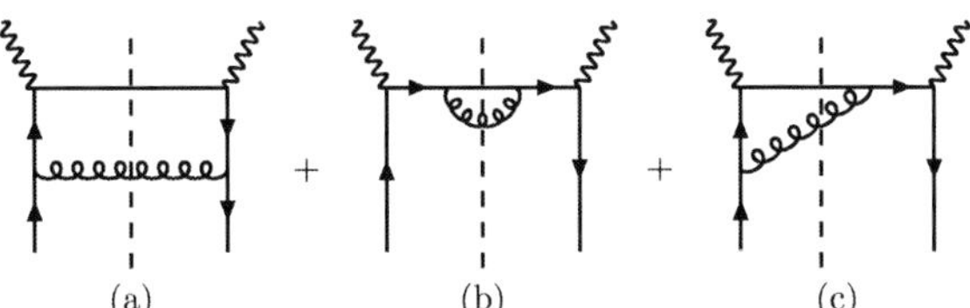

FIGURE 1. One-loop real diagrams for SIDIS.

where F is the spin-independent structure function. In the above equation, we have omitted other terms contributions which may depend on the spin of the hadrons and their factorizations are similar [6].

The one-loop real corrections to the structure function F are shown in Fig. 1. There is no contribution to the hard scattering kernel from any of these diagrams. In Fig. 1a, the soft-gluon radiation generates a transverse-momentum for the struck quark. There is no contribution from the fragmentation function because the contribution from the final state with a gluon in the n^μ direction and a soft quark is power suppressed. Therefore, the diagram must be factorizable into the parton distribution. Similarly for Fig. 1b, which again can be reproduced by the factorization formula with the one-loop fragmentation function and the soft factor S, and the tree-level parton distribution and the hard part. For Fig. 1c and its hermitian conjugate, we find three distinct contributions: where the first term corresponds to a gluon collinear to the initial quark, the second term a gluon collinear to the final state quark, and the third term a soft gluon. All these terms are reproduced by the factorization formula with one-loop parton distribution, fragmentation function, and the soft factor. Similarly, we can show that the virtual diagrams can also be factorized into different pieces in the factorization formula. From the vertex correction to SIDIS cross section, we get the hard factor as

$$H^{(1)}(Q^2,\mu^2,\rho) = \frac{\alpha_s}{2\pi}C_F\left[\left(1+\ln\rho^2\right)\ln\frac{Q^2}{\mu^2} - \ln\rho^2 + \frac{1}{4}\ln^2\rho^2 + \pi^2 - 4\right] . \quad (6)$$

Therefore we conclude that at the one-loop level, the general factorization formula Eq.(1) holds.

All Order Argument and Discussions

For arguments toward a factorization to all orders, we follow the discussions in [1, 6]. The procedure for this argument is the following. First, for any high order Feynman diagrams, using the power counting rules identifies the leading region contributions [8]. The leading regions clearly separate the soft, collinear, and hard gluons' contributions to the cross section (the cut diagram), where the soft gluons are only attached to the jet functions (parton distributions and/or fragmentation functions); hard gluons are included in the hard part; collinear gluons attached the jet functions to the hard part. On top of that, we can further use the Grammer-Yennie approximation to factorize out the soft factor, which can be expressed as matrix element of Wilson lines [1, 6], as defined above in

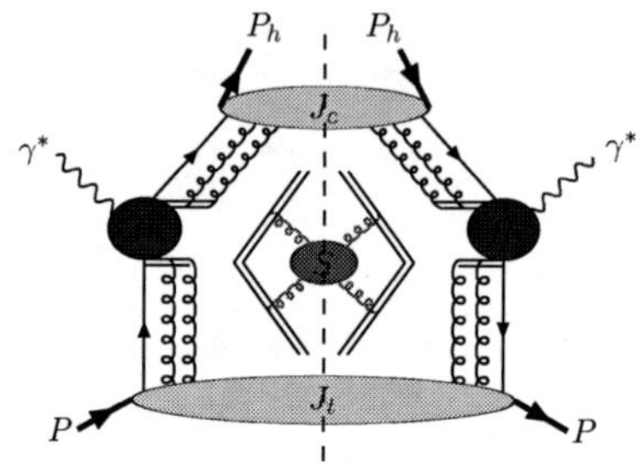

FIGURE 2. All order factorization for SIDIS.

Eq. (4). The Ward Identity will be used to further factorize the collinear gluons from the hard part, which results in a Wilson line (gauge link) association in the definition of the jet functions. The variation of the gauge link gives the Collins-Soper evolution equation for the jet functions [1]. After these procedures, the hard part only contains hard gluons, which can be calculated from perturbative QCD. Once all these being done, we will arrive at the factorization formula for SIDIS as in Eq. (1), and shown in Fig. 2.

Similar factorization formulas can be obtained for other semi-inclusive processes, including back-to-back di-hadron production in e^+e^- annihilation [1], low transverse momentum Drell-Yan [9], and di-jet or di-hadron correlation at hadron colliders [10]. The common feature of these semi-inclusive processes is that they all depend on the TMD parton distributions and/or fragmentation functions. The global analysis of all these processes will definitely provide us a unique picture about the nucleon structure, and reveal the relevant parton orbital motion in the nucleon.

The author is grateful to RIKEN, Brookhaven National Laboratory and the U.S. Department of Energy (contract number DE-AC02-98CH10886) for providing the facilities essential for the completion of his work.

REFERENCES

1. J. C. Collins and D. E. Soper, Nucl. Phys. B **193**, 381 (1981) [Erratum-ibid. B **213**, 545 (1983)]; Nucl. Phys. B **197**, 446 (1982).
2. J. C. Collins and D. E. Soper, Nucl. Phys. B **194**, 445 (1982).
3. S. J. Brodsky, D. S. Hwang and I. Schmidt, Phys. Lett. B **530**, 99 (2002).
4. J. C. Collins, Phys. Lett. B **536**, 43 (2002); Acta Phys. Polon. B **34**, 3103 (2003).
5. X. Ji and F. Yuan, Phys. Lett. B **543**, 66 (2002); A. V. Belitsky, X. Ji and F. Yuan, Nucl. Phys. B **656**, 165 (2003).
6. X. Ji, J. P. Ma and F. Yuan, Phys. Rev. D **71**, 034005 (2005); Phys. Lett. B **597**, 299 (2004); arXiv:hep-ph/0503015.
7. J. C. Collins and A. Metz, Phys. Rev. Lett. **93**, 252001 (2004).
8. J. C. Collins, D. E. Soper and G. Sterman, Adv. Ser. Direct. High Energy Phys. **5**, 1 (1988); in *Perturbative QCD* (A.H. Mueller, ed.) (World Scientific Publ., 1989).
9. J. C. Collins, D. E. Soper and G. Sterman, Nucl. Phys. B **250**, 199 (1985).
10. W. Vogelsang, F. Yuan, work in progress.

Large x Physics

Stanley J. Brodsky

Stanford Linear Accelerator Center, Stanford University, Stanford, CA, 94309

Abstract.
The large x_{bj} domain of deep inelastic lepton-proton scattering provides essential information on the structure of the proton when the struck quark is far off-shell and one must take into account inter-quark correlations, higher twist effects, and the breakdown of standard DGLAP evolution. I briefly review predictions from PQCD and AdS/CFT. I also discuss how intrinsic heavy quark Fock states lead to novel production mechanisms for heavy hadrons.

Keywords: Quantum Chromodynamics, Deep Inelastic Lepton-Hadron Interactions
PACS: 12.38.-t,12.38.Aw,13.60.Hb,11.25.Tq

THE FAR OFF-SHELL BEHAVIOR OF QCD WAVEFUNCTIONS

One of the most important arenas for testing QCD at a fundamental level is the large x_{bj} domain of deep inelastic structure functions as measured in inclusive lepton-proton scattering and other hard inclusive reactions. Predictions for the quark and gluon distributions at $x_{bj} \simeq 1$ can be made directly from the structure of the hadronic light-front wavefunctions from the near conformal behavior of QCD at short distances, using either perturbative expansions [1, 2, 3] or the AdS/CFT correspondence principle [4, 5] which maps conformal transformations to the behavior of hadronic wavefunctions in the fifth dimension of AdS space. These first-principle predictions depend in detail on flavor, spin, and angular momentum composition of the target hadron. Other issues include (a) the quenching of DGLAP evolution when the struck quark is off-shell; (b) duality between hard exclusive and inclusive channels; (c) the effects of dynamical higher-twist contributions from quark correlations within the hadron wavefunction. In addition, the distributions of intrinsic heavy quarks arising from the hadron bound state are peaked at high x_{bj}.

At leading twist $x_{bj} = Q^2/2p \cdot q$ can be equated to the light-cone momentum fraction $x = k^+/P^+ = (k^0 + k^z/P^0 + P^z)$ carried by the struck quark in the target hadron's light-front Fock state wavefunction $\Psi_n(x_i, k_{\perp_i})$. The mathematical limit $x \to 1$ at leading twist requires all of the light-cone momentum of the target to be carried by just one quark. Since $\sum_i^n x_i = 1$, all of the other (spectator) constituents are forced kinematically to have $x_i \to 0$, and the valence Fock state dominates. In fact, the requirement $k_i^+ = k_i^0 + k_i^z \to 0$ demands that $k_i^z \to -\infty$ for each spectator with nonzero mass or transverse momentum. Furthermore, the invariant mass of the Fock state $\mathscr{M}_n^2 = \sum_i^n((k_{\perp i}^2 + m_i^2)/x_i)$ becomes infinitely large. If one uses a covariant formulation of bound states, such as the Bethe-Salpeter formalism, one sees that the Feynman virtuality of the struck quark: $k_F^2 - m_q^2 = x_q(M^2 - \mathscr{M}_n^2)$ becomes infinitely spacelike. Thus measurements at the limit $x_{bj} \to 1$ test the extreme kinematics of the hadron bound state. It is thus not

CP792, *Deep Inelastic Scattering, DIS 2005*, edited by W. H. Smith and S. R. Dasu

surprising that, given asymptotic freedom, one can make first-principle predictions for the behavior of the parton distributions of hadrons at large x from the short-distance properties of QCD [6, 1, 2, 3]. The main power-law dependence at $x \sim 1$ is given by the minimal number of (vertical) gluon exchanges required to stop the hadronic spectator; *i.e.*, one iterates the interaction kernel of the hadron wavefunction to obtain the connected tree graphs, thus providing the minimal path for the momentum to flow to the struck constituent. For the case of the nucleon, the leading Fock state component is the $|qqq\rangle$ state, and two gluon exchanges with virtuality $k^2 \sim O\left((\vec{k}_\perp^2 + \widetilde{m}^2)/(1-x)\right)$ are required. The same basic QCD interactions which yield DGLAP evolution thus also predict the $x \to 1$ behavior. More generally, the nominal power-law prediction for the large x behavior for constituent a in hadron H is [7] $f_{a/H}(x) \sim (1-x)^{2n_s-1+2|\Delta S^z|}$, where n_s is the number of spectator constituents and $\Delta S^z = S^z_a - S^z_H$ is the difference of spin projections for the struck parton a and the hadron H. For example, for the valence quarks of the proton, $q_p^{\uparrow}(x) \sim (1-x)^3$ and $q_p^{\downarrow}(x) \sim (1-x)^5$ since $n_s = 2$ and $\Delta S^z = 0, 1$, respectively, whereas for the pion $q_\pi(x) \sim (1-x)^2$ since $n_s = 1$ and $2\Delta S^z = 1$. The power law must be even when one measures a fermion in a boson in order to have consistent crossing to the fragmentation functions at large z. The gluon distributions fall at least one power of $(1-x)$ faster than the valence quark distributions. The suppression of the antiparallel quarks reflects the fact that the internal orbital angular momentum of the quark is required for J_z angular momentum conservation. Helicity-flip terms from quark mass insertions are subleading at large x. The nominal powers for structure functions at large x can also be derived from the AdS/CFT approach [4, 5] without perturbation theory—reflecting the underlying conformal features of QCD. In general one predicts logarithmic corrections from the nonzero QCD β function as well as higher order diagrams.

If one assumes $SU(6)$ spin-flavor symmetry, then the number of quarks in the $S^z = +1/2$ proton valence state is normalized to $u^{\uparrow} : u^{\downarrow} : d^{\uparrow} : d^{\downarrow} = 5/3 : 1/3 : 1/3 : 2/3$. Thus one predicts $d(x)/u(x) \to d^{\uparrow}(x)/u^{\uparrow}(x) \to 1/5$ and $F_{2n}(x)/F_{2p}(x) \to 3/7$, at large x [6]. In contrast, the scalar diquark model predicts $d(x)/u(x) \to 0$. The proton structure function data suggest that a smooth, power-law behavior of the u-quark distribution, $\sim (1-x)^3$, works very well at moderate to large x. Evidently the reason why $d(x)$ is far from the pQCD prediction at moderate x is that the ratio of spin-antiparallel to parallel d quarks is 2:1 in the proton (for a spin-flavor symmetric wave function), whereas the ratio is 1:5 in the case of the u quarks. Measurements of the neutron to proton structure function ratio as proposed at Jefferson Laboratory with the 12-GeV upgrade will provide essential tests of these predictions.

MODIFICATION OF DGLAP EVOLUTION AND THE INCLUSIVE–EXCLUSIVE CONNECTION

QCD evolution predicts structure functions at large x of the form $F_2(x,Q^2) \propto (1-x)^{V+\widetilde{\xi}(Q^2,k^2)}$ where $(1-x)^V$ is the effective power-behavior at $Q^2 \sim O(k^2)$, and $\widetilde{\xi}(Q^2,k^2) = \frac{C_F}{\pi}\int_{k^2}^{Q^2} \frac{d\ell^2}{\ell^2}\alpha_s(\ell^2) \sim O(\log\log Q^2)$. If one uses this form for fixed

$W^2 = \mathscr{M}^2 = \frac{1-x}{x} Q^2$, then one obtains transition form factors which fall faster than any power, in direct contradiction to PQCD predictions. Thus the standard application of DGLAP evolution to the deep inelastic structure functions appears inconsistent with Bloom-Gilman duality at fixed missing mass W. This conflict is resolved if one takes into account the fact that the struck quark is far-off shell in the $x \to 1$ domain [1, 2]. The virtuality of the struck quark k_F^2 in fact sets the lower limit of the ℓ^2 integration in $\widetilde{\xi}$, severely truncating DGLAP evolution. One then finds that at fixed $\mathscr{M}^2$, $\widetilde{\xi} \to 0$, the inclusive and exclusive cross sections have the same fall-off. The "initial" or "starting" structure function is *no longer* unknown but is directly determined from QCD perturbation theory and distribution amplitude evolution in the large x domain. In a sense the most critical prediction from QCD is the nominal power law $(1-x)^3$ since the power 3 reflects the existence of a 3-quark Fock state as well as nearly scale-invariant QCD quark-quark interactions within the nucleon.

As one approaches the $x \to 1$ limit, dynamical higher-twist contributions from coherent scattering on more than one constituent becomes increasingly more probable [8]. For example, one can have subprocesses such as $qq\gamma^* \to qq$ where two quarks in the target are scattered coherently. These contributions are suppressed by powers of $1/Q^2$ but are enhanced at high x since the qq acts as a bosonic constituent with summed $x_{qq} = x_1 + x_2$. Such contributions, including the exclusive qqq subprocesses where all of the valence quarks scatter coherently, all contribute at leading order at fixed W^2, in agreement with Bloom-Gilman duality. Sufficient kinematic range in Q^2, as well as longitudinal-transverse separation, is required to cleanly separate the leading twist and higher twist contributions, all of which contribute to the measured structure functions.

INTRINSIC HEAVY QUARKS

It was originally suggested in Ref. [9] that there is a $\sim 1\%$ probability of IC Fock states in the nucleon; more recently, the operator product expansion has been used to show that the probability for Fock states in light hadron to have an extra heavy quark pair of mass M_Q decreases only as Λ_{QCD}^2/M_Q^2 in non-Abelian gauge theory [10]. In contrast, in the case of Abelian QED, the probability of an intrinsic heavy lepton pair in a light-atom such as positronium is suppressed by $\mu_{\rm bohr}^4/M_\ell^4$ where $\mu_{\rm bohr}$ is the Bohr momentum. The maximal probability for an intrinsic heavy quark Fock state occurs for minimal off-shellness; *i.e.*, at minimum invariant mass squared. Thus the dominant Fock state configuration is $x_i \propto m_{\perp i}$ where $m_{i\perp}^2 = m_i^2 + \vec{k}_{\perp i}^2$; *i.e.*, at equal rapidity, and the heaviest constituents carry the most momentum [9]. There are many experiments which confirm the presence of charm quarks at large x in the proton wavefunction beginning with the European Muon Collaboration (EMC) deep inelastic scattering experiment [11] which found a distinct excess of events in the charm quark distribution at $x_{bj} > 0.3$, at a rate at least an order of magnitude beyond lowest predictions based on gluon splitting and DGLAP evolution. The materialization of the intrinsic charm Fock state also leads to the production of open-charm states such as $\Lambda(cud)$ and $D^-(\bar{c}d)$ at large x_F through the coalescence of the valence and charm quarks which are co-moving with the same rapidity. The intrinsic charm model naturally accounts for the production of leading charm hadrons as observed at the ISR and Fermilab. The existence of the rare double IC Fock state such as $|uudc\bar{c}c\bar{c}\rangle$ leads to the production of two J/ψ's [12] or a double-charm baryon state at large x_F and small p_T. Double J/ψ events at a high combined $x_F \geq 0.8$ were in fact observed by NA3 [13]. The observation of the doubly-charmed baryon $\Xi_{cc}^{+}(3520)$ has been confirmed recently by SELEX at FNAL [14]; the presence of two charm quarks at large x_F thus has a natural interpretation within QCD. Another immediate consequence of intrinsic charm is the production of charmonium states at high $x_F = x_c + x_{\bar{c}}$ in a hadronic collision such as $pp \to J/\psi X$. The color octet $(c\bar{c})_{8_C}$ can be converted to a high x quarkonium state via gluon exchange with the target. Since the IC Fock state $|(uud)_{8_C}(c\bar{c})_{8_C}\rangle$ has a large color dipole moment which is strongly interacting, the production of quarkonium on a nucleus occurs at the front nuclear surface with an $A^{2/3}$ nuclear dependence. The IC mechanism also predicts that the strongest nuclear absorption appears when the J/ψ has minimum $p_T < 1$ GeV/c as well as large x_F. These features are precisely what has been observed by the NA3 Collaboration at CERN [15] in $pA \to J/\psi X$ and $\pi A \to J/\psi X$. Because of the existence of the IC plus usual A^1 contributions, the nuclear dependence of the total cross section is proportional to $A^{\alpha(x_F)}$ rather than the dependence predicted by leading-twist QCD $A^{\alpha(x_2)}$, where x_2 is the light-cone momentum fraction of the parton in the nucleus. This is in agreement with the NA3 and E866/NuSea [16] data which displays x_F shadowing, thus violating PQCD factorization, as emphasized in Ref. [17].

ACKNOWLEDGMENTS

I thank B. Kopeliovitch, I. A. Schmidt, and J. Soffer for helpful conversations. This work was supported by the Department of Energy, contract No. DE-AC02-76SF00515.

REFERENCES

1. S. J. Brodsky and G. P. Lepage, SLAC-PUB-2294
2. G. P. Lepage and S. J. Brodsky, Phys. Rev. D **22**, 2157 (1980).
3. X. d. Ji, J. P. Ma and F. Yuan, Phys. Lett. B **610**, 247 (2005) [arXiv:hep-ph/0411382].
4. J. Polchinski and M. J. Strassler, Phys. Rev. Lett. **88**, 031601 (2002) [arXiv:hep-th/0109174].
5. S. J. Brodsky and G. F. de Teramond, Phys. Lett. B **582**, 211 (2004) [arXiv:hep-th/0310227].
6. G. R. Farrar and D. R. Jackson, Phys. Rev. Lett. **35**, 1416 (1975).
7. S. J. Brodsky, M. Burkardt and I. Schmidt, Nucl. Phys. B **441**, 197 (1995) [arXiv:hep-ph/9401328].
8. E. L. Berger and S. J. Brodsky, Phys. Rev. Lett. **42**, 940 (1979).
9. S. J. Brodsky, P. Hoyer, C. Peterson and N. Sakai, Phys. Lett. B **93**, 451 (1980).
10. M. Franz, . V. Polyakov and K. Goeke, Phys. Rev. D **62**, 074024 (2000) [arXiv:hep-ph/0002240].
11. J. J. Aubert *et al.* [European Muon Collaboration], Nucl. Phys. B **213**, 31 (1983).
12. R. Vogt and S. J. Brodsky, Phys. Lett. B **349**, 569 (1995) [arXiv:hep-ph/9503206].
13. J. Badier *et al.* [NA3 Collaboration], Phys. Lett. B **114**, 457 (1982).
14. A. Ocherashvili *et al.* [SELEX Collaboration], arXiv:hep-ex/0406033.
15. J. Badier *et al.* [NA3 Collaboration], Z. Phys. C **20**, 101 (1983).
16. M. J. Leitch *et al.* [FNAL E866/NuSea collaboration], Phys. Rev. Lett. **84**, 3256 (2000) [arXiv:nucl-ex/9909007].
17. P. Hoyer, M. Vanttinen and U. Sukhatme, Phys. Lett. B **246**, 217 (1990).

Understanding the role of Cahn and Sivers effects in Deep Inelastic Scattering[1]

M. Anselmino*, M. Boglione*, U. D'Alesio†, A. Kotzinian**,‡, F. Murgia† and A. Prokudin*

*Dipartimento di Fisica Teorica, Università di Torino and INFN, Sezione di Torino, Via P. Giuria 1, I-10125 Torino, Italy
†INFN, Sezione di Cagliari and Dipartimento di Fisica, Università di Cagliari, C.P. 170, I-09042 Monserrato (CA), Italy
**Dipartimento di Fisica Generale, Università di Torino and INFN, Sezione di Torino, Via P. Giuria 1, I-10125 Torino, Italy
‡Yerevan Physics Institute, Alikhanian Brothers St. 2; AM-375036 Yerevan, Armenia; JINR, Dubna, 141980, Russia

Abstract. The role of intrinsic $k_\perp$ in semi-inclusive Deep Inelastic Scattering (SIDIS) processes ($\ell p \to \ell h X$) is studied with exact kinematics within QCD parton model at leading order; the dependence of the unpolarized cross section on the azimuthal angle between the leptonic and the hadron production planes (Cahn effect) is compared with data and used to estimate the average values of $k_\perp$ both in quark distribution and fragmentation functions. The resulting picture is applied to the description of the weighted single spin asymmetry $A_{UT}^{\sin(\phi_\pi-\phi_S)}$ recently measured by the HERMES collaboration at DESY; this allows to extract parameters for the quark Sivers functions. The extracted Sivers functions give predictions for the COMPASS measurement of $A_{UT}^{\sin(\phi_\pi-\phi_S)}$ in agreement with recent data, while their contribution to HERMES $A_{UL}^{\sin\phi_\pi}$ is computed and found to be small. Predictions for $A_{UT}^{\sin(\phi_K-\phi_S)}$ for kaon production at HERMES are also given.

Keywords: Single Spin Asymmetries, Sivers effect
PACS: 13.88.+e, 13.60.-r, 13.15.+g, 13.85.Ni

The role of intrinsic $k_\perp$ is known to be important in unpolarized SIDIS processes [1] and becomes crucial for the explanation of many single spin effects recently observed and still under active investigation in several ongoing experiments; spin and $k_\perp$ dependences can couple in parton distribution and fragmentation functions, giving origin to unexpected effects in polarization observables. One such example is the azimuthal asymmetry observed in the scattering of unpolarized leptons off polarized protons [2], [3] and deuterons [4].

A recent analysis of Single Spin Asymmetries (SSA) in $p^\uparrow p \to \pi X$ processes, with a separate study of the Sivers and the Collins contributions, has been performed respectively in Refs. [5] and [6], with the conclusion that the Sivers [7] mechanism alone can explain the data [8], while the Collins [9] mechanism is strongly suppressed.

We considered [11] the role of parton intrinsic motion in SIDIS processes within the QCD parton model at leading order. The average values of $k_\perp$ for quarks inside protons,

[1] Talk delivered by A. Prokudin at DIS05, Madison, WI, USA

CP792, *Deep Inelastic Scattering, DIS 2005*, edited by W. H. Smith and S. R. Dasu

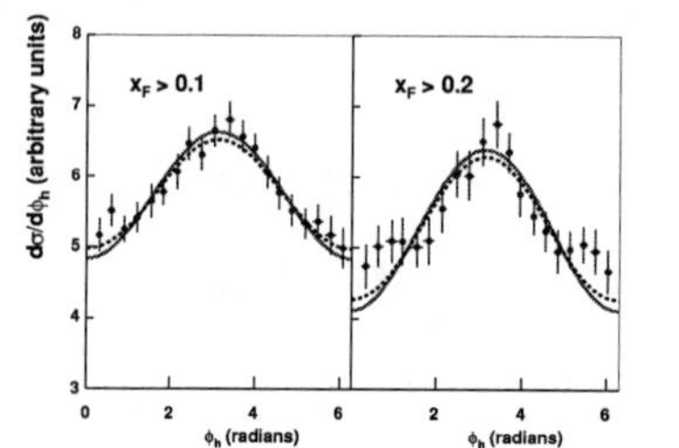

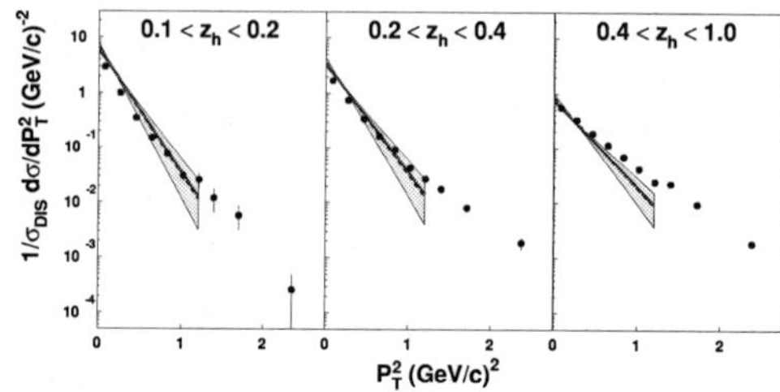

FIGURE 1. Description of the ϕ_h and P_T dependence of the cross section.

and $p_\perp$ for final hadrons inside the fragmenting quark jet, are fixed by comparison with data [10] on the dependence of the unpolarized cross section on the azimuthal angle between the leptonic and the hadronic planes and on P_T.

Within the factorization scheme, assuming an independent fragmentation process, the SIDIS cross section for the production of a hadron h in the current fragmentation region with the inclusion of all intrinsic motions can be written as [11]

$$\frac{d^5\sigma^{\ell p\to\ell hX}}{dx_B\,dQ^2\,dz_h\,d^2\mathbf{P}_T} = \sum_q e_q^2 \int d^2\mathbf{k}_\perp\, f_q(x,k_\perp)\,\frac{2\pi\alpha^2}{x_B^2 s^2}\,\frac{\hat{s}^2+\hat{u}^2}{Q^4} \times D_q^h(z,p_\perp)\,\frac{z}{z_h}\,\frac{x_B}{x}\left(1+\frac{x_B^2}{x^2}\frac{k_\perp^2}{Q^2}\right)^{-1}. \tag{1}$$

It is instructive, and often quite accurate, to consider the above equation in the much simpler limit in which only terms of $O(k_\perp/Q)$ are retained. In such a case $x \simeq x_B, z \simeq z_h$ and $\mathbf{p}_\perp \simeq \mathbf{P}_T - z_h\mathbf{k}_\perp$. In what follows we assume, both for parton densities and fragmentation functions, a factorized Gaussian $k_\perp$ and $p_\perp$ dependence.

In this way the $\mathbf{k}_\perp$ integration in Eq. (1) can be performed analytically, leading to the result, valid up to $O(k_\perp/Q)$:

$$\frac{d^5\sigma^{\ell p\to\ell hX}}{dx_B\,dQ^2\,dz_h\,d^2\mathbf{P}_T} \simeq \sum_q \frac{2\pi\alpha^2 e_q^2}{Q^4}\, f_q(x_B)\,D_q^h(z_h)\left[1+(1-y)^2 -4\,\frac{(2-y)\sqrt{1-y}\,\langle k_\perp^2\rangle\, z_h P_T}{\langle P_T^2\rangle\, Q}\cos\phi_h\right]\frac{1}{\pi\langle P_T^2\rangle}\,e^{-P_T^2/\langle P_T^2\rangle}\,, \tag{2}$$

where $\langle P_T^2\rangle = \langle p_\perp^2\rangle + z_h^2\langle k_\perp^2\rangle$. The term proportional to $\cos\phi_h$ describes the Cahn effect [1].

By fitting the data [10] on unpolarized SIDIS we obtain the following values of the parameters: $\langle k_\perp^2\rangle = 0.25\ (\mathrm{GeV/c})^2$, $\langle p_\perp^2\rangle = 0.20\ (\mathrm{GeV/c})^2$. The results are shown in Fig. 1.

Such values are then used to compute the SSA for $\ell p^\uparrow \to \ell hX$ processes. We considered the Sivers mechanism [7] alone. The unpolarized quark (and gluon) distributions inside a transversely polarized proton can be written as:

$$f_{q/p^{\uparrow}}(x,\mathbf{k}_{\perp}) = f_{q/p}(x,k_{\perp}) + \frac{1}{2}\Delta^N f_{q/p^{\uparrow}}(x,k_{\perp})\,\mathbf{S}\cdot(\hat{\mathbf{P}}\times\hat{\mathbf{k}}_{\perp})\,, \tag{3}$$

where $\mathbf{P}$ and $\mathbf{S}$ are respectively the proton momentum and transverse polarization vector, and $\mathbf{k}_{\perp}$ is the parton transverse momentum; transverse refers to the proton flight direction. Eq. (3) leads to non vanishing SSA, which can be calculated by substituting $f_{q/p}$ by $f_{q/p^{\uparrow}}$ in Eq. (1).

We parameterize, for each light quark flavour $q = u_v, d_v, u_s, d_s, \bar{u}, \bar{d}$, the Sivers function in the following factorized form: $\Delta^N f_{q/p^{\uparrow}}(x,k_{\perp}) = 2\,\mathrm{N}_q(x)\,h(k_{\perp})\,f_{q/p}(x,k_{\perp})$, where $\mathrm{N}_q(x) = N_q x^{a_q}(1-x)^{b_q}\,\frac{(a_q+b_q)^{(a_q+b_q)}}{a_q^{a_q} b_q^{b_q}}$, $h(k_{\perp}) = \sqrt{2e}\,\frac{k_{\perp}}{M'}\,e^{-k_{\perp}^2/M'^2}$.

Our fit [11] to the HERMES data on $A_{UT}^{\sin(\phi_\pi-\phi_S)}$ [3] is presented in the left panel of Fig. 2.

Having fixed all the parameters we can check the consistency of the model by computing $A_{UT}^{\sin(\phi_\pi-\phi_S)}$ for charged hadron production in COMPASS experiment [4]; our results are given in the right panel of Fig. 2, showing a very good agreement with the data.

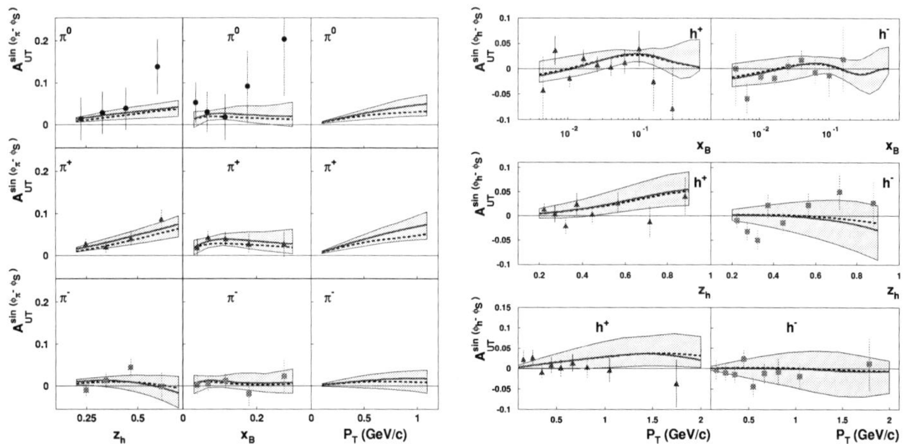

FIGURE 2. HERMES [3] (left) and COMPASS [4] (right) data on $A_{UT}^{\sin(\phi_\pi-\phi_S)}$ for scattering off a transversely polarized proton (deuterium) target and pion (hadron) production. The curves are the results of our fit to the HERMES data and description of the COMPASS data , with exact kinematics (dashed line) or keeping only terms up to $O(k_{\perp}/Q)$ (solid bold line). The shadowed region corresponds to the theoretical uncertainty due to the parameter errors.

We also compute $A_{UT}^{\sin(\phi_K-\phi_S)}$ for kaon production, which could be measured by HERMES. Our results are given in the right panel of Fig. 3.

Finally, we consider the HERMES data on $A_{UL}^{\sin\phi_\pi}$ obtained in the semi-inclusive electro-production of pions on a longitudinally polarized hydrogen target [2]. We have computed the Sivers contribution to this quantity again with our set of Sivers functions,

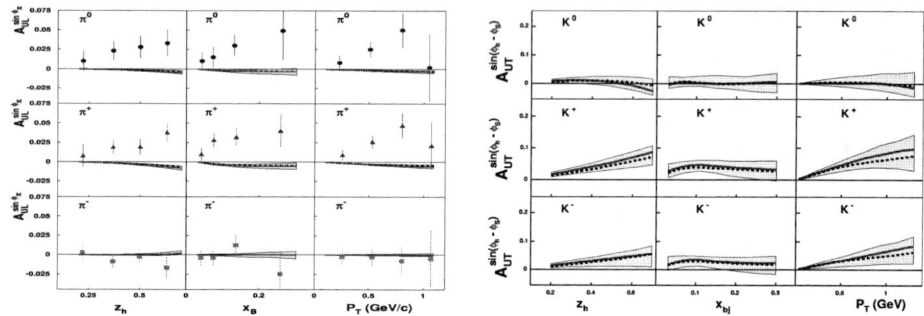

FIGURE 3. HERMES data on $A_{UL}^{\sin(\phi_\pi)}$ [2] for scattering off a longitudinally polarized proton target and pion production (left) and predictions of $A_{UT}^{\sin(\phi_K-\phi_S)}$ for kaon production at HERMES (right).

and compared with data (see left panel of Fig. 3). Notice that no agreement should be necessarily expected, as $A_{UL}^{\sin\phi_\pi}$ can be originated also (even dominantly) from the Collins mechanisms or higher-twist terms.

The HERMES data [3] clearly show a non zero Sivers effect; by a comparison with these data estimates of the Sivers functions for u and d (both valence and sea) quarks have been obtained. These functions not only describe well the HERMES data, but are also in agreement with COMPASS preliminary data [4].

A phenomenological study of SSA and azimuthal dependences, within a factorization scheme with unintegrated parton distribution and fragmentation functions, is now possible. SIDIS processes with measurements of the Cahn effect, and the various SSA $A_{UL}^{\sin\phi_h}$, $A_{UT}^{\sin(\phi_h-\phi_S)}$ and $A_{UT}^{\sin(\phi_h+\phi_S)}$ provide a rich ground to be further explored, both theoretically and experimentally.

REFERENCES

1. R.N. Cahn, *Phys. Lett.* **B78** 269 (1978), *Phys. Rev.* **D40** 3107 (1989)
2. HERMES Coll., A. Airapetian *et al.*, *Phys. Rev. Lett.* **84** 4047 (2000); *Phys. Rev.* **D64** 097101 (2001)
3. HERMES Collaboration, A. Airapetian *et al.*, *Phys. Rev. Lett.* **94** 012002 (2005), e-Print Archive: hep-ex/0408013
4. COMPASS Collaboration, V.Yu. Alexakhin et al., hep-ex/0503002
   COMPASS Collaboration, P. Pagano, talk delivered at the SPIN2004 Symposium, Trieste, Italy, October 10-16, 2004, e-Print Archive: hep-ex/0501035
   COMPASS Collaboration, R. Webb, talk delivered at BARYONS04, Oct 25-29 2004, Palaiseau, France, e-Print Archive: hep-ex/0501031
5. U. D'Alesio and F. Murgia , *Phys. Rev.* **D70** 074009 (2004)
6. M. Anselmino, M. Boglione, U. D'Alesio, E. Leader and F. Murgia, *Phys. Rev.* **D71** 014002 (2005)
7. D. Sivers, *Phys. Rev.* **D41** 83 (1990) ; **D43** 261 (1991)
8. D.L. Adams *et al.*, *Phys. Lett.* **B264** 462 (1991) ; A. Bravar *et al.*, *Phys. Rev. Lett* **77** 2626 (1996)
9. J.C. Collins, *Nucl. Phys.* **B396** 161 (1993)
10. EMC Coll., J.J. Aubert *et al.*, *Phys. Lett.* **B130** 118 (1983); EMC Coll., M. Arneodo *et al.*, *Z. Phys.* **C34** 277 (1987); Fermilab E665 Coll., M.R. Adams *et al.*, *Phys. Rev.* **D48** 5057 (1993)
11. M. Anselmino, M. Boglione, U. D'Alesio, A. Kotzinian, F. Murgia, A. Prokudin, *Phys. Rev.* **D71** 074006 (2005)

Forward π^0 Production from Transversely Polarized Protons at STAR

Steve Heppelmann for the STAR Collaboration

Penn State University, University Park Pa. 16802

Abstract. Forward electromagnetic calorimeters have been installed in the STAR detector at the Relativistic Heavy Ion Collider to enable the identification and reconstruction of photons from π^0's in the large rapidity region, $\eta>3$. First measurements of pp collisions with a transversely polarized proton beam at $\sqrt{s}=200GeV$ indicate that the π^0 production in the region of $X_F>0.3$ (Feynman X) exhibits a large single spin asymmetry A_N. These results are similar to those observed at lower energy, in Fermilab E704. Furthermore, for backward production, relative to the polarized beam, no significant asymmetry is observed.

Keywords: polarization, single spin asymmetries, RHIC spin, calorimeters
PACS: 13.85.Ni, 13.88.+e, 24.70.+s

Over the past 30 years, very large transverse asymmetries were observed in the production of pions in the interaction $p^{\uparrow}+p\rightarrow\pi+X$ within the forward kinematic region where the pion carries a large fraction X_F of the momentum of the incident proton. These results have long been a source of excitement and controversy. An early prediction of pQCD was that, at leading twist and with collinear factorization, the chiral properties of the theory would make the analyzing power A_N small for particles produced with transversely polarized proton beams [1] at high energy and with transverse momentum large compared to the hadronic scale. However from AGS energies [2] to Fermi Lab energies [3] and most recently at STAR [4] a large transverse spin analyzing power has been observed. The consistent trend is that the asymmetry in $p_{\uparrow}+p\rightarrow\pi^0+X$ increases rapidly for X_F above about 0.3. Transverse single spin asymmetries have also been observed in semi-inclusive DIS from polarized targets [5] and experimental studies of these spin effects is an active area of research.

While there are many possible phenomenological effects that can be identified as contributing sources for these large transverse asymmetries, they naturally divide into two types of contributions. One is the Sivers effect [6, 7], which depends upon an initial state correlation between the parton intrinsic transverse momentum k_T and the transverse spin of the incident nucleon. Within the Sivers framework, A_N is sensitive to the contribution of quark orbital angular momentum to the nucleon spin. Large transverse asymmetries are the result of a spin dependent p_T trigger bias favoring events where k_T is in the same direction as p_T.

As the Sivers effect connects A_N to the orbital angular momentum of quarks, a second effect, called the Collins effect [8,9], is directly sensitive to the transversity distribution function of the nucleon. Transversity is the transverse polarization of

CP792, *Deep Inelastic Scattering, DIS 2005,* edited by W. H. Smith and S. R. Dasu

quarks (and antiquarks) in a transversly polarized proton. In the Collins picture, the quark scatters preserving its transverse spin. It then fragments into pions and other hadrons with an azimuthal angular distribution reflecting the spin of the quark. The fragmentation function reveals the polarization of the fragmenting quark and thus the initial quark state. In this example, the asymmetry does not appear in the jet production directly, but only in the fragmentation. The jet axis would not show the transverse asymmetry, but a pion fragment would. It should be noted that even if the Collins effect makes a small contribution to the observed transverse spin asymmetry, the effect may still provide uniquely important information about the proton transversity structure.

If the Sivers effect is present, we can further characterize the effect with a measurement of the away side jet. Measurement of the difference in transverse momentum of these two back-to-back jets, or their surrogates, will depend upon k_T of the struck quark [10]. The asymmetry in this k_T measurement is exactly what the Sivers model predicts. If events are selected with a pair of π^0's separated by approximately 180º in azimuthal angle and produced within the very forward kinematic region, the momentum of an observed π^0 represents a very large fraction of the jet energy. The typical fragmentation fraction for these events is about 90%.

The STAR collaboration has established that these dramatic transverse spin asymmetries persist at the higher energies of the RHIC collider $\left(\sqrt{s} = 200GeV\right)$. With improving forward calorimetry, used in conjunction with the existing more central STAR detector systems, STAR will be able to distinguish the signatures of the Collins and Sivers effects.

The original plan for electromagnetic calorimetry in STAR emphasized coverage in the central rapidity region $(-1 < \eta < 2)$. Beginning in 2000, attempts were made to add electromagnetic coverage in the more forward region $(\eta > 2)$. Adding forward calorimetry to STAR proceeds in phases.

1. For the 2002 RHIC running period, a prototype calorimeter (**prototype FPD**) module was installed in the forward region. Asymmetry data from that run have been published [4].
2. Starting in the 2003 RHIC running period, arrays of lead glass photon detectors consisting of lead glass blocks of size 3.8 cm. x 3.8 cm. were installed left, right, above and below the beam. The left/right detectors consisted of 7x7 arrays of blocks and the top/bottom detectors consisted of 5x5 arrays. These detectors are referred to as the Forward Pion Detectors (**FPD**). Preliminary data from the FPD are shown as the $\eta = 3.8$ data in Figure 1.
3. STAR is in the process of increasing the scale of the forward electromagnetic coverage, extending to full forward coverage in azimuthal angle over the rapidity interval $2.5 < \eta < 4$. This new calorimeter, called the Forward Meson Spectrometer **FMS**, will consist of a single wall of lead glass. The inner blocks (nearest to the beam) will be of the same type as the FPD. The outer cells will be made from larger glass blocks. When the FMS is combined with the central STAR detector, it will provide nearly continuous coverage for the identification

and reconstruction of π^0's over the rapidity range $-1 < \eta < 4$. This coverage will allow us to distinguish between the signatures of Collins effect or Sivers effect.

From inclusive forward π^0 production with transversely polarized proton collisions $p^\uparrow + p \to \pi + X$, measurements have been obtained for the invariant cross sections and for the transverse asymmetry within the kinematic region defined by

$$1\frac{GeV}{c} < p_T < 2.5\frac{GeV}{c} \quad \textbf{and} \quad 3.3 < \eta < 4.1.$$

The transverse asymmetry is defined with the usual definition,

$$A_N = \frac{d\sigma^\uparrow - d\sigma^\downarrow}{d\sigma^\uparrow + d\sigma^\downarrow},$$

where $d\sigma^\uparrow$ represents a differential cross section for incident proton spin aligned along a transverse measurement direction and $d\sigma^\downarrow$ represents the corresponding differential cross section for the spin anti-aligned with that direction. The observed events are chosen to scatter in the plane transverse to the vertical axis (axis of spin quantization). Transverse asymmetry measurements with the prototype FPD (2002) and preliminary measurements with the FPD (2003), are shown in Figure 1.

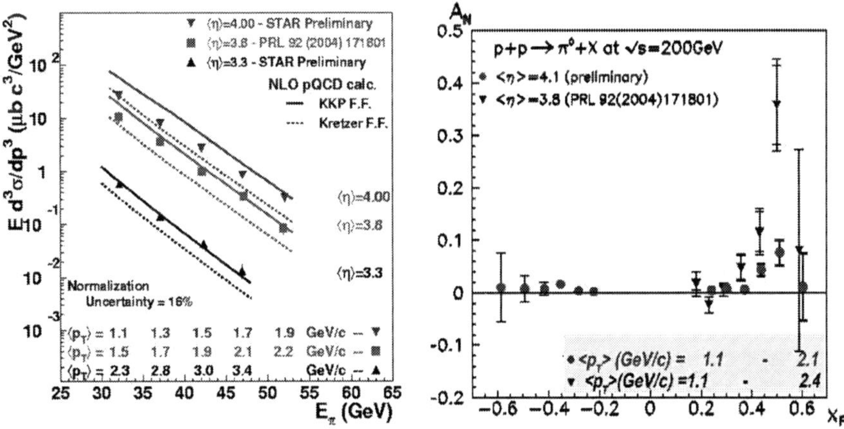

FIGURE 1. (Left) Invariant cross sections for inclusive π^0 production at large rapidity in p+p collisions at $\sqrt{s}$=200 GeV [4] compared to NLO pQCD calculations [11-13]. (Right) The corresponding STAR measurements of single spin asymmetry (A_N) for production of π^0's at the same energy. The η=4.1 data, shown with round markers, are preliminary. The published [4] η=3.8 data are shown with triangles. The equivalent p_T scale for these data is also indicated.

It is seen that the asymmetry measured has a form similar to that seen at much lower energy. The measurement of these cross sections is in good agreement with next to leading order perturbative calculations. It is noted that while the asymmetries observed are similar to those seen at lower energy, the cross sections at lower energy did not agree with perturbative calculations. We believe that this new RHIC result represents an interesting convergence of perturbative interpretations and asymmetry mechanisms which sets it apart from lower energy measurements, severely constraining the theoretical interpretation of the asymmetries.

With the FMS in place, STAR will be able to measure not only single inclusive π^0's but also correlated pairs of π^0's. These pairs can be generally associated with either a single jet (same side π^0's with small azimuthal separation angle) or with pairs of jets (π^0 pairs with nearly back to back azimuthal angles). These two topologies also correspond to regimes where Collins effect or Sivers effect can be identified and quantitatively studied.

By looking at pairs of same-side neutral pions, we can measure the asymmetry as a function of the two pion kinematics. With the FMS we will separately measure the contributions to this asymmetry that comes from the jet axis vs. that which comes from the jet structure. Many theory papers have studied this problem: however, the need for data is great. The FMS STAR experiments on transverse polarization will provide to theorists the necessary input to determine the relative contributions from the Sivers effect and the Collins effect.

The performance of the polarized RHIC collider has been steadily improving. Transverse polarization data sets from 2003 and 2004 were both obtained with less than ½ Picobarn^{-1} of beam. The average proton beam polarization in 2003 was less than 20%. In current running (2005), online polarization values in excess of 50% have been seen. Integrated luminosity for longitudinal running in 2005 will be about an order of magnitude greater than that corresponding to data presented in Figure 1. A much higher statistics measurement of these transverse asymmetries can be expected in the near future.

ACKNOWLEDGMENTS

This research is supported by the National Science Foundation, U.S. Department of Energy as well as funding agencies from 13 other nations.

REFERENCES

1. G.L. Kane, J. Pumplin and W. Repko, Phys. Rev. Lett. **41** (1978) 1689.
2. R.D. Klem *et al.*, Phys. Rev. Lett. **36** (1976) 929; W.H. Dragoset *et al.*, Phys. Rev. **D18** (1978) 3939; S. Saroff *et al.*, Phys. Rev. Lett. **64** (1990) 995; B.E. Bonner *et al.*, Phys. Rev. **D41** (1990) 13; K. Krueger *et al*, Phys. Lett. **B459** (1999); C.E. Allgower *et al.*, Phys. Rev. **D65** (2002) 092008.
3. B.E. Bonner *et al.* Phys. Rev. Lett. **61** (1988) 1918; D.L. Adams *et al.*, Phys. Lett. **B261** (1991) 201; Phys. Lett. **B264** (1991) 462; Z. Phys. **C56** (1992) 181; A. Bravar *et al.* Phys. Rev. Lett. **77** (1996) 2626.
4. J. Adams *et al.* (STAR collaboration), Phys. Rev. Lett. **92** (2004) 171801 [hep-ex/0310058].
5. A. Airapetian *et al.* (HERMES collaboration), Phys. Rev. Lett. **94** (2005) 012002.
6. D.Sivers, Phys. Rev. **D41** (1990) 83; **43** (1991) 261.
7. M. Anselmino, M. Boglione and F. Murgia, Phys. Lett. B362 (1995) 164; M. Anselmino and F. Murgia, Phys. Lett. B442 (1998) 470.
8. J. Collins, Nucl. Phys. **B396** (1993) 161; J. Collins, S.F. Heppelmann and G.A. Ladinsky, Nucl. Phys. **B420** (1994) 565.
9. M. Anselmino, M. Boglione and F. Murgia, Phys. Rev. **D60** (1999) 054027.
10. Daniel Boer and Werner Vogelsang, Phys. Rev. **D69** (2004) 094025 [hep-ph/0312320].
11 F. Aversa *et al.*, Nucl Phys. **B327** (1989) 105; B. Jager *et al.*, Phys. Rev. **D67** (2003) 054005; and D. de Florian, Phys. Rev. **D67** (2003) 054004.
12. B.A. Kniehl, G. Kramer and B. Poetter, Nucl Phys. **B597** (2001) 337.
13. S. Kretzer, Phys. Rev. **D62** (2000) 054001.

Transversity Physics Results from PHENIX

M. Chiu for the PHENIX Collaboration

University of Illinois
Urbana, IL 61801
U.S.A.
email: chiu@uiuc.edu

Abstract. During the 2001-2002 proton run at RHIC, PHENIX collected an integrated luminosity of 0.15 pb^{-1} of transversely polarized proton collisions at a $\sqrt{s} = 200GeV/c$. With this dataset the transverse single-spin asymmetry A_N for π^0 and non-identified charged hadrons at $x_F = 0$ has been measured up to $p_\perp = 5$ GeV/c. Transverse Single Spin Asymmetries (SSA) are thought to come from at least three mechanisms, and in the future PHENIX will be able to decouple contributions to SSA from the Sivers function by measuring asymmetries in the back-to-back correlation in azimuthal angle of two high-$p_\perp$hadrons.

Keywords: SSA,Single Spin Asymmetry,Transverse Spin,Transversity,Sivers Function
PACS: 14.20.Dh

Measurements by E704 of non-zero single-spin asymmetries (SSA) in inclusive particle production at fixed target energies [1, 2] have generated much interest since leading-twist collinear factorized pQCD predicts that these asymmetries should be small. This interest has recently grown with the discovery by the STAR collaboration that these asymmetries persist even to collider energies [3], where a perturbative treatment of the data seems validated by the agreement between data and pQCD calculations of the inclusive cross-section [3, 4]. Much of the rising interest at RHIC in single-spin asymmetries comes from the realization that a full QCD description of the proton requires the inclusion of transverse momentum, and that perhaps the three-dimensional structure of the proton may be probed with spin effects in hard collisions.

Currently, single inclusive transverse spin asymmetries are thought to come from some combination of the following three possible effects:

- An asymmetry in the k_T dependent parton distribution function, called the Sivers function [5].
- Non-zero transversity distributions in combination with a spin-dependent chiral-odd fragmentation function, such as the Collins function or an interference fragmentation function [6, 7].
- Higher Twist Initial-State or Final-State interactions [8, 9].

While a measurement of $A_N = 0$ implies some constraint on these contributions, future progress will require measurements which unambiguously separate these various effects. In the future PHENIX will be able to reduce these ambiguities in inclusive single-spin asymmetries by looking for the effects of the k_T asymmetry on the azimuthal distribution between hadrons fragmented from back-to-back jets [10].

CP792, *Deep Inelastic Scattering, DIS 2005,* edited by W. H. Smith and S. R. Dasu

INCLUSIVE SINGLE SPIN ASYMMETRY AT $X_F = 0$

The production of π^0 and charged hadrons were measured in the central arms of the PHENIX detector, which consist of two 90° spectrometer arms covering $|\eta| < 0.35$[11]. Charged hadrons are tracked using a drift chamber and pad chamber, with a veto in the RICH to eliminate electrons. The major source of background comes from weak decays of short lived particles that have their momentum mis-reconstructed since all tracks are assumed to come from the origin. The electron and decay background is less than 1% and 5%, respectively.

Neutral pions are reconstructed from their decay to two photons using either the Lead-Glass or Lead-Scintillator Electromagnetic Calorimeters, and identified by their invariant mass. The asymmetry from the combinatoric background is handled by the following:

$$A_N^{\pi^0} = \frac{A_N^{\text{peak}} - rA_N^{\text{bg}}}{1-r}, \sigma_{A_N^{\pi^0}} = \frac{\sqrt{\sigma^2_{A_N^{\text{peak}}} + r^2\sigma^2_{A_N^{\text{bg}}}}}{1-r} \tag{1}$$

That is, the background asymmetry is estimated by calculating the asymmetry in regions on both sides of the π^0 mass peak, and subtracted with a weight r, which is the fraction of the background under the π^0 peak.

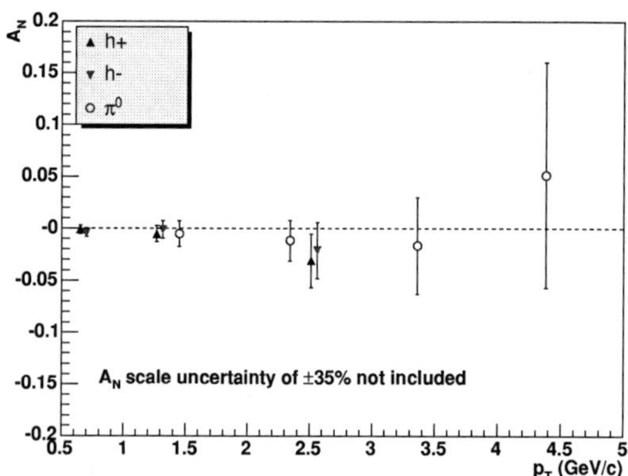

FIGURE 1. Inclusive A_N for π^0 and charged hadrons vs. $p_\perp$at $x_F = 0$. Positive hadrons have been shifted down by 50 MeV/c to improve readability [12].

The asymmetry is determined in two different ways. In one technique, the single-spin asymmetry is calculated using the formula

$$A_N = \frac{1}{P_b}\left(\frac{N^\uparrow - \mathscr{R}N^\downarrow}{N^\uparrow + \mathscr{R}N^\downarrow}\right), \tag{2}$$

where P_b is the beam polarization, $N^\uparrow(N^\downarrow)$ is the yield in both arms for π^0 or charged hadrons in positively (negatively) polarized $p^\uparrow + p$ collisions, and $\mathscr{R} = \mathscr{L}^\uparrow/\mathscr{L}^\downarrow$ the relative integrated luminosity of bunches that are positively versus negatively polarized.

Alternatively, we can calculate the asymmetry using

$$A_N = \frac{1}{P_b}\left(\frac{\sqrt{N_L^{\uparrow}N_R^{\downarrow}} - \sqrt{N_L^{\downarrow}N_R^{\uparrow}}}{\sqrt{N_L^{\uparrow}N_R^{\downarrow}} + \sqrt{N_L^{\downarrow}N_R^{\uparrow}}}\right), \tag{3}$$

where N_L (N_R) is the yield in the left (right) side of the polarized beam, and the arrow indicates the polarization of the beam. The first technique requires good knowledge of the beam luminosity but little knowledge of detector acceptance, while the second technique requires neither but is an approximation (but more than good enough for our purposes). Since the largest systematic error comes from the relative luminosity, the second technique provides a strong cross-check.

The asymmetry is determined for each fill and then averaged over all fills. The asymmetry is shown in fig. 1. There is an overall scale uncertainty of $\pm 35\%$ from uncertainty in the absolute polarization, and an absolute uncertainty of 0.2% on all points, neither of which are shown. The asymmetry at mid-rapidity is consistent with zero for all $p_\perp$ for both the π^0 and charged hadrons.

BACK-TO-BACK AZIMUTHAL CORRELATIONS OF DI-HADRONS

The current measurement of A_N at mid-rapidity provides some constraint on the gluon Sivers function, and the current data is sensitive to other contributions as well which weakens the constraint. Since the single inclusive SSA can come from many effects, different techniques must be used to deconvolute them. An asymmetry in the azimuthal correlation of back-to-back jets was suggested by Boer and Vogelsang [10] as a way of measuring effects from the Sivers function and which would be relatively insensitive to other effects.

Since the acceptance of the PHENIX central arms is too small to reconstruct jets explicitly, the back-to-back correlation must be measured using di-hadrons pairs. PHENIX has measured angular correlations between photons with a $p_\perp > 2.25 GeV/c$ and charged hadrons [13], where the photon comes dominantly from the decay of a π^0. These γ-hadron angular correlations show all the expected features of a jet, with a peak on the side near the decay photon due to the jet fragmentation, and a wider away side peak due to fragmentation plus k_T effects. In other words, the photon provides a good proxy for the underlying jet axis, and can be used as a substitute with some reduced resolution due to the jet fragmentation.

The asymmetry calculated by Boer and Vogelsang assumes that the di-jets are reconstructed. Since we use fragments from the jet, the asymmetry is smeared by the width of the jet fragmentation. In fig. 2, we plot the di-jet back-to-back A_N under the assumption that the gluon Sivers function, $\Delta_g^N(x, k_T) \sim g(x)\mathcal{N}_g(x)$, where $\mathcal{N}_g(x) = \mathcal{N}_d(x)$ and is taken from the parameterization of the down quark Sivers function determined from a fit to the SSA measured by the E704 collaboration [14]. The dashed line includes the effect from measuring with di-hadrons versus di-jets. The di-hadron smearing is done assuming a gaussian fragmentation, where the size of the gaussian is measured from

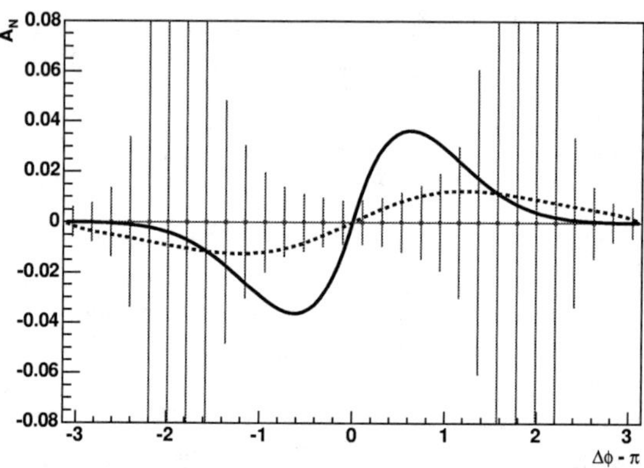

FIGURE 2. Estimated statistical errors in PHENIX for A_N with 0.35 pb^{-1} and 50% polarization. The solid line corresponds to an ansatz provided by Boer and Vogelsang [10]. The dashed line is the same ansatz but with realistic expectations for polarization and with di-hadron smearing included.

data. Additionally, the dashed line includes a 50% reduction in sensitivity to reflect the polarization at RHIC. The error bars represent the estimated sensitivity of PHENIX, and are estimated from 0.35 pb^{-1} of $p+p$ collisions taken during 2002-2003.

CONCLUSIONS

The single inclusive transverse spin asymmetry A_N has been measured by PHENIX for π^0 and non-identified charged hadrons at mid-rapidity and found to be consistent with 0 to the few percent level. This provides some constraint for the magnitude of any gluon Sivers function. PHENIX will improve upon this constraint by the measurement of back-to-back di-hadron azimuthal correlations.

REFERENCES

1. D. L. Adams, et al., *Phys. Lett.*, **B264**, 462–466 (1991).
2. D. L. Adams, et al., *Phys. Rev.*, **D53**, 4747–4755 (1996).
3. J. Adams, et al., *Phys. Rev. Lett.*, **92**, 171801 (2004).
4. S. S. Adler, et al., *Phys. Rev. Lett.*, **93**, 202002 (2004).
5. D. W. Sivers, *Phys. Rev.*, **D41**, 83 (1990).
6. J. C. Collins, *Nucl. Phys.*, **B396**, 161–182 (1993).
7. J. C. Collins, S. F. Heppelmann, and G. A. Ladinsky, *Nucl. Phys.*, **B420**, 565–582 (1994).
8. J.-w. Qiu, and G. Sterman, *Phys. Rev.*, **D59**, 014004 (1999).
9. Y. Kanazawa, and Y. Koike, *Phys. Lett.*, **B478**, 121–126 (2000).
10. D. Boer, and W. Vogelsang, *Phys. Rev.*, **D69**, 094025 (2004), `hep-ph/0312320`.
11. K. Adcox, et al., *Nucl. Instrum. Meth.*, **A499**, 469–479 (2003).
12. S. S. Adler (2005), `hep-ex/0507073`.
13. M. Chiu, *Nucl. Phys.*, **A715**, 761–764 (2003), `nucl-ex/0211008`.
14. M. Anselmino, U. D'Alesio, and F. Murgia, *Phys. Rev.*, **D67**, 074010 (2003), `hep-ph/0210371`.

Single Spin Asymmetries in the BRAHMS Experiment

F.Videbæk for the BRAHMS collaboration

Physics Department, Brookhaven National Laboratory

Abstract. The BRAHMS experiment at RHIC has the capability to measure the transverese spin asymmetries in polarized pp induced pion production at RHIC. The first results from a short run show a signaificant asymmetry for π^+ and π^- at moderate x_F. The trend of the data is in agreement with lower energy data while the absolute value are surprisingly large.

Keywords: Polarized protons, single spin asymmetries
PACS: 13.85Ni,13.88+e,12.38Qk

INTRODUCTION

In the last decade or so, measurements of transverse single spin asymmetries in pp collisions with polarized beams have attracted much theoretical and experimental interest. Results at moderate beam energies[1] show a sizeable asymmetry up to 30% at relative large Feynman-x (x_F) and at moderate p_T. It was expected, naively, from lowest order QCD estimates that the cross sections should have little spin dependence. In order to get a non-zero value both spin-flip amplitudes, a phase difference in the intrinsic states as well as a non zero scattering angle is necessary. This makes it a higher order effect that can be either in the initial state or in the final state parton scattering. The asymmetry or analyzing power A_N is defined as $(\sigma^+ - \sigma^-)/(\sigma^+ + \sigma^-)$, where $\sigma^{+(-)}$ is a spin dependent cross section for the scattering $pp \rightarrow \pi X$, and with the spin direction oriented up or down transversely to the beam momentum.scattering plane. The target is either un-polarized or the cross sections are averaged over polarization states. The experiments [1] has shown that $A_N(\pi^+) > A_N(\pi^0) > 0 > A_N(\pi^-)$. A recent result from STAR [2] shows a positive A_N for π^0 at large x_F in pp collisions at 200 GeV. The BRAHMS experiment at RHIC is primarily designed and operated to make measurements of semi-inclusive spectra of identified hadrons over a wide range in rapidity and p_T. The PID coverage for pions up to momenta of 40 GeV/c and the option to measure at 2.3 degrees ($\eta \approx 4$) makes it well suited to study Single Spin Asymmetries for identified pions at moderate x_F. The present contribution presents the first preliminary measurements of A_N for π^+ and π^- at moderate values of x_F in pp collisions at 200 GeV at RHIC.

RESULTS

The BRAHMS forward spectrometer consists of 4 dipole magnets, 5 tracking chambers, two Time-Of-Flight systems and a Ring Imaging Chrenkov Detector (RICH) for particle

CP792, *Deep Inelastic Scattering, DIS 2005*, edited by W. H. Smith and S. R. Dasu

identification. The angular coverage of the spectrometer is from 2.3 to 15 degrees, and the solid angle 0.8 msr. Details of experimental setup can be found in [3]. For transverse spin measurements the kinematic variables of interest are x_F and p_T. Shown in Fig. 1 is the BRAHMS acceptance for the data taken at $\theta = 2.3$ degrees at the maximum field setting of 7.2 Tm in the spectrometer. The momentum resolution $\delta p/p$ is estimated to be 1% at momenta of 22 GeV/c. There is an approximate linear correlation between x_F and p_T. It should be pointed out that the acceptance does not corresponds to a that of a fixed angle. Scattering angles of 2.3° and 4° are shown on the figure. Thus care should be taken when comparing to both other experiments (STAR) and to theory.

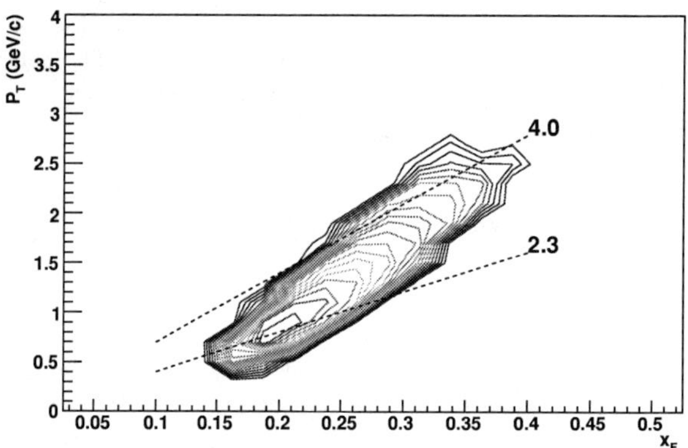

FIGURE 1. Acceptance in the BRAHMS experiment for pions at the nominal setting of 2.3 degrees in p_T vs. x_F. The dashed lines indicates the p_T- x_F correlations for fixed angles of 2.3° and 4°, respectively.

Tracks were reconstructed from measurements in at least 4 of the 5 chambers, and its momentum from 3 independent measurements. The tracks are required to project cleanly through the spectrometer. An approximate vertex can be determined from the timing measurements in sets of symmetricaly placed scintillator counters (INL) around the beam pipe at 1.5, 4.15 and 6.7 meters [4]. The position resolution of the vertex determination is about 10 cm from these measurements. In addition vertex positions and live rates are obtained from a set of Cherenkov Counters (BB) with limited acceptance at ± 2.15m and a pair of Zero Degree Calorimetres (ZDC) placed at ± 18m. The tracks accepted in the spectrometer are requiered to point backward to these measurements with an accuracy of 30 cm and to be within a narrow range of (-40,20) cm of the nominal interaction point. Due to the measuring angle of 2.3 degree, the spectrometer tends to accepts track weighted towards negative vertex positions.

The particle identification of the pions is done exclusively from the RICH. It is required that the calculated radius for pion is within .25 cm from the measured radius, and at the same time more than .30 cm away from the estimated radius assuming the track is from a kaon. This corresponds to about a 2 and 2.5 σ cuts, respectively. Figure 2 shows the radius of rings determined in the RICH for all events, and for those identified as pions using the above mentioned cuts in the momentum range p <35 GeV/c. The contamination of kaons into the pion sample is estimated to be less than a few percent.

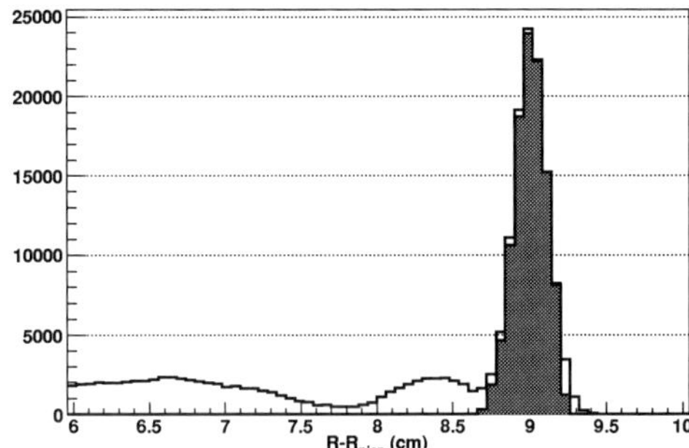

FIGURE 2. Distribution of radia of the rings in the RICH for accpeted particle in the spectrometer. The solid histogram represents those particles identified as pions.

In the RHIC accelerator the transverse spin polarization is altered between the 56 bunches of polarized protons that forms the beam in each of the two rings. Thus most experimental time-dependent effects originating from the spectrometer and the vertex determination cancel out when constructing the raw asymmetries

$$\varepsilon = (N^+ - L * N^-)/(N^+ + L * N^-)$$

The $N^{+(-)}$ represents the yield of pions in a given kinematic bin where the beam spin direction is up or down relative to the reaction plane determined by $k_{beam} \times k_{out}$. The factor L is the ratio of the luminosity of bunches with positive polaraization to those of negative polarization thus accounting for non-uniform bunch intensities. The luminosity ratio is determined independently from the spectrometer data using several measures of collision rates from the INEL, BB, and ZDC detector systems. It is estimated that the systematic error from the relative luminosity measurements is in order 0.5%.

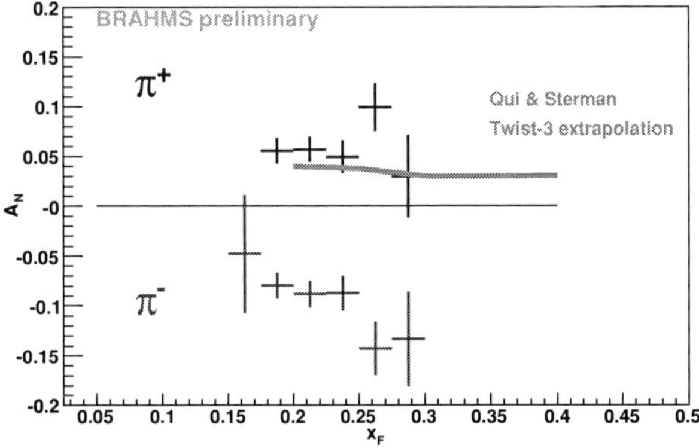

FIGURE 3. Analyzing power A_N for π^+ and π^- .

The asymmetry is the in turn determined from $A_N = \varepsilon/P$. The polarization (P) as determined from the CNI measurements [5] is $\approx 42\%$ for the π^+ measurements and $\approx 38\%$ for the for π^- measurements in the stores used. The systematic error on beam polarizations is $\approx 15\%$ and represents a scaling error on the values of A_N. This error is expected to be reduced after the final analysis of CNI data. The measured raw asymmetries corrected for the beam polarization is shown in Fig. 3 for π^- and π^+ . The π^+ asymmetries are positive while the π^- are negative i.e. the same sign dependence as seen in the E704 data at lower energy.

Several theorectical models have been worked out for the single spin asymmetries to clarify the importance of inital vs. final state effects as put forward through the Sivers and Collins effects. In this contribution we compare the data vs. extrapolations of twist 3 (initial state) calculations by Qiu and Sterman [6]. The pQCD calculations are apriori not valid at the lower values of p_T covered in the present measurements. Never the less it gives a good estimate how kinematic cuts may effect predictions as to give rise to a near constant A_N in a limited range of x_F. Both the magnitude and the x_F dependence is in reasonable agreement with the data.

During run-5 RHIC has delivered much increased integrated luminosity and with a larger beam polarization of $45-55\%$. BRAHMS has added new vertex detector that will provide a global vertex resolution of ≈ 2 cm. Data have been recorded for π^+ and π^- in the x_F range of 0.15-0.35 with about 10-20 times the statistics in the data presented here.

In summary, the first data from polarized proton data from BRAHMS were obtained in the RHIC Run-4 and shows a finite A_N for π^+ and π^- with sign ordering as observed previously in E704 at FNAL. In addition the protons are found to have $A_N \approx 0$. Data from the ongoing RHIC Run-5 will give an order of magnitude better statistics and should enable BRAHMS to make comparisons to several theoretical models

This work is suppord by the Division of Nuclear Physics of the Office of Science of the U.S. Department of energy under contract DE-AC02-98-CH10886, the Danish Natural Science Research Council, the Research Council of Norway, the Jagiellonian University Grants and the Romanian Ministry of Education and Research. I will like to thank J.Qui for supplying the calculations for the twist-3 extrapolations, and Brendan Fox for advice, calculations and help with the experimental setup for the first attempt to collect spin sorted data with BRAHMS spectrometer in 2003, a setup that was used in the brief 2004 run that resulted in the present data.

REFERENCES

1. D. L. Adams et al *Phys. Lett.*, **B264** 1991, 462
2. J. Adams et. al *Phys. Rev. Lett*, **92** 2004, 171801
3. J.Adamczyk et.al. *Nucl. Instr. Meth.* **A499** 2003, 437
4. I. Arsene at.al. *Phys. Rev. Letter*, **94** 2005, 032301
5. O. Jinnouchi et. al , Contribution to SPIN2004, nucl-ex/04122053
6. J. Qiu and G. Sterman *Phys. Rev.* **D59** 1998, 014064

Next-to-leading Order QCD Corrections to A_{TT} for Single-Inclusive Hadron Production

A. Mukherjee*, M. Stratmann† and W. Vogelsang**

*Instituut-Lorentz, University of Leiden, 2300 RA Leiden, The Netherlands
†Institut für Theoretische Physik, Universität Regensburg, D-93040 Regensburg, Germany
**Physics Department and RIKEN-BNL Research Center, Brookhaven National Laboratory, Upton, New York 11973, U.S.A.

Abstract. We report on a calculation of the next-to-leading order QCD corrections to the partonic cross sections contributing to single-inclusive high-p_T hadron production in collisions of transversely polarized hadrons. We give some predictions for the double spin asymmetry A_{TT}^{π} for the proposed experiments at RHIC and at the GSI.

Keywords:
PACS: PACS numbers: 12.38.Bx, 13.85.Ni, 13.88.+e

INTRODUCTION

The leading-twist partonic structure of a spin-$\frac{1}{2}$ hadron is given in terms of the unpolarized parton distribution functions $f(x,Q^2)$, the helicity distributions $\Delta f(x,Q^2)$, and the transversity distributions $\delta f(x,Q^2)$. Transversity describes the number density of a parton with the same transverse polarization as the nucleon, minus the number density for opposite polarization. Among the various parton distributions, the $\delta f(x,Q^2)$ are the ones about which we have the least knowledge. They are at present the focus of much experimental activity. Transversity will be probed by double-transverse spin asymmetries in transversely polarized pp collisions at the BNL Relativistic Heavy Ion Collider (RHIC) [1]. The most promising reaction is the Drell-Yan process, which offers the largest spin asymmetries, but whose main drawback is the rather moderate event rate [2]. Other relevant processes include high p_T prompt photon [3, 4] and jet production [3]. However, for these the asymmetry is expected to be much smaller because of the absence of gluon-initiated subprocesses in the transversely polarized cross section. Recently, it has also been investigated whether one could extract transversity from measurements of A_{TT} for the Drell-Yan process in transversely polarized $\bar{p}p$ collisions at the planned GSI-FAIR facility [5, 6, 7, 8, 9, 10]. In this note (for details, see Ref. [11]), we consider the spin asymmetry in single-inclusive production of pions at large transverse momentum p_T as a possible means for determining transversity at RHIC or the GSI. We report on the calculation of the next-to-leading order (NLO) corrections for this reaction, and also present some phenomenological results.

CP792, *Deep Inelastic Scattering, DIS 2005*, edited by W. H. Smith and S. R. Dasu

PROJECTION TECHNIQUE

It is known that the NLO QCD corrections are required in order to have firm theoretical predictions for hadronic scattering. Only with their knowledge can one reliably extract information on the partonic (spin) structure of nucleons. Apart from this motivation, also interesting new technical questions arise beyond leading order (LO) in the calculations of cross sections with transverse polarization. Unlike the longitudinally polarized case, where the spin vectors are aligned with the momenta, the transverse spin vectors specify extra spatial directions and, as a result, the cross section has non-trivial dependence on the azimuthal angle of the observed particle. For A_{TT} this dependence is of the form [4]

$$\frac{d^3\delta\sigma}{dp_T d\eta d\Phi} \equiv \cos(2\Phi) \left\langle \frac{d^2\delta\sigma}{dp_T d\eta} \right\rangle , \tag{1}$$

for a parity conserving theory with vector coupling. Here, the spin vectors are taken to point in the $\pm x$ direction. We furthermore consider the scattering in the center-of-mass frame of the initial hadrons and use their momenta to define the z axis. Because of the $\cos(2\Phi)$ dependence, integration over all azimuthal angles is not appropriate. This makes it difficult to use the standard techniques developed for NLO calculations of unpolarized and longitudinally polarized processes, because these techniques usually rely on the integration over the full azimuthal phase space and also on the choice of particular reference frames that are related in complicated ways to the center-of-mass frame of the initial hadrons. In [4] a general technique was introduced that facilitates NLO calculations for transverse polarization by conveniently projecting on the azimuthal dependence of the cross section in a covariant way. The projector

$$F(p, s_a, s_b) = \frac{s}{\pi t u}\left[2\,(p\cdot s_a)\,(p\cdot s_b) + \frac{tu}{s}\,(s_a\cdot s_b)\right] \tag{2}$$

reduces to $\cos(2\Phi)/\pi$ in the center-of-mass frame of the initial hadrons. Here p is the momentum of the observed particle in the final state and the s_i are the initial transverse spin vectors. The squared matrix element for the partonic process is multiplied with this projector and integrated over the full azimuthal phase space. Integrations of terms involving the product of the transverse spin vectors with the final-state momenta can be performed using a tensor decomposition. After this step, no scalar products involving the s_i are left in the squared matrix element. For the remainder of the phase space integrations, one can now use techniques familiar from the unpolarized and longitudinally polarized cases. This method is particularly convenient at NLO, where one uses dimensional regularization and the phase space integrations are performed in $n \neq 4$ dimensions.

APPLICATION TO SINGLE-INCLUSIVE HADROPRODUCTION : PHENOMENOLOGICAL RESULTS

Next, we give some phenomenological results for transversely polarized pp collisions at RHIC ($\sqrt{S} = 200$ and 500 GeV) and $\bar{p}p$ collisions at the GSI-FAIR facility in an

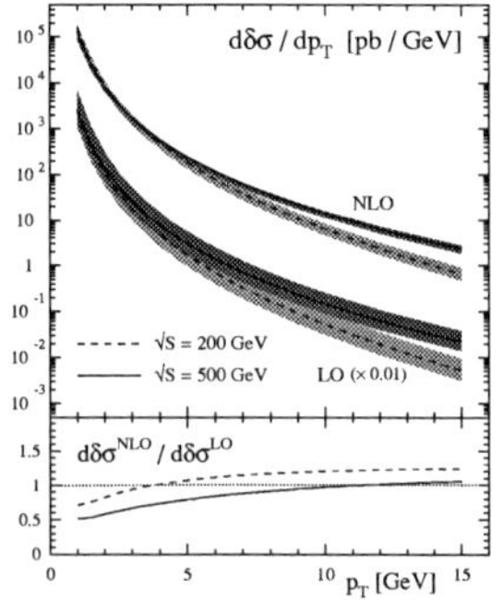

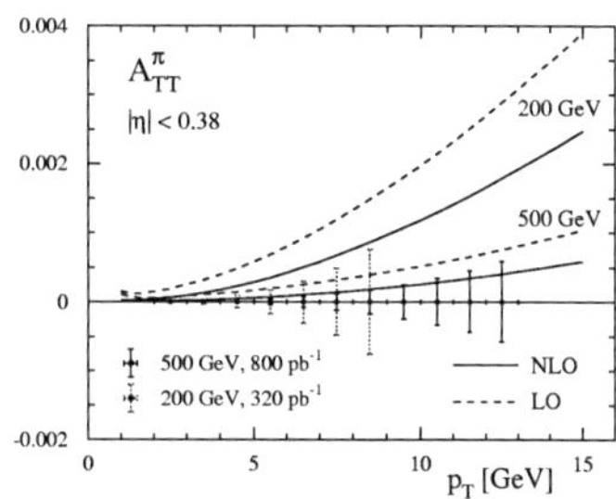

FIGURE 1. Predictions for the transversely polarized single-inclusive pion production cross sections at LO and NLO at RHIC (left), and for the transverse double-spin asymmetry $A_{\rm TT}^{\pi^0}$ (right). The shaded bands represent the range of predictions when the scale μ is varied in the range $p_T \leq \mu \leq 4p_T$. The lower panel on the left shows the ratios of the NLO and LO results (the "K-factors").

asymmetric collider mode with proton and antiproton energies of 3.5 GeV and 15 GeV, respectively [11]. For our numerical predictions, we model the transversity distributions by saturating the Soffer inequality [12] at some low input scale $\mu_0 \approx 0.6$ GeV, using the NLO (LO) GRV [13] and GRSV ("standard scenario") [14] densities $q(x,\mu_0)$ and $\Delta q(x,\mu_0)$, respectively.

Figure 1 (left) shows our predictions for the transversely polarized single-inclusive pion production cross sections at LO and NLO for the two different c.m.s. energies at RHIC. We also display the scale uncertainty. For the K-factor shown in the lower panel the scale choice is $\mu = 2p_T$. A significant decrease of scale dependence is observed when going from LO to NLO. Figure 1 (right) shows the asymmetry $A_{\rm TT}^{\pi^0}$ defined as the ratio of the polarized and unpolarized cross sections. Here, the scale is set to $\mu = p_T$. We also display the statistical errors that may be achievable in experiment. We have calculated these using

$$\delta A_{\rm TT} \simeq \frac{1}{P_1 P_2 \sqrt{\mathscr{L}\,\sigma_{\rm bin}}}, \tag{3}$$

where $P_1 = P_2 = 0.7$ are the transverse polarizations of the proton beams, $\mathscr{L}$ the integrated luminosity of the collisions indicated in the figure, and $\sigma_{\rm bin}$ the unpolarized cross section integrated over the p_T-bin for which the error is to be determined. The asymmetry is very small.

Figure 2 (left) shows the corresponding cross sections in transversely polarized $\bar{p}p$ collisions at $\sqrt{S} = 14.5$ GeV in an asymmetric collider mode at the GSI-FAIR facility. We have assumed beam polarizations of 30% and 50% for the antiprotons and protons, respectively. At GSI energies, the scale dependence does not really improve from LO to NLO. A resummation of the double-logarithmic corrections to the partonic cross sections [15] would be very desirable for the future, along with a study of power corrections. Figure 2 (right) shows the predicted transverse double-spin asymmetry at the GSI, which is much larger than at RHIC.

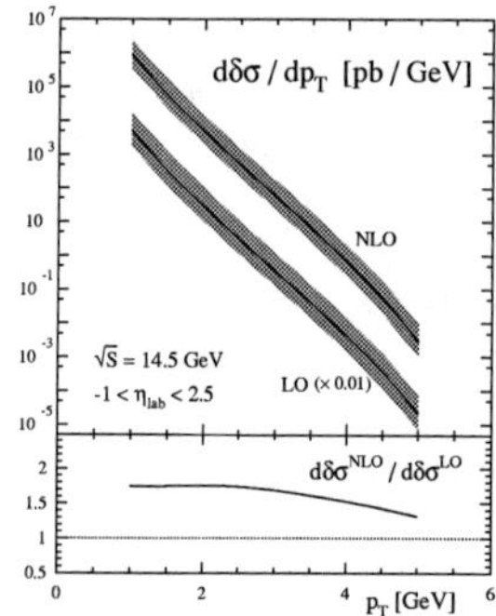

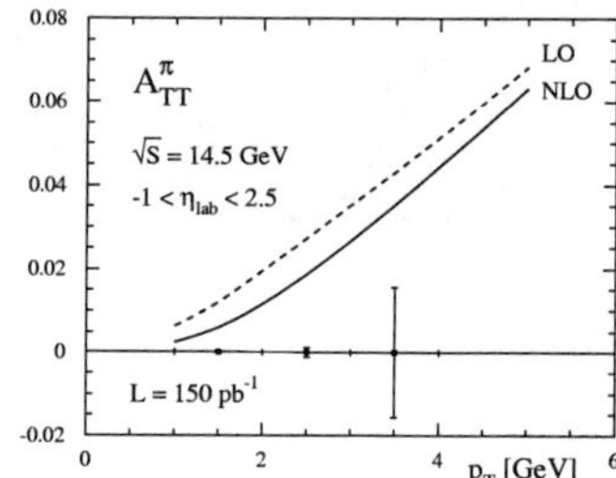

FIGURE 2. LO and NLO predictions for the cross section (left) and the transverse spin asymmetry $A_{TT}^{\pi^0}$ (right) for single-inclusive pion production in $\bar{p}p$ collisions at the GSI. The shaded bands represent the range of predictions if the scale μ is varied in the range $p_T \leq \mu \leq 4p_T$. The lower panel (left) shows the ratios of the NLO and LO results (the "K-factors").

ACKNOWLEDGMENTS

A.M. thanks the organizers of DIS 2005 for a wonderful conference and the invitation. W.V. is grateful to RIKEN, Brookhaven National Laboratory and the Department of Energy (contract number DE-AC02-98CH10886) for providing the facilities essential for the completion of this work. This work is supported in part by the "Bundesministerium für Bildung und Forschung (BMBF)" and by FOM, The Netherlands.

REFERENCES

1. See, for example: G. Bunce, N. Saito, J. Soffer, and W. Vogelsang, Annu. Rev. Nucl. Part. Sci. **50**, 525 (2000).
2. O. Martin, A. Schäfer, M. Stratmann, and W. Vogelsang, Phys. Rev. **D57**, 3090 (1998); **D60**, 117502 (1999).
3. J. Soffer, M. Stratmann, and W. Vogelsang, Phys. Rev. **D65**, 114024 (2002).
4. A. Mukherjee, M. Stratmann, and W. Vogelsang, Phys. Rev. **D67**, 114006 (2003).
5. P. Lenisa and F. Rathmann [the PAX Collaboration], `hep-ex/0505054` and `http://www.fz-juelich.de/ikp/pax/`
6. M. Maggiora [the ASSIA Collaboration], `hep-ex/0504011`; GSI-ASSIA Technical Proposal, Spokesperson: R. Bertini, `http://www.gsi.de/documents/DOC-2004-Jan-152-1.ps`
7. M. Anselmino, V. Barone, A. Drago, and N. N. Nikolaev, Phys. Lett. **B594**, 97 (2004).
8. A. V. Efremov, K. Goeke, and P. Schweitzer, Eur. Phys. J. **C35**, 207 (2004).
9. H. Shimizu, G. Sterman, W. Vogelsang, and H. Yokoya, Phys. Rev. **D71**, 114007 (2005).
10. A. Bianconi and M. Radici, `hep-ph/0504261`.
11. A. Mukherjee, M. Stratmann, and W. Vogelsang, `hep-ph/0506315`.
12. J. Soffer, Phys. Rev. Lett. **74**, 1292 (1995); D. Sivers, Phys. Rev. **D51**, 4880 (1995).
13. M. Glück, E. Reya, and A. Vogt, Eur. Phys. J. **C5**, 461 (1998).
14. M. Glück, E. Reya, M. Stratmann, and W. Vogelsang, Phys. Rev. **D63**, 094005 (2001).
15. D. de Florian and W. Vogelsang, Phys. Rev. **D71**, 114004 (2005).

Double helicity asymmetry measurements with PHENIX detector at RHIC

Abhay Deshpande for PHENIX collaboration

Department of Physics & Astronomy, SUNY-Stony Brook, NY 11794-3800 &
Riken-BNL Research Center, Brookhaven National Lab, Upton, NY 11973-5000

Abstract. Relativistic heavy ion collider (RHIC) at Brookhaven National Laboratory (BNL) is capable of colliding high energy polarized proton beams. These collisions provide a new technique to explore and understand the origin of nucleon spin. PHENIX detector at RHIC has started making measurements of double spin asymmetries that would eventually lead to the polarized gluon distribution. We present the results from these measurements made in 2003 and 2004 and present an outlook based on the data collected in 2005.

Keywords: spin, RHIC, PHENIX, nucleon spin, polarized gluon distribution, spin puzzle.
PACS: 12.38.Aw

MEASUREMENTS WITH A POLARIZED COLLIDER

Most of what we know so far about the nucleon spin structure comes from deep inelastic scattering (DIS) experiments performed with polarized leptons scattering off polarized fixed targets [1]. This is about to change. The Relativistic Heavy Ion Collider (RHIC), shown schematically in Figure 1 with components crucial for the spin program, was commissioned in 2002 at the Brookhaven National Laboratory (BNL) [2]. It is the first collider capable of colliding *polarized* proton beams. The center of mass (CoM) energies from ~60 GeV up to 500 GeV are possible, with varying luminosities, however major operations for physics are planned at 200 GeV and 500 GeV in CoM. Each ring of the collider (at RHIC they are called the "blue" and the "yellow") is filled with bunches of polarized protons (~10^{11}/bunch) from a chain of injectors consisting of the polarized source, the LINAC (200 MeV/c), the booster (~2 GeV/c), the alternating gradient synchrotron (AGS) (~24 GeV/c). In RHIC polarized protons have been accelerated to 100 GeV/c and collided routinely for physics measurements at Sqrt(s)=200 GeV in 2003, 2004 and 2005 (Runs 3,4,5 respectively). In future there are plans to accelerate them to 250 GeV/c. Techniques based on proton-Carbon and proton-proton elastic scattering in the coulomb nuclear interference (CNI) kinematic region are used to make measurements of beam polarization [3] in RHIC.

A polarized collider allows exquisite control over the experimental systematic uncertainties normally associated, and typically a cause of significant concern, with any spin related experiment. The principal concern always comes from "false asymmetries" of may kinds, principally, the time variation of detector performance

CP792, *Deep Inelastic Scattering, DIS 2005*, edited by W. H. Smith and S. R. Dasu

w.r.t. the different orientations of spin directions of the probe and the target involved in scattering [4]. At RHIC, each bunch is prepared with a specific (up or down) polarization. The typical time interval between the bunch-crossings is of the order of 100s of ns. As such, all false asymmetries associated with time dependence of detector operation and efficiencies become irrelevant or negligibly small for spin asymmetry calculations. A second order effect due to uneven filling of the differently polarized bunches with polarized protons remains, but has been (Runs 3 and 4) shown to be smaller than $2x10^{-4}$ [5].

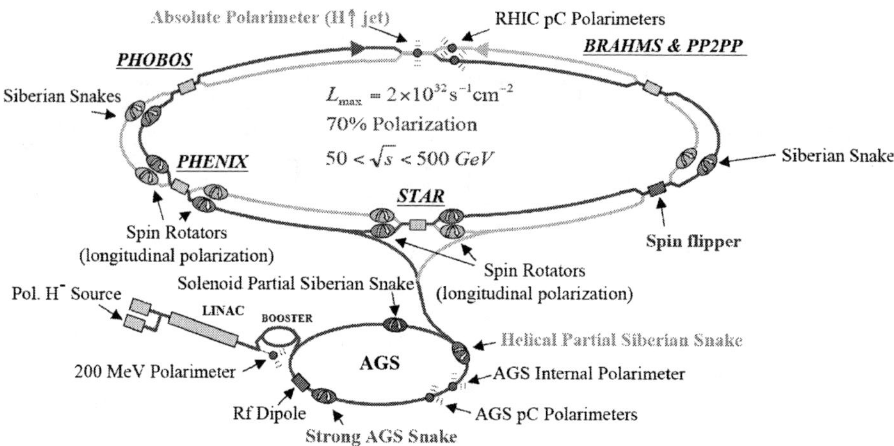

FIGURE 1: Layout of the RHIC as a polarized proton collider. Of significant importance to spin program are the components: the polarized proton source, the partial and strong AGS Siberian snakes, the AGS polarimeters, the RHIC Siberian snakes, the RHIC polarimeters (pC and hydrogen jet), and the spin rotator magnets around the PEHNIX and STAR detectors in both rings.

The spin rotator magnets (SRM) located on both sides of the experiment rotate (turn) the vertically oriented proton spins to longitudinal before the beam enters the experimental area, and turn them back to vertical orientation before the proton beams leave the experimental area. As such depending on whether these magnets are on or off, the experiments pursue programs of single or double longitudinal (helicity) asymmetry measurement or single and double transverse spin asymmetries.

The first stage of the RHIC Spin program witlCoM collisions at 200 GeV has started producing results. The first measurements of double spin asymmetry in RHIC will be discussed in this paper. Measurement of polarized gluon distribution is one of the major goals of the RHIC spin program [6]; the others being a) polarized quark and anti-quark distributions, separated by their flavor, using parity violating W boson production in polarized pp scattering and b) exploration of transverse spin effects in the nucleon [6,7].

In the present paper we present the results on the double spin asymmetry A_{LL} for inclusive π^0 production in longitudinally polarized protons-proton collisions corresponding to 0.22 nb^{-1} in Run-3 (2003) [5] and 0.75 nb^{-1} in Run-4 (2004) integrated with the PHENIX detector. The PHENIX experiment has reported the unpolarized cross section for π^0 production at mid-rapidity [8] for p_T = 1-14 GeV/c, which is described extremely well by the next-to-leading-order perturbative QCD (NLO pQCD) calculations over eight orders of magnitude. This becomes the basis on which we believe that the measured asymmetries will be interpretable in terms of polarized gluon distribution in the pQCD formalism.

THE PHENIX DETECTOR

The PHENIX detector [9] at RHIC is one of the two large collider detectors at RHIC used for the study of both, the heavy ion and polarized proton-proton collision studies. The design philosophy of PHENIX, dictated initially & mainly by the study of heavy ion collisions, was to emphasize high resolution measurements of electromagnetic final states and remnants of heavy ion collisions at the cost of not having full acceptance. With minimal additions to the original plan, the detector was made suitable for spin physics measurements. The design focuses on measurements of rare events in high rate environment, and as such, has highly efficient trigger. The experimental acceptance of PHENIX broadly divided in to two: 1) the central spectrometer coupled with fine grained electromagnetic calorimetry, tracking chambers and excellent particle identification devices, and 2) forward and backward muon spectrometers. The central spectrometer covers a pseudo rapidity region of –0.35 to 0.35 and an approximate azimuthal coverage of –45° to +45° w.r.t. the horizontal plane on both sides (east and west) of the beams. It consists of a central spectrometer magnet, Time of Flight (TOF) detectors, Pad Chambers (PC), Drift Chamber (DC), Ring Imaging Cerenkov Counter (RICH), Lead Glass(PbGl) and Lead Scintillator (PbSc). The forward and backward muon spectrometers consist of a radial field north and south muon tracking magnets backed by muon identification systems. The muon spectrometer and mu-ID system cover 2π in azimuthal acceptance and an η range of 1.2 to 2.4. In addition to these specialized detector systems, there exist two global detector systems: the Beam-Beam Counters (BBC) covers 2π acceptance in azimuth and 2.8 to 3.2 in η, and a Zero Degree Calorimeter (ZDC) at the end of the interaction region, between the two beam pipes when they separate and accepts neutrons scattered in the region +/- 2 mrad. The global detectors were designed for multiplicity measurements. The BBCs are used as collision counters and the ZDCs are used to catch the neutrons produced or released in the collisions. In heavy ion collisions the combination of BBC and ZDC hit multiplicity comparison is used to judge the centrality of nuclear collisions, while in polarized proton-proton collisions, the ratio of hits in the two as a function of bunch number indicates the relative luminosity uncertainties we may have in our double asymmetry calculation. The ZDC along with a Shower Max Detector (SMD) is also used as a local polarimeter to monitor the direction of the spin vector of the colliding proton bunches in the experimental area [10].

THE DOUBLE SPIN ASYMMETRY A_{LL} IN π^0 PRODUCTION

In perturbative QCD A_{LL} in π^0 production is directly sensitive to the polarized gluon distribution function in the proton through gluon-gluon and gluon-quark scattering sub-processes [11]. The double spin asymmetry in π^0 production is given by

$$A_{LL} = \frac{\sigma_{++} - \sigma_{+-}}{\sigma_{++} + \sigma_{+-}}$$

where σ_{++} and σ_{+-} are the cross sections of this reaction when the two colliding particles have the same and opposite helicity, respectively (neglecting parity violating difference in the cross sections between (++) and (--) and (+-) and (-+) beam helicity configurations). This can also be written as,

$$A_{LL} = \frac{1}{P_{blue} \cdot P_{yellow}} \frac{N_{++} - RN_{+-}}{N_{++} + RN_{+-}}; \qquad R = \frac{L_{++}}{L_{+-}}$$

where $P_{yellow/blue}$ are the beam polarizations and R is the ratio of luminosities of the protons colliding with like (++) to unlike(+-) helicities.

The analysis presented in the paper of the double spin asymmetry inπ^0 production will use the multiplicity in BBC, the ZDC for relative luminosity issues, and the central arm calorimeters for identification of π^0 event selection. The run 3 results of double spin asymmetry in π^0 production have already been published [5]. We calculate the A_{LL} in the two-photon invariant mass range of +/- 25 MeV around the π^0 mass peak. We define this region as signal. Then correct for the A_{LL} of the background to extract the $A_{LL}(\pi^0)$:

$$A_{LL}(\pi^0) = \frac{A_{LL}^{raw} - rA_{LL}^{BG}}{1-r}; \quad \Delta A_{LL}(\pi^0) = \frac{\sqrt{(\Delta A_{LL}^{raw})^2 + r^2(\Delta A - LL^{BG})^2}}{1-r}$$

where r is the fraction of the background in the signal region of $M_{\gamma\gamma}$. The measurements were performed in p_T range 1-5 GeV/c separated in 4 bins (1-2, 2-3,3-4 and 4-5 GeV/c). The background constituted 31%, 13%, 7% and 5% fraction of the total number of events seen in each of the p_T bins.

The uncorrelated bunch-to-bunch or fill-to-fill systematic uncertainty in A_L was evaluated using the "bunch-shuffling" method. In this we assign random polarization orientation to the beam bunches and evaluate the double spin asymmetries. We repeat this exercise multiple number of times and see the results of the center and the width of distributions of asymmetries. Bunch spin shuffling should result in a zero asymmetry, and the width of the distribution should be smaller or equal to the statistical uncertainty estimated in the asymmetry. Either of these conditions not being satisfied indicates problems with the asymmetry measurement, systematic in nature, that would need more attention. In our case no such effort was needed. All systematic sources of uncertainty estimated were significantly smaller than the statistical ones.

RESULTS

Figure 2 shows the combined Run-3 and Run-4 results for π^0 asymmetry result and their statistical uncertainties. The curves shown along with the data points are calculations of A_{LL} vs. p_T for next to leading order perturbative QCD fits performed on the polarized deep inelastic scattering data [12]. The GRSV-best corresponds to the best-fit gluon distribution from such an exercise and corresponds to $\int_0^1 \Delta g(x)dx = 0.7$ at Q^2=1 GeV^2. The GRSV-max corresponds to a hypothesis by the authors that the maximum value the polarized gluon density ΔG can take is that of the unpolarized gluon G itself. The NLO pQCD fits to the DIS data are unable to distinguish between the best fit and maximal gluon distribution scenarios [13].

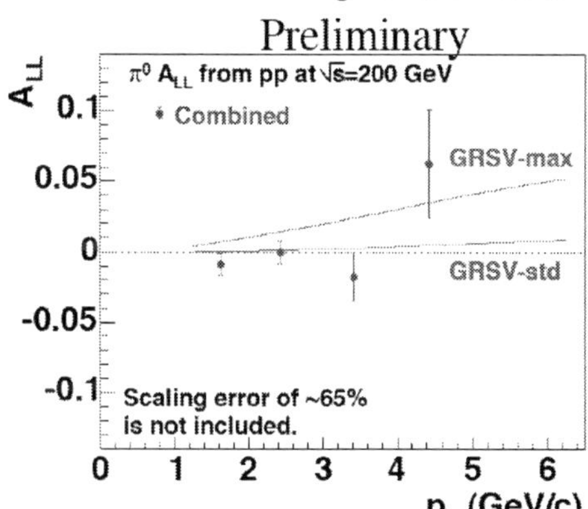

FIGURE 2 The PHENIX measurement of double spin asymmetry in π^0 production in polarized pp collisions from Run-3 and Run-4. The data are combined data sets weighted by the statistical errors of the two measurements. The theoretical curves of GRSV-std and GRSV-max are evaluated from the best fit at NLO to the world sample of DIS data from polarized fixed target experiments.

DISCUSSION OF RESULT & OUTLOOK

The data are consistent with ΔG=0 and also with GRSV-std curve with about 20-25% confidence level. Compared with this, the GRSV-max curve is inconsistent with the data with a χ^2 confidence level of ~0.2-1%. While one is inclined to infer that the larger value of ΔG is ruled out, presently we do not do that. The main reason being the uncertainties of theoretical nature (uncertainty due to scale variation in the GRSV NLO pQCD analysis, those due to the variation of α_S , the assumptions of the low x behavior of the spin structure functions) have not been included in this analysis. We feel the statistical uncertainty so far is too large to be concerned with these issues for the moment. We also note that the uncertainties due to experimental systematic such as the beam polarization measurement are also large, 65% at the time of release of these data. They consist mainly of the unknown absolute calibration of the p-Carbon CNI polarimeter. As this paper goes to print, this has been reduced to 20% [3]. These uncertainties are scale uncertainties which do not affect the *statistical significance* of any of the data points shown.

RHIC Run-5 recently ended. Compared to the data presented in this paper, an improvement of the order of ~30-50 in the figure of merit has been achieved, due to significantly enhanced beam polarization (average 26% and 39% in run Runs 3 and 4 increased to average 50% in Run-5) and a larger statistical sample (~350 nb^{-1} in Runs 3&4 compared to 3.6 pb^{-1} in Run-5). PHENIX expects to make the first definitive measurement of the $\Delta G/G$ in the accessible kinematic region using this method.

In Run-6, a cold super conducting Siberian Snake, recently installed in AGS is expected to be operational for physics. This is the last hardware related to the RHIC Spin program to go in, and is expected to boost the beam polarization in AGS (and hence in to RHIC) to above 60%. It is also expected to improve the beam aperture in to the RHIC and hence the beam intensity in to RHIC, allowing higher luminosity collisions in RHIC. The two together, with reasonably long polarized proton runs will allow measurements of polarized gluon distribution using rarer probes of polarized gluon distribution in pp collisions in the next few years.

ACKNOWLEDGMENTS

The authors (PHENIX collaboration) thank the BNL CAD for their dedication to improvement of the polarized beams in RHIC continually since the beginning of this program. AD thanks the organizers and conveners of the DIS2005 and Spin WG for the invitation of this talk. This work was supported by RBRC and US DOE.

REFERENCES

1. S. D. Bass, The Spin Structure of The Proton, *To appear in Reviews of Modern Physics*, hep-ph/0411005
2. T. Roser et al., Accelerating and colliding polarized protons in RHIC using Siberian Snakes, *Proceedings of EPAC 2002 Conference*
3. S. Bravar, Polarimetry at RHIC, these proceedings
4. B. Adeva et al., [Spin Muon Collaboration, (SMC)], *Physical Review* **D56** (1997) 5330
5. S. Adler et al., PHENIX Collaboration, *Physical Review Letters*, **93** 202002 (2004)
6. G. Bunce et al., Prospects of Spin Physics at RHIC, *Annual Review of Particle and Nuclear Science* **50** (2000) 525-575; also X. Wei, these proceedings
7. M. Chui, Single transverse spin measurements by PHENIX colloboration, these proceedings ; S. Heppelman, Single Spin Asymmerty Measurements by STAR collaboration, these proceedings
8. S. Adler et al., PHENIX Collaboration, *Physical Review Letters*, **91**, 251803 (2003)
9. S. Adcox et al., Overview of the PHENIX Detector, NIM **A499**, (2002) 469
10. M. Togawa, Phenix Local Polarimeter, *Proceedings of Spin 2004 Symposium*
11. W. Vogelsang, QCD Spin Physics : status and prospects for RHIC, *Pramana* **63** (2004) 1251
12. M. Gluck et al., *Physical Review* **D63**, 094005 (2001)
13. B. Jager et al., *Physical Review* **D67**, 054005 (2003)

The Longitudinal Spin Program at STAR

Robert V. Cadman for the STAR Collaboration

High Energy Physics Division, Argonne National Laboratory, Argonne, Illinois 60439

Abstract. The STAR Collaboration has nearly completed installation of electromagnetic calorimetry covering the full azimuth and pseudorapidity range $-1 < \eta < 2$. These calorimeters are essential for the STAR spin program, which requires efficient triggers for high-p_T jets, mesons, and photons. The results of this program will constrain Δg, the gluon contribution to the proton spin.

Keywords: polarization, nucleon spin structure, RHIC spin, calorimeters
PACS: 13.88.+e, 14.20.Dh, 29.40.Vj

For more than two decades the spin structure of the nucleon has been studied intensely. Interest was piqued by a result from the European Muon Collaboration [1] which violated the Ellis-Jaffe Sum Rule (EJSR) [2]. This result, suggesting that the quarks accounted for less of the proton spin than had been expected, led to the so-called "Spin Crisis". Further measurements have confirmed the violation of the EJSR, and have generally agreed that the spins of quarks account for between 20% and 30% of the total proton spin [3].

Another possible contribution to the proton spin comes from the spin of the gluons. The gluon polarization has been studied in deeply inelastic lepton scattering (DIS) [4]. Complementary measurements are also possible in the polarized proton-proton collisions at the Relativistic Heavy Ion Collider (RHIC). At RHIC, the hard scattering of quarks and gluons will be used to constrain the spin-dependent parton distribution functions. Using assumptions including partonic $k_T = 0$ in the initial state, the momentum fraction x of the incident partons can be determined from a two-body final state (di-jets or γ+jet) on an event-by-event basis. Within these assumptions, it will be possible to map out $\Delta g(x)$ over a range of x, yielding a more detailed understanding of the proton's spin structure.

Although the primary purpose of RHIC is to collide heavy nuclei, an important program of polarized proton collisions has been made possible with the installation of spin rotators and Siberian Snakes in RHIC and in the Alternating Gradient Synchrotron. During the 2005 run, RHIC delivered collisions at $\sqrt{s} = 200$ GeV with polarizations $P > 0.4$ for each beam. Polarized proton collisions at $\sqrt{s} = 410$ GeV were briefly achieved. The machine is eventually expected to reach $\sqrt{s} = 500$ GeV.

The two large experiments at RHIC, PHENIX [5] and the Solenoidal Tracker At RHIC (STAR), are able to measure longitudinal spin observables due to spin rotators installed in each beam on both sides of each experiment. Both experiments have a goal of measuring Δg, the gluon contribution to the proton spin. The STAR Collaboration is primarily interested in measuring the partonic level processes $q+g \rightarrow q+g$, $g+g \rightarrow g+g$, and $q+g \rightarrow q+\gamma$. The double-spin longitudinal analyzing power in the pQCD limit is in each case proportional to the product of the polarizations of the incident partons. The

CP792, *Deep Inelastic Scattering, DIS 2005,* edited by W. H. Smith and S. R. Dasu

analyzing powers of the partonic subprocesses are known from NLO pQCD. STAR will measure final state jets and high-p_T π^0's, the observable manifestations of the outgoing quarks and gluons. Initially these will be inclusive measurements. Di-jet coincidences have already been observed at STAR [6]. As the machine luminosity increases, measurements of spin observables for di-jets will be feasible. Eventually STAR plans to measure photon-jet coincidences, a channel which is a relatively clean signature of a gluon in the initial state. The STAR Endcap Electromagnetic Calorimeter was explicitly designed to obtain a clean signal from the direct photons produced in this channel with good suppression of the π^0 background.

Central to the baseline STAR experiment is a large Time Projection Chamber [7] which tracks charged particles in the pseudorapidity range $|\eta| < 1.4$. Additional detectors are required to trigger on the rare events of interest to the STAR spin program. These detectors include the Endcap and Barrel Electromagnetic Calorimeters [8], upgrades to STAR which will be completed before the 2006 run. The Barrel acceptance covers $|\eta| < 1.0$, while the Endcap covers $1.1 < \eta < 2.0$

The calorimeter electronics are designed to trigger when the largest single-tower energy deposition exceeds a programmable threshold and when any of 18 non-overlapping patches of $\eta \times \phi \approx 1 \times 1$ record an energy greater than another threshold. This second condition, the "jet-patch" trigger, is STAR's most efficient trigger for high-p_T jets. The single-tower trigger is sensitive primarily to single photons and π^0's. During the 2005 run, STAR took data with $> 50\%$ live time with unprescaled trigger thresholds set to $E_T \approx 3.5$ GeV for single towers and $E_T \approx 7$ GeV for the jet-patch trigger.

Both calorimeters are composed of alternating layers of lead and scintillator. The scintillators are cut into tiles covering $\eta \times \phi \approx 0.05 \times 0.05$ for the Barrel and $\eta \times \phi \approx 0.1 \times 0.1$ for the Endcap. The tiles are read out with wavelength-shifting (WLS) fibers. Both calorimeters have shower maximum detectors (SMD) at a depth of approximately 5 radiation lengths. The Barrel SMD is a wire proportional counter read out through strips on a printed circuit board. A 15×15 matrix of strips reads the ionization signal within an area corresponding to four towers. The Endcap has a novel SMD composed of triangular scintillator strips read out through WLS fibers. The Endcap strips are glued into 30 degree modules and are oriented at a 45 degree angle to the central radius of the module, as shown in Figure 1.

Installation of the Endcap was completed prior to the 2005 run. One-half of the Barrel (covering the full azimuth and $0 < \eta < 1$) was fully operational during the 2005 run. All detector components were installed. None of the $\eta < 0$ half of the Barrel was included in the trigger, and readout electronics were available for only one-half of the azimuth on the $\eta < 0$ side. The remaining electronics will be installed during the Summer 2005 shutdown, and the Barrel is expected to be fully operational for the following run.

The addition of these two calorimeters will allow STAR to trigger on jets and high-p_T π^0's, and the measurements allowed by these triggers will constrain Δg, the gluon contribution to the proton spin. Figure 2 shows estimates of the statistical precision for measurements at STAR of A_{LL} for inclusive jets with an integrated luminosity of 7 pb^{-1} and $P = 0.4$ for both beams [9]. These parameters reflect the goals of the 2005 run. The theory band in the figure shows the full physically allowed region. The central curve indicates the central value from fits to DIS data. The cusp on the lower edge of the theory band at $p_T \approx 18$ GeV/c reflects a change in the partonic channel which

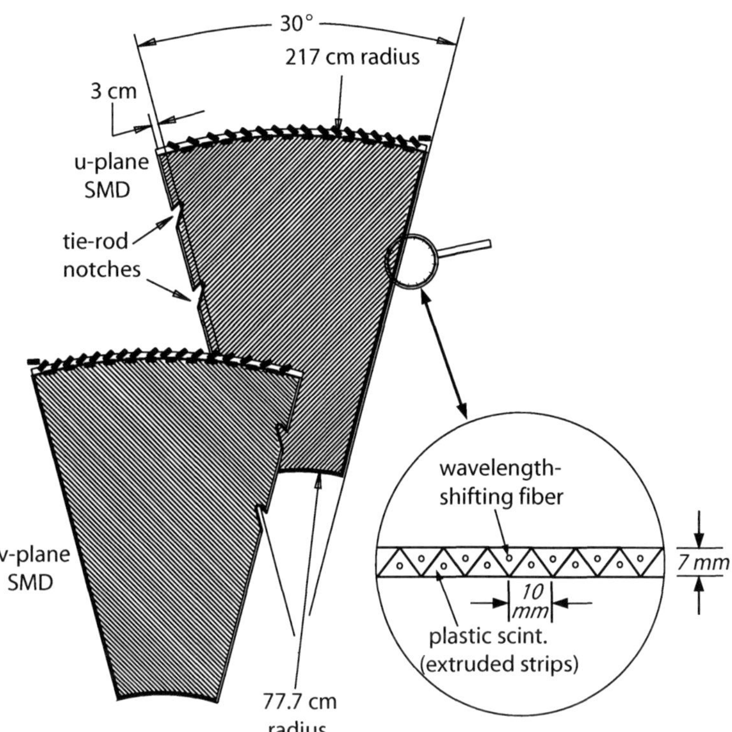

FIGURE 1. The Endcap Shower Maximum Detector. Two planes are shown as they would be installed in a single 30-degree sector, with the planes oriented such that the strips are perpendicular to those in the other plane, enabling reconstruction of a two-dimensional shower profile. A magnified view of the edge of a module shows the individual triangular scintillator strips.

dominates jet production, from $g+g$, which is insensitive to the sign of Δg, at lower p_T, to $q+g$ at higher p_T. The estimated experimental precision is based on realistic triggering and reconstruction efficiencies observed prior to the 2005 run, and assumes completed Barrel and Endcap Calorimeters. A significant fraction of the luminosity goal was achieved in the recently completed 2005 run, although only one-half of the Barrel was included in the trigger. The STAR Collaboration's eventual goals for $\sqrt{s}=200$ GeV are to record more than 100 pb^{-1} of integrated luminosity with $P=0.7$, improving the statistical precision of the measurements by two orders of magnitude compared to the projections in Figure 2. Ultimately, when integrated luminosity goals are reached at both $\sqrt{s}=200$ GeV and $\sqrt{s}=500$ GeV, STAR will measure $\Delta g(x)$ over the kinematic range $0.009 < x < 0.3$.

In summary, the STAR Collaboration has nearly completed calorimeter upgrades which will permit measurements sensitive to Δg. Initially STAR will measure A_{LL} for inclusive jets and π^0's at $\sqrt{s}=200$ GeV. The ultimate goal is to study the x-dependence of $\Delta g(x)$.

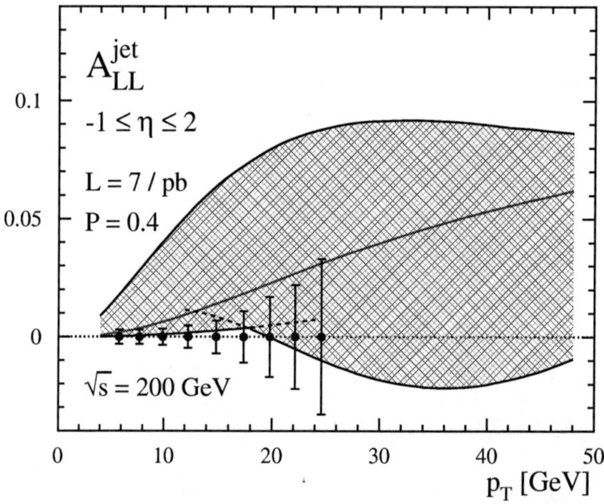

FIGURE 2. Projected precision of A_{LL} for inclusive jets at STAR with full calorimeter coverage over $-1 < \eta < 2$ and the full integrated luminosity requested for the 2005 run [9]. The band indicates the physically allowed range of A_{LL}. The central value from fits to DIS data is indicated by the central (red) curve.

ACKNOWLEDGMENTS

This work was supported by the U.S. Department of Energy under Contract W-31-109-ENG-38. The STAR Collaboration is grateful for the assistance of the RHIC Operations Group and the RHIC Computing Facility and for the support of funding agencies from 13 nations.

REFERENCES

1. J. Ashman *et al.* (European Muon Collaboration), *Phys. Lett.*, **B206**, 364 (1988); *Nucl. Phys.*, **B328**, 1 (1989).
2. J. Ellis and R. Jaffe, *Phys. Rev. D*, **9**, 1444 (1974); **10**, 1669 (1974).
3. E. W. Hughes and R. Voss, *Annu. Rev. Nucl. Part. Sci.*, **49**, 303 (1999).
4. C. Bernet, these proceedings.
5. A. Deshpande, these proceedings.
6. T. Henry (STAR Collaboration), *J. Phys. G*, **30**, S1287 (2004).
7. K. H. Ackermann *et al.*, *Nucl. Instrum. Meth. Phys. Res.* **A499**, 624 (2003); M. Anderson *et al.*, *Nucl. Instrum. Meth. Phys. Res.* **A499**, 659 (2003).
8. M. Beddo *et al.*, *Nucl. Instrum. Meth. Phys. Res.* **A499**, 725 (2003); C. E. Allgower *et al.*, *Nucl. Instrum. Meth. Phys. Res.* **A499**, 740 (2003).
9. C. Aidala *et al.*, "Research Plan for Spin Physics at RHIC", Brookhaven National Laboratory report BNL-73798-2005, February 2005 (unpublished).

Recent measurement of $\Delta G/G$ at COMPASS

C. Bernet, on behalf of the COMPASS collaboration

colin.bernet@cern.ch

Abstract. We present a preliminary measurement of the gluon polarization $\Delta G/G$ in the nucleon, based on the spin asymmetry of quasi-real photoproduction events for which a pair of large transverse momentum hadrons is produced. The data were obtained by the COMPASS experiment at CERN using a 160 GeV polarized muon beam scattered on a large polarized ^{6}LiD target. The preliminary helicity asymmetry for the selected events is $A_{\parallel}/D = 0.002 \pm 0.019(\text{stat.}) \pm 0.003(\text{exp.syst.})$. From this value, a leading order analysis based on the model of PYTHIA leads to the gluon polarization in the nucleon $\Delta G/G(x_g = 0.095, \mu^2 = 3\ \text{GeV}^2) = 0.024 \pm 0.089(\text{stat.}) \pm 0.057(\text{syst.})$. This value is consistent with parameterizations obtained from QCD fits to the g_1 data, with a first moment $\Delta G \equiv \int_0^1 \Delta G(x)dx \simeq 0.5$, at the same scale.

Keywords: gluons polarization nucleon
PACS: 13.60.Hb

The decomposition of the nucleon spin in the contributions from its constituents has been a central topic of investigation in polarized lepton-nucleon scattering in the last 20 years. The EMC measurement of the proton spin structure [1] has shown that only 20 to 30 % of the proton spin could be attributed to the total quark spin $\Delta\Sigma$, in contrast to the 60 % expected in the naive quark-parton model. In inclusive lepton-nucleon scattering the contribution of the gluon spin ΔG to the nucleon spin can only be measured indirectly by studying the Q^2 dependence of the polarized spin-structure functions in QCD. Fits at next-to-leading order (NLO) provide evaluations of ΔG which are of the order of 0.50. The precision of these fi ts is strongly limited by the small Q^2 range covered by the data at any value of x, a situation resulting from the lack of a polarized lepton-nucleon collider. In addition, the shape of $\Delta G(x)$ at a fi xed Q^2 used as reference, has to be provided as an input parameterization. It varies considerably between the different analyses and is only poorly constrained by the results of the fi ts. A direct measurement of the gluon polarization $\Delta G(x)/G(x)$ can be obtained from the helicity asymmetry of the photon-gluon fusion (PGF, $\gamma^* g \to q\bar{q}$) cross-section, which constitutes an important part of the experimental program of COMPASS

The COMPASS experiment [2] is located at the M2 beam line of the CERN SPS, which provides a 160 GeV μ^+ beam, with a natural polarization of $-76 \pm 5\%$. The target consists in an upstream and a downstream cell, longitudinally polarized in opposite directions. Typical target polarizations of $50.0 \pm 2.5\%$ are obtained. The forward spectrometer is divided in two stages allowing the reconstruction of the scattered muon and of the produced hadrons in broad momentum and angular ranges. The trigger system provides effi cient tagging down to $Q^2 = 0.002\ \text{GeV}^2$.

The present analysis focuses on the data collected during the 2002 and 2003 runs. We only consider quasi-real photoproduction events ($Q^2 < 1\ \text{GeV}^2$), in which at least two charged hadrons are associated to the primary vertex in addition to the incident

CP792, *Deep Inelastic Scattering, DIS 2005*, edited by W. H. Smith and S. R. Dasu

and the scattered muons. The fraction of PGF is enhanced by asking the two leading hadrons to have a large transverse momentum: $p_T^{h1} > 0.7$ GeV, $p_T^{h2} > 0.7$ GeV and $(p_T^{h1})^2 + (p_T^{h2})^2 > 2.5$ GeV^2. In total, around 350,000 events are selected. For this *high* p_T sample, the measured helicity asymmetry (defined as in [3]) is

$$\frac{A_{\parallel}}{D} = 0.002 \pm 0.019(stat) \pm 0.003(exp.syst), \tag{1}$$

where the quoted systematic error accounts for the false asymmetries related to the apparatus. Other sources of systematic errors, including the error on the beam and target polarizations, are only a few percents of the (small) measured asymmetry, and are therefore negligible. The 2004 data are currently under analysis, and represent the same amount of data as 2002 and 2003 altogether.

A Monte-Carlo simulation is needed to extract the gluon polarization from the high p_T asymmetry. The selected sample of high p_T events covers the transition region ranging from photoproduction ($Q^2 \approx 0$) up to DIS ($Q^2 \approx 1$ GeV^2). For this reason, we chose PYTHIA as an event generator because it provides a model for the lepton-nucleon interactions [4] at low Q^2. Two different kinds of processes are generated. In the so-called *direct processes*, the virtual photon takes part in the hard partonic interaction. In the *resolved processes*, it fluctuates to a hadronic state, from which a parton is extracted. This parton then interacts with a parton from the nucleon. At $Q^2 < 1$, the resolved processes constitute half of the high p_T sample. Their contribution falls to about 10% for $Q^2 > 1$ GeV^2 and becomes negligible for $Q^2 > 2$ GeV^2. Note that the analysis requires the factorization between the hard and soft parts of the reaction, hence the presence of a hard scale μ^2. As $Q^2 < 1$ GeV^2, the scale is provided by the transverse momentum of the partons involved in the reaction. Events for which no hard scale can be found are classified as *low* p_T.

The generated events are tracked through a GEANT description of the COMPASS spectrometer, and processed using the same reconstruction program as for real data. Then, the Monte-Carlo sample of high p_T events is selected through the same cuts.

Only one parameter of PYTHIA had to be changed to reach a good agreement with the data: the width of the Gaussian distribution of intrinsic transverse momentum of partons within the resolved virtual photon (PARP(99)) was decreased from 1 GeV/c to 0.5 GeV/c. Fig. 1 presents a comparison between the simulated and real data samples of high p_T events. Fig. 2 shows how the Monte-Carlo sample of high p_T events divides into the various PYTHIA subprocesses. The high p_T asymmetry can be written in terms of the contributions of the different processes:

$$\frac{A_{\parallel}}{D} = R_{PGF} \left\langle \hat{a}_{LL}^{PGF}/D \right\rangle \frac{\Delta G}{G} + R_{QCDC} \left\langle \hat{a}_{LL}^{QCDC}/D \right\rangle A_1 + \sum_{f,f'=u,d,s,\bar{u},\bar{d},\bar{s},g} R_{ff'} \left\langle \hat{a}_{LL}^{ff'} \left(\frac{\Delta f}{f}\right)^d \left(\frac{\Delta f'}{f'}\right)^{\gamma} \right\rangle. \tag{2}$$

Note that we have neglected the small contributions of the leading process $\gamma^* q \to q$ and of the low p_T scattering events because there is no hard scale allowing a perturbative treatment of these subprocesses (low transverse momentum, $Q^2 < 1$ GeV^2 events). As can be seen on Fig. 2, the fraction of photon-gluon fusion events is $R_{PGF} = 0.31$. The analyzing power $\hat{a}_{LL}^{PGF}$ is the helicity asymmetry of the $\mu g \to \mu' q\bar{q}$ scattering cross-section, $\hat{a}_{LL}^{PGF} \equiv d\Delta\sigma_{PGF}^{\mu g}/d\sigma_{PGF}^{\mu g}$. It is calculated from the kinematic variables of the partonic reaction for each PGF event in the high p_T Monte-Carlo sample. Averaging over the PGF events, we obtain $\langle \hat{a}_{LL}^{PGF}/D \rangle = -0.933$. The contribution of the PGF process to the high p_T asymmetry is thus $-0.292 \times \frac{\Delta G}{G}$. The contribution of the QCD Compton events is calculated in the same way to be 0.0063, using a fit on the world data for the virtual-photon deuteron asymmetry A_1^d.

Resolved photon subprocesses involve either a quark or a gluon from the nucleon. In the latter case, they are sensitive to the gluon polarization $\Delta G/G$, and contribute to the signal. The analyzing powers $\hat{a}_{LL}^{ff'}$ are calculated in pQCD at leading order [5], and are positive for all relevant channels. The polarizations $(\Delta f/f)^d$ of the u, d and s quarks in the deuteron are calculated using the unpolarized parton distribution functions from GRV98, and the polarized parton distribution functions from GRSV2000 [6], at leading order. The polarizations of quarks and gluons in the virtual photon $(\Delta f/f)^\gamma$ are unknown as the polarized PDFs of the virtual photon have not yet been measured. Nevertheless, theoretical considerations provide a minimum and a maximum value for each Δf^γ [7], called the *minimal* and *maximal scenarios*. The total contribution of the resolved photon processes to the high p_T asymmetry, which ranges between $0.000 + 0.012 \times \Delta G/G$ and $0.002 + 0.078 \times \Delta G/G$, will be taken into account in the systematic error on $\Delta G/G$.

The tuning of the PYTHIA parameters relevant to this analysis is an important source of systematic errors. This error was estimated by scanning these parameters independently over a range where the agreement between the simulation and real data remains reasonable. This resulted in several values for $\Delta G/G$, all based on the same high p_T asymmetry, Eq. (1). The value of $\Delta G/G$ appears to depend predominantly on the width of the intrinsic transverse momentum distribution for the partons in the photon. For instance, varying this parameter between 0.1 and 1 GeV/c results in a 30% variation of the fraction of photon-gluon fusion R_{PGF}. NLO effects seem to be small: varying the scale and (de)activating parton showers does not affect the result.

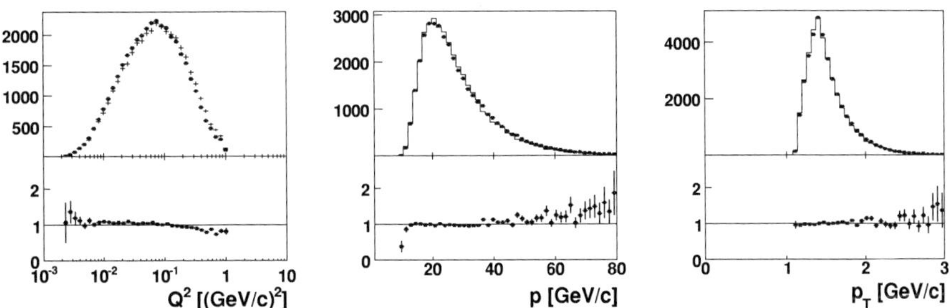

FIGURE 1. The upper part of these plots show a comparison between the simulated (histogram) and real data (points) samples of high p_T events, normalized to the number of events. The lower part shows the corresponding data/simulation ratio. p (p_T) is the total (transverse) momentum of the leading hadron. A similar agreement is obtained for the next-to-leading hadron.

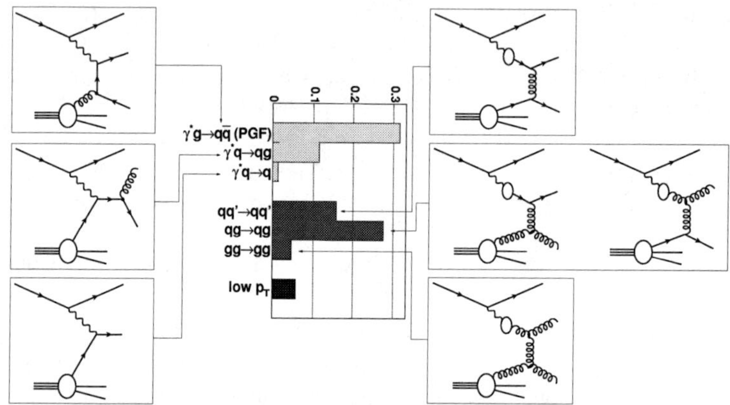

FIGURE 2. Contribution of each PYTHIA subprocess to the Monte-Carlo sample of high p_T events. On the left: direct processes (photon-gluon fusion, QCD Compton, and leading process); on the right: resolved processes; note that $q\bar{q} \to q'\bar{q}'$, $q\bar{q} \to gg$ and $gg \to q\bar{q}$ are neglected, as they altogether represent 0.6% of the sample.

Using Eq. (2) to extract the gluon polarization from the high p_T asymmetry, Eq. (1), we finally get:

$$\frac{\Delta G}{G}(x_g = 0.095, \mu^2 = 3 \text{ GeV}^2) = 0.024 \pm 0.089(stat.) \pm 0.057(syst.). \quad (3)$$

This result was compared to the recent distributions of $\Delta G(x)/G(x)$ from AAC [8] and LSS [9]. These two distributions, which strongly differ in shape, are almost equal at $x_g = 0.095$ and are compatible with our result within 1.5 σ. The first moments of ΔG are equal to 0.50 ± 0.35 and 0.68 ± 0.32 for the AAC and LSS fits, respectively. The fraction of PGF can also be enhanced by selecting events with charmed hadrons (D^0 and D^*), instead of high p_T hadrons. The results obtained with this method, which is much less model-dependent but suffers from low statistics, will be produced soon.

REFERENCES

1. J. Ashman, et al., *Phys.Lett. B*, **206**, 364 (1988).
2. G. K. Mallot, *Nucl. Instrum. Meth.*, **A518**, 121–124 (2004).
3. E. Ageev, et al., *Phys.Lett. B*, **612**, 154–164 (2005), `hep-ex/0501073`.
4. C. Friberg, and T. Sjostrand, *JHEP*, **09**, 010 (2000), `hep-ph/0007314`.
5. C. Bourrely, J. Soffer, F. M. Renard, and P. Taxil, *Phys. Rept.*, **177**, 319 (1989).
6. M. Gluck, E. Reya, M. Stratmann, and W. Vogelsang, *Phys. Rev.*, **D63**, 094005 (2001), `hep-ph/0011215`.
7. M. Gluck, E. Reya, and C. Sieg, *Eur. Phys. J.*, **C20**, 271–281 (2001), (*Note that our definition of the minimal scenario corresponds to a maximum negative polarization of the gluons in the VMD part of the photon*), `hep-ph/0103137`.
8. M. Hirai, S. Kumano, and N. Saito, *Phys. Rev.*, **D69**, 054021 (2004), `hep-ph/0312112`.
9. E. Leader, A. V. Sidorov, and D. B. Stamenov, *Eur. Phys. J.*, **C23**, 479–485 (2002), `hep-ph/0111267`.

The Effect of Positivity Constraints on Polarized Parton Densities

E. Leader*, A.V. Sidorov,† and D.B. Stamenov**

*Imperial College London, London WC1E 7HX, England
†Bogoliubov Theoretical Laboratory, Joint Institute for Nuclear Research, 141980 Dubna, Russia
**Institute for Nuclear Research and Nuclear Energy, 1784 Sofia, Bulgaria

Abstract. The impact of positivity constraints on the polarized parton densities has been studied. Special attention has been paid to the role of positivity constraints in determining the polarized strange quark and gluon densities, which are not well determined from the present data on inclusive polarized DIS.

Keywords: QCD, polarized parton densities, positivity constraints
PACS: 13.60.Hb, 12.38-t, 13.88.+e, 14.20.Dh

Spurred on by the famous European Muon Collaboration (EMC) experiment [1] at CERN in 1987, there has been a huge growth of interest in the partonic spin structure of the nucleon, *i.e.*, how the nucleon spin is built up out from the intrinsic spin and orbital angular momentum of its constituents, quarks and gluons. Our present knowledge about the spin structure of the nucleon comes from polarized inclusive and semi-inclusive DIS experiments at SLAC, CERN, DESY and JLab, polarized proton-proton collisions at RHIC and polarized photoproduction experiments. The determination of the longitudinal polarized parton densities in QCD is one of the important aspects of this knowledge. Many analyses [2, 3] of the world data on inclusive polarized DIS have been performed in order to extract them. It was shown that if the convention of a flavor symmetric sea is used[1] the polarized valence quarks are well determined, while the polarized strange sea and polarized gluon densities are weakly constrained[2].

In this talk we will discuss the effect of positivity constraints on the polarized parton densities and will demonstrate their importance in determining the strange and gluon densities, especially at high x.

The polarized parton densities have to satisfy the positivity condition, which in LO QCD implies:

$$|\Delta f_i(x,Q^2)| \leq f_i(x,Q^2), \quad |\Delta \bar{f}_i(x,Q^2)| \leq \bar{f}_i(x,Q^2). \tag{1}$$

The constraints (1) are the consequence of a probabilistic interpretation of the parton densities in the naive parton model, which is still valid in LO QCD. Beyond LO the parton densities are not physical quantities and the positivity constraints on the polarized parton densities are more complicated. They follow from the positivity condition for the

[1] In the absence of polarized charged current neutrino experiments a flavor decomposition is not possible.
[2] About the situation in semi-inclusive DIS see the talk by R. Sassot at this Workshop [4].

CP792, *Deep Inelastic Scattering, DIS 2005*, edited by W. H. Smith and S. R. Dasu

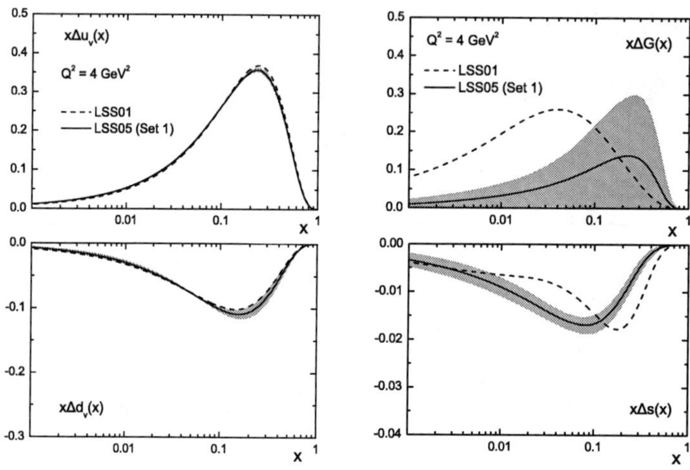

FIGURE 1. Comparison between our two sets of NLO($\overline{MS}$) polarized parton densities, LSS'01 and LSS'05(Set 1), at $Q^2 = 4\, GeV^2$.

polarized lepton-hadron cross-sections $\Delta\sigma_i$ in terms of the unpolarized ones ($|\Delta\sigma_i| \leq \sigma_i$) and include also the Wilson coefficient functions. It was shown [5], however, that for all practical purposes it is enough, at the present stage, to consider LO positivity bounds for LO as well as for for NLO parton densities, since NLO corrections are only relevant at the level of accuracy of a few percent. Note that, if the positivity constraints (1) are imposed at some Q_0^2, they are satisfied at any $Q^2 > Q_0^2$ [6]. So, in order to control easily the positivity conditions (1) it is enough to impose them for the minimum value of $Q^2 = Q_0^2$ in the data set used in the QCD analysis.

Let us consider how the use of different positivity constraints influences the results on the polarized parton densities. In Fig. 1 we compare LSS'05(Set 1) NLO($\overline{MS}$) polarized parton densities [3] with LSS'01 parton densities [7] presented on the HEPDATA web site. Both sets are determined from the data by the same method but using different positivity constraints. While the LSS'05 polarized PD are compatible with the positivity bounds (1) imposed by the MRST'02 unpolarized parton densities [8], those of the LSS'01 set are limited by the Barone et al. unpolarized parton densities [9]. As seen from Fig. 1, the valence quark densities Δu_v and Δd_v of the two sets are close to each other, while the polarized strange sea quark and gluon densities are significantly different. This comparison is a good illustration of the fact that the present inclusive polarized DIS data allow a much better determination of the valence quark densities (if SU(3) symmetry of the flavour decomposition of the sea is assumed) than the polarized strange quarks $\Delta s(x,Q^2)$ and the polarized gluons $\Delta G(x,Q^2)$. This is especially true for the high x region, where the values of $\Delta s(x,Q^2)$ and $\Delta G(x,Q^2)$ are very small and the precision of the data is not enough to extract them correctly. That is why different unpolarized sea quark and gluon densities (see Fig. 2) used on the RHS of the positivity constraints (1) are important and crucial in determining $\Delta s(x,Q^2)$ and $\Delta G(x,Q^2)$ in

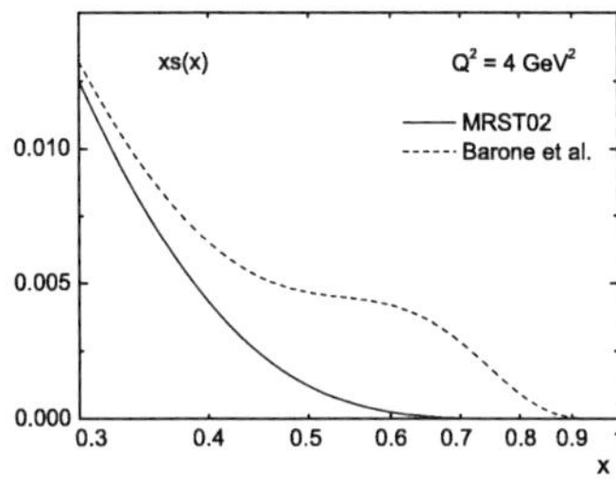

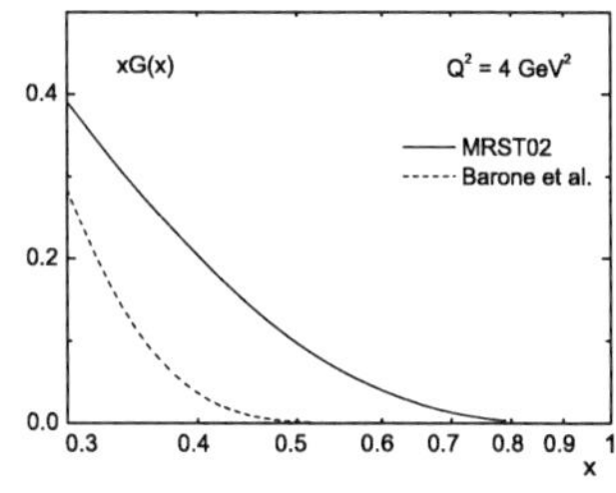

FIGURE 2. Comparison between the NLO($\overline{\text{MS}}$) unpolarized strange quark sea and gluon densities determined by MRST'02 [8] and Barone at al. [9].

this region. The more restrictive $s(x,Q^2)_{\text{MRST}'02}$ at high x leads to a smaller value of $|\Delta s(x,Q^2)|_{\text{LSS}'05}$ in this region, while the smaller $G(x,Q^2)_{\text{Bar.et.al}}$ provides a stronger constraint on $\Delta G(x,Q^2)_{\text{LSS}'01}$ (see Fig. 1).

To illustrate this fact once more, we compare the LSS'05 (Set 1) PPD at $Q^2 = 4\,GeV^2$ with those [2] obtained by GRSV, Blumlein, Bottcher and the Asymmetry Analysis Collaboration (AAC) using almost the same set of data. Note that all these groups have used the GRV unpolarized parton densities [10] for constraining their polarized parton densities at large x. In this x region the unpolarized GRV and MRST'02 gluons are practically the same, while the magnitude of the unpolarized GRV strange sea quarks is much smaller than that of MRST'02. Therefore, the GRV unpolarized strange sea quarks provide a stronger constraint on the polarized ones (see Fig. 3). The impact on the determination of the polarized strange sea density is demonstrated in Fig. 3. As a result, the magnitude of our polarized strange sea density $x|\Delta s(x,Q^2)|$ is larger in the region $x > 0.1$ than those obtained by the other groups. Note also that the magnitude of $x\Delta s$ obtained by the GRSV and BB is smaller than that determined by AAC. We consider the GRSV result to be a consequence of the fact that in their analysis, the GRV positivity constraint is imposed at lower value of Q^2: $Q^2 = \mu^2_{NLO} = 0.4\,GeV^2$, while AAC has used the same requirement at $Q^2 = 1\,GeV^2$. Finally, the different positivity conditions on Δs influence also the determination of the polarized gluon density for larger Q^2 because the evolution in Q^2 mixes the polarized sea quarks and gluons.

To end this discussion, we would like to emphasize that for the adequate determination of polarized strange quarks and gluons at large x, the role of the corresponding unpolarized densities is very important. That is why the latter have to be determined with good accuracy at large x in the preasymptotic (Q^2, W^2) region too. Usually the sets of unpolarized parton densities, presented in the literature, are extracted from the data on DIS using cuts in Q^2 and W^2 chosen in order to minimize the higher twist effects. In order to use the densities for constraining the polarized parton densities they have to be continued to the preasymptotic (Q^2, W^2) region. It is not obvious that the continued unpolarized parton densities would coincide well with those obtained from the data in the region ($Q^2 > 1\,GeV^2$, $W^2 > 4\,GeV^2$) in the presence of the HT corrections to unpolarized structure functions F_1 and F_2. So, a QCD analysis of the unpolarized world data

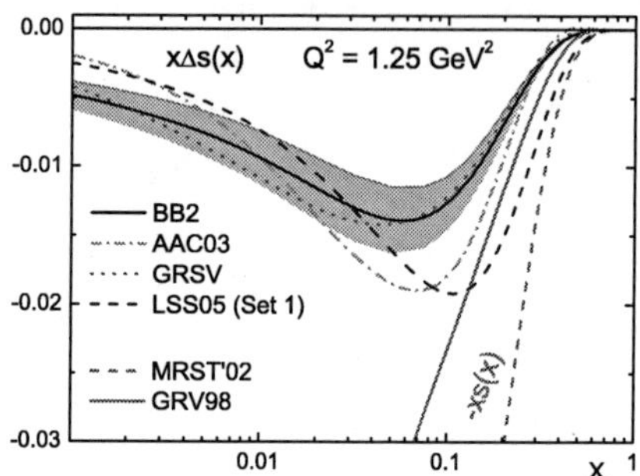

FIGURE 3. Comparison between our NLO($\overline{\text{MS}}$) polarized strange sea quark density (Set 1) [3] at $Q^2 = 1.25\, GeV^2$ with those [2] obtained by GRSV ('standard scenario'), BB (ISET=4 or BB2) and AAC (AAC03). The unpolarized MRST02 and GRV98 strange sea quark densities are also shown.

including the preasymptotic $(Q^2,\, W^2)$ region and taking into account HT corrections is needed in order to extract correctly the unpolarized parton densities in the preasymptotic region. Our arguments for the need for a precise determination of the unpolarized densities of strange quarks and gluons in both the asymptotic and preasymptotic regions in Q^2 and W^2, coming from spin physics, could be considered as additional to those discussed in the recent paper [11].

ACKNOWLEDGMENTS

This research was supported by the JINR-Bulgaria Collaborative Grant, by the RFBR (No 05-01-00992, 03-02-16816), by the Bulgarian National Science Foundation under Contract Ph-1010 and by the Royal Society of Edinburgh Auber Bequest.

REFERENCES

1. EMC, J. Ashman et al., *Phys. Letters* **B 206**, 364 (1988) ; *Nucl. Phys.* **B 328**, 1 (1989).
2. M. Glück, E. Reya, M. Stratmann, and W. Vogelsang, *Phys. Rev.* **D 63**, 094005 (2001); J. Blumlein, H. Bottcher, *Nucl. Phys.* **B 636**, 225 (2002); AAC, M. Hirai et al., *Phys. Rev.* **D 69**, 054021 (2004).
3. E. Leader, A.V. Sidorov and D.B. Stamenov, *J. High Energy Phys.* **06**, 033 (2005).
4. R. Sassot, this Workshop; D. de Florian, G.A. Navarro, and R. Sassot, *Phys. Rev.* **D 71**, 094018 (2005).
5. G. Altarelli, S. Forte, and G. Ridolfi, *Nucl. Phys.* **B 534**, 277 (1998); S. Forte, M. L. Mangano, and G. Ridolfi, *Nucl. Phys.* **B 602**, 585 (2001).
6. C. Bourrely, J. Soffer, and O.V. Teryaev, *Phys. Lett.* **B 420**, 375 (1998).
7. E. Leader, A.V. Sidorov, and D.B. Stamenov, *Eur. Phys. J.* **C 23**, 479 (2002).
8. A.D. Martin, R.G. Roberts, W.J. Stirling, and R.S. Thorne, *Eur. Phys. J.* **C 28**, 455 (2003).
9. V. Barone, C. Pascaud, and F. Zomer, *Eur. Phys. J.* **C 12**, 243 (2000).
10. M. Glück, E. Reya and A. Vogt, *Eur. Phys. J.* **C 5**, 461 (1998).
11. S. Forte, hep-ph/0502073.

New Results on Testing Duality in Spin Structure from Jefferson Lab

Nilanga Liyanage

University of Virginia, Charlottesville, VA, USA

Abstract. The Bloom-Gilman duality has been experimentally demonstrated for spin independent structure functions. Duality is observed when the smooth scaling curve at high momentum transfer is an average over the resonance bumps at lower momentum transfer, but at the same value of scaling variable x. Signs of quark-hadron duality for the spin Dependant structure function g_1 of the proton has been recently reported by the Hermes collaboration. Experimental Halls A, B and C at Jefferson lab have recently measured spin structure functions in the resonance region for the proton and the neutron. Data from these experiments combined with Deep-Inelastic-Scattering data provide a precision test of quark-hadron duality predictions for spin structure functions for both the proton and the neutron. This will be one of the first precision tests of spin and flavor dependence of quark-hadron duality.

Keywords: Quark-hadron duality, nucleon, spin

Thirty years ago Bloom and Gilman [1] made the observation that the scaling curve seen at high momentum transfer is an accurate average over the resonance bumps at lower momentum transfer but at the same value of x. This duality between the resonance region, which is best described by constituent quark models, and the scaling region, which is well described by pQCD, hints a common origin for both regions. Recent data from Jefferson Lab Hall C [2] have further confirmed that Bloom-Gilman duality holds to a few percent level down to small values of Q^2. The striking agreement shown by these data for unpolarized structure functions between the resonance and the scaling regions raises the question whether duality holds for polarized structure functions as well. Hermes collaboration has recently reported observing global quark-hadron duality for the spin structure function g_1 of the proton [3]. For a review of the published duality data and theoretical models, see [4].

Recently Jefferson lab Experimental Halls A, B and C have conducted precision experiments to measure spin structure functions in the resonance region for both the proton and the neutron (polarized ^{3}He and deuteron).

Jefferson Lab Hall A experiment E01-012 used Hall A polarized ^{3}He target for a precision extraction of the neutron spin structure functions g_1^n, g_2^n and the virtual photon asymmetries A_1^n, A_2^n in the resonance region up to $Q^2 = 4(\text{GeV}/c)^2$. Both Hall A High Resolution spectrometers (HRS) were used in a symmetric configuration in electron detection mode to measure the inclusive $^3\vec{H}e(e,e')X$ reaction. Three beam energies, 3 GeV, 4 GeV and 5 GeV were used with spectrometer angles of 25° and 32°. At each kinematic setting parallel and perpendicular cross sections and asymmetries were measured with the target spin parallel and perpendicular to the electron beam respectively.

CP792, *Deep Inelastic Scattering, DIS 2005,* edited by W. H. Smith and S. R. Dasu

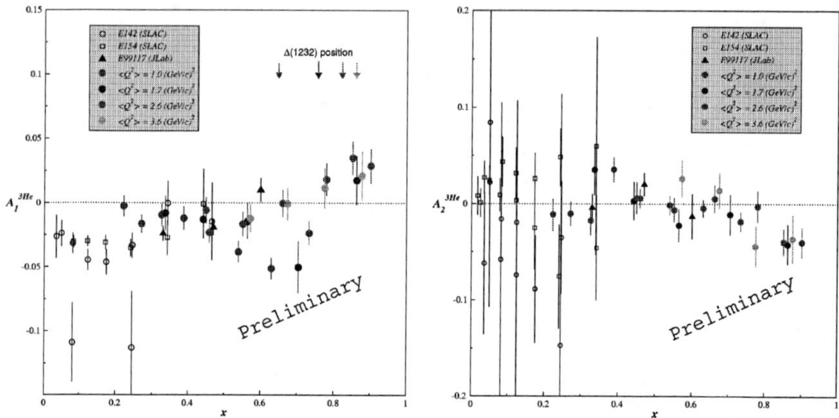

FIGURE 1. Preliminary results from Hall A experiment E01-012 for A_1^{3He} (left) and A_2^{3He} (right), compared A_1^{3He} and A_2^{3He} measured in the DIS region from Hall A experiment E99-117. The arrow indicates the location of the $\Delta(1232)$ resonance.

Figure 1 (left) shows preliminary results for A_1^{3He} in the resonance region at the four Q^2 values, compared to A_1^{3He} in the DIS region from Hall A experiment E99-117 [5]. The position of the $\Delta(1232)$ resonance is indicated by an arrow. The error bars shown are statistical only. The most noticeable feature of the plot is the negative contribution due to $\Delta(1232)$ at the two low Q^2 settings ($Q^2 < 2\ (GeV/c)^2$). It has been noted that quark-hadron duality for spin structure functions is not expected in the Delta resonance region at this low Q^2. For the two higher Q^2 settings, A_1^{3He} at the location of $\Delta(1232)$ is positive. It is also interesting to note that the results from these two settings ($Q^2 > 2\ (GeV/c)^2$) agree perfectly with each other, showing little or no Q^2 dependence, as expected in the scaling region. Furthermore, Our data seem to indicate that with increasing x_{Bj}, A_1 goes from negative to positive showing the same trend as indicated by the DIS data from Experiment 99-117. The behavior of A_1^n becoming positive at high x_{Bj} has been predicted for DIS data by relativistic constituent quark models and by pQCD inspired models [6].

Jefferson Lab Hall B CLAS collaboration's EG1 measured polarized structure functions for the proton and the deuteron in the resonance region covering a kinematic range of $0.05 < Q^2 < 4.5\ (GeV/c)^2$ and $0.2 < x < 0.8$. Hall B Longitudinally polarized ammonia targets were used for this experiment and the scattered electrons were detected in the CEBAF Large Acceptance Spectrometer (CLAS). A detailed description of the experimental setup can be found in references [7, 8]. The preliminary results for g_1^p are shown in Figure 2 compared to a parametrization of world's DIS data at $Q^2 = 10\ (GeV/c)^2$. For the settings where $Q^2 < 2\ (GeV/c)^2$, g_1^p is negative at the $\Delta(1232)$ resonance, clearly deviating from the DIS parametrization. However, at the higher Q^2 settings resonance data appear to be in closer agreement with the DIS parametrization indicating the onset of quark-hardron duality.

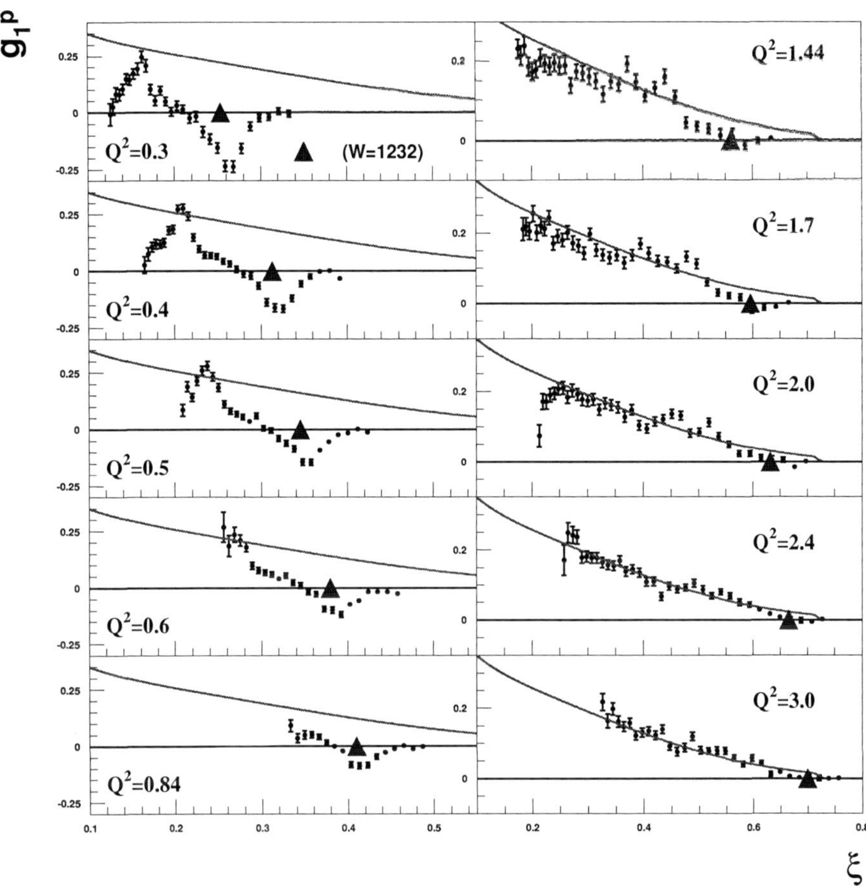

FIGURE 2. Preliminary results for g_1^p from Hall B EG1b experiment. The lines represent a parametrization of world's DIS data at $Q^2 = 10\ (GeV/c)^2$. The arrow indicates the location of the $\Delta(1232)$ resonance.

Jefferson lab Hall C experiment E01-006, Resonance Spin Structure (RSS), preformed a precision measurement of spin asymmetries A_1 and A_2 and spin structure functions g_1 and g_2 for the proton and the deuteron at $Q^2 \approx 1.3\ (GeV/c)^2$ and $0.8 < W < 2GeV$. This experiment used the UVa-Jefferson lab polarized ammonia target which allowed the measurement of both parallel and transverse asymmetries. The scattered electrons were detected in the Hall C High Momentum Spectrometer (HMS). Figure 3 shows preliminary results for g_1 and g_2 for the proton. The g_1^p results, measured at $Q^2 = 1.3\ (GeV/c)^2$, are compared to a parametrization of g_1^p in the DIS region evaluated at $Q^2 = 1.3\ (GeV/c)^2$. At the $\Delta(1232)$ resonance g_1^p is negative, dipping well below the DIS scaling curve, while at the higher resonance regions the resonance data appear to oscillate around the scaling curve. These observations are in qualitative agreement with

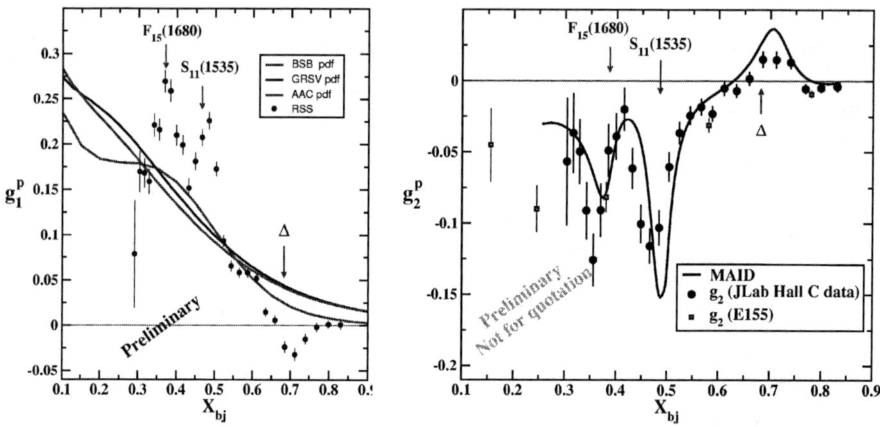

FIGURE 3. Preliminary results for g_1^p and g_2^p from Hall C RSS experiment. The lines on the g_1^p plot represent parameterizations of world's DIS data at $Q^2 = 1.3\ (GeV/c)^2$. The line shown on the g_2^p is from the MAID model.

Hall A and Hall B results taken at $Q^2 < 2\ (GeV/c)^2$ settings.

Summary and Outlook

Polarized experiments in Jefferson lab Halls A, B and C provide precision spin structure data in the resonance region for both the proton and the neutron. These data will provide a comprehensive test of the spin and flavor dependence of quark hadron duality. Preliminary results from all three experiments indicate that for Q^2 less than approximately 1.5 $(GeV/c)^2$, quark-hadron duality is violated for g_1 structure function, mostly due to the strong negative contribution of $\Delta(1232)$ resonance. Data taken at $Q^2 > 2\ (GeV/c)^2$ provide qualitative indications of global quark-hadron duality for g_1 and A_1.

The data analysis of all three experiments are in the final stages. Quantitative tests of global and local quark-hadron duality are being carried out with these data. Final results are expected soon.

REFERENCES

1. E. D. Bloom and F. J. Gilman, Phys. Rev. Lett. **25**, 1140 (1970); Phys. Rev. **D 4**, 2901 (1971).
2. I. Niculescu *et al.*, Phys. Rev. Lett. **85**, 1186 (2000); I. Niculescu *et al.*, Phys. Rev. Lett. **85**, 1182 (2000).
3. A. Airapetian *et al.*, Phys. Rev. Lett. **90**, 092002 (2003).
4. W. Melnitchouk, R. Ent and C. Keppel, Phys.Rept. 406, 127-301 (2005).
5. X. Zheng *et al.*, Phys. Rev. C **70**, 065207 (2004); X. Zheng *et al.*, Phys. Rev. Lett. **92**, 012004 (2004).
6. See the discussion in X. Zheng *et al.*, Phys. Rev. C **70** and references therein.
7. Vipuli Dharmawardane, Ph.D Thesis, Old Dominion University (2004)
8. Yelena Prok, Ph.D Thesis, University of Virginia (2004).

PAX: Polarized Antiproton eXperiments

P. Lenisa on behalf of the PAX Collaboration

Universitá di Ferrara and INFN, 44100 Ferrara, Italy

Abstract.
Polarized antiprotons produced by spin filtering with an internal polarized gas target provide access to a wealth of single– and double–spin observables, thereby opening a window to physics uniquely accessible with the HESR at FAIR. This includes a first measurement of the transversity distribution of the valence quarks in the proton, and a first measurement of the moduli and the relative phase of the time–like electric and magnetic form factors $G_{E,M}$ of the proton. In polarized and unpolarized $p\bar{p}$ elastic scattering open questions like the contribution from the odd charge–symmetry Landshoff–mechanism at large $|t|$ and spin–effects in the extraction of the forward scattering amplitude at low $|t|$ can be addressed.

Keywords: Polarization fenomena in reactions, Polarized beams, Polarized Targets
PACS: 29.27.Hj, 24.70.+s, 29.25.Pj

INTRODUCTION

The polarized antiproton–proton interactions at the High Energy Storage Ring (HESR) at the future Facility for Antiproton and Ion Research (FAIR) will provide unique access to a number of new fundamental physics observables, which can be studied neither at other facilities nor at HESR without transverse polarization of protons and antiprotons.

PHYSICS CASE

The transversity distributions is the last leading–twist missing piece of the QCD description of the partonic structure of the nucleon. It describes the quark transverse polarization inside a transversely polarized proton [1]. Unlike the more conventional unpolarized quark distribution $q(x,Q^2)$ and the helicity distribution $\Delta q(x,Q^2)$, the transversity $h_1^q(x,Q^2)$ can neither be accessed in deep–inelastic scattering of leptons off nucleons nor can it be reconstructed from the knowledge of $q(x,Q^2)$ and $\Delta q(x,Q^2)$. It may contribute to some single–spin observables, but always coupled to other unknown functions. The transversity distribution is directly accessible uniquely via the **double transverse spin asymmetry** A_{TT} in the Drell–Yan production of lepton pairs. The theoretical expectations for A_{TT} in the Drell–Yan process with transversely polarized antiprotons interacting with a transversely polarized proton target at HESR are in the 0.3–0.4 range [2, 3]; with the expected beam polarization achieved using a dedicated low–energy antiproton polarizer ring (APR) of $P \approx 0.3$ and the luminosity of HESR, the PAX experiment is uniquely suited for the defi nitive observation of $h_1^q(x,Q^2)$ of the proton for the valence quarks.

CP792, *Deep Inelastic Scattering, DIS 2005*, edited by W. H. Smith and S. R. Dasu

The origin of the unexpected Q^2–dependence of the ratio of the magnetic and electric form factors of the proton as observed at the Jefferson laboratory [4] can be clarified by a measurement of their relative phase in the time–like region, which discriminates strongly between the models for the form factor. This phase can only be measured via SSA in the annihilation $\bar{p}p^{\uparrow} \to e^+e^-$ on a transversely polarized target [5, 6]. The double–spin asymmetry will allow independently the $G_E - G_M$ separation and serve as a check of the Rosenbluth separation in the time–like region which has not been carried out so far.

Arguably, in $p\bar{p}$ elastic scattering the hard scattering mechanism can be checked beyond $|t| = \frac{1}{2}(s - 4m_p^2)$ accessible in the t–u–symmetric pp scattering, because in the $p\bar{p}$ case the u–channel exchange contribution can only originate from the strongly suppressed exotic dibaryon exchange. Consequently, in the $p\bar{p}$ case the hard mechanisms [7, 8, 9] can be tested at t almost twice as large as in pp scattering.

THE PAX EXPERIMENT

The possibility to test the nucleon structure via double spin asymmetries in polarized proton–antiproton reactions at the HESR ring of FAIR at GSI in Darmstadt (Germany) has been suggested by the PAX collaboration with a Letter–of–Intent submitted on January 15, 2004. The physics program of PAX has been positively reviewed by the QCD Program Advisory Committee (PAC) on May 14–16, 2004. Following the QCD–PAC report and the recommendation of the Chairman of the committee on Scientific and Technological Issues (STI) and the FAIR project coordinator, the PAX collaboration has optimized the technique to achieve a sizable antiproton polarization and the proposal for experiments at GSI with polarized antiprotons [10]. The work has been collected in a Technical Proposal submitted on January 15, 2005 [11].

The PAX collaboration proposes an approach that is composed of two phases. During these the major milestones of the project can be tested and optimized before the final goal is approached: An asymmetric proton–antiproton collider, in which polarized protons with momenta of about 3.5 GeV/c collide with polarized antiprotons with momenta up to 15 GeV/c. These circulate in the HESR, which has already been approved and will serve the PANDA experiment. The overall machine setup of the HESR complex is schematically depicted in Fig. 1. Its main features are:

1. An Antiproton Polarizer (APR) built inside the HESR area with the crucial goal of polarizing antiprotons, to be accelerated and injected into the other rings. The polarization method is based on spin-filtering of the circulating beam by an internal target to the storage ring. This technique has been successfull demonstrated with protons [12] and tests are foreseen to optimize it for an antiproton beam.
2. A second Cooler Synchrotron Ring (CSR, COSY–like) in which protons or antiprotons can be stored with a momentum up to 3.5 GeV/c. This ring shall have a straight section, where a PAX detector could be installed, running parallel to the experimental straight section of HESR.
3. By deflection of the HESR beam into the straight section of the CSR, both the collider or the fixed–target mode become feasible.

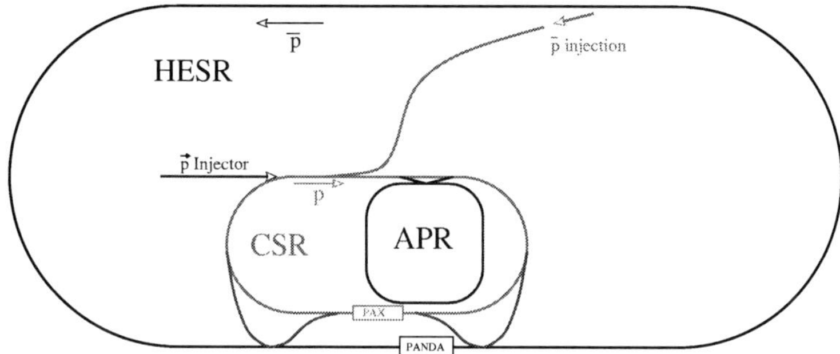

FIGURE 1. The proposed accelerator set–up at the HESR (black), with the equipment used by the PAX collaboration in Phase I: CSR (green), APR, beam transfer lines and polarized proton injector (all blue). In Phase II, by adding two transfer lines (red), an asymmetric collider is set up. It should be noted that, in this phase, also fi xed target operation at PAX is possible.

It is worthwhile to stress that, through the employment of the CSR, effectively a second interaction point is formed with minimum interference with PANDA. The proposed solution opens the possibility to run two different experiments at the same time.

The physics program should be pursued in two different phases.

Phase I A beam of unpolarized or polarized antiprotons with momentum up to 3.5 GeV/c in the CSR ring, colliding on a polarized hydrogen target in the PAX detector. This phase is independent of the HESR performance.
This fi rst phase, at moderately high energy, will allow for the fi rst time the measurement of the time–like proton form factors in single and double polarized $\bar{p}p$ interactions in a wide kinematical range, from close to threshold up to $Q^2 = 8.5$ GeV^2. It would enable to determine several double spin asymmetries in elastic $\bar{p}^\uparrow p^\uparrow$ scattering. By detecting back scattered antiprotons one can also explore hard scattering regions of large t: In proton–proton scattering the same region of t requires twice the energy. There are no competing facilities at which these topical issues can be addressed. For the theoretical background, see the PAX Technical Proposal [11] and the recent review paper [13].

Phase II This phase will allow the fi rst ever direct measurement of the quark transversity distribution h_1, by measuring the double transverse spin asymmetry A_{TT} in Drell–Yan processes $p^\uparrow\bar{p}^\uparrow \to e^+e^-X$ as a function of Bjorken x and Q^2 (= M^2)

$$A_{TT} \equiv \frac{d\sigma^{\uparrow\uparrow} - d\sigma^{\uparrow\downarrow}}{d\sigma^{\uparrow\uparrow} + d\sigma^{\uparrow\downarrow}} = \hat{a}_{TT} \frac{\sum_q e_q^2 h_1^q(x_1, M^2) h_1^{\bar{q}}(x_2, M^2)}{\sum_q e_q^2 q(x_1, M^2) \bar{q}(x_2, M^2)},$$

where $q = u, \bar{u}, d, \bar{d} \ldots$, M is the invariant mass of the lepton pair and $\hat{a}_{TT}$, of the order of one, is the calculable double–spin asymmetry of the QED elementary process $q\bar{q} \to e^+e^-$. The most promising scenario forsees a beam of polarized antiprotons from 1.5 GeV/c up to 15 GeV/c circulating in the HESR, colliding on a beam of polarized protons with momenta up to 3.5 GeV/c circulating in the CSR. Deflection of the HESR beam to the PAX detector in the CSR is necessary (see Fig. 1). By proper variation of the energy of the two colliding beams, this setup would allow a measurement of the transversity distribution h_1 in the valence region of $x > 0.05$, with corresponding $Q^2 = 4 \dots 100$ GeV^2 (see Fig. 2). A_{TT} is predicted to be larger than 0.3 over the full kinematic range, up to the highest reachable center–of–mass energy of $\sqrt{s} \sim \sqrt{200}$. The cross section is large as well: With a luminosity of $5 \cdot 10^{30}$ cm$^{-2}s^{-1}$ about 2000 events per day can be expected[1]. For the transversity distribution h_1, such an experiment can be considered as the analogue of polarized DIS for the determination of the helicity structure function g_1, i.e. of the helicity distribution $\Delta q(x, Q^2)$; the kinematical coverage (x, Q^2) will be similar to that of the HERMES experiment.

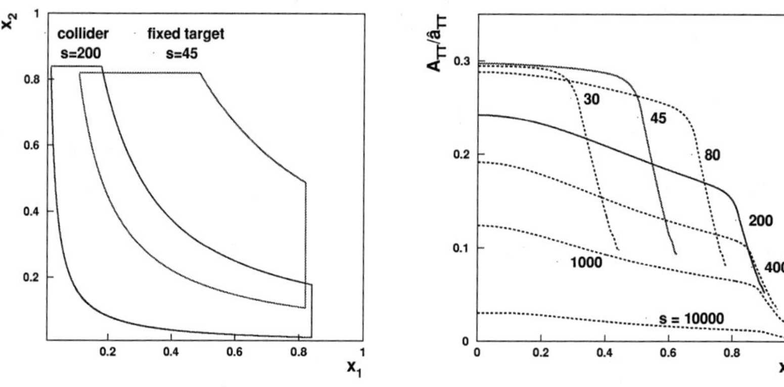

FIGURE 2. Left: The kinematic region covered by the h_1 measurement at PAX in phase II. In the asymmetric collider scenario (blue) antiprotons of 15 GeV/c impinge on protons of 3.5 GeV/c at c.m. energies of $\sqrt{s} \sim \sqrt{200}$ GeV and $Q^2 > 4$ GeV^2. The fi xed target case (red) represents antiprotons of 22 GeV/c colliding with a fi xed polarized target ($\sqrt{s} \sim \sqrt{45}$ GeV). Right: The expected asymmetry as a function of Feynman x_F for different values of s and $Q^2 = 16$ GeV^2.

DETECTOR AND SIGNAL ESTIMATES

An extensive program of studies has been started to investigate different options for the PAX detector confi guration, aiming at an optimization of the achievable performance. The primary goal of the PAX experimental program is to carry out a direct measurement of the h_1 transversity distribution. The proposed detector, described in the PAX techni-

---

[1] A first estimate indicates that in the collider mode luminosities in excess of 10^{30} cm^{-2}s^{-1} could be reached. We are presently evaluating the influence of intra–beam scattering, which seems to be one of the limiting factors.

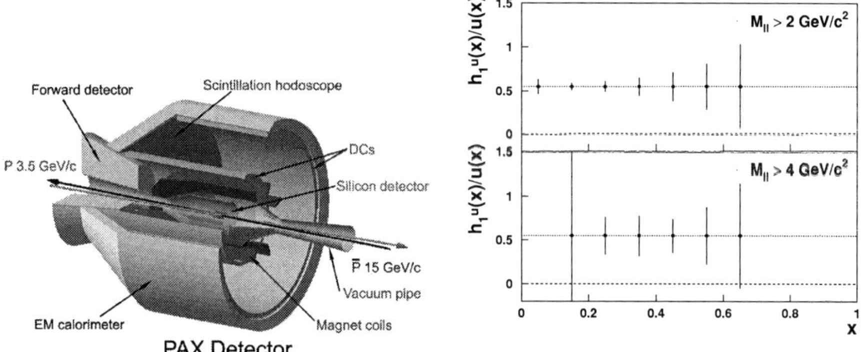

FIGURE 3. Left: Conceptual design of the PAX detector employed to estimate the performance and to show the feasibility of the transversity measurement in the asymmetric antiproton-proton collider at PAX. The artistic view is prodced by GEANT. Right: Expected precision of the $h_1^u(x)$ measurement for one year of data taking in the collider mode at PAX. A luminosity of $2 \cdot 10^{30}$ cm^{-2}s^{-}1 and a polar angle acceptance between 20° and 120° were assumed. The top panel shows the precision achievable within the full $Q^2 > 4GeV^2$ kinematic range, whereas the bottom panel shows the precision achievable in the restricted $Q^2 > 16GeV^2$ range.

cal proposal and shown in Fig. 3, is well-suited to provide large invariant-mass e^+e^- pair detection, from both Drell-Yan reactions and $\bar{p}p$ annihilations. In addition, such a detector is able to effi ciently detect secondaries in two body reactions, like elastic scattering events, where the over-constrained kinematic simplifi es the event reconstruction and reduces the particle identifi cation requirements. Alternative detector scenarios, e.g. with $\mu^+\mu^-$ Drell-Yan pair detection capability, with an instrumented forward section or with extended hadron particle identifi cation, are also under study. The present detector concept fulfi lls the following driving principles:

- Large acceptance. Good azimuthal coverage and symmetry are needed to be sensitive to the dependence of the observables on the angle between production plane and target spin orientation.
- Sensitivity to electron pairs. The overhelming hadronic background requires excellent lepton identifi cation.
- Use of a toroidal magnet. The spectrometer magnet should not affect the transverse spin orientation of the beam and provide an environment to ensure the operation of the Čerenkov detector. The toroid has almost neglible fringe-fi elds outside the active volume, both internally along the beam line and externally inside the tracking volume.

A detailed Monte Carlo study has been started to test the feasibility of the Drell-Yan measurement with the proposed detctor layout. The achievable precision of the ratio between the transverse h_1^u and the well known unpolarized u(x) distributions of the proton, in different intervals of the Bjorken-x and after one year of data-taking is shown in Fig. 3.

The situation is even more favorable for the measurements of electromagnetic form factors and $p\bar{p}$ elastic scattering foreseen in Phase I, as, in these cases, luminosity and cross-sections are such to guarantee high rates [11].

CONCLUSION

The PAX Collaboration has presented a rich and innovative physics program to be realized in the upcoming FAIR hadron facility. The storage of polarized antiprotons at HESR will open unique possibilities to test QCD in hitherto unexplored domains and make of FAIR a facility without competitors.

REFERENCES

1. A comprehensive review paper on the transverse spin structure of the proton can be found in: V. Barone, A. Drago and P. Ratcliffe, *Phys.Rep.* **359**, 1 (2002).
2. M. Anselmino, V. Barone, A. Drago and N. Nikolaev, *Phys. Lett. B* **594**, 97 (2004).
3. A. Efremov, K. Goecke and P. Schweitzer, *Eur. Phys. J. C* **35**, 207(2004).
4. M. K. Jones et al., [Jefferson Lab Hall A Collaboration], *Phys. Rev. Lett.* **84**, 1398 (2000). O. Gayou et al., [Jefferson Lab Hall A Collaboration], *Phys. Rev. Lett.* **88**, 092301 (2002).
5. A. Z. Dubnickova, S. Dubnicka, and M. P. Rekalo, *Nuovo Cimento* **109**, 241 (1966).
6. S.J. Brodsky et al., *Phys.Rev. D* **69**, 054022 (2004).
7. V. Matveev et al., *Lett. Nuovo Cimento* **7**, 719 (1972).
8. S. Brodsky and G. Farrar, *Phys. Rev. Lett.* **31**, 1153 (1973) and *Phys. Rev. D* **11**, 1309 (1973).
9. M. Diehl, T. Feldmann, R. Jakob and P. Kroll, *Phys. Lett. B* **460**, 204 (1999).
10. F. Rathmann et al., *Phys. Rev. Lett.* **94**, 014801 (2005)
11. *PAX Technical Proposal*, Spokespersons P. Lenisa and F. Rathmann http://www.fz-juelich.de/ikp/pax
12. F. Rathmann et al., *Phys. Rev. Lett.* **71**, 1379 (1993).
13. S.J. Brodsky *Testing Quantum ChromoDynamics with Antiprotons* arXiv:hep-ph/0412206 (2004)

Study Quark and Antiquark Contribution to Proton Spin Structure at RHIC

Wei Xie

Riken-BNL Research Center, Brookhaven National Lab, Upton, NY 11973

Abstract. Relativistic Heavy Ion Collider (RHIC) provides a unique opportunity for the direct measurement of quark and antiquark spin in proton utilizing the parity-violating feature of W boson production. Through probing the decay leptons from W bosons in the central and forward rapidity region, RHIC can dissociate contributions from different flavors of quarks to the proton spin with very high accuracy. The capabilities of the spin flavor dissociation at RHIC with current detector configuration and future upgrades are described.

Keywords: spin, RHIC, PHENIX, STAR, quark, upgrade, W boson, asymmetry, trigger, tracking, lepton.
PACS: 12.38.Aw

INTRODUCTION

In the late 80s, measurements from DIS experiment showed that the quarks and antiquarks contribute only ~ 20% of the proton spin [1]. To solve this "crisis", one needs direct measurement of the spin of different flavor of quarks and antiquarks. On the other hand, the unpolarized Fermilab E866 experiment discovered the flavor symmetry breaking in the quark sea [2]. It would be very interesting to see if similar flavor asymmetry exists in the polarized cases. The Semi-inclusive DIS experiment like HERMES [3] has made the first progress in this direction. The measurements are based on the assumption that the leading hadrons most likely come from the quarks hit by the virtual probe photons. The limitation is the low probe Q^2 where the quark fragmentation functions are not well measured and one has to rely on theoretical model calculations. RHIC can directly measure the quark sea flavor asymmetry *via* the decay leptons from W^+ and W^- bosons [4] based on the first principle and no model dependence. RHIC spin collaboration has the multi-year plan to do the W measurement with great sensitivity in $\sqrt{s} = 500\,GeV$ polarized p-p collisions. The capabilities of the measurements at RHIC with current detector configuration and future upgrades are described

SENSITIVITY OF QUARK AND ANTIQUARK SPIN MEASUREMENT AT RHIC

In the standard model, a W boson is produced *via* the V-A interaction and is directly coupled to the parent quark and antiquark with known helicity. Through studying the single longitudinal spin asymmetry of W-decay leptons, one can measure

CP792, *Deep Inelastic Scattering, DIS 2005,* edited by W. H. Smith and S. R. Dasu

the participating quark and antiquark spin inside protons at large Q^2 ($\sim M_w^2$) without any knowledge of fragmentation functions [4]. RHIC has the muti-year plan to measure W boson in polarized p-p collision at $\sqrt{s} = 500\,GeV$ .The projected integrated luminosity is 800 pb^{-1}. One expect to have 8000 decay muons from W^+ and 8000 decay muons from W^- in PHENIX muon arms, 15000 decay electrons from W^+ and 2500 decay electrons from W^- in PHENIX central arms . Recently P.M. Nadoski and C.P. Yuan made predictions on the W boson production in polarized p-p collision at RHIC based on the resummation calculation [5]. They proposed the approach of measuring quark and antiquark spin *via* studying W-decay lepton asymmetry. Fig.1 shows the prediction of the lepton asymmetry as a function of lepton rapidity with 800 pb^{-1} p-p collisions at $\sqrt{s} = 500\,GeV$. A 20GeV p_T cut is applied to reduce the backgrounds. The error bars on the solid point is statistical only. One can see the sensitivity is different in central and forward rapidity region. RHIC experiment will be able to cover both regions.

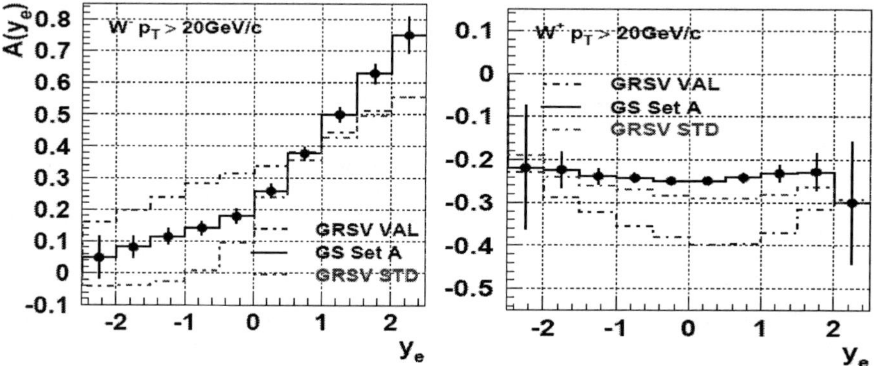

FIGURE 1. Asymmetries of decay leptons from W^- boson (left) and W^+ boson (right) predicted by resummation calculations using RhicBos [6]. Results shown are from GRSV VAL (blue) [7], GS Set A (black) [8] and GRSV STD (red) [9].

EXPERIMENTAL CAPABILITIES OF MEASURING QUARK AND ANTIQUARK SPIN AT RHIC

In order to successfully measure the quark and antiquark spin through W-decay leptons at RHIC, the experiments are required to have:

- Good charge sign identification at very high momentum. One can see from fig.1 that the decay lepton asymmetry from W^+ and W^- are quite different. Mixing them together will significantly deteriorate the sensitivity of the measurement.
- High performance level-1 trigger. In the $\sqrt{s} = 500\,GeV$ p-p runs at RHIC, the expected collision rate is 12 MHz. We need a single lepton level-1 trigger with very high rejection and efficiency to collect all the W-decay leptons through the limited trigger bandwidth in the RHIC experiments.
- Good hadron rejection factor or being able to measure the charged hadron asymmetry that can be subtracted to achieve the asymmetry of W-decay leptons.

Fig.2 shows the PYTHIA simulation on the p_T distribution of decay leptons from different sources in the central and forward rapidity region inside PHENIX acceptance. Keeping in mind the large uncertainties of the estimate, one can see the dominant background is the charged hadrons. We need either to reject them or be able to measure its asymmetry and subtract it later to get the W-decay lepton asymmetry. A rejection factor of at least 1000 is needed to bring down the charged hadrons to the level comparable to other backgrounds.

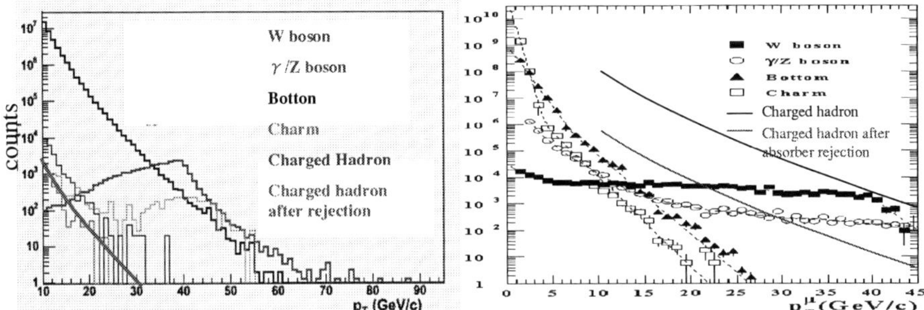

FIGURE 2. Yields of decay leptons from various sources in $\sqrt{s} = 500\,GeV$ p-p collisions with 800 pb^{-1} integrated luminosity inside PHENIX central arm (left) and muon arm (right) acceptance. Both raw charged hadron yield and the one after rejection are shown in the plots. For muon arm, the result from absorber rejection alone is shown and further rejection power can be achieved from shower profile cut.

Charged Hadron Rejection

For an experiment with calorimeters and tracking chambers, the usual ways to reject hadrons are through the cut on the ratio of energy over momentum (E/p cut) and shower profiles. In the case of measuring W bosons, its high p_T decay leptons should be isolated from charged hadrons and therefore an isolation cut is effective to reject hadrons. On the other hands, the p_T of W boson is much smaller than its decay leptons, an event containing a W boson decay must have very high p_T since the decay neutrino can not be detected. One therefore can reject any events with missing large p_T.

The STAR experiment [10] contains Electromagnetic Calorimeters (EMCal) and Time Projection Chambers (TPC) covering both central ($|\eta|<1$) and forward rapidity ($1<|\eta|<2$) region. The total hadron rejection factor is more than 10^3 and it comes equally from the E/p, isolation and large missing p_T cut. STAR is also equipped with a pre-shower detector in its forward calorimeter. A longitudinal shower profile cut should produce more hadron rejection powers.

The PHENIX [11] central arm detector contains EMCal and a drift chamber for track reconstruction covering $|\eta|<0.35$ region. The total rejection factor is between 2000 to 10^4 where E/p cut contributes ~40 to 200 depending on different arm and different tracking algorithm, shower profile and isolation cut contribute another factor of 50. A TEC/TRD detector is equipped in the east central arm and will be used for high p_T electron identification. It is expected to produce an additional rejection factor of 30 once its operation is stable.

The PHENIX Muon arm detector contains a muon tracking chambers (MuTr) and muon identification chambers (MuID) covering $1.2<|\eta|<2.4$ region. The total hadron rejection of ~100 comes mainly from the thick steel absorber located in front of MuTr. Further rejection power can be achieved *via* shower profile cut at MuID. Future PHENIX forward calorimeter upgrade will help reject hadrons through isolation and large missing p_T cut. The possibilities of measuring the large p_T hadron asymmetry that can be used for subtraction to get the W-decay lepton asymmetry are also being studied.

Charge Sign Identification

Both the PHENIX central and muon arm have very good charge sign identification capabilities. Only a few percent of charge sign misidentification is observed in p_T ~ 40GeV region with current detector setup. Further improvement is expected when the ongoing silicon vertex detector upgrade is finished.

STAR TPC has coarse pad readout granularity in the region of $|\eta|>1.0$. The charge sign misidentification is up to ~15% at p_T ~ 40GeV as shown from the simulation neglecting the huge event pile-up effect in the coming $\sqrt{s} = 500\,GeV$ high luminosity p-p run. The event pile-up effect is expected to further seriously deteriorate charge sign identification. An ongoing forward tracking upgrade project will be able to solve the problem and will be described briefly in one of the following sections.

Level-1 Trigger Performance

PHENIX central arm has built a powerful electron trigger using EMCal and Cerenkov detectors. The rejection factor can go well beyond 10^4 at reasonable trigger threshold where the efficiency is high. This satisfies the requirement for trigger performance in the high luminosity 500GeV run for PHENIX experiment.

STAR can trigger on electrons efficiently using the barrel and endcap EMCal by setting a certain energy threshold. The rejection factor is expected to be high enough to meet their requirement for detecting W bosons in the high luminosity 500GeV run.

PHENIX muon arm has a single muon trigger that has been used to successfully trigger on the muons decay from heavy flavor mesons. However, the rejection factor is about 20-50 times lower than that is required for W boson measurement at RHIC. A trigger upgrade project is ongoing to build a powerful momentum sensitive muon trigger to solve the problem. The project will be described briefly in one of the following sections.

FUTURE DETECTOR UPGRADE FOR QUARK AND ANTIQUARK SPIN MEASUREMENT AT RHIC

As we see from previous sections, to successfully measure the quark and antiquark spin at RHIC, STAR experiment needs a forward tracking upgrade to solve the charge sign identification problems and PHENIX need to upgrade the muon trigger to adapt to the coming high luminosity p-p collisions.

Forward Tracking Upgrade for STAR Experiment

The forward tracking upgrade will include 4 silicon disks with 50μm resolution in the inner region and a 3-layer GEM detector with 100μm resolution between endcap EMCal and TPC pad readout plane as shown in fig.3 (a). Fig.3 (b) shows the simulated results of new tracker performance with p_T=30GeV electrons. One can see the new tracking system alone has very little charge misidentification and much better momentum resolution compared to the TPC. The performances are comparable with and without adding the TPC. The proposal is planned to be submitted after FY06 run and the detectors are to be installed in the summer of 2009. Simulation and R&D efforts are ongoing actively.

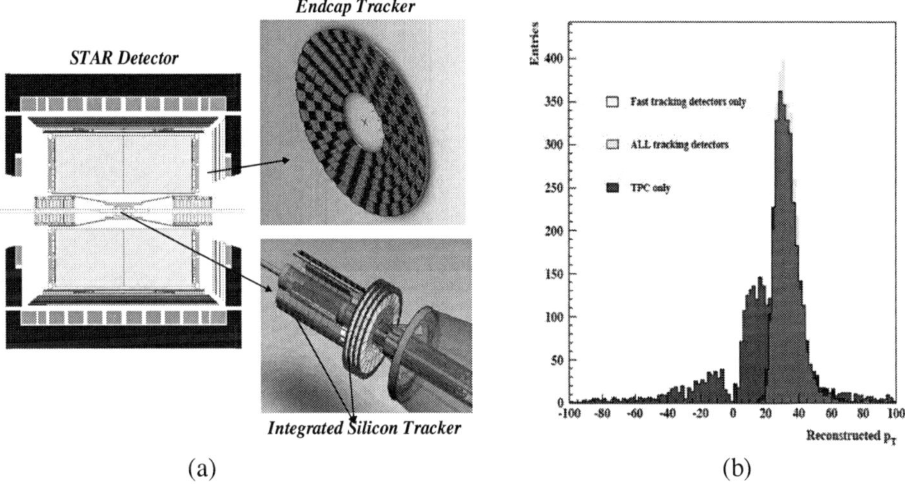

FIGURE 3. STAR forward tracking upgrade detectors (a) and performance (b). Right hand side plot shows reconstructed momentum of p_T=30GeV electrons with TPC only (red), TPC plus new forward tracking system (yellow) and forward tracking system only (brown).

Muon Level-1 Trigger Upgrade for PHENIX Experiment

There are two proposals for trigger upgrade. One is to install three resistive plate chambers (RPC) inside the muon arm. The trigger requires that the RPC hits are associated with the muon candidates of the existing muon trigger and apply an angle cut between hits in the two front RPCs to reject low momentum muons. The principle is illustrated in fig.3 (b). A proposal of two million dollars has been submitted to NSF. The other one is similar but MuTr hits are used for the angle cut and the association with muon candidates from current trigger and it requires upgrade of MuTr front end electronics. Efforts are being made to seek funding in Japan. Both proposals provide sufficient rejection factor ($> 10^4$) for W boson measurement in the coming high luminosity p-p run. If both proposals succeed in getting funding, their combination will produce an even more powerful trigger that can be used in the future higher luminosity RHICII environment. The installation of new RPC trigger detectors and electronics is planned to be finished before early 2009. Simulation and R&D work are ongoing actively on both fronts.

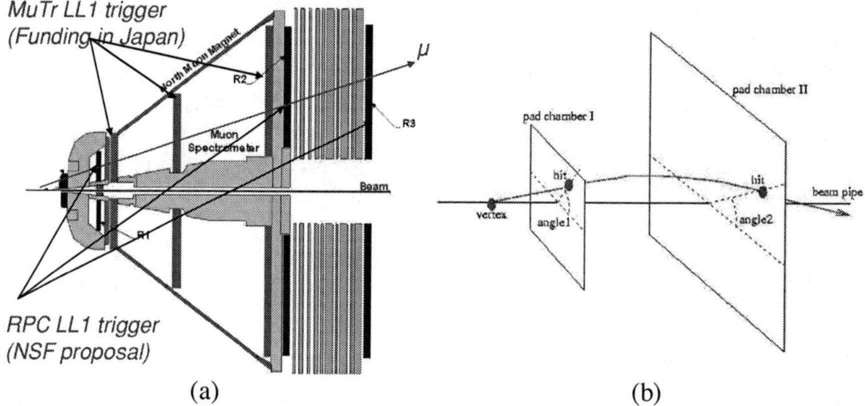

FIGURE 4. PHENIX forward muon trigger upgrade detectors (a) and working principle (b). In the left hand side plot, the proposal of RPC trigger is illustrated in back color and the proposal of MuTr trigger is illustrated in pink color.

SUMMARY

RHIC provides a unique opportunity for the direct measurement of quark and antiquark spin in proton utilizing the parity-violating feature of W boson production. With the successful upgrade of forward tracking system in STAR and forward muon trigger in PHENIX, RHIC will be able to dissociate contributions to proton spin from different flavors of quarks and antiquarks with very high accuracy through probing the decay leptons from W bosons in the central and forward rapidity region,

ACKNOWLEDGMENTS

The author expresses full thankfulness to B. Sorrow and A. Bazilevsky for providing many materials for the proceeding.

REFERENCES

1. Ashman J, et al [European Muon Collaboration (EMC)]. *Phys. Lett.* B206:364 (1988)
2. FNAL E866/NuSea Collaboration, E.A. Hawker *et al.*, Phys. Rev. Lett. 80 (1998) 3715
3. HERMES Collaboration, Phys.Rev. D71 (2005) 012073
4. G. Bunce *et al.*, Annu. Rev. Nucl. Part. Sci. 2000. 50:525-75
5. P.M. Nadolsky *et al.*, Nuclear Physics B 666 (2003) 31-55
6. P.M. Nadolsky *et al.*, http://hep.pa.msu.edu/people/nadolsky/RhicBos/index_frames.html
7. M. Gluck *et al.*, Phys. Rev. D 63 (2001) 094005
8. T. Gehrmann *et al.*, Phys. Rev. D 53 (1996) 6100
9. M. Gluck *et al.*, Phys. Rev. D 53 (1996) 4775
10. STAR Collaboration, NIM A 499 (2003) 624-813
11. PHENIX Collaboration, NIM A 499 (2003) 469-602

Electron Polarimetry: Status and Prospects

E. Chudakov

Thomas Jefferson National Accelerator Facility, Newport News, VA 23606, USA

Abstract. Polarized electron beams are used widely for DIS, parity violation, and other experiments. I discuss the methods and instrumentation used to measure the electron beam polarization, as well as the prospects for the future facilities. A number of recent achievements and projects are discussed. More details on the subject can be found in a 1998's review[1].

Keywords: electron beam polarimetry polarimeter Compton Møller Mott
PACS: 29.27.Hj, 29.27.Fh

Polarized electron beams in an energy range of $\sim 0.2-50$ GeV have been used in double-spin experiments, including studies of polarized structure functions in DIS, and in single-spin experiments, including SM tests in parity-violation (PV) scattering. In the double-spin experiments the requirements for the electron beam polarimetry have been typically modest - on the level of 2-3%, since the target (nucleon) polarimetry accuracy have been the limiting factor. In the single-spin experiments a higher accuracy, of about 1%, is often required.

Methods Used for Electron Polarimetry

Various methods to measure the electron beam polarization have been applied, in an energy range from $\sim$ 10 eV to 50 GeV. At low energy (up to several MeV), polarimetry is needed for developing and running the polarized guns and injectors.
Atomic Absorption. In a new low energy polarimeter[2], a 50 keV beam from the gun is decelerated to 13 eV and absorbed in argon. The degree of the circular polarization of the emitted fluorescence (811.5 nm) is related to the electron beam polarization. Potentially, this method could provide a 1% systematic accuracy, a 1% statistical accuracy in a 20 s measurement, and be non-invasive. For the moment, due to problems with handling the very low energy beam, the device has provided only relative polarization measurements and works at low beam currents (40 nA). It is used for invasive, but fast monitoring of the relative beam polarization.
Spin-Orbital Interaction. The standard polarimetry method used in the polarized electron injectors, is to measure the spacial (left-right) asymmetry of the Mott scattering of a transversely (vertically) polarized electron beam off heavy nuclei (see a review in [3]).The typical systematic error of the Mott polarimeters at 50 keV - 5 MeV was not better than $\sim$3%, however a still unpublished work[4] indicated that a higher accuracy of $\sim$1% can be obtained at 5 MeV. Mott polarimetry is invasive and typically uses a few μA beam.The measurements are done at conditions different from the experiment (energy, beam current, spin precession).
Spin-Spin Interaction. Electron scattering off polarized targets may provide large

CP792, *Deep Inelastic Scattering, DIS 2005,* edited by W. H. Smith and S. R. Dasu

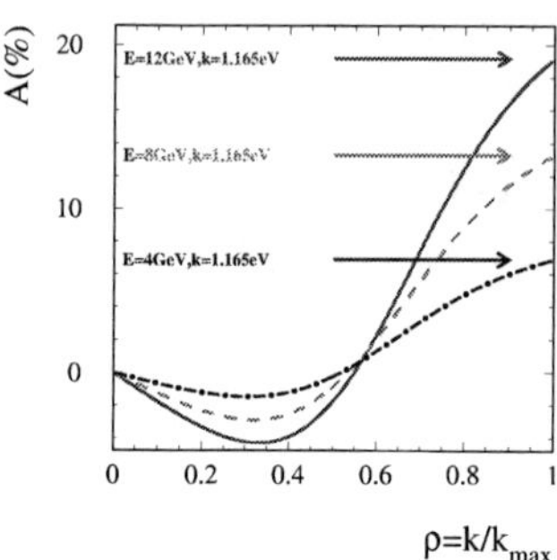

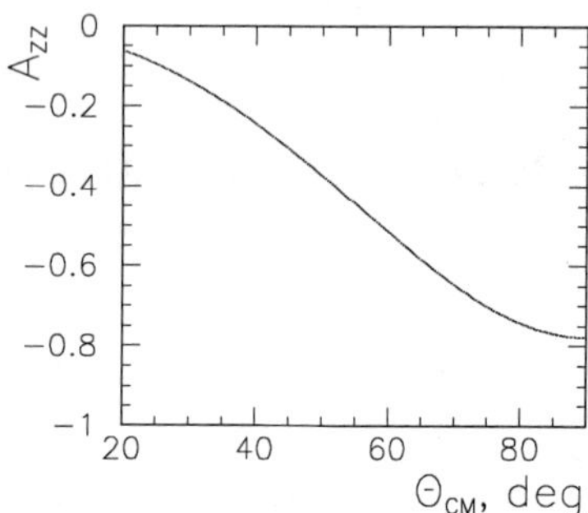

FIGURE 1. Left: the Compton analyzing power dependence on the scattered photon energy, for three beam energies: 4, 8 and 12 GeV. Right: the Møller analyzing power dependence on the scattering angle in CM.

polarization-driven asymmetries at high energies (let us consider the longitudinal polarizations only): $\frac{\sigma_{\uparrow\uparrow}-\sigma_{\uparrow\downarrow}}{\sigma_{\uparrow\uparrow}+\sigma_{\uparrow\downarrow}} = A \cdot \mathscr{P}_b \mathscr{P}_t$, where $\sigma_{\uparrow\uparrow}$ and $\sigma_{\uparrow\downarrow}$ are the cross sections for the parallel and anti-parallel spins of the beam and the target, $\mathscr{P}_b$ and $\mathscr{P}_t$ are the beam and target polarizations, and A is the analyzing power of the process. Two processes are widely used for polarimetry - Møller scattering on polarized electrons and Compton backscattering on circularly polarized laser light. The analyzing power of both processes have been calculated up to second order in QED. The deviation from the Born approximation for the typical polarimeter acceptances is about 0.3% for Møller scattering[5, 6], while for Compton scattering it is about 0.1%[7].

The Møller analyzing power (see Fig.1, right) does not depend on the beam energy and reaches -7/9 at 90° scattering in CM. The typical polarimeter accepts particles, scattered around 90° in CM, in the area of almost constant analyzing power. In contrast, the Compton analyzing power (see Fig.1, left) depends sharply on the scattered photon energy, well within the acceptance of the typical polarimeter, and its maximum value A_{max}, reached at the Compton edge, depends on the beam energy E_b and the initial photon energy k as $A \propto kE_b$ at $E_b < 20$ GeV. The figure of merit $\mathscr{F}$ (inversely proportional to the time needed to reach a certain statistical accuracy) is proportional to the counting rate and A^2. For Compton polarimeters $\mathscr{F} \propto E_b^2 k^2$, while for Møller polarimeters $\mathscr{F}$ does not depend on the energy. Additionally, the maximum backscattered photon energy (at the Compton edge) depends on the initial energies as: $k'_{max} = 4k(E_b/m_e)^2(1+4kE_b/m_e^2)^{-1}$, or as low as ~ 20 MeV for E_b=1 GeV and 1064 nm laser wavelength, which makes detection of Compton scattering harder at low energies. However, Compton polarimetry uses a nearly 100% polarized target (light) and is typically non-invasive. Møller polarimetry so far has used only one source of polarized electrons for the target - a magnetized ferromagnetic. Such targets have an average electron polarization of ~8%. The scattering on the internal shells of a heavy atom is smeared kinematically, which leads to a change of the effective target polarization, depending on the polarimeter acceptance (the Levchuk effect[8]). The ferromagnetic targets have to be thick enough ($> 1\ \mu$m), which makes polarimetry invasive. These targets can not stand continuous beam currents higher than several μA because of their heating and depolarization. All this excludes their application in storage rings. In extracted beams, Møller polarimetry normally uses beam currents lower than the experiments. Extrapolation of the results to the experimental conditions introduces a systematic error, which is sometimes hard to evaluate, in particular in case of CEBAF where the beam in a certain hall may contain a leakage from a differently polarized beam for another hall[9].

Development in Møller and Compton Polarimetry

A number of polarimeters used ferromagnetic foils, magnetized in relatively low fields of 0.01-0.03 T. The field was typically oriented along the beam and the foil was oriented at an angle of $\sim 20^\circ$ to the beam. The polarimeters detected either one scattered electron or two electrons in coincidence. The typical systematic error was about $2-4\%$ relative[10, 11], largely dominated by uncertainties in the foil polarization. A breakthrough in the accuracy was reached by magnetizing the foil to full saturation in a magnetic field of about 4 T[12]. The pure iron foils, 1-10 μm, thick were positioned perpendicularly to the beam. The absolute value of the foil magnetization has not been measured, but was taken from published results of magnetic measurements with the bulk material. A systematic error of 0.5% is claimed, dominated by the foil polarization uncertainty and the Levchuk effect. No extrapolation to different beam currents has been included at this stage. There are efforts to run the polarimeter at $\sim 100\ \mu$A, reducing the heating by introducing an effective duty cycle. In general, the limitation of the Møller polarimeter's accuracy comes from the beam current limits, the target polarization, the Levchuk effect and the invasiveness of the measurements. A possibility to improve the accuracy is to use electron-spin-polarized atomic hydrogen for the target. A jet of polarized deuterium atoms[13] with a density of about $5 \cdot 10^{11}$ cm^{-2} was used in a storage ring at ~ 100 mA. The polarization of the atoms was about 100% (the accuracy has not been discussed). So far, a 1% statistical accuracy could be reached in ~ 100 h. Another possibility discussed[14] is to use an ultra-cold trap for 100% polarized atomic hydrogen. This method could be used for CW beams up to 200 μA in a non-invasive way. It would remove the main systematic errors of, say [12], reducing the error to $\sim$0.2%.

The best systematic accuracy for Compton polarimetry of 0.5% relative has been achieved at a $\sim$46 GeV pulsed beam at the SLC[15]. The beam pulse contained $3.5 \cdot 10^{10}$ electrons. A 532 nm pulsed laser used at a 17 Hz repetition rate provided 50 mJ per pulse. The crossing angle was 10 mrad. Both the scattered electron and the backscattered photon were detected. The main contributions to the systematic error came from the analyzing power (0.4%), detector linearity (0.2%) and electronic noise (0.2%). Compton polarimeters work well at storage rings, such as the longitudinal polarimeter at HERA[16]. The HERA pulse contained $\sim 3 \cdot 10^{10}$ electrons, and a 532 nm pulsed laser (< 100 Hz, 100 mJ per pulse) was used. The systematic error was $\sim 1.6\%$. Another polarimeter at HERA[17] measured the transverse beam polarization, by measuring the left-right asymmetry of the backscattered photons. A 532 nm CW laser was used. A systematic error of about 1.9% was obtained, by comparing the measurements from both polarimeters. With CW electron beams the pulsed lasers are inefficient. At low energies the luminosity is particularly important. At JLab[18], a Fabry-Pérot cavity was used in order to amplify the light from a 1064 nm, 0.24 W CW laser by a factor of about 4000. Both the scattered electrons and photons were detected. This allowed to obtain a 1% statistical accuracy in $\sim$ 30 min with a 40 μA, 4.5 GeV beam. The systematic accuracy was about 1.2%. The accuracy is worse at lower energies, and becomes several % at 1.5 GeV. In order to extend the beam range to $\sim$0.8 GeV, a new cavity for a 532 nm laser is being developed. At HERA, a new longitudinal polarimeter, using a 1064 nm cavity has been built, which allows to measure the polarization of a single bunch.

At the ILC, the polarization has to be measured upstream and downstream of the interaction point, in order to take into account a considerable depolarization in the process of interaction. The method should be non-invasive. At the EIC, the polarization in a storage ring has to be measured, which also requires a non-invasive method. At the ILC, an improved SLC Compton polarimeter design[19] may provide a systematic accuracy of about 0.25%. At the EIC, an accuracy of about 1% is considered sufficient and can be achieved using the Compton polarimetry experience at HERA[20]. While at the moment the best accuracy achieved is about 0.5-1%, there is a motivation to reach 0.1-0.2% in the future. The experience with Møller and Compton polarimetry at SLAC (an initial discrepancy of nearly 10%) demonstrates that more than one method is needed to make reliable measurements in a new range of accuracy. At the ILC, there is a way to measure the effective polarization in the interaction point, using production of $W^{\pm}$ and other processes. At other machines, a development of non-invasive Møller polarimetry for high beam currents might be useful.

This work was supported by DOE contract DE-AC05-84ER40150 Modification No. M175, under which the Southeastern Universities Research Association (SURA) operates the Thomas Jefferson National Accelerator Facility.

REFERENCES

1. C. K. Sinclair, *AIP Conf. Proc.*, **451** (1998). 1
2. B. Collin, et al., *Nucl. Instrum. Meth.*, **A534**, 361–370 (2004). 1
3. T. Gay, and F. Dunning, *Rev. Sci. Instrum.*, **63**, 114–130 (1992). 1
4. J. S. Price, et al. (1998), prepared for 13th International Symposium on High-Energy Spin Physics (SPIN 98), Protvino, Russia, 8-12 Sep 1998. 1
5. V. A. Mosolov, N. M. Shumeiko, and J. G. Suarez, *Int. J. Mod. Phys.*, **A18**, 2807–2815 (2003). 2
6. A. Ilyichev, and V. Zykunov (2005), hep-ph/0504191. 2
7. A. Denner, and S. Dittmaier, *Nucl. Phys.*, **B540**, 58–86 (1999), hep-ph/9805443. 2
8. L. G. Levchuk, *Nucl. Instrum. Meth.*, **A345**, 496–499 (1994). 2
9. J. M. Grames, et al., *Phys. Rev. ST Accel. Beams*, **7**, 042802 (2004). 3
10. K. B. Beard, et al., *Nucl. Instrum. Meth.*, **A361**, 46–52 (1995). 3
11. H. R. Band, G. Mitchell, R. Prepost, and T. Wright, *Nucl. Instrum. Meth.*, **A400**, 24–33 (1997). 3
12. M. Hauger, et al., *Nucl. Instrum. Meth.*, **A462**, 382–392 (2001), nucl-ex/9910013. 3
13. M. V. Dyug, et al., *Nucl. Instrum. Meth.*, **A536**, 338–343 (2005). 3
14. E. Chudakov, and V. Luppov, *IEEE Trans. Nucl. Sci.*, **51**, 1533–1540 (2004). 3
15. M. Woods (1996), talk given at Workshop on High-energy Electron Polarimeters (Pre-symposium for SPIN 96), Amsterdam, Netherlands, 9 Sep 1996., hep-ex/9611005. 3
16. M. Beckmann, et al., *Nucl. Instrum. Meth.*, **A479**, 334–348 (2002), physics/0009047. 3
17. J. Bohme, *Eur. Phys. J.*, **C33**, s1067–s1069 (2004). 3
18. M. Baylac, et al., *Phys. Lett.*, **B539**, 8–12 (2002), hep-ex/0203012. 3
19. P. C. Rowson, and M. Woods (2000), contributed to 5th International Linear Collider Workshop (LCWS 2000), Fermilab, Batavia, Illinois, 24-28 Oct 2000., hep-ex/0012055. 4
20. W. Lorenzon, *AIP Conf. Proc.*, **698**, 797–800 (2004). 4

PROTON POLARIMETRY AT RHIC

Alessandro Bravar [1]

Brookhaven National Laboratory Upton, NY 11973, USA [2]

Abstract. The techniques used to measure precisely the polarization of the proton beams at RHIC are presented and discussed. Fast polarization measurements are performed using polarimeters based on *pC* elastic scattering. The absolute normalization is provided by a polarized hydrogen gas jet target. During the 2004 polarized proton run a relative precision on the beam polarization $\Delta P_{beam}/P_{beam}$ of 6.6% has been achieved.

Keywords: Polarization, Elastic Scattering
PACS: 13.85.Dz and 13.88.+e

INTRODUCTION AND METHOD

The RHIC Spin program [1] aims to determine the spin asymmetries with one or both beams polarized for a variety of processes with high precision, such to allow significant comparison with theoretical predictions and possibly unveil new physics. The proton polarimeters provide the normalization of the spin asymmetries measured by the RHIC experiments, and polarization measurements for the accelerator setup. A crucial requirement is the knowledge of the absolute polarization of the RHIC proton beams to 5% of its value and its constant monitoring.

The major requirements for a good and efficient polarimeter are a simple and well understood scattering process with a large cross section and known analyzing power A_N. The chosen *polarimetric* processes are *pC* and *pp* elastic scattering at very low momentum transfer t. The analyzing power A_N is defined as the left-right asymmetry of the cross section in the scattering plane normal to the beam polarization. Until recently, A_N^{pC} for *pC* scattering at RHIC energies was not well know. During the 2004 run a measurement and calibration of this process using a polarized hydrogen gas jet target has been performed to a precision better than 15%.

In high energy *pp* and *pA* elastic scattering at very low momentum transfer t, A_N originates from the interference between the real electromagnetic (Coulomb) spin-flip amplitude, which is generated by the proton's anomalous magnetic moment, and the imaginary hadronic (Nuclear) spin-nonflip amplitude (CNI = Coulomb Nuclear Interference), and thus provides important information on the spin dependence of the interaction. A_N has a maximum value of about 4-5% around $t \simeq -3 \times 10^{-3}$ $(\text{GeV}/c)^2$ [2].

[1] For the RHIC Polarimetry group: I. Alekseev, A. Bravar, G. Bunce, S. Dhawan, R. Gill, H. Huang, W. Haeberli, G. Igo, O. Jinnouchi, K. Kurita, Y. Makdisi, I. Nakagawa, A. Nass, H. Okada, N. Saito, H. Spinka, E. Stephenson, D. Svirida, C. Whitten, T. Wise, J. Wood, A. Zelenski

[2] Supported under Prime Contract between Brookhaven Science Associates and the Department of Energy No. DE-AC02-98CH10886.

CP792, *Deep Inelastic Scattering, DIS 2005*, edited by W. H. Smith and S. R. Dasu

PROTON-CARBON POLARIMETERS

Fast beam polarimetry at RHIC and AGS is based on *pC* elastic scattering at very small proton scattering angles. *pC* elastic scattering events are identified by detecting the recoil carbon ions. For very small angle scattering the elastic reaction dominates and the measurement of the recoil ions gives predominantly elastic events with very small backgrounds. Typical 4-momentum transfers squared are $-t \sim 0.01 - 0.02\ \text{GeV}^2/c^2$, corresponding to recoil carbon ion kinetic energies of $T_R \sim 0.4 - 1$ MeV. In this t region the analyzing power for *pC* elastic scattering A_N^{pC} is small, around 1%. The figure of merit, however, is high, since the *pC* elastic cross section is very large. The small value of A_N^{pC} makes it necessary to collect large data samples, of the order of 2×10^7 events per measurement. Event rates are relatively high, of the order of 10^5 elastic events / polarimeter channel / sec. A typical measurement in RHIC lasts around 10 sec.

The slow recoil carbon nuclei emerge at almost 90° w.r.t. the incident beam and are detected with six silicon detectors at different azimuthal angles (see Figure 1), which provide energy and time of flight (ToF) information. Each silicon detector consists of 12 strips, for a total of 72. The strips are oriented along the beam direction, thus each strip covers the same polar angle and detects the same physical process in the same kinematical region. The detectors are located inside the accelerator vacuum system at ~ 15 cm from the RHIC beam. On the basis of the T_R – ToF correlation, carbon recoils are identified and selected. Typical flight times are between 30 and 80 nsec. An ultra thin carbon ribbon of $3 - 5\ \mu$g/cm^2 and less than 10 μm wide is used as target and is inserted into the beam during the measurement. Two such devices are installed in RHIC for each beam, and one in AGS. The silicon detectors are readout with waveform digitizer, which provide a deadtimeless DAQ system with *on-board* event analysis, thus allowing us to handle the high rates. An algorithm in conjunction with on board FPGAs in the WFD units is used to extract energy and time information from the recorded waveforms.

Each detector channel covers a different azimuthal angle and can be viewed as an independent polarimeter. Figure 1 shows the asymmetry in the event yields for events from bunches with *up* polarization vs. bunches with *down* polarization, normalized with the relative luminosities, for each polarimeter channel. In Figure 1 these asymmetries are fitted with a sinusoidal function. The amplitude of this distribution, normalized with the corresponding A_N^{pC} gives P_{beam}. This is one method of extracting P_{beam}. These measurements are scattered around the fitted curve statistically. That indicates that all the measurements from each polarimeter channel are self consistent and that the systematic errors associated with each asymmetry measurement are small.

The largest systematical uncertainty in extracting P_{beam} comes from the absolute energy scale in determining the energy of the recoil carbon ions E_C. Since $A_N^{pC}(t)$ depends on E_C, an uncertainty is introduced in the extraction of P_{beam} from the measured asymmetries. The difficulty of the measurement resides in the fact that only a fraction of E_C is observed, part of this energy being deposited in the entrance window of the detector, and part being lost due to charge collection efficiencies. An energy correction term is estimated by requiring that the mass of the detected recoils correspond to the mass of carbon ions and is added to the measured energy on an event by event basis. At present this effect is estimated to be around 5% of the measured P_{beam} value.

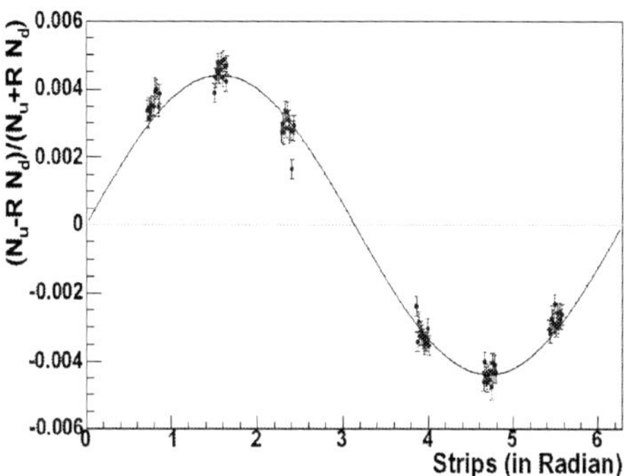

FIGURE 1. Asymmetry in the event yields for events from bunches with *up* polarization vs. bunches with *down* polarization, normalized with the relative luminosities, for each polarimeter channel. The amplitude of this distribution is directly related to P_{beam}. See text for more details.

THE POLARIZED GAS JET TARGET

An absolute polarimeter, based on *pp* elastic scattering, using a polarized hydrogen gas jet target, has been installed in 2004, and provides the absolute normalization for the fast *pC* polarimeters. The transverse spin asymmetry in *pp* elastic scattering of a polarized beam on an unpolarized target is identical to the unpolarized beam – polarized target one in the same kinematical region: $A_N^{p\uparrow p} = A_N^{pp\uparrow}$. This symmetry relation, which holds for the elastic scattering of spin 1/2 identical particles only, permits the direct transfer of the target polarization P_{target} to the beam polarization P_{beam}:

$$P_{beam} = P_{target} \times \frac{\varepsilon_{beam}}{\varepsilon_{target}} \quad , \tag{1}$$

where ε_{beam}, ε_{target} are the *left - right pp* scattering asymmetries obtained by averaging over the target polarization and beam polarization states, respectively (see Figure 2), using events from the same data set. This procedure is also referred to as the *self-calibration* method, and is independent of theoretical assumptions.

In the CNI region recoil protons from *pp* elastic scattering emerge close to 90° with respect to the incident beam direction. The recoil protons were detected using an array of silicon detectors. The scattered beam protons did not exit the beam pipe and they were not detected. In the covered *t* region, however, the elastic process is fully constrained by the recoil particle only, thus the detection of the scattered beam proton is not mandatory. Recoil protons were identified on the basis of the T_R – ToF non-relativistic relation $T_R = \frac{1}{2}M_p(\text{dist}/\text{ToF})^2$ and selected on the basis of the $\vartheta_R - T_R$ relation $T_R \simeq 2M_p\vartheta_R^2$ (ϑ_R is the polar angle). On the basis of the $\vartheta_R - T_R$ correlation one can reconstruct the mass of the scattered particle (so called missing mass M_X). For a *pp* elastic scattering process $M_X = M_p$. We found that $M_X \simeq M_p$ with little background below the elastic peak (figure not shown), confirming indeed that we were selecting *pp* elastic scattering events. The estimated background in the selected *pp* elastic scattering sample was less than 5%.

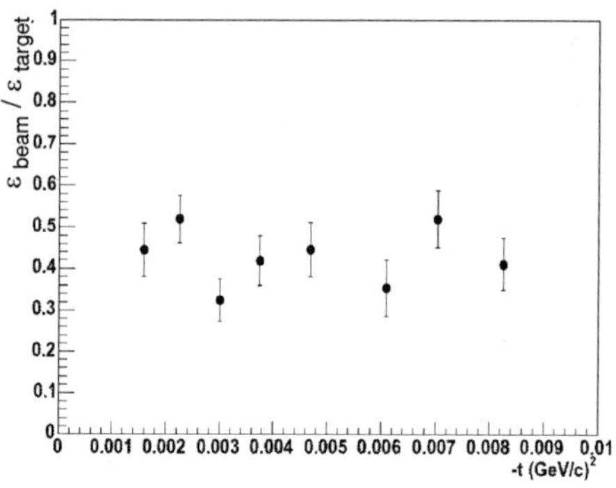

FIGURE 2. $\varepsilon_{beam}/\varepsilon_{target}$ as a function of t. For defi nitions see text.

The polarized hydrogen gas jet target crossed the RHIC beams from above with its polarization directed vertically. The polarized target is a free atomic beam jet. The state-of-the-art atomic polarized source delivered polarized protons with a polarization of 0.924 ± 0.018 (the dilution from molecular hydrogen is included in this figure), a density in excess of 10^{12} p/cm^2 in its center, and a FWHM profile of less than 6 mm. The target polarization was reversed every 5 to 10 minutes. The target polarization was constantly monitored with a Breit-Rabi polarimeter. For more details see [3].

During the 2004 polarized pp run, P_{beam} has been measured with an absolute precision of 2.7% using the *self-calibrating* method described above. The preliminary result on the average P_{beam} and the error from the past run is given here below:

$$P_{beam} = 0.392 + 0.021(\text{stat}) + 0.008(\Delta P_{target}) + 0.014(\text{sys}) \quad . \tag{2}$$

The second term in the systematic error comes mainly from backgrounds in the selected pp elastic sample. The relative precision of this measurement $\Delta P_{beam}/P_{beam}$ is 6.6%. For the 2005 polarized pp run a relative precision on P_{beam} better than 5% is expected.

ACKNOWLEDGMENTS

We would like to thank the Instrumentation Division at BNL for their work on the silicon detectors and electronics. The research reported here has been performed under the auspices of the U.S. DOE contract Nos. DE-AC02-98CH10886 and W-31-109-ENG-38, DOE grant No. DE-FG02-88ER40438, NSF grant PHY-0100348, and with support from RIKEN, Japan.

REFERENCES

1. G. Bunce *et al.*, Annu. Rev. Nucl. Part. Sci. **50**, 525 (2000).
2. N.H. Buttimore *et al.*, Phys. Rev. D **59**, 114010 (1999);
   B.Z. Kopeliovich and T.L. Trueman, Phys. Rev. D **64**, 034044 (2001).
3. A. Zelenski *et al.*, PST 2003, Nucl. Inst. and Meth. A **536**, 248 (2005).

FUTURE OF DIS PRESENTATIONS

Deeply Inelastic Scattering: Achievements and Needs

Johannes Blümlein

DESY, Platanenallee 6, D-15738 Zeuthen, Germany

Abstract. We discuss the present status of deeply inelastic scattering and possible developments needed to further complete the picture on the short–distance structure of nucleons.

Keywords: Deeply Inelastic Scattering
PACS: 13.60.-r

After a historic introduction to the field of deeply inelastic scattering, its contributions to the present understanding of the Standard Model are discussed. The experimental and theoretical achievements reached are outlined in detail both for unpolarized and polarized deep-inelastic scattering. Finally, the open questions of the field are addressed concerning future important measurements and the solution of a series of theoretical problems. Possible new measurements at present and upcoming facilities are discussed. There is preference for *high luminosity* facilities in optimal ranges of Q^2 and x to improve the current experimental accuracies considerably and to allow for the measurement of a series of observables, which cannot be accessed otherwise. Precision results from deeply inelastic scattering will play a key role in understanding the physics at LHC. Details of the contribution are given in Ref. [1].

REFERENCES

1. J. Blümlein, Proceedings of the *First Workshop on Quark-Hadron Duality and the Transition to pQCD*, Frascati, Italy, June 2005.

CP792, *Deep Inelastic Scattering, DIS 2005*, edited by W. H. Smith and S. R. Dasu

eRHIC: The Electron Ion Collider at BNL and Its Spin Physics Program

Abhay Deshpande

Department of Physics & Astronomy, SUNY at Stony Brook, NY 11794-3800 &
RIKEN BNL Research Center, Brookhaven National Laboratory, Upton NY 11973-5000

Abstract. We motivate the need for a future high luminosity, high energy polarized collider capable of deep inelastic scattering of polarized nucleons and nuclei. We propose that building a state of the art electron beam facility near the Relativistic Heavy Ion Collider (RHIC) complex to use one of its beams for such a facility would be the most cost effective way to achieve the physics goals. After a brief introduction to eRHIC project at BNL we present the spin physics program at eRHIC. The unpolarized e-A physics program and technical status of the collider design and detector ideas are discussed in other talks.

Keywords: eRHIC, nucleon spin, RHIC, heavy ion, low-x, gluon polarization, weak structure functions, Bjorken sum rule, strong coupling constant, confinement, color glass condensate
PACS: 21.10.Hw, 23.20.En, 24.85.+p, 25.30.-c

A (VERY) BRIEF HISTORY & STATUS OF QCD

Understanding the fundamental structure of matter is one of the central goals of scientific research through the ages. Towards the end of the last century scientists developed the theory of strong interactions, Quantum Chromodynamics (QCD), which explains all strongly interacting matter in terms of point-like spin half particles, called quarks, exchanging gauge bosons called gluons. During the last 30 years, experiments have verified QCD quantitatively in collisions involving exchange of large momenta between participants. Quarks and gluons are always confined, yet at the large momentum exchange regime, at distances scales smaller or comparable to the proton size, they seem to behave as if they are free. The discovery of proton's substructure in the late 1960s at SLAC led to a Nobel prize in 1990s to Friedman, Kendall and Taylor. The phenomena that quarks and gluons are quasi-free at short distances, follows from a fundamental property of QCD known as asymptotic freedom. For identifying and understanding this unique characteristic of QCD, Gross, Politzer and Wilczek received the 2004 Nobel prize in physics.

When the interaction distance between quarks and gluons becomes comparable to or larger than the typical hadronic size, the fundamental constituents of the nucleon are no longer free. They are bound by a strong force that does not allow for the observation of colored objects. Most hadronic matter exists in this regime where the symmetries of the quark-gluon interaction are hidden. The QCD calculations of hadrons in terms of dynamic properties of quarks and gluons is as yet impossible. The

CP792, *Deep Inelastic Scattering, DIS 2005,* edited by W. H. Smith and S. R. Dasu

only hope seems to be to carry out *ab initio* QCD calculations in strong QCD regime using Monte Carlo techniques on large scale computers.

Experimentation in the field of QCD has been the only way to gain insights and new knowledge so far. This has been going on for decades at SLAC, FNAL, CERN, DESY and BNL accelerator facilities. Measurements of structure function including its spin information, verification of many QCD predictions, determination of the strong coupling constant and its scale dependence, all have an experimental foundation.

COMPELLING QUESTIONS IN QCD AND A NEXT GENERATION COLLIDER

QCD has been accepted as the Standard Model of the strong interactions. Despite its great success in describing the strong interactions a detailed understanding of how QCD works at a detailed level is still a mystery. We broadly divide the compelling and un-understood questions in to the following three:

- What is the gluon distribution in an atomic nucleus?
- How does the spin structure of the nucleon understood in terms of its quark and gluon constituents?
- How does the process of hadronization exactly evolve starting from a soup of quarks and gluons?

Since we do not seem to be able to explain these phenomena quantitatively, it is clear that we still do not understand the subtleties of the theory of QCD. It should be the main motivation for detailed, comprehensive precision tests in experimental QCD.

We propose that a next generation lepton-nucleon(-nucleus) collider facility with high luminosity and capable of polarizing both the lepton and the hadron beam be built to study, test and understand QCD [1]. With precision studies in mind we propose a lepton (electron/positron beam), as QED is as yet the best understood interaction in nature and by judiciously choosing the experimental beam parameters, one can maximize sensitivity to different regions of quark and gluon distributions. To fully explore the kinematic regions dominated by quark(high x) *and* gluons(low x) we propose that this facility have a *variable* center-of-mass (CoM) energy (CoM Energy range from ~30-100 GeV). Up to 30 has been explored in fixed target experiments. Since the QED interaction between the lepton probe and the hadron target is weak, we propose a high luminosity (~10^{33} $cm^{-2}sec^{-1}$). To explore the spin properties of the nucleons protons and neutrons, we will need polarized protons and either deuteron or polarized ^3He). To understand the QCD in nuclei, where the gluon densities are expected to be extremely high perhaps forming a new phase of matter known as color glass condensate, one would need (heavy) nuclear beams as well. A large variation in A (the atomic number) would be highly desirable. And finally, to supplement such a facility with measurements of the complete event characteristics, a comprehensive detector capable of high rates, large acceptance, good particle identification, hadronic as well as electromagnetic calorimetry, efficient precision-tracking in the inner detector. It would be also highly desirable to have high efficiency and acceptance for the very large rapidity regions.

We believe that all reasonable analyses of the physics goals coupled with the present fiscal reality will lead to a proposal as we are making: the full utilization of existing RHIC in this proposal, i.e. to build an electron beam facility next to RHIC to collide with one of the RHIC rings. We call this facility the eRHIC [2].

The recent excitement in the field of understanding of the parton distributions in the nuclei comes from RHIC (d-A) collisions, and will be described as a motivation for the e-A physics at eRHIC [3]. Ideas of the detector presently under consideration are described in [4]. In this article we briefly describe eRHIC and then we go on to describe highlights of the polarized e-p physics program at eRHIC.

E-RHIC AT BNL

Relativistic Heavy IonCollider (RHIC) was commissioned in 2001. Since then it has not only realized every expectation of its design performance, but in astoundingly many instances, exceeded them. It has reached and exceeded its instantaneous luminosity goals for both heavy ion and proton collisions. It has exceeded its design goals for polarization at this stage of the program. Every year these performance parameters have been achieved in shorter and shorter time since machine turn on. We hence propose that a 10 GeV polarized electron/positron beam facility with high current be added to the RHIC complex. It will allow in addition to the presently possible polarized proton-proton and heavy ion collisions, polarized e-p and e-A collisions. This will make RHIC a unique QCD laboratory in the next decade.

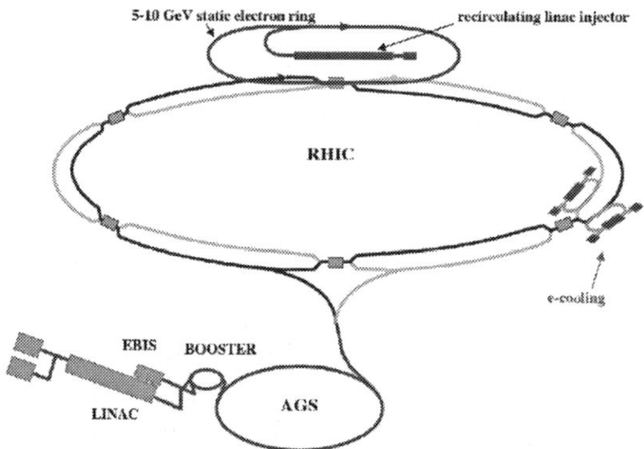

FIGURE 1. The layout of the eRHIC, the ring-ring design as proposed in the Zero-th Design Report (eRHIC-ZDR). A 10 GeV LINAC will inject in to an electron ring to be built next to the RHIC, and will have one interaction region. The linac injection energy range will be from 5 GeV/c to 10 GeV/c, with most optimal polarizaton conditions between 7-10 GeV/c electron/positron beam energy.

Figure 1 shows one of the two design layouts of the eRHIC as proposed in the Zero-th Design Report (ZDR) submitted to the BNL management in April 2004 [5]. Other options including a 10 GeV Energy Recovery Linac (ERL) colliding with one of the RHIC rings. A possible layout of the LINAC design which enable more than one IR has also been described in the appendix of the ZDR [5].

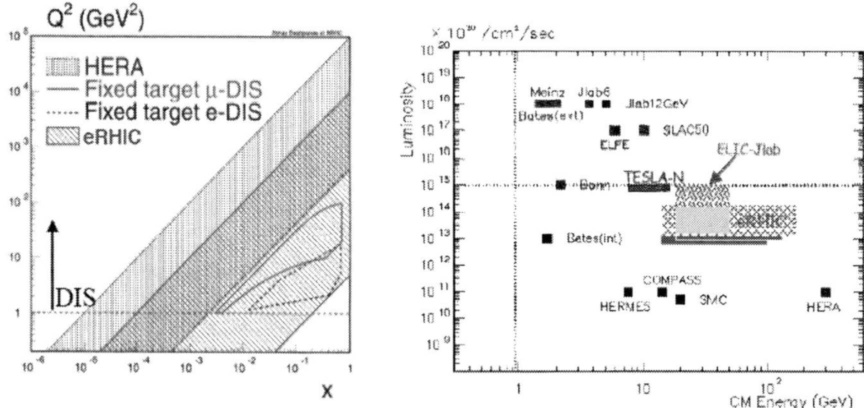

FIGURE 2. (a) The x-Q^2 range in the kinematic region that will be explored by the eRHIC (region between the two magenta lines) with 50-250 GeV/c variable proton beam energy and 5-10 GeV/c electron beam energy. Both beams will be polarized. (b) The center of mass vs. luminosity that eRHIC facility is planning to explore (shown in red). Notice the large range in the CoM possible, uniquely at eRHIC as against other DIS fixed target and collider facilities in the world. The Tesla-N and ELIC at Jlab are other future facilities envisioned.

POLARIZED E-P SCATTERING AT ERHIC

Polarized deep inelastic scattering has only been performed so far in fixed target experiments at SLAC, CERN and DESY [2] and were limited to a CoM energy of ~27 GeV. Figure 2 shows the regions explored by the fixed target facilities. A large x-Q^2 region is being explored by HERA at DESY, but with unpolarized proton beams. eRHIC will enable ~100 GeV in CoM with 250 GeV/c polarized proton beams of RHIC and ~10 GeV/c electron ring facility we propose to build. This will allow low x measurements of the spin structure functions up to 10^{-4} in Bjorken-x in a region of $Q^2 > 1$ GeV^2 (See Figure 2a). In terms of luminosity the design we hope to achieve is 100 times more than the HERA, and the other fixed target facilities (SMC and COMPASS at CERN, and HERMES at DESY). Other possible future facilities being discussed around the world are the TESLA-N (fixed target facility with the TESLA electron beam on a polarized target) and ELIC (Electron Light Ion Collider) being contemplated at Jefferson Laboratory [6].

While it is impossible to give a complete description and significance of the polarized DIS measurements that would be possible at eRHIC, we give a list of highlights and comment on some of them later. It is assumed that a new detector with at least the functionality of one of the HERA Detectors (H1 or ZEUS) will be built [4]. Other detector ideas for a dedicated forward physics measurement are also being discussed [7]. A selection of measurements we could make at eRHIC are [1,2] below:

- Measurement of the spin structure function of the proton and neutron at lower Bj-x than any other present or future facility (low x for $Q^2 > 1$ would be around 10^{-4}.

    - The neutron spin structure function will be measured using polarized ^{3}He beams. Measurement of proton and neutron to low x will allow the test of one of the most fundamental relations of QCD: the Bjorken Spin Sum Rule.
    - The extension of measurement of the Bjorken integral to low x will further allow the evaluation of the strong coupling constant α_S
- Measurement of polarized gluon distribution over the largest range of x: using two systematically different methods:
    - The scaling violation of g1 spin structure function
    - Using asymmetry produced in 2-Jet events produced in photon-gluon fusion in e-p scattering
- Measurement of the spin structure of the "resolved" virtual photon in terms of its quark/gluon components
- The parity violating spin structure function g_5 which will allow us to directly access the spin contributions from heavy quarks and anti-quarks through virtual W exchange
- Measurement of exclusive processes such as Deeply Virtual Compton Scattering (DVCS) will allow first look in to the low x behavior of the generalized parton distributions (GPD). GPDs hold a promise for future measurement of the orbital angulr momentum in the nucleon.
- Transversity and other transverse spin effects which have recently been discovered at fixed target ep and pp experiments and their connection to the orbital angular momentum

The list of such measurements is long and exciting. Special mention should be made of the diffractive measurements in the spin DIS which has not been explored to any significant level so far, however, recent surprises in the unpolarized sector begs us to explore this as yet unexplored region.

SUMMARY & OUTLOOK

In summary, we propose that building eRHIC at BNL in not too a distant future is the most cost effective way to study and understand QCD to its fullest extent. In addition the DIS technique, RHIC is a facility in which hadron-hadron scattering between nucleons and nuclei is on going. Adding a possibility to do DIS only enhances this facility. We believe this will allow us to have a very unique QCD facility in the world.

The eRHIC collaboration is forming. Detector simulation projects are starting up [4]. Physics interests are being explored. The basic design of the accelerator complex and the interaction region now exist in form of a ZDR. This work will continue in the next few years and the collaboration is expected to seek the blessing of the Nuclear Science Advisory Committee (NSAC) in their next Long Range Planning (LRP) exercise. This is expected to happen some time around 2007.

ACKNOWLEDGMENTS

The author thanks the organizers of this meeting, on behalf of eRHIC collaboration, for allowing the eRHIC collider presentations in the plenary session in three separate but complementary talks. This work was supported by RBRC.

REFERENCES

1. A White-Paper submitted to long range planning meeting of the NSAC, *The Electron Ion Collider*, BNL-**68933-02/07-REV**
2. A. Deshpande et al., *Study of the Fundamental Structure of Matter with An Electron Ion Collider,* to be published, **Volume 55**, Ann. Rev. of Part. & Nucl. Sci. December 2005, hep-ph/0506148
3. J. Jalilian-Marion, *d-A Physics at RHIC a motivation for the e-A collider at BNL*, these proceedings
4. B. Surrow, *eRHIC Accelerator & Detector Design*, these proceedings
5. L. Ahrens et al, eRHIC Zeroth Design Report, Edited by V. Ptitsyn & M. Farkhondeh; **http://www.bnl.gov/eic**
6. C. Keppel, *The Jlab upgrade plans,* these proceedings
7. I. Abt et al., *A Detector for Forward Physics at eRHIC: Feasibility Study*, hep-ph0407053

Heavy Ion Physics in eRHIC

Jamal Jalilian-Marian

Institute for Nuclear Theory, University of Washington, Box 351550, Seattle, WA, USA

Abstract. We review the physics of gluon saturation in heavy ions at small x and consider the applications of Color Glass Condensate formalism to Deep Inelastic Scattering (DIS) of leptons on nuclei and discuss the overlapping physics between high energy heavy ion collisions at RHIC and DIS in eRHIC.

Keywords: Deep Inelastic Scattering, structure functions, small x, Color Glass Condensate
PACS: 24.85.+p, 25.75.-q

DEEP INELASTIC SCATTERING

Fundamental constituents of hadron and nuclei can be studied in Deep Inelastic Scattering (DIS) lepton-hadron (nucleus) experiments where one the virtual photon radiated by the lepton probes the quark and gluon content of the target hadron or nucleus. The cross section for the lepton-hadron scattering can be written as a product of the leptonic tensor $L_{\mu\nu}$, which describes the electromagnetic part of the process, and the hadronic tensor $W^{\mu\nu}$ which contains all the information about the target hadron, written as

$$\frac{d\sigma}{dx\,dQ^2} \sim L_{\mu\nu} W^{\mu\nu} \tag{1}$$

where $Q^2 \equiv -q^2, x \equiv \frac{Q^2}{2q\cdot P}$ and k, q, P are the momenta of the incoming lepton, exchanged virtual photon and target hadron respectively. The leptonic tensor is well known and rather trivial. The hadronic tensor can be written in terms of two structure functions (for unpolarized processes and ignoring parity violating contributions) F_1 and F_2 as

$$\frac{d\sigma}{dx\,dQ^2} = \frac{2\pi\,\alpha_{em}^2}{Q^4}\left\{\left[1+(1-y)^2\right] F_2(x,Q^2) - y^2 F_L(x,Q^2)\right\} \tag{2}$$

where $y \equiv \frac{q\cdot P}{k\cdot P}$ and the longitudinal structure function F_L is defined as $F_L \equiv F_2 - 2xF_1$. The structure function F_2 is typically written in terms of quark and anti-quark distribution functions $q(x,Q^2), \bar{q}(x,Q^2)$ as

$$F_2(x,Q^2) = \sum_f e_f^2 \left[x q_f(x,Q^2) + x\bar{q}_f(x,Q^2)\right] \tag{3}$$

where the sum runs over all flavors. The quark and anti-quark distributions above then can then be used in a collinearly factorized cross section in order to, for example, calculate single inclusive particle production in hadron-hadron collisions via

$$E\frac{d\sigma^{h_1 h_2\to hX}}{d^3p} \sim f_{h1}(x,Q^2)\otimes f_{h2}(x,Q^2)\otimes\frac{d\sigma}{dt}\otimes D_{f/h}(x,Q^2) \tag{4}$$

CP792, *Deep Inelastic Scattering, DIS 2005*, edited by W. H. Smith and S. R. Dasu

The usefulness of a relation like (4) lies in the fact that the non-perturbative ingredients of this relation, such as the parton distribution functions $f_{h1,h2}$ and $D_{f/h}$ are universal, i.e. process independent and can be measured in other processes, such as in DIS while the hard scattering cross section is process dependent, but can be calculated perturbatively.

It is important to note that the relation (4) is not exact and receives corrections which are suppressed by the hard scale in the problem, Q^2. This may be taken to mean that these corrections are small and can be neglected if one is at sufficiently high momentum. However, these corrections may be enhanced by effects which are energy or A (nucleon number) dependent which could make them quite large for sufficiently high energies and/or large nuclei. One such effect is the large longitudinal phase space which becomes available when the center of mass energy squared s is much larger than the hard scale Q^2 such that $\alpha_s \ln 1/x \sim \alpha_s \ln(s/Q^2)$ is large. This would lead to breakdown of (4) since it would necessitate re-summation of the $\alpha_s \ln 1/x$ terms which can not be readily included in (4). Furthermore, it is known that the parton distributions defined in (3) are modified in nuclei. This modification of nuclear parton distributions is quite non-trivial and would generically signify the inappropriateness of a relation like (4) for calculating particle production in high energy heavy ion collisions.

DIS at small x

At Small x and in the rest frame of the target, virtual photon-hadron (nucleus) interaction can be separated into two independent parts, first the virtual photon fluctuates into a quark anti-quark pair and in the second stage, this pair multiply interacts with the target hadron (nucleus). The virtual photon-target cross section can be written as

$$\sigma_{T,L}^{\gamma^* p} = \int d^2b_t\, d^2r_t \int dz\, |\psi_{T,L}(z,r_t,Q^2)|^2\, \sigma_{q\bar{q}N}(x,r_t,b_t) \tag{5}$$

where $|\psi|^2$ is the probability of a virtual photon splitting into a quark anti-quark pair, with the quark carrying z and the anti-quark carrying $(1-z)$ fraction of the virtual photon energy. The quark anti-quark target cross section $\sigma_{q\bar{q}N}$includes all the multiple scatterings of the pair on the target as well as the x evolution of the target wave function and is given by

$$\sigma(x,r_t,b_t) \equiv \frac{1}{N_c} \mathrm{Tr}_c \left\langle 1 - V(b_t + \frac{r_t}{2}) V^\dagger (b_t - \frac{r_t}{2}) \right\rangle_\rho . \tag{6}$$

and V is a matrix in the fundamental representation of $SU(N_c)$ which goes along the light cone and sums the non-Abelian phases of the quark propagating through the color field of the nucleus. The cross section σ satisfies the non-linear evolution equations known as the JIMWLK equation. In the large N_c limit, the JIMWLK equations simplify and can be written in a closed form, known as the BK equation which for the dipole cross section, takes the form

$$\frac{dN_F}{d\ln 1/x} \sim \alpha_s\, \kappa_{BFKL} \left\{ N_F - N_F^2 \right\} \tag{7}$$

where $N_F \sim \int d^2b_t\, \sigma(x, r_t, b_t)$. The Wilson line V (and its adjoint counterpart U) are the universal ingredients on which any cross section involving hadrons or nuclei in high energy depend. For example, in single and double particle production cross section in DIS and proton-nucleus collisions. The difference between a hadron and a nucleus target in this formalism is in the initial condition of the JIMWLK equations which the dipole cross section satisfies.

PROTON-NUCLEUS COLLISIONS

In high energy proton-nucleus collisions can be formulated as a collision of a dilute system of partons, quarks and gluons in a proton, with the classical field of a nucleus if the center of mass energy is high enough and if we are not too far in rapidity from the proton. In case of super high energies and rapidities not too close to the proton, one may have to treat the proton as a classical field as well. However, this is probably not the case for the current and proposed collider experiments such as RHIC and LHC and at not too low transverse momenta.

Single inclusive hadron production cross section in proton-nucleus collisions can be written as

$$\frac{d\sigma^{pA\to hX}}{dY\, d^2P_t\, d^2b} = \frac{1}{(2\pi)^2}\int_{x_F}^{1} dx\, \frac{x}{x_F}\left\{ f_{q/p}(x,Q^2)\, N_F[\frac{x}{x_F}P_t, b]\, D_{h/q}(\frac{x_F}{x},Q^2) + f_{g/p}(x,Q^2)\, N_A[\frac{x}{x_F}P_t, b]\, D_{g/h}(\frac{x_F}{x},Q^2)\right\} \quad (8)$$

where Y and P_t are the rapidity and transverse momentum of the produced hadron and x_F is its Feynman-x while f and D are the proton parton distribution and parton-hadron fragmentation functions. Again, the cross section is written in terms of the (fundamental and adjoint) dipole cross sections. This makes it possible to use DIS on nuclear targets in order to extract or constrain the dipole cross section and use it in proton-nucleus collisions. This demonstrates how the Color Glass Condensate restores the universality of the ingredients of QCD cross sections in a dense environment, which is lost in a collinear factorization approach such as (4). Eq. (8) can be used to describe the recent RHIC data on charged particle production in the forward rapidity region where a good agreement with the charged hadron transverse momentum spectrum is found.

One can also investigate two particle production and the azimuthal correlation between the two produced particles in a proton-nucleus collision. In case of production of two hadrons, the cross section involves higher point functions of Wilson lines, such as three and four point functions. These expression are highly complicated and difficult to evaluate analytically even though one can investigate some interesting limits in transverse momentum and rapidity. Alternatively, one can use approximates solutions in different regions of interest and study the behavior of the correlations semi-quantitatively such as in where it has been shown that azimuthal correlations tend to get washed away due to x evolution, given there is a sufficiently large rapidity separation between the produced hadrons or jets.

Hadronization by its nature is non-perturbative and not well understood in QCD. One side-steps the question of the dynamics of hadronization by the use of fragmentation functions in perturbative QCD. However, there are theoretical uncertainties as well as experimental difficulties and limitation on the kinematics where these fragmentation functions can be measured. One can avoid these issues by investigating electromagnetic processes such as photon and dilepton production in proton nucleus collisions. Here, show the dilepton production cross section in proton nucleus collision using the Color Glass Condensate formalism. Again, the basic ingredients of the cross section are the Wilson lines which make it possible to use our knowledge from DIS. The cross section is given by

$$\frac{d\sigma^{qA\to ql^+l^-X}}{dz\,d^2b_t\,d^2k_t\,d\log M^2} = -\frac{2\alpha_{\rm em}^2}{3\pi}\int\frac{d^2l_t}{(2\pi)^4}\sigma(x,l_t,b_t,)$$
$$\left\{\left[\frac{1+(1-z)^2}{z}\right]\frac{z^2\,l_t^2}{[k_t^2+M^2(1-z)][(k_t-zl_t)^2+M^2(1-z)]}\right.$$
$$\left.-z(1-z)\,M^2\left[\frac{1}{[k_t^2+M^2(1-z)]}-\frac{1}{[(k_t-zl_t)^2+M^2(1-z)]}\right]^2\right\}$$

Nucleus-Nucleus Collisions

The knowledge gained from DIS and proton nucleus collisions will be invaluable for the pursuit of the so called Quark Gluon Plasma in high energy heavy ion collisions, such as those at RHIC and in the near future at LHC. A detailed knowledge of the nucleus wave function will be extremely useful in helping determine the initial conditions of a high energy heavy ion collisions, which is essential in determining the subsequent evolution of the produced hot and dense system. One can use the Color Glass Condensate formalism in order to determine the initial number of partons produced in the collisions and follow their further interactions after the collision until a possible thermalization or isotropization. One can determine the initial "temperature" and energy density of the produced system and use those as the input for kinetic theory or hydrodynamics type approaches. This avenue is of particular interest to heavy ion community and Deep Inelastic Scattering experiments at eRHIC will be able to explore kinematic regions which will be reached at LHC and where there is practically no information available on the structure of target nuclei.

In summary, there are strong indications from RHIC that the Color Glass Condensate has been observed in the forward rapidity region. The properties of the Color Glass Condensate can be precisely measured in eRHIC in a kinematic region desperately needed for the LHC experiment.

REFERENCES

1. For a review of Color Glass Condensate formalism and an extensive list of references, see J. Jalilian-Marian and Y. V. Kovchegov, arXiv:hep-ph/0505052.

eRHIC - Accelerator and Detector Design

B. Surrow

Massachusetts Institute of Technology, 77 Massachusetts Avenue, Cambridge, MA 02139

Abstract. An electron-proton/ion collider facility (eRHIC) is under consideration at Brookhaven National Laboratory (BNL). Such a new facility will require the design and construction of a new optimized detector profiting from the experience gained from the H1 and ZEUS detectors operated at the HERA collider at DESY. The details of the design will be closely coupled to the design of the interaction region, and thus to the machine development work in general. An overview of the accelerator and detector design concepts will be provided.

Keywords: Document processing, Class file writing, $\LaTeX\,2\varepsilon$
PACS: 43.35.Ei, 78.60.Mq

ERHIC ACCELERATOR DESIGN

The high energy, high intensity polarized electron/positron beam ($5-10\,\mathrm{GeV}/10\,\mathrm{GeV}$) facility (eRHIC) which is under consideration at Brookhaven National Laboratory will collide with the existing RHIC heavy ion (100 GeV per nucleon) and polarized proton beam ($50-250\,\mathrm{GeV}$). This facility will allow to significantly enhance the exploration of fundamental aspects of Quantum Chromodynamics (QCD), the underlying quantum field theory of strong interactions [1, 2]. A detailed report on the accelerator and interaction region (IR) design of this new collider facility has been completed based on studies performed jointly by BNL and MIT-Bates in collaboration with BINP and DESY [3]. The main design option is based on the construction of a 10 GeV electron/positron storage ring intersecting with one of the RHIC hadron beams. The electron beam energy will be variable down to 5 GeV with minimal loss in luminosity and polarization. The electron injector system will consist of linacs and recirculators fed by a polarized electron source. A study has shown that an ep luminosity of $4\times10^{32}\mathrm{cm}^{-2}\mathrm{s}^{-1}$ can be achieved for the high-energy mode (10 GeV on 250 GeV), if the electron beam facility is designed using today's state-of-the-art accelerator technology without an extensive R&D program. For electron-gold ion collisions (10 GeV on $100\,\mathrm{GeV}/u$), the same design results in a luminosity of $4\times10^{30}\mathrm{cm}^{-2}\mathrm{s}^{-1}$. The potential to go to higher luminosities at the level of $10\times10^{32}\mathrm{cm}^{-2}\mathrm{s}^{-1}$ (high-energy ep mode) by increasing the electron beam intensity will be explored in the future. A polarized positron beam of 10 GeV energy and high intensity will also be possible using the the process of self-polarization. A possible alternative design for eRHIC has been presented on the basis of an energy recovery superconducting linac (ERL) [4]. This option would be restricted to electrons only. Preliminary estimates suggest that this design option could produce higher luminosities at the level of $10^{34}\mathrm{cm}^{-2}\mathrm{s}^{-1}$ (high-energy ep mode). Significant R&D efforts for the polarized electron source and for the energy recovery technology is required.

CP792, *Deep Inelastic Scattering, DIS 2005,* edited by W. H. Smith and S. R. Dasu

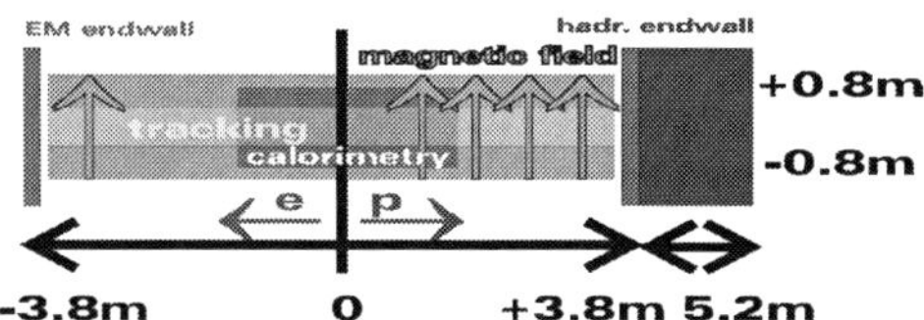

FIGURE 1. Conceptual detector layout focusing on forward physics with a 7 m long dipole field and an interaction region without machine elements extending from -3.8 m to $+5.8$ m.

ERHIC DETECTOR DESIGN

The following minimal requirements on a future eRHIC detector can be made:

- Measure precisely the energy and angle of the scattered electron (Kinematics of DIS reaction)
- Measure hadronic final state (Kinematics of DIS reaction, jet studies, flavor tagging, fragmentation studies, particle ID)
- Missing transvere energy measurement (Events involving neutrinos in the final state, electro-weak physics)

In addition to those demands on a central detector, the following forward and rear detector systems are crucial:

- Zero-degree photon detector to control radiative corrections and measure Bremsstrahlung photons for luminosity measurements
- Tag electrons under small angles (Study of the non-perturbative/perturbative QCD transition region and luminosity measurements from Bremsstrahlung ep events)
- Tagging of forward particles (Diffraction and nuclear fragments)

Optimizing all the above requirements is a challenging task. Two detector concepts have been considered so far. One, which focuses on the forward acceptance and thus on low-x/high-x physics which emerges out of HERA-III detector studies [5]. This detector concept is based on a compact system of tracking and central electromagnetic calorimetry inside a magnetic dipole field and calorimetric end-walls outside. Forward produced charged particles are bent into the detector volume which extends the rapidity coverage compared to existing detectors. A side view of the detector arrangment is shown in Figure 1. The required machine element-free region amounts to roughly $\pm$5m. This clearly limits the achieveable luminosity in a ring-ring configuration.

The second design effort focuses on a wide acceptance detector system similar to the current HERA collider experiments H1 and ZEUS to allow for the maximum possible Q^2 range. The physics program demands high luminosity and thus focusing machine elements in a ring-ring configuration have to be as close as possible to the IR while preserving good central detector acceptance. This will be discussed in more detail in the next section. A simulation and reconstruction package called ELECTRA has been developed to design a new eRHIC detector at BNL [6, 7]. Figure 2 shows a side view of a GEANT detector implementation of the above requirements on a central detector.

The hermetic inner and outer tracking system including the electromagnetic section of the barrel calorimeter is surrounded by an axial magnetic field. The forward calorimeter is subidvided into hadronic and electromagnetic sections based on a conventional lead-scintillator type. The rear and barrel electromagnetic consists of segmented towers, e.g. a tungsten-silicon type. This would allow a fairly compact configuration. Other options based on a crystal rear and barrel electromagnetic calorimeter are under study. The inner most double functioning dipole and quadrupole magnets are located at a distance of $\pm$3m from the IR. An initial IR design assumed those inner most machine elements at $\pm$1m. This would significantly impact the detector acceptance. More details on the IR design can be found in [3].

The bunch crossing frequency amounts to roughly 30 MHz. This sets stringent requirements on the high-rate capabililty of the tracking system. This makes a silicon-type detector for the inner tracking system (forward and rear silicon disks together with several silicon barrel layers) together with several GEM-type outer tracking layers a potential choice. The forward and rear detector systems have not been considered so far. The design and location of those detector systems has to be worked out in close collaboration to accelerator physicists since machine magnets will be potentially employed as sepectrometer magnetes and thus determine the actual detector acceptance and ultimately the final location. It is understood that demands on optimizing the rear/forward detector acceptance might have consequences on the machine layout and is therefore an iterative process.

CONSIDERATIONS ON THE ACCELERATOR/DETECTOR INTERFACE

The following section provides an overview of some aspects of the detector/machine interface. The specification of those items has only recently been started.

The direct synchrotron radiation has to pass through the entire IR before hitting a rear absorber system. This requires that the geometry of the beam pipe is designed appropriately with changing shape along the longitudinal beam direction which includes besides a simulation of the mechanical stress also the simulation of a cooling system of the inner beam pipe. The beam pipe design has to include in addidition the requirement to maximize the detector acceptance in the rear and forward direction. Furthermore the amount of dead material has to be minimized in particular to limit multiple scattering (track reconstruction) and energy loss for particles under shallow angles (energy reconstruction). The distribution of backscattered synchrotron radiation into the acutal detector volume has to be carefully evaluated. An installation of a collimator system has to be worked out. Those items have been started in close contact to previous experience at HERA [8].

The demand of a high luminosity ep/eA collider facility requires the installaton of focusing machine elements as close as possible to the central detector. An IR design with machine elements as close $\pm$1 m to the IR which has been presented in [3] would significantly limit the achievable detector acceptance. A new scheme has been presented in [9] which provides a machine-element free region of $\pm$3 m at the expense of approximatley half the luminosity for the IR design presented in [3]. A linac-ring option would not be limited by beam-beam effects compared to a ring-ring configuration. Even larger lumi-

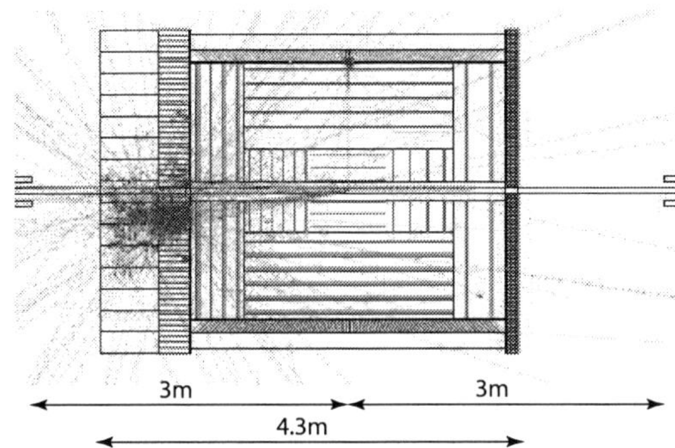

FIGURE 2. Side view of the GEANT dector implementation as part of the ELECTRA simulation and reconstruction package [6]. A deep-inelastic scattering event resulting from a LEPTO simulation is overlayed with $Q^2 = 361\,\text{GeV}^2$ and $x = 0.45$.

nosities could be achieved with a machine-element free region of approximatley $\pm 5\,\text{m}$. This scheme has been presented in [4].

The need for acceptance of scattered electrons beyond the central detector acceptance is driven by the need for luminosity measurements through ep/eA Bremstrahlung and photo-production physics. Besides that a calorimeter setup to tag radiated photons from inital-state radiation and Bremsstrahlung will be necessary. The scattered electrons will pass through the machine elements and leave the beam pipe through special exist windows. The simulation of various small-angle calorimeter setups has been started. This will require a close collaboration with the eRHIC machine design efforts to aim for an optimal detector setup.

The forward tagging system beyond the central detector will play a crucial role in diffractive ep/eA physics. The design of a forward tagger system based on forward calorimetry and Roman pot stations is foreseen. Charged particles will be deflected by forward machine elements. This effort will require as well a close collaboration with the eRHIC machine design efforts to ensure the best possible forward detector acceptance.

REFERENCES

1. A. Deshpande, these proceedings.
2. J. Jalilian-Marian, these proceedings.
3. M. Farkhondeh et al., 'eRHIC Zeroth-Order Design Report', BNL internal note, http://www.agsrhichome.bnl.gov/eRHIC/.
4. V. Litvinenko, Published contribution at PAC05, May 2005, Knoxville.
5. I. Abt, A. Caldwell, X. Liu and J. Sutiak, hep-ex/0407053.
6. A. Deshpande, J. Pasukonis and B. Surrow, Published contribution at PAC05, May 2005, Knoxville.
7. J. Pasukonis and B. Surrow, http://starmac.lns.mit.edu/~erhic/electra/.
8. J. Beebe-Wang et al., Published contribution at PAC05, May 2005, Knoxville.
9. C. Montag et al., Published contribution at PAC05, May 2005, Knoxville.

Prospects of DIS with fixed targets

Ewa Rondio

A. Soltan Institute for Nuclear Studies
00-681 Warsaw, Poland, Hoza 69
ewar@fuw.edu.pl

Abstract. Some prospects for important measurements in Deep Inelastic Scattering on the fixed targets are discussed. The subjects selected are parity violating asymmetries and deeply virtual Compton scattering. The last subject is related to the determination of generalized parton distribution. Possibilities for such measurements in the Compass experiment at CERN with high intensity proton beam are presented.

Keywords: DIS, parity violating asymmetries, deeply virtual Compton scattering
PACS: 13.88.+e, 13.60.Le

INTRODUCTION

There is no need to explain how important was the lepton hadron scattering for studies of nucleon structure. Some subjects in this field are still waiting for more precise or additional data. Presently the scene of such experiment is shrinking drastically. Some of the effort is going into collider experiments which can extend kinematic coverage, but some places are simply closed. Presently on the map we have CERN with Compass experiment, DESY with Hermes and in the US: Fermilab, SLAC and Jefferson Laboratory. The present and future plans of Jefferson Laboratory will be discussed in more details in separate presentation.

For a long time the main goal for DIS experiments was determination of the nucleon structure and in particular parton distribution functions. This was the main subject for first generation experiments and now the precision is given by extended range provided by collider measurements (HERA). For spin dependent distributions the fixed target data are still the only one and plans for step toward collider were presented at this conference. Subject still requiring precision data is a test of assumed symmetries like $u_p = d_n$ or $s = \bar{s}$ and measurements of the behavior at high x. The new data are needed for different approach to the nucleon structure - determination of generalized parton distributions (GPD) and here place for fixed target seems to exist again, as the statistics required is very high.

PARITY VIOLATING ASYMMETRIES

After the discovery of parity violation in beta decay, Zel'dovich predicted an analogous parity violating neutral current interaction [1]. The asymmetry should manifest in the measurements of cross section for leptons with spin parallel or anti-parallel to the momentum:

CP792, *Deep Inelastic Scattering, DIS 2005,* edited by W. H. Smith and S. R. Dasu

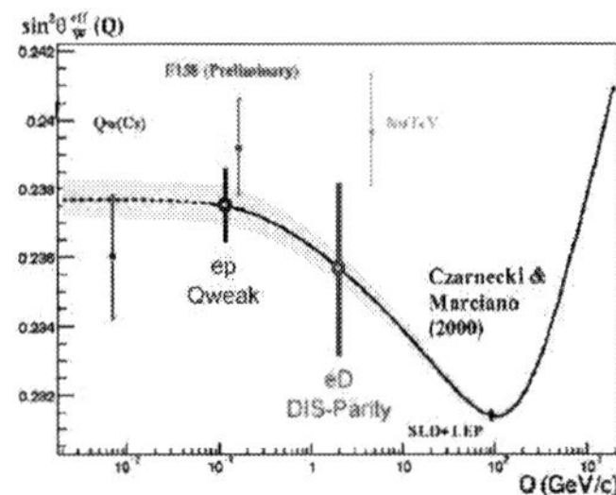

FIGURE 1. The running of $sin^2\theta_W$ in the $\bar{MS}$ scheme, showing measurement from atomic parity violation on Cs (APV Cs) [2], NuTeV result from ν DIS [3], combined measurements at Z^0 pole. Precision of proposed measurements by Q-Weak and SLAC E-158 Moller scattering as well as DIS parity violating asymmetries on deuteron target possible at SLAC or J-Lab after 12 GeV upgrade (DIS-Parity). the picture originates from [4].

$$A_{PV} = \frac{\sigma_R - \sigma_L}{\sigma_R + \sigma_L} \approx \frac{|A_Z|}{|A_\gamma|} \approx \frac{GQ^2}{4\sqrt{2}\pi\alpha} \tag{1}$$

and appears due to the interference between the weak and electromagnetic amplitudes. Present experiments allow measurements of asymmetries which are at the level of parts per million expected is such processes. The measurements of such asymmetries will allow verification of important questions. Within the Standard Model, the asymmetry is sensitive to the parameter $sin^2\theta_W$. Extensions of the Standard Model may modify the couplings, changing the apparent value of $sin^2\theta_W$ extracted from asymmetry measurement. An electron axial-vector coupling and vector coupling of target particle allow probing of nucleon structure. In the case when vector coupling of the target is very well known, or the ratio of vector to electromagnetic amplitude does not depend on the hadron structure such measurement is probing new physics.

Example of a problem still waiting for explanation is the so called NuTeV anomaly. In fig.1 presents measurements of $sin^2\theta_W$ at different scales (Q^2), where the measurement done with neutrino and anti neutrino beams shows significant deviation from the expectations [6]. Experimentally the precision is very well controlled and seems to be no reason to doubt the result. The only possible explanation is that one of the assumptions made to derive $sin^2\theta_W$ from the cross section ratio is not fulfilled. These assumptions are: isospin invariance (u quark in the proton is as d quark in the neutron), sea momentum symmetry for c and s quarks and nucleon effects common for W and Z boson exchange. The measurements of A_{PV} in DIS scattering at high x done on isoscalar and proton targets can bring enough information to verify the isospin invariance.

Additionally such measurements will constrain parton distribution function at high x and nucleon wave functions. Behavior of ratio u/d quarks for $x \to 1$ is predicted quite differently by SU(6) symmetric, SU(6) breaking models and by QCD [5]. A lot of such measurements are planed at Jefferson Laboratory with presently available beams and after 12 GeV upgrade. Measurements at higher energies could be done at SLAC, but there are no plans for such activity.

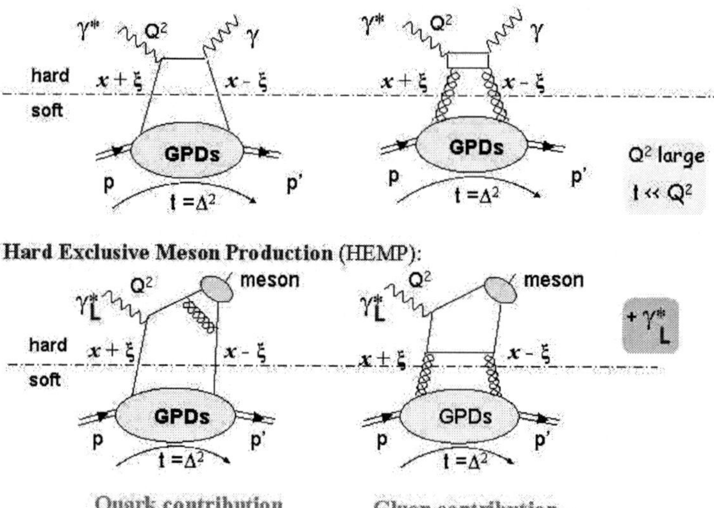

FIGURE 2. Handbag diagrams for the Compton scattering amplitudes (upper diagrams) and meson production amplitudes (lower diagrams). Diagrams with quark contributions are shown on the left and with gluons on the right.

THREE DIMENSIONAL PICTURE OF THE PARTONIC NUCLEON STRUCTURE

The interpretation of inclusive deep inelastic scattering results in terms of parton distributions allow precise description of the nucleon structure in terms of parton distribution functions (PDF). The PDF give one dimensional picture in which we know partons momenta. More precise information can be obtained from exclusive reactions, from which one can obtained three dimensional picture. The qualitatively new information from measurements of exclusive processes and their dependence on the momentum transfer can be translated to the information on distributions perpendicular to the direction of motion, giving three dimensional picture of the nucleon. In such approach Generalized Parton Distributions [9] are introduced.

There are different possibility to get information about such objects. One of them, which at present looks most promising, is Deeply Virtual Compton Scattering (DVCS). The connection between DIS and DVCS comes from optical theorem.

The process would be virtual photon scattering on the nucleon with real photon and nucleon in the final state. The inclusive meson production can be related to GPD's by the quark-parton duality. The handbag diagrams for these processes are presented schematically in fig.2

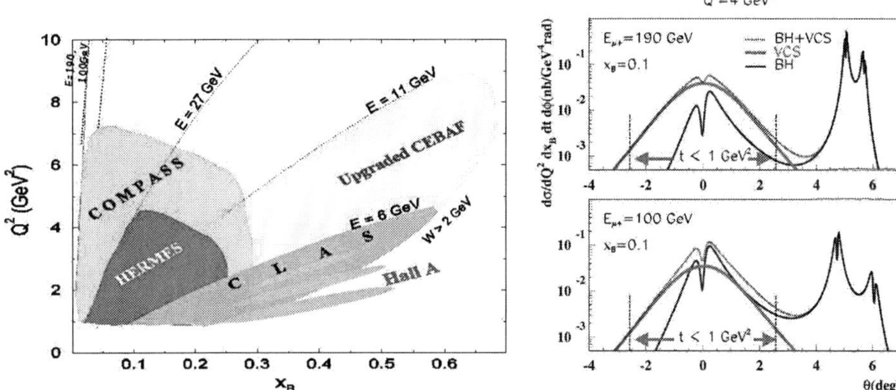

FIGURE 3. Presentation of the DVCS measurements: left figure - kinematic regions covered by present and planned experiments, right - for Compass experiment the expected contributions from Bethe-Heitler process and from genuine DVCS process, which produce identical final states. Upper figure is for incoming muon energy of 100 GeV and lower for 200 GeV.

Several theoretical publications [7],[8],[10] have shown that DVCS scattering can be factorized into hard-scattering part and non-perturbative nucleon part. The factorization is valid when Δ (see fig.2) is small compared to Q^2. In the diagrams shown in fig.2 the lower blobs represent the soft structure of the nucleon in terms of GPD's, conserving quark helicity): $H, \tilde{H}, E$ and $\tilde{E}$. Second moments of PDG's (with momentum transfer going to zero) allow the determination of angular momentum (spin+orbital) carried by quarks from the Ji sum rule [7]

$$2J_q = \int x|H+E|(x,\xi,0)dx \tag{2}$$

Several experiments are measuring DVCS processes presently, first results were already published from Hera, Hermess and J-Lab [11]. The complementarity of these results and the measurement planned in the Compass experiment at CERN is shown in fig.3.

The Compass spectrometer has to be equipped with recoil detector. Assuming 6 month of data taking with presently available muon intensity and with 2.5m liquid hydrogen target, the Q^2 is limited by 7.5 GeV^2. The increase of luminosity by factor two will increase the Q^2 coverage up to 11 GeV^2 and by factor 4 to 17 GeV^2. Such increase can be possible if the proton intensity upgrade plans at the CERN accelerator complex become reality. The upgrade will require modifications of injection to the PS, design and construction of new linac and possibly superconducting proton linac (SPL). The detailed discussion about this solution can be found in [13].

The measurement of DVCS with high energy muon beam gives a possibility to control the relative contributions of this process and Bethe-Heitler scattering. For energy about 200 GeV the DVCS cross section is dominant, while at 100 GeV both processes contribute with compatible strength and the interference term will provide DVCS amplitude. This allows access to real and imaginary parts of the amplitudes. From the sum and difference of the cross sections measured with positive and negative muon beams the DVCS and BH contributions can be separated. For more details and estimation of the sensitivity for such measurement see [12].

SUMMARY

In summary we can say that there are interesting measurements waiting for the fixed target experiments of DIS type. Some of the most interesting possibilities were presented in this talk. One of them, the detailed studies of the parity violating effects, gives possibility of unique tests of the standard model and searches for physics at the TeV scale. Measurements of this type will be performed at Jefferson Laboratory, but data at higher beam energies will also be extremely useful. The other measurements discussed here, determination of Generalized Parton Distributions in DVCS processes, can give access to three dimensional picture of the nucleon interior. For this kind of studies wide kinematic coverage is very important. Presently realized measurements at DESY and JLab can be complemented by future measurements in Compass experiment at CERN.

ACKNOWLEDGMENTS

The author would like to thank N.d'Hose for help in preparation of the part of this talk related to DVCS studies and K.Kurek for very useful discussions. It is a pleasure to thank the organizers for the stimulating meeting and providing very good opportunity for useful discussions. This work was supported by KPN grant 621/E-78/SPB/CERN/P-03/DWM576/2003-2006.

REFERENCES

1. Ya. B. Zel'dovich, *J. Exptl. Theoret. Phys. (USSR)*, **36**, 1959, 964.
2. S. C. Bennett and C. E. Wieman, *Phys. Rev. Lett.*, **82**, 1999, 2484.
3. G. P. Zeller *et al.*, Phys. Rev. Lett., ***88**, 2002, 091802.*
4. *P. E. Reimer, CP721, Neutrino Factories and Superbeams:*
5. *K. S. Kumar, CP675, Spin 2002: 15'th Spin Physics Symposium on Polarized Electron Sources and Polarimenters, AIP 2003, 142.*
6. *J. T. Londergan, contribution to 2nd workshop Structure of Nucleon at high Bjorken x, France, AIP Conf.Proc.747, 2005, 205.*
7. *X. Ji,* Phys. Rev. Lett. ***78**, 1997, 610;*Phys. Rev. ***D55**, 1997, 7114.*
8. *A. V. Radyushkin,*Phys. Rev. ***D56**, 1997, 5524.*
9. *M. Diehl, DESY-thesis-2003-018, hep-ph/0307382 and ref. therein.*
10. *J. C. Collins, L. Frankfurt, M. Strikmann* Phys. Rev. ***D56**, 1997, 2982.*
11. *H1, C. Adloff et al.,* Phys. Lett., ***B517**, 2001, 47; ZEUS, S. Chekanov et al.,* Phys. Lett., ***B573**, 2003, 46; CLAS, S. Stepanyan et al.,* Phys. Rev. Lett., ***87**, 2001, 182002; HERMESS, A. Airapetian et al.,* Phys. Rev. Lett., ***87**, 2001, 182001.*
12. *N. d'Hose contribution to SPSC Villars Meeting on a Future Fixed Target Programme at CERN*
13. *contributions to the workshop: Physics with a Multi-MW Proton Source, CERN, Geneva, May 25-27,2004.*

Physics at HERA and Beyond

Max Klein

DESY, 15 738 Zeuthen, Platanenallee 6
klein@ifh.de

Abstract. About two years prior to the planned termination of the operation of HERA, a short summary is given of the status of HERA, the physics goals of data taking until 2007, of the proposed experimental programme termed HERA III, and of ideas as to how deep inelastic scattering may conquer the TeV scale of energy.

Keywords: QCD, DIS, proton structure, electron-proton scattering, strong interactions
PACS: 12.38.-t,29.20.Dh

INTRODUCTION

This talk was given in the session devoted to the future of deep inelastic scattering (DIS). In 2001, the DESY directorate announced that HERA operation would be terminated in 2006 since one had promised to give PETRA to the synchrotron light physics community for building a 3rd generation brilliant light source and because DESY had hoped to have begun already with the construction of TESLA at Hamburg. Thus since the end of 2001, with a workshop promptly held at Durham, the HERA community had to think about its future, and, more generally, about the future of deep inelastic scattering. While this had been a more and more pressing question, which is not resolved today, most of the efforts naturally went into the realisation of the luminosity upgrade of HERA, the challenge caused by the modified interaction regions (IR) of H1 and ZEUS and the excitement real data and results cause.

In the years 1992-2000 the collider experiments took about $110\,\mathrm{pb}^{-1}$ of positron proton data and nearly $20\,\mathrm{pb}^{-1}$ with electrons. The mean specific luminosity of the HERA I phase was about $0.4 \cdot 10^{29}\mathrm{cm}^{-2}\mathrm{s}^{-1}/\mathrm{mA}^2$. The $e^{\pm}$ beams were unpolarised for H1 and ZEUS, and to about 50-60% longitudinally polarised at HERMES. In 2001/02 the IR's were equipped with superconducting focussing quadrupoles near the vertex region thereby reducing the accessible low Q^2 range. It took time, to the fall of 2003, until the machine in its new configuration was successfully commissioned: initially too high backgrounds were observed which in a series of systematic studies were traced back to the collisions of proton beam particles with restgas molecules steming from synchrotron radiation interaction with beam pipe and collimator elements. Modifications of the collimators at about -1.5 m including additional pumps were required as was an understanding of the role and use of the cold magnet surfaces near the IR. HERA II is since then running with specific luminosities increased by a factor of 3 (e^+) to 5 (e^-), lepton beam polarisations using new spin rotators around H1 and ZEUS of up to 50% and peak luminosity values of up to $5 \cdot 10^{31}\mathrm{cm}^{-2}\mathrm{s}^{-1}$, nearly as designed. At the Madison conference, May 2005, H1 and ZEUS had taken about half the HERA I e^+p

CP792, *Deep Inelastic Scattering, DIS 2005,* edited by W. H. Smith and S. R. Dasu

luminosity and about tripled the previous e^-p luminosity values. Utilising the beam polarisation rather promptly, a long awaited, now classic DIS measurement could be done establishing the linear dependence of the charged current (CC) cross section on the lepton beam polarisation and thus finding the electroweak standard model (SM) to be succesful yet another time, see Fig.1. Many more new results obtained by the HERA collaborations were presented at this conference by ZEUS [1], H1 [2] and HERMES [3].

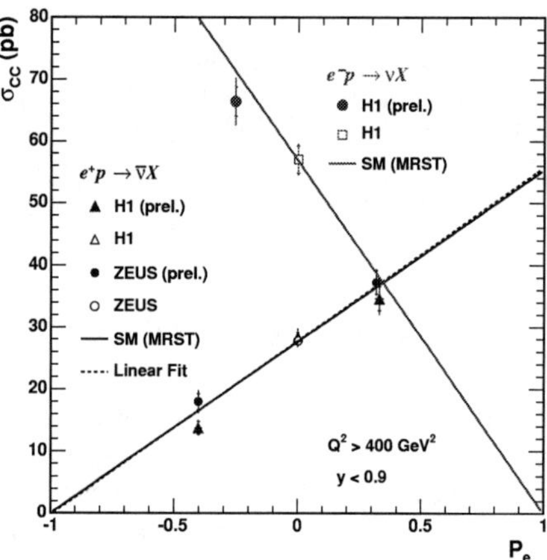

FIGURE 1. Measurements by ZEUS and H1 of the total cross section in deep inelastic charged current polarised electron and positron scattering as a function of the helicity of the $e^\pm$ beam.

This report presents a selection of physics subjects of HERA which can be expected to be investigated with enlarged statistics and higher precision than hitherto achieved. In a further section the motivation is summarised for a continued operation of HERA in a third phase. Finally ideas are presented for how deep inelastic scattering may be performed at TeV energies as is required for pursuing DIS physics further and for contributing to high energy physics as a whole at the new energy frontier which will be opened by the LHC.

PHYSICS WITH HERA II

Physics at HERA (II), as has been described in [4, 5, 6], is a wide field comprising classic inclusive charged and neutral current DIS, detailed tests of QCD and parton radiation, searches for new phenomena at the energy frontier and electroweak physics. So far, HERA found a number of new phenomena, in particular the rise of the proton structure function $F_2(x,Q^2)$ towards low x, which opened the field of high density QCD, hard diffraction which is still not explained, generalised parton distributions which lead to parton correlation measurements and exotic phenomena, specifically the observations of 5 quark states and of a few peculiar events with isolated leptons and large missing transverse momentum. It is not possible to describe this field in this contribution in any exhaustive way, which is why only a few possible highlights and particular challenges of physics at HERA II are briefly presented here. If the integrated luminosity gets strongly increased as is hoped for, new phenomena may become visible and rather surely new ideas will be developed in forthcoming analyses.

Increase of Precision

HERA is a laboratory for precision physics because of the very nature of deep inelastic scattering. It represents the cleanest microscope at high resolution of the world. The systematic measurement accuracy is much higher than in pp colliders because the kinematics is overconstrained allowing the calorimeter energy scales to be fixed at the per cent level. The tracking and calorimetric information are combined for optimum reconstruction and the tracking and trigger efficiencies determined from the data. The luminosity is measured to 1-2% accuracy. Further improvements of the experimental accuracy are still desirable, not only because of lacking statistics at highest Q^2 and x, but also since most QCD effects governed by gluon radiation are logarithmic, i.e. slowly varying, and some quantities like the gluon density are determined only by derivatives of cross sections. Three major methods are being pursued to increase the accuracy: i) the luminosity is increased: at Madison one could expect HERA II to deliver about $500\,\mathrm{pb}^{-1}$ of polarised $e^{\pm}$ data per experiment for analysis on tape, which would exceed the HERA I luminosity by a factor of 4 registered in about half the running time. A machine hardware improvement program had been endorsed including the exchange of all 32 “BU" magnets, large proton beam deflecting dipole magnets which exhibited aging effects, also upgrades and replacements in the proton rf., diagnostics and cryogenic systems. In order to deliver more luminosity, DESY decided to prolong the HERA life time by 6 months, until mid 2007; ii) the H1 and ZEUS detectors were upgraded substantially, in particular the forward trackers, Silicon trackers and triggers; iii) new ideas are being developed in order to cross calibrate the H1 and ZEUS measurements and thus by combining the HERA collider data significantly gaining in precision and reliability of the results [7]. Precision is the road to a success of HERA, its first and second phase.

The Strong Coupling Constant

The coupling constant α_s of the strong interaction is the worst measured fundamental coupling constant. Extrapolations of α_{elm}, G_F and α_s to the Planck scale are uncertain, not only in principle but also due to the large error of $\alpha_s(M_Z^2)$ which is usually taken to be 0.003. The determination of $\alpha_s(M_Z^2)$ from HERA [8] is experimentally already more precise, with a mean of $0.1186 \pm 0.0011(exp)$ in NLO QCD. A few comments are in order here: i) the data entering this measurement are from HERA I and improvements of about a factor of two should come with higher statistics and thus enabled increased calibration accuracy and reduced systematics; ii) while this mean value seems to just agree with the world average, there has been a peculiarity which deserves final clarification at HERA. While the inclusion of jet data leads to a large value of $\alpha_s(M_Z^2)$, near 0.120, the values from just the inclusive cross section analyses are systematically below: H1 data alone yield 0.115, BCDMS alone 0.110, ZEUS alone 0.110, H1 and BCDMS combined 0.115, see Fig. 2, furthermore H1 and NMC combined yield 0.116 [9] and the non-singlet (NNLO) analysis of [10] a value of $\alpha_s(M_Z^2)$=0.1135. The value of $\alpha_s(M_Z^2)$ is known to be particularly strongly related to the behaviour of the gluon distribution at large $x > 0.3$ where inclusive DIS has little sensitivity. The contradiction, if it was confirmed at higher accuracy, may thus be due to lacking input as to how xg approaches $x = 1$. Yet, there may also be other reasons, as for example $\alpha_s(M_Z^2)$ is related to the mass of gluinos [11]; iii) to the experimental uncertainty a large, so called theoretical error is

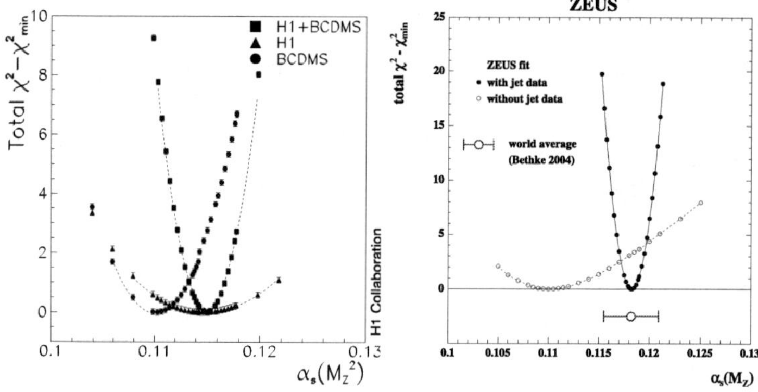

FIGURE 2. Determinations of the strong coupling constant α_s in NLO QCD fits to the BCDMS, the H1 and the combined H1+BCDMS data (left) and to the inclusive ZEUS and the combined ZEUS inclusive+jet data (right).

attached which is 0.005 in the HERA NLO analyses and stems essentially from a variation of the renormalisation (and factorisation) scale from $2\sqrt{Q^2}$ to $\sqrt{Q^2}/2$. This variation is an *ad hoc* prescription the size of which is in no way supported by the inclusive DIS analyses of H1 and ZEUS. Therefore it deserves serious reconsideration instead of applying such variations blindly in all analyses. The calculation of NNLO splitting and coefficient functions will reduce the scale uncertainties soon, yet, the genuine influence of the scales needs to be clarified nevertheless.

Heavy Flavour Physics

Heavy flavour physics at HERA explores the charm and beauty contents of the proton and of the photon, the production and fragmentation of charmed and beauty particles. The charm contribution to the NC cross section at low x is about 20% while beauty contributes to only 2% beyond threshold. Beauty physics at HERA II will play the role of charm physics at HERA I both becoming more accurately measured. At Madison first data were presented of the beauty structure function measured at low Q^2 and x. These and previous data at higher Q^2, both obtained with an impact parameter enrichment method based on H1's central Silicon tracker, represent the first ever measurement of $F_2^{b\bar{b}}$. The data, at the present level of accuracy, are well described by NLO QCD predictions, which determine the beauty quark density solely by the amount of scaling violations of the inclusive F_2 data and a prescription as to how beauty emerges near the threshold $Q^2 \sim m_b^2$. Since at the LHC all heavy quark distributions are of size comparable to those of the light quarks, accurate measurements of $F_2^{c\bar{c}}$ and $F_2^{b\bar{b}}$ are important for the prediction of hard scale processes such as W or Z production. These distributions can directly be measured only at HERA owing to the probing character of DIS.

Interesting observations which have been made by H1 and ZEUS such as the excess of beauty particle production in the forward direction as compared to NLO QCD, the trend of beauty particle cross sections to exceed the NLO predictions by perhaps a factor of 1.5-2, further, the measurements of the charm fragmentation function and fractions which examine the universality of charm particle formation in e^+e^- versus ep reactions and other measurements [1, 2], of charm jets for example, all will be performed at still increased accuracy at HERA II. This physics relies largely on the H1 and ZEUS Silicon strip detectors which have been installed as new or upgraded devices for the high luminosity phase of HERA. While the MVD of ZEUS and the CST of H1 are currently taking data, the forward and backward Silicon trackers of H1 are being repaired and upgraded with radiation hard electronics to be reinstalled in the shutdown 2005/06.

Electroweak Physics

HERA explores electroweak physics in the spacelike region and thus is complementary to LEP. At this workshop, the first HERA data were presented of the weak neutral current couplings of light quarks to the Z^0, see Fig. 3. These resolve the sign ambiguity

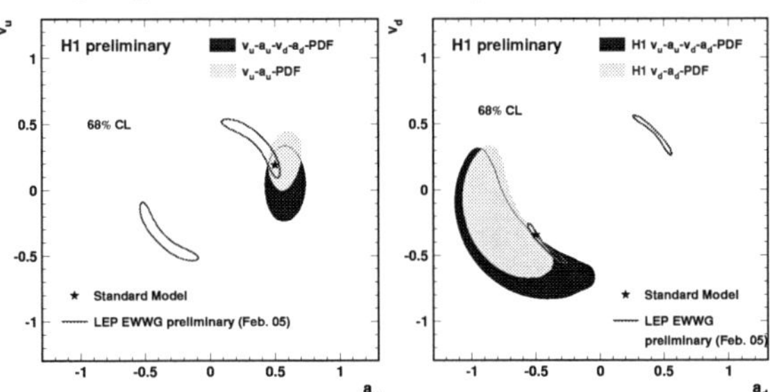

FIGURE 3. The first determinations at HERA of the weak neutral current vector and axial-vector couplings (v and a) of light quarks to the Z_0 boson compared with preliminary LEP Z resonance data.

which is inherent in the LEP measurements at resonance. The result uses only the HERA I data, in particular $15\,\text{pb}^{-1}$ of electron data only, and represents a first important step only. Higher accuracy can be expected from the HERA II data with enlarged luminosity and lepton polarisation. At HERA parton distributions have to be known to determine electroweak parameters. One may also use electroweak interactions to probe the strong interaction: a measurement of improved precision of the γZ interference structure function $xG_3 = xF_3^{\gamma Z}$ will allow to access the parton distributions, i.e. the valence quark combination $2u_V + d_V$ and a possible quark-antiquark distribution difference, $q - \overline{q}$, down to $x \sim 0.002$ and determining as well the sign of the quark electric charges.

Deeply Virtual Compton Scattering

In DVCS a real photon is produced in DIS and momentum transferred between the virtual and the real photons which allows parton correlations to be accessed. Experimentally, as much as phenomenologically [17], this field is new and developing rather rapidly with the first measurements being performed at HERA [1, 2, 3] and elsewhere. DVCS interferes with Bethe-Heitler scattering and thus parton amplitudes, or generalised parton distributions, can be measured from beam charge and helicity asymmetries. In the

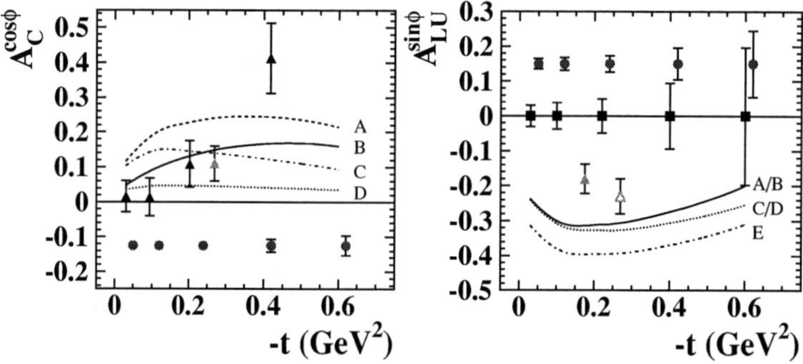

FIGURE 4. Projected accuracy for the t dependence of charge (left) and beam spin asymmetries (right, assuming a polarisation of 40%) in DVCS as expected to be measured differentially, for x between 0.04 and 0.4 and Q^2 between 1 and 8 GeV^2, by the HERMES experiment in 2006/07 using a new fibre-Silicon detector to tag the recoiling proton. The curves are different predictions by Vanderhaegen et al.

forward limit, via the 'Ji sum rule', these may be related to the unmeasured angular momentum contribution of the quarks to the spin of the proton. HERMES intends to dedicate the final phase of HERA II to DVCS measurements, see Fig. 4, employing a newly built detector to measure the recoiling proton in elastic DVCS. New cross section and HERA II asymmetry data from H1 and ZEUS will study DVCS in a different kinematic region at low x.

Low x Physics

HERA has established low x physics, which is characterised by the dynamics of the QCD vacuum rather than primarily the structure of the proton. This field is new, the appropriate theoretical description not necessarily identical to the conventional DGLAP approach, and it is related to superhigh energy neutrino physics, to nuclear effects as studied at RHIC and ALICE and to forward physics at the LHC. Three major contributions of HERA to this field have not been completed yet deserving further data and more unified analysis techniques. These can be sketched by the following questions:

Are the indications for a breakdown of DGLAP QCD to NLO, as observed in forward jet production, in association with dijets, and in the azimuthal decorrelation effects, true to the extent that the description of parton radiation needs to be modified in the direction of the CCFM or BFKL prescriptions? In particular which role do unintegrated parton distributions play which offer the advantage of incorporating transverse momenta, i.e the radiation kinematics, correctly?

What is the gluon distribution in the proton at low x? The gluon is determined by the $\ln Q^2$ derivative of F_2, which will be measured with yet increased precision. More generally, is the longitudinal structure function, which still needs to be measured at HERA, consistent with the predictions from scaling violations in QCD? The measurement of F_L is a necessary part of the HERA II programme which tests QCD at low x at higher orders. It requires a few months of data taking and highest possible precision at large inelasticities $y = Q^2/sx$. Additional information on xg is obtained from the heavy flavour structure functions and perhaps from inelastic J/Ψ production which is subject to theoretical challenges.

How should hard diffraction be integrated in low x theory? If it was an absorptive correction to be subtracted prior to a *pdf* QCD analysis [12], all distributions and their extrapolations to the LHC would be different from conventional wisdom, the desired extrapolation accuracy being 1% for determining the parton luminosity at the LHC. Moreover, the question isn't fully answered what the diffractive parton densities really are and to which extent they may be used for understanding pp interactions and in particular the diffractive production of exotic states at the LHC [13]. Behind these questions are strong attempts to measure diffraction at HERA accurately, including the new Very Forward Proton Spectrometer of H1, and is the need for intimate collaboration with theory, as for most of HERA physics.

Clarification of Puzzling Observations

Much increased luminosity is required for clarification of the unexpected observations at HERA. ZEUS has reported the observation of a strange pentaquark, θ^+, in the $K_s^0 p$ channel which is not confirmed by H1 though the preliminary limits derived by H1 do not contradict the ZEUS findings. H1 has observed a resonance in the D^*p, the minimal quark content of which is *cuudd*, at an acceptance corrected rate [14] of $D^*p/D^* = 1.59 \pm 0.33(sta)^{+0.33}_{-0.45}(sys)\%$, which may be about compared with a limit of 0.35%, quoted for $K\pi\pi$ and $K\pi\pi\pi$ D^* decays and a bit different cuts for the DIS data [15], resulting from the non-observation of such a resonance by ZEUS. These observations are being investigated with new data but still require higher statistics to be taken and the modified tracking apparatus to be carefully calibrated regarding the inclusion of Silicon detector information and material and dE/dX refinements.

Since many years H1 is observing an excessive production of events with isolated leptons in positron proton scattering. For missing transverse momentum $p_t^X > 25\,\text{GeV}$, and an integrated luminosity of 192 pb^{-1}, 17 events are observed and 6 expected to be compared with ZEUS finding 6 events at an expectation of 5.7, based on HERA I data alone. There wasn't enough electron data taken by the time of this workshop to distinguish the e^-p from the e^+p data, recent results, however, indicating a much less significant rate in electron scattering.

HERA OPTIONS

The future of HERA could involve an extended programme [16] including three major parts, two requiring limited luminosity while a spin collider needs very high luminosity and a proton or deuteron beam polarised:

i) The dedicated, high precision measurement of the transition from deep inelastic scattering to soft hadron collisions. The behaviour of the rise of F_2 towards low x is indicative for this transition to take place for Q^2 corresponding to an inner dimension of 0.3 fm. In the proton rest frame one may view DIS as proceeding by the photon splitting into a colour dipole of variable size $2/Q$ which interacts with the proton at an energy of $W^2 \simeq sy = Q^2/x$. The distance at which the splitting occurs, the coherence length (Gribov) is large as $L \propto 1/x$. Thus a scan of proton's structure, using both virtual photon polarisation states, would have been possible in a transition region of fundamental interest.

ii) The measurement of electron-nucleus collisions, *ed* and *eA*. Every DIS charged lepton scattering experiment, prior to HERA, used deuterons to study the neutron structure. HERA has discovered the rise of $F_2 \propto x(4\bar{u}+\bar{d})$ towards low x. There are non-perturbative models, like the chiral quark soliton model, which predict up and down anti-quarks not to be equal. While this has been indeed confirmed at larger $x \sim 0.1$, common wisdom assumes $(\bar{d}-\bar{u}) \rightarrow 0$ at low x but may yet be wrong. Deuteron data measure a singlet structure function and would allow the non-singlet and singlet evolutions at low x to be separated thus contributing to the development of QCD at low x. The measurement accuracy of $\alpha_s(M_Z^2)$ would be improved by about a factor of two. An *ed* programme at HERA would be much more powerful than that at fixed target experiments because, by tagging the spectator proton, *en* scattering could be measured nearly free of Fermi motion corrections. Moreover, a diffractive programme could be pursued on p, n and d which also would allow the shadowing corrections to be quantified. Extending this programme to heavier nuclei would determine nuclear parton distributions, as is required for ALICE at the LHC and for RHIC. Decisive tests are possible of the predictions of a 'black body limit' of DIS, see [16], in which $F_2 \rightarrow Q^2/\ln(\delta/x)$ and the diffractive scattering component reaches up to 50%. Since the gluon density is amplified in nuclei as $A^{-1/3}$, one would exceed the unitarity limit $\sim Q^2/\alpha_s$ and be able to study saturation phenomena which in *ep* scattering have not been observed unambiguously.

iii) With polarised protons and deuterons scattered at high luminosity off polarised leptons, HERA has the potential to make major qualitative progress in the investigation of nucleon spin components and the develoment of the QCD of polarised partons. Erraneously spin physics is sometimes considered to be nuclear physics while what it lacks is a polarised collider of high luminosity and variable beam energies such as eRHIC and HERA.

The proposed HERA III programme was widely supported and positively reviewed but finally ranked to be of less importance than the physics potential of the linear collider with which the programme did not intend to compete, neither in time nor financially. The result of this development is unfortunate. Building eRHIC is of interest. Not to use HERA further is wrong. The cost gained by reusing PETRA is negligible compared to the investment HERA represents. Running HERA for part time a year could have offered a way to solve manpower problems and still collect enough luminosity for the first two parts of HERA III, high precision low x physics and eN, both requiring integrated luminosities of the order of $10\,\mathrm{pb}^{-1}$only. For the coming years, beyond 2007, DESY intends to keep the HERA proton and electron rings together which in principle offers a chance to return to HERA which still represents one of the biggest investments in particle physics.

DEEP INELASTIC SCATTERING AT ENERGIES BEYOND HERA

Particle physics is preparing for a new level of energies. Both at energies of order 10 GeV and at the Fermi scale, explored by the colliders TeVatron, LEP and HERA, deep inelastic scattering has been making major contributions to understanding its own physics, the structure of matter and the strong interactions, and to interpreting and complementing the results from e^+e^- and from hadron-hadron scattering. It was a precision measurement in polarised ep scattering which in 1978 established the electron to be a right handed coupling singlet which was the final breakthrough for the $\mathrm{SU}(2)_L\mathrm{xU}(1)$ GWS model to become standard. With HERA's pdf's, to quote another example, one is able to interpret and calculate hard scale cross sections at the TeVatron. Electroweak cross sections require LEP's precision measurements and the top quark mass prediction wouldn't have been believed without finding the top at the TeVatron. The question arises how the symbiosis of ee, pp and ep scattering can be ensured at the new energy scale, or specifically, how may DIS conquer the TeV scale of collision energies and why should that be a high priority task for particle physics?

Towards a TeV ep Collider

An ep collider is cost effective when realised combining existing and/or future accelerators. Owing to the time structure of the cold ILC and the standing wave type cavities, it may be combined with HERA, as considered for the THERA study [19], or the TeVatron, and both arms of the linear collider may be used to realise a symmetric energy ep collider of cms energy up to 2 TeV. The luminosities for THERA were estimated to range between $4\cdot10^{30}\mathrm{cm}^{-2}\mathrm{s}^{-1}$, for $E_e = 250\,\mathrm{GeV}$ and $E_p = 1\,\mathrm{TeV}$, up to $2.5\cdot10^{31}\mathrm{cm}^{-2}\mathrm{s}^{-1}$, for $E_e = E_p = 500\,\mathrm{GeV}$ and using dynamic focussing. A study was made [20] for a 'QCD explorer' using a CLIC prototype of 75 GeV and the LHC proton ring of 7 TeV with a modified, 'super bunch' structure and an estimated luminosity of $1.1\cdot10^{31}\mathrm{cm}^{-2}\mathrm{s}^{-1}$. Both THERA and the 'Explorer' would have no problems to access a much extended low x region at rather large Q^2, see Fig. 5. Large rates close to the kinematic limits, however, are difficult to achieve with linac-ring combinations.

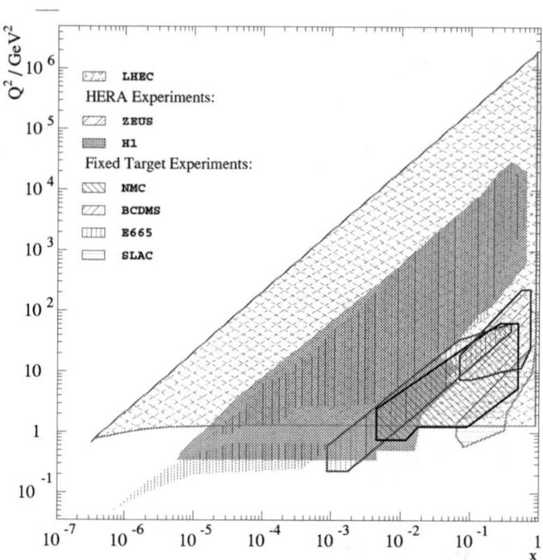

FIGURE 5. Kinematic range in Bjorken x and momentum transfer squared Q^2 as covered by fixed target experiments, by H1 and ZEUS at HERA and by a new ep collider at the LHC at CERN or using the ILC and a 1 TeV proton beam, from HERA at Hamburg or the TeVatron at Fermilab.

Higher luminosities in ep scattering at TeV energies appear possible combining the LHC with an electron ring [1]. Early studies, first considering LEP with the LHC mounted on top [21] and then the LHC with LEP mounted on top [22], lead to estimated luminosities of $1-2\cdot 10^{32}\text{cm}^{-2}\text{s}^{-1}$. For this workshop a tentative lattice study was made [23] of a new ep collider, LHEC, with a new electron ring mounted above the LHC. For standard LHC parameters, i.e. energy, bunch spacing, current, emittance, and a 75 GeV electron ring of 49 mA current using TESLA cavities, a luminosity was estimated of $2.4\cdot 10^{32}\text{cm}^{-2}\text{s}^{-1}$. This is achieved with strong focussing magnets near the IR limiting the acceptance range to polar angles between 10° and 170°. Without these magnets, the luminosity is lower but the acceptance enables low Q^2 and x physics to be studied. This design is being developed further regarding aspects as the luminosity, the effect of synchrotron radiation, crossing angle, the possibility of parasitic ep operations at the LHC, as well as infrastructure aspects and the feasibility to mount and close the electron ring in the LHC tunnel.

A symmetric energy machine requires an extended detector in the e beam direction to measure the scattered electron at low x. An asymmetric machine requires special detection efforts for the forward going hadronic final state to access the region of large x. Detectors for an ep collider at TeV energies may not pose extraordinary difficulties given the HERA experience and current technology developments.

---

[1] If the ILC was built at CERN, the possibility would appear of an ep collider of about 5 TeV energy with a maximum Q^2 of $2.8\cdot 10^7$ GeV^2.

DIS Physics at the TeV Scale

The physics of deep inelastic scattering at cms energies of the order of $1-2\,\text{TeV}$ was discussed early on [18] and much extended in a very detailed study [19], as part of the TESLA TDR. This can not be summarised here. It comprises searches for new states of matter, particles as leptoquarks, see Fig. 6, which in *ep* can be singly produced like SUSY particles in RP violating production. Searches can be performed for substructure with highest resolution, which is the classic field of *ep* scattering, and the partonic contents of the proton and the photon be investigated. The physics at low x may qualitatively differ at the LHEC from HERA, as then the rise of the gluon density is expected to be damped by unitarity. As sketched above, there are fundamental questions which may be studied with deuterons and heavier nuclei and which are relevant for *AA* scattering at the LHC.

For the LHC, the *pdf*'s measured at HERA are the basis of predicting cross sections and describing the QCD processes. The extrapolations from HERA, at least at the edges of the Bjorken x range, may yet be doubtful: at low x, in the new domain of high density QCD, gluon saturation effects possibly occuring in the LHC range may alter all HERA extrapolations, while at high x, the region related to the genuine, high mass discovery range at the LHC, resummation effects may similarly question such extrapolations. With new strongly interacting particles as gluinos possibly occuring in the LHC kinematic range these extrapolations would need to be altered significantly [11]. Discoveries at the LHC may as well be subtle, and high precision be required which is difficult to ensure with extrapolations over 1-2 orders of magnitude. Clearly the LHC will help defining the future of deep inelastic scattering.

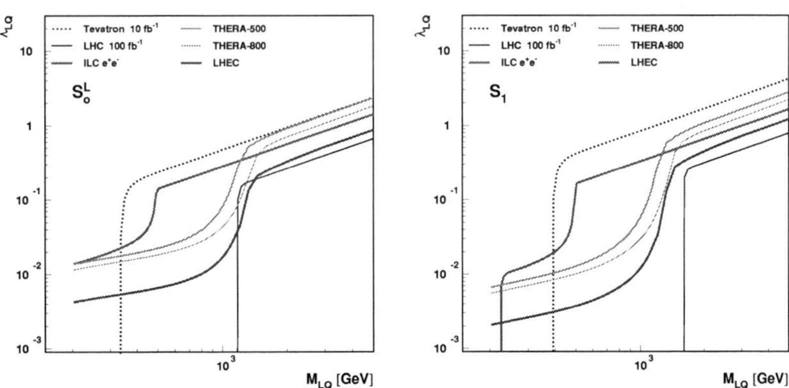

FIGURE 6. Comparison of accessible ranges for $S_0^L = e_L u,..$ and $S_1 = e_L d,..$ leptoquark production, from [24]. As for *ep* these particles can be singly produced and with changed initial conditions (charge and polarisation) their spectroscopy be studied, similarly for SUSY particles. Direct searches are limited by the cms energy while indirect searches may lead to higher masses. If particles are produced in the directly accessible range, the LHEC will be of much use to study particles which would be expected to be discovered at the LHC. Due to the production mechanism the next *ep* machine has a higher range in mass than the ILC.

ACKNOWLEDGMENTS

HERA and its physics results are due to a large community of motivated machine, experimental and theoretical physicists. I would like to thank Elke Aschenauer, Delia Hasch, Elisabetta Gallo, Rik Yoshida and many H1 colleagues for help with this talk. I thank the organisers of the session, Wesley Smith, Dave Soper and Uta Stösslein, together with Alan Caldwell, John Dainton, Tim Greenshaw, Aharon Levy, Daniel Pitzl and Ferdinand Willeke with whom and more colleagues the excitement and concern is shared for the present and the future of deep inelastic scattering.

REFERENCES

1. E. Gallo, these proceedings.
2. O. Behnke, these proceedings.
3. R. Ryckbosch, these proceedings.
4. Physics at HERA, Workshop at Ringberg, October 2003, Proceedings World Scientific, eds. G. Grindhammer, B. Kniehl and G. Kramer.
5. F. Eisele, Talk at the HERA LHC Workshop, March 2005, DESY.
6. R. Yoshida, Talk at the HERA LHC Workshop, March 2005, DESY.
7. A. Glazov, these proceedings.
8. C. Glasman, these proceedings.
9. H1 Collaboration, C. Adloff et al., Eur. Phys. J. **C21**,33 (2001) [hep-ex/0012053].
10. J. Blümlein, H. Böttcher and A. Guffanti, Nucl. Phys. Proc. Suppl. **135** (2004) 152 [hep-ph/0407089].
11. E. Berger et al., Phys. Rev. **D71**, 014007 (2005) [hep-ph/0406143].
12. A.D. Martin, M.G. Ryskin and G. Watt, Phys. Rev. D **70**, 091502 (2004) [hep-ph/0406225].
13. V. Khoze, these proceedings.
14. K. Daum, these proceedings.
15. Y. Eisenberg, these proceedings.
16. T. Alexopoulos *et al.*, eD Scattering with H1, A Letter of Intent, DESY 03-194;
H. Abramowicz *et al.*, A New Experiment for HERA, MPP-2003-62;
F. Willeke and G. Hoffstaetter, Talks at the Workshop on the Future of DIS, Durham 2001, unpublished;
http://hep.ph.liv.ac.uk/~green/HERA3/.
17. see e.g. V. Belitsky and D. Müller, Nucl. Phys. **A711**, 118 (2002) [hep-ph/0206306].
18. J. Feltesse and R. Rückl, Talks at the LHC Workshop, Aachen 1990, CERN 90-10 (1990), Proceedings Vol.1 p.219, 229, eds. C. Jarlskog and D. Rein.
19. The THERA Book, *ep* Scattering at $\sqrt{s} \sim 1$ TeV, DESY 01-123F, eds. U. Katz, M. Klein, A. Levy and S. Schlenstedt.
20. D. Schulte and F. Zimmermann, CLIC Note 608, CERN-AB-2004-079, unpublished;
see also L. Gladilin et al., hep-ex/0504008.
21. A. Verdier, LHC Workshop, Aachen 1990, CERN 90-10 (1990), Proceedings Vol.2 p.890, eds. C. Jarlskog and D. Rein.
22. E. Keil, LHC Project Report 93 (1997).
23. F. Willeke, private communication.
24. A. Zarnecki, private communication.

Using Neutrinos as a Probe of the Strong Interaction

Jorge G. Morfín

Fermi National Accelerator Laboratory, P.O. Box 500, Batavia, IL, 60510

Abstract. Neutrino scattering experiments have been studying QCD for over 30 years. From the Gargamelle experiments in the early 70's, through the subsequent bubble chamber and electronic detector experiments in the 80's and 90's, neutrino scattering experiments have steadily accumulated increasing statistics and minimized their systematic errors. While the most recent study of QCD with neutrinos is from the TeVatron neutrino beam (the NuTeV experiment with results presented by Martin Tzanov at this Workshop), near-future studies will shift to the Main Injector based NuMI facility also at Fermilab. The NuMI Facility at Fermilab provides an extremely intense beam of neutrinos making it an ideal place for high statistics (anti)neutrino-nucleon/nucleus scattering experiments. The MINERνA experiment at Fermilab is a collaboration of elementary-particle and nuclear physicists planning to use a fully active fine-grained solid scintillator detector to measure absolute exclusive cross-sections and nuclear effects in ν - A interactions as well as a systematic study of the resonance-DIS transition region and DIS with an emphasis on the extraction of high-x_{Bj} parton distribution functions. Further in the future an intense proton source, the Fermilab Proton Driver, will increase neutrino interaction rates by a further factor of 5 - 20.

Keywords: <neutrino, scattering, nucleus>
PACS: <13.15>

1. INTRODUCTION

The results of the latest study of QCD using neutrino scattering comes from the NuTeV experiment [1]. The NuTeV experiment accumulated over 3 million ν and $\overline{\nu}$ events in the energy range of 20 to 400 GeV and the details of the analysis can be found in Tzanov's contribution and contained references. The main points are that the NuTeV cross section agrees with the CCFR values (obtained using the same detector) for values of $x_{Bj} \leq 0.4$ but is systematically higher for larger values of x_{Bj} culminating at x_{Bj} = 0.65 where the NuTeV result is 20% higher than the CCFR result. NuTeV agrees with charged lepton data for $x_{Bj} \leq 0.5$ but there is increasing disagreement for higher values. Although NuTeV F_2 and xF_3 agree with theory for medium x, they find a different Q^2 behavior at small x and are systematically higher than theory at high x

These results can be summarized in four main questions to ask subsequent neutrino experiments:

- At low x, how does shadowing with incoming neutrinos differ from shadowing with incoming charged leptons?
- At high x, what is the behavior of the valence quarks as $x \to 1.0$?

CP792, *Deep Inelastic Scattering, DIS 2005*, edited by W. H. Smith and S. R. Dasu

- At low W, what is happening in the transition region between resonance production and the DIS regions?
- At all x and Q^2, what is yet to be learned if we can measure all six ν and $\overline{\nu}$ structure functions to yield maximal information on the parton distribution functions?

At Fermilab, these questions will be addressed using the recently commissioned NuMI neutrino facility, designed for the MINOS neutrino oscillation experiment, and based on the Main Injector (MI) accelerator. The neutrino beams from the MI yield several orders of magnitude more events per kg of detector and year of running than the higher energy Tevatron neutrino beam. With such a facility, one can now perform statistically significant experiments with much lighter targets than the massive iron, marble and other high-A detector materials used in the past. That these facilities are designed to study neutrino oscillations points out the second advantage of NuMI neutrino experiments; an excellent knowledge of the neutrino beam will be required to reduce the beam-associated systematics of the oscillation result. This knowledge of the neutrino spectrum will also reduce the beam systematics in the measurement of neutrino scattering phenomena.

2. THE MINERνA NEUTRINO SCATTERING EXPERIMENT

To take advantage of these major improvements, a collaboration of both elementary particle and nuclear physics institutions named MINERνA (Main Injector ExpeRiment: ν A) [2, 3] has been formed, received Stage I approval and is in the process of organizing funding with Fermilab's support.

The MINERνA neutrino scattering experiment in the NuMI beam offers a unique opportunity to study a broad spectrum of physics topics. Several have not previously been studied in any systematic way, while others have only few results that are compromised by large statistical and systematic errors. A summary of these topics where MINERνA will break fresh ground include:

- Precision measurement of the quasi-elastic neutrino–nucleus cross-section, including its E_ν and q^2 dependence, and study of the nucleon axial form factors. Figure 1 shows the predictions for the cross section measurement assuming a 4-year MINERνA run. Figure 2 shows the extraction of the axial-vector form factor from the quasi-elastic event sample accumulated over a 4-year MINERνA run. The data points are plotted as a ratio of F_A/F_A(Dipole). Also shown are the currently available values of F_A from early experiments.
- Determination of single- and double-pion production cross-sections in the resonance production region for both neutral-current and charged-current interactions, including a study of isospin amplitudes, measurement of pion angular distributions, isolation of dominant form factors, and measurement of the effective axial-vector mass.
- Clarification of the W ($\equiv$ mass of the hadronic system) transition region wherein resonance production merges with neutrino deep-inelastic scattering, including

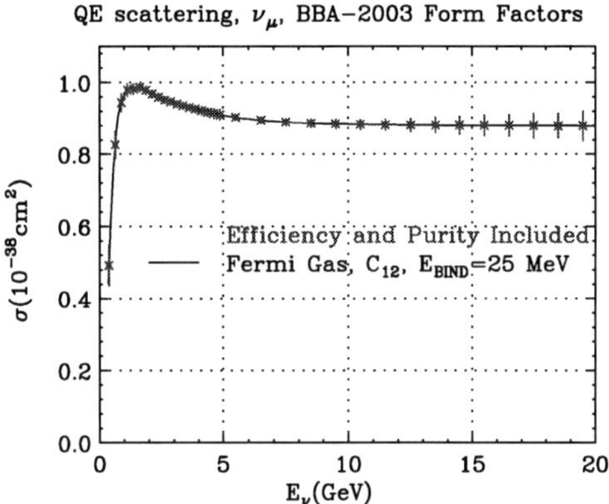

FIGURE 1. Cross section for MINERνA assuming a 4 year run with M_A =1.00 GeV, and the Fermi gas model.Statistical errors only are shown

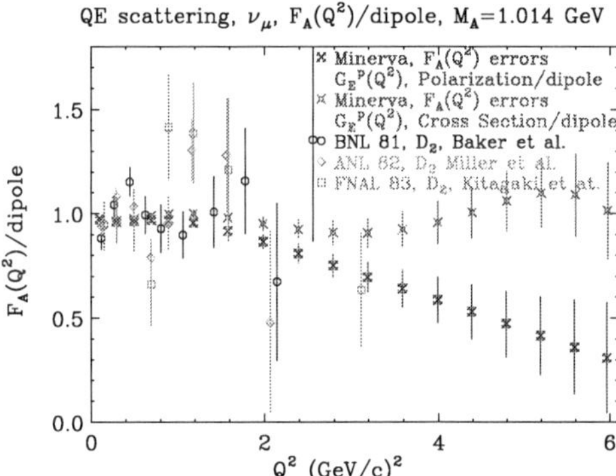

FIGURE 2. Extracted values of $F_A(q^2)/dipole$ for deuterium bubble chamber experiments Baker *et al.* [4] and Kitagaki *et al.* [5]. For MINERνA the projected results are shown for two different assumptions: F_A/dipole=G_E^p/dipole from cross section and F_A/dipole=G_E^p/dipole from polarization. The MINERνA errors are for a 4 year run.

tests of phenomenological characterizations of this transition such as quark/hadron duality.

- Precision measurement of coherent single-pion production cross-sections, with particular attention to target A dependence. Coherent π^0 production, via the neutral-current, is a significant background for next-generation neutrino oscillation experiments seeking to observe $\nu_\mu \to \nu_e$ oscillation. Figure 3 shows the expected precision (statistical errors only) of the MINERνA measurement of the charged current coherent pion production cross-section as a function of neutrino energy. Here it is assumed that the measured value is that predicted by Rein-Seghal. Also plotted are the only currently available measurements showing their total errors. Figure 4 illustrates the broad range in A to be covered by MINERνA's measurement of the

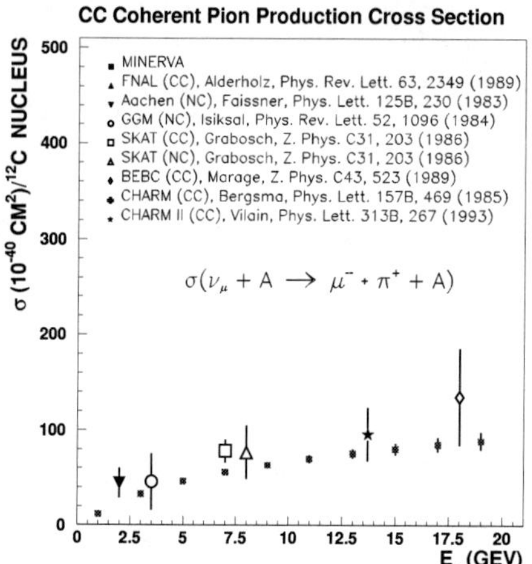

FIGURE 3. Coherent cross-sections as measured by MINERνA compared with existing published results. MINERνA errors here are statistical only.

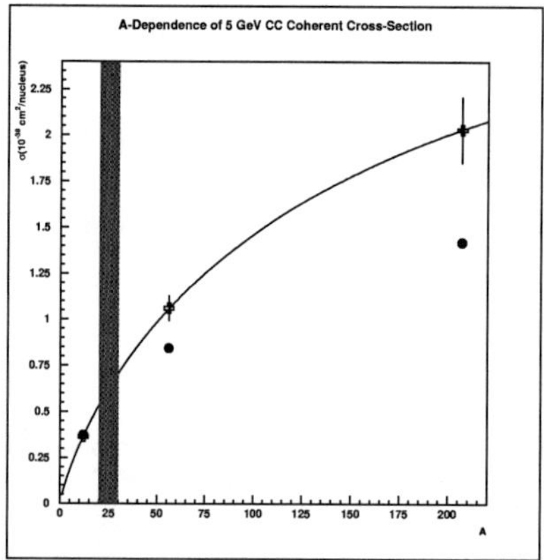

FIGURE 4. The range of A-dependent measurements of the coherent-pion cross-section to be measured by the MINERνA experiment

coherent pion cross-section. The shaded band is the range in A covered by existing experiments.

- Examination of nuclear effects in neutrino-induced interactions including energy loss and final-state modifications in heavy nuclei. These nuclear effects play a significant role in neutrino oscillation experiments that measure ν_μ disappearance as a function of E_ν. With sufficient $\overline{\nu}$ running, a study of quark flavor-dependent nuclear effects will also be performed.
- Clarification of the role of nuclear effects as they influence the determination of

$\sin^2\theta_W$ via measurement of the ratio of neutral-current to charged-current cross-sections off different nuclei.

- Much-improved measurement of the parton distribution functions will be possible using a measurement of all six ν and $\overline{\nu}$ structure functions (with sufficient $\overline{\nu}$ running).
- Examination of the leading exponential contributions of perturbative QCD.
- Precision measurement of exclusive strange-particle production channels near threshold, thereby improving knowledge of backgrounds in nucleon-decay searches, determination of V_{us}, and enabling searches for strangeness-changing neutral-currents and candidate pentaquark resonances. Measurement of hyperon-production cross-sections, including hyperon polarization, is feasible with exposure of MINERνA to $\overline{\nu}$ beams.
- Studies of nuclear physics for which neutrino reactions provide information complementary to JLab studies in the same kinematic range.

In addition to being significant fields of study in their own right, improved knowledge of many of these topics is essential to minimizing systematic uncertainties in neutrino-oscillation experiments.

2.1. The MINERνA Detector

For MINERνA to meet its physics goals, the detector must break new ground in the design of high-rate neutrino experiments. With final states as varied as high-multiplicity deep-inelastic reactions, coherent single-π^0 production and quasi-elastic neutrino scattering, the detector is a hybrid of a fully-active fine-grained detector and a traditional calorimeter. A complete description of MINERνA is found in the proposal [3]. The essential features are described here.

The MINERνA detector is made up of a number of sub-detectors with distinct functions in reconstructing neutrino interactions. The fiducial volume for most analyses is the inner "Active Target" shown in Figure 5, where all the material of the detector is the scintillator strips themselves. The scintillator detector does not fully contain events due to its low density and low Z, and therefore, the MINERνA design surrounds the scintillator fiducial volume with sampling detectors. To construct these sampling detectors, the scintillator strips are intermixed with absorbers. For example, the side, upstream (US) and downstream (DS) electromagnetic calorimeters (ECALs) have lead foil absorbers. Surrounding the ECALs are the US and DS hadronic calorimeter (HCAL) where the absorbers are steel plates. The US ECAL and HCAL serve a dual purpose as high-A targets for the study of nuclear effects. On the side of the detector, it is the outer detector (OD) that plays the role of the HCAL. Upstream of the detector is a veto of steel and scintillator strips to shield MINERνA from incoming soft particles produced upstream in the hall.

The core active element will be triangularly-shaped extruded scintillator strips readout *via* wavelength-shifting fibers, Readout of the fibers will be done with multi-anode photomultiplier tubes (MAPMTs), connected to the wavelength shifting fibers *via* an

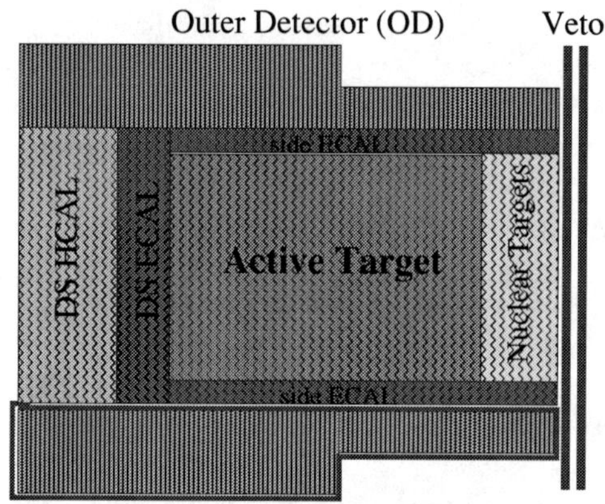

FIGURE 5. A side view schematic of the MINERνA detector

optical cable system and housed in a light tight "optical box" mounted on the outer detector. There are three distinct orientations of strips in the inner detector and veto, separated by 60°, and labeled X, U, V. A single module of MINERνA has two X layers to seed two-dimensional track reconstruction, and one each of the U and V layers to reconstruct three-dimensional tracks.

The MINERνA detector will be placed directly upstream of the MINOS near detector. Forward going muons exiting MINERνA can be measured in the MINOS near detector either by range or curvature.

3. A FUTURE NEUTRINO DETECTOR FOR NUMI

In a subsequent stage it would be most beneficial, for all physics topics of interest, to have a low-A target, preferably H_2 or D_2. An investigation of the technical and safety challenges of such a target is currently underway at Fermilab and a report [7] indicates that there are no real technical challenges in fabricating or efficiently operating a large LH_2 or LD_2 target. The main effort (and expense) for such a facility would be in satisfying safety requirements. For a fiducial volume with r = 80 cm. and l = 150 cm. we would expect 560 K CC events in LH_2 and 1280 K CC events in LD_2 per year of he-ν running.

An alternative proposal for a H_2 or D_2 target suggests that modern bubble chamber techniques, including recently-developed exceptionally smooth walls to eliminate extraneous bubbles and the use of CCD readout to allow electronic automatic scanning of events, makes a fully active liquid H_2 or D_2 target quite reasonable.

4. THE FERMILAB PROTON DRIVER

It has become increasingly clear that future long baseline neutrino experiments will require further factors of 5-10 improvements in proton luminosity. Such experiments at Fermilab are only feasible if a major proton source upgrade is undertaken. The Proton Driver project would replace the Booster with a new 8-GeV accelerator with 0.5-2 MW beam power, a factor of 15-60 more than the current Booster. It would also make the modifications needed to the Main Injector to upgrade it to simultaneously provide 120 GeV beams of 2 MW. A Proton Driver would bring with it other advantages. It would have the capacity to support a vigorous 8-GeV fixed-target program while providing 2 MW Main Injector beams. A Fermilab report [8] summarizes the physics potential of such a program.

A Proton Driver can also serve as a stepping-stone to future accelerators, both as an R&D test bed and as an injector, with connections to the Linear Collider, Neutrino Factories, and a VLHC.

A design study for a Superconducting Linac-based Proton Driver has been completed [9]. The simplicity of design should make it simpler to operate than booster/linac combinations. Limited emittance growth in a linac means that it can deliver the high brightness, low halo beams needed for running the Main Injector at high intensity with acceptable losses. The short MI "fill time" could deliver the full 2 MW of beam power at any energy from 40 to 120 GeV, and improvements to the MI ramp time could further increase the average proton intensity. There are many technical overlaps between the development and construction of such a machine and a cold technology Linear Collider. The 8 GeV superconducting linac could also accelerate electrons and could serve as a stepping-stone to future accelerators, both as an R&D test bed and as an injector, with connections to the Linear Collider, Neutrino Factories, and a VLHC.

Such a proton driver would allow an 8 GeV proton based neutrino program with a factor of ≈ 20 increase in intensity compared to the current MiniBooNe experiment and a factor 5 increase in the 120 GeV proton based Main Injector neutrino program compared to MINOS.

REFERENCES

1. M. Tzanov, 'NuTeV Structure Function Measurement", in these Proceedings.
2. The MINERνA Collaboration consists of groups from the following institutions: U Athens, U California/Irvine, U Dortmund, Fermilab, Hampton U, IL Inst. Tech., Inst. for Nuc. Research - Moscow, James Madison U, Jefferson Lab, N. Illinois U, U Pittsburgh, U Rochester, Rutgers U, Tufts U, William and Mary U.
3. D. Drakoulakos *et al.* [The MINERνA Collaboration], 'Proposal to Perform a High-Statistics Neutrino Scattering Experiment Using a Fine-grained Detector in the NuMI beam", December 2003 [rev. January 2004]. Submitted to the FNAL PAC. `http://www.pas.rochester.edu/minerva/`
4. N. J. Baker *et al.*, Phys. Rev. **D23**, 2499 (1981).
5. T. Kitagaki *et al.*, Phys. Rev. **D26**, 436 (1983).
6. P. Rubinov, FNAL-TM-2226. P. Rubinov, FNAL-TM-2227. P. Rubinov, FNAL-TM-2228.
7. J. Kilmer and T. J.Sarlina, 'Technical and Safety Considerations of Large Cryogenic Liquid Targets at the NuMI Near Hall" contribution to the Workshop on New Initiatives for the NuMI Neutrino Beam, Fermilab, 2002
8. S. Geer, Nucl.Phys.Proc.Suppl. **147**, 124-127 (2005)
9. R. D. Kephart,, Nucl.Phys.Proc.Suppl. **147**, 41-44 (2005)

Deep Inelastic Scattering at the Amplitude Level

Stanley J. Brodsky

Stanford Linear Accelerator Center, Stanford University, Stanford, CA, 94309

Abstract. The deep inelastic lepton scattering and deeply virtual Compton scattering cross sections can be interpreted in terms of the fundamental wavefunctions defined by the light-front Fock expansion, thus allowing tests of QCD at the amplitude level. The AdS/CFT correspondence between gauge theory and string theory provides remarkable new insights into QCD, including a model for hadronic wavefunctions which display conformal scaling at short distances and color confinement at large distances.

Keywords: Quantum Chromodynamics, Deep Inelastic Scattering, Gauge/String Duality
PACS: 12.38.Aw,13.60.Hb,11.25.Tq

WAVEFUNCTION REPRESENTATION OF DIS AND DVCS

The primary goal of deep inelastic lepton scattering is to resolve the fundamental structure of the nucleon. In fact, by combining measurements of DIS with measurements of deeply virtual Compton scattering, elastic lepton-hadron scattering, and other hard exclusive channels, it is possible to obtain information on the fundamental form of quark and gluon bound-state wavefunctions. Thus, for the first time, we have the potential to test QCD at the amplitude level.

If one quantizes QCD at fixed light-front time $x^+ = x^0 + x^3$, the bound state hadronic solutions $|\Psi_H\rangle$ are eigenstates of the light-front Heisenberg equation $H_{LF}\,|\Psi_H\rangle = M_H^2\,|\Psi_H\rangle$ [1]. The spectrum of QCD is given by the eigenvalues M_H^2. The projection of each hadronic eigensolution on the free Fock basis: $\langle n|\Psi_H\rangle \equiv \psi_{n/H}(x_i, \vec{k}_{\perp i}, \lambda_i)$ defines the LF Fock expansion in terms of the quark and transversely polarized gluon constituents in $A^+ = 0$ light-cone gauge. The light-front wavefunctions are frame-independent functions of the constituent light-cone fractions x_i, relative transverse momenta $\vec{k}_{\perp i}$, and spin projections $S_i^z = \lambda_i$. Observables in DIS and DVCS can be calculated directly from the hadron LFWFs. For example, the quark and gluon distributions measured in DIS are defined from the squares of the LFWFS summed over all Fock states n. Form factors, exclusive weak transition amplitudes [2] and the generalized parton distributions [3] measured in DVCS are overlaps of the initial and final LFWFS with $n = n'$ and $n = n' + 2$. The resulting distributions obey DGLAP, BFKL, and ERBL evolution as a function of the maximal invariant mass, thus providing a physical factorization scheme [4]. It is important to note that at large x where the struck quark is far-off shell, DGLAP evolution is quenched [5], so that the fall-off of the DIS cross sections in Q^2 satisfies inclusive-exclusive duality at fixed W^2. The gauge-invariant distribution amplitude $\phi_H(x_i, Q)$ defined from the integral over the transverse momenta $\vec{k}_{\perp i}^2 \le Q^2$ of the valence (smallest n) Fock state provides a fundamental measure of the hadron at the amplitude level [6, 7]; they are the nonperturbative input to the factorized form of hard

CP792, *Deep Inelastic Scattering, DIS 2005,* edited by W. H. Smith and S. R. Dasu

exclusive amplitudes and exclusive heavy hadron decays in PQCD. The front form provides a consistent definition of relative orbital angular momentum and J^z conservation. Fundamental sum rules such as Ji's measure of orbital angular momentum [8], and the vanishing of the "anomalous gravitomagnetic moment" $B(0)$ [9] are immediate properties of the LF Fock wavefunctions [10]. One can perform Fourier transforms of the Fock state wavefunctions in impact space $b_\perp$ and in $x^- = x^0 - x^3$ space to obtain the spatial form of Fock wavefunctions in coordinate space. See also: [11, 12] The DVCS amplitudes also enter the two-photon exchange contribution to elastic electron-proton scattering, which in turn produces a significant correction to the Rosenbluth method used for separating form factors [13, 14].

The E791 experiments at Fermilab [15] has shown how one can measure the valence LFWF directly from the diffractive di-jet dissociation of a high energy pion $\pi A \to q\bar{q}A'$ into two jets, nearly balancing in transverse momentum, leaving the nucleus intact. The measured pion distribution in x and $(1-x)$ is similar the form of the asymptotic distribution amplitude. The E791 experiment also find that the nuclear amplitude is additive in the number of nucleons when the quark jets are produced at high $k_\perp$, thus giving a dramatic confirmation of "color transparency", a fundamental manifestation of the gauge nature of QCD [16, 17] The LFWFs display other novel features, such as asymmetric sea-quark distributions $\bar{u}(x) \neq \bar{d}(x)$, $\bar{s}(x) \neq s(x)$, and intrinsic heavy-quark Fock states [18] of the proton $|uudQ\bar{Q}\rangle$ in which the heavy constituents carry the largest moment fractions. One can use the OPE to show that the probability of such states scales as $1/M_Q^2$ in contrast to $1/M_\ell^4$ fall-off of abelian theory [19]. The remarkable observations of the SELEX experiment of the double-charm baryon Ξ_{ccd} in $pA \to \Xi_{ccd}X$ and $\Sigma^- A \to \Xi_{ccd}X$ at large x_F [20] provides compelling evidence for double-charm intrinsic Fock states in the proton. The coherence of multi-particle correlations within the Fock states leads to higher-twist bosonic processes such as $e(qq) \to e'(qq)'$; although suppressed by inverse powers of Q^2, such subprocesses are important in the duality regime of fixed W^2, particularly in σ_L [21]. In the case of nuclei, one must include non-nucleonic "hidden color" [22] degrees of freedom of the deuteron LFWF.

Contrary to parton model expectations, the rescattering of the quarks in the final state in DIS has important phenomenological consequences, such as leading-twist diffractive DIS [23] and the Sivers single-spin asymmetry [24]. The Sivers asymmetry depends on the same matrix elements which produce the anomalous magnetic moment of the target nucleon as well as the phase difference of the final-state interactions in different partial waves. The rescattering of the struck parton generates dominantly imaginary diffractive amplitudes, giving rise to an effective "hard pomeron" exchange and a rapidity gap between the target and diffractive system, while leaving the target intact. This Bjorken-scaling physics, which is associated with the Wilson line connecting the currents in the virtual Compton amplitude survives even in light-cone gauge. Thus there are contributions to the DIS structure functions which are not included in the light-front wave functions computed in isolation and cannot be interpreted as parton probabilities [23]. DDIS in turn leads to nuclear shadowing at leading twist as a result of the destructive interference of multi-step processes within the nucleus. In addition, multi-step processes involving Reggeon exchange leads to antishadowing. In fact, because Reggeon couplings are flavor specific, antishadowing is predicted to be non-universal, depending on the type of current and even the polarization of the probes in nuclear DIS [25]. Another particularly interesting consequence of QCD is the Q^2-independent "$J = 0$ fixed-pole" contribution $M(\gamma^* p \to \gamma p) \sim s^0 F(t)$ to the real part of the DVCS amplitude, reflecting the effective contact interaction of the transverse currents [26]. DVCS can also be studied in the timelike domain from $e^+e^- \to H^+H^-\gamma$; the lepton charge asymmetry and single-spin asymmetries allow measurements of the relative phase of timelike form factors and the $\gamma^* \to H^+H^-\gamma$ amplitude [27].

ADS/CFT PREDICTIONS FOR HADRON PHYSICS

The AdS/CFT correspondence [28], between strongly-coupled conformal gauge theory and weakly-coupled string theory in the 10-dimensional $AdS_f \times S^5$ space is now providing a remarkable new insight into hadron wavefunctions of QCD. Although QCD is not conformal, it is nearly conformal in the asymptotic freedom domain and it resembles a strongly-coupled conformal theory at relatively soft momenta if the QCD coupling has an infrared fixed point [29]. Deur *et al.* have also shown that the effective charge $\alpha^s_{g_1}(Q^2)$ defined from the radiative corrections to the Bjorken sum rule also approaches a constant at low momentum [30]. The near-constancy of the effective QCD coupling at small scales helps explain the empirical success of dimensional counting rules for the power law fall-off of form factors and fixed angle scaling. The string/gauge theory duality provides a framework for predicting QCD phenomena based on the conformal properties of the AdS/CFT correspondence. For example, the dimensional counting rules for the power-law fall-off of hard exclusive hadron-hadron scattering amplitudes at large momentum transfer can be derived [31, 32, 33] without the use of perturbation theory. Polchinski and Strassler [31] have also derived counting rules for deep inelastic structure functions at $x \to 1$ in agreement with perturbative QCD predictions [34]

Recently Teramond and I [35, 36, 37] have shown how to compute the hadronic spectrum of light $q\bar{q}, qqq$ and gg bound states in a holographic dual of QCD defined on $AdS_5 \times S^5$. Specific hadrons are identified by the correspondence of string modes with the dimension of the interpolating operator of the hadron's valence Fock state, including orbital angular momentum excitations. Since only one parameter, the QCD scale Λ_{QCD}, is introduced, the agreement with the pattern of physical states is remarkable. In particular, the ratio of Δ to nucleon trajectories is determined by the ratio of zeros of Bessel functions We have also shown how one can use the extended AdS/CFT space-time theory to obtain a model for the form of hadron LFWFs. The model wavefunctions display confinement at large inter-quark separation and conformal symmetry at short distances. In particular, the scaling and conformal properties of the LFWFs at high relative momenta agree with perturbative QCD [38]. These AdS/CFT model wavefunctions could be used as an initial ansatz for a variational treatment of the light-front QCD Hamiltonian.

ACKNOWLEDGMENTS

This work was supported by the Department of Energy, contract No. DE-AC02-76SF00515.

REFERENCES

1. S. J. Brodsky, H. C. Pauli and S. S. Pinsky, Phys. Rept. **301**, 299 (1998) [arXiv:hep-ph/9705477].
2. S. J. Brodsky and D. S. Hwang, Nucl. Phys. B **543**, 239 (1999) [arXiv:hep-ph/9806358].
3. S. J. Brodsky, M. Diehl and D. S. Hwang, Nucl. Phys. B **596**, 99 (2001) [arXiv:hep-ph/0009254].
4. G. P. Lepage and S. J. Brodsky, Phys. Rev. D **22**, 2157 (1980).
5. S. J. Brodsky and G. P. Lepage, SLAC-PUB-2294
6. G. P. Lepage and S. J. Brodsky, Phys. Lett. B **87**, 359 (1979).
7. A. V. Efremov and A. V. Radyushkin, Phys. Lett. B **94**, 245 (1980).
8. X. D. Ji, Phys. Rev. Lett. **78**, 610 (1997) [arXiv:hep-ph/9603249].
9. O. V. Teryaev, arXiv:hep-ph/9904376.
10. S. J. Brodsky, D. S. Hwang, B. Q. Ma and I. Schmidt, Nucl. Phys. B **593**, 311 (2001) [arXiv:hep-th/0003082].
11. X. Ji, Ann. Rev. Nucl. Part. Sci. **54**, 413 (2004).
12. M. Burkardt, arXiv:hep-ph/0505189.
13. A. V. Afanasev, S. J. Brodsky, C. E. Carlson, Y. C. Chen and M. Vanderhaeghen, arXiv:hep-ph/0502013.
14. Y. C. Chen, A. Afanasev, S. J. Brodsky, C. E. Carlson and M. Vanderhaeghen, Phys. Rev. Lett. **93**, 122301 (2004) [arXiv:hep-ph/0403058].
15. D. Ashery [E791 Collaboration], arXiv:hep-ex/0205011.
16. G. Bertsch, S. J. Brodsky, A. S. Goldhaber and J. F. Gunion, Phys. Rev. Lett. **47**, 297 (1981).
17. S. J. Brodsky and A. H. Mueller, Phys. Lett. B **206**, 685 (1988).
18. S. J. Brodsky, P. Hoyer, C. Peterson and N. Sakai, Phys. Lett. B **93**, 451 (1980).
19. M. Franz, . V. Polyakov and K. Goeke, Phys. Rev. D **62**, 074024 (2000) [arXiv:hep-ph/0002240].
20. J. Engelfried [SELEX Collaboration], Nucl. Phys. A **752**, 121 (2005).
21. S. J. Brodsky, E. L. Berger and G. P. Lepage, SLAC-PUB-3027
22. S. J. Brodsky, C. R. Ji and G. P. Lepage, Phys. Rev. Lett. **51**, 83 (1983).
23. S. J. Brodsky, P. Hoyer, N. Marchal, S. Peigne and F. Sannino, Phys. Rev. D **65**, 114025 (2002) [arXiv:hep-ph/0104291].
24. S. J. Brodsky, D. S. Hwang and I. Schmidt, Phys. Lett. B **530**, 99 (2002) [arXiv:hep-ph/0201296].
25. S. J. Brodsky, I. Schmidt and J. J. Yang, Phys. Rev. D **70**, 116003 (2004) [arXiv:hep-ph/0409279].
26. S. J. Brodsky, F. E. Close and J. F. Gunion, Phys. Rev. D **5**, 1384 (1972).
27. A. V. Afanasev, S. J. Brodsky, C. E. Carlson, (in prepration).
28. J. M. Maldacena, Adv. Theor. Math. Phys. **2**, 231 (1998) [Int. J. Theor. Phys. **38**, 1113 (1999)] [arXiv:hep-th/9711200].
29. For a review and references see: S. J. Brodsky, arXiv:hep-ph/0408069.
30. A. Deur, et al. (This conference)
31. J. Polchinski and M. J. Strassler, Phys. Rev. Lett. **88**, 031601 (2002) [arXiv:hep-th/0109174].
32. R. C. Brower and C. I. Tan, Nucl. Phys. B **662**, 393 (2003) [arXiv:hep-th/0207144].
33. O. Andreev, Phys. Rev. D **67**, 046001 (2003) [arXiv:hep-th/0209256].
34. S. J. Brodsky, M. Burkardt and I. Schmidt, Nucl. Phys. B **441**, 197 (1995) [arXiv:hep-ph/9401328].
35. G. F. de Teramond and S. J. Brodsky, Phys. Rev. Lett. **94**, 201601 (2005) [arXiv:hep-th/0501022].
36. G. F. de Teramond and S. J. Brodsky, arXiv:hep-th/0409074.
37. S. J. Brodsky and G. F. de Teramond, Phys. Lett. B **582**, 211 (2004) [arXiv:hep-th/0310227].
38. X. d. Ji, J. P. Ma and F. Yuan, Phys. Rev. Lett. **90**, 241601 (2003) [arXiv:hep-ph/0301141].

Participant List

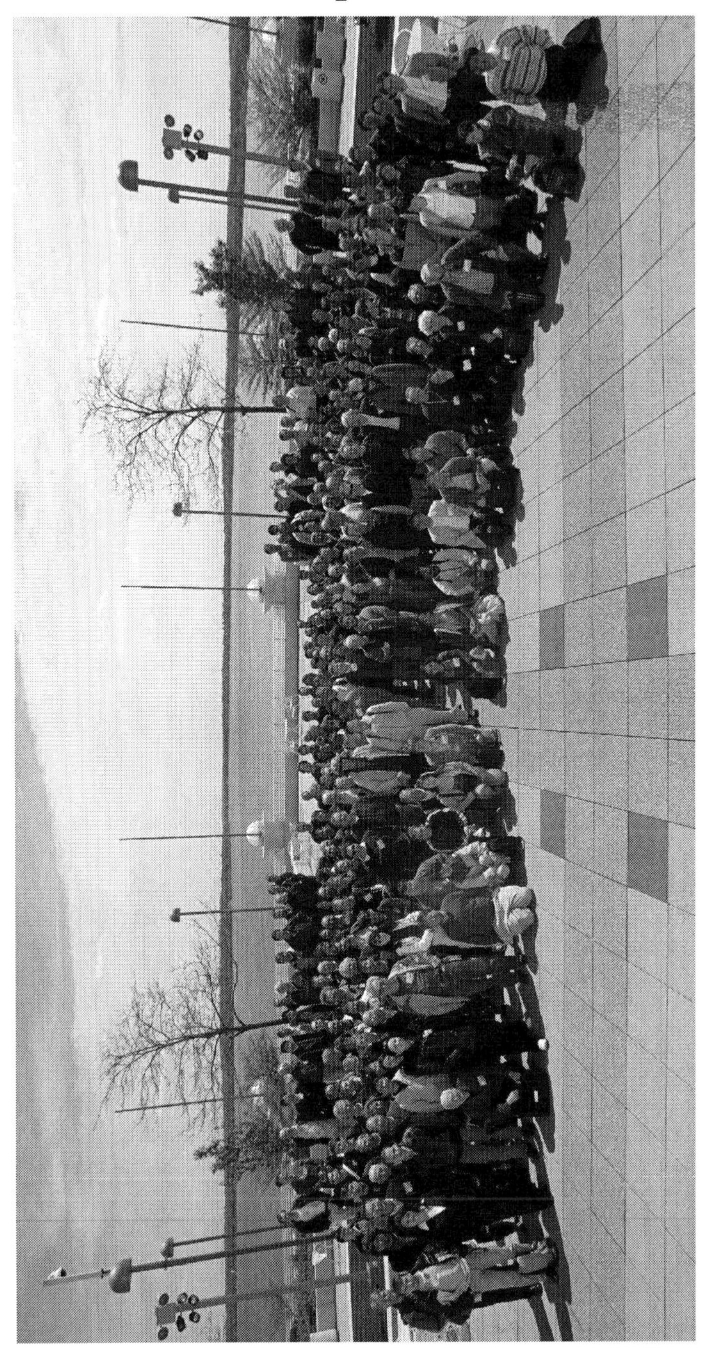

Halina Abramowicz.. Tel Aviv University
Leszek Adamczyk .... University of Science and Technology Cracow
Andrei Afanasev.. Jefferson Lab
Gayane Aghuzumtsyan..Bonn University
Avetik Airapetian .... DESY
Michael Albrow ....Fermilab
Johan Alwall ....Uppsala University
Jeppe Andersen .... University of Cambridge
Aaron Andrus............................. University of Illinois at Urbana Champaign
Tinne Anthonis...University of Antwerp
Alexei Antonov .... MEPHi / DESY
Jean-Francois Arguin..University of Toronto
Elke-Caroline Aschenauer .... DESY
Anatoli Astvatsatourov .... Vrije Universiteit Brussel
Harut Avagyan .... Jefferson Lab
Terry Awes..Oak Ridge National Laboratory
Howard Baer .... Florida State University
Csaba Balazs .... Argonne National Laboratory
Giuseppe Barbagli ....INFN-Firenze
Marco Battaglieri.. INFN
Matthew Beckingham... DESY
Olaf Behnke .... University of Heidelberg
Alexander Belyaev .... Michigan State University
Colin Bernet .... CERN
Robert Bernstein...Fermilab
Johannes Bluemlein... DESY
Arie Bodek ....University of Rochester
Juraj Bracinik .... Max-Planck Institute for Physics Munchen
Robert Bradford ....University of Rochester
Andrew Brandt.. University of Texas, Arlington
Alessandro Bravar .... Brookhaven National Laboratory
Stanley Brodsky .... SLAC, Stanford University
Duncan Brown .... UTA
Armen Bunyatyan .... MPI-Heidelberg / Yerevan
Peter Bussey... University of Glasgow
Robert Cadman.. Argonne National Laboratory
Allen Caldwell .... Max Planck Institute for Physics
Qinghong Cao .... Michigan State Universtiy
Vladimir Chekelian .... Max-Planck-Institut fuer Physik
Jian-ping Chen .... Jefferson Lab
Chunhui Chen .... University of Maryland
Mickey Chiu.................................. University of Illinois, Urbana-Champaign
Frank Chlebana ....Fermilab
Eugene Chudakov .... Jefferson Lab
Yeonsei Chung .... University of Rochester
Joanne Cole .... University of Bristol

Gennaro Corcella... CERN
Massimo Corradi .... INFN Bologna
Andrew Cottrell.. Oxford University
Brian Cox .... University of Manchester
Georges Cozzika ....CEA/DAPNIA/SPP
John Dainton .... University of Liverpool
Sridhara Dasu.. University of Wisconsin
Karin Daum... Wuppertal University/DESY
Brian Davies..Lancaster Universtiy U.K
Albert de Roeck... Univeristy of Manchester
Eddi De Wolf .... University of Antwerpen
Malcolm Derrick ....ANL
Abhay Deshpande .... SUNY-Stony Brook & RIKEN-BNL Research Center
William Detmold.. University of Washington
Alexandre Deur .... Jefferson Lab
Nicole D'Hose ....CEA Saclay
Pasquale Di Nezza... INFN Frascati
Markus Diefenthaler... Physikalisches Institut II
Lance Dixon...SLAC
Stanislav Dubnicka...Slovak Academy of Sciences
Anna Dubnickova...Comenius University
Eric Eckhart .... Colorado State University
Karsten Eggert... CERN
Yehuda Eisenberg .... Weizmann Institute
RikardEnberg .... Lawrence Berkeley National Lab
Boris Ermolaev.. Ioffe Phys. Tech. Institute
Marc Escalier .... LPNHE Paris
Adam Everett .... University of Wisconsin
Jozef Ferencei .... Institut of Experimental Physics
Rick Field... University of Florida
Lars Finke .... University of Hamburg
Gero Flucke.. DESY
Brian Foster... University of Oxford
Rainer Fries.. University of Minnesota
Stefano Frixione .... INFN
Anna Galas. Institute of Nuclear Physics Polish Academy of Sciences (DESY)
Hugh Gallagher ....Tufts University
Elisabetta Gallo ....INFN Firenze, Italy
Leonard Gamberg.. Penn State Berks
Gagik Gavalian... Old Dominion University
Walter Giele ....Fermilab
Claudia Glasman ....Universidad Autonoma de Madrid
Sasha Glazov .... DESY
Shrihari Gopalakrishna .... Northwestern University
Konstantin Goulianos .... The Rockefeller University
Guenter Grindhammer...MPI for Physics, Munich

Alberto Guffanti .... DESY
Wlodek Guryn .... Brookhaven National Lab
Claire Gwenlan .... University of Oxford
Carl Gwilliam .... The University of Manchester
Richard Hall-Wilton .... University College London
Tao Han .... University of Wisconsin
Jurgen Hannappel .... Mainz University
Amnon Harel .... Wuppertal University
Jonathan Hays .... Northwestern University
Chris Hays .... Duke University
Beate Heinemann .... University of Liverpool
Carsten Hensel .... University of Kansas
Steven Heppelmann .... Penn State University
Matthew Herndon .... The Johns Hopkins University
Rolf-Dieter Heuer .... DESY
Claus Horn .... DESY
Ulrich Husemann .... University of Rochester
Robert Illingworth .... Fermilab
Donald Isenhower .... Abilene Christian University
Robert Jaffe .... MIT
Jamal Jalilian-Marian .... University of Washington
Xavier Janssen .... DESY - Hamburg
Jiangyong Jia .... Columbia University
Rainer Joosten .... University of Bonn
Cynthia Keppel .... Hampton University / Jefferson Lab
Valeri Khoze .... IPPP, University of Durham
Nikolaos Kidonakis .... Kennesaw State University
Christian Kiesling .... Max-Planck-Institute for Physics
Michael Klasen .... LPSC Grenoble
Max Klein .... DESY
Spencer Klein .... Lawrence Berkeley National Lab.
Tatsiana Klimkovich .... DESY Hamburg
Thomas Kluge .... DESY
Bernd Kniehl .... University of Hamburg
Albert Knutsson .... Lund University
Takanori Kohno .... University of Oxford
Vesa Kolhinen .... University of Jyvaskyla
Mikhail Kopytin .... DESY
Yuri Kovchegov .... The Ohio State University
Alexander Kovner .... University of Connecticut
Henri Kowalski .... DESY
Peter Kroll .... Fachbereich Physik, Universitaet Wuppertal
Sebastian Kuhn .... Old Dominion University
Krishna Kumar .... University of Massachusetts
Albrecht Kyrieleis .... University of Manchester
Jean-Philippe Lansberg .... University of Liege

Francisco Larios .... Michigan State University
Paul Laycock .... University of Liverpool
Regis Lefevre .... IFAE - Barcelona
Paolo Lenisa .... Universita' di Ferrara and INFN - Sez. Ferrara
Boris Levchenko .... DESY
Eugene Levin .... Tel Aviv University
Aharon Levy .... Tel Aviv University
Heuijin Lim .... DESY
Loren A. Linden-Levy .... University of Illinois
Linus Lindfeld .... Uni Zuerich
Simonetta Liuti .... University of Virginia
Nilanga Liyanage .... University of Virginia
Ewelina Lobodzinska .... DESY Zeuthen
Andrea Longhin .... INFN and Padova University
Bino Maiheu .... University of Ghent
Nick Malden .... Manchester University
Cyrille Marquet .... CEA/Saclay
Alan Martin .... IPPP, Durham
Maria Martisikova .... DESY
David Mason .... University of Oregon
Stephen Maxfield .... University of Liverpool
Kajari Mazumdar .... Tata Institute of Fundamental Research
Andrew Mehta .... Liverpool
Thomas Meyer .... Iowa State University
Alexander Mitov .... University of Hawaii
Jorge Morfin .... Fermilab
Matthias Mozer .... Physikalisches Institut Universitaet Heidelberg
Asmita Mukherjee .... Lorentz Institute, University of Leiden
Luiz Mundim .... UERJ/Dzero
Pavel Nadolsky .... Argonne National Laboratory
Ioana Niculescu .... James Madison University
Andrei Nikiforov .... Max-Planck-Institut fuer Physik
Yujin Ning .... Columbia University
Carlo Oleari .... University Milano-Bicocca
Fred Olness .... SMU
Nurcan Ozturk .... University of Texas at Arlington
Paolo Pagano .... Istituto Nazionale di Fisica Nucleare - Trieste
Nirmalya Parua .... SUNY@Stony Brook
Emmanuelle Perez .... CEA-Saclay / DESY
Adrian Perieanu .... DESY - Hamburg
Alexey Petrukhin .... ITEP / DESY
Krzysztof Piotrzkowski .... UCL-Catholic University of Louvain
Benjamin Portheault .... Laboratoire de l'Accelerateur Lineaire
Alexey Prokudin .... The University of Turin
Jon Pumplin .... Michigan State University
Jianwei Qiu .... Iowa State University

Voica Radescu.. University of Pittsburgh
Greg Rakness ....Penn State University/Brookhaven National Lab
Kirti Ranjan... Fermilab and University of Delhi
Kenneth Read ....Oak Ridge National Laboratory
Don Reeder ....University of Wisconsin
Davide Reggiani .... University of Erlangen
Zhenhai Ren ....DEYS ZEUS Columbia
Roger Renner .... Physikalisches Institut, University of Bonn
Dru Renner...University of Arizona
Peter Renton ....University of Oxford
Jose Repond ....Argonne National Laboratory
Michael Rijssenbeek..Stony Brook University
Christiane Risler .... DESY
Eram Rizvi ....Queen Mary, University of London
Juan Rojo .... Facultat de Fisica, Universitat de Barcelona
Ewa Rondio... Soltan Institute for Nuclear Studies
Robert Roosen .... VUB
Michele Rosin ....University of Wisconsin
Christophe Royon...CEA Saclay / Fermilab
Joshua Rubin ....University of Illinois Urbana-Champaign
Zuzana Rurikova ....Max-Planck-Institut fuer Physik
Marta Ruspa .... U. Eastern Piedmont and INFN Turin
Dirk Ryckbosch..University of Gent
Rodolfo Sassot .... Universidad de Buenos Aires
David Saxon... University of Glasgow
Ingo Schienbein... DESY
Gerrit Schierholz .... DESY
William Schmidke...Columbia University
Andre' Schoening ....ETH Zurich
Ralf Seidl ....RBRC/University of Illinois at Urbana-Champaign
Tony Signal... Institute of Fundamental Sciences
Dennis Sivers .... University of Michigan
Peter Skands...Fermilab
Wesley Smith ....University of Wisconsin
Davison Soper ....University of Oregon
Paul Souder .... Syracuse University
David South .... DESY
Dimiter Stamenov ....Institute for Nuclear Research & Nuclear Energy
Jeffrey Standage .... York University
Anna Stasto ....BNL
Uta Stoesslein... DESY
Marco Stratmann ....University of Regensburg
Daniel Stump...Michigan State University
Bernd Surrow .... MIT
Adam Szczepaniak ....Indiana University
Dorota Szuba...INP-PAS

Tim Tait .... Argonne National Lab
Alex Tapper .... Imperial College London
Toshinari Tawara.. University of Tokyo
Juan Terron .... Universidad Autonoma de Madrid
Thomas Teubner.. University of Liverpool
Paul Thompson.. Birmingham University
Robert Thorne .... University of Cambridge
Katsuo Tokushuku.. KEK
Kirill Tuchin.. Brookhaven National Laboratory
Wu-Ki Tung .... Michigan State University
Teresa Tymieniecka .... Warsaw University
Martin Tzanov.. University of Pittsburgh
Artur Ukleja .... Warsaw University
Paul van der Nat .... NIKHEF
Arne Vandenbroucke.. Universiteit Gent
Christian Veelken .... Desy
Flemming Videbaek .... Brookhaven National Laboratory
Ivan Vitev .... Los Alamos National Lab
Nikolay Vlasov.. Freiburg University
Horst Wahl .... Florida State University
Roberval Walsh .... McGill University
Xie Wei.. Riken-BNL Research Center
Christian Weiss .... Jefferson Lab
Sebastian White.. Brookhaven National Lab
Jim Whitmore.. Penn State U./National Science Foundation
Matthew Wing.. DESY/UCL
Bruce Yabsley .... Virginia Polytechnic Institute and State U.
Yuji Yamazaki .... KEK
Un-ki Yang.. University of Chicago
Rik Yoshida .... Argonne National Laboratory
C.-P. Yuan.. Michigan State University
Chun Zhang.. Oak Ridge National Lab
John Zhou .... Rutgers University
Nikolay Zotov .... Moscow State University
Xiaomin Zu .... Florida State University

AUTHOR INDEX

A

B

C

D

E

F

G

H

I

J

K

L